结构抗风减振理论与实践

——湖南大学风工程试验研究中心成立 20 年论文精选

| 陈政清　华旭刚　主编

湖南大学出版社
·长沙·

内容简介

2024年是湖南大学HD-2风洞建成20周年，HD-3新风洞将于2024年底落成，编者团队将近20年来发表的代表性论文成果系统整理，编撰成集，影印出版。论文集共收集已发表论文65篇，包括8个主题，其中，电涡流阻尼减振技术8篇、桥梁多阶模态涡振10篇、索杆结构风致振动与控制8篇、新能源结构9篇、桥梁颤振与抖振8篇、振动控制理论与方法8篇、人致振动与控制3篇、健康监测与其他主题11篇。为方便读者检索，也考虑到中英文人名翻译的差异，在目录中，各英文篇目标注了英文名并括注了中文名。

本论文集可为工程界的抗风设计和振动控制提供直接的借鉴方案与经验，也可供国内风工程研究人员参考。

图书在版编目（CIP）数据

结构抗风减振理论与实践 ：湖南大学风工程试验研究中心成立20年论文精选 / 陈政清，华旭刚主编 . 长沙 ：湖南大学出版社，2024. 9. -- ISBN 978-7-5667-3873-8

Ⅰ. TU352. 2-53

中国国家版本馆CIP数据核字第2024DF0654号

结构抗风减振理论与实践
——湖南大学风工程试验研究中心成立20年论文精选
JIEGOU KANGFENG JIANZHEN LILUN YU SHIJIAN
——HUNAN DAXUE FENGGONGCHENG SHIYAN YANJIU ZHONGXIN CHENGLI 20 NIAN LUNWEN JINGXUAN

主　　编：陈政清　华旭刚
策划编辑：卢　宇
责任编辑：卢　宇　张佳佳
印　　装：湖南省众鑫印务有限公司
开　　本：880 mm × 1240 mm　1/16　　印　　张：57
字　　数：1766千字
版　　次：2024年9月第1版　　印　　次：2024年9月第1次印刷
书　　号：ISBN 978-7-5667-3873-8
定　　价：298.00元

出 版 人：李文邦
出版发行：湖南大学出版社
社　　址：湖南·长沙·岳麓山　　邮　　编：410082
电　　话：0731-88821559（营销部）88821315（编辑部）88821006（出版部）
传　　真：0731-88822264（总编室）
网　　址：http://press.hnu.edu.cn

2024 年建成的湖南大学风工程试验研究中心 HD-3 风洞大楼

2004 年建成的湖南大学风工程试验研究中心 HD-2 风洞大楼

刘光栋教授主持风工程与桥梁工程湖南省重点实验室第一届学术委员会第一次全体会议

刘光栋教授与青年学子交流

◎序　言

自20世纪80年代以来，我国大规模兴建基础设施，建设了众多大跨度桥梁、超高层建筑和特种结构。近年来，以风力机、光伏光热等新能源结构为代表的新基建也日益增多。许多已建和在建工程在规模与技术挑战方面均位居世界前列。对于这些大跨轻柔结构，设计阶段的抗风减振设计和运营阶段的状态监测与振动控制成为确保其结构安全与良好服役性能的关键技术。

湖南大学是我国最早从事风工程研究的高校之一。早在20世纪70年代，学校就建成了HD-1风洞，并在“211工程”和“985工程”的支持下，于2002年成立了湖南大学风工程试验研究中心。2004年，又建成并投入使用了大型边界层风洞HD-2，2012年在HD-2风洞的基础上增加开口试验段，2024年底即将建成包含宽18 m、高3.2 m试验段的大型边界层风洞HD-3。经过20年的不懈努力，湖南大学现已成为我国风致振动与控制研究的重要基地。回顾这一发展历程，我们要特别感谢以同济大学项海帆院士为代表的众多专家学者的帮助和支持。湖南大学风工程团队从创立之初的几人发展到目前的近20人，研究方向也从最初的抗风理论与实验技术，扩展到融合多学科交叉的电涡流阻尼器、振动监测与控制、桥梁抗风、建筑抗风、传统电力结构抗风、新能源结构抗风、索结构抗风等。

在陈政清院士的带领下，湖南大学学术团队已从事结构风致振动与控制研究20余年，并屡次获得国家重点研发计划和国家自然科学基金项目的资助。在风致振动理论、风洞试验技术和结构振动控制等方面取得了一系列原创性的科研成果。这些成果在国内外学术刊物上发表后广受好评，并被频繁引用。团队注重工程应用，完成了多项重大工程结构的抗风设计和振动控制，开发了以电涡流阻尼器为代表的多项振动控制技术，解决了实际工程中的诸多风致振动问题，取得了显著的经济效益和社会效益。项目成果得以广泛推广，获得了国家技术发明奖一等奖，并多次获得国家科技进步奖二等奖。

今年恰逢湖南大学 HD-2 风洞建成 20 周年以及 HD-3 新风洞落成，我建议将近 20 年来团队成员发表的代表性论文成果系统整理，编撰成集。希望该论文集不仅能为国内日益增多的风工程研究人员提供宝贵的学习参考资料，也能为工程界的抗风设计和振动控制提供直接的借鉴和经验。

刘光栋

湖南大学风工程试验研究中心

2024 年 8 月

◎ 目　录 /contents

人致振动与控制 ··· (709)

健康监测与其他主题 ··· (745)

电涡流阻尼减振技术

Vol.11, No.3 EARTHQUAKE ENGINEERING AND ENGINEERING VIBRATION September, 2012

Earthq Eng & Eng Vib (2012) 11: 391-401 **DOI**: 10.1007/s11803-012-0129-x

Feasibility study of a large-scale tuned mass damper with eddy current damping mechanism

Wang Zhihao[1,2†], Chen Zhengqing[1‡] and Wang Jianhui[1§]

1. Wind Engineering Research Center, Hunan University, Changsha 410082, China

2. School of Civil Engineering and Communication, North China University of Water Resources and Electric Power, Zhengzhou 450011, China

Abstract: Tuned mass dampers (TMDs) have been widely used in recent years to mitigate structural vibration. However, the damping mechanisms employed in the TMDs are mostly based on viscous dampers, which have several well-known disadvantages, such as oil leakage and difficult adjustment of damping ratio for an operating TMD. Alternatively, eddy current damping (ECD) that does not require any contact with the main structure is a potential solution. This paper discusses the design, analysis, manufacture and testing of a large-scale horizontal TMD based on ECD. First, the theoretical model of ECD is formulated, then one large-scale horizontal TMD using ECD is constructed, and finally performance tests of the TMD are conducted. The test results show that the proposed TMD has a very low intrinsic damping ratio, while the damping ratio due to ECD is the dominant damping source, which can be as large as 15% in a proper configuration. In addition, the damping ratios estimated with the theoretical model are roughly consistent with those identified from the test results, and the source of this error is investigated. Moreover, it is demonstrated that the damping ratio in the proposed TMD can be easily adjusted by varying the air gap between permanent magnets and conductive plates. In view of practical applications, possible improvements and feasibility considerations for the proposed TMD are then discussed. It is confirmed that the proposed TMD with ECD is reliable and feasible for use in structural vibration control.

Keywords: tuned mass damper; eddy current damping; permanent magnets; conductive plate; modal damping ratio

1 Introduction

In recent years, considerable attention has been given to research and development of structural control systems, with particular emphasis on suppressing the dynamic response of structures subjected to strong winds and earthquake excitations. Soong and Spencer (2002) have grouped structural control systems into three broad areas, namely, base isolation, passive energy dissipation, and active/hybrid/semi-active control systems. In particular, passive supplemental damping strategies, including viscoelastic dampers, viscous dampers, and tuned mass dampers (TMD), are well understood and widely accepted by the engineering community as a means for mitigating the effects of dynamic loading on structures (Spencer and Nagarajaiah, 2003).

Early applications of TMDs have been adopted to mitigate the stationary response of structures, such as wind-induced vibration (Gu *et al.*, 1998; Gu *et al.*, 2001; Xu *et al.*, 1992; Brownjohn *et al.*, 2010), human-induced vibration on footbridges (Dallard *et al.*, 2001; Wilkinson and Knapton, 2006; Chen and Liu, 2009) as well as building floors (Setareh and Hanson, 1992; Setareh *et al.*, 2007). There is also research related to the effectiveness of TMDs in reducing structural response subjected to seismic excitations (Li, 2000; Li and Zhu, 2006; Li and Qu, 2006; Marano *et al.*, 2007; Hoang *et al.*, 2008; Lin *et al.*, 2010) and impact loads (Abe and Igusa, 1996; Wang *et al.*, 2002).

One typical TMD comprises a lumped mass, a spring element and a damping element, which can form a single degree-of-freedom system with a specific resonant frequency tuned to the main structure and a target damping ratio. Traditionally, damping components in a TMD are provided by viscous dampers. However, viscous fluid may leak over time, which is the main concern for this type of damper. Furthermore, it is hard or expensive to change the damping ratio of TMD after the viscous dampers are already equipped in a TMD.

As a non-contacting damping strategy, eddy current damping (ECD) is one possible alternative solution to this problem. Eddy currents (also known as Foucault currents) can be generated when a conductor is exposed to a varying magnetic field. These circulating currents

Correspondence to: Chen Zhengqing, Wind Engineering Research Center, Hunan Univ., Changsha 410082, China
Tel: +86 731 88821424; Fax: +86 731 88823923
E-mail: zqchen@hnu.cn
†Assistant Professor; ‡Professor; §Associate Professor

Supported by: State Key Program of Natural Science Foundation of China Under Grant No. 50738002
Received November 27, 2011; **Accepted** June 4, 2012

can be induced either by relative motion between the conductor and the source of an external magnetic field or by varying the strength of the magnetic field around the conductor, which are called motional eddy currents and transformer eddy currents, respectively (Ebrahimi *et al.*, 2009a and Graves *et al.*, 2000). They create electromagnets with magnetic fields that oppose the change in an external magnetic field, causing a repulsive force proportional to the relative velocity between the magnetic field and conductor and making the ECD act the same way as viscous damping.

Sodano and Bae (2004) have thoroughly reviewed the research into various types of ECD mechanisms. It is found that three main research areas are eddy current braking, magnetic damping of rotor vibration, and ECD for suppressing structural vibration. Among them, eddy current braking is the most widely developed in both theoretical formulation and practical application. Mitigating the structural vibration with ECD is another research field that also has received significant attention.

Bae *et al.* (2004) have developed one simple ECD system to control the vibration of a cantilever beam, and good agreement between theoretical predictions and test results was observed. In a similar research, Sodano *et al.* (2005) have theoretically analyzed and experimentally investigated the control performance of a cantilever beam with ECD, where a cylindrical permanent magnet (PM), magnetized in the same direction as the beam motion, is fixed with a distance away from the tip of the beam, and a copper plate is also attached to the beam tip. The results show that the proposed ECD can increase the damping ratio of the beam by 150 times and provide sufficient damping force to quickly suppress vibrations of the beam. Later, Sodano *et al.* (2006) put another cylindrical PM on the other side of the beam tip so that the magnetic field around the conductor plate could be strengthened, and applied an image method to consider the boundary condition of zero eddy current density, which was assumed to be ignored in their earlier research. Recently, Sodano and Inman (2008) have continued and extended their passive ECD to active ECD. In this configuration, they employed an electromagnet with an eternal power supply to replace the commonly used PM. Consequently, an adjustable magnetic field around the conductor can be achieved by modifying the current flowing through the coils, thus the damping force due to generated eddy currents can be actively controlled.

In addition to the direct application of ECD, several studies have also focused on developing TMDs with an ECD mechanism. Kienholtz *et al.* (1994) developed several eddy current TMDs to suppress different vibration modes of a solar array simulator in the frequency range of 0.1–1 Hz. Similarly, one small scale TMD with adjustable ECD is constructed to control the wind-induced vibration of a full-bridge aerodynamic model in the wind tunnel test by Larose *et al.* (1995).

However, in this previous research, the performance of TMDs with eddy currents were mostly evaluated or tested using a small scale model, thus the performance of a larger TMD with eddy current in the practical implementation is still not reported. Therefore, the main focus of this paper is the feasibility study of a large-scale eddy current TMD with both theoretical analysis and experimental investigation.

In this paper, the theoretical formulation of the ECD in a horizontal TMD is derived first, which demonstrates that the ECD behaves like a viscous damper, followed by a numerical calculation of the magnetic flux due to rectangular PMs. A large-scale horizontal TMD with ECD is then constructed and experimentally investigated. Finally, possible improvement and feasibility considerations of the new TMD are discussed.

2 Modeling of eddy current damping

2.1 Eddy currents model

Figure 1 depicts the configuration of the ECD system adopted in the TMD, which consists of two fixed conductive plates and eight rectangular PMs that can move simultaneously with the mass components of TMD. Due to the relative movement of the PMs and conductive plates, the eddy currents are induced. If the surface charges in the conductive plate are assumed to be ignored, then the eddy currents in the conductive plate can be formulated as

$$\boldsymbol{J} = \sigma(\boldsymbol{v} \times \boldsymbol{B}) \tag{1}$$

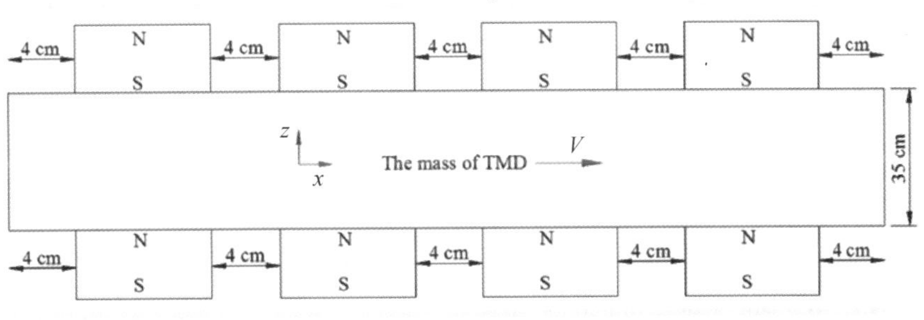

Fig. 1 Configuration of eddy current damping in a horizontal TMD

where σ is electrical conductivity of the conductive plate, $\boldsymbol{v}$ is the relative velocity between the PMs and conductive plates, and $\boldsymbol{B}$ is the magnetic flux density due to the PMs. $\boldsymbol{v}$ and $\boldsymbol{B}$ can be further expressed as follows:

$$\boldsymbol{v} = v_x\boldsymbol{i} + 0\boldsymbol{j} + 0\boldsymbol{k} \tag{2}$$

$$\boldsymbol{B} = B_x\boldsymbol{i} + B_y\boldsymbol{j} + B_z\boldsymbol{k} \tag{3}$$

where the velocity $\boldsymbol{v}$ is only in the x direction and the magnetic flux density $\boldsymbol{B}$ is spatially distributed.

Substituting Eqs. (2) and (3) into Eq. (1), the induced eddy currents can be obtained as

$$\boldsymbol{J} = \sigma v_x(-B_z\boldsymbol{j} + B_y\boldsymbol{k}) \tag{4}$$

The above equation shows that B_x representing x directional component of the magnetic field has no effect on the induced eddy current, and only the components of magnetic flux density in the y and z directions need to be considered.

Once the eddy currents are generated, they circulate in such a way that they induce a magnetic field with opposite polarity as the applied field, causing a repulsive force (Sodano and Bae, 2004). This electromagnetic force acting on the conductive plate can be written as

$$F = \int_V \boldsymbol{J} \times \boldsymbol{B}\mathrm{d}V \tag{5}$$

Substitution of Eqs. (3) and (4) into Eq. (5) allows the electromagnetic force to be further calculated as

$$F = \int_V \boldsymbol{J} \times \boldsymbol{B}\mathrm{d}V = \sigma v_x \int_V ((-B_z^2 - B_y^2)\boldsymbol{i} + (B_y B_x)\boldsymbol{j} + (B_z B_x)\boldsymbol{k})\mathrm{d}V \tag{6}$$

Hence, the damping force due to the eddy currents is given by:

$$F = -\sigma v_x \int_V (B_z^2 + B_y^2)\mathrm{d}V \tag{7}$$

It can be seen that the induced damping force is proportional to the relative velocity between the PMs and conductive plates, and the negative sign means that the force and velocity has exactly the opposite direction. Moreover, the damping force is only dependent on the components of magnetic flux density in the y and z directions, which shares the same characteristics with the induced eddy currents. As such, the PMs should be magnetized in the y or z directions to achieve larger damping force for a given size of damping device.

Finally, a linear damping coefficient can be defined as follows:

$$C = \sigma \int_V (B_z^2 + B_y^2)\mathrm{d}V \tag{8}$$

Assuming the mass in the proposed TMD is denoted as m, and the natural frequency of the TMD is ω, the modal damping ratio of TMD can be determined as

$$\xi = \frac{C}{2m\omega} \tag{9}$$

2.2 Magnetic field distribution of rectangular PMs

Before the ECD is theoretically calculated with the formulas presented in the previous section, the magnetic field distribution due to the rectangular PMs should be derived first. Based on the Biot-Savart Law, the analytical expressions for calculating the magnetic field distribution of rectangular PMs magnetized in one direction have been derived by Gou *et al.* (2004). For the completeness of this research, the formulas developed by Gou *et al.* (2004) are briefly given here, where the coordinate definition of the rectangular magnet is shown in Fig. 2. The magnetic flux density components due to a PM can be calculated as:

$$B_x = \int_0^h \mathrm{d}B_x = -\frac{K}{2}[\Gamma(a-x,y,z) + \Gamma(a-x,b-y,z) - \Gamma(x,y,z) - \Gamma(x,b-y,z)]\Big|_0^h \tag{10}$$

$$B_y = \int_0^h \mathrm{d}B_y = -\frac{K}{2}[\Gamma(b-y,x,z) + \Gamma(b-y,a-x,z) - \Gamma(y,x,z) - \Gamma(y,a-x,z)]\Big|_0^h \tag{11}$$

$$B_z = \int_0^h \mathrm{d}B_z = -K[\Phi(y,a-x,z) + \Phi(b-y,a-x,z) + \Phi(x,b-y,z) + \Phi(a-x,b-y,z) + \Phi(b-y,x,z)] + \Phi(y,x,z) + \Phi(a-x,y,z) + \Phi(x,y,z)]\Big|_0^h \tag{12}$$

where K, $\Gamma(\gamma_1,\gamma_2,\gamma_3)$ and $\Phi(\varphi_1,\varphi_2,\varphi_3)$ are defined as

$$K = \frac{\mu_0 J_{\mathrm{M}}}{4\pi} \tag{13}$$

$$\Gamma(\gamma_1,\gamma_2,\gamma_3) = \ln\frac{\sqrt{\gamma_1^2 + \gamma_2^2 + (\gamma_3 - z_0)^2} - \gamma_2}{\sqrt{\gamma_1^2 + \gamma_2^2 + (\gamma_3 - z_0)^2} + \gamma_2} \tag{14}$$

$$\Phi(\varphi_1,\varphi_2,\varphi_3) = \begin{cases} \arctan\left[\dfrac{\varphi_1}{\varphi_2}\dfrac{\varphi_3 - z_0}{\sqrt{\varphi_1^2 + \varphi_2^2 + (\varphi_3 - z_0)^2}}\right], & y \neq 0 \\ 0 & y = 0 \end{cases} \tag{15}$$

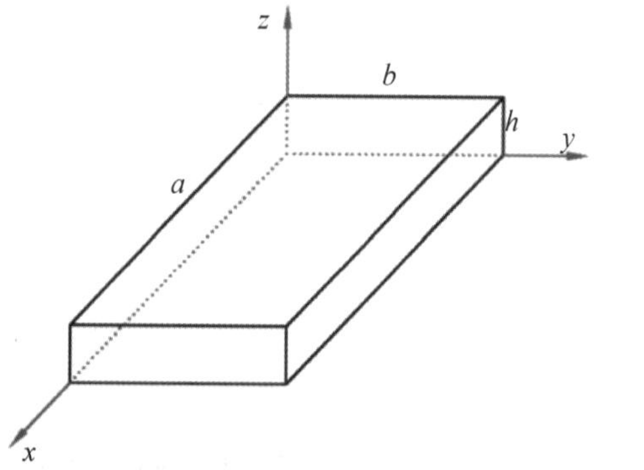

Fig. 2 Coordinate definition of a rectangular permanent magnet

where $[\bullet]|_0^h$ denotes the subtraction between the value of the function $[\bullet]$ at $z_0=h$ and $z_0=0$, $\mu_0 = 4\pi\times10^{-7}$ is the magnetic permeability in the air, and J_M represents the equivalent current density due to the PM.

From the above formulations, it can be seen that the magnetic flux density at any spatial location can be obtained with only one parameter J_M or K. To simplify the solution process, B_z at point $P(a/2, b/2, z)$ is measured to obtain J_M with the following formula:

$$J_M = -\frac{B_z(a/2, b/2, z)\pi}{\mu_0\left[\Phi(b/2, a/2, z)+\Phi(a/2, b/2, z)\right]\Big|_0^h} \tag{16}$$

The rare earth PMs made from alloys of rare earth elements, specifically NdFeB grade N35, were chosen due to their high remanent magnetic flux density (1.2 T), coercivity (8.7×10^5 Am^{-1}) , and the density of magnetic energy (2.8×10^5 Jm^{-3}) as well as availability in this research. The NdFeB PMs with a commercial size of 100 mm×100 mm×50 mm are adopted to reduce the manufacturing cost, where the height with 50 mm is the magnetized direction. The value of B_z at $P(a/2, b/2, 0)$ is 0.4 T measured by one magnetic flux density sensor, and then J_M is further determined as 9.55×10^5 A/m^2 according to Eq. (16).

Since Eqs. (10) – (12) can not be analytically integrated, the numerical integration is adopted with MATLAB software. The magnetic field density distribution in both the z and y directions at y = 0.25b 0-0.25b with different air gaps between the conductor and PM is shown in Fig. 3. Since B_x has no effect on the induced eddy current damping force in this research, its distribution is not given. From Fig. 3, it can be concluded that B_z with the same direction as the magnetization of PMs is the dominant magnetic flux density, which should also have a significant effect on the induced ECD.

To further evaluate the accuracy of the above analytical formulas, the results are compared with those obtained by the three-dimensional bar magnet calculator from Integrated Engineering Software. Table 1 lists these comparisons, where "AS" means analytical solution and "FEM" represents the finite element method. It can be seen that the two results are quite consistent with each other. Therefore, the precision of the analytical formulas are adequately confirmed, and are directly adopted to calculate the ECD force in the subsequent section.

2.3 Numerical results of eddy current damping

Since the damping force due to ECD is difficult to measure directly, the modal damping ratio of the TMD that can be easily identified from the free vibration test is taken as the evaluation value. For all the cases, the mass of TMD m is equal to 550 kg, and natural frequency ω is 5.84 rad/s. Due to high electrical conductivity of the aluminum and copper, they are chosen as the materials of the conductive plates, and their electrical conductivities are 3.6×10^7 $Sm^{-1}$and 5.6×10^7 Sm^{-1}, respectively.

On the basis of the results discussed in the above two sections, the modal damping ratio of the TMD configured as shown in Fig.1 can be theoretically predicted, and the corresponding results are shown in Fig. 4. Note that there are two curves for each conductive material: one is calculated with a finite conductor, and the other with an infinite conductor. As expected, the latter values are always larger than the former values, and ECD employing a copper plate can induce a larger damping ratio than with an aluminum plate with the same magnet gap. Figure 4 also illustrates that damping ratios drop quickly as the magnet gap increases, and when the magnet gap approaches the height of the magnet, the two curves for the same conductive material have the tendency to coincide with each other. Moreover, it can be seen that the predicted damping ratio can reach as much as 50% when the copper plate is quite close to the surface of the PM. This value is actually much larger than that needed by the traditional TMD. However, it provides two theoretical implications: the proposed TMD may be able to drive a larger scale TMD, and the damping ratio of the TMD can be varied in a border area. In summary, the design flowchart for the damping components of a TMD based on ECD is given in Fig. 5, which is a semi-analytical and semi-empirical method.

3 Performance tests of TMD

3.1 TMD development

To avoid the friction damping that is commonly experienced in the TMD with springs and guiding devices, four identical cantilever beams are adopted as the stiffness components of the proposed horizontal TMD, whose size is designed based on the principle of the same stress level along its length. As a result, the height of the beam section is gradually reduced with the same width along the length of the beam. Finally, the mass of the TMD is connected to the tip of the cantilever beams by four universal joint bearings.

To compare the damping characteristics with different electrical conductivity of conductors, a TMD equipped with two types of conductive plates is constructed and shown in Fig. 6. One accelerator is set on the top of the mass of the TMD to obtain the free vibration response of the TMD, and then the modal damping ratio can be identified with a logarithmic decrement of the test data. During the tests, the mass in the TMD is released from the same location every time.

3.2 Experimental results

Figure 7 shows several free response time histories of TMD, where the air gap is 5 mm as shown in both Fig. 7(b) and Fig. 7(c). From Fig.7, it can be seen that on one hand the intrinsic damping of TMD is quite small, which is identified as about 0.5 %, and on the other hand,

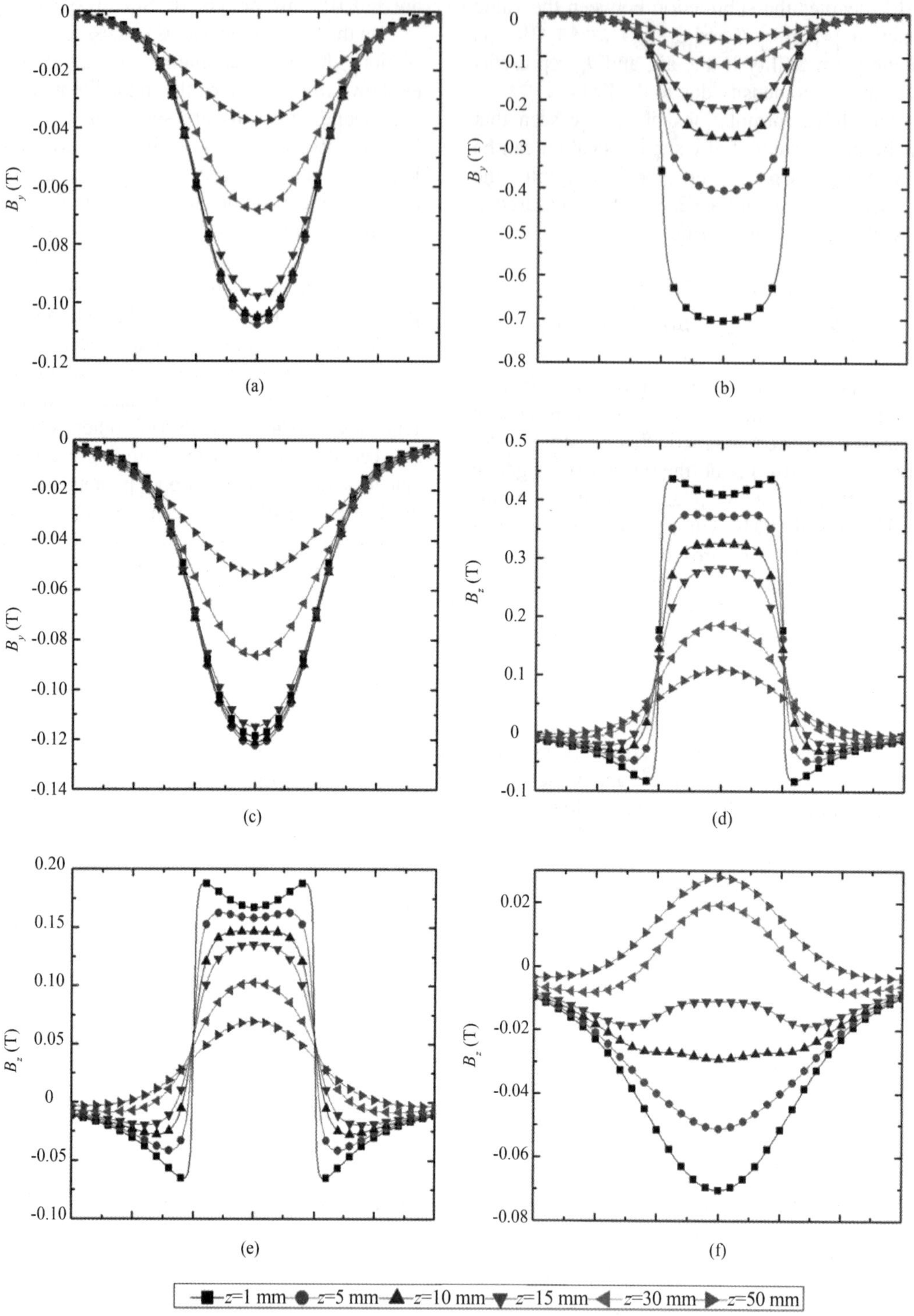

Fig. 3 Magnetic field distribution of a permanent magnet with analytical formulas. (a)-(c): B_y at $y=b/4, 0, -b/4$; (d)-(f) B_z at $y=b/4, 0, -b/4$

the ECD with both copper and aluminum plates can significantly increase the damping ratios of TMD.

Table 2 further gives the comparisons between the predicted and identified experimental results, where the experimental values have removed the contribution from the intrinsic structural damping ratio, which is identified as about 0.5% according to Fig. 7(a), and the predicted damping ratio is estimated with a finite conductor since an actual configuration of TMD is closer to a finite conductor. From Table 2, it is found that the experimental damping ratios are always about 50% smaller than the predicted damping ratios. Note that there is a similar error in Sodano *et al.* (2005). This error is mainly due to the following three factors. First, the edge effect of eddy

Table 1 Comparisons of the components of magnetic flux density due to a rectangular permanent magnet between the analytical formula and the FE analysis

x	y	z	B_x(T)	B_y(T)	B_z(T)	
0.5a	0.5b	h+1mm	0	0	0.393	AS
			0	0	0.394	FEM
0.55a	b	1.1h	0.010	0.406	0.159	AS
			0.010	0.406	0.159	FEM
0.55a	b	1.2h	0.010	0.284	0.147	AS
			0.010	0.284	0.147	FEM

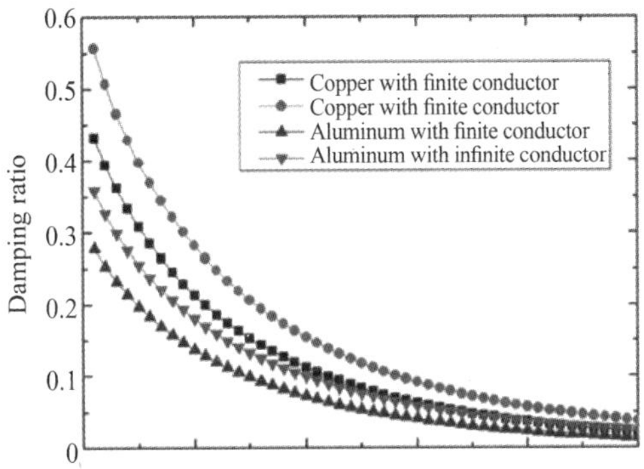

Fig. 4 Predicted eddy current damping ratios of TMD

currents on the conductive plate is neglected; second, the surface charge in the conductive plate is assumed to be ignored; and lastly, the steel mass attached behind the PMs is not considered in the theoretical analysis of magnetic field distribution. In general, the first and last factors can lead to overestimation of the magnetic flux density around the conductor, while the second factor can lead to overestimation of the eddy current in the conductor. In other words, all three factors contribute to overestimating the ECD.

Nonetheless, the predicted model can provide good directions for preliminary design of the TMD employing ECD. Furthermore, since the damping ratio can be easily adjusted by varying the magnet gap between the PMs and conductive plates, a target damping ratio can be realized

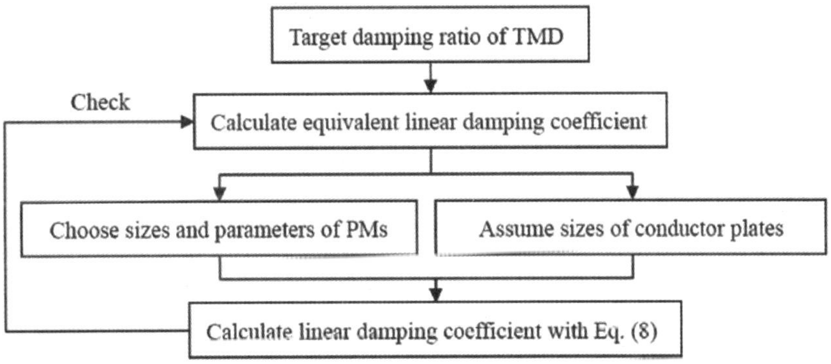

Fig. 5 Design flowchart of TMD based on eddy current damping

(a) No conductive plate (b) With pure copper plate (c) With aluminium plate

Fig. 6 Proposed TMDs with different conductive plates

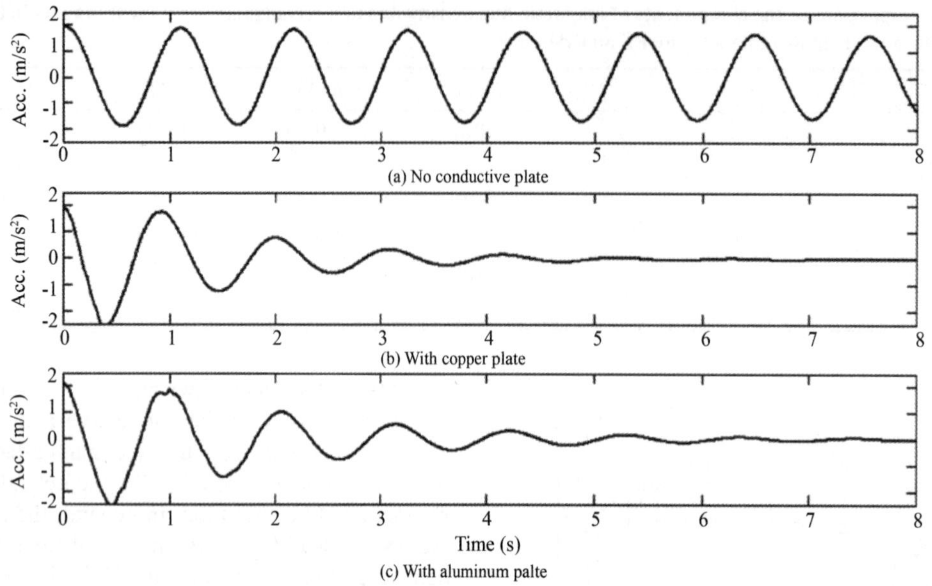

Fig. 7 Free vibration time histories of the TMD's acceleration

Table 2 Comparisons of experimental and predicted eddy-current damping ratios

Conductive plate	Magnet gap (mm)	Test values (%)	Predicted value (%)
Copper	5	14.5	30.8
	15	6.48	15.3
Aluminum	5	9.90	19.8
	15	4.30	9.8

as long as the maximum damping ratio that eddy current currents can achieve is beyond the desired value.

4 Feasibility considerations

4.1 Magnetic circuit design

The damping coefficient of ECD is shown to be proportional to the square of magnetic flux density. If the magnetic flux density can be strengthened, a larger damping ratio or smaller configuration size of the eddy current damping device can be achieved. The magnetic flux density is largely dependent on the PMs used to generate the field, the arrangement of PMs, and the design of the magnetic circuit. In general, there are three methods: more powerful PMs, optimum layout of PMs and another extra plate with high permeability behind each of the conductive plates. The effectiveness of the

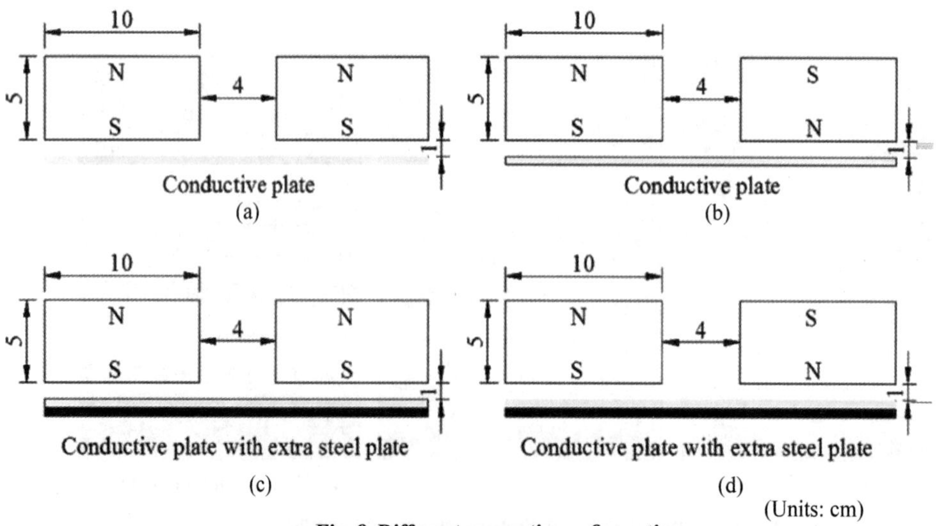

Fig. 8 Different magnetic configurations

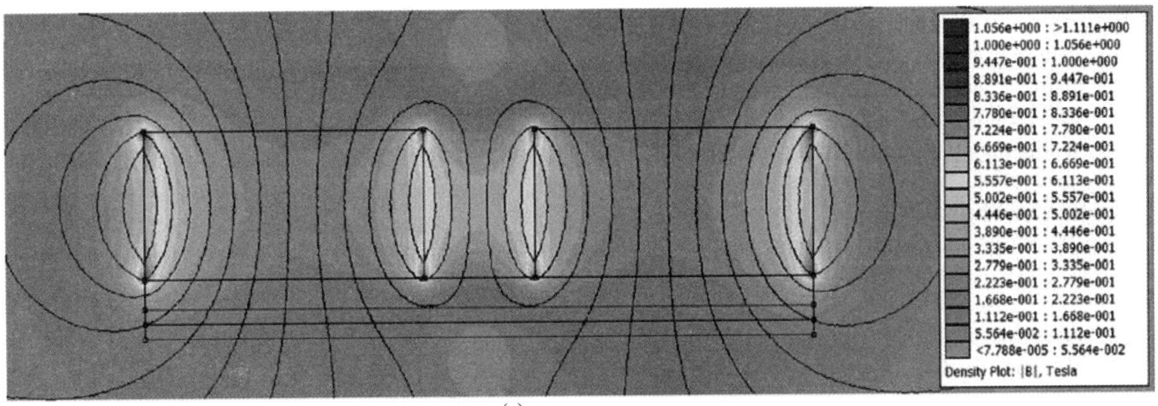

(a)

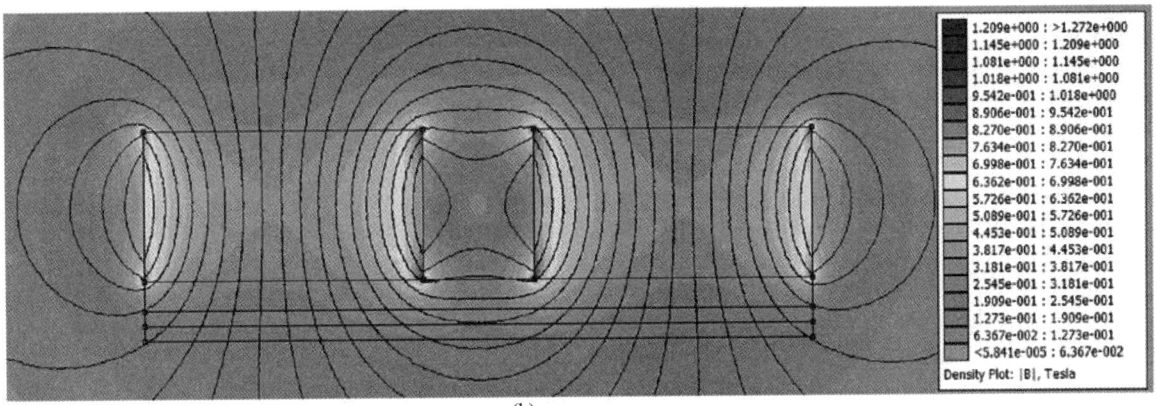

(b)

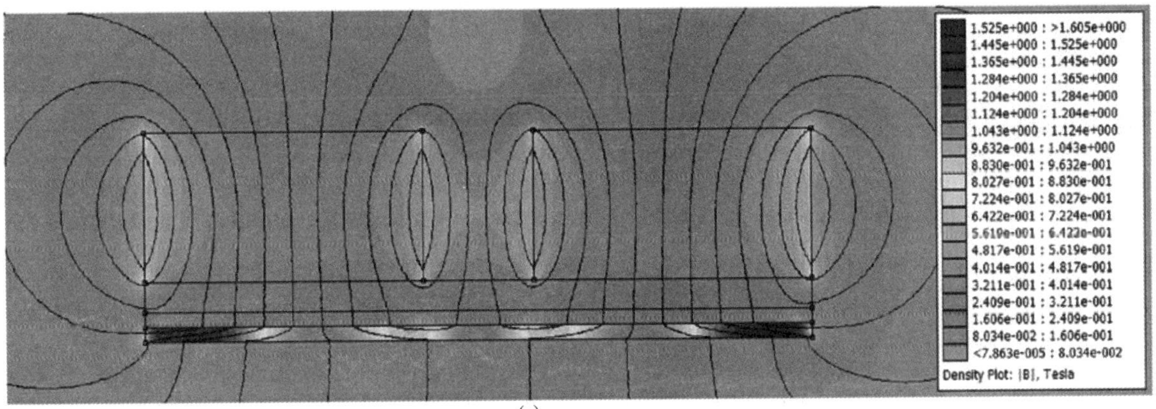

(c)

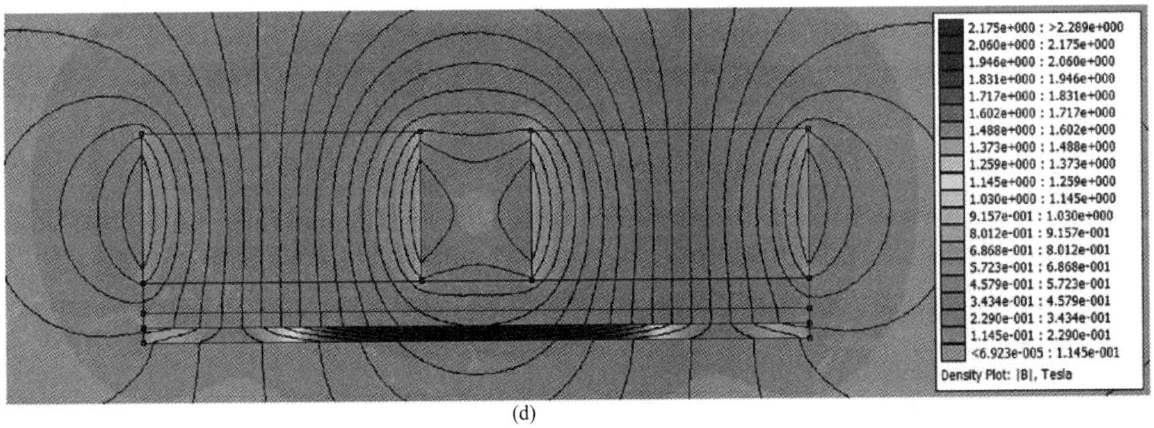

(d)

Fig. 9 Magnetic flux density with different magnetic configurations

first way is obvious at the cost of higher price. The improvements of the other two methods are demonstrated by finite element analysis, where the FEMM software is quite suitable for the two-dimensional planar problem adopted in this paper. To simplify the analysis without loss of generality, only two PMs are considered. The four cases illustrated in Fig. 8 are modeled and analyzed. The sizes and magnetic parameters of the PMs are the same as described in the previous section. High magnetically permeable Steel 1018 plates are attached behind the copper plate as shown in Fig. 7(c) and 7(d). Both the thickness of the copper plate and steel plate is 0.5 cm, and the magnet gap between the copper plate and magnets is set as 1 cm.

Figure 9 shows four diagrams of the magnetic flux density obtained with FEMM, and the calculation results displayed in Fig. 9(a)–(d) correspond to the configurations shown in Fig. 8(a)–(d), respectively. From Fig. 9, two main conclusions can be drawn. First, as expected, the alternative configuration of magnet poles is helpful to form a smaller magnetic circuit, while the adjacent magnets work independently when the magnetic poles of the two magnets are in the same direction. Second, extra steel plates can effectively pull the magnetic flux away from the magnet stack and further increase the magnetic flux density around the copper plates.

Figure 10 further reflects the variation of the magnetic flux density components in the magnets' magnetization direction along the two boundary lines of the copper plate, where Fig. 10(a) shows the density on the interior boundary line, and Fig. 10(b) shows the density on the exterior line. In the figure, cases 1–4 represent the ECD configurations depicted in Fig. 8(a)-(d), respectively. Note that the left magnet falls in the range of x (0–0.1m), while the second magnet corresponds to x (0.14–0.24m). It can be seen that B_z for all the cases is symmetrical along the axis of x=0.12, which is exactly at the middle of the two magnets' clearance. In the area of PMs, it is found that Case 4 has the largest envelop of B_z, followed by Case 3, Case 2 and Case 1, while in the area of gap between the magnets, the order of envelop is Case 3, case 1 and Case 2 or 4. It is also noted that in Case 3 and Case 4, there is almost no difference between the interior and exterior line for the B_z, while a significant difference can be seen from Case 1 and Case 2, especially Case 1. In conclusion, it is observed that the use of extra steel plates with the alternating magnetic pole configurations of magnets is the optimum topology of the ECD, which implies that a significant improvement of the ECD developed in the former section can be achieved with the optimized configuration.

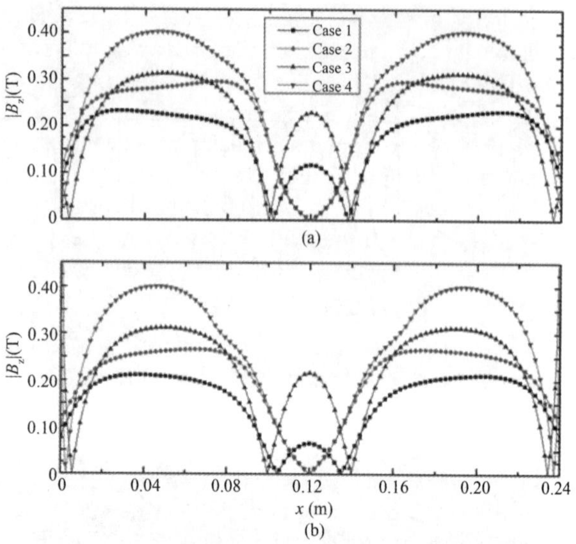

Fig. 10 Comparisons of different magnetic configurations

4.2 Durability and demagnetization effect of PMs

NdFeB is the most recently commercialized magnet material since the 1980s, and it is known as a rare earth magnet because it is composed of materials from the lanthanide series within the rare earth group of elements (Priya and Inman, 2009). Although the NdFeB can achieve the maximum energy product, this material suffers from poor corrosion resistance. However, exterior surface treatments such as nickel, zinc or tin plating and epoxy resin coatings can provide useful protection from corrosion. Moreover, NdFeB is extremely brittle, so it should not be subjected to any impact load during the assembly of the ECD.

Additionally, another concern about the rare earth PMs is a demagnetization effect where the magnetic property may reduce or degrade after a long time. Generally, factors that affect the magnet stability include time, temperature and adverse fields (Ebrahimi *et al.*, 2009b). With modern rare earth PMs, the time effect is quite small because of its high coercivity. Since the PMs are not in as compact a configuration as the PM motor, the induced heat can be easily dissipated. Consequently, its temperature should not be too high to influence the performance of PMs. An external magnetic field that is opposite to that generated by the PMs also has adverse effects. Two possible sources of such a field are induced eddy currents and adjacent magnets. The fundamental frequency of a typical civil structure is around 1 Hz or below, the same as the proposed TMD for a long-span footbridge as described in this paper, which should not have much influence on the strong magnetic flux density generated from the PMs, since as shown in Eq. (4), the induced eddy current is proportional to the relative velocity. On the other hand, with proper distance between the magnets and optimum magnetic loop, the second factor is also not a problem. In conclusion, the durability and reliability of the proposed TMD with ECD is confirmed.

4.3 Economic analysis

The feasibility of any product with new technologies for practical application is first restricted by its cost. Both raw materials and production contribute to the cost of the TMD, but for this product, the rare earth PMs are the dominant cost. In fact, the cost of the proposed large-scale TMD with ECD is mainly related to the high price of NdFeB PM, which is also the most significant cost of the passive electromagnetic damping developed by Palomera-Arias *et al.* (2008). However, as also pointed out by Palomera-Arias *et al.* (2008), it is believed that the cost of PMs will hopefully be significantly reduced with the development of PM technology, making the ECD a cost-effective solution for structural vibration control.

5 Conclusions

To eliminate the low durability problem of classical TMDs, this study has provided both the theoretical formulation and experimental evaluation of a large-scale TMD employing ECD. In general, the advantages of ECD used in TMDs include: (1) it is simple and mechanically robust; (2) it can be constructed entirely of metallic materials, thus avoiding concerns about deterioration with age; and (3) it creates a non-contacting device without any friction. Compared with TMDs based on viscous dampers, TMDs based on ECD have shown excellent performance in several areas. First, no fluids or seals are needed, and thus problems such as oil leakage are naturally avoided; second, the damping ratios can be easily adjusted without extra cost; and third, no extra stiffness will be induced from ECD at any working frequency of TMD (its working frequency can be quite wide, ranging from 0.1 Hz to over 10 Hz), while viscous dampers generate supplemental stiffness at 3 Hz or above, which leads to detuning of the TMD that can dramatically reduce control performance.

A theoretical model for estimating damping ratios due to eddy currents is proposed first to provide the preliminary design of the TMD. On this basis, one large-scale TMD is then constructed. Both the analytical study and test results show that the ECD is quite promising for TMD usage. The ECD ratio can achieve the desired value for structural vibration control; and it also can be easily adjusted using several different methods, such as changing the conductive plate with different conductivity, or varying the air gap between the conductors and PMs. Based on the theoretical study and experimental results, a feasibility analysis is conducted to illustrate the durability and economic viability of the proposed eddy-current TMD. Moreover, achievable strategies aimed at improving the ECD design are given in detail.

However, a more refined and improved analytical model is still needed to accurately predict the damping coefficient of ECD. Furthermore, the dynamic response of the proposed TMD subjected to forced vibration is another issue that needs further investigation. Moreover, the control performance of the proposed TMD on scaled or full-scale structures needs to be further confirmed by laboratory tests and/or field investigations. All the above issues will be addressed by the authors in future work.

Acknowledgement

The authors wish to express great thanks for the partial support received from Natural Science Foundation of China under Grant number 50738002.

Reference

Abe M and Igusa T (1996), "Semi-active Dynamic Vibration Absorbers for Controlling Transient Response," *Journal of Sound and Vibration*, **198**: 547–569.

Bae JS, Kwak MK and Inman DJ (2004), "Vibration Suppression of Cantilever Beam Using Eddy Current Damper," *Journal of Sound and Vibration*, **284**: 805–824.

Brownjohn JMW, Carden EP, Goddard CR and Oudin G (2010), "Real-time Performance Monitoring of Tuned Mass Damper System for a 183 m Reinforced Concrete Chimney," *Journal of Wind Engineering and Industrial Aerodynamics*, **98**(3): 169–179.

Chen ZQ and Liu GD (2009), "Pedestrian-induced Vibration Theory and Dynamic Design of Footbridges," *Engineering Mechanics*, **26**: 148–159. (In Chinese)

Dallard P, Fitzpatrick AJ, Flint A, Low A and Ridsdill-Smith, RM (2001), "The Millennium Bridge London: Problems and Solutions," *Structure Engineer*, **79**(8): 15–17.

Ebrahimi B, Khamesee MB and Golnaraghi F (2009a), "A Novel Eddy Current Damper: Theory and Experiment," *Journal of Physics D: Applied Physics*, **42**: 075001(1–6).

Ebrahimi B, Khamesee MB and Golnaraghi F (2009b), "Eddy Current Damper Feasibility in Automobile Suspension: Modeling, Simulation and Testing," *Smart Materials and Structures*, **18**: 015017(1–12).

Gou XF, Yang Y and Zheng XJ (2004), "Analytical Expressions of Magnetic Field Distribution of Rectangular Permanent Magnets," *Applied Mathematics and Mechanics*, **25**(3): 297–306.

Graves KE, Toncich D and Iovenitti PG (2000), "Theoretical Comparison of Motional and Transformer EMF Device Damping Efficiency," *Journal of Sound and Vibration*, **233**: 441–453.

Gu M, Chang CC, Wu W and Xiang HF (1998), "Increase of Critical Flutter Wind Speed of Long-span Bridges Using Tuned Mass Dampers," *Journal of Wind Engineering and Industrial Aerodynamics*, **73**(2): 111–123.

Gu M, Chen SR and Chang CC (2001), "Parametric Study on Multiple Tuned Mass Dampers for Buffeting Control of Yangpu Bridge," *Journal of Wind Engineering and Industrial Aerodynamics*, **89**(11): 987–1000.

Hoang N, Fujino Y and Warnitchai P (2008), "Optimal

Tuned Mass Damper for Seismic Applications and Practical Design Formulas," *Engineering Structures*, **30**: 707–715.

http://wwwintegratedsoftcom/papers/benchmark/bar_magnet/calculatorasp.

Kienholtz DA, Pendleton SC, Richards KE and Morgenthaler DR (1994), "Demonstration of Solar Array Vibration Suppression," *Proceedings of SPIE's Conference on Smart Structures and Materials*, Orlando, FL, 2193, 59–72.

Larose GL, Larsen A and Svensson E (1995), "Modeling of Tuned Mass Dampers for Wind-tunnel Tests on a Full-bridge Aeroelastic Model," *Journal of Wind Engineering and Industrial Aerodynamics*, **54/55**: 427–437.

Li CX (2000), "Performance of Multiple Tuned Mass Dampers for Attenuating Undesirable Oscillations of Structures under the Ground Acceleration," *Earthquake Engineering and Structural Dynamics*, **29**: 1405–1421.

Li CX and Qu WL (2006), "Optimum Properties of Multiple Tuned Mass Dampers for Reduction of Translational and Torsional Response of Structures Subject to Ground Acceleration," *Engineering Structures*, **28**(4): 472–494.

Li CX and Zhu BL (2006), "Estimating Double Tuned Mass Dampers for Structures under Ground Acceleration Using a Novel Optimum Criterion," *Journal of Sound and Vibration*, **298**(1-2): 280–297.

Lin CC, Chen CL and Wang JF (2010), "Vibration Control of Structures with Initially Accelerated Passive Tuned Mass Dampers under Near-fault Earthquake Excitation," *Computer-Aided Civil and Infrastructure Engineering*, **25**: 69–75.

Marano GC, Greco R, Trentadue F and Chiaia B (2007), "Constrained Reliability-based Optimization of Linear Tuned Mass Dampers for Seismic Control," *International Journal of Solids and Structures*, **44**(22–23), 7370–7388.

Palomera-Arias R, Connor JJ and Ochsendorf JA (2008), "Feasibility Study of Passive Electromagnetic Damping Systems," *Journal of Structural Engineering*, **134**: 164–170.

Setareh M and Hanson RD (1992), "Tuned Mass Dampers to Control Floor Vibrations from Humans," *Journal of Structural Engineering*, **118**(3): 741–762.

Priya S and Inman DJ (2009), *Energy Harvesting Technologies*, Springer.

Setareh M, Ritchey JK, Murray TM *et al.* (2007), "Semiactive Tuned Mass Damper for Floor Vibration Control," *Journal of Structural Engineering*, **133**(2): 242–250.

Sodano HA and Bae JS (2004), "Eddy Current Damping in Structures," *The Shock and Vibration Digest*, **36**(6): 469–478.

Sodano HA, Bae JS, Inman DJ and Belvin WK (2005), "Concept and Model of Eddy Current Damper for Vibration Suppression of a Beam," *Journal of Sound and Vibration*, **288**(4-5): 1177–1196.

Sodano HA, Bae JS, Inman DJ *et al.* (2006), "Improved Concept and Model of Eddy Current Damper," *Journal of Vibration and Acoustics*, **128**: 294–302.

Sodano HA and Inman DJ (2008) "Modeling of a New Active Eddy Current Vibration Control System," *Journal of Dynamic Systems, Measurement, and Control*, **130**(2): 021009(1–12).

Soong TT and Spencer BF, (2002), "Supplemental Energy Dissipation: State of Art and State of Practice," *Engineering Structures*, **22**: 243–259.

Spencer BF and Nagarajaiah S (2003), "State of the Art of Structural Control," *Journal of Structural Engineering*, **129**(7): 845–856.

Wang SQ, Li HJ, Ji CY and Jiao GY (2002), "Energy Analysis for TMD-structure Systems Subjected to Impact Loading," *China Ocean Engineering*, **16**(3): 301–310.

Wilkinson SM and Knapton J (2006), "Analysis and Solution to Human-induced Lateral Vibrations on a Historic Footbridge," *Journal of Bridge Engineering*, **11**(1): 4–12.

Xu YL, Kwok KCS and Samali B (1992), "Control of Wind-induced Tall Building Vibration by Tuned Mass Dampers," *Journal of Wind Engineering and Industrial Aerodynamics*, **40**(1): 1–32.

14th World Conference on Seismic Isolation,
Energy Dissipation and Active Vibration Control of Structures
September 9-11 2015 San Diego, Ca USA

Inerter-Damper-Spring Passive Vibration Control: Its System and Realization with Eddy-Current Apparent Mass Damper

ZhengQing Chen
Hunan University, Wind Engineering Research Centre,
1 Lushannan Road, 410082 Changsha Hunan, China
zqchen@hnu.edu.cn

ZhiWen Huang
Hunan University, Wind Engineering Research Centre,
1 Lushannan Road, 410082 Changsha Hunan, China
zwhuang213@hnu.edu.cn

XuGang Hua
Hunan University, Wind Engineering Research Centre,
1 Lushannan Road, 410082 Changsha Hunan, China
cexghua@hotmail.com

ABSTRACT - *The idea of structural vibration control by supplying external inertial forces may date back to the 1980s, however, its theoretical formulation is attributed to the work of Prof. Smith at the Cambridge University who explicitly proposed the inertial-force-producing component called the inerter and its potential use for vibration control in mechnical engineering in 2002. The inerter-based technolgy quickly moved to structural engineering and has become the third type of structural vibration control method. With the application of inerter, the traditional damper-spring passive vibration control system has been extended to the inerter-damper-spring(IDS) passive vibration control system which provide simultaneously damping force, elastic force and the inertial forces and shows great promise for structural applications.*

Eddy-current damping has many attractive properties and operational performance. By employing the ball screw mechanism, a novel type of eddy-current apparent mass damper has been developed by the authors. which is capable of producing sufficiently large damping coefficients and apparent mass. In this paper, the basic priciple, mechanical behaviour and experimental performance of a full-scale eddy-current apparent mass damper were reported.

Keywords: *passive vibration control, inerter, eddy-current damping, apparent mass, ball screw*

1 INTRODUCTION

The idea of structural vibration control by supplying external inertial forces can date back to the 1980s, however, its theoretical formula is attributed to the work of Prof. Smith at the Cambridge University who clearly proposed the inertial-force-producing unit called the *inerter* in 2002 [Smith, 2002]. He also illustrated a variety of vibration control schemes based on different combinations of elastic unit, damping unit and inertial unit for potential use in mechanical engineering. The inertial-force-producing unit produces an inertial force in proportion to the relative acceleration of its two ends, working like an extra mass added to the structure, so the constant of the proportionality also calls apparent mass. By employing the inerter, the traditional damper-spring passive vibration control system has been extended to the inerter-damper-spring (IDS) passive vibration control system which provides simultaneously damping force, elastic force and the inertial forces and shows great promise for structural applications.

The IDS passive vibration control theory has been investigated and used in vehicle suspension systems [Smith and Wang, 2004; Chen, *et al.*, 2009]. In 2007 Wang *et al.* [Wang, *et al.*, 2007] discussed the performance benefits of building suspension design employing inerters, offering new possibility for traditional passive structural vibration control. In recent years, the application of inertial force unit on seismic response control of building structures has been investigated in Japan. Kurota *et al.* [Kurota *et al.*, 2000; Ikago and Inoue, 2013] developed a viscous damper called RDT with an apparent mass using a ball screw amplifying mechanism for vibration control in buildings. Based on this device, Saito *et al.* [Saito, *et al.*, 2008] developed the tuned viscous mass damper (TVMD), which is a TMD-like system consisting of a RDT mounted in series to a supporting spring element. Ikago *et al.* [Ikago, *et al.*, 2012; Ikago and Inoue, 2013] verified its effectiveness for seismic control by shaking table test and developed simple design method for TVMD control system. Besides the RDT-based IDS passive vibration control systems, Matsuoka *et al.* [Matsuoka, *et al.*, 2006] developed an electromagnetic damper using apparent mass and a generator employing a ball screw mechanism. Recently, shaking table test and numerical simulations were conducted to demonstrate its seismic control performance [Nakamura, *et al.*, 2014].

In this paper, we develop a novel vibration control device, eddy current apparent mass damper (ECAMD), for IDS control system. Eddy current damping has been used in place of traditional fluid damping device. By employing a ball screw amplifying mechanism, this damper is capable of generating sufficiently large eddy current damping force and inertial force that are suitable for structural control. The ECAMD with ideal bi-linear constitutive model of force-speed relationship can be manufactured to meet various vibration control requirements. On one hand, the first part of the damping force-speed curve is steep, so it functions to suppress the vibration induced by vehicles and wind. On the other hand, the second part is quite flat to meet the seismic control requirement. Moreover, the ECAMD is a full-metal device without viscous fluid, so it requires little maintenance and has longer fatigue life. Performance tests were conducted on two types of ECAMD, disk-type ECAMD and cup-type ECAMD, to verify their work principle and industrial applicability.

2 MECHANICAL BEHAVIOUR OF THE ECAMD

Two types of ECAMD were developed by the authors. One is the disk-type ECAMD, and the other is the cup-type ECAMD. In both types, the ball screw amplifying mechanism is used to convert the translational movement of damper rods to rotation of mechanical components insider the dampers. The disk-type ECAMD produces eddy current damping and inertial force by rotating a conductive disk in the magnetic field created by the permanent magnets, while the cup-type ECAMD produced eddy damping and inertial force by rotating the cylindrical metal cup. Their configurations and work principles will be discussed in the next.

2.1 *Configuration and work principle of the disk-type ECAMD*

The configuration of a disk-type ECAMD is shown in Fig.1. The disk-type ECAMD is composed of a ball screw, a ball nut, a conductive disk, several pairs of permanent magnet, two iron backing and several columns. When a relative motion is given between the both rod ends, the ball nut and the conductive disk connected to it are rotated by the ball screw. The conductive disk rotates through the magnetic fluxes, so that the eddy currents are generated in the conductive disk and interact with the applied magnetic field, producing a damping torque that is proportional to the angular speed to resist the motion. At the same time, inertial force torque is also produced by the rotating disk. Utilizing the ball screw amplifying mechanism, the damping force torque and the inertial force torque are converted to a large axial force applied on the ball screw and the structural element attached to it.

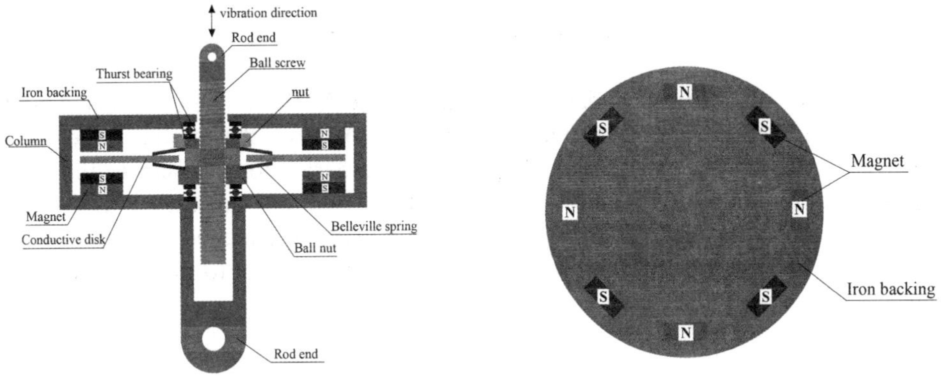

(a) (b)

Figure.1 Schematic representation of the Disk-type ECAMD

Similar with the EIMD proposed by Nakamura *et al.* [Nakamura, *et al.*, 2014], the total axial force generated by the disk-type ECAMD is given by the sum of eddy current damping force, inertial force and friction force, which is usually negligible,

$$F = m_e \ddot{u} + c_e \dot{u} + f_0 \cdot \mathrm{sign}(\dot{u}). \tag{1}$$

where,

$$m_e = \frac{1}{\eta}\left(\frac{2\pi}{L}\right)^2 I_f. \tag{2}$$

$$c_e = \frac{1}{\eta}\left(\frac{2\pi r}{L}\right)^2 C. \tag{3}$$

where m_e and c_e denotes the equivalent inertial mass and equivalent damping coefficient of the disk-type ECAMD, respectively. u is the axial displacement of the damper, f_0 is the friction force, η is the rotary efficiency of the ball screw, L is the ball screw lead, r is the distance between the magnet and the center of the ball screw, I_f is the moment of inertia of the rotating components, C is the damping coefficient calculated from the eddy current damping force and tangential speed of the conductive disk. In preliminary design the parameter C can be estimated as follows,

$$C = \frac{n(\eta_b B)^2 t}{8\rho} \bullet \frac{(r+a)^4 - r^4}{r^2} \bullet \frac{b}{r+a}. \tag{4}$$

where n is the number of magnet pair, B is the magnetic flux density at the surface of the conductive disk, t is the thickness of the conductive disk, a and b is the width and length of the magnet, respectively, ρ is the resistivity of the conductive plate, η_b is the dimensionless modification coefficient for B, which can be obtained from experiments or electromagnetic finite element simulation.

It is apparent from Eq.4 that the magnitude of magnetic flux density has a great influence on the equivalent damping coefficient. According to the electromagnetic theory, the magnetic flux density in the conductive plate is mainly influenced by the magnetic circuit, where the conductive plate is working. So the magnetic circuit should be well designed to improve the equivalent damping coefficient of the damper. In the disk-type ECAMD, the iron backing and column are made from iron in order to reduce the reluctance of the magnetic circuit and increase the magnetic flux density in the conductive plate. In addition, the magnetic flux density drops quickly as the air gap between the magnets increases. Therefore, it is reasonable to keep the air gap as small as possible.

From the Eq.1 the mechanical model of the disk-type ECAMD can be established as shown in Fig.2(a), in which the friction force is neglected. The Fig.2(b) shows the inertial force produced by the apparent mass cause negative stiffness. Compared with the traditional viscous fluid damper, the disk-type ECAMD produces larger and larger axial force without saturation as its speed and acceleration increases, so the device itself or the structural element attached to it may be damaged. So it is necessary to implement force-limiting component in the device to meet the requirement of seismic control.

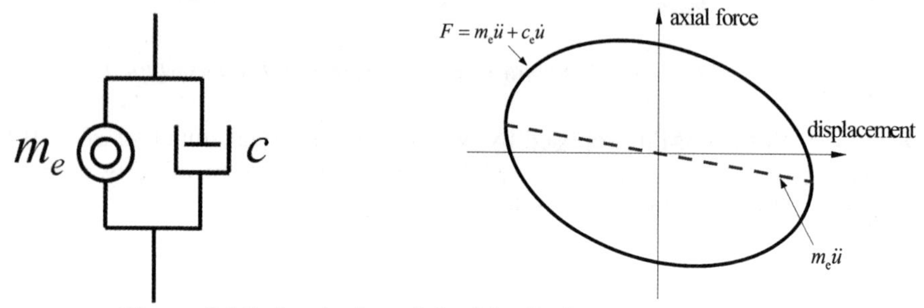

Figure.2 Mechanical model of the Disk-type ECAMD

2.2 *Disk-type ECAMD with force-limiting component*

Fig.3 shows the configuration of the disk-type ECAMD with force-limiting component (FR-D-ECAMD). To limit the maximum axial force, two Belleville springs are used to deliver the damping force torque and inertial force torque of the conductive disk to the ball nut by the friction force. The magnitude of the friction force can be adjusted by the nut at the top of the Belleville springs.

The force-limiting mechanism take effect only if the speed or acceleration of the ball screw is so large that the total torque produced by the nominal damping force and inertial force exceeds the friction force torque on the conductive disk. At this condition, there is a relative rotation between the Belleville springs and the conductive disk.

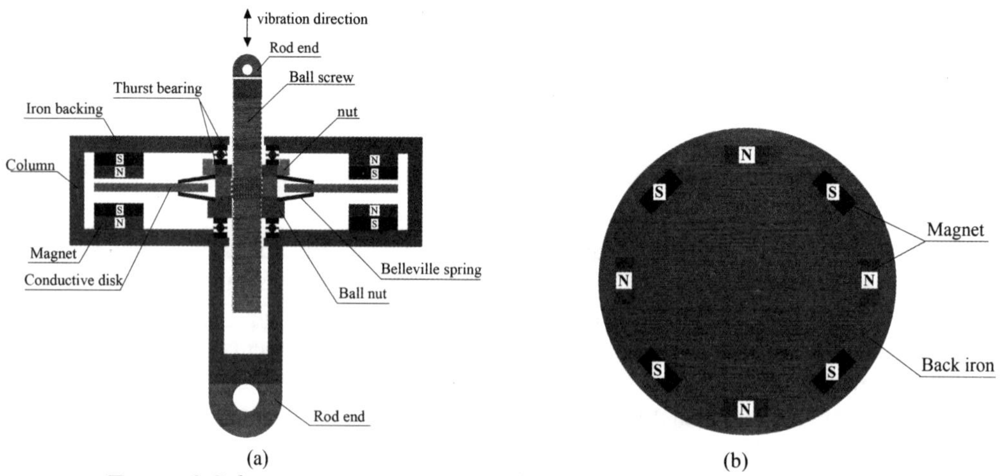

Figure.3 Schematic representation of the force-limiting Disk-type ECAMD

2.3 *Configuration and work principle of the Cup-type ECAMD*

Fig.4 shows the configuration of the Cup-type ECAMD. The magnets are installed on the stator, while the outer cup is connected to the ball nut by a thrust bearing and rotates with it. Several nuts are installed in the outer cup, and they contact with the conductive cylinder by friction force to restrict the maximum output axial force. Compared to the disk-type ECAMD, the cup-type ECAMD of the same mass has much larger moment of inertia and apparent mass. In addition, the conductor is cylindrical, so its area can be made with best efficiency to generate eddy current damping force. As a result, the cup-type ECAMD can produce sufficiently large axial force as the traditional viscous fluid dampers and it is suitable for structural vibration control.

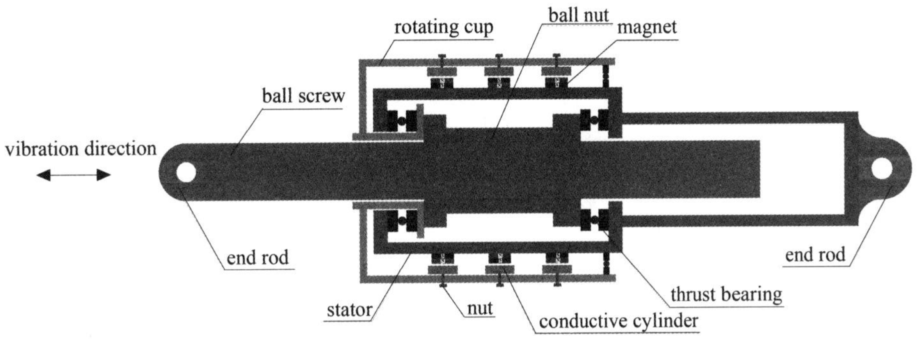

Figure.4 Configuration of the Cup-type ECAMD

The total axial force of the cup-type ECAMD is similar with that of the disk-type ECAMD. With considering of the force-limiting mechanism, the total axial force is given as follows,

$$F=\begin{cases} m_{\mathrm{e}}\ddot{u}+c_{\mathrm{e}}\dot{u}+f_0\cdot\mathrm{sign}(\dot{u}), & |F_n|\le f_r \\ f_r, & |F_n|>f_r \end{cases}. \tag{5}$$

where, f_{r} is design value of the maximum axial force of the cup-type ECAMD, F_{n} is the nominal total axial force calculated without considering the friction restriction mechanism. Fig.5 gives the mechanical model of the cup-type ECAMD, in which the force-limiting mechanism is represented by a friction damper connected to the apparent mass and eddy current damper in series. It can be seen from Fig.5 that the performance of the cup-type ECAMD changes with the input of the displacement, so it can both meet the control requirement of wind-induced vibration and seismic induced vibration.

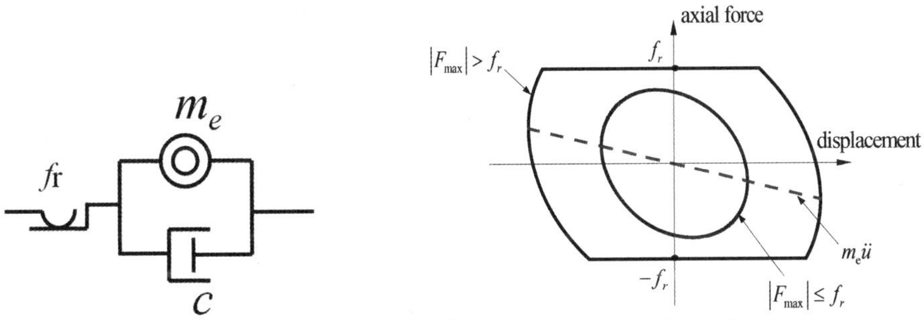

Figure.5 Mechanical model of the Cup-type ECAMD

3 PERFORMANCE TEST OF THE ECAMD

To verify the mechanical behaviour of the ECAMD, prototypes of the disk-type ECAMD and the cup-type ECAMD were manufactured, respectively. The results of the tests are discussed.

3.1 *Performance test of the prototype of the disk-type ECAMD with force restriction*

Table.1 gives the design parameters of the disk-type ECAMD with force-limiting component (FL-D-ECAMD). The number of the magnet pair and the type of the magnet are varied to obtain different equivalent damping coefficients. However, only two pairs of magnets were used in this performance test.

Table.1 Design parameters of the prototype of FL-D-ECDSD

Ball screw	Lead	5mm
	Diameter	16mm
	Rotary efficiency	0.94
	Stroke	50mm
Rare-earth magnet	Material	Nd-Fe-B
	Long	46mm
	Wide	46mm
	Thickness	23mm
	Magnet pairs	2
	Remanent flux density	1.4T
conductive disk	Material	copper
	outer radius	160mm
	Inner radius	68mm
	Thickness	8mm
	Resistivity	1.56×10-8
Air gap between magnet and conductive disk		9mm
Distance between magnet and ball screw		120mm

The FL-D-ECAMD was subjected to sinusoidal displacements using the MTS test machine. Fig.6 shows the prototype of the FL-D-ECAMD, and Fig.7 shows the loading setup of the FL-D-ECAMD. The top of the ball screw were fixed at the MTS, while the rod end at the bottom of the damper moved with the drive end of the MTS. The test conditions were summarized in Table.2. Four different tests called cs1, cs2, cs3 and cs4, respectively, were conducted by varying the frequency or the amplitude of the displacement.

Table.2 FL-D-ECAMD test conditions

	cs1	cs2	cs3	cs4
Frequency /Hz	0.2	0.2	0.2	0.5
amplitude/mm	10	20	30	20

Fig.8 shows the experimental relationship between the axial displacement and the axial resisting force of the FL-D-ECAMD, and Fig.9 shows the experimental relationship between the axial speed and the axial force. For test conditon cs1 and cs2, their hysteresis curve were nearly elliptical, while the force-speed curve were almost an obilique line. These indicated that the FL-D-ECAMD functions like a linear viscous damper when the force-limiting mechanism is not activated. However, for cs3 and cs4, the hysteresis curve and force-speed curve became very flat when the excitation velocities were larger than 40mm/s. At these conditions the friction-limiting mechanism takes effect and therefore restricts the output of the maximum axial force. Moreover, it can be seen from Fig.8 and Fig.9 that the maximum axial force is about 9KN. In summary, it is reasonable to conclude that the results of the performance test agrees well with the theory.

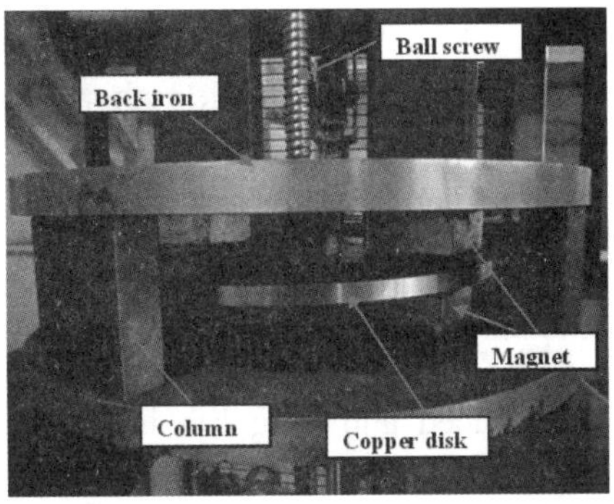

Figure.6 FL-D-ECAMD

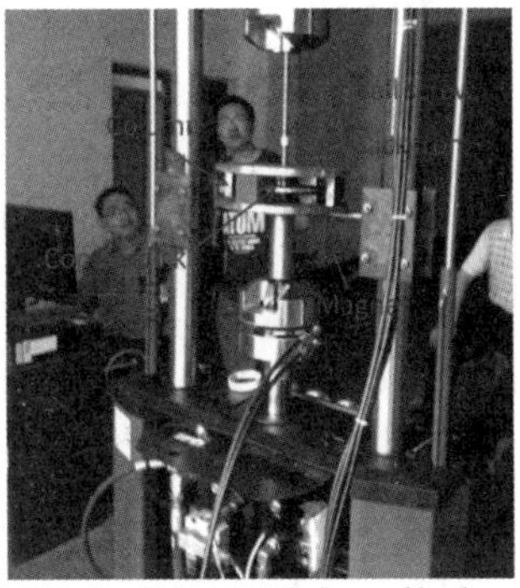

Figure.7 FL-D-ECAMD installed on the MTS

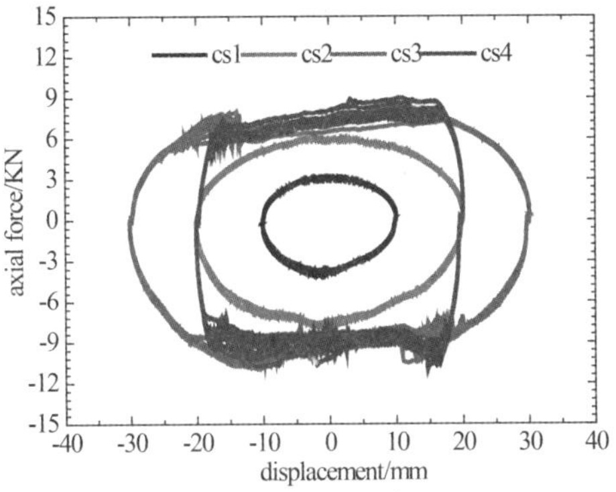

Figure.8 Hysteresis curve

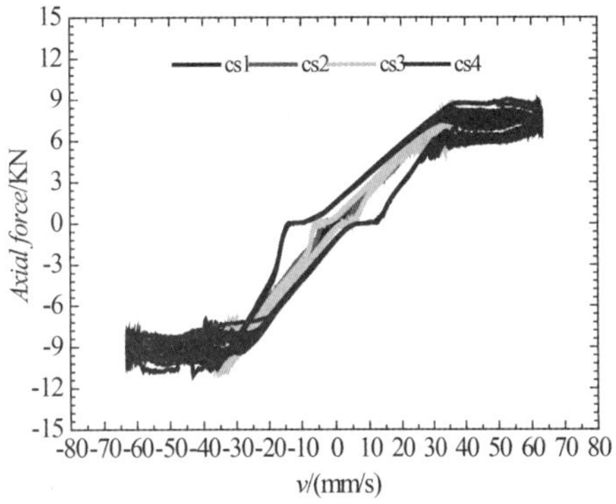

Figure.9 force-speed curve

3.2 *Performance test of the cup-type ECAMD*

The prototype of the cup-type ECAMD was designed to be able to provide the force up to 500KN. The performance tests were conducted on the hydraulic servo actuator at *Earthquake Engineering Research & Test Center of Guangzhou University*. Fig.10 shows the prototype of the cup-type ECAMD installed on the actuator. The cup-type ECAMD was subjected to sinusoidal displacement. The frequency and amplitude of the displacement is 0.24Hz and 60mm, respectively. Fig.11 shows the relationship between the axial displacement and the axial force produced by the cup-type ECAMD. It can be seen that the experimental result agrees well with the theoretical curve. The slope of the experimental hysteresis curve can be observed in Fig.11, which demonstrates the cup-type ECAMD has apparent mass and generates inertial force. In Fig.11, a fluctuation at the maximum displacement can also been seen, this may owe to the connection defect at the rod ends of the damper.

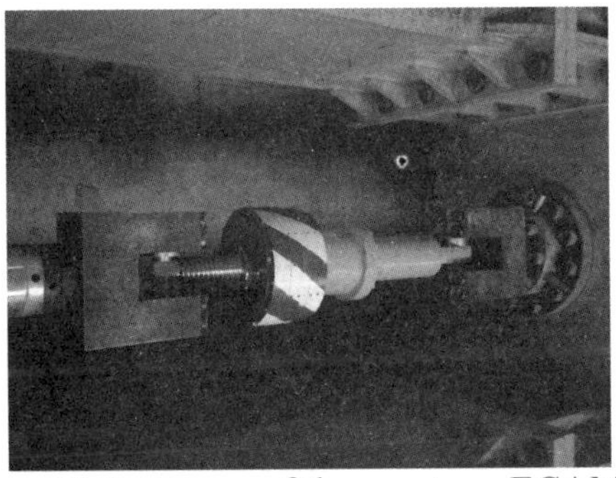

Figure.10 Prototype of the cup-type ECAMD

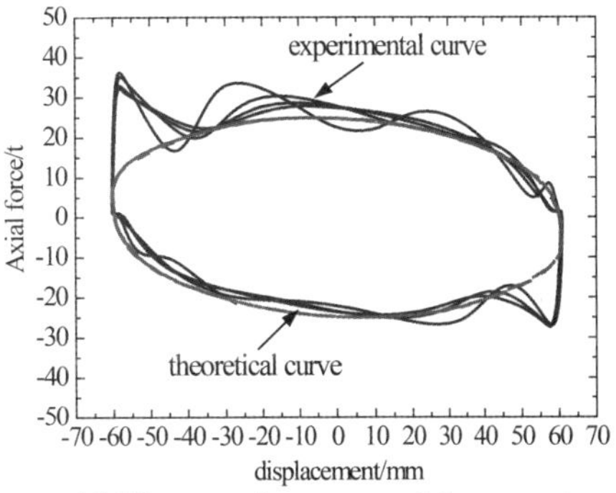

Figure.11 Hysteresis curve of the cup-type ECAMD

4 CONCLUSIONS

Two types of eddy current apparent mass damper were developed by the authors, the disk-type ECAMD and the cup-type ECAMD. Compared to the disk-type ECAMD, the energy dissipation ability of the cup-type ECAMD is much larger, producing force up to 500kN. A friction-based mechanism based on friction is implemented to limit the maximum output force of the damper at larger velocities, which is helpful for seismic control. The eddy current damping force is linear, which can be accurately predicted by finite-element simulation. In addition, the eddy current damping force is generated by a non-contact mechanism without working fluid. Therefore, the maintenance requirement of the ECAMD is largely reduced and long life is expected for the ECAMD.

The performance tests of the disk-type ECAMD and the cup-type ECAMD verify the mechanical behaviour of the ECAMD. In according to the theory, the ECAMD generates large linear viscous damping and apparent mass by employing the ball screw mechanism. Moreover, by employing the friction restriction mechanism, the maximum axial resisting force of the ECAMD can be restricted, so the ECAMD can be used to suppress the structural vibration induced by vehicles, winds and earthquakes.

References

Chen, M. Z., Papageorgiou, C., Scheibe, F., Wang, F. C., Smith, M. C. [2009] "The missing mechanical circuit element". *Circuits and Systems Magazine, IEEE*, *9*(1), 10-26.

Ikago. K., Inoue. N. [2013] "Seismic control of building using apparent mass dampers with rotational amplifying mechanisms: A review of the state of the art." *Proc. Of the 13th world conference on seismic isolation, energy dissipation and active vibration control of structures*, Sendai, Japan

Ikago, K., Saito, K., Inoue, N. [2012] "Seismic control of single - degree - of - freedom structure using tuned viscous mass damper." *Earthquake Engineering & Structural Dynamics*, *41*(3), 453-474.

Kuroda, H., Arima, F., Baba, K., Inoue, Y. [2000]. "Principles and characteristics of viscous damping devices (gyro-damper), the damping forces which are highly amplified by converting the axial movement to rotary one." *Proc. of the 12th World Conference on Earthquake Engineering, Auckland, New Zealand.*

Matsuoka, T., Hiramoto, K., Sunakoda, K., Ohtake, T., Roschke, P. N.[2006] "Numerical simulation of 3-story structure with vibration control device." *Proc. of the 4th International Conference on Earthquake Engineering*, Taipei, Taiwan.

Nakamura, Y., Fukukita, A., Tamura, K., Yamazaki, I., Matsuoka, T., Hiramoto, K., Sunakoda, K. [2014]. "Seismic response control using electromagnetic inertial mass dampers." *Earthquake Engineering & Structural Dynamics*, *43*(4), 507-527.

Smith, M. C. [2002] "Synthesis of mechanical networks: the inerter." *Automatic Control, IEEE Transactions on*, *47*(10), 1648-1662.

Smith, M. C., and Wang, F. C. [2004] "Performance benefits in passive vehicle suspensions employing inerters." *Vehicle System Dynamics*, *42*(4), 235-257.

Saito K., Sugimura Y, Nakaminami S, Kida, H., Inoue, N. [2008] "Vibration Tests of 1-Story Response Control System Using Inertial Mass and Optimized Soft Spring and Viscous Element." *Proc. of the 13th World Conference on Earthquake Engineering*. Beijing, China.

Wang, F. C., Chen, C. W., Liao, M. K., Hong, M. F. [2007] "Performance analyses of building suspension control with inerters." *Proc. of the 46th IEEE Conference on Decision and Control,* New Orleans, USA, 3786-3791.

Modeling, Testing, and Validation of an Eddy Current Damper for Structural Vibration Control

Z. W. Huang[1]; X. G. Hua[2]; Z. Q. Chen[3]; and H. W. Niu[4]

Abstract: Eddy current damping has been widely applied in mechanical engineering; however, its adaptation for civil engineering applications remains rather limited due to its low density of energy dissipation. This study presents the development of a new type of eddy current damper and theoretical and experimental characterization of its damping properties. In addition to rectangular permanent magnets and conductive plate, both primary back iron and secondary back iron are considered in the proposed eddy current damper, which represents an improved design over previous configurations. The analytical method based on the charge model of permanent magnets is extended to simulate the extra damping effect of the primary back iron and secondary back iron, and the magnet field and damping properties of the dampers are evaluated analytically. The use of a back iron in the proposed damper is shown to increase the damping coefficients by a factor of up to 5. The analytical results are compared with those obtained with the finite-element analysis in terms of accuracy in magnetic field distributions, eddy current distributions, and damping coefficients. The linear assumption, which implies the damping force is linearly proportional to velocity in the analytical model, is also examined for different motional velocities and relative permeability of the permanent magnet. It is shown that the linear damping assumption in the analytical model is only valid for a limited range of low velocity, and this velocity range can be increased by reducing the thickness or conductivity of the conductive plate. In addition, a modification is needed for the analytical model if the relative permeability of the permanent magnet significantly deviates from the assumed value of 1.0. Finally, the accuracy of the analytical model and the finite-element model is verified by experiment of a prototype damper mounted in a laboratory steel frame. **DOI: 10.1061/(ASCE)AS.1943-5525.0000891.** © *2018 American Society of Civil Engineers.*

Author keywords: Eddy current damping; Structural vibration control; Tuned mass dampers; Permanent magnet.

Introduction

Structural vibration control by using passive energy-dissipation devices is critical to improving severability and limiting damage to bridge and building structures when subjected to dynamic loadings such as earthquakes and windstorms. A number of energy-dissipation devices have been developed, which include friction dampers (Mualla and Belev 2002), metallic dampers (Vargas and Bruneau 2007), viscoelastic dampers (Kim et al. 2006), fluid viscous dampers (Weber et al. 2009; Chung et al. 2013; Lavan 2015), magnetorheological dampers (Xu et al. 2016), and pounding tuned mass dampers (TMDs) (Zhang et al. 2016; Song et al. 2016), among others. When properly designed, these devices can dissipate a large amount of energy imparted on structures and therefore have found wide applications in various engineering structures (Soong and Dargush 1997; Soong and Spencer 2002; Symans et al. 2008).

Eddy current damping (ECD) is another effective mechanism for dissipating kinetic energy (Palomera-Arias et al. 2008). In its simplest form, an ECD device consists of a conductive plate and magnets. Eddy currents are generated when the conductive plate moves through a stationary magnetic field or the strength of the magnet field varies. The interaction of the eddy currents and the applied magnetic field induces a drag force that resists the relative motion between the conductive plate and the magnets. Thus the drag force is also called an eddy current damping force. The moving magnet and conductive plates behave like a damper. Compared with other damping devices, such as friction damping and viscous fluid damping, the eddy current damping has many advantages (Sodano and Bae 2004). A notable advantage is that the eddy current damper may operate in outer space with extreme temperature conditions. Additionally, there is no fluid inside the damper and the generation of damping is independent of friction, potentially increasing eddy current damper longevity and lowering maintenance requirements. The fluid dampers may suffer from the problem of fluid degradation and leakage as has recently emerged in several bridges (Konstantinidis et al. 2011), and are also sensitive to temperature changes, which may detune mass dampers from optimum damping (Weber and Feltrin 2010). However, the energy-dissipation density of the eddy current damping system is usually very low, and this disadvantage does restrict its application (Zuo et al. 2011). In recent years, significant improvements have been made in the magnetic strength and thermal stability of the permanent magnet (Thompson 2009), and eddy current damping has been successfully used in auxiliary brake systems for heavy trucks and buses (Sakamoto et al. 1997; Gay 2005). In addition, the eddy current damping is also used to control the vibration of vehicle suspension systems (Ebrahimi et al. 2009).

For large-scale massive structures, the required damping will be several orders of magnitude larger than that in eddy current brakes

[1]Postdoctoral Researcher, College of Civil Engineering, Hunan Univ., Lushannan Rd. No. 2, 410082 Changsha, China. Email: zwhuang123@hnu.edu.cn

[2]Professor, Key Laboratory for Wind and Bridge Engineering of Hunan Province, College of Civil Engineering, Hunan Univ., Lushannan Rd. No. 2, 410082 Changsha, China (corresponding author). Email: cexghua@hnu.edu.cn

[3]Professor, College of Civil Engineering, Hunan Univ., Lushannan Rd. No. 2, 410082 Changsha, Hunan, China. Email: zqchen@hnu.edu.cn

[4]Senior Engineer, College of Civil Engineering, Hunan Univ., Lushannan Rd. No. 2, 410082 Changsha, Hunan, China. Email: niuhw@hnu.edu.cn

Note. This manuscript was submitted on February 6, 2018; approved on March 9, 2018; published online on June 13, 2018. Discussion period open until November 13, 2018; separate discussions must be submitted for individual papers. This paper is part of the ***Journal of Aerospace Engineering***, © ASCE, ISSN 0893-1321.

when eddy current dampers are directly mounted in large-scale structures for mitigation of dynamic vibrations. It is therefore more practical and economical to apply the eddy current damper as a damping element for a resonant-type absorber or TMD. The auxiliary mass weights of a TMD are just a small fraction (commonly 0.5–2%) of the controlled modal mass of the primary structure, and the damping required to mitigate the vibration of the auxiliary mass of a TMD is greatly reduced. Actually, the use of eddy current tuned mass dampers (ECTMDs) to reduce vibration of laboratory-scale structures has a long history. Larose et al. (1995) demonstrated the effectiveness of eddy current TMDs for mitigating the wind-induced vibrations of the bridge aeroelastic model in wind tunnel testing. Bae et al. (2012) analytically evaluated eddy current damping mounted in a TMD, and they also showed that the extra mass of a magnet increases the mass ratio of TMD, therefore enhancing the robustness of tuned mass dampers in the case of imperfect tuning. Wang et al. (2012) experimentally investigated eddy current tuned mass dampers with mass weight up to 3 t and verified that the optimum damping value can be achieved with affordable magnets. Chen et al. (2017) demonstrated the robustness of eddy current tuned mass dampers for vibration control of a cantilever beam. Several full-scale applications of eddy current tuned mass dampers in bridges and buildings have been reported for control of wind- or human-induced vibrations (Hua et al. 2016; Lu et al. 2017; Ao and Reynolds 2017).

Previous studies on ECTMD can mainly be divided into two categories: (1) modeling of the eddy current damper by analytical method or finite-element analysis to predict the performance of the ECTMD; and (2) development of novel eddy current dampers with high energy-dissipation density and compactness for ECTMD while minimizing cost and weights. Analytical modeling of eddy current dampers haw drawn much attention in previous studies since it is fast and flexible compared with finite-element analysis. Sodano et al. (2006) presented electromagnetic analysis of an eddy current damper and the damping coefficients can be significantly enhanced by including an additional permanent magnet. Bourquin et al. (2014) obtained an analytical expression for the eddy current damping force based on the Maxwell receding image method, and it yields the nonlinear dependence of the eddy current damping force on the velocity. Pluk et al. (2014) proposed a three-dimensional (3D) semianalytical modeling method to calculate the damping force in an eddy current damper with permanent magnets and a conductive plate of finite dimensions. By utilizing the method of images in two dimensions, the modeling accuracy was significantly increased. From a practical point of view, it is of great importance to increase the damping density of the eddy current damper. To achieve this purpose, high-grade NdFeB magnets can be used and optimization of the configuration of the eddy current damper is also necessary. Amjadian and Agrawal (2018) exploited the use of friction and eddy current damping to enhance energy dissipation. Zuo et al. (2011) proposed an eddy current damper in which the permanent magnets are arranged as an array and their experimental results showed that its damping density is as much as five times higher than the traditional configuration. However, the array configuration is not convenient for its use in a TMD. Wang et al. (2012) compared four different configurations of an eddy current damper that consisted of permanent magnets, a conductive plate, and an extra steel plate. The theoretical results indicated that the configuration with alternative magnetic poles and an extra steel plate behind the conductive plate outperforms the other three.

In this study, a novel eddy current damper was developed, which is composed of rectangular permanent magnets, conductive plate, primary back iron, and secondary back iron. The proposed damper represents an improvement over the previous one investigated by Wang et al. (2012), where the primary back iron was not considered. An analytical model that takes into account the effect of the primary back iron was formulated to calculate the magnetic field and linear damping coefficient based on the charge model of permanent magnets. The analytical results of the magnetic field and damping coefficients were compared with those obtained from the finite-element analysis. The comparisons were also made for the low-velocity range and high-velocity range of conductive plates to investigate the nonlinear dependence of damping force on the velocity. Finally, the accuracy of the analytical model and finite-element model (FEM) was further verified by experiment of a prototype of the planar eddy current damper.

Analytical Modeling of the Novel Eddy Current Damper

Configuration of Eddy Current Damper

Tuned mass dampers are very popular passive damping devices for vibration control of structures. It is a single degree of freedom (SDOF) system that consists of a mass, a spring, and a damper. If the TMD is tuned in resonance with the primary structure but in an opposite phase, a large amount of vibration energy is transferred to the TMD and then dissipated by damping in the TMD when the primary structure is subjected to harmonic excitations. The optimal design of the TMD for minimizations of dynamic response of a structure subject to random or harmonic excitations has been well established (e.g., Soong and Dargush 1997; Warburton 1982), and is not repeated here.

The damping in the TMD is provided by the eddy current damper. Fig. 1 shows the configuration of the proposed novel eddy current damper. The damper is composed of permanent magnets, conductive plate, primary back iron, and secondary back iron. The primary back iron, which is not considered in Wang et al. (2012), was introduced to optimize the magnetic field and enhance the energy-dissipation capacity, as shown subsequently. In general, the auxiliary mass of the ECTMD can be used as the primary back iron if it is made of steel or other high-permeability materials. Then in turn, the permanent magnets become parts of the auxiliary mass. The number of the permanent magnets is determined by the damping coefficient of the ECTMD and not limited to a pair as sketched in Fig. 1. In the following, the analytical method based on the charge model of permanent magnets is presented to exclusively include the key components of the damper responsible for the generation of the eddy damping effect, and in particular to take account of the primary back iron.

Assumptions

To derive the analytical model of the proposed eddy current dampers, the following assumptions were made:

1. The relative velocity between the permanent magnets and the conductive plate is low. In this case, the secondary magnetic field due to eddy currents in the conductive plate can be approximately ignored, and the skin effect inside the conductive plate is negligible.
2. Permeability of the primary back iron and the secondary back iron are infinite. In this case, the effects of the boundaries of the back iron on the magnetic field can be replaced by a series of images. This assumption is always valid since the primary and secondary back irons must be thick enough to avoid magnetic saturation in design.
3. The conductivities of the primary back iron and the secondary back iron are zero. This assumption really holds for the primary

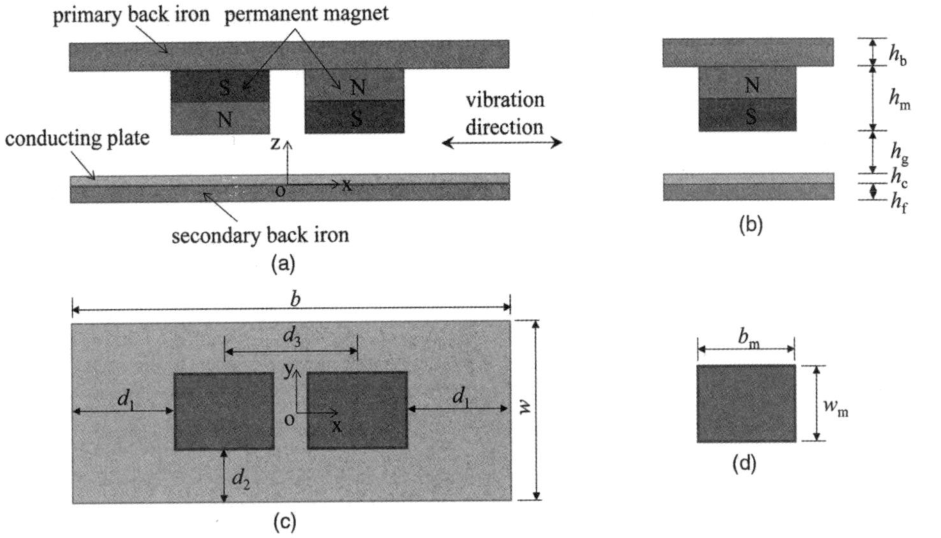

Fig. 1. Topology of the planar eddy current damper: (a) front view; (b) side view; and (c and d) top view.

back iron because no eddy current is induced in it. On the other hand, it is just an approximation for the secondary back iron if it is not laminated.

4. The permanent magnets are uniformly and linearly magnetized, and their relative permeability is 1.0. In this condition, the permanent magnets can be replaced by two surface charges sheets.
5. The reluctance of the conductive plate is neglected. This is true since the vibration frequencies of the large-scale civil engineering structures are really quite low.

Analytical Modeling

The magnetic field due to the positive surface charge of a single rectangular permanent magnet in the air is given by Furlani (2001)

$$B_x(x,y,z) = \frac{B_r}{4\pi}\sum_{m=0}^{1}\sum_{n=0}^{1}(-1)^{m+n}\ln(-Q+T) \tag{1}$$

$$\bar{B}_y(x,y,z) = \frac{B_r}{4\pi}\sum_{m=0}^{1}\sum_{n=0}^{1}(-1)^{m+n}\ln(-P+T) \tag{2}$$

$$\bar{B}_z(x,y,z) = \frac{B_r}{4\pi}\sum_{m=0}^{1}\sum_{n=0}^{1}(-1)^{m+n}\arctan\left(\frac{QP}{RT}\right) \tag{3}$$

where

$$P = x - x_n \tag{4}$$

$$Q = y - y_m \tag{5}$$

$$R = z - z_k \tag{6}$$

$$T = \sqrt{P^2 + Q^2 + R^2} \tag{7}$$

where x, y, and z = coordinates of any point outside the permanent magnet. This right-handed Cartesian system is shown in Fig. 1, fixed with respect to the permanent magnet. The coordinates of the corners of the permanent magnet are x_n, y_m, and z_k. The remanent flux density of the permanent magnet is B_r.

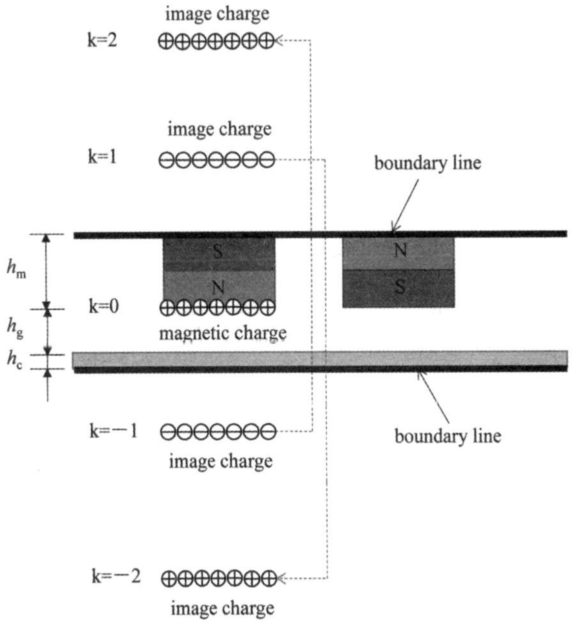

Fig. 2. Concept of method of images.

According to the study by Nasar and Xiong (1988), the influence of the primary back iron and the secondary back iron on the magnetic field of permanent magnets can be taken into account by the method of images, as shown in Fig. 2. The desired field of a single permanent magnet is then given by the sum of the applied field generated by its magnetic charges and their image fields, and the total field can be obtained by taking the superposition of the magnetic field solution for all the individual permanent magnets

$$B_x(x,y,z)=\frac{B_r}{4\pi}\sum_{j=0}^{n_m-1}\sum_{k=-\infty}^{+\infty}\sum_{m=0}^{1}\sum_{n=0}^{1}(-1)^{m+n+k+n_m}\ln(-Q+T) \tag{8}$$

$$B_y(x,y,z)=\frac{B_r}{4\pi}\sum_{j=0}^{n_m-1}\sum_{k=-\infty}^{+\infty}\sum_{m=0}^{1}\sum_{n=0}^{1}(-1)^{m+n+k+n_m}\ln(-P+T) \tag{9}$$

$$B_z(x,y,z)=\frac{B_r}{4\pi}\sum_{j=0}^{n_m-1}\sum_{k=-\infty}^{+\infty}\sum_{m=0}^{1}\sum_{n=0}^{1}(-1)^{m+n+k+n_m}\arctan\left(\frac{QP}{RT}\right) \tag{10}$$

where k and j = kth image and jth pole, respectively. Generally, the range of k may be taken to be $-r$ to $r(r=5\text{–}10)$.

The analytical expression for the x- and y-component of the eddy current density in the conductive plate can be obtained as follows (Furlani 2001):

$$J_x^p=v_x\sigma\frac{B_r}{4\pi}\sum_{j=0}^{n_m-1}\sum_{k=-\infty}^{+\infty}\sum_{m=0}^{1}\sum_{n=0}^{1}(-1)^{m+n+k+n_m}\ln(R+T) \tag{11}$$

$$J_y^p=v_x\sigma\frac{B_r}{4\pi}\sum_{j=0}^{n_m-1}\sum_{k=-\infty}^{+\infty}\sum_{m=0}^{1}\sum_{n=0}^{1}(-1)^{m+n+k+n_m}\arctan\left(\frac{-QR}{PT}\right) \tag{12}$$

where v_x and σ = moving velocity and conductivity of the conductive plate, respectively. The z-component of eddy current density is assumed to be zero in the deduction of Eqs. (11) and (12). This assumption is always valid for an ECTMD that works at low frequency and low speed since the thickness of the conductive plate is relatively smaller than the skin depth in this case.

Eqs. (11) and (12) are derived for an infinite conductive plate. To obtain the solution for a finite conductive plate, the method of image can be used as shown in Fig. 3, and the analytical expression for every image can be easily obtained based on the symmetry condition. For example, the x- and y-components for image M (1, 1) in Fig. 3 are given by

$$J_x^{M(1,1)}(x,y)=-J_x^p(2x_1-x,2y_1-y),$$
$$J_y^{M(1,1)}(x,y)=-J_y^p(2x_1-x,2y_1-y) \tag{13}$$

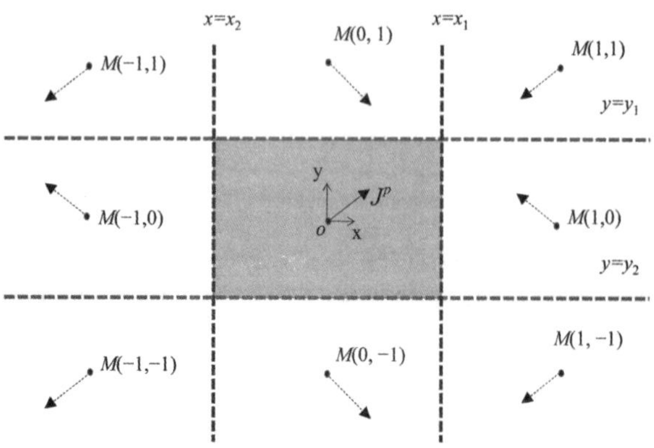

Fig. 3. Primary and imaginary electric field components.

where x_1 and y_1 = x-coordinate of the right edge of the rectangular conductive plate and the y-coordinate of the front edge, respectively. The actual current density in the finite conductive plate is then the summation of the primary and all imaginary current sources, that is

$$J_x=\sum_{i=-1}^{1}\sum_{j=-1}^{1}J_x^{M(i,j)},\qquad J_y=\sum_{i=-1}^{1}\sum_{j=-1}^{1}J_y^{M(i,j)} \tag{14}$$

in which

$$J_x^{M(0,0)}=J_x^p,\qquad J_y^{M(0,0)}=J_y^p \tag{15}$$

Finally, the x-component of the eddy current damping force can be calculated from the Lorentz force equation

$$F=\iiint_v J_yB_z dV \tag{16}$$

where V = volume of the conductive plate. Accordingly, the damping coefficient is calculated by

$$C=F/v_x \tag{17}$$

Comparison with 3D Finite-Element Model

A stationary 3D FEM was established in COMSOL Multiphysics to verify the accuracy of the proposed analytical model. The 3D FEM can consider the reaction field created by the motion-induced eddy currents, the actual geometry of the planar eddy current damper, and the soft ferromagnetic property of the back iron. Fig. 4 shows the mesh of the finite-element model, which has been refined to ensure the mesh independence of the numerical results. The parameters of the planar eddy current damper are given in Table 1. The velocity of the conductive plate used in the 3D FEM was v_x = 0.5 m/s. The variation of magnitude flux density B with magnetic field strength H (H-B curve) of the back iron (1010 steel) is plotted in Fig. 5.

Flux Density Distribution

Fig. 6 shows the flux density distribution along the x-direction at $z=h_c/2$ (middle of the conductive plate) and $y=0$. As can be seen, the analytical model predicted the flux density very well. However, there was a slight phase shift of B_z calculated by the 3D FEM. This difference was due to the counteracting magnetic field induced by the eddy current in the copper conductive plate. The flux density distribution along the y-direction at $z=h_c/2$ and $x=-(d_3+b_m)/2$ is shown in Fig. 7. Again, the analytical method shows a good agreement with the numerical simulation. Fig. 8 gives a more intuitive description of the flux density (z-component)

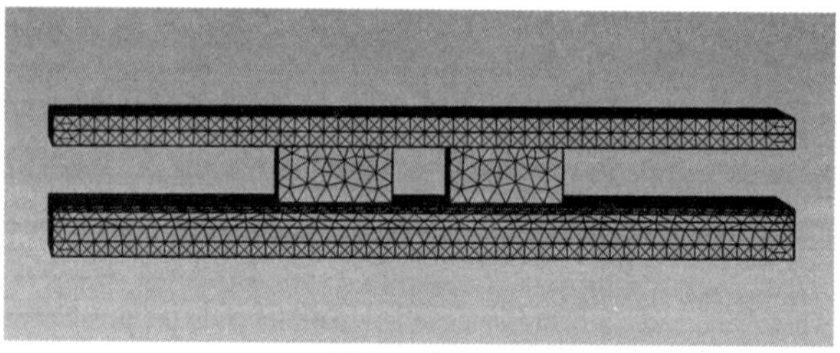

Fig. 4. Mesh of the model in COMSOL.

Table 1. Parameters of planar eddy current damper

Symbol	Description	Value
b_m	Length of the PM	40 mm
w_m	Width of the PM	40 mm
h_m	Thickness of the PM	20 mm
B	Length of the conductive plate	260 mm
W	Width of the conductive plate	120 mm
h_c	Thickness of the conductive plate	5 mm
h_f	Thickness of the secondary back iron	10 mm
h_b	Thickness of the primary back iron	10 mm
h_g	Air-gap length	2 mm
d_3	Pole pitch	60 mm
d_1	End-to-end distance of the PM and conductive plate in the x-direction	80 mm
d_2	End-to-end distance of the PM and conductive plate in the y-direction	40 mm
B_r	Remanent flux density of the PM (NdFeB)	1.40 T
μ_m	Relative permeability of the PM	1.0
σ_c	Conductivity of the conductive plate (copper)	5.8×10^7 S/m
σ_b	Electrical conductivity of the back iron	2×10^6 S/m
p	Pole-pair number	1

Note: PM = permanent magnetic.

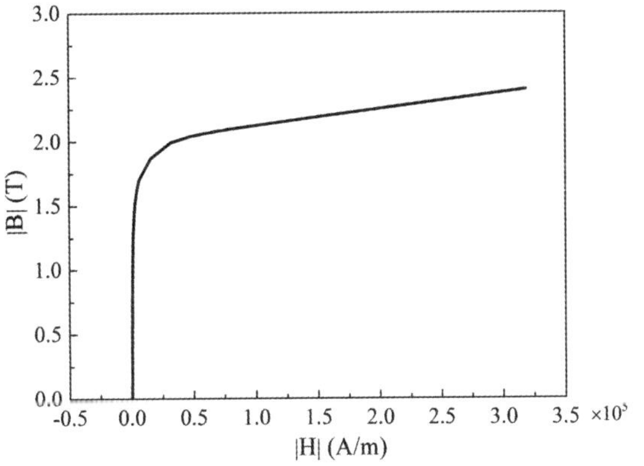

Fig. 5. H-B curve of the 1010 steel.

distribution in the middle of the conductive plate obtained with the 3D FEM. It can be seen that the maximum value of B_z was around 0.93 T.

Eddy Current Density Distribution

The eddy current distribution in the middle of the conductive plate ($z = h_c/2$) obtained with the 3D FEM is plotted in Fig. 9. The squares represent the position of permanent magnets. It can be seen that the induced currents flowed along three different closed paths inside the copper disk. Moreover, nearly all the induced currents flowed in the y-direction inside the useful regions corresponding to the projection of the magnet poles.

A comparison of the eddy current density distribution along the x-direction at $y = 0$ obtained from the 3D FEM and the proposed analytical formulas is shown in Fig. 10. It is clear that J_x and J_y were well predicted by the analytical model. Fig. 11 shows the eddy current distribution along the y-direction at $x = -(d_3 + b_m)/2$. The results of the 3D FEM verify that the z-component of the current density was nearly zero, as assumed in the analytical model.

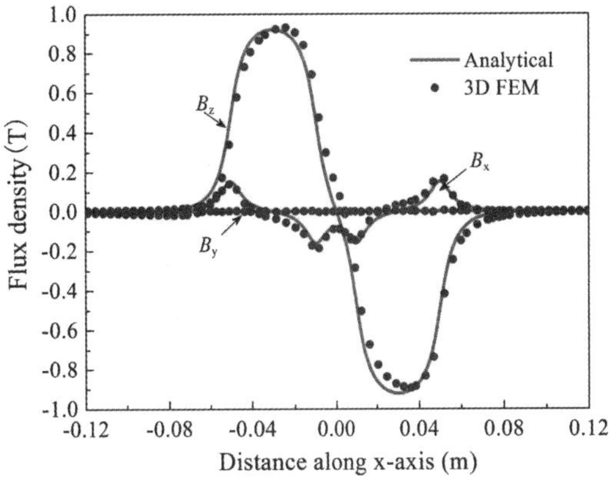

Fig. 6. Flux density distribution along the x-direction at $z = h_c/2$ and $y = 0$.

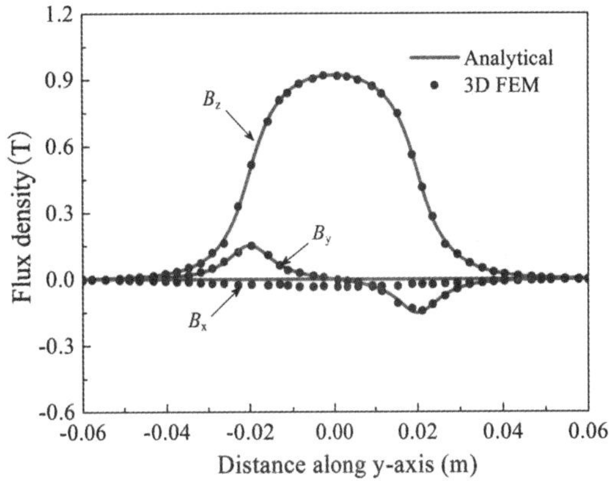

Fig. 7. Flux density distribution along the y-direction at $z = h_c/2$ and $x = -(d_3 + b_m)/2$.

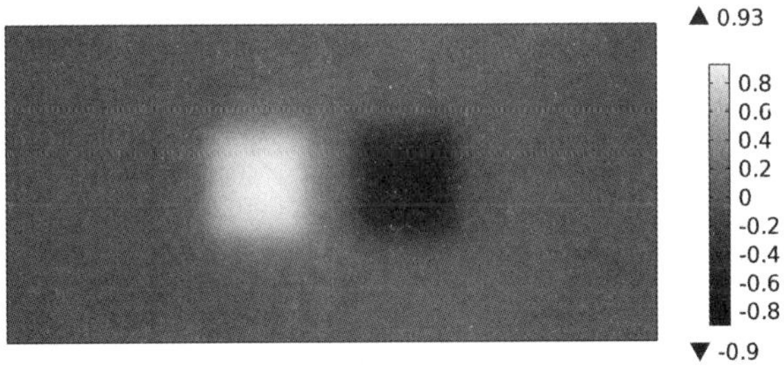

Fig. 8. Flux density distribution (z-component) contour at $z = h_c/2$ obtained by 3D FEM.

Damping Coefficient

Fig. 12 shows the variation of damping coefficient versus the air-gap distance. It can be seen that the damping coefficient was well predicted by the analytical formula, and the errors between the analytical and the numerical results were below 2%. As expected, the damping coefficient decreased with the air-gap distance since the reluctance of the magnetic circuit of the planar eddy current damper was dominated by the air gap and increased with it.

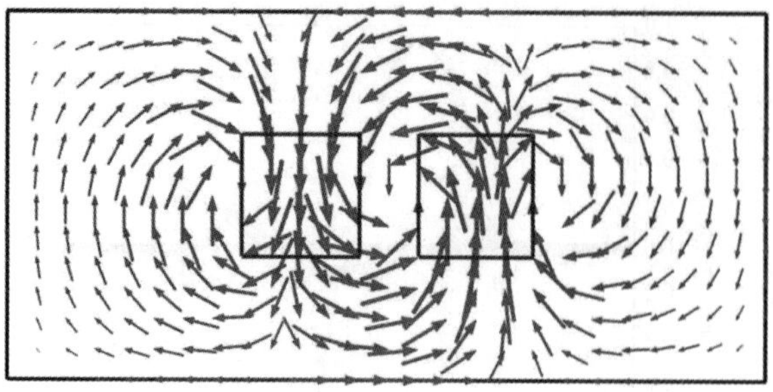

Fig. 9. Eddy current density distribution at $z = h_c/2$ (arrows).

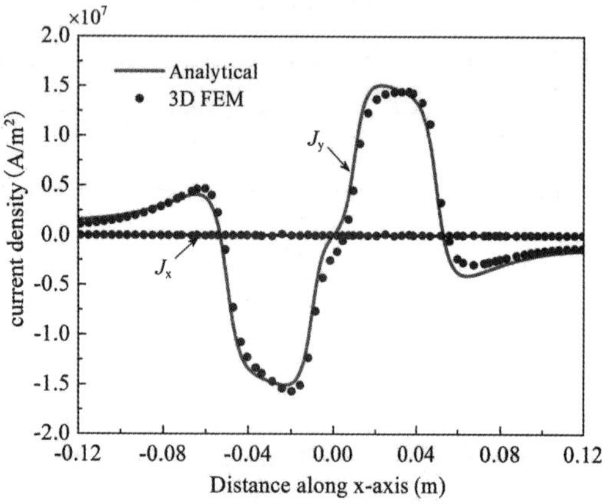

Fig. 10. Eddy current density distribution (J_x and J_y) along the x-direction at $z = h_c/2$ and $y = 0$.

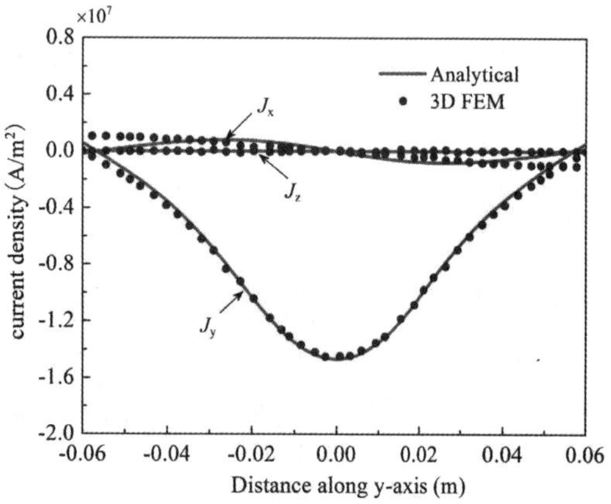

Fig. 11. Eddy current density distribution along the y-direction at $z = h_c/2$ and $x = -(d_3 + b_m)/2$.

In order to demonstrate the significance of back irons on the damping effect, the damping coefficients of the damper with or without the primary back iron or the secondary back iron are provided in Fig. 13. These results clearly indicate that the secondary back iron played the most important role in increasing the damping coefficient of the damper, and the primary back iron played the secondary role. Take the example of $h_g = 2$ mm: the damping coefficient of the eddy current damper was 321.34 $\mathrm{N \cdot s/m}$, while for

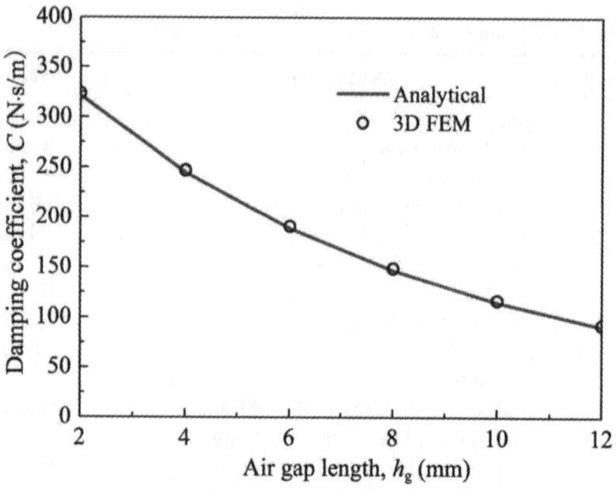

Fig. 12. Damping coefficient versus air-gap distance.

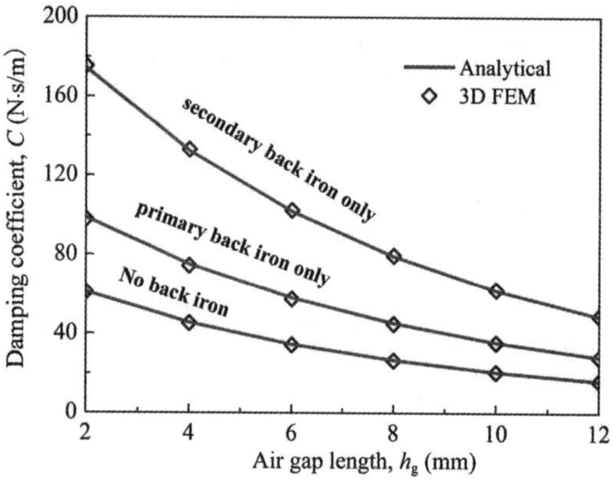

Fig. 13. Influence of back iron on damping coefficient obtained with 3D FEM.

the case of secondary back iron only, primary back iron only, and no back iron the damping coefficient was 175.3, 97.8, and 60.9 $\mathrm{N \cdot s/m}$, respectively. Notice that the average computation time was 18 s and 6 min for the analytical method and 3D FEM, respectively, on a Dell desktop (Dell, Xiamen, China) with an Intel i7-4790 eight core processor with 8.0 GB RAM, so the analytical method is more suitable for the parametric analysis and optimization.

Influence of the Velocity of the Conductive Plate

The proposed analytical model is only valid for a low velocity range. As the velocity increases, the error of the analytical model becomes larger and larger, and the eddy current damper cannot be assumed as a linear viscous damper any more. Therefore, the influence of the velocity on the damping forces and damping coefficients are studied in this section.

The damping force versus the velocity obtained with the 3D FEM is plotted in Fig. 14 with a 10% error bar, and is compared with the results calculated by the analytical formula. It can be seen that the error in the damping force obtained by the analytical formula was less than 10% in the velocity range of 0–1.0 m/s, so $v_x = 1.0$ m/s may be defined as the critical velocity below which the linear damping assumption holds for the studied planar eddy current damper. The damping coefficient versus the velocity is given in Fig. 15. The results clearly show that the damping coefficient

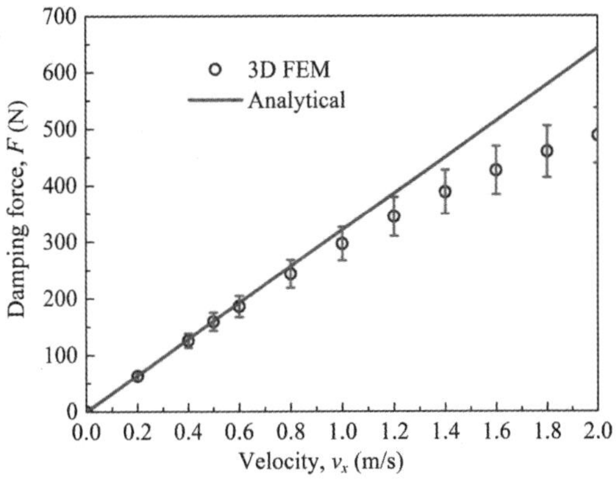

Fig. 14. Variation of damping force with velocity for copper plate with $h_c = 5$ mm.

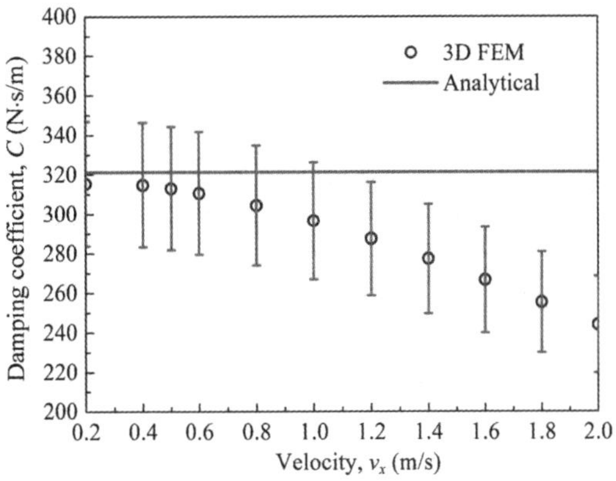

Fig. 15. Variation of damping coefficient with velocity for copper plate with $h_c = 5$ mm.

decreased with the velocity due to the magnetic field of the induced eddy current. For $v_x = 2.0$ m/s, the error of analytical method reached 31.7%, which is unacceptable for the design of ECTMD.

The thickness and conductivity of the conductive plate have a great influence on the force-velocity curve of the planar eddy current damper (Gay and Mehrdad 2006). If thinner conductive plate or conductive plate with lower conductivity is used, the critical velocity for the linear damping assumption can be increased dramatically. Figs. 16 and 17 show the damping coefficient versus velocity for copper plate with thickness $h_c = 2$ mm and aluminum plate with thickness $h_c = 5$ mm, respectively. It can be observed that the error in the damping coefficient of the analytical method remains less than 10% up to the velocity of 2.0 m/s for the 2-mm-thick copper plate, while the critical velocity for the aluminum conductive plate is about 1.6 m/s.

Influence of the Relative Permeability of the Permanent Magnets

In the previous section, the relative permeability of permanent magnets was specified as 1.0 in the 3D FEM simulation, which is the same as the assumption in the analytical model. In fact, the relative permeability of practical permanent magnets is always larger than

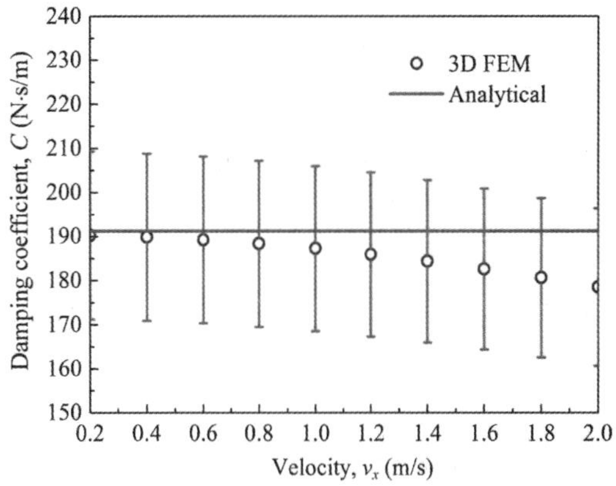

Fig. 16. Damping coefficient versus velocity for copper plate with $h_c = 2$ mm.

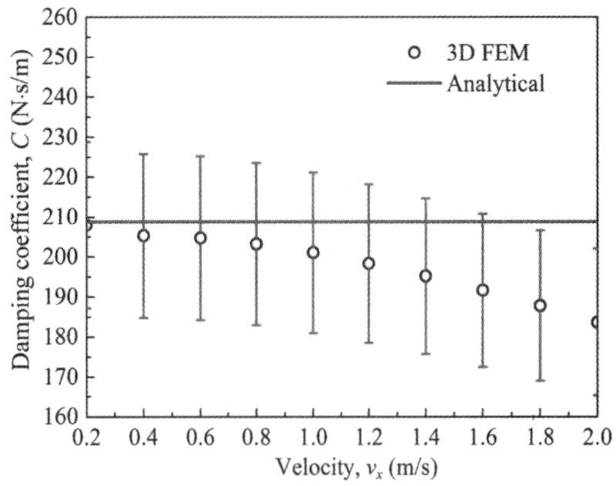

Fig. 17. Damping coefficient versus velocity for aluminum plate with $h_c = 5$ mm.

1.0, and may even reach 1.3 (Rovers et al. 2013). So in this section, the influence of the relative permeability of permanent magnet on the damping coefficient is discussed.

Fig. 18 shows the damping coefficient versus the relative permeability of permanent magnets obtained by the 3D FEM. The results indicate that the damping coefficient tends to decrease linearly as the relative permeability of permanent magnets increases. The error of the analytical formula versus the relative permeability of permanent magnet is given in Fig. 19. It can be observed that the error is beyond 15% when the relative permeability of permanent magnets reaches 1.3. Reducing the remanent flux density of a permanent magnet is a possible way of compensating the bias of the relative permeability of the magnets for analytical method, that is

$$B_{rm} = \frac{2B_1}{1+\mu_{mr}} \tag{18}$$

where B_{rm} = reduced remanent flux density of magnets; B_r = real remanent flux density of magnets; and μ_{mr} = relative permeability of the magnets. By replacing B_r by B_{rm} in Eqs. (1)–(17), we can obtain the relationship between the reduced damping coefficient C_{rm} and the original damping coefficient C

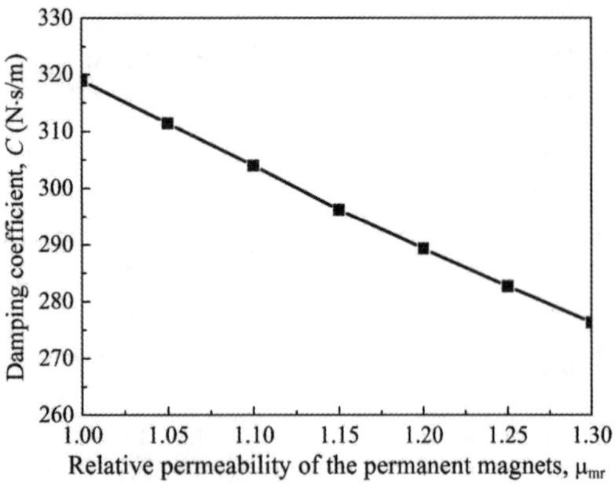

Fig. 18. Damping coefficient versus the relative permeability of the magnets.

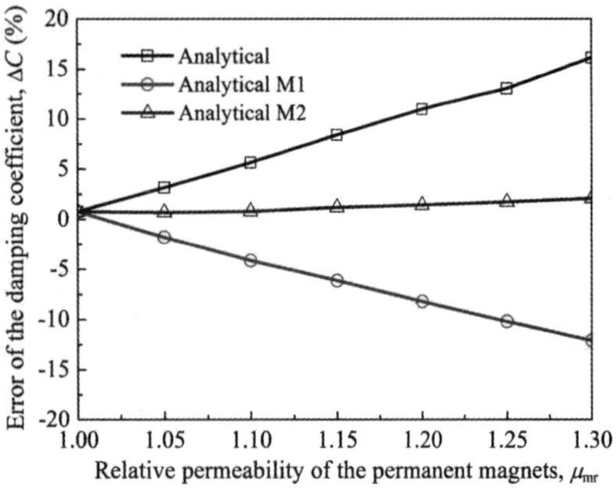

Fig. 19. Error in the damping coefficient as a function of the relative permeability of the magnets.

$$C_{rm} = \left(\frac{2}{1+\mu_{mr}}\right)^2 C \tag{19}$$

For simplicity, the damping coefficient calculated from Eq. (19) denotes the analytical M1. The error in damping coefficient for the analytical M1 is displayed Fig. 19. It can be seen that the analytical M1 severely underestimates the damping coefficient, indicating that Eq. (19) is not suitable for the planar eddy current damper. This may be due to the existence of high-permeability material such as the primary back iron and secondary back iron used in the planar eddy current damper. However, the comparison of the analytical method and the analytical M1 shows that averaging the damping coefficient obtained from the analytical method and the analytical M1 can significantly reduce the error, so an empirical formula can be given to modify the damping coefficient C, that is

$$C_m = \frac{1}{2}\left[1+\left(\frac{2}{1+\mu_{mr}}\right)^2\right] C \tag{20}$$

where C_m = modified damping coefficient, which we call the analytical M2 for simplicity. As shown in Fig. 19, the largest error of the analytical M2 is only 2.1% when $\mu_{mr} = 1.30$.

Experimental Verification

Description of Damper and Testing Rig

A prototype of the planar eddy current damper was constructed to validate the analytical model. The damper is shown in Fig. 20. It was made with four rectangular NdFeB magnets (grade N52), whose remanent flux density and relative permeability were 1.45 T and 1.05, respectively. The materials of the conductive plate were copper, and its conductivity was 5.80×10^7 S/m. The primary back iron and the secondary back iron were both made of 1010 steel, whose H-B curve is shown in Fig. 5. The geometrical parameters of the model are given in Fig. 20.

A special test rig was designed to determine the damping coefficient of the damper, as shown in Fig. 21. The test apparatus was a single-story steel frame, which can be well approximated by an SDOF system vibrating horizontally. The primary back iron and the permanent magnets were screwed on the top of the steel frame, while the conductive plate and the secondary back iron were installed on a stationary platform above the magnets.

Experimental Procedure

From the damped free vibration of the steel frame, the damping ratio with and without eddy current damping were determined. Then the damping coefficient induced by the eddy current damper can be determined as follows:

$$C_m = 2M_t\omega_t\zeta_t - C_0 \tag{21}$$

where M_t, ω_t, and ζ_t = modal mass, natural frequency, and natural damping ratio of the steel frame incorporated with the eddy current damper, respectively; and C_0 = natural damping coefficient of the

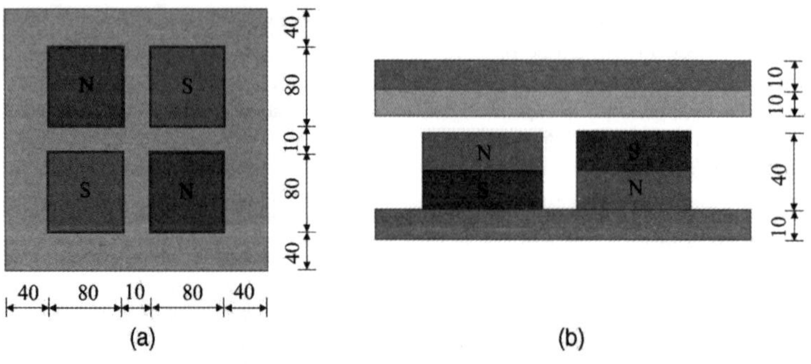

Fig. 20. Configuration of prototype eddy current damper (unit: millimeters): (a) top view; and (b) front view.

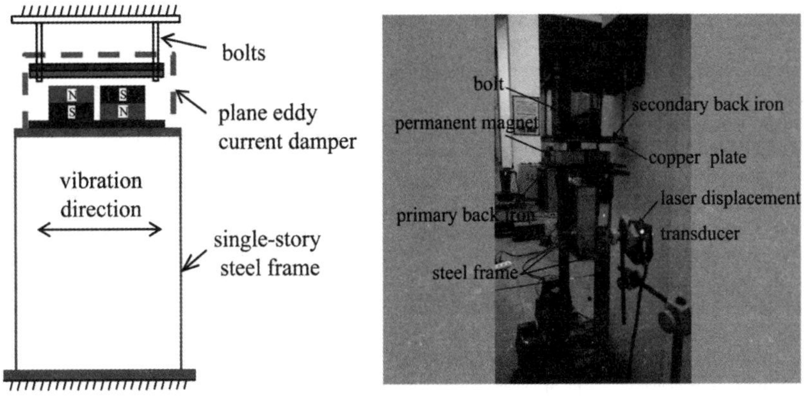

Fig. 21. Description of the test rig of a steel frame.

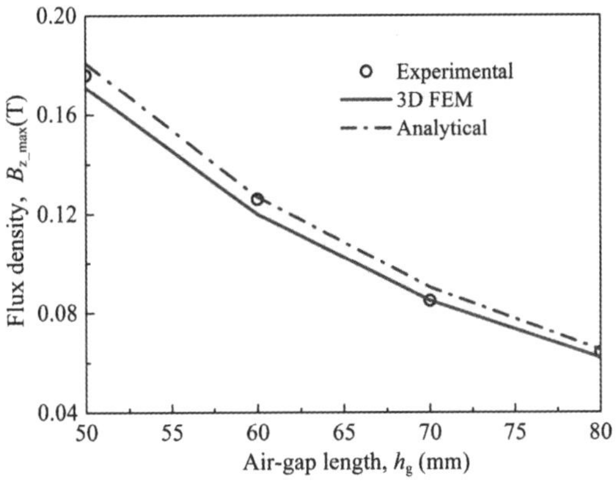

Fig. 22. Maximum flux density versus air-gap distance.

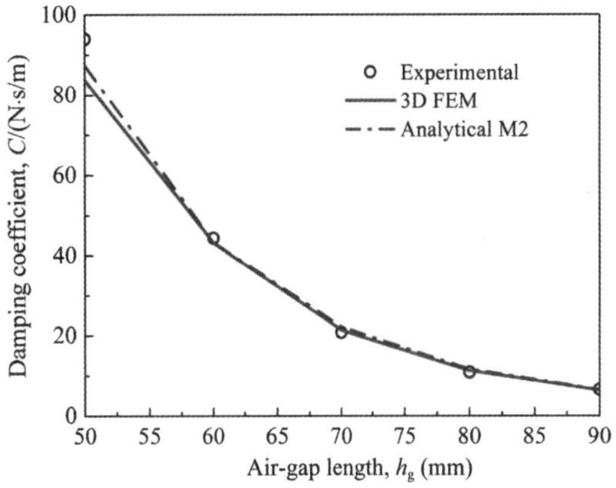

Fig. 23. Damping coefficient versus air-gap distance.

original vibration system including the steel frame, the permanent magnets, and primary back iron. In this experiment, M_t was 19.7 kg and C_0 was 0.53 N · s/m, corresponding to the generalized damping coefficient of the steel frame without the eddy current damper. The modal frequency for the fundamental mode was 1.553 Hz, therefore the maximum velocity of the frame top was relatively low.

Results and Comparison

During the experiment, the air-gap distance between the permanent magnets and the conductive plate could be varied by adjusting the height of four bolts shown in Fig. 21, thus different damping coefficients can be provided. Fig. 22 shows the maximum z-component of flux density at the bottom surface of the conductive plate versus the air-gap distance. In the tests, the flux density was measured by a tesla meter. It can be seen that the analytical model and the 3D FEM can both predict the experimental results very well. The flux density obtained by the analytical model was a little larger than that of the 3D FEM because the relative permeability of the permanent magnet in the analytical model was 1.0. The damping coefficient versus the air-gap distance is displayed in Fig. 23. The results obtained from the analytical M2 and the 3D FEM agree well with those of the experiments when the air-gap distance is larger than 50 mm. For example, when the air-gap distance $h_g = 70$ mm, the measured damping coefficient was 20.8 N · s/m, while the damping coefficients calculated from the analytical M2 and the 3D FEM, respectively, were 22.3 and 21.4 N · s/m. The errors of the analytical method and the 3D FEM were 7.2 and 2.9%, respectively. The error of the measured damping coefficient was large when the air-gap distance decreased to 50 mm because in this case the free damped vibration stopped only after about three cycles, making the accurate calculation of the natural damping ratio and natural damped frequency very difficult. Moreover, the remanent flux density and relative permeability of permanent magnets and the conductivity of the copper plate were not validated by test in our experiments, so the errors of the analytical method and the 3D FEM from the experimental measurement were unavoidable.

Conclusion

In this paper, a conceptual design of an eddy current damper was developed. The analytical model for the eddy current damper had been established and validated by 3D FEM and experimental measurements. The analytical expression of the eddy current damping coefficient was derived from the charge model of the permanent magnets, so it is able to reflect the influence of the geometric parameters and remanent of the permanent magnets. By using the method of images, both the magnetic boundary conditions in the back irons and the electric boundary conditions in the conductive plate were well approximated, ensuring the accuracy and simplicity of the analytical expression. If the velocity is sufficiently low and the relative permeability of the permanent magnets is 1.0, the difference in the damping coefficient between the analytical model and the 3D FEM is less than 2%. The proposed analytical method can be easily extended to calculate the damping coefficient of an eddy current damper with a different configuration by modifying the magnetic images, if the permanent magnets and the conductive plate are rectangular and their relative motion is perpendicular to the magnetized direction of the permanent magnets.

The analytical method and 3D FEM showed that the inclusion of the primary back iron and the secondary back iron can significantly increase the damping coefficient of the planar eddy current damper. For the studied modeling parameters, the damping coefficient of the eddy current damper with both primary back iron and secondary back iron was 321.34 $N \cdot s/m$, while for the case of secondary back iron only, primary back iron only, and no back iron, the damping coefficients were 175.3, 97.8, and 60.9 $N \cdot s/m$, respectively. This indicates that the eddy current damper has a much higher damping density than that of the previous ones without secondary back iron or primary back iron. Moreover, compared with the dampers configured in array of permanent magnet and conductive plate to achieve high energy-dissipation density, the proposed damper is simple in geometry, and especially suitable for damping components in a TMD.

Rigorously speaking, the eddy current damping is nonlinear due to the influence of the reaction field of the eddy current in the conductive plate. However, the eddy current damper can still be approximated by a viscous damper at low velocity, and the critical velocity defined for this linear damping assumption can be increased by reducing the thickness or conductivity of the conductive plate. The accuracy of the analytical model is greatly influenced by the relative permeability of the permanent magnets, and modification of the damping coefficient is needed when permanent magnets with high relative permeability are used.

Acknowledgments

This study is sponsored by the National Science Foundation of China under Program No. 51422806. The support from the State's Key Project of Research and Development Plan (Nos. 2016YFE0127900; 2017YFC07036002) is also greatly acknowledged.

References

Amjadian, M., and A. K. Agrawal. 2018. "Modelling, design and testing of a proof-of-concept prototype damper with friction and eddy current damping effect." *J. Sound Vibr.* 413: 225–249. https://doi.org/10.1016/j.jsv.2017.10.025.

Ao, W., and P. Reynolds. 2017. "Analytical and experimental study of eddy current damper for vibration suppression in a footbridge structure." In *Proc., Dynamics of Civil Structures*, 131–138. Cham, Switzerland: Springer.

Bae, J. S., J. H. Hawng, J. H. Roh, J. H. Kim, M. S. Yi, and J. H. Lim. 2012. "Vibration suppression of a cantiliver beam using megnitically tuned-mass-damper." *J. Sound Vibr.* 331 (26): 5669–5684. https://doi.org/10.1016/j.jsv.2012.07.020.

Bourquin, F., G. Caruso, M. Peigney, and D. Siegert. 2014. "Magnetically tuned mass dampers for optimal vibration damping of large structures." *Smart Mater. Struct.* 23 (8): 085009. https://doi.org/10.1088/0964-1726/23/8/085009.

Chen, J., G. Lu, Y. Li, T. Wang, W. X. Wang, and G. B. Song. 2017. "Experimental study on robustness of an eddy current-tuned mass damper." *J. Appl. Sci.* 7 (9): 895. https://doi.org/10.3390/app7090895.

Chung, L. L., L. Y. Wu, and C. S. Yang. 2013. "Optimal design formulas for viscous tuned mass dampers in wind-excited structures." *Struct. Control Health Monit.* 20 (3): 320–336. https://doi.org/10.1002/stc.496.

Ebrahimi, B., M. B. Khamesee, and F. Golnaraghi. 2009. "Eddy current damper feasibility in automobile suspension: Modeling, simulation and testing." *Smart Mater. Struct.* 18 (1): 015017. https://doi.org/10.1088/0964-1726/18/1/015017.

Furlani, E. P. 2001. *Permanent magnet and electromechanical devices: Materials, analysis, and applications*. San Diego, CA: Academic Press.

Gay, S. E. 2005. "Contactless magnetic brake for automotive applications." Ph.D. thesis, Dept. of Electrical Engineering, Texas A&M Univ.

Gay, S. E., and E. Mehrdad. 2006. "Parametric analysis of eddy current brake performance by 3-D finite element analysis." *IEEE Trans. Magn.* 42 (2): 319–328. https://doi.org/10.1109/TMAG.2005.860782.

Hua, X. G., Q. Wen, Z. Q. Chen, Y. Yang, and H. W. Niu. 2016. "Control of human-induced vibrations of a curved cable-stayed bridge: Design, implementation and field validation." *J. Bridge Eng.* 27 (1): 04016028. https://doi.org/10.1061/(ASCE)BE.1943-5592.0000887.

Kim, J., J. Ryu, and L. Chung. 2006. "Seismic performance of structures connected by viscoelastic dampers." *Eng. Struct.* 28 (2): 183–195. https://doi.org/10.1016/j.engstruct.2005.05.014.

Konstantinidis, D., J. M. Kelly, and N. Makris. 2011. *In-situ monitoring of the force output of fluid dampers: Experimental investigation*. PEER Rep. 2011/103. Berkeley, CA: Pacific Earthquake Engineering Research Center.

Larose, G. L., A. Larsen, and E. Svensson. 1995. "Modelling of tuned mass dampers for wind-tunnel tests on a full-bridge aeroelastic model." *J. Wind Eng. Ind. Aerodyn.* 54–55 (Feb): 427–437. https://doi.org/10.1016/0167-6105(94)00058-L.

Lavan, O. 2015. "Optimal design of viscous dampers and their supporting members for the seismic retrofitting of 3D irregular frame structures." *J. Struct. Eng.* 141 (11): 04015026. https://doi.org/10.1061/(ASCE)ST.1943-541X.0001261.

Lu, X. L., Q. Zhang, S. Wang, S. A. Mahin, Q. Ding, and F. Qian. 2017. "Improving performance of a super tall building using a new eddy-current tuned mass damper." *Struct. Control Health Monit.* 24 (3): e1882. https://doi.org/10.1002/stc.1882.

Mualla, I. H., and B. Belev. 2002. "Performance of steel frames with a new friction damper device under earthquake excitation." *Eng. Struct.* 24 (3): 365–371. https://doi.org/10.1016/S0141-0296(01)00102-X.

Nasar, S., and G. Xiong. 1988. "Determination of the field of a permanent-magnet disk machine using the concept of magnetic charge." *IEEE Trans. Magn.* 24 (3): 2038–2044. https://doi.org/10.1109/20.3398.

Palomera-Arias, R., J. J. Connor, and J. A. Ochsendorf. 2008. "Feasibility study of passive electromagnetic damping systems." *J. Struct. Eng.* 134 (1): 164–170. https://doi.org/10.1061/(ASCE)0733-9445(2008)134:1(164).

Pluk, K. J. W., T. A. V. Beek, J. W. Jansen, and E. A. Lomonova. 2014. "Modeling and measurements on a finite, rectangular, conducting plate in an eddy current damper." *IEEE Trans. Ind. Electron.* 61 (8): 4061–4072. https://doi.org/10.1109/TIE.2013.2279364.

Rovers, J., J. Jansen, and E. Lomonova. 2013. "Modeling of relative permeability of permanent magnet material using magnetic surface charges." *IEEE Trans. Magn.* 49 (6): 2913–2929. https://doi.org/10.1109/TMAG.2012.2236103.

Sakamoto, H., K. Araki, A. Ishida, S. Kobayashi, and T. Kuwahara. 1997. *Design of permanent magnet type compact ECB retarder.* SAE Technical Paper 973228. Warrendale, PA: SAE.

Sodano, H. A., and J. S. Bae. 2004. "Eddy current damping in structures." *Shock Vibr. Digest* 36 (6): 469–478. https://doi.org/10.1177/0583102404048517.

Sodano, H. A., J. S. Bae, D. J. Inman, and W. K. Belvin. 2006. "Improved concept and model of eddy current damper." *J. Vibr. Acoust. ASME* 128 (3): 294–302. https://doi.org/10.1115/1.2172256.

Song, G. B., P. Zhang, L. Y. Li, and M. Singla. 2016. "Vibration control of a pipeline structure using pounding tuned mass damper." *J. Eng. Mech.* 142 (6): 04016031. https://doi.org/10.1061/(ASCE)EM.1943-7889.0001078.

Soong, T. T., and G. F. Dargush. 1997. *Passive energy dissiptaion systems in structural engineering*. New York, NY: Wiley.

Soong, T. T., and B. F. Spencer. 2002. "Supplemental energy dissipation: State-of-the-art and state-of-the-pratice." *Eng. Struct.* 24 (3): 243–259. https://doi.org/10.1016/S0141-0296(01)00092-X.

Symans, M. D., F. A. Charney, A. S. Whittaker, M. C. Constantinous, C. A. Kircher, M. W. Johnson, and R. J. McNamara. 2008. "Energy dissipation systems for seismic applications: Current practice and recent developments." *J. Struct. Eng.* 134 (1): 3–21. https://doi.org/10.1061/(ASCE)0733-9445(2008)134:1(3).

Thompson, M. T. 2009. "Practical issues in the use of NdFeB permanent magnets in maglev, motors, bearings, and eddy current brakes." *Proc. IEEE* 97 (11): 1758–1767. https://doi.org/10.1109/JPROC.2009.2030231.

Vargas, R., and M. Bruneau. 2007. "Effect of supplement viscous damping on the seismic response of structural systems with metallic dampers." *J. Struct. Eng.* 133 (10): 1434–1444. https://doi.org/10.1061/(ASCE)0733-9445(2007)133:10(1434).

Wang, Z. H., Z. Q. Chen, and J. H. Wang. 2012. "Feasibility study of a large-scale tuned mass damper with eddy current damping mechanicsm." *Earthquake Eng. Eng. Vibr.* 11 (3): 391–401. https://doi.org/10.1007/s11803-012-0129-x.

Warburton, G. W. 1982. "Optimum absorber parameters for various combinations of response and excitation parameters." *Earthquake Eng. Struct. Dyn.* 10 (3): 381–401. https://doi.org/10.1002/eqe.4290100304.

Weber, B., and G. Feltrin. 2010. "Assessment of long-term behavior of tuned mass dampers by system identification." *Eng. Struct.* 32 (11): 3670–3682. https://doi.org/10.1016/j.engstruct.2010.08.011.

Weber, F., G. Feltrin, M. Maslanka, W. Fobo, and H. Distl. 2009. "Design of viscous dampers targeting multiple cable modes." *Eng. Struct.* 31 (11): 2797–2800. https://doi.org/10.1016/j.engstruct.2009.06.020.

Xu, Z. D., H. F. Xu, and X. Chen. 2016. "Intelligent vibration isolation and mitigation of a platform by using MR and VE devices." *J. Aerosp. Eng.* 29 (4): 04016010. https://doi.org/10.1061/(ASCE)AS.1943-5525.0000604.

Zhang, P., L. Li, D. Patil, M. Singla, H. N. Li, Y. L. Mo, and G. Song. 2016. "Parametric study of pounding tuned mass damper for subsea jumpers." *Smart Mater. Struct.* 25 (1): 015028. https://doi.org/10.1088/0964-1726/25/1/015028.

Zuo, L., X. Chen, and S. Nayfeh. 2011. "Design and analysis of a new type of electromagnetic damper with increased energy density." *J. Vibr. Acout. ASME* 133 (4): 041006. https://doi.org/10.1115/1.4003407.

第33卷 第11期 2020年11月
中国公路学报
China J. Highw. Transp.
Vol. 33 No. 11 Nov. 2020

文章编号:1001-7372(2020)11-0083-18

永磁电涡流阻尼新技术及其在土木工程中的应用

陈政清*[1,2],华旭刚[1,2],牛华伟[1,2],黄智文[1,2],张弘毅[1,2],陈谨林[1,2]

(1. 湖南大学 振动与冲击技术研究中心,湖南 长沙 410082; 2. 湖南大学 风工程与桥梁工程省重点实验室,湖南 长沙 410082)

摘要:针对永磁电涡流阻尼减振的理论与技术进展进行了综述,主要包括:永磁电涡流调谐质量减振器、大吨位的滚珠丝杠式轴向电涡流阻尼器等新型减振装置;板式电涡流阻尼及滚珠丝杠式电涡流阻尼器的性能仿真与设计方法;桥梁涡激共振、斜拉索振动控制的电涡流阻尼三元减振理论;以及基于电涡流阻尼力非线性特性的减震性能研究。介绍了湖南大学新建的大型阻尼器特性综合试验平台及主要功能,以及永磁电涡流阻尼技术的典型工程应用。最后指出了值得进一步研究的若干问题。

关键词:桥梁工程;电涡流阻尼器;综述;性能测试;振动控制;惯质

中图分类号:U44 **文献标志码:**A

Technological Innovations in Eddy Current Damping and Its Application in Civil Engineering

CHEN Zheng-qing*[1,2], HUA Xu-gang[1,2], NIU Hua-wei[1,2], HUANG Zhi-wen[1,2], ZHANG Hong-yi[1,2], CHEN Jin-ling[1,2]

(1. Vibration and Shock Technology Research Center, Hunan University, Changsha 410082, Hunan, China; 2. Key Laboratory for Wind and Bridge Engineering of Hunan Province, Hunan University, Changsha 410082, Hunan, China)

Abstract: The paper presents a summary review of the theoretical and technical innovations of eddy current damping (ECD) technology. The main contents include: ① development of two kinds of eddy current dampers, namely eddy current tuned mass dampers (EC-TMD) and ball-screw-based axial eddy current dampers with large output force; ② development and validation of the linear/nonlinear mathematical models for eddy current damping forces of EC-TMD and axial ECD; ③ vibration control of stay cables and bridge deck subjected to VIV with inerter-based damping devices, and seismic control of SDOF system considering nonlinear damping forces of axial ECD. The main function of newly built large-scale test platform for eddy current dampers is briefly described; A few typical examples of engineering applications of the ECD technology are also introduced. At last, some topics deserving further research are highlighted.

Keywords: bridge engineering; eddy current damper; review; performance test; vibration control; inerter

Received 05 July 2020

Funding: National Natural Science Foundation of China (50738002, 51422806, 51808210)

收稿日期:2020-07-05

基金项目:国家自然科学基金项目(50738002,51422806,51808210)

***作者简介:**陈政清(1947-),男,湖南邵东人,教授,博士研究生导师,中国工程院院士,工学博士,E-mail:zqchen@hnu.edu.cn。

*Corresponding author. *E-mail address*: zqchen@hnu. edu. cn.
https://doi. org/10. 19721/j. cnki. 1001-7372. 2020. 11. 009

0 引 言

地震和风灾是威胁人类生命和财产安全的 2 种最主要的自然灾害。如何抑制或减小土木工程结构在地震、强风及其他激振源作用下的动力响应,提高土木工程结构的防灾抗灾能力,一直都是土木工程学科及结构动力学的前沿课题[1]。结构振动控制技术发展至今已有 50 多年的历史,按照是否有外部能源输入,可以分为被动控制、主动控制、半主动控制和混合控制[2]。被动控制主要包括基础隔震和被动耗能减振技术,其中被动耗能减振技术通过在结构中设置各类阻尼装置,利用阻尼装置耗散结构振动的部分能量,可以降低结构的动力响应,显著提高结构抵御地震及其他灾害的能力[3]。被动耗能减振技术无需外部能源输入,具有较高的可靠性,在美国和日本等多地震发达国家已有广泛应用,近 20 多年在中国的发展也非常迅速[4]。很多应用被动耗能减振技术的结构都经受住了大地震和强风的考验,从而进一步推动了被动耗能减振技术的发展。

结构被动控制的核心技术之一在于开发可靠性高、耐久性好的阻尼装置。土木工程领域常用的被动阻尼装置包括摩擦阻尼器、金属阻尼器、黏弹性固体阻尼器、黏弹性流体阻尼器和黏滞流体阻尼器(以油阻尼器为代表)等[5-6]。油阻尼器可以满足大行程要求,20 世纪 80 年代末开始在土木工程领域应用。美国在这方面的研究与应用起步较早,中国油阻尼器产品经过 10 多年的发展也取得了很大的技术进步,已经在实际工程中应用。为了实现高耗能密度,土木工程使用的油阻尼器与机械工程的有源液压器件和内外管式油阻尼器不同,它采用了单管密封高压油(可达 100 MPa)结构[7]。然而,这种结构也带来 2 个难以克服的缺点:密封摩擦力大、减振灵敏度低。一旦磨损漏油就立即失去阻尼力,只能更换[8-9]。因此,目前使用的很多油阻尼器达不到减振设计要求,甚至 3～5 年就要更换,类似问题也存在于很多重型装备上的减振缓冲装置中。

电涡流阻尼(Eddy Current Damping, ECD)依据电磁感应原理产生阻尼,没有摩擦,也没有工作流体,具有结构简单、可靠性高、耐久性好、阻尼系数易调节等优点,是一种理想的阻尼形式[10]。电涡流阻尼在仪表、机械和外层空间技术中已有应用。早在 1936 年,Andre 等[11]首次提出将电涡流阻尼技术应用到汽车制动上,并申请了发明专利,该专利后来发展成为了法国 Telma 公司生产的电涡流制动器的雏形。目前,电涡流制动器已经在重载汽车和商用客车上得到了普遍应用。除了在重载汽车上作为辅助刹车装置外,电涡流制动器也在高速列车上使用,例如,20 世纪 80 年代日本在 100 系动车组上首创了盘形电涡流制动机制;2000 年,德国铁路 ICE3 型高速列车首次使用了平动涡流制动技术[12]。此外,电涡流阻尼技术还被用来控制旋转机械在超临界工作速度下的自激振动[13]。这些装置都是采用电磁铁,以间歇工作方式服役,工作速度大,阻尼系数小(一般小于 10 $kN \cdot s \cdot m^{-1}$)。对于需要连续工作长达数十年的大型土木工程结构,其减振宜采用永磁体。

与车辆、旋转机械的高速运动相比,土木工程结构的振动速度低几个数量级,在 1～1 000 $mm \cdot s^{-1}$ 之间,而且结构质量巨大,因此大质量低频土木工程结构的减振阻尼器需要巨大的阻尼系数,即便是斜拉桥拉索阻尼器,阻尼系数也在 100 $kN \cdot s \cdot m^{-1}$ 左右,抗震阻尼器的阻尼系数还要高 1～2 个数量级。从工程实用角度还必须考虑阻尼器的效率指数,其是指额定工作速度下,阻尼器阻尼力与其自身重量之比。大型阻尼器的效率指数必须在 100 以上才有工程应用价值,以前只有油阻尼器可以满足这个要求。电涡流阻尼本身耗能密度远低于黏滞流体阻尼,要实现与油阻尼器相当的效率是一个很大的难题,这是电涡流阻尼长期不能应用于大型土木工程的主要原因。例如,美国泰勒设备公司(Taylor Devices)制造的黏滞流体阻尼器的效率指数达到 200 左右,而他们 2002 年试制的一个电涡流阻尼器的效率指数只有 0.5,于是泰勒设备公司放弃了开发电涡流阻尼器的计划[14]。日本在 1990 年代也提出过一些电涡流阻尼器的专利,但是都没有达到工程应用要求。例如,文献[15]所述的日本专利不仅没有注意提高电涡流耗能效率,而且需要用机械方式连接,完全失去了电涡流阻尼无摩擦的本质优势。日本也开发过永磁电涡流制动器,也因阻力矩小,自重大,没有得到推广。

在中国国家自然科学基金等多个项目资助下,结合工程实际问题,笔者从提高电涡流阻尼效率和发挥电涡流阻尼无摩擦优势2个方面入手，进行了长期深入的研究，先后发明了配置复合板式永磁体电涡流阻尼单元的调谐质量减振器、大吨位的滚珠丝杠式轴向电涡流阻尼器等新型阻尼装置，建立了永磁电涡流阻尼器在低速及高速运动下线性与非线性阻尼力的数学模型，研究了桥梁涡激共振、斜拉索振动控制的电涡流阻尼三元被动控制理论，以及基于电涡流阻尼非线性特性的结构减震性能，并在工程应用实践中探索了相关的动力试验技术。基于上述研究成果开发的电涡流调谐质量阻尼器及轴向电涡流阻尼器，已在中国多座大跨度桥梁和多种建筑结构减振工程中得到应用，如湖南岳阳洞庭湖悬索桥和张家界玻璃桥、上海中心大厦、摩洛哥太阳能热电站及北京大兴国际机场等。

1　电涡流调谐质量减振器

调谐质量减振器（TMD）是一种应用广泛的吸能减振装置，它由质量单元、弹性刚度单元和阻尼单元构成，还必须有支撑和导向结构。当TMD的振动频率调谐与结构受控模态的固有频率一致时，TMD可以显著耗散结构振动的能量，从而达到减振的目的。TMD中的阻尼单元通常采用油阻尼器。由于油阻尼器的密封系统复杂、启动摩擦力大等原因，采用油阻尼的传统调谐质量减振器存在减振灵敏度不高、阻尼调节不便、服役寿命短等缺点。

1.1　基本结构

电涡流阻尼是基于电磁感应原理的一种物理现象，在低速时是一种理想的线性黏滞阻尼。如图1所示，当导体板在局部磁场中运动时，导体板内的磁通量发生变化，由电磁感应原理，在导体板内马上产生感应电流，并自动在板内形成涡电流，这就是电涡流效应。涡电流磁场又与原磁场相互作用，产生一个阻碍导体板运动的力，即电涡流阻尼力。在这一过程中，导体板与磁场相对运动的动能先转变为电能，再在导体板内转化为热能消耗于大气中，于是起到耗能减振的作用。由于导体板不与磁体接触，这是一种没有摩擦和磨损的理想阻尼方式，不会产生油阻尼器因密封带来的一系列问题。

然而，图1所示的传统电涡流阻尼生成方式有诸多缺点：①由于两磁体分置于导体板的两侧，漏磁大，有2道空气间隙导致磁阻偏大，是耗能密度低的重要原因；②导体板位于两磁体中间，在空间上受到很大约束，导致阻尼调节很不方便，采用永磁体时安装十分困难。图2给出了前言所述的电涡流调谐质量减振器日本专利[15]，可以看到，由于该专利也是采用传统的电涡流阻尼生成方式，所以阻尼调节也非常不便。

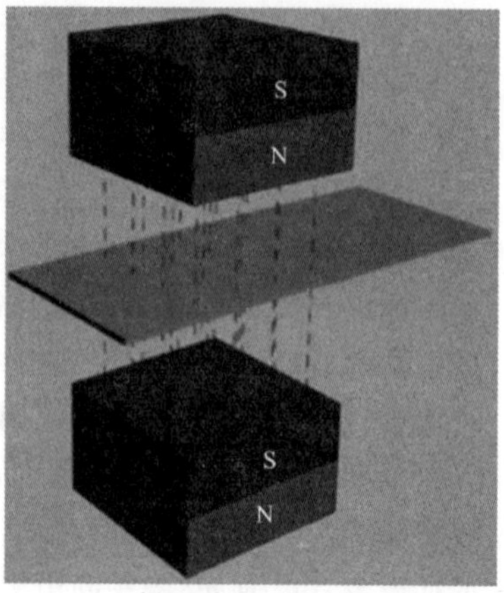

图1　电涡流阻尼生成的传统模式

Fig. 1　Traditional Method of Producing Eddy Current Damping

1.2　板式电涡流阻尼

笔者团队从2006年开始研究电涡流阻尼应用于大型土木工程减振的可行性。针对传统电涡流阻尼生成方式耗能效率低，导体板与磁体在空间上相互约束的问题，研究改进措施，于2010年发明了复合板式电涡流阻尼单元[16-19]，如图3(a)所示。它将两磁体置于同一侧，直接吸附于导磁背铁上，导体板固定在另一块背铁上，置于磁体下方。与传统电涡流方式相比，复合板式电涡流单元具有两大优点。

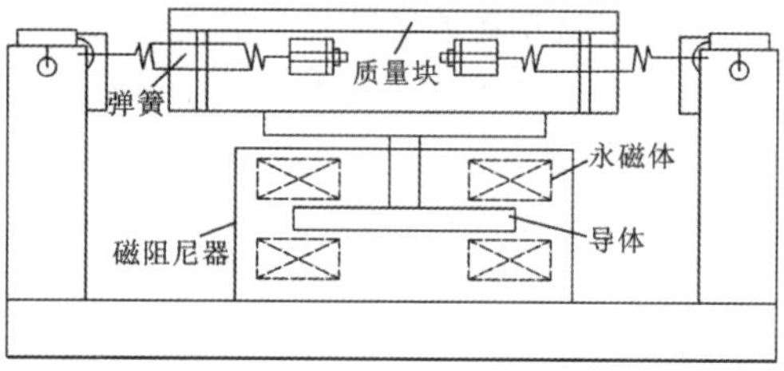

图2　1998年日本电涡流阻尼器专利[15]

Fig. 2　Schematic of an Eddy Current Tuned Mass Damper in a Japanese Patent (1988)[15]

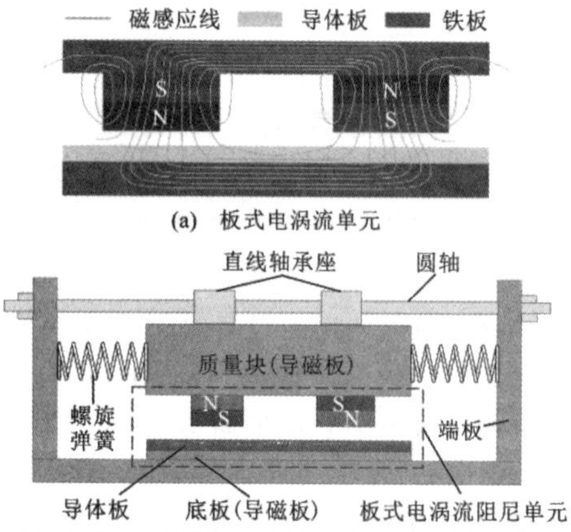

(a)　板式电涡流单元

(b)　采用板式电涡流单元的调谐质量阻尼器(TMD)

图3　板式电涡流阻尼单元及电涡流调谐质量减振器

Fig. 3　Schematic of the Planar Eddy Current Damping Element and the Eddy Current TMD

(1)实现了最短磁路设计,最大限度地减少漏磁和磁阻,并且导体板2次切割磁力线,这些措施大幅提高了电涡流阻尼耗能效率。表1给出了电涡流阻尼的传统模式、空间阵列式及板式3种结构的耗能密度对比,其中耗能密度按国际惯例定义为阻尼系数与阻尼单元轮廓体积之比。从表1看出,在相同体积及永磁体配置下,板式电涡流阻尼的耗能密度是传统模式的6倍多,是2011年佐治亚理工大学左磊教授等[20]研发的高密度空间阵列式电涡流阻尼的1.3倍。

表1 电涡流阻尼3种生成方式的耗能密度对比

Table 1 Comparison of the Energy Dissipation Density of Three Different Methods

电涡流阻尼生成方式	传统模式[1]	空间阵列式[20]	板式
阻尼系数体积密度比/($kN \cdot s \cdot m^{-4}$)	200	1 061	1 310

(2)形成了几何开放结构,磁场部分与导体部分在空间上互不约束。采用板式电涡流单元的调谐质量阻尼器示意图如图3(b)所示,两大部分分别安装在质量块和基座上,相互之间没有机械连接及由此产生的静摩擦力,阻尼调节方便,永磁体安装难度低。将图3(b)和图2对比可知,板式电涡流单元的构造优点十分明显,也优于左磊教授的阵列式。

基于板式电涡流阻尼单元,笔者团队研制了水平滑轨式、竖向支撑式、摆式等多种类型的调谐质量减振器(TMD),如图4所示,可以适应不同类型结构的减振。

1.3 低速电涡流阻尼力仿真与试验技术

1.3.1 低速电涡流阻尼力仿真

结构减振设计与分析需要建立板式电涡流阻尼单元阻尼力的数学模型。板式电涡流阻尼单元的阻尼力影响因素众多,主要包括导体板的尺寸和电磁特性、永磁体的形状、尺寸、磁化特性以及空间布置、导体板与永磁体的气隙等。板式电涡流阻尼器作为TMD的阻尼单元主要在低速下工作,为此,基于电磁场理论分析及三维有限元分析建立了低速下板式电涡流阻尼系数分析方法,研究了各设计参数对电涡流阻尼系数的影响,并开发了电涡流阻尼系数的测试装置,形成了电涡流阻尼的仿真与试验成套技术[21-24]。

(a) 水平滑轨式　　(b) 竖向支撑式　　(c) 摆式

图4 三种系列电涡流调谐质量减振器(TMD)

Fig. 4 Three Series of Eddy Current TMD

对于如图3(a)所示的板式电涡流阻尼器,当导体板发生运动时,可以利用洛伦兹力公式计算导体板受到的电涡流阻尼力 $\boldsymbol{F}$

$$\boldsymbol{F} = \iiint_{V_c} \boldsymbol{J}^s \boldsymbol{B} \mathrm{d}V \tag{1}$$

式中:$\boldsymbol{B}$ 为磁感应强度,$\boldsymbol{B} = B_x \boldsymbol{e}_i + B_y \boldsymbol{e}_j + B_z \boldsymbol{e}_k$,其中 x,y 均为导体板平面方向,z 为竖向;$\boldsymbol{J}^s$ 为导体板中电涡流强度,$\boldsymbol{J}^s = J_x^s \boldsymbol{e}_i + J_y^s \boldsymbol{e}_j + J_z^s \boldsymbol{e}_k$。磁感应强度和电涡流强度都是空间位置和时间的函数。

忽略沿 z 轴方向的电涡流(即 $J_z = 0$),可得沿导体板运动方向的阻尼力 F_x 为

$$F_x = \iiint_{V_c} J_y^s B_z \mathrm{d}V \tag{2}$$

$$J_y^s = \int \sigma v_x \frac{\partial B_y}{\partial x} \mathrm{d}z - \sigma v_x B_z - \int \sigma v_x \frac{\partial B_y}{\partial y} \mathrm{d}z \tag{3}$$

式中:σ 为介质的电导率。从式(2)可以看到,对阻尼力 F_x 影响最大的是与导体板运动方向垂直的电涡流分量 J_y^s 以及与导体板平面垂直的磁感应强度分量 B_z。从式(3)看出,电涡流阻尼力为速度的线性函数,因此板式电涡流阻尼在低速状态是一种理想的线性黏滞阻尼。考虑到低速条件下电涡流阻尼系数与速度无关,所以可以用阻尼系数为指标来衡量板式电涡流阻尼器的阻尼性能。

图5给出了采用理论公式和电磁有限元方法计算得到的电涡流阻尼系数随气隙大小的变化规律。可以看到,无论气隙大小,理论公式与有限元计算的结果都非常接近。通过调节永磁体与导体板之间的气隙大小,即可调节阻尼系数及电涡流调谐质量减振器的阻尼比,因此使用非常方便。图6展示了永

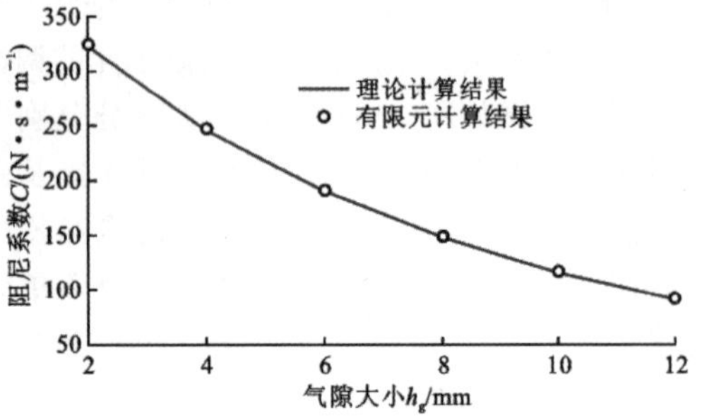

图5 电涡流阻尼系数随气隙大小的变化规律

Fig. 5 Damping Coefficient Versus Air-gap Distance

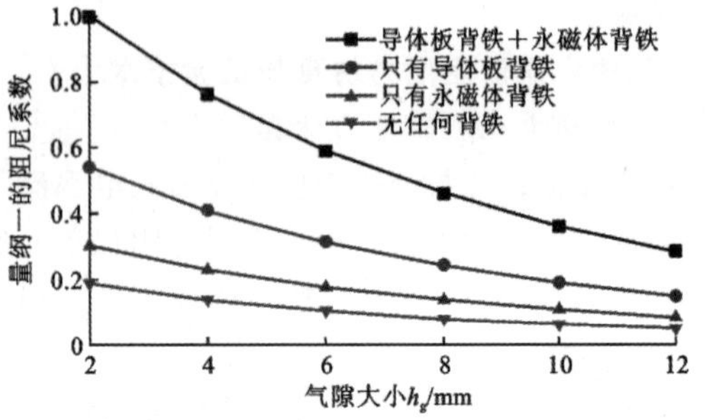

图6 背铁对阻尼系数的影响规律

Fig. 6 Influence of Back Iron on Damping Coefficient

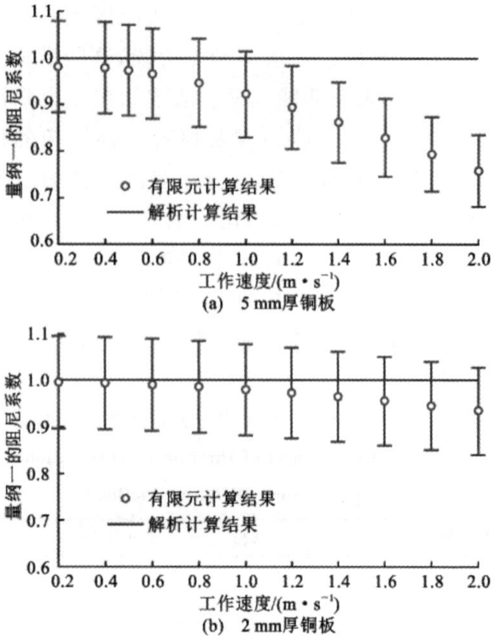

图7 量纲一的电涡流阻尼系数随工作速度的变化曲线

Fig. 7 Variation Curve of the Dimensionless Eddy Current Damping Coefficient with the Working Velocity

磁体背铁和导体板背铁对电涡流阻尼系数的影响，其中阻尼系数均以气隙大小为2 mm的板式电涡流阻尼器进行量纲一化。可以看出，导体板背铁和永磁体背铁都对阻尼系数有非常显著的影响，而且导体板背铁的影响更大，在导体板下增设背铁形成复合板，是提高电涡流阻尼耗能效率的一个关键措施。

上述理论推导中忽略了电涡流的感应磁场对永磁体磁场的影响，笔者采用有限元分析进一步评估了这一影响。图7分别给出了板式电涡流阻尼器采用5 mm厚铜板和2 mm厚铜板时阻尼系数随速度的变化情况，其中纵坐标所示的阻尼系数均以线性理论假定的计算结果进行了量纲一化。可以看到，随着工作速度的增加，阻尼系数逐渐降低。图7中还给出了阻尼系数的±10%误差棒，以评价线性阻尼假定适用的工作速度范围。可以看到，对于5 mm厚铜板，只要工作速度小于等于1.0 m·s⁻¹，线性假定的计算误差都在10%以内，此时可以认为线性阻尼假定是成立的。而对于2 mm厚铜板，即使工作速度达到2 m·s⁻¹，线性假定的计算误差仍在10%以内，其线性阻尼工作范围大幅增加。综上所述，板式电涡流阻尼器的阻尼特性受导体板厚度和工作速度的影响，在电涡流调谐质量减振器的常用工作速度范围，线性阻尼假定往往是成立的，电涡流阻尼在低速时的线性阻尼特性使其非常适合用于调谐质量减振器。

1.3.2 低速电涡流阻尼力测试技术

笔者团队研发了一套板式电涡流阻尼测定装置，可以用来测试不同工况的准确阻尼系数，校核理论公式及有限元计算结果。如图8所示，振动测试装置是一个强梁弱柱的单层框架，它在水平方向上的自由振动以一阶模态为主。铜板和背铁安装在框架上方，它们与永磁体之间的气隙可以通过图8中的高度调节螺杆进行调节，从而研究气隙大小对阻尼系数的影响。通过比较铜板安装前后振动测试装置阻尼比的变化情况来确定板式电涡流阻尼器的阻尼系数。

针对某典型工况，图9给出了改变相邻磁体间隙时电涡流阻尼系数随气隙大小的变化曲线。可以看到，测定装置能够准确测试气隙大小为2.5～30 mm之间的板式电涡流阻尼器的阻尼系数。试验发现：一方面，随着气隙的增大，电涡流阻尼系数快速减小；另一方面，随着相邻磁体间距的增大，电涡流阻尼系数先增大后减小，故存在最优的磁体间隙使阻尼系数达到最大值。

2 滚珠丝杠式轴向电涡流阻尼器

虽然永磁体板式电涡流阻尼单元的耗能效率有较大提高，并可将其用作调谐质量减振器TMD的阻尼单元，但对于大型结构直接耗能减振的轴向阻尼器，还需要进一步提高电涡流阻尼器的耗能密度。

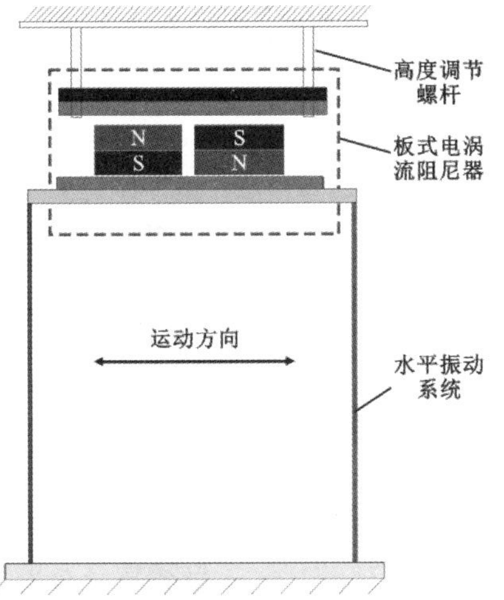

图 8 板式电涡流阻尼器的试验测试原理

Fig. 8 Description of the Test Rig of a Steel Frame

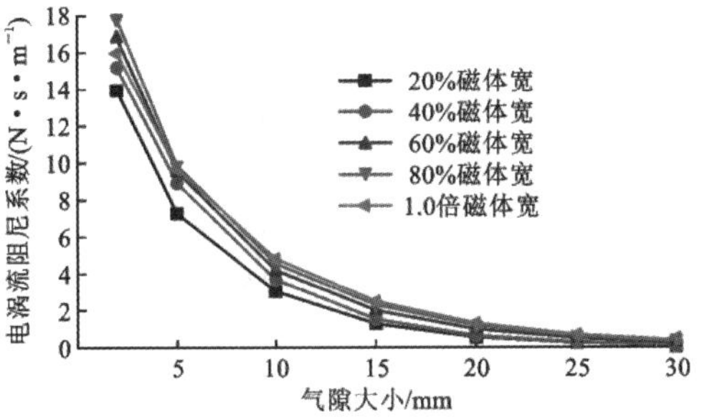

图 9 不同磁体间距下电涡流阻尼系数随气隙的变化规律

Fig. 9 Variations of ECD Coefficient with Air Gap Under Different Magnet Spacing

利用螺旋传动装置，将轴向阻尼器两端的直线相对运动转化为阻尼器内部的高速转动，可以制造出更大阻尼力的电涡流阻尼器，达到与油阻尼器相当的耗能密度，同时又避免了油阻尼器初始摩擦力大、减振灵敏度不高、服役周期短等缺点。齿轮齿条和滚珠丝杠是 2 类最为常用的螺旋传动机构。滚珠丝杠传动具有传动效率高、初始摩擦小及工作平稳等优点。滚珠丝杠式轴向电涡流阻尼器不仅能够用于对控制力和工作速度要求都很高的地震响应控制，而且非常适用于像风振和车致振动这一类频繁、往复的低速、小位移的振动控制。另外，它还具有惯质特性，可以发挥三元减振控制理论的优势。

2.1 基本结构

从 2013 年以来，笔者团队相继开发了基于滚珠丝杠的多种形式的轴向电涡流阻尼器[25-27]，最大出力超过 200 t。图 10 给出了开发的 3 类大吨位的轴向电涡流阻尼器，分别是盘式轴向电涡流阻尼器、外旋转管式轴向电涡流阻尼器和内旋转管式轴向电涡流阻尼器。以盘式轴向电涡流阻尼器为例，滚珠丝杠在结构构件相对运动的驱使下作直线运动，螺母通过推力轴承与阻尼器的外筒相连，在滚珠丝杠的驱动下带动导体圆盘一起做旋转运动，从而使盘式轴向电涡流阻尼器像汽车中使用的轴向电涡流制动器一样工作。轴向电涡流阻尼器中圆盘上产生的阻尼力矩经过滚珠丝杠副的逆传动作用又转化为轴向电涡流阻尼力施加到结构上。与板式电涡流阻尼器一样，永磁体在背铁沿圆周向按照正负磁极交替的方式进行布置，上下背铁对应位置上的永磁体极化方向相同。由于圆盘转动还产生惯性力，通过滚珠丝杠放大后可以形成远比圆盘物理质量大的惯性力。

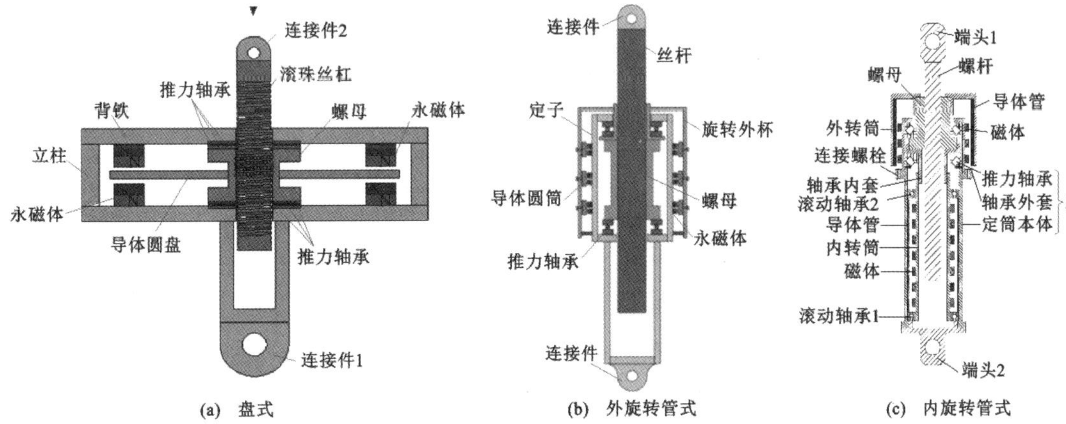

图 10 滚珠丝杠式电涡流轴向阻尼器

Fig. 10 Ball Screw Type Eddy Current Axial Damper

2.2 电涡流高速性能仿真与设计方法

2.2.1 力学模型

如图 10(a)所示，当滚珠丝杠式轴向电涡流阻尼器的两端发生相对位移 u 时，滚珠丝杠的轴向位移也为 u，螺母和导体圆盘的旋转角度 θ 可由滚珠丝杠副的几何特性得到

$$\theta=(2\pi)u/L \tag{4}$$

式中：L 为滚珠丝杠的导程，即螺母旋转一周滚珠丝杠的轴向位移。

往复转动的圆盘产生惯性力及阻尼力，阻尼器内部滚珠丝杠及连接件之间还有一定的机械摩擦力，阻尼器两端的轴向力可以描述为惯性力、阻尼力和摩擦力之和，即

$$F=m_e\ddot{u}+c_e\dot{u}+f_0\operatorname{sign}(\dot{u}) \tag{5}$$

$$\left.\begin{aligned} m_e&=\frac{4\pi^2}{\eta L^2}J \\ c_e&=\frac{4\pi^2}{\eta L^2}C \end{aligned}\right\} \tag{6}$$

式中：m_e 为滚珠丝杠式轴向电涡流阻尼器的表观质量，对应于圆盘转动产生的惯性力；c_e 为滚珠丝杠式轴向电涡流阻尼器的轴向等效阻尼系数，对应于磁场中圆盘转动受到的电涡流阻尼力，可以采用电磁场有限元分析得到；f_0 为摩擦力，可以通过试验测得；J 为圆盘的总转动惯量；C 为圆盘上轴向电涡流阻尼器的阻尼系数；η 为滚珠丝杠副的逆传动效率，一般在 0.85～0.90 之间。滚珠丝杠的导程一般是几十毫米量级，可以看出表观质量和等效电涡流阻尼系数都可以放大成千上万倍。

2.2.2　电涡流高速性能仿真技术[28-30]

由于采用滚珠丝杠的放大作用，圆盘转动的切线速度可以达到 10 m·s^{-1} 甚至更高，此时电涡流引起的感应磁场和永磁体的原磁场会相互干扰，其影响不可忽略，最终使电涡流阻尼力具有非线性特性。滚珠丝杠式轴向电涡流阻尼器的非线性特性是指轴向电涡流阻尼力随滚珠丝杠轴向运动速度的变化规律。针对特定的滚珠丝杠式轴向电涡流阻尼器，可以采用 COMSOL 的准静态分析模块 AC/DC 下的稳态分析方法对轴向电涡流阻尼器进行有限元分析，得到电涡流阻尼力矩随导体圆盘转速的变化曲线，再按式(5)就可以换算得到等效的电涡流阻尼系数。

图 11 给出了轴向电涡流阻尼力随滚珠丝杠轴向速度的变化曲线，其中的横坐标和纵坐标分别以最大阻尼力及对应的临界速度进行量纲一化[30]。当轴向速度较小时，轴向电涡流阻尼力与轴向速度近似成线性关系。随着速度增大，轴向电涡流阻尼力的增长逐渐减缓，当到达临界速度时阻尼力达到峰值。超过临界速度后，阻尼力随速度的增大而逐渐减小。轴向电涡流阻尼力随轴向速度的变化规律是由导体圆盘中电涡流形成的感应磁场对永磁体磁

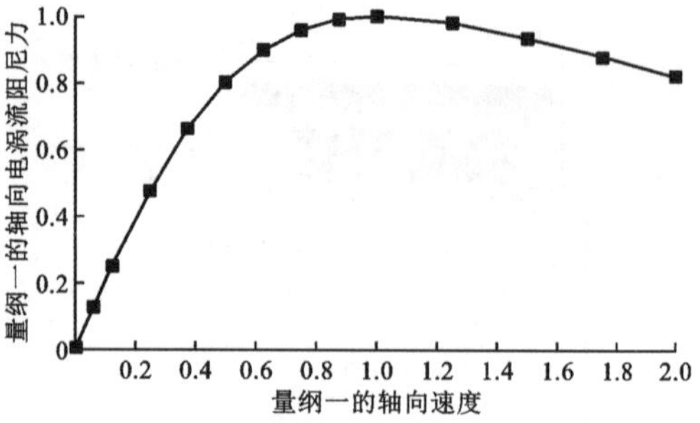

图 11　轴向电涡流阻尼力随轴向速度的变化

Fig. 11　Axial Eddy Current Damping Force Versus Axial Velocity

场的削弱造成的。因此，经过速度放大的轴向电涡流阻尼器与油阻尼器在阻尼力的高速特性方面截然不同。

导体圆盘的电导率和厚度、气隙大小、永磁体对数和形状是影响滚珠丝杠式轴向电涡流阻尼器非线性阻尼性能的主要因素，这里重点讨论临界速度和最大轴向电涡流阻尼力随上述参数的变化情况。图 12(a)以量纲一的参数的形式给出了轴向电涡流阻尼力的速度曲线随导体圆盘厚度的变化情况，其中横坐标所示的轴向速度和纵坐标所示的轴向电涡流阻尼力分别以厚度 2 mm 导体圆盘的临界速度和最大阻尼力进行量纲一化。可以看到，随着导体圆盘厚度增大，最大轴向电涡流阻尼力和临界速度都逐渐减小，而且临界速度与导体板厚度成反比。图 12(b)给出了相同厚度导体圆盘分别采用铜、铝和锌 3 种不同电导率的材料时，轴向电涡流阻尼力随轴向速度的变化情况。可以看到，随着导体圆盘电导率的降低，临界速度近似成比例增大，但最大轴向电涡流阻尼力保持不变。综上所述，导体圆盘的厚度和电导率对滚珠丝杠式轴向电涡流阻尼器的阻尼力特性有非常重要的影响。

2.3　非线性阻尼力的数学模型

从电涡流阻尼的非线性特性看，当运动速度小于临界速度时，电涡流阻尼力随着速度的增加而增加；一旦超过某临界速度，电涡流阻尼力随着速度的增加而减小。针对这一特性，不同学者提出了不同的数学模型。以 Wouterse 模型[31]为例，电涡流阻尼器阻尼力与速度的关系可以描述为

$$F(v)=F_{\max}\left[2\left(\frac{v_{cr}}{v}+\frac{v}{v_{cr}}\right)^{-1}\right] \tag{7}$$

上述模型虽然满足电涡流阻尼力的极限特性，但在整个工作速度范围内，并不能很好地拟合非线性阻尼力特性。为此，对 Wouterse 模型进行了修

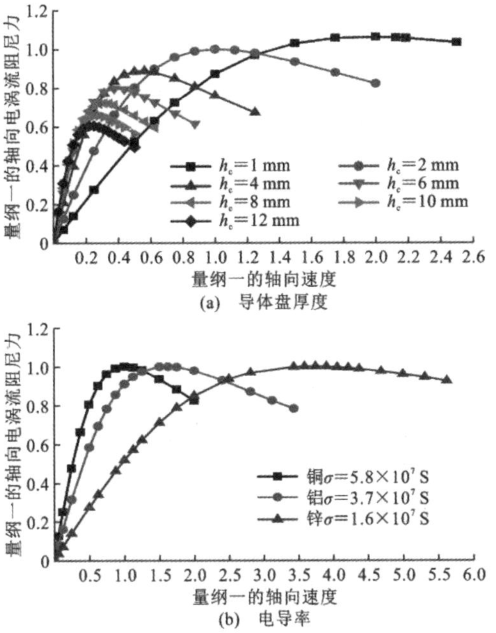

图 12 阻尼力速度曲线的变化

Fig. 12 Change of Damping Force Velocity Curve

正,使之具有更高的拟合精度,能够作为滚珠丝杠式轴向电涡流阻尼器的数学模型。修正的 Wouterse 模型[29]如下所示

$$F(v)=\frac{\alpha v}{v^2+\beta|v|+\gamma^2} \tag{8}$$

式中:α、β 和 γ 均为待拟合的参数,其中 β 和 γ 具有速度的量纲,α 具有功率的量纲。与 Wouterse 模型相比,修正 Wouterse 模型的分子中增加了速度的一次项,构成了一个完整的二次多项次。此外,考虑到有限元计算和试验都只能得到临界速度及最大轴向电涡流阻尼力的近似值,所以在修正的 Wouterse 模型中所有的参数都通过拟合得到。对修正 Wouterse 模型进一步分析可知,γ 代表了滚珠丝杠式轴向电涡流阻尼器的临界速度,α/γ^2 代表了初始阻尼系数,$\alpha/(2\gamma+\beta)$ 代表了最大轴向电涡流阻尼力。图 13 比较了修正的 Wouterse 模型、Wouterse 模型、Anwar 模型和板式电涡流阻尼器模型对有限元数据的拟合结果。可以看到,修正的 Wouterse 模型拟合结果最好。

基于大型阻尼器试验平台测定了旋转式永磁轴向电涡流阻尼器的工作性能。图 14 为某最大阻尼力为 850 kN 滚珠丝杠式轴向电涡流阻尼器及性能测试试验,试验中阻尼器两端分别固定于试验台,并测定不同激振频率及振幅下的阻尼器出力。图 15 为 2 个典型试验工况(频率相同,振幅不同)时一个周期内的滞回曲线试验值及理论曲线的比较。从图

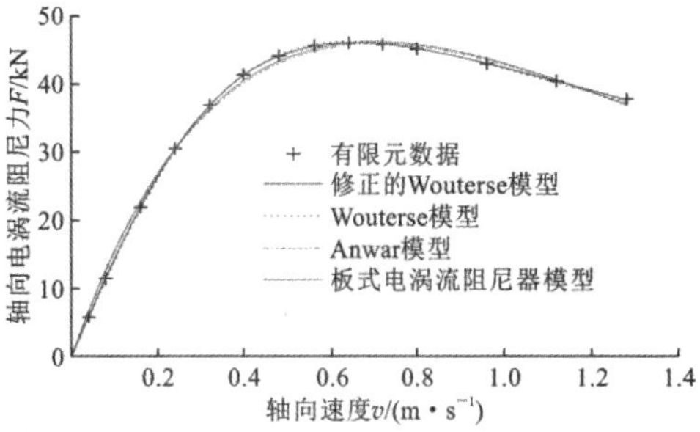

图 13 不同数学模型的拟合结果对比

Fig. 13 Fitting Results of Different Mathematical Models of Eddy Current Damping Force

15 看出,阻尼器滞洄曲线的试验值与理论值相比十分接近。滞回曲线中的“负刚度”现象是阻尼器的惯性质量引起的,而图 15(b)中滞回曲线中部向内凹,则是阻尼力在高速下略有下降引起的。图 16 给出了由滞回曲线得到的阻尼力随轴向速度的变化,可以看出理论计算结果与有限元结果几乎重合,与试验阻尼力差别也很小。综上分析,电涡流阻尼器无论是高速还是在低速运动状态下阻尼力均可以精确预测,这是电涡流阻尼器相比于油阻尼器的显著优势。

图 14 某滚珠丝杠式轴向电涡流阻尼器及性能测试

Fig. 14 A Ball Screw Type Axial Eddy Current Damper and Its Performance Test

3 基于电涡流阻尼的结构被动控制

结构的集总参数模型包括 3 种单元:质量单元、弹簧单元和阻尼单元。弹簧单元和阻尼单元是结构被动控制中经常使用的 2 种减振元件,只有调谐质量减振器中含有质量单元。2002 年 Smith 教授提出了惯质的概念[32]。惯质是一种两节点质量单元,理论上能够与阻尼单元、弹簧单元进行任意组合,实现振动控制所需的任意机械阻抗。笔者团队在中国土木工程领域较早注意到惯质的概念,并把基于弹簧单元、阻尼单元和惯质单元的被动控制理论称为三元被动控制理论[33-34]。三元被动控制理论为结构振动控制提供了新的方法,近年来已成为土木工

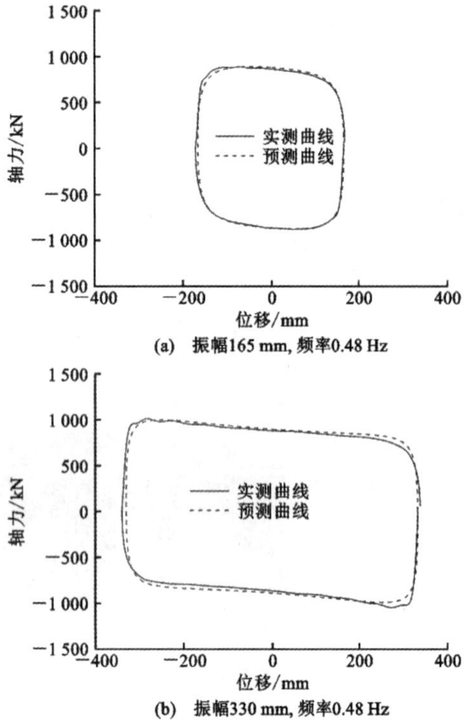

图 15 电涡流阻尼器样机理论滞回曲线与试验结果的比较

Fig. 15 Comparison of Theoretical Hysteretic Curves and Experimental Results of Prototype Eddy Current Damper

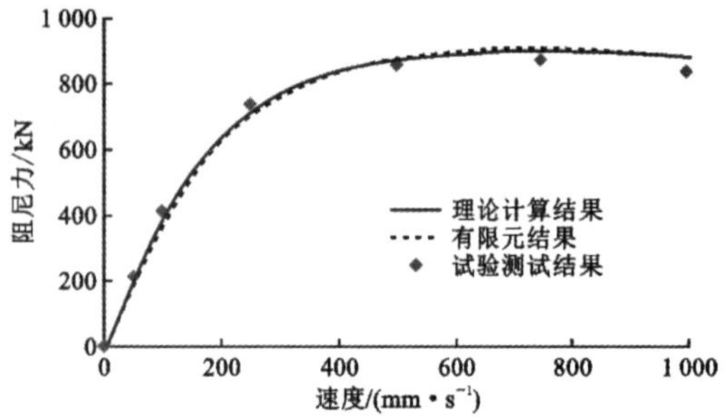

图 16 轴向电涡流阻尼器的阻尼力-速度曲线

Fig. 16 Damping Force-velocity Curves of Axial Eddy Current Damper

程领域研究的一个热点[35-39]。

滚珠丝杠式轴向电涡流阻尼器是一种复合减振单元，其力学模型是电涡流阻尼单元和惯质单元的并联体(其内部摩擦力很小，可以忽略)，如图 17 所示。基于滚珠丝杠式轴向电涡流阻尼器，对三元减振理论在结构振动控制领域的应用进行了深入研究。主要包括：①基于调谐黏滞质量阻尼器的桥梁竖向涡激共振控制；②基于轴向电涡流阻尼器的拉索多模态振动控制；③基于电涡流阻尼力非线性特性的减震性能。本文只对基本理论及主要结果作简要介绍，具体细节可参见已发表的相关论文。

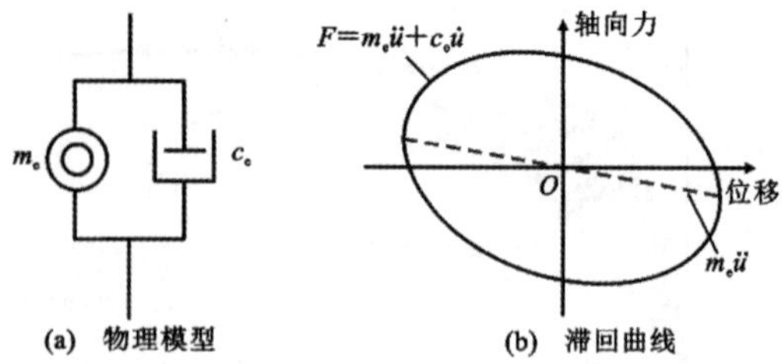

图 17 滚珠丝杠式轴向电涡流阻尼器的力学模型

Fig. 17 Mechanical Model of Ball Screw Type Axial Eddy Current Damper

3.1 基于 TVMD 的桥梁竖向涡振控制

把轴向电涡流阻尼器和弹簧串联可以构成调谐黏滞质量阻尼器(TVMD)。已有很多学者研究了 TVMD 对建筑结构的减震控制效果，但在桥梁风致振动控制领域对惯质阻尼器的研究非常稀少，本节从 Scanlan 经验非线性涡激力模型出发，研究了基于 TVMD 的悬索桥竖向涡振控制方法[29]。

3.1.1 TVMD 的参数优化

对于多跨连续全漂浮式悬索桥，可在桥塔处设置牛腿，并把 TVMD 安装于牛腿和主梁之间，用于主梁的竖向涡振控制(如图 18 所示)。此外，因为桥梁竖向涡振往往表现为单一模态的振动，所以涡激力作用下桥梁—TVMD 系统可以简化为一个两自由度系统，如图 19 所示，对应的运动方程可表示为

$$\left.\begin{aligned}&M(\ddot{y}+2\omega_n\xi_n\dot{y}+\omega_n^2 y)=\rho UD\left[\int_0^L Y_1(1-\varepsilon\phi^2(z)y^2)\phi^2(z)\mathrm{d}z\right]\dot{y}+m_1\omega_1^2(x-y)\\&m_1[\ddot{x}+2\omega_1\xi_1\dot{x}+\omega_1^2(x-y)]=0\end{aligned}\right\}\tag{9}$$

式中：z 为主梁沿桥跨方向的坐标；$M=\int_S m_s\phi^2(s)\mathrm{d}s$，为桥梁涡振的模态质量，$m_s$ 为桥梁各构件单位长度的质量；$\phi(\cdot)$为桥梁涡振模态的振型位移；ξ_n 和 ξ_1 分别为桥梁的模态阻尼比及 TVMD

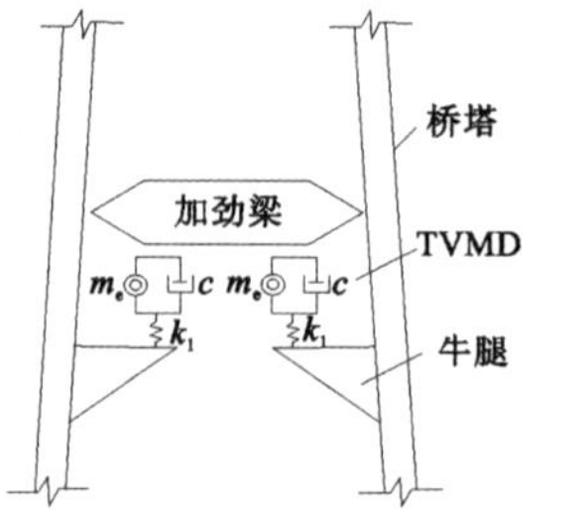

图 18 TVMD 在主梁与桥塔之间的安装示意

Fig. 18 TVMD Installed Between Main Beam and Bridge Tower

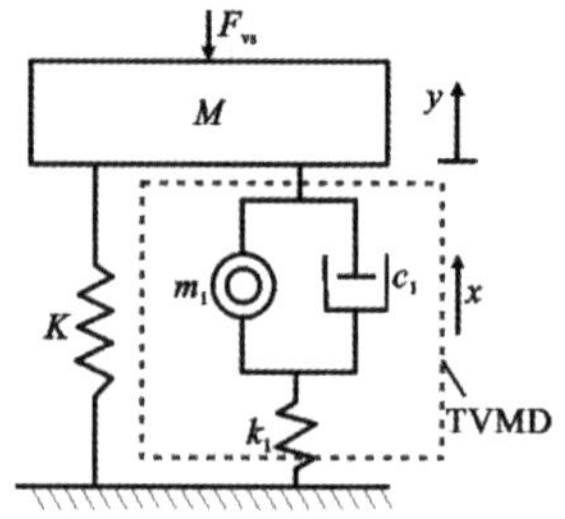

图 19 涡激力作用下主梁-TVMD 系统

Fig. 19 Girder-TVMD System Under Excitation of Vortex-induced Force

的阻尼比；$m_1=m_e\phi^2(z_0)$，为 TVMD 的模态质量，其中 m_e 为 TVMD 的表观质量，$\phi(z_0)$ 为 TVMD 安装位置处主梁的振型位移值；ω_n 和 ω_1 分别为桥梁涡振的模态频率和 TVMD 的固有圆频率。

式(9)是一个非线性自治系统，而且涡激振动的气动负阻尼往往与结构固有阻尼为同一数量级，所以只要 TVMD 的参数设计合理，就可以保证主梁-TVMD 系统存在稳定的零解，从而抑制主梁涡振的发生。鉴于此，本文以桥梁-TVMD 系统保持零解稳定所能承受的最大气动负阻尼作为优化目标，得到 TVMD 的最优参数如下所示

$$\left.\begin{aligned}\alpha_{opt}&=\frac{1}{\sqrt{1-\mu}}\\ \xi_{1\_opt}&=\sqrt{\frac{1-\sqrt{1-\mu}}{2}}\\ \beta_{max}&=\sqrt{\frac{1-\sqrt{1-\mu}}{2(1-\mu)}}\end{aligned}\right\}\tag{10}$$

式中：$\beta_{max}=\frac{\rho UDY_1}{2M\omega_n}\int_0^L\phi^2(z)\mathrm{d}z-\xi_n$，为主梁保持零解稳定所能承受的最大气动负阻尼；$\mu=m_1/M$，为 TVMD 的质量比；$\alpha_{opt}=\omega_1/\omega_n$，为 TVMD 的最优频率比。

3.1.2 TVMD 和 TMD 控制效果比较

采用相同的方法可以推导 TMD 的优化参数，图 20 对比了 TVMD 和 TMD 的参数优化结果。从图 20(a)可以看到：TVMD 的最优频率比总是略大于 1，而且随着质量比的增加而近似线性增加；TMD 的最优频率比总是略小于 1，而且随着质量比的增加而近似线性减小。从图 20(b)可以看到：TVMD 和 TMD 的最优阻尼比近似相等，而且随质量比的变化趋势相同。从图 20(c)可以看到：当质量比相同时，主梁-TVMD 系统保持零解稳定容许的最大气动负阻尼与主梁-TMD 系统近似相等，且前者略大于后者。

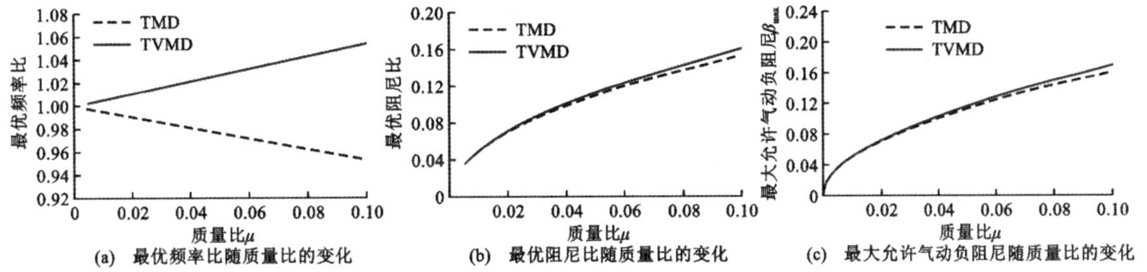

(a) 最优频率比随质量比的变化　(b) 最优阻尼比随质量比的变化　(c) 最大允许气动负阻尼随质量比的变化

图 20 TVMD 和 TMD 参数优化结果比较

Fig. 20 Optimization Parameters of TVMD and TMD

值得说明的是，与 TMD 相比，TVMD 的主要优点在于可以显著减小装置的物理质量和弹簧单元的静伸长。TVMD 和 TMD 的静伸长都可以采用公式(11)计算[40]

$$\Delta_s=\frac{mg}{k_1}=\frac{m_e g}{\lambda\phi_d^2 k_1}=\frac{g}{4\lambda\phi_d^2\pi^2 f_b^2}\approx\frac{0.25}{\lambda\phi_d^2 f_n^2}\tag{11}$$

式中：m 为 TMD 的质量或者 TVMD 中惯质的物理质量；m_e 为惯质的表观质量；ϕ_d 为 TVMD 安装处主梁受控模态的振型位移，对于 TMD，$\phi_d=1$；λ 为惯质的表观质量与其物理质量的比值，这里简称为质量放大系数，对于 TMD，$\lambda=1$；f_b 和 f_n 分别为 TVMD 和主结构的固有频率。

以大跨度桥梁竖向涡振为例说明上述静伸长问题。假定桥梁的竖弯涡振频率 $f_n=0.23$ Hz，模态质量 $M=20\ 000$ t，TVMD 和 TMD 的质量比 $\mu=0.5\%$，频率比 $\alpha=1$。对于 TMD，可以布置在主梁振型位移的峰值处，故所需总质量 $m_1=100$ t，弹簧静伸长 $\Delta=4.79$ m。可以看到，由于弹簧静伸长过长，TMD 在扁平钢箱梁中的实施难度很大。对于单个 TVMD，假定其安装处主梁的振型位移 $\phi_d=\phi(z_0)=0.4$，那么所需的表观质量为 $m_e=625$ t。利用按式(6)计算的滚珠丝杠的惯性放大效应，一个总质量约 $m_d=500$ kg 的旋转式轴向电涡流阻尼器的表观质量可以达到 $m_e=686$ t；再结合式(11)可计算得到 TVMD 的弹簧静伸长 $\Delta_1=0.024$ m。可以看到，TVMD 的总质量和弹簧静伸长都只有 TMD 的 1/200，从而具备了在大跨度悬索桥上应用的可行性。

3.2 基于轴向电涡流阻尼器的拉索多模态控制

已有研究表明，黏滞质量阻尼器能够显著提高拉索单一模态的阻尼比，但对于拉索的多模态控制尚不明确。为此，笔者团队开展了基于轴向电涡流阻尼的拉索多模态控制研究，以模态阻尼比为指标研究了轴向电涡流阻尼器对拉索前 4 阶模态的减振控制效果[41]。

3.2.1 自由振动方程

如图21所示的张紧弦结构，拉索-轴向电涡流阻尼器系统的量纲一的自由振动方程可表示为

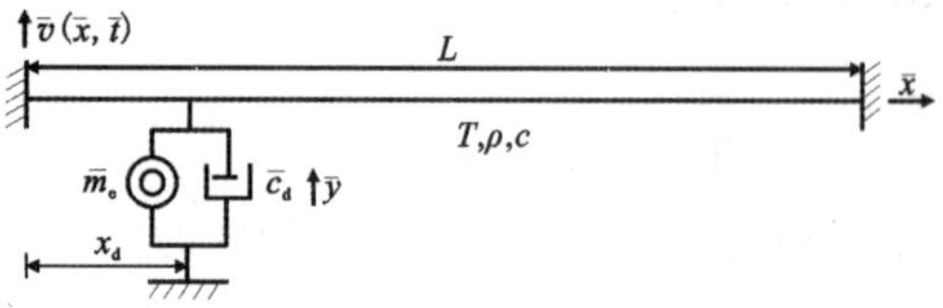

图21　拉索-轴向电涡流阻尼器系统示意

Fig. 21　Taut Cable with Axial Eddy Current Damper

$$\left.\begin{aligned}&\ddot{v}(x,t)-\frac{1}{\pi^2}v''(x,t)=F_d\delta(x-x_d)\\&F_d=-m_e\ddot{v}(x_d)-c_d\dot{v}(x_d)\end{aligned}\right\}\tag{12}$$

式中：量纲一的参数由以下参数定义

$$\left.\begin{aligned}&t=\omega_0\bar{t},x=\bar{x}/L,v(x,t)=v(\bar{x},\bar{t})/L\\&\delta(x-x_d)=L\delta(\bar{x}-\bar{x}_d),f(x,t)=L\bar{f}(\bar{x},\bar{t})/(\pi^2T)\\&F_d=\bar{F}_d(\bar{t})/(\pi^2T),m_e=\frac{\bar{m}_e}{m},c_d=\frac{\bar{c}_dL\omega_0}{\pi^2T}\end{aligned}\right\}\tag{13}$$

式中：ρ 为拉索每延米的质量；L 为拉索长度；T 为拉索的水平张力；F_d 为阻尼力；x_d 为拉索与邻近固定端的距离；$\omega_0=(\pi/L)/\sqrt{T/\rho}$表示无阻尼器时拉索的基频；$m_e$ 为轴向电涡流阻尼器的表观质量；$\bar{m}_e$ 和 $\bar{c}_e$ 分别为轴向电涡流阻尼器的表观质量和阻尼系数；m 为拉索的总质量。

通过引入拉索的 m 阶模态坐标，可以把式(12)离散化，进而得到拉索—轴向电涡流阻尼器系统的状态空间运动方程

$$\begin{bmatrix}\dot{\boldsymbol{q}}\\\ddot{\boldsymbol{q}}\end{bmatrix}=\begin{bmatrix}\boldsymbol{0}&\boldsymbol{I}\\-\boldsymbol{M}^{-1}\boldsymbol{K}&-\boldsymbol{M}^{-1}\boldsymbol{C}\end{bmatrix}\begin{bmatrix}\boldsymbol{q}\\\dot{\boldsymbol{q}}\end{bmatrix}\tag{14}$$

式中，$\boldsymbol{q}(t)=[q_1,q_2,\cdots,q_j,\cdots,q_m]^{\mathrm{T}}$，为模态坐标系下系统的广义自由度；$\boldsymbol{I}$ 为单位矩阵；$\boldsymbol{M}$、$\boldsymbol{K}$ 和 $\boldsymbol{C}$ 分别为质量矩阵、刚度矩阵和阻尼矩阵，其计算式如下所示

$$\left.\begin{aligned}M_{ij}&=\int_0^1\phi_i(x)\phi_j(x)\mathrm{d}x+m_e\phi_i(x_d)\phi_j(x_d)\\C_{ij}&=c_d\phi_i(x_d)\phi_j(x_d)\\K_{ij}&=\frac{1}{\pi^2}\int_0^1\phi_i'(x)\phi_j'(x)\mathrm{d}x\end{aligned}\right\}\tag{15}$$

通过求解式(14)对应的特征方程，可以得到安装轴向电涡流阻尼器后拉索各阶模态的频率和阻尼比，进而评价阻尼器的拉索减振效果。

3.2.2 多模态减振效果分析

为了不影响桥梁美观，拉索阻尼器的尺寸应尽可能小，因此本文仅考虑轴向电涡流阻尼器的量纲一的表观质量在0.5以内时，其对拉索前4阶模态的减振控制效果。

图22给出了量纲一的表观质量 $m_e=0.1$ 和 $m_e=0.5$ 时，拉索阻尼器系统的前4阶模态阻尼比随阻尼器量纲一的阻尼系数的变化曲线。可以看到，当阻尼器的表观质量一定时，存在不同的阻尼系数使拉索各阶模态的阻尼比达到最大；各阶模态所能达到的最大阻尼比及对应的阻尼系数，随阻尼器表观质量的变化而变化。实际工程中，拉索前4阶模态都有发生风雨振的可能，所以阻尼器的参数设计也应针对其进行优化。图22中的交叉点代表了拉索前4阶模态阻尼比极小值的峰值，对应的阻尼系数可以作为拉索多模态控制的最优参数。图23(a)和图23(b)分别给出了轴向电涡流阻尼器的量纲一的表观质量在0～0.5之间变化时，量纲一的最优阻尼系数及拉索前4阶模态阻尼比的变化情况，其中$m_e=0$代表了传统黏滞阻尼器的优化结果。可以看到，随着表观质量的增加，阻尼器的最优阻尼系数呈先增加后减小的变化规律，但总体变化不大，均处于2.4～2.8之间。另一方面，与黏滞阻尼器的控制效果相比，表观质量的增加使得拉索第2和第3阶模态的优化阻尼比有较大提高，但第1和第4阶模态的阻尼比改善较小。总体来看，表观质量的存在提高了轴向电涡流阻尼器对拉索多个模态的振动控制效果。

3.3 基于电涡流非线性阻尼力的减震性能

基于非线性阻尼力的 Wouterse 模型，研究了轴向电涡流阻尼器的能量耗散特征、参数优化及其对单自由度系统在谐波和地震基底作用下的减震性能[42]。

3.3.1 简谐运动下的耗能分析及参数优化

对于流体黏滞阻尼器，其阻尼力-速度曲线可以用指数形表达式

$$F_{\mathrm{FVD}}=c_dv^{\alpha}\tag{16}$$

它在简谐运动 $u=u_0\sin(\omega t)$ 一个周期内的耗能可以通过积分求解为

$$E_{\mathrm{FVD}}=4\int_0^{u_0}F_{\mathrm{FVD}}\mathrm{d}u=\left[2c_d\sqrt{\pi}u_0^{\alpha+1}\omega^{\alpha}\Gamma\left(\frac{\alpha}{2}+1\right)\cdot\left[\Gamma\left(\frac{\alpha+3}{2}\right)\right]^{-1}\right.\tag{17}$$

对于用 Wouterse 模型表示的轴向电涡流阻尼器，它在同样的简谐运动 $u=u_0\sin(\omega t)$ 作用下一个周期内的耗能为

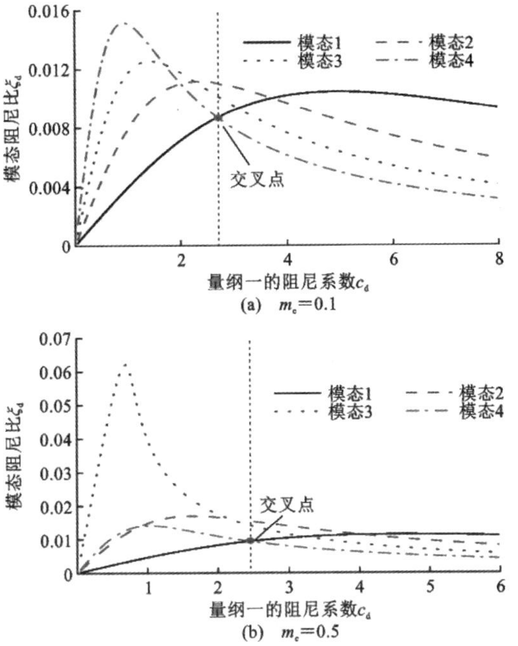

图 22 拉索-阻尼器系统的前 4 阶模态阻尼比随量纲一的阻尼系数的变化曲线($x_d=0.02$)

Fig. 22 Modal Damping Ratio Curves of the Cable-damper System with Nonoptimal Apparent Mass, $x_d=0.02$

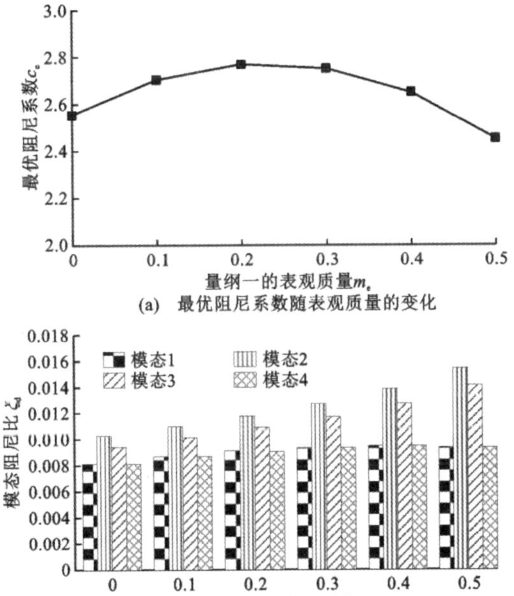

图 23 轴向电涡流阻尼器的最优参数及多模态控制效果

Fig. 23 Optimal Parameters of Axial Eddy Current Damper for Multi-modal Vibration Control

$$E_{\mathrm{ECD}}=4\int_0^{u_0}2F_{\max}\cdot\left(\frac{v}{v_{\mathrm{cr}}}+\frac{v_{\mathrm{cr}}}{v}\right)^{-1}\mathrm{d}u=\frac{4v_{\mathrm{cr}}F_{\max}\pi}{\omega}\left(1-\frac{v_{\mathrm{cr}}}{\sqrt{v_{\mathrm{cr}}^2+u_0^2\omega^2}}\right)\tag{18}$$

将式(18)和式(17)相除,可以得到电涡流阻尼和流体黏滞阻尼的耗能之比为

$$\frac{E_{\mathrm{ECD}}}{E_{\mathrm{FVD}}}=2\sqrt{\pi}\Gamma\left(\frac{\alpha+3}{2}\right)\cdot\left[\Gamma\left(\frac{\alpha}{2}+1\right)\right]^{-1}\cdot X\left[1-\left(\sqrt{1+\frac{1}{X^2}}\right)^{-1}\right]\tag{19}$$

式中:$X=v_{\mathrm{cr}}/(u_0\omega)$为量纲一的速度比,它是电涡流临界速度和简谐运动最大速度的比值。

基于式(19),图 24 给出了耗能比与速度比的关系。从图 24 中看出,当流体黏滞阻尼的指数 $\alpha>0.2$时,电涡流阻尼总能找到某一速度比区间,使它的耗能大于流体黏滞阻尼的耗能。另一方面,也可以看出,该耗能比随着速度比的增加先升高后降低,那么必存在一个最优速度比值。通过对式(19)进行两次求导,可以得到当速度比 $X=0.786$ 时,电涡流的耗能最大,详细推导过程见已发表的论文。

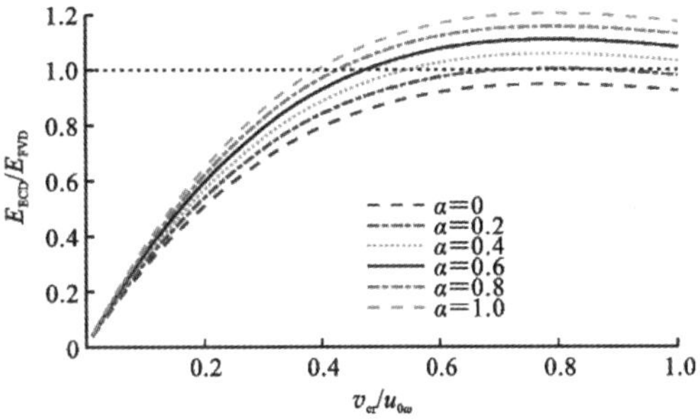

图 24 耗能比与速度比的关系

Fig. 24 Relation of Energy Consumption Ratio and Velocity Ratio

3.3.2 基底简谐激励下单自由度-电涡流阻尼系统响应分析

笔者团队研究了线弹性单自由度-电涡流阻尼系统在基底简谐激励下的响应,基底简谐激励的形式为 $\ddot{u}_g=\ddot{u}_{g0}\sin(\omega t)$,结构响应包括位移响应和加速度响应。定义位移响应比 $R_d=u_0/u_{st0}$ 为最大动态位移响应幅值与静态位移响应的比值,加速度传递比 $TR=\ddot{u}_0^t/\ddot{u}_{g0}$ 为最大动态绝对加速度幅值与基底峰值加速度的比值,频率比 ω/ω_n 为外界激励频率与系统固有频率的比值。位移响应比随频率比的变化关系如图 25。从图 25 看出:当频率比位于 1 附近(即在系统的共振频率附近)时,单自由度-电涡流阻尼系统的位移响应要比单自由度-流体黏滞阻尼系统的位移响应小,而且随着流体黏滞阻尼指数 α 的增加,这种优势越来越明显。然而需要指出,随着频率比的进一步增大,在十分远离共振频率的范围内,单自由度-电涡流阻尼系统的位移响应基本和单自由度-流体黏滞阻尼系统的位移响应相近,且有时

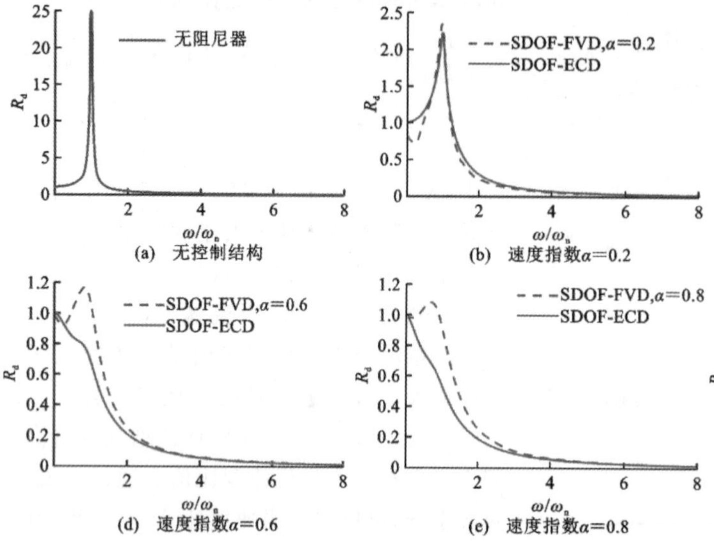

图 25 位移响应比与频率比的关系

Fig. 25 Relation of Displacement Response Ratio and Frequency Ratio

候会略大一些。这种现象也在后面的地震响应分析中得到了验证。加速度传递比基本也遵循相似的规律,在系统共振频率附近,单自由度-电涡流阻尼系统的加速度传递比要比单自由度-流体黏滞阻尼系统的加速度传递比小。

3.3.3 地震激励下电涡流阻尼系统减震效果

为进一步研究实际地震作用下的减震效果,选取了 8 条真实的地震波:Elcentro 波、Irpinia 波、Chichi 波、Kobe 波、Northridge 波、Niigat 波、Loma Prieta 波和 Iwate 波,其中 Kobe 波和 Loma Prieta 波为长周期地震波,对单自由度-电涡流阻尼系统进行响应分析。定义减震率为

$$\rho=(A_o-A_d)/A_o \tag{20}$$

式中:A_o 为没有附加阻尼器的峰值位移响应或者峰值加速度响应;A_d 为有附加阻尼器的峰值位移响应或者峰值加速度响应。ρ 值越大,表示附加阻尼器的减震性能越好。

图 26 给出了电涡流阻尼器和黏滞阻尼器对单自由度体系的位移减震率对比。可以看出,在大多数情况下,电涡流阻尼的减震性能要优于流体黏滞阻尼的减震性能。需要指出的是在长周期地震波 Kobe 波和 Loma Prieta 波的作用下,某些情况电涡流阻尼的减震性能要略低于流体黏滞阻尼的减震性能。这是因为如上节所述,在某些情况下远离共振频率一定范围或者更远的区域,电涡流阻尼的频响函数值会略高于流体黏滞阻尼,从而造成了某些情况下电涡流阻尼的减震性能会略低于流体黏滞阻尼的减震性能。

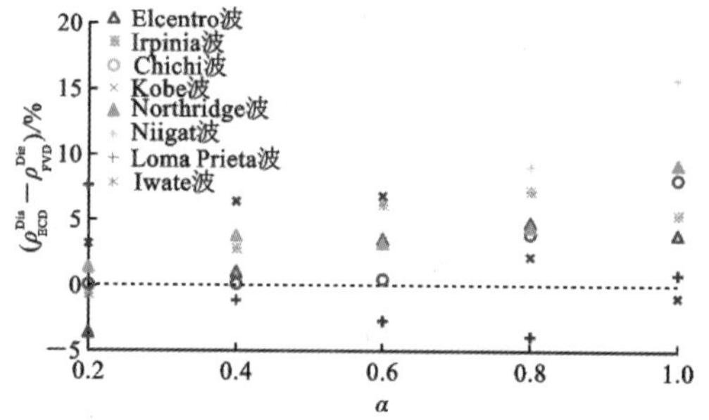

图 26 不同地震波作用下电涡流阻尼与流体黏滞阻尼位移减震率对比

Fig. 26 Comparison of Displacement Reduction Ratio Between Eddy Current Damping and Fluid Viscous Damping Under Different Seismic Waves

4 大型阻尼器特性综合试验平台

在湖南大学"双一流"建设经费支持下,建成了大型阻尼器特性综合试验平台(图 27),并开发了若干高速缓冲等特殊用途的试验装置,为深入研究电涡流阻尼器的动态性能奠定了基础。

整个试验平台共分为 10 个部分,包括:双通道全数字控制系统、阻尼器试验机自反力机架台体、水平动态电液伺服作动器、电动行走龙门吊车、油源、蓄能器组、管路系统、监视系统、四立柱试验台、竖向电液伺服作动器,如图 27(a)所示。其中自反力机架台体由土建基础、地锚栓、前端梁、纵梁、中间梁、辅助梁、丝杠、后端梁、动横梁、阻尼器安装夹具、动横梁移动机构、丝杠间隙消除机构、龙门吊等构成,见图 27(b)。

主试验平台为采用液压驱动的卧式结构,实物

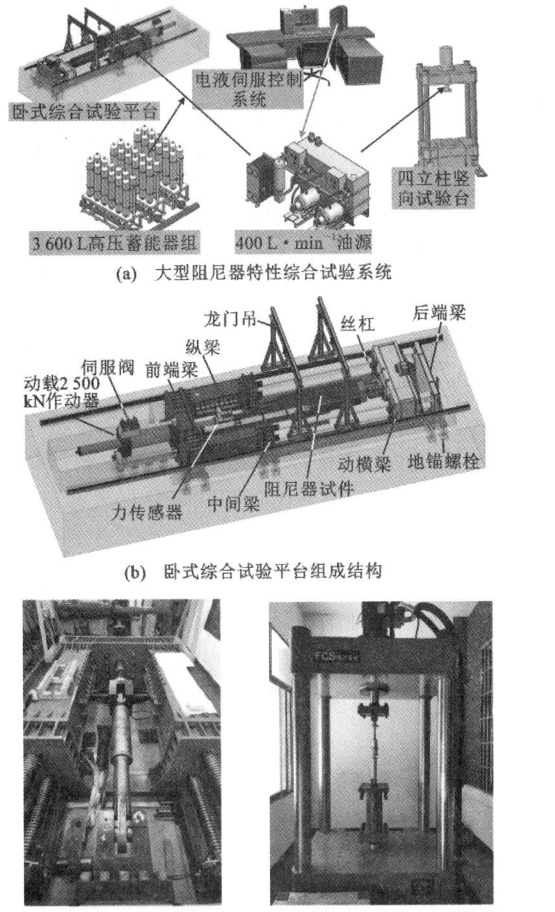

(a) 大型阻尼器特性综合试验系统

(b) 卧式综合试验平台组成结构

(c) 卧式主试验平台　(d) 四立柱竖向动态加载装置

图 27　大型阻尼器特性综合试验平台

Fig. 27　Comprehensive Experimental Platform for the Characteristics of Large-scale Dampers

照如图 27(c)所示。该试验平台的最大静载 4 000 kN,最大动载 2 500 kN,最大测试速度 1 m·s^{-1},工作行程±800 mm,油源流量为 400 L·min^{-1},油源动力系统功率控制在 220~250 kW 之间,3 600 L 蓄能器组为瞬时快速循环试验补充流量。可以进行 0.001~10 Hz 频率范围内的阻尼器性能动态试验。采用 Instron 8 800 mL 控制器,电液伺服闭环控制,可以进行载荷控制和位移控制,且可以平滑切换。系统精度:力精度±0.5%FS,位移精度±0.5%FS。能够安装的试件最大尺寸为长 6 m,直径 1 m。平台还配有 20~1 000 kN 力传感器,可以用于小吨位阻尼器的精确测试。测试平台框架在设计时除考虑承受的轴向荷载外,还考虑承受的弯矩荷载,模拟各种复杂工况。对于常规的两铰型阻尼器,其一端的铰可以上下移动,实现有竖向仰角的测试工况。

平台还包含一套四立柱竖向加载装置,见图 27(d)。最大静载 300 kN,最大动载 100 kN,最大测试速度 0.6 m·s^{-1},工作行程±100 mm,能够安装的试件最大高度为 1.2 m。

5　工程应用

经过 10 余年的研究,电涡流阻尼减振技术在土木工程领域已全面实现了工程应用。电涡流调谐质量减振器首先在高耸输电塔减振工程中进行了试验验证,取得了很好的减振效果[43]。绵阳三江大桥(原绵阳一号斜拉桥)是世界上首座全面应用电涡流调谐质量减振器的斜拉桥,全桥共安装了 38 套电涡流调谐质量减振器以控制大桥的横向及竖向人致振动[44-47]。电涡流调谐质量阻尼器目前已在上海中心大厦[48]、世界最高光热电站塔[49]、张家界玻璃桥[40]、榕江大桥[50-51]等一系列代表性工程中应用,电涡流轴向阻尼器也已经在杭瑞高速洞庭湖大桥、石首大桥等多座新建大跨度桥梁中成功应用。本节主要阐述具有代表性的工程应用以及在减振工程实施中开发的动力试验技术。此外,电涡流阻尼的相关设计与试验规范制定工作也在开展之中[52]。

5.1　电涡流调谐质量减振器

5.1.1　绵阳三江大桥

绵阳三江大桥是第 1 座全面采用电涡流阻尼减振技术的斜拉桥工程。该桥主跨为 200 m,在混凝土主梁的车行桥上空,悬挂一座钢箱梁人行桥,梁宽 6 m,高 2 m,为当时中国跨度最大的人行桥,且是曲线桥,如图 28 所示。

人致振动防治是保证该桥正常使用的关键。为提高行人舒适性,提出了采用连杆提高整体刚度和调谐质量减振器提高阻尼的联合减振方案[44],如图 28 所示。在主跨和边跨设置连杆后,人行桥的一阶侧向振动频率从原来的 0.48 Hz 提高了 0.723 Hz,人致振动敏感模态数量降低一半。此外,全桥安装了 38 套电涡流 TMD,用于控制 1 个侧向振动模态及 4 个竖向振动模态。人群动力试验表明:安装减振装置前,在不到 150 人行走时就激发横向锁定现象,并发生大幅横向振动,横向加速度达到 0.2 m·s^{-2};行人增加到 200 人时,加速度进一步增大,并出现行走困难现象;安装电涡流减振装置后,竖弯模态的等效模态阻尼比从原来的 0.8%提高到了 4.0%以上,400 人大步行走时无不舒适感,侧向振动加速度的实测最大值小于 0.08 m·s^{-2}。

准确测定动力参数对保证 TMD 减振效果极为重要。为准确识别大桥受控模态的模态质量,便于

(a) 大桥 (b) 连杆系统 (c) 电涡流TMD

图28 绵阳三江大桥及减振系统

Fig. 28 Mianyang Sanjiang Bridge and Its Vibration Control System

TMD优化设计与精准调试，研发了正弦激振、宽带扫频的模态全参数识别方法[45]。与现行技术相比，率先实现了模态质量的准确识别，为TMD设计提供了准确参数。

5.1.2 上海中心大厦

上海中心大厦建筑高度为632 m，是中国目前最高的建筑，如图29所示。大厦采用巨型框架-核心筒-伸臂桁架抗侧力体系，具有高宽比较大、结构自振周期(9.09 s)接近风荷载卓越周期的特点。依据抗风试验结果，大楼需要安装一台惯性质量为1 000 t的摆式TMD，这也是目前世界上最大的TMD。在强风或地震发生时，要求该台TMD的阻尼比随着振幅增大而增加，起到减轻强风和地震危害以及限制TMD行程不会过大的双重效果。

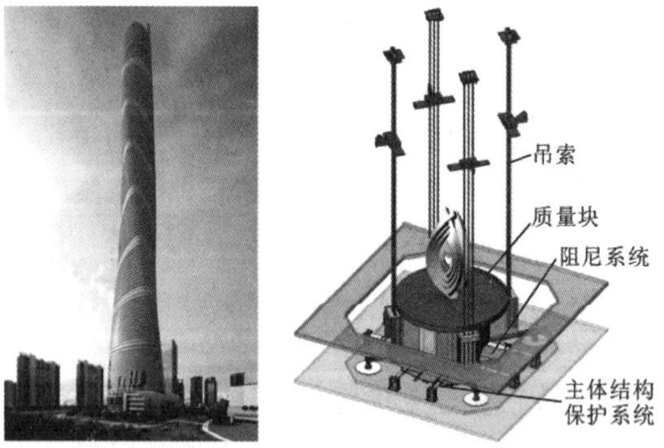

(a) 上海中心大厦 (b) 电涡流调谐质量减振器

图29 上海中心大厦及电涡流调谐质量阻尼器

Fig. 29 Shanghai Tower and Its Eddy Current TMD

如图29所示，上海中心阻尼器由吊索、1 000 t的质量块以及阻尼系统、主体结构保护系统组成，通过12根吊索悬挂在第125～131层之间。通过调节索的摆动长度即可使摆动频率与主体结构一致，达到降低结构加速度，提高建筑舒适度的作用。在抑制风致振动时，TMD工作在小振幅状态，所需阻尼比相对较小；当遭遇地震时，阻尼比会自动加大，设计最大阻尼比为148%，进入过阻尼状态，可以保证不会发生因地震激发TMD的有害振动。采用改变导体板与磁体的间距以及改变导体板厚度这2种措施，实现了电涡流阻尼比随振幅变化的要求，完全满足了TMD的设计要求。

目前，电涡流阻尼器已成功助力上海中心大厦抵抗台风“安比”和“利奇马”的考验，保证了大厦在强风作用下的舒适性。据观测，在强台风“安比”的作用下，阻尼器的振幅接近50 cm，由此推定结构最大动位移小于10 cm。

5.2 轴向电涡流阻尼器

杭瑞高速岳阳洞庭大桥是杭州-瑞丽高速公路上的一座两跨钢桁架悬索桥，如图30所示。大桥主跨1 480 m，钢桁加劲梁全宽35.4 m，梁高9 m，主缆垂跨比为1/10，主梁为桁板结合梁，桥塔为门式框架结构，由上塔柱、下塔柱、上横梁和下横梁组成。

图30 杭瑞高速岳阳洞庭湖悬索桥

Fig. 30 Yueyang Dongting Lake Suspension Bridge

经过减震分析，确定了在杭瑞高速岳阳洞庭湖大桥上需安装8台外杯旋转式轴向电涡流阻尼器，该阻尼器的设计最大阻尼力为1 500 kN、行程为±1 120 mm、最大速度为0.5 m·s^{-1}。图31为动力性能试验机上的轴向电涡流阻尼器及现场照片。经检测，轴向电涡流阻尼器的启动摩擦力小于最大阻尼力的1%，其速度相关性及频率相关性均满足规范要求。

6 总结与展望

本文主要介绍了笔者团队近十余年来在永磁电涡流阻尼减振技术方面的研究成果，包括电涡流阻尼器件的发明与产品研发、电涡流阻尼分析模型与仿真设计方法、结构风致振动与地震控制的三元减

(a) 轴向电涡流阻尼器性能试验 (b) 安装在实桥上的轴向电涡流阻尼器

图 31 轴向电涡流阻尼器性能试验及实桥安装

Fig. 31 Performance Test of Axial Eddy Current Damper and Installation

振理论、工程应用概况、标准与规范的制定等,形成了电涡流阻尼减振成套技术。研究成果已应用于上海中心大厦、绵阳三江大桥、洞庭湖二桥、摩洛哥世界最高光热电塔等一系列减振工程,实现了土木工程应用领域的全覆盖,使中国成为最先在工程结构减振领域应用电涡流阻尼的国家。自中国率先在大型土木工程中应用电涡流阻尼减振技术以来,德国、英国、西班牙和韩国学者也相继进行了电涡流阻尼的跟踪研究。

需进一步研究的课题如下:

(1)重型滚珠丝杠制造工艺及疲劳性能:本项目开发的永磁式轴向电涡流阻尼器,价格已明显低于国外进口的相同功能油阻尼器。重型滚珠丝杠副是关键构件,对于 200 t 的轴向阻尼器,重型滚珠丝杠副还需要进口。为进一步降低成本,亟需实现重型滚珠丝杠的国产化,需要从材料、制造工艺及疲劳性能方面开展技术攻关。

(2)基于电涡流阻尼的半主动和主动控制技术:现有的永磁电涡流阻尼减振技术采用永磁体,其产生的磁场及电涡流阻尼力无法调节。在实际工程环境中,结构在不同激励荷载及运营工况下,其振动特性可能存在显著差别;另外,对于超高耸结构及大跨度桥梁,其施工过程中的结构参数会不断变化。针对上述工况,需要给出变阻尼系数的电涡流阻尼技术。采用电磁铁的电涡流阻尼器可以实现上述要求,通过调节电流强度实现半主动控制,满足不同施工状态或工作环境下的控制要求。

参考文献:

References:

[1] HOUSNER G W, BERGMAN L A, CAUGHEY T K, et al. Structural Control: Past, Present, and Future [J]. Journal of Engineering Mechanics, 1997, 123 (9): 897-971.

[2] SOONG T T, DARGUSH G F. Passive Energy Dissipation Systems in Structural Engineering [M]. New York: John Wiley & Sons, 1997.

[3] 周福霖. 工程结构减震控制[M]. 北京:地震出版社, 1997.
ZHOU Fu-lin. Engineering Structure Seismic Control [M]. Beijing: Seismological Press, 2003.

[4] 滕 军. 结构振动控制的理论、技术和方法[M]. 北京:科学出版社, 2009.
TENG Jun. Theory, Technique and Method of Structural Vibration Control [M]. Beijing: Science Press, 2009.

[5] SPENCER B F, NAGARAJAIAH S. State of the Art of Structural Control [J]. Journal of Structural Engineering, 2003, 129 (7): 845-856.

[6] SYMANS M D, CONSTANTINOU M C. Semi-active Control Systems for Seismic Protection of Structures: A State-of-the-art Review [J]. Engineering Structures, 1999, 21 (6): 469-487.

[7] 陈永祁, 马良喆. 黏滞阻尼器在实际工程应用中相关问题讨论[J]. 工程抗震与加固改造, 2014, 36(3): 7-13.
CHEN Yong-qi, MA Liang-zhe. Some Discussion Related to Viscous Dampers Application in the Actual Project [J]. Earthquake Resistant Engineering and Retrofitting, 2014, 36 (3): 7-13.

[8] 陈永祁. 工程结构用液体黏滞阻尼器的漏油原因分析[J]. 钢结构, 2008, 23(9): 53-58.
CHEN Yong-qi. Leaking Analysis of Fluid Viscous Damper for Engineering Structure [J]. Steel Construction, 2008, 23 (9): 53-58.

[9] KONSTANTINIDIS D, KELLY J M, MAKRIS N. In-situ Monitoring of the Force Output of Fluid Dampers, Experimental Validation [R]. Berkeley: Pacific Earthquake Engineering Research Center, 2011.

[10] SODANO H A, BAE J S. Eddy Current Damping in Structures [J]. Shock and Vibration Digest, 2004, 36 (6): 469-478.

[11] ANDRE L, RAOUL S. Electrical Device for the Absorption of Mechanical Energy: U. S., 2068, 820 [P]. 1937-01-26.

[12] 陈爱芬. ICE3 率先采用轨道涡流制动运营[J]. 国外铁道车辆, 2001, 38(4): 37-39.
CHEN Ai-fen. ICE 3 Pioneers Application of Eddy-current Rail Brakes [J]. Foreign Rolling Stock, 2001, 38 (4): 37-39.

[13] 祝长生. 径向电涡流阻尼器对柔性转子系统振动的控制[J]. 机械工程学报, 2007, 43(12): 120-126.
ZHU Chang-sheng. Vibration Control of a Radial Journal Eddy Current Damper on Flexible Rotor Systems [J]. Chinese Journal of Mechanical Engineering, 2007, 43 (12): 120-126.

[14] 马良喆，陈永祁. 电涡流调频质量阻尼器应用分析[J]. 工业建筑，2015，45(增1)：434-440.
MA Liang-zhe, CHEN Yong-qi. Effect Analysis of Tuned Mass Damper (TMD) in High-rise Structure [J]. Building Structure, 2015, 45 (S1): 434-440.

[15] TOSHIO O, HIROSHI M, YASUHIKO T, et al. Vibration Control Device for Buildings: Japan, No 8-161890 [P]. 1998-01-13.

[16] 陈政清. 永磁电涡流调谐质量阻尼器：中国，ZL 201010022003.7[P]. 2011-08-10.
CHEN Zheng-qing. Permanent Eddy Current Tuned Mass Dampers: China, ZL 201010022003.7 [P]. 2011-08-10.

[17] WANG Z H, CHEN Z Q, WANG J H. Feasibility Study of a Large-scale Tuned Mass Damper with Eddy Current Damping Mechanism [J]. Earthquake Engineering and Engineering Vibration, 2012, 11 (3): 391-401.

[18] 汪志昊，陈政清. 永磁式电涡流调谐质量阻尼器的研制与性能试验[J]. 振动工程学报，2013，26(3)：374-379.
WANG Zhi-hao, CHEN Zheng-qing. Development and Performance Tests of an Eddy-current Tuned Mass Damper with Permanent magnets [J]. Journal of Vibration Engineering, 2013, 26 (3): 374-379.

[19] 汪志昊. 自供电磁流变阻尼器减振系统与永磁式电涡流TMD的研制及应用[D]. 长沙：湖南大学，2011.
WANG Zhi-hao. Development and Application of the Self-powered Magnetorheological Damper and TMDs Using Eddy Current Damping [D]. Changsha: Hunan University, 2011.

[20] ZUO L, CHEN X, NAYFEH S. Design and Analysis of a New Type of Electromagnetic Damper with Increased Energy Density [J]. ASME Journal of Vibration and Acoustics, 2011, 133: 041006.

[21] HUANG Z W, HUA X G, CHEN Z Q, NIU H W. Modeling, Testing, and Validation of an Eddy Current Damper for Structural Vibration Control [J]. Journal of Aerospace Engineering, 2018, 31 (5), 04018063.

[22] 陈政清，田静莹，黄智文，等. 板式电涡流阻尼系数的计算与试验修正方法[J]. 中国公路学报，2016，29(10)：46-53.
CHEN Zheng-qing, TIAN Jing-ying, HUANG Zhi-wen, et al. Calculation and Test Correction Method of Plan Type Eddy Current Damping Coefficient [J]. China Journal of Highway and Transport, 2016, 29 (10): 46-53.

[23] 陈政清，张弘毅，黄智文. 板式电涡流阻尼器有限元仿真与参数优化[J]. 振动与冲击，2016，35(18)：123-127.
CHEN Zheng-qing, ZHANG Hong-yi, HUANG Zhi-wen. FEM Simulation and Parameter Optimization of a Planar Eddy Current Damper [J]. Journal of Vibration and Shock, 2016, 35 (18): 123-127.

[24] 陈政清，黄智文. 一种板式电涡流阻尼器的有限元模拟及试验分析[J]. 合肥工业大学学报：自然科学版，2016，39(4)：499-502.
CHEN Zheng-qing, HUANG Zhi-wen. Finite Element Simulation and Experimental Test of a Plane-type Eddy Current Damper [J]. Journal of Hefei University of Technology: Natural Science, 2016, 39 (4): 499-502.

[25] 陈政清，黄智文，王建辉，等. 一种基于螺旋传动方式的轴向电涡流阻尼器：中国，ZL 201410106148.3[P]. 2015-05-20.
CHEN Zheng-qing, Huang Zhi-wen, Wang Jian-hui et al. Axial eddy Current Damper Based on Spiral Transmission: China, ZL 201410106148.3 [P]. 2015-05-20.

[26] 陈政清. 一种外杯旋转式轴向电涡流阻尼器：中国，ZL 201410475528.4[P]. 2015-07-08.
CHEN Zheng-qing. An Uuter Cup Rotating Axial Eddy Current Damper: China, ZL 201410475528.4 [P]. 2015-07-08.

[27] 陈政清. 一种外杯旋转式轴向电涡流阻尼器：中国，ZL 201811375220.7[P]. 2019-09-13.
CHEN Zheng-qing. An Eddy Current Axial Damper: China, ZL 201811375220.7 [P]. 2019-09-13.

[28] HUANG Z W, CHEN Z Q, HUA X G. Numerical Modeling and Experimental Testing of an Eddy Current Apparent Mass Damper [C] // ZHOU F L, BENZONI G. 14th World Conference on Seismic Isolation, Energy Dissipation and Active Vibration Control of Structures. San Diego: The Anti-seismic Systems International Society (ASSISi), 2015: 1-9.

[29] 黄智文. 电涡流阻尼器的理论研究及其在桥梁竖向涡振控制中的应用[D]. 长沙：湖南大学，2016.
HUANG Zhi-wen. Theoretical Study of Eddy Current Damper and Its Application in Vertical Vortex-induced Vibration Control of Bridge [D]. Changsha: Hunan University, 2016.

[30] ZHANG H Y, CHEN Z Q, HUA X G, et al. Design and Dynamic Characterization of a Large-scale Eddy Current Damper with Enhanced Performance for Vibration Control [J]. Mechanical Systems and Signal Processing, 2020, 145: 1-21.

[31] WOUTERSE J H. Critical Torque and Speed of Eddy Current Brake with Widely Separated Soft Iron Poles [J]. IEE Proceedings B — Electric Power Applications, 1991, 138 (4): 153-158.

[32] SMITH M C. Synthesis of Mechanical Networks: the Inerter [J]. IEEE Transactions on Automatic Control, 2002, 47 (10): 1648-1662.

[33] 陈政清.三元被动减振理论与它的一种实现方式—电涡流轴向阻尼器[M].北京:高等教育出版社,2016.
CHEN Zheng-qing. Inerter-based Damping Devices and Its Realization Mode-eddy Current Axial Damper [M]. Beijing: Higher Education Press, 2016.

[34] CHEN Z Q, HUANG Z W, HUA X G. Inerter-damper-spring Passive Vibration Control: Its System and Realization with Eddy-current Apparent Mass Damper [C] // ZHOU F L, BENZONI G. 14th World Conference on Seismic Isolation, Energy Dissipation and Active Vibration Control of Structures. San Diego: The Anti-seismic Systems International Society (ASSISi), 2015: 1-8.

[35] IKAGO K, SAITO K, INOUE N. Seismic Control of Single-degree-of-freedom Structure Using Tuned Viscous Mass Damper [J]. Earthquake Engineering & Structural Dynamics, 2012, 41: 453-474.

[36] WEN Y, CHEN Z Q, HUA X G. Design and Evaluation of Tuned Inerter-based Dampers for the Seismic Control of MDOF Structures [J]. Journal of Structural Engineering, 2016, 143 (4): 04016207.

[37] 张瑞甫,曹嫣如,潘 超.惯容减震(振)系统及其研究进展[J].工程力学,2019,36(10):8-27.
ZHANG Rui-fu, CAO Yang-ru, PAN Chao. Inerter System and Its State-of-the-art [J]. Engineering Mechanics, 2019, 36 (10): 8-27.

[38] ZHU H P, LI Y M, SHEN W A, et al. Mechanical and Energy-harvesting Model for Electromagnetic Inertial Mass Dampers [J]. Mechanical Systems and Signal Processing, 2019, 120: 203-220.

[39] LI J Y, ZHU S Y, SHI X, et al. Electromagnetic Shunt Damper for Bridge Cable Vibration Mitigation: Full-scale Experimental Study [J]. Journal of Structural Engineering, ASCE, 146 (1): 04019175.

[40] HUANG Z W, CHEN Z Q, HUA X G, et al. Optimal design of TVMD with Linear and Nonlinear Viscous Damping for SDOF Systems Subjected to Harmonic Excitations [J]. Structural Control and Health Monitoring, 26 (10): 2413.

[41] HUANG Z W, HUA X G, CHEN Z Q, et al. Performance Evaluation of Inerter-based Damping Devices for Structural Vibration Control of Stay Cables [J]. Smart Structures and Systems, 2019, 31 (6): 615-626.

[42] LIANG L T, FENG Z Q, CHEN Z Q. Seismic control of SDOF Systems with Nonlinear Eddy Current Dampers [J]. Applied Sciences, 2019, 9 (16): 3427.

[43] 陈政清,黄智文,王建辉,等.桥梁用 TMD 的基本要求与电涡流 TMD[J].湖南大学学报:自然科学版,2013,40(8):6-10.
CHEN Zheng-qing, HUANG Zhi-wen, WANG Jianhui, et al. Basic Requirement of Tuned Mass Damper for Bridge and the Eddy Current TMD [J]. Journal of Hunan University: Natural Science, 2013, 40 (8): 6-10.

[44] WEN Q, HUA X G, CHEN Z Q, et al. AMD-based Random Decrement Technique for Modal Identification of Structures with Close Modes [J]. Journal of Aerospace Engineering, ASCE, 2018, 31 (5): 04018063.

[45] WEN Q, HUA, X G, CHEN, Z Q, et al. Modal Parameter Identification of a Long-span Footbridge by Forced Vibration Experiments [J]. Advances in Structural Engineering, 2017, 20 (5): 661-673.

[46] WEN Q, HUA X G, CHEN Z Q, et al. Control of Human-induced Vibrations of a Curved Cable-stayed bridge: Design, Implementation, and Field Validation [J]. Journal of Bridge Engineering, 2016: 04016028.

[47] 汪志昊,华旭刚,陈政清,等.基于微型永电磁式涡流阻尼 TMD 的人行桥模型减振试验研究[J].振动与冲击,2014,33(20):129-132,139.
WANG Zhi-hao, HAU Xu-gang, CHEN Zheng-qing, et al. Experimental Study on Vibration Control of a Model Footbridge by a Tiny Eddy-current Tuned Mass Damper with Permanent Magnets [J]. Journal of Vibration and Shock, 2014, 33 (20): 129-132, 139.

[48] LU X L, ZHANG Q, WENG D G, et al. Improving Performance of a Super Tall Building Using a New Eddy Current Tuned Mass Damper [J]. Structural Control and Health Monitoring, 2017, 24 (3): 1-17.

[49] LI S Y, LIU M, LI H X, et al. Effects of Structural Damping on Wind-induced Responses of a 243-meter-high Solar Tower Based on a Novel Elastic Test Model [J]. Journal of Wind Engineering and Industrial Aerodynamics, 2018, 172: 1-11.

[50] 汪志昊,陈政清,王建辉.钢拱桥刚性细长吊杆减振用调谐质量阻尼器的试验研究[J].科学技术与工程,2013,13(19):5555-5560.
WANG Zhi-hao, CHEN Zheng-qing, WANG Jianhui. Experimental Study on Tuned Mass Dampers for Vibration Control of Rigid Slender Suspenders in Steel Arch Bridges [J]. Science Technology and Engineering, 2013, 13 (19): 5555-5560.

[51] NIU H W, CHEN Z Q, HUA X G, et al. Mitigation of Wind-induced Vibrations of Bridge Hangers Using Tuned Mass Dampers with Eddy Current Damping [J]. Smart Structures and Systems, 2018, 22 (6): 727-741.

[52] GB 50463—2019,工程隔振设计标准[S].
GB 50463—2019, Standard for Design of Engineering Vibration Isolation [S].

Mechanical Systems and Signal Processing 145 (2020) 106879

Contents lists available at ScienceDirect

Mechanical Systems and Signal Processing

journal homepage: www.elsevier.com/locate/ymssp

Design and dynamic characterization of a large-scale eddy current damper with enhanced performance for vibration control

H.Y. Zhang [a], Z.Q. Chen [a,b], X.G. Hua [a,b,*], Z.W. Huang [a], H.W. Niu [a]

[a] Key Laboratory for Wind and Bridge Engineering of Hunan Province, Hunan University, China
[b] Vibration and Shock Technology Research Center, College of Civil Engineering, Hunan University, China

ARTICLE INFO

Article history:
Received 13 August 2019
Received in revised form 5 April 2020
Accepted 8 April 2020

Keywords:
Vibration control
Eddy current dampers
Rotary dampers
Damping density
Force-speed characteristics
Seismic control

ABSTRACT

Eddy current dampers (ECD) are promising devices for vibration control of dynamic systems, but their application in civil structures is very limited due to a low damping density. This study describes the development and dynamic characterization of a large-scale rotary ECD with enhanced performance for structural applications. The rotary ECD consists of a conductive rotor, a stator with permanent magnets, and a set of ball screw which converts the linear motions across the damper into rotations of rotor. The torque at different rotating speeds is evaluated by using a 2-D analytical method considering the reaction magnetic field caused by eddy currents. Laboratory experiments are conducted to validate the numerical results obtained for a simple eddy current torque generator and a large-scale ECD. It is shown that the analytical model agrees with nonlinear finite-element simulations, but is higher than the test results. The large-scale rotary ECD achieves a high damping density up to 24.9 MN·s/m^4 at a low speed. Moreover, the damping force of rotary ECD is highly nonlinear and self-limiting as a function of rotating speeds; it first increases with speeds and after reaching a maximum at a critical speed it decreases for higher speeds. Numerical simulation on seismic control of a SDOF system and a 5-DOF system showed that the rotary ECD performs as well as the fluid viscous dampers having the same secant damping coefficients. Finally, effects of changing design parameters on force-speed curve are studied in details.

1. Introduction

Mitigation of vibration of civil structures induced by earthquake and strong wind is an important task for enhancing the structural functionality and safety. A typical way is to increase the energy dissipation capability of structures by installing damping devices [1–3]. During the past decades, various damping devices have been developed by exploiting the frictional sliding, yielding of metals, shear or compression deformation of viscoelastic solids/fluids, and fluid orificing as source of damping. All these types of dampers are able to supply additional damping to a structure, and some of them such as viscoelastic dampers provide both additional damping and stiffness to a structure. A considerable research has been devoted to seismic control of newly-built or existing buildings and bridges by employing these damping devices [4–10].

* Corresponding author at: Key Laboratory for Wind and Bridge Engineering of Hunan Province, Hunan University, China.
E-mail address: cexghua@hnu.edu.cn (X.G. Hua).

https://doi.org/10.1016/j.ymssp.2020.106879
0888-3270/

Nomenclature

A	Magnetic vector potential
B	Magnetic flux density
E	Electric field intensity
M	Magnetization density
μ_0	Vacuum permeability
μ_r	Relative permeability
μ_{rm}	Recoil permeability of the PMs
J	Total current density
$\mathbf{J}_f$	Electric current density
$\mathbf{J}_b$	Magnetization current density
σ	Electrical conductivity
p	Number of pole-pairs
n	Order of harmonics
ω	Angular speed of the Rotor
n_r	Rotational speed of the Rotor
R_i	Radius of the Stator's inner surface
R_s	Radius of the Stator's outer surface
R_m	Radius of the PMs' outer surface
R_c	Radius of the Conductor's inner surface
R_r	Radius of the Rotor's inner surface
R_o	Radius of the Rotor's outer surface
$I_{np}(\ )$, $K_{np}(\ )$	Modified Bessel functions
B_{rem}	Remanence flux density of PMs
α	Magnet-arc to pole-pitch ratio
L	Axial length
Ld	Lead of the ballscrew assembly
v	Axial velocity of the damper
P	Energy dissipation power
A	Displacement amplitude

The eddy current damper (ECD) is another effective mechanism for dissipating kinetic energy in dynamical systems [11–16]. The attractive merit of ECD includes no mechanical contact, limited dependency on temperature, truly linear viscous damper at low speed, and its long-term stability over time. In its simplest form, ECD consists of conductive sheet and permanent magnets (PMs), and eddy currents are generated in the conductor when relative motion between conductive sheet and PMs occurs. As the interaction between eddy currents and magnetic field of PMs creates a repulsive force (Lorentz force) that is proportional to relative velocity, the moving conductor and PMs behave like a viscous damper. Besides the plate-like configuration, ECD could also be made in tube-like configuration and produce axial force [13,17,18]. ECD has been successfully used in auxiliary brake systems for heavy trucks and buses, and vehicle suspension systems [19–21], among others. It should be noted that the electromagnetic dampers, which employs the coil as a substitution for solid conductive sheet, can also acts as both dampers and generators for vibration mitigation and energy harvesting [22]. The purpose of this study is only focused on ECD functions as a damper for structural application, the discussion on energy harvesting is beyond scope of this study.

The major disadvantage of ECD is associated with the low damping density, as it provides very limited forces compared to its size and weight. A few studies have been focused on optimal design of ECD for increasing the flux density in air gap by selecting proper materials and topology configuration [12,13,16], however linear ECD can hardly be applied to large-scale, heavy (5000 ton or more), low-frequency (less than 1 Hz) structures under earthquake and low-frequency wind excitations [18]. Another way to increase the damping density of ECD is to integrate eddy current damping and certain motion magnification mechanism, which is capable of amplifying the working speed and damping force of ECD. A commonly-used approach is to transform the linear motion to rotary motion by using rack pinion and ball screw mechanisms. Suda and Shiiba (1996) studied the energy generation and vibration control using rack pinion harvester for vehicle suspension systems who claimed the sufficient forces can be achieved for active vibration control [23]. Nakano (2004) studied a new self-powered regenerative suspension system by using ball screw, and the author showed that the energy recovered is adequate for active control without the need of external power [24]. Iiyama et al. (1998) proposed a new hybrid system for seismic isolation in which a ball-screw mechanism and rotational generator is implemented for active control [25]. A few recent studies on the feasibility and field tests of electromagnetic dampers have been conducted for active suspension system

[26–31]. By using gearbox, Ellis, Fink, and Rich (1989) [32] developed a high torque capacity eddy current damper with four planetary stages which achieved a 1600:1 gear ratio, and they used as a rate limiting device for a large solar array deployment mechanism.

In contrast to active vehicle suspension system, the integration of ECD and electromagnetic dampers with ball-screw amplifying mechanism for passive control and energy harvesting in civil engineering application is relatively new [33]. Ikago et al. proposed a new seismic control device called tuned viscous mass damper (TVMD), which consists of fluid viscous dampers, flywheel and ball screw amplifying mechanism [34,35]. They carried out parameter optimization of TVMD under harmonic loading, and the control efficiency of a small-scale TMVD is experimentally compared with those of FVD and viscous dampers. Chen et al. developed a disk-type eddy current damper consisting of solid conductive disk and PMs. In addition to the damping force, the rotating conductive disk may provide very large apparent inertial mass and the flywheel is unnecessary [36]. Perez-Diaz et al. employed linear magnetic gear to amply the working speed and developed an eddy current damper called Z-damper of vibration mitigation for aerospace turbines [37]. Nakamura et al. [38] proposed a novel electromagnetic inertial mass damper (EIMD) consisting of an inerter and an electromagnetic damper for seismic control of buildings. They developed a mathematical model for portraying the mechanical behavior of the EIMD, and the superior control performance under seismic loadings was demonstrated. A full-scale EIMD was manufactured and tested by employing shaking table. Wang et al. experimentally evaluated the control performance of a small-scale EIMD in stay cables [39]. Zhu et al. presented an improved model for EIMD and they also examined the performance of EIMD for both structural control and energy harvesting [40]. Amjadian and Agrawal developed a smart electromagnetic friction damper with smooth hysteretic behavior and applied it to seismic control of buildings [41]. More recently, Li et al. presented some new designs of eddy current dampers with different architectures to enhance the damping density of ECDs and EIMDs, laboratory experiments on small-scale damper prototypes indicated a remarkably high damping coefficient were achieved for certain design [42]. In these previous studies, the eddy current damper is treated as linear viscous damping regardless of rotating speeds, and nonlinearity of damping forces with respect to speeds is omitted. Actually, at high velocity range, the magnetic field induced by the eddy current counteracts the original magnetic field by PMs, therefore leading to nonlinearity in damping forces [43], as will be shown later.

The purpose of this study is to characterize the mechanical properties of a large-scale rotary ECD by using numerical and experimental approaches, and the speed range of the damper is large enough to highlight the nonlinear and limiting features of damping forces. The rotary ECD consists of stator with PMs, cylindrical conductor rotor and a ball-screw mechanism which converts the linear relative motion between damper ends into rotational motion of conductor. An analytical approach is adopted to compute the eddy current and the generated torque in the conductor at different rotating speeds, in which the magnetic field induced by the eddy current is considered. The numerical results are verified by nonlinear finite-element simulations and laboratory experiments for an eddy current torque generator and a full-scale rotary ECD with ball screw. It has been shown that the damping forces can be highly nonlinear at high velocity range, which is not observed in previous studies [36–40,42]. The achieved damping density for the full-scale ECD is compared to that of fluid viscous dampers. The control performances of a SDOF system and a 5-DOF system under seismic excitations are studied for linear FVD, nonlinear FVD and rotary ECD, respectively. The influence of design parameters on the force-speed curve is addressed.

2. 2D analytical evaluation of dynamic properties of rotary ECD

2.1. Principle and configuration of rotary ECD

Fig. 1 shows the sketch of a kind of rotary eddy current dampers (R-ECD) with enhanced performance. Fig. 2 illustrates the explored view of R-ECD. It is mainly composed of the rotor, the stator, and a set of ball screw. The working principle of an R-ECD is to transform the linear relative motion between damper's ends into the rotation of the rotor, through the ballscrew assembly and bearings. The rotor of an R-ECD is made of a conductor tube and a steel tube. The stator consists of PMs and a steel holder. As soon as the rotor rotates around PMs, an eddy current torque (damping force) will be generated and transformed into an axial force through the ballscrew assembly. The rotational speed is usually dozens of times larger than the

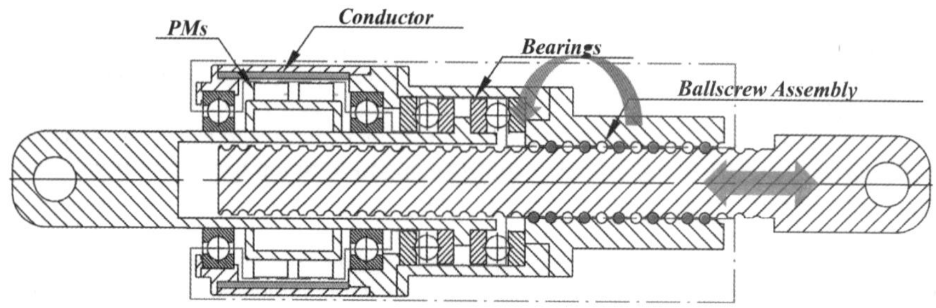

Fig. 1. Rotary eddy current damper (R-ECD).

Fig. 2. Explored view of rotary eddy current damper.

linear velocity due to the amplifying mechanism of the ballscrew, and therefore R-ECD is potential to produce large-capacity damping forces. When the eddy current damping force is amplified by the ball screw mechanism, the inertial force produced by rotational motion of the rotary part is also amplified [37,38]. So the total force produced by an R-ECD consists of a damping force related to velocity, an inertial force proportional to acceleration, and a friction force.

Considering the system architecture, rotary ECD is very similar to the drum permanent magnet retarder (DPMR) [44], and the slotless surface-mounted permanent magnet motor (SSPMM) [45,46]. The major difference among them is the current region. A DPMR usually generates eddy currents in the steel stator, while a SSPMM usually rotates the PMs, and generates current in the coil. For R-ECD, the eddy current is generated in the rotating conductor.

Tonoli investigated the dynamic characteristics of rotary eddy current dampers and couplers, and evaluated the influence of design parameters on forcing characteristics of the devices [47]. Therefore, it is important to investigate the dynamic characteristics of ECD in rotary configuration. In deriving the torque of the rotary ECD, the calculation of eddy current in conductor and its interaction with original magnetic field are much complicated than the electromagnetic dampers [35,38,40]. The following assumptions are made.

- End effects are ignored, that is, the cylinder is assumed to have infinite axial length compared to the distance between rotor and stator. The problem is therefore simplified to a 2D case. A rotor fixed reference axis is defined as shown in Fig. 3(a). The magnetic flux density has only radial component $B_r(r,\theta,t)$ and tangential component $B_\theta(r,\theta,t)$, while the electric field intensity has only axial component $E_z(r,\theta,t)$. In other words, the magnetic vector potential has only axial component $A_z(r,\theta,t)$, and the electric potential is 0 everywhere [45].
- The electromagnetic field can be treated as a quasi-static field [48], ignoring the displacement current, due to $\omega \ll \sigma/\varepsilon$ and $R \ll c/\omega$ where ω is the angular speed of the rotor, R is the character length of the study region, σ is the electrical conductivity, ε is the permittivity, and c is the speed of light in vacuum.
- All materials are isotropic. The permeability and conductivity are constant. The magnetic saturation and hysteresis loss are neglected. The steel has infinitely large relative permeability, $\mu_r^{\mathrm{I}} = \mu_r^{\mathrm{V}} = \infty$, and infinitely small conductivity. The relative permeability of airgap and the conductor is 1, $\mu_r^{\mathrm{III}} = \mu_r^{\mathrm{IV}} = 1$. Air and PMs in region II are assumed to have the same permeability. The relative recoil permeability of a magnet made of NdFeB is about 1.02 ~ 1.05.

2.2. Governing PDEs

An electromagnetic field can be fully described by the magnetic vector potential **A** as well as the electric potential ϕ (a scalar field) [48]:

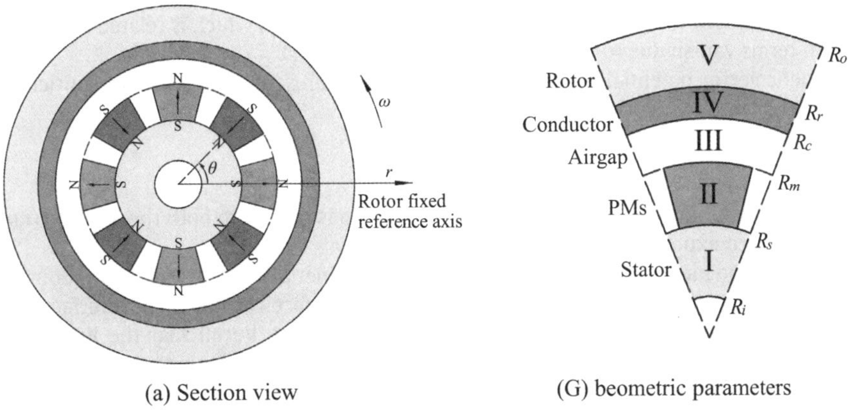

Fig. 3. Cross-section and geometric parameters of the R-ECD (Not to scale).

$$B = \nabla \times A$$
$$E = -\nabla\phi - \frac{\partial \mathbf{A}}{\partial t} \quad (1)$$

where **B** is the magnetic flux density, **E** is the electric field intensity.

By the potential field approach, Faraday's law and Gauss's law for magnetism reduce to identities. According to the assumption $\phi = 0$ and quasi-static condition, Ampère's law with Maxwell's addition reduces to:

$$\nabla^2 \mathbf{A} = -\mu_0(\sigma \mathbf{E} + \nabla \times \mathbf{M}) \quad (2)$$

where μ_0 is the vacuum permeability, σ is the electrical conductivity, and $\mathbf{A} = (0, 0, A_z(r, \theta, t))$ in a cylindrical coordinate (r, θ, z).

As illustrated in Fig. 3, the cylinder is divided into five regions: stator (region I), PMs (region II), air gap (region III), conductor (region IV) and rotor (region V). Free current only exists in the conductor. The magnetic hysteresis is ignored in rotor and stator, which are made of steel. Therefore, equations in each region are listed below:

$$\begin{cases} \nabla^2 \mathbf{A}^{\mathrm{I}} = 0 \\ \nabla^2 \mathbf{A}^{\mathrm{II}} = -\mu_0 \nabla \times M \\ \nabla^2 \mathbf{A}^{\mathrm{III}} = 0 \\ \nabla^2 \mathbf{A}^{\mathrm{IV}} = \mu_0 \sigma \frac{\partial \mathbf{A}}{\partial t} \\ \nabla^2 \mathbf{A}^{\mathrm{V}} = 0 \end{cases} \quad (3)$$

2.3. Magnetization density vector and boundary conditions

With the z-component ignored, the magnetization density vector of the permanent magnets is expressed by its radial and tangential components as

$$\mathbf{M} = M_r \mathbf{r} + M_\theta \boldsymbol{\theta} \quad (4)$$

where r and θ are the unit vectors in the radial and tangential directions.

Two different magnetization patterns are usually used in eddy current dampers: radial and parallel [43]. For the radial magnetization pattern, the tangential component of magnetization density vanishes and the n-th harmonic of radial magnetization density is given as

$$M_{rn} = \frac{4B_{rem} \sin\left(\frac{n\pi\alpha}{2}\right)}{\mu_0 n\pi} \quad (5)$$

where B_{rem} is the remanence flux density of PMs, α is the magnet-arc to pole-pitch ratio.

By using the analytic representation, the radial component of the magnetization density vector can be represented by a sum of Fourier series in exponential form [43,44]:

$$M_r(\theta, t) = \mathrm{Re}\left[\vec{M}_r(\theta, t)\right] = \mathrm{Re}\left[\sum_{n=1,3,5\cdots}^{\infty} M_{rn} e^{inp(\theta+\omega t)}\right] \quad (6)$$

where p is the number of pole-pairs, ω is the rotational angular velocity. The phasors $\vec{M}_r$ denoted by superscript arrows is called the analytic representation. The argument $np(\theta + \omega t)$ is called the phase, which is related to both the angular location θ and time t. The even terms vanish due to alternating arrangement of PMs.

Similarly, the magnetic vector potential generated by PMs can be represented by a sum of Fourier series, as follows:

$$A_z(r, \theta, t) = \mathrm{Re}\left[\vec{A}_z(r, \theta, t)\right] = \mathrm{Re}\left[\sum_{n=1,3,5\cdots}^{\infty} A_{zn}(r) e^{i\left[np(\theta+\omega t) - \frac{\pi}{2}\right]}\right] \quad (7)$$

where the phase of each term is $[np(\theta + \omega t) - \pi/2] + \mathrm{Arg}[A_{zn}(r)]$, which is related to both the location (r,θ) and time t. So far, we've got a variable-separated expression for magnetic vector potential.

At the interface between two media, the radial component of the magnetic flux density is continuous. The tangential component of the magnetic field intensity is also continuous as there is no surface current at the interface. For the present problem in study, at the boundary $r = R_i$ and $r = R_o$, magnetic insulation is set. Recall that the linear relation between the magnetic flux density and the magnetic field intensity exists for diamagnets and paramagnets, that is $\mathbf{B} = \mu_r \mu_0 \mathbf{H}$. According to the third assumption, $\mu_r^{\mathrm{I}} = \mu_r^{\mathrm{V}} = \infty$ for steel, $\mu_r^{\mathrm{III}} = \mu_r^{\mathrm{IV}} = 1$for airgap and conductor, and μ_{rm}^{II} for air and PMs in region II is about $1.02 \sim 1.05$. The boundary conditions can be expressed as follows:

$$
\begin{cases}
\left|B_r^{\mathrm{I}}(r,\ \theta,t)\right|_{r=R_i}\left|=\right|0 \\
\left|B_r^{\mathrm{I}}(r,\ \theta,t)\right|_{r=R_s}\left|=\right|\left|B_r^{\mathrm{II}}(r,\ \theta,t)\right|_{r=R_s} \\
\frac{1}{\mu_{rm}\mu_0}\left|B_\theta^{\mathrm{II}}(r,\ \theta,t)\right|_{r=R_s}\left|=\right|0 \\
\left|B_r^{\mathrm{II}}(r,\ \theta,t)\right|_{r=R_m}\left|=\right|\left|B_r^{\mathrm{III}}(r,\ \theta,t)\right|_{r=R_m} \\
\frac{1}{\mu_{rm}\mu_0}\left|B_\theta^{\mathrm{II}}(r,\ \theta,t)\right|_{r=R_m}\left|=\right|\frac{1}{\mu_0}\left|B_\theta^{\mathrm{III}}(r,\ \theta,t)\right|_{r=R_m} \\
\left|B_r^{\mathrm{III}}(r,\ \theta,t)\right|_{r=R_c}\left|=\right|\left|B_r^{\mathrm{IV}}(r,\ \theta,t)\right|_{r=R_c} \\
\frac{1}{\mu_0}\left|B_\theta^{\mathrm{III}}(r,\ \theta,t)\right|_{r=R_c}\left|=\right|\frac{1}{\mu_0}\left|B_\theta^{\mathrm{IV}}(r,\ \theta,t)\right|_{r=R_c} \\
\left|B_r^{\mathrm{IV}}(r,\ \theta,t)\right|_{r=R_r}\left|=\right|\left|B_r^{\mathrm{V}}(r,\ \theta,t)\right|_{r=R_r} \\
\frac{1}{\mu_0}\left|B_\theta^{\mathrm{IV}}(r,\ \theta,t)\right|_{r=R_r}\left|=\right|0 \\
\left|B_r^{\mathrm{V}}(r,\ \theta,t)\right|_{r=R_o}\left|=\right|0
\end{cases}
\tag{8}
$$

In the static condition, the magnetic potential is time-independent. The part on the right side of the equal sign of Eq. (3d) turns into 0, and the static field could be derived with Eq. (3) and the boundary condition Eq. (8). The derivation process is attached on the Appendix.

2.4. Induced current distribution and dissipation power

If the reaction magnetic field due to the eddy currents in Region IV is taken into account, the magnetic field will be more complicated. Fig. 4 shows a distribution sketch of the magnet's field B_0, the induced magnetic field B_i, and the induced current J. At the trailing edge of the magnet (left side), by the right hand rule the counterclockwise current creates a magnetic field pointed up, which is in the same direction as the magnet's field B_0 and enhances the magnetic field. In contrast, at the leading edge (right side), the clockwise current causes a magnetic field pointed down, attenuating the PMs' magnetic field. The induced current J is proportional to the total magnetic field $B_0 + B_i$.

The magnetic field and eddy current affect each other, as manifested in Eq. (3d). Combining Eq. (3d) with Eq. (7), we can get the governing equation for region IV (equations for all regions and the full solving progress are given in the appendix):

$$
\frac{\partial^2 A_{zn}^{\mathrm{IV}}}{\partial r^2}+\frac{1}{r}\frac{\partial A_{zn}^{\mathrm{IV}}}{\partial r}-\frac{(np)^2}{r^2}A_{zn}^{\mathrm{IV}}=inp\omega\sigma\mu_0 A_{zn}^{\mathrm{IV}} \tag{9}
$$

Eq. (9) is a Bessel's differential equation. The general solution could be written as:

$$
A_{zn}^{\mathrm{IV}}=C_7^{\mathrm{I}} I_{np}\left(\sqrt{inp\omega\sigma\mu_0}r\right)+C_8^{\mathrm{I}} K_{np}\left(\sqrt{inp\omega\sigma\mu_0}r\right) \tag{10}
$$

where $I_{np}(\ )$ and $K_{np}(\ )$ are *Modified Bessel functions* [49], consisting a fundamental set of solution to Eq. (9), C_7^{I} and C_8^{I} are arbitrary constants.

Substituting Eq. (1a) and Eq. (10) into the boundary conditions Eq. (8), one obtains C_7^{I} and C_8^{I}. Substitute them back into Eq. (1a) and Eq. (10), and an expression for radial component of magnetic flux density in Region IV is obtained:

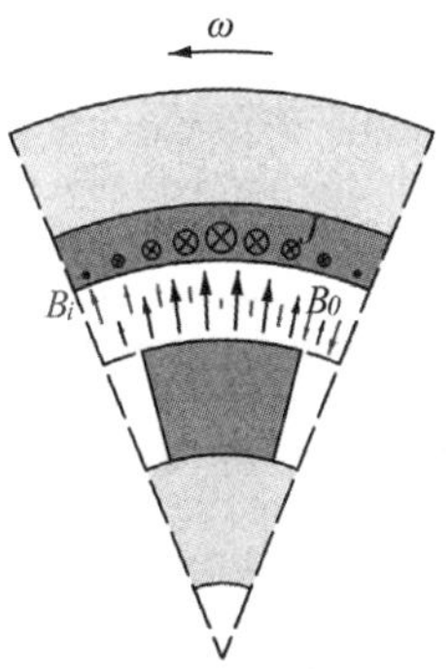

Fig. 4. A distribution sketch of the magnetic flux density and current density.

$$\vec{B}_r^{IV}(r,\theta,t) = \sum_{n=1,3,5\cdots}^{\infty} \frac{4B_{rem}\sin\left(\frac{n\pi\alpha}{2}\right)p}{\pi}\xi_n\frac{R_m}{r}\frac{I'_{np}\left(\frac{R_r}{\delta_n}\right)K_{np}\left(\frac{r}{\delta_n}\right)-K'_{np}\left(\frac{R_r}{\delta_n}\right)I_{np}\left(\frac{r}{\delta_n}\right)}{\psi_n}e^{inp(\theta+\omega t)} \tag{11}$$

The expressions for ξ_n and ψ_n are given in the appendix. It is worth to mention that Eq. (11) cannot be directly used to derive the static field, for which $\omega = 0$.

Similarly, the induced current distribution in conductor can be written as:

$$\vec{J}_z(r,\theta,t)\& = \sum_{n=1,3,5\cdots}^{\infty} -\sigma\frac{\partial}{\partial t}\left(A_{zn}^{IV}(r)e^{i\left[np(\theta+\omega t)-\frac{\pi}{2}\right]}\right) = \sum_{n=1,3,5\cdots}^{\infty} J_{zn}(r)e^{inp(\theta+\omega t)} \tag{12}$$

where

$$J_{zn}(r) = \omega\sigma\frac{4B_{rem}\sin\left(\frac{n\pi\alpha}{2}\right)p}{\pi}\xi_n R_m\frac{I'_{np}\left(\frac{R_r}{\delta_n}\right)K_{np}\left(\frac{r}{\delta_n}\right)-K'_{np}\left(\frac{R_r}{\delta_n}\right)I_{np}\left(\frac{r}{\delta_n}\right)}{\psi_n} \tag{13}$$

The Joule heating equation gives the dissipation power of eddy currents:

$$P = \iiint_{RegionIV}\frac{1}{\sigma}\mathrm{Re}\left(\vec{J}_z\right)^2 dV = \frac{L}{\sigma}\int_{R_c}^{R_r}\int_0^{2\pi}\mathrm{Re}\left(\vec{J}_z\right)^2 rd\theta dr \tag{14}$$

where L is the conductor length along z-axis.

Based on the orthogonality and periodicity of the Fourier terms, the dissipation power can be expressed as:

$$P = \sum_{n=1,3,5\cdots}^{\infty} P_n \tag{15}$$

where

$$P_n = \frac{\pi L}{\sigma}\int_{R_c}^{R_r}|J_{zn}(r)|^2 rdr \tag{16}$$

For an eddy current damper, a power balance equation is given by:

$$F\times v = T\times\omega = P \tag{17}$$

where T is the torque generated by eddy currents, F is the axial force transferred from the torque by the ball screw assembly, and v is the axial velocity of the damper.

3. Experimental validation

In this section, laboratory experiments of two rotary eddy current dampers are performed. The first damper is actually a simple torque generator consisting of cylindrical conductor rotating around PMs. In this example, the motor is used to drive a small-scale rotator and the generated torque is measured by force transducers. The second example is experimental tests of a full-scale rotary ECD with ball screw. In both cases, the eddy current torque is evaluated by the analytical method and finite element simulations, and is compared to the experimental results.

3.1. Simple torque generator

Fig. 5 shows the test setup for the simple eddy current torque generator. The testing device is mainly composed of three parts, a 3 kW three-phase asynchronous motor, the eddy-current torque generator and a torque sensor. The torque sensor measures torque by strain gages and has a rated torque of 20 N·m. The output voltage signal is measured by the computer through an A/D board. The eddy-current torque generator has the same topology as that given in Fig. 3(b). The rotor is driven by the rotational motor through a key. When it rotates, the eddy current will be generated in the conductor, which is linked to the rotor by bolts. The stator is fixed with a torque sensor. NdFeB PMs with alternating poles are positioned side by side along circumferential directions of the stator. The rotor and stator are linked by two deep groove ball bearings, which provide radial support and keep the airgap at a fixed value. When the motor starts, a torque will be transferred to the sensor.

Three groups of tests are conducted, and the corresponding test parameters (Test 1A, Test 1B, and Test 1C) are listed on Table 1. The main differences of the three groups of tests are the thickness of the conductor tube, though the manufacturing tolerance leads to other tiny differences in size. The motor's rotational speed is adjusted by a variable-frequency driver. In each test, a particular input AC frequency is set, and the motor will keep rotating at the specified speed. The torque data is record by the computer, and the mean speed is measured by a handheld tachometer. The friction torque has been removed from the data by an additional group of tests with no PMs.

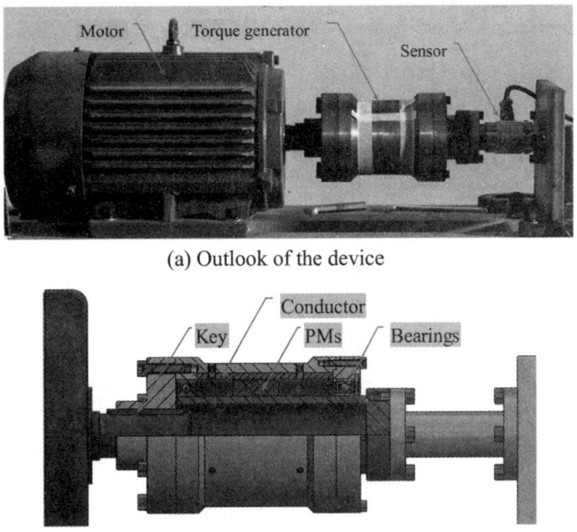

(a) Outlook of the device

(b) Cutaway view of the device

Fig. 5. Sketch of simple torque generator driven by motor.

The remanence magnetic flux density for NdFeB (N28) type magnet is 1.05 T. The theoretical electrical conductivity for copper is 58MS/m, while the copper we used in this experiment is about 53.5 MS/m tested by an electrical conductivity measurement.

From Eqs. (15)–(17), the torque at different ω is calculated by:

$$T(\omega) = \frac{1}{\omega} \sum_{n=1,3,5\cdots}^{\infty} \frac{\pi L}{\sigma} \int_{R_c}^{R_r} |J_{zn}(r,\omega)|^2 r dr \tag{18}$$

The first four Fourier terms (n = 1,3,5,7) are considered, and the effect of different order of harmonic on the generated torque will be shown later.

A 2D finite element model is also established by COMSOL software for comparison. The parameters and boundary conditions adopted in the 2D FEM are exactly the same as those in the analytical model, except that the relative permeability of the air in Region II is set 1 instead of μ_{rm}. Fig. 6 shows the magnetic flux density stream at different speed obtained from 2D FEM.

Fig. 7 shows the comparison of magnetic flux density distribution at $r = R_c$ obtained from 2D FEM and analytical method for null rotational speed and rotational speed of 3600RPM. At null rotational speed, the magnetic field is actually the static field due to PMs, which is symmetric. At n_r = 3600 RPM, the magnetic field become asymmetric due to the enhancing or attenuating effects of reaction field caused by the eddy current. A tiny difference occurs at the interface between the PMs and air in Region II. Generally, the results from 2D FEM and analytical method are quite close.

Fig. 8 shows the torque-speed curves obtained by 2D FEM, analytical method and the experiments. At low speed, the analytical results fit the test data well. Deviation between analytical results and testing values increases as the speed increases, especially in Test 1A. A possible reason is that the 2D FEM and analytical model used in this study neglects the loss of magnetic flux density in the z-axis. However, in the present case the axial length L is much larger than the air gap indicating that the three-dimensional effect is small. The main reason for the discrepancy is probably attributed to the rising temperature of

Table 1
Geometrical parameters and material properties of the tests.

	Unit	Test 1A	Test 1B	Test 1C	Test 2
B_{rem}	T	1.05	1.05	1.05	1.39
σ	MS/m	53.5	53.5	53.5	24.3
μ_{rm}	1	1.05	1.05	1.05	1.05
p	1	4	4	4	9
α	1	0.367	0.367	0.367	0.675
R_s	mm	25	25	25	99
R_m	mm	35	35	35	119
R_c	mm	37	36	36	120.2
R_r	mm	38.5	40	44	124
L	mm	30	30	30	4×40
Ld	mm				40

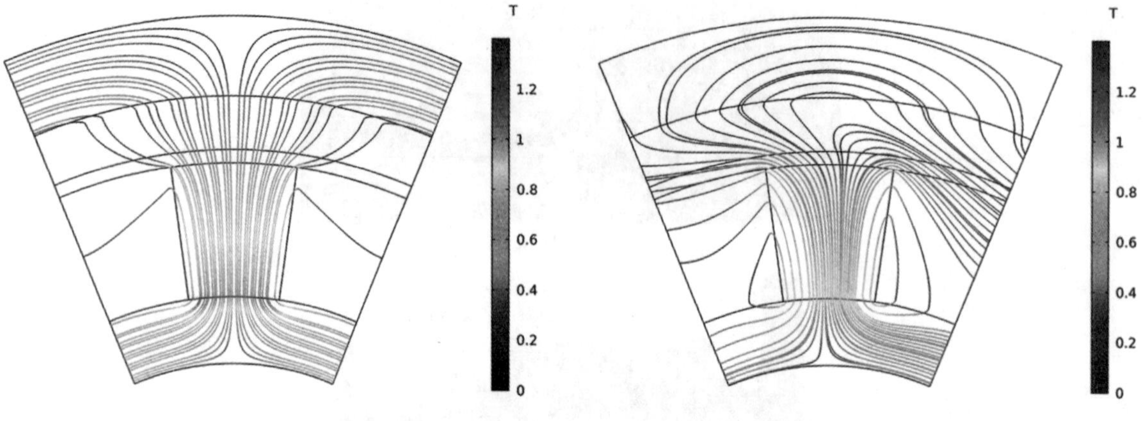

(a) 0 RPM (b) 3600 RPM

Fig. 6. Magnetic flux density stream resulted from 2D FEM (Test 1B).

copper with rotating speeds. Table 2 shows the recorded temperature at cylinder surface for several rotating speeds. As shown in Table 2, in the case of higher rotating speed, more time is taken by the rotor to reach stable, as a result a higher temperature rise in the rotor is expected. Especially, the temperature rise in copper for Test 1A is more significant than the others due to its smaller volume and higher power. It is expected that the copper's temperature will be slightly higher.

Temperature rise always results in a reduction of the electrical conductivity of copper, and the relationship between the electrical conductivity and the temperature rise can be approximately described by the formula below [50],

$$\sigma(T) = \frac{\sigma_0}{1 + K(T - T_0)} \tag{19}$$

where σ_0 is the electrical conductivity of copper at the reference temperature T_0, as shown in Table 1; K is the temperature correction coefficient, and for copper or aluminum alloy, $K = 0.004$.

To take into account the temperature rise on prediction of the experiment results, the measured temperatures of the rotor shown in Table 2 are interpolated to estimate the variation of the temperature with rotating speed. The actual electrical conductivity of the copper in different cases of rotating speed is calculated according to Eq. (19), and then they are employed to calculate the torque based on Eq. (15). The analytical results considering the thermal effect are also shown in Fig. 8 for comparison. It can be seen that the analytical expression can predict the experiment results pretty well for all the test cases after considering the thermal effect. Moreover, both the analytical and experimental peak torques are almost the same for Test 1B (or Test 1C). It is noted that temperatures of PMs also rise, which could cause the reduction in magnetization of the PMs. However, temperature rise in PMs should be much lower than that in copper since there is no direct contact between the PMs and the conductor.

3.2. Full-scale rotary ECD

In this section, a full-scale rotary ECD with ball screw is tested in a large-scale facility. The full-scale ECD is designed to have a comparable damping density as that of fluid viscous dampers. Four rotary ECDs with maximum damping forces up to 500kN each have been implemented on the Zhangjiajie Grand Canyon Glass Bridge in China, as shown in Fig. 9 (a). The dampers are fixed between the girder and the abutment to control the longitudinal vibration of the bridge due to seismic loads and pedestrian loadings, as shown in Fig. 9 (b). The design parameters for the damper are summarized in the last column of Table 1. The PMs are NdFeB (N50) type. They are arranged in four circles, and each magnet is 40 mm long. The conductor is made of 6061-T6 aluminum alloy. The maximum diameter of the R-ECD is 260 mm and the mid-stroke length is 1360 mm when the stroke is set to ±250 mm.

Experiments are conducted for one of four full-scale ECDs to validate the analytical method and characterize dynamic behavior of the damper. The test is conducted in the electro-hydraulic servo testing system for energy dissipating elements in the Earthquake Engineering Research & Test Center of Guangzhou University, China. One end of the rotary ECD is connected to actuator and the other side is connected to the frame, shown in Fig. 10. The electro-hydraulic servo actuator with a ±250 mm stroke is controlled by displacement commands through a PID controller, with feedback from an LVDT displacement sensor. The working frequency of this system ranges from 0.001 Hz to 5 Hz, and the maximum dynamic force capacity is 1000 kN. This testing system is particularly designed for large dampers used in civil engineering. The control instrumentation consists of a PID controller and a power amplifier that drives the servo-valve. There are sensors that provide measurements of the ram displacement x, the input current I_v to the servo-valve, and the load force F_L through the mechanical specimen that is being tested.

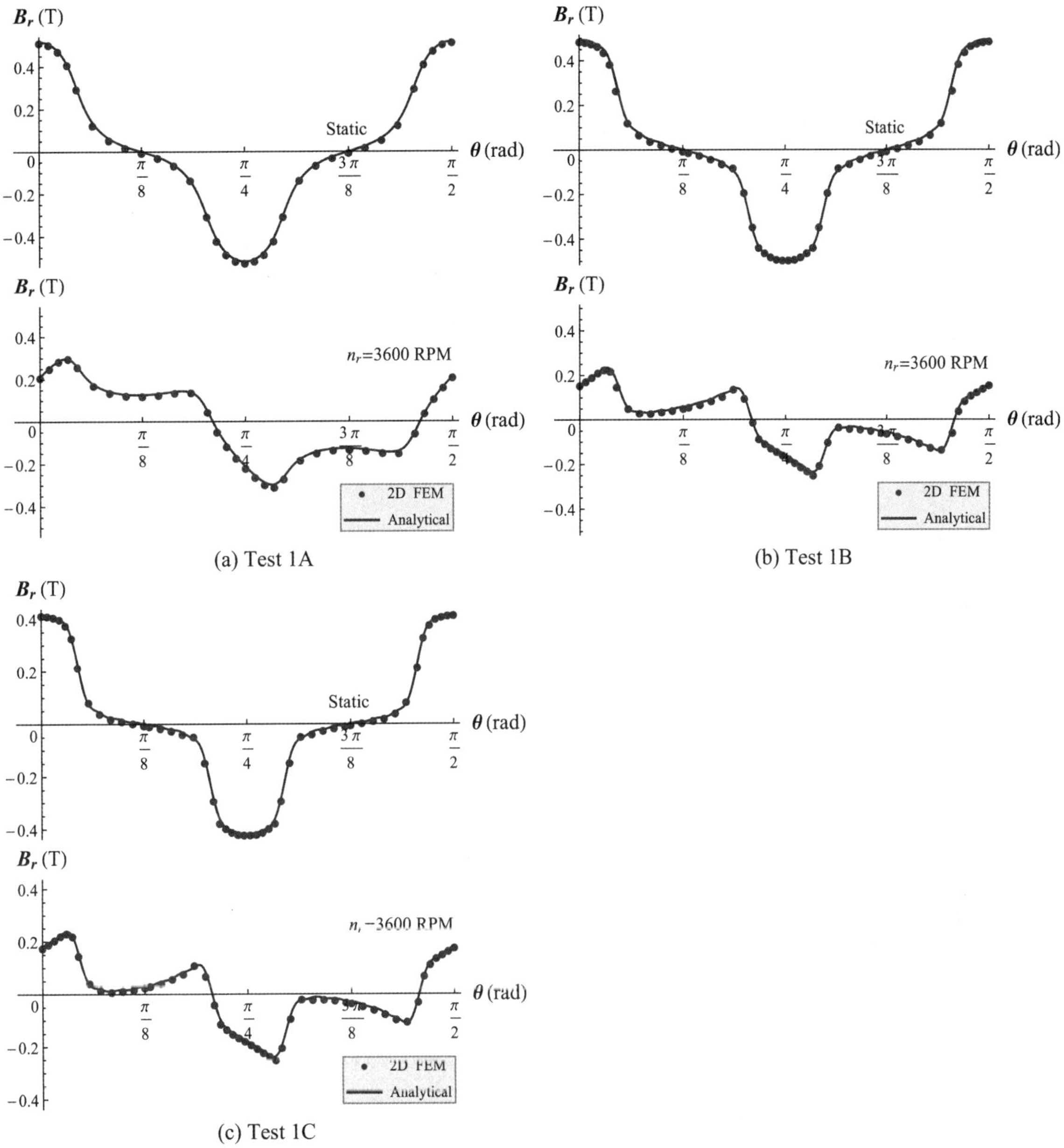

Fig. 7. Comparison of magnetic flux density distribution at $r = R_c$ from 2D FEM and analytical method.

During tests, the axial velocity of the damper is changed by increasing the displacement amplitude A between 20 mm and 200 mm, while the excitation frequency is maintained constant $f = 0.4$ Hz. Therefore, the range of axial velocity tested is between 50.3 mm/s and 503 mm/s. Fig. 11 shows the typical hysteretic loop of the full-scale ECD measured at two different amplitudes with the same excitation frequency $f = 0.4$ Hz. which looks like those of fluid viscous dampers. The output force from the rotary ECD consists of three parts: the damping force due to eddy currents, the inertial force due to rotational inertia of rotor, and friction force within the damper. The inertial force is proportional to the acceleration of rotor, and it vanishes when the velocity is maximum. The equivalent mass of this damper is about 18 tons, calculating from the rotary inertia and the lead of ball-screw [34], and the maximum inertial force is about 23 kN for $A = 200$ mm and $f = 0.4$ Hz. A friction force of 3.5 kN is measured by an additional testing case at a constant low velocity of 1 mm/s. Therefore, the damping force and the corresponding velocity are determined from the intersection between hysteric loop and y-axis, followed by the removal of friction force. In this way, we can obtain the tested *F-V* curve of a rotary ECD. The difference between tested and predicted forces is mainly caused by the temperature change in tests, as will be elaborated later.

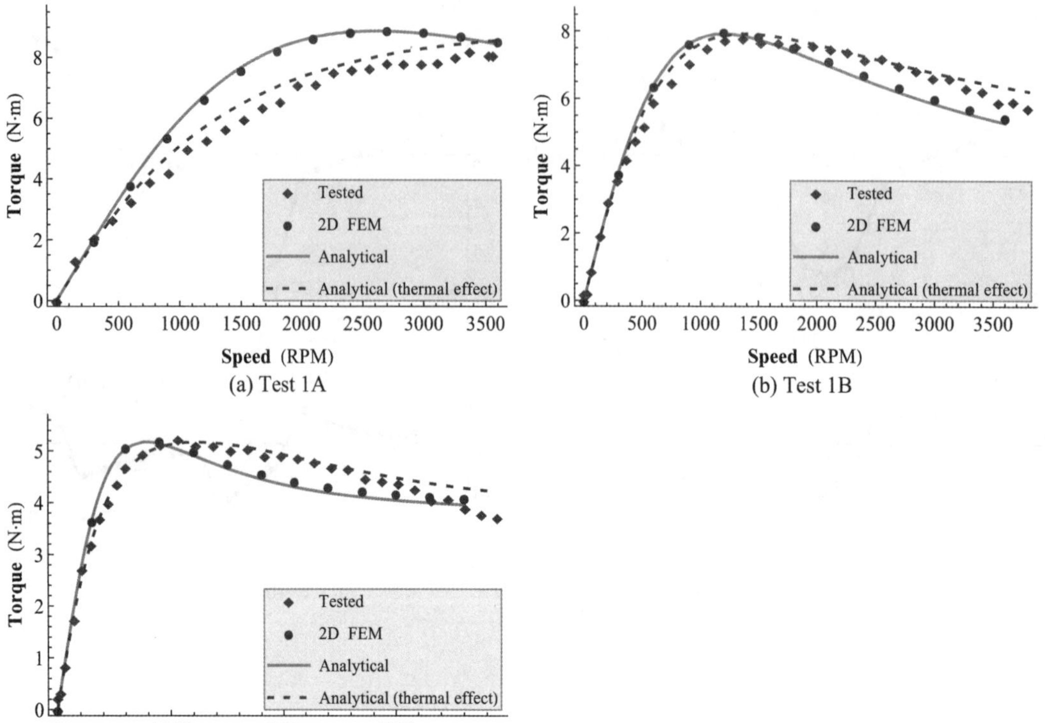

(a) Test 1A (b) Test 1B (c) Test 1C

Fig. 8. Comparison of eddy current torque for simple torque generator.

Table 2
Temperature measured at different rotating speeds for Tests 1 (℃).

Speed* (RPM)	0	145	750	1370	1970	2550	3100	3600
Test 1A	24	30	49	65	82	101	124	143
Test 1B	24	28	37	48	61	70	75	86
Test 1C	24	32	66	84	99	102	98	99

(*accurate speed of each test may vary 3% of nominal rotating speed).

(a) Panoramic view of the bridge (b) Rotary ECD fixed between girder and abutment

Fig. 9. Zhangjiajie Grand Canyon Glass Bridge and one of the full-scale rotary ECDs.

Fig. 10. Rotary ECD in test.

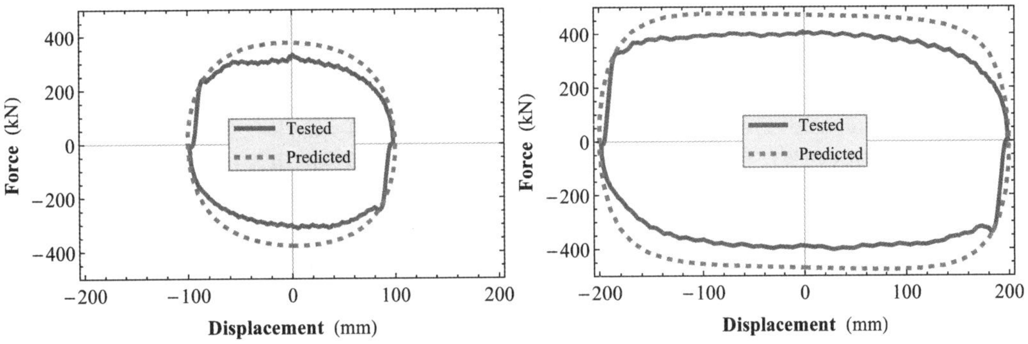

(a) Amplitude=100mm, v_{max}=251mm/s

(b) Amplitude=200mm, v_{max}=503mm/s

Fig. 11. Hysteretic loop of rotary ECD at excitation frequency f = 0.4 Hz.

For large-scale ECD with ball screw, the relationship between angular velocity ω and axial linear velocity v is $= v \times 2\pi/Ld$, where Ld is the lead of the ball screw assembly. From Eq. (15)–(17), the axial damping force at different axial velocity v is calculated by:

$$F(v) = \frac{1}{v}\sum_{n=1,3,5}^{\infty}\frac{\pi L}{\sigma}\int_{R_r}^{R_r}|J_{zn}(r,v)|^2 rdr \tag{20}$$

Fig. 12 compares the force-speed curve obtained from the analytical method (Eq. (20)) with those from the nonlinear finite-element simulations and experiments. The predicted output force is obtained by summation of damping force and inertial force, as previously shown in Fig. 11. Again, analytical method generally fits the 2D FEM well. The force tested is still smaller than that from calculations. In addition to temperature effect and neglect of loss of magnetic flux density in the z-axis mentioned in Section 3.1, there is one more non-negligible reason for this size of damper. The conductor expands with

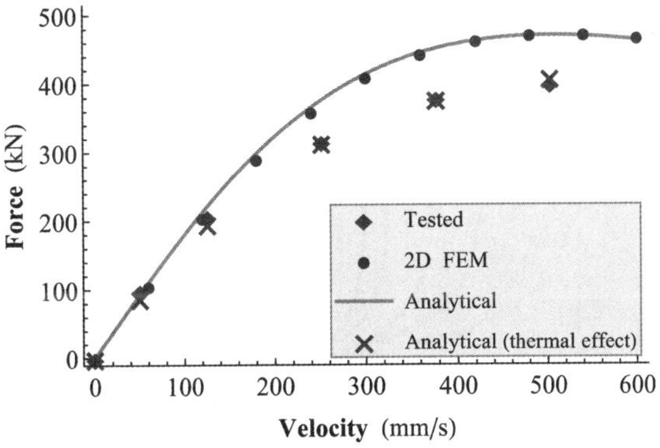

Fig. 12. Eddy current force calculated with different methods for full-scale rotary ECD.

the rise of temperature, which results in an increase of airgap between PMs and cylindrical conductor. The surface temperature of R-ECD is recorded to be 145 ℃ at the last test case (A = 200 mm, v = 503 mm/s), and the room temperature is about 20 ℃. Temperature of other cases is simply assumed to be linear with the axial velocity, as shown in Table 3. The actual conductivity of each testing case could be estimated according to Eq. (19), while change of the radius of conductor tube is followed by the equation, $R = R(1+\alpha_L)(T-T_0)$, where α_L is the coefficient of thermal expansion. For aluminum alloy, $\alpha_L = 23.6\times10^{-6}/°C$, and for steel, $\alpha_L = 13.0\times10^{-6}/°C$. For instance, at the maximum velocity, a 125℃ rise in temperature results in a 34% decrease in electrical conductivity and a 0.37 mm increase in air gap. The result of damping forces predicted by the analytical method considering thermal effects are also shown in Fig. 12, it can be seen that it agrees well with the experimental result for all the tested cases.

Additional tests are conducted on large-scale rotary ECD at different frequency between 0.2 Hz and 0.8 Hz, while maintaining the maximum velocity constant. The obtained hysteretic loops of the rotary ECD at different frequency are shown in Fig. 13. The R-ECD reaches a maximum velocity and a zero acceleration at $u = 0$ of each loop. This testing result reveals that the damping force of an R-ECD is frequency-independent at frequency range between 0.2 Hz and 0.8 Hz. The testing data for higher frequency are not available, and it will be addressed in the future research.

Compared with the traditional ECD, the developed rotary ECD with ball screw could achieve much higher damping density, which is defined as the ratio of damping coefficient to damper volume. An eddy current damper without any amplifying device can provide a damping density ranging from 1 ~ 3 MN·s/m^4 [12,18]. For eddy current damper with linear magnetic gear, a damping density of 8.4 MN·s/m^4 is reported [36]. Using the ball-screw, Nakamura et al. obtained a design damping density of 22.4 MN·s/m^4 for a full-scale electromagnetic damper with ±250 mm stroke [37], and the actual damping density is not provided. In this study, the experimental damping coefficient of the full-scale R-ECD at low velocity is 1.8 MN·s/m, which indicates a damping density of 24.9 MN·s/m^4. At a working velocity of 0.5 m/s, the damping density reduces to 11.1 MN·s/m^4. For comparison, a typical fluid viscous damper capable of providing a 500 kN force and ±250 mm stroke, is about 150 mm in diameter and 1820 mm in length, and the damping density can be estimated as 31.1 MN·s/m^4 at 0.5 m/s. Therefore, it can be concluded that the R-ECD has a damping density comparable to that of fluid viscous damper. In contrast to that of viscous dampers, the force-speed curve of rotary ECD can be tailored by properly choosing design parameters as will be discussed later, offering greater flexibility for vibration control.

It is interesting to note why previous studies in civil engineering didn't observe the nonlinear and force-limiting features of eddy current dampers. For simple eddy current torque generator in Section 3.1, the maximum rotational speed in Fig. 8 is 3900 RPM, and the corresponding tangential speed at conductor is $3900/60\times2\pi Rc$ = 14.7 m/s. The critical tangential speeds in Fig. 8 could be calculated to be 10.5 m/s for 1.5 mm thick copper, 4.5 m/s for 4 mm thick copper and 3.0 m/s for 8 mm thick copper. Similarly, the maximum axial velocity of the large-scale R-ECD is 600 mm/s, and the corresponding tangential speed is $600/Ld\times2\pi Rc$ = 11.3 m/s. The critical tangential speed for the large-scale R-ECD is about 9.4 m/s for the 3.8 mm thick 6061-T6 aluminum alloy. Nakamura et al. [38] has made some tests of electromagnetic dampers. However, the maximum tangential velocity of the rotor is only about 1.25 m/s. At a speed of that low, the damping force is almost proportional to the velocity and the nonlinearity is almost invisible.

Table 3
Parameters of R-ECD at different estimated temperatures.

Amplitude (mm)	20	50	100	150	200
Frequency (Hz)	0.4	0.4	0.4	0.4	0.4
Max velocity (mm/s)	50	126	251	377	503
Temperature (℃)	32.5	51.25	82.5	113.75	145
σ (MS/m)	23.2	21.6	19.5	17.7	16.2
R_c (mm)	120.24	120.29	120.38	120.47	120.56
R_r (mm)	124.02	124.05	124.10	124.15	124.20

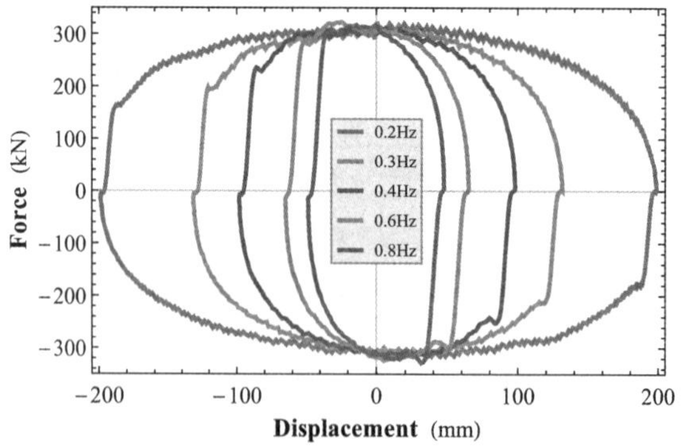

Fig. 13. Hysteretic loops of R-ECD at different excitation frequency.

4. Seismic control of structures

In this section, the passive control performance of the rotary ECD for two simplified systems subjected to seismic excitation is numerically investigated. For comparison of seismic control performance, the fluid viscous dampers are selected as they are widely used in civil engineering structures.

4.1. SDOF simplified system

In this section, numerical simulation of seismic control of a linear SDOF simplified system is carried out by using the rotary ECD, linear FVD and nonlinear FVD. The parameters of the SDOF system are as follows: modal frequency $f = 0.1486$ Hz, modal mass $M = 2904$ t, and $\zeta = 0.02$. The above parameters actually correspond to the first fundamental mode of the Zhangjiajie Grand Bridge.

Consider a SDOF system subjected to a ground motion as shown in Fig. 14. The governing equation of motion for the system is

$$M\ddot{u} + C\dot{u} + Ku + F_{damper}(\dot{u}) = -M\ddot{z} \tag{21}$$

where u is the displacement of the mass relative to the ground; and z is the ground motion.

El-centro wave is selected as ground motion and the peak ground acceleration is set to be 0.30 g, as shown in Fig. 15.

As a contrast, a SDOF system with four FVDs is also analyzed. The force–velocity curve of FVD can be described by

$$F = F_d\left(\frac{\dot{u}}{v_{ref}}\right)^{\alpha} \mathrm{sgn}(\dot{u}) \tag{22}$$

In this case, each FVD is set to have the damping force $F_d = 375$ kN at the reference velocity $v_{ref} = 0.25$ m/s which is the same as the rotary ECD. Different exponent α varies from 0.2 to 1.0 is considered. The comparison of the *F-V* curves between rotary ECD and different FVDs can be seen in Fig. 16. The main difference between these two kinds of *F-V* curves is that the damping force provided by an ECD has a maxima at a critical speed, while the force of an FVD will grow continuously with speed. This characteristic of the ECD may be beneficial because it could protect the structure and dampers from overload.

The response of this SDOF system is calculated using the Runge-Kutta algorithm. The peak value and RMS (root mean square) of displacement responses and the response reduction ratio ρ are shown in Table 4. As can be seen, both rotary ECD and fluid viscous dampers reduce the displacement response quite effectively. The reduction ratio of structure using

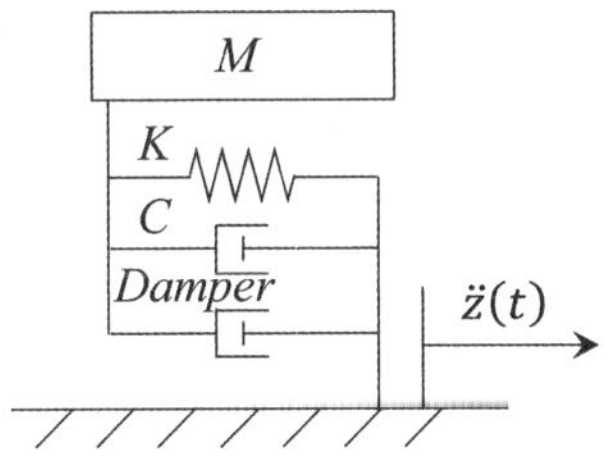

Fig. 14. A SDOF system with supplemental dampers subjected to ground motion.

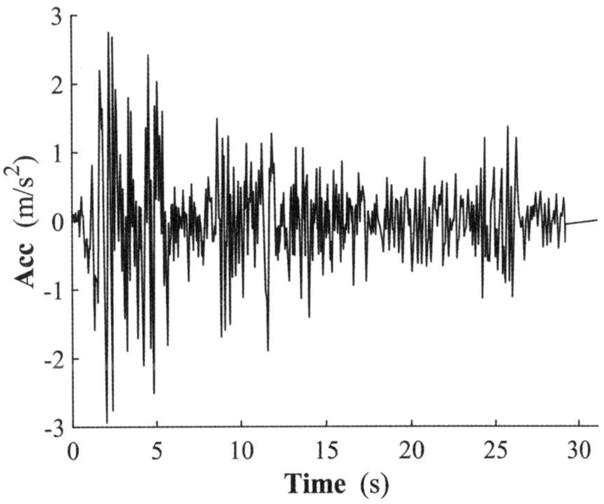

Fig. 15. Time histories of ground motion.

eddy current dampers is very close to that of the structure using fluid viscous dampers, especially the one with $\alpha = 0.8$. Either is the maximum damping force. Time histories of displacement response are shown in Fig. 17 and the time histories of velocity and damping force of the damper are shown in Fig. 18. ECD offers a new alternative for vibration control of structures in civil engineering.

4.2. 5-DOF simplified system

In this section, numerical simulation of seismic control of a base isolated 5-storey frame building is carried out. The building is simplified into a 5-DOF system. Dampers are set up between the 1st floor and the ground, as shown in Fig. 19. The parameters of the 5-DOF system are as follows: mass $M_1 = M_2 = \ldots = M_5 = 2.07 \times 10^5$t, horizontal stiffness $K_1 = 2043$ kN/mm, $K_2 = K_3 = \ldots = K_5 = 1.68 \times 10^5$ kN/mm, and $\zeta = 0.02$. The 1st modal frequency could be obtained to be 0.2219 Hz. The number of dampers is 152, instead of 4 in Section 4.1.

Consider the 5-DOF system subjected to a ground motion as shown in Fig. 19. The governing equation Eq. (21) turns into a matrix form. Rayleigh damping is used to obtain the damping matrix *C*. Likewise, El-centro wave is selected as ground motion and the peak ground acceleration is set to be 0.30 g, as shown in Fig. 15. Comparison is also conducted between the ECDs and FVDs, whose *F-V* curves are shown in Fig. 16.

The peak value and RMS of the 1st floor displacement responses and the response reduction ratio ρ are shown in Table 5. Similarly, the reduction ratio of structure using eddy current dampers is very close to that of the structure using fluid viscous dampers, especially the one with $\alpha = 0.8$. Time histories of displacement response are shown in Fig. 20 and the time histories of velocity and damping force of the damper are shown in Fig. 21. It's worth noting that the maximum damping force of ECD is always smaller than those of the FVD except for the case of $\alpha = 0.2$. The most distinct difference between ECD and FVD is the self-limiting characteristic of the eddy current damping force. Since the damping force is self-limiting, overloaded case induced by unanticipated excessive loadings may not occur to the ECD. Obviously, this feature improves the security and reliability of ECD.

5. Parametric analysis

5.1. Effect of different orders of harmonic fields

The influence of number of terms in Fourier series on damping forces is investigated. It is obvious that accuracy increases when much number of harmonic magnetic fields is used, but at a cost of much higher computation time to evaluate the high-order Bessel functions. Therefore, it's worthy of evaluating the number of harmonic fields to be used.

Fig. 22 shows the contribution of different order of harmonics on dissipating power. It is observed that proportion of the first two orders of dissipating power at different speed or velocity. The calculation parameters are taken from Table 1. Obvi-

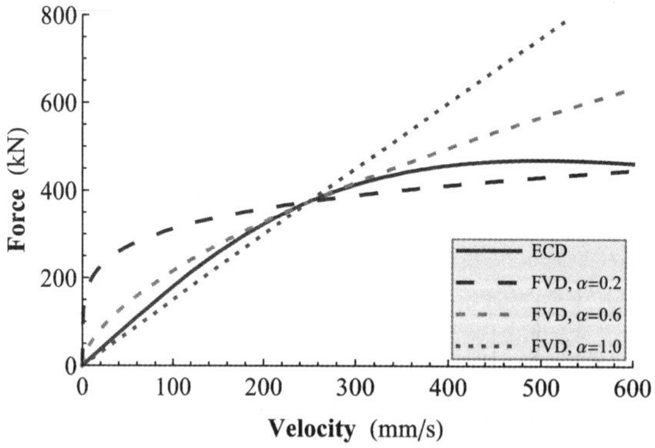

Fig. 16. Comparison of *F-V* curves between ECD and FVDs.

Table 4
Peak value and RMS of displacement responses of SDOF system.

Dampers	α	Peak value (mm)	ρ	RMS (mm)	ρ	Max damping force (kN)
None		448	0%	234	0%	
ECD		71	84.2%	22	90.7%	374
FVD	0.2	61	86.3%	16	93.0%	374
	0.4	62	86.1%	17	92.9%	374
	0.6	65	85.4%	18	92.1%	373
	0.8	72	84.0%	21	91.0%	373
	1.0	76	83.1%	25	89.5%	372

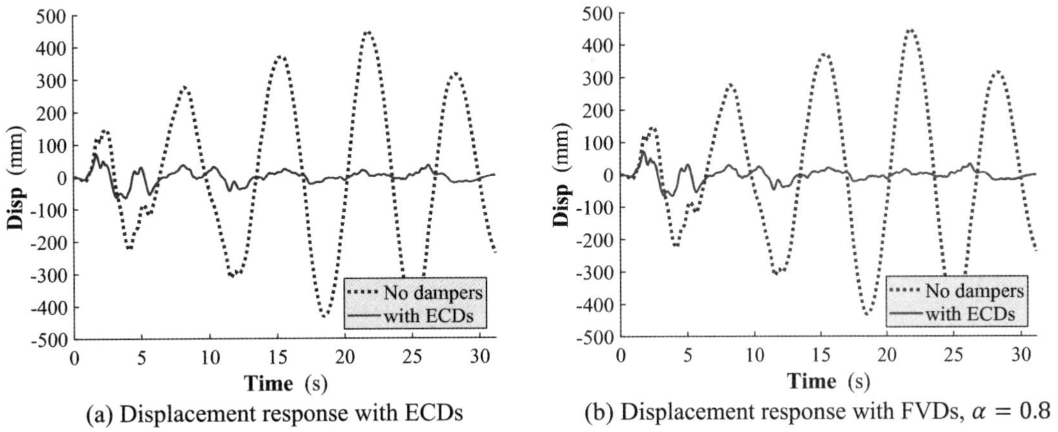

Fig. 17. Time histories of displacement response.

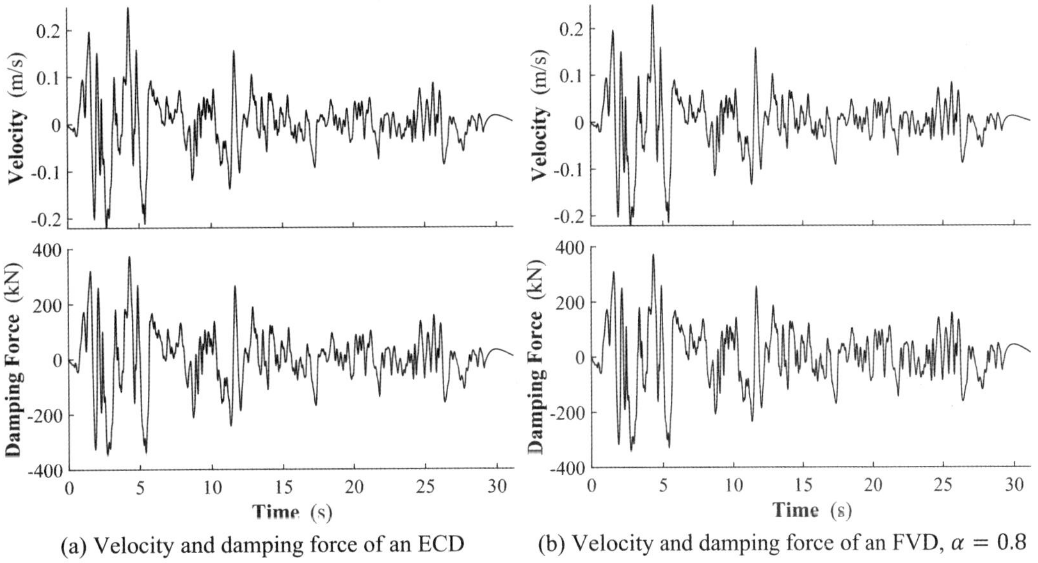

Fig. 18. Time histories of velocity and damping force.

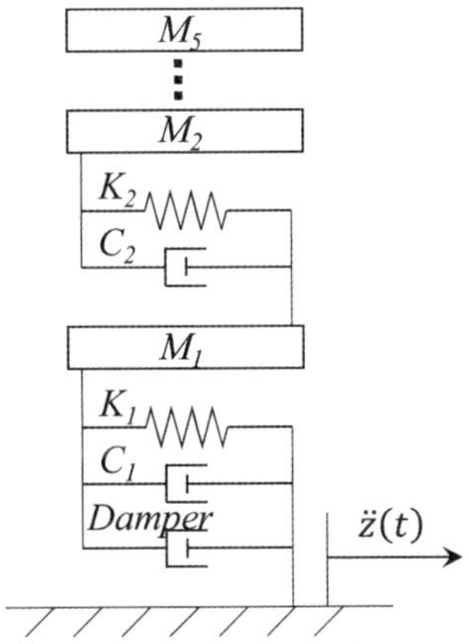

Fig. 19. A 5-DOF system with supplemental dampers subjected to ground motion.

Table 5
Peak value and RMS of 1st floor displacement responses of 5-DOF system.

Dampers	α	Peak value (mm)	ρ	RMS (mm)	ρ	Max damping force (kN)
None		244	0%	71	0%	
ECD		205	16.1%	46	35.6%	470
FVD	0.2	203	17.0%	40	43.0%	429
	0.4	203	17.0%	42	40.6%	490
	0.6	203	16.9%	43	38.4%	559
	0.8	203	16.9%	45	36.6%	638
	1.0	203	16.9%	46	34.8%	728

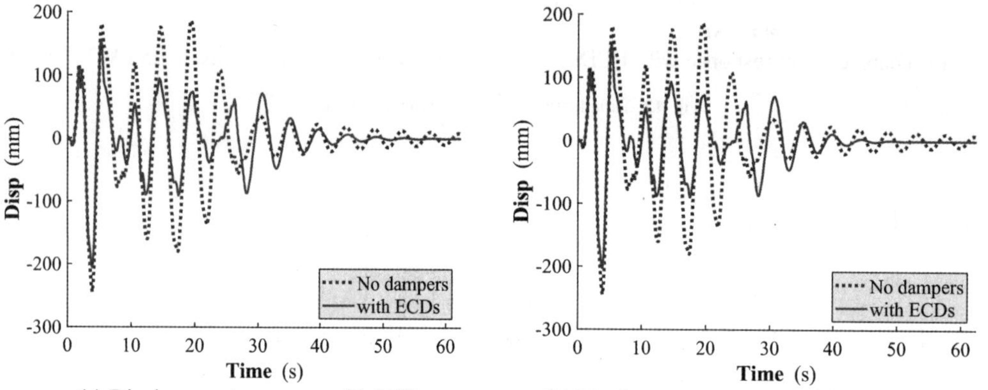

(a) Displacement response with ECDs (b) Displacement response with FVDs, $\alpha = 0.8$

Fig. 20. Displacement time histories of 1st floor for 5-DOF system under seismic excitation.

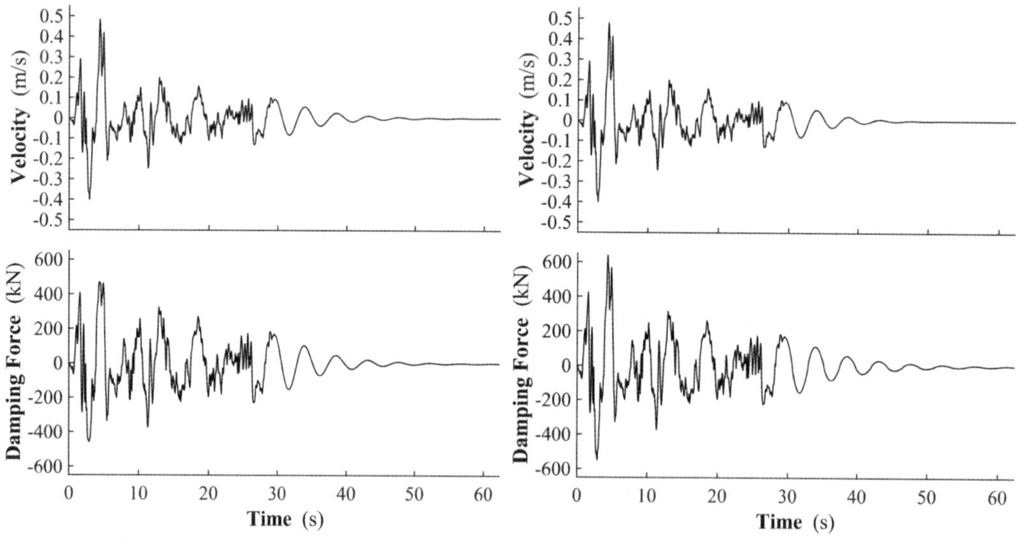

(a) Velocity and damping force of an ECD (b) Velocity and damping force of an FVD, $\alpha = 0.8$

Fig. 21. Time histories of velocity and damping force.

ously, the total proportion of P_1 and P_3 is more than 97% at different conditions in test1. Indeed, the proportion of P_1 is over 99% at different velocityfor full-scale rotary ECD. It is concluded that a sum of the first two orders of harmonic fields is enough for most engineering applications of eddy current damper.

5.2. Influence of design parameters

Several design parameters will be studied in this section to see how they affect the torque-speed curve. The parametric analysis is conducted only for the full-scale R-ECD model with ball screw, considering the thickness of each region in Test 2 is much smaller than the radius and the change of thickness would only make a little difference in the value of radius.

Fig. 23 shows the effect of the thickness t of the conductor on the force-speed curve. Obviously, as the conductor becomes thicker, the critical velocity becomes smaller as well as the critical force. However, the linear damping coefficient at low velocity C_{linear} will not keep increasing as the thickness of conductor increases. Actually, C_{linear} grows with t increasing from 1 mm to 4 mm, while almost remains unchanged with t increasing from 4 mm to 12 mm. In brief, thinner conductor could

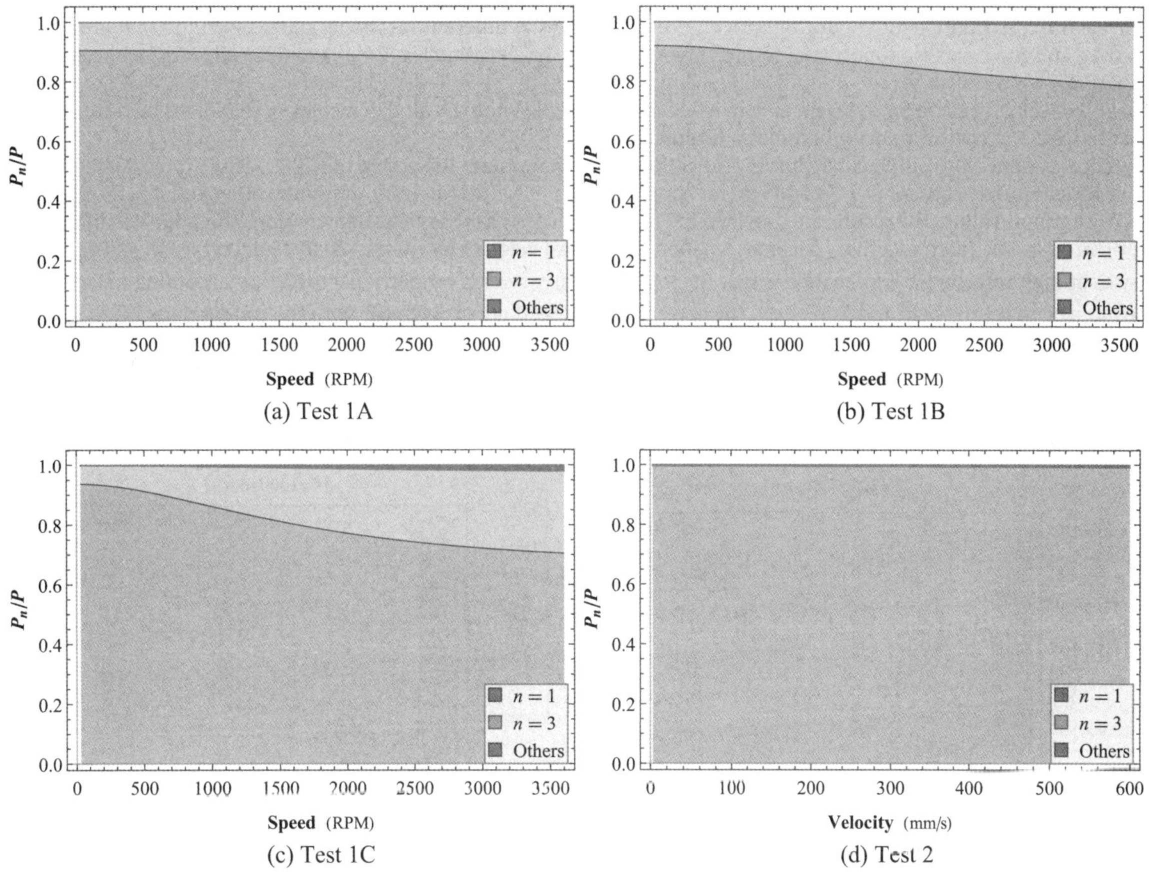

Fig. 22. Proportion of the first two orders of dissipating power at different speed/velocity.

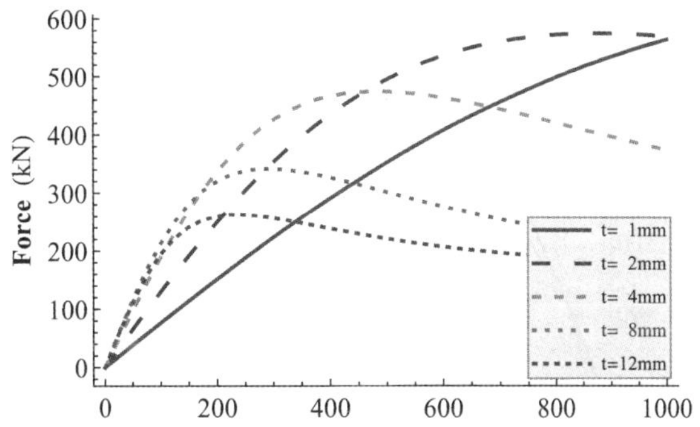

Fig. 23. Force-velocity curves for different thicknesses of the conductor.

generate larger force at high velocity, while appropriate thicker conductor could generate larger damping coefficient at low velocity. Similar phenomenon could be seen in paper [17], though the magnetic configurations are not exactly the same as that in this paper.

At low velocity, the thickness of conductor tube influences the damping force in three aspects which could be seen in Fig. 24(a). Firstly, thicker copper provides more energy consumption due to the larger volume, according to Eq. (16). Secondly, thicker copper leads to further distance between the outer surface of PMs and inner surface of the steel rotor, which could result in an overall reduction of magnetic flux density. Thirdly, "skin effect" causes the reduction of eddy currents along the thickness direction. As can be seen in Fig. 23, damping force at a low speed, such as 50 mm/s, increases with copper thickness in the range of 1 mm-8 mm. While the force at 50 mm/s even goes a bit lower for further increase of copper thickness. At high speed, "skin effect" could be more serious. Moreover, the induced magnetic field causes a reduction of the total magnetic field, which is more serious for thicker conductor. Fig. 24 shows the amplitude square of the first harmonic of eddy current density distribution for conductor with different thickness at different rotational speeds. Remind that area between each curve and the t-axis is proportional to the power. Comparison between Fig. 24(a) and SPS:refid::fig24(b) supports the explanation.

The force–velocity curve for different air gap is shown in Fig. 25. When the air gap increases, the force generated by the R-ECD decreases. The critical velocity becomes a bit lower.

Electrical conductivity of the conductor is also an important parameter for design of R-ECD. Three typical values of electrical conductivity are chosen: $\sigma = 53.5$ MS/m for copper, $\sigma = 35.9$ MS/m for 1060 aluminum alloy and $\sigma = 24.3$ MS/m for 6061-T6 aluminum alloy. The results are provided in Fig. 26. It is seen that the critical force keeps the same for different conductor materials. As the conductivity becomes smaller, the critical velocity increases, the force decreases at low velocity and increases at high velocity. An explanation is that the Joule heating power is proportional to $|J_{zn}|^2/\sigma$ according to Eq. (16). Due to the induced magnetic field and skin effect, the current density will not grow linear with the conductivity. So $|J_{zn}|^2/\sigma$ could be even smaller for larger σ at high velocity.

Fig. 27 shows the force-speed diagram for different number pole-pairs p of PMs. The size of PMs remains unchanged, while α changes with p. The purpose of this analysis is to identify the optimum placement of PMs to maximize the damping

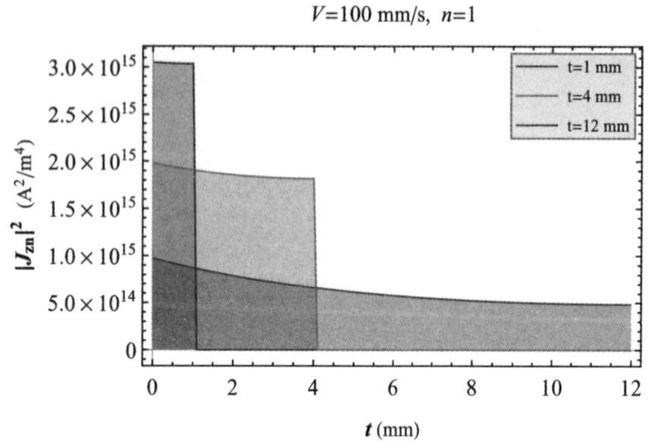

(a) Distribution at 100mm/s

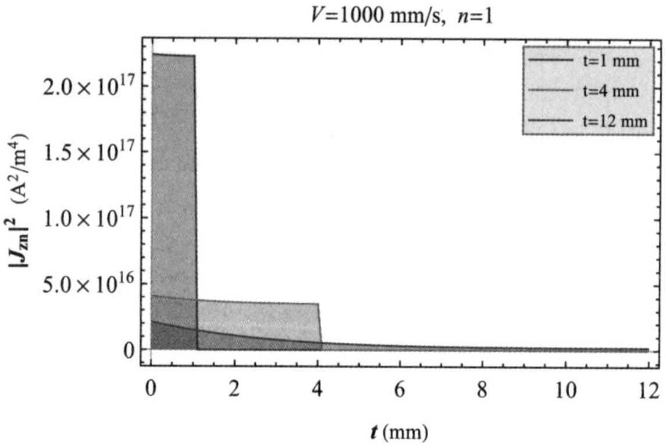

(b) Distribution at 1000mm/s

Fig. 24. Amplitude square of the first order harmonic of current density distribution.

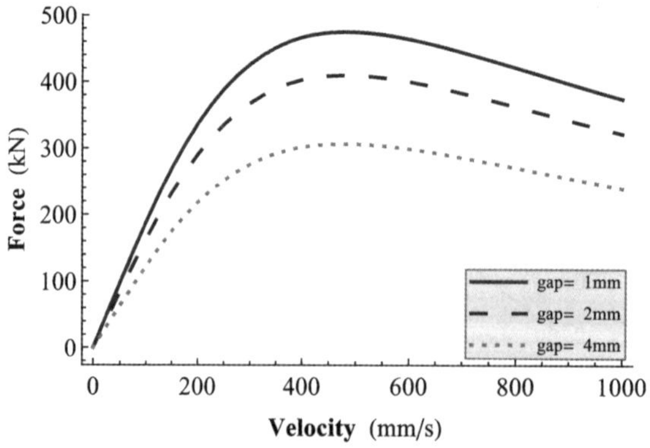

Fig. 25. Force-velocity curves for different size of airgap.

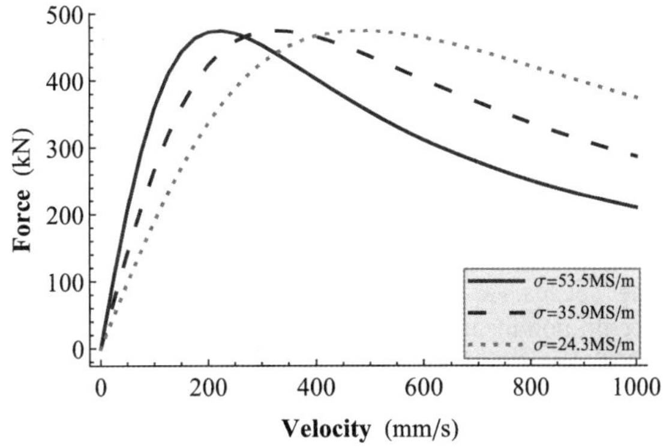

Fig. 26. Force-velocity curves for different electrical conductivity of the conductor.

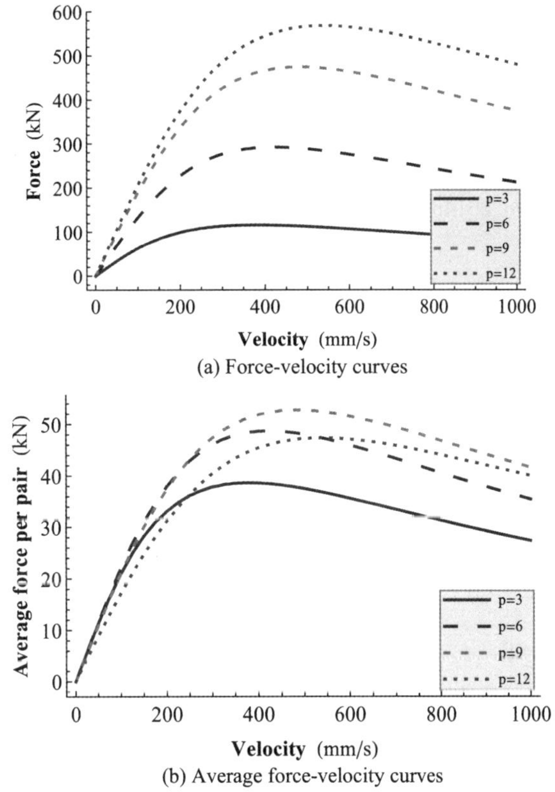

Fig. 27. Force-speed curve for different number of pole-pairs.

forces for every single PM. As observed in Fig. 27(a), the force generated by the R-ECD certainly increases with the number of pole-pairs *p*. However, as shown when *p* exceeds a critical value, the generated force per pair will decrease. That means, there is an optimized *p* value, which could make full use of each PM. This value would be an integer between 6 and 12, according to the plot. This is because, the distance of PMs has an influence on the distribution of magnetic flux density. Besides, the critical velocity increases slightly with *p*.

6. Conclusions

In this study, the design and dynamic characterization of rotary eddy current damper are described. Both 2D analytical method and nonlinear finite element simulations are used to evaluate the eddy current torque (i.e. damping force) at various rotating speeds of conductor. Laboratory experiments are conducted to validate the numerical results for a simple eddy current torque generator and a full-scale rotary ECD with ball screw, respectively. Results indicate that the torques obtained from analytical method and finite element simulations are consistent, but are higher than the experimental values. For the large-scale rotary ECD, an increase of 125℃ was recorded in the conductor after 5 cycles of tests, which may explain the decreased damping performance than numerical prediction. When the temperature rise is considered, numerically predicted damping forces agrees well with the experimental values.

The full-scale rotary ECD achieves an experimental damping density up to 24.9 MN·s/m^4 at small velocity, which is comparable to that of fluid viscous dampers (FVD). More notably, the damping force of rotary ECD is highly nonlinear and has limiting feature. The damping force first increases with rotating speeds and having reached a maximum at a critical speed it then decreases for higher speeds. Parametric study indicates that the conductivity and thickness of conductor and air gap have a notable influence on nonlinearity and limiting feature of force–velocity curves of rotary ECD. Seismic control of a SDOF system and a 5-DOF system is carried out by using fluid viscous dampers and the rotary eddy current damper, respectively. It has been shown both dampers can effectively reduce the peak and RMS displacement response of the system.

48of Competing Interest

The authors declare that they have no known competing financial interests or personal relationships that could have appeared to influence the work reported in this paper.

Acknowledgements

This project is supported by the National Key Research and Development Program of China (Grant No.2016YFE0127900, and No.2017YFC0703604), and the Open Research Fund Program of Guangdong Key Laboratory of Earthquake Engineering and Application Technology (Grant No. 2014B0303010 75-02).

Appendix

Considering the reaction magnetic field due to the induced eddy current in Region IV, the governing equations in each region are listed below:

$$\begin{cases} \frac{\partial^2 A_{zn}^{\mathrm{I}}}{\partial r^2}+\frac{1}{r}\frac{\partial A_{zn}^{\mathrm{I}}}{\partial r}-\frac{(np)^2}{r^2}A_{zn}^{\mathrm{I}}=0 \\ \frac{\partial^2 A_{zn}^{\mathrm{II}}}{\partial r^2}+\frac{1}{r}\frac{\partial A_{zn}^{\mathrm{II}}}{\partial r}-\frac{(np)^2}{r^2}A_{zn}^{\mathrm{II}}=-\mu_0\frac{np}{r}M_{rn} \\ \frac{\partial^2 A_{zn}^{\mathrm{III}}}{\partial r^2}+\frac{1}{r}\frac{\partial A_{zn}^{\mathrm{III}}}{\partial r}-\frac{(np)^2}{r^2}A_{zn}^{\mathrm{III}}=0 \\ \frac{\partial^2 A_{zn}^{\mathrm{IV}}}{\partial r^2}+\frac{1}{r}\frac{\partial A_{zn}^{\mathrm{IV}}}{\partial r}-\frac{(np)^2}{r^2}A_{zn}^{\mathrm{IV}}=inp\omega\sigma\mu_0A_{zn}^{\mathrm{IV}} \\ \frac{\partial^2 A_{zn}^{\mathrm{V}}}{\partial r^2}+\frac{1}{r}\frac{\partial A_{zn}^{\mathrm{V}}}{\partial r}-\frac{(np)^2}{r^2}A_{zn}^{\mathrm{V}}=0 \end{cases} \tag{A1}$$

The general solution could be written as:

$$\begin{cases} A_{zn}^{\mathrm{I}}=C_1^1\cosh\left(np\ln\frac{r}{R_c}\right)+C_2^1\sinh\left(np\ln\frac{r}{R_c}\right) \\ A_{zn}^{\mathrm{II}}=C_3^1\cosh\left(np\ln\frac{r}{R_c}\right)+C_4^1\sinh\left(np\ln\frac{r}{R_c}\right)-\frac{\mu_0 npr}{(np)^2-1}M_{rn} \\ A_{zn}^{\mathrm{III}}=C_5^1\cosh\left(np\ln\frac{r}{R_c}\right)+C_6^1\sinh\left(np\ln\frac{r}{R_c}\right) \\ A_{zn}^{\mathrm{IV}}=C_7^1 I_{np}\left(\sqrt{inp\omega\sigma\mu_0}r\right)+C_8^1 K_{np}\left(\sqrt{inp\omega\sigma\mu_0}r\right) \\ A_{zn}^{\mathrm{V}}=C_9^1\cosh\left(np\ln\frac{r}{R_c}\right)+C_{10}^1\sinh\left(np\ln\frac{r}{R_c}\right) \end{cases} \tag{A2}$$

where $C_1^1 \sim C_{10}^1$ are arbitrary constants.

Since $\delta_n=\frac{1}{\sqrt{inp\omega\sigma\mu_0}}=\frac{1-i}{\sqrt{2np\omega\sigma\mu_0}}$ is a complex variable, $A_{zn}^{\mathrm{IV}}(r)$ could also be a complex, whose amplitude and argument vary with r position. This means both the magnitude and phase of the electromagnetic field changes along the radial in Region IV.

Substituting Eq. (A2) into the boundary conditions Eq. (8), one obtains C_1^1 C_{10}^1. Expressions for C_7^1 and C_8^1 are listed below:

$$\begin{cases} C_7^1 = -\frac{4B_{rem}\sin\left(\frac{n\pi\alpha}{2}\right)}{n\pi}\xi_n R_m \frac{K'_{np}\left(\frac{R_r}{\delta_n}\right)}{\psi_n} \\ C_8^1 = \frac{4B_{rem}\sin\left(\frac{n\pi\alpha}{2}\right)}{n\pi}\xi_n R_m \frac{I'_{np}\left(\frac{R_r}{\delta_n}\right)}{\psi_n} \end{cases} \tag{A3}$$

where

$$\xi_n = \frac{\frac{R_s}{R_m} - \cosh\left(np\ln\frac{R_m}{R_s}\right) + np\sinh\left(np\ln\frac{R_m}{R_s}\right)}{(np)^2 - 1} \tag{A4}$$

and

$$\psi_n = \frac{R_c}{np}\left[\frac{\mu_{rm}+1}{2}\cosh\left(np\ln\frac{R_s}{R_c}\right) + \frac{\mu_{rm}-1}{2}\cosh\left(np\ln\frac{R_m^2}{R_cR_s}\right)\right]\left[I'_{np}\left(\frac{R_c}{\delta_n}\right)K'_{np}\left(\frac{R_r}{\delta_n}\right) - K'_{np}\left(\frac{R_c}{\delta_n}\right)I'_{np}\left(\frac{R_r}{\delta_n}\right)\right]$$

$$+\left[\frac{\mu_{rm}+1}{2}\sinh\left(np\ln\frac{R_s}{R_c}\right) + \frac{\mu_{rm}-1}{2}\sinh\left(np\ln\frac{R_m^2}{R_cR_s}\right)\right]\left[I_{np}\left(\frac{R_c}{\delta_n}\right)K'_{np}\left(\frac{R_r}{\delta_n}\right) - K_{np}\left(\frac{R_c}{\delta_n}\right)I'_{np}\left(\frac{R_r}{\delta_n}\right)\right] \tag{A5}$$

And the modified Bessel functions' derivative with respect to r is:

$$\begin{cases} I'_{np}\left(\frac{r}{\delta_n}\right) = \frac{\partial}{\partial r}I_{np}\left(\frac{r}{\delta_n}\right) = \frac{1}{2\delta_n}\left[I_{np-1}\left(\frac{r}{\delta_n}\right) + I_{np+1}\left(\frac{r}{\delta_n}\right)\right] \\ K'_{np}\left(\frac{r}{\delta_n}\right) = \frac{\partial}{\partial r}K_{np}\left(\frac{r}{\delta_n}\right) = -\frac{1}{2\delta_n}\left[K_{np-1}\left(\frac{r}{\delta_n}\right) + K_{np+1}\left(\frac{r}{\delta_n}\right)\right] \end{cases} \tag{A6}$$

If $\omega = 0$, the magnetic field turns into static, which means time-independent. PDEs in each region can be written as below:

$$\begin{cases} \frac{\partial^2 A_{zn}^{\mathrm{I}}}{\partial r^2} + \frac{1}{r}\frac{\partial A_{zn}^{\mathrm{I}}}{\partial r} - \frac{(np)^2}{r^2}A_{zn}^{\mathrm{I}} = 0 \\ \frac{\partial^2 A_{zn}^{\mathrm{II}}}{\partial r^2} + \frac{1}{r}\frac{\partial A_{zn}^{\mathrm{II}}}{\partial r} - \frac{(np)^2}{r^2}A_{zn}^{\mathrm{II}} = -\mu_0\frac{np}{r}M_{rn} \\ \frac{\partial^2 A_{zn}^{\mathrm{III}}}{\partial r^2} + \frac{1}{r}\frac{\partial A_{zn}^{\mathrm{III}}}{\partial r} - \frac{(np)^2}{r^2}A_{zn}^{\mathrm{III}} = 0 \\ \frac{\partial^2 A_{zn}^{\mathrm{IV}}}{\partial r^2} + \frac{1}{r}\frac{\partial A_{zn}^{\mathrm{IV}}}{\partial r} - \frac{(np)^2}{r^2}A_{zn}^{\mathrm{IV}} = 0 \\ \frac{\partial^2 A_{zn}^{\mathrm{V}}}{\partial r^2} + \frac{1}{r}\frac{\partial A_{zn}^{\mathrm{V}}}{\partial r} - \frac{(np)^2}{r^2}A_{zn}^{\mathrm{V}} = 0 \end{cases} \tag{A7}$$

General solution of Eq. (A7) is:

$$\begin{cases} A_{zn}^{\mathrm{I}} = C_1^0\cosh\left(np\ln\frac{r}{R_c}\right) + C_2^0\sinh\left(np\ln\frac{r}{R_c}\right) \\ A_{zn}^{\mathrm{II}} = C_3^0\cosh\left(np\ln\frac{r}{R_c}\right) + C_4^0\sinh\left(np\ln\frac{r}{R_c}\right) - \frac{\mu_0 npr}{(np)^2-1}M_{rn} \\ A_{zn}^{\mathrm{III}} = C_5^0\cosh\left(np\ln\frac{r}{R_c}\right) + C_6^0\sinh\left(np\ln\frac{r}{R_c}\right) \\ A_{zn}^{\mathrm{IV}} = C_7^0\cosh\left(np\ln\frac{r}{R_c}\right) + C_8^0\sinh\left(np\ln\frac{r}{R_c}\right) \\ A_{zn}^{\mathrm{V}} = C_9^0\cosh\left(np\ln\frac{r}{R_c}\right) + C_{10}^0\sinh\left(np\ln\frac{r}{R_c}\right) \end{cases} \tag{A8}$$

where $C_1^0 \sim C_{10}^0$ are arbitrary constants to be determined from the boundary conditions.

After substituting Eq. (A8) into the boundary conditions Eq. (8), the radial component of the magnetic flux density in Region IV is obtained:

$$\overrightarrow{B}_{0r}^{\mathrm{IV}}(r,\theta) = \sum_{n=1,3,5\cdots}^{\infty}\frac{4B_{rem}\sin\left(\frac{n\pi\alpha}{2}\right)p}{\pi}\xi_n\frac{R_m}{r}\frac{\cosh\left(np\ln\frac{r}{R_r}\right)}{\varphi_n}e^{inp\theta} \tag{A9}$$

where

$$\varphi_n = \frac{\mu_{rm}+1}{2}\sinh\left(np\ln\frac{R_r}{R_s}\right) + \frac{\mu_{rm}-1}{2}\sinh\left(np\ln\frac{R_rR_s}{R_m^2}\right)$$

As can be seen, $\overrightarrow{B}_{0r}^{\mathrm{IV}}$ is independent of R_i and R_o, which is caused by the assumption that rotor and stator have infinite permeability.

References

[1] T.T. Soong, G.F. Dargush, Passive Energy Dissipation Systems in Structural Engineering, Wiley, New York, 1997.
[2] T.T. Soong, B. Spencer Jr, Supplemental energy dissipation: state-of-the-art and state-of-the-practice, Eng. Struct. 24 (2002) 243–259.
[3] M. Symans, F. Charney, A. Whittaker, M. Constantinou, C. Kircher, M. Johnson, R. McNamara, Energy dissipation systems for seismic applications: current practice and recent developments, *J. Struct Eng.*, ASCE 134 (2008) 3–21.
[4] S. Gur, S.K. Mishra, K. Roy, Stochastic seismic responses of building with super-elastic damper, Mech. Syst. Sig. Process. 72–73 (2016) 642–659.
[5] I.H. Mualla, B. Belev, Performance of steel frames with a new friction damper device under earthquake excitation, Eng. Struct. 24 (3) (2002) 365–371.
[6] X.B. Nguyen, T. Komatsuzaki, Y. Iwata, H. Asanuma, Modelling and semi-active fuzzy control of magnetorheological elastomer-based isolator for seismic response reduction, Mech. Syst. Sig. Process. 101 (2016) 449–466.
[7] W.X. Wang, X.G. Hua, X.Y. Wang, G.B. Song, Z.Q. Chen, Numerical modeling and experimental study on a novel pounding tuned mass damper, J. Vib. Control 24 (17) (2018) 4023–4036.
[8] S.J. Dyke, B.F. Spencer, J.D. Carlson, Modeling and control of magnetorheological dampers for seismic response reduction, Smart Mater. Struct. 5 (1996) 565–575.
[9] D.I. Narkhede, R. Shnha, Behavior of nonlinear fluid viscous dampers for control of shock vibrations, J. Sound Vib. 333 (1) (2014) 80–98.
[10] X. Shi, S.Y. Zhu, Dynamic characteristics of stay cables with inerter dampers, J. Sound Vib. 432 (2019) 287–305.
[11] H.A. Sodano, J.-S. Bae, D.J. Inman, W.K. Belvin, Concept and model of eddy current damper for vibration suppression of a beam, J. Sound Vib. 288 (2005) 1177–1196.
[12] L. Zuo, X. Chen, S. Nayfeh, Design and analysis of a new type of electromagnetic damper with increased energy density, ASME *J. Vib. Acoust.* 133 (2011) 041006.
[13] B. Ebrahimi, M.B. Khamesee, F. Golnaraghi, A novel eddy current damper: theory and experiment, J. Phys. D Appl. Phys. 42 (2009) 075001.
[14] Q. Wen, X.G. Hua, Z.Q. Chen, Y. Yang, H.W. Niu, Control of human-induced vibrations of a curved cable-stayed bridge: design, implementation, and field validation, J. Bridge Eng. 21 (2016) 04016028.
[15] Z. Lu, B. Huang, Q. Zhang, X.L. Lu, Experimental and analytical study on vibration control effects of eddy-current tuned mass dampers under seismic excitations, J. Sound Vib. 421 (2008) 153–165.
[16] Z.W. Huang, X.G. Hua, Z.Q. Chen, H.W. Niu, Modeling, testing, and validation of an eddy current damper for structural vibration control, J. Aerosp. Eng. 31 (2018) 04018063.
[17] S.E. Gay, Contactless Magnetic Brake for Automotive Applications PhD thesis, Texas A & M University, 2010.
[18] R. Palomera-Arias, J. Connor Jerome, A. Ochsendorf John, Feasibility study of passive electromagnetic damping systems, J. Struct. Eng., ASCE 134 (1) (2008) 164–170.
[19] B. Ebrahimi, M.B. Khamesee, F. Golnaraghi, Eddy current damper feasibility in automobile suspension: modeling, simulation and testing, Smart Mater. Struct. 18 (2009) 015017.
[20] J. Meins, L. Miller, W. Mayer, The high speed Maglev transport system TRANSRAPID, IEEE Trans. Magn. 24 (1988) 808–811.
[21] B. Ebrahimi, M.B. Khamesee, M.F. Golnaraghi, Design and modeling of a magnetic shock absorber based on eddy current damping effect, J. Sound Vib. 315 (2008) 875–889.
[22] M.A.A. Abdelkareem, L. Xu, M.K.A. Ali, A. Elagouz, J. Mi, S.J. Guo, Y.L. Liu, L. Zuo, Vibration energy harvesting in automotive suspension system: a detailed review, Appl. Energy 229 (2018) 672–699.
[23] Y. Suda, T. Shiiba, A new hybrid suspension system with active control and energy regeneration, Veh. Syst. Dyn. 25 (S1) (1996) 641–654.
[24] K. Nakano, Combined type self-powered active vibration control of truck cabins, Veh. Syst. Dyn. 41 (6) (2004) 449–473.
[25] F. Iiyama, K. Sunakoda, K. Suzuki, Hybrid control of seismic isolation system using mechatro damper, Proceedings of the ASME/JSME Joint Pressure Vessels and Piping Conference, San Diego 379 (1998) 157–161.
[26] Y. Kawamoto, Y. Suda, H. Inoue, T. Kondo, Modeling of electromagnetic damper for automobile suspension, J. System Design Dyn. 1 (3) (2017) 524–533.
[27] P. Zhang, Design of electromagnetic shock absorbers for energy harvesting from vehicle suspensions Master thesis, Mechanical Engineering, Stony Brook University, 2010.
[28] Z. Li, L. Zuo, Electromagnetic energy-harvesting shock absorbers: design, modeling and road tests, IEEE Trans. Veh. Technol. (2013) I62.
[29] L. Zuo, B. Scully, J. Shestani, Y. Zhou, Design and characterization of an electromagnetic energy harvester for vehicle suspensions, Smart Mater. Struct. 19 (4) (2010) 045003.
[30] L. Zuo, P. Zhang, Energy harvesting, ride comfort, and road handling of regenerative vehicle suspensions, ASME J. Vib. Acoust. 135 (1) (2013) 011002.
[31] Y. Pei, Y. Liu, L. Zuo, Multi-resonant electromagnetic shunt in base isolation for vibration damping and energy harvesting, J. Sound Vib. 423 (2018) 1–7.
[32] Ellis, Fink, and Rich (1989) Eddy current damper. NASA, Marshall Space Flight Center, The 23rd Aerospace Mechanism Symposium, Honeywell Space and Aviation Systems. Durham, NC. US.
[33] S.Y. Zhu, W.A. Shen, Y.L. Xu, Linear electromagnetic devices for vibration damping and energy harvesting: modeling and testing, Eng. Struct. 34 (1) (2012) 198–212.
[34] G. Ikago, N. Inoue, Seismic control of buildings using apparent mass dampers with rotational amplifying mechanisms: are view of the state of the art, Proceedings of the 13th World Conference on Seismic Isolation, Energy Dissipation and Active Vibration Control of Structures, Sendai, Japan, ASSISi, 2013, p. 8.
[35] K. Saito, K. Ikago, Inoue, Seismic control of single-degree-of-freedom structure using tuned viscous mass damper, Earthquake Eng. Struct. Dyn. 41 (2012) 453–474.
[36] J.L. Perez-Diaz, I. Valiente-Blanco, C. Cristache, J. Sanchez-García-Casarubios, F. Rodriguez, J. Esnoz, E. Diez-Jimenez, A novel high temperature eddy current damper with enhanced performance by means of impedance matching, Smart Mater. Struct. 28 (2019) 025034.
[37] Z.Q. Chen, Z.W. Huang, X.G. Hua, Numerical modelling and experimental testing of an eddy current apparent mass damper, Proceedings of 14th World Conference on Seismic Isolation, Energy Dissipation and Active Vibration Control of Structures, September 9–11 2015, 2015, San Diego USA.
[38] Y. Nakamura, A. Fukukita, K. Tamura, I. Yamazaki, T. Matsuoka, K. Hiramoto, K. Sunakoda, Seismic response control using electromagnetic inertial mass dampers, Earthquake Eng. Struct. Dyn. 43 (4) (2014) 507–527.
[39] Z.H. Wang, Y.W. Xu, H. Guo, Z.Q. Chen, K. Xu, B. Zhao, Vibration control of a stay cable with a rotary electromagnetic inertial mass damper, Smart Struct. Syst. 23 (6) (2019) 627–638.
[40] H.P. Zhu, Y.M. Li, W.A. Shen, H.P. Zhu, Mechanical and energy-harvesting model for electromagnetic inertial mass dampers, Mech. Syst. Sig. Process. 120 (2019) 203–220.
[41] M. Amjadian, A.K. Agrawal, Seismic response control of multi-story base-isolated buildings using a smart electromagnetic friction damper with smooth hysteretic behavior, Mech. Syst. Sig. Process. 130 (2019) 409–432.
[42] J.-Y. Li, S. Zhu, J. Shen, Enhance the damping density of eddy current and electromagnetic dampers, Smart Struct. Syst. 24 (1) (2019) 15–26.
[43] J.H. Wouterse, Critical torque and speed of eddy current brakes with widely separated soft iron poles, IEE Proceedings B – Electric Power Applications 138 (4) (1991) 153–158.
[44] L. Ye, D. Li, Y. Ma, B. Jiao, Design and performance of a water-cooled permanent magnet retarder for heavy vehicles, IEEE Trans. Energy Convers. 26 (2011) 953–958.
[45] A. Rahideh, T. Korakianitis, Analytical magnetic field distribution of slotless brushless permanent magnet motors–Part I. armature reaction field, inductance and rotor eddy current loss calculations, IET Electr. Power Appl. 6 (2012) 628–638.
[46] T. Rahideh, A. Korakianitis, Analytical magnetic field distribution of slotless brushless PM motors. Part 2: open-circuit field and torque calculations, IET Electr. Power Appl. 6 (2012) 639–651.

[47] A. Tonoli, Dynamic characteristics of eddy current dampers and couplers, J. Sound Vib. 301 (2007) 576–591.
[48] H.A. Haus, J.R. Melcher, Electromagnetic Fields and Energy, vol. 107, Prentice Hall, Englewood Cliffs, NJ, 1989.
[49] L.C. Andrews, L.C. Andrews, Special Functions of Mathematics for Engineers, McGraw-Hill, New York, 1992, p. 407.
[50] S.O. Kasap, Principles of Electronic Materials and Devices (4th ed.). Mc-Graw Hill, 2018, p. 136.

Received: 22 August 2021 | Revised: 31 March 2022 | Accepted: 31 May 2022
DOI: 10.1002/stc.3022

RESEARCH ARTICLE

WILEY

A novel 500 kN axial eddy current damper using rack and gear mechanism: Design, testing, and evaluation

Shouying Li | **Yafeng Li** | **Weiyang Mao** | **Jianzhong Wang** | **Zhengqing Chen**

Key Laboratory for Wind and Bridge Engineering of Hunan Province, College of Civil Engineering, Hunan University, Changsha, China

Correspondence
Yafeng Li, Key Laboratory for Wind and Bridge Engineering of Hunan Province, College of Civil Engineering, Hunan University, Changsha 410082, China.
Email: leeliyf@163.com

Funding information
National Key Research and Development Program of China, Grant/Award Numbers: 2017YFC0703604, 2017YFC0703600

Summary

A novel axial eddy current damper by using rack and gear mechanism (RGA-ECD) was proposed. A 500 kN RGA-ECD prototype was manufactured and tested, and its nonlinear damping performance was systematically studied. First, the configuration of the RGA-ECD was introduced in detail, and its axial force was theoretically derived. Second, an RGA-ECD prototype with a target damping force of 500 kN at the velocity of 0.5 m/s was designed and manufactured, and a series of tests on the prototype were conducted. In the test, the maximum damping force of this prototype is 507 kN at the velocity of 0.509 m/s, which achieves the design goal. It appears that the tested results are in good agreement with the numerical results obtained from COMSOL Multiphysics when the loading velocity is less than 0.377 m/s. When the loading velocity is greater than 0.377 m/s, the maximum relative difference between the tested and numerical results is within 10.8%. The hysteresis curves of the prototype are relatively smooth, and the repeatability is also excellent, which indicates that the mechanical properties of the RGA-ECD are considerably stable and reliable. Third, the effectiveness of the RGA-ECD in mitigating seismic responses of base isolated structures was investigated by comparing it with fluid viscous damper. The results show that the RGA-ECD can be adopted as an alternative to the fluid viscous damper for seismic vibration control, and there is a force-limiting characteristic for the RGA-ECD, which can prevent the damage of the damper and controlled structure.

KEYWORDS

eddy current damper, force-limiting characteristic, nonlinear damping, rack and gear, vibration control

1 | INTRODUCTION

Energy dissipation devices, such as fluid viscous dampers (FVDs),[1,2] pounding dampers,[3,4] shape memory alloy dampers,[5,6] metallic dampers,[7,8] magnetorheological dampers,[9–11] and electromagnetic dampers,[12,13] have been widely adopted in civil engineering to mitigate structural vibration induced by earthquakes or strong winds. The eddy current damper (ECD) is another effective device for dissipating kinetic energy, which uses magnetic induction to generate eddy current damping. Compared with traditional viscous damping and friction damping, the eddy current damping has lots of advantages including no fluid, non-contact, less maintenance, long service life, high sensitivity, and strong reliability.[14]

Nowadays, the ECD has been widely adopted in industrial field to mitigate harmful vibrations, such as vehicle suspension system,[15] aerospace,[16] and high-speed train.[17] However, the application of the ECD in civil engineering is still in its infancy. The main reason for this situation may be that the vibration velocity of civil engineering structures is relatively small and the mass of them is considerably large. This results in insufficient energy dissipation efficiency for eddy current damping to effectively reduce vibration. In order to increase the flux density in air gap, many studies have been carried out on the optimal design of the ECD by selecting appropriate materials and configurations.[18–20] For example, Chen[21] invented a novel plate-type ECD with permanent magnets (PMs), in which the use of back irons increases the damping coefficients by more than five times.[22] This ECD has been widely used to reduce the wind-induced and human-induced vibrations of several civil structures.[23–27]

Another way to improve the energy dissipation efficiency of the ECD is to combine the eddy current damping with motion amplification mechanisms, which can significantly amplify the working velocity and damping force of the ECD. At present, the motion amplification mechanism commonly used in civil engineering is ball screw mechanism,[28] which converts linear motion into high velocity rotational motion. Ikago et al.[29] presented a tuned viscous mass damper (TVMD), which combines the ball screw mechanism, a flywheel, and a FVD. Nakamura et al.[30] proposed a novel electromagnetic inertial mass damper (EIMD) composed of the ball screw mechanism and an electric generator, and Zhu et al.[31] presented a mechanical model to predict the nonlinear characteristic of the EIMD. Wang et al.[32] proposed a rotary electromagnetic inertial mass damper (REIMD) by using the ball screw mechanism and the flywheel to reduce the vibration of stay cables. Liang et al.[22] and Zhang et al.[33] studied the dynamic characteristics of a large-scale ECD with the ball screw mechanism, and the results show that the large-scale ECD has enhanced performance for vibration control.

The ball screw mechanism has a high motion amplification efficiency and a stable transmission performance. However, the difficulty and cost for manufacturing the ball screw mechanism are significantly increased if the available maximum force of the damper increases (e.g., over 1000 kN). The rack and gear mechanism, a mature industrial product with low cost and strong reliability, can be used to replace the ball screw mechanism to achieve similar motion amplification. Li et al.[34,35] proposed a new type of damper that combines eddy current damping with the rack and gear mechanism, which can be installed between the superstructure and foundation to reduce the seismic responses of base isolated structures. Furthermore, the required inertia force realized by the apparent mass can also be achieved by the rack and gear mechanism, which provides a novel way for structural vibration control.

In this study, a novel axial ECD using rack and gear mechanism (RGA-ECD) was firstly presented. The configuration of the RGA-ECD was introduced in detail, and its axial forces were theoretically derived. A prototype of the RGA-ECD with a target damping force of 500 kN at the velocity of 0.5 m/s was manufactured, and its nonlinear damping performance was systematically investigated by using prototype tests and numerical simulations. In addition, the effectiveness of the RGA-ECD in mitigating seismic responses of base isolated structures was studied by comparing it with FVDs.

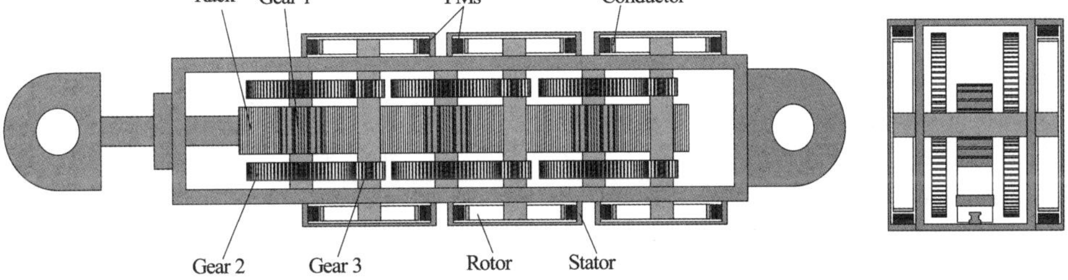

FIGURE 1 Configuration of the proposed ECD-RGD

2 | CONFIGURATION OF THE RGA-ECD

Sketch map of the proposed RGA-ECD, which consists of two main parts, is shown in Figure 1. The first part is a displacement transmission assembly consisting of rack, gears, transmission shafts, bearings, rotors, and so on. Among them, the rack meshes with Gear 1, Gear 1 and Gear 2 are coaxially connected, Gear 2 meshes with Gear 3, and Gear 3 drives the transmission shaft, which then drives the rotors and the conductors to rotate. The second part is an energy dissipation assembly by using eddy current damping, which consists of PMs, conductors, and stators. Among them, the PMs are fixed on the stators, and the stators are fixed on the gear box. Note that the PMs are arranged along the radial direction of the conductors, which is different from recent studies of the authors.[34,35] This arrangement is beneficial to reduce the deformation of the rotors and improve the manufacturing accuracy of dampers. The stators and rotors are both made of low-carbon steel materials, which also act as the back irons of the PMs and conductors, respectively. The cross section of the energy dissipation assembly is shown in Figure 2, and the detailed design parameters of the ECD are shown in Table 1. The grade of NdFeB PMs is N52, and the measured conductivity of aluminum alloy as conductor is 30.2 mS/m.

3 | THEORETICAL ANALYSIS OF THE AXIAL FORCE OF THE RGA-ECD

The relative velocity between the two nodes of the RGA-ECD is represented by $\dot{u}$, which is the axial velocity of the rack. Then, the rotating angular velocity of the rotors and conductors, $\dot{\theta}$, can be determined by

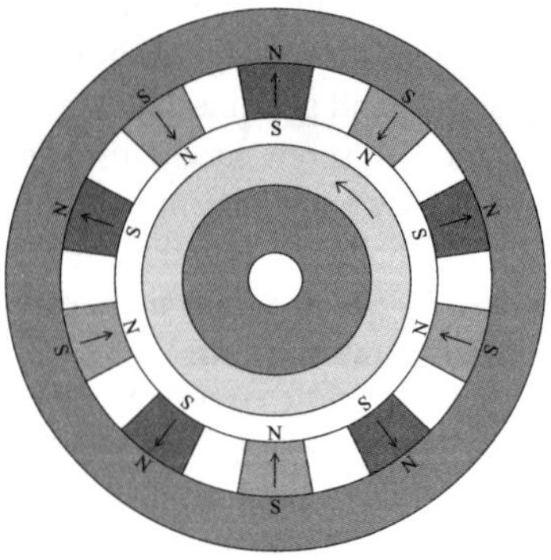

(a) Cross section of energy dissipation assembly

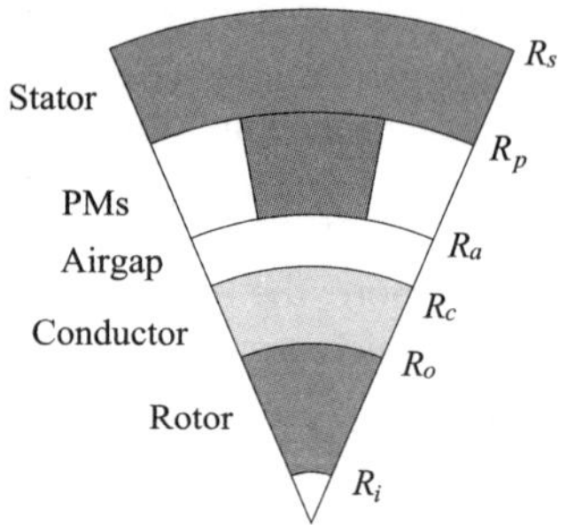

(b) Structural parameters

FIGURE 2 Cross section of energy dissipation assembly and corresponding structural parameters (non-prototype scale). (a) Cross section of energy dissipation assembly. (b) Structural parameters

TABLE 1 Detailed structural parameters of the prototype

Parameters	Notation	Values
Magnetic flux density (T)	B_r	1.42
Electrical conductivity of conductor (mS/m)	σ	30.2
Number of permanent magnets (pairs)	n	10
Radius of the rotor's outer surface (mm)	R_o	163
Radius of the conductor's outer surface (mm)	R_c	167
Radius of the air gap's outer surface (mm)	R_a	168
Radius of the PMs's outer surface (mm)	R_p	188
Radius of the stator's outer surface (mm)	R_s	200

$$\dot{\theta}=\frac{r_2}{r_1 r_3}\dot{u}, \tag{1}$$

where r_1, r_2, and r_3 are, respectively, the radius of Gears 1, 2, and 3.

The total moment T_s exported by the rotors and conductors can be obtained by

$$T_s=J\ddot{\theta}+T_e(\dot{\theta}), \tag{2}$$

in which $J\ddot{\theta}$ is the moment produced by the inertia of the rotors and conductors and T_e is the moment produced by the eddy current damping, which is a function of $\dot{\theta}$.

According to the energy conservation law, the input power should be equal to the output power:

$$F\dot{u}=\eta T_s\dot{\theta}, \tag{3}$$

where F is the axial force of the rack and η is the transmission efficiency of the rack and gear. Considering two-stage transmission for the gears, η is generally between 0.90 and 0.96.

Substituting Equations (1) and (2) into Equation (3), F can be described as

$$F=J\left(\frac{r_2}{r_1 r_3}\right)^2\eta\ddot{u}+T_e(\dot{\theta})\frac{r_2}{r_1 r_3}\eta. \tag{4}$$

The inertia force of the RGA-ECD Q_i can be written as

$$Q_i=J\left(\frac{r_2}{r_1 r_3}\right)^2\eta\ddot{u}=m_e\ddot{u}, \tag{5}$$

where m_e is the apparent mass of the RGA-ECD.

The damping force of the RGA-ECD Q_d can be well characterized by the Wouterse model[36] as follows:

$$Q_d=Q_{d,\max}\frac{2}{\frac{\dot{u}_{cr}}{\dot{u}}+\frac{\dot{u}}{\dot{u}_{cr}}}, \tag{6}$$

where $Q_{d,\max}$ is the maximum damping force of the RGA-ECD and $\dot{u}_{cr}$ is the critical velocity, which is the velocity corresponding to the maximum damping force.

Equation (4) can be rewritten as

$$F=m_e\ddot{u}+Q_{d,\max}\frac{2}{\frac{\dot{u}_{cr}}{\dot{u}}+\frac{\dot{u}}{\dot{u}_{cr}}}. \tag{7}$$

In addition, mechanical friction force will be unavoidably appeared in the RGA-ECD, which can be expressed as

$$Q_f=f_0\cdot\text{sign}(\dot{u}), \tag{8}$$

where f_0 is the magnitude of the mechanical friction force and sign(·) represents the direction of $\dot{u}$.

Therefore, the total axial force F can be obtained by

$$F = m_e\ddot{u} + Q_{d,\max}\frac{2}{\frac{\dot{u}_{cr}}{\dot{u}} + \frac{\dot{u}}{\dot{u}_{cr}}} + f_0 \cdot \mathrm{sign}(\dot{u}). \tag{9}$$

Figure 3a gives the mechanical model for the ECD-RGD, which includes the inertial force Q_i, the damping force Q_d, and the friction force Q_f. If the nonlinearity of eddy current damping is not considered, the relationship between Q_i, Q_d, and Q_f can be shown in Figure 3b.

4 | EXAMINATION FOR THE PROTOTYPE OF THE RGA-ECD

A prototype of the RGA-ECD with a target damping force of 500 kN at the velocity of 0.5 m/s was manufactured, and its photo is shown in Figure 4. All parts in the prototype except the gear box are products with standard sizes, which can be conveniently bought. The length and height of the gear box are 2350 and 400 mm, respectively, and the total width of gear box and stator is 340 mm. The test of this prototype was carried out on a testing system with a maximum force of 4000 kN in Hunan University. This testing system adopts an Instron 8800 ml controller and an electro-hydraulic servo actuator, which can perform displacement or force loading in the frequency range of 0.001 to 10 Hz.

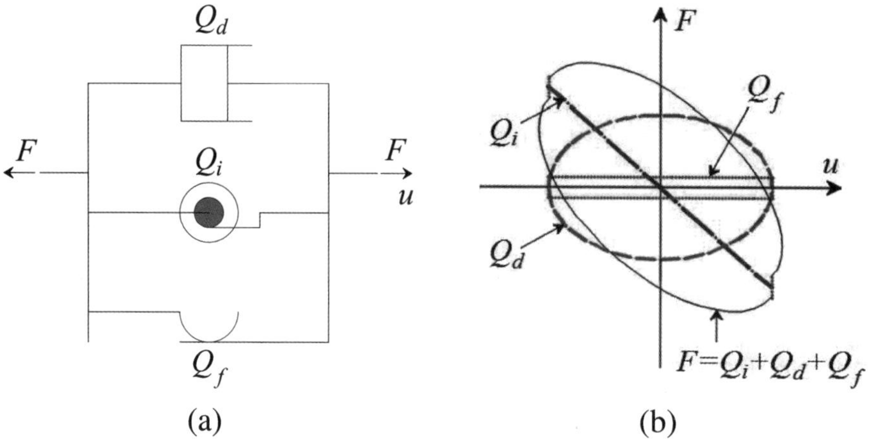

FIGURE 3 Sketch map of the axial force of the ECD-RGD. (a) Mechanical model of the RGA-ECD. (b) Relationship between Q_i, Q_d, and Q_f

FIGURE 4 Prototype of the RGA-ECD

The maximum displacement amplitude and maximum dynamic force capacity of this testing system are 800 mm and 2500 kN, respectively.

In the experiment, the testing system executed a harmonic motion with a constant amplitude of $A = 150$ mm, and its loading velocities can be achieved by changing the loading frequency of $f = 0.02–0.54$ Hz. Therefore, the axial maximum velocity amplitude of the prototype in the test is 0.019–0.509 m/s. Figure 5 presents the hysteretic curves of the prototype measured at four different loading frequencies with the same displacement amplitude $A = 150$ mm. It can be found from Figure 5 that the hysteresis curves in all cases are relatively smooth, and the repeatability is also excellent. This indicates that the performance of the prototype is considerably stable and reliable. It can be observed from Equation (9) that the axial force of the RGA-ECD consists of three parts: the damping force generated by the eddy currents, the inertial force generated by the rotors and conductors, and the friction force within the damper. The inertial force is proportional to the acceleration of the rotors and conductors, and it will disappear when the axial velocity of the RGA-ECD is at its maximum value. Therefore, the sum of the damping force and the friction force is determined when $u = 0$ mm. In the test, a friction force of 6.1 kN was measured at a constant low velocity of 1 mm/s. When $f = 0.54$ Hz and $A = 150$ mm ($\dot{u} = 0.509$ m/s), the damping force and the inertial force are 507 and 52 kN, respectively. Note that the predicted values in Figure 5 are obtained by adding the Wouterse model of eddy current damping force established by the numerical results in Figure 7 to the measured inertial force and friction force. In the Wouterse model established by the numerical results, the critical velocity and maximum damping force are 0.507 m/s and 567.99 kN, respectively. It appears that the tested results are in good agreement with the predicted values when $f \leq 0.32$ Hz. However, the differences between the tested and predicted results become significant at the loading frequency of 0.54 Hz, which might be caused by the temperature change during the test. This will be further discussed in the following.

The COMSOL Multiphysics, an electromagnetic finite element software, was adopted to numerically investigate the relationship between the moment produced by the eddy current damping, T_e, and the rotating angular velocity of the rotors, $\dot{\theta}$.[34] Then, the damping force induced by eddy current can be calculated by Equation (4). Figure 6 shows the electromagnetic finite element model of RGA-ECD, which is similar to the recent study of the authors,[34] except for

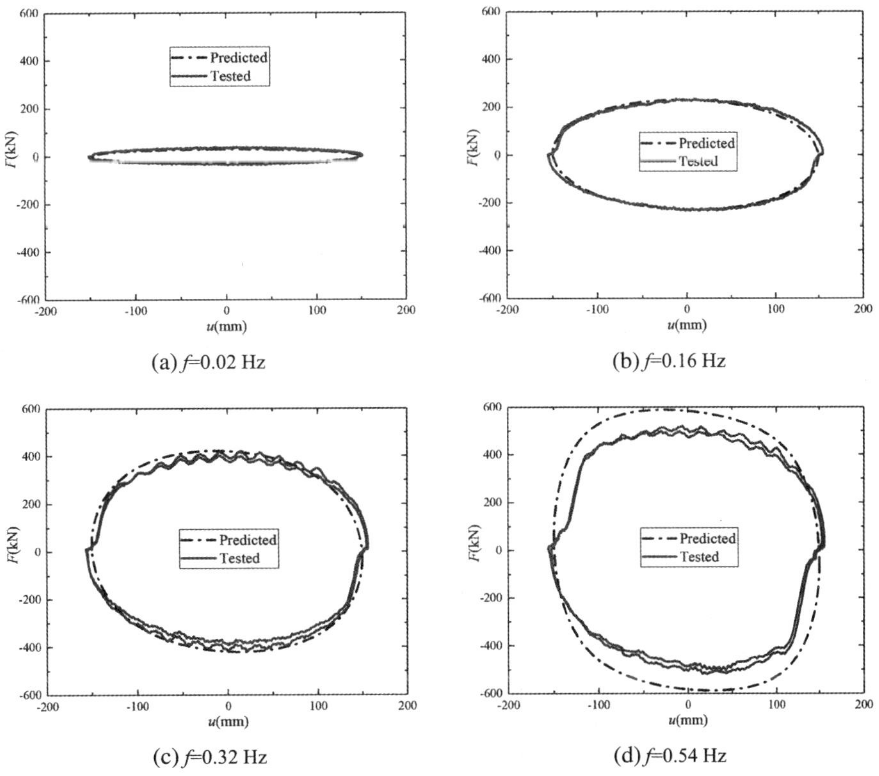

FIGURE 5 Hysteretic curves of the ECD-RGD. (a) $f = 0.02$ Hz. (b) $f = 0.16$ Hz. (c) $f = 0.32$ Hz. (d) $f = 0.54$ Hz

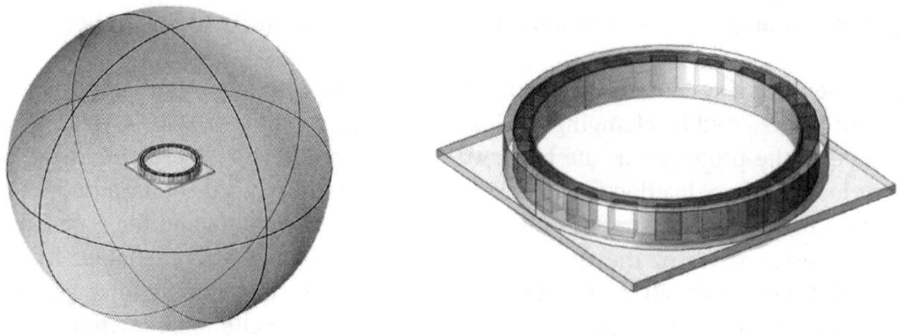

(a) Computational domain (b) Model of RGA-ECD

FIGURE 6 Electromagnetic finite element model. (a) Computational domain. (b) Model of RGA-ECD

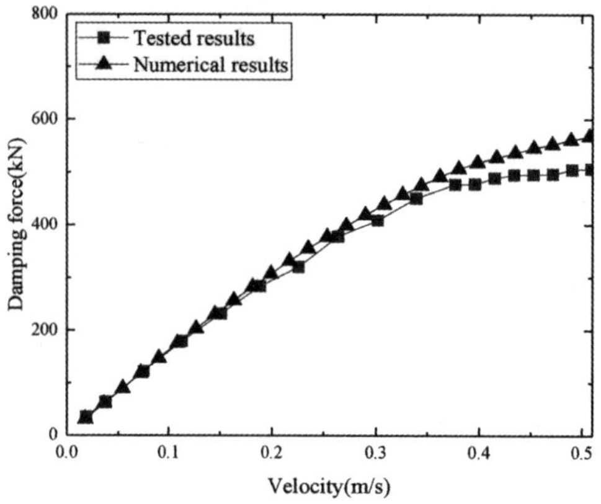

FIGURE 7 Comparison between tested and numerical results of the damping force of the RGA-ECD

the relative position of PMs and conductor. Note that the effect of temperature change upon the conductivity of conductor is ignored in the numerical simulation. Figure 7 presents the variation of the numerical damping force with the velocity, together with tested results. It can be found from Figure 7 that the tested results are in good agreement with the numerical results when the velocity is lower than 0.377 m/s. When the velocity is greater than 0.377 m/s, the relative differences between the tested and numerical results gradually increase with the increase of the velocity, but the maximum is not more than 10.8%. The reason for this may be that the increase in temperature reduces the conductivity of the conductor. When the loading velocity is low, the temperature of conductor can be quickly reduced to room temperature after each loading due to the heat dissipation holes on the stators of the RGA-ECD. However, a higher loading velocity causes the temperature of the conductor to rise significantly, and its temperature cannot drop to room temperature at the beginning of the next loading. In addition, when the loading velocities are 0.151 and 0.509 m/s, the effect of temperature on the eddy current damping force is observed within two cycles, as shown in Figure 8. The surface temperature of conductor is recorded to be 90.3°C when the loading velocity is 0.509 m/s, and the room temperature is about 10°C.

It appears in Figure 7 that the damping force keeps approximately linear with the velocity if the velocity is lower than 0.377 m/s. However, when the velocity is greater than 0.377 m/s, the tested results show that there is a significant nonlinear relationship between the damping force and the velocity. The nonlinear relationship is resulted from the interaction between the magnetic field of the PMs and the induced magnetic field generated by the eddy current in the conductor. When the velocity is lower than 0.377 m/s, the intensity of the induced magnetic field generated by eddy current is weak, and has little effect on the magnetic field of the PMs. At this stage, the relationship between the

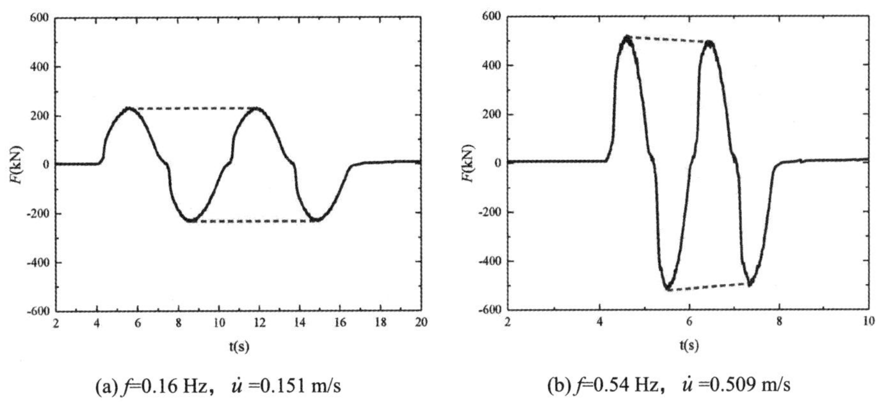

FIGURE 8 Time history results of axial force of the RGA-ECD. (a) $f = 0.16$ Hz, $\dot{u} = 0.151$ m/s. (b) $f = 0.54$ Hz, $\dot{u} = 0.509$ m/s

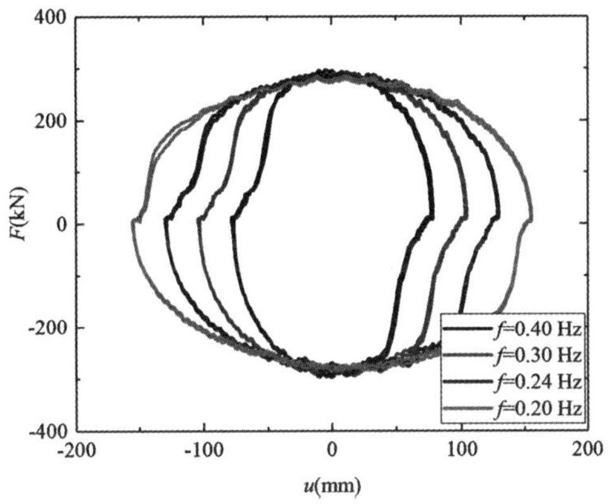

FIGURE 9 Hysteretic curves of RGA-ECD at different frequencies

damping force and the velocity is approximately linear. However, the induced magnetic field generated by the eddy current will increase when the velocity increases and significantly reduces the total magnetic intensity. At this stage, the relationship between the damping force and the velocity is nonlinear.

It should be noted that the damping force of the ECD has a maximum value at a certain velocity, that is, the critical velocity. Over this velocity, the damping force decreases with the increase of the velocity. Although the tested results of the damping force in Figure 7 do not cover the descending stage, it can be found from Figure 7 that the damping force is nearly unchanged with the increase of the velocity when the velocity is greater than 0.377 m/s. In addition, previous studies[22,33] have shown that the eddy current damping force will decrease at higher velocities. The characteristic of the force limiting of the RGA-ECD is beneficial to prevent the damage of the damper and the controlled structure when a large relative velocity is applied to the damper.

Another series of tests on the RGA-ECD were carried out at different frequencies from 0.20 to 0.40 Hz when the velocity amplitudes were maintained at 0.188 m/s, and the results are given in Figure 9. It can be found from Figure 9 that the maximum damping force of the RGA-ECD is not changed coincidently when the velocity amplitude is constant. Furthermore, the tests of the RGA-ECD at different displacement amplitudes with the same frequency of 0.50 Hz were also conducted, and the corresponding hysteresis curves are shown in Figure 10. It can be found from Figure 10 that the performance of RGA-ECD is still very stable when the displacement amplitude varies.

5 | SEISMIC CONTROL OF BASE ISOLATED STRUCTURE BY USING THE RGA-ECD

5.1 | Structural parameters of 5-DOF system

In this section, the seismic control of a base isolated structure with supplemental dampers was investigated by numerical simulation. As shown in Figure 11, the supplemental dampers are installed between the first layer and the foundation. The structural parameters of the base isolated structure simplified by Zhang et al.[33] are as follows: mass $m_1 = m_2 = m_3 = m_4 = m_5 = 2.07 \times 10^8$ kg, horizontal stiffness $k_1 = 2.043 \times 10^9$ N/m, $k_2 = k_3 = k_4 = k_5 = 1.68 \times 10^{11}$ N/m, and the Rayleigh damping is adopted for the inherent damping of the structure, and the damping ratio of the structure is 2%. Four typical earthquake records[37]: El Centro and Hachinohe, examples of far field earthquakes, Northridge and Kobe, examples of near fault earthquakes, were used as ground motions in the numerical simulations.

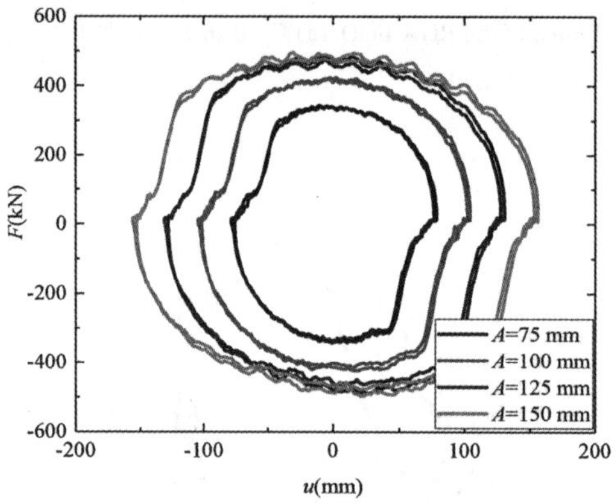

FIGURE 10 Hysteretic curves of RGA-ECD at different displacement amplitudes

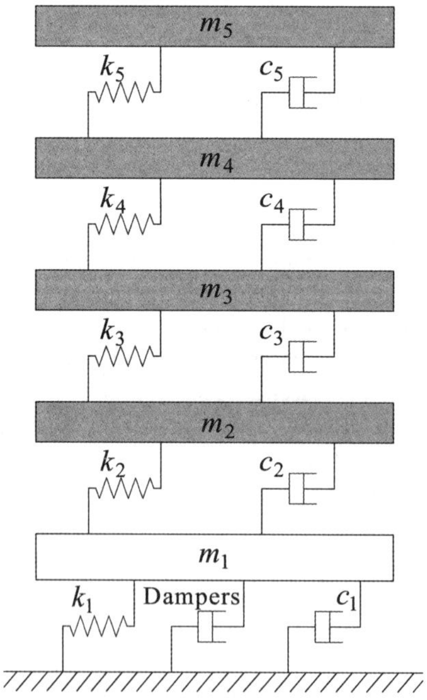

FIGURE 11 A 5-DOF system with supplemental dampers

These data were downloaded from the Peer Ground Motion Database. In order to evaluate the damping performance of the RGA-ECD more objectively, the conventional FVD widely used in civil engineering was also adopted in the numerical simulations for comparison.

If the supplemental dampers are the RGA-ECDs, the relationship between the damping force and the velocity is still described by the Wouterse model,[36] as shown in Equation (6).

If the supplemental dampers are the FVDs, the relationship between the damping force and the velocity can be described as

$$Q_{d,\mathrm{FVD}}(\dot{u}) = c_d \dot{u}^{\alpha}, \tag{10}$$

where c_d is the damping coefficient of the FVDs and α is a constant, which characterizes the nonlinearity of the damping force, and its range is often between 0 and 1.

Since the tested results of the damping force do not cover the descending stage, the maximum tested velocity of 0.509 m/s is determined as the critical velocity of the RGA-ECD, and the corresponding maximum damping force is 507 kN. The damping force of the FVD is determined to be equal to that of the RGA-ECD at the velocity of 0.509 m/s, and the α is considered to be 0.2, 0.4, 0.6, 0.8, and 1.0, respectively. Figure 12 presents the tested results of the RGA-ECD and the corresponding Wouterse model, together with the damping force–velocity curves of the FVD. It appears that the damping force of RGA-ECD has a maximum value, while the damping force of the FVD increases continuously with the increase of velocity.

5.2 | Seismic effectiveness evaluation

The response reduction ratios $\rho_{\max}$ and ρ_{rms} related to maximum value and root mean square (rms) value, respectively, are defined as

$$\rho_{\max} = \frac{A_{0,\max} - A_{d,\max}}{A_{0,\max}}, \tag{11}$$

$$\rho_{\mathrm{rms}} = \frac{A_{0,\mathrm{rms}} - A_{d,\mathrm{rms}}}{A_{0,\mathrm{rms}}}, \tag{12}$$

where A_0 and A_d represent the seismic responses of the 5-DOF system without or with supplemental dampers, respectively.

Two types of seismic responses, including the displacement of the first layer u relative to the foundation and the base shear f_b, were specially discussed. The response reduction ratios $\rho_{\max}$ and ρ_{rms} are both shown in Table 2. It can be found from Table 2 that the RGA-ECD is very effective in reducing the seismic responses, except for the maximum base

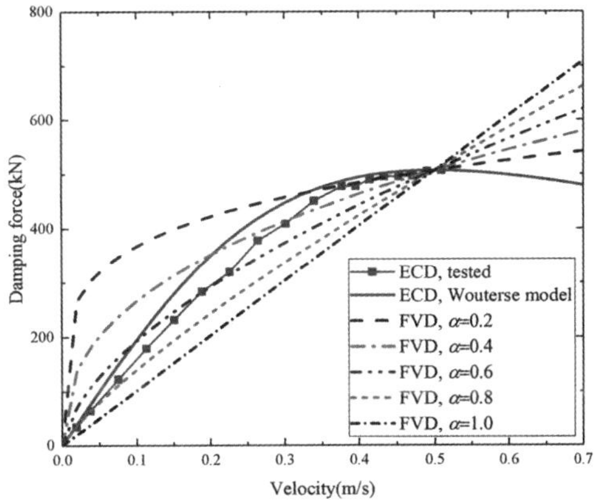

FIGURE 12 Comparison of damping force–velocity curves between the ECD-RGD and the FVD

TABLE 2 Effectiveness comparison between ECD-RGD and FVDs

Ground motion	Supplemental dampers	ρ_{max} (%)		ρ_{rms} (%)	
		u	f_b	u	f_b
El Centro	RGA-ECD	15.4	6.4	33.8	30.6
Uncontrolled	FVD, $\alpha = 0.2$	14.1	-1.7	42.7	33.4
$u_{max} = 0.176$ m	FVD, $\alpha = 0.4$	13.3	1.7	37.7	32.9
$f_{bmax} = 3.59 \times 10^8$ N	FVD, $\alpha = 0.6$	12.3	3.6	33.3	30.3
	FVD, $\alpha = 0.8$	11.4	4.7	29.3	27.3
	FVD, $\alpha = 1.0$	10.4	5.6	25.6	23.8
Hachinohe	RGA-ECD	13.5	6.9	29.3	25.6
Uncontrolled	FVD, $\alpha = 0.2$	21.2	9.9	36.3	28.6
$u_{max} = 0.225$ m	FVD, $\alpha = 0.4$	15.7	7.4	32.8	28.0
$f_{bmax} = 4.62 \times 10^8$ N	FVD, $\alpha = 0.6$	12.3	5.8	28.8	25.6
	FVD, $\alpha = 0.8$	9.9	4.9	24.8	22.5
	FVD, $\alpha = 1.0$	8.1	4.3	21.2	19.5
Northridge	RGA-ECD	3.6	-2.1	20.3	19.1
Uncontrolled	FVD, $\alpha = 0.2$	4.7	-1.9	20.9	17.6
$u_{max} = 0.524$ m	FVD, $\alpha = 0.4$	4.9	-1.4	21.9	19.6
$f_{bmax} = 1.08 \times 10^9$ N	FVD, $\alpha = 0.6$	5.2	-1.0	22.3	20.2
	FVD, $\alpha = 0.8$	5.5	-0.5	22.4	20.3
	FVD, $\alpha = 1.0$	5.9	-0.1	22.4	20.1
Kobe	RGA-ECD	34.6	27.8	58.4	56.9
Uncontrolled	FVD, $\alpha = 0.2$	35.9	28.5	61.0	58.5
$u_{max} = 0.431$ m	FVD, $\alpha = 0.4$	35.2	28.8	58.6	56.9
$f_{bmax} = 8.81 \times 10^8$ N	FVD, $\alpha = 0.6$	34.6	29.1	56.2	54.9
	FVD, $\alpha = 0.8$	34.2	29.3	53.9	52.6
	FVD, $\alpha = 1.0$	33.9	29.6	51.7	50.6

shear under Northridge earthquake. For example, the supplement of the RGA-ECDs, respectively, reduces the displacement relative to the foundation and the base shear by 58.4% and 56.9% when the structure is subjected to the Kobe earthquake. It seems that the effectiveness of the RGA-ECDs in reducing seismic responses is very close to those of the FVD, especially in the case of $\alpha = 0.6$. Figure 13 presents the displacement responses of the base isolated structure with the RGA-ECDs or the FVDs ($\alpha = 0.6$) under the Kobe earthquake, and Figure 14 presents the time history results of the normalized dissipated energy of the RGA-ECDs and the FVDs ($\alpha = 0.6$) under the Kobe earthquake. In Figure 14, the input energy of seismic excitations for the base isolated structure will ultimately be dissipated by the structural inherent damping and the RGA-ECDs or the FVDs ($\alpha = 0.6$). It can be considered from Figures 13 and 14 that the RGA-ECD has the same efficiency as the FVD ($\alpha = 0.6$) in reducing the seismic response of the base isolated structures.

5.3 | Force-limiting characteristic of the RGA-ECD

Table 3 lists the maximum damping forces of the RGA-ECD and the FVD in the time history analyses. It can be seen from Table 3 that the maximum damping forces of the RGA-ECD under four earthquakes are all lower than its design maximum value (507 kN). However, the maximum damping forces of the FVD in all α cases under the near fault earthquakes of Northridge and Kobe are considerably greater than 507 kN. This indicates that the FVD in the application may be damaged due to the unanticipated loads exceed its design capacity, while this shortcoming does not occur in

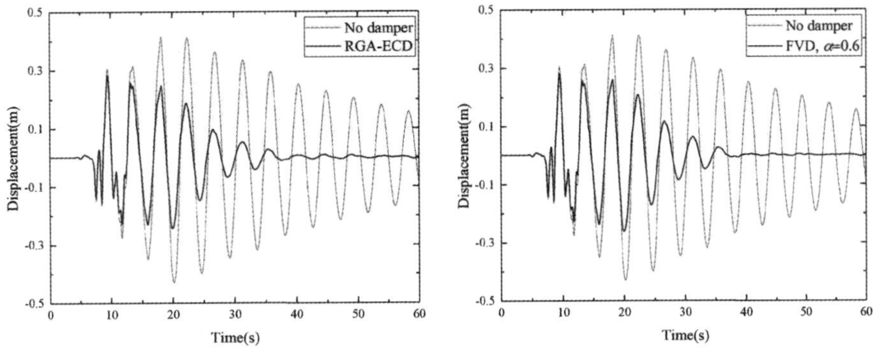

(a) The RGA-ECDs (b) The FVDs(α=0.6)

FIGURE 13 The displacement responses u under the Kobe earthquake. (a) The RGA-ECDs. (b) The FVDs (α=0.6)

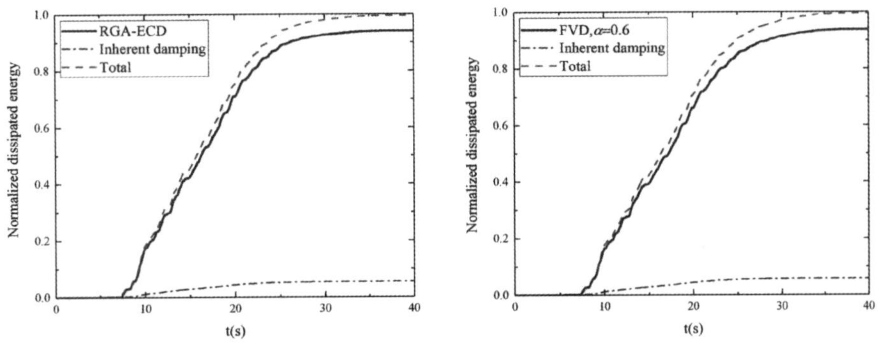

(a) The RGA-ECDs (b) The FVDs(α=0.6)

FIGURE 14 The time history results of the normalized dissipated energy under the Kobe earthquake. (a) The RGA-ECDs. (b) The FVDs (u – 0.6)

TABLE 3 Maximum damping force of RGA-ECD and FVD

	Maximum damping force (kN)			
Supplemental dampers	**El Centro**	**Hachinohe**	**Northridge**	**Kobe**
RGA-ECD	492.9	481.3	507.0	507.0
FVD, $\alpha = 0.2$	488.4	479.3	616.5	555.3
FVD, $\alpha = 0.4$	465.5	450.4	751.7	608.2
FVD, $\alpha = 0.6$	441.3	422.9	917.5	670.6
FVD, $\alpha = 0.8$	417.5	397.9	1119.6	740.4
FVD, $\alpha = 1.0$	394.6	375.2	1363.9	817.9

the RGA-ECD. The force-limiting characteristic improves the security and reliability of the dampers and the controlled structure.

Figure 15 presents the time histories of the damping force and the velocity of the RGA-ECD under the Kobe earthquake, together with the time histories of the FVD. It can be found from Figure 15 that the velocity of the RGA-ECD from 7.70 to 12.94 s is greater than its critical velocity of 0.509 m/s (the maximum velocity is 0.811 m/s), while the damping force at all times is not more than the design maximum value of 507.0 kN. It seems that when the velocity of RGA-ECD is low, such as $t < 6.90$ s or $t > 38.40$ s, the damping force of the RGA-ECD is significantly smaller than that

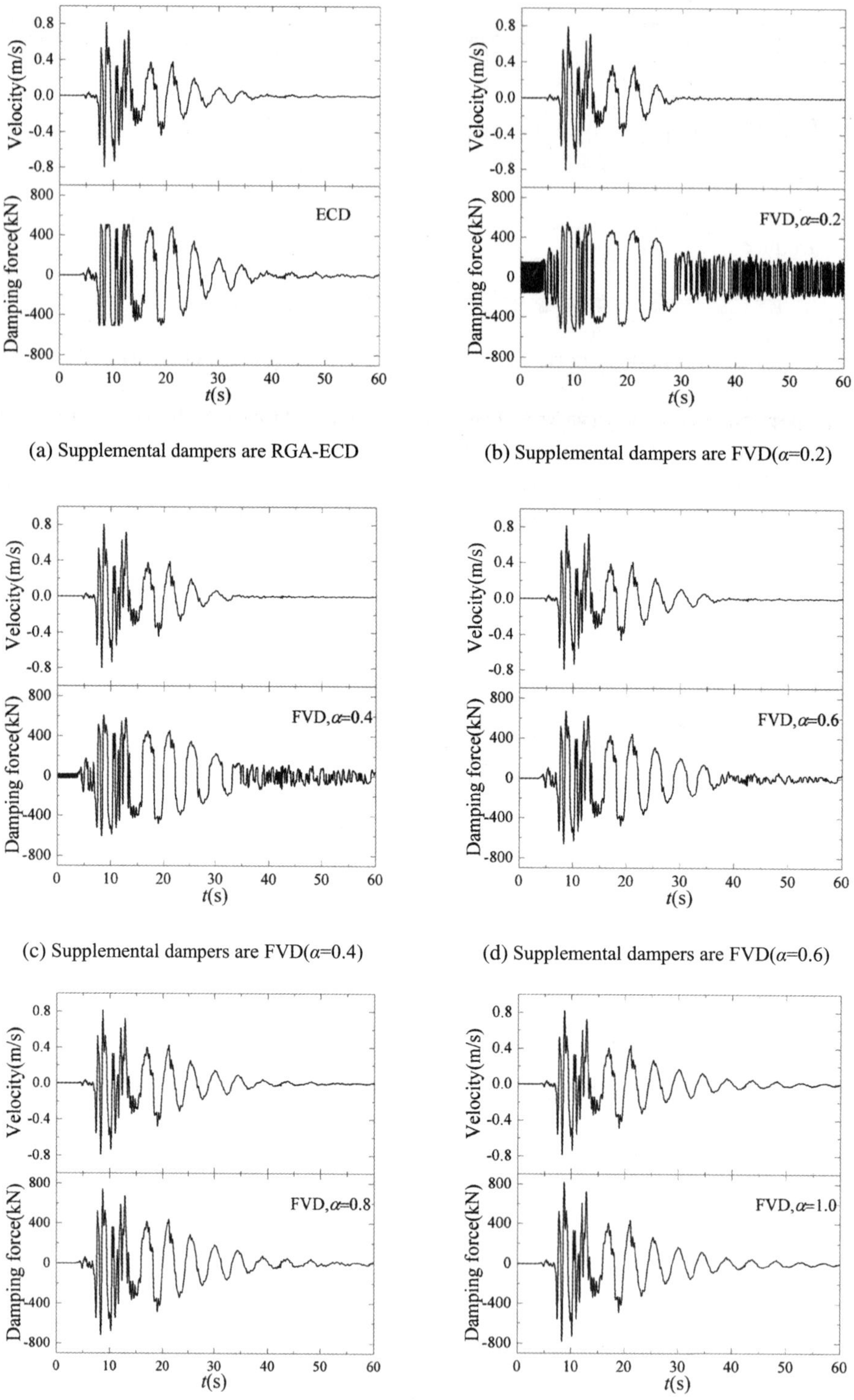

(a) Supplemental dampers are RGA-ECD

(b) Supplemental dampers are FVD(α=0.2)

(c) Supplemental dampers are FVD(α=0.4)

(d) Supplemental dampers are FVD(α=0.6)

(e) Supplemental dampers are FVD(α=0.8)

(f) Supplemental dampers are FVD(α=1.0)

FIGURE 15 Time histories of damping force and velocity under the Kobe earthquake. (a) Supplemental dampers are RGA-ECD. (b) Supplemental dampers are FVD ($\alpha = 0.2$). (c) Supplemental dampers are FVD ($\alpha = 0.4$). (d) Supplemental dampers are FVD ($\alpha = 0.6$). (e) Supplemental dampers are FVD ($\alpha = 0.8$). (f) Supplemental dampers are FVD ($\alpha = 1.0$)

of the FVD with $\alpha = 0.2$. This situation can be conveniently improved by changing the thickness or material of the conductor to adapt to different structures.

6 | CONCLUDING REMARKS

A novel RGA-ECD was proposed and manufactured in this study, and the nonlinear performance of eddy current damping was systematically investigated by numerical simulation and prototype testing. First, the configuration of the RGA-ECD was introduced in detail, and its axial force was derived theoretically. Second, a prototype of the RGA-ECD with a target damping force of 500 kN at the velocity of 0.5 m/s was manufactured, and the testing of this prototype was carried out. In the test, the maximum damping force of this prototype is 507 kN at the velocity of 0.509 m/s. It appears that the tested results are in good agreement with the numerical results obtained from COMSOL Multiphysics when the loading velocity is less than 0.377 m/s. When the loading velocity is greater than 0.377 m/s, the relative differences between the tested and numerical results gradually increase with the increase of the velocity, but the maximum is not more than 10.8%. The hysteresis curves of the prototype are relatively smooth, and the repeatability is satisfactory. This indicates that the mechanical properties of the RGA-ECD are considerably stable and reliable. Third, the effectiveness of RGA-ECD in mitigating seismic responses of base isolated structures was numerically studied by comparing it with the traditional FVD. The results show that the RGA-ECD with excellent performance can be adopted as an alternative to the FVD for vibration control, and there is a force-limiting characteristic for the RGA-ECD, which can prevent the damage of damper and controlled structure caused by excessive impact load.

ACKNOWLEDGMENT

This project is supported by the National Key Research and Development Program of China (Grants 2017YFC0703600 and 2017YFC0703604).

AUTHOR CONTRIBUTIONS

Shouying Li contributed significantly to analysis and manuscript preparation. Yafeng Li performed the data analyses and wrote the manuscript. Yafeng Li, Weiyang Mao, and Jianzhong Wang performed the experiment. Zhengqing Chen helped perform the analysis with constructive discussions.

DATA AVAILABILITY STATEMENT

Data are available on request from the authors.

ORCID

Shouying Li https://orcid.org/0000-0003-4516-4378
Yafeng Li https://orcid.org/0000-0001-9064-8246

REFERENCES

1. Lin W-H, Chopra AK. Earthquake response of elastic SDOF systems with non-linear fluid viscous dampers. *Earthq Eng Struct Dynam.* 2010;31(9):1623-1642.
2. Ahmadizadeh M. On equivalent passive structural control systems for semi-active control using viscous fluid dampers. *Struct Control Health Monit.* 2010;14(6):858-875.
3. Li LY, Song GB, Singla M, Mo Y. Vibration control of a traffic signal pole using a pounding tuned mass damper with viscoelastic materials (II): experimental verification. *J Vib Control.* 2013;21(4):670-675.
4. Wang WX, Wang XY, Hua XG, Song GB, Chen ZQ. Vibration control of vortex-induced vibrations of a bridge deck by a single-side pounding tuned mass damper. *Eng Struct.* 2018;173(OCT.15):61-75.
5. Olsen JS, Eijk C, Zhang ZL. Numerical analysis of a new SMA-based seismic damper system and material characterization of two commercial NiTi-alloys. *Smart Struct Syst.* 2008;4(2):137-152.
6. Casciati F, Faravelli L, Saleh RA. An SMA passive device proposed within the highway bridge benchmark. *Struct Control Health Monit.* 2009;16(6):657-667.
7. Guo W, Ma C, Yu Y, Bu D, Zeng C. Performance and optimum design of replaceable steel strips in an innovative metallic damper. *Eng Struct.* 2019;205:110118.
8. Lee CH, Ryu J, Kim DH, Ju YK. Improving seismic performance of non-ductile reinforced concrete frames through the combined behavior of friction and metallic dampers. *Eng Struct.* 2018;172(OCT.1):304-320.

9. Zapateiro M, Karimi HR, Luo N, et al. Real-time hybrid testing of semiactive control strategies for vibration reduction in a structure with MR damper. *Struct Control Health Monit.* 2010;17(4):427-451.
10. Jiménez R, Luis LI. LuGre friction model for a magnetorheological damper. *Struct Control Health Monit.* 2010;12(1):91-116.
11. Yang MG, Chen ZQ, Hua XG. An experimental study on using MR damper to mitigate longitudinal seismic response of a suspension bridge. *Soil Dynam Earthq Eng.* 2011;31(8):1171-1181.
12. Shen W, Zhu S, Zhu H. Unify energy harvesting and vibration control functions in randomly excited structures with electromagnetic devices. *J Eng Mech.* 2018;145(1):04018115.
13. Amjadian M, Agrawal AK. A passive electromagnetic eddy current friction damper (PEMECFD): theoretical and analytical modeling. *Struct Control Health Monit.* 2017;24(10):1-23.
14. Cheah SK, Sodano HA. Novel eddy current damping mechanism for passive magnetic bearings. *J Vib Control.* 2008;14(11):1749-1766.
15. Ebrahimi B, Khamesee MB, Golnaraghi F. Eddy current damper feasibility in automobile suspension: modeling, simulation and testing. *Smart Mater Struct.* 2009;18(1):015017.
16. Pan Q, He T, Xiao D, Liu X. Design and damping analysis of a new eddy current damper for aerospace applications. *Latin Am J Solids Struct.* 2016;13(11):1997-2011.
17. Kitanov S, Podolskii A. Analysis of eddy-current and magnetic rail brakes for high-speed trains. *Open Transport J.* 2008;2(1):19-28.
18. Zuo L, Chen X, Nayfeh S. Design and analysis of a new type of electromagnetic damper with increased energy density. *J Vib Acoust.* 2011;133(4):041006.
19. Ebrahimi B, Khamesee MB, Golnaraghi F. A novel eddy current damper: theory and experiment. *J Phys D.* 2009;42(7):075001.
20. Huang ZW, Hua XG, Chen ZQ, Niu HW. Modeling, testing, and validation of an eddy current damper for structural vibration control. *J Aerosp Eng.* 2018;31(5):04018063.
21. Chen ZQ. (2011) Permanent magnet eddy current tuned mass damper. Chinese patent 101033628. (in Chinese).
22. Liang LT, Feng ZQ, Chen ZQ. Seismic control of SDOF systems with nonlinear eddy current dampers. *Appl Sci.* 2019;9(16):3427.
23. Wen Q, Hua XG, Chen ZQ, Yang Y, Niu HW. Control of human-induced vibrations of a curved cable-stayed bridge: design, implementation, and field validation. *J Bridge Eng.* 2016;21(7):04016028.
24. Lu XL, Zhang Q, Weng D, et al. Improving performance of a super tall building using a new eddy-current tuned mass damper. *Struct Control Health Monit.* 2017;24(3):e1882.
25. Niu HW, Chen ZQ, Hua XG, Zhang W. Mitigation of wind-induced vibrations of bridge hangers using tuned mass dampers with eddy current damping. *Smart Struct Syst.* 2018;22(6):727-741.
26. Liu M, Li SY, Chen ZQ. Mitigation of wind-induced responses of cylinder solar tower by a tiny eddy current tuned mass damper based on elastic wind tunnel tests. *Smart Struct Syst.* 2020;26(5):619-629.
27. Liu M, Li SY, Wu T, Li YF, Meng HC, Chen ZQ. Eddy-current tuned mass dampers for mitigation of wind-induced response of the Noor III Solar Tower: design, installation, and validation. *J Struct Eng.* 2021;147(12):05021009.
28. Asai T, Araki Y, Ikago K. Structural control with tuned inertial mass electromagnetic transducers. *Struct Control Health Monit.* 2018; 25(2):e2059.
29. Ikago K, Saito K, Inoue N. Seismic control of single-degree-of-freedom structure using tuned viscous mass damper. *Earthq Eng Struct Dynam.* 2012;41(3):453-474.
30. Nakamura Y, Fukukita A, Tamura K, et al. Seismic response control using electromagnetic inertial mass dampers. *Earthq Eng Struct Dynam.* 2014;43(4):507-527.
31. Zhu H, Li Y, Shen W, Zhu S. Mechanical and energy-harvesting model for electromagnetic inertial mass dampers. *Mech Syst Signal Process.* 2019;120:203-220.
32. Wang ZH, Xu YW, Gao H, Chen ZQ, Xu K, Zhao SB. Vibration control of a stay cable with a rotary electromagnetic inertial mass damper. *Smart Struct Syst.* 2019;23(6):627-639.
33. Zhang HY, Chen ZQ, Hua XG, Huang ZW, Niu HW. Design and dynamic characterization of a large-scale eddy current damper with enhanced performance for vibration control. *Mech Syst Signal Process.* 2020;145:106879.
34. Li YF, Li SY, Wang JZ, Chen ZQ. A new type of damper combining eddy current damping with rack and gear. *J Vib Control.* 2021; 27(9-10):1087-1097.
35. Li SY, Li YF, Wang JZ, Chen ZQ. Theoretical investigations on the linear and nonlinear damping force for an eddy current damper combining with rack and gear. *J Vib Control.* 2022;28(9-10):1035-1044.
36. Wouterse JH. Critical torque and speed of eddy current brake with widely separated soft iron poles. *IEE Proc B Electr Power Appl.* 1991; 138(4):153-158.
37. Pietrosanti D, Angelis MD, Basili M. Optimal design and performance evaluation of systems with Tuned Mass Damper Inerter (TMDI). *Earthq Eng Struct Dynam.* 2017;46(8):1367-1388.

第36卷 第7期
2023年7月
中 国 公 路 学 报
China J. Highw. Transp.
Vol. 36 No. 7
July 2023

文章编号:1001-7372(2023)07-0067-13

Halbach阵列板式电涡流阻尼单元参数分析与优化设计

黄智文[1,2],蒲怡达[1,2],张弘毅*[1,2],华旭刚[1,2],陈政清[1,2]
(1. 湖南大学 风工程与桥梁工程湖南省重点实验室,湖南 长沙 410082;
2. 湖南大学 振动与冲击技术研究中心,湖南 长沙 410082)

摘要:为了提高板式电涡流阻尼单元的耗能效率,在板式电涡流阻尼单元中引入了Halbach永磁阵列,并开展了参数分析和优化设计研究。首先制作了一个Halbach阵列板式电涡流阻尼单元,并通过试验验证了采用三维电磁有限元分析计算其阻尼系数的可靠性。然后基于三维电磁有限元分析方法,研究了永磁体厚度、磁极数、磁极对数、极长比及永磁体背铁对Halbach阵列板式电涡流阻尼单元阻尼系数的影响。最后提出了二维电磁有限元分析与MATLAB遗传算法相结合的板式电涡流阻尼单元永磁阵列参数优化方法,并通过算例分别对有背铁Halbach阵列、无背铁Halbach阵列、垂向极化阵列和纵向极化阵列等4种方案进行了优化设计。研究结果表明:在一定范围内增加永磁体厚度能显著提高Halbach阵列板式电涡流阻尼单元的阻尼系数;在永磁体阵列总长度不变的情况下,阻尼系数随着Halbach阵列磁极数量的增大而增大,但增速逐渐减慢;设置永磁体背铁可提高Halbach阵列板式电涡流阻尼单元的阻尼系数,但永磁体厚度较大或磁极数较多时,其影响逐渐减弱;存在最优极长比使阻尼系数最大,其中有背铁时约为0.65,无背铁时约为0.5。针对优化设计算例,从耗能能力来看,有永磁体背铁的Halbach阵列大于垂向极化永磁体阵列大于无永磁体背铁的Halbach阵列大于纵向极化永磁体阵列;通过参数优化,可大幅提升板式电涡流阻尼单元的耗能能力。

关键词:桥梁工程;振动控制;参数优化;电涡流阻尼器;Halbach阵列
中图分类号:U441.3 **文献标志码:**A

Parametric Analysis and Optimal Design of Planar Eddy Current Damping Element with Halbach Array Magnets

HUANG Zhi-wen[1,2], PU Yi-da[1,2], ZHANG Hong-yi*[1,2],
HUA Xu-gang[1,2], CHEN Zheng-qing[1,2]
(1. Key Laboratory for Wind and Bridge Engineering of Hunan Province, Hunan University, Changsha 410082, Hunan, China; 2. Vibration and Shock Technology Research Center, Hunan University, Changsha 410082, Hunan, China)

Abstract: In order to increase the energy dissipation capacity of a planar eddy current damping element (PECDE), the Halbach array magnets are employed, and parametric analysis and optimization are conducted for the planar eddy current damping element with Halbach array (PECDE-HA). Firstly, a PECDE-HA was manufactured and tested, verifying the reliability of

收稿日期:2022-07-03
基金项目:湖南省自然科学基金项目(2020JJ5061);国家自然科学基金项目(51808210)
作者简介:黄智文(1986-),男,湖南怀化人,副教授,博士研究生导师,工学博士,E-mail:zwhuang213@hnu.edu.cn。
****通讯作者:**张弘毅(1991-),男,江苏无锡人,助理研究员,工学博士,E-mail:zhyzzb25@hnu.edu.cn。

the three-dimensional electromagnetic finite element analysis in calculating the damping coefficient. Secondly, based on the three-dimensional electromagnetic finite element analysis, the effects of several main factors, including the thickness of the permanent magnets, magnetic pole number, magnetic pole pair number, pole length ratio, and the back-iron, on the damping coefficient of the PECDE-HA were investigated in detail. Finally, an efficient optimal method, which combines the two-dimensional electromagnetic finite element and the genetic algorithm, was proposed to determine the optimal parameters of the PECDE with four different configurations of permanent magnets, including the Halbach arrays with back iron, Halbach arrays without back iron, vertically magnetized arrays (VA) and longitudinally magnetized arrays (LA). Results show that an increase in the thickness of permanent magnets can significantly increase the damping coefficient of PECDE-HA within a certain range; If the total length of the permanent magnet keeps unchanged, the damping coefficient of the PECDE-HA increases with the increasing magnetic pole number, but the increment decreases gradually; The permanent magnet back iron usually has a positive impact on the performance of PECDE-HA, but the impact attenuates as the thickness or pole number of the permanent magnet grows. An optimal pole length ratio can be obtained for the two-pole Halbach array, which is approximately 0.65 for the case with back-iron and 0.5 for the case without back-iron, respectively. In the optimal design of the PECDE, the optimal parameters and the maximum achievable damping coefficients for the four schemes are obtained, respectively. It can be seen that the Halbach arrays with back-iron performs best in terms of energy dissipation density, which is followed by the Halbach arrays without back-iron, the vertically magnetized arrays and longitudinally magnetized arrays.

Keywords: bridge engineering; structural vibration control; parametric optimization; eddy current damper; Halbach array

Received 03 July 2022

Funding: Natural Science Foundation of Hunan Province (2020JJ5061); National Natural Science Foundation of China (51808210)

*Corresponding author. *E-mail address*: zhyzzb25@hnu.edu.cn.

https://doi.org/10.19721/j.cnki.1001-7372.2023.07.005

0 引 言

电涡流阻尼具有非接触、无机械摩擦损耗、无工作流体、耐久性好等优点[1]，在载重汽车缓速控制[2]、列车制动[3]等机械领域有悠久的研究历史。近年来，电涡流阻尼装置在大型土木工程结构振动控制领域的研究逐渐引起关注[4-5]，例如大跨度桥梁的涡激共振[6]和人致振动[7]控制，高层建筑的风致振动[8-9]控制。目前已实现工程应用的电涡流阻尼装置主要有2种类型：①采用板式电涡流阻尼单元作为耗能元件的调谐质量阻尼器[10]；②作为直接耗能元件安装于结构构件之间的电涡流轴向阻尼器[11]，其基本组成也是板式电涡流阻尼单元。对于板式电涡流阻尼单元或电涡流轴向阻尼器，耗能密度通常可表示为阻尼装置的阻尼系数与其总体积的比值。对于总体积相同的阻尼装置，则可直接采用阻尼系数作为衡量耗能密度的相对指标，因此阻尼系数体现了阻尼器的耗能能力。要提高电涡流阻尼装置的耗能能力主要有2种方法：①在电涡流阻尼装置中引入速度和阻尼力放大机构；②优化电涡流阻尼装置的磁路结构[12]。第1种方法在电涡流轴向阻尼器中应用较多，它主要采用滚珠丝杠副或齿轮齿条等机械装置对阻尼器的工作速度和控制力进行放大[13]。第2种方法可提高各类电涡流阻尼装置的耗能能力，这也是其他机电装置优化设计的重要内容。电涡流阻尼装置中使用的板式电涡流阻尼单元主要采用永磁体垂向极化的单边复合板式结构[14]，其耗能能力受永磁体尺寸及其排列方式、导体板尺寸、背铁等众多因素的影响，要实现磁路参数的全局优化难度较大。已有文献通常采用参数分析的方法确定相对较优的磁路参数[15-18]，例如永磁体长宽比、间距和排布方式等。

Halbach阵列是Halbach在1980年提出的一种新型永磁体排布方式[19]，它是将不同极化方向的永磁体按照某一特定规律组合而成，使得阵列一侧磁力线增强，另一侧磁力线减弱，从而获得较为理想的单边正弦分布磁场。在20世纪90年代，Halbach阵列在高能物理领域中得到了首次应用[20]。由于工业发展对高效轻便永磁装置的迫切需求，Halbach阵列在永磁电机[21]、涡流制动[22]等领域得到了快速发展。例如在涡流制动方面，常用的旋转型电涡流阻尼缓速器按照永磁体极化方向不同可以分为径向极化阵列[23]、切向极化阵列[24]和Halbach阵列。而在直线型电涡流阻尼器中，常用垂向极化阵列和纵向极化阵列来描述前2种永磁体排布方式。纵向极化阵列产生对称分布的磁场，磁极之间需要导磁材料但不需要永磁体背铁，在气隙较小时，永磁体利用率高，但气隙较大时漏磁严重，影响阻尼器性能。垂向极化阵列是较为常用的方式，但阻尼性能受磁极间隙和永磁体背铁影响显著，同时永磁体背铁增大了结构质量和布置空间。常用的Halbach阵列是垂向极化和纵向极化的组合[25]，在磁场增强的一侧布置导体，可以达到提高阻尼器耗能能力的效果，在磁场较弱的一侧则降低了对磁体背铁的厚度要求，甚至可以取消背铁。

Halbach阵列的几何参数主要包括磁体厚度、磁极数、磁极对数以及不同极化方向磁体长度比等。Halbach阵列的优化设计目标根据其应用领域不同而变化。目前Halbach阵列的优化设计研究主要集中在永磁电机[26-27]和磁悬浮[28-29]领域，其优化目标主要是提高气隙处的磁场强度。在涡流制动方面，Jang等[30]较早地将Halbach阵列运用到电涡流制动器中，并通过参数分析比较了不同永磁体极化阵列制动器的性能，但Jang采用的是二维模型，且只关注了永磁体厚度、导体板厚度等厚度方向几何参数的优化。Bissal等[31]以高压直流断路器中的电涡流阻尼器为背景，分析了磁路结构对圆筒形电涡流阻尼器阻尼性能的影响，并指出Halbach阵列可提供最大阻尼系数。由于作者并没有对各种磁路进行优化，所以相关结论有很大的局限性。

本文在板式电涡流阻尼单元中引入了Halbach永磁阵列，并开展了Halbach永磁阵列板式电涡流阻尼单元的参数分析和优化设计研究。首先制作了一个Halbach阵列板式电涡流阻尼单元，并通过试验验证了三维电磁有限元稳态计算的可靠性。然后基于三维电磁场有限元分析方法，研究了永磁体厚度、磁极数、磁极对数、极长比及永磁体背铁对Halbach阵列板式电涡流阻尼单元阻尼系数的影响规律。最后，提出了二维电磁有限元分析与MATLAB遗传算法相结合的板式电涡流阻尼单元永磁阵列参数优化方法，并通过算例分别对有背铁Halbach阵列、无背铁Halbach阵列、垂向极化阵列和纵向极化阵列等4种方案进行了优化设计。

1 Halbach阵列板式电涡流阻尼器的基本构造

板式电涡流阻尼单元作为阻尼器发生器，是电涡流调谐质量阻尼器和电涡流轴向阻尼器的重要组成部分，图1为在各类电涡流阻尼装置中广泛使用的垂向极化板式电涡流阻尼单元[10]。图2为本文研究的Halbach阵列板式电涡流阻尼单元，其永磁体阵列由垂向极化和纵向极化的永磁体相间布置组成。Halbach阵列板式电涡流阻尼单元的工作原理与常用的垂向极化阵列板式电涡流阻尼单元相同，导体板和导体背铁可作为定子部分，永磁体阵列则作为动子部分，当动子部分相对于定子部分发生运动时，导体板切割磁力线，产生电涡流阻尼。与垂向极化阵列不同的是，按照合适的永磁体排布，Halbach阵列可增强导体板中的磁通密度，进一步增大阻尼力。

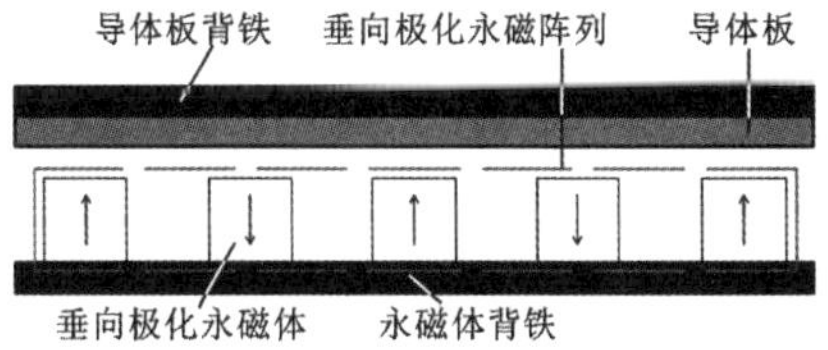

图1 垂向极化阵列板式电涡流阻尼单元

Fig. 1 Facade Arrangement of PECDE-VA

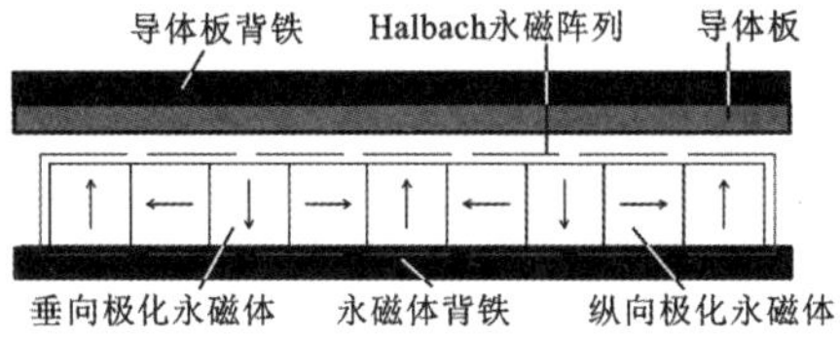

图2 Halbach阵列板式电涡流阻尼单元

Fig. 2 Facade Arrangement of PECDE-HA

2 Halbach阵列板式电涡流阻尼单元的电磁有限元分析及可靠性验证

2.1 阻尼特性的电磁有限元计算方法

本文主要通过电磁有限元方法计算Halbach阵列板式电涡流阻尼单元的阻尼特性。当板式电涡流阻尼单元作为调谐质量阻尼器的阻尼单元时，它的工作速度通常较小，阻尼特性接近线性黏滞阻尼，因此可以采用阻尼系数作为其阻尼特性或耗能密度的描述指标。本文的电磁有限元分析采用 COMSOL 进行，图 3 给出了在 COMSOL 中建立的 Halbach 阵列板式电涡流阻尼单元及其网格划分情况。

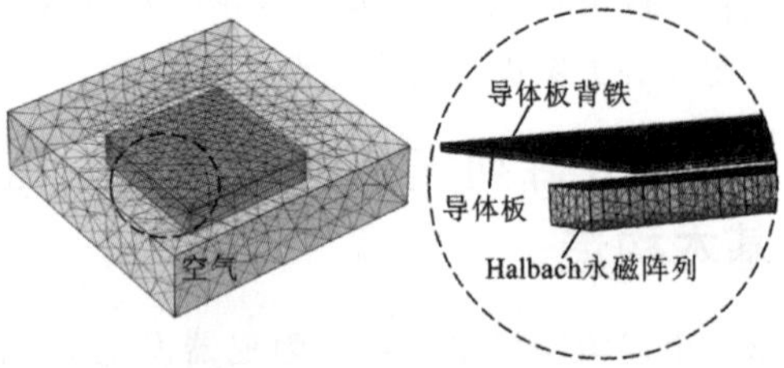

图 3　三维有限元模型

Fig. 3　3D Finite Element Model

通过在 COMSOL 的 AD/CD 模块下进行三维稳态磁场分析，可获得任意指定工作速度下板式电涡流阻尼单元的电磁场分布和阻尼力大小。有限元法的关键求解过程如下：由麦克斯韦方程组引入磁势量 $\boldsymbol{A}$ 和电标量 φ，再考虑动子部分的稳态电流项导出不规范的方程，使用弱形式的有限元法求解微分方程得到磁势量 $\boldsymbol{A}$，再进一步求得磁通密度 $\boldsymbol{B}$、电涡流密度 $\boldsymbol{J}$，然后将洛伦兹力在体积 $\boldsymbol{V}$ 中积分得到阻尼力 $\boldsymbol{F}$，见式(1)。用阻尼力 $\boldsymbol{F}$ 除以计算指定的速度 v 即可得到板式电涡流阻尼单元的阻尼系数 C_m。

$$\boldsymbol{F}=\int(\boldsymbol{J}\times\boldsymbol{B})\mathrm{d}V \tag{1}$$

$$C_m=\boldsymbol{F}/v \tag{2}$$

2.2　阻尼系数测试试验

为了保证电磁有限元分析结果的可靠性，首先制作了一个 Halbach 阵列板式电涡流阻尼单元，然后通过试验测试了其阻尼系数，并与三维电磁有限元分析结果进行了对比。图 4 是 Halbach 阵列板式电涡流阻尼单元的实物照片，该 Halbach 阵列主要由 5 块垂向极化永磁体和 4 块纵向极化永磁体构成。图 5 给出了几何参数示意图，表 1 列出了 Halbach 阵列式板式电涡流阻尼单元的关键设计参数和电磁有限元计算参数。

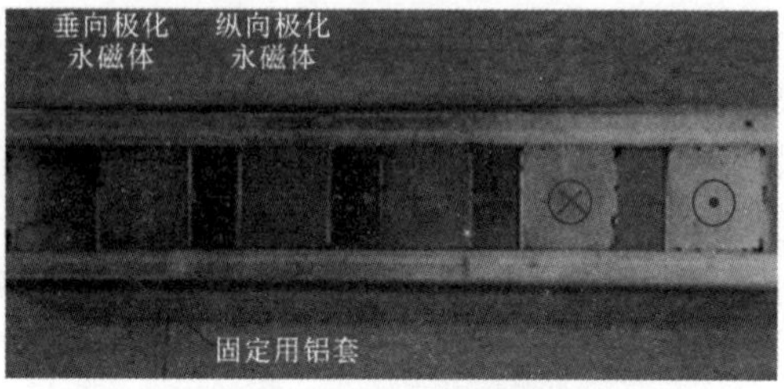

图 4　试验使用的 Halbach 板式电涡流阻尼单元

Fig. 4　PECDE-HA Used in Test

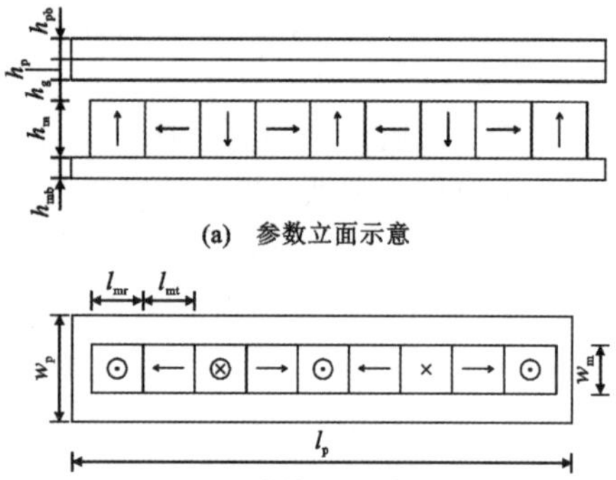

(a)　参数立面示意

(b)　参数平面示意

图 5　几何参数示意

Fig. 5　Geometrical Parameter Schematic

表 1　模型参数

Table 1　Model Parameters

名称	符号	值	名称	符号	值
导体板长度(平行运动方向)/mm	l_p	250	垂向极化永磁体厚度/mm	h_{mr}	20
导体板宽度(垂直运动方向)/mm	w_p	250	永磁体剩余磁感应强度/T	B_r	0.7
导体板厚度/mm	h_p	4	永磁体阵列磁极数	n_{mp}	2
导体板背铁厚度/mm	h_{pb}	9	永磁体阵列磁极对数	n_{mpp}	2/4
纵向充磁永磁体长度(平行运动方向)/mm	l_{mt}	15	永磁体与导体板间隙/mm	d	15～40
纵向充磁永磁体宽度(垂直运动方向)/mm	w_{mt}	50	导体板电导率/(MS·m^{-1})	σ_c	58
纵向极化永磁体厚度/mm	h_{mt}	20	空气、导体板相对磁导率	u_r	1
垂向极化永磁体长度(平行运动方向)/mm	l_{mr}	30	永磁体相对磁导率	u_{rm}	1.05
垂向极化永磁体宽度(垂直运动方向)/mm	w_{mr}	50	计算运动速度/(m·s^{-1})	v	0.05

为了测试 Halbach 阵列板式电涡流阻尼单元的阻尼系数，参照文献[10]的做法，可将它安装在如图 6 所示的单自由度水平振动框架上，通过水平框架的自由衰减振动识别板式电涡流阻尼器安装前后系统频率和阻尼比的变化，进而计算 Halbach 阵列板式电涡流阻尼单元的阻尼系数 C_m，如式(3)所示。永磁体和导体板之间的气隙可由螺杆调节，从而得到阻尼系数随气隙的变化规律。

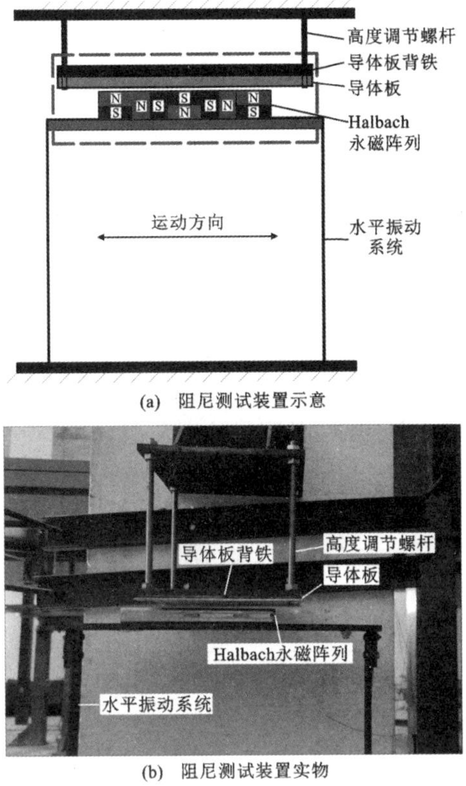

(a) 阻尼测试装置示意

(b) 阻尼测试装置实物

图 6 试验测试装置

Fig. 6 Experimental Test Setup

$$C_m = 2M_t\omega_t\zeta_t - C_0 \tag{3}$$

式中：M_t 为振动系统第一阶水平振动模态的模态质量；ω_t、ζ_t 和 C_0 分别为铜板安装后结构的固有圆频率、固有阻尼比、铜板及背铁安装前结构的阻尼系数。ω_t、ζ_t 和 C_0 均可通过水平振动系统的自由衰减振动试验得到。

图 7 给出了磁极对数为 4、气隙分别为 15、20、30、40 mm 时振动系统的自由衰减位移时程信号，对应图 8(b)中不同气隙下的阻尼系数。从中可以看出不同气隙下 Halbach 阵列式电涡流阻尼单元的阻尼系数相差较大，因此附加阻尼比相差较大，振动衰减曲线有明显差别。需要说明的是，图中还包括激励段时程，但本文仅关注自由衰减段时程。

2.3 电磁有限元计算的可靠性分析

试验选取 2 种不同磁极对数的 Halbach 阵列板式电涡流阻尼单元，测试了永磁体与导体板气隙 h_g 在 15～40 mm 之间变化时阻尼系数的变化情况，结果如图 8 所示。由图 8 可以看到，无论是磁极对数 $n_{mpp}=2$ 还是 $n_{mpp}=4$ 的 Halbach 永磁体阵列，随着气隙的增加，阻尼系数呈指数衰减。总体来看，当气隙较大、磁极数较多时三维电磁有限元计算结果与

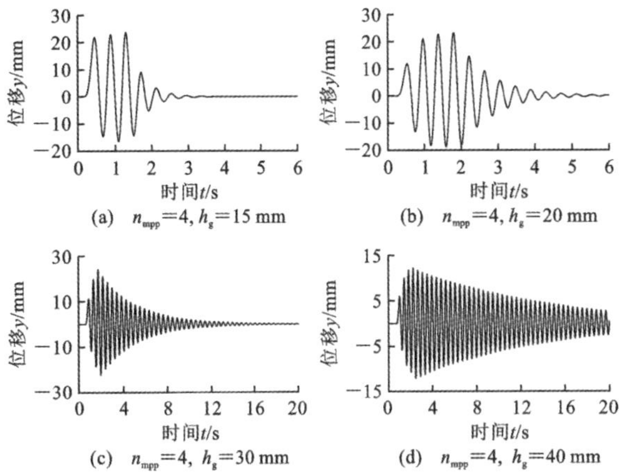

(a) $n_{mpp}=4$, $h_g=15$ mm (b) $n_{mpp}=4$, $h_g=20$ mm

(c) $n_{mpp}=4$, $h_g=30$ mm (d) $n_{mpp}=4$, $h_g=40$ mm

图 7 不同气隙下试验系统的自由振动衰减曲线

Fig. 7 Free Vibration Decay Curves for Test System with Different Air Gaps

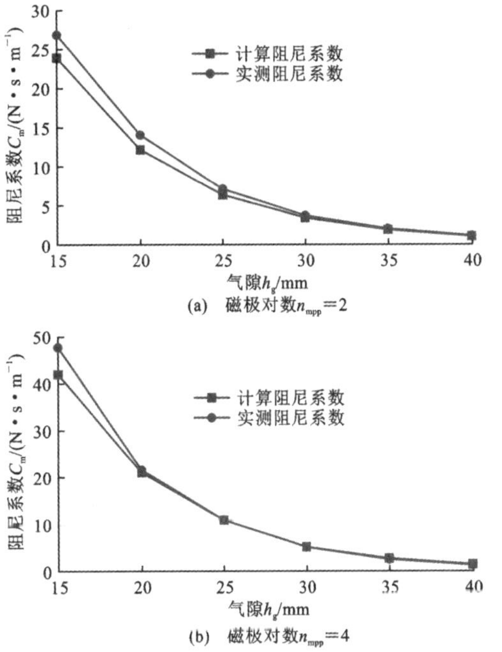

(a) 磁极对数$n_{mpp}=2$

(b) 磁极对数$n_{mpp}=4$

图 8 有限元计算及试验测试的电涡流阻尼系数

Fig. 8 Eddy Current Damping Coefficients for Finite Element Calculations and Experimental Tests

试验结果吻合很好，因此，采用电磁有限元方法开展 Halbach 阵列板式电涡流阻尼单元的参数分析和优化设计是合适的。随着气隙减小，有限元分析和试验结果之间的偏差逐渐增大，主要原因可能有如下几点：①气隙较小时振动系统接近临界阻尼状态，如图 7(a)所示，此时由自由衰减振动数据求出的阻尼比和阻尼系数容易产生较大误差；②气隙尺寸由游标卡尺手工测量，气隙较小时容易产生测量误差。图 9 还给出了磁极对数为 4 时，气隙分别为 15、20 mm 时计算的和实测的振动系统自由衰减位移时程

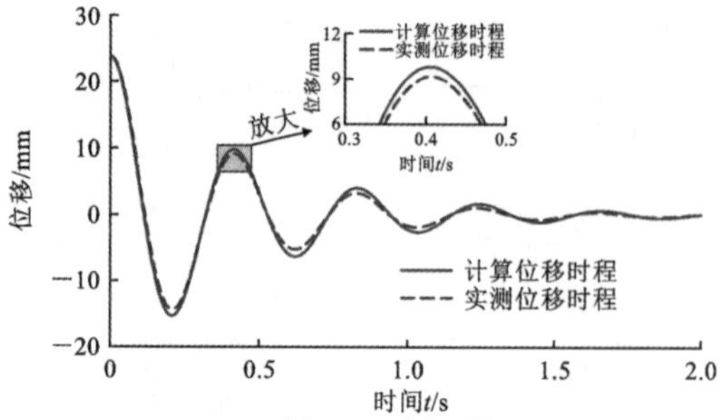

(a) $n_{mpp}=4, h_g=15$ mm

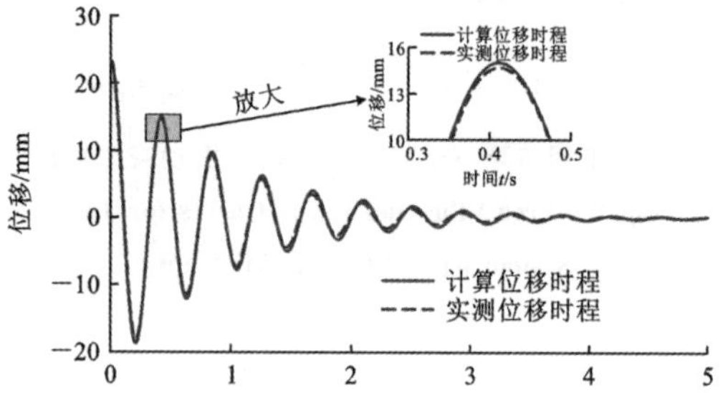

(b) $n_{mpp}=4, h_g=20$ mm

图 9 计算及试验测试的位移时程曲线

Fig. 9 Time Histories for Calculations and Experimental Tests

信号。由图9可以看到,实测和计算的位移时程比较吻合,气隙较小时,误差稍大,与图8中阻尼系数的计算规律一致。

3 Halbach阵列板式电涡流阻尼单元的参数分析

Halbach阵列板式电涡流阻尼单元的阻尼特性主要受永磁体阵列参数、导体板参数和背铁等因素的影响。本文主要关注Halbach永磁体阵列的参数分析和优化问题,导体板和气隙等因素对阻尼特性的影响与传统板式电涡流阻尼器基本相同,这里不做深入研究。永磁体厚度h_m、磁极数n_{mp}、磁极对数n_{mpp}和极长比α是Halbach永磁体阵列的4个关键参数,其物理意义如图10所示。参数分析选取在电涡流调谐质量阻尼器中常用的导体板和永磁体尺寸,如表2所示。在基本参数中,垂直极化和纵向极化的永磁体厚度都为$h_m=20$ mm,沿运动方向的长度都为25 mm,即极长比$\alpha=0.5$。实际工程中,永磁体的安装空间一般是有限的,因此本节永磁体阵列沿运动方向的总长度保持不变,取$L=225$ mm。

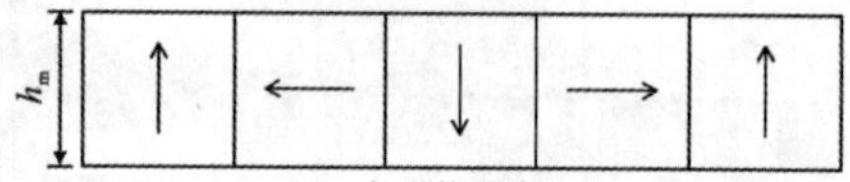

(a) 永磁体厚度h_m

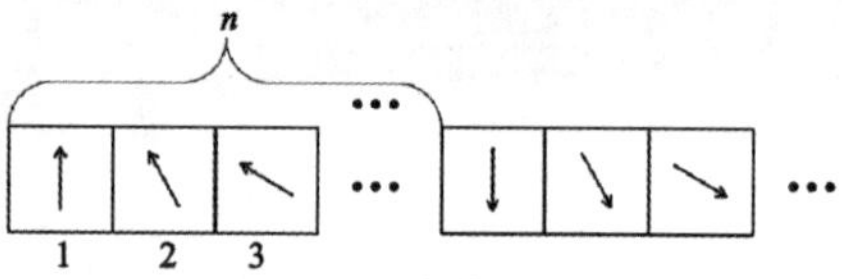

(b) 磁极数n_{mp}

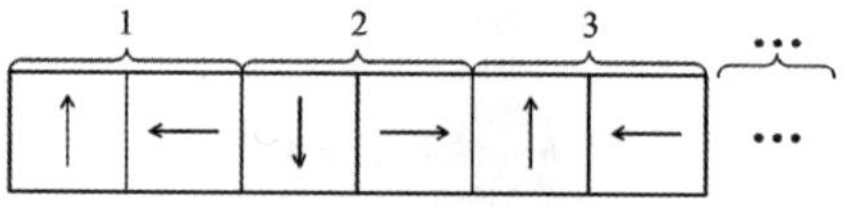

(c) 磁极对数n_{mpp}

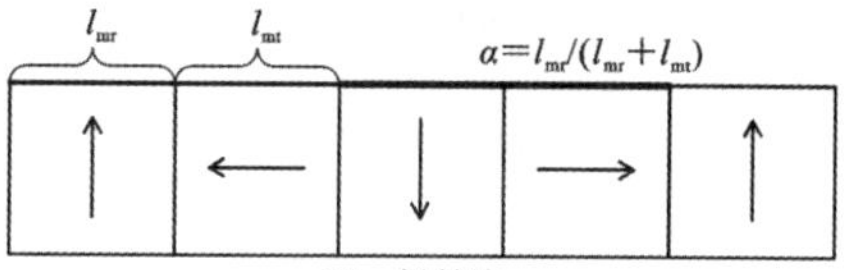

(d) 极长比α

图 10 Halbach阵列几何参数示意

Fig. 10 Schematic Representation of Geometric Parameters of Halbach Array

表 2 模型参数

Table 2 Model Parameters

名称	符号	值	名称	符号	值
导体板长度(平行运动方向)/mm	l_p	300	垂向极化永磁体厚度/mm	h_{mr}	20
导体板宽度(垂直运动方向)/mm	w_p	100	永磁体剩余磁感应强度/T	B_r	1.4
导体板厚度/mm	h_p	10	永磁体阵列磁极数	n_{mp}	2
导体板背铁厚度/mm	h_{pb}	10	永磁体阵列磁极对数	n_{mpp}	4
纵向极化永磁体长度(平行运动方向)/mm	l_{mt}	25	永磁体与导体板间隙/mm	d	10
纵向极化永磁体宽度(垂直运动方向)/mm	w_{mt}	40	导体板电导率/(MS·m^{-1})	σ_c	58
纵向极化永磁体厚度/mm	h_{mt}	20	空气、导体板相对磁导率	u_r	1
垂向极化永磁体长度(平行运动方向)/mm	l_{mr}	25	永磁体回复磁导率	u_{rm}	1.05
垂向极化永磁体宽度(垂直运动方向)/mm	w_{mr}	40	计算相对运动速度/(m·s^{-1})	v	0.05

由式(1)可知,在导体板几何尺寸、电磁属性以及相对运动速度不变的情况下,电涡流阻尼力的大小主要由导体板内垂直于导体板方向的磁通密度分量$\boldsymbol{B}_z$或磁通分量Φ_z以及电涡流密度决定。图11(a)为某参数下导体板下表面磁通密度分量$\boldsymbol{B}_z$的分布云图,由此可以看到在永磁体阵列投影区域的

磁通密度分量较大，占整个导体板的主要部分。图11(b)为永磁体阵列投影区域磁通密度分量 $\boldsymbol{B}_z$ 的平面分布，由此可以看到：在运动方向（x 轴方向）磁通密度分量 $\boldsymbol{B}_z$ 呈正弦分布；在垂直运动方向（y 轴方向），磁通密度分量 $\boldsymbol{B}_z$ 在中点最大，往两侧逐渐衰减。因此，可以用永磁体阵列投影区域沿运动方向中线的磁通密度分量 $\boldsymbol{B}_z$，或磁通量 Φ_z 的代表值定性解释各参数对阻尼系数的影响。$\boldsymbol{B}_z$ 和 Φ_z 的关系为

$$\Phi_z = \int \boldsymbol{B}_z \cdot \mathrm{d}l \tag{4}$$

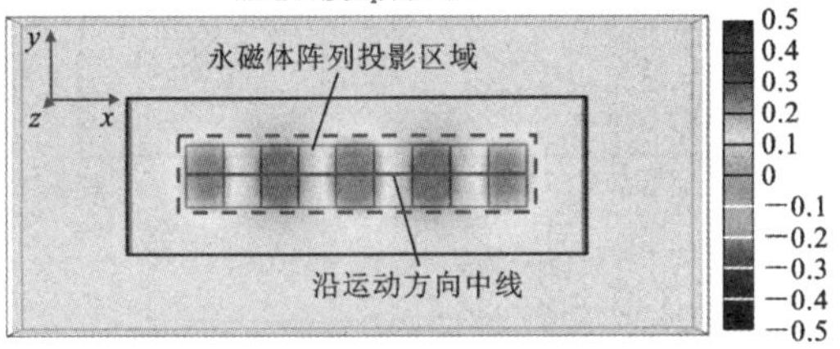

(a) 导体板下表面磁通密度$\boldsymbol{B}_z$分布

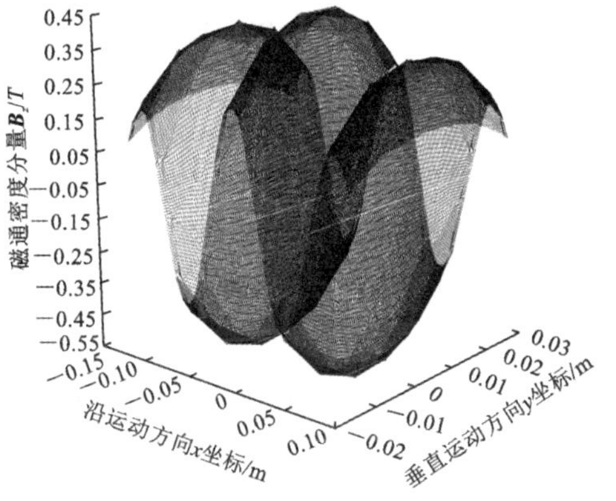

(b) 永磁体投影区域的$\boldsymbol{B}_z$二维分布

图 11 磁通密度分布

Fig. 11 Magnetic Flux Density Distribution

3.1 磁极数的影响

第 2 节的验证试验采用的是磁极数为 2 的 Halbach 永磁体阵列，本节采用电磁仿真方法进一步研究磁极数对阻尼系数的影响规律。固定永磁体阵列总长度 $L=225$ mm，保持磁极对数和极长比不变，研究了磁极数变化对阻尼系数的影响，结果如图12(a)所示。由此可以看到，无论是否有永磁体背铁，阻尼系数都随着磁极数 n_{mp} 的增加而先增加再逐渐趋于稳定。对于有背铁的工况，当磁极数 $n_{mp}=2$ 时，阻尼系数 $C_m=327$ N·s·m^{-1}，当磁极数增加到 $n_{mp}=3$ 时，阻尼系数增大 14%；当阻尼系数增至 $n_{mp}=10$ 时；阻尼系数增大 25%。当磁极数 $n_{mp}=2$ 时，无背铁工况的阻尼系数 $C_m=306$ N·s·m^{-1}，相对有背铁降低了 7%，远小于垂向极化板式电涡流阻尼单元中背铁的影响[10]，从而说明了 Halbach 阵列的自屏蔽作用。

为了分析磁极数对阻尼系数的影响机理，在图12(b)中计算了磁通量代表值 Φ_z 随磁极数的变化情况。由此可以看到，Φ_z 随磁极数的变化规律与 C_m 基本一致，从而说明磁极数引起的导体板内磁通量变化是引起板式电涡流阻尼器阻尼系数变化的主要原因。值得说明的是，针对永磁电机的研究也发现在 Halbach 阵列中采用多个磁极能有效改善气隙处磁通密度的波形[26]，从而提高电机的工作效率。但采用多磁极会大大增加永磁体的制造难度和加工成本问题，因此常用的 Halbach 阵列以两磁极式为主。

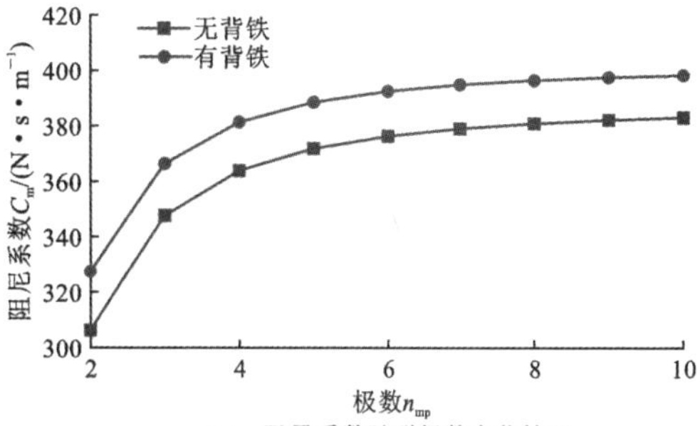

(a) 阻尼系数随磁极数变化情况

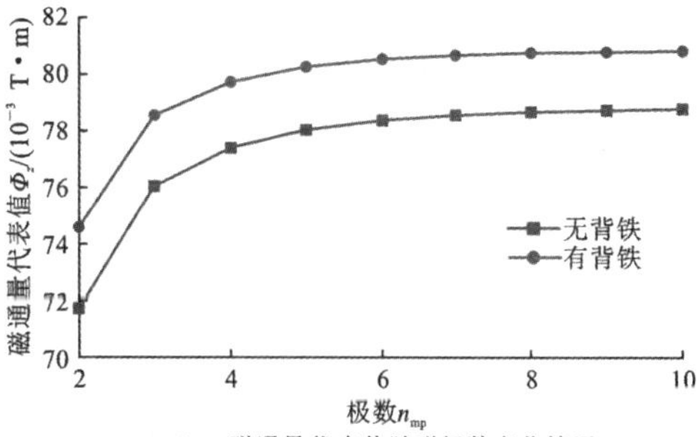

(b) 磁通量代表值随磁极数变化情况

图 12 磁极数的影响

Fig. 12 Effect of Number of Magnetic Poles

3.2 磁极对数的影响

固定永磁体阵列总长度 $L=225$ mm，保持磁极数 $n_{mp}=2$，极长比 $\alpha=0.5$，研究磁极对数变化对阻尼系数的影响，结果如图 13(a)所示。由此可以看到，无论是否有永磁体背铁，阻尼系数都随着磁极对数的增加而先增大再减小，当磁极对数 $n_{mpp}=3$ 时，永磁体构成一个半周期 Halbach 阵列，阻尼系数取到最大值。对比有无背铁的工况可以发现，在磁极对数 $n_{mpp}=1$ 时安装永磁体背铁可使阻尼系数增大60%左右，当磁极对数大于等于 4 时，背铁的影响几乎可以忽略不计。这主要是因为磁极对数较小时，

永磁体尚未构成完整的 Halbach 阵列，无背铁时产生严重漏磁。当磁极对数增多，构成完整的 Halbach 阵列以后，其磁场自屏蔽特性才能实现。

为了解释磁极对数对阻尼系数的影响机理，进一步分析了磁通量代表值 Φ_z 随磁极对数的变化规律，结果如图 13(b)所示，可以看到：当没有永磁体背铁时，Φ_z 随磁极对数的变化规律与阻尼系数变化规律基本一致；当有永磁体背铁时，Φ_z 随磁极对数的增大而逐渐减小，与阻尼系数的变化规律并不完全一致。为此，在图 14(a)、(b)中进一步给出了磁极对数分别为 1 和 3 时导体板下底面中线上的电涡流密度和磁通密度沿坐标轴的变化规律。由此可以看到：当磁极对数 $n_{mpp}=3$ 时，磁通密度和电涡流密度的波形都接近正弦波，且导体板左右两侧的边缘效应较小；当磁极对数 $n_{mpp}=1$ 时，磁通密度和电涡流密度的波形都严重偏离正弦波，电涡流密度的峰值也比 $n_{mpp}=3$ 小，且导体板左右两侧的边缘效应较强，从而导致 $n_{mpp}=1$ 时阻尼系数相对较小。

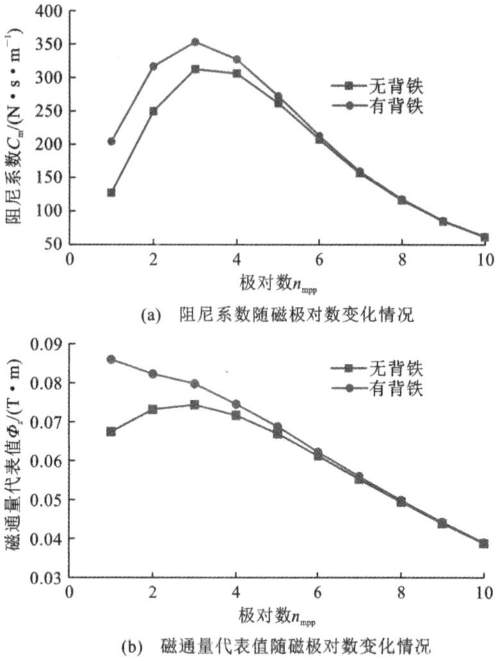

(a) 阻尼系数随磁极对数变化情况

(b) 磁通量代表值随磁极对数变化情况

图 13 磁极对数的影响

Fig. 13 Effect of Number of Magnetic Pole Pairs

3.3 极长比的影响

磁极数为 2 的 Halbach 阵列是垂向极化永磁体和纵向极化永磁体的组合，极长比是垂向极化永磁体长度与磁极对总长度的比值。固定永磁体阵列总长度 $L=225$ mm，固定单位磁极对长度 $L_p=50$ mm，图 15(a)展示了阻尼系数随极长比的变化规律，由图 15(a)可以看到，无论是否添加永磁体背

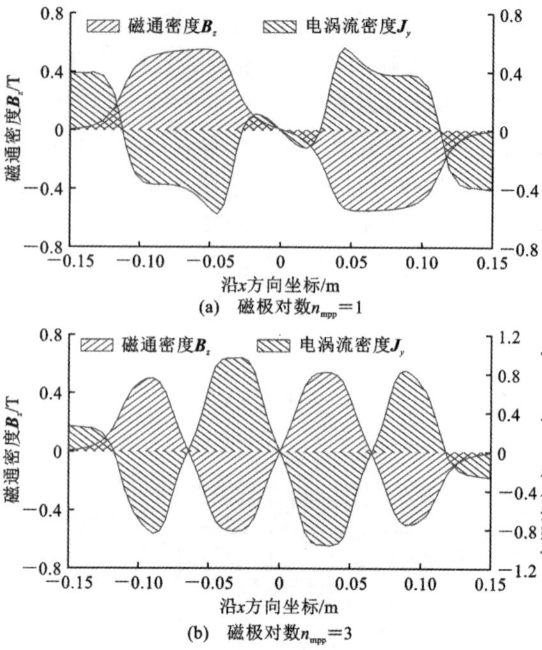

(a) 磁极对数$n_{mpp}=1$

(b) 磁极对数$n_{mpp}=3$

图 14 磁通密度及电涡流密度分布

Fig. 14 Magnetic Flux Density and Eddy Current Density Distribution

铁，阻尼系数 C_m 都是随着极长比 α 的增加而先增大再减小，但两者分别在极长比 $\alpha=0.5$ 和 0.7 时取得最大值。从图 15(b)可以发现，单位长度磁通量 Φ_z 与阻尼系数 C_m 的变化规律基本一致，但对于无背铁的工况，阻尼系数与磁通量的峰值对应的极长比分别为 0.5 和 0.6，存在微小差别。这一方面是因为，如式(1)～(2)所示，阻尼系数不仅与导体板上的磁通密度有关，还受到电涡流密度大小和分布情况的影响，因此，某些工况中磁通量的变化规律与阻尼系数不完全一致。另一方面，极长比的步长选取对阻尼系数与磁通量的峰值点也有一定影响。

通过对比图 15(a)中有无背铁的工况可以发现，在极长比 $\alpha=0.1\sim0.4$ 时，无背铁 Halbach 阵列的阻尼系数较大。这是因为当极长比 α 较小时，Halbach 阵列特性接近纵向极化阵列，此时永磁体背铁将原本是对称分布的磁场过多地引导到远离导体板的一侧，导致磁通密度降低，进一步导致阻尼系数降低。在极长比 $\alpha=0.4\sim0.9$ 时，有背铁 Halbach 阵列的阻尼系数较大。这是因为当极长比较大时，Halbach 阵列特性接近垂向极化阵列，永磁体背铁起到很好的导磁作用，使导体板中磁通密度增大，阻尼系数也随之增大。

3.4 永磁体厚度的影响

一般增大永磁体厚度能够有效地增大气隙磁通密度，进而提高阻尼系数。文献[11]在垂向极化阵

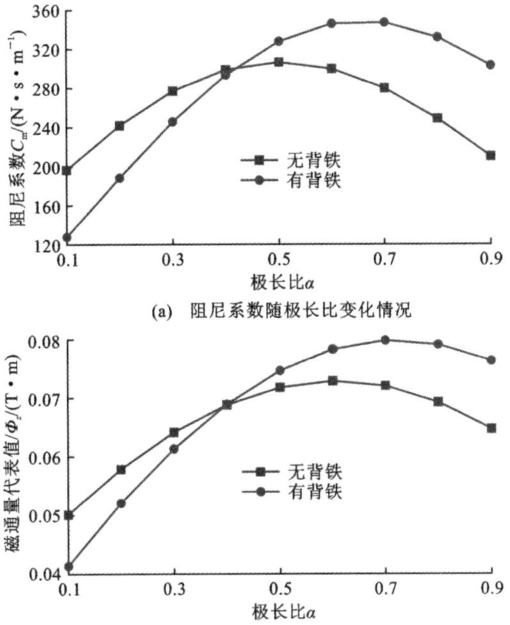

(a) 阻尼系数随极长比变化情况

(b) 磁通量代表值随极长比变化情况

图 15 极长比的影响

Fig. 15 Effect of Pole Length Proportion

列的板式电涡流阻尼单元中发现阻尼系数随着永磁体厚度的增大而逐渐增大至稳定。值得注意的是，永磁体厚度的增加意味着永磁阵列及整个板式电涡流阻尼单元体积的增大，进而可能引起单位体积的耗能能力降低。因此，以永磁体厚度 $h_m=5$ mm 为基准高度 h_0，引入等效阻尼系数来衡量耗能能力，其中等效阻尼系数计算公式如下

$$C_{me}=C_m\frac{h_0}{h_m} \tag{5}$$

图 16(a)给出了阻尼系数随永磁体厚度的变化情况，可以看到，无论是否添加永磁体背铁，阻尼系数 C_m 都是随着永磁体厚度 h_m 的增加而先增加再趋于稳定，背铁对阻尼系数的影响较小。图 16(b)给出了磁通量代表值 Φ_z 随永磁体厚度 h_m 的变化情况，可以发现 Φ_z 与 C_m 的变化情况基本一致。为了更准确地描述耗能能力，图 16(c)给出了等效阻尼系数 C_{me}随永磁体厚度的变化情况，可以看到，随着永磁体厚度的不断增大，等效阻尼系数先增大后减小，对于有、无背铁的方案，等效阻尼系数分别在永磁体厚度为 15 mm 和 20 mm 时达到最大值。因此，选择合适的永磁体厚度有利于提高 Halbach 板式电涡流阻尼单元的耗能能力。

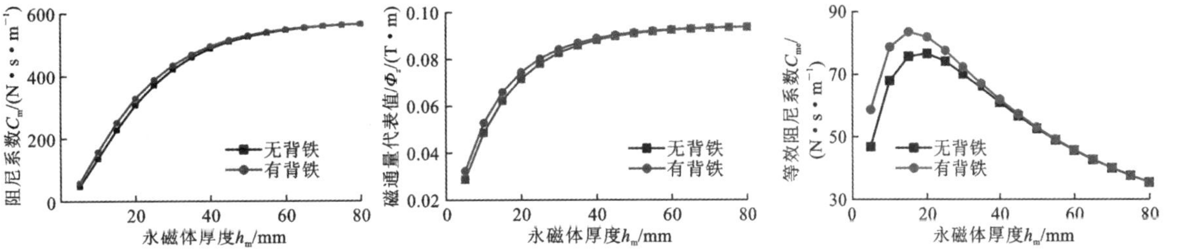

(a) 阻尼系数随永磁体厚度变化情况 (b) 磁通量代表值随永磁体厚度变化情况 (c) 等效阻尼系数随永磁体厚度变化情况

图 16 永磁体厚度的影响

Fig. 16 Effect of Permanent Magnet Height

4 案例优化设计

利用三维电磁有限元分析可获得 Halbach 永磁阵列的关键设计参数对板式电涡流阻尼单元阻尼性能的影响规律，但通过单参数分析难以获得最优的参数组合。另一方面，三维电磁有限元分析的网格数较多，计算时间较长，直接用于 Halbach 永磁体阵列的参数优化对计算机的算力要求非常高。为了提高多参数组合优化的计算效率，本节将二维电磁有限元分析与 MATLAB 遗传算法结合起来，以阻尼系数最大化为目标，对有背铁的 Halbach 阵列板式电涡流阻尼单元(方案 1)、无背铁的 Halbach 阵列板式电涡流阻尼单元(方案 2)、垂向极化阵列板式电涡流阻尼单元(方案 3)和纵向极化阵列板式电涡流阻尼单元(方案 4)等 4 种方案进行永磁阵列参数优化。

图 17(a)总结了上述 4 种方案的一般立面布置图，它们在各区域的材料特性如表 3 所示。区域Ⅰ的长度 l_{mr}、区域Ⅱ的长度 l_{mt}和磁极对数 n_{mpp}为待优化参数，永磁体阵列总长度 $l_l\leqslant 225$ mm，永磁体厚度考虑 $h_m=20$ mm，30 mm 和 50 mm 三种情况，导体板、气隙等其他设计参数的取值与表 2 相同。值得说明的是，采用二维平面电磁有限元方法计算阻尼系数假定了永磁体的宽度为无限长，而忽略了永磁体宽度有限引起的边缘效应。二维平面电磁有限元计算中的网格划分如图 17(b)所示。实际应用中，在垂直于相对运动方向上同列永磁体按相同极性布置以提高耗能能力[11,13][图 17(c)]，因此当永磁体沿垂直运动方向的总长度较大，或永磁体数量较多时，基于二维平面电磁有限元分析开展参数优化是可行的。

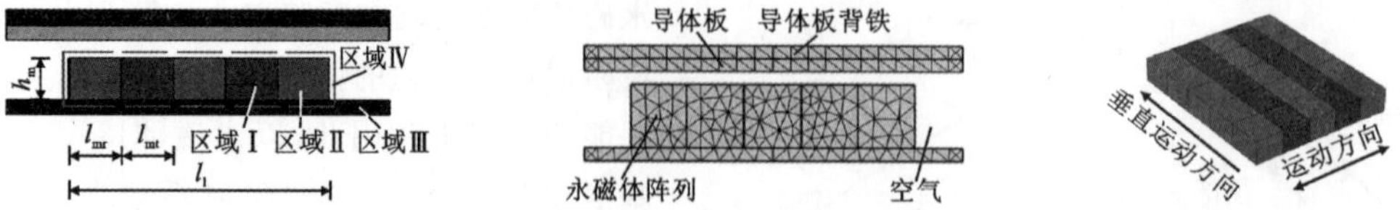

(a) 板式电涡流阻尼单元立面布置 (b) 板式电涡流阻尼单元二维网格划分 (c) 永磁体阵列布置示意

图 17 板式电涡流阻尼单元二维计算模型

Fig. 17 Two-dimensional Calculation Model for Eddy Current Damping Element

表 3 板式电涡流阻尼单元的 4 种设计方案

Table 3 Four Design Strategies for Eddy Current Damping Element

方案	区域Ⅰ	区域Ⅱ	区域Ⅲ
方案 1	纵向极化永磁体	垂向极化永磁体	高导磁钢材
方案 2	纵向极化永磁体	垂向极化永磁体	不导磁钢材
方案 3	空气	垂向极化永磁体	高导磁钢材
方案 4	纵向极化永磁体	导磁钢材	空气

优化计算的基本流程如图 18 所示，其中遗传算法的基本参数设置如表 4 所示，参数优化的数学模型可概括为方程(6)，MATLAB 遗传算法和电磁有限元计算软件 COMSOL 之间通过程序自动交互数据。

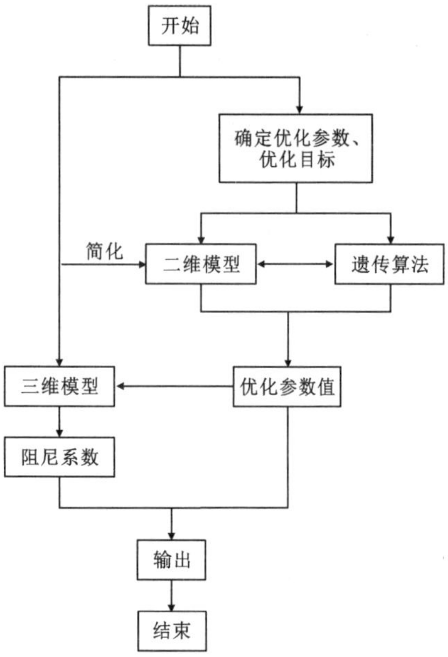

图 18 优化流程

Fig. 18 Optimization Flow Chart

$$\left.\begin{aligned}&\max(C_m(X))\\&X=(l_{mr},l_{mt},n_{mpp})\\&l_{mr}>0\\&l_{mt}>0\\&n_{mpp}=1,2,\cdots\\&l_{mr}(n_{mpp}+1)+l_{mt}n_{mpp}\leqslant l_1\end{aligned}\right\}\qquad(6)$$

假定单块永磁体宽度为 50 mm，每列 15 块永磁体在垂直于运动方向上紧密布置，即永磁体总宽度为 750 mm。针对该算例，先通过二维模型获得各方案的最优参数，再根据永磁体的实际总宽度建立三维电磁有限元模型，计算板式电涡流阻尼单元的准确阻尼系数，结果如表 5 所示。此外，图 19 以 h_m=20 mm 的永磁体为例给出了 4 种方案的布置立面图。

表 4 遗传算法参数设置

Table 4 Genetic Algorithm Parameter Settings

参数选项	设置值	参数选项	设置值
种群规模	40	杂交率	0.8
迭代数限值	300	突变方法	高斯分布法
选择方法	随机均匀分布	适应度尺度	排序法
直接传递率	0.05		

从阻尼系数来看，4 种方案的最优阻尼系数都随着永磁体厚度增大而逐渐增大，且有方案 1 大于方案 3 大于方案 2 大于方案 4，即有背铁的 Halbach 阵列板式电涡流阻尼单元的阻尼系数最大，而纵向极化板式电涡流阻尼单元的阻尼系数最小。特别地，有背铁 Halbach 阵列的最优磁极对数为 1，如图 19(a)所示，此时尚未构成完整的 Halbach 阵列。从参数优化结果来看：方案 1 和方案 2 的磁极对数优化值始终保持不变，但垂向极化和纵向极化永磁体的最优长度则随永磁体厚度不断变化；方案 3 的最优参数不随永磁体厚度变化，但方案 4 的最优磁极对数和永磁体长度都随永磁体厚度变化。值得说明的是，永磁阵列的优化参数与导体板和气隙厚度有关，同时也受永磁体总长度和单块永磁体最大长度等因素的影响，因此由本案例得到的参数优化结果不一定适用于其他约束条件。实际应用中可对参数优化的约束条件进行调整，获得更合理的优化解。

与参数分析相比，通过参数优化设计可显著提升各类板式电涡流阻尼单元的耗能能力。以有背铁的 Halbach 阵列板式电涡流阻尼单元为例，对于永磁体厚度 h_m=20 mm，磁极数 n_{mp}=2，永磁体阵列总长度 L=225 mm 的工况，取第 3 节参数分析得到最优极对数 n_{mpp}=3，最优极长比 α=0.5，计算得到其阻尼系数 C_m=17.7 kN·s·m^{-1}，相对方案1的

表 5　参数优化结果

Table 5　Results of Parametric Optimization

方案	h_m/mm	20		30		50		
	参数名称	优化参数	阻尼系数/(kN·s·m^{-1})	优化参数	阻尼系数/(kN·s·m^{-1})	优化参数	阻尼系数/(kN·s·m^{-1})	耗能密度/(kN·s·m^{-4})
方案Ⅰ	l_{mr}/mm	104	29.4	100	41.7	94	58.2	3.832
	l_{mt}/mm	17		25		37		
	n_{mpp}	1		1		1		
方案Ⅱ	l_{mr}/mm	43	15.4	43	25.9	33	42.2	2.779
	l_{mt}/mm	48		48		63		
	n_{mpp}	2		2		2		
方案Ⅲ	l_{mr}/mm	112	28.3	112	38.8	112	50.5	3.325
	l_{mt}/mm	0		0		0		
	n_{mpp}	1		1		1		
方案Ⅳ	l_{mr}/mm	3	9.2	7	16.4	9	30.4	2.002
	l_{mt}/mm	71		102		99		
	n_{mpp}	3		2		2		

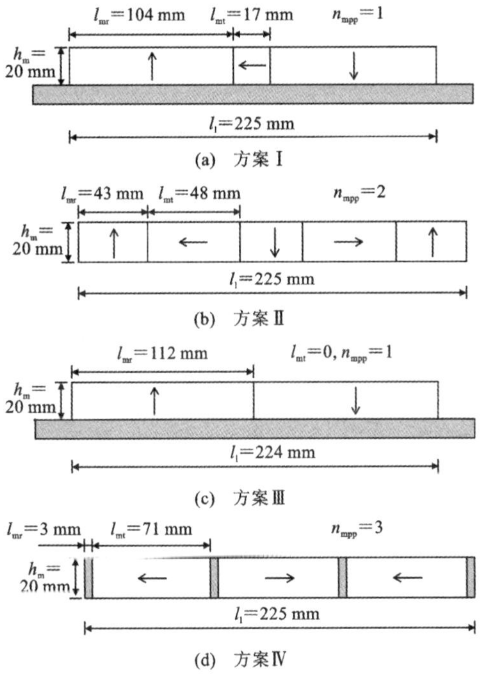

图 19　不同方案优化永磁阵列参数示意(h_m=20 mm)

Fig. 19　Schematics of Optimal Magnet Array Configuration for Different Eddy Current Damping Elements (h_m=20 mm)

优化结果小了 39.8%。需要说明的是，为了减小二维模型的计算误差，保证阻尼系数的计算精度，表 5 所示的最优阻尼系数是在使用二维模型确定最优参数后，通过三维有限元模型计算得到的。二维有限元计算的阻尼系数为 43.742 kN·s·m^{-2}，乘永磁体阵列宽度 750 mm，得到的三维阻尼系数为 32.807 kN·s·m^{-1}，比直接将优化参数代入三维计算的阻尼系数 29.4 kN·s·m^{-1}大 10.4%，误差在可接受范围，说明了简化计算的可行性。

5　结　语

本文分析了磁极数、磁极对数、极长比和永磁体厚度对 Halbach 阵列板式电涡流阻尼器阻尼系数的影响规律，建立了基于二维电磁场有限元分析与 MATLAB 遗传算法相结合的板式电涡流阻尼单元永磁阵列参数优化算法，可进一步提高板式电涡流阻尼单元的耗能能力，减小电涡流阻尼器的体积和质量，进而提升其在大跨度桥梁等土木工程结构中的应用范围。主要结论如下：

(1)永磁体阵列总长度不变的情况下，阻尼系数随着 Halbach 阵列磁极数量的增大而增大，但增速逐渐减慢。在一定范围内增加永磁体厚度能显著提高 Halbach 阵列板式电涡流阻尼单元的耗能能力。

(2)设置永磁体背铁可提高 Halbach 阵列板式电涡流阻尼单元的阻尼系数，但永磁体厚度较大或磁极数较多时，其影响逐渐减弱。

(3)从耗能能力来看，有背铁 Halbach 阵列板式电涡流阻尼单元大于垂向极化永磁阵列板式电涡流阻尼单元大于无背铁 Halbach 阵列板式电涡流阻尼单元大于纵向极化永磁阵列板式电涡流阻尼单元。

(4)与参数分析相比，结合二维电磁有限元分析与 MATLAB 遗传优化算法，可大幅提升板式电涡流阻尼单元的耗能能力。

为了得到阻尼力的快速计算方法以及考察各个

参数对阻尼力的影响，后续研究可以基于磁荷镜像法等各种方法推导 Halbach 阵列的板式电涡流阻尼单元的解析计算公式。此外，在优化设计中，可以根据实际应用需求对约束条件进行调整，如限制永磁体的最大最小尺寸，以获得更合理的优化解。

参考文献：

References：

[1] SODANO H A, BAE J S. Eddy current damping in structures [J]. Shock and Vibration Digest, 2004, 36 (6): 469-478.

[2] SHIN H J, CHOI J Y, CHO H W, et al. Analytical torque calculations and experimental testing of permanent magnet axial eddy current brake [J]. IEEE Transactions on Magnetics, 2013, 49 (7): 4152-4155.

[3] DANOEL V, IKER M, INIGO A, et al. Modeling eddy current brake emissions for electromagnetic compatibility with signaling devices in high-speed railways [J]. IEEE Transactions on Vehicular Technology, 2017, 66 (11): 9743-9752.

[4] SODANO H A, BAE J S, INMAN D J, et al. Concept and model of eddy current damper for vibration suppression of a beam [J]. Journal of Sound Vibration, 2005, 288 (4-5): 1177-1196.

[5] 陈政清，华旭刚，牛华伟，等. 永磁电涡流阻尼新技术及其在土木工程中的应用[J]. 中国公路学报，2020，33(11)：83-100.
CHEN Zheng-qing, HUA Xu-gang, NIU Hua-wei, et al. Technological innovations in eddy current damping and its application in civil engineering [J]. China Journal of Highway and Transport, 2020, 33 (11): 83-100.

[6] 华旭刚，黄智文，陈政清. 大跨度悬索桥的多阶模态竖向涡振与控制[J]. 中国公路学报，2019，32(10)：115-124.
HUA Xu-gang, HUANG Zhi-wen, CHEN Zheng-qing. Multi-mode vertical vortex-induced vibration of suspension bridges and control strategy [J]. China Journal of Highway and Transport, 2019, 32 (10): 115-124.

[7] WEN Q, HUA X G, CHEN Z Q, et al. Control of human-induced vibrations of a curved cable-stayed bridge: Design, implementation, and field validation [J]. Journal of Bridge Engineering, 2016, 21 (7): 04016028.

[8] WANG L K, NAGARAJAIAH S, SHI W X, et al. Study on adaptive-passive eddy current pendulum tuned mass damper for wind-induced vibration control [J]. The Structural Design of Tall and Special Buildings, 2020, 20: e1793. 1-e1793. 23.

[9] LU X L, ZHANG Q, WENG D G, et al. Improving performance of a super tall building using a new eddy-current tuned mass damper [J]. Structural Control and Health Monitoring, 2016, 24 (3): e1882. 1-e1882. 17.

[10] HUANG Z W, HUA X G, CHENG Z Q, et al. Modeling, testing, and validation of an eddy current damper for structural vibration control [J]. Journal of Aerospace Engineering. 2018, 31 (5): 04018063.

[11] ZHANG, H Y, CHEN Z Q, HUA X G, et al. Design and dynamic characterization of a large-scale eddy current damper with enhanced performance for vibration control [J]. Mechanical Systems and Signal Processing, 2020, 145: 106879.

[12] 陈政清，张弘毅，黄智文. 板式电涡流阻尼器有限元仿真与参数优化[J]. 振动与冲击，2016，35(18)：123-127.
CHEN Zheng-qing, ZHANG Hong-yi, HUANG Zhi-wen. FEM simulation and parameter optimization of a planar eddy current damper [J]. Journal of Vibration and Shock, 2016, 35 (18): 123-127.

[13] 李亚峰，李寿英，王健钟，等. 齿轮齿条式电涡流阻尼器的力学性能研究[J]. 土木工程学报，2020，53(3)：44-50.
LI Ya-feng, LI Shou-ying, WANG Jian-zhong, et al. Study on mechanical properties of eddy current damping-rack and gear damper [J]. China Civil Engineering Journal, 2020, 53 (3): 44-50.

[14] 陈政清，田静莹，黄智文，等. 板式电涡流阻尼系数的计算与试验修正方法[J]. 中国公路学报，2016，29(10)：46-53.
CHEN Zheng-qing, TIAN Jing-ying, HUANG Zhi-wen, et al. Calculation and test correction method of plane type eddy current damping coefficient [J]. China Journal of Highway and Transport, 2016, 29(10): 46-53.

[15] 黄智文. 电涡流阻尼器理论研究及其在桥梁竖向涡振控制中的应用[D]. 长沙：湖南大学，2016.
HUANG Zhi-wen. Theoretical study of eddy current damper and its application in vertical vortex-induced vibration control of bridges [D]. Changsha: Hunan University, 2016.

[16] 汪志昊，李国豪，周佳贞，等. 竖向 TMD 用板式电涡流阻尼器磁路对比分析[J]. 振动与冲击，2019，38(7)：233-239.
WANG Zhi-hao, LI Guo-hao, ZHOU Jia-zhen, et al. Contrastive analysis for magnetic circuit of a planar eddy current damper used in a vertical TMD [J]. Journal of Vibration and Shock, 38 (7): 233-239.

[17] 汪志昊，李国豪，田文文，等. 多层板式电涡流阻尼器构型与磁路优化分析[J]. 振动与冲击，2019，38(19)：103-108.
WANG Zhi-hao, LI Guo-hao, TIAN Wen-wen, et al. Configuration and magnetic circuit optimization of a multi-layer planar eddy current damper [J]. Journal of Vibration and Shock, 2019, 38 (19): 103-108.

[18] 王梁坤，施卫星，鲁　正. 电涡流 TMD 阻尼特性的有限元模拟与参数优化[J]. 结构工程师，2019，35(4)：98-104.
WANG Liang-kun, SHI Wei-xing, LU Zheng. Damping characteristics finite element simulations and parameters optimization of the eddy current TMD [J]. Structural Engineers, 2019, 35 (4): 98-104.

[19] HALBACH K. Design of permanent multipole magnets with oriented rare earth cobalt material [J]. Nuclear Instruments and Methods, 1980, 169 (1): 1-10.

[20] HALBACH K. Application of permanent magnets in accelerators and electron storage rings [J]. Journal of Applied Physics, 1984, 57: 3605-3608.

[21] ATALLAH K, HOWE D. The application of Halbach cylinders to brushless AC servo motors [J]. IEEE Transactions on Magnetics, 1998, 34 (4): 2060-2062.

[22] JANG S M, JEONG S S, CHA S D. The Application of linear Halbach array to eddy current rail brake system [J]. IEEE Transactions on Magnetics, 2001, 37 (4): 2627-2629.

[23] LUBIN T, REZZOUG A. Steady-state and Transient performance of axial-field eddy-current coupling [J]. IEEE Transactions on Industrial Electronics, 2014, 62 (4): 2287-2296.

[24] WANG J, LIN H Y, FANG S H, et al. A general analytical model of permanent magnet eddy current couplings [J]. IEEE Transactions on Magnetics, 2013, 50 (1): 1-9.

[25] JIAN L N, CHAU K T. A coaxial magnetic gear with Halbach permanent-magnet arrays [J]. IEEE Transactions on Energy Conversion, 2010, 25 (2): 319-328.

[26] 朱德明，严仰光. 离散 Halbach 永磁电动机气隙磁通密度特点及其空载电动势波形优化[J]. 电工技术学报，2008，23(11)：6. ZHU De-ming, YAN Yang-guang. Features of air-gap flux density in segmented Halbach permanent magnet synchronous motor and Its no-load EMF waveform optimization [J]. Transactions of China Electrotechnical Society, 2008, 23 (11): 6.

[27] SHEN Y, LIU G Y, XIA Z P, et al. Determination of maximum electromagnetic torque in pm brushless machines having two-segment Halbach array [J]. IEEE Transactions on Industrial Electronics, 2013, 61(2): 718-729.

[28] BIRD J Z, LIPO T A. Characteristics of an electrodynamic wheel using a 2-D steady-state model [J]. IEEE Transactions on Magnetics, 2007, 43 (8): 3395-3405.

[29] ZHANG J H, ZENG Y W, CHENG J, et al. Optimization of permanent magnet guideway for HTS maglev vehicle with numerical methods [J]. IEEE Transactions on Applied Superconductivity, 2008, 18 (3): 1681-1686.

[30] JANG S M, LEE S H, CHO S K, et al. Comparison of three types of permanent magnet linear eddy-current brakes according to magnetization pattern [J]. IEEE Transactions on Magnetics, 2003, 39 (5): 3004-3006.

[31] BISSAL A, SALINAS E, MAGNUSSON J, et al. On the design of a linear composite magnetic damper [J]. IEEE Transactions on Magnetics, 2015, 51 (11): 1-5.

Engineering Structures 299 (2024) 117102

Contents lists available at ScienceDirect

Engineering Structures

journal homepage: www.elsevier.com/locate/engstruct

Study on enhanced performance of CuFe alloy's eddy current damping and its advantages in multi-mode vibration control of stay-cable

Wenbin Yang [a,b], Hongyi Zhang [a,b,*], Zhengqing Chen [a,b], Xugang Hua [a,b], Huawei Niu [a,b], Shouying Li [a,b]

[a] Key Laboratory for Wind and Bridge Engineering of Hunan Province, Hunan University, Changsha 410082, China
[b] Vibration and Shock Technology Research Center, College of Civil Engineering, Hunan University, Changsha 410082, China

ARTICLE INFO

Keywords:
CuFe alloy
Eddy current damper
Bilinear characteristic
Mitigation vibration
Multi-mode control

ABSTRACT

This study presents an innovative solution to address the performance degradation of eddy current dampers (ECD) at high speeds. It replaces the traditional copper conductor with a magnetically and electrically conductive CuFe alloy. The damping force equation for this CuFe alloy-based ECD (CuFe-ECD) is derived and validated through rotor experiments and a simplified finite element model. Results confirm that CuFe alloy enhances damping performance, with CuFe20 exhibiting outstanding quasi-bilinear characteristics, particularly at higher speeds. CuFe10, CuFe15, and CuFe30 also demonstrate similar behavior. The study explores the impact of CuFe alloy properties (conductivity, permeability, and thickness) on the bilinear damping curve. CuFe alloys notably reduce magnetic reluctance, especially in thicker conductors, mitigating skin effect and armature reaction in the electromagnetic field. Finally, the CuFe-ECD is utilized for multi-mode vibration control using the long cable of the Sutong Bridge. The results indicate that the modal damping ratios of modes 1st-20th can reach 5‰, which is superior to other schemes. In high-order modes, the damping ratio of the 35th-40th modes still stabilizes at 3.72–3.90‰. It can be concluded that the bilinear damper has significant advantages in multi-mode vibration control: compared with the single linear damper scheme, it can improve the vibration reduction effect of multi-mode control, and compared with the dual linear damper scheme, it can reduce the number of dampers and achieve better economic benefits.

1. Introduction

In recent decades, the development of lightweight and high-strength materials has led to civil structures that exhibit low-frequency and low-damping vibration characteristics. While these structures meet aesthetic and spanning requirements, they impose higher demands on structural security and serviceability. This evolving development has rendered structures increasingly susceptible to severe vibrations under the influence of strong earthquakes and wind loads, ultimately culminating in potential engineering accidents [1–4]. Incorporating damping devices into structures or their components for energy dissipation has gradually become a well-established engineering practice for vibration mitigation in the past decades. Damping mechanisms that hinge on sliding friction, material yielding, viscous fluid, magnetorheological fluid, and particle collision have been subject to extensive inquiry and application within actual buildings and bridges [5–10]. These damping techniques exhibit their respective strengths, limitations, and domains of applicability, all of which merit further exploration.

The eddy current damper (ECD) represents a highly effective passive vibration control measure that has undergone significant development and application in recent years [11–15]. According to Faraday's law of electromagnetic induction, changes in magnetic flux passing through a closed circuit give rise to induced eddy currents within conductors. Low-resistivity conductors, which act as carriers of these eddy currents, transform them into heat, and the greater the eddy current power, the higher the efficiency of heat generation. From a mechanical standpoint, eddy current damping force impedes changes in magnetic flux, and its product with displacement yields the kinetic energy consumed by damping force. By applying the law of conservation of energy, the consumed kinetic energy equates to the heat energy generated by conductors, thereby achieving unity between mechanics and electromagnetics. Currently, ECD technology has been widely applied across various industrial fields, such as in the domains of vehicles, rails, and aviation.

* Corresponding author at: Key Laboratory for Wind and Bridge Engineering of Hunan Province, Hunan University, Changsha 410082, China.
E-mail address: zhyzzb25@hnu.edu.cn (H. Zhang).

https://doi.org/10.1016/j.engstruct.2023.117102
Received 12 May 2023; Received in revised form 5 October 2023; Accepted 28 October 2023
Available online 15 November 2023
0141-0296/

In civil engineering research, the initial approach involved directly installing ECDs on target structures as the primary method [16–19]. Sodano et al. [20] explored the application of an ECD to mitigate vibrations in a cantilever beam. They established an analytical model to assess the device's effectiveness in vibration control, followed by a model test that confirmed its exceptional performance. Larose et al. [21] developed a unilateral eddy current-tuned mass damper to control vortex-induced vibrations in a bridge aeroelastic model. To enhance the energy dissipation density of ECD, researchers have conducted topology and parameter optimization of the magnetic circuit [22–25]. Zhang et al. [14] achieved a 140–2200-fold increase in the damping coefficient using a ball screw device, addressing the low energy dissipation density issue of ECD. ECDs offer advantages such as contactless, velocity-dependent operation, effective vibration suppression without altering structural frequency characteristics, durability, stability, sensitivity to adjustments, and precise control of the damping coefficient through magnetic field strength adjustments, owing to the absence of working fluids, seals, and contact friction.

Industrial copper is a prevalent choice for ECD due to its excellent electrical and thermal conductivity. However, traditional conductor materials like copper face a significant constraint – magnetic resistance. This means that increasing conductor thickness to boost the eddy current damping coefficient leads to higher magnetic resistance and reduced magnetic flux density. This limitation hampers the progress of ECD's energy consumption density. To tackle this issue, this study explores CuFe alloy as an alternative conductor material for eddy current dampers in vibration control. Enhancing copper's strength while preserving its electrical conductivity is a key challenge, achieved through various strengthening methods. Incorporating magnetic conductivity into the conductor tube not only amplifies damping at low speeds but also mitigates skin and armature effects at high speeds. The skin effect refers to the distribution of eddy currents within a conductor, where the current density decreases from the surface to the interior of the conductor. The higher the velocity, the more non-uniform this distribution becomes. The armature effect, on the other hand, relates to an increase in the reverse magnetic field generated by eddy currents with increasing velocity. This effect weakens the magnetic field within the air gap and introduces a phase difference between this reverse magnetic field and the static magnetic field, leading to field distortion. Both the skin effect and the armature effect collectively result in the nonlinear behavior of eddy current damping forces at high speeds. The manifestation of skin and armature effects results in the rapid attainment of critical damping force values and a steep decline at high speeds, leading to a marked reduction in vibration mitigation efficacy. Furthermore, the low stiffness of copper conductors at high operating temperatures leads to significant degradation of material properties in application domains that require high material strength, such as buffer and brake systems [26–28].

CuFe alloys constitute a significant emerging industrial material that has found wide-ranging applications in aerospace, energy, and machinery due to their exceptional mechanical and high-temperature corrosion-resistant properties. Existing research has made significant strides in characterizing Cu-Fe alloy properties, including particle size and conductivity, the correlation between lattice strain and conductivity, and the influence of composition and powder size on magnetic properties [29–31]. CuFe alloys possess not only high strength, elasticity, and wear resistance but also superior plasticity, thermal conductivity, electrical conductivity, enormous magnetostriction effects, and unique physical properties. Notably, the introduction of the ferromagnetic material iron can significantly alter the magnetic circuit characteristics of ECDs, and the damping force–velocity constitutive characteristics of the conductor tube may exhibit more stable characteristics [32–35].

In the realm of cable vibration control, traditional linear dampers possess the merits of simple design and manufacturing. They excel in parameter adjustments based on classical linear vibration theory [36–38]. Nevertheless, their fixed damping characteristics limit adaptability to external load variations, resulting in restricted effectiveness in multi-mode control and a finite number of controllable orders. Hoang and Fujino [39] explored the combined influence of two linear dampers, revealing that appropriate positioning leads to a damping effect nearing the sum of individual damper contributions. While combined dampers offer increased controllable orders, they come at a higher economic cost compared to single dampers. Nonlinear dampers, tailored to structural serviceability, excel in managing large amplitude vibrations, outperforming linear counterparts. Liu et al. [40] delved into a nonlinear viscous-shear damper with viscosity and stiffness coefficients dependent on frequency and amplitude. Nonetheless, currently employed nonlinear dampers, such as viscous fluid and magneto-rheological dampers, contend with issues like fluid leakage and diminishing effectiveness. Wang et al. [13] introduced a novel ECD system to concurrently suppress cable vibration responses. The study demonstrated ECD's capacity to provide adequate modal damping ratios for multi-mode, high-order vibrations in ultra-long cables, effectively curbing vibration responses. Thus, this research elevates ECD's nonlinear attributes through CuFe alloys, augmenting its performance in multi-modal cable control and validating its efficacy against alternative damper schemes.

The highlight of this article is the proposal of a CuFe-ECD with enhanced damping efficiency and its use for multimodal vibration control with a bilinear constitutive curve. Section 2 establishes a damping force theory calculation formula compatible with CuFe-ECD, based on the theory of time-varying electromagnetic fields. In Section 3, a comprehensive experimental comparison assesses CuFe alloy conductors with varying thickness and iron proportions against copper and aluminum alloy conductors. This evaluation is performed by testing the damping-speed constitutive curve. Section 4 introduces a simplified finite element model (FEM) of ECD, comparing results from three damping force calculation methods: FED, experimental data, and theoretical formula. Additionally, a parameter analysis investigates CuFe alloy conductivity, magnetic conductivity, and thickness effects on the bilinear damping constitutive curve. Section 5 employs this approach to control multimodal vibration on Sutong Bridge's long cable, employing a CuFe alloy eddy current damper with bilinear properties. Section 6 critically discusses the study's significance and implications for future research.

2. Theoretical formula of damping force for CuFe-ECD

2.1. Assumptions of the simplified model

Given the introduction of iron in the conductor material, the CuFe alloy may exhibit a distinct damping force–velocity constitutive curve. It is imperative to investigate its nonlinear properties and potential engineering applications. Based on the basic electromagnetic field assumption, a two-dimensional theoretical formula for the damping force of CuFe-ECD has been developed to analyze its dynamic characteristics. This endeavor holds pragmatic importance for the design of damper parameters.

Building upon the author's prior work [14], the mechanical transmission scheme of CuFe-ECD persists in utilizing an external rotational system comprised of a composite tube, permanent magnets (PMs), a magnetic holder, and a ball screw to augment energy efficiency and operational stability, as depicted in Fig. 1. The rotor-like composite tube encompasses an internal conductor tube and an external steel casing. Fashioned from magnetic materials, the magnetic holder secures PMs in alternating positions encircling the N and S poles, constituting the stator of the damping mechanism. An air gap is present between the PMs and the inner wall of the conductor tube, which is maintained by the radial bearing. The fundamental operating principle involves converting axial motion into rotary motion via the ball screw. As the composite tube revolves around the stator, the conductor tube traverses magnetic field lines, generating eddy currents and damping torque. Subsequently, the

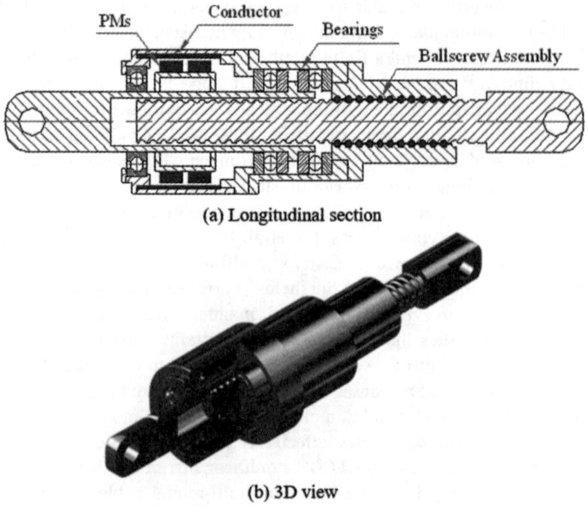

(a) Longitudinal section

(b) 3D view

Fig. 1. Structural drawing of ball screw type external rotating eddy current axial damper [14].

transmission system imparts axial damping forces to the vibration system. This study preserves the established ECD structural framework while integrating diverse experiments, structural computation models, and numerical methodologies.

The CuFe alloy, with iron mass content varying from 5% to 50%, exhibits remarkable attributes encompassing strength, thermal conductivity, and bending resistance. When the iron content exceeds 20%, the alloy demonstrates exceptional magnetic and electromagnetic shielding characteristics. Its low thermal expansion coefficient and elevated wear resistance make it an advantageous material for ECD conductors. Due to its electrical and magnetic conductivity, the dynamic behaviors of CuFe alloy conductor materials within a rotating electromagnetic field are inherently unique. To accurately compute damping forces, an electromagnetic structure model is indispensable. The interplay between the magnetic field from the PMs and the counteractive eddy current-induced anti-magnetic field in the alloy conductor can be intricate in practical scenarios. Thus, simplifying the model is essential to enable electromagnetic field analysis and derivation of the damping force calculation formula for the rotating CuFe-ECD system.

The electromagnetic model's axial consistency allows for a 2-D simplification of the damping generation device. This simplified model focuses on the electromagnetic field distribution in a single section, as illustrated in Fig. 2(a). To characterize electromagnetic properties, the damping component section is divided into four distinct radial domains (I-steel holder, II-PMs, III-air gap, and IV-conductor tube), progressing from the interior to the exterior, as depicted in Fig. 2(b). The following assumptions are made:

- Low carbon steel exhibits a relative magnetic permeability of 200–800 when PMs' remanence intensity is 0.002 T, rendering it effectively infinite compared to air. Neglecting magnetic hysteresis and saturation in the PM material, the steel holder in domain I is assumed to have infinite relative permeability (i.e. $\mu_r^{I} = \infty$).
- All materials in each region are considered homogeneous and isotropic.
- In domain II, two materials coexist PMs and air. Air has a relative permeability of 1, lower than PMs' relative recoil permeability (1.02–1.05) made of NdFeB. For simplicity, the relative permeability is averaged between these materials, disregarding current in PMs.

2.2. Governing equation of CuFe-ECD

Since the damping component of ECD is designed as a cylinder, therefore, the cylindrical coordinate system with the coordinate (r, θ, z) is used to establish the equations in this study. According to the above assumptions in Section 2.1, all physical quantities are independent of the axial coordinate z in the two-dimensional model as shown in Fig. 2. Within the research scope of this study, the influence of displacement current can be ignored, that is, there is an instantaneous relationship between source and field in electricity and magnetism. The relationship between the source and field at each instantaneous time is similar to that in the static field. Therefore, this field is called the quasi-static field. The magnetic intensity $\boldsymbol{B}$ and the magnetization intensity $\boldsymbol{M}$ have only in-plane component: $\boldsymbol{B} = (B_r(r,\theta,t), B_\theta(r,\theta,t), 0)$, and $\boldsymbol{M} = (M_r(r,\theta,t), M_\theta(r,\theta,t), 0)$. For the quasi-static field, the Ampère-Maxwell law in media under Coulomb gauge fixing condition can be simplified as follows:

$$\nabla \times \boldsymbol{B} = \mu_{rc}\mu_0 \boldsymbol{J} \tag{1}$$

$$\boldsymbol{J} = \sigma \boldsymbol{E} + \nabla \times \boldsymbol{M} \tag{2}$$

where $\boldsymbol{J}$ and $\boldsymbol{E}$ represent the total current density and electric field strength, respectively. Meanwhile, μ_0 denotes the vacuum permeability, μ_{rc} is the relative permeability, and σ pertains to conductivity. By introducing the magnetic vector potential $\boldsymbol{A}$, the magnetic intensity $\boldsymbol{B}$ and the electric field intensity $\boldsymbol{E}$ can be expressed as follows:

$$\boldsymbol{B} = \nabla \times \boldsymbol{A} \tag{3}$$

$$\boldsymbol{E} = -\frac{\partial \boldsymbol{A}}{\partial t}. \tag{4}$$

Then, according to Eq. (1) and (3), the following formula can be obtained:

$$\mu_{rc}\mu_0\sigma\frac{\partial \boldsymbol{A}}{\partial t} = \left(0, 0, \frac{\partial B_\theta}{\partial r} + \frac{1}{r}\left(B_\theta - \frac{\partial B_r}{\partial \theta}\right)\right) + \left(0, 0, \frac{\partial M_\theta}{\partial r} + \frac{1}{r}\left(M_\theta - \frac{\partial M_r}{\partial \theta}\right)\right). \tag{5}$$

The magnetic vector potential $\boldsymbol{A}$ has only an out-plane component:

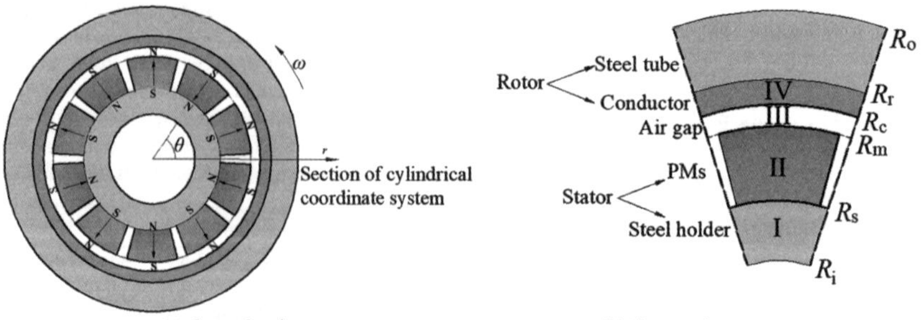

(a) Cross Section

(b) Geometric parameter

Fig. 2. Electromagnetic model of CuFe-ECD.

$\boldsymbol{A}$=(0, 0, $A_z(r,\theta,t)$). The Laplace operator of the magnetic vector potential $\boldsymbol{A}$ can be written as follows:

$$\nabla^2 \boldsymbol{A} = \left(0, 0, \frac{\partial^2 A_z}{\partial r^2} + \frac{1}{r}\frac{\partial A_z}{\partial r} + \frac{1}{r^2}\frac{\partial^2 A_z}{\partial \theta^2}\right). \tag{6}$$

Finally, according to Ampère-Maxwell law in potential form, the governing equations of I-IV calculation domains can be obtained:

$$\begin{cases} \nabla^2 \boldsymbol{A}^{\mathrm{I}} = 0 \\ \nabla^2 \boldsymbol{A}^{\mathrm{II}} = -\mu_0 \nabla \times M \\ \nabla^2 \boldsymbol{A}^{\mathrm{III}} = 0 \\ \nabla^2 \boldsymbol{A}^{\mathrm{IV}} = \mu_{rc}\mu_0 \sigma \dfrac{\partial \boldsymbol{A}^{\mathrm{IV}}}{\partial t} \end{cases}, \tag{7}$$

where A^{I}-A^{IV} represents the magnetic vector potential function of each calculation domain. In the calculation domains, namely domain I, and domain III, the governing equation reduces to the homogeneous formula since both the conductivity and magnetization are zero. On the other hand, in domain II, only the bound current term is represented on the right side of the governing equation as per the assumption, due to the conductivity being zero. Similarly, in domain IV, since there is no magnetization, only the free conduction current term is exhibited on the right side of the governing equation.

2.3. *Magnetization of PMs and boundary conditions of calculation domains*

The magnetization vector $\boldsymbol{M}_{re}$ of PMs can be resolved into its radial component $\boldsymbol{M}_r$ and tangential component $\boldsymbol{M}_\theta$:

$$\boldsymbol{M}_{re} = M_r \boldsymbol{e}_r + M_\theta \boldsymbol{e}_\theta, \tag{8}$$

where $\boldsymbol{e}_r$ and $\boldsymbol{e}_\theta$ represent radial and tangential unit vectors. Due to the uniform distribution of the PMs along the cycle and the alternating arrangement of N and S poles, the radial component of magnetization can be decomposed through Fourier transform. Here is the index representation of the aforementioned transformation:

$$\left[\vec{M}_r(\theta, t)\right] = Re\left[\sum_{n=1,3,5\cdots}^{\infty} M_{rn} e^{inp(\theta+\omega t)}\right], \tag{9}$$

where p represents the number of pole pairs, ω indicates the angular velocity, and $\vec{M}_r$ is a plural. Due to the presence of PMs with opposite poles in each cycle, the magnetization function takes on an odd form, leading to the Fourier series having only an odd term. The phase is expressed as $e^{inp(\theta+\omega t)}$, while the phase angle is denoted by $np(\theta + \omega t)$, both of which are linked to the spatial coordinates θ and time t. Performing a Fourier decomposition of the exponential form of Eq. (9) yields an expression for magnetization intensity containing only odd terms. This study employs a radial magnetization scheme for arc-shaped PMs, wherein each order component of magnetization is denoted as M_{rn} and $M_{\theta n}$, and is articulated as follows:

$$\begin{aligned} M_{rn} &= \frac{4B_{re}\sin\left(\dfrac{n\pi\alpha}{2}\right)}{\mu_0 n\pi}, \\ M_{\theta n} &= 0 \end{aligned} \tag{10}$$

where, α denotes the ratio of the arc length of the PMs to the arc length of the magnetic pole distance, which is commonly known as the pole arc coefficient.

The 2-D model in Fig. 2(b) illustrates that electric displacement (***D***) and magnetic induction (***B***) maintain normal continuity, while electric field (***E***) and magnetic field (***H***) exhibit tangential continuity at internal region interfaces. Due to the nearly infinite permeability of the magnetic steel holder and magnetic conducting tube, as well as the research focusing on the dynamic characteristics of the CuFe alloy conductor, the boundary conditions are disregarded in domain I and domain V. Boundary conditions for other interfaces and assumptions regarding conductivity and permeability are categorized as follows:

$$\begin{cases} \dfrac{1}{\mu_0} B_\theta^{\mathrm{II}}(r,\theta,t)\big|_{r=R_s} = 0 \\ B_r^{\mathrm{III}}(r,\theta,t)\big|_{r=R_m} = B_r^{\mathrm{II}}(r,\theta,t)\big|_{r=R_m} \\ \dfrac{1}{\mu_0} B_\theta^{\mathrm{III}}(r,\theta,t)\big|_{r=R_m} = \dfrac{1}{\mu_{rm}\mu_0} B_\theta^{\mathrm{II}}(r,\theta,t)\big|_{r=R_m} - \dfrac{1}{\mu_{rm}} M_{\theta n} \\ B_r^{\mathrm{III}}(r,\theta,t)\big|_{r=R_c} = B_r^{\mathrm{IV}}(r,\theta,t)\big|_{r=R_c} \\ \dfrac{1}{\mu_0} B_\theta^{\mathrm{III}}(r,\theta,t)\big|_{r=R_c} = \dfrac{1}{\mu_{rc}\mu_0} B_\theta^{\mathrm{IV}}(r,\theta,t)\big|_{r=R_c} \\ \dfrac{1}{\mu_{rc}\mu_0} B_\theta^{\mathrm{IV}}(r,\theta,t)\big|_{r=R_r} = 0 \end{cases}. \tag{11}$$

2.4. *Expression of damping force*

For a time-varying field, the CuFe alloy conductor in domain IV will generate an induced electric field, which in turn produces an induced magnetic field that is superimposed onto the original magnetic field. The interaction between the time-varying electric and magnetic fields results in a more intricate phenomenon compared to the static magnetic field. The induced magnetic field causes the original static magnetic field to fluctuate in strength at various positions, with these magnetic field distortions becoming even more pronounced during high-speed rotation. The superimposed magnetic field has an impact on the induced eddy current and subsequently on the damping force. Hence, it is imperative to acknowledge the distortion of the original magnetic field.

The equations governing each calculation domain can be derived from Eq. (7), akin to the static magnetic field. However, it is imperative to take into account the induced electric field term in calculation domain IV. Thus, the governing equation in this domain is articulated as follows:

$$\frac{\partial^2 A_{zn}^{\mathrm{IV}}}{\partial r^2} + \frac{1}{r}\frac{\partial A_{zn}^{\mathrm{IV}}}{\partial r} - \frac{(np)^2}{r^2} A_{zn}^{\mathrm{IV}} = inp\omega\sigma\mu_{rc}\mu_0 A_{zn}^{\mathrm{IV}}. \tag{12}$$

In comparison to the stationary magnetic field, the non-homogeneous terms are incorporated on the right-hand side of the governing equations for calculation domain IV, while the governing equations of other calculation domains remain unaltered. Eq. (12) represents a conventional Bessel differential equation, and its overall solution can be articulated through the Bessel function. Hence, the general solution can be presented as follows:

$$A_{zn}^{\mathrm{IV}} = C_7^1 I_{np}\left(\sqrt{inp\omega\sigma\mu_{rc}\mu_0}\, r\right) + C_8^1 K_{np}\left(\sqrt{inp\omega\sigma\mu_{rc}\mu_0}\, r\right), \tag{13}$$

where C_7^1 and C_8^1 are undetermined coefficients, and I_{np} and K_{np} represent the first and second kinds of modified Bessel functions respectively, which constitute the basic solution system of the Bessel equation. C_7^1 and C_8^1 can be obtained by substituting the relationship between the magnetic field intensity and the magnetic vector potential into the boundary condition in Eq. (11).

$$\begin{cases} C_7^1 = \dfrac{4B_{re}\sin\left(\dfrac{n\pi\alpha}{2}\right)}{n\pi} \xi_n R_m \dfrac{K'_{np}(kR_r)}{\psi_n} \\ C_8^1 = -\dfrac{4B_{re}\sin\left(\dfrac{n\pi\alpha}{2}\right)}{n\pi} \xi_n R_m \dfrac{I'_{np}(kR_r)}{\psi_n} \end{cases}, \tag{14}$$

where

$$\xi_n = \frac{\frac{R_s}{R_m} + np\sinh\left(npln\frac{R_m}{R_s}\right) - \cosh\left(npln\frac{R_m}{R_s}\right)}{(np)^2 - 1}, \tag{15}$$

and

$$\begin{aligned}\psi_n = {} & \frac{R_c}{np}\left[\frac{\mu_{rm}+1}{2}\cosh\left(npln\frac{R_c}{R_s}\right) + \frac{\mu_{rm}-1}{2}\cosh\left(npln\frac{R_cR_s}{R_m^2}\right)\right]\times \\ & \left[I'_{np}(kR_c)K'_{np}(kR_r) - K'_{np}(kR_c)I'_{np}(kR_r)\right] \\ & -\mu_{rc}\left[\frac{\mu_{rm}+1}{2}\sinh\left(npln\frac{R_c}{R_s}\right) + \frac{\mu_{rm}-1}{2}\sinh\left(npln\frac{R_cR_s}{R_m^2}\right)\right]\times \\ & \left[I_{np}(kR_c)K'_{np}(kR_r) - K_{np}(kR_c)I'_{np}(kR_r)\right]\end{aligned} \tag{16}$$

The expression for the radial component of magnetic induction intensity in domain IV may be derived in this manner:

$$\overrightarrow{B}_r^{IV}(r,\theta,t) = \sum_{n=1,3,5\cdots}^{\infty} \frac{4B_{re}\sin\left(\frac{n\pi\alpha}{2}\right)p}{\pi}\xi_n\frac{R_m}{r}\frac{I'_{np}(kR_r)K_{np}(kr) - K'_{np}(kR_r)I_{np}(kr)}{\psi_n}e^{inp(\theta+\omega t)}, \tag{17}$$

where the derivative of the modified Bessel function to polar coordinate r is written as follows:

$$\begin{cases} I'_{np}(kr) = \frac{\partial}{\partial r}I_{np}(kr) = \frac{k}{2}\left[I_{np-1}(kr) + I_{np+1}(kr)\right] \\ K'_{np}(kr) = \frac{\partial}{\partial r}K_{np}(kr) = -\frac{k}{2}\left[K_{np-1}(kr) + K_{np+1}(kr)\right] \end{cases}. \tag{18}$$

The conduction current density in domain IV is expressed as follows:

$$\overrightarrow{J}_z(r,\theta,t) = \sum_{n=1,3,5\cdots}^{\infty}\overrightarrow{J}_{zn}(r)e^{inp(\theta+\omega t)}, \tag{19}$$

where $\overrightarrow{J}_{zn}$ represents the n-th order of the current density phasor and it is shown as follows:

$$\overrightarrow{J}_{zn}(r) = -\omega\sigma\mu_{rc}\frac{4B_{re}\sin\left(\frac{n\pi\alpha}{2}\right)p}{\pi}\xi_nR_m\frac{I'_{np}(kR_r)K_{np}(kr) - K'_{np}(kR_r)I_{np}(kr)}{\psi_n}, \tag{20}$$

where μ_{rc} is the relative permeability of the CuFe alloy conductor tube in domain IV.

The damping force (or torque) may be derived using the Joule heating law.

$$T = \frac{\pi n_m L_m}{\sigma\omega}\sum_{n=1,3,5\cdots}^{\infty}\int_{R_c}^{R_r}\left|\overrightarrow{J}_{zn}(r,\omega)\right|^2 rdrA_{zn}^{IV}. \tag{21}$$

$$F = \frac{\pi n_m L_m}{\sigma v}\sum_{n=1,3,5\cdots}^{\infty}\int_{R_c}^{R_r}\left|\overrightarrow{J}_{zn}(r,v)\right|^2 rdr, \tag{22}$$

where v is the axial velocity of the damper.

The ultimate representation of the eddy current damping force takes the form of a Fourier series sum. Despite the convergence of such an infinite series, only a finite number of terms can be taken when performing numerical calculations. In ferromagnetic materials, such as CuFe alloys or other substances, the relationship between magnetization and external magnetic field strength shows strong nonlinearity due to the relative permeability. The CuFe alloy materials imply a relative permeability >1, which affects the expressions for the magnetic field strength, control equations, and boundary conditions throughout the derivation of the theoretical formula. The impact of permeability on the sum of the electrical current density vectors of each order, as well as their phases, is ultimately reflected in the nonlinear variation of the damping force in the ECD. Following Ohm's law, the free current density is directly proportional to the electric field strength, with their ratio being the conductivity of an isotropic medium. The influence of conductivity on current density is not consistently linear, particularly for CuFe conductors, as the "armature effect" and "skin effect" become stronger with increasing speed, causing the relationship between current density and conductivity to no longer be linear. Thus, at high speeds, low-conductivity conductors may yield greater damping forces than their high-conductivity counterparts.

3. Experimental verification

This chapter outlines the creation of a testing platform that comprises a rotating apparatus and a torque sensor. The platform aims to confirm the damping force and nonlinear traits of CuFe-ECD. Subsequently, utilizing finite element software, the magnetic induction intensity distribution and damping force–velocity curve of the FEM is calculated. The experimental findings are then compared with the theoretical formula and FEM calculation results.

3.1. *Experimental device and test*

The experimental platform consists of three distinct components, namely the driving device, damping generating device, and signal acquisition device, which are depicted in Fig. 3. The driving device is controlled by a mechanical transmission CA6250A/B lathe of the floor standing type, which outputs rotational velocity in the range of 10–1600 RPM. Separating the variables of the eddy current generator and the transmission components for testing offers a direct means of measuring induced torque. This approach eliminates the influence of the transmission process on damping forces, yielding more precise torque measurements solely attributable to eddy currents. The rotational velocity is measured through the employment of a velocity sensor, which is positioned at the left extremity of the damping generation device. The damping generating device comprises PMs, a magnetic steel holder, and a composite tube comprising both a conductor and an external steel holder tube. The detailed components are elaborated upon in Fig. 4.

The left-hand connector is affixed to the composite tube through a meticulously arranged sequence of bolts. The internal steel holder, replete with a ring of PMs that are held in place with screws, is connected to the composite tube by way of a rolling bearing which is fitted with a sealing ring. Adequate amounts of lubricating oil have been generously applied to the bearing assembly to facilitate smooth and uninterrupted rotation whilst simultaneously mitigating frictional torque. The steel holder shaft is anchored to the right operation platform through the employment of a three-directional clamp, as illustrated in Fig. 4. The principal dimensional parameters of the composite tube, magnetic steel, and all other pertinent components are comprehensively listed in Table 1.

When the lathe commences its operation, it will cause the composite tube to rotate around the steel holder. Furthermore, the conductor tube will incise the magnetic lines of force, resulting in the generation of eddy

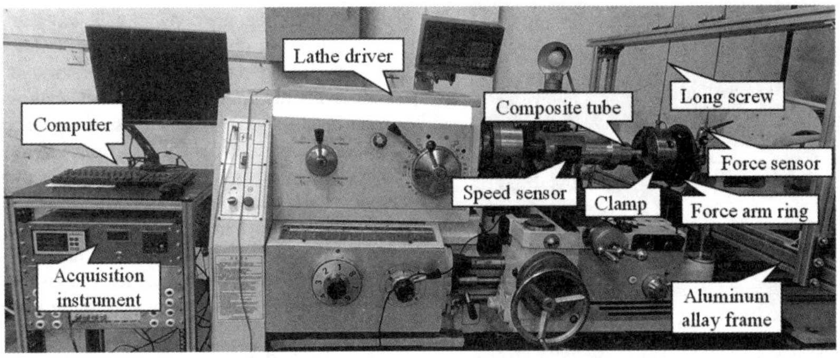

Fig. 3. Experimental equipment and scenario.

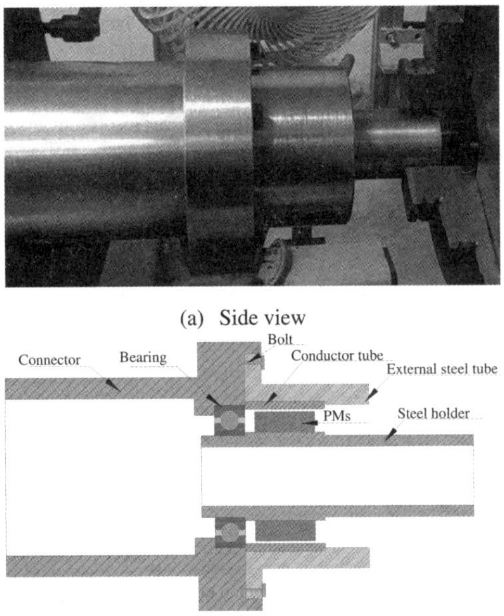

(a) Side view

(b) Sectional view

Fig. 4. ECD generator with CuFe-alloy conductor.

Table 1
Parameters of composite tube and PMs.

Geometrical parameter	Symbol	Value
Inner diameter of inner tube	r_0 (mm)	14
Inner diameter of conductor	r_1 (mm)	32.5
Air gap	t_{gap} (mm)	0.5
MPs thickness	t_{mag} (mm)	10
Arc angle of MPs	ang_{mag} (°)	30
Thickness of conductor	t_c (mm)	3 or 7
Thickness of iron tube	t_{iron} (mm)	9
Number of magnets	N	10
Remnant magnetization	Br (T)	1.42
Calculate air thickness	t_{air} (mm)	5
MPs length	h_{mag} (mm)	30
Number of pole pairs	p	5

current and the damping torque. Based on Newton's third law, the interior steel holder will also experience an equivalent and opposite reaction torque. As demonstrated in Fig. 3, the force sensors on both sides are pre-tightened on the aluminum alloy frame via a force arm ring with a radius of 0.15 m, and the steel holder is connected to the force arm using a clamp. The reaction torque bends the long screw that fixes the force sensors, with one side of the screw being pulled and the other being compressed. The tension and pressure sensors are employed to capture the reaction force signal, and the damping torque of the conductor tube can be computed by multiplying the sum of the absolute values of these two forces by the force arm. The conductor tube is fashioned from the materials copper, CuFe, and Al alloy, with the thickness of these materials being divided into 3.5 mm and 7 mm for comparative purposes. Since the existence of friction torque has an impact on the test results, after taking out the PMs from the interior steel holder, the friction torque was tested in advance. The final torque of ECD can be obtained by subtracting the friction torque from the test results with PMs.

Metallic materials' electromagnetic properties result from various factors, including composition, crystal structure, heat treatment, and mechanical traits. Key aspects defining their electrical performance include conductivity and permeability. The International Annealed Copper Standard (IACS) establishes the benchmark, rating standard annealed copper at 58 MS/m (at 20 °C) as 100% IACS [41–43]. Relative electrical conductivity is commonly expressed concerning this copper standard. This study utilizes relative permeability, the ratio of permeability μ to vacuum permeability μ_0, a non-dimensional constant. Thus, selecting CuFe alloy materials aligns with specific requirements, as depicted in Table 2, highlighting electromagnetic and mechanical parameters. The CuFe alloy conductor's base material is shown in Fig. 5.

3.2. *Experimental results*

Copper and aluminum alloys are common conductor materials in ECDs and both have electrical conductivity but not magnetic conductivity. The CuFe alloy material, as a new conductor material that is both electrically and magnetically conductive, may make a significant difference to the coupled magnetic field, producing new damping properties. The damping torque of 3.5 mm copper is higher than that of the

Table 2
Electromagnetic and Mechanical Parameters of Conductor Materials.

Material	Relative conductivity (IACS)	Relative permeability	Tensile strength (Mpa)
Aluminum alloy	40–42	1.0	205
Copper	100	1.0	210
CuFe5	65	1.2	500–600
CuFe10	62	1.3	550–650
CuFe15	58	1.6	580–630
CuFe20	53	2.3	600–650
CuFe30	42	3.0	750–850
CuFe45	34	4.2	900–1000

*CuFe5 denotes a CuFe alloy with an iron mass proportion of 5%, while other CuFe alloys are designated similarly.

Fig. 5. CuFe alloy conductor tube material.

CuFe45 alloy and aluminum alloy materials in the high-speed stage, as shown in Fig. 6(a). The damping torque curve of CuFe45 is between copper and aluminum alloy, and it has a higher damping torque than aluminum alloy in Fig. 6(b). Despite the high iron mass proportion (45%) in the CuFe45 alloy, the damping torque exhibited by the CuFe45 conductor tube with a thickness of 7 mm is close to that of a conventional copper conductor at 1600 RPM. Additionally, it is anticipated that the damping torque of the CuFe45 will surpass that of the copper conductor in the next velocity band. The damping torque-velocity variation curves obtained for CuFe45 imply that CuFe alloy materials have promising damping characteristics as conductor materials for ECDs, with the indication that they may exhibit enhanced damping properties in high-speed sections. The context of other CuFe alloy conductors with varying iron proportions warrants further investigation to fully understand and exploit their damping characteristics for practical applications. These findings provide insight into the potential application of CuFe alloys in ECDs and highlight the need for further research into their damping behavior and its underlying mechanisms.

Conventional copper conductors experience stiffness reduction at elevated temperatures, along with diminishing damping force in high-speed conditions. These challenges necessitate alternative materials with superior damping characteristics. Although aluminum alloys can partially compensate for copper's limitations, their lower electrical conductivity (33%) and reduced strength at high temperatures restrict their suitability for highly specialized applications. Notably, CuFe20 alloy, 7 mm thick, surpasses copper in damping torque output, with CuFe10, CuFe15, and CuFe30 exhibiting similar high-speed damping properties (Fig. 7 (b)). Fig. 7 shows that damping torque increases and then decreases with higher iron content for both 3.5 mm and 7 mm conductors. CuFe20 stands out as the best performer, offering maximum damping torque. CuFe alloys exhibit sustained growth, particularly in the second linear phase, making them excellent for multimodal

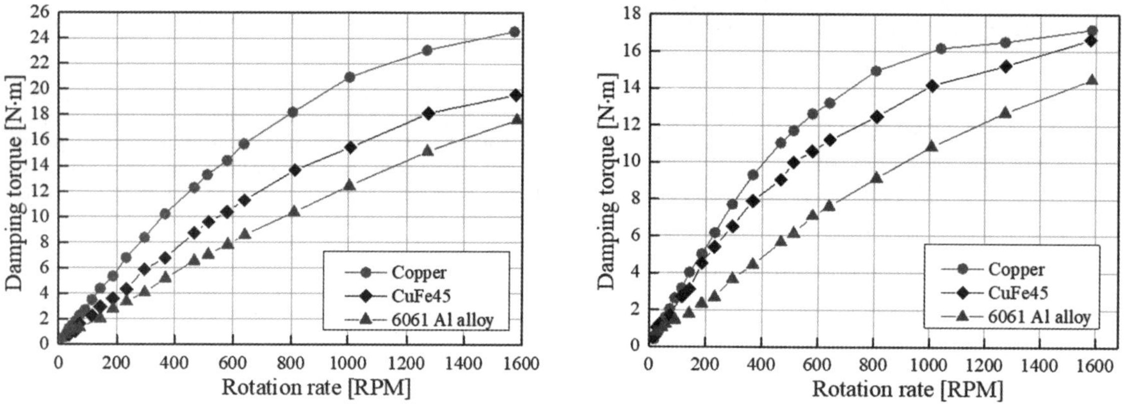

Fig. 6. Damping torque of the conductor tube with different materials.

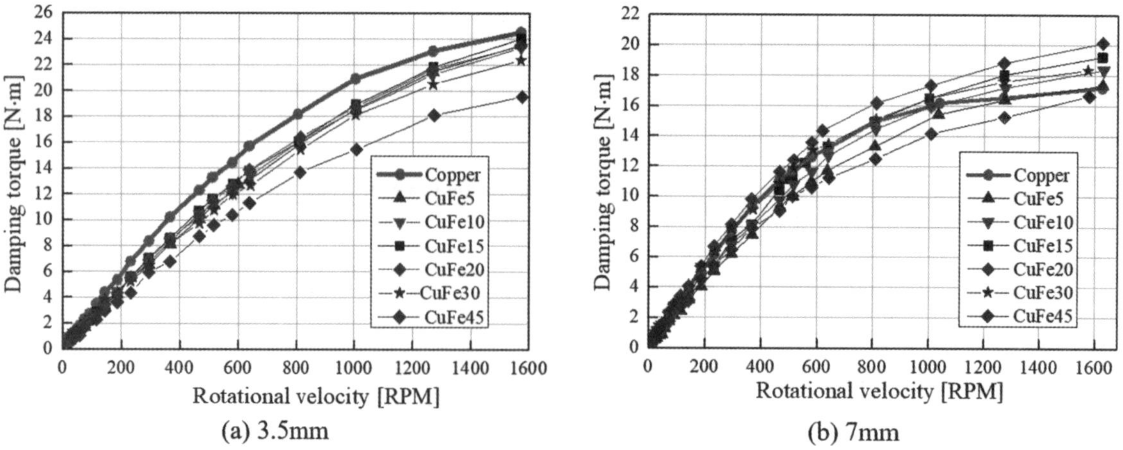

Fig. 7. Damping torque of the conductor tube with different iron mass proportions.

vibration control. This behavior compensates for copper's damping limitations, making CuFe alloys a promising choice for high-performance systems requiring effective damping control.

4. Analysis of experimental results and bilinear constitutive features

In this section, the influence of specific parameters on the damping force of CuFe-ECD as a conductor tube is investigated by using other invariant parameters. The characteristics of the damping force–velocity curve are studied. In addition, the values of other geometric and material parameters are shown in Table 1.

4.1. 3D-FEM and results comparison

The 3-D FEM of the damping torque generator is established in this section, in which the conductivity and relative permeability adopt the material parameters of the CuFe45 conductor tube. Simulation and analysis were conducted utilizing the AC/DC module within COMSOL Multiphysics software. This module is suitable for handling static and low-frequency electromagnetic fields, enabling the investigation of electromagnetic fields, electromagnetic compatibility, and interference phenomena through the resolution of Maxwell's equations. In the 3D-FEM framework, an axial expansion of the 2D model was performed to account for the real distribution of the PMs along the axial dimension, employing solid elements for modeling. Hexahedral meshing was applied to partition the model, and second-order shape functions were used for both magnetic vector potential and electric potential. The magnetization model of the exterior steel tube and steel holder is considered a linear relationship because it is difficult for the iterative calculation to converge if the nonlinear magnetization model is adopted in the 3D model. The relative permeability of the low-carbon steel is taken as $\mu_r = 800$. 3D-FEM considers the actual distribution of PMs in the axial direction, and its cross-section is shown in Fig. 2. The damping torque generator can be simplified to a $1/(2p)$ model by using the anti-symmetric periodic condition that is shown in Eq. (23). The periodic boundary is marked with blue dashed lines within the 2D radial plane in Fig. 2(b). In the 3D model, anti-symmetrical periodic boundary conditions were manually assigned to the two boundary domains. The computational freedom of the model is reduced and the computational efficiency is greatly improved by using the periodic condition to simplify the model.

$$\begin{cases} V_{\text{dst}} = -V_{\text{src}} \\ \boldsymbol{A}_{\text{dst}} = -\boldsymbol{A}_{\text{src}} \end{cases} \tag{23}$$

where V_{dst} and V_{src} represent electric potential; $\boldsymbol{A}_{dst}$ and $\boldsymbol{A}_{scr}$ represent the magnetic vector potential, on the left and right boundaries, respectively.

According to the geometric parameters and material properties of ECD, the simplified 3D-FEM based on the assumption and the boundary condition mentioned above is established. Fig. 8 shows the distribution of magnetic flux density in multiple planes according to the steady-state solution of the electromagnetic field in finite element software. The magnetic flux density is small in the calculation space and gradually decreases from the center of the PMs to the surrounding space when the rotation speed is 0 RPM as shown in Fig. 8(a). The eddy current in the conductor tube can produce a strong anti-magnetic field at a higher rotation speed, which makes the magnetic flux density higher than that near PMs. Therefore, Fig. 8(b) shows that the magnetic flux density reaches the maximum near the upper edge of the conductor tube when the rotation speed is 1600 RPM.

The experimental methodology for this study involved a specially designed damping rotor and a force transducer to measure the ECD's damping torque. In tandem, FEM was employed to explore electromagnetic fields and address electromagnetic compatibility concerns using Maxwell's equations. The magnetic vector potential and electric potential (scalar) were utilized as dependent variables, suitable for steady-state and frequency-domain calculations. Fig. 9 illustrates comparative damping moment curves for CuFe15 and CuFe20 conductors using three methods: experimental, FEM, and theoretical calculations. Numerical methods aligned closely with experimental measurements, showing a maximum error of only 3.8%, validating the accuracy of both FEM and theoretical formulations. Both FEM and electromagnetic theory formulas yield similar damping torque values, underlining the correction made to the original damping force calculation formula by considering the relative permeability of the CuFe alloy conductor tube material. However, discrepancies between numerical and experimental results arise due to assumptions regarding air's relative permeability between the PMs, the absence of considerations for thermal expansion coefficients, and temperature effects on conductivity and permeability. The theoretical formulation's accuracy concerning damping torque is further challenged by neglecting hysteresis loops and magnetic saturation in the PMs. Future research must delve deeper into these factors for a more comprehensive understanding of the damper's operation and design optimization.

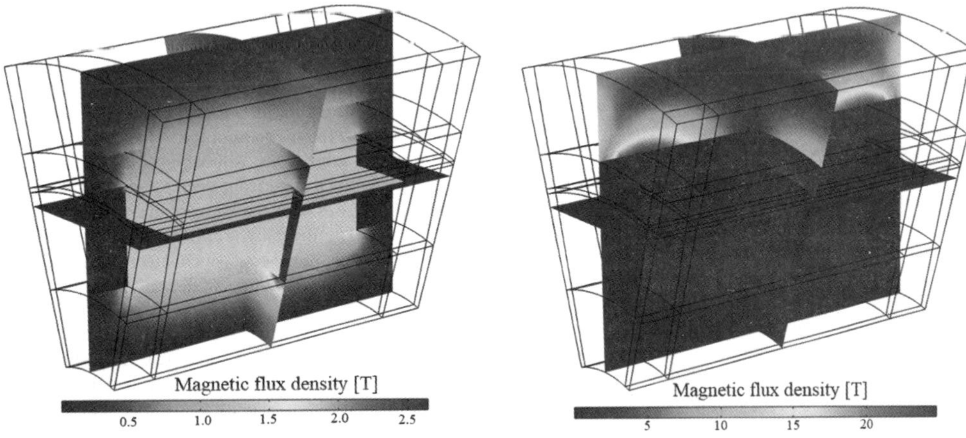

(a) ω=0 RPM (b) ω=1600 RPM

Fig. 8. Magnetic flux density of $1/2p$ FEM in multi-plane.

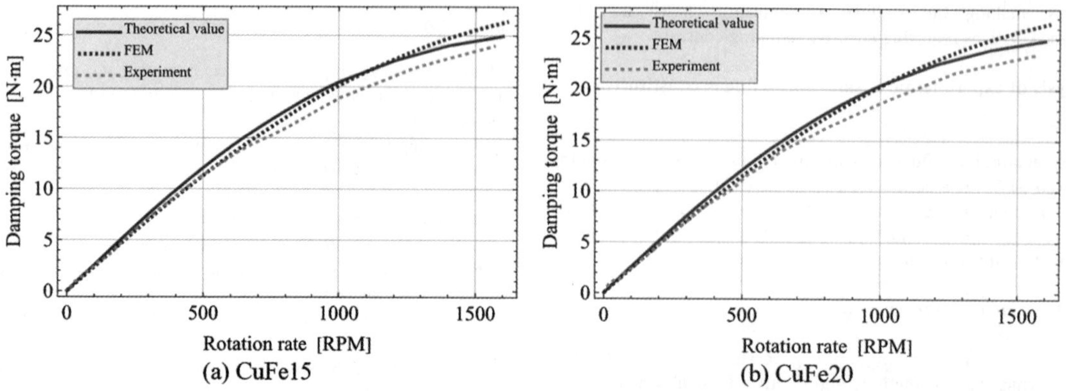

Fig. 9. Comparison of damping torque under different methods.

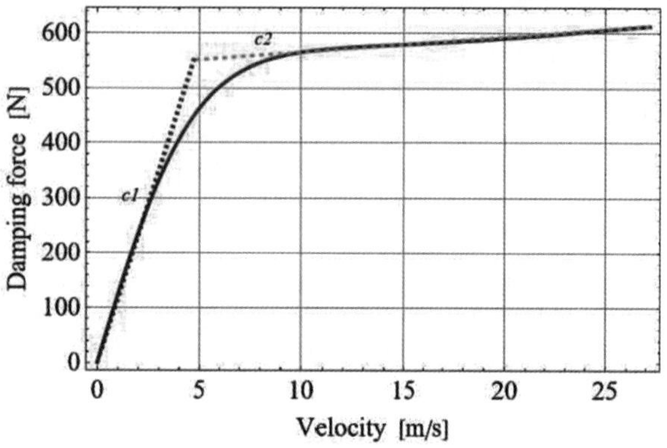

Fig. 10. Bilinear simplified model of the damping force–velocity constitutive curve.

4.2. Analysis of material parameters affecting bilinear characteristic

The study reveals a three-stage pattern in the damping force–velocity constitutive curve of the CuFe-ECD. The initial stage demonstrates a nearly linear increase in damping force with velocity, followed by a transitional second stage bridging the first and third stages. In the third stage, there's a gradual linear rise in damping force with a small damping coefficient. This complex curve embodies quasi-bilinear behavior, allowing us to approximate it as such with slopes denoted as c_1 and c_2, as illustrated in Fig. 10. The point where these lines intersect, defined as the "critical point," and its associated velocity, known as the critical velocity, encapsulate the bilinear damper's inherent attributes, serving as a pivotal design goal. Additionally, this section delves into conductor tube parameters, including thickness, electrical conductivity, and magnetic permeability, which significantly shape damping force output and constitutive properties in a stable electromagnetic environment. This analysis yields valuable insights for CuFe-ECD design and application.

According to the experimental and numerical results, it can be observed that the damping force–velocity constitutive curve of the CuFe-ECD exhibits a three-stage variation pattern. In the first stage, the damping force approximately linearly increases with the velocity. The third stage shows a gradual linear increase in damping force with a small damping coefficient, while the second stage acts as a transition between the first and third stages. The variation of the damping force–velocity curve exhibits a linear trend in the high and low-velocity regimes, with a relatively small transitional region in between. Therefore, to more accurately depict the quasi-bilinear nature of the CuFe alloy, the damping force–velocity constitutive curve can be approximated as a bilinear relationship with slopes (damping coefficients) c_1 and c_2, as depicted in Fig. 10. In this study, the point of intersection between two lines is defined as the "critical point," and the velocity at this point is taken as the critical velocity, which reflects the inherent characteristics of the bilinear damper and is one of the important design goals. The material parameters of the conductor tube play a fundamental role in determining the damping force output and the constitutive characteristics in a stable electromagnetic field. Thus, this section analyzes the parameters of the conductor tube, such as thickness, electrical conductivity, and magnetic permeability, to obtain valuable information on their variations, which provide practical guidance and a basis for the design and application of the CuFe-ECD.

4.2.1. Conductor tube thickness

The thickness of the conductor is a fundamental parameter affecting the ECD damping force. Thickness affects the eddy current density, which in turn affects damping force. Once the conductor thickness exceeds a critical value, it leads to a decrease in eddy current density, resulting in a decrease in damping force. Using CuFe20 as a research subject with a relative conductivity of 33% IACS and a relative magnetic permeability of 2.3, the numerical calculation results of the ECD damping force and damping coefficient are shown in Fig. 11. It can be observed that as the thickness of the conductor increases, c_1 first increases and then decreases, while c_2 first decreases and then increases. Once the conductor thickness exceeds 4 mm, the curve will no longer have a decreasing section, indicating that the damping coefficient $c_2 > 0$. Different from the copper material, the damping force of the CuFe alloy conductor with a certain thickness will not decrease after the velocity exceeds the critical point. It is actually that the damping force continues to increase with the increase of the velocity although the growth rate slows down in most cases. Although the damping force at the critical point experiences a significant decrease, the slope of the second stage gradually increases with the increase of the thickness of the conductor tube. It is worth noting that the critical speed decreases with the increase of the conductor thickness, which means that thickness can be an important indicator for adjusting the critical speed. Different thickness conductor damping force curves gradually converge to a small c_2 in the high-speed section, which is verified by the final convergence of the damping coefficient curve in Fig. 11(b). No matter how the thickness continues to increase, the damping force eventually tends to maintain a stable value in the high-velocity stage. This can be explained by the fact that greater conductor thickness results in stable magnetic resistance in the magnetic circuit. The introduction of the magnetic conductive metal element (iron) causes the CuFe alloy conductor to exhibit distinct magnetic circuit and magnetic resistance characteristics.

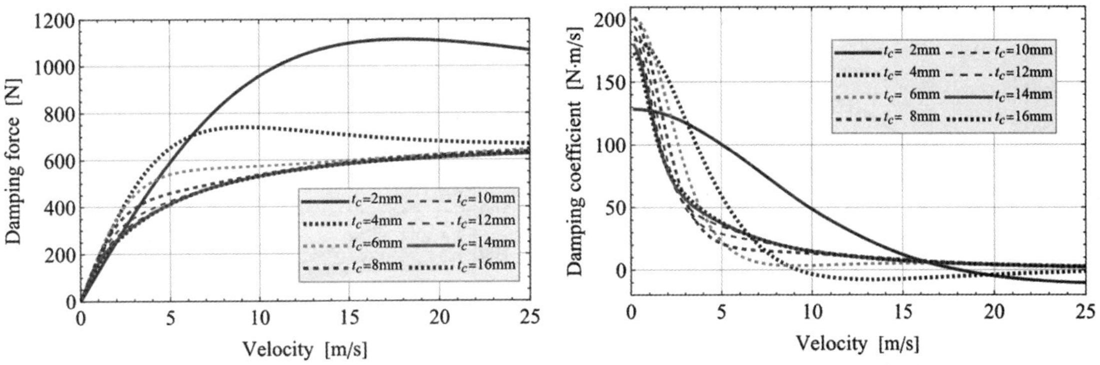

(a) Damping force (b) Damping coefficient

Fig. 11. Damping characteristic variation trend under different conductor tube thicknesses.

The damping-generating device used in the experiment maintains an air gap of 0.5 mm. By fine-tuning the thickness of the CuFe alloy conductor tube, the parameters c_1 and c_2 are varied within the range of 130–200 N·s/m and 0–10 N·s/m, respectively. Notably, the damping coefficient in the c_2 segment can reach a maximum value equivalent to 7.7% of c_1. Adjusting the air gap allows for the attainment of a broader variation of damping coefficients. In practical engineering applications, where conductor tube and PMs dimensions tend to be larger, and the magnet assembly exhibits increased length, coupled with mechanical systems like ball screw mechanisms, further optimization of CuFe-ECD's damping coefficients becomes viable. The damping coefficient of the eddy current generation device in this experiment can be amplified to 10 kN·s/m using the ball screw with a lead of 20 mm.

The bilinearity property of CuFe-ECD may facilitate multi-modal control of nonlinear dampers. The concurrent conductive and magnetic characteristics of CuFe alloys impart excellent eddy current damping enhancement properties. Furthermore, the unique combination of conductive and magnetic properties exhibited by CuFe alloys makes them ideal candidates for ECD. This property allows for the dissipation of kinetic energy as a result of induced magnetic fields, thereby reducing vibrations and enhancing damping performance. This feature is particularly beneficial in compensating for the limited damping force of traditional copper conductor ECDs under high-speed and high-temperature conditions. As a result, CuFe alloys hold tremendous promise for a wide range of applications, from industrial manufacturing to civil engineering and beyond.

4.2.2. Relative conductivity and permeability

Velocity-damping force curves, ranging from 5% to 100% IACS in relative conductivity, are shown in Fig. 12(a), while maintaining a constant relative permeability of 4.2 (CuFe45). Initially, at lower velocities, a decline in conductivity leads to a decrease in the damping coefficient. Lower conductivity results in a gradual rise in the critical velocity, with damping forces remaining at approximately 550 N at these critical points. This highlights that using higher-conductivity alloy materials initially yields slightly greater damping forces but indicates an upper limit to achievable damping. At high velocities, the damping force's growth rate diminishes, and the damping coefficient becomes notably smaller than in the initial stage. According to Eq.(22), damping force is directly proportional to $\left|\overrightarrow{J}_{zn}(\sigma)\right|^2/\sigma$, illustrating non-linear conductivity-damping force characteristics.

The reduced velocity is defined as:

$$v_r = v \times \sigma \tag{24}$$

where σ is the relative conductivity with IACS. The damping force-reduced velocity curve under varying conductivities is shown in Fig. 12(b), and it illustrates that other curves almost coincide with the curve at σ = 100%IACS when compressed horizontally. This indicates that the impact of conductivity on damping force is analogous to modifying velocity.

The magnetic permeability of a conductor is a crucial parameter that signifies its responsiveness to an external magnetic field. When subjected to a permanent magnetic field, moving conductors with high

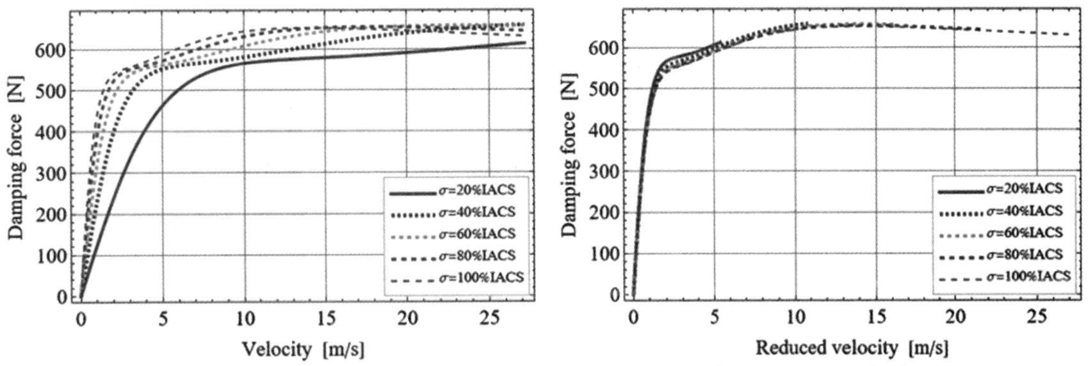

(a) Linear velocity (b) Reduced velocity

Fig. 12. Damping force–velocity curve under different conductivity.

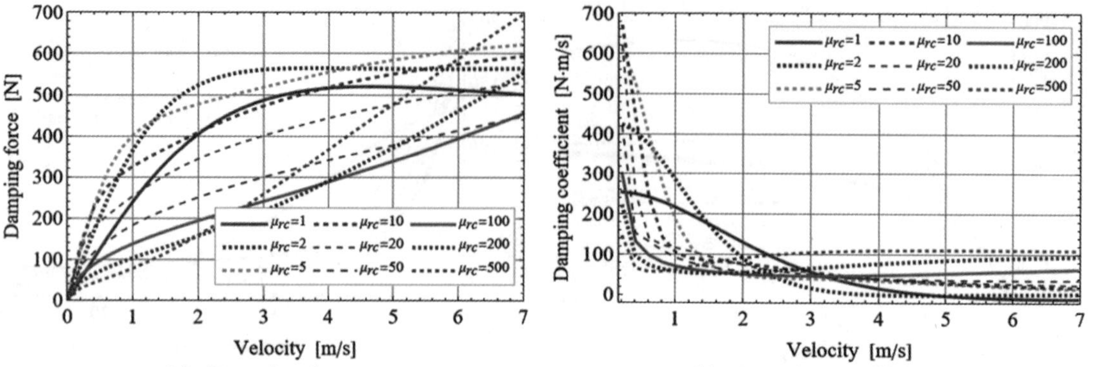

(a) Damping force (b) Damping coefficient

Fig. 13. Damping characteristic variation trend under different relative permeability.

electrical conductivity and no magnetic induce additional counteracting magnetic fields via eddy currents, influencing the original magnetic field's distribution and strength. Hence, conductors with appropriate magnetic permeability have the potential to enhance damper efficiency and overall performance. However, it's essential to consider potential drawbacks, including inductive coupling, energy loss, and electromagnetic interference, which can be exacerbated by excessive magnetic permeability, ultimately diminishing the ECD's energy dissipation capacity.

At a constant relative conductivity of 100% IACS, the damping force–velocity relationship for a 7 mm thick conductor is depicted in Fig. 13, while varying the relative permeability from 1 to 500. The curve at $\mu_{rc} = 1$ represents the damping force of copper to velocity. Notably, as velocity surpasses a critical value, the damping force starts decreasing, as supported by the negative damping coefficient in Fig. 13(b). For $\mu_{rc} > 1$, the damping force–velocity curve exhibits an apparent quasi-bilinear pattern. When relative permeability approaches infinity, the entire curve approximates linearity. As permeability rises, c_1 initially increases and then decreases, while c_2 consistently rises with permeability. Taking the examples of $\mu_{rc} = 2$ and $\mu_{rc} = 5$, where both the critical and maximum damping forces exceed those of copper conductors with $\mu_{rc} = 1$. This highlights the superior performance of high-conductivity, low-permeability CuFe alloy materials over copper conductors. Additionally, it's noteworthy that the critical velocity decreases with increasing relative permeability in a quasi-bilinear curve. Thus, relative permeability not only determines the damping coefficient magnitude but also the proportions of the two linear segments, serving as a vital indicator for delineating the bilinear damper behavior.

5. Case studies

The primary bridge of the Sutong Bridge is a cable-stayed bridge featuring a double cable plane and two towers with a length of 1088 m. The cable system and height of the bridge are illustrated in Fig. 14(a). This bridge holds the distinction of being the first cable-stayed bridge in China with a span exceeding one kilometer, and it was also the world's largest cable-stayed bridge with the longest cable. The length of the cables on the Sutong Bridge varies significantly, with the shortest cable measuring 153.15 m and the longest cable stretching 576.77 m. The ultra-long cable's fundamental frequency is less than 0.22 Hz, and its flexibility is much greater than that of standard cables. Sutong Bridge utilizes parallel steel wire stay cables with a tensile strength of 1770Mpa. There are eight cable planes, each consisting of 34 cables, and the lower ends of the cables are anchored 1.296 m vertically below the bridge deck. In terms of vibration control for the inner and outer spans of the cable on the Sutong Bridge, a configuration that involves installing two ECDs in a triangular arrangement on the cables has been utilized.

Table 3
Long cable parameters.

L (m)	Diameter (m)	Angle (°)	Cable force (N)	m (kg/m)	Modal frequency (Hz)
506	0.152	24.3	5.5×10^6	93	$0.24 \times n$

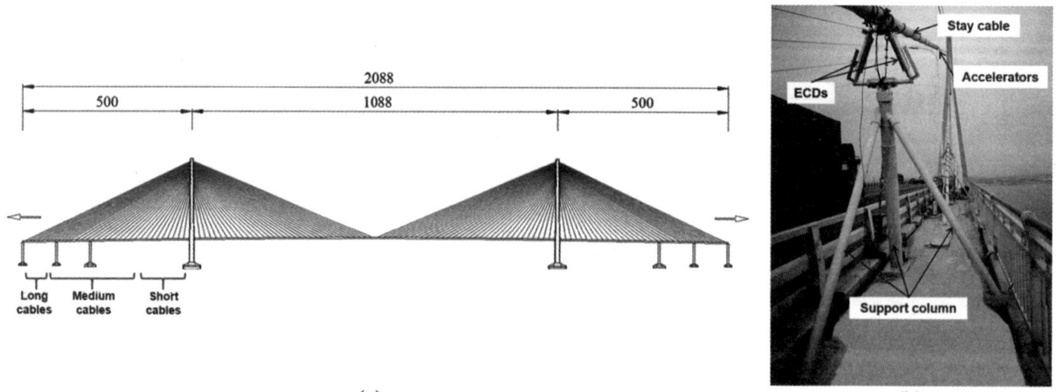

(a) (b)

Fig. 14. Cable layout of the Sutong Bridge.

This configuration ensures that any oscillation in either direction of the cables results in a change in the damper's stroke, thereby generating damping forces. This arrangement achieves simultaneous control for both the inner and outer spans, as illustrated in Fig. 14(b). Due to the increased flexibility of the long cable, it is susceptible to significant wind and rain vibration, which poses safety hazards to the structure. As a result, this study not only mitigates the bridge's vibration through ECD but also validates the CuFe-ECD conductor's vibration control effect on the long cable through numerical analysis. This investigation is based on the prior detection data of three long cables on the cable surface (i.e., near the upstream cable surface on the left bank), and the parameters of the long cable are detailed in Table 3.

5.1. *Bilinear damper design*

Transient and frequency response analyses of the cable, including characteristic frequencies and damping features, were performed using the Structural Mechanics module in COMSOL Multiphysics software. Stochastic vibration analysis was employed to accurately simulate the cable's response to wind-induced loads. Two-dimensional truss elements were used to simulate the pre-tensioned cables for conducting analyses related to natural frequencies and responses. The cables, constructed from steel wires, possess a high yield limit, allowing for a wide range of pre-tension. The transverse stiffness of the truss element is proportional to the tension, and the cable's stiffening due to stress is implemented in the FEM. The cables are anchored and fixed at both ends, with a point load applied at one end equal to the cable's force. The initial displacement was set to zero. The functions illustrating the variation of damping force with velocity, and the time-dependent behavior of the dynamic load, were predefined. Within the truss component, nonlinear dampers and dynamic loading were incorporated by calling the defined functions. Dynamic loading was applied to the cable in the form of a unit-length line load to excite the system. In the characteristic frequency solver, the relative tolerance was set to 1.0×10^{-6}, with a maximum iteration count of 300, and the ARPACK computational method [44] was employed. For computing the transient response of the cable, the generalized α-implicit method [45] was applied, with a time step of 0.01 and a tolerance factor of 0.1.

Typically, the damper is installed at a specific distance from the anchor end of the cable, as depicted in Fig. 15. Through FEM steady state calculation, Fig. 16 illustrates the variation of the mode damping ratio of the cable with the damper damping coefficient. The findings suggest the presence of an optimal damping coefficient for each mode, whereby the mode damping ratio initially increases and then decreases with the damper's damping coefficient. Notably, the optimal damping coefficient decreases as the mode number increases. The optimal damping coefficient for the damper at $a_1 = 5.51$ m for the second mode

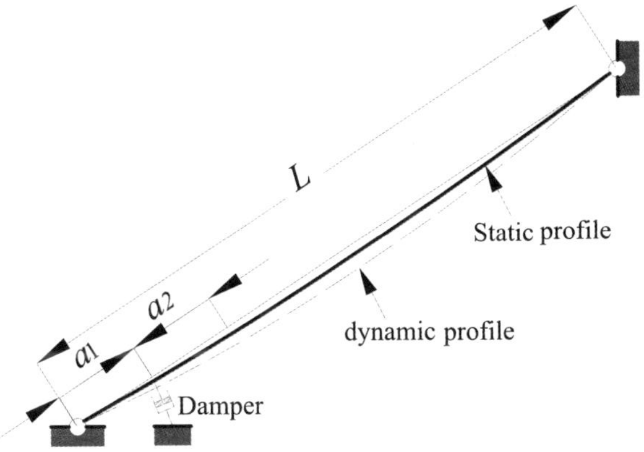

Fig. 15. Installation of the single damper on the cable.

is 219.3 kN·s/m, whereas for the 40th mode, it is 15.5 kN·s/m. These two damping coefficients form the fundamental parameters for the multimode control of bilinear dampers, and they serve as the basis for designing bilinear dampers to achieve multimode control.

Sutong Bridge monitoring results reveal that the long cables demonstrate characteristic frequency clustering between 2 and 13 Hz under wind speeds from 1 to 17 m/s, aligning with the 40th cable mode [46,47]. Notably, the influence of first-order vortex and rain-wind-induced vibrations in long cables is minimal, leading to the exclusion of first-order mode damping control. Leveraging experimental and theoretical insights into CuFe alloy, a bilinear damper was designed to facilitate multi-mode cable control, and subsequently validated for mitigation efficacy. The first section of the damping force–velocity constitutive curve employs the optimal linear damping coefficient from the second mode to govern low-order cable vibrations. The second section utilizes the optimal damping coefficient from the 40th mode for high-order vibration control. The transition between the two lines has been smoothly connected using the continuous first derivative, and the complete curve has been illustrated in Fig. 17. For cables with specific control requirements on higher-order modes, it is possible to design bilinear CuFe-ECD damping coefficients that meet these control demands and provide superior performance in controlling higher-order modes.

5.2. *Verification of vibration control effect*

This study aims to validate the CuFe-ECD's bilinear vibration mitigation during vortex-induced resonance in low-damping systems. In such systems, the maximum steady-state response amplitude occurs at a frequency ratio slightly less than 1 between the external load and the structural system. Due to minimal cable damping, the frequency ratio is taken to be 1, satisfying resonance conditions where the applied load's frequency matches the cable's natural frequency in undamped free vibration. Therefore, utilizing the method of structural dynamics, the modal damping ratio of the cable is computed via the dynamic amplification factor by recourse to the solution of the steady-state displacement time-history response. In FEM, a user-defined element and subroutine describe the nonlinear constitutive relation of the bilinear ECD. The eigenfrequency solver is employed to identify the vibration shape and eigenfrequencies of each mode. The cable's mode is excited by an external load adjusted to match its resonant frequency, following a sinusoidal load form:

$$F_L(x,t) = p(x) \bullet \sin(2\pi ft) \tag{25}$$

where $p(x)$ represents the distributed load amplitude, in N/m. f is the mode eigenfrequency and t represents the instantaneous time. According to the dynamic and static steady-state displacement response, the dynamic amplification coefficient of the displacement amplitude of the cable can be obtained, which is defined as follows:

$$\beta = \frac{v_d}{v_s}, \tag{26}$$

where v_d signifies dynamic cable displacement amplitude, and v_s represents static cable displacement amplitude.

Not only the cable itself is the nonlinear element, but the bilinear damper is bound to show nonlinear characteristics when the displacement amplitude exceeds a certain threshold. Therefore, the steady-state displacement amplitude excited by the natural frequency, based on the linear complex mode theory, may not be maximum at the actual resonance frequency of the cable. To address this issue caused by the bilinear damper's nonlinearity, the actual resonant dynamic amplitude (v_d) is ascertained via the Nelder-Mead optimization algorithm, wherein a tolerance coefficient of 0.001 is set and the optimization objective is to maximize the steady-state displacement amplitude at the remote point of the damper. The equivalent modal damping ratio is estimated based

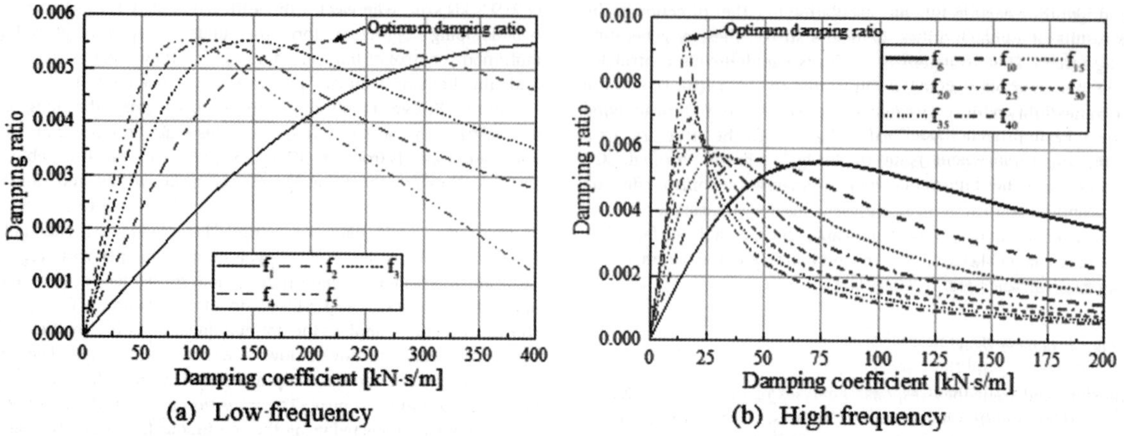

Fig. 16. Variation of the linear damping ratio of the cable with a damping coefficient of ECD damper at a = 5.51.

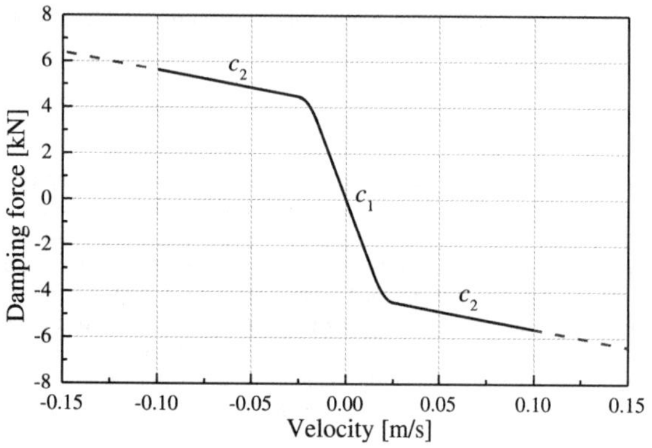

Fig. 17. Constitutive relation design of CuFe-ECD.

on the dynamic and static peak displacements, as illustrated below:

$$\xi = \frac{1}{2\beta_{max}}, \tag{27}$$

where β_{max} represents the maximum dynamic amplification factor of the ith mode at the point far from the damper installation location.

Fig. 18 illustrates the relationship between the cable's mode damping ratio and its vibration amplitude. For cable mode orders $n \leq 6$, as vibration amplitude increases, the mode damping ratio decreases. The first linear stage of the damper's constitutive curve is optimized for low modes, achieving maximum mode damping at minimal vibration, as shown in Fig. 18(a). With rising excitation amplitude, the damper transitions into the bilinear nonlinearity, reducing damping in low-order modes. In contrast, for $n > 6$, mode damping initially rises, then falls, achieving a peak damping ratio. As the damper enters the bilinear stage, damping coefficients approach optimal values for the current mode, leading to a gradual rise in mode damping, as shown in Fig. 18(b). However, if the amplitude continues to increase beyond this point, the bilinear damper deviates from the optimal damping coefficient of the current control mode, leading to a decrease in the mode damping ratio.

In practical engineering, assessing the impact of wind speed and vibration type on cable amplitude under wind load is crucial. To maintain structural integrity, the cable's allowable amplitude is limited to less than $L/1700$. Observations reveal that the amplitude of higher modes during vortex-induced resonance is usually consistent. To accurately determine the bilinear damper's effectiveness in vibration reduction, it is essential to determine cable amplitude and modal damping during vortex-induced resonance. Researchers have put forth empirical models for estimating structural vortex-induced vibration amplitude, aiding in engineering design parameter determination.

The Vickery & Basu model [48], which is grounded on linear random

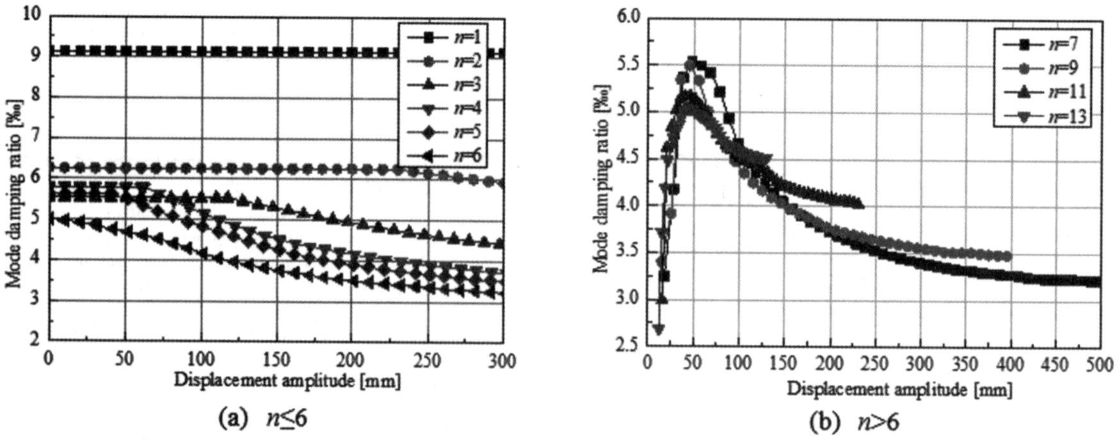

Fig. 18. Variation of modal damping ratio with displacement amplitude.

vibration theory, accounts for the impact of the vortex-induced force triggered by structural vibration interfering with the flow field. This model provides a framework for forecasting vortex-induced vibration amplitude and serves as an effective basis for designing vibration control measures. Building on the Vickery & Basu model, Grala et al. [49] have introduced an improved model with more accurate coefficients for predicting the displacement response of cylindrical structures. The aforementioned model represents an advancement in the field, and its formula is as follows:

$$\frac{x_d}{D_c} = \sqrt{\frac{\alpha^2}{2}(1-K) + \sqrt{\left[\frac{\alpha^2}{2}(1-K)\right]^2 + \frac{C_c \rho D_c^3 \alpha^2}{mLK_{a0}S_t^4}}}, \quad (28)$$

where x_d/D_c represents the normalized standard deviation of the structural displacement amplitude; x_d is the amplitude of vortex-induced resonance; d is the diameter of the stay cable; α is the normalized limit displacement amplitude, which is 0.4 for stay cables. K_{a0} is the aerodynamic damping parameter, and $K_{a0} = 5$ for Reynolds number Re $< 2 \times 10^5$, and $K_{a0} = 1.2$ for Re $> 2 \times 10^5$, according to the research of Vickery et al. [48,50] on this aerodynamic parameter of circular cross-section structures; the damping parameter $K = S_c/(4\pi \bullet K_{a0})$, in which Scruton number $S_c = m\xi/(\rho D_c^2)$; the coefficient C_c for circular cross-section structures is 0.0554 for the starting wind speed of vortex-induced resonance $v_{cr} < 11$ m/s; ρ is the air density, which is generally taken as 1.25 kg/m^3; S_t is the Strouhal number, which is generally taken as 0.2 for cylindrical members.

The modal damping ratio curve in Fig. 18, derived from FEM, can be combined with the empirical formula Eq. (28) to compute vortex-induced resonance amplitude. By utilizing an iterative computation process, the modal damping ratio and cable displacement amplitude can be determined in the presence of the bilinear damper control system during vortex-induced resonance. Previous research conducted by the authors' team introduced two optimization strategies for damper parameters, establishing a dynamic cable-damper system that yielded optimal vibration control outcomes across multiple modes. For a comprehensive comparison with the damper in this paper, Table 4 provides the parameters for the four damper cases.

Fig. 19 depicts the mode damping ratios of four damper mitigation cases. The data for Cases 1, 2, and 3 were obtained from a previous investigation [46]. The current Sutong Bridge dual-damper setup implements ECDs with aluminum alloy conductors. It's essential to note that both copper and aluminum alloys can function as suitable conductor materials for cable dampers. Whether using copper or aluminum alloy conductors, ECDs can be systematically designed to exhibit an optimal damping force–velocity constitutive relationship within their linear segments, yielding identical curves. Cases 1 and 2 employed linear dampers to optimize control for low-order or high-order modes, respectively. However, this simplified approach sacrifices control effectiveness for some corresponding modes. Notably, in high-order mode control (Case 2), damping for modes above the 25th order is maximized, but low-order modes have limited damping. Additionally, it's imperative to consider that low-order modes can lead to wind-rain-induced vibrations, potentially posing safety risks. Conversely, Case 1 exhibits inadequate damping for higher-order modes, implying ineffective control for potential higher-order vortex-induced vibrations. To address these limitations, a dual-damper system was introduced, effectively overcoming the drawbacks of Cases 1 and 2, demonstrating excellent control in multi-mode cable vibrations through optimized parameter design. Nevertheless, the implementation of a multi-damper system is costlier and poses installation challenges.

The damping ratio associated with a single bilinear damper surpasses that of two linear dampers within the first 20 mode orders. The modal damping ratios of modes 1st-20th can reach 5‰, which is superior to other schemes. Despite a gradual decline in mode damping ratio at high-order modes, the damping ratio between 35 and 40 orders eventually stabilizes at a range of 3.72–3.90‰. While the control effect of Case 4 in high-order modes is inferior to that of Case 3, it still provides superior and more stable control than Cases 1 and 2. In conclusion, bilinear dampers confer notable benefits in controlling vibrations across multiple modes, particularly in situations where the requirements for high-order mode vibration control are less exacting.

The long-term monitoring of the Sutong Bridge has revealed that during its operational period, the main cables frequently experience vortex-induced vibrations and wind-induced vibrations, involving multiple modes of oscillation, notably within the 6–10 Hz range. Consequently, this study employed a wind load comprising multiple harmonics to excite the main cables. Fig. 20 illustrates the displacement amplitudes and displacement spectrum of the long cable under four distinct control schemes. Under the influence of a single linear damper, the damping performance of an ECD designed for low-order modes proved less effective in mitigating vibrations dominated by high-order modes compared to an ECD designed for high-order modes. Spectral analysis affirmed that a low-order ECD is more suitable for controlling vibrations within its respective frequency range, while a high-order ECD is better suited for higher modes. In the displacement–time history depicted in Fig. 20(b), a single bilinear ECD achieved a root mean square (RMS) displacement amplitude of 0.0298, closely approaching the 0.0234 achieved by the two linear dampers scheme. The displacement spectrum indicates that both bilinear and dual damper control strategies effectively suppress vibrations across a range of modes. Despite being slightly inferior to the two ECDs scheme for simultaneous control of high and low-order vibrations, the bilinear damper approach offers superior economic value due to its lower damper count.

6. Discussion and conclusions

To address the limitations of traditional copper conductor ECDs, such as reduced damping force at high speeds and insufficient performance, this study introduces a CuFe ECD utilizing a bilinear constitutive relationship. The damping performance of this CuFe-ECD is validated through theoretical, experimental, and FEM approaches. A theoretical formula for damping force applicable to CuFe alloy conductors is formulated. Parameter analysis explores the effects of various CuFe alloy parameters, including conductivity, permeability, and thickness, on the bilinear damping constitutive curve. Case studies on the Sutong Bridge cable reveal that the bilinear damper demonstrates excellent mitigation under vortex-induced resonance, offering advantages over single-mode and multi-mode control methods. This study yields the following conclusions:

- The damping force–velocity constitutive relation of CuFe-ECD with a certain thickness conductor shows a two-stage growth characteristic.

Table 4
Parameters of the four damper cases.

Cases	Description	Damper	a/L	a (m)	c (N•s/m)
Case 1	Low mode control	Damper#1	0.02	11.02	1.8×10^5
Case 2	High mode control	Damper#1	0.02	11.02	2.4×10^4
Case 3	All mode control	Damper#1	$a_1/L = 0.01$	$a_1 = 5.51$	3.8×10^5
		Damper#2	$(a_2\text{-}a_1)/L = 0.01$	$a_2\text{-}a_1 = 5.51$	2.4×10^4
Case 4	Bilinear Damper	Damper#1	0.01	5.51	1.55×10^4-2.19×10^5

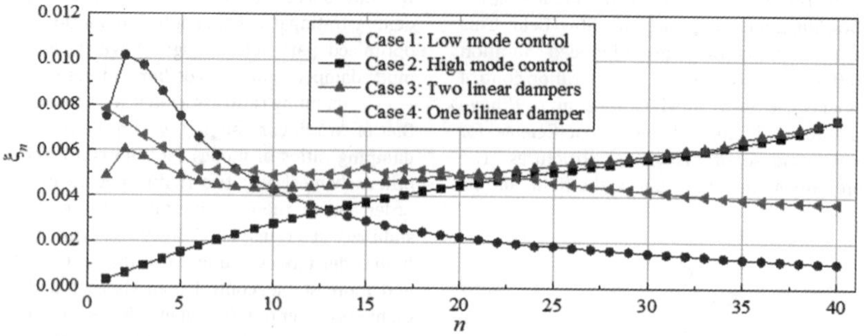

Fig. 19. Damping ratios of 40 modes with different damper cases.

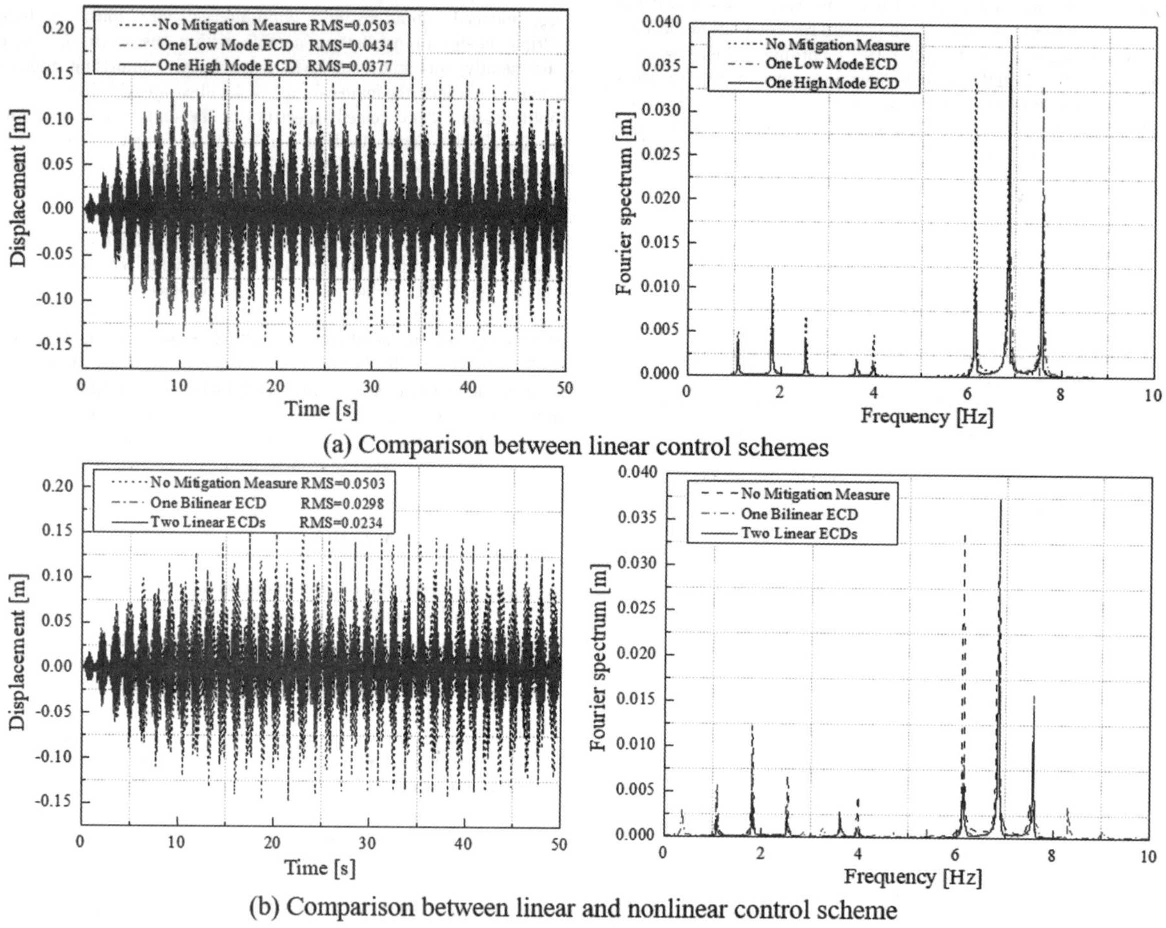

Fig. 20. Cable mid-position displacement response under various damper schemes.

The constitutive curve can be approximated as a bilinear relation with different slopes. The relative permeability represents a critical indicator that demarcates the bilinear damper into two linear stages, playing a pivotal role not only in determining the magnitude of the damping coefficient but also in reflecting the proportion of the two linear segments.

- CuFe alloys with 30–40% IACS conductivity already provide better damping performance than aluminum alloys. CuFe10, CuFe15, CuFe20, and CuFe30 demonstrate superior damping enhancement properties compared to copper, particularly in the high-speed range where this performance advantage is more pronounced. Notably, CuFe20 exhibits the highest damping force among CuFe alloys.
- The numerical damping torque results align with experimentally measured data, with a maximum error of less than 3.8%. This validation confirms the accuracy of the adjusted damping torque calculation formula, which now includes the relative permeability of CuFe alloy conductor materials.

- The cable's control performance reveals that a single bilinear damper outperforms the dual linear damper control in the first 20 mode orders, achieving modal damping ratios of up to 0.5%. Even in higher-order modes, although the damping ratio gradually decreases, it stabilizes between 3.72 and 3.90‰ for modes 35–40. This suggests that the bilinear damper has good advantages in multi-mode control and is suitable for situations where stringent high-order mode control is not required. Compared to the dual linear damper approach, the bilinear damper offers economic advantages by reducing the number of dampers needed for multi-mode control.

Due to the significant demand for conductor materials in ECD and the often prohibitive cost of industrial copper, the current study aims to investigate the damping performance of the more economical CuFe-ECD and explore its potential for mitigating vibrations. The focus of this research revolves around copper alloy materials containing an iron mass fraction ranging from 5% to 45%. The bilinear damping characteristics of these CuFe alloys exhibit significant potential for practical applications. The findings of the bilinear damping characteristics of CuFe alloys contribute to the ongoing efforts aimed at developing effective and efficient vibration control measures in civil engineering. The utilization of other alloy materials as conductors may enrich the variety of ECDs, and provide engineers with more diverse damping performance, which has broad prospects for applications in the field of structural vibration control.

In practical engineering applications, the load-bearing capacity of the ball screw assembly can be designed much higher than the required value to meet durability requirements. The uncertainties lie in variations in transmission efficiency and friction among different transmission components. Addressing these factors can be a focus of dedicated research before engineering applications. To control variables, the experiment exclusively investigated reference groups with different conductor materials and conductor thicknesses for ECD damping characteristics. More comprehensive experiments, encompassing parameters such as high-strength and high-conductivity materials, transmission efficiency, and friction, among others, will be pursued in future studies. Generally, the parameters of ECD are pre-optimized according to the structural control requirements. If there are changes in the dynamic characteristics of the controlled structure or adjustments are needed based on on-site measurements, damping force can be manually adjusted through various mechanically adjustable methods. Dampers based on electromagnetic field control hold the potential for real-time adjustment of damping coefficients, representing a prospective avenue for active control measures.

CRediT authorship contribution statement

Wenbin Yang: Software, Validation, Formal analysis. **Hongyi Zhang:** Conceptualization, Methodology, Supervision. **Zhengqing Chen:** Supervision, Visualization, Validation. **Xugang Hua:** Supervision, Validation. **Huawei Niu:** Supervision, Validation. **Shouying Li:** Supervision, Validation.

Declaration of Competing Interest

The authors declare that they have no known competing financial interests or personal relationships that could have appeared to influence the work reported in this paper.

Data availability

Data will be made available on request.

Acknowledgments

The authors gratefully acknowledge the financial support from the National Science Foundation of China under Grant numbers U2141242 and 52208478.

References

[1] Andersen H, Hommel D, Veje E. Emergency rehabilitation of the Zárate -Brazo Largo Bridges. Sweden: Malmo; 1996. p. 698–706.

[2] Dallard P, Fitzpatrick T, Flint A, Low A, Roche M. London millennium bridge: pedestrian-induced lateral vibration. J Bridge Eng 2001;6:412–7.

[3] Matsumoto M, Shirato H, Yagi T, Shijo R, Eguchi A, Tamaki H. Effects of aerodynamic interferences between heaving and torsional vibration of bridge decks: the case of Tacoma Narrows Bridge. J Wind Eng Ind Aerodyn 2003;91: 1547–57.

[4] Soong Jr TT, B.. Supplemental energy dissipation: state-of-the-art and state-of-the-practice. Eng Struct 2002;24:243–59.

[5] Ga A, Pa A, Ih B. Analysis of bilinear hysteretic structures with nonlinear fluid viscous dampers using modified stochastic linearization technique. Eng Struct 2022.

[6] Li S, Xiu S, Song W, Sun C, Yang H. Experimental study of surface roughness on improving the tribological performance of sealing pairs in magnetorheological damper. Smart Mater Struct 2022;31.

[7] Lu Z, Zhang J, Wang D. Energy analysis of particle tuned mass damper systems with applications to MDOF structures under wind-induced excitation. J Wind Eng Ind Aerodyn 2021;218:104766.

[8] Paronesso M, Lignos DG. Experimental study of sliding friction damper with composite materials for earthquake resistant structures. Eng Struct 2021.

[9] Shen S, Pan P, Sun J, Gong R, Wang H, Li W. Development of a double-sliding friction damper (DSFD). Smart Struct Syst 2017;20:151–62.

[10] Zhou Y, Sebaq MS, Xiao Y. Energy dissipation demand and distribution for multi-story buildings with fluid viscous dampers. Eng Struct 2022.

[11] Sang N, Zhang C, Chen J, Lin S, Yang G. A dual-sided hybrid excitation eddy current damper for vibration suppression in low damping linear motor system. IEEE Trans Ind Electron 2021;68:9897–907.

[12] Satpute N, Iwaniec M, Narina R, Presswala I, Arias LM. Design and Analysis of Eddy Current Damper. 2021 IEEE XVIIth Int. Conf. Perspect. Technol. Methods MEMS Des. MEMSTECH; 2021.

[13] Wang Y, Chen Z, Yang C, Liu Z, He J, Feng Z. A novel eddy current damper system for multi-mode high-order vibration control of ultra-long stay cables. Eng Struct 2022;262.

[14] Zhang HY, Chen ZQ, Hua XG, Huang ZW, Niu HW. Design and dynamic characterization of a large-scale eddy current damper with enhanced performance for vibration control. Mech Syst Signal Process 2020;145.

[15] Zhu Q, Yang W, Zhang Q, Du Y. A hybrid vibration mitigation method based on the crowd flow control and tuned mass damper on a footbridge. Eng Struct 2021;245: 112972.

[16] Niu H, Chen Z, Hua X, Zhang W. Mitigation of wind-induced vibrations of bridge hangers using tuned mass dampers with eddy current damping. Smart Struct Syst 2018;22:727–41.

[17] Sodano HA. Eddy current damping in structures. Shock Vib Dig 2004;36:469–78.

[18] Sodano HA, Inman DJ. Non-contact vibration control system employing an active eddy current damper. J Sound Vib 2007;305:596–613.

[19] Yan B, Wang Z, Ma H, Bao H, Wu CY. A novel lever-type vibration isolator with eddy current damping. J Sound Vib 2021;494:115862.

[20] Sodano HA, Bae JS, Inman DJ, Belvin WK. Concept and model of eddy current damper for vibration suppression of a beam. J Sound Vib 2005;288:1177–96.

[21] G. L, Larose, and, A, Larsen, and, et al. Modelling of tuned mass dampers for wind-tunnel tests on a full-bridge aeroelastic model. J Wind Eng Ind Aerodyn; 1995.

[22] Cheng NL, Shan J, Shi Z, Chang CC. Approximate analysis of eddy-current force under time-varying velocity motion for structural control. J Sound Vib 2020;475: 115295.

[23] Van T, Pluk KK, Jansen JH, Lomonova EE. Optimization and measurement of eddy current damping applied in a tuned mass damper. IET Electr Power Appl 2014;10: 609–15.

[24] Xiao X, Chen Y, Shen W, Zhu H. Vibration control of stress ribbon bridges subjected to moving vehicles. Struct Control Health Monit 2021;2021:e2835.

[25] Zuo L, Chen X, Nayfeh S. Design and analysis of a new type of electromagnetic damper with increased energy density. J Vib Acoust 2011;133.

[26] Wouterse JH. Critical torque and speed of eddy current brake with widely separated soft iron poles. IEE Proc Part B Electr Power Appl 1991;138:P.153-158.

[27] Lean MH. Dual simple-layer source formulation for two-dimensional eddy current and skin effect problems. J Appl Phys 1985;57:3844–6.

[28] Edwards JD, Jayawant BV, Dawson WRC, Wright DT. Permanent-magnet linear eddy-current brake with a non-magnetic reaction plate. IEE Proc Electr Power Appl 1999.

[29] Abbas S, Farhat P, Kyoung-Tae K, et al. Effect of composition and powder size on magnetic properties of rapidly solidified copper-iron alloys. J Alloys Compd Interdiscip J Mater Sci Solid State Chem Phys 2018.

[30] Abbas, Sardar, Farhat, Kim, Taek-Soo. Effect of lattice strain on the electrical conductivity of rapidly solidified copper-iron metastable alloys. J Alloys Compd Interdiscip J Mater Sci Solid-State Chem Phys 2018.

[31] Abbas SF, Seo SJ, Park KT, Kim BS, Kim TS. Effect of grain size on the electrical conductivity of copper–iron alloys. J Alloys Compd 2017:8–16.

[32] Liu K, Sheng X, Li Q, Zhang M, Atrens A. Microstructure and strengthening model of Cu–Fe in-situ composites. Materials 2020;13:3464.

[33] Schmidt S, Sathiaraj GD, Kumar SS, Sulkowski B, Skrotzki W. Effect of rolling and annealing temperature on the mechanical properties of CrMnFeCoNi high-entropy alloy. Mater Chem Phys 2021;124830.
[34] Yamashita T, Koga N, Kawasaki T, Morooka S, Tomono S, Umezawa O, et al. Work hardening behavior of dual phase copper–iron alloy at low temperature. Mater Sci Eng A 2021;819:141509 -.
[35] Lu T, Chen C, Li P, Zhang C, Han W, Zhou Y, et al. Enhanced mechanical and electrical properties of in situ synthesized nano-tungsten dispersion-strengthened copper alloy. Mater Sci Eng Struct Mater Prop Misrostructure Process 2021:799.
[36] Anne J. Impact of damper stiffness and damper support stiffness on the efficiency of a linear viscous damper in controlling stay cable vibrations. American Society of Civil Engineers; 2014.
[37] Huang Z, Jones NP. Damping of taut-cable systems: effects of linear elastic spring support. J Eng Mech 2011;137:512–8.
[38] Main JA, Jones NP. Free vibrations of taut cable with attached damper. I: linear viscous damper. J Eng Mech 2002;128:1062–71.
[39] Hoang N, Fujino Y. Combined damping effect of two dampers on a stay cable. J Bridge Eng 2008;13:299–303.
[40] Liu M, Yang W, Chen W, Xiao H, Li H. Experimental investigation on vortex-induced vibration mitigation of stay cables in long-span bridges equipped with damped crossties. J Aerosp Eng 2019;32:04019072.
[41] Weng GJ. Some elastic properties of reinforced solids, with special reference to isotropic ones containing spherical inclusions. Int J Eng Sci 1984;22:845–56.
[42] Davis JR. Copper and copper alloys. ASM international; 2001.
[43] El-Hadek MA, Kaytbay S. Mechanical and physical characterization of copper foam. Int J Mech Mater Des 2008;4:63–9.
[44] Lehoucq RB, Sorensen DC, Yang C. ARPACK users' guide: solution of large-scale eigenvalue problems with implicitly restarted Arnoldi methods. SIAM 1998.
[45] Aydi H, Jellali M, Karapınar E. Common fixed points for generalized α-implicit contractions in partial metric spaces: consequences and application. Rev Real Acad Cienc Exactas Físicas Nat Ser Matemáticas 2015;109:367–84.
[46] Chao Y, Zc A, Ww A, Xh A, Yw A. Optimal design of two viscous dampers for multi-mode control of a cable covering broad frequency range. Eng Struct 2021;245.
[47] Yang C, Chen Z, Hua X, Li X, Wang W. An explicit solution of modal-damping ratios for higher modes of a cable with an external damper. J Eng Mech 2022;148.
[48] Vickery BJ, Basu R. Simplified approaches to the evaluation of the across-wind response of chimneys. J Wind Eng Ind Aerodyn 1983;14:153–66.
[49] Grala P, Loredo-Souza AM, Rocha MM. A method to predict vortex shedding response based on Vickery and Basu model: a proposal for the new Brazilian Wind Code. Eng Struct 2019;198:109469 .1.
[50] Vickery BJ, Basu RI. Across-wind vibrations of structures of circular cross-section. Part I. Development of a mathematical model for two-dimensional conditions. J Wind Eng Ind Aerodyn 1983;12:49–73.

桥梁多阶模态涡振

Higher Modes' Vertical Vortex-induced Vibrations of Suspension Bridges—Part 2: Aeroelastic Model Study

Zhengqing Chen[a], Wen Chen[a], Xugang Hua[a], and Zhiwen Huang[a]

[a] *Wind Engineering Research Center, Hunan University, Changsha, China, zqchen@hnu.edu.cn*

Abstract

Part 1 of the companion papers presents experimental results regarding multi-mode vortex-induced vibrations (VIV) of steel-box girder suspension bridges based on sectional model study, which is built theoretically upon the two-dimensionality of both incoming flow and oscillating lift forces. In order to investigate the span-wise correlations of oscillating lift forces under different modes of vibration as well as to directly measure VIV amplitudes for different vertical modes, an aeroelastic model that can simulate the closely-spaced frequency characteristics as well as the mode shape in a suspension bridge is necessary. In this study, a novel multi-elastically supported aeroelastic model of shallow beam is proposed and is capable of reproducing as many as closely-spaced vertical modes of a real suspension bridge. The aeroelastic model is 8.1m long, 4cm high and 24cm wide, and the Froude number is followed in the model design. The stiffness coefficients of multiple elastic supports are determined by the assumed mode shape method. Steady sinusoidal excitations are used to confirm the modal frequencies, mode shapes and modal damping ratios of the aeroelastic model. The verified aeroelastic model is tested in smooth flow to measure the VIV amplitudes under different modes of vibrations. The 1:1 sectional model of the rectangular section is also tested in wind tunnel with the same vibration frequencies as in the aeroelastic model and the results are compared to those obtained on aeroelastic model.

1 Introduction

Multi-mode vertical vortex-induced vibrations (VIV) have been observed on several steel box girder suspension bridges under moderate wind and the vibration frequencies are usually between 0.2Hz~0.4Hz. The aim of the present study is to investigate the characteristics of higher modes' VIV and the relationship of VIV vibration amplitudes among different modes by using sectional model and aeroelastic model studies. Part 1 of the companion papers describes the sectional model experimental results regarding variation of VIV amplitude with model vibration frequency and model length scale for a shallow section with side ratio B/D=6 (Hua *et al.* 2013), which is built its theoretical base upon the two-dimensionality of both incoming flow and oscillating lift forces. However, vortex-induced pressures or the *resultant* lift forces between two cross-sections do not fully correlated in particular at small vibration amplitudes. For example, the correlation length, defined as the span-wise integration of correlation coefficient, is about $6D$ for a stationary square cross-section in smooth flow (Vickery, 1966), and the correlation length for oscillating square cross-section is greatly increased (Bearman and Obasaju, 1982). Therefore lack of knowledge of correlation of vortex-induced lift forces could lead to considerable discrepancy between predictions and experiments (Ehsan and Scanlan, 1990). Rigid sectional models are commonly used to determine the correlation between vortex-induced pressure/forces over a range of reduced wind velocity on stationary sectional models or dynamic sectional models (either spring-mounted models or forced vibration ones), and the variation of correlation with vibration amplitude can also be obtained when using dynamic models. While these experiments can delineate the span-wise correlation of pressures/forces caused by the three-dimensional flow around the model, the three-dimensional effect of mode shapes on the correlation has been omitted due to the uniform amplitudes of vibration at different parts of the model.

To study the three-dimensional effects of mode shapes on correlation of vortex-induced forces, a kind of flexible aeroelastic model whose natural frequencies are within the expected vortex-shedding frequencies and mode shapes coincides with the realistic modes of vibration will be necessary. A scaled aeroelastic model of the full bridge may satisfy such requirement, but it is practically infeasible due to the small scale ratio, usually at the order of 1/200, and can only simulate the first two or three lower vertical modes. The taut-strip model is capable of reproducing a pair of vertical and torsional modes of suspension bridges and can be used to investigate the flutter characteristics. The stiffness of the taut-strip model is provided by two internal parallel taut strings, and the vertical frequencies of the taut-strip models therefore increase linearly. For example, if the first vertical frequency is f_1, the modal frequencies of other vertical modes will be $2f_1$, $3f_1$, $4f_1$, ... Such frequency distribution is not the case for the vertical modes of suspension bridges. The vertical modes of a suspension bridge have the two salient features. They are closely-spaced such that there may exists more than ten modes in the frequency ranges of interests. The second feature is that the equivalent mass per unit length for every mode is nearly the same and approximately equals to the mass sum of stiffening girder and two main cables when the mode shape is nearly sinusoidal. As an example, Table 1 lists the modal characteristics of the first 8 vertical modes for a typical steel box suspension bridge with main span of about 800m. There are as many as 8 modes in the frequency range from 0.12Hz to 0.69Hz with an average increment of 0.07Hz, and the frequency distribution characteristics of suspension bridges differs from that of simply-supported beams (increase in a squared law) and that of taut strings (increase in a linear law).

Table 1 Modal parameters of the first 8 vertical modes for a suspension bridge with a main span of 800m

Mode number	Frequency (Hz)	Mode shape	Equivalent modal mass[1] (t/m)	Onset velocity of VIV[2] (m/s)
2	0.1181	1st asymmetrical vertical	37.1	3.44
3	0.1710	1st symmetrical vertical	33.8	4.98
5	0.2354	2nd symmetrical vertical	33.9	6.85
6	0.2645	2nd asymmetrical vertical	35.4	7.72
13	0.3588	3rd symmetrical vertical	33.8	10.5
18	0.4648	3rd asymmetrical vertical	46.0	13.6
26	0.5621	4th symmetrical vertical	33.8	16.3
34	0.6900	4th asymmetrical vertical	34.1	20.1

[1] The mass of stiffening box girder and the two main cables are 27.7t/m and 5.9t/m, the equivalent modal mass for every mode equals to the sum of stiffening girder and the two main cables except for mode 18.

[2] Evaluated according on height of stiffening girder D=3.5m and a nominal value of Strouhal number St=0.12

In this study, a novel elastically multi-supported aeroelastic model of a shallow beam is proposed as simplified representation of a suspension bridge, and it is capable of reproducing as many as closely-spaced vertical modes of suspension bridges. Only the stiffening girder is included so as to eliminate the possible interference from cables and bridge towers, and the model girder is elastically supported by springs at multiple locations. The appropriate numbers of elastic supports and the support stiffness are determined with the assumed mode shape method. The aeroelastic model is 8.1m long, 4cm high and 24cm wide, and the Froude number is followed in the model design. The modal frequencies and mode shapes of the elastically multi-supported aeroelastic model have been tested by steady sinusoidal excitations to extract modal parameters as well as to confirm the design method. The aeroelastic model is then tested in wind tunnel for two sets of support stiffness to investigate the relationship of VIV amplitudes among different vertical modes.

2 Elastically Multi-supported Aeroelastic Model

2.1 Model design

A suspension bridge with main span of 1620m is used as prototype for design and constructing the aeroelastic model. The stiffening girder is assumed to have a rectangular cross-section with side ratio *B/D*=6, as also used in the sectional model study (Hua *et al.*, 2013). Only the asymmetrical vertical modes with a stationary node at mid-span are considered in this study and thus half bridge span will be adequate in designing aeroelastic model. The modeling of asymmetrical vertical modes only also ensures sufficient frequency spacing to eliminate the potential interference of VIV of two neighboring modes. The length scale of the aeroelastic model is taken as 1/100. Table 2 shows the main parameters for the prototype and the aeroelastic model. It is assumed that frequency of the fundamental vertical mode is 0.15Hz and the frequency spacing between two consecutive vertical modes is 0.09Hz.

Table 2 Main parameters for the prototype and aeroelastic model

Parameter	Unit	Scale ratio	Prototype	Model
Length *L*	M	λ_L=1/100	810 (half span)	8.10
Width *B*	m	λ_L=1/100	24	0.24
Height *D (H)*	m	λ_L=1/100	4	0.04
Girder stiffness *EI*	N m	$\lambda_{EI}=\lambda_L^2=1/100^5$	$2.1\times10^{11}\times4=8.4\times10^{11}$	84
Equivalent mass	kg/m	$\lambda_m=\lambda_L^2=1/100^2$	30000	3
Fundamental Frequency *f*	Hz	$\lambda_f=\lambda_V/\lambda_L=10$	0.15	1.5

In model design, the stiffness of the stiffening girder is modeled by thin steel plate and the stiffness of main cables is represented by a series of flexible supports evenly distributed along the span. For suspension bridges, the mode shape of stiffening girder is very close to a sinusoidal one, and therefore has a form of $\phi_r(x)=\sin r\pi x/L$. By assuming that the elastic supports is evenly distributed and stiffness of the *i*th elastic supports is K_i, the generalized stiffness and mass for the *r*-th mode of the aeroelastic model is given as

$$K_r^* = \int_0^L EI\left(\phi_r''\right)^2 dx + \sum_{i=1}^{N} K_i\left[\phi_r\left(iL/N+1\right)\right]^2 \quad m_r^* = \int_0^L \rho A\phi_r^2 dx + \sum_{i=1}^{N} m_i\left[\phi_r\left(iL/N+1\right)\right]^2 \tag{1}$$

where N is the number of flexible supports which is taken as 17 in this study, and m_i is the effective mass of elastic supports. The aeroelastic model is intended to simulate the first eight to ten vertical modes of a suspension bridge, therefore the system of equation for solving support stiffness K based on equation (1) remains underdetermined. Therefore the stiffness of elastic support is assumed to symmetric and a number of 9 stiffness coefficients K_i (i=1, 2, …, N/2+1) are going to be determined, which can be solved by the first 9 target modal frequencies of the aeroelastic model. The cantilever beams of circular section are used to implement the required support stiffness, and the stiffness may be adjusted by simply changing the beam length. The detailed information about cantilever beams is shown in Figure 1.

Figure 2 shows the 8.1m long aeroelastic model in wind tunnel. The stiffness of the aeroelastic model is provided by a thin steel plate with 80mm wide and 4mm height, the outer configuration of the model is made of ABS plates which has a width of 240mm and height of 40mm (*B*/*D*=6). Thin steel plate and ABS plate are bolted together and foam materials are also filled inside to maintain the geometric configuration.

Additional small mass blocks are attached inside the model to satisfy the mass similarity. At both ends of the beam, the translational degrees of freedom are fixed and rotational degrees of freedom are allowed.

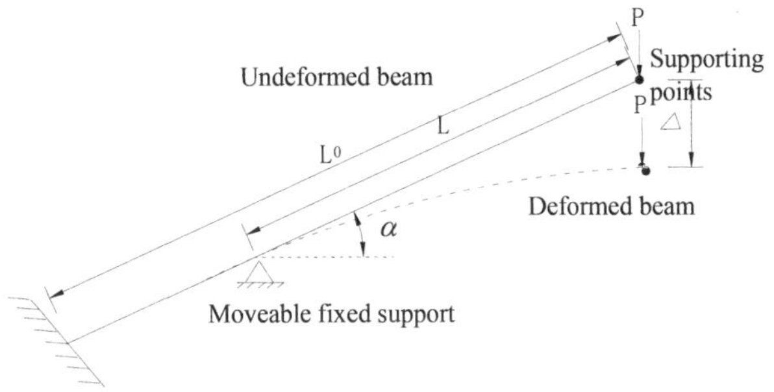

Figure 1: Cantilever beam with variable length for simulation of elastic supports, solid line-the undeformed state of the beam; dash line-the deformed state of the beam subject to weight of the model

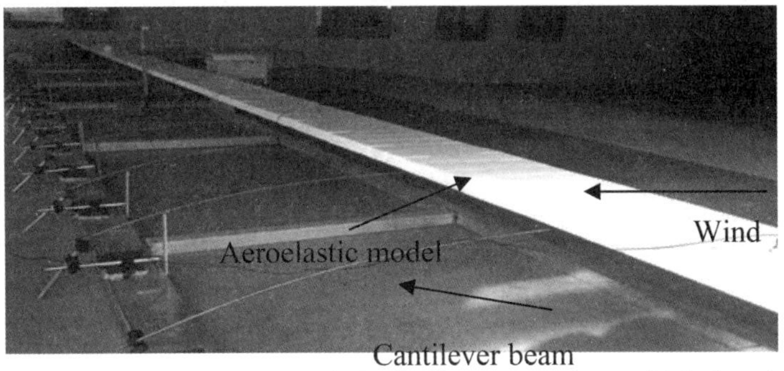

Figure 2: New aeroleastic model elastically-supported at multiple locations

2.2 Dynamic tests

Before wind tunnel test, dynamic tests are carried out to confirm the design method of aeroelastic model. The aeroelastic model is instrumented with 9 mini-accelerators evenly distributed along the span, as shown in Figure 3. A small vibrational exciter placed approximately at quarter span is used to derive the model into resonance with amplitude of 0.5mm, and the vibration amplitude at a particular resonant frequency is used to determine the mode shapes. Two types of methods, namely the forced vibration method and free decay method, are used to extract the modal frequencies and damping ratios of the aeroelastic model. In the forced vibration method, the FRF of the model is obtained by exciting the model with variable frequencies and a least squares fitting to FRF at the neighbourhood of resonant region is used to extract modal frequencies and modal damping ratios. In the free decay method, the model is derived into resonance mode by mode and after shutting down the exciter the free decay response is used to identify the modal frequencies and damping ratios. As for the mode shape, the magnitudes of measured acceleration at different locations are normalized to give mode shape with maximum amplitude being unit for both methods. Table 3 summarizes the identified modal frequencies and damping ratios from the two methods. Figure 3 shows the FRF of the first four modes obtained with forced vibration method. The measured FRF for the first three modes is relatively smooth but those for the remaining modes is quite scatter, and therefore no attempt has been made to identify the higher mode in the cases of forced vibration method. The identified modal frequencies are similar by the two methods. The damping ratios identified with forced vibration is about 50% larger than those by free decay method, which may be attributed by the relative large amplitude in forced vibration. Figure 4 shows the first six mode shapes of the aeroelastic model obtained by forced vibration method, which is in reasonable agreement with the assumed sinusoidal mode shape.

Table 3: The measured modal frequency and damping ratio for the aeroelastic model

Mode	Modal frequency (Hz)		Modal damping ratio (%)	
	Free vibration	Forced vibration	Free vibration	Forced vibration
1st mode (symmetrical)	1.51	1.51	0.50	0.82
2nd mode (asymmetrical)	2.43	2.40	0.71	1.13
3rd mode (symmetrical)	3.42	3.38	0.88	1.30
4th mode (asymmetrical)	4.41	4.34	0.72	1.16
5th mode (symmetrical)	5.42	—	0.82	—
6th mode (asymmetrical)	6.31	—	0.71	—
7th mode (symmetrical)	7.60	—	0.73	—
8th mode (asymmetrical)	9.03	—	0.70	—

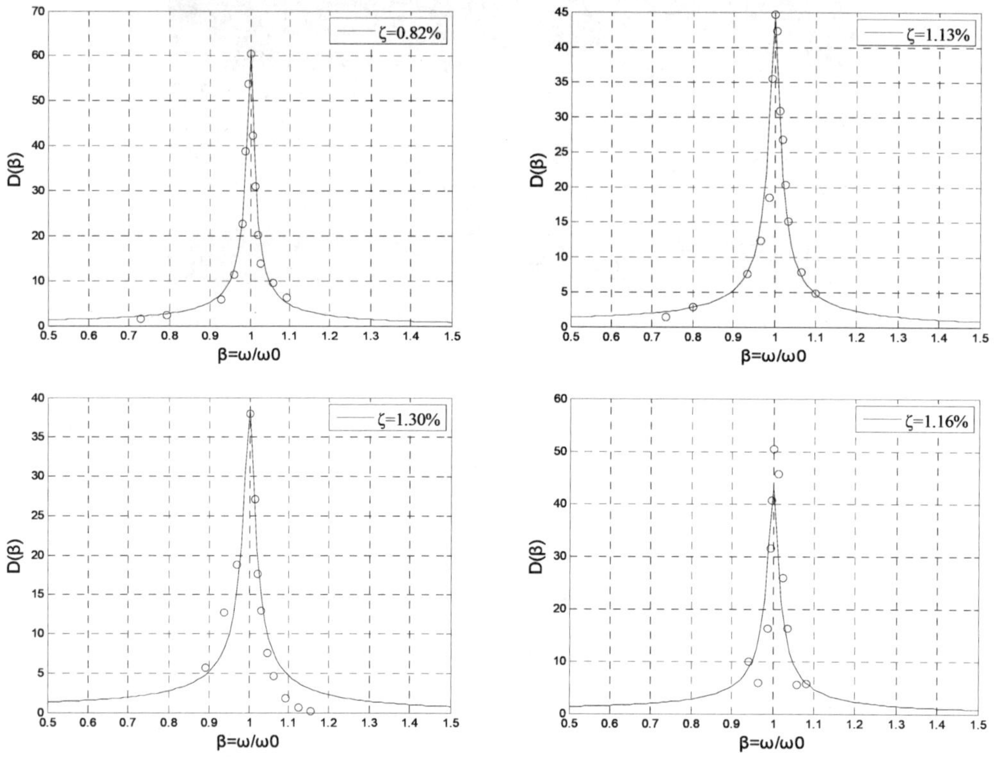

Figure 3: The identified FRF for the first four modes of vibration

For the aeroelastic model with fundamental frequency of 1.51Hz, the VIV onset velocities for the first four modes will be lower than 2m/s, so the free-stream turbulent intensity may be high enough to suppress VIV for these modes. Therefore another case (Case II) is designed by doubling the fundamental frequency but keeping the frequency spacing unchanged. Table 4 shows the identified modal frequencies and damping ratios for the case II.

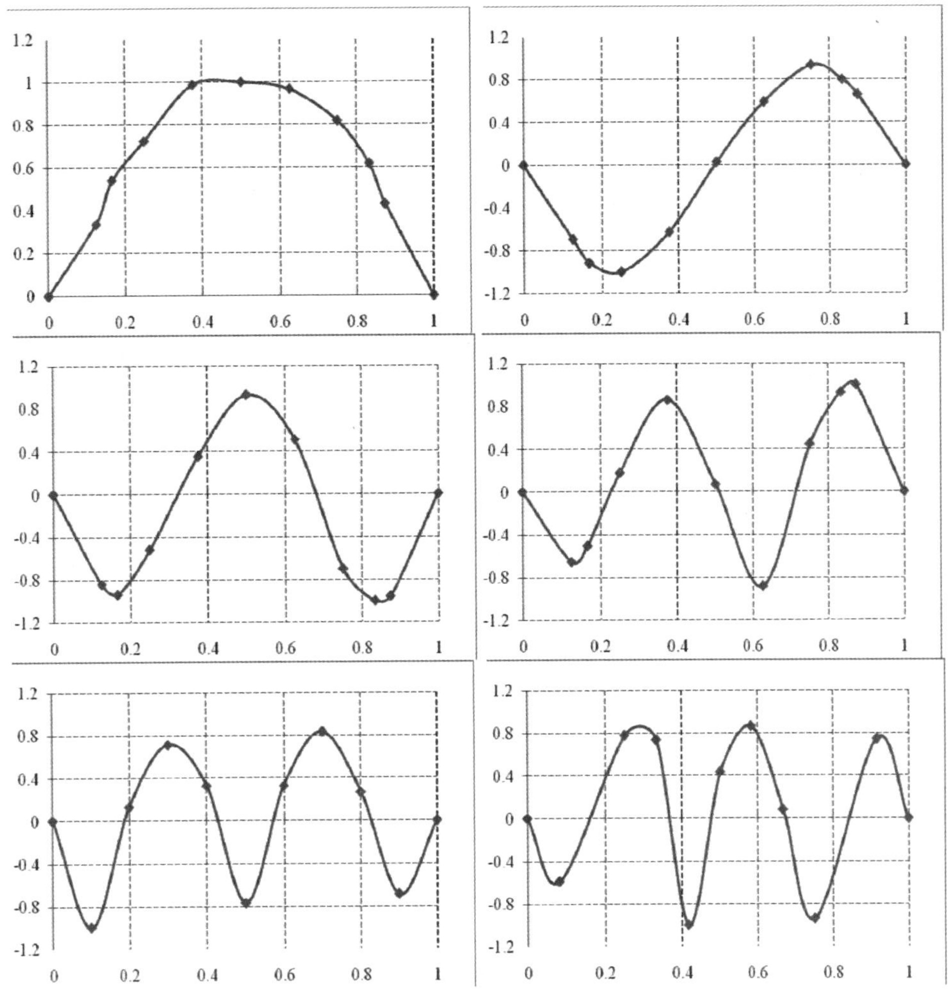

Figure 4: Mode shape of the first six modes

Table 4: The identified modal frequency and damping ratio for case II

Mode	Modal frequency (Hz)	Modal damping ratio (%)	Mode	Modal frequency (Hz)	Modal damping ratio (%)
1st mode	2.73	0.39	5th mode	5.86	0.61
2nd mode	3.61	0.55	6th mode	6.74	0.63
3rd mode	4.49	0.63	7th mode	7.86	0.63
4th mode	5.18	0.70	8th mode	9.18	0.60

3 VIV Test of the Aeroelastic Model

3.1 *Experimental results*

The wind tunnel experiments are carried out in the new 8.5m(W)×2m(H) ×15m(L) test section of the HD-2 wind tunnel. The non-uniformity of mean wind velocity is less than 2% and the turbulent intensity is generally less than 2% for wind velocity larger than 3m/s. The aeroelastic model is tested in wind tunnel with smooth flow for both cases. For case I, the first four modes do not develop VIV due to its low VIV onset velocity and high turbulence intensity, and modes 5~9 develop VIV continuously with increase of wind velocity. For case II, all of the nine modes develops VIV with increase of wind velocity. Table 5 gives the experimental results of vertical vortex-induced vibrations of the aeroelastic model. The *St* number for VIV of every mode is evaluated with the VIV onset velocity and the results are also shown in Table 5. The St number from aeroelastic model is very close to those of the second VIV in the sectional model tests (*Hua et al.* 2013). Figure 5 shows the variation of VIV amplitudes for different vertical modes for both cases. The VIV amplitudes does not decrease with increase of modal frequency.

Table 5: Experimental results of VIV for Case I and Case II

Case	Mode No.	Frequency (Hz)	Damping ratio (%)	VIV velocity U(m/s)	I_U(%)	*Re* number	y_{max}/D	*St* number
I	5	5.42	0.81	2.28~2.79	4.75~3.02	6080~7440	0.014	0.096
	6	6.31	0.71	2.81~3.32	2.98~2.28	7493~8853	0.022	0.090
	7	7.60	0.73	3.30~3.58	2.26~1.92	8800~9547	0.027	0.093
	8	9.03	0.70	3.88~4.39	1.90~1.38	10347~11707	0.029	0.094
	9	10.69	—	4.53~5.46	1.35~1.00	12080~14560	0.030	0.094
II	1	2.73	0.39	1.14~1.42	12.7~10.6	3040~3787	0.017	0.096
	2	3.61	0.55	1.48~1.79	8.77~5.89	3947~4773	0.027	0.098
	3	4.49	0.63	1.89~2.26	5.61~5.10	5040~6027	0.033	0.095
	4	5.18	0.70	2.24~2.53	3.53~3.19	5973~6747	0.034	0.093
	5	5.86	0.61	2.43~2.91	3.34~2.48	6880~7760	0.037	0.096
	6	6.74	0.63	2.91~3.30	2.24~1.96	7760~8800	0.037	0.093
	7	7.86	0.63	3.37~3.90	1.89~1.47	8687~10400	0.042	0.093
	8	9.18	0.60	3.92~4.65	1.39~1.10	10453~12400	0.042	0.094
	9	10.86	—	4.72~5.44	1.13~1.00	12587~14507	0.042	0.092

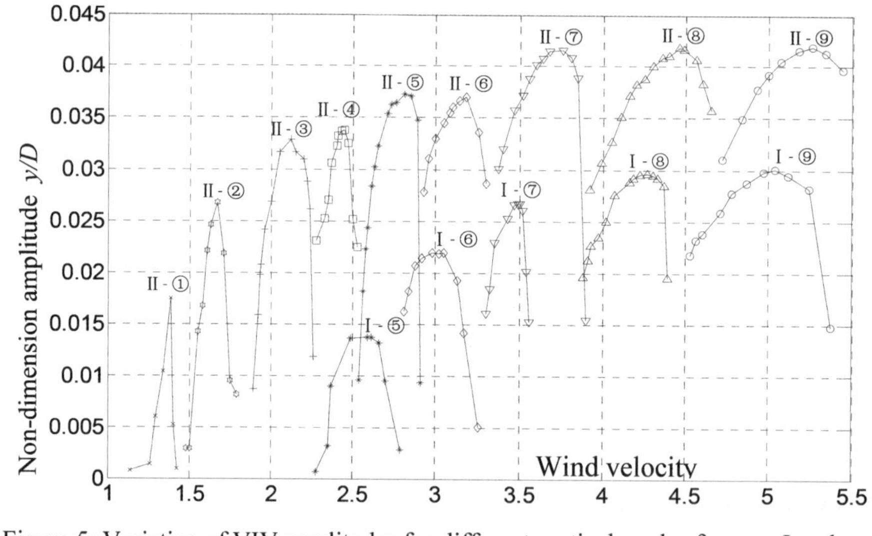

Figure 5: Variation of VIV amplitudes for different vertical modes for case I and case II

3.2 *Comparison with* 1:1 sectional model tests

The 1:1 sectional model of the rectangular section is also tested in wind tunnel with the same vibration frequencies and modal damping ratios as in the aeroelastic model for case II. The comparison between sectional model tests and aeroelastic model tests are shown in Figure 6 for modes 5~7. It is confirmed again that the sectional model tests produce the same results of VIV amplitudes regardless of model vibration frequency, as shown in the companion paper (Hua *et al.* 2013). And the VIV amplitudes from aeroelastic model tests give results consistently 30% larger than those from sectional model tests. Such difference may be explained by the mode shape correction factor (Macdonald *et al.*, 2002).

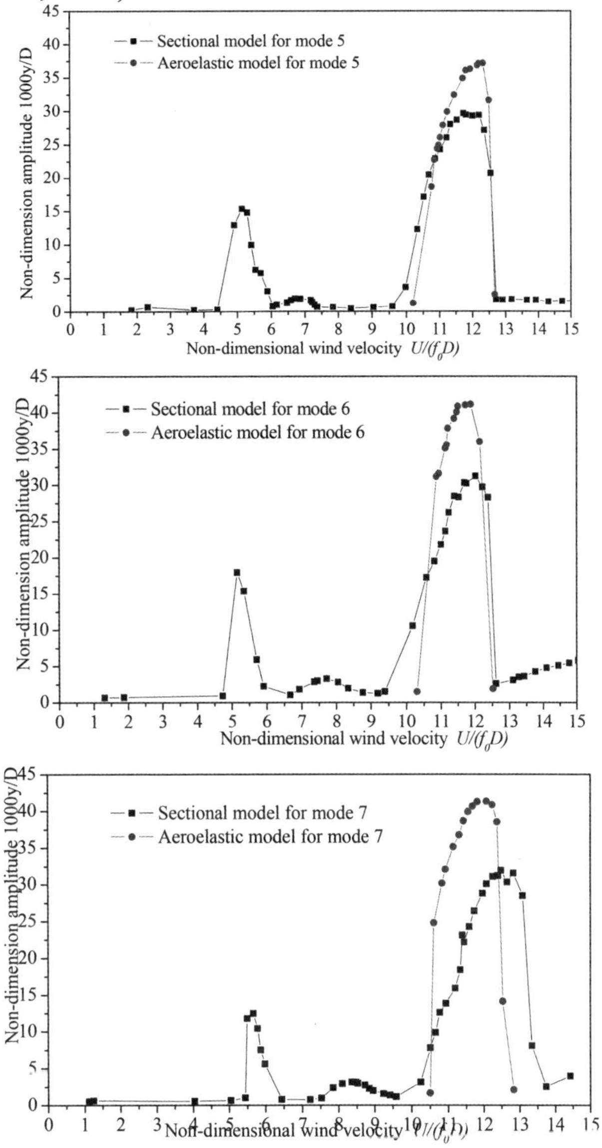

Figure 6: Comparison of VIV amplitude between aeroelastic model and 1:1 sectional model

4 Conclusions

An elastically multi-supported aeroelastic model has been proposed to directly measure the VIV amplitude of different vertical modes as well as to investigate the relationship of VIV amplitudes among different modes of long-span suspension bridges. The method for design such aeroelastic model is validated through dynamic tests. VIV of every important vertical mode is excited one by one with increase of wind velocity and VIV amplitude does not decrease with increase of wind velocity. It is also found that the VIV amplitudes from aeroelastic model tests give results consistently 30% larger than those from 1:1 sectional model tests with the same modal frequencies and damping ratios.

Measurement of span-wise pressures in the model undergoing different modes of vibration is underway.

5 Acknowledgements

The work is supported by the Natural Science Foundation of China (NSFC 51278189), and a key project from Ministry of Transportation of China.

References

Bearman P.W. and Obasaju E.D. 1982. An experimental study of pressure fluctuations on fixed and oscillating square-section cylinders, *Journal of Fluid Mechanics*, **119**, 297-321.

Ehsan F., and Scanlan R.H. 1990. Vortex-induced vibrations of flexible bridges, *Journal of Engineering Mechanics*, ASCE, 116, 1392-1411.

Hua X.G., Chen Z.Q., Chen W, and Huang Z.W. 2013. Higher modes' vertical vortex-induced vibrations of suspension bridge –Part 1: sectional model study, Submitted to the six European-Africa Conference on Wind Engineering, Cambridge, UK.

Macdonald, J. H. G. Irwin P. A. and Fletcher M. S. 2002. Vibrations of the Second Severn Crossing cable-stayed bridge—full-scale and wind tunnel, Paper 12723, *Proceedings of the Institution of Civil Engineers*: Structures & Buildings, 152(2): 123–134

Irwin P.A. 2008. Bluff body aerodynamics in wind engineering, *Journal of Wind Engineering and Industrial Aerodynamics*, **96**, 701-712.

Vickery B.J. 1966. Fluctuating lift and drag on a long cylinder of square cross-section in a smooth flow and in a turbulent flow. *Journal of Fluid Mechanics*, 25, 481-494.

第 28 卷 第 9 期
2015 年 9 月

中 国 公 路 学 报
China Journal of Highway and Transport

Vol. 28 No. 9
Sept. 2015

文章编号:1001-7372(2015)09-0030-08

大跨度桥梁竖弯涡振限值的主要影响因素分析

陈政清,黄智文
(湖南大学 风工程试验研究中心,湖南 长沙 410082)

摘要:为了合理确定大跨度桥梁的竖弯涡振限值,首先综述了其规范值和取值依据,讨论了其主要影响因素,包括人体舒适性和行车安全;然后分别依据人体承受全身振动的评价方法和行车视距的要求,分析了考虑人体舒适性和行车安全影响的大跨度桥梁竖弯涡振限值。研究结果表明:人体舒适性是确定桥梁竖弯涡振限值时必须考虑的因素;大跨度桥梁竖弯涡振频率低,桥上工作人员可能因涡振产生极不舒适的运动病感觉,其对应的涡振限值也最严格;大跨度桥梁的高阶模态涡振可能会对行车视距造成干扰,此时行车安全成为确定涡振限值的重要因素。

关键词:桥梁工程;竖弯涡激共振;理论分析;涡振限值;人体舒适性;行车安全

中图分类号:U441.3　　**文献标志码**:A

Analysis of Main Factors Influencing Allowable Magnitude of Vertical Vortex-induced Vibration of Long-span Bridges

CHEN Zheng-qing, HUANG Zhi-wen
(Wind Engineering Research Center, Hunan University, Changsha 410082, Hunan, China)

Abstract: In order to reasonably determine the allowable magnitude of vertical vortex-induced vibration (VVIV) of long-span bridges, firstly its normative values and the corresponding premises were reviewed, and the main influencing factors were discussed, including human-body comfort and vehicle safety. Then, the allowable magnitude of VVIV of long-span bridges with consideration of human-body comfort and vehicle safety was analyzed according to the evaluation methods for whole-body vibrations and the required sight distance for vehicles on road respectively. The results indicate that human-body comfort should always be taken into account for determining the allowable magnitude of VVIV for all types of bridges. The frequencies of VVIV of long-span bridges are very low, so people who work there are susceptible to the uncomfortable motion sickness resulting from VVIV, and the corresponding allowable magnitudes of VVIV are the strictest. Higher-order modes of VVIV for long-span bridges may interrupt the sight distance of drivers, and at this situation vehicle safety becomes crucial for determining the allowable magnitude of VVIV.

Key words: bridge engineering; vertical vortex-induced vibration; theoretical analysis; allowable magnitude of vortex-induced vibration; human-body comfort; vehicle safety

---

收稿日期:2014-10-25
基金项目:国家自然科学基金项目(91215302,51278189)
作者简介:陈政清(1947-),男,湖南湘潭人,教授,博士研究生导师,工学博士,E-mail:zqchen@hnu.edu.cn。

0 引 言

主梁涡激共振是大跨度桥梁在常遇风速下容易发生的一种风致振动现象，具有自激和自限幅的特性。涡激共振虽然不像颤振一样会在短时间内导致整体结构的毁灭性破坏，但过大的振幅和加速度会严重影响桥梁的使用功能，频繁发生的涡振还可能引起构件的疲劳破坏。因此对于在设计风速内有涡激共振隐患的桥梁应首先根据节段模型风洞试验或全桥气弹模型试验预测实桥的涡振响应，然后根据涡振响应的计算结果和参考的评价指标来评估桥梁的涡振性能，对于涡振性能不满足要求的桥梁，应采用气动措施或机械阻尼措施把涡振幅值降低到评价指标允许的范围。在桥梁设计规范中涡振限值通常以位移或加速度的形式给出，而本文把与桥梁涡振性能评价指标对应的涡振幅值允许值称为涡振限值。

涡振限值是桥梁涡振性能的直接评价指标，也是确定桥梁涡振控制目标的主要依据。准确评价桥梁的涡振性能，探索正常使用极限状态下桥梁涡振限值的合理取值，可以避免涡振限值过于严格而导致的桥梁建造困难，或者涡振限值过于宽松而影响桥梁的正常使用。目前国内外很多重要的桥梁设计规范都给出了桥梁涡振性能的评价指标，但对涡振限值的取值和影响涡振限值的具体因素缺乏深入分析。文献[1]～[3]中均要求从疲劳极限状态对桥梁涡振响应进行评估，但没有阐述正常使用极限状态下桥梁涡振性能的评估方法；文献[4]、[5]中规定了桥梁的涡振容许振幅，但主要用于跨径 200 m 以下的桥梁；此外，文献[6]中规定从行人舒适性、疲劳极限状态和承载能力极限状态方面对桥梁涡振性能进行评估，并给出了行人舒适性的评价指标和涡激共振等效荷载的计算方法。实际上，已有的工程实践表明桥梁涡激共振一般不会引起结构的强度破坏，造成构件疲劳破坏的可能性也较小，桥梁涡振限值通常是由其正常使用极限状态决定的[7-10]。

本文首先对国内外规范中推荐的涡振限值进行了综述，然后从人体舒适性及行车安全 2 个方面分析了涡振限值的取值并与规范值进行了比较，最后对桥梁涡振限值的确定提出了几点建议。需要说明的是，本文只考虑涡激共振的单独作用对桥梁正常使用极限状态的影响，涡振条件下考虑车辆-桥-风耦合的行车舒适性和行车安全问题不属于本文讨论的范围。此外，因为一般情况下大跨度桥梁主梁的扭转频率要比竖弯频率高，桥梁在常遇风速下发生扭转涡振的可能性较小，所以本文讨论的内容仅限于主梁竖弯涡振。

1 桥梁涡振限值的规范值

目前在国内外的桥梁设计规范中只有文献[4]（简称《中国规范》）、文献[5]（简称《日本指南》）以及文献[6]（简称《英国法则》）对主梁涡振限值做了直接规定。此外，Irwin[11] 在 RWDI 公司的技术报告中介绍了在北美和英国通常采用的主梁涡振限值。

《中国规范》采用和《日本指南》相同的涡振限值，本文统称为中日涡振限值，其规定主梁涡振容许振幅为

$$h_a=\frac{0.04}{f_b} \tag{1}$$

式中：h_a 为主梁涡振容许振幅；f_b 为主梁竖弯涡振频率。

因为桥梁涡振一般表现为单一模态的简谐振动，所以竖向涡振位移响应可以用正弦函数或余弦函数近似表示，相应的加速度幅值容许值则可由位移幅值容许值乘以涡振圆频率的平方得到，即

$$A_a=\frac{0.04}{f_b}\omega_b^2=\frac{0.04}{f_b}(4\pi^2 f_b^2)=1.58f_b \tag{2}$$

式中：A_a 为主梁涡振容许加速度幅值；ω_b 为主梁竖弯涡振圆频率。

根据《日本指南》的规定，式(1)主要用于跨度不大于 200 m 的桥梁，对跨度更大的桥梁建议研究专门的设计指标。陈政清在文献[12]中介绍了日本东京大学 Yagi 教授对式(1)的解释，可以看到，式(1)是在同时满足桥梁竖向振动加速度限值等于 1.0 $m\cdot s^{-2}$，竖弯基频采用近似公式 $f_b=100/L$ 计算，取桥梁跨度 $L=160$ m 这 3 个条件下推导出来的。为了定性描述涡振限值与桥梁跨度的关系，把上述基频计算公式代入式(2)可得

$$A_a=\frac{158}{L} \tag{3}$$

可以看到，主梁涡振容许加速度幅值与桥梁跨度成反比，当 $L>160$ m 时，$A_a<1.0\ m\cdot s^{-2}$，因此从满足桥梁竖向振动加速度限值等于 1.0 $m\cdot s^{-2}$ 的角度看，把式(1)用于计算跨度远大于 160 m 的大跨度桥梁是偏于严格的，但这一限值本身对低频振动的大跨度桥梁是否严格则是下文将要讨论的另一个问题。

《英国法则》用动力敏感参数 $K_D=y_m f_b^2$ 来评价桥梁涡振的荷载效应和对行人舒适度的影响，其中

K_D 与行人舒适度等级之间的对应关系如表 1 所示。K_D 的单位为 $mm \cdot s^{-2}$；y_m 表示涡振振幅，单位为 mm。作为补充，本文给出了 K_D 和主梁加速度幅值 A_D 之间的换算关系，即

$$A_D=\frac{y_m f_b^2 4\pi^2}{1\ 000}=0.039K_D \tag{4}$$

表 1 涡激振动效应评价[6]

Tab. 1 Assessment of Vortex-induced Vibration Effect[6]

$K_D/(mm \cdot s^{-2})$	$A_D/(m \cdot s^{-2})$	人体表征
≥100	3.90	可能存在不舒适
50	1.95	
30	1.17	使人感到不愉悦
20	0.78	
10	0.39	
5	0.20	可忍受
3	0.12	
2	0.08	可接受
1	0.04	稍微有感觉

注：运动舒适性评价仅用于涡振临界风速 $V_{cr}<20\ m \cdot s^{-1}$ 的情况。

《英国法则》认为当桥梁某阶模态涡振的临界风速小于 $20\ m \cdot s^{-1}$ 且 $K_D>30\ mm \cdot s^{-2}$ 时就可能导致行人不舒适，应对设计进行修正；因此 $K_D=30\ mm \cdot s^{-2}$（即 $A_D=1.17\ m \cdot s^{-2}$）可以理解为《英国法则》涡振限值，但实际工程中可以根据设计要求选择更高的行人舒适度等级和更严格的涡振限值。

Irwin[11] 在 RWDI 公司的技术报告中认为，对于竖向涡振北美通常采用的设计指标是当风速小于 $13\ m \cdot s^{-1}$ 时，桥梁竖向加速度幅值不超过 $0.5\ m \cdot s^{-2}$；当风速大于 $13\ m \cdot s^{-1}$ 时，桥梁竖向加速度幅值不超过 $1.0\ m \cdot s^{-2}$，本文称为 RWDI 北美涡振限值。英国则采用当风速小于 $20\ m \cdot s^{-1}$ 时主梁竖向加速度幅值不超过 $0.4\ m \cdot s^{-2}$ 作为设计指标，对涡振临界风速大于 $20\ m \cdot s^{-1}$ 的工况没有规定，本文称为 RWDI 英国涡振限值。尽管 Irwin 并没有介绍上述指标的文献出处，但从表 1 中可以看到，主梁加速度幅值等于 $0.4\ m \cdot s^{-2}$ 正好对应《英国法则》定义的“使人感到不愉悦”的下限值，这从侧面说明了在英国的桥梁抗风设计实践中倾向于采用比 $A_D=1.17\ m \cdot s^{-2}$ 更为严格的涡振限值，其中英国第二塞文桥的涡振性能评估就是典型的例子[7]。RWDI 北美涡振限值与风速大小直接相关，可以理解为考虑了涡振临界风速的发生概率和风速大小对涡振限值取值的影响，这与日本东京湾航道桥[8] 涡振容许振幅的取值思路是一致的。

为了更加直观地比较上述几种涡振限值，图 1(a)、(b)分别以加速度和位移的形式给出了上述涡振限值规范值随桥梁涡振频率变化的规律。其中，以加速度形式表示的涡振限值更适合用于行人或车辆乘员的振动舒适性评价，而以位移形式表示的涡振限值则更加直观，也便于工程设计使用。从图 1(a)容易看出：《英国法则》和 RWDI 都以不随频率变化的振动加速度作为桥梁涡振性能的评价指标，而中日涡振限值则是涡振频率的函数。在同一涡振频率下，各规范值之间差异显著，例如当涡振频率 $f_b=0.1$ Hz 时，中日规范中的加速度幅值允许值为 $0.16\ m \cdot s^{-2}$，大约只有《英国法则》涡振限值的 1/7，RWDI英国涡振限值的 1/2.5。当 $0.1\ Hz<f_b<0.25$ Hz时，中日涡振限值最严格；而当 $0.25\ Hz<f_b<1.0$ Hz 时，RWDI 英国涡振限值最严格。可以推测，上述差异必然会给设计阶段大跨度桥梁的涡振性能评估带来困难。由图 1(b)可以看到：当涡振频率 $f_b=0.1$ Hz 时，《英国法则》涡振限值为 3.0 m，从保障行车安全和行人安全的角度看这显然是不能接受的。

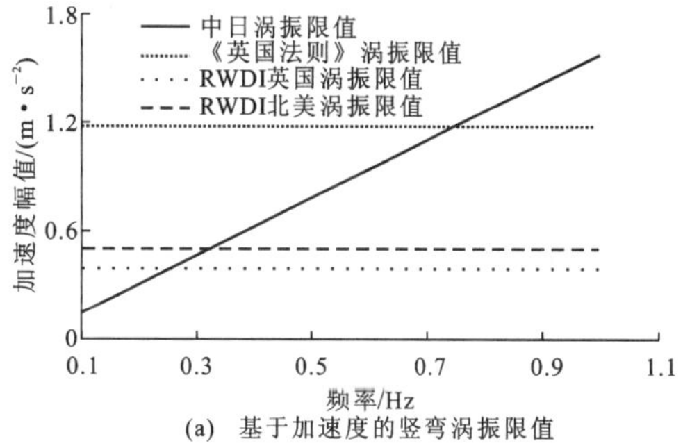

(a) 基于加速度的竖弯涡振限值

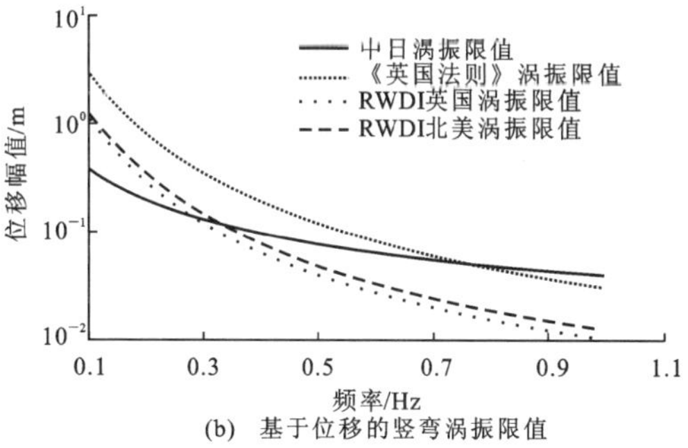

(b) 基于位移的竖弯涡振限值

图 1 不同规范竖弯涡振限值比较

Fig. 1 Comparison of Allowable Magnitudes of VVIV in Different Specifications

值得注意的是，上述涡振限值规范值并没有充分考虑振型以及振动持续时间等其他因素对涡振限值取值的影响。虽然涡振频率可以在一定程度上综合反映振型的差异和涡振风速的不同，但以涡振频

率一个参数来描述涡振限值显然是不全面的。从下文的分析中可以看到，除振动频率外，振型和振动持续时间都可能对涡振条件下的人体舒适性或行车安全造成影响，从而影响涡振限值的取值。

2 桥梁涡振对人体舒适性的影响

车辆乘员舒适性和桥上人员舒适性是考察桥梁正常使用极限状态的2个重要方面，它们都属于人体承受全身振动的舒适性问题。人体对振动的感知和评价是一个非常复杂的问题，它受振动幅值和频率、人体姿势以及振动持续时间等众多因素的影响，而且还有很大的主观性[13]。总体而言，人体承受振动时的反应包括心理反应和生理反应，其中“不舒适”等不好的心理反应源自不期望的振动，而不良的生理反应（例如晕船）则来源于低频、大振幅振动[14]。研究桥梁涡振对车辆乘员舒适性和桥上人员舒适性影响的理想方法是现场实测或在实验室中模拟桥梁涡振的低频、大振幅特性以及行车和行人环境。然而这样的试验条件是很难实现的，目前也没有文献通过试验方法专门研究过桥梁涡振对车辆乘员舒适性或桥上人员舒适性的影响。文献[15]～[17]系列标准是由国际标准化协会颁布的，用于评价通过立姿人的脚，坐姿人的臀部、背部和脚以及躺卧人的支撑区域传递到人体的全身振动，被广泛用于车辆乘员和船舶行驶等情况下人体承受全身振动的舒适性评价[18-20]。本节将基于该系列标准分析大跨度桥梁竖弯涡振对包括车辆乘员舒适性、桥上工作人员舒适性和行人舒适性在内的人体振动舒适性的影响，并讨论相应的涡振限值的取值。

2.1 对车辆乘员舒适性的影响

本文所指的车辆乘员舒适性只考虑车辆静止于桥面或车辆在桥上缓慢通行时桥梁涡激共振对车辆乘员（驾驶人和乘客）的影响。涡振是否加剧车桥耦合振动及车辆乘员的不舒适性不属于本文讨论的范围。实际上，Oehler[21]在研究车辆通过高速公路桥梁时的车辆乘员舒适性和行人舒适性时指出，运动车辆中的人对桥梁振动并不敏感，主要是静止车辆中的人或桥上行人对桥梁振动有不适反应。当车辆静止于桥面或行驶速度缓慢时，车辆乘员承受全身振动的幅值和频率与其所处位置的涡振幅值和频率相同，因而可以直接从桥梁涡振的幅值和频率出发来评价车辆乘员的振动舒适性，反之也可通过相应舒适度指标直接确定该工况下的桥梁涡振限值。

在最新的国际标准（ISO-2631-1—1997）中考虑振动对人体健康、舒适与感知可能影响的频率范围为0.5～80 Hz，对人体运动病可能影响的频率范围为0.1～0.5 Hz。运动病是对晕车、晕船等类似病症的总称，它可能导致人体出现出汗、反胃和呕吐等不良生理反应[13]。表2为国内外一部分有现场实测记录的桥梁竖弯涡振信息，表中 f_i 表示主梁第 i 阶竖弯模态涡振。由表2可以看到，大部分大跨度桥梁的竖弯涡振频率都在0.5 Hz以下，最大竖弯涡振频率也不超过0.75 Hz，由此可以推断桥梁涡振最有可能导致静止或缓慢行驶车辆中的人产生运动病。鉴于ISO-2631-1—1997只提供了运动病发病率的计算方法而没有给出与运动病相关的振动限值，本文将采用专门针对竖向振动导致的运动病评价标准ISO-2631-3—1985来分析这个问题。图2为ISO-2631-3—1985给出的可能导致运动病的人体振动加速度均方根限值，称为“极度不舒适限定值”。可以看到，“极度不舒适限定值”与人体承受振动的时间及振动频率有关。承受振动的时间分为3个等级，分别为30 min，2 h及8 h。因为实际情况中车辆在桥上可能停留的时间通常要小于30 min，因此可以保守地选择“30 min运动病限值”作为考虑车辆乘员舒适性影响的桥梁竖弯涡振限值。

从图2可以看到：运动病考虑的振动频率范围为0.1～0.63 Hz，其中当 $0.1\ \mathrm{Hz}<f_b<0.315\ \mathrm{Hz}$ 时，“30分钟运动病限值”对应的加速度均方根值等于 $1.0\ \mathrm{m\cdot s^{-2}}$；而当 $0.315\ \mathrm{Hz}<f_b<0.63\ \mathrm{Hz}$ 时，加速度均方根限值按照每倍频程增加10 dB的规律从 $1.0\ \mathrm{m\cdot s^{-2}}$ 增加到 $3.15\ \mathrm{m\cdot s^{-2}}$。对比图2和图1(a)可以发现：当前各国规范中的涡振限值都比“30 min运动病限值”严格，因此大跨度桥梁涡振导致车辆乘员运动病的可能性不大；但个别对低频振动特别敏感的驾驶人，有可能因桥梁涡振而影响正常驾驶。

2.2 对桥上工作人员舒适性的影响

大跨度桥梁的涡振频率一般都在运动病的敏感频率范围之内，桥梁在通车前后都存在施工或维护人员在桥上长时间工作的情况，宜按图2的“8 h运动病限值”曲线评价涡振时桥上工作人员的振动舒适性。由图2可以看到：需要求涡振加速度均方根值低于 $0.25\ \mathrm{m\cdot s^{-2}}$，即加速度幅值低于 $0.35\ \mathrm{m\cdot s^{-2}}$，桥上工作人员的舒适性才有保障。大跨度桥梁往往在正式通车前就可能发生涡激共振现象，并导致桥上的施工人员感到不舒适而影响正常施工，他们可能还会把这种不期望的桥梁振动及不舒

表 2　国内外大跨度桥梁竖弯涡振实测数据

Tab. 2　Information of VVIV of Long-span Bridges at Home and Abroad from Field Measurements

桥梁名称	主跨度/m	桥型	涡振频率/Hz	涡振风速/($m \cdot s^{-1}$)
日本东京湾航道桥[7]	240+240	连续钢箱梁桥	f_1=0.330 0	13.00～18.00
			f_2=0.470 0	>23.50
英国第二塞文桥[8]	456	叠合梁斜拉桥	f_1=0.326 0	15.00～20.00
挪威 Osterϕy Bridge[9]	595	钢箱梁悬索桥	f_2=0.300 0	5.40～6.10
			f_3=0.400 0	6.70～8.10
			f_4=0.560 0	9.00～13.00
			f_5=0.750 0	14.00～16.70
巴西里约尼泰罗伊大桥[22]	200+300+200	连续钢箱梁桥	f_1=0.340 0	15.00～16.50
英国科索克大桥[23]	240	叠合梁斜拉桥	f_1=0.510 0	22.00～25.00
丹麦大带东桥[24]	1 624	钢箱梁悬索桥	f_3=0.130 0	4.00～6.00
			f_5=0.209 0	6.50～9.70
			f_6=0.242 0	7.50～11.30
中国西堠门大桥[25-26]	1 650	中央开槽钢箱梁悬索桥	f_3=0.182 8	6.10～7.30
			f_4=0.229 7	8.30～8.90
			f_5=0.276 7	8.56～10.21
			f_6=0.326 5	9.74～12.64

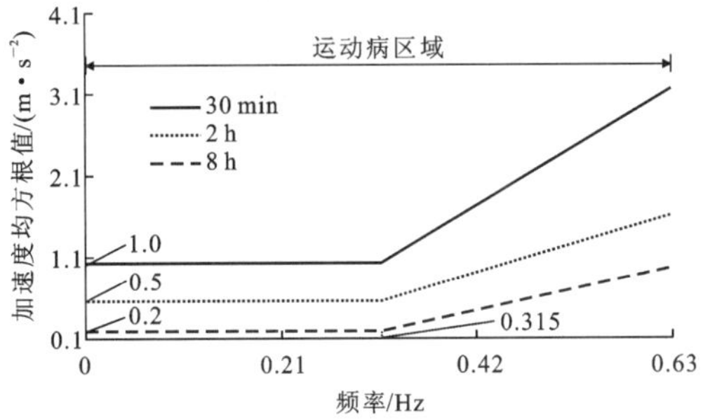

图 2　ISO-2631-3-1985 极不舒适限值

Fig. 2　ISO-2631-3-1985 Severe Discomfort Boundaries

适感与桥梁的设计不合理甚至可能垮塌联系起来，从而对桥梁运营造成不良的社会影响。日本东京湾大桥[7]、丹麦大带东桥[24]以及中国西堠门大桥[25]都是在正式通车前就观测到了显著的涡激共振。其中丹麦大带东桥的涡激共振正是由施工人员首先观察到的，而且从 Frandsen 在文献[27]中的描述可以看到，大带东桥在通车前的涡激共振已使工人感觉非常不舒适。根据表 2 给出的数据不难计算，即使该桥涡振振幅满足中日涡振限值的计算[式(1)]，表 2 中丹麦大带东桥各阶模态涡振对应的加速度均方根值也在 0.145～0.270 $m \cdot s^{-2}$ 之间，可能使桥上工作人员处于“极不舒适”的运动病状态。由此可见，桥上工作人员对涡振反应最为敏感，最为强烈，基于桥上工作人员舒适性的涡振限值要求也就最为严格。

2.3　对行人舒适性的影响

目前桥梁振动对行人舒适度的影响在人行桥人致振动领域研究最多，因此可以参考人行桥设计规范中的行人舒适度评价标准来分析与桥梁涡振相关的行人舒适性问题。在众多的人行桥设计规范中，文献[28]、[29]是当前国际上最新的人行桥设计指南，它们对人行桥行人舒适度的评价标准相同，如表 3 所示。由表 3 可以看出，定义的 4 个舒适度等级仅与主梁加速度幅值有关，而与振动频率无关。然而实际上人的步行特性决定了人行桥人致竖向振动的频率范围为 1.25～2.3 Hz。与桥梁的涡振频率相比，前者属于 ISO-2631 系列标准定义的引起人体不舒适的频率范围，后者属于运动病的频率范围，因此能否把人行桥行人舒适度标准用于确定桥梁涡振限值还有待进一步研究。

表 3　人行桥行人舒适度指标定义[28-29]

Tab. 3　Comfort Index Definition of Pedestrians on Footbridges[28-29]

$m \cdot s^{-2}$

等级	舒适度	竖向加速度	横向加速度
1	最大舒适度	<0.50	<0.10
2	中等舒适度	0.50～1.00	0.10～0.30
3	最小舒适度	1.00～2.50	0.30～0.80
4	不可忍受	>2.50	>0.80

作为一种参考，上述人行桥设计规范中舒适度等级的确定思路对桥梁涡振限值的取值具有更重要的借鉴意义。以文献[29]为例，它建议针对不同的设计工况选择不同的舒适度级别。例如，对于桥梁的竣工典礼，使用期内仅出现一次，对应的舒适度级别为 3 级；对经常往返者，是每天碰到的设计工况，

因此建议的舒适度级别为1级。按照上述舒适度等级的选择方法，在确定桥梁不同竖弯模态涡振的限值时应该结合桥址处的风速统计资料，对发生频率较高的涡振采用“最大舒适度”等级；而对于发生频率较小的则可选择“中等舒适度”。具体的加速度限值可以再结合满足桥梁正常使用极限状态的其他方面和制涡措施的经济代价来确定。

除上述人行桥设计规范外，图2所示的ISO-2631-3—1985运动病评价标准也可用于桥梁涡振的行人舒适度评价。与车辆相比，行人在桥上感受到涡振的时间会更长，结合桥梁涡振可能持续的时间，选择“2 h运动病限值”作为考虑行人舒适性影响的涡振限值可能更加合适。可以看到，在0.1～0.315 Hz的频率范围，“2 h运动病限值”对应的加速度均方根值为0.5 m·s^{-2}，相应的加速度幅值限值为0.7 m·s^{-2}。它属于德国人行桥行人舒适度指标中的中等舒适度，而且相比《英国法则》涡振限值要严格，但比中日涡振限值和RWDI涡振限值都要宽松。

3 桥梁涡振对行车安全的影响

Larsen等[24]在评价丹麦大带东桥的涡振性能时指出：主梁涡振对驾驶人视线的影响会使驾驶人分心并对桥上行车安全造成潜在威胁。图3为大带东桥发生涡振时摄像机在同一地点不同时刻拍摄的2张照片。照片里的圈中车辆为停车状态，由图3可以看到，在波峰时圈中车辆被桥面遮挡了一部分，而当振动回到波谷时摄像机对圈中车辆的视线又恢复了正常。由此可以预见，当涡振振幅增加到一定程度时摄像机位置对圈中车辆的视线将会被桥面全部遮挡。

桥梁竖向涡振之所以会对驾驶人的行车视线造成干扰，主要是因为涡振时桥面会形成凹凸起伏的曲线。为了保证汽车在竖曲线上的行驶安全，《公路工程技术标准》(JTG B01—2003)[30]根据停车视距的要求规定了不同设计车速下竖曲线应满足的最小半径和最小长度；但是仅要求涡振形成的竖曲线满足停车视距是不够的，更严格地要求涡振不干扰驾驶人的行车视线可能更加合理。因为涡振引起的视线干扰不但可能会使驾驶人分心而对行车安全造成潜在威胁，更重要的是它还可能会放大驾驶人对桥梁振动的感受，使他们把这种视线干扰及桥梁振动与桥梁的设计不安全联系起来，从而对桥梁的运营造成不良的社会影响。

(a) 振动的波峰时刻

(b) 振动的波谷时刻

图3 丹麦大带东桥涡振

Fig. 3 Vortex-induced Vibration of Great Belt Suspension Bridge in Denmark

桥梁发生竖向涡振时桥面竖曲线因涡振振型而异，且随振动做周期变化。如果涡振振型是数个波谷和波峰的组合，那么当车辆位于波谷时驾驶人对前方邻近波谷的视线最容易受到前方波峰的阻挡。以有3个半波的涡振振型为例(图4)，用几何作图法对涡振位移达到最大时的行车视线做了简单分析。图4中A_2为车辆所处的波谷位置，A_1为驾驶人的视线高度，假设路面障碍物的高度为h_2，那么可以看到在车辆前进方向的邻近波谷区域存在一个视觉盲区，在这个视觉盲区内高度小于h_2的障碍物对驾驶人都是不可见的。

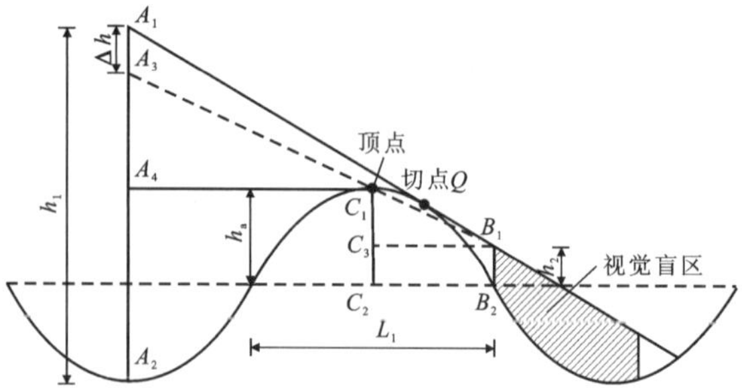

图4 桥梁竖向涡振条件下的驾驶人视距

Fig. 4 Sight Distance of Driver on Bridge Experiencing VVIV

假定桥梁的振型为简谐波形式，且相邻2个驻点之间的间距为L_1。根据图4可以近似计算满足桥梁涡振时桥面行车不存在视觉盲区的涡振振幅限值 h_a。图 4 中驾驶人的目高 $h_1=1.2$ m，障碍物高度 $h_2=0.1$ m[30]。由 $\triangle C_1B_1C_3$ 和 $\triangle A_3C_1A_4$ 的相似关系可得

$$\frac{\overline{C_3B_1}}{\overline{C_1C_3}}=\frac{\overline{A_4C_1}}{\overline{A_3A_4}} \tag{5}$$

式中：$\overline{C_3B_1}=L_1/2$；$\overline{A_4C_1}=L_1$；$\overline{C_1C_3}=h_a-h_2$；$\overline{A_3A_4}=h_1-2h_a-\Delta h$。

代入式(5)计算可得

$$h_a=\frac{h_1+2h_2-\Delta h}{4} \tag{6}$$

如果取 $\Delta h=0$，则可以得到 h_a 的一个估算值 h_{ae}，即

$$h_{ae}=\frac{h_1+2h_2}{4}=0.35\ \text{m} \tag{7}$$

因为 $h_{ae}=0.35$ m 总是要大于满足上述行车视线要求的涡振限值 h_a，所以可以把 h_{ae} 看作是 h_a 的一个上限值。

需要说明的是，当竖弯涡振振型仅为 1 个或者 2 个半波时，波峰对驾驶人视线的阻挡要明显弱于 3 个半波的情形，此时与涡振引起的人体舒适性问题相比保证行车视线对涡振限值的约束较小，因此可以不予考虑；当涡振振型存在 3 个以上的半波时，除前方邻近的波峰会对驾驶人的视线造成阻挡外，更远处的波峰会对驾驶人更远处的视线造成阻挡。考虑到在行车过程中驾驶人总是最关注前方离自己较近的路况，因此对于存在 3 个以上半波的涡振仍然可以取 $h_{ae}=0.35$ m 作为满足行车视线要求的涡振限值。从图 1(b)可以看到，当涡振频率为 0.1 Hz 时，中日涡振限值为 0.4 m，RWDI 英国限值为 1.01 m，而《英国法则》涡振限值达到了 3.0 m。因此它们都可能导致桥上行车存在视觉盲区，从保证行车安全的角度看是不能接受的。

4 结　语

(1)各国规范推荐的桥梁涡振限值存在较大差别，侧重点各有不同。大跨度桥梁的涡振限值应综合考虑人体舒适性和行车安全性的要求，仅满足单个规范要求可能是不够的。

(2)大跨度桥梁涡振频率低，满足人体舒适性的加速度限值随频率降低和持续时间增加而减少，宜以图 2 所示避免运动病的加速度限值作为舒适性指标。车上乘员、桥上行人和桥上工作人员可分别按 30 min，2 h，8 h 的运动病舒适度指标评价桥梁涡振性能和确定涡振限值。由于桥上工作人员的舒适度要求很高，大跨度桥梁一旦发生涡振，桥上工作人员就很容易产生很大反应，是导致桥梁涡振特别受到公众关注的重要原因。

(3)由于大跨度桥梁涡振频率和行人步行频率范围的差异，能否采用人行桥的行人舒适性指标进行桥梁涡振行人舒适度设计还有待试验研究。

(4)为了避免桥梁竖向涡振对驾驶人的行车视线造成影响，对于具有 3 个或者 3 个以上半波的高阶模态涡振，涡振振幅至少应限制在 0.35 m 以下，此时行车视距可能成为确定涡振限值的控制因素。

(5)本文研究成果可为大跨度桥梁的竖弯涡振性能评估和涡振控制指标的确定提供重要的参考依据。桥梁涡振引起的振动舒适性问题非常复杂，本文的分析以理论为主，进一步的结论还有待试验研究。此外，大跨度桥梁扭转涡振限值的取值方法也还有待分析。

参考文献：

References：

[1] EN 1991-1-4：2004，Eurocode 1：Actions on Structures—General Actions Part—1-4：Wind Actions[S].

[2] CAN/CSA-S6-06，Canadian Highway Bridge Design Code[S].

[3] AASHTO LRFD-2012，Bridge Design Specifications[S].

[4] JTG/T D60-01—2004，公路桥梁抗风设计规范[S]. JTG/T D60-01—2004，Wind-resistant Design Specification for Highway Bridges[S].

[5] SATO H. Wind Resistant Design Manual for Highway Bridges in Japan[J]. Journal of Wind Engineering and Industrial Aerodynamics，2003，91(12)：1499-1509.

[6] BD 49/01，Design Rules for Aerodynamic Effects on Bridges[S].

[7] MACDONALD J H G，IRWIN P A，FLETCHER M S. Vortex-induced Vibrations of the Second Severn Crossing Cable-stayed Bridge-full-scale and Wind Tunnel Measurements[J]. Structures & Buildings，2002，152(2)：123-134.

[8] FUJINO Y，YOSHIDA Y. Wind-induced Vibration and Control of Trans-Tokyo Bay Crossing Bridge[J]. Journal of Structural Engineering，2002，128(8)：1012-1025.

[9] ISAKSEN B，STRØMMEN E，GJERDING S K. Sup-

pressing of Vortex Shedding Vibrations at Osterøy Suspension Bridge[C]//KROKEBORG J. Strait Proceeding of the Fourth Symposium on Strait Crossings. London:Swets & Zeitlinger,2001:99-105.

[10] LARSEN A. Aerodynamic Aspects of the Final Design of the 1 624 m Suspension Bridge Across the Great Belt[J]. Journal of Wind Engineering and Industrial Aerodynamics,1993,48(2):261-285.

[11] IRWIN P A. Motion Criteria[R]. Ontario:Technical Notes of the Rowan Williams Davies & Irwin Inc.,1999.

[12] 陈政清.工程结构的风致振动、稳定与控制[M].北京:科学技术出版社,2013.
CHEN Zheng-qing. Wind-induced Vibration,Stability and Control of Engineering Structures[M]. Beijing:Science Press,2013.

[13] GRIFFIN M J. Handbook of Human Vibration[M]. London:Academic Press,1990.

[14] GAUNT J T,SUTTON C D. Highway Bridge Vibration Studies[R]. Indiana:Indiana Department of Transportation and Purdue University,1981.

[15] ISO-2631:1978,Guide for the Evaluation of Human Exposure to Whole-body Vibration[S].

[16] ISO-2631-3:1985,Evaluation of Human Exposure to Whole-body Vibration—Part 3:Evaluation of Exposure to Whole-body Z-axis Vertical Vibration in the Frequency Range 0.1 to 0.63 Hz[S].

[17] ISO-2631-1:1997,Mechanical Vibration and Shock—Evaluation of Human Exposure to Whole-body Vibration—Part 1:General Requirements[S].

[18] XU Y L,GUO W H. Effects of Bridge Motion and Crosswind on Ride Comfort of Road Vehicles[J]. Journal of Wind Engineering and Industrial Aerodynamics,2004,92(7):641-662.

[19] 韩万水,陈艾荣.风环境下大跨度斜拉桥上的车辆驾驶舒适性评价[J].中国公路学报,2008,21(2):54-60.
HAN Wan-shui,CHEN Ai-rong. Ride Comfort Assessment of Vehicles Running on Long-span Cable-stayed Bridge under Crosswind[J]. China Journal of Highway and Transport,2008,21(2):54-60.

[20] SARIÖZ K,NARLI E. Effect of Criteria on Seakeeping Performance Assessment[J]. Ocean Engineering,2005,32(10):1161-1173.

[21] OEHLER L T. Bridge Vibration—Summary of Questionnaire to State Highway Departments[R]. Michigan:Michigan State Highway Departments,1970.

[22] BATTISTA R C,PFEIL M S. Control of Wind Oscillations of Rio-Niterói Bridge, Brazil[J]. Structures and Buildings,2010,163(SB2):87-96.

[23] OWEN J S,VANN A M,DAVIES J P,et al. The Prototype Testing of Kessock Bridge:Response to Vortex Shedding[J]. Journal of Wind Engineering and Industrial Aerodynamics,1996,60(4):91-108.

[24] LARSEN A,ESDAHL S,ANDERSEN J E,et al. Storebaelt Suspension Bridge-vortex Shedding Excitation and Mitigation by Guide Vanes[J]. Journal of Wind Engineering and Industrial Aerodynamics,2000,88(2):283-296.

[25] LI H,LAIMA S,OU J,et al. Investigation of Vortex-induced Vibration of a Suspension Bridge with Two Separated Steel Box Girders Based on Field Measurements[J]. Engineering Structures,2011,33(6):1894-1907.

[26] LI H,LAIMA S,ZHANG Q,et al. Field Monitoring and Validation of Vortex-induced Vibrations of a Long-span Suspension Bridge[J]. Journal of Wind Engineering and Industrial Aerodynamics,2014,124(1):54-67.

[27] FRANDSEN J B. Simultaneous Pressures and Accelerations Measured Full-scale on the Great Belt East Suspension Bridge[J]. Journal of Wind Engineering and Industrial Aerodynamics,2001,89(1):95-129.

[28] Sétra:2006,Footbridges—Assessment of Vibrational Behaviour of Footbridges Under Pedestrian Loading Practical Guidelines[S].

[29] RFS2-CT-2007-00033,Design of Footbridges Guideline[S].

[30] JTG B01—2003,公路工程技术标准[S].
JTG B01—2003,Technical Standard of Highway Engineering[S].

Engineering Structures 141 (2017) 263–271

Contents lists available at ScienceDirect

Engineering Structures

journal homepage: www.elsevier.com/locate/engstruct

Experimental investigation of correction factor for VIV amplitude of flexible bridges from an aeroelastic model and its 1:1 section model

S. Zhou, X.G. Hua *, Z.Q. Chen, W. Chen

Key Laboratory for Wind and Bridge Engineering of Hunan Province, College of Civil Engineering, Hunan University, 410082 Changsha, Hunan, China

ARTICLE INFO

Article history:
Received 17 January 2017
Revised 14 March 2017
Accepted 15 March 2017
Available online 24 March 2017

Keywords:
Suspension bridges
Vortex-induced vibration
Aeroelastic model
Section model
Spanwise correlation

ABSTRACT

Evaluation of the maximum amplitudes of vortex-induced vibrations (VIV) in suspension bridges is mainly relied upon section model wind tunnel tests of bridge decks. Extrapolation to prototype responses requires addition correction to take into account the differences both in structural mode shape and in spanwise correlation of excitation forces between the section model and the prototype. The paper presents the results of VIV amplitudes measured on an aeroelastic model and on its section model with the same size of cross-section. The aeroelastic model is a new, elastically multi-supported one which is capable of reproducing the closely-spaced vertical modes of a suspension bridge. The vortex-induced vibrations for a number of vertical modes were successfully measured from the aeroelastic model test in smooth flow. The spring-mounted section model having the same size of cross-section as the aeroelastic model was also tested in smooth flow at the same mass, damping ratio and vibration frequency. It is shown that both the VIV onset wind velocity and lock-in range for each mode of vibration are in good agreement between the aeroelastic model and the section model. The aeroelastic model consistently produces a roughly 30% higher maximum amplitude than the section model for the observed modes of vibration. It is confirmed that the combined three-dimensional (3D) effects due to mode shape and imperfect correlation of excitation forces amplify the VIV amplitudes and disregarding these effects may underestimate the VIV amplitudes for section model tests. A correction factor of 1.3 for the VIV amplitude of vertical modes in suspension bridges may be adopted for practical use when extrapolating the section-model results to their corresponding full-scale values.

1. Introduction

The regular vortex shedding behind bridge decks placed in a flow induces periodic cross wind lift forces which can set the bridge decks into resonant vibrations if vortex-shedding frequency coincides with one of natural frequencies of the bridge. The vortex shedding frequency of the flow is generally proportional to the wind velocity as described by the Strouhal number $St = fD/U$. Once the bridge decks are set into resonant vibrations, the vortex-shedding frequency of the flow becomes locked onto the vibration frequency of bridge decks such that vortex-induced vibrations sustain over a range of wind velocity. Typically, a suspension bridge is very flexible and has closely-spaced bending modes in vertical direction. Hence, a number of vertical modes of a bridge may be selectively excited in turn with increase of the wind velocity, which was first observed for the first Tacoma Narrows Bridge and continues to occur for several steel box girder suspension bridges. For example, vortex-induced vibrations for the eigenmodes 3, 5, and 6 were observed on the Great Belt Bridge [1]; and vortex-induced vibrations for the eigenmodes 6, 7, 8, and 9 were observed on the Xihoumen Bridge [2]. The vortex-induced vibration may discomfort the users and even result in fatigue failure [3]. The problem of VIV is also a serious concern for marine structures and there have been numerous investigations within the broad field of vortex-induced vibrations, as well discussed in the comprehensive reviews [e.g. 4–8].

Due to the complexity of nonlinear fluid and structure interaction, wind tunnel tests on the scaled models remains the primary tool for evaluating the susceptibility of bridge decks to vortex-induced vibrations, and experimental results are commonly expressed in terms of the lock-in ranges of wind velocity and the maximum vibration amplitudes [9,10]. Full-bridge aeroelastic model and elastically mounted section model of bridge decks are commonly employed for wind tunnel tests. The former is a multi degree-of-freedom dynamic model of the prototype bridge allowing for the full three-dimensionality of the flow and the structure and are supposed as reproducing actual bridge responses to wind

* Corresponding author.
E-mail address: cexghua@hnu.edu.cn (X.G. Hua).

flow. Due to their small scale, the aeroelastic models are generally able to simulate only the first two or three lower-order vertical modes, which is inadequate for the higher-order vortex-induced vibrations of long-span suspension bridges. Moreover, the vortex-induced vibration responses are very sensitive to geometry details; the small scale may bring about the inadequacy in reproducing the bridge deck details such as barriers, which could leave large uncertainties on experimental results.

On the contrary, spring-mounted rigid section models merely simulate a limited length of whole bridge decks and therefore can be produced with large scale. The section model wind tunnel tests are based on the assumptions that the shed vortices are two dimensional and the excitation forces due to vortex shedding are fully correlated along the length of the model. However, vortex-induced lift forces over the entire span in prototype bridges do not fully correlated in particular at small amplitudes of vibration, and can vary case by case for different bridge deck cross-sections [11–13]. Lack of knowledge on the correlation of excitation forces could lead to discrepancy between section model experiments and full scale [14]. Another limitation of section model tests is that the model oscillates with the uniform amplitude and does not simulate the mode shapes in a bridge. The mode shape correction factor, also known as modal response factor [15], is invoked to account for the difference in mode shapes between the section model and the prototype bridge [16–18]. In practice, it is generally accepted that the correlation effect and mode shape effect cancel each other so that the overall correction factor for the vibration amplitudes is 1.0 [17].

Several full-scale measurements of vortex-induced vibrations in long-span bridges were carried out. Larsen et al. [1] and Frandsen [19] made independent measurements of wind velocity, acceleration, deck surface pressures during vortex-induced vibrations in the Great Belt Bridge. While the observed lock-in range of wind velocity are consistent with those from section model tests, taut-strip model and full aeroelastic model, the maximum VIV amplitude is larger than the upper bound of wind tunnel results for two damping levels [19]. Fujino and Yoshida [10] carried out the VIV measurements in a ten-span continuous steel box-girder bridge with varying depth of bridge deck. It is indicated the full-scale VIV amplitude is about 50% larger than the section model results. Macdonald et al. [17] reported that the VIV amplitude measured on the Second Severn Bridge is larger than those from section model tests, and showed that the mode shape correction factor may explain the difference. The field measurement on the Xihoumen suspension bridge by Li et al. [20] showed that the full-scale maximum VIV amplitudes are significantly larger than those of section model tests.

The purpose of the present study is to experimentally evaluate the correction factor for the maximum VIV amplitude when extrapolating section-model results to their prototype. It is ideal to directly compare the section model tests with full-scale measurements; however VIV amplitudes from full-scale measurements can be influenced by many factors such as wind direction, turbulence intensity and structural damping. Wind tunnel tests on full aeroelastic model and its section model in well controlled conditions provide a better way for comparison. As stated previously, a traditional full-bridge aeroelastic model of the suspension bridge usually simulates the first two or three lower-order vertical modes. Moreover, the frequencies of taut-strip aeroelastic model developed by Davenport et al. [21] increase linearly, which does not simulate the modal frequency characteristics for the vertical modes of suspension bridges. In this study, a novel elastically multi-supported aeroelastic model is introduced to investigate the vortex-induced vibrations of suspension bridges. The new model consists of a shallow beam with rectangular cross-section and a number of evenly distributed flexible springs to support the beam. Wind tunnel experiments are carried out for the new aeroelastic model and its elastically-mounted section model with the same size at the same flow and structural conditions. The results are presented of the maximum VIV amplitudes measured on the aeroelastic model and on the section model for the same mode of vibration. The experimental correction factors for the VIV amplitudes are also calculated for the observed modes, and they are compared to those obtained using the semi-empirical models of vortex-induced vibrations.

2. Theoretical background

In this section, the semi-empirical linear model for vortex-induced vibrations is briefly presented for discussion and comparison of experimental results, more details on other semi-empirical models can be found in literatures e.g. [9,22–24]. The vortex shedding excitation of a bridge deck usually involves in only one mode of vibration at a time. Therefore the equation of motion for a bridge deck with a length L undergoing vortex-induced vibrations may be expressed as

$$M_r\ddot{Y}_r(t) + 2\omega_r\zeta_r M_r\dot{Y}_r(t) + \omega_r^2 M_r Y_r(t) = \int_0^L \varphi_r(x)P(t)dx \tag{1}$$

where Y_r is the modal displacement; M_r, ω_r, ζ_r and φ_r are the effective mass, the natural circular frequency, damping ratio and mode shape for the rth mode of vibration, respectively; $P(t)$ is the vertical lift force due to vortex shedding.

The lift force can be approximately represented as a simple harmonic excitation, as

$$P(t) = \frac{1}{2}\rho U^2 D\tilde{C}_L \sin(\omega t + \psi) \tag{2}$$

where ρ is the air density; U is the flow velocity; D is the cross-wind dimension of a cross-section; $\tilde{C}_L$ is the lift coefficient of vortex shedding; and ω is the vortex shedding frequency.

Substituting Eq. (2) into Eq. (1), one obtains

$$\begin{aligned} & M_r\ddot{Y}_r(t) + 2\omega_r\zeta_r M_r\dot{Y}_r(t) + \omega_r^2 M_r Y_r(t) \\ & = \int_0^L \varphi_r(x)\frac{1}{2}\rho U^2 D\tilde{C}_L \sin(\omega t + \psi)dx \end{aligned} \tag{3}$$

When the bridge is excited into resonance by vortex excitation, vortex shedding frequency ω will be locked to natural frequency, ω_r, of the bridge for a certain range of wind velocity. The solution for the amplitude of modal displacement Y_r is given as

$$Y_{r,\max} = \frac{\frac{1}{2}\rho U^2 DR\tilde{C}_L \int_0^L \varphi_r(x)dx}{M_r\omega_r^2}\frac{1}{2\zeta_r} \tag{4}$$

where the factor R is employed accounting for imperfect correlation of vortex shedding forces over the bridge deck length [22].

The normalized/non-dimensional vibration amplitude can be rewritten as

$$\begin{aligned} \bar{Y}_{3D} & = \varphi_{r,\max}\frac{Y_{r,\max}}{D} = \varphi_{r,\max}\frac{\frac{1}{2}\rho U^2 R\tilde{C}_L \int_0^L \varphi_r(x)dx}{m\int_0^L \varphi_r^2(x)dx\cdot(2\pi f_r)^2}\cdot\frac{1}{2\zeta_r} \\ & = \frac{1}{4\pi}\cdot\frac{1}{S_C}\cdot\frac{1}{St^2}\cdot R\cdot\tilde{C}_L\cdot\frac{\varphi_{r,\max}\int_0^L \varphi_r(x)dx}{\int_0^L \varphi_r^2(x)dx} \end{aligned} \tag{5}$$

where m is the equivalent mass per unit length of the bridge deck; S_C is the mass-damping parameter (Sc number), $Sc = 4\pi m\zeta_r/\rho D^2$; St is the Strouhal number of bridge deck cross section, $St = f_r D/U$; the last term in right hand side of Eq. (5) represents the mode shape correction factor for the vibration amplitude, which is dependent on mode shape. Eq. (5) is one of the VIV maximum amplitude estimation formula used in Euro Code [16,25].

By Eq. (5), it is seen that the structural and aerodynamic parameters including Sc number, St number, lift coefficient and its spanwise correlation, and mode shape affect the maximum amplitude of vortex-induced vibrations. As mentioned earlier, it is common to employ section model test to predict the maximum amplitudes of vortex-induced vibration in prototype bridges. However, the section model test does not consider the mode shape and imperfect spanwise correlation of vortex shedding forces and may lead to unreliable results. In this study, wind tunnel tests are conducted with a novel aeroelastic model and its elastically-mounted section model with the same size. Under the same geometric scale, the discrepancies in cross-section dimension, Reynolds number, mass-damping parameter, and natural frequencies are eliminated, the difference in the maximum VIV amplitudes between these two models are purely caused by the three dimensional effects, namely mode shape and spanwise correlation of vortex shedding forces. By this way, the correction factor for the maximum VIV amplitudes between the section model and the aeroelastic model is evaluated.

3. Experimental procedure

3.1. Elastically multi-supported aeroelastic model

In this study, a novel elastically multi-supported aeroelastic model is proposed for studying the vortex-induced vibrations of a suspension bridge. As shown in Fig. 1, the model consists of a shallow beam and a number of evenly distributed flexible springs to support the beam. The shallow beam is made up of inner steel component for simulating vertical stiffness of a bridge deck and cladding component for representing the geometry of a bridge deck. The advantage of this new model over the traditional full aeroelastic model is its capability in reproducing as many as closely-spaced vertical modes of a suspension bridge, as will be shown later. The key in developing the new aeroelastic model is to determine the appropriate number of elastic supports as well as their stiffness by using the assumed mode shape method.

A suspension bridge design scheme with a main span of 1620 m is used as prototype for designing and constructing the aeroelastic model. For simplification, a rectangular cross-section with a side ratio $B/D = 6$ is adopted for the bridge deck. Only asymmetrical vertical modes with a stationary node at mid-span are considered in this study, which implies that half bridge span will be adequate for design of the aeroelastic model. The geometric scale of aeroelastic model is taken as 1:100. Table 1 shows the main structural parameters for the prototype bridge and its aeroelastic model. Following structural and modal parameters are employed: the effective mass of bridge deck (including those of cables) 3×10^4kg/m; vertical flexural stiffness of bridge deck $EI = 8.4 \times 10^{11}$N m; natural frequency of the fundamental vertical mode is 0.12 Hz and frequency step Δf is 0.08 Hz.

The stiffness of bridge deck is modeled by a thin steel plate and the stiffness of main cables is represented by a number of flexible supports evenly distributed along the span. For suspension bridges in particular for single-span one, mode shape of bridge deck is very close to a sinusoidal one, and therefore it is reasonable to employ a form of $\phi_r(x) = \sin r\pi x/L$. By assuming that the elastic supports is evenly distributed and stiffness of the ith elastic supports is K_i, the generalized stiffness and mass for the rth mode of the aeroelastic model are given in Eqs. (6) and (7).

$$K_r^* = \int_0^L EI(\phi_r'')^2 dx + \sum_{i=1}^{N} K_i[\phi_r(iL/N+1)]^2 \tag{6}$$

$$m_r^* = \int_0^L \rho A \phi_r^2 dx + \sum_{i=1}^{N} m_i[\phi_r(iL/N+1)]^2 \tag{7}$$

where N is the number of flexible supports which is taken as 17 in this study, EI is the model-scale flexural stiffness of bridge deck; ρA is the model-scale mass of bridge deck; and m_i is the effective mass of elastic supports, which is about 1/5.5 of the mass of the cantilever rod. The aeroelastic model is intended to simulate the first eight to ten vertical modes of a suspension bridge, therefore the system of equations for solving support stiffness K_i based on Eq. (6) remains underdetermined. For the verification of method of design such aeroelastic model, all the flexible support stiffness is assumed to have the same stiffness K. Therefore the stiffness of flexible support K is determined by using only the first modal frequency.

Fig. 2 shows the 8.1 m-long aeroelastic model in wind tunnel. The stiffness of the aeroelastic model is provided by an inner thin steel beam, while the cladding of the model is made of plastic materials, a kind of thin, light and rigid plastic called "ABS". As shown in Fig. 2, stiffness of the evenly-distributed elastic springs is realized by cantilever rods with variable length, and it is convenient to adjust the rods' overall vertical stiffness by simply changing their cantilever length. Extra small mass blocks are attached inside the model to satisfy the mass similarity. Fig. 3 illustrates the layout of cladding of the aeroelastic model. The cladding of the model is divided into 108 segments and each segment is 75 mm long. A net spacing of 2 mm between two adjacent segments is left to reduce structural damping. Each segment of cladding is fixed to the steel beam by two screws, and foam is also used to ensure the surface evenness of the model.

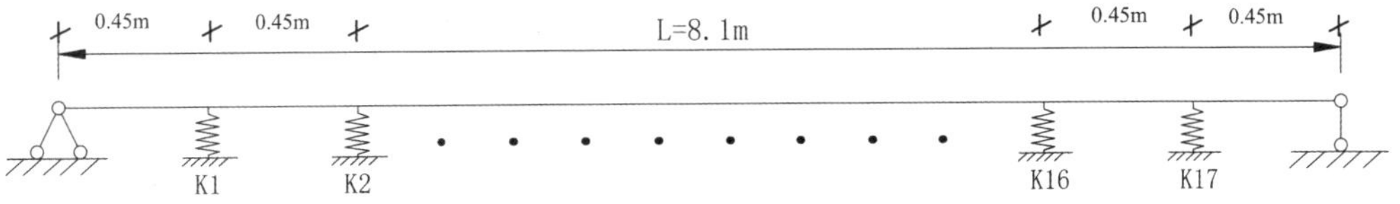

Fig. 1. Elastically multi-supported aeroelastic model.

Table 1
Structural parameters for prototype bridge and aeroelastic model.

Parameter	Unit	Scale ratio	Prototype	Model
Length L	m	$\lambda_L = 1/100$	810 (half of span)	8.10
Width B	m	$\lambda_L = 1/100$	24	0.24
Height H	m	$\lambda_L = 1/100$	4	0.04
EI	N m	$\lambda_{EI} = \lambda_L^5 = 1/100^5$	$2.1 \times 10^{11} \times 4 = 8.4 \times 10^{11}$	84
Mass	kg/m	$\lambda_m = \lambda_L^2 = 1/100^2$	30,000	3
Frequency f	Hz	$\lambda_f = \lambda_V/\lambda_L = 10$	0.12	1.2

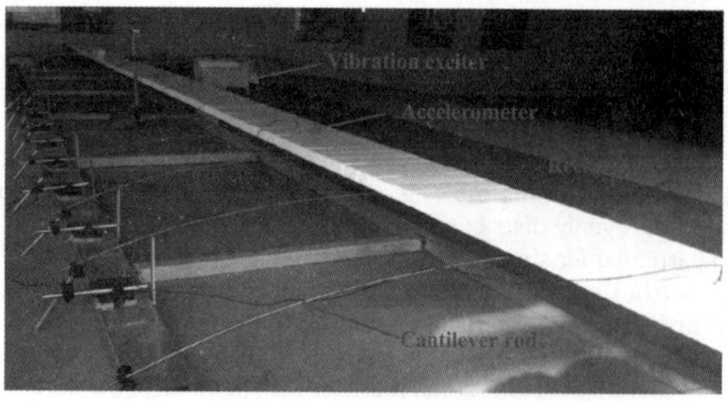

Fig. 2. Elastically multi-supported aeroelastic model in wind tunnel.

Prior to wind tunnel tests, dynamic tests are carried out to confirm modal properties of the model. A small-scale vibration exciter is used to drive the model into forced vibration, and the vibration amplitude at a particular resonant frequency is used to determine the mode shapes. Fig. 4 shows measurement locations of vertical acceleration at $3L/24$, $4L/24$, $6\ L/24$, $9L/24$, $12L/24$, $15L/24$, $18L/24$, $20L/24$, $21L/24$ respectively. For higher-order modes, more measurement locations are used to provide high-resolution mode shape.

Table 2 summarizes the modal frequency and damping ratio for the aeroelastic model obtained by free vibration data and forced vibration data, respectively. For forced vibration test, frequency response function (FRF) for each vibration mode is obtained by exciting aeroelastic model with varying frequencies and a least-squares curve fit to the measured FRF near resonant region is used to extract modal frequencies and damping ratios. In free vibration test, the model is derived into resonance mode by mode and after shutting down the exciter the free-decay response is used to identify modal frequencies and damping ratios. It is shown that the modal frequencies obtained with the two methods are almost identical, but the damping ratio deviates to some extent. Fig. 5 shows the mode shape of the first six modes for the aeroelastic model, which is in reasonable agreement with the assumed sinusoidal mode shape.

According to Table 2, the natural frequency of the first vertical mode is about 1.51 Hz, which is larger than the design value, 1.20 Hz. This is not unexpected since the cantiliver rods in the aeroelastic model have an inclination angle while those assumed in the model design is horizontal. Additionally, the static deformation of the cantilever rods due to model weight is sufficiently large to alter the original stiffness of the undeformed one. After accounting for the combined effects of cantiliver rod inclination and its nonlinearity, the predicted frequency of the first vertical mode is 1.49 Hz which is quite close to the experimental value, implying that the design method for aeroelastic model, as given by Eq. (6) and Eq. (7), is correct.

3.2. 1:1 Section model

Fig. 6 shows the elastically-mounted section model of rectangular cross-section in wind tunnel. The model is 1.51 m long, and has the same size of cross-sections as the aeroelastic model. The section model is made of a kind of rigid poly glass material, and there is a rigid steel rod inside the model to make sure the stiffness of the model itself. The section model surface is covered by the same material to guarantee the surfaced roughness is comparable with the aeroelastic model; moreover, two rectangular end plates are used to provide the 2D flow condition around the model.

Eight vertical coil springs are used to provide the required vibration frequency of the section model in vertical direction, and their stiffness can be adjusted to simulate different vibration frequencies. The torsional stiffness of the suspension system is high enough to eliminate the torsional mode oscillation. Additionally damping may be supplied by magnetic damping if required. The acceleration responses are measured by four mini accelerometers mounted vertically on the model surface. The vertical bending mode response is calculated by averaging the data measured from four accelerometers.

3.3. Test facilities and model parameters

Wind tunnel experiments were performed in the HD-2 wind tunnel laboratory at the Hunan University, China, all in smooth flow. The wind tunnel has three working sections for experiments. The aeroelastic model test is carried out in it's 15 m-long, 8.5 m-wide and 2.0 m-high low-speed working section with a maximum wind speed of 18 m/s; and the section model test is performed in it's 18 m-long, 3.0 m-wide and 2.0 m-high high-speed working section with a maximum wind speed of 56 m/s.

The testing parameters employed for the aeroelastic model and 1:1 section model are summarized in Table 3. The mass parameter

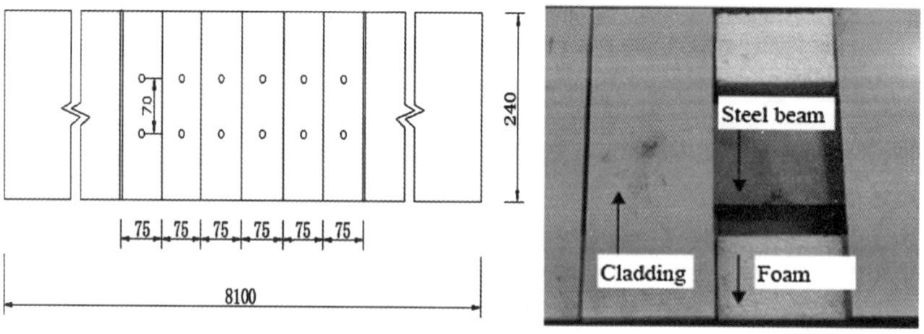

Fig. 3. Layout of cladding of aeroelastic model (unit: mm).

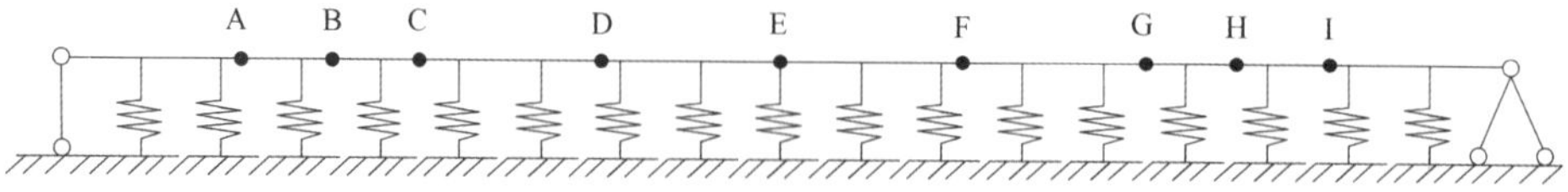

Fig. 4. General layout of accelerometer s for aeroelastic model.

Table 2
Measured modal parameters for aeroelastic model.

Mode	Modal frequency (Hz)		Modal damping ratio (%)	
	Free vibration	Forced vibration	Free vibration	Forced vibration
1st mode (symmetrical)	1.51	1.51	0.72	1.00
2nd mode (asymmetrical)	2.37	2.30	0.91	1.50
3rd mode (symmetrical)	3.36	3.30	1.38	1.78
4th mode (asymmetrical)	4.37	4.30	0.89	1.26
5th mode (symmetrical)	5.86	–	0.61	–
6th mode (asymmetrical)	6.74	–	0.63	–
7th mode (symmetrical)	7.86	–	0.63	–

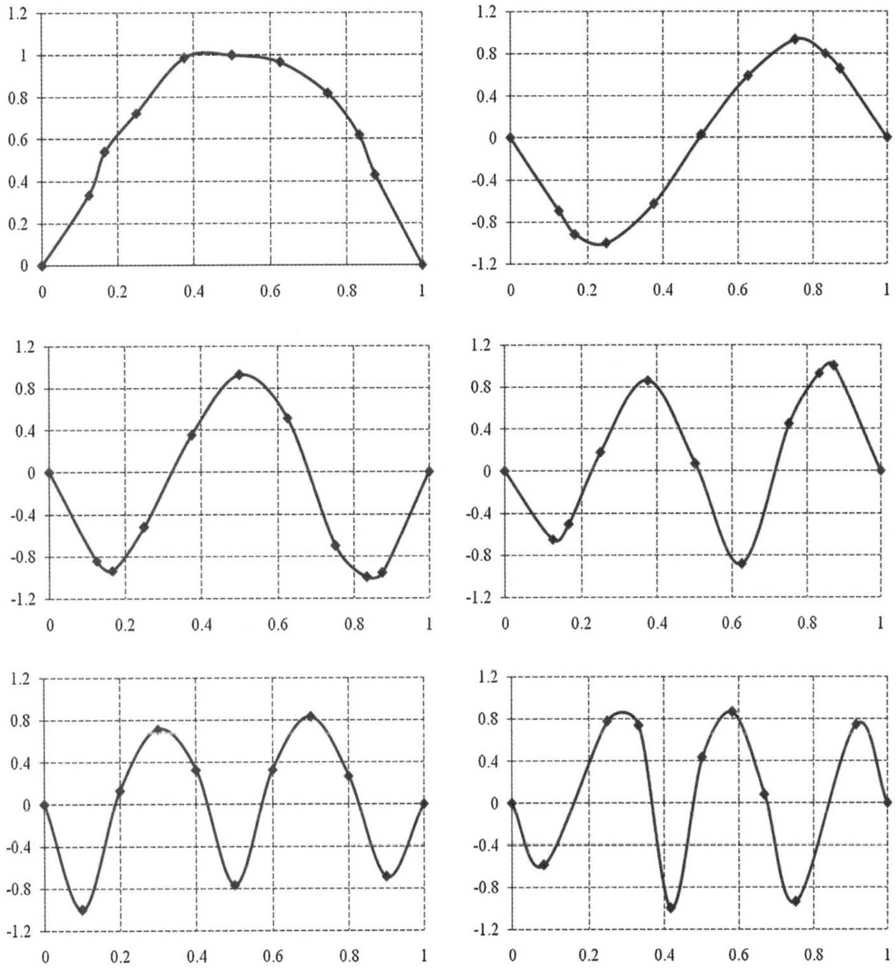

Fig. 5. Mode shape of the first six modes for aeroelastic model.

for the section model test is the equivalent mass of the model-suspension bridge including the section model, the effective mass of the elastically-mounted test rig. The comparison of vortex-induced vibrations between the aeroelastic model and the section model is only made for the 5th, 6th, and 7th vertical modes. It is noted that the vortex-induced vibrations for the first four vertical modes of the aeroelastic model develop for wind velocity below 2 m/s, the flow condition is actually not quite steady. Therefore they are excluded for comparison. While, the VIV lock-in for the 5th, 6th, and 7th vertical modes mode of the aeroelastic model are over 2 m/s, the uniform flow condition of both the aeroelastic model and the section model tests can be assured, the measured turbulence intensities are around 1%. By either increasing the stiffness of the model or increasing the depth of the rectangular cross-sections, VIV wind velocity range for lower-order modes can be promoted.

4. Results and discussions

4.1. Wind tunnel tests on aeroelastic model

The wind velocity range for the aeroelastic model wind tunnel test is 0.0 m/s ~ 6.0 m/s. Vortex-induced vibrations for the first

Fig. 6. Spring-mouted rigid section model in wind tunnel.

seven vertical modes of aeroelastic model are observed, and the acceleration data are recorded. For VIV lock-in regime of each vertical mode, the frequency spectrums of acceleration are characterized by one dominant peak except for very few transition points where two close modes are intermittently excited by vortex shedding. As mentioned before, the VIV onset wind velocity for the first four modes are below 2 m/s, the uniform flow condition cannot be guaranteed. Therefore only the measurement results of vortex-induced vibrations for the 5th, 6th, and 7th vertical modes are used for further analysis.

Fig. 7 shows the variation of VIV amplitudes with wind velocity for the 5th, 6th, and 7th modes of aeroelastic model. It is seen that the maximum VIV amplitudes for the three vertical modes are almost identical, and the value is around 1.0 mm. It is worth noting that there are a small range of overlap between the latter part of 5th VIV lock-in and the former part of 6th VIV lock-in, which is due to the mode transition between these two modes, and two peaks were also observed in frequency spectrum at these intermittent points. Fig. 8 shows the non-dimensional amplitudes for the three vertical modes. For each mode, the lock-in oscillations start at a reduced wind velocity of about 10.3, and cease at a reduced velocity of about 12.5. The Strouhal number is calculated to be about 0.1 based on the VIV onset wind velocity, which agrees well with other work [11,26]. The non-dimensional maximum amplitudes expressed in form of $1000A/D$ for the 5th, 6th, and 7th vertical modes are 37.0‰, 41.1‰ and 41.3‰, respectively.

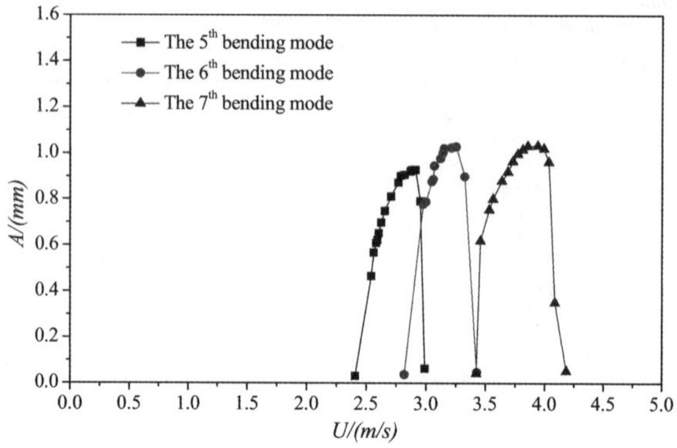

Fig. 7. Variation of VIV amplitudes for the 5th, 6th, and 7th modes of aeroelastic model.

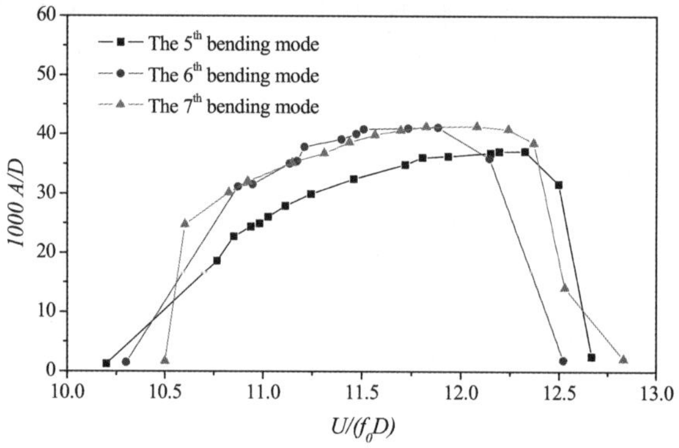

Fig. 8. Normalized amplitudes for the 5th, 6th, and 7th modes of aeroelastic model.

4.2. Wind tunnel tests on 1:1 section model

Wind tunnel tests on the section model are also performed for the 5th, 6th, and 7th vertical mode under the same testing conditions. The experimental parameters of the section model wind tunnel tests have been already provided in Table 3. The wind velocity

Table 3
Model parameters for aeroelastic model and section model.

Mode	Parameters	Symbol	Unite	Aeroelastic model	Section model	Deviation
5th, 6th, 7th mode	Length	L	mm	8100	1530	–
	Height	D	mm	40	40	0
	Width	B	mm	240	240	0
	Wind velocity	U	m/s	<6	<6	0
	Turbulence intensity	I	–	1%	1%	–
	Mass	m	kg/m	3.10	3.25	4.73%
5th mode	Natural frequency	$F_{0,5}$	HZ	5.86	5.90	0.68%
	Damping ratio	ζ_5	–	0.0061	0.0058	−4.92%
	Scruton number	Sc_5	–	121	121	−0.42%
	Reynolds number	Re	–	<6250	<6293	0.69%
6th mode	Natural frequency	$F_{0,6}$	HZ	6.74	6.84	1.48%
	Damping ratio	ζ_6	–	0.0063	0.0061	−3.17%
	Scruton number	Sc_6	–	125	127	1.41%
	Reynolds number	Re	–	<7189	<7296	1.49%
7th mode	Natural frequency	$F_{0,7}$	HZ	7.86	8.15	3.69%
	Damping ratio	ζ_7	–	0.0063	0.0062	−1.59%
	Scruton number	Sc_7	–	125	129	3.07%
	Reynolds number	Re	–	<8383	<8693	3.70%

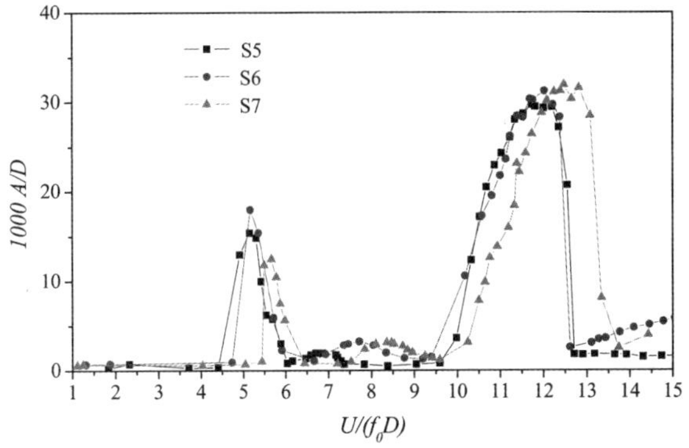

Fig. 9. Dimensionless VIV amplitude responses of section model tests.

Table 5
Mode shape correction factors for linear and nonlinear VIV models.

VIV model	Section model	Actual bridge	Correction factors (the sinusoidal mode shape)
Linear model	$\frac{A}{D}$	$\lambda_L \cdot \frac{A}{D} \cdot \varphi_{r,\max}(x) \frac{\int_0^{L_D} \lvert\varphi_r(x)\rvert dx}{\int_0^{L_D} \varphi_r^2(x) dx}$	$\lambda_L \cdot \frac{4}{\pi}$
Scanlan nonlinear model	$\frac{A}{D}$	$\lambda_L \cdot \frac{A}{D} \cdot \varphi_{r,\max}(x) \frac{\int_0^{L_D} \varphi_r^2(x) dx}{\int_0^{L_D} \varphi_r^4(x) dx}$	$\lambda_L \cdot \frac{2\sqrt{3}}{3}$

varies from 0.0 m/s to 6.0 m/s and the step size is 0.1 m/s. Fig. 9 shows the normalized VIV amplitudes of section model for the three vertical modes where the acronym "S5", "S6" and "S7" represent the three modes involved. For "S5", "S6" and "S7", two separate lock-in regimes of VIV were observed for each mode. The lock-in oscillations associated with the first regime (secondary regime) occur at a starting reduced velocity of 4.8, and have relatively small amplitude of vibrations. The lock-in oscillations associated with the second regime (primary regime) occur at higher reduced velocity of about 10.0, and are characterized by larger vibration amplitudes. The phenomena of two lock-in regimes were not observed in the aeroelastic model tests. The primary VIV

Table 4
Experimental parameters and measurement results for section model.

Parameters	Symbols	Section model/S5	Section model/S6	Section model/S7
Cross wind dimension	D/(mm)	40	40	40
Along wind dimension	B/(mm)	240	240	240
Equivalent mass	m/(kg/m)	3.25	3.25	3.25
Natural frequency	f/(HZ)	5.90	6.84	8.15
Damping ratio	ζ	0.0058	0.0061	0.0062
Scruton number	Sc	121	127	129
The first VIV lock-in	Strouhal number	0.22	0.21	0.19
	$1000A/D$	15.4	18.0	12.5
The second VIV lock-in	Strouhal number	0.10	0.11	0.10
	$1000A/D$	29.7	31.2	31.9

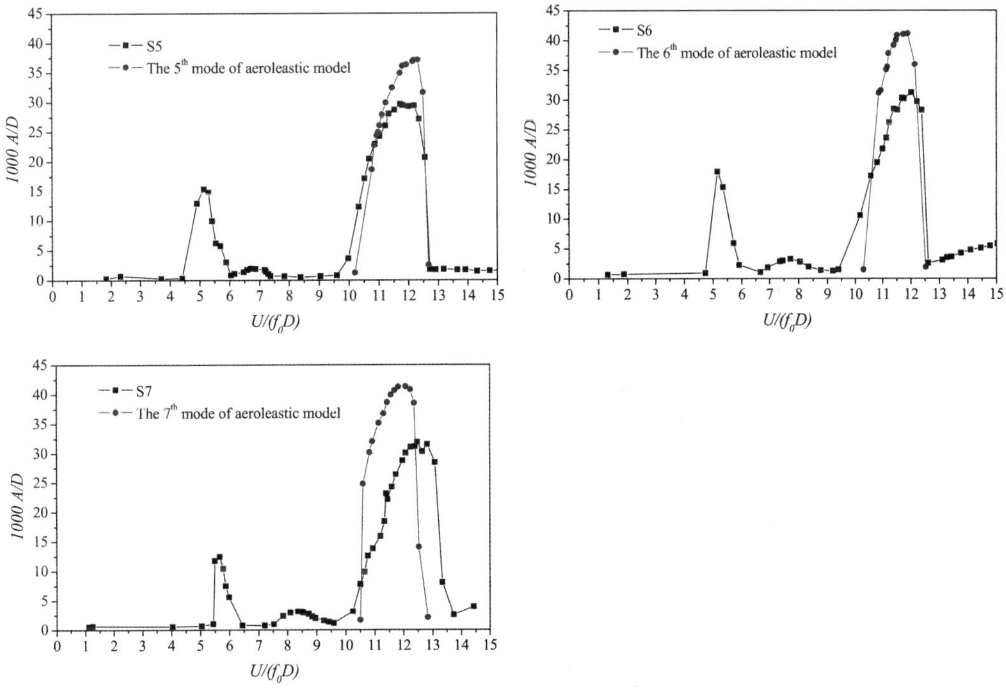

Fig. 10. Comparison of the normalized VIV amplitudes for the three modes: a) Mode 5; b) Mode 6; and c) Mode 7.

Table 6
Comparison of correction factors for VIV amplitude between aeroelastic model and section model.

Mode	Experiment 1000*A*/*D*	Correction factors		
		Experiment	Prediction with Linear model	Prediction with Non-linear model
The 5th mode of aeroelastic model/S5	37.2 29.7	1.25	1.38	1.20
The 6th mode of aeroelastic model/S6	41.1 31.2	1.32	1.24	1.15
The 7th mode of aeroelastic model/S7	41.3	1.29	1.37	1.20

regime can take control the secondary VIV regime due to the closely-spaced vertical modes in the aeroelastic model, and therefore the secondary VIV regime may be suppressed. The phenomena are also seldom observed at full scale. One exception is the Fred Hartman cable-stayed Bridge with a twin-deck section, where the starting reduced wind velocities for the first and second lock-in regimes are about 6.4 and 20.4, respectively [27]. The Strouhal numbers of the first lock-in regime are calculated to be around 0.2, while the second lock-in regime is about 0.1. The phenomenon of multiple lock-in regimes at a single vibration frequency has been reported in literature, and may be attributed to sub-harmonic and super-harmonic resonance in view of highly nonlinear vortex shedding forces and complicated vortex structures involving vortex shedding, vortices reattachment and movement on the surface of bluff bodies [16,26,28].

Table 4 shows the experimental parameters and results for section model tests. Referring to the “S5”, “S6” and “S7”, the observed normalized maximum amplitudes of the first lock-in regime are respectively for 15.4‰, 18.0‰ and 12.5‰, while those of the second lock-in regime are 29.7‰, 31.2‰ and 31.9‰. The VIV amplitudes of the second lock-in regime are selected for comparison with aeroelastic model tests results, because the second lock-in regime in section model tests has almost the same VIV reduced velocity range as the aeroelastic model.

4.3. Comparison and discussions

Figs. 10 show the comparison of the normalized vibration amplitudes between the aeroelastic model and its section model for the three modes. The VIV lock-in regime of the aeroelastic model generally coincides with the second VIV lock-in regime of the section model. Most importantly it is shown that the aeroelastic model consistently produce larger maximum vibration amplitudes than the section model tests for all the three modes tested. Fujino and Yoshida [10] reported that section model test, full aeroelastic model test, and full-scale measurements on the Tokyo Bay Bridge with varying depth of deck cross-section. A first comparison indicated that VIV amplitudes from aeroelastic model and full-scale measurements are about 50% larger than those from section model tests. For the second Severn crossing cable-stayed bridge, the vibration amplitude from the actual bridge is also larger than that measured from section model tests [18].

Both the mode shape and imperfect spanwise correlations of vortex shedding forces may cause difference in the maximum vibration amplitudes between section model and aeroelastic model (and also the prototype bridge). While the imperfect spanwise correlation of vortex shedding forces is hard to characterize and may vary case by case, the mode shape effect has been addressed by introducing a mode shape correction factor when extrapolating the section-model results to their prototype e.g., [16–18,29]. It is noted that different vortex-shedding force models lead to different mode shape correction factors. Table 5 lists the mode shape correction factors when using the linear model and the Scanlan's nonlinear model for vortex-induced vibrations. It is noted that the mode shape factors for linear and nonlinear VIV model are 1.27 and 1.15 for the ideal sinusoidal mode shape. Table 6 summarizes the detailed comparison of normalized vibration amplitudes for the 5th, 6th, and 7th vertical modes. The experimentally-measured correction factor the maximum vibration amplitudes for the 5th, 6th and 7th modes are 1.25, 1.32 and 1.29, respectively. If the fully-correlated excitation forces are assumed, the predicted correction factors considering only the mode shape effect are 1.38, 1.24 and 1.37 for linear VIV model, and are 1.20, 1.15 and 1.20 for nonlinear VIV model when using the mode shapes measured from aeroelastic model. It is noted that the experimental correction factors take into account both effects of mode shape and imperfect spanwise correlation of vortex shedding forces while those predicted from analytic models only consider the mode shape effect. By comparison, it seems that the mode shape effect is the primary source of VIV amplitude difference between section model and aeroelastic model and a full-correlated excitation force can be reasonably adopted.

5. Conclusion

This paper described the measurement results of VIV amplitude from an aeroelastic model and its 1:1 section model to assess the correction factor between them. The elastically multi-supported aeroelastic model used in this study is a new concept, and can be designed with the assumed mode shape method. The validity of the aeroelastic model in reproducing the vertical modes of an assumed example bridge is illustrated by modal tests. The measured 5th, 6th, and 7th vertical mode VIV responses from elastically multi-supported aeroelastic model are compared with those obtained from section models test under the same testing conditions. The experimental correction factors for the maximum VIV amplitude are 1.25, 1.32 and 1.29, respectively for the three modes. The present study confirms that the combined three-dimensional effects due to mode shape and imperfect correlation of excitation forces contribute positively and disregarding the effects may underestimates the maximum VIV amplitudes for section model tests. A correction factor of 1.3 for the VIV amplitude may be adopted for practical use when extrapolating the section-model results to their full-scale values.

As mentioned earlier, both structural mode shape and spanwise correlation of lift forces contribute to the correction factor for VIV amplitude. The span-wise correlation of lift forces may vary with different cross-sections and vibration amplitudes, which was not addressed in the present study. In order to achieve more general correction factors, further investigation of vortex-induced vibrations of some typical bridge cross-sections as well as the measurement of span-wise lift correlations under different VIV vibration amplitude are desirable.

Acknowledgments

This study is sponsored by the National Science Foundation of China (project No. 51278189 and 51422806) and the State's Key

Project of Research and Development Plan (Project No. 2016YFE0127900), which are acknowledged.

References

[1] Larsen A, Esdahl S, Andersen JE, Vejrum T. Storebalt suspension bridge-vortex shedding excitation and mitigation by guide vanes. J Wind Eng Ind Aerodyn 2000;88:283–96.
[2] Li H, Laima SJ, Ou JP, Zhao XF, Zhou WS, Yu Y, et al. Investigation of vortex-induced vibration of a suspension bridge with two separated steel box girders based on field measurements. Eng Struct 2011;33:1894–907.
[3] Chen XZ. Analysis of crosswind fatigue of wind-excited structures with nonlinear aerodynamic damping. Eng Struct 2014;74:145–56.
[4] Bearman PW. Vortex shedding from oscillating bluff bodies. Ann Rev Fluid Mech 1984;16:195–222.
[5] Parkinson GV. Phenomena and modeling of flow-induced vibration of bluff bodies. Prog Aerosp Sci 1989;26:169–224.
[6] Blevins RD. Flow-induced vibrations. New York: Van Nostrand Reinhold; 1990.
[7] Sarpkaya T. A critical review of the intrinsic nature of vortex-induced vibrations. J Fluids Struct 2004;19:389–447.
[8] Williamson CHK, Govardhan R. Vortex-induced vibrations. Ann Rev Fluid Mech 2004;36:413–55.
[9] Simiu E, Scanlan RH. Wind effects on structures: fundamentals and applications to design. New York: Wiley; 1996.
[10] Fujino Y, Yoshida Y. Wind-induced vibration and control of trans-tokyo bay crossing bridge. J Struct Eng, ASCE 2002;128:1012–25.
[11] Bearman PW, Obasaju ED. An experimental study of pressure fluctuations on fixed and oscillating square-section cylinders. J Fluid Mech 1982;119:297–321.
[12] Ramberg SE, Griffin OM. Velocity correlation and vortex spacing in the wake of a vibrating cable. J Fluids Eng, ASME 1976;98:10–8.
[13] Hann FL, Wu T, Kareem A. Correlation structures of pressure field and integrated forces on oscillating prism in turbulent flows. J Eng Mech, ASCE 2016;142:14. 04016017.
[14] Ehsan F, Scanlan RH. Vortex-induced vibrations of flexible bridges. J Eng Mech, ASCE 1990;116:1392–411.
[15] Griffin OM, Koopmann GH. The vortex-excited lift and reaction forces on resonantly vibrating cylinders. J Sound Vib 1977;54:435–48.
[16] Blevins RD, Burton TE. Fluid forces induced by vortex shedding. J Fluid Eng 1976;98:19–24.
[17] Irwin PA. The role of wind tunnel modeling in the prediction of wind effects on bridges. In: Larsen A, Esdahl S, editors. Bridge aerodynamic-proceedings of the international symposium on advances in bridge aerodynamics. Rotterdam: A. A. Balkema; 1998. p. 99–117.
[18] Macdonald J, Irwin PA, Fletcher M. Vortex-induced vibrations of the Second Severn Cross cable-stayed bridge-full-scale and wind tunnel measurements. Proc ICE: Struct Build 2002;152:123–34.
[19] Frandsen JB. Simultaneous pressures and accelerations measured full-scale on the great belt east suspension bridge. J Wind Eng Ind Aerodyn 2001;89:95–129.
[20] Li H, Laima SJ, Zhang QQ, Li N, Liu ZQ. Field monitoring and validation of vortex-induced vibrations of a long-span suspension bridge. J Wind Eng Ind Aerodyn 2014;124:54–67.
[21] Davenport AG, King JPC, Larose GL. Taut strip model tests. In: Proceedings of the 1st International Conference on the Aerodynamics of Large Bridges. Rotterdam: Balkema; 1992. p. 113–24.
[22] Caracoglia L, Noe S, Sepe V. Nonlinear computer model for the simulation of lock-in vibration on long-span bridges. Comput Aided Civil Infrastruct Eng 2008;24:130–44.
[23] Irwin PA. Bluff body aerodynamics in wind engineering. J Wind Eng Ind Aerodyn 2008;96:701–12.
[24] Wu T, Kareem A. An overview of vortex-induced vibrations of bridge decks. Front Civil Struct Eng 2012;6:335–47.
[25] Ruscheweyh H. Practical experience with wind induced vibrations. J Wind Eng Ind Aerodyn 1990;33:211–8.
[26] Matsumoto M. Vortex shedding of bluff bodies: a review. J Fluids Struct 1999;13:791–811.
[27] Mashnad M, Jones NP. A model for vortex-induced vibration analysis of long-span bridges. J Wind Eng Ind Aerodyn 2014;134:96–108.
[28] Garrett JL. Flow-induced vibration of elastically supported rectangular cylinders. Iowa State University; 2003. PhD thesis.
[29] Zhang ZT, Ge YJ, Chen ZQ. Vortex-induced oscillations of bridges: theoretical linkage between sectional model tests and full bridge responses. Wind Struct: Int J 2014;19:233–47.

Engineering Structures 173 (2018) 61–75

Contents lists available at ScienceDirect

Engineering Structures

journal homepage: www.elsevier.com/locate/engstruct

Vibration control of vortex-induced vibrations of a bridge deck by a single-side pounding tuned mass damper

Wenxi Wang[a], Xiuyong Wang[b], Xugang Hua[a,*], Gangbing Song[c], Zhengqing Chen[a]

[a] *Key Laboratory for Bridge and Wind Engineering of Hunan Province, College of Civil Engineering, Hunan University, Changsha, Hunan 410082, China*
[b] *Hunan Provincial Key Laboratory of Structures for Wind Resistance and Vibration Control, College of Civil Engineering, Hunan University of Science and Technology, Xiangtan, Hunan 411201, China*
[c] *Department of Mechanical Engineering, University of Houston, Houston, TX 77204, USA*

ARTICLE INFO

Keywords:
Vibration control
Pounding tuned mass damper
Vortex-induced vibration
Wake oscillator model

ABSTRACT

This paper proposes a new method to mitigate vortex-induced vibrations (VIVs) of a bridge deck using a single-side pounding tuned mass damper (SS-PTMD). The SS-PTMD is a passive control device and comprises an un-damped tuned mass with a pounding boundary covered with viscoelastic (VE) materials layer. A nonlinear force model for describing impact behavior of VE materials is used to simulate the response of a single degree of freedom (SDOF) system controlled by a SS-PTMD. The free pounding experiments are performed to determine the model parameters of impact force and validate the simulation method. The optimal design of SS-PTMD for SDOF system subjected to sinusoidal excitation is carried out by numerical optimization, and the optimized SS-PTMD is applied to control the vertical VIVs of a bridge deck. The control performance is experimentally examined by elastically mounted section model tests in wind tunnel. In addition, the classical wake oscillator model is used to predict the behavior of the coupled fluid-structure system under VIV and explore the control performance of the SS-PTMD. The experimental results show that the maximum response of the bridge deck model was reduced by 94% when a SS-PTMD with mass ratio of 2% was applied. It is also shown that non-linearity in vortex shedding forces has little influence on control performance of SS-PTMD optimized under sinusoidal excitation.

1. Introduction

Vortex-induced vibrations (VIVs) of the bridge deck happens if the frequency of vortex-shedding matches with one of the natural frequencies of the bridge [1–4]. It is a resonant-type vibration dominated by a single mode of vibration, and has been observed on many bridges. For the long-span bridges with small damping and more flexible configurations, large amplitude VIV may occur in a relatively low wind speed range about 5–10 m/s. In another word, long-span bridges may suffer from long-term vibration with considerably large amplitudes, which discomforts users and greatly shorten the life of bridges.

Tuned mass dampers (TMDs), as a low-cost and simple passive control device, have been applied to vibration control of flexible and low-damping structures under different excitation inputs, such as vehicle [5–8], wind [9–12] and earthquake excitations [13–16]. The effectiveness of using TMD to control VIVs of flexible structures, such as pipelines [17,18], high-rise buildings [19–21] and long-span bridges [22,23], were reported. The conventional TMD, which is firstly proposed by Frahm [24], only consists of a mass-spring system. When the frequency of the mass is exactly tuned to the frequency of the controlled structure, the vibration energy of the controlled structure will be transferred to the tuned mass. Den Hartog incorporated a linear viscous damper into TMD and the optimal parameters of TMDs to control an undamped system was derived [25,26]. Some other solutions to supply linear or nonlinear damping for TMDs were studied, such as particle dampers [27–29], Magneto-rheological (MR) dampers [30–33], friction dampers and eddy current dampers [34–38].

Recently, Song et al. [39] introduced the pounding damping into a TMD system to enhance the control performance and the combined system is called pounding tuned mass damper (PTMD), as shown in Fig. 1(a). In their design, two pounding boundaries are added in the either sides of the tuned mass and this type of damper is termed as a double-side PTMD. When the relative vibration displacement exceeds the gap between the tuned mass and the pounding boundaries, the tuned mass will pound on the boundary to provide supplemental energy dissipation. To enhance the energy dissipation during poundings, viscoelastic (VE) material layer is attached on the pounding boundary. Several applications of the double-side PTMD for vibration control of a

* Corresponding author.
E-mail address: cexghua@hnu.edu.cn (X. Hua).

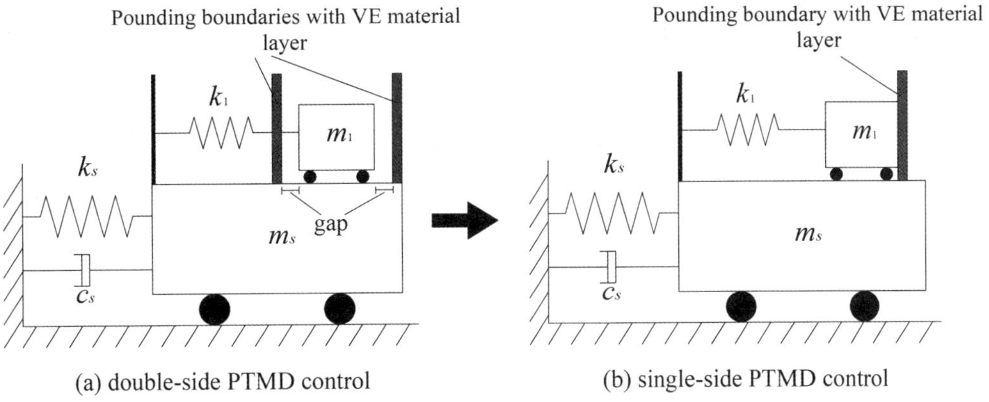

Fig. 1. A SDOF system with conventional PTMD and SS-PTMD.

traffic pole [40,42] and subsea pipelines [39,41,43] showed that the double-side PTMD is more robust and effective than the conventional TMD [42–44]. Furthermore, the effectiveness of the double-side PTMD for the seismic vibration protection of multiple degree of freedom (MDOF) structures was also investigated [45,46]. Based on the double-side PTMD design, Wang et al. [47,48] proposed a novel single-side pounding tuned mass damper (SS-PTMD) to control the free vibration and forced vibration of a low-damping structure, as shown in Fig. 1(b). Compared to the conventional double-side PTMD, only one pounding boundary is added next to the equilibrium position of the spring-mass system to provide additional damping for the TMD system. This modification eliminates the gap between the mass and pounding boundary as shown in Fig. 1(a) and simplifies the parametric design. The optimal design of the SS-PTMD for controlling a SODF system subjected to sinusoidal excitation has been studied [49,50].

In many situations, the installation space and stroke of tuned mass dampers is a practical constraint and the SS-PTMD provides a superior solution. Moreover, the energy dissipation provided by VE materials during poundings is not changed too much even after 350,000 poundings [51]. Although the effectiveness of the novel SS-PTMD to control the free vibration and the forced vibration under harmonic loadings were demonstrated, application to control the VIVs of bridge girders is still an interesting topic and has not been reported yet. Unlike sinusoidal excitation, vortex shedding forces responsible for VIV of bridge deck involves structure-fluid interaction and is highly nonlinear. In this study, the effectiveness of SS-PTMD for control of VIVs of a bridge deck is investigated experimentally and numerically. An impact force model which considers the nonlinear impact behavior of VE materials is used to simulate the response of a single degree of freedom (SDOF) system equipped with a SS-PTMD. The optimal parameters of the SS-PTMD are obtained for a SDOF structure subjected to sinusoidal excitations. Moreover, the optimized SS-PTMD is applied to control the vertical VIV of a bridge girder sectional model. The effectiveness of the proposed SS-PTMD to control VIVs is numerically studied by the wake oscillator model and examined by the wind tunnel experiments.

2. Modeling of single-side pounding tuned mass damper (SS-PTMD)

2.1. Dynamic properties of SS-PTMD

The mechanical model of a single degree of freedom (SDOF) structure equipped with the novel SS-PTMD to suppress vertical vibrations is shown in Fig. 2. The SS-PTMD has a pounding boundary covered by VE material layers beneath the tuned mass when the mass is in the equilibrium position. The natural frequency f_d and the equivalent damping ratio ζ_d of the novel SS-PTMD can be expressed as [48],

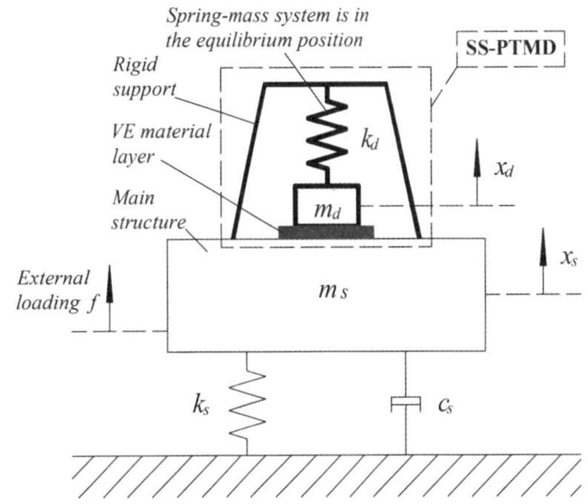

Fig. 2. A vertical vibration system controlled by the proposed SS-PTMD.

$$f_d = \frac{1}{\pi}\sqrt{\frac{k_d}{m_d}} \tag{1}$$

$$\zeta_d = \frac{1}{2\pi}\ln\left(\frac{1}{e}\right) \tag{2}$$

where k_d and m_d are the stiffness and mass of the SS-PTMD, e is the coefficient of restitution (COR) during impacts between tuned mass and the VE material layer. Contrast to a general spring-mass system, it is evident that the frequency of the SS-PTMD is doubled. This is because the pounding boundary prevents the mass move into the lower half stroke, as shown in Fig. 2. Since Eq. (1) ignores the pounding duration, the actual frequency should be slightly larger than that provided by Eq. (1). Moreover, it should be noted that Eq. (2) is an approximate expression for damping ratio calculation and it has good accuracy when e is larger than 0.4 [47].

2.2. Pounding force modeling

The conventional impact theory, employing the impact velocity exchange, neglects the pounding duration and cannot provide the specific value of the impact force [49]. In this paper, an impact force model, which considers the nonlinear behavior of poundings between VE material layer and steel, is utilized to simulate the response of the SS-PTMD. The pounding force between the VE materials layer and steel is given by [49],

$$F_p(t) = \begin{cases} k\delta^n(t) + \zeta\delta^n(t)\dot{\delta}(t) & \delta_{\max} \geqslant \delta(t) > 0, \quad \text{(approaching period)} \\ & \dot{\delta}(t) \geqslant 0 \\ f_e\left[\frac{\delta(t) - e_1\delta_{\max}}{(1-e_1)\delta_{\max}}\right]^n & \delta_{\max} > \delta(t) \geqslant \delta_e, \quad \text{(restitution period)} \\ & \dot{\delta}(t) < 0 \\ 0 & \delta_e > \delta(t) > 0, \\ & \dot{\delta}(t) < 0 \end{cases} \tag{3}$$

where k and ζ are the pounding stiffness and damping, respectively. $\delta(t)$ and $\dot{\delta}(t)$ are the indentation and relative velocity of the colliding bodies during pounding. n is the pounding nonlinear coefficient, $\delta_{\max}$ is the maximum indentation during pounding. e_1 and f_e are the remaining surface deformation ratio and the maximum elastic impact force, they can be expressed as [49],

$$e_1 = \frac{\delta_e}{\delta_{\max}} \tag{4}$$

$$f_e = k\delta_{\max}^n \tag{5}$$

where δ_e is the remaining surface deformation of the VE materials layer when the colliding bodies are just separated.

In the impact force model, the pounding stiffness k is determined by letting the maximum simulated impact force be equal to the measured value from free pounding experiments. The remaining surface deformation ratio e_1 also can be obtained from the pounding experiments by Eq. (4). The maximum indentation $\delta_{\max}$ can be collected during the pounding simulation. The pounding damping ζ, which represents the energy loss during one pounding, can be expressed as the following explicit formula [49],

$$\frac{1}{2}(1-e^2)\dot{\delta}_0^2 = (1-e_1)\left(\frac{k^2}{\zeta^2}\ln\left|\frac{\dot{\delta}_0 + \frac{k}{\zeta}}{\frac{k}{\zeta}}\right| - \frac{k}{\zeta}\dot{\delta}_0\right) + \frac{\dot{\delta}_0^2}{2} \tag{6}$$

where $\dot{\delta}_0$ is the relative pre-impact velocity of the colliding bodies. It should be noted that Eq. (6) can be solved by Newton iterative method to obtain the pounding damping for each pounding.

2.3. Free pounding experiment

In Eq. (2), the equivalent damping ratio of the SS-PTMD is only governed by the coefficient of restitution e. Therefore, a free pounding experiment was performed to study the coefficient of restitution of different VE materials under various pre-impact velocities, as shown in Fig. 3. In the experimental setup, a cantilever beam with a mass block at its end was fixed by a rigid support. The VE materials layer was placed upon a force sensor. The spherical striker was just contacted to the VE materials layer when the cantilever beam-mass system was in the equilibrium position. A laser displacement sensor (optoNCDT14, manufactured by $\mu\varepsilon$ Company) was used to measure the vertical displacement of the mass block, and the force sensor (PWSK-300N, manufactured by Shanghai Instruments Company) was used to collect the impact force during poundings. In each test, an initial vertical displacement x_{in} was given to the mass block and then let it freely pound on the VE materials layer. Based on the law of energy conservation, the pre-impact velocity V_{pre} and post-impact velocity V_{post} are given by,

$$V_{pre} = \sqrt{\frac{k_1}{m_1}}x_{in} \tag{7}$$

$$V_{post} = \sqrt{\frac{k_1}{m_1}}x_{re} \tag{8}$$

where x_{in} and x_{re} are respectively the initially released displacement and the maximum rebounded displacement after one pounding. m_1 is the weight of the mass block and k_1 is the stiffness provided by the cantilever beam which can be easily obtained from frequency test. Therefore, the coefficient of restitution e can be obtained by,

$$e = \frac{V_{post}}{V_{pre}} = \frac{x_{re}}{x_{in}} \tag{9}$$

The pre-impact velocity can be alternated by changing the initially released displacement. Two VE materials, as shown in Fig. 4, high energy dissipation rubber (HEDR, which was manufactured by Zhuzhou Times New Materials Technology Company) and very high bond (VHB, which was manufactured by 3M Company) were used in the experiment. Three different VE material layers were divided, namely 12 mm VHB layer, 12 mm HEDR layer and the composite layer composed of 3 mm VHB and 9 mm HEDR layer. In each case the different VE material layers with the same thickness of 12 mm was tested.

The relationship between the pre-impact velocity and the coefficient of restitution is shown in Fig. 5. It can be seen that the coefficient of restitution of each case is almost a constant under various pre-impact velocities. The average coefficients of restitution of the VHB materials and the HEDR are 0.3694 and 0.6296, respectively. The average coefficient of restitution of the composite layer composed of 3 mm VHB and 9 mm HEDR is 0.5008, which is somewhere in between the mean value of the case of VHB layer and the case of HEDR layer. This phenomenon indicates that the conception of using composite VE material layer can realize the adjustability of the equivalent damping ratio of the SS-PTMD in a certain range.

2.4. Validation of pounding force modeling

When a initially released displacement $x_{10} = 0.04$ m is given to the mass block and the 12 mm DEDR pounding layer is used in the free pounding experiment, the history of vertical displacement of the mass block and the impact force during poundings are shown in Fig. 5. To verify the accuracy of the impact force model, a simulation of the free

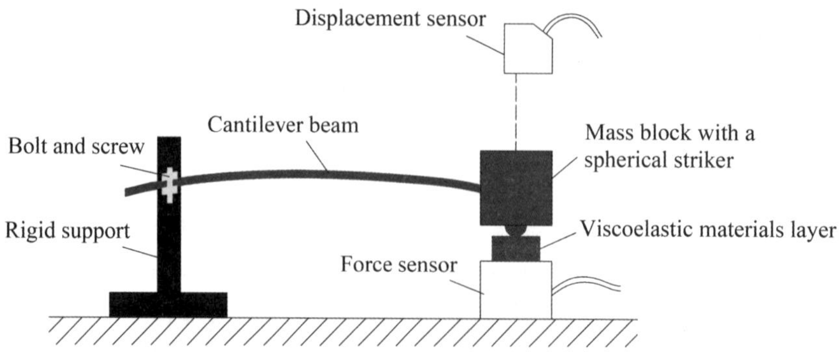

Fig. 3. Setup of the free pounding experiment.

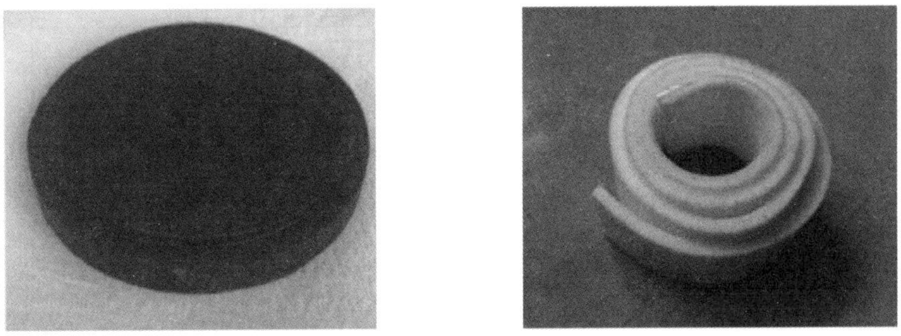

(a) HEDR (b) VHB

Fig. 4. Two kinds of VE materials.

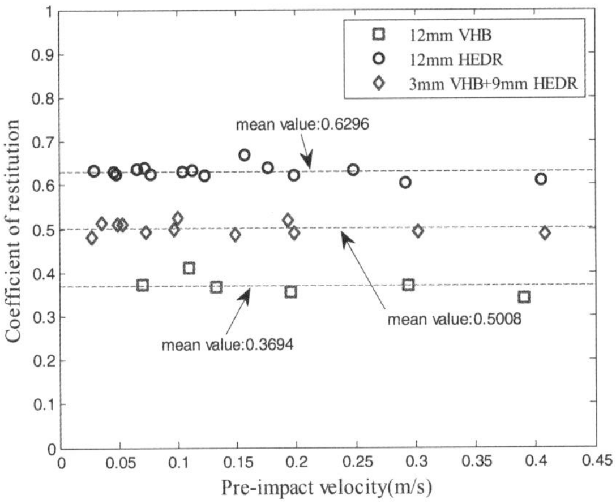

Fig. 5. Relationship between coefficient of restitution and different pre-impact velocities.

pounding experiment is carried out to compare the history of the displacement and impact force measured from the experiment. For the free pounding experiment, as shown in Fig. 3, the governing equation of motion is,

$$m_1\ddot{x}_1(t) + k_1x_1(t) = F_p(t) \tag{10}$$

where $\ddot{x}_1(t)$ and $x_1(t)$ are the vertical acceleration and displacement of the mass block, respectively. m_1 is the weight of the mass block, and k_1 is stiffness provided by the cantilever beam. $F_p(t)$ is the pounding force and it remains zero when $x_1(t) < 0$. $F_p(t)$ can be computed by substituting Eq. (3) into Eq. (10) when $x_1(t) > 0$ and the indentation $\delta(t)$ in Eq. (3) should be defined as,

$$\delta(t) = x_1(t) \tag{11}$$

According to the previous study on the pounding property of VE materials, the nonlinear pounding coefficient n is set to 1.5 [48] and the remaining surface deformation ratio e_1 is set to 0.5 [49]. The pounding stiffness is determined to be $6.4 \times 10^5\ \mathrm{N/m^{1.5}}$ by letting the maximum force value 67.7 N of the first pounding from experiment be equal to that from simulation [52,53]. The Runge-Kutta fourth-order method was employed to solve Eq. (10) and the time interval was 1×10^{-4} s. The initial condition of the simulation was set to $x_1 = 0.04$ m and $\dot{x}_1 = 0$. The weight of mass block is 0.86 kg and the stiffness provided by the cantilever beam is 85.5 N/m. The average coefficient of restitution of 12 mm HEDR layer is 0.6296.

The comparison of experimental and simulated results is shown in Fig. 6. It is clear that the simulated displacement and pounding force agree well with the experimental results, which demonstrates the accuracy of the impact force model. Since the friction force between the striker and the VE materials layer is not considered in Eq. (10), the displacement and impact force are not accurate enough during the fourth pounding when the friction force may account for a huge part of the contact force.

3. Optimum design of SS-PTMD

Based on the validated impact force model, a numerical optimization based on the H_∞ optimization theory was performed to obtain the optimal parameters of the SS-PTMD, including optimal tuning frequency ratio and the optimal coefficient of restitution. The simplified design formulas for the SS-PTMD were given and further discussed.

3.1. Equations of motion of a SDOF system controlled by a SS-PTMD

As the mechanical model shown in Fig. 2, the equations of motion of a SDOF structure controlled by a SS-PTMD are given by,

$$\begin{cases} m_s\ddot{x}_s(t) + c_s\dot{x}_s(t) + k_sx_s(t) + k_d(x_s(t) - x_d(t)) = f(t) + F_p(t) \\ m_d\ddot{x}_d(t) + k_d(x_d(t) - x_s(t)) = -F_p(t) \end{cases} \tag{12}$$

where m_s, c_s, and k_s are the mass, damping coefficient and stiffness of the main structure, respectively. m_d and k_d are the mass and stiffness of the SS-PTMD. $x_s(t)$ and $x_d(t)$ are the displacement of the main structure and the SS-PTMD, respectively. f is the external loading and $F_p(t)$ is the pounding force. The pounding force $F_p(t)$ is equal to zero when the relative displacement $x_s(t) - x_d(t) < 0$. When the relative displacement $x_s(t) - x_d(t) > 0$, the pounding force $F_p(t)$ can be computed by Eq. (3). The indentation $\delta(t)$ in Eq. (3) should be substituted by,

$$\delta(t) = x_s(t) - x_d(t) \tag{13}$$

3.2. Numerical optimization based on H_∞ theory

The H_∞ optimization theory [54,55], which can minimize the maximum response of the structure in a wide frequency range, is considered in the numerical optimization. The optimization objective based on the H_∞ optimization theory is given by [48],

$$\min_{f_i\ e_i}[\max_{\omega}(\mathrm{DMF}(\omega))] \tag{14}$$

where $f_i = \omega_d/\omega_s$ represents the frequency tuning ratio of the SS-PTMD. ω_s is the circular frequency of the main structure and ω_d is the circular frequency of the SS-PTMD according to Eq. (1). e_i is the coefficient of restitution of the VE materials layer. DMF(ω), the dynamic magnification factor (DMF) of the main structure, is given by,

$$\mathrm{DMF}(\omega) = A_s(\omega)k_s/f_0 \tag{15}$$

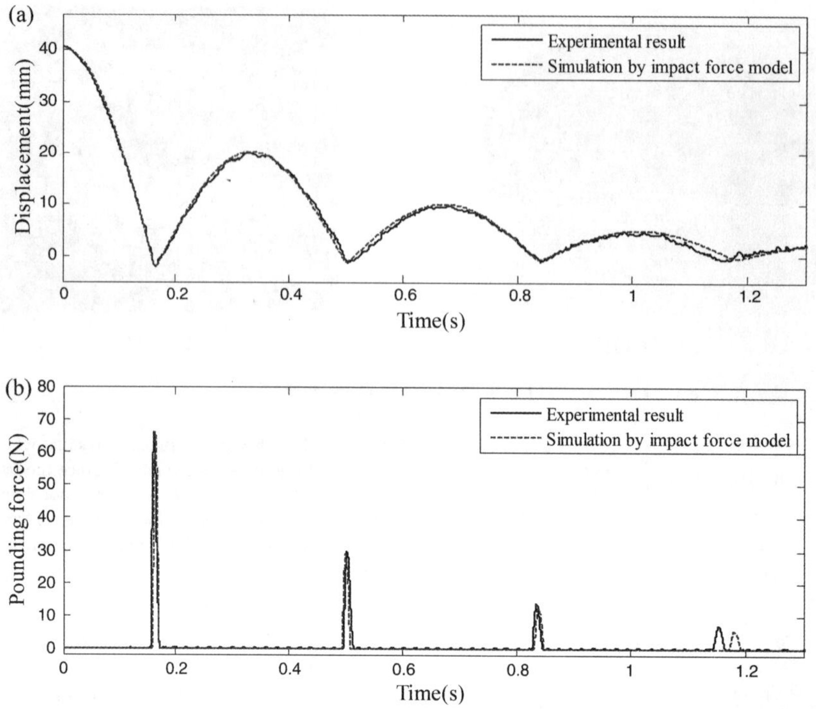

Fig. 6. Comparison of experimental and simulation results at an initial displacement 0.04 m: (a) Vertical displacement; (b) pounding force.

where $A_s(\omega)$ is the steady-state response amplitude of the main structure under a harmonic excitation force $f(t) = f_0\sin(\omega t)$.

When a certain mass ratio of the SS-PTMD and the damping ratio of the main structure are given, the control performance of the SS-PTMD only depends on the frequency tuning ratio and the coefficient of restitution e of VE material layer. A range of frequency tuning ratio $f_i \in [0.94, 1.23]$ and the coefficient of restitution $e_i \in [0.3, 0.9]$ was specified in the numerical optimization. The SS-PTMD is usually designed to enhance the damping and reduce the large amplitude vibrations for low-damping structures, such as long-span bridges and high-risen buildings. These flexible structures have a low value of damping ratio, even less than 0.5–1%. Therefore, the damping ratio of the main structure is set to 0.5% in the optimization. The mass ratio of SS-PTMD is set from 0.01 to 0.05 with an increment of 0.01 and the excitation frequency ω of the harmonic loading $f(t) = f_0\sin(\omega t)$ is varied from $0.8\omega_s$ to $1.2\omega_s$. Eq. (12) can be numerically solved by the Runge-Kutta fourth-order method to obtain the maximum DMF(ω) under different excitation frequencies, and the time interval of the solution is 1×10^{-4} s. Figs. 7 and 8 show the maximum DMF of the main structure when the frequency tuning ratio and the coefficient of restitution are varied in the search range. It is shown that the control performance of the SS-PTMD is insensitive to the derivation of the coefficient of restitution to the optimal value.

The optimal value of the frequency tuning ratio and the coefficient of restitution can be obtained by finding the point with the minimum value of the maximum DMF. The optimal frequency tuning ratio from the numerical optimization is plotted in Fig. 9(a) and the quadratic fitting curve is also shown. The optimal design formula for the frequency tuning ratio f_{opt} is given by,

$$f_{opt} = -28.6\mu^2 + 3.11\mu + 1.01 \tag{16}$$

where μ is the mass ratio of the SS-PTMD. It is shown from Fig. 8(a) that the optimal frequency tuning ratio for the SS-PTMD increases with the mass ratio. However for a conventional TMD system, the optimal frequency tuning ratio decreases with the increase of the mass ratio. This is because the pounding duration will be enlarged due to the increase of the pounding mass, and consequently the frequency of the SS-PTMD decreases.

Fig. 9(b) shows the optimal coefficient of restitution e_{opt} from the numerical optimization, and the optimum design formula proposed by Wang et al. [48] is also plotted. It can be seen that the optimal coefficient of restitution from the optimization is slightly larger than the value from the formula. Considering the damping detuning of TMD system will not greatly affect the control performance, the optimum formula proposed by Wang et al. [48] is recommended in the design, and the formula is given by,

$$e_{opt} = \frac{1}{\exp\left(2\pi\sqrt{\frac{3\mu}{8(1+\mu)}}\right)} \tag{17}$$

It should be noted that the wind loading such as the turbulence-induced buffeting force is generally a broad-band excitation which induces the multimode responses of a structure. For vortex-induced vibration in the present study, the excitation force is caused by the regular, periodic shedding of vortices in the wake of bridge decks. Although the nonlinear vortex shedding force is different to harmonic excitations, the vortex-induced vibration is generally dominated by a single mode [56], as also reflected from wind tunnel results of a bridge deck later. Therefore, the optimal parameters of the SS-PTMD obtained by applying a harmonic excitation on the primary structure are applicable to control the structural response under VIVs.

4. Control of a bridge deck model by a SS-PTMD under VIV

A sectional model test in wind tunnel is the most convenient way to study the various wind-induced responses, such as flutters and VIVs, of actual flexible bridges subjected to wind loadings. In the sectional model test, the dynamic properties of a flexible bridge are first determined from eigenvalue analysis, and the sectional model of the

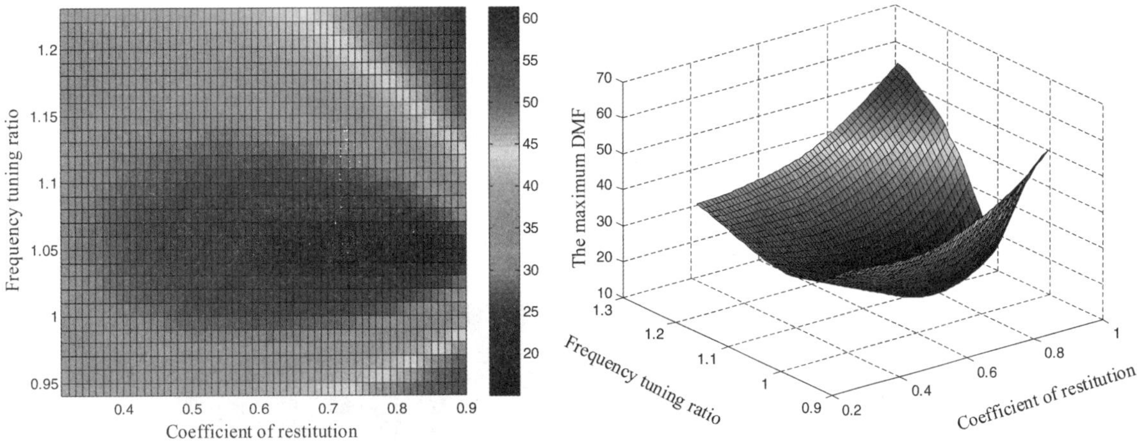

(a) Top view (b) 3D view

Fig. 7. Maximum DMF of main structure for $\mu = 0.01$: (a) Top view; (b) 3D view.

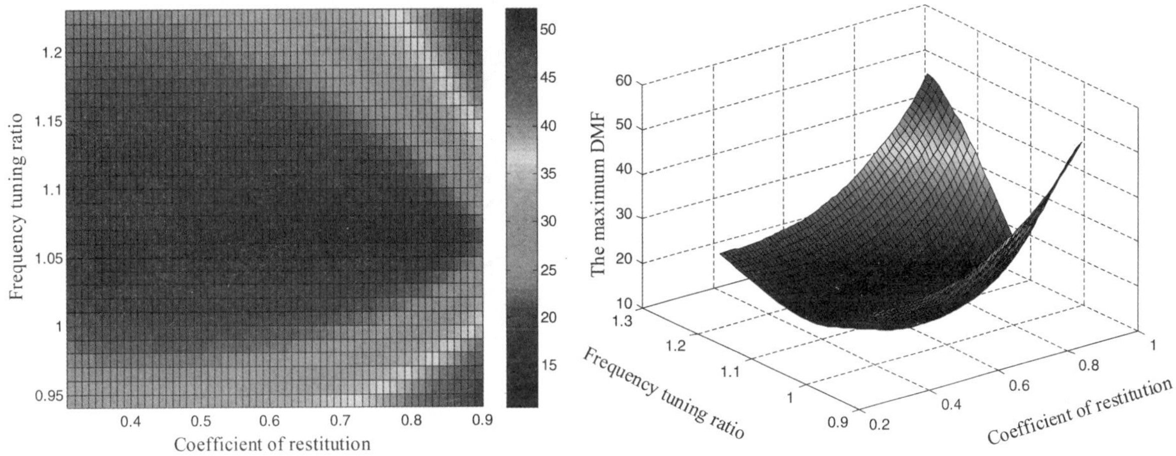

(a) Top view (b) 3D view.

Fig. 8. Maximum DMF of main structure for $\mu = 0.02$: (a) Top view; (b) 3D view.

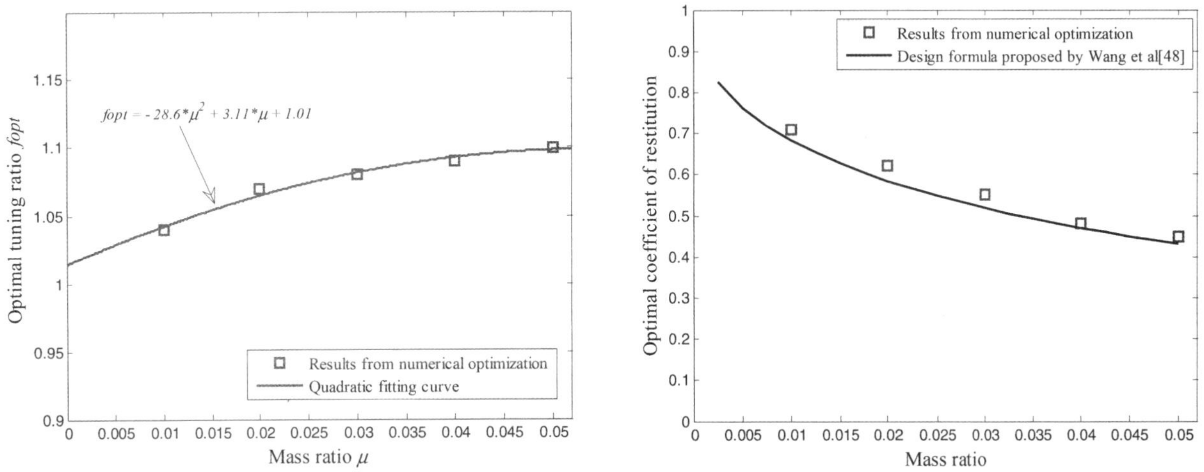

Fig. 9. Optimal parameters of SS-PTMD: (a) Frequency tuning ratio; (b) coefficient of restitution.

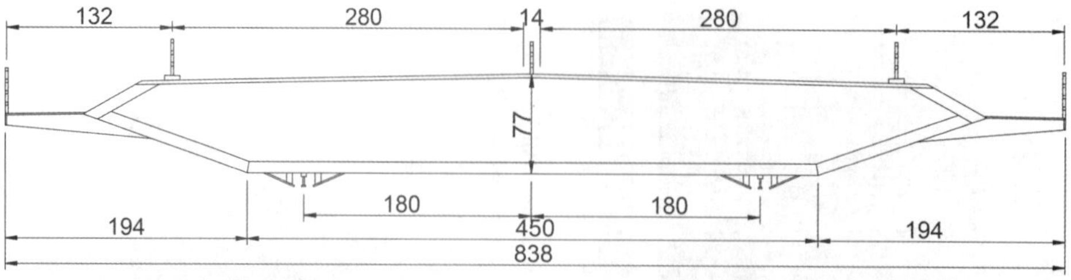

Fig. 10. Cross section of the tested bridge deck model (mm).

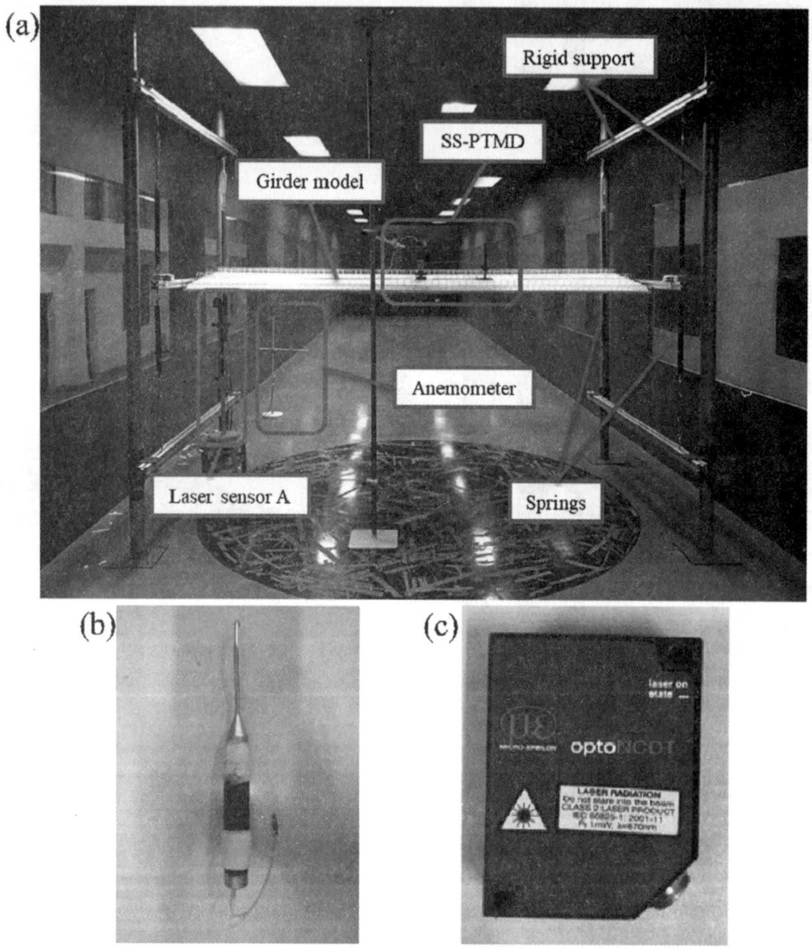

Fig. 11. Experimental setup of VIV tests in a wind tunnel: (a) Overview; (b) Cobra Probe; (c) laser displacement sensor.

Table 1
Dynamic parameters of bridge deck sectional model.

Mass m_m (kg)	Stiffness k_m (N/m)	Frequency f_m (Hz)	Damping coefficient c_m (N·m/s)	Damping ratio ζ_m
43.2	14343.0	2.90	11.02	0.70%

bridge deck is designed based on the modal properties of fundamental vertical mode and torsional mode. The sectional model is elastically mounted by coil springs in wind tunnel to simulate the target modal properties, and is tested in wind tunnel at varying wind speeds. Therefore, only a bridge deck model is used to numerically and experimentally study the effectiveness of a SS-PTMD for control of vortex-induced vibrations. The emphasis has been placed on considering nonlinear features of vortex shedding forces.

4.1. Modeling of a bridge deck model with a SS-PTMD under VIV

The vortex shedding force differs from the sinusoidal force due to its nonlinear feature of fluid and dependence on vibration amplitude. A semi-empirical model [57], based on van der Pol oscillator, was used to model the vortex-shedding vibration [57,58] in this section. Consider a section model with a height D and length L. According to Facchunetti et al. [57], the motion equations of elastically mounted bridge deck model immersed in flow is given by,

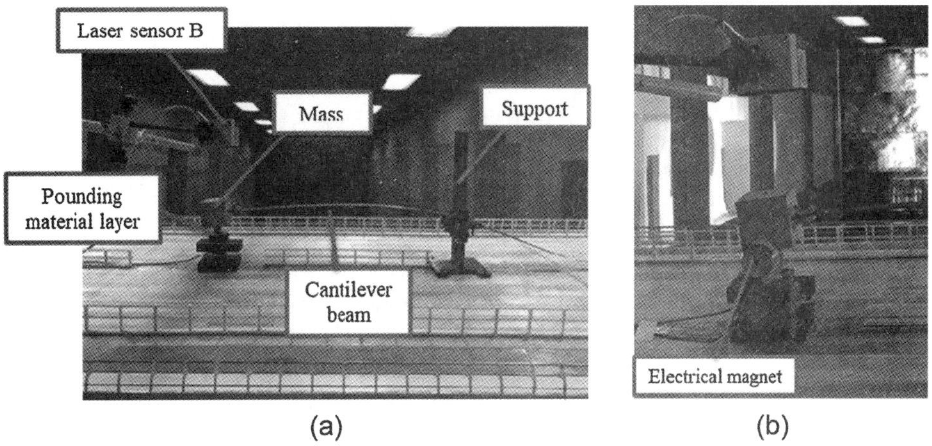

Fig. 12. Details of wind tunnel test: (a) SS-PTMD device; (b) locking device using an electrical magnet.

Table 2
Dynamic parameters of SS-PTMD.

Case	Mass m_d (kg)	Mass ratio	Stiffness k_d (N/m)	Frequency f_d (Hz)	Frequency tuning ratio	COR e
Optimal value	0.86	0.02	80.36	3.077	1.061	0.5835
Experimental value			81.57	3.100	1.069	0.6296

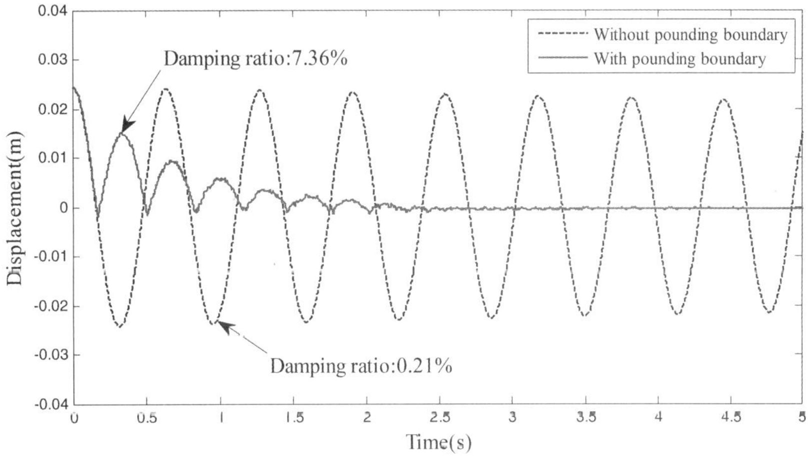

Fig. 13. Free vibration responses of the SS-PTMD with and without HEDR pounding boundary.

$$m_m\ddot{x}_m(t) + c_m\dot{x}_m(t) + k_m x_m(t) = F_L \tag{18}$$

$$F_L = (\gamma\omega_f\rho D^2 L)\dot{x}_m(t) + \left(\frac{1}{4}\rho U^2 D C_{L0} L\right)q(t) \tag{19}$$

$$\ddot{q}(t) + \varepsilon\omega_f[q^2(t)-1]\dot{q}(t) + \omega_f^2 q = [\omega_f(A/D)]\dot{x}_m(t) \tag{20}$$

where m_m, c_m and k_m are the mass, damping coefficient and stiffness of the spring-mounted bridge deck, respectively. $x_m(t)$ and $q(t)$ are the crosswind response of the model and the wake variable. F_L represents the nonlinear vortex shedding lift force based on wake oscillator model. ω_f is the frequency of vortex shedding, which is linear proportional to flow velocity outside lock-in range. ρ is the air density. γ is a stall parameter and C_{L0} is the magnitude of the lift coefficient caused by vortex. A is the velocity coupling coefficient and ε is the van der Pol coefficient. The four unknown model parameters (γ, C_{L0}, A, ε) in the wake oscillator model can be identified from vortex-induced vibrations of the cross section model in a wind tunnel test, as will be shown later.

When the bridge deck model is controlled by the SS-PTMD, the coupling system of the deck model and wake oscillator expressed by Eqs. (18) and (20) are given as,

$$m_m\ddot{x}_m(t) + c_m\dot{x}_m(t) + k_m x_m(t) = F_L + F_p(t) + k_d(x_d(t)-x_m(t)) \tag{21}$$

$$\ddot{q}(t) + \varepsilon\omega_f[q^2(t)-1]\dot{q}(t) + \omega_f^2 q = [\omega_f(A/D)]\dot{x}_m(t) \tag{22}$$

$$m_d\ddot{x}_d(t) + k_d(x_d(t)-x_m(t)) = -F_p(t) \tag{23}$$

where Eqs.(21)–(23) represent the motion equations of the deck model, the wake oscillator and the SS-PTMD. $F_p(t)$ is the pounding force, and it can be computed by Eq. (3) when the pounding condition is satisfied. The procedures of the proposed simulation method to compute the dynamic response of the bridge deck model are summarized below:

(1). Measure the VIV response of the bridge deck model which is elastically supported in a wind tunnel under different wind speeds.

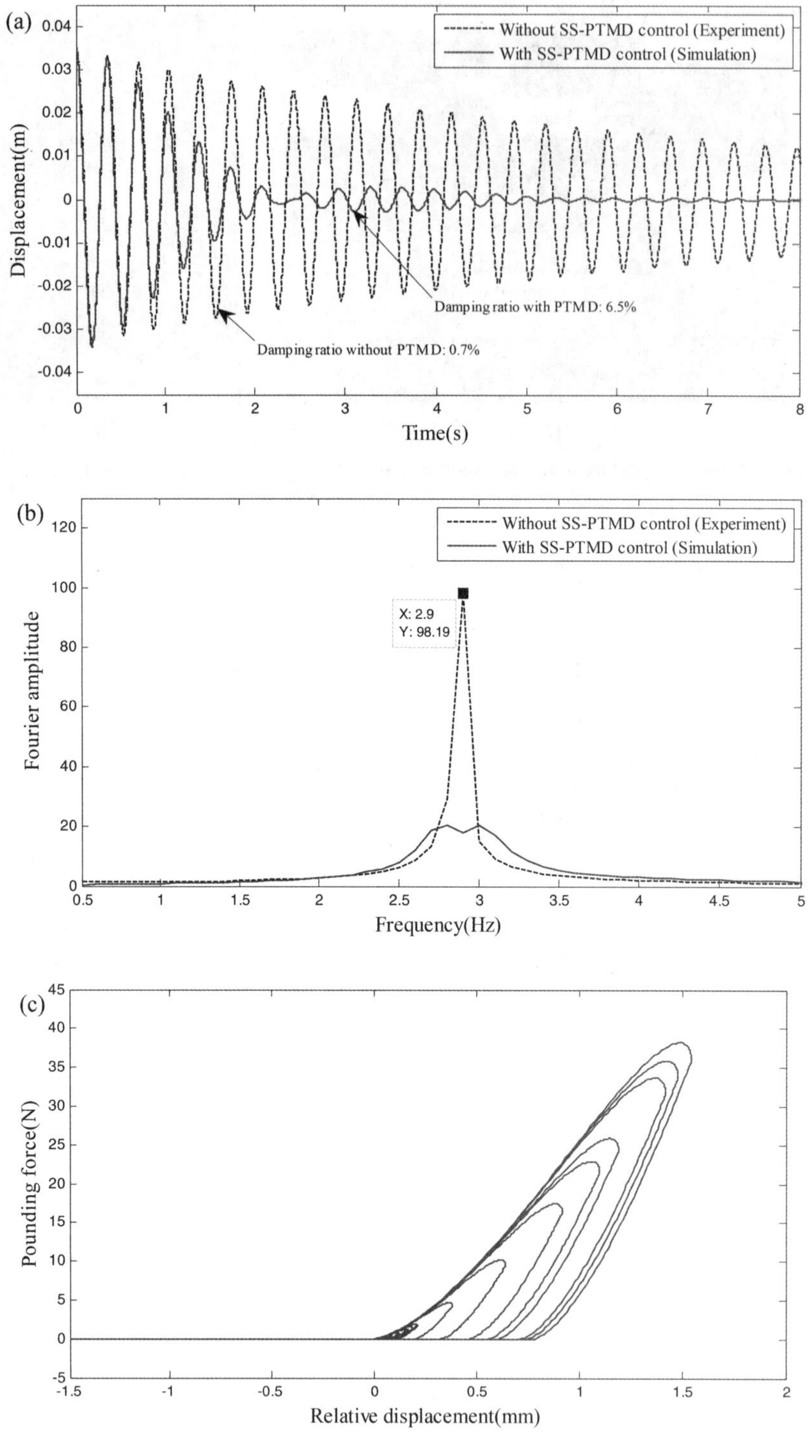

Fig. 14. Free vibration response of deck sectional model: (a) Time history; (b) amplitude-frequency response; (c) pounding force-relative displacement loop; (d) spring force-relative displacement loop.

(2). Use the Runge-Kutta fourth order method to compute the displacement of the bridge deck model when the unknown parameters (γ, C_{L0}, A, ε) in Eqs. (18)–(20) is assumed to certain values. Due to the existence of the negative damping, the VIV displacement of the bridge deck model will eventually converge to a stable-state response even if a random initial condition is given to Eqs. (18) and (20). Then a wind speed-VIV amplitude curve of the bridge deck can be obtained from the numerical simulation. The optimal value of the four unknown parameters (γ, C_{L0}, A, ε) to describe vortex shedding lift force should be identified through a numerical

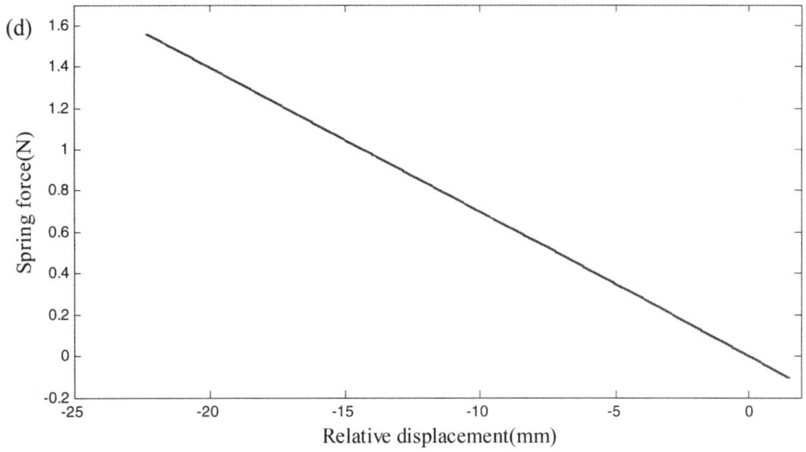

Fig. 14. *(continued)*

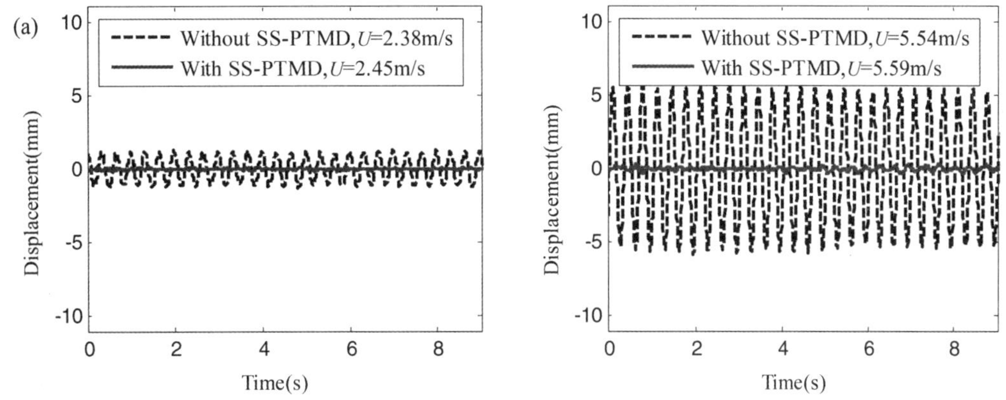

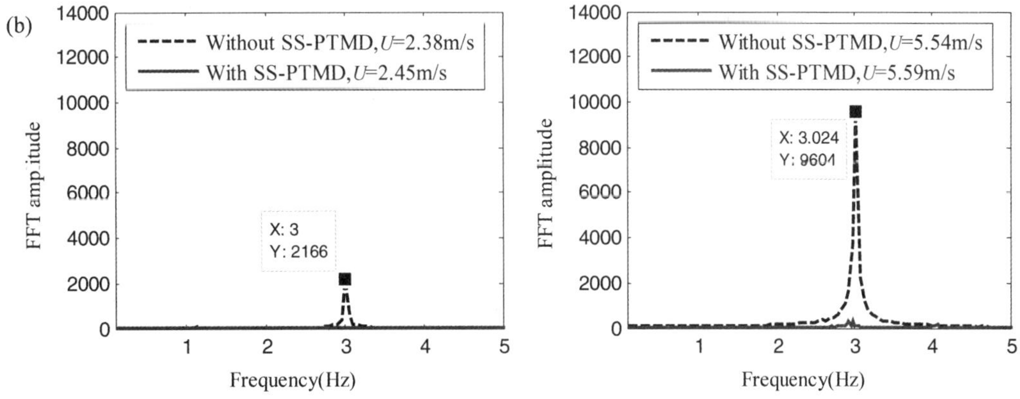

Fig. 15. Displacement of deck model and the corresponding frequency responses under different wind speeds: (a) Time history; (b) frequency response.

optimization program, letting the normalized error of the wind speed-VIV amplitude curve from simulation and wind tunnel tests to be minimized.

(3). Substitute the identified model parameters and the parameters of the SS-PTMD into Eqs. (21)–(23) and compute the wind speed-displacement amplitude curve of the bridge deck model with a SS-PTMD under VIV.

4.2. Experimental setup and dynamic properties of the model

A streamlined cross section shown in Fig. 10 was selected to study the VIV and the control effectiveness of the SS-PTMD for VIV control. It should be noted that the cross section of bridge deck studied in this paper is a typical one, and similar cross section has been adopted in some long span bridges such as the Tsurumi Bridge [59] and the Great Belt East Bridge [60]. As vortex-induced vibration is characterized by a single mode, one single vertical mode is simulated in wind tunnel test.

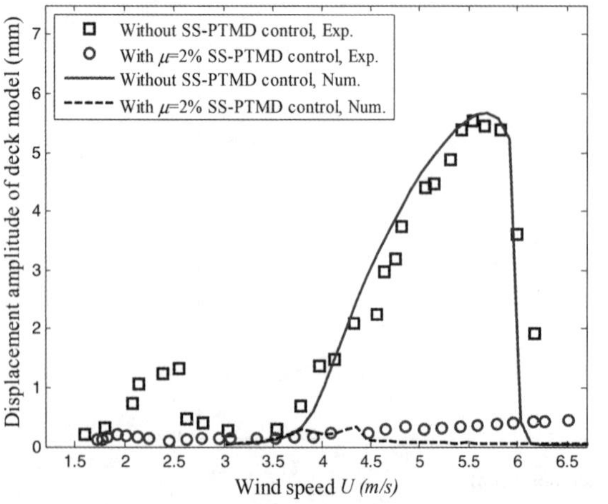

Fig. 16. Displacement amplitude of deck model under different wind speeds.

Table 3
Identified parameters for wake oscillator model.

Parameter	Stall parameter γ	Magnitude of lift coefficient C_{L0}	Velocity coupling coefficient A	van der Pol coefficient ε
Value	6.949	0.181	38.8	0.072

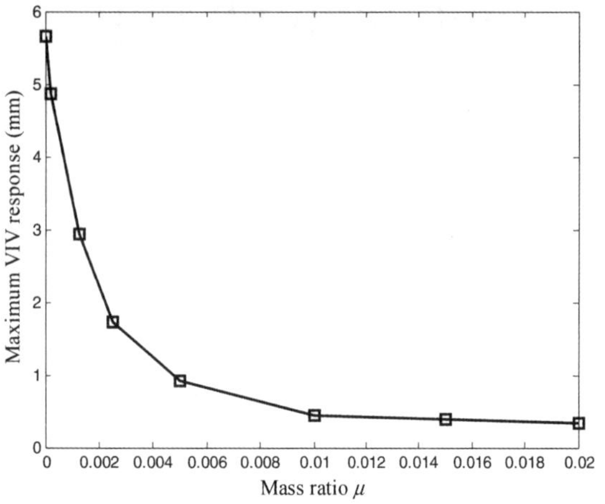

Fig. 17. Maximum VIV response of the deck model with a SS-PTMD at different mass ratios.

Therefore, both in numerical and experimental study, a scaled-down bridge deck model elastically supported by springs, which can capture the natural frequency of one vibration mode of the basic structure, are used to consider the dynamic behavior of bridge deck under VIVs.

The experimental setup of the wind tunnel test is shown in Fig. 11 and Table 1 shows the dynamic properties of the section model. The wind tunnel test was conducted in the KD-1 Wind Tunnel at Hunan University of Science and Technology. The deck model with a cross-flow length of $L = 3.0$ m was suspended by 8 linear springs in the wind tunnel and the effective mass of the bridge deck model is $m_m = 43.2$ kg. The laser displacement sensor A (model: optoNCDT14, which was manufactured by $\mu\varepsilon$ Company) was used to measure the vertical displacement of the model, and a pressure anemometer (model: TFI - Series 100 Cobra Probe, which was manufactured by Turbulent Flow Instrumentation Company) was placed in the front of the model to collect the wind speed during the wind tunnel test.

4.3. Design of SS-PTMD

A SS-PTMD device was designed to place on the deck mode suppress the response of the deck model under VIV, as showr Fig. 12(a). The frequency of the SS-PTMD can be adjusted by chan the length of the cantilever beam and the equivalent damping r depends on the coefficient of restitution of the used VE materials la The laser displacement sensor B (model: optoNCDT14, manufactu by $\mu\varepsilon$ Company) was used to record the displacement of the SS-PTI The existence of the SS-PTMD device may affect the aerodyna property of the bridge deck model. To keep the consistence on aerodynamic property of the tested model under controlled and controlled case, a locking device using an electrical magnet, as show Fig. 12(b), was employed to lock and release the SS-PTMD during wind tunnel test. When the uncontrolled case is tested, a 20 V DC po is applied to the magnet and the SS-PTMD will move together with model due to magnetic force. When the power supply for the magn cut down, the SS-PTMD will be released to absorb the vibration ene from the deck model and the vibration energy will be dissipated poundings.

The mass of the SS-PTMD installed on the deck model is 0.86 kg consequently the mass ratio of the SS-PTMD is $\mu = 0.02$. Accordin Eqs. (16) and (17), the optimal frequency tuning ratio is 1.061 and optimal coefficient of restitution of the VE material layer is 0.5 From Eq. (1), the optimal stiffness provided by the cantilever bean achieve the optimal frequency 3.077 Hz should be 80.36 N/m. W the length of the cantilever beam is set to 45.6 cm, the stiffness vided by the beam is 81.57 N/m and consequently the frequency of SS-PTMD is 3.1 Hz according to Eq. (1). It should be noted that stiffness provided by the cantilever beam can be calculated from free vibration tests of the SS-PTMD when the pounding boundar removed. Since the effectiveness is insensitive to the detuning of coefficient of restitution, the 12 mm HEDR layer with a coefficien restitution $e = 0.6296$, was used as the pounding boundary. The timal and experimental parameters of the SS-PTMD are summarize Table 2. It can be seen from Figs. 7 and 8, the deviation of the re tution coefficient to the optimal value only have litter influence on performance. Although the coefficient of restitution of the HI pounding boundary is not quite close to the optimal value, the ployed SS-PTMD device still can be regarded as an optimized one.

Fig. 13 shows the free vibration responses of the designed SS-PT when the 12 mm HEDR pounding boundary is installed and uninstal It is evident from Fig. 13 that the period of the case with the pounc boundary becomes half to that without the boundary and the damp ratio of the SS-PTMD is significantly enhanced from 0.21% to 7.36% poundings.

With the parameters listed in Tables 1 and 2, the free vibra response of the bridge deck model with SS-PTMD is obtained by merically solving Eq. (12) and the result is plotted in Fig. 14(a) and to compare with the uncontrolled result obtained from the experim It can be seen that the damping ratio of the model increases from 0.7 to 6.5% after the installation of the SS-PTMD, and the amplitude- quency curve also indicates that the vibration amplitude is gre suppressed. Fig. 14(c) plots the hysteretic loop between the pounc force and the relative displacement of the deck model and SS-PTMI can be seen that the energy dissipation of the deck-SS-PTMD sys only happens when the relative displacement is greater than zero when poundings between the deck and SS-PTMD are occurr Fig. 14(d) plots the hysteretic loop between the spring force and relative displacement of the deck model and SS-PTMD. It is shown the spring force does not dissipate the vibration energy at all.

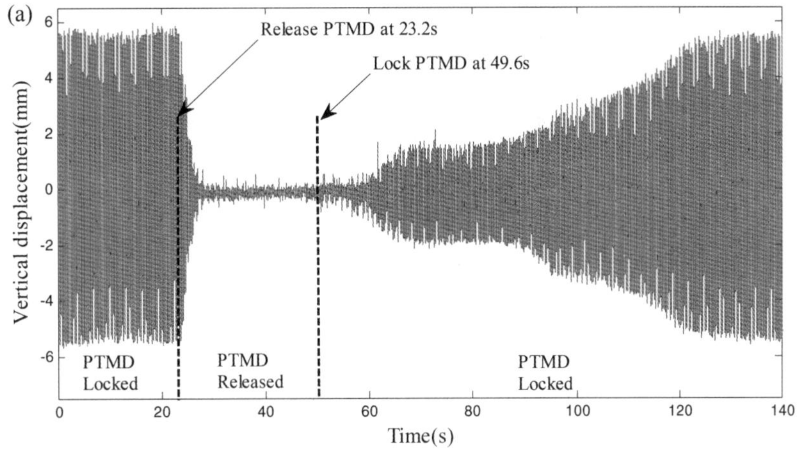

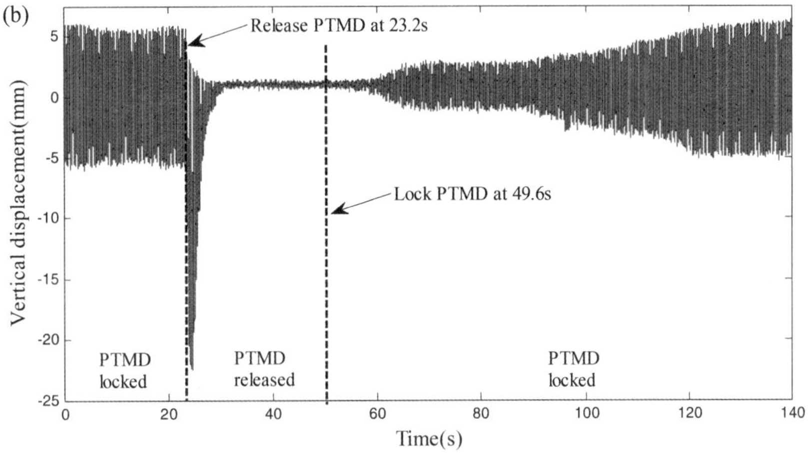

Fig. 18. Lock-release experiment under wind velocity $U = 5.54$ m/s: (a) Response of the bridge deck model; (b) response of SS-PTMD.

4.4. Wind tunnel results

The sectional model test in the wind tunnel was carried out to study the effectiveness of SS-PTMD for VIV control. The displacement histories of the deck model and the corresponding frequency response in the wind speed around 2.4 m/s, 4.6 m/s, 5.5 m/s and 6.0 m/s are shown in Fig. 15. It is shown that the vibration frequencies of the bridge deck model in the uncontrolled case under VIV are slightly larger than its natural frequency measured in the static air. The displacement amplitude of the bridge deck model without and with the optimized SS-PTMD in the wind speed range from 1.5 m/s to 6.5 m/s are shown in Fig. 16. It can be seen from the uncontrolled case that there are two lock-in regions of VIVs, which are $U = [1.8\text{ m/s}, 2.8\text{ m/s}]$ and $U = [3.8\text{ m/s}, 6.5\text{ m/s}]$, respectively. The maximum amplitude of VIV of 5.56 mm occurred in the second lock-in region at the wind speed of 5.54 m/s. When the SS-PTMD was activated, the large amplitude VIV of the model disappeared and the maximum VIV response of the model is suppressed by 94% compared to the uncontrolled case at the wind speed of 5.54 m/s. It is also shown that the large amplitude VIV responses in both the two lock-in regions are controlled.

4.5. Numerical results

The second VIV region $U = [3.8\text{ m/s}, 6.5\text{ m/s}]$ from the experimental result of the uncontrolled case, as plotted in Fig. 16, were utilized to identify the four unknown parameters (γ, $C_{L0,A}$, ε) in the wake oscillator model. Eqs. (18) and (19) can be numerically solved by the fourth order Runge-Kutta method to obtain the VIV responses under different wind speeds. The optimal values of the unidentified parameters were obtained by minimizing the normalized error between experimental and numerical results. The expression of the normalized error is given by,

$$err = \sum_{i=1}^{n} \sqrt{(E_i - N_i)^2} \tag{24}$$

where E_i and N_i are respectively the experimental and the numerical values of the vortex-induced response amplitude at the ith wind speed. An initial condition for the numerical solution of Runge-Kutta method is set to $[x_{m0} = dis_{i-1}, \dot{x}_{m0} = 0, q_0 = 0, \dot{q}_0 = 0]$, where dis_{i-1} is the steady-state vibration amplitude of the deck model from the last time computation [57]. The time step of the solution is 1×10^{-4} s and the wind speeds used in the numerical solution was set as same as that measured from wind tunnel experiments. The optimal values of the parameters are listed in Table 3 through the numerical identification program. The uncontrolled case from the simulation using the identified parameters is plotted in Fig. 16. It can be seen that the numerical results can well reflect the VIV response of the deck model when the SS-PTMD is deactivated. The identified parameters summarized in Table 3 will be continuously used to analyze the case with SS-PTMD control. In this paper, the four unknown parameters in the wake oscillator model

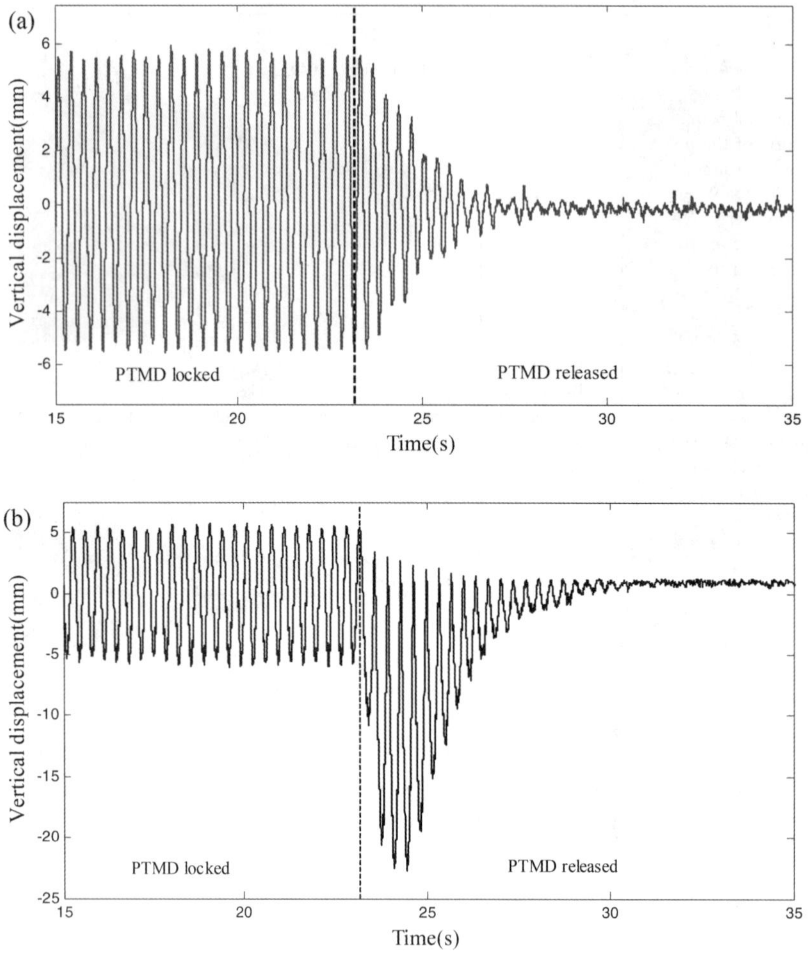

Fig. 19. Details of lock-release experiment from 15 s to 35 s: (a) Response of the bridge deck model; (b) response of SS-PTMD.

are identified from the numerical identification method following the previous studies on VIVs of rectangular section model [61]. Actually, the vortex lift coefficient can be obtained from wind tunnel tests by fixing the structure subjected to vortex shedding and measure the lift force. However due to the nonlinear feature of VIVs, the vortex lift force as well as C_{L0} are dependent on the vibration amplitude of the bridge deck under VIVs. Therefore C_{L0} is also taken as an unknown parameter in the identification.

The VIV response of the deck model controlled by a SS-PTMD with mass ratio of 2% was numerically computed by Eqs. (21)–(23) and the numerical results are also shown in Fig. 16 to compare with experimental results. It is shown that the numerical result matches well with the experimental data at the lower wind speed of the lock-in region, and however the error between experimental and numerical results at the higher wind speed of the lock-in region become greater and greater. This is because the van der Pol oscillator model only can take the self-excited vibration into account and the buffeting response of the deck model increases with the increase of the wind speed during wind tunnel tests.

Moreover, the deck model controlled by the SS-PTMD with different mass ratios was computed. Fig. 17 plots the maximum VIV response of the deck model controlled by the SS-PTMD with various mass ratios. It is shown that at the lower value of mass ratio such as less than 0.5%, the maximum response dramatically reduces with mass ratio, and however the performance of the SS-PTMD will not be enhanced too much with the increase of the mass ratio when the mass ratio is greater than 1%.

4.6. A lock-release experiment under the maximum VIV

A lock-release experiment was carried out to reveal the dynamic characteristics of the SS-PTMD to control the large amplitude VIV under wind speed $U = 5.54$ m/s, where the maximum vortex-induced response was occurred. The vertical responses of the model and the SS-PTMD during the lock-release experiment are shown in Fig. 18. It is shown that the large amplitude VIV occurred during the first 23.2 s when the SS-PTMD was locked by the electromagnet. After that the SS-PTMD was released by cutting off the DC power to the electromagnet, and the vibration amplitude rapidly deceased from 5.56 mm to 0.33 mm with a reduction of 94%. After locking the SS-PTMD at 49.6 s, the large amplitude VIV occurred again and the response amplitude returned to the value as same as that during the first 23.2 s. Moreover, the displacement of the SS-PTMD suddenly increased from the released moment, and the vibration energy was dissipated by the poundings between the SS-PTMD and the HEDR layer in the next few seconds. The zoomed view of the lock-release experiment between 15 s and 35 s is shown in Fig. 19. It should be noted that the lock-release experiment not only strongly demonstrates the effectiveness of the SS-PTMD for suppression of VIVs, but also reveals the dynamic characteristics of the SS-PTMD when is released.

5. Conclusions

This paper proposes a new method to control the vortex-induced vibration (VIV) of a bridge deck model by using singled-sided PTMD (SS-PTMD) with one VE material pounding boundary. An impact force model with the consideration of the poundings between VE materials layer and steel was employed to simulate the dynamic response of the SS-PTMD and was then validated by free pounding experiments. The coefficient of restitution of different VE material layers was also studied. Simplified design formulas for the PTMD were obtained from a numerical optimization based on the verified impact force model. The effectiveness of the proposed SS-PTMD was studied by wind tunnel tests and numerical simulations.

The coefficient of restitution of VE material layer is stable under various impact velocities and can be adjusted in a certain range by using a composite layer of VE materials. The results from wind tunnel tests show that there are two lock-in regions appeared when the PTMD is deactivated. The designed PTMD with mass ratio of 2% can reduce the maximum amplitude of VIVs by 94%. The performance of the PTMD may not be enhanced too much with the increase of the mass ratio, when mass ratio of the PTMD is larger than 1%. The lock-release experiment can well demonstrate the effectiveness of the PTMD to control VIVs resulted from wind excitations.

Acknowledgements

This research work was supported by the National Key Basic Research Program of China ('973 Project') (grant number 2015CB057702) and the National Natural Foundation of China (grant numbers 51422806; 51378203).

References

[1] Li H, Laima S, Ou J, Zhao X, Zhou W, Yu Y, et al. Investigation of vortex-induced vibration of a suspension bridge with two separated steel box girders based on field measurements. Eng Struct 2011;33(6):1894–907.
[2] Yuan W, Laima S, Chen W, Li H, Hu H. Investigation on the vortex-and-wake-induced vibration of a separated-box bridge girder. J Fluids Struct 2017;70:145–61.
[3] Li H, Laima S, Zhang Q, Li N, Liu Z. Field monitoring and validation of vortex-induced vibrations of a long-span suspension bridge. J Wind Eng Ind Aerodyn 2014;124(7):54–67.
[4] Seo JW, Kim HK, Jin P, Kim KT, Kim GN. Interference effect on vortex-induced vibration in a parallel twin cable-stayed bridge. J Wind Eng Ind Aerodyn 2013;116(116):7–20.
[5] Miguel LFF, Lopez RH, Torii AJ, Miguel LFF, Beck AT. Robust design optimization of TMDs in vehicle–bridge coupled vibration problems. Eng Struct 2016;126:703–11.
[6] Zhou D, Li J, Hansen CH. Suppression of the stationary maglev vehicle–bridge coupled resonance using a tuned mass damper. J Vib Control 2013;19(19):191–203.
[7] Shi X, Cai CS. Suppression of vehicle-induced bridge vibration using tuned mass damper. J Vib Control 2008;14(7):1037–54.
[8] Lin CC, Wang JF, Chen BL. Train-induced vibration control of high-speed railway bridges equipped with multiple tuned mass dampers. J Bridge Eng 2005;10(4):398–414.
[9] Wu WJ, Cai CS. Theoretical exploration of a taut cable and a TMD system. Eng Struct 2007;29(6):962–72.
[10] Gerges RR, Vickery BJ. Wind tunnel study of the across-wind response of a slender tower with a nonlinear tuned mass damper. J Wind Eng Ind Aerodyn 2003;91(8):1069–92.
[11] Venanzi I, Ubertini F, Materazzi AL. Optimal design of an array of active tuned mass dampers for wind-exposed high-rise buildings. Struct Control Health Monitor 2013;20(6):903–17.
[12] Abdel-Rohman M, John MJ. Control of wind-induced nonlinear oscillations in suspension bridges using a semi-active tuned mass damper. J Vib Control 2006;12(10):1049–80.
[13] Li C. Optimum multiple tuned mass dampers for structures under the ground acceleration based on DDMF and ADMF. Earthq Eng Struct Dyn 2002;31(4):897–919.
[14] Spencer Jr. BF, Nagarajaiah S. State of the art of structural control. J Struct Eng 2003;129(7):845–56.
[15] Hoang N, Fujino Y, Warnitchai P. Optimal tuned mass damper for seismic applications and practical design formulas. Eng Struct 2008;30(3):707–15.
[16] Lin CC, Ueng JM, Huang TC. Seismic response reduction of irregular buildings using passive tuned mass dampers. Eng Struct 2000;22(5):513–24.
[17] Bi K, Hao H. Using pipe-in-pipe systems for subsea pipeline vibration control. Eng Struct 2016;109:75–84.
[18] Nikoo HM, Bi K, Hao H. Passive vibration control of cylindrical offshore components using pipe-in-pipe (PIP) concept: an analytical study. Ocean Eng 2017;142:39–50.
[19] Bani-Hani KA. Vibration control of wind-induced response of tall buildings with an active tuned mass damper using neural networks. Struct Control Health Monitor 2007;14(1):83–108.
[20] Strommen E, Hjorth-Hansen E. On the use of tuned mass dampers to suppress vortex shedding induced vibrations. Wind Struct 2001;4(1):19–30.
[21] Giaralis A, Petrini F. Wind-induced vibration mitigation in tall buildings using the tuned mass-damper-inerter. J Struct Eng 2017;143(9):04017127.
[22] Fujino Y, Yoshida Y. Wind-induced vibration and control of trans-Tokyo bay crossing bridge. J Struct Eng 2002;128(8):1012–25.
[23] Mokrani B, Tian Z, Alaluf D, Meng F, Preumont A. Passive damping of suspension bridges using multi-degree of freedom tuned mass dampers. Eng Struct 2017;153:749–56.
[24] Frahm H. Device for damping vibrations of bodies. April, 18 1911. US Patent 989, 958; 1911.
[25] Den Hartog JP. Mechanical vibrations. McGraw-Hill; 1956.
[26] Rana R, Soong TT. Parametric study and simplified design of tuned mass dampers. Eng Struct 1998;20(3):193–204.
[27] Lu Z, Lu X, Masri SF. Studies of the performance of particle dampers under dynamic loads. J Sound Vib 2010;329(26):5415–33.
[28] Lu Z, Masri SF, Lu X. Parametric studies of the performance of particle dampers under harmonic excitation. Struct Control Health Monitor 2011;18(1):79–98.
[29] Lu Z, Masri SF, Lu X. Studies of the performance of particle dampers attached to a two-degrees-of-freedom system under random excitation. J Vib Control 2011;17(10):1454–71.
[30] Weber F, Boston C, Maślanka M. Adaptive TMD based on the emulation of positive and negative stiffness with MR damper. Smart Mater Struct 2010;20(1):015012.
[31] Weber F, Maslanka M. Frequency and damping adaptation of a TMD with controlled MR damper. Smart Mater Struct 2012;21(5):55011–27. (17).
[32] Weber F. Semi-active vibration absorber based on real-time controlled MR damper. Mech Syst Sig Process 2014;46(2):272–88.
[33] Cai CS, Asce M, Wu WJ, et al. Cable vibration control with a TMD-MR damper system: experimental exploration. J Struct Eng 2007;133(5):629–37.
[34] Wang Z, Chen Z, Wang J. Feasibility study of a large-scale tuned mass damper with eddy current damping mechanism. Earthq Eng Struct Vib 2012;11(3):391–401.
[35] Bae JS, Hwang JH, Kwag DG, Park J, Inman DJ. Vibration suppression of a large beam structure using tuned mass damper and eddy current damping. Shock Vib 2014;1:1–10.
[36] Wang W, Dalton D, Hua X, Wang X, Chen Z, Song G. Experimental study on vibration control of a submerged pipeline model by eddy current tuned mass damper. Appl Sci 2017;7(10):987.
[37] Chen J, Lu G, Li Y, Wang T, Wang W, Song G. Experimental study on robustness of an eddy current-tuned mass damper. Appl Sci 2017;7(9):895.
[38] Bae JS, Hwang JH, Roh JH, Kim JH, Yi MS, Lim JH. Vibration suppression of a cantilever beam using magnetically tuned-mass-damper. J Sound Vib 2012;331(26):5669–84.
[39] Song G, Zhang P, Li L, Singla M, Patil D, Li H, et al. Vibration control of a pipeline structure using pounding tuned mass damper. J Eng Mech 2016;142(6):04016031.
[40] Li L, Song G, Singla M, Mo YL. Vibration control of a traffic signal pole using a pounding tuned mass damper with viscoelastic materials (II): experimental verification. J Vib Control 2013;21(4):670–5.
[41] Zhang P, Li L, Patil D, Singla M, Li H, Song G. Parametric study of pounding tuned mass damper for subsea jumpers. Smart Mater Struct 2016;25(1):015028.
[42] Jiang J, Zhang P, Patil D, Li H, Song G. Experimental studies on the effectiveness and robustness of a pounding tuned mass damper for vibration suppression of a submerged cylindrical pipe. Struct Control Health Monitor 2017;24:e2027.
[43] Li H, Zhang P, Song G, Patil D, Mo YL. Robustness study of the pounding tuned mass damper for vibration control of subsea jumpers. Smart Mater Struct 2015;24(9):095001.
[44] Xue Q, Zhang J, He J, Zhang C. Control performance and robustness of pounding tuned mass damper for vibration reduction in SDOF structure. Shock Vib 2016;2016(2):1–15.
[45] Zhang P, Song G, Li H, Lin Y. Seismic control of power transmission tower using pounding TMD. J Eng Mech 2013;139(10):1395–406.
[46] Xue Q, Zhang J, He J, Zhang C, Zou G. Seismic control performance for Pounding Tuned Massed Damper based on viscoelastic pounding force analytical method. J Sound Vib 2017;411:362–77.
[47] Wang W, Hua X, Wang X, Chen Z, Song G. Numerical modeling and experimental study on a novel pounding tuned mass damper. J Vib Control 2017;2017(2). 107754631771871.
[48] Wang W, Hua X, Wang X, Chen Z, Song G. Optimum design of a novel pounding tuned mass damper under harmonic excitation. Smart Mater Struct 2017;26(5):055024.
[49] Wang W, Hua X, Wang X, Chen Z, Song G. Advanced impact force model for low-speed pounding between viscoelastic materials and steel. J Eng Mech 2017;143(12):04017139.
[50] Xue Q, Zhang C, He J, Zou G, Zhang J. An updated analytical structural pounding force model based on viscoelasticity of materials. Shock Vib 2016;2016(3):1–15.
[51] Zhang P, Huo L, Song G. Impact fatigue of viscoelastic materials subjected to pounding. Appl Sci 2018;8(1):117.
[52] Jankowski R. Non-linear viscoelastic modelling of earthquake-induced structural pounding. Earthq Eng Struct Dyn 2010;34(6):595–611.
[53] Anagnostopoulos SA. Pounding of buildings in series during earthquakes. Earthq Eng Struct Dyn 1988;16(3):443–56.
[54] Zuo L, Nayfeh SA. Minimax optimization of multi-degree-of-freedom tuned-mass

dampers. J Sound Vib 2004;272(3):893–908.
[55] Du D, Gu X, Chu D, Hua H. Performance and parametric study of infinite-multiple TMDs for structures underground acceleration by H_∞ optimization. J Sound Vib 2007;305(4–5):843–53.
[56] Zhou S, Hua XG, Chen ZQ, Chen W. Experimental investigation of correction factor for VIV amplitude of flexible bridges from an aeroelastic model and its 1:1 section model. Eng Struct 2017;141:263–71.
[57] Facchinetti ML, Langre ED, Biolley F. Coupling of structure and wake oscillators in vortex-induced vibrations. J Fluids Struct 2004;19(2):123–40.
[58] Xu K, Ge Y, Zhang D. Wake oscillator model for assessment of vortex-induced vibration of flexible structures under wind action. J Wind Eng Ind Aerodyn 2015;136:192–200.
[59] Mashnad M, Jones NP. A model for vortex-induced vibration analysis of long-span bridges. J Wind Eng Ind Aerodyn 2014;134:96–108.
[60] Larsen A, Esdahl S, Andersen JE, Vejrum T. Storebælt suspension bridge – vortex shedding excitation and mitigation by guide vanes. J Wind Eng Ind Aerodyn 2000;88(2–3):283–96.
[61] Marra AM, Mannini C, Bartoli G. Van der Pol-type equation for modeling vortex-induced oscillations of bridge decks. J Wind Eng Ind Aerodyn 2011;99(6–7):776–85.

第 32 卷 第 10 期
2019 年 10 月

中 国 公 路 学 报
China J. Highw. Transp.

Vol. 32 No. 10
Oct. 2019

文章编号：1001-7372(2019)10-0115-10

大跨度悬索桥的多阶模态竖向涡振与控制

华旭刚，黄智文，陈政清

（湖南大学 风工程试验研究中心，湖南 长沙 410082）

摘要：大跨度悬索桥具有多个竖向模态密集分布的特性。在常遇风速范围内，从低到高的各阶竖向模态随风速升高而逐个发生涡振，这就是大跨度悬索桥的多阶模态涡振问题。针对这一问题开展深入研究，讨论中国公路桥梁抗风设计规范中竖向涡振容许振幅的合理性；阐述了利用节段模型风洞试验和理论分析综合预测实桥多阶模态竖向涡振响应的基本方法，得到了各模态阻尼比相等时悬索桥各阶模态竖向涡振振幅基本相等的结论，并通过特殊设计的悬索桥竖向等效气弹模型和塔科马桥涡振实测资料，验证了这一结论；指出在既有桥梁上追加气动措施或安装调谐质量减振器抑制悬索桥多阶模态涡振都有很大的难度，进而提出了在加劲梁与桥塔之间安装直接耗能阻尼器的设想，并进行了气弹模型试验验证；讨论了采用电涡流阻尼器进行半主动涡振控制的可行性。研究结果表明：在相同阻尼比条件下大跨度悬索桥各阶竖弯模态的最大涡振振幅基本相等；依据最大加速度幅值按频率比的平方增加的原理，满足人体振动舒适性的高阶竖弯模态的容许振幅必然小于低阶模态，因此要更加重视起振风速在容许行车风速（25 m·s^{-1}）以内的高阶竖弯模态涡振；对于漂浮体系悬索桥，在加劲梁与对应桥塔之间设置阻尼器可有效抑制多阶模态涡振。

关键词：桥梁工程；悬索桥；风洞试验；涡振；振动控制；阻尼器

中图分类号：U441 **文献标志码**：A DOI：10.19721/j.cnki.1001-7372.2019.10.011

Multi-mode Vertical Vortex-induced Vibration of Suspension Bridges and Control Strategy

HUA Xu-gang, HUANG Zhi-wen, CHEN Zheng-qing

(Wind Engineering Research Center, Hunan University, Changsha 410082, Hunan, China)

Abstract: Long-span suspension bridges contain numerous closely spaced vertical modes, and these modes, in order of natural frequency, may develop vortex-induced vibrations (VIV) in turn with increase of wind velocity. This multi-mode vertical VIV problem has become a challenge in the development of suspension bridges. In this study, the problem has been addressed from several aspects. First, the rationality of the allowable amplitude of vertical VIV for different vibrational modes is discussed in reference to the wind-resistant design specifications for highway bridges in China. Second, the analytical method for predicting multi-mode vertical VIV is formulated on the basis of section model tests and theoretical analysis. The results indicate that when the same damping ratio is used, the maximum VIV amplitudes are almost the same for different vertical modes of suspension bridges. This is further verified by wind tunnel tests using a specially-designed novel aeroelastic model of suspension bridges and by existing measurement data of the Old Tacoma Narrows Bridge. Third, because it is quite difficult to apply aerodynamic

收稿日期：2018-11-13
基金项目：国家自然科学基金项目(51422806)
作者简介：华旭刚(1978-)，男，浙江义乌人，教授，博士研究生导师，工学博士，E-mail：cexghua@hnu.edu.cn。

countermeasures or tuned mass dampers on existing bridges experiencing VIV, a new vibration control strategy is proposed to suppress multi-mode VIV by using the direct energy dissipation devices installed between the tower and girders. The efficacy of the control strategy is preliminarily evaluated by conducting experiment susing the novel aeroelastic model, and the feasibility of semi-active control of multi-mode VIV by employing an eddy current damper is discussed. Through above study, it is concluded that the maximum amplitudes of VIV for different vertical modes were almost the same if their damping ratios were also the same; thus, their acceleration amplitudes increase in proportion to the square of natural frequency. In order to achieve the same allowable maximum acceleration amplitudes, the allowable VIV amplitudes for higher modes must be smaller than those for lower modes, which implies that VIV at higher modes should be considered with priority in wind-resistant designs of suspension bridges. Further, the direct energy dissipation device installed between the pylon and floating girder is shown to effectively suppress multi-mode VIV in multi-span suspension bridges without immediate support.

Key words: bridge engineering; suspension bridge; wind tunnel test; vortex-induced vibration; vibration control; damper

0 引 言

现有工程实践证明,大跨度悬索桥的颤振问题已经解决,按颤振理论进行抗风设计的桥梁都具有足够的抗风稳定性[1]。大跨度悬索桥的另一个特点是具有多个竖向模态密集分布的特性。在常遇风速范围内,从低到高的各阶竖向模态随风速升高而逐个发生涡振,这就是目前大跨度悬索桥发展中必须重视的多阶模态涡振问题。例如中国的西堠门大桥[2]和丹麦的大带东桥[3]等均在成桥以后观测到了多个模态涡振,加劲梁振幅超过 20 cm。悬索桥竖向涡振是气流流经加劲梁断面时产生流动分离和周期性旋涡脱落,使加劲梁上、下表面出现交替变化的正负压力,导致加劲梁发生横风向有限振幅的振动,即竖向振动。与颤振相比,大跨度悬索桥的竖向涡振具有如下特点:①涡激振动是一种非线性自限幅振动,它虽然不像颤振一样会在短时间内对结构造成灾难性破坏,但涡振振幅过大会影响桥上行车的舒适和安全,对桥梁的正常运营造成不利影响,频繁发生的涡振还可能引起构件的疲劳破坏;②大跨度悬索桥的竖向涡振通常表现为单一模态的振动,但大跨度悬索桥模态密集,在桥梁正常运营风速限值 25 m · s^{-1}内,固有频率在 0.6 Hz 以内的竖弯模态都有发生涡振的可能,这就是本文所指的多阶模态涡振问题。以西堠门大桥为例,其在 0.5 Hz 以下分布有 8 阶竖向模态,按梁高 $H=3.50$ m,斯托罗哈数 $S_t=0.11$ 估算,前 8 阶竖弯模态对应的涡振起振风速都在 15 m · s^{-1}以下。因此,确保大跨度悬索桥各阶模态的涡振振幅满足规范要求是工程人员和桥梁运营单位高度关注的工程问题。

从桥梁抗风设计的角度看,大跨度悬索桥的竖向涡振研究需要解决如下几个关键问题:①如何预测实桥多个不同模态的最大涡振振幅和涡振锁定风速区间,并验证其合理性;②不同模态竖向涡振的振幅容许值应该如何规定;③当实桥多个模态的最大涡振振幅超过容许值时,如何有效地进行振动控制。

对于第 1 个问题,目前主要采用弹性悬挂节段模型风洞试验加理论分析的预测方法。涡振振幅与气动参数、阻尼比及质量等模态参数有关。弹性悬挂节段模型风洞试验主要用于预测涡振锁定风速区间和节段模型最大涡振振幅、识别涡激力数学模型各参数和测定涡激力沿桥梁跨向的相关性。理论分析则是以涡激力数学模型为基础建立并求解实桥的涡振运动方程,最后得到计入模态振型影响的实桥涡振最大振幅。国内外学者已经对涡激力数学模型[4-6]、涡激力的跨向相关性[7-8]等进行了大量研究,建立了多种涡激力数学模型和涡激力跨向相关模型。一些学者对静止及振动状态下分离钢箱钝体断面的流场结构[9-10]进行了研究,有助于解释涡激共振的致振机理。现有的涡激力数学模型都具有很强的经验性,通常能够较好地重现用于数学模型参数识别的一些试验工况,但在预测其他试验工况时却容易出现较大误差。其中最突出的一个问题就是在某个 Scruton 数下识别的涡激力数学模型,难以精确预测 Scruton 数发生改变时结构的涡振响应[11],这无疑大幅增加了弹性悬挂节段模型风洞试验的工作量，也降低了实桥涡振响应的预测精度。另一方面，基于弹性悬挂节段模型风洞试验和理论分析的涡振振幅预测结果很难通过现场实测或常规气弹模型验证与评价。例如，朱乐东[12]、张志田等[13]分别从理论上推导了由节段模型涡振振幅到实桥涡振振幅的振型修正系数，但上述理论值很难通过现场实测或常规气弹模型试验进行验证。首先，实桥来流条件非常复杂，且实桥模态阻尼比存在较大的不确定性，这些因素难以在节段模型风洞试验中精确模拟，而桥梁涡振振幅又对来流条件和阻尼比非常敏感；其次，现有多阶模态的涡振试验技术也有较大的局限性：①常规气弹模型试验的缩尺比小，制作复杂、干扰因素众多，难以精确模拟多个竖向模态；②拉条模型虽然能模拟很多阶竖向模态，但其频率分布按自然数增加，因此无法模拟悬索桥的实际频率分布特性。鉴于此，本文作者开发了一种新型的悬索桥等效气弹模型，并开展了多阶模态竖向涡振的风洞试验研究。

对于第 2 个问题，中国《公路桥梁抗风设计规范》(JTG/T D60-01—2004)中规定桥梁竖弯涡振振幅容许值与涡振频率成反比[14]，但是没有说明上述涡振容许值的取值依据。此外，对于多个模态都可能发生涡激振动的大跨度悬索桥，是否应该分别对每一阶模态按其频率来计算相应的涡振振幅容许值尚不明确。各国桥梁抗风设计中对涡振容许限值的规定也不尽相同，但主要从结构疲劳、人体舒适性及承载能力极限状态 3 个方面对涡振容许幅值加以限定[15-18]。为了探明中国规范中涡振振幅容许值的取值依据，陈政清等[19]进行了深入调研及分析，也研究了基于行车视距的涡振容许值的取值方法。

对于第 3 个问题，目前桥梁涡振控制措施主要有气动措施、调谐质量减振器(TMD)措施和阻尼器措施 3 类。气动措施是控制涡振最常用的方法，主要包括导流板、抑流板、风嘴、稳定板和抑涡格栅等。以导流板对分离式箱梁的涡振控制为例，Yang 等[20-22]进行了大量研究，结果表明导流板的涡振控制效果不仅与原始断面外侧风嘴的角度、内侧梁底倒角形式、中央开槽率、风攻角等加劲梁气动外形有关，也受到导流板自身尺寸、安装位置的显著影响，还要特别关注雷诺数效应。Larsen 等[3]详细报道了丹麦大带桥采用导流板有效抑制了实桥发生的竖向模态涡振，导流板实施工期约为 2 个月。为提高传统气动措施的鲁棒性和通用性问题，国内外一些学者也提出了利用主动气动措施进行桥梁风致振动控制的方法[23-25]，目前都尚处于探索研究阶段。另一方面，对于在运营阶段发生大幅涡振的既有桥梁追加气动措施有一定难度，气动措施的施工过程可能影响桥梁的正常运营，引起某些负面社会影响；对于防颤振要求很高的桥梁，附加气动措施还可能影响桥梁的颤振性能，因此很难被甲方接受。此外，在连续梁桥涡振控制中使用较多的 TMD 用于大跨度悬索桥的涡振控制也有 2 个难题：①TMD 的参数优化设计主要针对单一模态控制，多个模态控制需要安装多套 TMD，导致桥面荷载增加过多；②TMD 弹簧的静伸长量由调谐频率单一控制，$\Delta=g/(2\pi f)^2$，其中 g 为重力加速度，f 为 TMD 频率。若假设加劲梁某阶次的 TMD 调谐频率为 $f=0.2$ Hz，则弹簧静伸长量可达 6.25 m，不仅制作困难，而且远大于悬索桥加劲梁的安装空间。因此，采用 TMD 控制大跨悬索桥的多阶竖向涡振控制基本上不具备可行性。综上所述，为解决既有大跨度悬索桥的涡振问题，亟需探索和研究新的涡振控制方法。

本文围绕大跨度悬索桥的多阶模态竖向涡振问题，首先讨论了中国《公路桥梁抗风设计规范》中竖向涡振容许振幅的合理性，然后阐述了利用节段模型风洞试验和理论分析综合预测实桥多阶模态竖向涡振响应的基本方法。为验证上述方法的正确性，本文提出、设计和制作了一种新型的悬索桥等效气弹模型，并开展了多阶模态竖向涡振的风洞试验研究；整理了旧塔科马桥多模态竖向涡振的现场实测数据，总结了实桥多模态涡振幅值响应的特点；最后针对多跨连续漂浮体系悬索桥，提出了在加劲梁与桥塔之间安装直接耗能阻尼器来控制多模态涡振的设想并进行试验验证，讨论采用电涡流阻尼器进行半主动涡振控制的可行性。

1 大跨度悬索桥多阶模态竖向涡振的振幅容许值

各国桥梁抗风设计中对涡振容许限值的规定不尽相同，但都是主要从结构疲劳、人体舒适性及承载能力极限状态 3 个方面对竖向涡振容许幅值加以限定[15-18]。Larsen 等[3]在评价丹麦大带东桥的涡振性能时指出主梁涡振对驾驶人视线的影响会使驾驶人分心，并对桥上行车安全造成潜在威胁。陈政清等[19]在分析已有文献的基础上，对中国规范中涡振振幅容许值的取值依据进行了讨论，主要从人体舒适性及行车视距 2 个方面讨论涡振容许幅值问题。

中国《公路桥梁抗风设计规范》(JTG/T D60-

01—2004)采用与日本高速公路桥梁抗风设计指南相同的涡振振幅容许值,它与竖弯模态频率成反比,如式(1)所示

$$h_a=0.04/f_b \tag{1}$$

式中:h_a 为竖弯涡振容许振幅;f_b 为桥梁竖向涡振频率。

如引言所述,大跨度悬索桥的多阶竖弯模态在容许通行风速范围内都有发生涡激振动的可能,所以式(1)应该标明模态阶数,即

$$h_{a,i}=0.04/f_{b,i} \tag{2}$$

式中:$h_{a,i}$为桥梁第 i 阶竖向涡振的容许振幅;$f_{b,i}$为桥梁第 i 阶竖向涡振的固有频率;i 为涡振模态阶数。

显然式(2)中模态阶数越高对应的容许振幅就越小,所以在大跨度悬索桥抗风设计阶段应该更关注高阶模态的涡振响应。

对于日本规范采用式(1)作为桥梁竖向涡振容许值的原因,日本京都大学 Tommi Yagi 教授援引文献[16]给出了解释。由桥梁竖弯振动基频频率的近似公式 $f=100L^{-1}$ 和涡振时桥梁加速度不宜超过 1 $\mathrm{m\cdot s^{-2}}$的要求,可得对应的涡振最大容许速度 V_a 为

$$V_a=\frac{1}{2\pi f}=\frac{L}{200\pi} \tag{3}$$

再由振幅与速度的关系得到最大容许振幅 h_a' 为

$$h_a'=\frac{V_a}{2\pi f}=\frac{L}{400\pi^2 f} \tag{4}$$

因为规范适用范围为 200 m 跨度以下桥梁,所以在式(4)中取代表性跨度 $L=160$ m 就得到容许振幅[式(1)]。从上述推导过程来看,式(1)主要适用于小跨度桥梁。从另一方面看,如果直接采用1.0 $\mathrm{m\cdot s^{-2}}$作为桥梁涡振的允许加速度,那么在假定涡激振动为简谐振动的前提下,对应的涡振容许振幅 $A_{a,i}$可表示为

$$A_{a,i}=0.025/f^2_{b,i} \tag{5}$$

图 1 对比了式(2)和式(5)计算的涡振振幅容许值随竖弯频率的变化规律。由图 1 可以看到,当竖弯频率小于 0.625 Hz 时,按规范式(2)计算的振幅容许值要小于式(5),而且竖弯频率越小,规范公式就相对更加严格。因此按规范计算竖弯涡振容许值,对于大跨度悬索桥是偏于安全的。

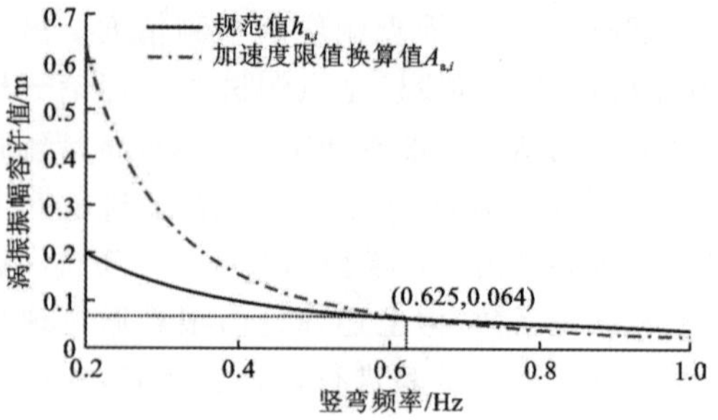

图 1 不同模态的竖弯涡振振幅容许值随频率的变化

Fig. 1 Variation of Allowable Amplitude of Vortex-induced Vibrations for Different Vertical Modes

2 多阶模态竖向涡振的节段模型风洞试验预测方法

弹性悬挂节段模型风洞试验与理论分析相结合的方法是预测悬索桥涡振性能的常用方法。实桥涡振响应预测主要包括预测涡振锁定风速区间、涡振振幅随风速的变化曲线以及最大涡振振幅。其中涡振锁定风速区间主要用于判断涡振是否会发生在桥梁的正常运营风速范围,而涡振振幅随风速的变化曲线主要用于确定最大涡振振幅,从而为是否需要进行涡振控制提供判据。由于涡振振幅和锁定风速区间都与雷诺数有关,因此在抗风设计阶段,主要采用雷诺数尽可能大的弹性悬挂节段模型风洞试验和理论分析相结合的方法预测涡振响应。为了得到尽可能大的雷诺数,大节段模型缩尺比可达 1/20~1/30。此外,湍流对桥梁涡振有显著影响。尽管规范中给出了各类地貌下的湍流度名义值,很多实测表明,对于跨海桥,实测湍流度可以低至 3%[2-23,27],不到名义值的 1/5。为此,节段模型风洞试验一般在均匀流下进行,以便得到偏于保守的试验结果。节段模型试验的基本思路是先选取某一低阶模态进行节段模型弹性悬挂风洞试验,测定断面的 S_t 数、发生涡激共振的锁定区间以及最大涡振振幅,再由该阶模态的风洞试验结果推算其余模态的结果。下面首先推导节段模型竖向涡振振幅响应的理论解,再阐述预测多阶模态涡振响应的具体步骤。

2.1 节段模型竖向涡振幅值响应的理论解

由于涡振通常表现为单模态振动形式,因此只需考虑桥梁单个模态的涡激振动运动方程

$$m_i(\ddot{y}+2\omega_i\zeta_i\dot{y}+\omega_i^2 y)=F(y,\dot{y},\ddot{y},U,t) \tag{6}$$

式中:ζ_i 为桥梁第 i 阶模态固有阻尼比;ω_i 为桥梁第 i 阶模态的固有频率;F 为作用在加劲梁上每延米的涡激力;U 为来流风速;y,$\dot{y}$ 和 $\ddot{y}$ 分别为位移,速度和加速度;t 为时间;m_i 为实桥第 i 阶模态涡振时加劲梁每延米的质量。

在节段模型风洞试验阶段,无法考虑振型的影响,y,$\dot{y}$ 和 $\ddot{y}$ 简化为与坐标 x 无关的单变量,而 m_i 须按缩尺比换算为加劲梁模型每延米的等效质量

$m_{e,i}$，其表达式为

$$m_{e,i}=\frac{1}{\lambda_L^2}\frac{1}{\int_0^L\varphi_i^2(x)\mathrm{d}x} \tag{7}$$

式中：λ_L 为节段模型相对实桥的几何缩尺比；$\varphi_i(x)$ 为桥梁第 i 阶模态的质量归一化振型函数。

式(6)的右边是涡激力，具有自激特性，常表述为 y、$\dot{y}$ 的显式。Scanlan 经验线性涡激力数学模型的表达式如下

$$F=\frac{1}{2}\rho U^2(2D)\left[Y_1(K_1)\frac{\dot{y}}{U}+Y_2(K_1)\frac{y}{D}+\frac{1}{2}C_L(K_1)\sin(\omega_n t+\phi)\right] \tag{8}$$

式中：ρ 为空气密度；D 为结构迎风特征尺寸；$K_1=\omega_n D/U$，为折算频率；ω_n 为节段模型固有频率；$Y_1(K_1)$、$Y_2(K_1)$ 和 $C_L(K_1)$ 分别为气动阻尼参数、气动刚度参数和强迫力参数，它们均为 K_1 的函数；ϕ 为相位差。

定义量纲一振幅 $\eta=y/D$，由式(6)～(8)可解得稳态振幅 η 为

$$\eta=\rho D^2 C_L\left[2m_{e,i}\sqrt{(K_0^2-K_1^2)^2+(2\gamma K_0K_1)^2}\right]^{-1} \tag{9}$$

$$K_0^2=K_1^2-\frac{\rho D^2}{m_{e,i}}Y_2 \tag{10}$$

$$\gamma=\frac{1}{2K_0}\left(2\zeta K_1-\frac{\rho D^2}{m_{e,i}}Y_1\right) \tag{11}$$

式(9)可比拟为一个量纲一振动频率为 K_0，阻尼比为 γ 的单自由度体系受频率为 K_1 的强迫振动的解。在涡振锁定区间，K_1 随着风速 U 的增加而减小；当 $K_0=K_1$ 时，节段模型达到最大涡振振幅。气动参数 $Y_1(K_1)$、$Y_2(K_1)$和 $C_L(K_1)$可由节段模型涡振风洞试验得到的 $K_1\sim\eta$ 曲线识别。

由式(9)～(11)描述的涡振振幅是气动参数 $Y_1(K_1)$、$Y_2(K_1)$ 和 $C_L(K_1)$ 的隐式表达式，把式(10)、(11)代入式(9)可得

$$\eta=\frac{C_L}{2\sqrt{Y_2^2+(2K_1^2S_c-Y_1)^2}} \tag{12}$$

式中：$S_c=4\pi m_{e,i}\zeta_i/\rho D^2$，表示质量阻尼参数 Scruton 数($S_c$ 数)。如果令 $Y_1=Y_2=0$，式(11)则退化为

$$\eta_L=\frac{C_L}{16\pi^2S_t^2S_c} \tag{13}$$

式中：η_L 为简谐涡激力模型作用下节段模型的最大涡振振幅响应；$S_t=f_nD/U_c$，为断面的 Strouhal 数，对于闭口钢箱梁 S_t 约为 0.1。很明显，最大涡振振幅与 S_t 及 S_c 成反比。

2.2 实桥多阶模态竖向涡振响应的预测方法

实桥各阶竖向模态的区别主要体现在振型、频率、阻尼比和模态质量的差异。依据某一阶竖弯模态的试验结果，可以进一步预测实桥多阶模态竖向涡振的振幅和涡振风速的锁定区间，具体步骤如下：

(1)由节段模型涡振振幅预测实桥涡振振幅时，除了乘以缩尺比外，还必须引入振型修正系数，即

$$y_p=\lambda_L\kappa_i\eta D \tag{14}$$

式中：y_p 为实桥的最大涡振振幅；κ_i 为振型修正系数。

对于式(8)所示的 Scanlan 经验线性涡激力数学模型，κ_i 的具体表达式为[12-13]

$$\kappa_i=\frac{\int_0^L|\varphi_i(x)|\mathrm{d}x}{\int_0^L\varphi_i^2(x)\mathrm{d}x} \tag{15}$$

大跨度悬索桥特别是单跨悬索桥的各阶竖向模态的振型一般可以用简谐波近似表示，即 $\varphi_i(x)=\sin(i\pi x/L)$。把上式代入式(15)积分可得，$\kappa_i=4/\pi$，由此可见 κ_i 与模态阶数无关，不同竖向模态涡振的区别只须考虑频率、阻尼比和模态质量的影响。

(2)很多研究表明，阻尼比和质量对桥梁涡振振幅的影响规律一致，可以采用 S_c 数合并考虑。例如，当采用简谐涡激力数学模型时，桥梁的最大涡振振幅与 S_c 数成反比，如式(13)所示。对大跨度悬索桥而言，各阶竖向模态的加劲梁每延米等效质量也基本相等。准确确定桥梁的各阶模态阻尼比非常困难，目前中国规范假定各阶模态阻尼比相同。实桥测试的阻尼比具有很大的离散性，一些研究表明，阻尼比随模态阶数升高而有降低的趋势[27]。若假定各模态阻尼比也相等，可以认为悬索桥各阶竖向模态的最大涡振振幅也基本相等。

(3)起振风速主要由 S_t 数确定。对于悬索桥而言，可认为 S_t 数不随风速而改变，于是各阶竖向模态的起振风速与频率成正比，即

$$U_{j,s}=U_{i,s}\frac{f_j}{f_i} \tag{16}$$

式中：$U_{j,s}$，$U_{i,s}$ 分别为第 j 阶和第 i 阶模态的涡振起振风速；f_j，f_i 分别为第 j 阶和第 i 阶模态的频率。

3 悬索桥的新型等效气弹模型及多阶模态涡振的风洞试验

3.1 多点弹性支承连续梁气弹模型

为了验证基于弹性悬挂节段模型风洞试验和理论分析的涡振响应预测结果，必须设计一个可以模拟多个模态密集分布的气弹模型，并通过风洞试验来验证。为了避开塔和缆索系统干扰，单纯研究加劲梁的涡激力跨向相关性与高阶模态振型变化之间

的关系，本文作者在文献[28]中提出了一种新型的竖向等效气弹模型——多点弹性支承连续梁气弹模型，如图2所示。在该气弹模型上可以直接测定低阶至高阶涡振的振幅，再结合节段模型风洞试验就可以研究和验证节段模型与气弹模型涡振振幅之间的转化关系[28]。

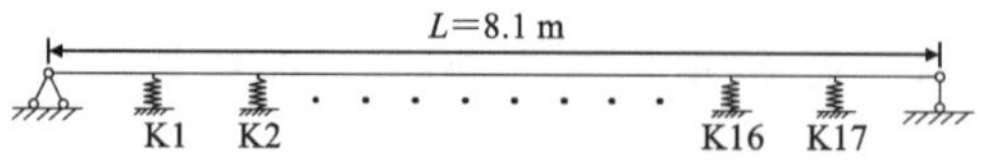

图2　多点弹性支承连续梁气弹模型结构示意

Fig. 2　Structural Diagram of Elastically Multi-supported Aeroelastic Beam Model

新型气弹模型设计的基本思路是通过一个等效的多点弹性支承连续梁模拟大跨度悬索桥多阶竖向模态，其中的关键步骤是通过假定振型法计算出等效的多点支承连续梁的支承弹簧刚度和支承点数量，从而模拟所希望的多个模态的频率和振型。按照这一方法以某悬索桥为背景，设计了加劲梁截面尺寸为240 mm(宽)×40 mm(高)，长度 L=8.1 m(半跨)的多点弹性支承连续梁气弹模型，如图3所示。等效模型刚度由一条薄钢板提供，外形用泡沫塑料板制作，如图4所示，弹性支承由位于模型气流后方的细钢筋悬臂梁实现，对气流无干扰。包括芯梁、外衣和悬臂梁有效振动质量在内的模型等效质量约为3 kg·m^{-1}。本文主要针对单跨悬索桥进行研究，其振型函数可以用简谐波描述。对于两跨及多跨悬索桥，可以按桥梁的实际振型来设计实现气弹模型[28-29]。

图3　多点弹性支承连续梁气弹模型

Fig. 3　Elastically Multi-supported eroelastic Beam Model in Wind Tunnel

通过在模型布置加速度传感器，测定了初始激励下的多点弹性支承连续梁气弹模型的加速度响应，并据此识别了动力特性。实测的各模态振型非常接近假定的正弦曲线，限于篇幅本文不再赘述[28]。气弹模型频率及阻尼比结果如表1所示。从表1可以看出，第3阶以及第5～8阶竖弯模态的阻尼比基本相等。总体而言，气弹模型的阻尼比略大，可以通过改善连接、减少摩擦及合理布置数据采集线来降低模态阻尼比。另外，由动力特性分析可知，气弹模型各阶模态的等效质量均相同，均为3.0 kg·m^{-1}。

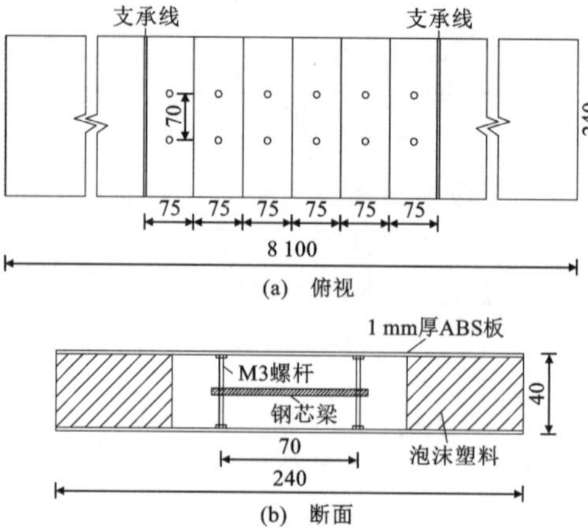

图4　气弹模型内支架和外衣示意(单位:mm)

Fig. 4　Skelton and Coat for Aeroelastic Model (Unit: mm)

表1　气弹模型动力特性测试结果

Table 1　Experimental Results of Dynamic Characteristics of Aeroelastic Model

竖弯模态	频率/Hz	阻尼比/%	竖弯模态	频率/Hz	阻尼比/%
第1阶	2.73	0.39	第5阶	5.86	0.61
第2阶	3.61	0.55	第6阶	6.74	0.63
第3阶	4.49	0.63	第7阶	7.86	0.63
第4阶	5.18	0.70	第8阶	9.18	0.60

3.2　风洞试验结果

风洞试验在湖南大学HD-2试验段开展。试验时从0开始逐渐增加风速，从而激发气弹模型发生从第1阶至高阶的各阶模态涡激振动，同时采用加速度传感器测量气弹模型的涡振响应。加速度传感器布置在模型上表面的中轴线上，根据高阶振型的分布特征，沿模型展向在 $3L/24$，$4L/24$，$6L/24$，$9L/24$，$12L/24$，$15L/24$，$18L/24$，$21L/24$ 处一共布置8个传感器。测振试验过程中，不用移动加速度传感器的位置即可以获得第1～9阶竖弯模态的加速度响应。

图5给出了气弹模型各阶模态涡振的量纲一振幅随试验风速的变化曲线。由图5可以看到，气弹模型的涡振振幅随模态频率的升高没有下降的趋势，特别是第7～9阶模态的最大涡振振幅基本相等，从而进一步说明了固有频率对最大涡振振幅没有影响，而且当各模态的 S_c 数相同时，它们的最大涡振振幅也相同。图5中低阶模态的涡振振幅相对较小，这主要是由于低阶模态涡振的起振风速非常小(3 m·s^{-1}以下)，此时风洞来流的紊流度较大，降低了涡振振幅。

3.3　大跨度悬索桥多阶模态竖向涡振的现场观测

现场实测是检验大跨度悬索桥多阶模态涡振风洞试验和理论预测结果的重要手段。尽管现场实测

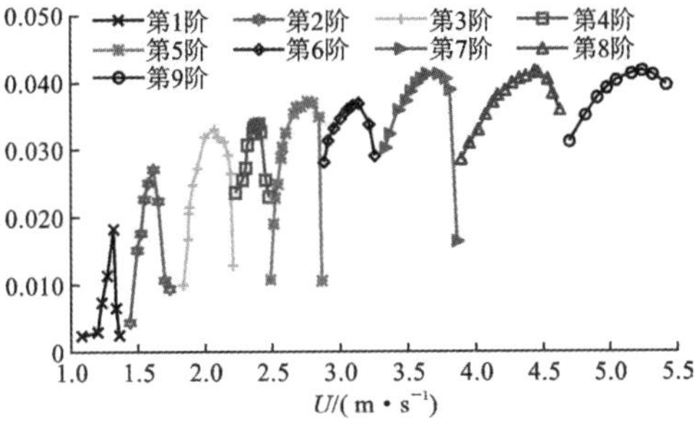

图 5　气弹模型各阶模态涡振的量纲一振幅随试验风速的变化

Fig. 5　Variation of Dimensionless Amplitude of VIV for Different Vertical Modes of Aeroelastic Model with Wind Velocity

的来流条件远比风洞试验复杂，实桥的固有阻尼比也存在不确定性，但基于大量现场实测的统计结果依然可以反映大跨度悬索桥多阶模态竖向涡振的本质特征。美国旧塔科马桥[30]、丹麦大带东桥[3]和中国西堠门大桥[2]的现场实测中都观测到了多阶模态竖向涡振。对于中国西堠门大桥及丹麦大带东桥，由于安装的传感器很有限，从测点位移估算各模态最大位移在某些情况下会产生很大的误差。表 2 整理了美国旧塔科马桥在风毁事故前多阶模态竖向涡振的现场实测数据[30]，并列出了具有代表性的各阶竖向涡振的振型、频率和最大单边涡振振幅。由表 2 可以看到，加劲梁第 1～3 阶正对称竖弯振型和第 2～3 阶反对称竖弯振型都发生了显著的涡激振动，而且随着模态阶次的增加，最大涡振振幅略增加，其中第 3 阶反对称竖弯涡振的最大单边振幅达到了 46 cm。

美国旧塔科马桥是左右边跨完全对称的三跨悬索桥，从表 2 所示的振型示意可以判断：第 1，2 阶正对称竖弯振型的等效质量较大，而第 2，3 阶反对称竖弯振型和第 3 阶正对称竖弯振型的等效质量相对较小，且基本相等。假定该桥的各阶模态阻尼比基本不变化，因此从理论上可以预测：随着旧塔科马桥竖弯模态阶次的增加，各阶模态的 S_c 数将基本保持不变，最大涡振振幅也基本不变。前 2 阶模态的涡振振幅偏小，可能是湍流度的影响。因此，可以认为上节的理论预测与表 2 所示的振幅实测结果是吻合的。由于各振型的不同，表 2 中的振幅没有考虑振型修正系数，除了第 2 阶模态外，其余 4 阶的涡振振型修正系数基本相等。

4　大跨度悬索桥多模态竖向涡振的阻尼器控制

引言部分已论述了治理既有桥梁多阶涡振的常用方法，指出增加气动措施和调谐质量减振器对于既有桥梁均存在较大的困难。基于这一背景，本文提出了一种直接耗能阻尼器的方法。

表 2　美国旧塔科马桥多模态竖向涡振的现场实测数据[30]

Table 2　Field Measurements of VIV of Different Vertical Modes of the Old Tacoma Bridge[30] in USA

编号	振型(跨径/m)	频率/Hz	最大单边振幅/cm
1	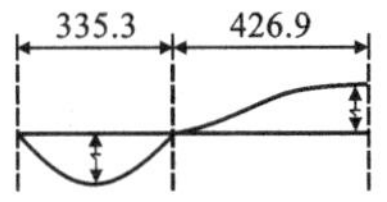	0.133	30.5
2	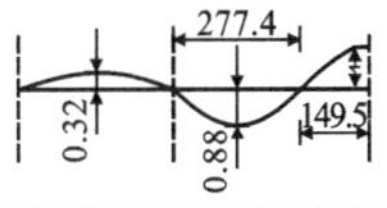	0.200	40.5
3	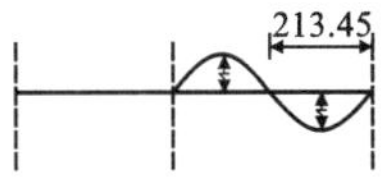	0.275	45.0
4	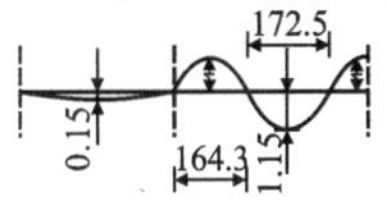	0.350	45.5
5	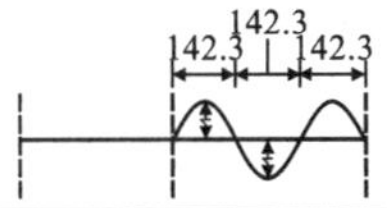	0.400	46.0

4.1　采用直接耗能阻尼器进行涡振控制的策略

对于两跨及多跨悬索桥，一般采用全漂浮体系以减小桥塔处主梁的负弯矩。例如中国西堠门大桥(主跨 1 650 m)、丹麦大带东桥(主跨 1 624 m)、土耳其 Izmit Bay 大桥(主跨 1 550 m)以及正在建设的土耳其 1915 大桥(主跨 2 023 m)均采用两跨连续或三跨连续全漂浮体系。对于采用漂浮体系的大跨度悬索桥，加劲梁与桥塔间不设置竖向支座。当加劲梁发生竖向涡振时，绝大多数模态都在加劲梁与桥塔之间的交汇处产生明显的相对竖向位移。例如，对于采用两跨连续的西堠门大桥，振动频率为 0.325 Hz 的加劲梁第 6 阶竖弯模态在北塔处的归一化振型位移(桥塔处加劲梁竖向模态位移与最大竖向模态位移的比值)就接近 0.4，如图 6 所示。因此，在加劲梁与桥塔之间设置直接耗能阻尼器，可以显著抑制全桥加劲梁多个模态的竖向与扭转振动。

图 7 显示了竖向直接耗能阻尼器在西堠门大桥上的设置方案：在北塔增设下横梁或牛腿，并与加劲梁之间设置竖向直接耗能阻尼器。经过初步计算，阻尼器在工作时的最大出力约为 200 kN，比常用的桥梁纵向阻尼器小了一个数量级左右。定性地看，采用竖向直接耗能阻尼器进行涡振控制不影响桥梁

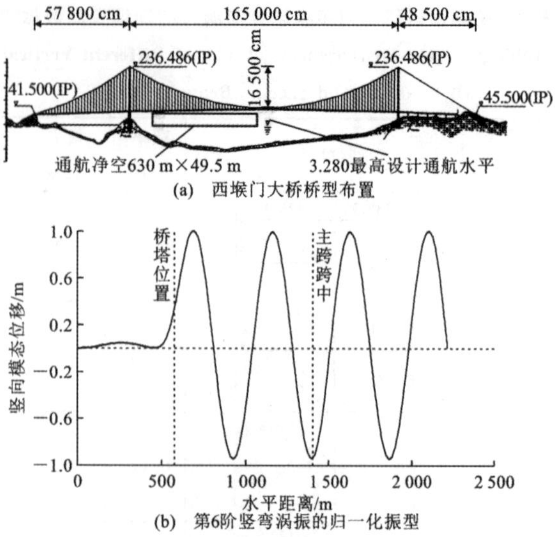

(a) 西堠门大桥桥型布置

(b) 第6阶竖弯涡振的归一化振型

图 6 中国西堠门大桥第 6 阶竖弯涡振的归一化振型

Fig. 6 Normalized Mode Shapes of the 6th Modes of the Xihoumen Bridge in China

正常运营，结构简单，实施方便，能有效控制大跨度漂浮体系悬索桥多个模态的涡振，大大降低了减振控制措施的实施难度，提高了减振控制措施的经济性，可靠性好，鲁棒性高，能够在结构参数和来流条件存在不确定因素的情况下，仍然保证较高的减振效率。

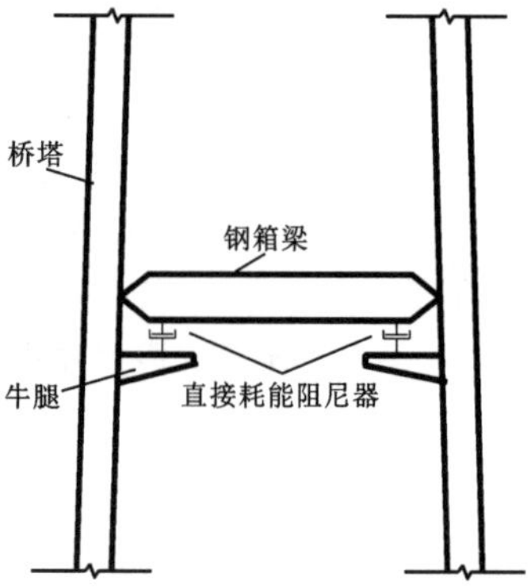

图 7 竖向阻尼器在加劲梁与桥塔之间的设置方案

Fig. 7 Installation Scheme of Vertical Dampers Between Tower and Girder

4.2 试验验证多点弹性支承连续梁气弹模型

为了初步验证采用直接耗能阻尼器控制漂浮体系悬索桥多模态涡振的有效性，采用多点弹性支承连续梁气弹模型研究了在漂浮加劲梁与桥塔之间设置电涡流阻尼器对桥梁各阶模态固有阻尼比的影响。

图 8 电涡流阻尼器在多点弹性支承连续梁气弹模型上的安装情况

Fig. 8 Installation of Eddy Current Damper on Elastically Multi-supported Aeroelastic Model

试验中通过板式电涡流阻尼器对气弹模型施加阻尼力，如图 8 所示。板式电涡流阻尼器由铜板和永磁体构成，其中铜板固定在气弹模型主梁 1/7 跨径处，永磁体通过刚性支架与气弹模型的底座相连。试验中采用自由振动方法测试了气弹模型在有、无阻尼器 2 种工况下频率和阻尼比的变化情况，结果如表 3 所示。由表 3 可以看到，安装电涡流阻尼器后气弹模型的各阶模态阻尼比都有大幅提高，其中第 3 阶模态的阻尼比提高了 4 倍；对于第 7 阶模态，电涡流阻尼器正好位于振型函数的驻点处，所以没有控制效果。此外，可以发现安装电涡流阻尼装置后，多点弹性支承连续梁气弹模型的固有频率几乎没有变化，这间接说明了电涡流阻尼装置几乎不提供附加刚度。

表 3 气弹模型各阶模态频率及有无阻尼器的阻尼比

Table 3 Modal Frequencies and Damping Ratios of Aeroelastic Models with and Without Dampers

模态阶次	无电涡流阻尼装置		有电涡流阻尼装置	
	频率/Hz	阻尼比/%	频率/Hz	阻尼比/%
1	2.73	0.39	2.73	1.03
2	3.61	0.55	3.61	2.26
3	4.44	0.63	4.40	2.96
4	5.08	0.70	5.08	1.85
5	5.71	0.61	5.71	1.82
6	6.64	0.63	6.64	1.01
7	7.81	0.63	7.86	
8	9.18	0.60	9.18	0.79

需要指出的是，上述气弹模型并未模拟西堠门悬索桥的实际振动模态，阻尼器安装在 1/7 跨径，但能基本验证本方案对抑制涡振的有效性。杨兴龙[31]研究了梁-线性黏滞阻尼器耦合系统的动力特性及阻尼器参数优化问题，得到了耦合系统频率和阻尼比的解析解与渐近解；在此基础上，得到了梁端附近位置安装阻尼器时梁-阻尼器系统的最大阻尼比及对应的最优阻尼器参数；并针对西堠门大桥的实际振动模态，得到了漂浮式悬索桥加劲梁附加竖向线性黏滞阻尼器的阻尼器优化参数。研究表明，由于阻尼器对于各阶模态安装相对位置的不同，控制各阶模态涡振的最优阻尼相差很大。为了使多个模态涡振控制都达到最优效果，可以采用基于电磁式电涡流阻尼器的半主动控制。电磁式电涡流阻尼器具有以下优点：①附加刚度小，通电后开始工作，不通电时对结构无任何干扰；②阻尼器参数可由电流强度调节实现半主动控制，抑制不同频率模态的涡振；③相对运动部件之间没有摩擦，所以使用寿命很长；④容易实现阻尼器顺桥向自由移动和竖向静挠度，而且桥塔处供电比较有保证。

5 结 语

(1)本文总结了悬索桥多阶竖向模态的涡振问题,包括涡振振幅的限值、多阶模态涡振振幅的预测与试验技术、多阶模态涡振的单一阻尼器控制方法。针对大跨度桥梁发生多阶涡振的现实可能性,需要对车辆通行风速内的各阶模态进行涡振性能评估,仅以加劲梁第1阶竖弯模态的涡振响应来评价桥梁的涡振性能是偏于不安全的。

(2)理论分析、基于多点弹性支承连续梁的新型气弹模型风洞试验以及美国旧塔科马桥的现场实测都表明,在各模态阻尼比相等的前提下,悬索桥各阶竖弯模态的最大涡振振幅基本相等,最大加速度幅值按频率比的平方增加,为此需更加重视高阶竖弯模态涡振问题。为抑制多阶模态的竖向涡振,本文提出了采用单一阻尼器的控制方法,并采用多点弹性支撑连续梁气弹模型验证了本方案的有效性。为了使多个模态涡振控制都达到最优效果,可以采用基于电磁式电涡流阻尼器的半主动控制。

(3)本文的一个重要假定是悬索桥各阶竖弯模态阻尼比相等,这也是中国桥梁抗风设计规范中沿用的假定,其合理性有待进一步研究。准确测定悬索桥的模态阻尼比,特别是高阶模态的阻尼比是一个极为困难的问题,一是缺少激振手段,二是阻尼比的识别也具有较大的离散性。为此,一个重要的研究方向是基于大量的实测数据,测定及评价若干重要代表性桥梁的低阶模态和高阶模态的阻尼比,可为桥梁抗风研究与设计提供基础资料。对于本文提出的阻尼器抑制涡振方案,需进一步完善优化参数设计及量化减振效果。

参考文献:

References:

[1] 项海帆. 现代桥梁抗风理论与实践[M]. 北京:中国人民交通出版社,2005.
XIANG Hai-fan. Modern Theory and Practice on Bridge Wind Resistance [M]. Beijing: China Communications Press, 2005.

[2] LI H, LAIMA S J, OU J P, et al. Investigation of Vortex-induced Vibration of a Suspension Bridge with Two Separated Steel Box Girders Based on Field Measurements [J]. Engineering Structures, 2011, 33 (6): 1894-1907.

[3] LARSEN A, ESDAHL S, ANDERSEN J E, et al. Storebaelt Suspension Bridge-vortex Shedding Excitation and Mitigation by Guide Vanes [J]. Journal of Wind Engineering and Industrial Aerodynamics, 2000, 88 (2): 283-296.

[4] EHSAN F, SCANLAN R H. Vortex-induced Vibrations of Flexible Bridges [J]. Journal of Engineering Mechanics ASCE, 1990, 116 (6): 1392-1411.

[5] ZHU L D, MENG X L, GUO Z S. Nonlinear Mathematical Model of Vortex-induced Vertical Force on a Flat Closed-box Bridge Deck [J]. Journal of Wind Engineering and Industrial Aerodynamics, 2013, 122 (11): 69-82.

[6] XU K, ZHAO L, GE Y J. Reduced-order Modeling and Calculation of Vortex-Induced Vibration for Large-span Bridges [J]. Journal of Wind Engineering and Industrial Aerodynamics, 2017, 167 (8): 228-241.

[7] SUN Y G, LI M S, LIAO H L. Nonlinear Approach of Vortex-induced Vibration for Line-like Structures [J]. Journal of Wind Engineering and Industrial Aerodynamics, 2014, 124 (7): 1-6.

[8] ZHU Q, CHEN B Y, ZHU L D. Investigation on Characteristics and Span-wise Correlation of Vortex-induced Forces on a Twin-box Deck Using Newly-developed Wind-tunnel Test Technique [J]. Journal of Wind Engineering and Industrial Aerodynamics, 2017, 164 (5): 69-81.

[9] LAIMA S, LI H, CHEN W, LI F C. Investigation and Control of Vortex-induced Vibration of Twin Box Girders [J]. Journal of Fluids and Structures, 2013, 39: 205-221.

[10] CHEN W L, LI H, HU H. An Experimental Study on the Unsteady Vortices and Turbulent Flow Structures Around Twin-box-girder Bridge Deck Models With Different Gap Ratios [J]. Journal of Wind Engineering and Industrial Aerodynamics, 2014, 132: 27-36.

[11] MARRA A M, MANNINI C, BARTOLI G. Measurements and Improved Model of Vortex-induced Vibration for an Elongated Rectangular Cylinder [J]. Journal of Wind Engineering and Industrial Aerodynamics, 2015 (11), 147: 358-367.

[12] 朱乐东. 桥梁涡激共振试验节段模型质量系统模拟与振幅修正方法[J]. 工程力学,2005,22(5):204-208.
ZHU Le-dong. Mass Simulation and Amplitude Conversion of Bridge Sectional Model Test for Vortex-excited Resonance [J]. Engineering Mechanics, 2005, 22 (5): 204-208.

[13] 张志田,陈政清. 桥梁节段与实桥涡激共振幅值的换算关系[J]. 土木工程学报,2011,44(7):77-82.
ZHANG Zhi-tian, CHEN Zheng-qing. Similarity of

Amplitude of Sectional Model to That of Full Bridge in the Case of Vortex-induced Resonance [J]. China Civil Engineering Journal, 2011, 44 (7): 77-82.

[14] JTG/T D 60-01—2004,中国公路桥梁抗风设计规范[S]. 2004.
JTG/T D 60-01—2004, Wind-resistant Design Specification for Highway Bridges [S].

[15] European Committee for Standardization. Eurocode 1: Actions on Structures—General Actions—Part 1-4: Wind Actions [S].

[16] SATO H. Wind-resistant Design Manual For Highway Bridges In Japan [J]. Journal of Wind Engineering and Industrial Aerodynamics, 2003, 91 (12): 1499-1509

[17] Highway Agency, Design Rules for Aerodynamic Effects on Bridges [S].

[18] IRWIN P A. Motion Criteria, Technical Notes of the Rowan Williams Davies and Irwin Inc. [R]. Guelph: RWDI, 1999.

[19] 陈政清,黄智文. 大跨度桥梁竖弯涡振限值的主要影响因素分析[J]. 中国公路学报,2015,28(9):30-37.
CHEN Zheng-qing, HUANG Zhi-wen. Analysis of Main Factors Influencing Allowable Magnitude of Vertical Vortex-induced Vibration of long-span Bridges [J]. China Journal of Highways and Transport, 2015, 28 (9): 30-37.

[20] YANG Y X, ZHOU R, GE Y J. Experimental Studies on Viv Performance and Countermeasures for Twin-box Girder Bridges With Various Slot Width Ratios [J]. Journal of Fluids and Structures, 2016, 66 (10): 476-489.

[21] 孙延国,廖海黎,李明水. 基于节段模型试验的悬索桥涡振抑振措施[J]. 西南交通大学学报,2012,(2):218-223,264.
SUN Yan-guo, LIAO Hai-li, LI Ming-shui. Mitigation Measures of Vortex-induced Vibration of Suspension Bridge Based on Section Model Test [J]. Journal of Southwest Jiaotong University, 2012 (2): 218-223, 264

[22] LARSEN A, SAVAGE M, LAFRENIERE A, et al. Investigation of Vortex Response of a Twin Box Bridge Section at High and Low Reynolds Numbers [J]. Journal of Wind Engineering and Industrial Aerodynamics, 2008, 96 (6): 934-944.

[23] XIN D B, ZHANG H, AND OU J P. Experimental Study on Mitigating Vortex-induced Vibration of a Bridge by Using Passive Vortex Generators [J]. Journal of Wind Engineering and Industrial Aerodynamics, 2018, 175 (4): 100-110.

[24] 赵 林,葛耀君,郭增伟,等. 大跨度缆索承重桥梁风振控制回顾与思考——主梁被动控制效果与主动控制策略[J]. 土木工程学报,2015,48(12):91-100.
ZHAO Lin, GE Yao-jun, GUO Zeng-wei, et al. Reconsideration of Wind-induced Vibration Mitigation of Long-span Cable Supported Bridges: Effects of Passive Control and Strategy of Active Control [J]. China Civil Engineering Journal, 2015, 48 (12): 91-100.

[25] LI K, GE Y J, GUO Z W, et al. Theoretical Framework of Feedback Aerodynamic Control of Flutter Oscillation for Long-span Suspension Bridges by the Twin-winglet System [J]. Journal of Wind Engineering and Industrial Aerodynamics, 2015, 145: 166-177.

[26] BAKIS K N, MASSARO M, WILLIAMS M S, et al. Aeroelastic Control of Long-span Suspension Bridges With Controllable Winglets [J]. Structural Control and Health Monitoring, 2016, 23 (12): 1417-1441.

[27] 华旭刚,孙瑞丰,温 青,等. 基于新奇检测技术的桥梁涡激共振自动识别研究[J]. 振动工程学报,2018,31(6):948-956.
HUA Xu-gang, SUN Rui-feng, WEN Qin, et al. Automatic Detection of Vortex-induced Vibration Events by Novelty Detection [J]. Journal of Vibration Engineering, 2018, 31 (6): 948-956.

[28] ZHOU S, HUA X G, CHEN Z Q, et al. Experimental Investigation of Correction Factor for Viv Amplitude of Flexible Bridges From an Aeroelastic Model and Its 1: 1 Section Model [J]. Engineering Structures, 2017, 141: 263-271.

[29] 温 青,大跨柔性桥梁高阶竖弯模态涡振振幅预测方法研究[D]. 长沙:湖南大学,2017.
WEN Qin. Amplitude Prediction of Vortex-induced Vibrations for high-order Vertical Modes of Long-span Flexible Bridges [D]. Changsha: Hunan University, 2007.

[30] FARQUHARSON F B, VINCENT G S. Aerodynamic Stability of Suspension Bridges with Special Reference to the Tacoma Narrows Bridge: A Report of an Investigation [M]. Seattle: University of Washington Press, 1950.

[31] 杨兴龙. 漂浮式悬索桥加劲梁—阻尼器系统动力特性与阻尼系数优化[D]. 长沙:湖南大学,2017.
YANG Xing-long. Dynamic Propeties and Damping Optimization of Girder-damper System of Floating Suspension Bridges [D]. Changsha: Hunan University, 2017.

Journal of Fluids and Structures 109 (2022) 103480

Contents lists available at ScienceDirect

Journal of Fluids and Structures

journal homepage: www.elsevier.com/locate/jfs

Effects of end condition and aspect ratio on vortex-induced vibration of a 5:1 rectangular cylinder

Chaoqun Wang [a], Qing Wen [b,*], Shuai Zhou [a,c], Xugang Hua [a,*], Zhiwen Huang [a], Zhengqing Chen [a]

[a] *Key Laboratory for Wind and Bridge Engineering of Hunan Province, College of Civil Engineering, Hunan University, Changsha 410082, China*
[b] *School of Civil Engineering, Hunan University of Science and Technology, Xiangtan 411201, China*
[c] *China Construction Fifth Engineering Division Corp. Ltd, Changsha 410004, China*

ARTICLE INFO

Article history:
Received 7 June 2021
Received in revised form 10 October 2021
Accepted 16 December 2021
Available online 6 January 2022

Keywords:
Vortex-induced vibration
Wind tunnel test
Rectangular cylinder
End condition
Aspect ratio
Sectional model

ABSTRACT

A series of wind tunnel tests were carried out to study the end condition and aspect ratio effects of a sectional model on the vortex-induced vibration (VIV) of a 5:1 rectangular cylinder. Five aspect ratios varying from 9 to 32 as well as eight endplates with different sizes were examined in the tests. Static and elastically mounted sectional models were employed to measure the surface pressure and VIV response of the cylinder, respectively. Sectional fluctuating lift forces, shedding frequency, as well as their spanwise distribution and correlation on the sectional model were obtained and discussed. Results indicated that the aspect ratio had a significant effect on VIV amplitudes, and small aspect ratios tended to remarkably underestimate the VIV amplitudes. There existed an end interference region with a length of 6–8 *D*, where the sectional lift forces could be, on average, half the value for the non-interference region. Besides, the aspect ratio was positively related to the spanwise correlation of sectional lift forces. Both end condition and VIV had little effect on the length of the end interference region. In the VIV 'lock-in' range, with the increasing reduced wind velocity the fluctuating lift force first increased and reached a maximum value earlier than the VIV amplitude, and after that further increase of reduced wind velocity led to decreased fluctuating lift force (toward a value even smaller than that of the stational model). This complicated variation of sectional lift forces implied that vortex shedding was highly nonlinear and self-limiting. Finally, an empirical method based on the test results was explored to account for the influence of aspect ratio on VIV amplitude.

1. Introduction

Slender structures such as bridges girders and cable structures in engineering are prone to various wind-induced vibrations. Due to the complexity of fluid–structure interaction, wind tunnel test remains the primary tool for load evaluation and response prediction of slender structures under dynamic wind action (Hansen et al., 2015; Mannini et al., 2015, 2016; Kwok et al., 2012). Compared to full aeroelastic bridge models, sectional models are more widely used due to the lower cost and larger scale ratio. They can be fixed or elastically mounted by springs in the wind tunnel, and the

* Corresponding authors.
E-mail addresses: cewenq@hnust.edu.cn (Q. Wen), cexghua@hnu.edu.cn (X. Hua).

wind loads as well as aeroelastic responses of bridge structures can be obtained through static and dynamic tests (Lee et al., 2014; Zhou et al., 2017; Wang et al., 2021b,c).

VIV is a typical wind-induced vibration which has been observed on several long-span bridges, such as the Greatbelt suspension bridge, Xihoumen suspension bridge, Humen bridge, etc. It is a significant challenge for long-span bridges in terms of structural safety (e.g., structural fatigue failure), traffic safety and comfort as well as social consequence (Wang et al., 2021a). VIV characteristics of bridge decks have been studied in depth through wind tunnel tests based on sectional models (Belloli et al., 2014; Larsen et al., 2008; Li et al., 2016; Zhu et al., 2017). Research emphases of these studies were usually paid to the effect of structural parameters (such as mass or damping) and aerodynamic modifications on VIV. Little attention has been paid to the effect of aspect ratios and end conditions of the sectional model on experimental results.

A sectional model only represents a finite length of an elongated structure, and the wake features along the sectional model may be quite different from the realistic situations due to end effects (Nakamura and Nakashima, 1986; Toy and Fox, 1986; Wang et al., 2021a). Therefore, endplates were usually attached to two ends of a sectional model to weaken the three-dimensional end effect of the flow (Mannini et al., 2014; Li et al., 2017; An et al., 2016; Hua et al., 2020; Wen et al., 2018). Generally, variation of size or shape for the endplate could significantly affect the characteristics of the flow field (Cowdrey, 1963; Lee, 1979). Based on a series of experimental study on a cylinder with two different aspect ratios (i.e., $L/D = 8$ and 16, where L and D denote the length and heigh of the model respectively) of the sectional model, Stansby (1974) suggested the basic requirements of mean base pressure for design of optional endplate sizes. Indeed, many studies involving circular cylinders have been performed using the end plate sizes suggested by Stansby. Subsequently, more researches were carried out concerning the effect of endplate size for sectional models in wind tunnel tests (Kubo et al., 1989; Stäger and Eckelmann, 1991; Szepessy, 1993). These researches indicated that the 2-dimensional characteristics of the flow field around the sectional model can be well improved by adopting an appropriate endplates size. However, the flow field close to the endplate was interfered by the endplates, which was mainly manifested by the boundary layer close to the endplate and inhibition of axial flow (Fox and West, 1990; Kubo and Kato, 1986; Namiranian, 1985; Inagaki et al., 2014; Norberg, 1994). Besides, Morse et al. (2008) found that the end condition could significantly change the VIV performance of a circular cylinder.

In general, aerodynamic forces acting on different cross sections of a sectional model are not completely correlated, and the spanwise correlation decreases with the increasing separation (Bearman, 1984; Blackburn and Melbourne, 1997; Matsumoto et al., 2003; Ricciardelli, 2010; West and Apelt, 1982, 1997; Yan et al., 2016). Using end plates, the equivalent fluctuating lift per unit length increases with the decrease of aspect ratio below about $L/D = 10$ for Reynolds number $Re = 10^4$ and below about $L/D = 6$ for $Re = 10^5$ (Norberg, 2003). Besides, in transitional and turbulent shedding flows, the vortex shedding does not occur in phase over the whole span of a sectional model. By affecting the fluctuating lift forces, aspect ratio could change the VIV behavior of a sectional model. In order to study the Reynolds number effect on VIV, sectional models with different scale ratios were usually employed in wind tunnel tests (Hansen et al., 2015; Hui et al., 2008; Larsen et al., 2008; Matsuda et al., 2001). Most researchers indicated that the VIV amplitudes were quite different between the sectional models with different scale ratios (Hansen et al., 2015; Larsen et al., 2008). In general, researchers were inclined to attribute the discrepancy of VIV amplitude to Reynolds number effect, whereas the effect of aspect ratios was not taken into full consideration.

End condition and aspect ratio of a sectional model have profound influences on the flow field around the model, leading to interferences on wind pressure distribution, vortex shedding frequency and the aerodynamic forces of the model (Inagaki et al., 2014; Lee and Budwig, 1991; Szepessy and Bearman, 1992). As a result, the VIV performance tends to be changed (Morse et al., 2008; Wooton, 1969). In this study, the effect of end condition and aspect ratio of a section model on VIV was studied through a series of wind tunnel tests, employing both static and elastically mounted sectional models. The rectangular cylinder with a width to height ratio of $B/D = 5$ (where B and D are the width and height of the cross section of the cylinder respectively, as shown in Fig. 2 below) was employed for the tests because this cross-section was generally considered to be representative of a generic bridge deck geometry and has been widely studied including the "Benchmark on the Aerodynamics of a Rectangular 5:1 Cylinder (BARC)". This paper is organized into five sections. In the second section, wind tunnel tests were introduced in detail. In the third and fourth sections, the effects of end condition and aspect ratio on the aerodynamic characteristics were analyzed and discussed. In the fifth section, the effects of end conditions and aspect ratios on VIV amplitude of the sectional model were discussed.

2. Wind tunnel tests

Both static and dynamic wind tunnel tests were performed on the rigid sectional models with different end conditions and aspect ratios for the 1:5 rectangular. The sectional model has a width of $B = 300$ mm and a height of $D = 60$ mm.

2.1. Sectional model

The sectional model employed for static tests was 1920 mm long and was composed of an inner rigid aluminum skeleton and surface coat made of poly glass plates. The model was instrumented by 14 arrays of pressure taps as shown in Fig. 1. Six of these taps spaced at uniform intervals of 60 mm (1D) were located near to one end of the model to study the end condition effect on vortex shedding forces. The remaining pressure taps were distributed with various distances

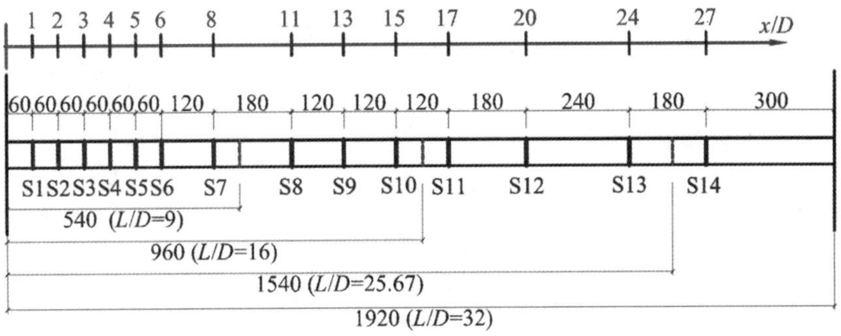

Fig. 1. Arrangement of spanswise pressure taps on the section model (Unit: mm).

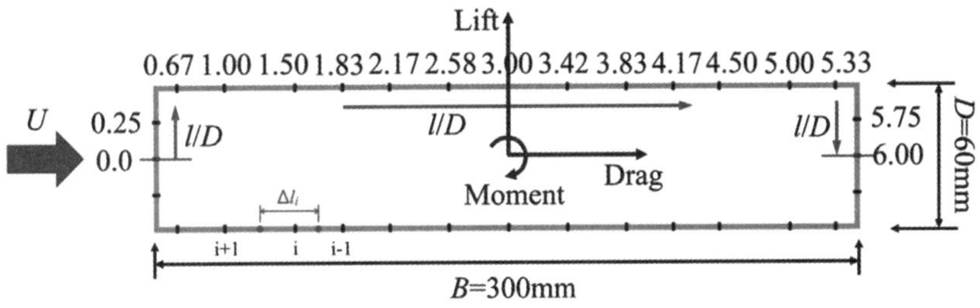

Fig. 2. Layout of pressure taps along the circumference of cross sections.

varying from 120 mm (2*D*) to 300 mm (5*D*). For each array, there are 32 pressure taps distributed circumferentially to measure the sectional fluctuating lift and drag forces, as shown in Fig. 2. In total, there were 448 pressure taps instrumented on the model. All the pneumatic tubes for the taps had a same length of 40 cm to ensure that the transfer function of the tubing was quite flat up to the frequency of 100 Hz and no additional time lag was induced between different taps. In dynamic test, the original model used in static tests (Fig. 3) was cut off for several times to change the aspect ratio, and this will be described detailly in Section 2.3. All the sectional models used in static and dynamic tests were sealed in good condition (including the connection positions between the model and endplates, as well as the possible surrounding gaps of the pressure probes on the model surface, etc.) to avoid unwanted test errors.

2.2. Outline of static tests

In static tests, the sectional model was rigidly connected to two vertical circular steel columns, as shown in Fig. 3. Various end conditions and aspect ratios were examined to study their effects on aerodynamic characteristics of the sectional model. Specifically, a total of eight end conditions were selected, including seven cases with various endplate sizes and one case without endplate (i.e., E11, E21, E22, E42, E44, E64 and E66 as shown in Table 1). The aspect ratio for these cases were consistent (i.e., $L/D = 32$, which is large enough to exclude the aspect ratio effect). All the endplates were rectangular wood plates with rounded corners and their geometric parameters are listed in Table 1. The thickness of the endplates was 5 mm, and the plates were very rigid to avoid unwanted vibration in the tests. In both static and dynamic tests, the large endplates were firmly fixed to the model by some temporary supports on the lateral side, which could ensure the stability of the endplates. Therefore, no obvious vibration of the endplates was observed in the wind tunnel test.

Besides, five aspect ratios varying from 9 to 32 were selected, as shown in Table 2. Aspect ratios of the model were altered by moving the endplate of E64 from one end toward the other end of the model, as shown in Fig. 3(b). Endplate E64 was employed for all the five cases, which was large enough to eliminate the interference between the two sections separated by the endplate. All the static tests were performed at wind incidence angle of 0° with three wind velocities of 4 m/s, 8 m/s and 12 m/s, corresponding to three different Reynolds numbers ($Re = \rho UD/\mu$, where U denotes the wind velocity of incoming flow, ρ and μ were the density and dynamic viscosity of the air, respectively) of 1.6×10^4, 3.2×10^4 and 4.8×10^4, respectively.

Both static and dynamic wind tunnel tests were carried out in the closed-circuit high-speed test section of wind tunnel HD-2 at Hunan University. The dimension of the test section is 3 m (width) × 2.5 m (height) × 17 m (length) and the wind velocity can be varied continuously up to 58 m/s. The longitudinal turbulence intensity in the bare tunnel is less than 0.5% for U > 2 m/s. Wind velocity of incoming flow was measured by a Cobra probe mounted at about 2.0 m ahead of the model in the windward side. Fluctuating pressures on the pressure taps of the static model were simultaneously recorded by seven pressure scanners (DMS 3400) placed inside of the model.

Fig. 3. Static sectional models in wind tunnel: (a) without endplate; (b) with endplate E64.

Table 1
Cases of static tests with different end conditions.

Case	Endplate width (mm)	Endplate height (mm)
E00	0	0
E66	1020(6*D*+*B*+6*D*)	780(6*D*+*D*+6*D*)
E64	1020(6*D*+*B*+6*D*)	540(4D+D+4*D*)
E44	780(4*D*+*B*+4*D*)	540(4*D*+*D*+4*D*)
E42	780(4*D*+*B*+4*D*)	300(2*D*+*D*+2*D*)
E22	540(2*D*+*B*+2*D*)	300(2*D*+*D*+2*D*)
E21	540(2*D*+*B*+2*D*)	180(1*D*+*D*+1*D*)
E11	420(1*D*+*B*+1*D*)	180(1*D*+*D*+1*D*)

Table 2
Cases of static tests with different aspect ratios.

Case	Effective length of model (mm)	Aspect ratio (*L*/*D*)
L32	1920	32
L26	1540	25.67
L16	960	16
L12	720	12
L9	540	9

2.3. Outline of dynamic tests

In dynamic tests, the sectional model was elastically mounted by eight coil springs through two rigid bar connected to the ends of the model (as shown in Fig. 4), hence it was allowed to oscillate in heaving and torsional directions. The lateral motion was restrained by two horizontal stay ropes connected to the two ends of the model. Three end conditions and three aspect ratios were employed in the tests, hence a total of nine dynamic test cases were performed, as outlined in Table 3. Parameter m in Table 3 denotes the equivalent mass per unit length of the sectional model, in which the mass of the two end plates as well as the pressure scanner inside the model is included. To model the different aspect ratios (i.e., $L/D = 26$, 16 and 9), the original sectional model ($L = 1920$ mm, $L/D = 32$) employed in static tests (Fig. 3) was firstly cut into a shorter one with a length of 1540 mm for the test cases with $L/D = 26$ (i.e., C2600, C2622 and C2611), and then it was cut again into a new one with a length of 960 mm for the cases with $L/D = 16$ (i.e., C1600, C1622 and C1611), finally, it was cut into a shortest one with the length of 540 mm for the cases with $L/D = 9$ (i.e., C0900, C0922 and C0911). The three cutting positions on the model were marked by red lines in Fig. 1. In this way, the total length of the sectional models in VIV tests for different aspect ratios varied with each other (e.g., Fig. 4 shows the sectional model with a length of 1540 mm for $L/D = 26$), which was different from that of the static tests (as shown in Fig. 3). For different aspect ratios, there remained different numbers of arrays of pressure taps on the model. Although the model length and endplate size were different between the nine cases, the equivalent mass per unit length of the model was kept consistent (namely $m = 10$ kg/m) by adjusting the added-mass provided by the attached mass blocks at the ends of the model.

In order to avoid the potential interference of VIV of heaving mode and torsional mode, the torsional frequency was kept much larger than the heaving frequency f_h. Detailed dynamic parameters of the elastically mounted system were listed in Table 3. Scruton number Sc in Table 3 was defined as $Sc = 4\pi m\xi_h/\rho D^2$, where ξ_h denotes the heaving damping ratio, $\rho = 1.225$ kg/m^3 was the air density. In VIV tests, the displacement responses of the model at upstream and downstream locations of the model were recorded by two laser displacement transducers (optoNCDT Model 2300) under the model. The two laser displacement transducers were symmetrically arranged at two sides of the model (namely upstream side and downstream side) with a horizontal distance $d = 240$ mm. In this way, both the heaving and torsional displacements of the model can be obtained by the synchronous data of the two laser displacement transducers (as shown

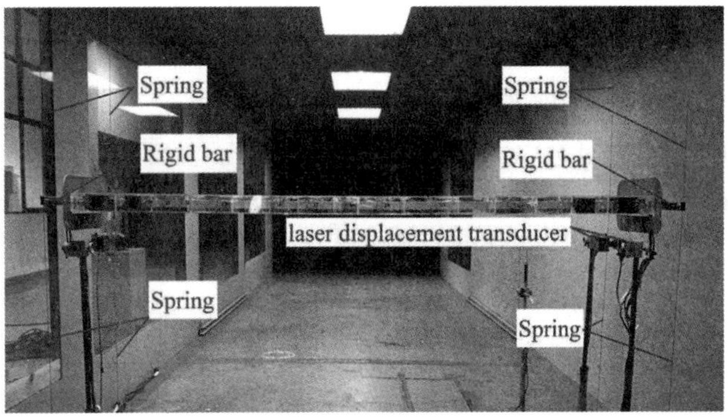

Fig. 4. Elastically-suspended sectional model in wind tunnel *(L/D = 26).*

Table 3
List of dynamic sectional models with different end condition and aspect ratios.

Case	L (mm)	L/D	Endplate	m (kg/m)	f_h (Hz)	ξ_h (%)	Sc
C2600	1540	26	E00	10	6.35	0.35	99.7
C2622	1540	26	E22	10	6.35	0.35	99.7
C2611	1540	26	E11	10	6.35	0.35	99.7
C1600	960	16	E00	10	6.50	0.35	99.7
C1622	960	16	E22	10	6.50	0.35	99.7
C1611	960	16	E11	10	6.50	0.35	99.7
C0900	540	9	E00	10	6.79	0.35	99.7
C0922	540	9	E22	10	6.79	0.35	99.7
C0911	540	9	E11	10	6.79	0.35	99.7

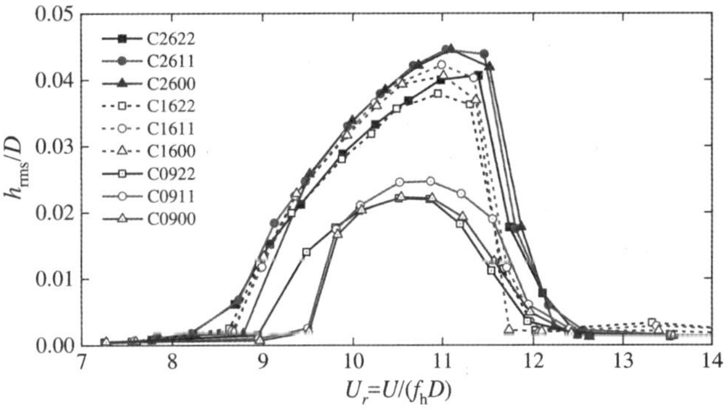

Fig. 5. Variation of the non-dimensional vertical amplitude with reduced velocity.

in Eq. (7) in Section 2.4). Meanwhile, the fluctuating pressures on the pressure taps of the model were simultaneously recorded by several pressure scanners (DMS 3400) fixed inside of the model.

With increasing wind velocity, two 'lock-in' ranges for heaving VIV were observed in each test case. The first 'lock-in' range, compared with the second one concerned in this study, occurs at a very small wind velocity range with lower wind velocity and very small VIV amplitude. The exact mechanism of this the first 'lock-in' range was not clear since the corresponding reduced frequency f_hD/U of it did not coincide with the Strouhal number of the cylinder obtained through static pressure measurement. In view of this, only the second normal 'lock-in' range closely related to the fluctuating aerodynamic forces caused by vortex-shedding was chosen for discussion. The reduced wind velocity–amplitude curves of VIV for the second 'lock-in' range are shown in Fig. 5, in which the reduced wind velocity is defined as $U_r = U/(f_hD)$, and h_{rms} denotes the root-mean-square (RMS) value of the heaving displacement. The maximum VIV amplitude for $L/D = 26$ was slightly larger than that for $L/D = 16$, but significantly larger than that for $L/D = 9$. That is, aspect ratio of the sectional model had remarkable effect on the VIV response. However, the effect of end condition on VIV response of the sectional model was not obvious, for there were slightly differences of the velocity–amplitude curves between the test cases with different end conditions and identical aspect ratio. These results will be discussed in detail in the following sections.

2.4. Data processing

The measured surface pressure $p_i(t)$ at the ith tap can be expressed as:

$$p_i(t) = p_{i,\text{mean}} + p_{i,fluc}(t) \tag{1}$$

where $p_{i,\text{mean}}$ is the mean wind pressure and $p_{i,\text{fluc}}$ is the fluctuating wind pressure as a function of time t. The mean wind pressure coefficient $Cp_{i,\text{mean}}$ and the fluctuating pressure coefficient $Cp_{i,\text{fluc}}$ (namely the RMS value of the instantaneous pressure coefficient) are written as:

$$Cp_{i,mean} = \frac{p_{i,mean}}{0.5\rho U^2} = \frac{1}{0.5\rho U^2}\frac{1}{N_f}\sum_{j=1}^{N_f} p_i(t_j) \tag{2}$$

$$Cp_{i,fluc} = \frac{1}{0.5\rho U^2}\sqrt{\frac{1}{N_f}\sum_{j=1}^{N_f}\left[p_{i,fluc}(t_j)\right]^2} = \frac{1}{0.5\rho U^2}\sqrt{\frac{1}{N_f}\sum_{j=1}^{N_f}\left[p_i(t_j) - p_{i,mean}\right]^2} \tag{3}$$

where U is the wind velocity of incoming flow; ρ is the air density; $N_f = 10000$ is the number of data points.

Lift coefficient $C_L(t)$, drag coefficient $C_D(t)$ and moment coefficient $C_M(t)$ were calculated by integrating the measured surface pressure coefficients, as follows:

$$C_L(t) = \frac{1}{0.5\rho U^2 B}\sum_{i=1}^{N_p} p_i(t)\Delta l_i \cos(\alpha_i) \tag{4}$$

$$C_D(t) = \frac{1}{0.5\rho U^2 D}\sum_{i=1}^{N_p} p_i(t)\Delta l_i \sin(\alpha_i) \tag{5}$$

$$C_M(t) = \frac{\sum_{i=1}^{N_p} p_i(t)\Delta l_i \sin(\alpha_i) d_{iy} + \sum_{i=1}^{N_p} p_i(t)\Delta l_i \cos(\alpha_i) d_{iz}}{0.5\rho U^2 B^2} \tag{6}$$

where Δl_i is the effective length corresponding to the i-th tap (as shown in Fig. 2, where the boundary points of Δl_i are the two midpoints between ith tap and the two adjacent taps respectively); α_i is the angle between the surface tangent at the ith tap and the direction of the free-stream wind, namely, it is 270°, 0°, 90° and 180° for the windward, upside, leeward and downside faces, respectively. The d_{iy} and d_{iz} are the horizontal and vertical distances between the ith tap and the center of the model.

During the dynamic tests, the vertical and torsional displacements were calculated by the two displacements $u_1(t)$ and $u_2(t)$ (measured by the two displacement transducers shown in Fig. 4) as follows:

$$h(t) = \frac{u_1(t) + u_2(t)}{2}, \quad \alpha(t) = \frac{u_1(t) - u_2(t)}{d} \tag{7}$$

where $d = 240$ is the distance between two measurement points of laser displacement transducers.

2.5. Validation of measurement data

For validation of the experimental data, the measured mean pressure coefficients and fluctuating pressure coefficients at the midspan cross section were compared with the existing data (Bruno et al., 2014; Nguyen et al., 2018; Liu et al., 2013; Mannini et al., 2017; Bronkhorst et al., 2011). For such a symmetric rectangular cylinder, the surface pressure distribution of the cross section should be symmetric, which could be a verification standard for the accuracy of the test results. Hence the pressure distributions on two sides of the cross section have been compared for all the test cases, and they all showed good agreement, both for mean pressure and fluctuating pressure. Therefore, only half part of the cross section was adopted for pressure plot (Figs. 6–8, Fig. 13 and Figs. 15–17).

The section model with an aspect ratio of 32 and endplate E64 were selected for comparison with other studies, as shown in Fig. 6. The measured mean pressure coefficients agreed well and the RMS values of fluctuating pressure coefficients measured in this study coincided well with the results measured by Bronkhorst et al. (2011) and Liu et al. (2013). The surface roughness, the length and end condition of the model, the flow condition, etc., may lead to the deviation of measured RMS of the fluctuating pressure coefficient. The measured mean drag coefficient was about 1.0, and the measured mean lift and moment coefficients were almost zero, which were in good agreement with the available wind tunnel data. The RMS lift coefficient obtained by means of wind pressure measurement was about 0.155 at Reynolds numbers of 4.8×10^4. It was larger than 0.08 at Reynolds numbers of 4.0×10^4 (Schewe, 2013) and 0.115 at Reynolds numbers of 5.2×10^4 (Nguyen et al., 2018) obtained by means of force measurement, and was close to 0.146 and 0.130 estimated by means of CFD. The Strouhal number estimated from the static test was 0.11, being in good agreement with the work of Bruno et al. (2014), Mannini et al. (2017) and Nguyen et al. (2018). In general, the measured data were reliable.

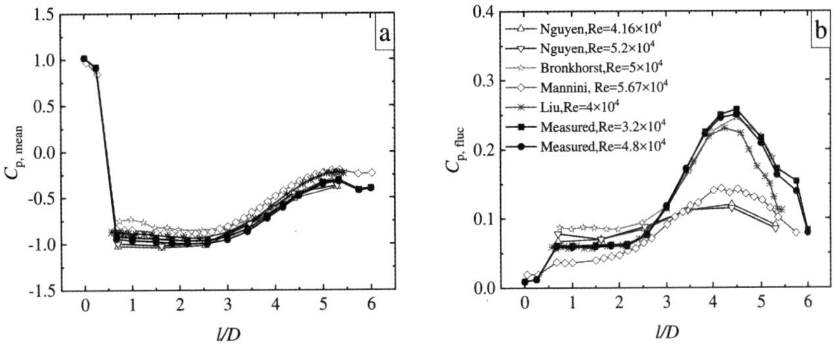

Fig. 6. Comparison of surface pressure distribution: (a) mean pressure coefficient; (b) fluctuating pressure coefficient.

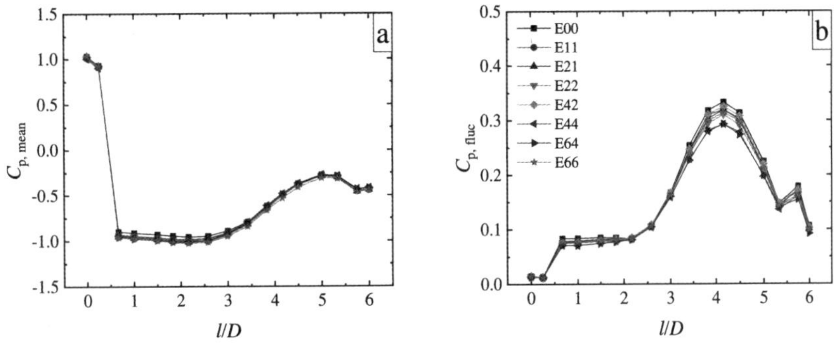

Fig. 7. Surface pressure distributions at midspan position for different endplates (L/D = 32): (a) mean pressure coefficient; (b) fluctuating pressure coefficient.

3. Results of static model tests

3.1. Effects of end condition

In this section, the end condition effects on the vortex-shedding parameters were studied based on static pressure measurement of cases E00~E66 (listed Table 1) with different endplates. Fig. 7 shows the circumferential surface pressure distribution at midspan of the sectional model for cases E00~E66. The mean surface pressure distributions of cases E00~E66 agreed very well with each other. In general, the fluctuating surface pressure distributions of cases E00~E66 also agreed well, except for the slight discrepancies in the flow reattachment region $3.5<x/D<5.0$.

Fig. 8 shows the circumferential surface pressure distribution at the selected cross sections S1~S8 (shown in Fig. 1) of the section model with endplate E22. For the mean circumferential surface pressure of the $C_{p,mean}$, measurement data at different cross sections showed good agreement except for those at the S1 and S2 cross sections which are close to the endplates. Therefore the mean surface pressures were slightly affected by the three-dimensional effects introduced by the endplates. Regarding the fluctuating surface pressure $C_{p,fluc}$, it is shown that the measurement data at the S3~S6 cross sections, especially those in the flow reattachment region, are significantly different from those at cross section S8. With the increasing distance away from the endplate, the surface pressure $C_{p,fluc}$ firstly increased and finally approached the values at the S8 cross section, which is almost the same as those at midspan. Close to the endplate, a turbulent boundary layer was formed. The interference from the boundary layer led to the increased fluctuating surface pressures at the flow separation region and the decreased fluctuating surface pressures at the flow reattachment region. The aerodynamic interference length of the endplate was about 2*D* for the mean pressure and was about 6-8*D* for the fluctuating pressure. In the interference region, fluctuating pressures in the reattachment region were greatly affected by the end effects, as a result of three-dimensional flow field caused by the endplates.

Fig. 9 shows the spanwise variation of fluctuating force coefficients (namely the RMS value of the instantaneous force coefficient) for different endplate sizes. It was observed that the aerodynamic interference length caused by the end condition was about 6~8*D*. Outside the interference region, the RMS value of fluctuating force coefficients at different cross sections was almost a constant value of about 0.16, while in the interference region it linearly increased with the separation measuring from the model end. Moreover, the trend in spanwise variation of fluctuating force coefficients was similar for different endplate sizes. However, comparing the fluctuating force coefficients and the mean pressure coefficients (Fig. 9a and Fig. 7a), it was obvious that the effect of endplates was more significant for fluctuating force coefficients. It was interesting that the maximum RMS values of fluctuating lift/moment coefficients occurred on the

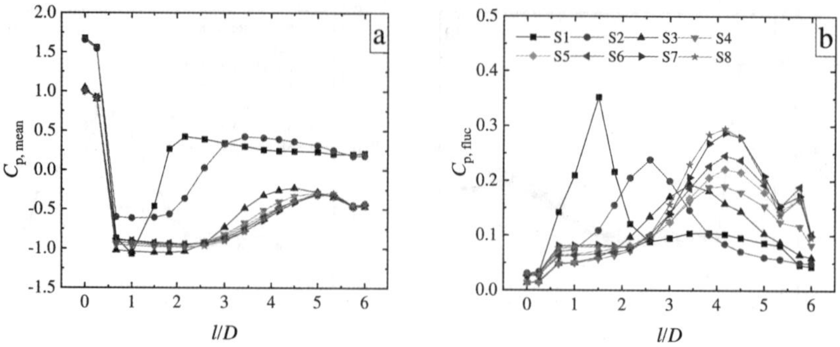

Fig. 8. Spanwise surface pressure distributions for different cross-sections (L/D = 32, endplate E22): (a) mean pressure coefficient; (b) fluctuating pressure coefficient.

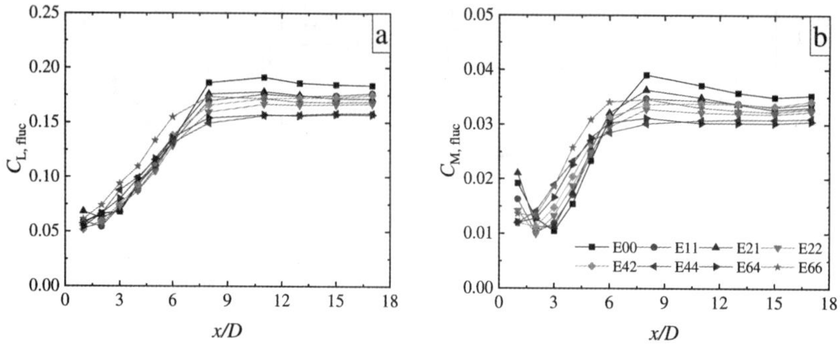

Fig. 9. Spanwise distributions of fluctuating force coefficients for different endplates (L/D = 32): (a) fluctuating lift coefficient; (b) fluctuating moment coefficient.

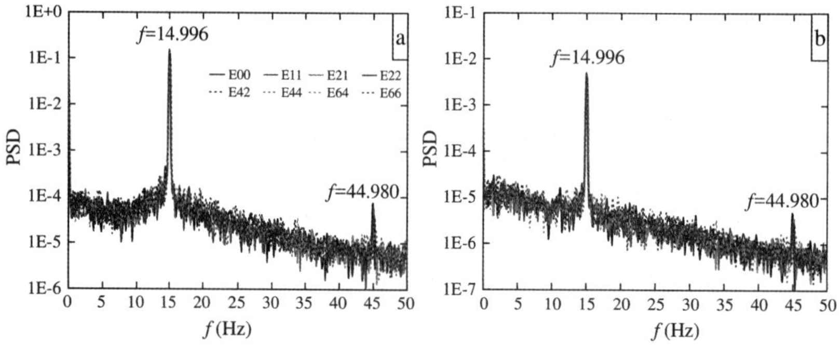

Fig. 10. PSD of lift and moment coefficients for different endplates at midspan (L/D = 32): (a) lift coefficient; (b) moment coefficient.

section model without endplate. Similar observation was reported in the study of Morse et al. (2008), in which larger VIV amplitude occurred when no endplate was employed for a long cylinder.

In general, the relationship between vortex shedding frequency (f_s) and the flow velocity (U) is described by Strouhal number, that is $St=f_sD/U$. The above sections have shown that the fluctuating surface pressures and the lifts were affected by end conditions. Consequently, it was speculated that the vortex shedding frequency might also be affected. Therefore, spectral analysis was performed on the lift and moment coefficients. Fig. 10 shows the power spectral densities (PSD) of the instantaneous lift and moment coefficients, based on measurement data at midspan cross section for sectional models with various end conditions at the reduced velocity $U/f_hD = 20.5$. Two extremal peak frequencies (i.e., $f = 14.996$ Hz and 44.98 Hz) were observed for each case. The dominant frequency was 14.996 Hz for all the cases, and the corresponding Strouhal number was St=0.11. The minor extremal frequency was about 44.98 Hz (namely three times of the dominant frequency) and was probably caused by the nonlinearity of fluctuating forces. The results indicated that the end condition had a negligible effect on vortex shedding frequency for cross section far away from model end.

The effect of end condition on spanwise vortex shedding frequency was also examined. Fig. 11 shows the PSDs of instantaneous lift and moment coefficients obtained at cross sections S1~S8 of the sectional model with the endplate E22.

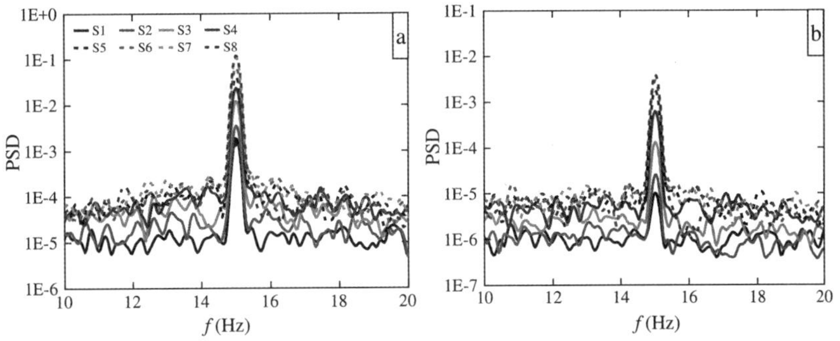

Fig. 11. PSD of force coefficients at different cross-sections (L/D = 32, endplate E22): (a) lift coefficient; (b) moment coefficient.

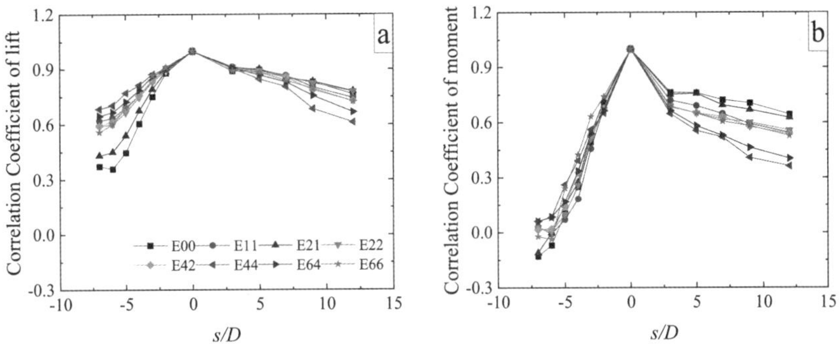

Fig. 12. Correlation coefficients of fluctuating lift and moment for different endplates (L/D = 32): (a) lift; (b) moment.

While all the peak frequencies were similar to those at midspan, the amplitudes of PSD for the peak frequencies were substantially smaller than those at midspan as described in Fig. 10. It was concluded that the regular vortex shedding became weaker for cross sections close to the endplates.

The spanwise correlation of fluctuating aerodynamic forces acting on the stationary and oscillating rectangular cylinders has been studied by some researchers. The spanwise correlation coefficients of aerodynamic forces for the stationary rectangular cylinders may be expressed as

$$R_{af}(s) = \exp(-s/\Lambda) \tag{8}$$

where s is the spanwise distance between the two measurement points of fluctuating forces, and Λ is the spanwise correlation length. The correlation coefficients of aerodynamic lifts and moments for the sectional models with the same aspect ratio $L/D = 32$ but various endplates are shown in Fig. 12. All correlation coefficients were calculated by taking the S7 cross section as the fixed reference point. It was observed that the correlation coefficients of fluctuating lifts outside the interference region (positive s/D) were much stronger than those within the interference region (negative s/D). In the interference region, the correlation coefficients of aerodynamic lifts sharply decreased with the increasing spanwise separation. The end condition had little effects on the correlation coefficients of aerodynamic lift outside the interference region. However, the end condition had significant effects on the correlation coefficients of aerodynamic moments. Comparing Fig. 9 and Fig. 12, it was concluded that the flow with stronger vortex shedding also had enhanced spanwise correlation.

3.2. *Effects of aspect ratios*

Wind tunnel tests were performed on the sectional models with various aspect ratios but with the same endplate of E64 to study the effect of aspect ratio on aerodynamic force characteristics. A total of five aspect ratios, namely L/D=9, 12, 16, 25.6 and 32, were considered. Fig. 13 shows the distribution of mean and fluctuating pressures at midspan of the sectional model for the five aspect ratios. The distributions of mean pressure coefficients were almost coincident for the five aspect ratios, implying that the steady aerodynamic forces were less affected by aspect ratios. In contrast, there were obvious differences in fluctuating pressure coefficients for different aspect ratios. The fluctuating pressure coefficients in the reattachment region of the cross-sections for the sectional models with $L/D = 32$ and $L/D = 25.6$ were much larger than those for other aspect ratios. The decreased fluctuating pressure coefficients in the reattachment region of cross-section for small aspects ratios was probably attributed to the interactions of regular vortex shedding with

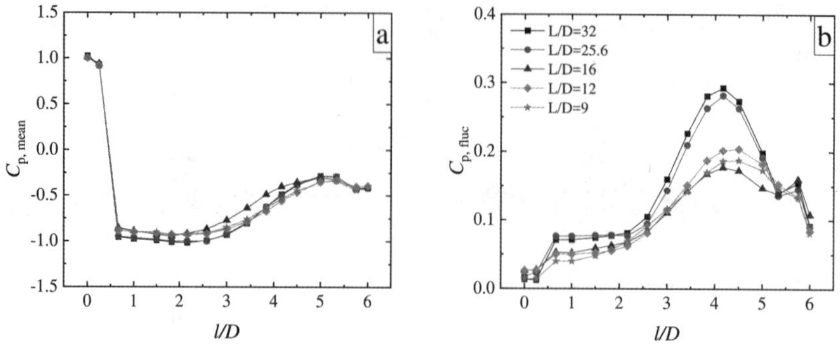

Fig. 13. Surface pressure distributions at midspan for different aspect ratios: (a) mean pressure coefficient; (b) fluctuating pressure coefficient.

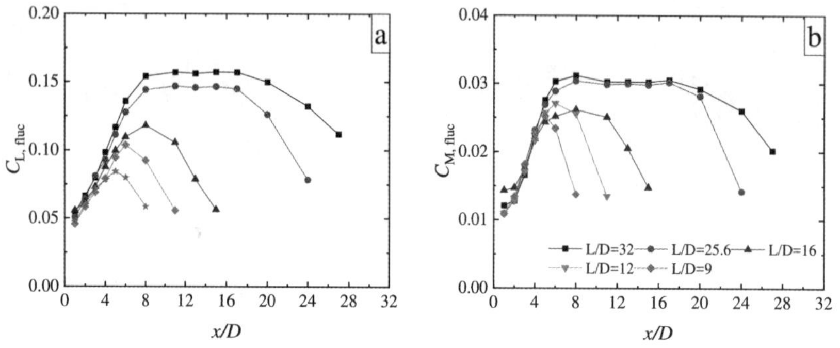

Fig. 14. Spanwise distributions of aerodynamic forces for different aspect ratios: (a) lift coefficient; (b) moment coefficient.

boundary layer created by the endplates. In other words, the aerodynamic forces were greatly affected because of the end interference for smaller aspect ratios. Consequently, the sectional model may lead to unreliable results if its length is smaller than double of the length of the end-interference region.

In addition to the pressure coefficients, Fig. 14 shows more clearly how the aspect ratio affects the fluctuating lift/moment along the span length. It was observed that the maximum RMS values of fluctuating lift coefficient showed a continuous reduction with decreasing aspect ratios, which was generally similar to the behavior of the aerodynamic pressures discussed above. The maximum RMS values of fluctuating lift were about 0.09 and 0.15 for L/D=9 and 25.6 respectively, indicating that the RMS values of the sectional fluctuating lifts were significantly underestimated for small aspect ratios. Increasing the aspect ratio from 25.6 to 32, no significant change in maximum fluctuating aerodynamic forces was observed. On the contrary, in the end-interference region (i.e., x/D<8 approximately) of the sectional model, the fluctuating lifts collapsed toward an almost constant value near the endplates. Nearby the endplates, the regular shedding pattern from the sectional model diminished significantly.

4. Results of dynamic tests

Measurement results of the stationary section model indicated that the fluctuating forces and their spanwise distribution could be significantly affected by the end conditions and aspect ratios. Therefore, sectional models with the same endplate E11 and several aspect ratios (namely cases C2611, C1611 and C0911) were employed for VIV tests. The results of heaving VIV responses were shown in Fig. 5 above.

4.1. Surface pressure distributions

The surface pressure distributions at midspan of the sectional model undergoing VIV for the three cases were obtained and shown in Fig. 15, along with that of the stationary model. For the dynamic model, two reduced wind speeds were selected for discussion: $U_r = \text{U1}$ at which the RMS peaks of the fluctuating lift coefficients were obtained, as well as $U_r = \text{U2}$ at which the amplitude peaks of VIV were observed (as shown in Fig. 15a). Besides, the surface pressure distribution of a dynamic model ($B/D = 5$, $L/D = 38.7$) undergoing torsional flutter (with an RMS torsional amplitude of 1.87° and an RMS heaving amplitude of 0.0078) measured by Ricciardelli and Marra (2008) was also presented in Fig. 15 for comparison.

According to Fig. 15, mean pressure distributions for the dynamic models in heaving VIV or torsional flutter were similar to that of the stationary model, both in magnitude and distribution law. Particularly, the mean pressure of

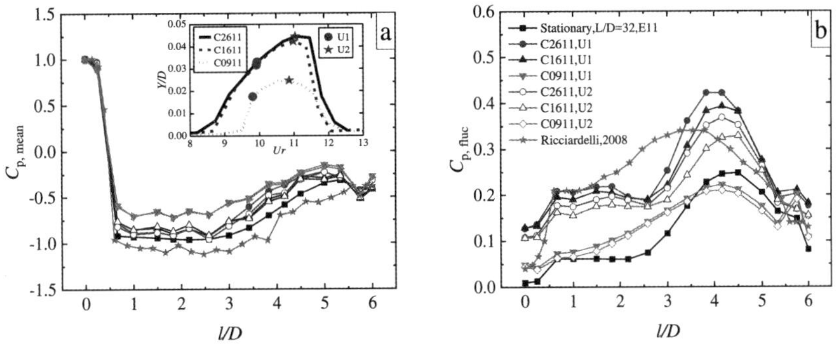

Fig. 15. Surface pressure distribution at midspan position for dynamic section model: (a) mean pressure coefficient; (b) fluctuating pressure coefficient.

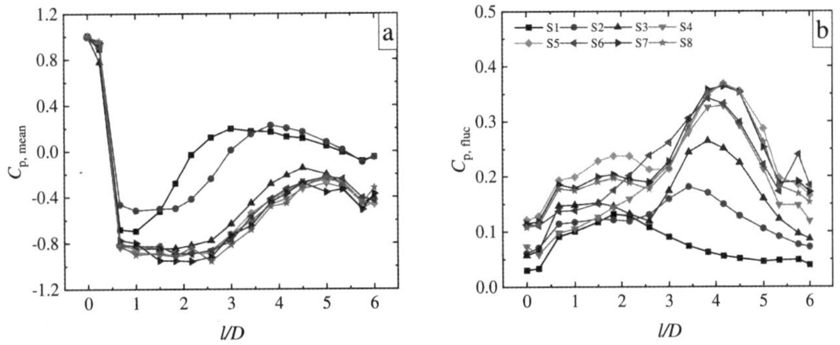

Fig. 16. Surface pressure distributions at different cross sections for case C2611 at Ur = U2: (a) mean pressure coefficient; (b) fluctuating pressure coefficient.

the stagnation point remained at 1.0 for all the cases. Regarding the fluctuating pressure distribution, things became complicated: firstly, the dynamic model with a small aspect ratio (i.e., case C0911 with $L/D = 9$) presented a special pressure distribution law comparing with the dynamic models with large aspect ratios (i.e., cases C0911 and C2611 with $L/D = 16$ and $L/D = 26$ respectively); distribution laws of the fluctuating pressure for the dynamic models with large aspect ratios in heaving VIV (namely cases C0911 and C2611) were similar to that for the stationary model, but the magnitudes of the fluctuating pressure were larger than those for the stationary model; the pressure distribution law of a dynamic model in heaving VIV was almost unchanged between U_r = U1 and U_r = U2, although there existed a small divergence in magnitude; there were large differences between the models in heaving VIV and torsional flutter, both in distribution law and magnitude.

The spanwise surface pressure distributions on the oscillating models of C2611 and C0911 at U_r=U2 were shown in Fig. 16 and Fig. 17, respectively. Similar to the stationary model, the surface distributions of mean pressure at the first two cross sections nearby the endplate were considerably different from those at midspan due to the end interference. Likewise, the distributions of fluctuating pressure coefficients obtained from the interference region tend to increase with the increasing distance from the end of model. The pressure fluctuation was sharply reduced by the end interference if the model length was less than double length of the end interference region. Obviously, end interference also suppressed the pressure fluctuations of the oscillating model in the interference region.

4.2. Properties of aerodynamic forces

When the heaving VIV occurred, the vortex shedding was 'locked' to the vibration of the section model. Hence, the main frequency component of the fluctuating lift force obtained at every cross-section of the oscillating model was identical to the heaving natural frequency, as shown in Figs. 18 and 19, indicating that the vortex shedding frequency f_s of any cross-section of the oscillating model was not disrupted by the end interference.

The spanwise variations of fluctuating lift and moment coefficients of the oscillating model were similar to those of the stationary model, as shown in Figs. 20 and 21. A "two-dimensional" pattern of fluctuating aerodynamic coefficients was observed outside the interference region, provided that the model was twice longer than the end interference length. On the other hand, in the end interference region, the "non-two-dimensional" spanwise distribution of the fluctuating aerodynamic forces was observed no matter how long the section model was, as shown in Figs. 20 and 21. The length of

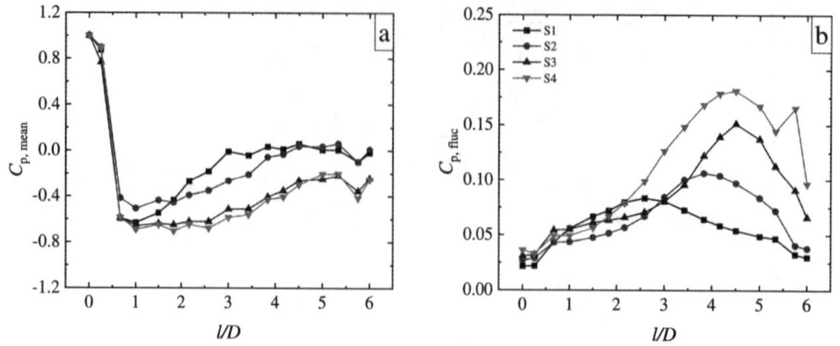

Fig. 17. Surface pressure distributions at different cross sections for case C0911 at Ur = U2: (a) mean pressure coefficient; (b) fluctuating pressure coefficient.

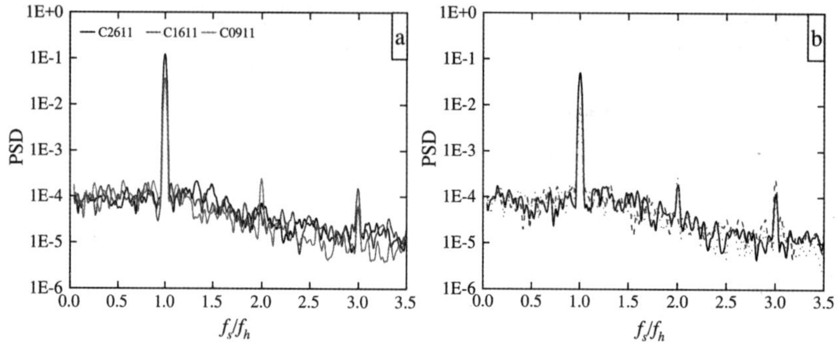

Fig. 18. PSD of aerodynamic lift at midspan for three oscillating models: (a) $U_r = U1$; (b)$U_r = U2$.

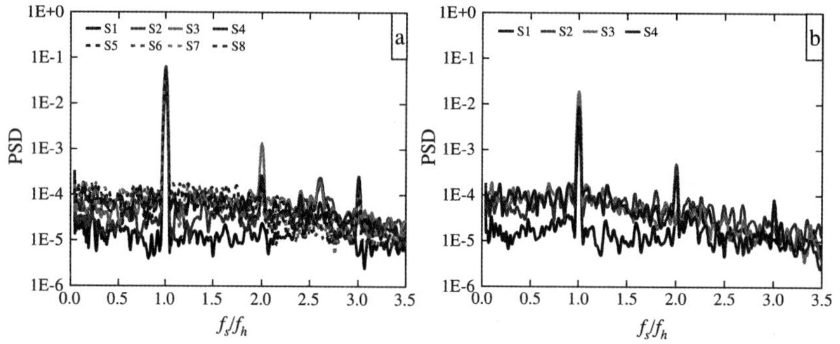

Fig. 19. PSD of aerodynamic lift at various cross sections of two oscillating models: (a) C2611; (b) C0911.

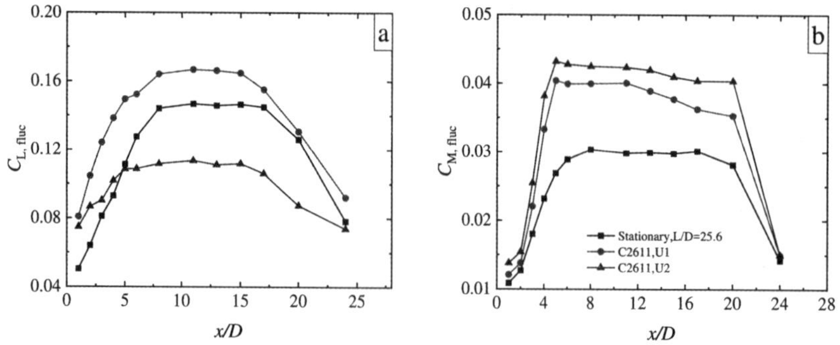

Fig. 20. Spanwise distribution of fluctuating aerodynamic coefficients for C2611 at two reduced speeds: (a) fluctuating lift coefficient; (b) fluctuating moment coefficient.

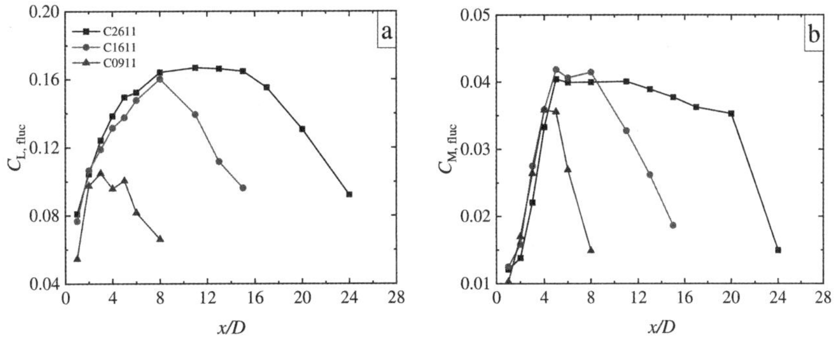

Fig. 21. Spanwise variation of fluctuating aerodynamic coefficients at U1: (a) fluctuating lift coefficient; (b) fluctuating moment coefficient.

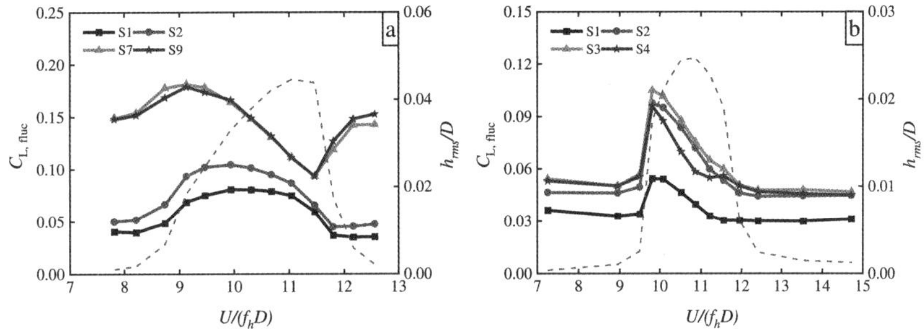

Fig. 22. Variation of fluctuating lift coefficients with reduced wind velocity for different cross sections: (a) C2611; (b) C0911.

end interference region for the oscillating model was close to that for the stationary model (namely 8*D*), indicating that VIV did not attenuate the end interference length of sectional modes.

Fig. 22 shows the variations of fluctuating lift coefficients with the reduced velocity during the VIV 'lock-in', in which the dot line represents the variation of VIV amplitudes with the reduced wind velocity. During the VIV 'lock-in', the fluctuating lift coefficients increased firstly with increasing wind velocity, and after reaching a maximum at a certain reduced velocity it decreased for higher reduced velocities; while outside of the 'lock-in', one can see that the fluctuating lift coefficients for both models approached to those for the stationary models. The complicated variation of fluctuating lifts implies that vortex shedding had a highly nonlinear and self-limiting feature. At the initialization stage of VIV, the strength of vortices shed at both sides of cross section enhanced with increasing VIV amplitudes, leading to the enhanced fluctuating lift coefficients. Under larger VIV amplitude, the vortices may become unstable such that the 'lock-in' behavior diminished, and further increase of wind velocity resulted in decreased responses. It was interesting to note that the maximum fluctuating lift was obtained at a wind velocity near to the onset wind speed, rather than the reduced speed corresponding to the maximum VIV amplitude, as shown in Fig. 22. This phenomenon was also reported by Nguyen et al. (2018).

5. Discussion

According to the results of the static and dynamic test, it was concluded that the aspect ratio considerably affected the magnitudes of fluctuating lift forces as well as its spanwise variation in the end interference region, confirming the need of a large aspect ratio for VIV tests. However, a large aspect ratio usually leads to a very small scale ratio of the sectional model due to the limited size of the wind tunnel, which may cause significant Reynolds number effects for some engineering structures. Therefore, the VIV results of a sectional model usually need to be correlated in consideration of the tridimensional effect of aerodynamic forces (Marra et al., 2017). In the following, a simple formula was developed to account for the aspect ratio effects on amplitude of VIV.

The semi-empirical linear model of VIV was employed to estimate the vibration amplitude of VIV. The lift force was represented as a simple harmonic excitation, as follows

$$F_L(x,t) = \frac{1}{2}\rho U^2 D C_L(x)\sin(2\pi f_s t + \theta) \tag{9}$$

where θ is the phase between the lift force and the structural displacement response. The VIV amplitude for a section model can be derived as

$$\frac{h_{rms}(U_r)}{D} = \frac{1}{4\pi}\cdot\frac{1}{Sc}\cdot U_r^2\cdot C_{LE,\,fluc}\cdot\sin(\theta) \tag{10}$$

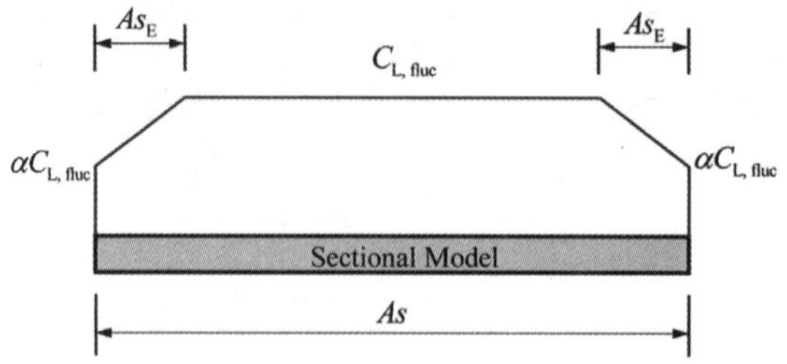

Fig. 23. K_L as the function of aspect ratio.

where $C_{LE,fluc}$ is the RMS value of the equivalent lift per unit length. It accounts for both the spanwise correlation of lift forces and the spanwise variation of lift forces due to the end interference. Herein, $C_{LE,fluc}$ was defined as

$$C_{LE,fluc} = C_{L,fluc} \cdot K_R \cdot K_E \tag{11}$$

where K_R is an equivalent coefficient of the lift per unite length accounting for the spanwise correlation, K_E is an equivalent coefficient of the lift per unite length due to the end interference.

The lift force was random with a narrow bandwidth, hence the sectional fluctuating lift forces between two cross sections with a certain spanwise distance s were not fully correlated. The correlation function of the fluctuating lift on the stationary model can be expressed as an exponential function. However, the correlation function of the fluctuating lift on the oscillating model undergoing VIV can be approximately written as

$$R_{LL}(s) = (1 - a_1)\exp(-s/a_2) + a_1, (s \geq 0) \tag{12}$$

where $a_1 = 0.7$ and $a_2 = 6.0$ are two factors (referencing Liu et al., 2017). K_R can be written as

$$K_R = \frac{\sqrt{2\int_0^L (L-s)R_{LL}(s)ds}}{L} \tag{13}$$

Substituting Eq. (12) into Eq. (13), K_R was

$$K_R = \sqrt{2(1-a_1)(\frac{a_2}{As} + \frac{a_2^2 e^{-As/a_2} - a_2^2}{As^2}) + a_1} \tag{14}$$

where $As=L/D$ is the aspect ratio.

Due to the end effect, the distribution of aerodynamic lift acting on the oscillating model was not homogeneous, and its spanwise variation on the model can be approximately expressed by a piecewise linear function as shown in Fig. 23. Consequently, the equivalent coefficient K_E of the lift per unite length can be obtained as

$$K_E = \begin{cases} 1 - \dfrac{(1-\alpha)As_E}{As}, & As \geq 2As_E \\ (1 - \frac{1-\alpha}{2})\dfrac{As}{2As_E}, & As < 2As_E \end{cases} \tag{15}$$

where $As_E = 8$ is the non-dimensional length of end interference region (as discussed in Section 4), $\alpha = 0.5$ is a reduction coefficient of fluctuating lift coefficient at the model end (according to Figs. 14, 20 and 21).

According to Eq. (14), K_R decreased slowly with the increasing aspect ratio. Eq. (15) indicated that when $As<16$ (namely As was less than twice the value of As_E), K_E decreased sharply with the decreasing aspect ratio, and when $As>16$, K_E increased slowly from 0.69 at $As = 16$ to 0.79 at $As = 40$. The product of K_R and K_E showed the similar trend with K_E. The variations of K_E, K_R, and $K_E K_R$ with respect to As are shown in Fig. 24.

Given the measured VIV amplitude $h_{rms,m}(U_r)$ from a finite-length sectional model, the predicted VIV amplitude $h_{rms,p}(U_r)$ of the model accounting for the end interference effect can be derived by

$$\frac{h_{rms,p}(U_r)}{D} = \frac{h_{rms,t}(U_r)}{D} \cdot \frac{1}{K_R \cdot K_E} \tag{16}$$

The predicted VIV amplitudes for test cases C2611, C1611 and C0911 were calculated based on Eq. (16) and shown in Table 4. Results indicated that the predicted VIV amplitudes (i.e., $h_{rms,p}(U_r)$=0.059, 0.061 and 0.062 for cases C2611, C1611 and C0911 respectively) derived from the measured VIV amplitudes for different aspect ratios were in good agreement, indicating that the semi-empirical linear model proposed above was reliable.

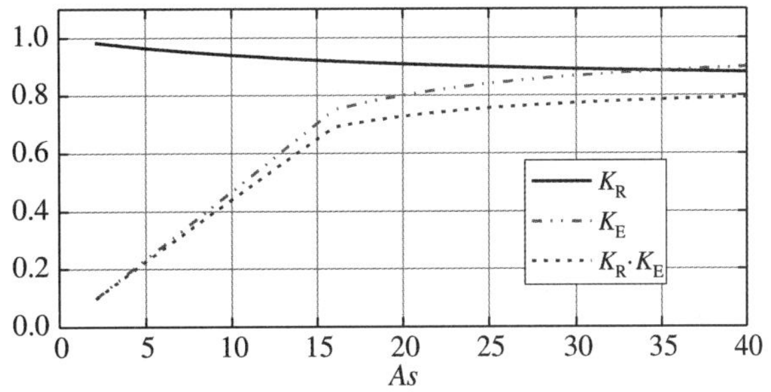

Fig. 24. K_R and K_E as functions of aspect ratio.

Table 4
Predicted VIV amplitudes for the three cases accounting for end interference effect.

Case	As	$K_R K_E$	Measured VIV amplitude	Predicted VIV amplitude
C2611	26.7	0.76	0.045	0.059
C1611	16	0.69	0.042	0.061
C0911	9	0.40	0.025	0.062

6. Conclusion

A series of wind tunnel tests were carried out by employing both static and elastically mounted sectional models to study the effects of end condition and aspect ratio of a sectional model on VIV performance for a 5:1 rectangular cylinder. Main conclusions are presented as follows: (1) there exists an aerodynamic interference region with a length of 6-8*D* close to the end of a sectional model, in which the end condition considerably interferes the flow field and resulting in a 'non-two-dimensional' spanwise distribution of surface pressure concerning various aspect ratios of the sectional model. Nearby the endplates, the regular shedding pattern from the sectional model disappeared basically. (2) Regular vortex shedding and fluctuating aerodynamic forces in the interference region are suppressed by the end effect, leading to a decrease of VIV amplitude, especially for the model with a length less than double of the interference region. (3) End condition has little effect on the pressure distribution of the model outside the interference region. Additionally, the VIV response of the model is almost independence of the end condition in the test range of this study (i.e., $L/D \geq 9$). (4) To obtain reliable VIV responses in wind tunnel tests by employing a sectional model, an aspect ratio no less than $L/D = 16$ is suggested for the model.

Based on the preliminary findings of this study, a flow field visualization technology such as Particle Image Velocimetry (PIV) could be helpful to further reveal the deeper mechanism of the end effect for the sectional model with a limited length. And this work could be performed for a further study.

CRediT authorship contribution statement

Chaoqun Wang: Investigation, Writing – review & editing. **Qing Wen:** Methodology, Experiment, Writing – original draft. **Shuai Zhou:** Investigation. **Xugang Hua:** Supervision, Conceptualization, Writing – review & editing. **Zhiwen Huang:** Investigation. **Zhengqing Chen:** Supervision, Conceptualization.

Declaration of competing interest

The authors declare that they have no known competing financial interests or personal relationships that could have appeared to influence the work reported in this paper.

Acknowledgments

This study is sponsored by the National Science Foundation of China with the program number Nos. 51708208 and 52025082, the Project of Scientific Research Department of Education of Hunan Province, China (No. 19B194), which are greatly acknowledged.

References

An, Y.H., Wang, C.Q., Li, S.L., Wang, D.W., 2016. Galloping of steepled main cables in long-span suspension bridges during construction. Wind Struct. 23 (6), 595–613.
Bearman, P.W., 1984. Vortex shedding from oscillating bluff bodies. Annu. Rev. Fluid Mech. 16, 195–222.

Belloli, M., Fossati, F., Giappino, S., Muggiasca, S., 2014. Vortex induced vibrations of a bridge deck: Dynamic response and surface pressure distribution. J. Wind Eng. Ind. Aerodyn. 133, 160–168.
Blackburn, H.M., Melbourne, W.H., 1997. Sectional lift forces for an oscillating circular cylinder in smooth and turbulent flows. J. Fluids Struct. 11, 413–431.
Bronkhorst, A.J., Geurts, C.P.W., van Bentum, C.A., 2011. Unsteady pressure measurements on a 5:1 rectangular cylinder. In: Proceedings of the Thirteenth International Conference on Wind Engineering. Amsterdam, The Netherlands.
Bruno, L., Salvetti, M.V., Ricciardelli, F., 2014. Benchmark on the aerodynamics of a rectangular 5:1 cylinder: An overview after the first four years of activity. J. Wind Eng. Ind. Aerodyn. 126, 87–106.
Cowdrey, C.F., 1963. A note on the use of end plates to prevent three-dimensional flow at the ends of bluff cylinders. In: Aeronautical Research Council, Current Paper (683) HMSO, London.
Fox, T.A., West, G.S., 1990. On the use of end plates with circular cylinders. Exp. Fluids 9, 237–239.
Hansen, S.O., Srouji, R.G., Isaksen, B., Berntsen, K., 2015. Vortex-induced vibrations of streamlined single box girder bridge decks. In: 14th International Conference on Wind Engineering.
Hua, X.G., Wang, C.Q., Li, S.L., Chen, Z.Q., 2020. Experimental investigation of wind-induced vibrations of main cables for suspension bridges in construction phases. J. Fluids Struct. 93, 102846.
Hui, M., Zhou, Z., Chen, A., Xiang, H., 2008. The effect of Reynolds numbers on the steady state aerodynamic force coefficients of the Stonecutters Bridge deck section. Wind Struct. 11, 179–192.
Inagaki, A., Mino, Y., Iwamoto, M., Yamada, H., 2014. Aspect ratio and end wall effects on the surface pressure coefficient of a circular cylinder. J. Fluid Sci. Technol. 9 (3), JFST0062-JFST0062.
Kubo, Y., Kato, K., 1986. The role of end plates in two dimensional wind tunnel tests. Doboku Gakkai Ronbunshu 368, 179–186.
Kubo, Y., Miyazaki, M., Kato, K., 1989. Effects of end plates and blockage of structural members on drag forces. J. Wind Eng. Ind Aerodyn. 32, 329–342.
Kwok, K.C.S., Qin, X.R., Fok, C.H., Hitchcock, P.A., 2012. Wind-induced pressures around a sectional twin-deck bridge model: Effects of gap-width on the aerodynamic forces and vortex shedding mechanisms. J. Wind Eng. Ind. Aerodyn. 110, 50–61.
Larsen, A., Savage, M., Lafrenière, A., Hui, M.C.H., Larsen, S.V., 2008. Investigation of vortex response of a twin box bridge section at high and low Reynolds numbers. J. Wind Eng. Ind. Aerodyn. 96, 934–944.
Lee, B.E., 1979. On the effects of end plates on the mean forces on square sectioned cylinders. J. Wind Eng. Ind. Aerodyn. 5 (1–2), 187–188.
Lee, T., Budwig, R., 1991. A study of the effect of aspect ratio on vortex shedding behind circular cylinders. Phys. Fluids. A 3, 309–315.
Lee, S., Kwon, S.-D., Yoon, J., 2014. Reynolds number sensitivity to aerodynamic forces of twin box bridge girder. J. Wind Eng. Ind. Aerodyn. 127, 59–68.
Li, S.L., An, Y.H., Wang, C.Q, Wang, D.W., 2017. Experimental and numerical studies on galloping of the flat-topped main cables for the long span suspension bridge during construction. J. Wind Eng. Ind. Aerodyn. 163, 24–32.
Li, M., Li, S., Liao, H., Zeng, J., Wang, Q., 2016. Spanwise correlation of aerodynamic forces on oscillating rectangular cylinder. J. Wind Eng. Ind. Aerodyn. 154, 47–57.
Liu, X., Cui, Y., Liu, Q., 2013. Wind tunnel study on spanwise correlation of aerodynamic forces on a 5: 1 rectangular cylinder. In: Eighth Asia-Pacific Conference on Wind Engineering, Chennai, India.
Liu, Z.W., Huang, L.K., Chen, Z.Q., 2017. Experimental study on spanwise correlation of aerodynamic forces of VIV of the rectangular section girder. J. Vibration Eng. 30, 422–431, (in Chinese).
Mannini, C., Belloli, M., Marra, A.M., Bayati, I., Giappino, S., Robustelli, F., Bartoli, G., 2016. Aeroelastic stability of two long-span arch structures: A collaborative experience in two wind tunnel facilities. Eng. Struct. 119, 252–263.
Mannini, C., Marra, A.M., Bartoli, G., 2014. Viv-galloping instability of rectangular cylinders: review and new experiments. J. Wind Eng. Ind. Aerodyn. 132, 109–124.
Mannini, C., Marra, A.M., Bartoli, G., 2015. Experimental investigation on VIV-galloping interaction of a rectangular 3:2 cylinder. Meccanica 50 (841).
Mannini, C., Marra, A.M., Pigolotti, L., Bartoli, G., 2017. The effects of free-stream turbulence and angle of attack on the aerodynamics of a cylinder with rectangular 5:1 cross section. J. Wind Eng. Ind. Aerodyn. 161, 42–58.
Marra, A.M., Mannini, C., Bartoli, G., 2017. Wind tunnel modeling for the vortex-induced vibrations of a yawed bridge tower. J. Bridge Eng. 22 (5), 04017006.
Matsuda, K., Cooper, K.R., Tanaka, H., Tokushige, M., Iwasaki, T., 2001. An investigation of Reynolds number effects on the steady and unsteady aerodynamic forces on a 1:10 scale bridge deck section model. J. Wind Eng. Ind. Aerodyn. 89, 619–632.
Matsumoto, M., Shirato, H., Araki, K., Haramura, T., Hashimoto, T., 2003. Spanwise coherence characteristics of surface pressure field on 2-D bluff bodies. J. Wind Eng. Ind. Aerodyn. 91, 155–163.
Morse, T.L., Govardhan, R.N., Williamson, C.H.K., 2008. The effect of end conditions on the vortex-induced vibration of cylinders. J. Fluids Struct. 24, 1227–1239.
Nakamura, Y., Nakashima, M., 1986. Vortex excitation of prisms with elongated rectangular, H and h cross-sections. J. Fluid Mech. 163, 149–169.
Namiranian, F., 1985. Fluctuating Lift on Cylinders of Rectangular Cross Section in Smooth and Turbulent Flows. University of British Columbia..
Nguyen, D.T., Hargreaves, D.M., Owen, J.S., 2018. Vortex-induced vibration of a 5:1 rectangular cylinder: A comparison of wind tunnel sectional model tests and computational simulations. J. Wind Eng. Ind. Aerodyn. 175, 1–16.
Norberg, C., 1994. Experimental investigation of the flow around a circular cylinder: Influence of aspect ratio. J. Fluid Mech. 258, 287–316.
Norberg, C., 2003. Fluctuating lift on a circular cylinder: review and new measurements. J. Fluids Struct. 17, 57–96.
Ricciardelli, F., 2010. Effects of the vibration regime on the spanwise correlation of the aerodynamic forces on a 5:1 rectangular cylinder. J. Wind Eng. Ind. Aerodyn. 98, 215–225.
Ricciardelli, F., Marra, A.M., 2008. Sectional aerodynamic forces and their longitudinal correlation on a vibrating 5:1 rectangular cylinder. In: Proceedings 6th International Colloquium on Bluff Bodies Aerodynamics and Applications. Milan, Italy.
Schewe, G., 2013. Reynolds-number-effects in flow around a rectangular cylinder with aspect ratio 1:5. J. Fluids Struct. 39, 15–26.
Stäger, R., Eckelmann, H., 1991. The effect of endplates on the shedding frequency of circular cylinders in the irregular range. Phys. Fluids A 3, 2116–2121.
Stansby, P.K., 1974. The effects of end plates on the base pressure coefficient of a circular cylinder. Aeronaut. J. 78.
Szepessy, S., 1993. On the control of circular cylinder flow by end plates. Eur. J. Mech. B/Fluids 12, 217–244.
Szepessy, S., Bearman, P.W., 1992. Aspect ratio and end plate effects on vortex shedding from a circular cylinder. J. Fluid Mech. 234, 191–217.
Toy, N., Fox, T.A., 1986. The effect of aspect ratio of end plate separation upon base pressures recorded on a square bar. Exp. Fluids 4 (266).
Wang, C.Q., Hua, X.G., Feng, Z.Q., Xiao, H.Z., Chen, Z.Q., Zhang, J.B., 2021a. Experimental investigation on vortex-induced vibrations of a triple-box girder with web modification. J. Wind Eng. Ind. Aerodyn. 218, 104783.
Wang, C.Q., Hua, X.G., Huang, Z.W., Tang, Y., Chen, Z.Q., 2021b. Post-critical behavior of galloping for main cables of uspension bridges in construction phases. J. Fluids Struct. 101 (03205).
Wang, C.Q., Hua, X.G., Huang, Z.W., Wen, Q., 2021c. Aerodynamic characteristics of coupled twin circular cylinders with near wake interference in cross flow. Appl. Sci. 11 (4189).

Wen, Q., Hua, X.G., Lei, X., Chen, Z.Q., Niu, H.W., 2018. Experimental study of wake-induced instability of coupled parallel hanger ropes for suspension bridges. Eng. Struct. 167, 175–187.
West, G.S., Apelt, C.J., 1982. The effects of tunnel blockage and aspect ratio on the mean flow past a circular cylinder with Reynolds numbers between 104 and 105. J. Fluid Mech. 114, 361–377.
West, G.S., Apelt, C.J., 1997. Fluctuating lift and drag forces on finite lengths of a circular cylinder in the subcritical reynolds number range. J. Fluids Struct. 11, 135–158.
Wooton, L.R., 1969. The oscillations of large circular stacks in wind. Ice Proc. 43, 573–598.
Yan, L., Zhu, L.D., Flay, R.G.J., 2016. Span-wise correlation of wind-induced fluctuating forces on a motionless flat-box bridge deck. J. Wind Eng. Ind. Aerodyn. 156, 115–128.
Zhou, S., Hua, X.G., Chen, Z.Q., Chen, W., 2017. Experimental investigation of correction factor for VIV amplitude of flexible bridges from an aeroelastic model and its 1:1 section model. Eng. Struct. 141, 263–271.
Zhu, Q., Chen, B.Y., Zhu, L.D., Xu, Y.L., 2017. Investigation on characteristics and span-wise correlation of vortex-induced forces on a twin-box deck using newly-developed wind-tunnel test technique. J. Wind Eng. Ind. Aerodyn. 164, 69–81.

第35卷 第11期
2022年11月

中国公路学报
China J. Highw. Transp.

Vol. 35 No. 11
Nov. 2022

文章编号:1001-7372(2022)11-0133-14

流线型钢箱梁涡激振动机理与气动控制措施

刘志文*[1],江智俊[1],黎晓刚[2],邵超逸[2],王灿东[3],陈政清[1]

(1. 湖南大学 风工程与桥梁工程湖南省重点实验室,湖南 长沙 410082; 2. 广州市高速公路有限公司,广东 广州 510290; 3. 中交公路规划设计院有限公司,北京 100120)

摘要:以某主跨390 m的独塔流线型钢箱梁斜拉桥为工程依托,采用风洞试验与计算流体动力学(Computational Fluid Dynamics, CFD)相结合的方法对流线型钢箱梁涡激振动机理与气动控制措施进行研究。首先,采用几何缩尺比为1:30的主梁节段模型进行主梁涡振性能与气动控制措施优化研究;其次,采用CFD方法对主梁涡振响应进行流固耦合计算,将Newmark-β算法嵌入ANSYS Fluent用户自定义函数(User Defined Functions, UDFs)实现主梁结构振动响应求解,同时结合动网格技术实现主梁断面流固耦合分析;并根据判断条件来检索箱梁壁面上的网格单元,以获得主梁断面振动过程中的表面压力,然后结合主梁结构振动响应、表面压力以及流场特征等对主梁涡激振动机理进行分析。结果表明:该桥主梁原设计方案存在涡激共振现象,将梁底检修车轨道内移120 cm可有效抑制主梁涡振响应;主梁涡激振动响应的数值模拟结果与风洞试验结果吻合较好;检修车轨道内移120 cm后主要改变了箱梁下表面平均压力系数分布特性,且箱梁表面各测点脉动压力卓越频率不一致,有效减小了主梁涡激振动响应;流线型箱梁靠近迎风侧的"被动区域"对结构涡振响应贡献较小,背风侧"驱动区域"发生周期性旋涡脱落是影响流线型箱梁涡振的主要因素。

关键词:桥梁工程;涡激振动机理;计算流体动力学;流线型钢箱梁;风洞试验;气动控制措施

中图分类号:U448.27 **文献标志码:**A

Study on Mechanism and Aerodynamic Control Measures for Vortex-induced Vibration of a Streamlined-box Steel Girder

LIU Zhi-wen*[1], JIANG Zhi-jun[1], LI Xiao-gang[2], SHAO Chao-yi[2], WANG Can-dong[3], CHEN Zheng-qing[1]

(1. Hunan Provincial Key Laboratory for Wind and Bridge Engineering, Hunan University, Changsha 410082, Hunan, China; 2. Guangzhou Expressway Co. Ltd., Guangzhou 510290, Guangdong, China; 3. CCCC Highway Consultants Co. Ltd., Beijing 100120)

Abstract: Based on a single-pylon streamlined steel box girder cable-stayed bridge with a main span of 390 m, the vortex-induced vibration (VIV) performance and aerodynamic control measures of streamlined steel box girder were studied using wind tunnel tests and computational fluid dynamics (CFD), respectively. First, wind tunnel tests of the bridge deck sectional model with a geometry scale of 1:30 were conducted to investigate the VIV responses and aerodynamic control measures. Second, the CFD method was used to calculate the VIV response of the main

收稿日期:2021-09-18
基金项目:国家自然科学基金项目(51778225,52178475)
*作者简介:刘志文(1975-),男,山西阳高人,教授,博士研究生导师,工学博士,E-mail:zhiwenliu@hnu.edu.cn。

deck, and the Newmark-β algorithm was embedded in the user-defined functions (UDFs) of ANSYS Fluent to solve the vibration responses of the main deck. Moreover, dynamic grid technology was applied to analyze the fluid-structure interactions of the main deck. According to the checked conditions, the grid elements on the surface of the main deck were retrieved to obtain the pressure of the oscillating main deck section. Finally, the VIV mechanism of the main deck was analyzed based on the vibration responses, surface pressure, and flow characteristics of the main deck. The results show that the original design of the main deck exhibits the phenomenon of VIV, which can be effectively suppressed by moving the inspection rails at the bottom of the main deck inward. The numerical results for the VIV response of the main deck are in good agreement with the wind tunnel test results. The average pressure distribution characteristics on the lower surface of the main deck are mainly affected by moving the inspection rails inward by 120 cm, and the dominant frequency of fluctuating pressures at different points on the surfaces of the main deck is not consistent with each other, which effectively suppressed the VIV responses of the main deck. Furthermore, it was found that the "passive area" of the streamlined box girder near the windward side has a lower contribution to the VIV response. However, the periodic vortex shedding in the "driving area" on the leeward side of the streamlined box girder s the main factor affecting the VIV response of the streamlined box girder.

Keywords: bridge engineering; mechanism of VIV; computational fluid dynamics; streamlined-box girder; wind tunnel test; aerodynamic control measure

Received 18 Sept. 2021

Funding: National Natural Science Foundation of China (51778225, 52178475)

*Corresponding author. *E-mail address*: zhiwenliu@hnu. edu. cn.

https://doi. org/10. 19721/j. cnki. 1001-7372. 2022. 11. 013

0 引 言

大跨度桥梁具有自振频率低、阻尼比低、质量相对较轻等特点，在风作用下容易发生较为明显的振动现象，如颤振、涡激振动、驰振、抖振及拉索振动等。涡激振动是气流经过结构断面时产生周期性旋涡脱落，从而对结构产生周期性激励作用，当其卓越频率与结构某一阶固有频率接近时所产生的一种风致振动现象，具有自激、限幅特征；若不有效控制则会引起行车舒适性与结构疲劳等问题[1]。国内外多座大跨度桥梁发生了较为明显的涡激共振现象，如英国赛文桥[2]、丹麦大带东桥[3]、日本东京湾大桥[4]以及中国西堠门大桥[5]、武汉鹦鹉洲大桥和广东虎门大桥等，引起社会广泛关注。国内外许多学者对大跨度桥梁涡激振动特征、控制措施及机理进行了大量研究。

目前，针对涡振的抑制，主要措施有结构措施、气动措施和机械措施。结构措施是指增加结构刚度以提高结构固有频率，从而提高结构涡振起振风速。气动控制措施则通过改变结构断面气动外形来减小其涡振响应[6]。Larsen 等[7]采用风洞试验和热线法结合进行了流体可视化研究，认为底板和下斜腹板保持角度小于 15°能有效抑制涡振，并在挪威的 Hålogaland 桥和土耳其的 Izmit 海湾桥验证了该措施的有效性。工程实践中仅仅通过修正主梁断面气动外形尚不能完全抑制主梁涡振响应，还需要通过调整附属设施和附加气动措施的组合形式来优化涡振性能。赵林等[8]基于大跨度桥梁中常见的主梁类型，系统地总结了涡激振动气动控制方法。许福友等[9]通过风洞试验测压法研究发现，在栏杆扶手上加装抑流板，减弱了上表面气流在前缘的分离和上表面的风压脉动程度，能够抑制涡振的发生。管清海等[10]针对栏杆透风率展开研究，结果表明提高栏杆的阻风率能有效抑制涡振。朱思宇等[11]在 +7°大攻角下进行主梁涡振试验研究，发现内移检修车轨道位置能够有效地抑制加劲梁涡激振动响应，背风侧检修车轨道是引起加劲梁涡激振动的主要影响因素；同时发现内外侧的导流板会对涡振产生不同的影响。此外，由于调谐质量阻尼器(Tuned Mass Dampers, TMDs)针对单一模态振动的涡激振动具有非常好的抑制效果，在气动外形优化效果受限时还可以采用机械措施进行控制。陈政清等[12]研发的电涡流调谐质量阻尼器已经应用于榕江大桥和绵阳三江大桥等一系列桥梁工程中，并已经取得较好的减振效果。

由于结构涡振影响因素多且较为复杂，其发生机理尚未完全明确。Paidoussis 等[13]归纳了涡振在不同断面下的发生机理，指出桥梁主梁断面涡振主要与 4 种典型涡脱有关：顶部涡脱、尾部涡脱、交错涡脱以及顶部碰撞涡脱。Larsen 等[7]研究了流线型箱梁断面涡振发生机制，发现下底板处产生的分离旋涡是涡振发生的主要驱动因素。Hu 等[14]基于流线型闭口箱梁断面研究了模型各区域气动力对涡激力的贡献值，从气动力的角度揭示了涡振发生机理。另外，有研究表明雷诺数对主梁断面气动特性的影响也不可忽略。Schewe 等[15]在丹麦大贝尔特东桥的主梁断面风洞试验研究中发现，断面斯特罗哈数会随着雷诺数缓慢增加。金挺等[16]、李加武等[17]的研究也都表明扁平钢箱梁的斯特罗哈数具有明显的雷诺数效应。

此外，现场实测也是桥梁风工程研究的重要方法之一。Li 等[18]通过压力监测的手段对西堠门大桥开展了涡振现场实测研究，并采用镭射激光片、烟雾发生器和 CCD 摄像机进行现场可视化试验，从流场特征的角度对涡激振动机理进行了一定程度上的解释。同时，随着计算机技术的发展，计算流体动力学(Computational Fluid Dynamics，CFD)因具有计算效率高、可视化强等优点，被广泛应用于结构风工程[19-21]。刘钥等[22]对箱梁断面静风力系数进行 CFD 数值模拟，模拟结果与试验结果吻合较好；祝志文等[23]对带防撞护栏的扁平箱梁进行数值模拟，能有效预测高阶模态涡振锁定区间和涡振振幅。但目前基于 CFD 流固耦合的涡振控制机理研究相对较少。

综上所述，尽管国内外许多学者针对大跨度桥梁主梁涡激振动已经开展了大量的研究工作，然而由于桥梁主梁断面涡激振动的复杂性，对于主梁涡激振动的机理尚未形成明确的认识。为此，本文采用几何缩尺比为 1/30 的大比例主梁节段模型进行涡激振动特征与气动控制措施研究。借助动网格技术和 UDF 二次开发功能，实现了主梁断面流固耦合振动与压力监测。根据主梁结构振动响应、表面压力以及流场特征等对主梁涡激振动机理进行研究。

1 工程概况

主桥为某主跨 390 m 独塔双索面半漂浮体系钢箱梁斜拉桥，跨径布置为 66 m＋390 m＋324 m＋2×66 m＝912 m；主梁采用扁平流线型整幅式钢箱梁，主梁宽 B＝43.2 m，主梁中心处梁高 H＝3.80 m，主梁标准断面如图 1 所示。

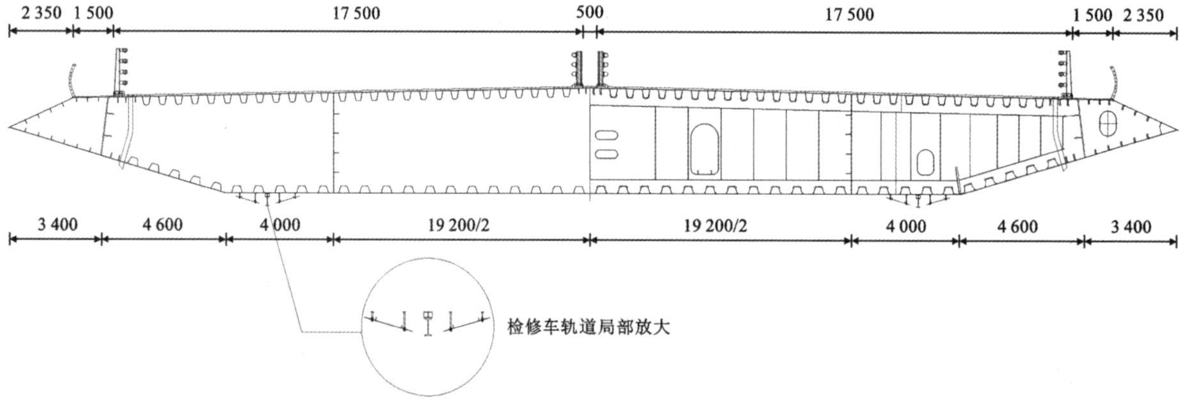

图 1　主梁标准断面(单位：mm)

Fig. 1　Standard Section of the Main Deck (Unit: mm)

根据该桥桥位处周边气象站观测数据及风参数专题研究成果可知，大桥桥位基本风速 V_{10}＝36.8 m·s^{-1}，平均风剖面指数 α＝0.12，成桥状态桥面高度处设计基准风速 V_d＝44.0 m·s^{-1}。采用 ANSYS 软件对主桥成桥状态进行结构动力特性计算，表 1 给出成桥状态前 10 阶频率与振型描述。

2 主梁大比例节段模型涡激振动试验

2.1 试验装置与工况

主梁大比例节段模型试验在湖南大学 HD-2 号风洞第三试验段进行，该试验段尺寸为 8.5 m(宽)×2 m(高)×15 m(长)，试验段风速范围为 0～

表 1　大桥成桥状态动力特性结果

Table 1　Dynamic Characteristics of the Bridge Structure in-service Stage

序号	频率/Hz	振型描述
1	0.248 3	主梁纵飘＋竖弯
2	0.388 2	一阶主梁纵飘
3	0.440 1	桥塔侧摆
4	0.479 3	一阶主梁正对称竖弯
5	0.599 2	一阶主梁反对称侧弯
6	0.603 5	主梁竖弯
7	0.733 7	二阶主梁竖弯
8	0.826 3	桥塔侧弯＋主梁扭转
9	0.848 5	三阶主梁竖弯
10	0.894 5	一阶主梁对称扭转

$15\ \mathrm{m\cdot s^{-1}}$，均匀流场紊流度小于 1.0%。由于主梁涡激共振响应对主梁断面几何外形十分敏感，为了尽可能真实地模拟主梁细部构造，同时考虑主梁断面雷诺数效应，综合考虑主梁几何外形、质量以及风洞试验条件等因素，最终确定大桥主梁大比例节段模型的几何缩尺比 $\lambda_L=1:30$。主梁模型长度取为 $L=3\ 000$ mm，宽度 $B=1\ 440$ mm，高度 $H=127$ mm，模型长宽比 $L/B=2.08$。主梁模型骨架采用不锈钢框架制作而成，外衣采用优质木材制作，保证几何外形的相似。模型两端采用铝板作为端板，以保证主梁断面附近气流的二元特性。主梁上的防撞护栏和检修道护栏分别采用亚克力板和钢丝制作，并模拟了防撞护栏的形状与透风率。检修车轨道及检修车轨道两侧的导流板采用 ABS 制作。在靠近主梁节段模型两端的正下方布置 4 个激光位移计，位移计间距为 0.53 m，以测试主梁断面竖向、扭转振动位移响应，激光位移计采样频率 200 Hz。主梁大比例节段模型试验参数见表 2。图 2 为大比例主梁节段模型风洞试验照片。表 3 为具体试验工况汇总。

2.2　试验结果

2.2.1　允许振动响应根方差

根据《公路桥梁抗风设计规范》(JTG/T 3360-01—2018)第 8.2.9 条规定，成桥状态主梁竖弯涡激共振及扭转涡激共振允许振幅分别为

$$[h_a]=0.04/f_b=0.04/0.479\ 3=0.083\ 5\ \mathrm{m} \tag{1}$$

$$[\alpha_a]=4.56/(Bf_t)=4.56/(43.2\times 0.894\ 5)=0.118^\circ \tag{2}$$

式中：$[h_a]$、$[\alpha_a]$分别为主梁竖弯及扭转涡激共振振幅允许值；f_b、f_t 分别为主梁竖弯、扭转频率。

表 2　1∶30 几何缩尺比主梁节段模型试验参数

Table 2　Experimental Parameters of the Main Deck Section Model with Length Scale of 1∶30

参数	实桥值	缩尺比	模型	
			设计值	实现值
主梁单位长度质量 $m/(\mathrm{kg\cdot m^{-1}})$	36 062	$1/30^2$	40.069	40.382
主梁等效质量惯性矩 $I_m/(\mathrm{kg\cdot m^2\cdot m^{-1}})$	6 587 010	$1/30^4$	8.132	8.057
一阶对称竖弯频率 f_b/Hz	0.479 3	7.39	3.595	3.540
一阶对称扭转频率 f_t/Hz	0.894 5	7.45	6.709	6.665
扭弯频率比 ε	1.866	1	1.866	1.883

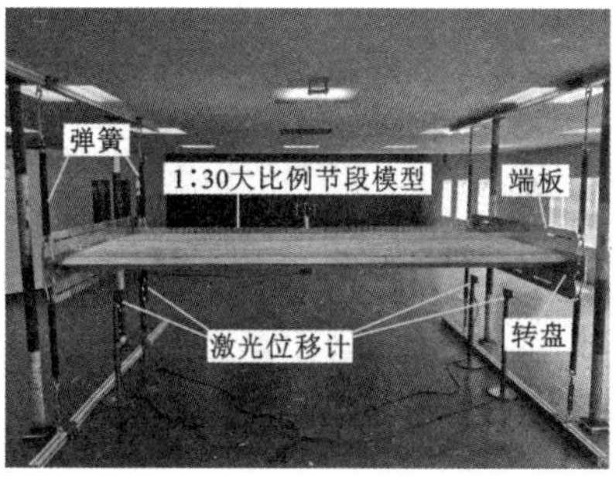

图 2　1∶30 几何缩尺比主梁节段模型

Fig. 2　Main Deck Section Model with Length Scale of 1∶30

表 3　试验工况汇总

Table 3　Summary of Test Cases

工况	措施	竖弯阻尼比/%	扭转阻尼比/%
1	原始断面	0.248	0.168
2	1.2 m 高中央稳定板	0.248	0.168
3	高阻尼	0.430	0.406
4	1.5 m 长水平分离板	0.273	0.193
5	中阻尼	0.322	0.275
6	移动检修车轨道 1.2 m	0.297	0.271

2.2.2　主梁断面原设计方案

图 3 所示为主梁断面原设计方案成桥状态在不同风攻角下竖向、扭转的量纲一振动响应根方差随折算风速的变化曲线。由图 3 可知，在桥梁运营风速范围内(即桥面高度处风速 $V_{deck}\leqslant 25\ \mathrm{m\cdot s^{-1}}$)，在风攻角为 0°时，折算风速区间 $V_{red}=0.57\sim0.65$(实桥 $V_{deck}=22\sim25\ \mathrm{m\cdot s^{-1}}$)主梁断面发生扭转涡激共振现象，对应振幅超限，不满足规范要求；其余风攻角下涡激共振振幅均满足规范要求。

超过桥梁运营风速范围(即桥面高度处风速 $V_{deck}>25\ \mathrm{m\cdot s^{-1}}$)，风攻角为＋3°时，在折算风速区

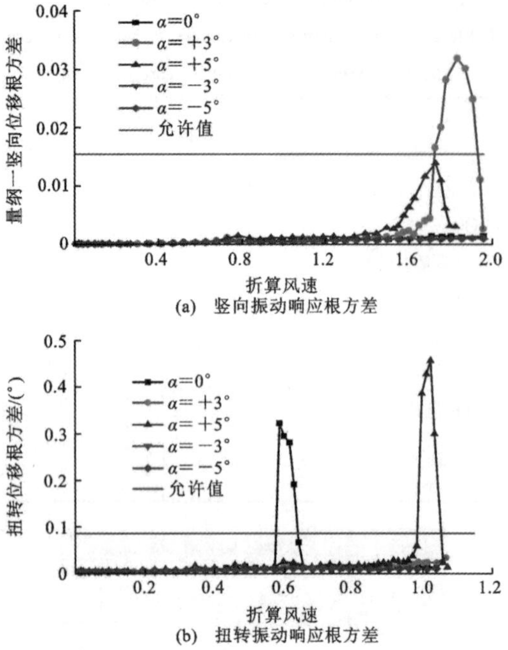

(a) 竖向振动响应根方差

(b) 扭转振动响应根方差

图 3 主梁原设计方案成桥状态量纲一根方差随折算风速变化曲线

Fig. 3 Non-dimensional RMS of the Vortex-induced Vibration Responses of the Original Section of the Main Deck in-service Stage vs. Reduced Wind Speed

间为 $V_{red}=1.74\sim2.00$(实桥 $V_{deck}=36\sim42\ \text{m}\cdot\text{s}^{-1}$)时，主梁发生了竖向涡激共振现象，最大振幅超过规范要求；风攻角为+5°时，在折算风速区间为 $V_{red}=1.59\sim1.84$(实桥 $V_{deck}=33\sim38\ \text{m}\cdot\text{s}^{-1}$)时，发生了明显的竖向涡激共振现象；在折算风速区间为 $V_{red}=0.96\sim1.05$(实桥 $V_{deck}=37\sim40\ \text{m}\cdot\text{s}^{-1}$)时，主梁断面发生了扭转涡激共振现象，且最大振幅超过规范要求。

为进一步减小或消除主梁断面在设计风速范围内的涡激振动现象，对主梁涡激振动气动控制措施进行研究。

2.2.3 不同抑振措施试验结果

针对主梁断面原设计方案成桥状态观测到的明显涡激振动现象，分别进行了中央稳定板(高度 $H=1.20$ m)、水平分离板(长度 $L=1.50$ m)和移动检修车轨道、增加主梁节段模型系统阻尼比等控制措施试验，不同气动控制措施细部构造如图 4 所示。试验模型阻尼比调整是通过在弹性悬挂弹簧上缠绕胶带实现的。

图 5 所示为主梁成桥状态不同气动控制措施下，竖向、扭转量纲一振动响应根方差随折算风速的变化曲线。由图 5(a)、(b)可知，工况 2(增设高度为

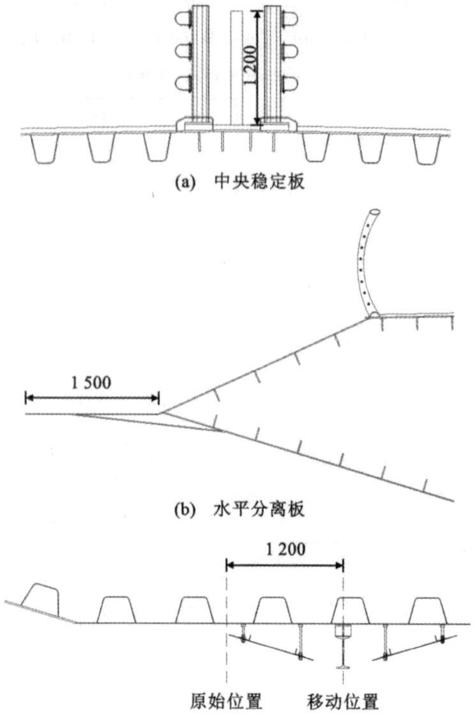

(a) 中央稳定板

(b) 水平分离板

(c) 移动检修车轨道

图 4 不同气动控制措施细部构造示意(单位:mm)

Fig. 4 Schematic Diagram of the Details of Different Aerodynamic Control Measures (Unit: mm)

1.20 m 中央稳定板)在桥梁运营风速范围内时，当风攻角为 0°时，折算风速区间为 $V_{red}=0.6\sim0.7$(实桥 $V_{deck}=23\sim26\ \text{m}\cdot\text{s}^{-1}$)时，出现了明显的扭转涡激共振现象，且扭转涡激共振振幅超限，不满足规范要求。工况 3(增加主梁模型系统阻尼比)在桥梁运营风速范围内，未出现竖向和扭转涡激共振振幅超限。

由图 5(c)、(d)可知，工况 4(原方案风嘴处加设长度为 1.50 m 的水平分离板)在桥梁运营风速范围内，所有风攻角下涡激共振振幅均满足规范要求；但超过桥梁运营风速后，当风攻角为+5°时出现了“锁定”折算风速区间为 $V_{red}=0.70\sim0.84$(实桥 $V_{deck}=27\sim32\ \text{m}\cdot\text{s}^{-1}$)的扭转涡激共振现象，最大振幅超过规范要求。需要说明的是，该风速区间已超过桥梁正常运营的风速限制值 $V_{deck}=25\ \text{m}\cdot\text{s}^{-1}$，且实际情况中桥位处出现 $V=27\ \text{m}\cdot\text{s}^{-1}$ 以上的风速的时段并不长，故对与该“锁定”风速区间下的涡激振动认为可以不予控制，同时考虑到现行《公路桥梁抗风设计规范》(JTG/T 3360-01—2018)中规定仅针对−3°～+3°度范围内进行涡振检验，故本桥可不对+5°攻角进行涡激共振响应振幅的检验。因此可认为加设水平分离板方案的主梁成桥状态涡激

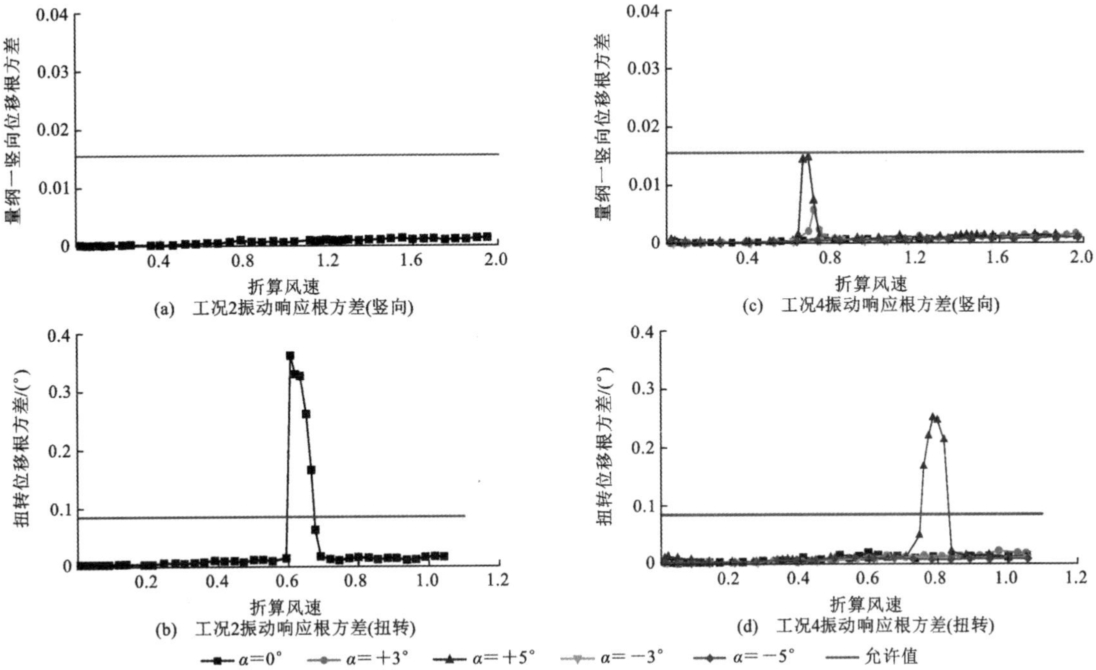

图 5　不同工况主梁成桥状态下量纲一根方差随折算风速变化曲线

Fig. 5　Non-dimensional RMS of the Vortex-induced Vibration Responses of the Main Deck in-service Stage vs. Reduced Wind Speed for Different Cases

共振响应风洞试验结果满足规范要求；但是值得注意的是，在+3°攻角时出现了原方案下未发生的竖向涡激振动，分析可能是水平分离板改变主梁断面气动外形所带来了不明显的负作用，因此该方案可作为一个备选方案。

图 6 所示为工况 5（阻尼比分别为 $\xi_b=0.322\%$、$\xi_t=0.275\%$）对应的主梁竖向、扭转振动量纲一根方差随折算风速变化曲线。由图 6 可知，在桥梁运营风速范围内所有风攻角下涡激共振振幅均满足规范要求。图 7 所示为主梁断面原设计方案施工状态在中等阻尼比条件下，对应的主梁竖向、扭转振动量纲一根方差随折算风速变化曲线。由图 7 可知：当风攻角为 0°时，在折算风速区间为 $V_{red}=0.33\sim0.40$（实桥 $V_{deck}=12.5\sim15.0\ m\cdot s^{-1}$）时，出现扭转涡激共振现象，对应的最大振动响应根方差为 0.15°，超过规范允许振动响应根方差为 0.112°；当风攻角为+3°时，在折算风速区间为 $V_{red}=0.30\sim0.40$（实桥 $V_{deck}=6.0\sim8.0\ m\cdot s^{-1}$）时，主梁断面发生了明显的竖向涡振现象，对应的最大量纲一竖向振动根方差为 0.009 m，小于规范允许值 0.020 m；在折算风速区间为 $V_{red}=0.33\sim0.40$（实桥 $V_{deck}=12.5\sim15.0\ m\cdot s^{-1}$）下出现扭转涡激共振现象，对应的最大振动响应根方差为 0.08°，小于规范允许值 0.112°。

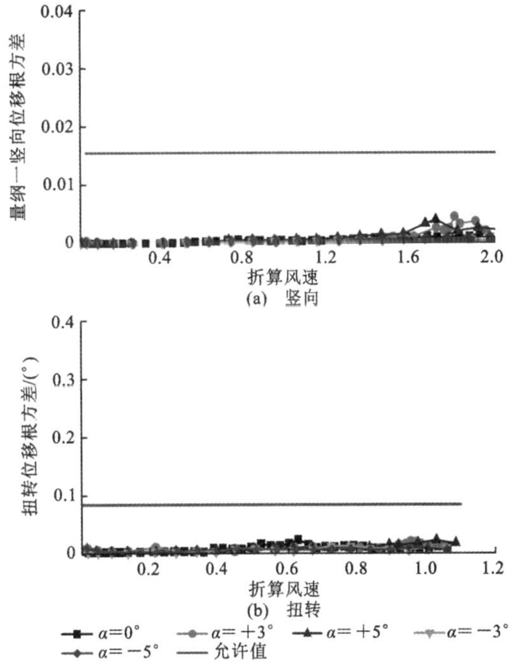

图 6　工况 5 主梁成桥状态下量纲一根方差随折算风速变化曲线

Fig. 6　Non-dimensional RMS of the Vortex-induced Vibration Responses of the Main Deck In-service Stage vs. Reduced Wind Speed for Case 5

综合考虑闭口流线型箱梁涡振性能对附属设施比较敏感。刘志文等[24]、李永乐等[25]验证了检修车

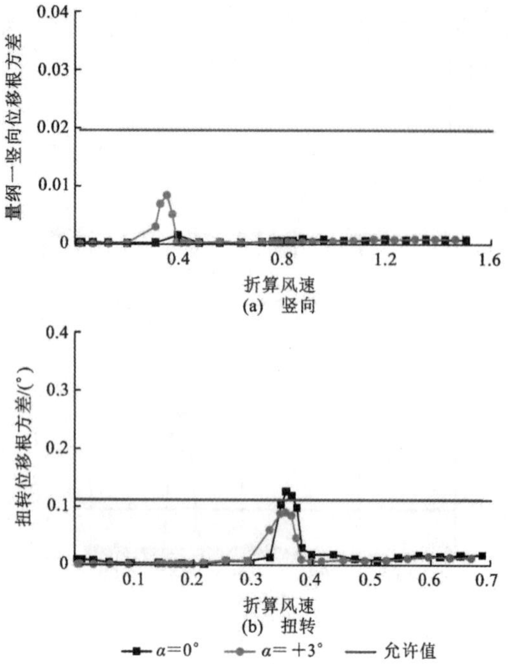

图7 工况5主梁施工状态下量纲一根方差随折算风速变化曲线

Fig. 7 Non-dimensional RMS of the Vortex-induced Vibration Responses of the Main Deck at Construction Stage vs. Reduced Wind Speed for Case 5

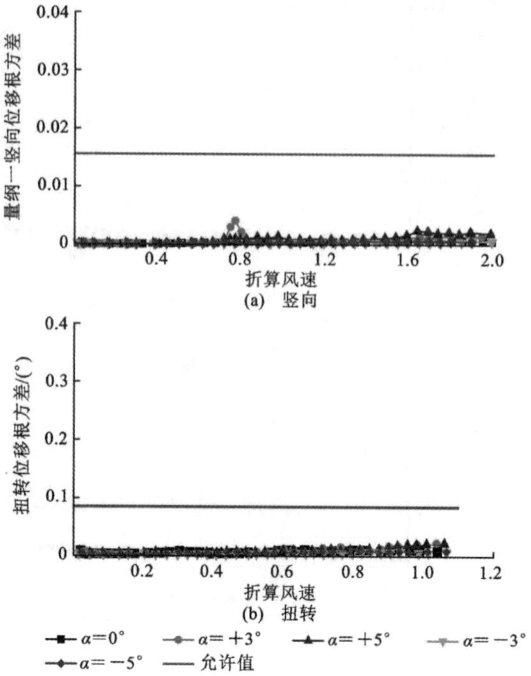

图8 工况6主梁成桥状态下量纲一根方差随折算风速变化曲线

Fig. 8 Non-dimensional RMS of the Vortex-induced Vibration Responses of the Main Deck In-service Stage vs. Reduced Wind Speed for Case 6

轨道内移能有效抑制涡振，故在主梁成桥状态原方案基础上将检修轨道向内移动120 cm，试验中竖弯阻尼比$\xi_b=0.297\%$和扭转阻尼比$\xi_t=0.271\%$，即为工况6。图8所示为工况6（检修车轨道内移120 cm）不同攻角下主梁竖向、扭转振动响应量纲一根方差随折算风速的变化曲线。由图8可知，当风攻角为+3°时，主梁断面发生了小幅竖向涡激共振现象，对应锁定折算风速区间为$V_{red}=0.33\sim0.40$（实桥$V_{deck}=15.0\sim17.0\ \mathrm{m\cdot s^{-1}}$），最大量纲一竖向振动根方差为0.004 m，远小于规范允许值0.016 m，满足规范要求，相比于原始断面+3°攻角下的位移响应降低了85%；其余风攻角下未观测到明显的涡激振动现象。

综上所述，增加主梁阻尼比可有效抑制主梁涡激共振响应，将检修车轨道由原设计位置向内侧移120 cm也可有效控制主梁涡激共振响应。综合考虑推荐采用内移检修车轨道的措施进行主梁涡激共振响应控制。

3 流固耦合数值模拟

3.1 流固耦合计算方法

主梁结构风致振动响应可用两自由度模型来描述，即主梁竖向和扭转两自由度振动方程为

$$m(\ddot{h}+2\xi_b\omega_b\dot{h}+\omega_b^2h)=L \tag{3}$$

$$I(\ddot{\alpha}+2\xi_t\omega_t\dot{\alpha}+\omega_t^2\alpha)=M \tag{4}$$

式中：m为主梁单位长度质量；I为单位长度质量惯性矩；h、α分别为竖向和扭转位移；ξ_b为竖向振动阻尼比；ω_b为竖向振动圆频率；ξ_t为扭转振动阻尼比；ω_t为扭转振动圆频率；L、M分别为升力与力矩。

在直角坐标系下，基于雷诺平均的黏性不可压流体的Navier-Stokes方程可表示为

$$\frac{\partial \bar{u}_i}{\partial x_i}=0 \tag{5}$$

$$\frac{\partial \bar{u}_i}{\partial t}+\frac{\partial}{\partial x_j}(\bar{u}_i\bar{u}_j)=-\frac{1}{\rho}\frac{\partial \bar{p}}{\partial x_j}+\frac{\partial}{\partial x_j}\left(\nu\frac{\partial \bar{u}_i}{\partial x_j}-\overline{u_i'u_j'}\right) \tag{6}$$

式中：ρ、ν分别为空气密度和运动学黏性系数；$\bar{p}$为平均压强；u_i、u_i'分别为直角坐标系x_i方向下的平均速度分量和脉动速度分量；t为时间。

为了同步监测主梁断面在振动过程中的表面压力，本文采用基于ANSYS Fluent的用户自定义函数（UDF）进行二次开发，实现主梁断面流固耦合同步测振测压分析。具体实现方法为：首先根据主梁断面两自由度振动方程确定主梁断面测压点坐标，再通过设置判定条件检索主梁表面上距离压力测点

最近的网格单元，进而通过宏命令 DEFINE_EXECUTE_AT_END 和 F_P 获得每一时间步对应的压力信息，并将压力数据写入文件。主梁断面流固耦合计算流程如图 9 所示。

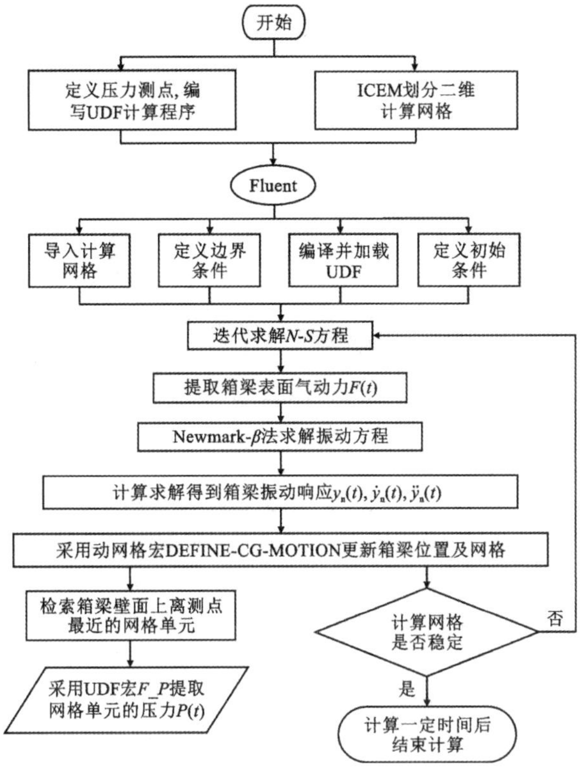

图 9　主梁流固耦合计算及表面压力监测流程

Fig. 9　Flow Chart of Fluid-structures Interaction and Pressure Monitoring of the Main Deck

3.2　计算域及边界条件

以主梁断面成桥状态 0°风攻角为例，进行主梁断面流固耦合数值模拟。结合经验确定计算域为长 $28B$、宽 $16B$ 的矩形区域，主梁模型距离计算域上游边界及上下侧边界均为 $8B$，距离计算域下游侧边界为 $20B$。主梁模型的阻塞率为 $\delta=0.6\%$。计算域边界条件：计算域左侧边界设置为速度入口边界，湍流强度为 0.5%，湍流黏性比为 2.0。计算域右侧边界设置为压力出口边界，对应的湍流强度和湍流黏性比分别为 0.5%和 2.0。计算域上、下侧设置为对称边界。图 10 为主梁断面计算域及网格划分区域分布示意图；图 11 为主梁断面网格划分图，对应的网格总数为 356 834。

采用大型商业软件 ANSYS Fluent 进行流场求解，采用剪应力输运湍流模型（SST k-ω），选择有限体积法求解控制方程，通过任意拉格朗日-欧拉（ALE）描述来适应动网格。空间离散采用二阶迎风格式，时间离散则采用二阶隐式时间积分；采用 SIMPLEC 算法求解压力-速度耦合方程。时间步长统一取为 0.001 s，收敛残差控制在 1×10^{-6}。

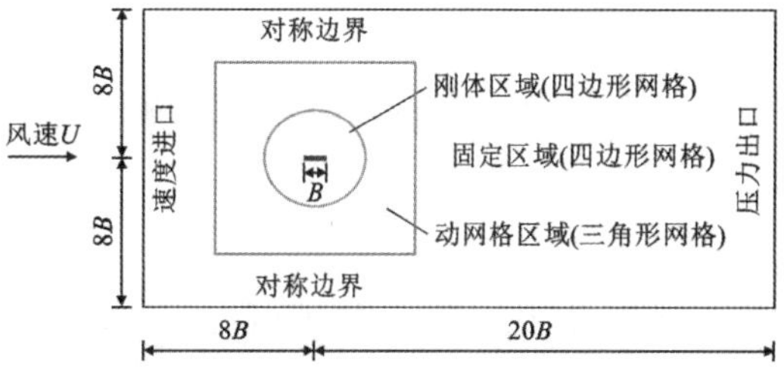

图 10　计算域示意

Fig. 10　Computational Domain

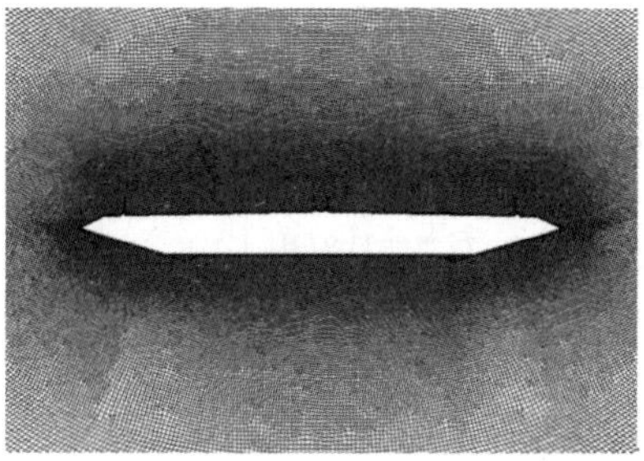

图 11　箱梁整体网格划分

Fig. 11　Global Meshing of Box-girder

3.3　计算结果

3.3.1　静止绕流

在进行流固耦合计算前，先对闭口流线型箱梁断面在风速 $U=10\ \text{m}\cdot\text{s}^{-1}$ 时，进行静止网格三分力计算，得到箱梁断面的静气动力时程，如图 12 所示；再对其进行傅里叶变换得到气动扭矩卓越频率，如图 13 所示；然后，通过式(7)先求得主梁断面 St 数，进而得到原始箱梁断面的量纲一化起振风速为 $V_{\text{red}}=0.585$。

$$St=\frac{fH}{V} \tag{7}$$

式中：St 为主梁断面的斯特罗哈数；f 为涡脱频率；V 为来流风速；H 为梁高。

3.3.2　流固耦合

本文流固耦合参数与风洞试验工况 1（原始断面）试验参数保持一致：模型宽度 B 和高度 H 分别为 1.44 m 和 0.127 m，风攻角 $\alpha=0°$，一阶竖弯频率 3.54 Hz，对应的模态质量 $m=40.382\ \text{kg}\cdot\text{m}^{-1}$，阻尼比 $\xi_b=0.248\%$；一阶扭转频率为 6.66 Hz，对应的模态质量惯性矩 $I_m=8.057\ \text{kg}\cdot\text{m}^2\cdot\text{m}^{-1}$，阻尼比 $\xi_t=0.168\%$。

针对折算风速区间 $V_{\text{red}}=0.0\sim1.05$ 进行了同步测振测压流固耦合数值模拟。图 14 为主梁断面

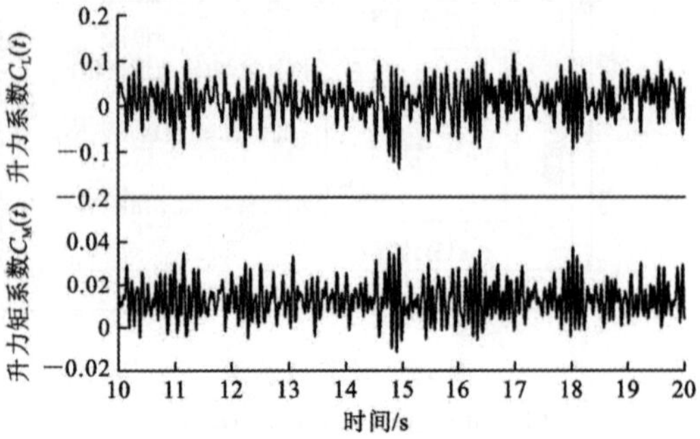

图12 主梁断面气动升力系数、扭矩系数时程曲线

Fig. 12 Time Histories Curves of Aerodynamic Lift and Torque Coefficients of the Main Deck Section

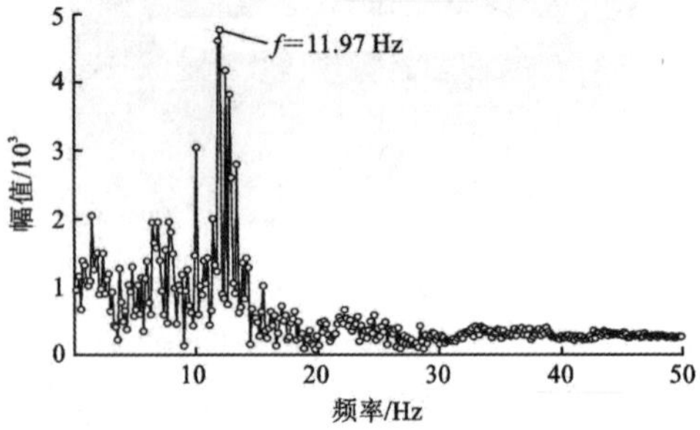

图13 主梁断面气动扭矩系数频谱

Fig. 13 Amplitude Spectrum of the Aerodynamic Torque Coefficients of the Main Deck Section

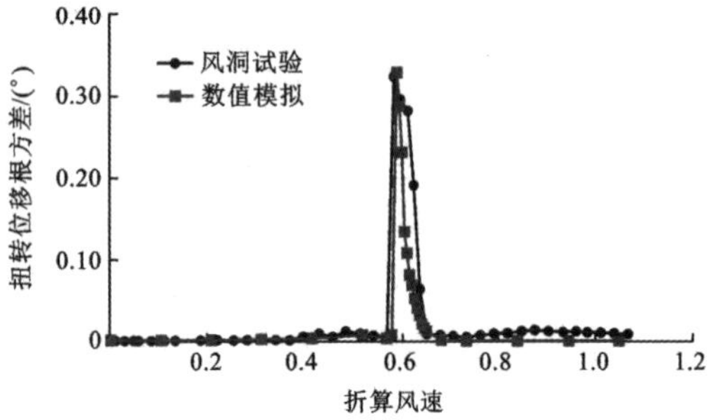

图14 主梁断面扭转振动响应根方差随折算风速变化曲线

Fig. 14 RMS of the Rotational Angles of the Main Deck vs. the Reduced Wind Velocity

成桥状态风攻角为0°时，扭转振动响应根方差随折算风速的变化曲线。由图14可知：闭口流线型主梁断面成桥状态扭转涡振响应锁定区间及最大振幅CFD计算结果与风洞试验结果吻合较好；在折算风速区间为$V_{red}=0.58\sim0.65$时，发生了明显的扭转涡激共振现象，对应的最大振动响应根方差为0.328。图15所示为折算风速$V_{red}=0.588$时，主梁断面扭转振动响应时程曲线与功率谱曲线的CFD数值模拟与风洞试验结果对比，由图15可知两者吻合较好。

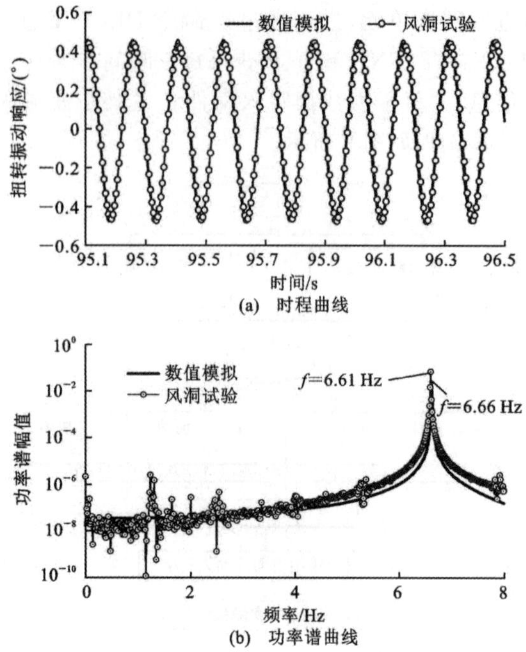

图15 折算风速$V_{red}=0.588$主梁断面扭转振动响应数值模拟和风洞试验时程与功率谱

Fig. 15 Numerical and Experimental Results of Time Histories and Power Spectrum of the Rotational Angles of the Main Deck for Reduced Wind Velocity of $V_{red}=0.588$

4 气动措施机理分析

4.1 宏观流场分析

为了进一步研究闭口流线型主梁断面涡激振动的机理，分别针对主梁断面原设计方案与内移检修车轨道120 cm，在折算风速$V_{red}=0.588$(对应振幅最大)时进行流固耦合数值模拟。图16所示为主梁断面原设计方案、内移检修车轨道120 cm对应的一个振动周期内不同时刻的瞬时涡量演变。

一般而言，具有较高涡量的漩涡是影响箱梁断面流场的主导漩涡，其次就是具有较大尺度的漩涡。由图16(a)可知，主梁断面原方案中气流遇到护栏和检修车轨道都发生了气流分离，均在不连续处产生了高涡量的漩涡脱落，后伴随着一段狭长的涡量削落但尺度较大的漩涡脱落，同时在尾流区域出现交替脱落的“卡门涡街”。值得注意的是，在背风侧风嘴附近还存在着一个涡量较大的漩涡脱落区域。对比一个周期内4个不同时刻的瞬时涡量图，迎风侧前缘的涡量变化程度不大，主要变化区域集中在背风侧的顶板、风嘴和斜腹板位置。在0时刻，气流经过背风侧检修车轨道后产生一个高涡量且尺度较大的漩涡A，在背风侧风嘴尖部出现一个尺度较小但是涡量较高的漩涡B；随着振动的发展和流场的发育，在 $T/4$ 和 $T/2$ 时刻，漩涡 A 的尺度持续变大但涡量削弱，漩涡 B 逐渐向尾部移动，尺度基本保持不变但涡量同样减弱；在 $3T/4$ 时刻，漩涡 A 即将脱落，漩涡 B 则汇入一个附近脱落的狭长漩涡后脱落。

由图 16(b)可以发现：梁底检修车轨道内移 120 cm 后断面的一个周期内，不同时刻的瞬时涡量图基本一致，气流经过背风侧检车轨道之后没有再附着到斜腹板上，而是向尾流水平移动形成一个狭长的漩涡。在背风侧没有发现周期性的漩涡脱落现象，不会对箱梁背风侧下斜腹板产生交替冲击的力，因此涡振消失。由此推断在背风侧发生的周期性的漩涡脱落是诱发箱梁涡振的驱动因素。从流场的宏观角度可以看出：箱梁产生扭转涡振的条件是背风侧的风嘴和斜腹板上出现周期性的漩涡脱落。

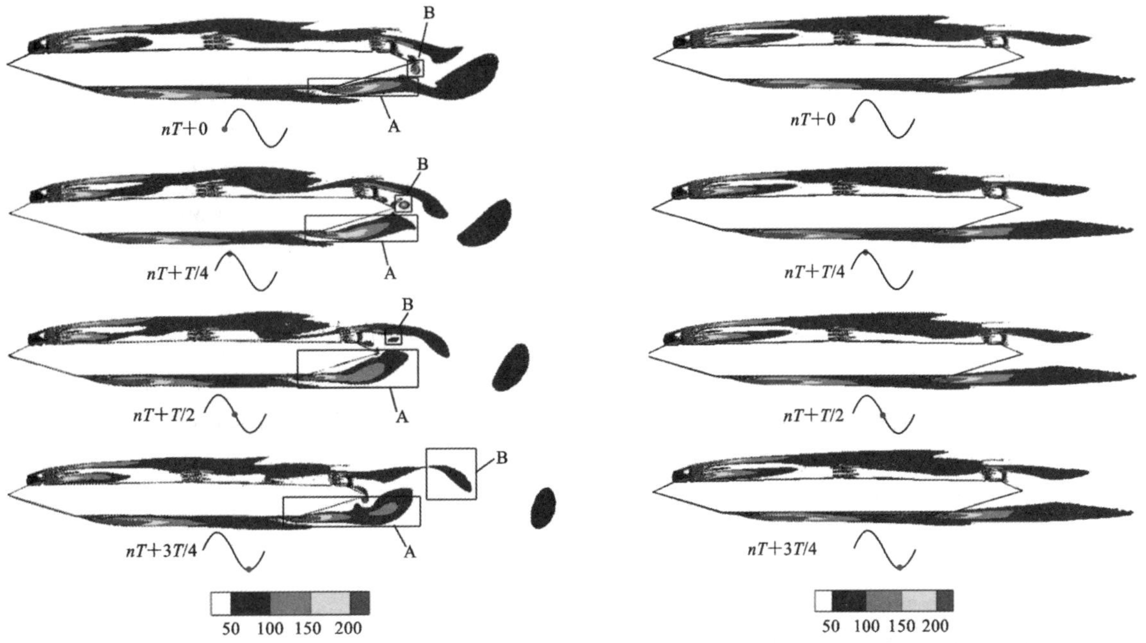

(a) 原始断面瞬时涡量演变

(b) 移动检修车轨道后断面瞬时涡量演变

图 16 移动检修车轨道前后瞬时涡量演变

Fig. 16 Evolution Chart of Instantaneous Vorticity Before and After Moving the Inspection Rails

4.2 气动力与位移响应相位差

采用 CFD 方法可以方便地获得主梁断面气动力与位移响应，本节重点分析主梁断面位移响应与总气动力之间的相位差，从能量的角度进行主梁断面涡振机理解释。图 17 所示分别为主梁断面原设计方案、检修车轨道内移 120 cm 后，在折算风速为 $V_{red}=0.588$ 时主梁断面气动力与位移的相位差。由图 17 可知：主梁断面原设计方案气动扭矩和扭转位移之间相位差为 21°，气动扭矩对结构做正功；而移动检修车轨道后，气动扭矩与扭转位移之间的相位角接近 0°，气动扭矩基本不做功。

对主梁断面原设计方案涡振的发展过程进行同样的相位差分析，具体分为上升段、限幅段和稳定段 3 个阶段进行分析，如图 18 所示。图 19 所示为主梁断面扭矩和扭转位移相位差随着振动发展的演变过程。从图 19 可以看出，上升段、限幅段和稳定段 3 个阶段的相位差分别为 39°、28°和 17°，表明主梁断面涡激共振响应随时间的增加气动力做功效率逐渐减小，直至振幅稳定。

4.3 表面压力分析

主梁断面总气动力能够反映结构整体变化的过程，而分析表面压力能够反映气动力的局部变化过程，还能够反映断面气体的绕流状态，从更为细观的角度来研究涡振特性及其抑制机理。目前，主梁断面表面压力一般通过风洞试验压力采集装置获得，但是存在诸如测点布置空间受限、不同测点采集延迟等现象，而采用 CFD 则可以很好地解决这些问题。箱梁原始断面和内移检修车轨道 120 cm 后的表面测压点布置仅在检修车轨道位置发生变化，其余位置压力测点编号均未改变，具体测点布置如图 20 所示。

主梁断面第 i 测点压力系数定义为

$$C_{pi}(t)=\frac{p_i(t)}{0.5\rho U^2} \tag{8}$$

式中：$C_{pi}(t)$ 为测点压力系数时程；$p_i(t)$ 为测点压力时程；U 为来流风速。

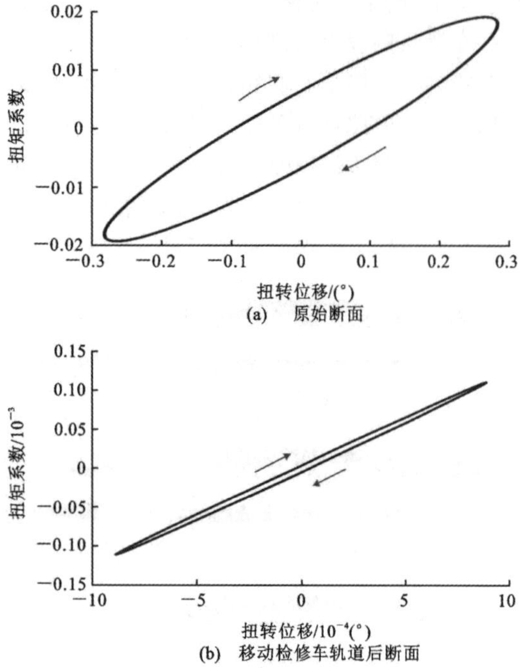

图 17 主梁断面移动检修车轨道前后相位差

Fig. 17 Phase-difference Diagram of Torque and Rotational Angle of the Main Deck Before and After the Moving the Inspection Rails

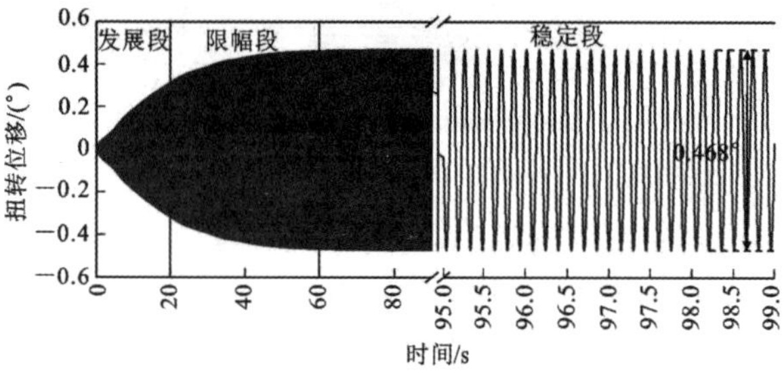

图 18 折算风速 $V_{red}=0.588$ 时主梁断面成桥状态扭转振动位移时程

Fig. 18 Time Histories of Rotational Angle of the Main Deck for Reduced Wind Velocity of Reduced Wind Velocity of $V_{red}=0.588$

图 21 所示为折算风速 $V_{red}=0.588$ 下，移动检修车轨道前后 2 个断面各测点的平均压力系数分布。位于箱梁内部的压力轮廓线表示指向箱梁的正压力，位于箱梁外部的压力轮廓线表示背向箱梁的负压力，即遵循“内正外负”的原则。从图 21 可以看出，移动检修车轨道前后 2 个断面的平均压力系数的空间分布变化趋势基本一致，压力正值只分布在迎风侧风嘴处且出现正压峰值，随着来流气流流动，压力正值逐渐减小。在主梁断面原设计方案中，上表面气流遇到迎风侧检修道护栏的阻挡发生气流分离，在转角处急剧变化由正变负，后又遇到外侧防撞护栏的阻挡，发生更加强烈的气流分离，在此处出现一个负压峰尖，经过迎风侧的防撞护栏后的压力负值绝对值减小，气流表现出再附的趋势。在中间防撞护栏处压力出现一小段波动上升后，又表现为气流再附的趋势，压力负值绝对值减小，分离活动降低。当气流经过背风侧的防撞护栏和检修道护栏之后，保持稳定的压力系数。下表面气流在遇到迎风侧检修车轨道后，压力系数出现波动，且伴随出现一个负压峰尖，代表着此处气流分离程度较大，底板区域的气流也表现出再附的趋势，呈现负压绝对值减小，经过背风侧检修车轨道后压力系数保持稳定。

相比于主梁断面原设计方案，将梁底检修车轨道内移 120 cm 后，主梁断面表面平均压力系数绝对值在迎风侧斜腹板与底板交界处急剧下降，表明此处气流发生再附，经过迎风侧检修车轨道后出现剧烈波动。同时，梁底迎风侧、背风侧检修车轨道之间距离减小，使得底板处气流的再附程度增大，背风侧处的气流分离活动提前发生，导致背风侧下斜腹板的压力系数幅值出现明显的减小，分离强度降低。对比移动检修车轨道前后主梁断面附近压力特征可以发现，区别只存在于底板和背风侧下斜腹板，说明移动检修车轨道后主要改变了主梁下表面的平均压力系数分布特性，进而导致主梁涡振响应减小。

结构表面压力频谱可反映压力脉动的频率特征，卓越频率则反映了压力变化的主导频率。数值模拟计算结果显示，原始断面各测点压力卓越频率均与涡激振动频率一致；而移动检修车轨道后，结构表面不存在一个统一的卓越频率，无法激发整体结构涡振效应，故振动消失。

为探究主梁涡激振动与其表面压力特性之间的联系，特对主梁表面压力演变过程进行分析。图 22(a)为原始断面涡振振幅稳定后一个周期内不同时刻的压力系数，可以明显看到原始断面同一周期不同时刻的压力分布有较大差距，而移动检修车轨道后断面不同时刻的压力分布基本相同，压力脉动值非常小，且不存在一个固定的周期。内移检修车轨道 120 cm 后某一时刻的压力分布如图 22(b)所示。由图 22(a)、(b)可知：在迎风侧的风嘴、顶板以及底板的区域，同一周期下不同时刻的压力系数变化规律基本趋于均匀变化，整体的趋势和移动检修车轨道后断面的压力分布相似，仅在数值上产生差异，可认为此区域是由运动状态决定，称为“被动区域”；而在背风侧的风嘴和斜腹板区域不同时刻的压力系数有出现交替变化的现象，可以看到背风侧斜腹板位

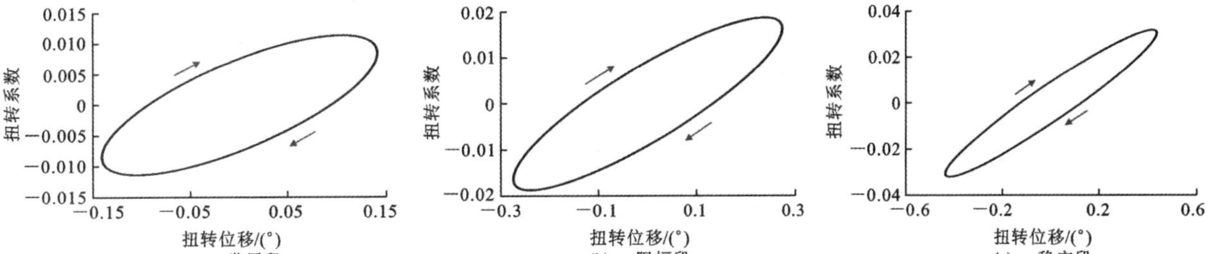

(a) 发展段　(b) 限幅段　(c) 稳定段

图 19　主梁断面原设计方案扭矩与扭转位移相位差演变

Fig. 19　Evolution of the Phase-difference Between Torque and Rotational Angle of the Original Main Deck

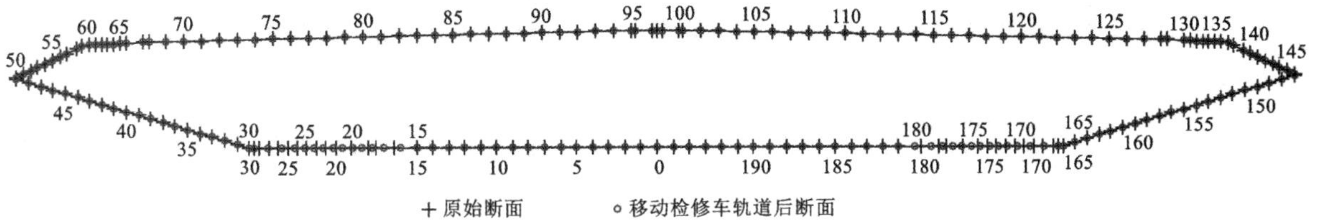

图 20　主梁断面原始断面和移动检修车轨道后断面的压力测点布置

Fig. 20　Layout of Pressure Measuring Points for the Original Section of the Main Girder and the Section of Moving the Inspection Rails

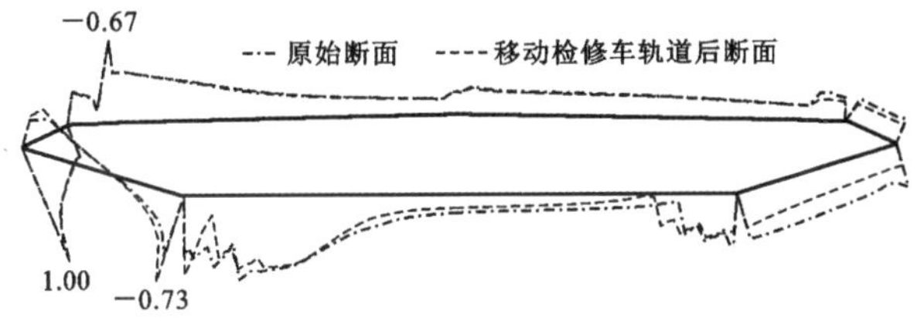

图 21　移动检修车轨道前后断面平均压力系数空间分布

Fig. 21　Spatial Distribution of Average Pressure Coefficient Before and After Moving the Maintenance Rails

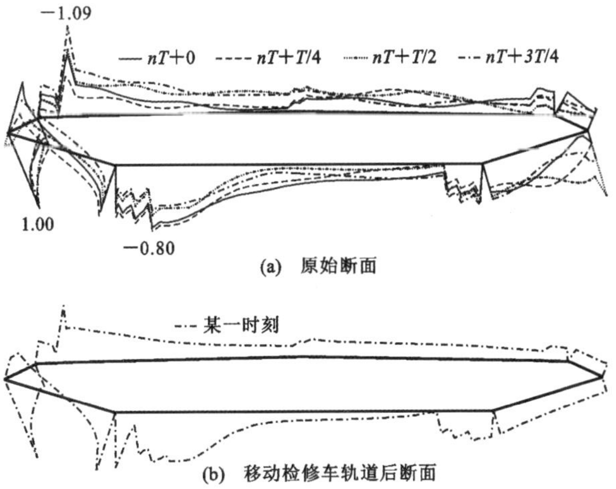

图 22　移动检修车轨道前后断面一个周期内不同时刻压力系数空间分布

Fig. 22　Spatial Distribution of Pressure Coefficient at Different Time in a Cycle of Original Section Before and After Moving the Inspection Rails

置表现出明显的不均匀性，具有较大的脉动压力，可推断是涡振的“驱动区域”。

综上所述，箱梁扭转涡振发生的内在机理是：原始断面在背风侧斜腹板与底板交界处正好设有与斜腹板同样角度的检修车轨道导流板，能够使底板处分离的漩涡平滑过渡向“驱动区域”移动，与其他漩涡相互碰撞而产生周期性脱落的漩涡，打乱了此区域分布气动力的均匀性，对箱梁背风侧下斜腹板产生了交替冲击作用。移动检修轨道之后，导流板也相应向内移动导致结构的气动高度降低，使箱梁更趋于扁平化，在“驱动区域”的气流分离活动减弱，交界处出现较大的转角变化导致漩涡在此处没法向“驱动区域”转移，而是向下游开阔区域移动，因此就不会产生周期性脱落的漩涡，所以涡振消失。

5　结　语

针对大跨度桥梁涡激振动问题，以实际工程为依托，采用风洞试验与 CFD 数值模拟相结合的方法，对闭口流线型钢箱梁涡激振动特征、气动控制措施及机理进行了研究，得到如下主要结论：

(1)闭口流线型主梁断面原设计方案在风攻角分别为 0°、+3°和 +5°时出现了较为明显的涡激共振现象，将主梁断面梁底检修车轨道内移 120 cm 后其涡振性能满足规范要求；

(2)结合动网格技术与 ANSYS Fluent 用户自定义函数(UDF)，建立了主梁断面流固耦合数值模拟方法，实现了主梁断面涡激振动响应与主梁表面压力同步计算；

(3)闭口流线型箱梁原设计断面背风侧风嘴和斜腹板上出现周期性的漩涡脱落是其发生扭转涡振的主要原因，将梁底检修车轨道内移 120 cm 后，有效减小了主梁断面背风侧风嘴和斜腹板上出现周期性的漩涡脱落，从而有效抑制了闭口流线型箱梁断面涡振响应。

(4)本文仅采用几何缩尺比 1∶30 的主梁节段模型进行了主梁涡激振动响应特征、气动控制措施和机理研究，考虑因素较为单一；考虑到主梁涡激振动影响因素较为复杂，如来流风特性、模型缩尺比、结构阻尼比等，今后将结合现场实测、大比例节段模型测振、测压试验以及 CFD 数值模拟等方法进一步开展研究。

参考文献：

References：

[1] 陈政清，黄智文．大跨度桥梁竖弯涡振限值的主要影响因素分析[J]．中国公路学报，2015，28(9)：30-37.
CHEN Zheng-qing，HUANG Zhi-wen. Analysis of main factors influencing allowable magnitude of vertical vortex-induced vibration of long-span bridges [J]. China Journal of Highway and Transport，2015，28 (9)：30-37.

[2] MACDONALD J，IRWIN P A，FLETCHER M S. Vortex-induced vibrations of the second severn crossing cable-stayed bridge-full-scale and wind tunnel measurements [J]. Structures & Buildings，2003，152 (2)：123-134.

[3] LARSEN A，ESDAHL S，ANDERSEN J E，et al. Storebaelt suspension bridge-vortex shedding excitation and mitigation by guide vanes [J]. Journal of Wind Engineering and Industrial Aerodynamics，2000，88 (2)：283-296.

[4] FUJINO Y，YOSHIDA Y. Wind-induced vibration and control of trans -Tokyo bay crossing bridge [J]. Journal of Structural Engineering，2002，128 (8)：1012-1025.

[5] LI H，LAIMA S J，OU J P，et al. Investigation of vortex-induced vibration of a suspension bridge with two separated steel box girders based on field measurements [J]. Engineering Structures，2011，33 (6)：1894-1907.

[6] 葛耀君，赵　林，许　坤．大跨桥梁主梁涡激振动研究进展与思考[J]．中国公路学报，2019，32(10)：1-18.
GE Yao-jun，ZHAO Lin，XU Kun. Review and reflection on vortex-induce vibration of main girders of long-span bridges [J]. China Journal of Highway and Transport，2019，32 (10)：1-18.

[7] LARSEN A，LAROSE G L. Dynamic wind effects on suspension and cable-stayed bridges [J]. Journal of Sound and Vibration，2015，334：2-28.

[8] 赵　林，李　珂，王昌将，等．大跨桥梁主梁风致稳定性被动气动控制措施综述[J]．中国公路学报，2019，32(10)：34-48.
ZHAO Lin，LI Ke，WANG Chang-jiang，et al. Review on passive aerodynamic countermeasures on main girders aiming at wind-induced stabilities of long-span bridges [J]. China Journal of Highway and Transport，2019，32 (10)：34-48.

[9] 许福友，林志兴，李永宁，等．气动措施抑制桥梁涡振机理研究[J]．振动与冲击，2010，29(1)：73-76.
XU Fu-you，LIN Zhi-xin，LI Yong-ning，et al. Study on the mechanism of aerodynamic measures to suppress bridge vortex vibration [J]. Journal of Vibration and Shock，2010，29 (1)：73-76.

[10] 管青海，李加武，胡兆同，等．栏杆对典型桥梁断面涡激振动的影响研究[J]．振动与冲击，2014，33(3)：150-156.
GUAN Qing-hai，LI Jia-wu，HU Zhao-tong，et al. Effects of railings on vortex-induced vibration of a bridge deck section [J]. Journal of Vibration and Shock，2014，33 (3)：150-156.

[11] 朱思宇，李永乐，申俊昕，等．大攻角来流作用下扁平钢箱梁涡振性能风洞试验优化研究[J]．土木工程学报，2015，48(2)：79-86.
ZHU Si-yu，LI Yong-le，SHEN Jun-xin，et al. Optimization of vortex-induced vibration of flat steel box girders at large attack angle by wind tunnel test [J]. China Civil Engineering Journal，2015，48 (2)：79-86.

[12] 陈政清，华旭刚，牛华伟，等．永磁电涡流阻尼新技术及其在土木工程中的应用[J]．中国公路学报，2020，33(11)：83-100.
CHEN Zheng-qing，HUA Xu-gang，NIU Hua-wei，et al. Technological innovations in eddy current damping and its application in civil engineering [J]. China Journal of Highway and Transport，2020，33 (11)：83-100.

[13] PAIDOUSSIS M P，PRICE S J，LANGRE E D. Fluid-structure interactions：Cross-flow-induced instabilities [M]. Cambridge：Cambridge University Press，2011.

[14] HU C X，ZHAO L，GE Y J. Mechanism of suppression of vortex-induced vibrations of a streamlined closed-box girder using additional small-scale components [J]. Journal of Wind Engineering and Industrial Aerodynamics，2019，189：314-331.

[15] SCHEWE G，LARSEN A. Reynolds number effects in the flow around a bluff bridge deck cross section [J]. Journal of Wind Engineering and Industrial Aerodynamics，1998，74 (2)：829-838.

[16] 金　挺，林志兴．扁平箱形桥梁断面斯特罗哈数的雷诺数效应研究[J]．工程力学，2006，23(10)：174-179.
JIN Ting，LIN Zhi-xing. Reynolds number effect on Strohal number of flat-box girder bridge decks [J]. Engineering Mechanics，2006，23 (10)：174-179.

[17] 李加武．桥梁断面雷诺数效应及其控制研究[D]．上海：同济大学，2003.
LI Jia-wu. Reynolds number effect and its control on bridge decks [D]. Shanghai：Tongji University，2003.

[18] LI H，LAIMA S J，ZHANG Q Q，et al. Field monitoring and validation of vortex-induced vibrations of a long-span sus-

pension bridge [J]. Journal of Wind Engineering and Industrial Aerodynamics, 2014, 124: 54-67.

[19] CHEN X Y, WANG B, ZHU L D, et al. Numerical study on surface distributed vortex-induced force on a flat-steel-box girder [J]. Engineering Applications of Computational Fluid Mechanics, 2017, 12 (1): 1-16.

[20] ZHANG Z T, ZHANG X X, GE Y J. Motion-induced vortex shedding and lock-in phenomena of a rectangular section [J]. Nonlinear Dynamics, 2020, 102 (3): 1-14.

[21] LAIMA S J, LI H, CHEN W L, et al. Investigation and control of vortex-induced vibration of twin box girders [J]. Journal of Fluids and Structures, 2013, 39 (5): 205-221.

[22] 刘 钥,陈政清,张志田.箱梁断面静风力系数的CFD数值模拟[J].振动与冲击,2010,29(1):133-137.
LIU Yao, CHEN Zheng-qing, ZHANG Zhi-tian. CFD numerical simulation of static wind coefficient of box girder section [J]. Journal of Vibration and Shock, 2010, 29 (1): 133-137.

[23] 祝志文,石亚光,颜 爽.带防撞栏杆扁平箱梁高阶模态涡激振动的CFD研究[J].振动与冲击,2021,40(10):228-234.
ZHU Zhi-wen, SHI Ya-guang, YAN Shuang. Investigation on vortex-induced high-mode vibration of flat box girders with crash barriers based on CFD [J]. Journal of Vibration and Shock, 2021, 40 (10): 228-234.

[24] 刘志文,洪 涵,梁立农,等.广东江顺大桥抗风性能试验研究[J].湖南大学学报(自然科学版)2015,42(3):112-119.
LIU Zhi-wen, HONG Han, LIANG Li-nong, et al. Experimental investigation of the wind-resistance performance of jiang-shun bridge in Guangdong [J]. Journal of Hunan University (Natural Science), 2015, 42 (3): 112-119.

[25] 李永乐,候光阳,向活跃.大跨度悬索桥钢箱梁主梁涡振性能优化风洞试验研究[J].空气动力学学报,2011,29(6):702-708.
LI Yong-le, HOU Guang-yang, XIANG Huo-yue. Optimization of the vortex-induced vibration for steel box girder of long-span suspension bridges by wind tunnel test [J]. Acta Aerodynamica Sinica, 2011, 29 (6): 702-708.

Mechanism and Aerodynamic Countermeasures of Vortex-Induced Vibration of a Cable-Stayed Bridge with Narrow Π-Shaped Girder Sections

Zhiwen Liu[1]; Fawei He[2]; Aiguo Yan[3]; Zhenbiao Liu[4]; Tao Yin[5]; and Zhengqing Chen, M.ASCE[6]

Abstract: In this study, the investigations on the mechanism and aerodynamic countermeasures (ACs) of vortex-induced vibration (VIV) for a narrow Π-shaped girder (NG) section with the aspect ratio of $B/D = 5$ (B is the width and D is the height of the main deck) are conducted. First, the wind tunnel tests on VIV of the NG section with/without ACs are carried out. Furthermore, numerical simulations of VIV of the NG section with/without AC-VII, namely combined AC of modified fairings and inclined deflectors, for wind attack angle of +3° are conducted via fluid–structure interaction, respectively. Moreover, the mechanism of VIV of the NG section is analyzed from the characteristics of vorticity, pressure, and vortex-induced force (VIF) distributions, respectively. The research results show that the original NG section exhibits obvious vertical and torsional VIV for wind attack angles of +3° and +5°, respectively. Moreover, the ACs of low vertical stabilizers and horizontal separators, which are commonly applied to suppress the VIV responses of the wide Π-shaped girder section with aspect ratio $B/D = 10$–14, cannot effectively suppress the VIV responses of the NG section with $B/D = 5$. However, the AC-VII can effectively suppress the VIV responses. For the original NG section, the shear layer separation vortices that are generated on the windward side will extend to the leeward of the section and form the periodic shedding vortex, which is the mechanism of the VIV of the NG section. However, the AC-VII effectively suppresses the scale and vorticity intensity of the upper dominant vortex, avoiding the appearance of the vortex acceleration region at the leeward of the NG section. Therefore, the vorticity intensity of the upper and lower dominant vortexes is reduced during the shedding process, and the peak value of the fluctuating pressure in the VIV's driving region is reduced by 86%. **DOI: 10.1061/JBENF2.BEENG-6285.**

Author keywords: Narrow Π-shaped girder section; Vortex-induced vibration (VIV); Mechanism; Aerodynamic countermeasures; Wind tunnel tests; Fluid–structure interaction.

Introduction

The composite steel–concrete Π-shaped girders, which consist of two longitudinal steel plate girders, diaphragms, and concrete slabs, are widely used for cable-supported bridges because of their excellent mechanical performance and efficiency. The composite steel–concrete girder was adopted as the main deck of the Annacis Bridge in Canada, which was built in 1986 (Taylor and Torrejon 1987). Now, the composite steel–concrete girders have become the most widely used girder for the main deck of long-span cable-supported bridges, such as Nanpu Bridge in Shanghai, and Parrot Island Yangtze River bridge in Wuhan of China, Baytown Bridge in Texas, USA, and Rion-Antirion Bridge in Greece (Scanlan and Jones 1990; Zhu et al. 2000; Oliveira and Reis 2016). However, the aerodynamic performance of the Π-shaped girder section is slightly worse than that of the streamlined-box girder. Moreover, vortex-induced vibration (VIV) is prone to occur and the flutter critical wind velocity is relatively low compared with the streamlined-box girder section (Owen et al. 1996; Macdonald et al. 2002; Li et al. 2018b; Huang et al. 2022).

Several investigations have been conducted for mechanism and aerodynamic countermeasures (ACs) of VIV of the wide Π-shaped girder (WG) sections with aspect ratios of $B/D = 10$–14. The results show that the large-scale separation vortices are formed on the windward side of the WG section when the air flow passes through the WG section, which cause the obvious VIV responses of the WG section. Furthermore, many ACs were investigated to suppress the VIV responses of the WG section. The fairings proposed by Wardlaw (1972) can better improve the aerodynamic shape of the main deck section, which accelerate the velocity of the airflow and reduce the flow separation. Therefore, the fairings can effectively weaken the scale of the shear layer separation vortex (Meng et al. 2011; Li et al. 2022d), and they were used to suppress the VIV responses of the Longs-Creek bridge in Canada. The vertical edge plates can also improve the aerodynamic shape of the main deck section, allowing the incoming flow pass through the main deck section more smoothly and reducing the scale and intensity of the vortex (Zhang et al. 2020; Bai et al. 2021), and they were applied to suppress the VIV responses of the main deck of the Ting

[1]Hunan Provincial Key Lab for Wind and Bridge Engineering, College of Civil Engineering, Hunan Univ., Changsha 410082, China (corresponding author). ORCID: https://orcid.org/0000-0002-0900-6420. Email: zhiwenliu@hnu.edu.cn
[2]College of Civil Engineering, Hunan Univ., Changsha 410082, China. Email: faweihe@hnu.edu.cn
[3]China Railway Siyuan Survey and Design Group Co., Ltd., Wuhan 430063, China. Email: 1261309610@qq.com
[4]China Railway Siyuan Survey and Design Group Co., Ltd., Wuhan 430063, China. Email: liuzhenb@sohu.com
[5]China Railway Siyuan Survey and Design Group Co., Ltd., Wuhan 430063, China. Email: 95363938@qq.com
[6]Hunan Provincial Key Lab for Wind and Bridge Engineering, College of Civil Engineering, Hunan Univ., Changsha 410082, China. Email: zqchen@hnu.edu.cn
Note. This manuscript was submitted on January 14, 2023; approved on August 10, 2023; published online on September 22, 2023. Discussion period open until February 22, 2024; separate discussions must be submitted for individual papers. This paper is part of the ***Journal of Bridge Engineering***, © ASCE, ISSN 1084-0702.

Kau Bridge, Hong Kong, China. The concave region between the side longitudinal steel plate girders of the WG provides enough space for the formation and development of a vortex, and the lower stabilizers can effectively break up the large-scale vortex on the lower surface of the WG into a small-scale vortex to reduce the contribution of the pressure on the lower surface of the main deck to the vortex-induced force (VIF) (Katsuchi et al. 2009; Li et al. 2022a), and the lower stabilizers have been applied on the Yumenkou Yellow River Bridge in China and Quincy Bridge in USA. Setting horizontal separators inside the side longitudinal steel plate girders can also effectively suppress the development of a vortex in the concave region and reduce the pressure pulsation on the lower surface of the main deck (Li et al. 2019), and the horizontal separators on the outside of the side longitudinal steel plate girders can effectively suppress the formation of a *secondary vortex* (Li et al. 2022c). However, with the construction of railway cable-stayed bridges, more and more narrow Π-shaped girders (NGs) will appear in the future, and the ACs that have a better VIV suppression effect on WGs may not necessarily suppress the VIV of NGs, so it is necessary to conduct an in-depth study on the mechanism and ACs of VIV of NGs.

As a typical section, the aerodynamic performances of the rectangular section have been investigated by many researchers, and the results show that the aerodynamic performances of rectangular sections with various aspect ratios are quite different, which has important reference values for investigations on the mechanism and ACs of VIV of the NG section. Shiraishi and Matsumoto (1983) studied the aerodynamic characteristics of the rectangular sections with different aspect ratios. They found that the separated vortex at the leading edge of a rectangular section with a small aspect ratio will wrap the entire section and generate a *Von Kármán vortex street* at the trailing edge. However, the separation vortex at the leading edge of a rectangular section with a large aspect ratio generated a *secondary vortex* at the trailing edge. Nakamura and Nakashima (1986) conducted a study on the VIV characteristics of ├-shaped section prisms and rectangular prisms with different aspect ratios. The results showed that the VIV on the section with a small aspect ratio had a higher critical onset speed. Kubo et al. (2001) investigated the VIV characteristics of a Π-shaped section with different opening sizes (C/D, C is the distance from the center of the side girder to the edge of the main deck and D is the height of the main deck) using the flow trace display technology, and they found that the upper separation bubble is smaller for the section with a large opening size and it is prone to vibrate when the two plate girders are placed near the ends of the floor. Overall, the aerodynamic performances of rectangular sections with different aspect ratios are quite different. Furthermore, it is reasonable to infer that the Π-shaped girder sections with different aspect ratios will also have obvious different aerodynamic characteristics. Therefore, it is necessary to study the aerodynamic performance of the NG section.

The analysis of flow field and surface pressure characteristics is widely applied to study the mechanism of VIV responses of structures in wind engineering, as evidenced by the works of Yuan et al. (2017), Ma et al. (2019), Li and He (2020), and Wang et al. (2022). In particular, Li et al. (2011) conducted a mechanism analysis of full-scale vortex-induced vibration of a bridge with twin box girders based on bridge deck pressure distribution data obtained from field measurements. Furthermore, they utilized computational fluid dynamics (CFD) to conduct a detailed analysis of the pressure distributions, energy transfer between the flow and motion of the body, and evolution of flow patterns over the vortex-induced vibration process of the twin box girder and proposed corresponding vibration suppression measures, and an automatic identification of bridge VIVs from huge monitoring data sets was first achieved by clustering analysis (Laima et al. 2013; Li et al. 2017b). Therefore, in this paper, CFD will be utilized to obtain the flow field around the structure and the surface pressure of the structure to investigate the VIV mechanism of the NG section.

In this study, the investigation on the mechanism and ACs of VIV for an NG section with an aspect ratio of $B/D = 5$ is conducted. First, the wind tunnel tests of an NG section for different wind attack angles are carried out. Then, the ACs for VIV of the NG section are investigated. Furthermore, the numerical simulation and mechanism analysis of the VIV and its ACs of the NG section are conducted, respectively. Finally, a discussion and a summary of the results are presented.

Engineering Background

Jiya waterway railway bridge (JYB) is a cable-stayed bridge with steel–concrete composite girders for central span and concrete girders for side spans under construction, which crosses the Jiya waterway in Guangzhou, China. The span arrangement of the JYB is (50 + 70 + 70 + 408 + 70 + 70 + 50) $m = 788$ m. The Π-shaped steel–concrete composite girder is 20.0 m wide and 4.0 m high. Moreover, the heights of the pylons of JYB are 135.6 and 139.6 m, respectively. The main deck is supported by 128 stay cables of Φ7 mm galvanized parallel steel wire bundles, which are arranged in four inclined planes of 16 stay cables at each of two H-shaped pylons. The anchorage spacings of the stay cables at the main deck and pylon are 12.0 and 2.0–2.5 m, respectively. The Π-shaped steel–concrete composite girder is used for the central span of the JYB, which is 20.0 wide and 4.0 m high. Considering that the Π-shaped section is prone to VIV, the original fairings are designed on both sides of the main deck. The overall layout and standard cross section of the main deck of JYB are shown in Figs. 1 and 2, respectively. The basic wind speed at the bridge site is 38.5 m/s, and the design reference wind speed at the main deck height of the JYB in-service stage is 44.5 m/s.

The structural dynamic characteristics of the JYB in-service stage are calculated by commercial software ANSYS (version 15.0). The main deck and piers are modeled by BEAM188 and BEAM4, respectively, and the pylons and stay cables are modeled by BEAM188 and LINK10, respectively. The mass of diaphragms, rails, and other ancillary facilities is modeled by MASS21. The structural constraints of the main bridge are set as follows: the pylons and piers are fixed at the bearing platform, the main deck is vertically supported on two sides by the bridge pylons and piers, and laterally supported on one side by the bridge pylons and piers, as shown in Fig. 3. The finite-element method (FEM) model of the JYB in-service stage is shown in Fig. 3. Fig. 4 shows the first symmetry and antisymmetry vertical bending, lateral bending, and torsional mode shapes of the main deck of the JYB in-service stage. As shown in Fig. 4, the frequencies of the first symmetry vertical, lateral bending, and torsional modes of the main deck are 0.3936, 0.6126, and 0.7841 Hz, respectively. The equivalent mass per unit length of the ith mode of the bridge structure is shown as follows:

$$m_i{}^* = \frac{\int_B m(s)\phi_i^2(s)ds}{\int_g \phi_i^2(s)ds} \tag{1}$$

where m_i^* = equivalent mass per unit length of the ith mode; $\int_B$ = integration of the full bridge structure, including the pylons, stay-cables, and piers; $\int_g$ = integral of the main deck; $m(s)$ = mass per unit length of the full bridge structure, including the main deck,

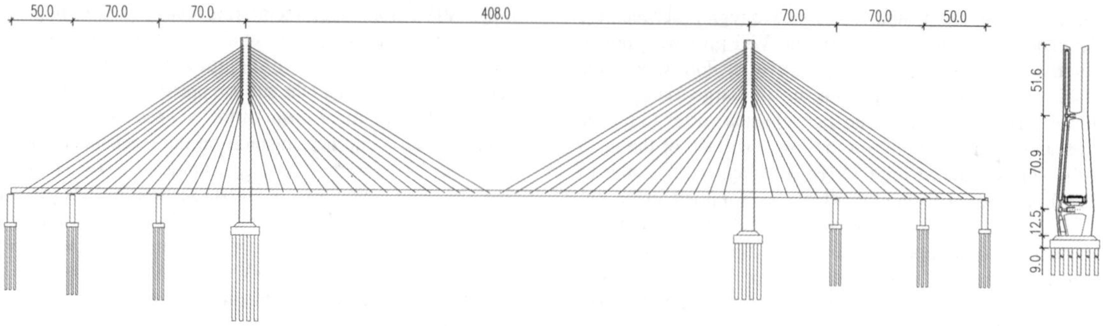

Fig. 1. Elevation layout of JYB (unit: m).

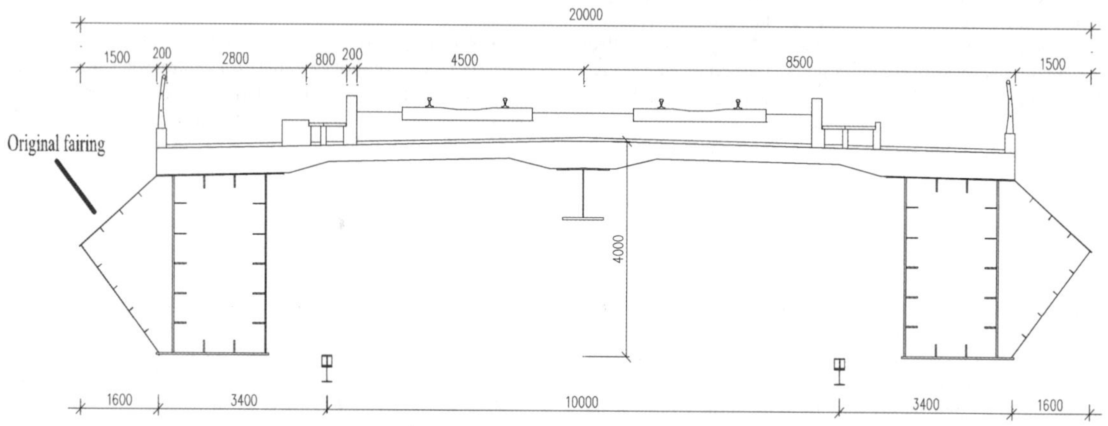

Fig. 2. Standard cross section of a Π-shaped girder for the main span of JYB (unit: mm).

pylons, stay-cables, and piers; and $\phi_i(s) = i$th mode shape function of the bridge structure.

Experimental Study of VIV

Experimental Setup

The wind tunnel tests of the main deck section model are carried out in Section I of the HD-2 wind tunnel in Hunan University, which is 3.0 m wide, 2.5 m high, and 17.0 m long. Furthermore, the turbulence intensity and the maximum wind velocity in the empty wind tunnel are approximately 0.5% and 58.0 m/s, respectively. Considering the Reynolds number effect, model blocking ratio, model geometry, and other factors, the geometric scaling ratio of the section model is finally determined as $\lambda_L = 1{:}60$. The blockage ratios of the model in the wind tunnel test at the angle of attack of 0°, ±3°, and ±5° are 2.67%, 3.43%, and 4.40%, respectively. The length, width, and height of the section model are $L = 1{,}733$ mm, $B = 333$ mm, and $D = 67$ mm, respectively. Two laser displacement sensors (IL-300) supplied by KEYENCE CO., Ltd. are set below the center of the section model with a spacing of 230 mm, and one laser displacement sensor (IL-300) is set near the end of the section model. The F.S. of the laser displacement sensor IL-300 is ±20 mm, and the linear error of IL-300 is 0.25%, F.S. = ±0.05 mm. The data are measured with a DH5920 multichannel signal acquisition system developed by Donghua Testing Technology Co., Ltd. The sampling frequency of the displacement sensor is 1,000 Hz, and the sampling time is approximately 60 s. The experimental photo of the main deck section model is shown in Fig. 5, and the specific design parameters and measured parameters are shown in Table 1. The experimental cases are given in Table 2. Fig. 6 shows the arrangement of ACs.

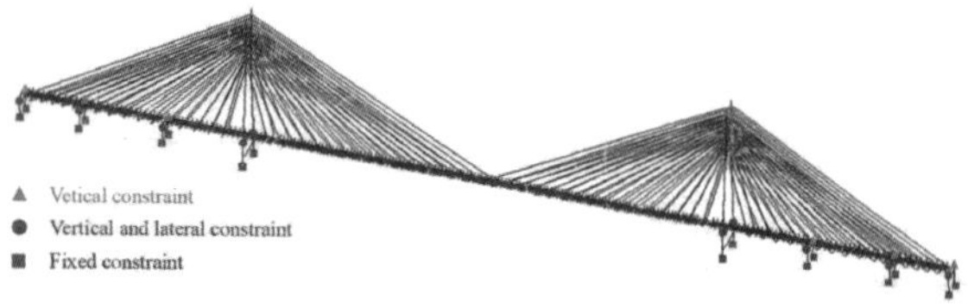

Fig. 3. FEM model of the JYB in-service stage.

VIV Response of the Original Main Deck

Allowable Amplitudes

According to the Chinese code of "Wind-Resistant Design Specifications for Railway Bridges" (Draft edition for comments), the allowable amplitudes of the vertical and torsional VIV of the main deck of the JYB in-service stage are determined as follows:

$$[h_v]_{\text{railway}} = \frac{0.01}{f_v^2 + \gamma_n\left(\frac{V}{L}\right)^2} = \frac{0.01}{0.3936^2 + 0.5 \times \left(\frac{69.44}{408}\right)^2} = 0.0590\ \text{m} \tag{2}$$

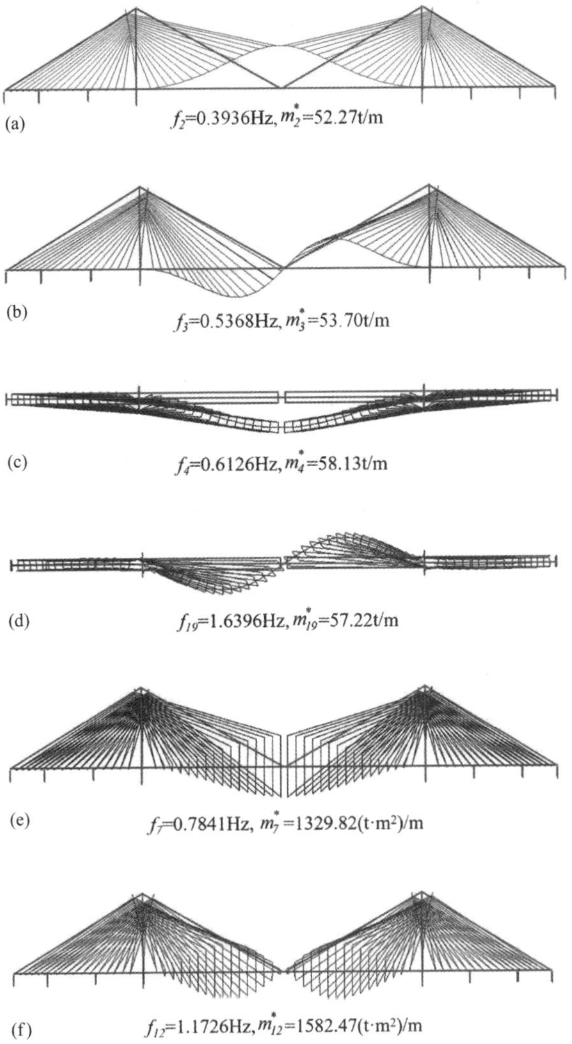

Fig. 4. Typical mode shapes of the main deck of the JYB in-service stage: (a) 1-S-V of the main deck; (b) 1-A-V of the main deck; (c) 1-S-L of the main deck; (d) 1-A-L of the main deck; (e) 1-S-T of the main deck; and (f) 1-A-T of the main deck. S and A = symmetry and antisymmetry mode shape, respectively, and V, L, and T = vertical, lateral, and torsional mode shapes, respectively.

$$[\theta_t]_{\text{railway}} = \frac{1}{f_t^2 + \gamma_n\left(\frac{v}{L}\right)^2} \cdot \frac{1.14}{B} = \frac{1}{0.7841^2 + 0.5 \times \left(\frac{69.44}{408}\right)^2} \times \frac{1.14}{20} = 0.1139^\circ \quad (3)$$

In addition, the corresponding root mean square (RMS) of the vertical and torsional VIV displacements are as follows:

$$[\sigma_{va}]_{\text{railway}} = \frac{0.0590}{\sqrt{2}} = 0.0417\,\text{m} \quad (4)$$

$$[\sigma_{ta}]_{\text{railway}} = \frac{0.1139^\circ}{\sqrt{2}} = 0.0805^\circ \quad (5)$$

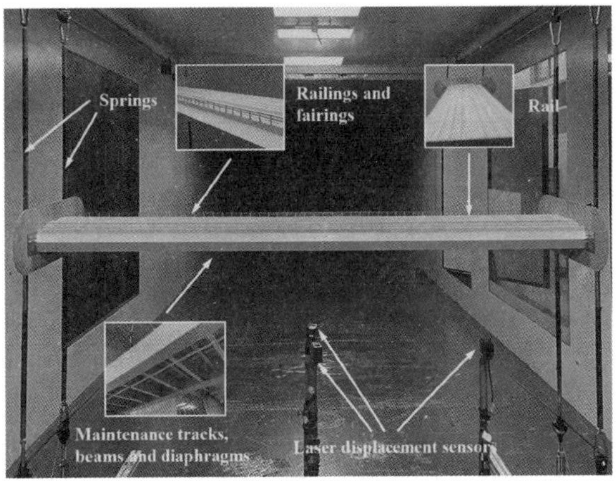

Fig. 5. Experimental photo of the main deck model supported by springs.

where $[h_v]$ and $[\theta_t]$ = allowable amplitudes of the vertical and torsional VIV displacements, respectively; $[\sigma_{va}]$ and $[\sigma_{ta}]$ = allowable RMS of the vertical and torsional VIV displacements; f_v and f_t = vertical and torsional frequencies of the main deck of JYB (Hz); B = width of the cross section (m); L = length of the main span (m); v = speed of the train (m/s); and γ_n = VIV mode shape coefficient, and the first-order symmetrical mode shape is taken as 0.5.

VIV Performance

In order to study the VIV performance of the original NG section, the wind tunnel tests of the original NG section model are conducted in uniform flow for attack angles of −5°, −3°, 0°, +3°, and +5°, respectively. The RMS of the vertical and torsional displacements of the original NG section versus reduced wind velocity for different wind attack angles are shown in Fig. 7. It should be noted that in Fig. 7, the abscissa represents the reduced wind velocity $U^* = U/f_vB$ or $U^* = U/f_tB$, and the ordinate represents the RMS of the vertical displacement $\sigma^* = \sigma/D$ or RMS of the torsional displacement φ, where U is the incoming wind velocity, B and D are the width and height of the NG section model, respectively. f_v and f_t are the vertical and torsional frequencies of the NG section model, respectively. As shown in Fig. 7, when the wind attack angles are +3° and +5°, the vertical VIV amplitude exceeds the limits of the railway bridge specification, and when the wind attack angle is +5°, the torsional VIV amplitude exceeds the limit of railway bridge specification.

Aerodynamic Countermeasure Effects

Many research results have shown that the large-scale vortex shedding on the upper and lower surfaces is an important reason for VIV of the WG section. The lower stabilizers (Murakami et al. 2002; Chen et al. 2019), the lower horizontal separators (Zhang et al. 2019; Li et al. 2022b), the fairings (Wardlaw 1990; Huang et al. 2021), and the deflectors (Hu et al. 2020; Li et al. 2022e) are commonly used to control the vortex shedding at the lower and upper parts of the WG section. Therefore, the lower stabilizers, horizontal separators, modified fairings, and inclined deflectors are applied to suppress the VIV responses of the NG section in this study. Fig. 8 shows the RMS of the vertical and torsional displacements of the NG section with ACs versus reduced wind velocity for Cases 6–11. As shown in Figs. 8(a and b), setting one central lower stabilizer (Case 6) or two lower stabilizers (Case 7) on the NG section

Table 1. Experimental parameters of the main deck section model

Parameter	Unit	Prototype values	Scale ratio	Model values		
				Design values	Measured values	Relatively error (%)
Width, B	m	20.0	1:60	0.33	0.33	0.00
Height, D	m	4.0	1:60	0.067	0.067	0.00
Mass per unit length, M	kg/m	52270	1:60^2	14.52	14.72	1.38
Mass moment of inertia per unit length, I_m	kg·m^2/m	1329820	1:60^4	0.103	0.101	1.94
Vertical frequency, f_v	Hz	0.3936	1:12	4.72	4.64	1.69
Torsional frequency, f_t	Hz	0.7841	1:12	9.41	9.52	1.17
Vertical damping ratio, ζ_b	%	0.65	1:1	0.65	0.64–0.67	−1.5 to −4.69
Torsional damping ratio, ζ_t	%	0.65	1:1	0.65	0.64–0.68	−1.5 to 6.25

cannot effectively suppress the vertical VIV responses of the NG section. However, the torsional VIV response of the main deck can be effectively controlled. From Figs. 8(c and d), it can be seen that setting one horizontal separator (Case 8) cannot effectively suppress the vertical and torsional VIV responses of the NG section; however, this increases the maximum RMS value of torsional displacement from 0.6° to 0.7°. Moreover, setting of two horizontal separators (Case 9) on the NG section can effectively control the torsional VIV responses, while it cannot effectively suppress the vertical VIV responses. It can be seen from Figs. 8(e and f) that the modified fairings (Case 10) can effectively suppress the vertical and torsional VIV responses of the NG section, and the maximum RMS of the vertical and torsional VIV responses are reduced from 0.02 and 0.07° to 0.007 and 0.04°, respectively. Moreover, the inclined deflectors (Case 11) can also effectively control the vertical and torsional VIV responses of the main deck, and the maximum RMS values of the vertical and torsional VIV responses are reduced from 0.02 and 0.07° to 0.006 and 0.02°, respectively.

To further investigate the VIV control effect of the AC-V and AC-VI, the wind tunnel tests of the NG section for wind attack angles of −5° to +5° are conducted. Fig. 9 shows the RMS of vertical and torsional displacements of the NG section with AC-V and AC-VI versus reduced wind velocity under wind attack angles of −5° to +5°. It can be seen from Figs. 9(a and b) that setting AC-V can effectively suppress the vertical and torsional displacements of the NG section for wind attack angles of −5° to +3°. However, the RMS of the vertical and torsional VIV responses of the NG section exceed the limit value by the railway specification under a wind attack angle of +5°. As shown in Figs. 9(c and d), the control effect of AC-VI on VIV responses of the NG section is more significant than that of AC-V, and the RMS values of the vertical and the torsional VIV of the NG section under wind attack angles of −5° to +5° are less than the limit value by the railway specification. However, it should be noted that the RMS value of torsional VIV is relatively large for a wind attack angle of 5°. Therefore, the wind tunnel tests of the main deck with the combined ACs of *modified fairings and inclined deflectors* (AC-VII) are conducted. The RMS of the vertical and torsional displacements of the main deck with AC-VII under wind attack angles of −5° to +5° versus reduced wind velocity are shown in Figs. 9(e and f). From Figs. 9(e and f), it can be seen that the vibration suppression effect of AC-VII is better than that of a single AC. Therefore, AC-VII is adopted as the recommended AC for the NG section of JYB.

Based on the investigation on the aerodynamic countermeasures for VIV suppression on the NG section, the following conclusions can be drawn: (1) According to scholars' research on the vibration suppression effect of flow deflectors, the typical inclination angle

Table 2. Test cases of the NG section model w/o ACs

Case no.	Deck w/o AC	Wind angle of attack
1	Without AC	−5°
2	Without AC	−3°
3	Without AC	0°
4	Without AC	+3°
5	Without AC	+5°
6	One lower stabilizer (AC-I)	+3°
7	Two lower stabilizers (AC-II)	—
8	One horizontal separator (AC-III)	—
9	Two horizontal separators (AC-IV)	—
10	Modified fairings (AC-V)	—
11	Inclined deflectors (AC-VI)	—
12–15	Modified fairings (AC-V)	−5°, −3°, 0°, +5°
16–19	Inclined deflectors (AC-VI)	−5°, −3°, 0°, +5°
20–24	Modified fairings and inclined deflectors (AC-VII)	−5°, −3°, 0°, +3°, +5°

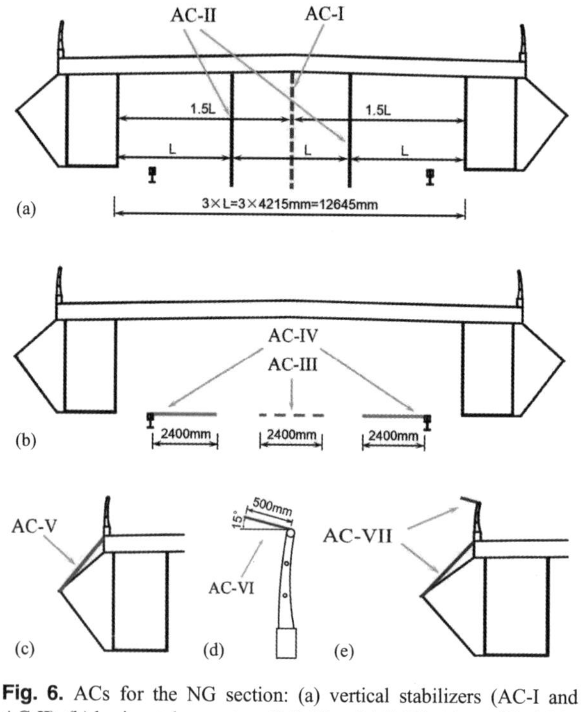

Fig. 6. ACs for the NG section: (a) vertical stabilizers (AC-I and AC-II); (b) horizontal separators (AC-III and AC-IV); (c) modified fairings (AC-V); (d) inclined deflectors (AC-VI); and (e) modified fairings and inclined deflectors (AC-VII).

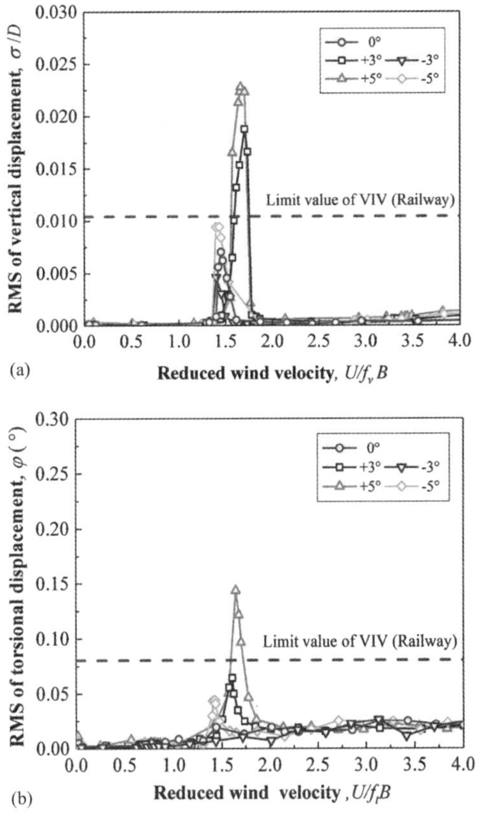

Fig. 7. RMS of VIV responses of the original NG section (Cases 1–5): (a) vertical VIV response; and (b) torsional VIV response.

of deflectors is 15°, and the width of flow deflectors ranges from 0.3 to 2.0 m. Generally, a wider width leads to a better vibration suppression effect, and the specific value can be determined based on the actual situation (Fang et al. 2017; Li et al. 2017a; Li et al. 2022e); and (2) when there is a significant step at the intersection of the fairing and the bridge deck, modifying the fairing to cover the existing step can improve the aerodynamic shape of the main deck section.

Numerical Simulation and Mechanism Analysis

Numerical Model

The numerical simulations of the NG section with AC-VII under a wind attack angle of +3° are conducted to study the mechanism of the vertical VIV and its AC of the NG section. The numerical simulation parameters are summarized in Table 3.

The computational domain is selected as a rectangular region with a length of $30B$ and a width of $20B$, where B is the width of the NG section. The blocking ratio of the NG model is $\delta = 1.0\%$, which is less than the required value of 5.0% (Saathoff and Melbourne 1987). The boundary conditions of the computational domain are defined as follows: the left and lower boundaries of the computational domain are set as the velocity inlet, in which the turbulent intensity and viscosity ratios are set to 0.5% and 2.0, respectively. The right and upper boundaries of the computational domain are set as the pressure outlet, and the NG section model is set as a no-slip wall condition. The block meshing method is adopted to categorize the computational domain into, namely the rigid domain, dynamic grid domain, and static grid domain. Fig. 10 shows the computational domain and boundary conditions of the NG section model. The structural grids are adopted to mesh the computational domain, and the Y^+ is less 5 near the main deck section. Fig. 11 shows the grids near the NG section model, and the total number of corresponding meshes is 433156. The shear-stress transportation k-ω turbulence model is adopted to elucidate the Navier–Stokes equations in the time domain (Menter 1994). Moreover, the adaptive dynamic grid is described by the arbitrary Lagrangian–Eulerian. The SIMPLEC second-order implicit algorithm is adopted for spatial discretization of pressure and momentum, respectively, and the convergence residual is set as 10^{-5}.

The NG section model is simplified as a single-degree-of-freedom (SDF) system, which is mounted by springs with viscous dashpots appropriate to the structural mode of vibration under consideration. The structure equation of motion for the dynamic behavior of a damped linear vibration system with SDF is as follows:

$$m \cdot \ddot{y}(t) + c \cdot \dot{y}(t) + k \cdot y(t) = F_y(t) \tag{6}$$

where m, c, and k = mass, damping, and stiffness of the NG section per unit length in vertical bending mode, respectively; $\ddot{y}(t)$, $\dot{y}(t)$, and $y(t)$ = acceleration, velocity, and displacement of the NG section, respectively; and $F_y(t)$ = VIF on the NG section. The Newmark-β method was applied to solve Eq. (10). The fluid–structure interaction (FSI) numerical simulation is carried out for the reduced wind velocity range $U^* = V/f_vB = 0.0$–4.0.

To further study the mechanism of the VIV and its control mechanism of the NG section, the pressure distribution on the surface of the NG section with/without AC-VII should be analyzed. However, it is difficult to measure the pressure distribution of the NG section via wind tunnel tests because of the difficulty of the pressure taps' layout. Therefore, the pressure distribution on the surface of the NG section with/without AC-VII is monitored via numerical simulation based on commercial software ANSYS Fluent. The grids on the surface of the NG section are retrieved by checking the condition, and the surface pressures of the NG section during the vibration are monitored via user-defined functions of the ANSYS Fluent (version 15.0) (Liu et al. 2022a). Fig. 12 shows the distribution of the 648 measuring taps on the surface of the NG section with AC-VII. Moreover, the surface of the NG section is divided into 10 regions (A–J). As shown in Fig. 12, p_i is the pressure at a measuring tap, and F is the VIF of the NG section, where p_{ih} and p_{iv} represent the horizontal and vertical components of tap pressure, respectively; the range of each region has been distinguished by letters (A–J).

Numerical Results of VIV of the Original Main Deck Section

The independency tests of the gird size and time step are performed first for incoming wind velocity of 7.0 m/s and +3 wind angle of attack, which is consistent with the parameters of wind tunnel tests. Tables 4 and 5 show the numerical results of the main deck section for different meshes and time steps, respectively. From Tables 4 and 5, it can be seen that the aerostatic coefficients of the main deck section corresponding to Grid 2 and Grid 3 are generally close. Moreover, the horizontal and vertical force coefficients of the main deck section corresponding to the three time steps are relatively close; however, there are obvious differences for the vertical force coefficient for the three time steps. Considering that the

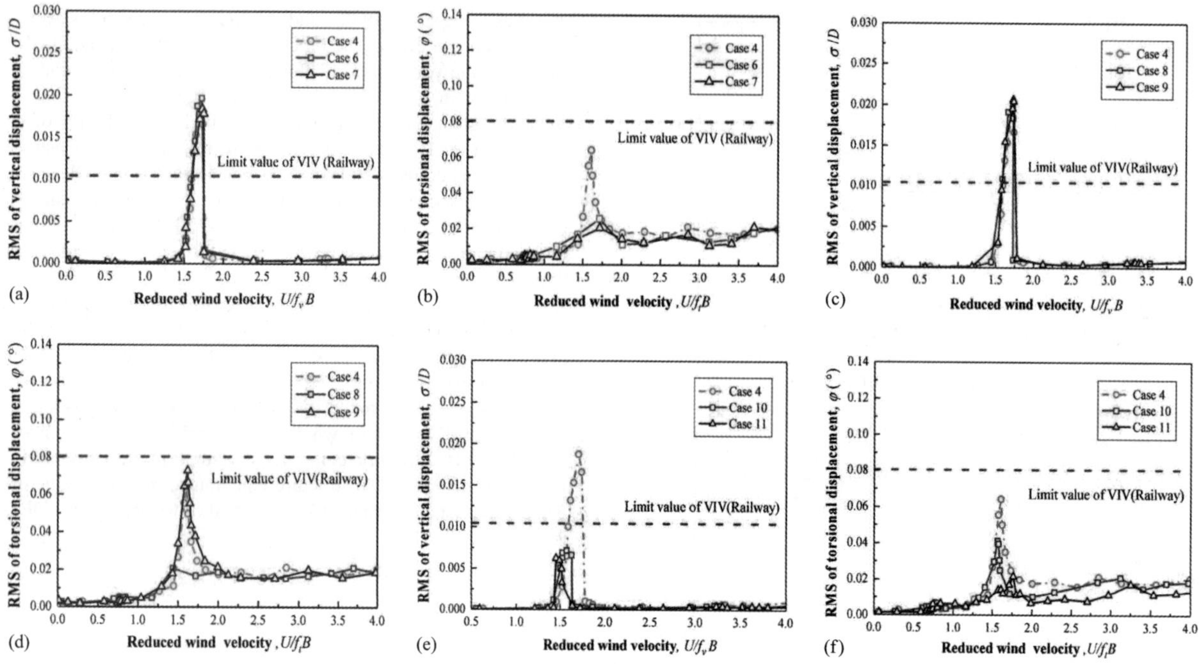

Fig. 8. RMS of the VIV responses of the NG section with AC-I to AC-VI: (a) vertical VIV response for AC-I and AC-II; (b) torsional VIV response for AC-I and AC-II; (c) vertical VIV response for AC-III and AC-IV; (d) torsional VIV response for AC-III and AC-IV; (e) vertical VIV response for AC-V and AC-VI; and (f) torsional VIV response for AC-V and AC-VI.

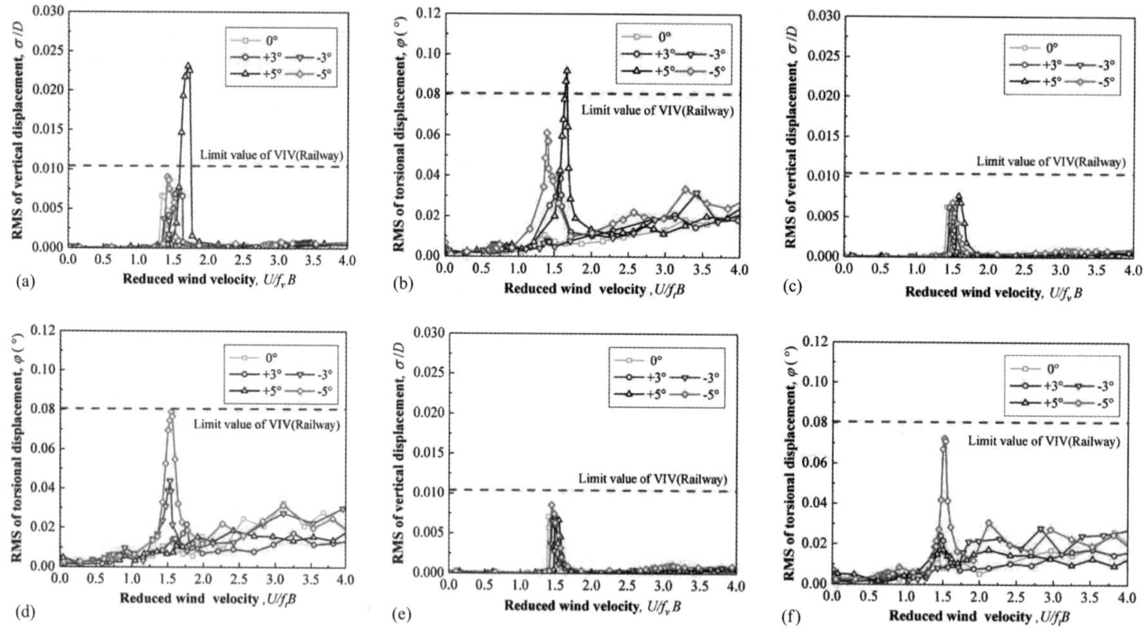

Fig. 9. RMS of VIV responses of the NG section with AC-V to AC-VII vs. reduced wind velocity: (a) vertical VIV response for AC-V; (b) torsional VIV response for AC-V; (c) vertical VIV response for AC-VI; (d) torsional VIV response for AC-VI; (e) vertical VIV response for AC-VII; and (f) torsional VIV response for AC-VII.

numerical results of the vertical force coefficient of the main deck section for a time step of 0.0004 s are in good agreement with the experimental results, it is determined to adopt Grid 2 and a time step of 0.0004 s to conduct the numerical simulation of flow around the stationary main deck section. For the subsequent fluid–structure interaction calculation, the time step of 0.001 s is adopted.

To validate the pressure monitoring method utilized in this study, the aerostatic coefficients of the NG section with AC-VII

Table 3. Numerical simulation parameters of the NG section

Parameter	Unit	Prototype value	Scale ratio	Model value
Width, B	m	20.0	1:60	0.3333
Height, D	m	4.0	1:60	0.0667
Mass per unit length, M	kg/m	52270	$1:60^2$	14.724
Vertical frequency, f_v	Hz	0.3936	$1:60^4$	4.639
Vertical damping ratio, ξ_v	%	0.65	1:1	0.64

are computed for a wind attack angle of +3° and a wind velocity of 7 m/s at the time step of 0.0004 s. Table 6 presents the numerical and experimental results of aerostatic coefficients of the NG section with AC-VII. As presented in Table 6, the numerical results of the aerostatic coefficients based on pressure integration agree well with the FLUENT results, which also agree well with the experimental results of aerostatic coefficients of the NG section except pitch force coefficient. There is a certain difference between the experimental results and the numerical results of the pitch coefficients. The main reason may be that the pitch coefficient of the main deck is relatively small, and there are certain experimental errors, such as low amplitude vibration and wind attack angle errors, which may result in a certain difference between the experimental results and the numerical results.

Figs. 13(a–c) show the nondimensional displacement time history of the main deck for reduced wind velocities of 1.39, 1.68, and 1.84, which correspond to the onset, peak, and ending reduced wind velocity of the lock-in region of the main deck with a geometric scale ratio of 1:60, respectively. Fig. 13(d) shows the power spectrum of the displacement of the experimental ($U^* = 1.48$) and numerical results ($U^* = 1.68$) of the main deck with a geometric scale ratio of 1:60. The RMS of the vertical displacement of the original NG section of the numerical and experimental results are given in Fig. 13(e).

As shown in Fig. 13(e), the numerical results of the RMS of the vertical displacement of the original NG section are in good agreement with the experimental results. Nevertheless, the amplitude of VIV experiences a gradual increase followed by a rapid decrease for CFD simulations. Conversely, the wind tunnel tests indicate a distinct trend. The possible reason may be that the VIV amplitude of the NG section with AC-VII is generally small, and some inevitable test errors, such as damping ratio, incoming turbulence, and wind attack angle of the sectional model, may cause certain differences between the experimental results and CFD calculation results.

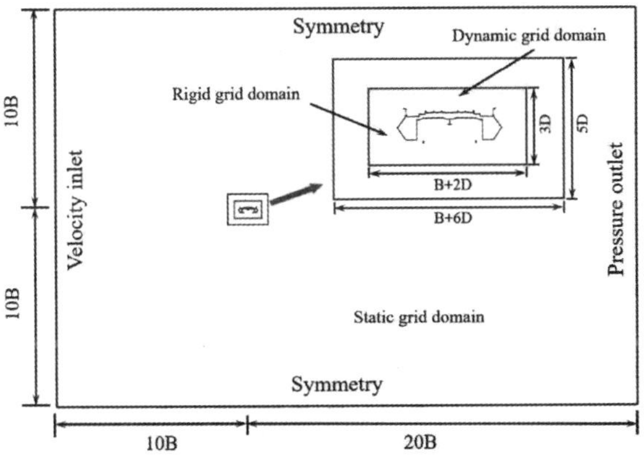

Fig. 10. Computational domain of the NG section.

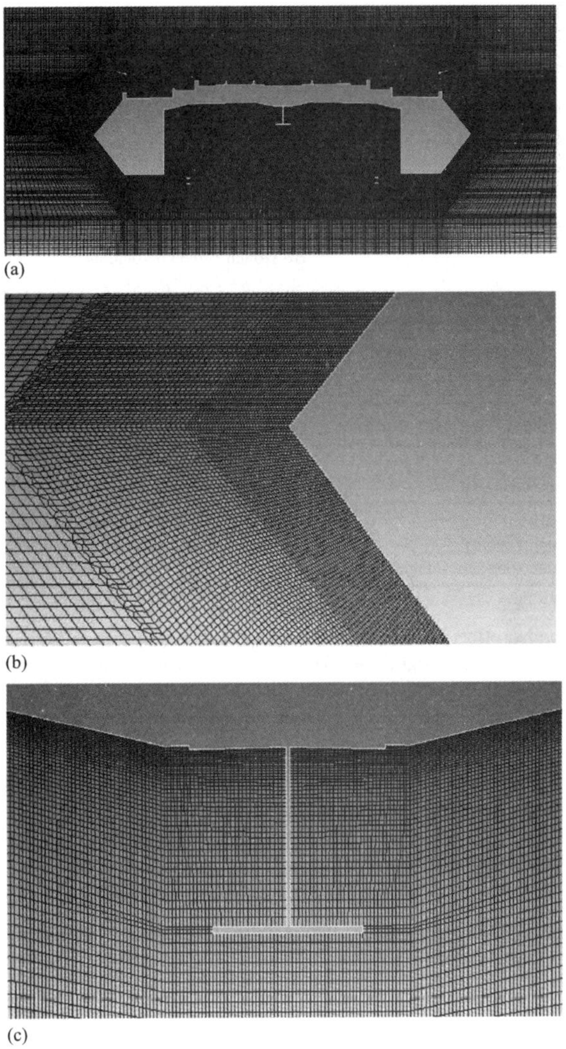

Fig. 11. Computational grids near the NG section model: (a) overall grid; (b) grid near fairing; and (c) grid near the center stringer.

Additionally, to study the Reynolds number effects on the VIV responses of the NG section with AC-VII, the VIV responses of the NG section with AC-VII of a geometric scale ratio of 1:30 are also calculated, and the numerical results are shown in Fig. 13(e). As shown in Fig. 13(e), the numerical results of the lock-in wind velocity and maximum amplitude of the NG section with AC-VII for a geometrical scale ratio of 1:30 are close to that of the NG section with AC-VII for a geometrical scale of 1:60, and the Reynolds

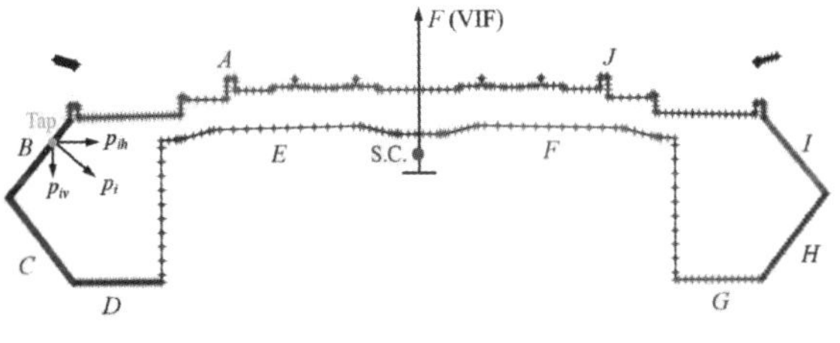

Fig. 12. Distribution of measurement taps on the NG section surface.

Table 4. Numerical results for aerostatic coefficients at the time step of 0.0004 s

Grid	Grid number	Horizontal force coefficients, $C_H=F_H/0.5\rho V^2D$	Vertical force coefficients, $C_V=F_V/0.5\rho V^2B$	Pitch coefficients, $C_M=M/0.5\rho V^2B^2$
Grid 1	349513	1.0771	−0.1036	0.0345
Grid 2	445243	1.0838	−0.0768	0.0365
Grid 3	543068	1.0849	−0.0613	0.0388

Table 5. Numerical results for aerostatic coefficients for Mesh 2

Time step/s	Horizontal force coefficients, $C_H=F_H/0.5\rho V^2D$	Vertical force coefficients, $C_V=F_V/0.5\rho V^2B$	Pitch coefficients, $C_M=M/0.5\rho V^2B^2$
0.0003	1.1865	−0.0323	0.0522
0.0004	1.0838	−0.0768	0.0365
0.0005	1.0650	−0.1571	0.0254

Table 6. Numerical results of aerostatic coefficients of the NG section with AC-VII

Methods	Horizontal force coefficients, $C_H=F_H/0.5\rho V^2D$	Vertical force coefficients, $C_V=F_V/0.5\rho V^2B$	Pitch coefficients, $C_M=M/0.5\rho V^2B^2$
Exp.	0.9445	−0.0756	0.0073
Num. (fluent)	1.0838	−0.0768	0.0365
Num. (pressure integration)	1.0445	−0.0776	0.0378

number effect on VIV responses of the NG section with AC-VII in this study is not significant. Therefore, it is feasible to use the main deck sectional model with appropriate geometrical scale to investigate the VIV responses of the main deck section with wind tunnel tests or numerical simulations.

Mechanism Analysis of VIV

Flow Field Characteristics

To further study the mechanism of the VIV and its aerodynamic countermeasures of the NG section, the FSI numerical simulation of the NG section with/without AC-VII is conducted for the peak reduced wind velocity. Figs. 14(a and b) show the flow traces near the NG section with/without AC-VII, and Fig. 14(c) shows the nondimensional displacement time history of the two sections under peak reduced wind velocity. It can be seen from Figs. 14(a and b) that there is a large-sized vortex on the upper surface of the original NG section. However, the size of the vortex on the upper surface of the main deck section with AC-VII is significantly suppressed.

There are two types of vortex shedding on a bridge cross section: alternating tail vortex (Matsumoto et al. 1993) and shear layer separation vortex (Matsumoto 1999). Fig. 15 shows the evolution of the instantaneous vorticity near the main deck section without [Fig. 15(a)] and with [Fig. 15(b)] AC-VII in one vibration cycle. The value in parentheses is the vorticity value in the center of the vortex. The vorticity W is expressed as the curl of the fluid velocity vector as follows:

$$W=\left(\frac{\partial w}{\partial y}-\frac{\partial v}{\partial z}\right)i+\left(\frac{\partial u}{\partial z}-\frac{\partial v}{\partial x}\right)j+\left(\frac{\partial v}{\partial x}-\frac{\partial u}{\partial y}\right)k \tag{7}$$

From Fig. 15, it can be seen that because of the influence of the blunt cross section, shear layer instability occurs at the windward side railing and side box girder, resulting in shear layer separation vortices A_0 and B_0. A_0 forms a periodic shedding vortex A_1 on the leeward side of the NG section, and B_0 forms a periodic shedding vortex B_1 after attaching to the bottom surface of the box girder on the leeward side (the secondary vortex). Therefore, it can be inferred that vortices A_1 and B_1 shedding periodically at the tail causes the vertical VIV of the original NG section. Therefore, reducing the vorticity of the dominant vortices A_1 and B_1 or destroying the periodic shedding of the vortices is the key for suppressing VIV. It is worth noting that the motion shape of the dominant vortex B_0 is inconsistent with the research results of Li et al. (2018a) and Liu et al. (2022b) on the WG sections with aspect ratios of 11.48 and 8.44, respectively. The vortex B_0 does not enter the concave region between the side box girders after forming at the lower surface of the side box girder but directly crosses this region to form a secondary vortex at the trailing edge of the cross section. This is also the reason why the lower stabilizers and the lower horizontal separators have negligible effects on VIV suppression of the NG section. It can be further inferred that if the cross section aspect ratio continues to decrease, the separation vortices A_0 and B_0 will wrap the entire cross section, which is consistent with the research conclusions based on small aspect ratio blunt cross sections (Shiraishi and Matsumoto 1983; Nakamura and Nakashima 1986).

For the NG section with AC-VII, it can be seen from the instantaneous vorticity evolution diagram that the original separation vortex A_0 with a large size and high vorticity is replaced by high-frequency vortex A_0' with low vorticity and small size. For the upper dominant vortex, the dominant vortices A_1 and A_1' are about to shed at time nT, and the vorticities of A_1 and A_1' are 50 and 29. That is, the vorticity of A_1 is weakened by 42% with AC-VII. At $(n+1/4)T$, A_1 and A_1' have shed, and the vorticity of both is around 27. That is to say, the vorticity change of the upper dominant vortex during the shedding process becomes smaller for the NG section with AC-VII. For the lower dominant vortex, the vorticity change trend of dominant vortex B_1 is the same as that of dominant vortex A_1. The vorticity of vortex B_1 is weakened, and the vorticity change during the shedding process decreases. It should be noted that the size reduction of vortex A_1 causes this change. In the original cross section, because of the large size of vortex A_1, there is an *acceleration region*. (It is the collision region of the upper and lower dominant vortices. The two vortices rotate in opposite directions, and at the collision point, their airflow directions are aligned. This causes an acceleration of the vortex rotation and an increase in vorticity between vortices A_1 and B_1 caused by the collision of vortices.) The central vorticity value

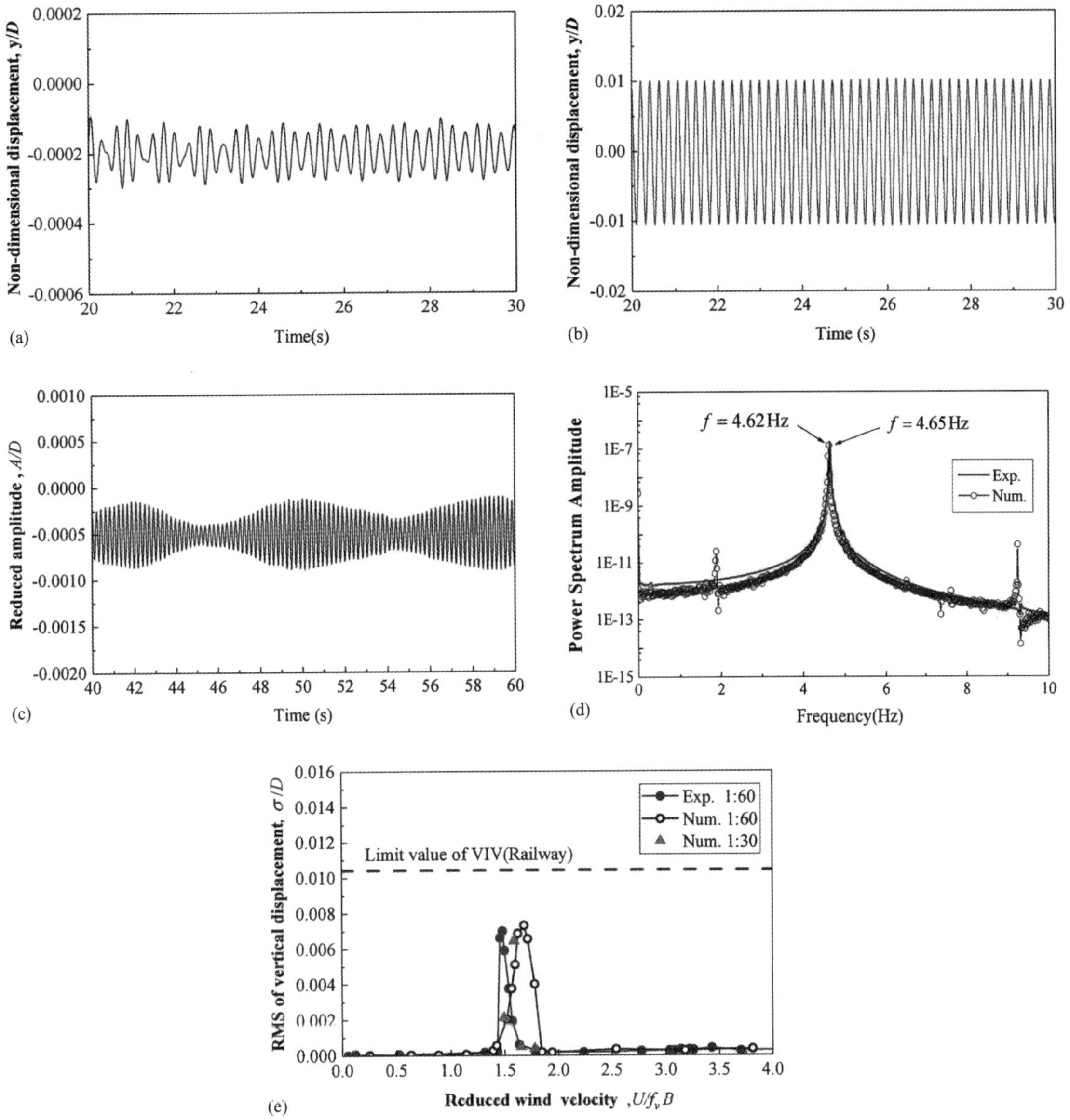

Fig. 13. Numerical and experimental results of the NG section with AC-VII: (a) $U^*=1.39$; (b) $U^*=1.68$; (c) $U^*=1.84$; (d) power spectrum of vertical displacement of the NG section with AC-VII; and (e) RMS of the VIV responses of the NG section with AC-VII.

of B_1 reaches 393 in the formation stage. However, for the NG section with AC-VII, the size of A_1 is reduced, and the *acceleration region* disappears, so the vorticity of B_1 is also reduced.

To sum up, because of the small aspect ratio of the NG section, the separation vortex B_0 generated by the leading edge has crossed the concave region, so the lower stabilizers and the horizontal separators have minor vibration suppression effect. It can be inferred that the alternately shedding vortices at the tail cause the VIV of the original NG section. The mechanism of AC-VII to suppress VIV is to weaken the size and vorticity of the separation vortex A_0, which in turn reduces the size and vorticity of the upper dominant vortex A_1 and avoids the *acceleration region*, thus reducing the vorticity change during the dominant vortex shedding process.

Surface Pressure Evolution Characteristics

Mean Pressure Coefficient and Pulsation Pressure Coefficient

The spatial distributions of the mean pressure coefficient of the NG section with/without AC-VII are shown in Fig. 16, respectively. It should be noted that, for the convenience of display, the NG section is simplified. Moreover, the positive pressure is drawn on the inside of the main deck section, and the negative pressure is drawn on the outside of the NG section. In Fig. 16, the values in/outside the brackets are the mean pressure coefficients with/without AC-VII, respectively. As shown in Fig. 16, AC-VII mainly interferes with the flow field at the NG section. The absolute values of the mean pressure coefficient in Region A are significantly reduced for the NG section with AC-VII, by which it can be inferred that AC-VII restricts the vortices away

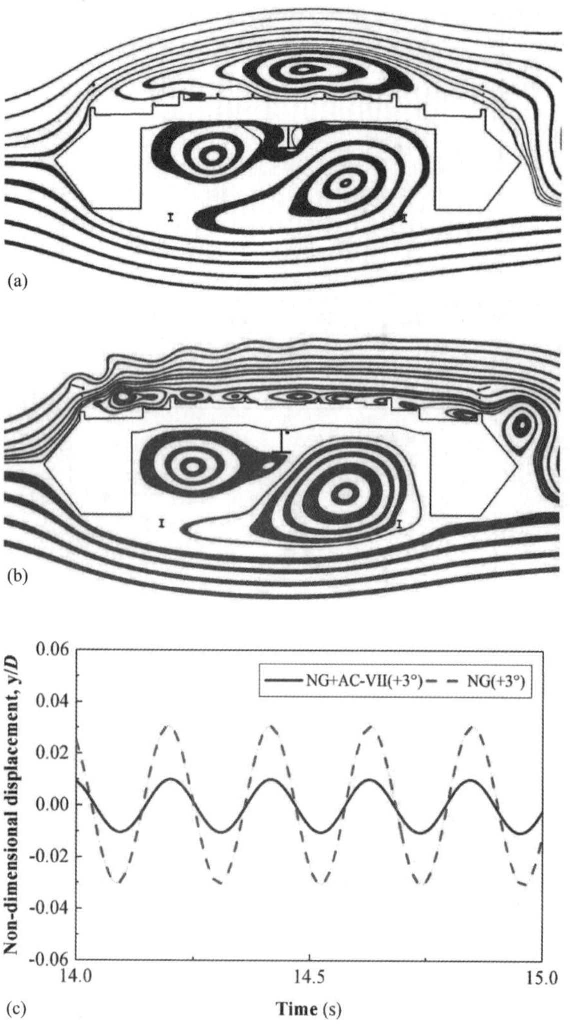

Fig. 14. Flow trace near the NG section with/without AC-VII for peak reduced wind velocity: (a) the original NG section; (b) the NG section with AC-VII; and (c) the nondimensional displacement time history of the two sections.

from the NG section, thereby reducing the suction force on the NG section.

Fig. 17 shows the spatial distribution of the RMS of the fluctuating pressure coefficient on the surface of the NG section. It can be seen from Fig. 17 that the RMS of the fluctuating pressure coefficient on the surface of the original NG section is relatively large and not uniformly distributed, and the fluctuating pressure coefficients in J, G, and H regions are relatively large. However, the RMS of the fluctuating pressure coefficient in A and B local regions of the NG section with AC-VII is slight larger than that of the NG section without AC-VII, and the RMS of the fluctuating pressure coefficient in the other regions of the NG section with AC-VII are reduced significantly, which is consistent with the results of Xu et al. (2010) on the vibration suppression mechanism of the inclined deflector. Moreover, the RMS of the fluctuating pressure coefficient in the J region has the most significant decrease, with a decrease of 86% at the peak value. It should be noted that although AC-VII is not designed for the lower dominant vortex B_0, the RMS of the fluctuating pressure coefficients in the G and H regions are reduced because of the disappearance of the *acceleration region* caused by the vortices' collision.

Additionally, to exclude the effect of different displacement amplitudes on the results of the pressure coefficient, the surface pressure distribution of the original NG section with the same amplitude as that of the AC-VII section has been calculated and is depicted in Fig. 17. As shown in Fig. 17, when the displacement amplitude of the original NG section is the same as that of the NG section with AC-VII, the pressure fluctuation coefficient on its surface decreases; however, the characteristics of the surface pressure distribution are not affected by the amplitude. In other words, the influence of displacement amplitude on the pressure coefficient can be excluded.

Surface Pressure Spectrum Characteristics

Fig. 18 shows the predominant frequency distribution at each pressure measurement tap. It can be seen from Fig. 18 that the predominant frequency of each pressure measurement tap on the original NG section is consistent with the frequency of the vertical VIV of the original NG section. However, the large separation vortices generated at the railing of the NG section with AC-VII are broken up into small vortices with high frequencies, which cause the change of the predominant frequencies in A, B, and C regions of the NG section. Furthermore, the RMS fluctuating pressure coefficient in A and B regions increases significantly. Moreover, Fig. 19 shows the spectrum analysis of pressure sampling points on the surface of upstream inclined deflector, region J and region B of the VII section.

Vortex-Induced Force Evolution Characteristics

Correlation between Pressure and Vortex-Induced Force

The correlation coefficient between the vertical component of the pressure at tap i (p_i) and the VIF (F) of the NG section can be expressed as follows (Liu et al. 2021):

$$\rho_i = \frac{\mathrm{Cov}(F, p_i)}{\sqrt{D(F)}\sqrt{D(p_i)}} \tag{8}$$

where ρ_i = correlation coefficient between the pressure and VIF at the measuring tap i; $D(F)$ and $D(p_i)$ = variance of the VIF and the pressure at measuring tap i; and $Cov(F, p_i)$ = covariance of the VIF and pressure at measuring tap i. The correlation coefficient ranges from −1 to 1; $\rho_i = 1$ indicates that the pressure is positively correlated with the VIF; $\rho_i = -1$ indicates that the pressure is inversely correlated with the VIF.

Fig. 20 shows the spatial distribution characteristics of the correlation coefficient between the pressure and the VIF on the surface of the NG section. It can be seen from Fig. 20 that the distribution of correlation coefficients on the surface of the main deck is not uniform, and the value of the correlation coefficient on the leeward side of the main deck is larger than that on the windward side. However, the correlation coefficients of Regions A, B, and C of the NG section with AC-VII have been significantly reduced because the high-frequency vortices shedding off the inclined deflector disrupt the consistency of the pressure and vortex pulsations in the A, B, and C regions, but the regions with large correlation coefficients such as J, F, H, and G regions have not changed significantly. Combined with the analysis of the spectral characteristics of the pressure, it can be concluded that the position of the VIV *driving region* has not been changed by the AC-VII.

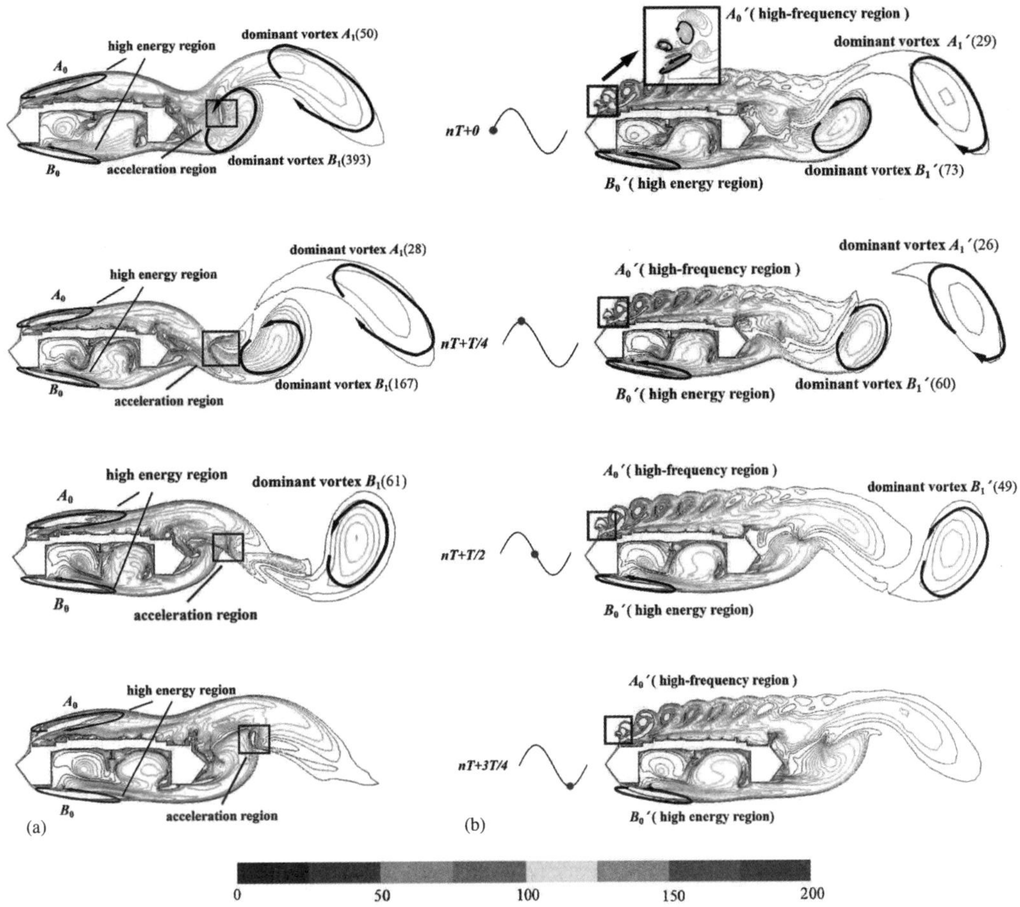

Fig. 15. Instantaneous vorticity of the NG section with/without AC-VII for reduced wind velocity corresponding to the maximum amplitude of VIV responses at a +3° angle of attack: (*a*) the original NG section; and (*b*) the NG section with AC-VII.

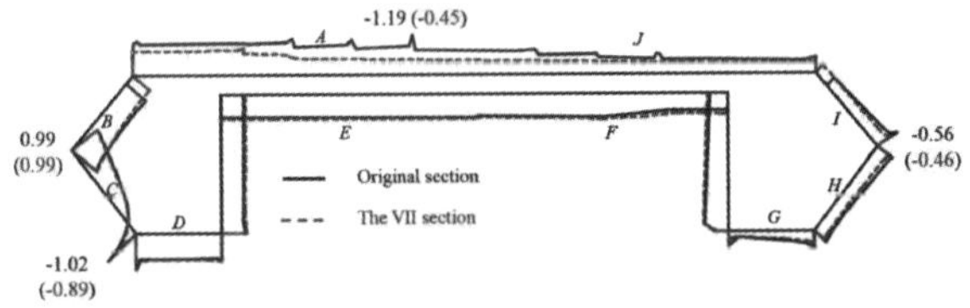

Fig. 16. Mean pressure coefficients of the NG section.

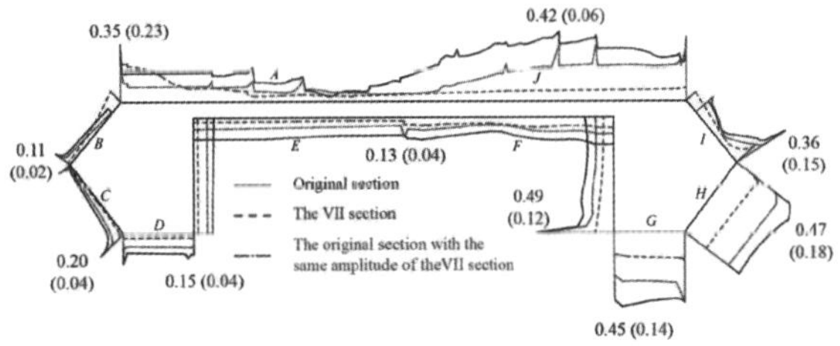

Fig. 17. RMS of fluctuating pressure coefficients of the NG section.

Contribution of Pressure at Taps

In order to identify the *driving region* for the VIV of the NG section, it is necessary to analyze the contribution of the pressure at each measuring tap to the VIF of the NG section. The correlation between the pressure at the *i*th measuring tap (p_i) and the VIF (F) of the NG section is as follows:

$$C_{\text{aero}-i} = \frac{\sigma_i b_i \rho_i}{0.5\rho U^2 B} \tag{9}$$

where $C_{\text{aero-}i}$ = dimensionless contribution value; σ_i = RMS of the fluctuating pressure at the measuring tap i; b_i = bridge width B (for vertical VIV); and ρ_i = correlation coefficient between the p_{iv} and the F at the measuring tap i. When $C_{\text{aero-}i}$ is positive, the pressure of the measurement tap i enhances the F; when $C_{\text{aero-}i}$ is negative, the p_i of the i measurement tap inhibits the F.

Fig. 21 shows the spatial distribution of the contribution of the p_i to the F. It can be seen from Fig. 21 that the contribution of the pressure of the original NG section to the VIF is not uniform and the leeward contribution of the NG section is significantly larger than on the windward side of the NG section. Moreover, the contributions of the J, H, G, and F regions of the original NG section to the VIF are more significant than for other regions. Therefore, it can be concluded that the J, H, G, and F regions are the VIV's *driving region* of the NG section. However, by setting the AC-VII, the contributions of the pressure of the J, H, G, and F regions on the VIF of the NG section are significantly reduced. Furthermore, the size and vorticity of the upper dominant vortex A_1 are

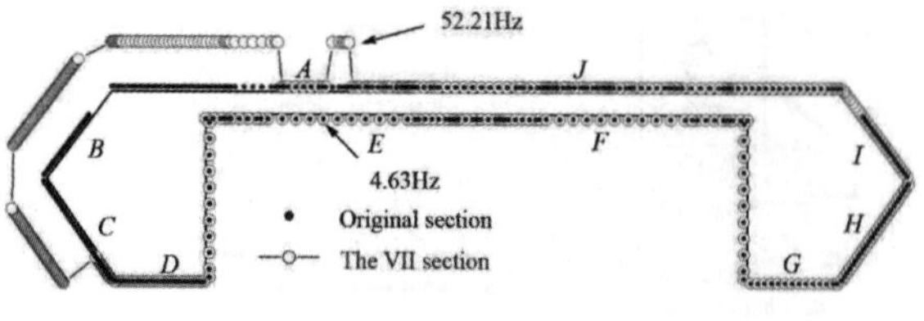

Fig. 18. Predominant frequency of various pressure taps.

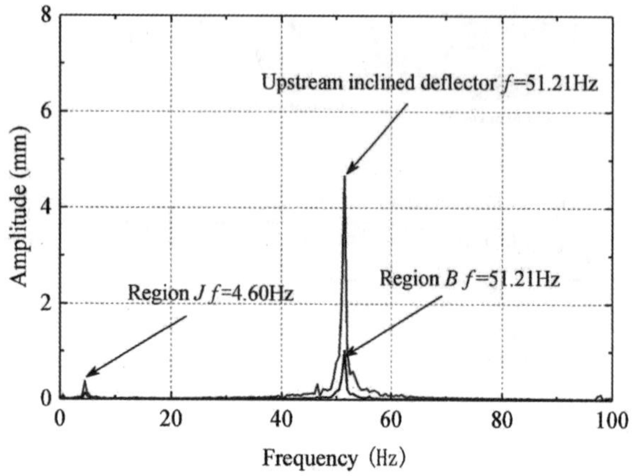

Fig. 19. Spectrum analysis of typical regions.

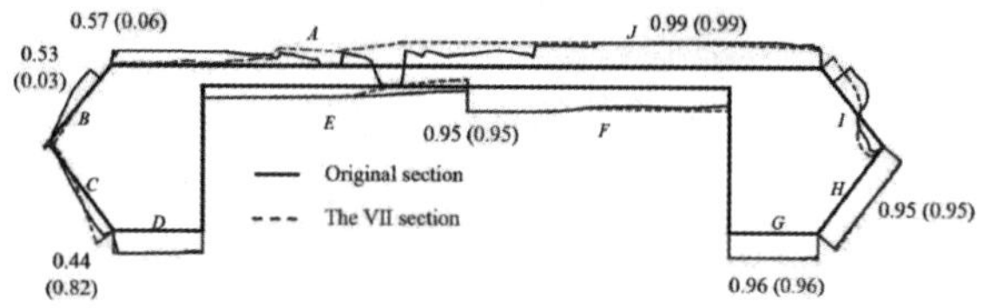

Fig. 20. Characteristics of the correlation coefficient between the pressure and the VIF.

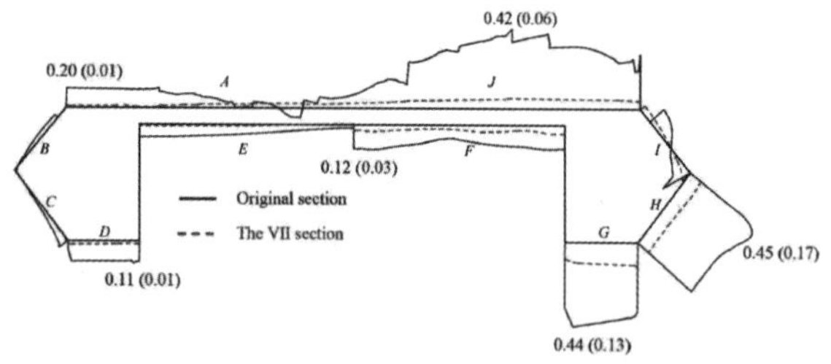

Fig. 21. Contribution of the pressure to the VIF.

significantly reduced, and the maximum contribution value of the J region drops from 0.42 to 0.06. Meanwhile, because of the disappearance of the *accelerated region*, the contribution values of the H and G regions are also significantly reduced, and the maximum contribution values of the H and G regions drop by 62% and 70%, respectively. In summary, setting the AC-VII, the contribution of

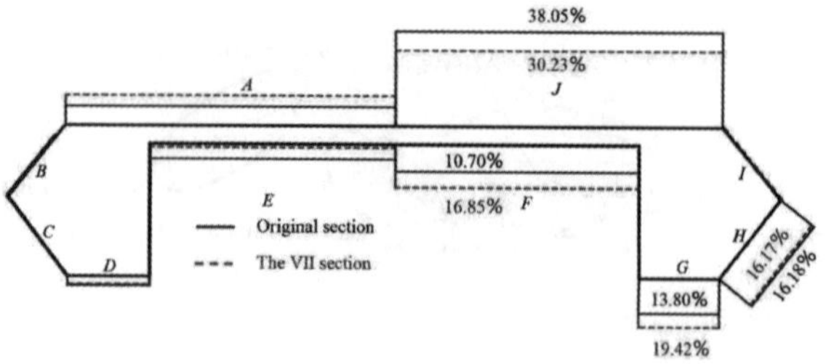

Fig. 22. Contribution of each region to the VIF in the total contribution of the section.

the surface pressure on the VIF of the NG section is significantly reduced.

Contribution Ratio of the Driving Region to Vortex-Induced Force

In order to further investigate the mechanism of the VIV of the NG section, the spatial distribution of the contribution of the aerodynamic force coefficient at each region to the aerodynamic force coefficient in the whole NG section is defined as follows:

$$P_k = \frac{C_{\text{areo-k}}}{C_{\text{areo-section}}} \tag{10}$$

where P_k = percentage of the contribution of the aerodynamic force coefficient at the k region to the aerodynamic force coefficient of the whole NG section; $C_{\text{aero-}k}$ = aerodynamic force coefficient in the k region; and $C_{\text{aero-section}}$ = aerodynamic force coefficient of the whole NG section.

Fig. 22 shows the contribution of aerodynamic force at each region to the VIF of the whole NG section. It can be seen from Fig. 22 that the P_{region} value in the J, H, G, and F regions are higher than for other regions. For the convenience of subsequent analysis, the four regions are divided into the following two categories: *upper driving region* (J) and *lower driving region* (H, G, and F).

To better describe the difference in contribution of the upper and lower driven regions of the NG section to the VIF, the contribution ratio is defined in this study. The contribution ratio of the *upper driving region* to the *lower driving region* on the original NG section is 38.05%/(10.70% + 13.80% + 16.18%) = 0.94, indicating that under such condition, the high-vorticity *acceleration region* [Fig. 15(a)] controls both the upper and lower dominant regions, and there is no significant difference between them. After applying AC-VII, the VIV amplitude of the main deck is significantly reduced, and the pressure fluctuation amplitude in the driving region of the main deck also decreases. In the new VIV process, the contribution ratio of the *upper driving region* to the *lower driving region* is 30.23%/(16.85% + 19.42% + 16.17%) = 0.58, indicating that the *lower driving region* plays a dominant role in the VIV process at this time. This is consistent with the phenomenon shown in Fig. 15 (b), where the vortex intensity is larger in the lower dominant vortex than in the upper dominant vortex. Therefore, the VIV mode of the main deck changes from the original *double-sided driving mode* to *one-sided dominant driving mode.*

Conclusions

The experimental and numerical investigations on VIV and its AC mechanisms of an NG section with the aspect ratio of $B/D = 5$ are conducted. The following conclusions can be drawn:

1. An obvious VIV occurs for the original NG section with the aspect ratio of $B/D = 5$ for a wind attack angle of −5° to +5°. The

ACs of low vertical stabilizers or low horizontal separators, which are commonly applied to suppress the VIV responses of the WG section with the aspect ratio of $B/D = 10–14$, cannot effectively suppress the VIV responses of the NG section with the aspect ratio of $B/D = 5$. However, the VIV responses can be significantly suppressed by the AC of the combination of the modified fairings and the inclined deflectors.

2. Because of the small aspect ratio of the narrow Π-shaped section, the shear layer separation vortex formed at the windward side railing of the main deck crosses the upper surface of the main deck and then falls off periodically at the tail of the main deck (upper dominant vortex), and the shear layer separation vortex formed at the windward side girder crosses the concave region and generates a *secondary vortex* at the bottom surface of the downstream girder (lower dominant vortex). The *Karman vortex street* that the upper and lower dominant vortices alternately shed at the leeward of the main deck is the mechanism of the VIV in the original NG section.
3. The AC-VII weakens the size and vorticity of the upper dominant vortex, avoiding the generation of an *acceleration region* at the leeward of the main deck, and further reduces the vorticity of the lower dominant vortex. Furthermore, the vorticity change of the upper and lower dominant vortices of the NG section with AC-VII during the shedding process is reduced, thereby reducing the VIF of the NG section caused by the vortex shedding. The peak value of the pressure fluctuation in the *driving region* of the VIV on the surface of the main deck is reduced by 86%, the contribution of the surface pressure to the VIF is significantly reduced, and so the VIF on the main deck is reduced and the VIV response is suppressed.

Data Availability Statement

Some data that support the findings of this study are available from the corresponding author upon reasonable request, such as the detailed information of the experimental and numerical results of the main deck.

Acknowledgments

This research was financially supported by the National Natural Science Foundation of China (Grant Number 51778225), Scientific Research and Development Program of China National Railway Group Co., Ltd. (N2021G034), Scientific Research Program of China Railway Construction Corporation Limited (2021-B17), and Scientific and Technological Research and Development Project of China Railway Siyuan Survey and Design Group Co., Ltd. (2018K001, 2021K015), for which the authors are grateful.

References

Bai, H., R. Li, G. Xu, and A. Kareem. 2021. "Aerodynamic performance of Π-shaped composite deck cable-stayed bridges including VIV mitigation measures." *J. Wind Eng. Ind. Aerodyn.* 208: 104451. https://doi.org/10.1016/j.jweia.2020.104451.

Chen, Y., Y. Zhou, Y. X. Yang, and Y. J. Ge. 2019. "Vortex-induced vibration control for twin box girder bridges with vertical central stabilizers." [In Chinese.] *J. Tongji Univ. (Nat. Sci.)* 47 (5): 617–626. https://doi.org/10.11908/j.issn.0253-374x.2019.05.004.

Fang, G. S., Y. X. Yang, Y. J. Ge, and Z. Y. Zhou. 2017. "Vortex-induced vibration performance and aerodynamic countermeasures of semi-open separated twin-box deck." [In Chinese.] *China Civ. Eng. J.* 50 (3): 74–82. https://doi.org/10.15951/j.tmgcxb.2017.03.009.

Hu, C. X., L. Zhao, Z. Y. Zhou, and Y. J. Ge. 2020. "Suppressing mechanism of spoilers on vortex-induced vibrations around a streamlined closed-box girder based on characteristics of aerodynamics forces and flow field." [In Chinese.] *J. Vib. Eng.* 33 (1): 1–11. https://doi.org/10.16385/j.cnki.issn.1004-4523.2020.01.001.

Huang, L., J. H. Dong, Q. Wang, Z. G. Li, G. Gao, and S. W. Li. 2021. "Vortex-induced vibration performance of a cable-stayed railway bridge with rectangular steel box girder and its aerodynamic countermeasure." [In Chinese.] *J. Vib. Shock* 40 (6): 23–32. https://doi.org/10.13465/j.cnki.jvs.2021.06.003.

Huang, L., J. H. Dong, Q. Wang, H. L. Liao, and Z. G. Li. 2022. "Influence of the inclination angle of the deflector on the vortex vibration performance of the Π-shaped composite beam." [In Chinese.] *J. Vib. Eng.* 1–12. https://doi.org/kns.cnki.net/kcms/detail/32.1349.TB.20220817.1020.004.html.

Katsuchi, H., H. Yamada, and H. Ohashi. 2009. "A study on aerodynamic characteristics of edge-girder deck of composite cable-stayed bridge." *J. Wind Eng.* 34 (4): 103–110. https://doi.org/10.5359/jwe.34.103.

Kubo, Y., K. Sadashima, E. Yamaguchi, K. Kato, Y. Okamoto, and T. Koga. 2001. "Improvement of aeroelastic instability of shallow π cross-section." *J. Wind Eng. Ind. Aerodyn.* 89 (14–15): 1445–1457. https://doi.org/10.1016/S0167-6105(01)00151-9.

Laima, S., H. Li, W. Chen, and F. Li. 2013. "Investigation and control of vortex-induced vibration of twin box girders." *J. Fluids Struct.* 39: 205–221. https://doi.org/10.1016/j.jfluidstructs.2012.10.009.

Li, C. G., Y. Mao, H. B. Yan, Y. Han, and X. H. Zhou. 2022a. "Influence mechanism of stabilizer height on vortex induced vibration performance of side girder cable stayed bridge with deck conveyor." [In Chinese.] *J. Chang'an Univ. (Nat. Sci.)* 42 (2): 56–67. https://doi.org/10.19721/j.cnki.1671-8879.2022.02.006.

Li, C. G., Y. Mao, H. B. Yan, A. H. Liang, and Y. Han. 2022b. "Experimental study on vortex-induced vibration performance and countermeasures for side girder beam with conveyer." [In Chinese.] *J.S.W. Jiaotong Univ. (Nat. Sci.)* 54 (4): 886–893. https://doi.org/10.3969/lj.issn.0258-2724.20210224.

Li, C. G., J. Zhang, Y. B. Fan, and X. G. Hua. 2017a. "Study of aerodynamic optimization measures for vortex-induced vibration performance of wide streamlined steel box girder." [In Chinese.] *Bridge Const.* 47 (1): 35–40. https://doi.org/10.3969/j.issn.1003-4722.2017.01.007.

Li, H., and X. H. He. 2020. "Lateral aerodynamic interference between an interior train and a flat box bridge-deck." *Exp. Therm Fluid Sci.* 117: 110115. https://doi.org/10.1016/j.expthermflusci.2020.110115.

Li, H., X. H. He, H. F. Wang, M. T. Liu, and S. Peng. 2018a. "Wind tunnel tests for vortex-induced vibration control measures of a super high cable-stayed bridge with π-cross cross-section." [In Chinese.] *J. Vib. Shock* 37 (7): 62–68. https://doi.org/10.13465/j.cnki.jvs.2018.07.010.

Li, H., S. Laima, J. Ou, X. Zhao, W. Zhou, Y. Yu, N. Li, and Z. Liu. 2011. "Investigation of vortex-induced vibration of a suspension bridge with two separated steel box girders based on field measurements." *Eng. Struct.* 33 (6): 1894–1907. https://doi.org/10.1016/j.engstruct.2011.02.017.

Li, J., P. Wu, J. Hao, and H. Pan. 2022c. "Experimental and numerical studies on the two 'lock-in' regions characteristic of vertical vortex-induced vibration of Π-shaped composite bridge deck." *J. Wind Eng. Ind. Aerodyn.* 228: 105122. https://doi.org/10.1016/j.jweia.2022.105122.

Li, J. W., M. J. Xu, Z. J. Wang, and S. Xing. 2022d. "Effect of geometric parameters of wind fairing on vortex-induced vibration performance of two-side box type Π-shaped girder." [In Chinese.] *J. Arch. Civ. Eng.* 39 (5): 74–83. https://doi.org/10.19815/j.jace.2022.06030.

Li, K., G. W. Qian, Y. J. Ge, L. Zhao, and J. Di. 2019. "Control effect and mechanism investigation on the horizontal flow-isolating plate for PI shaped bridge decks' VIV stability." *Wind Struct.* 28 (2): 99–110. https://doi.org/10.12989/was.2019.28.2.099.

Li, S., S. Laima, and H. Li. 2017b. "Cluster analysis of winds and wind-induced vibrations on a long-span bridge based on long-term field monitoring data." *Eng. Struct.* 138: 245–259. https://doi.org/10.1016/j.engstruct.2017.02.024.

Li, S., S. Laima, and H. Li. 2018b. "Data-driven modeling of vortex-induced vibration of a long-span suspension bridge using decision tree learning and support vector regression." *J. Wind Eng. Ind. Aerodyn.* 172: 196–211. https://doi.org/10.1016/j.jweia.2017.10.022.

Li, Y. X., C. Y. Ma, H. Y. Li, and Y. J. Tian. 2022e. "Experimental study on vortex-induced vibration of large-span pedestrian suspension bridge." [In Chinese.] *J. Vib. Meas. Diag.* 42 (3): 483–489. https://doi.org/10.16450/j.cnki.issn.1004-6801.2022.03.010.

Liu, S., L. Zhao, G. Fang, C. Hu, and Y. Ge. 2021. "Investigation on aerodynamic force nonlinear evolution for a central-slotted box girder under torsional vortex-induced vibration." *J. Fluids Struct.* 106: 103380. https://doi.org/10.1016/j.jfluidstructs.2021.103380.

Liu, Z. W., Z. J. Jiang, X. G. Li, C. Y. Shao, C. D. Wang, and Z. Q. Chen. 2022a. "Study on mechanism and aerodynamic control measures of vortex-induced vibration of a streamlined-box steel girder." [In Chinese.] *China J. Highway Trans.* 35 (11): 133–146. https://doi.org/10.19721/j.cnki.1001-7372.2022.11.013.

Liu, Z. W., H. Xiao, L. Wang, J. Sheng, and Z. Q. Chen. 2022b. "Vortex-induced vibration of a Π-shaped steel–concrete composite girder and its aerodynamic countermeasures." [In Chinese.] *J. Hunan Univ. (Nat. Sci.)* 49 (3): 68–78. https://doi.org/10.13465/j.cnki.jvs.2018.07.010.

Ma, K., C. Hu, and Z. Zhou. 2019. "Investigation on vortex-induced vibration of twin rectangular 5:1 cylinders through wind tunnel tests and POD analysis." *J. Wind Eng. Ind. Aerodyn.* 187: 97–107. https://doi.org/10.1016/j.jweia.2019.02.014.

Macdonald, J. H. G., P. A. Irwin, and M. S. Fletcher. 2002. "Vortex-induced vibrations of the second severn crossing cable-stayed bridge—Full-scale and wind tunnel measurements." *Proc. Inst. Civ. Eng. Struct. Build.* 152 (2): 123–134. https://doi.org/10.1680/stbu.2002.152.2.123.

Matsumoto, M. 1999. "Vortex shedding of bluff bodies: A review." *J. Fluids Struct.* 13 (7–8): 791–811. https://doi.org/791-811.10.1006/jfls.1999.0249.

Matsumoto, M., N. Shiraishi, H. Shirato, S. Stoyanoff, and T. Yagi. 1993. "Mechanism of, and turbulence effect on vortex-induced oscillations for bridge box girders." *J. Wind Eng. Ind. Aerodyn.* 49 (1–3): 467–476. https://doi.org/10.1016/0167-6105(93)90041-L.

Meng, X., L. Zhu, and Z. Guo. 2011. "Aerodynamic interference effects and mitigation measures on vortex-induced vibrations of two adjacent cable-stayed bridges." *Front. Archit. Civ. Eng. China* 5 (4): 510–517. https://doi.org/10.1007/s11709-011-0129-3.

Menter, F. R. 1994. "Two-equation eddy-viscosity turbulence models for engineering applications." *AIAA J.* 32 (8): 1598–1605. https://doi.org/10.2514/3.12149.

Murakami, T., K. Takeda, M. Takao, and R. Yui. 2002. "Investigation on aerodynamic and structural countermeasures for cable-stayed bridge with 2-edge I-shaped girder." *J. Wind Eng. Ind. Aerodyn.* 90 (12–15): 2143–2151. https://doi.org/10.1016/S0167-6105(02)00330-6.

Nakamura, Y., and M. Nakashima. 1986. "Vortex excitation of prisms with elongated rectangular, H and [vdash] cross-sections." *J. Fluid Mech.* 163: 149–169. https://doi.org/10.1017/S0022112086002252.

Oliveira Pedro, J. J., and A. J. Reis. 2016. "Composite cable-stayed bridges: State of the art." *Proc. Inst. Civ. Eng. Bridge Eng.* 169 (1): 13–38. https://doi.org/10.1680/bren.14.00005.

Owen, J. S., A. M. Vann, J. P. Davies, and A. Blakeborough. 1996. "The prototype testing of Kessock Bridge: Response to vortex shedding." *J. Wind Eng. Ind. Aerodyn.* 60: 91–108. https://doi.org/10.1016/0167-6105(96)00026-8.

Saathoff, P. J., and W. H. Melbourne. 1987. "Freestream turbulence and wind tunnel blockage effects on streamwise surface pressures." *J. Wind Eng. Ind. Aerodyn.* 26 (3): 353–370. https://doi.org/10.1016/0167-6105(87)90005-5.

Scanlan, R. H., and N. P. Jones. 1990. "A minimum design methodology for evaluating bridge flutter and buffeting response." *J. Wind Eng. Ind. Aerodyn.* 36: 1341–1353. https://doi.org/10.1016/0167-6105(90)90130-5.

Shiraishi, N., and M. Matsumoto. 1983. "On classification of vortex-induced oscillation and its application for bridge structures." *J. Wind Eng. Ind. Aerodyn.* 14 (1–3): 419–430. https://doi.org/10.1016/B978-0-444-42341-2.50052-0.

Taylor, P., and J. E. Torrejon. 1987. "Annacis bridge." *Concrete Int.* 9 (7): 13–22.

Wang, C., Z. Huang, X.. Hua, X. He, S. Zhou, and Z. Chen. 2022. "Aerodynamic mechanism of triggering and suppression of vortex-induced vibrations for a triple-box girder." *J. Wind Eng. Ind. Aerodyn.* 227: 105051. https://doi.org/10.1016/j.jweia.2022.105051.

Wardlaw, R. L. 1972. *Some approaches for improving the aerodynamic stability of bridge road decks.* Ottawa: National Research Council Press.

Wardlaw, R. L. 1990. "Wind effects on bridges." *J. Wind Eng. Ind. Aerodyn.* 33 (1–2): 301–312. https://doi.org/10.1016/0167-6105(90)90045-E.

Xu, F. Y., Z. X. Lin, Y. N. Li, and W. J. Lou. 2010. "Vortex resonance depression mechanism of a bridge deck with aerodynamic measures." [In Chinese.] *J. Vib. Shock* 29 (1): 73–76. https://doi.org/10.13465/j.cnki.jvs.2010.01.028.

Yuan, W., S. Laima, W. Chen, H. Li, and H. Hu. 2017. "Investigation on the vortex-and-wake-induced vibration of a separated-box bridge girder." *J. Fluids Struct.* 70: 145–161. https://doi.org/10.1016/j.jfluidstructs.2017.01.015.

Zhang, T. Y., Y. G. Sun, M. S. Li, and H. L. Liao. 2019. "Experimental study on vortex-induced vibration performance and aerodynamic countermeasures for a wide-width double-box composite beam." [In Chinese.] *China J. Highway Trans.* 32 (10): 107–168. https://doi.org/10.19721/j.cnki.1001-7372.2019.10.010.

Zhang, T. Y., Y. Sun, M. Li, and X. Yang. 2020. "Experimental and numerical studies on the vortex-induced vibration of two-box edge girder for cable-stayed bridges." *J. Wind Eng. Ind. Aerodyn.* 206: 104336. https://doi.org/10.1016/j.jweia.2020.104336.

Zhu, L. D., H. F. Xiang, and Y. L. Xu. 2000. "Triple-girder model for modal analysis of cable-stayed bridges with warping effect." *Eng. Struct.* 22 (10): 1313–1323. https://doi.org/10.1016/S0141-0296(99)00077-2.

Journal of Wind Engineering & Industrial Aerodynamics 243 (2023) 105612

Contents lists available at ScienceDirect

Journal of Wind Engineering & Industrial Aerodynamics

journal homepage: www.elsevier.com/locate/jweia

Discrete viscous dampers for multi-mode vortex-induced vibration control of long-span suspension bridges

Yiwen Cao [a,b], Zhiwen Huang [a,b,*], Hongyi Zhang [a,b], Xugang Hua [a,b], Zhengqing Chen [a,b], Tianbao Wan [c]

[a] State Key Laboratory of Bridge Engineering Safety and Resilience, Hunan University, Changsha, 410006, China
[b] Key Laboratory for Wind and Bridge Engineering of Hunan Province, Hunan University, Changsha, 410006, China
[c] China Railway Major Bridge Reconnaissance & Design Institute Co. Ltd., Wuhan, 430056, China

ARTICLE INFO

Keywords:
Discrete viscous dampers
Long-span suspension bridges
Vortex-induced vibration
Multi-mode control

ABSTRACT

Multi-mode vertical vortex-induced vibrations have been observed on several long-span suspension bridges, where several vertical modes of stiffening girder are excited in turn with increasing of wind velocity. This paper proposed a discrete viscous dampers solution and its optimization targeting multi-mode vortex-induced vibration control for long-span suspension bridges. The minimal modal damping ratio was used to evaluate the performance of the damper system. The wind tunnel test verified that the modal damping ratio is an alternative to the resonance response amplitude and can be obtained by complex modal analysis. The ternary search algorithm was adopted to find the optimal damper parameter for the Yingwuzhou Yangtze River Bridge, which is a suspension bridge that suffered excessive vortex-induced vibrations. Using linear damper elements to connect the bridge tower and the bridge girder directly turned out to be a feasible method of increasing multiple modal damping ratios. Four torsional eddy current dampers have been installed in the Yingwuzhou Yangtze River Bridge, based on the numerical optimization results calculated in this paper. Although the measured modal damping ratios were smaller than the numerical results, the structural damping increased by three times after dampers were installed, and the requirement of the current specification was satisfied.

1. Introduction

Long-span bridges are susceptible to wind-induced vibration, such as flutter, buffeting, and vortex-induced vibration (VIV), due to their low frequencies of oscillation and low mechanical damping (Simiu and Scanlan, 1996). The VIV occurs when the vortex shedding frequency coincides with a natural structural frequency (Ehsan and Scanlan, 1990). For long-span suspension bridges, there exist several closely spaced vertical modes, and these modes, in order of natural frequency, may be excited by VIV in turn with an increase of wind velocity. This is the multi-mode vertical VIV problem, which usually occurs at a low wind speed that is commonly encountered with a high-frequency mode (Diana et al., 2006). Multi-mode VIV events have been reported in many suspension bridges in actual use worldwide. For example, the Xihoumen Bridge experienced vertical VIV events with frequencies of 0.183 Hz, 0.276 Hz, 0.327 Hz, 0.378 Hz, 0.435 Hz, and 0.495 Hz (Li et al., 2017). The Great Belt Bridge experienced vertical VIV events with frequencies of 0.130 Hz, 0.209 Hz, and 0.242 Hz (Frandsen, 2001). The Humen Bridge experienced vertical VIV events with frequencies of 0.13 Hz, 0.23 Hz, 0.27 Hz, and 0.37 Hz (Zhao et al., 2022). Though the VIV is self-limiting and the response amplitude is usually moderate, it can affect serviceability or cause fatigue problems at a low wind speed even below 10 m/s. Therefore, VIV mitigation is necessary in practice and has drawn extensive attention in the field of bridge wind engineering during the last decades.

Countermeasures to suppress VIV can be divided into two types: aerodynamical and mechanical. Theoretically, aerodynamical countermeasures can fundamentally eliminate the VIV by interfering the flow field around the main girder to prevent the formation and shedding of the periodic vortex. To achieve this aim, configurations of some non-structural components (e.g., crash barriers (Kubo et al., 2002) and maintenance rails (Laima et al., 2018)) can be adjusted. Alternatively, installing additional components such as winglets, guide vanes, and deflectors has also been proven to suppress VIV (Larsen et al., 2000; Bai et al., 2020). Due to the intrinsic complexity of fluid-structure interaction existing in the VIV, the performance of aerodynamical

* Corresponding author. State Key Laboratory of Bridge Engineering Safety and Resilience, Hunan University, Changsha, 410006, China.
E-mail address: zwhuang213@hnu.edu.cn (Z. Huang).

countermeasures is highly sensitive to the configuration of the main girder (Yang et al., 2016). As a result, for each long-span suspension bridge, the effective aerodynamic countermeasure is generally obtained from sectional model wind tunnel tests through a tedious trial-and-error procedure. Besides, simplifications and model scaling in wind tunnel experiments have been reported to have unexpected influences on aerodynamic characteristics (Schewe and Larsen, 1998; Jones et al., 1995; Larose and D'Auteuil, 2006). Further examination is required to check whether the aerodynamical countermeasures can maintain the same effect in the actual structure. Moreover, for bridges in service, the construction process of applying aerodynamical measures is complex and expensive, which may also interrupt normal operation and cause widespread social concern.

Since the VIV of bridges is highly sensitive to the structural damping level, mechanical countermeasures are also frequently used to suppress VIV of long-span bridges by installing certain energy dissipation devices to supply additional modal damping ratios. For long-span continuous steel box-girder bridges, tuned mass damper (TMD) is the most popular and effective mechanical countermeasure to control the excessive vibration due to VIV, and it has been employed in many bridges worldwide, such as the Trans-Tokyo Bay Crossing Bridge in Japan (Fujino and Yoshida, 2002), the Rio-Niterói Bridge in Brazil (Battista and Pfeil, 2000), the Chongqi Bridge in China (Sun et al., 2020) and the Volgograd Bridge in Russia (Weber and Maślanka, 2012). A traditional TMD works by transferring the vibration energy of the host structure to an additional mass and dissipating it, which has to be finely tuned to a particular control mode and typically requires an additional mass between 0.5% and 2.0% of the controlled modal mass in applications. It is relatively easy to achieve this goal for TMD utilized on a continuous steel box-girder bridge, where the VIV event is often dominated by only one or two modes. However, there exist some inherent limitations for long-span suspension bridges. Firstly, the control bandwidth of TMD is narrow, and multiple TMDs should be used to control multi-mode vertical VIV. The additional mass of TMD required to control VIV in a single mode generally weighs hundreds of tons for long-span suspension bridges. As a result, a total mass of TMDs weighing thousands of tons is needed to solve the multi-mode vertical VIV problem of long-span suspension bridges. Secondly, due to the low frequency of long-span suspension bridges, the tuned TMD stiffness is extremely low, resulting in a large static deformation of the springs, which is difficult to manufacture and install on existing bridges. The static deformation of TMD springs can be estimated by $\Delta \approx 0.25/f_n^2$ (f_n is the modal frequency of the bridge to be controlled) (Huang et al., 2019). For an example, the static deformation corresponding to the frequency 0.2 Hz is 6.25 m, which could be too large.

To reduce the large static deformation of the spring of the classical hanging-type TMD, one way is to use the lever-type TMD (Gu et al., 2002). The other is to employ the TMDI, in which an inerter is added to a traditional TMD between the secondary mass and the structure (Giaralis and Petrini, 2017). The static stretching of the spring of TMDI is much smaller than that of the traditional TMD, due to the dynamic negative stiffness contributed by the inerter (Xu et al., 2020; Weber and Maślanka, 2012). However, compared with the TMD, the control efficiency of TMDI could be reduced (Xu et al., 2019) unless the inerter is installed at a position far from the mass stock (Dai et al., 2021), which may face practical challenges. Moreover, in consideration of the nonlinearity of the inerter, the effectiveness of the TMDI would be further reduced (Xu et al., 2022).

Apart from the dynamic absorbers such as TMD and TMDI, the vibration of civil engineering structures such as cables and beams can also be controlled by discrete dampers directly. Krenk (2004) established a state space framework for calculating the dynamic characteristics of continuous structures with discrete dampers. Then, Main and Krenk (2005) developed an approximate solution to estimate the complex modal of a discrete system with discrete dampers and examined the damper effect on a 10-story frame model. Yang et al. (2021) applied two viscous dampers to control the cable vibration and verified the multi-mode control performance of the discrete dampers. Møller et al. (2019) proposed a damping system for flutter control of a long-span suspension bridge, which is composed of a spring and a viscous damper in parallel, connecting the tower with bridge main cables via a pre-tensioned cable. Recently, Zoltowski et al. (2022) applied an extended viscous damper to control pedestrian vibrations. However, to the authors' best knowledge, the study on the performance of discrete dampers for multi-mode VIV control of suspension bridges has not yet begun. The question of how to optimize damper parameters to control multi-mode vibrations remains unclear.

Since VIV control of long-span suspension bridges is still a challenging issue. This paper hence proposed a discrete viscous dampers solution for multi-mode VIV control of long-span bridges. The solution is suitable for VIV control of almost all suspension bridges without the additional mass. With the help of the complex modal analysis, the optimal parameters can be easily obtained. The construction of the discrete damper system for a long-span suspension bridge was discussed in Section 2. The governing equation of the complex modal analysis and a multi-mode optimization process of structure with discrete dampers was proposed in Section 3. In Section 4, the controlled performance of the discrete damper system was verified by the application in the VIV control of the Yingwuzhou Yangtze River Bridge. Finally, the conclusions and discussions were briefly addressed in Section 5.

2. Installation scheme of the damper system

Similar to the conception of the vibration control of stay cable using a damper attached near the anchorage, a feasible option for a long-span suspension bridge is connecting the bridge girder and the bridge tower. For the floating suspension bridge, there is a relative modal displacement between the girder and the tower. Damping effects can be achieved by installing the damper directly at the intersection of the girder and tower, as shown in Fig. 1. When there is no relative displacement at the intersection, such as in simply-supported bridges, the position of the damper needs to be far away from the support to ensure the relative displacement, as shown in Fig. 2. A rigid corbel is required when the damper is located beyond the bridge tower. For in-serve bridges, dampers are suggested to be placed under the local stiffeners, such as the diaphragms and the suspender's junctions. Otherwise, additional local stiffeners need to be installed at the damper's junction.

Though fluid viscous damper has been widely used to reduce the dynamic response of large-scale structures (Møller et al., 2019; Chen et al., 2020; Yang et al., 2021; Zoltowski et al., 2022), special considerations should be taken when a viscous damper is employed to supply vertical damping force for a long-span suspension bridge. Besides controlling the vertical VIV, the viscous damper needs to work under different loading conditions, such as the traffic load, the temperature fluctuation, and the seismic loading. Under the action of traffic load or temperature fluctuation, the suspension bridge not only has vertical deflection but also longitudinal horizontal motion, the amplitude of which is usually much greater than that of vertical VIV at the damper position. Under strong earthquakes, the longitudinal motion of long-span suspension bridges may be even larger than the former ones, so a viscous damper with a length of up to several meters may be needed to meet the stroke requirement, which may be difficult to manufacture, hard to maintain, and expensive. In other words, it is preferred to reduce the influence of the longitudinal movement of the girder on the working of the damper. The damper scheme and the stroke capability employed in an actual bridge are shown in Fig. 3. A torsional eddy current damper is developed, which utilizes the crank-connecting rod mechanism to convert the vertical vibration of the bridge into a rotational motion of the torque damper. With the help of the crank-connecting rod mechanism, the longitudinal motion of the long-span suspension bridges can be well met without much increase in the stroke of the damper.

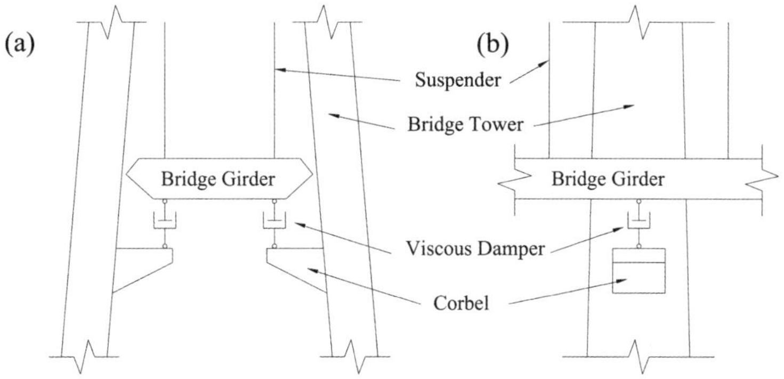

Fig. 1. Schematic of the damper arrangement of the floating suspension bridge at (a) transverse direction and (b) longitudinal direction of the bridge.

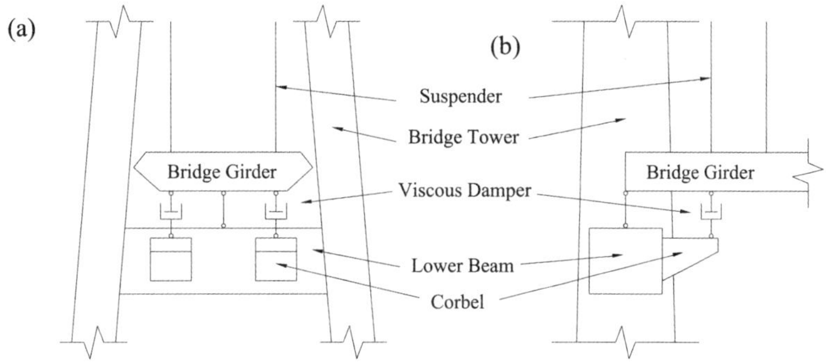

Fig. 2. Schematic of the damper arrangement of the simply-supported suspension bridge at (a) transverse direction and (b) longitudinal direction of the bridge.

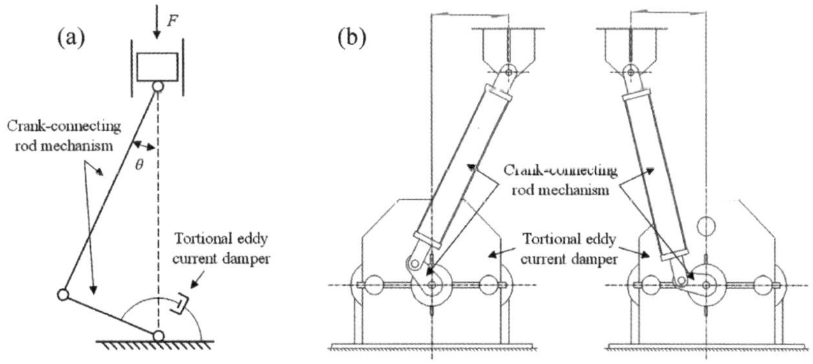

Fig. 3. Illustrations of (a) schematic and (b) stroke capability of the damper mechanism.

Additionally, a torsional eddy current damper does not require sealing and has a longer service life than that of the fluid viscous damper.

3. Governing equations and optimization

3.1. Complex modal analysis and damper's contribution

Based on the complex modal theory, the n-DOF linear motion equation with linear viscous damping can be described in the state-space form:

$$\begin{bmatrix} \boldsymbol{C} & \boldsymbol{M} \\ \boldsymbol{M} & \boldsymbol{0} \end{bmatrix} \begin{Bmatrix} \dot{\boldsymbol{X}} \\ \ddot{\boldsymbol{X}} \end{Bmatrix} + \begin{bmatrix} \boldsymbol{K} & \boldsymbol{0} \\ \boldsymbol{0} & -\boldsymbol{M} \end{bmatrix} \begin{Bmatrix} \boldsymbol{X} \\ \dot{\boldsymbol{X}} \end{Bmatrix} = \begin{Bmatrix} \boldsymbol{F} \\ \boldsymbol{0} \end{Bmatrix} \tag{1}$$

where, submatrices $\boldsymbol{K}$, $\boldsymbol{C}$, and $\boldsymbol{M}$ respectively represent the stiffness, damping, and mass of the linear system; $\{\boldsymbol{F}\}$ is the external force; vector $\{\boldsymbol{X}, \dot{\boldsymbol{X}}\}^T$ is known as the state vector; the dot marks the differentiation of time.

The complex modes are defined by the solution of the homogenous linear equation, that is $\{F\} = 0$, which is independent of the characteristic of the load and can be expressed as:

$$X = \boldsymbol{\Psi} e^{\lambda t}, \dot{X} = \boldsymbol{\Psi} \lambda e^{\lambda t} \tag{2}$$

where, λ is the eigenvalue of Equation 1, and $\{\boldsymbol{\Psi}, \lambda\boldsymbol{\Psi}\}^T$ is the corresponding eigenvector. It has n pairs of conjugate eigenvalues and eigenvectors. The r-th eigenvalue λ_r and its conjugate λ_r^* are usually rewritten by two real parameters, the modal damping ratio ζ_r and the modal frequency ω_r, that is:

$$\lambda_r, \lambda_r^* = -\zeta_r\omega_r \pm \sqrt{1\text{-}\zeta_r^2}\,\omega_r i \tag{3}$$

where, $i = \sqrt{-1}$; $\zeta_r = \text{-Re}\{\lambda_r\}/|\lambda_r|$; $\omega_r = |\lambda_r|$. ω_r is an artificially defined frequency that is equal to the undamped frequency ω_{0r} only if the damping matrix $\boldsymbol{C}$ meets the following Caughey-O'Kelly condition (Caughey and O'Kelly, 1965):

$$\boldsymbol{C}\boldsymbol{M}^{-1}\boldsymbol{K} = \boldsymbol{K}\boldsymbol{M}^{-1}\boldsymbol{C} \tag{4}$$

For discrete viscous dampers system, the damping matrix is defined as:

$$\boldsymbol{C} = \sum_j c_j \boldsymbol{w}_j \boldsymbol{w}_j^T \tag{5}$$

where, c_j is the viscous damping parameter of the j-th damper, and vector $\boldsymbol{w}_j$ represents the attachment position and orientation. For example, the vector $\boldsymbol{w}_1$ corresponds to a damper connecting 3rd and 4th DOF of a 5-DOF system is defined by:

$$\boldsymbol{w}_1 = [0\ \ 0\ \ \text{-}1\ \ 1\ \ 0]^T \tag{6}$$

It is obvious that the damping matrix does not strictly match the Caughey-O'Kelly condition. The damped system is still linear, but its eigenvectors are complex, indicating that the state of the damped system is not only determined by displacements but also by velocities of discrete nodes.

When the imaginary part of the eigenvector $\boldsymbol{\Psi}_r$ is negligible, $\omega_r \approx \omega_{0r}$, the damped system is approximately equivalent to a combination of n SDOF systems. Furthermore, VIV can be simplified as a single-modal resonance vibration. The resonance response $\boldsymbol{X}$ can be approximated as:

$$\boldsymbol{X} \approx \frac{\boldsymbol{\Psi}_r \boldsymbol{\Psi}_r^T \boldsymbol{F}}{2\zeta_r M_r \omega_r^2} \tag{7}$$

where, $M_r = |\boldsymbol{\Psi}_r^T \boldsymbol{M} \boldsymbol{\Psi}_r|$ is the modal mass, and usually set to 1 to normalize the modal shape vector $\boldsymbol{\Psi}_r$ and $\boldsymbol{F}$ is the excitation force. It indicates that the VIV response is inversely proportional to ζ_r. The generally considered optimization objective of the smallest resonance amplitude can be simplified to the calculation of the maximum complex modal damping ratio ζ_r.

The r-th modal damping ratio ζ_r of a structure with discrete viscous dampers can be approximated as (Main and Krenk, 2005):

$$\zeta_r \approx \frac{\Delta\omega_{\infty r}\eta_r}{\omega_{0r}\left(1+\eta_r^2\right)} \tag{8}$$

where, $\Delta\omega_{\infty r}$ is the frequency increment between undamped system ($c_j = 0$) and constrained system ($c_j = \infty$); η_r is the dimensionless viscous damping parameter defined as:

$$\eta_r = \frac{\boldsymbol{\Psi}_{0r}^T \boldsymbol{C} \boldsymbol{\Psi}_{0r}}{2\Delta\omega_{\infty r}} = \frac{\sum_j c_j \psi_{0r,j}^2}{2\Delta\omega_{\infty r}} \tag{9}$$

where, $\boldsymbol{\Psi}_{0r}$ is the r-th modal shape of the undamped system, and $\psi_{0r,j}$ represents the relative modal displacement between the two positions where the j-th damper connects. When the position of dampers is determined, $\Delta\omega_{\infty r}$ is constant and proportional to $\psi_{0r,j}$ (Main and Krenk, 2005). Fig. 4 shows that ζ_r increases then decreases when η_r increases.

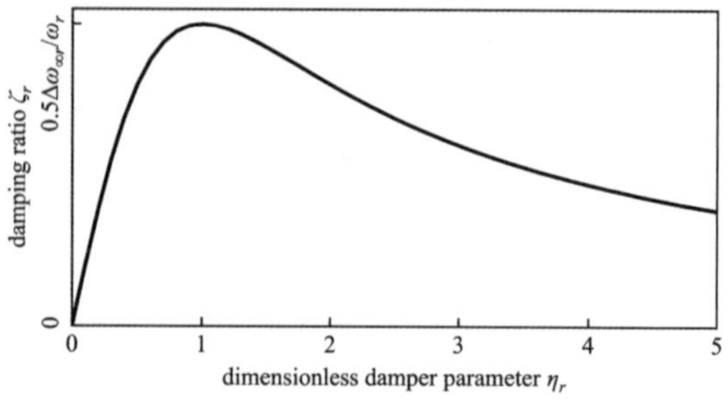

Fig. 4. Unimodal characteristic of ζ_r versus η_r.

When $\eta_r = 1$, ζ_r takes the maximum value. A damper placed in a location with larger $\psi_{0r,j}$ has a larger optimal ζ_r and requires a smaller optimal c_j.

3.2. *Parameter contribution and multi-mode optimization*

A long-span bridge has dense low-frequency modes, and VIV resonance may occur in multiple modes. For each selected damper location, ζ_r is an unimodal function of the damper parameter c_j. The optimal damper parameter for a single-mode $c_{j,opt}$ can be approximately obtained as:

$$c_{j,opt} = \frac{2\Delta\omega_{\infty r}}{\psi_{0r,j}^2} \tag{10}$$

However, the optimal damper parameter varies from mode to mode. In practice, the most crucial factor is the minimum ζ_r of the considered modes. The optimal objective is to maximize the minimum ζ_r, that is:

$$\text{maximize}\{\min\{\zeta_1, \zeta_2 \ldots \zeta_n\}\} \tag{11}$$

Fig. 5 shows that the minimum function maintains the unimodal characteristics when the derivatives of all the functions are strictly monotonic. In addition, the minimum function is a piecewise function, and the maximum value is located at the functions' intersection or a maximum point of one of the functions, which makes it difficult to obtain the analytical solution.

Ternary search is an iterative algorithm for finding the peak value of such an unimodal function. The algorithm starts with a limited open interval (l, r), takes two points (a, and b) to divide the interval into three parts, and after comparing the function values at a and b, reduces the interval to the two sub-intervals around the maximum, then moves to the next iteration. An optimized version of the ternary search algorithm proposed by Kiefer (1953) is utilized as shown in the dashed box in Fig. 6. The golden ratio, $(\sqrt{5}\text{-}1)/2$, is introduced to divide the interval so that the function value can be reused in the next iteration. The algorithm continually narrows the interval proportionally during the iteration and

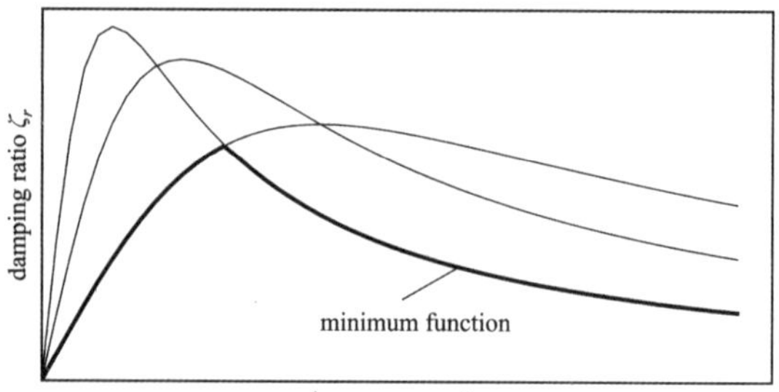

Fig. 5. Unimodal characteristic of the minimum function.

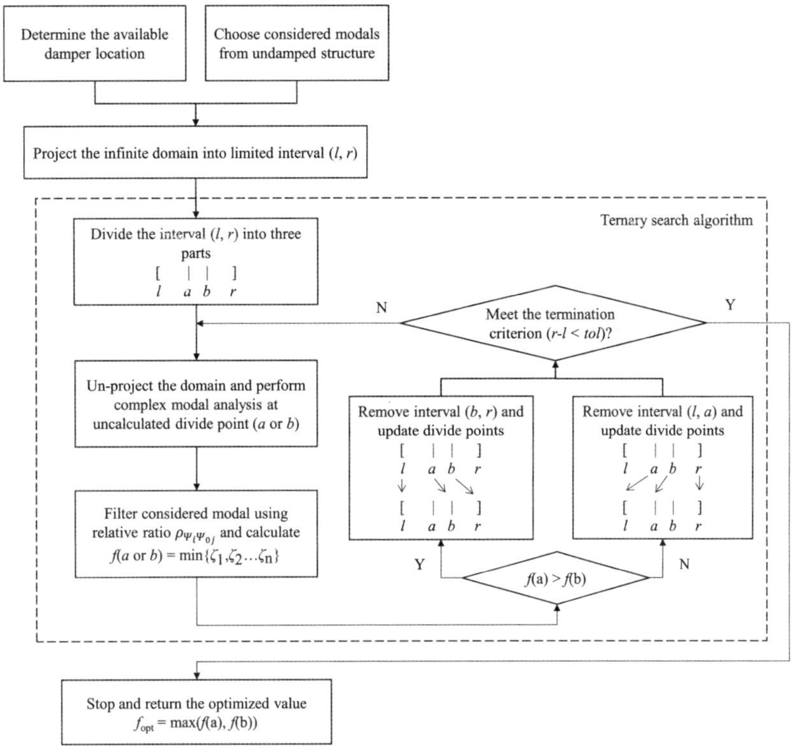

Fig. 6. Flowchart of the multi-mode optimization process for a single parameter.

has been proved to provide the best performance of the ternary search algorithm.

In this study, the argument of the unimodal function is the damper parameter c_j. The initial interval is determined by the available range of c_j. If the available range is the infinite interval $(0, +\infty)$, a substituted parameter can be used to project it to a limited interval, for instance:

$$c_j = c_0 \cdot \tan x \tag{12}$$

where, x is the substituted parameter defined in range $(0, \pi/2)$. The scale parameter c_0 can be set to the single-mode optimal damper parameter by calculating Equation 10.

To programmatically implement the iterative process, the program needs to be able to automatically extract the desired modes. In this paper, we assume that the difference in the modal shapes between the damped and undamped system is negligible, and the relative ratio can be used to quantify the similarity between modes:

$$\rho_{\Psi_i \Psi_{0j}} = \left| \frac{\Psi_i^T \Psi_{0j}}{\Psi_{0j}^T \Psi_{0j}} \right| \tag{13}$$

where, Ψ_{0j} denotes the j-th modal shape of the undamped system. A damped mode with the largest $\rho_{\Psi_i \Psi_{0j}}$ is considered to correspond to the j-th undamped mode.

Fig. 6 summarizes the multi-mode optimization process for a single damper parameter. The available locations should be first determined based on the conception described in Section 2. Then, the considered modes can be determined based on the modal analysis of the undamped structure. Before performing the iteration, the infinite domain of the damper parameter should be projected into a limited interval using a substituted parameter. For each iteration, complex modal analysis is performed at one of the uncalculated divide points. Then, the considered modes are filtered, the minimum damping ratio is calculated, and the iteration interval is updated based on the ternary search algorithm. When the interval is smaller than the termination threshold, the maximum value is returned as the optimal damper parameter. In addition, the unimodal characteristic of the minimum function can be considered to simplify the optimization algorithm for multiple damper parameters tunning, and the mutual influence between dampers should be studied in future investigations.

4. Case study

4.1. Structural properties of the Yingwuzhou Yangtze River Bridge

The Yingwuzhou Yangtze River Bridge is located on the Yangtze River in Wuhan, Hubei, China. It has three towers and 4 spans in an arrangement of 200 + 850+850 + 200 m, as shown in Fig. 7. The bridge girder is simply supported on the lower beam of the tower. Fig. 8 illustrates the girder section is 3 m high and 38 m wide, consisting of a 20 cm thick concrete deck and steel longitudinal beams. An I-shaped steel diaphragm is installed to connect the longitudinal beams every 3 m along the bridge.

The Yingwuzhou Yangtze River Bridge opened on December 28, 2014. Two perceptible vertical vibrations occurred on April 26, 2020, causing discomfort to the drivers and arousing public attention. During the vibration, the monitoring data showed that the average wind speed was between 7 and 7.6 m/s with small fluctuation (I_u = 5.8–7.0%), the measured maximum acceleration amplitude was 0.75 m/s^2 located at 3/8 of the main span on the Hanyang side, and the main frequency of the monitored vibration was 0.2441 Hz (Zhu et al., 2021). Since VIV of a bridge always has a quasi-sinusoidal displacement and acceration, the corresponding maximum amplitude of the displacement can be obtained, $A = 0.75/(2\pi \times 0.2441)^2 = 0.32$ m. To further investigate the vibration characteristic, the first 23 vertical modes of the undamped structure were calculated by the commercial finite element software

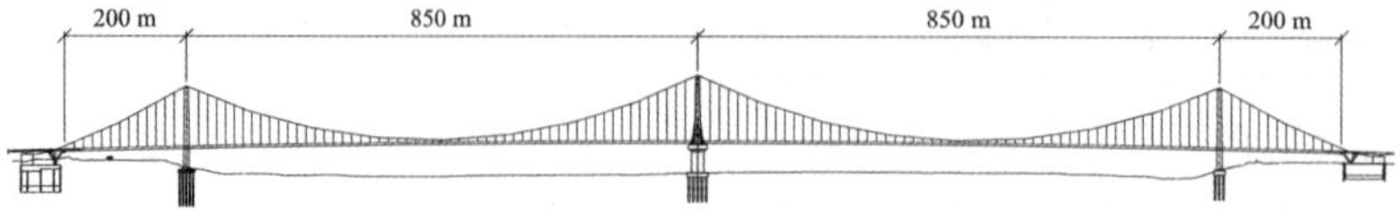

Fig. 7. The elevation view of the Yingwuzhou Yangtze River Bridge.

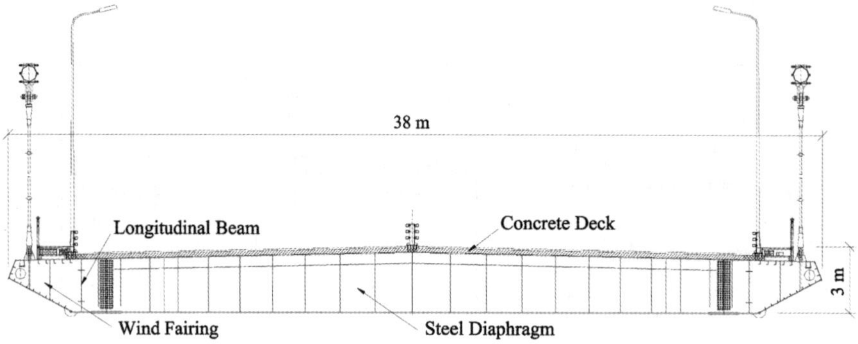

Fig. 8. The section of the bridge girder.

ANSYS and listed in Table 1. The 7th modal frequency ($f_{v7} = 0.2421$ Hz) is close to the vibration frequency, and the monitored vibration shape was close to the 7th vertical modal shape of the structure, in which the amplitude was more significant in 1/8, 3/8, 5/8, and 7/8 span of the two middle spans as shown in Table 1. It was confirmed that the vibration was vortex-induced.

The Strouhal number (St number) is a non-dimensional parameter that indicates the frequency-related VIV starting wind speed. The number corresponding to the monitored VIV event is $St = f_{v7} \times H/U = 0.2441 \times 3/7 = 0.105$. The predicted VIV starting wind speed of each mode was listed in Table 1, and the first 23 modes covered all the modes that possibly suffered VIV events when the wind speed was lower than 25 m/s. The equivalent mass ($m_{eq} = \varphi_g^T\varphi_g L/M_r$) calculated for 2-DOF test prediction was also addressed in Tables 1 and it was similar in different modes except for some modes coupling the side span vibration. It should be noted that the frequency of the 8th mode was extremely close to that of the 7th mode, so it could be probably excited at a similar wind condition. It seems that more than one mode may be excited by the vortex-induced force at a certain wind velocity since their frequencies are so close. However, due to the phenomenon of lock-in originating from the nonlinearity of VIV, generally only one mode will be excited as observed from field measurement. At present, the VIV lock-in regimes of long-span suspension bridges with closely spaced vertical modes are still unclarified. It may be influenced by many factors, including the time history of wind, the distribution of wind speeds and wind attack along the bridge, the modal characteristics of adjacent modes, the vortex shedding mechanism, and so on. It is still difficult to accurately predict which mode of the bridge the vortex-induced vibration will be lock-in to under a certain wind velocity. To reduce the occurrence of vortex-induced vibration of a long-span suspension bridge, multiple modes should be taken into consideration in VIV control.

4.2. *Damper design and parameter optimization*

After comprehensive consideration of the structural characteristics and the construction difficulty, 4 dampers were decided to be placed on the two sides of the middle tower. Fig. 9(a) shows the schematic of the damper arrangement. A total of five locations were considered, and the distance between the dampers and the center of the middle tower (D) was 5, 8, 11, 17.5, and 20 m, respectively. The location where $D = 20$m corresponds to the position of the suspenders closest to the middle tower. Due to the symmetrical arrangement, the damper parameters were consistent ($c_j = c$). The contribution of the dampers was modeled by the longitudinal spring-damper elements "Combin14", which forms the location-related term in the damping matrix. One end of the damper element was attached to the bridge girder, and the other end to the rigid beam extended out of the tower, as shown in Fig. 9(b).

The single-mode optimization was conducted first to examine the influence of the damper parameter c and damper location D on each vertical mode. The iterative optimization algorithm introduced in Section 3 was used to find the optimal parameter. Fig. 10(a) shows the change in the optimal damper parameter c_{opt} with the increase of the vertical modal order, and the change of the corresponding optimal additional modal damping ratio ζ_{opt} were illustrated in Fig. 10(b). It was inspiring that c_{opt} of high-order modal are close, indicating that a set of optimal dampers can simultaneously control multi-mode vibration. Moreover, c_{opt} decreased and ζ_{opt} increased with the increase of the modal order, except for the 4, 6, and 9th modes. The inefficiency of the 4, 6, and 9th modes is because the side spans participated in the modal shape. Under the considered cases, the modal amplitude at the damper location increased with D. When D increased from 5m to 20m, c_{opt} became over ten times smaller and ζ_{opt} increased by a factor of 1.5 times.

In the multi-mode optimization, the target modes were chosen based on the following considerations. For modes with low frequency, the lock-in wind speed of VIV is low and narrow. Since high turbulence intensity is unavoidable for low wind speed, VIV rarely occurs in low frequency modes. From the field monitoring of VIVs of the Xihoumen suspension bridge, VIV is more likely to occur at a wind velocity above 5 m/s (Li et al., 2017). On the other hand, for modes with high frequency, the lock-in wind speed of VIV is high, and their occurrence probability is considerably low. Moreover, the operational wind velocity for a long-span bridge is usually below 25 m/s, and it is suggested by the Wind-resistant Design Specification for Highway Bridge of China that only large-amplitude VIV happens below the wind speed of 25 m/s needs to be controlled. Therefore, it is reasonable to choose the target modes corresponding to the VIV wind speed within the range of 5–25 m/s.

For the case of Yingwuzhou Yangtze River Bridge, it can be seen from

Table 1
The first 23rd vertical modes of the undamped structure.

Mode	Modal shape	Frequency (Hz)	Equivalent mass (t/m)	VIV starting wind speed (m/s)
v1		0.1008	40.58	2.9
v2		0.1277	41.07	3.7
v3		0.1280	41.17	3.7
v4		0.1607	41.49	4.6
v5		0.1827	40.85	5.3
v6		0.2105	52.97	6.1
v7		0.2421	40.94	7.0
v8		0.2421	40.85	7.0
v9		0.2888	107.81	8.4
v10		0.3086	40.83	8.9
v11		0.3162	46.20	9.1
v12		0.3779	40.94	10.9
v13		0.3780	40.94	10.9
v14		0.4526	41.06	13.1
v15		0.4532	41.08	13.1
v16		0.5325	41.30	15.4
v17		0.5325	41.29	15.4
v18		0.6191	41.57	17.9
v19		0.6192	41.57	17.9
v20		0.7126	41.98	20.6
v21		0.7126	41.95	20.6
v22		0.8140	42.42	23.5

(continued on next page)

Table 1 (*continued*)

Mode	Modal shape	Frequency (Hz)	Equivalent mass (t/m)	VIV starting wind speed (m/s)
v23		0.8140	42.42	23.5

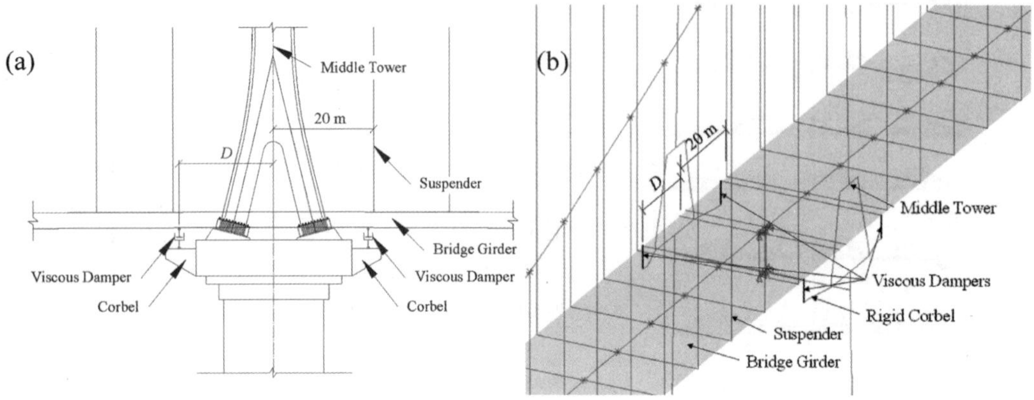

Fig. 9. The (a) Schematic and (b) finite element model of the damper arrangement.

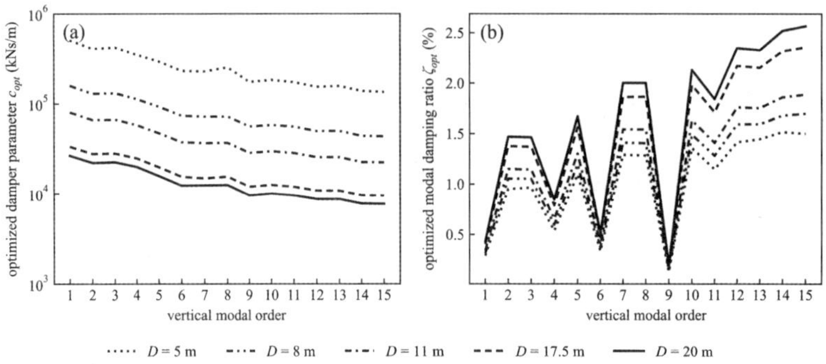

Fig. 10. Optimal (a) damper parameters and (b) modal damping ratio for each vertical mode and damper location.

Table 1 that the vertical bending mode 5 to mode 23 corresponds to the wind speed range of 5–25 m/s. In the design of the damper system, special consideration was given to the 7th mode that experienced VIV during the field monitoring. It was set to be the lower bound in multi-mode optimization to ensure that its additional damping ratio is large enough. For higher modes, more attention was paid to those corresponding to VIV wind speed below 15 m/s because the monitoring data of wind speed in recent years showed that the wind speed on the bridge deck is mainly below 15 m/s. Moreover, the optimal damper parameters and additional damping ratios for modes higher than the 15th are close based on the parameter analysis. In addition, the modal displacement at the damper position for the 9th vertical mode is so small that dampers showed little efficiency in controlling it, so this mode was not considered in the optimization. Based on the above consideration, multi-mode

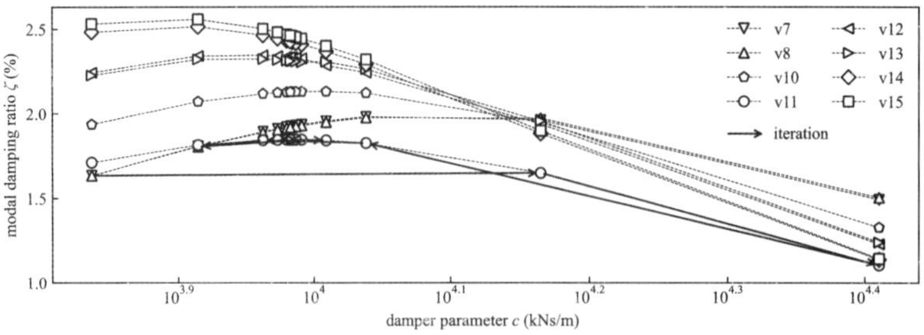

Fig. 11. Iteration of the multi-mode optimization.

optimization was performed for the 7, 8, and 10-15th vertical modes in the present study.

The dampers were placed under the closest suspenders ($D = 20$ m). The algorithm took thirteen iterations to make the relative accuracy reach 1%, as shown in Figs. 11 and 12. The arrow line illustrates that the iteration started from the left point and converged at the optimal point. In this study, c_{opt} was equitant to the optimal parameter for the 11th mode. Moreover, the additional damping for each considered mode had little variance near the optimal point, suggesting that the optimal damper parameter was stable and robust.

4.3. Wind tunnel test on the effect of damping ratio

To examine the contribution of the damping ratio to the VIV characteristics, a wind tunnel test was conducted in the HD-2 wind tunnel at Hunan University. The structure was simplified to a 2-DOF rigid suspension model based on the similarity of the first vertical and torsional modes. The rigid model was made up of a steel skeleton and wooden cover. A total of eight springs hung the model on the aluminum frame, as shown in Fig. 13. The structure parameters of the model and real bridge are listed in Table 2.

In the present study, the influence of the damping ratio was examined under a wind attack angle of +3°, which was the worst condition in the previous experiment. Three laser displacement sensors were adopted to measure the vibration displacements. The incoming wind speed was measured by an ultrasonic wind speed indicator, which is located 2 m in front of the model to eliminate the influence on the wind field. The effect of damper location is ignored because the damper was located near towers in the present study with no change in the sectional characteristic of the bridge girder. The structure damping ratio was adjusted by two plate-typed eddy current dampers, whose damping force can be precisely adjusted by tunning the air gaps (Hua et al., 2022), and the amplitude-related damping characteristic of each damping state is shown in Fig. 14. A total of 6 damping states were examined in the present study. The damping ratio was observed to increase slightly with the vibration amplitude. It should be noted that the damping characteristic was different between the sectional model and the real bridge. The value corresponding to the peak VIV amplitude was used as the nominal damping ratio in the present study which overestimated the structural damping, and the resulting response was expected located under the safety side.

Fig. 15(a) illustrates the vertical response amplitude of the wind speed under different damping ratios. The effect of the modal shape was considered by multiplying a correction factor of 1.15 based on the previous studies (Zhang et al., 2014). A dashed line was appended to indicate the maximum allowed amplitude of the 7th mode by the Chinese specification (Ministry of Communications of the People's Republic of China, 2018). Two VIV lock-in interval was observed in a lower

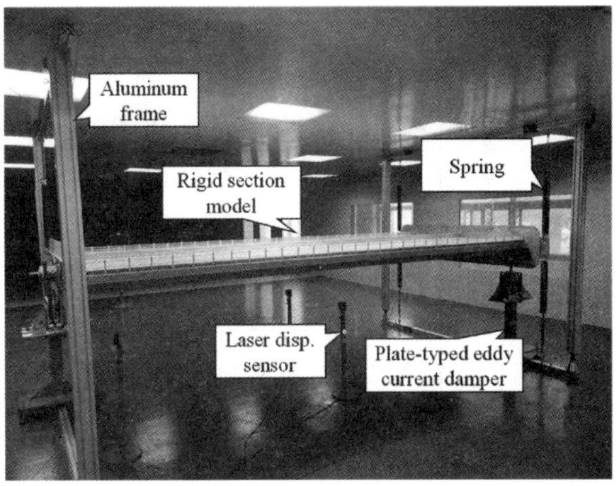

Fig. 13. Photo of the wind tunnel experiment.

Table 2
Parameter comparison between the model and actual bridge.

Parameter	Actual	Scale ratio	Model
Rigid model length L(m)	/	/	3.78
Girder width B(m)	38	1/25	1.52
Girder height H(m)	3	1/25	0.12
Equivalent mass m_{eq}(kg/m)	40581.3	$1/25^2$	64.93
Equivalent inertia momentum I_{eq}(kg·m^2/m)	5759780	$1/25^4$	14.75
Frequency of the 1st vertical mode f_{v1}(Hz)	0.101	1/23.9	2.411

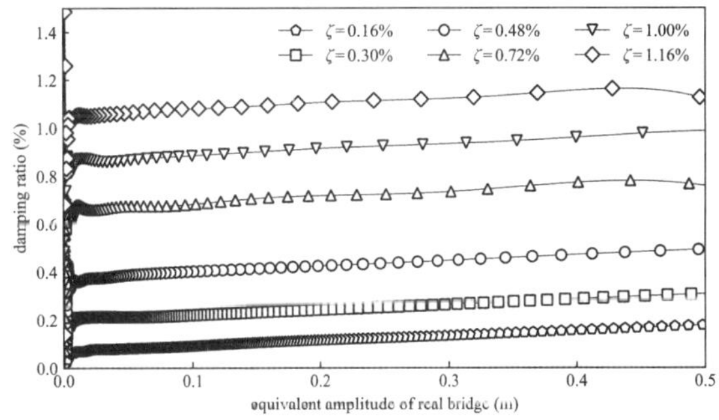

Fig. 14. The amplitude-related damping characteristics of the sectional model.

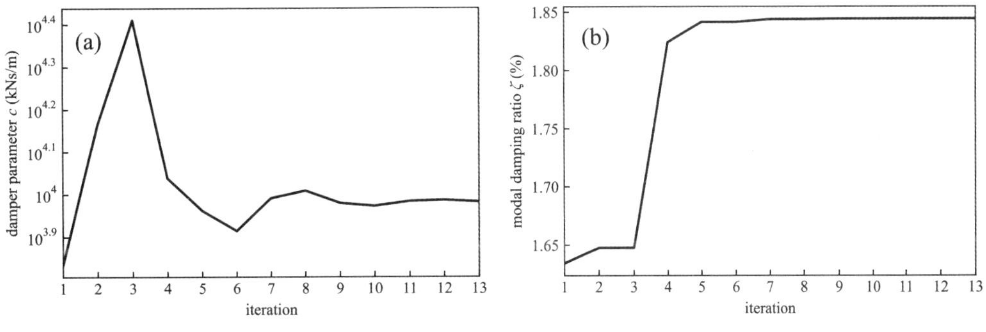

Fig. 12. Multi-mode optimization convergence curve of (a) the damping parameter and (b) the minimum modal damping ratio.

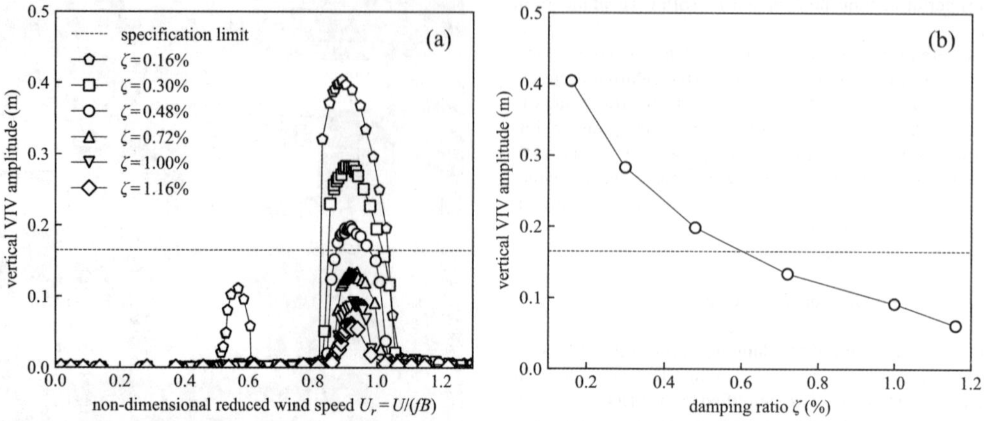

Fig. 15. The vertical VIV amplitude versus (a) wind speed and (b) damping ratio.

damping ratio, representing two St numbers, $St_1 = 0.152$ and $St_2 = 0.098$. The small interval with lower wind speed quickly disappeared when the damping ratio increased. The other interval with higher wind speed had a similar St number with the monitored VIV event ($St = f_{v7} \times H/U = 0.2441 \times 3/7 = 0.105$), and the VIV amplitude had an approximately inversely proportional relation to the damping ratio, as shown in Fig. 15(b). To satisfy the specification requirement, the damping ratio must be larger than 0.6% in the real bridge.

4.4. *Field installation and measurement*

Based on the aboved numerical examination, the dampers' designed parameter c was set to 1×10^4kNs/m and the dampers were placed under the first suspenders (Hua et al., 2023). The additional damping ratio of each vertical mode is listed in Table 3. All the mode with VIV starting wind speed lower than 25 m/s was validated. Similar to that in the cable's vibration control, the higher mode is expected to have a better control effect for its higher additional damping ratio.

It should be addressed that the modal characteristics are slightly different between structures with or without dampers. For the 7th vertical mode in which VIV occurred, the modal frequency increases from 0.2421 Hz to 0.2456 Hz, and the difference in the modal shape is illustrated in Fig. 16. The modal shape of the damped structure consists of real and imaginary parts. The real part is similar to the undamped modal shape and turning points exist at the damper locations ($D = \pm 20$ m). Different from the structure without dampers, there is no zero-magnitude stationary point in the modal shape of the damped structure due to the contribution of the imaginary part of the complex modal shape. The difference of modal frequency and modal shape was ignored in the present study and the modal damping ratio is considered as the main factor for field measurement.

Table 3
Numerical predicted additional damping ratio of each vertical mode.

Mode	Numerical predicted additional damping ratio	Mode	Numerical predicted additional damping ratio	Mode	Numerical predicted additional damping ratio
v1	0.282%	v9	0.191%	v17	2.490%
v2	1.079%	v10	2.131%	v18	2.497%
v3	1.064%	v11	1.842%	v19	2.533%
v4	0.667%	v12	2.295%	v20	2.512%
v5	1.487%	v13	2.317%	v21	2.544%
v6	0.496%	v14	2.389%	v22	2.493%
v7	1.944%	v15	2.422%	v23	2.537%
v8	1.939%	v16	2.469%		

In this project, a type of torsional eddy current damper (TECD) is adopted. Fig. 17 shows the construction and figure of the damper. The connecting rods translate the vertical vibration and release the constraint of the lateral and torsional displacement. The translated vertical vibration is then amplified by the gear system and then dissipated by the torsional damper. Key parameters of the TECD are listed in Table 4. The TECD only weighs about 600 kg, which is a thousand times less than the additional mass required for TMD (calculated by 1% of the modal mass of the main spans, that is, 1% × 40 t × 1700 m = 680 t). Compared to the traditional oil damper, TECD can provide a large damping coefficient under low vibration speed. Its metal structure requires low maintenance costs and has no oil leakage problems.

The structural dynamic characteristics were measured under the condition of closed traffic for five nights from July 24 to July 28, 2021. The environment wind speed was relatively small, no VIV event happened, and the contribution of the aerodynamic damping was ignored. The measurement compared the structural damping ratio under the conditions with or without dampers. To enlarge the response amplitude and improve the measurement accuracy, the continuous vehicle bumping excitation method was used for excitation (Hua et al., 2017). Since the frequency of the symmetry and anti-symmetry modes were close, the monitors were installed on the half side of the bridge. A total of 36 wireless acceleration sensors were installed to monitor the vibration characteristics. Fig. 18 shows a velocity time history obtained in an excitation experiment without or with dampers. The background vibration was suppressed after the damper was installed. The stochastic subspace identification (SSI) technique was adopted to extract the modal frequencies and the modal damping ratios. The SSI method has a complete theory and powerful algorithm and is one of the most effective modal parameter recognition methods based on environmental vibration or free vibration. The extracted results were averaged over repeated experiments of different excitation modes and listed in Table 5. The extracted modes were labeled by the closest anti-symmetry modes in the finite element model. The extracted damping ratio of the structure without dampers was between 0.25% and 0.38%, and the theoretical VIV amplitudes of all the modes exceeded the specification limits. For the bridge with dampers, the modal damping ratio is between 0.64% and 1.11%. For the 7th vertical model in which VIV occurred, the additional damping ratio increased from 0.34% to 1.00% and the predicted amplitude of VIV met the requirements of the specification. In addition, the VIV locking interval was reduced to half of the original one based on the result of the wind tunnel test, and the probability of VIV occurrence should be greatly reduced. Meanwhile, the structure health monitoring system has been updated after the VIV event, and three years have passed since the dampers were installed, and no VIV has occurred again

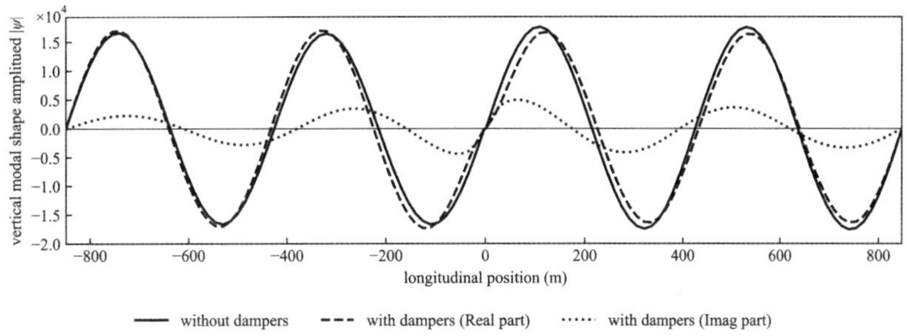

Fig. 16. The modal shape amplitude difference of structures with or without dampers.

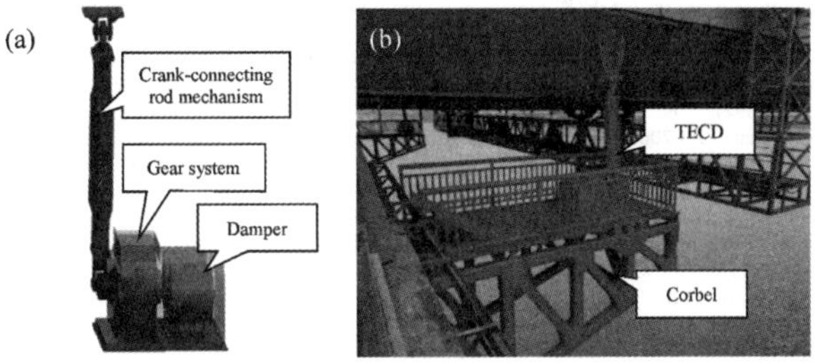

Fig. 17. The (b) construction and (c) photo of the torsional eddy current damper.

Table 4
Key parameters of the TECD.

Size (mm)	Weight (kg)	Working stroke (mm)	Maximum damping force (kN)	Damping Coefficient (kNs/m)
1600 × 1200	600	±80	500	10000

on the bridge. Though a further track should be conducted to validate the control performance of the dampers, the solution is considered a feasible approach in VIV controlling for long-span suspension bridges. However, the measured damping ratio was much smaller than the numerical results, especially for the high-order modes. Similar results have been observed and reported by Zoltowski et al. (2022). One possible reason was that the structural gaps and frictions of the dampers and the bridge supports made the structural system inconsistent with the numerical model. In addition, the vibration amplitude excited by the jump load was too small to fully unleash the performance of the dampers. For example, the static friction may lock the support's rotation and the bridge pin support will change to a fixed support. The modal shape will hence change and reduce the modal displacement at the dampers' location. It suggests that the local characteristic between dampers and bridge supports plays a significant role in the dampers' performance. A further examination is required to explore the inconsistency.

5. Conclusions

In the present study, a discrete linear viscous damper solution is proposed to suppress the multi-mode VIV response of the long-span suspension bridges. The construction concept and the optimization process of the damping system are fully addressed, and an application on the Yingwuzhou Yangtze River Suspension Bridge is presented in detail as an illustration. The main conclusions are as follows:

1. Viscous dampers near the support of girder are a viable method to increase the modal damping of long-span suspension bridges and suppress vertical VIV, provided there is enough distance from the support to the point of attachment of the damper. The maximum possible modal damping increases with this distance.
2. For a specified location, the maximum possible damping can be obtained by employing an optimal damper size, which varies with the mode number. Fortunately, the difference in optimal damper size gradually decreases as the increasing mode number, which indicates that the viscous damper can robustly control multi-mode VIV

Table 5
The monitored modal frequency and damping ratio without or with dampers.

Mode	Frequency (Hz)	Damping ratio without dampers (%)	Damping ratio with dampers (%)	Additional damping ratio (%)	Numerical predicted additional damping ratio (%)
v1	0.105	0.37	0.74	0.37	0.282%
v3	0.135	0.29	0.65	0.36	1.064%
v5	0.187	0.25	0.64	0.39	1.487%
v7	0.245	0.34	1.00	0.66	1.944%
v11	0.318	0.38	1.11	0.73	1.842%
v13	0.390	0.31	0.72	0.41	2.317%

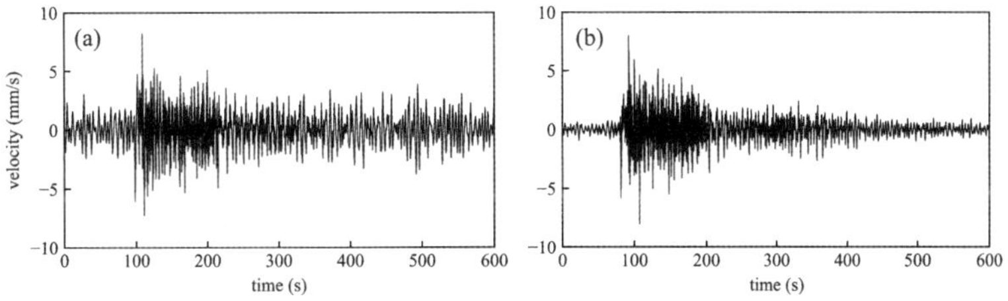

Fig. 18. Velocity time history during excitation (a) without or (b) with dampers.

simultaneously. However, a damper size optimized for the first several modes may result in ineffectiveness for controlling VIV in higher-order modes.

3. A design procedure for viscous dampers for damping of multiple modes, corresponding to vertical VIVs that occur in the wind speed range of 5~25 m/s, has been presented. In this method, the optimal damper size is obtained by maximizing the minimum modal damping ratio of all target modes. Such an optimization is able to provide enough damping capacity for all assigned modes of the suspension bridge.
4. The damping system has been applied successfully to the Yingwuzhou Yangtze River Suspension Bridge.The measured damping ratios for some critical modes are increased by three times after installation of the dampers. As for the mode that experienced VIV, its modal damping ratio has been raised high enough to suppress the VIV response according to the sectional model wind tunnel test results and the requirement of Wind-resistant Design Specification for Highway Bridges of China. Though a long-term monitoring should be conducted to validate damper's performance, three years have passed since the dampers were installed, and no VIV event has occurred again to the bridge.
5. It should be noted that there is still a gap between the additional damping ratios obtained from the field measurement and those from the theretical analysis. Many factors may influence this gap, including the support constraints in real bridge, a mechanical gap in the connection of the damper, and so on. Meanwhile, the sectional modal test cannot consider the contribution of the modal shape and 3D effect of the real bridge. Aeroelastic model testing should be examined in the future to validate and study the different performances between the model and the real bridge and the effect on the different wind conditions.

CRediT authorship contribution statement

Yiwen Cao: Validation, Investigation, Data curation, Writing – original draft, Writing – review & editing, Visualization. **Zhiwen Huang:** Methodology, Formal analysis, Validation, Visualization, Writing – review & editing. **Hongyi Zhang:** Validation, Investigation, Data curation, Writing – review & editing. **Xugang Hua:** Conceptualization, Methodology, Formal analysis, Validation, Writing – review & editing. **Zhengqing Chen:** Conceptualization, Methodology, Validation, Writing – review & editing, Supervision. **Tianbao Wan:** Validation, Data curation, Writing – review & editing.

Declaration of competing interest

The authors declare that they have no known competing financial interests or personal relationships that could have appeared to influence the work reported in this paper.

Data availability

Data will be made available on request.

Acknowledgment

This study was supported by the National Key R&D Program of China (2022YFC3005300) and the National Natural Science Foundation of China (No. 52278499). The authors would like to thank the teams of Prof. Shun Weng and Associate Prof. Qing Wen for providing the data of field tests, and Dr. Chao Yang for his help in establishing the methodology.

References

Bai, H., Ji, N., Xu, G., Li, J., 2020. An alternative aerodynamic mitigation measure for improving bridge flutter and vortex induced vibration (VIV) stability: sealed traffic barrier. J. Wind Eng. Ind. Aerod. 206, 104302 https://doi.org/10.1016/j.jweia.2020.104302.

Battista, R.C., Pfeil, M.S., 2000. Reduction of vortex-induced oscillations of Rio–Niterói bridge by dynamic control devices. J. Wind Eng. Ind. Aerod. 84, 273–288. https://doi.org/10.1016/S0167-6105(99)00108-7.

Caughey, T.K., O'Kelly, M.E.J., 1965. Classical normal modes in damped linear dynamic systems. J. Appl. Mech. 32, 583–588. https://doi.org/10.1115/1.3627262.

Chen, L., Sun, L., Xu, Yuyuan, Di, F., Xu, Yingmei, Wang, L., 2020. A comparative study of multi-mode cable vibration control using viscous and viscoelastic dampers through field tests on the Sutong Bridge. Eng. Struct. 224, 111226 https://doi.org/10.1016/j.engstruct.2020.111226.

Dai, J., Xu, Z.-D., Gai, P.-P., Hu, Z.-W., 2021. Optimal design of tuned mass damper inerter with a Maxwell element for mitigating the vortex-induced vibration in bridges. Mech. Syst. Signal Process. 148, 107180 https://doi.org/10.1016/j.ymssp.2020.107180.

Diana, G., Resta, F., Belloli, M., Rocchi, D., 2006. On the vortex shedding forcing on suspension bridge deck. J. Wind Eng. Ind. Aerod. 94, 341–363. https://doi.org/10.1016/j.jweia.2006.01.017.

Ehsan, F., Scanlan, R.H., 1990. Vortex-induced vibrations of flexible bridges. J. Eng. Mech. 116, 1392–1411. https://doi.org/10.1061/(ASCE)0733-9399(1990)116:6(1392).

Frandsen, J.B., 2001. Simultaneous pressures and accelerations measured full-scale on the Great Belt East suspension bridge. J. Wind Eng. Ind. Aerod. 89, 95–129. https://doi.org/10.1016/S0167-6105(00)00059-3.

Fujino, Y., Yoshida, Y., 2002. Wind-induced vibration and control of Trans-Tokyo Bay crossing bridge. J. Struct. Eng. 128, 1012–1025. https://doi.org/10.1061/(ASCE)0733-9445(2002)128:8(1012).

Giaralis, A., Petrini, F., 2017. Wind-induced vibration mitigation in tall buildings using the tuned mass-damper-inerter. J. Struct. Eng. 143, 04017127 https://doi.org/10.1061/(ASCE)ST.1943-541X.0001863.

Gu, M., Chen, S.R., Chang, C.C., 2002. Control of wind-induced vibrations of long-span bridges by semi-active lever-type TMD. J. Wind Eng. Ind. Aerod. 90, 111–126. https://doi.org/10.1016/S0167-6105(01)00165-9.

Hua, X., Ma, W., Huang, Z., Chen, Z., Wan, T., Sun, Y., Liu, S., 2022. Effect of Sc number on vertical vortex-induced vibration of bridges based on precise damping adjustment. J. Hunan Univ., Nat. Sci. 49, 17–25. https://doi.org/10.16339/j.cnki.hdxbzkb.2022045.

Hua, X., Zhang, J., Wan, T., Sun, Y., Chen, Z., 2023. A tower-girder damping system to control multi-mode vertical vortex-induced vibrations of suspension bridge. Bridge Constr. 53, 16–24. https://doi.org/10.20051/j.issn.1003-4722.2023.04.003.

Hua, X., Zhou, Y., Yang, K., Wen, Q., Chen, Z., Li, Y., 2017. Damping identification of long-span bridges from continuous vehicle bumping excitations. J. Railw. Sci. Eng. 14, 1664–1673. https://doi.org/10.19713/j.cnki.43-1423/u.2017.08.013.

Huang, Z., Hua, X., Chen, Z., Niu, H., 2019. Optimal design of TVMD with linear and nonlinear viscous damping for SDOF systems subjected to harmonic excitation. Struct. Control Health Monit. 26 https://doi.org/10.1002/stc.2413.

Jones, N.P., Scanlan, R.H., Sarkar, P.P., Singha, L., 1995. The effect of section model details on aeroelastic parameters. J. Wind Eng. Ind. Aerodyn., Third Asian-Pacific Sympos. Wind Eng. 54–55, 45–53. https://doi.org/10.1016/0167-6105(94)00028-C.

Kiefer, J., 1953. Sequential minimax search for a maximum. Proc. Am. Math. Soc. 4, 502–506. https://doi.org/10.1090/S0002-9939-1953-0055639-3.

Krenk, S., 2004. Complex modes and frequencies in damped structural vibrations. J. Sound Vib. 270, 981–996. https://doi.org/10.1016/S0022-460X(03)00768-5.

Kubo, Y., Kimura, K., Sadashima, K., Okamoto, Y., Yamaguchi, E., Kato, K., 2002. Aerodynamic performance of improved shallow p shape bridge deck. J. Wind Eng. Ind. Aerod. https://doi.org/10.1016/S0167-6105(02)00328-8.

Laima, S., Li, H., Chen, W., Ou, J., 2018. Effects of attachments on aerodynamic characteristics and vortex-induced vibration of twin-box girder. J. Fluid Struct. 77, 115–133. https://doi.org/10.1016/j.jfluidstructs.2017.12.005.

Larose, G.L., D'Auteuil, A., 2006. On the Reynolds number sensitivity of the aerodynamics of bluff bodies with sharp edges. J. Wind Eng. Ind. Aerodyn., Eighth Ital. Nat. Conf. Wind Eng. IN-VENTO-2004 94, 365–376. https://doi.org/10.1016/j.jweia.2006.01.011.

Larsen, A., Esdahl, S., Andersen, J.E., Vejrum, T., 2000. Storebælt suspension bridge – vortex shedding excitation and mitigation by guide vanes. J. Wind Eng. Ind. Aerodyn., Int. Conf. wind Eng. 88, 283–296. https://doi.org/10.1016/S0167-6105(00)00054-4.

Li, S., Laima, S., Li, H., 2017. Cluster analysis of winds and wind-induced vibrations on a long-span bridge based on long-term field monitoring data. Eng. Struct. 138, 245–259. https://doi.org/10.1016/j.engstruct.2017.02.024.

Main, J.A., Krenk, S., 2005. Efficiency and tuning of viscous dampers on discrete systems. J. Sound Vib. 286, 97–122. https://doi.org/10.1016/j.jsv.2004.09.022.

Ministry of Communications of the People's Republic of China, 2018. Wind-resistant Design Specification for Highway Bridges. China Communications Press, Beijing, China. JTG/T 3360-01-2018.

Møller, R.N., Krenk, S., Svendsen, M.N., 2019. Damping system for long-span suspension bridges. Struct. Control Health Monit. 26 https://doi.org/10.1002/stc.2448.

Schewe, G., Larsen, A., 1998. Reynolds number effects in the flow around a bluff bridge deck cross section. J. Wind Eng. Ind. Aerod. 74–76, 829–838. https://doi.org/10.1016/S0167-6105(98)00075-0.

Simiu, E., Scanlan, R.H., 1996. Wind Effects on Structures: Fundamentals and Applications to Design, third ed. John Wiley, New York.
Sun, Z., Zou, Z., Ying, X., Li, X., 2020. Tuned mass dampers for wind-induced vibration control of Chongqi bridge. J. Bridge Eng. 25, 05019014 https://doi.org/10.1061/(ASCE)BE.1943-5592.0001510.
Weber, F., Maślanka, M., 2012. Frequency and damping adaptation of a TMD with controlled MR damper. Smart Mater. Struct. 21, 055011 https://doi.org/10.1088/0964-1726/21/5/055011.
Xu, K., Bi, K., Ge, Y., Zhao, L., Han, Q., Du, X., 2020. Performance evaluation of inerter-based dampers for vortex-induced vibration control of long-span bridges: a comparative study. Struct. Control Health Monit. 27 https://doi.org/10.1002/stc.2529.
Xu, K., Bi, K., Han, Q., Li, X., Du, X., 2019. Using tuned mass damper inerter to mitigate vortex-induced vibration of long-span bridges: analytical study. Eng. Struct. 182, 101–111. https://doi.org/10.1016/j.engstruct.2018.12.067.
Xu, K., Dai, Q., Bi, K., Fang, G., Ge, Y., 2022. Closed-form design formulas of TMDI for suppressing vortex-induced vibration of bridge structures. Struct. Control Health Monit. https://doi.org/10.1002/stc.3016.
Yang, C., Chen, Z., Wang, W., Hua, X., Wang, Y., 2021. Optimal design of two viscous dampers for multi-mode control of a cable covering broad frequency range. Eng. Struct. 245, 112830 https://doi.org/10.1016/j.engstruct.2021.112830.
Yang, Y., Zhou, R., Ge, Y., Zhang, L., 2016. Experimental studies on VIV performance and countermeasures for twin-box girder bridges with various slot width ratios. J. Fluid Struct. 66, 476–489. https://doi.org/10.1016/j.jfluidstructs.2016.08.010.
Zhang, Z., Ge, Y., Chen, Z., 2014. Vortex-induced oscillations of bridges: theoretical linkages between sectional model tests and full bridge responses. Wind Struct. Int. J. 19, 233–247. https://doi.org/10.12989/was.2014.19.3.233.
Zhao, L., Cui, W., Shen, X., Xu, S., Ding, Y., Ge, Y., 2022. A fast on-site measure-analyze-suppress response to control vortex-induced-vibration of a long-span bridge. Structures 35, 192–201. https://doi.org/10.1016/j.istruc.2021.10.017.
Zhu, J., Xiong, Z., Xiang, H., Huang, X., Li, Y., 2021. Ride comfort evaluation of stochastic traffic flow crossing long-span suspension bridge experiencing vortex-induced vibration. J. Wind Eng. Ind. Aerod. 219, 104794 https://doi.org/10.1016/j.jweia.2021.104794.
Zoltowski, K., Banas, A., Binczyk, M., Kalitowski, P., 2022. Control of the bridge span vibration with high coefficient passive damper. Theoretical consideration and application. Eng. Struct. 254, 113781 https://doi.org/10.1016/j.engstruct.2021.113781.

Journal of Wind Engineering & Industrial Aerodynamics 250 (2024) 105782

Contents lists available at ScienceDirect

Journal of Wind Engineering & Industrial Aerodynamics

journal homepage: www.elsevier.com/locate/jweia

Experimental and numerical study on vibration response, VIV mechanism and aerostatic coefficients of triple-separated decks under different vertical distances

Han Xiao [a,b], Zhiwen Liu [a,b,*], Zhengqing Chen [a,b], Zhenbiao Liu [c], Aiguo Yan [c], Ou Yang [a,b]

[a] State Key Laboratory of Bridge Engineering Safety and Resilience, Hunan University, Changsha, Hunan, 410082, China
[b] College of Civil Engineering, Hunan University, Changsha, Hunan, 410082, China
[c] China Railway Siyuan Survey and Design Group Co., LTD., Wuhan, Hubei, 430063, China

ARTICLE INFO

Keywords:
Triple-separated decks
Aerodynamic interference effects
Vibration response
Vortex-induced vibration mechanism
Aerostatic coefficient
Wind tunnel tests
Numerical simulations

ABSTRACT

In this study, aerodynamic interference effects (AIE) on the vibration response, aerostatic coefficient (AC) and vortex-induced vibration (VIV) mechanism of the triple-separated decks under different incoming wind directions and vertical distances were investigated via wind tunnel tests and numerical simulations, respectively. The test results show that the AIE increases the displacement of VIV, and the VIV responses reaches maximum value under S/H = −1.0 (S is the vertical distance between the centers of RB and HB decks, and H is the height of RB deck) when the RB deck is on the windward side (the maximum displacements of VIV are 193.86% and 37.30% higher than the case of single RB deck and S/H = 0.0, respectively). The pressure distribution and flow structure show that the scale of vortex that alternately shed off from the tail of the deck becomes stronger and the fluctuating pressure coefficient increases under AIE, which increase the VIV amplitude of the RB deck and cause the VIV of HB decks. The experimental and numerical results indicated that the aerostatic coefficients (AC) of the leeward girder are more affected by the AIE. The interference factor (IF) of drag coefficient (C_D) is far less than 1.0 under the case of F-5 (S/H = 0.0).

1. Introduction

In order to meet the demand of the increasing of the traffic volume, more and more parallel bridges or bridges with two separated main decks were built in past years over the world, such as the Kansai International Airport Access Bridge, Onomichi bridge and Meiko Nishi bridge in Japan (Okubo and Enami, 1972; Honda et al., 1990; Suzuki et al., 1999); Tacoma Narrows Bridge and Fred Hartman Bridge in the USA (Svensson and Lovett, 1990; Irwin et al., 2005); Jindo Bridge in South Korea (Seo et al., 2013; Kim et al., 2013); Haihe Bridge, Yongjiang Bridge, Hongdao Bridge, and Pingsheng Bridge in China (Liu et al., 2008; Liu et al., 2009; Meng et al., 2011; Chen, 2016). Existing researches show that the aerodynamic interference effect (AIE) between the separated decks cannot be ignored. Liu et al. (2008) and (2009) studied the AIE between the twin-parallel rectangular decks, streamlined decks and Π-shaped decks through wind tunnel tests and numerical simulations, respectively. The results show that the AIE not only affects the aerostatic coefficients, but also affects the wind-induced vibration (WIV) of these twin-parallel decks. Moreover, Seo et al. (2013), Kim et al. (2013) and Park et al. (2017) investigated the AIE of VIV responses of the two separated decks of Jindo Bridge in South Korea via wind tunnel tests. The results show that the AIE can affect the VIV performance and change the frequencies of the sectional models, and the VIV responses of the parallel decks are mainly affected by the horizontal net distances between them.

As the AIE becomes more and more obvious between parallel decks, researchers have conducted many studies on the AIE under different factors, such as incoming wind direction, wind attack angle and horizontal net distance. Argentini et al. (2015) and Liu et al. (2022) found that different types of decks have different aerodynamic shapes, frequencies and qualities, which can cause the WIV performance under the AIE in different incoming wind directions. For the studies of horizontal net distance, which is one of the most common factors, an earlier study has found that the aerodynamic interference could be significant even with a horizontal net distance seven times as large as the width of girder (Kimura et al., 2008). Seo et al. (2013), Kim et al. (2013) and Park et al.

* Corresponding author. State Key laboratory of Bridge Engineering Safety and Resilience, Hunan University, Changsha, Hunan, 410082, China.
E-mail address: zhiwenliu@hnu.edu.cn (Z. Liu).

https://doi.org/10.1016/j.jweia.2024.105782
Received 21 December 2023; Received in revised form 9 May 2024; Accepted 22 May 2024
Available online 29 May 2024

(2017) also verified this result when conducting research on Jindo Bridge in South Korea. The results show that for small D/B of 0.7–1.3, only the VIV of the downstream deck occurs, and when the D/B increases to 2.1–5.4, the vortex at the tail of the upstream deck circulates and amplifies between the two decks, resulting in vibrations in both the upstream and downstream decks. When the D/B increases to more than 7.4, the VIV disappears, and the AIE between parallel decks can be considered to disappear. For separated decks with different types of aerodynamic shapes, the existence of vertical distance between separated decks is common, such as the Kansai International Airport Access Bridge in Japan, the Fenghua River Bridge (Zeng, 2014) and the Yongjiang Bridge in China. According to existing research, the AIE under different vertical distance not only affects the VIV, but also affects the aerostatic coefficient (Zhou et al., 2017).

Based on the above research, current research on the AIE mostly focuses on the incoming wind direction, wind attack angle and horizontal net distance, while few studies on vertical distance. Moreover, when studying the AIE on aerodynamic performance, most of the research focuses on the aerostatic coefficients and VIV. However, existing studies have shown that AIE can also affect flutter stability (Zhu et al., 2010). The author has conducted a detailed study on the VIV performance of two parallel railway- (RB) and highway (HB) bridges with triple-separated decks under AIE and the effects of different incoming wind directions, horizontal net distance and vertical distance were considered Xiao et al., (2024). In this study, the highway bridge (HB) with two separated decks and railway Bridge (RB) with single deck over Quanzhou Bay projects are also taken as the engineering background, and the AIE on vibration responses of the two bridges are further conducted by wind tunnel tests under different incoming wind directions and vertical distances. Moreover, the AIE on aerostatic coefficient (AC) and VIV mechanism are conducted by wind tunnel tests and numerical simulations under different cases. The organization of the paper is as follows: The introduction of RB and HB, the experimental setup adopted for this study, and the test cases are presented in Section 2. Moreover, Section 3 describes the experimental results, including the vibration responses of RB and HB under different incoming wind directions and vertical distances. Furthermore, the pressure distribution of the RB is obtained through the pressure-measurement tests and the AIE

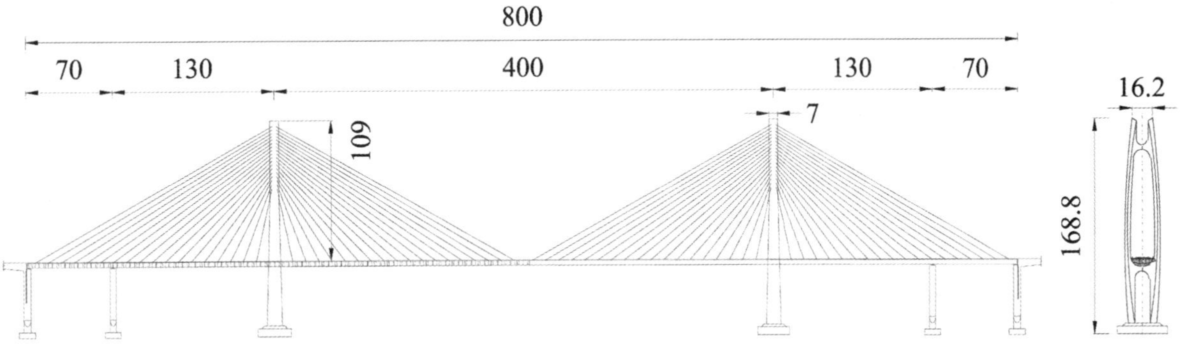

(a)

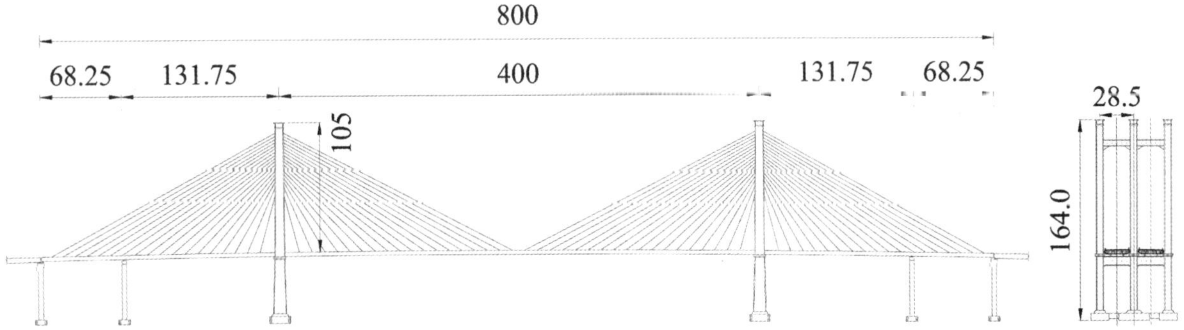

(b)

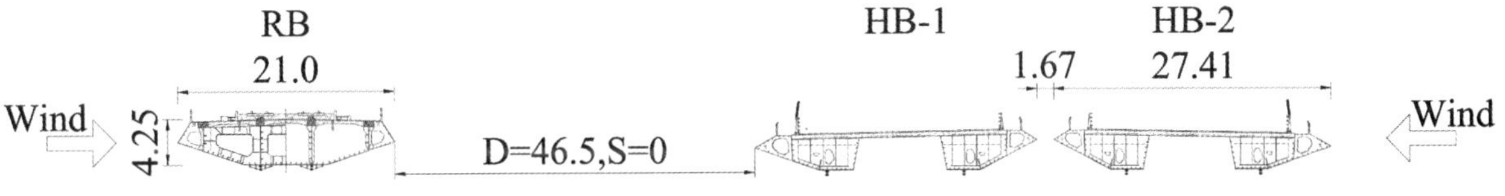

(c)

Fig. 1. Elevation layout of RB and HB, (a) Elevation layout of RB, (b) Elevation layout of HB, and (c) Main girder sections of RB and HB (unit: m).

was analyzed. Then, Section 4 describes the ACs of the RB and HB obtained from force-measurement tests and numerical simulations, respectively. Moreover, based on the flow structures obtained from numerical simulations, the mechanism of AIE is analyzed combined with the pressure distributions in Section 4. Finally, the main conclusions are summarized in Section 5.

2. Experimental setup

2.1. *Engineering background*

The newly built RB with single deck and the existing HB with twin-separated decks over Quanzhou Bay are taken as the background engineering. The RB is a cable-stayed bridge with a steel-concrete composite girder. The width and height of the streamlined-box steel-concrete composite girder of the RB are B = 21.0m and H = 4.25m, respectively. The HB is a cable-stayed bridge with two separated P–K steel-concrete composite girder sections. The width and height of the single main deck of HB are B = 27.41 m and H = 3.5 m, respectively, and the horizontal net distance of the two separated parallel decks is D = 1.67 m. Moreover, the horizontal net distance and vertical distance between the main decks of the two bridges are D = 46.5 m, and S = 0.0 m, respectively. The elevation, pylons, and main decks of RB and HB are shown in Fig. 1, respectively. The reference basic wind speed of the RB and HB site is 34.0 m/s, and the design wind speed at the main deck level of RB and HB is 51.8 m/s.

2.2. *Experimental set-up*

2.2.1. *Sectional model*

In order to meet the requirements of the *Wind-resistant Design Specifications for Highway Bridges (JTG/T 3360–01 2018)*, the geometry scale of the sectional model is selected as λ_L = 1:60, and the blocking ratio of the main decks of RB and HB in the wind tunnel is 2.83% and 2.33%, respectively. The AIE on aerodynamic characteristic of RB and HB under different vertical distances are investigated in the first test section of HD-2 wind tunnel, Hunan University. The main parameters of the elastic-supported sectional models of the main decks of RB and HB are shown in Table 1. It should be noted that the damping ratios of the main deck model of RB are less than the recommended values by *Wind-resistant Design Specifications for Highway Bridges (JTG/T 3360–01 2018)*.

The vibration responses of the main deck of RB and HB are measured by 8 Keyence laser displacement meters. The DH5920 multi-channel signal acquisition system is used to collect the time varying displacement signals with a sampling frequency of 200 Hz, and a sampling time of 60 s. The average wind speed monitoring in smooth flow is measured by the R. M. Young Model 81000 ultrasonic anemometer with an accuracy of 0.05 m/s, and collected with CR1000 provided by Campbell Scientific with a sampling frequency of 10.0 Hz, and a sampling time of 60s. The setup of the vibration-measurement tests is illustrated in Fig. 2.

According to existing research, when the vertical distance reached S/H = 2/3, the AIE still existed (S is the vertical distance between the centers of main decks of the RB and HB, and H is the height of the main deck of RB). Moreover, the aerodynamic shapes of RB and HB are different, so when the RB deck is located at windward or leeward side, the vibration responses would be different under different incoming wind directions. Therefore, the cases of the single RB deck and twin-separated HB decks, 2 test cases of different incoming wind directions (i.e. the RB deck is located at the windward side and the RB deck is located at the leeward side) and 3 test cases of different vertical distances (i.e. S/H = 0.0, −1.0 and 1.0) were conducted to investigate the AIE on vibration responses of the triple-separated decks at wind angle of attack of α = 0°. The test cases are shown in Table 2.

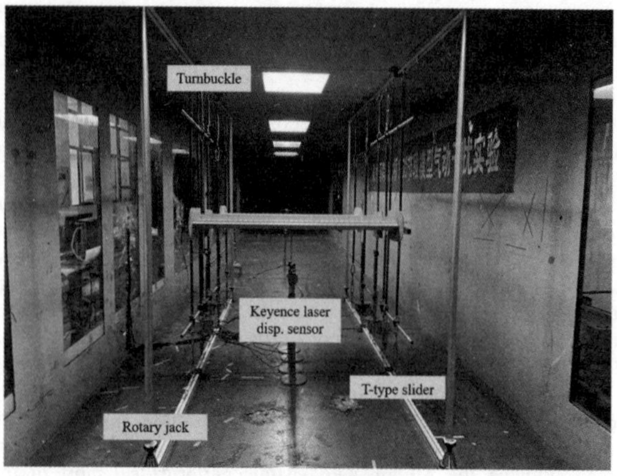

Fig. 2. Setup of the vibration-measurement tests.

2.2.2. *Pressure-measurement experiments*

Because the RB is under construction, the study of its wind stability is more important than that of the HB that has been completed in 2015. Moreover, the existing studies have shown that VIV do not occur in HB when there is no railway bridge next to it, so the pressure-measurement test is only conducted on the deck of RB. The pressure taps on the surface of RB are evenly arranged on both sides 10 cm away from the central axis of the main deck, with 54 taps set on each side and a total of 108 taps set on both sides. It should be noted that the pressure taps are only arranged on the surface of the main deck, and are not arranged on the maintenance rail, handrail and railway ballast. The layout position of the pressure taps and detailed pressure-measurement tests setup of RB are shown in Fig. 3. The pressure measurement and data collection use the DTC Initium measurement specialties pressure system, which includes a pressure measurement device fixed inside the model, a data collection PC device and software. The sampling frequency of the pressure system is 330Hz, the number of collected data is set to 10000, and the static pressure and total pressure values in the wind tunnel are

Table 1
Main parameters of the main deck sectional models of RB and HB.

Parameters	RB			HB		
	prototype bridge	Scale ratio	section mode	prototype bridge	Scale ratio	section mode
Length, L (m)	/	1:60	1.60	/	1:60	/
Width, B (m)	21.0	1:60	0.35	27.41	1:60	0.457
Height, H (m)	4.25	1:60	0.071	3.50	1:60	0.058
Equivalent mass, m_{eq} (kg/m)	5.21×10^4	$1:60^2$	14.472	3.565×10^4	$1:60^2$	9.903
Equivalent mass inertial moment, J_{meq} (kg.m^2/m)	2.130×10^6	$1:60^4$	0.1644	1.850×10^6	$1:60^4$	0.1427
Vertical frequency, f_h (Hz)	0.3946	10:1	3.946	0.3353	10:1	3.353
Torsional frequency, f_α (Hz)	1.3504	5.93:1	8.008	0.9497	6.30:1	5.981
Vertical damping ratio, ξ_h (%)	0.30	1:1	0.14	0.30	1:1	0.22
Torsional damping ratio, ξ_α (%)	0.30	1:1	0.17	0.30	1:1	0.31

Table 2
Vibration-measurement test cases.

Case	Configuration	Windward deck	Horizontal distance ratio (D/B)	Vertical distance ratio (S/H)	Vertical damping ratio, ξ_h (%)	Torsional damping ratio, ξ_α (%)
V-1	RB	/	/	/	$\xi_{h1}=0.14$	$\xi_{\alpha1}=0.17$
V-2	HB-1+HB-2	/	/	/	$\xi_{h2}=0.22$ $\xi_{h3}=0.22$	$\xi_{\alpha2}=0.31$ $\xi_{\alpha3}=0.31$
V-3	RB + HB-1+HB-2	RB	2.2	0.0	$\xi_{h1}=0.14\xi_{h2}=0.22$	$\xi_{\alpha1}=0.17\xi_{\alpha2}=0.31$
V-4				−1.0	$\xi_{h3}=0.22$	$\xi_{\alpha3}=0.31$
V-5				1.0		
V-6	HB-2+HB-1+RB	HB-2	2.2	0.0		
V-7				−1.0		
V-8				1.0		

Note: V indicates vibration-measurement test case. RB indicates the railway bridge deck. HB-1 and HB-2 indicate the highway bridge deck-1 and highway bridge deck-2, respectively. All cases are conducted under wind angle of attack of $\alpha = 0°$.

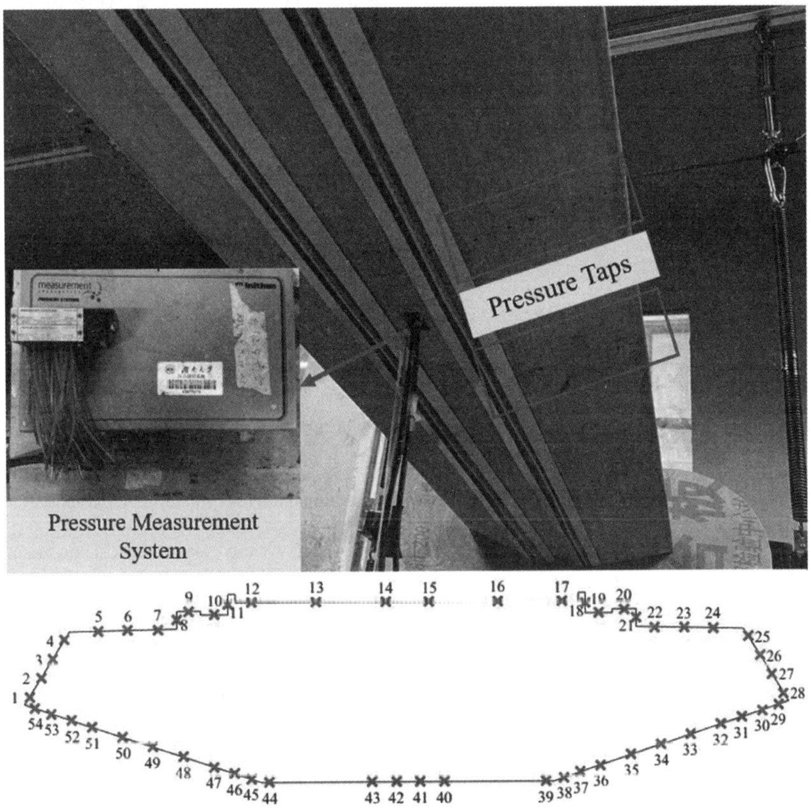

Fig. 3. Setup of pressure-measurement tests and locations of pressure taps for sectional model of RB.

measured through the pitot tube. The pressure-measurement tests and the vibration-measurement tests are conducted simultaneously, so that the surface pressure distribution under different vibration response conditions can be obtained. Therefore, the pressure-measurement test cases are consistent with the cases of vibration-measurement tests.

2.2.3. Force-measurement experiments

In order to obtain the ACs of the main deck under AIE, a special system shown in Fig. 4 was designed for force-measurement tests. In the force-measurement test system, a transverse fixation device is used to connect the sectional model of RB to a six-dimensional force-measurement device (KD4000) and the external control system of the device can also realize changes in different wind attack angles. The sectional models of HB are arranged in the wind tunnel through a self-made fixed bracket and the stability of the bracket can meet the needs of the wind tunnel test. Changing the position of the bracket can realize different incoming wind directions and moving the slide adjustment device up and down can realize different vertical distances between the separated decks. The force-measurement tests also just focus on the main deck of RB and the main decks of HB play the role of interference. Therefore, based on the cases of vibration-measurement tests, all 7 test cases were conducted to investigate the AIE on the AC of the main deck of RB and the test cases are shown in Table 3.

3. Experimental results

Based on the existing studies of the AIE on VIV performances (Xiao et al., 2024), there are significant differences in VIV of the triple-separated decks of RB and HB under different wind incoming wind directions, horizontal net distances and vertical distances. In order to

Fig. 4. Setup of force-measurement tests.

Table 3
Force-measurement test cases.

Case	Configuration	Windward deck	Horizontal distance ratio(D/B)	Vertical distance ratio(S/H)
F-1	RB	/	/	/
F-2	RB + HB-	RB	2.2	0.0
F-3	1+HB-2			−1.0
F-4				1.0
F-5	HB-2+HB-	HB-2	2.2	0.0
F-6	1+RB			−1.0
F-7				1.0

Note: F indicates force-measurement test case. RB indicates the railway bridge deck. HB-1 and HB-2 indicate the highway bridge deck-1 and highway bridge deck-2, respectively. All cases are conducted under wind angle of attack of $\alpha = 0°$.

study the more detailed wind-induced vibration responses and AIE mechanism on VIV performances, the vibration-measurement experiments of single RB deck, twin HB decks and triple RB and HB decks and the pressure-measurement experiments of RB deck under different cases were conducted, respectively. Considering there is currently little research on the effect of the vertical distance on AIE, this study focuses more on the effect of vertical distance under different incoming directions on the AIE.

3.1. Results of the main deck of RB

3.1.1. Vibration performance

Fig. 5 shows the RMS of vertical and torsional displacements of the main deck of RB vs. reduced wind velocity for different vertical distances when the RB is located at the windward, respectively. As shown in Fig. 5, for case of single RB deck (V-1), there is an obvious vertical VIV lock-in interval of the main deck of RB for reduced wind velocity V_{rh} = 2.52–3.90, and the maximum RMS of non-dimensional vertical displacements is σ_h/H = 0.0114, corresponding to the reduced wind velocity of V_{rh} = 3.10. For cases of triple-separated decks under different S/Hs when the RB deck is on the (V-3~5), affected by the AIE, the vibration response of RB increases, and existing of the vertical distance further changes the vibration response: (1) For the case of triple-separated decks under S/H = 0.0 when the RB deck is on the windward side (V-3), compared with the case of V-1, the maximum displacement of VIV increased by 114.03%, and the lock-in interval did not change under the AIE. As the wind velocity increases, the vibration response of RB deck is larger than that under the case of V-1. (2) For the case of triple-separated decks under S/H = −1.0 when the RB deck is on the windward side (V-4), compared to the case of V-1, the maximum displacement of VIV increased by 193.86%, and compared to S/H = 0.0, the maximum displacement of VIV increased by 37.30%. Moreover, as the wind velocity increases, the vibration response increases and becomes more unstable. (3) For the case of triple-separated decks under S/H = 1.0 when the RB deck is on the windward side (V-5), compared to the case of V-1, the maximum displacement of VIV increased by 159.65%, and compared to the case of S/H = 0.0, the maximum displacement of VIV increased by 21.31%. As the wind velocity increases, the vibration response is more stable than that of the cases of S/H = 0.0 and S/H = −1.0, while when the reduced velocity reaches 2.4, the torsional angle is greater than that of the cases of S/H = 0.0 and S/H = −1.0.

Fig. 6 shows the RMS of vertical and torsional displacements of the main deck of RB vs. reduced wind velocity for different vertical distances when the RB is located at the leeward, respectively. As shown in Fig. 6, no obvious VIV occurred and the main deck of RB showed an unsteady vibration response under the AIE: (1) For the case of triple-separated

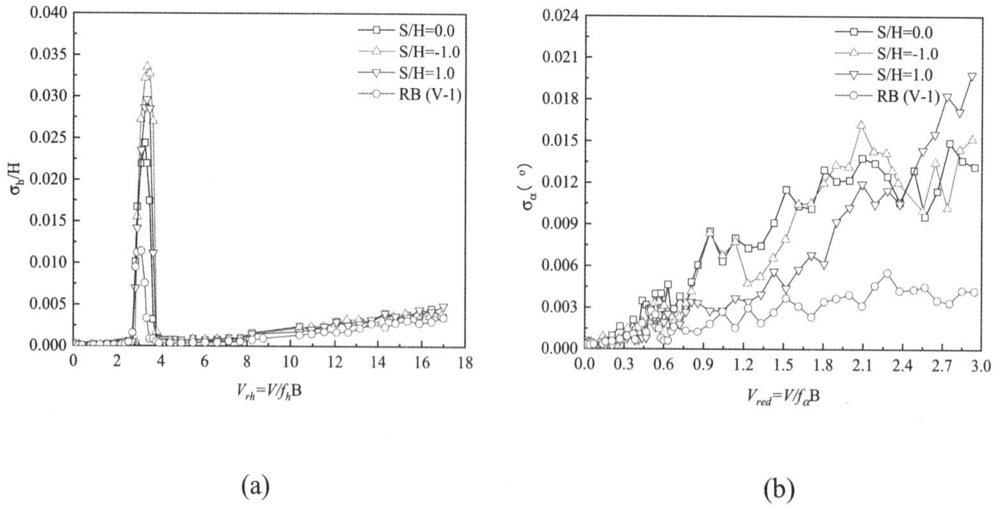

Fig. 5. RMS of displacement of main deck of RB vs. reduced wind velocity at different S/Hs when the RB deck is on the windward side: (a) Vertical displacements, and (b) torsional angle.

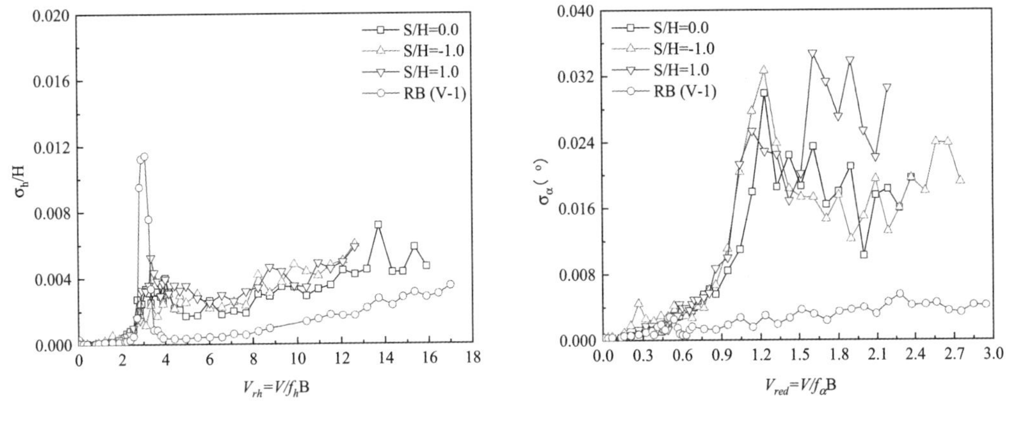

(a) (b)

Fig. 6. RMS of displacement of main deck of RB vs. reduced wind velocity at different S/Hs when the RB deck is on the leeward side: (a) Vertical displacements, and (b) torsional angle.

decks under S/H = 0.0 when the RB deck is on the leeward side (V-6), compared to the case of V-1, there is no obvious VIV in the main deck of RB and the vibration response becomes larger and unstable affected by the AIE. As the wind velocity increases, the vibration displacement firstly increases, especially for the torsional angle, and reaches the maximum value of $\sigma_\alpha = 0.0299°$ at the $V_{ah} = 1.2342$. Then, the vibration responses become unstable, showing a trend of first increasing and then decreasing. (2) For the case of triple-separated decks under S/H = −1.0 when the RB deck is on the leeward side (V-7), compared to the cases of V-1 and S/H = 0.0, there is still no obvious VIV in the main deck of RB and the vibration becomes more unstable. As the wind velocity increases, the torsional angle reaches the maximum value of $\sigma_\alpha = 0.0033°$ at $V_{ah} = 1.2342$ and the unstable vibration responses at high wind velocities become larger. (3) For the case of triple-separated decks under S/H = 0.0 when the RB deck is on the leeward side (V-8), no obvious VIV occurred. Compared to cases of S/H = 0.0 and −1.0, the maximum torsional angle decreases to $\sigma_\alpha = 0.0253°$, and as the wind velocity increases, the vibration response becomes more unstable and the unstable vibration responses at high wind velocities become larger.

3.1.2. Pressure coefficients

Based on the pressure distribution on the surface of the main deck, which is obtained by the pressure-measurement tests carried out simultaneously with the vibration-measurement tests, the vibration characteristics of the main deck of RB can be analyzed in more detail. In general, the mean pressure coefficients can reflect the separation and reattachment of flow on the deck, and the fluctuating pressure coefficients can reflect the dynamic load acting on the surface of deck when the deck is vibrating. The pressure on the deck under the wind field is usually expressed by the dimensionless pressure coefficient:

$$C_{pi}(t) = p_i(t)\text{-}p_O/0.5\rho U^2 \quad (1)$$

$$C_{pi,mean} = \sum C_{pi}(t)/N \quad (2)$$

$$C_{pi,rms} = \sum (C_{pi}(t_n)\text{-}C_{pi,mean})/N \quad (3)$$

Where, $C_{pi}(t)$ and $p_i(t)$ are the pressure coefficient and pressure at point i on the surface of the main deck of RB, respectively; ρ is the air density, $\rho = 1.225$ kg/m^3; $C_{pi,mean}$ and $C_{pi,rms}$ are the mean and fluctuating pressure coefficients, respectively; P_O is the static pressure; U is the mean wind speed, and N is the total number of data.

When the RB is located at the windward, the surface mean pressure coefficients and fluctuating pressure coefficients at the maximum value of the VIV are selected to explain the VIV characteristic under the AIE from the flow structures to some extent. As shown in Fig. 7, the existence of the AIE causes significant changes in the fluctuating pressure coefficients, and after changing the vertical distance, the mean pressure coefficients further change: (1) For the case of single RB deck (V-1), the mean pressure coefficient at the handrail on the upper surface changes from positive to negative, indicating that flow was strongly separated due to the obstruction of the handrail and has weak adhesion at this time, and then reaches the peak of negative mean pressure coefficients at the railway track position. As the flow continues to move downstream, the mean pressure coefficients increase, showing a tendency to reattach, and changes to positive value at the downstream wind fairing. The mean pressure coefficient on the lower surface of the windward wind fairing is positive, indicating that the flow attaches to the main deck surface and moves downstream. The mean pressure coefficient becomes negative when flow moves through the maintenance rail, and changes to positive value again after flow leaving this region. (2) For S/H = 0.0, the existence of the AIE does not cause a large change in the mean pressure coefficient. It is worth noting that the mean pressure coefficient at the tail railway track position on the upper surface changes from positive to negative, indicating that affected by the AIE, the flow separation at the tail becomes stronger and the adhesion weakens. Moreover, the fluctuating pressure coefficient increases and becomes more unstable, which may be the main reason for the increase in the maximum value of VIV under the AIE. (3) For S/H = −1.0, compared to the cases of single RB deck (V-1) and S/H = 0.0, the absolute value of the mean pressure coefficient at the handrail position increased, indicating that the flow separation becomes stronger and the adhesion weakens. As the flow moves downstream, the mean pressure coefficient increases but remains negative. The mean pressure coefficient of the lower surface also becomes negative at the position of the maintenance rail, and the absolute value increases. The mean pressure coefficient increases after leaving this region, but still remains negative. The fluctuating pressure coefficient further increases and the fluctuation becomes stronger. (4) For S/H = 1.0, compared to S/H = −1.0, the change trend of the mean pressure coefficient is the same, but the value of the fluctuating pressure coefficient decreases. Compared to S/H = 0.0, the fluctuating pressure coefficient is smaller but it becomes more unstable. Moreover, the mean pressure coefficient at the tail decrease, indicating that the flow separation becomes stronger.

When the RB is located at the leeward, there is no obvious VIV in the main deck of RB. In order to explain the reasons for the disappearance of

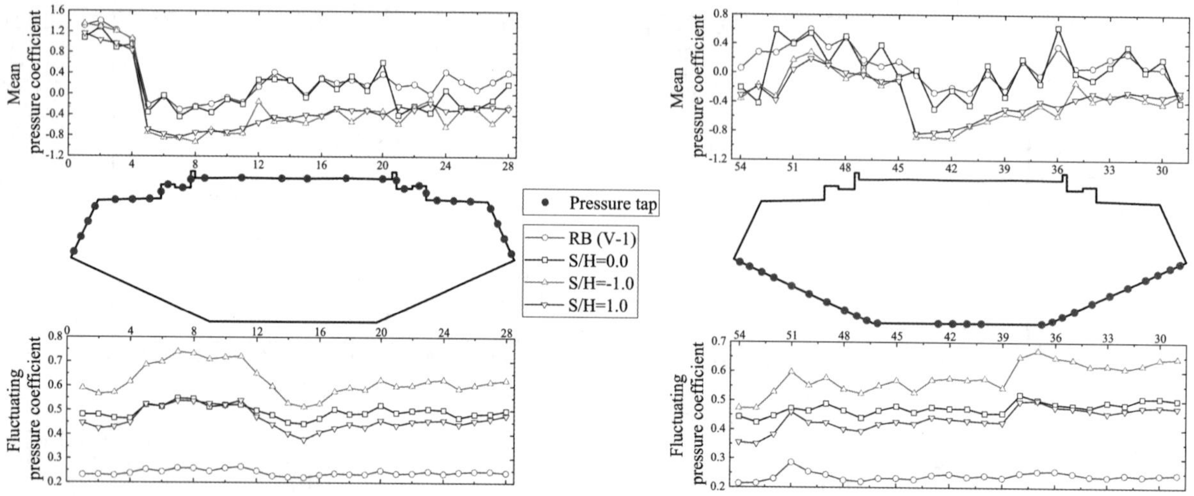

(a) (b)

Fig. 7. Mean and fluctuating pressure coefficient with different S/Hs (The RB deck is on the windward side): (a) Upper surface, (b) Lower surface.

VIV and the vibration response under the AIE from the flow pattern, the surface mean pressure coefficient and fluctuating pressure coefficient at the same wind velocity when the RB is located upstream were selected for analysis. As shown in Fig. 8, when the RB is located at the leeward, the mean pressure coefficient at the windward wind fairing of the upper and lower surfaces decreases, indicating that the incoming flow weakens. The mean pressure coefficient on the main deck surface is mostly negative, indicating that the flow separation is stronger and has low adhesion. Moreover, the fluctuating pressure coefficient is larger and more unstable than that of the case of single RB deck (V-1): (1) For S/H = 0.0, the mean pressure coefficient at the windward wind fairing approaches 0, indicating that due to the shielding effect of the upstream HB, the incoming flow to the downstream RB is weakened. The mean pressure coefficient at the handrail on the upper surface and at the maintenance rail on the lower surface changes from positive to negative. As the flow moves downstream, the surface mean pressure coefficient gradually increases but remains negative. Moreover, the fluctuating pressure coefficient is significantly increased and becomes more unstable compared to the case of single RB deck (V-1). However, compared to the case of S/H = 0.0 when the RB is on the windward side (V-3), the fluctuating pressure coefficient decreases, but becomes more unstable. The fluctuating pressure coefficient reaches the maximum value on the windward side of the lower surface and as the flow moves downstream, the fluctuating pressure coefficient gradually decreases. (2) For S/H = −1.0, compared to the case of S/H = 0.0, the mean pressure coefficient at the windward wind fairing of the upper surface increases, indicating that the incoming flow become stronger and still changes to negative value at the position of the handrail and maintenance rail. The

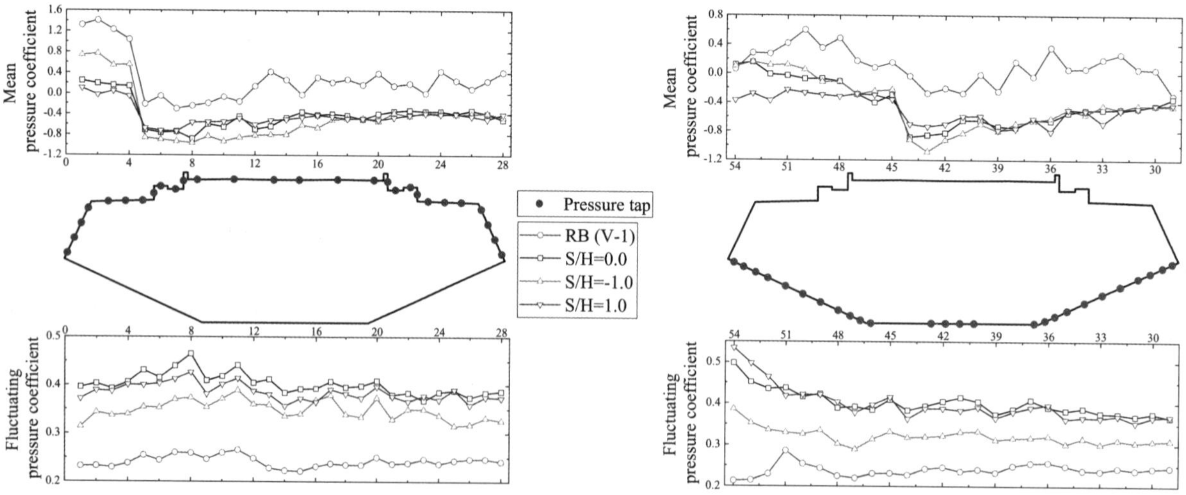

(a) (b)

Fig. 8. Mean and fluctuating pressure coefficient with different S/Hs (The RB deck is on the leeward side): (a) Upper surface, (b) Lower surface.

fluctuating pressure coefficient decreases but becomes more unstable at the flow separation regions, especially at the railway bridge track position on the upper surface. (3) For S/H = 1.0, compared to the case of S/H = 0.0, the mean pressure coefficient of the upper surface has no significant change, but the mean pressure coefficient of the lower surface is all negative, indicating that the flow separation becomes stronger and the adhesion is weaker, especially on the windward wind fairing. Since the RB is located above the HB, the fluctuating pressure coefficient on the windward side of the lower surface further increases, which may be related to flow falling off the end of the deck of HB.

In summary, the different flow separation and attachment conditions on the upper and lower surfaces may contribute to the occurrence of VIV in the RB to some extent. Moreover, for the case of V-1, the mean pressure coefficient at the tail of the main deck becomes negative, indicating that there is flow separation and falling off at this time, and whether this is a Karman vortex that periodically sheds requires further study. Affected by the AIE, the fluctuating pressure coefficient becomes larger and more unstable, which contributes to the increase in the maximum value of the VIV. When the deck of RB is located at the leeward, affected by shielding effect of the two-separated decks of HB, the incoming flow and the adhesion of the flow of the deck of RB is weakened, which results in insufficient energy to cause VIV in the deck of RB. Meanwhile, the downstream deck is also affected by the wake of the upstream deck. The flow falling off the end of the deck is not as stable as the incoming wind, and this unstable incoming flow can contribute to the unstable vibration responses at high wind velocities. When the vertical distance between the separated decks S/H increases, the incoming flow on the deck of RB becomes stronger and can also be affected by the wake of the upstream deck, which further makes the unstable vibration responses at high wind velocities larger.

3.2. *Results of the main deck of HB-1*

Fig. 9 shows the RMS of vertical and torsional displacements of the main deck of HB-1 vs. reduced wind velocity for different vertical distances when the deck of RB is located at the windward, respectively. As shown in Fig. 9, for case of twin-separated HB deck (V-2), there is no obvious VIV within the test wind velocity and when the $V_{ah} = 2.23$, the torsional angle suddenly increased. Affected by the AIE, the vertical and torsional VIV both occurred for the main deck of HB-1 and changing vertical distance have obvious effects on the vibration responses: (1) For the case of triple-separated decks under S/H = 0.0 when the RB deck is on the windward side (V-3), affected by the AIE, there is a torsional VIV lock-in interval for $V_{ah} = 0.51$–1.09 and a vertical VIV lock-in interval for $V_{rh} = 1.09$–2.79. The maximum torsional angle of $\sigma_\alpha = 0.1098°$ and the maximum vertical displacement of $\sigma_h/H = 0.0039$ were reached at $V_{ah} = 0.76$ and $V_{rh} = 3.10$, respectively. Compared to case of twin-separated HB deck (V-2), the VIV performance of the main deck of HB-1 becomes worse, while the flutter stability becomes better. As the wind velocity increases, the vibration displacement gradually increases from $V_{ah} = 2.64$, while the sudden increase does not occur. (2) For the case of triple-separated decks under S/H = −1.0 when the RB deck is on the windward side (V-4), compared to the case of S/H = 0.0, the maximum amplitude of torsional and vertical VIV decreased by 28.05% and 10.25%, respectively. As the reduced wind velocity increases, the vibration amplitude gradually increases from $V_{ah} = 2.19$, and compared to S/H = 0.0, the vibration response becomes larger. (3) For the case of triple-separated decks under S/H = 1.0 when the RB deck is on the windward side (V-5), compared to case of S/H = 0.0, the maximum angle of torsional VIV increased by 16.94% and the maximum displacement of vertical VIV decreased by 10.25%. As the reduced wind velocity increases, the vibration amplitude gradually increases from $V_{ah} = 2.64$, and compared to S/H = −1.0, the vibration response becomes smaller, but still larger than that of S/H = 0.0.

Fig. 10 shows the RMS of vertical and torsional displacements of the main deck of HB-1 vs. reduced wind velocity for different vertical distances when the RB is located at the leeward, respectively. For the cases of triple-separated decks under different S/Hs when the RB deck is on the leeward side (V-6~8), the HB-1 is located downstream of the HB-2, and there is no obvious VIV in the main deck of HB-1 within the range of test wind velocity. Compared to the case of twin-separated HB decks, a sudden increase in amplitude corresponds to a higher wind velocity affected by the AIE, which in turn becomes lower due to the existence of vertical distances.

3.3. *Results of the main deck of HB-2*

Fig. 11 shows the RMS of vertical and torsional displacements of the main deck of HB-2 vs. reduced wind velocity for different vertical distances when the RB is located at the windward, respectively. As shown in Fig. 11, there are also torsional VIV and vertical VIV in the main deck of HB-2. Compared to HB-1, the lock-in interval of VIV does not change, while the maximum torsional angle and vertical displacement of VIV both decreases. Moreover, as the reduced wind velocity increases, the

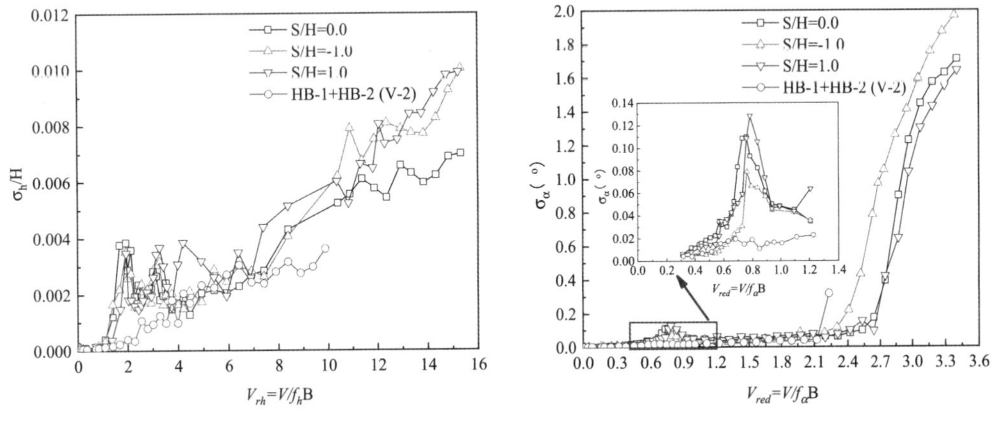

(a) (b)

Fig. 9. RMS of displacement of main deck of HB-1 vs. reduced wind velocity at different S/Hs when the RB deck is on the windward side: (a) Vertical displacements, and (b) torsional angle.

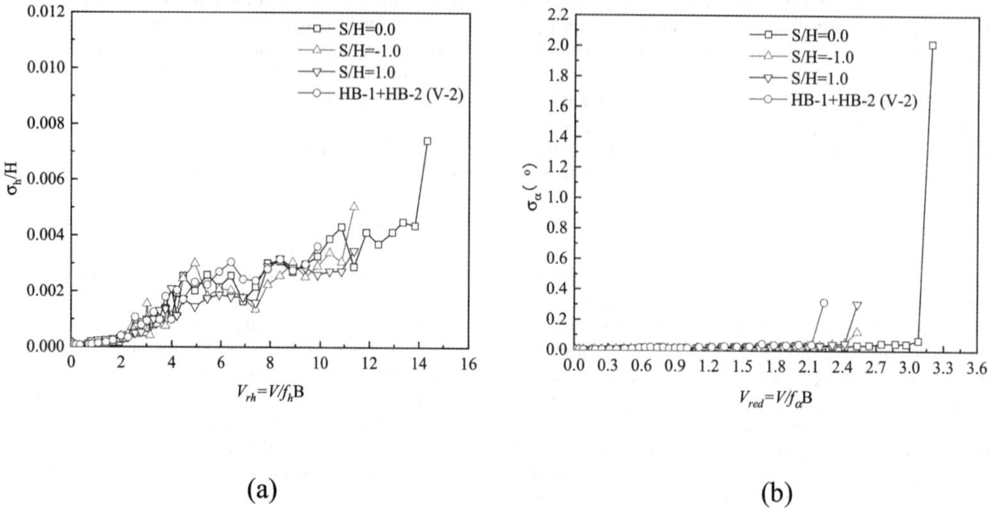

Fig. 10. RMS of displacement of main deck of HB-1 vs. reduced wind velocity at different S/Hs when the RB deck is on the leeward side: (a) Vertical displacements, and (b) torsional angle.

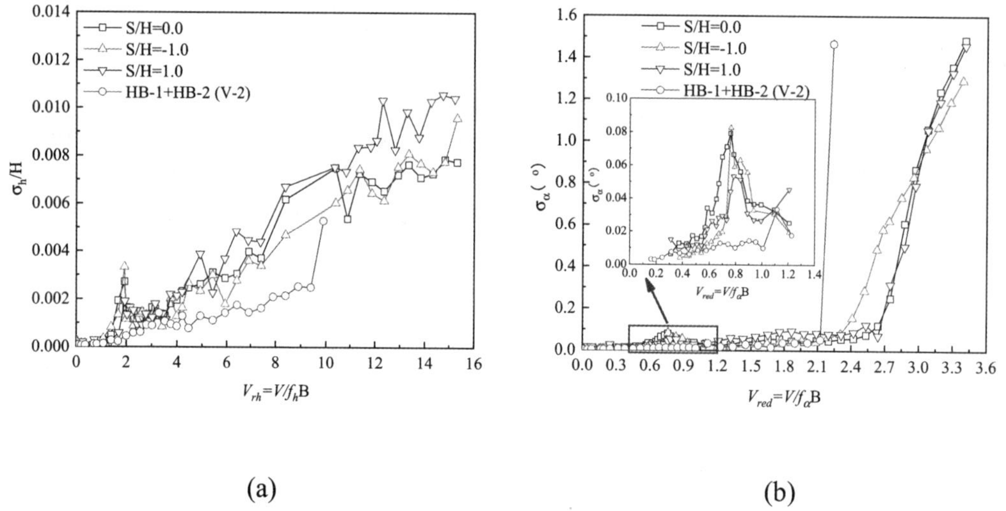

Fig. 11. RMS of displacement of main deck of HB-2 vs. reduced wind velocity at different S/Hs when the RB deck is on the windward side: (a) Vertical displacements, and (b) torsional angle.

vibration response of HB-1 becomes larger: (1) For the case of triple-separated decks under S/H = 0.0 when the RB deck is on the windward side (V-3), the maximum angle of torsional VIV is $\sigma_\alpha = 0.071^\circ$ (about 35.34% lower than HB-1) and the maximum displacement of vertical VIV is $\sigma_h/H = 0.0027$ (about 30.77% lower than HB-1). (2) For the case of triple-separated decks under S/H = −1.0 when the RB deck is on the windward side (V-4), compared to the case of S/H = 0.0, the maximum amplitude of torsional and vertical VIV decreased by 15.49% and 22.22%, respectively. (3) For the case of triple-separated decks under S/H = 1.0 when the RB deck is on the windward side (V-5), compared to case of S/H = 0.0, the maximum angle of torsional VIV increased by 25.35% and the maximum displacement of vertical VIV decreased by 29.63%.

Fig. 12 shows the RMS of vertical and torsional displacements of the main deck of HB-2 vs. reduced wind velocity for different vertical distances when the RB is located at the leeward, respectively. As shown in Fig. 12, only vertical VIV occurred in the main deck of HB-2 and the existence of vertical distance suppress this VIV phenomenon: (1) For the case of triple-separated decks under S/H = 0.0 when the RB deck is on the leeward side (V-6), there is a vertical VIV lock-in interval for V_{rh} = 2.43–3.02. The maximum vertical displacement of $\sigma_h/H = 0.0017$ were reached at $V_{rh} = 2.79$. When the $V_{\alpha h} = 3.1685$ (41.86% higher than the case of twin-separated HB decks), the torsional angle suddenly increased. (2) For the case of triple-separated decks under S/H = −1.0 when the RB deck is on the leeward side (V-7) and S/H = 1.0 (V-8), there is no obvious VIV occurring on the main deck of HB-2 and as the wind velocity increases, the torsional angle suddenly increased at $V_{\alpha h} = 2.53$ (20.11% lower than that of S/H = 0.0).

In summary, for VIV performance, affected by the AIE, the VIV performance of the triple-separated decks of RB and HB becomes worse, especially for the cases of triple-separated decks when the RB deck is on the windward side, which also confirms that the AIE between the separated decks cannot be ignored. Moreover, even when the vertical distance reaches S/H = 1.0 or −1.0, the influence of aerodynamic

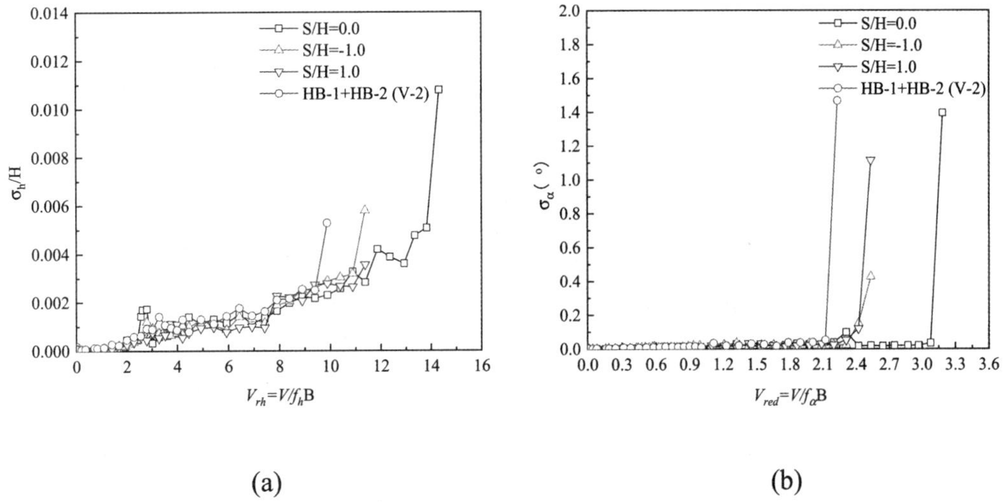

Fig. 12. RMS of displacement of main deck of HB-2 vs. reduced wind velocity at different S/Hs when the RB deck is on the leeward side: (a) Vertical displacements, and (b) torsional angle.

interference still exists. Therefore, in actual bridge engineering, the vertical distance will be less than L/H = 1.0 or −1.0, which also shows that the influence of vertical distance needs to be considered.

4. Numerical simulation

Due to the limitation of wind tunnel test, the experimental studies on the ACs under the AIE just focus on the deck of RB, and other separated decks just play the role of interference. Since the wind tunnel test of the separated decks involves many detail facilities, it is difficult to ensure test accuracy. Therefore, computational fluid dynamic (CFD) is becoming an increasingly attractive tool for the investigation of the AIE on the separated decks in recent years (Liu et al., 2010; Alvarez et al., 2018; Jeong et al., 2019; He et al., 2021).

4.1. Numerical model

A schematic diagram of the computational domain arrangement for the cases when the RB deck is on windward side (F-1) is shown in Fig. 13, which is inspired by Liu, 2011). For the cases when the RB deck is on windward side (F-2~4), the left inlet boundary is $10B_1$ away from the shearing center of main deck of RB, and the right outlet boundary is 20 B_1 away from the shearing center of the main deck of RB, where B_1 is the width of RB deck. The upper and lower boundaries are 10 B_1 away from the shear center of the main deck of RB respectively. The cases when the RB deck is on leeward side (F-5~7) are also considered, and its computational domain arrangement is consistent with the cases when the RB deck is on windward side (F-2~4). For the cases when the RB deck is on leeward side (F-5~7), the left inlet boundary is 10 B_2 away from the shearing center of the main deck of HB-2, and the right outlet boundary is 20 B_2 away from the shearing center of the main deck of HB-2, where B_2 is the width of HB-2 deck. The upper and lower boundaries are 10 B_2 away from the shear center of main deck of HB-2 respectively. For the boundary conditions, the main deck is set to the non-slip wall boundary, and the upper and lower boundaries are both set as symmetry boundaries. The inlet boundary of the computational domain is set as the velocity-inlet boundary, and the inlet velocity set to 10.0 m/s. Moreover, the inlet boundary turbulence intensity is set to 0.5%, and the turbulent viscosity ratio is set to 2.0. The outlet boundary is set as the pressure-outlet boundary.

The overall and detailed views of the computational mesh for the case of triple-separated decks are shown in Fig. 14. In order to ensure

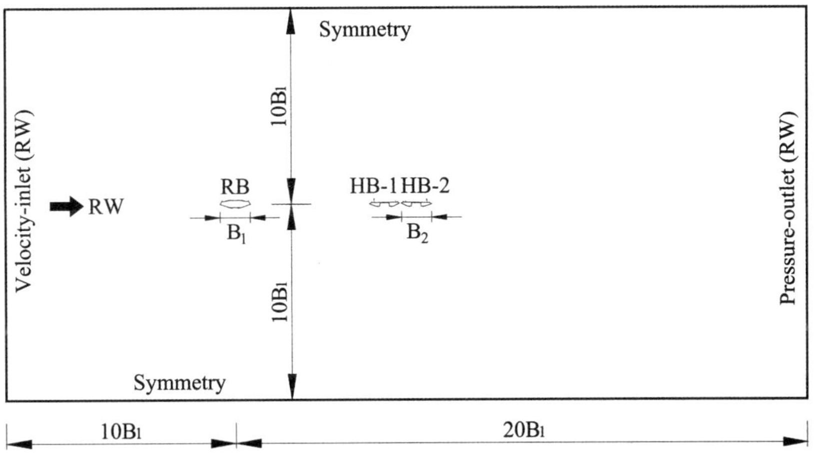

Fig. 13. Computational domain for cases of triple-separated decks when RB deck is on the windward side.

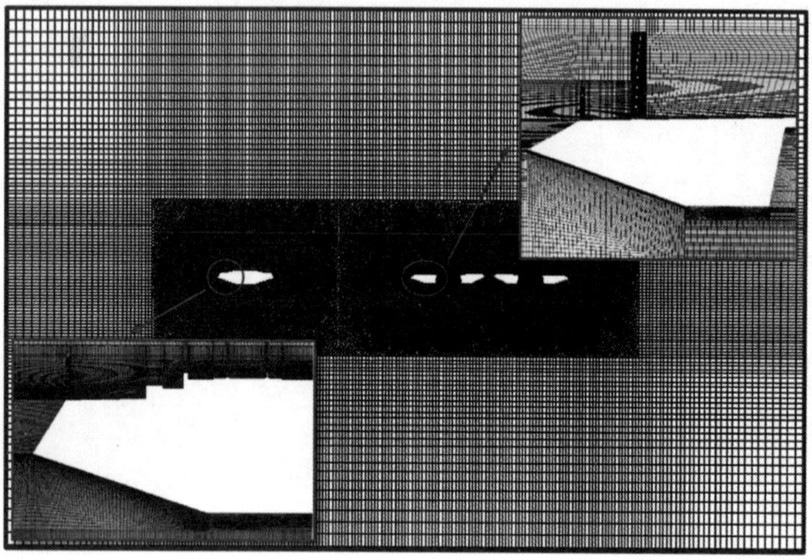

Fig. 14. Computational mesh.

calculation accuracy and efficiency, the meshes in different areas are divided into blocks for drawing. Among them, the mesh is densified in the area close to the main deck (the first height of the first layer δ_h = 0.0001B1), and as the distance from the main deck increases (the grid change ratio $\lambda = 1.2$), the mesh gradually becomes sparse. The turbulence model uses the Large Eddy Simulation (LES) model. To achieve better stability of the numerical solution, the second-order implicit scheme and second-order upwind scheme were applied to perform temporal and spatial discretization, respectively. Moreover, the coupling of the pressure and velocity was handled by the SIMPLEC algorithm.

4.2. *Validation of computational results*

The mesh density of the computational domain and the time step play the vital role in determining the accuracy of the results, so it is necessary to conduct study on the independence checks of mesh density and time step. Three different cell numbers and three different time steps are chosen to verify the suitability of the grid and the rationality of the computational model for single RB deck, single HB deck, twin HB decks and triple-separated decks under different incoming wind directions via the aerodynamic coefficients of RB and HB deck. The aerodynamic coefficients are defined as follows:

$$C_D = F_D / 0.5\rho U^2 H \tag{4}$$

$$C_L = F_V / 0.5\rho U^2 B \tag{5}$$

$$C_M = M / 0.5\rho U^2 B^2 \tag{6}$$

Where C_D, C_L and C_M are the aerodynamic coefficients of drag, lift and moment; ρ is the air density, respectively; U is the wind velocity; F_D, F_V and M are the drag force, lift force and lift moment, respectively; H is the height of the main deck of RB or HB; B is the width of the main deck of RB or HB.

Tables 4–7 show the ACs of each deck under three different cell numbers and three different time steps for wind attack angle of 0°. As shown in Tables 4–7, the ACs of each deck have almost no changes. Thus, for different cases, the medium cell numbers of the computational domain can provide precise enough results at a lower computational source and it was chosen for all the simulation cases of single deck, twin decks and triple decks. Since the LES method is time-consuming, to achieve reasonable accuracy and acceptable time, the time step of 0.0002s is chosen for all the simulations to minimize the computational time.

Table 5
Independence checks of cell number and time step size for twin-separated decks of HB.

Cell Number	△T (Unit: s)	HB-1			HB-2		
		C_D	C_L	C_M	C_D	C_L	C_M
304000	0.0002	0.939	−0.447	0.021	0.655	−0.120	0.045
428000		0.933	−0.448	0.02	0.658	−0.121	0.043
553000		0.932	−0.445	0.022	0.652	−0.125	0.046
428000	0.0001	0.931	−0.440	0.021	0.657	−0.124	0.046
	0.0002	0.933	−0.448	0.02	0.658	−0.121	0.043
	0.0005	0.934	−0.443	0.022	0.659	−0.120	0.040

Table 4
Independence checks of cell number and time step size for single RB deck and single HB deck.

Cell Number	△T (Unit: s)	SRB			Cell Number	△T (Unit: s)	SHB		
		C_D	C_L	C_M			C_D	C_L	C_M
238000	0.0002	0.804	0.127	0.0013	323000	0.0002	1.202	−0.302	0.006
367000		0.804	0.128	0.0013	417000		1.207	−0.308	0.006
513000		0.802	0.123	0.0014	539000		1.203	−0.307	0.006
367000	0.0001	0.804	0.127	0.0013	417000	0.0001	1.204	−0.307	0.006
	0.0002	0.804	0.128	0.0013		0.0002	1.207	−0.308	0.006
	0.0005	0.808	0.128	0.0012		0.0005	1.201	−0.304	0.006

Note: SRB indicates the single RB deck; SHB indicates the single HB deck.

Table 6
Independence checks of cell number and time step size for triple-separated decks when RB deck is on the windward side.

Cell Number	△T (Unit: s)	RB			HB-1			HB-2		
		C_D	C_L	C_M	C_D	C_L	C_M	C_D	C_L	C_M
314000	0.0002	0.799	0.086	0.0025	0.906	−0.216	−0.015	0.504	−0.158	0.0115
436000		0.796	0.087	0.0025	0.904	−0.214	−0.015	0.507	−0.152	0.011
569000		0.795	0.085	0.0026	0.903	−0.215	−0.015	0.502	−0.155	0.0111
436000	0.0001	0.794	0.089	0.0026	0.905	−0.215	−0.015	0.503	−0.153	0.0111
	0.0002	0.796	0.087	0.0025	0.904	−0.214	−0.015	0.507	−0.152	0.011
	0.0005	0.795	0.082	0.0026	0.907	−0.212	−0.016	0.504	−0.154	0.0115

Table 7
Independence checks of cell number and time step size for triple-separated decks when RB deck is on the leeward side.

Cell Number	△T (Unit: s)	RB			HB-1			HB-2		
		C_D	C_L	C_M	C_D	C_L	C_M	C_D	C_L	C_M
327000	0.0002	0.515	0.252	0.0036	1.012	−0.393	−0.023	0.665	−0.136	0.0349
435000		0.514	0.255	0.0035	1.013	−0.397	−0.022	0.660	−0.137	0.0340
563000		0.511	0.251	0.0038	1.019	−0.396	−0.023	0.667	−0.131	0.0358
435000	0.0001	0.514	0.256	0.0036	1.016	−0.399	−0.023	0.667	−0.133	0.0345
	0.0002	0.514	0.255	0.0035	1.013	−0.397	−0.022	0.660	−0.137	0.0340
	0.0005	0.512	0.254	0.0037	1.017	−0.39	−0.023	0.668	−0.131	0.0358

In order to analyze quantitatively the influence of AIE on the ACs of upstream and downstream decks, the interference factors (IF) of the ACs are defined as follows:

$$IF = C_{AIE}/C_S$$

Where, C_{AIE} is the AC of each deck under the AIE, and C_S is the AC of the single deck.

In order to further verify the accuracy of the numerical simulation, the experimental (EXP) and numerical results comparison of the AC and IF of RB deck are shown in Table 8. Due to the limitations of the wind tunnel test, the test only obtained the AC of RB deck. As shown in Table 8, the C_L and C_M are in good agreement, but C_D obtained from numerical simulation is smaller than that from the wind tunnel test.

For the differences between the results of numerical simulation and wind tunnel test, there are many detail facilities involved in the test, and a tiny change may affect results. Moreover, the flow field between the triple-separated decks is so complex that numerical simulation may ignore some changes. In summary, although there are differences in the C_D obtained from numerical simulations and wind tunnel tests, the changing trends are in good agreement. Thus, the results of the numerical simulations by LES model can be considered to accurately reflect the AIE on the ACs and accurately reflect the flow structures between the triple-separated decks.

4.3. Results of aerostatic coefficients

In order to obtain the ACs of the triple-separated decks under the AIE with different vertical distances, the numerical simulations of the single deck of RB (SRB), single deck of HB (SHB), twin-separated decks of HB-1 and HB-2 (THB), and triple-separated decks of RB and HB were conducted, respectively. Moreover, the effects of different incoming wind directions and different vertical distances were also considered for the case of triple-separated decks. Fig. 15,Fig. 19 show the simplified CFD models and all the numerical simulation cases.

4.3.1. RB deck is on the windward side

Fig. 16 shows the mean values and IFs of C_D for different decks under different cases of triple-separated decks when the RB deck is on the windward side. For RB deck, the drag coefficient of the case of single RB deck is 0.804. Under the AIE with different S/Hs, the IFs of C_D are all close to 1, and changing vertical distances does not cause obvious changes. For HB-1 deck and HB-2 deck, compared to the case of single HB deck (the drag coefficient is 1.207), the drag coefficients are all smaller in both cases of triple-separated decks and twin-separated HB decks. Under the case of S/H = 0.0, the IFs of C_D are 0.75 and 0.42 for HB-1 deck and HB-2 deck respectively, and the existence of vertical distance makes the IFs larger. For HB-1 deck, the IFs increase to 0.84 and 0.76 for S/H = −1.0 and 1.0 respectively. For HB-2 deck, the IFs increase to 0.46 and 0.43 for S/H = −1.0 and 1.0 respectively.

Fig. 17 shows the mean values and IFs of C_L for different decks under different cases of triple-separated decks when the RB deck is on the windward side. For RB deck, compared with the case of single RB deck (the lift coefficient is 0.128), the lift coefficient becomes smaller, and the IF of C_L is 0.66 under the AIE when S/H = 0.0. Changing vertical distances does not cause obvious changes. For HB-1 deck, the lift coefficient of the case of single HB deck is −0.308, and the lift coefficient shows an opposite trend in the case of triple-separated decks and the case of twin-separated HB decks. Under the case of S/H = 0.0, the IF of C_L is 0.7, and the existence of vertical distance increases the IFs to 0.76 and 0.77 for S/H = −1.0 and 1.0 respectively. For HB-2 deck, compared with the case of single HB deck, the lift coefficients increase in both cases of triple-separated decks and twin-separated HB decks, but still remain

Table 8
Comparison of ACs and IFs of RB deck obtained from EXP and CFD under different cases.

Case	Configuration	S/H	C_D-EXP	C_D-CFD	C_L-EXP	C_L -CFD	C_M-EXP	C_M-CFD
F-1	RB	/	1.015	0.804	0.122	0.128	0.0011	0.0013
F-2	RB + HB-1+HB-2	0.0	0.932	0.796	0.100	0.087	0.0013	0.0015
F-3		−1.0	0.937	0.798	0.096	0.085	0.0009	0.0010
F-4		1.0	0.929	0.798	0.097	0.086	0.0013	0.0016
F-5	HB-2+HB-1+RB	0.0	0.656	0.514	0.255	0.255	0.0025	0.0025
F-6		−1.0	0.861	0.756	0.25	0.254	0.0025	0.0027
F-7		1.0	0.635	0.616	0.231	0.243	0.0025	0.0022

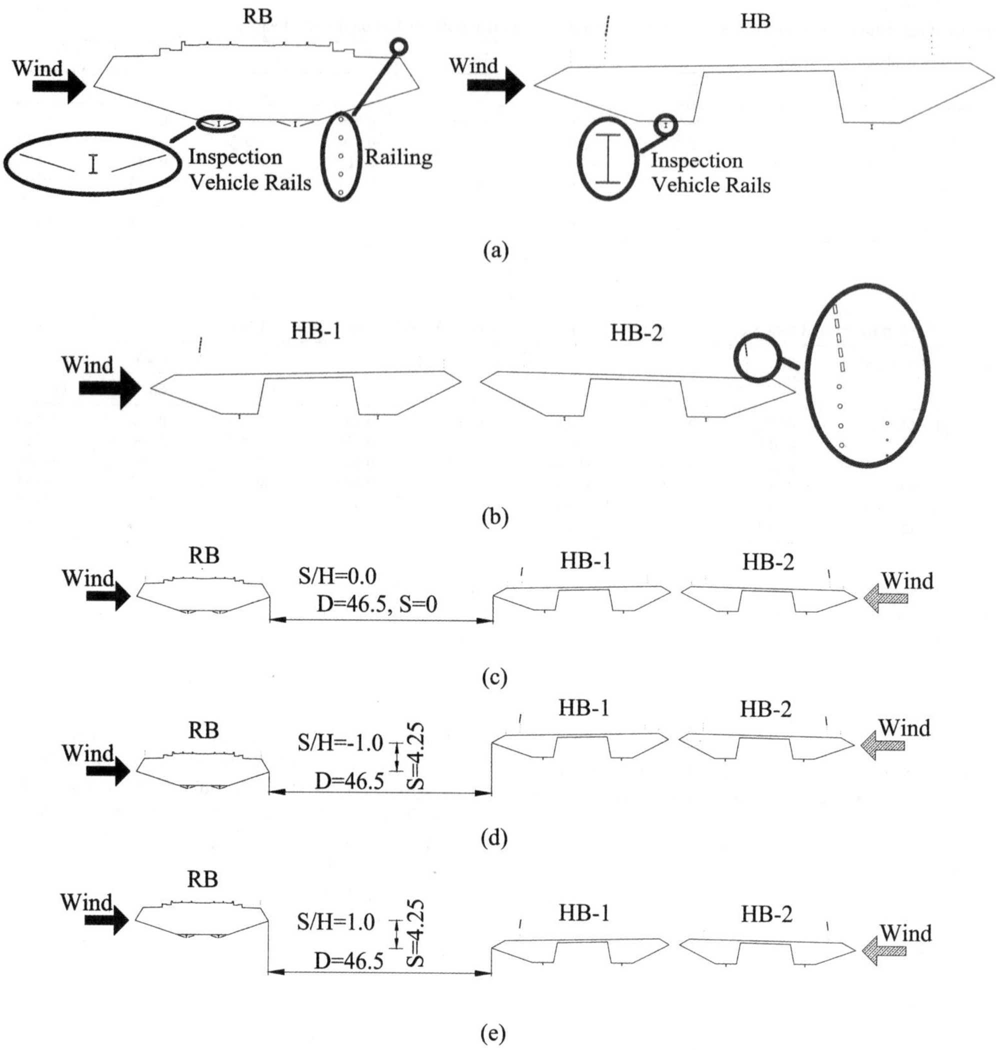

Fig. 15. Simplified CFD models of different cases: (a) cases of single RB and single HB, (b) case of twin-separated HB, (c) cases of triple-separated decks under S/H = 0.0, (d) cases of triple-separated decks under S/H = −1.0, (e) cases of triple-separated decks under S/H = 1.0.

negative. Under the case of S/H = 0.0, the IF of C_L is 0.49 and the existence of vertical distance increases the IFs of C_L to 0.52 and 0.53 for S/H = −1.0 and 1.0 respectively.

Fig. 18 shows the mean values and IFs of C_M for different decks under different cases of triple-separated decks when the RB deck is on the windward side. For RB deck, the lift moment coefficient of the case of single RB deck is 0.0013. Under the AIE with different S/Hs, the lift moment coefficients are still close to 0. For S/H = 0.0, the IF of C_M is 1.92 and changes to1.85 and 2 for S/H = −1.0 and 1.0 respectively. For HB-1 deck, under different cases, the lift moment coefficient is close to 0.0 (the lift moment coefficient of the case of single HB deck is 0.006), but there are great changes in the IF of C_M. Under the cases of triple-separated decks with different S/Hs, the lift moment coefficient becomes negative. The IF of C_M is −2.5 and changes to −4.2 and −2.2 for S/H = −1.0 and 1.0 respectively. For HB-2 deck, the lift moment coefficient is also close to 0, and there are also great changes in the IFs of C_M. Under the case of S/H = 0.0, the IF of C_M is 1.8 and the existence of vertical distance increases the IF of C_M to 3.0 and 2.3 for S/H = −1.0 and 1.0 respectively.

4.3.2. *RB deck is on the leeward side*

Fig. 19 shows the mean values and IFs of C_D for different decks under different cases of triple-separated decks when the RB deck is on the leeward side. For RB deck under AIE with S/H = 0.0, compared with the case of single RB deck (the drag coefficient is 0.804), the drag coefficient is significantly decreased, and the IF of C_D is 0.64. Changing vertical distance increases the IFs of C_D to 0.98 and 0.77 for S/H = −1.0 and S/H = 1.0 respectively. For HB-1 deck, compared to the case of single HB deck (the drag coefficient is 1.207), the drag coefficient decreases and the IF of C_D is 0.55 under AIE with S/H = 0.0. The existence of vertical distance decreases the IFs to 0.52 and 0.51 for S/H = −1.0 and 1 respectively. For HB-2 deck, compared to the case of single HB deck, the drag coefficient decreases and the IF of C_D is 0.84 under AIE with S/H = 0.0. The existence of vertical distance increases the IFs to 0.86 and 0.87 for S/H = −1.0 and 1.0 respectively.

Fig. 20 shows the mean values and IFs of C_L for different decks under different cases of triple-separated decks when the RB deck is on the leeward side. For RB deck (the lift coefficient of the case of single RB deck is 0.128), the IF of C_L is 1.99 under the AIE with S/H = 0.0, and the existence of vertical distance decreases the IFs to 1.98 and 1.90 for S/H

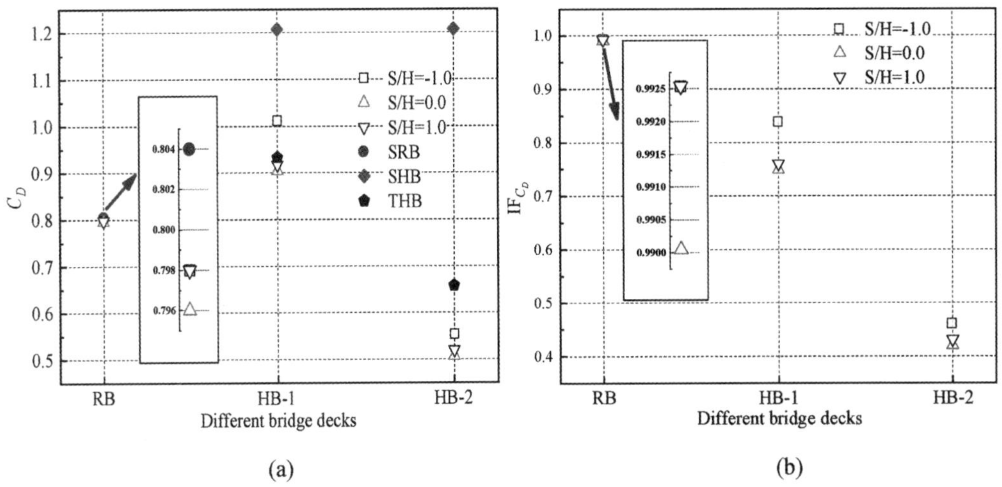

Fig. 16. Mean values and IFs of C_D for different decks under different cases: (a) C_D, (b) IFs.

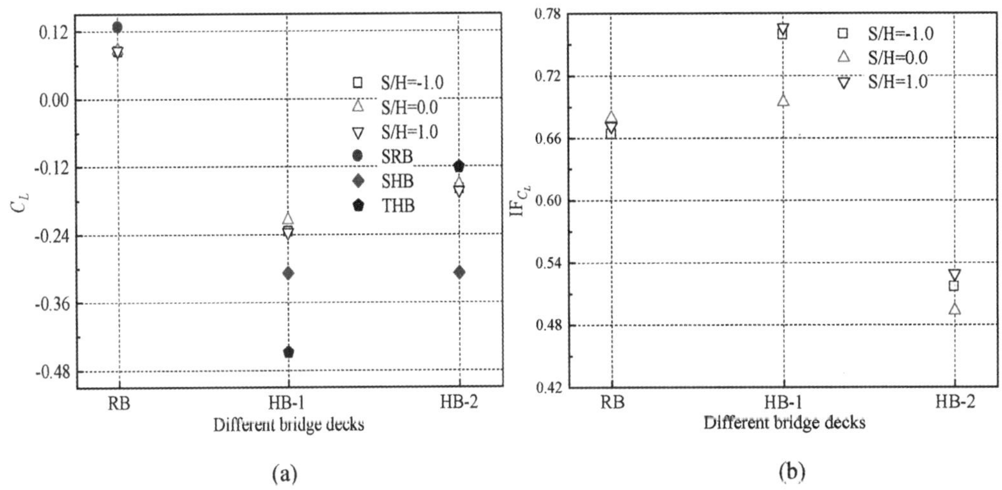

Fig. 17. Mean values and IFs of C_L for different decks under different cases: (a) C_L, (b) IFs.

$= -1.0$ and 1.0, respectively. For HB-1 deck (the lift coefficient of the case of single HB deck is -0.308), the lift coefficient is still negative and the IF of C_L is 0.45 under the AIE with S/H = 0.0. The existence of vertical distance makes the IF of C_L increase to 0.47 for both S/H = -1.0 and 1.0. For HB-2 deck, the lift coefficient is still negative and the IF of C_L is 1.29 under S/H = 0.0. The existence of vertical distance decreases the IF of C_L to 1.25 and 1.19 for S/H = -1.0 and S/H = 1.0, respectively.

Fig. 21 shows the mean values and IFs of C_M for of different decks under different cases of triple-separated decks when the RB deck is on the leeward side. For RB deck, compared with the case of single RB deck (the lift moment coefficient is 0.0013), the lift moment coefficients are still close to 0.0 under different S/Hs, while there are large changes in IFs. The IF of C_M is 2.69 under AIE with S/H = 0.0, and the existence of vertical distance changes the IFs of C_M to 2.85 and 2.46 for S/H = -1.0 and 1.0 respectively. For HB-1 deck, compared with the case of single HB deck (the lift moment coefficient is 0.006), the lift moment coefficient is still close to 0.0, but the IF of C_M is 5.67 under the AIE with S/H = 0.0. The existence of vertical distance decreases the IFs of C_M to 4.83 and 4.00 for S/H = -1.0 and 1.0 respectively. For HB-2 deck under AIE with different S/Hs, the lift moment coefficient is still close to 0, but becomes negative. The IF of C_M is -3.67 under AIE with S/H = 0.0, and the existence of vertical distance changes the IFs of C_M to -4.33 and -4.5 for S/H = -1.0 and S/H = 1.0, respectively.

4.4. *VIV mechanism analysis*

The VIV mechanism of the RB and HB decks under AIE are analyzed based on the results of pressure-measurement tests and the static flow structures.

4.4.1. *RB deck is on the windward side*

The vorticity distribution and magnitude of flow structures for different cases of the triple-separated decks when the RB deck is on the windward side and comparison with cases of single RB deck (SRB), single HB deck (SHB) and twin-separated HB decks (THB) are shown in Fig. 22, which can analyze the AIE in combination with the pressure distributions. For the case of single RB deck, the flow on the upper surface has been experiencing flow separation due to the influence of ancillary facilities, and shows weak adhesion. The flow on the lower surface only separates at the position of maintenance rails, and mostly adheres to the deck surface and moves. The fluid structure around the deck is consistent with the results of pressure-measurement tests, which

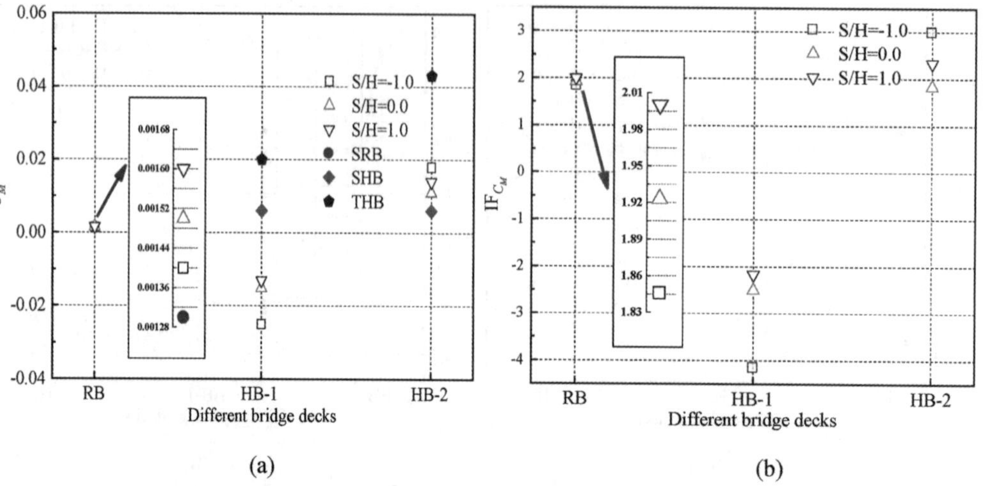

Fig. 18. Mean values and IFs of C_M for different decks under different cases: (a) C_M, (b) IFs.

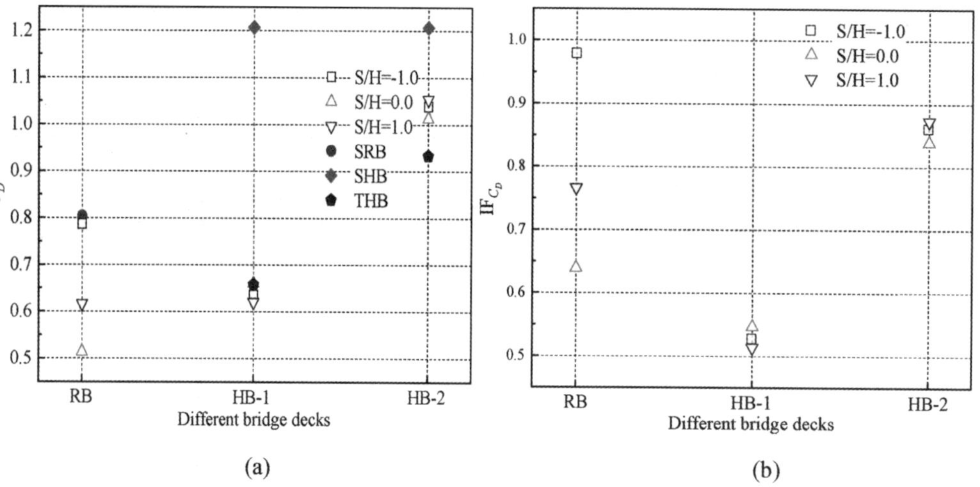

Fig. 19. Mean values and IFs of C_D for different decks under different cases: (a) C_D, (b) IFs.

also illustrates the accuracy of the numerical simulation and the results can be used to describe the fluid structure around the deck. It is worth noting that, combined with the results of the pressure-measurement tests, there are vortex that alternately fall off at the tail of the deck, which further shows that when the deck vibrates, the periodic pressure generated by the vortex that alternately fall off at the tail causes VIV. Moreover, the difference in the scale of vortex between the upper and lower surfaces of the main deck can lead to a pressure difference, which also contributes to the occurrence of VIV. For the case of single HB deck, due to the existence of the anti-throw net on the upper surface of the deck, the airflow does not move away from the deck, and the flow on the lower surface is mostly concentrated in the notches of the deck. Therefore, there is no large pressure difference between the upper and lower surfaces, and the flow at the tail of the deck does not show an alternating shedding trend at this time. For the case of twin-separated HB decks, due to the close distance between the twin-separated decks, the flow at the tail of the upstream deck is further destroyed, and due to the shielding effect of the upstream deck, the incoming wind acting on the downstream deck is weaker. According to the current research, the vortex that periodically sheds from the tail is the main cause of VIV in the P–K section (Liu et al., 2023). Therefore, no VIV occurred in the main twin-separated decks of HB during the tests.

For the case of triple-separated decks under S/H = 0.0 when the RB deck is on the windward, the alternate shedding of vortex at the tail of the RB deck is more obvious. The results of the pressure-measurement test show that the mean pressure coefficient at the tail of the deck decreases, further indicating that the scale of the vortex alternately shedding at the tail becomes larger. Moreover, the fluctuating pressure coefficients become larger and more unstable, which all contributes to the increase in the VIV amplitude of the main deck of RB. Due to the shielding effect of RB, the incoming flow acting on the downstream HB is significantly weakened, and the scale of vortex around the deck becomes smaller. However, the vortex shed from the tail of the RB deck will move downstream and act on the HB-1, which may contribute to the vibration of the main deck of HB. Although the scale of vortex on the surface of the HB deck is small at this time, the phenomenon of alternate shedding occurs at the tail of the twin-separated decks, which is one of the reasons for the VIV in the main deck of HB.

For the case of triple-separated decks under S/H = −1.0 when the RB deck is on the windward, the alternate shedding of vortex at the tail of RB deck still exists. Combined with the results of pressure-measurement test, it is shown that the increase in the fluctuating pressure coefficient is

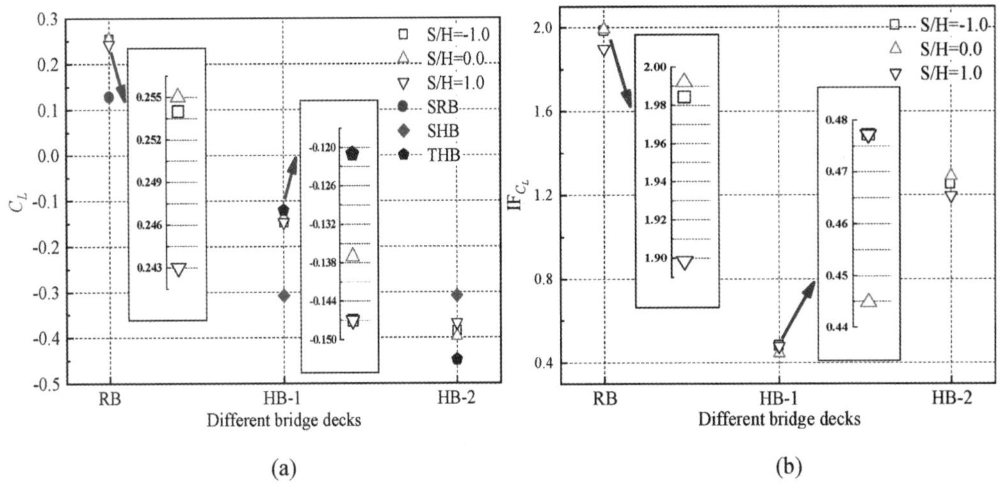

Fig. 20. Mean values and IFs of C_L for different decks under different cases: (a) C_L, (b) IFs.

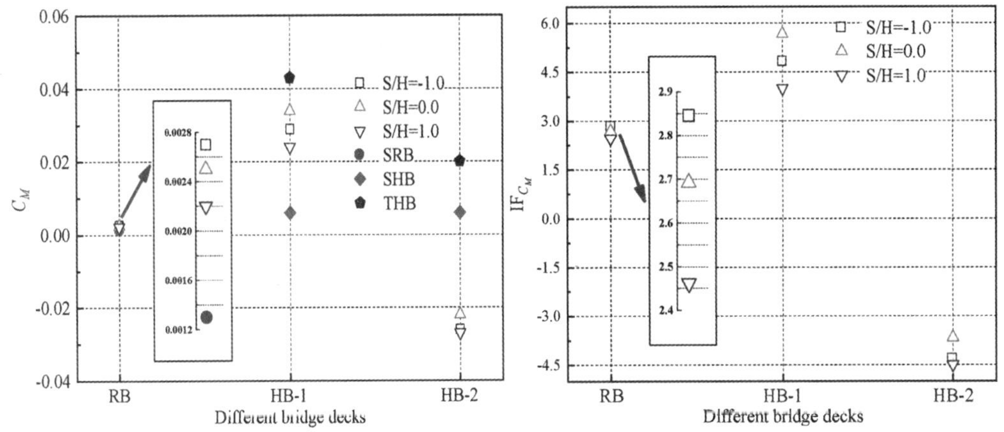

Fig. 21. Mean values and IFs of C_M for different decks under different cases: (a) C_M, (b) IFs.

the main reason for the further increase in the amplitude of VIV. Moreover, due to the existence of vertical distance, not only the incoming wind acting on HB is strengthened, but also the impact of the vortex shed from the tail of RB is enhanced. The scale of vortex on the surface of the HB-1 deck increases significantly, and the alternating vortex shedding at the tail is weakened at this time, resulting in a reduction in the amplitude of VIV. The HB-2 deck is blocked by HB-1, and the incoming wind is still weak, resulting in no obvious change in VIV.

For the case of triple-separated decks under S/H = 1.0 when the RB deck is on the windward, the increase of fluctuating pressure coefficient is still the main reason for the increase of VIV amplitude. Due to the existence of the vertical distance, the incoming flow acting on the HB deck is enhanced, and the vortex shed from the tail of RB separate on the surface of the HB-1 deck and moves downstream along the deck surface. The increase in the scale of vortex on the upper and lower surfaces, especially the upper surface, further increases the scale of vortex that alternately fall off the tail of the HB-1 deck, so the VIV amplitude of the main deck of HB-1 increases at this time. Moreover, the scale of vortex around the HB-2 deck also increased, which can contribute to the increase in the VIV amplitude of the main deck of HB-2.

4.4.2. RB deck is on the leeward side

The vorticity distribution and magnitude of flow structures for different cases of the triple-separated decks when the RB deck is on the leeward side and comparison with cases of single RB deck, single HB deck and twin-separated HB decks are shown in Fig. 23. For the case of triple-separated decks under S/H = 0.0 when the RB deck is on the leeward, due to the shielding effect of the upstream HB deck, the incoming flow acting on RB is weakened, which is consistent with the mean pressure coefficient of the windward surface of the RB deck being close to 0. Moreover, the wake vortex generated by HB are distributed around the RB deck, which also results in an increase in the fluctuating pressure coefficient on the RB deck surface. The weakening of the incoming flow and the absence of alternating shedding of vortex at the tail result in no VIV on the main deck of RB, while the effect of the irregular wake of HB contributes to the unstable vibration responses at high wind velocities of RB. At this time, compared with cases of twin-separated HB decks and single HB deck, the vortex movement of the windward HB-2 deck is closer to the deck surface, and the scale of vortex around the leeward HB-1 deck surface becomes smaller.

For the case of triple-separated decks under S/H = −1.0 when the RB deck is on the leeward, the shielding effect on the RB deck is weakened,

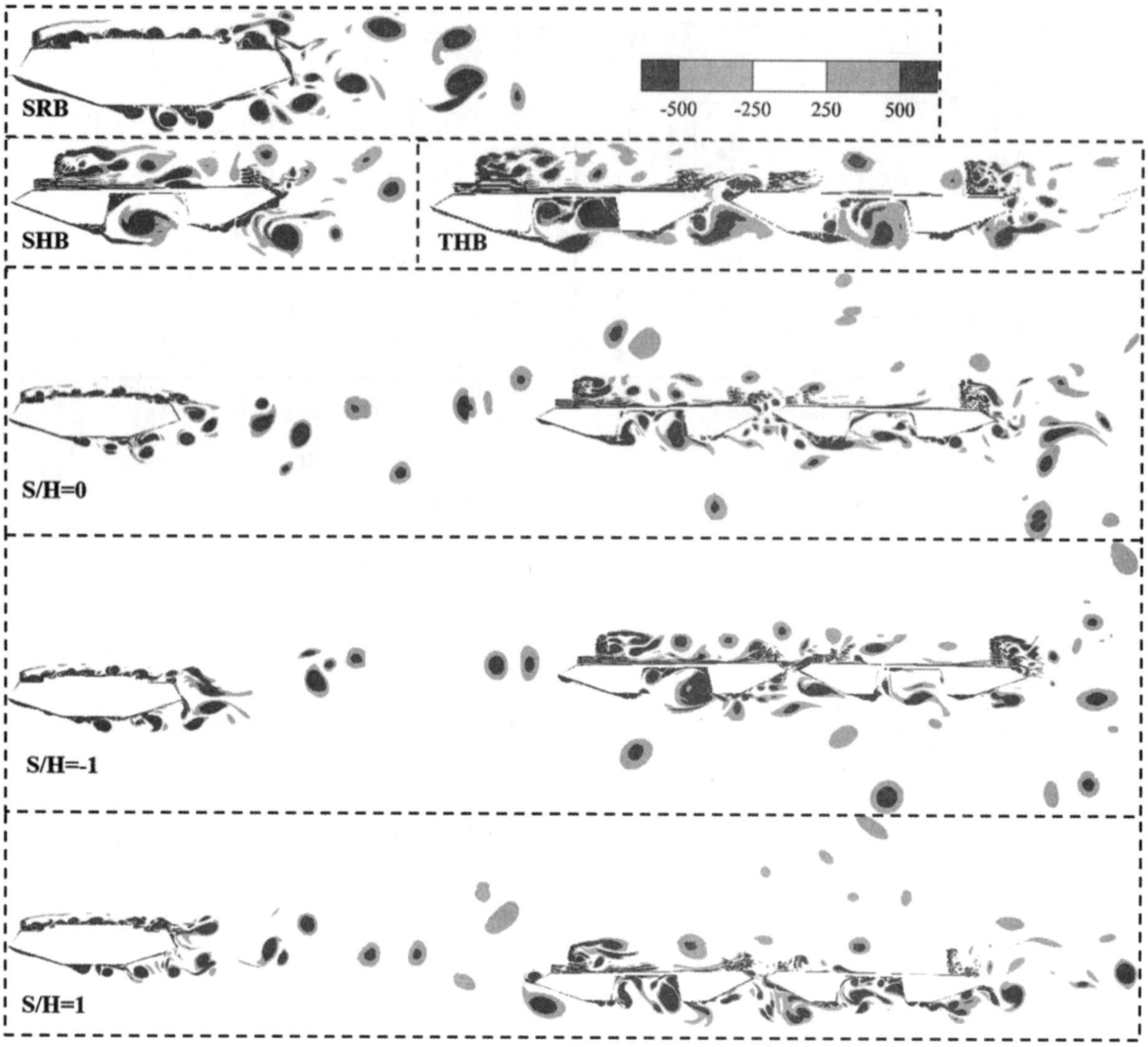

Fig. 22. Vorticity distribution and magnitude of flow structures when RB deck is on the windward side and comparison with cases of SRB, SHB and THB.

and the incoming flow becomes stronger, which is consistent with the increase in the mean pressure coefficient on the windward side of the RB deck. Moreover, the scale of wake vortex shed from the tail of HB deck becomes smaller, and the fluctuating pressure coefficient of the RB deck decreases, while the fluctuations are stronger. The scale of vortex shed from the tail of the RB deck increases, which may be the cause of unstable torsional vibration. The scale of vortex on the surface of the HB deck does not change significantly, but compared with S/H = 0.0, the distance between the vortex on the surface of the HB-2 deck and the deck surface increases.

For the case of triple-separated decks under S/H = 1.0 when the RB deck is on the leeward, combined with the pressure distribution, the vortex at the tail of HB deck acts on the lower surface of windward wind fairing of the RB deck, causing its fluctuating pressure coefficient to increase. The flow structure shows that no large-scale vortex appears on the lower surface of the RB deck except at the windward wind fairing, and indicate that the large-scale vortex at the wind fairing separates and moves away from the deck, which is consistent with the mean pressure coefficient becoming negative. Moreover, there are no alternating shedding vortex at the tail of the RB deck, so no VIV occurs. It is worth noting that compared with S/H = −1.0, the vortex on the surface of the HB-2 deck are closer to the deck surface, and there is more vortex distributed at the tail of the HB-1 deck at this time.

In summary, the flow distribution between RB and HB with triple-separated decks is more complex. Although the existence of vertical distance weakens the shielding effect of the upstream bridge deck, it can make the interference between separated decks more concentrated on the upper or lower surface of the bridge deck. Based on the research results, for RB and HB, the AIE can make the scale of vortex that alternately shed off the tail of the deck stronger and increase the fluctuating pressure coefficient on the deck surface, causing the amplitude of VIV to increase, or the unstable vibration responses at high wind velocities occur. For the influence of vertical distance, compared with S/H = 0.0 and S/H = −1.0, the flow structure on the deck surface is more stable at S/H = 1.0, so the wind-induced vibration of each deck is also more stable.

This study focused on the AIE between the triple-separated decks under different incoming wind directions and vertical distances, which is relatively rare in the study of AIE. The wind-induced vibration, aerostatic coefficient and VIV mechanism under different incoming

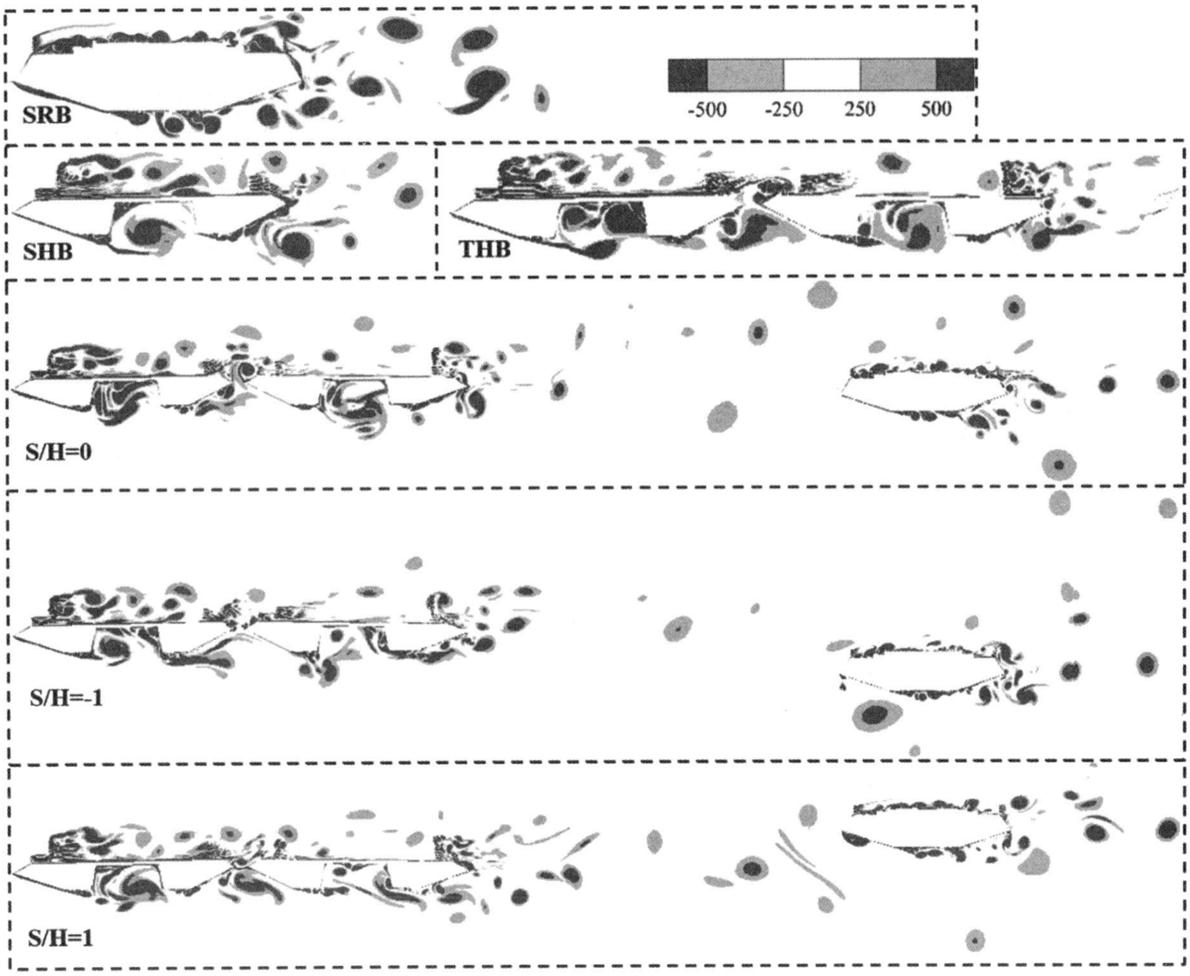

Fig. 23. Vorticity distribution and magnitude of flow structures when RB deck is on the leeward side and comparison with cases of SRB, SHB and THB.

wind directions and vertical distances have important reference value for the design of the parallel bridges with separated decks. However, due to the limitations of wind tunnel test, the critical value of vertical distance was not obtained in the study, that is, when S/H reaches the critical value, the AIE can be considered to disappear, which need to be considered more cases. Moreover, for VIV mechanism, this study only analyzes the flow structure around stationary triple-separated decks, and the flow distribution around the moving decks would provide a more detailed explanation of the AIE mechanism, which will also be studied in detail in subsequent research. For flutter stability, although studies have shown that the AIE effect can not only decrease the flutter critical velocity (Tang et al., 2020), but also change the flutter type, little attention has been paid to the AIE on flutter stability. In this study, the AIE increases the flutter critical wind velocity, and the flutter type changed to soft flutter when the RB deck is on the windward side. However, compared with $S/H = 0.0$, the existence of vertical distance decreases the flutter critical wind velocity, which also show that the vertical distance between separated decks cannot be ignored. According to the existing study (Tang et al., 2020), the AIE on the ACs of the triple-separated decks may be the cause of changes in flutter critical wind velocity and flutter type. Because flutter mechanisms corresponding to different deck types are different, the impact mechanisms of aerodynamic interference on different decks are also different. This study does not summarize the AIE on the flutter stability, and more detailed studies are needed to analyze it later.

5. Conclusion

The aerodynamic interference effect (AIE) between two bridges with triple-separated parallel decks on aerodynamic characteristic with different incoming wind directions and vertical distances are investigated in this study. The main conclusions are summarized as follows.

1. For parallel bridges with separated decks, the AIE can affect the vibration responses, and the effects of vertical distance cannot be ignored. The VIV response of the RB deck reaches the maximum value under $S/H = -1.0$ when the RB deck is on the windward side (the maximum displacements of VIV are 193.86% and 37.30% higher than the case of single RB deck and $S/H = 0.0$, respectively). Compared to the case of twin-separated HB decks, VIV occurs in the HB decks under the AIE.
2. The drag coefficient of the deck is most affected by AIE, especially when the deck is located at the leeward. Due to the shielding effect of the upstream deck, the drag coefficient of the leeward deck is significantly reduced and the existence of vertical distance can weaken the influence of AIE on the ACs.
3. Due to the interaction of flows between the separated decks under the AIE, the larger scale of vortex that alternately shed off from the

tail of the deck and the larger fluctuating pressure coefficient are the reasons for the change in the VIV responses of triple-separated decks. The existence of vertical distance can make the interaction of flow more concentrated on the upper or lower surface, which makes the fluctuating pressure coefficient more unstable and then increase the VIV amplitude.

CRediT authorship contribution statement

Han Xiao: Writing – original draft, Investigation, Formal analysis, Data curation. **Zhiwen Liu:** Writing – review & editing, Validation, Project administration, Methodology, Conceptualization. **Zhengqing Chen:** Supervision, Project administration. **Zhenbiao Liu:** Resources, Funding acquisition. **Aiguo Yan:** Resources, Funding acquisition. **Ou Yang:** Resources, Funding acquisition.

Declaration of competing interest

The authors declare that they have no known competing financial interests or personal relationships that could have appeared to influence the work reported in this paper.

Data availability

Data will be made available on request.

Acknowledgements

This research was financially supported by the National Natural Science Foundation of China (grant numbers 52178475 and 51778225), National Key Research and Development Projects (grant number 2019YFC1511101), Scientific Research and Development Program of China National Railway Group Co., Ltd (N2021G034), Scientific Research Program of China Railway Construction Corporation Limited (2021-B17), and Scientific and technological research and development project of China Railway Siyuan Survey and Design Group Co., Ltd. (2018K001, 2021K015), for which the authors are grateful.

References

Argentini, T., Rocchi, D., Zasso, A., 2015. Aerodynamic interference and vortex-induced vibrations on parallel bridges: the Ewijk bridge during different stages of refurbishment. J. Wind Eng. Ind. Aerod. 147, 276–282. https://doi.org/10.1016/j.jweia.2015.07.012.

Alvarez, A.J., Nieto, F., Kwok, K.C.S., Hernandez, S., 2018. Aerodynamic performance of twin-box decks: a parametric study on gap width effects based on validated 2D URANS simulations. J. Wind Eng. Ind. Aerod. 182, 202–221. https://doi.org/10.1016/j.jweia.2018.09.004.

Chen, P., 2016. Influence of aerodynamic interference effects of adjacent three separated deck bridges on wind stability of bridges. Bridge Constr. 46 (5), 6 (In Chinese). https://CNKI:SUN:QLJS.0.2016-05-010.

Honda, A., Shiraishi, N., Motoyama, S., 1990. Aerodynamic stability of Kansai international airport access bridge. J. Wind Eng. Ind. Aerod. 33, 369–376. https://doi.org/10.1016/0167-6105(90)90052-E.

He, X., Kang, X., Yan, L., Flay, R., Ren, P., Wu, T., 2021. Numerical investigation of flow structures and aerodynamic interference around stationary parallel box girders. J. Wind Eng. Ind. Aerod. 215, 104610 https://doi.org/10.1016/j.jweia.2021.104610.

Irwin, P., Stoyanoff, S., Xie, J., Hunter, M., 2005. Tacoma narrows 50 years later-wind engineering investigations for parallel bridges. J. I. Coast Archaeol. 1, 3–17. https://doi.org/10.1080/1573248042000274551.

Jeong, W., Liu, S., Bogunovic Jakobsen, J., Ong, M.C., 2019. Unsteady RANS simulations of flow around a twin-box bridge girder cross section. Energies 12, 2670. https://doi.org/10.3390/en12142670.

Kimura, K., Shima, K., Sano, K., Kubo, Y., Kato, K., Ukon, H., 2008. Effects of separation distance on wind-induced response of parallel box girders. J. Wind Eng. Ind. Aerod. 6, 954–962. https://doi.org/10.1016/j.jweia.2007.06.021.

Kim, S., Kim, H., Calmer, R., Park, J., Kim, S., Lee, K., 2013. Operational field monitoring of interactive vortex-induced vibrations between two parallel cable-stayed bridges. J. Wind Eng. Ind. Aerod. 123, 143–154. https://doi.org/10.1016/j.jweia.2013.10.001.

Liu, Z., Chen, Z., Hu, J., He, S., 2008. Aerodynamic interference effects on of large-span bridge with twin separated decks. J. Chang'an Univ. (Nat. Sci. Ed.) 28 (6), 55–59. https://doi.org/10.3969/j.issn.1003-4722.2007.06.003 (in Chinese).

Liu, Z., Chen, Z., Gao, L., Shao, X., 2009. Experimental study of aerodynamic interference effects on aerostatic coefficients of twin deck bridges. Front. Architect. Civ. Eng. China 3 (3), 292–298. https://doi.org/10.1007/s11709-009-0048-8.

Liu, Z., Cai, S., Chen, Z., 2010. Numerical Simulation of Aerodynamic Interference Effects on Aerostatic Coefficients of Two Bluff Sections in Tandem. The 5th International Symposium on Computational Wind Engineering (CWE2010). Chapel Hill, North Carolina, USA.

Liu, X., 2011. Study on Aerodynamic Interferences of Long-Span Bridges with Twin Separate Parallel Decks. Hunan University (In Chinese). https://10.7666/d.y1908787.

Liu, L., Zhou, Y., He, X., Yang, J., Wang, Z., Guo, D., 2022. Experimental investigation on vortex-induced vibration of a long-span rail-cum-road bridge with twin separated parallel decks. J. Wind Eng. Ind. Aerod. 228, 105086 https://doi.org/10.1016/j.jweia.2022.105086.

Liu, Z., He, F., Yan, A., Liu, Z., Yin, T., Chen, Z., 2023. Mechanism and aerodynamic countermeasures of vortex-induced vibration of a cable-stayed bridge with narrow Π-shaped girder sections. J. Bridge Eng. 28 (12), 04023090 https://doi.org/10.1061/JBENF2.BEENG-6285.

Meng, X., Zhu, L., Guo, Z., 2011. Aerodynamic interference effects and mitigation measures on vortex-induced vibrations of two adjacent cable-stayed bridges. Front. Architect. Civ. Eng. China 5 (4), 510–517. https://doi.org/10.1007/s11709-011-0129-3.

Okubo, T., Enami, Y., 1972. Results of wind tunnel model tests and field observations of wind effects on the Onomichi bridge. In: Publication of: Precast Concrete UK, Precast Concrete, Public Works Research Institute, Tokyo, pp. 249–260. http://trid.trb.org/view/1972/C/102781.

Park, J., Kim, S., Kim, H., 2017. Effect of gap distance on vortex-induced vibration in two parallel cable-stayed bridges. J. Wind Eng. Ind. Aerod. 162, 35–44. https://doi.org/10.1016/j.jweia.2017.01.004.

Svensson, H., Lovett, T., 1990. The twin cable-stayed composite bridge at Bay-town, Texas. In: IABSE Symposium (Brussels): Mixed Structures, Including New Materials, pp. 317–322. https://doi.org/10.5169/seals-46499.

Suzuki, Y., Mizuguchi, K., Ueda, T., 1999. Field observation on aerodynamic response of meiko west bridge. In: IABSE Conference (Malmö): Cable-Stayed Bridges: Past, Present and Future, pp. 149–150. https://doi.org/10.5169/seals-62168.

Seo, J., Kim, H., Park, J., Kim, K., Kim, G., 2013. Interference effect on vortex-induced vibration in a parallel twin cable-stayed bridge. J. Wind Eng. Ind. Aerod. 116, 7–20. https://doi.org/10.1016/j.jweia.2013.01.014.

Tang, B., Cao, J., Yang, Y., Tan, X., Ge, Y., 2020. Interference effect on flutter performance of long span bridges with parallel twin decks. J. Tongji Univ. 48 (4), 490–497. https://doi.org/10.11908/j.issn.0253-374x.19188 (In Chinese).

Xiao, H., Liu, Z., He, F., Chen, Z., Liu, Z., Yan, A., 2024. Aerodynamic interference effects on vortex-induced vibration of two parallel railway– and highway–cable-stayed bridges with triple-separated decks. J. Bridge Eng. 29 (2), 04023112. https://ascelibrary.org/doi/abs/10.1061/JBENF2.BEENG-6380.

Zhu, L., Zhou, Q., Guo, Z., Meng, X., 2010. Aerodynamic interference effects on flutter and vortex-excited resonance of bridges with twin-separate parallel box decks. J. Tongji Univ. 38 (5), 7. https://doi.org/10.3969/j.issn.0253-374x.2010.05.002 (In Chinese).

Zeng, H., 2014. Aerodynamic Interference and Wind Environment of Parallel Bridges. Southwest Jiaotong. University (In Chinese). https://read.cnki.net/web/Dissertation/Article/10613-1014254510.nh.html.

Zhou, R., Yang, Y., Zhang, L., Ge, Y., 2017. Interference effect on stationary aerodynamic performance between parallel bridges. Int. J. Struct. Stabil. Dynam. 17 (2) https://doi.org/10.1142/s0219455417500171.

索杆结构风致振动与控制

Flutter, Galloping, and Vortex-Induced Vibrations of H-Section Hangers

Z. Q. Chen, M.ASCE[1]; M. G. Liu[2]; X. G. Hua[3]; and T. M. Mou[4]

Abstract: Hangers in arch bridges and other long vertical bars in truss bridges are typically slender members and often adopt an aerodynamically unfavorable H-section due to the conveniences in manufacture, construction, and maintenance. In contrast to the very narrow range of wind attack angles of horizontal members, hangers may suffer wider wind attack angles, possibly from 0° to 360°, and are therefore more prone to wind-induced vibrations. In August 2006, large torsional vibration with severe damage was observed on the 13 longest hangers of the Dongping arch bridge in China during a strong wind. While vibrations of hangers were usually caused by galloping, and vortex shedding excitation, the present case was likely to be a kind of torsional flutter instability. Therefore an in-depth investigation on the hangers' aerodynamic performances in forms of flutter, galloping, and vortex shedding has been conducted through a series of wind tunnel tests. First, with section model and aeroelastic models tests of the longest hanger in the bridge, the observed field vibration is confirmed as torsional flutter under large attack angles from 15° to 25°, and the experimental onset velocity coincides well with the field observation; the flutter derivative A_2^* becomes positive at a low reduced wind velocity, which further implies that the H-section is prone to flutter instability. Then, the influences of web perforation on flutter, galloping, and vortex shedding are studied with four section models having different web perforation ratios but the same depth-to-width ratio $D/B = 0.416$ (flange depth D to web width B). It is found that the web perforation may increase the galloping critical velocity to some extent but have no obvious effects on flutter instability and the Strouhal numbers, at least for the shallow H-section with $D/B = 0.416$. Next, a total of 16 H-section models with different D/B ratios, web perforation ratios, and flange perforation ratios are tested to investigate their effects on aerodynamic behaviors of hangers. A comparison of the experimental results with previous work is made, which may explain why the flutter instability of H-shaped hangers under large attack angles was not treated by earlier investigators. **DOI: 10.1061/(ASCE)BE.1943-5592.0000268.** *© 2012 American Society of Civil Engineers.*

CE Database subject headings: Arch bridges; Hangars; Vibration; Wind forces; Wind tunnels.

Author keywords: Bridges; Arch bridges; Hangers; H-sections; Wind-induced vibration; Flutter; Vortex-excited vibration; Galloping; Wind tunnel test.

Introduction

Long and slender vertical steel members with H- or box-sections were widely adopted as hangers or suspenders in arch bridges recently built in China. For example, the H-section hanger in the Dongping arch bridge in Guangdong, China, has a maximum length of 40.2 m and a slenderness ratio of 342; the hanger in the Dashengguan bridge in Jiangsu, China, has a maximum length of 56 m and a slenderness ratio of 120. In comparison with the steel-rope or steel-strand vertical hangers with circular sections, the hanger of this kind possess an aerodynamically unfavorable shape and is prone to a variety of wind-induced vibrations, which may potentially inducing severe fatigue damage, especially in the connections between hanger and bridge girder. From the early 1970's, there have been many reports about the large-amplitude wind-induced vibrations of arch bridge hangers or truss bridge bars. For example, Ulstrup (1980) listed nine episodes of wind-induced vibrations on slender bridge components. In China, the first case was the Jiujiang Yangtze River Bridge, built in the 1990's, in which most of its hangers experienced vortex-induced vibrations at moderate wind shortly after completion of the bridge, and tuned mass dampers were installed as remedy measures to reduce the vibration (Gu et al. 1994; Yu et al. 1996). Unfortunately, this lesson does not appear to be well-learned by bridge designers. More recently, large-amplitude wind-induced vibrations of the H-section hangers were observed in several steel arch bridges. In August 2006, typhoon Prapiroon attacked the Dongping bridge, a steel arch bridge that was nearly completed at that time, as shown in Fig. 1. On August 4, 2006, Foushan Diary reported the ground gust wind on the Beaufort wind force scale of 8–10, which corresponds to a wind velocity of about 25 m/s at bridge deck level ("Typhoon" 2006). More than 20 hangers oscillated simultaneously in a torsional direction for 24 h, with maximum vibration amplitude reaching 35°, and the 13 longest hangers were damaged with severe cracks on the flanges near the hanger-to-girder connections, as shown in Fig. 2. While galloping and vortex shedding excitation were more common for hangers, the present case was likely to

[1]Professor, Director of Wind Engineering Research Center, Hunan Univ., Changsha 410082, Hunan, China (corresponding author). E-mail: zqchen@hnu.cn

[2]Assistant Professor, College of Civil Engineering and Transportation, South China Univ. of Technology, Guangzhou, P.R. China; formerly, PhD Student, Wind Engineering Research Center, Hunan Univ., Changsha 410082, Hunan, P.R. China.

[3]Associate Professor, Wind Engineering Research Center, Hunan Univ., Changsha 410082, Hunan, China.

[4]Senior Engineer, Planning, Survey and Design Institute of Highways of Dept. of Transportation of Shichuan Province, Chengdu 610041, Shichuan, China.

Note. This manuscript was submitted on October 11, 2010; approved on May 4, 2011; published online on May 6, 2011. Discussion period open until October 1, 2012; separate discussions must be submitted for individual papers. This paper is part of the ***Journal of Bridge Engineering***, Vol. 17, No. 3, May 1, 2012. ©ASCE, ISSN 1084-0702/2012/3-500–508/$25.00.

Fig. 1. The Dongping bridge in Guangdong, China (before completion)

Fig. 2. Wind-induced crack at the connection between hanger and girder

a kind of flutter instability, based on the careful scrutiny of the video recorded in situ (Chen et al. 2010).

Significant effort has been made to study the aerodynamic characteristics of the H-sections used on arch bridge hangers and truss bridge bars over the past decades. Wardlaw is an early investigator who published several papers regarding wind-induced vibration and its mitigation countermeasures for the I-section truss bars and road deck hangers (Wardlaw 1990). Chi et al. (1977) conducted wind tunnel tests of three typical H-sections with aspect ratios being approximately 1:1, 3:4, and 1:2, and they determined that the Strouhal numbers for these sections are in the neighborhood of 0.125 ~ 0.134. Chi and Vossoughi (1985) predicted the vortex-induced response of H-section cylinders by a modified Hartlen-Currie model, which needs to be calibrated with wind tunnel data. Maher and Wittig (1980) investigated the vibration of H-section members of the Commodore Barry bridge, and its possible remedy countermeasures included modifying H-sections into box-sections and the use of cable ties and perforated web. They concluded that vortex-excited vibration in heaving modes and torsional modes may build up for the H-sections studied, and the vibration in the Commodore Barry Bridge is very likely a kind of traditional Karman vortex shedding. Kubo et al. (1980) studied the vortex-excited oscillation and galloping of H-section cylinders of Langer bridge, Japan, in both laminar and turbulent flow. They clarified that both vortex-induced vibrations and galloping may develop in the laminar flow, but in the turbulent flow the former seldom takes place and the oscillation became mainly a galloping one. Ruscheweyh (1990, 1996) developed a mathematical model for predicting the vortex-excited vibration, formulated by taking into account the increasing correlation length of the exciting force with increasing vibration amplitude. Using this model, Ruscheweyh predicted the amplitudes of the vortex-excited bending vibration of several steel stacks and bridge hangers. The predicted amplitudes had been verified with full-scale measurements and the agreement is very good. This model has been included into Eurocode ENV "Wind Action." Ma et al. (2005) studied the characteristics of galloping and vortex-induced vibration of H-section hangers, and they pointed that there were apparent effects on the galloping stability by changing the size/shape of web perforation in hangers. However, none of the above investigators have pointed out the dangerous flutter instability possibly encountered for H-sections. Matsumoto et al. (1992, 2008) studied the torsional flutter mechanism of H-section cylinders. Their research is connected with the horizontal bridge girders like the Old Tacoma Bridge under zero wind attack angle. The H-section girders in their study have a width-to-depth (web width to flange depth) ratio of 2–15 or an equivalent depth-to-width (flange depth D to web width B) ratio of 0.067 to 0.5. For the vertical hangers, the worst scenario may be not in parallel with strong/weak-axis of H-section and the wind attack angle may vary from 0° to 90°. In addition, the width-to-depth ratio of H-sections adopted in hangers is usually less than 2, which is not covered by Matsumoto et al. (1992, 2008).

Although H-sections are aerodynamically unstable, they are convenient to manufacture, and connection and maintenance are frequently applied in practice. It is therefore very important to study the aerodynamic characteristics of H-sections from the point of view of its wind-resistant design. Flutter, galloping, and vortex-induced vibrations of H-section hangers is examined in this study by section model and aeroelastic model tests. The longest hanger in the Dongping arch bridge which experienced strong torsional vibration and severe structural damage after storm was selected as prototype.

First, the flutter instability of the H-sections is studied with a section model and full aeroelastic model test for the prototype hanger. The critical flutter wind speed for the hanger is directly measured by a free vibration test of the elastically mounted section model and is further verified by the aeroelastic model test. It is shown that the experimental critical speed is quite close to the in-site wind speed. The flutter derivative A_2^* of the H-sections is identified with section model test data for wind attack angles of 15°, 20°, and 25°, respectively. The critical wind velocity predicted with A_2^* agrees with the directly measured one in wind tunnel. It is concluded that the observed prototype vibration is a kind of torsional flutter instability.

Second, the influences of web perforation on the performances of flutter, galloping, and vortex-excited vibration are studied using four section models with different web perforation ratios but the same depth-to-width ratio $D/B = 0.416$. For galloping, the measured critical wind speed is compared with that predicted from the Den Hartog's galloping criteria. The performance of vortex-excited vibration of the H-sections is assessed by their Strouhal numbers and lock-in wind speed ranges.

In view of the fact that web perforations has little impact on improving flutter stability and vortex-excited vibration, a new research scheme has been launched to investigate aerodynamic performances of the H-sections with different depth-to-width ratios and web and flange perforation ratios.

Finally, a comparison of the above results with previous works is made, which may explain why the flutter instability of H-section hangers under large wind attack angles was not treated by earlier investigators.

Torsional Flutter of H-Section Hangers under Large Wind Attack Angles

Prototype and Model

The longest hanger in the Dongping arch bridge (Fig. 1) was selected as a prototype for wind tunnel study of aerodynamic characteristics of the H-sections. The prototype hanger is 40.212 m long and the section is so positioned such that the web plane is in the transverse direction of the bridge. The web is 1.2 m wide and 10 mm thick, and the flange is 0.5 m deep and 18 mm thick. The depth-to-width ratio of the H-section is $D/B = 0.5/1.2 = 0.416$, so it is a shallow H-section. The web is perforated mainly for improving the aerodynamic performance and the web perforation ratio is 27% (the perforation ratio is defined as the ratio of perforation area to the whole web area before perforation). The dynamic modes of the hanger were analyzed with a finite element model using shell elements, and the minimum axial force applied on the prototype was considered as the initial stress effect. A 1:4 section model of the hanger was made and named A-3, whose geometry and size are provided in Fig. 3. A 1:16 aeroelastic model was also designed without conforming to the Froude number similarity. Table 1 summarizes the main parameters of the prototype hanger, the 1:4 section model, and the 1:16 aeroelastic model.

Section Model Tests

The section model was elastically suspended in a wind tunnel with 8 springs in such a way that the ratio of the torsional frequency to the weak-axis-bending frequency was nearly the same ratio as that of the prototype's, as shown in Fig. 4. Each spring was connected to a dynamometer to measure the vibration. The vibration data were recorded with a sampling frequency of 100 Hz. The definition of wind attack angle α is illustrated in Fig. 5. The wind direction normal to the flange is defined as $\alpha = 0°$ and the direction perpendicular to the web is defined as $\alpha = 90°$.

The first test was to measure the critical flutter wind speed of the section model under different wind attack angles in uniform flow. The tests were carried out for wind attack angles in the range of 0° ~ 90° with increments of 5°. The torsional instabilities developed for wind attack angles of 15°, 20°, and 25°. Fig. 6 provides the variation of torsional amplitudes with prototype full-scale wind speed for the three wind attack angles. The critical wind speed was 26.9 m/s at a wind attack angle of 15°. The criterion for flutter instability was taken as the torsional amplitude being large than 5°. This threshold is based on Chinese Guideline for Wind-resistant Design of Highway Bridges (1996), where it is indicated that a maximum torsional response of 1° ~ 5° during wind tunnel tests indicates occurrences of flutter instability. By further considering the effect of large wind attack angles associated with vertical hangers, a maximum torsional response of 5° was taken as the threshold. The critical wind speed obtained in the section model test was found to be in good agreement with that reported in Foushan Diary on August 4, 2006 ("Typhoon" 2006). Since the amplitude continuously increases with wind speed, these cases

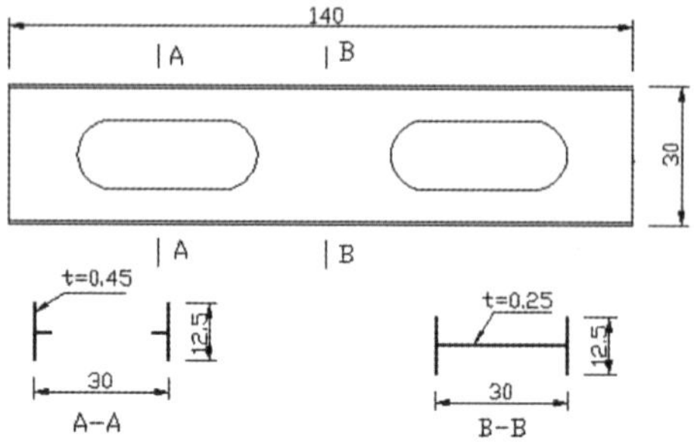

Fig. 3. H-shaped section model (dimensions: cm)

Table 1. Parameters of Prototype and Models

Parameters	Prototype	Section model A-3	Aeroelastic model
Length (m)	40.212	1.400	2.513 (1:16)
Flange depth D (m)	0.5	0.125 (1:4)	0.0312 (1:16)
Web width B (m)	1.2	0.3 (1:4)	0.075 (1:16)
Mass (kg/m)	211.6	13.2 (1:4^2)	0.852 (1:16^2)
Mass moment (kg – m^2/m)	63.06	0.25 (1:4^4)	9.76×10^{-4} (1:16^4)
Weak-axis bending (Hz)	2.001	1.953 (0.976:1)	14.258 (7.125:1)
Strong-axis bending (Hz)	5.936	1.953 (0.345:1)	41.992 (7.074:1)
Torsional frequency (Hz)	2.251	2.148 (0.954:1)	17.188 (7.635:1)
Damping ratio	0.002	0.0033 for bending 0.0017 for torsion	0.007 for bending 0.005 for torsion
Wind speed ratio		1:4.098 weak-axis bending 1:11.5 strong-axis bending 1:4.192 torsion	1:2.246 weak-axis bending 1:2.262 strong-axis bending 1:2.096 torsion

Note: Figures in parenthesis are scale ratios, and all of the scale ratios are defined as models to prototypes.

Fig. 4. Section model elastically suspended in wind tunnel

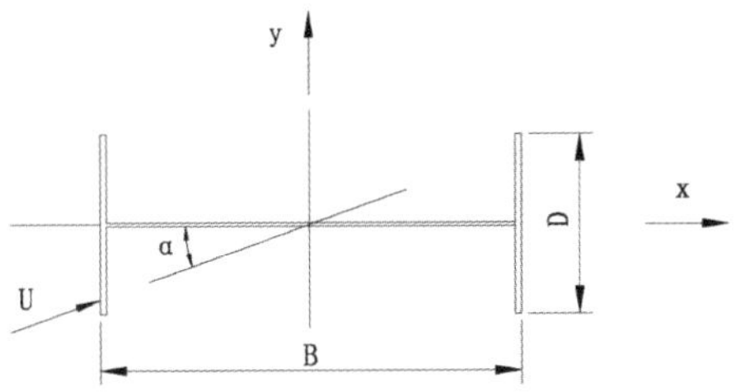

Fig. 5. Definition of wind attack angle

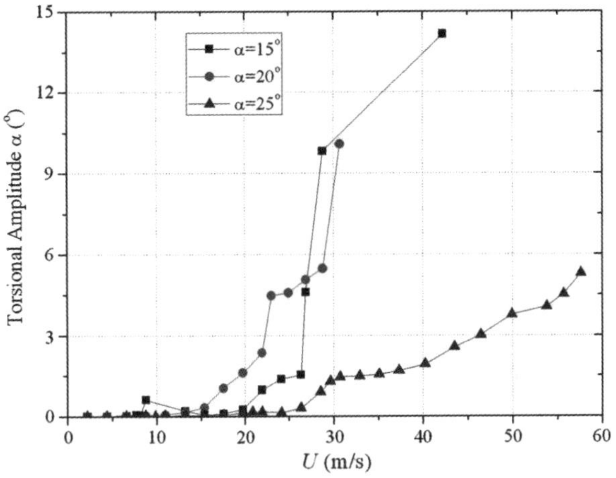

Fig. 6. Variation of vibration amplitude of model A-3 with wind velocity

should be classified as torsional flutter. Additionally, because the Strouhal number, St, for the H-sections is approximately 0.12, the hanger would torsionally resonate due to vortex shedding at a wind velocity of approximately 9.0 m/s, which is well below the field wind velocity. Therefore it is not likely the vibration is caused by Karman vortex shedding.

Fig. S1 shows the variation of the equivalent damping ratio with the reduced velocity U/fB for the three attack angles along with wind attack wind angles of 0°. When the testing wind velocity approached the flutter critical threshold, torsional response built up very rapidly for wind attack angles of 15° and 20°, as observed in Fig. 6. There was concern that the rapidly divergent response may potentially damage the wind tunnel facilities, thus no effective data was recorded for some cases. An extrapolation of experimental net damping curves was adopted to estimate the critical reduced wind velocity initiating flutter instability for wind attack angels of 15° and 20°. The curves for wind attack angles of 15° were the first ones to cross zero at a reduced velocity of $U/fB = 7.5$. Since the vibration frequency and width of the section are 2.25 Hz and 1.2 m, respectively, the onset full-scale flutter wind velocity is $V_{cr} = 7.5 \times 1.2 \times 2.25 = 20$ m/s, which is slightly lower than the measured critical wind speed of 26.9 m/s. The discrepancy is mainly caused by taking a torsional amplitude large than 5° as the threshold of flutter instability.

Flutter is an aerodynamic instability with torsional vibration. There are two kinds of flutter instability: classical flutter and separated flow flutter (Simiu and Scanlan 1996). Separated flow flutter is a purely torsional vibration and can be analyzed by a flutter equation with single degree of freedom (SDOF) as follows:

$$I_m \ddot{\alpha} + c\dot{\alpha} + k\alpha = \rho U^2 B^2 \left(K A_2^* \frac{B\dot{\alpha}}{U} + K^2 A_3^* \alpha \right) \tag{1}$$

where I_m, c, and k = inertial, damping, and stiffness parameters of the model, respectively; α = torsional displacement of the model; A_2^* and A_3^* = flutter derivatives, which are functions of the reduced wind frequency, $K = Bf/U$; ρ = mass density of air; U = wind velocity; and B = width of the H-section. The right-hand-side (RHS) term of Eq. (1) represents the unsteady self-excited moment. Moving the right term to left side, the equation becomes:

$$I_m \ddot{\alpha} + (c - \rho U B^3 K A_2^*)\dot{\alpha} + (k - \rho U^2 B^2 K^2 A_3^*)\alpha = 0 \tag{2}$$

The coefficient in the bracket of the second term is the equivalent damping of the model. It is apparent that the hanger may exhibit torsional flutter when the flutter derivatives of A_2^* is positive. Therefore, the same model was tested by the forced vibration method to identify the flutter derivatives A_2^* for different wind attack angles. In the forced vibration method, the motion of the section model is driven by the numerically controlled external exciter fixed at two ends of the model; the aerodynamic forces acting on the model for a particular wind velocity and the response were measured and were used to identify the flutter derivatives (Chen et al. 2005). The identified flutter derivatives A_2^* for wind attack angles of 0°, 15°, 20°, and 25° are given in Fig. S2. It is indicated that positive values of A_2^* is most notable for wind attack angles of 15° and 20°, which is consistent with the results in Figs. 6 and S1. The flutter derivatives A_2^* curve for wind attack angle 15° has the largest positive value, which implies the lowest critical wind velocity.

Aeroelastic Model Test

A full aeroelastic model of the above prototype hanger was further tested in uniform flow to verify the torsional flutter observed in section model test. The geometrical scale of the aeroelastic model was 1:16. As the gravity effect is of minor importance for hangers, the Froude number similarity was not respected. The aeroelastic model consisted of three welded thin steel plates and its ends were fixed to the bottom and upper floors of the wind tunnel, as shown in Fig. 7. To keep the mass similarity, additional mass blocks were

Fig. 7. Aeroelastic model in wind tunnel

attached to the thin steel plates. The parameters of the aeroelastic model are listed in Table 1. The model was tested in uniform flow for wind attack angles of 0°~90°, with increments of 5°. The structural damping ratio of torsional mode of the aeroelastic model was about 0.5%, which is larger than that used in the section model test, 0.17%, because of the difficulty in firmly fixing the added mass blocks to the thin steel plates.

At a wind attack angle of 15°, torsional flutter occurs at the prototype full-scale wind velocity of 29 m/s. The wind velocity in this case is a little higher than the critical flutter wind speed of 26.9 m/s that was obtained for the section model. Because the damping ratio of the torsion mode for the aeroelastic model is higher than that for the section model, this result is reasonable and acceptable.

As a remedy countermeasure to reduce the torsional vibration, the midspan of the aeroelastic model was connected to two columns with steel wires, as shown in Fig. 7. The horizontal distance from the aeroelastic model to each column was the same as that of two suspenders on an actual bridge (15.35 m); the stiffness of steel wires was also expected to follow the same similarity principle used by the aeroelastic model, but the stiffness of the columns was not modeled in this study. The use of steel wires increased the torsional and weak-axis bending modal frequencies and promoted the lowest flutter critical velocity exceeding 60 m/s, well above the design requirement. This suppression measure was applied on the actual bridge according to the above experimental results. Fig. 8 shows the design scheme actually implemented on the bridge. For each of three main arches in the bridge, two wind-resistant cables, whose size was equal to the full-scale value of the steel wire used in the test, were connected through the flanges of hangers and the cable ends were fixed to the main arch. The size of each cable was of $4\phi 15.2$ steel strands. No attempt was made to optimize the size of steel wire in wind tunnel tests.

Influences of Web Slot on Aerodynamic Features of H-Section Hangers

Section Models

To study the influence of web perforation on flutter, galloping, and vortex-excited vibration of the H-section hangers, four section models were used. The four section models had the same depth-to-width ratio as the model A-3 ($D/B = 0.416$) but with different web perforation ratios. The web perforation ratios of these models were 0%, 14%, 27%, and 38%, and these four models were denoted as A-1, A-2, A-3, and A-4, respectively. Model A-3 with the perforation ratio of 27% was the one actually used on the Dongping arch bridge. The mass and stiffness of the hangers varied slightly for the model as a result of different perforation ratios. Table 2 provides the modal frequencies of the longest hanger with different web perforation ratios. It was found that the strong-axis-bending frequency for the model A-4 decreased appreciably due to web perforation. The aerodynamic performance of flutter, galloping, and vortex-excited vibration of these section models were tested for wind attack angles from 0° to 90° with increments of 5°. The damping ratios during the test were nearly the same of about 0.36% in bending and 0.17% in torsion.

Table 2. Prototype Mass and Frequencies of the Four Models ($D/B = 0.416$)

Model	Mass (kg/m)	Mass moment ($kg.m^2/m$)	Frequency (Hz) Strong-axis bending	Weak-axis bending	Torsion
A-1	235.553	63.986	6.057	1.901	2.175
A-2	223.342	63.829	6.112	1.952	2.173
A-3	211.635	63.055	5.936	2.001	2.251
A-4	202.709	61.512	5.469	2.042	2.409

Flutter Performance

The flutter performances of the four models were directly tested by an elastically suspended rig, as described previously. Torsional flutter instability was observed for all four models under large wind attack angles. Table 3 summarizes the ranges of wind attack angle and the prototype critical wind velocities for the four models under the criterion of torsional displacement exceeding 5°. The results show that the web perforation does not promote the flutter critical wind speed of H-sections, and actually tends to decrease flutter performance of H-sections. For the shallow H-section with $D/B = 0.416$, the flutter critical wind speed is lower than the wind-resistant design requirements in most parts of China. The critical wind velocity for a particular web perforation listed in Table 3 is the lowest one for all wind attack angles; such presentation is convenience for bridge designers in order to clearly understand which kind of perforation is best in connection with flutter

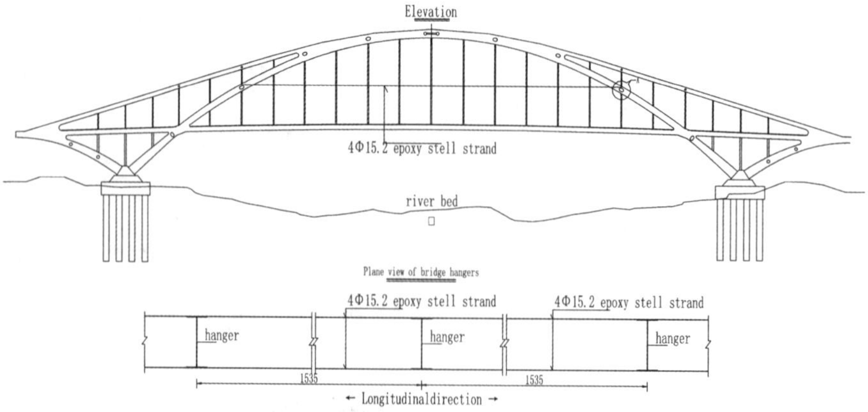

Fig. 8. Design scheme of wind-resistant cables for hangers

Table 3. Torsional Flutter of the Models with Different Web Perforation Ratios ($D/B = 0.416$)

Section model	Wind attack angle	Prototype critical wind speed	Reduced wind speed (V/fB)
A-1	30° ~ 35°	36.9 m/s	14.1
A-2	20°	24.3 m/s	9.32
A-3	15° ~ 25°	26.9 m/s	9.96
A-4	15° ~ 25°	23.1 m/s	7.99

performance. However, in order to study the underlying mechanism of improvement of flutter performance, it is necessary to change only one parameter while keep the remaining parameters fixed, such as by keeping wind attack angles unchanged and varying the web perforation ratio.

Galloping Performance

Galloping of slender members is a kind of bending, wind-induced, self-excited, and unstable oscillation. According to the quasi-steady theory of galloping, the galloping stability for a certain section can be evaluated by its Den Hartog coefficient, $C_L' + C_D$. Following this theory, the critical wind speed of galloping is calculated as

$$V_{cg} = -\frac{4m\omega\zeta}{\rho B(C_L' + C_D)} \tag{3}$$

where ζ = modal damping ratio. Obviously, a necessary condition for galloping is to have a negative Den Hartog coefficient. This method is based on assumption of quasi-steady and linearization and is suitable for engineering purposes. More advanced methods considering the nonlinearity in aerodynamic galloping loads will be explored in future (Yu et al. 1993; Luongo and Piccardo 1998).

To evaluate the Den Hartog coefficient, the section model was connected with a force balance and the assembly was then fixed vertically on a turntable in the wind tunnel to measure the steady aerodynamic lift coefficient C_L and drag coefficient C_D at wind attack angles from 0° to 90°, with increments of 2°. The test was conducted in uniform flow. The measured drag and lift coefficients are shown in Fig. S3 with the web width B being the characteristic length. Fig. S4 shows the variations of Den Hartog's galloping coefficients with wind attack angles. The negative slope for the C_L is obvious at wind attack angles close to 0° and to 90° for model A-1 without web perforation and the Den Hartog coefficients for the model A-1 are also negative, indicating that the galloping instability may occur in the vicinity of the two wind attack angles. For the section models with web perforations, namely A-2, A-3, and A-4, the negative slope is obviously weakened. For models A-2 and A-3, with perforation ratios of 14% and 27%, respectively, the negative slope is not clear at a wind attack angle around 0°. It is concluded that modest web perforations have a positive effect in stabilizing the galloping of H-sections.

Because the Den Hartog coefficients vary considerably with wind attack angles, which value should be selected to calculate the galloping critical wind speed from Eq. (3)? Taking model A-1 as an example, the values of the Den Hartog coefficient at wind attack angles of 0, 2, 4, 6, and 8° are −1.8, −2.1, −3.0, −3.5, and +1.3, respectively. Following the symmetry, the Den Hartog coefficients of model A-1 for wind attack angle within ±8° are shown in Fig. 9. It is seen that the attack angle with a steady state of galloping is within ±6°. It is unreasonable to use the minimum Den Hartog coefficient, i.e., $C_L' + C_D = -3.5$ to calculate the critical wind speed of galloping. This is because the effective wind attack angle

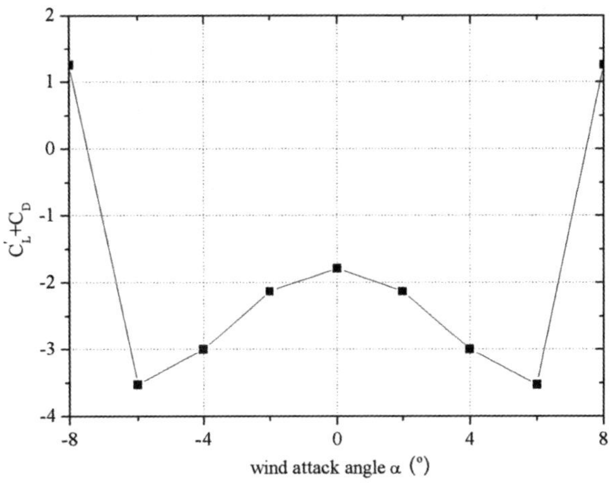

Fig. 9. Den Hartog coefficient of model A-1 at angle interval ±8°

relative to the structure varies during vibration. The structure will withdraw rapidly from the galloping state when the effective wind attack angle is larger than 6°. It is suggested in this study to use the average values of Den Hartog coefficients between 0° and 6° to calculate the critical galloping wind speed. The average galloping coefficient of the model A-1 along the weak-axis is −2.6. Table 4 shows the Den Hartog coefficients calculated with this method and galloping critical wind speeds of the prototype for the four section models. The parameters used in Eq. (3) are given in Table 2. As seen in Table 4, the galloping performance for the weak-axis of H-sections is much improved by web perforation. For model A-4 with a slot ratio of 38%, the galloping coefficient increases more than that of model A-1 by a factor of 5. However, the galloping coefficient at the strong axis changes little with the web perforation and is approximately −0.5.

As verification, wind tunnel experiments were also carried out in uniform flow for the four section models freely suspended with the same test rig as that used for flutter instability. The galloping critical speed, along with wind attack angles, are shown in Table 5. As consistent with those in Table 4, only models A-1 and A-4 developed galloping. For model A-1, a galloping vibration normal to the weak-axis at wind attack angles of 0° and 5° is observed at a

Table 4. Den Hartog Coefficient of H-Sections with $D/B = 0.416$

Model	Galloping coefficient		Galloping critical velocity (m/s)	
	Strong axis	Weak axis	Strong axis	Weak axis
A-1	−0.6	−2.6	146.4	11.5
A-2	−0.5	+	140.0	Stable
A-3	−0.5	+	128.9	stable
A-4	−0.4	−0.5	113.7	69.4

Table 5. Critical Galloping Wind Velocity for Different Web Perforations

Model	A-1 (web peroration 0%)		A-4 (web peroration 38%)		
Wind attack angle[a] (°)	0	5	80	85	90
Critical galloping speed (m/s)	26.2	34.7	137.9	122.4	92.2

[a]The increment of wind attack angle is 5°.

full-scale wind velocity of 26.2 m/s and 34.7 m/s, respectively. For model A-4, galloping with vibration normal to strong axis occurs at attack angles of 80°, 85°, and 90°. However, the critical wind speed is 92.2 m/s, which is well above the design requirements. The galloping for model A-4 is mainly attributed to the notable decrease in modal frequency due to large web perforation. Additionally, the galloping critical wind speed for model A-1 calculated by Eq. (3) is approximately half of the experimental values, and the quasi-steady theory of galloping is therefore conservative. These experimental results conclude that a modest web perforation can promote the galloping performance, especially normal to the weak-axis.

Vortex-Excited Vibration Performances

Vortex-excited resonance takes place when the frequency of vortices regularly shed from the model coincides with one of the model vibration frequencies. The vortex shedding frequency is given as

$$f_v = St\frac{V}{L} \tag{4}$$

where L is a characteristic dimension of the body projected on a plane normal to the mean flow velocity, defined as $L = D$ for $0° \le \alpha \le 45°$ and $L = B$ for $45° \le \alpha \le 90°$; V = velocity of the oncoming flow; and St = Strouhal number. The use of this definition is based on the two points. The first is that wind-induced vibrations for wind directions between 0° and 45° are normal to the web plate or in the direction of the weak axis, so the flange depth D is used for L. Wind-induced vibrations for wind directions between 45° and 90° are normal to the flange plate or in the direction of strong axis, so the web width B is used for L. The second is the same characteristic length ($L = D$ Or $L = B$), which is convenient for bridge designers.

Table S1 shows the vortex-induced vibration characteristics for the four H-section models for wind attack angles from 0° to 90°. The wind speed and amplitude of vortex shedding response have been converted to full-scale values.

From Table S1, it is shown that (1) torsional vortex-induced vibration with small amplitude at low wind speed is observed for section models A-2 and A-3. Among the wind velocities of vortex-induced vibration at various wind attack angles, the minimum one is less than 10 m/s, and the torsional amplitude is lower than 1.5°; (2) for section models A-1 and A-4, the onset wind speed of vortex-induced vibrations at different attack angles is usually around 10 ~ 30 m/s. However, the torsional amplitude is large and generally exceeds 3°; (3) torsional motions of vortex-induced vibrations dominate for the four models, and models A-2 and A-4 may develop bending motions at a certain wind attack angle; (4) the maximum Strouhal numbers for model A-1 are 0.077, whereas the Strouhal numbers for other models generally exceed 0.11, with the largest one being 0.236 for model A-4, indicating that the H-section members with web slots are more easily vulnerable to vortex shedding. Thus, it is concluded that web perforation hardly improves the vortex shedding response of H-sections.

Effects of Three Sectional Parameters of H-Section Hangers

In view of the fact that web perforation has little effect on improving the flutter stability and vortex shedding, a new research scheme was launched to investigate the aerodynamic performances of H-shaped hangers with different sectional parameters, namely, depth-to-width ratios, web perforation ratios, and flange perforation ratios. As a first step, a group of 16 section models were designed with the orthogonal test method. In these models, the depth-to-width ratios were 1/2.4, 1/1.6, 1/1, and 1/0.75, i.e., $D/B = 0.416$, 0.625, 1.0, and 1.33, respectively; the web perforation ratios were 0%, 14%, 27%, and 38%, respectively; the flange perforation ratios were 0%, 11%, 20.1%, and 28.6%, respectively. The flange perforation was of a rectangular hole, as shown in Fig. 10. The selection of these parameters was mainly based on the general design dimensions of hangers in arch bridges. For example, the depth-to-width ratio D/B varies from 0.416 to 1.33, which means the variation of an H-section from a shallow one to a deep one.

The sectional parameters of the 16 section models are provided in Table S2. The same hanger in Table 1 was taken as the structural parameters of these models. The geometrical scale of the section model was still taken as 1/4, but the spring-mounted test rig improved with a bending damping ratio of 0.15% and a torsional damping ratio of 0.10%. The wind-induced vibration characteristics of torsional flutter, galloping, and vortex-induced vibration were tested in uniform flow for wind attack angles of 0° ~ 90°, with increments of 5°. Table S3 shows some selected results of the full-scale critical wind speeds for torsional flutter, galloping, and vortex-induced vibrations, as well as the maximum amplitude of vortex-induced vibration for each kind of H-sections.

The results listed in Table S3 reveal some general trends of the effects of the three sectional parameters on the aerodynamic characteristics of H-sections: (1) increasing the depth-to-width ratio D/B notably improves the aerodynamic performances, implying that the deep H-section is better than the shallow one; (2) flange perforation can enhance the aerodynamic capability against flutter and vortex shedding; (3) web perforation has positive effects merely on galloping; (4) for the 16 section models, section model 10, with a model with depth-to-width ratio of 1.0, a

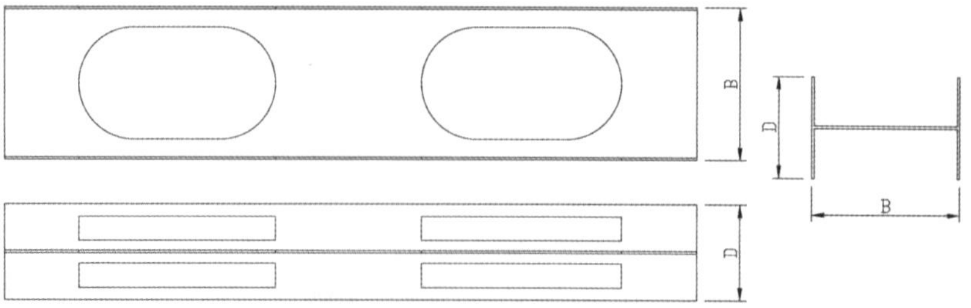

Fig. 10. The shape on web and flange perforation of H-sections

web perforation ratio of 14%, and a flange perforation ratio of 28.6%, gives the best aerodynamic performance and has only vortex-induced oscillation at the prototype wind velocity of 70 m/s, which is larger than design requirement.

To obtain the quantitative description of the effects of these parameters, more experiments and analyses are needed. Therefore, this investigation is still in progress. The results will be reported elsewhere.

Comparison with Previous Works

After the severe wind-induced vibration of tensile members happened on the Commodore Barry Bridge during the construction, Maher and Wittig (1980) conducted a careful study of the H-section model by section model tests. In their tests, the range of attack angles was $0° \sim 90°$, with increments of 10°. However, there was no flutter instability found. The studied prototype member of the Commodore Barry Bridge was a deep H-section member with $D/B = 1.28$, a flange depth of 910 mm, and a web width of 710 mm. The prototype wind speed range in the tests did not exceed 50 m/s. From Table S3, the estimated flutter speed of this member exceeded 60 m/s, therefore they missed the flutter in the tests. The prototype hanger used in the present study was a shallow H-section with $D/B = 0.416$, and the onset flutter velocity is much lower than that of a deep one. Therefore it was easy to find in the test.

Ruscheweyh (1996) made a valuable contribution on the wind-induced vibration of slender elements. He presented a mathematical model for predicting the amplitudes of vortex-excited vibration, called "The correlation length model." He also studied the critical speed of galloping. Besides the formula for bending galloping, he also presented a similar equation to estimate the critical speed of "torsional galloping." It should be pointed out that "torsional galloping" is an inadequate concept, because if a self-excited instability is torsional oscillation, it should be classified as flutter according to the modern theory of wind-induced vibrations. Ruscheweyh's formula for galloping was based on the quasi-steady theory; its applicability to flutter that is caused by the unsteady self-excited force is questionable.

Matsumoto et al. (1992, 2008) studied aerodynamic behaviors of the H-sections only for a wind attack angle of 0°. They concluded that the model is flutter-stable for $B/D < 3.4$ or equivalently $D/B > 0.3$. As consistent with Matsumoto's results, the H-sections with $D/B = 0.416$ is flutter-stable at a wind attack angle of 0°, as given in Table 4. However, torsional flutter instability of H-sections easily develops for large wind attack angles even at relatively low wind velocity as far as hangers and other vertical members are concerned.

Conclusions

A series of experimental studies of wind-induced vibrations of vertical flexible hangers with H-sections is carried out using both section models and an aeroelastic model. It is highlighted that the vertical hangers with H-sections are featured by large wind attack angles and have more complicated and unfavorable aerodynamic characteristics than the horizontal members. The main conclusions from this study are as follows: (1) the torsional flutter of vertical H-shaped hangers under large wind attack angle has been confirmed, especially for shallow H-sections, which has not been pointed out by previous investigators; (2) flange perforation can enhance the capability against flutter and vortex-induced vibration, while web perforation usually increases the critical wind velocity of galloping; and (3) the increase of D/B significantly improves the aerodynamic performance, which implies that the deep H-section is better than the shallow one in resisting wind-induced vibrations.

A total of 16 section models with different depth-to-width ratios, web perforation ratios, and flange perforation ratios have been studied. A more complete investigation of the effect of the above three parameters on the aerodynamic performance of H-sections is underway.

Acknowledgments

This study is sponsored by the National Science Foundation of China with the key research program number 50738002, 2006BAJ03B04, and is partially supported by research fund from Department of Transportation of Sichuan Province (Grant No. 2008.2-4-3).

Supplemental Data

Figs. S1–S4 and Tables S1–S3 are available online in the ASCE Library (www.ascelibrary.org).

References

Chen, Z. Q., Liu, M. G., Liu, G. D., and Jin, Z. J. (2010). "Wind-induced vibration and wind-resistant design of H-shaped suspenders under large attack angle." *China Civ. Eng. J.*, 43(2), 1–11.

Chen, Z. Q., Yu, X. D., Yang, G., and Spencer, B. F. (2005). "Wind-induced self-excited loads on bridges." *J. Struct. Eng.*, 131(12), 1783–1793.

Chi, M., Neal, E., and Dennis, B. G. (1977). "Determination of strouhal characteristics and power spectrum for elastically restrained H-shape sections." FHWA-RD-78-26, Federal Highway Administration, Office of Research and Development, Washington, DC.

Chi, M., and Vossoughi, J. (1985). "Response of slender structural members in self-excited oscillation." *J. Sound Vib.*, 101(1), 75–83.

Gu, J. H., Zhao, Y. C, and Shao, K. H. (1994). "Application of new TMD to suppressing vortex-shedding vibrations of hangers of Jiujiang bridge over Yangtze river." *China Civ. Eng. J.*, 27(3), 3–13 (in Chinese).

Kubo, Y., Sakurai, K., and Azuma, S. (1980). "Aerodynamic characteristics of H-shaped section cylinder in laminar and turbulent flows." *Mem. Kyushu Inst. Technol. Eng.*, 10, 1–14.

Luongo, A, and Piccardo, G. (1998). "Nonlinear galloping of sagged cables in 1:2 internal resonance." *J. Sound Vib.*, 214(5), 915–940.

Ma, C., Liao, H., Zheng, S., and Li, J. (2005). "Wind tunnel experiment on the aerodynamic performances of H-shaped booms." *Zhongguo Tiedao Kexue*, 26(4), 42–46.

Maher, F. J., and Wittig, L. E. (1980). "Aerodynamic response of long H-sections." *J. Struct. Div.*, 106(1), 183–198.

Matsumoto, M., Shirato, H., and Hirai, S. (1992). "Torsional flutter mechanism of 2-D H-shaped cylinders and effect of flow turbulence." *J. Wind Eng. Ind. Aerodyn.*, 41–44(1–3), 687–698.

Matsumoto, M., Shirato, H., Mizuno, K., Shijo, R., and Hikida, T. (2008). "Flutter characteristics of H-shaped cylinders with various side-ratios and comparisons with characteristics of rectangular cylinders." *J. Wind Eng. Ind. Aerodyn.*, 96(6–7), 963–970.

Ruscheweyh, H. (1990). "Practical experiences with wind-induced vibrations." *J. Wind Eng. Ind. Aerodyn.*, 33(1–2), 211–218.

Ruscheweyh, H. (1996). "Vortex-excited vibrations and galloping of slender elements." *J. Wind Eng. Ind. Aerodyn.*, 65(1–3), 347–352.

Simiu, E., and Scanlan, R. H. (1996). *Wind effects on structures: Fundamentals and applications to design*, 3rd Ed., Wiley, New York.

"Typhoon Prapiroon blows through Foushan city with strong wind and rain." (2006) *Foshan Diary*, Aug. 4, A4, ⟨http://epaper.citygf.com/szb/history/html/2006-08/04/content_56449.htm⟩ (in Chinese).

Ulstrup, C. C. (1980). "Aerodynamic lessons learned from individual bridge members." *Ann. N.Y. Acad. Sci.*, 352(1), 265–281.

Wardlaw, R. L. (1990). "Wind effects on bridges." *J. Wind Eng. Ind. Aerodyn.*, 33(1–2), 301–312.

Xiang, H. F., et al. (1996). *Chinese guideline for wind-resistant design of highway bridges*, China Communications Press, Beijing, China (in Chinese).

Yu, L., Gu, J., and Wang, Z. (1996). "Experimental study of vortex-induced oscillation for the booms on large bridge." *J. Mech. Strength*, 18(4), 16–20.

Yu, P., Desai, Y. M., Shah, A. H., Popplewell, N. (1993). "Three-degrees-of-freedom model for galloping: formulation and solutions." *J. Eng. Mech.*, 119(12), 2404–2425.

Rain-Wind-Induced In-Plane and Out-of-Plane Vibrations of Stay Cables

Shouying Li[1]; Zhengqing Chen, M.ASCE[2]; Teng Wu, S.M.ASCE[3]; and Ahsan Kareem, Dist.M.ASCE[4]

Abstract: A theoretical model of in-plane and out-of-plane rain-wind-induced vibrations (RWIVs) of stay cables is developed in this paper. The proposed scheme models the cable, akin to sectional models of a bridge deck, as a segmental rigid element with two degrees of freedom (DOF) (in plane and out of plane) and the rivulet as a single DOF oscillator. The interaction between the cable surface and the water rivulet is first analyzed by using a wetting theory, which suggests a combined application of linear and Coulomb damping forces between the rivulet and the cable. Based on the model, the response of the combined cable-rivulet system is numerically evaluated. The results show that the in-plane and out-of-plane DOF of the cable are coupled. For the equilibrium of the rivulet on the cable surface, the gravitational and aerodynamic forces on the rivulet have similar time-averaged values with opposite signs. The interaction between the cable surface and the rivulet leads to energy dissipation, which helps to stabilize the rivulet. Like galloping, RWIV appears to be triggered by a sudden drop in the mean aerodynamic force coefficient. However, it is not a divergent type of vibration because the rivulet oscillates on the cable surface and its equilibrium position changes with wind velocity. Finally, the turbulence effect on the RWIVs is investigated. As the longitudinal turbulence intensity approaches 15%, the RWIV abates concomitantly. **DOI: 10.1061/(ASCE)EM.1943-7889.0000612.** © *2013 American Society of Civil Engineers.*

CE Database subject headings: Cable-stayed bridges; Wind; Rainfall; Vibration; Turbulent flow.

Author keywords: Stay cables; Rain-wind-induced vibrations; Theoretical model; Turbulent flow.

Introduction

The rain-wind-induced vibration (RWIV) of the stay cables of cable-stayed bridges can seriously impact the safety of the cables, and eventually the safety of the bridges (Watson and David 1988). This type of vibration was first noticed in Japan in the 1980s on the MeikoNishi Bridge. Hikami and Shiraishi (1988) conducted field observations to understand this behavior and a series of wind tunnel tests to reproduce this phenomenon. The results showed that the type of cable vibration experienced on the MeikoNishi Bridge could not be explained by the established aerodynamic theories. Hikami and Shiraishi (1988) named these RWIVs. Since that time, many investigations concerning the mechanism of RWIV have been performed involving field observations (e.g., Hikami and Shiraishi 1988; Matsumoto et al. 2003; Chen et al. 2004), wind tunnel tests (e.g., Hikami and Shiraishi 1988; Yamaguchi 1990; Flamand 1995; Gu et al. 2002; Gu and Lu 2001; Cosentino et al. 2003a; Gu and Du 2005), and theoretical analyses (e.g., Yamaguchi 1990; Gu and Lu 2001; Peil and Nahrath 2003; van der Burgh and Hartono 2004; Xu and Wang 2003; Wilde and Witkowski 2003; Cosentino et al. 2003b; Cao et al. 2003; Gu et al. 2009). In recent years, with the development of modern computer hardware and advances in computational fluid dynamics, numerical simulations have been used to investigate RWIVs (e.g., Rocchi and Zasso 2002; Li and Gu 2006; Li et al. 2010; Taylor and Robertson 2011). Thus far, the complexity involved in RWIVs has precluded the development of a theoretical basis concerning their mechanism.

Despite the significance of the RWIVs of stay cables on the safety of cable-stayed bridges, thus far neither the underlying mechanism has been fully understood nor has a well-recognized model been established. Currently, two types of explanations have been advanced. The first, shared by most investigators, postulates that the rivulet formed on the upper surface of the cable, which alters the flow near the cable, is the key factor in the onset and sustenance of RWIVs. The second viewpoint is advanced by Matsumoto et al. (1995), which is based on the field observations of the role of axial flow along the cable axis on RWIVs. These vibrations were observed to take place at even higher reduced wind speeds of 20, 40, 60, and 80.

The first theoretical analysis model of RWIV was carried out by Yamaguchi (1990), in which both the natural frequency of the rivulet and modal aerodynamic damping of the cable with the rivulet were analyzed. In this model, the cable and the rivulet were treated as a unit. Therefore, it was not necessary to separate the wind forces acting on the rivulet and the cable, and the interaction between the cable and the rivulet was not needed. Accordingly, this simulation did not include the oscillation of the rivulet on the cable surface. There was little progress in theoretical analysis of RWIVs in the 1990s. To simplify the over complexity, some researchers assumed sinusoidal oscillations of the rivulet for analyzing RWIVs (Xu and Wang 2003; Wilde and Witkowski 2003; van der Burgh and Hartono 2004; Li and Gu 2007). The range of velocities over which the RWIVs take place was established by invoking an additional assumption concerning the relationship between the rivulet equilibrium

[1]Associate Professor, Wind Engineering Research Center, Hunan Univ., Yuelu Hill, Changsha 410082, China (corresponding author). E-mail: shyli@hnu.edu.cn

[2]Professor, Wind Engineering Research Center, Hunan Univ., Yuelu Hill, Changsha 410082, China. E-mail: zqchen@hnu.edu.cn

[3]Ph.D. Candidate, NatHaz Modeling Laboratory, Dept. of Civil Engineering and Geological Sciences, Univ. of Notre Dame, Notre Dame, IN 46556.

[4]Professor, NatHaz Modeling Laboratory, Dept. of Civil Engineering and Geological Sciences, Univ. of Notre Dame, Notre Dame, IN 46556.

Note. This manuscript was submitted on December 28, 2011; approved on February 22, 2013; published online on February 25, 2013. Discussion period open until May 1, 2014; separate discussions must be submitted for individual papers. This paper is part of the ***Journal of Engineering Mechanics***, Vol. 139, No. 12, December 1, 2013. ©ASCE, ISSN 0733-9399/2013/12-1688–1698/$25.00.

position and the wind velocity. However, the assumption of a rivulet oscillating sinusoidally was unable to explain exactly the mechanism involved in RWIVs (Li and Gu 2007).

Gu and Lu (2001) established equations of motion for both the cable and the upper rivulet by assuming that the aerodynamic force on the rivulet was proportional to the torsional force on the cable with the rivulet, while the interaction force between the cable surface and rivulet was set to zero. Peil and Nahrath (2003) put forward a theoretical model including the in-plane and out-of-plane degrees of freedom (DOF) of the cable and a tangential DOF of the rivulet. They assumed the interaction between the cable surface and rivulet to be modeled by a linear damping force. However, the focus of their study was only on the nonlinear stability and bifurcation of the system (Cosentino et al. 2003b). Peil et al. (2003) extended their study by including the lower rivulet. Cao et al. (2003) treated the rivulet oscillations as a stochastic process and assumed the interaction of cable surface and rivulet by a restoring force and a linear damping force. Gu et al. (2009) carried out wind tunnel tests to obtain the aerodynamic forces acting on the cable and upper rivulet and established a theoretical model in which the in-plane DOF of the cable and the tangential DOF of the rivulet were taken into account. In this model, the interaction between the cable surface and rivulet was assumed to be a combination of a Coulomb damping force and a linear damping force, while the coupled oscillations of the cable and upper rivulet were discussed in detail to explain the RWIV (Li et al. 2008).

Concerning the role of turbulence, the fluctuating component of the approach flow has been generally considered to dampen the RWIV (Matsumoto et al. 1992). However, during field observations, RWIVs were experienced at turbulence intensities of 9–11% (Chen et al. 2004). Uniform flow fields have been used in previous wind tunnel tests and theoretical analyses (Hikami and Shiraishi 1988; Yamaguchi 1990). Therefore, the influence of turbulence in the approach flow has not been systematically delineated.

In this context, the interactions between the cable surface and the rain-induced rivulet of water, and the coupling between the cable in-plane and out-of-plane vibrations, require further examination. Additionally, the effect of turbulence in the approach flow may also play an important role in the cable vibrations. In this paper, the interaction between the cable surface and rivulet is first analyzed based on a wetting theory, and then a new model for the RWIV of cables is proposed, which is based on the initial work in Gu et al. (2009).

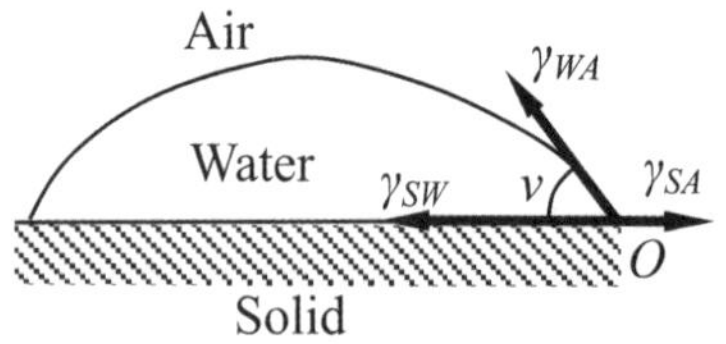

Fig. 1. Forces acting on a small droplet of water over a smooth surface

Simple spring-mass damper oscillators representing the cable and the upper rivulet define a system with three DOF; i.e., the in-plane and out-of-plane motions of the cable and the tangential (to the cable surface) motion of the upper rivulet. The coupled response of the cable and the rivulet is analyzed in detail by solving equations numerically by a fourth-order Runge-Kutta scheme. The results show that the out-of-plane DOF of the cable may play an important role in the RWIV of cables. Finally, the influence of various levels of inflow turbulence in the approach flow on the cable vibration is examined.

Interaction between Cable Surface and Rivulet

The shape of a droplet of water on a solid surface is determined by the properties of the solid and liquid interface, as indicated in Fig. 1. According to Young's condition (de Gennes 1985), the relationship between the contact angle (ν) and the surface free energy per unit area (γ_{WA}, γ_{SA}, and γ_{SW}) in equilibrium can be expressed as

$$\gamma_{WA}\cos\nu=\gamma_{SA}-\gamma_{SW} \tag{1}$$

where γ_{WA}, γ_{SA}, and γ_{SW} are in J/m^2, which can also be rewritten as N/m. Therefore, γ_{WA}, γ_{SA}, and γ_{SW} can be treated as distributed forces along triple line O (normal to the plane) shown in Fig. 1. In many practical situations—for example, on a rough solid surface (de Gennes 1985)—triple line O will be pinned and immobile. As a result, the contact angle will lie within a finite interval around ν

$$\nu_r<\nu<\nu_a \tag{2}$$

where ν_r and ν_a = advancing and receding angles, respectively. Eq. (2) indicates the well-documented phenomenon of contact angle hysteresis (de Gennes 1985) in wetting theory. For instance, if a droplet of water has a tendency to move from right to left, its shape will be changed, as indicated in Fig. 2. Two frictions, f_a and f_r, will appear at triple lines A and R, respectively, to satisfy the force balance in the horizontal direction (de Gennes 1985). Then, the force equilibriums in the horizontal direction at triple lines A and R can be written as

$$\gamma_{WA}\cos\nu_a=\gamma_{SA}-\gamma_{SW}-f_a \tag{3}$$

$$\gamma_{WA}\cos\nu_r=\gamma_{SA}-\gamma_{SW}+f_r \tag{4}$$

Subtracting Eq. (3) from Eq. (4) yields

$$\gamma_{WA}(\cos\nu_r-\cos\nu_a)=f_a+f_r=F_0 \tag{5}$$

Thus, as indicated by Eq. (5), a force F_0, somewhat similar to static friction, should be overcome before the water droplet moves. For water, air, and polyethylene, the surface tension is γ_{WA} = 0.072 N/m, while the contact angle is ν = 96.8° at a temperature

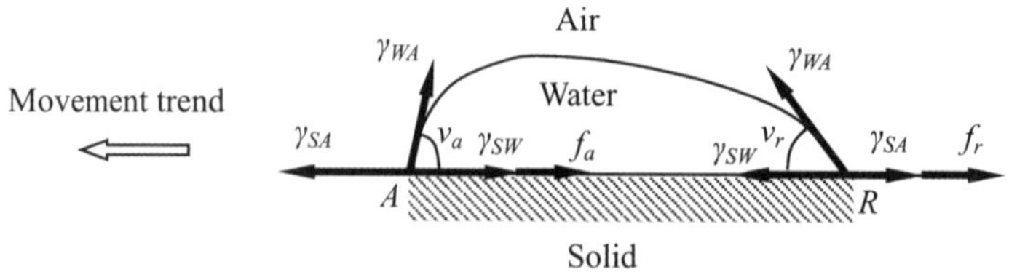

Fig. 2. Shape of a droplet of water on a rough surface with a tendency to move

of 25°C. Letting $\Delta\theta = \nu_a - \nu_r$ and assuming $\nu_a - \nu = \nu - \nu_r = \Delta\theta/2$, the relationship between F_0 and $\Delta\theta$ can be calculated as shown in Fig. 3. For a solid surface that has not been specifically prepared, interval $\Delta\theta$ may be 10° or more (de Gennes 1985). In this paper, $\Delta\theta$ is set equal to be 16°, corresponding to $F_0 = 0.02\,\mathrm{N/m}$.

The interaction between the cable surface and rivulet can be expressed as a linear combination of a Coulomb damping force $-F_0\,\mathrm{sign}(\dot{\theta})$ and a linear damping force $-c_r R\dot{\theta}$ as follows:

$$F_c = -F_0\,\mathrm{sign}(\dot{\theta}) - c_r R\dot{\theta} \tag{6}$$

where $\dot{\theta}$ = relative velocity of the rivulet on the cable; R = radius of the cable; and c_r = linear damping coefficient between the cable surface and rivulet. The sign function is defined as

$$\mathrm{sign}(\dot{\theta}) = \begin{cases} 1 & \dot{\theta} > 0 \\ 0 & \dot{\theta} = 0 \\ -1 & \dot{\theta} > 0 \end{cases} \tag{7}$$

Computational Model

The following assumptions are made in this study: (1) the lower rivulet has no effect on RWIVs, therefore it is not included here; (2) the shape of the rivulet is an arc, and its mass remains invariant when oscillating; (3) the quasi-steady assumption; (4) restoring force does not exist between the cable surface and the water rivulet; and (5) the effect of axial flow is neglected.

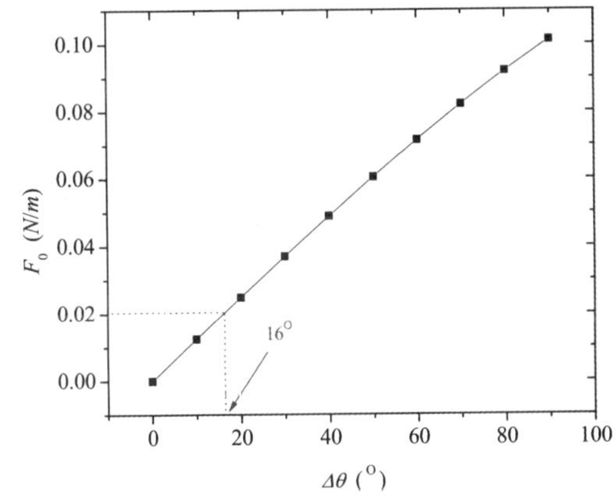

Fig. 3. Relationship between static friction F_0 and $\Delta\theta$

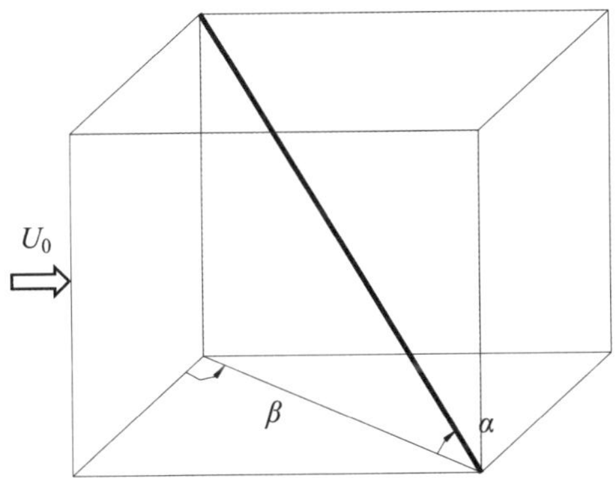

Fig. 4. Orientation of the cable

Equations of Motion of the Cable

The orientation of the stay cable can be defined by angle of inclination α and yaw angle β, as indicated in Fig. 4. The approaching wind velocity is U_0, while the component of U_0 perpendicular to the cable, U, can be expressed as follows (Wilde and Witkowski 2003):

$$U = U_0\sqrt{\cos^2\beta + \sin^2\alpha\sin^2\beta} = U_0\sqrt{\sin^2\alpha + \cos^2\alpha\cos^2\beta} \tag{8}$$

Accordingly, angle γ between U and a horizontal line is

$$\gamma = \arcsin\left(\frac{\sin\alpha\sin\beta}{\sqrt{\cos^2\beta + \sin^2\alpha\sin^2\beta}}\right) \tag{9}$$

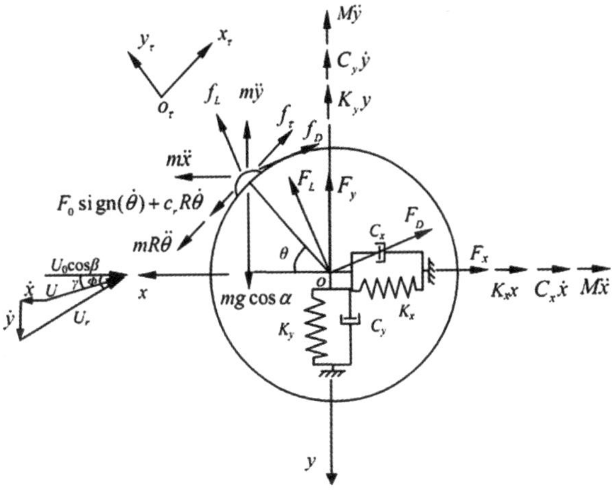

Fig. 5. Forces acting on the cable and the rivulet

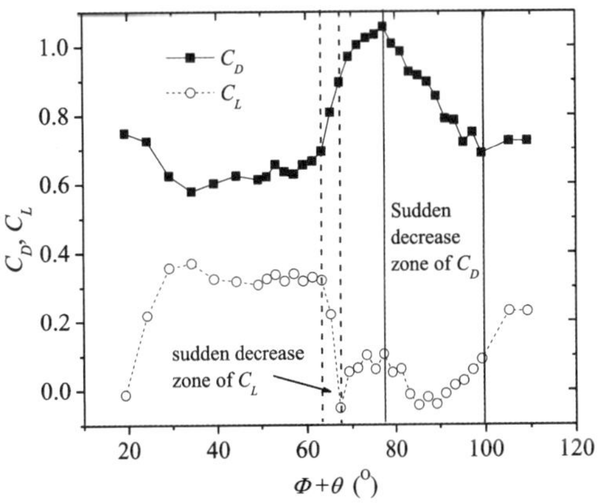

Fig. 6. Aerodynamic coefficients C_D and C_L of the cable for $\alpha = 30°$ and $\beta = 35°$ [data from Gu et al. (2009)]

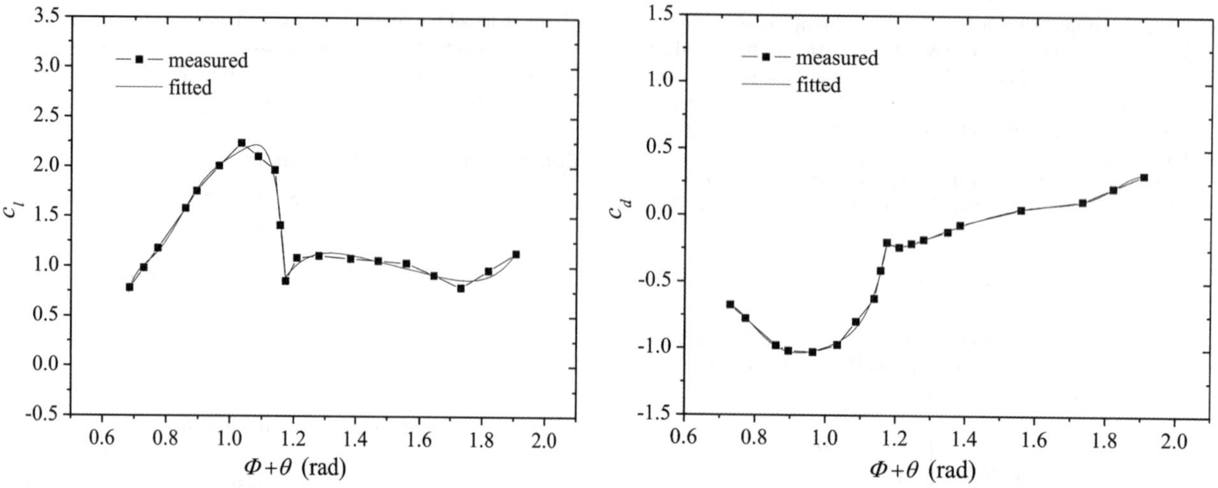

Fig. 7. Aerodynamic coefficients c_d and c_l of the rivulet for $\alpha = 30°$ and $\beta = 35°$ [data from Gu et al. (2009)]

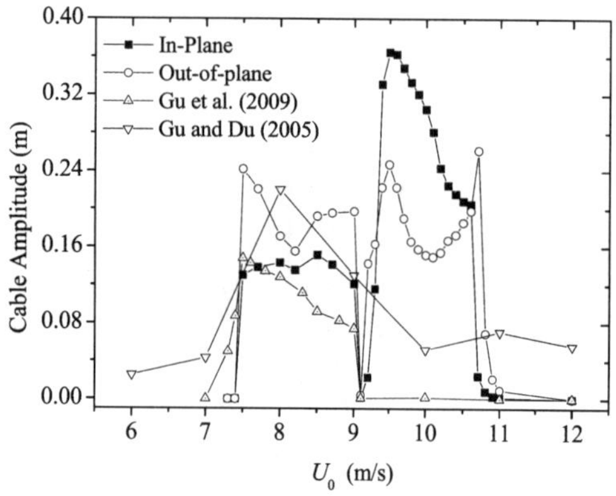

Fig. 8. Cable amplitude (peak-to-peak value) versus wind velocity

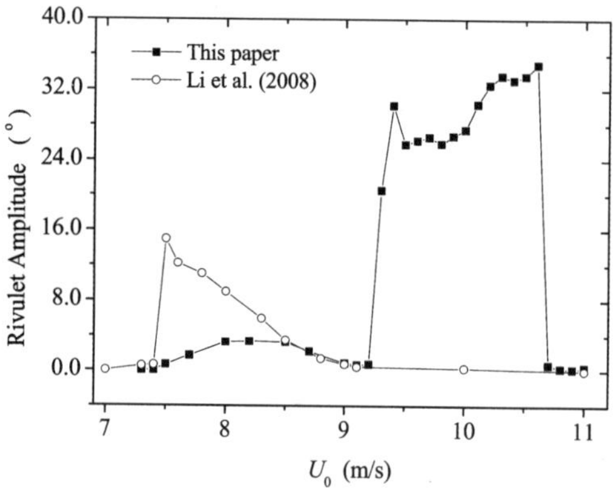

Fig. 9. Relationship between the rivulet amplitude (peak-to-peak value) and the wind velocity

All forces acting on the cable and the water rivulet are schematically shown in Fig. 5, in which two coordinate systems, oxy and $o_T x_T y_T$, are defined for the cable and the water rivulet, respectively. The forces acting on the cable include aerodynamic forces F_x and F_y, restoring forces $K_x x$ and $K_y y$, damping forces $C_x \dot{x}$ and $C_y \dot{y}$, inertia forces $M\ddot{x}$ and $M\ddot{y}$, and the forces induced by the water rivulet. Because of the small weight of the water rivulet with respect to the cable, the forces induced by the rivulet can be neglected. The in-plane and out-of-plane equations of motion of the cable can be written as

$$\ddot{x} + 2\xi_x \omega_x \dot{x} + \omega_x^2 x = \frac{-F_x}{M} \tag{10}$$

$$\ddot{y} + 2\xi_y \omega_y \dot{y} + \omega_y^2 y = \frac{-F_y}{M} \tag{11}$$

where ξ_x and ξ_y = in-plane and out-of-plane damping ratios; ω_x and ω_y = in-plane and out-of-plane natural frequencies; M = mass of the cable oscillator; F_x and F_y = in-plane and out-of-plane aerodynamic forces; and the overdots = differentiation with respect to

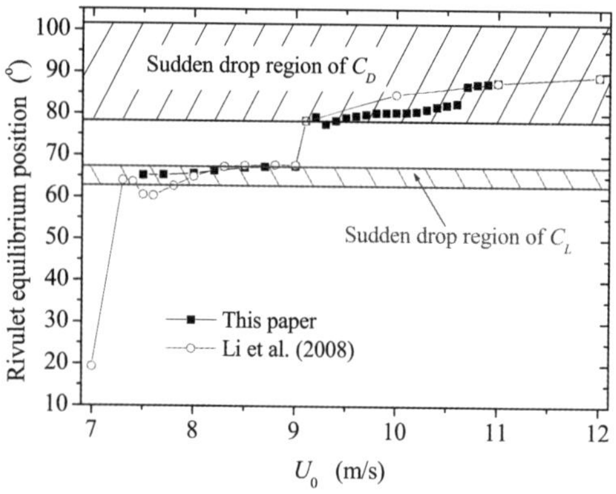

Fig. 10. Rivulet equilibrium position versus wind speed

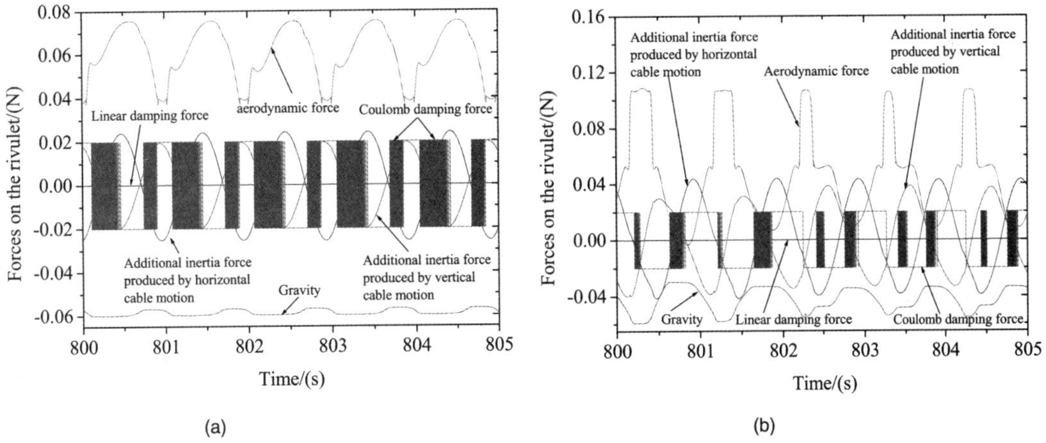

Fig. 11. Time histories of force acting on the rivulet in the tangential direction with respect to the cable cross section: (a) $U_0 = 8.0\,\mathrm{m/s}$; (b) $U_0 = 9.5\,\mathrm{m/s}$

time. On the basis of quasi-steady theory, F_x and F_y can be obtained from the following expressions:

$$F_x = F_D \cos\Phi - F_L \sin\Phi = (1/2)\rho[U_0(t)]^2 \times D[C_D(\Phi+\theta)\cos\Phi - C_L(\Phi+\theta)\sin\Phi] \tag{12}$$

$$F_y = F_L \cos\Phi + F_D \sin\Phi = (1/2)\rho[U_0(t)]^2 \times D[C_L(\Phi+\theta)\cos\Phi + C_D(\Phi+\theta)\sin\Phi] \tag{13}$$

where ρ = air density; D = diameter of the cable; C_D and C_L = aerodynamic drag and lift coefficients of the cable, respectively; Φ = angle between U_r and the horizontal direction; U_r = relative wind velocity between the approaching wind and the oscillating cable, and θ = rivulet position, measured through the central angle. Here, Φ can be expressed as

$$\Phi = \arctan\left(\frac{\dot{y} + U\sin\gamma}{\dot{x} + U\cos\gamma}\right) \tag{14}$$

The aerodynamic drag and lift coefficients of the cable and the rivulet measurements reported in Gu et al. (2009) are adopted in this study. The aerodynamic coefficients of the cable, C_D and C_L, with an inclined angle of 30° and a yaw angle of 35°, are shown in Fig. 6.

Equation of Motion of the Rivulet

The forces acting on the rivulet in the tangential direction of the cable surface include the gravitational force ($mg\cos\alpha\cos\theta$), inertia force ($mR\ddot{\theta}$), Coulomb damping force and linear damping force [$F_0\,\mathrm{sign}(\dot{\theta}) + c_r R\dot{\theta}$], aerodynamic force ($f_\tau$), and additional inertia force induced by the oscillation of the cable ($m\ddot{y}\cos\theta - m\ddot{x}\sin\theta$). The equation of motion of the rivulet in the tangential direction of the cable can be written as

$$mR\ddot{\theta} + F_0\,\mathrm{sign}(\dot{\theta}) + c_r R\dot{\theta} = f_\tau + m\ddot{y}\cos\theta - m\ddot{x}\sin\theta - mg\cos\alpha\cos\theta \tag{15}$$

where the rivulet aerodynamic force f_τ is

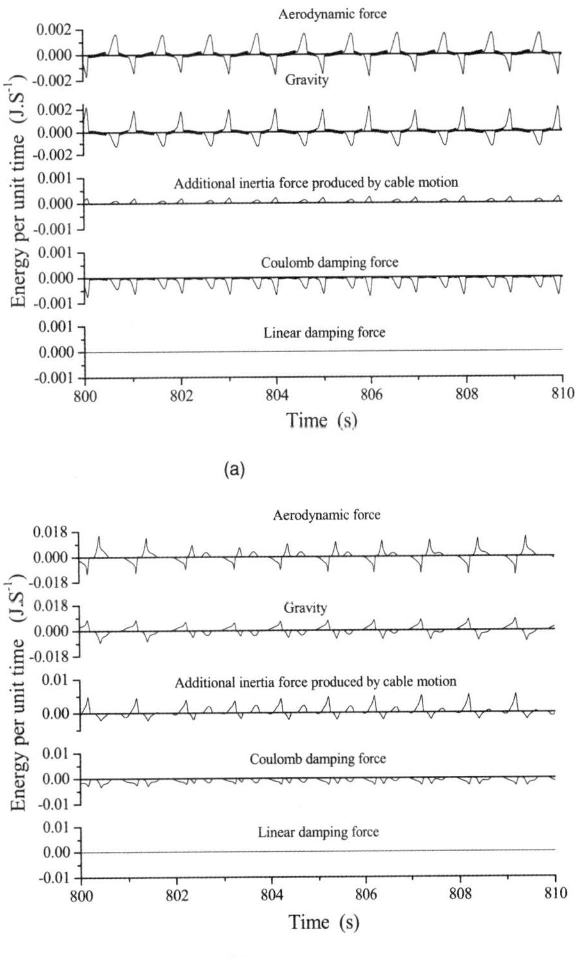

Fig. 12. Energy per unit time of forces acting on the rivulet: (a) $U_0 = 8.0\,\mathrm{m/s}$; (b) $U_0 = 9.5\,\mathrm{m/s}$

$$f_\tau = f_D\sin(\Phi+\theta) + f_L\cos(\Phi+\theta) = (1/2)\rho[U_0(t)]^2 \times B[c_d(\Phi+\theta)\sin(\Phi+\theta) + c_l(\Phi+\theta)\cos(\Phi+\theta)] \quad (16)$$

where c_d and c_l = aerodynamic drag and lift coefficients of the rivulet, respectively; and B = radius of the arc of the rivulet. Here, c_d and c_l are based on measurements with an inclined angle of 30° and a yaw angle of 35° (as shown in Fig. 7). Coupled Eqs. (10), (11), and (15) are solved using the fourth-order Runge-Kutta method.

Numerical Experiments

Uniform Flow

For the numerical experiment, the approach flow was assumed to be uniform. The cable diameter was set to be 120 mm; the mass per unit length of the cable was 6 kg/m; the in-plane and out-of-plane natural frequencies were both 1 Hz; the in-plane and out-of-plane damping ratios were 0.1%; the inclined angle α of the cable was 30°; and the yaw angle β of the cable was 35°. The diameter, width, and height of the arc rivulet were 13.5, 9, and 1.7 mm, respectively. The mass per unit length of the rivulet, calculated in terms of the aforementioned size, was 0.01 kg. The Coulomb damping force per unit length between the cable surface and the rivulet, F_0, was $0.02\,\mathrm{N/m}$, as discussed previously. The linear damping coefficient between the cable surface and the rivulet, c_r, was $0.0008\,\mathrm{N{\cdot}S/m^2}$, corresponding to a damping ratio of approximately 0.1%, provided that the natural frequency of the rivulet is 1 Hz. The initial displacement and initial velocity of the cable were both set to zero, the initial velocity of the rivulet was also set to zero, and the initial position of the rivulet was set to 1 rad (≅57°). The previous parameters of the cable were exactly the same as those used in the wind tunnel test in Gu and Du (2005), whereas the rivulet parameters were matched with those of the theoretical analysis in Gu et al. (2009).

Fig. 8 shows the variations of cable amplitude (peak-to-peak value) with respect to the wind velocity, together with the results of Gu and Du (2005) obtained from wind tunnel tests, as well as those obtained from the analysis in Gu et al. (2009). In both of these studies, only the in-plane DOF of the cable was considered. By comparing the results from these studies in Fig. 8, two differences from the present analysis are noted. First, there are two identifiable velocity zones in which large cable vibration amplitude takes place in the present investigation. One velocity zone is $U_0 = 7.3\text{–}9.0\,\mathrm{m/s}$, the other is $U_0 = 9.2\text{–}10.8\,\mathrm{m/s}$. This is in contrast to Gu and Du (2005) and Gu et al. (2009), where there is only one velocity zone, $U_0 = 7.3\text{–}9.0\,\mathrm{m/s}$. Second, large out-of-plane amplitude of the cable was found in the present investigation. These results suggest that the range of the wind velocity zone in which RWIVs take place is extended when taking into account the out-of-plane DOF of the cable.

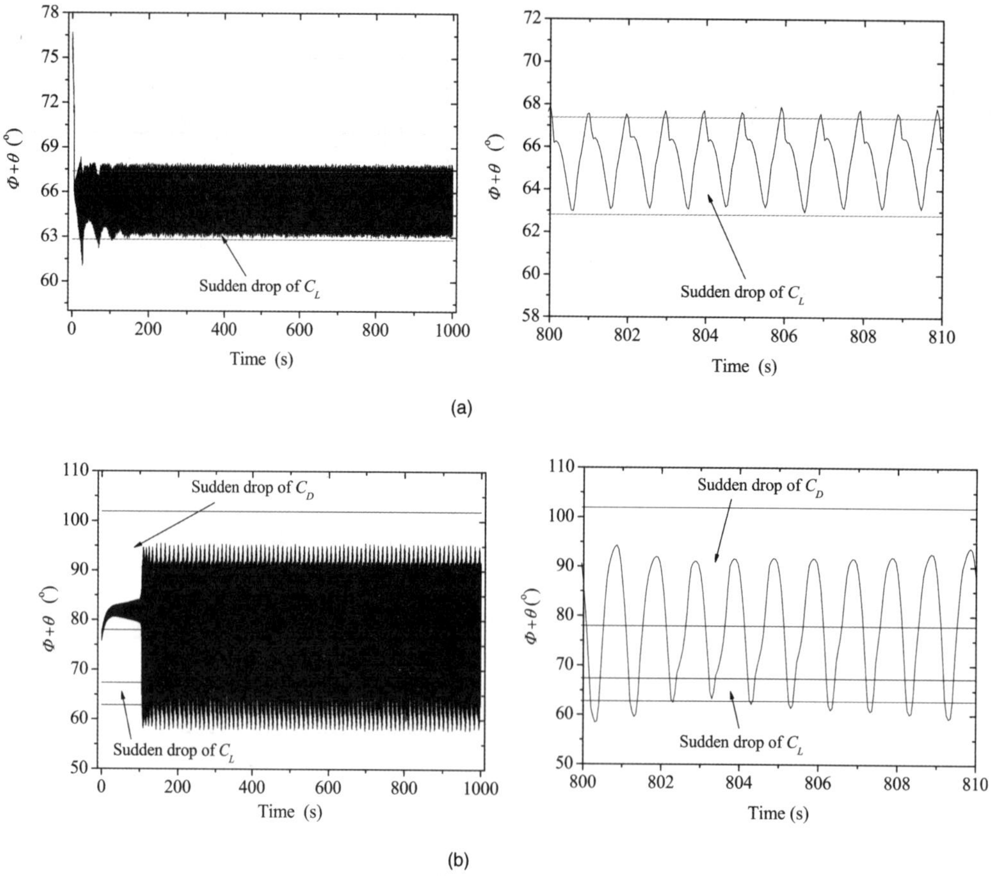

Fig. 13. Variation of wind angle of attack $\Phi+\theta$ with time t: (a) $U_0 = 8.0\,\mathrm{m/s}$; (b) $U_0 = 9.5\,\mathrm{m/s}$

The relationship between the rivulet amplitude (peak-to-peak value) and wind velocity is shown in Fig. 9, together with the numerical result from Li et al. (2008). It can be seen in Fig. 9 that the largest rivulet amplitude (3.3°) in the first velocity zone ($U_0 = 7.3-9.0\,\mathrm{m/s}$) is much smaller than the 15° computed by Li et al. (2008). However, in the second velocity zone ($U_0 = 9.2-10.8\,\mathrm{m/s}$), the largest rivulet amplitude (35°) is much greater than that from Li et al. (2008), which is nearly 0°.

Fig. 10 shows the variation of the rivulet equilibrium position with wind velocity, together with the result from Li et al. (2008). The rivulet equilibrium position is defined as the mean value of $\Phi + \theta$. From Fig. 10, it is seen that the rivulet equilibrium position in this study is nearly the same as in Li et al. (2008). This is because the balance of the rivulet on the cable surface is mainly affected by the aerodynamic forces and gravity of the rivulet in the tangential direction of the cable cross section, which are not affected by the out-of-plane DOF. It should be noted that the rivulet equilibrium position locates itself within the region of sudden decrease in C_L, $\Phi + \theta = 62.8-67.4°$, when $U_0 = 7.3-9.0\,\mathrm{m/s}$. Similarly, when U_0 is between 9.2 and 10.8 m/s, the rivulet equilibrium position falls into the region of sudden decrease in C_D; i.e., $\Phi + \theta = 78.0-102.0°$.

The oscillations of the rivulet on the cable surface are extremely significant in explaining the mechanism of RWIV. Fig. 11 shows the time histories of the forces acting on the rivulet in the tangential direction of the cross section of the cable when the wind velocities are $U_0 = 8.0\,\mathrm{m/s}$ (in the first velocity zone) and $U_0 = 9.5\,\mathrm{m/s}$ (in the second velocity zone). It can be seen from Fig. 11 that the mean value of the aerodynamic force acting on the rivulet is approximately equal to that of gravity acting on the rivulet (about 0.06 and 0.04 N/m for $U_0 = 8.0$ and 9.5 m/s, respectively) but with opposite signs. The phase difference between the additional inertia forces produced by the motion of the cable in the horizontal and vertical directions is about 180° with similar peak values. Because of the relatively small tangential velocity of the rivulet, the linear damping force acting on it is very small. The Coulomb damping force, which is 0.02 N, changes its direction when the direction of motion of the rivulet changes. The amplitudes of the aerodynamic and gravitational forces acting on the rivulet in the second velocity zone ($U_0 = 9.5\,\mathrm{m/s}$) are larger than those in the first velocity zone ($U_0 = 8.0\,\mathrm{m/s}$). This is because the amplitude of motion of the rivulet in the second velocity zone is larger than that in the first zone, as indicated in Fig. 9.

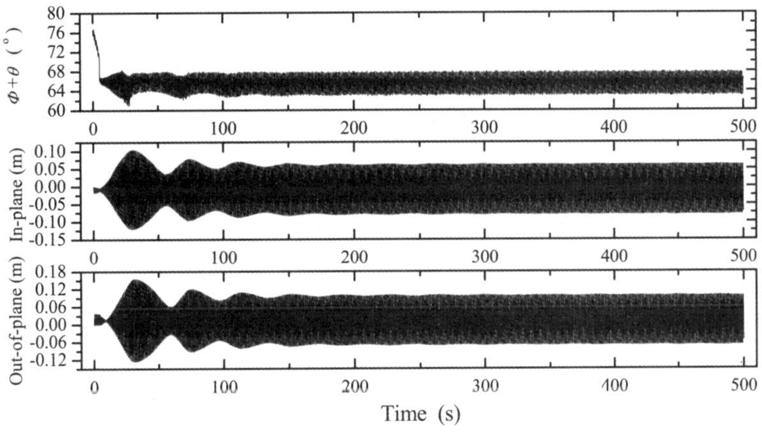

(a)

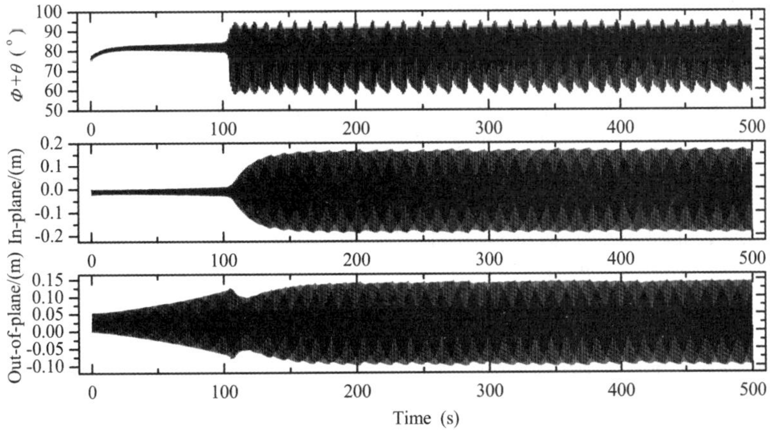

(b)

Fig. 14. Coupled nature of the cable and rivulet vibration: (a) $U_0 = 8.0\,\mathrm{m/s}$; (b) $U_0 = 9.5\,\mathrm{m/s}$

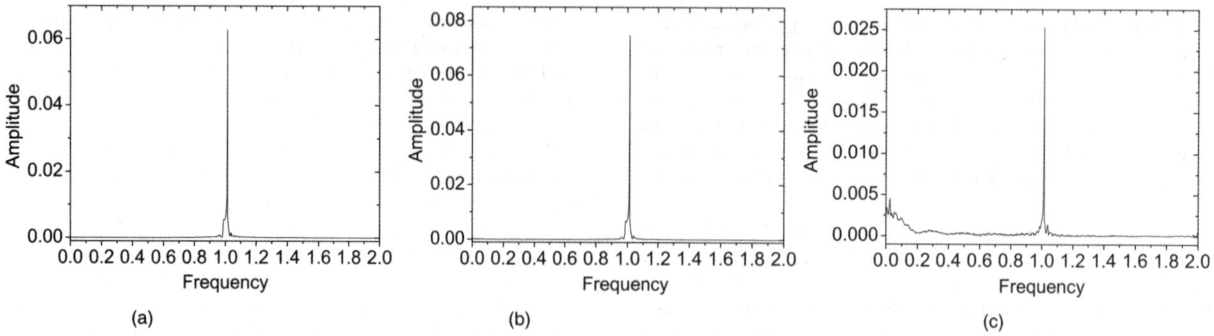

Fig. 15. Cable and rivulet displacement power spectral densities in uniform flow ($U_0 = 8.0\,\mathrm{m/s}$): (a) in plane of the cable; (b) out of plane of the cable; (c) rivulet

The forces acting on the rivulet during its oscillations may be better examined in terms of energy budget per unit time (forces on the rivulet multiplied by the rivulet velocity). Fig. 12 reports the energy per unit time of aerodynamic forces, gravity, additional inertial forces produced by the cable motion, and the Coulomb and linear damping forces when the wind velocities are $U_0 = 8.0$ and $9.5\,\mathrm{m/s}$. Three interesting points can be made based on the information presented in Fig. 12. First, the energy per unit time in the aerodynamic and gravitational forces are the highest among those of all forces acting on the rivulet while their signs are opposite. Second, the energy per unit time of the Coulomb and linear damping forces is always negative, which means that these two forces always dissipate the vibration energy of the rivulet. Furthermore, the energy per unit time of the Coulomb damping force is much higher than the linear damping force. Third, the energy per unit time of the additional inertial forces produced by the cable motions is always positive when $U_0 = 8.0\,\mathrm{m/s}$. At $U_0 = 9.5\,\mathrm{m/s}$, negative energy per unit time is sometimes noticed; however, the mean remains positive. Therefore, the cable motion imparts energy to the motion of the rivulet.

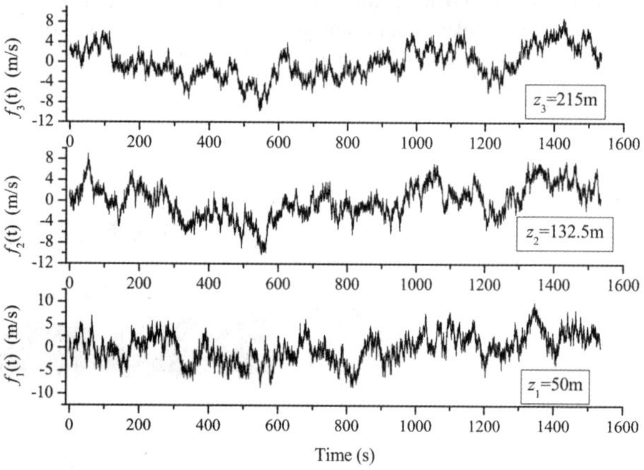

Fig. 16. Sample of simulated wind velocities

Discussion of Results in Uniform Flow

When the wind velocity is small, water on the cable surface falls off because of gravity. With an increase in the wind velocity (for example, $U_0 \geq 7.3\,\mathrm{m/s}$), the rivulet can be sustained on the cable surface under the combined action of the aerodynamic and gravitational forces. If the rivulet equilibrium position falls into the sudden decrease region of the lift aerodynamic force on the cable (C_L), as indicated in Fig. 13(a) for $U_0 = 8.0\,\mathrm{m/s}$, negative aerodynamic damping will be produced. This will lead to large in-plane oscillation of the cable. Then, because of the energy input by the aerodynamic force and additional inertia force produced by the cable oscillation, the rivulet will oscillate on the cable surface around the equilibrium position. The oscillation of the rivulet will result in a periodic force on the cable, which will trigger cable motion in the out-of-plane DOF. The coupled oscillations of the rivulet and the cable when $U_0 = 8.0\,\mathrm{m/s}$ are shown in Fig. 14(a).

Fig. 13(b) provides the variations of wind angle of attack $\Phi + \theta$ with time t when wind velocity U_0 is $9.5\,\mathrm{m/s}$, which places it within the second velocity zone. Fig. 14(b) shows the coupled oscillations of the rivulet and the cable when $U_0 = 9.5\,\mathrm{m/s}$. If only the in-plane DOF of the cable is considered (Li et al. 2008), the cable vibration amplitude is very small for $U_0 = 9.5\,\mathrm{m/s}$ because the rivulet equilibrium position is far away from the sudden decrease region of C_L. However, in the present paper the in-plane and out-of-plane DOF are both considered. In the second velocity zone ($U_0 = 9.2 - 10.8\,\mathrm{m/s}$)—for example, $U_0 = 9.5\,\mathrm{m/s}$ as shown in Fig. 13(b)—the rivulet equilibrium position falls into the sudden decrease region of C_D, which will lead to a large out-of-plane oscillation of the cable [see Fig. 14(b), $t = 0 - 100\,\mathrm{s}$]. It can be seen how the rivulet amplitude increases with the increase of the out-of-plane cable amplitude. Finally, the rivulet moves into the sudden decrease region of C_D, which leads to a large in-plane oscillation of the cable [see Fig. 14(b), $t = 100 - 500\,\mathrm{s}$]. Therefore, the sudden decrease of C_L and C_D can both produce considerable RWIV with in-plane and out-of-plane coupled oscillations.

Just as in galloping, the RWIV of stay cables is activated by the sudden decrease of the aerodynamic force coefficients of the cable with the rivulet. Because of the oscillations of the rivulet, the RWIV falls in the classification of being amplitude and velocity restricted, whereas galloping is a divergent-type motion. Fig. 15 presents the power spectral densities of the cable and the rivulet vibrations when the wind velocity is $8.0\,\mathrm{m/s}$, where it can be seen that the energy of the cable and the rivulet vibration are centered at the natural frequency of the cable, i.e., 1 Hz.

Turbulent Flow Field

The turbulent wind flow was simulated through the spectral representation method (e.g., Deodatis 1996; Li and Kareem 1993). The power spectral density of wind fluctuations proposed by Kaimal

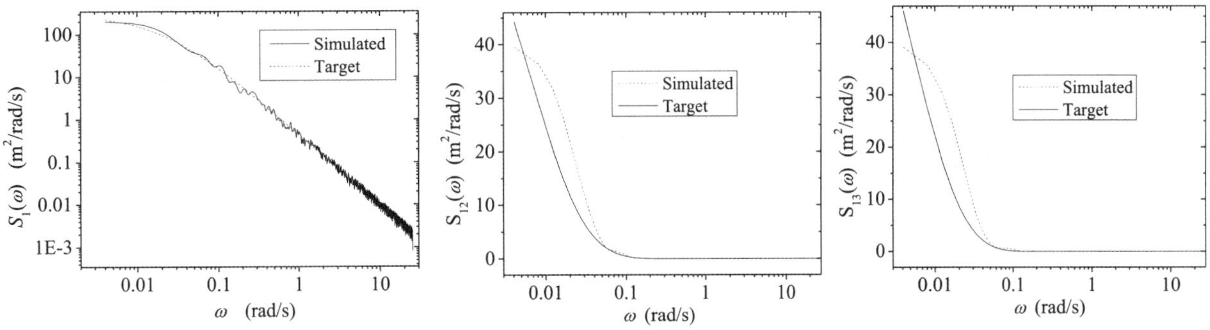

Fig. 17. Comparison between simulated and target power spectral densities

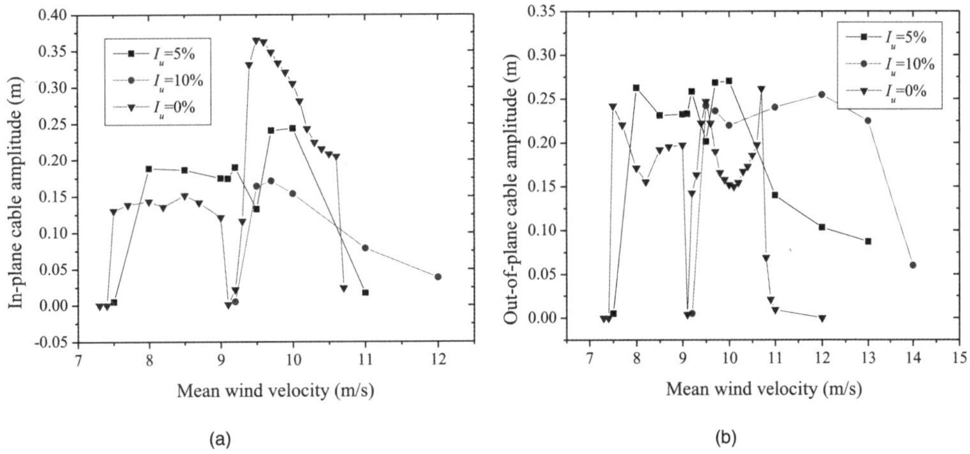

Fig. 18. Cable amplitudes under simulated approach flows at various turbulence intensities: (a) in plane; (b) out of plane

et al. (1972) and the coherence function proposed by Davenport (1968) were used in this simulation (e.g., Simiu and Scanlan 1996). The mean wind velocity at 50 m was set at 8.0 m/s and the mean wind velocity profile was described by a power law with exponent $\alpha = 0.16$. The target turbulence intensities at these locations were 0.152, 0.133, and 0.125. The simulated turbulence intensities were found to be 0.155, 0.137 and 0.126. Fig. 16 shows the results of a simulated three-point wind velocity sequence at the three heights. Fig. 17(a) presents a comparison of the simulated and target power spectral densities at 50-m height and Figs. 17(b and c) show the cross-spectral densities between 50 and 132.5 m (S_{12}) and between 50 and 215 m (S_{13}), respectively.

Results and Discussions in Simulated Turbulent Flows

The cable and rivulet parameters were kept the same as in the previous section. Three levels of turbulence intensities (i.e., 5, 10, and 15%) were considered here, which reflect conditions observed frequently at most bridge locations.

The results of the numerical simulation show that when the turbulence intensity equals 15% the rivulet will oscillate strongly, causing it to fall off of the cable surface and leaving the cable fairly stable. Fig. 18 presents the relationship between the cable amplitudes and mean wind velocity under simulated turbulent fluctuations when $I_u = 0, 5$, and 10%. In Fig. 18 it can be seen that the wind velocity zone within which RWIVs take place when $I_u = 5\%$

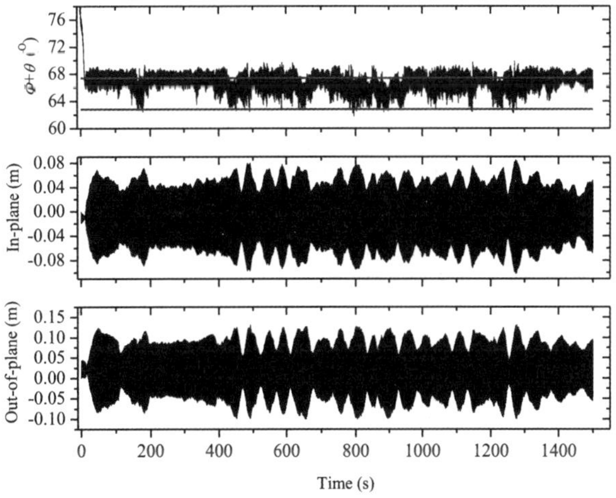

Fig. 19. Time histories of the cable and rivulet motions under simulated flow ($I_u = 5\%, \overline{U} = 8.5$ m/s)

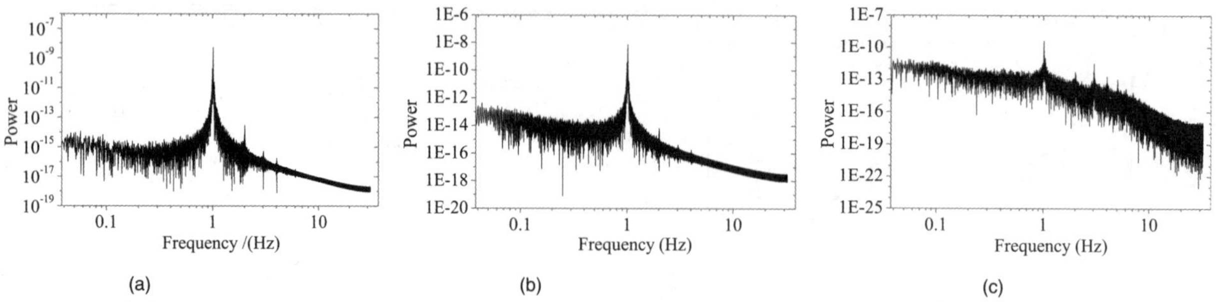

Fig. 20. Cable and rivulet displacement power spectral densities under simulated flow ($I_u = 5\%$, $\overline{U} = 8.5$ m/s): (a) in plane of the cable; (b) out of plane of the cable; (c) rivulet

are nearly the same as that of uniform flow while the cable amplitudes do not obviously decrease. When $I_u = 10\%$, the in-plane cable amplitudes decrease significantly and the wind velocity zone under which RWIVs take place increases by about 2 m/s. The buffeting responses induced by the fluctuating wind velocity are included in Fig. 18.

Fig. 19 shows the time histories of the cable and rivulet as a result of the simulated turbulent approach flow with $I_u = 5\%$ and $\overline{U} = 8.5$ m/s. Fig. 20 gives the corresponding power spectral densities of the cable and rivulet displacements. It can be seen in Fig. 19 that the cable and rivulet amplitudes change with time. This coincides with the observed characteristics in field observations. However, the cable and rivulet amplitudes remain nearly constant in the theoretical analysis in uniform flow, as indicated in Fig. 14. On the other hand, in simulated turbulent flow large cable amplitude will not take place until the rivulet moves into the zone of sudden decrease of the aerodynamic force coefficients. In Fig. 20 the resonant peaks at the cable's natural frequency in the power spectral densities of the cable and the rivulet are clearly evident, as well as the fairly significant background response induced by a rather broad banded fluctuating flow.

Concluding Remarks

This paper presents a model for the description of in-plane and out-of-plane RWIVs of stay cables on long-span, cable-stayed bridges.

1. In particular, the interaction between the cable surface and the rivulet was first analyzed by a wetting theory, which suggests the use of combined Coulomb and linear damping forces.
2. The proposed model for the RWIVs of stay cables is comprised of a coupled two-mass oscillator with three DOF, which represents the in-plane and out-of-plane DOF of the cable. The mechanical coupling between the in-plane and out-of-plane DOF of the cable leads to a coupled vibration response, which results in a wider range of wind velocity over which the RWIVs take place compared with a model that considers only the in-plane DOF for the cable.
3. The gravity and aerodynamic forces on the rivulet have approximately similar time-averaged values; however, their signs are opposite for the gravity and aerodynamic forces that satisfy the rivulet equilibrium on the cable surface. This results in the upward movement of the equilibrium position of the rivulet on the cable surface as the wind velocity increases. At certain wind velocities, the rivulet enters in the region that corresponds to the sudden decrease in the aerodynamic force coefficients. Consequently, it leads to negative aerodynamic damping, thus triggering RWIVs. This behavior suggests that the RWIVs can be considered as a special type of cable galloping with a moving rivulet and explains the important velocity-restricted characteristic of RWIV.
4. The interaction forces between the cable surface and the rivulet dissipate rivulet energy, thereby stabilizing it, while the additional inertial forces resulting from the cable oscillations add energy to the rivulet.
5. Finally, the effects of turbulence were investigated using simulated turbulent fluctuations. The results show that the RWIVs seem to diminish when the turbulence intensity reaches 15% for the parameters of the cable and the rivulet used in this study. This warrants additional investigation because the RWIVs of a stay cable have been reproduced under nearly 20% turbulence intensity in wind tunnel tests.

The study offers a model that captures the salient features of the RWIVs of a segmental representation of a stay cable both from inception to sustenance in various velocity regimes and demonstrates the influence of turbulence, which mirrors some of the field observations.

Acknowledgments

This project was jointly supported by the National Natural Science Foundation (50708035) and the Chinese Postdoctoral Science Foundation (20060400873). This study was also supported in part by U.S. National Science Foundation Grant No. 0928282.

References

Cao, D. Q., Tucker, R. W., and Wang, C. (2003). "A stochastic approach to cable dynamics with moving rivulets." *J. Sound Vib.*, 268(2), 291–304.

Chen, Z. Q., Wang, X. Y., Ko, J. M., Ni, Y. Q., Yang, G., and Hu, J. H. (2004). "MR damping system for mitigating wind-rain induced vibration on Dongting Lake cable-stayed bridge." *Wind Struct.*, 7(5), 293–304.

Cosentino, N., Flamand, O., and Ceccoli, C. (2003a). "Rain-wind induced vibration of inclined stay cables. Part I: Experimental investigation and physical explanation." *Wind Struct*, 6(6), 471–484.

Cosentino, N., Flamand, O., and Ceccoli, C. (2003b). "Rain-wind induced vibration of inclined stay cables. Part II: Mechanical modeling and parameter characterization." *Wind Struct.*, 6(6), 485–498.

Davenport, A. G. (1968). "The dependence of wind load upon meteorological parameters." *Proc., Int. Research Seminar on Wind Effects on Buildings and Structures*, University of Toronto Press, Toronto, 19–82.

de Gennes, P. G. (1985). "Wetting: Statics and dynamics." *Rev. Mod. Phys.*, 57(3), 827–863.

Deodatis, G. (1996). "Simulation of ergodic multivariate stochastic processes." *J. Eng. Mech.*, 122(8), 778–787.

Flamand, O. (1995). "Rain-wind induced vibration of cables." *J. Wind Eng. Ind. Aerodyn.*, 57(2–3), 353–362.

Gu, M., and Du, X. Q. (2005). "Experimental investigation of rain-wind-induced vibration of cables in cable-stayed bridges and its mitigation." *J. Wind Eng. Ind. Aerodyn.*, 93(1), 79–95.

Gu, M., Du, X. Q., and Li, S. Y. (2009). "Experimental and theoretical simulations on wind–rain-induced vibration of 3-D rigid stay cables." *J. Sound Vib.*, 320(1–2), 184–200.

Gu, M., Liu, C. J., Xu, Y. L., and Xiang, H. F. (2002). "Response characteristics of wind excited cables with artificial rivulet." *Appl. Math. Mech.*, 23(10), 1176–1187.

Gu, M., and Lu, Q. (2001). "Theoretical analysis of wind-rain induced vibration of cables of cable-stayed bridges." *Proc., 5th Asia-Pacific Conf. on Wind Engineering*, Japan Association for Wind Engineering, Kyoto, Japan, 125–128.

Hikami, Y., and Shiraishi, N. (1988). "Rain-wind induced vibrations of cables stayed bridges." *J. Wind Eng. Ind. Aerodyn.*, 29(1–3), 409–418.

Kaimal, J. C., Wyngaard, J. C., Izumi, Y., and Coté, O. R. (1972). "Spectral characteristics of surface-layer turbulence." *J. R. Meteorolog. Soc.*, 98(417), 563–589.

Li, H., Chen, W. L., Xu, F., Li, F. C., and Ou, J. P. (2010). "A numerical and experimental hybrid approach for the investigation of aerodynamic forces on stay cables suffering from rain-wind induced vibration." *J. Fluids Struct.*, 26(7–8), 1195–1215.

Li, S. Y., Chen, Z. Q., and Gu, M. (2008). "Coupled motion between stay cables and rivulets in rain-wind induced vibration." *J. Vib. Shock*, 27(10), 1–5 (in Chinese).

Li, S. Y., and Gu, M. (2006). "Numerical simulations of flow around stay cables with and without fixed artificial rivulets." *Proc., 4th Int. Symp. on Computational Wind Engineering*, Japan Association for Wind Engineering, Yokohama, Japan, 307–310.

Li, S. Y., and Gu, M. (2007). "A theoretical model of rain-wind induced vibration of three-dimensional continuous stay cable with quasi-moving rivulet." *Proc., 12th Int. Conf. on Wind Engineering*, Australasian Wind Engineering Society, Carns, Australia, 887–894.

Li, Y., and Kareem, A. (1993). "Simulation of multivariate random processes: Hybrid DFT and digital filtering approach." *J. Eng. Mech.*, 119(5), 1078–1098.

Matsumoto, M., Saitoh, T., Kitazawa, M., Shirato, H., and Nishizaki, T. (1995). "Response characteristics of rain-wind induced vibration of stay-cables of cable-stayed bridges." *J. Wind Eng. Ind. Aerodyn.*, 57(2–3), 323–333.

Matsumoto, M., Shiraishi, N., and Shirato, H. (1992). "Rain-wind induced vibration of cables of cable-stayed bridges." *J. Wind Eng. Ind. Aerodyn.*, 43(1–3), 2011–2022.

Matsumoto, M., Shirato, H., Yagi, T., Goto, M., Sakai, S., and Ohya, J. (2003). "Field observation of the full-scale wind-induced cable vibration." *J. Wind Eng. Ind. Aerodyn.*, 91(1–2), 13–26.

Peil, U., and Nahrath, N. (2003). "Modeling of rain-wind induced vibrations." *Wind Struct.*, 6(1), 41–52.

Peil, U., Nahrath, N., and Dreyer, O. (2003). "Modeling of rain-wind induced vibrations." *Proc., 11th Int. Conf. on Wind Engineering*, American Association of Wind Engineering, Lubbock, TX, 389–396.

Rocchi, D., and Zasso, A. (2002). "Vortex shedding from a circular cylinder in a smooth and wired configuration: Comparison between 3D LES simulation and experimental analysis." *J. Wind Eng. Ind. Aerodyn.*, 90(4–5), 475–489.

Simiu, E., and Scanlan, R. H. (1996). *Wind effects on structures*, 3rd Ed., Wiley, New York.

Taylor, I. J., and Robertson, A. C. (2011). "Numerical simulation of the airflow-rivulet interaction associated with the rain-wind induced vibration phenomenon." *J. Wind Eng. Ind. Aerodyn.*, 99(9), 931–944.

van der Burgh, A. H. P., and Hartono (2004). "Rain-wind-induced vibrations of a simple oscillator." *Int. J. Non-linear Mech.*, 39(1), 93–100.

Watson, S. C., and David, S. (1988). "Cables in trouble." *Civ. Eng.*, 58(4), 38–41.

Wilde, K., and Witkowski, W. (2003). "Simple model of rain-wind-induced vibrations of stayed cables." *J. Wind Eng. Ind. Aerodyn.*, 91(7), 873–891.

Xu, Y. L., and Wang, L. Y. (2003). "Analytical study of wind-rain-induced cable vibration: SDOF model." *J. Wind Eng. Ind. Aerodyn.*, 91(1–2), 27–40.

Yamaguchi, H. (1990). "Analytical study on growth mechanism of rain vibration of cables." *J. Wind Eng. Ind. Aerodyn.*, 33(1–2), 73–80.

Smart Structures and Systems, *Vol. 22, No. 6 (2018) 727-741*
DOI: https://doi.org/10.12989/sss.2018.22.6.727

Mitigation of wind-induced vibrations of bridge hangers using tuned mass dampers with eddy current damping

Huawei Niu[*1], Zhengqing Chen[1a], Xugang Hua[1b] and Wei Zhang[2c]

[1]*Key Laboratory for Wind and Bridge Engineering of Hunan Province, College of Civil Engineering, Hunan University, Changsha 410082,, Hunan, China*
[2]*Department of Civil and Environmental Engineering, University of Connecticut, Storrs, CT 06269, USA*

(Received April 7, 2018, Revised August 26, 2018, Accepted November 14, 2018)

Abstract. To mitigate vibrations, tuned mass dampers (TMD) are widely used for long span bridges or high-rise buildings. Due to some durability concerns, such as fluid degradation, oil leakage, etc., the alternative solutions, such as the non-contacted eddy current damping (ECD), are proposed for mechanical devices in small scales. In the present study, a new eddy current damping TMD (ECD-TMD) is proposed and developed for large scale civil infrastructure applications. Starting from parametric study on finite element analysis of the ECD-TMD, the new design is enhanced via using the permanent magnets to eliminate the power need and a combination of a copper plate and a steel plate to improve the energy dissipation efficiency. Additional special design includes installation of two permanent magnets at the same side above the copper plate to easily adjust the gap as well as the damping. In a case study, the proposed ECD-TMD is demonstrated in the application of a steel arch bridge to mitigate the wind-induced vibrations of the flexible hangers. After a brief introduction of the configuration and the installation process for the damper, the mitigation effects are measured for the ambient vibration and forced vibration scenarios. The results show that the damping ratios increase to 3% for the weak axis after the installation of the ECD-TMDs and the maximum vibration amplitudes can be reduced by 60%.

Keywords: eddy current damping tuned mass damper (ECD-TMD); arch bridges; hanger; tuned mass damper; wind-induced vibration; vibration control

1. Introduction

Large vibrations of slender steel hangers have been one of the major concerns for the design of long-span arch bridges or suspension bridges, which could be critical for the bridge safety, serviceability, and durability. Many large amplitude vibrations of hangers were reported since 1980s (Ulstrup 1980). In the 1990's, most of the H section hangers of the Jiujiang Yangtze River Bridge, China were found vibrating with large amplitudes at a low wind speed of 8 m/s (Gu *et al.* 1994). More recently in 2006, large torsional wind-induced vibrations of H-shaped steel hangers of Dongping Bridge, China, were observed. More than 20 hangers vibrated violently for over 20 hours leading to severe cracks on the flanges near the hanger-girder connections for all the 13 longest hangers (Chen *et al.* 2010, 2012). Many mitigation plans were proposed, such as using perforations on the web or the flange (Ma *et al.* 2005), wind-resistant cables (Chen *et al.* 2012), tuned mass dampers (Gu *et al.* 1994, Li *et al.* 2012) etc. However, these mitigation measures generated new problems. For example, perforations on the web or the flange may increase the galloping critical velocity, and no obvious effects were found for flutter instability and vortex shedding excitation. Connecting the hangers using wind-resistant cables is effective to increase the critical velocity of hangers. However, the adverse effects on bridge aesthetics limited its wider applications.

In addition to these vibration control measures, the turned mass dampers (TMDs) (Tamura 2008, Fujino 2002, Spencer and Nagarajaiah 2003, Lin *et al.* 2010, Pourzeynali and Esteki 2009, Zhu *et al.* 2016), as one type of passive mitigation method, have been widely used for many civil infrastructures, as well, especially for long span bridges to suppress the wind-induced vibrations (Koshimura *et al.* 1994, Fujino and Yoshida 2002, Larsen *et al.* 1995, Chen and Kareem 2003, Chang *et al.* 2003). A typical TMD include a lumped mass, a spring element, and damping elements to dissipate the vibration energy (Soong *et al.* 1997). Traditionally, the damping in a TMD is provided by the viscous dampers, oil tanks or frictions. However, some researchers found the fluid dampers were susceptible to fluid degradation and sealing over time (Wang *et al.* 2012; Sodano and Bae 2004, Amjadian and Agrawal 2017). For example, the San Francisco–Oakland Bay Bridge and the Vincent Thomas Bridge were found with the leakage problems for their dampers during their inspections (Konstantinidis *et al.* 2011). Meanwhile, with a large start-up force caused by friction inside the fluid or friction

*Corresponding author, Associate Professor
E-mail: niuhw@hnu.edu.cn
[a] Professor
E-mail: cexghua@hnu.edu.cn
[b] Professor
E-mail: cexghua@hnu.edu.cn
[c] Assistant Professor
E-mail: zqchen@hnu.edu.cn

dampers, many TMDs does not work for small amplitudes of vibrations of the building, such as the TMD in the Taibei 101 Tower. Nevertheless, the temperature-sensitive damping properties of fluids increased the uncertainty of the damping effects for different seasons, which endanger the control performance of the TMD (Weber and Feltrin 2010).

In parallel, eddy current tuned mass damper system have been developed and used in many fields other than large scale civil engineering projects, such as the magnetic braking system and in the vibration control devices for the rotating machinery (Sodano and Bae 2004, Anantha Krishna and Sathish Kumar 2018), and wind tunnel testing, such as in the BLWT Laboratory at University of Western Ontario, Canada and Hunan University, China. In an ECD system, the circulating eddy currents are generated in a conductor when there is a relative motion between the conductor and magnetic field. A resistive force will cause energy dissipation due to the electrical resistance. The resistive force induced by the eddy currents is proportional to the relative velocity. Therefore, the damping system can be described as a form of viscous damping, as well. After development of this damping mechanism, many theoretical simulations and experimental investigations were carried out, which show their excellent control performances of ECD (Sodano *et al.* 2005, 2008). With many followed studies and applications (Wen *et al.* 2016, Lu *et al.* 2018, Amjadian and Agrawal 2018, David Saige *et al.* 2017), the feasibility of the non-contact strategy for mitigations of wind-induced vibrations of flexible bridge elements have demonstrated their strong potential especially at a very low wind speed due to the non-frictions of the ECD-TMD. Without liquids and seals, a better durability is expected comparing with the traditional fluid-based TMDs. To apply such ECD-TMD concept from a small and lab scale to a scale for the real long-span bridge, lots of experimental testing and sensitivity analysis were carried out in the present study. Therefore, seeking for a better alternative to the current ECD technique is the motivation of the present study.

The present study focused on the development of an innovative passive TMD with eddy current damping induced by permanent magnets. This paper is organized as the following. After introduction, basic theories for ECD and TMD parameter will be briefly introduced. Secondly, the structure and main elements of the designed ECD-TMD as well as numerical analysis for ECD configuration are presented. After demonstration of the application of the ECD-TMD to suppress the wind-induced vibration of long hangers, and the implementation process of these ECD-TMDs on the site, the control effects of the ECD-TMDs were discussed based on the field experiments.

2. Modelling of ECD-TMD

A hanger with a TMD subjected to harmonic excitation can be described by a two-degree-of-freedom system, as shown in Fig. 1. The dynamic motion equations for this system can be written as (Soong and Dargush 1997)

$$\begin{bmatrix} m_p & 0 \\ 0 & m_\alpha \end{bmatrix}\begin{bmatrix} \ddot{x}_p(t) \\ \ddot{x}_\alpha(t) \end{bmatrix}+\begin{bmatrix} c_p+c_\alpha & -c_\alpha \\ -c_\alpha & c_\alpha \end{bmatrix}\begin{bmatrix} \dot{x}_p(t) \\ \dot{x}_\alpha(t) \end{bmatrix}+ \begin{bmatrix} k_p+k_\alpha & -k_\alpha \\ -k_\alpha & k_\alpha \end{bmatrix}\begin{bmatrix} x_p(t) \\ x_\alpha(t) \end{bmatrix}=\begin{bmatrix} F_0 \\ 0 \end{bmatrix}\sin(\omega t) \tag{1}$$

in which $x_p(t),\dot{x}_p(t),\ddot{x}_p(t)$ and $x_\alpha(t),\dot{x}_\alpha(t),\ddot{x}_\alpha(t)$ are the vibration displacement, velocity, acceleration of the hanger and the TMD; m_p,k_p,c_p and $m_\alpha,k_\alpha,c_\alpha$ are the modal mass, stiffness, damping coefficients of the hanger and the TMD respectively; F_0 and ω are the amplitude and circular frequency of the excited force.

For the hanger-TMD system, the dynamic magnification factor (DMF) by displacement R can be determined as

$$R=\sqrt{\frac{\left(\beta^2-\alpha^2\right)^2+(2\xi\beta)^2}{\left[(\beta^2-\alpha^2)(1-\alpha^2)-\mu\beta^2\alpha^2\right]^2+[(2\xi\beta)(1-\beta^2-\mu\beta^2)]^2}} \tag{2}$$

where μ is the mass ratio between the added mass of TMD and the hanger; ξ is the damping ratio, $\xi=c_\alpha/(2m_\alpha\omega_\alpha)$; α is the frequency ratio of the TMD to the hanger; β is the frequency ratio of the excitation force to the natural frequency of the hanger.

Derived from the fixed points of DMF curve, the TMD optimal frequency ratio α_{opt} and the optimal damping ratio ξ_{opt} can begiven as (Warburton 1982, Rana and Soong 1998)

$$\alpha_{opt}=\frac{1}{1+\mu} \tag{3}$$

$$\xi_{opt}=\sqrt{\frac{3\mu}{8(1+3\mu)}} \tag{4}$$

When the damping is generated by an ECD system, such as shown in Fig. 2, the damping force and the damping coefficients in the velocity direction due to the eddy currents can be obtained (Sodano *et al.* 2006, Wang *et al.* 2012)

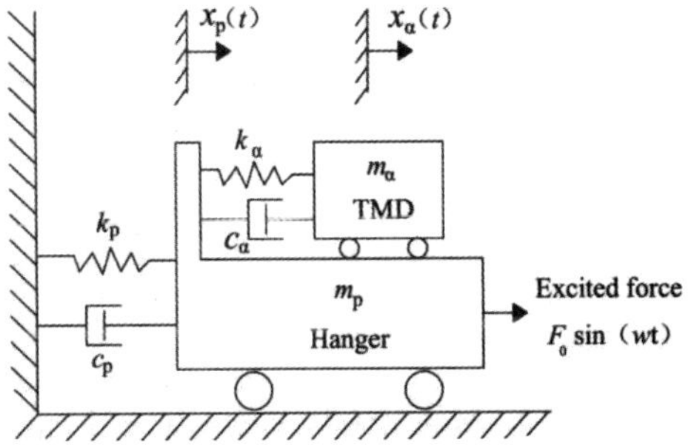

Fig. 1 Two degree-of-freedom modelling of the hanger and TMD system

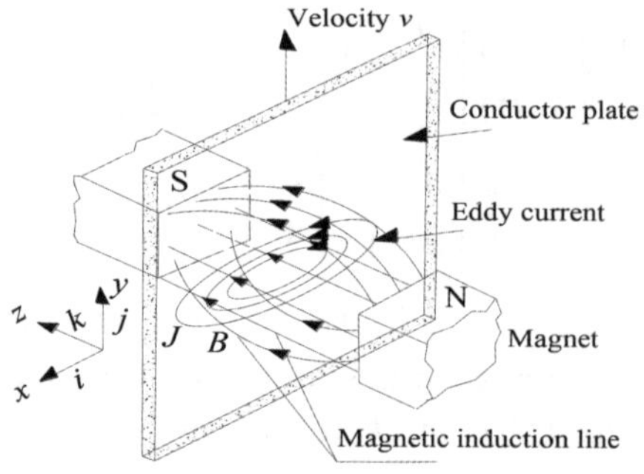

Fig. 2 Schematic of generated eddy current by the conductor plate moving with a magnetic field

$$F_{vy} = -\sigma v_y \int_v (B_X^2 + B_Z^2) dV \tag{5}$$

$$C_{vy} = -\sigma \int_v (B_X^2 + B_Z^2) dV \tag{6}$$

where σ is the electrical conductivity of the conductive plate; V is the relative velocity between the permanent magnets and the conductive plate; and B is the magnetic flux density due to the permanent magnets. It can be seen that the induced damping force and damping coefficient are proportional to the relative velocity between the permanent magnets and the conductive plate, and the damping force and damping coefficient in y direction is only dependent on the components of the magnetic flux density in x direction and z direction. Therefore, the permanent magnets should better be magnetized in x direction or z direction to achieve larger damping coefficient with a given size of damping device.

For an ECD-TMD, if the added muss m_α and circular frequency ω_α are obtained previously, the damping ratio of TMD can be determined as

$$\xi_{TMD} = \frac{c_{vy}}{2m_\alpha \omega_\alpha} \tag{7}$$

3. Design of the ECD-TMD

The proposed eddy current damping TMD (ECD-TMD) consists of an outer frame, a cantilever beam, two lumped mass blocks and an ECD system, as shown in Fig. 3. Here, the plates to fix the nylon columns in the real model are omitted in the sketch to show the main parts more clearly. Compared with the aforementioned references, the ECD-TMD has two improvements: (1) a new configuration of ECD system that the magnets are placed in one side of the conductive plates which could easily adjust the damping through changing the gap; (2) a cantilever beam with uniform strength, which improves the anti-fatigue performance of the stiffness element of the TMD.

3.1 Design of ECD system

For an ECD system in the TMD used on a bridge, three key aspects have to be considered to optimize the design: (1) zero power consumption; (2) higher energy dissipation efficiency especially in the low frequency range (the flexible bridge elements usually have a lower frequency and a lower moving speed, while the energy dissipation amount of the ECD system is related to vibration velocity); (3) easy assembling and adjustment especially in the field. Given this, a new configuration with two rectangular permanent magnets was proposed in the ECD system, as illustrated in Fig. 4, in which the permanent magnets rather than electromagnets were used to avoid any power supply, and the two magnets were arranged at the same side above the conductor plate so that the damping can be adjusted conveniently by changing the gap between permanent magnets and conductor plate. During its operations, the two magnets mounted on the turned mass by the fixed plate will move and induce eddy currents in the conductor plate. In the ECD system, a concentrating flux plate was used to improve the efficiency of energy dissipation, which was further investigated by finite element analysis using COMSOL Multiphysics Software.

The positive effects of a concentrating plate were previously studied by Lequesne *et al.* (1997) to solid-rotor eddy-current couplers and brakes, and by Wang *et al.* (2012) to a large-scale TMD, which proved the feasibility using the concentrating flux plate to improve the damping of a damper.

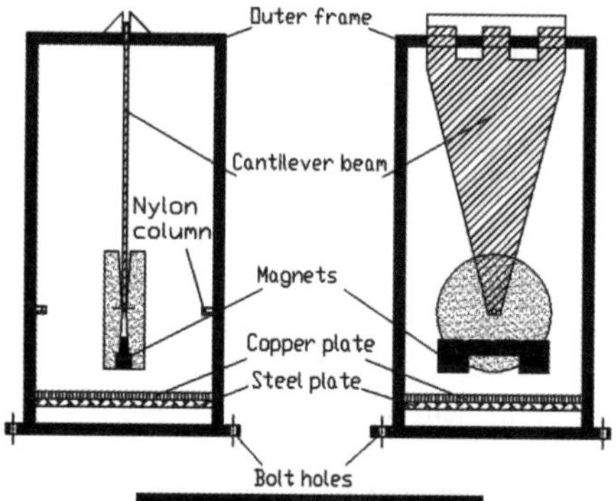

Fig. 3 Structure schematic and prototype picture of the ECD-TMD

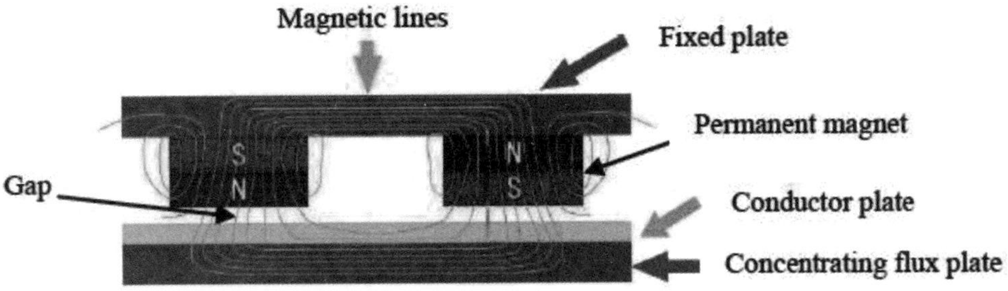

Fig. 4 Sketch of the ECD system in TMD

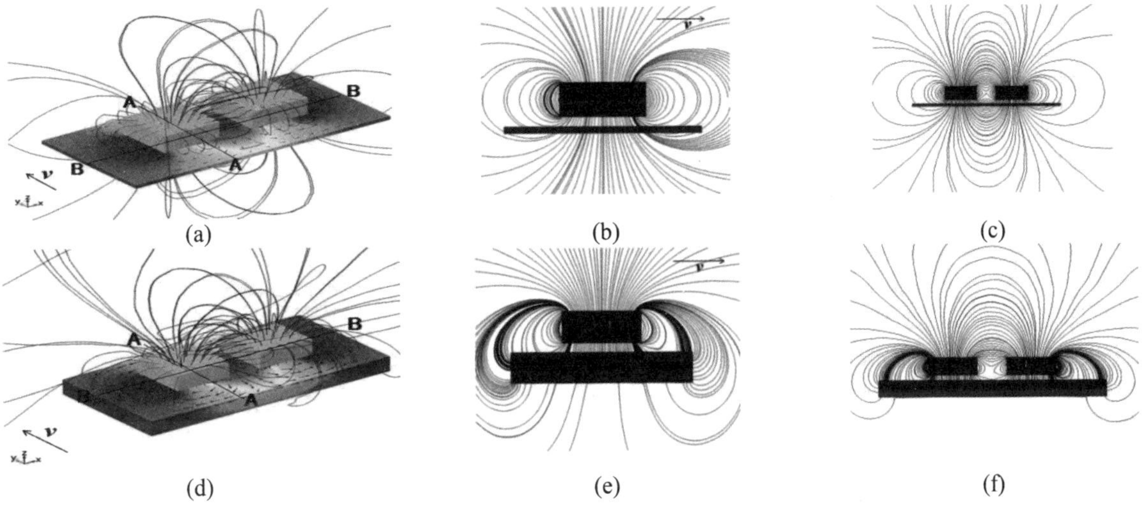

Fig. 5 Simulated magnetic induction lines with two permanent magnets and copper plate. (a) without steel plate, (b) A-A cross-section, without steel plate, (c) B-B cross-section, without steel plate, (d) with steel plate, (e) A-A cross-section, with steel plate and (f) B-B cross-section, with steel plate

To further investigate the effects, two rectangular permanent magnets and a common steel plate used as concentrating flux plate, as shown in Fig. 4, were simulated by COMSOL software. At first, mesh sizes were optimized by trial and error with different element mesh sizes until the calculated results were in accordance with the theoretical values (Gou *et al.* 2004, Wang *et al.* 2012). The simulated results with and without the steel plate are compared in Fig. 5. The gathering effects produced by the steel plate are shown clearly by the cross-section along two directions A-A and B-B, and shown as closed circles of magnetic induction lines at the sides of the magnets. Based on the configuration with the steel plate, the damping coefficients C_v or damping forces F_v were investigated. In the calculations, the thickness of the steel plate was defined as 5 mm, and the magnetization intensity M of the magnets was assigned as 1×10^6 A/m referring to the real parameter of the used magnets in TMD, as well as a copper plate was used as the conductor plate.

As shown in Fig. 6, an optimal interval can be found between two magnets corresponding to the maximum damping coefficient C_v. The damping coefficient decreases with the increasing magnetic gap, which implies the damping ratio of the ECD-TMD can be easily adjusted by varying this gap. In addition, C_v increases first and then decreases with the added thickness of conductor plate, which suggests the necessity of the optimization for the plate thickness. The linear relationship is found for the damping force and relative speed in most conditions. However, the nonlinear relationship becomes dominant with the increases of thicknesses or relative speeds. The calculated results serve as the design guidance of the ECD system for TMDs described hereinafter.

Referring to the previously calculated results, the ECD system was designed to include two permanent magnets, one copper plate, and one steel plate. The sketch of the ECD system is shown in Fig. 4. The thicknesses of the copper plate and steel plate are both 5 mm. To easily adjust the damping ratio of TMD, the copper plate is attached on the upper surface of the steel plate. Therefore, the gap between the magnets and the copper plate can be changed from 0 to 15 mm. The two magnets were embedded at the bottom of the lumped mass blocks and the interval between them is 25 mm.

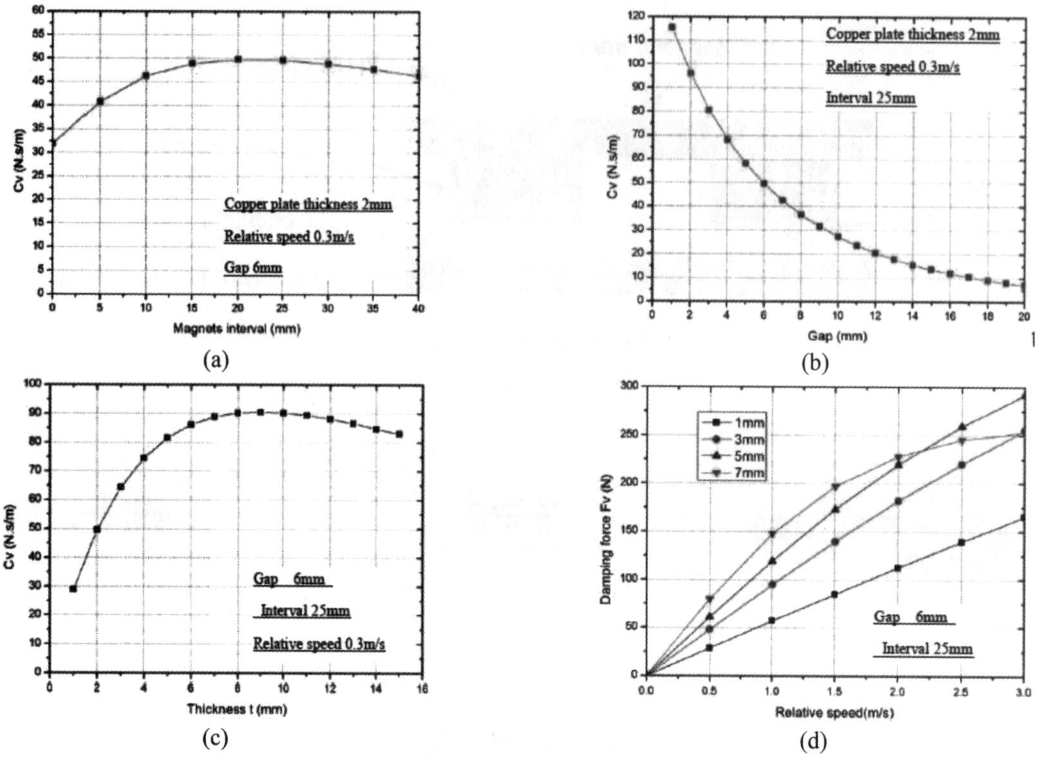

Fig. 6 Calculated eddy current damping coefficients or damping forces with different parameters. (a) interval between two magnets, (b) gaps between magnets and copper plate, (c) copper plate thickness and (d) relative speeds and copper plate thicknesses

3.2 Design of stiffness element

To improve fatigue performances, the cantilever beam with uniform strength is adopted as stiffness components, which is effective to avoid the root damage due to stress fatigue. According to the theory of the uniform strength beam, the width $b(x)$ of the cantilever beam with a rectangular section can be given by

$$b(x) = 6k_{\alpha}\delta_{\max}x / \sigma_{\max}h^2 \tag{8}$$

where k_{α} is the effective stiffness of the beam, $k_{\alpha} = m_{\alpha}\omega_{\alpha}^2$; $\sigma_{\max}$ is the allowable stress corresponding to the steel fatigue; x is the distance from the designed point to the tip of the cantilever beam; $\delta_{\max}$ is allowable maximum displacement of the lumped mass blocks; m_{α} are the moving masses including the cantilever beam and the lumped mass blocks with two magnets; ω_{α} are the circular frequency of the TMD; h is the thickness of the steel plate used for the cantilever beam. The preliminary parameters of the cantilever beam can be calculated, and the final parameters were evaluated and optimized based on the FEA results. In addition, in order to prevent the overload conditions that may cause fatigue issues, two nylon short columns are mounted on the outer frame to limit the maximum displacement of the cantilever beam.

As illustrated in Fig. 3, the cantilever beam was riveted at the top of the outer frame of which the foot would be anchored on the central diaphragm plate of hangers. To meet the requirement of adjusting the frequency deviation derived from the difference between the FEM analysis and the field test, the top end of the cantilever beam is reserved as 100 mm long which can be used to adjust the TMD frequency according to the tested frequency of a hanger.

3.3 Design of the tuned mass

As illustrated in Fig. 3, the lumped mass blocks were fixed at the tip of the cantilever beam so that it enables the frequency of the TMD to be tuned easily. To ease frequency adjustments, additional reserved space is reserved for a steel plate to adjust the total lumped mass at the two sides of the mass blocks.

Due to the strength and space limitation of the hangers, the total weight is strictly limited under 50 kg for an easy move by one or two workers, considering future maintenance and repair. According to the pre-defined mass ratio μ, the required TMD masses may still surpass this limitation. In this circumstance, the required concentrated masses can be divided into two or more ECD-TMDs in the real application.

Fig. 7 General view of the Rongjiang Steel Arch Bridge

Table 1 Basic parameters of all types of the hanger

Hanger Parameter/No.	G13A13	G14A14	G15A15	G16A16	G17A17	G18A18	G19A19	G20A20
Length (m)	10.313	19.250	26.813	33.000	37.813	41.250	43.313	44.000
λ_{across}	12	23	32	39	45	49	51	52
λ_{along}	16	29	41	50	57	63	66	67
Total mass (kg)	1773.5	3535	4191.2	5320.5	6072.9	6634.3	6963.3	7066.3

4. Case study

4.1 Description of the Rongjiang Steel Arch Bridge

Rongjiang Steel Arch Bridge is across the Rongjiang River, located in Shantou, China, as shown in Fig. 7. The bridge's total length is 660 m consisting of two main spans of 220 m and two side spans of 110 m. A steel truss girder was used, which comprises of two main trusses with an interval and height of both being 15 m. The rise-to-span ratio of the two arches is 1/5. The arch follows a second-degree parabola line. The heights of two arch tops over the upper chord of truss girder are 44 m. To keep the smooth running of the high-speed train, a total of 60 rigid hangers are used for the two arches. These hangers are grouped into eight types by their lengths, as shown in Table 1. All the hangers have the same rectangular cross-sections of 1.082 m (across bridge)×0.824 m (along bridge), and the longest hanger is 44 m with the slenderness ratio across bridge 52 and along bridge 67.

4.2 Wind-resistant stability assessment of the hangers

For the bridge, the vertical steel hangers are the vital structural elements. With a large slenderness ratio and the typical rectangular-shaped cross section, these hangers are prone to wind induced vibrations, especially for the wind-induced vortex shedding and galloping. In addition, the bridge locates in a typhoon prone region and the bridge site is just 30 km away from Shantou Bay. The design wind speed at 10 m height is 39.4 m/s. With an additional elevation difference of 57 m between the arch foot and the design water level, the design wind speed for hangers corresponding to half height of the arches reaches 54.5 m/s. Therefore, it is critical for the wind-resistant stability design of the hangers. At the initial design phase of the bridge, wind tunnel tests were carried out to understand wind structure interactions.

Since cross sections of all hangers are the same, the longest hanger numbered G20A20 was selected for further investigation of wind-induced vibrations due to its slenderness. A rigid 1/5-scale sectional model was designed for the experiment. The cross section is 0.216×0.165 m, and the length of the model is 1.53 m. The experimental wind flow is uniform, and the turbulence intensity is less than 1%. Both the static aerodynamic force coefficients and wind-induced vibration tests were performed in HD-2 Wind Tunnel of Hunan University, China. The tested aerodynamic force coefficients and wind-induced vibrations for the longest hanger are illustrated in Fig.8. Since the hanger is vertical, the yaw angle for the bridge deck turns into the wind attack angle for the hanger. In the wind tunnel tests, the wind attach angles β were measured from 0~90°. Specifically, β=0°is corresponding to across wind direction for the bridge deck, which means that the wind direction is along the strong axis of the hanger. Meanwhile, the β=90°means the wind direction is along the weak axis of the hanger.

Based on the experimental results, the main instability problem for the hangers is related to the coupling vortex-induced vibration and galloping (Niu *et al.* 2015), which needs to be further studied. In the present work, a simple

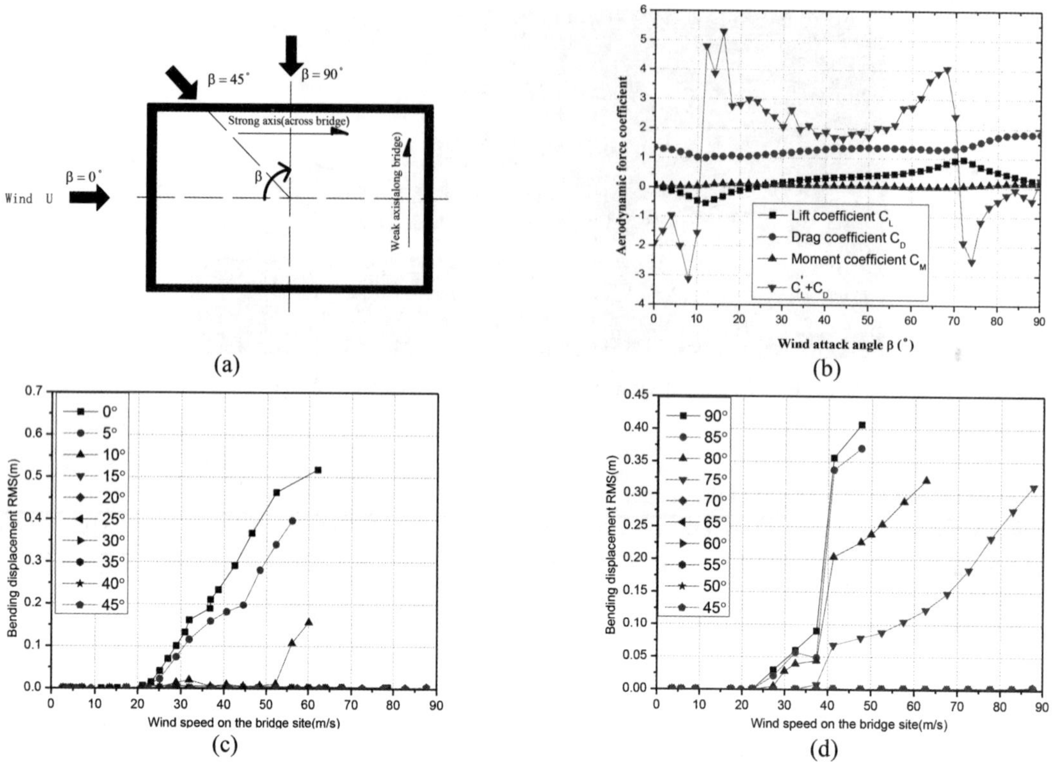

Fig. 8 The wind tunnel test results of aerodynamic force coefficients and wind-induced bending responses (a) definition of the wind attack angle β, (b) aerodynamic force coefficients, (c) wind-induced responses along weak axis and (d) wind-induced responses along strong axis

Table 2 Evaluated Critical Wind Speeds of the Hangers Numbered G_iA_i *(i=16~20)*

Hanger Number	Axis	G13A13	G14A14	G15A15	G16A16	G17A17	G18A18	G19A19	G20A20	Checking speed (m/s)
Vortex-induced resonance onset wind speed (m/s)	Strong	352	132	71	**43**	**31**	**25**	**25**	**24**	*54*
	Weak	309	115	56	**37**	**29**	**25**	**23**	**22**	
Galloping wind speed (m/s)	Strong	960	360	194	116	84	69	67	**64**	*65*
	Weak	347	129	68	**42**	**32**	**27**	**25**	**24**	

method is employed to assess the instability of the hangers, which separately evaluates the onset wind speeds of the vortex-induced vibration and the critical wind speeds of the galloping through measured Strouhal number and galloping coefficient. The onset wind speed of vortex-induced vibration U_{cr} and the critical wind speed of galloping V_{cg} can be determined as (Simiu and Scanlan 1996)

$$U_{cr}=\frac{f_v D}{St} \tag{9}$$

$$V_{cg}=-\frac{4m\omega\xi}{\rho D}\cdot\frac{1}{C_L'+C_D} \tag{10}$$

where D is the length of the side of the rectangular cross-section normal to mean flow velocity; f_v is the vibration frequency of the hanger, which is calculated by finite element method; St is the Strouhal number that is calculated by the measured onset wind velocity through hanger G20A20 model at 0° and 90°, 0.112 and 0.163 respectively; m is mass per unit length of the hanger; ω is the circular frequency of the hanger; ξ is modal damping ratio of the hanger; ρ is air density; $C_L'+C_D$ is Den Hartog coefficient, as illustrated in Fig. 8(b)). Considering the natural wind direction varying with turbulence, the mean values of Den Hartog coefficients of -1.85 and -0.85 at 0°~10°and 72°~90° are used to evaluate the critical wind speed of galloping. The evaluated results of critical wind speeds are listed in

Table 2, which shows the critical wind speeds of the hangers numbered G16A16~G20A20 cannot meet the wind-resistant demands in both direction of strong and weak axis.

To investigate the effects of damping on the suppression of wind-induced vibrations, the sectional model was further tested by varying damping ratios, and the results were illustrated in Fig. 9. It was found that the necessary damping ratios to control the wind-induced vibration of the long hanger are 0.01 at strong axis and 0.03 at weak axis respectively, which provided the reference data for ECD-TMD design and vibration control performance assessment.

4.3 Design parameters for an ECD-TMD

To get the target mass ratio μ, the complex eigenvalue analysis to the model of coupled hanger-TMD system was carried out. Corresponding to the aforementioned necessary damping ratios of the system, the mass ratio μ of 0.006 and 0.01 was obtained corresponding to the strong axis and weak axis. Then the damping of TMD ξ_{opt} can be calculated by Eq. (4) and will be realized by the previous ECD system. Considering the total weight and easy installation, the lumped mass for each axis of the hangers numbered G16A16~G20A20 was designed to divide into 2 ECD-TMDs. Therefore, a total of 144 ECD-TMDs will be installed in the entire bridge, as shown in Table 3.

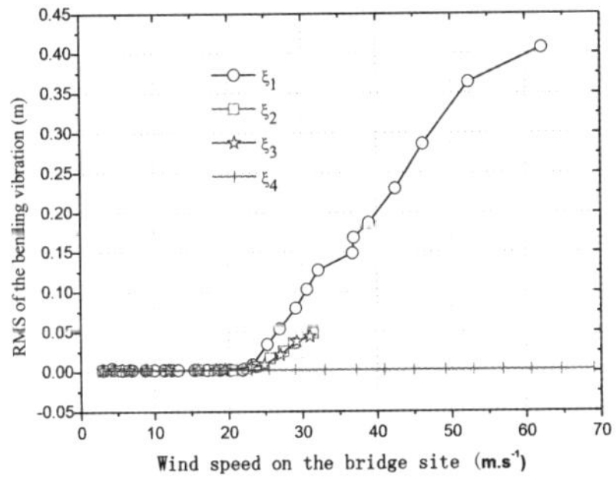

(a)

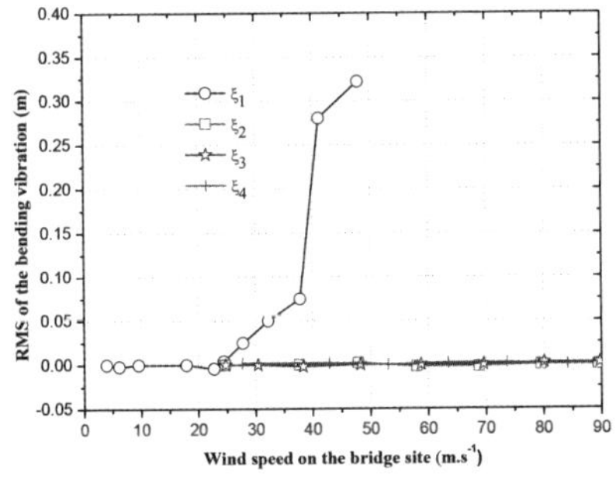

(b)

Fig. 9 Tested results of the longest hanger G20A20 with different damping $\xi1=0.002$, $\xi2=0.01$, $\xi3=0.02$, $\xi4=0.03$. (a) weak axis, 0° and (b) strong axis, 90°

Table 3 Concentrated mass parameters and related numbers for the ECD-TMDs

Hangers Type	Quantity	Needed concentrated mass(kg)		Total number of designed TMD	
		Strong axis	Weak axis	Strong axis	Weak axis
G16A16	8	31.92	63.85	16	16
G17A17	8	36.44	60.73	16	16
G18A18	8	39.8	66.34	16	16
G19A19	8	41.78	69.63	16	16
G20A20	4	42.4	70.66	8	8

In addition, the frequency is related to the stiffness element of TMD. Therefore, the control efficiency of the TMD is very sensitive to the frequency ratio between the TMD and the main structure. During the design of the ECD-TMD, the finite element analysis with a full bridge model and multi-segment hangers was carried out to obtain the frequencies of these hangers. Since the real frequencies of hangers are usually different from the calculated ones obtained from finite element analysis (FEA), in-situ measurements were carried out to find the real frequencies before the design. The onsite implementations of TMDs were divided into several steps to keep up with the construction progress of the two arches and hangers. In the designed construction steps, the main truss girder would be joined first, followed by building the two arches with the hangers one by one. Finally the temporary piers were removed and the concrete decks with tracks were paved. Based on the FEA, the construction stage when one arch with hangers had been completed was selected to obtain the frequencies of the hangers on the field. The frequency of each TMD was fine-tuned to the optimal target value. After that, the fine adjusted ECD-TMDs were installed into the hangers one by one. Finally, when the two arches were both closed, the frequencies and damping ratios of hangers were tested on the site to check and evaluate the control performance of the CED-TMD system.

Considering the symmetry of this bridge, the eleven hangers at the upstream side of the arch to Shenzhen, namely Arch1, were measured using accelerometers, as shown in Fig. 10. During the tests, two accelerometers were installed at the midpoint of each hanger for monitoring the ambient vibrations for the strong axis and the weak axis. The recorded typical acceleration signals and their frequency spectrum of the longest hanger G20A20 are illustrated in Fig. 11, and the vibration frequencies can be obtained easily from the frequency spectrum. The measured frequencies were demonstrated in Table 4 with the calculated frequencies by FEA. Here, different from the beam model that a hanger is simulated by a beam of fixed ends, the full bridge model simulates the boundary conditions of the hangers more accurately by taking account of the main girder and the main arches of the bridge. To investigate the change of these frequencies, these hangers were measured again when the Arch2 were completed and the temporary supports were released, as well as the ECD-TMDs had been installed into these hangers, the results were also listed in Table 4.

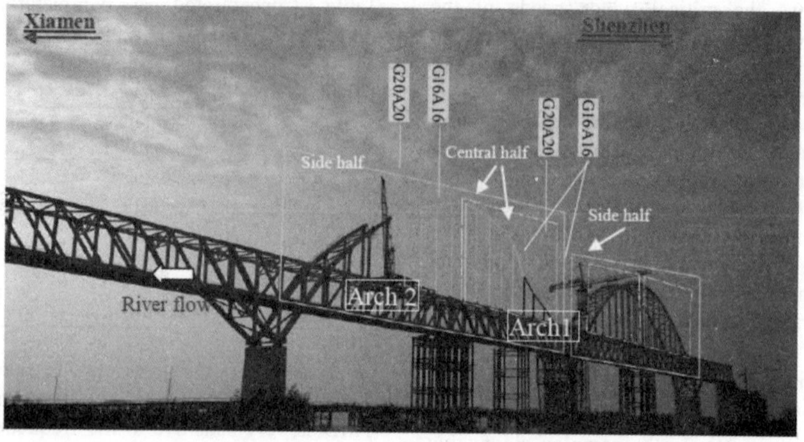

Fig. 10 Position diagram for the hangers and the frequency tests

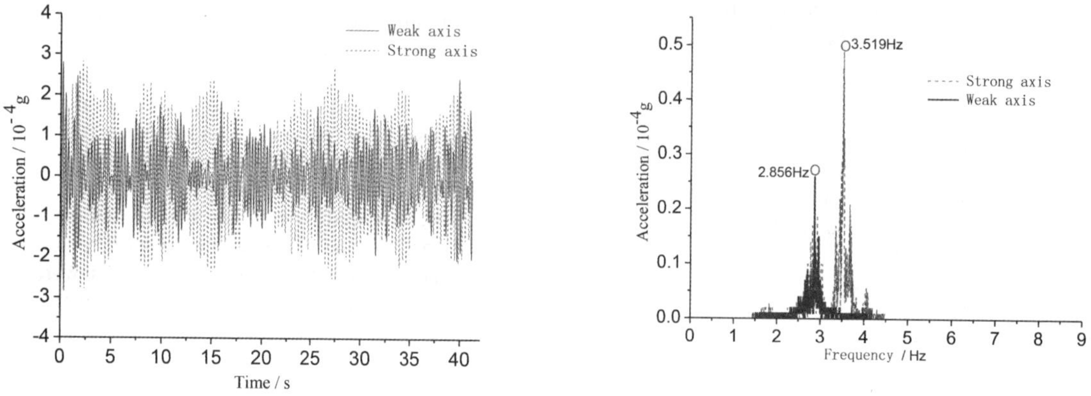

Fig. 11 Acceleration time-history under ambient excitation and the corresponding spectrum of hanger G20A20

Table 4 Frequencies of hangers tested at different positions and situations

Axis	Position & situation description	Hangers No.				
		G16A16	G17A17	G18A18	G19A19	G20A20
Weak axis	Arch1-central half- 1st	5.24	3.78	3.20	2.97	2.86
	Arch1-side half-1st	5.25	3.71	3.18	2.93	2.86
	Arch1-central half-2nd	/	3.89	3.22	2.98	2.90
	Arch2-central half-2nd	5.27	/	3.23	3.17	2.76
	Beam model	5.57	4.29	3.63	3.31	3.21
	Full bridge model	5.05	3.96	3.35	3.07	2.97
Strong axis	Arch1-central half- 1st	5.89	4.30	3.74	3.33	3.52
	Arch1-side half-1st	5.81	4.27	3.72	3.31	3.52
	Arch1-central half-2nd	/	4.54	3.95	3.68	3.47
	Arch2-central half-2nd	5.69	/	3.98	3.71	3.41
	Beam model	7.22	5.55	4.70	4.27	4.15
	Full bridge model	6.45	4.68	3.82	3.75	3.55

*1st test corresponding to the situation only Arch1 was completed, the hanger without TMD; 2nd test corresponding to the situation both Arch1 and Arch2 were completed, the hanger with TMD; "/" refers to no effective data

Table 5 Final Parameters of the ECD-TMDs

	Hanger No.	G16A16	G17A17	G18A18	G19A19	G20A20
Strong axis	Target frequency (Hz)	5.83	4.25	3.70	3.29	3.48
	Tested TMD frequency (Hz)	5.76	4.23	3.69	3.26	3.50
	TMD total weight (kg)	23	30	33	31	33
	Target dampingratio (%)	4.7-6.0	4.7-6.0	4.7-6.0	4.7-6.0	4.7-6.0
	TestedTMD dampingratio (%)	6.05	5.80	5.94	6.25	6.61
Weak axis	Target frequency (Hz)	5.19	3.75	3.16	2.94	2.83
	Tested TMD frequency (Hz)	5.15	3.75	3.16	2.95	2.85
	TMD total weight (kg)	30	38	46	45	46
	Target dampingratio (%)	6-7	6-7	6-7	6-7	6-7
	TestedTMD dampingratio (%)	7.30	6.96	7.01	6.88	7.15

4.4 Installation of ECD-TMDs on the site

To install the ECD-TMDs on these hangers, a diaphragm plate with reserved bolt holes was designed. The plate is welded together with each hanger at the middle point in factory. For the hangers numbered G_iA_i(i=16~20), four ECD-TMDs were installed on the diaphragm plates and two ECD-TMDs would vibrate along the strong axis or the weak axis at the diagonal position, as shown in Fig. 12. In fact, two different installation processes were adopted when installing the ECD-TMDs on the hangers in Arch1 and Arch2. For Arch1, the arch with hangers had been completed on the site, therefore, the TMDs had to be lifted by the tower crane and to be dragged into the hanger from the reserved manhole. After that, the TMDs were then mounted on the diaphragm plate; as for Arch2, since the hangers and TMDs were constructed at the same time. The TMDs were pre-mounted on the diaphragm plate and then hoisted together with the hanger to be docked with the arch. Fig. 13 demonstrates the typical situations on the field during the installation process.

4.5 Vibration control performance assessment

Similar to the control performance evaluation method for a cable vibration control system (Chen *et al.* 2004), the modal damping ratios are used to assess the control performance of the supplemental ECD-TMDs. To measure the damping ratio of the hanger with ECD-TMD, two types of experiments were performed on the field, namely, the ambient vibration tests and the forced vibration tests. As aforementioned method to acquire the frequencies of these hangers, the recorded signals of ambient vibrations were also used to identify the needed damping ratio by RDT techniques (Yang 2004, Carlos 2016). Hence the same cases as Table 4 for Arch2 and Arch1 are used to mainly demonstrate the changes of damping ratios with and without ECD-TMDs, as listed in Table 6. For the 1st tests without TMDs in the hangers, the results indicate most of the damping ratios are larger than 0.5%.

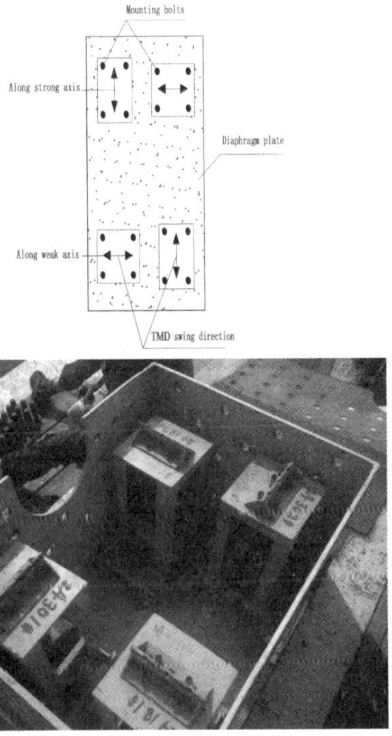

Fig. 12 Installation layout and real picture of four ECD-TMDs on the diaphragm plates

It is worthy to mention that the damping ratios may be related to the influence of the temporary wire rope bracings between two hangers, which are served as a wind-resistance countermeasure during the construction. In addition, the influences derived from the wind-induced aerodynamic damping and the errors of identification method based on ambient excitations are also not neglectable.

To better identify the damping ratios of the hangers, the forced sinusoidal excitations through an eccentric wheel system driven by an electric motor, of which the frequency can be adjusted by a frequency converter, was applied to excite the hanger to vibrate steadily when the excited frequency was very close to its natural frequency. Due to the limitation of the construction time, only the two longest hangers, numbered G19A19 and G20A20, in Arch2 were tested, as shown in Fig. 14.

In each test, the vibration exciter was first mounted on the surface of the hanger by an electromagnet. Then the frequency converter was adjusted gradually to excite the hanger to a steady vibration. Finally, the exciter was suddenly separated from the hanger by turning off the electricity to the electromagnet. Therefore, the hanger went into a free decay vibration at this time, and the corresponding acceleration time history was recorded to estimate the damping ratio by the logarithmic decrement method. However, due to the interaction of the hanger and the TMD which have two closely-spaced frequencies, the free vibration exhibits beating phenomena. Considering the vibration of the hanger itself was obviously dominant in the signals, the free decay vibration curves were fitted by 1-DOF nonlinear least square method based on the vibration frequency of the hanger. The average value of the damping ratios in the five tests was used as the measured damping ratio for the experimental case. The recorded typical acceleration time histories of the decay vibration are shown in Fig. 15, and the acquired damping ratios are summarised in Table 6.

Fig. 14 Vibration exciter for forced vibration tests of the damping ratios of hanger-TMD System

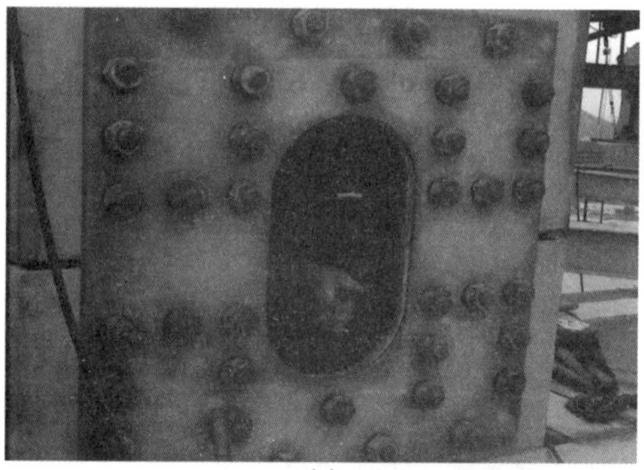

(a)

(b)

Fig. 13 Typical situations for the TMD installation on the field: (a) installation through the manhole, for Arch1 and (b) pre-installation of ECD-TMDs, for Arch2

Compared to the results from ambient vibration tests, the damping ratios acquired by forced vibration tests have about 10% differences. The differences between the identified damping values can mostly be attributed to three aspects: 1) different approaches, ambient vibration and free decay vibration; 2) different aerodynamic damping due to different wind speed when the tests were performed; 3) different amplitudes, which induced the different signal-to-noise ratio and the damping deviation corresponding to the nonlinear damping-amplitude relationship. However, as shown in the table, the identified damping ratios without ECD-TMDs by the ambient vibration tests are less than the same hanger with ECD-TMDs, and most of the damping ratios with ECD-TMDs are improved more than 1% and 2.5% corresponding to the strong axis and weak axis respectively. Furthermore, the damping ratios acquired by forced vibration tests affirm the beneficial effects of ECD-TMDs, and the results assure the damping ratios of longer hangers being beyond the desired values verified by wind tunnel tests, especially for the weak axis over 3%. Meanwhile, oscillation of the lumped mass blocks of the ECD-TMDs was observed with 3~6 mm amplitude during the forced excitation on the hanger, which proved the ECD-TMDs work in good conditions. To further assess the control effects of the devices, two acceleration sensors along the strong and weak axis were fixed in one of the longest hangers of Arch-1 during the construction stage, and mounted on the diaphragm plates where the ECD-TMDs were mounted. Meanwhile, a propeller anemometer was fixed on the top of the truss girder, where the height above the water level is about 55 m. The recorded typical signals without and with the ECD-TMDs are presented in Fig. 16, in which two wind speed cases are compared.

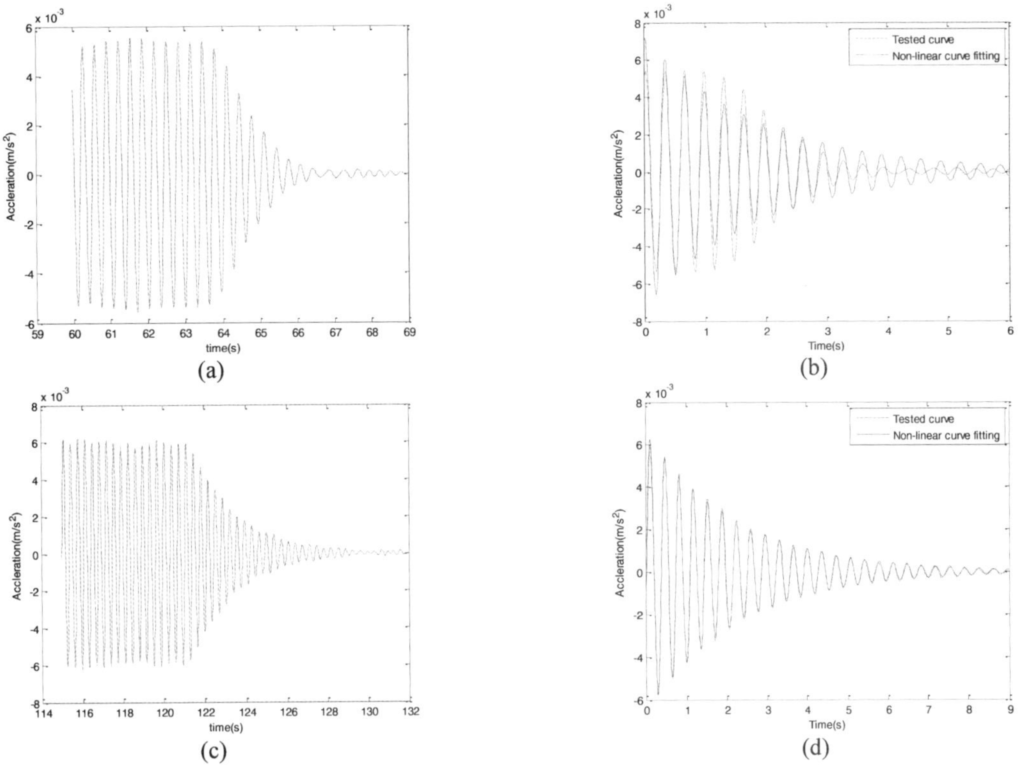

Fig. 15 Typical acceleration time history of the hangers during forced vibration tests, weak axis (a) G19A19, recorded curve, (b) G19A19, fitting curve of the decaying free motion, (c) G20A20, recorded curve and (d) G20A20, fitting curve of the decaying free motion

Table 6 Identified Equivalent Modal Damping Ratios of the Hangers through Field Tests

Axis		Position & situation	G16A16	G17A17	G18A18	G19A19	G20A20
Weak axis	Ambient vibration test	Arch1-central half- 1st	0.0075	0.0065	0.0067	0.0071	0.007
		Arch1-side half-1st	0.0062	0.0084	0.0049	0.0058	
		Arch1-central half 2nd	/	0.0321	0.0308	0.0342	0.0282
		Arch2-central half-2nd	0.0277	/	0.0316	0.0354	0.027
	Forced vibration test	***Arch2-central half-2nd***	*/*	*/*	*/*	***0.0324***	***0.0297***
Strong axis	Ambient vibration test	Arch1-central half- 1st	0.0083	0.0088	0.0074	0.0072	0.0077
		Arch1-side half-1st	0.0092	0.0094	0.0058	0.005	
		Arch1-central half-2nd	/	0.0309	0.0249	0.0228	0.0252
		Arch2-central half-2nd	0.0195	/	0.0275	0.0271	0.0286
	Forced vibration test	***Arch2-central half-2nd***	*/*	*/*	*/*	***0.0206***	***0.0304***

*1st test corresponding to the hanger without TMD; 2nd test corresponding to the hanger with TMD; "/" refers to no effective data

As shown in the figure, the wind induced responses reduced more than 60% after the ECD-TMDs were installed in the long hanger. Moreover, when the Monster Super Typhoon USAGI, reported maximum speed 52 m/s, landed 40km away from the bridge in September 2013, no visible vibrations of these hangers were observed by telescope on the bank. Following that, the bridge also underwent the tropical storm Hagibis in 2014 as well as the severe tropical storm Linfa in 2015, and no special vibrations were reported by the owners. These experiences also proved the feasibility of the ECD-TMDs to suppress the wind-induced vibrations of the slender hangers.

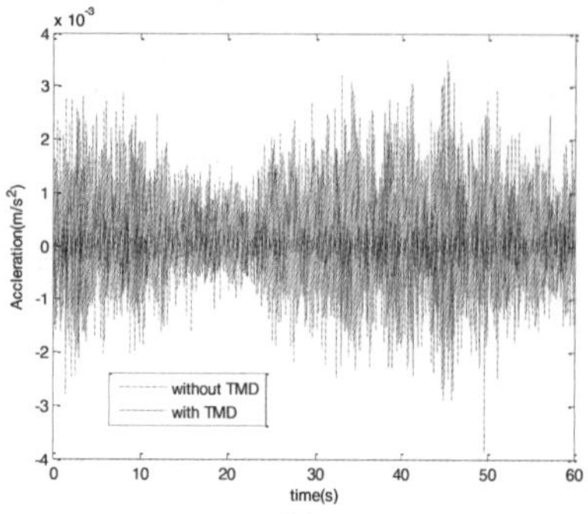

(a)

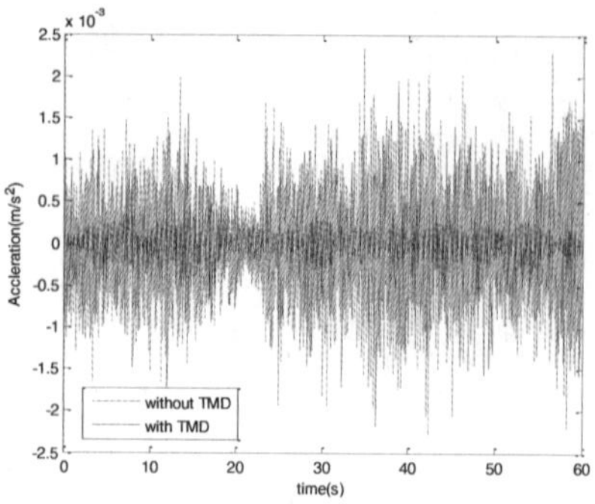

(b)

Fig. 16 Recorded typical acceleration time history of the hangers without ECD-TMDs (wind speed, 11.7m/s) and with ECD-TMD (wind speed, 10.6m/s), G20A20: (a) weak axis and (b) strong axis

5. Conclusions

In the present study, a new TMD is proposed that uses the non-contact ECD system to generate damping and dissipate energy for large scale civil infrastructures. A uniform strength cantilever beam is used to mitigate the potential fatigue issues of the beam. Without any viscous damper, oil tank or friction element, the new ECD-TMD is expecting a better durability. Due to its none-friction feature, the new ECD-TMD could also begin to work even at a very low wind speed compared with typical TMDs. With the application of the permanent magnets rather than electromagnets, power supply is not necessary. In addition, the new design of the copper plate with a steel plate has been implemented to improve the energy dissipation efficiency. Nevertheless, the two permanent magnets were arranged at the same side above the copper plate to adjust the damping more conveniently by changing the gap between the magnets and the copper plate. After careful introduction of the ECD-TMD device including the design of the uniform strength cantilever beam, a case study is presented when the ECD-TMD was used to suppress the wind-induced vibrations of the flexible hangers of a steel arch bridge. The TMD parameters, especially the frequencies of ECD-TMDs on the field, were discussed, and the installation process of these ECD-TMDs was introduced, as well. Based on the data from onsite measurements using the ambient vibration and forced vibration method, the wind vibration mitigation were evaluated. The results show that the damping ratios increased more than 1% and 2.5% in the strong axis and weak axis, which is beyond the desired damping ratios, especially for the weak axis over 3%. The ECD-TMDs are also proven to be effective to mitigate the vibrations in the recent several super typhoons. In addition, the proposed ECD-TMD in the bridge not only has a smaller size compared with a conventional oil tank but also has 15% lower cost compared with a conventional oil tank. Meanwhile, the life cycle costs are obviously lower than the conventional oil damper TMD due to the absence of the liquid and the seals. The chances of possible friction issues or fatigue issues are also greatly reduced, which could help improve the durability of these ECD-TMDs.

Acknowledgements

The authors would like to acknowledge support from Natural Science Foundation of China under grant number51478181, U1534206, 50908085 and the support rom China Scholarship Council (No.201506135039).The authors also would like to express great thanks to our collaboration of the China Railway 13th Construction Bureau, LTD. and China Railway Siyuan Survey and Design Group Co. LTD., in particular senior engineer Yu Wang and Jie Zhang, is of special mention. Our colleague Associate Prof. Jianhui Wang and Ph.D student Mr. Xu Lei and Jinlin Chen who provided great assistance on the field are also acknowledged.

References

Amjadian, M. and Agrawal, A.K. (2017), "A passive electromagnetic eddy current friction damper (PEMECFD): Theoretical and analytical modelling", *Struct Control Health Monit.*, 2017, e1978, 23 pages, https://doi.org/10.1002/stc.1978.

Amjadian, M. and Agrawal, A.K. (2018), "Modeling, design, and testing of a proof-of-concept prototype damper with friction and eddy current damping effects", *J. Sound Vib.*, **413**, 225-249.

Anantha Krishna, G.L. and Sathish Kumar, K.M. (2018), "Experimental investigation of influence of various parameters on permanent magnet eddy current braking system", Materials Today: Proceedings, 5, 2575-2581.

Bae, J.S., Hwang, J.H., Kwag, D.G., Park, J. and Inman, D.J. (2014), "Vibration suppression of a large beam structure using tuned mass damper and eddy current damping", *Shock Vib.*, Article ID 893914, 10 pages, doi:10.1155/2014/893914.

Bae, J.S., Hwang, J.H., Park, J.S. and Kwag, D.G. (2009), "Modeling and experiments on eddy current damping causedby a permanent magnet in a conductive tube", *J. Mech. Sci. Technol.*, **23**, 3024-3035.

Bae, J.S., Hwang, J.H., Roh, J.H., Kim, J.H., Yi, M.S. and Lim, J. H. (2012), "Vibration suppression of a cantilever beam using magnetically tuned-mass-damper", *J. Sound Vib.*, **331**, 5669-5684.

Carlos, A.P.R., Juan, P.A.S., Hojjat, A., Martin, V.R., David, C.M. and Rene, J.R.T. (2016), "New methodology for modal parameters identification of smart civil structures using ambient vibrations and synchrosqueezed wavelet transform", *Eng. Appl. Artif. Intel.*, **48**, 1-12.

Chang, C.C., Gu, M. and Tang, K.H. (2003), "Tuned mass dampers for dual-Mode buffeting controlof bridges", *J. Bridge Eng.*, **8**(4), 237-240.

Chen, X.Z. and Kareem, A. (2003), "Efficacy of tuned mass dampers for bridge flutter control", *J. Struct. Eng.*, **129**(10), 1291-1300.

Chen, Z.Q., Liu, M.G., Hua, X.G. and Mou, T. M. (2012), "Flutter, galloping, and vortex-Induced vibrations of H-section hangers", *J. Bridge Eng.*, **17**(3), 500-508.

Chen, Z.Q., Liu, M.G., Liu G.D. and Jin, Z.J. (2010), "Wind-induced vibration and wind-resistant design of H-shaped suspenders under large attack angle", *China Civ. Eng. J.*, **43**(2), 1-11 (in Chinese).

Chen, Z.Q., Wang, X.Y., Ko, J.M., *et al.* (2004), "MR damping system for mitigating wind-rain induced vibration on Dongting Lake Cable-Stayed Bridge", *Wind Struct.*, 7(5), 293-304.

David Saige, W.I., Engelhardt, I.J. and Sebastian Katz, I. (2017), "Application of eddy current damper technology for passive tuned mass damper systems within footbridges", *Procedia Eng.*, **199**, 1804-1809.

Den Hartog, J.P. (1956), Mechanical vibrations, 4th Ed., McGraw-Hill, New York.

Fujino, Y. (2002), "Vibration, control and monitoring of long-spanbridges—recent research, developments andpractice in Japan", *J. Constr. Steel Res.*, **58**, 71-97.

Fujino, Y. and Yoshida, Y. (2002), "Wind-induced vibration and control of Trans-Tokyo BayCrossing Bridge", *J. Struct. Eng.*, **128**(8), 1012-1025.

Gou, X.F., Yang, Y. and Zheng, X.J. (2004), "Analytical expressions of magnetic field distribution of rectangular permanent magnets", *Appl. Math. Mech.*, **25**(3), 297-306.

Gu, J.J., Zhao, Y.C. and Shao, K.H. (1994), "Application of new TMD to suppressing vortex-shedding vibrations of hangers of Jiujiang Bridge over Yangtze river", *China Civ. Eng. J.*, **27**(3), 3-13 (in Chinese).

Keller, P., Higgins, C. and Lovejoy, S. (2014), "Evaluation of torsional vibrations in steel truss bridge members", *J. Bridge Eng.*, 10.1061/(ASCE)BE.1943-5592.0000688, 04014102.

Konstantinidis, D., Kelly, J.M. and Makris, N. (2011), "In-situ monitoring of the force output of fluid dampers, experimental validation, PEER Report 2011/103", Pacific Earthquake Engineering Research Center, University of California, Berkeley.

Koshimura, K., Tatsumi, M. and Hata, K. (1994), "Vibration control of the main towers of the Akashi Kaikyo Bridge", In: Proc. of 1st Int. Conf. on Structural Control, TP3, vol. 2, 98-106.

Larose, G.L., Larsen, A. and Svensson, E. (1995), "Modeling of tuned mass dampers for wind-tunnel tests on a full-bridge aeroelastic model", *J. Wind Eng. Ind. Aerod.*, **54-55**, 427-437.

Larsen, A., Svensson, E. and Andersen, H. (1995), "Design aspects of tuned mass dampersfor the Great Belt East Bridge approach spans", *J. Wind Eng. Ind. Aerod.*, **54-55**, 413-426.

Lequesne, B., Liu, B.Y. and Nehl, T.W. (1997), "Eddy current machines with permanent magnets and solid rotors", *IEEE T. Ind. Appl.*, **33**(5), 1289-1294.

Li, R.Q., Zhu, S.F. and Li, D.C. (2012), "Application of a new type of suspender shock absorber (TLMD) for the nanjing daShengguan yangtze river bridge", *World Bridges*, **40**(6), 68-72 (in Chinese).

Lin, C.C., Lu, L.Y., Lin, G.L. and Yang, T.W. (2010), "Vibration control of seismic structures using semi-active friction multiple tunedmass dampers", *Eng. Struct.*, **32**, 3404-3417.

Lu, Z., Huang, B., Zhang, Q. and Lu, X. (2018), "Experimental and analytical study on vibration control effects of eddy-current tuned mass dampers under seismic excitations", *J. Sound Vib.*, **421**, 153-165.

Ma, C., Liao, H., Zheng, S. and Li, J. (2005), "Wind tunnel experiment on the aerodynamic performances of H-shaped booms", *China Railway Science*, **26**(4), 42-46 (in Chinese).

Niu, H.W., Zhou, S., Chen, Z.Q. and Hua, X.G. (2015), "An empirical model for amplitude prediction on VIV-galloping instability of rectangular cylinders", *Wind Struct.*, **21**(1), 85-103.

Paul, P., Ingale, C. and Bhattacharya, B. (2014), "Design of a vibration isolation systemusing eddy current damper", *J Mech. Eng. Sci.*, **228**(4), 664-675.

Pourzeynali, S. and Esteki, S. (2009), "Optimization of the TMD parameters to suppress the vertical vibrations of suspension bridges subjected to earthquake excitations", *IJE Transactions B: Appl.*, **22**(1), 23-34.

Rana, R. and Soong, T.T. (1998), "Parametric study and simplified design of tuned mass dampers", *Eng. Struct.*, **20**, 193-204.

Ruscheweyh, H. (1996), "Vortex-excited vibrations and galloping of slender elements", *J. Wind Eng. Ind. Aerod.*, **65**(1-3), 347-352.

Shen, W.A., Zhu, S.Y. and Xu, Y.L. (2012), "An experimental study on self-powered vibration control and monitoring systemusing electromagnetic TMD and wireless sensors", *Sensor. Actuat. A*, **180**, 166-176.

Simiu, E. and Scanlan, R.H. (1996), *Wind effects on structures: Fundamentals and applications to design*, 3rd Ed., Wiley, New York.

Sodano, H.A. and Bae, J.S. (2004), "Eddy current damping in structures", *Shock Vib. Dig.*, **36**(6), 469-478.

Sodano, H.A., Bae, J.S., Inman, D.J. and Belvin, W.K. (2005), "Concept and model of eddy current damper for vibration suppression of a beam," *J. Sound Vib.*, **288**(4-5), 1177-1196.

Sodano, H.A., Bae, J.S., Inman, D.J. and Belvin, W.K. (2006), "Improved concept and model of eddy current damper", *J. Vib. Acoust.*, **128**, 294-302.

Sodanoa, H.A. and Inman, D.J. (2007), "Non-contact vibration control system employing an active eddycurrent damper", *J. Sound Vib.*, **305**, 596-613.

Sodano, H.A. and Inman, D.J. (2008), "Modeling of a new active eddy current vibration control system", *J. Dyn. Syst. Meas. Control*, **130**(2), 021009(1-12).

Soong, T.T. and Dargush, G.F. (1997), Passive energy dissipation systems in structural engineering, John Wileys & Sons, New York.

Spencer, B.F. and Nagarajaiah, S. (2003), "State of the art of structural control", *J. Struct. Eng.*, **129**(7), 845-856.

Tamura, Y. (1998). "Application of damping devices to suppress wind-induced responses of buildings", *J. Wind Eng. Ind. Aerod.*, **74-76**, 49-72.

Ulstrup, C.C. (1980), "Aerodynamic lessons learned from individual bridge members". *Ann. N.Y. Acad. Sci.*, **352**(1), 265-281.

Wang, Z.H., Chen, Z.Q. and Wang, J.H. (2012), "Feasibility study of a large-scale tuned mass damper with eddy current damping mechanism", *Earthq. Eng. Eng. Vib.*, **11**(3), 391-401.

Warburton, G.W. (1982), "Optimum absorber parameters for various combinations of response and excitation parameters", *Earthq. Eng. Struct. D.*, **10**, 381-401.

Weber, B. and Feltrin, G. (2010), "Assessment of long-term behavior of tuned mass dampers by system identification", *Eng. Struct.*, **32**, 3670-3682.

Wen, Q., Hua, X.G., Chen, Z.Q., Yang, Y. and Niu, H.W. (2016), "Control of human-induced vibrations of a curved cable-stayed bridge: design, implementation, and field validation", *J. Bridge Eng.*, 10.1061/(ASCE) 1084-0702(2016), **21**(7), 04016028,1-13

Yang, J.N., Lei, Y. and Huang, N. (2004), "Identification of natural frequencies and damping of in situ tall buildings using ambient wind vibration data", *J. Eng. Mech.*, **130**(5), 570-577.

Zhu, J., Zhang, W., Zheng, K.F. and Li, H.G. (2016), "Seismic design of a long span cable-stayed bridge with fluid viscous dampers", *ASCE Practice Periodical on Structural Design and Construction*, **21**(1):04015006. 10.1061/(ASCE)SC.1943-5576.0000262.

BS

Engineering Structures 245 (2021) 112830

Contents lists available at ScienceDirect

Engineering Structures

journal homepage: www.elsevier.com/locate/engstruct

Optimal design of two viscous dampers for multi-mode control of a cable covering broad frequency range

Chao Yang [a,b], Zhengqing Chen [a], Wenxi Wang [a,*], Xugang Hua [a], Yafei Wang [a]

[a] Key Laboratory for Bridge and Wind Engineering of Hunan Province, College of Civil Engineering, Hunan University, Changsha, Hunan 410082, China

[b] Liuzhou Orient Engineering Rubber Poducts Co., Ltd, Liuzhou, Guangxi 545025, China

ARTICLE INFO

Keywords:
Cables vibration
Viscous dampers
Modal damping
Higher mode vibration
Two dampers

ABSTRACT

For the conventional design of viscous dampers attached to cables, only the first few modes are of interest. However, a recent cable vibration event occurred on the Sutong Bridge highlights that the higher mode vibrations of cables also need to be controlled. To this end, a novel optimal design of two viscous dampers near to a cable anchorage is explored which can broaden the control frequency range covering both the lower and higher modes. In the proposed design scheme, the one damper nearest to the cable anchorage is targeted for lower-mode control and the other damper is targeted for higher-mode control. The mode shape characteristics of a cable with two viscous dampers, which is prerequisite for optimal damper design, are investigated. Based on that, the optimal design of the two dampers in a cable is simply reduced to two individual designs of one single viscous damper, which can lower the design complexity. The control performance of the proposed design both for lower and higher vibration modes is numerically validated by a cable on the Sutong Bridge and compared with the conventional design considering only lower vibration modes. Finally, the effects of parametric variations of the two dampers in the proposed design are discussed. It is shown that the proposed design using two dampers can both increase damping ratios of cables for lower and higher vibration modes, and the acceleration response of the cable is effectively reduced to around 60% of that using the conventional design.

1. Introduction

Stay cables are light-weight, low-damped structures, which may be easily excited into large amplitude vibrations under a certain condition of wind and rain. As a critical component in cable-stayed bridges, undesired cable vibrations may significantly shorten the expected life of the cable itself and also threaten the safety of the bridges [1,2]. The methods of vibration control of stay cables mainly include aerodynamic measures [3,4], structural measures [5–8] and damping measures [9–12]. Using one viscous damper attached to the lower end of stay cables is the most common way to suppress the large amplitude vibrations. Several applications are reported such as the Sutong Bridge in China and the Russky Island Bridge in Russia [13–19].

The damping effect of viscous dampers on a stay cable has been well studied in the past few decades. Kovacs et al. [20] used the semi-analytical method to investigate the damping characteristics of the cable-viscous damper system and obtained a maximum attainable modal damping ratio. In 1993, Pacheco et al. [21] proposed a "universal curve" for estimating the modal damping ratios after determining the damper installation location and the damping coefficient, which is useful in the practical design of a viscous damper for lower vibration modes. Krenk [22] developed an analytical equation for the cable-damper system by assuming the mode shape function as two sinusoidal curves, and obtained an asymptotic solution of modal damping ratios for the first few modes. Main et al. [23] assumed another form of mode shape functions and proposed the same solution as Krenk [22] gave before. Moreover, the effects of cable sags, bending stiffness and inclination angles on the damping optimization of the cable-damper system have been considered and well studied [24–28]. With the help of the above works, the optimal damping coefficient and the maximum achievable damping ratio of a viscous damper for a specific cable mode can be easily obtained. Recently, the method of using two viscous dampers for cable control is proposed, and the damping effect of a cable with two viscous dampers at different locations has also been studied [29–31]. In particular, Xu and Yu [29] discussed the enhancement of the modal damping ratio of the 1st in-plane mode. Caracoglia and Jones [30] developed the closed-form solutions for small frequency shifts and discussed the evolution of the modal solutions for large frequency shifts. Hoang et al. [31] provided

* Corresponding author.
E-mail address: wxwang@hnu.edu.cn (W. Wang).

https://doi.org/10.1016/j.engstruct.2021.112830
Received 20 February 2021; Received in revised form 2 July 2021; Accepted 6 July 2021
Available online 23 July 2021

Table 1
Parameters of the cable.

Length (m)	Diameter (m)	Tension (N)	Mass per length (kg/m)	Calculated flexural frequency of nth mode (Hz)
471	0.142	5.2×10^6	78.5	$0.273 \times n$ (n greater than 1)

the asymptotic solution for lower modes when two dampers are both close to a support.

Most studies of damping enhancement of cables are currently focused on the first several vibration modes, such as achieving considerable damping ratios for the first three or five modes through utilizing a viscous damper [32–35]. However, more and more evidences show that the higher-modes vibrations of cables also need to be controlled [36–39]. As early as 2002, Main et al. [23,40] pointed out that the higher mode vibrations of cables should also be focused. In 2010, Zuo et al. [41] observed that the cable vibrations of up to 10 vibration modes frequently occur on the Fred Hartman Bridge. In particular, a field study to monitor the vortex-induced vibrations of stay cables of a bridge located in the East China Sea was reported by Chen et al. [42], and it is found that the larger vibration amplitudes were mainly located in the vibration mode range of 18th to 40th. The frequency of higher vibration modes of a cable may vary from 3 to 10 Hz. Obviously, the inertial force of higher modes is quite larger than that of lower modes when the displacement amplitude of each mode is identical. Therefore, suppressing the higher mode vibration is an urgent task for bridge designers.

The traditional design of using one viscous damper may not be able to control the lower and higher mode vibrations of cables at the same time. The previous studies of using two viscous dampers for cable control are only focused on enhancing the damping ratio for a certain vibration mode, and the damping effects for higher modes has not been investigated. In this study, the optimal design of two viscous dampers for multi-model control of cables is investigated, aiming at extending the control frequency range covering both the lower and higher modes. Firstly, a recent cable vibration event of higher modes with considerable amplitude on the Sutong Bridge is analyzed and the limitation of using one single viscous damper for multi-modes control of cables is clarified. As a prerequisite for optimal damper design, the effect of adding two viscous dampers in a cable on the mode shape is analyzed and verified by the computational results from the finite element software ANSYS. Based on the results of the mode shape analysis, a simplified estimation method is proposed and modified to precisely calculate the damping ratios of a cable controlled by two viscous dampers. Moreover, the optimal design of the two dampers in a cable is simply reduced to two individual designs of one single viscous damper, which can lower the design complexity. The control performance of the proposed design both for lower and higher modes is verified by a cable from the Sutong Bridge and compared with the conventional design only for lower vibration modes. Finally, the effects of canceling the limitation of the furthest installation location and exchanging the damping coefficients of the two dampers in the proposed design are discussed.

2. Limitation of multi-mode control using one single viscous damper

2.1. Higher mode vibrations of stay cables on the Sutong Bridge

Recently, higher mode vibrations of stay cables were observed on the Sutong Bridge (China) under a strong wind event, even viscous dampers are installed to control the first 5 modes of those cables. The in-plane acceleration response of a cable with length of 471 m was recorded and the wind speed of the bridge site was also measured. The parameters of the measured cable are listed in Table 1. The flexural frequency for nth mode is calculated by the following equation,

$$f = \frac{n}{2l}\sqrt{\frac{T}{m}} \tag{1}$$

where T is the cable force, l is the length of the cable, and m is the mass per unit length.

The measured cable on the Sutong Bridge and the layout of the sensors are shown in Fig. 1.The accelerometer (model: DH1A202E, Donghua Tech. Company) with the sampling frequency of 100 Hz is installed at the height of 15 m vertically above the bridge deck and two 2-dimensional anemometers (model: CFF3D-1, Zhixiangyu Instrument Equipment Company) with sampling frequency of 1.0 Hz were installed at the level of deck and tower top to measure the real-time wind speed. The measurement results in Fig. 2 show that the wind speed at the top of the tower is around 30 to 40 m/s and that on the bridge deck is about 10 to 20 m/s. Moreover, the maximum acceleration amplitude is close to 15 m/s^2 and the corresponding dominant frequencies are located around 2.46 to 6.00 Hz, which are the 9th to 22nd in-plane vibration modes. The cable vibrations with higher frequencies may rapidly shorten the lifecycle of the structural members, and result in severer damages between the cable and its anchorage, as shown in Fig. 3.

2.2. Computation of damping ratios for higher modes

To analyze the higher modal damping ratios of the cable with a viscous damper, the conventional design curve [21] and asymptotic formula [22] may be inapplicable due to the assumption of lower vibration modes. Therefore, the iterative method proposed by Krenk [22] is used to compute the cable damping ratios of both lower and higher modes. The mechanical model of the cable with a viscous damper is plotted in Fig. 4. The cable force is T, the length of the cable is l, and the mass per length is m. The viscous damper is installed at the distance a away from the anchorage.

The partial differential equation of the cable with a viscous damper is expressed as,

$$T\frac{\partial^2 v}{\partial x^2} - m\frac{\partial^2 v}{\partial t^2} = c\frac{\partial v}{\partial t}\cdot\delta(x-a) \tag{2}$$

where v is the transverse displacement of the cable, c is the damping coefficient of the damper, and δ is the Dirac delta function.

As proposed by Krenk [22], the following complex frequency equation can be obtained when the assumed mode shapes are divided into two sinusoidal functions respectively for the left and right part of the damper location,

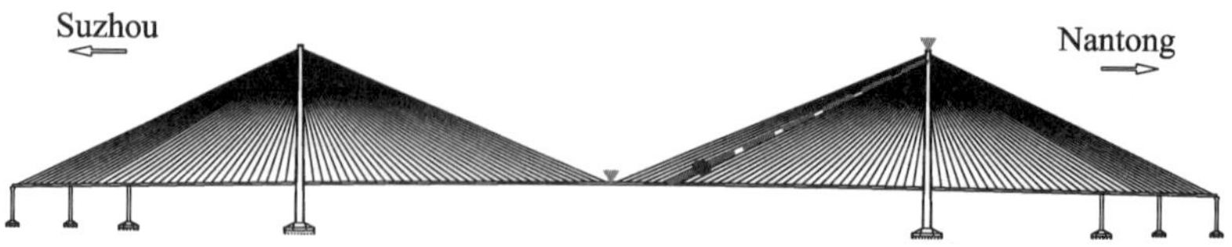

Fig. 1. The measured cable and distribution of sensors on the Sutong Bridge.

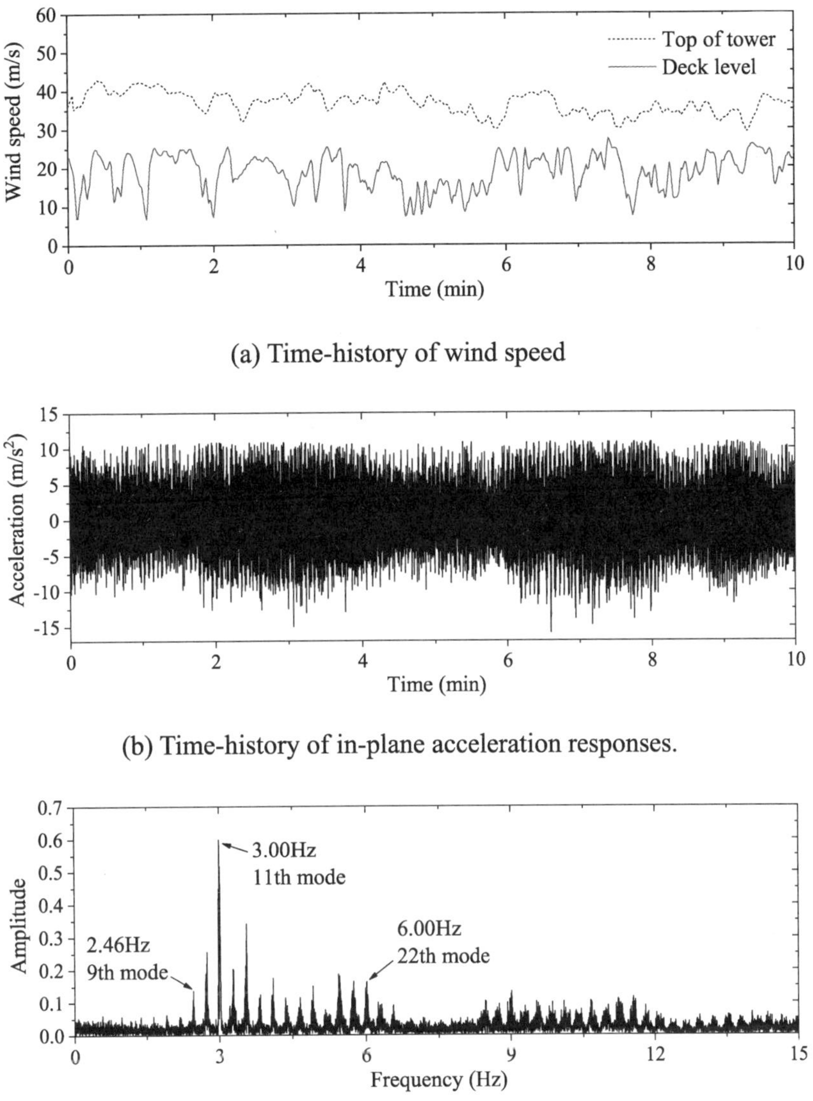

(a) Time-history of wind speed

(b) Time-history of in-plane acceleration responses.

(c) FFT spectrum of in-plane acceleration responses.

Fig. 2. Field measurement data of a cable vibration event of higher modes on the Sutong Bridge. (a) Time-history of wind speed, (b) Time-history of in-plane acceleration responses, (c) FFT spectrum of in-plane acceleration responses.

$$tan(\beta l) = \frac{i\eta sin^2(\beta a)}{1 + i\eta sin(\beta a) sin(\beta a)} \tag{3}$$

where $\beta_n = \omega_n(m/T)^{0.5}$ represents the wave number, $\eta = c/(Tm)^{0.5}$ is the dimensionless damping coefficient, and i represents the complex unit. ω_n is the complex eigen-frequency and it can be expressed as,

$$\omega_n = |\omega_n|\left(\sqrt{1-\zeta_n^2} + i\zeta_n\right) \tag{4}$$

where $|\omega_n|$ is the magnitude of the angular frequency, and ζ_n is the nth modal damping ratio. Eq. (3) can be solved by an iteration process. An initial value of β_n can be substituted into the right hand of Eq. (3) to obtain an iterative value at the left hand. The entire process can be expressed as,

$$\beta_n^{j+1} l = n\pi + arctan\left(\frac{i\eta sin^2(\beta_n^j a)}{1 + i\eta cos(\beta_n^j a) sin(\beta_n^j a)}\right), \ j = 0, 1, \cdots \tag{5}$$

As suggested by Krenk [22], the initial value of β_n can be set to $\beta_n^0 = n\pi/l$. The damping ratio ζ_n of nth mode can be solved from the ratio of Im(β_n)/Re(β_n).

2.3. Optimal damping coefficients of lower and higher modes

If the result of the first iteration βn^1 is regarded as the final iteration result, the optimal damping coefficient $\eta_{opt\ n}$ to achieve the maximum modal damping ratio for nth mode can be determined by the following equation,

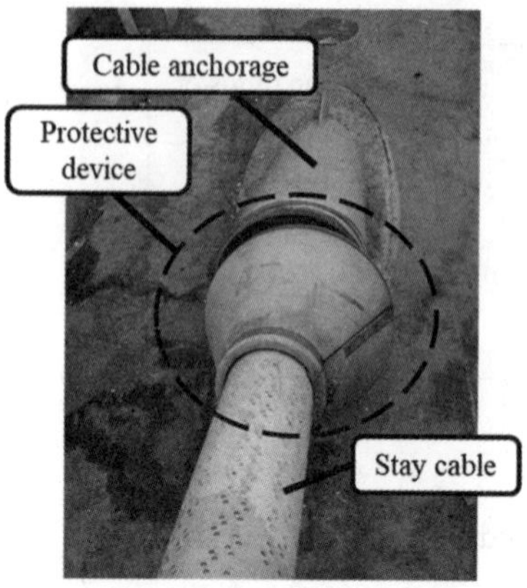

Fig. 3. Damage of protective device for the measured stay cable at the anchored end.

$$\frac{\partial(Im(\beta_n^1 l))}{\partial(\eta)}=0 \tag{6}$$

By solving Eq. (6), the expression of the optimal dimensionless damping coefficient targeted to nth mode can be obtained as,

$$\eta_{\text{opt } n}=\frac{1}{\left|\sin\left(\frac{n\pi a}{l}\right)\right|} \tag{7}$$

Eq. (7) is similar to the expression of the optimal dimensionless damping coefficient proposed by Krenk [22]. The Krenk's equation has a form of $1/\sin(n\pi a/l)$ rather than $1/|\sin(n\pi a/l)|$. Therefore, different from the Krenk's equation, using Eq. (7) to calculate the optimal dimensionless damping coefficient for higher modes will not result in a negative value. It should be noted that Eq. (7) is an approximation of the optimal dimensionless damping coefficient determined from the first iteration βn^1, and the accuracy for the modes higher than 5 has not been examined in the previous studies. To verify the accuracy of Eq. (7), a numerical study is performed to obtain the relation between η and ζ_n through Eqs. (3) to (5), as shown in Fig. 5. The convergence condition of the iteration is set to $(\beta_n^{j+1}-\beta_n^j)/\beta_n^j<1\%$. Moreover, the optimal damping coefficient targeted to each mode calculated by Eq. (7) is also plotted in Fig. 5. The inherent damping of the cable for each mode is set to 0. In the numerical simulation, a/l is set to 0.01. It can be seen that the calculated result from Eq. (7) agrees well with the peaks of the curves obtained

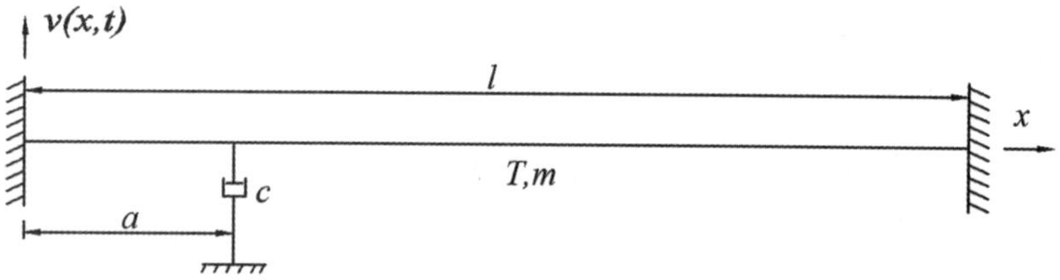

Fig. 4. Mechanical model of a cable with a viscous damper.

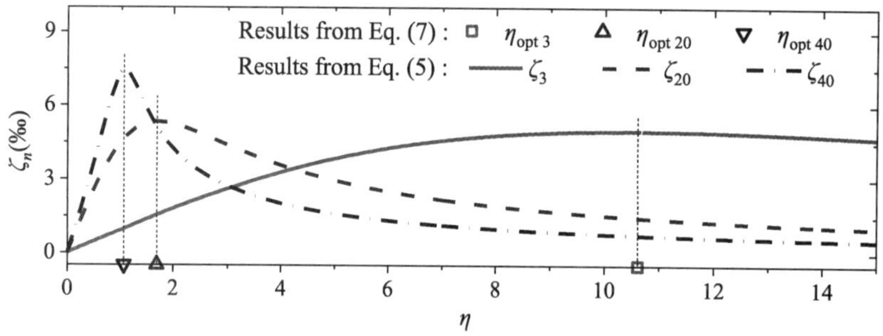

Fig. 5. Damping ratios of several modes for $a/l=0.01$ against different dimensionless damping coefficients.

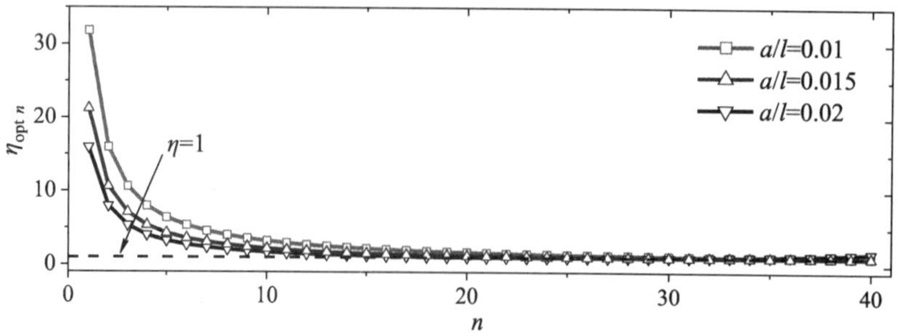

Fig. 6. Optimal dimensionless damping coefficients of each mode for different value of a/l.

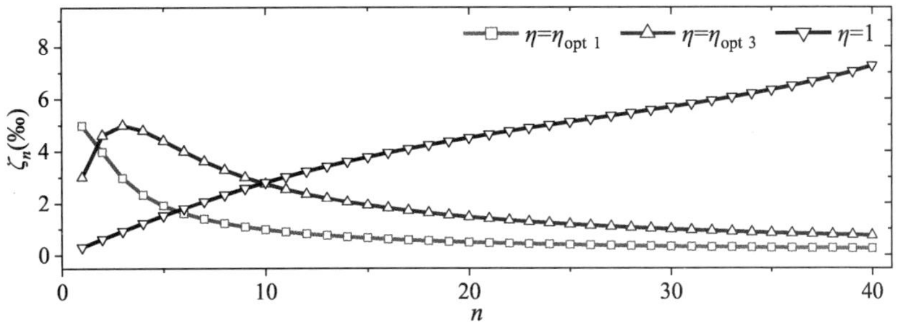

Fig. 7. Modal damping ratios of the first 40 modes for $a/l = 0.01$.

from multiple iterations. It indicates that although Eq. (7) is obtained by using the result of the first iteration β_n^1 in Eq. (5), it is also accurate enough to calculate the optimal damping coefficient for each mode, including lower and higher modes. Additionally, it is important to note that the optimal damping coefficients are different respected to different vibration modes. The damping ratio of *n*th mode can only reach a relatively high level when the dimensionless damping coefficient η is set approaching to $\eta_{opt\ n}$ which is the optimal dimensionless damping coefficient targeted to *n*th mode.

According to Eq. (7), the optimal dimensionless damping coefficients $\eta_{opt\ n}$ respectively tuned to *n*th mode for optimal control is shown in Fig. 6. It can be seen that only the values of $\eta_{opt\ n}$ targeted to the first several modes are relatively large and decrease rapidly with the increase of the controlled mode. However, the values of $\eta_{opt\ n}$ targeted to higher modes are approaching to 1 even for different modes. Therefore, it indicates that an identical value of $\eta_{opt\ n} = 1$, can be designed to control the most of higher modes for different values of a/l.

2.4. *Modal damping characteristics of a cable with one viscous damper*

In this section, modal damping ratios of a cable with a damper installed close to the anchorage are obtained to study the damping contribution of the damper to each mode. Modal damping ratios of the first 40 modes for different η of a cable with one single viscous damper are shown in Fig. 7. The installation location a/l is set to 0.01, and the inherent damping of the cable for each mode is assumed to 0. The dimensionless damping coefficient η of the damper is approximately set to $\eta_{opt\ 1}$ and $\eta_{opt\ 3}$ respectively tuned to the 1st and 3rd mode. Moreover, the damping ratios of the case $\eta = 1$ targeted to higher modes, as suggested by Fig. 6, is also plotted in Fig. 7. It can be found that only the first few modal damping ratios are significantly increased if the damper is optimally designed for the 1st and 3rd mode. However, the damping ratios of the modes higher than 10th are not effectively improved and only remain at a relatively low value. The results indicate that only considering the lower modes may be the main reason why the damper system installed on the Sutong Bridge fails to suppress the large amplitude vibrations of higher modes. Additionally for the case of $\eta = 1$, the damping ratios of higher modes in a broad range are significantly improved. However, the damping ratios of lower modes, such as the first 5 modes, are not significantly increased, even smaller than 0.2%.

3. Novel design of two viscous dampers for multi-mode control

The conventional optimal design of a viscous damper for cable control aims at enhancing the damping ratios of the lower modes, such as the first 5 modes. Therefore, the dimensionless damping coefficient η is normally tuned to $\eta_{opt\ 3} \sim \eta_{opt\ 1}$ to guarantee the damping ratios of the first 5 modes achieving a relatively high level [32,33]. However, it is found that the conventional design cannot effectively control higher modes. In such a case, one more damper should be added to enhance the damping ratios and mitigate the responses of higher modes.

In this section, we propose a methodology for optimal design of two viscous dampers which can extend the control frequency range covering both the lower and higher modes of a cable. The mechanical model of a cable with two viscous dampers is plotted in Fig. 8. The damper 1 is installed at the location of a_1 away from the left anchorage for controlling the lower modes, and the damper 2 installed at the location of a_2 away from the left anchorage is deployed for controlling the higher modes. In this study, we have $a_2 > a_1$, and the dimensionless damping coefficients for the two dampers can be rewritten as $\eta_i = c_i/(Tm)^{0.5}$, for $i = 1, 2$. Therefore, the dimensionless damping coefficients of the two dampers in the proposed design should be set as

$$\begin{cases} \eta_1 = \eta_{1,\ opt\ 3} - \eta_{1,\ opt\ 1} \\ \eta_2 \approx 1 \end{cases} \tag{8}$$

where $\eta_{i,opt\ n}$ is the optimal dimensionless damping coefficient targeted to *n*th mode when the damper *i* is installed alone on the cable.

3.1. *Mode shape characteristics of lower and higher modes*

The mode shapes of a cable with two dampers shown in Fig. 8 are different from that with one damper. The selection of the damping coefficients (or the dimensionless damping coefficients) will dramatically affect the mode shapes. Therefore, in this section, the mode shapes of a cable with two dampers suggested as the proposed design are studied. As mentioned by Krenk [22], the mode shapes of a cable with one damper

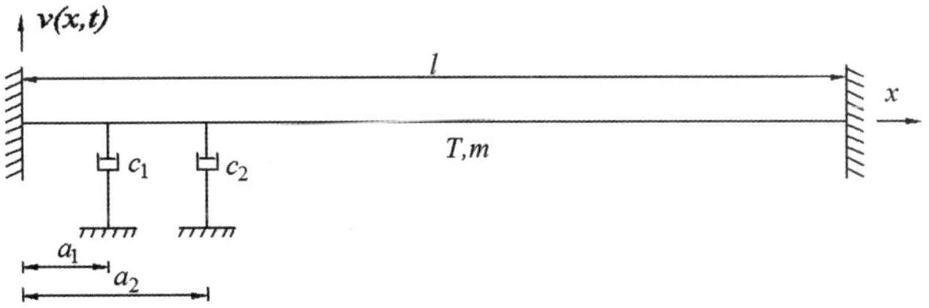

Fig. 8. Mechanical model of a cable with two viscous dampers near the anchorage.

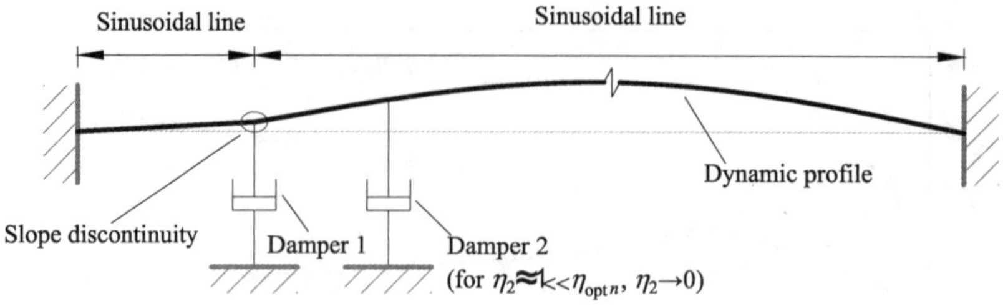

(a) Mode shapes of lower modes

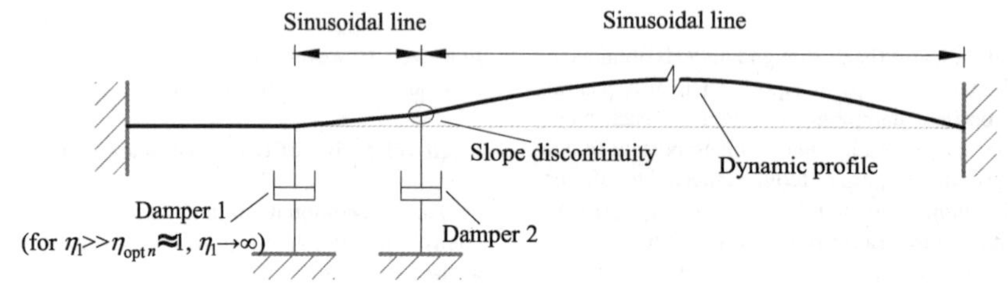

(b) Mode shapes of higher modes

Fig. 9. Mode shapes simplified by separately considering lower modes and higher modes. (a) Mode shapes of lower modes. (b) Mode shapes of higher modes.

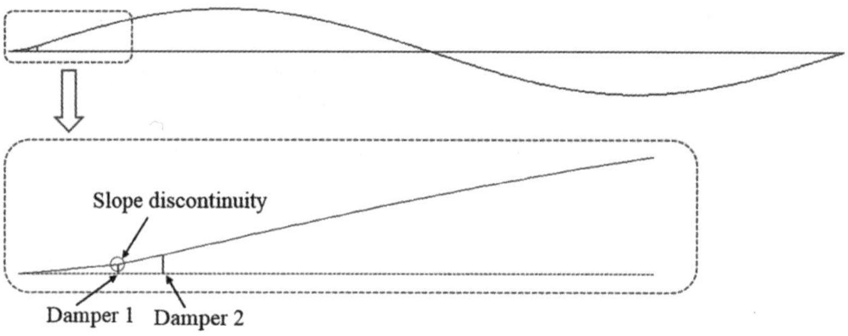

(a) 2nd mode shape.

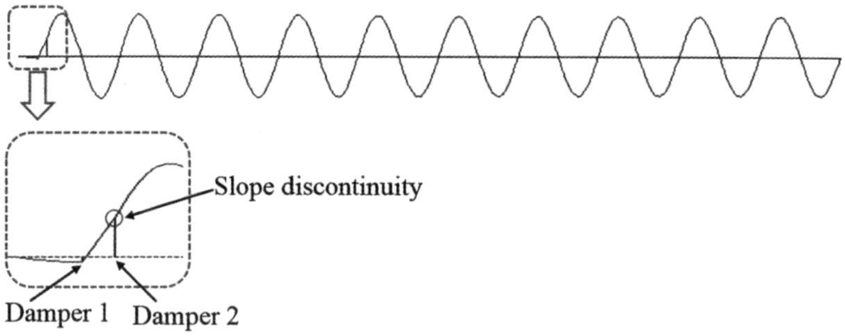

(b) 20th mode shape.

Fig. 10. Results of mode shape analysis by ANSYS. (a) 2nd mode shape, (b) 20th mode shape.

suggested as the conventional proposed design can be divided into two different sinusoidal curves, respectively in the left and right sides of the installation location. Obviously, there is a slope discontinuity on the mode shape at the installation location of the damper. The mode shapes of a cable with two dampers are even more complicated. In order to simplify the study, the authors assume that $\eta_i \to \infty$ if η_i is much larger

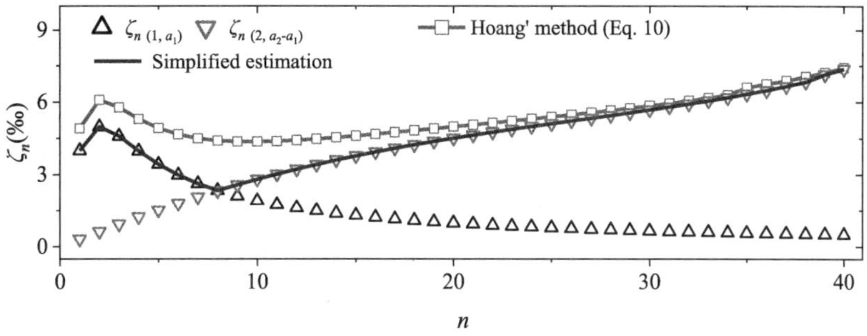

Fig. 11. The first 40 Modal damping ratios through the proposed simplified estimation and the iteration method.

than $\eta_{i,\text{opt}\ n}$. On the contrary, it is assumed that $\eta_i \to 0$ if η_i is much smaller than $\eta_{i,\text{opt}\ n}$. With the help of the above assumption, we can discuss the mode shapes of a cable with two dampers suggested as the proposed design by the following two cases,

Considering mode shapes of lower modes: Since the dimensionless damping coefficient of the damper 2 $\eta_2 \approx 1$ is much smaller than the optimal dimensionless damping coefficients of lower modes, we can assume $\eta_2 \to 0$ in this case. The mode shapes of the cable with the two dampers are the same as that of the cable only with the damper 1, as shown in Fig. 9(a).

Considering mode shapes of higher modes: Because that damper 1 is optimally designed for a lower mode, η_1 is much greater than $\eta_{1,\text{opt}\ n}$ of the higher modes which are approaching to 1. Therefore, we can assume $\eta_1 \to \infty$. In this case, the damper 1 is regarded as a fixed support, and the mode shapes of the cable with two dampers for higher modes can be simplified into that only with the damper 2 at the location of a_2-a_1, as shown in Fig. 9(b).

The mode shapes of a cable with two viscous dampers are studied to verify the aforementioned hypothesis by a finite element software ANSYS. The cable parameters are listed in Table 1. The damper 1 and 2 are installed at the position of 11 m and 17 m away from the left anchorage, respectively. The damping coefficient of the damper 1 is set to $c_1 = (Tm)^{0.5} \cdot \eta_{1,\text{opt}\ 2} = 138200$ N·s/m, which is exactly targeted to the 2nd mode. The damping coefficient of the damper 2 is set to $c_2 = 20000$ N·s/m that gives rise to $\eta_2 \approx 1$ for higher mode control. The link element (LINK 8) in the ANSYS is used to model the cable. The entire cable is uniformly divided into 1000 elements and each one is about 4.71 m. Two damper elements (COMBIN 14), which can simulate the viscous damper by inputting the damping coefficient, are attached to the cable. The two anchorages are fixed, and the complex modal analysis is performed. The mode shapes of 2nd and 20th modes are respectively plotted in Fig. 10. It is shown that the results from the ANSYS software agree well with the aforementioned assumption.

3.2. *Simplified estimation of modal damping ratios*

Based on the characteristics of mode shapes with two dampers respectively for lower and higher modes control, the modal damping ratios of a cable with two dampers can be simply estimated by separately considering lower modes and higher modes. As shown in Fig. 9(a), we consider $\eta_2 \to 0$. Therefore, the effect of the damper 2 can be neglected while calculating the modal damping ratios of lower modes. The modal damping ratios of lower modes are mainly provided by the damper 1 installed at a_1. Moreover, as shown in Fig. 9(b), we consider $\eta_1 \to \infty$. The damper 1 is regarded as an approximate support. The mechanical model in such a case can be regarded as a cable of length l-a_1 only with the damper 2 installed at a_2-a_1 away from the installation position of the damper 1. However, the change of the cable length can be ignored due to $l \gg a_1$, we still consider that the length of the cable is l. Therefore, based on the damping ratios of a cable with one damper, the modal damping ratios of a cable equipped with two dampers can be estimated by

$$\zeta_n \approx \begin{cases} \zeta_{n\ (1,\ a_1)} & \text{for lower modes} \\ \zeta_{n\ (2,\ a_2-a_1)} & \text{for higher modes} \end{cases} \tag{9}$$

where $\zeta_{n(1,\ a1)}$ repents the modal damping ratios when only the damper 1 is installed at a_1, $\zeta_{n(1,\ a2\text{-}a1)}$ repents the modal damping ratios when only the damper 2 is installed at a_2-a_1.

With the help of Eq. (9), the damping ratio of a cable with two dampers can be estimated by that with only one damper. The computation of damping ratio for one damper case can be computed by the iteration method. The proposed simplified estimation for damping ratios of a cable with two dampers is verified with the comparisons to the solutions proposed by Hoang et al. [31]. In this paper, the iteration method from Hoang et al. [31] is employed to solve the complex eigenfrequency function, and we have the iteration equation as,

$$\beta_n^{j+1} l = n\pi + \arctan \left(\frac{-\eta_1\eta_2 \sin\beta_n^j a_1 \sin\beta_n^j a_2 \sin\beta_n^j (a_2 - a_1) + \mathrm{i}(\eta_1 \sin^2\beta_n^j a_1 + \eta_2 \sin^2\beta_n^j a_2)}{1 - \eta_1\eta_2 \sin\beta_n^j a_1 \cos\beta_n^j a_2 \sin\beta_n^j (a_2 - a_1) + \mathrm{i}(\eta_1 \sin\beta_n^j a_1 \cos\beta_n^j a_1 + \eta_2 \sin\beta_n^j a_2 \cos\beta_n^j a_2)} \right), \quad j = 0, 1, \cdots \tag{10}$$

The initial iteration value for Eq. (10) is still set to $\beta_n^0 = n\pi/l$. The convergence condition of the iteration is set to $(\beta_n^{j+1}$-$\beta_n^j)/\beta_n^j < 1\%$. The modal damping ratios ζ_n of the cable with two dampers will be solved from the ratio of $\mathrm{Im}(\beta_n)/\mathrm{Re}(\beta_n)$. The damping ratios of the cable with the damper 1 installed at a_1 and the damper 2 installed at a_2-a_1 are also computed separately, and the result is plotted in Fig. 11. The damper 1 is installed at $a_1/l = 0.01$ and η_1 is set to $\eta_{1,\text{opt}\ 2}$ for optimal control of the 2nd mode. The damper 2 is installed at $(a_2$-$a_1)/l = 0.01$ and η_2 is set to 1 for higher modes control. Based on the conception of the damping estimation from Eq. (9), a specific expression for damping estimation of each mode is proposed as,

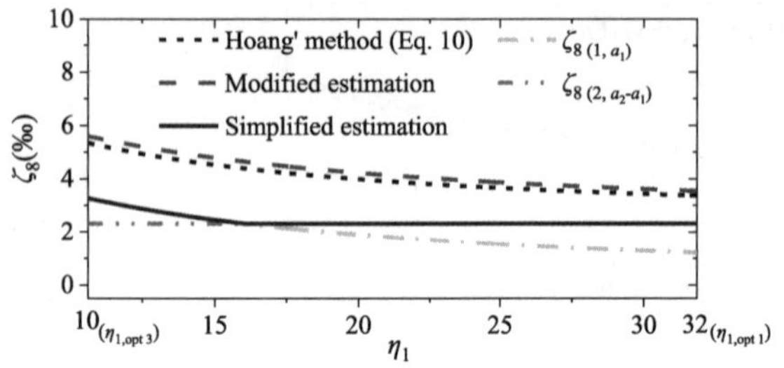

Fig. 12. Damping ratio of the 8th mode computed by different methods.

$$\zeta_n \approx \max\left(\zeta_{n(1,\ a_1)}, \zeta_{n(2,\ a_2-a_1)}\right) \tag{11}$$

By using Eq. (11), the simplified estimation curve is plotted in Fig. 11, and the results from Hoang's method are also plotted as a reference. Fig. 11 shows the modal damping ratios computed by Hoang's method, are larger than that from the proposed simplified estimation. It is indicated that the proposed method can be used to conservatively calculate the damping ratios of a cable with two dampers.

3.3. Modification of the simplified estimation method

It is shown in Fig. 11 that there are some errors in calculating modal damping ratios by the simplified estimation Eq. (11). In particular, the estimated modal damping ratios of the intermediate modes, such as the 8th mode, are significantly lower than the accurate results from the

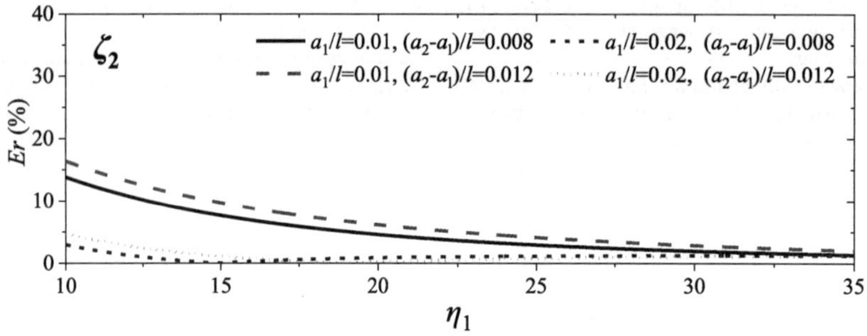

(a) 2nd mode.

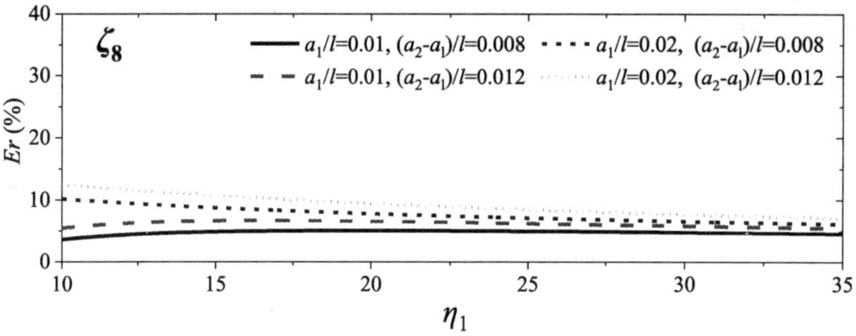

(b) 8th mode.

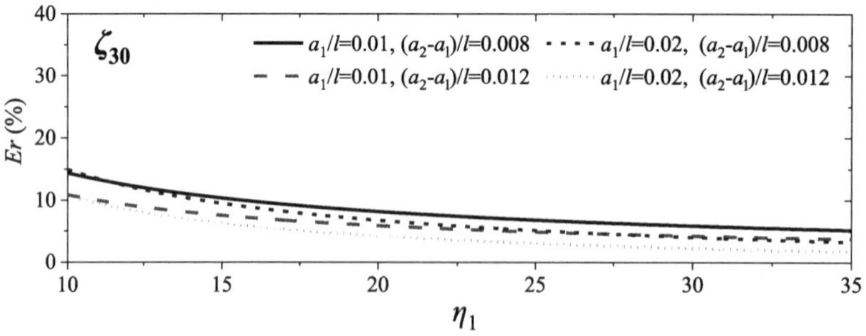

(c) 30th mode.

Fig. 13. Relative errors of modified estimation in calculating damping ratios of several modes when the two dampers are installed in different positions for η_1 varies from 10 to 35. (a) 2nd mode, (b) 8th mode, (c) 30th mode.

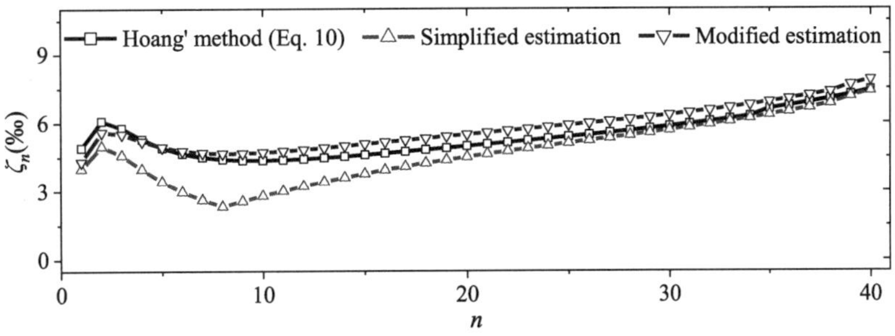

Fig. 14. Damping ratios of the first 40 modes computed by different methods.

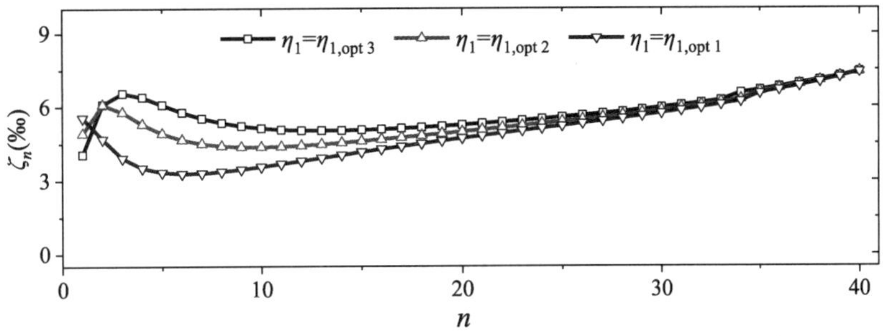

Fig. 15. Differences of the modal damping ratio distributions for different η_1.

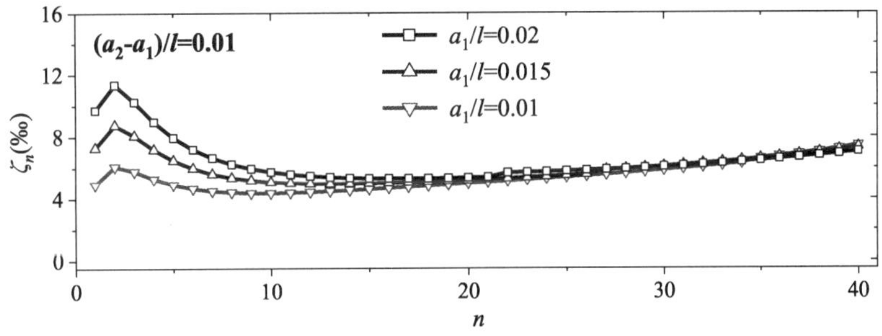

(a) Modal damping ratios for different values of a_1/l under $(a_2$-$a_1)/l$=0.01.

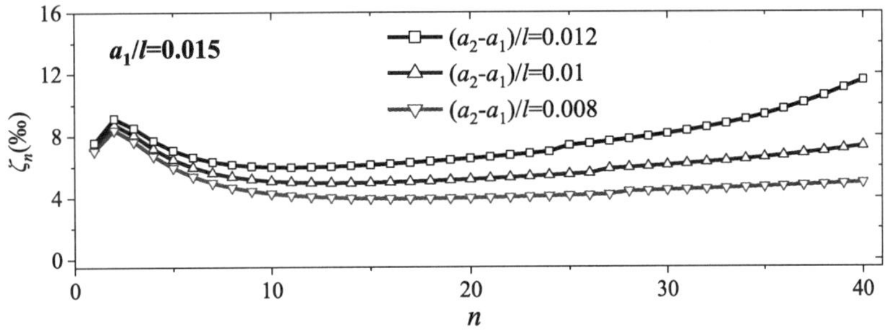

(b) Modal damping ratios for different values of $(a_2$-$a_1)/l$ under a_1/l=0.015.

Fig. 16. Modal damping ratios computed by Eq. (13) for different locations of the two dampers. (a) Modal damping ratios for different values of a_1/l under $(a_2\text{-}a_1)/l = 0.01$, (b) Modal damping ratios for different values of $(a_2\text{-}a_1)/l$ under $a_1/l = 0.015$.

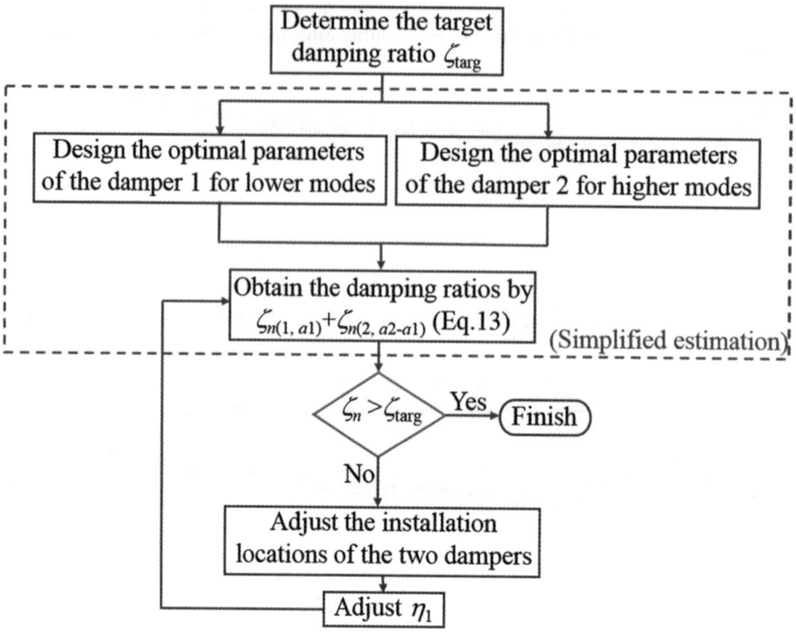

Fig. 17. Flowchart of parametric optimization for a cable with two dampers using the proposed design.

Hoang's method. It is indicated that the ignorance of the damping contribution of the damper 2 to lower modes and that of the damper 1 to higher modes may significantly affect the estimation results. If the aforementioned reasons are considered, a modified expression of Eq. (9) can be written as,

$$\zeta_n \approx \begin{cases} \zeta_{n\,(1,\,a_1)} + \zeta_{n\,(2,\,a_2-a_1)} & \text{for lower modes} \\ \zeta_{n\,(2,\,a_2-a_1)} + \zeta_{n\,(1,\,a_1)} & \text{for higher modes} \end{cases} \tag{12}$$

Obviously, Eq. (12) can be unified into a simple form as

$$\zeta_n \approx \zeta_{n\,(1,\,a_1)} + \zeta_{n\,(2,\,a_2-a_1)} \tag{13}$$

The damping ratio of the 8th mode, which is the mode with the largest error in Fig. 11, is re-calculated by the modified estimation Eq. (13) in Fig. 12 when η_1 is changed from 10 to 35 with an interval of 0.01. The results from the simplified estimation Eq. (11) and the Hoang's method from Eq. (10) are also plotted in Fig. 12. The other parameters of the two dampers are set as $a_1/l = 0.01$, $(a_2\text{-}a_1)/l = 0.01$ and $\eta_2 = 1$. It is shown from Fig. 12 that the modified estimation can reach a good accuracy even for different values of η_1 compared to the Hoang's method.

A relative error index for quantitatively evaluating the estimation accuracy can be defined as

$$Er = \frac{|\text{estimated value} - \text{exact value}|}{\text{exact value}} \tag{14}$$

Fig. 13 shows the relative errors of the modified estimation from Eq. (13) in calculating ζ_2, ζ_8 and ζ_{30} when the two dampers are installed in different positions for η_1 varying from 10 to 35. It can be seen that relative errors are basically within 15% when η_1 greater than 10 in Fig. 13. Moreover, since we assume that η_1 is much larger than $\eta_2 = 1$, Er decreases with the increase of η_1. For example in Fig. 13(a), Er is about 10% when $\eta_1 = 15$, and decreases to 7% when $\eta_1 = 20$.

The modified estimation Eq. (13) is applied to calculate the first 40 modal damping ratios, and the results are shown in Fig. 14. The parameters of the two dampers are the same as that for calculation of Fig. 11. It is obvious that the results from the modified estimation are more accurate than that from the simplified estimation. Therefore, the modified estimation Eq. (13) is suggested to be used to calculate the damping ratios for both lower and higher modes.

3.4. Parametric optimization

In order to reasonably design the parameters of the two dampers, the damping performance under different parameters of the two dampers is discussed. The damper parameters include the dimensionless damping coefficient of the damper 1 and installation location of the two dampers. It should be noted that the dimensionless damping coefficient of the damper 2 remains to $\eta_2 = 1$ for optimal control of higher modes according to Fig. 6.

3.4.1. *Damping performance under different η_1*

As shown in Eq. (8), the dimensionless damping coefficient of the damper 1 may vary from $\eta_{1,\text{opt }3}$ to $\eta_{1,\text{opt }1}$. The differences of the distributions for the first 40 modal damping ratios are calculated for $a_1/l = (a_2\text{-}a_1)/l = 1$, and the results are shown in Fig. 15. The inherent damping of the cable for each mode is still set to 0 in the numerical simulation. It is shown that the damping ratio of each mode is quite similar and remains at a relatively high value when η_1 is set to $\eta_{1,\text{opt }2}$ or $\eta_{1,\text{opt }3}$. It is worthy to notice that the change of η_1 has limited effects on the damping ratio of higher modes.

3.4.2. *Damping performance under various installation locations*

The installation locations of the two dampers also affect the performance of the proposed design with two dampers. Therefore, the modal damping ratios for different installation locations are compared in Fig. 16. The dimensionless damping coefficient η_1 of the damper 1 is set to $\eta_{1,\text{opt }2}$ for optimal control of the 2nd mode considering different installation positions, and η_2 is still set to 1 for higher modes control. Fig. 16(a) shows the increase of the damping ratios of lower modes can be achieved by increasing the installation location of the damper 1, and increasing a_1 only slightly affect the damping ratios of the higher modes. However, as shown in Fig. 16(b), increasing the distance between the damper 2 and damper 1 can improve the damping ratios of higher modes, and the variation of $a_2\text{-}a_1$ only has limited effects on the damping ratios of lower modes.

Table 2
Parameters of the two designs.

Scheme		a/l	a (m)	η	c (N·s/m)
Conventional design		0.02	9.42	7.3	1.47×10^5
Proposed design	Damper 1	$a_1/l=0.01$	$a_1=4.71$	15.9	3.20×10^5
	Damper 2	$(a_2\text{-}a_1)/l=0.01$	$a_2\text{-}a_1=4.71$	1	2.01×10^4

3.4.3. Optimization procedures

Generally, the damping ratios of a cable with two dampers can be optimized through changing the installation locations and the damping coefficients of the two dampers. In the proposed estimation, the optimization process of a cable with two dampers can be changed into that of two individual cables respectively with one single viscous damper. The proposed estimation reduces the computation complexity of the optimization process, which is formerly an optimization with four varied parameters and now changed into two optimization processes respectively with only two varied parameters.

The framework of parametric optimization for a cable with two dampers is illustrated in Fig. 17, aiming at extending the control frequency range covering both the lower and higher modes. Firstly, a target damping ratio ζ_{targ} for all interested modes can be determined based on the engineering requirements. Secondly, the parameters of the damper 1 can be optimized by the previous method, such as the "universal curve" from Pacheco [21] or the asymptotic solution from Krenk [22].The damping ratios of the lower modes, such as the first 5 modes, should be optimized larger than ζ_{targ}. Meanwhile, an appropriate value of $(a_2\text{-}a_1)/l$ can be selected to make the damping ratios of higher modes larger than ζ_{targ} by the iteration solutions from Krenk [22] when η_2 is set approaching to1. Thirdly, the total modal damping ratios can be obtained by letting $\zeta_{n(1,a1)}$ plus with $\zeta_{n(2,a2\text{-}a1)}$ through Eq. (13). Finally, the

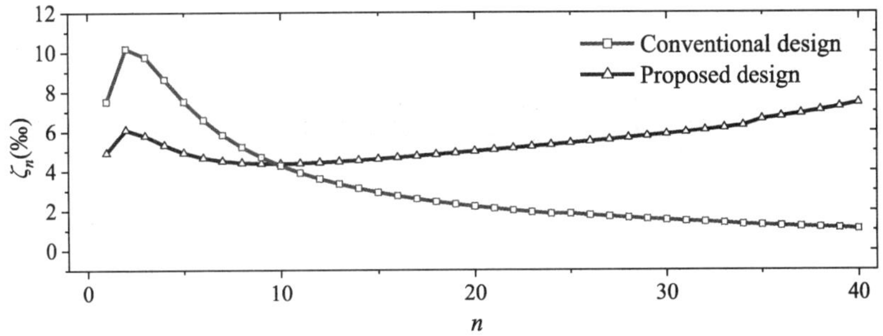

Fig. 18. Modal damping ratios computed by Eq. (13) of the first 40 modes controlled by the proposed and the conventional designs.

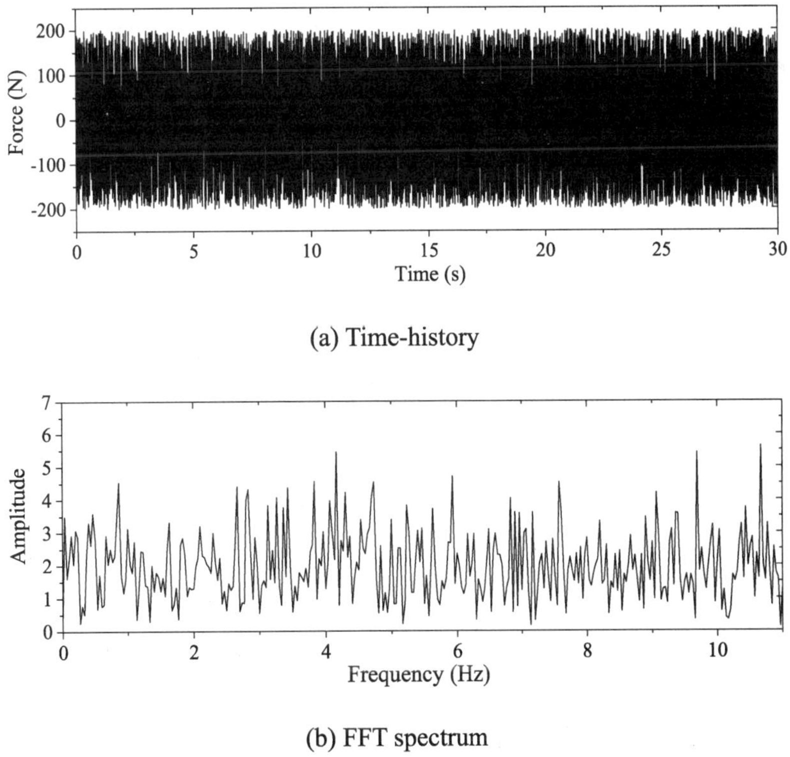

Fig. 19. A typical white noise excitation (a) Time-history, (b) FFT spectrum.

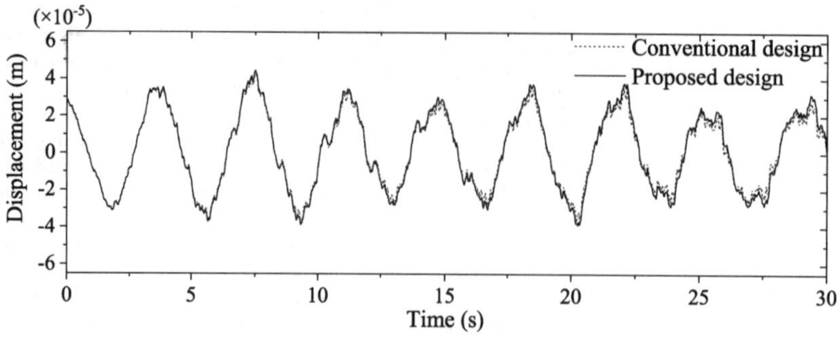

(a) Time-history of displacement responses.

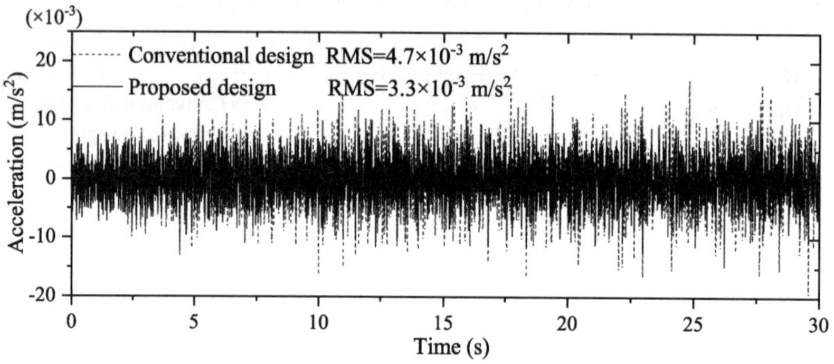

(b) Time-history of acceleration responses.

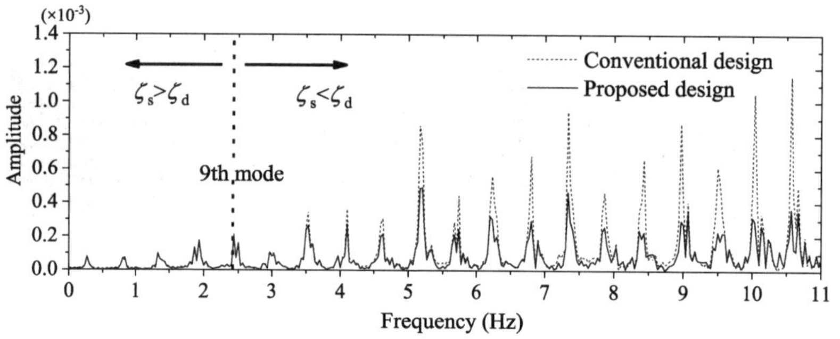

(c) FFT spectra of the acceleration responses.

Fig. 20. Responses under a typical white noise excitation controlled by the proposed and the conventional designs. (a) Time-history of displacement responses, (b) Time-history of acceleration responses, (c) FFT spectra of the acceleration responses.

modal damping ratios of all interested modes need to be judged whether they are greater than ζ_{targ}. If $\zeta_n \geq \zeta_{targ}$, the parametric optimization are finished. If $\zeta_n < \zeta_{targ}$, the installation locations of the two dampers should be adjust. η_1 is also need to be updated correspondingly. The total modal damping ratios should be recalculated by letting the adjusted value $\zeta_{n(1,a1)}$ plus with the $\zeta_{n(2,a2\text{-}a1)}$.

4. Performance comparison between the proposed and conventional design

The control performance of the proposed control design with two dampers both for lower and higher modes is compared with the conventional design only using one single damper for lower modes. The parameters of the cable used in simulations are listed in Table 1. For the conventional design using one single damper, the installation position is set to $a/l = 0.02$. To control all of the first five mode vibrations, the damping coefficient of the damper is chosen according to the equation as $\zeta_1 = \zeta_5$ as mentioned by Weber et al. [32]. In such a design, c is approaching to $c_{opt\ 2}$, which is the optimal damping coefficient targeted to 2nd mode. The damping ratios of the first 5 modes can achieve a considerable level at the same time. For the proposed design using two dampers, the damper 1 is installed at the position of $a_1/l = 0.01$, and the damper 2 is installed at $(a_2\text{-}\ a_1)/l = 0.01$. The damping coefficient of damper 1 is still set to $c_{1,opt\ 2}$, and the dimensionless damping coefficient of the damper 2 η_2 is set to 1 for higher modes control. The detailed parameters of the dampers for the two designs are respectively listed in

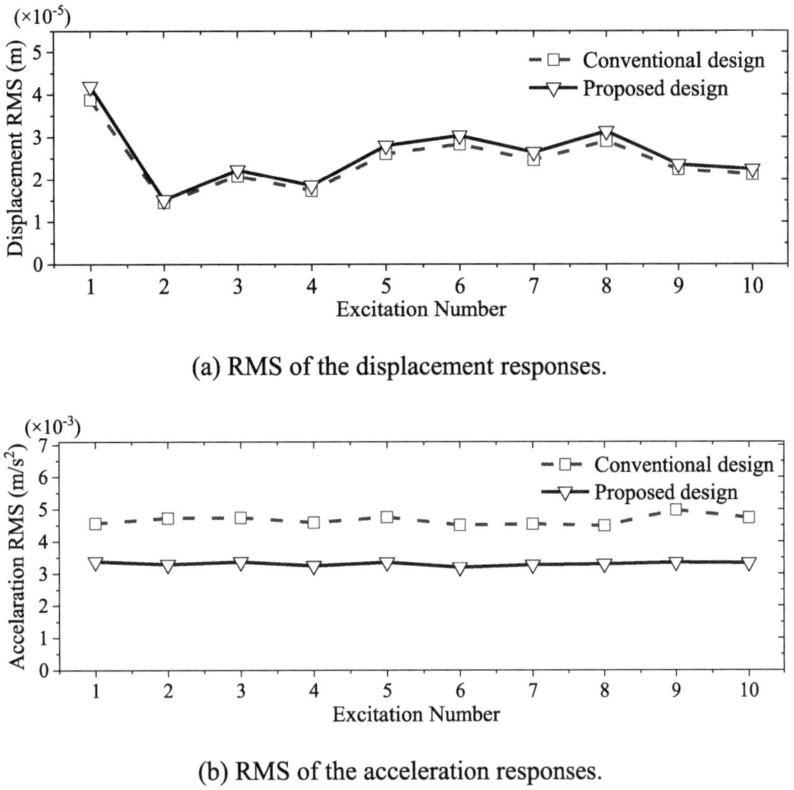

Fig. 21. RMS of the responses under 10 different records of the random white noise excitations. (a) RMS of the displacement responses, (b) RMS of the acceleration responses.

Table 2. The first 40 modal damping ratios of the two designs computed by Eq. (13) are shown in Fig. 18. The inherent damping ratio of the bare cable for each mode is set to 0. It can be seen that the damping ratios for the all 40 modes are larger than 0.4% with the proposed design using two dampers. However, the conventional design using one single damper fails to provide enough damping for the higher modes.

4.1. Responses under white noise excitation

The displacement and acceleration response of the cable controlled by the proposed and the conventional designs is studied under white noise excitations. The time history and Fourier spectrum of a typical white noise are shown in Fig. 19. The mode shapes of a cable can be considered as sin $(n\pi a/l)$. The mid-span point of the cable is the location of the largest modal displacement for all the symmetrical modes. Therefore, the white noise excitations are applied at the mid-span point, and the time interval in the simulations is set to 1/300 s to precisely compute the cable response. The mode-superposition method (considering the first 40 modes) is selected to obtain the dynamic responses of the cable. The modal damping ratio of each mode in the conventional and proposed designs is inputted according to Fig. 18.

The responses at the mid-span point under a typical white noise excitation are shown in Fig. 20. It can be seen from Fig. 20(a) that the displacement responses respectively controlled the two designs are almost identical. However, the proposed design can effectively control the acceleration response from Fig. 20(b). Moreover, it is shown in Fig. 20(a) that the displacement response for the two designs is dominated by the lower vibration modes of the cable. Since the modal mass is the same for each mode and the frequency of the nth mode is equal to n times of the fundamental frequency, the modal stiffness of the 1st mode is only $1/n^2$ times to that of the nth mode. The displacement response of each mode is determined by the modal force of the external load and the modal stiffness. Therefore, it is clearly seen that the dominant vibration mode of the response history is the fundamental mode from Fig. 20(a), and it also indicates that the proposed design has the almost same damping capacity compared to the conventional design for lower modes. The Fourier spectra of the acceleration response of the cable respectively controlled by the two designs are shown in Fig. 20(c). It can be seen that the acceleration response of the first 9 modes is controlled and the both FFT amplitude remains at a relatively low level by the two designs. However, the conventional design cannot effectively control the cable vibration of higher modes, and meanwhile the proposed design is more effective in controlling the vibration of the higher modes.

Fig. 21 compares the root of mean square (RMS) of displacement and acceleration at the mid-span point under 10 different records of the random white noise excitations. It is repeatedly demonstrated that the proposed and the conventional design has the almost identical performance in controlling the displacement response. However, it is shown in Fig. 21(b) that the RMS of acceleration responses controlled by the proposed design is reduced to 70% compared to that controlled by the conventional design.

4.2. Responses under the combined load with multiple sinusoidal excitations

In this section, a combined load composing of 40 sinusoidal waves with excitation frequencies targeted to the first 40 modes of the cable is used to excite the resonant vibrations of the first 40 modes. The results is employed to examine the control performance of the proposed design under the multiple resonances. The expression of the combined load is

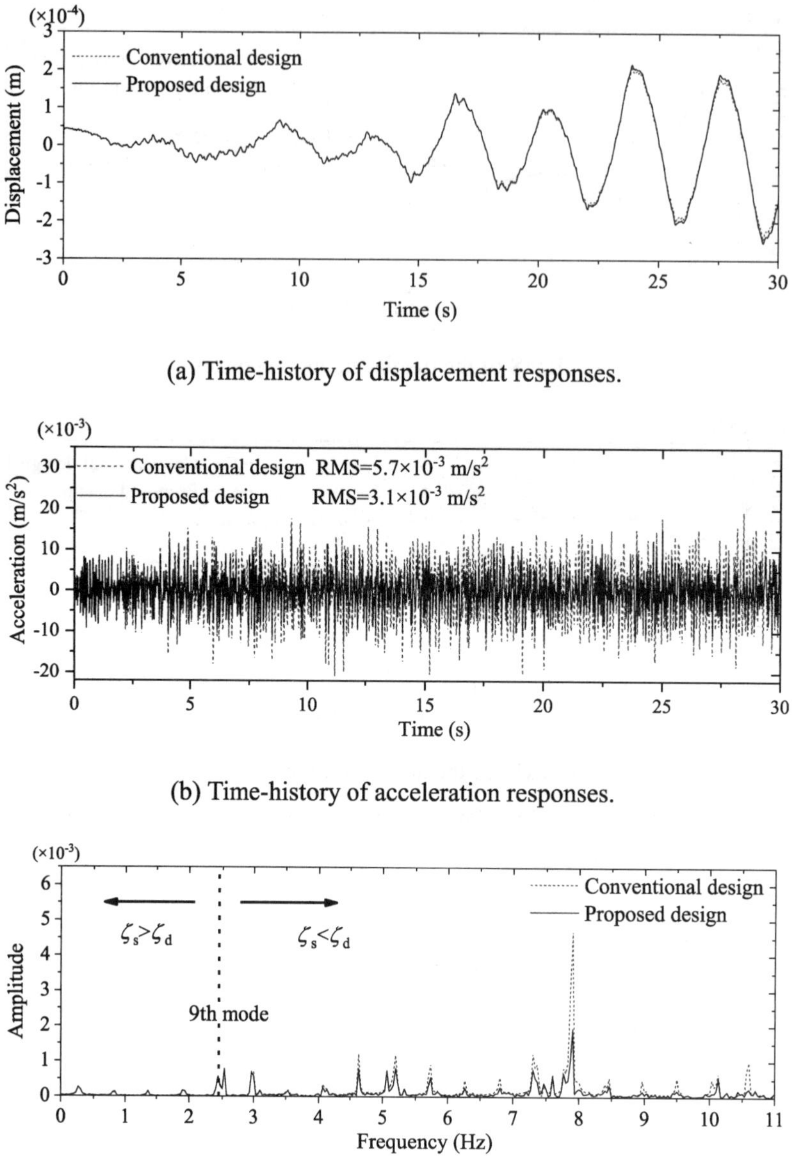

Fig. 22. Responses under the multiple sinusoidal excitation controlled by the proposed and the conventional designs (a) Time-history of displacement responses, (b) Time-history of acceleration responses, (c) FFT spectra of the acceleration responses.

written as,

$$F = \sum_{n=1}^{40} A\cdot sin(n\omega_n t) \tag{15}$$

where A represents the amplitude of sinusoidal load, and it equals to 200 N for each mode. The multiple sinusoidal excitations are also applied at the mid-span point, and the time interval in the simulations is also set to 1/300 s. The comparison respectively controlled by the two designs under the combined load is shown in Fig. 22. The displacement and the acceleration histories in Fig. 22 are recorded at the mid-span point of the cable. It can be seen from Fig. 22(a) that the displacement response controlled by the proposed design is almost identical to that controlled by the conventional design. It is shown from Fig. 22(b) that the acceleration controlled by the proposed design is dramatically decreased compared to that controlled by the conventional design. From Fig. 22(c), it is found that the control performance of the proposed design in frequency domain outweighs that of the conventional design.

5. Discussions of the proposed design

5.1. *Effect of canceling the limitation of the furthest installation location*

The control performances of the proposed and conventional designs has been adequately compared when the furthest installation location of the dampers are both set to 0.02. It is found that the damping ratios of the all 40 modes are larger than 0.4% from the proposed design.

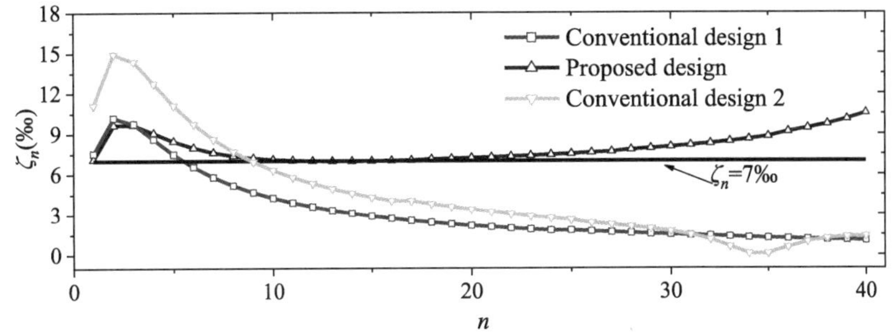

Fig. 23. Modal damping ratios of the three designs.

Table 3
Parameters of the three designs.

Scheme		a/l	a (m)	η	c (N·s/m)
Conventional design 1		0.02	9.42	7.3	1.47×10^5
Proposed design	Damper 1	$a_1/l = 0.018$	$a_1 = 8.48$	8.0	1.61×10^5
	Damper 2	$(a_2\text{-}a_1)/l = 0.011$	$a_2\text{-}a_1 = 5.18$	1	2.01×10^4
Conventional design 2		0.029	13.66	5.0	1.01×10^5

However, the damping ratios of the first 9 modes from the proposed deign are smaller than that from the conventional design. If the proposed design is aimed to enhance the damping ratios of the first 40 modes to 0.7% for both lower and higher modes, the parameters of the two dampers can be chosen based on the flowchart illustrated in Fig. 17 by setting the value of the target damping as 0.7%. As a comparison, the conventional design 1 proposes installing one damper at $a/l = 0.02$ to increase the damping ratios of the first 5 modes to 0.7%. The modal damping ratios with the two designs are respectively calculated in Fig. 23. It is seen that the damping ratios of the first 5 modes for the conventional design 1, and all of the 40 modes for the proposed design are both higher than the target value. The detailed parameters of the two designs are respectively listed in Table 3. It can be seen that the installation location of the damper 1 in the proposed design is slightly smaller than that of the damper in the conventional design 1. However, the furthest installation location is increased about 1.45 times to that in the conventional design 1. Moreover, as another comparison, the conventional design 2 proposes installing the damper at $a/l = 0.029$ as the same location as the damper 2 in the proposed design. The modal damping ratios of cable for the conventional design 2 are also calculated and plotted in Fig. 23. It can be seen that although the damping ratios of the first 8 modes are greatly enhanced, the damping ratios of higher modes are still<0.7%.

Fig. 24 shows the acceleration response of the cable controlled by the proposed design with the parameters listed in Table 3 under a typical white noise excitation as shown in Fig. 19. The results are also compared

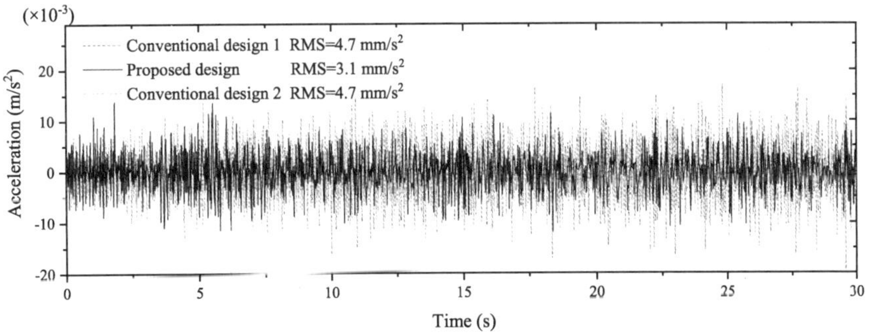

Fig. 24. Time-history of acceleration responses under a typical white noise excitation when 0.7% is set as the target damping ratio.

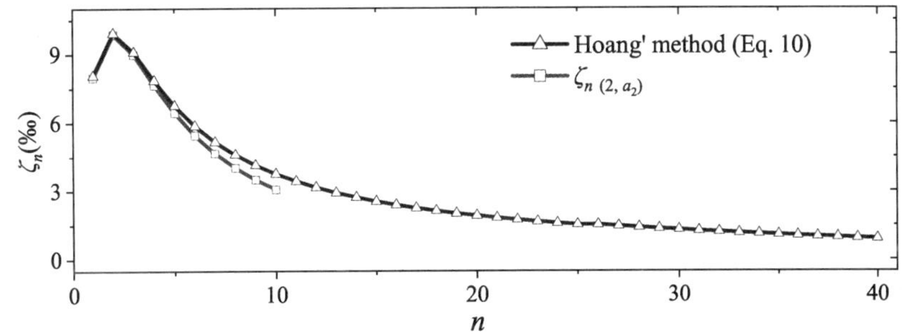

Fig. 25. The first 40 modal damping ratios of a cable with two dampers designed by Eq. (16) and a cable only with the damper 2.

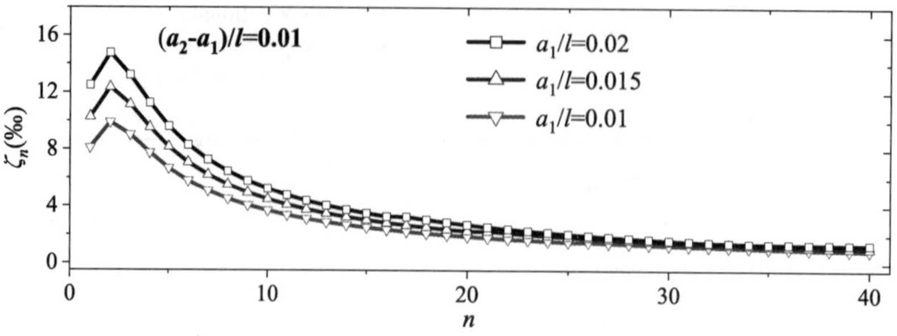

(a) Modal damping ratios for different values of a_1/l under $(a_2$-$a_1)/l$=0.01.

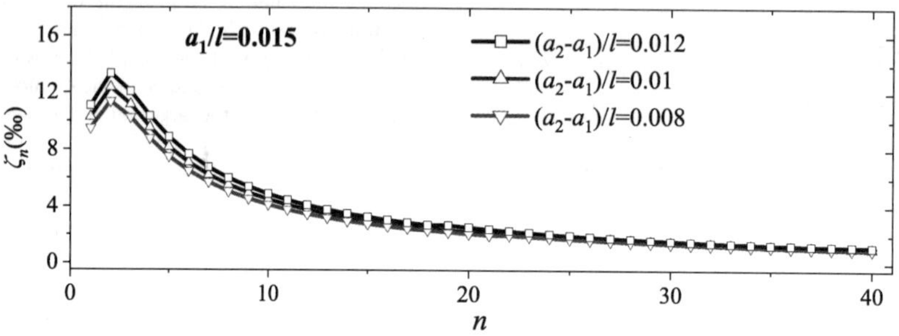

(b) Modal damping ratios for different values of $(a_2$-$a_1)/l$ under a_1/l=0.015.

Fig. 26. Modal damping ratios of a cable with two dampers designed by Eq. (11) for different installation locations(a) Modal damping ratios for different values of a_1/l under $(a_2\text{-}a_1)/l = 0.01$, (b) Modal damping ratios for different values of $(a_2\text{-}a_1)/l$ under $a_1/l = 0.015$.

with that controlled by the two conventional designs in Fig. 24. It is shown that the amplitude of acceleration responses controlled by the proposed design is reduced to 65% compared to that controlled by the conventional designs. However, the damper performance from the conventional design 2 is not enhanced and the acceleration RMS is almost identical with the conventional design 1.

5.2. *Effect of exchanging the damping coefficients of the two dampers*

For the proposed design, the damper 1 is used to control the cable vibrations of lower modes, and the damper 2 is applied to enhance the modal damping ratios of higher modes. However, another case, using the damper 1 for higher mode control and the damper 2 for lower mode control, is also worthy to be discussed. In such a case, the dimensionless damping coefficients of the two dampers η_1 and η_2 can be set as

$$\begin{cases} \eta_1 \approx 1 \\ \eta_2 = \eta_{2,\ \mathrm{opt}\ 3} - \eta_{2,\ \mathrm{opt}\ 1} \end{cases} \tag{16}$$

Similarly, with the help of the assumption in section 3.1, we can simply estimate the modal damping ratios by separately considering lower modes and higher modes. Since the dimensionless damping coefficient of the damper 1 $\eta_1 \approx 1$ for higher modes is much smaller than the optimal dimensionless damping coefficients for lower modes, we can assume $\eta_1 \to 0$. In such a case, the modal damping ratios of lower modes are mainly provided by the damper 2 installed at a_2. However, considering the modal damping ratios of higher modes, the damper 2 can be considered as a fixed support since η_2 is much greater than $\eta_{2,\mathrm{opt}\ n}$ of the higher modes. Moreover, the effect of the damper 1 will be limited by the damper 2 which can be regarded as a barrier in such a case. The two dampers only provide limited damping effect for higher modes of the cable and, therefore, the modal damping ratios can be estimated by

$$\zeta_n \approx \begin{cases} \zeta_{n\ (2,\ a_2)} \text{ for lower modes} \\ 0 \text{ for higher modes} \end{cases} \tag{17}$$

where $\zeta_{n(2,\ a2)}$ repents the modal damping ratios when only the damper 2 is installed at a_2.

The modal damping ratios of the first 10 modes of a cable only with the damper 2 installed at a_2 on the cable are shown in Fig. 25, and the results of the first 40 modes of a cable with two dampers designed by Eq. (16) are also plotted in the figure. The inherent damping ratios of each mode are set to 0, and the parameters of the two dampers are set as $a_1/l = 0.01$, $(a_2\text{-}a_1)/l = 0.01$, $\eta_1 = 1$, and $\eta_2 = \eta_{2,\mathrm{opt}\ n}$. It can be seen that the damping ratios of lower modes, especially the first 5 modes, of the cable with two dampers are almost identical to that only with the damper 2 installed at a_2. Moreover, the damping ratios of higher modes remain at a relatively low level.

The modal damping ratios for different installation locations are shown in Fig. 26. The dimensionless damping coefficient of the damper 1 η_1 is set to 1, and η_2 is set to $\eta_{2,\mathrm{opt}\ n}$ considering different installation positions. It is found that the damping ratios of the modes higher than 10th cannot be effectively enhanced wherever the two dampers installed. It also indicates that the design discussed in this subsection cannot be employed for simultaneously controlling lower and higher modes.

6. Conclusions

In this paper, the higher mode vibrations of cables occurred on the Sutong Bridge is reported and the limitation of using one single viscous damper for multi-modes control of cables is clarified. A novel framework that designs two dampers near the cable anchorage for controlling lower and higher mode vibrations simultaneously is proposed. The damper 1 near the cable anchorage with a relatively large damping coefficient is designed to control the lower mode vibrations. Since the damper 1 can be regarded as an approximate support in the higher modes due to the large damping coefficient, the damper 2 with a dimensionless damping coefficient $\eta_2 \approx 1$ is installed farther than the damper 1 away from the cable anchorage to control the higher mode vibrations. The effect of adding the two viscous dampers in a cable on the mode shapes is analyzed and verified by the computational results from the finite element software ANSYS. Based on the calculation of damping ratio for a cable with one viscous damper, a simplified estimation method is proposed and modified to precisely evaluate the damping ratios of a cable controlled by two viscous dampers. A flowchart of parametric optimization for a cable with two dampers is summarized. The control effectiveness of the proposed design is validated and compared with the conventional design. Finally, the effects of canceling the limitation of the furthest installation location and exchanging the damping coefficients of the two dampers in the proposed design are discussed.

The measurement results of a cable on the Sutong Bridge show that the maximum acceleration amplitude of higher mode vibrations is up to 15 m/s^2. The cable vibrations with higher frequencies result in severer damages between the cable and its anchorage. The conventional design using one single viscous damper is not able to control the lower and higher modes of cables simultaneously. The mode shapes of a cable with two viscous dampers can be simply described by the way of separately considering the lower and higher modes. The damping ratios of a cable controlled by two viscous dampers in the proposed design can be precisely estimated by a proposed simplification. The damping ratios of both lower and higher modes can be significantly enhanced by the optimal design of two dampers. The optimal design of the two dampers in a cable can be simply reduced to two individual designs of one single viscous damper, which can lower the design complexity. The proposed design is effective to control the acceleration responses of higher modes vibrations, both under white noise and a combined load mixed with multiple sinusoidal excitations.

Declaration of Competing Interest

The authors declare that they have no known competing financial interests or personal relationships that could have appeared to influence the work reported in this paper.

Acknowledgments

The financial supports from the National Key Research and Development Program of China (grant number 2019YFC1511101), the National Science Fund for Distinguished Young Scholars (grant number 52025082), the National Natural Foundation of China (grant number 51908210), and the Natural Foundation of Hunan Province (grant number 2020JJ5074) are greatly appreciated.

References

[1] Cai CS, Wu WJ, Araujo M. Cable Vibration Control with a TMD-MR Damper System: Experimental Exploration. J Struct Eng 2007;133(5):629–37.
[2] Shi X, Zhu S, Nagarajaiah S. Performance Comparison between Passive Negative-Stiffness Dampers and Active Control in Cable Vibration Mitigation. J Bridge Eng 2017;22(9):04017054.
[3] de Sá CE. Cable vibrations in cable-stayed bridges. IABSE 2007.
[4] McTavish S, D'Auteuil A, Raeesi A. Effect of cable surface characteristics and flow turbulence on the aerodynamic behaviour of stay cables in dry conditions. J Wind Eng Ind Aerodyn 2020;207:104414.
[5] Zhou HJ, Wu YH, Li LX, Sun LM, Xing F. Free vibrations of a two-cable network inter-supported by cross-links extended to ground. Smart Structures and Systems 2019;23(6):653–67.
[6] Sun L, Hong D, Chen L. In-plane free vibrations of shallow cables with cross-ties. Structural Control and Health Monitoring 2019;26(10):e2421.
[7] Jing H, He X, Zou Y, Wang H. In-plane modal frequencies and mode shapes of two stay cables interconnected by uniformly distributed cross-ties. J Sound Vib 2018;417:38–55.
[8] Liu M, Yang W, Chen W-L, Xiao H, Li H. Experimental Investigation on Vortex-Induced Vibration Mitigation of Stay Cables in Long-Span Bridges Equipped with Damped Crossties. J Aerosp Eng 2019;32(5):04019072.
[9] Jing H, He X, Li H. Initial condition and damping effects on rain–wind induced cable vibration. J Wind Eng Ind Aerodyn 2018;175:376–83.
[10] Lu L, Duan Y-F, Spencer B, Lu X. Inertial mass damper for mitigating cable vibration. Structural Control and Health Monitoring 2017;24(10):e1986.
[11] Shi X, Zhu S, Spencer B. Experimental Study on Passive Negative Stiffness Damper for Cable Vibration Mitigation. J Eng Mech 2017;143(9):04017070.
[12] Huang P, Wang X, Wen Q, Wang W, Sun H. Active Control of Stay Cable Vibration Using a Giant Magnetostrictive Actuator. J Aerosp Eng 2018;31(5):04018074.
[13] Chen Z, Wang X, Ko J, Ni Y, Spencer B, Yang G, et al. MR damping system for mitigating wind-rain induced vibration on Dongting Lake Cable-Stayed Bridge. Wind & structures 2004;7(5):293–304.
[14] Wang W, Hua X, Wang X, Wu J, Sun H, Song G. Mechanical behavior of magnetorheological dampers after long-term operation in a cable vibration control system. Struct Control Health Monit 2019;26(1):e2280.
[15] Li H, Liu M, Li J, Guan X, Ou J. Vibration control of stay cables of the shandong binzhou yellow river highway bridge using magnetorheological fluid dampers. J Bridge Eng 2007;12(4):401–9.
[16] Javadinasab Hormozabad S, Ghorbani-Tanha AK. Semi-active fuzzy control of Lali Cable-Stayed Bridge using MR dampers under seismic excitation. Frontiers of Structural and Civil Engineering 2020;14(3):706–21.
[17] Huber P, Benicke O. The testing campaign on seismic and wind protection systems of the worlds largest cable stayed bridge. The Russkij Bridge at Vladivostok 2012.
[18] Larsen A, Larose GL. Dynamic wind effects on suspension and cable-stayed bridges. J Sound Vib 2015;334:2–28.
[19] An Y, Wang Z, Ou G, Pan S, Ou J. Vibration mitigation of suspension bridge suspender cables using a ring-shaped tuned liquid damper. J Bridge Eng 2019;24(4):04019020.
[20] Kovacs I. Zur frage der seilschwingungen und der seildämpfung. Bautechnik 1982;59(10).
[21] Pacheco BM, Fujino Y, Sulekh A. Estimation curve for modal damping in stay cables with viscous damper. J Struct Eng 1993;119(6):1961–79.
[22] Krenk S. Vibrations of a taut cable with an external damper. J Appl Mech 2000;67(4):772–6.
[23] Main JA, Jones NP. Free vibrations of taut cable with attached damper. I: Linear viscous damper. J Eng Mech 2002;128(10):1062–71.
[24] Main JA, Jones NP. Vibration of tensioned beams with intermediate damper. I: Formulation, influence of damper location. J Eng Mech 2007;133(4):369–78.
[25] Sun L, Shi C, Zhou H, Cheng W. A full-scale experiment on vibration mitigation of stay cable. IABSE Symposium Report 2004;88:154.
[26] Fujino Y, Hoang N. Design formulas for damping of a stay cable with a damper. J Struct Eng 2008;134(2):269–78.
[27] Huang Z, Jones NP. Damping of taut-cable systems: Effects of linear elastic spring support. J Eng Mech 2011;137(7):512–8.
[28] Raftoyiannis IG, Michaltsos GT. Movable anchorage systems for vibration control of stay-cables in bridges. Eng Struct 2016;112:162–71.
[29] Xu YL, Yu Z. Mitigation of three-dimensional vibration of inclined sag cable using discrete oil dampers—II. Application. Journal of Sound and Vibration. 1998;214(4):675–93.
[30] Caracoglia L, Jones NP. Damping of taut-cable systems: Two dampers on a single stay. J Eng Mech 2007;133(10):1050–60.
[31] Hoang N, Fujino Y. Combined damping effect of two dampers on a stay cable. J Bridge Eng 2008;13(3):299–303.
[32] Weber F, Feltrin G, Maślanka M, Fobo W, Distl H. Design of viscous dampers targeting multiple cable modes. Eng Struct 2009;31(11):2797–800.
[33] PTI. (Post-Tensioning Institute). Recommendations for stay cable design, testing and installation. Phoenix, AZ: PTI; 2001.
[34] Wang XY, Ni YQ, Ko JM, Chen ZQ. Optimal design of viscous dampers for multi-mode vibration control of bridge cables. Eng Struct 2005;27(5):792–800.
[35] Gao H, Wang H, Li J, Wang Z, Liang R, Xu Z, et al. Optimum design of viscous inerter damper targeting multi-mode vibration mitigation of stay cables. Eng Struct 2021;226:111375.
[36] Chen L, Sun L, Xu Y, Di F, Xu Y, Wang L. A comparative study of multi-mode cable vibration control using viscous and viscoelastic dampers through field tests on the Sutong Bridge. Eng Struct 2020;224:111226.
[37] Liu Z, Shen J, Li S, Chen Z, Ou Q, Xin D. Experimental study on high-mode vortex-induced vibration of stay cable and its aerodynamic countermeasures. J Fluids Struct 2021;100:103195.
[38] Chen L, Nagarajaiah S, Sun L. A unified analysis of negative stiffness dampers and inerter-based absorbers for multimode cable vibration control. J Sound Vib 2021;494:115814.
[39] Chen L, Sun L, Nagarajaiah S. Cable vibration control with both lateral and rotational dampers attached at an intermediate location. J Sound Vib 2016;377:38–57.
[40] Main JA, Jones NP. Evaluation of Viscous Dampers for Stay-Cable Vibration Mitigation. J Bridge Eng 2001;6(6):385–97.
[41] Zuo D, Jones NP. Interpretation of field observations of wind-and rain-wind-induced stay cable vibrations. J Wind Eng Ind Aerodyn 2010;98(2):73–87.
[42] Chen W-L, Gao D, Laima S, Li H. A Field Investigation on Vortex-Induced Vibrations of Stay Cables in a Cable-Stayed Bridge. Applied Sciences 2019;9(21):4556.

Journal of Wind Engineering & Industrial Aerodynamics 215 (2021) 104685

Contents lists available at ScienceDirect

Journal of Wind Engineering & Industrial Aerodynamics

journal homepage: www.elsevier.com/locate/jweia

Galloping vibration of stay cable installed with a rectangular lamp: Field observations and wind tunnel tests

An Miao [a], Li Shouying [a,*], Liu Zhiwen [a], Yan Banfu [a], Li Longan [b], Chen Zhengqing [a]

[a] *Key Laboratory for Wind and Bridge Engineering of Hunan Province, College of Civil Engineering, Hunan University, Changsha, 410082, China*
[b] *Engineering, China Railway Major Bridge Reconnaissance & Design Institute Co., Ltd, Wuhan, 430050, China*

ARTICLE INFO

Keywords:
Galloping vibration
Stay cables
Rectangular lamps
Wind tunnel tests
Damping ratio
The Kuimen Bridge

ABSTRACT

In recent years, stay cables on cable-stayed bridges have been equipped with various kinds of lamps to lighten the cables at night. Large amplitude vibration of the stay cables of the Kuimen Bridge was observed after some rectangular lamps were installed. The mechanism and countermeasure of the cable vibration of this bridge were investigated in this paper. First, wind-induced responses of six stay cables were extracted from a video observed in the field. Second, according to Den Hartog's theory, a series of force measurement wind tunnel tests on stay cable attached with a rectangular box were carried out to test the possibility of galloping. Third, a series of vibration measurement wind tunnel tests based on 2-dimensional and 3-dimensional test models were carried out to reproduce the vibration of the cable-lamp test model. In the test, the features of wind-induced vibration were studied at the damping level of 0.1%, 0.3%, 0.6% and the inclination angles of the stay cable include 25°, 35°, 40°, and 45°. The results show that the rectangular aluminum alloy box of the lamp on stay cables can result in the unstable galloping vibration at a very low wind velocity. The critical wind velocity of the cable-lamp test model can be as low as 6.3 m/s (damping ratio $\zeta = 0.1\%$), which is a wind velocity often encountered. It appears that the critical wind velocity increases with the increase of the structural damping ratio, but is only 9–11 m/s when the damping ratio is 0.6%. It seems that increasing the damping ratio has little effect on suppressing the galloping of the stay cable. Finally, the rectangular aluminum alloy box of the lamp was removed and only an electrical wire was attached on the cable surface. The vibration measurement wind tunnel tests based on the 3-dimensional cable-wire test model indicate that the stay cable attached with an electrical wire is aerodynamically stable if the position angle of the wire is 0°, which is consistent with the actual situation.

1. Introduction

In the past two years, it has become a common phenomenon to install lamps on (stay) cables. There are some examples of the installation of lamps on stay cables or hangers of some bridges in China, including the Kuimen Bridge in 2019, the Second Changjiang River Bridge in Wuhan in 2019, the Kuipu Bridge in 2020 and the Gushan Bridge in 2020. Large amplitude vibration of the stay cables of the Kuimen Bridge in China was observed after some lamps were installed on them in May of 2019. There was also a small vibration in the process of installing lamps for the Gushan bridge in August of 2020. It is found that there is no standard for lamp designers to guide them to choose the appropriate aerodynamic shape.

Stay cables experienced several types of wind-induced vibrations, including rain-wind-induced vibration (Hikami and Shiraishi, 1988; Gu and Du, 2005, 2009; Jing et al., 2017; Li et al., 2013, 2014), galloping vibration (Li et al., 2014b; Ma et al., 2019), dry galloping (Cheng et al., 2008a, 2008b, 2008b; Matsumoto et al., 2010), vortex-induced vibration (Zuo et al., 2008; Matsumoto et al., 2001; Chen et al., 2015) and wake galloping (Acampora et al., 2014; Li et al., 2019).

Vortex-induced vibration (VIV) has limited-amplitude vibration, while galloping is a kind of self-excited vibration with a large amplitude and low frequency. VIV is caused by the vortices that alternatively separate from each side of a cylinder and are supposed to take place at the reduced velocity that is the reciprocal of the Strouhal number. The nondimensional amplitude A/D of the VIV (A and D are the amplitude and the diameter of the cable, respectively) of circular cables is usually less than 0.5 and it is considered not to have a significant impact on the

* Corresponding author.
E-mail addresses: anmiao324@hnu.edu.cn (A. Miao), shyli@hnu.edu.cn (L. Shouying), zhiwenliu@hnu.edu.cn (L. Zhiwen), yanbanfu@hnu.edu.cn (Y. Banfu), lila@brdi.com.cn (L. Longan), zqchen@hnu.edu.cn (C. Zhengqing).

https://doi.org/10.1016/j.jweia.2021.104685
Received 3 January 2021; Received in revised form 1 June 2021; Accepted 1 June 2021
Available online 18 June 2021
0167-6105/

safety of stay cables (Simiu and Yeo, 2019). Liu et al. (2021) found obvious high-mode VIV vibration on the stay cables of the Sutong Bridge, and the high-mode VIV vibration can be effectively mitigated if the damping ratio of the stay cable was as high as 0.48%.

Dry galloping occurs on no rainy days at higher wind velocity than that of the VIV (Matsumoto et al., 2010). It is extremely sensitive to wind direction and cable inclination angle, which makes it difficult to reoccur by using wind tunnel tests or numerical simulations. Many studies have focused on clarifying the mechanism of dry galloping vibration (Benidir et al., 2015; Jakobsen et al., 2012; Ma et al., 2019; Matteoni and Georgakis, 2015; Nikitas and Macdonald, 2015; Nikitas et al., 2012; Raeesi et al., 2014; Matsumoto et al., 2010 ; Cheng et al.,2008a, 2008b, 2008b), and several mechanisms have been proposed. The first explanation was proposed by Matsumoto et al. (2010). They believed that an axial flow behind the yaw and/or inclined cable was the key reason because the axial flow can mitigate the Karman vortex shedding. The second explanation was classical galloping or divergent-type galloping, which can be studied by the quasi-steady assumption (Cheng et al., 2008b; Ma et al., 2017a,b). Finally, Acampora et al. (2014) found that there was a significant drop in the aerodynamic damping in the critical Reynolds number range. They believed that it was to be related to the large amplitude cable vibrations observed on some bridges in dry conditions.

Stay cables may experience galloping vibration if some small components (or surface protrusion) are attached to the cable surface. For example, galloping vibration might take place if lamps are installed on the cable surface. Li et al. (2014b) investigated the aerodynamic effects of lamp installation on the stay cables of the Hedong Bridge in China based on wind tunnel tests and CFD simulations. It was found that galloping vibration took place at a wind velocity of around 18 m/s, which was also lower than the design wind velocity. Nguyen et al. (2015) studied the aerodynamic and aeroelastic behavior of a prismatic element with complex shapes by means of static and aeroelastic wind tunnel tests. It was found that Den Hartog's theory was able to predict the critical wind directions that give rise to galloping. However, it failed to predict the critical wind velocity associated with some angles of attack.

In the past, the galloping vibrations of horizontal members, for example, iced conductors or rectangular cylinders, were often investigated, and the wind attack angle was mostly within ±5°. Gao and Zhu (2017) established a nonlinear theoretical model with unsteady galloping force by using a 2:1 rectangular horizontal cylinder. Hua et al. (2020) experimentally investigated the galloping of main cables of suspension bridges under construction. In the test, the segmental model was placed horizontally with an attack angle of ±2°. However, the yaw and/or inclination angle of the stay cable has significant effects on the stay cable that can change the aerodynamics of the structure.

In this study, the mechanism and countermeasure of the cable vibration of the Kuimen Bridge were investigated. First, wind-induced responses of six stay cables of the Kuimen Bridge were extracted from a video taken by field observation. Then, a series of wind tunnel tests including force and vibration measurement wind tunnel tests on stay cable attached with a rectangular box was carried out to reproduce the cable vibration. In the test, the features of wind-induced vibration were studied at several damping levels of 0.1%, 0.3%, 0.6%, and the inclination angles of the stay cable included 25°, 35°, 40°, and 45°. Finally, the effectiveness of countermeasures was also experimentally investigated.

2. Outline of the Kuimen Bridge

The Kuimen Bridge, which is a cable-stayed bridge with a main span of 460 m, is across the Yangtze River and is located in Fengjie County, Chongqing, China. It has two bridge towers, three spans, and double cable planes, and there is an auxiliary pier on each side span. The elevation of the Kuimen Bridge is shown in Fig. 1. The Kuimen Bridge

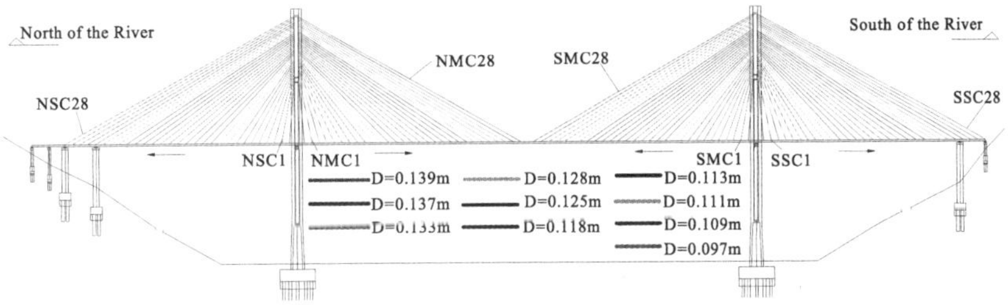

(a)

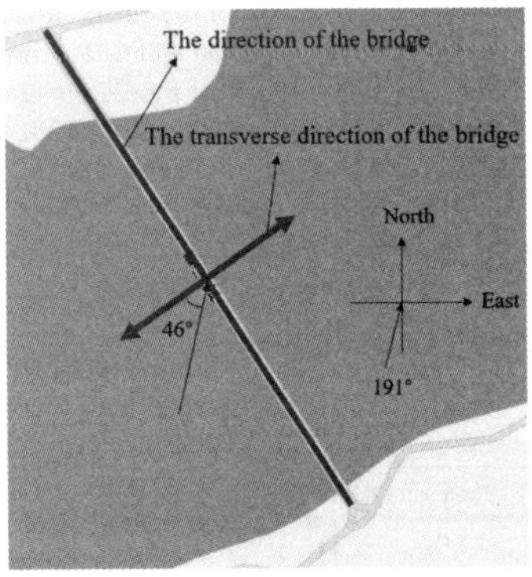

(b)

Fig. 1. Sketch of the Kuimen Bridge (a) Elevation of the Kuimen Bridge (b) Definition of wind direction.

has a total of 224 stay cables, and the cable number is also given in Fig. 1. The inclination angle of the cable varies from 23° to 75°. The standard cable distance along the main beam is 7.85m. The cable distance along the tower is 1.7 m. The fundamental frequency of the stay cables SMC27~22 is shown in Table .1.

3. Wind-induced vibration of stay cables attached with lamps

The Kuimen Bridge was completed in 2006, and no obvious vibration of the stay cables was found after the completion. A lighting project of this bridge was started on April 1, 2019, and all lamps were installed on the stay cables by the end of May of 2019. The circular section of the smooth stay cable was attached with a rectangular aluminum alloy box, which may lead the stay cable to become aerodynamically unstable. The photos of the stay cable attached with lamps are given in Fig. 2.

From the late May of 2019, obvious vibration of the stay cables of the Kuimen Bridge was observed several times. A typical example was around 5:00 p.m. on August 10, 2019. Almost all the stay cables experienced a large amplitude of vibration. The distance between the two stay cables varies from the bridge tower to the main beam, and the distance at the end of the bridge tower is the shortest. It can be seen from the video that the vibration of the outer long cable is more severe. Taking the seventh long cable as an example, the minimum distance between two cables perpendicular to the axis of the stay cable is 11.8*D*. With the increase of stay cable length, the distance between stay cables also increases. Li et al. (2019) found that wake galloping would occur when the distance of the hangers was less than 9D. Moreover, when wake-induced vibration occurs, the cable vibration is usually asynchronous. Obviously, the wake-induced vibration is not the main reason for the cable vibration of the Kuimen bridge. The maximum wind speed provided by Shapingba weather station located in Chongqing at the same time is 6.2 m/s, and the wind direction is 191°, which is 46° for the angle between the wind direction and the transverse direction of the bridge. The angle definition is given in Fig. 1(b). The wind direction of 0° indicates the wind coming from the north. The angle increases clockwise. The altitude of the weather station is 259.1m. The absolute elevation of the middle point of the cable is about 245m. It should be pointed out that the weather station is 329 km away from the bridge site, and the wind environment of the bridge site may be different from that of the station. Therefore, the wind speed value can only be used as a reference. Unfortunately, there was no health monitoring system on the stay cables of the bridge to record the exact wind-induced responses in time. However, some videos of the cable vibration have been taken, and the wind-induced responses of the cable vibration can be extracted from these videos.

Two images are taken from the video at 5:00 p.m. on August 10, 2019, as shown in Fig. 3. The time interval between the two images is 0.42 s. The displacement time histories of Cables SMC27~22 (as shown in Fig. 3) extracted from the video are given in Fig. 4, and their corresponding power spectral densities (PSDs) are presented in Fig. 5. Digital Image Correlation (DIC) algorithm integrating FTCC and IC-GN proposed by Yan et al. (2020) was used in the process. It can be found from Fig. 4 that the maximum peak to peak amplitude of the stay cable is more than 11*D*, *D* is the diameter of the stay cable. The displacement direction is perpendicular to the axis of the cable in the camera plane.

Table 1
Detailed information of the stay cables of the Kuimen Bridge.

Number	Diameter *D* (m)	Inclination angle θ (°)	Frequency *f*(Hz)
SMC27	0.137	23.9	0.54
SMC26	0.133	24.3	0.57
SMC25	0.133	24.7	0.58
SMC24	0.133	25.2	0.60
SMC23	0.128	25.7	0.63
SMC22	0.128	26.2	0.65

The fundamental frequency of Cables SMC27~22 varies from 0.54 Hz to 0.65Hz (as given in Table .1). The energy peaks in the PSDs are around 0.60 Hz, which seems to correspond to the fundamental frequency of the stay cables. It is worth noting that there is a certain difference between the frequency extracted from the video and the fundamental frequency given in Table 1. The difference of Cable SMC26 is the largest, which is 9.5%. There are three reasons for this difference: (1) The values listed in Table 1 are the design parameters. The exact values after the construction of the bridge may be different from those listed in Table 1. (2) The displacement extracted by Digital Image Correlation (DIC) largely depends on the quality of the video image. This video was not taken by professionals with professional equipment. The difference between the identified frequency and the actual structure may be due to the shooting quality of the video. (3) The aeroelastic effect is also an important factor for the frequency difference. In terms of amplitude and vibration frequency, galloping vibration might be the key reason for the cable vibration of the Kuimen Bridge because the rectangular aluminum alloy box changes the stable configuration of the smooth cable.

4. Force measurement wind tunnel tests

4.1. *Test models and test cases for force measurement*

There are ten types of cable diameters on the Kuimen Bridge, including 97, 109, 111, 113, 118, 125, 128, 133, 137, 139 mm. The size of the rectangular aluminum alloy box is 55 mm × 58 mm, as shown in Fig. 6. Since the aluminum alloy bracket is added to the surface of the stay cable later, there is a gap at the contact surface. But the gap is very small, only 1–2 mm, and this gap is usually filled with glue or silica gel in the construction process. Therefore, the gap size is not considered in the model design. To consider the effects of the diameter of stay cable on the aerodynamic forces, three test models with different diameters 109 mm, 125 mm, and 139 mm, were designed. The three test models are named Model 1, Model 2, and Model 3 respectively, as given in Fig. 6(a). Aerodynamic countermeasure might be the most economical way to reduce the wind-induced vibration of the Kuimen Bridge, so another three kinds of test models, named Models 1Ã3A as shown in Fig. 6 (b), were also designed to measure their aerodynamic forces. Models 1Ã3A are formed by adding baffles along the vertex of the rectangle and the tangent of the circle on Models 1–3. The geometric scale ratio of the test models is 1:1, and the length of the test models is 1.2 m.

Force measurement wind tunnel tests were carried out in the HD-3 Boundary Layer Wind Tunnel (HD-3BLWT) at Hunan University. It is an open-circuit-type wind tunnel with a maximum velocity of 20 m/s. The test section is 24.0 m in length by 3.0 m in width by 2.5 m in height.

High-frequency force balance (HFFB) was adopted to measure the aerodynamic forces of the test models. The HFFB was fixed at the center of the turntable and the test model was vertically installed at the top of the HFFB. To ensure a 2-dimensional flow state at the upper end of the test model, an end plate with a diameter of 550 mm was installed on a steel tube which was protruded from the roof of the test section at the upper part of the model. The gap between the end plate and the upper-end face of the test model is 3 mm. Fig. 7 shows the photo of the test model installed in the test section.

Force measurement wind tunnel tests were carried out in a uniform flow with a wind speed of 7.5 m/s, under which the stay cable of the Kuimen Bridge oscillated violently. The orientation of the test model was defined by the wind attack angle α, as shown in Fig. 8. Due to the symmetry of the test models, the aerodynamic forces were measured in the range of $\alpha = 0°–180°$ with an interval of 2°. The sampling frequency of the HFFB was 1000 Hz and the sampling time was 30 s.

4.2. *Definition of mean force coefficients*

In this study, only the average values of the aerodynamic forces were considered. The mean forces directly measured by the HFFB are F_x and

Fig. 2. Photos of stay cable attached with a rectangular aluminum alloy box.

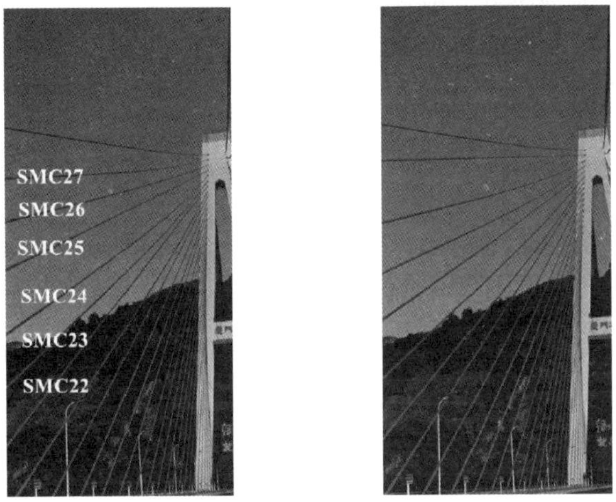

Fig. 3. Violent vibration images from the video at 5:00 p.m. on August 10, 2019.

F_y, which is corresponding to the body axis, as shown in Fig. 8. The mean drag and lift forces defined in the wind axis, F_D and F_L, can be determined by,

$$F_D = F_x\cos\alpha + F_y\sin\alpha \tag{1}$$

$$F_L = F_y\cos\alpha - F_x\sin\alpha \tag{2}$$

and the mean drag and lift coefficients, C_D and C_L, can be obtained from,

$$C_D = \frac{F_D}{0.5\rho U^2 DL} \tag{3}$$

$$C_L = \frac{F_L}{0.5\rho U^2 DL} \tag{4}$$

in which, U is the mean wind velocity of the approaching flow; D and L are the diameter and the length of the test model, respectively; ρ is the density of air. The effect of the boundary layer near the floor of the wind tunnel was taken into account to modify the aerodynamic force coefficients. According to classical galloping theory (Den Hartog, 1932), the galloping force coefficient C_g can be obtained from,

$$C_g = C_D + \frac{dC_L}{d\alpha} \tag{5}$$

The lift coefficient derivative is calculated by forward difference estimator and can be calculated by the following formula:

$$\frac{dC_L}{d\alpha} = \frac{C_L(\alpha + \Delta\alpha) - C_L(\alpha)}{\Delta\alpha} \tag{6}$$

in which, $\Delta\alpha$equals 2°.

According to Den Hartog's criterion, the galloping vibration of the stay cable maybe take place if the galloping force coefficient $C_g < 0$.

4.3. *Experimental results for force measurement wind tunnel tests*

Fig. 9 presents the variation of mean drag and lift coefficients with wind attack angle for the six test models. In the test, the Reynolds number of the smooth stay cables is between 5.64×10^4 and 7.19×10^4, within which the mean drag coefficient of smooth cable is about 1.2 and the mean lift coefficient is 0. It can be found from Fig. 9 that the

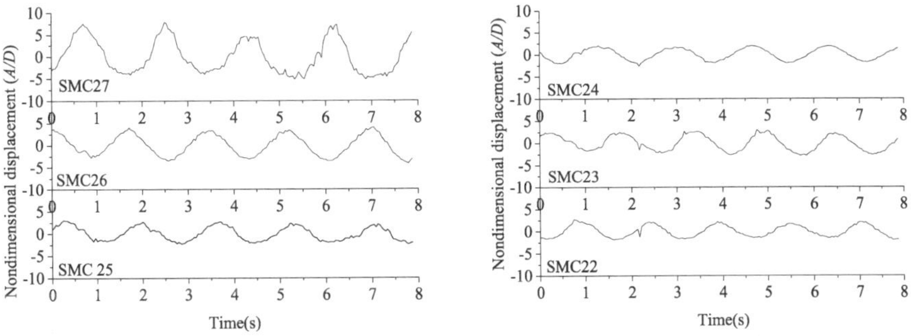

Fig. 4. Displacement time histories of Cables SMC27~SMC22.

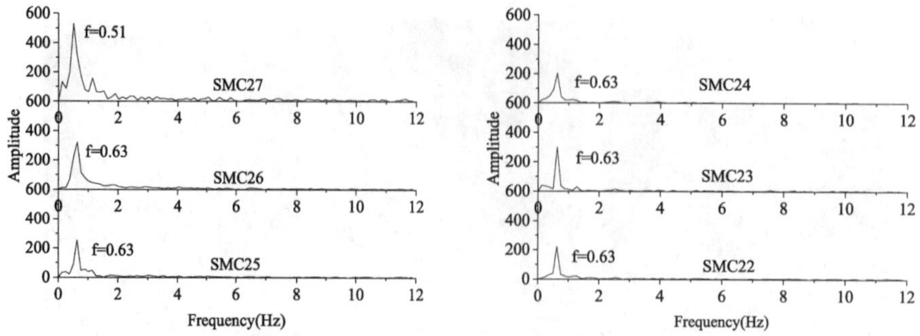

Fig. 5. PSDs of displacement time histories of Cables SMC27~SMC22.

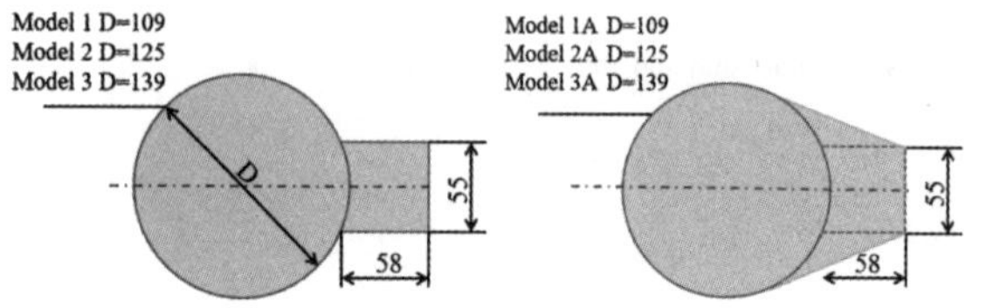

(a) Models 1~3 (b) Models 1A~3A

Fig. 6. Test models for force measurement wind tunnel tests (unit: mm).

Fig. 7. Photo of the test model installed in the test section.

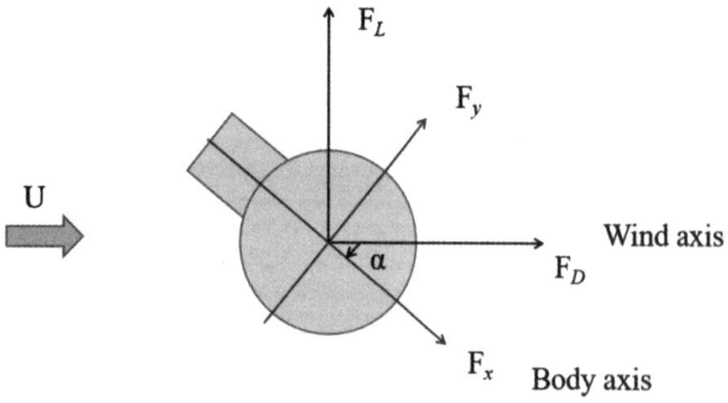

Fig. 8. Definition of wind attack angle α

maximum drag coefficient is 2.6 and the maximum lift coefficient is 1.38. The maximum drag coefficient occurs near the wind attack angle of 90° because of the increase of the windward area by the rectangular aluminum alloy box. It should be noted that there are two obvious sudden decreases of the mean lift force near the wind attack angle of 10° and 170° for all the six test models. Large negative galloping force coefficients are also observed near the above two wind attack angle, as shown in Fig. 10. This indicates the possibility of galloping vibration maybe take place near these wind attack angles. Comparing Figs. 9 (a) and Fig.10 (a) with Figs. 9 (b) and Fig.10 (b), it appears that the aerodynamic countermeasure as shown in Fig. 6 (b) has no effect on vibration suppression.

Table .2 shows the range of wind attack angle for the galloping force coefficient $C_g < 0$, together with the minimum value of C_g and the corresponding wind attack angle. It can be found from Fig. 10 and Table 2 that the maximum negative galloping force coefficient appears in the range of 10–20° and 160–170° for Models 1–3. The maximum negative values of C_g are −9.3, −10.1, and −12.9 for Models 1, 2, and 3, respectively, and the corresponding wind attack angles are respectively 10°, 12°, and 14°. It seems that the diameter of the stay cable has no obvious effect on the aerodynamic forces of the test models. Therefore, only the stay cable with a diameter of 139 mm was adopted in the following vibration measurement wind tunnel tests.

5. Vibration measurement wind tunnel tests

5.1. Similarity relationship of test models for vibration measurement

An elastic segmental test model of stay cable was used in the vibration measurement wind tunnel tests. Different from the test models used in the force measurement wind tunnel tests, the main structural parameters, including stiffness, mass, and damping, etc., should be carefully simulated in the test model. Three dimensionless parameters, elastic parameter (U/fD) inertia parameter ($m/\rho D^2$), and damping ratio, need to be satisfied. To ensure the similarity of Reynolds number, the model with a geometric scale ratio of 1:1 is usually used in the previous research on the cylinder, and the simulation of model mass is abandoned. In this way, it is difficult to convert the critical wind speed and amplitude between the prototype and the model. There are also a small number of studies using full-scale simulation, which is a huge challenge to the support system of the test device. The cross-section studied is the combination of rectangular and circular cross-sections. The Reynolds number effect is different from that of circular cross-section. From the field observation, the critical wind speed is low and the Reynolds number is in the subcritical region. To obtain more accurate critical wind speed, the geometric scale ratio less than 1:1 was adopted in the design of the elastic segmental test model. The target stay cable is Cable NSC27 with a diameter of 139 mm, a mass per unit length of 72.8 kg/m, and a fundamental frequency of 0.53 Hz. The geometric scale is 1:2.78, and the diameter of the cable model is 50 mm. The detailed similarity

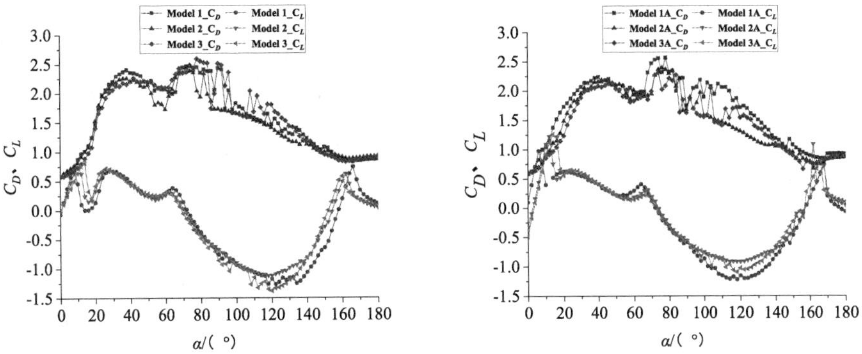

Fig. 9. Variation of mean drag and lift coefficients with wind attack angle.

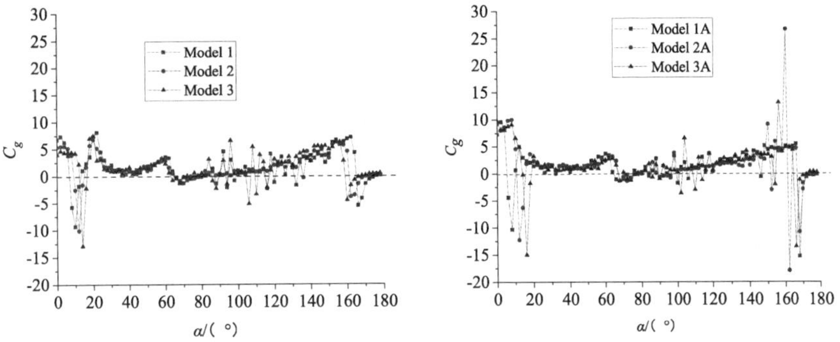

Fig. 10. Variation of galloping force coefficients with wind attack angle.

Table 2
Minimum galloping force coefficient and its corresponding wind attack angle.

Test models	Wind attack angles for $C_g < 0$ (°)	Minimum C_g	α for minimum C_g
Model 1	8–12、66–80、88、94、98、116、120、132、166–174	−9.3 (−5.5)	10° (166°)
Model 2	10–14、66–80、136、162–170	−10.1 (−3.9)	12° (162°)
Model 3	14、16、64–74、86、88、94、106、110、116、160–166、	−12.9 (−4.55)	14° (160°)

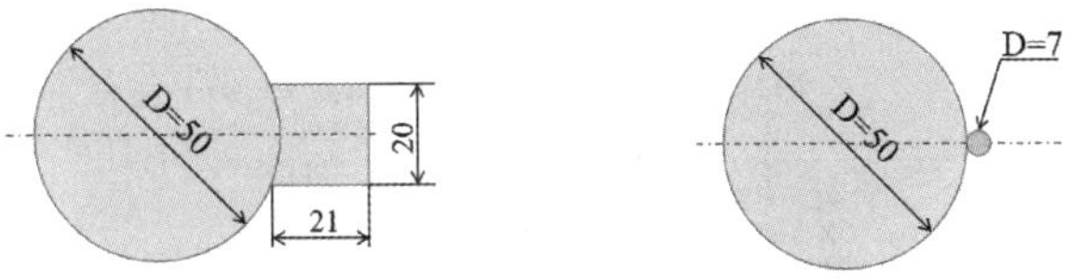

(a) Model A (b) Model B

Fig. 11. Cross sections of the elastic test models (unit: mm).

scales of the elastic test model are given in Table .3. Under this geometric scale, the static displacement of the test model is only 0.067 m. The test model including stay cable and the rectangular aluminum alloy box is named Model A, as shown in Fig. 11(a). On the other hand, the rectangular aluminum alloy boxes were removed from the stay cables of the Kuimen Bridge after the violent vibration was observed. There is only one electrical wire left on the surface of each stay cable, and the diameter of the electrical wire is 20 mm. Therefore, another type of test model, which is named Model B as shown in Fig. 11(b), was also designed. In Model B, the diameter of the electrical wire is 7 mm. The length of both models is 1.8 m. Table .3 shows the detailed similarity scales of the elastic test model.

Table 3
Similarity scales of the elastic test model.

Parameters	Test model	Prototype	Scales
Diameter (mm)	50	139	1/2.78
Frequency (Hz)	1.87	0.53	3.53/1
Wind speed	–	–	1.27/1
Mass per unit length (kg/m)	9.4	72.8	$1/2.78^2$
Damping ratio	0.1%, 0.3%, 0.6%	0.1%	1/1

5.2. Outline of vibration measurement wind tunnel tests

Vibration measurement wind tunnel tests were conducted in the high-speed test section of HD-2 Boundary Layer Wind Tunnel (HD-2BLWT) in Hunan University, Changsha, China. The HD-2BLWT is a closed-circuit-type wind tunnel. The high-speed test section is 2.5 m in height by 3.0 m in width by 17 m in length with a maximum wind velocity of 58 m/s. The definition of wind attack angle α of a 2-dimensional stay cable ($\beta = \theta = 0$) has been shown in Fig. 8. The wind attack angle α represents the spatial relationship between the approaching flow and the aluminum alloy box (or the electrical wire). The two-dimensional vibration measurement system is shown in Fig. 12(c).

The orientation of the 3-dimensional stay cable can be defined by three angles ϕ, β, and θ, as indicated in Fig. 12(a)and(b). The position angle ϕ represents the spatial relationship between the cable and the aluminum alloy box (or the electrical wire); The definition of $\phi = 0°$ and $\phi = 70°$ is shown in Fig. 12 (b), in which U_0 is the component of U perpendicular to the cable plane; θ is the inclination angle of stay cable;

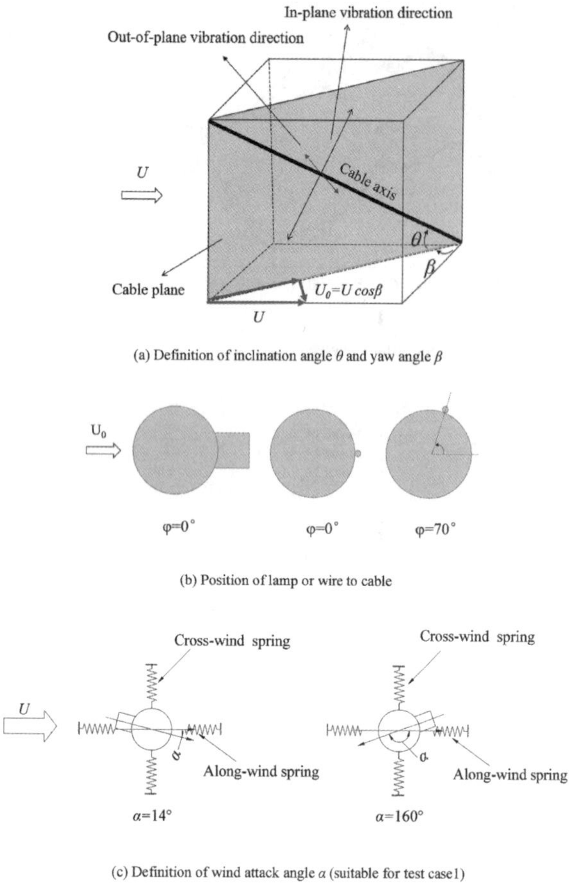

Fig. 12. Orientation of stay cable with the approaching flow.

β is the yaw angle of the cable plane relative to the approaching flow. The in-plane vibration direction refers to the direction perpendicular to the cable axis in the cable plane. The out-of-plane vibration direction refers to the direction perpendicular to the cable plane.

To conveniently adjust the inclination angle θ in the vibration measurement wind tunnel tests, a support system was specially designed, as shown in Fig. 13. The inclination angle of the stay cable can be changed by changing the fixed points on the four columns. The support system was installed on the turntable of the high-speed test section of the HD-2BLWT, and the yaw angle β can be conveniently adjusted by the turntable.

The stiffness of the elastic test model was provided by 8 springs connected to the two ends of test model. The natural frequency of the elastic test model was identified by free vibration decay method, together with the damping ratio. Fig. 14 and Fig. 15 present the time histories and its corresponding PSDs of the decrement displacement for Models A and B, respectively. To study the effect of structural damping on the wind-induced responses of the stay cable attached with aluminum alloy box, three levels of damping ratio, $\zeta = 0.1\%$, 0.3%, and 0.6%, were obtained by pasting adhesive tape on the springs. It can be found from Figs. 14 and 15 that the realized in-plane and out-of-plane natural frequencies are almost the same value. Natural frequency mainly depends on the spring stiffness and section model mass. The tape pasted on the spring affects the spring stiffness, which is the main reason for the difference in vibration frequency in the two vibration directions.

There are 5 types of test cases in the force measurement wind tunnel tests. In test case 1, Model A was horizontally suspended and the

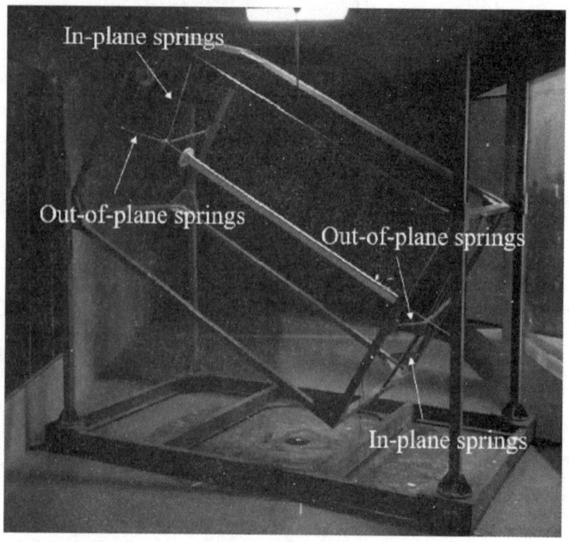

Fig. 13. Photo of the support system in the test section.

approaching wind was always perpendicular to the cable axis, i.e., $\theta =$ 0° and $\beta = 0°$. The wind attack angle α was fixed to be 14° and 160° which correspond to the maximum negative galloping coefficients as indicated in Table .2, and the wind velocity varied from 6.2 to 10.4 m/s with an interval of about 0.5 m/s. It should be pointed out that the wind speed interval of 0.5 m/s is a reference value. In the test, the value of wind speed is changed by changing the voltage. Affected by the air temperature and humidity, the wind speed interval can change slightly when changing the same voltage, which is about 0.5 m/s、and maybe 0.4 or 0.6. The damping ratios of 0.1%, 0.3%, and 0.6% were adopted. It should be noted that the condition of test case 1 was the same as that of the force measurement wind tunnel test, but was different from the prototype inclined cable. In test case 2, the inclination angle of the Model A was $\theta = 25°$, 30°, 35°, and 45°, and the position angle $\phi = 0°$, the yaw angle β was from 0° to 360° with an interval of 2°. The wind velocity in test case 2 was always 15 m/s. The dangerous range of yaw angle β can be found from test case 2 for the above-mentioned four inclination angles. In test case 3, the yaw angle was fixed in the dangerous range to obtain the effects of wind velocity and damping ratio. The wind velocity varied from 9.2 to 15.4 m/s and 6.3–15.6 m/s with an interval of about 0.5 m/s, and the damping ratios were 0.1%, 0.3% and 0.6%. Test case 4 was designed to find the dangerous range of yaw angle β for Model B. $\phi = 0°$ corresponded to the actual position of the wire on the stay cable and the selection of $\phi = 70°$ was based on the dangerous location of the upper rivulet in the rain-wind-induced vibration (Gu and Du, 2005). Han et al. (2021) studied the VIV response of a tapered light pole varying with the position of the weld. It was found that when the weld position angle was in the range of 55°–80°, the weld had a significant enhancing effect on the VIV of the light pole, and the displacements of the VIV were greater than those for the light pole without the weld. Alam et al. (2010) investigated the wake of a circular cylinder with two tripwires attached at azimuthal angle $a = \pm 10°–70°$. It was found that C_D, $C_{D,\ rms}$ and $C_{L,\ rms}$ were large at $a =$ 45–70°. The position angle $\phi = 70°$ seems to be a dangerous location. Test case 5 was used to study the effects of wind velocity for Model B. The detailed information for the test cases in vibration measurement wind tunnel tests is given in Table .4.

5.3. *Experimental results for 2-dimensional Model A (test case 1)*

Fig. 16 presents the variation of the amplitude of the horizontal

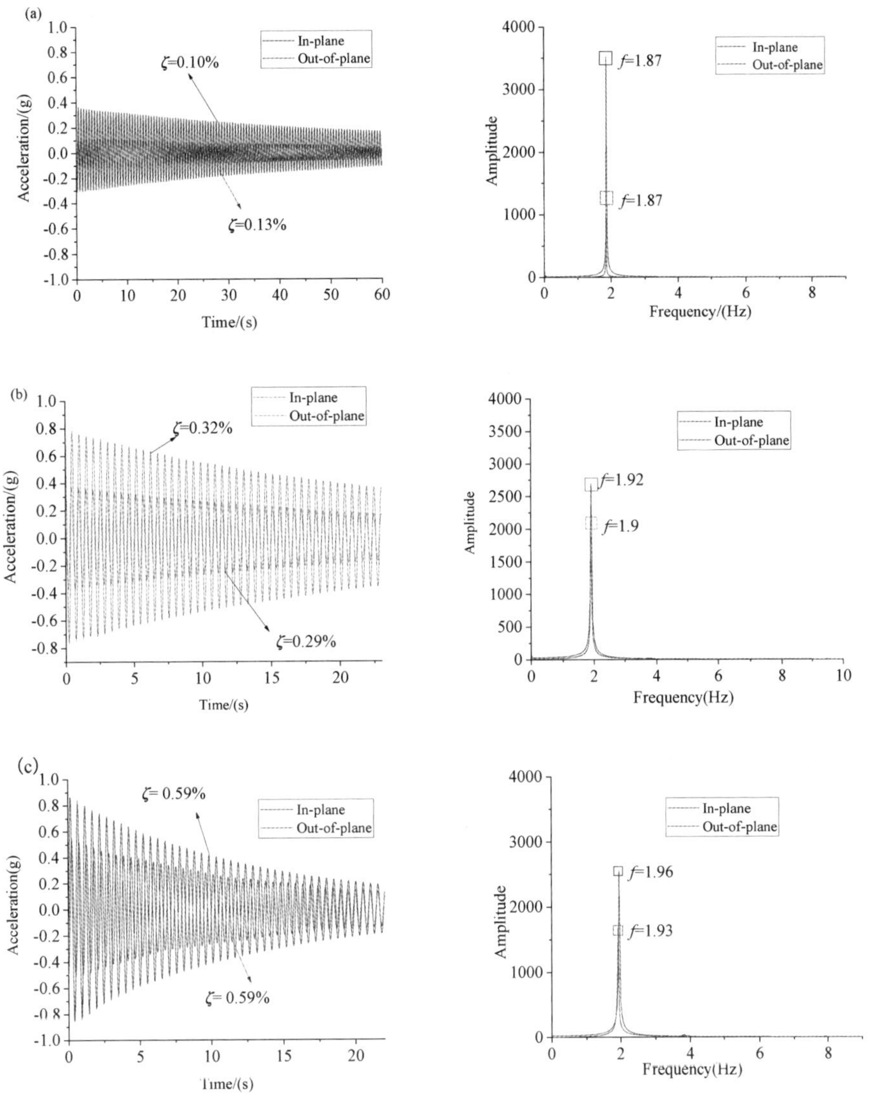

Fig. 14. Time histories and its corresponding PSDs of the decrement displacement of Model A for different structural damping: (a) $\zeta = 0.1\%$; (b) $\zeta = 0.3\%$; (b) $\zeta = 0.6\%$.

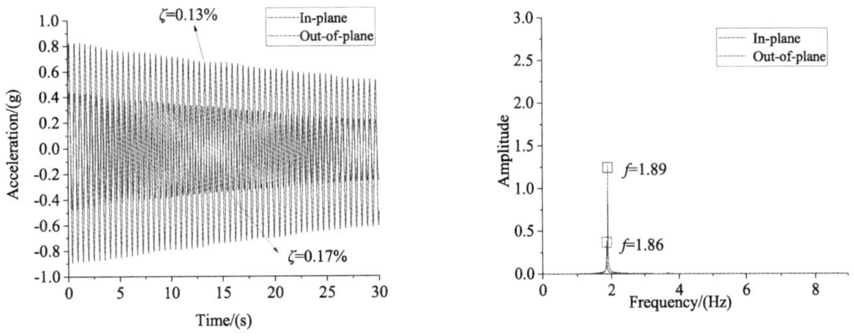

Fig. 15. Time histories and its corresponding PSDs of the decrement displacement of Model B for $\zeta = 0.1\%$.

Table 4
Test cases in the vibration measurement wind tunnel tests.

Test cases	Models	θ (°)	α (°)	ϕ (°)	β (°)	ζ (%)	Test wind velocity U (m/s)
1	Model A	0	14, 160	—	0	0.1, 0.3, 0.6	6.2–10.4, $\Delta U = 0.5$
2	Model A	25, 30, 35, 45	—	0	0–360, $\Delta\beta = 2$	0.1	15
3	Model A	25	—	0	38 154	0.1, 0.3,0.6	9.2–15.4, $\Delta U = 0.5$ 6.3–15.6, $\Delta U = 0.5$
4	Model B	30	—	0, 70	0–360, $\Delta\beta = 2$	0.1	15
5	Model B	30	—	70	206	0.1	5.6–11.6, $\Delta U = 1$

Model A with wind velocity under two dangerous wind attack angles (14° and 160°) for three levels of damping ratios (0.1%, 0.3%, and 0.6). It can be found from Fig. 16 that the galloping vibration can take place at a very low wind velocity. The critical reduced wind velocity is about 66.3 under the wind attack angle of 14°, which corresponds to the actual bridge wind velocity of 4.88 m/s. It appears that the amplitude in the cross-wind direction is significantly larger than that in the along-wind direction. The maximum nondimensional amplitude (A/D) in the cross-wind direction can reach about 1.4 for reduced wind velocity $U/fD = 105.9$. No obvious difference between the cable responses under the two wind attack angles of 14° and 160° is observed. It seems that the cable amplitude can be reduced a little for a given speed by increasing the structural damping ratio. However, the structural damping ratio seems to have little effect on the critical wind velocity.

According to classical galloping theory, critical wind velocity U_{cr} can be determined as

$$\frac{U_{cr}}{fD} = \frac{-4m\delta}{\rho D^2(C_L' + C_D)}$$

where C_D is the drag coefficient, C'_L refers to the derivate of lift coefficient with respect to wind incidence angle. According to the formula, there is a linear relationship between the critical reduced wind speed and the damping ratio. But from the experimental results, this linear relationship is not tenable. The critical reduced wind speed is the same when the damping ratio is 0.1% and 0.3% at $\alpha = 14°$. When the damping is increased from 0.3% to 0.6%, the damping is doubled. However, the critical reduced wind speed for galloping increased from 70 to 95, an increase of 0.35 times at $\alpha = 160°$. The Den Hartog criterion cannot effectively predict the relationship between critical reduced wind speed and damping.

5.4. *Experimental results for 3-dimensional Model A (test cases 2 and 3)*

The test model is horizontally arranged in test case 1, which is different from the prototype cable on cable-stayed bridges. The features of galloping vibration of an inclined (3-dimensional) cable might be significantly different from that of the horizontal one. Fig. 17 presents the variation of the amplitude of the inclined Model A with the yaw angle in the in-plane and out-of-plane directions for $U = 15$ m/s (test wind velocity), $\zeta = 0.1\%$, $\phi = 0°$, $\theta = 25°$, 35°, 40° and 45° (test case 2).

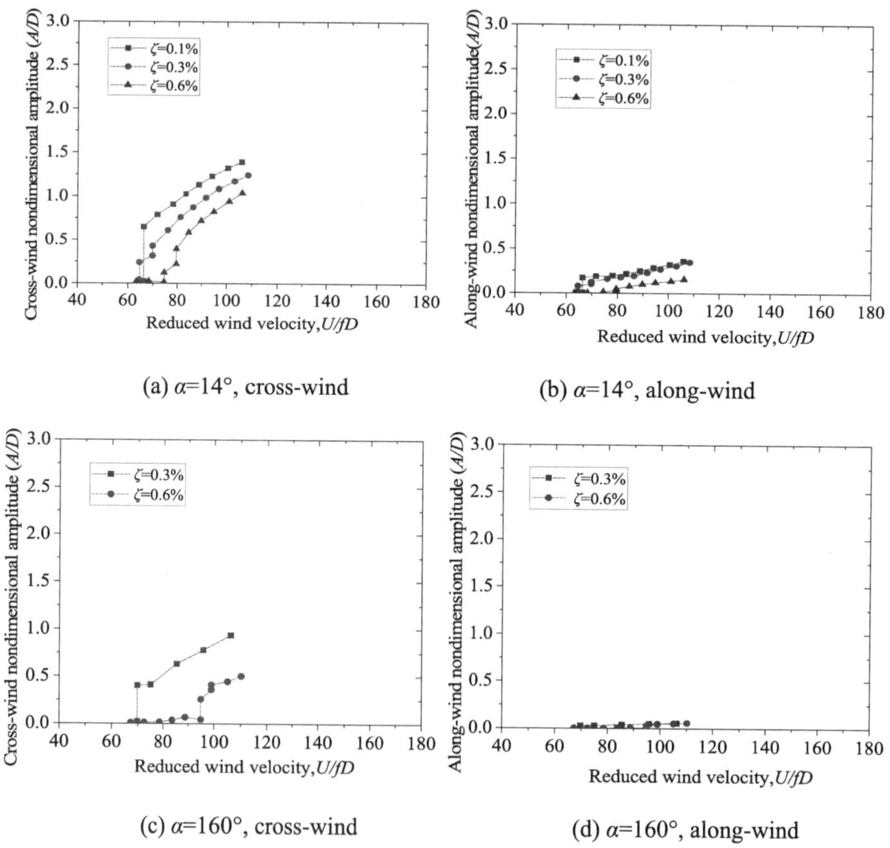

(a) α=14°, cross-wind

(b) α=14°, along-wind

(c) α=160°, cross-wind

(d) α=160°, along-wind

Fig. 16. Variation of the amplitude of the horizontal Model A with wind velocity (test case 1).

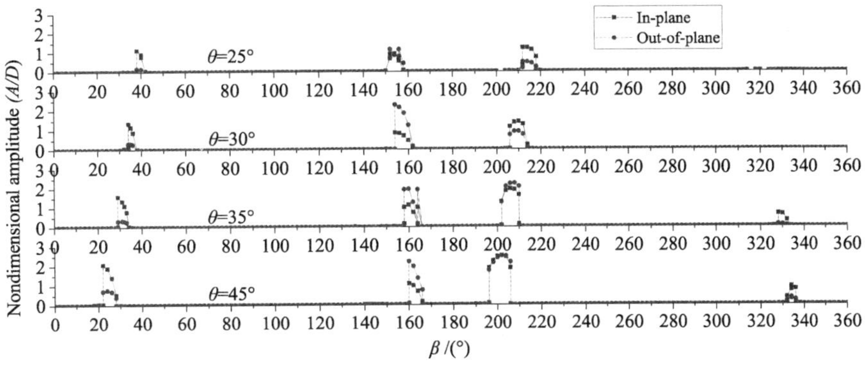

Fig. 17. Variation of the amplitude of the inclined Model A with the yaw angle in the in-plane and out-of-plane directions (test case 2).

The rectangular aluminum alloy box installed on the stay cables of the Kuimen Bridge corresponds to $\phi = 0°$. It can be found from Fig. 17 that the dangerous yaw angles of the 3-dimensional Model A are within the ranges of $\beta = 22°–42°$, $150°–166°$, $196°–220°$, and $332°–336°$. With the increase of inclination angle, the dangerous angle approaches 0° or 180°. It seems that the inclination angle of the test model has a significant effect on the range of the dangerous yaw angle.

Fig. 18 shows the variation of the amplitude of the inclined Model A in the in-plane and out-of-plane directions with the wind velocity for three levels of damping ratios (test case 4). The inclination angle is fixed to be 25° and the yaw angle is fixed to be 38° and 154°, which is in the dangerous range in Fig. 17. It can be found from Fig. 18 that the critical reduced wind velocities of the inclined Model A are respectively about 124 and 78 for $\beta = 38°$ and 154°, which corresponds to the actual bridge wind velocities of 9.13 m/s and 5.75 m/s. The measured critical wind velocity in the test is consistent with the results obtained from field observation, and the critical wind velocity is a little bit higher than that of the 2-dimensional test model, as shown in Fig. 18. This is a very low wind velocity that can be often encountered. It seems that the cable vibration can be reduced by increasing the damping ratio, which is the same as that of the 2-dimensional test model. However, the critical wind velocity is only 9–11 m/s when the damping ratio is 0.6%. This indicates that higher damping is needed to effectively mitigate the galloping vibration than the rain-wind induced vibration, for which 0.5% of damping ratio is enough (Li et al., 2013).

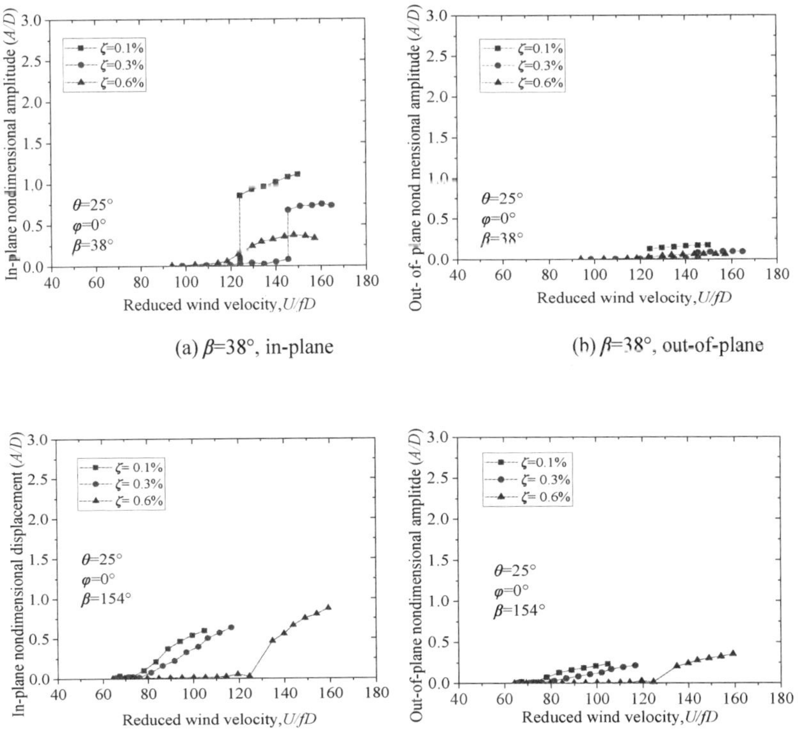

(a) β=38°, in-plane (b) β=38°, out-of-plane

(c) β=154°, in-plane (d) β=154°, out-of-plane

Fig. 18. Variation of the amplitude of the inclined Model A with the wind velocity (test case 3).

5.5. Experimental results for 3-dimensional Model B (test cases 4 and 5)

Fig. 19 shows the variation of the amplitude of the inclined Model B in the in-plane and out-of-plane directions with the yaw angle for $U = 15$ m/s, $\zeta = 0.1\%$, $\phi = 0°$, and $70°$, $\theta = 30°$ (test case 4). It can be found from Fig. 19 that when the position angle ϕ is equal to $0°$, the inclined Model B keeps calm for all the yaw angle β. However, the inclined Model B experiences large amplitude vibration when the position angle ϕ is $70°$ and the yaw angle β is near $206°$. Fig. 20 shows the variation of the amplitude of the inclined Model B with the wind velocity for $\phi = 70°$ and $\beta = 206°$ (test case 5). It can be found from Fig. 20 that the critical reduced wind speed is 59.3 and the corresponding actual bridge wind speed is 4.37 m/s. The nondimensional amplitude in the cross-wind direction reaches about 2.0 when the reduced wind velocity is 123. For the Kuimen Bridge, the electrical wire on the stay cables is at $\phi = 0°$. Therefore, the galloping vibration of the stay cable can be avoided by removing the aluminum alloy box from the aerodynamic point of view.

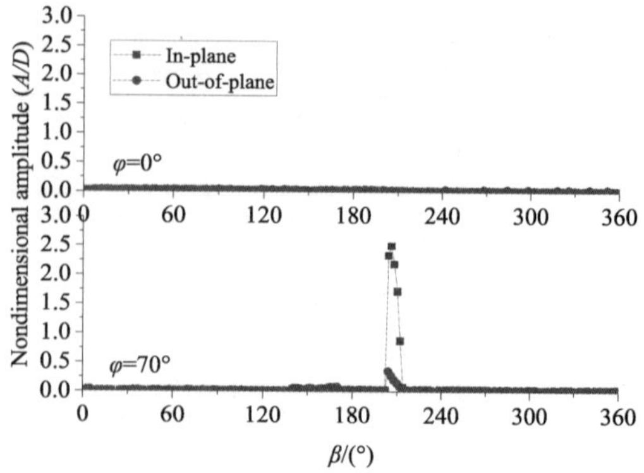

Fig. 19. Variation of the amplitude of the inclined Model B with the yaw angle (test case 4).

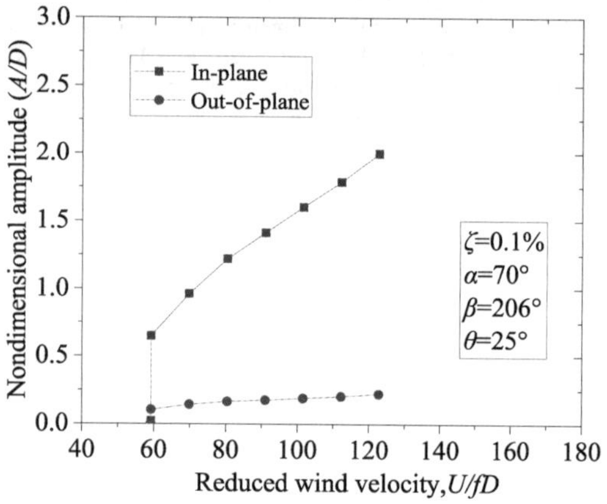

Fig. 20. Variation of the amplitude of the inclined Model B with the wind velocity for $\phi = 70°$ and $\beta = 206°$ (test case 5).

6. Concluding remarks

The vibration of stay cables caused by rectangular lamps is reported in the paper. The characteristic and countermeasures of the violent vibration of stay cables of the Kuimen Bridge, which were attached with a rectangular lamp, was studied by field observations and wind tunnel tests.

The energy peaks in the PSDs of the wind-induced responses extracted from the video are all around 0.60 Hz for the six stay cables, which corresponds to the fundamental frequency of these stay cables. The maximum peak to peak amplitude of the stay cable is more than $11D$, D is the diameter of the stay cable. It appears that the rectangular aluminum alloy box might be the key factor for the cable galloping. From the force test results of the 2-dimensional stay cable attached with lamps, it seems that the diameter of the stay cable has no obvious effect on the aerodynamic forces of the test models. The maximum negative values of the galloping force coefficient are as low as -12.9 near the wind attack angle of $14°$. This indicates that the galloping vibration of the stay cable attached with lamps might take place. The critical wind velocity obtained from 2-dimensional vibration tests is 4.88 m/s at $\zeta = 0.1\%$. The relationship between critical wind velocity and damping is nonlinear. It appears that the Den Hartog criterion cannot effectively predict the relationship between critical reduced wind speed and damping.

The dangerous yaw angles of the 3-dimensional Model A are within the ranges of $\beta = 22°–42°$, $150°–166°$, $196°–220°$, and $332°–336°$. With the increase of inclination angle, the dangerous angle approaches $0°$ or $180°$. The critical wind velocity of the 3-dimensional cable-lamp test model can be as low as 6.3 m/s (damping ratio $\zeta = 0.1\%$), which is a wind velocity often encountered. The critical wind velocity of galloping measured in the test is consistent with that observed in the field observations. It appears that the critical wind velocity increases with the increase of the structural damping ratio, but is only 9–11 m/s when the damping ratio is 0.6%. The vibration measurement wind tunnel tests based on the 3-dimensional cable-wire test model indicate that the stay cable attached with an electrical wire is aerodynamically stable if the position angle is $0°$, which is consistent with the actual situation. However, when the position angle is $70°$, it can also cause a large vibration of the stay cable. The position angle has an important influence on the vibration of stay cables, and the position of wires should be strictly controlled in practical engineering.

CRediT authorship contribution statement

An Miao: Conceptualization, Methodology, Formal analysis, Data curation, Investigation, Writing – original draft. **Li Shouying:** Conceptualization, Methodology, Resources, Writing – review & editing, Funding acquisition. **Liu Zhiwen:** Resources, Visualization, Methodology. **Yan Banfu:** Methodology, Software, Data curation. **Li Longan:** Validation, Investigation, Resources. **Chen Zhengqing:** Writing – review & editing, Supervision.

Declaration of competing interest

The authors declare that they have no known competing financial interests or personal relationships that could have appeared to influence the work reported in this paper.

Acknowledgements

This project is jointly supported by the National Key Research and Development Program of China (Grant No.2017YFC0703600 and No.2017YFC0703604) and the National Natural Science Foundation of China (Grant No.51578234).

Appendix A. Supplementary data

Supplementary data to this article can be found online at https://doi.org/10.1016/j.jweia.2021.104685.

References

Acampora, A., Macdonald, J., Georgakis, C.T., et al., 2014. Identification of aeroelastic forces and static drag coefficients of a twin cable bridge stay from full-scale ambient vibration measurements. J. Wind Eng. Ind. Aerod. 124 (1), 90–98.

Alam, M.M., Zhou, Y., Zhao, J.M., et al., 2010. Classification of the tripped cylinder wake and bi-stable phenomenon. Int. J. Heat Fluid Flow 31 (4), 545–560.
Benidir, A., Flamand, O., Gaillet, L., Dimitriadis, G., 2015. Impact of roughness and circularity-defect on bridge cables stability. J. Wind Eng. Ind. Aerod. 137, 1–13.
Chen, W.L., Zhang, Q.Q., Li, H., Hu, H., 2015. An experimental investigation on vortex induced vibration of a flexible inclined cable under a shear flow. J. Fluid Struct. 54, 297–311.
Cheng, S.H., Irwin, P.A., Tanaka, H., 2008b. Experimental study on the wind-induced vibration of a dry inclined cable-Part II: proposed mechanisms. J. Wind Eng. Ind. Aerod. 96 (12), 2254–2272.
Cheng, S.H., Larose, G.L., Savage, M.G., Tanaka, H., Irwin, P.A., 2008a. Experimental study on the wind-induced vibration of a dry inclined cable—Part I: Phenomena. J. Wind Eng. Ind. Aerod. 96 (12), 2231–2253.
Den Hartog, J.P., 1932. Transmission line vibration due to sleet. Trans. Am. Inst. Electr. Eng. 51 (4), 1074–1081.
Gao, G.Z., Zhu, L.D., 2017. Nonlinear mathematical model of unsteady galloping force on a rectangular 2:1 cylinder. J. Fluid Struct. 70, 47–71.
Gu, M., Du, X.Q., 2005. Experimental investigation of rain-wind-induced vibration of cables in cable-stayed bridges and its mitigation. J. Wind Eng. Ind. Aerod. 93, 79–95.
Gu, M., Du, X.Q., Li, S.Y., 2009. Experimental and theoretical simulations on wind-rain-induced vibration of 3-d rigid stay cables. J. Sound Vib. 320 (1–2), 184–200.
Han, Y., Zhou, X., Wang, L., et al., 2021. Experimental investigation of the vortex-induced vibration of tapered light poles. J. Wind Eng. Ind. Aerod. 211 (4), 104555.
Hikami, Y., Shiraishi, N., 1988. Rain-wind induced vibrations of cables in cable stayed bridges. J. Wind Eng. Ind. Aerod. 29 (1–3), 409–418.
Hua, X.G., Wang, C.Q., Li, S.L., Chen, Z.Q., 2020. Experimental investigation of wind-induced vibrations of main cables for suspension bridges in construction phases. J. Fluid Struct. 93, 102846.
Jakobsen, J.B., Andersen, T.L., Macdonald, J.H.G., Nikitas, N., Larose, G.L., Savage, M. G., et al., 2012. Wind-induced response and excitation characteristics of an inclined cable model in the critical Reynolds number range. J. Wind Eng. Ind. Aerod. 110, 100–112.
Jing, H.Q., Xia, Y., Li, H., Xu, Y.L., Li, Y.L., 2017. Excitation mechanism of rain-wind induced cable vibration in a wind tunnel. J. Fluid Struct. 68, 32–47, 2017.
Li, S.Y., Chen, Z.Q., Wu, T., et al., 2013. Rain-wind-induced in-plane and out-of-plane vibrations of stay cables. ASCE J. Eng. Mech. 139 (12).
Li, S.Y., Chen, Z.Q., Dong, G.C., Luo, J.H., 2014b. Aerodynamic stability of stay cables incorporated with lamps: a case study. Wind Struct. 18, 83–101.
Li, S.Y., Deng, Y.C., Huang, J., et al., 2019. Experimental investigation on aerodynamic interference of two kinds of suspension bridge hangers – science direct. J. Fluid Struct. 90, 57–70.
Li, Y.L., Xu, Y.L., Xiang, H.Y., Chen, B., Shum, K.M., Liao, H.L., 2014. Analysis of rain-wind induced cable vibration using spatially measured aerodynamic coefficients, Advances in. Struct. Eng. Earthq. Eng. 17 (7), 961–977.
Ma, W.Y., Macdonald, J.H.G., Liu, Q.K., 2017a. Aerodynamic characteristics and excitation mechanisms of the galloping of an elliptical cylinder in the critical Reynolds number range. J. Wind Eng. Ind. Aerod. 171, 342–352.
Ma, W.Y., Macdonald, J.H.G., Liu, Q.K., et al., 2017b. Galloping of an elliptical cylinder at the critical Reynolds number and its quasi-steady prediction. J. Wind Eng. Ind. Aerod. 168, 110–122.
Liu, Z.W., Shen, J.S., Li, S.Q., Chen, Z.Q., Ou, Q.B., Xin, D.B., 2021. Experimental study on high-mode vortex-induced vibration of stay-cable and its aerodynamic countermeasures. J. Fluid Struct. 100, 103195.
Ma, W.Y., Liu, Q.K., Matsumoto, M., 2019. Excitation of the large-amplitude vibrations of a circular cylinder under normal wind conditions in the critical Reynolds number range. J. Fluid Struct. 84, 318–328.
Matsumoto, M., Yagi, T., Hatsuda, H., Shima, T., Tanaka, M., Naito, H., 2010. Dry galloping characteristics and its mechanism of inclined/yawed cables. J. Wind Eng. Ind. Aerod. 98 (6), 317–327.
Matsumoto, M., Yagi, T., Shigemura, Y., Tsushima, D., 2001. Vortex-induced cable vibration of cable-stayed bridges at high reduced wind velocity. J. Wind Eng. Ind. Aerod. 89 (7–8).
Matteoni, G., Georgakis, C.T., 2015. Effects of surface roughness and cross-sectional distortion on the wind-induced response of bridge cables in dry conditions. J. Wind Eng. Ind. Aerod. 136, 89–100.
Nguyen, C.H., Freda, A., Solari, G., Tubino, F., 2015. Experimental investigation of the aeroelastic behavior of a complex prismatic element. Wind Struct. 20 (5), 683–699.
Nikitas, N., Macdonald, J.H.G., 2015. Aerodynamic forcing characteristics of dry cable galloping at critical Reynolds numbers. Eur. J. Mech. B Fluid 49, 243–249.
Nikitas, N., Macdonald, J.H.G., Jakobsen, J.B., Andersen, T.L., 2012. Critical Reynolds number and galloping instabilities: experiments on circular cylinders. Exp. Fluid 52 (5), 1295–1306.
Raeesi, A., Cheng, S.H., Ting, D.S.-K., 2014. A two-degree-of-freedom aeroelastic model for the vibration of dry cylindrical body along unsteady air flow and its application to aerodynamic response of dry inclined cables. J. Wind Eng. Ind. Aerod. 130, 108–124.
Simiu, E., Yeo, D.H., 2019. Wind Effects on Structures, fourth ed. WILEY Blackwell, Hoboken, USA.
Yan, B.F., Li, D.R., Chen, W.B., et al., 2020. Mode shape–aided cable force determination using digital image correlation. Struct. Health Monit. (6) 147592172095216.
Zuo, D., Jones, N.P., Main, J.A., 2008. Field observation of vortex- and rain-wind-induced stay-cable vibrations in a three-dimensional environment. J. Wind Eng. Ind. Aerod. 96 (6–7), 1124–1133.

Engineering Structures 234 (2021) 111985

Contents lists available at ScienceDirect

Engineering Structures

journal homepage: www.elsevier.com/locate/engstruct

Experimental investigation on wake-induced vibrations of the hangers of suspension bridges based on three-dimensional elastic test model

Deng Yangchen, Li Shouying *, Chen Zhengqing

Key Laboratory for Wind and Bridge Engineering of Hunan Province, College of Civil Engineering, Hunan University, Changsha 410082, China

ARTICLE INFO

Keywords:
Hangers
Wake-induced vibrations
3-D elastic test model
Wind tunnel tests
Structural damping
Helical wires
Spacers

ABSTRACT

Large amplitude wind-induced vibrations of the hangers have been observed on several suspension bridges, and the underlying mechanism is not clearly clarified until now. In this study, the mechanism of wake-induced vibration of the hanger was studied by using wind tunnel tests based on three-dimensional (3-D) elastic test model. First, a new method was proposed to design 3-D elastic test model to accurately simulate a long cable in wind tunnel with a limited height. Second, taking the No.2 hanger of the Xihoumen Bridge as the background, a 3-D elastic hanger model was designed and manufactured by using the new method, and a series of wind tunnel tests were carried out within the spatial region of $3 \le P \le 10$ and $0° \le \alpha \le 20°$ to obtain the wake-induced responses of the hanger. The results show that obvious vibrations of the downstream cable are observed under all wind velocities tested ($3 \sim 14$ m/s), and the unstable region of the downstream cable tends to increase with the increase of wind velocity. Third, the effects of the structural damping on the wake-induced vibrations of the hangers were studied in detail. The results show that the wake-induced cable vibration of the hangers is not totally mitigated even if the structural damping ratio is as high as 1.37%. Finally, the effectiveness of two kinds of countermeasures (helical wires and spacers) on the wake-induced vibrations of the hangers of suspension bridges was experimentally studied. The results show that the helical wire can suppress the wake-induced vibration of the test case of $P = 7$, $\alpha = 5°$, but have no effect for the test case of $P = 3.5$, $\alpha = 20°$. It seems that the spacer is a reasonable countermeasure to mitigate the wake-induced vibrations of the hangers of suspension bridges.

1. Introduction

The hangers of large-span suspension bridges are extremely prone to wind-induced vibrations due to its light weight, small structural damping and low natural frequency. Serious wind-induced vibrations of the hangers may cause fatigue problem, and even threaten the safety of the bridge. Large amplitude wind-induced vibrations of the hangers have been observed on several suspension bridges all over the world, for example, the Great Belt East Bridge in Denmark [1], the Akashi-Kaikyo Bridge in Japan [2], and the Xihoumen Bridge in China [3]. The engineers were often confused that effective countermeasures on different bridges had different effects. On the Great Belt East Bridge, several measures, including helical wires, cross ties, spacers and tuned liquid dampers (TLD), have proved ineffective. Until 2005, the hanger vibration was effectively mitigated by simultaneously using the spacers and hydraulic dampers [1]. On the Akashi-Kaikyo Bridge, the helical wires were used to successfully suppress the hanger vibration [2]. On the Xihoumen Bridge, the vibration of the hanger was successfully mitigated by rigid spacers [4,5]. This indicates that the mechanisms of the wind-induced vibration for different bridges are different.

Some researchers have studied the underlying mechanism of the hanger vibration. Laursen *et al.* [1] speculated that the ice accretions might lead to the vibration of the hanger based on a field observation on the Great Belt East Bridge. Kashima *et al.* [6] concluded that vortex-induced vibration and wake-induced flutter might be partly responsible for the hanger vibration by monitoring the Akashi-Kaikyo Bridge. Zhang *et al.* [7] proposed that the vibration of the main cable can result in large amplitude vibration of the hangers near the pylon based on numerical analyses. Li *et al.* [8] and Chen *et al.* [9–11] both found that the wake of the pylon might lead to large amplitude vibration of the hangers through wind tunnel tests. The authors conducted a series of field measurements on the No.2 hanger of the Xihoumen Bridge, and found that the oscillation amplitude of the downstream cables was obviously larger than that of the upstream ones, and the trajectory of the cable was close to a clockwise ellipse [3]. This indicates that the aerodynamic interference between the cables of the hanger might play a key

* Corresponding author.
E-mail addresses: ycdeng@hnu.edu.cn (D. Yangchen), shyli@hnu.edu.cn (L. Shouying), zqchen@hnu.edu.cn (C. Zhengqing).

https://doi.org/10.1016/j.engstruct.2021.111985
Received 27 November 2020; Received in revised form 14 January 2021; Accepted 23 January 2021
Available online 12 February 2021

role for the hanger vibration.

During the past few decades, the aerodynamic interference of two circular cylinders has been extensively studied, mainly focusing on the flow pattern, aerodynamic force, response characteristics and excitation mechanisms. Zdravkovich [12], Zdravkovich [13], Sumner [14], Bearman [15] and Zhou [16] conducted comprehensive reviews about the progress achieved over the last few decades. In recent years, some researchers have paid their attentions on the aerodynamic interference between the cables of the cable-supported bridges. Based on a series of wind tunnel tests to measure the mean drag and lift coefficients of a cable in the wake of another one, Li *et al.* [17] established a 2-D theoretical model for the wake-induced vibration of the hanger by using quasi-steady theory. They found that the hanger vibration can be evoked by the aerodynamic interference between the cables, and aerodynamic negative stiffness can be considered as the key factor for the hanger vibration. Deng *et al.* [18] extended the quasi-steady theoretical model of Li *et al.* [17] to a 2-D unsteady model, and found similar results. Li *et al.* [19] carried out a series of wind tunnel tests based on the 2-D elastic model to study the aerodynamic interference of the PWS and WR cables, and they found that the aerodynamic stability of WR cables is better than that of PWS cables. Similarly, based on the wind tunnel test of 2-D model, Jafari *et al.* [20] and Kim *et al.* [21] studied the aerodynamic interference of stay cables.

All of the above mentioned studies simplify the continuous cable into a segmental rigid element with two degree-of-freedoms, which cannot fully reflect the characteristics of the actual cable of the cable-supported bridges [22]. In addition, the wind-induced nonlinear behavior of the cable-supported bridge is significant [23,24], so it is necessary to carry out a 3-D elastic model wind tunnel test to study the aerodynamic interference of the hangers. However, it is very difficult to design a 3-D elastic test model of the cable to strictly simulate the prototype. The diameter of the cable model will be very small (obvious Reynolds number effects) for a small length scale, or the length of the cable model will be very long (unable to install in wind tunnel) for a large length scale. Hua *et al.* [5] designed a 3-D elastic hanger model with a small length scale of 1:36. At this time, the diameter of the hanger model was only 2.5 mm. Li *et al.* [25], Chen *et al.* [26], He *et al.* [27] and Jafari *et al.* [28] both designed a 3-D elastic stay cable model with a large length scale, but the length of the test model was far shorter than its target value. Similar 3-D elastic test models of the hangers were also designed by Li *et al.* [8] and Chen *et al.* [9].

In this study, a new method was firstly proposed to design a 3-D elastic test model to accurately simulate a long cable in wind tunnel with a limited height. Then, taking the No.2 hanger of the Xihoumen Bridge as the background, a 3-D elastic hanger model was designed and manufactured by using the new method, and a series of wind tunnel tests were carried out to obtain the wake-induced responses of the hanger. The effects of structural damping on the wake-induced vibrations of the hangers were studied in detail. Finally, the effectiveness of two kinds of countermeasures (helical wires and spacers) were experimentally investigated.

2. Similarity relationship of test model

The length of the hanger model was limited by the height of the test section of wind tunnel. Therefore, the diameter of the cable model will be extremely small if the length scales in the directions of length and diameter are the same. This may lead to extremely low test wind velocity and obvious Reynolds number effect. Furthermore, it is difficult to accurately measure the response of the small cable model in the wind tunnel test. The nth natural frequency of the cable is determined by [29],

$$f_n = \frac{n}{2l}\sqrt{\frac{T}{m}} \tag{1}$$

where, l, T and m are the length, the axial force and the mass per unit length of the cable, respectively. It could be found from Equation (1) that the cable frequency can be adjusted to its target value by changing the axial force T if the length of the cable l is not totally simulated in the test model. In this study, the length of the cable model is far shorter than it target value according to the length scale, but the natural frequencies of the cable model can be totally simulated by decreasing the axial force of the cable model.

Assuming that the length, diameter, axial force and mass per unit length of a prototype cable are l_p, D_p, T_p and m_p, respectively, its corresponding values of the cable model under the strict similarity ratios can be respectively l_m, D_m, T_m and m_m. Here we supposed that l_m is far larger than the height of the test section of wind tunnel. The motion equation of this cable model for the nth mode can be written as [30],

Table 1
Main similarity scales of the test model.

Scales	Values
Diameter	$\lambda_D = 1/4$
Velocity	$\lambda_U = 1/1$
Mass per unit length	$\lambda_M = 1/16$
frequency	$\lambda_f = 4/1$
Damping ratio	$\lambda_\xi = 1/1$

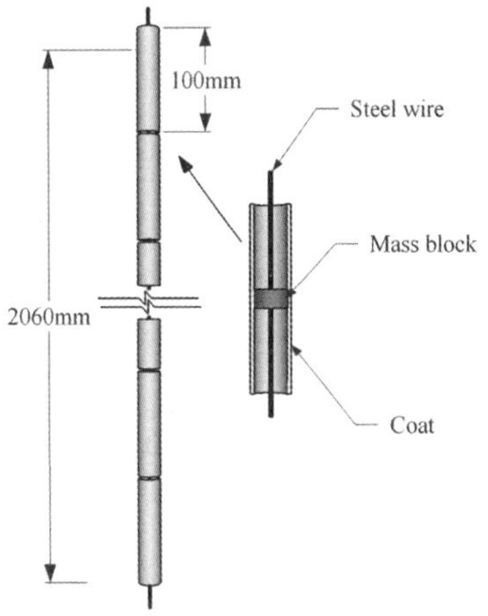

Fig. 1. Sketch map of cable models.

Fig. 2. Photo of cable models in the wind tunnel.

$$\ddot{Y}_n + 2\zeta_n\omega_n\dot{Y}_n + \omega_n^2 Y_n = F_n/M_n \tag{2}$$

where, Y_n, F_n and M_n are the generalized displacement, force and mass of the nth mode, respectively; ζ_n and ω_n are the damping ratio and circular frequency of the nth mode, respectively.

Assuming the nth mode shape $\varphi_n(x)$ is [29],

$$\varphi_n(x) = \sin n\pi x/l_m \tag{3}$$

then the generalized mass M_n can be determined by,

$$M_n = \int_o^{l_m} m_m \cdot \varphi_n^2(x)\mathrm{d}x = \frac{m_m l_m}{2} \tag{4}$$

Assuming a uniform force p_m is acting on the cable model, F_n can be determined by,

$$F_n = \int_o^{l_m} p_m \cdot \varphi_n(x)\mathrm{d}x = \left[1 + (-1)^{n-1}\right]\frac{p_m l_m}{n\pi} \tag{5}$$

Substituting Equations (4) and (5) into Equation (2),

$$\ddot{Y}_n + 2\zeta_n\omega_n\dot{Y}_n + \omega_n^2 Y_n = \left[1 + (-1)^{n-1}\right]\frac{2p_m}{n\pi m_m} \tag{6}$$

Then shorten the length of the cable model l_m to l_{ms}, but keep the diameter and mass per unit length of the cable model (D_m and m_m) unchanged. At the same time, the axial force T_m becomes smaller (T_{ms}) to satisfy the following relationship,

$$\sqrt{T_m}/l_m = \sqrt{T_{ms}}/l_{ms} \tag{7}$$

This indicates that the nth circular frequency of the shorter cable model is still ω_n, which is the same as that in Equation (6). The nth generalized mass and force of the shorter cable model, M_{ns} and F_{ns}, can be respectively written as,

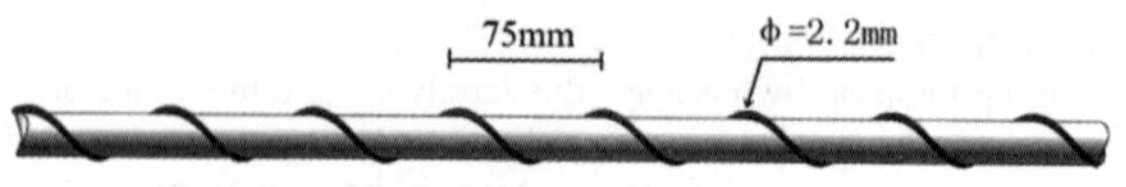

Fig. 4. Schematic diagram of the helical wire setting.

$$M_{ns} = \int_o^{l_{ms}} m_m \cdot \varphi_n^2(x)\mathrm{d}x = \frac{m_m l_{ms}}{2} \tag{8}$$

$$F_{ns} = \int_o^{l_{ms}} p_m \cdot \varphi_n(x)\mathrm{d}x = \left[1 + (-1)^{n-1}\right]\frac{p_m l_{ms}}{n\pi} \tag{9}$$

Substituting Equations (8) and (9) into Equation (2), the motion equations of the shorter cable model can be obtained, and is the same as Equation (6). This means that the long cable model can be reasonably represented by the shorter one.

3. Outline of wind tunnel tests

3.1. Design and manufacture of test model

The prototypes of the test model is the No.2 hanger of the Xihoumen Bridge, which is the second hanger near the bridge tower. The diameter, length and mass per unit length of the cables of the No.2 hanger are respectively 88 mm, 160 m and 33 kg/m, and the first natural frequency

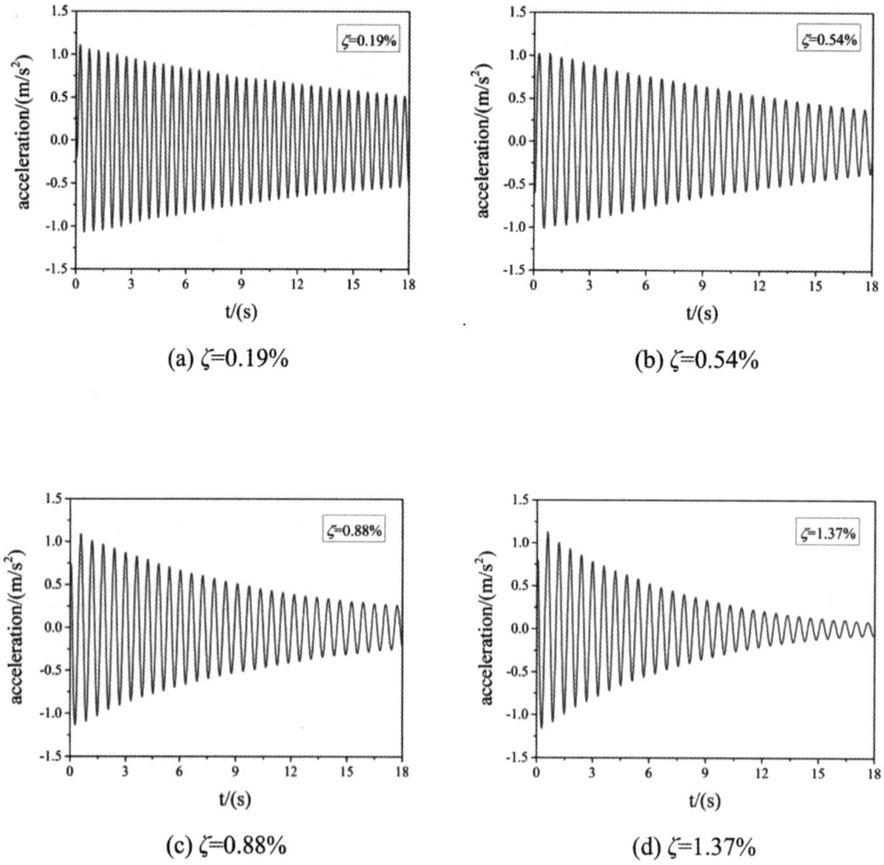

Fig. 3. Free-vibration decay curve of the acceleration for damping ratio of 0.19%, 0.54%, 0.88% and 1.37%

(a)Only on upstream cable

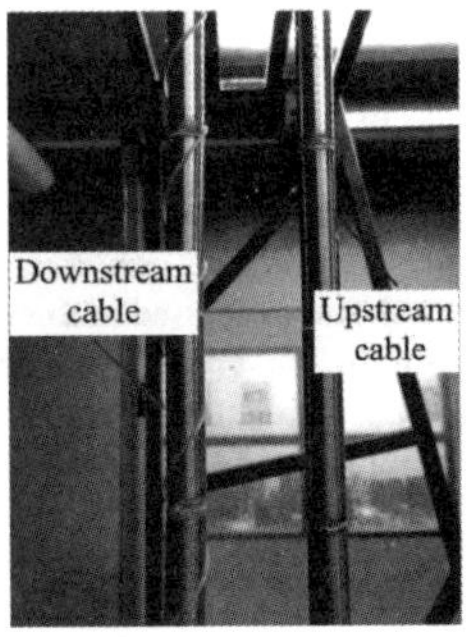

(b) Only on downstream cable

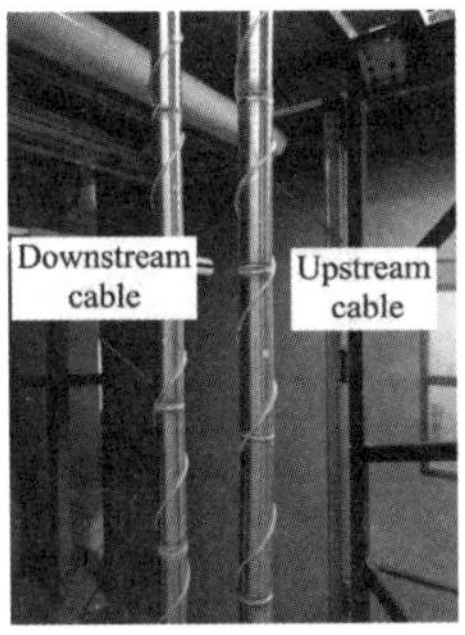

(c) On both the upstream and the downstream cables

Fig. 5. Photos of cable models attached with helical wire.

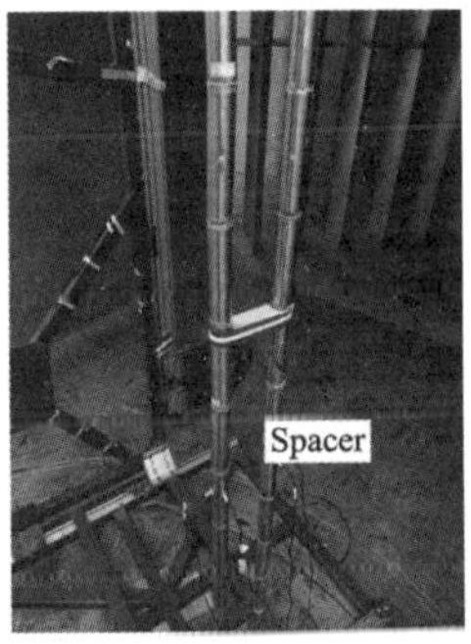

Fig. 6. Photo of cable models installed with a single spacer at the mid-span.

is 0.40 Hz. For the hangers of suspension bridges, the gravity is vertical, while the displacement is horizontal. This indicates that the Froude number is not necessary to accurately simulate. So the length scale in the diameter direction was 1/4, and the velocity scale was set to be 1/1. Table 1 gives the main similarity scales of the test model. As discussed in Section 2, the length of the cable model, which is 2.06 m, is far shorter than the target value (40 m).

This study focused on the hanger with two separate cables. Each cable model is made of core steel wire, mass block and coats, similar with that by He *et al* [27]. Core steel wires with a diameter of 2 mm were used in the test model to provide the stiffness of the cable model, and the shape of the cable was guaranteed by coats. A mass block was designed to connect the steel wire with each coat, and the mass block was also used as additional mass. Sketch map of the cable models is shown in Fig. 1. The length of each coat is 100 mm, and the gap between two

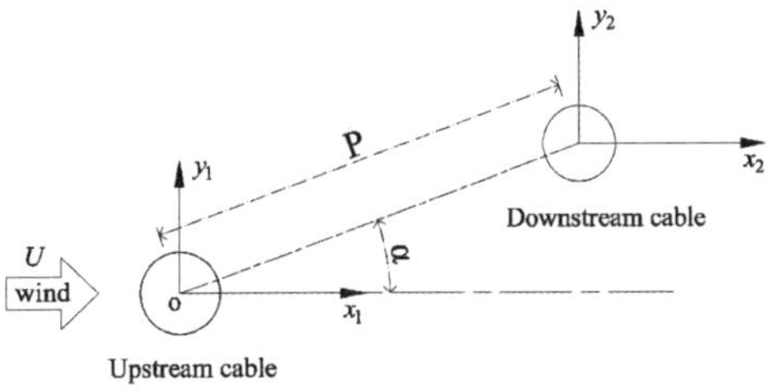

Fig. 7. Definition of the spatial position of the two cables.

adjacent coats is 3 mm. It should be noted that the surface of this cable model is smooth, while the hangers of the Xihoumen Bridge are made of wire ropes cable with rough surface. The two cable models were installed on a steel frame, which was fixed on the turntable in the test section to conveniently adjust its wind attack angle. Two chutes were arranged at the top and bottom of the steel frame to adjust the distance between the two cable models. The photo of the cable models installed on the steel frame in the wind tunnel is shown in Fig. 2. It should be noted that the axial force of the cable model was not directly measured by force sensors, but by controlling the cable frequency.

3.2. Structural damping ratio of cable model

The original structural damping ratio of the cable models was $\zeta = 0.19\%$, as shown in Fig. 3 (a). In order to study the effects of structural damping ratio on wake-induced vibrations of the hangers, the damping ratio of the cable models was adjusted by pasting different types of adhesive tapes upon the gaps between the coats, and the damping ratios were measured by means of free-vibration decay method, as shown in Fig. 3. Four levels of structural damping ratio of the cable models, including 0.19%, 0.54%, 0.88% and 1.37%, were obtained.

3.3. Helical wire

The countermeasure of helical wire was studied in the wind tunnel tests. The diameter of the helical wire was 2.2 mm and the pitch was 75 mm, as shown in Fig. 4. There were three kinds of test conditions, the first was to wound the helical wire only on the upstream cable, the second was to wrap the helical wire only on the downstream cable, and the third was to wound the helical wire on both the upstream and downstream cables. Fig. 5 shows the photos of cable models attached with helical wire.

3.4. Spacer

The countermeasure of spacer was studied in the wind tunnel tests. The spacer adopted in the test was made of high-density polystyrene foam, and is very light only with a mass of 11 g. Fig. 6 shows the photo of cable models installed with a single spacer at the mid-span.

3.5. Test cases

Wind tunnel tests were conducted in the HD-3 Wind Tunnel in Hunan

Table 2
Test cases of the wind tunnel tests.

Parameters	Values
Spatial position	$3 \le P \le 10$, $0^\circ \le \alpha \le 20^\circ$
Spatial interval	$\triangle P = 0.5$, $\triangle\alpha = 5^\circ$ $(3 \le P \le 6)$; $\triangle P = 1$, $\triangle\alpha = 5^\circ$ $(6 < P \le 10)$
Wind velocity, U (m/s)	$3 \sim 14$, $\triangle U = 1$
Reynolds number ($\times 10^4$)	$0.44 \sim 2.05$

University, Changsha, China, which is an open-circuit type of boundary layer wind tunnel with its test section of 2.5 m in height by 3.0 m in width and 17.0 m in length. The maximum wind velocity in this test section is 20 m/s.

Four accelerometers, which were installed at the height of 0.75 m over the lower end of the cable, were used to measure the along-wind and cross-wind responses of the cable models. The accuracy of the accelerometers is 0.001 g. The sampling frequency was 1000 Hz, and the sampling time was 60 s. The data were collected with a DH5920 multichannel signal acquisition system developed by Donghua Testing TechnologyCo., Ltd.

The relative spatial position of the two cables is defined by two parameters, P and α, as shown in Fig. 7. In which, P represents the ratio of the center-to-center spacing to the diameter D; α is the wind attack angle. In consideration of the dimensions of the prototype hangers on suspension bridges, the wind tunnel tests were carried out within the spatial region of $3 \le P \le 10$ and $0° \le \alpha \le 20°$. The interval is $\triangle P = 0.5$, $\triangle\alpha = 5°$ when $3 \le P \le 6$, and $\triangle P = 1$, $\triangle\alpha = 5°$ when $6 < P \le 10$. Uniform flow was used in the wind tunnel tests, and the wind velocity U was from 3 m/s to 14 m/s with an interval of $\triangle U = 1$ m/s, corresponding to Reynolds number of 0.44×10^4-2.05×10^4. Test cases of the wind tunnel test are shown in Table 2.

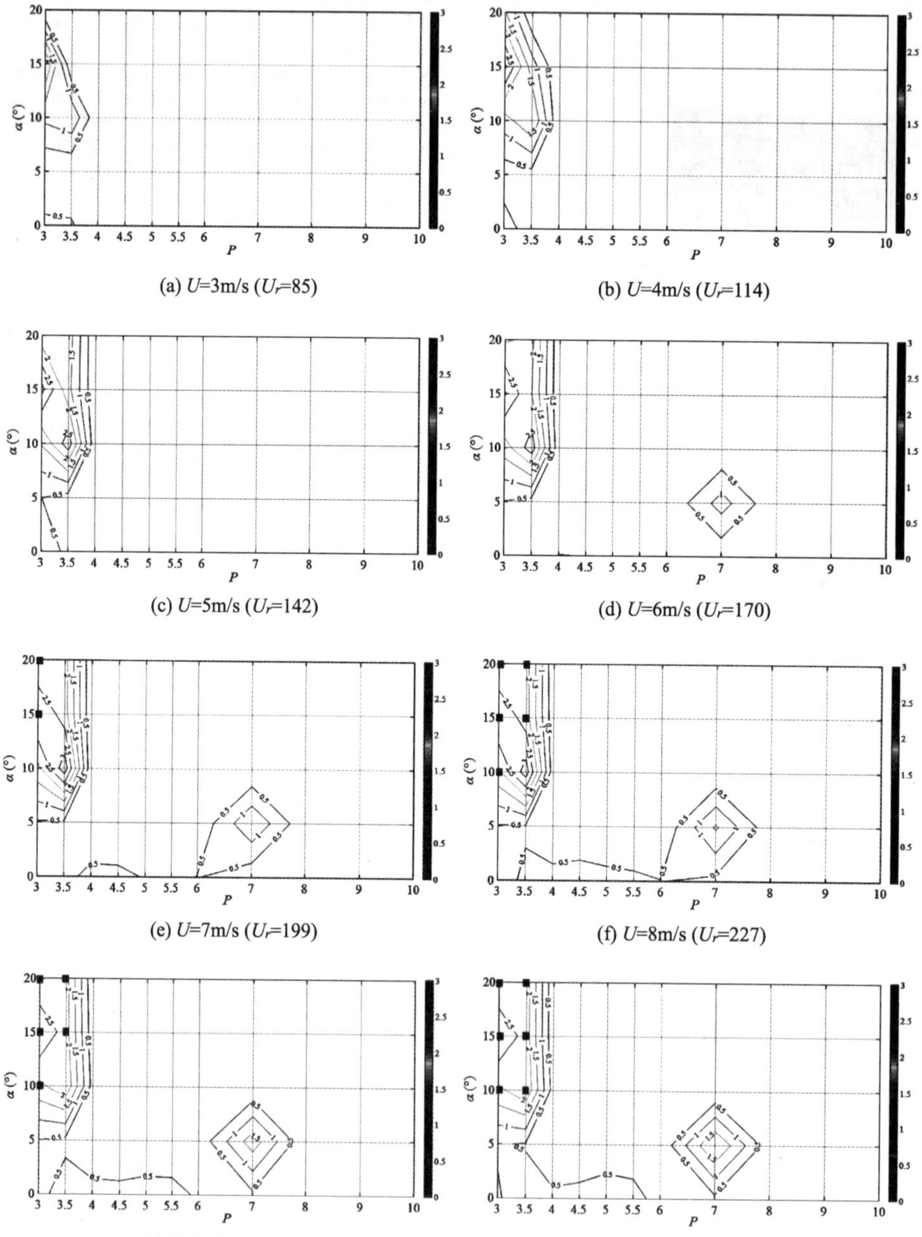

(a) U=3m/s (U_r=85)

(b) U=4m/s (U_r=114)

(c) U=5m/s (U_r=142)

(d) U=6m/s (U_r=170)

(e) U=7m/s (U_r=199)

(f) U=8m/s (U_r=227)

(g) U=9m/s (U_r=256)

(h) U=10m/s (U_r=284)

Fig. 8. Spatial distribution of the one-side resultant amplitude of the downstream cable for different wind velocities.

4. Responses of aeroelastic model

4.1. *Spatial distributions of cable amplitude*

Fig. 8 presents the spatial distributions of the maximum one-side resultant amplitude (A_{max}/D) of the downstream cable for different wind velocities of 3 ~ 14 m/s (Reduced wind velocity $U_r = U/fD$ from 85 to 398). It should be noted that the maximum value of the amplitude of the cable is the most important parameter affecting the safety of the hanger, which can be used to directly judge whether the cables will collide, so only the maximum value of the amplitude is given in this study. During the wind tunnel tests, cable collisions were found under some test cases and this test cases are marked by red points in Fig. 8. It can be found from Fig. 8 that obvious oscillation of the downstream cable is observed under all wind velocities tested from 3 to 14 m/s. Moreover, it appears that the unstable region tends to increase with the increase of wind velocity. When $U \leq 5$ m/s ($U_r \leq 142$), the obvious oscillation of the downstream cable mainly occurs at the space region of $3 \leq P \leq 4$ and $5° \leq \alpha \leq 20°$. With further increase of wind velocity, for example $6 \leq U \leq 12$ m/s ($170 \leq U_r \leq 341$), more unstable space region within $6 \leq P \leq 8$ and $0° \leq \alpha \leq 10°$ is found. When $U = 14$ m/s ($U_r = 369$), it seems that obvious vibration of the downstream cable is observed even if the center-to-center spacing of the two cables is as large as 10*D*. It is worth noting that the center-to-center spacing between cables of the hanger of the Xihoumen Bridge is 3.4*D* in the longitudinal direction and 6.8*D* in the transverse direction. Obvious hanger vibration was observed on the Xihoumen Bridge [3].

The upstream cable remains stationary in most of the test cases, except for the test case of $P = 3.5$, $\alpha = 0°$, which has a maximum amplitude about 1.2*D* for $U = 14$ m/s, as indicated in Fig. 9. This may be related to flow pattern of the shear layer reattachment, where the shear layer from the upstream cylinder reattaches on the surface of the downstream cylinder, and there is a strong backflow between the two cables, which will lead to the negative drag force for the downstream cable [3].

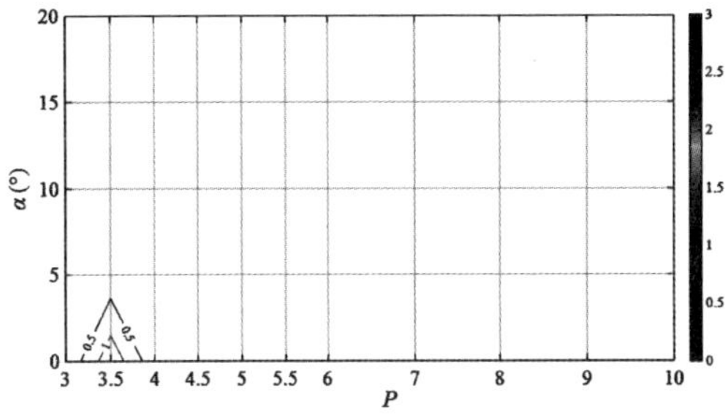

Fig. 9. Spatial distribution of the one-side resultant amplitude of the upstream cable for $U = 14$ m/s ($U_r = 398$).

4.2. *Variation of cable amplitude with wind velocity*

In consideration of the actual spacing between the cables of the No.2 hanger of the Xihoumen Bridge, three typical test cases ($P = 3.5$, $\alpha = 0°$; $P = 3.5$, $\alpha = 20°$; $P = 7$, $\alpha = 15°$) were taken to analyze the detailed responses. Fig. 10 presents the variation of the maximum along and cross-wind cable amplitudes with the wind velocity for the three typical cases.

For the test case of $P = 3.5$, $\alpha = 0°$, the upstream cable keeps calm with its maximum amplitude lower than 0.2*D* when $U \leq 11$ m/s ($U_r \leq 313$). When $U \geq 12$ m/s, the upstream cable starts to be evoked, and it appears that the upstream cable vibrates nearly in the along-wind direction, as shown in Fig. 10 (a). Compared with the upstream cable, the downstream cable is evoked at a very low wind velocity of 3 m/s ($U_r = 85$) under the test case of $P = 3.5$, $\alpha = 0°$, as indicated in Fig. 10 (b). An obvious decrease of cable amplitude is found when the wind velocity increases from 3 to 6 m/s. When $6 \leq U \leq 14$ m/s ($170 \leq U_r \leq 398$), the cable starts to vibrate again and its maximum amplitude is lower than 0.8*D*. It seems that the cross-wind amplitude is obviously larger than the along-wind one in the test case of $P = 3.5$, $\alpha = 0°$. Some studies have shown that the vibration at low reduced wind velocity ($U_r = 85$) might be vortex-induced vibration, and the vibration at higher wind velocity ($170 \leq U_r \leq 398$) might be wake galloping or wake-induced flutter

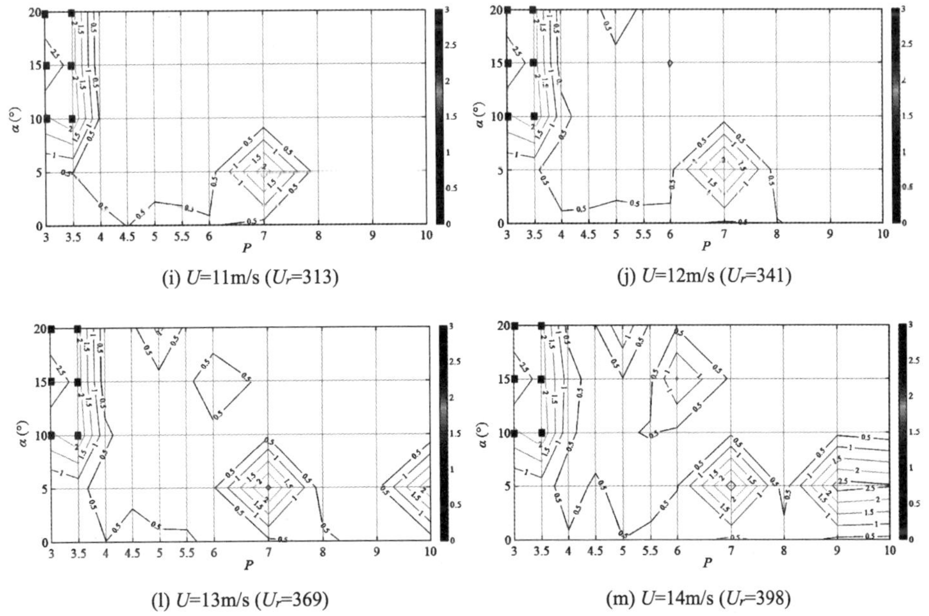

(i) U=11m/s (U_r=313)

(j) U=12m/s (U_r=341)

(l) U=13m/s (U_r=369)

(m) U=14m/s (U_r=398)

Fig. 8. (*continued*).

[31,32]. It can be seen from Fig. 10 (b) that the trajectory of the downstream cable under the test case of $P = 3.5$, $\alpha = 0°$ is approximately a straight line, which was also found in wind tunnel tests on sectional model [19].

For the test case of $P = 3.5$, $\alpha = 20°$, it is obvious that the onset wind velocity of the instability of the downstream cable is about $U = 5$ m/s ($U_r = 142$). When the wind velocity is higher than 5 m/s, the amplitude of the downstream cable increases rapidly with the increase of wind velocity. It seems that the along-wind amplitude is almost the same as that of the cross-wind, and the vibration trajectory is elliptical, as shown in Fig. 10 (c). It should be mentioned that the two cables collided when wind velocity reached 8 m/s. So the wind tunnel tests under this test case stopped at $U = 8$ m/s ($U_r = 227$).

For the test case of $P = 7$, $\alpha = 15°$, the amplitude of the downstream cable slowly increases with the increase of the wind velocity, as shown in Fig. 10 (d). It appears that there is no obvious onset wind velocity of the instability. Similar to the test case of $P = 3.5$, $\alpha = 20°$, cable amplitudes in along-wind and cross-wind directions of the test case of $P = 7$, $\alpha = 15°$ are almost the same, and its trajectory is also elliptical, as shown in Fig. 10 (d).

4.3. Power spectral density

Fig. 11 present the power spectral density (PSD) of the along-wind acceleration of the upstream and downstream cables at different reduced wind velocities for the test case of $P = 3.5$, $\alpha = 0°$. It can be found that the upstream cable vibrates in form of the first mode with very weak energy when $U_r \leq 313$ ($U = 11$ m/s). When $U_r = 369$ ($U = 13$ m/s, higher than onset wind velocity), obvious energy peaks are found at the first three natural frequency of the cable. On the other hand, the downstream cable vibrates in form of higher modes (even 10th mode), and it is not always that the first-order mode vibration dominates. The differences between the PSDs of $U_r = 85$ and $U_r \geq 142$ indicate that their vibration mechanisms are totally different, as discussed in Fig. 10 (b).

Fig. 12 present the PSD of along-wind acceleration of the downstream cable at different reduced wind velocities for the test cases of $P = 3.5$, $\alpha = 20°$ and $P = 7$, $\alpha = 15°$. For the test case of $P = 3.5$, $\alpha = 20°$, no obvious energy peaks are found when $85 \leq U_r \leq 114$. When wind velocity is higher than its onset velocity ($U_r = 142$), the PSD of the cable vibration has multiple energy peaks, but the first-order mode vibration completely dominates. For the test case of $P = 7$, $\alpha = 15°$, the PSD of the downstream cable vibration has multiple energy peaks with no obvious dominant mode when $170 \leq U_r \leq 227$. When the reduced wind velocity reaches 284, the energy near the third mode starts to increase and become obviously larger than that of other modes. The features of the PSD of cross-wind acceleration are nearly the same as those of along-wind one.

5. Suppression of wake-induced vibration

5.1. Effects of the structural damping

Figs. 13 and 14 respectively present the variation of the one-side resultant cable amplitudes with the reduced wind velocity for the four levels of structural damping under the two typical test cases of $P = 3.5$, $\alpha = 0°$ and $P = 3.5$, $\alpha = 20°$.

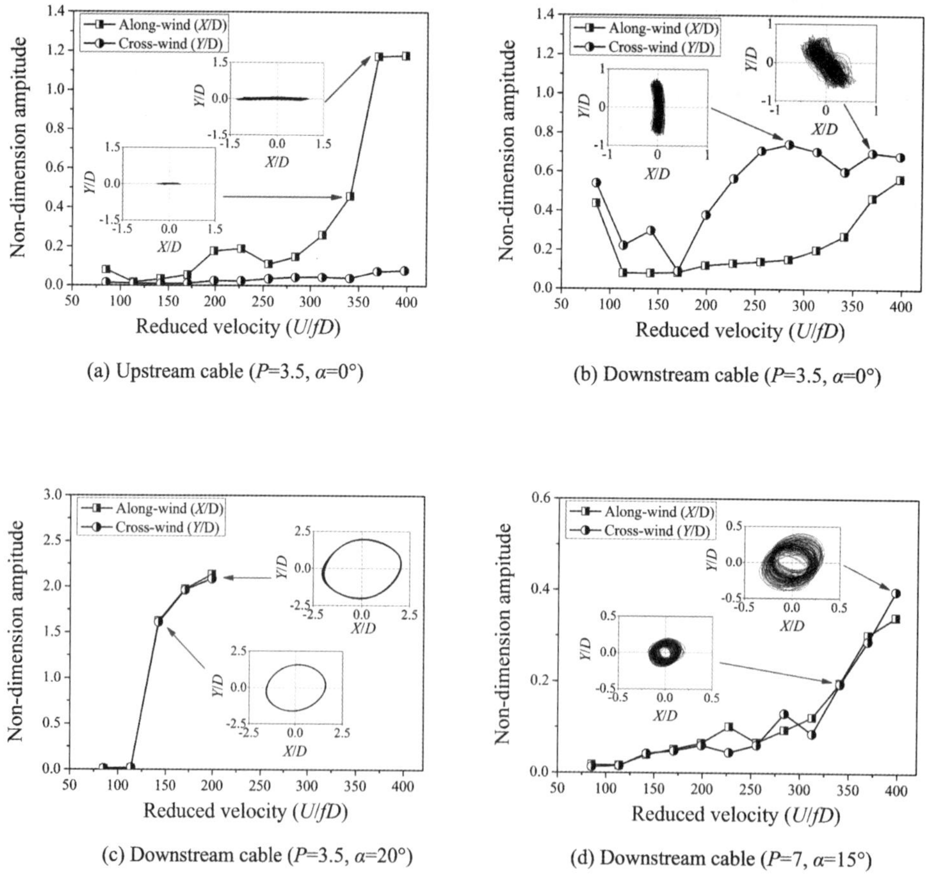

(a) Upstream cable (P=3.5, α=0°)

(b) Downstream cable (P=3.5, α=0°)

(c) Downstream cable (P=3.5, α=20°)

(d) Downstream cable (P=7, α=15°)

Fig. 10. The variation of the maximum amplitudes of cable with the reduced wind velocity.

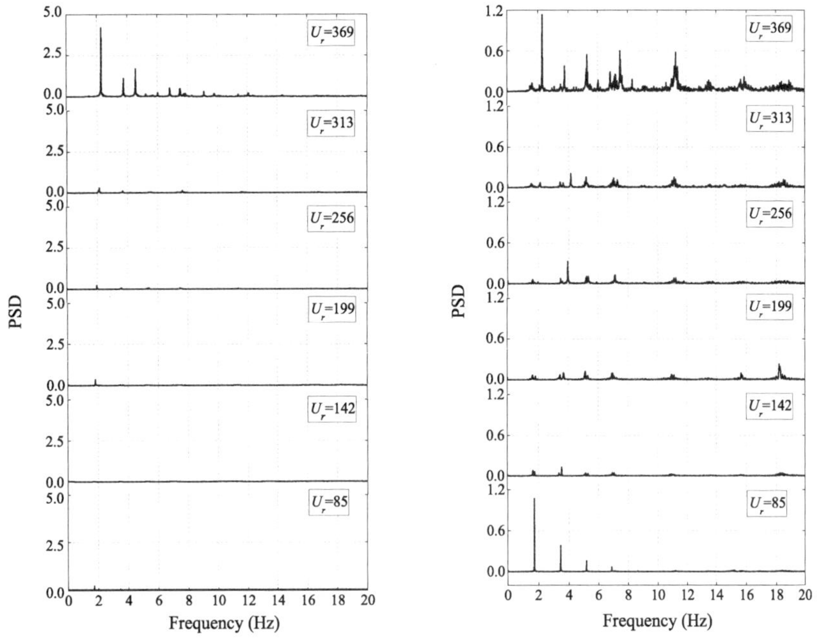

(a) Upstream cable (b) Downstream cable

Fig. 11. PSD of along-wind acceleration of the upstream and downstream cables for different wind velocity ($P = 3.5$, $\alpha = 0°$).

In the test case of $P = 3.5$, $\alpha = 0°$, the upstream cable vibration is well mitigated when the damping ratio increases to 0.88%, as indicated in Fig. 13 (a) and (b). It is strange that the upstream cable amplitude for $\zeta = 0.19\%$ is a little bit smaller than that for $\zeta = 0.54\%$. On the other hand, the downstream cable amplitude always decreases with the increase of the structural damping. However, it seems that the downstream cable vibration is not well mitigated even if the structural damping ratio reaches 1.37%. In the case of $P = 3.5$, $\alpha = 20°$, the downstream cable amplitude has no obvious change when the structural damping ratio increases from 0.21% to 0.88%. When the damping ratio reaches 1.37%, the downstream cable amplitude is well mitigated at low wind velocity ($U < 10$ m/s). It appears that the onset wind velocity increases from 4

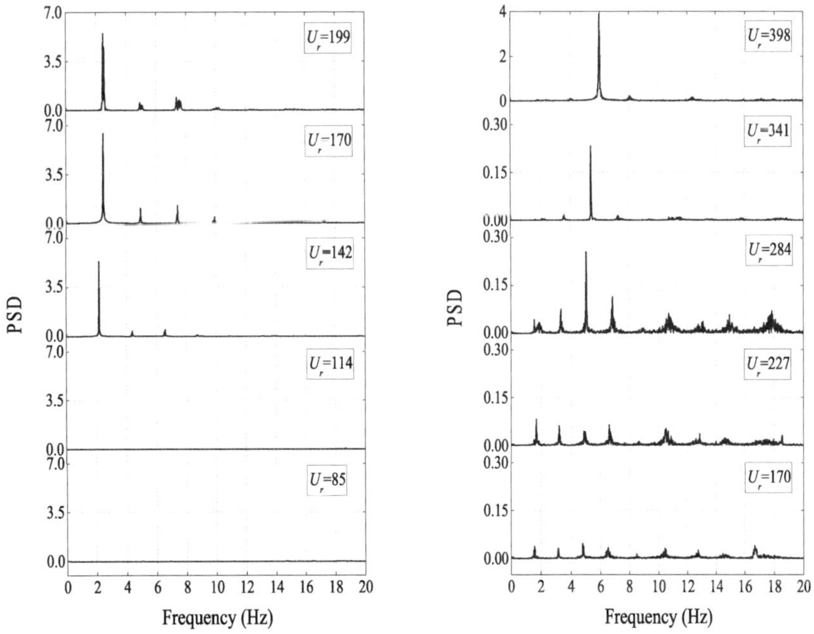

(a) P=3.5, α=20° (b) P=7, α=15°)

Fig. 12. PSD of along-wind acceleration of the downstream cable for different reduced wind velocities ($P = 3.5$, $\alpha = 20°$; $P = 7$, $\alpha = 15°$).

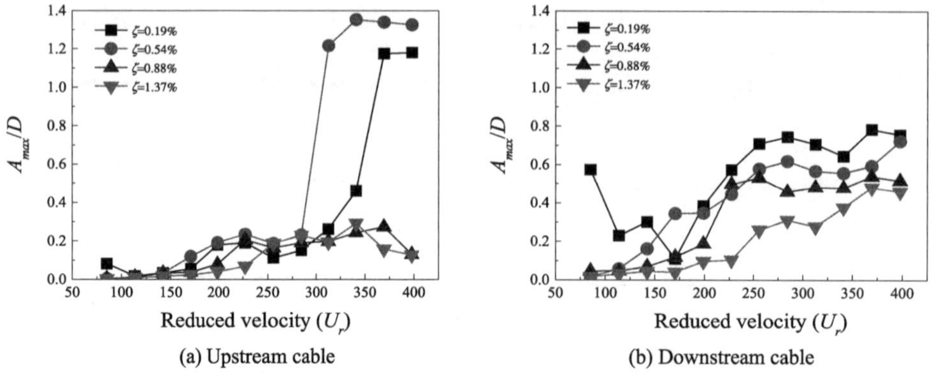

Fig. 13. Effects of structural damping on the cable amplitude for the test case of $P = 3.5$, $\alpha = 0°$.

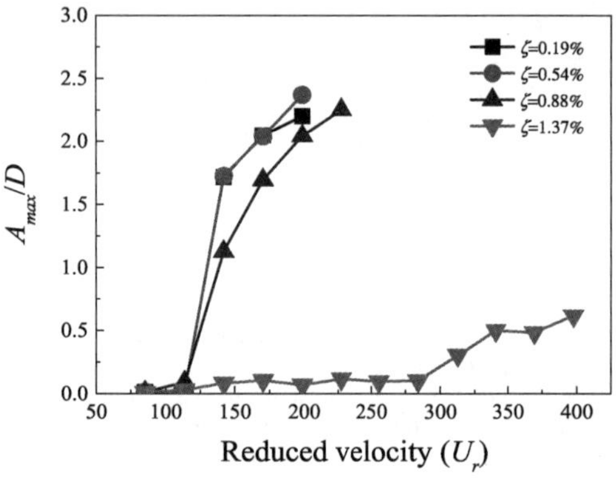

Fig. 14. Effects of structural damping on the downstream cable amplitude for the test case of $P = 3.5$, $\alpha = 20°$.

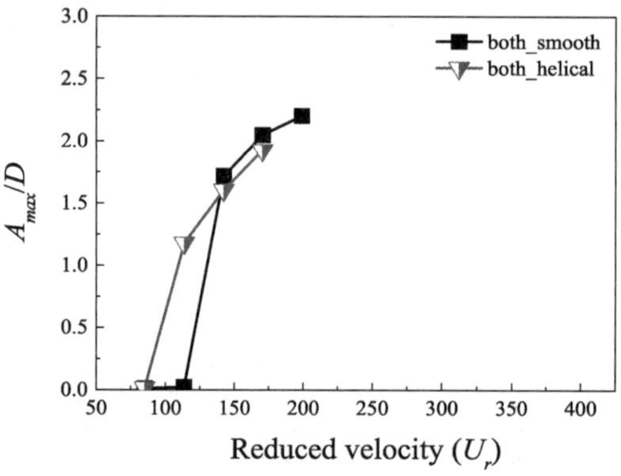

Fig. 16. Effects of helical wire on the downstream cable amplitude for the test case of $P = 3.5$, $\alpha = 20°$.

m/s ($U_r = 114$) to 10 m/s ($U_r = 284$) when the structural damping ratio increases from 0.88% to 1.37%. The one-side resultant cable amplitude is about 0.62D for $\zeta = 1.37\%$ and $U = 14$ m/s. The results show that the wake-induced cable vibration of the hangers is not totally mitigated even if the structural damping ratio is as high as 1.37%. The main reason might be related to the negative aerodynamic stiffness, which has been discussed in detail by Deng *et al.* [18] and Li *et al.* [17].

5.2. *Effects of the helical wire*

Fig. 15 ~ 17 present the effects of the helical wire on the one-side resultant cable amplitudes for the three typical test cases ($P = 3.5$, $\alpha = 0°$; $P = 3.5$, $\alpha = 20°$; $P = 7$, $\alpha = 5°$) with structural damping ratio of 0.19%. In the case of $P = 3.5$, $\alpha = 0°$, the results of the three kinds of test conditions are included, whereas only the test condition for both the upstream and the downstream cables with helical wire is discussed in the rest two test cases ($P = 3.5$, $\alpha = 20°$; $P = 7$, $\alpha = 5°$).

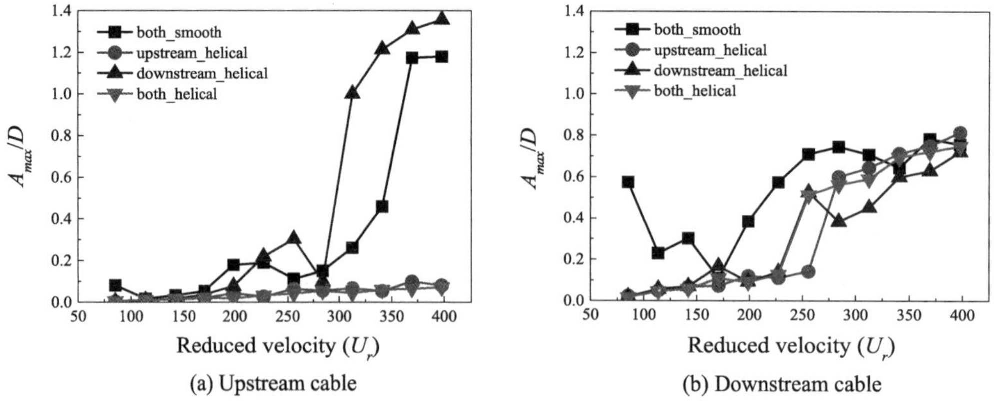

Fig. 15. Effects of helical wire on the cable amplitude for the test case of $P = 3.5$, $\alpha = 0°$.

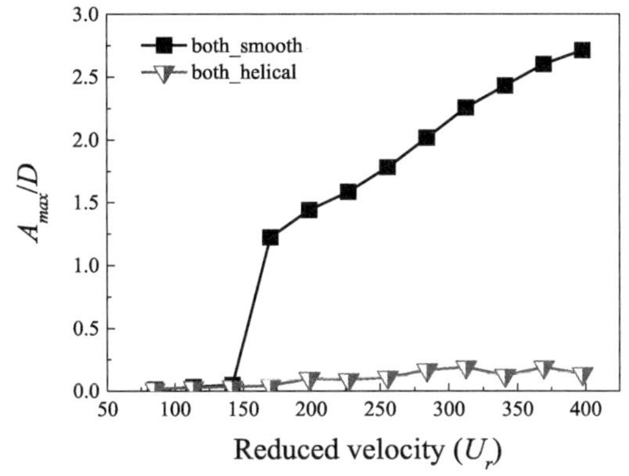

Fig. 17. Effects of helical wire on the downstream cable amplitude for the test case of $P = 7$, $\alpha = 5°$.

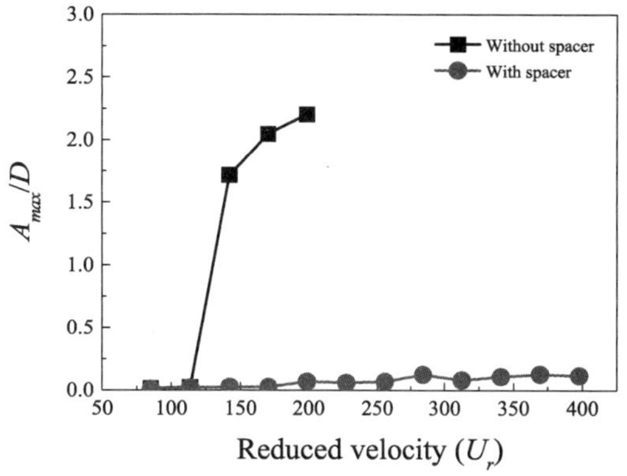

Fig. 19. Effects of spacer on the downstream cable amplitude for the test case of $P = 3.5$, $\alpha = 20°$.

In the test case of $P = 3.5$, $\alpha = 0°$, the upstream cable has no obvious vibration when the upstream cable is wounded by helical wire, as indicated in Fig. 15 (a). But it appears that the upstream vibration cannot be well mitigated by wrapping helical wire on the downstream cable. On the other hand, the helical wire has no effects on the downstream cable vibration under the three kinds of test conditions, as shown in Fig. 15 (b). In the test case of $P = 3.5$, $\alpha = 20°$, the helical wire also has no effects on the downstream cable vibration, as shown in Fig. 16. However, it seems that the downstream cable oscillation is totally mitigated by the helical wire for the test case of $P = 7$, $\alpha = 5°$, as indicated in Fig. 17. This indicates that the mechanisms of the wake-induced vibration of the hangers under the two test cases ($P = 3.5$, $\alpha = 20°$; $P = 7$, $\alpha = 5°$) are completely different.

5.3. Effects of the spacer

Figs. 18 and 19 present the effects of the spacer on the one-side cable amplitudes for the two typical test cases ($P = 3.5$, $\alpha = 0°$; $P = 3.5$, $\alpha = 20°$) with structural damping ratio of 0.19%. It can be found from Figs. 18 and 19 that the cable vibration can be obviously mitigated by installing a spacer at the mid-span under the two typical test cases. In the test case of $P = 3.5$, $\alpha = 0°$, the upstream cable amplitude decreases from 1.18D to 0.26D by installing the spacer at the wind velocity of 14 m/s, and the downstream cable amplitude decreases from 0.78D to 0.27D. In the test case of $P = 3.5$, $\alpha = 20°$, the downstream cable amplitude is very small under all the test wind velocities. It seems that the spacer is a reasonable countermeasure to mitigate the wake-induced vibration of the hangers of suspension bridges because the stiffness of the cable can be obviously enhanced by shortening its effective length. Several spacers were installed on the hangers of the Xihoumen Bridge in the summer of 2015. Since then, no significant hanger vibration was observed.

6. Concluding remarks

Wake-induced vibrations of twin-cable hanger of suspension bridge were investigated base on wind tunnel tests by using 3-D elastic test model. First, a new method was proposed to design a 3-D elastic test model to accurately simulate a long cable in wind tunnel with a limited height. Second, taking the No.2 hanger of the Xihoumen Bridge as the background, a 3-D elastic hanger model was designed and manufactured by using the new method, and a series of wind tunnel tests were carried out within the spatial region of $3 \le P \le 10$ and $0° \le \alpha \le 20°$ to obtain the wake-induced responses of the hanger. The results show that obvious vibrations of the downstream cable are observed under all wind velocities tested (3 ~ 14 m/s), and the unstable region of the downstream cable tends to increase with the increase of wind velocity. No obvious oscillation of the upstream cable is found except for the test case of $P = 3.5$, $\alpha = 0°$, which has a vibration in along-wind direction. Third, Four levels of structural damping ratio, including 0.19%, 0.54%, 0.88% and 1.37%, were achieved on the cable model, and the effects of the structural damping on the wake-induced vibrations of the hangers were studied in detail. The results show that the wake-induced cable vibrations of the hangers is not totally mitigated even if the structural damping ratio is as high as 1.37%. Finally, the effectiveness of two kinds of countermeasures (helical wires and spacers) on the wake-induced vibrations of the hangers of suspension bridges were

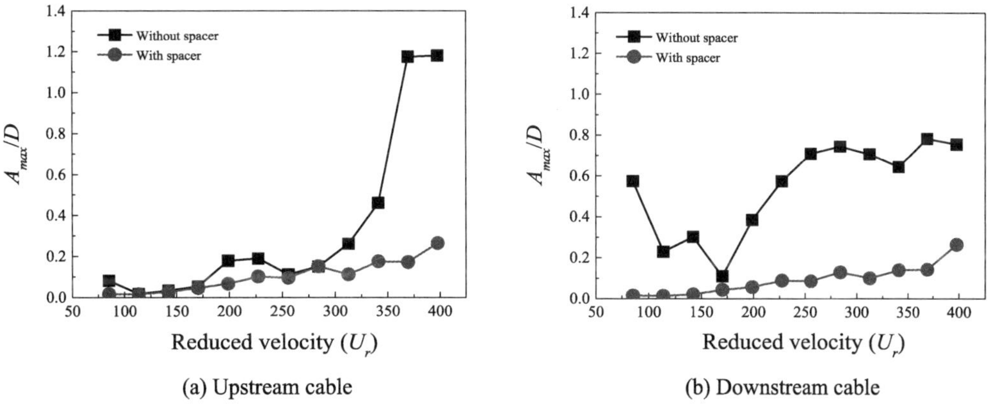

Fig. 18. Effects of spacer on the cable amplitude for the test case of $P = 3.5$, $\alpha = 0°$.

experimentally studied. The results show that the helical wire can suppress the wake-induced vibration of the test case of $P = 7$, $\alpha = 5°$, but have no effect for the test cases of $P = 3.5$, $\alpha = 20°$. It seems that the spacer is a more reasonable countermeasure to mitigate the wake-induced vibration of the hangers of suspension bridges.

Declaration of Competing Interest

The authors declared that there is no conflict of interest.

Acknowledgments

This project is jointly supported by the National Natural Science Foundation of China (51578234), the National Key Research and Development Program of China (2017YFC0703604) and Hunan Provincial Innovation Foundation For Postgraduate (CX20190290), and those supports are greatly appreciated by the authors.

References

[1] E. Laursen, N. Bitsch, J. E. Andersen, Analysis and mitigation of large amplitude cable vibrations at the Great Belt East Bridge, IABSE (2005), Copenhagen.
[2] Fujino Y, Kimura K, Tanaka H. Wind resistant design of bridges in Japan: Developments and Practices. New York: Springer; 2012.
[3] Li SY, Deng YC, Lei X, Wu T, Chen ZQ. Wake-induced vibration of the hanger of a suspension bridge: field measurements and theoretical modeling. Structural Engineering and Mechanics 2019;72(2):169–80.
[4] Wen Q, Hua XG, Chen ZQ, Niu HW. Experimental study of wake-induced instability of coupled parallel hanger ropes for suspension bridges. Eng Struct 2018;167(15): 175–87.
[5] Hua XG, Chen ZQ, Lei X, Wen Q, Niu HW. Monitoring and control of wind-induced vibrations of hanger ropes of a suspension bridge. Smart Structures and Systems 2018;23(6):125–41.
[6] Kashima S, Yanaka Y, Suzuki S, Mori K. Monitoring the akashi kaikyo bridge: first experiences. Structural Engineering International 2001;11(2):120–3.
[7] Zhang ZT, Wu XB, Chen ZQ, Ge YJ. Mechanism of hanger oscillation at suspension bridges: buffeting-induced resonance. ASCE J Bridge Engineering 2016;21(3): 04015066.
[8] Li YL, Tang HJ, Lin QM, Chen XZ. Vortex-induced vibration of suspenders in the wake of bridge tower by numerical simulation and wind tunnel test. J Wind Eng Ind Aerodyn 2017;164:164–73.
[9] Chen WL, Gao DL, Li H, Hu H. Wake-flow-induced vibrations of vertical hangers behind the tower of a long-span suspension bridge. Eng Struct 2018;169:188–200.
[10] Chen WL, Yang WH, Xu F, Zhan CG. Complex wake-induced vibration of aligned hangers behind tower of long-span suspension bridge. J Fluids Struct 2020;92. 102829.
[11] Chen WL, Huang YW, Meng H. Wake-induced vibration of a suspender cable in the rear of a bridge tower. J Fluids Struct 2020;99. 103166.
[12] Zdravkovich MM. Review of flow interference between two circular cylinders in various arrangements. J Fluids Struct 1977;99:618–33.
[13] Zdravkovich MM. The effects of interference between circular cylinders in cross flow. J Fluids Struct 1987;1:239–61.
[14] Sumner D. Two circular cylinders in cross-flow: a review. J Fluids Struct 2010;26: 849–99.
[15] Bearman PW. Circular cylinder wakes and vortex-induced vibrations. J Fluids Struct 2011;27:648–58.
[16] Y. Zhou Y , M. M. Alam, Wake of two interacting circular cylinders: A review, International Journal of Heat and Fluid Flow 62 (2016) 510–537.
[17] Li SY, Xiao CY, Wu T, Chen ZQ. Aerodynamic interference between the cables of the suspension bridge hanger. Adv Struct Eng 2019;19(5):481–503.
[18] Deng YC, Li SY, Chen ZQ. Unsteady theoretical analysis on the wake-induced vibration of the hangers of suspension bridges. J Bridge Eng 2019;24(2):04018113.
[19] Li SY, Deng YC, Huang J, Chen ZQ. Experimental investigation on aerodynamic interference of two kinds of suspension bridge hangers. J Fluids Struct 2019;90: 57–70.
[20] Jafari M, Sarkar PP. Wind tunnel study of wake-induced aerodynamics of parallel stay-cables and power conductor cables in a yawed flow. Wind Structures International J 2020;30(6):617–31.
[21] Kim S, Kim HK. Wake galloping phenomena between two parallel/unparallel cylinders. Wind Structures An International J 2014;18(5):511–28.
[22] Li SY, Chen ZQ, Li SK. Theoretical investigation on rain-wind induced vibration of a continuous stay cable with given rivulet motion. Wind Structures An International J 2014;19(5):481–503.
[23] Zhou R, Ge YJ, Yang YX, Liu SY, Du YL, Zhang LH. A nonlinear numerical scheme to simulate multiple wind effects on twin-box girder suspension bridges. Eng Struct 2019;183:1072–90.
[24] Zhou R, Ge YJ, Yang YX, Du YL, Zhang LH. Wind-induced nonlinear behaviors of twin-box girder bridges with various aerodynamic shapes. Nonlinear Dyn 2018;94: 1095–115.
[25] Li YL, Wu MX, Chen XZ, Wang T, Liang HL. Wind-tunnel study of wake galloping of parallel cables on cable-stayed bridges and its suppression. Wind and Structures An International J 2013;16(3):249–61.
[26] Chen WL, Zhang QQ, Li H, Hu H. An experimental investigation on vortex induced vibration of a flexible inclined cable under a shear flow. J Fluids Struct 2015;54: 297–311.
[27] He XH, Cai C, Wang ZJ, Jing HQ, Qin CW. Experimental verification of the effectiveness of elastic cross-ties in suppressing wake-induced vibrations of staggered stay cables. Eng Struct 2018;167:151–65.
[28] Jafari M, Sarkar PP. Wind-induced response characteristics of a yawed and inclined cable in ABL wind: Experimental-and numerical-model based study. Eng Struct 2020;214. 110681.
[29] Irvine HM. Cable Structure, The MIT Press. Cambridge: Massachusetts, and London, England; 1981.
[30] Clough RW, Penzien J. Dynamics of structures. New York: McGraw-Hill Inc.; 1975.
[31] Fujino Y, Siringoringo D. Vibration mechanisms and controls of long-span bridges: a review. Structural Engineering International 2013;23(3):248–68.
[32] Païdoussis MP, Price SJ, Langre ED. Fluid-structure interactions: cross-flow-induced instabilities. Cambridge University Press; 2011.

Experimental Investigation on High-Mode Vortex-Induced Vibration of a Flexible Stay Cable in Smooth Flow

Zhiwen Liu[1]; Shuqiong Li[2]; Lianhua Wang[3]; Banfu Yan[4]; Ruilin Zhang, Ph.D.[5]; and Zhengqing Chen, M.ASCE[6]

Abstract: With the increase in the main spans of cable-stayed bridges, the wind-induced vibration, especially the high-mode vortex-induced vibration (VIV) of the long stay cables has become a concern for many researchers. In this study, to investigate the high-mode VIVs of stay cables via wind tunnel tests, the mode order amplification factor (MOAF) of the stay cable was proposed to design the flexible stay cable model with a relatively large-scale ratio (λ_{Re}) of the Reynolds number (*Re*). Furthermore, the wind-induced vibration characteristics of the flexible stay cable model with a smooth surface for wind yaw angle (β) of 0° were investigated. In addition, the β effects on the wind-induced vibration characteristics of the stay cable were studied. Finally, double-helical fillets were applied to suppress the high-mode VIV responses of the flexible stay cable model for different β. The results showed that the flexible stay cable model that was designed by the proposed method of MOAF satisfied the similarity of the prototype stay cable and the damping ratios (ξ) of the first several modes of the flexible stay cable model were from approximately 0.021% to 0.064%. The stay cable exhibited five in-plane VIVs in the test wind velocity (V) range for $\beta = 0°$. Furthermore, the maximum amplitude of the displacement of the in-plane VIV of the flexible stay cable model decreased with the increase in the VIV mode order. The significant in-plane VIV responses of the stay cable were observed for β of 15°, 30°, −15°, −30°, and −45°, respectively, and the in-plane vibration acceleration responses were significantly larger than the out-of-plane vibration acceleration responses of the flexible stay cable model. The VIV responses of the stay cable could be effectively suppressed through the double-helical fillets with a diameter (d) of $d = 0.10\,D$ (where D = diameter of the stay cable) and pitch (P) of $P = 12\,D$. **DOI: 10.1061/(ASCE)BE.1943-5592.0001909.**

Author keywords: Stay cable; High-mode vortex-induced vibration; Flexible stay cable model; Mode order amplification factor; Helical fillets.

Introduction

Currently, the largest span cable-stayed bridge under construction is the Changzhou-Taizhou Yangtze River Bridge, Changzhou, Jiangsu, China, with a main span of 1,176 m and a longest stay cable length of 633 m (Qin et al. 2020). Moreover, some researchers conducted a feasibility study of the cable-stayed bridges with main spans from 1,200 to 1,500 m and a longest stay cable length from 670 to 800 m (Nagai et al. 2004; Zhu et al. 2011; Zhang and Zhu 2013; Hu et al. 2020). With the further increase of the stay cable length, the natural frequency (f_1) and damping ratio (ξ) of the stay cables will significantly decrease, and the wind-induced vibrations of the stay cable, such as rain–wind-induced vibration (RWIV), vortex-induced vibration (VIV), high reduced velocity VIV, and dry galloping (DG), will become more prominent, which might result in visual panic in drivers, damage to the stay cable dampers, and fatigue damage to the stay cables (Yamaguchi and Fujino 1998; Fujino 2002). The mechanism and countermeasures against the RWIV of the stay cables have been investigated via field measurements, wind tunnel tests, and theory analysis. However, significant VIVs in the stay cables were observed in some large-span cable-stayed bridges recently (Susumu and Masahiro 2014; Ge and Chen 2019; Liu et al. 2021).

To investigate the characteristics, mechanism, and countermeasures against VIVs and DG in stay cables, some researchers performed experimental investigations via rigid stay cable models (Shirakashi et al. 1986; Nakagawa et al. 1998; Sarkar and Gardner 2000; Matsumoto et al. 2001; Zuo and Jones 2009). Moreover, to satisfy the similarity of the Reynolds numbers (*Re*) of the stay cable model and the prototype stay cables, wind tunnel experiments on the rigid segment stay cable models with a geometric scale of 1:1 were performed to study the wind-induced vibration characteristics and the aerodynamic countermeasures for the stay cables (Duy et al. 2014; McTavish et al. 2018; Liu et al. 2021).

However, the prototype stay cable is a typical flexible three-dimensional structure, and some field measurement results for the VIV of the stay cables exhibited single or multimode VIV characteristics. Therefore, it is necessary to study the wind-induced vibration characteristics of the stay cables via the flexible stay cable model

[1]Hunan Provincial Key Lab for Wind & Bridge Engineering, College of Civil Engineering, Hunan Univ., Changsha 410082, China (corresponding author). ORCID: https://orcid.org/0000-0002-0900-6420. Emails: zhiwenliu@hnu.edu.cn; liuzhiwen757@126.com

[2]College of Civil Engineering, Hunan Univ., Changsha 410082, China. Email: lsq1995@hnu.edu.cn

[3]Hunan Provincial Key Lab for Wind & Bridge Engineering, College of Civil Engineering, Hunan Univ., Changsha 410082, China. Email: lhwang@hnu.edu.cn

[4]Hunan Provincial Key Lab for Wind & Bridge Engineering, College of Civil Engineering, Hunan Univ., Changsha 410082, China. Email: yanbanfu@hnu.edu.cn

[5]College of Civil Engineering, Hunan Univ., Changsha 410082, China. Email: ruilinzhang@hnu.edu.cn

[6]Hunan Provincial Key Lab for Wind & Bridge Engineering, College of Civil Engineering, Hunan Univ., Changsha 410082, China. Email: zqchen@hnu.edu.cn

Note. This manuscript was submitted on August 25, 2021; approved on April 11, 2022; published online on June 15, 2022. Discussion period open until November 15, 2022; separate discussions must be submitted for individual papers. This paper is part of the ***Journal of Bridge Engineering***, © ASCE, ISSN 1084-0702.

(Bartoli et al. 2006; Cluni et al. 2007; Chen et al. 2015; Han et al. 2017; Daniel Urdareanu et al. 2019; Liu et al. 2019; Gao et al. 2019; Jafari and Sarkar 2020). Bartoli et al. (2006) studied the wind-induced vibration characteristics of a cable model in laminar and turbulent flows, respectively. Moreover, the cable tension, the along-wind reaction at supports, and the displacements at the midpoint were analyzed. Chen et al. (2015) studied the VIV characteristics of a flexible cable in shear flow. The results showed that the displacement amplitudes of the single-mode VIVs were larger than those of multimode VIVs. Daniel Urdareanu et al. (2019) investigated the wind-induced vibration of a scaled stay cable model via wind tunnel tests and numerical simulations. Liu et al. (2019) conducted an experimental investigation into the mitigation effects of the damped crossties on VIVs of stay cables in long-span bridges. Their results showed that the proposed damped stay cable crossties could significantly mitigate the VIV responses of the stay cables. Gao et al. (2019) investigated the multimode VIVs and RWIVs of an inclined and yawed cable model in a wind tunnel. The results showed that the VIV mode increased from the first to third modes with the increase in wind velocity (V). Jafari and Sarkar (2020) studied the wind-induced vibration characteristics of an inclined stay cable in an atmospheric–boundary layer wind with a novel flexible cable model.

Similar to the high-mode VIV of stay cables, the experimental investigations on VIV responses and the countermeasures in long flexible risers were performed (Chaplin et al. 2005; Trim et al. 2005; Lie and Kaasen 2006; Gao et al. 2015). Chaplin et al. (2005) conducted experimental investigations on VIVs in a vertical tension riser in a stepped current. They obtained the frequencies and amplitudes of the VIV responses up to the 8th and 12th modes in the transverse and in-line directions, respectively. Trim et al. (2005) performed the experimental investigations into VIVs of a long marine riser model with the length-to-diameter (L/D) ratio of $L/D=1{,}400$, the tests were conducted for the riser model with and without various strakes, respectively. Lie and Kaasen (2006) applied the modal analysis method to analyze the high-mode VIV test data of a large-scale riser model in linearly sheared flow. Gao et al. (2015) performed experimental investigations on the VIV responses and fatigue damage in a long flexible riser that was fitted with strakes in uniform and linear sheared currents. They observed that the bare flexible riser might exhibit obvious multimode VIVs.

Many experimental investigations into the VIV of the stay cables have been conducted via rigid stay cable and flexible stay cable models, respectively. In summary, some previous investigations into the VIV or RWIV of the stay cable have been performed via flexible stay cable models. However, some stay cable models did not strictly satisfy the requirements of the similarity of the stay cable. In addition, the damping ratios of the high-order modes of the flexible cable models were relatively higher or difficult to be excited in the flexible stay cable model. In this case, the high-mode vibration phenomenon might be suppressed by the higher ξ. Therefore, it is necessary to investigate the vibration characteristics and the aerodynamic control measures in the high-mode VIV of stay cables via a novel flexible stay cable model, which might be designed based on prototype stay cables and with relatively low modal ξ, simultaneously.

In this study, the characteristics, and aerodynamic countermeasures against the high-mode VIV of the stay cables are investigated via a flexible stay cable model under different wind yaw angles (β). This study is organized as follows. First, the experimental setups and flexible stay cable model are introduced. Then, the characteristics of the wind-induced vibration responses, and wind yaw effects on the wind-induced vibration of the flexible stay cable model are analyzed. Finally, a discussion and a summary of the experimental results are presented.

Experiment Setup and Stay Cable Model

Experimental Setup

The wind tunnel experiments on the high-mode wind-induced vibration of a flexible stay cable model were conducted in the third work section of the HD-2 wind tunnel at Hunan University, Changsha, Hunan, China, which is 8.5 m wide, 2.0 m high, and 15.0 m long. Furthermore, the turbulence intensity and the maximum V in the empty wind tunnel were <1.0% and 15.0 m/s, respectively. The R. M. Young Model 81,000 V ultrasonic anemometer, which is a three-axis, no-moving-parts wind sensor with four synchronized voltage input channels and serial output, was applied to measure V in the wind tunnel. The accuracy of the ultrasonic anemometer was 0.05 m/s, and the sampling frequency was 10.0 Hz. Of note, the wind tunnel experiments were conducted in smooth flow. Therefore, the mean V was monitored through the ultrasonic anemometer.

Introduction of Sutong Bridge

Using the Suzhou–Nantong Yangtze River Highway Bridge (STB) as the engineering background, the experimental investigations into the high-mode VIV of stay cables were conducted in this study. The STB is a large-span cable-stayed bridge with a main span of 1,088 m, which crosses the Yangtze River between Suzhou and Nantong, Jiangsu, China, which was the largest span cable-stayed bridge in the world when it opened to traffic in 2008. The span arrangement of the STB is $100+100+300+1{,}088+300+100+100=2{,}088$ m, and the main deck of the STB is a streamlined

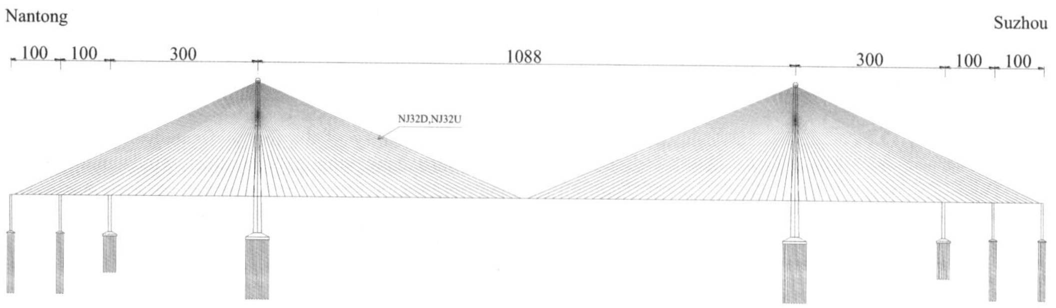

Fig. 1. Elevation layout of STB (m).

Table 1. Parameters of the prototype and model of the stay cable NJ32D of STB

Parameters	Unit	Prototype cable	FASM		SASM		
			Scale	Design value	Secondary scale	Design value	Realized value
L	m	545.539	1/5	109.1078	1/18.747	5.82	5.82
D	mm	152.00	1/5	30.4	1	30.4	30.0
α^a	°	21.131	1/1	21.131	1	15.8	15.8
T	N	5,684,498	$1/5^3$	45,475.98	$(1/18.747)^2 n_f^2$	1,409.11	1,354
m	kg/m	90.8	$1/5^2$	3.6320	1	3.6320	3.485
f_1	Hz	0.2293	$\sqrt{5}$	0.5127	n_f	1.6919	1.600
ξ	%	0.01–0.05	1	0.01–0.05	1	0.01–0.05	0.021–0.064^b
V	m/s	0–15.0	$1/\sqrt{5}$	0–6.71	1	0.00–6.71	0.00–6.71
$Re = \frac{UD}{v}$	—	$0–1.52 \times 10^5$	$1/5\sqrt{5}$	$0–1.34 \times 10^4$	1	$0–1.34 \times 10^4$	$0–1.34 \times 10^4$
y and z^v	m	—	1/5	—	1	—	—
Vibration mode order	n	—	1	—	n_f	—	—

[a]Due to the limitation of the size of the wind tunnel test section, α of the stay cable was not modeled in this study.
[b]The maximum ξ of most modes of the flexible stay cable model was from 0.021% to 0.064% except for the ξ of 0.162% of the first in-plane mode, as given in Table 2.

steel box girder that is 41.0 m wide and 4.0 m high. Fig. 1 shows the elevation layout of the STB. The main deck is supported by 272 stay cables, which are arranged in four inclined planes of 34 stays at each of two inverted Y-shaped pylons. The length of the

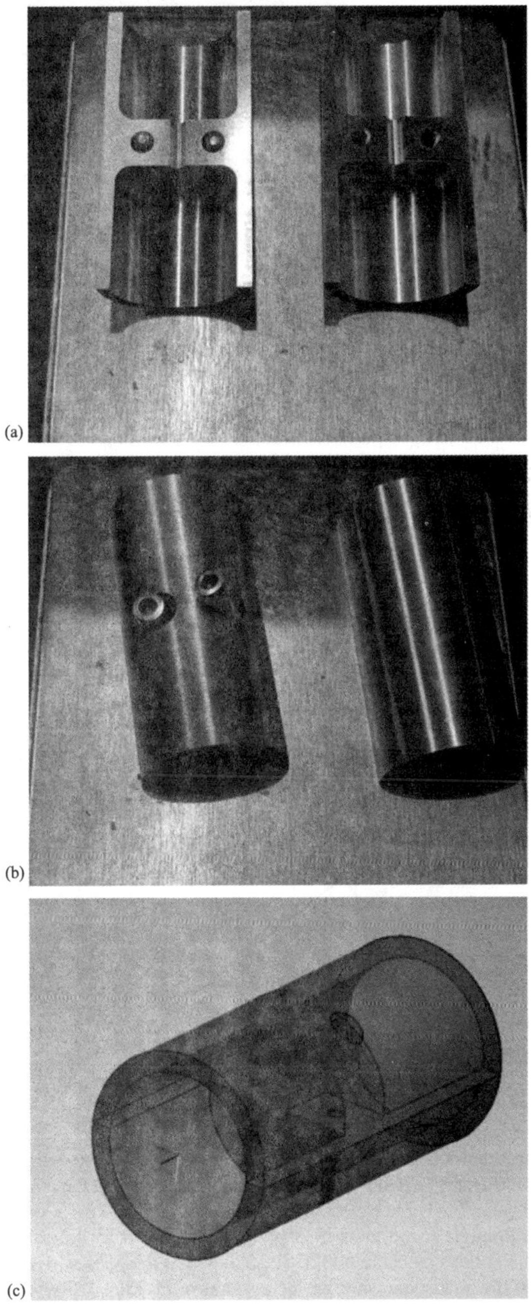

Fig. 2. Photos of the cylindrical copper tube: (a) inner side view of the copper tube; (b) outside view of the copper tube; and (c) assembly of copper tube.

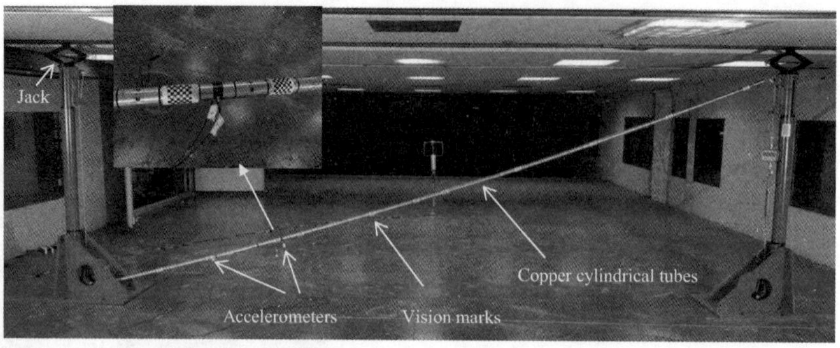

Fig. 3. Flexible stay cable model in wind tunnel.

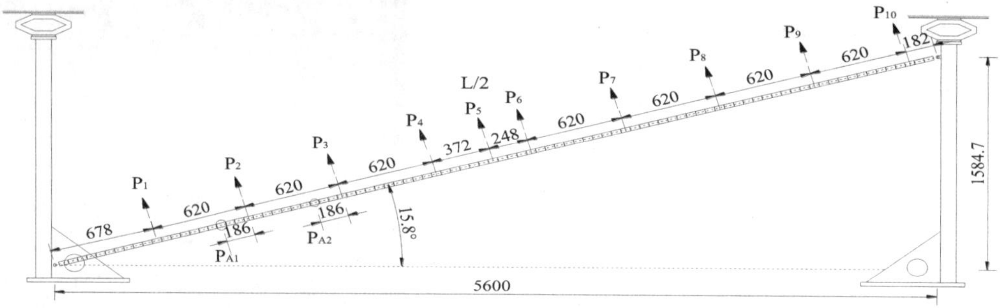

Fig. 4. Layout of measuring points for accelerometers and computer vision marks on stay cable model (mm). The computer vision marks were set at P_1 to P_{10}, of note, P_2 and P_3 were set near P_{A1} and P_{A2}.

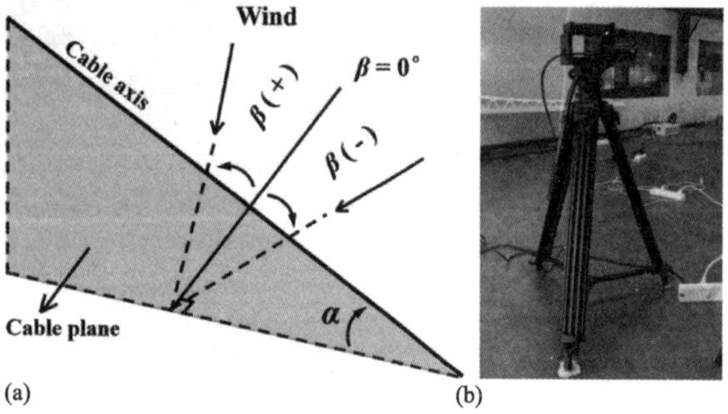

Fig. 5. Definition of β and the photo of VDS: (a) definition of β; and (b) the photo of VDS.

stay cables is from 153 to 577 m, with varying diameters from 116 to 161 mm. They are attached at 16 m intervals along the main deck (Zhang 2004; Zhang et al. 2006; Wang et al. 2013). To study the high-mode wind-induced vibration characteristics of the stay cable, the stay cable NJ32D (PES7-283) of the STB was selected as the prototype stay cable, as shown in Fig. 1. Where $L = 545.539$ m and $D = 152$ mm for the stay cable NJ32D of the STB.

Similarity Relationship of the Stay Cable Model

Considering that the width of the third work section of the wind tunnel at Hunan University and the L of the prototype stay cable of the STB, the geometry scale ratio (λ_L) of the full aeroelastic stay cable model (FASM) was $\lambda_L = 1/70$. Therefore, the L and D of the FASM of the stay cable NJ32D of the STB were approximately 7.793 m and 2.17 mm, respectively. Moreover, the scale ratio (λ_{Re}) of the Re of the FASM was $\lambda_{Re} = (1/70)^{3/2} = 1/585.66$, and the Re might significantly affect the VIV responses of the stay cable. Therefore, it was necessary to balance the relationship between the geometry scale and Re effects on the high-model VIV of the flexible stay cable model. To achieve this, a novel method was proposed to design the flexible stay cable model.

A two-step model design method was used to design the stay cable model in this study. First, according to the similarity criterion of the aeroelastic model, the length scale ratio (λ_{L1}) of FASM of the NJ32D stay cable of the STB was determined ($\lambda_{L1} = 1/5$), and the detailed similarity scales and design values for different parameters of the FASM of the NJ32D stay cable of the STB are listed in Table 1. However, due to the limitations of the size of the wind tunnel test section, the FASM of the NJ32D stay cable of the STB had to be further scaled (i.e., $\lambda_{L2} = 1/18.747$), which was the length of the shorter aeroelastic stay cable model (SASM) and was 5.82 m (Deng et al. 2021). Second, the mode order amplification factor [(MOAF) n_f] of the stay cable was proposed to consider the high-mode wind-induced vibration of the stay cable, which could be determined based on the wind tunnel test section size and the length of the prototype stay cable. Here, $n_f = 3.3$.

The ith f_1 of the stay cable with flat sag can be approximately calculated as follows (Irvine 1981):

$$f_i = \frac{i}{2L}\sqrt{\frac{T}{m}} \quad (i = 1, 2, \ldots\ldots n) \tag{1}$$

where T, L, and m = force, length, and mass per unit of the stay cable, respectively.

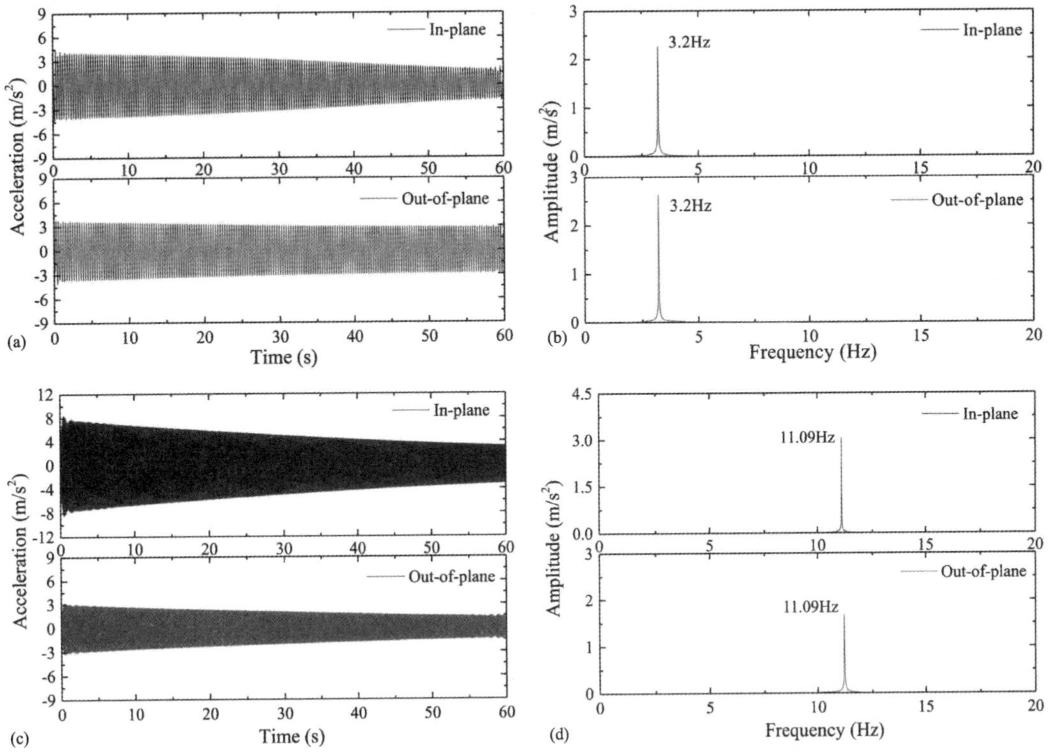

Fig. 6. Time histories and amplitude spectra of the in-plane and out-of-plane vibration acceleration responses at P_{A1} of the flexible stay cable model during the free vibration decay tests: (a) time histories of 2nd mode; (b) FFT of 2nd mode; (c) time histories of 7th mode; and (d) FFT of 7th mode.

Table 2. Dynamics parameters of the flexible stay cable model with or without helical fillets

Mode order	Vibration direction	Without helical fillets		With helical fillets ($d=0.10\ D$)	
		Frequency (Hz)	ξ (%)	Frequency (Hz)	ξ (%)
1	In-plane	2.183	0.162	2.175	0.140
	Out-of-plane	1.600	0.037	1.625	0.048
2	In plane	3.200	0.064	3.213	0.037
	Out-of-plane	3.200	0.030	3.225	0.034
3	In-plane	4.833	0.037	4.863	0.028
	Out-of-plane	4.816	0.042	4.838	0.031
4	In-plane	6.433	0.036	6.463	0.041
	Out-of-plane	6.417	0.055	6.450	0.034
5	In-plane	8.000	0.046	8.063	0.044
	Out-of-plane	8.000	0.032	8.038	0.032
7	In-plane	11.090	0.028	11.190	0.031
	Out-of-plane	11.090	0.021	11.150	0.025

Table 3. Wind tunnel experimental cases for the flexible stay cable model

Cases	α (°)	β (°)	With or without helical fillets	V (m/s)
1–9	15.8	0, ±15, ±30, ±45, ±60	without helical fillets	0–7.0
10		0	with double-helical fillets ($d=2.0$ mm, $P=12\ D$)	0–7.0
11–17		0, ±15, ±30, ±45	with double-helical fillets ($d=3.0$ mm, $P=12\ D$)	0–7.0

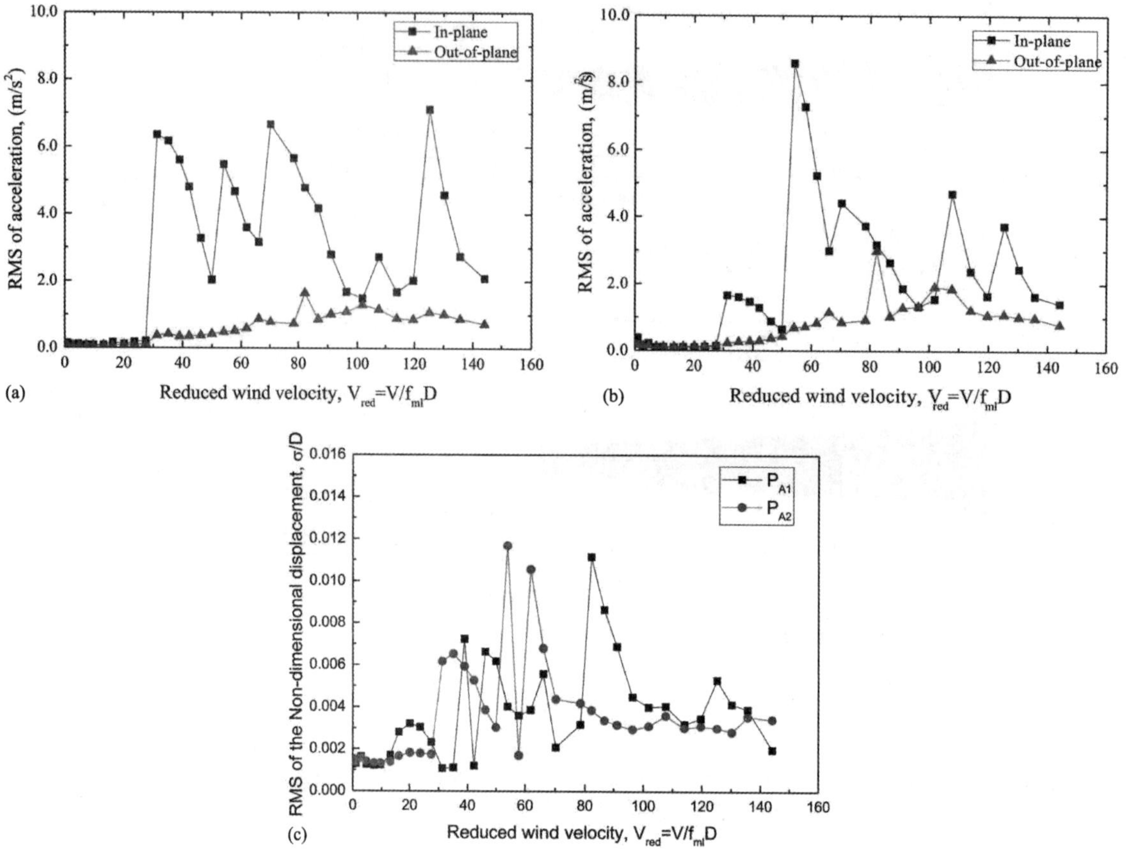

Fig. 7. RMS of acceleration and displacements at P_{A1} and P_{A2} of the SASM of the stay cable for $\beta = 0°$: (a) RMS of acceleration at P_{A1}; (b) RMS of acceleration at P_{A2}; and (c) RMS of displacements at P_{A1} and P_{A2}.

Since L of the FASM was further scaled and the fundamental frequency of the FASM was amplified by n_f to consider the high-order vibration of the stay cable, and maintaining the m of the FASM unchanged, to ensure that the frequencies of the FASM were similar, and the cable T was determined according to the frequency invariance criterion after the cable was enlarged. The T of the SASM should satisfy the following relationship:

$$T_{ms} = \lambda_{L2}^2 n_f^2 T_m \quad (2)$$

where T_m, and T_{ms} = force of the FASM and SASM, respectively. Of note, the fundamental frequency of the SASM of the stay cable was n_f times that of the FASM. Therefore, the SASM could only reflect the high-order vibration response characteristics of the stay cable with an interval of n_f times frequency. The main similarity relationship of the SASM is given in Table 1. As given in Table 1, the similarity relationship between the displacement of the stay cable and the prototype stay cable was, for instance, $\lambda_{y,z} = 1/5$. Since the L and m of the stay cable model were determined, the sag of the stay cable model is only related to the T and axial stiffness of the stay cable model.

Fabrication of Stay Cable Model

In this study, a steel wire rope with $d = 3.0$ mm and cylindrical copper tubes with $D = 30.0$ mm were used to simulate the mass and aerodynamic shape of the prototype stay cable. The L and mass of each cylindrical copper tube were 60.0 mm and 223.0 g, respectively. Two half copper tubes were fixed by clamping the steel wire rope with two screws, as shown in Fig. 2. Moreover, the mass per meter of the stay cable model was 3.485 kg/m, and 16 segmental cylindrical copper double tubes with a 2.0 mm gap between two adjacent tubes were used for the flexible stay cable model. Overall, the L and D of the flexible stay cable model were 5.82 m and 30.0 mm, respectively, and the aspect ratio of the stay cable model was $L/D = 194.0$. Of note, due to the limitations of the wind tunnel section, the inclination angle (α) of the flexible stay cable model was $\alpha = 15.8°$, which was $\alpha = 21.131°$ of the prototype stay cable. The flexible stay cable model is shown in Fig. 3.

Dynamic Characteristics of Stay Cable Model

Four accelerometers were applied to measure the in-plane and out-of-plane vibration responses at P_{A1} and P_{A2}, which were $3L/16$ and $7L/24$ to the lower fixed end (L = length of the flexible stay cable model), respectively. The sampling frequency was 1,000 Hz, and the sampling time of each case was 80 s. Moreover, the video displacement system (VDS) was used to monitor the in-plane vibration displacement responses at different monitoring points, for example, P_1 to P_{10} of the flexible stay cable model. The VDS developed by Yan et al. (2020) included a video acquisition device, a lighting device, and a video image processing system. The digital image correlation algorithm was used to analyze the video image to obtain the displacement responses at the

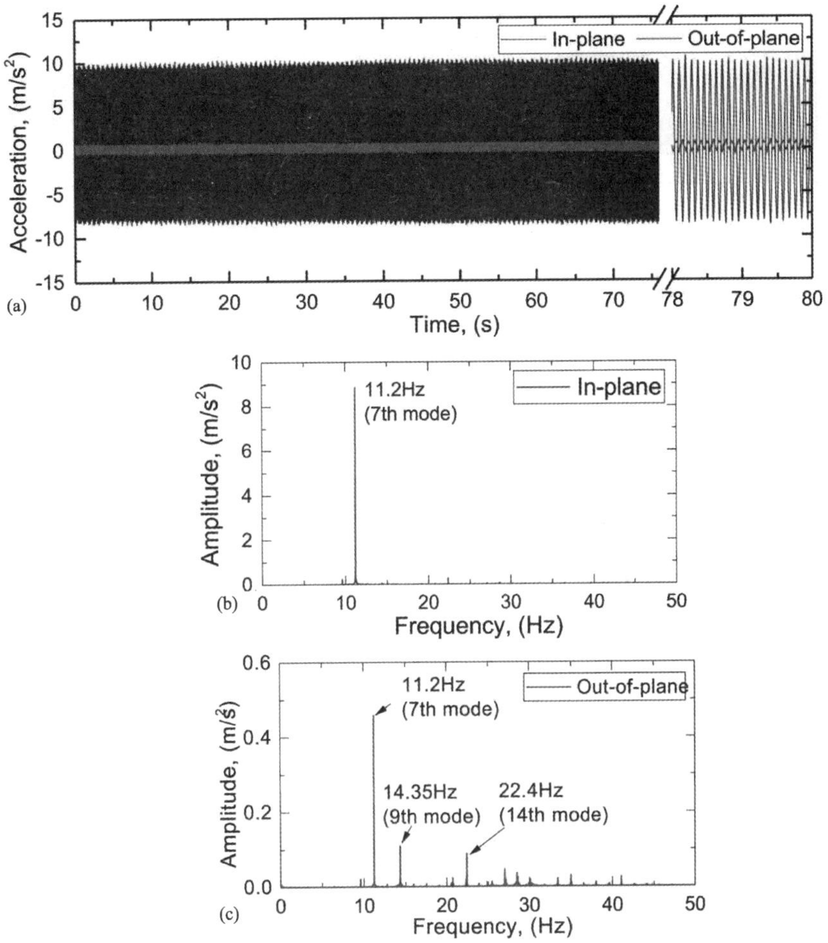

Fig. 8. Time histories and amplitude spectra of the acceleration responses at P_{A1} of the stay cable for $V_{red}=31.29$: (a) acceleration responses; (b) amplitude spectrum of in-plane acceleration; and (c) amplitude spectrum of out-of-plane acceleration.

measurement points of the stay cable model. The sampling frequency was 600 Hz, and the sampling time was 80 s. Fig. 4 shows the layout of the accelerometers and the VDS marks set on the flexible stay cable model, respectively. The definition of the β of the stay cable and the VDS are shown in Fig. 5, respectively.

The free vibration method was applied to identify the dynamic parameters of the flexible stay cable model with or without helical fillets in still air. The identification f_1 and ξ of each mode of the SASM of the stay cable was carried out by artificial excitation. For different modes of the SASM of the stay cable, the acceleration responses at P_{A1} and P_{A2} were different, and the measurement points with larger acceleration responses were analyzed. The f_1 was obtained from the acceleration responses at P_{A1} and P_{A2} of the flexible stay cable model using Fast Fourier Transform (FFT). Furthermore, the ξ of the ith mode of the stay cable model (ξ_i) was determined via the following equation:

$$\xi_i \approx \frac{1}{2\pi n}\ln\left(\frac{y_k}{y_{k+n}}\right) \tag{3}$$

where y_k and y_{k+n} = responses amplitude of the kth cycle and the $(k+n)$th cycle of the flexible stay cable model, respectively (Chopra 1995). Fig. 6 shows the time histories and amplitude spectra of the acceleration responses of the in-plane and out-of-plane vibration at P_{A1} of the flexible stay cable model during the free decay tests. Table 2 gives the frequencies and modal ξ of the first five modes and the seventh mode of the flexible stay cable model. As given in Table 2, the frequencies of the first in-plane and out-of-plane modes of the flexible stay cable model without helical wires were 2.183 and 1.60 Hz, respectively. Furthermore, the corresponding frequencies of the flexible stay cable model with helical fillets were 2.175 and 1.625 Hz, respectively. The ξ of the first in-plane vibration mode of the stay cable model was 0.16%, which was larger than the inherent ξ of the prototype stay cable (i.e., parallel cables) of 0.1% based on Eurocode 1: Actions on structures Part 1–4: Generation actions – Wind actions. However, the ξ of the high-order vibration modes of the stay cable model was approximately 0.021%–0.055%, and the ξ gradually decreased with the increase in the vibration mode of the flexible stay cable model, which was consistent with the field measurement results of the prototype stay cables (Yamaguchi and Fujino 1998). Therefore, in this study, the high-order modal ξ of the stay cable model was close to the ξ of the high-order vibration modes of the prototype stay cable. Furthermore, the corresponding Scruton number (S_c) of the flexible stay cable was $S_c = m\xi/\rho D^2 = 0.664$–$5.12$. To investigate the high-mode

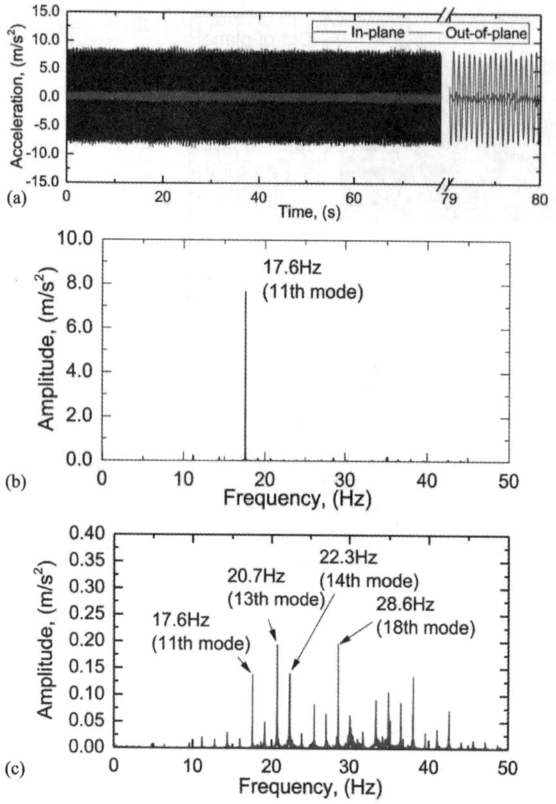

Fig. 9. Time histories and amplitude spectra of the acceleration responses at P_{A1} of the stay cable for $V_{red} = 54.02$: (a) acceleration responses; (b) amplitude spectrum of in-plane acceleration; and (c) amplitude spectrum of out-of-plane acceleration.

wind-induced vibration characteristics of the flexible stay cable, the wind tunnel experiments were conducted to investigate the wind-induced vibration in the flexible stay cable model with a smooth surface under different β. In addition, the effects of aerodynamic countermeasures on the high-mode wind-induced vibration of the flexible stay cable model were studied for different β. A total of 17 experimental cases were conducted in this study, and the wind tunnel experimental cases are listed in Table 3.

Experimental Results

High-mode VIVs of Flexible Stay Cable

Acceleration Responses

To describe the wind-induced vibration characteristics of the stay cable model, the reduced wind velocity (V_{red}) was defined as $V_{red} = V/f_{ml}D$, where V is the wind velocity, D is the diameter of the stay cable, and $f_{ml} = 1.60\ HZ$ is the f_1 of the flexible stay cable model. The root mean square (RMS) of the acceleration responses at P_{A1} and P_{A2} of the flexible stay cable model for $\beta = 0°$ were plotted against V_{red} as shown in Fig. 7. From Fig. 7, when the V_{red} increased from $V_{red} = 0.0$ to $V_{red} = 27.5$ the RMS of acceleration of the vibration responses of the stay cable were relatively small, and no obvious VIVs occurred. With the further increase in the V_{red}, the stay cable exhibited five in-plane VIVs, and the

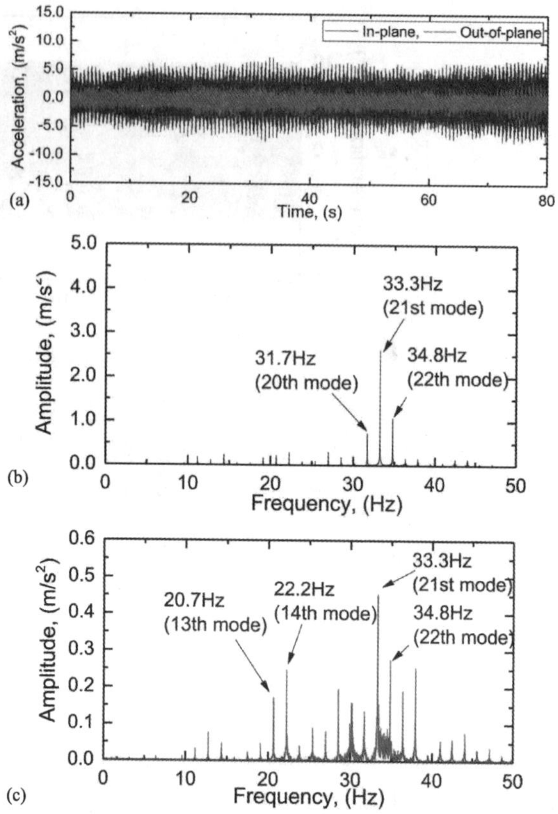

Fig. 10. Time histories and amplitude spectra of the acceleration responses at P_{A1} of the stay cable for $V_{red} = 107.68$: (a) acceleration responses; (b) amplitude spectrum of in-plane acceleration; and (c) amplitude spectrum of out-of-plane acceleration.

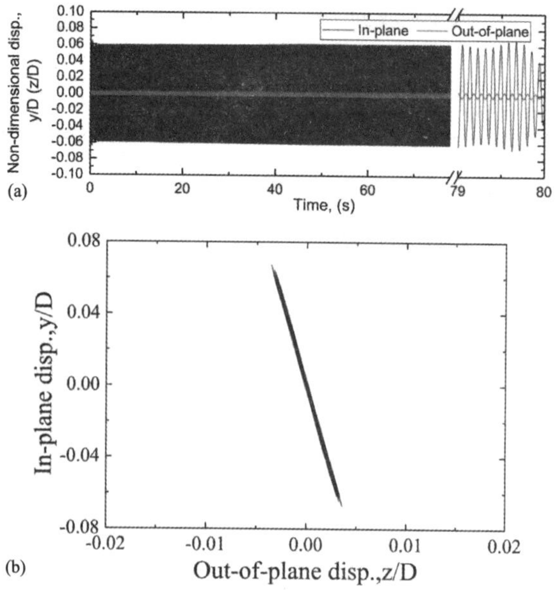

Fig. 11. Nondimensional displacement and trajectory at P_{A1} of the stay cable for $V_{red} = 31.29$: (a) nondimensional displacement; and (b) trajectory.

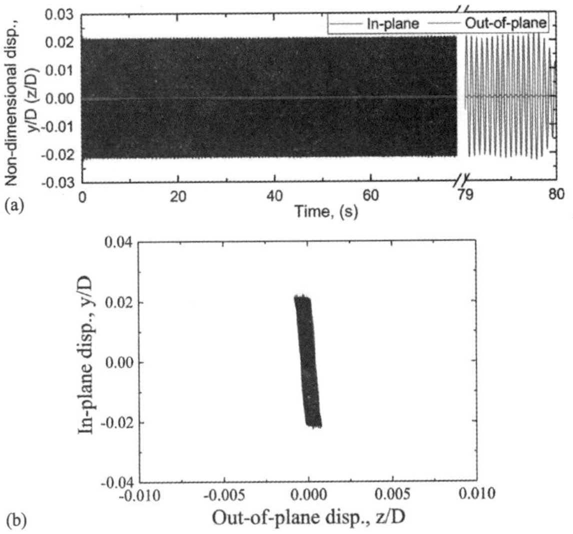

Fig. 12. Nondimensional displacement and trajectory at P_{A1} of the stay cable for $V_{red}=54.02$: (a) nondimensional displacement; and (b) trajectory.

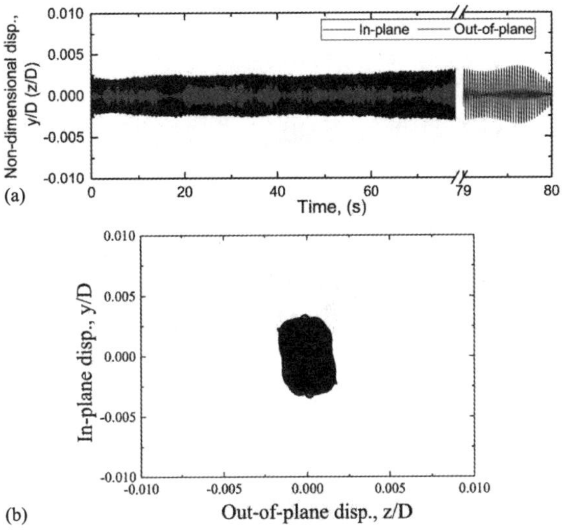

Fig. 13. Nondimensional displacement and trajectory at P_{A1} of the stay cable for $V_{red}=107.7$ (21st mode): (a) nondimensional displacement; and (b) trajectory.

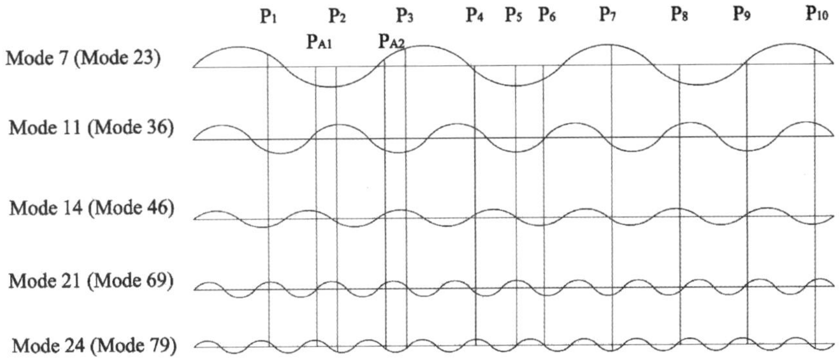

Fig. 14. Relative position of the VDS marks the different mode shapes of the flexible stay cable model. The vibration modes in parentheses correspond to the vibration modes of the prototype stay cable.

maximum RMS of acceleration of the lock-in at P_{A1} of the stay cable model were 6.0, 5.48, 6.67, 2.73, and 7.12 m/s^2, respectively, and the corresponding lock-in V_{red} ranges were $V_{red}=27.5$–50.0, $V_{red}=50.0$–66.0, $V_{red}=66.0$–96.4, $V_{red}=101.8$–114.0, and $V_{red}=114.0$–144.2, respectively. As shown in Fig. 7, the stay cable exhibits an out-of-plane VIV lock-in with a relatively small amplitude when $V_{red}=78.5$–86.8, and the maximum acceleration of the lock-in at P_{A1} of the stay cable model was 1.65 m/s^2. The similar wind-induced vibration characteristics at P_{A2} of the stay cable model are shown in Fig. 7(b). The in-plane vibration dominated the high-mode VIV response of the stay cable.

Time Histories and Amplitude Spectra

To further investigate the wind-induced vibration characteristics of the stay cable, the time histories and amplitude spectra of the acceleration responses were analyzed, which are shown in Figs. 8–12, respectively. From Fig. 8, when $V_{red}=31.29$, the stay cable exhibited in-plane single-mode VIV responses with a maximum VIV amplitude of 9.66 m/s^2, and the dominant frequency of the VIV responses was 11.2 Hz, which corresponded to the 7th vibration mode of the SASM of stay cable, which corresponded to the 23rd-order vibration mode of the prototype stay cable that considered the MOAF of the SASM (i.e., $n_f=3.3$). However, the out-of-plane vibration responses with a small amplitude of 0.98 m/s^2 of the stay cable model were observed, which was approximately 0.10 times the maximum amplitude of the in-plane vibration. Furthermore, the stay cable model exhibited multimode out-of-plane vibration, and the dominant frequencies were 11.2, 14.35, and 22.4 Hz, which corresponded to the 7th, 9th, and 14th vibration modes of the SASM of the stay cable, respectively. They corresponded to the 23rd, 30th, and 46th vibration modes of the prototype stay cable, respectively.

With the V_{red} further increased, the stay cable exhibited higher-order single-mode in-plane VIV responses, such as at the 11th, 14th, 21st, and 24th vibration modes of the SASM of the stay cable, respectively. The characteristics of the 11th, 14th, and 24th mode VIVs responses of the SASM of the stay cable were

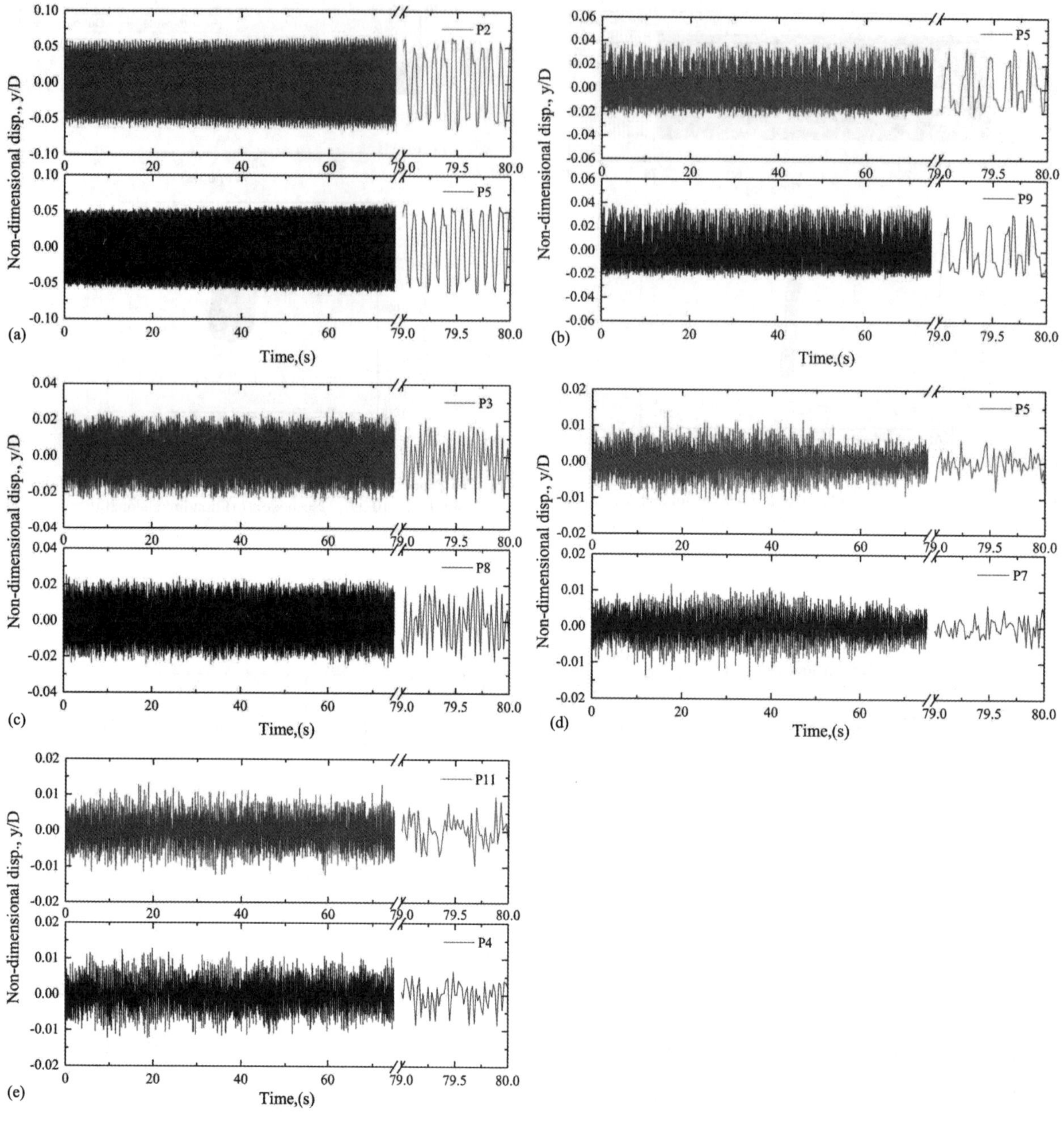

Fig. 15. Time histories of the nondimensional displacement near the maximum mode of corresponding VIV mode of the flexible stay cable model: (a) 7th mode; (b) 11th mode; (c) 14th mode; (d) 21st mode; and (e) 24th mode.

similar to that of the 7th mode VIVs responses of the SASM of the stay cable. In addition, the SASM of the stay cable exhibited multimode wind-induced vibrations in the out-of-plane direction for V_{red} of 54.02, 70.24, and 125.21, respectively. Fig. 9 shows the time histories and amplitude spectra of the acceleration responses of the SASM of the stay cable for $V_{red}=54.02$. However, as shown in Fig. 10, the multimode in-plane VIV responses of the SASM of the stay cable were observed when $V_{red}=107.68$, and the dominant frequencies were 31.7, 33.3, and 34.8 Hz, which corresponded to the 20th, 21st, and 22nd vibration modes of the SASM of the stay cable, respectively. They corresponded to the 66th, 69th, and 73rd vibration modes of the prototype stay cable. Moreover, the stay cable exhibited multimode out-of-plane vibration.

Displacements and Trajectories

To study the wind-induced vibration characteristics of the stay cable, the in-plane and out-of-plane vibration displacements at P_{A1} of the flexible stay cable model were determined based on the measured acceleration responses. To achieve this, the inverse Fourier transform was applied to calculate the displacement at the corresponding measuring points, and the Fourier component

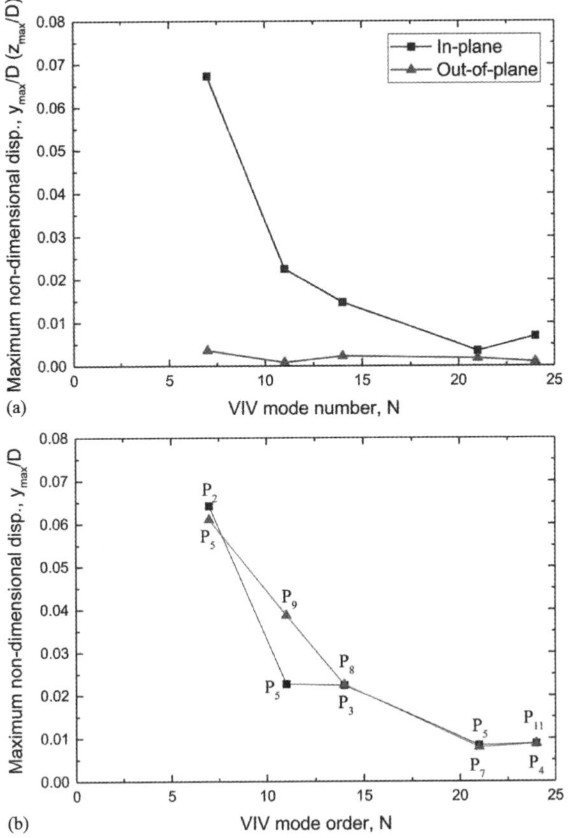

Fig. 16. Nondimensional displacement amplitudes versus the corresponding VIV mode order of SASM of the stay cable: (a) in-plane and out-of-plane nondisplacement amplitude based on acceleration at P_{A1}; and (b) nondimensional displacement by VDS.

of the acceleration response can be expressed as

$$a(t) = Ae^{j\omega t} \tag{4}$$

where $a(t)$ = Fourier component of the acceleration responses at a circular frequency (ω); A = coefficient of $a(t)$; and j = unit imaginary number, for instance, $j = \sqrt{-1}$. When the initial velocity equals zero, the velocity component can be obtained by integrating $a(t)$ of the acceleration responses

$$v(t) = \int_0^t a(\tau)d\tau = \int_0^t Ae^{j\omega\tau}d\tau = \frac{A}{j\omega}e^{j\omega t} = Ve^{j\omega t} \tag{5}$$

where $v(t) = a(t)$ of the velocity at ω. When the initial velocity and displacement all equal zero, the displacement component can be calculated by integrating $a(t)$ of the acceleration responses twice

$$x(t) = \int_0^t \left[\int_0^\tau a(\lambda)d\lambda\right]d\tau = \int_0^t Ve^{j\omega\tau}d\tau = \frac{V}{j\omega}e^{j\omega t} = -\frac{A}{\omega^2}e^{j\omega t} = Xe^{j\omega t} \tag{6}$$

where $x(t) = a(t)$ of the displacement at ω.

However, the displacement can be calculated by numerical integration of the discrete acceleration data in the frequency domain using the discrete Fourier transform (DFT) method

$$y(r) = \sum_{k=0}^{N-1} -\frac{1}{(2\pi k\Delta f)^2}H(k)X(k)e^{j2\pi kr/N} \tag{7}$$

where

$$H(k) = \begin{cases} 1 & (f_d \le k\Delta f \le f_u) \\ 0 & k\Delta f < f_d \text{ or } k\Delta f > f_u \end{cases} \tag{8}$$

where f_d and f_u = lower and upper cut off frequency, respectively; $X(k)$ = DFT of $x(r)$; and $x(r)$ = frequency resolution.

The time histories of nondimensional displacement and trajectories of in-plane and out-of-plane vibration responses at P_{A1} of the flexible stay cable model are shown in Figs. 11–13. As shown in Fig. 11(a), when $V_{red} = 31.29$ the stay cable model exhibited significant VIV responses with the 7th mode, and the maximum nondimensional displacement amplitude of in-plane and out-of-plane vibration responses at P_{A1} of the stay cable model were approximately $y_0/D = 0.067$ and $z_0/D = 0.0036$, respectively. As shown in Fig. 11(b) the VIV response of the stay cable was dominated by the in-plane vibration.

As the V_{red} further increased, the stay cable exhibited a higher-order VIV response. As shown in Fig. 12(a), when $V_{red} = 54.02$ the stay cable model exhibited significant VIV responses in the 11th mode, and the maximum nondimensional displacement amplitude of in-plane and out-of-plane vibration responses at P_{A1} of the stay cable model were $y_0/D = 0.0225$ and $z_0/D = 0.0008$, respectively. Moreover, the in-plane VIV responses of the stay cable were more significant than that of the out-of-plane vibration, as shown in Fig. 12(b).

However, as shown in Fig. 13(a), the in-plane vibration displacement was slightly larger than that of the out-of-plane vibration when $V_{red} = 107.7$. In addition, the maximum nondimensional displacement amplitude of in-plane and out-of-plane vibration responses at P_{A1} of the stay cable model was approximately $y_0/D = 0.0035$ and $z_0/D = 0.0018$, respectively.

In addition, the VDS was applied to measure the in-plane vibration displacements at different positions on the stay cable model. Fig. 14 shows the relative position of the video displacement marks for different mode shapes of the SASM of the stay cable. Of note, the vibration modes in the parentheses correspond to the vibration modes of the prototype stay cable. As shown in Fig. 14, according to the FFT of the acceleration response data for the SASM of the stay cable, the dominant vibration frequency of the SASM of the stay cable was determined. Furthermore, the displacement amplitude could be obtained based on the displacement measurement at the VDS marks near the maximum value of the vibration mode shape. Fig. 15 shows the time histories of displacements near the maximum mode of the corresponding VIV modes. As shown in Fig. 15, the nondimensional displacement near the maximum mode of the corresponding VIV mode of the SASM of the stay cable was close to each other.

Fig. 16 shows the maximum amplitude of the nondimensional displacement with the vibration mode order of the SASM of the stay cable. It can be seen from Fig. 16(a) that the in-plane vibration displacements of the stay cable were significantly larger than the out-of-plane vibration displacements response, and the in-plane vibration displacement amplitudes of the stay cable gradually decreased with the increase in the vibration mode order of the stay cable. Furthermore, it can be seen from Fig. 16 that the in-plane VIV nondimensional displacement amplitudes based on the acceleration were close to the measured values by VDS, and the maximum amplitude of the nondimensional displacement of the in-plane VIV of the flexible stay cable model decreased with the

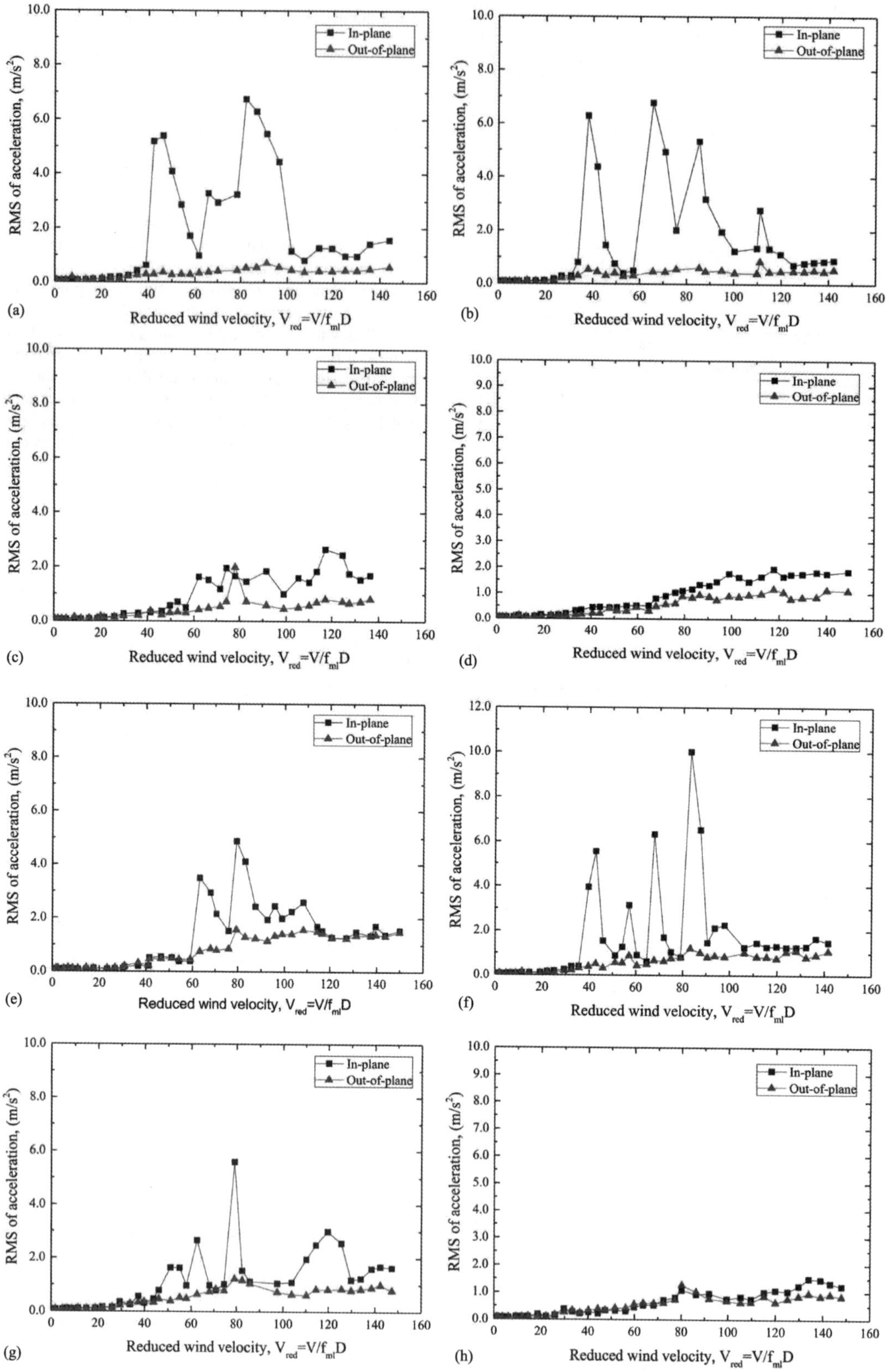

Fig. 17. RMS of acceleration at P_{A1} of the flexible stay cable model for different β: (a) $\beta=15°$; (b) $\beta=30°$; (c) $\beta=45°$; (d) $\beta=60°$; (e) $\beta=-15°$; (f) $\beta=-30°$; (g) $\beta=-45°$; and (h) $\beta=-60°$.

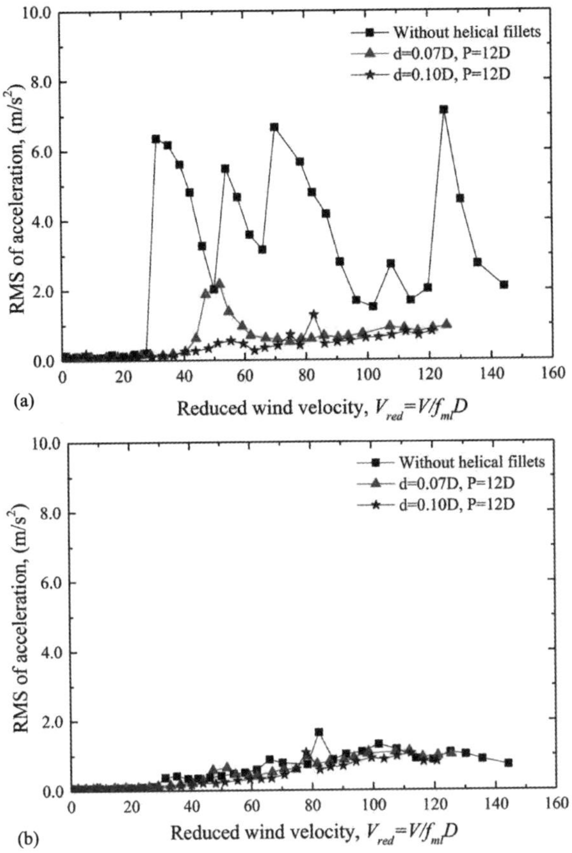

Fig. 18. RMS of acceleration at P_{A1} of the flexible stay cable model for $\beta=0°$ versus V_{red}: (a) in-plane vibration; and (b) out-of-plane vibration.

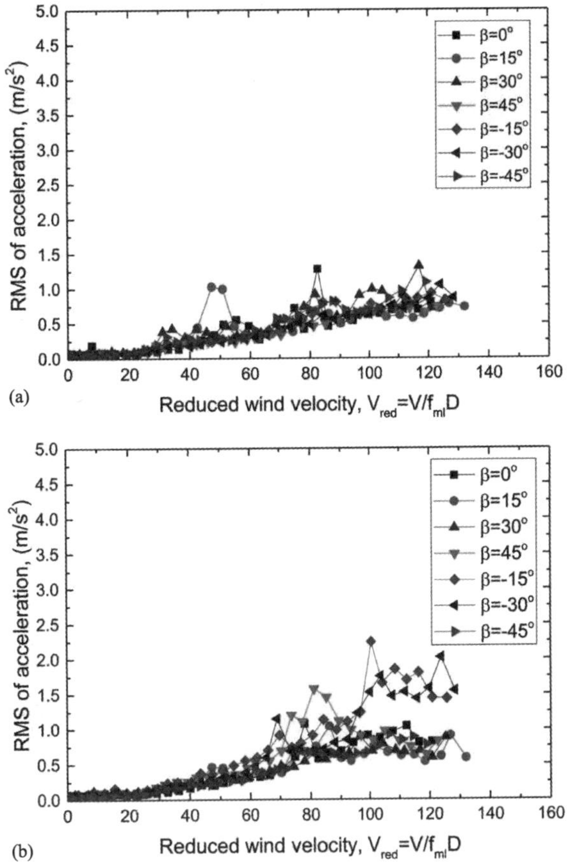

Fig. 19. RMS of acceleration at P_{A1} of the flexible stay cable model with double-helical wires of $d=0.10\ D$ and $P=12\ D$ for different β: (a) in-plane vibration; and (b) out-of-plane.

increase of the vibration mode order of the stay cable. However, the maximum amplitudes of the nondimensional displacement of the in-plane VIV varied slightly when the vibration mode order of the SASM of the stay cable was larger than the 21st, which corresponded to the 69th vibration mode order of the prototype stay cable. Furthermore, the maximum amplitude of the nondimensional displacement of the out-of-plane vibration varied slightly for all the vibration modes.

For the 7th vibration mode of the SASM of the stay cable, which corresponded to the 23rd vibration mode of the prototype stay cable, from Fig. 16(a) it can be seen that the nondimensional displacement amplitude based on acceleration data at P_{A1} was 0.0673, which was near the displacement measured point of P_2. Moreover, the nondimensional displacement amplitudes at P_2 and P_5 by VDS were 0.0642 and 0.0610, respectively. In addition, the maximum amplitudes of the nondimensional displacement of the VIV for mode order of the11th, 14th, 21st, and 24th VIV of the SASM of the stay cable were 0.0225, 0.0147, 0.00349, and 0.00688, respectively, which corresponded to the 36th, 46th, 69th, and 79th mode order of the prototype stay cable.

β Effects

The wind might come from different directions at actual bridge sites. Therefore, the β effects on the high-mode VIV responses of the stay cable should be investigated. Fig. 17 shows the RMS of in-plane and out-of-plane acceleration responses of the stay cable for different β. As shown in Fig. 17, the significant in-plane VIV responses of the stay cable were observed for $\beta=15°$, 30°, −15°, −30°, and −45°, respectively. Furthermore, the in-plane vibration acceleration responses were significantly larger than the out-of-plane vibration acceleration responses of the stay cable for $\beta=15°$, 30°, −15°, −30°, and −45°, respectively. As shown in Fig. 17(a), two lock-ins of the in-plane vibration of the stay cable were observed in the test wind speed range for $\beta=15°$, and the maximum RMS of acceleration of the lock-in at P_{A1} of the stay cable were 5.39 and 6.75 m/s^2, respectively. Figs. 17(c and d) show that no VIV response of the stay cable occurred when $\beta=45°$ and $\beta=60°$, respectively. As shown in Figs. 17(e–g), the stay cable exhibited significant VIV responses when $\beta=-15°$, −30°, and −45°. The stay cable might exhibit significant VIV responses when β was in the range of $\beta=-45°$–30°, and the VIV mode increased with the V_{red}.

Aerodynamic Countermeasures

To suppress the VIV response of the stay cable via aerodynamic measures, the experimental investigations into the flexible stay cable model with circular double-helical fillets with $d=2$ mm ($d=0.07\ D$, where D is the diameter of the SASM of the stay cable), $d=3$ mm ($d=0.10\ D$), and pitch (P) $P=12\ D$ were conducted for $\beta=0°$, respectively. Fig. 18 shows the RMS of acceleration at P_{A1} of the flexible stay cable model. As shown in Fig. 18,

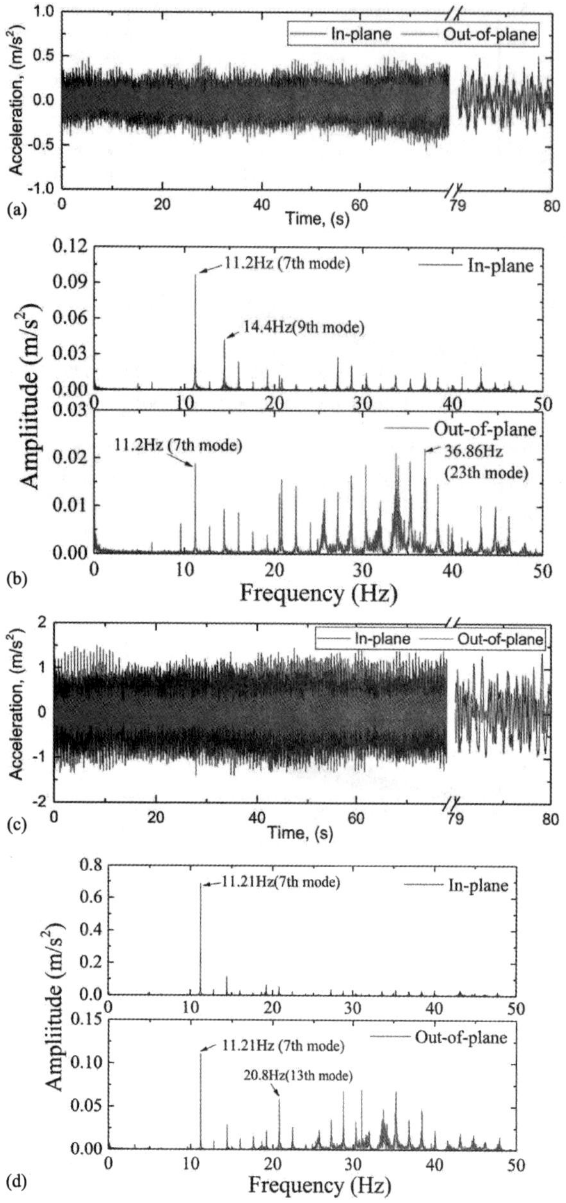

Fig. 20. Time histories and amplitude spectra of acceleration responses of the flexible stay cable model with double-helical fillets with $d=0.10\,D$ and $P=12\,D$: (a) time histories of acceleration responses for $V_{red}=32.9$; (b) amplitude spectra of acceleration responses for $V_{red}=32.9$; (c) time histories of acceleration responses for $V_{red}=55.4$; and (d) amplitude spectra of acceleration responses for $V_{red}=55.4$.

the VIV responses of the flexible stay cable model could be effectively suppressed by the double-helical fillets of $d=0.07\,D$ or $d=0.10\,D$ and $P=12\,D$. However, the control effects on the VIV responses of the double-helical fillets of $d=0.10\,D$ and $P=12\,D$ were better than that of the double-helical fillets with $d=0.07\,D$ and $P=12\,D$. Therefore, the double-helical fillets with $d=0.10\,D$ and $P=12\,D$ were applied to control the VIV responses of the flexible stay cable model for different β. As shown in Fig. 18(b), the out-of-plane vibration responses of the stay cable with double-helical fillets with $d=0.07$ or $d=0.10$ and $P=12\,D$ were close to that of the stay cable without helical fillets.

Fig. 19 shows the RMS of the acceleration of in-plane and out-of-plane vibration responses at P_{A1} of the flexible stay cable model with fillets of $d=0.10\,D$ for $\beta=0°$, 15°, 30°, 45°, −15°, −30°, −45°. As shown in Fig. 19, the VIV responses of the stay cable for different β can be effectively suppressed with the double-helical fillets with $d=0.10\,D$ and $P=12\,D$. The VIV responses of the stay cable could be effectively suppressed with the double-helical fillets of $d=0.10\,D$ and $P=12\,D$.

Fig. 20 shows the time histories and amplitude spectra of acceleration responses of in-plane and out-of-plane vibration at P_{A1} for $\beta=0°$. As shown in Fig. 20, the acceleration response amplitudes of the in-plane vibration of the flexible stay cable model with double-helical fillets were close to that of the out-of-plane vibration. In addition, the acceleration response amplitude spectrum included relatively more frequency components for the stay cable with double-helical fillets, and the stay cable exhibited the multifrequency random vibration. The aerodynamic force of the stay cable with double-helical fillets might be weakened in the spanwise direction, which might cause an increase in the frequency components of the in-plane and out-of-plane vibration responses of the stay cable.

Discussion and Conclusions

In this study, the MOAF of the stay cable was proposed to design the flexible stay cable model for the prototype stay cable that was based on the similarity criteria for aeroelastic model design. The ξ of the flexible stay cable model was approximately 0.021% to 0.064%, and the ξ gradually decreased with the increase in the vibration mode of the flexible stay cable model. From the identified frequencies and ξ of the first few modes of the flexible stay cable model, it might be feasible to design a flexible stay cable model via the proposed MOAF of the stay cable to simulate the high-mode vibration of the stay cables in this study. However, the fundamental f_1 of the flexible stay cable model was n_f times the f_1 of the flexible stay cable model, which was based on traditional similarity. Therefore, some high-order vibration modes might not be reproduced by the flexible cable model. Furthermore, if the aeroelastic model of the stay cable was designed according to the traditional method, the corresponding λ_{Re} of the Re was approximately 1/585.66 for the stay cable NJ32D of the STB. However, the corresponding λ_{Re} of the Re was approximately 1/11.2 for the flexible stay cable that was designed by the proposed method in this study, which could effectively reduce the effects of the Re on the high-order vibration responses of the stay cable.

In this study, the high-mode VIV characteristics and aerodynamic VIV control measures of the stay cable were investigated by applying the flexible stay cable model. The experimental results of the flexible stay cable with or without helical fillets were analyzed and compared, respectively. The following conclusions were drawn.

1. The novel design method could be used to design a flexible stay cable model with a relatively large λ_{Re} of the Re, which satisfied the similarity of the prototype stay cable.
2. The stay cable exhibited five in-plane VIVs in the test V range for $\beta=0°$, among them, four were single-mode in-plane VIVs with a relatively large amplitude, and one was a multimodal in-plane VIV with a relatively small amplitude. The maximum amplitude of the nondimensional displacement of the in-plane VIV of the flexible stay cable model decreased with the increase in the vibration mode.

3. The significant in-plane VIV responses of the stay cable were observed for $\beta = 15°$, $30°$, $-15°$, $-30°$, and $-45°$, respectively, and the in-plane vibration acceleration responses were significantly larger than the out-of-plane vibration acceleration responses of the flexible stay cable model.
4. The VIV responses of the stay cable could be effectively suppressed with the circular double-helical fillets with $d = 0.10\,D$ and $P = 12\,D$.

Data Availability Statement

Some data that support the findings of this study are available from the corresponding author upon reasonable requests, such as the detailed information on the stay cable model, and experimental results.

Acknowledgments

This research was financially supported by the National Natural Science Foundation of China (Grant Nos. 52178475, 51778225), for which the authors are grateful. The authors also gratefully acknowledge Sutong Bridge Co., Ltd. of China for its financial support.

References

Bartoli, G., F. Cluni, V. Gusella, and L. Procino. 2006. "Dynamics of cable under wind action: Wind tunnel experimental analysis." *J. Wind Eng. Ind. Aerodyn.* 94: 259–273. https://doi.org/10.1016/j.jweia.2006.01.002.

Chaplin, J. R., P. W. Bearman, F. J. Huera Huarte, and R. J. Pattenden. 2005. "Laboratory measurements of vortex-induced vibrations of a vertical tension riser in a stepped current." *J. Fluids Struct.* 21 (1): 3–24. https://doi.org/10.1016/j.jfluidstructs.2005.04.010.

Chen, W.-L., Q.-Q. Zhang, H. Li, and H. Hu. 2015. "An experimental investigation on vortex induced vibration of a flexible inclined cable under a shear flow." *J. Fluids Struct.* 54: 297–311. https://doi.org/10.1016/j.jfluidstructs.2014.11.007.

Chopra, A. K. 1995. *Dynamics of structures theory and applications to earthquake engineering*. Englewood Cliffs, NJ: Prentice-Hall A Simon & Schuster Company.

Cluni, F., V. Gusella, and F. Ubertini. 2007. "A parametric investigation of wind-induced cable fatigue." *Eng. Struct.* 29: 3094–3105. https://doi.org/10.1016/j.engstruct.2007.02.010.

Daniel Urdareanu, V., I. Radu Racanel, M. Degeratu, C. Ioan Cosoiu, C. Lucian Ghindea, R. Iuliu Cruciat, et al. 2019. "Experimental study of the dynamic response of a cable under wind flow." *Mater. Today: Proc.* 12: 446–454. https://doi.org/10.1016/j.matpr.2019.03.148.

Deng, Y. C., S. Y. Li, and Z. Q. Chen. 2021. "Experimental investigation on wake-induced vibrations of the hangers of suspension bridges based on three-dimensional elastic test model." *Eng. Struct.* 234: 111985. https://doi.org/10.1016/j.engstruct.2021.111985.

Duy, H. V., H. Katsuchi, H. Yamada, and M. Nishio. 2014. "Experimental study on dry-state galloping with various wind relative angles and its countermeasures." *J. Struct. Eng.* 60A: 428–435.

Fujino, Y. 2002. "Vibration, control and monitoring of long-span bridges—recent research, developments and practice in Japan." *J. Constr. Steel Res.* 58: 71–97. https://doi.org/10.1016/S0143-974X(01)00049-9.

Gao, D., W.-L. Chen, R.-T. Zhang, Y.-W. Huang, and H. Li. 2019. "Multi-modal vortex- and rain–wind- induced vibrations of an inclined flexible cable." *Mech. Syst. Sig. Process.* 118: 245–258. https://doi.org/10.1016/j.ymssp.2018.08.057.

Gao, Y., S. Fu, T. Ren, Y. Xiong, and L. Song. 2015. "VIV response of a long flexible riser fitted with strakes in uniform and linearly sheared currents." *Appl. Ocean Res.* 52: 102–114. https://doi.org/10.1016/j.apor.2015.05.006.

Ge, C. X., and A. R. Chen. 2019. "Vibration characteristics identification of ultra-long cables of a cable-stayed bridge in normal operation based on half-year monitoring data." *Struct. Infrastruct. Eng.* 15 (12): 1–16.

Han, Q., Y. Ma, W. Xu, Y. Lu, and A. Cheng. 2017. "Dynamic characteristics of an inclined flexible cylinder undergoing vortex-induced vibrations." *J. Sound Vib.* 394: 306–320. https://doi.org/10.1016/j.jsv.2017.01.034.

Hu, C., Z. Zhou, and J. Yan. 2020. "Wind-induced stability of a cable-stayed bridge with double main spans of 1,500 m and a twin-Box section." *J. Bridge Eng.* 25 (1): 04019135. https://doi.org/10.1061/(ASCE)BE.1943-5592.0001501.

Irvine, H. M. 1981. *Cable structures*. Cambridge, MA, London: The MIT Press.

Jafari, M., and P. P. Sarkar. 2020. "Wind-induced response characteristics of a yawed and inclined cable in ABL wind: Experimental- and numerical-model based study." *Eng. Struct.* 214: 110681. https://doi.org/10.1016/j.engstruct.2020.110681.

Lie, H., and K. E. Kaasen. 2006. "Modal analysis of measurements from a large-scale VIV model test of a riser in linearly sheared flow." *J. Fluids Struct.* 22: 557–575. https://doi.org/10.1016/j.jfluidstructs.2006.01.002.

Liu, M., W. Yang, W. Chen, H. Xiao, and H. Li. 2019. "Experimental investigation on vortex-induced vibration mitigation of stay cables in long-span bridges equipped with damped crossties." *J. Aerosp. Eng.* 32 (5): 04019072. https://doi.org/10.1061/(ASCE)AS.1943-5525.0001061.

Liu, Z., J. Shen, S. Li, Z. Chen, Q. Ou, and D. Xin. 2021. "Experimental study on high-mode vortex-induced vibration of stay cable and its aerodynamic countermeasures." *J. Fluids Struct.* 100: 103195. https://doi.org/10.1016/j.jfluidstructs.2020.103195.

Matsumoto, M., T. Yagi, Y. Shigemura, and D. Tsushima. 2001. "Vortex-induced cable vibration of cable-stayed bridges at high reduced wind velocity." *J. Wind Eng. Ind. Aerodyn.* 89: 633–647. https://doi.org/10.1016/S0167-6105(01)00063-0.

McTavish, S., A. Raeesi, A. D'Auteuil, K. Yamauchi, and H. Sato. 2018. "An investigation of the mechanisms causing large-amplitude wind-induced vibrations in stay cables using unsteady surface pressure measurements." *J. Wind Eng. Ind. Aerodyn.* 183: 19–34. https://doi.org/10.1016/j.jweia.2018.10.004.

Nagai, M., Y. Fujino, H. Yamaguchi, and E. Iwasaki. 2004. "Feasibility of a 1,400 m span steel cable-stayed bridge." *J. Bridge Eng.* 9 (5): 444–452. https://doi.org/10.1061/(ASCE)1084-0702(2004)9:5(444).

Nakagawa, K., K. Kishida, and K. Igarashi. 1998. "Vortex-induced oscillation and lift of yawed circular cylinders in cross-flow." *J. Fluids Struct.* 12: 759–777. https://doi.org/10.1006/jfls.1998.0157.

Qin, S. Q., W. Xu, Q. F. Lu, Q. G. Zheng, Z. G. Fu, R. A. Yan, and J. L. Sun. 2020. "Overall design and concept development for main navigation channel bridge of Changzhou-Taizhou Yangtze River bridge." [In Chinese.] *Bridge Constr.* 50 (3): 1–10.

Sarkar, P. P., and T. B. Gardner. 2000. "Model tests to study rain/wind-induced vibration of stay cables." In *Advanced Technology in Structural Engineering*. Reston, VA: ASCE.

Shirakashi, M., A. Hasegawa, and S. Wakiya. 1986. "Effect of the secondary flow on Karman Vortex shedding from a Yawed Cylinder." *Bull. JSME* 29 (250): 1124–1128. https://doi.org/10.1299/jsme1958.29.1124.

Susumu, F., and T. Masahiro. 2014. "Aerodynamic characteristics of indent stay cables of Tatara Bridge." In *IABSE Madrid Symp. Engineering for Progress, Nature and People*, 2749–2756. Madrid, Spain: IABSE.

Trim, A. D., H. Braaten, H. Lie, and M. A. Tognarelli. 2005. "Experimental investigation of vortex-induced vibration of long marine risers." *J. Fluids Struct.* 21: 335–361. https://doi.org/10.1016/j.jfluidstructs.2005.07.014.

Wang, H., A. Li, J. Niu, Z. Zong, J. Li, et al. 2013. "Long-term monitoring of wind characteristics at Sutong Bridge site." *J. Wind Eng. Ind. Aerodyn.* 115: 39–47. https://doi.org/10.1016/j.jweia.2013.01.006.

Yamaguchi, H., and Y. Fujino. 1998. "Stayed cable dynamics and its vibration control." In *Proc., Int. Symp. on Advances in Bridge Aerodynamics*, edited by A. Larsen and S. Esdahl, 235–253. Rotterdam, Netherlands: Balkema.

Yan, B. F., D. R. Li, W. B. Chen, L. Deng, and X. M. Jiang. 2020. "Mode shape-aided cable force determination using digital image correlation." *Struct. Health Monit.* (6). https://doi.org/10.1177/1475921720952163.

Zhang, H. J., and L. D. Zhu. 2013. "Influence of additional attack angle on flutter performance of a 1400m long cable stayed bridge." [In Chinese.] *J. Vibr. Shock* 32 (17): 95–99.

Zhang, X. G. 2004. "Conceptual design of Sutong Bridge." [In Chinese.] *Highway* 7: 1–11.

Zhang, X. G., H. Yuan, M. S. Pei, et al. 2006. "Key technologies research and innovation design of Sutong Bridge." [In Chinese.] In *Proc., National Bridge Academic Conf. of the Bridge and Structural Engineering Branch of the China Highway Society*, 3–6. Beijing: China Communications Press.

Zhu, L. D., H. J. Zhang, and X. H. Hu. 2011. "Test study of aerodynamic measures for flutter control of a proposed 1400m span steel box deck cable-stayed bridge." [In Chinese.] *Bridge Constr.* 41 (2): 9–12.

Zuo, D., and N. P. Jones. 2009. "Wind tunnel testing of yawed and inclined circular cylinders in the context of field observations of stay-cable vibrations." *J. Wind Eng. Ind. Aerodyn.* 97: 219–227. https://doi.org/10.1016/j.jweia.2009.06.009.

Journal of Wind Engineering & Industrial Aerodynamics 247 (2024) 105703

Contents lists available at ScienceDirect

Journal of Wind Engineering & Industrial Aerodynamics

journal homepage: www.elsevier.com/locate/jweia

Monitoring-based identification of nonlinear aerodynamic damping in multi-mode vortex-induced vibrations of a stay cable

Ruilin Zhang [a,b], Zhiwen Liu [a,b,c,*], Shuai Zhou [c], Yafei Wang [a,b], Zhengqing Chen [a,b,c]

[a] National Key Laboratory of Bridge Engineering Safety and Resilience, Hunan University, Changsha, Hunan, 410082, China
[b] Key Laboratory for Wind & Bridge Engineering of Hunan Province, Hunan University, Changsha, Hunan, 410082, China
[c] College of Civil Engineering, Hunan University, Changsha, Hunan, 410082, China

ARTICLE INFO

Keywords:
Stay cable
Vortex-induced vibration
Nonlinear aerodynamic damping
Sruton number criterion
System identification
Structural health monitoring

ABSTRACT

New issues of wind-induced cable vibration, i.e., multi-mode vibrations, increasingly occur in long-span bridges, which are closely associated with underlying damping features. In this study, an analytical method is proposed to identify nonlinear aerodynamic damping in multi-mode wind-induced cable vibrations in fields, involving the assessment of total and mechanical damping. The method is available to extract a series of mono-component resonances/decays from the measured signals based on the analytical mode decomposition (AMD) and random decrement technique (RDT), and then to determine the damping parameters using the Hilbert transform (HT) combined with the smoothing spline or linear least-square fitting. Furthermore, this method is applied to address the vortex-induced vibrations (VIV) of an ultra-long stay cable on the Sutong Bridge. Firstly, the vibrations are reported in terms of a typical event and one-year statistics. Subsequently, the damping parameters are identified to illustrate the in-situ aeroelasticity. The results show that the cable exhibits high-frequency vibrations primarily in the 41st to 43rd modes, and the aerodynamic damping exhibits significant amplitude-dependent nonlinearity. Finally, the damping identifications are employed to reveal the onset mechanism for the vibrations, and a Scruton number (Sc) criterion, i.e., Sc > 5.4, is suggested for mitigating the vibrations.

1. Introduction

Bridge stay cables are potentially excited into dramatic vibration of various types, including vortex-induced vibration (VIV) (Argentini et al., 2016; Urushadze and Pirner, 2018; He et al., 2022), rain-wind-induced vibration (RWIV) (Daniotti et al., 2021), and galloping (An et al., 2021). These vibrations, for shorter stay cables (e.g., span L < 200m), are typically dominated by a single mode of the first few during a specific vibration event. As stay cables continue to grow in length, however, high-order multi-mode vibrations are increasingly observed in recent years. These vibrations are not effectively controlled with the traditional countermeasures, resulting in more undue stress and thereby more severe fatigue damage. This emerging issue poses significant challenges to engineering designers.

Field monitoring is a powerful approach to investigating the multi-mode wind-induced vibrations of stay cables. The multi-mode VIVs on a stay cable (L = 323m) were observed by Chen et al. (2019), indicating that the vibrations primarily exhibited over the 18th to 40th modes under wind velocities ranging from 4 to 6 m/s. The vibration characteristics of a stay cable (L = 576.77m) were identified by Ge and Chen (2019), finding that the cable experienced RWIVs in the 13th mode and VIVs in the 43rd and 44th modes. The multi-mode VIVs of a stay cable (L = 349.2m) were reported by Kim et al. (2022), revealing the vibrations within the frequency ranging from 10 to 40 Hz in a wide range of wind velocities. The multi-mode vibrations of a stay cable (L = 512m) were presented by Wang et al. (2022a), summarizing that the VIVs and RWIVs occurred over frequency bands of 2.4–12.3 Hz and 2.2–4.1 Hz, respectively. The dynamic responses of stay cables (L = 337.4m, 541.1m) under extreme winds were analyzed by H. Zhang et al. (2022), showing the multi-mode vibrations with frequencies above 3 Hz. To sum up, these reports have revealed the vibration characteristics of longer cables, whereas the underlying aeroelastic mechanism remains unclear.

Multi-mode cable vibrations in fields can be addressed well using aeroelastic dynamics modeling, in which, aerodynamic damping, governing the energy transfer from air to specific cable modes, is of significance for predicting and mitigating cable vibrations. Great efforts have been experimentally/numerically made to identify aeroelastic

* Corresponding author. National Key Laboratory of Bridge Engineering Safety and Resilience, Hunan University, Changsha, Hunan, 410082, China.
E-mail address: zhiwenliu@hnu.edu.cn (Z. Liu).

https://doi.org/10.1016/j.jweia.2024.105703
Received 4 August 2023; Received in revised form 29 February 2024; Accepted 7 March 2024
Available online 26 March 2024
0167-6105/

parameters of sectional circular cylinders in terms of VIV (Lupi et al., 2018; R. Zhang et al., 2023), RWIV (Jing et al., 2018) and dry-cable galloping (Jafari and Sarkar, 2019). With these identifications, Jafari and Sarkar (2019) estimated the minimum damping required to prevent the dry-cable galloping up to the designed wind velocity; R. Zhang et al. (2023) predicted the VIV amplitudes of a circular cylinder under various mass-damping conditions. Nevertheless, the identifications are not directly applicable to in-situ cables due to involving multi-mode couplings, large aspect ratios, non-uniform turbulent winds, etc. In this sense, some attempts have been made to identify aerodynamic damping of stay cables based on field monitoring. Ni et al. (2007) estimated the segment-averaged aerodynamic damping of the first three modes of a stay cable during RWIVs. Hua and Zuo (2019) evaluated the van der Pol-type nonlinear aerodynamic damping of the second to fourth modes of stay cables under RWIVs. However, there are currently rare field reports on nonlinear aerodynamic damping identifications of stay cables, especially for high-order multi-mode vibrations.

An alternative scheme for identifying aerodynamic damping is to first derive the total damping by measuring the growing or decaying rate of the vibration evolution and then subtract the corresponding mechanical damping (Mashnad and Jones, 2014). However, two aspects of issues are encountered when identifying multi-mode aerodynamic damping in fields, namely the multi-mode coupling in the vibration signals and the unknown mechanical damping of in-situ cables. To address these issues, some advanced techniques in signal processing and system identification can be considered. For instance, the techniques, e.g., the wavelet transform, empirical mode decomposition (EMD) (Huang et al., 1998), and analytical mode decomposition (AMD) (Chen and Wang, 2012), are available for decomposing multi-frequency components; the methods, e.g., the random decrement technique (RDT) (Ibrahim, 1977) and natural excitation technique (NExT) (James et al., 1995), are feasible for modal identification. These techniques can be integrated to identify the multi-mode linear mechanical systems under operational conditions (He et al., 2011; Lardies and Ta, 2011; Wen et al., 2018). However, there is still a lack of attempts to identify multimode aeroelastic systems of nonlinearity, especially for closely-spaced modal frequency systems. Consequently, it is crucial to establish an analytical framework well-suited for identifying nonlinear aerodynamic damping in multi-mode wind-induced cable vibrations in field environments.

This study presents a monitoring-based investigation of the multi-mode VIVs of an ultra-long stay cable (L = 551.01m) by applying a proposed method for aerodynamic damping identification. The remainder of this paper is organized as follows. Section 2 introduces the details of the analytical method. Section 3 presents a case study on the Sutong Bridge. Finally, Section 4 summarizes the paper.

2. Methodology

Based on the dynamics of flexible cables and the modeling of wind loads, an analytical method is proposed to identify the multi-mode nonlinear aerodynamic damping of stay cables in the context of field monitoring, which is achieved by identifying the total and mechanical damping.

2.1. Theoretical background

For an inclined stay cable subjected to dynamic wind loads, the equations of motion, taking the in-plane vibration as an example, can be obtained by employing the Hamilton principle (Wang et al., 2005):

$$m\frac{\partial^2 y}{\partial t^2}+c\frac{\partial y}{\partial t}-T\frac{\partial^2 y}{\partial x^2}-\frac{\lambda^2 T}{L^3}\int_0^L y\mathrm{d}x=F(x,t) \tag{1}$$

where $y(x,t)$ is the dynamic displacement; m and c are the mass and damping coefficient per unit length of the stay cable, respectively; T is the initial tension; L is the cable span; $F(x,t)$ is the external wind loads on the stay cable; λ^2 is the dimensionless sag parameter, defined as follows:

$$\lambda^2=\frac{(mgL\cos\alpha)^2EA}{T^3L_e/L} \tag{2}$$

where g is the gravity acceleration; α is the inclined angle; E is Yong's modulus; A is the cross-section area; $L_e=L[1+(mgL\cos\alpha/T)^2/8]$ is the effective length of the stay cable in the static configuration. By neglecting the coupling between in-plane and out-of-plane motions, the wind loads on the oscillating cable at lock-in, i.e., the nonlinear self-excited aerodynamic forces, can be written as (Simiu and Scanlan, 1996; Jafari et al., 2020):

$$F(x,t)=\frac{1}{2}\rho U^2(2D)\left\{Y_1(K)\left[1-\varepsilon(K)\frac{y^2}{D^2}\right]\frac{\dot y}{U}+Y_2(K)\frac{y}{D}\right\} \tag{3}$$

where ρ is air density; U is the wind speed; D is the diameter of the stay cable; $K=\omega D/U$ is the reduced frequency with ω being the vibration frequency; Y_1^* and ε are the parameters associated with linear and nonlinear aerodynamic damping terms, respectively; Y_2^* is the parameter related to aerodynamic stiffness term. These parameters are functions of the reduced frequency and should depend on the yaw angle and turbulences of incoming winds in the fields. Strictly speaking, the self-excited force on the cable derives from not only in-plane but also out-of-plane motions, and the out-of-plane motion plays an important role in the galloping of cables (Jafari and Sarkar, 2020). Whereas, in Eq. (3), the out-of-plane motion's effects are neglected for convenience due to its relatively small participation in the VIVs of more concern herein.

According to the Galerkin method, the dynamic response $y(x,t)$ can be expressed in terms of a finite series (Wang and Zhao, 2009):

$$y(x,t)=\sum_{i=1}^{n}\varphi_i(x)q_i(t) \tag{4}$$

where $q_i(t)$ is the generalized coordinate and $\varphi_i(x)$ is the shape of the i th mode; n is the number of modes of concern. Substituting Eq. (4) into Eq. (1) and dropping the negligible aerodynamic stiffness term, one yields the following matrix equation of motion:

$$\mathbf{M}\ddot{\mathbf{q}}+(\mathbf{C}_s+\mathbf{C}_a)\dot{\mathbf{q}}+\mathbf{K}_s\mathbf{q}=\mathbf{0} \tag{5}$$

In Eq. (5), $\mathbf{M}=\mathrm{diag}[m_i]$, $\mathbf{C}_s=\mathrm{diag}[2m_i\xi_{si}\omega_{si}]$ and $\mathbf{K}_s=\mathrm{diag}[m_i\omega_{si}^2]$ are the generalized mass, damping and stiffness matrices of the stay cable, respectively, where $m_i=m\int_0^L\varphi_i^2\mathrm{d}x$ is the i th modal mass; ξ_{si} and ω_{si} are the i th modal mechanical damping ratio and natural frequency, respectively; $\mathbf{C}_a=\mathrm{diag}[2m_i\xi_{ai}\omega_{si}]$ is the generalized aerodynamic damping matrix, with ξ_{ai} being the i th modal aerodynamic damping ratio, which is defined as:

$$\xi_{ai}=-\frac{1}{2\mu K_{ref}}Y_{1,g}\left(1-\varepsilon_g\eta_i^2\right) \tag{6}$$

where $\mu=m/(\rho D^2)$ is the mass ratio of the stay cable to the placed air; $K_{ref}=\omega_{si}D/U_{ref}$ is the reduced frequency for reference with U_{ref} being a referenced wind velocity; $Y_{1,g}$ and ε_g are the generalized linear and nonlinear aerodynamic damping parameters, respectively, which are expressed as follows:

$$Y_{1,g}=\frac{\int_0^L\frac{U}{U_{ref}}Y_1\varphi_i^2\mathrm{d}x}{\int_0^L\varphi_i^2\mathrm{d}x},\ \varepsilon_g=\frac{\int_0^L\frac{U}{U_{ref}}Y_1\varepsilon\varphi_i^4\mathrm{d}x}{Y_{1,g}\int_0^L\varphi_i^2\mathrm{d}x} \tag{7}$$

and $\eta_i=q_i/D$ is the normalized generalized coordinate of the i th mode. Moreover, according to the harmonic balance method (Chen, 2013), the modal displacement-dependent aerodynamic damping in Eq. (6) can be equivalently transformed into the functions of modal displacement amplitude for convenience of parameter identification, as follows:

$$\xi_{ai} = -\frac{1}{2\mu K_{ref}} Y_{1,g}\left(1 - \frac{1}{4}\varepsilon_g A_{\eta_i}^2\right) \tag{8}$$

where A_{η_i} is the i th normalized modal displacement amplitude.

2.2. Analytical approach

For a given stay cable, the mass and stiffness matrices in Eq. (5) can be determined through the design data, whereas the damping parameters are unknown and require identification using particular approaches.

2.2.1. Signal decomposition

Stay cable vibrations involve multi-mode participation as well as quasi-static responses. In light of this, the AMD proposed by Chen and Wang (2012) is employed to extract a series of sub-signals of individual modes. According to the theorem, a real signal $y(t)$ consisting of n significant frequency components $(\omega_1, \omega_2, \cdots, \omega_n)$ can be decomposed into a series of time series whose Fourier spectra over mutually exclusive frequency ranges:

$$y(t) = \sum_{i=1}^{n} y_i(t) \tag{9}$$

where each time series $y_i(t)$ has a narrow frequency band and can be defined by:

$$y_1(t) = s_1(t), \cdots, y_i(t) = s_i(t) - s_{i-1}(t), \cdots, y_n(t) = y(t) - s_{n-1}(t) \tag{10}$$

In Eq. (10), the new series $s_i(t)$ could be determined as follows:

$$s_i(t) = \sin(\omega_{bi}t)H[y(t)\cos(\omega_{bi}t)] - \cos(\omega_{bi}t)H[y(t)\sin(\omega_{bi}t)] \tag{11}$$

where $H[\cdot]$ denotes the Hilbert transform operator, and ω_{bi} is the bisecting frequency between ω_i and ω_{i+1}. The selected frequency ranges should capture the target spectral bands, as illustrated in Fig. 1. For stay cable systems, the natural frequency approximately increases by multiple integers of the fundamental frequency, and hence the bisecting frequencies can be initially defined as the natural frequencies minus or plus half of the frequency interval, i.e., $\omega_{si} \pm \Delta\omega_s/2$. However, the spectral peaks of the measured responses may deviate slightly from the calculated natural frequencies. In such cases, the bisecting frequencies need to be redefined accordingly.

Moreover, the decomposed modal components $y_i(t)$ at the measured point $x = x_0$ can be scaled to yield the corresponding generalized modal responses according to the mode shapes:

$$q_i(t) = y_i(t) \,/\, \varphi_i(x_0) \tag{12}$$

In Eq. (12), the mode shape of a stay cable can be approximately expressed as: $\varphi_i(x_0) = \sin(i\pi x_0/L)$.

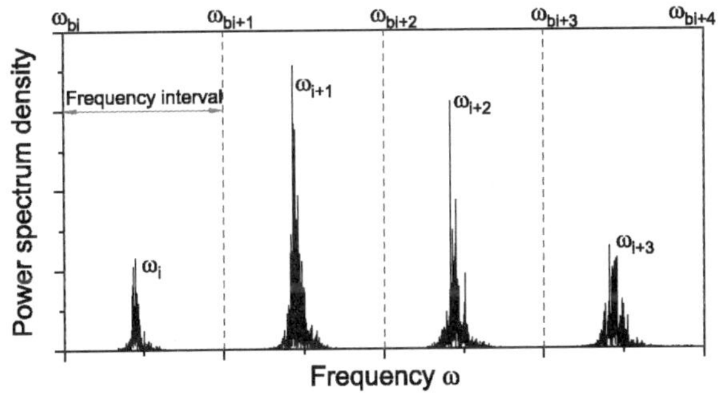

Fig. 1. Selection of bisecting frequencies in the analytical mode decomposition for stay cable systems.

2.2.2. Transient response

Two types of transient responses, i.e., wind-induced resonance and free-to-decay signature, are the basis for identifying total and mechanical damping, respectively. The evolution of the wind-induced resonance generally includes three stages, namely growth-to-resonance, fully-developed, and resonance-to-decay stages, as depicted in Fig. 2. This evolution could reflect the energy-trapping/dissipating properties of stay cables from/to their surrounding winds through the cable-wind interaction.

On the other hand, the RDT (Ibrahim, 1977) is adopted to determine the free-to-decay signatures from the ambient-excited cable responses in the time domain. The theoretical basis for this approach is described as follows. According to Duhamel's integral principle, the dynamic response $y_i(t)$ at a location $(x = x_0)$ for the i th mode of a linear system subjected to arbitrary loads consists of free decay and forced vibration, as follows:

$$y_i(t) = e^{-\xi_{si}\omega_{si}t}\left\{y_i(0)\left[\cos(\omega_{di}t) + \frac{\xi_{si}\omega_{si}}{\omega_{di}}\sin(\omega_{di}t)\right] + \frac{\dot{y}_i(0)}{\omega_{di}}\sin(\omega_{di}t)\right\}$$
$$+ \varphi_i(x_0)\int_0^t h_i(t-\tau)F_i(\tau)\mathrm{d}\tau \tag{13}$$

where $y_i(0)$ and $\dot{y}_i(0)$ are the initial displacement and velocity at $t = 0$, respectively; $\omega_{di} = \omega_{si}\sqrt{1-\xi_{si}^2}$ is the i th modal damped frequency; F_i is the generalized force, and $h_i(t-\tau)$ is the impulse response function of the i th mode. Supposing that the external force is a zero-mean stationary Gaussian process, the forcing convolution integral in Eq. (13) over time is thus expected to be zero. In this sense, the transient decay can be determined by averaging the ensemble data series under the specific initial condition $(y_i = a, \dot{y}_i = b)$, given as follows:

$$E\left[y_i(t+\tau)|_{y_i(t)=a, \dot{y}_i(t)=b}\right] = A_{y,i}e^{-\xi_{si}\omega_{si}\tau}\sin(\omega_{di}\tau + \varphi_i) \tag{14}$$

In Eq. (14), $A_{y,i}$ and φ_i are the initial amplitude and phase, respectively. Considering the limited duration of the signal, the random decrement signature $\delta_i(\tau)$ for the i th mode can be extracted in a new time coordinate system by averaging the selected N signal segments:

$$\delta_i(\tau) = \frac{1}{N}\sum_{k=1}^{N} y_i(t_k + \tau) \tag{15}$$

In practice, the initial conditions are typically given as the specific displacement or acceleration, and thus the free vibration due to the initial velocity will be averaged out. The process of RDT is illustrated in Fig. 3.

2.2.3. Parameter identification

Once the single-mode transient resonance/decay responses are extracted, the Hilbert transform (HT) is applied to identify the modal parameters of either linear or nonlinear systems. The HT of a time series $y_i(t)$ is defined as follows:

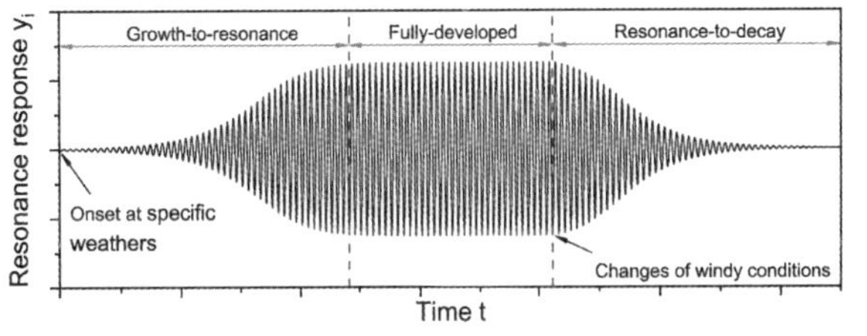

Fig. 2. Schematical illustration of wind-induced resonance evolution during a typical vibration event of a stay cable.

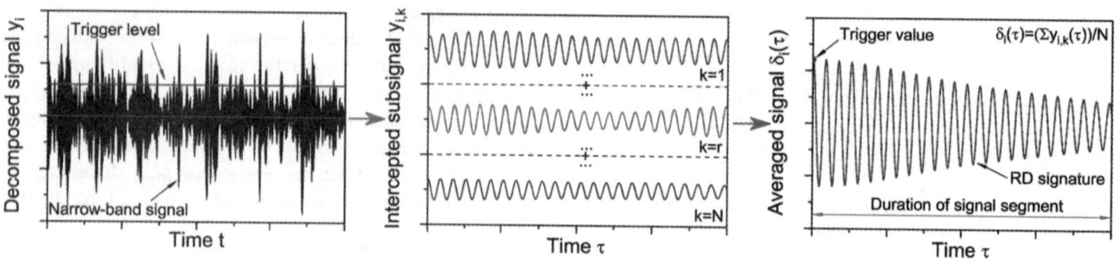

Fig. 3. Extraction of the free-to-decay response (random decrement signature) from the narrow-band ambient-excited signal of stay cables by the random decrement technique.

$$\widetilde{y}_i(t)=H[y_i(t)]=\frac{1}{\pi}\int_{-\infty}^{\infty}\frac{y_i(\tau)}{t-\tau}\mathrm{d}\tau \tag{16}$$

Then, an analytic signal associated with $y_i(t)$ can be determined as follows:

$$Y_i(t)=y_i(t)+\mathrm{i}\widetilde{y}_i(t)=A(t)e^{i\theta(t)} \tag{17}$$

where $A(t)$ and $\theta(t)$ are the instantaneous amplitude (i.e., the envelope) and phase angle of $y_i(t)$, respectively, and they are defined as:

$$A(t)=\left[y_i^2(t)+\widetilde{y}_i^2(t)\right]^{1/2},\theta(t)=\tan^{-1}[\widetilde{y}_i(t)/y_i(t)] \tag{18}$$

By applying the logarithmic and differential operators to Eq. (18), the instantaneous frequency and damping ratio can be determined from the following equations:

$$\omega(t)=\frac{\mathrm{d}\theta_{fit}(t)}{\mathrm{d}t},\xi(t)=-\frac{1}{\omega(t)}\frac{\mathrm{dln}A_{fit}(t)}{\mathrm{d}t} \tag{19}$$

where $\theta_{fit}(t)$ and $A_{fit}(t)$ are the instantaneous amplitude and phase after fitting, respectively. This fitting process aims to eliminate their local fluctuations due to the possible mixing among components of adjacent modes and the end effects of the HT. Specifically, smoothing spline fitting is employed to remove the spikes of the envelopes when identifying the total damping during wind-induced resonance, whereas linear least-square fitting is applied to the functions of $\theta(t)$ and $\ln A(t)$ when determining the natural frequency and mechanical damping ratio. Fig. 4 illustrates the two types of fitting in parameter identification.

The increasing/decreasing rate of wind-induced cable resonance is characterized as total damping, which is the sum of mechanical damping and aerodynamic damping (Main and Jones, 2001; Ni et al., 2007), and thus the aerodynamic damping is estimated as:

$$\xi_{ai}=\xi_{total,i}-\xi_{si} \tag{20}$$

where $\xi_{total,i}$ is the total damping ratio of the i th mode. Conceivably, wind-induced cable instability will occur if the total damping becomes zero or less than zero which is possible when the aerodynamic damping becomes negative with a magnitude equal to or exceeding that of the mechanical damping.

2.3. Processing procedure

By applying the above analytical approach, the multi-mode nonlinear damping parameters can be identified based on the monitoring data, which mainly includes the following steps.

(1) *Decomposing vibration signals*: The wind-induced resonances and/or ambient-excited responses are first selected from the monitoring data by detecting the response magnitude and wind conditions. Then, the Fourier spectra are employed to determine the modes of concern, along with the corresponding bisecting frequencies ω_{bi}. On this basis, the narrow-band signal can be extracted by performing the AMD on the concerned frequency ranges (see Eqs. (10) and (11)).
(2) *Extracting transient responses*: For the wind-induced resonances, the decomposed signals are directly taken as the transient responses available for the following analysis. Whereas, for the ambient-excited responses, the RDT is still required to extract the free-to-decay signatures. To achieve this, (i) the trigger condition and the segment length are first set; (ii) the segmented signals are then gathered when the trigger condition is met; (iii) the free-to-decay signatures are obtained by averaging the segments (see Eq. (15)).
(3) *Identifying damping parameters*: Based on the obtained two types of transient responses, the HT is first applied to determine their envelopes and instantaneous phases via Eq. (18). Then, the natural frequencies and mechanical damping ratios, as well as the time-varying total damping ratios, can be identified according to Eq. (19). Finally, the aerodynamic damping can be determined by subtracting the mechanical damping from the total damping (see Eq. (20)). The generalized nonlinear aerodynamic parameters are further obtained by fitting the amplitude-dependent aerodynamic damping (see Eq. (8)).

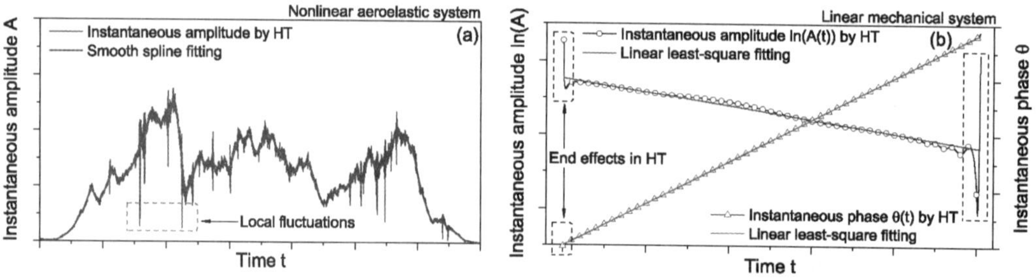

Fig. 4. Fitting of instantaneous amplitude and phase of the transient responses for parameter identification: (a) smoothing spline fitting; (b) linear least-square fitting.

3. Application to the Sutong Bridge

Based on the monitoring data of the Sutong Yangtze River Bridge in Jiangsu Province, China, the proposed method is applied to deal with the multi-mode vortex-induced vibrations of an ultra-long stay cable.

3.1. *Description of the bridge*

The Sutong Bridge is a cable-stayed bridge with a main span of 1088 m. The bridge deck is composed of a steel closed-box girder with a width of 41.0 m and a height of 4.0 m, which is supported by two inverted Y-shaped towers through a total of 272 stay cables arranged in two planes. The Sutong Bridge has been instrumented with a structural health monitoring (SHM) system after completion to evaluate the bridge's performances, as shown in Fig. 5. This monitoring system consists of 15 types of sensors with a total number of over one thousand. Particularly, four three-dimensional (3D) ultrasonic anemometers are installed at the bridge deck and the tower tops to record the wind velocity and direction. Furthermore, biaxial accelerometers are mounted at the bridge deck and the specific stay cables (depicted in Fig. 5) to measure their accelerations. In recent years, some stay cables of the bridge have experienced dramatic vibrations. To detect these vibrations, supplemental monitoring instruments are installed on this bridge (Yang et al., 2021; Wang et al., 2022a). Specifically, the accelerometers (DH1A202E, Donghua Tech.) are installed on the eight stay cables (denoted in Fig. 5) to record the in-plane and out-of-plane vibrations of the stay cables simultaneously. Also, it is noted that the anemometers are capable of measuring the wind velocity up to 45.0 m/s and wind direction ranging from 0° to 360° with a sampling frequency of 1.0 Hz, while the accelerometers have a wide measurement range of ± 4g and a much higher sampling frequency of 100.0 Hz, enabling the capture of possible high-frequency vibrations with greater precision.

This investigation focuses on an ultra-long stay cable of the Sutong Bridge, NJ32U, with a length of $L = 551.01$m and a diameter of $D = 0.152$m (aspect ratio $L/D = 3625$), as illustrated in Fig. 6. The stay cable is fabricated using parallel wires wrapped in polyethylene protective tubes covered with dimpled surfaces. More importantly, an MR damper is equipped in the plane of the stay cable to mitigate vibrations, and this damper has been thoroughly examined and repaired before the present monitoring. Also, high-damping rubber is attached to the cable anchorages at the bridge deck. An auxiliary steel wire rope is incorporated into the stay cable near the MR damper to partly participate the vibration control. More parameters of the stay cable are illustrated in Table 1.

3.2. *Overview of cable vibrations*

In 2019, the SHM system recorded over fifty high-frequency vibration events of the stay cable NJ32U, most of which occurred between May and September with the most endured and intense record taking place on September 13.

Fig. 7 shows the whole evolution of the typical vibration event of the stay cable observed on September 13, 2019. During this event, the mean wind velocities and directions are 7.1 m/s and 72.4° (yaw angle is $\beta = 28.2°$) at the deck level and 10.7 m/s and 84.8° (yaw angle is $\beta = 15.8°$) at the tower top, respectively (see Fig. 7a and b). The recorded winds are remarkably steady without significant variation in wind directions, and the mean turbulence intensity at the deck level and the tower top are only 4.5 and 2.9 percent, respectively. Fig. 7c and d presents the corresponding time histories of the in-plane and out-of-plane accelerations of the stay cable. The in-plane vibration is triggered at about 19:40 and gradually increases to its maximum amplitude of 36 m/s^2. Afterward, the vibration experiences rising and descending for about 2 h before eventually decaying out. Interestingly, the decay rate of the vibration is much larger than the growth rate. Specifically, it takes about 53 min from the onset to the maximum response, while the vibration completely decays in 30 min. On the other hand, the out-of-plane vibration develops similarly but with much smaller amplitudes. Furthermore, the evolutionary spectra of the stay cable accelerations are determined through the short-time Fourier transform (STFT), as illustrated in Fig. 7e and f. Both the in-plane and out-of-plane vibrations are characterized by three prominent frequencies around 10.4 Hz, corresponding to the 41st to 43rd modes of the stay cable, indicating obvious high-order multi-mode behaviors.

Subsequently, it is to elucidate the vibration mechanism of the stay cable based on the observations as follows. Due to the high-frequency features and dry conditions, the vibration is thus inferred to be excited by the Karman vortices in the cable wake. In this sense, the vortex-shedding frequency f_{vs} can be determined based on the Strouhal law:

$$f_{vs} = \frac{StU\cos\beta^*}{D} \tag{21}$$

In Eq. (21), St is the Strouhal number and is considered as 0.19 for the circular cylinder in the higher subcritical regime (R. Zhang et al., 2023); $\beta^* = \sin^{-1}(\sin\beta\cos\alpha)$ is the equivalent wind yaw angle (Jafari and Sarkar, 2019). Then, the time-averaged shedding frequencies f_{vs} at the deck level and tower top are 7.97 Hz and 12.94 Hz, respectively, and their mean (10.45 Hz) matches closely the observed dominant frequency of the stay cable vibration (10.4 Hz). Therefore, this event is attributed to VIV. To further prove this inference, Fig. 8 shows the variation of the vibration responses against the wind velocity during this event. The responses are observed so sensitive to incoming wind velocity that even a slight change can result in the onset or decay of the vibration. Furthermore, the dominant frequencies, corresponding to the maximum in the power spectrum density (PSD) function of the acceleration, are somewhat scattered at relatively low wind velocities but stabilize at about 10 Hz when the wind velocity exceeds 10 m/s, demonstrating the vortex lock-in behavior.

After analyzing the above vibration event, it is to make some statistics between the stay cable vibrations and windy conditions over the one-year monitoring data in 2019. The wind data from the tower-top anemometer is utilized as a reference due to its smaller disturbances. Moreover, the two turbulence parameters, i.e., turbulence intensity I_u and integral length L_u, are determined via the following equations (Simiu and Scanlan, 1996):

$$I_u = \frac{\sigma_u}{U}, L_u = \frac{U}{\sigma_u^2}\int_0^\infty R_u(\tau)d\tau \tag{22}$$

where σ_u is the standard deviation of the wind velocity; U is the mean

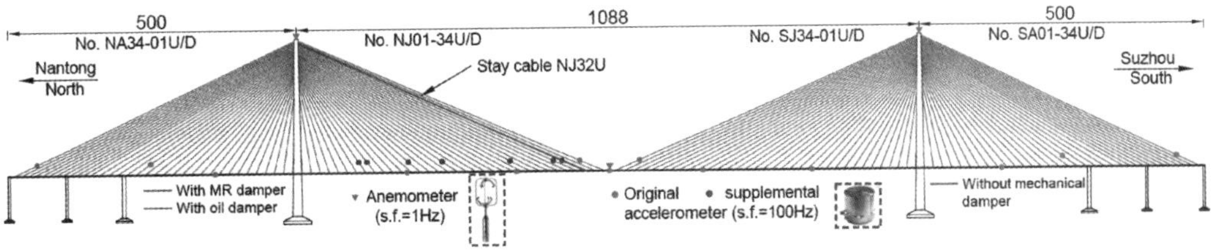

Fig. 5. Structural health monitoring system of the Sutong Yangtze River Bridge and partial instrumented sensors (unit: m).

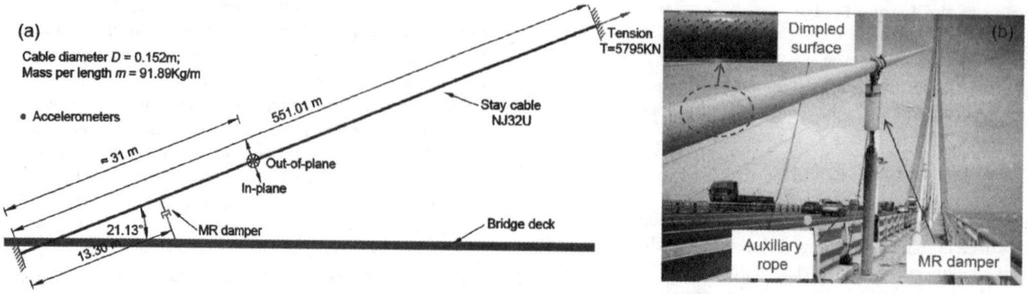

Fig. 6. Stay cable NJ32 of the Sutong Yangtze River Bridge: (a) technical illustration; (b) onsite photo.

Table 1
Parameters of the stay cable NJ32U of the Sutong Yangtze River Bridge.

L (m)	D (m)	α (°)	m (kg/m)	T (kN)	E (MPa)	A (m^2)	λ^2	$f_{s,1}^{in}$ (Hz)	$f_{s,1}^{out}$ (Hz)
551.01	0.152	21.13	91.89	5795	1.95×10^5	0.01022	2.20	0.259	0.241

Note that $f_{s,1}^{in}$ and $f_{s,1}^{out}$ are the in-plane and out-of-plane fundamental frequencies of the stay cable, respectively, and the measured in-plane fundamental frequency $f_{s,1}^{in}$ (0.259 Hz) is slightly larger than the frequency interval between modes (0.248 Hz) due to the cable sag.

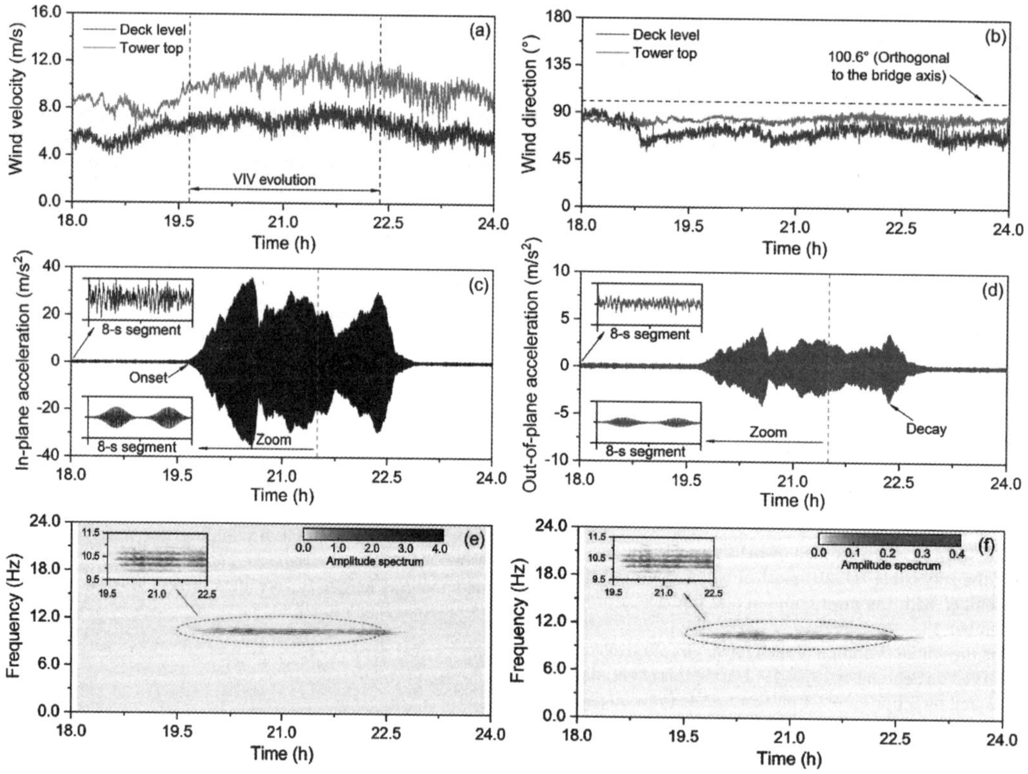

Fig. 7. Vortex-induced vibration event of the stay cable on September 13, 2019: (a) (b) wind velocities and directions; (c) (d) time histories of the accelerations; (e) (f) evolution of the acceleration spectra.

wind velocity; $R_u(\tau)$ is the autocorrelation function of the wind velocity. The analysis is taken with a basic interval of 10 min.

Fig. 9 illustrates the effects of windy conditions on the modal responses of the stay cable, which is obtained through the AMD and modal coordinate amplification (see Eqs. (10)–(12)). The results show that the vibrations predominantly occur in the 41st to 43rd modes, while the responses in other modes (e.g., the 40th and 44th modes) are considerably smaller. On the other hand, as depicted in Fig. 9a, the large-amplitude vibrations are mainly within a range of wind velocity U_{tower} from 7.0 to 14.0 m/s (or equivalently U_{deck} = 4.4–8.8 m/s). The maximum occurs at the wind velocity U_{tower} = 10.8 m/s (or equivalently U_{deck} = 6.8 m/s), corresponding to an averaged reduced velocity U/fD =

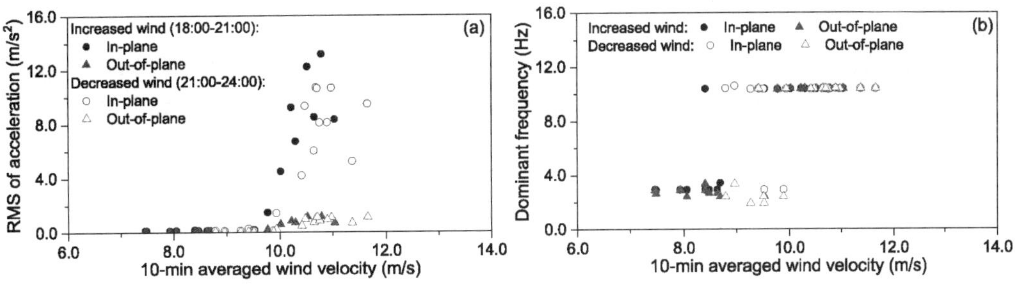

Fig. 8. Relationship between the tower-top wind velocity and the vibration response of the stay cable during the vortex-induced vibration event on September 13, 2019: (a) 10-min RMS (root mean square) of the acceleration; (b) dominant vibration frequency.

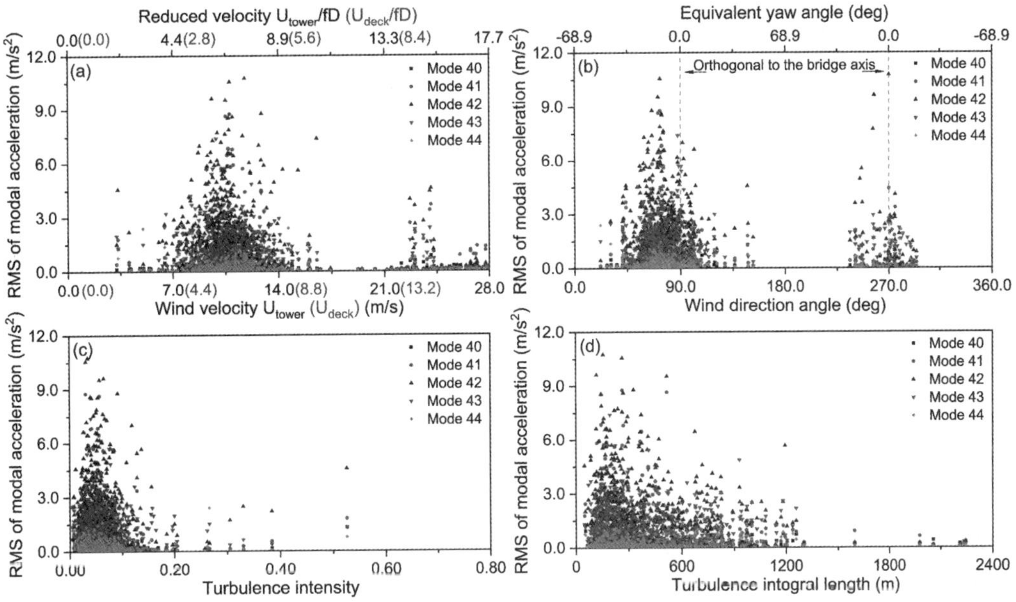

Fig. 9. Effects of the windy conditions on the 10-min RMS of the in-plane modal accelerations of the stay cable during the vortex-induced vibrations observed in 2019: (a) wind velocity; (b) relative wind direction; (c) turbulence intensity; (d) turbulence integral length. Note that, in (a), the deck-level wind velocities presented are determined based on the gradient wind coefficient referring to the tower-top wind velocities; the reduced velocity is determined by referring to $f = 10.4$ Hz, i.e., the natural frequency of the 42nd mode.

5.6 with $f = 10.4$ Hz, i.e., the natural frequency of the 42nd mode. As seen in Fig. 9b, the cable vibrations are prone to occur under the winds perpendicular to the bridge axis (or the stay cable plane), but they are not restricted solely to the orthogonal winds. Specifically, the VIV can be observed in two wide regions of the relative wind direction angle, i.e., 45°–115° and 240°–290°. This suggests that there still are noticeable vibration responses even at the wind yaw angle reaching 45°, which is similar to the experimental observations by Liu et al. (2021). Fig. 9c and d demonstrate the dependence of the cable vibrations on the incoming flow turbulence. The vibration events are found mostly linked with relatively smooth flows with turbulence intensity below 0.10. More significant vibrations occur at the turbulence intensity of approximately 0.05. Additionally, the vibrations can be observed over a wide range of turbulence integral lengths from 50 m to 1200m. These lengths are considerably greater than the diameter of the stay cable, which could potentially reduce the sensitivity of the vibrations to the turbulence integral length.

3.3. *Results of damping identification*

The above analysis focuses on the response characteristics of the stay cable during the VIVs. Whereas, the associated damping features deserve further investigation. The mechanical damping of the stay cable is first identified for determining the aerodynamic damping subsequently. To this end, the ambient-excited responses of the stay cable are selected at enough small wind velocities, i.e., $U_{tower} < 4.0$ m/s, to decrease the wind effects as much as possible (Li et al., 2011). To reduce the errors due to possible uncertainties, six sets of 4-h acceleration signals, recorded on different days in 2019, are herein employed. Fig. 10 displays the time histories and PSD functions of both in-plane and out-of-plane accelerations in a case. On this basis, the present method is utilized to identify the multi-mode mechanical damping ratios of the stay cable. The first fifty single-mode free-decay responses are determined by applying the AMD and RDT successively. In the RDT process, the trigger level is set as 1.5 times the standard deviation of the decomposed acceleration responses, as suggested by Wen et al. (2018), and the duration of the signal segment is about 25 cycles of the decays. Subsequently, the mechanical damping parameters are identified by

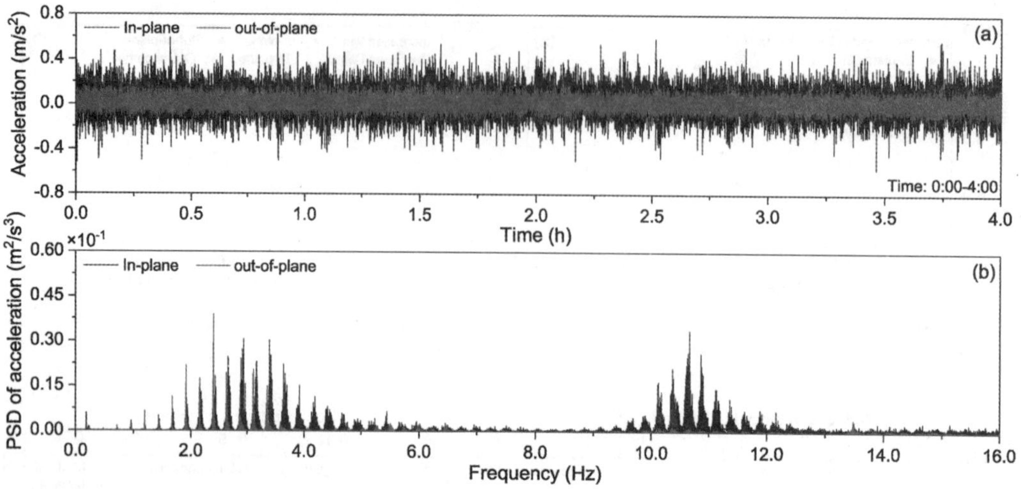

Fig. 10. Ambient-excited accelerations of the stay cable on April 25, 2019: (a) time histories; (b) power spectrum densities.

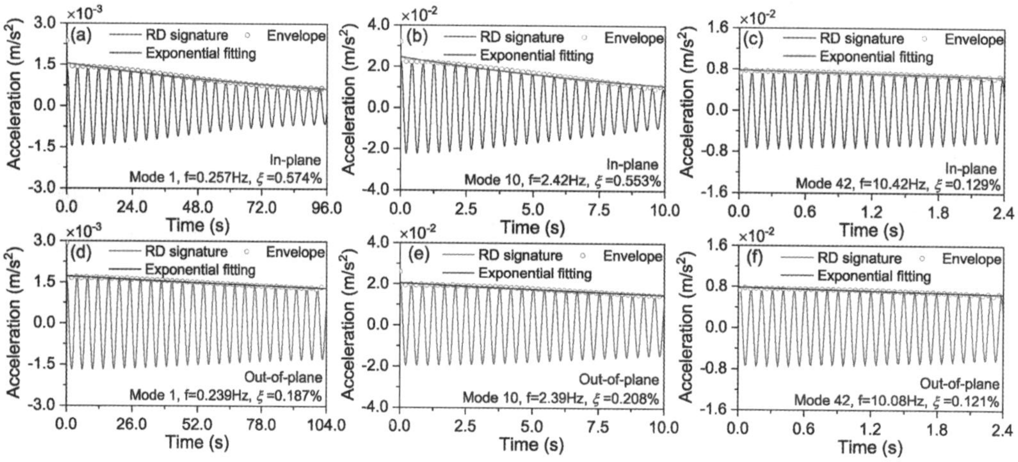

Fig. 11. Free decay responses and their envelopes of the three individual modes identified based on the measured acceleration of the stay cable on April 25, 2019.

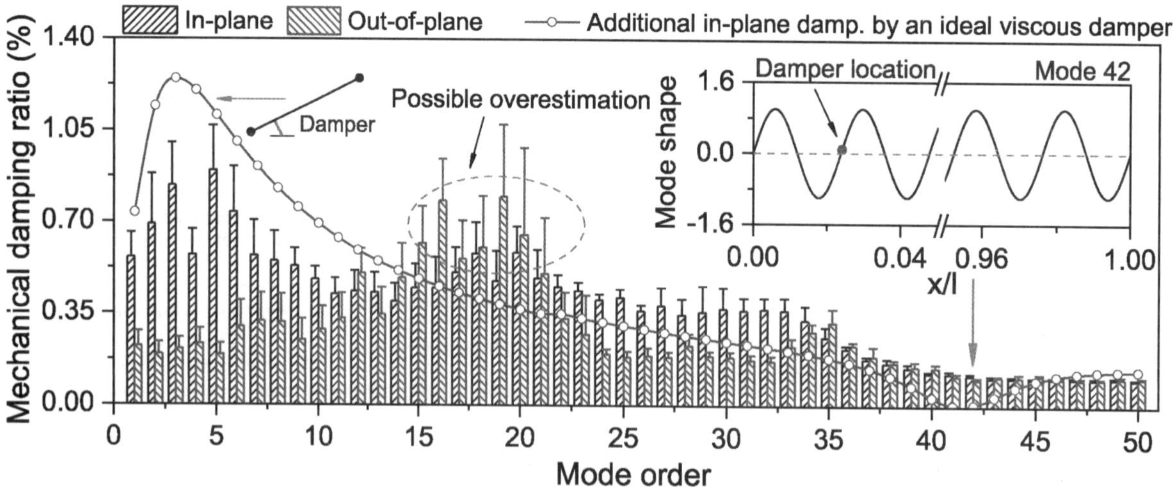

Fig. 12. Estimated mechanical damping ratios of the stay cable based on the measured response under ambient excitations.

combining the HT with the linear least-square fitting, as shown in Fig. 11. The satisfactory fits achieved illustrate that the stay cable can be approximately considered as a linear mechanical system at small amplitudes.

Accordingly, the identified mechanical damping ratios of the stay cable for the first fifty modes are summarized in Fig. 12, including the means and standard deviations. The in-plane and out-of-plane fundamental-mode mechanical damping ratios are estimated to be 0.56% and 0.22%, respectively. The in-plane mechanical damping ratio is significantly larger than the out-of-plane one, which is primarily ascribed to the MR damper equipped in the cable plane. The out-of-plane fundamental-mode mechanical damping ratio (without the MR damper's effect) is close to, but slightly greater than, the previously-reported results for other stay cables in the in-plane fundamental modes, e.g., 0.178% in Ni et al. (2007) and approximately 0.16% in Yamaguchi and Fujino (1998), confirming the reliability of the identified results. Furthermore, the in-plane mechanical damping ratio is generally decreased in higher modes. For the high modes above the 40th order, both in-plane and out-of-plane mechanical damping ratios are only approximately 0.1%. These mechanical dampings should be primarily attributed to the inherent damping properties of the stay cable itself and the MR damper as well as the high-damping rubber. Moreover, to characterize the effects of the external damper, the additional in-plane mechanical damping ratios by an ideal viscous damper optimized targeting the third mode, determined via the iterative scheme (Krenk, 2000), are also drawn in Fig. 12. It is observed that, beyond the optimal mode, the additional mechanical damping ratio gradually decreases as the mode order increases, which is consistent with the field-identified results. In addition, by comparing with the theoretical results, it is noticed that the mechanical damping ratios around the 18th mode may be overestimated due to the weak signals in these modes (see Fig. 10), while those in the 45th to 50th modes may be underestimated as the MR damper does not effectively detect the high-mode ambient-excited vibration with small displacements.

Subsequently, the total damping of the cable during the VIVs can be identified using the proposed method, as shown in Fig. 13. Based on the observed typical VIV event (see Fig. 7), multiple groups of individual-mode responses are firstly extracted from the overall signal. The attached spectra demonstrate that the single-mode signal is successfully obtained after using the AMD. This decomposition also shows that the vibrations manifest initially on the 42nd and 43rd modes, 12 min earlier than that of the 41st mode. In addition, the decay of the vibrations varies among the three modes, and the 42nd mode endures over 10 min longer than the other two modes. Therefore, the 42nd mode is believed to play a more dominant role in this VIV event. Moreover, based on the smoothing spline fittings of the modal displacement envelopes, the time-varying total damping ratios are determined in this figure. It is found that the maximum negative total damping ratios occur around the onset instants, which are identified as 0.095‰, 0.11‰, and 0.098‰ for the 41st, 42nd, and 43rd in-plane modes, respectively. As the vibration amplitudes increase, the negative total damping ratios generally decrease. In the fully-developed stage, the total damping ratios are relatively small with slight fluctuations. During the decaying stage, the positive total dampings increase as the vibration amplitudes decrease, and the maximum positive total dampings are usually accompanied by the end of the decays. Overall, the total damping ratios are relatively small, which results in the VIV development being so slow that it spans approximately 1 h. Moreover, the in-plane and out-of-plane motions have almost the same total dampings, indicating that the two motions share similar evolution patterns. More specifically, the out-of-plane motion follows the in-plane motion.

Based on the mechanical and total damping identified above, aerodynamic damping of the stay cable in the VIVs can be obtained to illustrate the in-situ aeroelasticity. Fig. 14 shows the aerodynamic damping against the modal displacement amplitude during the growth-to-resonance stage of the vibration. Interestingly, several closed loops

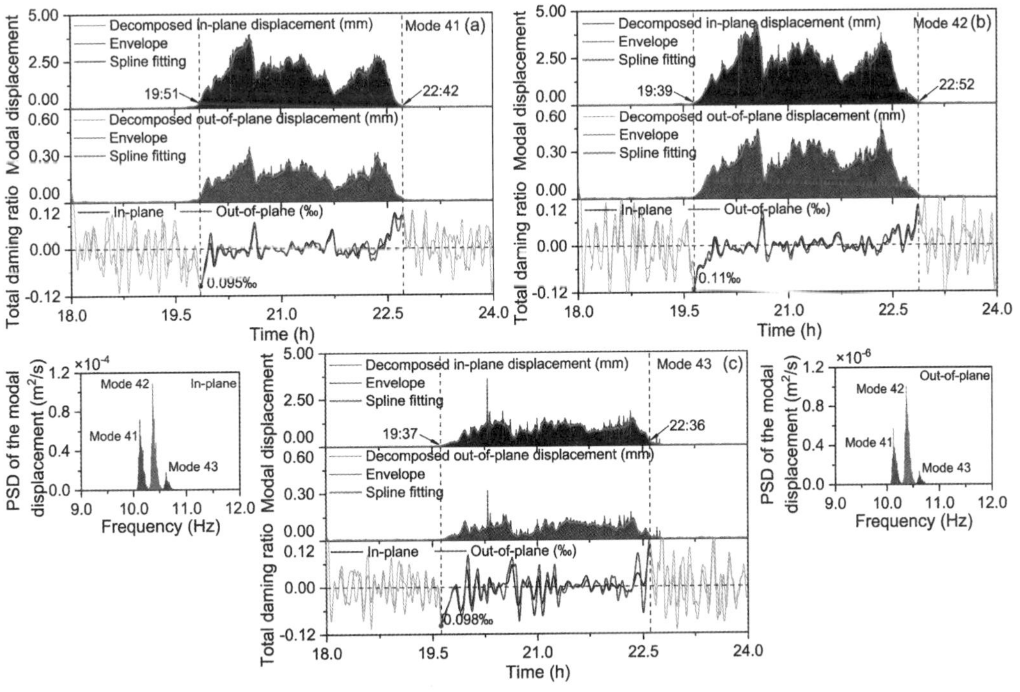

Fig. 13. Identification of the time-varying total damping ratios of the stay cable during the vortex-induced vibration event on September 13, 2019: (a) the 41st mode; (b) the 42nd mode; (c) the 43rd mode. Note that the measured accelerations have been numerically integrated to yield the displacements via the frequency-domain algorithm (Liu et al., 2022) before the identification.

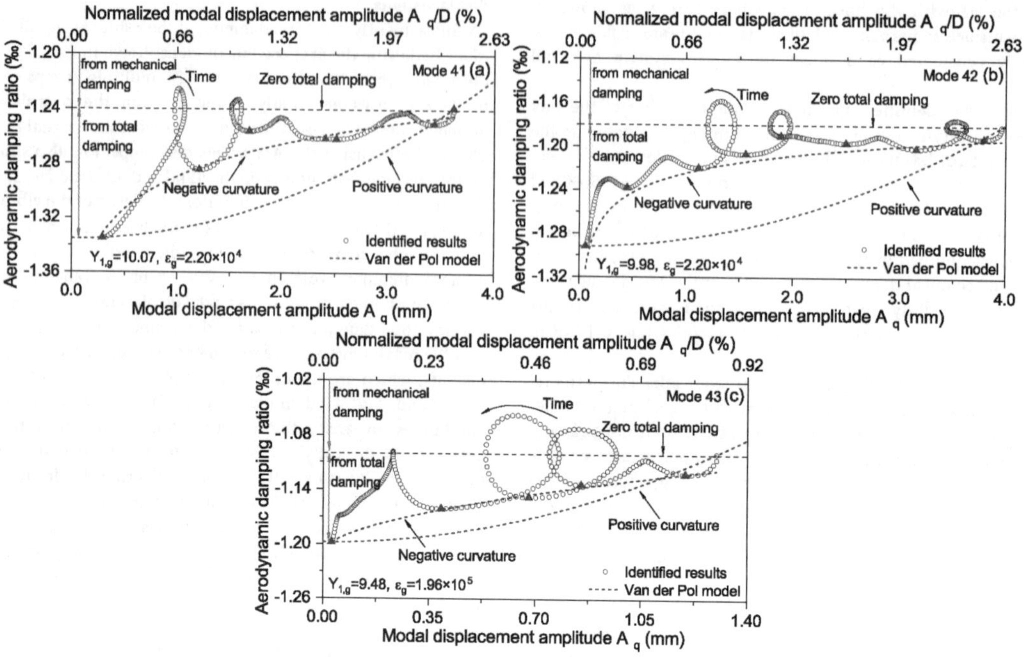

Fig. 14. Identified amplitude-dependent aerodynamic damping ratios during the growth-to-resonance stage of the vortex-induced vibration event on September 13, 2019: (a) the 41st mode; (b) the 42nd mode; (c) the 43rd mode. Note that in this figure, "curvature" is defined according to the aerodynamic damping (negative value) rather than the aerodynamic parameter (positive value), and thus its sign is opposite to these in Lupi et al. (2018), Guo et al. (2021) and M. Zhang et al. (2021).

are observed in these curves, resulting in multiple aerodynamic damping values for a given modal displacement amplitude. The direct cause for this phenomenon is that the vibration evolution may experience local downs in growth rate (i.e., total damping) and even decrease but then continue to increase progressively (see Fig. 13). Therefore, the aerodynamic damping, or more specifically total damping, may exhibit ups and downs over time, as denoted by the arrays in this figure. The inherent reason for this phenomenon is that the aerodynamic damping is not solely determined by the vibration amplitude, but varies with the windy condition, etc. Specifically, slight variations in wind velocity and/or direction may significantly alter the vibration evolution. In addition, multi-mode coupling may influence the surrounding flow fields and thereby fluctuate aerodynamic damping. Overall, the aerodynamic damping of an in-situ flexible cable is more complicated than that of a sectional cable model from experimental/numerical tests.

More importantly, if ignoring these loops, the aerodynamic damping can be roughly characterized by their local minima denoted in dots, which represent stronger exciting capability around this amplitude to some extent. In this sense, the negative aerodynamic damping is found to generally decrease as the displacement amplitude increases. This decrease is mainly due to the nonlinear aeroelastic effects that the vortex-induced forces are significantly dependent on the oscillation amplitudes (Lupi et al., 2018; R. Zhang et al., 2023). Moreover, the van der Pol-type polynomials are applied to model the variation of the aerodynamic damping ratio against the modal displacement amplitude, according to the principle of capturing the maximum aerodynamic damping at the initial small-amplitude vibration and the amplitude at zero total damping, as shown in Fig. 14. Accordingly, the linear and nonlinear generalized aerodynamic parameters are obtained as follows: $Y_{1g} = 10.07$, 9.98, and 9.48 and $\varepsilon_g = 2.20 \times 10^4$, 2.20×10^4 and 1.96×10^5 for the 41st to 43rd modes with $K_{ref} =$ 1.16, 1.19 and 1.22, respectively. Also, it is noticed that, in the van der Pol model, the negative aerodynamic damping decays faster at a larger vibration amplitude, characterized as a positive curvature ($d^2\xi_a/dA_q^2 > 0$), whereas, the field-identified aerodynamic damping seems to decay more slowly, i.e., a negative curvature ($d^2\xi_a/dA_q^2 < 0$). Similar findings regarding the curvature are also seen in the experimental and numerical investigations of circular cylinders (Lupi et al., 2018; Guo et al., 2021; M. Zhang et al., 2021; R. Zhang et al., 2023). Therefore, it is necessary to develop more advanced models for nonlinear aerodynamic damping in VIVs (Larsen, 1995; M. Zhang et al., 2020, 2021; Guo et al., 2021), but this aspect is not the focus herein.

3.4. *Discussion*

With the identified damping parameters, the VIVs of the stay cable can be re-analyzed, aiming to reveal the mechanism for the onset of the vibrations and to give a control strategy.

3.4.1. *On onset mechanism*

In general, wind-induced cable vibrations could occur if the negative aerodynamic damping, related to Karman vortices or wind-rain action, exceeds the corresponding mechanical damping. For VIVs, the Karman vortices could be locked mode-by-mode as the wind velocity increases, indicating potential excitations in all concerned modes. However, in the present observations, the VIVs only or primarily exhibit in the 41st to 43rd modes, rather than mode-by-mode excitations. The reasons are mainly two-fold: (i) The vortex-induced negative aerodynamic damping is relatively small, having a very weak capability to overcome the mechanical damping of the stay cable. For example, the maximum negative aerodynamic damping during the typical VIV event is only 1.3‰, approximately. (ii) The stay cable exhibits considerable mechanical damping, derived from the MR damper, the high-damping rubber and the cable itself, in most modes, even including the high-order modes (e.g., above the 45th mode), as demonstrated in the measured and analytical results in Fig. 12. However, several particular high-order modes are exceptions, due to these modes exhibiting a shape node close to the MR damper (see Fig. 12). Consequently, the additional mechanical damping by the damper is almost zero for these modes. Therefore, the VIVs are observed only within the particular frequency

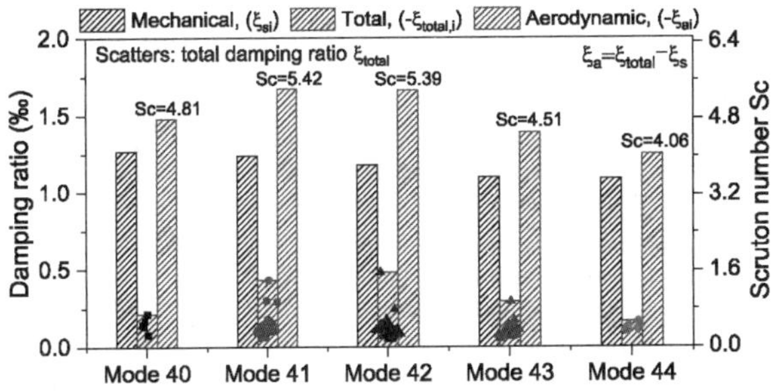

Fig. 15. Suggested Scruton number thresholds for mitigating the high-order multi-mode vortex-induced vibrations based on the identified damping parameters of the stay cable.

bands where the damper is located on the mode shape nodes. It is also expected that the VIVs could be observed in a wider frequency band without the MR damper.

In addition, it is interesting to discuss a phenomenon in practice that VIVs are prone to occur in long cables, whereas rarely observed in relatively short cables. An important factor is the shorter cables have a larger fundamental frequency, which results in the onset velocity for these "particular" modes being above the wind velocity encountered commonly at the bridge sites.

3.4.2. On control criterion

More importantly, the identified aerodynamic damping can be in turn used to determine the mechanical damping required for mitigating VIVs. To improve the generality, the maximum negative aerodynamic damping ratios are identified from the recorded VIVs in 2019, as summarized in Fig. 15. The scatters are primarily in the 41st to 43rd modes, with less in the 40th and 44th modes, because the vibrations are predominantly triggered in the 41st to 43rd modes. Moreover, for convenience in engineering practice, the extreme aerodynamic damping is further equivalently normalized to be the Scruton number (Sc) that unifies the aerodynamic damping ratio and the mass ratio:

$$Sc > \frac{m|\xi_a|}{\rho D^2} \tag{23}$$

Accordingly, the obtained Sc threshold is shown in Fig. 15, characterizing the extreme capability of the vortex-induced loads to excite the stay cable into vibrations. The Sc thresholds for the 40th to 44th modes are 4.81, 5.42, 5.39, 4.51, and 4.06, respectively. If the Sc number of the stay cable itself ($Sc = m\xi_s/\rho D^2$) exceeds these thresholds, VIVs will not occur. Overall, based on the damping identifications, this study suggests a Scruton number criterion of $Sc > 5.4$ for mitigating the VIVs of stay cables, which can be further applied used for the optimal design of damper systems (Di et al., 2021; Wang et al., 2022b). Also, note that the current Sc threshold, determined by the maximum negative aerodynamic damping at the onset of the VIVs, is capable of controlling VIVs completely, rather than reducing the VIV amplitude to a certain level. Partly due to this, the Sc threshold is larger than that derived from the aerodynamic parameters of sectional cylinders at normalized amplitudes of 0.03 ($Sc = 1.33$ (Lupi et al., 2018),) and 0.05 ($Sc = 1.16$ (R. Zhang et al., 2023),).

On the other hand, it is necessary to indicate that the criterion of Sc > 10 is widely adopted for controlling wind-induced cable vibration, particularly for RWIV (Irwin, 1997), which is much larger than the present results. It implies that the Karman vortices, compared with the rain-wind action, are weaker in exciting cables into vibrations. In other words, mitigating VIV needs a smaller mechanical damping. In addition, the present criterion for VIVs is different from that for dry-cable galloping where the Sc threshold increases with the critical velocity required (Jafari and Sarkar, 2019, 2020).

4. Conclusions

This study proposed an analytical method to identify the nonlinear aerodynamic damping associated with multi-mode wind-induced cable vibrations in field environments. For this method, a series of individual-mode transient wind-induced resonances and free-to-decay signatures are firstly extracted from the monitored signals through the AMD and RDT, and then the total and mechanical damping, as well as the corresponding aerodynamic damping, can be estimated using the HT combined with the smoothing spline and linear least-square fitting. Moreover, based on the one-year field monitoring on the Sutong Bridge, the method was applied to address the response characteristics, onset mechanism, and control strategies of the multi-mode VIVs of an ultra-long stay cable ($L = 551.01$m, $D = 0.152$m, $L/D = 3625$) with an external damper.

The monitoring results showed that the stay cable experienced high-frequency vibrations, mainly involving the 41st to 43rd modes (around 10.4 Hz). The vibrations were featured by the non-planar motions, with dominant in-plane motion and accompanied out-of-plane motion. The vibration mechanism was detected as VIV according to the Strouhal law. The statistical analysis revealed that the VIVs mainly occurred within the wind velocity range of $U_{tower} = 7.0–14.0$ m/s or equivalently $U_{deck} = 4.4–8.8$ m/s and across wide wind direction ranges (e.g., the yaw angle reaching 45°). The VIVs were mostly associated with relatively smooth flows with turbulence intensity below 0.10 but were not very sensitive to the turbulence integral length. Subsequently, by applying the proposed method, the mechanical and time-varying total damping ratios were identified based on the measured ambient-excited responses and the resonance evolution, respectively, yielding the corresponding aerodynamic damping ratios. The results showed that the high-mode mechanical damping ratios are notably small, i.e., approximately 1.0‰ for the modes above the 40th order. The aerodynamic damping exhibits a maximum negative value at the onset instant and then decreases generally with increasing oscillation amplitude. The amplitude-dependent aerodynamic damping was roughly described by the van der Pol model with nonlinear aerodynamic parameters. Finally, the mechanism for the onset was revealed, attributing to the weak mechanical damping for the specific modes with a shape node encountered by the external damper. More importantly, based on the identified extreme value of aerodynamic damping, a criterion of $Sc > 5.4$ was proposed for mitigating the VIVs of stay cables, which is helpful for the optimal design of damper systems.

This study only focused on applying the analytical method to investigate the multi-mode VIVs of a stay cable. In fact, the method can be extended to analyze other types of vibrations, such as RWIV and galloping of stay cables, as well as VIVs on the decks of cable-stayed and suspension bridges (Xu et al., 2019). In this extension, the mode orders of concern should be lower than those in the present case, and better results may be achieved due to larger frequency intervals.

CRediT authorship contribution statement

Ruilin Zhang: Writing – original draft, Visualization, Methodology, Investigation, Formal analysis, Data curation. **Zhiwen Liu:** Writing – review & editing, Validation, Project administration, Methodology, Investigation, Conceptualization. **Shuai Zhou:** Writing – review & editing, Formal analysis. **Yafei Wang:** Investigation, Formal analysis. **Zhengqing Chen:** Supervision, Project administration.

Declaration of competing interest

The authors declare that they have no known competing financial interests or personal relationships that could have appeared to influence the work reported in this paper.

Data availability

Data will be made available on request.

Acknowledgment

This research was financially supported by the National Natural Science Foundation of China (Grant Nos. 52178475, 51778225), for which the authors are grateful. The authors also gratefully acknowledge Jiangsu Sutong Bridge Co., Ltd., China, for the data support.

References

An, M., Li, S., Liu, Z., Yan, B., Li, L., Chen, Z., 2021. Galloping vibration of stay cable installed with a rectangular lamp: field observations and wind tunnel tests. J. Wind Eng. Ind. Aerod. 215, 104685.

Argentini, T., Rosa, L., Zasso, A., 2016. Wind-induced vibrations of the stay cables of a roundabout flyover: assessment, repair, and countermeasures. J. Perform. Constr. Facil. 30, 04016046.

Chen, G., Wang, Z., 2012. A signal decomposition theorem with Hilbert transform and its application to narrow band time series with closely spaced frequency components. Mech. Syst. Signal Process. 28, 258–279.

Chen, W.L., Gao, D., Laima, S., Li, H., 2019. A field investigation on vortex-induced vibrations of stay cables in a cable-stayed bridge. Appl. Sci. 9, 4556.

Chen, X., 2013. Estimation of stochastic crosswind response of wind-excited tall buildings with nonlinear aerodynamic damping. Eng. Struct. 56, 766–778.

Daniotti, N., Jakobsen, J.B., Snæbjörnsson, J., Cheynet, E., Wang, J., 2021. Observations of bridge stay cable vibrations in dry and wet conditions: a case study. J. Sound Vib. 503, 116106.

Di, F., Sun, L., Chen, L., 2021. Suppression of vortex-induced high-mode vibrations of a cable-damper system by an additional damper. Eng. Struct. 242, 112495.

Ge, C., Chen, A., 2019. Vibration characteristics identification of ultra-long cables of a cable-stayed bridge in normal operation based on half-year monitoring data. Structure and Infrastructure Engineering 15, 1567–1582.

Guo, K., Yang, Q., Liu, M., Li, B., 2021. Aerodynamic damping model for vortex-induced vibration of suspended circular cylinder in uniform flow. J. Wind Eng. Ind. Aerod. 209, 104497.

He, M., Liang, P., Wang, Y., Xia, Z., Wang, X., 2022. Online automatic monitoring of abnormal vibration of stay cables based on acceleration data from structural health monitoring. Measurement 195, 111102.

He, X.H., Hua, X.G., Chen, Z.Q., Huang, F.L., 2011. EMD-based random decrement technique for modal parameter identification of an existing railway bridge. Eng. Struct. 33, 1348–1356.

Huang, N.E., Shen, Z., Long, S.R., Wu, M.C., Shih, H.H., Zheng, Q., Yen, N.-C., Tung, C.C., Liu, H.H., 1998. The empirical mode decomposition and the Hilbert spectrum for nonlinear and non-stationary time series analysis. Proceedings of the Royal Society of London. Series A: Math. Phys. Eng. Sci. 454, 903–995.

Hua, J., Zuo, D., 2019. Evaluation of aerodynamic damping in full-scale rain-wind-induced stay cable vibration. J. Wind Eng. Ind. Aerod. 191, 215–226.

Ibrahim, S.R., 1977. Random decrement technique for modal identification of structures. J. Spacecraft Rockets 14, 696–700.

Irwin, P.A., 1997. Wind vibrations of cables on cable-stayed bridges. In: Proceedings of Structures Congress XV. Structural Engineering Institute/ASCE, Portland, Oregon, pp. 383–387.

Jafari, M., Hou, F., Abdelkefi, A., 2020. Wind-induced vibration of structural cables. Nonlinear Dynam. 100, 351–421.

Jafari, M., Sarkar, P.P., 2019. Parameter identification of wind-induced buffeting loads and onset criteria for dry-cable galloping of yawed/inclined cables. Eng. Struct. 180, 685–699.

Jafari, M., Sarkar, P.P., 2020. Wind-induced response characteristics of a yawed and inclined cable in ABL wind: experimental- and numerical-model based study. Eng. Struct. 214, 110681.

James, G.H., Carne, T.G., Lauffer, J.P., 1995. The natural excitation technique (NExT) for modal parameter extraction from operating structures. Int. J. Anal. Exp. Modal Anal. 10, 260–277.

Jing, H., He, X., Li, H., 2018. Initial condition and damping effects on rain-wind induced cable vibration. J. Wind Eng. Ind. Aerod. 175, 376–383.

Kim, S., Kim, S., Kim, H.K., 2022. High-mode vortex-induced vibration of stay cables: monitoring, cause investigation, and mitigation. J. Sound Vib. 524, 116758.

Krenk, S., 2000. Vibrations of a taut cable with an external damper. J. Appl. Mech. 67, 772–776.

Larsen, A., 1995. A generalized model for assessment of vortex-induced vibrations of flexible structures. J. Wind Eng. Ind. Aerod. 57, 281–294.

Lardies, J., Ta, M.-N., 2011. Modal parameter identification of stay cables from output-only measurements. Mech. Syst. Signal Process. 25, 133–150.

Li, H., Laima, S., Ou, J., Zhao, X., Zhou, W., Yu, Y., Li, N., Liu, Z., 2011. Investigation of vortex-induced vibration of a suspension bridge with two separated steel box girders based on field measurements. Eng. Struct. 33, 1894–1907.

Liu, Z., Shen, J., Li, S., Chen, Z., Ou, Q., Xin, D., 2021. Experimental study on high-mode vortex-induced vibration of stay cable and its aerodynamic countermeasures. J. Fluid Struct. 100, 103195.

Liu, Z., Li, S., Wang, L., Yan, B., Zhang, R., Chen, Z., 2022. Experimental investigation on high-mode vortex-induced vibration of a flexible stay cable in smooth flow. J. Bridge Eng. 27, 04022068.

Lupi, F., Niemann, H.-J., Höffer, R., 2018. Aerodynamic damping model in vortex-induced vibrations for wind engineering applications. J. Wind Eng. Ind. Aerod. 174, 281–295.

Main, J.A., Jones, N.P., 2001. Evaluation of viscous dampers for stay-cable vibration mitigation. J. Bridge Eng. 6, 385–397.

Mashnad, M., Jones, N.P., 2014. A model for vortex-induced vibration analysis of long-span bridges. J. Wind Eng. Ind. Aerod. 134, 96–108.

Ni, Y.Q., Wang, X.Y., Chen, Z.Q., Ko, J.M., 2007. Field observations of rain-wind-induced cable vibration in cable-stayed Dongting Lake Bridge. J. Wind Eng. Ind. Aerod. 95, 303–328.

Simiu, E., Scanlan, R.H., 1996. Winds Effects on Structures: Fundamentals and Applications to Design. Wiley, New York.

Urushadze, S., Pirner, M., 2018. Vortex-induced vibration of stay cables, verification on the footbridge. Experimental Vibration Analysis for Civil Structures 5, 497–504.

Wang, L., Zhao, Y., 2009. Large amplitude motion mechanism and non-planar vibration character of stay cables subject to the support motions. J. Sound Vib. 327, 121–133.

Wang, X.Y., Ni, Y.Q., Ko, J.M., Chen, Z.Q., 2005. Optimal design of viscous dampers for multi-mode vibration control of bridge cables. Eng. Struct. 27, 792–800.

Wang, Y., Chen, Z., Yang, C., Liu, Z., He, J., Feng, Z., 2022a. A novel eddy current damper system for multi-mode high-order vibration control of ultra-long stay cables. Eng. Struct. 262, 114319.

Wang, Y., Li, S., Qie, K., Zhou, S., Chen, Z., 2022b. Effectiveness of damping and inertance of two dampers on mitigation of multimode vibrations of stay cables by using the finite difference method. J. Vib. Control 0, 1–14.

Wen, Q., Hua, X.G., Chen, Z.Q., Niu, H.W., Wang, X.Y., 2018. AMD-based random decrement technique for modal identification of structures with close modes. J. Aero. Eng. 31, 04018057.

Xu, S., Ma, R., Wang, D., Chen, A., Tian, H., 2019. Prediction analysis of vortex-induced vibration of long-span suspension bridge based on monitoring data. J. Wind Eng. Ind. Aerod. 191, 312–324.

Yamaguchi, H., Fujino, Y., 1998. Stayed cable dynamics and its vibration control. Proceedings of the international symposium on advances in bridge aerodynamics 235–253.

Yang, C., Chen, Z., Wang, W., Hua, X., Wang, Y., 2021. Optimal design of two viscous dampers for multi-mode control of a cable covering broad frequency range. Eng. Struct. 245, 112830.

Zhang, H., Wang, H., Xu, Z., Zhang, Y., Tao, T., Mao, J., 2022. Monitoring-based analysis of wind-induced vibrations of ultra-long stay cables during an exceptional wind event. J. Wind Eng. Ind. Aerod. 221, 104883.

Zhang, M., Xu, F., Øiseth, O., 2020. Aerodynamic damping models for vortex-induced vibration of a rectangular 4:1 cylinder: Comparison of modeling schemes. J. Wind Eng. Ind. Aerod. 205, 104321.

Zhang, M., Yu, H., Ying, X., 2021. Incorporation of subcritical Reynolds number into an aerodynamic damping model for vortex-induced vibration of a smooth circular cylinder. Eng. Struct. 249, 113325.

Zhang, R., Liu, Z., Wang, L., Chen, Z., 2023. Component decomposition and spatial distribution of unsteady forces on a circular cylinder oscillated under various amplitudes at Re $= 2.0\times10^4$. Ocean Eng. 281, 114746.

新能源结构

建筑结构学报 Journal of Building Structures
第34卷 第6期 2013年6月
Vol. 34 No. 6 June 2013
文章编号：1000-6869(2013)06-0060-08

基于完全气动弹性模型的冷却塔风致响应风洞试验研究

邹云峰，牛华伟，陈政清
（湖南大学 风工程试验研究中心，湖南长沙 410082）

摘要：风荷载是冷却塔设计的控制荷载，完全气动弹性模型风洞试验是研究其风致响应的有效途径。基于此，推导了冷却塔完全气动弹性模型相似关系并用有限元分析方法予以验证，据此相似关系设计并制作了某核电站200 m高超大型冷却塔的1∶400完全气动弹性模型，并在风洞中模拟的B类风场对其风致响应进行测试，试验前对模型的动力特性进行检验。试验结果表明：完全气动弹性模型能较为精确模拟冷却塔结构的质量、刚度和阻尼相似；迎风面的风致变形较大，而其中又以喉部附近最大；背景响应占总响应的主导地位，动力放大效应不明显；风振系数值随高度的增加而减小，各控制点风振系数均值略小于DL/T 5339—2006《火力发电厂水工设计规范》规定值。

关键词：冷却塔；完全气动弹性模型；风洞试验；风致响应；风振系数
中图分类号：TU332 TU311.3 TU317.1 **文献标志码**：A

Wind tunnel test on wind-induced response of cooling tower based on full aero-elastic model

ZOU Yunfeng, NIU Huawei, CHEN Zhengqing
(Wind Engineering Research Center, Hunan University, Changsha 410082, China)

Abstract: Wind loading is the control load for the design of cooling tower, and full aero-elastic model wind tunnel test is an effective way to study wind-induced response of cooling tower. Based on this, similitude relations for cooling tower full aero-elastic model were derived and verified, according to which a full aero-elastic model with geometrical scale of 1∶400 was designed and processed for a 200 m high cooling tower of a nuclear power plant. Wind tunnel tests were carried out to measure the wind-induced response in the simulated wind field of terrain category B, before which the dynamic characteristics of the model was tested. The test result indicates full aero-elastic model can accurately simulate the similitude relation of mass, stiffness and damping of prototype cooling tower. Meanwhile, it is found that the maximum displacement occurring in windward side of the throat and the average displacement obtained with full aero-elastic model are almost the same as the result calculated with data measured through rigid model pressure-measurement test. It is also shown that background response plays a major role in the total displacement response and the dynamic amplification effect is not obvious. At the same time, test result shows that gust loading factor decreases with height increasing and the mean value is slightly less than the specified value.

Keywords: cooling tower; full aero-elastic model; wind tunnel test; wind-induced response; gust loading factor

基金项目：国家自然科学基金项目(50908085,51208067)。
作者简介：邹云峰(1984—)，男，湖南邵阳人，博士研究生。E-mail：67793657@qq.com
通信作者：牛华伟(1978—)，男，河南驻马店人，工学博士，工程师。E-mail：niuhw_hd@126.com
收稿日期：2012年7月

0 引言

获得结构物的风致响应有两种途径，即测试结构物的风荷载进而计算出结构的响应和直接测量结构物的风致响应。风洞试验中，风荷载可以通过刚性模型测压和高频底座天平测量广义气动力进行研究，但不能考虑结构物的气弹效应，而采用气弹模型可以直接测量结构物的风致响应。

热电厂大型发电机组的冷却塔塔体高、体型大、壁厚薄，对风荷载极为敏感，风荷载是其设计的控制荷载。1965 年英国渡桥电厂 3 座冷却塔的倒塌引起各国学者对冷却塔风荷载及风致振动的广泛研究，Isyumov[1]、Armitt[2] 等的研究表明，冷却塔的风致动态响应和静态响应属同一数量级，且共振响应随风速的增长速率远大于准静态响应，冷却塔结构的气弹效应显著。气弹模型风洞试验是测量气弹效应显著的结构物风致响应最直接的途径。美国 ASCE 的建筑风洞试验手册[3] 指出冷却塔的风致响应应通过“replica models”即完全气动弹性模型来测量，所谓“replica models”直译为“原型的复制品”，它是一个按冷却塔原型严格几何缩尺得到的连续壳体，通过选择合适的材料，可以实现质量与刚度连续分布的相似性，由于是连续体模型，也容易实现塔体各阶模态阻尼比的相似。但这种气弹模型制作和测试都非常困难，至今只有少数学者成功进行了冷却塔完全气动弹性模型风洞试验。Tschanz 等[4] 通过完全气动弹性模型风洞试验获得了原型塔的风致响应，开启了冷却塔风致动态响应研究；Armitt[2] 应用完全气动弹性模型对渡桥电厂事故进行研究，发现模型试验结果与事故发生时的风速及 3 个塔的倒塌顺序非常吻合，证明了完全气动弹性模型试验的可靠性；Niemann 等[5] 利用完全气动弹性模型研究了冷却塔的风致干扰，给出了不同塔间距下的干扰因子建议值，研究结果被德国冷却塔设计规范引用；陈凯等[6] 基于完全气动弹性模型对冷却塔风致变形进行了测试，并对通过由差分得到的加速度值获得的风振力进行分析，指出风振力沿环向基本不变，但以顶层最大。但以上研究都未对冷却塔完全气动弹性模型的相似关系作出阐述，且试验大多在上世纪 70 年代甚至更早进行，由于位移计精度的限制，只对冷却塔应变响应进行测量，没有对冷却塔的风致变形进行研究或得到的变形精度较差，因而无法得到位移风振系数，不能进一步了解位移响应特性。

本文对冷却塔完全气动弹性模型的相似关系进行推导，并通过有限元分析方法对相似关系进行验证，为冷却塔完全气动弹性模型的设计提供参考，据此相似关系设计和制作了某核电站 200 m 高双曲冷却塔的 1:400 完全气动弹性模型，通过高精度激光位移计对其风致位移进行了测试以进一步了解大型冷却塔结构的风致变形和响应特性，并将气弹模型试验结果与刚性模型测压试验风压时程有限元分析结果进行了比较。

1 相似关系推导及验证

1.1 相似关系推导

在气弹模型设计中，不仅要求模型在外形上与实际结构保持几何相似，还要保证模型的质量、刚度和阻尼与实际结构满足如柯西数、弗劳德数、斯托罗哈数等相似的要求。文献[3]指出，对于冷却塔这种自立式结构，完全气动弹性模型设计时可以忽略重力的影响，即可以不考虑弗劳德数，只要保证柯西数相似即可满足刚度相似要求，即要求式(1)恒成立。

$$E_{\mathrm{meff}}/(\rho_{\mathrm{m}}U_{\mathrm{m}}^2) = E_{\mathrm{peff}}/(\rho_{\mathrm{p}}U_{\mathrm{pp}}^2) \tag{1}$$

式中：E_{meff}、E_{peff} 分别为模型和原型的等效模量；ρ_{m}、ρ_{p} 分别为模型和原型所处环境空气密度；U_{m}、U_{p} 分别为模型和原型所处环境特征风速。

冷却塔通常采用旋转薄壳并由人字柱支承，需要考虑的相似刚度有壳体的弯曲刚度 D_1、拉伸刚度 D_2、剪切刚度 D_3、扭转刚度 D_4 与人字柱的拉伸刚度 D_5、抗弯刚度 D_6 及扭转刚度 D_7，各刚度及对应的等效模量表达式如表 1 所示，表中 E 为弹性模量，t 为壳体厚度，μ 为泊松比，A 为面积，I 为抗弯惯性矩，G 为抗扭惯性矩，L 为特征尺寸。

设风速比 $U_{\mathrm{m}}/U_{\mathrm{p}} = 1/m$，几何缩尺比 $L_{\mathrm{m}}/L_{\mathrm{p}} = 1/n$，由式(1)及表 1 中的等效模量表达式可得满足壳体弯曲刚度、拉伸刚度、剪切刚度与扭转刚度相似

表 1 冷却塔刚度及其等效模量表达式

Table 1 Expression of cooling tower stiffness and effective modulus

刚度类型	壳体弯曲刚度	壳体拉伸刚度	壳体剪切刚度	壳体扭转刚度	人字柱拉伸刚度	人字柱抗弯刚度	人字柱抗扭刚度
刚度表达式	$D_1=\dfrac{Et^3}{12(1-\mu^2)}$	$D_2=\dfrac{Et}{1-\mu^2}$	$D_3=\dfrac{Et}{2(1+\mu)}$	$D_4=\dfrac{Et^3}{12(1+\mu)}$	$D_5=EA$	$D_6=EI$	$D_7=EG$
等效模量表达式	$E_{\mathrm{eff1}}=\dfrac{D_1}{L^3}$	$E_{\mathrm{eff2}}=\dfrac{D_2}{L}$	$E_{\mathrm{eff3}}=\dfrac{D_3}{L}$	$E_{\mathrm{eff4}}=\dfrac{D_4}{L^3}$	$E_{\mathrm{eff5}}=\dfrac{D_5}{L^2}$	$E_{\mathrm{eff6}}=\dfrac{D_6}{L^4}$	$E_{\mathrm{eff7}}=\dfrac{D_7}{L^4}$

要求的条件表达式分别为:

$$\frac{t_{\mathrm{m}}}{t_{\mathrm{p}}}=\left(\frac{1}{n}\right)\left[\frac{E_{\mathrm{p}}}{E_{\mathrm{m}}}\left(\frac{1}{m}\right)^{2}\frac{1-\mu_{\mathrm{m}}^{2}}{1-\mu_{\mathrm{p}}^{2}}\right]^{1/3} \tag{2}$$

$$\frac{t_{\mathrm{m}}}{t_{\mathrm{p}}}=\left(\frac{1}{n}\right)\left[\frac{E_{\mathrm{p}}}{E_{\mathrm{m}}}\left(\frac{1}{m}\right)^{2}\frac{1-\mu_{\mathrm{m}}^{2}}{1-\mu_{\mathrm{p}}^{2}}\right] \tag{3}$$

$$\frac{t_{\mathrm{m}}}{t_{\mathrm{p}}}=\left(\frac{1}{n}\right)\left[\frac{E_{\mathrm{p}}}{E_{\mathrm{m}}}\left(\frac{1}{m}\right)^{2}\frac{1+\mu_{\mathrm{m}}}{1+\mu_{\mathrm{p}}}\right] \tag{4}$$

$$\frac{t_{\mathrm{m}}}{t_{\mathrm{p}}}=\left(\frac{1}{n}\right)\left[\frac{E_{\mathrm{p}}}{E_{\mathrm{m}}}\left(\frac{1}{m}\right)^{2}\frac{1+\mu_{\mathrm{m}}}{1+\mu_{\mathrm{p}}}\right]^{1/3} \tag{5}$$

以上壳体刚度相似要求应同时满足,即式(2)～式(5)之间应互等。由于材料泊松比一般在0.20左右,以上各式中含泊松比μ的比例因子接近于1,以本文模型为例,$\mu_{\mathrm{m}}=0.26$,$\mu_{\mathrm{p}}=0.17$,式(2)～(5)中含泊松比μ的比例因子分别为0.99、0.96、1.07、1.02,因此泊松比相似条件可以忽略。于是得到同时满足壳体各项刚度相似的条件为$t_{\mathrm{m}}/t_{\mathrm{p}}=1/n$,$1/m=\sqrt{E_{\mathrm{m}}/E_{\mathrm{p}}}$。同理,可得满足人字柱拉伸刚度、抗弯刚度及抗扭刚度相似要求的条件为$1/m=\sqrt{E_{\mathrm{m}}/E_{\mathrm{p}}}$。因此冷却塔完全气动弹性模型刚度相似要求可由模型制作材料的弹性模量和其它条件调节试验风速比完全实现。

除刚度相似条件外,质量分布相似由模型制作材料密度与混凝土密度接近直接实现,阻尼比相似可由制作工艺保证。于是冷却塔完全气动弹性模型与原型结构的相似参数见表2(其中λ_E为弹性模量相似比)。

表2 冷却塔完全气动弹性模型相似比

Table 2 Similitude ratio of full aero-elastic model of cooling tower

参数	相似系数值	相似要求
几何尺寸L	$\lambda_L=1/n$	几何相似
风速U	$\lambda_U=\sqrt{\lambda_E}$	柯西数
频率f	$\lambda_f=\lambda_U/\lambda_L=n\sqrt{\lambda_E}$	斯托罗哈数
时间t	$\lambda_t=1/\lambda_f=1/n\sqrt{\lambda_E}$	斯托罗哈数
单位面积质量m	$\lambda_m=\lambda_\rho\lambda_L^3=\lambda_L^3=1/n^3$	量纲
单位长度拉伸刚度EA	$\lambda_{EA}=\lambda_E\lambda_L^2=\lambda_E/n^2$	量纲
单位长度弯曲刚度EI	$\lambda_{EI}=\lambda_E\lambda_L^4=\lambda_E/n^4$	量纲
单位长度抗扭刚度GJ	$\lambda_{GJ}=\lambda_G\lambda_L^4=1/n^5$	量纲
阻尼比ξ	$\lambda_\xi=1$	阻尼比
位移S	$\lambda_S=\lambda_L=1/n$	量纲
加速度a	$\lambda_a=\lambda_U^2\lambda_L=\lambda_E/n$	量纲

1.2 相似关系验证

相似关系的推导结果表明,模型与原型结构的频率比λ_f等于风速比λ_U除以几何缩尺比λ_L。采用通用大型有限元分析软件ANSYS分别建立冷却塔原型结构和气弹模型的有限元分析模型,比较二者的模态分析结果以验证上述相似关系的正确性。风筒采用SHELL 63壳单元模拟,人字柱采用BEAM 188 Timoshenko梁单元模拟,支柱上端节点与壳体末节圆的有关节点位置保持一致,边界条件为人字柱底端固接。原型冷却塔材料为混凝土,气弹模型制作材料为密度与混凝土十分接近的环氧树脂(devcon)。模型与原型结构材料弹性模量之比$\lambda_E=1/5.6$,本文取几何缩尺比$\lambda_L=1/400$,则频率比$\lambda_f=168.78$,即气弹模型的频率为原型结构频率的168.78倍。动力特性分析结果如表3所示,从表中可以看出,模型频率实际值与设计值的偏差在±0.3%以内,设计的气弹模型频率与原型结构的频率符合所推导的频率相似关系,二者振型也一致,证明了上述推导的相似关系正确。

2 风洞试验概况

2.1 试验模型

某核电站拟建冷却塔塔顶标高200.20 m,喉部标高156.70 m,进风口标高12.59 m,塔顶直径96.60 m,喉部直径94.60 m,底部直径153.00 m,风筒采用分段等厚,最小厚度在喉部断面,壁厚0.25 m,最大厚度在下环梁位置,壁厚1.4 m,由均匀分布的52对人字柱支撑。风洞试验在湖南大学HD-2风洞的高速试验段进行,试验段长17.0 m,宽3.0 m,高2.5 m。试验模型采用密度与混凝土接近的环氧树脂devcon制作,devcon与混凝土材料特性的比较如表4所示,模型材料与原型结构材料的密度一致保证了质量相似。几何缩尺比为1∶400,严格保证外形与实际结构保持几何相似,几何尺寸精度由数控车床精加工保证,模型的最小厚度为0.63 mm,试验模型照片如图1所示。在模型表面0.15H、0.45H、0.65H、0.71H、0.79H、0.95H(H为模型高度)处布置位移测点,每层沿环向等间距布置12个测点,共计72个位移测点,测点布置及各断面测点编号如图2所示。考虑到冷却塔为轴对称结构,仅进行了1个风向角下的风洞试验。

图1 试验模型照片

Fig. 1 Full aero-elastic model of cooling tower

表 3　完全气动弹性模型与原型结构动力特性分析结果比较

Table 3　Comparison of dynamic characteristics between model and prototype

阶数	原型冷却塔		模型冷却塔		结果比较	
	振型	振型描述及频率	振型	振型描述及频率	模型频率理论值/Hz	误差
1,2	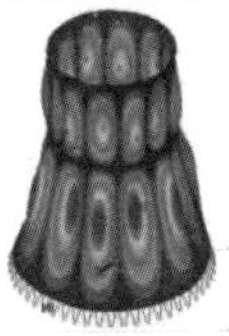	6 个环向谐波 2 个竖向谐波 $f_{1,2}$ =0.807 83 Hz	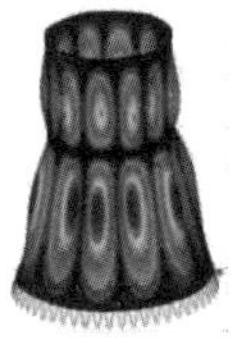	6 个环向谐波 2 个竖向谐波 $f_{1,2}$ =136.23 Hz	136.34	-0.08%
3,4	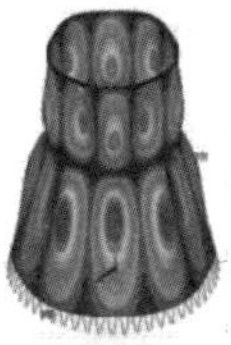	5 个环向谐波 2 个竖向谐波 $f_{3,4}$ =0.817 61 Hz	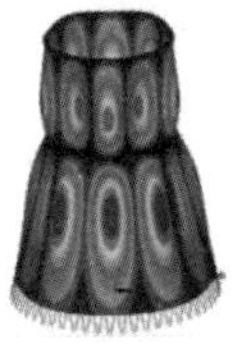	5 个环向谐波 2 个竖向谐波 $f_{3,4}$ =137.63 Hz	137.99	-0.26%
5,6	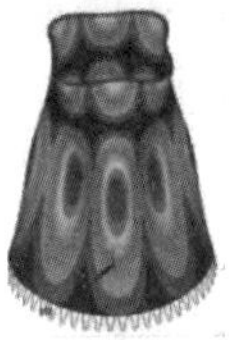	4 个环向谐波 2 个竖向谐波 $f_{5,6}$ =0.933 59 Hz		4 个环向谐波 2 个竖向谐波 $f_{5,6}$ =157.32 Hz	157.56	-0.15%
7,8	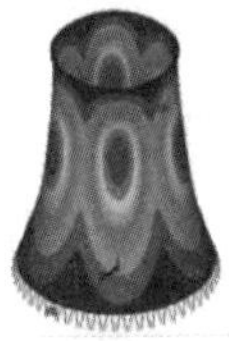	3 个环向谐波 1 个竖向谐波 $f_{7,8}$ =0.959 57 Hz	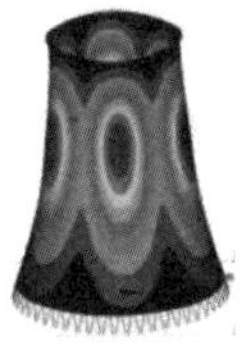	3 个环向谐波 1 个竖向谐波 $f_{7,8}$ =161.69 Hz	161.95	-0.16%
9,10	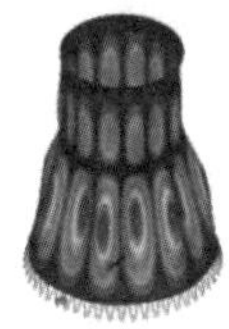	7 个环向谐波 2 个竖向谐波 $f_{9,10}$ =0.971 24 Hz		7 个环向谐波 2 个竖向谐波 $f_{9,10}$ =163.88 Hz	163.92	-0.02%

表 4　Devcon 与混凝土材料特性的对比

Table 4　Comparison of material properties between devcon and concrete

材料参数	容重/(kN·m^{-3})	弹性模型/GPa	泊松比
Devcon	24.3	5.8	0.26
混凝土	25.0	32.5	0.17

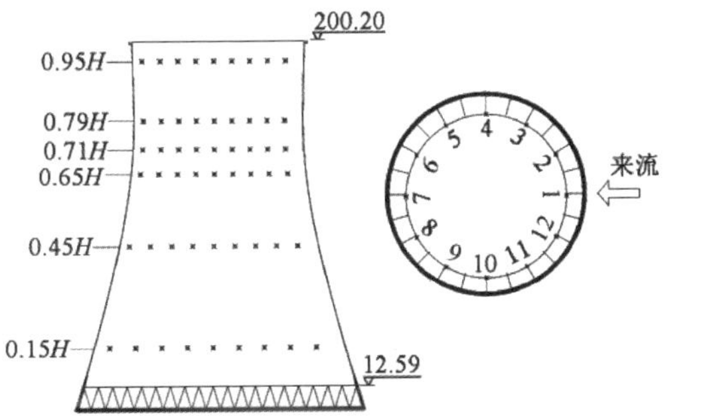

图 2　测点布置及各断面测点编号

Fig. 2　Measuring point arrangement

试验采用的位移计为 LK-G10 非接触式高精度激光位移计，测量不受色彩、表面材质或离散光线所影响，测量精度为 0.01 μm，动态响应频率 1 000 Hz 以上。根据基恩士激光位移传感器原理，本文所指的位移是指测点总位移在光路上的分量，即冷却塔的径向位移，正位移指的是在平衡位置沿半径向内的变形，在平衡位置沿半径向外的变形则为负位移。测量信号采样频率为 1 kHz，采样时长 20 s。

2.2　风场模拟

本项目厂址周边地形与 GB 50009—2001《建筑结构荷载规范》规定的 B 类地貌类似。模型的几何缩尺比由风洞中的紊流积分尺度与实际大气边界层的积分尺度之比来确定，一般取 1/300 ~ 1/600。本项目为满足阻塞率要求，选取模型的几何缩尺比为 1:400，因此在湖南大学 HD-2 风洞高速试验段采用"尖劈 + 粗糙元 + 格栅条"的方法模拟 B 类风场，转盘中心处的模拟结果如图 3 所示。从图 3a 可以看出，风洞中模拟的平均风速剖面与 GB 50009—2001《建筑结构荷载规范》规定的 B 类风场基本一致，模拟的边界层厚度约为 90 cm，相当于实际高度 360 m；对于湍流度剖面，我国现行规范未作规定，但从脉动系数可反推得到紊流度，同时可参考国外相关规范[7-10]，风洞中模拟的湍流度剖面与规范值比较如图 3b 所示，可以看出，湍流度剖面也与实际大气中的情况基本一致;图3c给出了转盘中心50 cm高处的顺风向脉动风谱,可以看出,模拟的顺风向脉动风谱与常用的von Karman、Kaimal、Davenport等理论谱基本一致。另外,风洞模拟的积分尺度约为0.40 m,相当于实际积分尺度为160 m,积分尺度与实际大气中的情况基本一致。

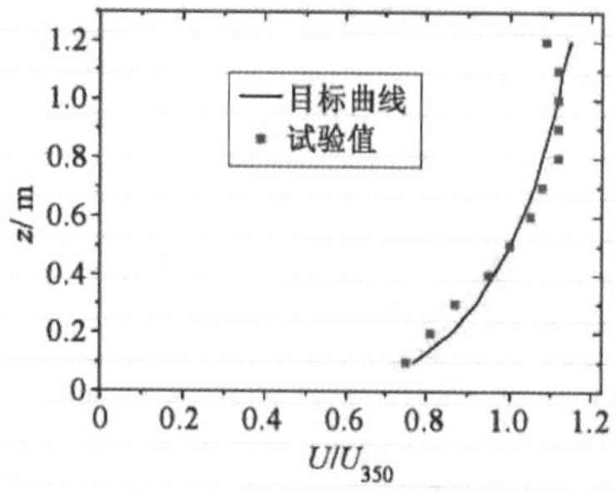

(a) 平均风速剖面

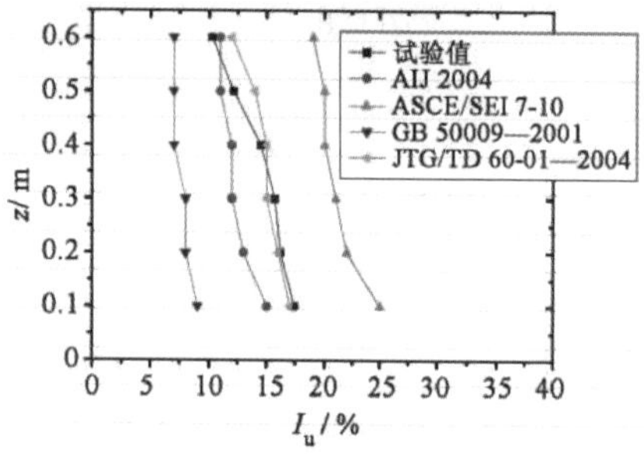

(b) 湍流度剖面

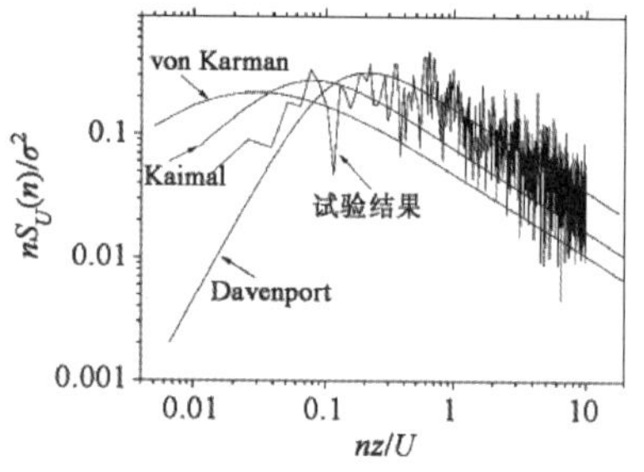

(c) 顺风向脉动风谱

图3 风洞中大气边界层的模拟

Fig. 3 Simulation of wind characteristics in BLWT

2.3 雷诺数效应模拟

风洞试验中由于采用缩尺模型导致试验的雷诺数比实际的雷诺数小2~3个数量级,且由于风洞试验中风速的限制,很难通过增加试验风速来提高雷诺数,而冷却塔等类圆柱结构风压大小及其分布与雷诺数密切相关,因此在风洞试验中必须采取一定的措施对雷诺数效应进行补偿。圆形截面结构表面的绕流特性除受雷诺数影响外,还受结构本身表面粗糙度的影响。实践证明[11],适当增大模型表面粗糙度可以在低雷诺数条件下模拟高雷诺数时的绕流特性,赵林等[12]、沈国辉等[13]采用在模型表面粘贴粗糙纸带以增大表面粗糙度的方法对模型试验的雷诺数效应进行补偿,本文试验也采用该方法提高模型表面的粗糙度。

已有研究表明[14],冷却塔中段各高度的风压分布曲线基本一致,且考虑到布置太多测压孔对气弹模型影响过大,仅选择在冷却塔中部0.55H高度处布置风压测点,将不同粗糙纸带厚度和试验风速下的测试结果与目标曲线比较,发现在模型表面粘贴36条子午向通长宽10 mm厚0.3 mm的粗糙纸带在U =18 m/s的试验风速下风压系数分布曲线与目标曲线(以该塔的设计曲线VGB-K1.2为目标曲线)最为接近(图4),且进一步提高风速时,风压分布曲线基本不变,表明在该粗糙度下模型表面的绕流特性已接近超临界状态而保持稳定状态,风压分布不再随雷诺数的增加而变化。为保证雷诺数模拟需求,以U =18 m/s为试验风速。

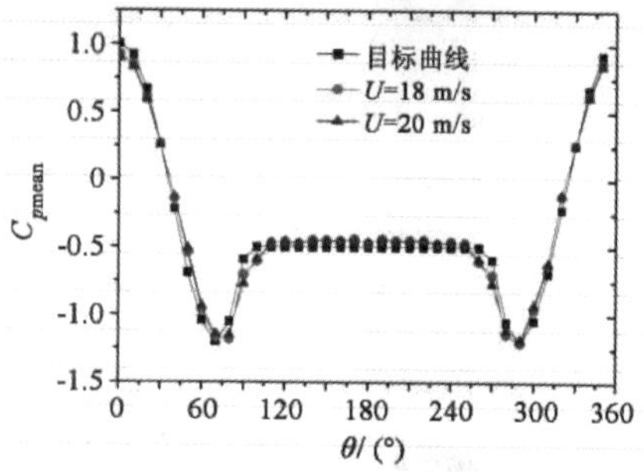

图4 模拟结果与目标曲线的比较

Fig. 4 Comparison of simulation results with target curves

2.4 模型动力特性测试

在进行风洞试验之前,需要对模型的动力特性进行测试检验,主要测定试验模型的频率、振型及阻尼比。但冷却塔结构模态频率密集,该气弹模型的1阶频率约为130 Hz,属高频结构,识别此类高频、密集模态结构的模态参数相当困难,为验证识别参数的准确性,本文采用两种方法进行模态参数识别:1)风荷载激励法:把模型固定在风洞内,用紊流风激励模型,采集模型的加速度响应,采样频率1 kHz,采样时长20~30 min;2)锤击法激振:应用单输入多输出模态参数识别方法获得该模型的模态参数,试验采用的敲击力锤为KD3002型石英力锤,加速度传感器信号与力锤信号同步采集,采样频率为1 kHz,采用DHMA模态分析软件进行模态参数识别。试验结果表明,两种方法识别的结果基本一致。由风荷载激励法得到的气弹模型频率试验值与理论值比较如表5所示,测试结果中考虑了雷诺数模拟时粘贴的粗糙纸带,从该表中可以看出,前10阶模型频率的测试值与设计值的误差都在3.5%以内,说明完全气动弹性模型能精确模拟冷却塔结构的质量、刚度相似。此外,由半功率带宽法得到模型阻尼比约为1.1%,与文献[15]对某大型冷却塔阻尼比实测结果1.0%接近。

表 5　冷却塔气弹模型频率理论值与实测值比较

Table 5　Comparison of theoretical and measured frequencies of full aero-elastic model

振型阶次	理论频率 f_t /Hz	测试频率 f_m /Hz	$(f_m - f_t)/f_t$
1、2	136.34	137.70	1.00%
3、4	137.99	142.58	3.33%
5、6	157.56	159.18	1.03%
7、8	161.95	163.09	0.70%
9、10	163.92	168.95	3.07%

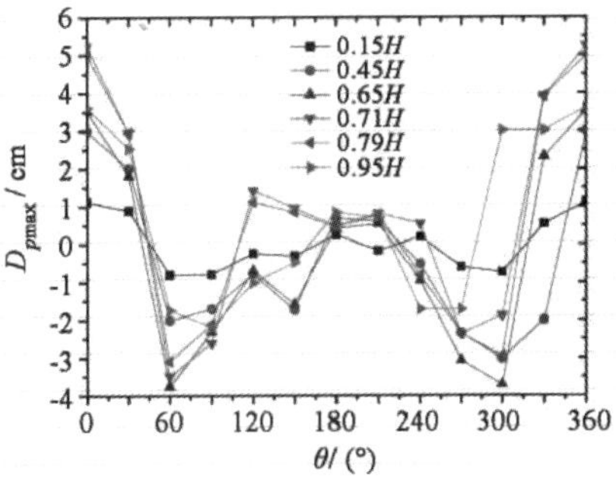

图 6　冷却塔极值位移沿环向分布

Fig. 6　Distribution of extreme circumferential displacement

3　试验结果及分析

3.1　风致变形

图 5 给出试验测试结果由相似比换算至原型基本风速 U_0 =26.8 m/s 风荷载作用下的冷却塔各高度平均位移沿环向的分布，从图中可以看出：冷却塔的位移与其平均风压分布比较类似，沿环向基本呈对称分布；迎风向(0°) 的正位移最大，且大于发生在最大负压(60°) 附近极值负位移的绝对值，尾流区的位移均值很小，在 0 附近波动；冷却塔的位移同平均风压分布一样，具有明显的"三维效应"，呈中段位移大、两端位移小的分布特征，其中以喉部附近 (z = 0.71H) 的位移最大，最大值为 3.03 cm。

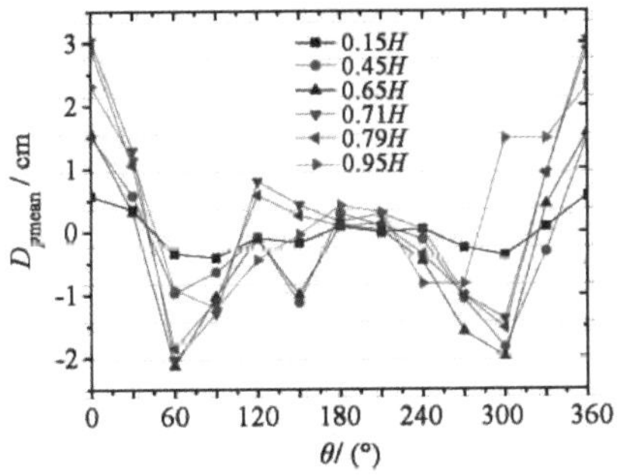

图 5　冷却塔平均位移沿环向分布

Fig. 5　Distribution of mean circumferential displacement

为考虑脉动位移的影响，极值位移 D_{pmax} 定义为：

$$D_{pmax} = D_{pmean} \pm g\sigma_D \tag{6}$$

式中：D_{pmean} 为平均位移；g 为峰值因子，取 g =4.0；σ_D 为位移根方差。

图 6 给出冷却塔各高度极值位移沿环向分布，可以看出，极值位移的分布趋势与平均位移类似，最大极值位移为 5.24 cm，发生在喉部附近 (z = 0.71H) 。总的来说，冷却塔的风致变形相当小(冷却塔风致变形示意见图 7)，最大极值位移也仅为喉部半径的 0.1%，约相当于喉部厚度的 20%。

3.2　位移响应特性

动力放大系数 φ 定义为：

$$\varphi = \frac{D_{pmean} \pm g\sigma_{Total}}{D_{pmean} \pm g\sigma_B} \tag{7}$$

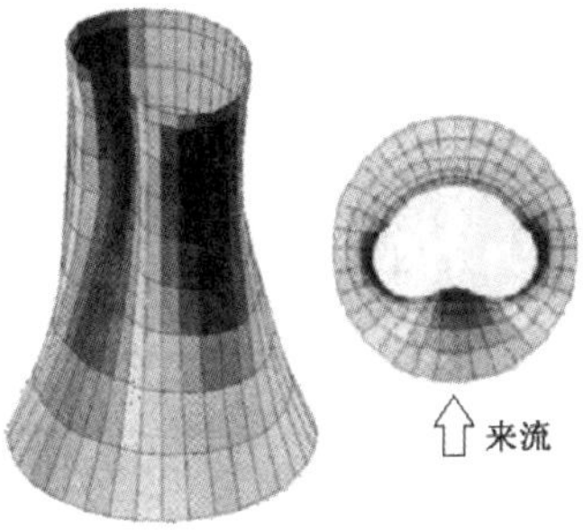

图 7　冷却塔风致变形示意图

Fig. 7　Displacement distribution of cooling tower

式中：σ_{Total} 为总风致动力响应的标准差；σ_B 为共振响应的标准差；g 为峰值因子，取 g =4.0。

图 8 为典型高度中最大正位移、极值负位移和背压区测点位移(换算至原型) 响应的自功率谱密度图，可以看出，冷却塔结构的风致动力响应主要由低阶振型贡献(0 ~ 1 Hz 频率范围包含了结构前 10 阶频率)，频率大于 1.5 Hz 的高阶振型对风致动力响应的贡献很小；共振响应随着高度的增大而愈加显著，同一高度中，以背压区测点的共振响应最为明显。

表 6 给出位移总动力响应、背景响应和共振响应标准差及动力放大系数，各测点中以测点 0.95H-7 (测点位于正背风面，图 2) 共振响应占总动力响应的比例最大，达 43.43%，该测点的动力放大系数也达到最大值 1.14，除测点 0.79H-7 共振响应比例接近 10.0%，动力放大系数为 1.04 之外，其它测点的共振响应比例都小于 5.0%，动力放大系数均在 1.0 左右；各个测点动力放大系数均值为 1.02，略小于按德国规范所得计算值 1.05，其原因在于规范计算值有一定的安全系数。

3.3　位移响应及风振系数

考虑到冷却塔顶部和下部位移较小，由此计算得到的风振系数会偏大甚至不合理，因此试验数据分析过程经试算比较最终设定原型结构平均位移为 1.5 cm 的阈值，当测点的平均位移超过该阈值时计入其风振系数的值。冷却塔风振系数计算结果如表 7 所示，可以看出，风振系数值随高度的增加而减小，

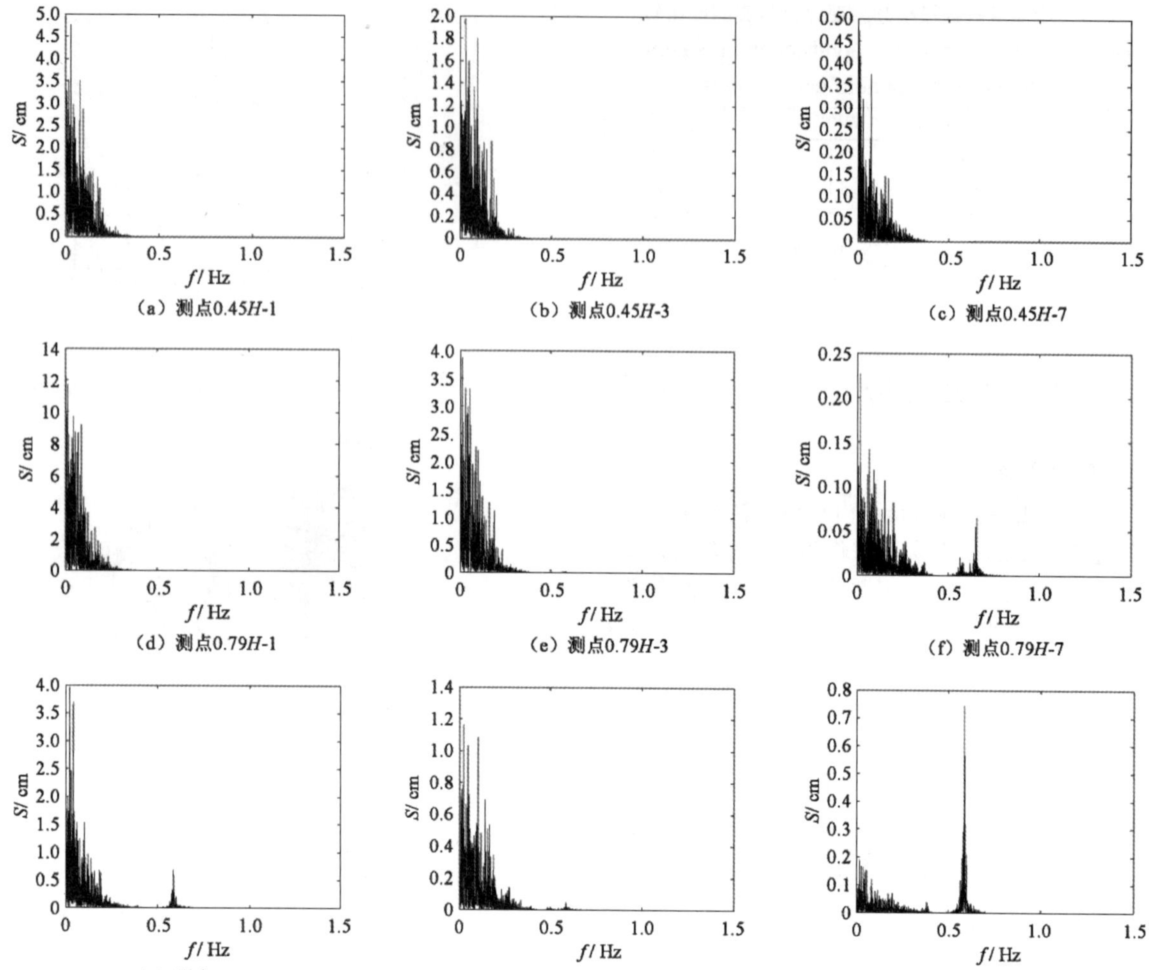

(a) 测点0.45H-1 (b) 测点0.45H-3 (c) 测点0.45H-7
(d) 测点0.79H-1 (e) 测点0.79H-3 (f) 测点0.79H-7
(g) 测点0.95H-1 (h) 测点0.95H-3 (i) 测点0.95H-7

图8 典型测点位移响应自功率谱密度

Fig. 8 Power spectral density of displacement of typical measurement points

表6 典型测点位移总响应、背景响应和共振响应标准差及动力放大系数

Table 6 Standard deviation of total, background and resonant responses of displacement and dynamic amplification factor

位移响应	测点编号								
	0.45*H*-1	0.45*H*-3	0.45*H*-7	0.79*H*-1	0.79*H*-3	0.79*H*-7	0.95*H*-1	0.95*H*-3	0.95*H*-7
σ_{Total} /cm	0.128 3	0.060 7	0.010 9	0.314 4	0.110 4	0.006 9	0.098 7	0.039 1	0.011 3
σ_B /cm	0.128 2	0.060 6	0.010 8	0.313 6	0.110 1	0.006 2	0.093 6	0.038 6	0.006 4
$(\sigma_{Total}-\sigma_B)$ /cm	0.000 1	0.000 1	0.000 1	0.000 8	0.000 3	0.000 7	0.005 1	0.000 5	0.004 9
$(\sigma_{Total}-\sigma_B)/\sigma_{Total}$	0.11%	0.16%	0.57%	0.25%	0.24%	9.88%	5.15%	1.21%	43.43%
φ	1.000 3	1.000 5	1.001 8	1.000 5	1.000 6	1.040 7	1.009 3	1.002 8	1.142 7
φ_{mean}					1.022 1				

表7 冷却塔位移响应及风振系数

Table 7 Displacement response and gust loading factor of cooling tower

测点高度	D_{pmean} /cm	D_{pmax} /cm	σ_D /cm	风振系数 η
0.45*H*	1.52	0.37	3.00	1.97
0.65*H*	1.57	0.33	2.89	1.84
0.71*H*	3.04	0.58	5.36	1.76
0.79*H*	2.90	0.52	4.98	1.72
0.95*H*	2.31	0.34	3.67	1.59
均值	2.27	0.43	3.98	1.75

注：D_{pmean}、D_{pmax}、σ_D 为位移响应的均值、极值和均方根。

这与我国近年来针对冷却塔结构风振系数所取得的研究成果规律性一致[16]；各高度控制点平均位移响应的均值为2.27 cm，根方差均值为0.43 cm，极值位移均值为3.98 cm，平均风振系数为1.75，略小于我国规范[17]规定的1.90。

4 结论

通过对冷却塔完全气动弹性模型刚度相似关系

的推导及验证、大型双曲冷却塔完全气动弹性模型的动力特性测试及其风洞试验结果的分析，得到如下主要结论：

1) 完全气动弹性模型能较精确模拟冷却塔结构的质量、刚度和阻尼相似，并能实现质量、刚度的连续分布；高精度的激光位移计是适合冷却塔完全气动弹性模型的位移响应测量仪器。

2) 冷却塔风致位移总体较小，平均位移沿环向分布与平均风压分布类似，沿子午向的分布有明显的三维效应，以喉部附近的位移最大。

3) 总风致动力响应中以背景响应占主导地位，动力放大效应不明显，得到的动力放大系数均值为1.02，略小于按德国规范计算得到的1.05。

4) 当测点的平均位移超过以设定原型结构平均位移1.5 cm的阈值时计入其风振系数值，风振系数值随高度的增加而减小，平均风振系数为1.75，略小于我国规范的1.90。

参 考 文 献

[1] Isyumov N, Abu-Sitta S H. Approaches to the design of hyperbolic cooling towers against the dynamic action of wind and earthquakes [J]. Bulletin of the International Association of Shell Structures, 1972, 48(3): 3-22.

[2] Armitt J. Wind loading on cooling towers [J]. Journal of the Structure Division, ASCE, 1980, 106(ST3): 623-641.

[3] ASCE. Manuals and reports on engineering practice: No. 67: wind tunnel models studies of buildings and structures [M]. Reston, VA: ASCE, 1987: 29-36.

[4] Tschanz T, Davenport A G. The base balance technique for the determination of dynamic wind loads [J]. Journal of Wind Engineering and Industrial Aerodynamics, 1983, 13(3): 429-439.

[5] Niemann H J, Kopper H D. Influence of adjacent buildings on wind effects on cooling towers [J]. Engineering Structures, 1998, 20(10): 874-880.

[6] 陈凯，魏庆鼎. 冷却塔风致振动试验研究 [C]//第十一届结构风工程学术会议. 上海: 同济大学出版社, 2003: 177-182. (CHEN Kai, WEI Qingding. Investigation of wind-induced vibration of cooling towers [C]//The Proceedings of 11th Structural Wind Engineering Conference. Shanghai: Tongji University Press, 2003: 177-182. (in Chinese))

[7] AIJ 2004 Recommendations for loads for building [S]. Tokyo, Japan: AIJ, 2004.

[8] ASCE/SEI 7-10 Minimum design loads for buildings and other structures [S]. Reston, VA: ASCE, 2010.

[9] GB 50009—2001 建筑结构荷载规范 [S]. 北京: 中国建筑工业出版社, 2001. (GB 50009—2001 Load code for the design of building structures [S]. Beijing: China Architecture & Building Press, 2001. (in Chinese))

[10] JTG/TD 60-01—2004 公路桥梁抗风设计规范 [S]. 北京: 人民交通出版社, 2004. (JTG/TD 60-01—2004 Wind-resistent design specification of highway bridges [S]. Beijing: China Communication Press, 2004. (in Chinese))

[11] Farell C, Guven O, Maisch F. Mean wind loading on rough-walled cooling towers [J]. Journal of the Engineering Mechanics Division, ASCE, 1976, 102 (EM6): 1059-1081.

[12] 李鹏飞，赵林，葛耀君，等. 超大型冷却塔风荷载特性风洞试验研究 [J]. 工程力学, 2008, 25(6): 60-67. (LI Pengfei, ZHAO Lin, GE Yaojun, et al. Investigation on wind load characteristics for super large cooling tower in wind tunnel [J]. Engineering Mechanics, 2008, 25(6): 60-67. (in Chinese))

[13] 沈国辉，余关鹏，孙炳楠，等. 大型冷却塔双塔干扰的风洞试验研究 [J]. 振动与冲击, 2011, 30(3): 109-114. (SHEN Guohui, YU Guanpeng, SUN Bingnan, et al. Study on interference effects of two large hyperbolic cooling towers by using wind tunnel test [J]. Journal of Vibration and Shock, 2011, 30(3): 109-114. (in Chinese))

[14] 孙天风，周良茂. 无肋双曲线型冷却塔风压分布的全尺寸测量和风洞研究 [J]. 空气动力学报, 1983, 12(4): 68-76. (SUN Tianfeng, ZHOU Liangmao. A full scale and wind tunnel study of wind pressure distribution around a ribless hyperbolic cooling tower [J]. Acta Aerodynamica Sinica, 1983, 12(4): 68-76. (in Chinese))

[15] Emilia J, Zdenek B, Ondrej F. Vibrations characters of a cooling tower shell [J]. Journal of Wind Engineering and Industrial Aerodynamics, 1983, 12(2): 145-154.

[16] 李佳颖，任春玲，黄志龙. 自然通风冷却塔的实验及有限元分析 [J]. 力学季刊, 2007, 28(3): 443-447. (LI Jiaying, REN Chunling, HUANG Zhilong. Experiment study and finite element analysis of a natural draft cooling tower [J]. Chinese Quarterly of Mechanics, 2007, 28(3): 443-447. (in Chinese))

[17] DL/T 5339—2006 火力发电厂水工设计规范 [S]. 北京: 中国电力出版社, 2006. (DL/T 5339—2006 Code for hydraulic design of fossil fuel power plants [S]. Beijing: China Electric Power Press, 2006. (in Chinese))

Wind and Structures, *Vol. 21, No. 1 (2015) 85-103*
DOI: http://dx.doi.org/10.12989/was.2015.21.1.085

An empirical model for amplitude prediction on VIV-galloping instability of rectangular cylinders

Huawei Niu[*1], Shuai Zhou[1,2a], Zhengqing Chen[1] and Xugang Hua[1]

[1]*Wind Engineering Research Center of Hunan University, Changsha, Hunan, China*
[2]*China Railway Eryuan Engineering Group Co.Ltd, Chengdu, Sichuan, China*

(Received October 28, 2014, Revised May 4, 2015, Accepted May 8, 2015)

Abstract. Aerodynamic forces of vortex-induced vibration and galloping are going to be coupled when their onset velocities are close to each other, which will induce the cross-wind amplitudes of the structures increased continuously with ever-increasing wind velocities. The main purpose of the present work is going to propose an empirical formula to predict the response amplitude of VIV-galloping interaction. Firstly, two typical mathematical models for the coupled oscillations, i.e., Tamura & Shimada model and Parkinson & Corless model are comparatively summarized. Then, the key parameter affecting response amplitude is determined through comparative numerical simulations with Tamura & Shimada model. For rectangular cylinders with the side ratio from 0.5 to 2.5, which are actually prone to develop the VIV and galloping induced interaction responses, an empirical amplitude prediction formula is proposed after regression analysis on comprehensively collected experimental data with the predetermined key parameter.

Keywords: rectangular cylinders; VIV-Galloping interaction; amplitude prediction; empirical model

1. Introduction

Rectangular cylinders are typical bluff bodies, which are prone to different types of aerodynamic instabilities, depending on the side ratio, such as vortex-induced vibration (VIV) and transverse galloping. VIV is caused by the resonance of the transverse force due to alternative vortex shedding, which is characteristic by amplitude limitation to the order of one cylinder diameter, and occur only in a restricted range of oncoming velocities. The onset velocity of the typical lock-in can be effectively determined by Strouhal law, and the amplitude of vibration depends on the mass ratio and structural damping. VIV is a heat point in the research fields of wind induced vibrations, large efforts on this topic had been done, which are comprehensively reviewed by Bearman (1984), Sarpkaya (1979, 2004), Matsumoto (1999), Williamson and Govardhan (2008). The researches on amplitude estimation are also widely expanded for decades, see e.g., Skop and Griffin (1973), Tamura (1983), Hemon (1999), Govardhan and Williamson (2006) and Marra *et al.* (2011).

By contrast, transverse galloping is caused by a self-excitation due to wind action. Den Hartog

*Corresponding author, Ph.D., E-mail: niuhw@hnu.edu.cn
[a] Ph.D. Student, E-mail: 344953076@qq.com

(Den Hartog 1932) provided an effective way to determine the onset velocity of the galloping instability. The quasi-steady galloping theory is reasonably established on the basis of neglecting unsteady effects, which means the different vibration periods of different amplitudes are taken as only the matters of the varying of relative wind attack angles between structures and approaching flow. Parkinson and Brooks (1961) used polynomials to fit the drag and lift force coefficients curves before expressing the galloping aerodynamic force with limited terms of Taylor series, and the experimental results were reported to be in good agreements with theoretical predictions. In the last five decades, galloping instability of rectangular cylinders were studied by Parkinson and Smith (1964), Parkinson (1965), Bearman *et al.* (1987), Hortmanns and Ruscheweyh (1997), Luo *et al.* (2003), Macdonald and Larose (2006), Gjelstrup and Georgakis (2011), Joly *et al.* (2012), Names *et al.* (2012), etc.

To rectangular cylinders, both VIV and galloping are possible, and there are two major differences between them. Firstly, the response amplitude of galloping would never limited only to be one diameter of cylinder as VIV, which is actually observed to be even 100 times more, for example the transmission lines covered with ice of certain shape. Secondly, the occurring wind velocity ranges of galloping are typically for all speeds above the critical velocity, whereas the VIV is well known to be limited in the lock-in. According to Eurocode 1 (EN 1991-1-4,2010), if the ratio of the galloping to the VIV onset wind speed is either larger than 1.5 or lower than 0.7, the two phenomena can be considered separately. Nevertheless, some experimental data challenged this statement (see e.g., Hansen 2013, Mannini *et al.* 2014). To the slender structural elements with rectangular cross sections, however, when the onset velocities of VIV and galloping are very close, the aerodynamic forces of VIV and galloping are prone to be coupled, therefore interaction between the two phenomena can be expected. Then structural response appears to be a novel form which is beyond the single VIV or galloping style. As reported by many investigations (Parkinson 1965, Borri *et al.* 2012, Mannini *et al.* 2013, Mannini *et al.* 2014, etc.), the oscillations will start at the onset velocities which are always decided by Strouhal law or quasi-steady galloping theory, the response amplitude increase almost linearly with wind velocity, no signs of setback.

In fact, the interaction cases of VIV and galloping are very common in the slender rectangular cylinders because of their bluff aerodynamic shapes. For the practical engineering applications, it will be very convenient if there is an empirical formula to predict the amplitude of the VIV-galloping instability, just like Griffin Plot for VIV (Skop and Griffin 1973). While many investigations have been carried out for studying the response amplitude due to combined VIV and galloping, a simple model is still absent for practical applications purpose even under a certain pre-determined conditions. So the main purpose of the present work is trying to build an empirical model, based on a lot of collected experimental data and wind tunnel tested data performed by present work, by curve fitting.

2. Materials and methods

2.1 Characteristics of the interaction effects

Fig. 1 is showing a group of experimental curves of rectangular cylinders in different aspect ratios, which performed by Parkinson and Brooks (1961). In this figure, the x-axis is the dimensionless wind velocity, while the ordinate is the dimensionless amplitude response. It is worth noting the response curves of rectangular cylinders whose aspect ratios *d/h* are above 0.75, the amplitudes are observed to be generally increased as linear fashion with wind velocities. It is also reported that cylinders are dominantly vibrating at the natural frequencies and the time history of each case are keeping steady in this whole experimental group.

Fig. 2 is showing another typical group of results on VIV and galloping interaction induced oscillations, which are presented by Corless and Parkinson (1988). Vibration curves of the same square cylinder at different damping levels are presented, and the experimental results coincide with numerical results relatively well. While the onset velocities are different for the three cases, the rate of change of response amplitude with respect to the wind velocity for each case are observed to be almost the same.

Many more results can be found that VIV and galloping interaction induced vibration curves of rectangular cylinders like Figs. 1 and 2 in the foregoing researches. The main characteristics of the oscillation curves maybe concluded as following two points, firstly, the onset velocities are always determined by Strouhal law or quasi-steady galloping theory; secondly, the shape of the curves are showing linearly increased behaviors, which means the rate of change of the amplitude with respect to wind velocity is constant after the starting point, although the values are varying with aspect ratios. That's to say, if we can appropriately the slope of the response curve, the amplitude prediction may be straightforwardly made. Therefore, we define a new parameter S_L as Eq. (1) to present this parameter.

$$S_L=\frac{\Delta Y}{\Delta U_r} \tag{1}$$

where, ΔY is the increment of amplitude; $Y=A/D$ is the dimensionless amplitude response; ΔU_r is the increment of wind velocity associate with increased amplitude; $U_r=V/(f_vD)$ is the normalized wind velocity; V is oncoming wind velocity; f_v is the dominant oscillating frequency.

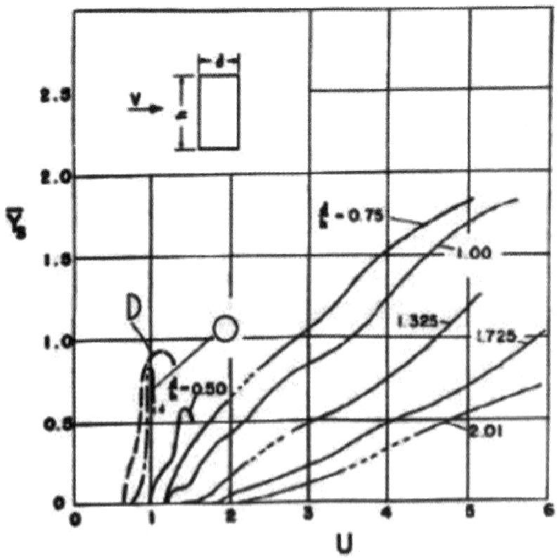

Fig. 1 Experimental curves of amplitude versus dimensionless wind velocity for different aspect ratio rectangular cylinders (Parkinson and Brooks 1961)

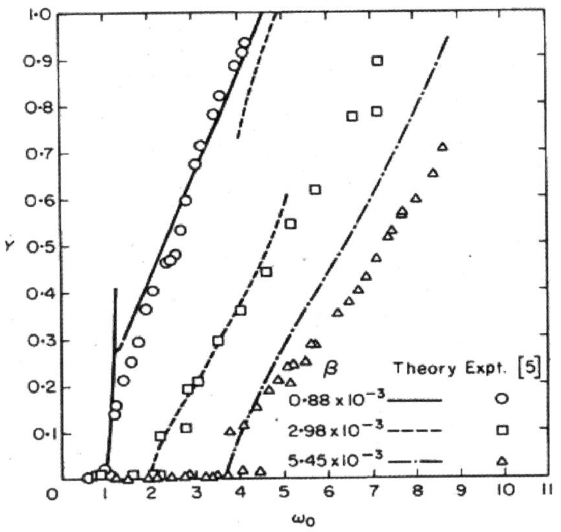

Fig. 2 Comparision on amplitude responses of three levels of system damping between simulation and experimental results (Corless and Parkinson, 1988, Bearman *et al.* 1987)

2.2 Mathematical models

The mathematical modeling on interaction effects between VIV and galloping could be classified into two main branches, one is proposed by Corless and Parkinson (1988), which is characterized by parametrical mathematical fitting on measured vibration results; the other one is Tamura and Matsui model (Tamura and Matsui 1979), which pay more attention on the physical meanings of parameters. And the common point of these two models is that both they are based on the quasi-steady galloping theory with one kind of VIV model to form the coupled VIV and galloping model.

2.2.1 Parkinson and Corless model

In order to take into account the effects of vortex shedding force and quasi-steady galloping force, Parkinson and Bouclin (1977) describe the integrated unsteady aerodynamic force acting on a square cylinder by adding the quasi-steady perturbation to the Hartlen-Currie fluid oscillator, then Eq. (2) is obtained.

$$\ddot{Y}+Y=nU^2(C_{FY}+C_L)$$

$$\ddot{C}_L\text{-}G\left\{C_{L0}^2\text{-}(4/3)\left(\dot{C}_L/k\right)^2\right\}\dot{C}_L+k^2C_L=H\dot{Y}$$

$$C_{FY}=A_1\cdot\left(\frac{\dot{Y}}{V}\right)^1+A_2\cdot\left(\frac{\dot{Y}}{V}\right)^2+A_3\cdot\left(\frac{\dot{Y}}{V}\right)^3+A_4\cdot\left(\frac{\dot{Y}}{V}\right)^4+A_5\cdot\left(\frac{\dot{Y}}{V}\right)^5+A_6\cdot\left(\frac{\dot{Y}}{V}\right)^6+A_7\cdot\left(\frac{\dot{Y}}{V}\right)^7 \quad (2)$$

where, G, k, H are the constants which are determined by mathematical fitting on the response curves; C_L is the lift coefficient; $\dot{Y}$ is the structural velocity response; n is the mass ratio, defined as the ratio of fluid mass over structural mass of the same unit; C_{FY} represents the quasi-steady galloping aerodynamic force, it is obtained by expanding Taylor series of drag and lift coefficient according to quasi-steady galloping theory, and up to 7th term is proved to be sufficient by many investigations. The good coincidences between experimental and simulation results demonstrate that this mathematical model can successfully consider the interaction effects of VIV and galloping.

Corless and Parkinson (1988) improved it by adding a new coupling term of $B\ddot{Y}$ into the differential equation in the expectation of considering some effects of cylinder acceleration on the wake vortices. And Facchinettia *et al.* (2004) proved that using acceleration as coupling term is indeed more effective than using displacement or velocity by comparative wind tunnel tests studies. According to Corless and Parkinson (1998), Eq. (2) is modified as

$$\begin{aligned}&\ddot{Y}+Y=nU^2(C_{FY}+C_L)\\&\ddot{C}_L-G\left\{C_{L0}^2-(4/3)\left(\dot{C}_L/k\right)^2\right\}\dot{C}_L+k^2C_L=H\dot{Y}+B\ddot{Y}\end{aligned}\tag{3}$$

The differential equations Eq. (1) can be solved by the method of multiple scales. The simulation obtained curves have a good agreement with the experimental results, which show that the effects of VIV could be simply added to the time-averaged position of shear layer to predict the aerodynamic force.

2.2.2 Tamura and Shimada model

There is a main branch in VIV mathematical models research field, they considered the alternatively shedding vortices as an independent oscillator, this oscillator could be interacted with structural vibration to form a combined motion equation, e.g., Birkhoff type VIV model. And the modified Birkhoff type VIV mathematical model on circular cylinders is proposed by Tamura and Matsui (1979), which is characteristically for has taken into account the effects of verified length of the wake-oscillator on aerodynamic forces. The coupled dynamic equation between wake oscillator and circular cylinder motion is expressed as Eq. (4)

$$\begin{aligned}&\ddot{\alpha}-2\zeta v\left\{1-\left(4f^2/C_{L0}^2\right)\alpha^2\right\}\dot{\alpha}+v\alpha=-m^*\ddot{Y}-vS^*\dot{Y}\\&\ddot{Y}+\left\{2\eta+n(f+C_D)v/S^*\right\}\dot{Y}+Y=-fnv^2\alpha/S^{*2}\\&C_L=-f\left(\alpha+S^*\dot{Y}/v\right)\end{aligned}\tag{4}$$

where, α is the angular displacement of wake oscillator; Y is the dimensionless transverse displacement of circular cylinder; V is oncoming wind velocity; ζ is wake oscillator damping ratio, $\zeta=1/(2\sqrt{2}*\pi^2)*f/Hr$; $v=U_r/U_s$ is the non-dimensional flow velocity; $U_s=f_vD/St$ is the Strouhal number determined from the critical wind velocity; η is the mechanical damping ratio of structure; f is the parameter which determined by the Magnus effect; Hr is the dimensionless width of the wake oscillator; C_{L0} is the static lift coefficient amplitude for circular cylinders; C_D is the drag coefficient for circular cylinders; m^* is the equivalent mass ratio, $m^*=1/(0.5+Hr)$; S^* is the equivalent Strouhal number. For circular cylinders, the aerodynamic parameters are determined by the previous experiments as following: $m^*=0.625$, $S^*=1.26$, $f=1.16$, $C_D=1.2$. The differential equation Eq. (4) can be solved by Runge-Kutta method, and the numerical results are reported to agree with the experimental results qualitatively and quantitatively (Tamura and Matsui 1979).

For square cylinders, when the onset velocities of VIV and galloping are very close, the oscillation curves are observed to be coupled, Tamura and Shimada (1987) proposed a coupling mathematical model to consider this interaction effects. Using previously proposed modified Birkhoff type wake oscillator model to consider the vortex shedding force, and quasi-steady perturbation model (Parkinson and Brooks 1961) for galloping aerodynamic force, the coupled mathematical model is established by the superimposing of these two discrete aerodynamic force parts. The detailed processes of constructing the coupled model is as following: expanding the quasi-steady galloping force C_{FY} by Taylor series and taking the first 7 orders as galloping force, combining the equivalent galloping force into the modified Birkhoff type VIV model, then the coupled mathematical is finally established as Eq. (5)

$$\ddot{\alpha} - 2\zeta v\{1 - (4f^2/C_{L0}^2)\alpha^2\}\dot{\alpha} + v\alpha = -m^*\ddot{Y} - vS^*\dot{Y}$$

$$\ddot{Y} + \left\{2\eta + n(f-A_1)v/S^* - nA_3S^*/v\dot{Y}^2 - nA_5(S^*/v)^3\dot{Y}^4 - nA_7(S^*/v)^5\dot{Y}^6 \ldots\right\}\dot{Y} + Y = -fnv^2\alpha/S^{*2}$$

$$C_L = -\left\{(f - A_1)S^*\dot{Y}/v - A_3(S^*\dot{Y}/v)^3 - A_5(S^*\dot{Y}/v)^5 - \ldots + f\alpha\right\} \tag{5}$$

where, A_1, A_3, A_5, A_7 are the coefficients of the Taylor series expansions, the meanings of other symbols are the same as mentioned above. The effectiveness of the Tamura-Shimada mathematical model on interaction effects between VIV and galloping is verified by good consistency between theoretical prediction and experimental results, and most importantly, all the parameters in this model have definite physical interpretations.

Since the parameters in the Hartlen-Currie model are mainly a kind of mathematical curves-fitting on experimental results; while all the parameters in the Tamura-Shimada mathematical model have their physical meanings, therefore the later will shed insight into the vibration mechanism. Allison and Corless (1995) also indicate that the mathematical model of the Tamura-Shimada model is more rationally derived than their Hartlen-Currie model.

2.3 The determination of key parameter

As defined before, S_L is the key parameter for predicting amplitude response of coupled VIV and galloping vibrations, here, in this part, we are going to further investigate by numerical simulations on which is the main influence factor of this key parameter.

Regarding which are the potential influencing factors on S_L, the steady vibration history of each wind velocity point should be carefully analyzed, the limit-cycle vibration status of each case means that the total energy absorbing from the oncoming flow are going to balance the energy dissipated by the system damping and maintain the kinetic energy of structure (Vio *et al.* 2007). Therefore, it is reasonable to assume that S_L should be associated with the aspect ratio of the cross section, because it is directly affecting the efficiency of structure on absorbing energy from the approaching flow. On the contrary, structural physical mass, system damping and vibration frequency are the main factor who are affecting the efficiency of energy dissipating, they are deserved to be regarded as potential factors on S_L. Moreover, the Reynolds number is also an important parameter that can't be neglected.

Consequently, the S_L may be assumed as the function of several factors which is shown in Eq. (6)

$$S_L = F\left(n, \eta, f_v, A_F, R, Re\right) \tag{6}$$

where, n is mass ratio; η is mechanical damping ratio of structure; f_v is dominant vibrating frequency, which is basically equaling to the structural natural frequency in VIV and galloping coupled vibrations, so it will be excluded in the dimensionless expressions; A_F is representing the influences of all the fluid parameters, such as f, C_{L0}, H_r .etc; $R=B/D$ is the aspect ratio of rectangular cylinders, the ratio of along wind dimension over cross wind dimension; Re, is Reynolds number, its effect may be combined into the aerodynamic coefficients, i.e., aspect ratio of rectangular cylinders. Therefore, S_L could be further simplified as Eq. (7)

$$S_L=F(n,\eta,A_F,R) \tag{7}$$

2.3.1 The effect of structural parameters

The Tamura-Shimada mathematical model (Eq. (5)) is selected to perform the comparative simulations in this part, because the physical interpretations of each variables could be checked and the whole equation group is more rational derived which is concluded in section 2.2.

First of all, a set of classical data on VIV and galloping interaction is taken here to verify the efficiency of Tamura-Shimada mathematical model. As shown in Fig. 2, the experimental data of square cylinder is provided by Bearman *et al.* (1987), Corless and Parkinson (1988) verified it by their mathematical model. The parametrical details of mass ratio, damping ratio and galloping coefficients are shown in Tables1 and 2. Some other parameters are needed before calculating with the Tamura-Shimada model, where St=0.12 is used for Strouhal number; the fluid parameters i.e. Hr and f are respectively assumed to 1.8 and 1.16, which are the same value as circular cylinder, and their effects will be separately investigated later; amplitude of lift coefficient on a stationary square cylinder C_{L0} is reported by many researchers, C_{L0}=0.7 is basically adopted for calculation. Eq. (5) is solved by Runge-Kutta method, time step is scaled to as small as 0.05, and the longest calculation time is up to 20000 to ensure all the obtained results are steady and reasonable.

Fig. 3 shows the comparison between calculation results and experimental measured data of Bearman .etc, at the low wind velocity range, there is a small range of VIV lock-in appearing in calculation processes but not provided in experimental curves. The differences might be attributed to the error of values of fluid parameters because of lacking the precise experimental results. However, the good agreements of calculation and experiments can be observed after the VIV and galloping interaction starting point. Once again, the Tamura-Shimada model is proved to be efficient on predicting such kind of interaction effects.

Table 1 is a list of parameters, from which we can see the only varying parameter is damping ratio, three different levels separately according to experimental data of Fig. 2. Similarly, we input these parameters into Tamura-Shimada model and solve it by Runge-Kutta method, the results are shown in Fig. 4. Different damping ratios give rise to different oscillation curves, more specifically, higher damping ratios postpone the oscillation starting point; and the same wind velocity point, the higher of damping ratios, the lower of amplitudes will be. However, the most concerned amplitude slope S_L of present work are observed to be basically the same even the significantly changed damping ratios, which means that S_L may not be sensitive with damping.

And in the comparative investigation of mass on amplitude slope S_L, as shown in Fig. 5, the only varying parameter mass ratio are adjusted by double, and S_L are basically the same. Therefore, the similar conclusion as damping ratio can be drawn: S_L is also not sensitive with mass.

Table 1 Investigation on damping ratio

Structural parameters		Fluid parameters			Aerodynamic parameters			
Mass ratio (n)	Damping ratio (η)	Wake oscillator parameter (Hr)	Magnus effect (f)	Lift coefficient (C_{L0})	A_1	A_3	A_5	A_7
4.7e-4	0.88e-3 2.98e-3 5.45e-3	1.8	1.16	0.70	4.87	421	1.70e4	1.94e5

Table 2 Investigation on mass ratio

Structural parameters		Fluid parameters			Aerodynamic parameters			
Mass ratio (n)	Damping ratio (η)	Wake oscillator parameter (Hr)	Magnus effect (f)	Lift coefficient (C_{L0})	A_1	A_3	A_5	A_7
0.5*4.7e-4 4.7e-4 2*4.7e-4	2.98e-3	1.8	1.16	0.70	4.87	421	1.70e4	1.94e5

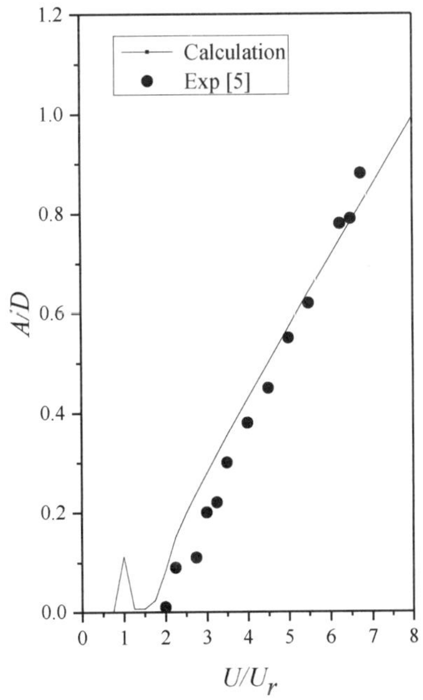

Fig. 3 Comparison between Bearman's experimental data and calculation results

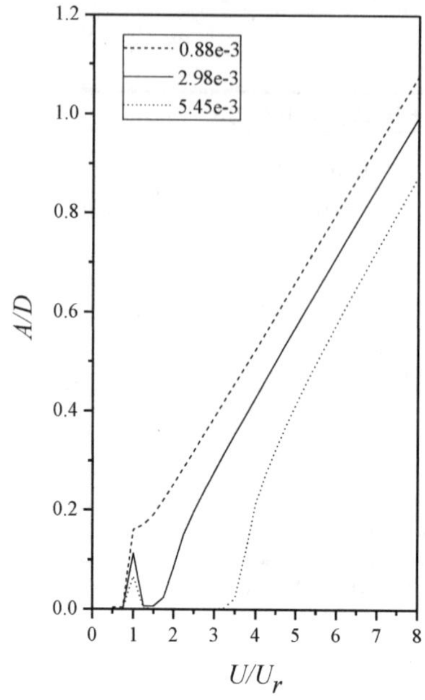

Fig. 4 The effects of damping on amplitude response

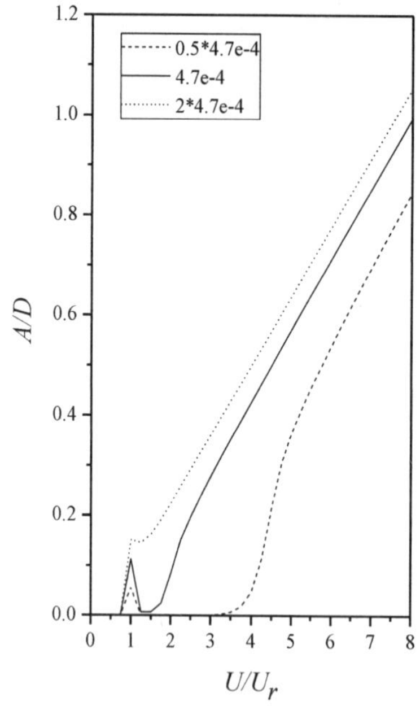

Fig. 5 The effects of mass on amplitude response

2.3.2 The effect of fluid parameters

Here, in this part, we are going to check the effects of fluid parameters (i.e., f, C_{L0}, H_r) on amplitude response parameter S_L by the Tamura-Shimada model. Three cases are formed due to the main principle of changing only one parameter in each case. The calculating parameters of each case are separately listed from Table 3 to Table 5. The way to vary parameter values in each case is unified as scaling 2 times and 0.5 times on the basis of a standard value, then three different values are obtained in each case.

Similarly, the Tamura-Shimada models are solved by Runge-Kutta method, simulation curves are obtained as shown in Fig. 6 to Fig. 8. It can be observed from Fig. 6 that the large extent of changing fluid parameter *Hr* did not bring any obvious changes on simulation results, simulation curves are almost coincided both in low wind range VIV and high wind interaction responses. The effects of fluid parameters f and C_{L0} on response amplitude are shown in Figs. 7 and 8 respectively. While different values of each parameter causes significant response variation in VIV response amplitude, the coupled VIV and galloping response are basically the same. Therefore it is concluded that w the fluid parameters f, C_{L0}, H_r are not the main affecting factors on response slope S_L.

Table 3 Investigation on *Hr*

Structural parameters		Fluid parameters			Aerodynamic parameters			
Mass ratio (n)	Damping ratio (η)	Wake oscillator parameter (Hr)	Magnus effect (f)	Lift coefficient (C_{L0})	A_1	A_3	A_5	A_7
4.7e-4	2.98e-3	0.5*1.8 1.8 2*1.8	1.16	0.70	4.87	421	1.70e4	1.94e5

Table 4 Investigation on f

Structural parameters		Fluid parameters			Aerodynamic parameters			
Mass ratio (n)	Damping ratio (η)	Wake oscillator parameter (Hr)	Magnus effect (f)	Lift coefficient (C_{L0})	A_1	A_3	A_5	A_7
4.7e-4	2.98e-3	1.8	0.5*1.16 1.16 2*1.16	0.70	4.87	421	1.70e4	1.94e5

Table 5 Investigation on C_{L0}

Structural parameters		Fluid parameters			Aerodynamic parameters			
Mass ratio (n)	Damping ratio (η)	Wake oscillator parameter (Hr)	Magnus effect (f)	Lift coefficient (C_{L0})	A_1	A_3	A_5	A_7
4.7e-4	2.98e-3	1.8	1.16	0.5*0.70 0.70 2*0.70	4.87	421	1.70e4	1.94e5

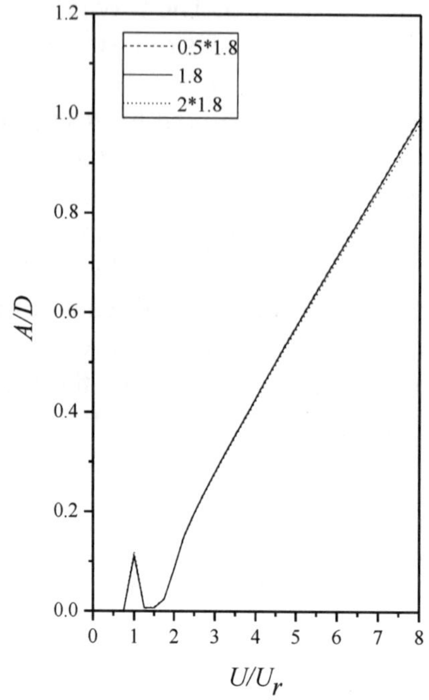

Fig. 6 The effects of wake oscillator parameter *Hr* on amplitude response

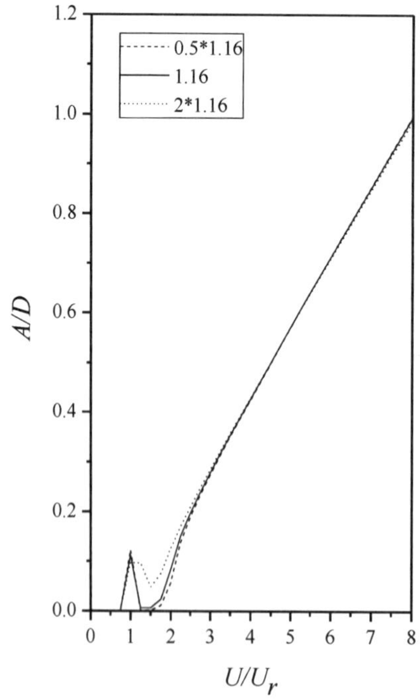

Fig. 7 The effects of Magnus parameter *f* on amplitude response

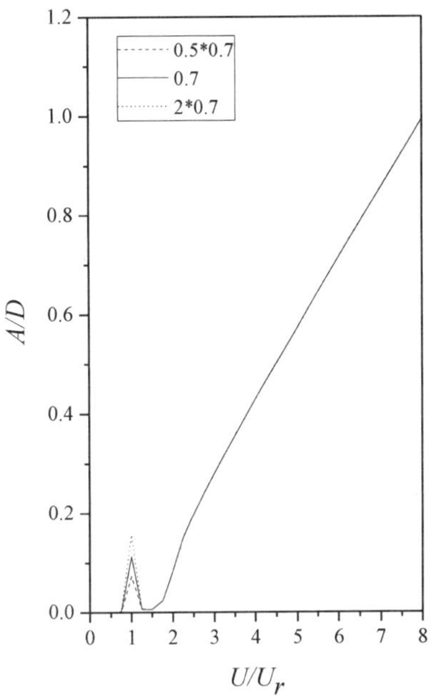

Fig. 8 The effects of lift coefficient C_{L0} on amplitude response

Table 6 Investigation on quasi-steady galloping coefficients

Structural parameters		Fluid parameters			Aerodynamic parameters			
Mass ratio (n)	Damping ratio (η)	Wake oscillator parameter (Hr)	Magnus effect (f)	Lift coefficient (C_{L0})	A_1	A_3	A_5	A_7
4.7e-4	2.98e-3	1.8	1.16	0.70	4.87 1.30	421 315	1.70e4 5.70e4	1.94e5 1.72e5

2.3.3 The effect of aerodynamic parameters

The effect of aerodynamic parameters A_1, A_3, A_5... A_N on amplitude response parameter S_L is numerically studied in this part. Two cases are selected, one is square cylinder, the aspect ratio of cross section is definitely to be 1.0, and the aerodynamic parameters A_1, A_3, A_5, A_7 are 4.87, 421, 1.70e4, 1.94e5 respectively according to the former researchers' report (Corless and Parkinson, 1988); the other one is a rectangular cylinder whose aspect ratio is 1.4, the aerodynamic parameters are correspondingly determined to be as 1.30, 315, 5.70e4, 1.72e5 by static force coefficients wind tunnel tests in present work.

Numerical simulations are performed under Tamura-Shimada model with the same processes as before, comparative results curves are plotted in Fig. 9. As we can see, two curves exhibit the obvious differences on interaction responses stages not only at starting point but for amplitude increasing slope S_L. The value of S_L of square cylinder is calculated to be 0.1385, while the result of the other rectangular cylinder is 0.0935, the deviation between them is over 40%. Therefore, we may conclude that aerodynamic parameters are the main affecting factors on amplitude slope S_L. It is worth noting that for rectangular cylinders, aerodynamic parameters are only determined by aspect ratio, so we can further conclude that the aspect ratio of rectangular cylinder is actually the key factor for S_L.

3. Results and discussions

The key parameter S_L is found to be only associated with aspect ratio ($R=B/D$) of rectangular cylinders through comparative numerical simulations in last part. For rectangular cylinders who are suffered from VIV and galloping coupled oscillations, S_L is going to be uniformly determined in this part by regression analysis on $R=B/D$ with comprehensively collected experimental data.

3.1 The collected data

Table 7 is the list of collected experimental data of rectangular cylinders on interaction effects between VIV and galloping, 26 cases in total. The corresponding vibration curves of 26 cases are separately presented in Fig. 10, the high magnitude of mathematical fitting deviation R^2 for all cases are clearly shown, and the obviously linear relationship between normalized wind velocities and dimensionless amplitudes are presented.

For each case, the aspect ratio, mass, damping, predicted onset velocity ratio of galloping over VIV, S_L and dimensionless onset velocity of interaction vibration are presented in Table 7, which shows the good representative for collected data: aspect ratio range is varying from 0.71 to 2.49; the predicted onset velocity ratio of galloping over VIV is distributing from 0.1 to 8.4; and most importantly, S_L , i.e., the slope of increased amplitude over increased wind velocity, obtained by the least square linear fitting on vibration curves of collected data (See Fig. 10), is ranging from 0.011 to 0.104, which is showing the promising signs for the establishment of empirical amplitude prediction formula.

Regarding the onset velocities of interaction effects, on the basis of Table 7, the general conclusion may be drawn as following: the dimensionless onset velocity is controlled by VIV when $0.2 < V_g/V_v < 5$, then U_0 should be floating at 1.0; and the onset velocity will be controlled by galloping when $V_g/V_v > 5$ or $V_g/V_v < 0.2$, U_0 appears to be determined by quasi-steady galloping theory.

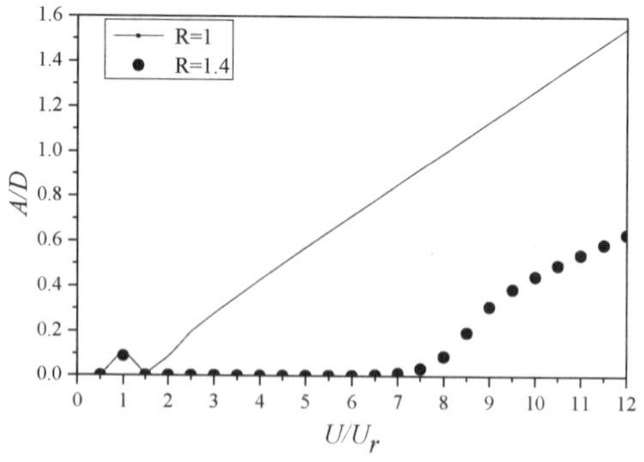

Fig. 9 The effects of aspect ratio on amplitude response

Table 7 Main parameters and fitting data

No.	*R=B/D*	*η/n*	V_g/V_v	S_L	U_0
Case 1	1.40	2.2	2.4	0.037	1.0
Case 2	1.40	2.7	3.0	0.029	1.0
Case 3	1.40	2.7	3.1	0.032	1.0
Case 4	1.40	3.7	4.1	0.030	0.8
Case 5	1.40	4.0	4.5	0.023	0.9
Case 6	1.40	2.0	2.2	0.026	1.1
Case 7	1.40	0.1	0.1	0.104	0.4
Case 8	1.40	1.6	3.0	0.024	0.7
Case 9	1.31	0.7	1.4	0.048	0.7
Case 10	2.00	1.0	3.0	0.017	1.2
Case 11	2.00	1.1	1.6	0.016	0.6
Case 12	2.00	1.7	2.6	0.019	0.6
Case 13	2.00	2.2	3.3	0.021	0.7
Case 14	2.00	2.4	3.6	0.019	0.6
Case 15	1.00	1.3	1.1	0.048	1.1
Case 16	1.00	1.8	1.5	0.045	1.0
Case 17	1.50	7.6	8.4	0.021	2.9
Case 18	1.50	1.3	1.4	0.024	2.0
Case 19	0.76	0.7	1.6	0.068	0.6
Case 20	0.71	1.6	5.1	0.075	4.7
Case 21	0.75	-	<1	0.064	0.9
Case 22	1.00	-	<1	0.026	0.9
Case 23	1.33	-	<1	0.019	0.9
Case 24	1.73	-	<1	0.015	0.9
Case 25	2.01	-	<1	0.011	0.9
Case 26	2.49	-	<1	0.016	0.9

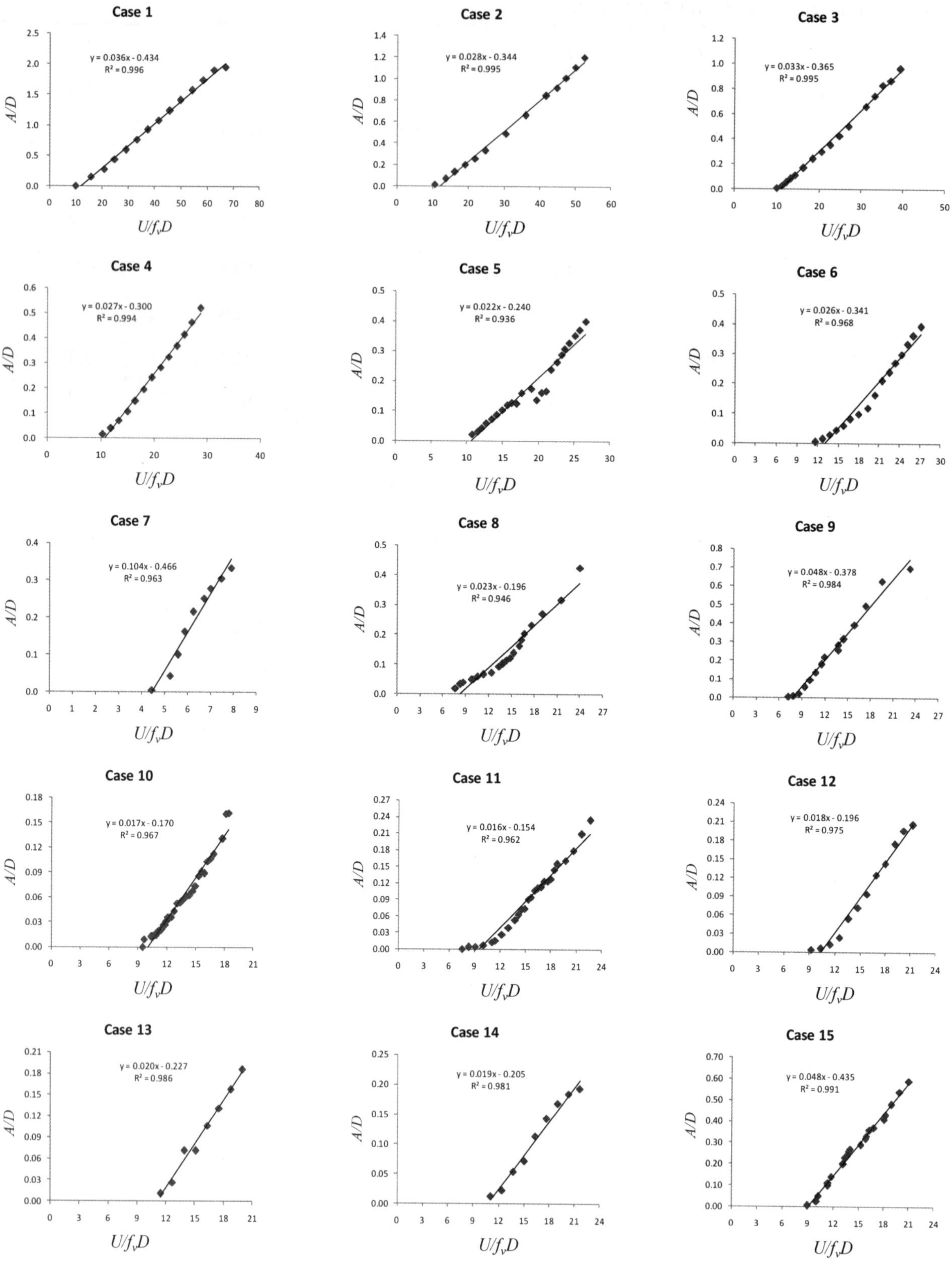
Case 1
y = 0.036x - 0.434
R² = 0.996
Case 2
y = 0.028x - 0.344
R² = 0.995
Case 3
y = 0.033x - 0.365
R² = 0.995
Case 4
y = 0.027x - 0.300
R² = 0.994
Case 5
y = 0.022x - 0.240
R² = 0.936
Case 6
y = 0.026x - 0.341
R² = 0.968
Case 7
y = 0.104x - 0.466
R² = 0.963
Case 8
y = 0.023x - 0.196
R² = 0.946
Case 9
y = 0.048x - 0.378
R² = 0.984
Case 10
y = 0.017x - 0.170
R² = 0.967
Case 11
y = 0.016x - 0.154
R² = 0.962
Case 12
y = 0.018x - 0.196
R² = 0.975
Case 13
y = 0.020x - 0.227
R² = 0.986
Case 14
y = 0.019x - 0.205
R² = 0.981
Case 15
y = 0.048x - 0.435
R² = 0.991
A/D
U/f_vD

Continued-

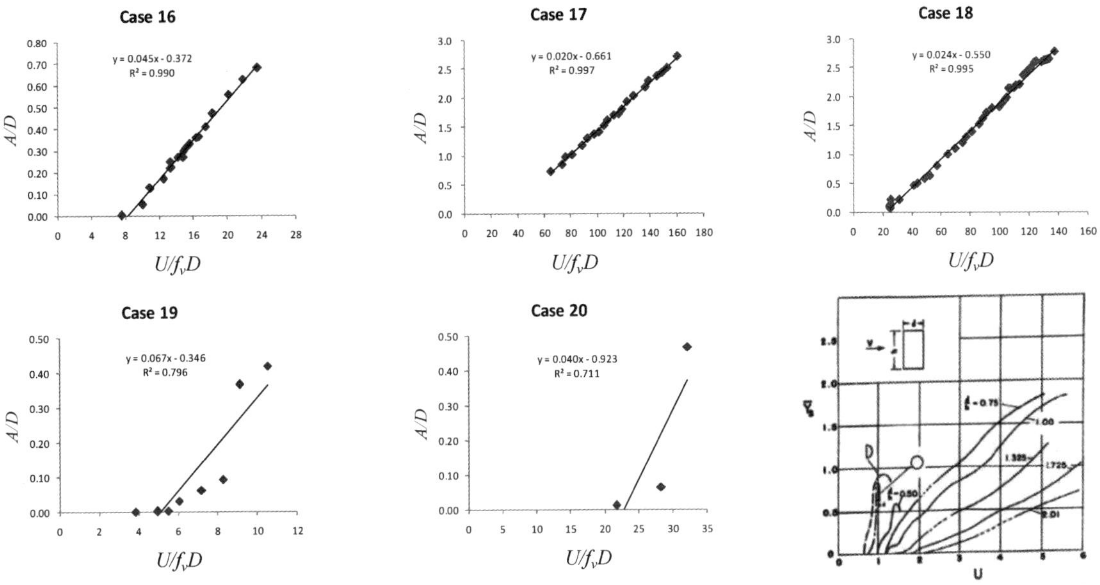

Fig. 10 The plot lists on the normalized wind velocity vs. dimensionless amplitude and the corresponding linear fitting equations

3.2 Empirical formula

The fitting curve between aspect ratio $R=B/D$ of rectangular cylinders and linear slope S_L of amplitude response is obtained by polynomial fitting method as shown in Fig. 11, from which we can get the mathematical expression of S_L on $R=B/D$.

Because of the onset velocity could be reasonably obtained by Strouhal law, then finally, the empirical formula to predict the amplitude of coupled VIV and galloping response can be obtained as Eq. (8)

$$\frac{A}{D}=\left(-0.016\cdot R^3+0.101\cdot R^2-0.222\cdot R+0.182\right)\cdot\left(U_r-U_0\right) \tag{8}$$

Where, A/D is the dimensionless amplitude response; R is the aspect ratio of rectangular cylinders; $U_r=V/(f_vD)$ is the normalized wind velocity where we want predict the amplitude response; $U_0=0.9/St$ is onset velocity according to the generally summarized data.

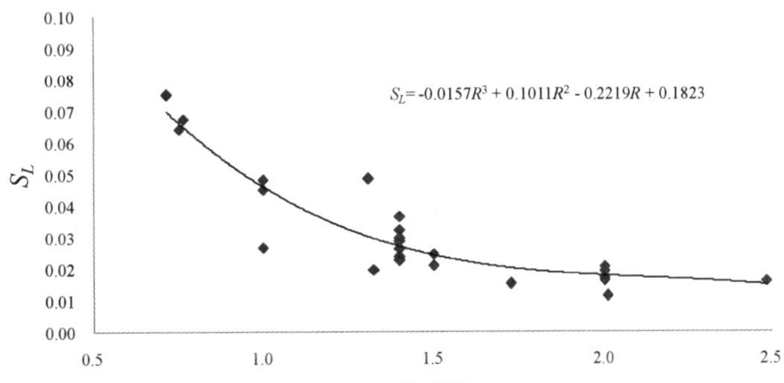

Fig. 11 The mathematical fitting between R and S_L based on the collected data

However, there are still several limitations for the use of this formula on amplitude prediction of rectangular cylinders should be noted:

Firstly, due to the limited collected data during mathematical fitting process, the aspect ratio of rectangular cylinders should be confined to the range of 0.5< $R=B/D$ < 2.5; secondly, there are few points in Fig. 11 observed to be scattered in the mathematical fitting between S_L and R, more data is required in the future investigations to improve the accuracy of this empirical curve; and more importantly ,this empirical formula is only applicable for the VIV-galloping coupled cases.

4. Conclusions

The coupled oscillations between VIV and galloping of rectangular cylinders are observed by many researchers; mathematical models which have taken into account the combined effects of VIV and galloping are also separately presented, mainly, Tamura & Shimada model and Corless & Parkinson model.

The interaction oscillations are characterized by: firstly, there will be an onset velocity which is decided by Strouhal law or quasi-steady galloping theory; secondly, the amplitude response curves are always observed to be linearly increased accompanied by wind velocities. Therefore, it is reasonable to propose an empirical amplitude estimation formula on the basis of careful research on onset velocity and amplitude increasing slope S_L .

Tamura & Shimada model is taken as the main model to perform comparative numerical simulations in present work, large amount of simulation works indicate that S_L is only associated with aspect ratio of rectangular cylinders. On the basis of comprehensively collected experimental data on interaction oscillations, S_L is determined by regression analysis, an empirical formula to predict the amplitude response of interaction oscillations of rectangular cylinders is finally proposed.

However, more efforts are need to improve the accuracy of this empirical formula, as several points are observed to be scattered during the mathematical fitting processes.

Acknowledgements

The authors are gratefully acknowledging the support of Natural Science Foundation of China under the Grant Nos. 51478181, 50908085 and 91215302. Special thanks to Prof. Tamura of Tokyo Polytechnic University and Prof. Borri of University of Florence for their patient and valuable comments on this paper.

References

Allison E, A.M. and Corless, R.M. (1995), "Prediction of closed-loop hysteresis with a flow-induced vibration model", *Proceedings of the 15th Canadian Congress of Applied Mechanics*, Victoria, Canada, May.

Bearman, P.W. (1984), "Vortex shedding from oscillating bluff bodies", *Annu. Rev. Fluid Mech.*, **16**(1), 195-222.

Bearman, P.W., Gartshore, I.S., Maull, D.J. and Parkinson, G.V. (1987), "Experiments on flow-induced vibration of a square section cylinder", *J. Fluids Struct.*, **1**(1), 19-34.

Borri, C., Zhou, S. and Chen, Z. (2012), "Coupling investigation on vortex-induced vibration and galloping of rectangular cylinders", *Proceedings of the Seventh International Colloquium on Bluff Body Aerodynamics and Applications*, Shanghai, China, September.

Corless, R.M. and Parkinson, G.V. (1988), "A model of the combined effects of vortex-induced oscillation and galloping", *J. Fluids Struct.*, **2**(3), 203-220.

Den Hartog, J.P. (1932), "Transmission-line vibration due to sleet", *AIEE*, **51**, 1074-1086.

EN 1991-1-4 (2010), Eurocode 1-Actions on Structures, Parts 1-4: General Actions-Wind Actions.

Facchinetti, M.L., Langrea, E.de. and Biolley, F. (2004), "Coupling of structure and wake oscillators in vortex-induced vibrations", *J. Fluids Struct.*, **19**(2) ,123-140

Garrett, J.L. (2003), *Flow-induced vibration of elastically supported rectangular cylinders*, Ph.D. Dissertation, Iowa State University, Ames.

Govardhan, R.N. and Williamson, C.H.K. (2006), "Defining the 'modified griffin plot' in vortex-induced vibration: revealing the effect of Reynolds number using controlled damping", *J. Fluid Mech.*, **561**,147-180.

Gjelstrup, H. and Georgakis, C.T. (2011), "A quasi-steady 3 degree-of-freedom model for the determination of the onset of bluff body galloping instability", *J. Fluids Struct.*, **27**(7), 1021-1034

Hansen, S.O. (2013), "Wind loading design codes", Fifty Years of Wind Engineering —Prestige Lectures from the Sixth European and African Conference on Wind Engineering, Cambridge, UK, July.

Hortmanns, M. and Ruscheweyh, H. (1997), "Development of a method for calculating galloping amplitudes considering nonlinear aerodynamic coefficients measured with the forced oscillation method", *J. Wind Eng. Ind. Aerod.*, **69-71**, 251-261.

Hemon, P. (1999), "An improvement of time delayed quasi-steady model for the oscillations of circular cylinders in cross-flow", *J. Fluids Struct.*, **13**(3), 291-307.

Joly, A., Etienne, S. and Pelletier, D. (2012), "Galloping of square cylinders in cross-flow at low Reynolds numbers", *J. Fluids Struct.*, **28**, 232-243.

Luo, S.C., Chew, Y.T. and Ng, Y.T. (2003), "Hysteresis phenomenon in the galloping oscillation of a square cylinder", *J. Fluids Struct.*, **18**(1), 103-118.

Macdonald, H.G. and Larose, G.L. (2006), "A unified approach to aerodynamic damping and drag/lift instabilities, and its application to dry inclined cable galloping", *J. Fluids Struct.*, **22**(2), 229-252.

Mannini, C., Marra, A.M., Massai, T. and Bartoli, G. (2013), "VIV and galloping interaction for a 3:2 rectangular cylinder", *Proceedings of the 6th European and African Conference on Wind Engineering*, Cambridge, UK, July.

Mannini C., Marra A.M. and Bartoli G. (2014), "VIV galloping instability of rectangular cylinders: Review and new experiments", *J. Wind Eng. Ind. Aerod.*, **132**, 109-124.

Marra, A.M., Mannini, C. and Bartoli, G. (2011), "Van der Pol-type equation for modeling vortex-induced oscillations of bridge decks", *J. Wind Eng. Ind. Aerod.*, **99**(6-7), 776-785.

Matsumoto, M. (1999), "Vortex shedding of bluff bodies: a review", *J. Fluids Struct.*, **13**(7-8), 791-811.

Names, A., Zhao, J., Lo Jacono, D. and Sheridan, J. (2012), "The interaction between flow-induced vibration mechanisms of a square cylinder with varying angles of attack", *J. Fluid Mech.*, **710**, 102-130.

Parkinson, G.V. and Bouclin, D. (1977), "Hydroelastic oscillation of square cylinders", International Research Seminar on Safety of Structures under Dynamic Loading, Trondheim, Norway, June.

Parkinson, G.V. and Brooks, N.P.H. (1961), "On the aeroelastic instability of bluff cylinders", *J. Appl. Mech. - TASME*, **28**(2), 252-258.

Parkinson, G.V. and Smith, J.D. (1964), "The square cylinder as an aeroelastic non-linear oscillator", *Quart. J. Mech. Appl. Math.*, **17**(2), 225-239.

Parkinson, G.V. (1965), "Aeroelastic galloping in one degree of freem", Wind Effects on Buildings and Structures: Proceedings of the Conference Held at the National Physical Laboratory, Teddington, UK, June.

Parkinson, G.V. and Wawzonak, M.A. (1981), "Some consideration of combined effects of galloping and vortex resonance", *J. Wind Eng. Ind. Aerod.*, **8**(1-2), 135-143.

Skop, R.A. and Griffin, O.M. (1973), "A model for the vortex-excited resonant response of bluff cylinders", *J. Sound Vib.*, **27**(2), 225-233.

Sarpkaya, T. (1979), "Vortex-induced oscillations: a selective review", *J. Appl. Mech. - T ASME*, **46**(2), 241-258.

Sarpkaya, T. (2004), "A critical review of the intrinsic nature of vortex-induced vibrations", *J. Fluids Struct.*, **19**(4), 389-447.

Tamura, Y. and Matsui, G. (1979), "Wake-oscillator model of vortex-induced oscillation of circular cylinder", *Proceedings of the 5th international conference on wind engineering*, Fort Collins, USA, July.

Tamura, Y. (1983), "Mathematical model for vortex-induced oscillations of continuous systems with circular cross section", *J. Wind Eng. Ind. Aerod.*, **14**, 431-442.

Tamura, Y. and Shimada, K. (1987), "A mathematical model for the transverse oscillations of square cylinders", *Proceedings of the 1st International Conference on Flow Induced Vibrations*, Bowness on Windermere, UK, May.

Vio, G.A., Dimitriadis, G. and Cooper, J.E. (2007), "Bifurcation analysis and limit cycle oscillation amplitude prediction methods applied to the aeroelastic galloping problem", *J. Fluids Struct.*, 23(7), 983-1011.

Williamson, C. and Govardhan, R. (2008), "A brief review of recent results in vortex-induced vibrations", *J. Wind Eng. Ind. Aerod.*, **96**(6-7), 713-735.

Received: 24 November 2017 | Revised: 19 March 2018 | Accepted: 22 May 2018
DOI: 10.1002/we.2218

RESEARCH ARTICLE

WILEY

Modelling damping sources in monopile-supported offshore wind turbines

Chao Chen | Philippe Duffour

Department of Civil, Environmental and Geomatic Engineering, University College London, London, UK

Correspondence
P. Duffour, Department of Civil, Environmental and Geomatic Engineering, University College London, London, UK.
Email: p.duffour@ucl.ac.uk

Abstract

Vibration damping in offshore wind turbines is a key parameter to predict reliably the dynamic response and fatigue life of these systems. Damping in an offshore wind turbine originates from different sources, mainly, aerodynamic, structural, hydrodynamic, and soil dampings. The difficulties in identifying the individual contribution from each damping source have led to considerable uncertainty and variation in the values recommended. This paper proposes simplified but direct modelling approaches to quantify the damping contributions from the aerodynamic, hydrodynamic, and soil interactions. Results from these models were systemically compared to published values and when appropriate with simulation results from the software package FAST. The range of values obtained for aerodynamic damping confirmed those available in the literature, and blade element modelling theory was shown to provide good results relatively efficiently. The influence of couplings between fore-aft and side-side directions on the aerodynamic damping contribution was highlighted. The modelling of hydrodynamic damping showed that this damping is much smaller than usually recommended and could be safely ignored for offshore wind turbines. Soil damping strongly depends on the soil specific nonlinear behaviour.

KEYWORDS
damping modelling, dynamic response, offshore wind turbine

1 | INTRODUCTION

Offshore wind farms (OWFs) are set to play a key role in achieving renewable energy targets around the world over the next decades. Many countries are making plans to develop wind generation offshore as the wind resource is stronger and steadier at sea.[1] As wind and waves are highly fluctuating loads and offshore wind turbines (OWTs) are slender structures, their dynamics must be considered carefully. Damping is critical in correctly predicting the dynamic behaviour of an OWT because it is the only factor that limits the amplitude of the response at resonance. The dynamic response in turn affects the severity of fatigue damage. Therefore, it is essential to quantify properly the damping from different sources to assess the fatigue damage of an OWT. There are 5 main sources of damping in OWTs: aerodynamic, hydrodynamic, structural, soil, and supplemental damping provided by mechanical dissipating devices.[2] As OWTs are lightly damped structures, it is usual to assume that the total damping factor in the system is simply the sum of the damping factors from different sources as expressed by Equation 1:

$$\zeta_{Total} = \zeta_{Aero} + \zeta_{Struc} + \zeta_{Hydro} + \zeta_{Soil} + \zeta_{Damper} \, , \tag{1}$$

where ζ_{Total}, ζ_{Aero}, ζ_{Struc}, ζ_{Hydro}, ζ_{Soil}, and ζ_{Damper} are respectively the total, aerodynamic, structural, hydrodynamic, soil, and supplemental damping factors as a percentage of critical damping. The aerodynamic damping is the largest and most variable contributor, so the other damping sources are sometimes lumped together under the term "additional damping." We shall call this the baseline damping in this paper to avoid any confusion between additional and supplemental. For consistency and ease of comparison, damping ratios are used throughout this paper (as opposed to logarithmic decrement), and all published damping values cited have been converted accordingly. The conversion of logarithmic decrement values to damping ratios (a division by 2π) as well the varying precision with which results are available in the literature has caused some difficulty in reporting damping values with a consistent level of significant figures. It should also be noted that the repeatability of damping measurements is usually quite poor (typically with 10%-20% of variability) rendering the precision of many damping values reported here somewhat spurious. However, this was deemed necessary for the sake of comparison.

Damping in offshore and onshore wind turbines has been researched fairly extensively; however, the variations in published values for most damping sources remain large, and this directly translates into uncertainty in the prediction of the dynamic response and fatigue life of the system. As a result, designs may be unsafe or overly conservative. Therefore, there is a need to better characterize each source of damping in OWT systems.

This paper initially started as a literature review on damping in OWTs intending to provide the offshore wind community with the most up-to-date damping values. This still constitutes the first part of this paper. However, because of the lack of consensus for many sources of damping found in the literature, it was decided to complement the literature review with direct estimations of the various damping contributions through simple but physics-based models. This is the second part of the paper. The aims of this second part are first to better characterize the values reported in literature but also to test the extent to which these simple models can reliably provide damping values. This may make the paper somewhat unusual, but we believe that the 2 parts complement each other usefully.

For the second part of the paper, a finite element (FE) model was developed and used in conjunction with the coupled Aerodynamic Modelling Package FAST provided by NREL.[3] As most OWTs installed so far are supported on monopiles, this study focuses on this type of support structure.[4]

Section 2 provides the comprehensive literature review highlighting areas of agreement as well the scatter in the published damping data. Section 3 presents the model and the general methodology used to estimate the damping. Sections 4, 5, and 6 describe the modelling approaches and results for the aerodynamics, hydrodynamic, and soil damping respectively. The paper concludes in Section 7.

2 | LITERATURE REVIEW

Various physical processes contribute to the damping of OWTs, and isolating them is not straightforward. Published analytical studies tend to concentrate on a particular damping source either by direct modelling or through back-calculation, ie, estimating one source by subtracting other contributions from the overall damping. By contrast, experimental studies tend to focus on the overall damping in the system either in parked or operational regime, as this can be measured more readily. As a rough guide, Windenergie recommend using a total damping ratio of 1% for parked turbines and 7% in operation.[5]

2.1 | Experimental measurements of damping in parked OWTs

Two main methods have been used to measure damping experimentally in OWTs depending on the type of excitation. Modal properties (including damping) have been obtained by operational modal analysis (OMA) when the system is excited by ambient sources (wind and waves), or standard time or frequency domain damping identification techniques when the dynamic response of the system is caused by a controlled measurable excitation. Applying a controlled excitation to an actual OWT presents a number of practical difficulties because of access and size. Loads from the environment could also influence the results. The so-called rotor stop test is the most commonly used technique (eg,[6-10]) and consists in suddenly turning the blades into feathered position so that the tower experiences a downward step in rotor thrust. "Overspeed test" is a similar technique that has also been used extensively (eg,[11-15]). Other types of excitations have also been used—for instance, Koukoura et al used boat impact at sea level.[16] These techniques intrinsically reduce the influence of aerodynamic damping by keeping the rotor speed low or at zero, so that they are better suited at estimating the baseline damping. As a result, the difference in damping between the fore-aft (FA) and side-side (SS) directions tends to be small for idle turbines.

Using these various techniques, a range of values have been reported. Tarp-Johansen et al[10] measured the baseline damping ratio in a 3.5 MW OWT using rotor stop tests and estimated it at 1.91%. Versteijlen et al[6] measured the baseline damping in 3.6 MW Siemens OWTs using spectra of the bending moment at the base of the tower. They obtained a damping ratio of 3% for the first bending mode of the turbine. Damgaard et al[7-9] used "rotor stop" tests and OMA on turbines subjected to ambient excitation in 4 wind farms. The rotor stop tests gave a first FA mode damping ratio in the range 2.39% to 2.55%. Results from ambient excitation tests gave fairly close results with damping factors in the range 2.55% to 2.86%. A series of comprehensive damping studies on 3 MW Vestas V90 OWTs in Belwind Wind farm were completed by Shirzadeh et al.[11-15] Using overspeed tests and OMA under ambient excitation, they measured the damping ratios of the first FA and SS modes at 1.05% and 1.27% respectively (with an installed tuned-mass damper kept inactive). Koukoura et al[16] studied the damping in both parked and operating turbines excited by a boat impact and ambient excitation. They measured the FA and SS damping ratios of 1.8% and 1.9% respectively for a parked turbine. Dampers were installed and active in their study, but their contribution was not specified.

In conclusion, the damping ratios for parked turbines have been found to vary between 1% and 3% with reasonable agreement between published results.

2.2 | Experimental measurements of damping in operating OWTs

Hansen et al[17] used two experimental methods to estimate the aerodynamic damping in operating OWTs; however, no measurement of the baseline damping was provided in their study, so their results are effectively overall damping values. One method estimated the damping from the decay in the OWT free response obtained after an artificial periodic pitching of the blades was stopped. The measured overall damping ratios obtained this way were extremely scattered, averaging around ζ^{FA}~8% in the FA direction and ζ^{SS}~2.4% in the SS direction. The other method used was OMA with wind excitation which gave an overall damping ratio of 13.2% for the FA mode and 8.0% in the SS mode. The difference between the results obtained from the two methods is large, and OMA gave much higher values than those usually found in the literature. Koukoura et al[16] found that the overall FA damping ratio for an operating OWT in a relatively high wind speed of 19 m/second is 10.35% and 4.77% in the SS direction. However, this again seems to include some unspecified supplemental damping. In conclusion, there is a lot of uncertainty on the measured overall damping in operating wind turbines, and it transpires that further work is needed to establish a reliable methodology to measure it.

2.3 | Aerodynamic damping

Aerodynamic damping is mostly generated by the air drag experienced by the rotating blades oscillating in the airflow because of tower bending vibration.[18] The vibrating tower also experiences some air drag even if the rotor is idle but to a much lesser extent. Aerodynamic damping values have been reported in the range 4% to 8% in the FA direction (after rounding off). This dissipative effect is much less effective at damping the SS motion. In this case, published damping factors range from 0.08% to 1.43%.

Aerodynamic damping comes in such a wide range because it is strongly influenced by the wind speed, the rotation speed, and the control system which changes the pitch angle of blades and the yaw angle of the rotor to optimize power output.[19,20] A few papers have proposed theoretical derivations for the aerodynamic damping in OWTs. Salzmann and Tempel[21] summarized several analytical solutions for this damping in constant-speed turbines, and they proposed their own analytical model for variable-speed turbines. Using blade element momentum (BEM) theory,[18] they obtained closed-form expressions that account for the relationship between wind speed and the motion of turbine. Comparing their numerical results to published experimental data, they found that simulation largely underestimates the aerodynamic damping when the wind speed is much higher than the rated wind speed. Valamanesh and Myers[19] proposed a semianalytical solution also based on BEM theory that predicts the aerodynamic damping for both fore-aft and side-side vibration. Their numerical results agree fairly well with results from FAST simulations. Tarp-Joahnsen et al[10] simulated the SS aerodynamic damping in a 3.5 MW OWT using HAWCStab and found it to range from 0.08% to 1.43%. Schafhirt and Muskulus[22] proposed a decoupled modelling process to model the dynamic behaviour of wind turbine in separate stages. Their approach includes a simplified derivation of the aerodynamic forces, but their damping coefficients were eventually obtained by matching the response to that of a fully coupled model.

Aerodynamic damping was studied experimentally by Ozbek and Rixen[23] on a 2.5 MW *onshore* turbine. The damping ratios were estimated using OMA from photogrammetry and laser vibrometry measurements. As onshore foundations are stiffer; onshore wind turbines are expected to experience much less soil damping and no hydrodynamic damping at all. Also, they have no transition piece as such so their structural damping is bound to be low. Therefore, the baseline damping is likely to be very small so that a good estimate of the aerodynamic damping can be obtained by measuring the overall damping in operation. They found 5% in the FA direction and around 0.5% in SS mode. These values were in good agreement with results from matching HAWCS simulations they carried out. Although no other study has measured the aerodynamic damping directly, the overall damping values provided in the previous section can be used for this assuming standard values for the baseline damping. Assuming 2% baseline damping component (*our* estimate), overall damping results from Hansen et al[17] give an estimated ζ_{Aero}^{FA}~6% and ζ_{Aero}^{SS}~0.4% for the aerodynamic contributions using the controlled excitation. This is in line with other studies. Subtracting 2% from their OMA overall damping ratios gives ζ_{Aero}^{FA}~11.2% and ζ_{Aero}^{SS}~6%. As Koukoura et al[16] did not specify their supplemental damping, it seems prudent not to use their results in this way.

Therefore, published values concur that the aerodynamic damping is the strongest contribution in operation; however, both predicted or measured values can vary widely. Also, there is currently not much published data for larger OWTs.

2.4 | Hydrodynamic damping

Hydrodynamic damping comes from 2 sources: (1) wave radiation damping and (2) viscous damping caused by hydrodynamic drag.[24] The wave radiation is proportional to wave velocities whereas viscous damping is proportional to the relative velocities squared.[11] Few studies have been conducted on hydrodynamic damping of OWTs. Windenergie[5] suggest the value of 0.15% for viscous damping, and 0.11% is proposed for the radiation damping in Cook and Vandiver.[25] A radiation damping ratio of 0.24% was suggested by Tarp-Johansen et al[10] after simulating the radiation/diffraction component using the wave modelling program WAMIT. Using HAWC2 time-domain simulations (implementing Morison's equation), Shirzadeh et al[11] estimated the viscous hydrodynamic damping ratio at around 0.004%. Again, the reported values for this source of damping vary widely, but for an OWT on monopile located in relatively shallow water, these damping values are consistently very low compared to other sources.

2.5 | Soil/foundation damping

Soil damping plays a more prominent role when the turbine is idle or when the side-side behaviour is considered, as the aerodynamic damping is much smaller in these cases. Damgaard[9] modelled the influence of soil damping directly by introducing hysteretic springs in a cantilever model of the turbine to account for the soil-structure interaction (SSI). From this, they found the contribution of soil damping to range between 0.8% and 1.3% which is in good agreement with an experimental estimate of soil damping (1%) they obtained by back-calculation. Carswell et al[26] used a similar method for converting hysteretic energy loss into a viscous rotational dashpot at mudline to represent OWT foundation damping in a lumped-parameter model. From numerical simulations of NREL 5 MW reference turbine, they found the soil damping ratio to be in the range of 0.17% to 0.28% when inferred from free vibration results. However, a higher damping value of 0.72% was obtained from time history analyses in response to extreme wind and wave conditions. It should be noted that back calculating soil damping this way tends to increase the final uncertainty by propagating the uncertainties from other damping sources. Modelling the soil as a viscoelastic soil block in Abaqus, Tarp-Johansen et al[10] found a soil damping ratio of 0.56%. They suggested that 0.80% should be used to account for the nonlinear behaviour of the soil. Their values are currently recommended by GL guidelines.[5] Although very few studies have attempted to model soil damping directly in OWT, the damping in a standard single pile in horizontal vibration has been the subject of many past papers (eg,[27-29]). These could form the basis for modelling soil dissipation in monopiles. Geotechnical earthquake engineers also have a long history of studying soil damping (eg,[30,31]).

2.6 | Structural damping

Structural damping in offshore wind turbines has received very little specific attention. It is usually assumed to follow the behaviour of standard steel structures for which damping values are available. For instance, Eurocode 1[32] recommends a value of 0.19% for unlined welded steel stacks without external thermal insulation. More specific values can be found in the offshore oil and gas literature.[33] Structural damping in *onshore* wind turbines was measured by Schaumann and Seidel,[34] and it was found to be between 0.2% and 0.5% (excluding soil damping). Ozbek and Rixen[23] measured it at 0.3% in the FA direction and between 0.3% and 0.9% in the SS direction. However, structural damping in OWTs could be larger (from 0.5% to 1.5% according to Shirzadeh[11]) as the damping from other structural parts like the grouted connection may have a significant contribution. Offshore wind farms also tend to have much larger turbines than those investigated in the studies reported in this section, so it is not clear how reliably their measurements can be generalized to current larger OWTs.

2.7 | Summary

The damping estimations from past papers are summarized in Table 1 for comparison. These values reflect the range and uncertainties in damping from different sources for different directions and operational regimes. Cells with a blue (dark grey) background represent values on which many papers agreed or when the range of proposed values is quite narrow thus indicating a higher degree of confidence than the yellow (light grey) cells for which these conditions are not met.

Structural damping for the whole structure could vary from 0.2% to 1.5%. Published values for hydrodynamic damping range from 0.11% to 0.39% which is low both in absolute terms and compared to other sources. Soil damping has been reported in the range 0.17% to 1.3%. Aerodynamic damping values are reported in a fairly narrow range for parked turbines (0.08%-0.24%), but for operating turbines, these values present large variations because of the influence of wind speed, rotor speed, and vibration direction (FA and SS). Given the large uncertainties in the available damping values, it is of interest to explore simplified but direct modelling approaches that allow the reported values to be tested.

The values reported in this section warrant a few words of caution. First, in principle, for a linear dynamic system, the damping is a property of the system itself and does not depend on the type of excitation used. This means that the damping values quoted for idle turbines should be the same regardless of the means of obtaining them (eg, boat impact or rotor stop). However, when aerodynamic damping is present, the applied wind force is coupled with damping mechanism (which makes it nonlinear), so the use of standard linear damping identification techniques is only a convenient expedient. Second, the way the data is processed to identify the damping may affect the results. For instance, if the force applied excites significantly more than 1 mode, then the identification technique used should involve some spectral analysis (eg, FFT or wavelet) or the damping estimations may be inaccurate. However in most of the published papers, the vibration response seems to be clearly dominated by the first bending mode, so the values reported above should be considered as first mode damping ratios.

The models proposed in the remainder of the paper also focus on prediction of the first bending mode damping ratio.

TABLE 1 Damping ratio range based on published results [Colour table can be viewed at wileyonlinelibrary.com]

Operation Condition	Direction	Overall Damping	Structural Damping	Aerodynamic Damping	Soil Damping	Hydrodynamic Damping
Parked	FA	1.0%-3.0%	0.2%-1.5%	0.08%-0.24%	0.17%-1.30%	0.11%-0.39%
Operational	FA	5.0%-10.0%	0.2%-1.5%	4.0%-8.0%	0.17%-1.3%	0.11%-0.39%
Parked	SS	1.0%-3.0%	0.2%-1.5%	0.08%-0.24%	0.17%-1.3%	0.11%-0.39%
Operational	SS	1.0%-4.77%	0.2%-1.5%	0.08%-1.43%	0.17%-1.3%	0.11%-0.39%

Yellow (ligth grey) cells indicate large uncertainties, while blue (dark grey) cells indicate smaller uncertainties.

3 | NUMERICAL MODELLING APPROACH

3.1 | Description of the OWT finite element model

In this second part, the behaviour of a monopile-supported OWT was modelled using a bespoke *decoupled* finite element model. "Decoupled" here means that the tower FE model will predict the dynamic response of the system but does not include the wind/rotor aerodynamic interaction simultaneously. Decoupled models in this sense have been used to save computational time (eg,[22,35]). The model is based on the widely used 5 MW Reference Offshore Wind Turbine published by NREL.[36] The schematic is shown in Figure 1, and the basic properties of this turbine are listed in Table 2. A lumped mass at the top of the tower represents the rotor nacelle assembly (RNA). The tower/monopile is modelled with an assembly of 39 2D Euler-Bernoulli elements. The discretization of the tower followed the segmentation in the data provided for the turbine, and a smaller mesh size was used for the monopile. A convergence study (results not shown) confirmed that this element size was adequate. The general expression for the system equation of motion is:

$$\mathbf{M}\ddot{\mathbf{u}}(t)+\mathbf{C}\dot{\mathbf{u}}(t)+\mathbf{K}\mathbf{u}(t)=\mathbf{F}(t), \tag{2}$$

where **M**, **K**, and **C** stand for the mass matrix, stiffness matrix, and damping matrix respectively, **u**(t) is the displacement vector, and **F**(t) is the external force vector. Given the section and material properties of tower over height, the mass and stiffness matrices were formed.

The damping calculations from each source will be described in detail in turn in the following sections, but there are broadly two approaches. For aerodynamic and hydrodynamic dampings, the dissipation models produced directly a modal damping factor which was calculated with a separate MATLAB program. In this case, the FE model was only used to compute the modal properties of the system (see Section 3.2), and the

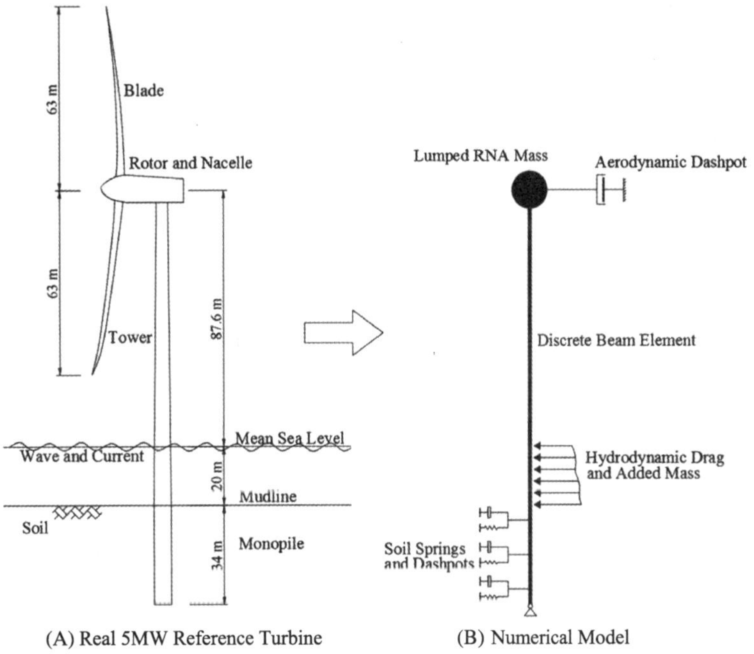

FIGURE 1 Schematics for (A) 5 MW reference turbine and (B) corresponding numerical model

TABLE 2 Basic properties of NREL 5 MW reference OWT

Rotor diameter, *R*	126 m
Hub height from MSL	87.6 m
Water depth, *h*	20 m
Tower diameter, *D*	3.87-6.00 m
Tower thickness, *t*	19-60 mm
Monopile diameter, D_{Pile}	6 m
Monopile thickness, t_{Pile}	90 mm
Lumped mass at top	3.5×10^5 kg
Natural frequency for model fixed at mudline	0.27 Hz
Natural frequency for model considering SSI	0.24-0.25 Hz

model was fixed at mudline to avoid having to choose a particular soil profile. This is consistent with usual practice in published studies. In the second approach, relevant for soil damping, nonlinear springs and dashpots in parallel were included below the mudline to model the SSI which contributed to the damping and stiffness matrices in Equation 2. The overall modal damping in the system was then extracted through time history analysis. This is described in more detail in Section 3.3.

Structural damping is made of many different mechanisms (material losses, joint friction, etc.) which cannot be modelled straightforwardly from first principles. Hence, this damping was not considered any further in this paper. However, any assumed value could easily be added to a finite element model for instance using a Rayleigh damping matrix in Equation 2.

3.2 | Modal analysis

Modal damping factors were related to the system properties by writing the equation of motion in modal coordinates. Although standard, this is briefly described here to introduce the notation which is used later. Premultiplying Equation 2 with the transpose of the ith mode shape vector Φ_i^T, the equation of motion becomes:

$$\overline{m}_i\ddot{\alpha}_i(t)+\overline{c}_i\dot{\alpha}_i(t)+\overline{k}_i\alpha_i(t)=\overline{f}(t), \tag{3}$$

where $\overline{m}_i$, $\overline{k}_i$, and $\overline{c}_i$ are respectively the modal mass, stiffness, and damping coefficients and $\alpha_i(t)$ is the ith modal coordinate. $\overline{f}_i(t)$ is a generalized external force defined as

$$\overline{f}_i(t)=\Phi_i^t\mathbf{F}(t) \tag{4}$$

Rearranging Equation 3 and introducing the modal frequency ω_i and the modal damping ratio ζ_i, the equation of motion for the ith mode becomes

$$\ddot{\alpha}_i(t)+2\zeta_i\omega_i\dot{\alpha}_i(t)+\omega_i^2\alpha_i(t)=\frac{\Phi_i^T\mathbf{F}(t)}{\overline{m}_i}. \tag{5}$$

In most cases, the contribution from higher modes is negligible, so the quantities of interest are the fundamental frequency ω_1, the corresponding damping ratio ζ_1, and the first modal mass $\overline{m}_1$. If a 3D model of the turbine is available, then the FA and SS modal properties might be slightly different as the two corresponding rotor moments of inertia are quite different. In this case, the relevant values for $\overline{m}_1$ and ω_1 should be used in the subsequent calculations.

3.3 | Time history analyses and damping identification

Whenever the damping models did not produce damping ratios directly (ie, the damping ratios were not calculated from equations based on modal analysis such as Equation 5), time domain analyses were conducted by implementing the numerical integration scheme HHT-α[37] which is a generalized version of the Newmark-β method. The nonlinear stiffness and damping coefficients of the soil springs were directly added to the stiffness and damping matrices at the beginning of each time step. The accuracy of this method was checked by comparing the model with a convergence algorithm (Newton-Raphson method), and good agreement was found.

To extract damping factors from the simulated response time histories, a decaying oscillation was triggered in the system by assigning an initial displacement or acceleration to the tower top causing a transient response superimposed on the steady-state behaviour. A sonogram FFT technique was then used on the transient decay so that the damping of the first mode could be isolated.[38] In principle, this damping identification technique assumes that the underlying dissipation mechanism is linear. This is not quite the case for the soil model used here; however, the shape of the decays obtained were close enough to those of a linear system to justify the use of the method.

4 | AERODYNAMIC DAMPING

4.1 | Theoretical derivation of the aerodynamic force

Aerodynamic damping is caused by the drag forces experienced by the oscillating rotor in the surrounding air flow. In our model, the rotor is represented by a lumped mass at the top of the beam representing the tower. The motion of the entire rotor is assumed to be consistent with that of the beam tip in translation but not rotation. In BEM theory, the resultant aerodynamic force applied to the hub can be obtained by introducing a relative wind speed experienced by each blade element and summing up the element thrust component. These resultant forces and relative speeds are different in the fore-aft and side-side direction, so these 2 cases must be dealt with separately.

4.1.1 | Fore-aft aerodynamic damping

It is assumed that the rotor is rotating in its own plane with the angular speed of ω and exposed to a uniform steady incoming wind field of average speed V_{Wx}, as shown in Figure 2. When the rotor does not oscillate, the three blade elements at radius r from the hub, with thickness dr, experience a total thrust $dT|_{V_x}$:

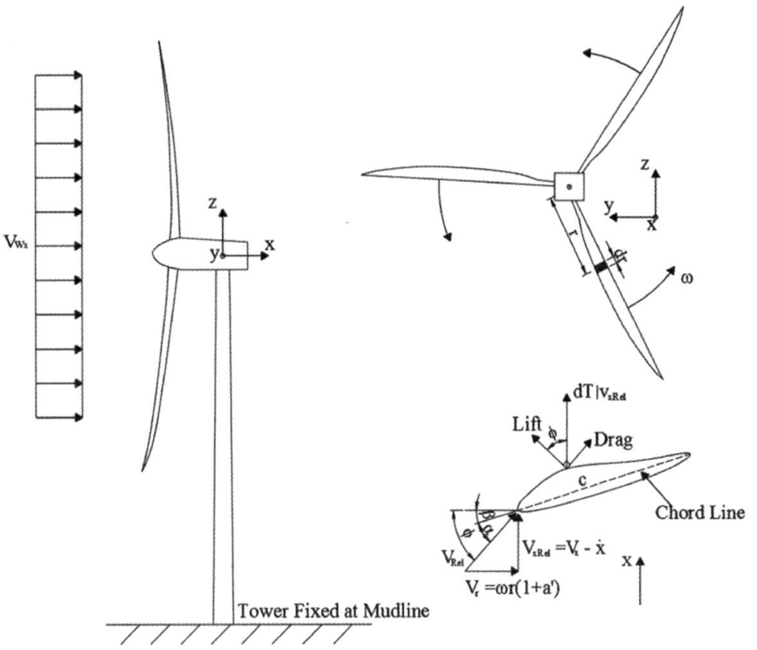

FIGURE 2 Blade element forces for fore-aft vibration

$$dT|_{V_x} = \frac{1}{2}\rho_a N_b \frac{V_x^2}{\sin^2\phi} cC_n dr, \tag{6}$$

where V_x is the axial component of the relative wind speed felt by the blade element, N_b is the number of blades, ρ_a is the air density, c is the chord length, $C_n = C_l \cos\phi + C_d \sin\phi$, with $C_{l/d}$ the lift/drag coefficients, and ϕ is the sum of the attack, pitch, and twist angles. If the tower top moves in the fore-aft direction with a velocity $\dot{x}$, the element now feels a relative velocity V_{xRel}:

$$V_{xRel} = V_x - \dot{x} \tag{7}$$

Assuming that $\dot{x}$ is a small compared to V_x and that it is the same throughout the rotor (the rotor remains parallel to itself), $dT|_{V_{xRel}}$, the thrust felt by the three blade elements under a relative speed of V_{xRel}, can be obtained from $dT|_{V_x}$ through a first-order Taylor expansion:

$$dT|_{V_{xRel}} = dT|_{V_x} + dT'\Big|_{V_x} \cdot (V_{xRel} - V_x) = dT|_{V_x} - \frac{d(dT)}{dV_x}\dot{x}. \tag{8}$$

From this derivation, it is clear that the aerodynamic damping will come into the equation of motion of the tower through the viscous term $\frac{d(dT)}{dV_x}\dot{x}$. The objective is to calculate this term and obtain the resultant thrust.

Assuming that dT is a function of V_x and ϕ, and that ϕ is a function of V_x, $\frac{d(dT)}{dV_x}$ can be calculated as

$$\frac{d(dT)}{dV_x} = \frac{\partial(dT)}{\partial V_x} + \frac{\partial(dT)}{\partial\phi}\frac{d\phi}{dV_x}. \tag{9}$$

In Equation 9, $\frac{\partial(dT)}{\partial V_x}$ and $\frac{\partial(dT)}{\partial\phi}$ can be obtained by differentiating Equation 6:

$$\frac{\partial(dT)}{\partial V_x} = \rho_a N_a \frac{V_x}{\sin^2\phi} cC_n dr, \tag{10}$$

$$\frac{\partial(dT)}{\partial\phi} = \frac{1}{2}\rho_a N_a V_x^2 \frac{\frac{dC_n}{d\phi}\sin^2\phi - 2\sin\phi\cos\phi C_n}{\sin^4\phi} cdr, \tag{11}$$

where

$$\frac{dC_n}{d\phi} = \frac{\partial C_l}{\partial\phi}\cos\phi + \frac{\partial C_d}{\partial\phi}\sin\phi + C_d\cos\phi - C_l\sin\phi, \tag{12}$$

$$\frac{dC_t}{d\phi}=\frac{\partial C_l}{\partial \phi}\sin\phi-\frac{\partial C_d}{\partial \phi}\cos\phi+C_l\cos\phi+C_d\sin\phi. \tag{13}$$

To obtain an expression for $\frac{d\phi}{dV_x}$, the following 2 equations can be used:

$$\tan\phi=\frac{V_x}{\omega r(1+a')}, \tag{14}$$

$$a'=\frac{1}{\frac{4\sin\phi\cos\phi}{\sigma C_t}-1}. \tag{15}$$

Differentiating Equation 14 with respect to V_x and Equation 15 with respect to ϕ gives:

$$\frac{d\phi}{dV_x}=\frac{1}{\omega r\left(\frac{1+a'}{\cos^2\phi}+\frac{da'}{d\phi}\tan\phi\right)}, \tag{16}$$

and

$$\frac{da'}{d\phi}=\frac{-4\left(2\cos 2\phi C_t-\frac{dC_t}{d\phi}\sin\phi\cos\phi\right)}{\sigma C_t^2\left(\frac{4\sin\phi\cos\phi}{\sigma C_t}-1\right)^2}. \tag{17}$$

Equations 10, 11, and 16 can be substituted into Equation 9, and the resulting expression for $\frac{d(dT)}{dV_x}$ can be used to obtain $dT|_{V_{xRel}}$ in Equation 8 which is the aerodynamic force subjected by the three blade elements. For simplicity, denote

$$A\cdot dr=\frac{\partial(dT)}{\partial V_x}, \tag{18}$$

$$B\cdot dr=\frac{\partial(dT)}{\partial\phi}\frac{d\phi}{dV_x}. \tag{19}$$

So, the thrust felt by the three blade elements is:

$$dT|_{V_{xRel}}=dT|_{V_x}-(A+B)dr\cdot\dot{x}. \tag{20}$$

The total thrust felt by the rotor is:

$$F_x=\int_0^R dT|_{V_{xRel}}=\int_0^R dT|_{V_x}-\int_0^R(A+B)dr\cdot\dot{x}, \tag{21}$$

where R is the radius of the rotor. Equation 21 shows that the total thrust can be divided in 2 parts: the first part $\int_0^R dT|_{V_x}$ represents the static force caused by the steady wind on a rigid structure; the second part $-\int_0^R(A+B)dr\cdot\dot{x}$ is the aerodynamic damping force proportional to the oscillating velocity of the entire rotor. Applying the total thrust on the top of the tower, the turbine model is able to account for the aerodynamic damping at a given rotor speed, wind speed, and pitch angles without coupling the tower and the rotor models. The damping part $\int_0^R(A+B)dr$ can be used to calculate the damping ratio with the following equation:

$$\zeta_{Aero\_FA}=\frac{\int_0^R(A+B)dr}{2\overline{m}_1\omega_1}. \tag{22}$$

4.1.2 | Side-side aerodynamic damping

Following a similar procedure, the total side-side aerodynamic force felt by the rotor can be expressed as:

$$F_y=\frac{N_b}{2}\int_0^R\left(A'+B'\right)dr\cdot\dot{y}, \tag{23}$$

where A' and B' are derived in Appendix 1 and $\dot{y}$ is the side-side velocity of the tower top. Using modal analysis, the side-side damping ratio can be expressed as:

$$\zeta_{Aero\_SS}=\frac{-\frac{N_b}{2}\int_0^R\left(A'+B'\right)dr}{2\overline{m}_1\omega_1}. \tag{24}$$

4.2 | Aerodynamic damping comparison

Equations 22 and 24 were implemented in MATLAB and coupled with the modal properties obtained from the FE model. These results are here compared to FAST simulations obtained with identical settings for validation purposes. For a range of steady wind speeds, FAST simulations were run, leaving the control module in FAST to alter the pitch angles and the rotating speed of rotor to achieve the desired power output. Figure 3 shows the relationship obtained this way between rotor speed, pitch, and wind speed for the same NREL 5 MW turbine.

For each average wind speed, an initial displacement of 1 m was given to the tower top and FAST was used to calculate time series of the tower top response. From these time series, damping ratios were estimated as described in Section 3.3.

Figure 4 shows the comparison of damping ratios in the fore-aft and side-side directions (respectively) using FAST and the proposed modelling approach for varying wind speeds. The FA and SS degrees of freedom were decoupled in the FAST simulations to mirror the BEM modelling assumptions. For example, when looking at the FA damping, the FA degree of freedom (DOF) was on, but the side-side DOF was off and vice versa. In the FA direction, good agreement is found between FAST and BEM results. Figure 4A confirms that for the turbine modelled here, FA aerodynamic damping increases from 5% at low wind speed to 7% at the rated speed after which it remains broadly constant. This range is in full agreement with the results reported in the literature. In the SS direction, the damping ratios obtained from BEM and FAST follow similar trends, and the damping ratios are much lower than in the FA direction, and they increase steadily from 0.2% to 1.2 to 1.6% at the top wind speed in the operating range. Blade element momentum results appear systematically higher than those from FAST by a fraction of a percent. Further research currently under way suggests that this difference is associated with the rotation of the top which is ignored here (but not in FAST). In fact, the tilting and sideway motions of the tower mean that the induced velocity seen by blade elements differs from the one we calculate and the wake flow angle is less than 90° to the rotor plane.[39] This phenomenon generates unbalanced resultant forces in the SS direction at the rotational frequency of the rotor which cause the tower to oscillate. In turn, the SS motion causes a harmonic excitation on the FA motion. FAST simulations not shown here suggest that this coupling does not change much the overall range of damping values obtained, but it can affect the dependence on the wind speed. In summary, the proposed aerodynamic model based on BEM theory is adequate to model the aerodynamic

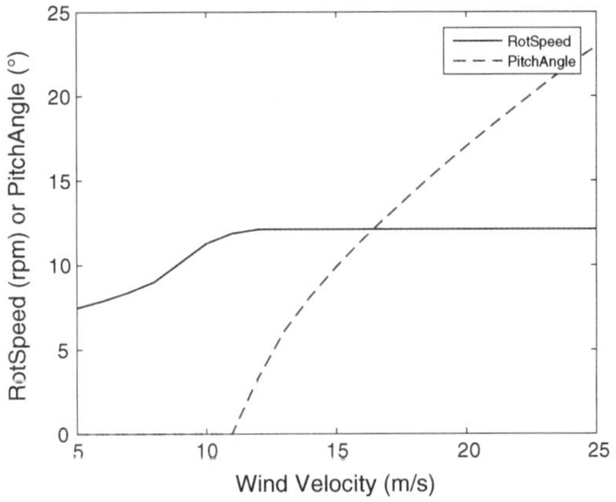

FIGURE 3 Steady-state pitch angles and rotor speeds for different wind speed in the FA direction [Colour figure can be viewed at wileyonlinelibrary.com]

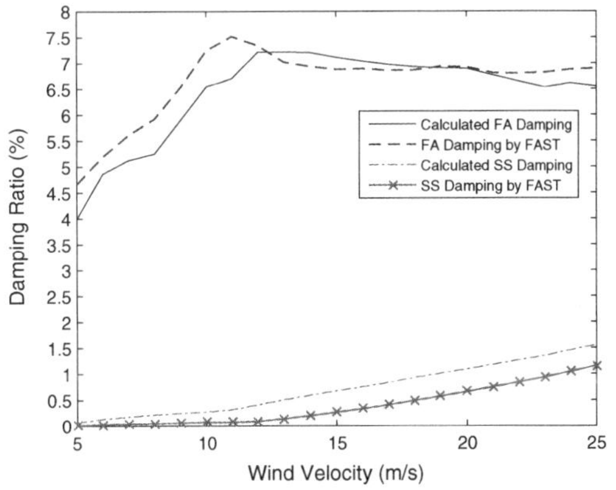

FIGURE 4 Comparison of FA and SS damping ratios for varying wind speeds obtained from BEM method and from FAST [Colour figure can be viewed at wileyonlinelibrary.com]

damping in the FA direction and gives a reasonable estimate in the SS direction. However, it cannot capture more detailed effects related to coupling which appear to affect the dependency of the damping ratio on the wind speed above the rated speed of the turbine.

5 | HYDRODYNAMIC DAMPING

In this section, hydrodynamic damping is quantified directly on the basis of classical theories and numerical simulations. As a simple but representative model, a flexible cantilevered cylinder immersed in water up to MSL is considered and subjected to wave excitation as shown in Figure 5. There are mainly 2 approaches to calculate the hydrodynamic forces resulting from the interaction between the cylinder and the seawater: (1) through Morison's equation or (2) potential flow theory. Morison's equation deals with the viscous effects when flow separation is significant, while potential flow theory accounts for the diffraction and radiation phenomena.

The physics of the water-cylinder interaction can be very different depending on the value of key nondimensional parameters. The most relevant ones here are the Keulegan-Carpenter number KC, the Stokes parameter β_{Stokes}, and the diffraction factor D/λ where λ is the wavelength and D the diameter of the pile. By definition, $KC = 2\pi\hat{x}/D$ where $\hat{x}$ is the amplitude of the sideway motion, and $\beta_{Stokes} = D^2/v_wT$ where v_w is the kinematic viscosity of water and T is the oscillation period. In the case of a large monopile submerged in water, the KC number is in the range of low values, around 10^{-2}, while the Stokes parameter and the diffraction factor are relatively large. These parameter values indicate that the flow separation tends to be small but not insignificant, while diffraction and radiation are the dominant phenomena.[40] This is supported by Johanning et al,[41] who found that the damping of an oscillating cylinder can be influenced by both the viscous damping and radiation damping caused by hydrodynamic forces. In this paper, hydrodynamic damping is estimated by modelling the viscous (Morison's equation) and wave (potential theory) contributions separately.

In addition to the two dissipative effects mentioned above, the added mass coming from the volume of water moving together with the cylinder can influence the mass distribution of the monopile under mean sea level (MSL). The added mass tends to reduce the damping ratio, but this effect has been found to be negligible.[42] The low relative acceleration in the submerged tower (because this part is near to the clamped end) results in a very small effect on the frequency and dynamic response of the whole system. Therefore, in the subsequent calculations, the modal mass does not account for the hydrodynamic added mass.

Both for viscous and radiation damping, the wave height and wave period were set to 6 m and 10 seconds for all calculations, and the wave velocities were calculated using linear wave theory.[43] This single combination of wave period and wave height was selected for simplicity, but is representative of the most common sea states.[44]

5.1 | Viscous damping

For an oscillating cylinder excited by waves, Morison's equation gives the forces on a strip of cylinder at depth z:

$$F_{\text{Morison}} = \frac{1}{2}\rho_w D C_d \left|\dot{u}_w - \dot{v}_{tower}\right|\left(\dot{u}_w - \dot{v}_{tower}\right) + \frac{\pi}{4}\rho_w D^2 C_a\left(\ddot{u}_w - \ddot{v}_{tower}\right) + \frac{\pi}{4}\rho_w D^2 \ddot{u}_w, \tag{25}$$

where u_w is the velocity of water particles, v_{tower} is the velocity of the vibrating tower under MSL, C_d is the drag coefficient, C_a is the added mass coefficient, and ρ_w is the density of water. The damping term is quadratic in the relative velocity and influenced by the drag coefficient C_d. The value of C_d is mainly dependent on the KC number and the Stokes parameter.[45,46] According to Bearman and Russell, for low values of KC number (up to 5) and high values of Stokes number (in the order of 10^5), C_d can be obtained by the following equation:

$$C_d = \frac{2\times 26.24}{KC\sqrt{\beta_{Stokes}}} + 0.08KC, \tag{26}$$

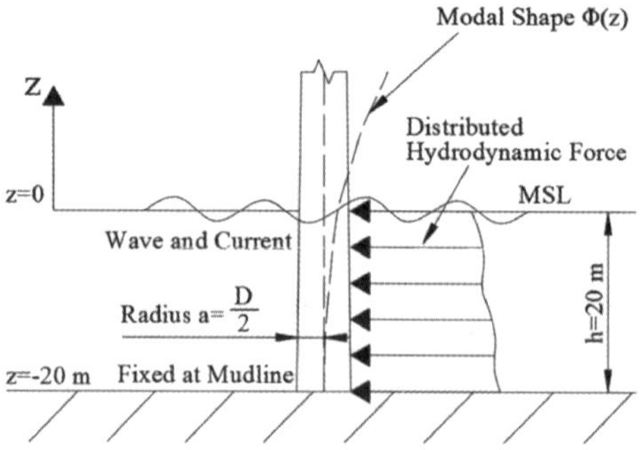

FIGURE 5 Schematic of hydrodynamic forces on OWT substructure

where the second term in this equation can be neglected for even smaller *KC* numbers. As *KC* depends on the vibration amplitude, C_d could be made to vary along the depth of the cylinder according to Equation 26. This was tested, and the damping did not show much sensitivity to C_d, so $C_d = 1$ was used throughout, based on the velocity amplitude of the pile at MSL obtained from simulation of the 5 MW reference turbine and recommendations from past study (eg,[11]).

The viscous drag term in Morison's equation varies along the immersed cylinder as a result of the nonuniform distribution of vibration accelerations and velocities. Following Cook[47] and after rearranging, the damping ratio for a single pile subjected to viscous drag can be written as:

$$\zeta_{Hydro,Vis} = \frac{\int_{-h}^{0} \frac{1}{2}\rho_w D C_d \sqrt{\frac{8}{\pi}}\sigma_r(z)\Phi^2(z)dz}{2\overline{m}\omega_n}, \tag{27}$$

where $\sigma_r(z)$ is the root mean square of the relative speed between the water wave particle and the pile (and other notation was defined in Section 3.2). For large OWTs, the velocity of the pile can be estimated by $2\pi\frac{\hat{x}}{T}\tilde{\ }0.1$ m/second. This is much smaller than the wave particle velocity (usually above 10 m/second), so $\sigma_r(z)$ can be replaced by the root mean square of the particle velocity. Usually, the wave profile can be generated using specific spectra, so $\sigma_r(z)$ could be calculated from these spectra.

5.2 | Radiation damping

According to linear potential flow theory, when only wave radiation is considered (ie, ignoring the fluid viscosity), wave-structure interaction produces two kinds of forces: those produced by the incoming wave field on a rigid cylinder and those because of the motion of the structure in a disturbance-free fluid. Only the latter gives rise to dissipation and can be expressed in Equation 5 by

$$\overline{f}(t) = -\int_{-h}^{0} C_{Rad} dz \cdot \dot{\alpha}(t), \tag{28}$$

where C_{Rad} is the radiation factor, the derivation of which is shown in Appendix 2 based on the study by Petrauskas.[48] Substituting Equation 28 into Equation 5 and rearranging gives a radiation damping ratio:

$$\zeta_{Hydro,Rad} = \frac{\int_{-h}^{0} C_{Rad} dz}{2\omega_1 \overline{m}}. \tag{29}$$

5.3 | Hydrodynamic damping comparison

For the model of the 5 MW reference OWT, the viscous damping ratio calculated by Equation 27 is 0.0006%, while the radiation damping ratio according to Equation 29 is 0.007%, giving a total hydrodynamic damping ratio of 0.0076%. This result is much smaller than what has been reported in past research papers such as Shirzadeh[11] which suggests that the viscous damping ratio is around 0.004% and radiation damping ratio is 0.12%. This difference is explained by the size of the structure which is much larger in our case than those considered in the literature. In Petrauskas[48] from which the 0.12% value for radiation damping value seems to originate, the diameter of the pile considered was 1.2 m (4 ft), and the structure was much shorter above MSL (so the dynamics of the structures were very different). When their dimensions are used in the equations presented in this section, 0.11% is obtained confirming the soundness of the calculation and the effect of the size. To explore this further, plausible dimensions for tower height, water depth, and pile thickness (based on the turbine scaling equations proposed in Jamieson[49]) were estimated so that damping values could be calculated for various turbine sizes. The results are shown in Figure 6, confirming that both viscous and radiation damping are always very small for large diameter rotors.

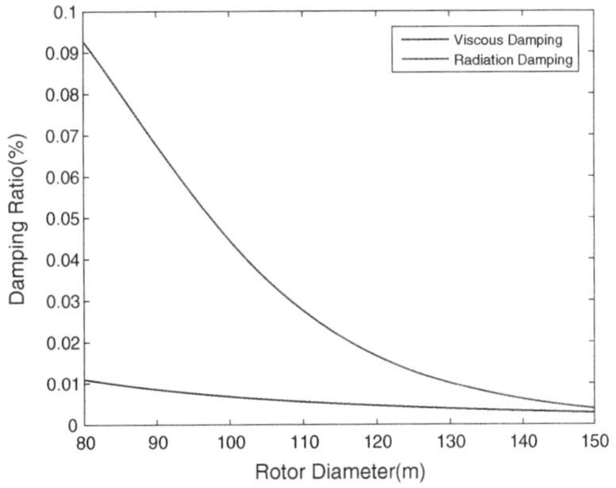

FIGURE 6 Variation of viscous and radiation damping ratios for different turbine sizes [Colour figure can be viewed at wileyonlinelibrary.com]

The conclusion from these models is that hydrodynamic damping in OWTs is much smaller than usually assumed and for all practical purposes could be safely ignored.

6 | SOIL DAMPING

The soil surrounding the monopile foundation can dissipate vibrational energy in two ways: (1) the motion of the pile can cause waves to radiate away from the pile into the ground (radiation damping) and (2) hysteretic material damping can occur as the soil is being cyclically stressed.[26,50] Radiation damping depends on the frequency ω of the external excitation. The nondimensional parameter $a_0 = \omega D/V_s$ can be used to assess the significance of the wave phenomenon. It compares the order of magnitude of the pile lateral velocity and the shear wave velocity of the soil stratum V_s. The shear wave velocity ranges from 150 to 300 m/second depending on the soil type (see Table 4 below). For a monopile-supported OWT, the forces experienced by the soil will usually have a low frequency around the first natural frequency (less than 1 Hz) as the response of the system is dominated by the first vibration mode. In these conditions, a_0 can be expected to be of the order of 10^{-2}, indicating that the behaviour of the soil is largely quasistatic so that radiation damping for a monopile-supported OWT can be ignored.[28]

6.1 | Soil model

6.1.1 | Stiffness and damping coefficients for soil springs and dashpots

Soil-pile interaction can be modelled using different approaches such as linear or nonlinear Winkler's springs or finite element solid modelling. For design purposes, *p-y* curves which capture the nonlinear displacement-force relationship in the soil are frequently used. In the present study, a combination of classic *p-y* curves and the study by Gazetas and Dobry[28] is adopted. The soil is represented by a series of horizontal springs and dashpots in parallel from the mudline to bottom of the pile. The stiffness of the springs was obtained from *p-y* curves, while the dashpot coefficients representing the soil material damping were calculated following Gazetas and Dobry[28]:

$$c_m = 2k\frac{\beta_m}{\omega}, \tag{30}$$

where β_m represents the material damping ratio, k is the secant modulus defined as the ratio between the static local soil reaction for a unit length of pile and the corresponding local pile deflection, and ω is the circular frequency of the excitation at the pile head. The magnitude of material damping ratio for a soil stratum is related to the shear strain in the soil, and can be influenced by the properties of soil such as plastic index and mean effective confining stress. Darendeli[31] proposed an analytical model to generate material damping curves for various values of the over consolidation ratio (OCR) and plasticity index (PI). This study adopts this method and uses the coefficients for sand and clay from Southern California provided by Darendeli as the parameters he uses are somewhat idiosyncratic and not available from standard data tables. However, Darendeli's data as well as other experimental studies in soil damping[30] show that the curves produced using his formulae are representative of many soil types.

Figure 7 shows the relationship between the material damping ratio and soil shear strain for clay in which the damping depends on the PI to some extent, so various curves were plotted for different PI values. The PI values used for the clay soils is shown in Table 3. Sandy soils (not shown) have a very similar relationship between material damping ratio and shear strain, but their PI should be 0. Soil material damping also depends slightly on the OCR, but in the context of offshore wind turbines, OCR should be 1, and this value was used for all soil types considered in this study.

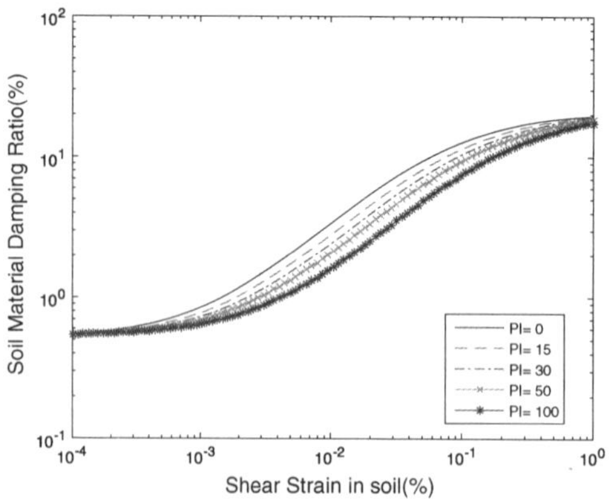

FIGURE 7 Relationship between material damping ratio and shear strain in clay for different PI values [Colour figure can be viewed at wileyonlinelibrary.com]

TABLE 3 Soil profile definition

Profile 1	Profile 2	Profile 3	Profile 4	Profile 5	Profile 6
Medium clay	Stiff clay	Medium sand	Dense sand	28-34 m soft clay 14-28 m medium clay 0-14 m stiff clay	28-34 m loose sand 14-28 m medium sand 0-14 m dense sand
PI = 20	PI = 20	PI = 0	PI = 0	PI = 20	PI = 0
OCR = 1	OCR = 1	OCR = 1	OCR = 1	OCR = 1	OCR = 1

Following Gazetas and Dobry,[28] the shear strain γ_e can be determined by

$$\gamma_e = \frac{1+v}{2.5D} y_{Amp}(z,t), \tag{31}$$

where $y_{Amp}(z)$ is the oscillating amplitude of the pile in the horizontal direction, v is Poisson's ratio for the soil stratum, and D is the diameter of the pile. k is determined from the nonlinear p-y curves:

$$k = \frac{p(z,t)}{y(z,t)}. \tag{32}$$

The soil around a wind turbine experiences some cyclic strain on top of a constant strain caused by the mean wind thrust. Whether the total amplitude or simply the dynamic component should be used for $y_{Amp}(z)$ is not obvious. In this study, the amplitude y_{Amp} used to calculate the strain (and subsequently the damping ratio) was half the range of oscillation. An alternative method commonly used in earthquake engineering[30] is to evaluate the damping constant using 65% of the maximum strain reached in the soil strata during a response time history. Earthquake ground motion does not usually have a static component, so it is still not clear how to implement this rule here, but this was also tested and found to give very similar results to those obtained using Equation 31.

6.1.2 | Soil properties and *p*-*y* curves

Here, 6 soil profiles are selected to investigate the soil damping contribution. Profiles 1 to 4 are single material soils made of medium clay, stiff clay, medium sand, and dense sand respectively. The low stiffness of soft clay and loose sand would significantly reduce the stiffness of the entire system and lead to unrealistic designs, so they were excluded as monolayered soils. Profiles 5 and 6 are layered soil profiles combining a mix of clay and sandy soils respectively. The layered clay profile is a combination of stiff clay, medium clay, and soft clay from bottom to pile head with the height of 14, 14, and 6 m respectively. The layered sandy profile is a combination of loose sand, medium sand, and dense sand from bottom to pile head with the height of 14, 14, and 6 m respectively which is a typical layered profile based on the data provided in Appendix 2 in Subramanian.[51] The different soil profiles used are summarized in Table 3, while the properties of the different soil types are listed in Table 4.

The SSI effect was modelled using static or cyclic p-y curves recommended in DNV 2014[52] and API standards,[53] and the steps to establish the p-y curves are detailed in Reese and Van Impe.[54] Although it would seem more sensible to use the cyclic curves to represent the SSI of an OWT under long-term excitations, both static and cyclic p-y curves were used to allow comparison with published results. Figure 8 shows the static and cyclic p-y curves for each soil types at the depth $z = -8.5$m (from mudline). These figures illustrate that, whether static or cyclic p-y curves are used, the forces from the soil plateau vary quickly in sandy soils as the pile moves sideway. By contrast, the behaviour of clay soils is more complex and variable as the lateral displacement of the pile increases. The cyclic p-y curves are recognizably similar to the static ones except that much larger forces can be reached in the cyclic case.

In this section, the finite element model of the OWT included soil springs and dashpots whose properties were obtained using Equations 30 and 32 respectively. As the response of the soil is strongly strain-dependent, the model was subjected to forces of increasing intensity. This was implemented using thrust time series obtained from FAST with mean speed increasing from 5 to 25 m/second by 1 m/second increments. To avoid the stochastic component of the wind adding noise to the response of the system and interfering with the damping calculation, these time series were averaged and the average was used as a static force applied at the tower top in the finite element model. Then, the soil damping was calculated from the vibration decay ensuing an applied initial condition of a magnitude consistent with the response of the system to the given force. For these simulations, the control system was kept on in FAST. This enables blade pitching and causes a decrease in aerodynamic force in the FA direction for wind speeds above the rated speed (12 m/second in this case).

TABLE 4 Mechanical properties of the soil profiles used in the model

	Soil Type	Density γ (1×10^3 kg/m^3)	Undrained Shear Strength c_u (kPa)	Poisson's Ratio v	Friction Angle φ	Shear Modulus G_s (MPa)	Shear Wave Velocity V_s (m/second)	Elastic Modulus E_s (MPa)
Clay	Soft clay	1.75	30	0.45	-	43	158	15
	Medium clay	1.90	80	0.45	-	87	214	30
	Stiff clay	2.00	125	0.45	-	145	269	50
Sand	Loose sand	1.75	-	0.30	33°	47	164	18
	Medium sand	1.90	-	0.30	36°	109	240	42
	Dense sand	2.07	-	0.30	38°	182	297	70

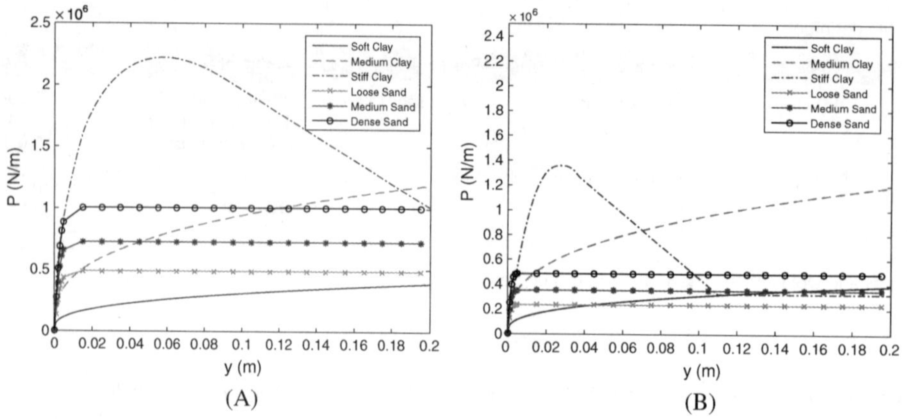

FIGURE 8 Static (A) and cyclic (B) *p-y* curves for different soil types [Colour figure can be viewed at wileyonlinelibrary.com]

6.2 | Soil damping results and discussion

Figure 9 shows the variation of the damping ratios in the various soils for mean wind speeds ranging from 5 to 25 m/second.

The plots in Figure 9 all show a similar behaviour. Soil damping ratios increase from 0.2% to 0.3% at low wind speeds up to around 1.3% at the rated wind speed 12 m/second. Thereafter, it decreases back to 0.3% to 0.6%. This range is in agreement with the published results reported in Section 2.5, and our results confirm that soil damping is quite variable and can become quite significant for some tower amplitudes.

To understand these results better, Figure 10A shows the average thrust for varying wind speeds, and Figure 10B shows the modal stiffnesses for the clay soils obtained with static *p-y* curves. The behaviour for sands is very similar, so the corresponding graph is not shown.

Figure 10A shows that up to the rated wind speed, the thrust increases. Then, the control system starts feathering the blades so that the overall thrust decreases despite increasing wind speeds. It can be seen that the shape of the damping curves in Figure 9 follows closely that of the thrust.

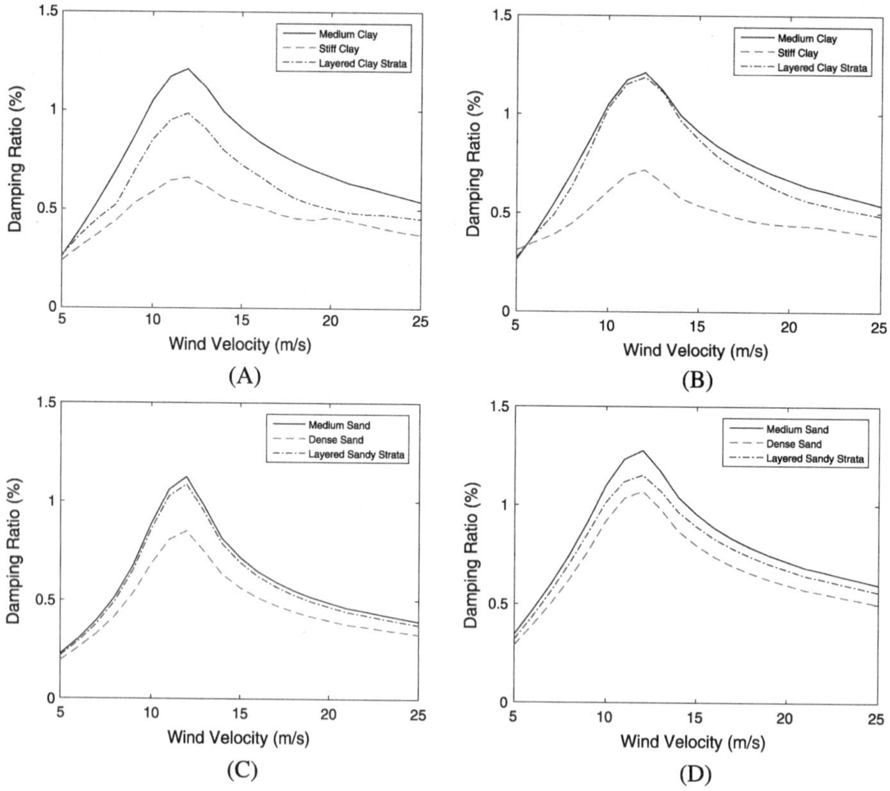

FIGURE 9 Soil damping in wind speed for various soil profiles and *p-y* curves. A, The clay profiles using static *p-y* curves. B, The clay profiles using cyclic *p-y* curves. C, The sand profiles using static *p-y* curves. D, The sand profiles using cyclic *p-y* curves [Colour figure can be viewed at wileyonlinelibrary.com]

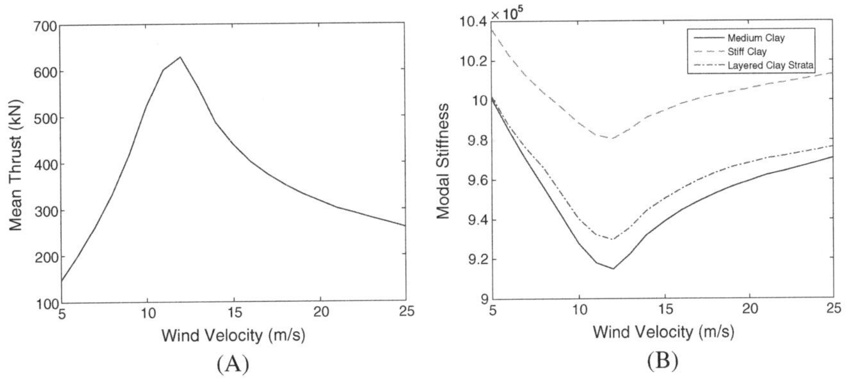

FIGURE 10 A, Wind speed thrust at tower top relationship. B, Modal stiffness (no physical unit) in wind speed using static *p-y* curve soil model [Colour figure can be viewed at wileyonlinelibrary.com]

This phenomenon is a consequence of two concomitant mechanisms: (1) the nonlinear softening of the soil as described by *p-y* curves (Figure 8) and (2) the increase in soil material damping with increasing strain described by Figure 7. When the deflection caused by the wind speed increases beyond some critical values (the abrupt change in slope in the *p-y* curves), the soil stiffness decreases sharply, which reduces the stiffness of the whole system. This stiffness reduction leads to an increase in the soil damping contribution. The drop in stiffness in Figure 10B occurs when the wind speed is around the rated wind speed. This is when the largest rotor thrust and thus the largest displacement amplitude are experienced. According to Equation 30, the increased displacement magnitude also results in increased soil material damping ratio. As the constant of each soil dashpot increases, the overall soil damping is further increased.

It should be noted that the above observations are based on the assumption that the soil behaviour is well described by *p-y* curves. Some soils actually stiffen because of cyclic loading (ratchetting behaviour), and in those cases, the dependence of soil damping on the wind speed may be very different. This should be the subject of further study.

7 | CONCLUSION

Vibration damping in offshore wind turbines is a critical parameter to predict reliably the dynamic response and fatigue life of the system. Although much research work has been done to characterize these sources of dissipation in OWTs, published values vary widely and a lot of uncertainty remain as to what should be used at the design stage, when actual measurements are not available. Following a detailed literature survey, this paper proposed a series of simple physics-based models that allowed the quantification of aerodynamic, hydrodynamic, and soil damping separately for monopile-supported OWTs. The dissipation models were implemented around a bespoke finite element beam model of the system based on the 5 MW NREL offshore wind turbine. The damping results in this study are for the first fore-aft or side-side mode, which dominates the OWT dynamic response. These damping values can be easily implemented in a FE model at the design stage either specifying the damping ratio for the first mode or by inputting Rayleigh damping coefficients so that the first mode damping reaches the desired level.

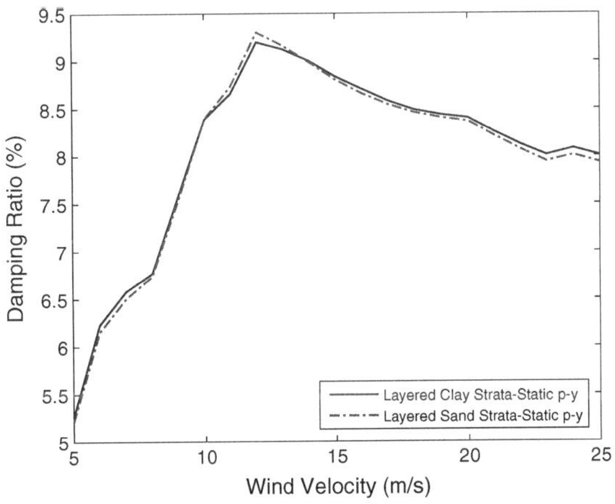

FIGURE 11 Overall FA damping ratio in a 5 MW offshore wind turbine in wind speed for 2 representative soil profiles [Colour figure can be viewed at wileyonlinelibrary.com]

Results showed that aerodynamic damping is in the range of 5% to 7% in the FA direction and 0.2% to 1.6% in the SS direction depending on the wind speed. Preliminary findings also pointed to the importance of the tilt angle and coupling between FA and SS directions although more work is needed to characterize this better.

The viscous and radiation components of hydrodynamics damping were modelled from first principle. Results showed that for the current size of offshore wind turbines, hydrodynamic damping is much lower than what is usually quoted in the literature and for all practical purposes could be safely ignored.

Soil damping was modelled in the form of dissipative springs distributed along the embedded length of the pile. The properties of these springs are based on relevant *p-y* curves, which are nonlinear, and as such, their overall dissipative effect depends on the amplitude of vibration. Soil damping was calculated for different types of soils and incoming wind speeds. Results showed that ζ_{soil} ranges from 0.2% to 1.3% with a strong dependence on the stiffness of the system. As the thrust increases, the springs weaken, so the amplitude increases relatively more. This increases the material damping and decreases the overall stiffness of the stiffness, thereby increasing the damping ratio. This effect has not previously been studied systematically.

Figure 11 combines the results from the various models by adding up the contribution from various damping sources in the fore-aft direction. ζ_{struct} = 1% was used. The similarity between the fore-aft damping curves in Figures 4 and 11 shows that the overall damping is completely dominated by the aerodynamic component. It increases sharply from 6% to 9% up to the rated wind speed and then decreases slowly down to 8%.

Outside the operative range of the turbine, the aerodynamic damping drops dramatically so that the overall damping is dominated by the structural and soil components. Therefore, a reasonable range for the total damping in a nonoperating turbine could be between 1% and 1.5%.

If the purpose of calculating the dynamic response of the system to design a currently nonexisting turbine, it is important to get the damping values in the correct range always bearing in mind that lower damping values will lead to safe but potentially overconservative designs. By testing published values against simple but robust models, this paper can hopefully contribute to more optimal OWT designs.

From this study, two issues emerge as particularly deserving of further investigation: (1) there is still no data available for the structural damping in large OWTs even though this contribution is probably not insignificant and (2) given the importance of the aerodynamic damping component, there is comparatively little reliable experimental data available to test numerical predictions. In fact, what seems to be needed first is a robust experimental methodology to measure and characterize this damping source properly.

ORCID

Chao Chen http://orcid.org/0000-0002-0441-8709
Philippe Duffour http://orcid.org/0000-0003-3551-1217

REFERENCES

1. European Wind Energy Association, "*The European offshore wind industry—key trends and statistics 2013*," 2014.
2. Fontana C, Carswell W, Arwade SR, DeGroot DJ, Myers AT. Sensitivity of the dynamic response of monopile-supported offshore wind turbines to structural and foundation damping. *Wind Eng*. 2015;39(6):609-628.
3. J. M. Jonkman and M. L. J. Buhl, "FAST user's guide—updated august 2005," Technical Report No.NREL/TP-500-38230. *National Renewable Energy Laboratory (NREL)*, Golden, CO, Oct. 2005.
4. Wind Europe, "*The European offshore wind industry—key trends and statistics 2016*," 2017.
5. Windenergie GL. Overall damping for piled offshore support structures. In: *Guideline for the Certification of Offshore Wind Turbines*; 2005.
6. Versteijlen WG, Metrikine AV, Hoving JS, Smid EH, De Vries WE. Estimation of the vibration decrement of an offshore wind turbine support structure caused by its interaction with soil. In: *Proceedings of the EWEA Offshore 2011 Conference*; 2011.
7. Damgaard M, Andersen JKF. Natural frequency and damping estimation of an offshore wind turbine structure. In: *The Twenty-second International Offshore and Polar Engineering Conference*; 2012.
8. Damgaard M, Andersen JKF, Ibsen LB, Andersen LV. Time-varying dynamic properties of offshore wind turbines evaluated by modal testing. In: *Proceedings of the 18th International Conference on Soil Mechanics and Geotechnical Engineering*; 2013:2343-2346.
9. Damgaard M, Ibsen LB, Andersen LV, Andersen JKF. Cross-wind modal properties of offshore wind turbines identified by full scale testing. *J Wind Eng Ind Aerodyn*. 2013;116:94-108.
10. Tarp-Johansen NJ, Andersen L, Christensen ED, Mørch C, Kallesøe B, Frandsen S. Comparing sources of damping of cross-wind motion. In: *The European Wind Energy Conference*. The European Wind Energy Association; 2009.
11. Shirzadeh R, Devriendt C, Bidakhvidi MA, Guillaume P. Experimental and computational damping estimation of an offshore wind turbine on a monopile foundation. *J Wind Eng Ind Aerodyn*. 2013;120, no SEPTEMBER:96-106.
12. Devriendt C, El-Kafafy M, De Sitter G, Guillaume P. Damping estimation of offshore wind turbines on a monopile foundation using state-of-the-art operational modal analysis techniques. *Proc. Int. Conf. Noise Vib. Eng. ISMA*. 2012;2012:2647-2662.
13. Devriendt C, Jordaens PJ, De Sitter G, Guillaume P. Damping estimation of an offshore wind turbine on a monopile foundation. *Proc Int Conf Noise Vib Eng ISMA*. 2012;2012:4411-4419.
14. Shirzadeh R, Weijtjens W, Guillaume P, Devriendt C. The dynamics of an offshore wind turbine in parked conditions: A comparison between simulations and measurements. *Wind Energy*. 2014;18(10):1685-1702.

15. Devriendt C, Jordaens PJ, Van Ingelgem Y, De Sitter G, Guillaume P. Monitoring of resonant frequencies and damping values of an offshore wind turbine. *IET Renew. Power Gener.* 2014;8(4):433-441.

16. Koukoura C, Natarajan A, Vesth A. Identification of support structure damping of a full scale offshore wind turbine in normal operation. *Renew Energy.* 2015;81:882-895.

17. M. H. Hansen, K. Thomsen, P. Fuglsang, and T. Knudsen, "two methods for estimating aeroelastic damping of operational wind turbine modes from experiments," no. January, pp. 179–191, 2006.

18. Hansen MOL. *Aerodynamics of wind turbines.* 2nd ed. Earthscan; 2008.

19. Valamanesh V, Myers AT. Aerodynamic damping and seismic response of horizontal axis wind turbine towers. *J Struct Eng.* 2014;140(11):1-9.

20. Ozbek M, Meng F, Rixen DJ. Challenges in testing and monitoring the in-operation vibration characteristics of wind turbines. *Mech Syst Signal Process.* 2013;41(1–2):649-666.

21. Salzmann DJC, Van Der Tempel J. Aerodynamic damping in the design of support structures for offshore wind turbines. *Proc Copenhagen Offshore Conf.* 2005;9.

22. Schafhirt S, Muskulus M. Decoupled simulations of offshore wind turbines with reduced rotor loads and aerodynamic damping. *Wind Energy Sci.* 2018;3(1):25-41.

23. Ozbek M, Rixen DJ. Operational modal analysis of a 2.5 MW wind turbine using optical measurement techniques and strain gauges. *Wind Energy.* 2013;17:367-381.

24. Faltinsen O. *Sea loads on ships and offshore structures.* Cambridge University Press; 1993.

25. Cook MF, Vandiver JK. Measured and predicted dynamic response of a single pile platform to random wave excitation. In: *Proceedings of 14th Annual Offshore Technology Conference in Texas*; 1982.

26. Carswell W, Johansson J, Løvholt F, et al. Foundation damping and the dynamics of offshore wind turbine monopiles. *Renew Energy.* 2015;80:724-736.

27. Novak M. Dynamic stiffness and damping of piles. *Canadian Geotechnical Journal.* 1974;11(4):574-598.

28. Gazetas G, Dobry R. Horizontal response of piles in layered soils. *J Geotech Eng.* 1984;110(1):20-40.

29. El Naggar MH, Bentley KJ. Dynamic analysis for laterally loaded piles and dynamic p-y curves. *Can Geotech J.* 2000;37(6):1166-1183.

30. H. B. Seed and Idriss I. M., "Soil moduli and damping factors for dynamic response analyses," Report NO. EERC 70–10. *California Univ Berkely Earthquake Engineering Research Centre,* 1970.

31. Darendeli MB. "Development of a new family of normalized modulus reduction and material damping curves," PhD Dissertation. In: *The University of Texas at Austin*; 2001.

32. EN 1991-1-4. Eurocode 1: Actions on structures—part 1–4: General actions—wind actions. *Eur Comm Stand.* 2005;4(2005):1-148.

33. Barltrop NDP, Adams AJ. *Dynamics of fixed marine structures.* Butterworth-Heinemann; 1991.

34. Schaumann P, Seidel M. Eigenschwingverhalten von windenergieanlagen-berechnungen und messungen. In: *Tagungsband DEWEK*; 2000:35-38.

35. Rezaei R, Fromme P, Duffour P. Fatigue life sensitivity of monopile-supported offshore wind turbines to damping. *Renew Energy.* 2018;123: 450-459.

36. J. M. Jonkman, S. Butterfield, W. Musial, and G. Scott, "Definition of a 5-MW reference wind turbine for offshore system development," Technical Report NREL/TP-500-38060 February 2009, *National Renewable Energy Laboratory (NREL),* Golden, CO., 2009.

37. H. P. Gavin, "Numerical integration in structural dynamics," *Department of Civil & Environmental Engineering Duke University.* Lecture Notes, *Department of Civil & Environmental Engineering Duke University.* 2016.

38. Hodges CH, Power J, Woodhouse J. The use of the sonogram in structural acoustics and an application to the vibrations of cylindrical shells. *J Sound Vib.* 1985;101(2):203-218.

39. Hansen MOL, Sørensen JN, Voutsinas S, Madsen HA. State of the art in wind turbine aerodynamics and aeroelasticity. *Prog Aerosp Sci.* 2006;42(4):285-330.

40. Sarpkaya T. *Wave forces on offshore structures.* Cambridge University Press; 2010.

41. Johanning L, Bearman PW, Graham JMR. Hydrodynamic damping of a large scale piercing circular in planar oscillatory motion. *J. Fluids Struct.* 2001;16(15):891-908.

42. B. Petersen, M. Pollack, B. Connell, D. Greeley, D. Davis, C. Slavik, and B. Goldman, "*Evaluate the effect of turbine period of vibration requirements on structural design parameters: Technical report of findings,*" 2010.

43. R. Salmon, "Introduction to ocean dynamics," Lecture Notes, University of California, San Diego, 2000.

44. J. Van Der Tempel, "Design of support structures for offshore wind turbines," PhD Dissertation, *Delft University of Technology,* 2006.

45. Kuiper GL, Metrikine AV, Battjes JA. A new time-domain drag description and its influence on the dynamic behaviour of a cantilever pipe conveying fluid. *J Fluids Struct.* 2007;23(3):429-445.

46. Chaplin JR. Hydrodynamic damping of a cylinder at $\beta \approx 10^6$. *J. Fluids Struct.* 2000;0:1101-1117.

47. M. F. Cook, "Damping estimation, response prediction and fatigue calculation of an operational single pile platform," Master Dissertation, *Massachusetts Institute of Technology,* 1978.

48. C. Petrauskas, "Hydrodynamic damping and "added mass" for flexible offshore platforms." Technical Paper No. 76–18, *California Univ Berkely Hydraulic Engineering Lab,* 1976.

49. Jamieson P. *Innovation in wind turbine design.* John Wiley & Sons; 2011.

50. Van Buren E, Muskulus M. Improving pile foundation models for use in bottom-fixed offshore wind turbine applications. *Energy Procedia.* 2012;24(1876):363-370.

51. Subramanian N. *Steel structures—design and practice.* Oxford University Press; 2011.

52. DNV, Design of offshore wind turbine structures, DNV-OS-J101, 2014.

53. API, *Recommended Practice for Planning*, Designing and constructing fixed offshore platforms—working stress design, API Recommended Practice 2A-WSD, 2000.

54. Reese L, Van Impe W, Holtz R. *Single piles and pile groups under lateral loading*. CRC Press; 2010.

How to cite this article: Chen C, Duffour P. Modelling damping sources in monopile-supported offshore wind turbines. *Wind Energy*. 2018;21:1121–1140. https://doi.org/10.1002/we.2218

APPENDIX 1 DERIVATION FOR SIDE-SIDE AERODYNAMIC DAMPING

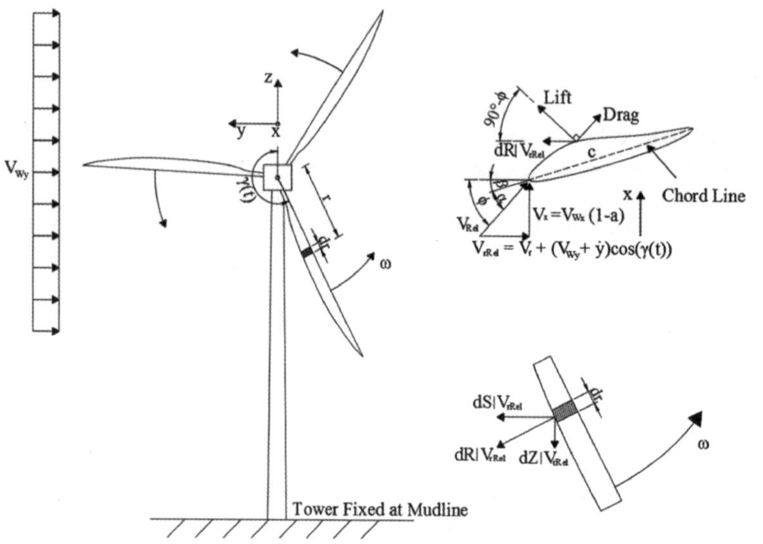

FIGURE A1 Blade element forces for side-side vibration

For the side-side direction, a similar approach can be followed. Calling V_r the speed component in the tangential direction, BEM theory gives $V_r = \omega r(1 + a') + V_{Wy} \cos(\gamma_i(t))$ [18] where V_{Wy} is the inflow wind speed in side-side direction and $\gamma_i(t)$ is the azimuth angle of the blade which is a function of time t when the turbine is in operation. The force in the y-direction felt by an element at distance r along the ith blade is

$$dS|_{V_r} = \frac{1}{2}\rho_a \frac{v_r^2}{\cos^2\phi} cC_t \cos(\gamma_i(t))dr, \tag{A1}$$

where $C_t = C_l \sin\phi - C_d \cos\phi$ and $\gamma_i(t)$ is the azimuth angle of the blade which is a function of time t when the turbine is in operation. It is noted that here, the elemental force in y direction is assumed to be a function of V_r and ϕ. If there is a small change in V_r because of the overall tower top motion, the relative speed of the element in tangential direction V_{rRel} becomes:

$$V_{rRel} = V_r + (V_{W_y} + \dot{y})\cos(\gamma_i(t)). \tag{A2}$$

In the following derivation, we can always assume that $V_{Wy} = 0$ because the turbine can be controlled and oriented so as to face the wind. Again, a first-order Taylor's expansion of the side-side force gives the expression of $dS|_{V_{rRel}}$ which is the tangential force felt by the blade element under a relative speed of V_{rRel}:

$$dS|_{V_{rRel}} = dS|_{V_r} + dS'\Big|_{V_r} \cdot (V_{rRel} - V_r) = dS|_{V_r} + \frac{d(dS)}{dV_r}\dot{y}\cos(\gamma_i(t)). \tag{A3}$$

Assuming that dS is a function of V_r and ϕ, and ϕ is a function of V_r, $\frac{d(dS)}{dV_r}$ can be calculated as:

$$\frac{d(dS)}{dV_r} = \frac{\partial(dS)}{\partial V_r} + \frac{\partial(dS)}{\partial\phi}\frac{d\phi}{dV_r}. \tag{A4}$$

$\frac{\partial(dS)}{\partial V_r}$ and $\frac{\partial(dS)}{\partial\phi}$ can be obtained by differentiating Equation A 1:

$$\frac{\partial(dS)}{\partial V_r}=\rho_a\frac{V_r}{\cos^2\phi}cC_t\cos(\gamma_i(t))dr, \tag{A5}$$

$$\frac{\partial(dS)}{\partial\phi}=\frac{1}{2}\rho_aV_r^2\frac{\frac{dC_t}{d\phi}\cos^2\phi+2\cos\phi\sin\phi C_t}{\cos^4\phi}c\cos(\gamma_i(t))dr. \tag{A6}$$

To obtain an expression for $\frac{d\phi}{V_r}$, the following 2 equations can be used:

$$\cot\phi=\frac{V_r}{V_{Wx}(1-a)}, \tag{A7}$$

$$a=\frac{1}{\frac{4\sin^2\phi}{\sigma C_n}+1}. \tag{A8}$$

Differentiating Equation A 7 with respect to V_r and Equation A 8 with respect to ϕ, resulting in

$$\frac{d\phi}{dV_r}=\frac{1}{V_{Wx}\left(\frac{1-a}{\sin^2\phi}-\frac{da}{d\phi}\cot\phi\right)}, \tag{A9}$$

and

$$\frac{da}{d\phi}=\frac{-4\left(\sin2\phi C_n-\frac{dC_n}{d\phi}\sin^2\phi\right)}{\sigma C_n^2\left(\frac{4\sin^2\phi}{\sigma C_n}+1\right)^2}. \tag{A10}$$

Equations A 5, A 6, and A 9 can be substituted into Equation A 4, and the resultant $\frac{d(dS)}{dV_r}$ can be used to obtain $dS|_{V_{rRel}}$ in Equation A 3 which is the side-side aerodynamic force subjected by the rotor element. For simplicity, let

$$A'\cos(\gamma_i(t))\cdot dr=\frac{\partial(dS)}{\partial V_r}, \tag{A11}$$

$$B'\cos(\gamma_i(t))\cdot dr=\frac{\partial(dS)}{\partial\phi}\frac{d\phi}{dV_r}. \tag{A12}$$

So, the side-side aerodynamic force for 1 element is

$$dS|_{V_{rRel}}=dS|_{V_r}+\left(A'+B'\right)dr\cdot\dot{y}\cos^2(\gamma_i(t)). \tag{A13}$$

The side-side aerodynamic force felt by the ith blade should be

$$F|_{\gamma_i}=\int_0^R dS|_{V_{rRel}}=\int_0^R dS|_{V_r}+\int_0^R\left(A'+B'\right)dr\cdot\dot{y}\cos^2(\gamma_i(t)). \tag{A14}$$

Following a similar procedure to that outlined in the fore-aft case, the total side-side aerodynamic force felt by the rotor can be expressed as:

$$F_y=\sum_{i=1}^{N_b}\int_0^R dS|_{V_r}+\sum_{i=1}^{N_b}\int_0^R\left(A'+B'\right)dr\cdot\dot{y}\cos^2(\gamma_i(t)). \tag{A15}$$

For a symmetric 3-blade rotor on steady condition without vibration and inflow wind in side-side direction, the total force on side-side direction, $\sum_{i=1}^{N_b}\int_0^R dS|_{V_{rRel}}$ is zero, so F_y can be simplified to

$$F_y=\frac{N_b}{2}\int_0^R\left(A'+B'\right)dr\cdot\dot{y}, \tag{A16}$$

because $\sum_{i=1}^{N_b}\int_0^R dS|_{V_r}=0$ and $\sum_{i=1}^{N_b}cos^2(\gamma_i(t))=\frac{N_b}{2}$. It should be noted that the term $A'+B'$ is negative.

APPENDIX 2 DERIVATION FOR RADIATION DAMPING COEFFICIENT FOR WAVE-CYLINDER INTERACTION

The radiation force applied on the cylinder highly depends on its oscillating frequency σ, and the well-known linear dispersion relation is important for the radiation problem. The linear dispersion relation can be written as

$$\sigma^2 = gk \cdot \tanh(kh) = -gk_n \cdot \tan(k_n h), \quad n = 1, 2, 3, \ldots$$

where g is the gravitational acceleration, k is the wavenumber, and h is the depth of the water. Assuming the distributed hydrodynamic force over the height of tower under water is $f_{Rad}(z, t)$, $\bar{f}(t)$ in Equation 5 excluding other damping terms can be written as

$$\bar{f}(t) = \int_{-h}^{0} f_{Rad}(z, t)\Phi(z)dz. \quad \text{(B1)}$$

According to Charles Petrauskas[48] who solves the PDEs for velocity potentials, $f_{Rad}(z, t)$ can be expressed by

$$f_{Rad}(z, t) = -\rho_w \pi a \sigma h G_0(kh) P_1(ka) \cosh(k(z+h)) \cdot \dot{\alpha}(t), \quad \text{(B2)}$$

where

$$G_0(kh) = \frac{2\int_{-h}^{0} \cosh(k(z+h))\Phi(z)dz}{h(\sinh(kh)\cosh(kh) + kh)}, \quad P_1(ka) = \frac{2}{\pi ka\left(J_1'(ka)^2 + Y_1'(ka)^2\right)}. \quad \text{(B3)}$$

$J_1'(ka)$ and $Y_1'(ka)$ are the derivatives of the Bessel functions of the first kind and second kind respectively. Denoting

$$C_{Rad} = \rho_w \pi a \sigma h G_0(kh) P_1(ka) \cosh(k(z+h))\Phi(z), \quad \text{(B4)}$$

$\bar{f}(t)$ can be written as

$$\bar{f}(t) = -\int_{-h}^{0} C_{Rad} dz \cdot \dot{\alpha}(t). \quad \text{(B5)}$$

Eddy-Current Tuned Mass Dampers for Mitigation of Wind-Induced Response of the Noor III Solar Tower: Design, Installation, and Validation

Min Liu[1]; Shouying Li[2]; Teng Wu, M.ASCE[3]; Yafeng Li[4]; Huachang Meng[5]; and Zhengqing Chen, M.ASCE[6]

Abstract: The Noor III solar tower of the Ouarzazate Solar Power Station in Morocco is a very slender structure sensitive to wind excitations. Hence, tuned mass dampers (TMDs) were employed to reduce its wind-induced response. Finite-element modeling and wind-tunnel test results were employed to design the parameters of the TMD system. Four 10-t TMDs with eddy-current damping (EC-TMD) were selected for implementation (via a customized steel frame) at the top of the solar tower (i.e., inside the receiver). The effectiveness of the designed TMDs in controlling the tower was numerically studied under free vibration and forced vibration. Then, considering design and modeling errors and the possible need for on-site adjustments of EC-TMD parameters, pendulum-type EC-TMDs were fabricated. Dynamic properties of the tower were identified based on field measurements. With the installation and adjustment of the four designed EC-TMDs, the damping ratio of the Noor III solar tower was enhanced from 0.72% to 3.78%. Field measurements of tower vibrations under strong winds with/without participation of the EC-TMDs showed that significant mitigation of wind-induced response was effectively achieved by the TMDs with eddy-current damping. In addition, the on-site acceleration measured on the prototype tower was compared with the acceleration obtained for the scale model in a wind tunnel to highlight the significant Reynolds number effects on the prediction of the wind-induced response of the solar tower structure. **DOI: 10.1061/(ASCE)ST.1943-541X.0003180.**

Introduction

Solar energy has been widely used in recent years all over the world due to its environmentally friendly nature. An efficient way to utilize solar energy is based on concentrated solar power (CSP) (González-Roubaud et al. 2017; Ozoegwu et al. 2017; Tahri et al. 2015). There are generally four CSP technologies, including solar power towers, parabolic troughs, linear Fresnel systems, and parabolic dish systems (González-Roubaud et al. 2017). For tower systems, a solar tower is centered in a field of mirrors that reflect the sun's rays toward a receiver at the top. A solar tower usually has a large aspect ratio and, hence, is extremely sensitive to wind excitations. Li et al. (2018) conducted a series of wind tunnel tests based on the Noor III solar tower of the Ouarzazate Solar Power Station in Morocco. They observed significant vortex-induced vibration (VIV) responses under winds far lower than the design wind speed. It was found that large wind-induced vibrations could be effectively reduced by increasing the damping ratio to over 1.0%.

Tuned mass dampers (TMDs) are one of the most frequently used countermeasures for increasing effective damping to mitigate wind-induced vibrations of civil structures. A number of successful applications on high-rise buildings have demonstrated the effectiveness of TMDs in reducing wind-induced vibrations (Vickery and Davenport 1971; Kwok 1983; Sacks and Swallow 1993; Tuan and Shang 2014; Xu et al. 2014). A TMD system is an energy absorber in nature and usually consists of a mass element, a stiffness element, and a damping element. The most widely-used damping element in TMD systems is a fluid viscous damper (Mcnamara and Taylor 2003; Narkhede and Sinha 2014). However, this type of damper has several shortcomings, such as oil leakage and limited service life. Moreover, the activation of fluid viscous dampers may be problematic with small-amplitude vibrations due to their large initial friction force.

Eddy-current dampers have been utilized as an effective alternative to overcome the aforementioned limitations of fluid viscous dampers. Eddy-current damping (ECD) is generated by moving a conductive plate in a magnetic field, and it has been actually adopted to reduce machine vibrations for several decades (Sakamoto et al. 1997; Ebrahimi et al. 2009). In the late 1980s, ECD was used on the cryogenic pumps and turbomachinery of space shuttles to suppress rotor vibrations (Gunter et al. 1983; Cunningham 1986). Simeu and Georges (1996) designed an eddy-current disc brake to control the speed of a rotating system in the presence of a randomly varied braking resistance torque. Several researchers found that ECD is also suitable for accurately providing damping in the scale-model tests (Larose et al. 1995; Kwak et al. 2003; Bae et al. 2005; Lou et al. 2019; Liu et al. 2020). Until recently, however,

[1]Ph.D. Student, Key Laboratory for Wind and Bridge Engineering of Hunan Province, College of Civil Engineering, Hunan Univ., Changsha 410082, China. Email: 464933874@qq.com

[2]Professor, Key Laboratory for Wind and Bridge Engineering of Hunan Province, College of Civil Engineering, Hunan Univ., Changsha 410082, China (corresponding author). ORCID: https://orcid.org/0000-0003-4516-4378. Email: shyli@hnu.edu.cn

[3]Associate Professor, Dept. of Civil, Structural and Environmental Engineering, Univ. at Buffalo, Buffalo, NY 14260. Email: tengwu@buffalo.edu

[4]Ph.D. Student, Key Laboratory for Wind and Bridge Engineering of Hunan Province, College of Civil Engineering, Hunan Univ., Changsha 410082, China. Email: 394296559@qq.com

[5]Engineer, Liuzhou Orient Engineering Rubber Products Co., Ltd., Putaoshan St., Yufeng District, Liuzhou 545000, China. Email: 314530293@qq.com

[6]Professor, Key Laboratory for Wind and Bridge Engineering of Hunan Province, College of Civil Engineering, Hunan Univ., Changsha 410082, China. Email: zqchen@hnu.edu.cn

Note. This manuscript was submitted on January 4, 2021; approved on July 20, 2021; published online on September 27, 2021. Discussion period open until February 27, 2022; separate discussions must be submitted for individual papers. This paper is part of the *Journal of Structural Engineering*, © ASCE, ISSN 0733-9445.

ECD technology has not been widely adopted for vibration control in prototype civil structures. The output damping force produced by a traditional eddy-current damper is not necessarily sufficient for vibration mitigation of civil structures, which usually possess heavy masses and relatively small vibration velocities (Palomera-Arias et al. 2008; Huang et al. 2018; Zhang et al. 2020). To this end, a new planar ECD element was developed by Wang et al. (2012). The damping force can be enhanced by about six times with this planar ECD element. This new ECD technology was recently used for the TMDs installed on the Mianyang First Bridge to mitigate human-induced vibrations (Wen et al. 2016) and on the Shanghai Tower to reduce the wind-induced response (Lu et al. 2017). Niu et al. (2018) investigated the effectiveness of eddy-current TMDs (EC-TMDs) on reducing the vibrations of bridge hangers. Another approach to enhancing the output damping force of an eddy-current damper by using a ball screw (Ohmata 1983; Ohmata et al. 1986) was applied for the mitigation of earthquake-induced bridge responses (Huang et al. 2015).

In this study, pendulum-type EC-TMDs based on the new planar eddy-current damper proposed by Z. Q. Chen ["Permanent magnet eddy current tuned mass damper," US Patent No. 101,033,628 (2011)] were designed and manufactured for the Noor III Solar Tower to reduce its wind-induced responses. First, numerical simulations were conducted to demonstrate that the developed EC-TMDs could significantly enhance the damping capacity of the eddy-current device. Then, the EC-TMDs were installed on the tower in November 2017, and their frequency was accurately adjusted to the target value after the completion of the tower in October 2018. The structural damping of the tower was identified with and without the EC-TMDs, and the effectiveness of the EC-TMDs was validated by a series of field measurements of the tower responses under a strong wind. Finally, the results of field measurements and wind tunnel tests were compared in order to investigate the effect of Reynolds number on wind-induced responses of the Noor III Solar Tower.

EC-TMD Design and Manufacture

Engineering Background

The Noor III solar tower (with a height of 243 m) is located 10 km northeast of Ouarzazate in the Kingdom of Morocco. The receiver at the top is a 43-m steel structure (with an outer diameter of 20.0 m) used to absorb solar energy reflected from thousands of surrounding mirrors. The 200-m reinforced concrete structure has a ring-shaped cross section, and its outer diameter varies from 23.0 m (bottom) to 20.0 m (top). Correspondingly, the thickness of the RC wall changes from 800 mm at the ground level to 450 mm at an elevation of 200 m.

Design Parameters

The new planar eddy-current damper proposed by Z. Q. Chen ["Permanent magnet eddy current tuned mass damper," US Patent No. 101,033,628 (2011)] was used to provide the damping force in the TMDs of the Noor III Solar Tower, as shown in Fig. 1.

The parameter central to TMD system design is the mass ratio. Once the mass ratio is determined, the optimal frequency and damping ratio can easily be determined from the classical theory (Den Hartog) that

$$eqf_{\mathrm{TMD}}/f_s = (1+\mu)^{-1} \tag{1}$$

$$\zeta_{\mathrm{TMD}} = [3\mu/8(1+\mu)]^{-0.5} \tag{2}$$

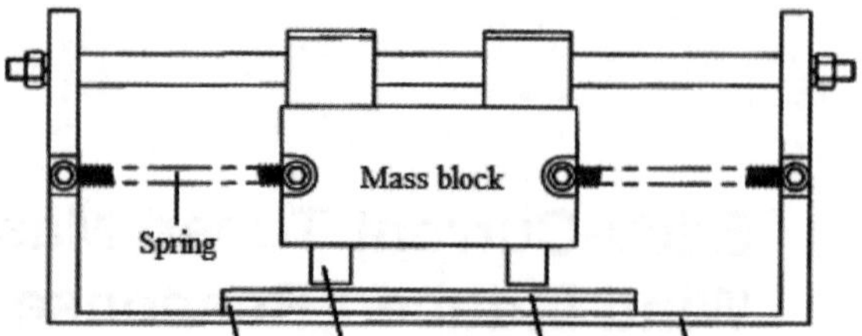

Fig. 1. New planar eddy-current damper.

where μ = mass ratio. In this study, a finite-element model of the Noor III solar tower was established based on ANSYS version 15.0 software. Modal analysis suggested that the fundamental frequency and generalized mass of the tower were 0.28 Hz and 4,270 t, respectively. Next, wind tunnel tests were carried out for the Noor III solar tower using an aeroelastic building model with a scaled-down TMD in the HD-2 wind tunnel at Hunan University in China (Li et al. 2018; Liu et al. 2019). The experimental results showed that the wind-induced responses of this tower result mainly from first-order mode contributions and that a 1% mass ratio (between TMD and tower) would be adequate to sufficiently mitigate the wind-induced vibration of the tower (corresponding to an optimal TMD damping ratio of 6.1%) (Liu et al. 2020). Accordingly, the mass of the prototype TMD was about 40 t (with a mass ratio of 1%), and the optimum frequency and damping ratio were 0.277 Hz and 6.1%, respectively. However, due to the existence of a salt container, there was not enough room to accommodate the 40-t TMD near the top of the Noor III solar tower, where the amplitude of wind-induced vibration is largest. Therefore, four separate TMDs, each with a mass of 10-t, were specially designed to be symmetrically arranged around the center line of the tower at a height of 237 m above the ground, as shown in Fig. 2. The configuration and design parameters for the four separate TMDs were identical (the optimum frequency and damping ratio were 0.277 Hz and 6.1%, respectively) for simplicity. The pendulum cable for each TMD had a length of about 3.23 m (with a frequency of 0.28 Hz), and considering design and modeling errors and possible need for on-site TMD parameter adjustment (especially for frequency), it was designed in such a way that the pendulum length could be conveniently modified. Specifically, the adjustable frequency range of each TMD was 0.25–0.35 Hz. The damping of each TMD was provided by an eddy-current damper, which consisted of six pairs of permanent magnets and a conductive plate. The conductive plates were made of copper and had a size of $1{,}300 \times 1{,}300 \times 10\ \mathrm{mm}^3$. The permanent magnets had dimensions of $80 \times 80 \times 25\ \mathrm{mm}^3$. By adjusting the gaps between the permanent magnets and the conductive plate, the damping ratio for each TMD could be adjusted in the range of 3%–10% (corresponding to gaps ranging from 33–10 mm).

Control Performance Analysis

The system was analyzed by modeling the dampers as a single 40-t TMD and then as four individual 10-t TMDs, with identical results that agreed well with the expected behavior of an oscillating system equipped with a TMD (Den Hartog 1956).

In engineering practice, it is very difficult to ensure that the dynamic parameters of four TMDs are identical with each other and agree well with the optimal design values because of errors caused by manufacture, transportation, installation, and so forth. Moreover, the frequency of a controlled structure may often change under different conditions due to structural complexity and nonlinearity. Therefore, the control performance of the four-TMD system with frequency and damping detuning was studied on the basis of a random simulation method. The maximum deviations for the stiffness and damping coefficients of the four TMDs were set to 5% and 25% of their optimal values, respectively. Fig. 3 presents the mean steady-state displacement response amplitude of the controlled tower under forced vibration with sinusoidal forces at different excitation frequencies (the amplitude of the sinusoidal force was set to 5,000 N, and 2,000 tests were performed for each case).

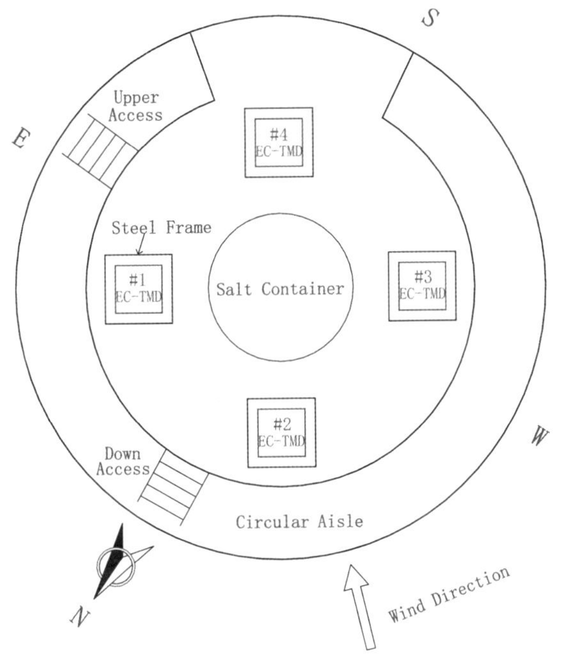

Fig. 2. Arrangements of four EC-TMDs at a height of 237 m.

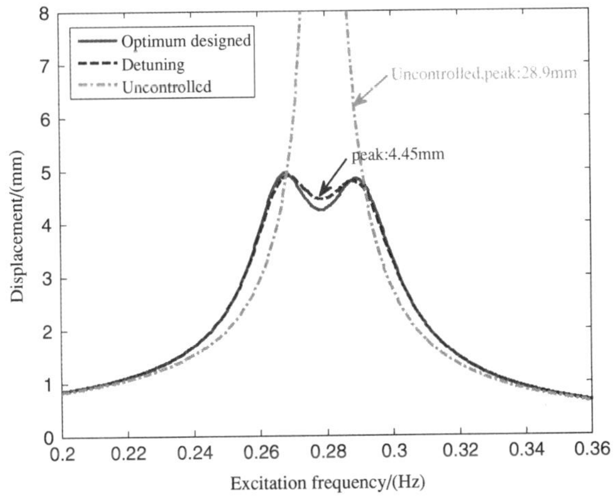

Fig. 3. Steady-state displacement response amplitudes of the solar tower under forced vibration with the four 10-t TMDs detuned.

The figure shows that although the control performance of the four 10-t TMDs degraded with frequency and damping detuning under/close to the resonant frequency, the displacement reduction was still more than 84.6% of the displacement amplitude in the uncontrolled case, which indicates that the designed TMD system would work well in suppressing vibration for the Noor III solar tower.

Manufacture of EC-TMDs

The designed EC-TMDs were manufactured by Liuzhou Orient Engineering Rubber Products (Liuzhou, China). Fabrication drawings of a 10-t TMD and the finished product are shown in Figs. 4(a and b), respectively. A customized steel frame was designed to facilitate the accommodation of the 10-t EC-TMDs in the tower. As shown in Fig. 4, it had four beams at various levels, and the permanent magnets were installed on Beam 3. The EC-TMD mass block was suspended by four cables, and the copper conductive plate was placed on top of the mass block. The upper end points of the four cables were fixed on Beam 1. The pendulum length

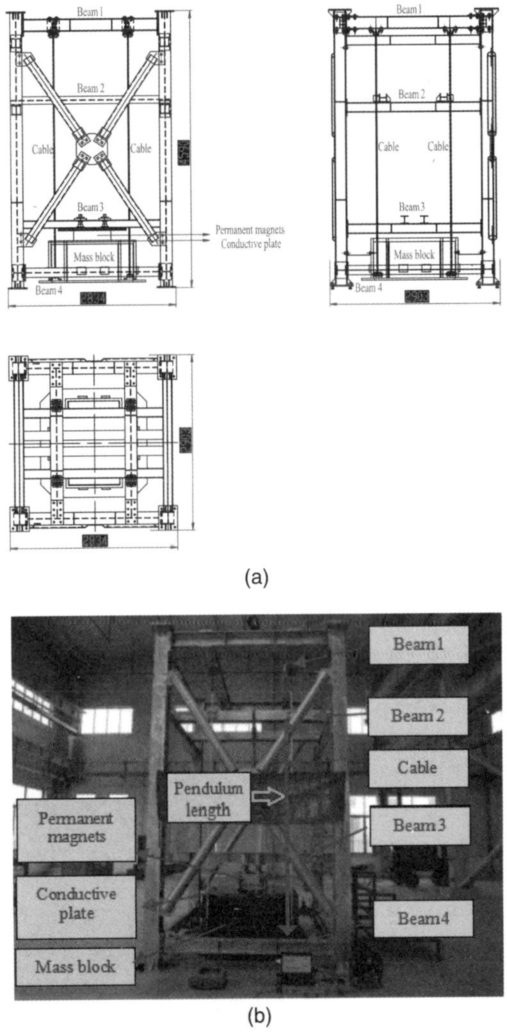

Fig. 4. The designed 10-t TMDs to be installed on site: (a) fabrication drawings; and (b) finished product.

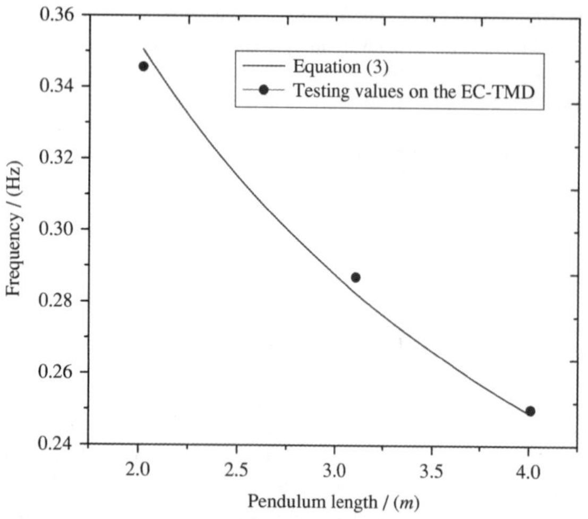

Fig. 5. Relationship between pendulum length and frequency.

could be conveniently adjusted by modifying the effective cable length through movable clamps on Beam 2; the pendulum length was calculated from the anchor points of the mass block to the bottom of Beam 2, as shown in Fig. 4(b). In addition, the permanent magnets could also be shifted up and down slightly to adjust the distance between them and the conductive plate (i.e., air gap) to refine the achievable damping ratio of the EC-TMD (about 6.1%). The air gap remains unchanged when the mass block is moving horizontally. Beam 4 was used to limit unexpected vibrations of the EC-TMD to protect the safety of the surrounding facilities.

After manufacture was completed, testing was conducted on the dynamic properties of the designed EC-TMDs, and some selected results from the testing are presented in this section. Note that the frequency for an idealized pendulum, in which the mass block is treated as a point, can be theoretically determined by

$$f_d = \frac{1}{2\pi}\sqrt{\frac{g}{l}} \tag{3}$$

where g = acceleration of gravity; and l = pendulum length. Fig. 5 shows the relationship between pendulum length and frequency measured on the prototype 10-t EC-TMD as well as the relationship determined theoretically. Fig. 5 shows that the test values agreed well with the theoretical values, indicating that the position of the anchor points of the cables on the mass block had little effect on the relationship between the pendulum length and frequency. Therefore, the pendulum lengths for the EC-TMDs could be adjusted roughly based on Eq. (3).

Identification of the Structural Parameters of the Tower

The four 10-t TMDs on the Noor III solar tower were installed on the tower in November 2017. The customized steel frame was fixed on the tower through the upper end points of its four columns. The pendulum lengths and the gaps between the magnets and the conductive plates were adjusted on the basis of the identified parameters of the tower at the time (the fundamental frequency was around 0.33 Hz). However, because the construction of the Noor III solar tower was not completed and some extra structural components were to be added, it was expected that the EC-TMD parameters according to the tower under construction would not necessarily be optimal. Testing on the dynamic properties of the tower (e.g., fundamental natural frequency and structural damping) was conducted again in October 2018, after completion of the Noor III solar tower, under ambient vibration and free vibration, as shown in Fig. 6; based on this testing, the EC-TMDs were retuned.

Coupled Dynamic Properties of EC-TMDs and the Structure

The tower and EC-TMD coupled system can be simplified to a five-degree-of-freedom system. Each TMD had the same mass, damping coefficient, and stiffness, as designed. Accordingly, the frequency of the main structure f_s, the frequency of each TMD f_d, and the total mass ratio of the TMDs to the main structure μ can be expressed as

$$f_s = \frac{1}{2\pi}\sqrt{\frac{k_s}{m_s}}, \qquad f_d = \frac{1}{2\pi}\sqrt{\frac{k_i}{m_i}}, \qquad \mu = \frac{\sum m_i}{m_s} \tag{4}$$

Then, the natural frequencies of the coupled structural system are

$$f_{1,2}^2 = \frac{f_s^2 + f_d^2(1+\mu)}{2}\left(1 \pm \sqrt{1 - \frac{4f_s^2 f_d^2}{[f_s^2 + f_d^2 \overline{1} + \mu]^2}}\right)$$
$$f_{3,4,5} = f_d \tag{5}$$

Fig. 7 shows the variations of coupled frequencies against the frequency of the main structure f_s and the TMDs f_d (mass ratio μ of 1.0%). Fig. 7 shows that the first and second natural frequencies of the coupled system (f_1 and f_2) significantly deviated from both

(a) (b)

Fig. 6. EC-TMDs on the tower in October 2018: (a) completed tower; and (b) EC-TMD installation. (Images by Shouying Li.)

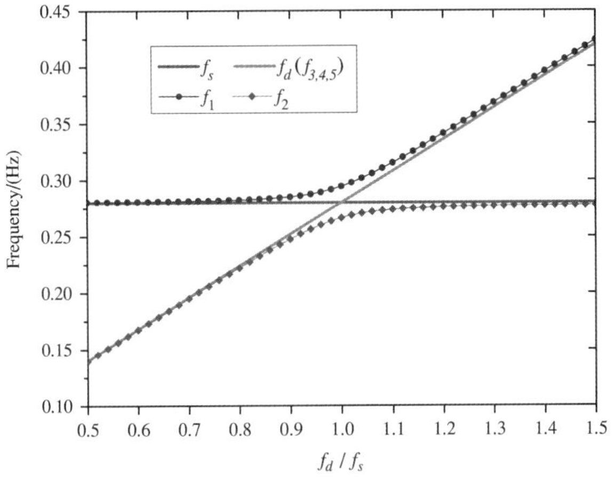

Fig. 7. Variation of coupled frequencies against f_s and f_d ($\mu = 1.0\%$).

f_s and f_d when $f_d/f_s = 1.0$. However, f_1 and f_2 approached f_s and f_d, respectively, with the decrease of f_d/f_s, and f_1 and f_2 approached f_d and f_s, respectively, with the increase of f_d/f_s. This observation indicates that the difference between the fundamental frequencies of the coupled system and the tower decreased as the frequency of the TMDs deviated from that of the tower.

Based on the foregoing discussion, the testing of the dynamic parameters for the tower in October 2018 was conducted by adjusting the pendulum length of the TMDs deviated from the optimal values as much as possible.

Ambient Vibration Testing

Accelerometers (model 2D001V, Donghua Testing Technology Company, Ltd., Jiangsu, China) with a resolution ratio for acceleration measurement of 3×10^{-6} m/s^2 (that is 3×10^{-4} milli-g) were installed near the top of the tower to measure the acceleration responses. The environmental excitations were mainly based on the wind loads on the tower, and Figs. 8(a and c) show the time histories of acceleration in two orthogonal directions, respectively, at the level of the EC-TMDs (237 m above the ground) induced by environmental excitations. Figs. 8(b and d) show the corresponding power spectrum densities (PSDs) of the acceleration. The four EC-TMDs were locked on the tower when the data in Fig. 8 was sampled. Figs. 8(b and d) show that the responses of the tower were primarily in the first fundamental sway mode of vibration, and the fundamental natural frequency in the two orthogonal directions was the same. The identified frequency after the tower was completed

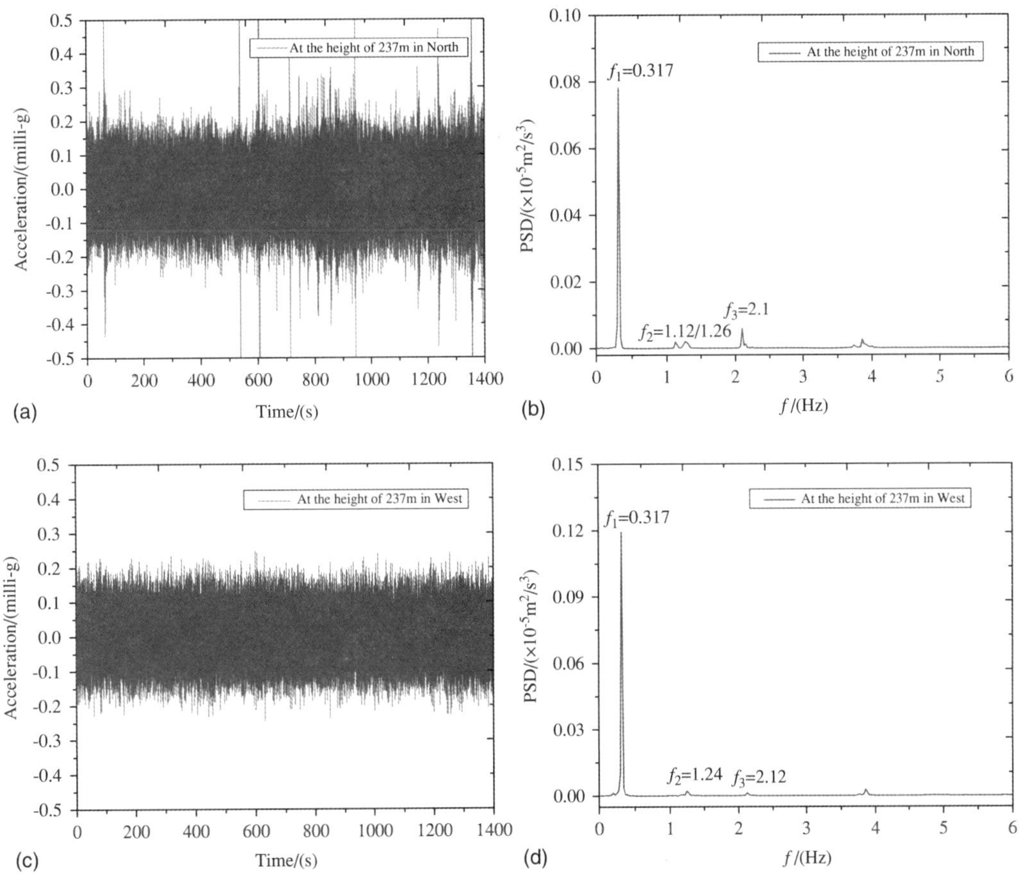

Fig. 8. Time history and PSD of acceleration at the height of the EC-TMDs induced by environmental excitations: (a) time history in north; (b) PSD; (c) time history in west; and (d) PSD.

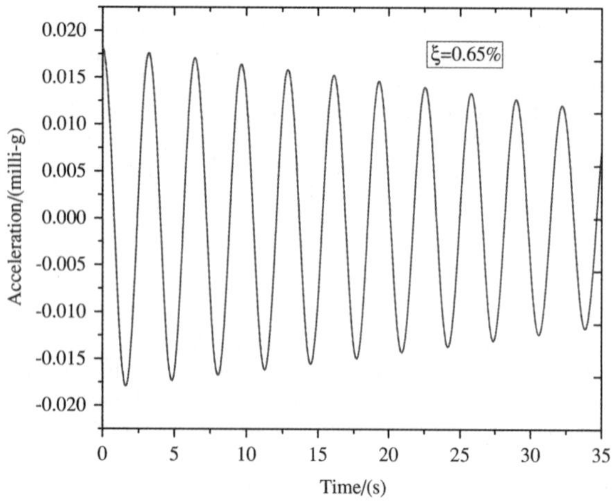

Fig. 9. Decay curve obtained from the time history of acceleration using the RDT method.

was 0.317 Hz, which was slightly smaller than the value measured in November 2017 (0.33 Hz). The random decrement technique (RDT) with a trigger value of the standard deviation was used to identify the structural damping of the tower from the environmental excitations. Fig. 9 presents the decay curve obtained from the time history of acceleration in Fig. 8(c) using the RDT method; the structural damping ratio of the tower calculated from Fig. 9 is 0.65%.

Free Vibration Testing

For the free vibration testing to identify the structural parameters, the initial excitations were obtained by manually swaying one of the EC-TMDs and then suddenly locking the swayed EC-TMD on the tower. Fig. 10 presents the decay response of the tower and the PSD of the acceleration of the tower at a height of 237 m. Fig. 10(b) shows that the fundamental frequency of the tower was 0.317 Hz, which agreed well with the results obtained from the ambient vibration testing. The structural damping ratio of the tower identified from Fig. 10(a) is 0.72%, a little higher than the value identified from the ambient vibration testing. The main reason for this obvious difference was that the acceleration amplitude in Fig. 10(a) was larger than that in Fig. 8(c). Overall, the damping measurements showed that the structural damping ratio of 1.0% to 4.0% suggested for the tower by the American Concrete Institute [ACI 307 (ACI 2008)] code at the design stage may have been an overestimation. Wind tunnel testing results for the Noor III solar tower showed that structural damping had an important effect on the wind-induced responses of the tower, especially in the cross-wind direction when the structural damping ratio was lower than 1.0% (Li et al. 2018).

Adjustment of EC-TMDs

The procedure for adjusting the pendulum lengths for the EC-TMDs installed on the tower was as follows: (1) identifying the fundamental natural frequency of the tower; (2) calculating the optimal frequency of the TMDs by Eq. (1); and (3) adjusting the pendulum lengths of the TMDs according to the relationship between pendulum length and frequency as expressed by Eq. (3). The damping of the EC-TMDs is not readjusted this time due to its independence of the frequency.

Based on the fundamental frequency of 0.317 Hz identified in the section "Identification of the Structural Parameters of the Tower," the pendulum lengths of the EC-TMDs were adjusted. The time histories of acceleration and the corresponding PSDs of the four EC-TMDs with three subsequent excitations are shown in Figs. 11–14. The identified damping ratios and frequencies for each EC-TMD are listed in Table 1.

Table 1 indicates that the identified frequencies for TMD Nos. 2 and 3 were found to be identical, with a value of 0.317 Hz that agreed well with the target value of 0.314 Hz. TMD Nos. 1 and 4 were found to be about 6.7% and 8.9% different from the target value, respectively. In addition, the damping ratios of the four TMDs (ranging from 6.3% to 7.37%) were a little higher than the target value of 6.1%, which will be good for the reduction of the maximum operating amplitude for the TMDs.

Field Validation

Free Vibration

The total structural damping ratio of the tower with the four EC-TMDs in working condition was identified using free vibration

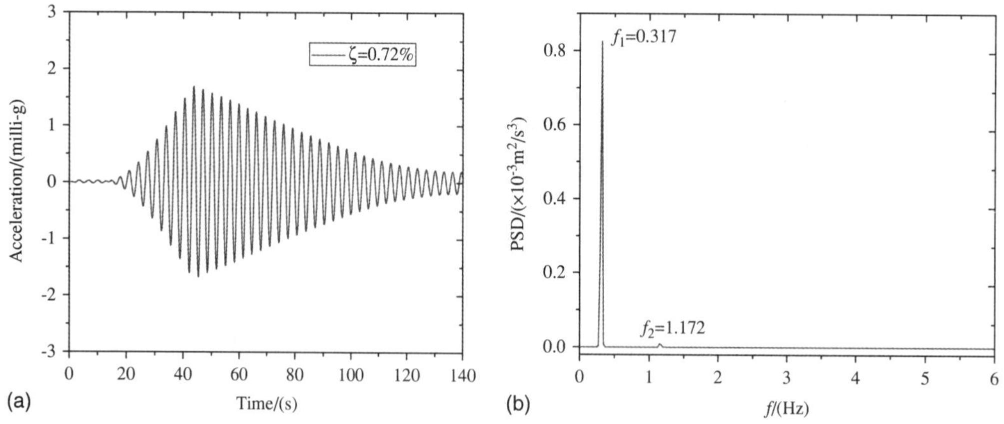

Fig. 10. Decay curve and PSD of the acceleration of the tower in free vibration testing: (a) decay curve; and (b) PSD.

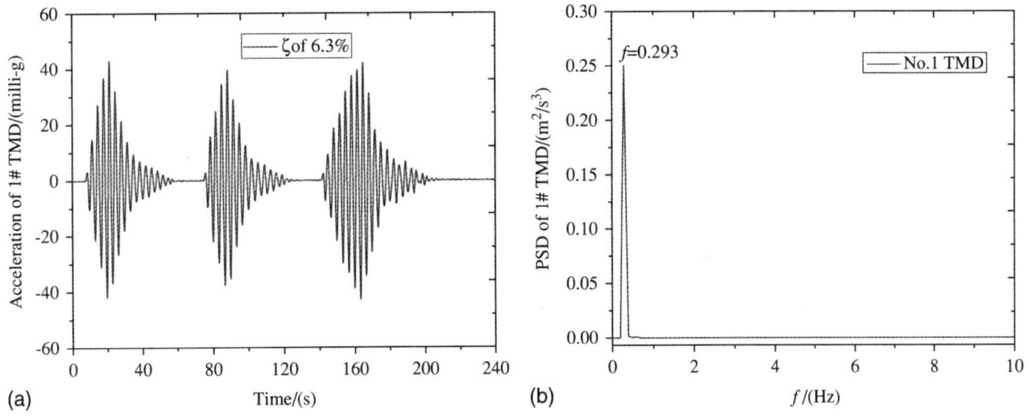

Fig. 11. Time history and PSD of acceleration of EC-TMD No. 1 induced by manual excitations: (a) time history; and (b) PSD.

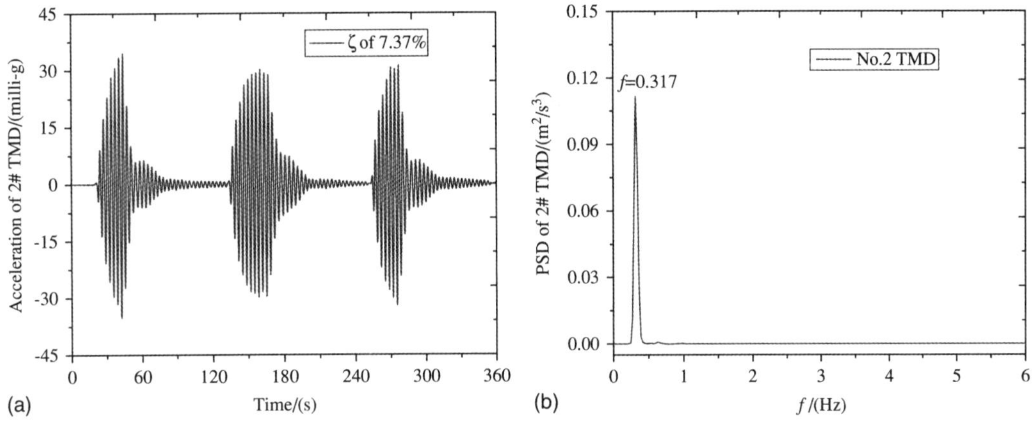

Fig. 12. Time history and PSD of acceleration of EC-TMD No. 2 induced by manual excitations: (a) time history; and (b) PSD.

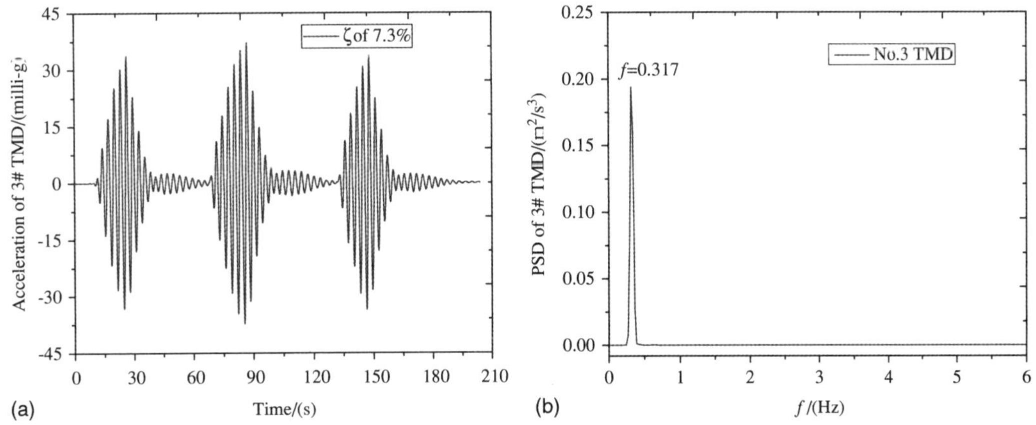

Fig. 13. Time history and PSD of acceleration of EC-TMD No. 3 induced by manual excitations: (a) time history; and (b) PSD.

testing. Fig. 15 shows the time histories of acceleration of the tower with two subsequent excitations. Fig. 15 shows that the total structural damping ratio of the tower with the four TMDs in working condition was about 3.78%, which agrees well with the values from both numerical simulation and experiments (Liu et al. 2020). This relatively high structural damping was adequate to sufficiently mitigate the wind-induced vibration of the tower, according to the results obtained from wind tunnel tests (Li et al. 2018).

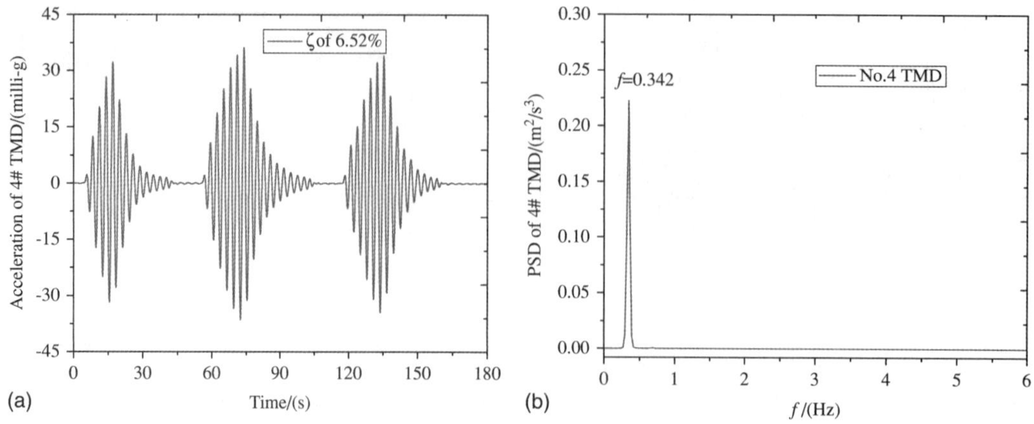

Fig. 14. Time history and PSD of acceleration of EC-TMD No. 4 induced by manual excitations: (a) time history; and (b) PSD.

Table 1. Frequency and damping ratio of each EC-TMD after adjustment

TMD number	No. 1	No. 2	No. 3	No. 4
Frequency (Hz)	0.293	0.317	0.317	0.342
Damping ratio (%)	6.30	7.37	7.30	6.52

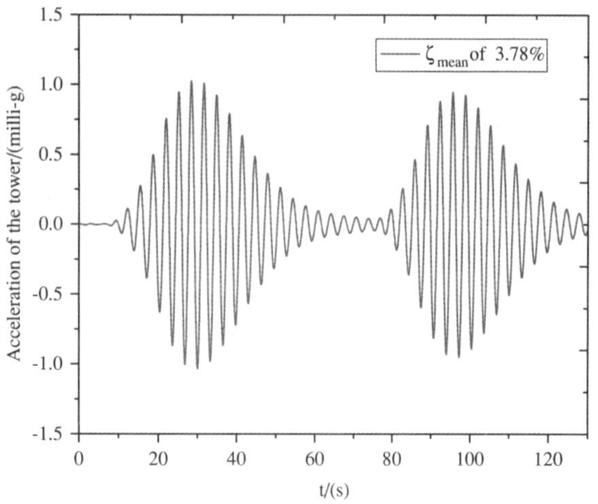

Fig. 15. Total structural damping ratios of tower with four EC-TMDs in working condition.

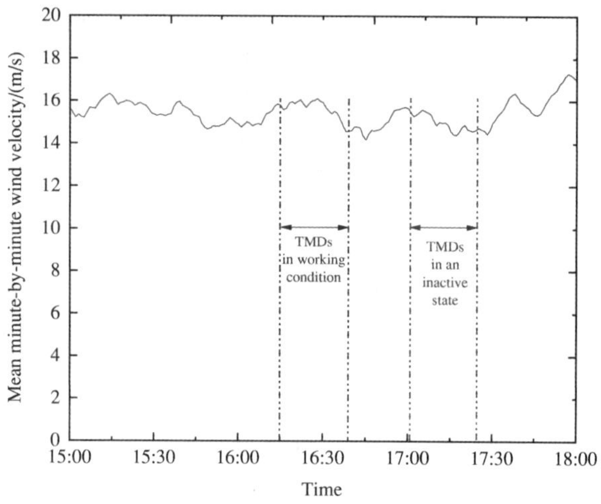

Fig. 16. Time history of minute-by-minute mean wind velocity from 15:00 h to 18:00 h on the afternoon of October 9, 2018.

Wind-Induced Vibration

There was a strong wind attacking the Noor III solar tower on the afternoon of October 9, 2018, immediately after the completion of its construction. Fig. 16 presents the time history of the minute-by-minute mean wind velocity measured by an anemometer installed at 15 m above ground near the tower (about 3 km away) from 15:00 h to 18:00 h on October 9, 2018. The minute-by-minute mean wind velocity (average velocity for one minute) varied between 14 and 17 m/s; this was significantly lower than the critical wind velocity of 23.2 m/s at a height of 10 m obtained based on wind tunnel tests (Li et al. 2018). Furthermore, the anemometer record indicates that the prevailing wind direction was always northwest. The four EC-TMDs were in working condition from 16:15 h to 16:39 h, and the corresponding acceleration responses in the direction of the wind were measured as shown in Fig. 17(a). Then, the four EC-TMDs were locked on the tower (i.e., in an inactive state), and the corresponding acceleration in the direction of the wind was sampled from 17:01 h to 17:25 h, as shown in Fig. 17(b). Fig. 17 shows that the four EC-TMDs installed on the Noor III solar tower significantly reduced the wind-induced vibration of the tower. The maximum accelerations of the tower with the EC-TMDs in working condition and in an inactive state were about 0.04 and 0.08 m/s^2, respectively, indicating a 50% reduction in acceleration response due to effective damping provided by the installed EC-TMDs.

Fig. 18 presents the PSD of the acceleration of the tower in the along-wind direction with and without contributions from EC-TMDs. Fig. 18 shows that the first two modes were excited under the action of the strong wind, but it is obvious that the component of the first mode was dominant. This observation is in agreement with the along-wind results obtained in wind tunnel tests using an aeroelastic building model (Li et al. 2018).

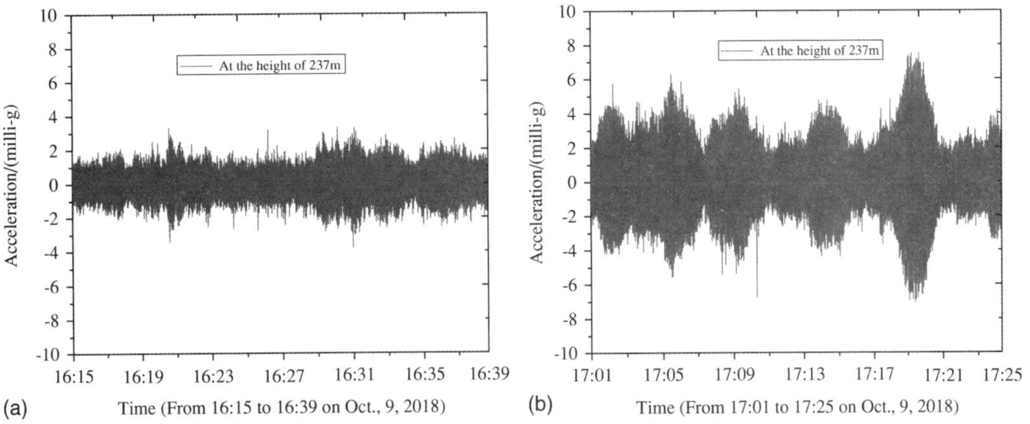

Fig. 17. Time histories of wind-induced tower acceleration with and without EC-TMDs.

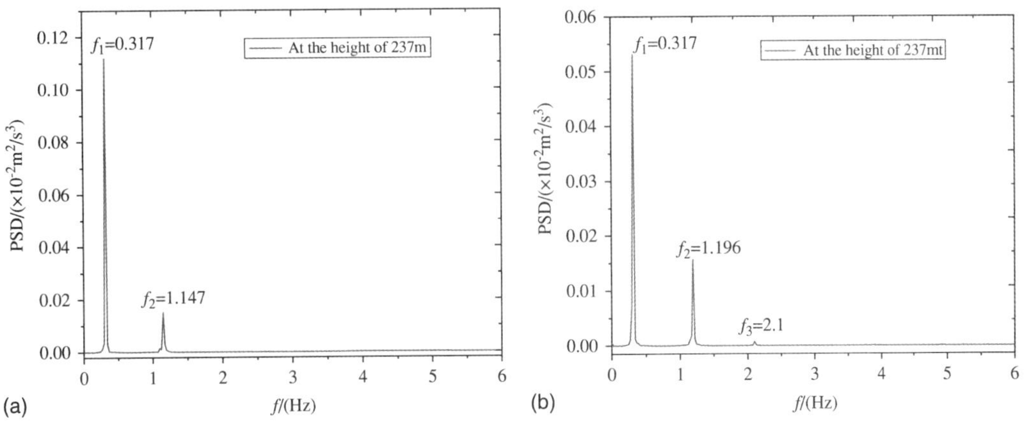

Fig. 18. PSDs of wind-induced tower acceleration in the direction of the wind with and without EC-TMDs: (a) in working condition; and (b) in an inactive state.

Fig. 19 compares the wind-induced accelerations of the tower and EC-TMD No. 2 in the direction of the wind. Fig. 19(a) shows that the maximum acceleration of EC-TMD No. 2 (about 0.15 m/s^2) with the EC-TMDs in working condition was far larger than that of the tower (about 0.04 m/s^2). This indicates that the kinetic energy of the tower was considerably absorbed by the EC-TMDs. However, the maximum accelerations of the tower and EC-TMD No. 2 with the EC-TMDs in an inactive state were almost the same. The slight differences in Fig. 19(b) were mainly due to local disturbances of the EC-TMDs.

Fig. 20 presents the decay curves obtained from the time history of the accelerations in Fig. 17 using the RDT method. The equivalent damping ratios of the tower were calculated to be 1.6% and 0.35% with the TMDs in working condition and in an inactive condition, respectively. This demonstrates that, in addition to absorbing the kinetic energy of the tower as mentioned previously, the EC-TMDs can also efficiently enhance the damping ratio of the tower. Moreover, the damping ratio of 0.35% shown in Fig. 20(b) with the TMDs in an inactive state was much lower than the ratio of 0.65% from gentle wind excitation shown in Fig. 9, indicating that negative aerodynamic damping may have been produced in the tower under the strong wind event.

Comparison with the Results of Wind Tunnel Tests

During the construction of the solar tower studied in this paper, a series of wind tunnel tests were conducted based on a 1:200-scale elastic model by Li et al. (2018). The results showed that the wind-induced responses were far higher than the values calculated from the design code. The Reynolds numbers of the model and the prototype tower were about 5.1×10^4 and 5.6×10^7, respectively. The Reynolds number effect was not considered in the study, which means that the test results were not necessarily representative of reality. As is well known, the response of circular structures is very sensitive to Reynolds number compared to the response of bluff bodies with sharp edges or corners. In wind tunnel tests, it is important for a scaled-down model to simulate the flow separation in disturbed regions of the prototype structure; as has been suggested in previous studies (e.g. Szechenyi 1975; Marukawa et al. 1984; Lawson 1982; Adachi 1997), altering the surface roughness of a test model can help to simulate this. Therefore, Liu et al. (2019) carried out wind tunnel tests to study the Reynolds number effect on the wind-induced responses of the tower by increasing the surface roughness of the elastic test model in Li et al. (2018).

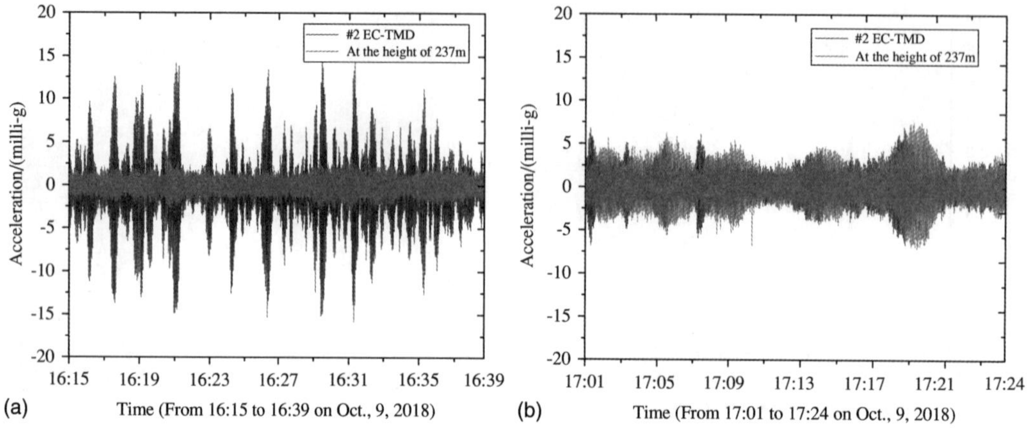

Fig. 19. Comparison of wind-induced responses of the tower and EC-TMD No. 2: (a) in working condition; and (b) in an inactive state.

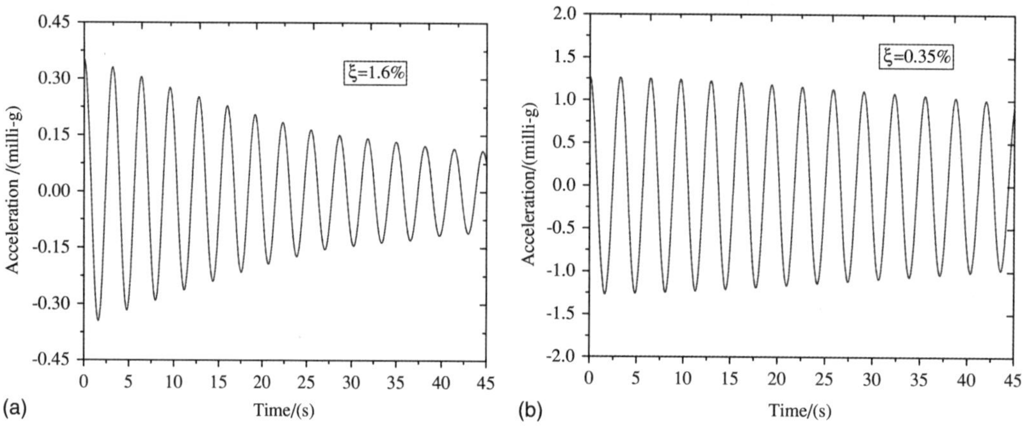

Fig. 20. Decay curve obtained from the time history of acceleration using the RDT method under strong wind events with TMDs: (a) in working condition; and (b) in an inactive state.

Fig. 21 presents the along-wind root-mean square (RMS) acceleration on the top of the tower obtained from wind tunnel tests using aeroelastic building model (Li et al. 2018; Liu et al. 2019), together with the field-measurement results of the prototype tower discussed in the section "Field Validation" (i.e., 0.007 and 0.016 m/s^2 for EC-TMDs in working condition and in an inactive state, respectively). The structural damping ratios of the test model in Li et al. (2018) and Liu et al. (2019) were 0.7%—almost the same as the prototype tower, as indicated in Fig. 10. Fig. 21 shows that the RMS acceleration of the prototype tower was significantly smaller than the values obtained from the wind tunnel tests. Moreover, the modification of the Reynolds number effect by increasing surface roughness can mitigate the Reynolds number effect and, hence, reduce the difference between full-scale and scaled-down structural responses under winds. Although increasing the surface roughness can help to achieve the flow state of the supercritical Reynolds number for a scaled elastic model in a wind tunnel test, the discrepancy between the response of the full-scale and scaled-down tower cannot be completely eliminated. The current study focused on simulating aerodynamic performance in the direction of the wind, but crosswind loads play an important role in structural

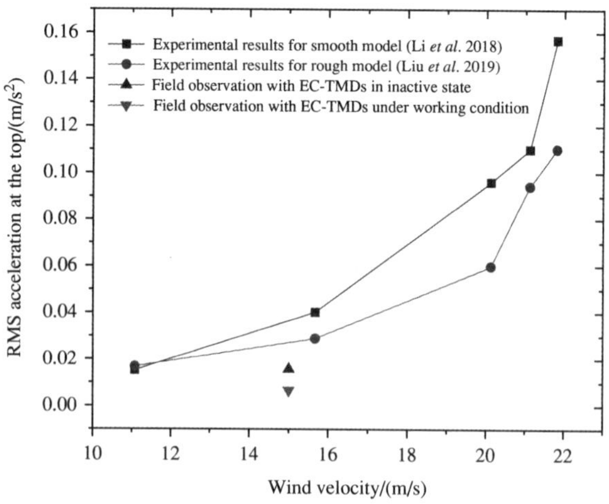

Fig. 21. Comparison between results of wind tunnel tests and field observations.

design for solar towers. And it would work well just under the circumstances that enough field measured data of similar structures could be referenced to for the determination of the surface roughness, such as the cool tower. However, there is little field-measured data for the solar tower to date. Therefore, a large amount of field-measured data for the solar tower in both along-wind and cross-wind directions is needed to further improve the modified method for wind tunnel tests.

Concluding Remarks

An eddy-current tuned mass damper with a mass ratio of 1.0% (about 40 t), determined by experiment, was designed and installed at the top of the Noor III solar tower to reduce its wind-induced responses. Due to limitations of space on the site, the single 40-t EC-TMD was divided into four 10-t EC-TMDs with identical structural parameters determined based on Den Hartog's optimum design formulas. A series of numerical simulations were conducted to validate the control performance of the TMDs. The results showed that even considering parameters (e.g., frequency and damping) detuning to the optimal values for the TMDs in engineering practice, the response reduction was no less than 84.6% with respect to the resonance responses under uncontrolled. This implies that the EC-TMDs were appropriately designed. The EC-TMDs were installed on the tower under construction and symmetrically arranged around the center line of the tower at the top. Structural parameters of the tower were identified by both ambient vibration testing and free vibration testing approaches. The fundamental frequency of the completed tower was 0.317 Hz, and the damping ratio was around 0.72% with an acceleration amplitude of 0.02 m/s^2. Based on the identified frequency of 0.317 Hz, the pendulum lengths of the four EC-TMDs were adjusted. Then, a series of measurements were carried out to validate the effectiveness of the EC-TMDs. The total structural damping ratio of the tower was enhanced to 3.78% due to the contributions of the four EC-TMDs under free vibration testing. The wind-induced responses of the tower under a strong wind event were significantly reduced with the EC-TMDs. The maximum tower accelerations with the EC-TMDs in working condition and in an inactive state were about 0.04 and 0.08 m/s^2, respectively, indicating a 50% reduction in the acceleration response due to effective damping provided by the installed EC-TMDs. In addition, the acceleration of the prototype tower was significantly smaller than the values obtained from wind tunnel tests. This indicates an important Reynolds number effect on the wind-induced response of the Noor III solar tower.

Data Availability Statement

Some or all data, models, or code generated or used during the study are proprietary or confidential in nature and may only be provided with restrictions.

Acknowledgments

This project was supported by the National Key Research and Development Program of China (Grant Nos. 2017YFC0703600 and 2017YFC0703604). The National Natural Science Foundation of China (Grant No. 51508184) and the Foundation of Hu'nan Province Education Department (Grant No. 19C1660) are also acknowledged.

References

ACI (American Concrete Institute). 2008. *Code requirements for reinforced concrete chimneys*. ACI 307. Farmington Hills, MI: ACI.

Adachi, T. 1997. "Effects of surface roughness on the universal Strouhal number over the wide Reynolds number range." *J. Wind Eng. Ind. Aerodyn.* 69–71 (Jul): 399–412. https://doi.org/10.1016/S0167-6105(97)00172-4.

Bae, J. S., M. K. Kwak, and D. J. Inman. 2005. "Vibration suppression of cantilever beam using eddy current damper." *J. Sound Vib.* 284 (3–5): 805–824. https://doi.org/10.1016/j.jsv.2004.07.031.

Cunningham, R. E. 1986. *Passive eddy-current damping as a means of vibration control in cryogenic turbomachinery*. NASA-TP-2562. Washington, DC: National Aeronautics and Space Administration.

Den Hartog, J. P. 1956. *Mechanical vibrations*. 4th ed. New York: McGraw-Hill.

Ebrahimi, B., M. B. Khamesee, and F. Golnaraghi. 2009. "Eddy current damper feasibility in automobile suspension: Modeling, simulation and testing." *Smart Mater. Struct.* 18 (1): 015017. https://doi.org/10.1088/0964-1726/18/1/015017.

González-Roubaud, E., D. Pérez-Osorio, and C. Prieto. 2017. "Review of commercial thermal energy storage in concentrated solar power plants: Steam vs. molten salts." *Renewable Sustainable Energy Rev.* 80 (Dec): 133–148. https://doi.org/10.1016/j.rser.2017.05.084.

Gunter, E. J., R. R. Humphris, and S. J. Severson. 1983. *Design study of magnetic eddy-current vibration suppression dampers for application to cryogenic turbomachinery*. NASA-CR-173273, Charlottesville, VA: Univ. of Virginia.

Huang, Z. W., Z. Q. Chen, and X. G. Hua. 2015. "Numerical modeling and experimental testing of an eddy current apparent mass damper." In *Proc., 14th World Conf. on Seismic Isolation Energy Dissipation and Active Vibration Control of Structures*. San Diego: Energy Dissipation and Active Vibration Control of Structures.

Huang, Z. W., X. G. Hua, Z. Q. Chen, and H. W. Niu. 2018. "Modeling, testing, and validation of an eddy current damper for structural vibration control." *J. Aerosp. Eng.* 31 (5): 04018063. https://doi.org/10.1061/(ASCE)AS.1943-5525.0000891.

Kwak, M. K., M. I. Lee, and S. Heo. 2003. "Vibration suppression using eddy current damper." *Korean Soc. Noise Vib. Eng.* 233 (10): 441–453.

Kwok, K. C. S. 1983. "Full-scale measurements of wind-induced response of Sydney Tower." *J. Wind Eng. Ind. Aerodyn.* 14 (1–3): 307–318. https://doi.org/10.1016/0167-6105(83)90033-8.

Larose, G. L., A. Larsen, and E. Svensson. 1995. "Modelling of tuned mass dampers for wind-tunnel tests on a full-bridge aeroelastic model." *J. Wind Eng. Ind. Aerodyn.* 54 (Feb): 427–437. https://doi.org/10.1016/0167-6105(94)00058-L.

Lawson, T. V. 1982. "The use of roughness to produce high Reynolds number flows around circular cylinders at lower Reynolds numbers." *J. Wind Eng. Ind. Aerodyn.* 10 (3): 381–387. https://doi.org/10.1016/0167-6105(82)90009-5.

Li, S. Y., M. Liu, H. X. Li, Y. Hui, and Z. Q. Chen. 2018. "Effects of structural damping on wind-induced responses of a 243-meter-high solar tower based on a novel elastic test model." *J. Wind Eng. Ind. Aerodyn.* 172 (Jan): 1–11. https://doi.org/10.1016/j.jweia.2017.10.027.

Liu, M., S. Y. Li, and Z. Q. Chen. 2020. "Mitigation of wind-induced responses of cylinder solar tower by a tiny eddy current tuned mass damper based on elastic wind tunnel tests." *Smart Struct. Syst.* 26 (5): 619–629.

Liu, M., S. Y. Li, H. X. Li, S. K. Li, and Z. Q. Chen. 2019. "Reynolds number effects on the wind-induced responses of a 243-meter-high solar tower in elastic wind tunnel tests." *J. Aerosp. Eng.* 32 (4): 04019053. https://doi.org/10.1061/(ASCE)AS.1943-5525.0001033.

Lou, W. J., Z. P. Wen, Y. Chen, and M. F. Huang. 2019. "Wind tunnel study on bidirectional vibration control of lattice towers with omnidirectional cantilever-type eddy current TMD." *Appl. Sci.* 9 (15): 2978. https://doi.org/10.3390/app9152978.

Lu, X. L., Q. Zhang, D. Weng, Z. Zhou, S. Wang, S. A. Mahin, S. Ding, and F. Qian. 2017. "Improving performance of a super tall building using a new eddy-current tuned mass damper." *Struct. Control Health Monit.* 24 (3): e1882. https://doi.org/10.1002/stc.1882.

Marukawa, H., Y. Tamura, S. Sanada, and O. Nakamura. 1984. "Lift and across-wind response of a 200 m concrete chimney." *J. Wind Eng. Trans. Jpn. Assoc. Wind Eng.* 19 (Apr): 37–52.

McNamara, R. J., and D. P. Taylor. 2003. "Fluid viscous dampers for high-rise buildings." *Struct. Des. Tall Spec. Build.* 12 (2): 145–154. https://doi.org/10.1002/tal.218.

Narkhede, D. I., and R. Sinha. 2014. "Behavior of nonlinear fluid viscous dampers for control of shock vibration." *J. Sound Vib.* 333 (1): 80–98. https://doi.org/10.1016/j.jsv.2013.08.041.

Niu, H. W., Z. Q. Chen, X. G. Hua, and W. Zhang. 2018. "Mitigation of wind-induced vibrations of bridge hangers using tuned mass dampers with eddy current damping." *Smart Struct. Syst.* 22 (6): 727–741.

Ohmata, K. 1983. "Vibration control of beams by ball screw type dampers (case in which dampers possess viscous damping)." *Bull. JSME* 26 (222): 2172–2177. https://doi.org/10.1299/jsme1958.26.2172.

Ohmata, K., H. Shimoda, and A. Hosoya. 1986. "Characteristics of ball screw type damper with magnetic damping." *Bull. JSME* 29 (253): 2255–2259. https://doi.org/10.1299/jsme1958.29.2255.

Ozoegwu, C. G., C. A. Mgbemene, and P. A. Ozor. 2017. "The status of solar energy integration and policy in Nigeria." *Renewable Sustainable Energy Rev.* 70: 457–471. https://doi.org/10.1016/j.rser.2016.11.224.

Palomera-Arias, R., J. Connor Jerome, and A. Ochsendorf John. 2008. "Feasibility study of passive electromagnetic damping systems." *J. Struct. Eng.* 134 (1): 164–170. https://doi.org/10.1061/(ASCE)0733-9445(2008)134:1(164).

Sacks, M. P., and J. C. Swallow. 1993. "Tuned mass dampers for towers and buildings." In *Proc., Symp. on Structural Engineering in Natural Hazards Mitigation*, 640–645. Reston, VA: ASCE..

Sakamoto, H., K. Araki, A. Ishida, S. Kobayashi, and T. Kuwahara. 1997. *Design of permanent magnet type compact ECB retarder.* SAE Technical Paper 973228. Warrendale, PA: Society of Automotive Engineers.

Simeu, E., and D. Georges. 1996. "Modeling and control of an eddy current brake." *Control Eng. Pract.* 4 (1): 19–26. https://doi.org/10.1016/0967-0661(95)00202-4.

Szechenyi, E. 1975. "Supercritical Reynolds number simulation for two-dimensional flow over circular cylinders." *J. Fluid Mech.* 70 (3): 529–542. https://doi.org/10.1017/S0022112075002170.

Tahri, M., M. Hakdaoui, and M. Maanan. 2015. "The evaluation of solar farm locations applying geographic information system and multi-criteria decision-making methods: Case study in southern Morocco." *Renewable Sustainable Energy Rev.* 51 (Nov): 1354–1362. https://doi.org/10.1016/j.rser.2015.07.054.

Tuan, A. Y., and G. Q. Shang. 2014. "Vibration control in a 101-storey building using a tuned mass damper." *J. Appl. Sci. Eng.* 17 (2): 141–156.

Vickery, B. J., and A. G. Davenport. 1971. *An investigation of the behaviour in wind of the proposed Centrepoint Tower.* Engineering Science Rep. No. BLWT-1-70. London: Univ. of Western Ontario.

Wang, Z. H., Z. Q. Chen, and J. H. Wang. 2012. "Feasibility study of a large-scale tuned mass damper with eddy current damping mechanism." *Earthquake Eng. Eng. Vib.* 11 (3): 391–401. https://doi.org/10.1007/s11803-012-0129-x.

Wen, Q., X. G. Hua, Z. Q. Chen, Y. Yang, and H. W. Niu. 2016. "Control of human-induced vibrations of a curved cable-stayed bridge: Design, implementation, and field validation." *J. Bridge Eng.* 21 (7): 04016028. https://doi.org/10.1061/(ASCE)BE.1943-5592.0000887.

Xu, H. B., C. W. Zhang, H. Li, P. Tan, J. P. Ou, and F. L. Zhou. 2014. "Active mass driver control system for suppressing wind-induced vibration of the Canton Tower." *Smart Struct. Syst.* 13 (2): 281–303. https://doi.org/10.12989/sss.2014.13.2.281.

Zhang, H. Y., Z. Q. Chen, X. G. Hua, Z. W. Huang, and H. W. Niu. 2020. "Design and dynamic characterization of a large-scale eddy current damper with enhanced performance for vibration control." *Mech. Syst. Signal Process.* 145 (Nov): 106879. https://doi.org/10.1016/j.ymssp.2020.106879.

Renewable Energy 198 (2022) 772–788

Contents lists available at ScienceDirect

Renewable Energy

journal homepage: www.elsevier.com/locate/renene

Analytical study on the aerodynamic and hydrodynamic damping of the platform in an operating spar-type floating offshore wind turbine

Qingshen Meng [a,b], Xugang Hua [a,b], Chao Chen [a,b,*], Shuai Zhou [a,b], Feipeng Liu [c], Zhengqing Chen [a,b]

[a] Key Laboratory for Wind and Bridge Engineering, Hunan University, Changsha, China
[b] College of Civil Engineering, Hunan University, Changsha, China
[c] Northwest Electric Power Design Institute Co., Ltd, Xi'an, China

ARTICLE INFO

Keywords:
Floating wind turbine
Aerodynamic damping
Hydrodynamic damping
Spar platform

ABSTRACT

Accurate evaluation of aerodynamic damping and hydrodynamic damping is critical to response prediction and vibration control of spar-type floating offshore wind turbines. This paper presents a comprehensive analytical study of these two damping sources with respect to platform motions. Aerodynamic damping is evaluated by a newly derived aerodynamic damping matrix based on linearisation of aerodynamic resultant forces at tower top. Both radiation and viscous drag effects are considered for hydrodynamic damping. The former is analytically expressed in the form of a radiation damping matrix, while the latter is derived based on Morison's equation and strip theory. A simplified model is established based on the analytical damping expressions and successfully verified against FAST and Aqwa. Under various operational conditions, the damping ratios for different degrees of freedom of the platform are estimated using complex modal analysis. It is found that the modal damping ratios arising from different sources can be very different for different vibration modes of the platform, and the tower flexibility is proved to have negligible impact on the platform damping. For a typical operational state, surge, sway, pitch and yaw motions are highly damped, with the roll motion intermediately damped and the heave motion nearly undamped.

1. Introduction

Offshore wind energy is becoming increasingly important when the world is faced with the energy crisis and climate-change problems [1]. Extending the buildable area in deep water for offshore wind turbines allows harvesting stronger and steadier wind resource, but it requires developing more advanced Floating Offshore Wind Turbine (FOWT) technology to reduce cost.

A typical horizontal-axis FOWT is a fairly complex system that combines mooring lines, a floating platform, a tower and a rotor, and it is subjected to highly fluctuating wind and wave loads. As a result, its dynamics must be considered carefully to avoid resonance and large vibration magnitude so that the safety of key components and normal electricity generation can be guaranteed. Damping is critical for accurately predicting the dynamic behaviour of a FOWT platform because it greatly limits vibration amplitude at resonance. Therefore, it is essential to properly quantify the damping from different sources. There are mainly four sources of damping in FOWTs: aerodynamic, hydrodynamic, structural damping and damping from mooring lines. From literature, the contribution from structural damping is relatively low and normally referred to offshore oil and gas literature [2] and standard steel structure code [3]. Aerodynamic damping and hydrodynamic damping have much larger contributions, but there are relatively large uncertainties as they highly depend on the stochastic environment and operational conditions [4]. The overall dynamic behaviour of a FOWT is mainly dependent on the oscillation of the platform, with vibrations of the tower and blades also worthwhile to study. To be specific and avoid confusion, the present study only focuses on the aerodynamic and hydrodynamic damping of the FOWT platform.

Aerodynamic damping has been studied extensively for bottom-fixed wind turbines. Experimental studies using time-domain and frequency-domain techniques were carried out to measure aerodynamic damping in a number of studies (e.g. Refs. [5–9]). Aerodynamic damping can also be quantified analytically. In early studies by Salzmann and Tempel [10], several closed-form expressions based on Blade Element Momentum (BEM) theory are summarised for aerodynamic damping in the Fore-Aft (FA) direction. Valamanesh and Myers [11] and Chen and

* Corresponding author. Key Laboratory for Wind and Bridge Engineering, Hunan University, Changsha, China.
E-mail address: steinchen@hnu.edu.cn (C. Chen).

https://doi.org/10.1016/j.renene.2022.07.126
Received 29 March 2022; Received in revised form 4 July 2022; Accepted 22 July 2022
Available online 14 August 2022

Nomenclature

ξ	Wave oscillating amplitude
ρ_w	Water density
$\Psi_j(t)$	Rotor azimuth angle
σ	Wave angular frequency
ω_n	Natural frequency of nth mode
ϖ	Platform motion angular frequency
Ω	Rated rotor speed
a	Floater cylinder radius
$\mathbf{C}$	Global damping matrix
C_A	Normalized added-mass coefficient
C_D	Normalized viscous-drag coefficient
h_c	Height of a merged cylinder
D	Diameter of the cylinder strip
$\mathbf{F}^0_{\text{Aero}}$	External aerodynamic force vector
$\mathbf{F}^0_{\text{Buoy}}$	External buoyancy force vector
$\mathbf{F}_{\text{Gra}}(t)$	External gravity force vector
$\mathbf{F}^0_{\text{Hydro}}(t)$	External hydrodynamic force vector
$\mathbf{F}^0_{\text{Moor}}$	External mooring force vector
f_{Tower}	Tower natural frequency
g	Gravity acceleration
H_s	Wave height
H_T	Length from tower top to spar top
h_R	Length from rotor centre to spar mass centre
h_T	Length from tower base to spar mass centre
$\mathbf{K}$	Global stiffness matrix
k	Wave number
$\mathbf{M}$	Global mass matrix
M_0	Lumped mass at tower top
N_b	Number of blades
R	Rotor radius
r	Radius of blade element
$\dot{\mathbf{U}}$	Platform velocity vector
$\dot{\mathbf{u}}$	Tower velocity vector
$\mathbf{V}_B$	Blades velocity vector in global coordinate system
V_0	Inflow wind of steady velocity
V_r	Wind speed in tangential direction
$\mathbf{x}(t)$	Displacement vector
Z_w	Water depth

Duffour [12] proposed semi-analytical solutions that calculate the aerodynamic damping in both FA and Side-Side (SS) directions. Semi-analytical expressions of aerodynamic damping in wind turbines were also derived in other studies [13]. For example, Chen et al. [14] derived an expression to calculate the FA aerodynamic damping ratio considering blade flexibility. However, aerodynamic damping expressions in the FA and SS directions were separately derived in these studies, so the damping coupling from motions in the FA and SS directions cannot be accounted. Further, Chen et al. [15,16] derived an aerodynamic damping matrix to model the aerodynamic damping coupling between the FA and SS directions so that the wind-rotor interaction can be better captured. The blades were assumed to be rigid, and the aerodynamic damping matrix describes the overall aerodynamic damping caused by the wind-rotor interaction to the tower. It can be implied that for a FOWT the aerodynamic damping coupling between motions in different directions should still exist and could have great impact on the dynamic responses.

Nevertheless, the studies on aerodynamic damping in FOWTs are still limited so far. Karimirad and Moan [17] derived a simplified expression of aerodynamic damping for platform rigid motion in a FOWT by BEM theory. This expression results in a relatively rough estimation of aerodynamic damping as it assumes fixed induction factors for the whole rotor. Jonkman [18] studied the influence of blade-pitch control actions on the pitch damping (including aerodynamic damping and hydrodynamic damping) of a wind turbine mounted on a floating barge using FAST. Cheng et al. [19] studied aerodynamic damping in a floating vertical-axis wind turbine based on simulated free-decay dynamic responses of pitch motion. Chen and Hu [20] studied the aerodynamic damping in a semi-submissible floating wind turbine analytically and experimentally. The aerodynamic damping derivation in this study can be regarded as an updated version of that by Karimirad and Moan [17] as it additionally considers the aerodynamic damping contributions from blade elements. Gueydon [21] also conducted an experimental study on the effect of the rotor thrust on the motions of a semi-submersible FOWT. It was found that the aerodynamic damping has strong effect on the resonance peak of surge and pitch motions. Thiagarajan et al. [22] introduced a one Degree-Of-Freedom (DOF) nonlinear vibration model to model the nonlinear decay motion of a floating wind turbine platform. It is concluded in this study that the static pitch angle caused by the steady wind force on the rotor clearly influences the overall platform damping. To obtain more accurate results than those based on the BEM theory, some attempt has been made to evaluate aerodynamic damping by combing a quasi-static blade-element method and an unsteady boundary-element method [23].

Hydrodynamic damping also plays an important role on dynamic responses of floating offshore wind turbines. The determination of hydrodynamic damping coefficients of offshore platforms is a topical issue [24] and a few studies on hydrodynamic damping in FOWTs have been presented. Karimirad and Moan [17] also studied the hydrodynamic damping in a spar-type FOWT based on Morison's equation. It was found that the quadratic drag coefficient in Morison's equation significantly influences the resonant response of low frequency surge motion. Subbulakshmi and Sundaravadivelu [25] studied experimentally and numerically the effects of heave plate on heave damping of a spar supporting a FOWT. Pegalajar-Jurado and Bredmose [26] studied the slow-drift forcing and damping of a semi-submissible platform in a FOWT through a numerical model and operational modal analysis on test data. It was found that the dynamic response of the platform can be generally well reproduced when the damping is calibrated for each sea state. Clement et al. [27] particularly focused on the effects of viscous damping on the dynamic response of a hybrid floating platform in a FOWT. Combining a potential-based model and a computational fluid dynamics model, the viscous damping of the platform was determined. Results show that the rotational motion of the hybrid platform is highly impacted by viscous damping. Yang et al. [28] investigated experimentally and numerically to study the nonlinear responses of a 6 MW spar-type FOWT. An equivalent linear damping model and a quadratic damping model were used to comparatively study the hydrodynamic damping of the platform. It was found that a better agreement with the free-decay experimental results can be obtained by the quadratic damping method.

The effect of the aerodynamic damping and hydrodynamic damping can also be evaluated from dynamic responses generated by mainstream FOWT modelling software such as FAST [29] and HAWC2 [30]. Aerodynamic and hydrodynamic damping effect is implicitly included in these softwares. As a result, it is unable to quantify these two damping sources directly. It is of interest to look at the physics-based characteristics of the aerodynamic and hydrodynamic damping in the FOWT system. In summary, most of previous studies focused on damping effect from either aerodynamic damping or hydrodynamic damping. On the other hand, to conveniently obtain the analytical damping expression, many simplifications in these studies could lead to omission of important mechanisms. For instance, in many studies the damping in surge and pitch motions caused by aerodynamic damping were investigated, but those in platform sway and roll motions were mostly ignored. Therefore, a comprehensive study is required to quantitatively analyse the contributions from aerodynamic damping and hydrodynamic damping in FOWTs.

The present work aims at studying the aerodynamic damping and hydrodynamic damping in spar-type FOWTs, through extending the authors' previous study for bottom-fixed wind turbines [15,16]. The aerodynamic damping and hydrodynamic damping are linearised with respect to the motion of the platform. Both damping sources are expressed in the form of damping matrix, from which the modal damping ratio are obtained by complex eigenvalue analysis. A quick evaluation of the damping levels of the platform is beneficial for the preliminary design and parameter optimisation of FOWTs. In addition, efficient time-domain and frequency-domain models can be created to conduct quick fatigue and reliability analysis with the proposed expressions of aerodynamic damping and hydrodynamic damping matrices. This paper is organised as follows. Section 2 proposes the FEM model and introduces the damping calculation method. Section 3 proposes the derivation of the aerodynamic damping model considering aerodynamic damping coupling. Section 4 provides the model to calculate the radiation damping and viscous damping caused by water-platform interaction. Section 5 verifies these two damping models against FAST. Section 6 shows and analyses the damping calculation results. Section 7 concludes the paper.

2. Simplified model description and damping calculation

2.1. FEM model

A spar-type OC3 Hywind model [31] was established in MATLAB to study the damping characteristics of FOWTs. This model is mainly composed of three blades, a nacelle, a hub, a steel tower, a spar-type platform, and three mooring lines, as shown in Fig. 1, with basic properties listed in Table 1. The definitions of the global and local coordinate systems are shown in Fig. 1. The global coordinate system origin is located at the gravity centre of the platform at its initial state. The tower local coordinate system origin is placed at the tower base, and the local coordinate system for each blade is defined as a rotating system fixed at the blade root. X, x, and x' axes are perpendicular to the rotor plane. The positive direction of Z-axis in the global coordinate system is vertically upward, while the positive directions of z and z' axes point to the tower top and blade tip respectively. Y, y, and y' axes are finally determined by right-hand rule.

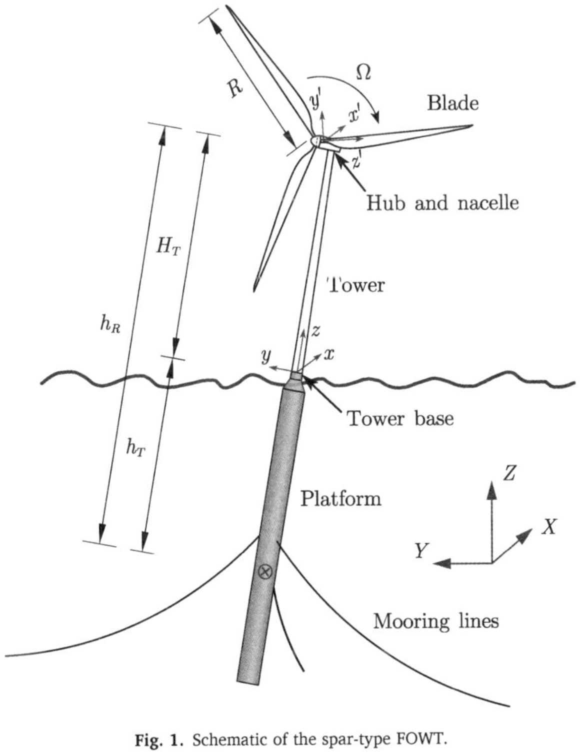

Fig. 1. Schematic of the spar-type FOWT.

The spar-type platform is modelled as a rigid body with three translational DOFs and three rotational DOFs. The translational DOFs correspond to surge, sway and heave respectively, while the rotational DOFs correspond to roll, pitch, and yaw respectively, as shown in Fig. 2.

The tower is modelled as a Euler-Bernoulli beam with 10 elements fixed at the top of the spar-type platform, and the Rotor-Nacelle-Assembly (RNA) is modelled as a lumped mass at the tower top. The beam nodes can translate and rotate in three perpendicular directions. x (FA), y (SS), and z (vertical) denote translations while θ_x, θ_y and θ_z denote rotations about the x, y and z axes respectively, as shown in Fig. 3. After assembling the elemental matrices of the beam elements and the matrix of the spar in one global matrix, the equation of motion of the FOWT model can finally be expressed by:

$$\mathbf{M}\ddot{\mathbf{x}}(t) + \mathbf{C}\dot{\mathbf{x}}(t) + \mathbf{K}\mathbf{x}(t) = \mathbf{F}^0_{\text{Aero}} + \mathbf{F}^0_{\text{Buoy}} + \mathbf{F}^0_{\text{Hydro}}(t) + \mathbf{F}^0_{\text{Moor}} + \mathbf{F}_{\text{Gra}}(t) \quad (1)$$

where **M**, **C**, and **K**, are the mass matrix, damping matrix, and stiffness matrix respectively. Added mass caused by water-platform interaction is included in the system mass matrix. Aerodynamic damping and hydrodynamic damping are included in the system damping matrix, while buoyancy stiffness and mooring line stiffness is included in system stiffness matrix. Especially, as the present paper focuses on the steady-state platform responses under a regular wave and the associated radiation damping, so the conventional convolution term for hydrodynamic radiation damping force is not used for simplicity. According to Jonkman [32], if the incident wave propagates at a single amplitude, frequency and direction (the incident wave is a regular wave) and the platform also oscillates at the same frequency as the incident wave, the radiation damping force can be expressed as a radiation damping coefficient multiplied by the platform velocity. Therefore, a linear radiation damping matrix is directly added into the damping matrix **C**. $\mathbf{x}(t)$ is the displacement vector. $\mathbf{F}^0_{\text{Aero}}$, $\mathbf{F}^0_{\text{Buoy}}$, $\mathbf{F}^0_{\text{Hydro}}(t)$ and $\mathbf{F}^0_{\text{Moor}}$ are the external aerodynamic, buoyancy, hydrodynamic and mooring force vectors. $\mathbf{F}_{\text{Gra}}(t)$ is the gravity force vector. It should be noted that $\mathbf{F}^0_{\text{Aero}}$, $\mathbf{F}^0_{\text{Buoy}}$, and $\mathbf{F}^0_{\text{Moor}}$ are time independent, since $\mathbf{F}^0_{\text{Aero}}$ represents the static thrust and torque under the steady wind. The time-dependent aerodynamic damping forces are successfully separated from the total aerodynamic

Table 1
Structural properties of the NREL-5MW spar-type FOWT.

Rotor Radius, R	63m
Rated Rotor Speed, Ω	1.267 rad/s
Length from Rotor Centre to Spar Mass Centre, h_R	168m
Length from Tower Top to Spar top, H_T	87.6m
Length from Tower Base to Spar Mass Centre, h_T	78m
Lumped Mass at Tower Top, M_0	3.5×10^5kg
Platform Diameter	9.4m
Platform Depth Below SWL	120m
Water Depth	320m
Length from Spar Bottom to Spar Mass Centre, h_B	42m
Platform Mass, including Ballast	7466330 kg
Additional Yaw Spring Stiffness	9.834×10^7Nm/rad
Number of Mooring Lines	3
Depth to Fairleads Below SWL	70.0m
Radius to Anchors from Platform Centreline	853.87m
Unstretched Mooring Line Length	902.2m

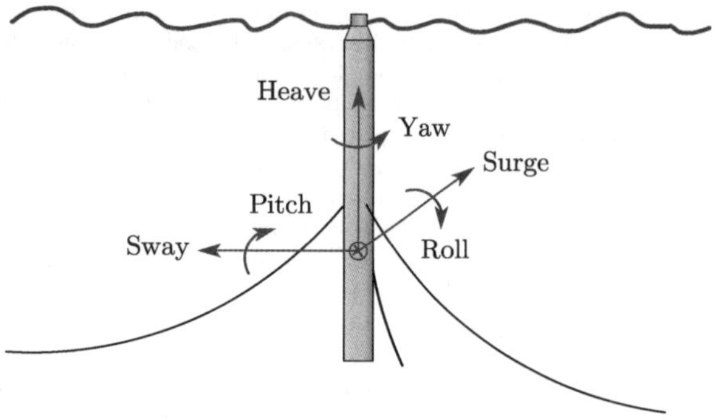

Fig. 2. 6 DOFs of the spar platform.

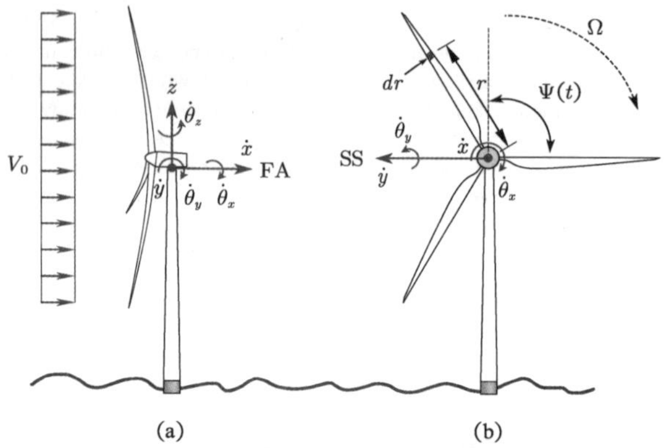

Fig. 3. Fore-aft (a) and side-side (b) motions of tower top.

loads, and then assembled into the damping matrix **C** as an aerodynamic damping matrix in this paper. Moreover, $\mathbf{F}^0_{\text{Buoy}}$ and $\mathbf{F}^0_{\text{Moor}}$ represent the static buoyancy and mooring loads respectively for the platform at the undisturbed position, which are also time independent.

The lumped mass representing the RNA was added to the relevant terms in the mass matrix. For simplicity, the moments of inertia of the RNA were set to zero. The mass and stiffness matrices of the tower can be easily formed for each element, using the material and geometric properties of the tower. The mass matrix of the platform can be obtained based on the structural properties provided in Ref. [31], while the platform stiffness matrix includes buoyancy stiffness defined in Ref. [32] and mooring line stiffness which will be defined in Section 2.2 later. Structural damping could easily be added as Rayleigh damping to this model, but zero structural damping was assumed throughout as this article aims at quantifying aerodynamic damping and hydrodynamic damping.

The FOWT is aerodynamically damped by wind-tower interaction and wind-rotor interaction. However, according to the literature, when a wind turbine is in operation the aerodynamic damping caused by wind-tower interaction is much smaller than that caused by wind-rotor interaction [12]. It can be implied that it is similar for a FOWT. Therefore, the aerodynamic damping caused by wind-tower interaction is ignored in this study.

The restrictive forces from the mooring lines are calculated by an open-source OpenMOOR program compiled in C++, which allows users to call this program in MATLAB [33]. The time integration algorithm used to solve the equation of motions in Eq. (1) in the time domain is HHT-α [34]–a generalized version of the widely-used Newmark-β method.

2.2. Linearised mooring lines

It is more straightforward to quantify the damping for a linear vibration system. However, calculating the real-time mooring forces by OpenMOOR [33] results in nonlinear stiffness and time-varying system stiffness matrix. Linearization of the mooring stiffness enables the direct calculation of aerodynamic and hydrodynamic damping ratios.

Mooring system stiffness linearization can be done in a theoretical way [31]. A more straightforward method was implemented in this study by directly fitting the stiffness terms with regard to different motion directions from simulated displacement-force relationships of the mooring lines. Six tests were conducted in the developed MATLAB model to obtain the displacement-force relationships. In each test, the spar-type platform was forced to move in a designated route, so mooring forces in six directions can be obtained. Six different routes result in 36 mooring displacement-force relationships. For example, the nonlinear relationship (red, solid) between the mooring force and spar displacement in the surge direction was first obtained by OpenMOOR, as shown in Fig. 4. Then a line (blue, dashed) was fitted to the nonlinear relationship by using a six-order polynomial, from which the stiffness term with respect to the "surge-surge" element in the platform stiffness matrix can be determined by the first-order derivative of the polynomial. The 6 × 6 linear mooring stiffness matrix accounting for the mooring effect can be finally determined via the above method, and relationships between the mean displacement and rotation of the platform and the mooring line stiffness in different motion directions are partly shown in Fig. 5. Moreover, this paper also investigates the central difference method for the stiffness determination [35], and the results (not shown) agree well with those in Fig. 5. The expression of the mooring stiffness matrix at the undisturbed position is .shown in Appendix A.

2.3. Damping calculation method

According to the derivations of aerodynamic damping and hydrodynamic which will be shown in Sections 3 and 4, the system damping matrix of the FOWT is asymmetric and nonproportional to the system mass and stiffness matrices. As a result, the determination of the damping ratios for the FOWT platform requires complex modal analysis [36]. With no external loads considered, the equation of motion (Eq. (1)) can be rewritten in the state space by

$$\mathbf{A}\dot{\mathbf{z}}(t) + \mathbf{B}\mathbf{z}(t) = \mathbf{0}, \tag{2}$$

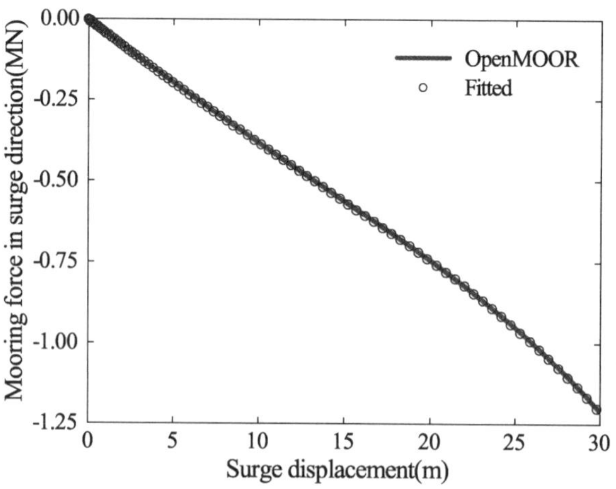

Fig. 4. The Real-time nonlinear (red, solid) and fitted (blue, dashed) relationships between the mooring force and spar displacement in the surge direction.

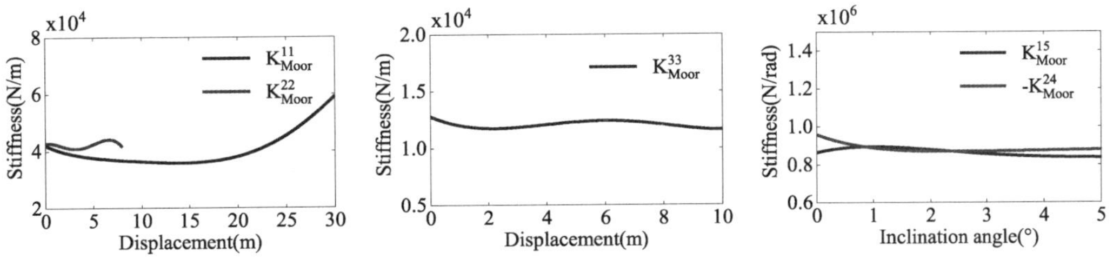

Fig. 5. Mooring system stiffnesses versus platform mean positions.

where

$$\mathbf{A}=\begin{bmatrix}\mathbf{C} & \mathbf{M}\\ \mathbf{M} & \mathbf{0}\end{bmatrix},\ \mathbf{B}=\begin{bmatrix}\mathbf{K} & \mathbf{0}\\ \mathbf{0} & -\mathbf{M}\end{bmatrix}. \tag{3}$$

The eigenvalues and eigenvectors of the state space model can be determined by the characteristic equation

$$\det(\lambda\mathbf{A}+\mathbf{B})=0. \tag{4}$$

The complex eigenvalue λ_n for the nth mode can be expressed by

$$\lambda_n=-\zeta_n\omega_n+i\omega_n\sqrt{1-\zeta_n^2}, \tag{5}$$

where ω_n is the natural frequency, and ζ_n is the damping ratio.

3. Aerodynamic damping derivation

3.1. Aerodynamic damping derivation considering tower bending

Aerodynamic damping caused by wind-rotor interaction originates from the fact that a moving blade experiences a drag force from the wind that impedes the movement of the blade. As the aerodynamic forces on a wind turbine blade depend on the relative wind velocity felt by the blade, analytically deriving aerodynamic damping for this blade first requires the expression of relative wind speed. The relative wind speed felt by a blade is not only related to inflow wind speed, but also related to the vibration of the blades and tower, platform motion and rotor rotation. It requires to introduce several assumptions so that the relative speeds felt by different blade elements can be conveniently determined. These assumptions are as follows.

(1) The connections between the tower top, the nacelle, and the rotor are considered rigid. This means that the motion of the rotor can be directly linked to the translational and rotational motions of the tower top. Vertical motion of the tower top is considered the same as that of the platform since the axial elastic deformation of the tower is negligible compared with the platform heave motion.
(2) The blades are assumed to be rigid, so blade flapwise and edgewise vibrations are ignored. This means the relative wind speed felt by a blade element is only influenced by the inflow wind speed, the tower top motion and the rotor rotation speed. This assumption has been proved to be reasonable in Ref. [15] for fixed-bottom wind turbines, and its feasibility will be examined for FOWTs later. The shaft tilt and blade precone angles were permanently set to zero to simplify the derivation.
(3) The vibration velocity of every blade element is much smaller than the corresponding inflow wind speed and the linear velocity caused by rotor rotation. This makes it possible for the aerodynamic loads to be linearised using first-order Taylor expansion near the inflow wind speed and rotor rotation speed.

With the assumptions made above, the vibration speed of a blade element can be divided into two parts. One part is the speed due to the platform motions, while another part is the speed due to tower top motions. Thus, the vibration velocity vector of a blade element can be written as

$$\mathbf{V}_\mathrm{B}=\mathbf{A}_\mathrm{p}\dot{\mathbf{U}}+\mathbf{A}_\mathrm{t}\dot{\mathbf{u}}. \tag{6}$$

The blades velocity vector in the global coordinate system is $\mathbf{V}_\mathrm{B}=\{V_{B,x}\ V_{B,y}\ V_{B,z}\ V_{B,\theta_x}\ V_{B,\theta_y}\ V_{B,\theta_z}\}^\mathrm{T}$. $\dot{\mathbf{U}}$ and $\dot{\mathbf{u}}$ represent the velocity vectors of the platform and the tower top respectively. These two velocity vectors are defined as

$$\dot{\mathbf{U}}=\{\dot{U}_1\ \dot{U}_2\ \dot{U}_3\ \dot{U}_4\ \dot{U}_5\ \dot{U}_6\}^\mathrm{T},\ \dot{\mathbf{u}}=\{\dot{x}\ \dot{y}\ \dot{z}\ \dot{\theta}_x\ \dot{\theta}_y\ \dot{\theta}_z\}^\mathrm{T}, \tag{7}$$

where the $\dot{U}_1$ to $\dot{U}_6$ represent the platform velocities in the surge, sway, heave, roll, pitch, and yaw directions respectively, as shown in Fig. 2. $\dot{\mathbf{u}}$ is defined in Fig. 3, in which $\dot{x}$, $\dot{y}$ and $\dot{z}$ represent the tower top translational velocities while $\dot{\theta}_x$, $\dot{\theta}_y$ and $\dot{\theta}_z$ represent the tower top rotational velocities. $\dot{z}$ is very small compared to $\dot{U}_3$, so only considering the vertical velocity of the platform is accurately enough to calculate the aerodynamic loads. Therefore, $\dot{z}$ will be set to zero in the following analysis. $\mathbf{A}_\mathrm{p}$ and $\mathbf{A}_\mathrm{t}$ are two transformation matrices through which the vibration speeds of blade elements can be linked to the platform velocities and tower top velocities. $\mathbf{A}_\mathrm{p}$ and $\mathbf{A}_\mathrm{t}$ are two similar 6-dimensional matrices. The matrix $\mathbf{A}_\mathrm{p}$ can be written as

$$\mathbf{A}_\mathrm{p}=\begin{bmatrix}1 & 0 & 0 & 0 & h_R+r\cos\Psi_j(t) & r\sin\Psi_j(t)\\ 0 & 1 & 0 & -(h_R+r\cos\Psi_j(t)) & 0 & 0\\ 0 & 0 & 1 & -r\sin\Psi_j(t) & 0 & 0\\ 0 & 0 & 0 & 0 & 0 & 0\\ 0 & 0 & 0 & 0 & 0 & 0\\ 0 & 0 & 0 & 0 & 0 & 0\end{bmatrix} \tag{8}$$

where r is the distance from the blade element to the rotor centre, $\Psi_j(t)$ is the azimuth of the jth blade ($j=1,2,3$). $\mathbf{A}_\mathrm{t}$ can be obtained from the expression of $\mathbf{A}_\mathrm{p}$ with h_R set to zero.

Assume wind turbine rotor experiences an inflow wind of steady velocity V_0 in the positive direction of the x axis. The rotor rotates positively clockwise around the x axis at a speed Ω, so a blade element at distance r along the blade length moves at a speed of $V_r=\Omega r$ in the direction tangential to the circular path at r. For an arbitrary blade element at distance r from the hub centre with azimuthal position $\Psi_j(t)$ of the jth blade, the relative wind speed in the normal direction, $V_{xj,Rel}$, and the relative wind speed in the tangential direction $V_{rj,Rel}$ can be written as

$$\begin{aligned}V_{xj,Rel}&=V_0-V_{B,x},\\ V_{rj,Rel}&=V_r-V_{B,y}\cos\Psi_j(t)-V_{B,z}\sin\Psi_j(t).\end{aligned} \tag{9}$$

Combining Eq. (6)~ Eq. (8), Eq. (9) can be rewritten as

$$\begin{aligned}V_{xj,Rel}&=V_0-(\dot{x}+\dot{U}_1+h_R\dot{U}_5)-(\dot{\theta}_y+\dot{U}_5)r\cos\Psi_j(t)-(\dot{\theta}_z+\dot{U}_6)r\sin\Psi_j(t),\\ V_{rj,Rel}&=V_r-(\dot{y}+\dot{U}_2-h_R\dot{U}_4)\cos\Psi_j(t)-\dot{U}_3\sin\Psi_j(t)+(\dot{\theta}_x+\dot{U}_4)r.\end{aligned} \tag{10}$$

A positive $V_{xj,Rel}$ is a velocity toward the positive direction of the x axis, while a positive $V_{rj,Rel}$ is a velocity having the opposite sign to the rotor rotation speed. Platform and tower top motions cause a variation of relative wind speed around V_0 and V_r.

With the expression of wind speed experienced in the normal and tangential directions, the aerodynamic forces applied on a single blade element can be calculated by the BEM theory. An explicit expression of the aerodynamic forces is unavailable but the forces can be regarded as the function of $V_{x,j,Rel}$ and $V_{r,j,Rel}$. Therefore, the forces in the normal and tangential directions can be written as $dT(V_{x,j,Rel}, V_{r,j,Rel})$ and $dS(V_{x,j,Rel}, V_{r,j,Rel})$. As mentioned earlier, platform and tower top velocities are small enough compared with the V_0 and V_r, which allows the Taylor's expansion at V_0 and V_r. Through Taylor's expansion, the aerodynamic force in the normal direction is given by

$$\begin{aligned} dT(V_{x,j,Rel}, V_{r,j,Rel}) = {} & dT(V_0, V_r) - \frac{\partial(dT)}{\partial V_0}(\dot{x}+\dot{U}_1+h_R\dot{U}_5) - \frac{\partial(dT)}{\partial V_0}(\dot{\theta}_y+\dot{U}_5) \\ & r\cos\Psi_j - \frac{\partial(dT)}{\partial V_0}(\dot{\theta}_z+\dot{U}_6)r\sin\Psi_j - \frac{\partial(dT)}{\partial V_r}(\dot{y}+\dot{U}_2-h_R\dot{U}_4) \\ & \cos\Psi_j - \frac{\partial(dT)}{\partial V_r}\dot{U}_3\sin\Psi_j + \frac{\partial(dT)}{\partial V_r}(\dot{\theta}_x+\dot{U}_4)r, \end{aligned} \tag{11}$$

where $dT(V_0, V_r)$ and $dS(V_0, V_r)$ are the forces in normal and tangential directions when the blade element experiences wind speeds of V_0 and V_r in x direction and the tangential direction. For simplicity, $dT(V_0, V_r)$, $dS(V_0, V_r)$ and $\Psi_j(t)$ are abbreviated as dT, dS and Ψ_j. Similarly, the force in the tangential direction is given by

$$\begin{aligned} dS(V_{x,j,Rel}, V_{r,j,Rel}) = {} & dS(V_0, V_r) - \frac{\partial(dS)}{\partial V_0}(\dot{x}+\dot{U}_1+h_R\dot{U}_5) - \frac{\partial(dS)}{\partial V_0}(\dot{\theta}_y+\dot{U}_5) \\ & r\cos\Psi_j - \frac{\partial(dS)}{\partial V_0}(\dot{\theta}_z+\dot{U}_6)r\sin\Psi_j - \frac{\partial(dS)}{\partial V_r}(\dot{y}+\dot{U}_2-h_R\dot{U}_4) \\ & \cos\Psi_j - \frac{\partial(dS)}{\partial V_r}\dot{U}_3\sin\Psi_j + \frac{\partial(dS)}{\partial V_r}(\dot{\theta}_x+\dot{U}_4)r. \end{aligned} \tag{12}$$

The moments about x, y and z axes are given by

$$\begin{aligned} dM_x(V_{x,j,Rel}, V_{r,j,Rel}) &= dS(V_{x,j,Rel}, V_{r,j,Rel})r \\ dM_y(V_{x,j,Rel}, V_{r,j,Rel}) &= dT(V_{x,j,Rel}, V_{r,j,Rel})r\cos\Psi_j \\ dM_z(V_{x,j,Rel}, V_{r,j,Rel}) &= dT(V_{x,j,Rel}, V_{r,j,Rel})r\sin\Psi_j \end{aligned} \tag{13}$$

The total forces and moments from the rotor applied to the tower top are obtained by summing the elemental forces and moments of the three blades. As $\sum_{j=1}^{N_b}\cos\Psi_j(t) = 0$ for any time t, the total force in the x (FA, Surge) direction is

$$\begin{aligned} F_x = {} & \sum_{j=1}^{N_b} \\ & \times \int_0^R dT(V_{x,j,rel}, V_{r,j,Rel}) = N_b\int_0^R dT(V_0, V_r) - N_b(\dot{x}+\dot{U}_1+h_R\dot{U}_5)\int_0^R \frac{\partial(dT)}{\partial V_0} \\ & + N_b(\dot{\theta}_x+\dot{U}_4)\int_0^R r\frac{\partial(dT)}{\partial V_r}, \end{aligned} \tag{14}$$

where N_b is the number of blades, and r is the radius of the blade. The total force in the y (SS, Sway) direction is

$$\begin{aligned} F_y = {} & -\sum_{j=1}^{N_b}\int_0^R dS(V_{x,j,Rel}, V_{r,j,Rel})\cos\Psi_j = \frac{N_b}{2}(\dot{\theta}_y+\dot{U}_5)\int_0^R r\frac{\partial(dS)}{\partial V_0} \\ & + \frac{N_b}{2}(\dot{y}+\dot{U}_2-h_R\dot{U}_4)\int_0^R \frac{\partial(dS)}{\partial V_r}. \end{aligned} \tag{15}$$

The negative sign on the left-hand side of the summation operator is due to the fact that for a rotor rotating positively, the blade at a position above the hub is subjected to a tangential force towards the negative y-direction. This is a consequence of the chosen coordinate system. In addition, it can be easily checked that $\sum_{j=1}^{N_b}\cos^2\Psi_j(t) = \frac{N_b}{2}$. The total force in the z direction is

$$\begin{aligned} F_z = {} & -\sum_{j=1}^{N_b}\int_0^R dS(V_{x,j,Rel}, V_{r,j,Rel})\sin\Psi_j(t) = \frac{N_b}{2}(\dot{\theta}_z+\dot{U}_6)\int_0^R r\frac{\partial(dS)}{\partial V_0} \\ & + \frac{N_b}{2}\dot{U}_3\int_0^R \frac{\partial(dS)}{\partial V_r}. \end{aligned} \tag{16}$$

The total moment about the x axis is

$$\begin{aligned} M_x = {} & \sum_{j=1}^{N_b} \\ & \times \int_0^R dM_x(V_{x,j,Rel}, V_{r,j,Rel}) = N_b\int_0^R dS(V_0, V_r)r - N_b(\dot{x}+\dot{U}_1+h_R\dot{U}_5)\int_0^R r\frac{\partial(dS)}{\partial V_0} \\ & + N_b(\dot{\theta}_x+\dot{U}_4)\int_0^R r^2\frac{\partial(dS)}{\partial V_r}. \end{aligned} \tag{17}$$

Similarly, the total moment about the y axis is

$$\begin{aligned} M_y = {} & \sum_{j=1}^{N_b}\int_0^R dM_y(V_{x,j,Rel}, V_{r,j,Rel}) = -\frac{N_b}{2}(\dot{\theta}_y+\dot{U}_5)\int_0^R r^2\frac{\partial(dT)}{\partial V_0} \\ & - \frac{N_b}{2}(\dot{y}+\dot{U}_2-h_R\dot{U}_4)\int_0^R r\frac{\partial(dT)}{\partial V_r}. \end{aligned} \tag{18}$$

Whereas the total moment about the z axis is

$$\begin{aligned} M_z = {} & \sum_{j=1}^{N_b}\int_0^R dM_z(V_{x,j,Rel}, V_{r,j,Rel}) = -\frac{N_b}{2}(\dot{\theta}_z+\dot{U}_6)\int_0^R r^2\frac{\partial(dT)}{\partial V_0} \\ & - \frac{N_b}{2}\dot{U}_3\int_0^R r\frac{\partial(dT)}{\partial V_r}. \end{aligned} \tag{19}$$

The partial derivatives in Eqs. (14) ~ (19) can be found in Appendix B of the authors' previous work [15].

3.2. *Aerodynamic load resultants at the tower top and platform*

The aerodynamic loads applied on the tower top have been derived using Eqs. (14) ~ (19). These load terms can be divided into static terms and terms related to blade vibration. The static terms $N_b\int_0^R dT(V_0, V_r)$ and $N_b\int_0^R dS(V_0, V_r)r$ are irrelevant to blade vibration state (displacement, velocity and accelerate), only depending on the inflow wind speed and rotor rotation speed. While other terms are "damping" terms as they are proportional to the tower and platform velocities. The static terms can be easily calculated using the BEM theory coded in MATLAB given the inflow wind speed and the rotational speed. The other terms in F_x, F_y, F_z, M_x, M_y and M_z appear as aerodynamic damping components since they depend linearly on the translational or angular velocities of the tower top and platform. Let a force vector $\mathbf{F}_{Aerot} = \{F_x\ F_y\ F_z\ M_x\ M_y\ M_z\}^T$ collects the aerodynamic resultants on the tower top. According to theorem of force transformation [37], the aerodynamic force applied on the platform denoted as $\mathbf{F}_{Aerop}$ can be obtained from $\mathbf{F}_{Aerot}$ by

$$\mathbf{F}_{Aerop} = \mathbf{A}_{pt}\mathbf{F}_{Aerot}, \tag{20}$$

in which $\mathbf{A}_{pt}$ is a transformation matrix. More specifically, the platform damping force $\mathbf{F}^{damp}_{Aerop}$ can be written as

$$\mathbf{F}^{damp}_{Aerop} = \mathbf{C}^{pp}_{Aero}\dot{\mathbf{U}} + \mathbf{C}^{pt}_{Aero}\dot{\mathbf{u}}. \tag{21}$$

It means that the damping force is composed of two parts, one is induced by the tower top velocities while another is the force caused by the platform motion velocities. The two platform damping matrices $\mathbf{C}^{pp}_{Aero}$, $\mathbf{C}^{pt}_{Aero}$ can be written according to Eqs. (14) ~ (19). $\mathbf{C}^{pp}_{Aero}$ is defined by:

$$\mathbf{C}^{pp}_{Aero} = \mathbf{A}_{pt}\mathbf{C}^{tp}_{Aero} = \begin{bmatrix} c_{xU_1} & 0 & 0 & c_{xU_4} & c_{xU_5} & 0 \\ 0 & c_{yU_2} & 0 & c_{yU_4} & c_{yU_5} & 0 \\ 0 & 0 & c_{zU_3} & 0 & 0 & c_{zU_6} \\ c_{\theta_x U_1} & -h_R c_{yU_2} & 0 & c_{\theta_x U_4} - h_R c_{yU_4} & c_{\theta_x U_5} - h_R c_{yU_5} & 0 \\ h_R c_{xU_1} & c_{\theta_y U_2} & 0 & c_{\theta_y U_4} + h_R c_{xU_4} & c_{\theta_y U_5} + h_R c_{xU_5} & 0 \\ 0 & 0 & c_{\theta_z U_3} & 0 & 0 & c_{\theta_z U_6} \end{bmatrix}. \tag{22}$$

$\mathbf{C}^{pt}_{Aero}$ can be given as follows:

$$\mathbf{C}^{pt}_{Aero} = \mathbf{A}_{pt}\mathbf{C}^{tt}_{Aero} = \begin{bmatrix} c_{xx} & 0 & 0 & c_{x\theta_x} & 0 & 0 \\ 0 & c_{yy} & 0 & 0 & c_{y\theta_y} & 0 \\ 0 & 0 & 0 & 0 & 0 & c_{z\theta_z} \\ c_{\theta_x x} & -h_R c_{yy} & 0 & c_{\theta_x\theta_x} & -h_R c_{y\theta_y} & 0 \\ h_R c_{xx} & c_{\theta_y y} & 0 & h_R c_{x\theta_x} & c_{\theta_y\theta_y} & 0 \\ 0 & 0 & 0 & 0 & 0 & c_{\theta_z\theta_z} \end{bmatrix}. \tag{23}$$

$\mathbf{C}^{pt}_{Aero}$ has a similar form with $\mathbf{C}^{pp}_{Aero}$, while the latter is more complex than the former. The detailed expressions of the coefficients in these two aerodynamic damping matrices are listed in Appendix B. $\mathbf{C}^{pp}_{Aero}$ has several relevant structural features:

(1) For a given combination of operation states including the inflow wind speed, rotation speed, and the blade pitch angle, the coefficients in $\mathbf{C}^{pp}_{Aero}$ are constant and can therefore be calculated using the steady BEM theory. Moreover, unlike the conventional proportional damping matrix, antisymmetric coefficients are included in the aerodynamic damping matrix which is proved common in rotating machinery [38].

(2) Platform translational DOFs (surge and sway) are not coupled directly since there are no relevant coefficients in the damping matrix while the translational and rotational DOFs are coupled through the off-diagonal terms.

(3) Rotational DOFs around X -axis and Y -axis are coupled directly, due to the submatrix:

$$\begin{bmatrix} c_{\theta_x U_4} - h_R c_{yU_4} & c_{\theta_x U_5} - h_R c_{yU_5} \\ c_{\theta_y U_4} + h_R c_{xU_4} & c_{\theta_y U_5} + h_R c_{xU_5} \end{bmatrix} \tag{24}$$

(4) c_{xU_4} introduces a coupling effect between surge and roll, while the roll motion is also coupled with the sway motion, resulting in an indirect coupling between the surge and sway motions. c_{yU_5} also has a similar effect.

(5) Heave and yaw motions are coupled, but these two motions are not coupled with any other platform DOFs.

The aerodynamic damping matrix $\mathbf{C}^{pp}_{Aero}$ derived in this section is helpful to provide a better understanding of the coupling between platform motions from aerodynamic damping.

4. Hydrodynamic damping derivation

Hydrodynamic effects may also introduce significant damping to a spar-type platform due to water-platform interaction. The platform can be regarded as an oscillating cylinder subjected to hydrodynamic forces categorised as added mass in phase with platform acceleration, radiation damping, and viscous drag both in phase with platform velocity and buoyancy related to the volume below water [39]. The relative importance of the inertial, drag, and diffraction/radiation wave forces depends on some key nondimensional parameters such as the Keulegan-Carpenter number and Stokes parameter. For large-size spar platform of FOWTs, it was reported in literature [40] that both viscous and radiation damping may be important depending on wave height. Therefore, it is worthwhile to quantify the hydrodynamic damping contributions of the spar platform originated from wave radiation and viscous effects. In the present study, radiation damping effect is considered by an analytical potential flow solution, while viscous damping effect is considered via Morison's equation. Taper on the upper part of the platform is ignored in this paper, and the platform is considered a simple cylinder instead, which can simplify hydrodynamic damping derivations.

4.1. *Radiation damping via analytical potential flow solution*

To quantify the dissipation effect from water radiation caused by an oscillating cylinder, numerical and analytical methods based on potential flow theory [41] are usually used. As this study aims at obtaining a quick and preliminary quantification of damping contribution, an analytical method recently developed by Li et al. [42] is used to calculate the hydrodynamic added mass and radiation damping coefficients of the spar platform. This analytical method solves water wave radiation by vertical truncated circular cylinders based on linear potential flow theory and multi-term Galerkin method, which has the merits of rapid convergence. A spar platform can be categorised as the Type 1 cylinder which is a surface-piercing floating cylinder in Li's method. Assuming the cylinder is oscillating with an angular frequency of ϖ, the added mass coefficient $c^{A}_{U_1U_1}$ and radiation damping coefficient $c^{rad}_{U_1U_1}$ with regard to surge motion can be determined by

$$c^{A}_{U_1U_1} + \mathrm{i}\frac{c^{rad}_{U_1U_1}}{\varpi} = -\xi\rho_w \int_{-h_c}^{0}\int_{0}^{2\pi} \varphi^{(1)}_1(a,\theta,\overline{Z})(-a\cos\theta)\mathrm{d}\theta\mathrm{d}\overline{Z}, \tag{25}$$

where $\mathrm{i} = \sqrt{-1}$, ξ is the platform oscillating amplitude, ρ_w is the water density, h_c is the height of the cylinder, $\varphi^{(1)}_1$ represents the radiation velocity potential with regard to surge motion in the exterior domain, which is a function of the radius position r_c (here $r_c = a$, a is the cylinder radius), angular position θ and vertical position $\overline{Z}$ (the coordinate origin is placed on Still Water Level (SWL)). The formulae to calculate the added mass coefficients and radiation damping coefficients with regard to the heave, pitch motions are introduced in Appendix C. Also, the calculation formulae of the hydrodynamic coefficients associated with coupling motions such as surge-pitch motion are provided with slight modifications to formulae in Ref. [42]. Radiation damping matrix can be expressed as follows:

$$\mathbf{C}_{Rad} = \begin{bmatrix} c^{rad}_{U_1U_1} & 0 & 0 & 0 & c^{rad}_{U_1U_5} & 0 \\ 0 & c^{rad}_{U_2U_2} & 0 & c^{rad}_{U_2U_4} & 0 & 0 \\ 0 & 0 & c^{rad}_{U_3U_3} & 0 & 0 & 0 \\ 0 & c^{rad}_{U_4U_2} & 0 & c^{rad}_{U_4U_4} & 0 & 0 \\ c^{rad}_{U_5U_1} & 0 & 0 & 0 & c^{rad}_{U_5U_5} & 0 \\ 0 & 0 & 0 & 0 & 0 & 0 \end{bmatrix} \tag{26}$$

To check the accuracy of the analytical method, the radiation damping coefficients analytically calculated were compared to those obtained by Aqwa, as shown in Fig. 6. It can be seen that the analytical method is able to calculate very similar damping coefficients for surge, pitch and surge-pitch motions compared to those from Aqwa. However,

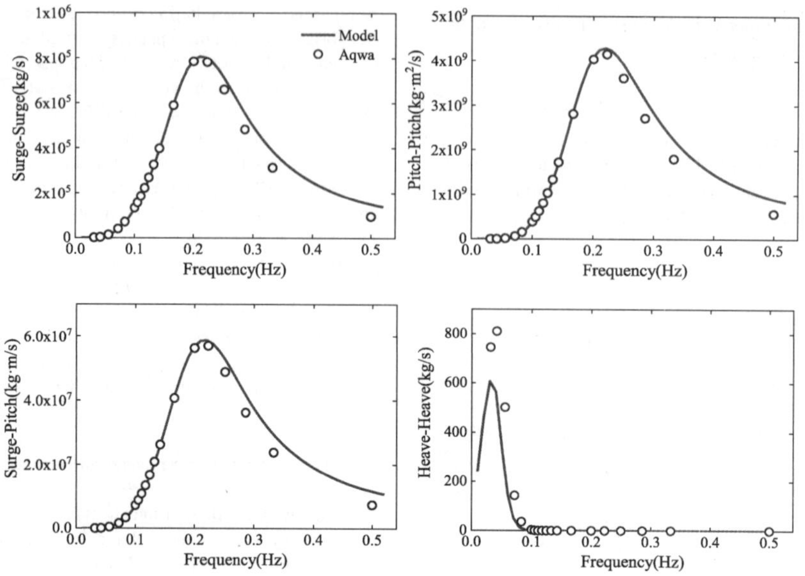

Fig. 6. Radiation damping of platform verification against Aqwa: platform height = 120m, water depth = 320m, platform diameter = 9.4m.

the damping coefficient for heave motion obtained by the analytical method is higher than those from Aqwa in the low frequency range. As the heave damping coefficients is much lower than other three coefficients, the less accurate heave damping coefficient only has insignificant impact.

4.2. *Viscous damping via Morison' equation*

Morison's equation is a widely used method to calculate the hydrodynamic forces applied on a vibrating structure. Morison's equation has widely been used in the analysis of fixed-bottom offshore wind turbines [43,44]. It has also been used directly in the analysis of FOWTs [45]. Morison's equation can be used to efficiently compute the linear or nonlinear wave loads, especially for slender vertical surface-piercing cylinders that extend to the seafloor. A structure is first split into several strips, the added-mass and viscous-drag forces applied on each strip can be determined, from which the overall hydrodynamic forces can be obtained by integrating the forces applied all strips. According to Morison's equation, for a fixed-bottom structure, at time t, hydrodynamic load applied on a platform strip located at heigh of Z can be written as

$$\mathrm{d}F_i^{Platform}(t,Z)=-C_A\rho_w\left(\frac{\pi D^2}{4}\mathrm{d}Z\right)\ddot{U}_i(Z)+(1+C_A)\rho_w\left(\frac{\pi D^2}{4}\mathrm{d}Z\right)a_i(t,Z)$$
$$+\frac{1}{2}C_D\rho_w(D\cdot\mathrm{d}Z)[v_i(t,Z)-\dot{U}_i(Z)]\left[(v_1(t,Z)-\dot{U}_1(Z))^2+(v_2(t,Z)-\dot{U}_2(Z))^2\right]^{\frac{1}{2}}, \tag{27}$$

where the coefficients C_A and C_D are the normalized hydrodynamic-added-mass and viscous-drag coefficients respectively. For a spar-type platform, C_A and C_D are taken as 0.97 and 0.6 [31], D is the diameter of the cylinder strip, dZ is the strip height, $\ddot{U}_i(Z)$ and $\dot{U}_i(Z)$ are the strip acceleration and velocity of the ith DOF of a platform, v_i and a_i are the components of the undisturbed fluid-particle velocity and acceleration in the direction of ith DOF. The quadratic drag force becomes very small if the platform moves slowly [45].

The drag force in Morison' equation causes viscous damping effect, from which the viscous damping of the platform can be quantified. However, the viscous damping ratio cannot be directly calculated due to the quadratic drag terms. It is required to linearise the viscous hydrodynamic damping for the damping ratio calculation. For a floating platform, both global translational and rotational vibrate velocities contribute to finial translational velocities for a platform strip. Therefore, the drag force for each platform strip can be linearised as follows [46]:

$$\begin{aligned}\mathrm{d}F_1^{Viscous}(t,Z)&=\frac{1}{2}C_D\rho_w D\mathrm{d}Z\times[v_1(t,Z)|v_1(t,Z)|-2(\dot{U}_1+\dot{U}_5Z)|v_1(t,Z)|]\\ \mathrm{d}F_2^{Viscous}(t,Z)&=\frac{1}{2}C_D\rho_w D\mathrm{d}Z\times[0-2(\dot{U}_2-\dot{U}_4Z)|v_1(t,Z)|]\\ \mathrm{d}F_4^{Viscous}(t,Z)&=-\mathrm{d}F_2^{Viscous}(t,Z)Z\\ \mathrm{d}F_5^{Viscous}(t,Z)&=\mathrm{d}F_1^{Viscous}(t,Z)Z.\end{aligned} \tag{28}$$

This linearization method was introduced for fixed bottom monopile platforms in Ref. [46]. When used for floating spar type platform simulation, its accuracy should be verified firstly. Given an unsteady wave generated from the JONSWAP spectrum, both the original Morison's equation (labeled Nonlinear) and the lineari ed method were used to calculate the platform responses. As shown in Fig. 7, the linear Morison's equation is accurate enough for the spar-type platform response simulation under the JONSWAP spectrum as the platform responses by these two methods are very close.

Viscous drag forces are related to fluid particle velocities and platform moving velocities in different directions, among which the terms proportional to the platform velocities can be defined as the viscous damping forces. Then, a hydrodynamic viscous damping matrix can be further defined from the viscous damping forces. The viscous damping force at a strip can be expressed by:

$$\begin{aligned}\mathrm{d}F_{1,damp}^{Viscous}(t,Z)&=-C_D\rho_w D\mathrm{d}Z|v_1(t,Z)|(\dot{U}_1+\dot{U}_5Z),\\ \mathrm{d}F_{2,damp}^{Viscous}(t,Z)&=-C_D\rho_w D\mathrm{d}Z|v_1(t,Z)|(\dot{U}_2-\dot{U}_4Z).\end{aligned} \tag{29}$$

The total viscous damping forces applied to the platform in the surge, sway, pitch, and roll directions can be further given by:

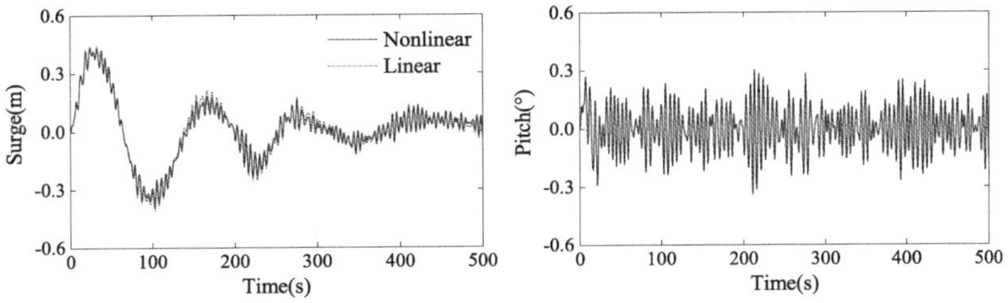

Fig. 7. Linearization of Morison's equation verification against the nonlinear model, surge and pitch responses under JONSWAP spectrum.

$$
\begin{aligned}
F_{1,damp}^{Viscous}(t,Z) &= -C_D\rho_w D\int_{-h_B}^{h_T}(\dot{U}_1+\dot{U}_5 Z)|v_1(t,Z)|\mathrm{d}Z,\\
F_{2,damp}^{Viscous}(t,Z) &= -C_D\rho_w D\int_{-h_B}^{h_T}(\dot{U}_2-\dot{U}_4 Z)|v_1(t,Z)|\mathrm{d}Z,\\
F_{4,damp}^{Viscous}(t,Z) &= C_D\rho_w D\int_{-h_B}^{h_T}(\dot{U}_2 Z-\dot{U}_4 Z^2)|v_1(t,Z)|\mathrm{d}Z,\\
F_{5,damp}^{Viscous}(t,Z) &= -C_D\rho_w D\int_{-h_B}^{h_T}(\dot{U}_1 Z+\dot{U}_5 Z^2)|v_1(t,Z)|\mathrm{d}Z.
\end{aligned}
\tag{30}
$$

Based on the linear wave theory, in deep water, the fluid-particle velocity at a given water depth Z at any time t can be obtained as follows [47].

$$
v_1(t,Z)=H_s\sigma e^{kZ}\sin\sigma t \tag{31}
$$

where H_s is the wave height, σ is the wave angular frequency, k is the wavenumber. For deep water, the dispersion relation is simplified as follows [48,49].

$$
\sigma^2=gk \tag{32}
$$

Thus, hydrodynamic viscous damping forces applied on the platform DOFs can be further developed by inserting Eq. (31) and Eq. (32) into Eq. (30). The expressions for linearised viscous damping forces are given as

$$
\begin{aligned}
F_{1,damp}^{Viscous}(t,Z) &= -C_D\rho_w DH_s\sigma|\sin\sigma t|\left[A_1 e^{k(Z-h_T)}\dot{U}_1+A_2 e^{k(Z-h_T)}\dot{U}_5\right]_{-h_B}^{h_T},\\
F_{2,damp}^{Viscous}(t,Z) &= -C_D\rho_w DH_s\sigma|\sin\sigma t|\left[A_1 e^{k(Z-h_T)}\dot{U}_2-A_2 e^{k(Z-h_T)}\dot{U}_4\right]_{-h_B}^{h_T},\\
F_{4,damp}^{Viscous}(t,Z) &= C_D\rho_w DH_s\sigma|\sin\sigma t|\left[A_2 e^{k(Z-h_T)}\dot{U}_2-A_3 e^{k(Z-h_T)}\dot{U}_4\right]_{-h_B}^{h_T},\\
F_{5,damp}^{Viscous}(t,Z) &= -C_D\rho_w DH_s\sigma|\sin\sigma t|\left[A_2 e^{k(Z-h_T)}\dot{U}_1+A_3 e^{k(Z-h_T)}\dot{U}_5\right]_{-h_B}^{h_T},
\end{aligned}
\tag{33}
$$

in which

$$
\begin{aligned}
A_1 &= \frac{1}{k}\\
A_2 &= \frac{Z}{k}-\frac{1}{k^2}\\
A_3 &= \frac{Z^2}{k}-\frac{2Z}{k^2}+\frac{2}{k^3}
\end{aligned}
\tag{34}
$$

A viscous damping matrix can be finally defined by:

$$
\mathbf{C}_{\mathrm{Vis}}^{\mathrm{Morison}}=\begin{bmatrix}
c_{U_1U_1}^{vis} & 0 & 0 & 0 & c_{U_1U_5}^{vis} & 0\\
0 & c_{U_2U_2}^{vis} & 0 & c_{U_2U_4}^{vis} & 0 & 0\\
0 & 0 & 0 & 0 & 0 & 0\\
0 & c_{U_4U_2}^{vis} & 0 & c_{U_4U_4}^{vis} & 0 & 0\\
c_{U_5U_1}^{vis} & 0 & 0 & 0 & c_{U_5U_5}^{vis} & 0\\
0 & 0 & 0 & 0 & 0 & 0
\end{bmatrix}
\tag{35}
$$

in which

$$
\begin{aligned}
c_{U_1U_1}^{vis} &= C_D\rho_w DH_s\sigma\tau A_1 e^{kZ}\Big|_{-h_B}^{h_T},\\
c_{U_1U_5}^{vis} &= C_D\rho_w DH_s\sigma\tau A_2 e^{kZ}\Big|_{-h_B}^{h_T},\\
c_{U_4U_4}^{vis} &= C_D\rho_w DH_s\sigma\tau A_3 e^{kZ}\Big|_{-h_B}^{h_T},\\
c_{U_2U_2}^{vis} &= c_{U_1U_1}^{vis},\ c_{U_5U_5}^{vis}=c_{U_4U_4}^{vis},\\
c_{U_2U_4}^{vis} &= c_{U_4U_2}^{vis}=-c_{U_1U_5}^{vis}=-c_{U_5U_1}^{vis}.
\end{aligned}
\tag{36}
$$

Damping terms in the viscous damping matrix are time-varying as $|\sin\sigma t|$ changes with time. A parameter τ is introduced to make the viscous damping matrix time invariant. Eq. (37) was proposed to average the viscous damping terms:

$$
\tau=\frac{\int_0^{\frac{2\pi}{\sigma}}|\sin\sigma t|\mathrm{d}t}{2\pi/\sigma}=\frac{2}{\pi}. \tag{37}
$$

Based on the derivations of aerodynamic damping and hydrodynamic damping, a simplified dynamic model can be established from Eq. (1) to rapidly calculate dynamic responses of FOWTs. Additionally, in contrast to conventional operational modal analysis which aims at identifying modal damping ratios for a structural system (e.g. Refs. [6, 48,49]), damping identification for an operating FOWT could aim to identify the aerodynamic damping matrix (e.g. using a time-frequency analysis such as that proposed by Chen et al. [50]).

5. Model verification against FAST

5.1. FAST model settings for verification

A FOWT model in FAST was used to verify the derived expressions of aerodynamic damping and hydrodynamic damping. A bulk of simplifications is made in the FAST model to make the settings in FAST are consistent with the derivations: (1) the RNA centre is moved to the tower top; (2) the moments of inertia of the RNA relative to the tower top is considered as zero; (3) the mode shapes coefficients are updated due to the tower top mass changes; (4) "classic BEM theory with the Prandtl and Glauert's corrections" were chosen in FAST's AeroDyn module for keeping consistent with the derivation in Section 3; (5) the mooring system calculation method is chosen as "MAP++". To consider the influence of the control system, the standard relationships between wind speed, rotor speed, and pitch angle were used, as shown in Fig. 8. In the following discussion, the inflow wind speed, pitch angle, and rotor speed are kept constant for one operational state.

5.2. Verification and general description of the response

With the expressions of aerodynamic damping and hydrodynamic damping previously derived, a simplified FOWT model described in Section 2.1 can be established for rapid dynamic response generation. In this FOWT model, the aerodynamic damping matrix, radiation damping matrix, and viscous damping matrix are added to the total damping matrix. To test the simplified model, it was first assumed that the FOWT is only subjected to a regular wave (wave height is 5m, and wave

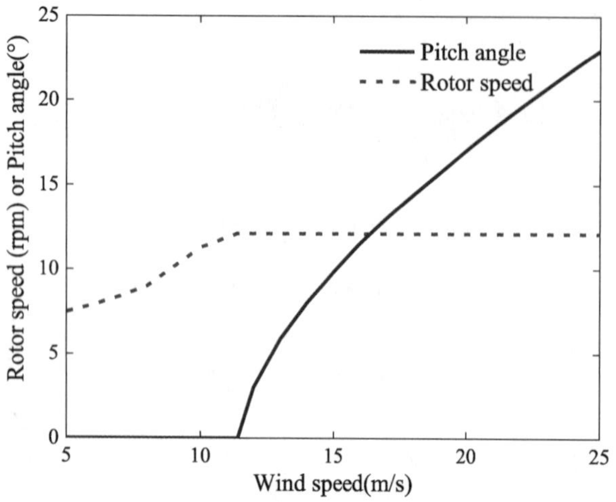

Fig. 8. Standard relationship between the rotor speed (dashed), pitch angle (solid), and inflow wind speed, based on [51].

frequency is 0.01Hz–0.45Hz). The resultant Response Amplitude Operators (RAOs), which represents the ratios of the platform motion amplitude to the wave amplitude [52], are compared with those from the FAST model with equivalent settings in Fig. 9. RAOs were obtained from the steady-state platform motions. It is observed the surge, roll and pitch motion RAOs induced by the regular wave from the simplified FOWT model and the FAST model agree well at most frequencies. Larger difference in RAOs exists around the resonance frequency as the response is more sensitive to the wave frequency in this frequency range. In this case, the roll motion from the simplified FOWT model is hardly excited as there is no coupling between the surge/pitch motion and the roll motion through the hydrodynamic radiation and viscous damping matrices. The above comparison results prove the linearised model with the hydrodynamic damping matrices previously defined can correctly generate platform RAOs under regular wave effect.

Then, the computations of the FOWT responses were performed by these two models under the regular wave and constant wind field. These computations were conducted to further verify the contribution of aerodynamics damping. Several combinations of constant wind speeds (5 m/s, 7 m/s, 9 m/s, 11.4 m/s, 15 m/s, 17 m/s, 20 m/s, 22 m/s, and 25 m/s) and regular wave frequencies from 0.01Hz–0.25Hz were selected and for each combination the RAOs of the platform were calculated. The resultant RAOs in different motion directions are compared, as shown in the waterfall plot in Fig. 10. It shows that the platform responses from these two models generally agree well with slight frequency shifts and small differences in vibration magnitude. A great agreement is obtained at all mean wind speeds. It should be noted that the surge response decreases when the wind speed grows until the rated wind speed, and the surge RAO keeps stable with the growth of wind speed between the rated and cut-out wind speed.

A similar and even more noticeable trend is observed for the pitch motion, while a different trend occurs for the roll motion. By comparing the roll motion RAOs in Fig. 9 and in Fig. 10, it can be found that the roll motions in the case with both wave and wind considered are excited, and this is because of the constant wind field and aerodynamic coupling effect caused by the non-zero off-diagonal elements in the aerodynamic damping matrix in Eq. (23).

Platform responses from the simplified FOWT model in the time domain are also compared with those from the FAST model for further verification. A typical operational condition with a wave height of 5m, wave frequency of 0.035Hz, and mean wind speed of 15 m/s was used for this comparison. Initial displacement of 40m and rotations of 5° and 10° were given for the surge, roll, and pitch motions respectively. The time domain responses for the surge, roll and pitch motions are compared in Fig. 11 and good agreements in amplitude and phase can be found between the responses from the simplified model and the FAST model for both the initial decays and the subsequent steady-state responses, proving the proposed linearised model is accurate in dynamic response generation.

6. Discussion of results

This section presents the damping ratio calculation results of the spar-type platform and studies the influence of different damping sources on FOWT dynamic responses. Damping ratios were calculated using the method in Section 2.3. The natural frequencies of the FOWT for primary vibration modes of the platform and tower are listed in Table 2. The calculation of the natural frequencies is based on the mooring stiffness corresponding to the undisturbed position of the platform.

6.1. Aerodynamic damping

6.1.1. Considering tower flexibility

Calculation results of aerodynamic damping ratios of the platform are present in this subsection, with aerodynamic forces calculated by Eq.

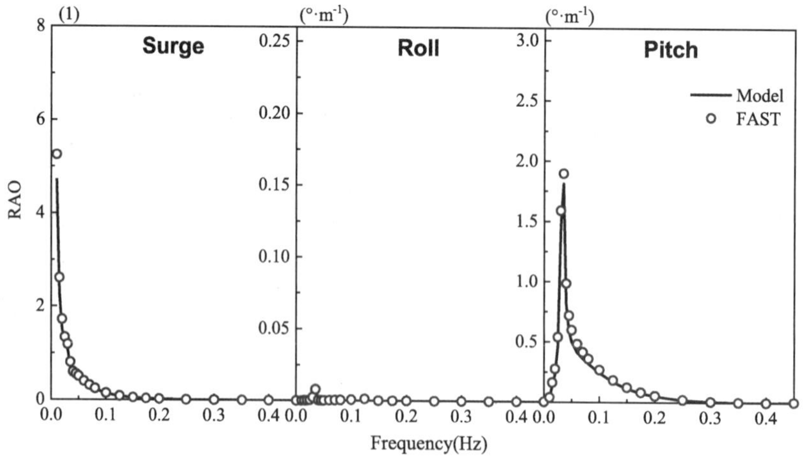

Fig. 9. Wave-induced platform motion RAOs (without wind effect, wave frequency is 0.01Hz–0.45Hz) from the simplified FOWT model and the FAST model.

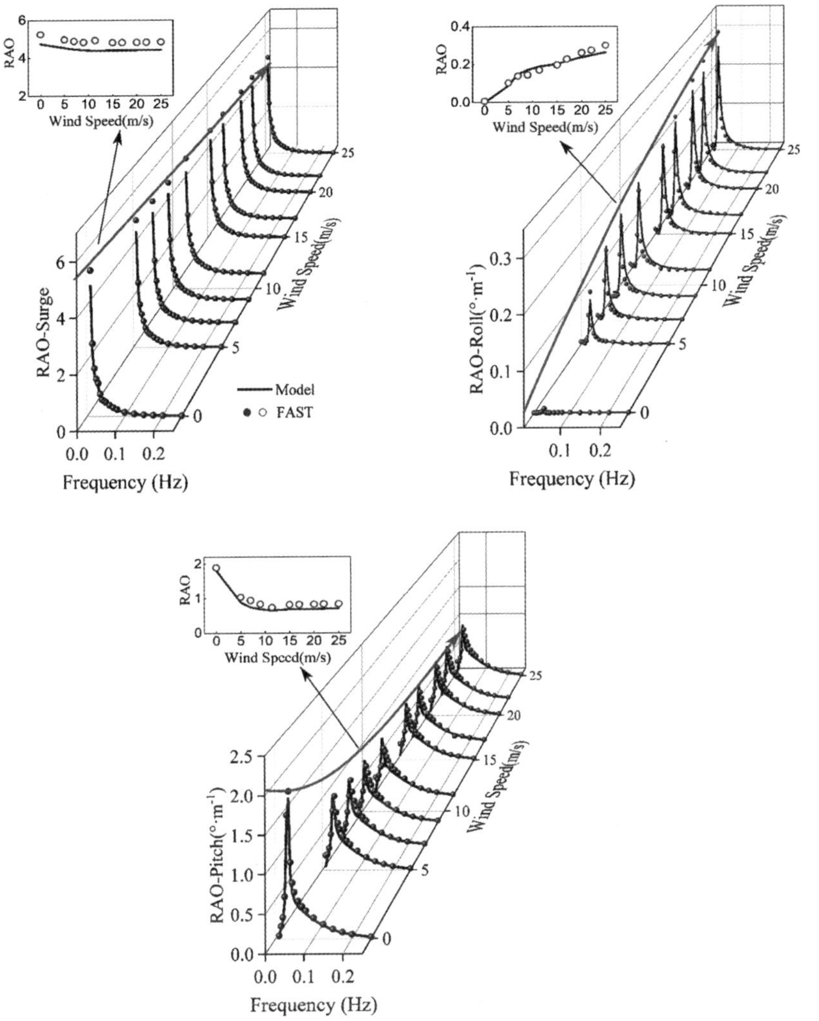

Fig. 10. Joint wind-wave-induced platform motion RAOs (constant wind speed from 0 to 25 m/s, wave frequency is 0.01Hz–0.25Hz) from the simplified FOWT model and the FAST model.

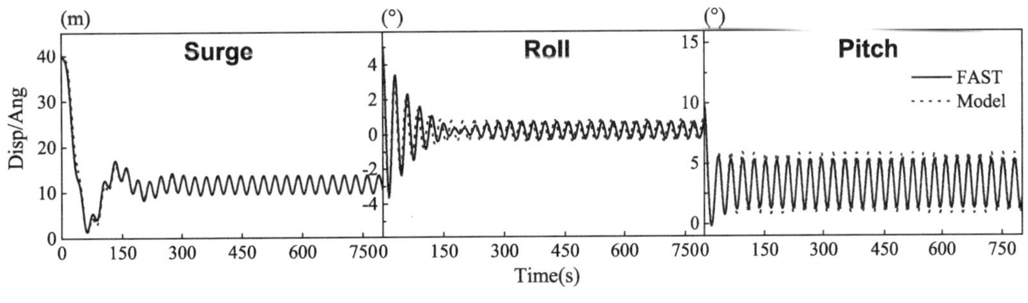

Fig. 11. Decay test with the regular wave and constant wind effect (wave height = 5m, wave frequency = 0.035Hz, wind speed = 15m/s).

(21) considering tower flexibility. When the mean wind speed increases from 5 m/s to 25 m/s, the calculated aerodynamic damping ratios are shown in Fig. 12. The aerodynamic damping ratio curves for the surge, pitch, and yaw motions share a similar trend: aerodynamic damping ratio increases with the wind speed when the wind speed is lower than the rated speed and then keep nearly constant when the wind speed higher than the rated speed. It can be seen that the aerodynamic damping ratios in the surge motion is generally around 5%, much lower than those in the yaw and pitch motions. The aerodynamic damping ratio curves in sway, heave, and roll motions also have a similar trend: aerodynamic damping ratio increases slowly with the wind speed when the wind speed is lower than the rated speed and then grows quickly

Table 2
Platform and tower frequencies.

Platform Frequencies (Hz)		Tower Frequencies (Hz)	
Surge	0.0082	1st FA	0.5320
Sway	0.0082	1st SS	0.5312
Heave	0.0325	2nd FA	4.2346
Roll	0.0339	2nd SS	4.2829
Pitch	0.0340		
Yaw	0.1223		

with the wind speed higher than the rated speed. The effect of pitch controller accounts for the changes of trend at the rated wind speed. It can be seen that aerodynamic damping ratios in the surge, pitch, and yaw motions are much larger than those in sway, heave, and roll motions, in which the maximal aerodynamic damping ratios in yaw and pitch motions are up to 28% and 18% respectively. This is because the wind-rotor interaction provides large drag forces in the direction perpendicular to the rotor plane but much smaller forces in the rotor plane when the FOWT is operating.

6.1.2. Rigid tower assumption

It is of interest to study the influence of the assumption with tower flexibility not considered, as ignoring tower flexibility greatly simplifies the derivation of the aerodynamic damping ratios for the platform. In the aerodynamic damping matrix derivation in Section 3, the blades are considered to be rigid, but the tower is treated as a flexible cantilever beam. If the flexibility of the tower is ignored (the whole tower moves with the platform but does not bend), Eq. (21) can be written as $\mathbf{F}_{Aerop}^{damp} = \mathbf{C}_{Aero}^{pp}\dot{\mathbf{U}}$ with $\mathbf{C}_{Aero}^{pt}$ omitted. With the rigid tower assumption, aerodynamic damping matrix of the platform becomes a simpler 6×6 matrix, making the aerodynamic damping forces are only related to the platform motions. For the platform motions except the roll and surge motions, the dynamic responses obtained from the model without tower flexibility are very close to those from the model with tower flexibility considered (responses comparison not shown). The aerodynamic damping ratios in different directions were also calculated using the model without tower flexibility considered. The relationship between the aerodynamic damping ratios and mean wind speeds for the model without tower flexibility (marked by dots) is very similar to those obtained from the model considering tower flexibility (solid curves), as shown in Fig. 13. It shows the aerodynamic damping can be effectively and accurately obtained by a simplified model with the rigid tower assumption.

6.2. Hydrodynamic damping

6.2.1. Hydrodynamic radiation damping

Calculated hydrodynamic radiation damping ratios as a function of wave frequency are plotted in Fig. 14. Surge/sway radiation damping ratio is larger than that for the pitch/roll motion, but they reach the peaks simultaneously at the wave frequency of about 0.21–0.22Hz. The peak value of surge/sway damping ratio is about 53.5%, while the peak pitch/roll damping ratio is 24.5%. The radiation damping ratios are generally quite large when the platform vibration frequency ranges from 0.1 to 0.4Hz, meaning that when the platform moves under this frequency range the radiation damping plays an important role in limiting the vibration magnitude. However, the damping ratios near platform vibration frequencies are much lower. It means when low-frequency motions are excited, the radiation damping effect can be ignored.

6.2.2. Viscous damping from Morison's equation

Viscous damping ratios were also calculated and analysed, as shown in Fig. 15. It can be known that the viscous damping ratio is related to the wave height and wave period as in Morison's equation. Therefore, the viscous damping ratios were calculated given different combinations of wave heigh and wave period. Fig. 15 shows the value of surge/sway viscous damping ratio becomes larger with the increase of wave height and wave period. The trend for the viscous damping ratio of the pitch/roll motion is similar. The viscous damping ratios for translational

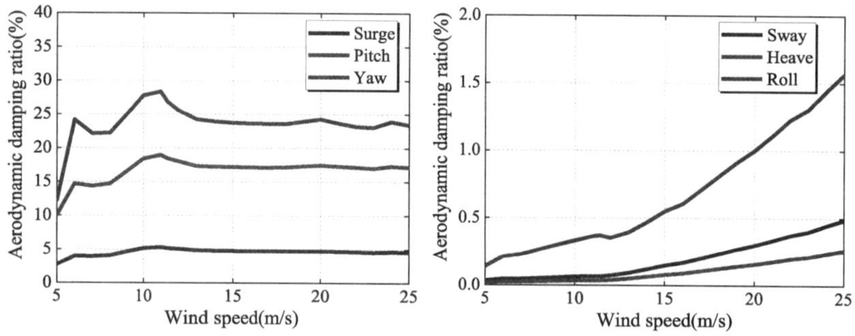

Fig. 12. Aerodynamic damping ratios for different platform DOFs when the mean wind speeds range from 5 to 25 m/s, with tower flexibility considered.

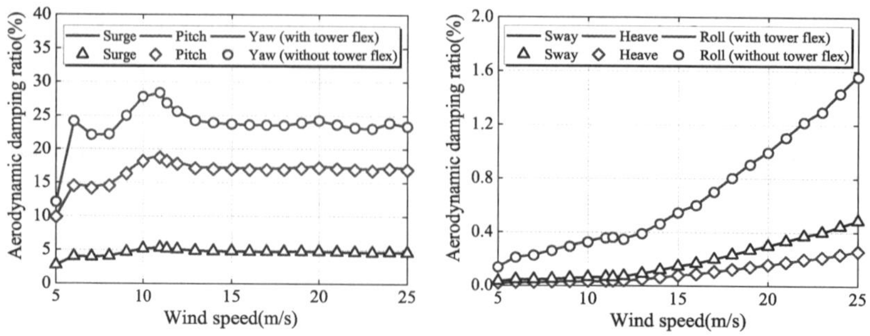

Fig. 13. Aerodynamic damping ratios for different platform DOFs versus wind velocity including (solid curves)/excluding (marked by dots) tower flexibility.

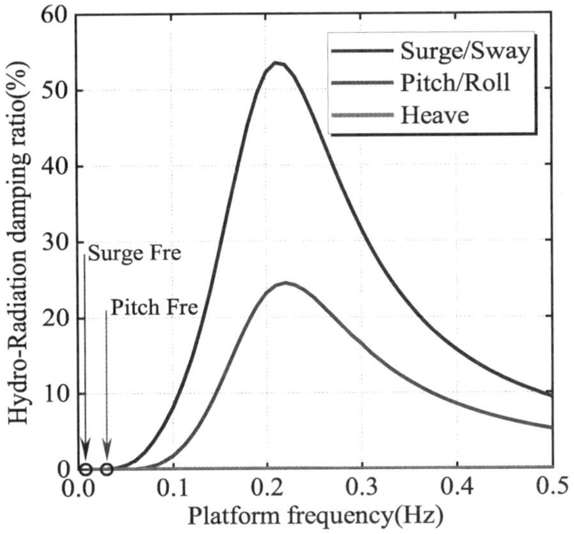

Fig. 14. Hydrodynamic radiation damping ratios.

motions are generally larger than those for the angular motions. For example, when the wave height is 10m and the wave period is 10s, the viscous damping ratio for the surge/sway motion is around 35%, while the value for the pitch/roll motion is about 10%. Moreover, the viscous damping ratios for the surge/sway motion and pitch/roll motion display significantly different sensitivities with regard to wave height and wave period. Compared with surge and sway motions, the viscous damping ratio of pitch and roll DOFs keeps stable when the wave height is equal to 10m. It should be noted that this hydrodynamic viscous damping model is developed with the assumption $\dot{U}_1(Z) \ll v_1(t,Z)$, which possibly causes less accurate damping estimation at extreme wave condition as $\dot{U}_1(Z)$ can be very large.

6.3. *Total damping analysis*

6.3.1. *Damping analysis for a special case*

Aerodynamic damping, hydrodynamic radiation damping, and hydrodynamic viscous damping have been analysed separately in previous sections. The overall damping analysis for the platform is given in this section for a special case. A representative operation condition should be proposed firstly. Here the inflow mean wind speed is set to the rated speed of 11.4 m/s, with the wave height of 5m and wave period of 9.6s, which corresponds to a typical ocean environment [53]. For this operation condition, the calculated aerodynamic damping, radiation damping, viscous damping and overall ratios are shown in Table 3. The overall damping values listed are directly calculated from the FOWT model including aerodynamic and hydrodynamic damping matrices, which are almost identical to the values obtained by adding all individual aerodynamic and hydrodynamic radiation and viscous damping ratios. The total damping ratios for the surge, sway, pitch and yaw motions are all over 25%, meaning that these motions are all highly damped. On the other hand, the damping ratio for the roll motion is at a medium level, compared to the negligible damping for the heave damping. It can be concluded that the pitch and yaw motions are mainly damped by the aerodynamic effect. Nevertheless, for the dissipation of sway and roll motions, hydrodynamic damping plays a much more important role than aerodynamic damping. Besides, aerodynamic damping and hydrodynamic damping have both important contributions to the dissipation of the surge motion.

The damping ratios for the roll and pitch motions obtained from this study generally agree well with the data in Ref. [54] in which the damping ratios of the OC3-Hywind FOWT were calculated. Other relevant studies [22,27] also presented calculated damping ratios of floating platforms. But these values cannot be directly compared to the results in this section, as the size and platform type in these studies are different from those used in this study.

6.3.2. *Influence of damping on platform RAOs*

The influences of different damping sources on the dynamic responses of the FOWT platform are present in this section. For simplicity, it is assumed that the FOWT is subjected to a 5m-height regular wave with a range of wave periods (2.857s–100s, corresponding to the wave frequency from 0.01Hz–0.35Hz) and an inflow wind field with a mean wind speed of 11.4 m/s. RAOs are evaluated at every frequency from the steady-state platform responses from both the simplified FOWT model and the FAST model. The platform RAOs with different single damping sources were calculated and compared with the RAOs with the overall damping sources in Fig. 16.

It can be seen in Fig. 16 (a) that the RAO of the surge motion is mainly influenced by the viscous damping and aerodynamic damping at the resonance frequency, but insensitive to the radiation damping. Aerodynamic damping can excite vibrations in roll (Fig. 16 (b)) and sway directions (not shown) because of the aerodynamic damping

Table 3
Damping values of aerodynamic damping, radiation damping, and Morison's viscous damping in the wind speed of 11.4 m/s, wave height = 5m, and wave period = 9.6s.

Damping Ratio Values (%)	Surge	Sway	Heave	Roll	Pitch	Yaw
Aerodynamic	5.16	0.07	0.04	0.36	18.17	26.83
Radiation	8.92	9.66	≈0	2.33	2.33	\
Morison's Viscous	16.21	17.54	\	5.25	5.26	\
Overall	32.93	25.35	0.04	7.92	26.03	26.83
Ref. [54]	\	\	\	<15	≈30	\

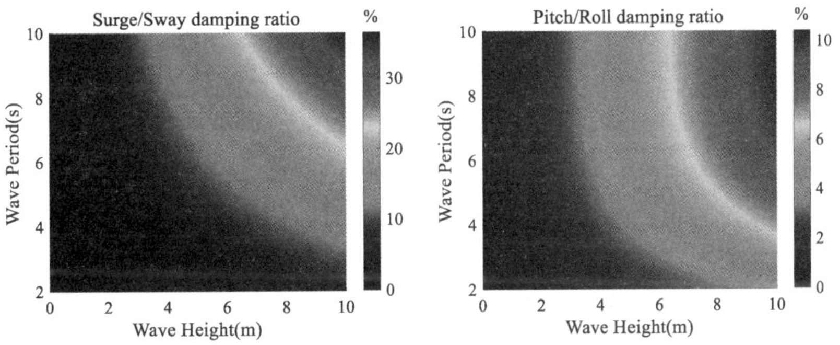

Fig. 15. Viscous damping rations from Morison's equation. (wave height = 0–10m, wave period = 2–10s).

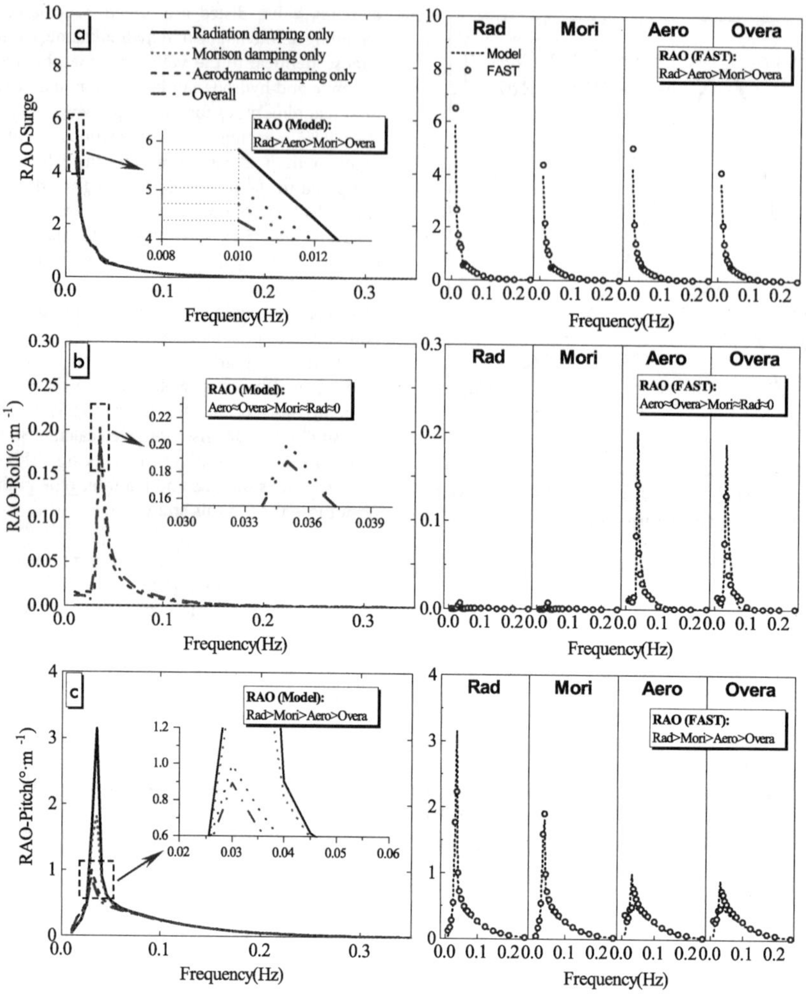

Fig. 16. RAOs for the surge (a), roll (b), pitch (c) motions with different single and overall damping sources, compared to FAST results. (wind speed = 11.4 m/s, wave height = 5m, wave frequency = 0.01Hz–0.35Hz).

coupling. The effect of aerodynamic damping coupling is clearly shown in Fig. 16 (b), as both the FAST model and the linearized model can only generate noticeable roll motions when the aerodynamic damping is considered by including the wind field. On the other hand, aerodynamic damping and viscous damping are much more significant than radiation damping for the pitch motion as shown in Fig. 16 (c), where the pitch motion is mostly damped via aerodynamic damping.

7. Conclusion

This paper proposes analytical solutions to quantify the aerodynamic damping and hydrodynamic damping for the FOWT platform. To represent the dissipation effect caused by wind-rotor interaction, an aerodynamic damping matrix is derived based on the BEM theory and the linearization of aerodynamic forces on the platform. An analytical method based on potential flow theory is adopted to form a radiation damping matrix from which the radiation damping can be quantified. Additionally, a hydrodynamic viscous damping matrix is derived based on Morison's equation and integrating the viscous drag forces applied on platform strips. From the derived analytical expressions of aerodynamic and hydrodynamic damping, a simplified state-space model of the FOWT can be established, from which the damping ratios for different DOFs of the platform are calculated via complex modal analysis. Then, the analytical expressions of damping and simplified model are successfully verified against FAST and Aqwa.

Aerodynamic damping and hydrodynamic damping ratios were calculated given different combinations of mean wind speeds, platform vibration frequency, wave height and wave period. For different platform DOFs, it is found that aerodynamic damping and hydrodynamic damping can have very different effects. The aerodynamic damping has great impact on the dissipations of the surge, pitch, and yaw motions, with the aerodynamic damping ratio for the yaw motion larger than 25%. However, the aerodynamic damping is much less for the sway, heave and roll motions, with values less than 2%. The radiation damping and viscous damping contribute larger damping ratios (up to 50% and 30% respectively) for the surge and sway motions but has relatively less impact (up to 20% and 10% respectively) on the pitch and roll motions. However, heave and yaw motions are nearly not damped by radiation and viscous effect. Overall, for a typical operation state, the platform is high damped for the surge, sway, pitch and yaw motions through the combination of aerodynamic and hydrodynamic damping effects, with total damping ratios larger than 25%. The roll motion is intermediately damped with a total damping ratio of 8%, while the damping for the heave motion is negligible.

The analytical expressions and simplified model proposed in the present study can be easily extended to quantify the damping contributions of different damping sources for FOWTs with different sizes, thus conveniently and effectively assisting the evaluation of the vibration amplitude of the FOWT platforms.

CRediT authorship contribution statement

Qingshen Meng: Formal analysis, Software, Validation, Writing – original draft. **Xugang Hua:** Methodology, Writing – review & editing, Funding acquisition, Supervision. **Chao Chen:** Conceptualization, Methodology, Software, Writing – original draft, Writing – review & editing, Funding acquisition, Supervision. **Shuai Zhou:** Data curation. **Feipeng Liu:** Investigation. **Zhengqing Chen:** Supervision.

Declaration of competing interest

The authors declare that they have no known competing financial interests or personal relationships that could have appeared to influence the work reported in this paper.

Acknowledgement

The financial supports from the National Natural Science Foundation of China (No. 52108280), the National Science Fund for Distinguished Young Scholars (No. 52025082) and Key Project of Strategy Consulting of Chinese Academy of Engineering (Grant No. 2021-XZ-37) are greatly appreciated.

Appendix A

The 6 × 6 linear mooring stiffness matrix accounting the mooring effect can be determined by (at undisturbed position)

$$\mathbf{K}_{\text{Moor}} = \begin{bmatrix} K_{Moor}^{11} & 0 & 0 & 0 & K_{Moor}^{15} & 0 \\ 0 & K_{Moor}^{22} & 0 & K_{Moor}^{24} & 0 & 0 \\ 0 & 0 & K_{Moor}^{33} & 0 & 0 & 0 \\ 0 & K_{Moor}^{42} & 0 & K_{Moor}^{44} & 0 & 0 \\ K_{Moor}^{51} & 0 & 0 & 0 & K_{Moor}^{55} & 0 \\ 0 & 0 & 0 & 0 & 0 & K_{Moor}^{66} \end{bmatrix},$$

where.

$$K_{Moor}^{11} = 4.21 \times 10^4 N/m,\ K_{Moor}^{22} = 4.27 \times 10^4 N/m$$
$$K_{Moor}^{15} = 9.07 \times 10^5 N/rad,\ K_{Moor}^{51} = 8.65 \times 10^5 N$$
$$K_{Moor}^{24} = -9.60 \times 10^5 N/rad,\ K_{Moor}^{42} = -9.21 \times 10^5 N$$
$$K_{Moor}^{33} = 1.28 \times 10^4 N/m,\ K_{Moor}^{44} = -5.48 \times 10^6 N\cdot m/rad$$
$$K_{Moor}^{55} = -7.59 \times 10^6 N\cdot m/rad,\ K_{Moor}^{66} = 1.16 \times 10^7 N\cdot m/rad$$

Appendix B

Elements in aerodynamic damping matrix $\mathbf{C}_{\text{Aero}}^{\text{pp}}$:

$$c_{xU_1} = N_b \int_0^R \frac{\partial(\mathrm{d}T)}{\partial V_0},\ c_{xU_4} = -N_b \int_0^R r\frac{\partial(\mathrm{d}T)}{\partial V_r},\ c_{xU_5} = N_b h_R \int_0^R \frac{\partial(\mathrm{d}T)}{\partial V_0},$$

$$c_{yU_2} = -\frac{N_b}{2} \int_0^R \frac{\partial(\mathrm{d}S)}{\partial V_r},\ c_{yU_4} = \frac{N_b}{2} h_R \int_0^R \frac{\partial(\mathrm{d}S)}{\partial V_r},\ c_{yU_5} = -\frac{N_b}{2} \int_0^R r\frac{\partial(\mathrm{d}S)}{\partial V_0},$$

$$c_{zU_3} = -\frac{N_b}{2} \int_0^R \frac{\partial(\mathrm{d}S)}{\partial V_r},\ c_{zU_6} = -\frac{N_b}{2} \int_0^R r\frac{\partial(\mathrm{d}S)}{\partial V_0},\ c_{\theta_x U_1} = N_b \int_0^R r\frac{\partial(\mathrm{d}S)}{\partial V_0},$$

$$c_{\theta_x U_4} = -N_b \int_0^R r^2\frac{\partial(\mathrm{d}S)}{\partial V_r},\ c_{\theta_x U_5} = N_b h_R \int_0^R r\frac{\partial(\mathrm{d}S)}{\partial V_0},\ c_{\theta_y U_2} = \frac{N_b}{2} \int_0^R r\frac{\partial(\mathrm{d}T)}{\partial V_r},$$

$$c_{\theta_y U_4} = -\frac{N_b}{2} h_R \int_0^R r\frac{\partial(\mathrm{d}T)}{\partial V_r},\ c_{\theta_y U_5} = \frac{N_b}{2} \int_0^R r^2\frac{\partial(\mathrm{d}T)}{\partial V_0},\ c_{\theta_z U_3} = \frac{N_b}{2} \int_0^R r\frac{\partial(\mathrm{d}T)}{\partial V_r},$$

$$c_{\theta_z U_6} = \frac{N_b}{2} \int_0^R r^2\frac{\partial(\mathrm{d}T)}{\partial V_0}$$

Elements in aerodynamic damping matrix $\mathbf{C}_{\text{Aero}}^{\text{pt}}$:

$$c_{xx}=N_b\int_0^R\frac{\partial(\mathrm{d}T)}{\partial V_0},c_{x\theta_x}=-N_b\int_0^R r\frac{\partial(\mathrm{d}T)}{\partial V_r},c_{yy}=-\frac{N_b}{2}\int_0^R\frac{\partial(\mathrm{d}S)}{\partial V_r},$$

$$c_{y\theta_y}=-\frac{N_b}{2}\int_0^R r\frac{\partial(\mathrm{d}S)}{\partial V_0},c_{z\theta_z}=-\frac{N_b}{2}\int_0^R r\frac{\partial(\mathrm{d}S)}{\partial V_0},c_{\theta_x x}=N_b\int_0^R r\frac{\partial(\mathrm{d}S)}{\partial V_0},$$

$$c_{\theta_x\theta_x}=-N_b\int_0^R r^2\frac{\partial(\mathrm{d}S)}{\partial V_r},c_{\theta_y y}=\frac{N_b}{2}\int_0^R r\frac{\partial(\mathrm{d}T)}{\partial V_r},c_{\theta_y\theta_y}=\frac{N_b}{2}\int_0^R r^2\frac{\partial(\mathrm{d}T)}{\partial V_0},$$

$$c_{\theta_z\theta_z}=\frac{N_b}{2}\int_0^R r^2\frac{\partial(\mathrm{d}T)}{\partial V_0}$$

Appendix C

According to Li et al. [42], for a Type 1 cylinder oscillating with an angular frequency of ϖ, the added mass coefficient $c^A_{U_3U_3}$ and radiation damping coefficient $c^{rad}_{U_3U_3}$ with regard to heave motion can be determined by

$$c^A_{U_3U_3}+\mathrm{i}\frac{c^{rad}_{U_3U_3}}{\varpi}=-\xi\rho_w\int_a^0\int_0^{2\pi}\varphi_2^{(1)}(r,\theta,-h_c)r\mathrm{d}\theta\mathrm{d}r,$$

where $\varphi_2^{(1)}$ represents the radiation velocity potential with regard to heave motion in the exterior domain. Other notations have been defined in Section 4.1. The added mass coefficient $c^A_{U_5U_5}$ and radiation damping coefficient $c^{rad}_{U_5U_5}$ with regard to pitch motion can be determined by

$$c^A_{U_5U_5}+\mathrm{i}\frac{c^{rad}_{U_5U_5}}{\varpi}=M_{yh}+M_{yd},$$

where

$$M_{yh}=-\xi\rho_w\int_{-c}^0\int_0^{2\pi}\varphi_1^{(1)}(a,\theta,Z)(\overline{Z}-\overline{Z}_0)(-a\cos\theta)\mathrm{d}\theta\mathrm{d}\overline{Z},$$

$$M_{yd}=\xi\rho_w\int_0^a\int_0^{2\pi}\varphi_2^{(1)}(r,\theta,-h_c)r^2\cos\theta\mathrm{d}\theta\mathrm{d}r.$$

In which the integral lower limit c represents the height from cylinder bottom to the SWL, and $\overline{Z}_0$ is the rotation center of roll and pitch motions.

The added mass coefficient $c^A_{U_5U_1}$ and radiation damping coefficient $c^{rad}_{U_5U_1}$ with regard to pitch-surge interaction can be determined by

$$c^A_{U_5U_1}+\mathrm{i}\frac{c^{rad}_{U_5U_1}}{\varpi}=-\xi\rho_w\int_{-h_c}^0\int_0^{2\pi}\varphi_1^{(1)}(a,\theta,\overline{Z})(-a\cos\theta)\mathrm{d}\theta\mathrm{d}\overline{Z}.$$

The radiation damping coefficients with regard to sway, roll and sway-roll interaction can be easily determined from coefficients previously defined due to symmetry, i.e. $c^{rad}_{U_2U_2}=c^{rad}_{U_1U_1}$, $c^{rad}_{U_4U_4}=c^{rad}_{U_5U_5}$, $c^{rad}_{U_1U_5}=c^{rad}_{U_5U_1}=c^{rad}_{U_2U_4}=c^{rad}_{U_4U_2}$.

References

[1] M. Karimirad, Modeling aspects of a floating wind turbine for coupled wave-wind-induced dynamic analyses, Renew. Energy 53 (2013) 299–305.
[2] N.D.P. Barltrop, A.J. Adams, Dynamics of Fixed Marine Structures, Butterworth-Heinemann, 2013.
[3] European Standard, EN 1991-1-4 Eurocode 1: Actions on Structures -Part 1-4: General Actions -Wind Actions, The European Union, Bruxelles, Belgium, 2005.
[4] R. Shirzadeh, C. Devriendt, M.A. Bidakhvidi, P. Guillaume, Experimental and computational damping estimation of an offshore wind turbine on a monopile foundation, J. Wind Eng. Ind. Aerod. 120 (2013) 96–106.
[5] M. Ozbek, D.J. Rixen, Operational modal analysis of a 2.5 MW wind turbine using optical measurement techniques and strain gauges, Wind Energy 17 (2013) 367–381.
[6] C. Koukoura, A. Natarajan, A. Vesth, Identification of support structure damping of a full scale offshore wind turbine in normal operation, Renew. Energy 81 (2015) 882–895.
[7] X. Dong, J. Lian, H. Wang, T. Yu, Y. Zhao, Structural vibration monitoring and operational modal analysis of offshore wind turbine structure, Ocean Eng. 150 (2018) 280–297.
[8] W. Weijtjens, Rasoul Shirzadeh, Gert De Sitter, C. Devriendt, Classifying resonant frequencies and damping values of an offshore wind turbine on a monopile foundation for different operational conditions, in: Proc. Eur. Wind Energy Assoc., Barcelona, Spain, 2014.
[9] W. Hu, S. Thöns, R.G. Rohrmann, S. Said, W. Rücker, Vibration-based structural health monitoring of a wind turbine system. Part I: resonance phenomenon, Eng. Struct. 89 (2015) 260–272.
[10] D.J.C. Salzmann, J. Van Der Tempel, Aerodynamic damping in the design of support structures for offshore wind turbines, Proc. Copenhagen Offshore Conf. (2005) 9.
[11] V. Valamanesh, A.T. Myers, Aerodynamic damping and seismic response of horizontal axis wind turbine towers, J. Struct. Eng. 140 (2014) 1–9.
[12] C. Chen, P. Duffour, Modelling damping sources in monopile-supported offshore wind turbines, Wind Energy 21 (2018) 1121–1140.

[13] X. Liu, C. Lu, G. Li, A. Godbole, Y. Chen, Effects of aerodynamic damping on the tower load of offshore horizontal axis wind turbines, Appl. Energy 204 (2017) 1101–1114.
[14] Y. Chen, D. Wu, Y. Yu, W. Gao, An improved theory in the determination of aerodynamic damping for a horizontal axis wind turbine (HAWT), J. Wind Eng. Ind. Aerod. 213 (2021) 1–19.
[15] C. Chen, P. Duffour, P. Fromme, Modelling wind turbine tower-rotor interaction through an aerodynamic damping matrix, J. Sound Vib. 489 (2020), 115667.
[16] C. Chen, P. Duffour, K. Dai, Y. Wang, P. Fromme, Identification of aerodynamic damping matrix for operating wind turbines, Mech. Syst. Signal Process. 154 (2021), 107568.
[17] M. Karimirad, Torgeir Moan, Effect of aerodynamic and hydrodynamic damping on dynamic response of a spar type floating wind turbine, Proc. Eur. Wind Energy Conf. 6 (2010) 4856–4866.
[18] J.M. Jonkman, Influence of control on the pitch damping of a floating wind turbine, Contract (2008) 1–15.
[19] Z. Cheng, H.A. Madsen, Z. Gao, T. Moan, Numerical study on aerodynamic damping of floating vertical axis wind turbines, J. Phys. Conf. Ser. 753 (2016), 102001.
[20] Jiahao Chen, Zhiqiang Hu, Study on aerodynamic damping of semi-submersible floating wind turbines, Chin. J. Theor. Appl. Mech. 51 (2019) 1255–1265.
[21] S. Gueydon, Aerodynamic damping on a semisubmersible floating foundation for wind turbines, Energy Proc. 94 (2016) 367–378.
[22] K.P. Thiagarajan, R. Urbina, W. Hsu, Nonlinear pitch decay motion of a floating offshore wind turbine structure, J. Offshore Mech. Arctic Eng. 137 (2015) 1–7.
[23] S. Salehyar, Q. Zhu, Aerodynamic dissipation effects on the rotating blades of floating wind turbines, Renew. Energy 78 (2015) 119–127.
[24] M.J. Rao, S. Nallayarasu, S.K. Bhattacharyya, CFD approach to heave damping of spar with heave plates with experimental validation, Appl. Ocean Res. 108 (2021), 102517.
[25] A. Subbulakshmi, R. Sundaravadivelu, Heave damping of spar platform for offshore wind turbine with heave plate, Ocean Eng. 121 (2016) 24–36.
[26] A. Pegalajar-Jurado, H. Bredmose, Reproduction of slow-drift motions of a floating wind turbine using second-order hydrodynamics and Operational Modal Analysis, Mar. Struct. 66 (2019) 178–196.
[27] C. Clement, S. Kosleck, T. Lie, Investigation of viscous damping effect on the coupled dynamic response of a hybrid floating platform concept for offshore wind turbines, Ocean Eng. 225 (2021), 108836.
[28] J. Yang, Y.P. He, Y.S. Zhao, Y.L. Shao, Z.L. Han, Experimental and numerical studies on the low-frequency responses of a spar-type floating offshore wind turbine, Ocean Eng. 222 (2021), 108571.
[29] J.M. Jonkman, M.L.J. Buhl, FAST User's Guide, National Renewable Energy Laboratory, Golden, CO, USA, 2005. Technical Report NREL/EL-500-38230.
[30] J. Torben, A. Melchior, How 2 HAWC2, the User's Manual, Technical Report Risø-R-1597(ver. 3-1)(EN), Risø National Laboratory, Denmark, 2007.
[31] J.M. Jonkman, Definition of the Floating System for Phase IV of OC3, Technical Report NREL/TP-500-47535, National Renewable Energy Laboratory.Golden, CO, USA, 2010.
[32] J.M. Jonkman, Dynamics Modeling and Loads Analysis of an Offshore Floating Wind Turbine, Technical Report NREL/TP-500-41958, National Renewable Energy Laboratory: Golden, CO, USA, 2007.
[33] L. Chen, B. Basu, S.R.K. Nielsen, A coupled finite difference mooring dynamics model for floating offshore wind turbine analysis, Ocean Eng. 162 (2018) 304–315.
[34] T. Belytschko, T.J.R. Hughes, Computational Methods for Transient Analysis, Elsevier, Amsterdam, North-Holland, 1983.
[35] A. Pegalajar-Jurado, M. Borg, H. Bredmose, An efficient frequency-domain model for quick load analysis of floating offshore wind turbines, Wind Energy Sci 3 (2018) 693–712.
[36] J. Roy R. Craig, A.J. Kurdila, Fundamentals of Structural Dynamics, second ed., John Wiley & Sons, Inc., Hoboken, New Jersey, 2006.
[37] D. Deleanu, Theoretical Mechanics. Theory and Applications, Nautica Publishing House, Constanta, Roumania, 2015.
[38] E.S. Gutierrez-wing, Modal Analysis of Rotating Machinery Structures, PhD Thesis, Imperial College London, 2003.
[39] H. Gu, P. Stansby, T. Stallard, E. Carpintero Moreno, Drag, added mass and radiation damping of oscillating vertical cylindrical bodies in heave and surge in still water, J. Fluid Struct. 82 (2018) 343–356.
[40] M. Karimirad, Q. Meissonnier, Z. Gao, T. Moan, Hydroelastic code-to-code comparison for a tension leg spar-type floating wind turbine, Mar. Struct. 24 (2011) 412–435.
[41] D.D. Bhatta, M. Rahman, On scattering and radiation problem for a cylinder in water of finite depth, Int. J. Eng. Sci. 41 (2003) 931–967.
[42] A. jun Li, Y. Liu, H. jun Li, New analytical solutions to water wave radiation by vertical truncated cylinders through multi-term Galerkin method, Meccanica 54 (2019) 429–450.
[43] M.J. Kuhn, Dynamics and Design Optimisation of Offshore Wind Energy Conversion Systems, Report No. 2001.002, DUWIND, Delft University Wind Energy Research Institute, Delft, The Netherlands, 2001.
[44] H.F. Veldkamp, J. Van Der Tempel, Influence of wave modelling on the prediction of fatigue for offshore wind turbines, Wind Energy 8 (2005) 49–65.
[45] J.E. Withee, Fully Coupled Dynamic Analysis of a Floating Wind Turbine System, PhD Thesis, Massachusetts Institute of Technology, 2004.
[46] B. Van Dao, J. Penzien, Comparison of treatments of non-linear drag forces acting on fixed offshore platforms, Appl. Ocean Res. 4 (1982) 66–72.
[47] R.G. Dean, R.A. Dalrymple, Water Wave Mechanics for Engineers and Scientists, first ed., World scientific publishing company, 1991.
[48] C. Devriendt, F. Magalhães, W. Weijtjens, G. De Sitter, Á. Cunha, P. Guillaume, Structural health monitoring of offshore wind turbines using automated operational modal analysis, Struct. Health Monit. 13 (2014) 644–659.
[49] A. Bajrić, J. Høgsberg, F. Rüdinger, Evaluation of damping estimates by automated Operational Modal Analysis for offshore wind turbine tower vibrations, Renew. Energy 116 (2018) 153–163.
[50] B. Chen, Z. Zhang, X. Hua, B. Basu, S.R.K. Nielsen, Identification of aerodynamic damping in wind turbines using time-frequency analysis, Mech. Syst. Signal Process. 91 (2017) 198–214.
[51] J.M. Jonkman, Dynamics of offshore floating wind turbines-model development and verification, Wind Energy 12 (2009) 459–492.
[52] A. Pecher, J.P. Kofoed, Handbook of Ocean Wave Energy, Springer Nature, 2017.
[53] C.E. Høeg, Z. Zhang, The influence of gyroscopic effects on dynamic responses of floating offshore wind turbines in idling and operational conditions, Ocean Eng. 227 (2021).
[54] S. Christiansen, T. Bak, T. Knudsen, Damping wind and wave loads on a floating wind turbine, Energies 6 (2013) 4097–4116.

Journal of Sound and Vibration 521 (2022) 116565

Contents lists available at ScienceDirect

Journal of Sound and Vibration

journal homepage: www.elsevier.com/locate/jsvi

Optimal calibration of a tuned liquid column damper (TLCD) for rotating wind turbine blades

Bei Chen [a,b], Zili Zhang [b,*], Xugang Hua [a,*], Biswajit Basu [c]

[a] *Key Laboratory for Wind and Bridge Engineering, Hunan University, 410082 Changsha, China*
[b] *Department of Engineering, Aarhus University, 8000 Aarhus, Denmark*
[c] *School of Engineering, Trinity College Dublin, Dublin 2, Ireland*

ARTICLE INFO

Keywords:
Tuned liquid column damper (TLCD)
Offshore wind turbine
Vibration control
Wind turbine blade
Edgewise vibration
Pole-placement calibration

ABSTRACT

Blade edgewise vibration is a growing concern for modern megawatt wind turbines, as it may significantly shorten the blade lifetime. In this paper, optimal calibration (tuning) of a blade-mounted tuned liquid column damper (TLCD) using closed-form formulas is investigated. A 2-DOF model for the blade-TLCD system is first established, acting as the basis for analytical derivation of the optimal tuning formulas. Two different frequency-tuning criteria are employed, one based on the equal dynamic amplification at two neutral frequencies and the other one based on the equal modal damping ratio of the two free vibration modes. Two optimal frequency-tuning (frequency ratio) formulas having similar expressions are derived based on these two criteria, and both formulas are dependent on the rotational speed, the mounting position, the blade edgewise eigenfrequency and the TLCD prescribed parameters. Next, an explicit expression for the optimal equivalent TLCD damping ratio is obtained by equalizing the dynamic amplification at three excitation frequencies, and it also exhibits dependence on the rotational speed and the edgewise eigenfrequency. A closed-form formula for the optimal TLCD head loss coefficient is then proposed, providing a straightforward method to determine the optimal value without resorting to iteration. Comparisons of the two optimal frequency-tuning formulas are conducted in terms of the dynamic amplification, the root loci of the 2-DOF system, and the parametric sensitivity, using data from the NREL 5MW reference wind turbine. The results indicate that these two formulas lead to very similar optimal design of the blade-mounted TLCD as well as the damper performance, although the 1st formula only guarantees the equal dynamic amplification while the 2nd one only guarantees the equal modal damping ratio. The set of closed-form formulas developed in this study facilitates a convenient and robust design of blade-mounted-TLCD for suppressing edgewise vibrations, and forms the basis for developing semi-active control algorithms.

1. Introduction

The translational vibration modes of the wind turbine blade are normally classified as flapwise and edgewise modes. The flapwise vibration (or the out-of-plane vibration) is highly aerodynamically damped under normal working conditions (i.e. the airflow is fully attached to the blade) [1], rendering the vibration to be quasi-static. On the other hand, the edgewise vibration (or the in-plane vibration), which is associated with low aerodynamic damping, is usually the severest problem [2]. Onset of instability might also take place due to negative edgewise aerodynamic damping. Large oscillation amplitude of the blade may trigger the generation and

* Corresponding authors.
E-mail addresses: chamber@hnu.edu.cn (B. Chen), zili_zhang@cae.au.dk (Z. Zhang), cexghua@hnu.edu.cn (X. Hua), basub@tcd.ie (B. Basu).

https://doi.org/10.1016/j.jsv.2021.116565
Received 15 June 2021; Received in revised form 24 September 2021; Accepted 18 October 2021
Available online 21 December 2021

Nomenclature

$a = \Phi(x_0)$	Auxiliary parameter related to the local displacement of the blade [m]
A	Cross-sectional area of vertical tube of the TLCD [m^2]
A_0	Cross-sectional area of horizontal tube of the TLCD [m^2]
$b = \Phi'(x_0)$	Auxiliary parameter related to the local rotation of the blade [-]
B	Horizontal length of the liquid inside the TLCD [m]
C	Excitation intensity factor [-]
C_w	Equivalent damping coefficient of the TLCD [kg/s]
C_0	Structural damping coefficient of the blade [kg/s]
$\hat{f} = Cm_0 g$	Amplitude of the external force [N]
$f(t) = \hat{f}e^{i\omega t}$	External harmonic force [N]
g	Gravitational acceleration [m/s^2]
H	Vertical length of the liquid inside the TLCD [m]
$i = \sqrt{-1}$	Imaginary unit [-]
$k_{ori,0}(\Omega, t)$	Original modal stiffness of the blade [kg/m]
$k_0(\Omega)$	Simplified modal stiffness of the blade [kg/m]
k_e	Elastic stiffness of the blade [kg/m]
k_1, k_2	Stiffness related parameters [kg/m], given in Eq. (9)
L_B	Length of the blade [m]
$L = B + 2H$	Overall length of the liquid inside the TLCD [m]
m_w	Overall mass of the liquid [kg]
m_0	Modal mass of the blade [kg]
m_1, m_2	Blade kinetic energy related parameters [kg], given in Eq. (6)
$q(t)$	Edgewise tip displacement of the blade [-]
R	Radius of the circular inversion [rad/s]
t	Time [s]
${}^0\mathbf{T}_1$	Transformation matrix between coordinate $(X_1, X_2, X_3)_0$ and $(x_1, x_2, x_3)_1$
${}^1\mathbf{T}_2$	Transformation matrix between coordinate $(x_1, x_2, x_3)_1$ and $(y_1, y_2, y_3)_2$
$u_{X_i,b}(t)$	X_i– component of the blade point 'p_b' displacement vector in the $(X_1, X_2, X_3)_0$-coordinate [m]
$u_{x_i,b}(t)$	x_i– component of the blade point 'p_b' displacement vector in the $(x_1, x_2, x_3)_1$-coordinate [m]
$u_{y_i,r}(t)$	y_i– component of the point 'p_r' displacement vector in the $(y_1, y_2, y_3)_1$-coordinate [m]
$u_{X_i,r}(t)$	X_i– component of the blade point 'p_r' displacement vector in the $(X_1, X_2, X_3)_0$-coordinate [m]
$\dot{u}_{y_i,l}(t)$	y_i– component of the liquid point 'p_l' velocity vector in the $(y_1, y_2, y_3)_1$-coordinate [m/s]
$v(t)$	Displacement of the liquid in the vertical tube [m]
$\hat{v}$	Displacement amplitude of the liquid in the vertical tube [m]
$v_0(t)$	Displacement of the liquid in the horizontal tube [m]
x_0	Mounting position of the TLCD [m]
$\alpha = A/A_0$	Cross-sectional area ratio of the vertical tube versus horizontal tube [-]
$\gamma = B/L$	Ratio of the horizontal length to overall length of the liquid column [-]
ζ_w	Equivalent damping ratio of the TLCD [-]
$\mu = m_w/m_0$	Ratio of the liquid mass to the modal mass of the blade [-]
ξ	Head loss coefficient due to the orifice [-]
ρ	Liquid density [kg/m^3]
$\Phi(x_3)$	Fundamental edgewise modal shape of the blade [m]
$\Psi(t)$	Azimuthal angle of the blade [rad]
$\omega_{ori,0}(\Omega, t)$	Original fundamental edgewise circular frequency of the blade [rad/s]
$\omega_0(\Omega)$	Simplified fundamental edgewise circular frequency of the blade [rad/s]

propagation of blade cracks (especially on the trailing edge), and finally result in damage of the blade [3,4]. The earliest edgewise instability case was reported by Stiesdal [5] in 1995 on a stall regulated wind turbine with a rotor diameter around 37 m. Later, eighteen 3700 LM Glasfiber 19m wind turbine blades were reported to be seriously damaged [6] due to edgewise vibrations (which caused serious longitudinal cracks on the trailing edge). For the megawatt-sized wind turbines with blade lengths ranging from 40 m to more than 80 m, the potential catastrophic blade failure due to edgewise vibration has also been experimentally demonstrated by

$\omega_{ori,w}(\Omega, t)$	Original circular frequency of the TLCD [rad/s]
$\omega_w(\Omega)$	Simplified circular frequency of the TLCD [rad/s]
ω	Excitation circular frequency [rad/s]
ω_1, ω_2	Roots of the characteristic equation Eq. (39) in the first quadrant of the complex plane [rad/s]
ω_*	Root of the characteristic equation Eq. (39) in the first quadrant of the complex plane with Eq. (57) and Eq. (56) inserted, i.e. the bifurcation point [rad/s]
Ω	Rotational speed of the rotor [rad/s]
Ω_0	Rated speed of the rotor [rad/s]

Malkin [7]. Although there have been no blade failure examples reported so far for the commercially utilized megawatt-sized wind turbines, this vibration issue will affect the power output of turbines (e.g. some wind turbines need to be shut down before reaching the cut-out speed under certain unfavorable working conditions in order to avoid the onset of edgewise vibrations, and this will in turn reduce the power output and finally result in an increase of the levelized cost of energy). The current trend in commercial wind turbine development shows a rapid increase in turbine dimensions [8]. As a result, the blades of modern large-scale wind turbines become slender and more flexible, making them more vulnerable to vibrations particularly in the edgewise direction. As also theoretically demonstrated by Chaviaropoulos [9], the edgewise vibration can occur on all rotor blades, and this problem will increase with blade size. Hence, there is a need to address the edgewise vibration problem for modern wind turbines in order to ensure the safe operation during their lifetime.

The severity of edgewise vibration depends on the sum of the inherent structural damping (normally very low for wind turbine blades) and the aerodynamic damping, and thus is highly related to the structural dynamic property as well as the blade airfoil characteristics [9,10]. In view of its, edgewise vibration can be addressed either by increasing the aerodynamic damping [11,12] or by introducing extra damping elements to the blade (such as attaching mechanical dampers inside the blade or integrating damping layers into the blade laminates) [10,13–22]. The first approach is usually referred to as the aerodynamic modification method. By slightly modifying the blade geometric shape (such as mounting stall strips along the blade or optimizing the airfoil shape), an improvement of edgewise aerodynamic damping can be achieved [11]. Such kind of solution (i.e. by optimizing the structural and aerodynamic design of the blade) is also commonly employed by current wind turbine companies. Despite its simplicity and efficiency, it may negatively affect the power output of the turbine. Furthermore, the design of those aerodynamic devices is rather cumbersome and highly dependent on load characteristics, and usually resort to wind tunnel experiments and numerical simulations [11,12]. However, both approaches cannot reflect the reality especially under complex flow conditions. By comparison, the second approach also known as the mechanical method would be more desirable, as it maintains the blade aerodynamic shape thus has not influence on the power output.This mechanical solution is also gradually used by companies due to its high vibration suppressing efficiency, simple configuration and lower manufacturing cost (particularly for the liquid-related damper such as the TLCD).

Active, semi-active and passive control devices have been widely employed by researchers for mitigating the blade edgewise vibration. The active and semi-active solutions have the advantage of being able to continuously adjust the control force or damper parameters, in reaction to the varying excitation and structural property. Huang et al. [13] employed semi-active tuned mass dampers (STMDs) to control the edgewise vibration of floating wind turbine blades, where the stiffness of each STMD can be online optimized using the short-time Fourier transform (STFT) based semi-active control strategy. Fitzgerald et al. [14] used active tuned mass dampers (ATMDs) to suppress the edgewise vibration, taking into account the influence of the tower-blade interaction. Later, an extra element (i.e. a pretension cable) was added to that ATMD resulting in an additional control force at the blade tip to further suppress edgewise vibrations [15]. Staino et al. [10] proposed a novel active controller composing of two linear actuators (active tendons) attached to the blade tip, which allows for a continuously adjustment of the control force to be applied in edgewise direction. However, such a controller requires a blade-length support frame that is in practically difficult to be realized. Chen et al. [16] refined Staino's device by replacing those linear actuators with semi-active magnetorheological (MR) dampers, and the corresponding support frames were also shorten. Despite encouraging results from these numerical studies, the complexity of the design, the high cost and power requirement barricade the possible extensive application of active and semi-active devices for wind turbine blades.

In comparison with the active and semi-active solutions, the passive damping devices are considered more feasible to be applied in practice, due to their cost-effectiveness and robustness. Various types of pendulum-analogous passive dampers were proposed for suppressing blade edgewise vibrations, including the roller damper [18], the circular liquid column damper (CLCD) [20] and the tuned liquid damper (TLD) [23]. The main advantage of these devices is the elimination of damper stroke issue. Moreover, the liquid dampers can be manufactured with a very low cost, making them suitable for large-scale applications. Friction damping constitutes the energy dissipation mechanism of the roller damper [18] while fluid damping is the dissipation mechanism for liquid dampers [20,23], both being inherently nonlinear. Consequently, nonlinear mathematical models for the blade–damper system were established where the blade vibration was modeled by one degree of freedom (DOF) representing the fundamental edgewise mode. These models enabled time-domain simulations of the coupled blade–damper system for performance evaluation. However, no explicit formulas for optimal design (calibration) of the dampers were derived due to the difficulties arising from the nonlinear equations of motion, and numerical optimization (numerical search) was used to optimize the dampers. More recently, Biglari et al. [17] proposed to use the genetic algorithm [24] to optimize a blade-mounted damper with shunt damping. Other numerical optimization techniques (e.g. particle swarm optimization [25]) can in principle also be used for optimal damper calibration. However, numerical optimization for blade-mounted dampers suffers from two shortcomings: 1) it can be computationally inefficient to run optimization algorithms, thus inconvenient for practical application; 2) more importantly, the varying operational conditions (e.g. varying rotational speed) of the wind turbine make it impossible to numerically determine the optimal damper design for all possible operational conditions.

To overcome these issues, a series of studies were conducted for developing closed-form optimal design formulas for various types of linear vibration absorbers, which were installed in a rotating blade for suppressing edgewise vibrations. These include TMD [26], tuned mass–damper–inerter (TMDI) [27] and rotational inertia double-tuned mass damper (RIDTMD) [28], all of which result in linear equations of motion of the blade–damper system due to the linear energy dissipation mechanism (linear viscous damper). The classic TMD was first investigated by Zhang [26], and two explicit frequency-tuning formulas were derived using two different criteria (the equal dynamic amplification [29] and the equal modal damping ratio [30]). The first formula is rotor-speed independent while the second one is dependent on the rotor speed, although the difference resulted from these two is insignificant for realistic rotor speeds [26]. Later, the inerter was utilized by Zhang and Fitzgerald [27] to enhance performance of the classic TMD, resulting in the blade-mounted TMDI. The optimal frequency ratio and optimal damping ratio formulas were analytically derived based on frequency response analysis, and both were dependent on the rotor speed and the blade edgewise eigenfrequency. Very recently, Zhang and Larsen [28] developed explicit optimal calibration formulas for the blade-mounted RIDTMD (3-DOF model with one extra resonance) using the pole-placement method, and all optimal parameters were dependent on the rotor speed and edgewise eigenfrequency. It turns out that the optimal TMDI outperforms the optimal TMD in terms of the reduced damper stroke [27], while the optimal RIDTMD outperforms the optimal TMD in terms of the better vibration control effect [28]. It should noted that closed-form design formulas have only been developed for linear dampers so far, and corresponding research on nonlinear dampers is still lacking.

In the present paper, closed-form optimal calibration formulas are to be derived for a weakly nonlinear device, the TLCD, installed inside a rotating blade. This will push the state-of-the-art one step forward from a theoretical point of view, and facilitate design and application of this very cost-effective device in wind turbine blades from a practical point of view. First, a 2-DOF model for the rotational blade-TLCD system is established through analytical dynamics method. The geometric stiffness due to the centrifugal force and the centripetal softening effect are both accounted for. Next, explicit expressions for the optimal frequency ratio and optimal equivalent TLCD damping ratio are derived using frequency response analysis by equalizing the dynamic amplification at three excitation frequencies. Subsequently, a closed-form formula for the optimal TLCD head loss coefficient is developed, without resorting to iteration. Furthermore, an alternative frequency-tuning formula is derived by means of the pole-placement method, ensuring that the modal damping ratios of the two free vibration modes of the 2-DOF system are equal. Comparison of the two slightly different frequency-tuning formulas are carried in terms of the resulting optimal damper parameters, the dynamic amplification, the root loci diagram, and the modal damping ratios. For the considered NREL 5MW wind turbine, the difference resulted from the two formulas turns out to be insignificant. Parametric study on various parameters (including the rotor speed, the mass ratio, etc.) is also performed. The present study proves the feasibility of developing closed-form design formulas for weakly-nonlinear devices installed in a rotating blade.

2. Theoretical model for the rotational blade-TLCD coupled system

2.1. Definition of the problem

The schematic diagram of a rotating wind turbine blade equipped with a TLCD is shown in Fig. 1. Assuming the three blades to be identical, only one blade is considered in the current study, as shown in Fig. 1(a). Three Cartesian coordinate systems (indicated by a subscript $i, i = 0, 1, 2$) are introduced to describe the motions of both the blade and liquid inside the TLCD tube. $(X_1, X_2, X_3)_0$-coordinate system corresponds to the global fixed coordinate system with the origin 'O_X' located at the center of the hub. The blade edgewise vibration is described in a rotating $(x_1, x_2, x_3)_1$-coordinate system fixed to the blade, with the original point O_x being coincided with O_X. The x_1-axis is unidirectional to the mean wind velocity, and the x_3-axis is placed along the undeformed blade axis orientated from the hub center towards the blade tip. The blade is modeled as a Bernoulli–Euler beam, and the mass per unit length and bending stiffness in the edgewise direction are denoted $\rho_b(x_3)$ and $EI(x_3)$, respectively. The liquid response inside the TLCD is described in a local $(y_1, y_2, y_3)_2$-coordinate system, as will be elaborated in Section 2.2. Assuming a constant rotational speed Ω of the blade, the rotational angle of the $(x_1, x_2, x_3)_1$-coordinate system with respect to the $(X_1, X_2, X_3)_0$-coordinate system is specified by the azimuthal angle, $\Psi(t)$, given by:

$$\Psi(t) = \Omega t \tag{1}$$

As shown in Fig. 1(a), 'p_b' represents a blade cross section located at x_3. When the blade is deformed, the position of 'p_b' in the local $(x_1, x_2, x_3)_1$-coordinate system can be determined by an augmented vector [31], given as:

$$\mathbf{u}_{x,b}(x_3, t) = \begin{bmatrix} u_{x_1,b}(x_3,t) \\ u_{x_2,b}(x_3,t) \\ u_{x_3,b}(x_3,t) \\ 1 \end{bmatrix} = \begin{bmatrix} 0 \\ -\Phi(x_3)q(t) \\ x_3 \\ 1 \end{bmatrix} \tag{2}$$

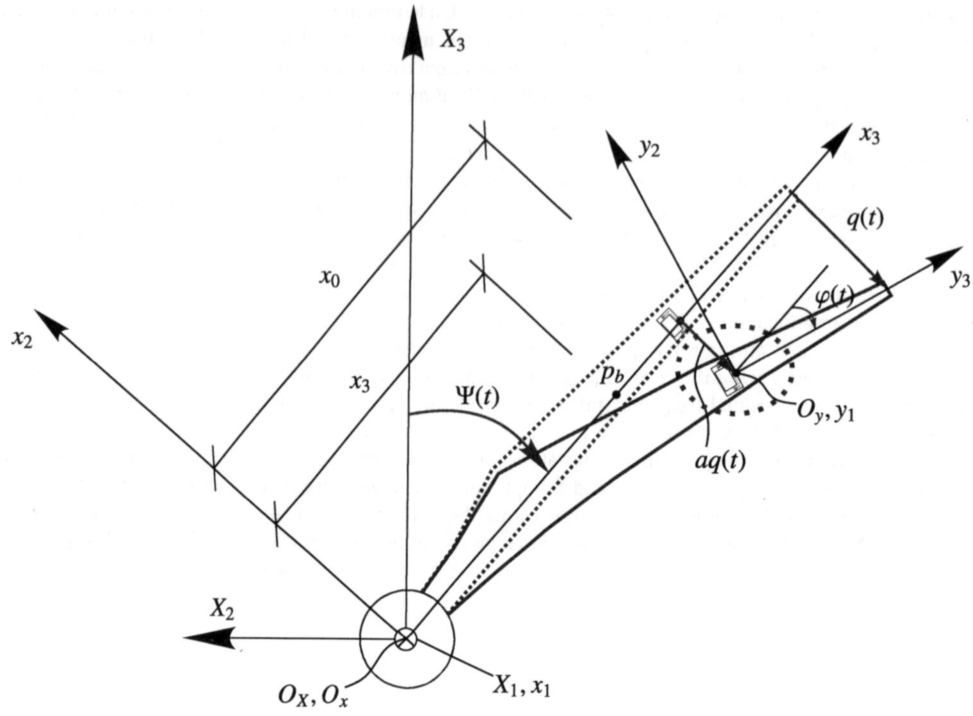

(a) Overview of a rotating blade quipped with a sealed TLCD

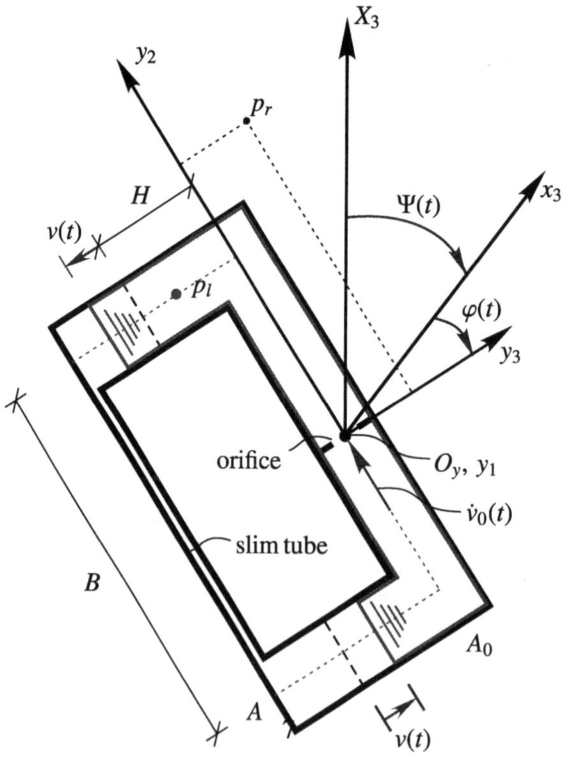

(b) Zoomed TLCD part

Fig. 1. Schematic diagram of the blade-TLCD coupled system. (a) Overview of a rotating blade equipped with a sealed TLCD (the TLCD part is indicated by a red circle); (b) zoomed TLCD part.

where $u_{x_i,b}(x_3,t), i = 1,2,3$ is the x_i–component of $\mathbf{u}_{x,b}(x_3,t)$. $\Phi(x_3)$ is the fundamental edgewise eigenmode of the blade, which is normalized to 1.0 at the blade tip; $q(t)$ is the edgewise DOF of the blade in the negative x_2–direction. The TLCD only aims at suppressing the fundamental mode of vibration, and thus the blade deflection in the x_2-direction is specified by a single degree of freedom (SDOF) $q(t)$. For the present study focusing on edgewise vibrations, the flapwise (x_1–direction) and axial (x_3–direction) deformations are not considered in Eq. (2).

The augmented position vector $\mathbf{u}_{X,b}(x_3,t)$ of point 'p_b' in the global fixed $(X_1, X_2, X_3)_0$-coordinate system can be obtained through the homogeneous transformation [31]:

$$\mathbf{u}_{X,b}(x_3,t) = {}^0\mathbf{T}_1\,\mathbf{u}_{x,b}(x_3,t) = \begin{bmatrix} u_{X_1,b}(x_3,t) \\ u_{X_2,b}(x_3,t) \\ u_{X_3,b}(x_3,t) \\ 1 \end{bmatrix} = \begin{bmatrix} 0 \\ -\sin(\Psi)x_3 - \cos(\Psi)\Phi q \\ \cos(\Psi)x_3 - \sin(\Psi)\Phi q \\ 1 \end{bmatrix} \tag{3}$$

where $u_{X_i,b}(x_3,t), i = 1,2,3$ is the X_i–component of $\mathbf{u}_{X,b}(x_3,t)$. The homogeneous transformation matrix ${}^0\mathbf{T}_1$ between the coordinate systems $(X_1, X_2, X_3)_0$ and $(x_1, x_2, x_3)_1$ is given by:

$${}^0\mathbf{T}_1 = \begin{bmatrix} 1 & 0 & 0 & 0 \\ 0 & \cos(\Psi) & -\sin(\Psi) & 0 \\ 0 & \sin(\Psi) & \cos(\Psi) & 0 \\ 0 & 0 & 0 & 1 \end{bmatrix} \tag{4}$$

The augmented velocity vector $\dot{\mathbf{u}}_{X,b}(x_3,t)$ of point 'p_b' in the global $(X_1, X_2, X_3)_0$-coordinate system becomes:

$$\dot{\mathbf{u}}_{X,b}(x_3,t) = \begin{bmatrix} \dot{u}_{X_1,b}(x_3,t) \\ \dot{u}_{X_2,b}(x_3,t) \\ \dot{u}_{X_3,b}(x_3,t) \\ 0 \end{bmatrix} = \begin{bmatrix} 0 \\ -\Phi\dot{q}\cos(\Psi) + \Omega\Phi q\sin(\Psi) - \Omega x_3\cos(\Psi) \\ -\Phi\dot{q}\sin(\Psi) - \Omega\Phi q\cos(\Psi) - \Omega x_3\sin(\Psi) \\ 0 \end{bmatrix} \tag{5}$$

The total kinetic energy of the blade T_{blade} can then be written as:

$$\begin{aligned} T_{\text{blade}} &= \int_0^{L_B} \frac{1}{2}\rho_b(x_3)\dot{\mathbf{u}}^{\mathrm{T}}_{X,b}(x_3,t)\dot{\mathbf{u}}_{X,b}(x_3,t)\mathrm{d}x_3 \\ &= \frac{1}{2}m_0\left(\dot{q}^2 + \Omega^2 q^2\right) + \Omega m_1\dot{q} + \frac{1}{2}\Omega^2 m_2 \end{aligned} \tag{6}$$

where L_B is the total length of the blade, superscript 'T' denotes the transpose operator, and $m_0 = \int_o^{L_B}\rho_b(x_3)\Phi^2(x_3)\mathrm{d}x_3$ is the modal mass of the blade. Further, two density-related parameters have been defined as: $m_1 = \int_o^{L_B}\rho_b(x_3)\Phi(x_3)x_3\mathrm{d}x_3$, $m_2 = \int_o^{L_B}\rho_b(x_3)x_3^2\mathrm{d}x_3$.

The total potential energy of the blade U_{blade} can be written as [17,19]:

$$U_{\text{blade}} = \frac{1}{2}k_{\text{ori},0}(\Omega,t)q^2 \tag{7}$$

where $k_{\text{ori},0}(\Omega,t)$ denotes the time-dependent modal stiffness of the blade including the contributions from the elastic deformation, centrifugal effect and gravity, given by:

$$k_{\text{ori},0}(\Omega,t) = k_e + k_1\Omega^2 - k_2\,g\cos(\Psi) \tag{8}$$

with g being the gravitational acceleration. k_e is the elastic modal stiffness of the non-rotational blade. The second term on the right side of Eq. (8) is related to the geometric stiffening due to centrifugal stiffening. The last term indicates the geometric softening caused by gravity. As seen, $k_{\text{ori},0}$ not only depends on the rotational speed Ω, but also varies with time t. For deriving closed-form formulas, the time-dependent term within $k_{\text{ori},0}(\Omega,t)$ (the third term in Eq. (8)) will be disregarded leading to a time-independent modal mass stiffness $k_0(\Omega)$, as will be elaborated in Section 3. The stiffness-related parameters (k_e, k_1 and k_2) in Eq. (8) are defined as:

$$\begin{cases} k_e = \displaystyle\int_0^{L_B} EI(x_3)\left(\frac{\mathrm{d}^2\Phi(x_3)}{\mathrm{d}x_3^2}\right)^2\mathrm{d}x_3 & \text{(a)} \\ k_1 = \displaystyle\int_0^{L_B} N_1(x_3)\left(\frac{\mathrm{d}\Phi(x_3)}{\mathrm{d}x_3}\right)^2\mathrm{d}x_3, \quad N_1(x_3) = \int_{x_3}^{L_B} z_3\rho_b(z_3)\mathrm{d}z_3 & \text{(b)} \\ k_2 = \displaystyle\int_0^{L_B} N_2(x_3)\left(\frac{\mathrm{d}\Phi(x_3)}{\mathrm{d}x_3}\right)^2\mathrm{d}x_3, \quad N_2(x_3) = \int_{x_3}^{L_B} \rho_b(z_3)\mathrm{d}z_3 & \text{(c)} \end{cases} \tag{9}$$

Thus, the time-dependent circular eigenfrequency of the fundamental edgewise mode becomes:

$$\omega_{\text{ori},0}(\Omega,t)=\sqrt{\frac{k_{\text{ori},0}(\Omega,t)}{m_0}} \tag{10}$$

2.2. Modeling of the TLCD

The layout of the U-shaped TLCD is presented in Fig. 1(b) with an orifice located in the middle of the horizontal tube. In order to remove the air spring effect [32] and avoid liquid leaking, an extra slim tube is added to connect the two vertical tubes. The lengths of the horizontal and vertical tubes are denoted by B and H, respectively. Thus, the overall length of liquid inside the TLCD is $L=B+2H$. The cross-sectional areas of the horizontal and vertical tubes are denoted A_0 and A, respectively. The overall mass of liquid inside the TLCD becomes:

$$m_w=\rho(A_0B+2AH) \tag{11}$$

with ρ being the liquid density.

Assuming the TLCD to be mounted at the coordinate x_0, the deflection $u_{x_2,b}(x_0,t)$ and the elastic rotation angle $\varphi(t)$ of the blade at this position in the local $(x_1,x_2,x_3)_1$-coordinate system can be obtained by setting $x_3=x_0$ in Eq. (2):

$$\left.\begin{aligned}u_{x_2,b}(x_0,t)&=-aq(t)\\ \varphi(t)&=\ \ bq(t)\end{aligned}\right\} \tag{12}$$

where two auxiliary parameters have been introduced for simplicity: $a=\Phi(x_0)$, $b=\Phi'(x_0)$, with the superscript ″′′ denoting the derivative with respect to x_3.

The local displacements of liquid inside the horizontal and vertical tubes are defined as $v_0(t)$ and $v(t)$, respectively. Assuming incompressible flow, the relationship between $v_0(t)$ and $v(t)$ can be derived based on the continuity equation:

$$\dot{v}_0(t)A_0=\dot{v}(t)A\Rightarrow\dot{v}_0(t)=\alpha\dot{v}(t) \tag{13}$$

in which $\alpha=A/A_0$ is the area ratio.

To describe the liquid response inside the TLCD, a local $(y_1,y_2,y_3)_2$-coordinate system is introduced with its origin O_y fixed at the middle point of the horizontal tube. The y_3-axis is placed on the symmetry line of the damper, and the time-varying angle between y_3- and x_3-axis is equal to $\varphi(t)$, as shown in Fig. 1(b). An auxiliary point, p_r (the black point in Fig. 1(b)), is defined in the local $(y_1,y_2,y_3)_2$-coordinate system by an augmented position vector:

$$\mathbf{u}_{y,r}(y_2,y_3)=\begin{bmatrix}u_{y_1,r}\\u_{y_2,r}\\u_{y_3,r}\\1\end{bmatrix}=\begin{bmatrix}0\\y_2\\y_3\\1\end{bmatrix} \tag{14}$$

where $u_{y_i,r}(x_3,t),i=1,2,3$ is the y_i-component of $\mathbf{u}_{y,r}(x_3,t)$. The augmented position vector, $\mathbf{u}_{X,r}(y_2,y_3,t)$, of point '$p_r$' in the global fixed $(X_1,X_2,X_3)_0$-coordinate system can be obtained through the homogeneous transformation [31]:

$$\begin{aligned}\mathbf{u}_{X,r}(y_2,y_3,t)&={}^0\mathbf{T}_1\,{}^1\mathbf{T}_2\,\mathbf{u}_{y,r}(y_2,y_3)\\&=\begin{bmatrix}u_{X_1,r}(y_2,y_3,t)\\u_{X_2,r}(y_2,y_3,t)\\u_{X_3,r}(y_2,y_3,t)\\1\end{bmatrix}\\&=\begin{bmatrix}0\\y_2\left(\cos(\Psi)-b\sin(\Psi)q\right)-y_3(\sin(\Psi)+b\cos(\Psi)q)\\-aq\cos(\Psi)-x_0\sin(\Psi)\\y_2\left(\sin(\Psi)+b\cos(\Psi)q\right)+y_3\left(\cos(\Psi)-b\sin(\Psi)q\right)\\-aq\sin(\Psi)+x_0\cos(\Psi)\\1\end{bmatrix}\end{aligned} \tag{15}$$

where $u_{X_i,r}(x_3,t),i=1,2,3$ is the X_i-component of $\mathbf{u}_{X,r}(x_3,t)$. Here the small deflection assumption for the blade is introduced to linearize the transformation matrix ${}^1\mathbf{T}_2$ relating the coordinate systems $(x_1,x_2,x_3)_1$ and $(y_1,y_2,y_3)_2$:

$${}^1\mathbf{T}_2=\begin{bmatrix}1&0&0&0\\0&1&-\varphi&-aq\\0&\varphi&1&x_0\\0&0&0&1\end{bmatrix} \tag{16}$$

Hence, the augmented velocity vector $\dot{\mathbf{u}}_{y,r}(y_2,y_3,t)$ of point 'p_r' in the local $(y_1,y_2,y_3)_2$-coordinate system can be obtained as:

$$\dot{\mathbf{u}}_{y,r}(y_2,y_3,t)={}^1\mathbf{T}_2^{-1}\,{}^0\mathbf{T}_1^{-1}\,\dot{\mathbf{u}}_{X,r}(y_2,y_3,t) \tag{17}$$

$$= \begin{bmatrix} 0 \\ \left(b^2 y_2 q - a - b y_3\right)\dot{q} - \Omega b(b y_3 + a) q^2 - \Omega(x_0 + y_3) \\ \left(b^2 y_3 q + abq + b y_2\right)\dot{q} + \Omega b^2 y_2 q^2 + \Omega(x_0 b - a) q + \Omega y_2 \\ 0 \end{bmatrix} \tag{18}$$

with the superscript $[\cdot]^{-1}$ being the inverse operator.

Furthermore, another auxiliary point p_l (indicated by the blue point) within the TLCD liquid is introduced for describing the liquid motion, as shown in Fig. 1(b). The corresponding augmented velocity vector $\dot{\mathbf{u}}_{y,l}\left(y_2, y_3, t\right)$ of point 'p_l' in the local $(y_1, y_2, y_3)_2$-coordinate system can be obtained by combining the relative liquid velocities $\dot{v}_0(t)$ and $\dot{v}(t)$ with $\dot{\mathbf{u}}_{y,r}(y_2, y_3, t)$, given as:

$$\dot{\mathbf{u}}_{y,l}\left(y_2, y_3, t\right) = \begin{cases} \dot{\mathbf{u}}_{y,l}\left(-\frac{B}{2}, y_3, t\right) = \dot{\mathbf{u}}_{y,r}\left(-\frac{B}{2}, y_3, t\right) + \begin{bmatrix} 0 \\ 0 \\ \dot{v} \\ 0 \end{bmatrix} & y_3 \in [-H + v, 0], \text{(Left)} \quad \text{(a)} \\ \dot{\mathbf{u}}_{y,l}\left(\frac{B}{2}, y_3, t\right) = \dot{\mathbf{u}}_{y,r}\left(\frac{B}{2}, y_3, t\right) + \begin{bmatrix} 0 \\ 0 \\ -\dot{v} \\ 0 \end{bmatrix} & y_3 \in [-H - v, 0], \text{(Right)} \quad \text{(b)} \\ \dot{\mathbf{u}}_{y,l}\left(y_2, 0, t\right) = \dot{\mathbf{u}}_{y,r}(y_2, 0, t) + \begin{bmatrix} 0 \\ \dot{v}_0 \\ 0 \\ 0 \end{bmatrix} & y_2 \in \left[-\frac{B}{2}, \frac{B}{2}\right], \text{(Horizontal)} \quad \text{(c)} \end{cases} \tag{19}$$

where $\dot{\mathbf{u}}_{y,l}\left(-B/2, y_3, t\right)$, $\dot{\mathbf{u}}_{y,l}\left(B/2, y_3, t\right)$ and $\dot{\mathbf{u}}_{y,l}\left(y_2, 0, t\right)$ represent the velocities of point 'p_l' located in the left, right and horizontal tube, respectively.

Thus, the kinetic energy T_{TLCD} and potential energy U_{TLCD} of the TLCD can be calculated:

$$\begin{aligned} T_{\text{TLCD}} = & \int_{-H+v}^{0} \frac{1}{2}\rho A \dot{\mathbf{u}}_{y,l}^{\text{T}}\left(-\frac{B}{2}, y_3, t\right)\dot{\mathbf{u}}_{y,l}\left(-\frac{B}{2}, y_3, t\right)\mathrm{d}y_3 \\ & + \int_{-H-v}^{0} \frac{1}{2}\rho A \dot{\mathbf{u}}_{y,l}^{\text{T}}\left(\frac{B}{2}, y_3, t\right)\dot{\mathbf{u}}_{y,l}\left(\frac{B}{2}, y_3, t\right)\mathrm{d}y_3 \\ & + \int_{-B/2}^{B/2} \frac{1}{2}\rho A_0 \dot{\mathbf{u}}_{y,l}^{\text{T}}\left(y_2, 0, t\right)\dot{\mathbf{u}}_{y,l}\left(y_2, 0, t\right)\mathrm{d}y_2 \end{aligned} \tag{20}$$

$$\begin{aligned} U_{\text{TLCD}} = & \int_{-H+v}^{0} \rho A g\, u_{X_3,r}\left(-\frac{B}{2}, y_3, t\right)\mathrm{d}y_3 \\ & + \int_{-H-v}^{0} \rho A g\, u_{X_3,r}\left(\frac{B}{2}, y_3, t\right)\mathrm{d}y_3 \\ & + \int_{-B/2}^{B/2} \rho A_0 g\, u_{X_3,r}\left(y_2, 0, t\right)\mathrm{d}y_2 \end{aligned} \tag{21}$$

The total kinetic energy T_{total} and total potential energy U_{total} of the blade-TLCD system can be obtained based on Eqs. (6), (7), (20) and (21):

$$T_{\text{total}} = T_{\text{blade}} + T_{\text{TLCD}} \tag{22}$$

$$U_{\text{total}} = U_{\text{blade}} + U_{\text{TLCD}} \tag{23}$$

The equations of motion (EOMs) for the 2-DOF model (one DOF $q(t)$ representing the blade response and the other DOF $v(t)$ representing the vertical liquid response) can be obtained using the Euler–Lagrange equation [33]:

$$\begin{aligned} & \frac{\mathrm{d}}{\mathrm{d}t}\left(\frac{\partial T_{\text{total}}}{\partial \dot{q}}\right) - \frac{\partial T_{\text{total}}}{\partial q} + \frac{\partial U_{\text{total}}}{\partial q} = f\left(\dot{q}, t\right) - C_0\dot{q} + f_g(t) \Rightarrow \\ & (m_0 + m_w a^2)\ddot{q} - \rho A B a \ddot{v} + \left(k_{ori,0}(\Omega, t) - \Omega^2 m_0 - \Omega^2 a^2 m_w\right) q \\ & + \Omega^2 \rho A B a v = f(\dot{q}, t) - C_0\dot{q} + m_3 g \sin(\Psi) + m_w g a \sin(\Psi) \end{aligned} \tag{24}$$

$$\begin{aligned} & \frac{\mathrm{d}}{\mathrm{d}t}\left(\frac{\partial T_{\text{total}}}{\partial \dot{v}}\right) - \frac{\partial T_{\text{total}}}{\partial v} + \frac{\partial U_{\text{total}}}{\partial v} = f_{\text{TLCD,nonlinear}}(\dot{v}, t) \Rightarrow \\ & -\rho A B a \ddot{q} + \rho A(2H + B\alpha)\ddot{v} + \frac{1}{2}\rho A \xi \alpha |\dot{v}|\dot{v} + \Omega^2 \rho A B a q \\ & + 2\rho A\left(\Omega^2 x_0 - \Omega^2 H - g\cos(\Psi)\right) v = -\rho A B g \sin(\Psi) \end{aligned} \tag{25}$$

where the elastic rotation $\varphi(t) = bq(t)$ of the blade is disregarded (due to its very small value [20]) leading to simplification of the EOMs. C_0 is the structural damping coefficient, $f(\dot{q},t)$ is the wind-induced modal load on the blade accounting for aerodynamic damping. $f_g(t)$ is the modal load due to gravity, given as:

$$f_g(t) = \sin(\Psi)g\int_0^{L_B}\rho_b(x_3)\Phi(x_3)\mathrm{d}x_3 = m_3 g\sin(\Psi) \tag{26}$$

in which $m_3 = \int_0^{L_B}\rho_b(x_3)\Phi(x_3)\mathrm{d}x_3$. $f_{\mathrm{TLCD,nonlinear}}(\dot{v},t) = -1/2\rho A\xi\alpha|\dot{v}|\dot{v}$ in Eq. (25) is the inherent nonlinear damping force of the TLCD, with ξ being the head loss coefficient that can be adjusted by varying the orifice area [34,35]. It is noteworthy that the nonlinear coupling and high-order terms related to q and v, such as qv, v^2 and q^2, in the derivation procedure are omitted due to their weak contributions. Thus, Eqs. (24) and (25) are the reduced form of the fully coupled EOMs (Eqs. (18) and (19), given in [23]).

To facilitate closed-form optimal calibration of the TLCD, the nonlinear damping force $f_{\mathrm{TLCD,nonlinear}}(\dot{v},t)$ is replaced by a linearized force $f_{\mathrm{TLCD,linear}} = -C_w\dot{v}$ under the assumption of harmonic external excitation [34]. C_w is the equivalent damping coefficient, and will be elaborated in Section 3.4.

The linearized EOMs can then be reformulated into the following matrix format:

$$\begin{aligned}&\begin{bmatrix} m_0+m_w a^2 & -\rho ABa \\ -\rho ABa & \rho A(2H+B\alpha)\end{bmatrix}\begin{bmatrix}\ddot{q}\\ \ddot{v}\end{bmatrix}+\begin{bmatrix} C_0 & 0\\ 0 & C_w\end{bmatrix}\begin{bmatrix}\dot{q}\\ \dot{v}\end{bmatrix}\\ &+\begin{bmatrix} k_{\mathrm{ori},0}(\Omega,t)-\Omega^2 m_0-\Omega^2 a^2 m_w & \Omega^2\rho ABa\\ \Omega^2\rho ABa & k_{\mathrm{ori},w}(\Omega,t)\gamma_A\end{bmatrix}\begin{bmatrix} q\\ v\end{bmatrix}\\ &=\begin{bmatrix} f(\dot{q},t)+m_3 g\sin(\Psi)+m_w ga\sin(\Psi)\\ -\rho ABg\sin(\Psi)\end{bmatrix}\end{aligned} \tag{27}$$

with $k_{\mathrm{ori},w}(\Omega,t)$ being the time-dependent equivalent stiffness of the rotating bladed-mounted TLCD, defined as:

$$k_{\mathrm{ori},w}(\Omega,t) = \frac{2\Omega^2\rho A\left(x_0 - H - g\cos(\Psi)\right)}{\gamma_A} \tag{28}$$

in which the non-dimensional parameter γ_A has been defined as:

$$\gamma_A = \frac{\alpha(1+\alpha\gamma-\gamma)}{\alpha-\alpha\gamma+\gamma} \tag{29}$$

Thus, the time-dependent circular eigenfrequency of the rotating blade-mounted TLCD can be obtained as:

$$\omega_{\mathrm{ori},w}(\Omega,t) = \sqrt{\frac{k_{\mathrm{ori},w}(\Omega,t)}{m_w}} = \sqrt{\frac{2\Omega^2 x_0 - \Omega^2(1-\gamma)L - 2\,g\cos(\Psi)}{L+(\alpha-1)\gamma L}} \tag{30}$$

with $\gamma = B/L$ being the horizontal length ratio. It turns out that $\omega_{\mathrm{ori},w}(\Omega,t)$ is dependent on L, x_0, Ω, γ and α, as well as time t (due to the azimuthal angle $\Psi(t)$).

3. Optimal calibration of the blade-mounted TLCD based on frequency response analysis

For deriving closed-form expressions for optimal TLCD calibration, three approximations are made here: (1) the primary structure has negligible damping, i.e. $C_0 = 0$. Such a simplification is reasonable because the overall damping (both structural and aerodynamic damping) for the blade in edgewise direction is low [9,12,36]. (2) The three time-dependent terms $m_3 g\sin(\Psi)$, $m_w ga\sin(\Psi)$ and $-\rho ABg\sin(\Psi)$ within the external load vector (the load vector in Eq. (27)) are numerically negligible compared with $f(\dot{q},t)$ [26]. (3) Similarly, the term $g\cos(\Psi)$ within $k_{\mathrm{ori},0}(\Omega,t)$ and $k_{\mathrm{ori},w}(\Omega,t)$ (in the stiffness matrix of Eq. (27)) is numerically negligible [19].

Thus, the linearized EOMs given in Eq. (27) are simplified to be:

$$\begin{aligned}&\begin{bmatrix} m_0+m_w a^2 & -\rho ABa \\ -\rho ABa & \rho A(2H+B\alpha)\end{bmatrix}\begin{bmatrix}\ddot{q}\\ \ddot{v}\end{bmatrix}+\begin{bmatrix} 0 & 0\\ 0 & C_w\end{bmatrix}\begin{bmatrix}\dot{q}\\ \dot{v}\end{bmatrix}\\ &+\begin{bmatrix} k_0-\Omega^2 m_0-\Omega^2 a^2 m_w & \Omega^2\rho ABa\\ \Omega^2\rho ABa & k_w\gamma_A\end{bmatrix}\begin{bmatrix} q\\ v\end{bmatrix}=\begin{bmatrix} f(t)\\ 0\end{bmatrix}\end{aligned} \tag{31}$$

where the original time-dependent blade modal stiffness $k_{\mathrm{ori},0}(\Omega,t)$ (Eq. (8)) and the equivalent TLCD stiffness $k_{\mathrm{ori},w}(\Omega,t)$(Eq. (28)), used in Eq. (27), are replaced by the time-independent terms denoted by $k_0(\Omega)$ and $k_w(\Omega)$, respectively:

$$\left.\begin{aligned} k_0(\Omega) &= k_e + k_1\Omega^2\\ k_w(\Omega) &= \frac{2\Omega^2\rho A(x_0-H)}{\gamma_A}\end{aligned}\right\} \tag{32}$$

Correspondingly, the time-dependent eigenfrequency of the blade and the TLCD ($\omega_o(\Omega, t)$ and $\omega_w(\Omega, t)$, presented in Eqs. (10) and (30)), are simplified to:

$$\left.\begin{aligned}\omega_0(\Omega) &= \sqrt{\frac{k_0(\Omega)}{m_0}} = \sqrt{\frac{k_e + k_1\Omega^2}{m_0}} \\ \omega_w(\Omega) &= \sqrt{\frac{k_w(\Omega)}{m_w}} = \Omega\sqrt{\frac{2(x_0 - H)}{L + (\alpha - 1)\gamma L}}\end{aligned}\right\} \tag{33}$$

For the classic non-rotating TLCD [37], ω_w is merely a function of L, γ and α. In the present case, however, ω_w is also dependent on Ω and x_0, which is in consistence with the rotating bladed-mounted TMD [19].

Eq. (31) is further nondimensionalized:

$$\begin{aligned}&\begin{bmatrix}1+\mu a^2 & -a\alpha\gamma_B\mu \\ -a\alpha\gamma_B\mu & \gamma_A\mu\end{bmatrix}\begin{bmatrix}\ddot{q}\\ \ddot{v}\end{bmatrix} + \begin{bmatrix}0 & 0 \\ 0 & \frac{C_w}{m_0}\end{bmatrix}\begin{bmatrix}\dot{q}\\ \dot{v}\end{bmatrix} \\ &+ \begin{bmatrix}\omega_0^2 - \Omega^2(1+\mu a^2) & \Omega^2 a\alpha\gamma_B\mu \\ \Omega^2 a\alpha\gamma_B\mu & \gamma_A\omega_w^2\mu\end{bmatrix}\begin{bmatrix}q\\ v\end{bmatrix} = \begin{bmatrix}\frac{f(t)}{m_0}\\ 0\end{bmatrix}\end{aligned} \tag{34}$$

with the normalized parameters introduced:

$$\mu = \frac{m_w}{m_0},\ \gamma_B = \frac{\gamma}{\alpha - \alpha\gamma + \gamma} \tag{35}$$

where μ is the mass ratio of the TLCD.

3.1. *Frequency response analysis of the 2-DOF system*

Assuming a harmonic external force with an angular frequency ω (i.e. $f(t) = \hat{f}e^{i\omega t} = Cm_0 g e^{i\omega t}$, with C being the excitation intensity factor and i being the imaginary unit), the steady-state responses of the blade (the 1st edgewise modal response) and the liquid can be written as:

$$q(t) = \hat{q}e^{i\omega t},\ v(t) = \hat{v}e^{i\omega t} \tag{36}$$

with $\hat{q}$ and $\hat{v}$ being the complex amplitudes.

By inserting Eq. (36) into Eq. (34), the normalized complex amplitudes of the responses can be obtained as:

$$\frac{\hat{q}}{\hat{f}/k_0} = \frac{\gamma_A\omega_0^2(\omega^2 - \omega_w^2) - 2i\zeta_w\omega_w\omega_0^2\omega}{\left(\begin{aligned}&\left[\mu a^2(\alpha^2\gamma_B^2 - \gamma_A) - \gamma_A\right]\omega^4 + \left[\Omega^2\mu a^2(2\alpha^2\gamma_B^2 - \gamma_A) + \gamma_A\omega_w^2(1+\mu a^2) + \gamma_A(\omega_0^2 - \Omega^2)\right]\omega^2 \\ &+ \left[\Omega^2\mu a^2(\Omega^2\alpha^2\gamma_B^2 + \gamma_A\omega_w^2) + \gamma_A\omega_w^2(\Omega^2 - \omega_0^2)\right] \\ &+ 2i\zeta_w\omega_w\omega\left[(1+\mu a^2)\omega^2 + \Omega^2(1+\mu a^2) - \omega_0^2\right]\end{aligned}\right)} \tag{37}$$

$$\frac{\hat{v}}{\hat{f}/k_0} = \frac{a\alpha\gamma_B\omega_0^2(\omega^2 + \Omega^2)}{\left(\begin{aligned}&\left[\mu a^2(\alpha^2\gamma_B^2 - \gamma_A) - \gamma_A\right]\omega^4 + \left[\Omega^2\mu a^2(2\alpha^2\gamma_B^2 - \gamma_A) + \gamma_A\omega_w^2(1+\mu a^2) + \gamma_A(\omega_0^2 - \Omega^2)\right]\omega^2 \\ &+ \left[\Omega^2\mu a^2(\Omega^2\alpha^2\gamma_B^2 + \gamma_A\omega_w^2) + \gamma_A\omega_w^2(\Omega^2 - \omega_0^2)\right] \\ &+ 2i\zeta_w\omega_w\omega\left[(1+\mu a^2)\omega^2 + \Omega^2(1+\mu a^2) - \omega_0^2\right]\end{aligned}\right)} \tag{38}$$

where ζ_w is the equivalent damping ratio of the TLCD ($C_w = 2\zeta_w\omega_w m_w$, cf.Section 3.4).

The characteristic equation of the 2-DOF system can be directly obtained from the denominator of Eq. (38) or Eq. (37), given as:

$$\begin{aligned}&\omega^4 + \left[\frac{\Omega^2\mu a^2(2\alpha^2\gamma_B^2 - \gamma_A) + \gamma_A\omega_w^2(1+\mu a^2) + \gamma_A(\omega_0^2 - \Omega^2)}{\mu a^2(\alpha^2\gamma_B^2 - \gamma_A) - \gamma_A}\right]\omega^2 \\ &+ \left[\frac{\Omega^2\mu a^2(\Omega^2\alpha^2\gamma_B^2 + \gamma_A\omega_w^2) + \gamma_A\omega_w^2(\Omega^2 - \omega_0^2)}{\mu a^2(\alpha^2\gamma_B^2 - \gamma_A) - \gamma_A}\right] + 2i\zeta_w\omega_w\omega\left[\frac{(1+\mu a^2)\omega^2 + \Omega^2(1+\mu a^2) - \omega_0^2}{\mu a^2(\alpha^2\gamma_B^2 - \gamma_A) - \gamma_A}\right] = 0\end{aligned} \tag{39}$$

The complex roots of Eq. (39) represent dynamic characteristics of the free vibration of the 2-DOF system, and will be used for optimal TLCD calibration in Section 4.

A special limiting case, $\zeta_w \Rightarrow \infty$, indicating infinite damping value of the TLCD (i.e. the orifice area is totally blocked), leads to a real root of Eq. (39):

$$\omega_\infty(\Omega) = \sqrt{\frac{\omega_0^2(\Omega)}{1+\mu a^2} - \Omega^2} \tag{40}$$

which is in perfect consistent with the blade-TMD system [26], and is different from the result of the conventional non-rotating structure-TLCD system [38]. This limiting frequency will be used as an important reference value in the subsequent calibration.

3.2. Optimal frequency tuning by equalized dynamic amplification

For classic SDOF-TMD system, two neutral frequencies (i.e. two fixed points) ω_M and ω_N exist in the frequency response curve of the structural response when structural damping is disregarded, where dynamic amplifications at these two frequencies are independent of the TMD damping ratio [29]. The same observation can be made for the non-rotating SDOF-TLCD model [38,39]. In the present case, this important feature still holds (as shown in Fig. 3), and will be theoretically and numerically demonstrated in the following. According to Den Hartog's theory [29], the optimal frequency tuning of the damper can be realized by equalizing the dynamic amplifications at the two neutral frequencies ω_M and ω_N. This will be the first criterion employed for the blade-mounted TLCD.

From Eq. (37), the dynamic amplification of the structure response can be expressed in a compact format [29]:

$$\frac{\hat{q}}{\hat{f}/k_0}=\frac{\tilde{A}+2i\zeta_w\tilde{B}}{\tilde{C}+2i\zeta_w\tilde{D}}\Rightarrow\left|\frac{\hat{q}}{\hat{f}/k_0}\right|^2=\frac{\tilde{A}^2+(2\zeta_w)^2\tilde{B}^2}{\tilde{C}^2+(2\zeta_w)^2\tilde{D}^2}\tag{41}$$

where $|\cdot|$ is the absolute value operator.

The condition of the dynamic amplification being independent of ζ_w leads to the following relationship:

$$\frac{\tilde{A}^2}{\tilde{C}^2}=\frac{\tilde{B}^2}{\tilde{D}^2}\Rightarrow\tilde{A}\tilde{D}=\pm\tilde{B}\tilde{C}\tag{42}$$

Substituting the expressions of $\tilde{A}$, $\tilde{B}$, $\tilde{C}$ and $\tilde{D}$ and realizing that the plus sign in Eq. (42) only leads to a trivial solution $\omega=0$, the following equation fulfilled by $\omega=\omega_M$ and $\omega=\omega_N$ is obtained:

$$\begin{aligned}&\left[2\gamma_A(1+\mu a^2)-a^2\alpha^2\gamma_B^2\mu\right]\omega^4\\&-\left[2\Omega^2\left(-a^2\alpha^2\gamma_B^2\mu+\gamma_A(1+\mu a^2)\right)+2\gamma_A\omega_0^2+2\gamma_A\omega_w^2(1+\mu a^2)\right]\omega^2\\&-2\gamma_A\omega_0^2\left(\frac{1}{1+\mu a^2}-\frac{\Omega^2}{\omega_0^2}-\frac{a^2\alpha^2\gamma_B^2\mu}{\gamma_A(1+\mu a^2)^2}\right)\left(\Omega^2(1+\mu a^2)-\omega_0^2\right)-\Omega^4a^2\alpha^2\gamma_B^2\mu=0\end{aligned}\tag{43}$$

from which the first equation for $\omega_M^2+\omega_N^2$ can be obtained directly:

$$\omega_M^2+\omega_N^2=\frac{2\Omega^2\left(-a^2\alpha^2\gamma_B^2\mu+\gamma_A(1+\mu a^2)\right)+2\gamma_A\omega_0^2+2\gamma_A\omega_w^2(1+\mu a^2)}{2\gamma_A(1+\mu a^2)-a^2\alpha^2\gamma_B^2\mu}\tag{44}$$

Since the dynamic amplification at ω_M and ω_N is independent of ζ_w, it can be expressed in a simple format with the limiting condition $\zeta_w\Rightarrow\infty$:

$$\left|\frac{\hat{q}}{\hat{f}/k_0}\right|_{\omega=\omega_M,\omega_N}=\pm\frac{\omega_0^2}{(1+\mu a^2)\omega^2+\Omega^2(1+\mu a^2)-\omega_0^2}\tag{45}$$

Then, the equal dynamic amplification criterion [29] leads to the second equation for $\omega_M^2+\omega_N^2$:

$$\omega_M^2+\omega_N^2=\frac{2\omega_0^2-2\Omega^2(1+\mu a^2)}{1+\mu a^2}\tag{46}$$

Equalizing Eq. (44) with Eq. (46) results in the explicit formula for the optimal frequency ratio of the blade-mounted TLCD:

$$\left(\frac{\omega_w}{\omega_0}\right)_{\text{opt}}=\sqrt{\frac{1}{1+\mu a^2}-\frac{\Omega^2}{\omega_0^2(\Omega)}-\frac{a^2\alpha^2\gamma_B^2\mu}{\gamma_A(1+\mu a^2)^2}}\tag{47}$$

As seen, the optimal frequency ratio of the blade-mounted TLCD exhibits nonlinear dependence on the effective mass ratio μa^2. Moreover, it also explicitly depends on the rotor rotational speed Ω and the edgewise eigenfrequency ω_0. This is different from the optimal frequency ratio of the blade-mounted TMD [26] based on the same criterion (equalized dynamic amplification), which is independent of Ω and ω_0. Furthermore, Eq. (47) is also quite different from the result of the classic non-rotating structure-TLCD system, in which $(\omega_w/\omega_0)_{\text{opt}}$ is merely a function of μ and γ [38,39].

Inserting Eq. (47) into Eq. (43), two roots can be obtained:

$$\omega_{M,N}^2(\Omega)=\omega_\infty^2\pm\frac{a\alpha\gamma_B\sqrt{\mu}\omega_0^2(\Omega)}{(1+\mu a^2)\sqrt{2\gamma_A(1+\mu a^2)-a^2\alpha^2\gamma_B^2\mu}}\tag{48}$$

where ω_∞ is the reference frequency given in Eq. (40). One may observe that the specific values of those two neutral frequencies (i.e. ω_M and ω_N) are implicitly dependent on Ω, and the distance between these two points $|\omega_N-\omega_M|$ (implying the bandwidth of the TLCD) also varies with Ω. Furthermore, Eq. (48) also implies that ω_∞ lies in the middle of ω_M and ω_N, and this property will be used for determining the optimal damping ratio in Section 3.3. The dynamic amplification at ω_M or ω_N under the optimal frequency tuning condition Eq. (47) can be subsequently obtained by inserting Eq. (48) into Eq. (45):

$$\left|\frac{\hat{q}}{\hat{f}/k_0}\right|_{\omega=\omega_M,\omega_N}=\sqrt{\frac{2\gamma_A(1+\mu a^2)}{\mu a^2\alpha^2\gamma_B^2}-1}\tag{49}$$

It is interesting to observe that the dynamic amplitude at the neutral frequencies (i.e. ω_M and ω_N) is independent of Ω when $(\omega_w/\omega_0)_{\text{opt}}$ in Eq. (47) is used. This finding is consistent with that from the rotating blade-TMD system [26].

3.3. Optimal equivalent damping ratio

The damping tuning criterion proposed by Krenk [30] leads to a fairly level behavior of the dynamic amplification of the relative TMD motion in comparison with Den Hartog's criterion [29]. Such an observation still holds for the non-rotating structure-TLCD system as demonstrated by our previous work [38]. Therefore, Krenk's damping tuning criterion is also employed in the present study, which concerns equalizing the dynamic amplification at three excitation frequencies (i.e. ω_M, ω_N and ω_∞) in the frequency response curve, as shown in Fig. 3.

Inserting Eqs. (40) and (47) into Eq. (37), the dynamic amplification at ω_∞ is obtained (not shown here for conciseness). Equalizing this dynamic amplification with Eq. (49), the optimal equivalent damping ratio $\zeta_{w,\text{opt}}$ becomes:

$$\zeta_{w,\text{opt}} = \frac{\sqrt{2}}{2}\alpha\gamma_B\omega_0^2\sqrt{\frac{\gamma_A\mu\left(\gamma_A(1+\mu a^2)-a^2\alpha^2\gamma_B^2\mu\right)}{(1+\mu a^2)\left(\Omega^2(1+\mu a^2)-\omega_0^2\right)\left(\gamma_A\Omega^2(1+\mu a^2)^2-\gamma_A\omega_0^2(1+\mu a^2)+a^2\alpha^2\gamma_B^2\mu\omega_0^2\right)}} \tag{50}$$

As seen, $\zeta_{w,\text{opt}}$ exhibits strong nonlinear dependency on the rotational speed Ω and the blade eigenfrequency ω_0, which is quite different from that of the non-rotating SDOF-TLCD system where $\zeta_{w,\text{opt}}$ only depends on γ and μ [38,39].

3.4. Optimal head loss coefficient

Having the equivalent damping ratio ζ_w is not sufficient for designing the TLCD. The head loss coefficient ξ needs be determined in order to specify the orifice opening area [37,40]. $\xi = 0$ represents the TLCD without damping element, and the opening ratio of the orifice is 100%. Conversely, $\xi \Rightarrow \infty$ represents the case where the orifice area is totally blocked, and the TLCD becomes a concentrated mass attached to the structure. Since a harmonic external excitation force ($f(t) = \hat{f}e^{i\omega t}$) is assumed in the frequency response analysis of the analogue 2-DOF system (as presented in Section 2.1), the equivalent damping coefficient C_w of TLCD in Eq. (27) can thus be obtained using the harmonic linearization method (see details in [34]), given as:

$$C_w = 2\zeta_w\omega_w m_w = \frac{4}{3\pi}\rho A\xi\omega|\hat{v}| \tag{51}$$

where ζ_w is the equivalent damping ratio, ω is the excitation frequency and $|\hat{v}|$ is the absolute value of the liquid response amplitude. Thus, the head loss coefficient can be directly obtained from Eq. (51):

$$\xi = \frac{3\pi\zeta_w\omega_w m_w}{\rho A\omega|\hat{v}|} = \frac{3\pi x_0\zeta_w\omega_w(\alpha-\alpha\gamma+\gamma)}{\alpha\omega|\hat{v}|\left((1+\alpha\gamma-\gamma)\frac{\omega_w^2}{\Omega^2}+1-\gamma\right)} \tag{52}$$

To obtain the optimal head loss coefficient, optimal values $\zeta_{w,\text{opt}}$ and $\omega_{w,\text{opt}}$ given in Eqs. (47) and (50) should be substituted in Eq. (52). The excitation frequency ω is chosen as $\omega = \omega_\infty$, in consistent with the optimal damping tuning criterion in Section 3.3. Obtaining $|\hat{v}|$ normally requires an iterative procedure. In the present study, however, we propose to obtain $|\hat{v}|$ directly from Eq. (38), by letting $\omega = \omega_\infty$ (given in Eq. (40)) and substituting optimal values $\omega_{w,\text{opt}}$ and $\zeta_{w,\text{opt}}$ into Eq. (38). As a result, the optimal head loss coefficient can be written as:

$$\xi_{\text{opt}} = \frac{3\pi x_0\zeta_{w,\text{opt}}\omega_{w,\text{opt}}(\alpha-\alpha\gamma+\gamma)}{\alpha\omega_\infty|\hat{v}|_{\omega=\omega_\infty}\left((1+\alpha\gamma-\gamma)\frac{\omega_{w,\text{opt}}^2}{\Omega^2}+1-\gamma\right)} \tag{53}$$

Therefore, based on the frequency response analysis of the blade-TLCD system, closed-form formulas for the optimal frequency ratio and optimal head loss coefficient of the blade-mounted TLCD are obtained, as given by Eqs. (47) and (53), respectively.

4. Optimal calibration of the blade-mounted TLCD using the pole-placement method

Once the values of Ω, μ, a, γ, α, x_0, ζ_w, ω_w and ω_0 are determined, four complex roots can be obtained from Eq. (39). Two roots of them (denoted by ω_1 and ω_2) are located in the first quadrant of the complex plane, and the other two symmetrically located roots ($-\bar{\omega}_1$ and $-\bar{\omega}_2$) are in the second quadrant, where the bar symbol ($\bar{}$) denotes complex conjugate. By varying the equivalent damping ratio ζ_w from zero to infinity (using a large value in numerical calculation), the so-called root loci diagram of the characteristic equation (Eq. (39)) can be obtained [30].

For classic non-rotating SDOF-TMD system, equal dynamic amplification at the two neutral frequencies is equivalent to the existence of a bifurcation point ω_* (a complex root of Eq. (39) indicated by the red asterisk point $*$ in Fig. 5) where two branches of the root locus intersect [30]. It should be note that the so-called bifurcation point here it is not a real bifurcation phenomenon of the mechanical system, but it is only with mathematical relevance. This will further ensure that the modal damping ratios of the two free vibration modes of the SDOF-TMD system are equal. However, this equivalence is lost for the present rotating blade-TLCD system, i.e., equal dynamic amplification at the neutral frequencies does not guarantee the existence of a bifurcation point as well as the equal modal damping ratios, as will be shown in the numerical example Section 5.2. In view of this inconsistency, a second criterion (i.e. the existence of a bifurcation point ω_*) for determining the optimal frequency ratio in additional to the equal dynamic amplification criterion is introduced in this section.

The existence of the bifurcation point ω_* means that the following equation holds [30]:

$$(\omega - \omega_*)^2(\omega - \bar{\omega}_*)^2 = 0 \tag{54}$$

This equation can be expanded into the following format:

$$\omega^4 - 4\mathrm{i}\,\mathrm{Im}[\omega_*]\omega^3 - \left(2|\omega_*|^2 + 4\,\mathrm{Im}[\omega_*]^2\right)\omega^2 + 4\mathrm{i}\,\mathrm{Im}[\omega_*]|\omega_*|^2\omega + |\omega_*|^4 = 0 \tag{55}$$

where Im[.] is the imaginary part operator.

Comparison of the coefficients of the linear and cubic terms in Eqs. (39) and (55) leads to the expression of $|\omega_*|$:

$$|\omega_*| = \sqrt{\frac{\omega_0^2}{1+\mu a^2} - \Omega^2} \Rightarrow |\omega_*| = \omega_\infty \tag{56}$$

The result $|\omega_*| = \omega_\infty$ in Eq. (56) indicates that the bifurcation point is located on a circle around the origin of the complex plane with a radius equal to $R = \omega_\infty$.

Next, comparison of the constant terms in Eqs. (39) and (55) provides the alternative optimal frequency ratio formula:

$$\left(\frac{\omega_w}{\omega_0}\right)_{\mathrm{opt}} = \sqrt{\frac{1}{1+\mu a^2} - \frac{\Omega^2}{\omega_0^2(\Omega)} - \frac{a^2\alpha^2\gamma_B^2\mu\left(2\Omega^2(1+\mu a^2) - \omega_0^2\right)}{\gamma_A(1+\mu a^2)^2\left(\Omega^2(1+\mu a^2) - \omega_0^2\right)}} \tag{57}$$

As seen, Eq. (57) exhibits a similar format as in Eq. (47), except for the additional factor $\left(2\Omega^2(1+\mu a^2) - \omega_0^2\right) / \left(\Omega^2(1+\mu a^2) - \omega_0^2\right)$ within the third term. This nonlinear factor will lead to a slightly lower value of the optimal frequency ratio.

Hence, two similar frequency-tuning formulas, i.e. Eqs. (47) and (57), are derived based on two different criteria. For the pole-placement-based method, a simplified approximation method based on the damping ratio at bifurcation point was suggested for determining the optimal equivalent damping ratio for a non-rotating SDOF-TLCD system, which is $\zeta_{w,\mathrm{opt}} = \sqrt{2}/2\zeta_*$ [38]. This method can in principle be employed in the present paper as well. However, for investigating the difference between the two frequency-tuning formulas, the damping tuning criterion is kept unchanged in the present study. This means that the optimal equivalent damping ratio provided in Eq. (50) will still be treated as the optimal value when Eq. (57) is used for the optimal frequency tuning.

Moreover, comparison of the coefficients in the cubic and quadratic term of Eqs. (39) and (55) provides the closed-form expression of the critical equivalent damping ratio:

$$\zeta_* = \frac{a\alpha\gamma_B\omega_0^2\sqrt{\gamma_A\mu\left(\gamma_A(1+\mu a^2) - a^2\alpha^2\gamma_B^2\mu\right)}}{\sqrt{(1+\mu a^2)\left(\gamma_A(1+\mu a^2)\left(\Omega^2(1+\mu a^2) - \omega_0^2\right)^2 + a^2\alpha^2\gamma_B^2\mu\omega_0^2\left(2\Omega^2(1+\mu a^2) - \omega_0^2\right)\right)}} \tag{58}$$

which corresponds to the equivalent damping ratio of the TLCD at the bifurcation point ω_*.

Finally, the modal damping ratio $\zeta_i (i = 1, 2)$ of the 2-DOF system is determined from the complex frequency ω_i (the two roots of Eq. (39) located in the first quadrant) [30]:

$$\zeta_i = \frac{\mathrm{Im}[\omega_i]}{|\omega_i|}, i = 1, 2 \tag{59}$$

Once the optimal frequency-tuning ratio $(\omega_w/\omega_0)_{\mathrm{opt}}$ and optimal head loss coefficient ξ_{opt} are determined, the resulting dynamic amplification curves of the 2-DOF system under the optimal tuning condition can be obtained through a simple iterative process in the frequency domain [39]. For clarity, a flowchart for determining the dynamic amplifications of the structural and liquid responses is given in Fig. 2.

5. Numerical example

The use of closed-form calibration formulas for the blade-mounted TLCD will be demonstrated by means of the NREL 5MW reference turbine [41]. The rated rotor speed is $\Omega_0 = 1.27$ rad/s and the rated mean wind velocity at the hub is 11.4 m/s. In the present study, the rotor speed Ω will be varied from Ω_0 to $3\Omega_0$ in order to highlight the rotational effect on the calibration results, although in reality Ω will modestly fluctuate around Ω_0 during the normal operational condition (unless the brake system brakes down) [42]. The choice of $3\Omega_0$ is of little practical implications (as such large a rotor overspeed only occurs in a catastrophe), and it is merely used in the numerical example to better highlight the influence of different rotor speeds on the optimal TLCD parameters as well as its performance. By doing so, a more comprehensive understanding of the parametric sensitivity can be achieved. The total length and overall mass of each blade are $L_B = 61.5$ m and 17740 kg, respectively. The edgewise stiffness $EI(x_3)$, the mass per unit length $\rho_b(x_3)$ and the fundamental modal shape in the edgewise direction are provided in [41]. The modal mass corresponding to the fundamental mode can be calculated as $m_0 = 1.354 \times 10^3$ kg. A possible preliminary design for a TLCD mounted on $x_0 = 45$ m was firstly conducted (with practical parameters $\mu = 0.05$, $\gamma = 0.6$ and $\alpha = 1$) to demonstrate the feasibility of installing the TLCD into the space-limited blade. The results show ed the TLCD only took up around 20% of the airfoil area (with possible reduction by adjusting γ and α), verifying its practicability.

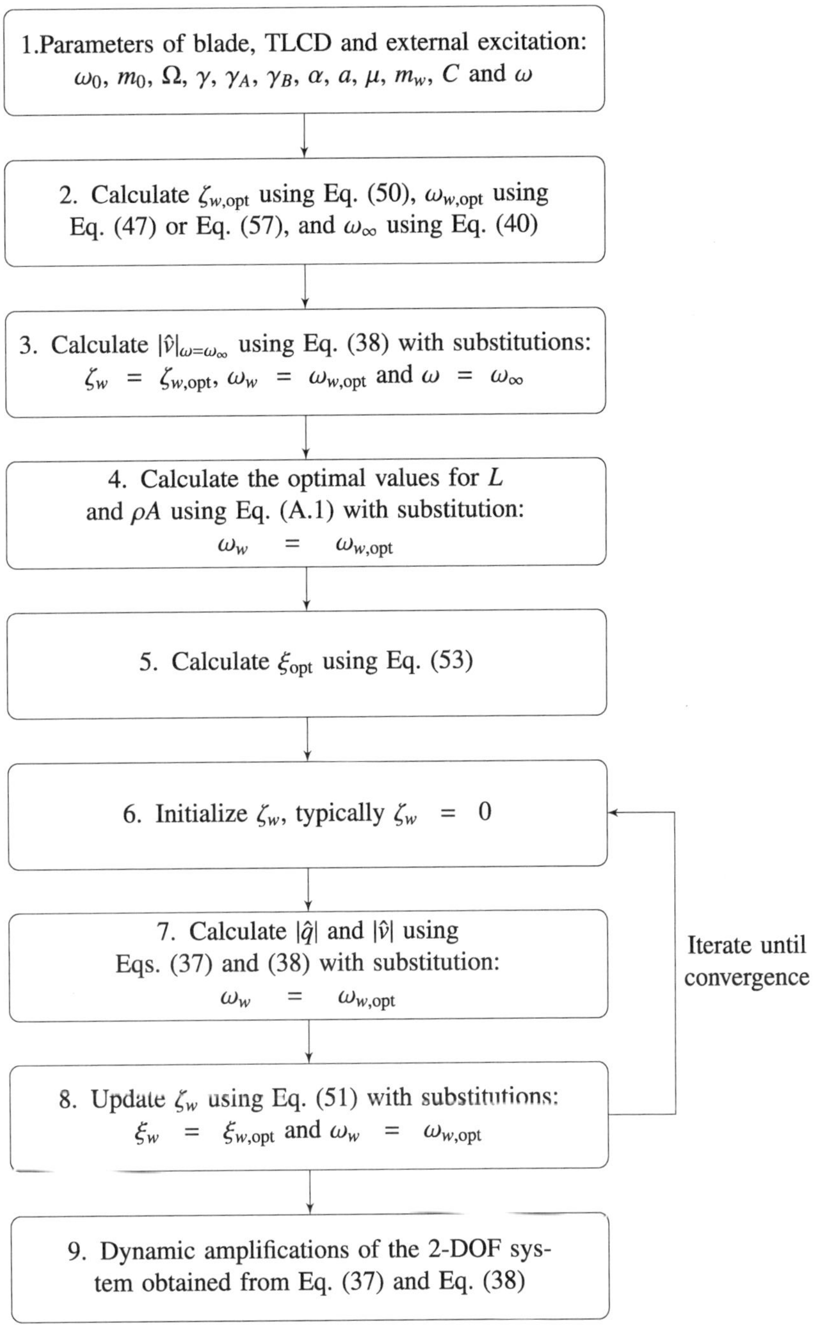

Fig. 2. Flowchat for determining the resulting dynamic amplifications of the 2-DOF system after the TLCD is optimally calibrated. $(\omega_w/\omega_0)_{\mathrm{opt}}$ is obtained from Eq. (47) or Eq. (57), and ξ_{opt} is obtained from Eq. (53).

In the following, the dynamic amplifications of the structural and liquid responses obtained from the two optimal frequency-tuning formulas, i.e. Eqs. (47) and (57), are compared. Subsequently, the root loci diagrams of the 2-DOF system obtained from the two frequency tuning formulas are also compared. Parametric studies in terms of the rotor speed Ω, the area ratio α, the mass ratio μ, etc. are then performed.

5.1. Dynamic amplification

Fig. 3 shows the dynamic amplifications of the structural and liquid responses versus the normalized excitation frequency ω/ω_0, using the optimal frequency-tuning formula Eq. (47). The prescribed parameters are $x_0 = 45$ m, $\mu = 0.05$, $\gamma = 0.6$, $\alpha = 1$ and

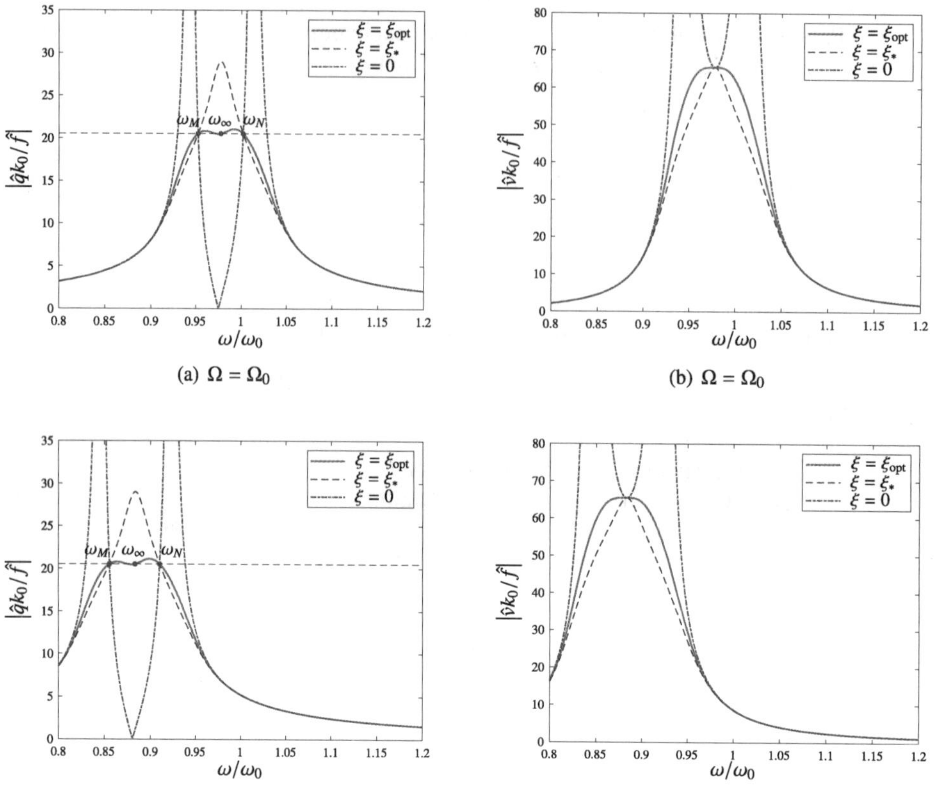

(a) $\Omega = \Omega_0$ (b) $\Omega = \Omega_0$

(c) $\Omega = 3\Omega_0$ (d) $\Omega = 3\Omega_0$

Fig. 3. Dynamic amplification curves of the 2-DOF system with the optimal frequency ratio $(\omega_w/\omega_0)_{opt}$ given by Eq. (47), for three different head loss coefficients. $x_0 = 45$ m, $\mu = 0.05$, $\gamma = 0.6$, $\alpha = 1$ and $C = 0.02$. (a)(c) Structural response, (b)(d) liquid response. (a)(b) $\Omega = \Omega_0$ (1.27 rad/s), (c)(d) $\Omega = 3\Omega_0$ (3.81 rad/s).

$C = 0.02$. Two rotational speeds are considered here (i.e. $\Omega = \Omega_0, 3\Omega_0$), excluding the limiting case $\Omega = 0$ that will render the TLCD nonfunctioning due to the loss of centrifugal force. For each Ω, three different head loss coefficients are considered, i.e. $\xi = \xi_{opt}$ given by Eq. (53), $\xi = \xi_*$ (numerically larger than ξ_{opt}, representing the over-optimal damping case) obtained by replacing $\zeta_{w,opt}$ with ζ_* (given by Eq. (58)) in Eq. (53), and the limiting case $\xi = 0$ (representing the undamped condition).

As shown in Figs. 3(a) and 3(c), the two neutral frequencies ω_M and ω_N, which are independent of ξ, do exist for both rotor speeds. Furthermore, the amplitudes at ω_M and ω_N are equal, in accordance with the employed equal dynamic amplification criterion in Section 3.2. The horizontal interval between these two points (i.e. $|\omega_M - \omega_N|$) is slightly broaden with increasing value of Ω, which can be revealed by the theoretical expression Eq. (48). When $\xi = \xi_{opt}$, the dynamic amplification at ω_∞ is equal to that at ω_M and ω_N, as indicated by the horizontal dashed line in Figs. 3(a) and 3(c). In addition, ξ_{opt} also leads to a fairly level behavior of the curve between these two fixed points. On the other hand, $\xi = \xi_*$ leads to a single peak between ω_M and ω_N, as indicated by the blue dashed line in Figs. 3(a) and 3(c). In the limit $\xi = 0$ (indicated by the purple line in Figs. 3(a) and 3(c)) corresponding to the undamped condition, the amplitude at the two natural frequencies of the 2-DOF system becomes infinite, and a zero amplitude is obtained at ω_w. Comparing Figs. 3(a) and 3(c), one can observe that a larger rotor speed leads to a shift of all the curves towards left. Figs. 3(b) and 3(d) show the corresponding results of the liquid response. It can be observed that the amplitude of the liquid response at ω_∞ is independent of ξ. Further, $\xi = \xi_{opt}$ (the green line) leads to a more level behavior of the frequency response curve. Any value of ξ larger than ξ_{opt} (such as $\xi = \xi_*$, indicated by the blue line) leads to a narrow peak in the curve. On the other hand, any value of ξ smaller than ξ_{opt} leads to two peaks exceeding the dynamic amplification at ω_∞ (such as the limiting case $\xi = 0$, indicated by the purple lines). It should be noted that the purple lines in Figs. 3(b) and 3(d) only provide a qualitative trend for the liquid response under $\xi = 0$ condition, because the actual dynamic response of liquid cannot reach infinity due to the constraint from the vertical tube length (i.e. $\hat{v} \leq H$) [37]. Similar to Figs. 3(a) and 3(c), the liquid response curves shift towards lower frequencies when Ω increases.

Fig. 4 shows the results when the alternative optimal frequency-tuning formula Eq. (57) is used. Unlike Figs. 3(a) and 3(c), the amplitudes at ω_M and ω_N are no longer equal (exhibiting a negative inclination), and this phenomenon becomes more significant for larger values of Ω, as shown in Fig. 4(c). Despite the unequal amplitudes at ω_M and ω_N, $\xi = \xi_{opt}$ still provides the optimal performance of the structural response. It is worth noting that due to the unbalanced feature of the curve, the maximal structural response will switches from the right peak to the left peak when Ω increases from Ω_0 to $3\Omega_0$, as shown in the zoomed parts of Figs. 4(a) and 4(c). This phenomenon will cause a nonlinear variation of the maximum structural response $\left|\hat{q}k_0/\hat{f}\right|_{max}$ versus Ω,

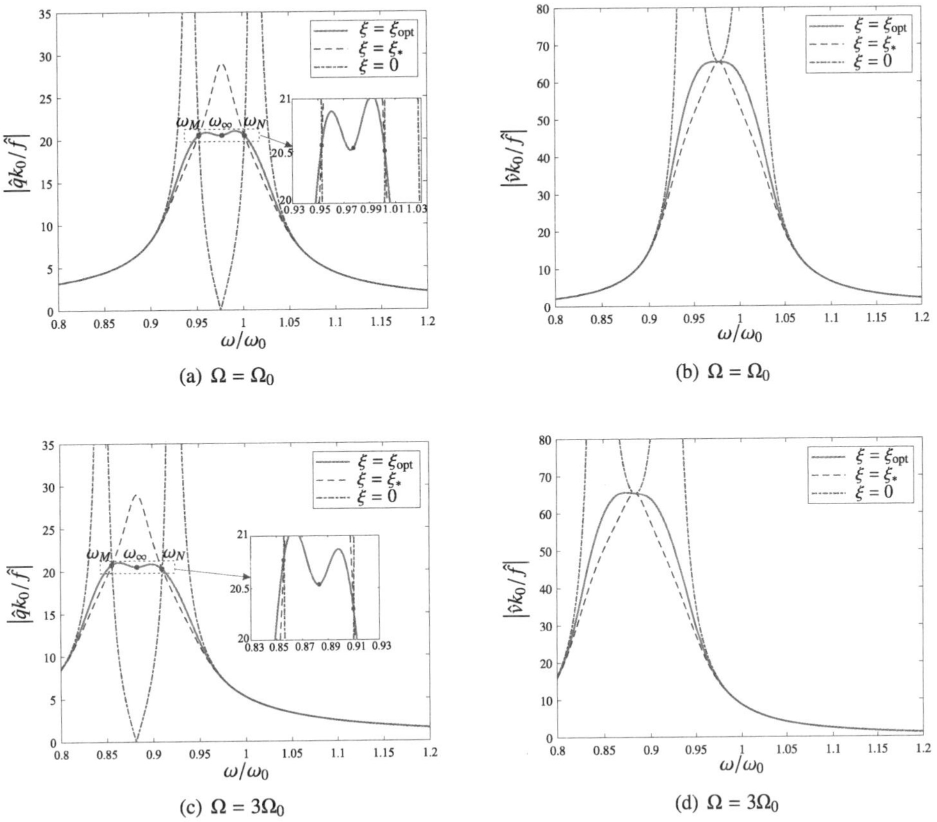

Fig. 4. Dynamic amplification curves of the 2-DOF system with the optimal frequency ratio $(\omega_w/\omega_0)_{\text{opt}}$ given by Eq. (57), for three different head loss coefficients ξ. $x_0 = 45$ m, $\mu = 0.05$, $\gamma = 0.6$, $\alpha = 1$ and $C = 0.02$. (a)(c) Structure response, (b)(d) liquid response. (a)(b) $\Omega = \Omega_0$ (1.27 rad/s), (c)(d) $\Omega = 3\Omega_0$ (3.81 rad/s).

as will be shown in Fig. 9(a). Similar to Figs. 3(b) and 3(d), $\xi = \xi_{\text{opt}}$ leads to a fairly level behavior of the liquid response curve, although the curve exhibits a slightly negative inclination. Both Figs. 3 and 4 show similar trend with that of the blade-TMD system (given in [26])

5.2. Root loci of the 2-DOF blade-TLCD system

It can be seen from Figs. 3 and 4 that the optimal frequency tuning criterion Eq. (47) is superior to Eq. (57) in terms of the balanced dynamic amplification curve. In this subsection, however, the superiority of Eq. (57) will be shown in terms of the resulting equal modal damping ratios.

Fig. 5 shows the root loci of the 2-DOF model in the first quadrant of the complex plane (the other two roots are symmetrically located in the second quadrant) with the two optimal frequency-tuning formulas Eqs. (47) and (57) employed. Two mounting positions ($x_0 = 25$ m and $x_0 = 45$ m) and two rotor speeds ($\Omega = \Omega_0$ and $3\Omega_0$) are considered. The zero equivalent damping ratio (i.e. $\zeta_w = 0$) leads to two roots (denoted by the red circles ∘) of the characteristic equation Eq. (39), which are located in the positive real axis. As ζ_w increases, these two roots corresponding to the optimal equivalent damping ratio $\zeta_{w,\text{opt}}$ move into the positive imaginary half plane with the evolution directions indicated by arrow symbols (◀ or ◀). Further, the limiting case $\zeta_w \Rightarrow \infty$ leads to two roots, one located at the positive image axis (not shown here) and the other one located at the positive real axis denoted by the red square symbols □(i.e. ω_∞). Figs. 5(a) and 5(c) show the results when the first optimal frequency-tuning formula Eq. (47) is applied. Taking the results of $x_0 = 45$ m for instance, the two local branches get closer (without intersecting) initially as ζ_w increases, but move away from each other after ζ_w exceeds the critical value ζ_*. Therefore, no bifurcation point exists. This property becomes more significant as the value of Ω increases, as shown in Fig. 5(c). In addition, two dashed lines separately connecting the origin and the two roots do not coincide with each other, implying that the modal damping ratios of the 2-DOF system are not equal. Similar observations can be made for the case of $x_0 = 25$m, but with smaller encircled areas. Further, a higher rotor speed leads to a shift of the root loci towards left.

Figs. 5(b) and 5(d) show the results when the alternative optimal frequency-tuning formula Eq. (57) is used. It is clearly observed that two branches of root locus intersect at the bifurcation point ω_* (denoted by the red asterisk ∗) when $\zeta_w = \zeta_*$. Further increase of ζ_w leads to a split of two branches in the opposite directions, one ending at the positive image axis and the other one ending at ω_∞. In addition, the two roots of Eq. (39) (denoted by the red dots) lies on the same dashed line passing the origin when $0 \leq \zeta \leq \zeta_*$,

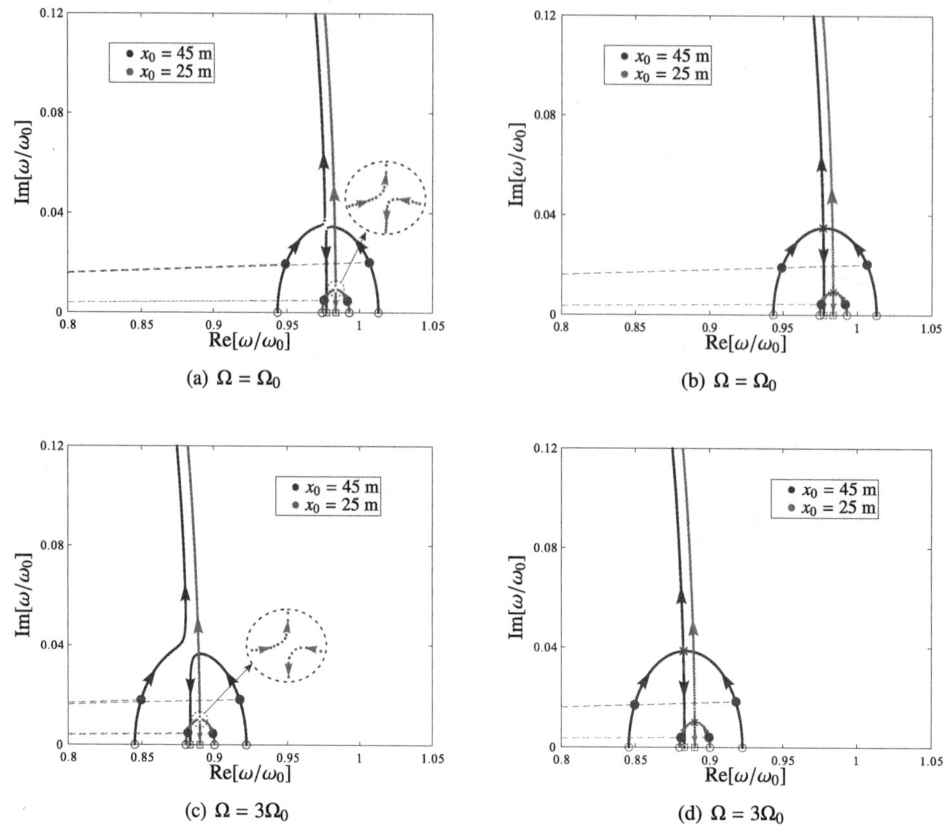

Fig. 5. Root loci of the 2-DOF system for two different mounting positions x_0, with $\mu = 0.05$, $\gamma = 0.6$, $\alpha = 1$ and $C = 0.02$. (a)(c) Optimal frequency-tuning formula Eq. (47) has been used, (b)(d) optimal frequency-tuning formula Eq. (57) has been used. (a)(b) $\Omega = \Omega_0$ (1.27 rad/s), (c)(d) $\Omega = 3\Omega_0$ (3.81 rad/s). The red points • denote the roots of Eq. (39) with $\zeta_{w,\mathrm{opt}}$; The red square symbols □ denote the root of Eq. (39) under the limit $\zeta_w \Rightarrow \infty$ (i.e. ω_∞); The red circles ∘ denote the roots of Eq. (39) when $\zeta_w = 0$; The red asterisks ∗ denote the root of Eq. (39) when $\zeta_w = \zeta_*$ (i.e. the bifurcation point ω_*). The arrows ◀ or ◀ denote the evolution direction of roots of Eq. (39) with increasing value of ζ_w. (For interpretation of the references to color in this figure legend, the reader is referred to the web version of this article.)

indicating that these two roots are inverse points with respect to a circle centered at the origin with a radius $R = \omega_\infty$. This feature also indicates that the two modal damping ratios of the 2-DOF system are identical.

The modal damping ratios ζ_1 and ζ_2 (obtained through Eq. (59)) of the 2-DOF system are presented in Fig. 6. Due to the nonlinear property of the TLCD damping element (as discussed in Section 3.4), ζ_i is a function of the external frequency ω. The non-linear dependence of modal damping ratios on ω can be observed in Figs. 6(a) and 6(c), where the first optimal frequency-tuning formula Eq. (47) is used. As seen, the highest damping values are achieved within the interval $[\omega_M, \omega_N]$. Furthermore, these two modal damping ratios are slightly different with each other, and this difference becomes more significant as Ω increases. When the alternative optimal frequency-tuning formula Eq. (57) is used, these two modal damping ratios are exactly the same, as shown in Figs. 6(b) and 6(d). This is in perfect agreement with the root loci results in Figs. 5(b) and 5(d) (i.e. two complex roots lie in the same dashed line). In addition, a higher rotor speed also leads to a shift of the damping ratio curve.

Therefore, Figs. 5 and 6 demonstrate that Eq. (57) is superior to Eq. (47) in terms of the equal modal damping ratio. This conclusion is in good agreement with that of the blade-TMD system [26]. However, these two optimal frequency-tuning formulas actually lead to very similar optimal TLCD parameters as well as the vibration control performance, and this will be further illustrated in the following parametric study.

5.3. Parametric study

Fig. 7 shows the variation of two optimal frequency ratios $(\omega_w/\omega_0)_{\mathrm{opt}}$ (i.e. Eqs. (47) and (57)) versus different parameters. As seen, these two formulas lead to similar results, i.e. the red and blue curves are almost on top of each other. Actually, Eq. (57) provides a slightly lower value due to the additional factor $\left(2\Omega^2(1+\mu\alpha^2)-\omega_0^2\right)/\left(\Omega^2(1+\mu\alpha^2)-\omega_0^2\right)$, already discussed in Section 4. In general, $(\omega_w/\omega_0)_{\mathrm{opt}}$ decreases with increasing values of these parameters, except for the area ratio α. The rotor speed Ω turns out to be the most predominant parameter, which causes a reduction of $(\omega_w/\omega_0)_{\mathrm{opt}}$ about 12% from Ω_0 to $3\Omega_0$, as shown in Fig. 7(a). Other parameters result in a reduction less than 1%, especially for α that leads to a non-linear variation of $(\omega_w/\omega_0)_{\mathrm{opt}}$ with a maximum

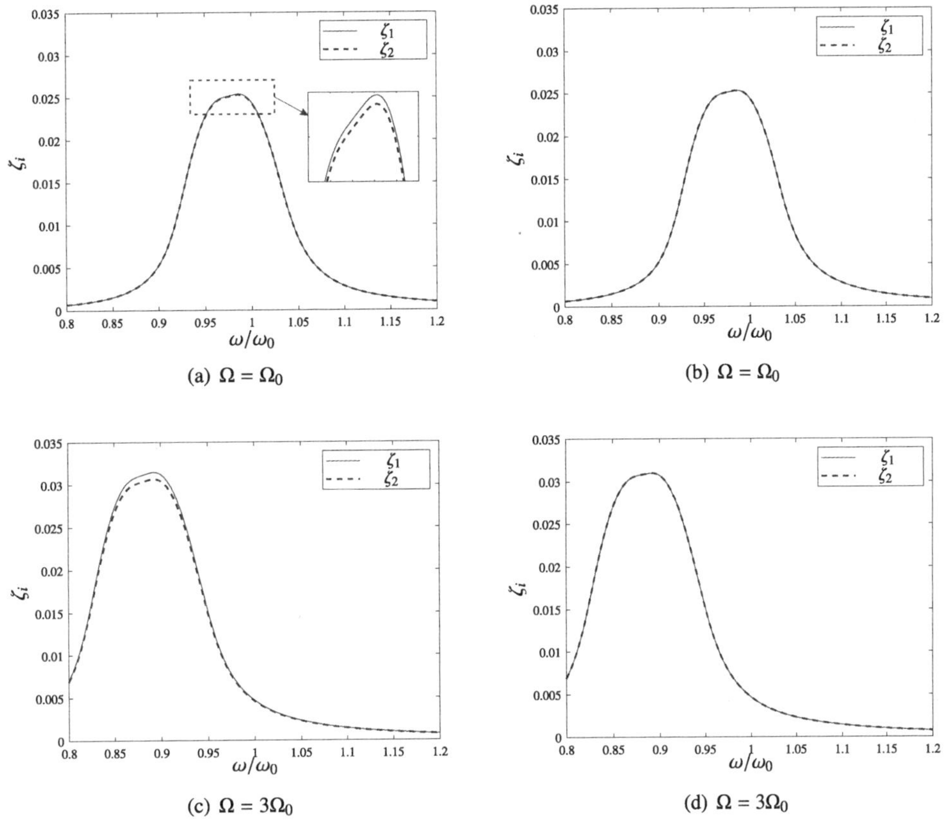

Fig. 6. Modal damping ratios ζ_i of the 2-DOF system versus the external excitation frequency ω, with $x_0 = 45$ m, $\mu = 0.05$, $\gamma = 0.6$, $\alpha = 1$ and $C = 0.02$. (a)(c) Optimal frequency-tuning formula Eq. (47) has been used, (b)(d) optimal frequency-tuning formula Eq. (57) has been used. (a)(b) $\Omega = \Omega_0$ (1.27 rad/s), (c)(d) $\Omega = 3\Omega_0$ (3.81 rad/s).

Table 1
Variation of the absolute difference between Eq. (47) and Eq. (57) with various parameters.

Ω/Ω_0	$\lvert\Delta(\frac{\omega_w}{\omega_0})_{opt}\rvert$	α	$\lvert\Delta(\frac{\omega_w}{\omega_0})_{opt}\rvert$	μ	$\lvert\Delta(\frac{\omega_w}{\omega_0})_{opt}\rvert$	γ	$\lvert\Delta(\frac{\omega_w}{\omega_0})_{opt}\rvert$	x_0 (m)	$\lvert\Delta(\frac{\omega_w}{\omega_0})_{opt}\rvert$
1	0.7×10^{-4}	0.5	7.0×10^{-5}	0.02	3.1×10^{-5}	0.4	0.3×10^{-4}	25	0.5×10^{-5}
1.5	1.7×10^{-4}	1	7.8×10^{-5}	0.03	4.7×10^{-5}	0.5	0.5×10^{-4}	30	1.2×10^{-5}
2	3.1×10^{-4}	1.5	7.5×10^{-5}	0.04	6.2×10^{-5}	0.6	0.7×10^{-4}	35	2.6×10^{-5}
2.5	4.9×10^{-4}	2	6.9×10^{-5}	0.05	7.8×10^{-5}	0.7	1.0×10^{-4}	40	4.7×10^{-5}
3	7.0×10^{-4}	2.5	6.4×10^{-5}	0.06	9.3×10^{-5}	0.8	1.3×10^{-4}	45	7.8×10^{-5}

* $\lvert\Delta(\frac{\omega_w}{\omega_0})_{opt}\rvert = \lvert(\frac{\omega_w}{\omega_0})_{opt}$ Eq. (47) $-(\frac{\omega_w}{\omega_0})_{opt}$ Eq. (57)$\rvert$.
* The reference values for those parameters are taken as $\Omega = \Omega_0$ (1.27 rad/s), $x_0 = 45$ m, $\mu = 0.05$, $\gamma = 0.6$, $\alpha = 1$ and $C = 0.02$, and only one parameter is varied in each column.

difference around 0.05%. $(\omega_w/\omega_0)_{opt}$ exhibits almost a linear relationship with the mas ratio μ (though theoretically not), which is similar to the results in the classic non-rotating SDOF-TLCD system [39]. The length ratio γ (as given in Fig. 7(d)) has a modest influence on $(\omega_w/\omega_0)_{opt}$, which is also in consistent with the classic non-rotating SDOF-TLCD system [38]. Further, $(\omega_w/\omega_0)_{opt}$ is insensitive to the mounting position x_0 when the TLCD is located near the hub (i.e. $x_0 \le 35$ m), but exhibits noticeable reduction when the TLCD moved towards the blade tip, as shown in Fig. 7(e).

The absolute difference $\left|\Delta(\omega_w/\omega_0)_{opt}\right|$ between the results from Eqs. (47) and (57) is summarized in Table 1. As the values of Ω, μ, γ and x_0 increase, the discrepancies between these two formulas become more significant, as shown in columns two, six, eight and ten. On the other hand, $\left|\Delta(\omega_w/\omega_0)_{opt}\right|$ does not have a monotonic variation with α, as presented in column four. Overall, the absolute difference between the optimal frequency ratios from the two formulas is quite insignificant (in the orders of magnitude of 10^{-5} to 10^{-4}).

Due to nonlinear property of the TLCD damping element (as discussed in Section 3.4), the TLCD damping is represented here by the head loss coefficient ξ, which in practice can be modified by adjusting the orifice area inside the horizontal tube [35,40]. The optimal values ξ_{opt} corresponding to the two optimal frequency-tuning formulas, can be obtained by inserting Eqs. (47) and (57) into Eq. (53), respectively. The variation of ξ_{opt} versus various parameters is shown in Fig. 8. Similar to the finding in Fig. 7,

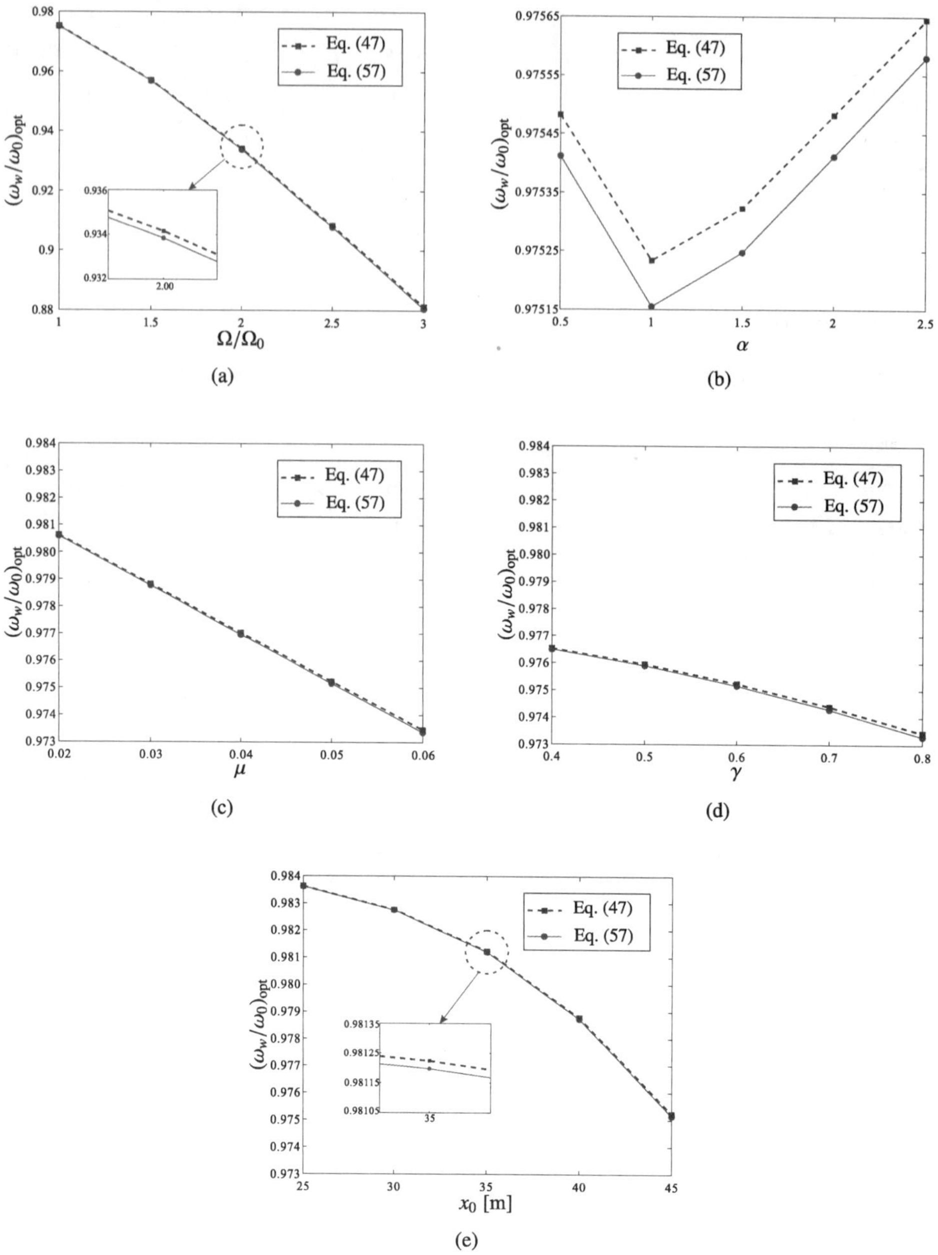

Fig. 7. Variation of the optimal frequency ratio $(\omega_w/\omega_0)_{opt}$ with (a) the rotational speed Ω, (b) the section ratio α, (c) the mass ratio μ, (d) the horizontal length ratio γ and (e) the mounting location x_0. The basic values for those parameters are taken as $\Omega = \Omega_0$ (1.27 rad/s), $x_0 = 45$ m, $\mu = 0.05$, $\gamma = 0.6$, $\alpha = 1$ and $C = 0.02$, and only one variable is considered in each subfigure. The blue dashed line – – – corresponds to $(\omega_w/\omega_0)_{opt}$ given by Eq. (47), the red solid line —— corresponds to $(\omega_w/\omega_0)_{opt}$ given by Eq. (57). (For interpretation of the references to color in this figure legend, the reader is referred to the web version of this article.)

Eqs. (47) and (57) lead to almost identical results, although Eq. (57) provides slightly lower value (i.e. the red lines are beneath the blue lines in all subfigures, as indicated in the zoomed parts of Fig. 8). As depicted in Fig. 8(a), the rotor speed Ω still plays a crucial role in determining ξ_{opt}, and ξ_{opt} becomes 10 times larger as Ω increases from Ω_0 to $3\Omega_0$. Regarding the area ratio α (Fig. 8(b)), ξ_{opt} increases notably with α when $\alpha < 1.5$, and starts to reach an asymptotic value as α increases further. As can be observed from Figs. 8(c) and 8(d), the increase of both the mass ratio μ and the length ratio γ lead to higher values of ξ_{opt}, and the impact from γ is a bit more prominent. This observation is different from that of the non-rotating SDOF-TLCD system [38], where γ has insignificant influence on ξ_{opt}. Regarding the mounting position x_0 (Fig. 8(e)), ξ_{opt} increases nonlinearly as x_0 increases (as TLCD is mounted closer to blade tip). Finally, Fig. 8(f) shows the relationship between the external force intensity factor C and ξ_{opt}. As seen, ξ_{opt} decreases as C increases, which is in good agreement with result of the classic non-rotating system [38,39].

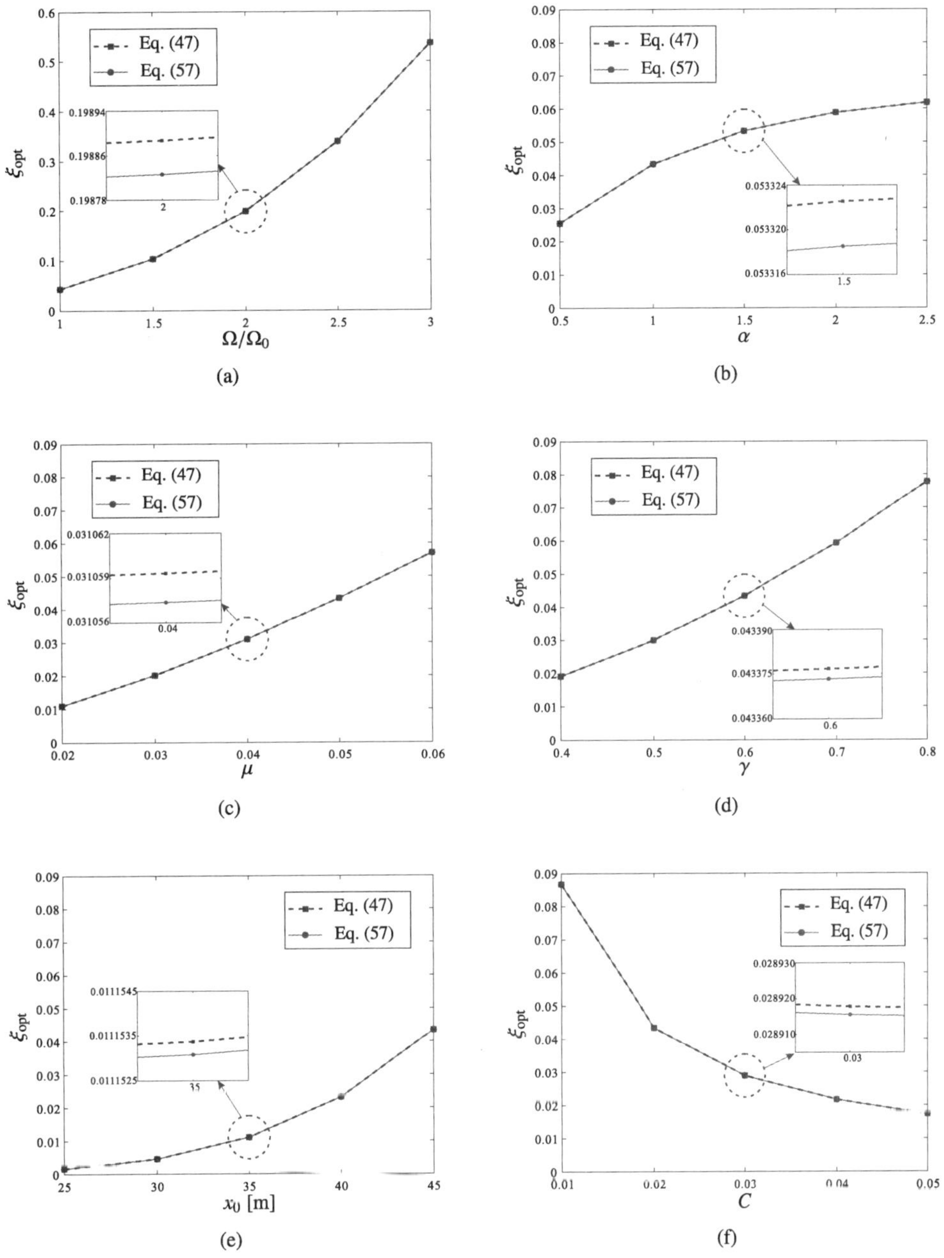

Fig. 8. Variation of the optimal head loss coefficient ξ_{opt} with (a) the rotational speed Ω, (b) the area ratio α, (c) the mass ratio μ, (d) the horizontal length ratio γ, (e) the mounting location x_0 and (f) the excitation intensity factor C. The reference values for those parameters are taken as $\Omega = \Omega_0$ (1.27 rad/s), $x_0 = 45$ m, $\mu = 0.05$, $\gamma = 0.6$, $\alpha = 1$ and $C = 0.02$, and only one variable is considered in each sub-figure. The blue dashed line – – – corresponds to ξ_{opt} when Eq. (47) is used, and the red solid line ——— corresponds to ξ_{opt} when Eq. (57) is used. (For interpretation of the references to color in this figure legend, the reader is referred to the web version of this article.)

The absolute difference of ξ_{opt} corresponding to Eqs. (47) and (57) is summarized in Table 2. It can be observed that $|\Delta \xi_{opt}|$ increases with increasing Ω, μ, γ and x_0. On the other hand, the results corresponding to the force intensity factor C shows an opposite trend, as seen from the last column. $|\Delta \xi_{opt}|$ exhibits nonlinear relationship with the area ratio α, as shown in column four. Nevertheless, the absolute difference of ξ_{opt} is very insignificant for all the cases considered.

The vibration control performance of the blade-mounted TLCD using the two optimal frequency-tuning formulas (Eqs. (47) and (57)) are compared in Fig. 9, in terms of the maximal dynamic amplification of the structural response $\left|\hat{q}k_0/\hat{f}\right|_{max}$. Similar to the observations in Figs. 7 and 8, Eqs. (47) and (57) lead to almost identical results of $\left|\hat{q}k_0/\hat{f}\right|_{max}$. In terms of the vibration control effect, Eq. (57) is slightly superior to Eq. (47), as the red lines lie a bit lower than the blue lines. It is well known that the closer the

Table 2

Variation of the absolute difference of ξ_{opt} (obtained by inserting Eq. (47) and Eq. (57) into Eq. (53), respectively) with parameters.

Ω/Ω_0	$\lvert\Delta\xi_{opt}\rvert$	α	$\lvert\Delta\xi_{opt}\rvert$	μ	$\lvert\Delta\xi_{opt}\rvert$
1	0.03×10^{-4}	0.5	1.76×10^{-4}	0.02	0.34×10^{-6}
1.5	0.18×10^{-4}	1	3.39×10^{-4}	0.03	0.95×10^{-6}
2	0.60×10^{-4}	1.5	4.04×10^{-4}	0.04	1.94×10^{-6}
2.5	1.57×10^{-4}	2	4.15×10^{-4}	0.05	3.39×10^{-6}
3	3.46×10^{-4}	2.5	4.02×10^{-4}	0.06	5.36×10^{-6}
γ	$\lvert\Delta\xi_{opt}\rvert$	x_0 (m)	$\lvert\Delta\xi_{opt}\rvert$	C	$\lvert\Delta\xi_{opt}\rvert$
0.4	0.07×10^{-5}	25	0.01×10^{-6}	0.01	6.79×10^{-6}
0.5	0.16×10^{-5}	30	0.06×10^{-6}	0.02	3.39×10^{-6}
0.6	0.34×10^{-5}	35	0.29×10^{-6}	0.03	2.26×10^{-6}
0.7	0.64×10^{-5}	40	1.09×10^{-6}	0.04	1.70×10^{-6}
0.8	1.10×10^{-5}	45	3.39×10^{-6}	0.05	1.36×10^{-6}

* $\lvert\Delta\xi_{opt}\rvert = \lvert(\xi_{opt}$ using Eq. (47) − ξ_{opt} using Eq. (57)$)\rvert$.
* The reference values for those parameters are taken as $\Omega = \Omega_0$ (1.27 rad/s), $x_0 = 45$ m, $\mu = 0.05$, $\gamma = 0.6$, $\alpha = 1$ and $C = 0.02$, and only one parameter is varied in each column.

TLCD is attached to the blade tip, the better control performance the damper can achieve. In practice, the optimal mounting position $x_{0,opt}$ is usually a tradeoff taking both the control efficiency and the space constraint within the blade into consideration. As shown in Fig. 9(e), about 55% reduction in structural response amplitude can be achieved by moving the TLCD from $x_0 = 25$ m to $x_0 = 40$ m. But the performance improvement becomes minor when $x_0 \geq 40$m. Furthermore, it can be observed from Fig. 9(a) that the use of both Eqs. (47) and (57) can guarantee a fairly consistent performance of the TLCD under different rotor speeds. In addition, the result from using Eq. (57) (indicated by the red line) exhibits a slightly nonlinear tendency due to the rotor-speed-dependency of the dynamic amplification curve, as discussed in relation to Fig. 4. Fig. 9(b) shows the influence of the area ratio α on $\left|\hat{q}k_0/\hat{f}\right|_{max}$. As seen, the best control performance when using Eqs. (47) and (57) is consistent, i.e. the minimum value of $\left|\hat{q}k_0/\hat{f}\right|_{max}$ is always achieved with uniform cross sections of the U-tube (i.e. $\alpha = 1$), regardless of the optimal frequency-tuning formulas. This observation is in a good agreement with the numerically optimized result of the blade-TLCD system [19]. Similar to the classic non-rotating SDOF-TLCD system, higher values of the mass ratio μ and the length ratio γ are favorable for the control performance, as shown in Figs. 9(c) and 9(d).

To further illustrate the efficiency of TLCD, the edgewise responses of the blade-TLCD system using those two optimal frequency-tuning formulas (Eqs. (47) and (57)) are displayed in Fig. 10. As a comparison, the uncontrolled blade responses are also provided, as indicated by blue dashed lines. The external generalized force (or the modal load $f(t)$, cf.Eq. (24)) exerted on the blade is obtained from the Ref. [19]. It is clearly show in Fig. 10(a), the response amplitude of the uncontrolled blade is effectively attenuated by TLCD, with a reduction in standard deviation values (STD) around 20%. Similar to the results in Fig. 9, the control performance corresponding to those two formulas (Eqs. (47) and (57)) is almost the same, as indicated by the overlapped green and red lines in the partial enlargement figure. Fig. 10(b) provides the corresponding Fourier spectrums. As seen, the amplitude at the blade edgewise fundamental frequency ω_0 (the second peak in Fig. 10(b)) is significantly reduced by the damper. By comparison, the lower frequency components seem to be unaffected, for instance, the frequency component corresponding to the rotor speed Ω (indicated by the first peak). Strictly speaking, Eq. (47) or Eq. (57) is not the optimal value for the rotational blade-mounted TLCD due to the assumptions employed in the calibration procedure, as illustrated in Section 3. More specifically, the time-dependent term $\Psi = \Omega t$ within the blade-TLCD system (cf. Eq. (27)), which is skipped in the derivation part, leads to time-dependent fundamental frequency $\omega_{ori,0}(\Omega, t)$ (as presented in Eq. (10)). In this regard, the optimal design parameters for the blade-mounted TLCD should vary with time. Nevertheless, the TLCD with time-independent parameters obtained from the simplified system (cf. Eq. (31)) can still achieve satisfactory control performance, as demonstrated in Fig. 10.

As discussed above, the optimal parameters (particularly the optimal frequency) of the blade-mounted TLCD are closely related to the rotor speed Ω. As known, the commercial wind turbines will go through four operating regions (divided by the cut-in speed, the rated speed and the cut-out speed) during normal working condition, and Ω will vary from 0 to Ω_0, accordingly. The Region III, corresponding to the region between the rated speed and the cut-out speed, is often regarded as the full load region where the rotor speed will be limited to the rated speed Ω_0 in order to ensure an almost constant power production (with some fluctuations due to turbulence). It is also the region where most of blade vibration issues, such as the edgewise vibration, will occur. In this regard, the rated speed Ω_0 is recommended to be chosen as the reference value for the TLCD design in real application. Further improvement for the blade-mounted TLCD design accommodating for the entire four regions can be achieved by installing active or semi-active devices which can continuously adjust the eigenfrequency of the TLCD (such as pressurized TLCD with air chambers [43]).

6. Conclusion

The present paper develops closed-form formulas for optimal calibration of the TLCD installed in a rotating blade for suppressing edgewise vibrations. A 2-DOF model has been established representing the blade-TLCD system, which acts as the basis for analytical derivation of the optimal frequency ratio and optimal damping ratio formulas. Two different criteria have been used for deriving the

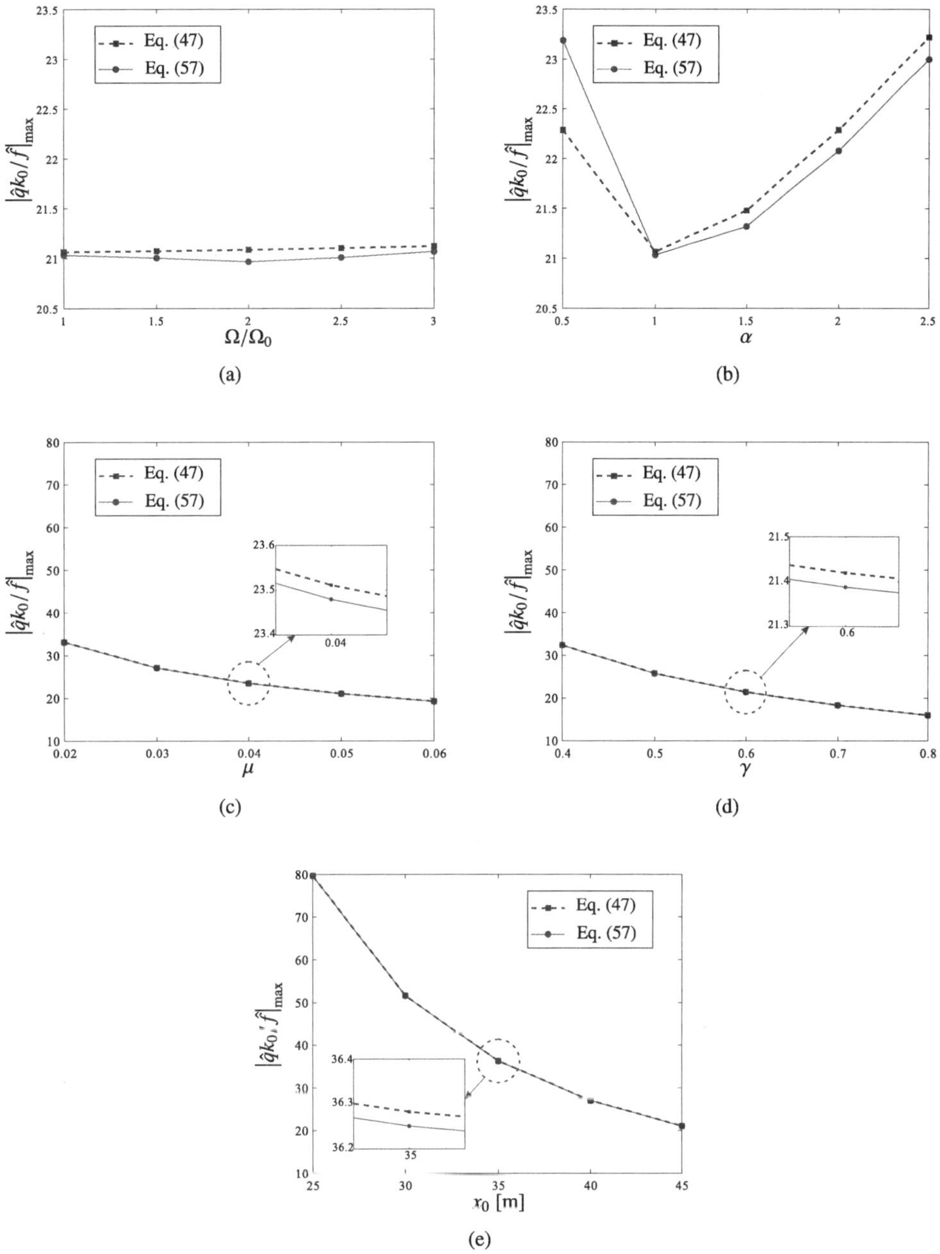

Fig. 9. Variation of the maximum dynamic amplification of the structural response $\left|\hat{q}k_0/\hat{f}\right|_{\max}$ with respect to (a) the rotational speed Ω, (b) the area ratio α, (c) the mass ratio μ, (d) the horizontal length ratio γ and (e) the mounting location x_0. The reference values for those parameters are taken as $\Omega = \Omega_0\,(1.27\,\text{rad/s})$, $x_0 = 45$ m, $\mu = 0.05$, $\gamma = 0.6$, $\alpha = 1$ and $C = 0.02$, and only one parameter is varied in each subfigure. The blue dashed line – – – corresponds to $\left|\hat{q}k_0/\hat{f}\right|_{\max}$ when Eq. (47) is used, and the red solid line —— corresponds to $\left|\hat{q}k_0/\hat{f}\right|_{\max}$ when Eq. (57) is used. (For interpretation of the references to color in this figure legend, the reader is referred to the web version of this article.)

optimal frequency ratio of the blade-mounted TLCD. The first one is the equal dynamic amplification criterion based on frequency response analysis, and the second one is the equal modal damping ratio criterion using the pole-placement method. Unlike the classic non-rotating SDOF-TLCD system where these two criteria result in an identical frequency-tuning formula, in the present study, two slightly different optimal frequency-tuning formulas (Eqs. (47) and (57)) have been developed based on the two criteria. The optimal frequency-tuning formula Eq. (47) obtained from the 1st criterion turns out to be dependent on the rotational speed Ω, the mounting position x_0, the blade eigenfrequency ω_0 and the prescribed TLCD parameters (mass ratio, length ratio and area ratio), which is different from the formula of a non-rotating SDOF-TLCD system that is only dependent on the prescribed TLCD parameters. It is also different from the blade-mounted TMD system, whose optimal frequency-tuning formula based on the 1st criterion is independent

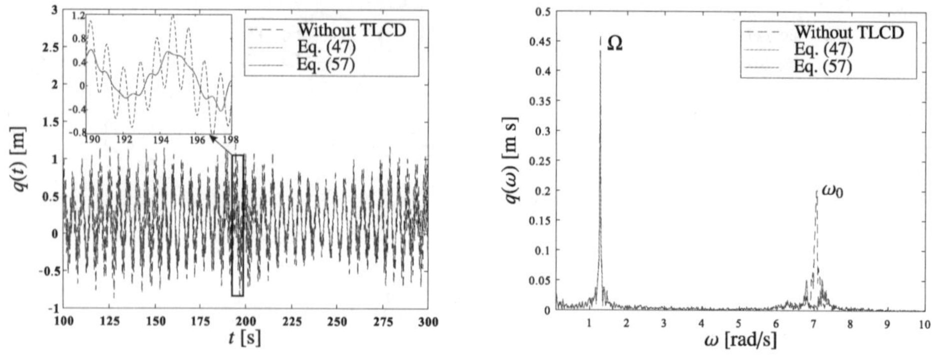

(a) (b)

Fig. 10. Blade edgewise responses with and without TLCD. The reference values are taken as $\Omega = \Omega_0\,(1.27\,\text{rad/s})$, $x_0 = 45$ m, $\mu = 0.05$, $\gamma = 0.6$, $\alpha = 1$. (a) Time series, (b) Fourier amplitude.

of Ω. The optimal frequency-tuning formula Eq. (57) obtained from the 2nd criterion possess a similar format as the first one, and results in a slightly lower frequency ratio due to an additional factor in the third term of the expression.

Equalizing the dynamic amplification at three frequencies, the explicit expression for optimal equivalent damping ratio of the TLCD Eq. (50) has been derived, which is also dependent on Ω, x_0 and ω_0. An closed-form formula for calibrating the optimal TLCD head loss coefficient has been developed based on Eq. (50) together with one of the optimal frequency-tuning formulas (Eq. (47) or Eq. (57)). As a result, a convenient procedure for optimal calibration of a rotating blade-mounted TLCD has been established with only closed-form formulas, eliminating the need of cumbersome iteration-based optimization methods. These formulas also form a basis for developing semi-active control algorithms.

For the considered NREL 5MW reference wind turbine, the difference of the optimal parameters resulted from the two optimal frequency-tuning formulas turns out to be insignificant. In addition, the rotor speed plays a predominant role for determining the optimal frequency ratio and optimal head loss coefficient, while other parameters such as the mass ratio, length ratio and area ratio, have less significant influence on the optimal values. Eq. (57) is slightly superior to Eq. (47) in terms of the vibration control performance. All in all, both Eqs. (47) and (57) can lead to similarly satisfactory performance of the TLCD in suppressing blade edgewise vibrations, even under extreme working conditions (such as $\Omega = 3\Omega_0$).

The present work focuses on the rigorous development of mathematical models of the coupled system and the analytical derivation of close-form optimal design formulas for the blade-mounted TLCD. Full-scale blade-TLCD tests (reduced-scale test will suffer from the different scale effects of the solid and the fluid, and thus may not reveal the true performance of the TLCD) will be conducted to further validate the feasibility and effectiveness of the blade-mounted TLCD in our future works.

CRediT authorship contribution statement

Bei Chen: Investigation, Software, Writing – original draft, Writing – review & editing. **Zili Zhang:** Conceptualization, Investigation, Software, Supervision, Writing – original draft, Writing – review & editing. **Xugang Hua:** Supervision, Writing – review & editing. **Biswajit Basu:** Supervision, Writing – review & editing.

Declaration of competing interest

The authors declare that they have no known competing financial interests or personal relationships that could have appeared to influence the work reported in this paper.

Acknowledgments

The authors gratefully acknowledge the financial support from the National Natural Science Foundation of China (No. 51908209), the Hunan Provincial Natural Science Foundation of China (No. 2020JJ5052), the International Postdoctoral Exchange Fellowship Program (No. PC2021019) and the supports of Aarhus University Research Foundation, Denmark under the AUFF Assistant Professor Starting Grant (AUFF-E-2017-7-20).

Appendix. Constant parameters

$$
\left.\begin{aligned}
m_0 &= \int_0^{L_B} \rho_b(x_3)\Phi^2(x_3)\mathrm{d}x_3 \\
m_1 &= \int_0^{L_B} \rho_b(x_3)\Phi(x_3)x_3\mathrm{d}x_3 \\
m_2 &= \int_0^{L_B} \rho_b(x_3)x_3^2\mathrm{d}x_3 \\
m_3 &= \int_0^{L_B} \rho_b(x_3)\Phi(x_3)\mathrm{d}x_3 \\
m_w &= \rho(A_0B + 2AH) = \rho A_0 L(\alpha - \alpha\gamma + \gamma) \\
k_e &= \int_0^{L_B} EI(x_3)\left(\frac{d^2\Phi(x_3)}{dx_3^2}\right)\mathrm{d}x_3 \\
k_1 &= \int_0^{L_B} N_1(x_3)\left(\frac{\mathrm{d}\Phi(x_3)}{\mathrm{d}x_3}\right)^2\mathrm{d}x_3, N_1(x_3) = \int_{x_3}^{L_B} z_3\rho_b(z_3)\mathrm{d}z_3 \\
k_2 &= \int_0^{L_B} N_2(x_3)\left(\frac{\mathrm{d}\Phi(x_3)}{\mathrm{d}x_3}\right)^2\mathrm{d}x_3, N_2(x_3) = \int_{x_3}^{L_B} \rho_b(z_3)\mathrm{d}z_3 \\
\gamma_A &= \frac{\alpha(1+\alpha\gamma-\gamma)}{\alpha-\alpha\gamma+\gamma} \\
\gamma_B &= \frac{\gamma}{\alpha-\alpha\gamma+\gamma} \\
L &= \frac{2x_0}{\frac{\omega_w^2}{\Omega^2}(1+\alpha\gamma-\gamma)+1-\gamma}\text{(derived from Eq. (33))} \\
\rho A &= \frac{m_w\alpha}{L(\alpha-\alpha\gamma+\gamma)}\text{(derived from Eq. (11))}
\end{aligned}\right\} \quad (A.1)
$$

References

[1] M.H. Hansen, Aeroelastic instability problems for wind turbines, Wind Energy 10 (6) (2007) 551–577.

[2] K. Thomsen, J.T. Petersen, E. Nim, S. Øye, B. Petersen, A method for determination of damping for edgewise blade vibrations, Wind Energy 3 (4) (2000) 233–246.

[3] O. Al-Khudairi, H. Hadavinia, C. Little, G. Gillmore, P. Greaves, K. Dyer, Full-scale fatigue testing of a wind turbine blade in flapwise direction and examining the effect of crack propagation on the blade performance, Materials 10 (10) (2017) 1152.

[4] L. Witek, Experimental crack propagation and failure analysis of the first stage compressor blade subjected to vibration, Eng. Fail. Anal. 16 (7) (2009) 2163–2170.

[5] H. Stiesdal, Extreme wind loads on stall regulated wind turbines, in: 16th Annual Conference of the British Wind Energy Association, Mechanical Engineering Publications Ltd., Stirling, United Kingdom, 1994, pp. 101–106.

[6] T. Moeller, Blade cracks signal new stress problem, WindPower Mon. 25 (1997).

[7] M. Malkin, D. Griffin, Edgewise vibration and wind turbine blade failure, 2016, Available at: www.windtech-international.com/editorial-features/edgewise-vibration-and-wind-turbine-blade-failure.

[8] C.C. Qin, E. Loth, D.S. Zalkind, L.Y. Pao, S. Yao, D.T. Griffith, M.S. Selig, R. Damiani, Downwind coning concept rotor for a 25 MW offshore wind turbine, Renew. Energy 156 (2020) 314–327.

[9] P. Chaviaropoulos, Flap/lead–lag aeroelastic stability of wind turbine blade sections, Wind Energy 2 (2) (1999) 99–112.

[10] A. Staino, B. Basu, S.r.R. Nielsen, Actuator control of edgewise vibrations in wind turbine blades, J. Sound Vib. 331 (6) (2012) 1233–1256.

[11] J. Thirstrup Petersen, K. Thomsen, H. Aagaard Madsen, Stall strips can control edgewise vibrations, Tech. rep., AED-RB No. AED-RB-6 [EN], 1998.

[12] C. Anderson, H. Heerkes, R. Yemm, The use of blade-mounted dampers to eliminate edgewise stall vibration, in: EWEC-CONFERENCE-, 1999, pp. 207–211.

[13] C. Huang, J. Arrigan, S. Nagarajaiah, B. Basu, Semi-active algorithm for edgewise vibration control in floating wind turbine blades, in: Earth and Space 2010: Engineering, Science, Construction, and Operations in Challenging Environments, 2010, pp. 2097–2110.

[14] B. Fitzgerald, B. Basu, S.r.R. Nielsen, Active tuned mass dampers for control of in-plane vibrations of wind turbine blades, Struct. Control Health Monit. 20 (12) (2013) 1377–1396.

[15] B. Fitzgerald, B. Basu, Cable connected active tuned mass dampers for control of in-plane vibrations of wind turbine blades, J. Sound Vib. 333 (23) (2014) 5980–6004.

[16] J. Chen, C. Yuan, J. Li, Q. Xu, Semi-active fuzzy control of edgewise vibrations in wind turbine blades under extreme wind, J. Wind Eng. Ind. Aerodyn. 147 (2015) 251–261.

[17] H. Biglari, V. Fakhari, Edgewise vibration reduction of small size wind turbine blades using shunt damping, J. Vib. Control 26 (3–4) (2020) 186–199.

[18] Z. Zhang, J. Li, S.r.R. Nielsen, B. Basu, Mitigation of edgewise vibrations in wind turbine blades by means of roller dampers, J. Sound Vib. 333 (21) (2014) 5283–5298.

[19] Z. Zhang, B. Basu, S.r.R. Nielsen, Tuned liquid column dampers for mitigation of edgewise vibrations in rotating wind turbine blades, Struct. Control Health Monit. 22 (3) (2015) 500–517.

[20] B. Basu, Z. Zhang, S.r.R. Nielsen, Damping of edgewise vibration in wind turbine blades by means of circular liquid dampers, Wind Energy 19 (2) (2016) 213–226.

[21] W.P. Howson, A. Zare, Exact dynamic stiffness matrix for flexural vibration of three-layered sandwich beams, J. Sound Vib. 282 (3–5) (2005) 753–767.
[22] S. Bornassi, H. Navazi, H. Haddadpour, Edgewise bending vibration analysis of a rotating sandwich beam with magnetorheological elastomer core, Int. J. Struct. Stab. Dyn. 18 (11) (2018) 1850134.
[23] Z. Zhang, S.r.R. Nielsen, B. Basu, J. Li, Nonlinear modeling of tuned liquid dampers (TLDs) in rotating wind turbine blades for damping edgewise vibrations, J. Fluids Struct. 59 (2015) 252–269.
[24] R.L. Haupt, S. Ellen Haupt, Practical Genetic Algorithms, second ed., Wiley Online Library, 2004.
[25] A.Y. Leung, H. Zhang, C. Cheng, Y. Lee, Particle swarm optimization of TMD by non-stationary base excitation during earthquake, Earthq. Eng. Struct. Dyn. 37 (9) (2008) 1223–1246.
[26] Z. Zhang, Optimal tuning of the tuned mass damper (TMD) for rotating wind turbine blades, Eng. Struct. 207 (2020) 110209.
[27] Z. Zhang, B. Fitzgerald, Tuned mass-damper-inerter (TMDI) for suppressing edgewise vibrations of wind turbine blades, Eng. Struct. 221 (2020) 110928.
[28] Z. Zhang, T.G. Larsen, Optimal calibration of the rotational inertia double tuned mass damper (RIDTMD) for rotating wind turbine blades, J. Sound Vib. 493 (2021) 115827.
[29] J.P. Den Hartog, Mechanical Vibrations, fourth ed., McGraw-Hill, New York, 1956.
[30] S. Krenk, Frequency analysis of the tuned mass damper, J. Appl. Mech. 72 (6) (2005) 936–942.
[31] J.J. Craig, Introduction To Robotics: Mechanics and Control, third ed., Pearson Education India, 2009.
[32] K. Shum, Y.L. Xu, W. Guo, Wind-induced vibration control of long span cable-stayed bridges using multiple pressurized tuned liquid column dampers, J. Wind Eng. Ind. Aerodyn. 96 (2) (2008) 166–192.
[33] L. Pars, A Treatise on Analytical Dynamics, Ox Bow Press Woodbridge, 1979.
[34] H. Gao, K. Kwok, B. Samali, Optimization of tuned liquid column dampers, Eng. Struct. 19 (6) (1997) 476–486.
[35] T. Balendra, C. Wang, H. Cheong, Effectiveness of tuned liquid column dampers for vibration control of towers, Eng. Struct. 17 (9) (1995) 668–675.
[36] B. Chen, B. Basu, X. Hua, Z. Feng, Z. Zhang, Z. Chen, S.r.R. Nielsen, Online DWT algorithm for identification of aerodynamic damping in wind turbines, Mech. Syst. Signal Process. 152 (2021) 107437.
[37] P. Hitchcock, K. Kwok, R. Watkins, B. Samali, Characteristics of liquid column vibration absorbers (LCVA)-I, Eng. Struct. 19 (2) (1997) 126–134.
[38] B. Chen, Z. Zhang, X. Hua, Closed-form optimal calibration of a tuned liquid column damper (TLCD) for flexible structures, Int. J. Mech. Sci. 198 (2021) 106364.
[39] K. Shum, Closed form optimal solution of a tuned liquid column damper for suppressing harmonic vibration of structures, Eng. Struct. 31 (1) (2009) 84–92.
[40] J.-C. Wu, Experimental calibration and head loss prediction of tuned liquid column damper, Tamkang J. Sci. Eng. 8 (4) (2005) 319–325.
[41] J. Jonkman, S. Butterfield, W. Musial, G. Scott, Definition of a 5-MW reference wind turbine for offshore system development, Tech. rep., National Renewable Energy Lab.(NREL), Golden, CO (United States), 2009.
[42] T. Burton, D. Sharpe, N. Jenkins, E. Bossanyi, Wind Energy Handbook, vol. 2, Wiley Online Library, 2001.
[43] C. Riascos, J. Marulanda, P. Thomson, Structural control of a grandstand using a semi-active pressurized tuned liquid column damper considering effects of passive crowds, Struct. Control Health Monit. 26 (10) (2019) e2426.

第43卷 第8期 2022年8月
太阳能学报 ACTA ENERGIAE SOLARIS SINICA
Vol. 43, No. 8 Aug., 2022

DOI:10.19912/j.0254-0096.tynxb.2020-1411 文章编号:0254-0096(2022)08-0216-08

多排槽式反射镜风压分布及干扰研究

康佳鑫[1], 牛华伟[1], 李红星[2], 何邵华[2]

(1. 湖南大学土木工程学院, 长沙 410082; 2. 中国电力工程顾问集团西北电力设计院有限公司, 西安 710075)

摘 要: 为研究多排槽式反射镜不同位置处的风荷载特性及其互相干扰,对单个反射镜及多排反射镜进行缩尺为1/40的刚性模型测压试验,试验变量为测压模型的风攻角、风向角以及其在群镜中的位置。首先测得各工况下镜面的风压分布数据,绘制镜面风压系数等值线图和反射镜力系数变化曲线,然后分析了反射镜互相干扰对镜面风荷载的影响。研究表明:中间列首排反射镜所受风荷载最大,其阻力系数是单镜的2倍,中间列其余排反射镜阻力系数为单镜的30%~70%;边缘列首排反射镜阻力系数为单镜的1.88倍,其余各排在单镜的1.20倍以上;建议将镜场分为4类区域分别进行结构设计。

关键字: 太阳能热发电; 槽式反射镜; 风洞试验; 风压分布; 干扰效应; 群镜

中图分类号: TU3; TK513.1 **文献标志码**: A

0 引 言

在槽式太阳能热发电站中建有若干槽式反射镜成阵列布置,其建设安装成本占发电站总投资的30%以上[1]。槽式反射镜属于风敏感结构,对其所受风荷载的精细化研究可降低其生产成本。中国在单个槽式反射镜风荷载方面有较多研究:邹琼等[2]通过风洞试验得到槽式反射镜镜面的平均风压和脉动风压分布特性,给出了镜面分区及各分区的峰值风压系数;付康维[3]、邓波[4]运用CFD技术对槽式反射镜风载特性进行研究,其结果与风洞试验结果相吻合。实际镜场中各反射镜之间存在互相遮挡,不同位置受到风荷载不同,邹琼[5]对3×3的反射镜阵列进行了风洞试验研究,得到中间反射镜的阻力系数小于单镜的84%;邓波[4]对3×3的反射镜阵列进行CFD数值模拟研究,给出了阵列中各位置反射镜的风压分布,但3×3规模的反射镜阵列与实际镜场的排布存在较大差异,其规律无法直接应用于工程实际。

槽式反射镜结构不同于常规的工业和民用建筑结构,中国规范对其风荷载无明确规定,其风荷载取值多依靠风洞试验。本文以开口宽8.6 m的槽式反射镜阵列为研究对象,进行缩尺为1/40的风洞测压试验,以单镜测压试验结果为基准,分别对比镜场边缘列、中间列反射镜风压分布和力系数,为镜场风荷载分区计算提供依据。

1 工程背景及风洞试验概况

本文研究的槽式反射镜结构如图1所示,开口宽B=8.6 m、轴长L=12.9 m、柱高h=4.55 m,在工作过程中结构最高点可达8.85 m。镜面由28块小镜面分4排组成。镜场中12个反射镜串联在一起形成1个基本单元。基本单元之间水平间距7.7 m,前后间距为槽式反射镜开口宽的3倍即25.8 m。

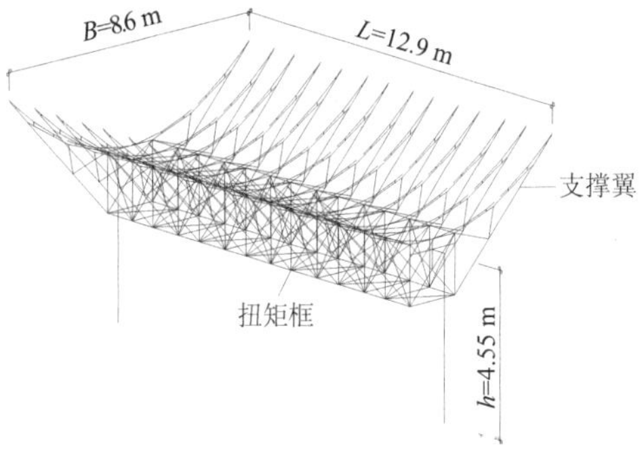

图1 槽式反射镜结构单线图

Fig. 1 Single line diagram of parabolic trough solar collectors

风洞试验在湖南大学风工程试验研究中心HD-2大气边界层风洞开口试验段进行,该试验段长15 m、宽8.5 m、高2.0 m,中间有直径6.5 m的大转盘。

根据《建筑结构荷载规范》[6]按A类地貌对大气边界层风场进行模拟[7],使用TFI公司的三维眼镜蛇探针进行风速剖面和湍流度剖面测量,图2为模拟风场风参数与荷载规范规定的标准曲线对比图,图中风速剖面横坐标为测试点风速与参考高度处风速的比值,湍流度剖面横坐标为湍流度,纵坐标均为模拟风场高度,由图可看出模拟风场风速和湍流度与A类地貌基本吻合。

收稿日期: 2020-12-28
通信作者: 牛华伟(1978—), 男, 博士、教授级高级工程师, 主要从事输电塔、冷却塔等电力工程结构风荷载、结构减振技术方面的研究。niuhw@hnu.edu.cn

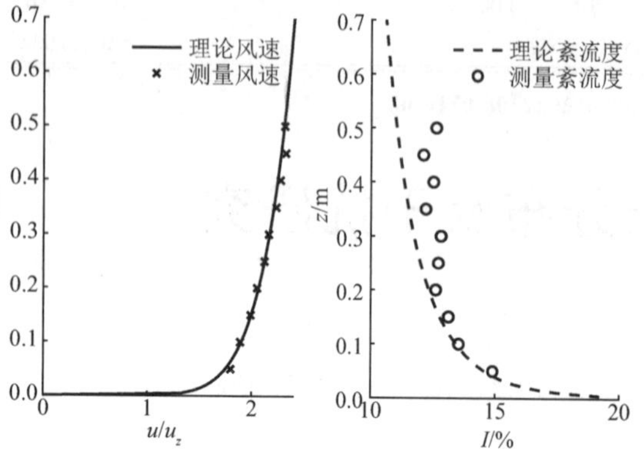

a. 风速剖面　　b. 湍流度剖面

图2　风速和湍流度剖面

Fig. 2　Profiles of velocity and turbulence intensity

1.1　试验模型

根据风洞尺寸及预模拟的镜场阵列范围确定试验模型的几何缩尺比为 1/40。模型按几何缩尺比进行制作，镜面、支撑翼和扭矩框选用 ABS 材料，支撑柱选用钢材，保证结构的强度及刚度。反射镜模型可绕扭矩框与柱交点沿竖直平面自由转动。试验模型共 108 个分为 2 类：一类为带有测压孔的刚性测压模型(6 个)，一类为不带测压孔的普通刚性模型(102 个)，用于形成反射镜阵列。测压模型的镜面在凹凸两侧均布置有测压点，两侧测点一一对应，每侧各 62 个，共 124 个(镜面测压点位置如图 3)。普通刚性模型除镜面外其余与刚性测压模型相同。

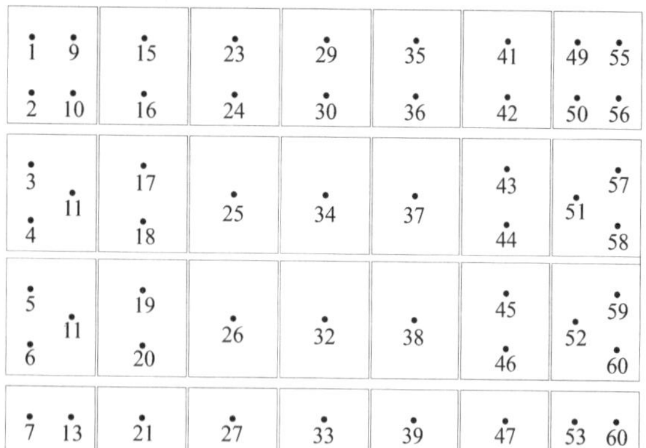

图3　镜面测压点位置示意图(凹凸两侧相同)

Fig. 3　Locations of pressure measurement points (both sides have same arrangement)

1.2　试验工况

试验包括单个反射镜(简称单镜)测压试验和模拟实际镜场排布的多排反射镜(简称群镜)测压试验。

试验风攻角、风向角定义如图 4 所示。风攻角 α 为来流风与反射镜焦线(镜面对称轴)在竖直方向的夹角，来流风与镜面垂直时为 0°，α 取 0°、30°、90°、150°、180°；风向角 β 为来流风与焦线在水平方向的夹角，取 0°~360°隔 15°测量一次。使用自主设计的角度测量装置确定风攻角，风洞自带转盘改变风向角。下文中用 XXX-XXX 对工况进行描述(XXX-XXX 表示 α 的度数-β 的度数)。

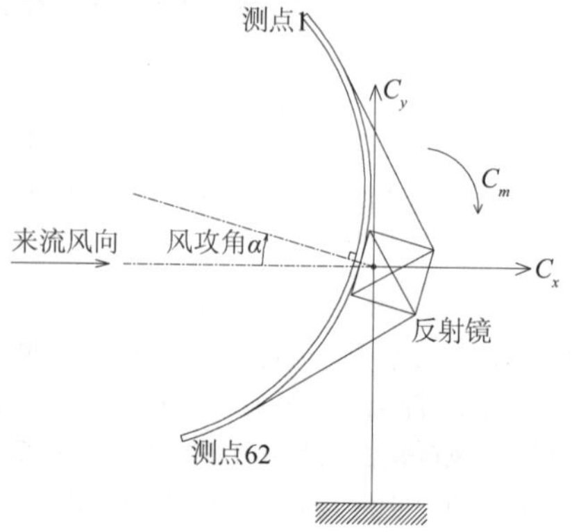

a. 风攻角 α 示意图

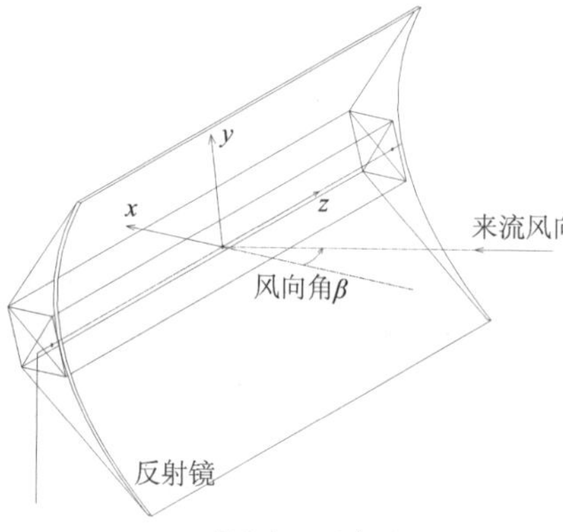

b. 风向角 β 示意图

图4　试验风攻角、风向角定义图

Fig. 4　Demonstrations of pitch angles α and wind directions β

群镜试验时将 108 个单镜模型分 6 排放置，将 6 个测压模型编号为 A、B、C、D、E、F 与 6 个普通模型进行串联作为 1 个基本单元参与试验。镜场中反射镜可分为边缘列和中间列，为涵盖所有位置将基本单元按图 5 所示的 12 种情况进行排布，镜场边缘列各排反射镜由 R1E~R6E 表示，中间列各排反射镜由 R1M~R6M 表示。风洞试验实际布置图如图 6。

1.3　风压系数及力系数定义

槽式反射镜受到的总风压用凹凸两侧的压力差来表示，定义各测点的净风压系数 C_{pi} 为：

$$C_{pi}=\frac{p_{iu}-p_{id}}{0.5\rho U^2} \tag{1}$$

式中：p_{iu}——试验测得的反射镜凹面第 i 测点处的压力均值，Pa；p_{id}——试验测得的反射镜凸面第 i 测点处的压力均值，Pa；ρ——空气密度，kg/m³；U——参考点风速(试验参考高度 0.25 m，对应实际高度 10 m)，kg/m³。由 C_{pi} 定义可知，C_{pi} 以垂直朝向凹面为正。

阻力系数 C_x、扭转系数 C_m 定义（统称力系数）如式（2）、式（3），正方向如图 4a 所示。

$$C_x=\frac{\sum_{i=1}^{n}C_{pxi}A_i}{BL} \tag{2}$$

$$C_m=\frac{\sum_{i=1}^{n}C_{pi}A_id_i}{B^2L} \tag{3}$$

式中：C_{pxi}——C_{pi} 在 X 方向的分系数；A_i——第 i 个测点的影响面积，m²；d_i——测点到旋转点的距离，m；B——镜面开口宽度，m；L——镜面长度，m。

2 反射镜镜面风压分布

镜场中不同位置处反射镜由于周边反射镜的干扰，所受风荷载不同，镜面风压分布规律也不同，中间区域反射镜因周围反射镜的遮挡所受风荷载将会大幅降低。考虑到文章篇幅，以 R1M-A~R6M-A、R1E-F~R6E-F 的结果为代表进行相关结果分析（X-X 表示：位置-测压模型编号）。

2.1 单镜典型工况下的风压分布

图 7 为 $\beta=0°$，$\alpha=0°$、30°和 90°时单镜的风压系数等值线图。由图 7 可得：工况 000-000 时镜面左右两侧风压系数对称，上下分布较均匀，从中间向两边有明显的梯度变化，此时阻力系数较大而扭矩系数较小；工况 030-000 时镜面风压沿短边方向出现变化，底端（近风侧）风压系数达到 1.5，此时

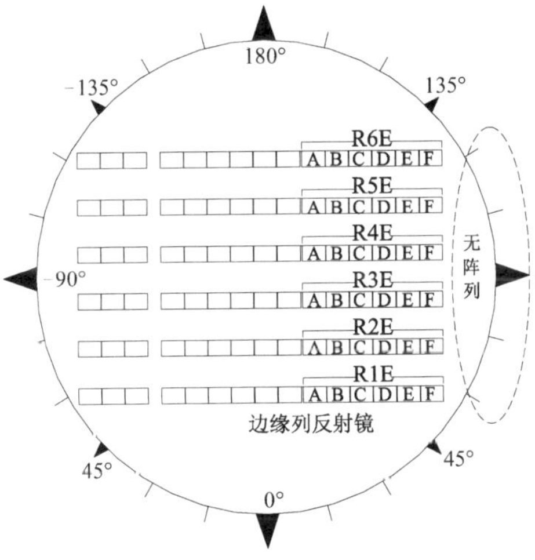

a. 边缘列反射镜 R1E~R6E 位置示意图

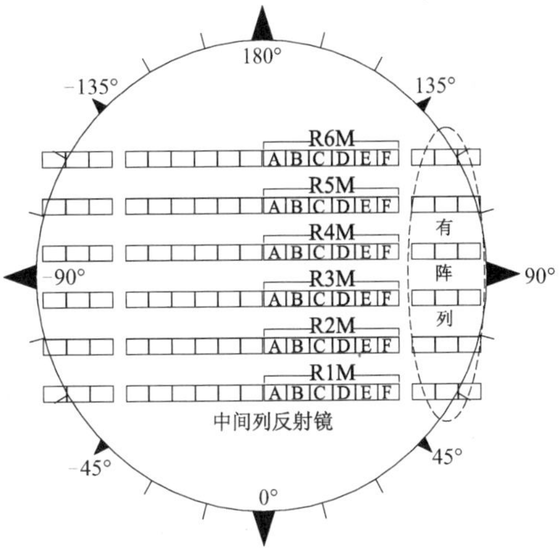

b. 中间列反射镜 R1M~R6M 位置示意图

图5 群镜各试验位置及名称定义示意图

Fig. 5 Investigated positions in array arrangement

图6 中间列反射镜风洞试验布置图

Fig. 6 Arrangement of collectors located in middle rows for wind-tunnel test

镜面风压系数大于工况 000-000，阻力系数增大；工况 090-000 时镜面远风侧边角处风压为正，其余位置风压系数为负值，镜面风压沿短边方向梯度分布，镜面近风侧受到的风荷载更大，其负风压由反射镜凹面受到的风吸力与凸面受到的风压力共同引起，最不利负压可达-1.1，远风侧的风压系数绝对值较小约在-0.2，此时因风压系数沿镜面短边方向分布不均匀使得扭矩系数大于其余工况。

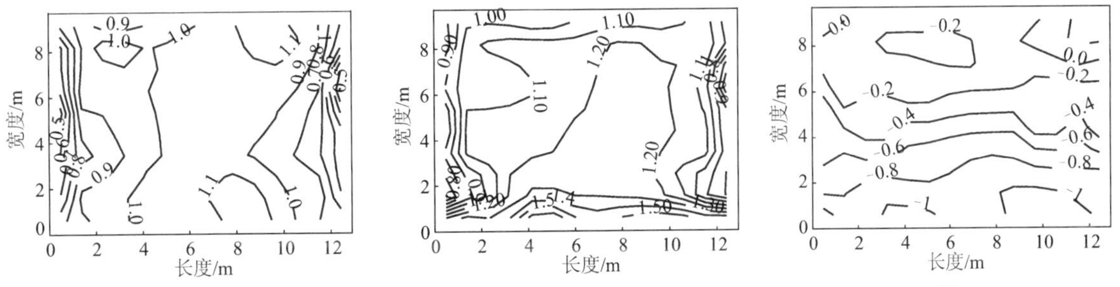

a. 工况 000-000　　b. 工况 030-000　　c. 工况 090-000

图7 单镜典型工况下净风压力系数等值线图

Fig. 7 Contour maps of net wind pressure coefficient under typical condition of single collector

2.2 群镜镜面风压分布

图8为边缘反射镜R1E-F~R6E-F在工况000-000时的净风压系数等值线图。前排反射镜首先与来流风接触，风压系数较大，后排反射镜因前排反射镜遮挡风压系数较小。工况000-000时因R1E-F正面受风，其风压系数远大于其余位置，达到1.8是单镜的1.20倍；气流在R1E-F镜面边缘产生分离和绕流，使得R2E-F镜面周围气流分布较复杂，风压系数在R2E-F处变为负值；随后气流逐渐趋于稳定，风压系数在R3E-F~R6E-F时逐渐转正，镜面风压分布有明显梯度变化，由图可看出反射镜R3E-F~R6E-F镜面风压向左下方逐渐减小，镜面右侧(外侧)所受风压较大，这是由于镜面右侧离气流近且气流在镜面右上角发生分离形成漩涡区，增大了镜面所受风荷载，此时反射镜结构绕短边方向会产生较大扭矩。

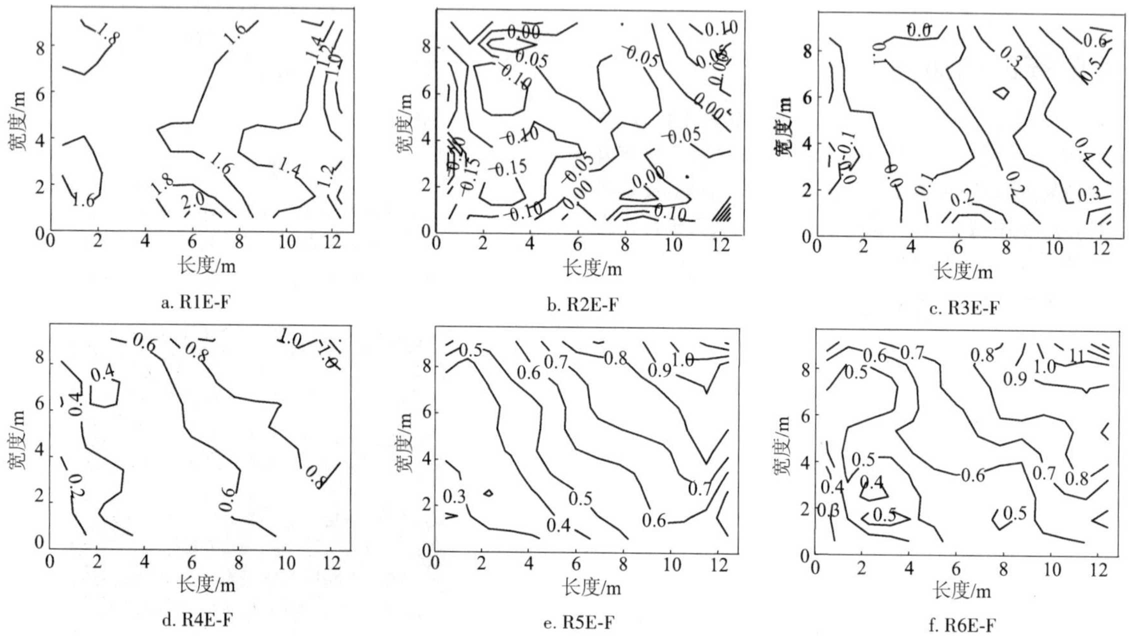

图8 R1E-F~R6E-F反射镜净风压系数等值线图(工况000-000)

Fig. 8 Contour maps of net wind pressure coefficient of collectors R1E-F~R6E-F($\alpha=0°$, $\beta=0°$)

图9为中间反射镜R1M-A~R6M-A在工况000-000时的净风压系数等值线图。阵列中间反射镜受前排遮挡更加明显，因反射镜两侧气流不通，各位置镜面气流情况更加复杂。在工况000-000时R1M-A的风压系数达到2.1是单镜的1.40倍，而R2M-A的风压系数仅为-0.2。中间列除首排外，其余各位置风压系数均小于边缘镜面。

2.3 群镜与单镜镜面风压分布对比

由以上分析可知反射镜阵列正面迎风(工况000-000)时，首排风荷载大于单镜，第2排受风吸力作用。在图10中给出了工况000-000时单镜、R1M-A和R2M-A凹凸两侧风压系数等值线分布图。对比单镜，R1M-A凹面的正压稍有增加，但凸面负压明显增大，约为单镜的2倍，R2E-F凹面风压为负且大于其凸面风压故其净风压系数为负值，这是开口宽度与前后排间距共同引起的，工况000-000时前后排之间出现了较明显的负压区使得首排风荷载突增。文献[8-9]的多排试验中出现了类似的现象，但首排风荷载值增加有限，反射镜截面不同对此可能有影响；除此之外前后排的间距对风荷载的分布也有较大影响。

3 反射镜力系数分析

3.1 单镜力系数分析

图11a为单镜在$\alpha=0°$~$90°$时C_x随风向角β的变化曲线，横坐标为风向角β，纵坐标为C_x系数值。同一风攻角下，C_x随着风向角的增大先正向减小再负向增大，其最小值约在$\beta=90°$，此时来流风与镜面平行；同一风向角下C_x在$\alpha=0°$时绝对值最大，在$\alpha=90°$时绝对值最小；C_x最大正值为0.89，出现在工况030-000时。图11b为单镜在$\alpha=0°$~$90°$时C_m随风向角的变化曲线。在$\alpha=0°$时C_m值较小，且随风向角变化不大；在$\alpha=30°$时，反射镜凸面迎风($\beta=90°$~$180°$)受到的扭矩比反射镜凹面迎风($\beta=0°$~$90°$)更大；在$\alpha=90°$时C_m值随风向角的变化起伏明显，在工况090-045时出现最大值，为0.155。

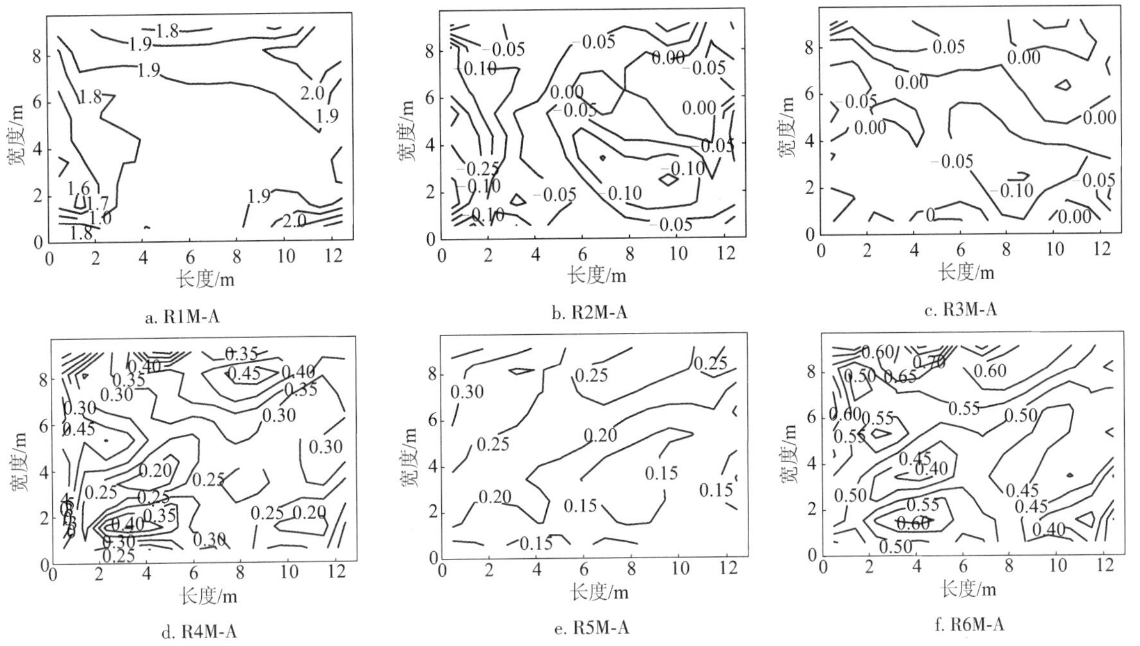

图9　R1M-A~R6M-A反射镜净风压系数等值线图(工况000-000)

Fig. 9　Contour maps of net wind pressure coefficient of collectors R1M-A-R6M-A($\alpha=0°$, $\beta=0°$)

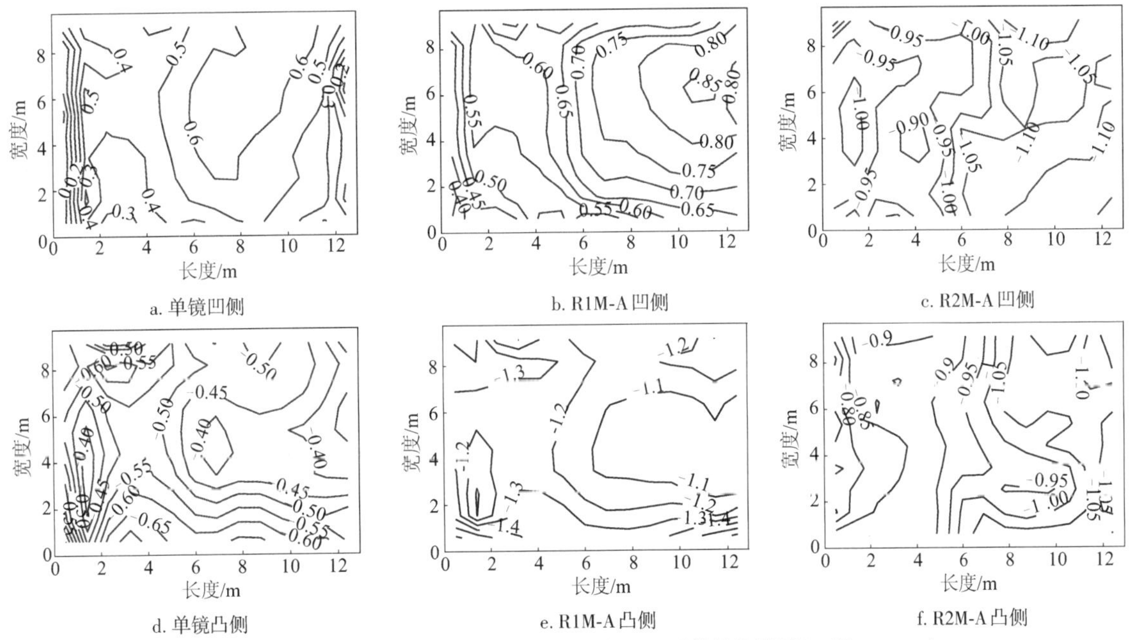

图10　R1M-A和R2M-A以及单镜凹凸两侧风压系数等值线图(工况000-000)

Fig. 10　Contour maps of wind pressure coefficient of both sides of R1M-A ,R2M-A and single collector($\alpha=0°$, $\beta=0°$)

单镜试验结果与邹琼[5]研究结果相对比，力系数变化趋势吻合，但因镜面尺寸以及开口宽度不同具体数值有差异，该数据作为基准具有可靠性。

3.2　群镜力系数分析及与单镜对比

3.2.1　边缘列反射镜力系数分析

边缘列反射镜位于阵列外围，直接承受来流风带来的影响。图12为 $\alpha=0°$ 时R1E~R6E位置处镜面F的阻力系数 C_x 随 β 的变化曲线，横坐标表示 β 的值(详见图5)。在 $\beta=0°$~90°时边缘列各排镜面实际迎风面积随风向的改变先增大后减小，C_x 也先增大后减小，与迎风面积正相关，$\beta=90°$~180°时R6E-F等后排反射镜由背风侧变为迎风侧 C_x 值有所增加。风向角 $\beta=0°$ 时前对后排风荷载的遮挡最明显，随着风

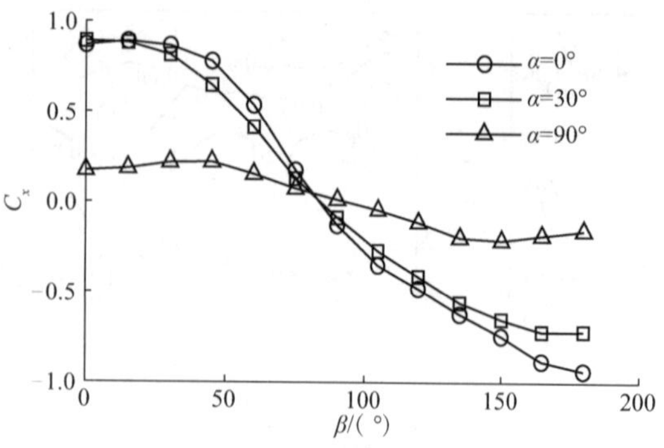

a. 单镜 C_x 系数变化曲线示意图

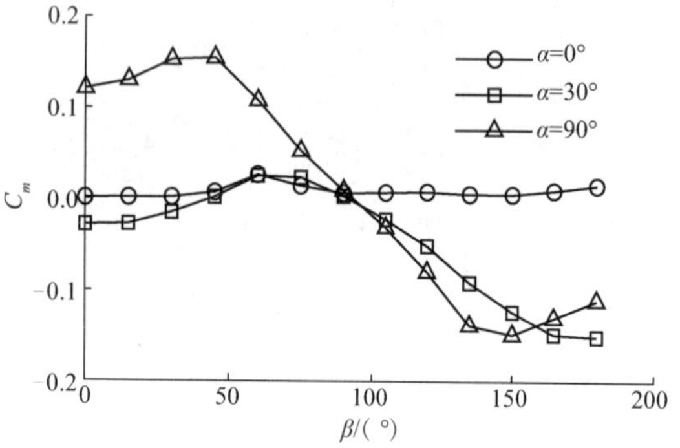

b. 单镜 C_m 系数变化曲线示意图

图11　单镜力系数变化曲线示意图

Fig. 11　Trend curve of force coefficient of single collector

向角增大，遮挡效应减弱。R1E-F 的最不利 C_x 出现在 β=15°，R2E-F~R6E-F 的最不利 C_x 均出现在 β=30°。图 13 为在 α=90°时 R1E~R6E 位置处镜面 F 扭矩系数 C_m 随风向角 β 的变化曲线。β=0°~90°时各位置 C_m 值随风向角的变化先增大后减小，各位置最不利值均出现在约风向角 30°、45°，后排反射镜 C_m 值更不利。

特殊工况下 R1E-F~R6E-F 的力系数与单镜对比绘于图 14 中，图中同时给出了力系数最大值随位置的变化趋势，

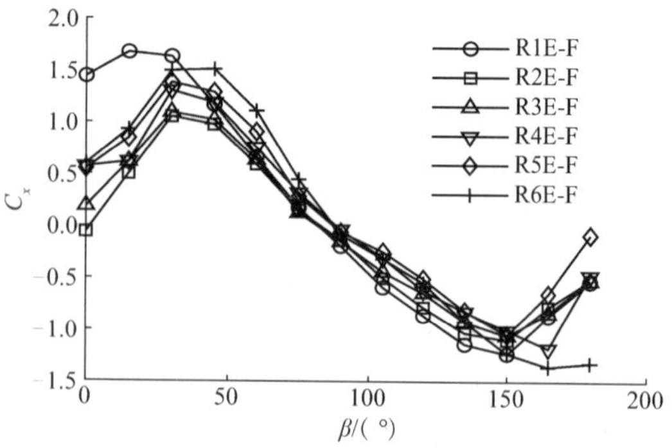

图12　α=0°时边缘各镜面 C_x 变化曲线

Fig. 12　Force coefficients C_x for collectors at edge in all rows with pitch angles of α=0°

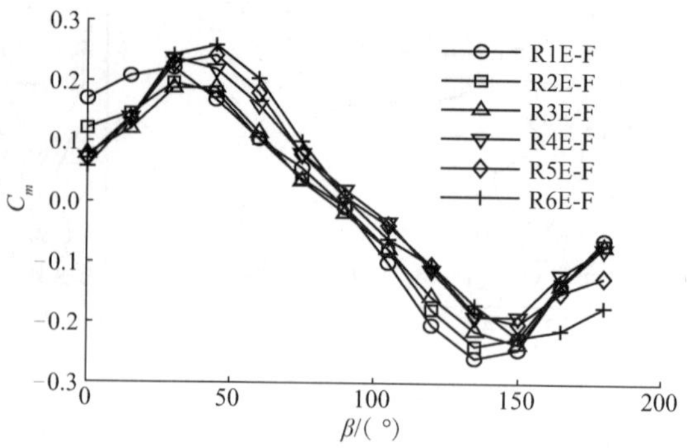

图13　α=90°时边缘各排镜面 C_m 变化曲线图

Fig. 13　Force coefficients C_m collectors at edge in all rows with pitch angles of α=90°

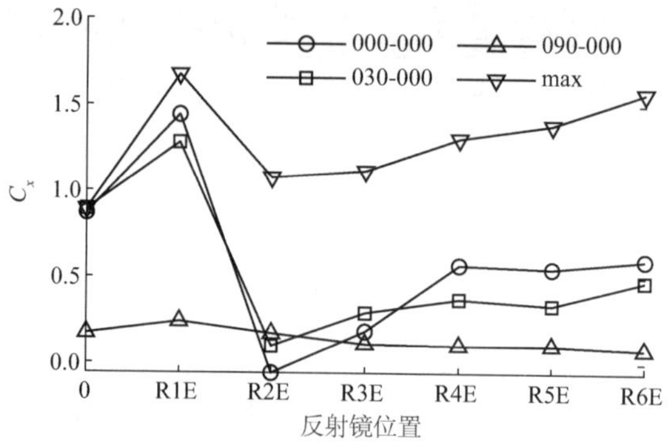

a. C_x

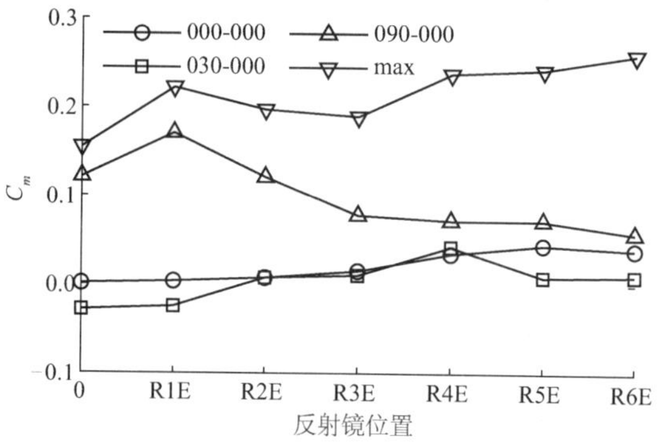

b. C_m

图14　边缘列各排镜面力系数与其所在位置的相关曲线

Fig. 14　Effect of row position along edge of array field

图中横坐标表示镜面 F 所在位置，原点 0 表示单镜状态，max 值表示各位置处所有工况下力系数的最大值。由图可得：工况 000-000 时，R1E-F 的 C_x 突增大于单镜，R2E-F 的 C_x 减小变为负值，R3E-F~R6E-F 的 C_x 逐渐增大，在 R4E-F 后基本不变；工况 030-000 时各排 C_x 的变化趋势与 000-000 相似；090-000 时 C_x 的值很小且变化幅度较小。C_x 值在 R1E 处突增以及在 R2E 处降低，是由于两者之间存在负压区引起的，这与前面分析一致。而 090-000 时 R1E-F、R2E-F 变化不明显是

因为 $\alpha=90°$时镜面与地面之间的空隙较大，同时镜面迎风面积减小，对风的遮挡减少，两者之间负压区较弱或不存在。边缘列反射镜 C_x 最大值出现在 R1E-F 处，为单镜的 1.88 倍，在 R2E-F 时最大值回落随后各排 C_x 最大值再次增大但均小于首排且大于单镜。C_m 在 $\beta=0°$时各工况变化不同，$\alpha=90°$时首排增大后逐排减小，$\alpha=0°$、30°时始终增大，值大于单镜。由最大值变化曲线可知 R6E-F 处 C_m 值最大，为单镜 1.68 倍。

3.2.2 中间列反射镜力系数分析

镜场中反射镜除了边缘几排外其余均属于中间区域，两侧有其余反射镜遮挡。图 15 为 $\alpha=0°$时 R1M~R6M 位置处镜面 A 的阻力系数 C_x 随风向角 β 的变化曲线。由图 15 可看出受两侧反射镜影响，中间列反射镜的 C_x 的变化趋势与单镜相比变化较大，且比边缘反射镜变化明显；在 $\beta=0°$~90°时 R1M-A 处的 C_x 值远大于其余位置，最不利值出现在 $\beta=15°$，值为 1.83；反射镜 R2M-A~R5M-A 因前后均有反射镜遮挡，风荷载始终较小，且随 β 变化不大，C_x 在±0.5 以内；R6M-A 在 $\beta=90°$~180°时变为迎风侧，C_x 开始逐渐增大。图 16 为 $\alpha=90°$时 R1M~R6M 位置处镜面 A 扭矩系数 C_m 随风向角 β 的变化曲线。在 $\beta=0°$~90°时，各位置镜面扭矩系数变化趋势一致随风向角的增大而减小，迎风侧反射镜 C_m 值大于背风侧，最不利值均出现在 $\beta=15°$时。

统计 $\beta=0°$时 R1M-A~R6M-A 力系数值及力系数最大值如图 17 所示，图中坐标含义同图 13。由图可看出 C_x 和 C_m 均在 R1M 时增大，max 值分别为单镜的 2.05 倍和 1.19 倍，在 R2M~R6M 时减小，C_x 为单镜的 30%~70%（R1M 与 R2M 处 C_x 的变化原因与 R1E、R2E 类似此处不再赘述）；C_m 为单镜的 40%~70%。中间列反射镜相较边缘反射镜受到的干扰影响大，除首排外其余各排力系数最大值明显减小。

综上，镜场各位置处反射镜风荷载有较大区别，边缘和中部各排风压系数有明显规律，根据力系数最不利值可将镜场分为 4 类区域，边缘列分为首排和中间排，中间列分为首排和中间排，边缘列与中间列的具体划分需进一步研究。

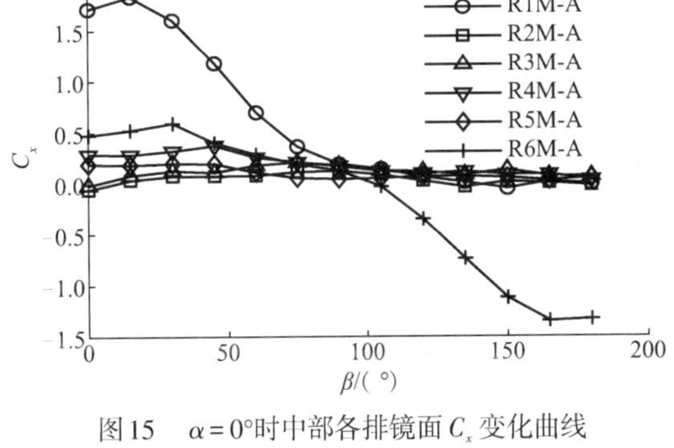

图15 $\alpha=0°$时中部各排镜面 C_x 变化曲线

Fig. 15 Force coefficients C_x for collectors at center in all rows with pitch angles of $\alpha=0°$

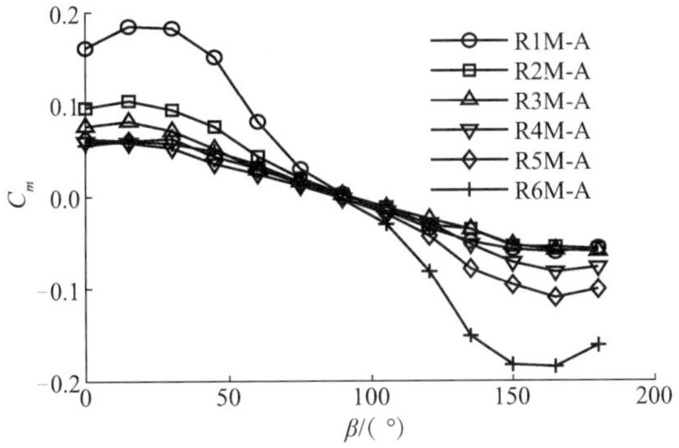

图16 $\alpha=90°$时中部各排镜面 C_m 变化曲线

Fig. 16 Force coefficients C_m for collectors at center in all rows with pitch angles of $\alpha=90°$

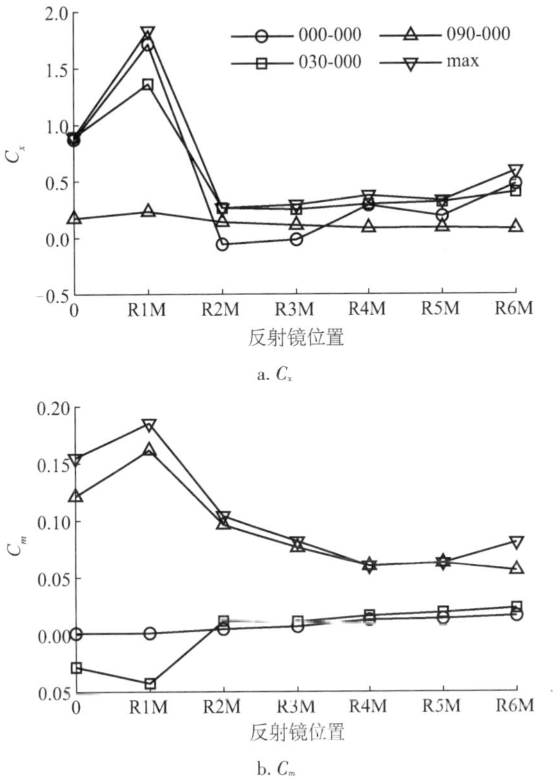

图17 镜场中间列各排镜面力系数与其所在位置的相关曲线

Fig. 17 Effect of row position along center of array field

4 结 论

本文通过单镜及群镜模型的风洞测压试验，给出了特殊工况下的镜面风压系数等值线图，分析了镜场不同位置处反射镜的风压分布以及力系数取值变化情况，对此类结构的相互干扰研究有借鉴作用，具体结论如下：

1）在工况 000-000 时，边缘列反射镜 R1E-R 局部风压系数达 1.8，是单镜的 1.20 倍；中间列反射镜 R1M-A 的局部风压系数达到2.1，是单镜的1.40倍。

2)镜场首排反射镜风压增大，主要是由于镜面凸面负压增大引起，故在首排和第2排之间有较强的负压区。

3)单个反射镜，C_x 最大正值0.89，出现在工况030-000，C_m 最大正值为0.155，出现在工况090-045。

4)边缘列反射镜的 C_x 及 C_m 的变化趋势与单镜类似，各排 C_x 最大值均大于单镜，R1E-F的值最大，为单镜的1.88倍；各位置处 C_m 最大值也大于单镜，R6E-F的值最大，为单镜1.68倍。

5)中间列反射镜受到周围反射镜干扰更明显，其 C_x 和 C_m 均在R1M时最不利，最不利值分别为单镜的2.05倍和1.19倍，在R2M~R6M时值较小，C_x 为单镜的30%~70%；C_m 为单镜的40%~70%。

[参考文献]

[1] SUN H H, GONG B, YAO Q, et al. A review of wind loads on heliostats and trough collectors[J]. Renewable & sustainable energy reviews, 2014, 32(1): 206-221.

[2] 邹琼，李正农，吴红华. 槽式聚光镜的脉动风压特性与极值风压分布[J]. 太阳能学报，2016，37(2)：407-414.
ZOU Q, LI Z N, WU H H. Extreme wind pressure distribution and fluctuation characteristics of trough concentrating mirror[J]. Acta energiae solaris sinica, 2016, 37(2): 407-414.

[3] 付康维. 槽式聚光镜风效应的CFD数值模拟[D]. 长沙：湖南大学，2014.
FU K W. CFD numerical simulation of curved mirror[D]. Changsha: Hunan University, 2014.

[4] 邓波. 槽式聚光镜抗风性能研究[D]. 重庆：重庆大学，2016.
DENG B. Study of wind resistant performance of parabolic condenser[D]. Chongqing: Chongqing University, 2016.

[5] 邹琼. 槽式聚光镜组系统的抗风性能研究[D]. 长沙：湖南大学，2016.
ZOU Q. Research on wind resistant performance of trough solar collector system[D]. Changsha: Hunan University, 2016.

[6] GB 50009—20112，建筑结构荷载规范[S].
GB 50009—2012, Load code for the design of building structures[S].

[7] 孙瑛，武岳，曹正罡. 建筑风洞实验指南[M]. 北京：中国建筑工业出版社，2011：25-29.
SUN Y, WU Y, CAO Z G. The wind tunnel experiment guide[M]. Beijing: China Building Industry Press, 2011: 25-29.

[8] HOSOYA N, PETERKA J A, GEE R C, et al. Wind tunnel tests of parabolic trough solar collectors[R]. NREL/SR-550-32282, 2008.

[9] WINKELMANN U, KAMPER C, HOFFER R, et al. Wind actions on large- aperture parabolic trough solar collectors: Wind tunnel tests and structural analysis [J]. Renewable energy, 2020, 146: 2390-2407.

STUDY ON WIND PRESSURE DISTRIBUTION AND INTERFERENCE OF MULTI-ROW PARABOLIC TROUGH SOLAR COLLECTORS

Kang Jiaxin[1], Niu Huawei [1], Li Hongxing[2], He Shaohua[2]

(1. *College of Civil Engineering, Hunan University, Changsha* 410082, *China*;

2. *Northwest Electric Power Design Institute Co.,Ltd., China Power Engineering Consulting Group, Xi'an* 710075, *China*)

Abstract: In order to study the wind pressure characteristics and mutual interference of multi-row parabolic trough solar collectors, a single trough solar collector and multi-row trough solar collectors were subjected to a 1/40 scale pressure measurement test. The pitch angles, wind direction angles and its position were selected as the experimental variables. Firstly, the wind pressure distribution on the collector surface was measured; then, the contour map of wind pressure coefficient and changing curve of force coefficient were produced; at last, the influences of mutual interference of trough solar collectors on the wind load of collector surface were analyzed. The results show that the trough solar collectors at the first row in the middle column subjected to the maximum wind pressure, and its drag coefficient is twice that of a single collector. The drag coefficients are 30% to 70% that of single collector for the rest of the mirrors in the middle column. The drag coefficient for the collector at edge column of first row is 1.88 times that of single collector, and the rest of the rows are more than 1.2 times that of a single collector. Hence, it is recommended to divide the collector field into four types of areas for structural design.

Keywords: solar thermal power; parabolic trough solar collectors; wind tunnels; pressure distribution; interference effects; solar collector fields

Journal of Wind Engineering & Industrial Aerodynamics 236 (2023) 105355

Contents lists available at ScienceDirect

Journal of Wind Engineering & Industrial Aerodynamics

journal homepage: www.elsevier.com/locate/jweia

ELSEVIER

Experimental study on critical wind velocity of a 33-meter-span flexible photovoltaic support structure and its mitigation

Jiaqi Liu [a], Shouying Li [a,*], Jing Luo [b], Zhengqing Chen [a]

[a] Key Laboratory for Wind and Bridge Engineering of Hunan Province, College of Civil Engineering, Hunan University, Changsha, 410082, China
[b] ATEC Energy & Environment CO., LTD, Shenzhen, 518057, China

ARTICLE INFO

Keywords:
Flexible photovoltaic modules support structures
Critical wind velocity
Wind-induced responses
Wind tunnel tests
Non-contact video displacment measuring system
Stability cables

ABSTRACT

Flexible photovoltaic (PV) modules support structures are extremely prone to wind-induced vibrations due to its low frequency and small mass. Wind-induced response and critical wind velocity of a 33-m-span flexible PV modules support structure was investigated by using wind tunnel tests based on elastic test model, and the effectiveness of three types of stability cables on enhancing the critical wind velocity of the flexible PV modules support structures was carefully examined. First, an elastic test model of the flexible PV modules support structure was designed and manufactured. Second, a series of wind tunnel tests based on the elastic test model were carried out to obtain the wind-induced responses of the 33-m-span PV modules support structure. Third, three types of stability cables (T1, T2 and T3) were installed on the test model, and a series of wind tunnel tests were carried out to examine the effectiveness of the three types of mitigation countermeasures. Finally, the wind load factor was discussed by using the displacement time histories measured from the wind tunnel tests. The results show that 180° and 0° are the two most unfavorable wind directions, and the critical wind velocity is only about 18.5 m/s for the deformation threshold of 1/100 span. The critical wind velocities for stability cables T1, T2 and T3 (1/100 of the span) are respectively 24.6, 29.3 and 36.7 m/s, which are significantly higher than the value without stability cable (18.5 m/s). The wind load factor at the windward side under wind directions of 0° and 180° are respectively 1.56 and 1.30 for the test case with stability cable T3.

1. Introduction

The society development is accompanied by a large amount of energy consumption. With the climate change and the increasing depletion of non-renewable energies, some clean energies should be vigorously developed, such as solar energy, wind energy and ocean energy. The solar energy is a typical clean energy, and has been widely used. In recent years, photovoltaic (PV) power stations have been built all over the world.

In the early years, the PV modules were often mounted on the building roofs. Radu et al. (1986, 1989) studied the wind loads on the PV modules mounted on the building roofs. They found that the wind loads on the leeward side were lower than those on the windward side due to the shading effect of the PV modules. Kopp et al. (2002) proposed a linearized model to predict the peak system torque of PV arrays. Some other researchers (Geurts et al., 2009; Velicu et al., 2010; Jubayer and Hangan, 2016) conducted wind tunnel tests on PV modules mounted on building roofs to measure the drag and lift forces of the PV modules. The effects of some structural parameters, including inclined angle and row spacing, on the wind loads of the PV modules mounted on building roofs were also systematically investigated (Kopp et al., 2012; Cao et al., 2013; Naeiji et al., 2017). Kopp and Banks (2013) carried out a series of wind tunnel tests to determine the design wind loads for roof-mounted PV modules.

Due to the limitation of roof area, the PV power stations on the ground, mostly fixed PV modules support structures, were gradually developed in recent years. Some studies were performed to determine the wind loadings on the PV modules mounted on the fixed support structures. Bitsuamlak et al. (2010) and Aly and Bitsuamlak (2013) carried out numerical simulations by using computational fluid dynamics (CFD) method to determine the aerodynamic effects of the prototype of PV modules mounted on the fixed support structures, and conducted wind tunnel tests on test models with different geometrical scales to systematically study the sensitivity of wind loads on the PV modules. Jubayer and Hangan (2016) numerically studied the wind loads and wind fields around the ground-mounted PV modules by using

* Corresponding author.
E-mail addresses: ljq940114@hnu.edu.cn (J. Liu), shyli@hnu.edu.cn (S. Li), maqingtao@atec-sz.com.cn (J. Luo), zqchen@hnu.edu.cn (Z. Chen).

https://doi.org/10.1016/j.jweia.2023.105355
Received 2 October 2022; Received in revised form 6 February 2023; Accepted 10 February 2023
Available online 10 March 2023
0167-6105/

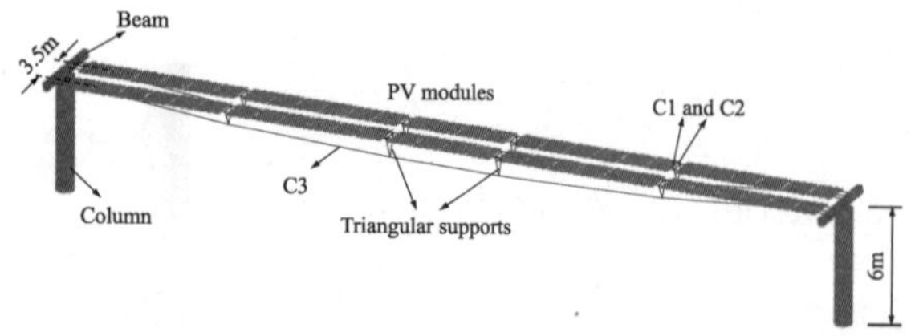

(a) Three-dimensional diagram of the structure of the flexible PV support

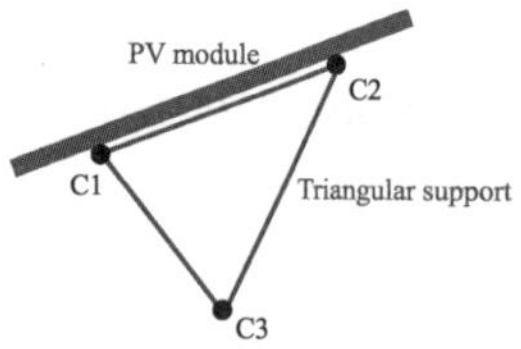

(b) Cross section of the PV module, cables and triangular supports

Fig. 1. Structure of the flexible PV support.

Table 1
Similarity relationships of the aeroelastic model.

Model parameters	Similarity ratio	Scale
Length(λ_L)	1:n	1:20
Velocity(λ_V)	1: $\sqrt{n}$	1: $\sqrt{20}$
Frequency(λ_f)	$\sqrt{n}$:1	$\sqrt{20}$:1
Force(λ_F)	1:n^3	1:20^3

Table 2
The parameters of aeroelastic model of flexible PV support.

		Scale	Prototype	Target value	Test model
PV module	Length/(mm)	1:20	2256	112.8	112
	Width/(mm)	1:20	1133	56.65	56
	Thickness/(mm)	1:20	35	1.75	2
	Mass/(g)	1: 20^3	32300	4.04	4
C1、C2	Diameter/(mm)	1:20	12.7	0.635	0.5
	Force/(N)	1: 20^3	21800	2.725	2.7
C3	Diameter/(mm)	1:20	12.7	0.635	0.5
	Force/(N)	1: 20^3	42100	5.2625	5.3
C4、C5	Diameter/(mm)	1:20	9.5	0.475	0.4
	Force/(N)	1: 20^3	0	0	0

CFD method, and the wind velocity around the PV modules was accurately predicted. Meroney and Neff (2010) numerically investigated the wind field around the PV modules by using different turbulence models, and the drag, lift forces and overturning moment of the PV modules were also obtained. The results showed that the windward row of the modules has larger wind load under the wind directions of 0° and 180°. The effects of some structural parameters, such as row spacing, module clearance and inclined angle, on the wind loads of the PV modules were systematically investigated by using CFD and wind tunnel tests (Shademan et al., 2014; Warsido et al., 2014; Abiola-Ogedengbe et al., 2015; López et al., 2019).

In recent years, the PV power stations were gradually built on mountains, Gobi and other complex terrains due to the shortage of land resources. In many areas, fishery-photovoltaic and agricultural-photovoltaic hybrid PV project also need to be considered. Under these cases, the span of the fixed PV support structures are too small, and it is necessary to develop a new type of PV modules support structure with larger span to adapt these complex terrains. Baumgartner et al. (2008, 2009, 2010) first proposed the concept of flexible PV modules support structure, in which the PV modules were mounted on the cables. Ma et al. (2021) investigated the effects of wind direction, inclination angle, spacing ratio and installation position on the wind loads of the flexible PV modules support structures. He et al. (2021) investigated the mechanical properties of a new flexible PV modules support structure with a span of 30 m, and discussed the effects of row spacing, inclination angle, initial cable force and cable diameter on the structural properties. As the most important part of the flexible PV modules support structures, the cable is prone to wind-induced vibrations due to its small mass and low frequency (Li et al., 2014, 2019; Li et al., 2019a,b; Deng et al., 2021). Tamura et al. (2015) investigated the flexible PV modules support structure with double cables by the wind tunnel tests. The effects of sag-span ratio, wind velocity and wind direction on the aeroelastic stability of the flexible PV modules support structures were systematically analyzed. Kim et al. (2018, 2020) studied the effect of the PV module shape on wind-induced vibrations of the flexible PV modules support structures under four wind environments through conducting a series of wind tunnel tests. He et al. (2020) proposed a new type of flexible PV modules support structure with three cables. The wind-induced responses and its mitigation of this type of structure were studied by using elastic wind tunnel tests. Du et al. (2022) studied the distribution of the average and fluctuating wind loads of a long-span PV modules support structure, and the wind-induced response was analyzed by finite element method. At present, the research on the control standard and critical wind velocity of flexible PV supports is not sufficient.

In the previous studies, the wind-induced responses were measured only on a few PV modules because the limitations on the number of sensors. Furthermore, little attentions were paid on the critical wind velocity of the flexible PV modules support structures. In this study, wind-induced response and critical wind velocity of a 33-m-span flexible PV support structure was experimentally studied by using a non-contact video displacement measuring system. An elastic test model of the flexible PV modules support structure was firstly designed and manufactured. Second, a series of wind tunnel tests based on the elastic test model were carried out to obtain the wind-induced responses of the 33-m-span PV modules support structure. The displacement time histories for 176 marked points on 88 PV modules were simultaneously sampled by using a non-contact video displacement measuring system. Third, three types of stability cables were installed on the test model, and a series of wind tunnel tests were carried out to examine the effectiveness of the three types of stability cables. The critical wind velocities of the three stability cables were obtained. Finally, the wind load factor was discussed by using the displacement time histories measured from the wind tunnel tests.

2. Outline of wind tunnel tests

2.1. Test model

The structure of the flexible PV support adopted in this study is shown in Fig. 1. The height of the columns is 6 m, and the center-to-center spacing between two adjacent rows of PV modules is 3.5 m. The span of the flexible PV support is 33 m, which is consisted of 28 PV modules. The inclination angle between the PV modules and the horizontal plane is 15°, and the PV modules are mounted on two steel cables C1 and C2. Furthermore, steel cable C3 is set to reduce the vertical deformation under the actions of wind and snow loadings. Four triangular supports are arranged between C1, C2 and C3 within each span. The triangular supports are made of hollow round steel pipes, the four triangular supports are placed at the 6th, 12th, 16th, and 22nd PV modules, respectively. The length, width and thickness of the PV module are respectively 2256 mm, 1133 mm and 35 mm, and the weight for each PV module is 32.3 kg.

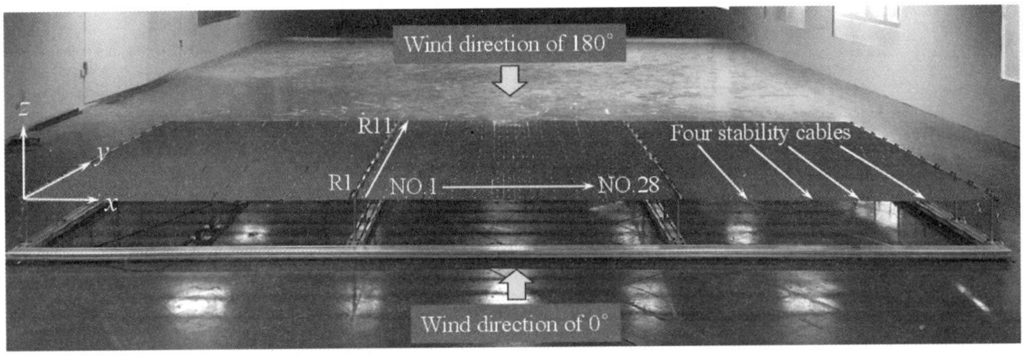

Fig. 2. Photo of the test model and the coordinate system.

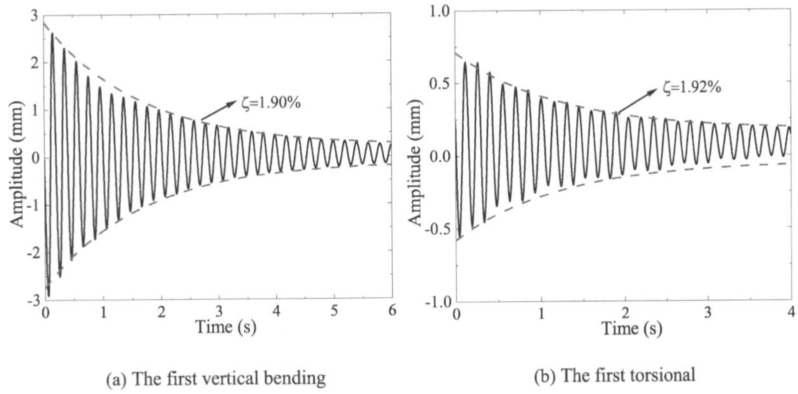

Fig. 3. Free vibration decay curves of the test model.

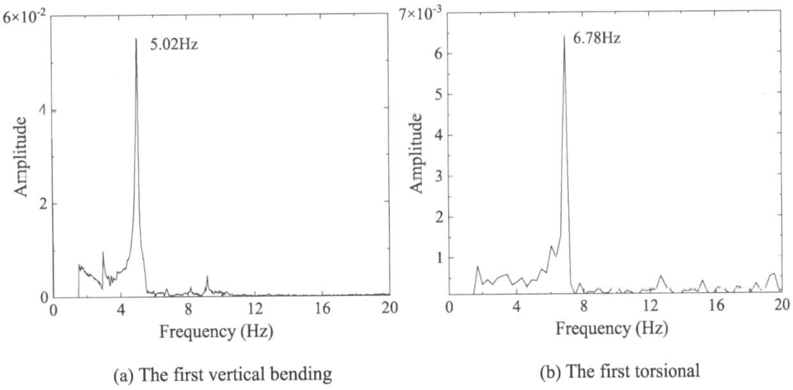

Fig. 4. Power spectrum densities of free vibration decay curves.

Considering the geometrical size and maximum wind velocity of the test section of wind tunnel used in this study, the length scale of the test model is $\lambda_L = 1:20$, and the maximum blockage ratio is about 3.4%. According to Strouhal and Froude numbers, other scales, including velocity, frequency and force (λ_V, λ_f and λ_F), can be determined, as shown in Table 1. The PV module model was made of Paulownia wood panel, and its density is about 300 kg/m^3, which is close to the density of PV module(360 kg/m^3). Therefore, Paulownia can simultaneously ensure the simulations of the thickness and the mass of the PV module. High strength piano wire with the elastic modulus of 2000Mpa was adopted to simulate the cables, which can ensure a better material properties under large tension. The models of the columns and the beams were made of aluminium bar. The main structural parameters of aeroelastic model of flexible PV support are shown in Table 2.

The test model was assembled according to the actual process. Cables C1 and C2 were first tensioned, then installed the PV modules on the cable C1 and C2, and cable C3 was finally tensioned. When the test model was assembled, the PV modules was basically located on a horizontal line. Due to the effects of geometric nonlinear, the tensions of cables C1 and C2 will significantly change when cable C3 was tensioned. Therefore, the cable tension needs to be adjusted several times to make sure the accuracy of the ultimate frequencies of the test model. In order

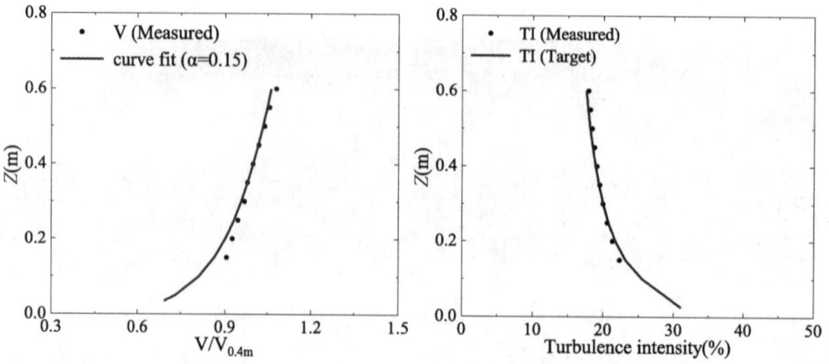

(a) Mean wind velocity profile

(b) Turbulence intensity profile

Fig. 5. Simulation of wind field in the test section.

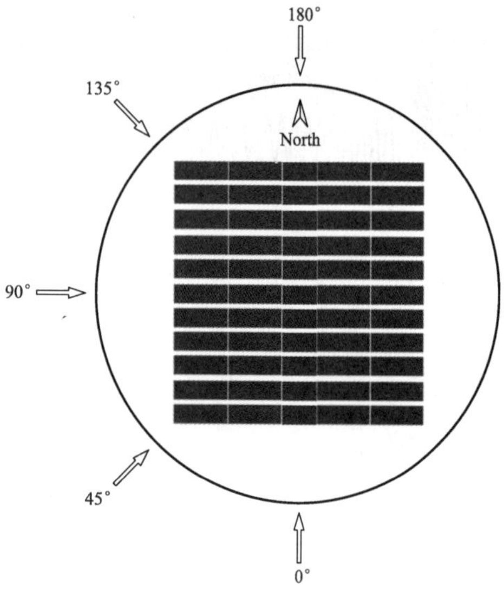

Fig. 6. Definition of wind directions.

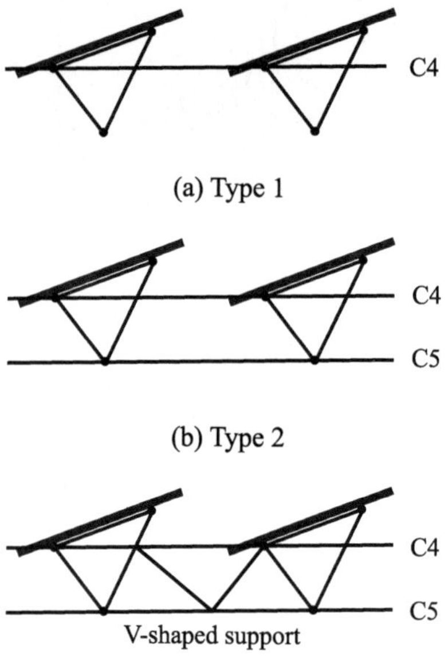

(a) Type 1

(b) Type 2

(c) Type 3

Fig. 7. Diagram of stability cables.

to considering the interference effect of the adjacent PV modules, the aeroelastic test model consists of 3 spans and 11 rows. The photo of the assembled elastic test model of the flexible PV support is shown in Fig. 2. A cordinate system *oxyz* is established, as shown in Fig. 2. The rows of the test model are defined as R1 to R11 along the positive direction of y axis, and the PV modules with each span are defined as No.1 to No.28 along the positive direction of x axis.

Software ANSYS was used to numerically analyze the dynamic characteristics of the prototype structure of the flexible PV support. The results showed that the first two frequencies are 1.07 Hz and 1.56 Hz, which respectively correspond to the modes of the first vertical bending and the first torsion. According to the frequency scale listed in Table 1, the first two frequency of the test model should be 4.79 Hz and 6.96 Hz.

The dynamic characteristics of the test model were identified by laser displacement sensors installed on the PV modules. The free vibration decay curves of the first vertical bending and the first torsional modes of the test model obtained by artificial excitations are respectively shown in Fig. 3 (a) and (b). The damping ratios of the first two modes are 1.90% and 1.92%, respectively. The power spectrum densities (PSDs) of the time histories in Fig. 3 are given in Fig. 4. It can be found from Fig. 4 that the first two frequencies of the test model are 5.02 Hz and 6.78 Hz. Comparing with the target values obtained by numerical simulations (4.79 Hz and 6.96 Hz), the deviation ratio is smaller than 5%.

Table 3
Test cases of the aeroelastic wind tunnel tests.

Test cases	Stability cables	Wind directions (°)	Wind velocity (m/s)
Test case 1	No	0°~180° with an interval of 15°	0–18.5
Test case 2	Type 1		0–31.2
Test case 3	Type 2		0–34.3
Test case 4	Type 3		0–40.6

2.2. Wind tunnel and wind field simulation

The aeroelastic wind tunnel tests were conducted in the open-circuit

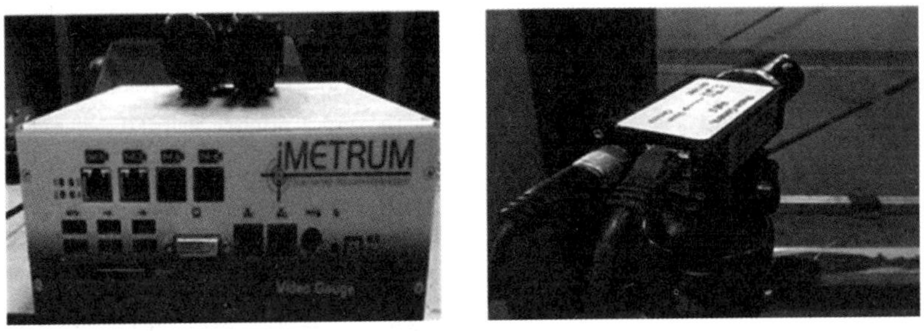

(a) Measurement and analysis system (b) Camera and lense

Fig. 8. Non-contact video displacement measuring system.

Fig. 9. Photo of the arrangement of marked points on the mid-span of the test model.

test section of the HD-2 atmospheric boundary layer wind tunnel in Hunan University, Changsha, China. The test section is 8.5 m in wide by 2.0 m in height by 15.0 m in length. The wind velocity ranges from 0 to 14 m/s, and the turbulence intensity can be controlled below 1%. Roughness elements and spires were adopted to produce the wind field in the test section. Terrain category B stipulated in Chinese code was simulated, and the mean wind velocity and turbulence intensity profiles are shown in Fig. 5 (a) and (b), respectively.

2.3. *Test cases*

The direction along the positive direction of x axis (i.e., from R1 to R11) is defined as 0°, and increases counterclockwise, as shown in Fig. 6. The wind-induced responses were recorded in the wind direction range of 0–180° with an interval of 15°. The wind velocity was gradually increased from 0 m/s and the test was stopped when a large amplitude of vibration was observed.

The flexible PV support structure introduced in Section 2.1 is prone to wind-induced vibrations. Necessary countermeasures should be put forward to reduce the wind-induced vibration and ensure the safety of the structure. In this study, four stability cables along the y axis direction within each span were arranged at the locations of the four triangular supports to enhance the wind resistance of the structure of the flexible PV support, as indicated in Fig. 2. Three types of stability cables, types 1–3, are adopted. For type 1 (T1), a longitudinal cable C4 was arranged to connect the cable C2 in the y axis direction, as shown in Fig. 7(a). For type 2 (T2), add another longitudinal cable C5 to connect the cable C3 on the basis of type 1, as shown in Fig. 7(b). For type 3 (T3), add a V-shaped rigid support between cables C4 and C5 on the basis of type 2, as shown in Fig. 7(c). Cables C4 and C5 are made of steel wires with a diameter of 0.4 mm, which have no pre-stress applied on them.

There are four types of test cases in this aeroelastic wind tunnel tests, as shown in Table 3. The maximum test wind velocities in test cases 1–4 are respectively 18.5, 31.2, 34.3 and 40.6 m/s, which were determined by the responses during the tests.

In order to measure the overall wind-induced responses of the PV modules on this test model, a non-contact video displacement measuring system (as shown in Fig. 8) made by the British company IMETRUM was used to record the vertical displacement of the PV modules in the mid-span. A marked point should be attached on the PV module before the displacement can be measured by the non-contact video displacment measuring system. White paper (1 cm × 1 cm) was adopted as the marked point because the PV module model is red. In order to simultaneously measure the vertical and torsional displacement, two marked points which were 35 mm from the center of PV module were arranged on the PV module model. The two white marks on the same PV module are named P1 and P2, respectively. For each row within the mid-span of the test model, marked points were arranged on 8 PV module models, including No.5, 9, 11, 14, 15, 18, 20 and 24 (16 marked points). That is, there were 176 marked points in total on the mid-span of the test model, as shown in Fig. 9.

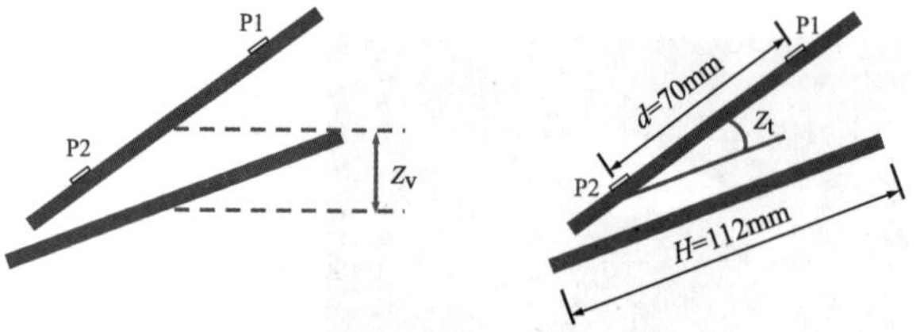

(a) Vertical and torsional displacements of the PV modules

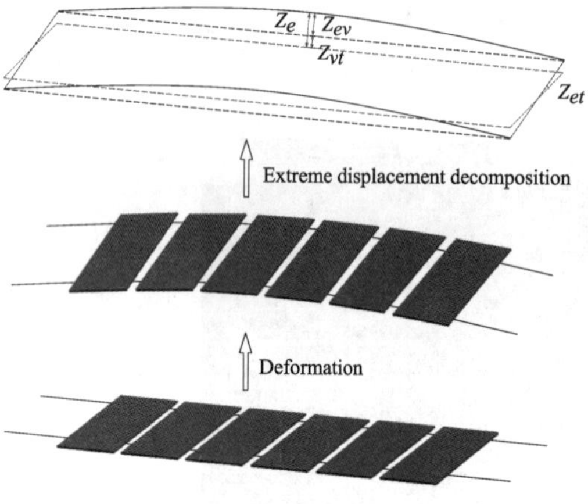

(b) Definition of extreme displacement

Fig. 10. Diagram of the displacement of the PV module.

2.4. *Definitions of structural responses*

The PV module has two main displacements, one is the vertical displacement of the PV module center Z_v along z axis, the other is the torsional displacement Z_t around x axis, as indicated in Fig. 10 (a). By using the vertical displacements measured from the two marked points, the vertical displacement of the PV module center Z_v and the torsional displacement Z_t can be obtained by,

$$Z_v = \frac{z_1 + z_2}{2} \tag{1}$$

$$Z_t = \frac{z_1 - z_2}{d} \tag{2}$$

where, z_1 and z_2 are the vertical displacements measured at the two marked points (i.e. P1 and P2, as indicated in Fig. 10(a)), for which the upward displacement is positive; For torsional displacement Z_t, the clockwise rotation is positive. d is the distance between the two marked points, $d = 70$ mm.

The average vertical and torsional displacements Z_{av} and Z_{at}, the standard deviation of vertical and torsional displacements σ_v and σ_t can be obtained from Z_v and Z_t calculated by Equations (1) and (2). The vertical fluctuating displacement at the PV module center Z_{fv} and the torsional fluctuating displacement Z_{ft} can be respectively calculated by,

$$Z_{fv} = g \times \sigma_v \tag{3}$$

$$Z_{ft} = g \times \sigma_t \tag{4}$$

where g is the peak factor, $g = 3.5$.

The extreme vertical displacement at the PV module center Z_{ev} and the extreme torsional displacement Z_{et} can be respectively obtained by,

$$Z_{ev} = max(|Z_{av} \pm Z_{fv}|) \tag{5}$$

$$Z_{et} = max(|Z_{at} \pm Z_{ft}|) \tag{6}$$

The vertical displacement at the edge of the PV module Z_{vt} produced by the torsional displacement Z_{et} can be calculated by,

$$Z_{vt} = Z_{et} \times \frac{H}{2} \tag{7}$$

where H is the length of the PV module model, $H = 112$ mm.

The extreme vertical displacement at the edge of the PV module, Z_e, can be obtained by (as indicated in Fig. 10(b)),

$$Z_e = max(|Z_{ev} \pm Z_{vt}|) \tag{8}$$

3. Experimental results for test case 1

During the test case 1, large amplitude vibration of the test model can be found when the wind velocity increased to 18.5 m/s, which is the maximum wind velocity in this test case.

3.1. *Average responses of the PV modules*

Fig. 11 and Fig. 12 respectively show the contour maps of the average vertical displacement at the PV module center Z_{av} and the average torsional displacement Z_{at} of the mid-span of the test model without stability cables under the wind directions of 0°, 30°, 150°, 180°. The test wind velocity is 18.5 m/s. It can be found from Figs. 11 and 12 that the largest values of Z_{av} and Z_{at} of the PV modulues both occur on the row R1 if the wind directions are 0° and 30°. Under these cases, the average wind pressure acting on the PV modules is downward. For the wind directions of 150° and 180°, the largest values of Z_{av} and Z_{at} of the PV modulues both occur on the row R9~R11. Under these cases, the average wind pressure acting on the PV modules is upward. It appears that the maximum values of the vertical displacement Z_{av} under the wind directions of 150° and 180° are significantly larger than those of the wind directions of 0° and 30°. However, for the torsional displacements Z_{at}, the maximum values under the wind directions of 0° and 30° are slightly larger than those of the directions of 150° and 180°. It also can be found from Figs. 11 and 12 that the largest average displacements (including Z_{av} and Z_{at}) take place at the locations of one-third span on the windward row of the test model.

Figss. 13 and 14 present the variations of the average vertical displacement Z_{av} and the average torsional displacement Z_{at} at the location of one-third of the mid-span for different wind directions with wind velocity. It can be found from Figss. 13 and 14 that the average vertical and torsional displacements of the PV modules both increase with the increase of wind velocity. Under each wind velocity, the maximum average vertical and torsional displacements appears at the first row on windward side, i.e., R1 for 0° and R11 for 180°. The maximum average vertical displacement (about 99 mm for wind velocity of 18.5 m/s) occurs at row R11 under the wind direction of 180°. The maximum average torsional displacement (about 0.04 rad for wind velocity of 18.5 m/s) occurs at row R1 and R11 under the wind directions of 0° and 180°, respecitvely.

Fig. 15 shows the variations of the average vertical displacements Z_{av} of rows R1~R11 at the locations of mid and one-third spans with wind directions for wind velocity of 18.5 m/s. It can be found from Fig. 15 that the average vertical displacements of all rows decrease with the increase of the wind direction from 0° to 90°, and increase with the increase of the wind direction from 90° to 180°. The smallest average vertical displacement occrus at the wind direction of 90°. The average vertical displacements of row R1 are significantly larger than other rows under the wind directions from 0° to 90°, while the average vertical displacements of R11 are significantly larger than other rows under the wind directions from 90° to 180°. It seems that the average vertical

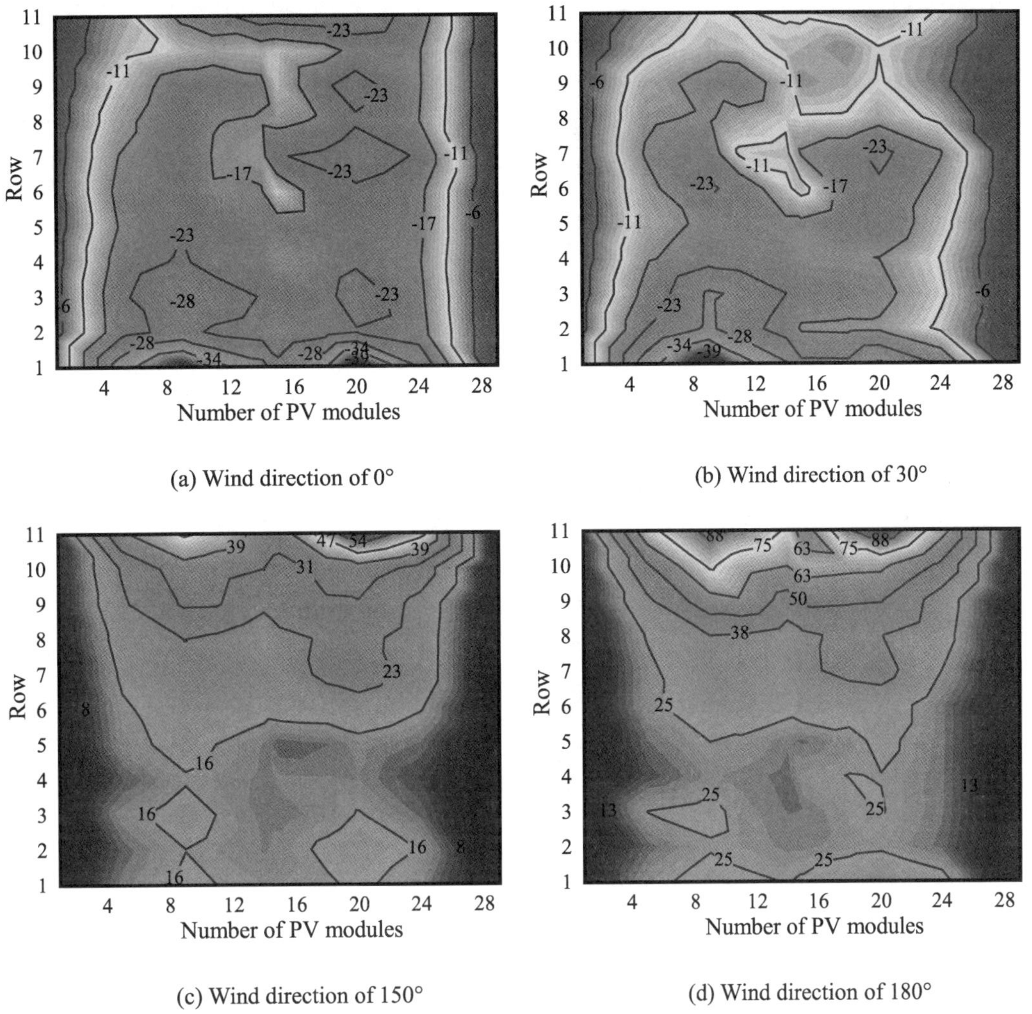

Fig. 11. Contour maps of the average vertical displacement at the PV module center under wind velocity of 18.5 m/s for different wind directions (unit: mm).

displacements at the one-third span are obviously higher than the values at the mid-span. The maximum average vertical displacement (about 99 mm) occurs at row R11 under the wind direction of 180°, which can also be found from Fig. 14.

Fig. 16 shows the variations of the average torsional displacements Z_{at} of rows R1~R11 at the locations of mid and one-third spans with wind directions for wind velocity of 18.5 m/s. It can be found from Fig. 16 that only the first row at the windward side has large values of the average torsional displacement, i.e., R1 for wind directions from 0° to 90° and R11 for wind directions from 90° to 180°. Different from the average vertical displacement, it appears that the average torsional displacements at the one-third span are obviously smaller than the values at the mid-span, and the maximum average torsion displacement of R1 under wind direction of 0° are almost the same as that of R11 under wind direction of 180°.

3.2. Fluctuating responses of the PV modules

Fig. 17 and Fig. 18 respectively present the contour maps of the fluctuating vertical displacement at the PV module center σ_v and the fluctuating torsional displacement σ_t of the mid-span of the test model without stability cables under the wind directions of 0°, 30°, 150°, 180°. The test wind velocity is 18.5 m/s. It can be found from Figs. 17 and 18 that large values of σ_v and σ_t are found at the first three rows on the windward side of the test model. However, the largest values of σ_v and σ_t occur at the second or the third rows, not the first row, which is different from the results of the average displacements as Figs. 11 and 12. Like the average results, the largest fluctuating vertical displacement (σ_v) takes place at the locations of one-third span on the windward rows of the test model. But for the fluctuating torsional displacement (σ_t), the largest value occurs at the location of mid-span.

Figss. 19 and 20 show the variations of the fluctuating vertical displacement σ_v and the fluctuating torsional displacement σ_t at the location of one-third span under different wind directions with wind velocity. It can be found from Figss. 19 and 20 that the fluctuating vertical and torsional displacements increase with the increase of the wind velocity, which is the same as the average results. It seems that the fluctuating values of the first three rows on the windward side are almost the same, and are higher than those of other rows.

Fig. 21 and Fig. 22 respectively show the variations of the fluctuating vertical and torsional displacements of rows R1~R11 at the locations of mid and one-third spans with wind directions for wind velocity of 18.5 m/s. It can be found from Figs. 21 and 22 that the fluctuating displacements decrease with the increase of wind direction from 0° to 90°,

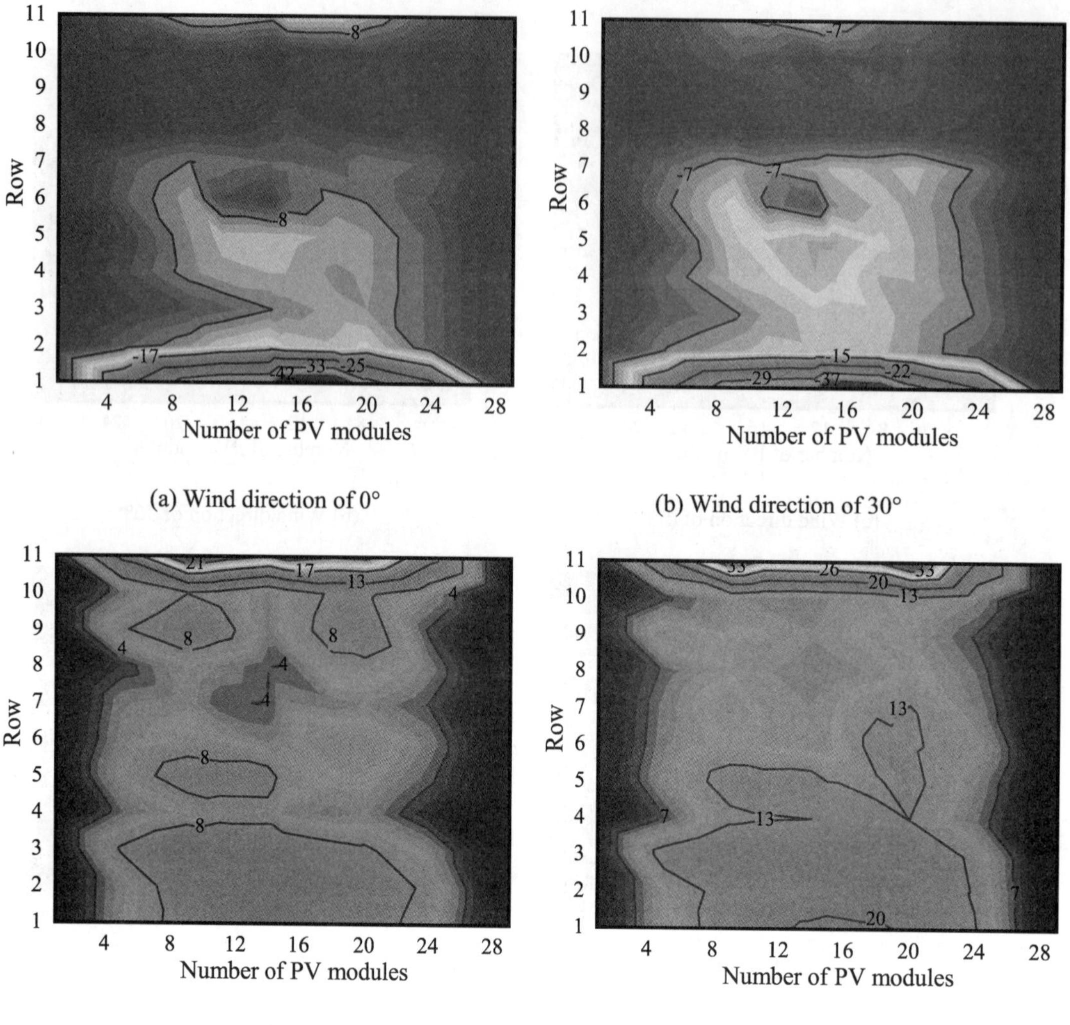

(a) Wind direction of 0°　(b) Wind direction of 30°

(c) Wind direction of 150°　(d) Wind direction of 180°

Fig. 12. Contour maps of the average torsional displacement under wind velocity of 18.5 m/s for different wind directions (unit: rad × 10^{-3}).

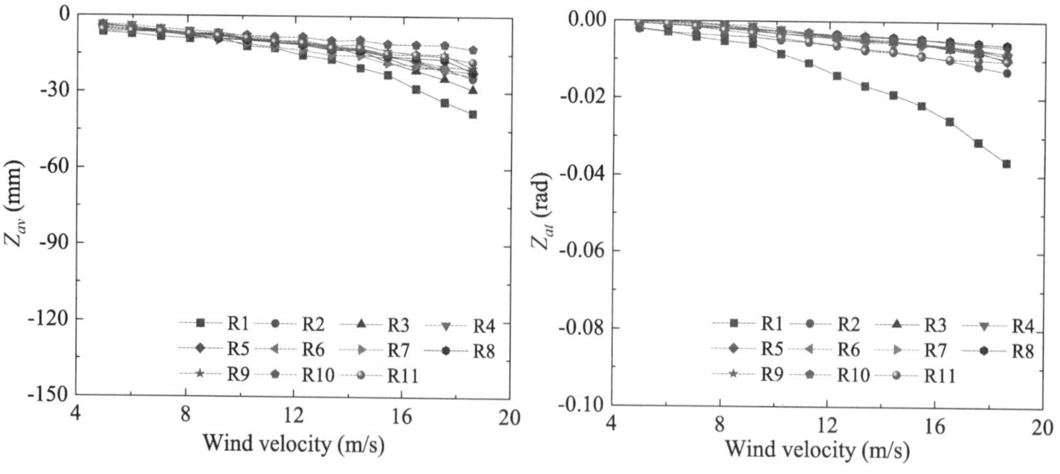

(a) Average vertical displacement　(b) Average torsional displacement

Fig. 13. Variations of average vertical and torsional displacements with wind velocity for wind direction of 0°.

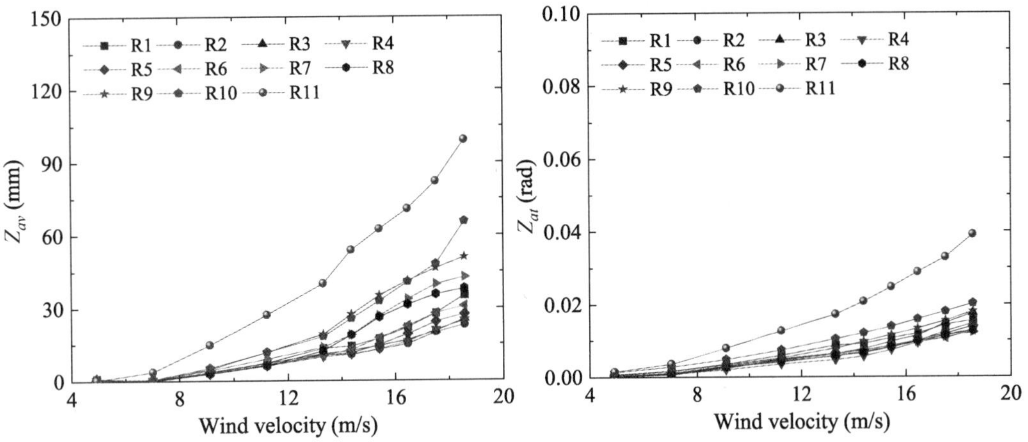

(a) Average vertical displacement (b) Average torsional displacement

Fig. 14. Variations of average vertical and torsional displacements with wind velocity for wind direction of 180°.

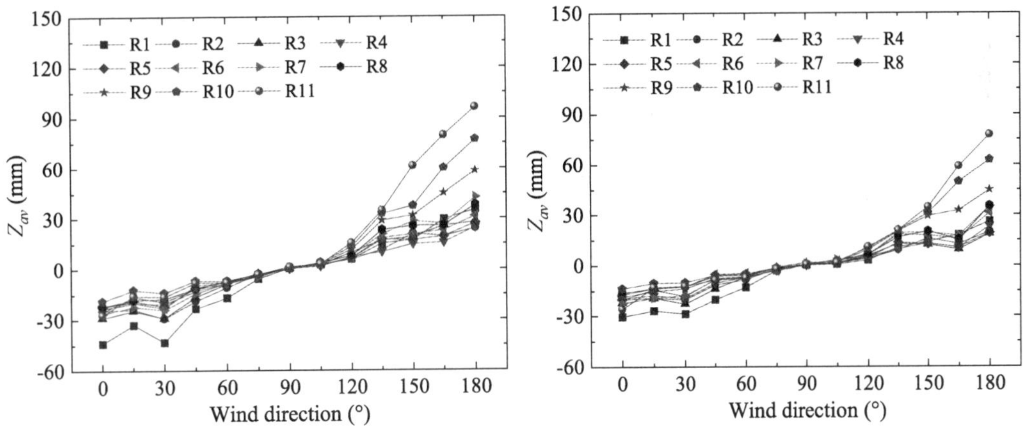

(a) One-third span (b) Mid-span

Fig. 15. Variations of the average vertical displacements with wind directions for wind velocity of 18.5 m/s.

and increase with the increase of wind direction from 90° to 180°. The smallest fluctuating displacement takes place at the wind direction of 90°. The distribution of the fluctuating displacement along wind direction is almost the same that of the average displacement, as indicated in Figs. 15 and 16.

3.3. *Extreme responses of the PV modules*

According to the previous discussions, largest wind-induced responses of the PV modules occur under the wind directions of 0° and 180°. Fig. 23 and Fig. 24 respectively present contour maps of the extreme vertical displacement at the PV module center Z_{ev} and the extreme torsional displacement Z_{et} under wind directions of 0° and 180° for wind velocity of 18.5 m/s. It can be found from Figs. 23 and 24 that large values of the extreme vertical and torsional displacements arise at the first three rows on the windward side of the test model for each wind direction.

Fig. 25 and Fig. 26 respectively show the variations of the extreme vertical and torsional displacements of rows R1~R11 at the locations of mid and one-third spans with wind directions for wind velocity of 18.5 m/s. It can be found from Figs. 25 and 26 that the highest Z_{ev} and Z_{et} appears under the wind direction of 180°, and the values are respectively 191 mm and 0.127 rad.

The extreme vertical displacement at the edge of the PV module, Z_e, can be calculated by Equation (8). Fig. 27 presents the contour maps of the extreme vertical displacement at the edge of the PV module Z_e for wind directions of 0° and 180° under wind velocity of 18.5 m/s. Fig. 28 shows the variations of the extreme vertical displacement at the edge of the PV module Z_e at the locations of mid and one-third spans with wind direction under wind velocity of 18.5 m/s.

It can be found from Figs. 27 and 28 that the largest value of the extreme vertical displacement at the edge of the PV module under wind velocity of 18.5 m/s occurs at row R11 under the wind direction of 180°, and is 333 mm. This is close to 1/100 of the span of the flexible PV support structure. Until now, there are no particular regulations on the deformation of the flexible PV supports. In the design of these structures, the extreme vertical displacement less than 1/100 of the span was often used. In this study, the threshold of the vertical deformation is determined as 1/100 of the span. Therefore, the critical wind velocity of this flexible PV modules support structure is 18.5 m/s. In Chinese code for design of photovoltaic modules support structures, 25-year return

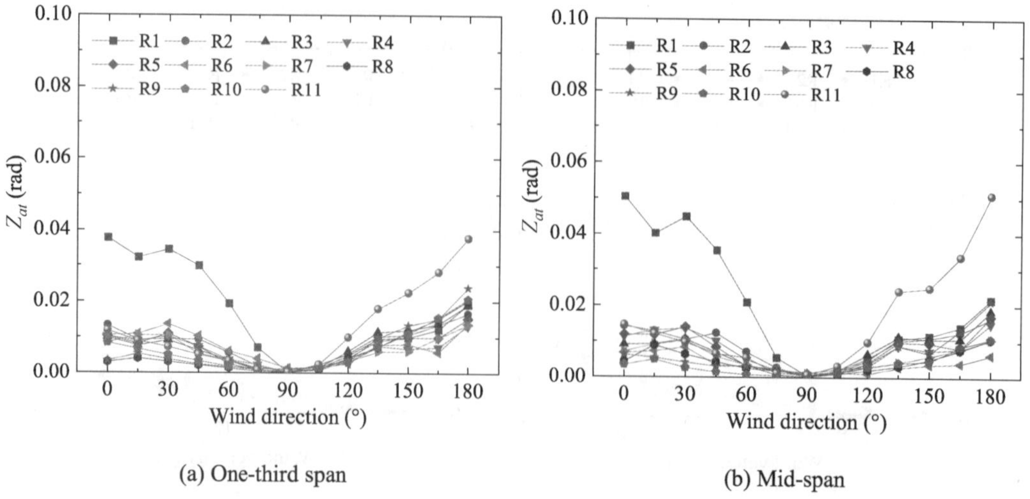

Fig. 16. Variations of the average torsional displacements with wind directions for wind velocity of 18.5 m/s.

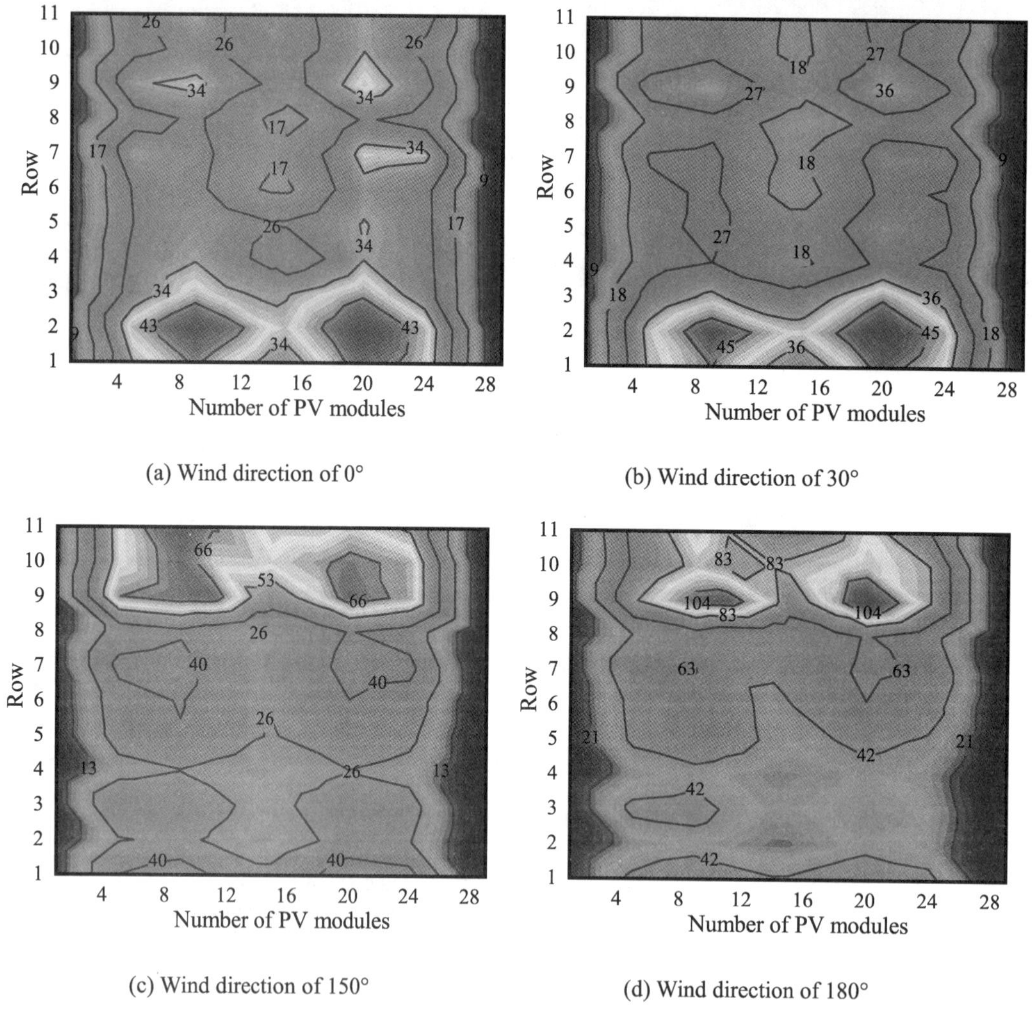

Fig. 17. Contour maps of the fluctuating vertical displacement at the PV module center under wind velocity of 18.5 m/s for different wind directions (unit: mm).

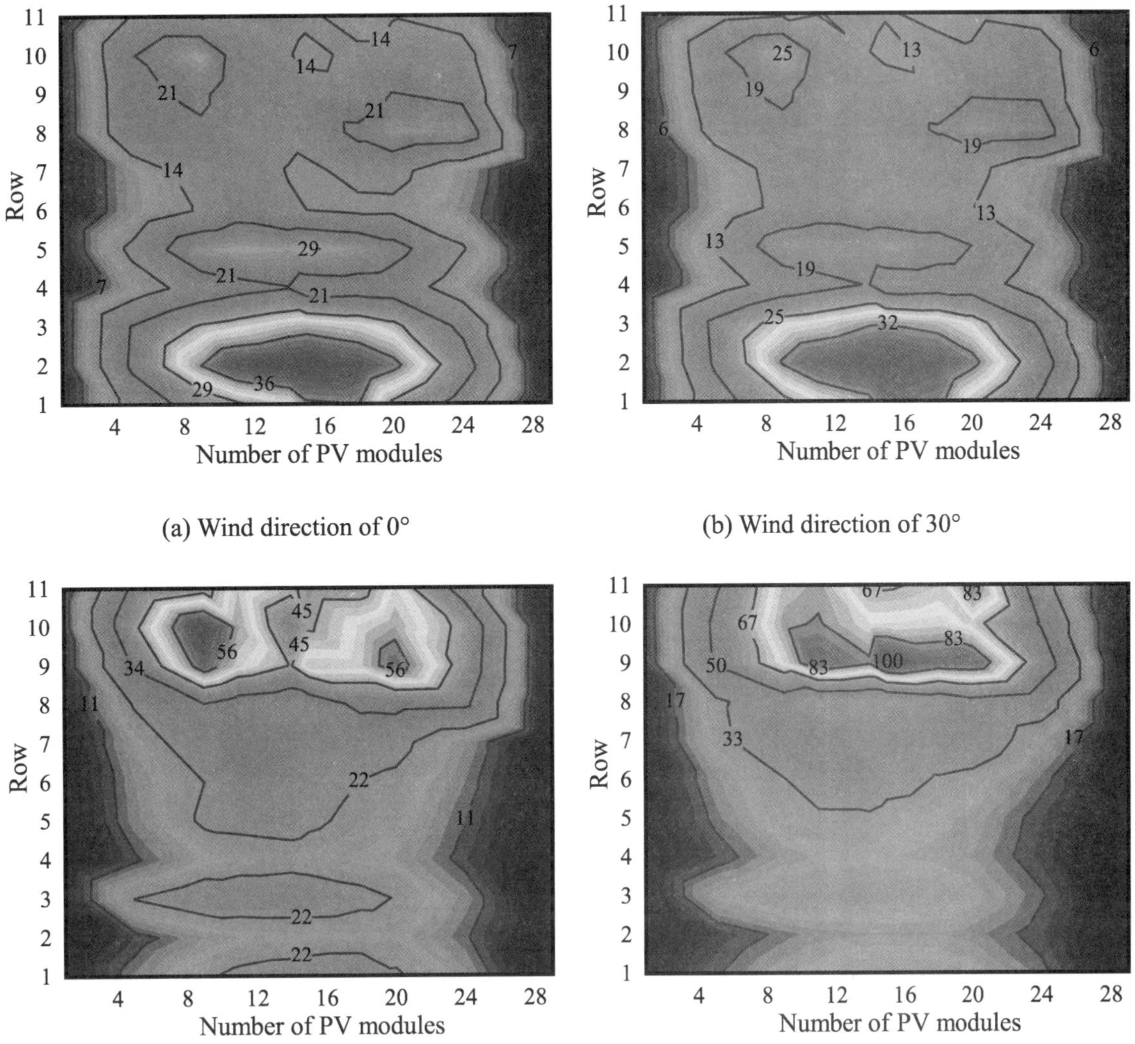

(a) Wind direction of 0°　　(b) Wind direction of 30°

(c) Wind direction of 150°　　(d) Wind direction of 180°

Fig. 18. Contour maps of the fluctuating torsional displacement under wind velocity of 18.5 m/s for different wind directions (unit: rad × 10^{-3}).

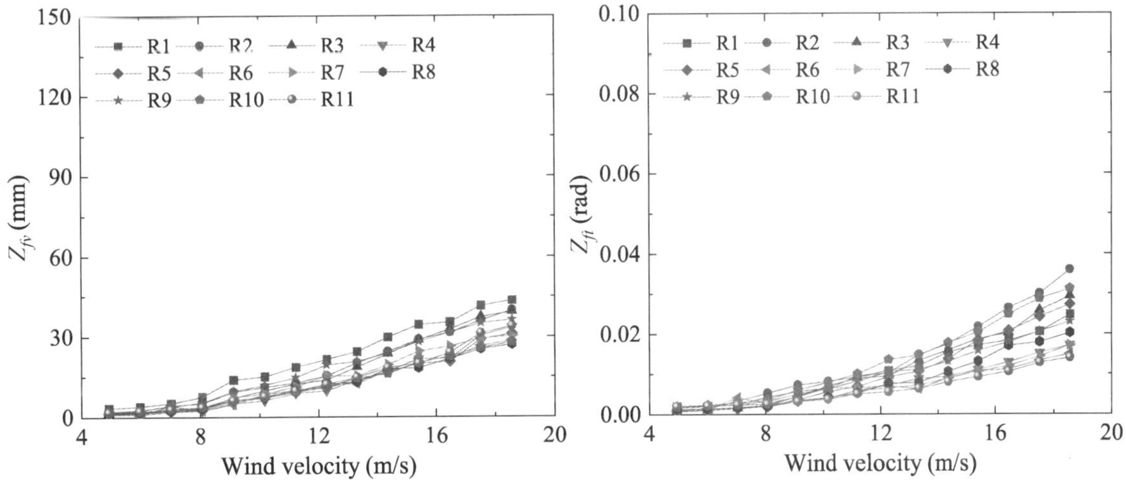

(a) Fluctuating vertical displacement　　(b) Fluctuating torsional displacement

Fig. 19. Variations of fluctuating vertical and torsional displacements with wind velocity for wind direction of 0°.

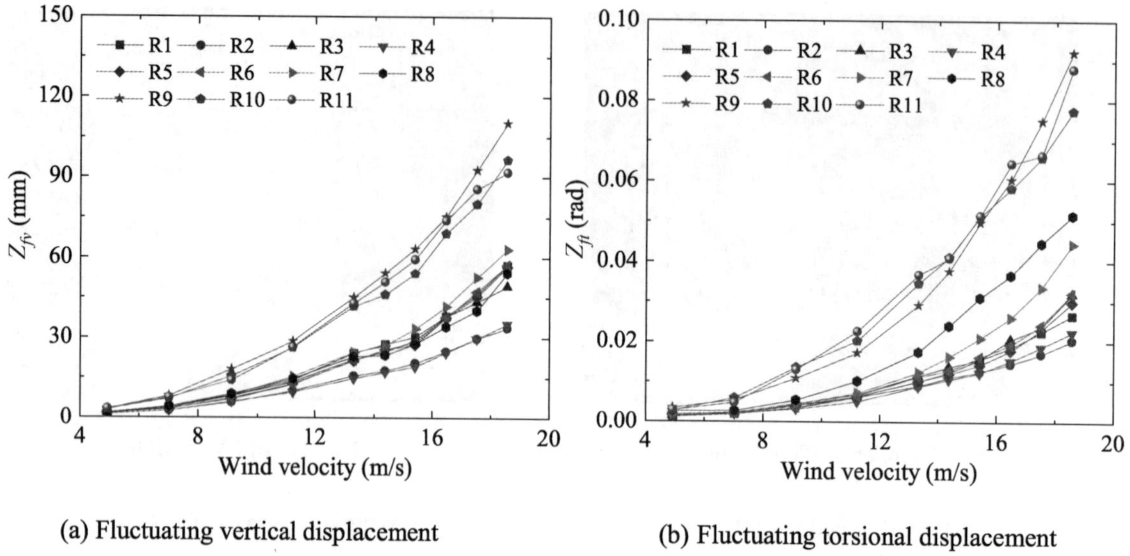

Fig. 20. Variations of fluctuating vertical and torsional displacements with wind velocity for wind direction of 180°.

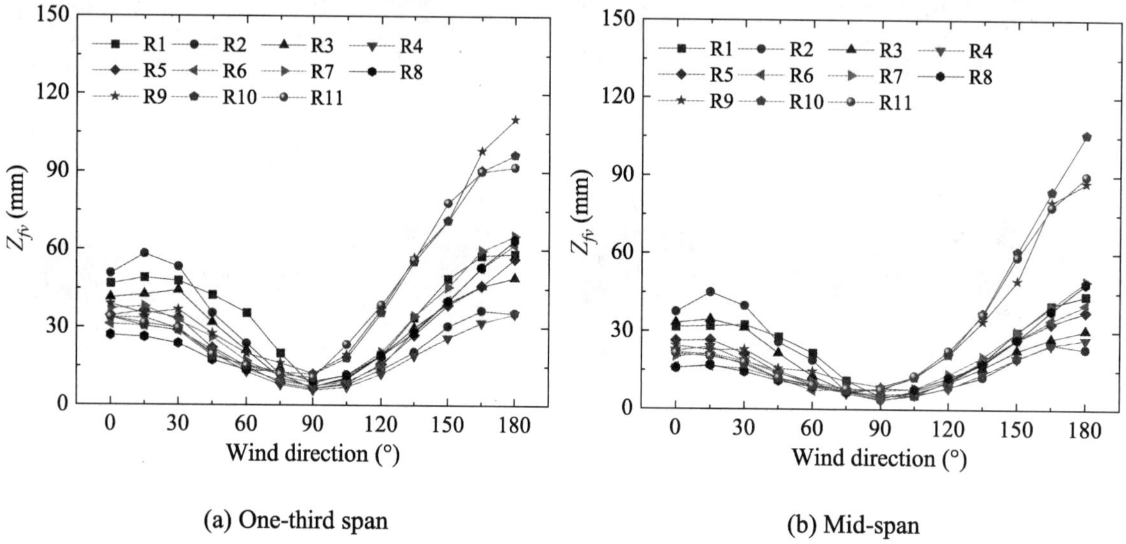

Fig. 21. Variations of the fluctuating vertical displacements with wind directions for wind velocity of 18.5 m/s.

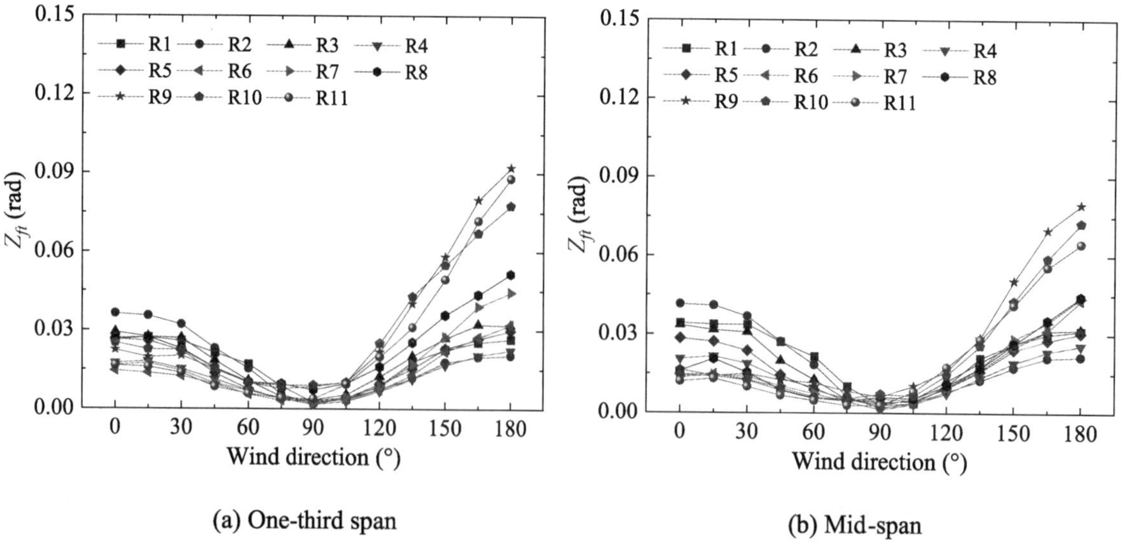

Fig. 22. Variations of the fluctuating torsional displacements with wind directions for wind velocity of 18.5 m/s.

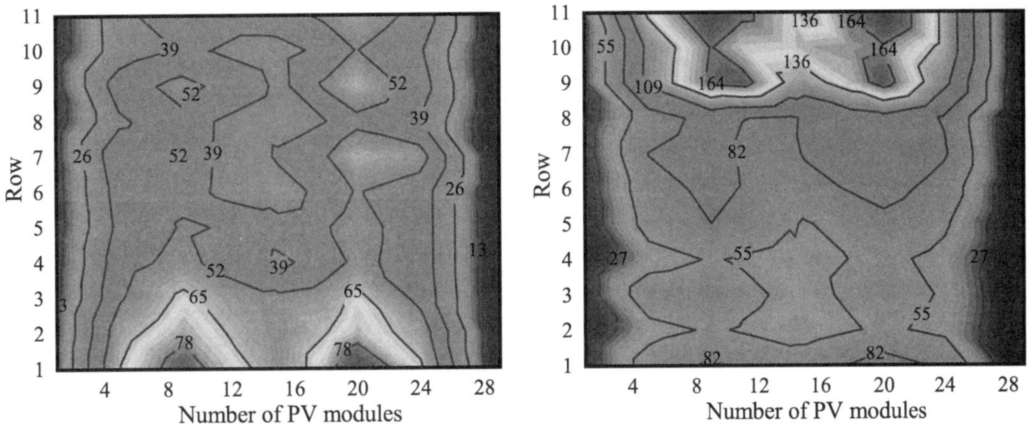

(a) Wind direction of 0°

(b) Wind direction of 180°

Fig. 23. Contour maps of the extreme vertical displacement at the PV module center under wind velocity of 18.5 m/s (unit: mm).

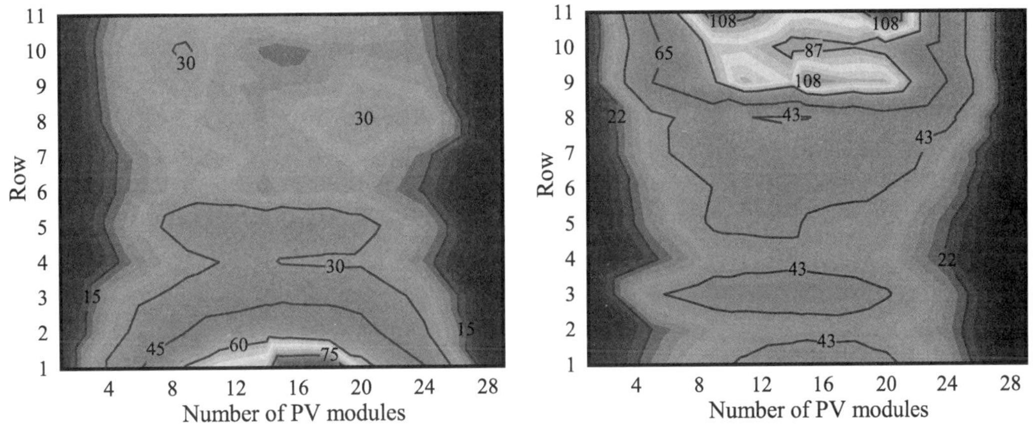

(a) Wind direction of 0°

(b) Wind direction fo 180°

Fig. 24. Contour maps of the extreme torsional displacement under wind velocity of 18.5 m/s (unit: rad × 10^{-3}).

period is used. The wind pressure corresponding to wind velocity of 18.5 m/s is only about 0.215 kN/m^2, which is significantly lower than basic wind pressure over the 25-year return period for most of China. Therefore, the flexible PV modules support structure introduced in Section 2.1 can not be used before some effective countermeasures are imposed.

4. Effectiveness of the three stability cables

Fig. 29 shows the maximum extreme displacement at the edge of the PV modules (Z_e) for each row of test model under the four test cases listed in Table 3. In which, the wind velocity is 18.5 m/s and the wind directions is 0° and 180°. Table 4 and Table 5 further summary the ratio (R_{ze}) between the maximum Z_e with stability cables (test cases 2, 3, and 4) and that without stability cable (test case 1) under wind direction of 0° and 180°, respectively, together with the average value of R_{ze} for each type of stability cable. It can be found from Fig. 29, Table 4 and Table 5 that the three types of stability cables all have significant effects on reducing the wind-induced vibration of the PV modules. It appears that the average values of R_{ze} gradually decreases with the stability cables from T1 to T3. For instance, the average R_{ze} is respectively 58%, 52% and 36% for stability cables T1, T2 and T3 under the wind direction of 0°, and is respectively 57%, 37% and 28% for stability cables T1, T2 and T3 under the wind direction of 180°. This indicates that stability cable T3 have the most effective effects on reducing the wind-induced vibration of the flexible PV modules support structure among the three stability cables used in the study.

Fig. 30 shows the variation of the extreme displacement Z_e at the position of one-third span with the wind velocity for the four test cases under wind direction of 180°. The defomation threshhold of 1/100 of span is used, and is plotted as horizontal lines in Fig. 30. It can be concluded from Fig. 30 that the critical wind velocities corresponding to the flexible PV modules support structures with stability cables T1, T2 and T3 are respectively 24.6 m/s, 29.3 m/s and 36.7 m/s, which are significantly higher than the value without stability cable (18.5 m/s). The basic wind pressures corresponding to the above three critical wind velocities are respectively 0.35 kN/m^2, 0.50 kN/m^2 and 0.80 kN/m^2. This indicates that the flexible PV modules support structure installed with stability cable T3 can be safely adopted in most parts of China (25-year return period).

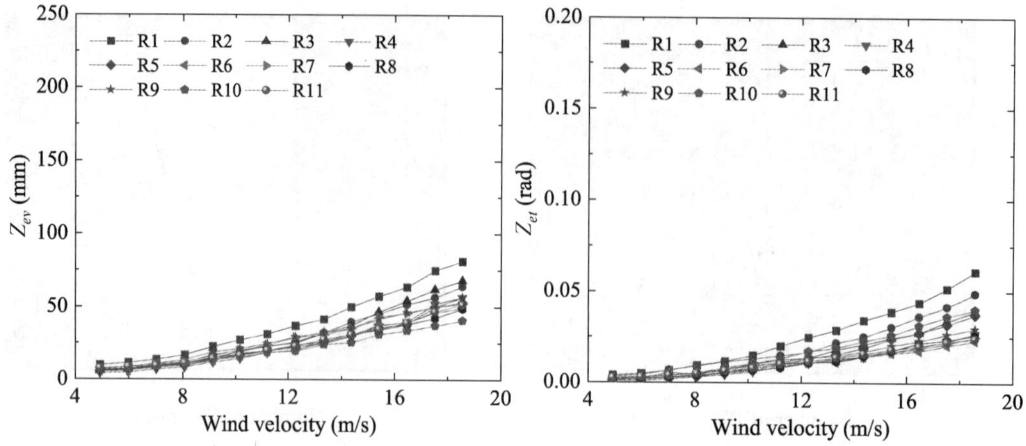

(a) Extreme vertical displacement (b) Extreme torsional displacement

Fig. 25. Variations of extreme vertical and torsional displacements with wind velocity for wind direction of 0°.

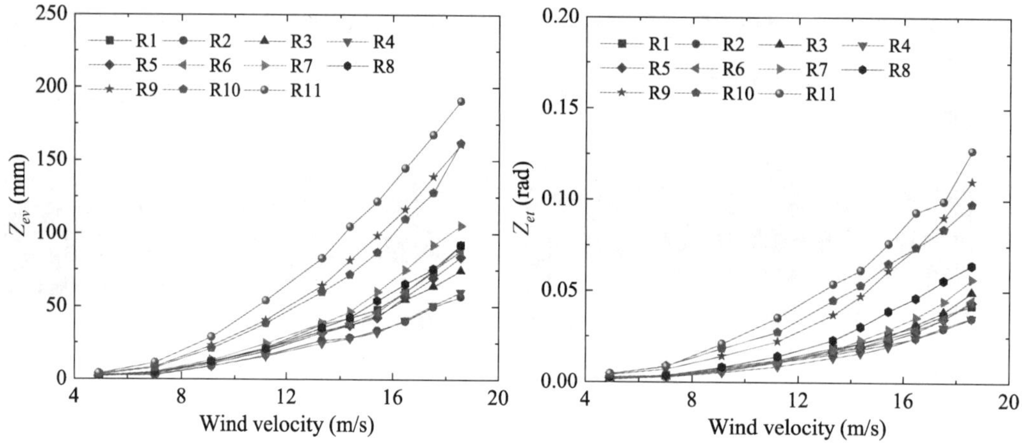

(a) Extreme vertical displacement (b) Extreme torsional displacement

Fig. 26. Variations of extreme vertical and torsional displacements with wind velocity for wind direction of 180°.

5. Wind load factor

Wind load factor is an important parameter for the structural design of the flexible PV support structure. The time history of the maximum displacement in each row of the test model is used to obtain the wind load factors, in which the peak factor is 2.5. Fig. 31 shows the variation of the wind load factor for rows R1 and R11 with wind velocity under wind direction of 180° for the four test cases. When the wind velocity is very low, the wind load factors are rather large due to the small average displacement. It can be found the wind load factors at the windward side (row R11) are a little bit lower than those at the leeward side (R1). For example, the wind load factor of row R11 for test case 4 under wind vleocity of 36.7 m/s is 1.3, whereas the value of row R1 is 1.63.

Table 6 and Table 7 respectively show the wind load factors of each row at wind directions of 0° and 180° under the maximum wind velocities obtained from Fig. 30. It can be found Table 6 and Table 7 that the wind load factors of test case 4 are obviously lower than those of test cases 2–3, which mean that the design wind load for the PV modules support structure installed with stability cable T3 is the smallest. The wind load factor of the outermost row at the windward side under wind directions of 0° and 180° are respectively 1.56 and 1.30 for test case 4, 1.99 and 1.55 for test case 3, 1.82 and 1.61 for test case 2, 1.76 and 1.76 for test case 1.

6. Concluding remarks

Wind-induced response of a type of flexible PV modules support structure with a span of 33 m was investigated by using wind tunnel tests based on elastic test model, and the effectiveness of three types of stability cables on reducing wind-induced vibration of the flexible PV modules support structures were carefully examined. First, an elastic test model of the flexible PV modules support structure including 3 spans and 11 rows was designed and manufactured. The frequency and damping ratio of the test model were identified by free vibration method. The results show that the vertical and torsional frequencies of the test model agree well with the target values, and the damping ratios are about 1.9%, which is close to the expected value. Second, a series of wind tunnel tests based on the elastic test model were carried out to obtain the wind-induced responses of the 33-m-span PV modules support structure without any mitigation countermeasures. The

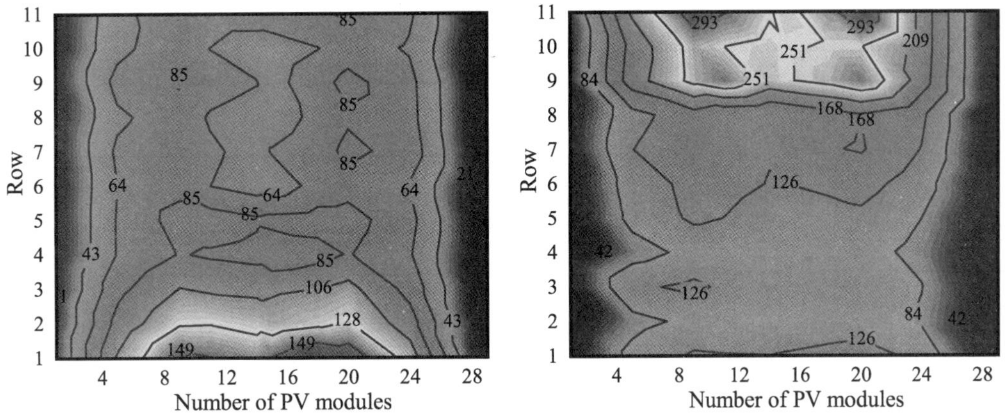

(a) Wind direction of 0°

(b) Wind direction of 180°

Fig. 27. Contour maps of the extreme vertical displacement at the edge of the PV module under wind velocity of 18.5 m/s (unit: mm).

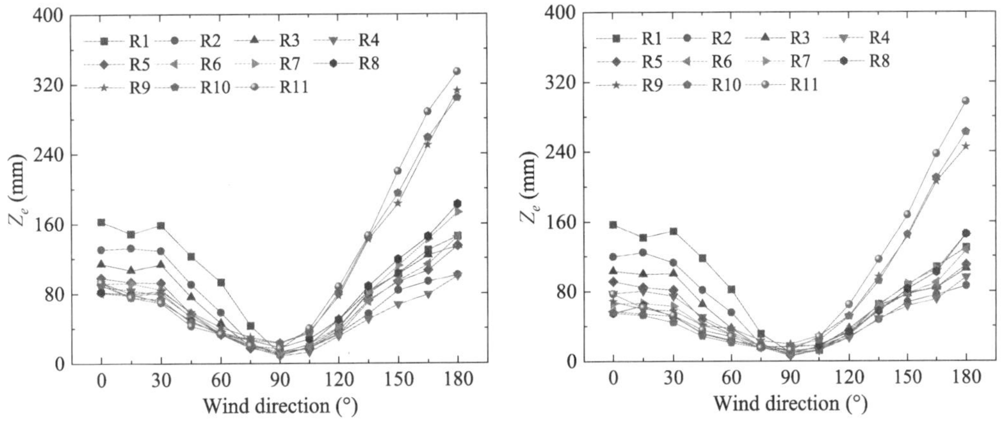

(a) One-third span

(b) Mid-span

Fig. 28. Variations of the extreme vertical displacement at the edge of the PV module with wind direction under wind velocity of 18.5 m/s.

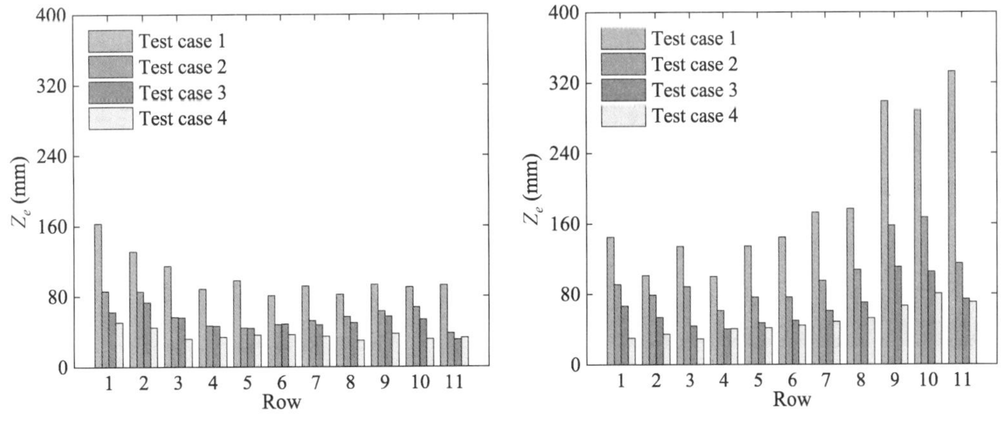

(a) Wind direction of 0°

(b) Wind direction of 180°

Fig. 29. The maximum extreme displacement at the edge of the PV modules for each row of the test model under the four test cases (wind velocity of 18.5 m/s, wind directions of 0° and 180°).

Table 4
Ratio (R_{ze}) between the maximum Z_e with stability cables and that without stability cable under wind velocity of 18.5 m/s and wind direction of 0°.

Rows	R1	R2	R3	R4	R5	R6	R7	R8	R9	R10	R11	Average
R_{ze} of T1 (%)	53	65	50	53	45	59	57	69	67	75	41	58
R_{ze} of T2 (%)	38	56	49	53	45	60	52	61	61	60	33	52
R_{ze} of T3 (%)	31	34	28	38	37	45	38	37	40	35	36	36

Table 5
Ratio (R_{ze}) between the maximum Z_e with stability cables and that without stability cable under wind velocity of 18.5 m/s and wind direction of 180°.

Rows	R1	R2	R3	R4	R5	R6	R7	R8	R9	R10	R11	Average
R_{ze} of T1 (%)	63	78	66	61	57	53	55	59	51	55	34	57
R_{ze} of T2 (%)	46	53	33	40	35	34	35	39	35	35	22	37
R_{ze} of T3 (%)	21	34	22	40	31	31	28	29	21	27	21	28

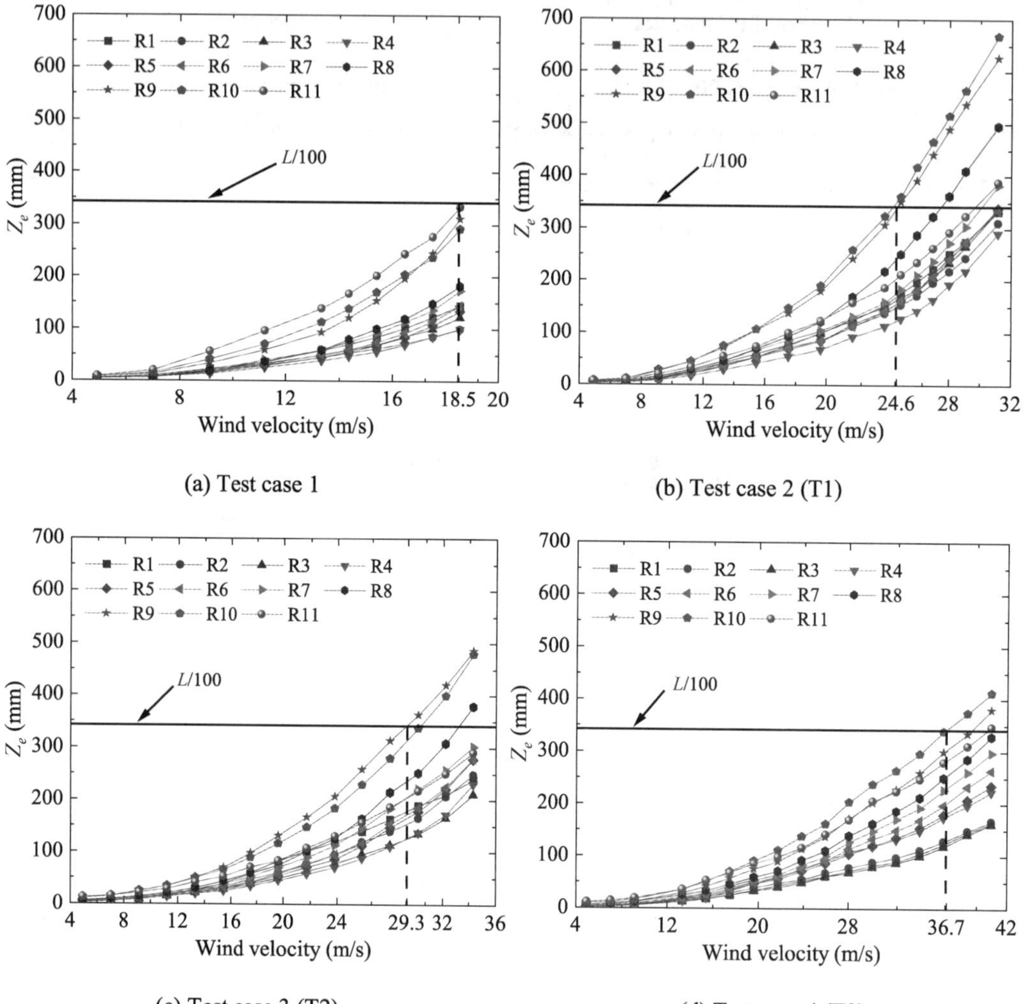

(a) Test case 1
(b) Test case 2 (T1)
(c) Test case 3 (T2)
(d) Test case 4 (T3)

Fig. 30. Variation of the extreme displacement at the position of one-third span with the wind velocity for the four test cases under wind direction of 180°.

displacement time histories for 176 marked points on 88 PV modules were simutaneously sampled by using a non-contact video displacement measuring system. The results show that 180° and 0° are the two most unfavorable wind directions, and the critical wind velocity is about 18.5 m/s if the deformation threshhold is set as 1/100 of the span. This indicates that this flexible PV modules support structure introduced in this study can not be used before some effective countermeasures are imposed. Third, three types of stability cables (T1, T2 and T3) were installed on the test model, and a series of wind tunnel tests were carried out to examine the effectiveness of this types of mitigation countermeasures. The results show that the three types of stability cables all have significant effects on reducing the wind-induced vibration of the

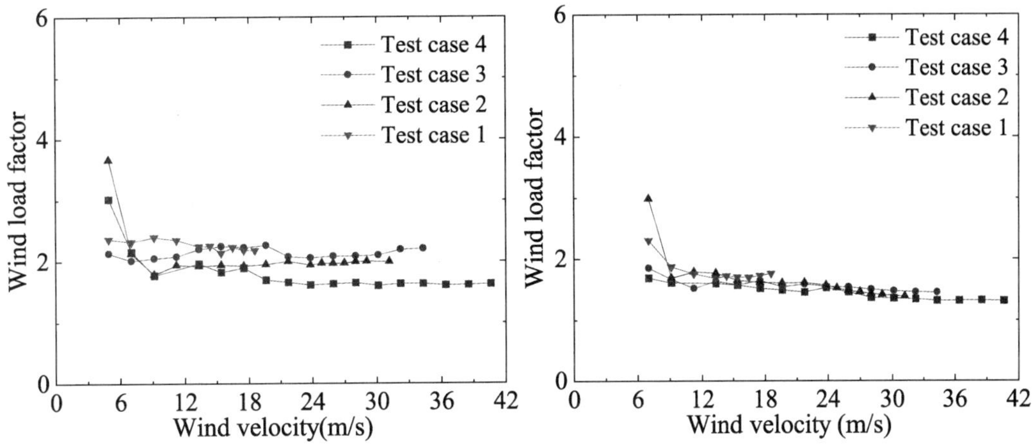

(a) R1 (b) R11

Fig. 31. Variation of the wind load factor for rows R1 and R11 with wind velocity under wind direction of 180° for the four test cases.

Table 6
Wind load factor of each row under wind direction of 0°.

	R1	R2	R3	R4	R5	R6	R7	R8	R9	R10	R11
Test case 1	1.76	2.48	2.03	2.03	2.11	2.04	2.01	1.88	2.10	2.30	1.89
Test case 2	1.82	1.55	1.72	1.65	1.64	1.66	1.66	1.59	2.11	2.02	1.85
Test case 3	1.99	1.81	1.79	1.67	1.71	1.65	1.69	1.78	1.99	2.15	1.90
Test case 4	1.56	1.61	1.52	1.44	1.58	1.5	1.48	1.6	1.56	1.8	1.66

Table 7
Wind load factor of each row under wind direction of 180°.

	R1	R2	R3	R4	R5	R6	R7	R8	R9	R10	R11
Test case 1	2.19	2.1	2.11	2.01	2.47	2.41	2.1	2.2	2.52	2.27	1.76
Test case 2	1.95	1.86	1.83	1.93	1.95	2.08	1.76	1.78	1.92	1.89	1.61
Test case 3	2.26	1.81	1.89	1.69	1.58	1.75	1.61	1.6	1.72	1.59	1.55
Test case 4	1.63	1.41	1.46	1.49	1.37	1.32	1.33	1.32	1.33	1.35	1.3

PV modules. The critical wind velocities for stability cables T1, T2 and T3 (1/100 of the span) are respectively 24.6, 29.3 and 36.7 m/s, which are significantly higher than the value without stability cable (18.5 m/s). The basic wind pressures corresponding to the above three critical wind velocities are respectively 0.35, 0.50 and 0.80 kN/m^2. This indicates that the flexible PV modules support structure installed with stability cable T3 can be safely adopted in most parts of China (25-year return period). Finally, the wind load factor was discussed by using the displacement time histories measured from the wind tunnel tests. The results show that the wind load factor of the outermost row at the windward side under wind directions of 0° and 180° are respectively 1.56 and 1.30 for test case 4, 1.99 and 1.55 for test case 3, 1.82 and 1.61 for test case 2, 1.76 and 1.76 for test case 1.

CRediT authorship contribution statement

Jiaqi Liu: Conceptualization, Methodology, Formal analysis, Data curation, Investigation, Writing – original draft. **Shouying Li:** Conceptualization, Methodology, Resources, Writing – review & editing, Funding acquisition. **Jing Luo:** Validation, Investigation, Resources. **Zhengqing Chen:** Writing – review & editing, Supervision.

Declaration of competing interest

The authors declare that they have no known competing financial interests or personal relationships that could have appeared to influence the work reported in this paper.

Data availability

Data will be made available on request.

Acknowledgements

This project is jointly supported by the National Key Research and Development Program of China (Grant No.2017YFC0703600 and No.2017YFC0703604) and the National Natural Science Foundation of China (Grant No.51578234).

References

Abiola-Ogedengbe, A., Hangan, H., Siddiqui, K., 2015. Experimental investigation of wind effects on a standalone photovoltaic (PV) module. Renew. Energy 78, 657–665.
Aly, A., Bitsuamlak, G., 2013. Aerodynamics of ground-mounted solar panels: test model scale effects. J. Wind Eng. Ind. Aerod. 123, 250–260.
Baumgartner, F.P., Büchel, A., Bartholet, R., 2008. Solar wings: a new lightweight PV tracking system. In: Proceedings of the 23rd European Photovoltaic Solar Energy Conference.
Baumgartner, F.P., Büchel, A., Bartholet, R., 2009. Experiences with cable-based solar wings tracking system and progress towards two-axis large scale solar systems. In: Proceedings of the 24th European Photovoltaic Solar Energy Conference, Hamburg, September, pp. 21–24.
Baumgartner, F.P., Büchel, A., Bartholet, R., 2010. Cable-based solar wings tracking system: two-axis system and progress of one-axis system. In: Proceedings of the 25th European Photovoltaic Solar Energy Conference, Valencia, September 6-9.

Bitsuamlak, G.T., Dagnew, A.K., Erwin, J., 2010. Evaluation of wind loads on solar panel modules using CFD. In: Proceedings of the Fifth International Symposium on Computational Wind Engineering. Chapel Hill, North Carolina, USA. May 23-27.
Deng, Y.C., Li, S.Y., Chen, Z.Q., 2021. Experimental investigation on wake-induced vibrations of the hangers of suspension bridges based on three-dimensional elastic test model. Eng. Struct. 234, 111985.
Du, H., Xu, H.W., Zhang, Y.L., Lou, W.J., 2022. Wind pressure characteristics and wind vibration response of long-span flexible photovoltaic support structure. J. Harbin Inst. Technol. 54 (10), 67–74.
Geurts, C.P.W., Steenbergen, R.D.J.M., 2009. Full scale measurements of wind loads on stand-off photovoltaic systems. In: Proceedings of the Fifth European and African Conference on Wind Engineering Florence, Italy.
He, X.H., Ding, H., Jing, H.Q., Zhang, F., Wu, X.P., Weng, X.J., 2020. Wind-induced vibration and its suppression of photovoltaic modules supported by suspension cables. J. Wind Eng. Ind. Aerod. 206.
He, X.H., Ding, H., Jing, H.Q., et al., 2021. Mechanical characteristics of a new type of cable-supported photovoltaic module system[J]. Sol. Energy 226, 408–420.
Kim, Y.C., Tamura, Y., Yoshida, A., Ito, T., Shan, W., Yang, Q.S., 2018. Experimental investigation of aerodynamic vibrations of solar wing system. Adv. Struct. Eng. 21 (15), 2217–2226.
Jubayer, C.M., Hangan, H., 2016. A numerical approach to the investigation of wind loading on an array of ground mounted solar photovoltaic (PV) panels. J. Wind Eng. Ind. Aerod. 153, 60–70.
Kim, Y.C., Shan, W., Yang, Q.S., et al., 2020. Effect of panel shapes on wind-induced vibrations of solar wing system under various wind environments[J]. J. Struct. Eng. 146 (6), 04020104.
Kopp, G.A., Banks, D., 2013. Use of the wind tunnel test method for obtaining design wind loads on roof-mounted solar arrays. J. Struct. Eng. 139, 284–287.
Kopp, G.A., Surry, D., Chen, K., 2002. Wind loads on a solar array. Wind & Structures An International Journal 5 (5), 393–405.
Kopp, G.A., Farquhar, S., Morrison, M.J., 2012. Aerodynamic mechanisms for wind loads on tilted, roof-mounted, solar arrays. J. Wind Eng. Ind. Aerod. 111, 40–52.
Li, S.Y., Chen, Z.Q., Li, S.K., 2014. Theoretical investigation on rain-wind induced vibration of a continuous stay cable with given rivulet motion[J]. Wind & Structures An International Journal 19 (5), 481–503.
Li, S.Y., Deng, Y.C., Huang, J., et al., 2019a. Experimental investigation on aerodynamic interference of two kinds of suspension bridge hangers[J]. J. Fluid Struct. 90, 57–70.
Li, S.Y., Deng, Y.C., Lei, X., et al., 2019b. Wake-induced vibration of the hanger of a suspension bridge: field measurement and theoretical modeling. Struct. Eng. Mech. 72 (2), 169–180.
López, A., Parnás, V.E., Cataldo, J., 2019. Wind tunnel experiments on ground-mounted photovoltaic solar panels[J]. Revista Ingenieria de Construction 34 (1), 15–24.
Ma, W.Y., Chai, X.B., Ma, C.C., 2021. Experimental study on wind load influencing factors of flexible support photovoltaic modules. Acta Energiae Solaris Sin. 42 (11), 9.
Meroney, R.N., Neff, D.E., 2010. Wind effects on roof mounted solar photovoltaic arrays: CFD and wind-tunnel evaluation. In: The Fifth International Symposium on Computational Wind Engineering (CWE2010). Chapel Hill, North Carolina, USA. May 23-27.
Naeiji, A., Raji, F., Zisis, I., 2017. Wind loads on residential scale rooftop photovoltaic panels. J. Wind Eng. Ind. Aerod. 168, 228–246.
Radu, A., Axinte, E., 1989. Wind forces on structures supporting solar collectors. J. Wind Eng. Ind. Aerod. 32, 93–100.
Radu, A., Axinte, E., Theohari, C., 1986. Steady wind pressures on solar collectors on flat-roofed buildings. J. Wind Eng. Ind. Aerod. 23, 249–258.
Shademan, M., Barron, R.M., Balachandar, R., Hangan, H., 2014. Numerical simulation of wind loading on ground-mounted solar panels at different flow configurations. Can. J. Civ. Eng. 41 (8), 728–738.
Tamura, Y., Kim, Y.C., Yoshida, A., 2015. Wind-induced vibration experiment on solar wing. In: 6th International Conference on Experimental Vibration Analysis for Civil Engineering Structures, Zurich, October 19-21.
Velicu, R., Moldovean, G., Scaletchi, I., Butuc, B.R., 2010. Wind loads on an azimuthal photovoltaic platform. Experimental study. In: Proceeding of International Conference on Renewable Energies and Power Quality, Granada, Spain, March 23-25.
Warsido, W.P., Bitsuamlak, G.T., Barata, J., 2014. Influence of spacing parameters on the wind loading of solar array. J. Fluid Struct. 48, 295–315.

第 44 卷 第 11 期 2023 年 11 月
建筑结构学报 Journal of Building Structures Vol. 44 No. 11 Nov. 2023
文章编号:1000-6869(2023)11-0153-09 DOI:10. 14006/j. jzjgxb. 2022. 0891

基于气弹模型风洞试验的柔性光伏支架气动稳定性及干扰效应研究

陈 权[1],牛华伟[1],李红星[2],姜 东[2]
(1. 湖南大学 土木工程学院,湖南长沙 410082;
2. 中国电力工程顾问集团西北电力设计院有限公司,陕西西安 710075)

摘要:为完善柔性光伏支架气动稳定性及阵列气弹干扰效应的研究,通过一倾角可调的柔性光伏支架气弹模型进行风洞试验,研究了不同倾角、不同风向角、不同均匀紊流风场下单柔性光伏支架气动稳定性;在此基础上,通过改变风向角、光伏支架数量和光伏支架之间连接(有、无连接),对 10°倾角下光伏支架干扰效应进行研究。结果表明:测试倾角中,10°倾角颤振稳定性最差;测试全风向角中,0°、15°和 30°风向角结构颤振稳定性较差;紊流可以抑制光伏支架的颤振,但会增大光伏支架的抖振响应,在 10°倾角、0°风向角工况下,紊流度从小于 0. 5% 提高到 5. 0%、10. 0%、15. 0% 的过程中,结构颤振临界风速分别提高 2. 92%、6. 10% 和 28. 38%,光伏支架的加速度响应(标准差)分别为均匀流场的 1. 78、4. 56 倍和 6. 12 倍;阵列时,来流方向第 2 行受尾流影响最为剧烈,尾流的影响随着距离的增大逐渐减弱,影响区域大致在 10 ~ 15 m 范围内;根据阵列风振响应情况,将阵列划分为尾流区、过渡区和遮挡区;在阵列柔性光伏支架之间布置连接件会减弱阵列尾流区与遮挡区结构振动响应(结构加速度标准差最多减少了 56. 38%),但对于过渡区结构较为不利(结构加速度标准差增大了 24. 43%)。

关键词:柔性光伏支架;风洞试验;气动稳定性;扭转振动;尾流致振;遮挡效应
中图分类号:TU352. 2 TU375. 4 **文献标志码**:A

Aerodynamic stability and interference effect on a flexible photovoltaic based on wind tunnel test with aeroelastic model

CHEN Quan[1], NIU Huawei[1], LI Hongxing[2], JIANG Dong[2]
(1. College of Civil Engineering, Hunan University, Changsha 410082, China;
2. Northwest Electric Power Design Institute Co., Ltd., of China Power Engineering Consulting Group, Xi'an 710075, China)

Abstract: In order to foster a better understanding of the aerodynamic stability of flexible photovoltaic and the aeroelastic interference effect of flexible photovoltaic array, the aerodynamic stability of single flexible photovoltaic in different inclination angles, wind directions and uniform turbulent wind fields was studied by wind tunnel tests with an aeroelastic model that can adjust angles. Furthermore, by changing the wind direction, the number of flexible photovoltaic and the connector conditions between flexible photovoltaics (with or without connection), the interference effect of flexible photovoltaic under 10° was studied. The results show that the flutter stability of 10° inclination angle is the worst; 0°, 15° and 30° wind direction leads to poorer performance on flutter stability. Turbulence can inhibit flutter but increase the buffeting response of flexible photovoltaic. Under the condition of 10° inclination angle and 0° wind direction, when turbulence increased from less than 0. 5% to 5. 0%, 10. 0% and 15. 0%, the flutter critical wind speed is increased by 2. 92%, 6. 10% and 28. 38%, respectively, and the acceleration buffeting response of flexible photovoltaic is 1. 78, 4. 56 and 6. 12 times of uniform flow field. In the case of array, the second row in the direction of incoming flow is most affected by wake, which gradually weakens with the increase of distance, and the affected area is roughly within the range of 10 to 15 m. According to the wind-vibration response of the array, the array is divided into wake zone, transition zone and shielding zone. The installation of connectors between the flexible photovoltaic array will weaken the structural vibration response in the wake zone and the shield zone of the array (the standard deviation of the structure acceleration is reduced by 56. 38% at most), but it is unfavorable to the structure in the transition zone (the standard deviation of the structure acceleration is increases by 24. 43%).

Keywords: flexible photovoltaic; wind tunnel test; aerodynamic stability; twisting vibration; wake induced vibration; shadowing

基金项目:国家自然科学基金面上项目(52178477)。
作者简介:陈权,出生于 1998 年,男,工学硕士,主要从事结构风工程研究。email:chen1998@ hnu. edu. cn
通信作者:牛华伟,出生于 1978 年,男,工学博士,教授,主要从事结构风工程与振动控制、电涡流阻尼新技术、新能源结构抗风理论与应用研究。email:niuhw@ hnu. edu. cn
收稿日期:2022 年 12 月

0 引言

光伏发电作为一种技术成熟的可再生能源产生途径，因兼具环保性和经济性而得到迅速发展。随着光伏发电规模的不断扩大，安装条件较好的平原、荒漠、山地等土地资源日益减少，一些地形起伏大、跨径较大的峡谷、水面、鱼塘等特殊地形已成为光伏发电领域的必然选择。但是传统形式光伏支架由于跨越能力、安装条件、经济性等条件限制，并不适用于这类场址区域。而由预应力拉索、光伏组件以及两端支撑构成的柔性光伏支架体系成为目前光伏支架体系发展的一个重要方向。由于柔性光伏支架体系对风荷载作用更加敏感，其风振性能与气动稳定性是建设中的关键性问题，因此开展柔性光伏支架的风振特性研究具有十分重要的工程价值。

目前，国内外学者已经对各类光伏支架体系开展了一系列研究工作，并取得了一定成果。诸如：Aly等[1]通过刚性模型测压风洞试验和CFD模拟，研究了不同几何尺度下风荷载对地面安装固定式光伏支架的敏感性，发现平均荷载不受模型尺寸的显著影响，但峰值荷载对来流的几何比例和频率成分都很敏感。Jubayer等[2-3]通过数值模拟，研究了25°倾角下单个固定式光伏支架周围的风荷载及流场，计算了其对应的三分力系数，并研究了5×3阵列、典型倾角与风向角下光伏支架周围的风荷载及流场分布特性。Abiola-ogedengbe等[4]通过固定式光伏支架刚性模型测压风洞试验研究了光伏组件表面风荷载分布情况以及倾角、风向角、组件之间的间隙、风场类型对风荷载的影响。楼文娟等[5]采用刚性模型测压风洞试验和CFD模拟研究了某超大型固定式光伏板阵列体型系数及群体遮挡效应，给出了遮挡效应影响因素，将光伏阵列分为边缘区、渐变区和稳定区，并给出各区体型系数取值建议。马文勇等[6-8]及李晓娜[9]、韩晓乐[10]通过固定式光伏支架刚性模型测压风洞试验，研究了倾斜角度、风向角、离地高度、流场类型、底部阻塞度、间距比和安装位置等因素对于光伏支架风荷载的影响；提出了光伏板均匀分布、梯形分布和偏心分布三种风荷载取值模型；分析了阵列光伏支架折减效应、遮挡效应。杜航等[11]基于刚性模型研究了某大跨柔性光伏支架光伏组件板面在不同风向角和倾角组合下的风压分布特性，并利用ANSYS软件进行有限元分析给出了该结构的风振系数建议值。Taylor等[12]通过刚性模型风洞试验提出了一种精确地结合了抖振力、惯性力和自激力的平单轴光伏支架设计风荷载取值方法。Strobel等[13]通过刚性模型对平单轴式光伏支架自激力和抖振力进行分析，评估典型光伏板的动态响应，发现结构的固有频率高于1 Hz，也应纳入光伏板共振荷载估算，并分析了结构的固有频率、阻尼等对风振特性的影响。Kim等[14]通过一系列气弹模型风洞试验，研究了不同风环境下不同面板形状柔性光伏支架的风致振动特性。He等[15]通过气弹模型风洞试验模拟了一种索支撑光伏组件的风致振动，并根据响应特性提出了抑制措施。唐俊福等[16]通过有限元模拟对比了斜拉索、钢斜柱、八字形钢斜柱三类不同的柔性光伏支架结构水平承载构件的受力特点，分析了水平力承载构件与地面倾角的变化对结构力学行为的影响。

综上，现有研究主要针对光伏支架光伏组件表面风荷载分布情况、风荷载影响因素及风荷载取值方法，而对于结构气动稳定性与阵列干扰工况下的风振效应研究较少，特别是柔性光伏支架的气动稳定性及多列干扰工况下的风振性能还需要深入研究。因此，本文中通过柔性光伏支架气弹模型风洞试验进行柔性光伏支架气动稳定性及多列干扰下的风振特性研究。

1 试验概况

1.1 原型简介

柔性光伏支架结构原型立柱间跨径L为20 m，支架运行倾角α为0~40°。支架结构由17块光伏组件组成，各组件之间间隔G为20 mm，单块组件平面尺寸$L_1 \times B \times T$(长度×宽度×厚度)为2 256 mm×1 133 mm×40 mm；两根直径15.2 mm的拉索通过锚具与光伏组件相连并固定在下部支架上；拉索为预应力索，施加预应力为115 kN。图1a为17块光伏组件并联而成的单排柔性光伏支架；图1b为柔性光伏支架倾角及风向角示意，风向角按照逆时针增加为正；阵列柔性光伏支架布置及部分参数示意见图1c。

1.2 气弹模型设计

1.2.1 有限元分析

为了保证气弹模型能够良好地模拟出结构原型的各项力学特性，确保研究结果的正确性，首先对柔性光伏支架进行有限元分析获得其动力特性参数。

柔性光伏支架由光伏组件、拉索和下部支架(立柱、斜撑、斜拉杆)组成。采用有限元软件ANSYS建立柔性光伏支架有限元模型，光伏组件采用SHELL181单元模拟，拉索采用LINK10单元模拟，下部支架均采用BEAM188单元模拟。每块光伏组件与拉索通过4个节点采用约束方程耦合[11]，拉索与支架顶部固接，支架底部与地面铰接。

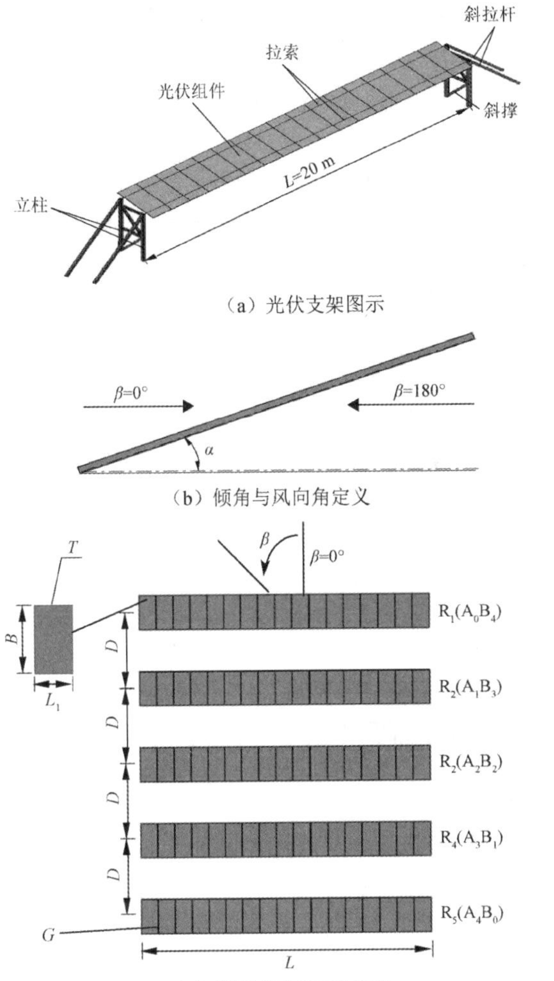

（a）光伏支架图示

（b）倾角与风向角定义

（c）阵列布置及参数定义

注：α 为倾角；β 为风向角；D 为阵列柔性光伏支架各行之间行中心距，$D=5$ m；$R_1 \sim R_5$ 为阵列柔性光伏支架 1～5 排编号；定义试验中不同工况下各位置处光伏支架编号为 A_xB_y，其中 A_x 代表该编号光伏支架上游方向光伏支架数量，B_y 代表该编号光伏支架下游方向光伏支架数量。例如，A_1B_2 代表阵列一共 4 行光伏支架，该支架位于第 2 行，该光伏支架上游方向有 1 行光伏支架，下游方向有 2 行光伏支架。

图 1　原型结构示意及试验参数定义

Fig. 1　Schematic diagram of prototype structure and test parameter definition

对不同倾角下柔性光伏支架结构进行模态分析，前 5 阶频率结果如表 1 所示，发现倾角的改变对柔性光伏支架结构的模态影响不大。不同倾角下柔性光伏支架振型一致，以 $\alpha=30°$ 为例，结构前 5 阶振动模态如图 2 所示。

1.2.2　气弹模型

为确保气动弹性模型的试验结果能真实反映实际结构在大气边界层的气动弹性效应，需要满足模型与原型气动外形相似，同时，还要求实际结构和风洞模型之间满足表 2 所列的无量纲参数的一致性条件[17]。

表 1　柔性光伏支架前 5 阶自振频率

Table 1　The first five natural vibration frequencies of flexible photovoltaic

α/(°)	f_1/Hz	f_2/Hz	f_3/Hz	f_4/Hz	f_5/Hz
0	2.13	4.18	4.29	6.23	7.57
10	2.13	4.18	4.30	6.24	7.59
20	2.13	4.19	4.35	6.24	7.61
30	2.12	4.19	4.40	6.24	7.61
40	2.12	4.18	4.31	6.24	7.60

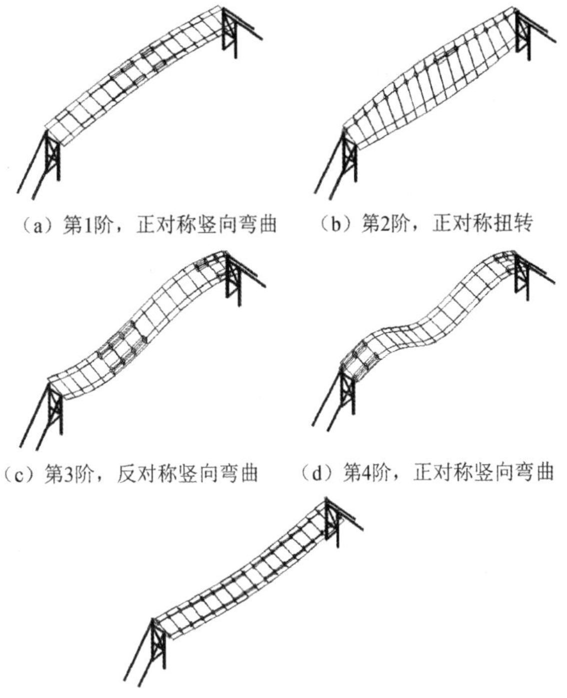

（a）第1阶，正对称竖向弯曲　（b）第2阶，正对称扭转

（c）第3阶，反对称竖向弯曲　（d）第4阶，正对称竖向弯曲

（e）第5阶，正对称竖向弯曲

图 2　柔性光伏支架振动模态（$\alpha=30°$）

Fig. 2　Vibration modes of flexible photovoltaic（$\alpha=30°$）

表 2　无量纲参数的相似要求

Table 2　Similar requirements for dimensionless parameters

参数	物理意义	模拟情况
Reynolds 数	气动惯性力与空气黏性力之比	钝体不模拟
Froude 数	气动惯性力与结构重力之比	严格相似
Strouhal 数	时间尺度	严格相似
Cauchy 数	结构弹性力与气动惯性力之比	严格相似
密度比	结构惯性力与气动惯性力之比	严格相似
阻尼比	每个周期耗能与振动总能量之比	严格相似

风洞试验模型（图 3）按照 1∶10 比例进行缩尺，最大阻塞度 3.87%，对应结构 40°倾角。模型总质量为1.92 kg，目标质量为 1.86 kg，两者之间误差为 3.23%。光伏组件由轻质木板制成，拉索采用高强度钢绞线模拟，光伏组件与拉索的连接锚具利用塑料等比例 3D 打印，质量均严格满足相似条件；拉索初始张力由索力传感器测试得到，其值与设计张力（115 N）之间最大相差 1.90%；立柱、斜撑、斜拉杆和连接件统一采用钢材作为芯梁，外部采用 ABS 塑料

板包裹以保证气弹模型几何外形相似。采用“桁架四棱锥式”的连接方式,即在各行光伏支架跨中位置布置1个四棱锥桁架(与拉索相连),并通过1根横杆将相邻的两行光伏支架连接在一起,最终将5排连接成整体,以提高结构扭转刚度。该连接件原型为Φ40×3的Q235B薄壁圆钢管,风洞试验时用钢材进行模拟。图3b为连接示意,图3c为连接风洞试验时所用连接件实物,图3d为布置连接件后5×1光伏支架阵列。

(a) 单光伏支架

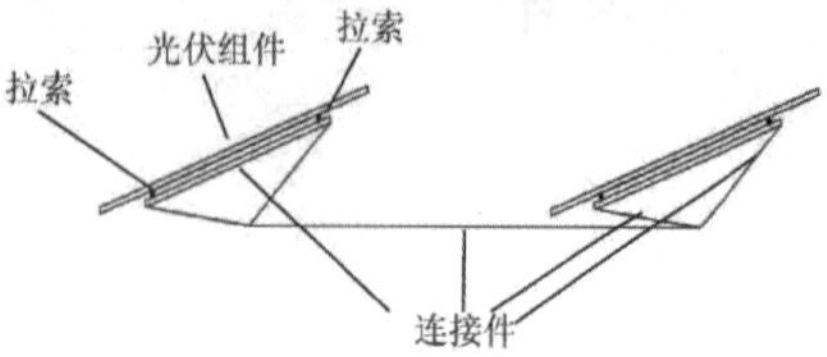

(b) 连接示意

(c) 连接件实物

(d) 布置连接件后,5×1支架阵列

(e) 阵列光伏支架

图3　风洞试验模型

Fig. 3　Wind tunnel test model

1.3　传感器布置及参数设置

风洞试验在湖南大学风工程试验研究中心HD-2大气边界层风洞的高速试验段进行。该试验段宽度3.0 m,高度2.5 m,试验风速在0～58 m/s之间连续可调。

模型倾角范围为0°～40°,测试角度间隔10°。试验时在每一行模型跨中布置1个LC0408T型压电式加速度计,加速度计测试频率范围1～18 000 Hz,单个传感器质量为6 g,灵敏度约为6PC/g(g为重力加速度,其值为9.81 m/s^2)。对于质量很小的柔性光伏支架,如果采用的测试传感器质量过大,将会对模型的模态质量及振型产生较大的影响,从而影响测试结果的真实性;本研究使用的微型加速度计质量仅占整体结构质量的0.63%,对试验测试结果不会产生影响。振动测试时采样频率为500 Hz,采样时间为60 s。

1.4　动力特性标定

为了检验柔性光伏支架模型的振型、频率及阻尼比,需进行模型动力特性标定。以倾角30°为例,给出模型前两阶(竖向弯曲和扭转模态)动力特性标定结果,标定中的加速度时程以及频谱如图4所示。

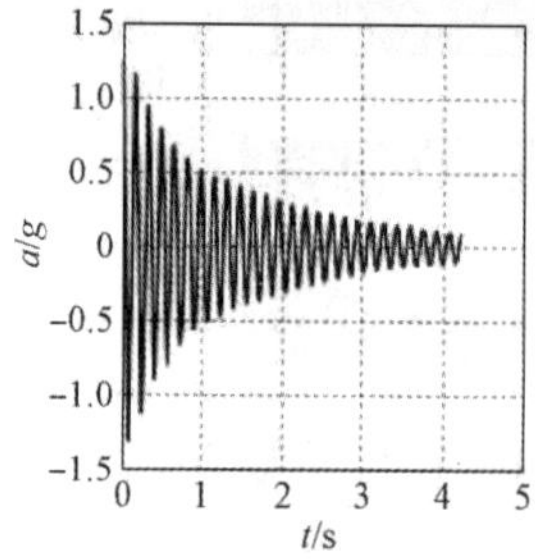

(a) 第1阶加速度时程

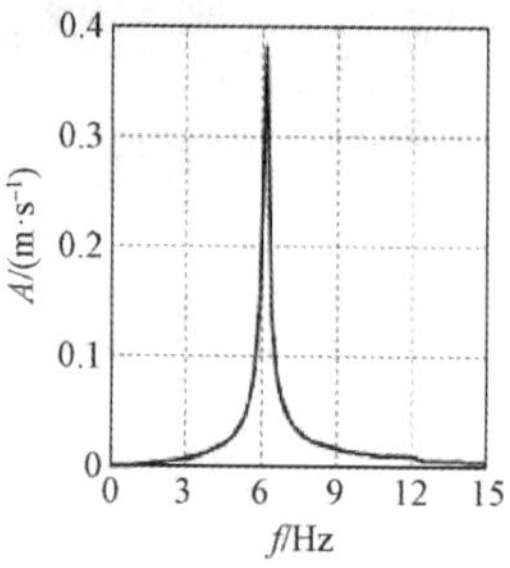

(b) 第1阶频谱

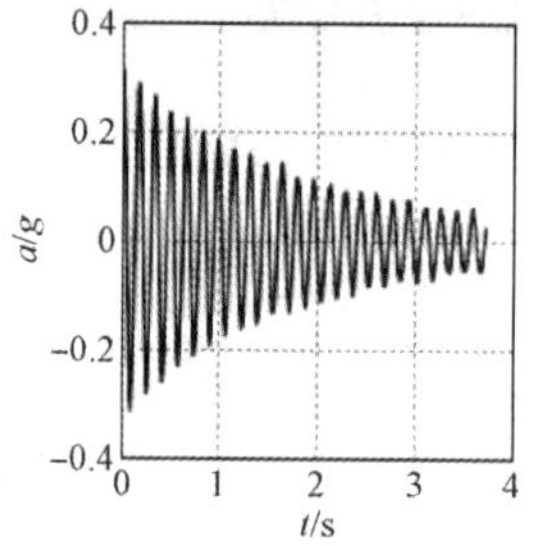

(c) 第2阶加速度时程

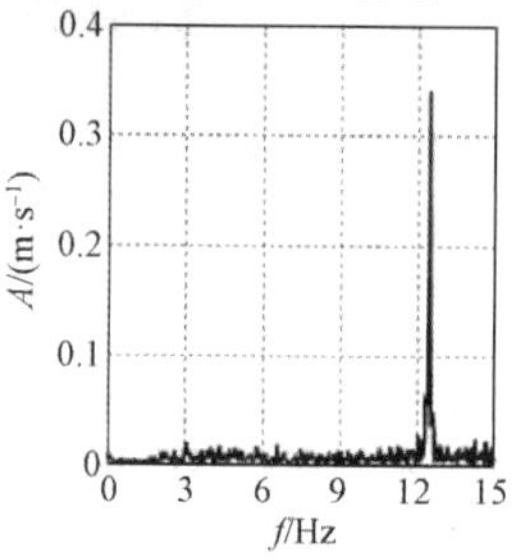

(d) 第2阶频谱

图4　动力特性标定结果

Fig. 4　Calibration results of dynamic characteristics

模型前两阶实测频率分别为6.18 Hz和12.45 Hz,与理论频率6.74 Hz和13.25 Hz相比误差分别为−8.30%和−6.04%;模型前两阶模态实测阻尼比分别为1.66%和1.60%。综上,气弹模型的设计和制作满足试验要求。根据动力特性标定结果最终确定模型相似系数,见表3。

表3　柔性光伏支架气弹模型相似系数

Table 3　Similarity coefficient of flexible photovoltaic

参数名称	相似系数值
几何缩尺比 λ_L	1/10.00
频率相似系数 λ_f	2.92
风速比 λ_U	1/3.42
加速度相似系数 λ_a	1/1.17

1.5 试验工况

试验时分为单支架试验与阵列支架试验，试验工况见表4。通过钢立柱与高强度木板搭接格栅紊流场，各木板以及钢立柱之间用绑扎带以及铁丝固定连接。通过调整格栅条木板宽度、格栅条中心距和格栅断面之间的相对距离得到A、B、C、D四类格栅紊流场，四类格栅紊流场紊流度 I_u 分别为小于0.5%（均匀流场，无格栅）、5.0%、10.0%和15.0%。

表4　试验工况

Table 4　Wind tunnel test case

试验类别	工况编号	α/(°)	β/(°)	流场类型	备注
单支架	1～13	30	0～180	均匀流	β 间隔15°
	14～16	0	0、15、30	均匀流	—
	17～19	10	0、15、30	均匀流	—
	20～22	20	0、15、30	均匀流	—
	23～25	40	0、15、30	均匀流	—
	26～28	10	0、15、30	5.0%均匀紊流	—
	29～31	10	0、15、30	10.0%均匀紊流	—
	32～34	10	0、15、30	15.0%均匀紊流	—
阵列支架	35	30	0	15.0%均匀紊流	5×1
	36	30	0	15.0%均匀紊流	4×1
	37	30	0	15.0%均匀紊流	3×1
	38	30	0	15.0%均匀紊流	2×1
	39	30	0	15.0%均匀紊流	5×1带连接件

1.6 数据处理

试验时风速 U 为0～15 m/s。无量纲风速 U^* 定义为

$$U^* = \frac{U}{fB} \tag{1}$$

式中：U 为风洞来流风速；f 为风致振动时的振动频率，静止状态时，f 按照结构的1阶扭转频率 f_t 取值；B 为结构宽度。

将风洞试验测试得到的模型加速度响应按照下式换算到原型，即

$$a_{\mathrm{std,p}} = \frac{a_{\mathrm{std,m}}}{\lambda_a} \tag{2}$$

式中：下标p、m分别表示原型和模型，a_{std} 为加速度标准差，λ_a 为加速度相似系数。

2　试验结果及分析

2.1 单支架试验

2.1.1　倾角及风向角对结构颤振性能影响

试验原型位于我国西北区域，该区域最常用支架结构倾角为30°。因此首先针对30°倾角柔性光伏支架结构开展风洞试验，取 β 为0°～180°且间隔15°，试验流场为均匀流风场（工况1～13）。

原型加速度标准差随风速变化曲线如图5所示，试验中发现 $\alpha=30°$ 时，β 为0°、15°和30°三个工况下结构发生了明显的颤振，表现为加速度标准差数值突然剧烈增加，对应颤振临界无量纲风速分别为4.05、3.65和3.91，其余风向角下结构颤振临界无量纲风速均大于4.40。

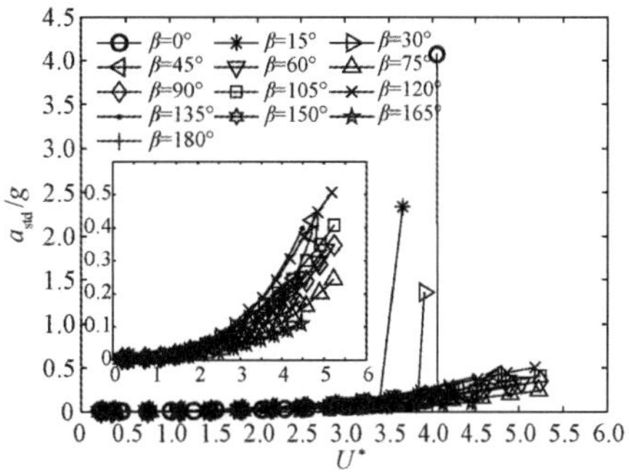

图5　$\alpha=30°$时结构加速度标准差

Fig. 5　Standard deviation of structural acceleration at $\alpha=30°$

针对颤振明显的 β 为0°、15°、30°三个风向角工况，对 α 为0°、10°、20°、40°四组变化倾角工况进行风洞试验（工况14～25），以研究倾角及风向角组合对结构颤振性能的影响，结果如图6所示。

由图5、6可见，不同倾角柔性光伏支架颤振稳定性能差距较大，5个测试倾角下，各倾角对应的颤振临界风速分别为3.33、1.89、2.19、3.65和大于5.50，其中 $\alpha=40°$ 时颤振临界风速是 $\alpha=10°$ 的2.91倍。可见，倾角10°的气动稳定性最差。$\alpha<30°$ 时，风向角越大，结构颤振性能越优越；$\alpha>30°$ 时，β 为15°、0°、30°工况下结构颤振性能依次变强，分析原因为大倾角下，来流在结构处绕流情况发生变化而产生此现象。

2.1.2　紊流对结构颤振性能影响

对结构颤振性能较差的10°倾角进行不同紊流强度测试（工况26～34），测试结果见图7。由图7可见，随着紊流度增大（紊流度5%、10%、15%），对比均匀流场情况，结构颤振临界风速分别提高2.92%、6.10%和28.38%；同时，结构加速度标准差分别为均匀流场的1.78、4.56和6.12倍（$\beta=0°$，$U^*=1.89$）。即紊流度可以抑制柔性光伏支架的颤振，同时紊流度增加，来流风的脉动成分会增大光伏支架颤振临界状态之前结构的抖振响应。

综上，倾角、风向角和紊流度对柔性光伏支架的气动稳定性和风振响应影响很大。在实际工程中应考虑结构所处位置的风参数条件选择合适的安装倾角，避免较不利情况的风向角和大紊流度出现。

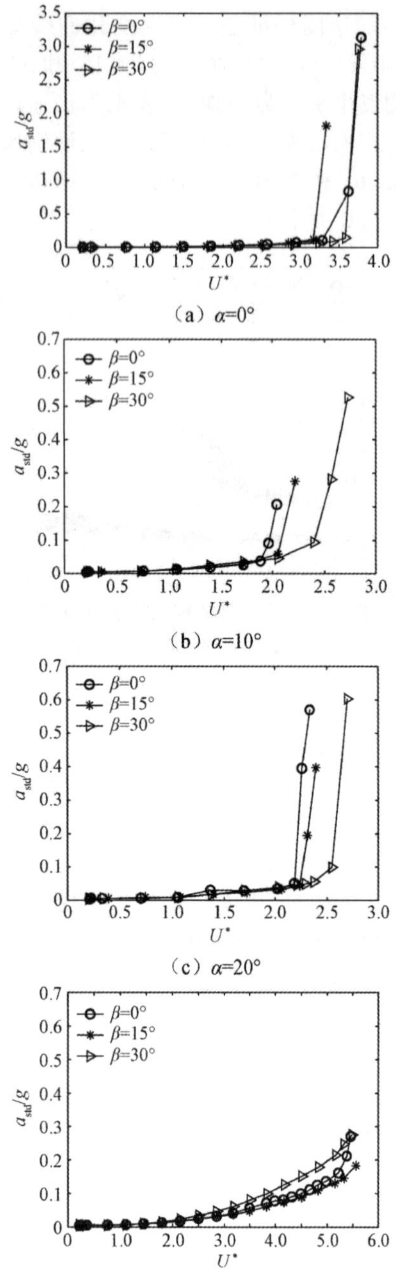

图6　不同倾角下结构加速度标准差

Fig. 6　Standard deviation of structural acceleration for different inclination angles

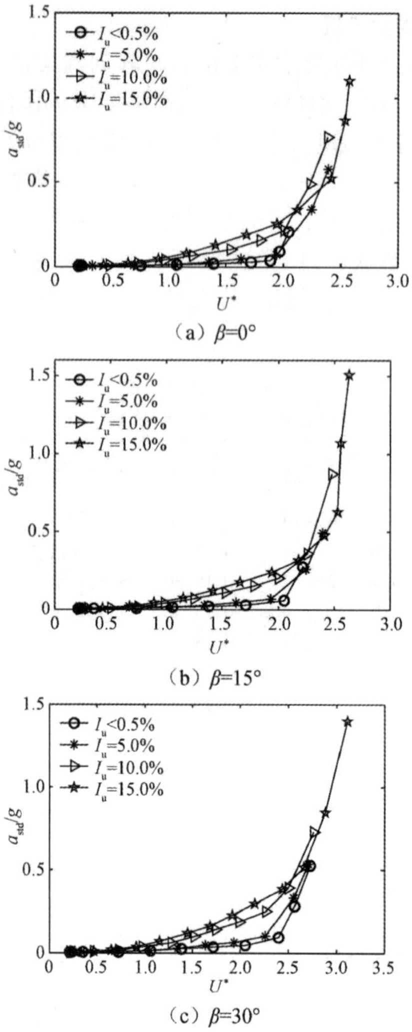

图7　当 $\alpha=10°$ 时不同紊流度下结构加速度标准差

Fig. 7　Standard deviation of structural acceleration for different turbulence degrees at $\alpha=10°$

2.2　阵列支架试验

当光伏支架以阵列的形式出现时，由于上游光伏支架的存在，往往会对下游光伏支架的风荷载产生影响，进而可能改变光伏支架振动情况，这种影响的本质是流场的影响[3,5]。

由2.1.2节内容可知，紊流对于光伏支架的颤振有一定的抑制作用，且结构原型处于B类地貌，因此，在15%均匀紊流场下进行光伏阵列支架风振干扰效应研究。测试过程中 $\alpha=30°$（西北地区最常用角度），$\beta=0°$（上游光伏支架对下游光伏支架遮挡效应最严重角度）。阵列工况为2×1、3×1、4×1和5×1（工况35~38），以考虑上下游不同数量光伏支架的干扰影响。

2.2.1　尾流致振现象及遮挡效应

5×1阵列（工况35）支架风振响应如图8所示，图中 R_1~R_5 分别代表阵列柔性光伏支架第1~5排（图1c）。由于上游光伏支架的影响，下游光伏支架会发生不同于单个光伏支架的振动。

由图8可见，R_2 和 R_3 振动响应显著大于 R_1，当无量纲风速为2.41时，R_2 和 R_3 振动响应分别为 R_1 的2.02和1.15倍，这是来流经过 R_1 产生的尾流造成的，初步分析尾流影响区域小于 $3D$（15 m）。R_4 和 R_5 振动响应小于 R_1，当无量纲风速为2.41时，R_4 和 R_5 振动响应分别为 R_1 的82%和71%，这一区域超过尾流作用区域，因为上游支架对下游支

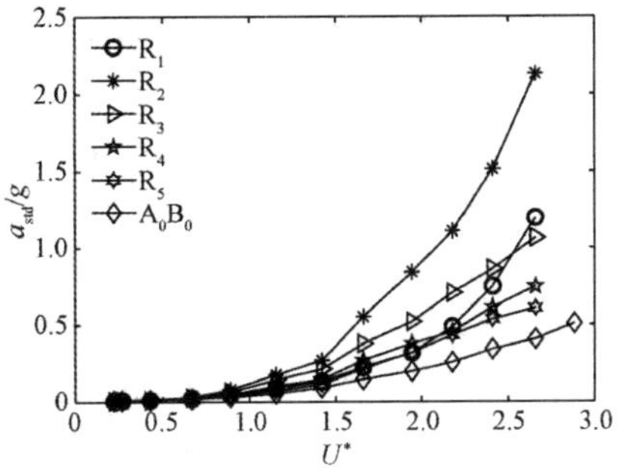

图 8　5×1 阵列与单支架的结构加速度标准差

Fig. 8　Standard deviation of structural acceleration for 5×1 arrys and single flexible photovoltaic

架的遮挡作用，下游支架(R_4 和 R_5)加速度响应小于上游支架(R_1)。同时，由于下游光伏支架的存在，R_1 振动响应大于单光伏支架，当无量纲风速为 2.41 时，R_1 振动响应为单光伏支架的 2.2 倍。

为研究其余阵列形式是否存在不同于单个光伏支架的振动情况，从 5 行光伏支架开始，依次拆除 R_1 ~ R_3 组成不同测试阵列，即 4×1、3×1 和 2×1 阵列(工况 36 ~ 38)。不同测试工况下结构振动响应如图 9 所示。由图 9 可知，由于上游光伏支架尾流的影响，下游光伏支架会发生明显的“尾流致振现象”，其中来流方向第 2 行受尾流影响最为剧烈，尾流的影响随着距离的增大逐渐减弱，在来流方向第 4 行基本趋于稳定。在来流方向第 4 行(含)以后由于上游光伏支架的遮挡作用，其振动响应小于其上游光伏支架响应。同时，由于下游光伏支架的存在，也会加剧上游首行光伏支架的振动。综上，可将光伏支架阵列划分为三个区域，即尾流区、过渡区、遮挡区，分区示意如图 10 所示。

2.2.2　尾流致振现象及遮挡效应影响因素

综合图 8，采用两排模型，分别调整模型中心距为 $2D$ 和 $3D$，研究结构尾流作用区域。不同中心距下结构振动加速度响应见图 11。由图 11 可见，光伏支架尾流影响区域大致在 $2D$ ~ $3D$(10 ~ 15 m)范围内。

连接方式(工况 39)对于阵列光伏支架尾流及干扰效应影响结果见图 12。由图可见，光伏支架之间布置连接件后，尾流区内光伏支架的振动均有一定的减弱效果，$U^*=2.88$ 时(下同)，R_1、R_2、R_3 的加速度标准差分别减小 56.38%、31.97%、28.92%；过渡区 R_4 的加速度标准差较无连接件时增大 24.43%，遮挡区 R_5 的加速度标准差较无连接件时减小 15.62%。

由于连接件的存在，减小了尾流区的振动响应；同时，连接件将 5 行结构连成整体，尾流区会带动过渡区一同振动，使得过渡区振动加剧；遮挡区由于连接件的存在振动响应也会增大，但受到上游光伏支架遮挡最严重，故振动响应总体呈减小趋势。

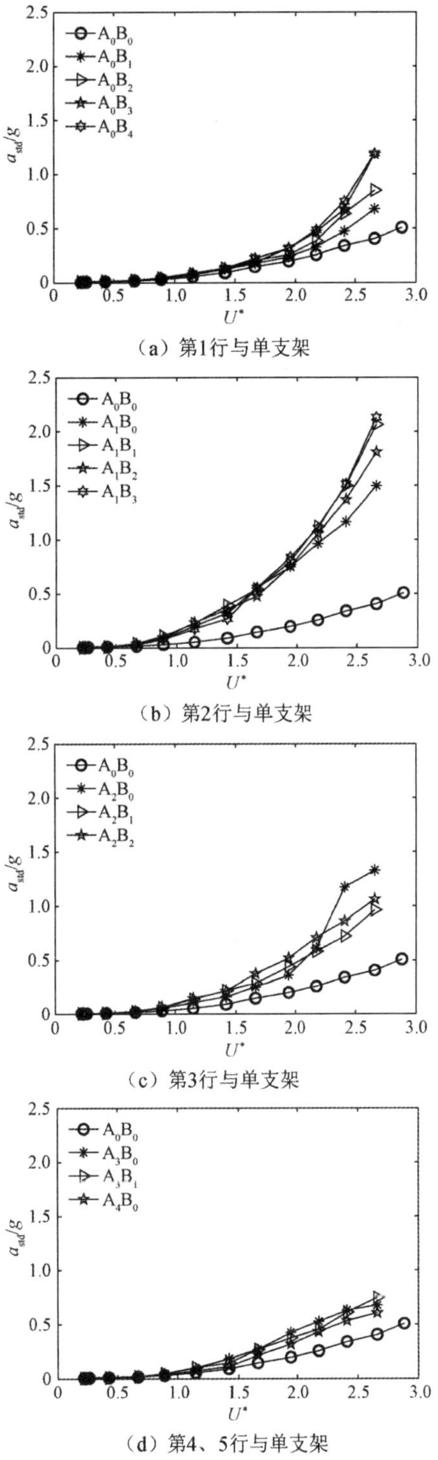

图 9　不同工况下来流方向结构加速度标准差

Fig. 9　Standard deviation of structural acceleration for different working conditions

综上，将结构连接成整体会减弱阵列尾流区与遮挡区结构振动响应，但对过渡区结构较为不利。

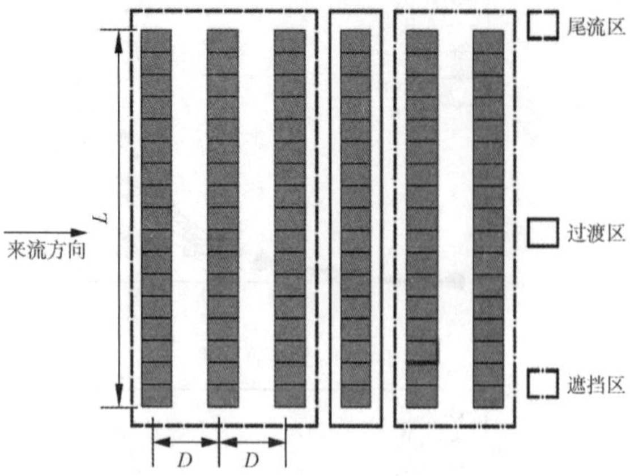

图10 分区示意

Fig. 10 Schematic diagram of partition

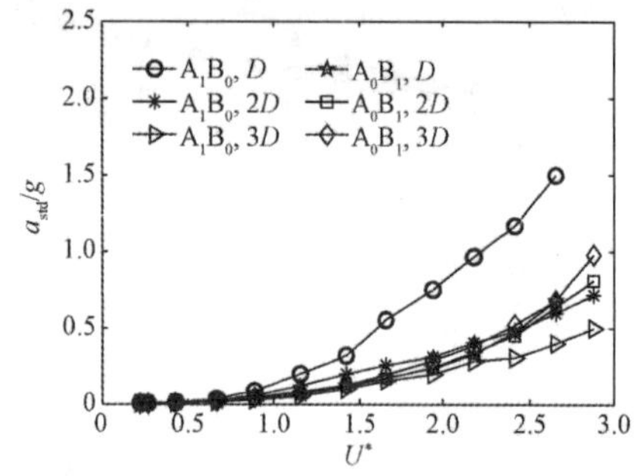

图11 不同中心距的结构加速度标准差

Fig. 11 Standard deviation of structural acceleration for different center distances

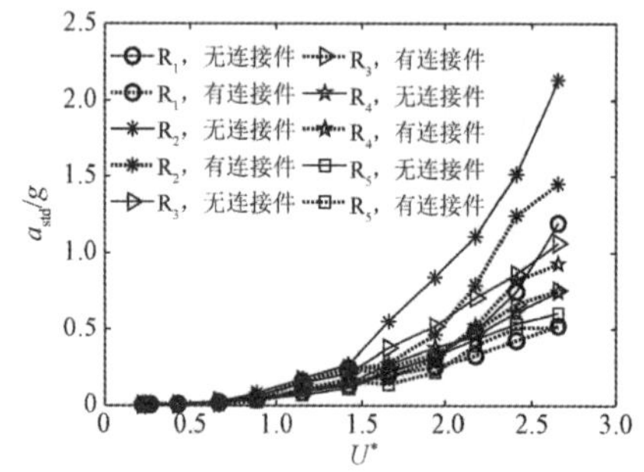

图12 有无连接件的结构加速度标准差

Fig. 12 Acceleration standard deviation of structure with or without connectors

对于实际工程，建议在各柔性光伏支架之间布置连接件减小结构风振响应，提高结构气动稳定性。

3 结论

1) 不同倾角和风向角柔性光伏支架颤振稳定性能差距较大，测试倾角中，10°倾角颤振稳定性最差；测试风向角中，0°、15°和30°风向角结构颤振稳定性较差；不同倾角与风向角下颤振临界风速最大比值超过2.91倍（$\alpha=40°$、$\beta=0°$与$\alpha=10°$、$\beta=0°$相比）。

2) 紊流度可以抑制光伏支架的颤振，同时，由于来流风的脉动成分，会增大光伏支架的抖振响应。紊流度从小于0.5%提高到5.0%、10.0%、15.0%的过程中，结构颤振临界风速分别提高2.92%、6.10%、28.38%，光伏支架的加速度标准差分别为均匀流场的1.78、4.56、6.12倍（$\alpha=10°$、$\beta=0°$、$U^*=1.89$）。

3) 由于上游光伏支架尾流的影响，下游光伏支架会发生明显的"尾流致振现象"，其中来流方向第2行受尾流影响最为剧烈（最大加速度响应超过单支架时的5.30倍），尾流的影响随着距离的增大逐渐减弱，在来流方向第4排（3D距离）基本趋于稳定。光伏支架尾流影响区域大致在10～15 m范围内。

4) 根据阵列柔性光伏支架风振响应情况，将阵列区域划分为尾流区、过渡区和遮挡区；在阵列柔性光伏支架之间布置连接件会减弱阵列尾流区与遮挡区结构振动响应，加速度标准差最多减小了56.38%，但对于过渡区结构较为不利，加速度标准差增大了24.43%。

参考文献

[1] ALY A M, BITSUAMLAK G. Aerodynamics of ground-mounted solar panels: test model scale effects [J]. Journal of Wind Engineering and Industrial Aerodynamics, 2013, 123: 250-260.

[2] JUBAYER C M, HANGAN H. Numerical simulation of wind effects on a stand-alone ground mounted photovoltaic (PV) system [J]. Journal of Wind Engineering and Industrial Aerodynamics, 2014, 134: 56-64.

[3] JUBAYER C M, HANGAN H. A numerical approach to the investigation of wind loading on an array of ground mounted solar photovoltaic (PV) panels [J]. Journal of Wind Engineering and Industrial Aerodynamics, 2016, 153: 60-70.

[4] ABIOLA-OGEDENGBE A, HANGAN H, SIDDIQUI K. Experimental investigation of wind effects on a standalone photovoltaic (PV) module [J]. Renewable Energy, 2015, 78: 657-665.

[5] 楼文娟，单弘扬，杨臻，等. 超大型阵列光伏板体型系数遮挡效应研究[J]. 建筑结构学报，2021，42(5)：47-54.（LOU Wenjuan, SHAN Hongyang, YANG Zhen, et al. Study of shielding effect on shape coefficient of super-large photovoltaic arrays [J]. Journal of Building Structures, 2021, 42 (5): 47-54. (in Chinese)）

[6] 马文勇，孙高健，刘小兵，等. 太阳能光伏板风荷载

分布模型试验研究[J]. 振动与冲击, 2017, 36(7): 8-13. (MA Wenyong, SUN Gaojian, LIU Xiaobing, et al. Tests for wind load distribution model of solar panels [J]. Journal of Vibration and Shock, 2017, 36(7): 8-13. (in Chinese))

[7] 马文勇, 马成成, 王彩玉, 等. 光伏阵列风荷载干扰效应风洞试验研究[J]. 实验流体力学, 2021, 35(4): 19-25. (MA Wenyong, MA Chengcheng, WANG Caiyu, et al. Wind tunnel experimental study on the wind load interference effect of solar panel arrays [J]. Journal of Experiments in Fluid Mechanics, 2021, 35(4): 19-25. (in Chinese))

[8] 黄伯城, 马文勇. 串列光伏板体型系数遮挡效应研究[C] //第25届全国结构工程学术会议. 北京: 中国力学学会结构工程专业委员会, 2016: 326-329. (HUANG Bocheng, MA Wenyong. Study on the effectof sheelding tandem photovoltaic panels shape coeffcient [C] // Proceedings of the 25th National Conference on Structural Engineering. Beijing: Structural Engineering Committee, Chinese Society of Theoretical and Applied Mechanics, 2016: 326-329. (in Chinese))

[9] 李晓娜. 太阳能光伏支架风荷载体型系数研究[D]. 石家庄: 石家庄铁道大学, 2015: 9-27. (LI Xiaona. Study on wind load of solar photovoltaic bracket [D]. Shijiazhuang: Shijiazhuang Tiedao University, 2015: 9-27. (in Chinese))

[10] 韩晓乐. 太阳能光伏阵列风荷载取值研究[D]. 石家庄: 石家庄铁道大学, 2016: 14-32. (HAN Xiaole. Study on wind load of solar photovoltaic array [D]. Shijiazhuang: Shijiazhuang Tiedao University, 2016: 14-32. (in Chinese))

[11] 杜航, 徐海巍, 张跃龙, 等. 大跨柔性光伏支架结构风压特性及风振响应[J]. 哈尔滨工业大学学报, 2022, 54(10): 67-74. (DU Hang, XU Haiwei, ZHANG Yuelong, et al. Wind pressure characteristics and wind vibration response of long-span flexible photovoltaic support structure [J]. Journal of Harbin Institute of Technology, 2022, 54(10): 67-74. (in Chinese))

[12] TAYLOR Z J, BROWNE M T L. Hybrid pressure integration and buffeting analysis for multi-row wind loading in an array of single-axis trackers [J]. Journal of Wind Engineering and Industrial Aerodynamics, 2020, 197: 104056.

[13] STROBEL K, BANKS D. Effects of vortex shedding in arrays of long inclined flat plates and ramifications for ground-mounted photovoltaic arrays [J]. Journal of Wind Engineering and Industrial Aerodynamics, 2014, 133: 146-149.

[14] KIM Y C, SHAN W, YANG Q S, et al. Effect of panel shapes on wind-induced vibrations of solar wing system under various wind environments [J]. Journal of Structural Engineering, 2020, 146(6): 04020104.

[15] HE Xuhui, DING Hao, JING Haiquan, et al. Wind-induced vibration and its suppression of photovoltaic modules supported by suspension cables [J]. Journal of Wind Engineering and Industrial Aerodynamics, 2020, 206: 104275.

[16] 唐俊福, 林建平, 霍静思. 柔性光伏支架结构特性分析及其优化设计[J]. 华侨大学学报(自然科学版), 2019, 40(3): 331-337. (TANG Junfu, LIN Jianping, HUO Jingsi. Structural characteristic analysis and optiomal design of flexible photovoltaic support structure [J]. Journal of Huaqiao University (Natural Science), 2019, 40(3): 331-337. (in Chinese))

[17] 陈政清. 桥梁风工程[M]. 北京: 人民交通出版社, 2005: 125-126. (CHEN Zhengqing. Bridge wind engineering [M]. Beijing: China Communications Press, 2005: 125-126. (in Chinese))

桥梁颤振与抖振

第20卷 第5期
2007年9月
中 国 公 路 学 报
China Journal of Highway and Transport
Vol. 20 No. 5
Sept. 2007

文章编号:1001-7372(2007)05-0041-07

基于ANSYS的桥梁全模态颤振频域分析方法

华旭刚,陈政清
(湖南大学 风工程试验研究中心,湖南 长沙 410082)

摘要:提出了一种在ANSYS中直接分析大跨桥梁三维全模态颤振的有限元模型和频域方法。该有限元模型采用ANSYS的自定义单元Matrix27来模拟桥面受到的自激气动力,而将单元Matrix27的刚度或阻尼矩阵描述成风速和振动频率的函数。通过阻尼特征值分析,确定风-桥耦合系统的各复特征值,其中复特征值的实部为阻尼而其虚部则为振动频率;并给出了求解系统各阶复模态特性随风速变化的迭代算法。最后以一简支梁桥和一悬索桥算例验证了颤振分析模型和方法的正确性和可靠性。该方法能使桥梁设计人员和研究人员可以直接利用ANSYS来分析大跨桥梁的耦合颤振。

关键词:桥梁工程;全模态颤振;有限元方法;频域分析
中图分类号:U441.3 **文献标志码**:A

Full-mode Flutter Frequency Domain Analysis Method of Bridge Based on ANSYS

HUA Xu-gang, CHEN Zheng-qing
(Wind Engineering Experiment Research Center, Hunan University, Changsha 410082, Hunan, China)

Abstract: Authors presented a finite element (FE) model and frequency analysis method for analyzing 3-D full-mode flutter of long-span bridges using ANSYS software. This model utilized a special kind of element Matrix27 in ANSYS to simulate the aeroelastic forces acting on bridges, wherein the stiffness or damping matrixes of Matrix27 elements were described as function of wind velocity and flutter derivatives. Using this FE model, damped complex eigenvalue analysis was then carried out to determine the complex eigenvalues of wind-bridge coupled system; the real part of complex eigenvalues was damp while the imaginary part was vibration frequency. The iteration algorithm for solving variation of system order complex mode property vs wind velocity was put forward. Validity of the method was verified by example of a simple-supported bridge and a suspension bridge. The proposed method enables the bridge designers and engineering practitioners to analyze flutter instability by using ANSYS.

Key words: bridge engineering; full-mode flutter; FEM; frequency domain analysis

0 引 言

颤振是一种由于气动不稳定引起的自激发散振动。这种振动一旦发生,将导致结构彻底破坏。自1940年美国的旧Tacoma大桥风毁以来,桥梁颤振问题的研究得到了普遍重视。现代桥梁日益向轻

收稿日期:2007-02-19
基金项目:国家自然科学基金项目(50478051);国家自然科学基金重大项目(59895410)
作者简介:华旭刚(1978-),男,浙江义乌人,副教授,工学博士,E-mail:cexghua@hotmail.com。

柔、大跨径方向发展,使桥梁的气动稳定问题日益突出,确定桥梁颤振临界风速已成为大跨度桥梁设计中的一个基本问题[1]。

经过近 50 a 的研究,桥梁的颤振分析方法,尤其是颤振频域分析方法已日趋成熟。颤振频域分析方法主要分为 2 类:第 1 类方法称为多模态方法,它将受气动自激力的桥梁结构运动方程用模态坐标表示,然后通过求解模态坐标下自激运动方程的广义复特征值问题来判断结构的气动稳定性[2-11];第 2 类方法通常称为全模态方法,它将风-桥梁作为耦合系统,直接在物理坐标中分析耦合系统的复模态特性来判断系统的气动稳定性[12-15]。由于这 2 类方法中系统的系数矩阵均为风速和振动频率的函数,所以这 2 类方法最终都归结为搜索使系统出现动力不稳定的最小风速和相应的频率,由此得到的最小风速和相应的频率即为颤振临界风速和颤振频率。颤振问题涉及到非对称矩阵的复特征值问题,因此以往颤振分析往往需要专用程序才能完成。

自 20 世纪 70 年代以来,随着计算机技术的飞速发展,国外相继出现了许多大型通用有限元分析程序,如 ANSYS、ABAQUS、ADINA 和 MSC/NASTRAN 等,这些程序具有良好的界面、方便的前后处理和强大的计算分析功能以及开放的二次开发系统。近 10 a 来,各大型通用有限元分析程序已在桥梁设计和研究单位中被广泛应用,其中可能以 ANSYS 的应用最为广泛。随着桥梁跨径不断增大,桥梁设计人员迫切需要在桥梁早期设计阶段估计桥梁的颤振临界风速,然而,各通用分析软件无法考虑结构分析的各个方面,当前尚缺乏直接采用大型通用程序进行桥梁三维颤振分析的方法。

基于笔者提出的在 ANSYS 中考虑自激气动力的有限单元模型[16],笔者[17]和胡峰强等[18]同时独立地提出了在 ANSYS 中实现全模态颤振频域分析的实用方法。本文中笔者给出了基于 ANSYS 的全模态颤振分析模型和频域分析算法。该模型采用 ANSYS 中自定义单元 Matrix27 来模拟桥面受到的自激气动力,而单元 Matrix27 的刚度或阻尼矩阵被描述成风速和振动频率的函数。通过阻尼特征值分析,可确定风-桥耦合系统的各个复特征值,其中特征值的实部为阻尼而其虚部则为阻尼振动频率,最后以一简支梁桥和一悬索桥算例验证了本颤振分析模型和方法。

1 基本理论

桥梁结构在均匀流中的运动方程可以描述为

$$\mathbf{M}\ddot{\mathbf{X}}+\mathbf{C}\dot{\mathbf{X}}+\mathbf{K}\mathbf{X}=\mathbf{F}_{se} \tag{1}$$

式中:$\mathbf{M}$、$\mathbf{C}$、$\mathbf{K}$ 分别为结构的质量、刚度和阻尼矩阵;$\mathbf{X}$、$\dot{\mathbf{X}}$、$\ddot{\mathbf{X}}$ 分别为节点的位移、速度和加速度;$\mathbf{F}_{se}$ 为桥梁受到的自激气动力。

根据 Scanlan 自激气动力表达式,单位长度主梁上受到的气动升力 L_{se}、阻力 D_{se} 和气动扭矩 M_{se} 可分别表示为竖向位移 h、水平位移 p 和扭转位移 α 的函数,如图 1 所示

$$L_{se}=\frac{1}{2}\rho U^2\times 2B\left(KH_1^*\frac{h}{U}+KH_2^*\frac{B\dot{\alpha}}{U}+K^2H_3^*\alpha+K^2H_4^*\frac{h}{B}+KH_5^*\frac{\dot{p}}{U}+K^2H_6^*\frac{p}{B}\right) \tag{2}$$

$$D_{se}=\frac{1}{2}\rho U^2\times 2B\left(KP_1^*\frac{\dot{p}}{U}+KP_2^*\frac{B\dot{\alpha}}{U}+K^2P_3^*\alpha+K^2P_4^*\frac{p}{B}+KP_5^*\frac{h}{U}+K^2P_6^*\frac{h}{B}\right) \tag{3}$$

$$M_{se}=\frac{1}{2}\rho U^2\times 2B^2\left(KA_1^*\frac{h}{U}+KA_2^*\frac{B\dot{\alpha}}{U}+K^2A_3^*\alpha+K^2A_4^*\frac{h}{B}+KA_5^*\frac{\dot{p}}{U}+K^2A_6^*\frac{p}{U}\right) \tag{4}$$

式中:U 为风速;ρ 为空气的质量密度;B 为桥面宽;K 为量纲一的频率,$K=B\omega/U$,ω 为振动圆频率;H_i^*、A_i^*、P_i^* 为量纲一的颤振导数,$i=1,2,\cdots,6$,是量纲一的风速 $\tilde{U}$ 的函数,$\tilde{U}=U/(fB)$,与桥梁截面的几何形状有关,颤振导数通常由节段模型风洞试验得到。

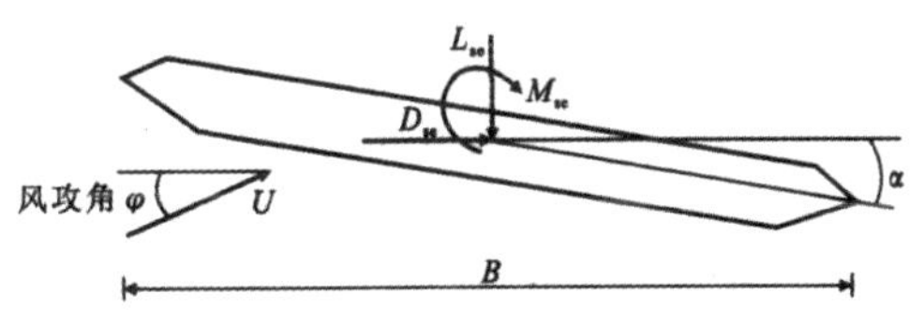

图 1 桥梁断面的气动力

Fig. 1 Acrodynamics of Bridge Section

式(2)~(4)为桥面单位长度上受到的气动力,将这些分布荷载转化为作用于单元两节点的集中荷载。作用于单元 e 两节点的等效自激力可表示为

$$\mathbf{F}_{ae}^e=\mathbf{K}_{ae}^e\mathbf{X}^e+\mathbf{C}_{ae}^e\dot{\mathbf{X}}^e \tag{5}$$

式中:$\mathbf{X}^e$、$\dot{\mathbf{X}}^e$ 分别为单元 e 的节点位移和节点速度向量;$\mathbf{K}_{ae}^e$、$\mathbf{C}_{ae}^e$ 分别为单元 e 的气动刚度矩阵和气动阻尼矩阵。类似于一致质量矩阵和集中质量矩阵的推导,容易导出单元的一致气动力矩阵和集中气动力矩阵。当采用集中气动力矩阵时,气动刚度和气动阻尼矩阵为

$$\mathbf{K}_{ae}^e=\begin{bmatrix}\mathbf{K}_{ae1}^e & \mathbf{0}\\ \mathbf{0} & \mathbf{K}_{ae1}^e\end{bmatrix},\quad \mathbf{C}_{ae}^e=\begin{bmatrix}\mathbf{C}_{ae1}^e & \mathbf{0}\\ \mathbf{0} & \mathbf{C}_{ae1}^e\end{bmatrix} \tag{6}$$

$$\mathbf{K}_{ae1}^{e}=a\begin{bmatrix}0&0&0&0&0&0\\0&P_6^*&P_4^*&BP_3^*&0&0\\0&H_6^*&H_4^*&BH_3^*&0&0\\0&BA_6^*&BA_4^*&B^2A_3^*&0&0\\0&0&0&0&0&0\\0&0&0&0&0&0\end{bmatrix}\tag{7}$$

$$\mathbf{C}_{ae1}^{e}=b\begin{bmatrix}0&0&0&0&0&0\\0&P_5^*&P_1^*&BP_2^*&0&0\\0&H_5^*&H_1^*&BH_2^*&0&0\\0&BA_5^*&BA_1^*&B^2A_2^*&0&0\\0&0&0&0&0&0\\0&0&0&0&0&0\end{bmatrix}\tag{8}$$

式中：$a=\rho U^2K^2L_e/2$；$b=\rho UBKL_e/2$，L_e为单元e的长度。需要指出的是，式(6)～(8)中各元素的具体位置还依赖于整体坐标系的方向。由式(6)～(8)可知，单元e的2个节点上受到的自激力只与各自的节点位移有关。

ANSYS的Matrix27单元是一种功能很强的单元。如图2所示，该单元具有2个节点，每个节点有6个自由度，其单元坐标系和整体坐标系平行；该单元没有固定的几何形状。跟其他结构分析单元不同的是，该单元可以通过实常数输入对称或不对称的质量、刚度或阻尼矩阵。为将式(6)～(8)中的自激力荷载在ANSYS中实现，笔者采用了如图3所示的计算方法(对某个桥面单元e)：在每个桥面主梁节点处添加1对Matrix27单元(包括1个刚度单元和1个阻尼单元)，该单元的一个节点为桥面节点i或j，另一个节点固定。1个Matrix27单元只能模拟气动刚度或者气动阻尼，而不能同时模拟两者。例如，在节点i处，单元e_1用于模拟气动刚度而单元e_3用于模拟节点i处的气动阻尼，单元e_1和e_3共用节点i和k。

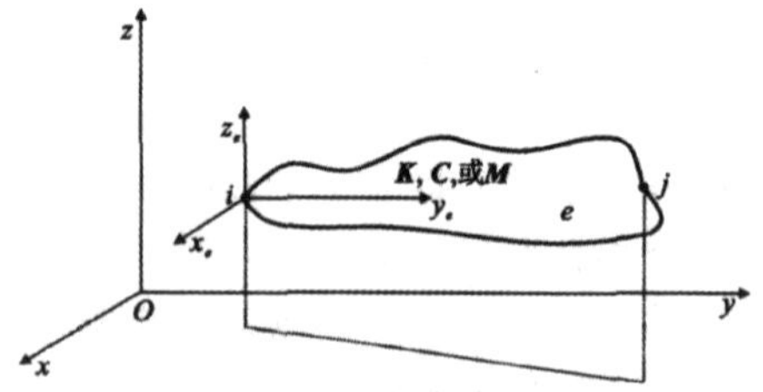

图2　Matrix27单元

Fig. 2　Element Matrix27

当各桥面单元长度相等时，各Matrix27单元的刚度矩阵或阻尼矩阵简化为

$$\mathbf{K}_{e_1}=2\mathbf{K}_{ae}^{e},\quad \mathbf{C}_{e_3}=2\mathbf{C}_{ae}^{e},\quad \mathbf{K}_{e_2}=2\mathbf{K}_{ae}^{e},\quad \mathbf{C}_{e_4}=2\mathbf{C}_{ae}^{e}$$

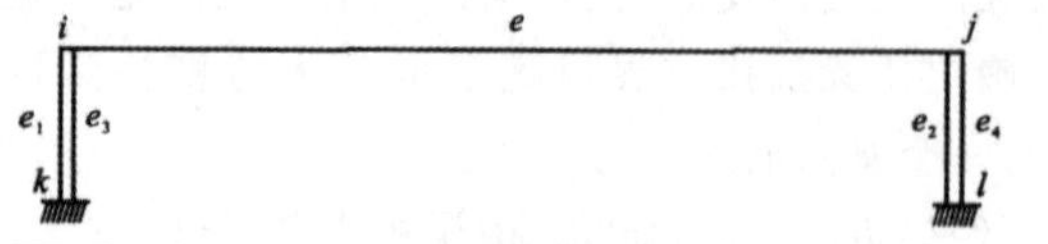

图3　ANSYS中的颤振分析计算模型

Fig. 3　Flutter Analysis Calculation Model in ANSYS

组装单元气动力矩阵后，即可以得到总体气动力矩阵

$$\mathbf{F}_{ae}=\mathbf{K}_{ae}\mathbf{X}+\mathbf{C}_{ae}\mathbf{X}\tag{9}$$

式中：$\mathbf{C}_{ae}$、$\mathbf{K}_{ae}$分别为总体气动阻尼矩阵和气动刚度矩阵。

将式(9)代入式(1)可得到风-桥梁系统的运动方程

$$\mathbf{M}\mathbf{X}+(\mathbf{C}-\mathbf{C}_{ae})\mathbf{X}+(\mathbf{K}-\mathbf{K}_{ae})\mathbf{X}=\mathbf{0}\tag{10}$$

式(10)描述了风-桥梁整合系统(简称系统)的参数化运动方程，其中风速和振动频率为系统参数，系统共有n对共轭特征值$\lambda_j=\sigma_j\pm \mathrm{i}\omega_j$。如果所有特征值的实部$\sigma$都小于0，那么系统是动力稳定的；如果存在至少1对特征值的实部大于0，那么系统是动力不稳定的。颤振的临界状态为在某风速下，系统有且只有1对特征值的实部为0。考虑到桥梁的最低颤振临界风速往往发生在低阶频率，所以一般只需要跟踪结构的低阶复特征值随风速的变化。

结构的阻尼通常以模态阻尼比的形式给出，在进行阻尼复特征值分析时，需要将模态阻尼比描述为瑞利阻尼形式

$$\mathbf{C}=\alpha\mathbf{M}+\beta\mathbf{K}\tag{11}$$

式中：α、β分别为瑞利阻尼系数。当只考虑两阶模态p和q时，瑞利阻尼系数与结构模态阻尼比ξ的关系可以描述为

$$\left.\begin{aligned}\alpha&=2\frac{\omega_p\omega_q}{\omega_q^2-\omega_p^2}(\omega_q\xi_p-\omega_p\xi_q)\\ \beta&=2\frac{\omega_p\omega_q}{\omega_q^2-\omega_p^2}\left(-\frac{\xi_p}{\omega_q}+\frac{\xi_q}{\omega_p}\right)\end{aligned}\right\}\tag{12}$$

当更多的模态阻尼比需要等效时，瑞利阻尼系数可以由式(11)最小二乘拟合得到。

当考虑瑞利结构阻尼后，需要对式(10)进行适当修改。在添加了Matrix27单元的系统中，系统的质量$\mathbf{M}$跟原来的桥梁相同，而系统刚度则变为$\mathbf{K}-\mathbf{K}_{ae}$而不是原桥梁刚度$\mathbf{K}$。因此，系统的瑞利阻尼矩阵为

$$\mathbf{C}=\alpha\mathbf{M}+\beta(\mathbf{K}-\mathbf{K}_{ae})\tag{13}$$

所以，考虑瑞利结构阻尼时式(10)所描述的运动方程描述变为

$$\mathbf{M}\mathbf{X}+(\mathbf{C}-\mathbf{C}_{ae}')\mathbf{X}+(\mathbf{K}-\mathbf{K}_{ae})\mathbf{X}=\mathbf{0}\tag{14}$$

式中：C'_{ae}为修改的气动阻尼矩阵，$C'_{ae}=C_{ae}-\beta K_{ae}$。

2 数值方法

从式(6)～(8)可以看出，Matrix27 单元的的系数矩阵依赖于 3 个参数：风速、频率和量纲一的频率。其实只有 2 个参数是独立的，因此桥梁颤振的识别需要对风速进行搜索及对频率进行迭代。笔者采用逐个模态跟踪法求解。

步骤 1：建立无 Matrix27 单元的初始有限元模型，计算前 m 阶频率 $\omega_i^0(i=1,2,\cdots,m)$。

步骤 2：建立有 Matrix27 单元的整合系统有限元模型，颤振导数以表的方式储存。

步骤 3：给定初始搜索风速 U_0 和风速增量 ΔU。

步骤 4：假定系统的的第 i 阶振动频率 ω_0 为其初始模态圆频率 ω_i^0。

步骤 5：根据当前风速和振动频率确定 Matrix27 的刚度矩阵和阻尼矩阵，进行阻尼特征值分析。

步骤 6：比较计算的第 i 对特征值 λ_i 的虚部和试算值 ω_0。如果$|(\mathrm{Im}\ \lambda_i-\omega_0)/\mathrm{Im}\ \lambda_i|>10^{-3}$，则令 $\omega_0=\mathrm{Im}\ \lambda_i$，重复步骤 5 和步骤 6；否则进入步骤 7。

步骤 7：对所有考虑的前 m 个复特征值重复步骤 4～6。如果所有特征值的实部小于 0，则令 $U_0=U_0+\Delta U$，重复步骤 4～7；否则停止。

上述计算过程很容易通过 ANSYS 提供的参数化设计语言 APDL 实现。

3 数值算例

3.1 具有理想平板断面的简支梁桥

为验证本文模型和求解方法的正确性，笔者先以理想平板断面的简支梁桥为例介绍 ANSYS 中全模态颤振频域分析的实现过程。简支梁桥长300 m，宽 40 m，两端的扭转自由度均固定。平板断面的竖向刚度和横向刚度分别为 2.1×10^6 MPa·m^4、1.8×10^7 MPa·m^4，扭转刚度为 4.1×10^5 MPa·m^4。每延米质量为20 t，质量惯性矩为 4.5×10^6 kg·m，空气密度为 1.248 kg·m^{-3}。

在结构的初始有限元模型中，对质量的考虑采用集中质量矩阵形式。桥面主梁采用 Beam4 单元，质量惯性矩采用 Mass21 单元模拟。模型有 30 个桥面梁单元和 29 个扭转质量单元，见图 4。表 1 给出了计算得到的前 10 阶自振频率。

图 4 桥梁结构的初始有限元模型

Fig. 4 Initial Finite Element Model for Bridge Structure

表 1 简支梁前 10 阶频率

Tab. 1 The First Ten Order Frequencies of Simple-supported Girder

阶 次	频率/Hz	振型描述	阶 次	频率/Hz	振型描述
1	0.178 8	对称竖弯	6	1.503 0	对称扭转
2	0.502 8	对称扭转	7	1.609 6	对称竖弯
3	0.523 6	对称侧弯	8	1.997 6	反对称扭转
4	0.715 4	反对称竖弯	9	2.094 4	反对称侧弯
5	1.004 3	反对称扭转	10	2.486 7	对称扭转

在颤振分析模型中，桥梁结构同样采用 30 个桥面梁单元和 29 个扭转质量单元。此外，采用 58 个 Matrix27 单元模拟桥面受到的自激气动力，其中 29 个 Matrix27 单元用于模拟气动刚度，另外 29 个单元用于模拟气动阻尼，颤振分析有限元模型见图 5。

图 5 用于颤振分析的有限元模型

Fig. 5 Finite Element Model for Flutter Analysis

在颤振分析时，假定结构阻尼为 0，采用理想平板颤振导数在风速范围 0～180 m·s^{-1}内分析系统的低阶复模态随风速的变化。图 6 给出了系统的前 10 对复特征值的实部和虚部随风速的变化。从图 6 可看出：对竖弯模态，其复特征值的虚部和实部，即振动频率和阻尼，都随风速的增大而减少；对侧弯模态，其振动频率和阻尼几乎不变；对于扭转模态，其振动频率随着风速增大而减少，阻尼则先减小后增大。从图 6(a)可看出，在风速约为 135.10 m·s^{-1}时，第 2 阶复特征值的实部几乎为 0，表示颤振即将发生，与此对应的振动频率为 0.394 0 Hz。从图 6 还可看出：当某阶频率对应的量纲一的风速超过 10.0 后，系统的该阶复模态将不再随风速的增大而改变，定性的解释和结果有待于进一步研究。

图 7 给出了第 2 阶复模态振型在竖弯和扭转 2 个方向的广义质量随风速的变化。从图 7 可看出：随着风速的增大，第 2 阶复模态逐渐从纯扭转模态转化为竖弯与扭转耦合的振动模态。

对此简单结构，可以采用时域线性颤振分析方法来验证频域分析方法的计算结果。表 2 给出了不同方法的结果，其中精确解由Bleich提出的二模态

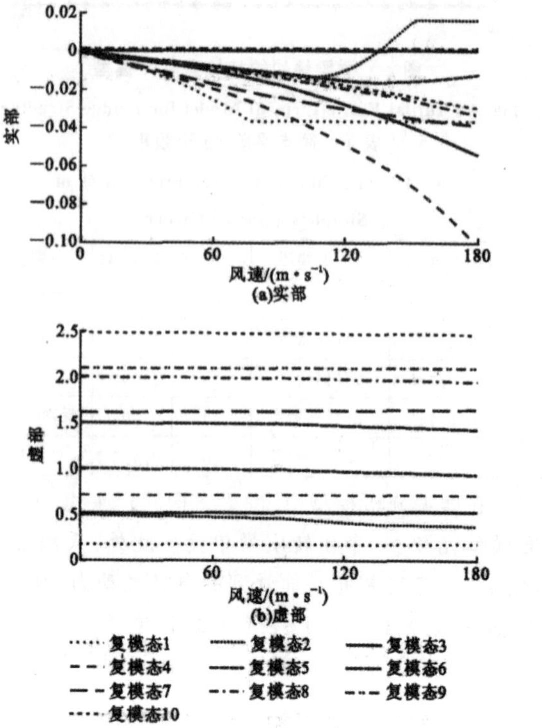

图 6　复特征值随风速的变化

Fig. 6　Variations of Complex Eigenvalues vs Wind Velocities

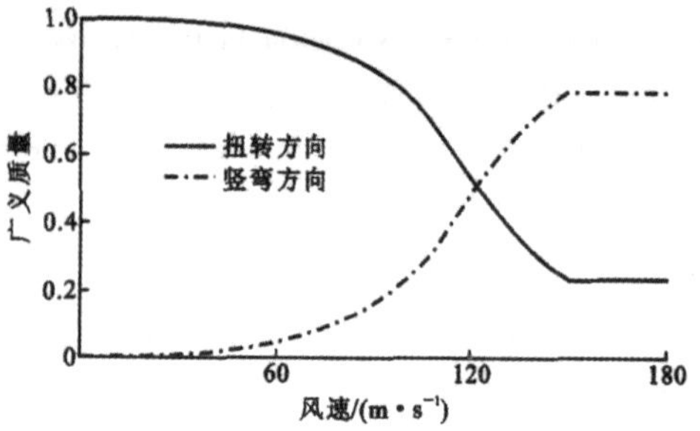

图 7　第 2 阶复模态广义质量随风速的变化

Fig. 7　Variation of Generalized Mass vs Wind Speed for the Second Order Complex Mode

表 2　颤振分析结果比较 1

Tab. 2　Comparison 1 of Flutter Analysis Results

方　法	颤振风速/(m·s^{-1})	颤振频率/Hz
本文方法(频域)	135.1	0.394 0
本文方法(时域)	134.1	0.394 0
M-S 法	134.3	0.393 6
精确解	136.3	0.391 4

注:本文精确解与文献[15]略有不同,可能是由于理想平板颤振导数差异引起的。

颤振理论得到;M-S 结果采用多模态参与单参数搜索法得到。从表 2 中可以看出本文方法与精确解及 M-S 法吻合良好。

3.2　虎门悬索桥

虎门悬索桥是主跨为 888 m 的单跨悬索桥,桥面宽 35.6 m。图 8 为虎门悬索桥的初始有限元模型,模型中共有 642 个节点和 736 个单元。表 3 给出了其前 10 阶频率。同样,在初始有限元模型基础上,通过在桥面主梁节点处添加 148 个 Matrix27 单元来模拟桥面受到的自激力,并形成颤振分析有限元模型,如图 9 所示。

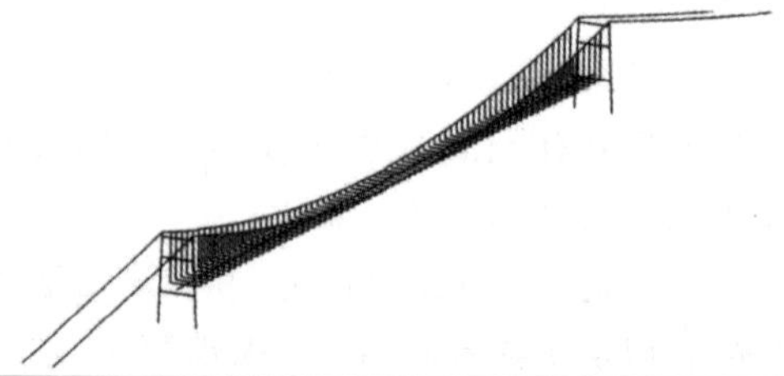

图 8　虎门悬索桥初始有限元模型

Fig. 8　Initial Finite Element Model of Humen Bridge

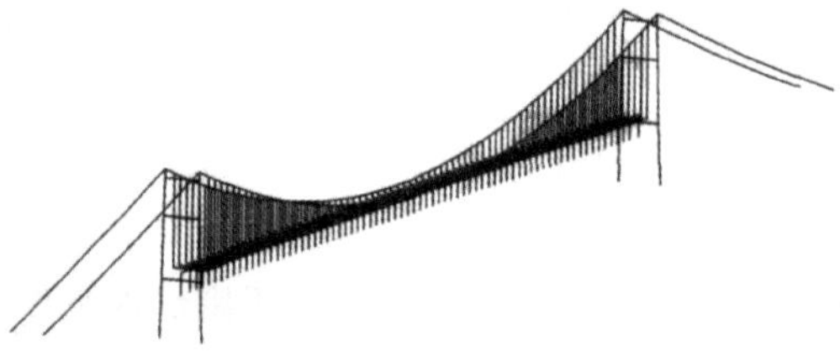

图 9　用于颤振分析的虎门悬索桥有限元模型

Fig. 9　Finite Element Model of Humen Suspension Bridge for Flutter Analysis

表 3　虎门悬索桥前 10 阶频率

Tab. 3　The First Ten Order Frequencies for Humen Suspension Bridge

阶　次	频率/Hz	振型描述	阶　次	频率/Hz	振型描述
1	0.090 1	对称侧弯	6	0.283 1	反对称侧弯
2	0.114 4	反对称竖弯	7	0.284 0	反对称竖弯
3	0.161 2	反对称竖弯	8	0.315 8	主缆振动
4	0.174 3	对称竖弯	9	0.326 1	主缆振动
5	0.227 2	对称竖弯	10	0.329 7	对称扭转

首先采用理想平板颤振导数并不考虑结构阻尼,在风速 0～120 m·s^{-1}内分析了前 10 对特征值随风速的变化,其结果见图 10。对于密频结构,随着风速的增加,振动频率可能出现交叉。为减小错误跟踪频率的可能性,采用振型相关系数 C 来判断前后级风速的模态配对

$$C=\frac{(\boldsymbol{\Phi}^{\mathrm{T}}\boldsymbol{\varphi})^2}{(\boldsymbol{\Phi}^{\mathrm{T}}\boldsymbol{\Phi})(\boldsymbol{\varphi}^{\mathrm{T}}\boldsymbol{\varphi})} \tag{15}$$

式中:$\boldsymbol{\Phi}$为在风速 U_{i-1}时的模态振型;$\boldsymbol{\varphi}$ 为在风速 U_i 时的模态振型。C 的数值越接近于 1,说明 2 个模

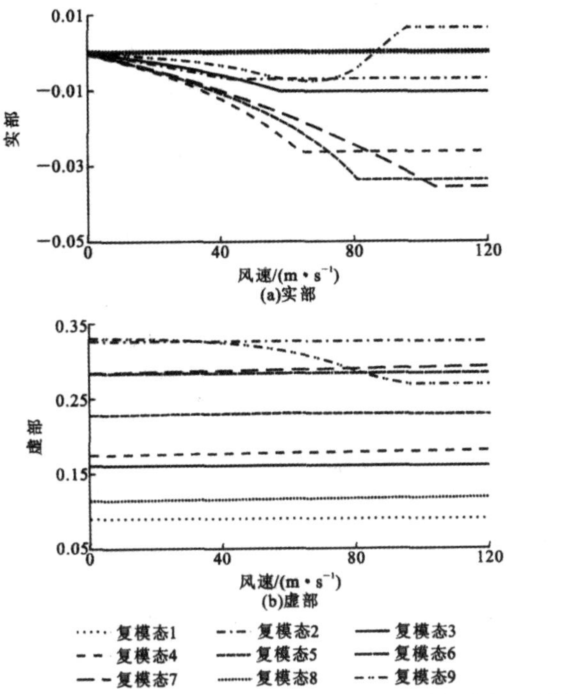

图 10 基于理想平板颤振导数的复特征值随风速的变化

Fig. 10 Variations of Complex Eigenvalues vs Wind Velocities Based on Flutter Derivatives of Ideal Flat

态为同一阶模态的可能性越大。从图 10 可以看出：当不考虑结构阻尼时预测的颤振临界风速和颤振频率分别为 85.90 $m \cdot s^{-1}$和 0.278 1 Hz；当结构各阶模态阻尼比为 0.5%时，其颤振临界风速和颤振频率分别为 92.23 $m \cdot s^{-1}$和 0.275 2 Hz。

在 0°和 3°攻角下实测颤振导数，并假定各阶模态阻尼比为 0.5%，重新计算各复模态随风速的变化，结果见图 11。表 4 为本文分析结果和 M-S 法计算值及风洞实测值的比较，结果表明本文方法和 M-S 法计算值及风洞实测结果非常接近，可以看出计算值和实测值相差小于 5%。

表 4 颤振分析结果比较 2

Tab. 4 Comparison 2 of Flutter Analysis Results

攻角/(°)	参 数	计算值		风洞实测值
		本文方法	M-S 法	
0	颤振临界风速/($m \cdot s^{-1}$)	89.36	89.55	88.00
	颤振频率/Hz	0.297 3	0.300 9	
3	颤振临界风速/($m \cdot s^{-1}$)	75.20	71.68	72.00
	颤振频率/Hz	0.305 0	0.308 9	

4 结 语

(1)笔者提出了直接采用 ANSYS 在物理坐标中分析大跨度桥梁结构三维耦合颤振的有限元模型和频域方法。以具有理想平板断面的简支梁桥和虎门悬索桥为例，采用理想平板理论颤振导数和实测颤振导数，在 ANSYS 中分析了颤振临界风速和颤振频率。数值算例表明，本文计算结果与精确解、其他多模态方法计算值及风洞试验值吻合良好，说明了本文方法的正确性和可靠性。本文方法使得桥梁设计和研究人员可以直接采用 ANSYS 确定大跨桥梁的最低颤振临界风速。研究结果也表明：当某阶频率对应的量纲一的风速超过 10.0 以后，系统的该阶复模态将不随风速的增大而改变，定性的解释和结论有待于进一步研究。

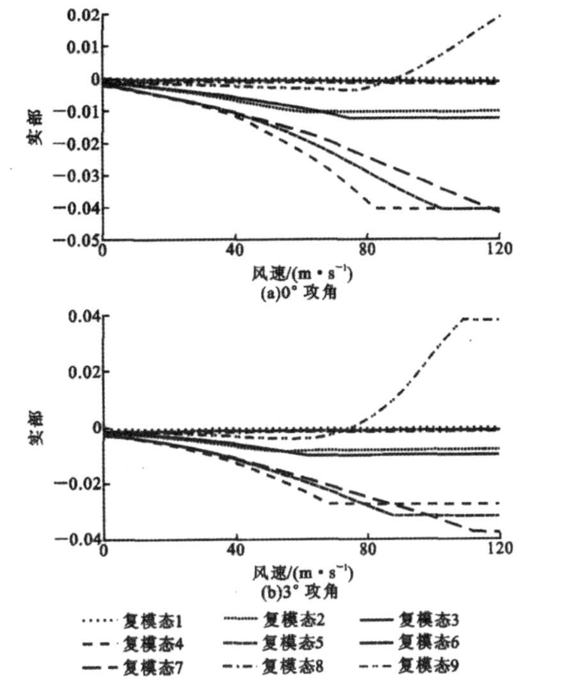

图 11 基于实测颤振导数的复特征值实部随风速的变化

Fig. 11 Variations of Real Part of Complex Eigenvalues vs Wind Velocities Based on Field Measured Flutter Derivatives

(2)本文方法属于颤振分析的精确方法或全阶方法，较常用的多模态分析方法计算费时。为减少计算量，可采用经验公式估算初始颤振临界风速，如 Van der Put 公式，事实上，不需要对所有复模态进行跟踪，只需要跟踪主要扭转模态的复特征值随风速的变化，因此计算量可以大大降低。

参考文献：

References：

[1] 项海帆，葛耀君. 悬索桥空气动力跨径的动力极限[J]. 土木工程学报，2005，38(1)：60-70.

XIANG Hai-fan, GE Yao-jun. On Aerodynamic Limits to Suspension Bridges[J]. China Civil Engineering Journal, 2005, 38(1): 60-70.

[2] BLEICH F. Dynamic Instability of Truss-stiffened Suspension Bridges Under Wind Action[J]. Transactions of American Society of Civil Engineering, 1948, 74(7): 1 269-1 314.

[3] 谢霁明,项海帆. 桥梁三维颤振分析的状态空间法[J]. 同济大学学报:自然科学版,1985,3(3):1-12.

XIE Ji-ming, XIANG Hai-fan. State-space Method for 3-D Flutter Analysis of Bridge Structures[J]. Journal of Tongji University: Natural Science, 1985, 3(3): 1-12.

[4] AGAR T J A. Aerodynamic Flutter Analysis of Suspension Bridges by a Modal Technique[J]. Engineering Structures, 1989, 11(2): 75-82.

[5] NAMINI A, ALBRECHT P, BOSCH H. Finite Element-based Flutter Analysis of Cable-suspended Bridges[J]. Journal of Structural Engineering, 1992, 118(6): 1 509-1 526.

[6] 张新军,项海帆,陈艾荣. 大跨度悬索桥三维颤振非线性分析[J]. 土木工程学报,2002,35(5):42-46.

ZHANG Xin-jun, XIANG Hai-fan, CHEN Ai-rong. Three-dimensional Nonlinear Flutter Analysis of Long-span Suspension Bridges[J]. China Civil Engineering Journal, 2002, 35(5): 42-46.

[7] CHEN Z Q. Finite Element Based Flutter Analysis of Cable-suspended Bridges[J]. Journal of Structural Engineering, 1994, 120(3): 1 044-1 046.

[8] KATSUCHI H, JONES N P, SCANLAN R H. Multi-mode Coupled Flutter and Buffeting Analysis of the Akashi-Kaikyo Bridge[J]. Journal of Structural Engineering, 1999, 125(1): 60-70.

[9] 华旭刚,陈政清. 桥梁风致颤振临界状态的全域自动搜索法[J]. 工程力学,2002,19(2):68-72.

HUA Xu-gang, CHEN Zheng-qing. Wind-induced Flutter Analysis of Long-span Bridges with Fully Automatic Searching Method[J]. Engineering Mechanics, 2002, 19(2): 68-72.

[10] 丁泉顺,陈艾荣,项海帆. 大跨度桥梁多模态耦合颤振的自动分析[J]. 土木工程学报,2002,35(4):52-58.

DING Quan-shun, CHEN Ai-rong, XIANG Hai-fan. Automatic Analysis for Multimode-coupled Flutter of Long-span Bridges[J]. China Civil Engineering Journal, 2002, 35(4): 52-58.

[11] 刘 高,强士中,周述华. 悬索桥颤振的多模态耦合及机理研究[J]. 土木工程学报,2001,34(3):59-66.

LIU Gao, QIANG Shi-zhong, ZHOU Shu-hua. Study on Flutter Stability for Suspension Bridges and Mechanism of Multi-mode Coupling[J]. China Civil Engineering Journal, 2001, 34(3): 59-66.

[12] MIYATA T, YAMADA H. Coupled Flutter Estimate of a Suspension Bridge[J]. Journal of Wind Engineering and Industrial Aerodynamics, 1990, 33(1): 341-348.

[13] DUNG N N, MIYATA T, YAMADA M N. Flutter Responses in Long Span Bridges with Wind Induced Displacement by the Mode Tracing Method[J]. Journal of Wind Engineering and Industrial Aerodynamics, 1998, 77(8): 367-379.

[14] GE Y J, TANAKA H. Aerodynamic Analysis of Cable-supported Bridge by Multi-model and Full-mode Approaches[J]. Journal of Wind Engineering and Industrial Aerodynamics, 2000, 86(3): 123-153.

[15] 丁泉顺,陈艾荣,项海帆. 大跨度桥梁耦合颤振的全阶分析方法[J]. 振动工程学,2002,15(4):436-441.

DING Quan-shun, CHEN Ai-rong, XIANG Hai-fan. Coupled Flutter Analysis of Long-span Bridges by Full-order Approach[J]. Journal of Vibration Engineering, 2002, 15(4): 436-441.

[16] 华旭刚,陈政清,祝志文. 一种在ANSYS中实现颤振时域分析的方法[J]. 中国公路学报,2002,15(4):32-34.

HUA Xu-gang, CHEN Zheng-qing, ZHU Zhi-wen. Approach of Time-history Analysis of Flutter in ANSYS[J]. China Journal of Highway and Transport, 2002, 15(4): 32-34.

[17] 华旭刚,陈政清. 一种基于ANSYS的颤振频域分析方法[C]//中国土木工程学会桥梁结构分会. 第12届全国结构风工程会议论文集. 北京:人民交通出版社,2005:517-525.

HUA Xu-gang, CHEN Zheng-qing. A Flutter Frequency Domain Analysis Method Based on ANSYS[C]//Bridge Structure Branch of China Civil Engineering Society. The 12th National Structural Wind Engineering Conference Symposium. Beijing: China Communications Press, 2005: 517-525.

[18] 胡峰强,陈艾荣. 基于ANSYS实现全模态颤振分析的实用方法[C]//中国土木工程学会桥梁结构分会. 第12届全国结构风工程会议论文集. 北京:人民交通出版社,2005:506-509.

HU Feng-qiang, CHEN Ai-rong. Practical Method of Full-mode Flutter Analysis Based on ANSYS[C]//Bridge Structure Branch of China Civil Engineering Society. The 12th National Structural Wind Engineering Conference Symposium. Beijing: China Communications Press, 2005: 506-509.

第22卷 第6期 2009年11月
中 国 公 路 学 报
China Journal of Highway and Transport
Vol.22 No.6 Nov. 2009

文章编号：1001-7372(2009)06-0053-07

中央稳定板提高桁架梁悬索桥颤振稳定性的气动机理

陈政清，欧阳克俭，牛华伟，华旭刚
(湖南大学 风工程试验研究中心，湖南 长沙 410082)

摘要：为了研究中央稳定板的气动作用机理，首先采用三自由度耦合振动时域识别的强迫振动法，一次识别出桁架断面在有无中央稳定板时的18个气动导数，然后以识别的气动导数作为输入参数，利用自主研发的NACS软件对矮寨大桥进行了颤振分析；最后通过能量原理对中央稳定板的作用机理进行了证明。结果表明：中央稳定板对气动性能的影响主要体现在气动导数 A_2^* 降低了气动负阻尼；中央稳定板的作用机理是提高了颤振发生时竖弯自由度的参与程度和扭弯耦合程度，使得颤振形态由单自由度扭转振动向弯扭耦合振动转移，从而降低了颤振频率，提高了颤振临界风速。

关键词：桥梁工程；中央稳定板；强迫振动法；颤振稳定性；能量原理
中图分类号：U448.25　　**文献标志码**：A

Aerodynamic Mechanism of Improvement of Flutter Stability of Truss-girder Suspension Bridge Using Central Stabilizer

CHEN Zheng-qing, OUYANG Ke-jian, NIU Hua-wei, HUA Xu-gang
(Wind Engineering Experiment Research Center, Hunan University, Changsha 410082, Hunan, China)

Abstract: In order to analyze the aerodynamic mechanism of central stabilizers, firstly, a data processing method for identification of the bridge aerodynamic derivatives in time domain was presented based on three degrees of freedom coupling forced vibration method and 18 aerodynamic derivatives were identified at truss-girder section with and without central stabilizer. Regarded identified aerodynamic derivatives as input parameters, a multimode-based flutter analysis for predicting the lowest critical flutter velocity of Aizhai Bridge was then carried out by using the software NACS. Aerodynamic mechanism of central stabilizer was verified by energy principle. The results show that influences of central stabilizer on aerodynamic property mainly incarnate that aerodynamic derivatives A_2^* decreases negative aerodynamic damping. The extent of mode coupling between vertical-bending mode and torsional mode increases when using the central stabilizers. It makes flutter mode shift from single degree of freedom torsional vibration to coupling vibration between vertical-bending mode and torsional mode so that flutter frequency is decreased and flutter critical wind velocity is increased.

Key words: bridge engineering; central stabilizer; forced vibration method; flutter stability; energy principle

---

收稿日期：2009-01-05
基金项目：国家自然科学基金重点项目(50738002)
作者简介：陈政清(1947-)，男，湖南湘潭人，教授，博士研究生导师，工学博士，E-mail:zqchen@hnu.cn。

0 引 言

桁架加劲梁具有较大的抗扭刚度和较高的透风率等众多优点，且方便在交通运输困难的峡谷地区施工，因此成为了中国跨越山区峡谷的悬索桥建设中一种被广泛采用的加劲梁。例如主跨 1 176 m 的湖南湘西矮寨大桥、位于黔西地区的主跨 1 088 m 的坝陵河大桥、主跨 900 m 的湖北四渡河峡谷大桥等均采用了桁架加劲梁悬索桥。由于山区峡谷地区风环境在时间和空间上变化快而且复杂，如何确保大桥的气动安全就成为控制因素之一。采用适当的气动控制措施满足颤振稳定性要求成为了一种广泛采用的方法[1-2]。

风洞试验证明，在主梁断面上设置中央稳定板是提高桥梁颤振稳定性的一种有效方法。日本学者 Ueda 等[3]认为，在明石海峡悬索桥的桁梁桥面下设置中央稳定板能有效提高颤振临界风速。中国的项海帆等[4]发现润扬悬索桥在箱梁上表面设置中央稳定板之后，颤振临界风速从 62 $m \cdot s^{-1}$ 提高到了 79 $m \cdot s^{-1}$。Matsumoto 等[5]采用分部求解的方法对一些桥梁断面的颤振机理进行了一系列研究。Larsen[6]基于 CFD 中的离散涡方法对 Tacoma 桥的颤振机理进行了研究。杨詠昕等[7]利用二维三自由度耦合颤振分析方法对几个典型桥梁断面和一些气动控制措施的颤振机理进行了研究。尽管不少学者对颤振机理进行了研究，但是对气动控制措施作用下气动机理的研究相对较少。

本文中以提高湖南矮寨大桥颤振稳定性为背景，研究了中央稳定板对桁梁加劲的悬索桥气动性能的影响。首先通过弹性悬挂二维节段模型试验得到了节段模型不同工况下的颤振临界风速，然后利用三自由度耦合振动的风洞试验强迫振动法[8-10]，一次识别了节段模型的 18 个气动导数。以这些气动导数作为输入参数，应用笔者自主研发的大跨度桥梁静动力有限元程序 NACS 软件[11]中的多模态参与单参数搜索的颤振分析方法，对矮寨大桥颤振临界状态进行了分析。

1 工程概况

矮寨大桥为湖南吉首至茶洞公路段的双塔两主缆地锚式悬索桥，主跨跨径 1 176 m，以桥隧相连方式，一跨越过跨度超过 1 000 m 的矮寨峡谷，跨中桥面距谷底高度 340 m，采用钢桁加劲梁，梁宽 27 m，梁高 7.5 m(上下弦杆形心距离)。桥面原设计方案为中央开槽分离车道的混凝土桥面板方案，达不到颤振稳定性要求。为此湖南大学风工程试验研究中心对矮寨大桥各设计阶段进行了抗风性能研究。通过节段模型风洞试验对钢桁加劲梁进行了气动性能优化措施研究，试验了桥面板中央开槽、设中央稳定板、设水平稳定板等多种可能的气动措施，最终确定了方便有效的封闭原方案的中央开槽带、设置上下中央稳定板的措施[12]。其中上稳定板即桥面中央高 1 m 的不透风中央防撞栏杆，无需另行设置。下稳定板高 1 m，紧贴桥面板下表面设置。图 1 为矮寨大桥最终方案的标准横断面。

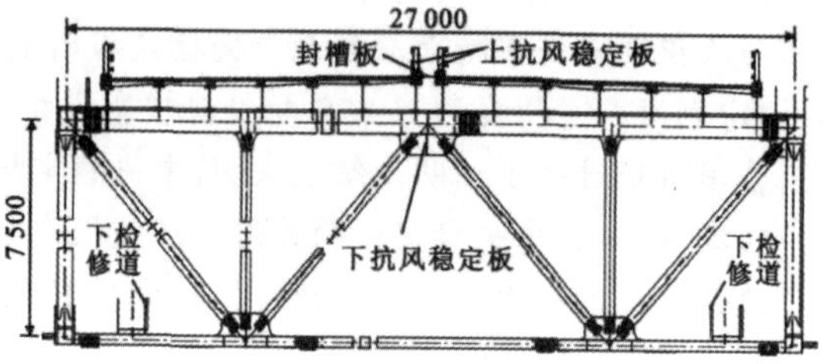

图 1 标准横断面(单位:mm)

Fig.1 Standard Cross Section (Unit:mm)

节段模型颤振试验是在湖南大学风工程试验研究中心 HD-2 号风洞第 1 试验段边界层风洞中进行的。节段模型采用洞内支架弹性悬挂，试验参数如表 1 所示，整个节段模型测振装置如图 2 所示。

表 1 颤振临界风速试验的节段模型主要参数

Tab.1 Main Parameters of Section Model for Critical Flutter Wind Velocity Test

参数	实桥值	模型值
主桁架长度 L/m	87.0	1.74
主桁架宽度 B/m	27.0	0.54
主桁架高度 H/m	7.5	0.15
等效质量 m_{eq}/($kg \cdot m^{-1}$)	30 165	12.066
等效质量惯性矩 $J_{m,eq}$/($kg \cdot m^2 \cdot m^{-1}$)	3 085 095	0.493 6
竖弯基频 f_h/Hz	0.160 5	1.758
扭转基频 f_t/Hz	0.294 5	3.32
扭弯频率比 ε	1.835	1.835

2 节段模型气动导数的强迫振动法识别

2.1 气动导数的基本定义

根据 Scanlan 线性自激气动力表达式，单位长度上受到的气动升力 L_{se}、阻力 D_{se} 和气动扭矩 M_{se} 是竖向位移 h、水平位移 p 和扭转位移 α 的函数

图 2　节段模型测振试验装置

Fig. 2　Vibration Test Device for Section Model Testing

$$
\left.
\begin{aligned}
L_{se} &= \frac{1}{2}\rho U^2(2B)\Big(KH_1^* \frac{\dot{h}}{U} + KH_2^* \frac{B\dot{\alpha}}{U} + \\
&\quad K^2 H_3^* \alpha + K^2 H_4^* \frac{h}{B} + KH_5^* \frac{\dot{p}}{U} + K^2 H_6^* \frac{p}{B}\Big) \\
D_{se} &= \frac{1}{2}\rho U^2(2B)\Big(KP_1^* \frac{\dot{p}}{U} + KP_2^* \frac{B\dot{\alpha}}{U} + \\
&\quad K^2 P_3^* \alpha + K^2 P_4^* \frac{p}{B} + KP_5^* \frac{\dot{h}}{U} + K^2 P_6^* \frac{h}{B}\Big) \\
M_{se} &= \frac{1}{2}\rho U^2(2B^2)\Big(KA_1^* \frac{\dot{h}}{U} + KA_2^* \frac{B\dot{\alpha}}{U} + \\
&\quad K^2 A_3^* \alpha + K^2 A_4^* \frac{h}{B} + KA_5^* \frac{\dot{p}}{U} + K^2 A_6^* \frac{p}{U}\Big)
\end{aligned}
\right\} \quad (1)
$$

式中：U 为风速；ρ 为空气的质量密度；B 为桥面宽；K 为量纲一的频率，$K=B\omega/U$；ω 为振动圆频率；H_i^*，A_i^*，P_i^*（$i=1,2,\cdots,6$）为量纲一的气动导数，它们是量纲一的风速 $\bar{U}=U/(fB)$ 或量纲一的频率的函数，与桥梁截面的几何形状有关，f 为振动频率。

2.2　气动导数识别的强迫振动法

湖南大学风工程试验研究中心开发了一套完全数控的三自由度强迫振动装置。该装置两端与节段模型刚性连接，一端置于风洞外，置于风洞内的一端外包流线形的整流罩。强迫振动法利用专用的机械装置驱动模型作频率、振幅及相位可控的简谐振动，直接测量作用在模型上的气动力和模型运动的位移或加速度等振动信号，然后利用参数识别方法得到气动导数。与自由振动相比，强迫振动法识别过程简单、信号的信噪比高、气动导数的量纲一的频率识别范围更广且有更高的识别精度，便于研究气动导数的影响参数和气动力特性。图 3 为强迫振动装置两端驱动的原理。图 4 为安装在强迫振动装置上的节段模型。

2.3　节段模型设计

由于强迫振动装置要求模型长度小于 1.54 m

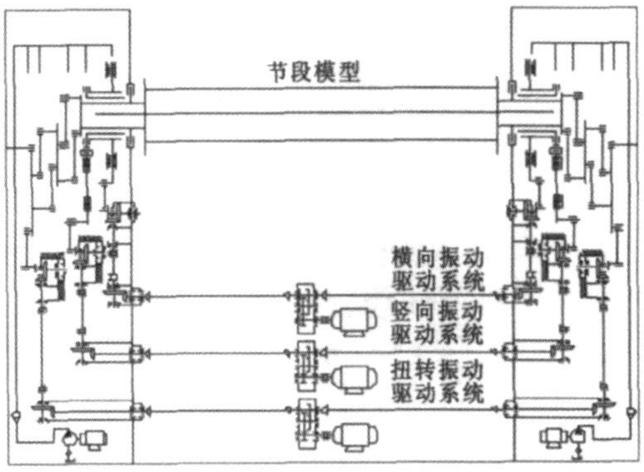

图 3　强迫振动驱动原理

Fig. 3　Driving Principle of Forced Vibration

图 4　固定在强迫振动装置上的节段模型

Fig. 4　Section Model Fixed on Forced Vibration Device

且要尽量轻以减少惯性力，又特别制作了第 2 个节段模型。节段模型 2 的外框采用铝合金材料制作而成，内部桁架用 ABS 材料经电脑雕刻机雕刻而成，内包方形木棍，尽量减轻模型质量，以减小节段模型惯性力对强迫振动装置测力的影响，其中主桁各弦杆及桥面行车道系、栏杆、检修轨道等都按缩尺比精确模拟，以确保气动外形与实桥成相似关系。桥面与主桁采用刚性连接以确保刚性节段模型的风洞试验要求。模型缩尺比为 1/50，长度 1.53 m，质量为 7.5 kg。

2.4　气动导数识别的强迫振动时域法

湖南大学风工程试验研究中心开发的全数控三自由度强迫振动系统通过编码器的反馈脉冲获取模型运动的位移信号，再由位移信号经数值微分得到速度信号。与记录模型振动加速度的方法相比，直接获取位移时程信号的方法避免了通过运动加速度信号积分得到速度和位移信号的困难过程，提高了时域方法的识别精度。

本文中选用三自由度耦合强迫振动一次识别 18 个气动导数，模型三自由度的振动试验参数如

表2所示。依据记录的三自由度位移时程信号和模型振动的竖向力、水平力和扭转时程信号，采用基于最小二乘原理的时域识别法识别出所有18个气动导数，强迫振动时域识别法的原理与计算公式见文献[13]。

表2　节段模型强迫振动试验参数

Tab.2　Test Parameters of Section Model for Forced Vibration Method

竖向振动频率/Hz	1.6	侧向振动频率/Hz	1.1
扭转振动频率/Hz	2.6	竖向振动幅值/m	0.012
侧向振动幅值/m	0.012	扭转振动幅值/(°)	2
采样频率/Hz	200	采样点数	12 400
风攻角/(°)	0,3		

2.5　气动导数识别结果与分析

本文中对0°和3°两种主要攻角的气动导数进行了识别，结果见图5～7。

对于0°攻角，首先分析直接导数 A_2^*。A_2^* 表示扭转速度 $\dot{\alpha}$ 对气动升力矩的贡献，表现为一种气动阻尼效应，若 A_2^* 为负，相当于正阻尼，反之则为负阻尼。有无稳定板的2组 A_2^* 都出现了由负转正的情况，但从图5(b)可以看出，设置了稳定板的 A_2^* 出现由负转正的折减风速较没有设置稳定板的要高很多，而且有稳定板 A_2^* 的绝对值相对无稳定板的要小，也就是说由 A_2^* 所对应的气动负阻尼也就小。

有无稳定板的导数 H_2^* 的变化趋势相同，但有稳定板的 H_2^* 绝对值较无稳定板的要大3倍以上。

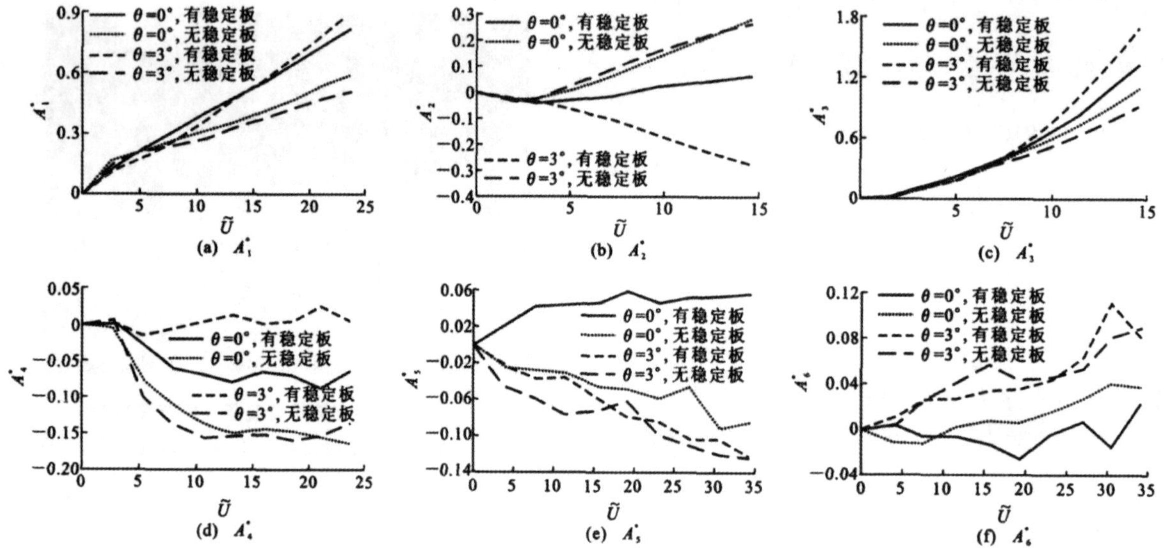

图5　不同攻角有无稳定板节段模型气动导数 A_i^* 对比

Fig.5　Comparisons of Aerodynamic Derivatives of Section Model with and Without Central Stabilizers Under Different Attack Angles

H_1^*、H_3^*、H_4^*、A_1^*、A_3^* 这5种气动导数的相应数值相差很小，变化趋势一致。其他数值较小的次要导数变化趋势相同，只是在数值上有所差异。

对于3°攻角，有无稳定板的所有导数中 A_2^* 差别最为明显，两者变化趋势几乎相反。无稳定板的 A_2^* 在相对较低的折减风速出现了由负转正，有稳定板的 A_2^* 一直为负值，绝对值随着折减风速的增大逐渐增大。2组导数 H_4^* 都出现了由负转正的情况，但有稳定板的 H_4^* 由负转正的转折点相对无稳定板靠前，在折减风速相对较小时就出现了正值。其他导数的变化情况类似于0°攻角。在3°攻角，有无稳定板对气动外形的影响主要体现在 A_2^*、H_2^*、H_4^* 变化规律的改变上。

综合比较2种攻角的气动导数，变化最为明显的是 A_2^*，A_2^* 值为正所对应的气动负阻尼是造成颤振发生的主要原因之一。2种攻角的气动导数最为接近的是 H_3^*，不管是否设置了稳定板，或者改变攻角，2组导数的数值接近。

3　全桥三维颤振稳定性分析

3.1　M-S法简介

多模态参与的单参数搜索(M-S)法是一种全桥的三维颤振分析方法，该法最后归结为一个 m 阶的复特征值问题，在量纲一的风速范围内直接对量纲一的风速 $\bar{v}$ 进行搜索，应用于中国虎门桥、江阴桥等多座大桥颤振分析，均能与全桥模型试验基本吻合。经改进后的M-S法实现了量纲一的风速范围内多个可能颤振形态的全域自动搜索，构造了考虑结构

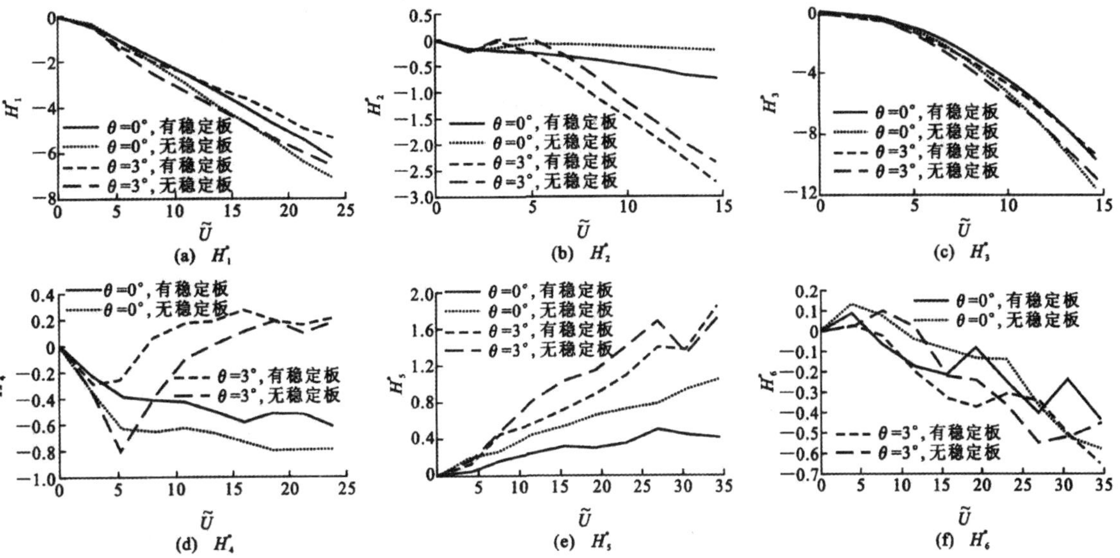

图 6　不同攻角有无稳定板节段模型气动导数 H_i^* 对比

Fig. 6　Comparisons of Aerodynamic Derivatives H_i^* of Section Model with and Without Central Stabilizers Under Different Attack Angles

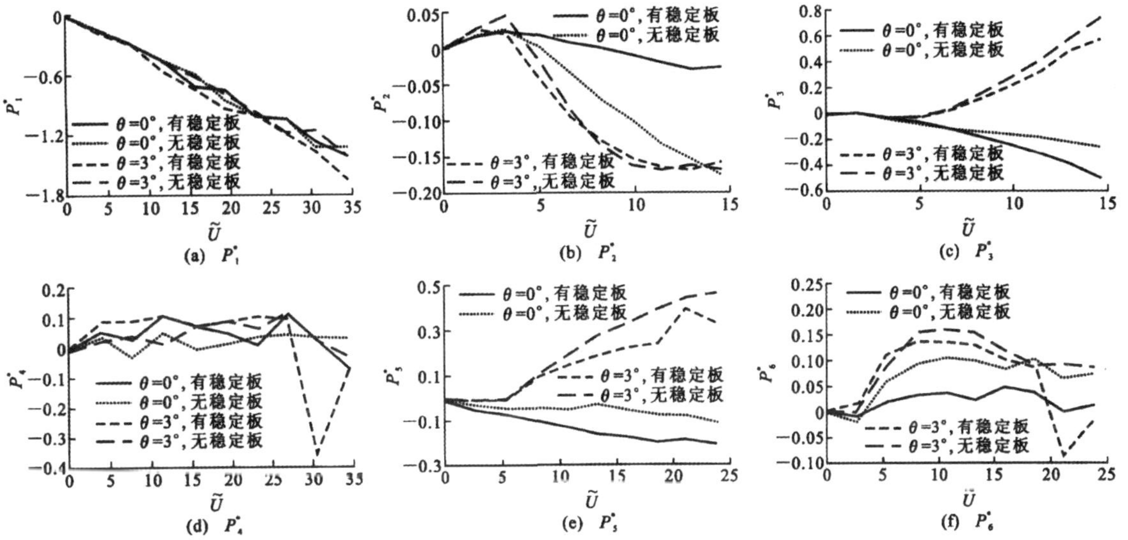

图 7　不同攻角有无稳定板节段模型气动导数 P_i^* 对比

Fig. 7　Comparisons of Aerodynamic Derivatives P_i^* of Section Model with and Without Central Stabilizers Under Different Attack Angles

阻尼的自动迭代方法,全部计算过程已在笔者开发的大跨桥梁空间非线性静动力分析程序 NACS 中实现,有关 M-S 法的计算原理见文献[11]～[15]。

3.2　实例分析

根据笔者自主研发的 NACS 程序,应用其中的自振计算模块计算出该桥前 30 阶振型,颤振分析按 NACS 程序多模态参与单参数全域自动搜索法进行,利用攻角 0°、3°时的气动导数进行三维颤振分析,各阶振型阻尼比取为 0.005,应用其中的颤振振型分析模块,根据能量原理得到了各个颤振参与振型的能量百分比。计算结果与节段模型试验结果见表 3。

其中第 12 阶振型为正对称扭转振型,第 4 和第 5 阶振型都为正对称竖弯振型。从表 3 可以看出:对于未加稳定板的断面形式,多模态耦合颤振的主要参与振型均为一阶正对称扭转,且几乎所有的能量都集中到一阶正对称扭转振型上,这说明桁架加劲梁形式的悬索桥颤振的发生,仍然是纯扭转振动颤振。加上中央稳定板之后,提高了颤振发生时竖弯自由度的参与程度和扭弯耦合程度,而这种强的

表3 矮寨大桥颤振分析结果

Tab.3 Results of Flutter Analyses of Aizhai Bridge

颤振检验风速/(m·s⁻¹)	攻角/(°)	有无稳定板	折减风速	颤振频率/Hz	NACS临界风速/(m·s⁻¹)	参与振型	能量百分比/%	节段模型试验临界风速/(m·s⁻¹)
51.5	0	有	8.844	1.891	71.85	12、4、5	90.2、4.9、4.7	75.30
		无	6.938	1.921	57.28	12	99.9	52.66
	3	有	19.063	1.631	133.59	5、4、12	43.5、28.7、27.1	105.00
		无	5.469	1.944	45.68	12	97.3	48.98

竖向自由度参与程度有利于结构颤振稳定性能。这证明中央稳定板的作用机理是增加竖弯自由度的参与程度，使颤振形态由单自由度扭转振动向弯扭耦合振动转移。

对于3°攻角有稳定板的情况，理论值和试验值差别较大，主要原因是理论计算的折减风速(19.063)超过了几项重要气动导数折减风速(H_2^*、H_3^*、A_2^*、A_3^*)的实际测量区间，超过区间范围里使用的气动导数值是试验值的线性外插值，存在一定误差，这将在一定程度上影响NACS的识别精度。总的来说，从试验和理论结果都能看出，中央稳定板对3°风攻角的颤振控制效果非常显著。

4 单个气动导数对颤振稳定性的影响

如果在颤振分析中将无稳定板情况下的1个或2个气动导数用有稳定板气动导数替换，可以考察单个气动导数对提高颤振稳定性的贡献。本文中分别替换气动导数A_2^*、H_2^*，以及同时变换A_2^*和H_2^*，进行了3次颤振分析，0°和3°攻角的各3次颤振分析结果如表4所示。

表4 改变单个气动导数的颤振分析结果

Tab.4 Results of Flutter Analyses of Changing Single Aerodynamic Derivative

攻角/(°)	改变的气动导数	折减风速	颤振频率/Hz	NACS临界风速/(m·s⁻¹)	参与振型	能量百分比/%
0	A_2^*	15.250	1.789	117.21	12,5,4	98.6,0.7,0.6
	H_2^*	7.031	1.921	58.03	12	99.8
	A_2^*，H_2^*	17.750	1.750	133.51	12,5,4	97,1.6,1.3
3	A_2^*	34.625	1.657	246.52	5,4,12	48.2,43.5,7.6
	H_2^*	5.531	1.652	46.20	5,4,12	97,1.5,1.3
	A_2^*，H_2^*	35.938	1.994	255.09	12,4,5	48.3,44.1,7

颤振分析结果表明，A_2^*和H_2^*两项气动导数对颤振分析结果都有影响。对于0°攻角，单独改变H_2^*只能将颤振临界风速提高1.3%，单独改变A_2^*则能将颤振临界风速提高104.6%，同时改变A_2^*和H_2^*能将颤振临界风速提高133.1%。对于3°攻角，单独改变H_2^*只能将颤振临界风速提高1.1%，单独改变A_2^*或者同时改变A_2^*和H_2^*都能很大程度提高颤振临界风速。计算结果表明：A_2^*是提高颤振临界风速的根本原因，H_2^*对颤振临界风速的影响较小，但是A_2^*和H_2^*共同作用的效果比A_2^*单独作用效果更好。比较表3、4还可以看到，仅替换A_2^*和H_2^*所获得的颤振临界风速比全部气动导数都替换为由稳定板的临界风速还高，这说明稳定的A_2^*和H_2^*有提高颤振临界风速的作用，而其他气动导数的作用则可能是降低颤振临界风速。

5 结 语

(1)通过强迫振动装置三自由度耦合振动法，一次识别了桁架节段模型的18个气动导数。

(2)中央稳定板对气动性能的影响主要体现在气动导数A_2^*、H_2^*、H_4^*变化规律的改变上，其中以2个攻角状态的A_2^*变化最为显著。没有稳定板的A_2^*出现了由负转正的情况，形成了气动负阻尼，使结构总阻尼减小，颤振频率增大，颤振临界风速降低。

(3)用笔者自主研发的NACS程序计算了有无稳定板2组气动导数对颤振临界状态主要参数的影响，验证了设置中央稳定板对提高颤振临界风速的正确性，由能量法证明稳定板的作用机理是提高了颤振发生时竖弯自由度的参与程度和扭弯耦合程度，使颤振形态由单自由度扭转振动向弯扭耦合振动转移，从而降低颤振频率，提高颤振临界风速。

参考文献：

References：

[1] XIANG H F，GE Y J. State-of-the-art on Long-span Bridge Aerodynamics in China[J]. Structural Engin-

eering International,2005,15(4):240-247.

[2] 宋锦忠,林志兴,徐建英.桥梁抗风气动措施的研究及应用[J].同济大学学报:自然科学版,2002,30(5):618-621.
SONG Jin-zhong, LIN Zhi-xing, XU Jian-ying. Research and Appliance of Aerodynamic Measures About Wind-resistance of Bridges [J]. Journal of Tongji University:Natural Science,2002,30(5):618-621.

[3] UEDA O,TANAKA H,MATSUSHITA Y. Aerodynamic Stabilization for Super Long-span Suspension Bridges[C]//IABSE. Proc IABSE Symposium Long-span and High-rise Structures. Kode:IABSE,1998:721-728.

[4] 项海帆,葛耀君.悬索桥跨径的空气动力极限[J].土木工程学报,2005,38(1):60-70.
XIANG Hai-fan, GE Yao-jun. On Aerodynamic Limits to Suspension Bridges[J]. China Civil Engineering Journal,2005,38(1):60-70.

[5] MATSUMOTO M,DAITO Y,YOSHIZUMI F,et al. Torsional Flutter of Bluff Bodies[J]. Journal of Wind Engineering and Industrial Aerodynamics,1997,69-71:871-882.

[6] LARSEN A. Aerodynamics of the Tacoma Narrow Bridge: 60 Years Later [J]. Structural Engineering International,2000,10(4):243-248.

[7] 杨詠昕,葛耀君,项海帆.中央稳定板颤振控制效果和机理研究[J].同济大学学报:自然科学版,2007,35(2):149-155.
YANG Yong-xin,GE Yao-jun,XIANG Hai-fan. Flutter Controlling Effect and Mechanism of Central Stabilizer[J]. Journal of Tongji University: Natural Science,2007,35(2):149-155.

[8] 于向东,陈政清.桥梁主梁断面颤振导数的强迫振动识别法[J].中国公路学报,2001,14(2):36-39.
YU Xiang-dong,CHEN Zheng-qing. Forced Vibration Method for Identifying Flutter Derivatives of Bridges Sectional Model [J]. China Journal of Highway and Transport,2001,14(2):36-39.

[9] 郭震山.桥梁断面气动导数识别的三自由度强迫振动法[D].上海:同济大学,2006.
GUO Zhen-shan. Three Degree-of-freedom Forced Vibration Method for Identification of Aerodynamic Derivatives of Bridge Decks [D]. Shanghai: Tongji University,2006.

[10] 牛华伟.气动导数识别的三自由度强迫振动法及颤振机理研究[D].长沙:湖南大学,2008.
NIU Hua-wei. The Research on Three Degree-of-freedom Forced Vibration Method for Identification of Aerodynamic Derivatives and Flutter Mechanism[D]. Changsha: Hunan University,2008.

[11] 陈政清.结构风工程研究新进展与应用[M].上海:同济大学出版社,1993.
CHEN Zheng-qing. New Development and Application of a Structural Wind Engineering[M]. Shanghai: Tongji University Press,1993.

[12] 湖南大学风工程试验研究中心.矮寨大桥悬索桥抗风与抗震研究之二:节段模型风洞试验及气动措施研究报告[R].长沙:湖南大学,2006.
Wind Engineering Experiment Research Center of Hunan University. Research on Wind Resistance and Earthquake Resistance of Aizhai Bridge: The Second Report of Wind Surroundings Bridge State of Aizhai Bridge[R]. Changsha: Hunan University,2006.

[13] 陈政清,胡建华.桥梁颤振导数识别的时域法与频域法对比研究[J].工程力学,2005,22(6):127-133.
CHEN Zheng-qing, HU Jian-hua. A Comparative Study Between Time-domain Method and Frequency-domain Method for Identification of Bridge Flutter Derivatives[J]. Engineering Mechanics,2005,22(6):127-133.

[14] 华旭刚,陈政清.桥梁风致颤振临界状态的全域自动搜索法[J].工程力学,2002,19(2):68-72.
HUA Xu-gang, CHEN Zheng-qing. Wind-induced Flutter Analysis of Long-span Bridges with Fully Automatic Searching Method[J]. Engineering Mechanics,2002,19(2):68-72.

[15] 陈政清.斜拉索风雨振现场观测与振动控制[J].建筑科学与工程学报,2005,22(4):5-10.
CHEN Zheng-qing. On-site Observation of Wind-rain Induced Vibration of Stay Cables and Its Control[J]. Journal of Architecture and Civil Engineering,2005,22(4):5-10.

Wind and Structures, Vol. 13, No. 3 (2010) 293-307

New estimation methodology of six complex aerodynamic admittance functions

Y. Han[1,2], Z.Q. Chen[1]* and X.G. Hua[1]

[1]*Wind Engineering Research Center, College of Civil Engineering, Hunan University, Changsha 410082, Hunan, P. R. China*
[2]*School of Civil Engineering and Architecture, Changsha University of Science & Technology, Changsha 410076, Hunan, P. R. China*

(Received January 7, 2009, Accepted February 3, 2010)

Abstract. This paper describes a new method for the estimation of six complex aerodynamic admittance functions. The aerodynamic admittance functions relate buffeting forces to the incoming wind turbulent components, of which the estimation accuracy affects the prediction accuracy of the buffeting response of long-span bridges. There should be two aerodynamic admittance functions corresponding to the longitudinal and vertical turbulent components, respectively, for each gust buffeting force. Therefore, there are six aerodynamic admittance functions in all for the three buffeting forces. Sears function is a complex theoretical expression for the aerodynamic admittance function for a thin airfoil. Similarly, the aerodynamic admittance functions for a bridge deck should also be complex functions. This paper presents a separated frequency-by-frequency method for estimating the six complex aerodynamic admittance functions. A new experimental methodology using an active turbulence generator is developed to measure simultaneously all the six complex aerodynamic admittance functions. Wind tunnel tests of a thin plate model and a streamlined bridge section model are conducted in turbulent flow. The six complex aerodynamic admittance functions, determined by the developed methodology are compared with the Sears functions and Davenport's formula.

Keywords: admittance functions; Sears function; thin plates; buffeting forces; long-span bridges.

1. Introduction

The advancements in modeling of the buffeting forces are generally accompanied by the developments of the aerodynamic admittance functions that relate buffeting forces to the incoming wind turbulent components. The accuracy of predicting buffeting response of a long-span bridge depends largely on the accurate estimations of the aerodynamic admittance functions. Sears function, which is a theoretical expression for the aerodynamic admittance function for an idealized thin airfoil, is conventionally applied to the gust response analysis of long-span bridges (Mendes and Branco 2001, Denoel and Degee 2006). Thin airfoil dynamic force theories depend on unique circulation functions (Theodorsen 1935, Sears 1941, Liepmann 1952). However, the circulation functions cannot be realized for bluff bodies that experience flow separation. Extending the theoretical thin-airfoil force coefficients to the bluff bodies presented by bridge deck sections is

* Corresponding Author, Professor, E-mail: zqchen@hnu.cn

basically incorrect from both mathematical and physical viewpoints. The common use of the Sears function in the context of bridge deck buffeting can be viewed only as a suggested approximation. Davenport (1962) also derived a formula for aerodynamic admittance function of drag force in relation to horizontal turbulence component.

Since the thin airfoil theory does not apply to bluff bodies, aerodynamic admittance functions of bridge deck sections cannot be defined analytically. They are usually determined experimentally by wind tunnel tests or derived from flutter derivatives as proposed by Scanlan and other researchers (Scanlan 1993, Scanlan and Jones 1999, Scanlan 2000, Chen and Kareem 2002, Tubino 2005) and the relationships are verified by Hatanaka and Tanaka (2002, 2008). Jancauskas and Melbourne (1986) and Kawatani and Kim (1992) studied the aerodynamic admittance functions for rectangular sections by experiments and found that they are different from Sears function. An experimental comparison of the aerodynamic admittance functions for a thin plate and different streamlined bridge decks was performed by Larose and Livesey (1997). Larose and Mann (1998) analytically obtained the aerodynamic admittance functions for a thin airfoil considering two wave-numbers and compared the analytical results with experimental ones obtained for a family of streamlined deck sections. Zhou and Kareem (2003) developed a procedure that uses high frequency base balance to estimate aerodynamic admittance functions and noted some discrepancies between experimentally obtained and theoretical aerodynamic admittance functions in the high frequency range.

The methods used in the literature reviewed above are usually direct estimation methods which mean the modul square of aerodynamic admittance function is estimated by the ratio of the power spectral density functions of fluctuating forces and the wind velocity fluctuations. However, since the aerodynamic admittance functions for a bridge deck are complex functions like Sears function (Sears 1941), they cannot be obtained by this approach. Diana, *et al.* (2002, 2004) developed the concept of complex aerodynamic admittance functions and identified three complex aerodynamic admittance functions for realistic bridge sections associated with vertical gusts by using an active turbulence generator. However, there are six aerodynamic admittance functions for gust forces corresponding to the longitudinal and vertical turbulent components. This paper presents a separated frequency-by-frequency methodology for estimating the six complex aerodynamic admittance functions using an active turbulence generator (Han 2007). The six complex aerodynamic admittance functions for a thin plate were identified by using the developed methodology and compared with Sears function and Davenport's formulae.

2. Theoretical basis of the new estimation methodology

The aerodynamic lift force, drag force, and pitching moment of a bridge deck with a length of D can be defined as (Chen and Kareem 2002)

$$L_b(t) = \frac{1}{2}\rho U^2 BD\left[2C_L\chi_{Lu}\frac{u(t)}{U} + (C'_L + C_D)\chi_{Lw}\frac{w(t)}{U}\right] \tag{1a}$$

$$D_b(t) = \frac{1}{2}\rho U^2 BD\left[2C_D\chi_{Du}\frac{u(t)}{U} + (C'_D - C_L)\chi_{Dw}\frac{w(t)}{U}\right] \tag{1b}$$

$$M_b(t) = \frac{1}{2}\rho U^2 B^2 D\left[2C_M\chi_{Mu}\frac{u(t)}{U} + C'_M\chi_{Mw}\frac{w(t)}{U}\right] \tag{1c}$$

where ρ is the air density; B is the deck width; U is the mean longitudinal wind velocity; $u(t)$ and $w(t)$ are the longitudinal and vertical turbulence wind components, respectively; C_D, C_L, C_M are the drag force, lift force, and pitching moment coefficients, respectively; C_D', C_L', C_M' are the derivates of C_D, C_L and C_M with respect to wind attack angle, respectively; and χ_{Rk} ($R = L, D, M$; $k = u, w$) are aerodynamic admittance functions.

To explain the concept, suppose that $u(t)$ and $w(t)$ are time-varying harmonic turbulent wind components expressed as:

$$u(t) = A\cos(\omega_1 t + \varphi_1) = A\cos(2\pi k_1 t/T + \varphi_1) \tag{2a}$$

$$w(t) = B\cos(\omega_2 t + \varphi_2) = B\cos(2\pi k_2 t/T + \varphi_2) \tag{2b}$$

where A and B are the amplitude of harmonic functions $u(t)$ and $w(t)$, respectively; ω_1 and ω_2 are vibration circular frequency of $u(t)$ and $w(t)$, respectively, and $\omega_1 \neq \omega_2$; T is the total duration and $T = 2\pi k_1/\omega_1 = 2\pi k_2/\omega_2$; k_1 and k_2 are random positive whole numbers corresponding to ω_1 and ω_2; and φ_1 and φ_2 are initial phase angle of $u(t)$ and $w(t)$, respectively.

By using Fast Fourier Transform (FFT) algorithm, Eqs. (2a-2b) are transformed as:

$$F[L_b(t)] = \frac{1}{2}\rho U^2 BD\left[2C_L\chi_{Lu}\frac{F[u(t)]}{U} + (C'_L + C_D)\chi_{Lw}\frac{F[w(t)]}{U}\right] \tag{3a}$$

$$F[D_b(t)] = \frac{1}{2}\rho U^2 BD\left[2C_D\chi_{Du}\frac{F[u(t)]}{U} + (C'_D - C_L)\chi_{Dw}\frac{F[w(t)]}{U}\right] \tag{3b}$$

$$F[M_b(t)] = \frac{1}{2}\rho U^2 B^2 D\left[2C_M\chi_{Mu}\frac{F[u(t)]}{U} + C'_M\chi_{Mw}\frac{F[w(t)]}{U}\right] \tag{3c}$$

where the sign $F[\ ]$ denotes the operation of FFT algorithm.

The FFTs of $u(t)$ and $w(t)$ are expressed as:

$$\begin{aligned} F[u(t)] &= \int_{-T/2}^{T/2} u(t)e^{-i\omega t}dt = \int_{-T/2}^{T/2} A\cos(2\pi k_1 t/T + \varphi_1)e^{-i(2\pi kt/T)}dt \\ &= \frac{AT}{2}e^{i\varphi_1}\frac{\sin\pi(k_1 - k)}{\pi(k_1 - k)} + \frac{AT}{2}e^{-i\varphi_1}\frac{\sin\pi(k_1 + k)}{\pi(k_1 + k)} \\ &= \frac{AT}{2}e^{i\varphi_1}\Theta(\omega_1 - \omega) + \frac{AT}{2}e^{-i\varphi_1}\Theta(\omega_1 + \omega) \end{aligned} \tag{4a}$$

$$\begin{aligned} F[w(t)] &= \int_{-T/2}^{T/2} w(t)e^{-i\omega t}dt = \int_{-T/2}^{T/2} B\cos(2\pi k_2 t/T + \varphi_2)e^{-i(2\pi kt/T)}dt \\ &= \frac{BT}{2}e^{i\varphi_2}\frac{\sin\pi(k_2 - k)}{\pi(k_2 - k)} + \frac{BT}{2}e^{-i\varphi_2}\frac{\sin\pi(k_2 + k)k}{\pi(k_2 + k)k} \\ &= \frac{BT}{2}e^{i\varphi_2}\Theta(\omega_2 - \omega) + \frac{BT}{2}e^{-i\varphi_2}\Theta(\omega_2 + \omega) \end{aligned} \tag{4b}$$

where

$$\Theta(\omega_1-\omega)=\frac{\sin(\omega_1-\omega)T/2}{(\omega_1-\omega)T/2},\ \Theta(\omega_1+\omega)=\frac{\sin(\omega_1+\omega)T/2}{(\omega_1+\omega)T/2} \tag{5a}$$

$$\Theta(\omega_2-\omega)=\frac{\sin(\omega_2-\omega)T/2}{(\omega_2-\omega)T/2},\ \Theta(\omega_2+\omega)=\frac{\sin(\omega_2+\omega)T/2}{(\omega_2+\omega)T/2} \tag{5b}$$

Suppose that the FFTs of $L_b(t)$, $D_b(t)$ and $M_b(t)$ are defined as:

$$F[L_b(t)]=L_b(\omega) \tag{6a}$$

$$F[D_b(t)]=D_b(\omega) \tag{6b}$$

$$F[M_b(t)]=M_b(\omega) \tag{6c}$$

Replacing Eqs. (4a-4b) and (6a-6c) into Eq. (3) provides:

$$\begin{aligned}L_b(\omega)&=\frac{1}{2}\rho U^2BDC_L\frac{2}{U}\chi_{Lu}\frac{AT}{2}(e^{i\varphi_1}\Theta(\omega_1-\omega)+e^{-i\varphi_1}\Theta(\omega_1+\omega))\\&+\frac{1}{2}\rho U^2BD\frac{(C'_L-C_D)}{U}\chi_{Lw}\frac{BT}{2}(e^{i\varphi_2}\Theta(\omega_2-\omega)+e^{-i\varphi_2}\Theta(\omega_2+\omega))\end{aligned} \tag{7a}$$

$$\begin{aligned}D_b(\omega)&=\frac{1}{2}\rho U^2BDC_D\frac{2}{U}\chi_{Du}\frac{AT}{2}(e^{i\varphi_1}\Theta(\omega_1-\omega)+e^{-i\varphi_1}\Theta(\omega_1+\omega))\\&+\frac{1}{2}\rho U^2BD\frac{(C'_D-C_L)}{U}\chi_{Dw}\frac{BT}{2}(e^{i\varphi_2}\Theta(\omega_2-\omega)+e^{-i\varphi_2}\Theta(\omega_2+\omega))\end{aligned} \tag{7b}$$

$$\begin{aligned}M_b(\omega)&=\frac{1}{2}\rho U^2BDC_M\frac{2}{U}\chi_{Mu}\frac{AT}{2}(e^{i\varphi_1}\Theta(\omega_1-\omega)+e^{-i\varphi_1}\Theta(\omega_1+\omega))\\&+\frac{1}{2}\rho U^2B^2D\frac{C'_M}{U}\chi_{Mw}\frac{BT}{2}(e^{i\varphi_2}\Theta(\omega_2-\omega)+e^{-i\varphi_2}\Theta(\omega_2+\omega))\end{aligned} \tag{7c}$$

If $\omega=\omega_1$, Eqs. (7a-7c) turn into:

$$L_b(\omega_1)=\frac{1}{2}\rho U^2BDC_L\frac{2}{U}\chi_{Lu}\cdot\frac{AT}{2}e^{i\varphi_1} \tag{8a}$$

$$D_b(\omega_1)=\frac{1}{2}\rho U^2BDC_D\frac{2}{U}\chi_{Du}\cdot\frac{AT}{2}e^{i\varphi_1} \tag{8b}$$

$$M_b(\omega_1)=\frac{1}{2}\rho U^2B^2DC_M\frac{2}{U}\chi_{Mu}\cdot\frac{AT}{2}e^{i\varphi_1} \tag{8c}$$

According to Eq. (8), the complex aerodynamic admittance functions χ_{Lu}, χ_{Du} and χ_{Mu} corresponding to longitudinal turbulent component $u(t)$ can be easily derived as:

$$\chi_{Lu}(\omega_1) = \frac{L_b(\omega_1)}{\frac{1}{2}\rho U^2 BDC_L\frac{2}{U}\cdot\frac{AT}{2}e^{i\varphi_1}} \tag{9a}$$

$$\chi_{Du}(\omega_1) = \frac{D_b(\omega_1)}{\frac{1}{2}\rho U^2 BDC_D\frac{2}{U}\cdot\frac{AT}{2}e^{i\varphi_1}} \tag{9b}$$

$$\chi_{Mu}(\omega_1) = \frac{M_b(\omega_1)}{\frac{1}{2}\rho U^2 B^2 DC_M\frac{2}{U}\cdot\frac{AT}{2}e^{i\varphi_1}} \tag{9c}$$

If $\omega = \omega_2$, Eqs. (7a-7c) turn into:

$$L_b(\omega_2) = \frac{1}{2}\rho U^2 BD\frac{(C'_L + C_D)}{U}\chi_{Lw}\cdot\frac{BT}{2}e^{i\varphi_2} \tag{10a}$$

$$D_b(\omega_2) = \frac{1}{2}\rho U^2 BD\frac{(C'_D - C_L)}{U}\chi_{Dw}\cdot\frac{BT}{2}e^{i\varphi_2} \tag{10b}$$

$$M_b(\omega_2) = \frac{1}{2}\rho U^2 B^2 D\frac{C'_M}{U}\chi_{Mw}\cdot\frac{BT}{2}e^{i\varphi_2} \tag{10c}$$

From Eq. (10), the complex aerodynamic admittance functions χ_{Lw}, χ_{Dw} and χ_{Mw} corresponding to the vertical turbulent component $w(t)$ can be easily derived as:

$$\chi_{Lw}(\omega_2) = \frac{L_b(\omega_2)}{\frac{1}{2}\rho U^2 BD\frac{(C'_L + C_D)}{U}\cdot\frac{BT}{2}e^{i\varphi_2}} \tag{11a}$$

$$\chi_{Dw}(\omega_2) = \frac{D_b(\omega_2)}{\frac{1}{2}\rho U^2 BD\frac{(C'_D - C_L)}{U}\cdot\frac{BT}{2}e^{i\varphi_2}} \tag{11b}$$

$$\chi_{Mw}(\omega_2) = \frac{M_b(\omega_2)}{\frac{1}{2}\rho U^2 B^2 D\frac{C'_M}{U}\cdot\frac{BT}{2}e^{i\varphi_2}} \tag{11c}$$

Up to now, the complex formulations of the six aerodynamic admittance functions are given as Eqs. (9) and (11). If there is a special experimental setup which can generate the harmonic oscillating longitudinal and vertical velocity components with different frequencies simultaneously, the six complex aerodynamic admittance functions can be estimated at one time, frequency-by-frequency.

3. Experimental technique

An active turbulence generator is developed for generating simultaneously the harmonic oscillating longitudinal and vertical wind velocity components with different frequencies in the HD-

2 wind tunnel at Hunan University. The experimental set-up as shown in Fig. 1 is made of a series of wings linked together to produce the harmonic oscillating vertical wind velocity component and a series of motionless thin plates installed in the middle of wings to produce the harmonic oscillating longitudinal velocity component. The motions of all the wings are controlled by an AC servomotors and the motion frequency is controllable and does not depend on the mean wind velocity.

A number of tests are conducted to check the correlation between two points around the model in the wind field. The locations of measurement points cover the whole length and width of the sectional model, and are shown in Fig. 2. Two cobra probes are used to simultaneously measure the longitudinal/vertical turbulence components at two points as given in Fig. 2. Correlation coefficients are larger than 0.99 for both vertical and horizontal components for all the testing conditions; the incoming flow can be thus considered as a perfect 2D flow.

A force-measurement system as shown in Fig. 3 is set up and a five-component force balances at each support are used to measure the five-direction forces, allowing a calculation of the total drag force, lift force, and pitching moment. Because one side of the force-measurement equipment is set inside the wind tunnel, a fairings is designed to avoid the disturbance around the side of the force-measurement equipment inside the wind tunnel. By using the measured aerodynamic forces and the measured turbulence components, the six complex aerodynamic admittance functions, relating the

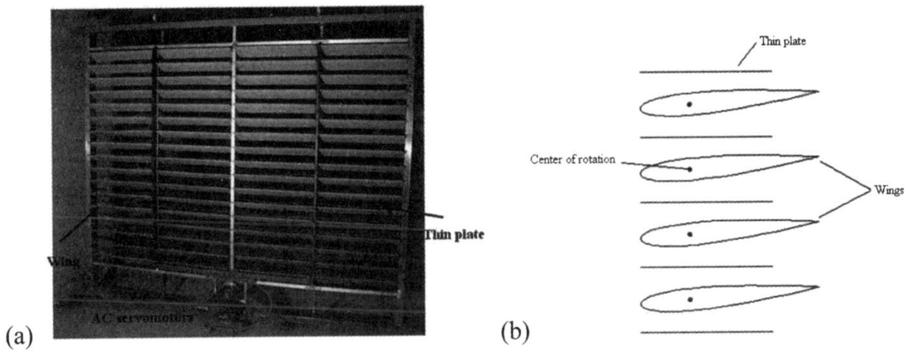

Fig. 1 An active turbulence generator: (a) actual deployment, (b) sketch

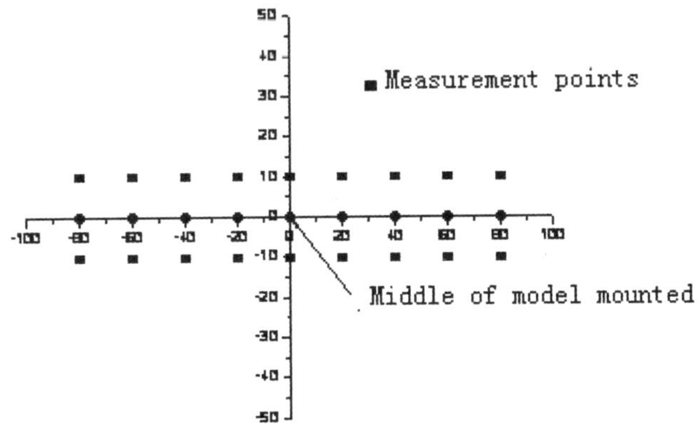

Fig. 2 Locations of measurement points of wind velocity (units: cm)

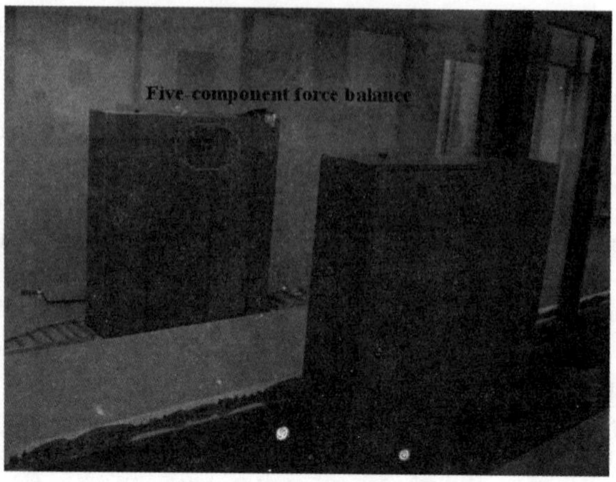

Fig. 3 The force-measurement equipment (with the left side inside the wind tunnel)

three buffeting forces to the two turbulent wind components, are then identified using Eqs. (9) and (11) by sweeping the mean wind velocity but keeping the vibration frequency of wings unchanged. It should be noted that the active turbulence generator developed by Diana, *et al.* (2002, 2004) can only produce the vertical turbulent wind component and therefore restricts itself to identification of the three aerodynamic admittance functions related to turbulent wind component *w*.

4. Turbulence flow characterization

Examples of experimentally generated time histories of turbulence components, obtained by driving the generator with a frequency of 1 Hz, are shown in Fig. 4. The mean horizontal wind speed is, for this test, 8 m/s. Fig. 5 shows the amplitude spectra of the longitudinal and vertical

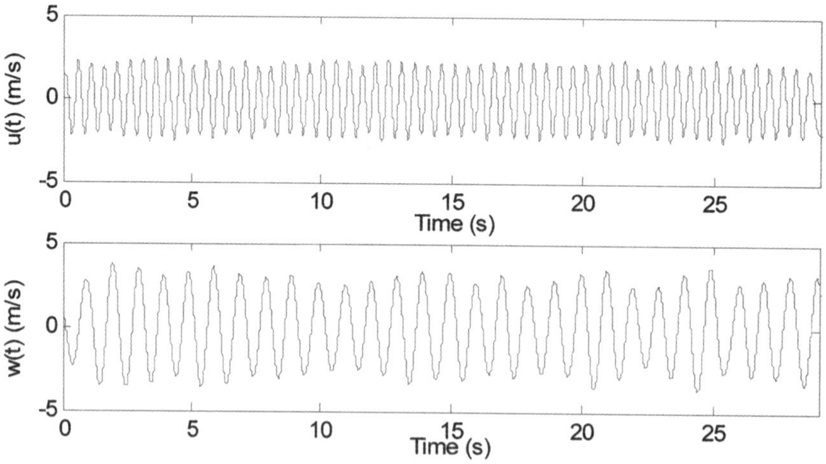

Fig. 4 The longitudinal and vertical turbulent components time histories

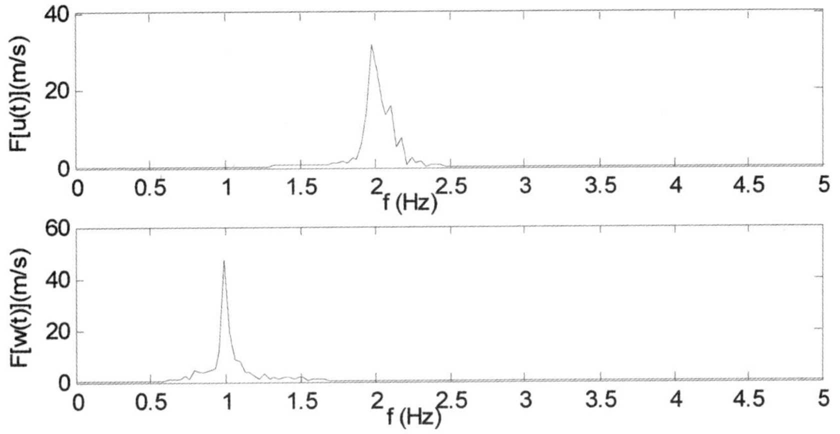

Fig. 5 The amplitude spectra of the longitudinal and vertical turbulent components

turbulent components. From Figs. 4 and 5, it is observed that the two turbulence components are mainly made up of just one harmonic component and the frequency of the vertical component is different from the frequency of the longitudinal one, as desired. They also show that the frequency of the vertical turbulence is identical with the frequency of the wings (1 Hz) and the frequency of the longitudinal turbulence (2 Hz) is twice the frequency of the wings.

5. Determination of complex aerodynamic admittance functions

As discussed earlier, the experimental procedure to obtain the complex aerodynamic admittance functions is based on the use of the developed active generator. Suppose that the generated turbulences have a vertical frequency ω_w and horizontal one ω_u. Similar to Eqs. (9) and (11), the six complex admittance functions can be obtained for the aerodynamic drag, lift and moment as:

$$\chi_{Lu}(\omega_u) = \frac{F[L(t)]}{\frac{1}{2}\rho U^2 BDC_L \frac{2}{U} \cdot F[u(t)]} \tag{12a}$$

$$\chi_{Du}(\omega_u) = \frac{F[D(t)]}{\frac{1}{2}\rho U^2 BDC_D \frac{2}{U} \cdot F[u(t)]} \tag{12b}$$

$$\chi_{Mu}(\omega_u) = \frac{F[M(t)]}{\frac{1}{2}\rho U^2 B^2 DC_M \frac{2}{U} \cdot F[u(t)]} \tag{12c}$$

$$\chi_{Lw}(\omega_w) = \frac{F[L(t)]}{\frac{1}{2}\rho U^2 BD \frac{(C'_L + C_D)}{U} \cdot F[w(t)]} \tag{13a}$$

$$\chi_{Dw}(\omega_w) = \frac{F[D(t)]}{\frac{1}{2}\rho U^2 BD\frac{(C_D - C_L)}{U}\cdot F[w(t)]} \tag{13b}$$

$$\chi_{Mw}(\omega_w) = \frac{F[M(t)]}{\frac{1}{2}\rho U^2 B^2 D\frac{C_M}{U}\cdot F[w(t)]} \tag{13c}$$

where $\omega_u = 2\omega_w$.

Wind tunnel tests of a thin plate model and a streamlined bridge section model (as shown in Fig. 6) are conducted in turbulent flow. The thin plate section (length × width × depth = 155 cm × 45 cm × 2 cm) is made of ABS plates and there is an aluminum bar inside to improve the stiffness of the model. The streamlined section model is made of wood plates and there also is an aluminum bar inside. The section models are fixed on the force-measurement equipment firmly (as shown in Fig. 7) and the natural frequencies of the whole experimental system which is composed of the section model and the equipment are listed in Table 1. Form Table 1, it is observed that the natural frequencies of the whole experimental equipment system are large enough to satisfy the need of stiffness.

Before identifying the aerodynamic admittance functions, the static aerodynamic coefficients of the sections must be measured. Figs. 8 and 9 show the drag force, lift force, and pitching moment coefficients of the thin plate section model and the streamlined section model, respectively. In this

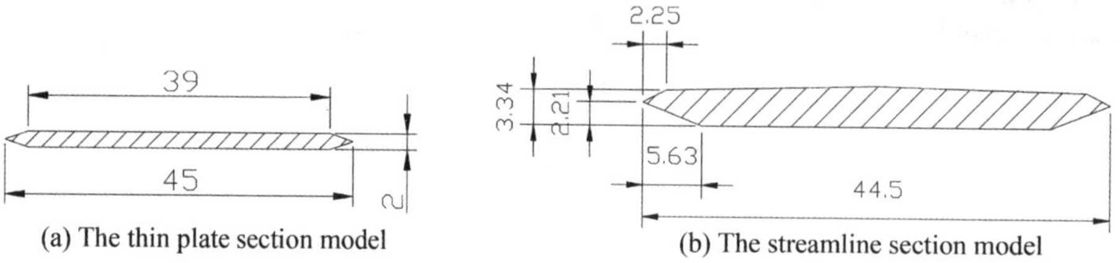

(a) The thin plate section model (b) The streamline section model

Fig. 6 The section models (unit. cm)

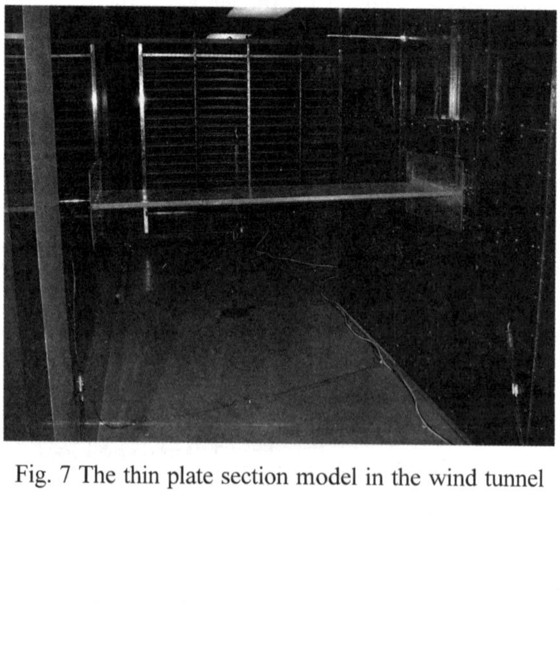

Fig. 7 The thin plate section model in the wind tunnel

Table 1 Frequencies of the whole experimental system

Motion direction	Frequencies (Hz)	
	The thin plate section model case	The streamlined section model case
Vertical bending	21.2	36.4
Lateral bending	45.6	58.7
Pitching moment	82.5	86.5

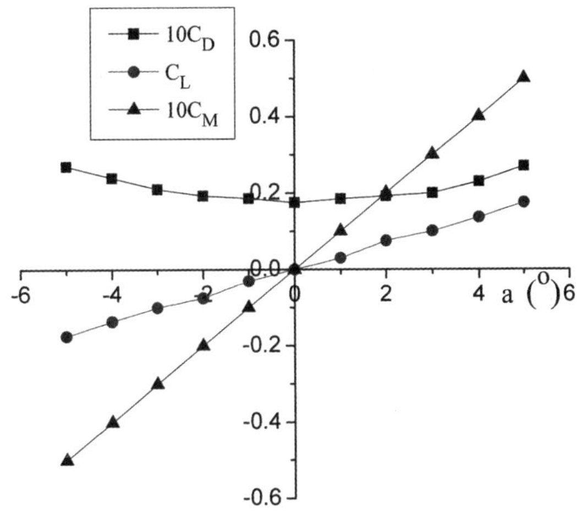

Fig. 8 Steady aerodynamic coefficients of the thin plate section model

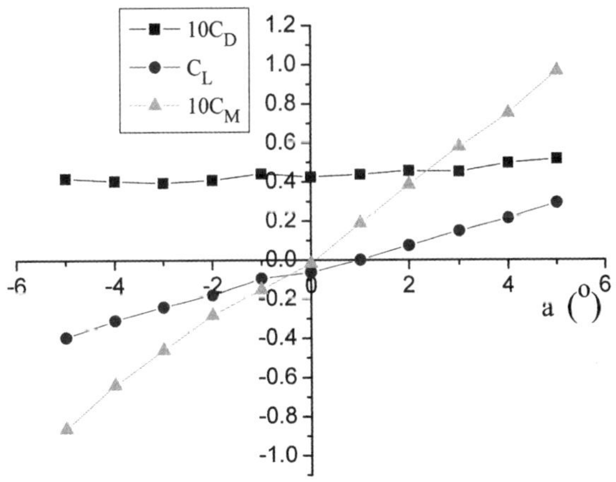

Fig. 9 Steady aerodynamic coefficients of the streamlined section model

test, the vibration frequency of the wings is 1 Hz and unchanged, and the mean wind velocity is changeable and the changing range is 0 m/s ~ 12 m/s.

Fig. 10 and Fig. 11 show the variation of the squared amplitudes and the phases of three identified complex aerodynamic admittance functions versus the reduced frequency for the thin plate section at 0° wind attack angle. The aerodynamic admittance functions χ_{Dw}, χ_{Lu} and χ_{Mu} for the

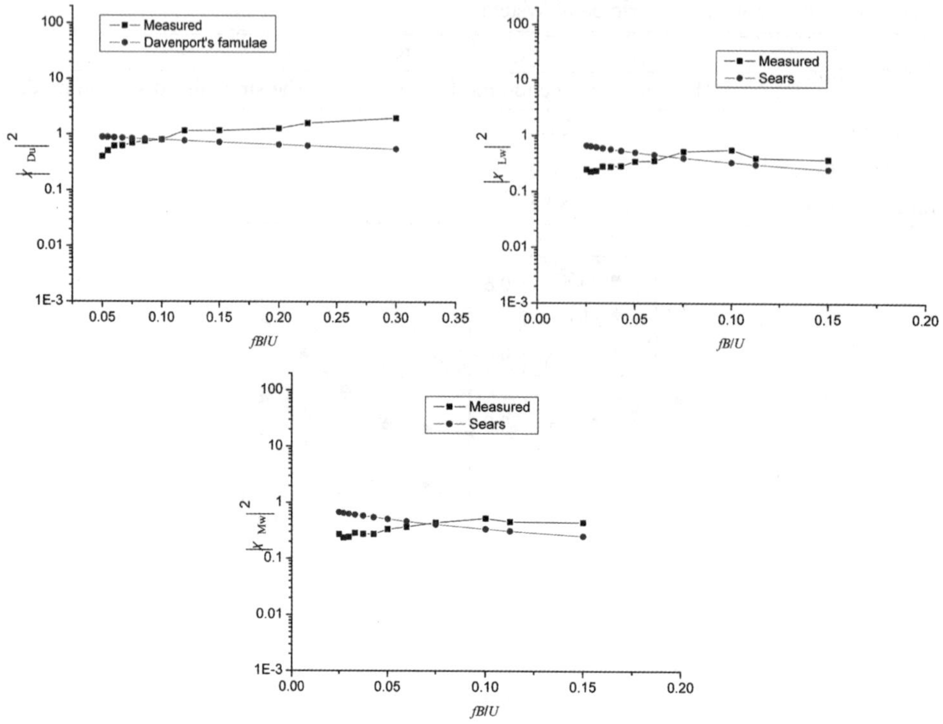

Fig. 10 The squared amplitude of CAAFs for the thin plate section at 0° attack angle

thin plate section are nonexistent because the section is symmetrical about the horizontal axis and its static aerodynamic coefficients C_D', C_L, C_M are close to zero, as given in Fig. 8. Thus χ_{Dw}, χ_{Lu} and χ_{Mu} for the thin plate section are not shown in Figs. 10 and 11. Similarly, the variation of the squared amplitudes and the phases of the six complex aerodynamic admittance functions versus the reduced frequency for the streamlined section at 0° wind attack angle are shown in Figs. 12 and 13. The measured aerodynamic admittance function χ_{Du} is compared to Davenport's formulae (Davenport 1962). The remaining aerodynamic admittance functions are compared with Sears formulae (Sears 1941).

From Figs. 10 and 12, it is seen that the measured magnitude of aerodynamic admittance functions deviate from those of the Sears function or Davenport's formulae, especially the aerodynamic admittance functions corresponding to the longitudinal component and the admittance functions for drag force. Sears function is the aerodynamic admittance functions of a thin aerofoil associated with the vertical force and in fact caused by the vertical turbulent component. Davenport's formula is the aerodynamic admittance function of drag forces associated with horizontal turbulence component. In addition, the admittance functions corresponding to the longitudinal component are different from those corresponding to the vertical component. Therefore, it is necessary to estimate all the six admittance functions and not just the three equivalent admittance functions. In other words, we cannot simply assume that the admittance functions corresponding to the longitudinal component are equal to those corresponding to the vertical component. Unlike Sears function, some of the identified aerodynamic admittance functions increase with the increase of the reduced frequency. Actually,

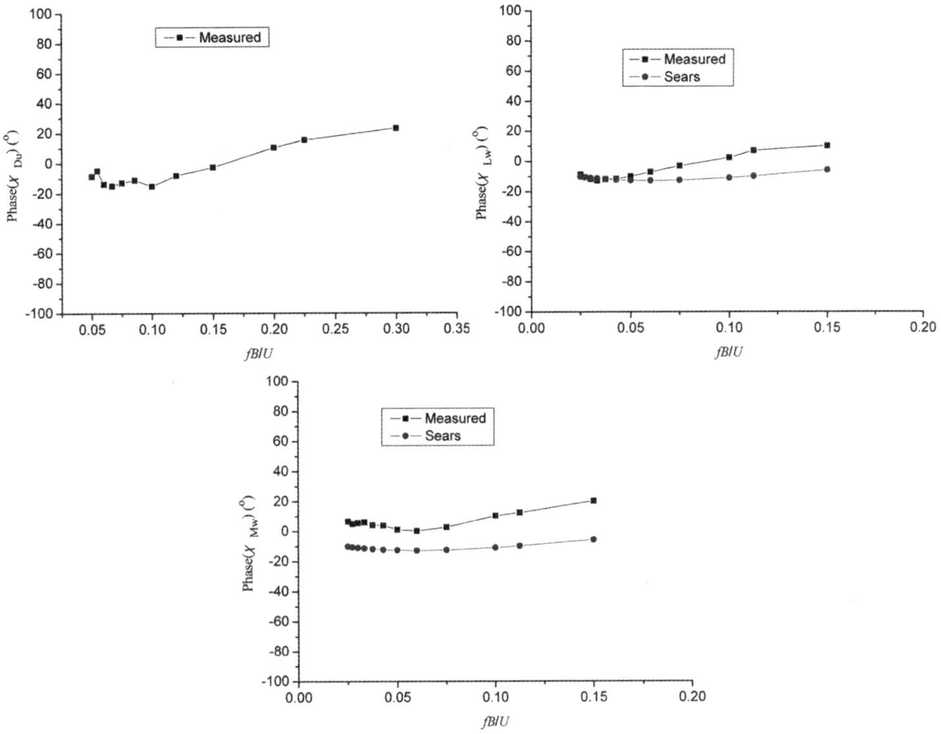

Fig. 11 The phase of CAAFs for the thin plate section at 0° attack angle

similar observations of certain increase trend with the reduced frequency were also made by Scanlan (2000) and Diana, *et al.* (2004) for bridge sections. The value of the aerodynamic admittance functions may also exceed unity for some cases in this study, which was also reported by several other researchers (Larose and Livesey 1997, Larose, *et al.* 1998, Scanlan 2000, Diana, *et al.* 2004). From Figs. 11 and 13, it can be seen that the phases of the estimated admittance functions increase with the increase of the reduced frequency just like Sear functions.

Large vortex shedding will occur and quasi-steady theory may not be applicable in the range of attack angle beyond ±10°. However, for long-span bridges constructed with mountain area with deep valley, such large attack angle may exist even in strong winds. The admittance function identified in such condition may be explained as 'equivalent' admittance functions in conjunction with quasi-steady theory. However, these 'equivalent' admittance functions may deviate significantly from the Sears function or Davenport's formulae. Nevertheless, the characterization of the aerodynamic admittance functions deserves further exploration in the future.

6. Conclusions

A new methodology of estimating the six complex aerodynamic admittance functions is developed by first through theoretical derivations and then by developing experimental techniques. Based on the results of this limited study, the following conclusions are drawn: (1) Drag-force admittance

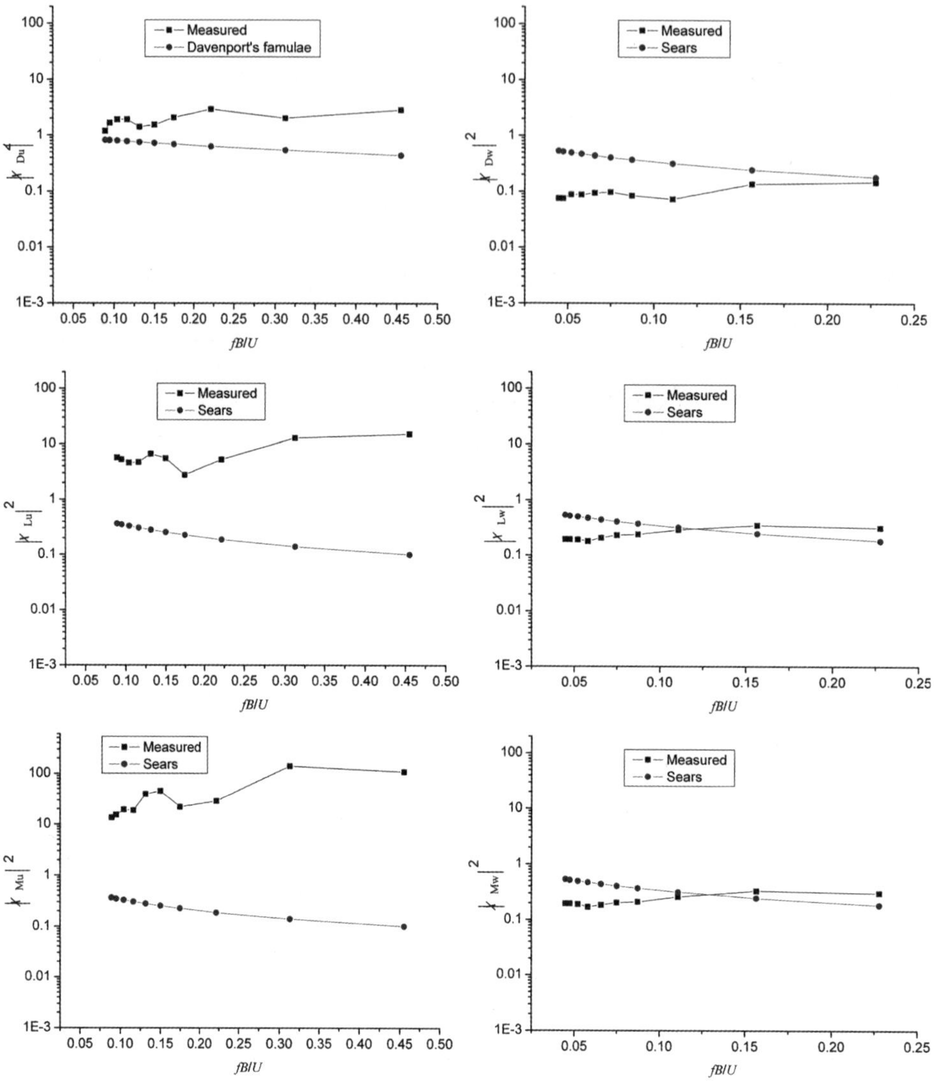

Fig. 12 The squared amplitude of CAAFs for the streamlined section at 0° attack angle

functions and admittance functions corresponding to the longitudinal component derivate significantly from the Sears function. The Sears function can only be employed as an approximation of the lift-force and pitching-moment aerodynamic admittance functions corresponding to the vertical turbulence component when the experimentally determined admittance functions are not available. (2) The admittance functions corresponding to the longitudinal component are different from those corresponding to the vertical component and it is necessary to estimate all the six admittance functions. (3) Some of the identified aerodynamic admittance functions increase with the increase of the reduced frequency. (4) The phases of the estimated admittance functions increase with the increase of the reduced frequency just like Sear functions.

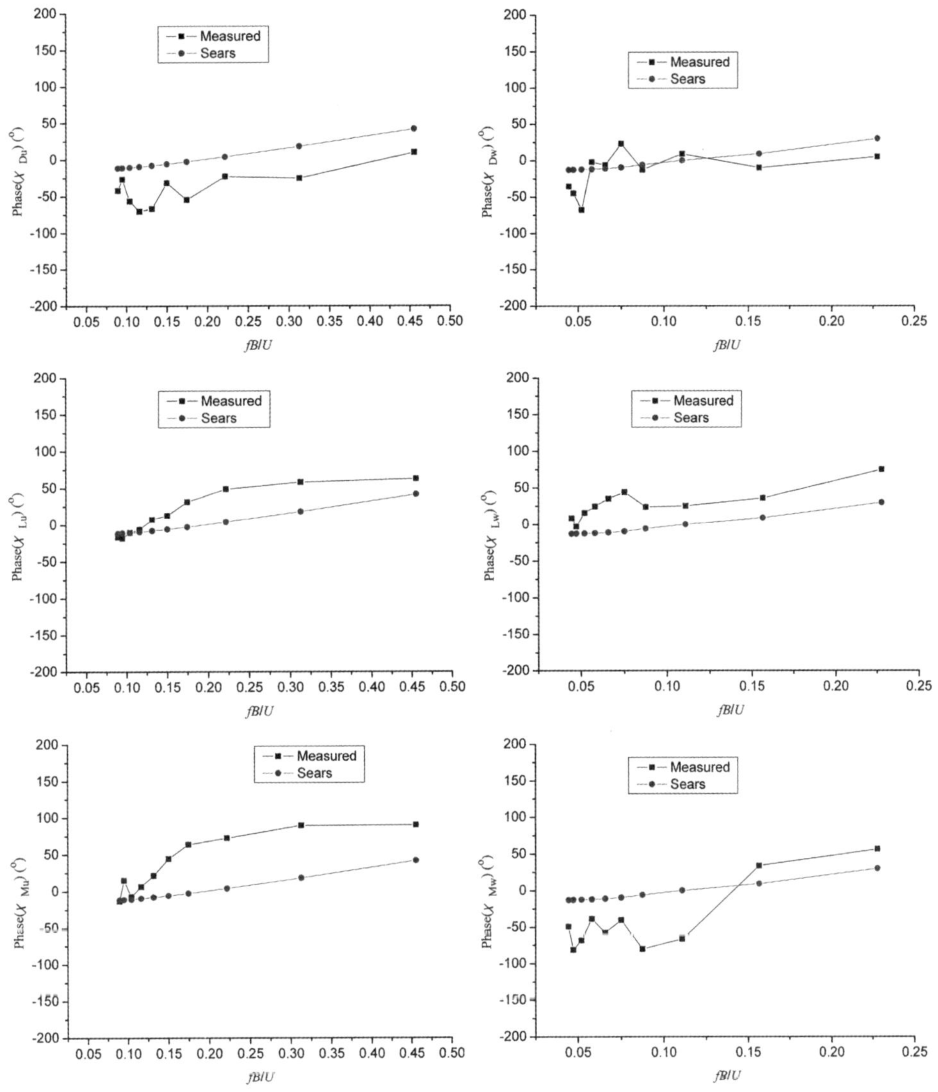

Fig. 13 The phase of CAAFs for the streamline section at 0° attack angle

It is also recommended that more research be performed for the aerodynamic admittance functions to predict the buffeting responses of long-span bridges more accurately.

Acknowledgements

The authors would like to gratefully acknowledge the supports from the National Science Foundation of China (Project Nos. 50478051 and 50738002) and from the Wind Engineering Research Center of Hunan University and the School of Civil Engineering and Architecture of

Changsha University of Science and Technology in China.

References

Chen, X. and Kareem, A. (2002), "Advances in modeling aerodynamic forces on bridge decks", *J. Eng. Mech.-ASCE*, **128**(11), 1193-1205.

Davenport, A.G. (1962), "Buffeting of a suspension bridge by storm winds", *J. Struct. Div., ASCE*, **88**(3), 233-268.

Denoël, V. and Degée, H. (2006), "Influence of the non-linearity of the aerodynamic coefficients on the skewness of the buffeting drag force", *Wind Struct.*, **6**, 457-471.

Diana, G., Bruni, S., Cigada, A. and Zappa, E. (2002), "Complex aerodynamic admittance function role in buffeting response of a bridge deck", *J. Wind Eng. Ind. Aerod.*, **90**, 2057-2072.

Diana, G., Resta, F., Zasso, A., Belloli, M. and Rocchi, R. (2004), "Forced motion and free motion aeroelastic tests on a new-concept dynamometric section model of the Messina suspension bridges", *J. Wind Eng. Ind. Aerod.*, **92**, 441-462.

Han, Y. (2007), *Study on complex aerodynamic admittance functions and refined analysis of buffeting response of bridges*, Ph.D. Thesis, College of Civil Engineering, Hunan University, China. (In Chinese).

Hatanaka, A. and Tanaka, H. (2002), "New estimation method of aerodynamic admittance function", *J. Wind Eng. Ind. Aerod.*, **90**, 2073-2086.

Hatanaka, A. and Tanaka, H. (2008), "Aerodynamic admittance functions of rectangular cylinders", *J. Wind Eng. Ind. Aerod.*, **96**, 945-953.

Jancauskas, E.D. and Melbourne, W.H. (1986), "The aerodynamic admittance of two-dimensional rectangular section cylinders in smooth flow", *J. Wind Eng. Ind. Aerod.*, **23**, 395-408.

Kawatani, M. and Kim, H. (1992), "Evaluation of aerodynamic admittance for buffeting analysis", *J. Wind Eng. Ind. Aerod.*, **41-44**, 613-624.

Larose, G.L. and Livesey, F.M. (1997), "Performance of streamlined bridge decks in relation to the aerodynamics of a thin plate", *J. Wind Eng. Ind. Aerod.*, **69-71**, 851-860.

Larose, G.L. and Mann, J. (1998), "Gust loading on streamlined bridge decks", *J. Fluid. Struct.*, **12**, 511-536.

Larose, G.L., Tanaka, H., Gimsing, N.J. and Dyrbye, C. (1998), "Direct measurement of buffeting wind forces on bridge decks", *J. Wind Eng. Ind. Aerod.*, **74-76**, 809-818.

Liepmann, H.W. (1952), "On the application of statistical concepts to the buffeting problem", *J. Aeronaut. Sci.*, **19**, 793-800.

Mendes, P.A. and Branco, F.A. (2001), "Unbalanced wind buffeting efforts on bridges during double cantilever erection stages", *Wind Struct.*, **4**, 45-62.

Sears, W.R. (1941), "Some aspects of non-stationary airfoil theory and its practical application", *J. Aeronaut. Sci.*, **8**, 104-108.

Scanlan, R.H. (1993), "Problematics in formulation of wind-force model for bridge deck", *J. Eng. Mech.-ASCE*, **119**, 1353-1375.

Scanlan, R.H. and Jones, N.P. (1999), "A form of aerodynamic admittance for use in bridge aeroelastic analysis", *J. Fluid. Struct.*, **13**, 1017-1027.

Scanlan, R.H. (2000), "Motion-related body-force functions in two-dimensional low-speed flow", *J. Fluid. Struct.*, **14**, 49-63.

Theodorsen, T. (1935), *General theory of aerodynamic instability and the mechanism of flutter*, NACA Report, No. 496, Langley: U.S. National Advisory Committee for Aeronautics.

Tubino, F. (2005), "Relationships among aerodynamic admittance functions, flutter derivatives and static coefficients for long-span bridges", *J. Wind Eng. Ind. Aerod.*, **93**, 929-950.

Zhou, Y. and Kareem, A. (2003), "Aerodynamic admittance functions of tall buildings", *Proc. of the 11th Int. Conf. on Wind Engineering*, June, Lubbock, TX.

Dynamic Performance of a Slender Truss Bridge Subjected to Extreme Wind and Traffic Loads Considering 18 Flutter Derivatives

Huawei Niu[1]; Jin Zhu, M.ASCE[2]; Zhengqing Chen[3]; and Wei Zhang, M.ASCE[4]

Abstract: The aerodynamic performance of long-span bridges under extreme environmental conditions, such as under strong winds and heavy traffic, is critical for bridge design. Either free vibrational or forced vibrational tests are usually carried out to obtain the flutter derivatives, either traditional eight flutter derivatives from two-degree-of-freedom (2-DOF) model tests or 18 flutter derivatives from 3-DOF model tests, to evaluate structural dynamic performance. In the present study, an experimental facility is developed based upon a digitally controlled 3-DOF forced vibration system to enable the extraction of 18 flutter derivatives of bridge decks through wind-tunnel tests. After validating the proposed forced vibration test facility using typical section types, i.e., a thin-plate section and a truss-girder section, the 18 flutter derivatives for the truss-girder section are extracted and used for bridge dynamic performance assessment considering combined loads from traffic and wind. The case study of a truss bridge shows that ignorance of the additional 10 flutter derivatives can result in a significant overestimation of the bridge lateral responses, especially under high wind speeds. However, ignorance of the additional 10 flutter derivatives on the bridge vertical and torsional responses, as well as on the dynamic responses of the running vehicles, are not significant. Similar observation were also applied to the stress responses at the critical location of the bridge deck. **DOI: 10.1061/(ASCE)AS.1943-5525.0001068.** *© 2019 American Society of Civil Engineers.*

Author keywords: Flutter derivative; Forced vibration; Identification method; Wind-tunnel test; Long-span bridges; Vehicle–bridge–wind system.

Introduction

Owing to high flexibility and low structural damping, long-span bridges are potentially susceptible to a variety of types of wind-induced instabilities, among which flutter is the most critical and may serve as a governing criterion for the bridge design and construction. As essential parameters for flutter estimations of wind sensitive long-span bridges, flutter derivatives are commonly determined using free vibration and forced vibration techniques, i.e., wind-tunnel tests or computational fluid dynamics (CFD)-based numerical simulations (Xu 2015; Xu et al. 2012). During the last several decades, wind-tunnel tests of bridge section models, including free vibrational tests and forced vibrational tests, were used to obtain the flutter derivatives. With an easy experimental setup and being less demanding of testing facilities, free vibration techniques are more popular. In comparison, because the flutter derivatives are extracted directly from aerodynamic force and motion signals, forced vibration techniques are believed to provide better estimates of flutter derivatives (Chen and Kareem 2004; Xu et al. 2012).

Starting from the thin-plate theory proposed by Theodorsen (1935), researchers have continued to obtain self-excited forces of different types of bridge sections. For bridges with typical bluff cross sections, Scanlan and Tomko (1971) improved the thin-plate theory, in which the six flutter derivatives were extracted through two-stage free vibration tests on both one-degree-of-freedom (1-DOF) and coupled 2-DOF section models. With further expansion of self-excited forces using eight flutter derivatives, i.e., H_i^* and A_i^* ($i = 1$–4) (Simiu and Scanlan 1996), more methods were proposed to extract the eight flutter derivatives using coupled 2-DOF free vibration tests, such as the modified Ibrahim time domain (MITD) method (Sarkar et al. 1992, 1994), extended Kalman filter method (Iwamoto and Fujino 1995, Yamada et al. 1992), uni fying least-squares (ULS) method (Ding et al. 2010; Gu et al. 2000; Zhu et al. 2002), stochastic subspace method (Qin and Gu 2004), neural network method (Chen 2003; Jung et al. 2004), and the subsection extended-order iterative least-square (SEO-ILS) algorithm (Chen et al. 2010), among others.

Most research on flutter derivative identification of the bridge deck in the past have only focused on the flutter derivatives related to the self-excited lift force and torsional moment, and flutter derivatives related to the self-excited drag force have commonly been ignored (Xu et al. 2016). This is due to the assumption that the contribution of self-excited lift and torsional moment to the wind-induced vibration of the bridge are more significant than that of the self-excited drag force. However, such an assumption is questionable, and several reports have highlighted the importance of the self-excited drag force in the aeroelasticity analysis of long-span bridges under some conditions (Jones et al. 2003; Katsuchi et al. 1998; Miyata 1995).

[1]Associate Professor, Wind Engineering Research Center, Hunan Univ., Changsha 410082, China.

[2]Assistant Professor, Dept. of Bridge Engineering, Southwest Jiaotong Univ., Chengdu 610031, China; formerly, Ph.D. Candidate, Dept. of Civil and Environmental Engineering, Univ. of Connecticut, Storrs, CT 06269.

[3]Professor, Wind Engineering Research Center, Hunan Univ., Changsha 410082, China.

[4]Assistant Professor, Dept. of Civil and Environmental Engineering, Univ. of Connecticut, Storrs, CT 06269 (corresponding author). ORCID: https://orcid.org/0000-0001-8364-9953. Email: wzhang@engr.uconn.edu

Note. This manuscript was submitted on September 11, 2018; approved on April 1, 2019; published online on July 19, 2019. Discussion period open until December 19, 2019; separate discussions must be submitted for individual papers. This paper is part of the ***Journal of Aerospace Engineering***, © ASCE, ISSN 0893-1321.

In recognizing the potential effects from lateral motions on bridge aerodynamics, researchers attempted to extract the complete set of flutter derivatives (total number of 18) experimentally and numerically. Singh et al. (1996) proposed the MITD method with a 3-DOF suspension system aiming to identity the lateral flutter derivatives P_i^* ($i = 1$–6) for both streamlined and bluff bridge-deck sections using the free vibration technique. Due to the limitation of the test facility, however, only the flutter derivatives P_1^* and P_4^* related to the lateral motion were identified. Later on, the development of system identification techniques facilitated the extraction of 18 flutter derivatives in the 3-DOF coupled dynamic system (Chowdhury and Sarkar 2003; Mishra et al. 2006; Xu et al. 2012). For example, Sarkar et al. (2004) developed a 3-DOF suspension system to identify the 18 flutter derivatives of an airfoil and a bridge deck using the iterative least-square (ILS) algorithm. Additional efforts were also made by Chen et al. (2002), Mishra et al. (2006), and Xu et al. (2012), among others.

Alternatively, CFD-based techniques have also been employed as cost-effective and efficient approaches to extracting the flutter derivatives. Xu et al. (2014) proposed a CFD-based numerical simulation approach to synchronously extract the 18 aerodynamic derivatives of bridge deck using 3-DOF coupled forced vibration, and the accuracy and efficacy of the proposed numerical approach was verified. Later, through incorporating a nonlinear mathematical model, Xu et al. (2016) further extended the flutter derivative identification framework to evaluate the nonlinear features of self-excited drag forces induced by the vertical and torsional motions of both streamlined and bluff deck sections. Despite remarkable progress made in determination of flutter derivatives in recent decades, challenges still remain such that most current CFD-based approaches are applicable for platelike or boxlike deck sections, but are almost impossible for the trusslike deck section. In such cases, wind-tunnel tests are still the only reliable methods to identify the flutter derivatives.

Although significant progress has been achieved in extracting the complete set of flutter derivatives, challenges remain, particularly in accurately identifying the lateral flutter derivatives with high fidelity (Xu et al. 2016). Consequently, the obtained lateral flutter derivatives were usually scattered instead of in a typical continuous format, and further improvement is still needed, especially in eliminating the scatter and fluctuation of the obtained data. Therefore, there is a research need on the identification techniques for the 18 flutter derivatives and their effects on bridge dynamics (Chen and Kareem 2006; Jones et al. 2003). The present study aims to extract 18 flutter derivatives of a truss deck section and further evaluate their effects on bridge dynamics and fatigue damage accumulations. The least-square method is proposed to obtain the 18 flutter derivatives of bridge decks. After the brief introduction of the research background, the paper is organized as follows. First, the 3-DOF forced vibration system and proposed identification method are introduced. Next, the proposed identification method is applied for two bridge section models to obtain the flutter derivatives, i.e., a thin-plate section and a truss-girder section, using 1-DOF and 3-DOF forced vibration methods. The effectiveness and accuracy of the proposed flutter derivative identification framework are validated by comparing the results of the thin-plate model using the proposed method and Theodorsen's analytical results (Theodorsen 1935). Finally, to further evaluate the effects of the additional 10 flutter derivatives, both the bridge dynamic and stress analysis of a suspension bridge with truss girder by employing 18 and 8 flutter derivatives are performed in the context of a coupled vehicle–bridge–wind (VBW) system.

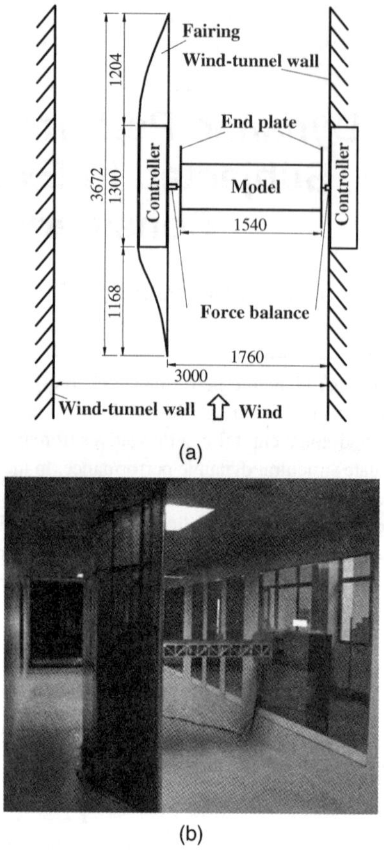

Fig. 1. Three-degree-of-freedom forced vibration facility in HD-2 wind tunnel: (a) plan view (unit: millimeters); and (b) three-dimensional view (image by Huawei Niu).

Three-DOF Forced Vibration System

Experimental Setup

Both the free vibration technique and forced vibration technique can be employed to experimentally extract the 18 flutter derivatives. Compared with the free vibration technique, the forced vibration technique provides more reliable estimates of flutter derivatives. Nevertheless, the forced vibration technique was not widely used when the facility was designed because it involves a complicated driving mechanism and measurement setups (Chen et al. 2005). In order to overcome such difficulties, the experimental facility was designed and manufactured in 2007 at the HD-2 Wind Tunnel of Hunan University, China, to enable better estimation of flutter derivatives using the forced vibration technique. Fig. 1(a) shows the plan view of the 3-DOF forced vibration system. As shown in Fig. 1(a), one side of the 3-DOF forced vibration system is built outside of the wind tunnel to increase the rigidity of the forced vibration system and avoid a possible large blockage rate of the wind tunnel. The other side is placed inside the wind tunnel to mount the section model. To improve the flow quality and reduce the perturbations due to the instruments of the bluff body shape inside the wind tunnel, a fairing is adopted during the wind-tunnel tests. The airfoil-shaped fairing is placed in front of the inside

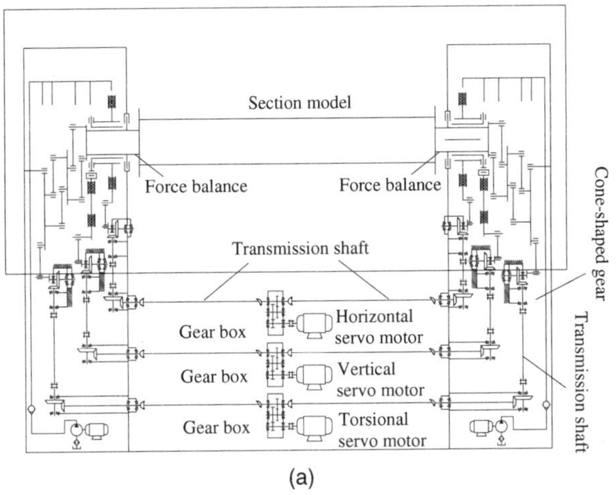

Fig. 2. Mechanical drive and transmission system: (a) overview; and (b) coupling mechanism.

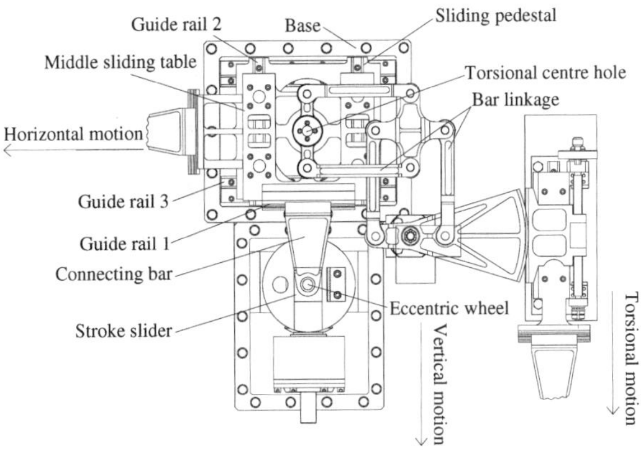

Fig. 3. Coupling setup with sinusoidal motion generation mechanism.

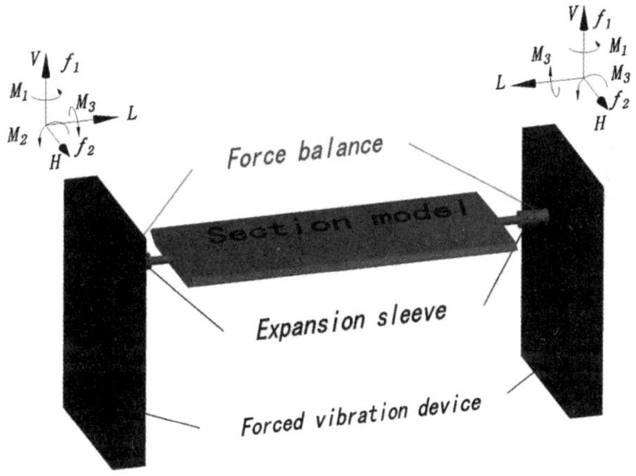

Fig. 4. Force measurement system.

supporting structures as shown in Fig. 1(b). To fulfill the requirement of wind-tunnel tests, the aerodynamic shape of the fairing has been carefully designed and validated using both the wind-tunnel tests and the CFD numerical simulations.

Schematic diagrams of the mechanical drive and transmission system of the 3-DOF forced vibration system are illustrated in Fig. 2. With this forced vibration system, three DOFs, i.e., torsional (T), vertical (V), and lateral (L) motions are generated. Each DOF is controlled by a servomotor and a digital servodrive system, which allows a purely sinusoidal oscillation. In addition, a coupling setup is also designed and implemented in the forced vibration system to realize the arbitrary DOF combinations among the 3-DOF motions, as shown in Fig. 3. Accordingly, a total of seven movement modes with different DOF combinations, i.e., 1-DOF (separate T, V, and L), coupled 2-DOF (T and V, T and L, and V and L), and coupled 3-DOF (T and V and L), can be actuated by the system. A specific mechanism is employed to generate pure sinusoidal motion for each DOF, as shown in Fig. 3. The implemented experimental facility used in the present study was carefully designed to achieve (1) fully automatic control of the motion in each of the 3-DOFs using a digital servodrive system; and (2) arbitrary DOF combinations among the 3-DOF motions using a coupling setup.

To obtain the aerodynamic lift, drag, and moment acting on a section model, two strain-gauged force balances are designed and implanted to facilitate the measurement of two forces f_1 and f_2 and three moments M_1, M_2, and M_3, as shown in Fig. 4. In order to reduce the mechanical interferences, one end of the force balance is rigidly connected to the model through an expansion sleeve, and the other end is fixed on the torsional center hole (Fig. 3) of the forced vibration device. As a result, the two force balances can vibrate together with the model. The expansion sleeves are specially designed to avoid overloaded damages to the force balances and to rotate the section model conveniently with a chosen wind attack angle.

In the forced vibration system, a photoelectric encoder and inductance proximity switch are used to control the model motions accurately. In order to get more feedback of electrical pulses in one motion period, a high-resolution increment photoelectric encoder was connected to the eccentric wheel by a small gear, as illustrated in Fig. 5(a). Thus, the displacement of each DOF can be calculated by the feedback pulses of the photoelectric encoder and mechanical amplitude according to the flowchart in Fig. 5(b) using the following equations:

$$\Phi = n\frac{2\pi}{N} \tag{1}$$

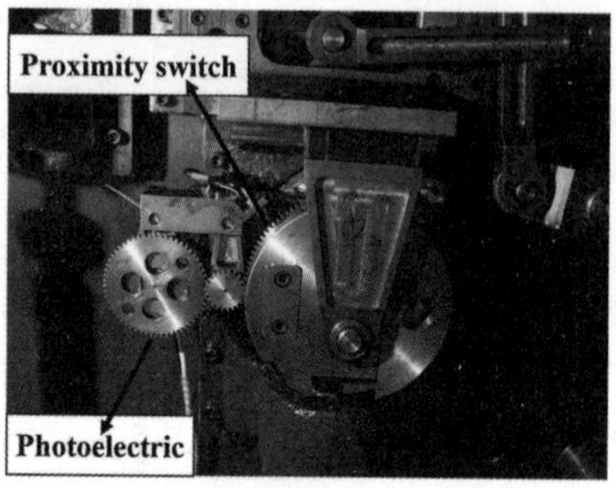

(a)

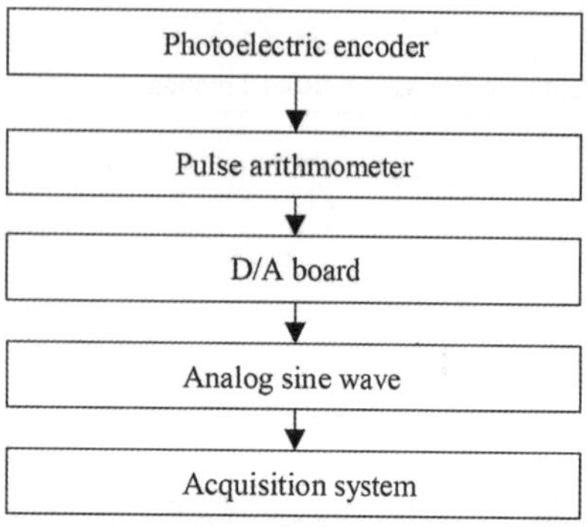

(b)

Fig. 5. Displacement measurement by electrical feedback pulses of the photoelectric encoder: (a) photoelectric encoder and proximity switch (image by Huawei Niu); and (b) flowchart to acquire displacement signals.

$$d = A \sin \Phi \tag{2}$$

where Φ = rotation angle of the eccentric wheel calculated from the original point reserved by the proximity switch; n = number of the feedback pulse; N = total number of the equally spaced pulses generated by the photoelectric encoder in one motion period (here, $N = 9{,}600$); A = set mechanical amplitude by the slider; and d = derived displacement of the section model.

The acquired displacements were compared with the recorded displacement of the middle part of the force balance by a laser displacement sensor. The differences in the amplitudes and the phase angles were less than 2%.

Bridge Deck Model

Experiments for a truss-girder section (based upon Aizhai Bridge, China) were performed for demonstration purposes. The section is shown in Fig. 6. The model is 1.54 m long, and the weight is 6 kg. Two plastic end plates were attached to the section model to reduce end effects.

Identification Method for 18 Flutter Derivatives

Self-excited forces for bridge decks can be expressed using the following equations using the 18 flutter derivatives (Chowdhury and Sarkar 2003; Sarkar et al. 1994):

$$\mathbf{L}_{se} = \rho U^2 B\left[K_h H_1^* \frac{\dot{h}}{U} + K_\alpha H_2^* \frac{B\dot{\alpha}}{U} + K_\alpha^2 H_3^* \alpha + K_h^2 H_4^* \frac{h}{B} + K_p H_5^* \frac{\dot{p}}{U} + K_p^2 H_6^* \frac{p}{B}\right] \tag{3}$$

$$\mathbf{D}_{se} = \rho U^2 B\left[K_p P_1^* \frac{\dot{p}}{U} + K_\alpha P_2^* \frac{B\dot{\alpha}}{U} + K_\alpha^2 P_3^* \alpha + K_p^2 P_4^* \frac{p}{B} + K_h P_5^* \frac{\dot{h}}{U} + K_h^2 P_6^* \frac{h}{B}\right] \tag{4}$$

$$\mathbf{M}_{se} = \rho U^2 B^2\left[K_h A_1^* \frac{\dot{h}}{U} + K_\alpha A_2^* \frac{B\dot{\alpha}}{U} + K_\alpha^2 A_3^* \alpha + K_h^2 A_4^* \frac{h}{B} + K_p A_5^* \frac{\dot{p}}{U} + K_p^2 A_6^* \frac{p}{B}\right] \tag{5}$$

where ρ = air density; U = mean wind velocity; B = width of the model; $K_i = B\omega_i/U$ $(i = h,\ p,\ \alpha)$ are the reduced frequencies; h, p, and α = forced vibration displacements along the vertical, lateral and torsional directions; the dot above the displacement denotes a derivative with respect to time; H_i^*, P_i^*, and A_i^* $(i = 1\text{–}6)$ = 18 flutter derivatives; and $\mathbf{L}_{se}$, $\mathbf{D}_{se}$, and $\mathbf{M}_{se}$ = self-excited aerodynamic lift, drag, and moment.

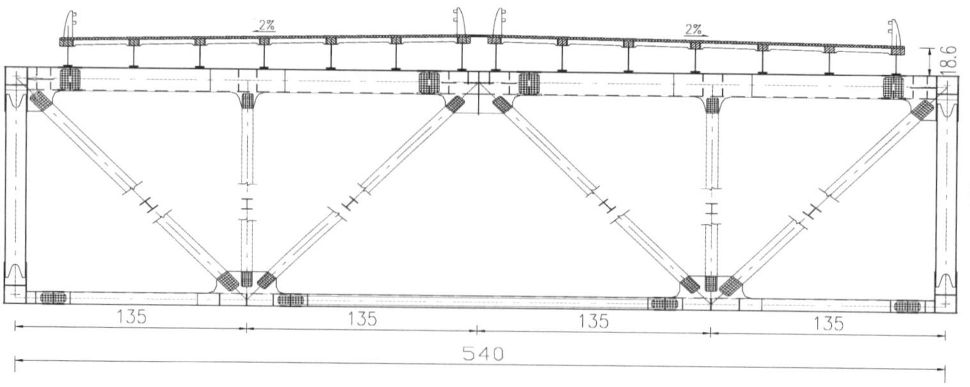

Fig. 6. Truss-girder section model (unit: millimeters).

In the present study, to extract the self-excited forces, a two-step experiment was carried out. First, a baseline reference test with forced vibration was carried out in still air (wind speed = 0 m/s), which was used to identify the inertial forces as well as some other mechanical interference forces; second, repeated tests with the same settings were carried out under various wind speeds. Because the motion parameters remain constant during the forced vibration tests, the inertia forces along with the other mechanical interference forces, namely the dynamic interference forces, are assumed to remain the same magnitude in the two-step experiments. Therefore, the aerodynamic forces can be obtained as the differences of the measured forces between the two sets of the experiments. The dynamic interference force can be expressed

$$\mathbf{F}_{\mathrm{dy},0} = \mathbf{X}_0\mathbf{C} \tag{6}$$

where $\mathbf{F}_{\mathrm{dy},0} = \{\mathbf{L}_{\mathrm{dy},0} \quad \mathbf{D}_{\mathrm{dy},0} \quad \mathbf{M}_{\mathrm{dy},0}\}$ is the dynamic interference force matrix in still air ($U = 0$); $\mathbf{L}_{\mathrm{dy},0}$, $\mathbf{D}_{\mathrm{dy},0}$, and $\mathbf{M}_{\mathrm{dy},0}$ = time histories of lift, drag, and moment; $\mathbf{X}_0 = \{\dot{h}_0 \quad h_0 \quad \dot{p}_0 \quad p_0 \quad \dot{\alpha}_0 \quad \alpha_0\}$ is the motion vector matrix at $U = 0$; and $\mathbf{C} = (\mathbf{X}_0^{\mathrm{T}}\mathbf{X}_0)^{-1}\mathbf{X}_0^{\mathrm{T}}\mathbf{F}_{\mathrm{dy},0}$ is the coefficient matrix between the dynamic interference forces and the motion vectors, which can be solved by the following least-square problem:

$$\mathbf{J} = \mathbf{e}^T\mathbf{e} = (\mathbf{F}_{\mathrm{dy},0} - \mathbf{X}_0\mathbf{C})^T(\mathbf{F}_{\mathrm{dy},0} - \mathbf{X}_0\mathbf{C}) \tag{7}$$

Therefore, the self-excited forces acting on the section model at wind speed U can be calculated

$$\mathbf{F}_{\mathrm{se,U}} = \mathbf{F}_{\mathrm{dy,U}} - \mathbf{X}_{\mathrm{U}}\mathbf{C} \tag{8}$$

where $\mathbf{F}_{\mathrm{se,U}} = \{\mathbf{L}_{\mathrm{se}} \quad \mathbf{D}_{\mathrm{se}} \quad \mathbf{M}_{\mathrm{se}}\}$ is the self-excited force matrix at wind speed U; $\mathbf{F}_{\mathrm{dy,U}} = \{\mathbf{L}_{\mathrm{dy,U}} \quad \mathbf{D}_{\mathrm{dy,U}} \quad \mathbf{M}_{\mathrm{dy,U}}\}$ is the measured dynamic total force matrix; and $\mathbf{X}_{\mathrm{U}} = \{\dot{h}_U \quad h_U \quad \dot{p}_U \quad p_U \quad \dot{\alpha}_U \quad \alpha_U\}$ is the motion vector matrix. Rewriting the Eqs. (3)–(5) leads to

$$\mathbf{L}_{\mathrm{se}} = \mathbf{SH} \tag{9}$$

$$\mathbf{D}_{\mathrm{se}} = \mathbf{RP} \tag{10}$$

$$\mathbf{M}_{\mathrm{se}} = \mathbf{SA} \tag{11}$$

where $\mathbf{S} = \{\dot{h}_U \quad \dot{\alpha}_U \quad \alpha_U \quad h_U \quad \dot{p}_U \quad p_U\}$ and $\mathbf{R} = \{\dot{p}_U \quad \dot{\alpha}_U \quad \alpha_U \quad p_U \quad \dot{h}_U \quad h_U\}$ are the motion vector matrices; and $\mathbf{H}$, $\mathbf{P}$, and $\mathbf{A}$ are the vectors of flutter derivatives given by

$$\mathbf{H} = \begin{Bmatrix} H_1 \\ H_2 \\ H_3 \\ H_4 \\ H_5 \\ H_6 \end{Bmatrix} = \rho U^2 B \begin{bmatrix} \frac{K_h}{U} & & & & & \\ & \frac{K_\alpha B}{U} & & & & \\ & & K_\alpha^2 & & & \\ & & & \frac{K_h^2}{B} & & \\ & & & & \frac{K_p}{U} & \\ & & & & & \frac{K_p^2}{B} \end{bmatrix} \begin{Bmatrix} H_1^* \\ H_2^* \\ H_3^* \\ H_4^* \\ H_5^* \\ H_6^* \end{Bmatrix} \tag{12}$$

$$\mathbf{P} = \begin{Bmatrix} P_1 \\ P_2 \\ P_3 \\ P_4 \\ P_5 \\ P_6 \end{Bmatrix} = \rho U^2 B \begin{bmatrix} \frac{K_p}{U} & & & & & \\ & \frac{K_\alpha B}{U} & & & & \\ & & K_\alpha^2 & & & \\ & & & \frac{K_p^2}{B} & & \\ & & & & \frac{K_h}{U} & \\ & & & & & \frac{K_h^2}{B} \end{bmatrix} \begin{Bmatrix} P_1^* \\ P_2^* \\ P_3^* \\ P_4^* \\ P_5^* \\ P_6^* \end{Bmatrix} \tag{13}$$

$$\mathbf{A} = \begin{Bmatrix} A_1 \\ A_2 \\ A_3 \\ A_4 \\ A_5 \\ A_6 \end{Bmatrix} = \rho U^2 B \begin{bmatrix} \frac{K_h}{U} & & & & & \\ & \frac{K_\alpha B}{U} & & & & \\ & & K_\alpha^2 & & & \\ & & & \frac{K_h^2}{B} & & \\ & & & & \frac{K_p}{U} & \\ & & & & & \frac{K_p^2}{B} \end{bmatrix} \begin{Bmatrix} A_1^* \\ A_2^* \\ A_3^* \\ A_4^* \\ A_5^* \\ A_6^* \end{Bmatrix} \tag{14}$$

Hence, the flutter derivatives can be identified by solving the corresponding least-square equations for the self-excited forces given by

$$\mathbf{J} = \mathbf{e}^T\mathbf{e} = (\mathbf{L}_{\mathrm{se}} - \mathbf{SH})^T(\mathbf{L}_{\mathrm{se}} - \mathbf{SH}) \tag{15}$$

$$\mathbf{J} = \mathbf{e}^T\mathbf{e} = (\mathbf{D}_{\mathrm{se}} - \mathbf{RP})^T(\mathbf{D}_{\mathrm{se}} - \mathbf{RP}) \tag{16}$$

$$\mathbf{J} = \mathbf{e}^T\mathbf{e} = (\mathbf{M}_{\mathrm{se}} - \mathbf{SA})^T(\mathbf{M}_{\mathrm{se}} - \mathbf{SA}) \tag{17}$$

The vectors of $\mathbf{H}$, $\mathbf{P}$, and $\mathbf{A}$ can be rewritten

$$\mathbf{H} = (\mathbf{S}^T\mathbf{S})^{-1}\mathbf{S}^T\mathbf{L}_{\mathrm{se}} \tag{18}$$

$$\mathbf{P} = (\mathbf{R}^T\mathbf{R})^{-1}\mathbf{S}^T\mathbf{D}_{\mathrm{se}} \tag{19}$$

$$\mathbf{A} = (\mathbf{S}^T\mathbf{S})^{-1}\mathbf{S}^T\mathbf{M}_{\mathrm{se}} \tag{20}$$

Finally, the 18 flutter derivatives can be obtained from the vectors $\mathbf{H}$, $\mathbf{P}$, and $\mathbf{A}$ according to Eqs. (12)–(14).

During the experiments, the time histories of the dynamic forces and displacements were directly measured by the proposed forced vibration system, and the velocities can be obtained as the first time derivative of the recorded displacement signals. Similarly, the procedures can be used for 1-DOF and coupled 2-DOF cases by only keeping the terms associated with the DOFs that are of consideration in these equations.

Extraction of Flutter Derivatives

Forced Vibration Parameters

1-DOF (T, V, and L) and coupled 3-DOF (T-V-L) experiments were carried out for the truss-girder section model. Table 1 gives the

Table 1. Forced vibration testing parameters for different DOF combinations

Parameter	Value
T frequency	2.5 Hz
T amplitude	2°
V frequency	1.8 Hz
V amplitude	12 mm
L frequency	1.1 Hz
L amplitude	12 mm
Sampling rate	204.8 Hz
Sampling number	12,288
Wind speed	0–25 m/s
Wind speed interval	2.5 m/s
Wind attack angle	0°
Model natural frequency	>10 Hz

Note: T = torsional; V = vertical; and L = lateral.

experimental parameters for all the vibration modes during the forced vibration tests. The reduced frequencies were obtained by varying the wind speed from 0 to 25 m/s with an increment of 2.5 m/s, whereas the vibration frequencies remain constant in all tests. In addition, all the experiments were performed in a uniform flow with the turbulence intensity $I < 0.01$.

Flutter Derivatives

The flutter derivatives were obtained and are shown in Figs. 7–9. To better demonstrate the results, the flutter derivatives for a thin plate [Theodorsen plate (TP)], the flutter derivatives of a rectangular cross section with a side ratio of $h/B = 0.07$ (referred to B1 hereafter) are also presented in Figs. 7–9 for comparison purposes. The flutter derivatives of the B1 section model were obtained using a 3-DOF elastic suspension system based on a free vibration technique (Chowdhury and Sarkar 2003). As shown in Fig. 7, the curves of H_i^* ($i = 1$–4) for both the TP section and truss girder of the Aizhai Bridge (AZ) section are all smooth with clear trends. In comparison with the TP section, H_1^* and H_3^* of the AZ section display similar trends as those of TP section, whereas H_2^* and H_4^* of the AZ section show distinguishable trends. This is because the AZ section is a truss girder (bluff body), which is different from the streamlined configuration of the TP section. As shown in Fig. 7, the identified H_2^* of the AZ section is found to decrease with the increase of the reduced wind speed, showing an opposite trend compared with that of the TP section. In addition, for the AZ section, the identified values of H_1^*, H_2^*, and H_3^* obtained from 1-DOF test are identical with those obtained from 3-DOF tests, but large variations can be found for H_4^* between the two tests when $V^* > 15$.

Different from the smoothness of H_i^* ($i = 1$–4), H_5^* and H_6^* curves for the AZ section model show more fluctuations, although

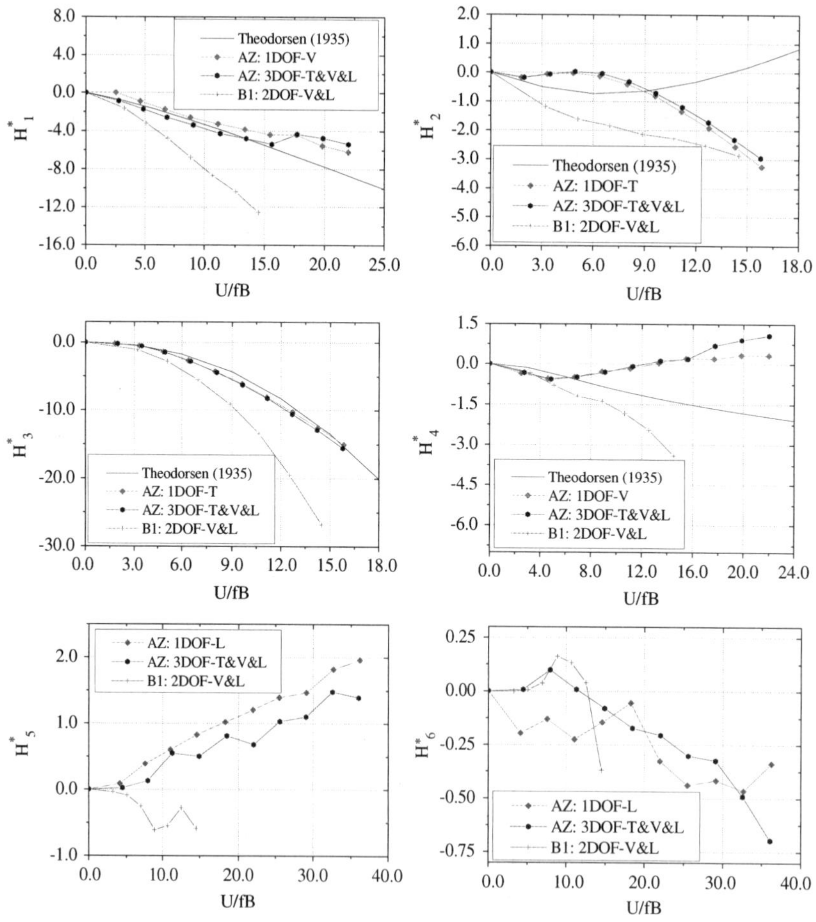

Fig. 7. Vertical flutter derivatives H_i^* ($i = 1$–6).

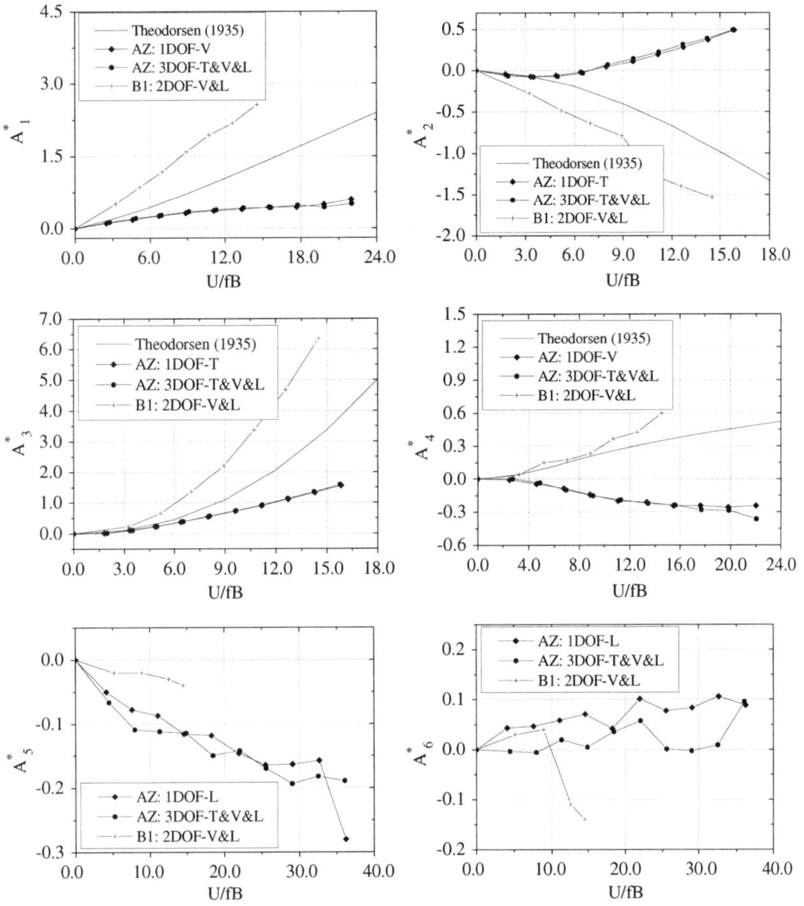

Fig. 8. Torsional flutter derivatives A_i^* (i = 1–6).

their general trends can still be found. Both the identified values of H_5^* and H_6^* show some variations between the 1-DOF test and 3-DOF test. Comparing the H_5^* and H_6^* curves of Section B1 with those of the AZ section reveals that the curves of Section B1 have more fluctuations with unclear trends. The comparison results indicate that the proposed forced vibration method can provide better prediction results with fewer fluctuations compared with the free vibration method.

Similar trends can be found for the torsional flutter derivatives, as shown in Fig. 8. The A_i^* (i = 1–4) of the AZ section has smooth curves with clear trends. In addition, similar to the H_i^* (i = 1–4), the A_i^* (i = 1–4) curves for the AZ section are also found to deviate from the TP curves, among which A_2^* and A_4^* curves have an opposite trend compared with the TP curves. Compared with A_i^* (i = 1–4), the curves of both A_5^* and A_6^* show more fluctuations, in which the A_5^* curve has a clear trend but the A_6^* curve has no clear trend.

Finally, the lateral flutter derivatives for the AZ section are shown in Fig. 9. All the lateral flutter derivatives P_i^* (i = 1–6) have smooth curves. The curves of P_1^*, P_2^*, P_3^*, and P_5^* have a clear trend, but no clear trend can be found for P_4^* and P_6^*. In addition, for all P_i^* (i = 1–6), the identified values obtained from 1-DOF test and multi-DOF test agree well with each other. Again, the curves of the flutter derivatives identified from the proposed forced vibration technique are smoother than those obtained from the free vibration technique.

In summary of the preceding discussions, the proposed forced vibration technique can be used to extract 18 flutter derivatives with a smoother and clearer trend than the free vibration technique. In addition, the proposed forced vibration technique also facilitates flutter derivative identification with different DOF modes, i.e., 1-DOF and 3-DOF tests for cross checking. The results obtained from 1-DOF and multi-DOF tests overall agree well with each other, indicating the accuracy and robustness of the proposed identification technique. The large variation in the identified values between the 1-DOF and multi-DOF tests only occur for those flutter derivatives with small values, i.e., H_5^*, A_5^*, H_6^*, A_6^*, and P_4^*. Such large variations are due to the relatively small ratios of the corresponding self-excited force component to the total force.

Numerical Simulation

Bridge Model

As discussed previously, the identification of additional flutter derivatives gives the opportunity to discuss the potential effects of these parameters on the bridge dynamic and stress performance.

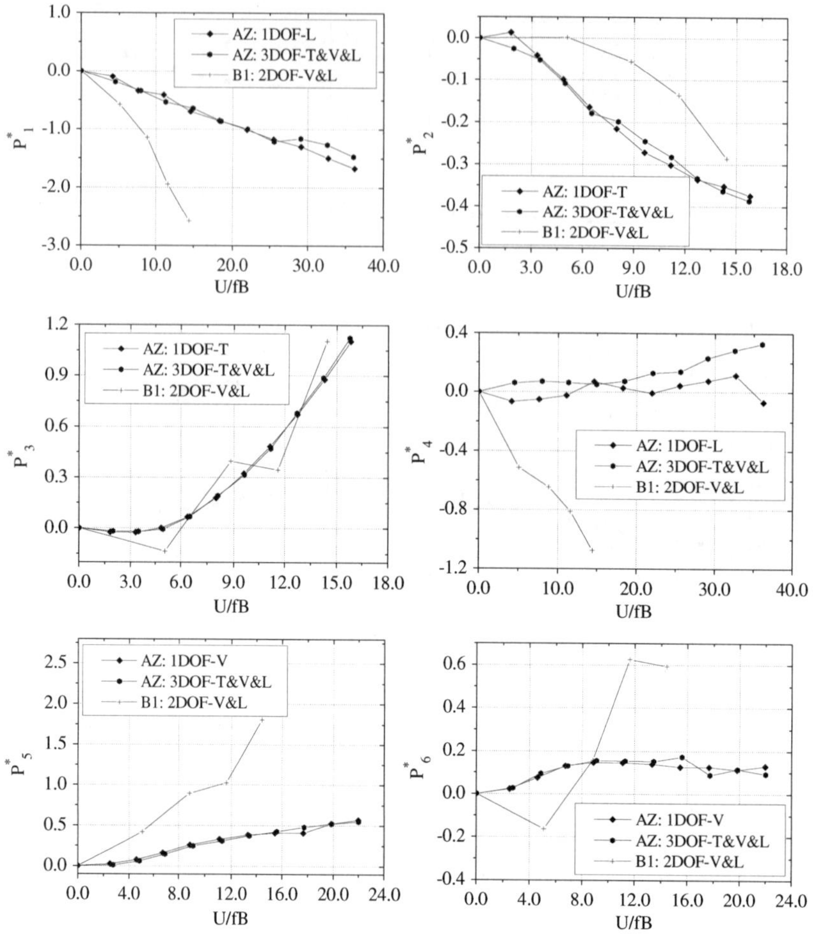

Fig. 9. Lateral flutter derivatives P_i^* ($i = 1$–6).

To this end, Aizhai Bridge is adopted in the present study to evaluate its dynamic responses and stresses for local details using the conventional eight flutter derivatives as well as the 18 flutter derivatives based upon the established vehicle–bridge–wind (VBW) numerical framework (Zhu et al. 2018a; Zhu and Zhang 2017). Located in Hunan Province, China, Aizhai Bridge is a suspension bridge across a deep river valley with a span arrangement of $242 + 1{,}176 + 116$ m, as shown in Fig. 10. Its main span is 1,000.5 m long and supported by hangers with an equal space of 14.5 m. The truss girder is 27 m wide and 7.5 m high, which supports four lanes of traffic (two in each direction). The height from the bridge midspan to the bottom of the valley is about 350 m. Table 2 presents the natural frequencies and corresponding mode shapes of the first 10 modes of Aizhai Bridge. The effects of the additional 10 flutter derivatives on the bridge dynamic and stress responses are discussed in the following subsections.

Effects on the Bridge Dynamic Responses under Wind Load Only

The effects of the 10 additional flutter derivatives on the bridge dynamic characteristics under only wind load were investigated. To generate the wind field, the fluctuating wind velocities along the bridge deck in both vertical and horizontal directions were generated using a simplified spectral representation method (Cao et al. 2000). A total of 139 wind simulation points, evenly distributed along the bridge deck axis with an equal interval of 7.25 m, were generated for the wind field around the bridge. The Kaimal's spectrum and Lumley-Panofsky's spectrum were adopted for the horizontal and vertical wind spectrum, respectively. Three key locations along the bridge span, i.e., 1/4 span, mid-span, and 3/4 span, were adopted in the present study to investigate the influence of different set of flutter derivatives on the bridge dynamic responses. Here and hereafter, the 18 flutter case and eight flutter case are used to represent the simulation case using 18 and 8 flutter derivatives, respectively, for the sake of brevity.

The total wind forces acting on the bridge were assumed to have three components, i.e., steady-state forces resulting from the mean wind speed, self-excited forces resulting from wind–bridge interactions, and buffeting forces resulting from wind turbulence. For the sake of brevity, a detailed derivation of the wind forces is not elaborated in the present study but can be found in the literature (Simiu and Scanlan 1996; Xu 2013; Wu et al. 2017). In addition, Rayleigh damping was adopted to construct the bridge damping matrix as a function of stiffness and mass matrices, in which the damping coefficient for the mass and stiffness matrices are obtained by assuming a damping ratio of 0.5% for the first two fundamental modes.

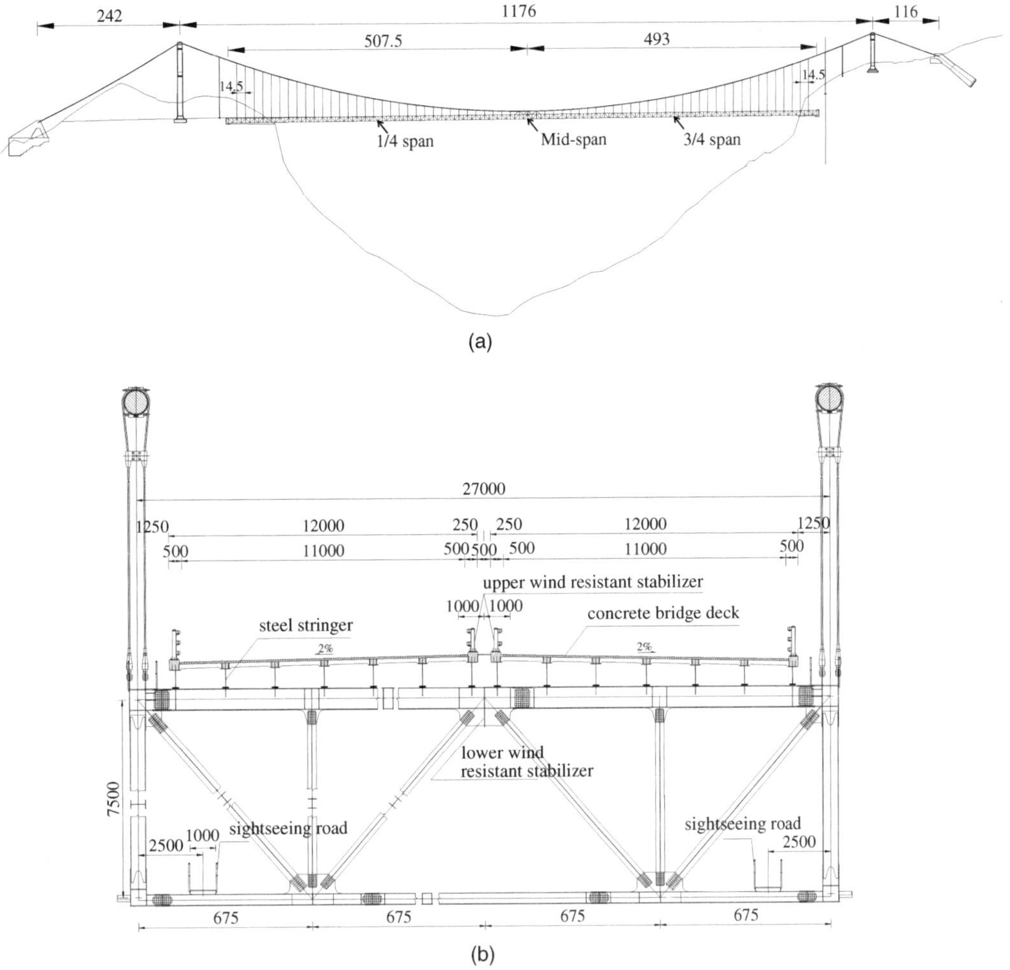

Fig. 10. Layout of Aizhai Bridge: (a) elevation view (unit: meters); and (b) cross-section (unit: millimeters).

Table 2. First 10 modes of Aizhai Bridge

Mode No.	Natural frequency (Hz)	Period (s)	Mode shape description
1	0.0564	17.7173	1. symmetric lateral, deck
2	0.1159	8.6259	1. asymmetric vertical, deck
3	0.1422	7.0309	1. asymmetric lateral, deck
4	0.1602	6.2418	1. symmetric vertical, deck
5	0.2138	4.6777	2. symmetric vertical, deck
6	0.2538	3.9404	2. asymmetric vertical, deck
7	0.2558	3.9101	1. symmetric lateral, cable
8	0.2585	3.8691	1. symmetric torsional, deck
9	0.2712	3.6877	2. symmetric lateral, cable
10	0.2769	3.6119	1. asymmetric lateral, cable

Fig. 11 shows the displacement responses at the midspan in vertical, lateral, and rotational directions under a crosswind speed of 30 m/s. As shown in Fig. 11, the 18 flutter case and eight flutter case have almost identical vertical and torsional displacements, but an obvious discrepancy can be found with regard to the lateral displacement between the two cases. The maximum absolute lateral displacements in the 8 and 18 flutter cases are 1.39 and 1.00 m. Therefore, the eight flutter case can result in 39.0% higher maximum lateral displacements than the 18 flutter case. Similar to the midspan, the maximum lateral displacements for 1/4 span and 3/4 span of the bridge deck using the eight flutter case are 36.8% and 41.4% higher than those using 18 flutter derivatives, respectively, as shown in Fig. 12. The possible explanation is as follows. The 18 flutter derivatives contain P_1^*, i.e., the lateral damping term associated with lateral motion, which is usually negative, as shown in Fig. 9. Therefore, the additional negative aerodynamic damping term P_1^* in the 18 flutter case can result in the reduction of the bridge lateral displacement responses compared with the eight flutter case.

In addition, Fig. 13 shows the maximum absolute lateral displacement responses at the three key locations along the bridge deck under different wind speeds using both 18 and 8 flutter derivatives in order to further investigate the effects of the wind speed on the bridge lateral dynamic responses. The results indicate that the discrepancy of the bridge lateral responses between the eight flutter case and the 18 flutter case increases significantly as the wind speed increases for all the three locations. In other words, using eight flutter derivatives can lead to significant overestimation of the bridge lateral displacement, especially under high wind speed, compared

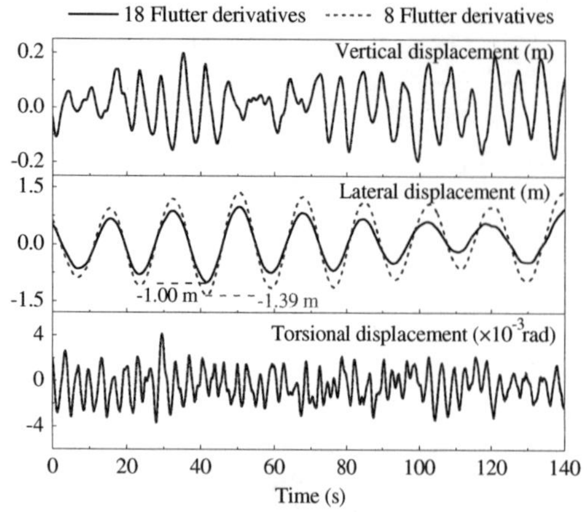

Fig. 11. Comparison of displacement time histories at the midspan using 18 and 8 flutter derivatives under crosswind $U = 30$ m/s.

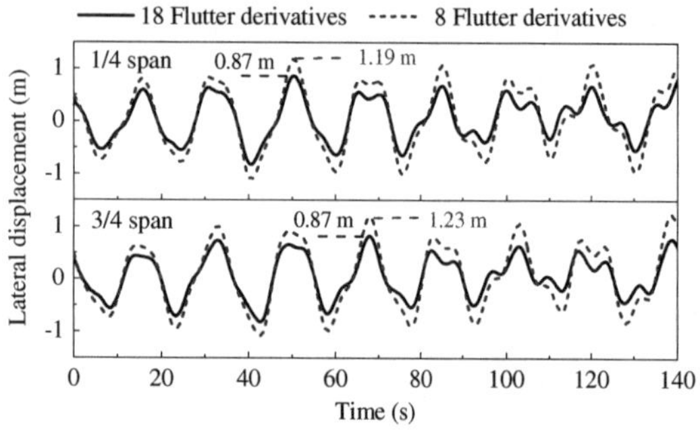

Fig. 12. Lateral displacement time histories at the 1/4 and 3/4 span of the bridge deck using 18 and 8 flutter derivatives under crosswind $U = 30$ m/s.

with the 18 flutter case for Aizhai Bridge. Taking the midspan as an example, the maximum absolute lateral displacements in the eight flutter case under the wind speed of 10, 15, 20, 25, and 30 m/s are 13.3%, 20.0%, 23.8%, 31.74%, and 39.0% higher, respectively, than those in the 18 flutter case. The vertical and torsional displacements are the same for both the 18 and 8 flutter cases for all the three locations.

To summarize, the results suggest that the ignorance of the additional 10 flutter derivatives will result in significant overestimation of the bridge lateral responses, especially under the high wind speed. However, the additional 10 flutter derivatives have negligible influences on both the vertical and torsional bridge responses.

Effects on the Dynamic Response under Combined Wind and Traffic Loads

To further evaluate the dynamic performance of the bridge–traffic system under windy environments, VBW analysis was carried out. The presence of traffic flow has an important primary impact on the aerodynamic configuration of bridge deck, which may influence the aerostatic and aerodynamic coefficients and in turn, affect the self-excited forces acting on the bridge. The aerostatic and aerodynamic coefficients adopted in the present study are original wind parameters of the truss-girder section without considering the wind interference between the bridge and moving traffic flow for the following reasons. Currently, it is still a challenging task to quantify the wind coefficients of the coupled system of bridge and moving traffic flow due to the lack of the state-of-the art techniques. In fact, a variety of factors such as the traffic pattern (e.g., vehicle position, vehicle configuration, and vehicle speed) and wind turbulence play a significant role in modifying the shape of the combined cross section for the bridge–traffic system. Consequently, it is very difficult to rationally and effectively quantity the wind coefficients experimentally if all the possible combinations of the influential factors are considered. Although there has been some limited studies on wind-tunnel testing on evaluating the flutter derivatives by considering the aerodynamic interference (Han et al. 2014), no comprehensive studies have been reported publicly in the literature so far.

In the present study, the mean crosswind at the deck level was taken as 20 m/s (44.74 mi/h). A free-flow traffic condition was adopted for the traffic simulation, in which approximately 20 vehicles with an equal distance of 50 m are traveling simultaneously on the outer lane in the windward side of the bridge deck at a driving speed of 20 m/s. A light truck was utilized as the representative vehicle in the present study. Detailed dimensional and dynamic properties of the light truck have been given by Zhu et al. (2018a, b). In addition, the road surface condition was assumed as good [ISO 8608 (ISO 1995)]. Figs. 14(a and b) show the vehicle displacement and acceleration responses of a selected vehicle when running on the bridge in both the 18 and 8 flutter cases. It is observed from Figs. 14(a and b) that the vehicle displacements and accelerations in both cases are almost identical, indicating that the additional 10 flutter derivatives have a negligible influence on the dynamic characteristics of the running vehicles.

In addition, Fig. 15 also shows the lateral displacement at the 1/4 span, midspan, and 3/4 span using both 18 and 8 flutter derivatives in the context of coupled VBW analysis. As shown in Fig. 15, the bridge lateral displacement in the eight flutter case at the 1/4 span, midspan, and 3/4 span are 24.3%, 28.9%, and 27.6% higher, respectively, than those in the 18 flutter case. Recall from the previous bridge–wind analysis that under the same wind speed of 20 m/s (Fig. 13), the percentage increases at the 1/4 span, midspan, and 3/4 span due to the use of eight flutter derivatives are 27.5%, 31.3%, and 28.6%, respectively. The percentage increases of the bridge lateral responses due to the replacement of the 18 flutter derivatives with eight flutter derivatives in the context of VBW analysis are very close to those in the bridge–wind analysis because the major traffic loading effects are in the vertical direction. This indicates that the vehicle–bridge coupling effect may have negligible influence on the discrepancy of the bridge lateral responses between the 18 and 8 flutter cases. Meanwhile, the additional 10 flutter derivatives also have negligible effects on the bridge's vertical and torsional responses.

Effects on the Stress Response of Bridge Deck under Combined Wind and Traffic Loads

To evaluate the fatigue performance, the effects of additional 10 flutter derivatives on the stress response of the bridge deck in the context of VBW analysis were investigated as well. In the present study, multiscale finite-element analysis (FEA) was performed to evaluate the maximum principal stress at critical locations of the bridge deck. For the sake of brevity, detailed procedures about the analysis scheme are not introduced in the present study, and more details have been given by Zhu and Zhang (2018). In the present study, the cross section of the bridge deck at the

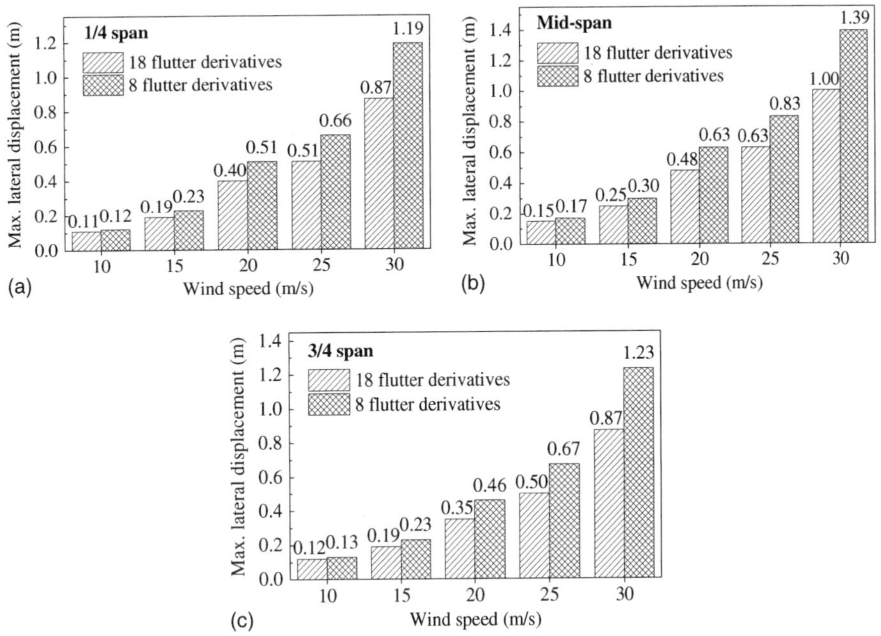

Fig. 13. Maximum absolute lateral displacement responses under different wind speeds at key locations along the bridge deck: (a) 1/4 span; (b) midspan; and (c) 3/4 span.

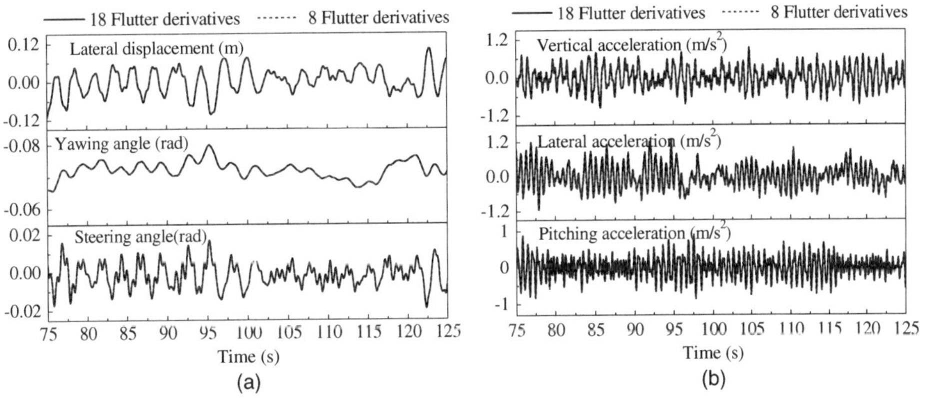

Fig. 14. Vehicle responses using 18 and 8 flutter derivatives ($U = 20$ m/s with one lane of traffic): (a) displacement; and (b) acceleration.

midspan was identified as the most critical section, in which the Element 1023 was identified as the most critical element, as shown in Fig. 16. Element 1023 as shown in dark gray in Fig. 16 is the bottom chord in the middle of the cross frame. The stress response at the middle section of critical Element 1023 under combined wind and traffic loads was adopted for the subsequent analysis.

The stress responses at the identified critical location of the bridge deck under only wind load were first investigated, and the results are shown in Fig. 17, in which the wind speed is 20 m/s. It is interesting to find that the stress response using 18 and 8 flutter derivatives are almost identical, with a maximum difference within 3%, indicating the additional 10 flutter derivatives have a negligible influence on the stress responses. Similar observations were also found under other wind speeds. The possible reason may be due to that the contribution of the bridge lateral motion on the resultant stress response is not significant compared with that due to bridge motion in other directions. Consequently, the additional 10 flutter derivatives can hardly affect the resultant stress response, considering that the additional 10 flutter derivatives only have impacts on the bridge lateral motion.

Furthermore, the stress responses at the critical location of the bridge deck under a wind speed of 20 m/s with simultaneous traffic flow was also investigated, and the results are displayed in Fig. 18. For simplicity, the traffic flow was assumed to be deterministic in which the running vehicles, i.e., the light truck, are uniformly distributed along the four traffic lanes and the distance between the two adjacent vehicles in the same lane is 50 m. It is observed that the vehicle-induced stress in terms of both stress

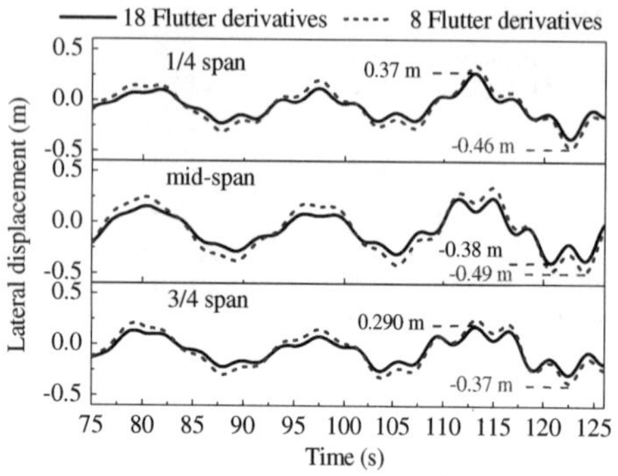

Fig. 15. Lateral displacement response at the 1/4 span, midspan, and 3/4 span along the bridge deck using 18 and 8 flutter derivatives ($U = 20$ m/s, one lane of traffic).

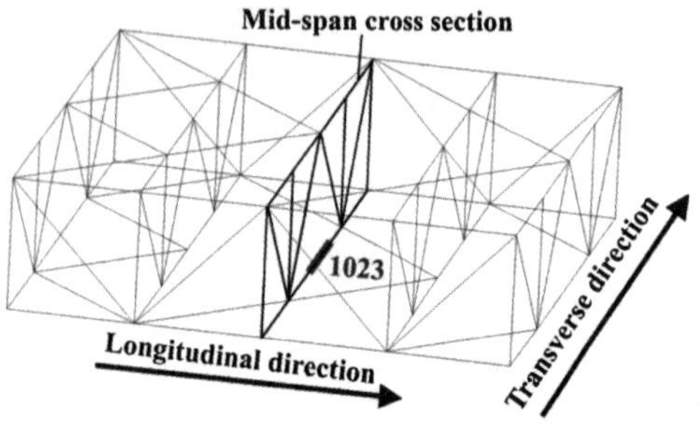

Fig. 16. Identified critical deck section and critical element.

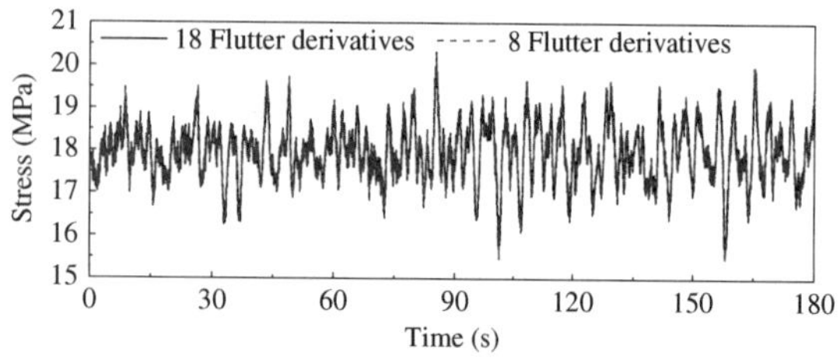

Fig. 17. Wind-induced stress responses at the critical location of bridge deck under crosswind speed of 20 m/s for a duration of 3 min.

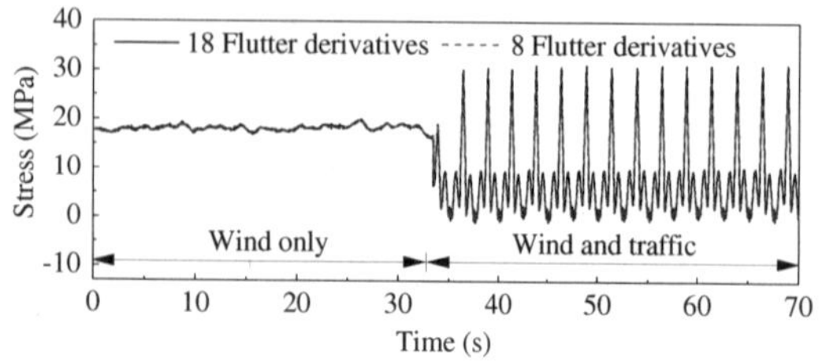

Fig. 18. Stress response segment at the critical location of bridge deck under combined wind and traffic loads ($U = 20$ m/s).

amplitude and stress range is much greater than that induced by wind, even for the wind speed of 20 m/s. The vehicle-induced stress is the major factor of the stress response at the critical location of the bridge deck in the context of VBW stress analysis. As a result, the effects of the additional 10 flutter derivatives on the stress response of the bridge deck under combined wind and traffic load is insignificant, as shown in Fig. 18.

To summarize, the study of the influence of the additional 10 flutter derivatives in the context of VBW stress analysis shows that ignoring the additional 10 flutter derivatives has negligible influence on the stress response at the selected critical locations of the bridge deck, which could possibly be explained by the following two reasons. First, although the additional 10 flutter derivatives can cause a large discrepancy of the resultant bridge lateral motion, the stress responses contributed by the bridge lateral motion is much less than those due to other components of the bridge motion. Second, both the stress amplitude and stress range induced by wind load is much less than that induced by the traffic load.

Conclusions

In this paper, an innovative 3-DOF forced vibration system was developed to experimentally extract 18 flutter derivatives of bridge decks. After validating the effectiveness and accuracy based on a thin-plate section model, the proposed flutter derivative identification technique was employed to extract 18 flutter derivatives of a streamlined box-girder section model and a truss-girder section model, respectively. The results show that for those flutter derivatives that are hardly recognized in the current state-of-the-art methods, the proposed method could provide the smooth curves with clear trends.

To further evaluate the effects of the additional 10 flutter derivatives, both the bridge dynamic and stress analysis of a long-span bridge with truss girders by employing 18 and 8 flutter derivatives were performed in the context of coupled VBW analysis. The VBW dynamic analysis shows that ignoring the additional 10 flutter derivatives will result in significant overestimation of the bridge lateral responses, especially under high wind speeds. This could possibly be explained by the existence of the additional negative aerodynamic damping term P_1^* in the 18 flutter case, which can result in suppression of the bridge lateral displacement responses. Different from the bridge lateral motion, the additional 10 flutter derivatives were found to have negligible effects on bridge motions in other directions.

The VBW stress analysis showed that ignoring the additional 10 flutter derivatives has a negligible influence on the stress response at critical locations of the bridge deck. One possible reason is that the contribution of the bridge lateral motion to the bridge stress response is insignificant compared with those due to the bridge motion in other directions, even though the additional 10 flutter derivatives can cause a large discrepancy in the resultant bridge lateral motion. The other reason is due to the fact that both the stress amplitude and stress range induced by only wind load are much less than that induced by the traffic load, leading to the negligible influence of the additional 10 flutter derivatives on the overall resultant stress response.

The present study provides some new insights on evaluating the role of 10 additional flutter derivatives on the dynamic and stress responses of a slender truss suspension bridge in the context of dynamic vehicle–bridge–wind system. Further investigations can be performed to explore the 10 additional flutter derivatives on the dynamic and fatigue performance of long-span bridges with different dynamic performances and other types of decks. In addition, due to the remaining technical challenges of wind-tunnel tests, the wind coefficients of the bridge–traffic system accounting for their aerodynamic interference were not considered in the present study. These aerodynamic interference effects can be incorporated in a future study to obtain more realistic wind coefficients for better predicting the wind-induced vibration of the bridge–traffic system.

Acknowledgments

The research is supported by Natural Science Foundation of China under Grant Nos. 51478181, U1534206, 50908085, and 91215302 and by the China Scholarship Council (No. 201506135039), which are gratefully acknowledged. The authors also would like to express great thanks to Professor Kun Luo and Professor Zhongyu Cheng, who provided great assistance with the development of the 3-DOF forced vibration system.

References

Cao, Y., H. Xiang, and Y. Zhou. 2000. "Simulation of stochastic wind velocity field on long-span bridges." *J. Eng. Mech.* 126 (1): 1–6. https://doi.org/10.1061/(ASCE)0733-9399(2000)126:1(1).

Chen, A., X. He, and H. Xiang. 2002. "Identification of 18 flutter derivatives of bridge decks." *J. Wind Eng. Ind. Aerodyn.* 90 (12–15): 2007–2022. https://doi.org/10.1016/S0167-6105(02)00317-3.

Chen, C. H. 2003. "Determination of flutter derivatives via a neural network approach." *J. Sound Vib.* 263 (4): 797–813. https://doi.org/10.1016/S0022-460X(02)01279-8.

Chen, X., and A. Kareem. 2004. "Efficacy of the implied approximation in the identification of flutter derivatives." *J. Struct. Eng.* 130 (12): 2070–2074. https://doi.org/10.1061/(ASCE)0733-9445(2004)130:12(2070).

Chen, X., and A. Kareem. 2006. "Revisiting multimode coupled bridge flutter: Some new insights." *J. Eng. Mech.* 132 (10): 1115–1123. https://doi.org/10.1061/(ASCE)0733-9399(2006)132:10(1115).

Chen, Z., Y. Han, Y. Luo, and X. Hua. 2010. "Identification of aerodynamic parameters for eccentric bridge section model." *J. Wind Eng. Ind. Aerodyn.* 98 (4–5): 202–214. https://doi.org/10.1016/j.jweia.2009.10.016.

Chen, Z. Q., X. D. Yu, G. Yang, and B. F. Spencer Jr. 2005. "Wind-induced self-excited loads on bridges." *J. Struct. Eng.* 131 (12): 1783–1793. https://doi.org/10.1061/(ASCE)0733-9445(2005)131:12(1783).

Chowdhury, A. G., and P. P. Sarkar. 2003. "A new technique for identification of eighteen flutter derivatives using a three-degree-of-freedom section model." *Eng. Struct.* 25 (14): 1763–1772. https://doi.org/10.1016/j.engstruct.2003.07.002.

Ding, Q., Z. Zhou, L. Zhu, and H. Xiang. 2010. "Identification of flutter derivatives of bridge decks with free vibration technique." *J. Wind Eng. Ind. Aerodyn.* 98 (12): 911–918. https://doi.org/10.1016/j.jweia.2010.09.005.

Gu, M., R. Zhang, and H. Xiang. 2000. "Identification of flutter derivatives of bridge decks." *J. Wind Eng. Ind. Aerodyn.* 84 (2): 151–162. https://doi.org/10.1016/S0167-6105(99)00051-3.

Han, Y., C. S. Cai, J. Zhang, S. Chen, and X. He. 2014. "Effects of aerodynamic parameters on the dynamic responses of road vehicles and bridges under cross winds." *J. Wind Eng. Ind. Aerodyn.* 134 (Nov): 78–95. https://doi.org/10.1016/j.jweia.2014.08.013.

ISO. 1995. *Mechanical vibration-road surface profiles—Reporting of measured data*. ISO 8608. Geneva: ISO.

Iwamoto, M., and Y. Fujino. 1995. "Identification of flutter derivatives of bridge deck from free vibration data." *J. Wind Eng. Ind. Aerodyn.* 54–55 (Feb): 55–63. https://doi.org/10.1016/0167-6105(94)00029-D.

Jones, N. P., J. D. Raggett, and E. Ozkan. 2003. "Prediction of cable-supported bridge response to wind: Coupled flutter assessment during retrofit." *J. Wind Eng. Ind. Aerodyn.* 91 (12–15): 1445–1464. https://doi.org/10.1016/j.jweia.2003.09.030.

Jung, S. M., J. Ghaboussi, and S. D. Kwon. 2004. "Estimation of aeroelastic parameters of bridge decks using neural networks." *J. Eng. Mech.* 130 (11): 1356–1364. https://doi.org/10.1061/(ASCE)0733-9399(2004)130:11(1356).

Katsuchi, H., N. P. Jones, R. H. Scanlan, and H. Akiyama. 1998. "Multimode flutter and buffeting analysis of the Akashi-Kaikyo Bridge." *J. Wind Eng. Ind. Aerodyn.* 77–78: 431–441. https://doi.org/10.1016/S0167-6105(98)00162-7.

Mishra, S. S., K. Kumar, and P. Krishna. 2006. "Identification of 18 flutter derivatives by covariance driven stochastic subspace method." *Wind Struct. Int. J.* 9 (2): 159–178. https://doi.org/10.12989/was.2006.9.2.159.

Miyata, T. 1995. "Full model testing of large cable-supported bridges." In *Proc., 9th ICWE, State-of-the-Art Volume*. New Delhi, India: Wiley.

Qin, X. R., and M. Gu. 2004. "Determination of flutter derivatives by stochastic subspace identification technique." *Wind Struct. Int. J.* 7 (3): 173–186. https://doi.org/10.12989/was.2004.7.3.173.

Sarkar, P. P., A. G. Chowdhury, and T. B. Gardner. 2004. "A novel elastic suspension system for wind tunnel section model studies." *J. Wind Eng. Ind. Aerodyn.* 92 (1): 23–40. https://doi.org/10.1016/j.jweia.2003.09.036.

Sarkar, P. P., N. P. Jones, and R. H. Scanlan. 1992. "System identification for estimation of flutter derivatives." *J. Wind Eng. Ind. Aerodyn.* 42 (1–3): 1243–1254. https://doi.org/10.1016/0167-6105(92)90131-S.

Sarkar, P. P., N. P. Jones, and R. H. Scanlan. 1994. "Identification of aeroelastic parameters of flexible bridges." *J. Eng. Mech.* 120 (8): 1718–1742. https://doi.org/10.1061/(ASCE)0733-9399(1994)120:8(1718).

Scanlan, R. H., and J. J. Tomko. 1971. "Airfoil and bridge deck flutter derivatives." *J. Eng. Mech. Div.* 97 (6): 1717–1737.

Simiu, E., and R. H. Scanlan. 1996. *Wind effects on structures: Fundamentals and applications to design*. 3rd ed. New York: Wiley.

Singh, L., N. P. Jones, R. H. Scanlan, and O. Lorendeaux. 1996. "Identification of lateral flutter derivatives of bridge decks." *J. Wind Eng. Ind. Aerodyn.* 60 (1–3): 81–89. https://doi.org/10.1016/0167-6105(96)00025-6.

Theodorsen, T. 1935. *General theory of aerodynamic instability and the mechanism of flutter*. NACA Rep. 496. Langley Field, VA: NASA Ames Research Center.

Wu, M. X., Y. L. Li, and W. Zhang. 2017. "Impacts of wind shielding effects of bridge tower on railway vehicle running performance." *Wind Struct.* 25 (1): 63–77. https://doi.org/10.12989/was.2017.25.1.063.

Xu, F. Y. 2015. "Direct approach to extracting 18 flutter derivatives of bridge decks and vulnerability analysis on identification accuracy." *J. Aerosp. Eng.* 28 (3): 04014080. https://doi.org/10.1061/(ASCE)AS.1943-5525.0000413.

Xu, F. Y., X. Z. Chen, C. S. Cai, and A. R. Chen. 2012. "Determination of 18 flutter derivatives of bridge decks by an improved stochastic search algorithm." *J. Bridge Eng.* 17 (4): 576–588. https://doi.org/10.1061/(ASCE)BE.1943-5592.0000295.

Xu, F. Y., T. Wu, X. Ying, and A. Kareem. 2016. "Higher-order self-excited drag forces on bridge decks." *J. Eng. Mech.* 142 (3): 06015007. https://doi.org/10.1061/(ASCE)EM.1943-7889.0001036.

Xu, F. Y., X. Y. Ying, and Z. Zhang. 2014. "Three-degree-of-freedom coupled numerical technique for extracting 18 aerodynamic derivatives of bridge decks." *J. Struct. Eng.* 140 (11): 04014085. https://doi.org/10.1061/(ASCE)ST.1943-541X.0001009.

Xu, Y. L. 2013. *Wind effects on cable-supported bridges*. Singapore: Wiley.

Yamada, H., T. Miyata, and H. Ichikawa. 1992. "Measurement of aerodynamic coefficients by system identification methods." *J. Wind Eng. Ind. Aerodyn.* 42 (1–3): 1255–1263. https://doi.org/10.1016/0167-6105(92)90132-T.

Zhu, J., and W. Zhang. 2017. "Numerical simulation of wind and wave fields for coastal slender bridges." *J. Bridge Eng.* 22 (3): 04016125. https://doi.org/10.1061/(ASCE)BE.1943-5592.0001002.

Zhu, J., and W. Zhang. 2018. "Probabilistic fatigue damage assessment of coastal slender bridges under coupled dynamic loads." *Eng. Struct.* 166 (7): 274–285. https://doi.org/10.1016/j.engstruct.2018.03.073.

Zhu, J., W. Zhang, and M. Wu. 2018a. "Coupled dynamic analysis of vehicle-bridge-wind-wave system." *J. Bridge Eng.* 23 (8): 04018054. https://doi.org/10.1061/(ASCE)BE.1943-5592.0001268.

Zhu, J., W. Zhang, and M. Wu. 2018b. "Evaluation of ride comfort and driving safety for moving vehicles on slender coastal bridges." *ASME J. Vib. Acoust.* 140 (5): 051012. https://doi.org/10.1115/1.4039569.

Zhu, L. D., Y. L. Xu, and H. F. Xiang. 2002. "Tsing Ma bridge deck under skew winds. Part II: Flutter derivatives." *J. Wind Eng. Ind. Aerodyn.* 90 (7): 807–837. https://doi.org/10.1016/S0167-6105(02)00159-9.

知己知彼知差距

——谈桥梁颤振研究的发展与挑战

陈政清

邓文中院士曾经讲过一件事。大意是：大家都以为中国的建桥速度很快，其实，美国金门大桥的建设速度并不比中国慢，而且当时的机械化程度没有现在高。他的话提醒我们要认真关注信息不对称问题。信息不对称有两个原因：一个是客观原因，就是信息传播的技术手段。我们现在的技术手段相当发达了，由客观原因引起的信息不对称已经非常少了。只要认真查阅，就能发现很多中文文献都有介绍：金门大桥于 1933 年开建，1937 年通车。还有一个是主观原因，就是知道但不说或不敢说，前者是掩耳盗铃，后者是皇帝的新衣。我个人认为，这确实是我们应该正视的问题。我们写论文作报告，先想一想：世界上其他国家是什么样的状态？我们是不是真的第一？总而言之，我们首先应该了解世界，才能够让世界了解我们，这就是本文的宗旨。

（一）颤振及颤振临界风速理论

桥梁的振动有多种类型，最明显的就是风振和地震，还有车桥振动和人致振动。每一类振动还可以再分为多种振动形式。本文只讨论风致颤振问题。

颤振是长的条带状结构在横向气流作用下发生的大幅振动并且具有以扭转振动为主的特征。最早报道的颤振是飞机机翼的失速颤振。如果飞机起飞的迎角太大，机翼会发生强烈振动甚至导致飞机因失速而坠毁。在桥梁上，最有名的一次颤振灾害是 1940 年美国旧塔科马桥的坠毁。

颤振是一种典型的气流与结构振动强烈耦合的效应，即气流导致结构振动，结构振动反过来又导致气动力增强，于是振动更加剧烈，最终导致振动发散结构毁坏。气动弹性力学就是专门研究这种强耦合气动效应的科学。桥梁颤振，必须要用气动弹性力学才能解释。依据气动弹性力学，桥梁颤振可以表示为

$$M\ddot{u} + C\dot{u} + Ku = f(u, \dot{u}, U, \omega) \qquad (1)$$

式（1）右端荷载项不仅与风速有关，也与自身的运动状态（即位移和速度）有关。所以称为气动自激力。颤振就是一种由气动自激力引起的自激振动。结构的运动状态是待求解的未知函数，式（1）是一种两边都含有未知函数的微分方程，它是很难精确求解的。

现有风洞实验技术已经可以精确再现颤振的全过程：在低风速下，振动幅度很小；高于某一特定风速后，振幅突然增大；风速再增加，振幅就会增加得更大，形成振动发散状态。这个特定风速就称为颤振临界风速。塔科马桥坠毁过程就是这样一种现象，临界风速在 18 m/s 左右。从抗风设计的角度来看，不一定需要分析颤振全过程，只要能确定颤振临界风速，并确保临界风速

高于桥址处的设计检验风速就可以了。这样一来，问题得到极大的简化，因为临界风速可以归结为方程（1）的特征值问题，求解相对容易。不同于静力稳定荷载，颤振临界风速是一个与动力稳定荷载相关的量。到目前为止，抗风设计一直是基于临界风速理论进行设计的。

（二）二维颤振临界风速理论的发展

旧塔科马桥事故，促使了桥梁风工程学科的创立和发展。若干力学大师经过努力，将航空领域的颤振理论加以改造，在20世纪50年代建立了二维两自由度颤振理论。这个理论把桥梁假定成无限长的等截面的一条带子，叫作“条带假定”。依据这一假定，只要研究桥梁截面的振动，就可以简化为二维平面问题。考虑到桥梁的宽度远大于高度，可进一步忽略横向振动，于是只要考虑竖向和扭转两个自由度，所以称为二维两自由度颤振理论，计算模式如图1所示。

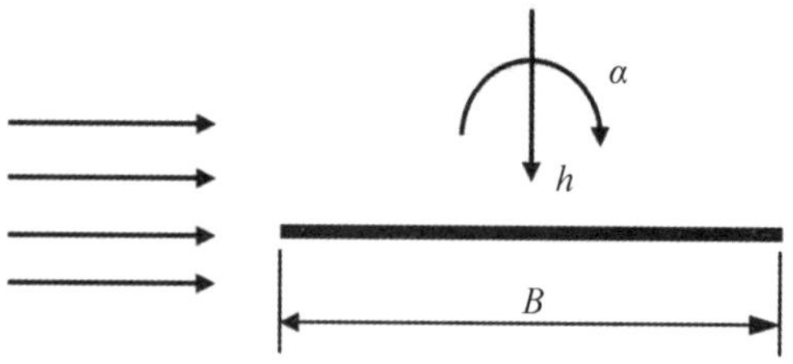

图1 二维两自由度的桥梁颤振计算模式

这样一来。方程（1）就可以简化为两个方程的组合。第一个方程是竖向振动，第二个方程是扭转振动。接下来的问题是如何确定竖向气动力和扭转气动力矩。1935年，希奥多尔森就把这个条带做了进一步的假定，假定这个截面的厚度无限薄，薄到没有厚度，只有宽度，称为“理想平板”，他采用复变函数的位势理论得到了理想平板自激气动力的理论解。理想平板模型有助于理解气动弹性力学特性。设想如果理想平板不动，气流沿平板宽度方向横向流过，由于理想平板的厚度为零，对流场没有任何干扰，也就不会对平板产生任何作用力，但是如果这个平板本身振动起来的话，气流就会对平板产生与运动相关的作用力了，所以它是一种气动弹性的力。

用理想平板的气动力来计算桥梁的颤振临界风速，准确度不高。因为桥梁的断面不是一个理想的平板，而是一个钝体，更具体地说，对于理想平板而言气流沿物体的轮廓流动，基本不产生分离和尾流，而气流流经钝体断面时会出现明显的流动分离，钝体后部会产生宽阔的尾流。真正的桥梁断面的气动力是没有理论解的。于是斯坎农教授等多位学者引进了颤振导数的概念，把这个气动力假定为位移和速度的一个线性组合。那么，这个位移和速度前面有一个组合系数，这个系数就叫作颤振导数，它随着无量钢风速的变化而变化。斯坎农同时提出了一种通过节段模型风洞试验确定桥梁颤振导数的方法，称为自由振动法。至此，一个能基本满足设计需要的二维颤振理论就基本建立了。

这个理论至今仍有很重要的指导性意义，它预示了两种基本的颤振方式：一种是钝体截面的纯扭转颤振，如旧塔科马桥颤振；第二种是当桥面很接近理想平板时出现的竖向和扭转耦合的颤振。现在我们仍然采用的节段模型颤振试验，就是以这个理论为依据的。

20世纪60年代，英国学者在运用二维颤振理论方面作出了杰出贡献。通过颤振研究，英国塞文桥在世界上第一次采用了扁平钢箱梁，而且计算出了施工全过程中的颤振临界风速变化曲线（图2）。这个曲线表明，当已安装梁段长度达到跨度的20%～30%的时候，这个状态是颤振临界风速最低，也是最危险的施工状态。以后随着拼接长度的增加，它的颤振临界风速也会增加。这个结论至今都对指导施工有重要意义。

我国现代意义上的桥梁风致振动研究，是从

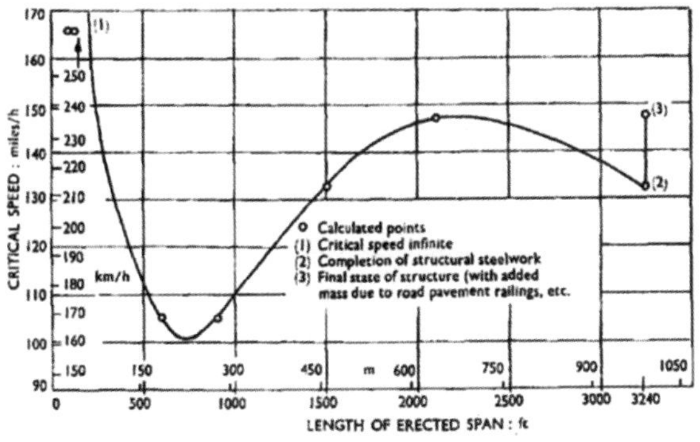

图 2　20 世纪 60 年代英国学者计算出塞文桥施工全过程的颤振临界风速变化曲线

（原载于 **1966** 年 **Aerodynamics Division, National Physical Laboratory**）

20 世纪 80 年代开始的。同济大学项海帆院士团队最早在国内实现了二维颤振理论的试验和分析技术，开发了识别颤振导数的自由振动法。笔者团队于 1996 年在国内开发了识别颤振导数的强迫振动法。与自由振动法比较，强迫振动法可以更为准确地在更宽的风速范围内测定 8 个颤振导数。西南交通大学和长安大学也是我国早期开展桥梁风致振动研究的单位。

（三）三维颤振理论的发展

二维颤振理论的局限性是很明显的，它至多考虑竖向和扭转两个纯模态的耦合，而很多桥梁（如斜拉桥等）存在模态不纯和多个模态参与颤振的现象，这就需要三维颤振理论。20 世纪 80 年代，计算机与有限元已经很发达了，在这样的情况下，利用三维有限元，把气动力荷载加进去，就可以进行三维颤振的数值分析。美国的萨卡和斯坎隆提出在三维颤振理论中，应该进一步考虑截面横向自由度运动的气动力，于是形成三维有限元模型和三个气动自由度结合的三维颤振理论。在这个理论框架内，需要识别 18 个颤振导数。

三维颤振分析有两类方法，一类叫模态坐标法，另一类叫物理坐标法。直接在三维有限元的模型上进行颤振分析，就是物理坐标法。这个方法最先由宫田敏男和山田仁在 1990 年提出。由于方程包含上千个甚至上万个自由度，以当时的计算机水平，难以推广。如果用模态缩减的办法，只考虑对颤振有贡献的低阶模态，也就是前面十几阶模态，就可以把这个方程压缩到十几阶，这就是模态坐标法。所以，在当时的条件下，最先发展的是模态坐标法，因为它的计算量相对少很多。早在 20 世纪 80 年代，谢霁明和项海帆、艾格、纳米尼等多人提出了各自的三维颤振的模态分析法。

笔者 1991 到 1992 年在英国学习期间，注意到前人模态坐标法的一个共同的缺点，就是它需要将风速和频率作为两个参数（因为风速和频率共同确定了描述气动力所需的颤振导数），在两个参数的平面区域内去搜索颤振解。搜索过程需要专业的判断，难以在设计人员中应用。在两个参数的一个平面上找到一个精确解，就好比要在一张大地图上找出某个地点，这是很困难的。

笔者在 1992 年提出了一种只需要对无量纲风速这一个参数搜索颤振解的模态分析法，称为 M-S 法。由于单参数法可以实现自动搜索，计算量和操作难度大大降低。1993 年，用这个方法计算出虎门悬索桥的施工全过程颤振临界风速，结果与同济大学气弹模型试验一致（见图 3）。与图 2 相比，曲线最低点位置基本一致，但以后一直上升，不再下降。1994 年，朱乐东又应用这一方法得到了江阴悬索桥施工全过程的颤振临界风速曲线。

2000 年，同济大学葛耀君教授在加拿大学习的时候，认为物理坐标法可以考虑全部模态，应该称为全模态法，并且用它计算了国内外多个大桥的施工全过程颤振临界风速。需要指出的是，这个全模态并不是桥梁结构真实的全模态，有限元模型中的全部模态中，只有前一部分模态可以模拟桥梁真实的模态，后面的高阶模态总会逐渐失真。

华旭刚在2006年提出了一个标准化的全模态方法。这个方法不需要自编计算程序，可以在现有的有限元商业软件（如ANSYS）上通过定义气动力单元来实现颤振分析。这个方法发表以后，受到学者欢迎，在其所发表的期刊中连续几个月下载量排名第一，并且有很多国家的学者直接采用了这个方法。

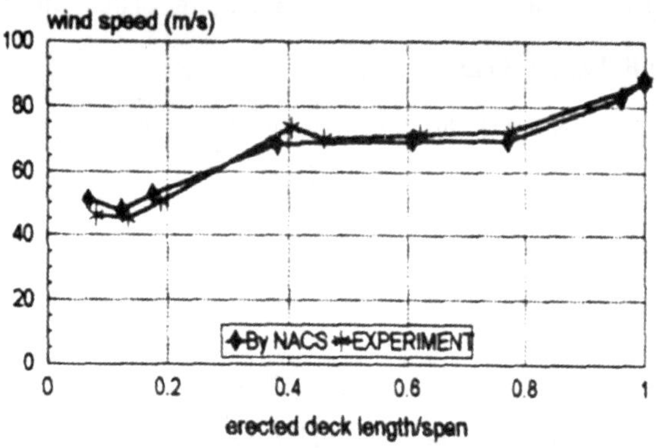

图3　虎门桥架梁全过程颤振临界风速变化曲线（陈政清1993）

三维颤振分析需要识别三个自由度的18个颤振导数，这一工作更加需要强迫振动法。笔者团队和同济大学团队合作，相继实现了18个颤振导数识别法，达到了国际先进水平。

最近，中铁大桥勘测设计院集团有限公司设计了一座跨度六百多米的人行桥，采用悬带桥结构，桥面不是平的，而是始终与主缆曲线保持平行。风洞试验证明其颤振形态是横向弯曲和扭转的耦合，颤振分析和试验都必须考虑横向自由度和横向颤振导数。这个实例说明，随着桥型的多样化，三维颤振分析的必要性会更加明显。

（四）提高颤振临界风速的主要途径

如果桥梁初步设计方案的颤振临界风速低于设计检验风速，就需要采取提高颤振临界风速的措施，包括结构强化措施和气动优化措施。结构强化措施通过提高影响颤振的结构扭转模态和其他模态的频率来提高颤振临界风速。气动优化措施通过改变主梁断面的外形来优化气动性能。借鉴飞行器设计的经验，在原有断面上附加各种导流或扰流装置，如导流板、扰流板、上下稳定板等，或者改变栏杆、检修轨道的位置、形状和尺寸，都具有气动优化效果。但目前优化过程没有定规，主要依靠经验和反复进行风洞试验，此外在优化颤振性能时，还要注意保持断面抗涡振的性能不至降低。

（五）抗风设计面临新的挑战

颤振研究的目的是为桥梁抗风设计服务。一个大跨度桥梁的抗风设计，不仅要满足颤振、涡振和抖振的多方面的要求，还要兼顾成本和施工技术等重要因素，因此需要评估—优化—再评估的多次反复。在提高抗风设计水平的道路上，我们还有很多工作要做。

1. **变单方案单目标抗风设计为多方案多目标优化设计**

由于各种原因，我国一些桥梁在选定一个方案后，尽管抗风评估出现了大的问题，也仍然对原方案情有独钟，不愿意作出大的改变，而是修修补补，甚至存在侥幸心理，认为实验室预测的风振在建成后不一定出现或可以补救。采用这种方式设计的桥梁在建成后，往往会留下遗憾。

日本明石海峡悬索桥的设计过程很值得我国桥梁设计者学习研究。明石海峡悬索桥主跨跨度1991 m，1998年建成直到今天一直保持着世界最大跨度记录。该桥在施工期遭遇阪神大地震但没有被破坏，建成后除出现过长吊索的局部振动外，没有其他风振问题。设计方花了非常长的时间来选择它的桥梁断面。选择的原则不仅要性能最优，还要造价低、施工方便。三个因素同时兼顾，这是一个很好的设计思想。

明石海峡悬索桥设计过程中一共研究了四种截面，如图 4 所示。第一种截面是分离双箱梁，它具有造价低、颤振临界风速容易满足要求的优点，但是即使增加稳定板，也会有涡振，于是坚决舍弃了这一方案。第二个方案是很扁平的桁架，加了一个流线型的风嘴后颤振临界风速还是达不到要求。第三个方案是单箱钢箱梁，该方案在靠近桥塔处用高的钢箱梁，在跨中部分用低的钢箱梁。两者结合也可以满足全部抗风要求，而且价格也不高，于是列为备选方案。第四个方案是高桁架方案，加设上、下稳定板后可以满足颤振要求，也没有涡振问题。第四个方案与第三个方案相比，造价相当，但是制造施工更为方便，成为最后的实施方案。第三个方案虽然没有实现，但是它表明，即使跨度达到两千米，单箱梁方案也是有可能满足抗风要求的。

后来，我国润扬大桥的钢箱梁也设置了上稳定板，这可能是我国第一座采用气动优化措施的悬索桥。湖南矮寨大桥也是桁梁悬索桥，加装了上下稳定板。目前在我国桥梁抗风设计中，气动优化措施已经得到普遍推广。

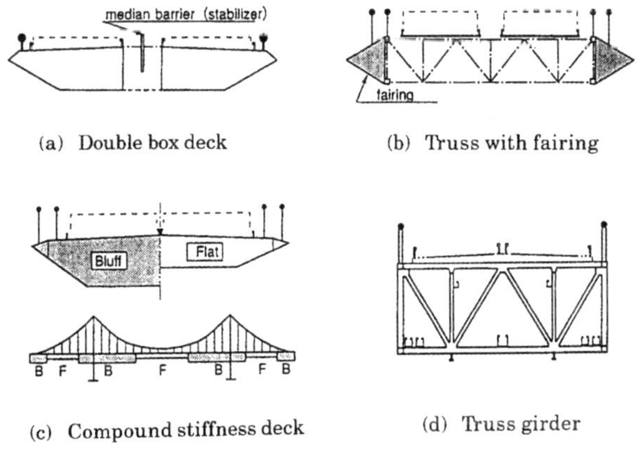

(a) Double box deck (b) Truss with fairing (c) Compound stiffness deck (d) Truss girder

图 4 日本明石海峡悬索桥抗风研究的四种加劲梁设计方案

2. 有必要进行超临界颤振研究

对于目前大量采用的钝体截面索支承桥，颤振风洞试验得到的风速-振幅曲线没有明显的临界风速点，具有软颤振特性。现有的基于特征值分析的颤振临界风速理论无法描述这种软颤振特性。另一方面，金门大桥 1950 年遇到了一次大风，实际上已经超过了它的颤振临界风速，进入了颤振后状态。但实测的风速-振幅曲线远低于风洞试验曲线。说明实际桥梁在风速超过临界风速后，振幅的增长速度并没有如节段模型风洞试验曲线增长得那么快。这种现象是可以理解的，因为颤振临界风速理论没有考虑到振幅增加后索支承桥的阻尼和刚度都可能增加，而气动自激力效应有可能下降等因素。如果要考虑上述因素，需要深入研究桥梁超临界状态的特性，建立超临界颤振理论。这是一个很难但是很有意义的课题。中国的桥梁研究工作者有得天独厚的条件和机会，应该在理论上有所突破，在世界桥梁发展历程中作出重要贡献，否则，我们就只是在某些技术层面上丰富和应用了前人的成果。

3. 积极慎重地开展颤振的主动和半主动控制研究

随着技术的进步，采用主动或半主动的振动控制手段抑制颤振也是有可能的。目前国内外也在这方面有所探索，应当积极鼓励。这种探索应该首先考虑所提出的方案在特大风速时和特大跨度桥梁上应用的可能性。例如，有一种模仿飞机在桥梁两侧加装水平可变翼板的方案，由于水平变翼板工作时必须按颤振频率不断改变姿态，结构刚度和技术要求比飞机襟翼还高。如果在千米以上的桥梁上安装水平变翼板，成本会非常高，而且可靠性很难保证。

（本文根据作者在 2019WTC 桥梁发展论坛上的演讲报告整理）

Journal of Fluids and Structures 88 (2019) 275–291

Contents lists available at ScienceDirect

Journal of Fluids and Structures

journal homepage: www.elsevier.com/locate/jfs

Experimental investigation of flutter characteristics of shallow *Π* section at post-critical regime

Y. Tang [a,b], X.G. Hua [a,*], Z.Q. Chen [a], Y. Zhou [c]

[a] *Key Laboratory of Wind and Bridge Engineering of Hunan Province, College of Civil Engineering, Hunan University, Changsha, Hunan 410082, PR China*
[b] *School of Civil Engineering and Architecture, Southwest Petroleum University, Chengdu, Sichuan 610500, PR China*
[c] *Division of Bridge Design and Research, Hunan Provincial Communications Planning, Survey & Design Institute, Co LTD, Changsha, Hunan 410008, PR China*

A R T I C L E I N F O

Article history:
Received 27 June 2018
Received in revised form 15 April 2019
Accepted 17 May 2019
Available online 31 May 2019

Keywords:
Aeroelastic flutter
Limit cycle oscillation
Post-critical behaviour
Wind tunnel tests
Shallow *Π* section
Bridges

A B S T R A C T

The aeroelastic response of a bridge deck at post-critical regime of flutter is featured by limit cycle oscillation (LCO) due to nonlinearities in both fluid and structural-dynamics. In this study, wind tunnel tests are conducted on an elastically-mounted rigid section model with shallow *Π* cross section, which is free to vibrate in vertical and torsional directions. The aim is to explore the response characteristics of shallow *Π* cross section at different initial angles of attack (AOA) for wind speeds above the flutter threshold. Firstly, the structural nonlinearities of the elastically-mounted system for section model are carefully characterized in still air to evaluate their effects on aeroelastic responses of the model. Then, post-critical responses of the model subjected to flutter instability are measured at different initial AOA. Experimental phenomena in terms of the aerostatic displacement, critical velocity, steady state amplitude, time history of oscillations, frequency characteristics, phase-angle shift between heaving and torsional motions, and damping evolution are discussed in detail as well as the influence of structural damping on post-critical responses. The results show that the oscillations of *Π* cross section at post-critical regime of flutter grow slowly with wind speed, instead of building up to large amplitude abruptly. Both the critical flutter velocity and the steady state amplitude of flutter oscillations are sensitive to initial AOA and structural damping. One degree of freedom (DOF) motion approximation is suitable for small amplitude flutter oscillations. For large amplitude oscillation description of coupled 2-DOF motion would be necessary for the bluff *Π* cross section. The torsional and vertical heaving oscillations tend to be synchronous at high velocity with large amplitude. Compared to slight nonlinear structural damping in the dynamic system, the nonlinear aerodynamic damping related to higher-order terms of displacement is more significant, implying that the aerodynamic nonlinearity is the main source leading to LCO.

1. Introduction

Aeroelastic flutter, a typical fluid–structure interaction, is an extremely important topic during design of long-span bridges (Simiu and Scanlan, 1996; Miyata, 2003). It occurs when a certain critical flow velocity is reached, resulting in

* Corresponding author.
E-mail address: cexghua@hnu.edu.cn (X.G. Hua).

https://doi.org/10.1016/j.jfluidstructs.2019.05.010
0889-9746/

very large amplitude of vibration or even collapse of bridges. Due to its catastrophic consequence, long-span bridges are commonly designed to warrant a safety margin with respect to the onset of flutter (Larsen and Larose, 2015).

At present the design of long span bridges with respect to flutter is governed by the linearized approaches. The most widely used analytical method for bridge flutter is established by Scanlan and co-workers (Scanlan and Tomoko, 1971; Scanlan, 1993, 1997a,b). In their work, the self-excited forces are expressed as linear functions of structural motions with the coefficients given by flutter derivatives, and flutter problem is thus treated as a linearized stability problem where the vanishing of the complex determinant of flutter characteristic equations at a critical speed indicates the onset of flutter. The linear approach assumes that the structural response goes unbounded beyond a critical velocity, and there is no need to deal with the actual aeroelastic responses at post-critical state. Extensive studies have been conducted focusing on the flutter onset in frequency domain (e.g. Náprstek and Pospíšil, 2012; Náprstek et al., 2015) or in time domain (e.g. Caracoglia and Jones, 2013).

Following the linearized approach, long-span bridges are designed with the critical flutter wind velocity well above a reference wind velocity specified in various design codes. The usefulness of the critical flutter velocity as a design metric for long-span bridges is well proved by design and construction of a number of landmark long-span suspension bridges such as the Akashi-Kaikyo Bridge, the Greatbelt Bridge and the Xihoumen Bridge, and structural failure due to the destructive flutter never occurs since the old Tacoma Bridge. It should be noted that in many situations the critical flutter velocity of a bridge must be improved by additional countermeasures to satisfy the stringent requirement of flutter check wind velocity, such as increasing the stiffness of structure (Kazama et al., 1995), optimizing the shape of the bridge deck (Diana et al., 1995), and setting extra plate near bridge deck as stabilizer (Ueda et al., 1998), among others. In the Akashi-Kaikyo Bridge, a 14 m-high deep stiffening truss, along with central stabilizer, was adopted to satisfy a flutter check velocity of 78 m/s; In the Xihoumen Bridge, a twin bridge girder spaced at open air gap of 6 m and connected structurally by heavy cross girders was utilized to raise the critical flutter velocity (Ge and Xiang, 2008).

Despite of the fact that flutter instability is strictly prohibited in current wind-resistant design, there is still of academic value to investigate the post-critical behaviours of bridge decks subjected to flutter instability. Moreover, even when a bridge is not intentionally designed to operate at wind velocity beyond linear flutter boundary, parameter uncertainty and/or large disturbances due to turbulence may push the system dynamics into post-critical state. The presence of nonlinear mechanisms may affect the system dynamics predicted by linear flutter. The deck of old Tacoma Narrows bridge had experienced about 70 min of oscillatory torsional motion with amplitudes of approximately $\pm 30°-35°$ before its final collapse (Larsen, 2000). Violent torsional oscillation with maximum double amplitude more than 12°, which is deemed to be flutter, also had been observed on the stiffening girder of the Golden Gate Bridge on 1 December 1951 (Vincent, 1958). In these published limited literatures, the actual aeroelastic response of bridges did not strictly behave as the linearized stability method predicted. The reason for these discrepancies would come down to neglect of nonlinearity. In fact, a few investigations have been focused on torsional or coupled vertical–torsional flutter-based wind harvester which implies that flutter can be well controlled in some situations, rather than being a simple dynamic divergence (Pigolotti et al., 2017a; Caracoglia, 2017, 2018).

The linear flutter is an idealized theoretical concept, and the aeroelastic response of a bridge deck at post-critical regime of flutter is featured by limit cycle oscillation (LCO) due to nonlinearities in both fluid and structural-dynamics. Long-span bridges, such as suspension bridges and cable stayed bridges are complicated systems with various nonlinearity mechanisms. The geometric nonlinearities originating from the suspension cables and the high flexural–torsional slenderness of stiffening girder play an important role in the dynamic response of bridge, in the meantime, the nonlinear effective stiffness of cables in its prestressed equilibrium configuration under dead loads can also remarkably influence the flutter behaviour (Arena and Lacarbonara, 2012). A new explanation on the sudden appearance of torsional oscillations during the old Tacoma Narrows Bridge collapse suggested by Arioli and Gazzola (2015) is turned out to be internal resonance which is a nonlinear structural instability depending on the bridge structure only. In addition, the equivalent modal damping ratio of bridge is considered as a constant parameter during design which shows considerable scatter in the actual measurement. A number of field investigations on bridges in service have indicated a tendency that the damping ratio increases with the increasing amplitude of vibration. In terms of aerodynamics, self-excited forces of the oscillating deck also contain nonlinear components due to flow separation which referred to the bluff configurations (e.g. Chen et al., 2005), large angles of attack (Chen and Kareem, 2001, 2003; Diana et al., 2010) and large amplitudes (Noda et al., 2003; Wang et al., 2013).

A few experimental studies recently have been carried out to investigate the nonlinear effects on the dynamic behaviours at the post-critical state. Matsumoto et al. (1992) revealed a non-divergent velocity restricted characteristic of torsional flutter when testing bluff bodies such as rectangular and H-shaped cylinders. Kubo et al. (2001) tested a Π section for a long span bridge on the critical velocity and vibration amplitude of post-critical flutter. Náprstek et al. (2007, 2008) tested a bridge type section in post-critical state and some characteristics of the nonlinear response were drawn. Zhu and Gao (2015) carried out a series of wind tunnel tests to investigate the aeroelastic responses of four typical cross-sections at post-critical flutter, and they observed that post-critical responses is a coupled vertical–torsional motion with a single frequency for all the tested four typical section models. Zhang et al. (2017) experimentally studied nonlinear flutter characteristics of a streamlined box section model, and concluded that the steady-state torsional and vertical amplitude of flutter almost increase linearly with the increase of reduced wind speed. These investigations provide beneficial experience for research on nonlinear flutter of bridge. Amandolese et al. (2013) experimentally evaluated low speed flutter and limit

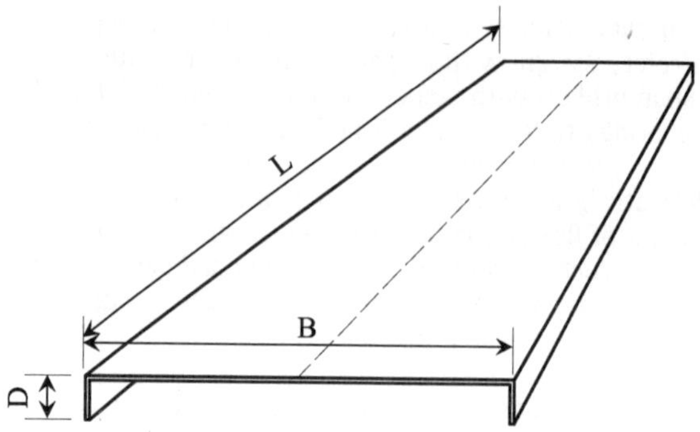

Fig. 1. Sketch overview of section model with Π cross section.

cycle oscillations of a two-degree-of-freedom flat plate in a wind tunnel. While most existing experimental studies of flutter are performed in smooth flow, turbulence may play an important role on both flutter threshold and post-critical responses. The turbulence effects on flutter manifests itself by changes in flutter derivatives (e.g. Huston, 1986; Larose et al., 1993; Scanlan, 1997a,b), loss of correction of spanwise self-excited forces (Scanlan, 1997a,b; Haan, 2000), and/or change in effective angle of attack (Chen and Kareem, 2003; Diana et al., 2010). However, a recent study by Pigolotti et al. (2017c) indicated the longitudinal turbulence intensity up to 15% have little influence on post-critical responses of a two-DOF airfoil undergoing the coupled flutter, where the mass flow separation at large amplitude of vibration overshadowed the turbulence of incoming flow.

Some advanced analytical models accounting for nonlinear aerodynamic forces were introduced to analyse the aeroelastic responses of bridges at post-critical regime for correlating measurement and prediction. These nonlinear models can be roughly classified into three types where the system responses at post-critical states are simplified as single 1-DOF torsional motion (e.g. Wang and Liao, 2012; Zhu and Gao, 2015; Zhang et al., 2017), coupled 2-DOF motion (Chen and Kareem, 2003; Náprstek et al., 2007; Diana et al., 2010; Wu and Kareem, 2011, 2014; Tang, 2015) and coupled 3-DOF oscillation (Xu and Cao, 2001; Prud'Homme et al., 2015). These nonlinear models of self-excited loads can either be quasi-steady or unsteady. The analytical models offer tools for exploring and understanding the underlying mechanisms of post-critical behaviours of aeroelastic systems subjected to flutter instability. Nevertheless, due to the complicated nature of actual aeroelastic systems, reliable analytical models are yet not available to describe the nonlinear oscillation of coupled flutter within a range of vibration amplitude.

In the present study, the elastically-mounted rigid section model tests are carried out in smooth flow to investigate the flutter and limit cycle oscillation of a shallow Π cross section for a range of initial angles of attack (AOA). The nonlinearities for the elastically-mounted system are carefully characterized to eliminate their effect on aeroelastic response of the model at post-critical state. Experiment results are described in terms of aerostatic displacement, critical velocity, steady state amplitude, frequency characteristics and time history of oscillation, phase shift between vertical heaving and torsional motions. The impact of structural damping on flutter responses is evaluated.

2. Wind tunnel tests

2.1. Experimental setup

The wind tunnel tests are carried out in the HD-2 boundary layer wind tunnel of Hunan University, China. This Prantl-type (closed-circuit) wind tunnel has a work section of 17.0 m (length) × 3.0 m (width) × 2.5 m (height). The background turbulence intensity is less than 0.5% in empty tunnel for wind velocity above 3 m/s. The model has a shallow Π cross section shown in Fig. 1. The length of the model is $L = 1.54$ m and the width-to-depth ratio of the cross section is $B/D = 10$, considered as a typical ratio for long-span bridge girder with this kind of cross section.

The model is made of aluminium alloy and manufactured by rolling a complete piece of raw material in plane. The thickness of model is 4 mm for flange and web which provides sufficient stiffness to prevent local distortion under flutter. The model is elastically supported by eight helical springs (with an individual nominal stiffness of 400 N/m and a maximum elongation of 800 mm), and is allowed to oscillate in the vertical heaving and torsional motions, as shown in Fig. 2. The lateral motion is restrained by two horizontal guy cables connected to the model ends. Two sizeable oval end plates made of light wood, with a maximum height 10 times larger than the depth of the model and a maximum width 3 times larger than the model width, is mounted at model ends to prevent the air outside the end plates from flowing into the central part of the section model, which ensures the two-dimensional flow conditions for the section model undergoing large-amplitude oscillation. Main parameters of the suspension model are listed in Table 1. The moment of inertia is evaluated with respect to the centre of gravity (C.G.) of the model, as well as the aerodynamic forces shown later.

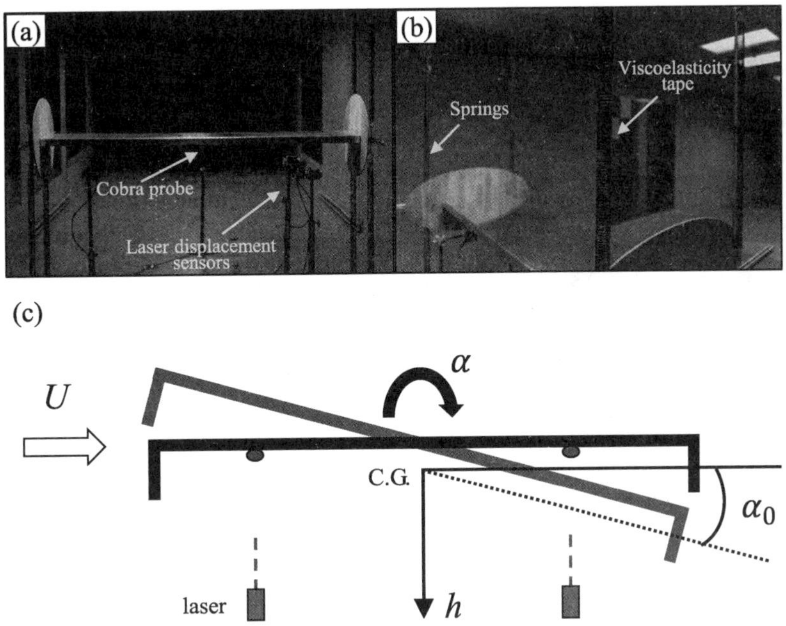

Fig. 2. Section model elastically mounted in wind tunnel.

Table 1
Parameter for shallow Π section model.

Parameter	Value
Length [L (m)]	1.54
Width [B (m)]	0.40
Height [H (m)]	0.04
Mass [m (kg/m)]	7.61
Moment of inertia [I (kg m^2/m)]	0.20

Three single-point laser displacement sensors (optoNCDT Model 2300), with an adjustable measuring range of 100–300 mm, are used to measure the upstream and downstream vertical displacements of the model. The horizontal distance between upstream and downstream laser points is 240 mm. The pure vertical and torsional displacements of the model can be calculated by a simple geometrical relationship based on signals of laser sensors. For laser triangulation system, apparent distance change between sensor and measuring object should be considered for large tilt angles. According to the instruction manual of optoNCDT 2300 (Micro-Epsilon Messtechnik GmbH, 2018), the apparent distance change is approximately 1.5 mm (0.5% of the measuring range 300 mm) for title angles between 15° and 30°, which gives rise to an maximum measurement error of 0.7° for torsional displacement. The measurement error is within the acceptable level. The coordinate of the test system is illustrated in Fig. 2c, where the vertical displacement h is positive downward and the torsional displacement α is positive clockwise. The sampling frequency of the displacement measurement is 200 Hz. The initial angle of attack α_0 has the same symbol definition with torsional displacement.

Force measurements are first performed to determine the aerodynamic drag (F_D), lift (F_L) and torsional moment (F_M) of the model per unit length from the section model rigidly attached on the five-component force balance, illustrated in Fig. 3a. The steady aerodynamic force coefficients C_D, C_L and C_M, defined in a wind-axis reference system, are obtained by normalizing the mean aerodynamic forces according to Eq. (1) below. Fig. 3b shows the variation of the aerodynamic force coefficients as functions of AOA, where the lift force is positive downward and the torsional moment is positive clockwise.

$$C_D = \frac{F_D}{0.5\rho U^2 DL} \quad C_L = \frac{F_L}{0.5\rho U^2 BL} \quad C_M = \frac{F_M}{0.5\rho U^2 B^2 L} \tag{1}$$

Flutter tests are carried out in smooth flow. The mean flow speed is measured by a three-axial Cobra probe located upstream of the model. During the test, the Reynolds number is in the range 80,000–320,000 ($Re = UB/\upsilon$). The maximum blockage ratio is about 5% when the torsional amplitude is about 20°.

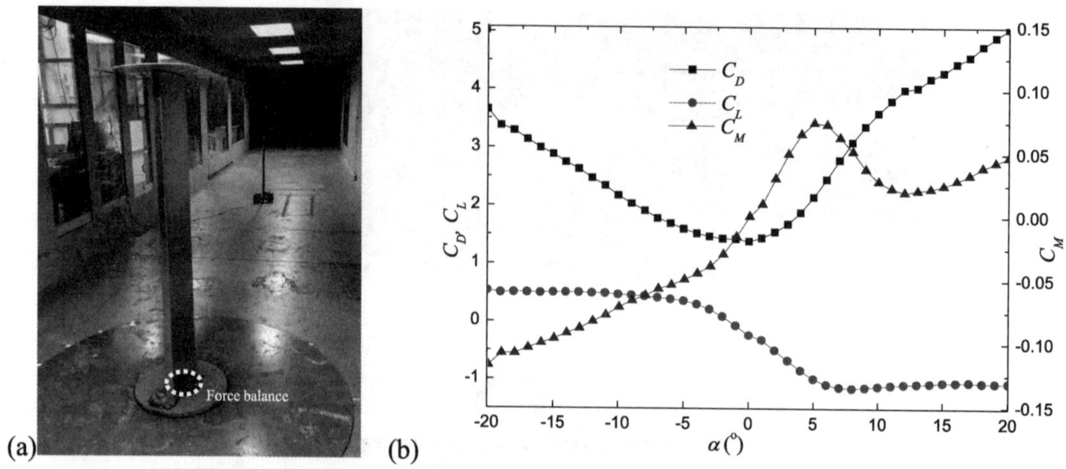

Fig. 3. Force measurement and the aerodynamic force coefficients.

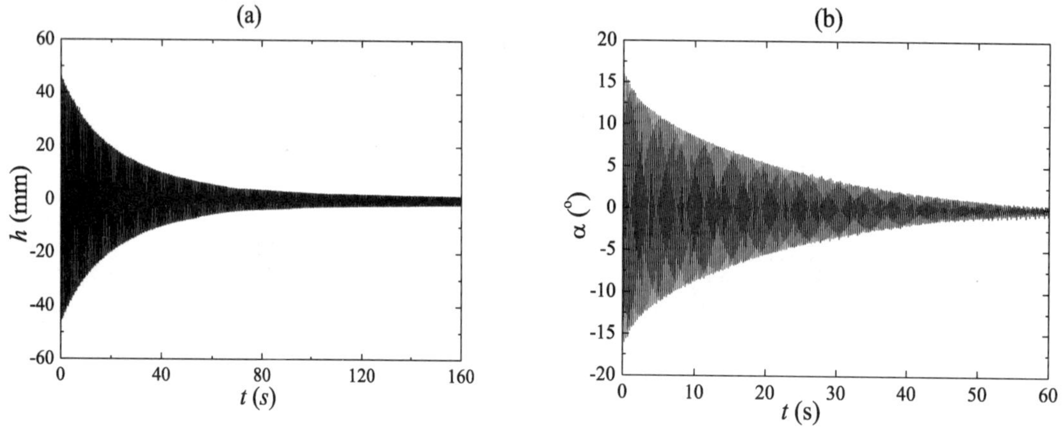

Fig. 4. Typical vertical heaving and torsional signals from the SDOF free-decay test.

2.2. Nonlinear stiffness of the elastically mounted model

The aerodynamic response of a wind-structure coupled dynamic system can be highly sensitive to the nonlinearities that stem from structural properties such as stiffness and damping, in addition to those from the aerodynamic origin (Dowell, 2014). Therefore the nonlinear characteristics should be carefully described and evaluated before flutter test. It should be noticed that for the Π cross-section there is difference between centre of gravity and centre of torsion, which should be accounted for during analytical and wind tunnel studies. In this study, the dynamic properties of the model as well as aerodynamic forces are evaluated with respect to the centre of gravity.

The stiffness of the spring-mounted model, both in vertical heaving and torsional degrees of freedom, is determined indirectly through dynamic tests. Large-amplitude free-decay tests of single degree of freedom are carried out in still air to measure the natural frequency and damping ratio of the suspension system with section model. Fig. 4 plots typical free-decay response samples in vertical heaving and torsional directions. The recorded vertical displacement is decayed from initial amplitude of 47.4 mm and the torsional displacement from 15.9°. In order to evaluate the nonlinearities in stiffness, instantaneous frequencies of oscillations are identified by time–frequency analysis with short time Fourier transform (using toolbox function tfrstft.m in MATLAB). Hamming window is used to minish the sidelobes of frequency spectrum, meanwhile, the instantaneous frequency is further estimated by interpolation of Fourier coefficients of weighted samples in each window for purpose of improving the identification accuracy.

Fig. 5 shows the change of vertical heaving and torsional frequencies as a function of time. One can see that the overall variation of frequency magnitudes in both vertical heaving and torsional modes is very little at the range of vibration amplitude tested. Fig. 6 illustrates the evolution of instantaneous stiffness with respect to amplitude. It is observed that the system stiffness decreases slightly along with the increasing amplitude of oscillation for both directions. From tiny oscillation to large amplitude oscillation, the variations of vertical stiffness and torsional stiffness are about 0.17% and

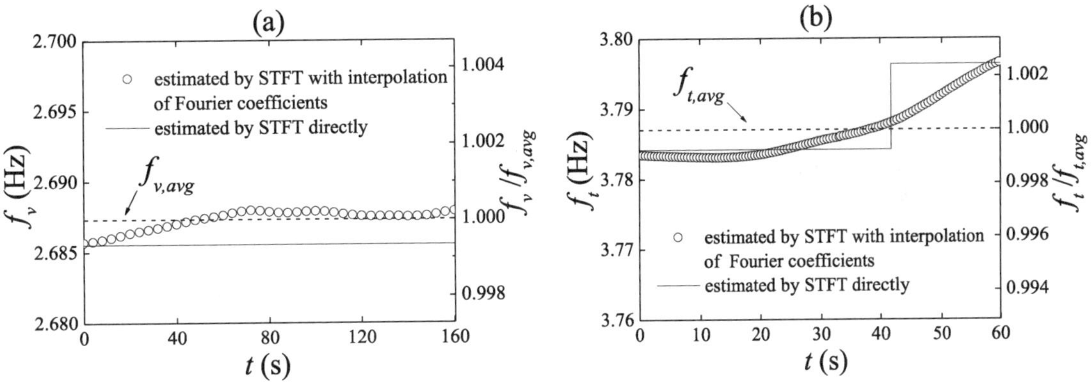

Fig. 5. Instantaneous frequencies extracted from free decay oscillation.

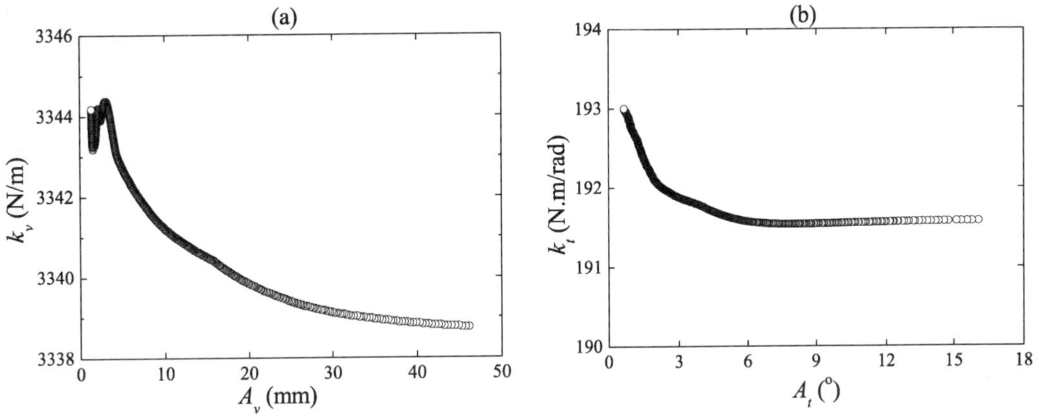

Fig. 6. Variation of instantaneous stiffness with vibration amplitude.

0.74% respectively. The decrement of vertical heaving and torsional stiffness with increasing vibration amplitude suggests a kind of softening nonlinearity, which may be mainly caused by the imperfect connection between the toggle and spring (see Fig. 2b). Several additional free-decay responses are analysed, and only little nonlinearity of the suspension system is observed. Therefore the stiffness nonlinearity is considered as negligible and the experimental set up for section model can be treated as linear.

2.3. Nonlinear damping of the elastically mounted model

In the case of viscously-damped linear dynamic system in still air, the damping ratio is assumed as constant and amplitude independent. However, the actual damping for the spring-mounted system is not completely viscous, and there inevitably also exist other damping mechanisms arisen due to imperfect connections between the toggle and spring (see Fig. 2b) as well as connection between model and springs. Nonlinear variation of structural damping ratio with respect to vibration amplitude has been reported in several free-decay tests on spring-mounted section model (Gao and Zhu, 2015; Pigolotti et al., 2017a). Therefore, local least squares fitting (LLSF) method is conducted here to identify the variation of damping ratio with vibration amplitude. Fig. 7 illustrates the flow diagram of LLSF for torsional oscillation time history. For a given moment t_n, a local time window of free-decay displacement signal is cut out firstly, defined by centre t_n and size $2\Delta t$ which corresponds to 20 periods of vibration for vertical and torsion modes. Then, negative signal in the local window is reversed into positive values with the purpose of making better use of the negative peak data to reduce fitting error. The damping ratio ζ_n with respect to instantaneous amplitude A_n at the specified moment t_n is estimated by applying conventional LSF in the local window. After that, the calculation is moved to next window centred at t_{n+1}. In the progress of free-decay testing, pure single DOF oscillation is difficult to obtain especially for small damping system. The sampled single DOF signal will be disturbed unavoidably which results in dispersion of the identified damping ratio; therefore a polynomial fit is requisite. For relevant more exquisite signal extraction of nonlinear damping, one can refer to the literature by Gao and Zhu (2015).

Fig. 8 shows the variation of damping ratios in vertical and torsional directions with vibration amplitudes. The vertical damping ratio (Fig. 8a) increases with the amplitude throughout the tested range. This curve can be roughly divided into

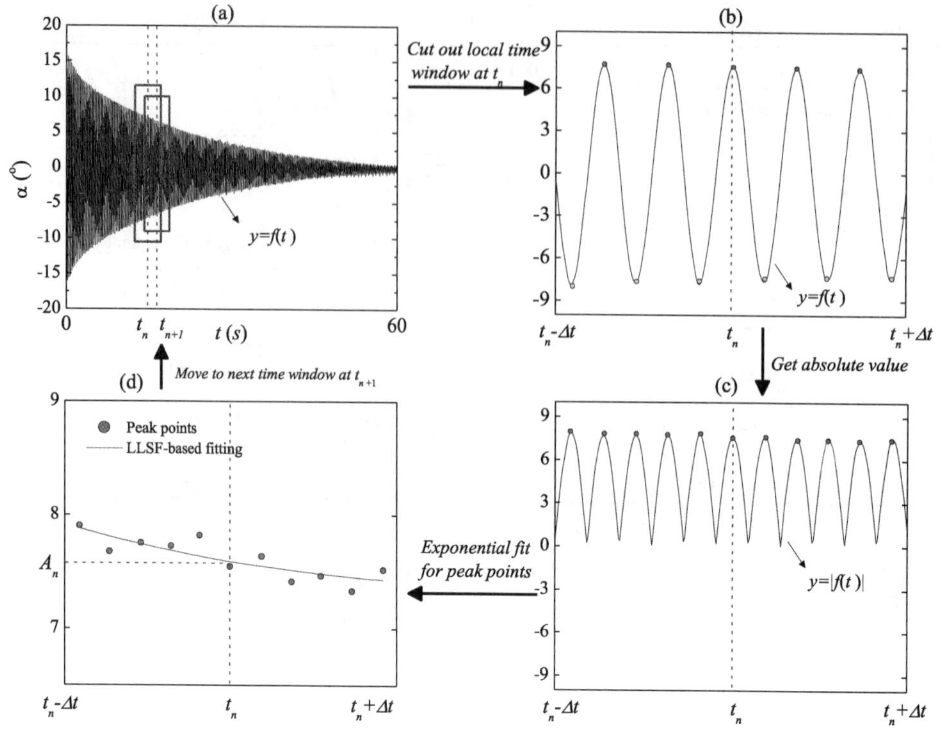

Fig. 7. Flow diagram for identification of nonlinear damping ratio.

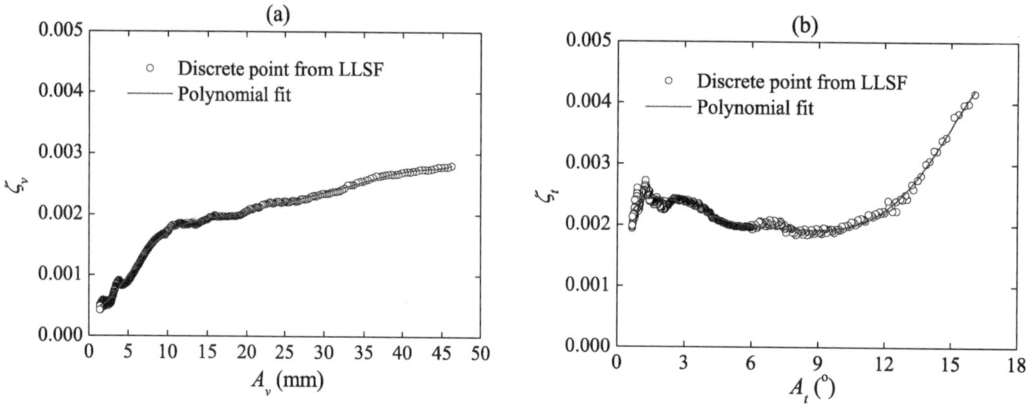

Fig. 8. Identified nonlinear damping ratio with respect to vibration amplitude.

two linear parts with different slops, accompanied by craggy slop for vertical amplitude smaller than 10 mm and gradual slop for larger amplitude. However, non-monotonic increase of the torsional damping ratio with amplitude is found in Fig. 8a. The torsional damping ratio increases a bit to a peak nearby the torsional amplitude of 1.5° at the early stage and then decreases slightly until 9°. After that the torsional damping ratio returns to increase with even more larger slop. These nonlinear behaviours of damping ratios have reflected the sophisticated mechanism of energy dissipation on the suspension system. The predominant contribution seems to be ascribable to frictions of the suspension system, although it may be partly rooted in the still-air aerodynamic damping. In this article, with neglect of still-air aerodynamic damping but without loss of universality, the identified nonlinear damping ratio of free-decay oscillation in Fig. 8 is used to stand for the structural damping ratio.

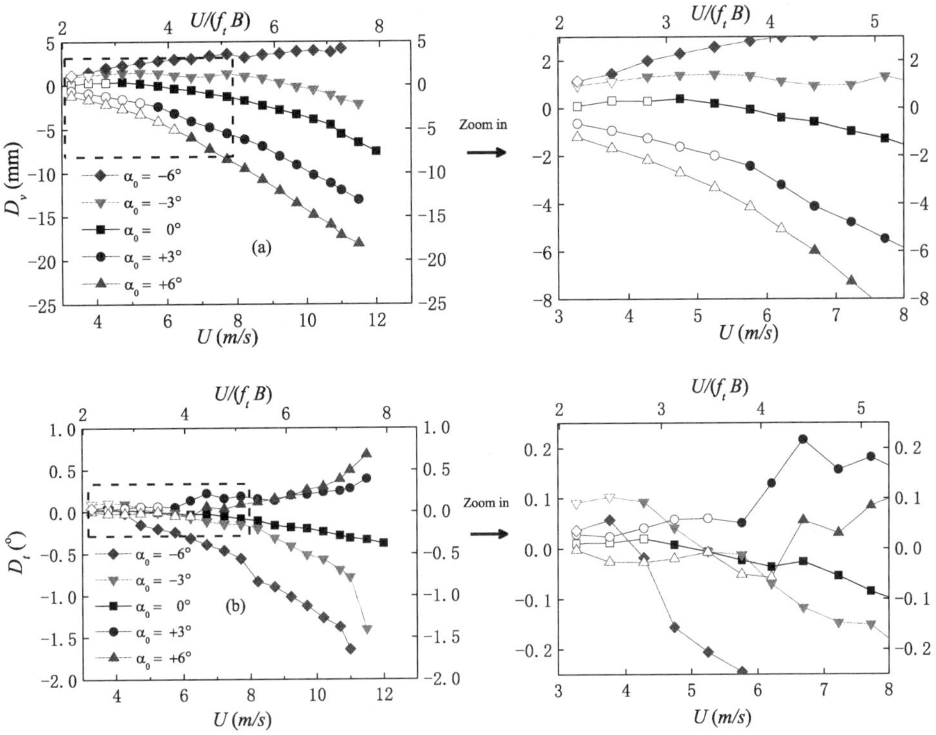

Fig. 9. Aerostatic displacement at different initial AOA.

3. Aeroelastic response characteristics

In this section, the flutter tests of the Π cross section are performed for five initial AOA, namely $\alpha_0 = 0°, \pm 3°, \pm 6°$ at the low structural damping, illustrated in Fig. 7. In addition, the structural damping is increased to two higher levels by intertwining viscoelasticity tape (see Fig. 2b) for the purpose of comparative study.

3.1. Aerostatic displacement

Displacement responses of flutter in smooth flow consist of the self-excited and aerostatic components. It has been found that the aerostatic component at high wind velocities close to the flutter onset can considerably affects the flutter response of bridge in the form of added steady AOA due to the mean wind loads (Chen and Kareem, 2003). Before the occurrence of flutter, the static deformation induced by mean wind loads is directly obtained from the displacement histories. At flutter critical condition and post-critical regime, the sinusoidal curve fitting is employed to identify the equilibrium position of oscillation, the steady-state amplitude, the steady-state frequency, the phase shift between bending and torsion motion.

The aerostatic displacements of vertical and torsional oscillations at different initial AOA are shown in Fig. 9. Solid markers refer to circumstance of flutter and empty markers refer to non-flutter cases. The reduced wind velocity drawn in figures as well as the actual wind velocity is defined by $U_r = U/f_t B$ for both vertical heaving and torsional components, where f_t is fixed at the torsional frequency of 3.784 Hz measured in still air.

For all five initial AOA cases it can be found from Fig. 9a that the aerostatic vertical heaving displacement of the Π cross section generally varies regularly with wind velocity. In addition, for initial AOA $\alpha_0 = 0°$, the monotonicity of vertical displacement–speed curve changes once before flutter, and for initial AOA $\alpha_0 = -3°$, it changes twice during post-critical flutter state. These behaviours would be due to the aerodynamic nonlinearities dominated by section configuration since the stiffness nonlinearities have been found to be insignificant already.

The aerostatic torsional displacement illustrated in Fig. 9b shows non-monotonic increase with respect to the wind speed, which is different form the linear result found on a box deck section by Zhang et al. (2017). The monotonicity of torsional displacement–speed curve has changed no less than once during the reduced velocity range of 2~5 for all test cases. As for $\alpha_0 = -6°$, the magnitude of torsional aerostatic displacement increases rapidly, and reaches its maximum at wind speed of 11.0 m/s, while the corresponding increase of vertical aerostatic displacement is more gently than that of other initial AOAs. This indicates that the added aerodynamic moment of π section is not totally offered by added aerodynamic lift, and it is quite different from thin airfoil which has an almost unchanged aerodynamic centre. These

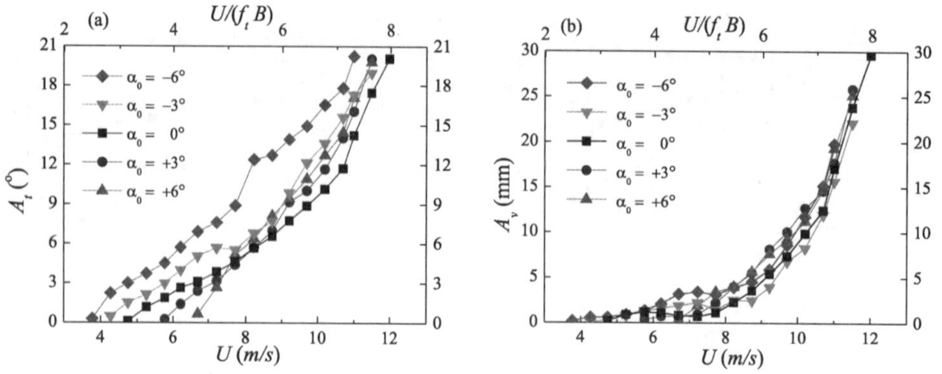

Fig. 10. Steady-state amplitude of Π section for low structural damping.

nonlinearities of added steady AOA have drawn attention in the aerostatic stability analysis (Boonyapinyo et al., 1994; Zhang et al., 2013).

It is noted that the torsional aerostatic displacement is no more than 0.1° prior to the onset of flutter at all initial AOA, thus the effect of added steady AOA on critical velocity is minor and limited. However, the maximum of torsional aerostatic displacement in the test have exceeded 1.5° for $\alpha_0 = -6°$, which would have non-negligible impact on post-critical flutter response.

3.2. *Critical velocity and steady-state amplitude of oscillation*

For coupled wind-structure system, linear flutter analysis can be carried out to predict the variation of complex eigenmodes with respect to wind velocity by using experimental flutter derivatives, and flutter occurs when the damping of one of eigenmodes vanishes at a certain critical velocity (e.g. Hua et al., 2007). For section model wind tunnel tests, damping of heaving and torsional modes can be extracted experimentally from the wind-excited oscillations. Due to small magnitude of damping of torsional mode at the neighbourhood of the onset of flutter, accurate identification of damping is difficult. For nonlinear flutter of a bluff body such as the shallow Π cross section, there may be no clear boundary compared to that of the classic linear flutter typically found on streamlined cross-sections. In this study, for the sake of convenience, flutter is assumed to occur when the torsional amplitude of the section model has reached 0.1°.

Fig. 10 shows the variation of aeroelastic self-excited responses of the section model with wind velocity. The starting points of curves correspond to the threshold of flutter. The critical velocity is obviously influenced by initial AOA (Fig. 10a). For $\alpha_0 = 0°$, the flutter critical velocity at which LCOs initialize is about 4.7 m/s. For negative initial AOA, the flutter critical velocity reduces. On the contrary, the positive initial AOA leads to higher critical velocity. The steady-state torsional amplitude increases stably with wind velocity and is also affected by initial AOA remarkably. When the wind velocity is no more than 7.5 m/s, the amplitude of LCOs could be reduced by positive initial AOA and enhanced by negative initial AOA. Once the test speed exceeds 8.5 m/s, the LCO amplitude for $\alpha_0 = 0°$ becomes smaller than that of any other initial AOAs. In this sense, the tested Π section at $\alpha_0 = 0°$ shows the better performance under this special condition, though the viewpoint would be different if evaluating on basis of critical flutter. And more remarkably, for the case of $\alpha_0 = +6°$ where the Π cross section is deemed to be less blunt with a higher critical velocity, the increase slop of torsional amplitude is the sharpest. The sensitivity to initial AOA indicates that the primary nonlinearity is rooted in the fluid rather than in the structure.

The steady state vertical-heaving amplitude is show in Fig. 10b for different initial AOA. Vertical amplitudes are relatively small (less than 4 mm or 0.01B) for $U < 8.0$ m/s, while the corresponding torsional amplitude has exceeded 5°. Once the wind velocity exceeds 8.5 m/s, and the rate of increase of response seems quite large. The coupled heaving-torsional responses of the model at higher velocity above flutter threshold indicate the necessary of a 2-DOF model for prediction of aeroelastic responses at post-critical regime. It is noteworthy that the shape of response-velocity curves at post-critical state depends on both the aerodynamic nonlinearity of the cross sections and the structural dynamic properties (Pigolotti et al., 2017b). Nevertheless, it is very difficult to numerically predict the variation of post-critical responses with wind velocity, as reliable nonlinear models of self-excited loads are yet not available for large amplitude of vibration.

3.3. *Displacement time history and vibration characteristic*

Since the aerostatic and oscillatory displacement are remarkably influenced by initial AOA, detailed comparison of time history of the measured response need to be drawn up. Fig. 11 illustrates a comparison of the time history responses for $\alpha_0 = 0°$ at four selected wind velocities, which corresponds to critical, small amplitude, medium amplitude and large

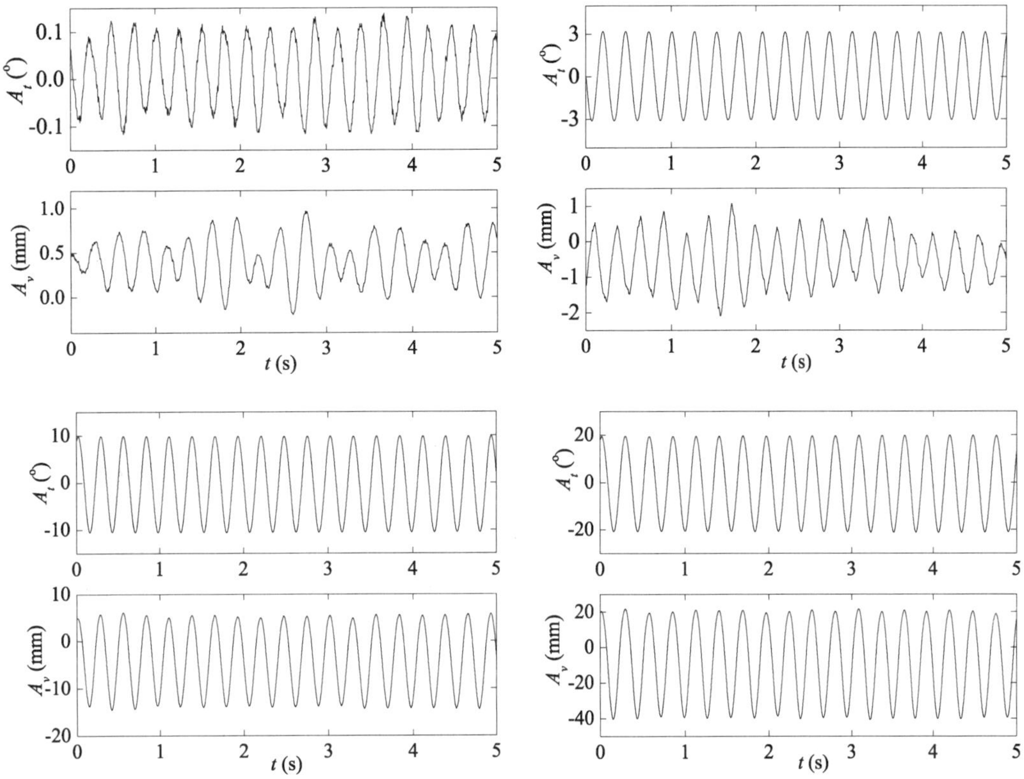

Fig. 11. Time history response of flutter for $\alpha_0 = 0°$ (i) onset ($U = 4.7$ m/s), (ii) small amplitude case ($U = 5.7$ m/s), (iii) medium amplitude case ($U = 10.2$ m/s), (iv) large amplitude case ($U = 12.0$ m/s).

amplitude of vibration. It can be found from the curve shape that the torsional displacement is always nearly simple harmonic during the flutter oscillation, enhancing from critical state to large amplitude. But for the vertical displacement, there exists strong signal interference at the very start of flutter and small amplitude oscillation. This interference gradually disappears with the increasing amplitude of oscillation, and becomes less prominent in the medium or large amplitude oscillations. The vertical displacement at critical flutter state is mainly caused by the signature turbulence, i.e. the turbulence caused by oscillating model.

Fast Fourier transform (FFT) is employed to analyse the time history of the flutter oscillation, and the frequency resolution is about 0.01 Hz. Fig. 12 shows the frequency characteristics for initial AOA $\alpha_0 = 0°$ at different wind velocities with torsional amplitudes being about 0.04°, 0.1°, 1.9°, 3.1°, 10.2°, and 20.1°, respectively. At the critical velocity of 4.7 m/s, the torsional oscillation is dominated by a single frequency close to the structural torsional frequency as already shown in Fig. 4b. The vertical response is dominated by the torsional mode.

Variation of the dominant frequencies of flutter oscillation is shown in Fig. 13 for different initial AOA, and the corresponding damping will be discussed later. In general, both frequencies for vertical displacement and torsional displacement are almost equal and reduce with the increasing wind velocity. The quantity of reduction is in the order of 10^{-1} Hz. Note that the reduction of frequency during the large amplitude free decay oscillation in still air (Fig. 4) is only in the order of 10^{-2} Hz. It can be concluded that the frequency reduction of flutter oscillation is mainly caused by the aerodynamic stiffness rather than the nonlinearities of structural property. The local variations of vertical frequency found at critical velocity for $\alpha_0 = 0°$ and $\alpha_0 = -3°$ are consistent with the frequency bifurcation phenomenon (Fig. 12), which come from the coupling behaviour of the involved degrees of freedom at the critical state of flutter.

During the oscillating of coupled flutter, in general, there is a shift of phase angle between vertical and torsional motions. The phase-angle shift plays crucial role in process of energy input where the flutter oscillation is driven by aerodynamic mechanism of coupling motion. Pure torsional flutter rarely happens to streamlined cross sections in single DOF if the initial AOA effect is excluded. For streamlined cross section with 2-DOF as illustrated by Dyrbye and Hansen (1997), the total energy input during a complete oscillating cycle is shown to be zero when the phase-angle shift of torsional to vertical oscillation is 0°, and turns into positive at the phase-angle shift of $-135°$.

The phase angles of each DOF oscillation are identified in sync by trigonometric function fitting. The resulting phase-angle shifts of torsional to vertical oscillation are presented in Fig. 14. For small initial AOA cases ($\alpha_0 = -3°, 0°$, and $+3°$), the phase-angle shift lies in range of 130°~150° at the critical state of flutter, then rises with the increase of wind speed,

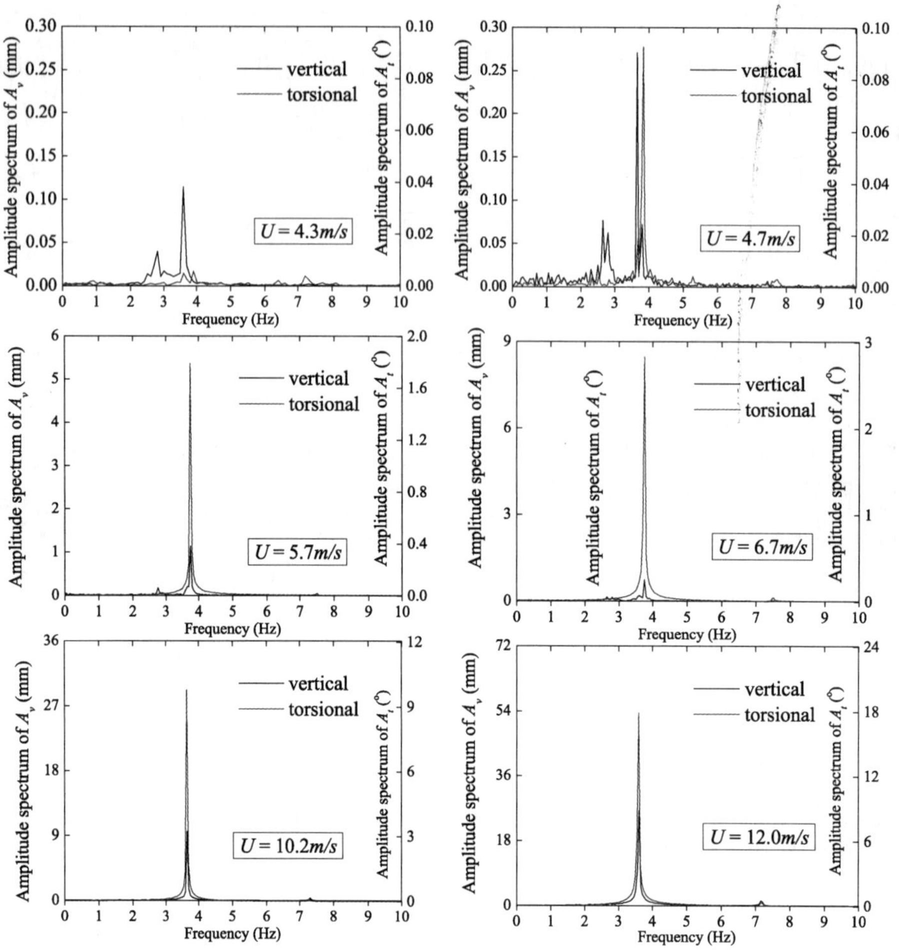

Fig. 12. Frequency characteristic of flutter oscillation for $\alpha_0 = 0°$.

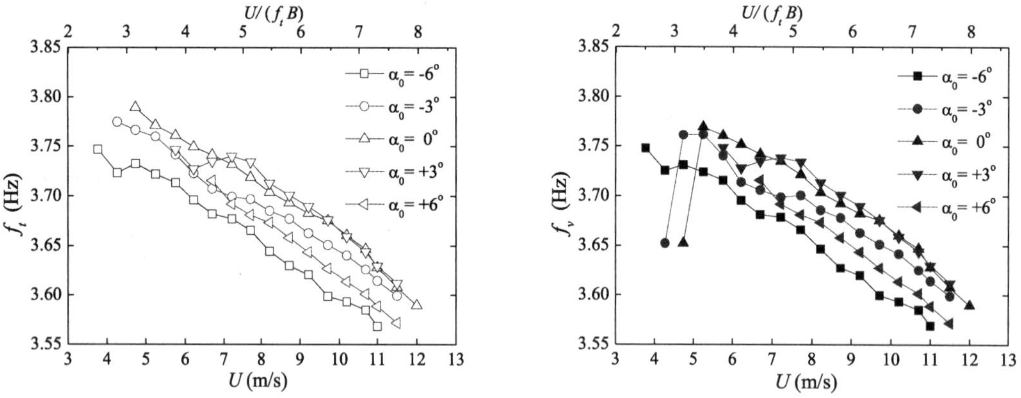

Fig. 13. Evolution of the dominant frequencies of flutter oscillation versus wind velocity.

finally approaches zero. For $\alpha_0 = -6°$, the phase shift starts from zero almost, then falls down to 90°, gradually recovers to zero in the end. For $\alpha_0 = +6°$, the phase shift waves nearby 0° all the time aside from the flutter critical state. Similar observations on limit cycles with phase shift close to 0° were also reported by Amandolese et al. (2013) and Pigolotti et al. (2017b).

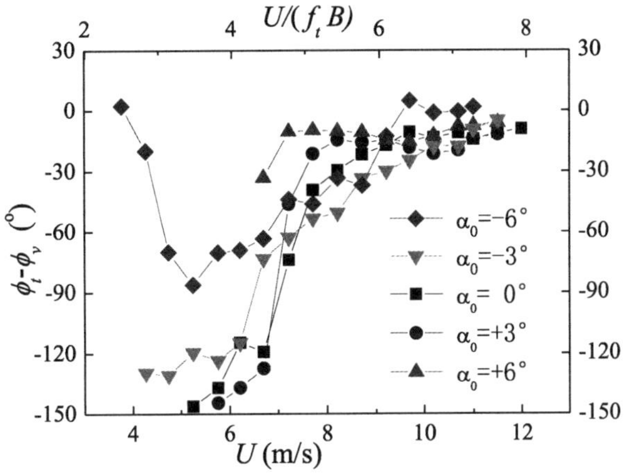

Fig. 14. Phase-angle shift of torsional to vertical oscillation.

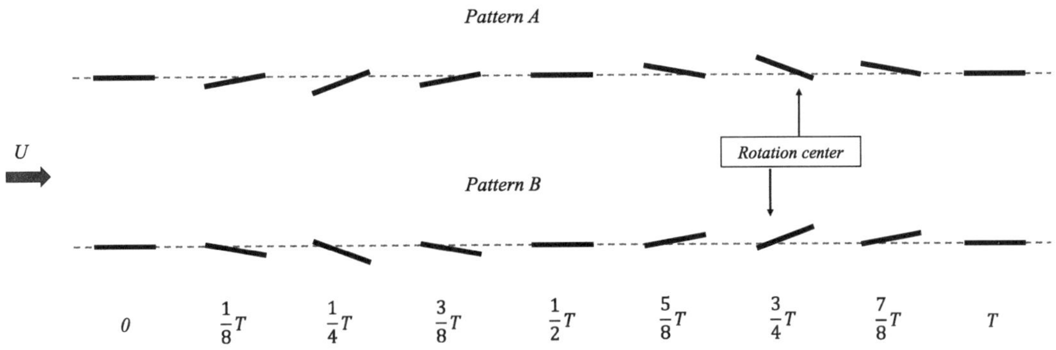

Fig. 15. Schematic illustration of the oscillating patterns.

3.4. *Motion in phase plane and pattern*

Two kinds of typical oscillating patterns are observed during the test. Fig. 15 presents the schematic diagram of motions in one whole cycle where the oscillation amplitude is overstated for illustrative purpose. Oscillations of pattern A are only found for $\alpha_0 = 0°$ at low speed with small amplitude response, where the rotation centre is located downstream. On the contrary, oscillation of pattern B occurs at high speed with large amplitude response for all initial AOA in the test. The corresponding rotation centre moves upstream.

More direct evidences come from the measuring instruments. For $\alpha_0 = 0°$, Fig. 16 shows the original signals from single point laser displacement sensors symmetrically placed under the section model with distance of 240 mm (refer to Fig. 3). At the wind speed of 5.2 m/s (Fig. 16a), the amplitude of displacement signal from leeward laser is about two thirds of the signal from windward laser. As the wind speed increases to 12.0 m/s (Fig. 16c), the leeward displacement becomes nearly three times of the windward displacement at the measured point. The transformation from pattern A to pattern B is a step by step process with the increasing wind speed. Then a special status (Fig. 16b), bordering upon pure torsional oscillation where the signal amplitude from both leeward and windward laser equal to each other, is observed at $U = 7.2$ m/s.

The phase diagram given in Fig. 17 shows the trajectory of 2-DOF steady state motions from critical state to large amplitude oscillation. These limit cycles in phase space are oval in shape with positive time direction of clockwise. Along with the increasing wind speed, the cycle size grows up and its length direction rotates counter-clockwise which indicates the variation of phase angle, as already shown in Fig. 14.

With regard to bluff body such as the present Π section, the accepted flutter mechanism is driven by negative damping of torsional motion. Therefore the observed vertical motions are more likely to dissipate energy and stabilize the coupled oscillation. The tendency that phase shift approaches to 0° for large amplitude vibration at all initial AOAs indicates some general character for nonlinear flutter of this kind. Moreover, the oscillations patterns (see Fig. 15) are affirmatively caused by phase-angle shift and it can be proved by using simple geometrical relationship. When the phase-angle shift is closing to $-180°$, the oscillation of the pattern A occurs. And when the phase-angle shift comes near 0°, the oscillation of the pattern B is observed.

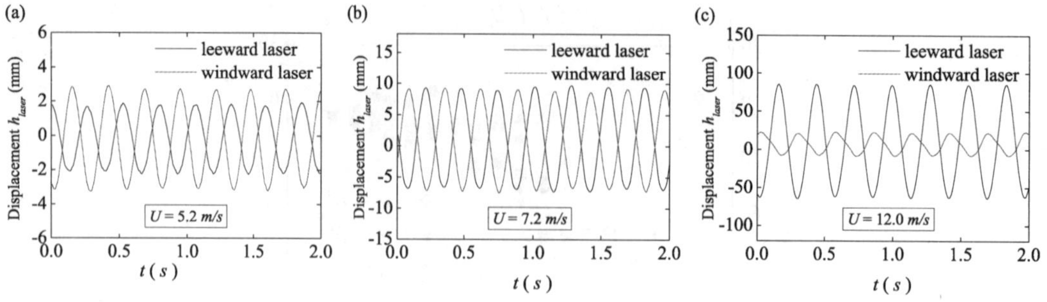

Fig. 16. Measured time history of single-point laser displacement sensor for $\alpha_0 = 0°$.

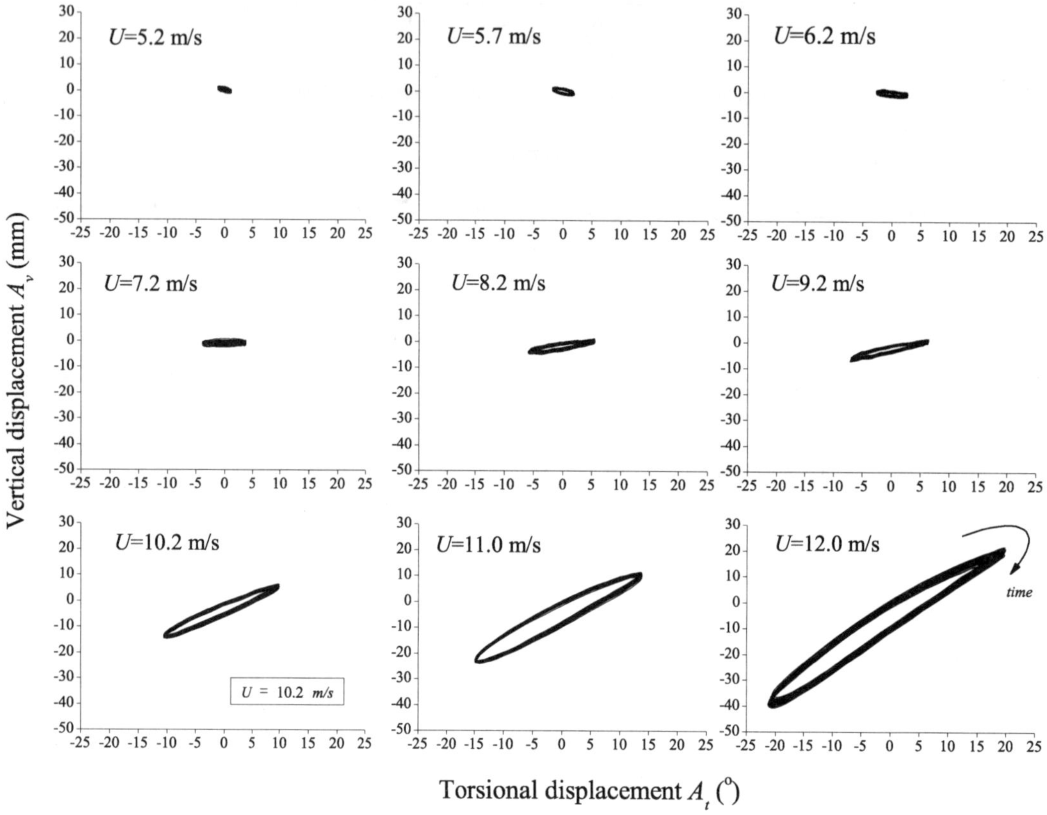

Fig. 17. Evolution of limit cycle versus wind speed for $\alpha_0 = 0°$.

3.5. Damping evolution of the oscillation from rest or large amplitude release

It has been reported in experiments that steady state amplitudes of coupled flutter are sensitive to initial condition (Amandolese et al., 2013; Pigolotti et al., 2017a). Tests with increasing and decreasing wind speed results in different oscillation amplitudes near critical velocity and generally larger amplitude is found in the latter case. However, this phenomenon is not observed in the present test. This might be due to the low level participation of vertical oscillation near critical velocity.

To further reveal the damping characteristics of the oscillating system in flutter, the test model is released from still/large amplitude at the wind speed region where flutter oscillations have been observed. Fig. 18a shows the entire process that the torsional oscillation released from still and enhances to violent vibration with amplitude of 20° for $U =$ 12 m/s. Fig. 18b shows the decaying process at the critical velocity that the torsional oscillation is released from large amplitude and reduces to tiny vibration.

At post-critical regime of flutter, the aeroelastic system is nonlinear and can be modelled by the Van der Pol oscillator, in which the aerodynamic damping consists of a linear part and a nonlinear part related to the 2nd or 4th degree of

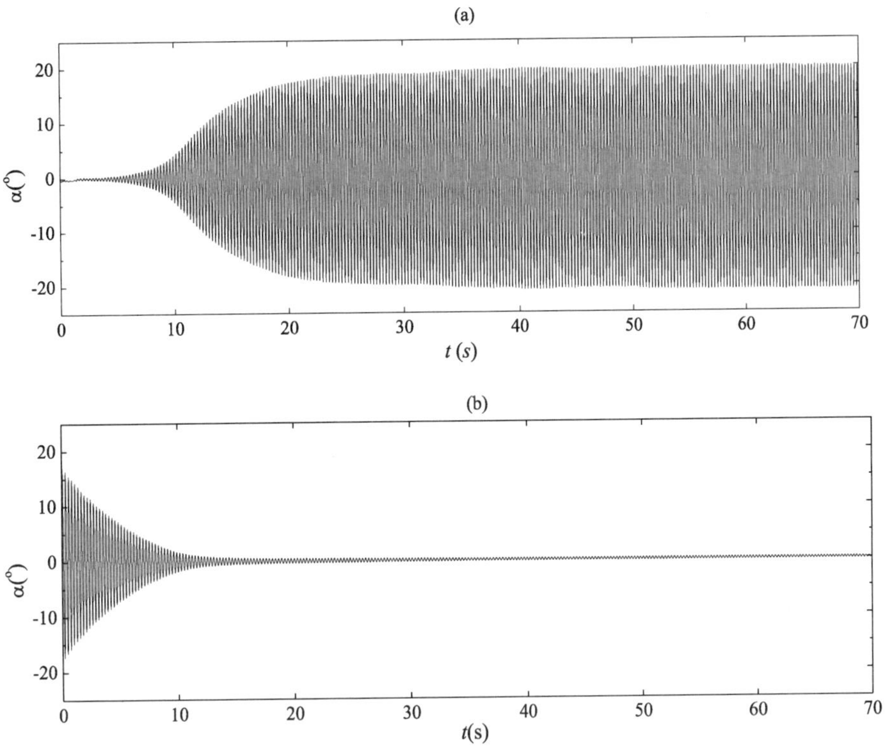

Fig. 18. Torsional response from different initial conditions for $\alpha_0 = 0°$ (a) release from still when $U = 12.0$ m/s (b) release from large amplitude when $U = 4.7$ m/s.

displacement. For small-amplitude vibration the linear damping term dominates and response increases exponentially; at large amplitude the nonlinear term take effects and tends to cancel the negative linear damping, therefore leading to a LCO.

The linear damping ratios of the torsional mode (not the structural damping ratio) at wind velocities after flutter onset are identified by LSF method at small amplitude oscillation (A_t varies from 0.2° to 0.5°) when the linear theory still applies. Results are shown in Fig. 19 by transparent bar. In general, the magnitude of the equivalent negative damping ratio for torsional oscillation increases with wind speed. In the meanwhile, the linear damping ratios for torsional mode before flutter onset are extracted from wind-excited random motions with the use of stochastic subspace identification (SSI) method and also drawn in Fig. 19 by solid bar. It should be noted that due to small magnitude of torsional damping at wind velocities close to flutter onset, accurate identification of damping is difficult. Therefore, it is expected that the critical flutter velocity determined from the zero torsional damping may differ from that determined from predefined threshold of vibration amplitude.

3.6. Effect of structural damping

To investigate the effect of structural damping ratio on responses of nonlinear flutter, the helical springs are additionally intertwined with viscoelasticity tape to add structural damping of the suspension model system. The original structural system of small damping are strengthened into medium damping system (MDS) and large damping system (LDS) as shown in Fig. 20. The damping ratio of MDS varies nearby 0.005, which is the recommended structural damping value for long span bridges with steel girder in the Chinese specification JTG/T D60-01-2004, Wind-resistant deign specification for bridges. The damping ratio of LDS varies nearby 0.01 that is the recommended structural damping value for long span bridges with steel–concrete composite girder (often with Π shape cross section) in Chinese specification JTG/T. The damping ratio curves respect to amplitude of both MDS and LDS show better monotony than the small damping system. This is caused by the damping mechanism differences between the original system and the added viscoelasticity tape.

Fig. 21 shows the comparison of steady-state LCO amplitudes at three different levels of structural damping. It can be seen that the damping ratio plays important role in responses of nonlinear flutter. The critical velocity shows clear rise along with the increase of damping ratio. When the original small damping system is strengthened into LDS, the critical velocity is raised from 4.7 m/s to 6.7 m/s. The structural damping also has remarkable suppression on the steady state

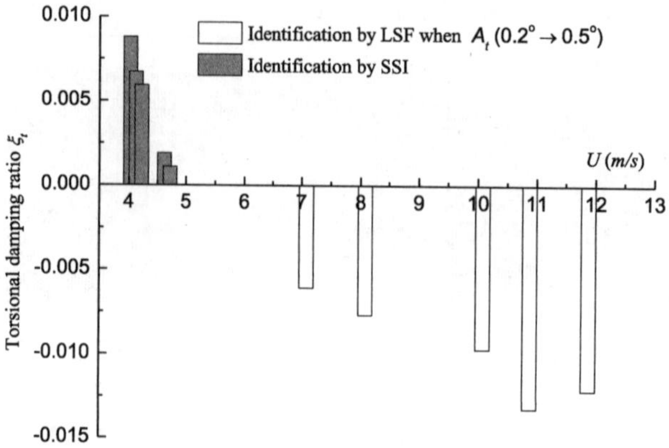

Fig. 19. Damping evolution of the torsional oscillation with wind speed for $\alpha_0 = 0°$.

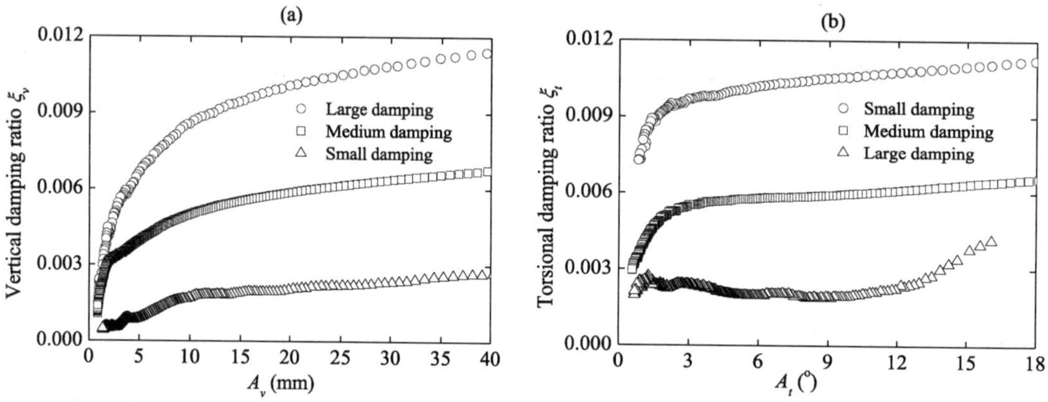

Fig. 20. Comparisons of nonlinear structural damping ratios versus vibration amplitude.

amplitude of flutter oscillation. At the wind speed of 12.0 m/s, for instance, the torsional amplitude of flutter oscillation is about 20° for small structural damping system. The torsional amplitude drops down to 5.6° for LDS. It can be found that the percentage of reduction on oscillation amplitude surpasses 70%.

The positive effects of structural damping on both flutter critical velocity and oscillation amplitude of Π section indicate that the flutter performance of bridge girder with similar shape can be controlled, at least to a great extent, by enhancing structural damping or adding mechanical damping. Consistent view can be found in other literature with theoretical analysis (Chen and Kareem, 2004). This provides alternative engineering measures either for wind-resistant design of newly-built bridge or for active bridges under threat from extreme climatic change where the critical design wind speed for flutter may be exceeded.

4. Conclusions

In the present study, post-critical behaviour of a shallow Π cross section subjected to flutter instability has been experimentally investigated by mean of elastically mounted section model under different initial angles of attack. The spring-suspended model has negligible nonlinearities in stiffness and slight nonlinearities in structural damping. Quantitative identifications of these nonlinear factors are necessary for the study of nonlinear flutter.

Flutter critical velocity of the cross section is obvious influenced by initial AOA. The flutter oscillations of Π section show gradual enhancement with wind velocity rather than abrupt instability during the tests. The participation of vertical degree within flutter oscillation is inconspicuous at low wind speed and noticeable at high wind speed. Therefore, flutter oscillation with small amplitude can be treated as 1-DOF motion approximately. But for large amplitude flutter, description on coupled 2-DOF motion would be necessary. Two kind of typical oscillating patterns are observed in flutter experiments. It is caused by the phase-angle shifts between vertical and torsional oscillations. For the tested cases of initial AOA, the phase-angle shifts of torsional to vertical oscillation are different from each other at flutter critical state. However, they all tend to be synchronous at high wind speed with large amplitude. The damping ratios of torsional motions of flutter are revealed which draw deep insight of the flutter oscillations. The nonlinear aerodynamic damping related to higher-order terms of displacement is the main source leading to LCO at the post-critical state.

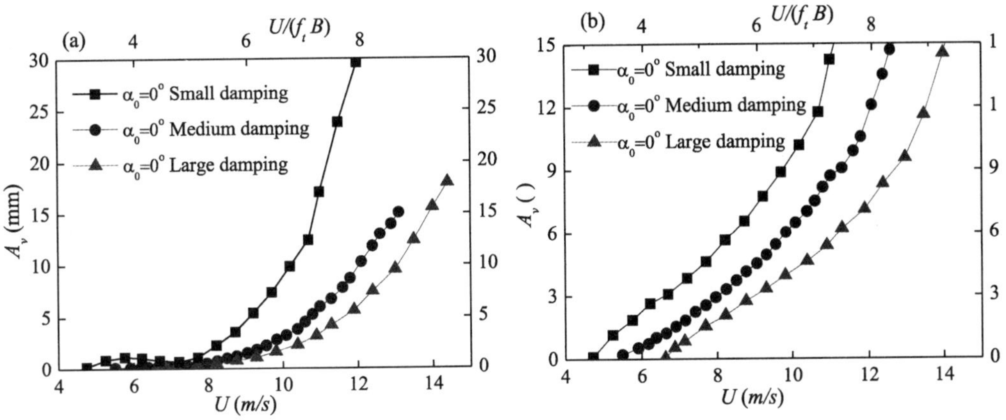

Fig. 21. Steady-state LCO amplitudes for different structural damping.

Acknowledgements

The authors gratefully acknowledge the support for the research work jointly provided by the National Natural Science Foundations of China (No. 51808470, 51422806).

References

Amandolese, X., Michelin, S., Choquel, M., 2013. Low speed flutter and limit cycle oscillations of a two-degree-of-freedom flat plate in a wind tunnel. J. Fluids Struct. 43, 244–255.
Arena, A., Lacarbonara, W., 2012. Nonlinear parametric modeling of suspension bridges under aeroelastic forces: torsional divergence and flutter. Nonlinear Dynam. 70 (4), 2487–2510.
Arioli, G., Gazzola, F., 2015. A new mathematical explanation of what triggered the catastrophic torsional mode of the Tacoma Narrows Bridge. Appl. Math. Model. 39 (2), 901–912.
Boonyapinyo, V., Yamada, H., Miyata, T., 1994. Wind-induced nonlinear lateral-torsional buckling of cable-stayed bridge. J. Struct. Eng. 120 (2), 486–506, ASCE.
Caracoglia, L., 2017. Numerical investigations on the operational regimes of a torsional flutter-based wind harvester. In: Proceedings of the X International Conference on Structural Dynamics, EURODYN 2017, Procedia Engineering, Vol. 199. Elsevier, Rome, Italy, pp. 3434–3439.
Caracoglia, L., 2018. Modeling the coupled electro-mechanical response of a torsional-flutter-based wind harvester with a focus on energy efficiency examination. J. Wind Eng. Ind. Aerodyn. 174, 437–450.
Caracoglia, L., Jones, N.P., 2013. Time domain vs. frequency domain characterization of aeroelastic forces for bridge deck sections. J. Wind Eng. Ind. Aerodyn. 91, 371–402.
Chen, X., Kareem, A., 2001. Nonlinear response analysis of long-span bridges under turbulent winds. J. Wind Eng. Ind. Aerodyn. 89 (14–15), 1335–1350.
Chen, X., Kareem, A., 2003. Aeroelastic analysis of bridges: effects of turbulence and aerodynamic nonlinearities. ASCE J. Eng. Mech. 129 (8), 885–895.
Chen, X., Kareem, A., 2004. Efficacy of tuned mass dampers for bridge flutter control. J. Struct. Eng. ASCE 129 (10), 1291–1300.
Chen, Z.Q., Yu, X.D., Yang, G.Q., Spencer, B.F., 2005. Wind-induced self-excited loads on bridges. J. Struct. Eng. ASCE 131 (12), 1783–1793.
Diana, G., Falco, M., Bruni, S., 1995. Comparisons between wind tunnel tests on a full aeroelastic model of the proposed bridge over Stretto di Messina and numerical results. J. Wind Eng. Ind. Aerodyn. s 54–55 (94), 101–113.
Diana, G., Rocchi, D., Argentini, T., Muggiasca, S., 2010. Aerodynamic instability of a bridge deck section model: Linear and nonlinear approach to force modeling. J. Wind Eng. Ind. Aerodyn. 98 (6), 363–374.
Dowell, E.H., 2014. A Modern Course in Aeroelasticity, fifth ed. Spring, New York.
Dyrbye, C., Hansen, S.O., 1997. Wind Loads on Structures. Wiley, New York.
Gao, G., Zhu, L., 2015. Nonlinearity of mechanical damping and stiffness of a spring suspended sectional model system for wind tunnel tests. J. Sound Vib. 355, 369–391.
Ge, Y., Xiang, H., 2008. Recent development of bridge aerodynamics in China. J. Wind Eng. Ind. Aerodyn. 96 (6–7), 736–768.
Haan, F.L., 2000. The Effects of Turbulence on Aerodynamics of Long-Span Bridges (Ph.D. thesis). Department of Aerospace and Mechanical Engineering, Notre Dame.
Hua, X.G., Chen, Z.Q., Ni, Y.Q., Ko, J.M., 2007. Flutter analysis of long-span bridges using ANSYS. Wind Struct. 10 (1), 61–82.
Huston, D.R., 1986. The Effects of Upstream Gusting on the Aeroelastic Behavior of Long Suspended-Span Bridges (Ph.D. thesis). Department of Civil and Environmental Engineering, Princeton University.
Kazama, K., Yamada, H., Miyata, T., 1995. Wind resistant design for long span suspension bridges. J. Wind Eng. Ind. Aerodyn. 54-55 (94), 65–74.
Kubo, Y., Sadashima, K., Yamaguchi, E., Kato, K., Okamoto, Y., Koga, T., 2001. Improvement of aeroelastic instability of shallow π section. J. Wind Eng. Ind. Aerodyn. 89 (14), 1445–1457.
Larose, G.L., Davenport, A.G., King, J.P.C., 1993. On the unsteady aerodynamic forces on a bridge deck in turbulent boundary lay flow. In: Proceedings of the 7th U.S. National Conference on Wind Engineering. pp. 373–382.
Larsen, A., 2000. Aerodynamics of the tacoma narrows bridge-60 years later. Struct. Eng. Int. 10 (4), 243–248.
Larsen, A., Larose, G.L., 2015. Dynamic wind effects on suspension and cable-stayed bridges. J. Sound Vib. 334, 2–28.
Matsumoto, M., Shirato, H., Hirai, S., 1992. Torsional flutter mechanism of 2-D H-shaped cylinders and effect of flow turbulence. J. Wind Eng. Ind. Aerodyn. 41 (1-3), 687–698.
Micro-Epsilon Messtechnik GmbH & Co. KG (2018). Instruction Manual optoNCDT 2300, https://www.micro-epsilon.com/download/manuals/man--optoNCDT-2300--en.pdf, accessed on 21 2018.

Miyata, T., 2003. Historical view of long-span bridge aerodynamic. J. Wind Eng. Ind. Aerodyn. 91, 1393–1410.
Náprstek, J., Pospíšil, S., 2012. Response types and general stability conditions of linear aero-elastic system with two degrees-of-freedom. J. Wind Eng. Ind. Aerodyn. 111, 1–13.
Náprstek, J., Pospıšil, S., Höffer, R., Sahlmen, J., 2008. Self-excited nonlinear response of a bridge-type cross section in post-critical state. In: Proceeding of the 6th International Colloquium on Bluff Bodies Aerodynamics and Applications (BBAA6), AnsaldoBreda, Milan, Italy, Vol. 2008. pp. 20–24.
Náprstek, J., Pospíšil, S., Hračov, S., 2007. Analytical and experimental modelling of non-linear aeroelastic effects on prismatic bodies. J. Wind Eng. Ind. Aerodyn. 95 (9–11), 1315–1328.
Náprstek, J., Pospíšil, S., Yau, J.D., 2015. Stability of two-degrees-of-freedom aero-elastic models with frequency and time variable parametric self-induced forces. J. Fluids Struct. 57, 91–107.
Noda, M., Utsunomiya, H., Nagao, F., Kanda, M., Shiraishi, N., 2003. Effects of oscillation amplitude on aerodynamic derivatives. J. Wind Eng. Ind. Aerodyn. 91 (1), 101–111.
Pigolotti, L., Mannini, C., Bartoli, G., 2017a. Critical and post-critical behaviour of two-degree-of-freedom flutter-based generators. J. Sound Vib. 404, 116–140.
Pigolotti, L., Mannini, C., Bartoli, G., 2017b. Experimental study on the flutter-induced motion of two-degree-of-freedom plates. J. Fluids Struct. 75, 77–98.
Pigolotti, L., Mannini, C., Bartoli, G., 2017c. Turbulence effects on the post-critical response of flutter-based generators. In: Proceedings of European and African Conference on Wind Engineering, Paper 167, 4-7 2017, Liege, Belgium.
Prud'Homme, S., Legeron, F., Laneville, A., 2015. Transient flutter analysis of bluff bodies. J. Wind Eng. Ind. Aerodyn. 145, 139–151.
Scanlan, R.H., 1993. Problematics in formulation of wind-force models for bridge decks. J. Eng. Mech. 119 (7), 1353–1375.
Scanlan, R.H., 1997a. Amplitude and turbulence effects on bridge flutter derivatives. J. Struct. Eng. ASCE 123 (2), 232–236.
Scanlan, R.H., 1997b. Some observations on the state of bluff-body aeroelasticity. J. Wind Eng. Ind. Aerodyn. 69, 77–90.
Scanlan, R.H., Tomoko, J.J., 1971. Airfoil and bridge deck flutter derivatives. J. Eng. Mech. Div. ASCE 97 (6), 1717–1737.
Simiu, E., Scanlan, R.H., 1996. Wind Effects on Structures: Fundamentals and Applications to Design, third ed. John Wiley, New York.
Tang, Y., 2015. Nonlinear Self-Excited Forces of Streamlined Box Deck and Nonlinear Flutter Response (Ph.D. thesis). Southwest Jiaotong University, Chengdu, China, (in Chinese).
Ueda, O., Tanaka, H., Matsushita, Y., 1998. Aerodynamic stabilization for super long-span suspension bridges. In: Proceeding of IABSE Symposium Long-Span and High-Rise Structures. Kobe, Japan, Vol. 1998. pp. 721–728.
Vincent, G., 1958. Golden gate bridge vibration studies. J. Struct. Div. 84 (6), 1–40.
Wang, Q., Liao, H., 2012. The nonlinear aerodynamic stability of long-span bridges: post flutter. In: Proceeding of the 7th International Colloquium on Bluff Body Aerodynamics and Applications (BBAA7), Shanghai, China, 2012.
Wang, Q., Liao, H., Li, M., Ma, C., 2013. Aerodynamic hysteresis of thin airfoil and streamline box girder under large amplitude oscillation. J. Exp. Fluid Mech. 27 (1), 32–37, (in Chinese).
Wu, T., Kareem, A., 2011. Modeling hysteretic nonlinear behavior of bridge aerodynamics via cellular automata nested neural network. J. Wind Eng. Ind. Aerodyn. 99 (4), 378–388.
Wu, T., Kareem, A., 2014. Simulation of nonlinear bridge aerodynamics: A sparse third-order volterra model. J. Sound Vib. 333 (1), 178–188.
Xu, X., Cao, Z., 2001. Linear and nonlinear aerodynamic theory of interaction between flexible structure and wind. Appl. Math. Mech. (Engl. Ed.) 22 (12), 57–65.
Zhang, Z.T., Ge, Y., Yang, Y.X., 2013. Torsional stiffness degradation and aerostatic divergence of suspension bridge decks. J. Fluids Struct. 40, 269–283.
Zhang, M., Xu, F., Ying, X., 2017. Experimental investigations on the nonlinear torsional flutter of a bridge deck. J. Bridge Eng. 22 (8), 04017048.
Zhu, L., Gao, G., 2015. Influential factors of soft flutter phenomenon for typical bridge deck sections. J. Tongji Univ. (Nat. Sci.) 43 (9), 1289–1294, (in Chinese).

第46卷 第3期 湖南大学学报(自然科学版) Vol.46,No.3
2019年3月 Journal of Hunan University(Natural Sciences) Mar. 2019

文章编号:1674-2974(2019)03-0001-09

大跨度流线型箱梁悬索桥颤振稳定性气动优化

刘志文[1†],谢普仁[1],陈政清[1],徐国平[2],徐军[2]
(1. 湖南大学 风工程与桥梁工程湖南省重点实验室,湖南 长沙 410082;
2. 中交公路规划设计院有限公司,北京 100010)

摘 要:以主跨为1 660 m流线型箱梁悬索桥为工程依托,采用风洞试验和CFD数值模拟相结合的方法对影响大跨度悬索桥颤振稳定性的主要因素(主缆空间形式、主梁气动外形和中央稳定板高度)进行了研究,并对气动控制措施机理进行了探讨.结果表明:主缆布置形式对桥梁结构颤振临界风速的影响主要表现为主缆布置形式导致桥梁结构扭转频率的改变,从而影响桥梁结构颤振临界风速;适当增加主梁断面宽高比可有效提高桥梁结构颤振临界风速;设置合适高度的中央稳定板可有效提高带水平分离板的流线型箱梁断面颤振临界风速.中央稳定板附近产生的涡会引起主梁断面竖向气动力增加,导致主梁断面竖向运动参与程度提高,抑制了主梁断面扭转运动,从而提高了流线型箱梁断面颤振稳定性.

关键词:大跨悬索桥;颤振稳定性;气动优化;风洞试验;数值模拟

中图分类号:U441.3 **文献标志码:**A

Aerodynamic Optimization of Flutter Stability for a Long-span Streamlined Box Girder Suspension Bridge

LIU Zhiwen[1†], XIE Puren[1], CHEN Zhengqing[1], XU Guoping[2], XU Jun[2]
(1. Key Laboratory for Wind and Bridge Engineering of Hunan Province, Hunan University, Changsha 410082, China;
2. China Highway Planning and Design Institute Consultants Inc, Beijing 100010, China)

Abstract: Taking a suspension bridge with a streamlined girder of 1 660 m span as engineering background and using the methods of wind tunnel test and numerical simulation of CFD, this paper studies three main factors affecting the flutter stability of long-span suspension bridges, namely the main cable space form, aerodynamic shape of the girder section and height of central stabilizer, and then the mechanism of aerodynamic countermeasures is discussed. The results show that the influence of main cable layout on the flutter critical wind speed of bridge structures is mainly caused by the change of the torsional frequency of bridge structure according to the layout of main cables, which affects the critical wind speed of bridge structures. Appropriately increasing the ratio of width to height of the main girder section can effectively improve the flutter critical wind speed of bridge structures. Setting an appropriate height of central stabilizer can effectively improve the flutter critical wind speed of streamlined girder section with horizontal separation plates. The major effect of vortices generated near the central stabilizer is to generate the vertical aerodynamic force of streamlined main beam, leading to the increase of vertical movement degree of the main

* 收稿日期:2018-01-26
基金项目:国家自然科学基金资助项目(51778225,51478180),National Natural Science Foundation of China(51778225,51478180)
作者简介:刘志文(1975—),男,山西阳高人,湖南大学副教授,工学博士
† 通讯联系人,E-mail:zhiwenliu@hnu.edu.cn

girder section and the suppression of torsional movement degree of the main girder section. Thus, the flutter stability of the streamlined girder section is improved.

Key words: long-span suspension bridges; flutter stability; aerodynamic optimization; wind tunnel tests; numerical simulation

---

颤振稳定性往往是大跨度悬索桥设计的控制因素之一,如何有效提高大跨度悬索桥的颤振稳定性是其抗风设计的重点. 影响桥梁结构颤振的主要因素有桥梁结构体系、加劲梁气动外形和气动措施三大类. 国内外许多学者从这三个方面对大跨度悬索桥的颤振稳定性进行了研究.

桥梁结构体系是影响大跨度悬索桥颤振稳定性的重要因素之一.Miguel[1]研究了多种超大跨径悬索桥主缆数量、结构形式及吊杆布置对颤振性能的影响,结果表明交叉布置吊杆是提高桥梁颤振稳定性最为简洁的方法之一.Michael 等[2]研究了扭转频率特性对箱梁断面悬索桥颤振临界风速的影响,结果表明桥梁结构扭转频率低于竖向频率,可有效避免颤振问题的出现.李翠娟等[3]研究了交叉吊索设置对 CFRP 主缆悬索桥颤振稳定性的影响, 结果表明在中跨 $L/3$ 和边跨跨中各设置一竖向交叉吊索,结构颤振稳定性最好.

加劲梁气动外形是影响大跨度悬索桥颤振稳定性最为重要的因素,也是桥梁结构颤振稳定性研究重点.Wang 等[4]详细研究了流线型箱梁断面气动外形对桥梁颤振和涡振性能的影响,研究表明闭口流线型箱梁断面斜腹板倾角为 15°时, 桥梁结构颤振临界风速可有效提高, 且涡振现象也可以消除. Larsen 等[5]结合机翼断面升力随攻角的变化规律,研究了流线型箱梁断面斜腹板倾角对涡振的影响,结果表明流线型箱梁断面斜腹板最优倾角为 15°,该角度条件下可消除箱梁涡振现象.刘志文等[6-7]针对闭口流线型箱梁的颤振性能和涡激力展向相关性进行了试验研究, 结果表明采用闭口流线型箱梁大跨度斜拉桥颤振稳定性较好,振动状态流线型主梁断面涡激力展向相关系数与振幅、锁定区风速等相关. 此外,Yoshinobu 等[8]对 Π 型断面气动外形优化进行了有益尝试.结果表明,改变 Π 型截面两板梁的间距可有效减小涡振振幅和提高颤振临界风速.

气动措施是改善桥梁断面颤振稳定性的重要措施.徐洪涛等[9]研究了桁架加劲梁桥面板中央开槽、裙板和气动翼板等措施对桥梁颤振临界风速的影响,结果表明桥面板中央开槽、裙板、气动翼板都能够改善桁架桥颤振稳定性.张宏杰等[10]针对流线型箱梁断面研究了中央稳定板、中央开槽、悬臂水平分离板对斜拉桥颤振稳定性的影响,结果表明在流线型箱梁断面两侧设置水平分离板能够显著改善桥梁颤振性能.徐爱军等[11]对板-桁分离式钢桁架桥和板-桁组合式钢桁架桥主梁断面进行了气动外形优化研究,结果表明采用下中央稳定板或中央开槽气动措施可有效改善板-桁分离式钢桁架桥颤振性能; 采用上中央稳定板和水平稳定板对板-桁组合式钢桁架桥颤振性能有明显改善.夏锦林等[12]研究了单侧和上下组合设置中央稳定板对单箱断面悬索桥颤振性能的影响,结果表明相比于单侧稳定板,同时设置上、下稳定板对于气动稳定性的改善效果更佳.Yang 等[13]对中央开槽双箱梁断面颤振性能进行了研究,结果表明中央稳定板可有效改善中央开槽双箱梁断面气动性能,中央稳定板颤振控制效果与双箱梁断面开槽宽度有关.Tang 等[14]对桥面中央开槽钢桁架主梁断面气动性能进行了研究,结果表明封闭桥面中央槽隙和设置竖向稳定板有利于改善主梁断面颤振性能.杨咏昕等[15]对薄平板断面中央稳定板气动控制措施的颤振控制效果及控制机理进行了研究,结果表明采用合适高度的中央稳定板可有效改善桥梁颤振稳定性;中央稳定板颤振控制机理是增加竖向自由度参与程度,改变耦合气动阻尼性质和发展规律,从而抑制系统扭转运动发散,使颤振形态转化为竖弯形式.陈政清等[16]对矮寨大桥有无中央稳定板时气动导数进行了颤振分析,结果表明中央稳定板提高了颤振发生时竖向自由度的参与程度和扭弯耦合程度,使颤振形态由单自由度扭转振动向弯扭耦合振动转移,从而提高颤振临界风速. 欧阳克俭等 [17] 通过 CFD 数值模拟及 PIV 风洞试验对中央稳定板提高桥梁结构颤振稳定性能的细观作用机理进行了研究, 结果表明中央稳定板可使桥梁断面在颤振中竖向自由度参与作用增强.

综上所述,国内外许多学者分别从桥梁结构体系、主梁断面气动外形以及气动措施等角度对大跨度桥梁颤振稳定性进行研究,并得到了一些具有重要工程应用价值的研究成果.对于大跨度悬索桥,当桥位颤振检验风速较高且采用流线型箱梁断面时有必要从桥梁结构体系、主梁断面气动外形和气动措施三个层面来提高桥梁结构颤振稳定性. 本文以主跨为 1 660 m 的伶仃洋悬索桥为工程背景，对主缆空间形式、主梁气动外形及中央稳定板高度对流线型整体钢箱梁的颤振控制效果及机理进行研究.

1 试验介绍

1.1 工程简介

拟建的深圳至中山跨江通道工程伶仃洋大桥初步设计阶段分别拟定了分离箱方案和整体箱方案,其中整体箱梁方案根据桥塔布置又分为钻石型桥塔空间缆方案和门型桥塔平行缆方案. 本文主要针对深中通道伶仃洋大桥整体箱空间缆和平行缆方案颤振稳定性问题进行研究. 深中通道伶仃洋大桥整体箱设计方案为主跨 1 660 m 的悬索桥方案，跨径布置为 530 m+1 660 m+530 m=2 720 m,桥塔塔高 266 m，图 1 为该桥整体箱方案总体布置示意图. 加劲梁采用闭口流线型钢箱梁,梁宽为 49.7 m(含风嘴、水平分离板),加劲梁中心线处梁高为 4.0 m.钻石型桥塔方案(图 1(c)所示)和 H 型桥塔方案(图 1(d)所示)对应主缆分别为空间缆和平行缆,吊杆布置与之相适应.表 1 给出了相应的动力特性分析结果.

表 1 伶仃洋大桥整体箱方案成桥状态结构动力特性

Tab.1 Dynamic characteristics of the Lingding suspension bridge with single box girder section in service stage

设计方案	一阶对称侧弯		一阶对称竖弯		一阶对称扭转	
	频率/Hz	等效质量/(10^3 kg·m^{-1})	频率/Hz	等效质量/(10^3 kg·m^{-1})	频率/Hz	等效质量惯性矩/(10^3 kg·m^2·m^{-1})
钻石型桥塔	0.058 3	39.0	0.096 8	41.5	0.244 4	16 100
H 型桥塔	0.056 5	38.9	0.100 0	42.4	0.222 8	9 840

1.2 试验装置与试验工况

综合考虑确定伶仃洋大桥主梁节段模型几何缩尺比为 $\lambda_L = 1/70$. 为减少节段模型端部三维流动的影响,主梁模型长度取 $L = 2\ 000$ mm,主梁宽度为 $B = 710$ mm,模型高度 $H = 57$ mm,模型长宽比 $L/B = 2.817$.节段模型骨架采用铝合金框架制作而成,外衣采用优质木材制作,以保证几何外形相似.模型两端采用胶合板作为端板,以保证主梁断面附近气流的二元特性. 主梁上的防撞护栏、人行道栏杆采用 ABS 板雕刻而成,并模拟了防撞护栏形状与透风率.检修车轨道及检修车轨道两侧的导流板采用 ABS 板制作. 加劲梁节段模型颤振稳定性试验在湖南大学 HD-2 边界层风洞中进行，该试验段高 2.5 m,宽 3.0 m,长 17 m.试验风速范围为 0.5~58 m/s,流场不均匀性小于 0.55%,紊流度小于 0.5%.悬挂于风洞中的节段模型照片如图 2 所示.

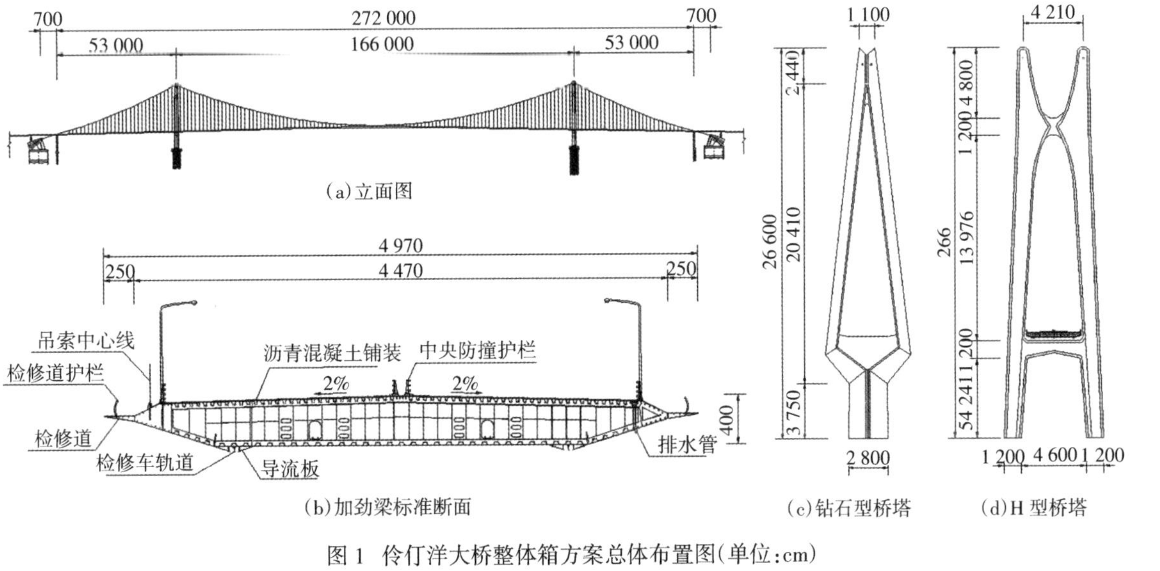

图 1 伶仃洋大桥整体箱方案总体布置图(单位:cm)

Fig.1 General layout of Lingding Bridge with mono-box girder (unit: cm)

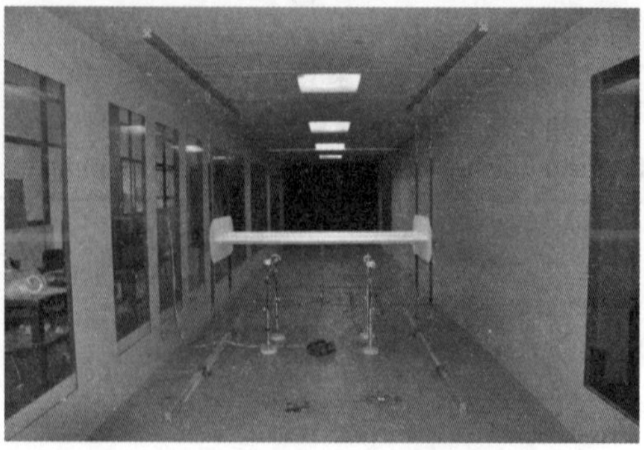

图2 HD-2风洞中的节段模型

Fig.2 Section model in HD-2 wind tunnel

首先分别针对钻石型桥塔、H型桥塔原加劲梁断面方案进行颤振稳定性节段模型试验研究；然后以H型塔方案为基础，分别进行主梁断面气动外形优化和中央稳定板气动措施颤振稳定性试验研究.详细试验工况见表2. 各工况下的动力特性见表1和表3.由表1可得，钻石型桥塔空间缆方案扭转频率较H型桥塔平行缆方案高9.7%；钻石型桥塔空间缆方案扭弯频率比为2.525，H型桥塔平行缆方案的扭弯频率比为2.228，钻石型桥塔空间缆方案扭弯频率较H型桥塔平行缆方案增大13.3%. 由表3可知，对于主跨1 660 m的悬索桥，加劲梁梁高和外形改变对桥梁结构动力特性的影响较小.

表2 伶仃洋大桥整体箱方案颤振稳定性气动优化试验工况

Tab.2 Test cases of aerodynamic optimizations for flutter stability of Lingding Bridge with mono-box girder section

工况编号	桥塔	梁高/m	气动措施	试验攻角/(°)
DG-4.0	钻石型塔	4	原断面	0、±3、±5
HG-4.0	H型塔	4	原断面	0、±3、±5
			原断面+CS(1.0 m)	0、±3
			原断面+CS(1.2 m)	
			原断面+CS(1.4 m)	
HG-4.5-1	H型塔	4.5	气动优化断面1	0、±3
HG-4.5-2	H型塔	4.5	气动优化断面2	0、±3
HG-4.5-3	H型塔	4.5	气动优化断面3	0、±3
HG-5.0	H型塔	5	气动优化断面	0、±3

注：DG-4.0表示钻石型桥塔+4 m梁高方案；HG-4.0表示H型桥塔+4 m梁高方案；其余工况编号含义类似.CS表示桥面中央稳定板.

表3 加劲梁气动优化方案关键动力特性参数

Tab.3 Key dynamic characteristics parameters of aerodynamic optimization of the stiffen girder sections

设计方案	一阶对称竖弯		一阶对称扭转	
	频率/Hz	等效质量/(10^3 kg·m^{-1})	频率/Hz	等效质量惯性矩/(10^3 kg·m^2 m^{-1})
HG-4.5-1	0.104 2	44.66	0.222 2	10 113.6
HG-4.5-2	0.104 4	44.65	0.224 6	10 440.4
HG-4.5-3	0.104 2	44.60	0.222 1	10 422.5
HG-5.0	0.100 4	44.74	0.223 0	10 900.0

2 试验结果

2.1 主缆布置对桥梁颤振稳定性的影响

钻石型桥塔和H型桥塔方案分别对应空间缆和平行缆，对应颤振临界风速试验结果汇总见表4.由表4可知，工况DG-4.0“钻石型桥塔+4.0 m梁高方案”对应最小颤振临界风速为79.5 m/s；工况HG-4.0“H型桥塔+4.0 m梁高方案”对应最小颤振临界风速为72.4 m/s.即“钻石型桥塔+4.0 m梁高方案”颤振临界风速较“H型桥塔+4.0 m梁高方案”颤振临界风速偏高约9.8%，钻石型桥塔空间缆的颤振稳定性要优于H型桥塔平行缆方案.主要原因是钻石型桥塔空间缆结构体系整体扭转刚度大于H型塔平行缆结构体系整体扭转刚度.实际上，钻石型塔方案对应的扭转频率较H型桥塔方案对应的扭转频率偏大9.7%.考虑到实际工程中钻石型桥塔空间缆施工难度较大，故仅针对平行缆方案进行进一步气动性能优化试验研究.

表4 钻石型桥塔和H型桥塔方案颤振临界风速结果汇总

Tab.4 Summary of critical flutter wind speeds of the bridge with diamond pylon and H-shaped pylon

工况编号	风攻角/(°)	颤振临界风速/(m·s^{-1})
DG-4.0	-5	>84.3
	-3	>86.7
	0	81.9
	+3	79.5
	+5	79.5
HG-4.0	-5	>92.3
	-3	79.9
	0	74.9
	+3	72.4
	+5	75.8

2.2 流线型箱梁断面外形对桥梁颤振稳定性的影响

流线型箱梁断面气动外形对桥梁结构颤振稳定性影响至关重要,决定流线型箱梁断面气动外形的主要参数包括加劲梁断面宽度、高度、斜腹板与底板夹角及风嘴参数等.参考已有文献研究成果[4-5],综合考虑流线型箱梁断面的梁高、梁宽、斜腹板与底板夹角、风嘴锐度及风嘴位置等参数对桥梁颤振稳定性的影响进行了试验研究,流线型箱梁断面气动外形具体参数汇总见表 5,箱梁断面外形及基本尺寸见图 3.

表 5 流线型箱梁断面几何外形参数汇总

Tab.5 Summary of geometry parameters of the streamlined-box girder sections

工况编号	梁高 H/m	梁宽 *B/m	梁宽 B_0/m	风嘴与底板高 h_1/m	斜腹板倾角 α/(°)	风嘴锐度 β/(°)
HG-4.0	4.0	49.70	44.70	2.48	17	41
HG-4.5-1	4.5	49.70	44.70	2.98	17	43
HG-4.5-2	4.5	49.70	44.70	2.25	16	56
HG-4.5-3	4.5	49.70	44.70	2.79	16	41
HG-5.0	5.0	51.32	46.32	3.10	15	40

注:工况编号含义同表 2;梁宽 * 表示含水平分离板.

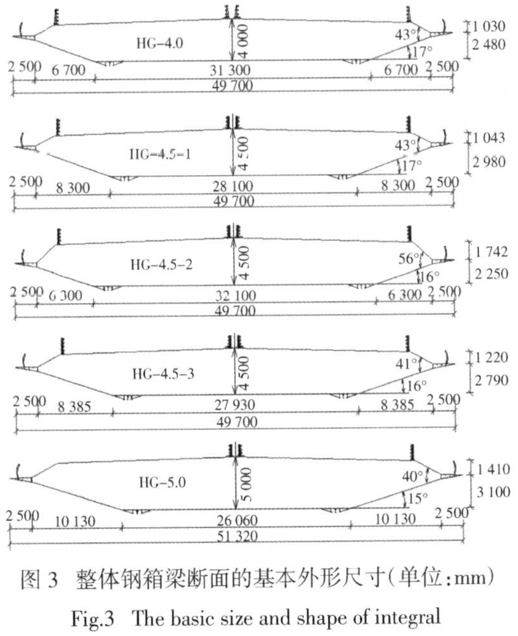

图 3 整体钢箱梁断面的基本外形尺寸(单位:mm)

Fig.3 The basic size and shape of integral steel box girder section (unit: mm)

为分析主梁断面关键气动参数对桥梁结构颤振稳定性的影响,对表 5 中各工况进行分组,尽量减小不同参数变化导致的相互影响. 首先进行箱梁断面宽高比 B/H 对桥梁结构颤振临界风速影响分析. 考虑到工况 HG-4.0 与工况 HG-4.5-1、工况 HG-4.5-3 与工况 HG-5.0 对应的风嘴外形变化较小,故将工况 HG-4.0 与工况 HG-4.5-1 作为组合 1,将工况 HG-4.5-3 与工况 HG-5.0 作为组合 2;然后进行风嘴参数对桥梁结构颤振临界风速影响分析,即对工况 HG-4.5-1、HG-4.5-2 及 HG-4.5-3 对应的试验结果进行分析.

表 6 所示为不同主梁宽高比 B/H 对应的桥梁结构颤振临界风速试验结果.由表 6 可知,对于组合 1,0°、+3°和-3°攻角条件下,工况 HG-4.0 对应颤振临界风速分别为 74.9 m/s、72.4 m/s 和 79.9 m/s;0°、+3°和-3°攻角条件下,工况 HG-4.5-1 对应颤振临界风速分别为 61.6 m/s、47.9 m/s 和 83.0 m/s. 即在 0°、+3°攻角条件下,工况 HG-4.5-1 对应颤振临界风速分别较工况 HG-4.0 减小 13.3 m/s、24.5 m/s;在-3°攻角条件下,工况 HG4.5-1 对应颤振临界风速较工况 HG-4.0 增加 3.1 m/s.工况 HG-4.0 对应最小颤振临界风速为 72.4 m/s;工况 HG-4.5-1 对应最小颤振临界风速为 47.9 m/s.工况 HG-4.5-1 颤振临界风速比工况 HG-4.0 颤振临界风速低 24.5 m/s,表明随主梁断面宽高比减小,主梁断面颤振临界风速会降低,且影响较为明显.

表 6 主梁宽高比对颤振临界风速的影响

Tab.6 Influence of height width ratio of the main girder on critical wind speed

组合	方案	H/m	B/H	颤振临界风速/(m·s^{-1})		
				-3°	0°	+3°
1	HG-4.0	4.0	12.43	79.9	74.9	72.4
	HG-4.5-1	4.5	11.04	83.0	61.6	47.9
2	HG-4.5-3	4.5	11.04	>84.4	80.4	65.9
	HG-5.0	5.0	10.26	83.0	82.2	62.1

由表 6 可知,对于组合 2,0°、+3°和-3°攻角条件下,工况 HG-4.5-3 对应颤振临界风速分别为 80.4 m/s、65.9 m/s 和大于 84.4 m/s;0°、+3°和-3°攻角条件下,工况 HG-5.0 对应颤振临界风速分别为 82.2 m/s、62.1 m/s 和 83.0 m/s. 即在 0°攻角下工况 HG-5.0 对应颤振临界风速较工况 HG-4.5-3 对应颤振临界风速增加 1.8 m/s;+3°和-3°攻角下,HG-5.0 对应颤振临界风速较工况 HG-4.5-3 对应颤振

临界风速分别下降了 3.8 m/s 和 1.4 m/s.对应最小颤振临界风速由工况 HG-4.5-3 对应的 65.9 m/s 降到了工况 HG-5.0 对应的 62.1 m/s.

综上可知，总体而言随着主梁断面宽高比 B/H 的减小，主梁断面颤振临界风速会降低，适当增加主梁断面宽高比 B/H，可有效提高桥梁结构颤振稳定性.

表 7 所示为相同梁高不同风嘴形式主梁断面颤振稳定性试验结果汇总. 由表 7 可知，工况 HG-4.5-1、HG-4.5-2 和 HG-4.5-3 对应各攻角下颤振临界风速随风嘴形状的改变而改变，相对而言，工况 HG-4.5-3 对应桥梁颤振临界风速是最高的，表明选择合适的风嘴形状可有效提高桥梁结构颤振稳定性. 此外，工况 HG-4.5-1、HG-4.5-2 及 HG-4.5-3 方案风嘴锐度、h_1/H 变化趋势与颤振临界风速的变化趋势不尽相同，表明风嘴锐度和 h_1/H 共同影响桥梁结构颤振临界风速. 当风嘴锐度为 41°、风嘴位置 h_1/H=0.62 时（HG-4.5-3）颤振临界风速最高.表 7 中工况 HG-4.0 对应风嘴锐度为 41°、风嘴位置 h_1/H 为 0.62，对应颤振临界风速试验结果也反映了这一规律.

表 7 风嘴对颤振临界风速的影响

Tab.7 Influence of wind fairing on flutter critical velocity

工况	风嘴锐度/(°)	h_1/H	颤振临界风速/(m·s⁻¹)		
			−3°	0°	+3°
HG-4.5-1	43	0.66	83.0	61.6	47.9
HG-4.5-2	56	0.50	84.8	64.2	56.5
HG-4.5-3	41	0.62	>84.4	80.4	65.9

2.3 中央稳定板对桥梁颤振性的影响

中央稳定板是提高桥梁颤振临界风速的有效气动措施之一. 以工况 HG-4.0 原主梁断面为基础，针对不同高度中央稳定板对带水平分离板流线型箱梁断面颤振临界风速的影响进行研究.综合考虑，确定中央稳定板的高度分别为 1.0 m、1.2 m 和 1.4 m，即分别为 0.25H、0.3H 和 0.35H.表 8 给出了工况 HG-4.0 原断面和增设不同高度中央稳定板对应桥梁颤振临界风速试验结果.图 4 所示为工况 HG-4.0 原断面增设不同高度中央稳定板后不同攻角条件下桥梁颤振临界风速随中央稳定板高度变化曲线.

表 8 中央稳定板高度对桥梁颤振临界风速的影响

Tab.8 Influence of central stabilizer heights on flutter critical velocity of the bridge

工况	颤振临界风速/(m·s⁻¹)		
	−3°	0°	+3°
HG-4.0	79.9	74.9	72.4
HG-4.0+CS1.0	84.1	81.8	91.6
HG-4.0+CS1.2	86.5	87.9	92.1
HG-4.0+CS1.4	87.4	90.2	94.9

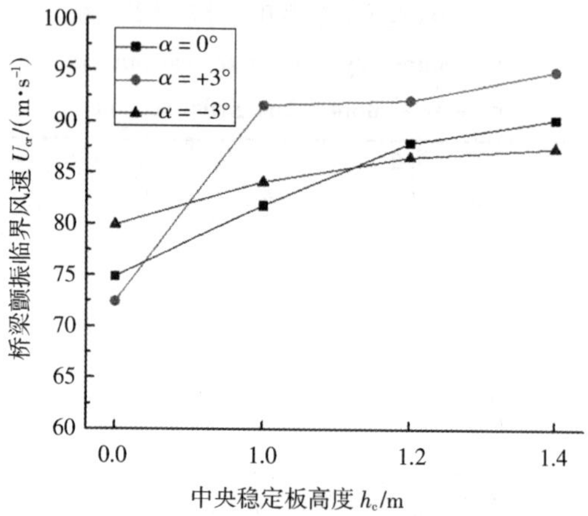

图 4 桥梁颤振临界风速随中央稳定板高度的变化曲线

Fig.4 Flutter critical velocity of the bridge vs. the central stabilizer height

由表 8 可知，0°、+3°和−3°攻角条件下，工况 HG-4.0 原断面对应颤振临界风速分别为 74.9 m/s、72.4 m/s 和 79.9 m/s；增设高度为 1.0 m 的中央稳定板后对应颤振临界风速分别为 81.8 m/s、91.6 m/s 和 84.1 m/s；增设高度为 1.2 m 的中央稳定板后对应的颤振临界风速分别为 87.9 m/s、92.1 m/s 和 86.5 m/s；增设高度为 1.4 m 中央稳定板后对应的颤振临界风速分别为 90.2 m/s、94.9 m/s 和 87.4 m/s.结果表明设置合适高度的中央稳定板可有效提高带水平分离板的流线型箱梁断面颤振临界风速.

由图 4 可知，各风攻角对应桥梁结构颤振临界风速均随中央稳定板高度的增加而逐渐增加；当中央稳定板高度超过 1.2 m(0.3H)后中央稳定板对桥梁颤振临界风速提高效果不太明显. 不同风攻角条件下，中央稳定板对桥梁颤振临界风速提高效果也存在一定的差异，+3°攻角中央稳定板对桥梁颤振临界风速提高效果最为明显；−3°攻角中央稳定板对桥梁颤振临界风速提高效果相对最小，这可能与流线型箱梁断面两侧设置水平分离板有一定关系.

3 流线型箱梁断面颤振控制机理探讨

由上述试验结果可知，主缆布置形式、箱梁断面气动外形及中央稳定板均会对桥梁颤振临界风速产生一定的影响. 主缆布置形式对桥梁结构颤振临界风速的影响主要表现为主缆布置形式导致桥梁结构扭转频率改变，从而影响桥梁结构颤振临界风速. 箱梁断面气动外形及中央稳定板高度对桥梁结构颤振临界风速的影响机理则较为复杂. 已有研究成果表明增设中央稳定板后能有效增加竖向自由度参与程度，改变耦合气动阻尼的性质和发展规律，从而抑制系统扭转运动的发散，提高原断面颤振稳定性[15-17].

现采用计算流体力学方法从流场结构角度对流线型箱梁断面中央稳定板颤振控制机理进行研究. 基于二维大涡模拟和动网格技术建立桥梁主梁断面流固耦合分析方法，以工况 HG-4.0 和 HG-4.0+CS1.4 为研究对象，分别进行 0°攻角主梁断面自由振动数值模拟，根据不同风速条件下主梁断面的振动响应曲线确定桥梁结构颤振临界风速. 为便于与试验结果进行比较，确定主梁断面几何缩尺比为 1/70，风速比为 1/5.计算域左侧边界距主梁断面 $4B$（B 为主梁断面宽度），计算域右侧边界距主梁断面 $10B$，计算域上、下侧距主梁断面中心线 $4B$，模型阻塞率为 1.0%，满足阻塞率小于 5%的要求.网格划分采用分块化思路，靠近主梁断面附近采用结构化网格进行划分，主梁断面附近变形区域采用三角形网格，计算域最外侧区域采用结构化网格，以提高网格划分质量.边界条件确定如下：计算域左侧为速度入口（Velocity inlet），计算域右侧为压力出口（Pressure outlet），计算域上、下侧为对称边界（Symmetry）.图 5 为主梁断面计算域、网格划分分块及边界条件示意图.图 6 为工况 HG-4.0 及工况 HG-4.0+CS1.4 主梁模型网格示意图.

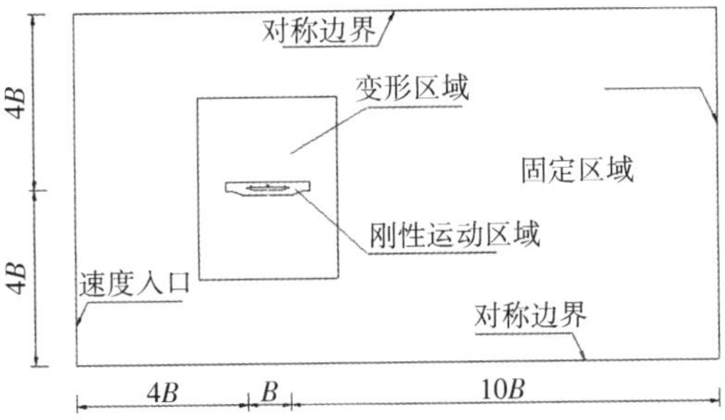

图 5 主梁断面计算域及边界条件

Fig.5 Computational domain and boundary conditions of the main deck section

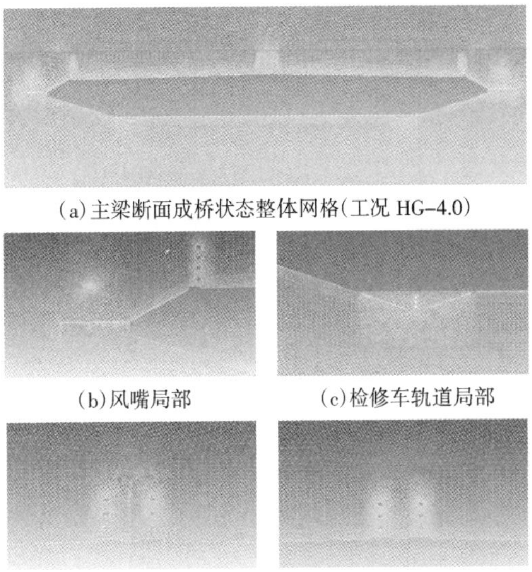

(a) 主梁断面成桥状态整体网格(工况 HG-4.0)

(b) 风嘴局部 (c) 检修车轨道局部

(d) 中央护栏(无中央稳定板) (f) 中央护栏+CS(1.4 m)

图 6 工况 HG-4.0 和工况 HG-4.0+CS1.4 主梁断面网格

Fig.6 Mesh of main deck of HG-4.0 case and HG-4.0+CS1.4 case

湍流模型选择二维大涡模拟（Large Eddy Simulation，LES）中的 Smagorinsky-Lilly 湍流模型，Smagorinsky 常数 C_s=0.10.具体求解参数设置如表 9 所示.0°攻角下工况 HG-4.0 及工况 HG-4.0+CS1.4 主梁断面在颤振发散前后位移响应计算结果如图 7、图 8 所示，数值模拟结果与试验结果的比较见表 10. 由图 7、图 8 可知，工况 HG-4.0 实际桥梁结构颤振临界风速约为 75~77.5 m/s，工况 HG-4.0+CS1.4 实际桥梁结构的颤振临界风速约为 80~82.5 m/s.

表 9 求解参数设置

Tab.9 Parameter setting of calculation

方法及参数		设置
速度-压力耦合求解		SIMPLEC
空间离散	梯度	Green-Gause Node Based
	压力	Second Order
	动量	Bounded Central Differencing
时间离散		Second Order Implicit
时间步/s		0.000 2

表 10 桥梁结构颤振临界风速计算结果与试验结果比较

Tab.10 Comparison of numerical and experimental results of flutter critical wind speeds of the bridge structures

工况	颤振临界风速/(m·s^{-1})		相对误差/%
	试验结果	数值模拟	
HG-4.0	74.9	75~77.5	0.1~3.5
HG-4.0+CS1.4	90.2	80~82.5	9.3~12.8

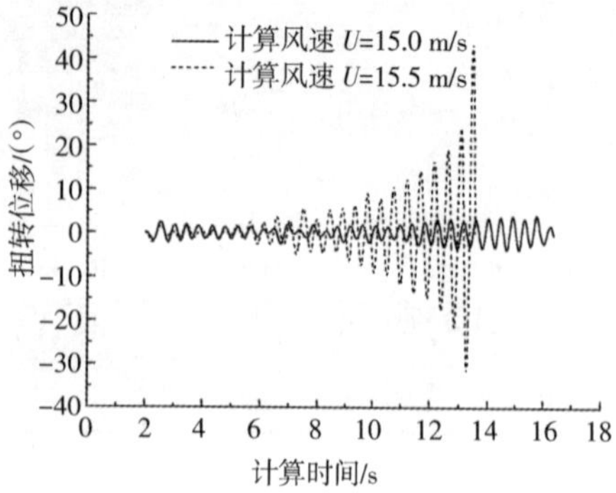

(a)主梁断面扭转位移响应

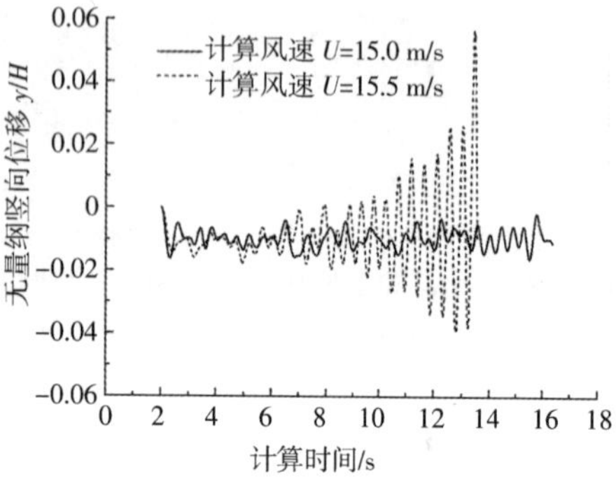

(b)主梁断面竖向位移响应

图7 工况 HG-4.0 主梁断面颤振前后位移响应

Fig.7 Displacements of the main deck of HG-4.0 case before and after flutter divergence

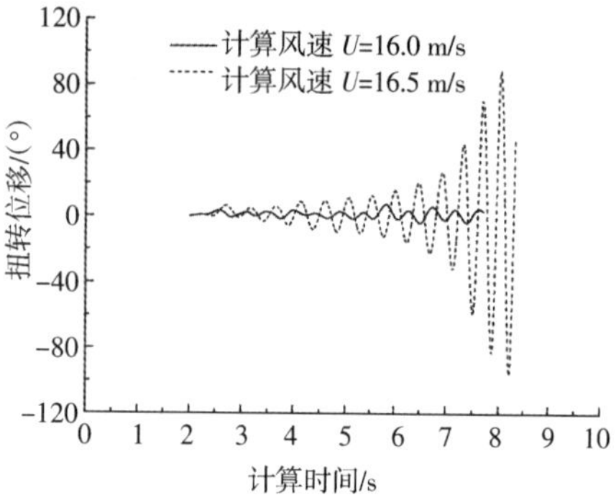

(a)主梁断面扭转位移响应

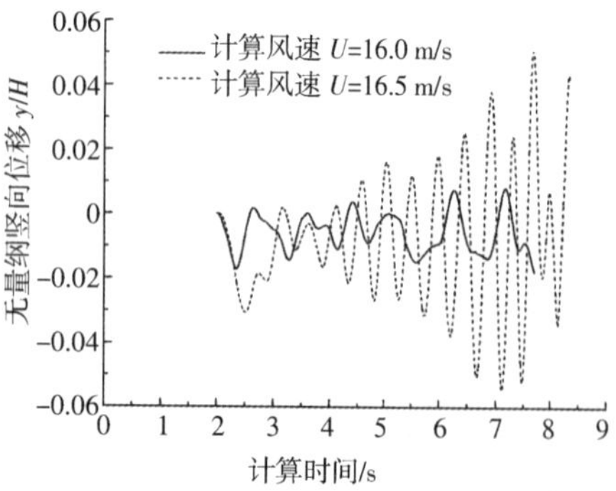

(b)主梁断面竖向位移响应

图8 工况 HG-4.0+CS1.4 主梁断面颤振前后的位移响应

Fig.8 Displacement of the main deck of HG-4.0+CS1.4 case before and after flutter divergence

由表10可知,0°攻角下工况 HG-4.0 和工况 HG-4.0+CS1.4 对应的实际桥梁结构颤振临界风速数值模拟结果与试验结果总体吻合较好,验证了本文数值模拟方法的精度.

为了方便分析设置中央稳定板后流场的变化,图9、图10给出颤振前后一个周期内断面的流场显示.从图9可以看出,当计算风速为15.5 m/s时,原主梁断面主要表现为扭转运动;主梁断面附近的涡主要位于主梁断面前后缘处,断面中央附近涡较少.

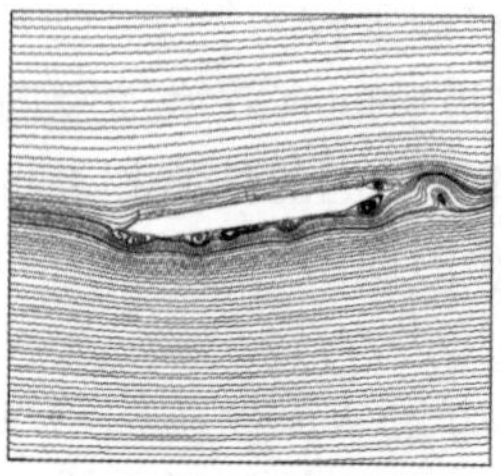

(a)$T/4$

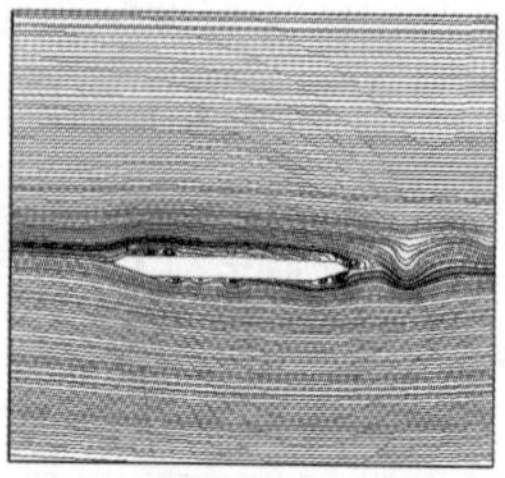

(b)$2T/4$

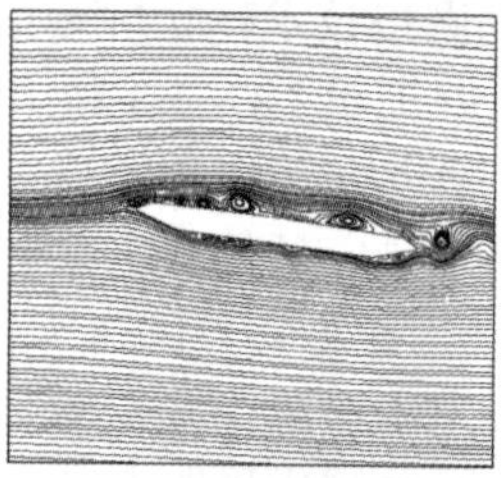

(c)$3T/4$

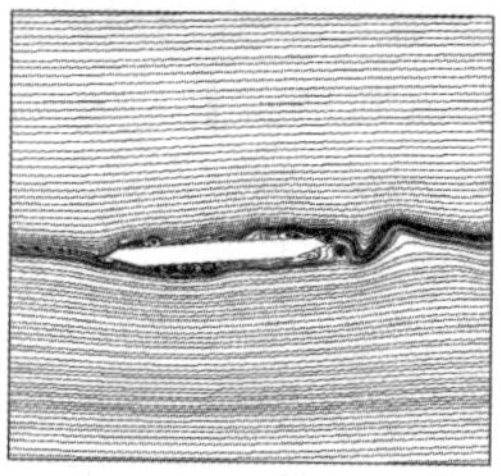

(d)$4T/4$

图9 无中央稳定板颤振发散后的流场结构(U=15.5 m/s)

Fig.9 The streamline of the flow field after flutter without central stabilizer (U=15.5 m/s)

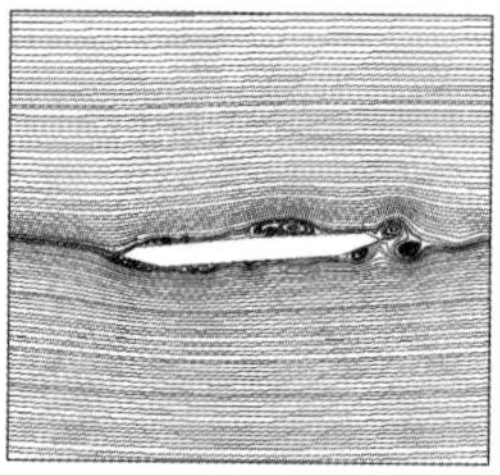

(a)$T/4$

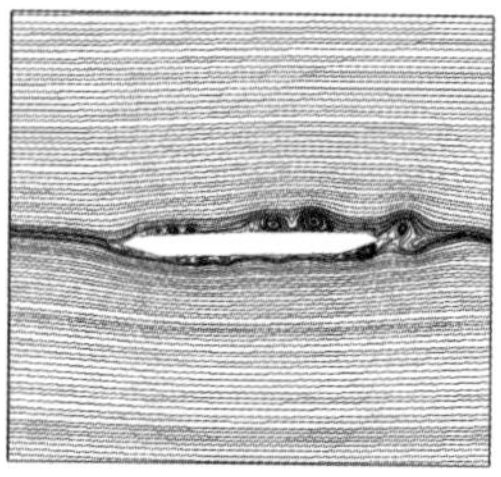

(b)$2T/4$

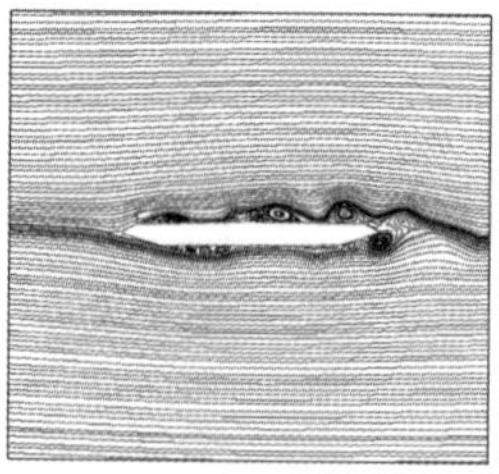

(c)$3T/4$

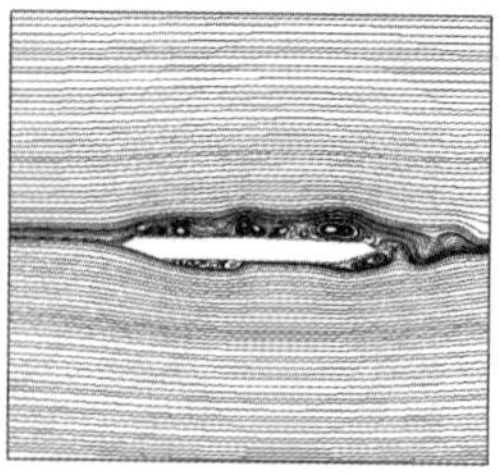

(d)$4T/4$

图10 有中央稳定板颤振发散前的流场结构(U=16.0 m/s)

Fig.10 The streamline of the flow field before flutter with 1.4 m central stabilizer (U=16.0 m/s)

从图 10 可以看出，当计算风速为 16.0 m/s 时，设置中央稳定板后主梁断面主要表现为竖向运动；主梁断面在中央稳定板附近存在明显的涡，从而对主梁产生竖向气动力，导致主梁断面竖向运动参与程度提高，抑制了主梁断面扭转运动.

4 结 论

以拟建的伶仃洋大桥为工程依托，对流线型箱梁断面的颤振稳定性影响因素和气动控制措施进行了研究，得到如下主要结论：

1) 主缆布置形式对桥梁结构颤振临界风速的影响主要表现为主缆布置形式导致桥梁结构扭转频率改变，从而影响桥梁颤振临界风速.

2) 适当增加主梁断面宽高比可有效提高桥梁结构颤振稳定性；设置合适高度中央稳定板可有效提高带水平分离板流线型箱梁断面颤振临界风速.

3) 中央稳定板附近产生的涡主要对流线型主梁断面产生竖向气动力作用，导致主梁断面竖向运动参与程度提高，抑制了主梁断面的扭转运动，从而提高了流线型箱梁断面颤振稳定性.

参考文献

[1] MIGUEL A A.Flutter stability of very long suspension bridges[J]. Journal of Bridge Engineering, 1998, 3(3):132—139.

[2] MICHAEL S A, JENS J, ANDERS B, *et al*. Aerodynamic stability of long span suspension bridges with low torsional natural frequencies [J]. Engineering Structures, 2016, 120:82—91.

[3] 李翠娟，李永乐，唐茂林，等.交叉吊索对超大跨 CFRP 主缆悬索桥颤振性能的改善作用[J].土木工程学报，2017，50(3):83—90.
LI C J, LI Y L, TANG M L, *et al*. Improvement of flutter stability of super long span suspension bridge with CFRP cable by crossed hangers[J]. China Civil Engineering Journal, 2017, 50(3):83—90. (In Chinese)

[4] WANG Q, LIAO H L, LI M S, *et al*. Influence of aerodynamic configuration of a streamline box girder on bridge flutter and vortex-induced vibration[J]. Journal of Modern Transportation, 2011, 19(4):261—267.

[5] LARSEN A, WALL A.Shaping of bridge box girders to avoid vortex shedding response[J]. Journal of Wind Engineering and Industrial Aerodynamics, 2012, 104/106:159—165.

[6] 刘志文，洪涵，梁立农，等.广东江顺大桥抗风性能试验研究[J].湖南大学学报(自然科学版)，2015，42(3):112—119.
LIU Z W, HONG H, LIANG L N, *et al*. Experimental investigation of the wind- resistant performance of Jiangshun Bridge in Guangdong [J]. Journal of Hunan University (Natural Sciences), 2015, 42(3):112—119.(In Chinese)

[7] 刘志文，龚慧星，陈政清. 流线型箱梁断面涡激力展向相关试验研究[J]. 湖南大学学报(自然科学版)，2017，44(5):80—87.
LIU Z W, GONG H X, CHEN Z Q. Experimental study on spanwise correlation of vortex-excited forces of a streamlined box girder[J]. Journal of Hunan University(Natural Sciences), 2017, 44(5):80—87.(In Chinese)

[8] YOSHINOBU K, KENJI S, EIKI Y, *et al*.Improvement of aeroelastic instability of shallow π section[J].Journal of Wind Engineering and Industrial Aerodymics, 2001, 89:1445—1457.

[9] 徐洪涛，何勇，邹隽珺，等.特大跨径桁架加劲梁桥颤振稳定性气动优化措施风洞试验研究[J].公路交通科技，2010(2):48—53.
XU H T, HE Y, ZOU J J, *et al*. Study of wins tunnel test of aerodynamic optimization measures for flutter stability of super long-span bridge with truss girder [J]. Journal of Highway and Transportation Research and Development, 2010(2): 48—53. (In Chinese)

[10] 张宏杰，朱乐东. 箱形主梁悬臂水平分离板的颤振控制效果与机理 [J]. 同济大学学报（自然科学版），2011，39(11):1569—1574，1640.
ZHANG H J, ZHU L D. Control effect and mechanism of cantilever horizontal splitting plates on flutter performance of a box deck[J]. Journal of TongJi University (Natural Science), 2011, 39 (11): 1569—1574, 1640. (In Chinese)

[11] 徐爱军，王凯，李明水，等. 板-桁组合式钢桁梁悬索桥颤振稳定性选型研究[J]. 实验流体力学，2015，29(4): 52—57.
XU A J, WANG K, LI M S, *et al*. Flutter stability selection study of a long-span steel truss suspension bridge with a combine deck plate[J]. Journal of Experiments in Fluid Mechanics, 2015, 29(4): 52—57.(In Chinese)

[12] 夏锦林，杨咏昕，葛耀君. 上、下组合中央稳定板对于箱梁颤振性能的影响[J]. 中国公路学报，2017，30(7):86—93.
XIA J L, YANG Y X, GE Y J. Influence of combined central stabilizing plates on flutter performances of box girder [J]. China Journal of Highway and Transport, 2017, 30(7): 86—93. (In Chinese)

[13] YANG Y X, ZHOU R, GE Y J, *et al*. Flutter characteristics of twin-box girder bridges with vertical central stabilizers [J]. Engineering Structures, 2017, 133:33—48.

[14] TANG H J, LI Y L, WANG Y F, *et al*. Aerodynamic optimization for flutter performance of steel truss stiffening girder at large angles of attack [J]. Journal of Wind Engineering & Industrial Aerodynamics, 2017, 168:260—270.

[15] 杨咏昕，葛耀君，项海帆. 中央稳定板颤振控制效果和机理研究[J]. 同济大学学报(自然科学版)，2007，35(2):149—155.
YANG Y X, GE Y J, XIANG H F. Flutter controlling effect and, mechanism of central stabilizer[J]. Journal of TongJi University (Natural Science), 2007, 35(2):149—155.(In Chinese)

[16] 陈政清，欧阳克俭，牛华伟，等. 中央稳定板提高桁架梁悬索桥颤振稳定性的气动机理[J]. 中国公路学报，2009，22(6): 53—59.
CHEN Z Q, OUYANG K J, NIU H W, *et al*. Aerodynamic mechanism of improvement of flutter stability of truss-girder suspension bridge using central stabilizer [J]. China Journal of Highway and Transport, 2009, 22(6): 53—59. (In Chinese)

[17] 欧阳克俭，陈政清. 中央稳定板提高颤振稳定性能的细观作用机理[J]. 振动与冲击，2016，35(1):11—16.
OUYANG K J, CHEN Z Q. Micro-mechanism of a central stabilizer for improving a bridge's flutter stability [J]. Journal of Vibration and Shock, 2016, 35(1):11—16. (In Chinese)

第36卷 第8期 2023年8月
中 国 公 路 学 报 China J. Highw. Transp.
Vol. 36 No. 8 Aug. 2023

文章编号:1001-7372(2023)08-0014-08

桥梁颤振导数识别及颤振分析的不确定性研究

封周权[*1,2],林 阳[1,2],华旭刚[1,2],陈政清[1,2]

(1. 湖南大学 风工程与桥梁工程湖南省重点实验室,湖南 长沙 410082;
2. 桥梁工程安全与韧性全国重点实验室(湖南大学),湖南 长沙 410082)

摘要:由于风洞试验和理论模型的各种不确定性,通过风洞试验获得的颤振导数及相应的颤振临界风速存在不确定性。为了量化这些不确定性,提出了一种创新的近似贝叶斯方法。该方法通过抽样和模拟来近似表达似然函数,从而实现颤振导数的准确识别和不确定性量化。同时,还研究了颤振导数不确定性在颤振分析中的传播情况。采用子集模拟技术与近似贝叶斯方法相结合,以提高参数后验样本的抽样效率。该方法不仅能够获得颤振导数和颤振临界风速的最优估计,还能获得其后验概率分布。通过理想平板数值模拟和实桥主梁断面风洞试验,验证了该方法的有效性,并将其与传统最小二乘法进行了比较。研究结果显示:该方法得到的颤振导数最优估计与最小二乘法结果非常接近;在低风速下,所有导数的不确定性都较小,而在中高风速情况下,大多数导数都具有较大的不确定性,尤其是接近颤振临界风速时,所有导数的不确定性均较大;颤振导数的不确定性会在颤振分析中传播,导致颤振临界风速也存在较大的不确定性。所提出的近似贝叶斯方法能够准确识别颤振导数,并量化其不确定性,从而实现桥梁颤振性能的概率性评价;为桥梁颤振分析提供了新的思路,为确保桥梁的抗风安全提供了有力支持。

关键词:桥梁工程;不确定性量化;近似贝叶斯计算;颤振导数;颤振分析;子集模拟
中图分类号:U441.3 **文献标志码:**A

Uncertainty Quantification on Flutter Derivative Identification and Flutter Analysis of Bridges

FENG Zhou-quan[*1,2], LIN Yang[1,2], HUA Xu-gang[1,2], CHEN Zheng-qing[1,2]

(1. Hunan Provincial Key Laboratory of Wind and Bridge Engineering, Hunan University, Changsha 410082, Hunan, China; 2. State Key Laboratory of Bridge Engineering Safety and Resilience (Hunan University), Changsha 410082, Hunan, China)

Abstract: Inherent uncertainties in wind tunnel testing and theoretical models are associated with flutter derivatives and critical flutter wind speeds obtained through wind tunnel testing. To quantify these uncertainties, we propose an innovative approximate Bayesian computational method. This method approximates the likelihood function through sampling and simulation, thereby achieving accurate identification and uncertainty quantification of flutter derivatives. In addition, we investigated the propagation of uncertainties in the flutter derivatives during flutter analysis. Subset simulation was combined with approximate Bayesian computation to improve the sampling efficiency of the posterior samples. This method provides optimal estimates for flutter

收稿日期:2023-02-26
基金项目:国家自然科学基金项目(52178284,51708203);湖南省自然科学基金项目(2021JJ30106);重庆市自然科学基金项目(2022NSCQ-MSX5727);长沙市政府采购研究项目(CSCG-202107060066)
* **作者简介:**封周权(1982-),男,湖南衡南人,副教授,博士研究生导师,工学博士,E-mail:zqfeng@hnu.edu.cn。

derivatives and critical wind speeds and yields their posterior probability distributions. We validated the effectiveness of our approach through numerical simulations of flat plates and wind tunnel tests on an actual bridge girder sectional model, and compared it with the traditional least squares method. The results demonstrate a high degree of consistency between the optimal estimates obtained using the proposed method and those obtained using the least-squares method. At low wind speeds, the uncertainties in all derivatives were relatively small; whereas, at medium to high wind speeds, most derivatives exhibited larger uncertainties, particularly near the critical wind speed of the flutter, where all derivatives became highly uncertain. The uncertainty of the flutter derivatives propagates in the flutter analysis, resulting in substantial uncertainty in the critical wind speed of the flutter. The proposed approximate Bayesian method accurately identifies flutter derivatives and quantifies their uncertainties, thereby enabling a probabilistic assessment of bridge flutter performance. This study provides new insights for bridge flutter studies and offers robust support for ensuring the wind-resistant safety of bridges.

Keywords: bridge engineering; uncertainty quantification; approximate Bayesian computation; flutter derivatives; flutter analysis; subset simulation

Received 26 Feb. 2023

Funding: Natural Science Foundation of China (52178284, 51708203); Natural Science Foundation of Hunan Province (2021JJ30106); Natural Science Foundation of Chongqing (2022NSCQ-MSX5727); Changsha Municipal Government Procurement Research Project (CSCG-202107060066)

* Corresponding author. *E-mail address*: zqfeng@hnu. edu. cn.

https://doi. org/10. 19721/j. cnki. 1001-7372. 2023. 08. 002

0 引 言

随着国民经济的日益发展，桥梁建设的规模和质量也在不断提高，特别是大跨径桥梁的数量逐年增加。然而，大跨径桥梁通常比较轻柔，极易在外荷载作用下发生振动。其中，风是引发桥梁振动的主要外荷载之一，对于大跨桥梁来说，风致振动是一个不可忽视的问题。风引起的桥梁振动包括颤振、驰振、抖振和涡激振动等。其中，颤振是一种可能发散的自激振动，严重时可能导致桥梁结构失稳甚至垮塌。自从旧塔科马大桥因颤振现象坍塌以来，桥梁的颤振问题开始被广泛关注。颤振导数是表征桥梁颤振性能和进行颤振临界风速预测的关键参数，准确识别颤振导数是进行颤振分析的前提。

识别颤振导数的方法可以分为自由振动法、强迫振动法和随机振动法三大类。自由振动法通过测量无风和有风下节段模型弯扭耦合自由振动响应，并从自由振动响应中提取颤振导数。强迫振动法则是指利用特殊的机械装置驱动模型做频率和振幅可控的谐波振动，然后直接测量模型上的振动信号（如气动力、位移或加速度），并进行频谱分析或时域分析，以获得颤振导数。上述 2 类方法都是在均匀流风场中进行的，忽略了湍流的影响。然而，湍流会改变截面周围的流动模式，从而影响颤振导数的识别。因此，随机振动法通过在格栅产生的紊流风场中进行节段模型风洞试验，以识别桥梁的颤振导数。强迫振动法识别颤振导数不确定性较小，但需要复杂的设备和严格的试验条件。随机振动法也需要专门的试验设备，并且识别算法相对复杂。相比之下，自由振动法虽然颤振导数的识别不确定性较大，但方法简单易实现，无需复杂设备，因此被广泛应用。

在自由振动法识别颤振导数方面，Scanlan 等[1]最早提出了颤振导数的分状态识别方法。该方法首先从竖弯和扭转的单自由度振动信号中分别识别出直接导数，然后进行耦合振动识别，得到交叉导数。Sarkar 等[2]提出了一种系统识别方法，旨在通过单个两自由度自由衰减振动测试同时识别 8 个颤振导数。Gu 等[3]提出了统一最小二乘方法（ULS），采用了竖弯和扭转运动的统一误差函数。为了平衡竖弯信号和扭转信号，丁泉顺等[4]率先在 ULS 方法中引入权重因子进行颤振导数识别。后来 Li 等[5]提出加权整体最小二乘法（WELS）来提取桥梁颤振导数。Bartoli 等[6]对 ULS 方法进行了改进，在统一误差函数中引入了适当的权重因子，增强了迭代求解过程。李加武等[7]进一步采用加权矩阵，提出了加权最小二乘法(WLS)，该方法具有更好的识别精度和计算效率。Xu 等[8]利用改进的随机搜索算法对 ULS 中的迭代方案进行了改进。

上述基于最小二乘原理的方法都是确定性方法。然而，在颤振导数识别的整个过程中存在着多种不确定性[9-11]。例如在结构建模时，由于计算模型与真实结构之间存在差异，会引起一定的不确定性。在试验过程中，仪器测量和外界环境干扰也会导致不确定性。此外，在数据处理和计算过程中，浮点数和舍入误差也会引入不确定性[12]。一般来说，数据量的不充分也是不确定性增大的原因。尽管这些不确定性可以在一定程度上减小，但是很难完全消除。采用自由振动方法识别颤振导数时，高折减风速下，由于气动阻尼的影响，振幅会很快衰减，导致高折减风速下颤振导数识别结果的离散性很大。因此，对颤振导数的不确定性进行量化十分必要，这也是量化颤振分析结果不确定性的前提[13]。在众多不确定性量化方法中，贝叶斯方法(Bayesian method)是一种广泛应用且十分有效的方法。通常情况下，将概率理解为大量重复样本出现的频次。然而，在贝叶斯方法中，概率被解释为对未知命题的信任程度[14]，这使其在参数识别方面更适用[15]。Beck 等[16]和 Katafygiotis 等[17]较早地建立了基于贝叶斯方法的结构系统识别理论框架。通过应用贝叶斯方法进行颤振导数识别，不仅能得到其最优值，还能给出颤振导数的后验概率分布，从而量化颤振导数的不确定性。

贝叶斯方法在参数识别中能够提供后验概率分布，但在处理复杂问题，尤其是具有多个待求参数和高维度的情况下，直接求解计算就比较困难。这些问题通常涉及高维积分的求解，从而引发维数灾难问题，使得传统的数值积分方法如高斯积分失效[18]。因此，在这种情形下，通常会采用近似贝叶斯计算(Approximate Bayesian Computation, ABC)[19-21]或马尔可夫蒙特卡洛方法(Markov Chain Monte Carlo, MCMC)[22]。在经典贝叶斯方法中，有时难以获得似然函数的精确表达式，这限制了贝叶斯方法的应用范围。近似贝叶斯计算通过采样的方式来近似似然函数，将样本与实测数据进行比较，从而筛选出合适的样本。通过这种方法，近似贝叶斯计算避免了对似然函数的直接计算，因此被称为“似然自由”(Likelihood-free)方法[23]。这种基于样本的逼近方法为复杂问题的贝叶斯推断提供了一种可行的选择。

为了应对确定性模型在颤振导数识别和颤振分析中所面临的不确定性挑战，本文首次引入了近似贝叶斯计算方法。首先，基于最小二乘原理构建样本与实测数据之间的误差函数[24]，从而实现了对似然函数的近似表达。这样一来，可以在不直接计算似然函数的情况下，通过近似来获得参数的后验分布。接着，为了获取颤振导数的后验分布，本文采用了子集模拟的方法进行采样。具体而言，对模型参数进行随机抽样，并利用这些抽样参数进行模拟计算，从而得到颤振导数的后验分布。通过这种方法，可以在复杂问题中避免对高维积分的直接计算，同时考虑到不确定性因素，为颤振导数提供了更全面准确的估计。在得到颤振导数的后验分布之后，本文进一步利用该信息来求解颤振临界风速的分布范围，从而为桥梁抗风设计提供更可靠的依据[9,13]。这种基于后验分布的分析方法能够更好地理解颤振行为，并能够更全面地考虑到不确定性对结构性能的影响，从而提高抗风设计的可靠性。

1 颤振导数识别及颤振分析

1.1 两自由度桥梁断面节段模型振动响应

两自由度桥梁节段模型自由振动方程为

$$m\ddot{h}+2m\zeta_h\omega_h\dot{h}+m\omega_h^2h=F_{\mathrm{L}} \tag{1}$$

$$I\ddot{\alpha}+2I\zeta_\alpha\omega_\alpha\dot{\alpha}+I\omega_\alpha^2\alpha=F_{\mathrm{M}} \tag{2}$$

式中：F_{L}、F_{M} 分别为自激升力和扭矩；h、α 分别为竖向位移和扭转角；$\dot{h}$、$\dot{\alpha}$ 分别为竖向速度和扭转角速度；$\ddot{h}$、$\ddot{\alpha}$ 分别为竖向加速度和扭转角加速度；m、I 分别为单位长度质量和质量惯性矩；ζ_h、ζ_α 分别为竖弯和扭转阻尼比；ω_h、ω_α 分别为竖弯和扭转圆频率。

其中自激升力和扭矩分别为

$$F_{\mathrm{L}}=\frac{1}{2}\rho U^2(2B)\left(KH_1^*\frac{\dot{h}}{U}+KH_2^*\frac{\dot{\alpha}B}{U}+K^2H_3^*\alpha+K^2H_4^*\frac{h}{B}\right) \tag{3}$$

$$F_{\mathrm{M}}=\frac{1}{2}\rho U^2(2B^2)\left(KA_1^*\frac{\dot{h}}{U}+KA_2^*\frac{\dot{\alpha}B}{U}+K^2A_3^*\alpha+K^2A_4^*\frac{h}{B}\right) \tag{4}$$

式中：ρ 为空气密度；U 为均匀来流风速；B 为桥梁宽度；$K=B\omega/U$ 为折算频率；ω 为振动圆频率；H_i^* 和 A_i^* $(i=1,2,3,4)$ 为颤振导数。

将式(1)、(2)代入式(3)、(4)中有

$$\ddot{\boldsymbol{y}}+\boldsymbol{C}\dot{\boldsymbol{y}}+\boldsymbol{K}\boldsymbol{y}=0 \tag{5}$$

式中：

$\boldsymbol{y}(t)=[h(t)\ \ \alpha(t)]^{\mathrm{T}}$

$$\boldsymbol{C}=\begin{bmatrix} 2\zeta_h\omega_h-\rho B^2\omega H_1^*/m & -\rho B^3\omega H_2^*/m \\ -\rho B^3\omega A_1^*/I & 2\zeta_\alpha\omega_\alpha-\rho B^4\omega A_2^*/I \end{bmatrix}$$

$$\boldsymbol{K}=\begin{bmatrix} \omega_h^2-\rho B^2\omega^2 H_4^*/m & -\rho B^3\omega^2 H_3^*/m \\ -\rho B^3\omega^2 A_4^*/I & \omega_\alpha^2-\rho B^4\omega^2 A_3^*/I \end{bmatrix}$$

上式可写为

$$\begin{bmatrix} \dot{\boldsymbol{y}} \\ \ddot{\boldsymbol{y}} \end{bmatrix}=\begin{bmatrix} 0 & \boldsymbol{E} \\ -\boldsymbol{K} & -\boldsymbol{C} \end{bmatrix}\begin{bmatrix} \boldsymbol{y} \\ \dot{\boldsymbol{y}} \end{bmatrix} \tag{6}$$

式(6)中 $\boldsymbol{E}$ 为单位矩阵。并令

$$\boldsymbol{A}=\begin{bmatrix} 0 & \boldsymbol{E} \\ -\boldsymbol{K} & -\boldsymbol{C} \end{bmatrix}$$

求解方程(6)即可得系统的响应。矩阵 $\boldsymbol{A}$ 具有复共轭特征值 λ_i、λ_i^*

$$\lambda_i=a_i+\mathrm{i}b_i,\lambda_i^*=a_i-\mathrm{i}b_i \tag{7}$$

式中：a_i、b_i，$(i=1,2)$是待求参数；i 为虚数单位。

由此可得节段模型的理论响应值

$$h(t)=\sum_{i=1}^{2}\mathrm{e}^{a_it}(c_i\cos(b_it)+d_i\sin(b_it)) \tag{8}$$

$$\alpha(t)=\sum_{i=1}^{2}\mathrm{e}^{a_it}(e_i\cos(b_it)+f_i\sin(b_it)) \tag{9}$$

式中：c_i、d_i、e_i，f_i $(i=1,2)$同样也是待求参数。

因此，在式(8)、(9)中，有 $a_i\sim f_i$ 共 12 个待求参数。求解出这 12 个待定参数，则可以得到系统的状态向量，进一步可得到颤振导数。

1.2 颤振分析

根据 Scanlan 颤振理论，对于式(1)、(2)中的气动力模型，颤振临界风速的计算可以通过同时求解以下 2 个方程的共同解得到

$$s_1X^4+s_2X^3+s_3X^2+s_4X+\gamma^2=0 \tag{10}$$

$$t_1X^4+t_2X^3+t_3X^2+t_4X=0 \tag{11}$$

上面两式中 $X=\omega/\omega_h$，因为竖弯振动圆频率 ω_h 为已知常量，因此 X 可以认为只与振动圆频率 ω 有关，而 ω 又与风速 U 有关。式中其他符号的定义如下

$$\gamma=\frac{\omega_\alpha}{\omega_h}$$

$$s_1=1+\frac{\rho B^2H_4^*}{m}+\frac{\rho B^4A_3^*}{I}+\frac{\rho B^2}{m}\frac{\rho B^4}{I}\cdot(-A_2^*H_1^*+A_1^*H_2^*+A_3^*H_4^*-A_4^*H_3^*)$$

$$s_2=2\zeta_\alpha\gamma\frac{\rho B^2}{m}H_1^*+2\zeta_h\frac{\rho B^4}{I}A_2^*$$

$$s_3=-\gamma^2-\gamma^2\frac{\rho B^2}{m}H_4^*-1-\frac{\rho B^4}{I}A_3^*-4\zeta_h\zeta_\alpha\gamma$$

$$s_4=0$$

$$t_1=\frac{\rho B^2}{m}H_1^*+\frac{\rho B^4}{I}A_2^*+\frac{\rho B^2}{m}\frac{\rho B^4}{I}\cdot(A_3^*H_1^*-A_1^*H_3^*+A_2^*H_4^*-A_4^*H_2^*)$$

$$t_2=-2\zeta_\alpha\gamma-2\zeta_\alpha\gamma\frac{\rho B^2}{m}H_4^*-2\zeta_h-2\zeta_h\frac{\rho B^4}{I}A_3^*$$

$$t_3=-\gamma^2\frac{\rho B^2}{m}H_1^*-\frac{\rho B^4}{I}A_2^*$$

$$t_4=2\zeta_h\gamma^2+2\zeta_\alpha\gamma$$

上述式中的符号与前面交代的符号意义相同，不再赘述。上式中独立的待求量为颤振临界风速 U_{cr}和颤振临界频率与竖弯频率比值 X_{cr}。通常，强非线性方程很难直接求解，因此常采用搜索方法解决问题。在求解颤振临界风速时，研究者常使用追赶法作为有效的求解方法[25]。其核心思想是将联立方程求解问题转化为优化问题，并通过不断迭代逼近方程的解。追赶法通过更新递推式中的参数，逐步逼近满足方程条件的解。这种方法能够有效处理非线性方程，并在颤振分析等领域得到广泛应用。

2 近似贝叶斯方法

2.1 贝叶斯推断

对于测量数据集 $y\in D$，根据贝叶斯公式，模型参数 $\theta\in\Theta$ 的后验概率分布可由下式求得

$$p(\theta|D)=\frac{p(D|\theta)p(\theta)}{p(D)} \tag{12}$$

式中：$p(D|\theta)$为似然函数，即测量数据集 $y\in D$ 在参数 $\theta\in\Theta$ 条件下的概率分布；$p(\theta)$为参数的先验分布，是由经验或人为假设确定；$p(D)$为参数集的全概率分布。

$p(D)$由下式确定

$$p(D)=\int_\theta p(D|\theta)p(\theta)\mathrm{d}\theta \tag{13}$$

在这里，式(13)的多维积分是一个常数，故式(12)可写为

$$p(\theta|D)=cp(D|\theta)p(\theta) \tag{14}$$

式中：$c=1/p(D)$为一个归一化的常数。

从上式可以看出，贝叶斯方法是根据当前样本 $y\in D$ 及其条件概率 $p(D|\theta)$来修正先验概率 $p(\theta)$，最终得到后验概率分布 $p(\theta|D)$。有时似然函数的构建比较困难，近似贝叶斯方法使用误差统计函数近似代替似然函数，可规避复杂似然函数的计算。

2.2 近似贝叶斯计算

在近似贝叶斯方法中，采用误差统计函数替代似然函数，对于从先验 $p(\theta)$ 中抽取的样本，计算仿真样本并与实测值进行对比，根据两者接近程度接受或拒绝该样本，直到获得足够量的样本，从而估计出 $p(\theta|D)$。因此需要定义接受和拒绝的条件。

设在给定度量 β，容差 ε 时观测数据的近似邻域 $N_\varepsilon(z)$ 定义为

$$N_\varepsilon(z)=\{x\in D:\beta(\boldsymbol{\eta}(x),\boldsymbol{\eta}(z))\leqslant\varepsilon\} \tag{15}$$

式中：$x\in D$ 为从分布 $p(\cdot|\theta)$ 中抽取的样本；$\boldsymbol{\eta}(\cdot)$ 为汇总统计信息用于比较 x 和 y 靠近程度的低维向量。

在此基础上近似的后验概率分布可表示为

$$p_\varepsilon(\theta,x|z)\propto P(x\in N_\varepsilon(z)|x)p(x|\theta)p(\theta) \tag{16}$$

式中：$P(x\in N_\varepsilon(z)|x)=\mathrm{II}_{N_\varepsilon(z)}(x)$。

其具体表达式为

$$P(x\in N_\varepsilon(z)|x)=\mathrm{II}_{N_\varepsilon(z)}(x)=\begin{cases}1 & \beta(\boldsymbol{\eta}(x),\boldsymbol{\eta}(z))\leqslant\varepsilon\\ 0 & \beta(\boldsymbol{\eta}(x),\boldsymbol{\eta}(z))>\varepsilon\end{cases} \tag{17}$$

上式即为样本的接受或拒绝规则。

令人感兴趣的通常是边缘后验分布，即

$$p_\varepsilon(\theta|z)\propto p(\theta)\int_D p(x|\theta)\,\mathrm{II}_{N_\varepsilon(z)}(x)\mathrm{d}x=P(x\in N_\varepsilon(z)|\theta)p(\theta) \tag{18}$$

从上述公式可以看出，度量 β 和容差 ε 的选择十分重要。当 $\varepsilon\to+\infty$ 时，得到的 $p(\theta|D)$ 就是 $p(\theta)$。当 $\varepsilon\to0$ 时，得到的 $p(\theta|D)$ 就趋于精确，相应的计算耗时就会更长。ABC 方法得到的后验分布质量取决于度量 β 和容差 ε 的选择[26]。

近似贝叶斯方法通常与 MCMC 方法相结合，本文采用一种基于子集模拟(Subset Simulation)的近似贝叶斯方法(ABC-SubSim)[26-27]。

本文采用基于最小二乘原理构建的度量函数，即 β 为

$$\beta(\boldsymbol{\eta}(y^h,y^a),\boldsymbol{\eta}(y_i^h,y_i^a))=\sum_{i=1}^{N_t}\left(w_h\sum_{j=1}^{N}[y_i^h(t_j)-y_i^h(t_j)]^2+w_a\sum_{j=1}^{N}[y_i^a(t_j)-y_i^a(t_j)]^2\right) \tag{19}$$

式中：y^h、y^a 分别为主梁竖弯和扭转的计算值；y^h、y^a 分别为主梁竖弯和扭转的实测值；w_h、w_a 分别为主梁竖弯和扭转的残差加权值。

ABC-SubSim 的求解步骤如下：

(1)使用 MCMC 方法从 $p(\theta)$ 中采样，得到 N 个样本；

(2)计算这 N 个样本的度量并按度量大小进行编号；

(3)从第(2)步中选择 NP_0 个样本作为种子，$P_0\in[0,1]$，生成长度为 $1/P_0$ 的马尔科夫链；

(4)对新生成的样本进行排序；

(5)检查停止准则，若满足则结束，若不满足，重复步骤(2)～(5)。

3 数值模拟

为了验证本文所提出方法的有效性，首先采用理想平板对该方法进行验证。理想平板的宽度为 0.45 m，单位长度质量为 11.25 kg·m^{-1}，单位长度质量惯矩为 0.282 8 kg·m^2·m^{-1}，板的竖弯和扭转频率分别为 1.927 4、3.023 9 Hz，竖弯和扭转阻尼比均为 0.5%。理想平板的自激力采用 Theodorson 理论解[28]，由该理论解生成自由振动曲线并添加 2%的白噪声。采用近似贝叶斯方法识别平板的颤振导数并量化其不确定性，如图 1 所示，并以误差棒的形式给出每个导数的不确定性。

从识别结果可以看出：近似贝叶斯方法得到的参数最优值与理论解能够很好地吻合；对于所有导数来说，在高风速情况下其不确定性相对较大。与其他导数相比，H_4^* 和 A_4^* 的不确定性在高风速情况下更大，但其均值仍能与理论值较好地吻合。理想平板仿真试验验证了基于子集模拟近似贝叶斯方法在颤振导数识别方面的有效性。

4 桥梁断面颤振导数识别

为了进一步验证本方法在桥梁断面颤振导数识别中的有效性，对实桥断面的颤振导数进行识别。自由振动风洞试验在意大利的 CRIACIV 实验室开展，断面的形状如图 2 所示，断面具体数据参见文献[13]，在试验中每个工况重复 10 次。

基于近似贝叶斯方法的实桥断面颤振导数识别结果如图 3 所示，并将识别结果与 MULS 法的结果进行对比[13]。从识别结果中可以看出，近似贝叶斯方法识别的均值能与 MULS 法较好的吻合，进一步验证了该方法在实桥断面颤振导数识别中的有效性。从识别结果中可以看出，对于全部 8 个导数来说，随着风速的增大，识别结果的离散度增大，即其不确定性增大。在所有导数中 A_2^*、A_3^*、H_3^* 的不确定性相对较小，仅在高风速表现出稍大的不确定性，但与其余导数的低风速相比，其不确定性仍较小。

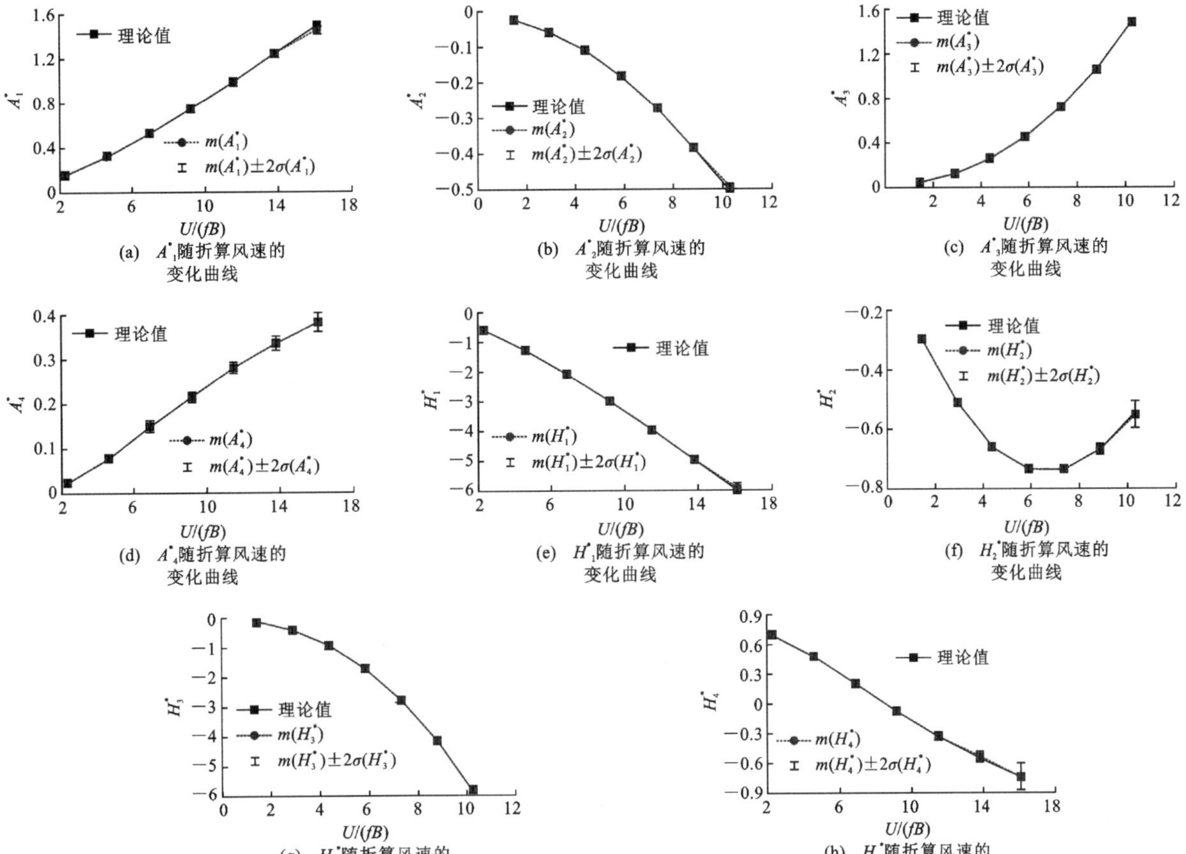

(a) A_1^*随折算风速的变化曲线

(b) A_2^*随折算风速的变化曲线

(c) A_3^*随折算风速的变化曲线

(d) A_4^*随折算风速的变化曲线

(e) H_1^*随折算风速的变化曲线

(f) H_2^*随折算风速的变化曲线

(g) H_3^*随折算风速的变化曲线

(h) H_4^*随折算风速的变化曲线

图 1　采用近似贝叶斯方法识别的理想平板的颤振导数

Fig. 1　Flutter Derivatives of an Ideal Plate Identified by Approximate Bayesian Method

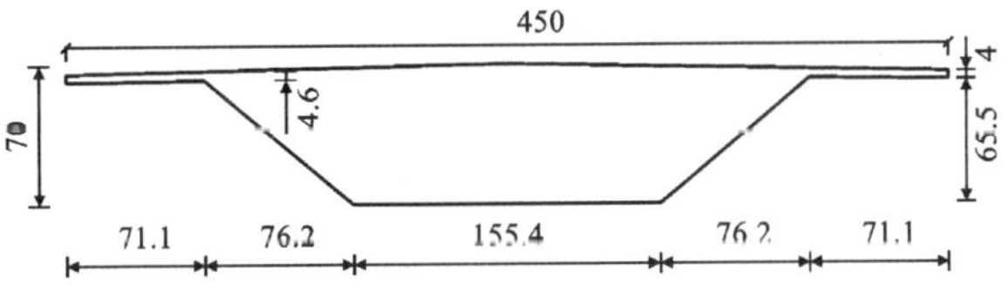

图 2　自由振动试验的模型断面(单位:mm)

Fig. 2　Model Bridge Section for Free Vibration Test (Unit: mm)

A_4^*、H_1^*、H_2^*、H_4^* 仅在较低风速下具有较小的不确定性,当风速开始增大时,其不确定性就开始增加,在中风速下就开始展现出较大的不确定性。对于 A_1^* 来说,其不确定性随着风速的增加开始逐步增大。从上述结果可以观察到,所有颤振导数中在低风速下呈现出较小的不确定性,而大多数导数在稍大的风速下都具有较大的不确定性。这种现象主要是由于在高风速下气动阻尼的影响,导致模型振幅迅速衰减,进而使得识别结果呈现出较大的离散性。因此,对颤振导数的不确定性进行量化是非常必要的。

基于近似贝叶斯方法,获得了颤振导数的后验概率分布,并以此为基础对该桥梁进行颤振分析。采用追赶法计算颤振临界风速,并与文献[13]的结果进行了对比,文献[13]中折算临界风速为 9.42,而本方法得到的折算临界风速的平均值为 9.65,两者相差 2.4%,结果显示了较好的一致性,从而验证了本方法的准确性。为了更好地理解颤振临界风速的不确定性,笔者绘制了折算颤振临界风速的累积分布函数(CDF)和分布直方图,如图 4 所示,从图 4 可以观察到,颤振导数的不确定性导致了临界风速的不确定性存在。因此,对颤振导数的不确定性进行量化显得十分必要。

5　结　语

(1)本文针对颤振导数识别和颤振分析过程中普遍存在的不确定性问题,采用近似贝叶斯方法进行颤振导数的不确定性量化研究,并在此基础上进行了颤振分析,得到了颤振临界风速的概率分布,为桥梁抗风设计提供更可靠的依据。近似贝叶斯方法能够避免对复杂似然函数的计算,从而扩大了贝叶

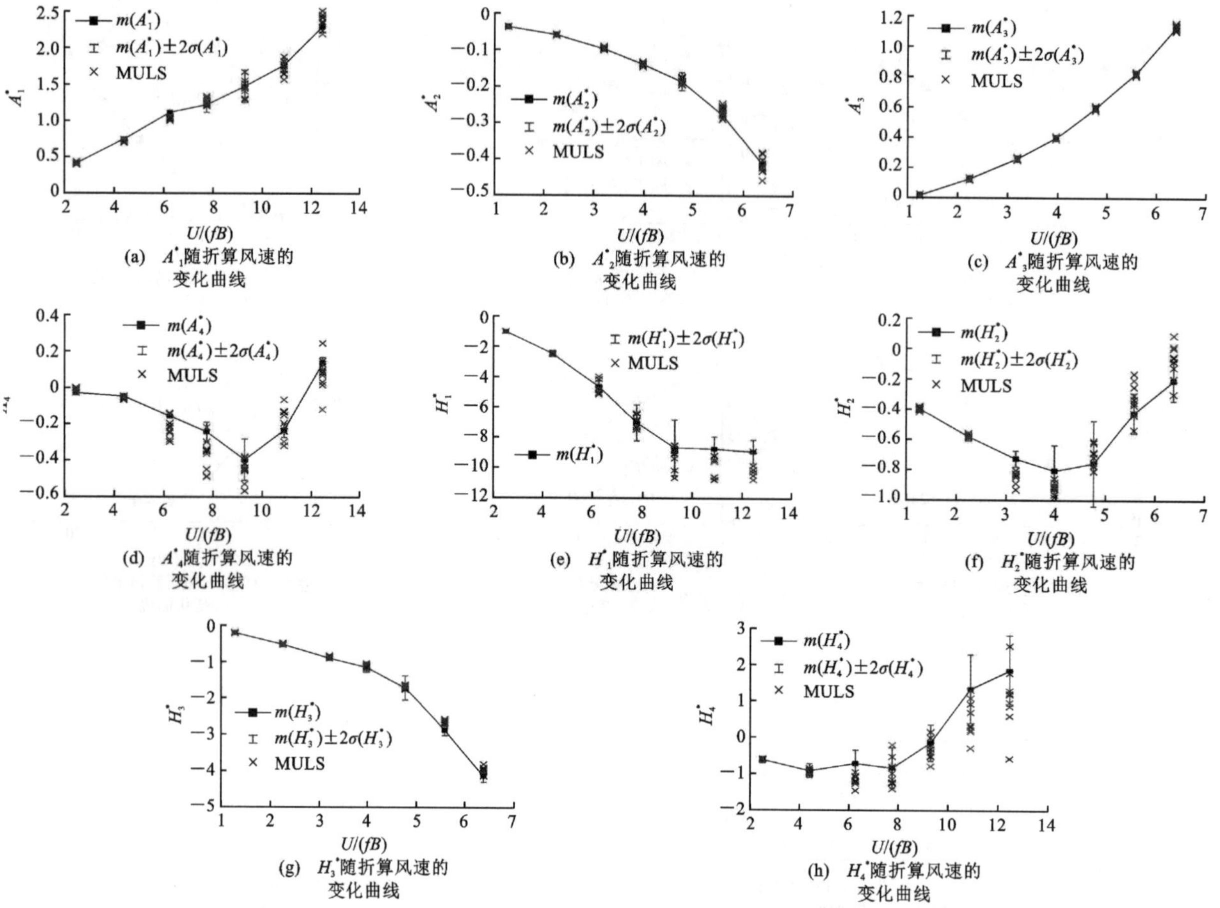

(a) A_1^*随折算风速的变化曲线

(b) A_2^*随折算风速的变化曲线

(c) A_3^*随折算风速的变化曲线

(d) A_4^*随折算风速的变化曲线

(e) H_1^*随折算风速的变化曲线

(f) H_2^*随折算风速的变化曲线

(g) H_3^*随折算风速的变化曲线

(h) H_4^*随折算风速的变化曲线

图 3　采用近似贝叶斯方法识别的实桥断面颤振导数

Fig. 3　Flutter Derivative of Real Bridge Cross-section Identified by Approximate Bayesian Method

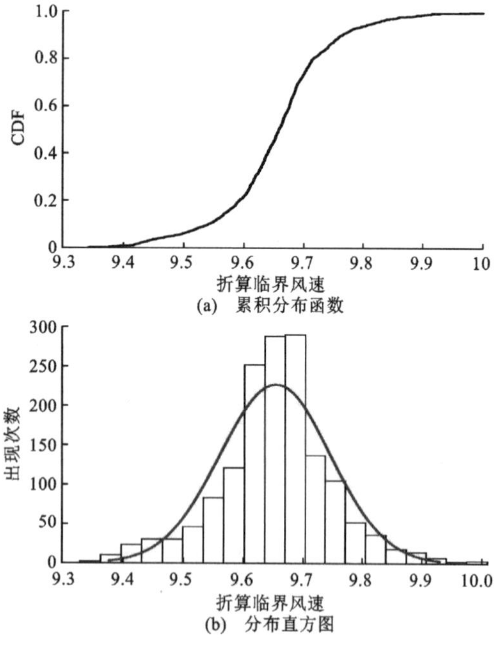

(a) 累积分布函数

(b) 分布直方图

图 4　折算颤振临界风速的累积分布函数和分布直方图

Fig. 4　Cumulative Distribution Function and Distribution Histogram of Non-dimensional Flutter Critical Wind Speed

斯方法的应用范围。本文将子集模拟技术与近似贝叶斯方法相结合，提高了采样效率，加快了近似贝叶斯的计算速度。

(2)通过理想平板仿真试验和实桥断面颤振导数识别的验证，证明了基于近似贝叶斯方法识别颤振导数的有效性，并量化了颤振导数的不确定性。研究结果显示，低风速下所有导数的不确定性相对较小，而在高风速下，所有导数的不确定性都较大。这些不确定性对颤振分析结果产生了显著影响。因此，本文的研究结果进一步证实了在颤振导数识别和颤振分析中考虑不确定性的必要性。

(3)然而，工程实际中存在多种不确定因素，准确地区分和建模各种不确定性因素是比较困难的。本文在颤振分析中仅考虑了由颤振导数引起的不确定性，而未考虑颤振分析模型的不确定性等其他因素。因此，未来的研究可以综合考虑颤振分析模型等因素引起的不确定性，以更全面地解决该问题。

(4)本研究为颤振导数识别和颤振分析领域做出了有益的探索，并为工程实践提供了有价值的参考。随着研究的进一步深入，相信能够推动颤振导数识别和颤振分析方法的不断发展，为桥梁抗风设计提供更加可靠和有效的解决方案。

参考文献：

References：

[1] SCANLAN R H，TOMKO J J. Airfoil and bridge deck flutter derivatives [J]. Journal of the Engineering Mechanics Division，1971，97 (6)：1717-1737.

[2] SARKAR P P，JONES N P，SCANLAN R H. System identification for estimation of flutter derivatives [J]. Journal of Wind Engineering & Industrial Aerodynamics，1992，42 (1-3)：1243-1254.

[3] GU M，ZHANG R，XIANG H. Identification of flutter derivatives of bridge decks [J]. Journal of Wind Engineering & Industrial Aerodynamics，2000，84 (2)：151-162.

[4] 丁泉顺，陈艾荣，项海帆. 桥梁断面气动导数识别的修正最小二乘法[J]. 同济大学学报(自然科学版)，2001，29(1)：25-29.
DING Quan-shun，CHEN Ai-rong，XIANG Hai-fan. Modified least-square method for identification of bridge deck aerodynamic derivatives [J]. Journal of Tongji University (Natural Science Edition)，2001，29 (1)：25-29.

[5] LI Y，LIAO H，QIANG S. Weighting ensemble least-square method for flutter derivatives of bridge decks [J]. Journal of Wind Engineering & Industrial Aerodynamics，2003，91 (6)：713-721.

[6] BARTOLI G，CONTRI S，MANNINI C，et al. Toward an improvement in the identification of bridge deck flutter derivatives [J]. Journal of Engineering Mechanics，2009，135 (8)：771-785.

[7] 李加武，张 斐，吴 拓. 桥梁断面颤振导数识别的加权最小二乘法[J]. 振动工程学报，2017，30(6)：993-1000.
LI Jia-wu，ZHANG Fei，WU Tuo. Weighted least square method for identification of flutter derivatives of bridge section [J]. Journal of Vibration Engineering，2017，30 (6)：993-1000.

[8] XU F Y，CHEN X Z，CAI C S，et al. Determination of 18 flutter derivatives of bridge decks by an improved stochastic search algorithm [J]. Journal of Bridge Engineering，2012，17 (4)：576-588.

[9] ABBAS T，MORGENTHAL G. Framework for sensitivity and uncertainty quantification in the flutter assessment of bridges [J]. Probabilistic Engineering Mechanics，2016，43：91-105.

[10] FANG G，CAO J，YANG Y，et al. Experimental uncertainty quantification of flutter derivatives for a pk section girder and its application on probabilistic flutter analysis [J]. Journal of Bridge Engineering，2020，25 (7)：1-17.

[11] FENG Z，LIN Y. Flutter derivatives identification and uncertainty quantification for bridge decks based on the artificial bee colony algorithm and bootstrap technique [J]. Applied Sciences，2021，11 (23)：11376.

[12] 颜王吉，曹诗泽，任伟新. 结构系统识别不确定性分析的Bayes方法及其进展[J]. 应用数学和力学，2017，38(1)：44-59.
YAN Wang-ji，CAO Shi-ze，REN Wei-xin. Uncertainty quantification for system identification utilizing the Bayesian theory and its recent advances [J]. Applied Mathematics and Mechanics，2017，38 (1)：44-59.

[13] MANNINI C，BARTOLI G. Aerodynamic uncertainty propagation in bridge flutter analysis [J]. Structural Safety，2015，52：29-39.

[14] 茆诗松. 贝叶斯统计[M]. 北京：中国统计出版社，1999.
MAO Shi-song. Bayesian statistics [M]. Beijing：China Statistics Press，1999.

[15] BECK J L. Bayesian system identification based on probability logic [J]. Structural Control and Health Monitoring，2010，17 (7)：825-847.

[16] BECK J L，KATAFYGIOTIS L S. Updating models and their uncertainties. I：Bayesian statistical framework [J]. Journal of Engineering Mechanics，1998，124 (4)：455-461.

[17] KATAFYGIOTIS L S，BECK J L. Updating models and their uncertainties. II：Model identifiability [J]. Journal of Engineering Mechanics，1998，124 (4)：463-467.

[18] 韦来生. 贝叶斯统计[M]. 北京：高等教育出版社，2016.
WEI Lai-sheng. Bayesian statistics [M]. Beijing：Higher Education Press，2016.

[19] TURNER B M，VAN ZANDT T. A tutorial on approximate Bayesian computation [J]. Journal of Mathematical Psychology，2012，56 (2)：69-85.

[20] MARIN J M，PUDLO P，ROBERT C P，et al. Approximate Bayesian computational methods [J]. Statistics and Computing，2012，22 (6)：1167-1180.

[21] WARNE D J，BAKER R E，SIMPSON M J. Multilevel rejection sampling for approximate Bayesian computation [J]. Computational Statistics & Data Analysis，2018，124：71-86.

[22] FIGUEIREDO E，RADU L，WORDEN K，et al. A Bayesian approach based on a Markov-chain Monte Carlo method for damage detection under unknown sources of variability [J]. Engineering Structures，2014，80：1-10.

[23] BEAUMONT M A. Approximate Bayesian computation [J]. Annual Review of Statistics and Its Application，2019，6：379-403.

[24] 林 阳，封周权，华旭刚，等. 基于自由振动响应识别桥梁断面颤振导数的人工蜂群算法[J]. 工程力学，2020，37(2)：192-200.
LIN Yang，FENG Zhou-quan，HUA Xu-gang，et al. Artificial be colony algorithm for flutter derivatives identification of bridge decks using free vibration records [J]. Engineering Mechanics，2020，37 (2)：192-200.

[25] 许福友，陈艾荣，王达磊. 搜索平板颤振临界风速的追赶法[J]. 工程力学，2005，22(5)：48-53.
XU Fu-you，CHEN Ai-rong，WANG Da-lei. Pursuit method for searching the critical flutter wind velocity of flat plate [J]. Engineering Mechanics，2005，22 (5)：48-53.

[26] CHIACHIO M，BECK J L，CHIACHIO J，et al. Approximate Bayesian computation by subset simulation [J]. SIAM Journal on Scientific Computing，2014，36 (3)：A1339-A1358.

[27] FENG Z，LIN Y，WANG W，et al. Probabilistic updating of structural models for damage assessment using approximate Bayesian computation [J]. Sensors，2020，20 (11)：3197.

[28] THEODORSON T. General theory of aerodynamic instability and mechanism of flutter [R]. Washington DC：NACA，1935.

振动控制理论与方法

Advanced Impact Force Model for Low-Speed Pounding between Viscoelastic Materials and Steel

Wenxi Wang[1]; Xugang Hua, Ph.D., M.ASCE[2]; Xiuyong Wang, Ph.D.[3]; Zhengqing Chen, Ph.D., M.ASCE[4]; and Gangbing Song, Ph.D.[5]

Abstract: Viscoelastic (VE) materials, which have a lower value for their coefficient of restitution, are widely used in vibration control and reduction of structural pounding. Even if an impact occurs at a low relative velocity, elastic aftereffect phenomenon may take place on the pounding surface of viscoelastic materials. In this situation, the surface deformation of viscoelastic materials is still unrecovered right before the separation of the colliding bodies, which can be regarded as the remaining surface deformation. Most existing viscoelastic impact force models do not consider the remaining surface deformation resulting from the elastic aftereffect phenomenon. In this paper, an advanced nonlinear viscoelastic impact force model considering the elastic aftereffect phenomenon is proposed to accurately model low-speed pounding between viscoelastic materials and steel. Energy dissipations from the material damping and remaining surface deformation are considered in this model. Pounding experiments are conducted to determine the unknown model parameters from the measured impact force and indentation. In addition, the stability verification of the proposed model for various precoefficients of restitution is presented. Comparative studies between the proposed model and other viscoelastic impact force models demonstrate that the proposed model performs much better than other viscoelastic impact force models in the impact simulation of a steel striker pounding on a viscoelastic materials layer. In addition, the proposed model can provide a reasonable explanation for the remaining surface deformation observed in the pounding experiments. **DOI: 10.1061/(ASCE)EM.1943-7889.0001372.**

Author keywords: Impact force model; Viscoelastic Materials; Elastic aftereffect; Pounding experiment; Impact force prediction.

Introduction

Impact-related topics, such as impact-induced vibration (Wagg and Bishop 2002; Wagg 2007; Bayly and Virgin 1993; Dimentberg and Iourtchenko 1999; Iourtchenko and Song 2006; Song and Iourtchenko 2006), pounding between adjacent buildings (Polycarpou et al. 2013; Maison and Kasai 1990; Kim and Shinozuka 2003; Malhotra 1998), and impact dampers (Al-Shudeifat et al. 2013; Xu et al. 2005; Lu et al. 2010, 2012; Zhang et al. 2013), have been studied. A key issue for impact-induced problems is the mathematical modeling of the impact force, which is highly nonlinear. There are two main methods to simulate impact-induced problems (Jankowski 2005). The first one is the stereomechanical method, which uses conservative laws of energy and momentum to update the postvelocity of contacting bodies (Masri and Caughey 1966; Cross and Jones 1993). However, it ignores the local stress and surface deformation during impacts. Furthermore, numerical simulations of structural pounding by this approach may result in an unrealistic response by neglecting the pounding duration.

The second method, based on the impact force model, uses a linear or nonlinear spring-damper element to directly model the pounding force. This method has been used to analyze structural pounding for insufficiently separated building structures and bridge girders under severe earthquakes (Muthukumar and DesRoches 2006; Jankowski et al. 1998; Anagnostopoulos 1988; Anagnostopoulos and Spiliopoulos 1992; Jankowski 2010). Impact force models include two important categories: elastic force models and viscoelastic force models (Machado et al. 2012). Elastic force models only contain an elastic term. If energy dissipation is considered, the coefficient of restitution shall be used to update the postimpact velocities (Dimentberg and Iourtchenko 2004). On the other hand, viscoelastic force models include an elastic term and a viscous term that models energy dissipation during impacts. To simulate impacts involving high damping materials, such as viscoelastic materials, viscoelastic impact force models are more straightforward and convenient.

Viscoelastic (VE) materials, which have a lower value coefficient of restitution and stiffness, can increase the energy dissipation and reduce the pounding force during an impact. Some novel applications involving pounding of viscoelastic materials have been studied. Song et al. (2016) proposed a pounding tuned mass damper (PTMDs) to mitigate structural vibration, as shown in Fig. 1(a). Viscoelastic materials are used to cover the pounding boundary of the PTMD to enhance the energy dissipation during pounding (Song et al. 2016; Li et al. 2015, 2013; Xue et al. 2016b). When the PTMD is tuned to the primary structure, the vibration energy of the primary structure will be transferred to the PTMD and is dissipated by pounding. Moreover, a viscoelastic materials bumper [Fig. 1(b)] has been applied in reduction of structural

[1]Ph.D. Candidate, Key Laboratory for Bridge and Wind Engineering of Hunan Province, College of Civil Engineering, Hunan Univ., Changsha, Hunan 410082, China. E-mail: wxwang@hnu.edu.cn

[2]Professor, Key Laboratory for Bridge and Wind Engineering of Hunan Province, College of Civil Engineering, Hunan Univ., Changsha, Hunan 410082, China. E-mail: cexghua@hnu.edu.cn

[3]Professor, College of Civil Engineering, Hunan Univ. of Science and Technology, Xiangtan, Hunan 411201, China. E-mail: cexywang@hnust.edu.cn

[4]Professor, Key Laboratory for Bridge and Wind Engineering of Hunan Province, College of Civil Engineering, Hunan Univ., Changsha, Hunan 410082, China. E-mail: zqchen@hnu.edu.cn

[5]Professor, Dept. of Mechanical Engineering, Univ. of Houston, Houston, TX 77204 (corresponding author). E-mail: gsong@uh.edu

Note. This manuscript was submitted on February 6, 2017; approved on June 12, 2017; published online on October 5, 2017. Discussion period open until March 5, 2018; separate discussions must be submitted for individual papers. This paper is part of the ***Journal of Engineering Mechanics***, © ASCE, ISSN 0733-9399.

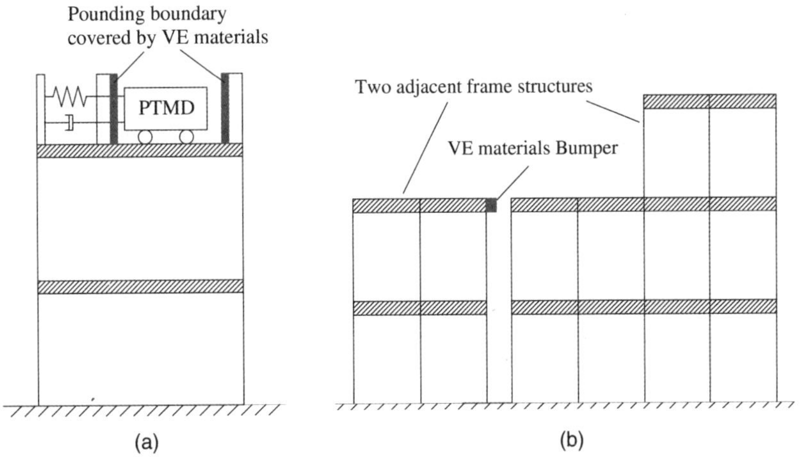

Fig. 1. Viscoelastic materials involved poundings in civil engineering: (a) PTMD with viscoelastic materials pounding boundary for vibration control; (b) viscoelastic materials bumper for mitigating structural pounding

pounding between adjacent buildings under a severe earthquake (Polycarpou et al. 2013; Anagnostopoulos 1988). An impact force model to accurately model the pounding involving viscoelastic materials should be developed.

Differing from the pounding between hard materials, a relaxation phenomenon (Xue et al. 2016a), called the elastic aftereffect, obviously occurs in pounding between a soft material, such as viscoelastic materials, and a hard material, such as steel. The elastic aftereffect phenomenon in the impact can be simply described as a nonzero indentation at the separation of two colliding bodies caused by the deformation of the contacting bodies always lagging behind the changes of impact force. The unrecovered deformation on the surface of the viscoelastic materials layer at the end of the pounding is called the remaining surface deformation, and the deformed surface of the viscoelastic materials layer will rapidly return to zero before the next impact comes. However, most existing impact force models cannot consider the remaining surface deformation resulting from the elastic aftereffect phenomenon, which may lead to error in the simulation of pounding and fail to provide a reasonable explanation of experimental observations of the elastic aftereffect phenomenon.

In this paper, the remaining surface deformation resulting from the elastic aftereffect phenomenon was observed at the end of poundings between a steel striker and a viscoelastic materials layer. A new impact force model considering the remaining surface deformation is proposed to simulate low-speed pounding between viscoelastic materials and steel. The main application area of the proposed model is the simulation of impact dampers with a pounding boundary covered by viscoelastic materials and the structural pounding between adjacent buildings involving a bump of viscoelastic materials. Numerical and experimental verifications are performed to show the effectiveness of the proposed impact force model. For the purpose of illustrating the enhanced performance of the proposed model in impact simulations, a comparative study of the proposed model with other viscoelastic impact force models has been carried out. Finally, experimental results are shown to validate the numerical prediction of the impact force made using the proposed model.

Existing Viscoelastic Impact Force Models

Consider the viscoelastic impact problem of two bodies with mass m_1 and m_2 approaching each other with respective velocities of v_{10} and v_{20}, as shown in Fig. 2. Generally, this can be represented by a single-degree-of-freedom (SDOF) system with an equivalent mass $m_e = m_1 m_2/(m_1 + m_2)$ and a relative approaching velocity of $\dot{\delta}_0 = v_{10} - v_{20} (v_{10} > v_{20})$. The coefficient of restitution e is expressed

$$e = -\frac{v_{11} - v_{21}}{v_{10} - v_{20}} \tag{1}$$

where v_{11} and v_{21} = separated velocities of the two bodies.

Because of the complexity of impact, which may involve local cracking and permanent deformation of the surface, many impact force models have been proposed to apply to numerical simulations. From experimental studies, some of them have been verified to have a good performance in pounding simulations.

Linear Viscoelastic Model

Investigation of impact force models started almost a century ago. First, an elastic force model, called the Hertz elastic model (Goldsmith 1960), was introduced to describe the contacting force of rigid impacts. A crucial drawback of this model was that it cannot simulate the energy dissipation during pounding. Therefore, a linear viscoelastic (LV) model (Kelvin-Voigt model) has been proposed to revise the Hertz elastic model by adding a damping term (Anagnostopoulos 1988, 2004; Jankowski et al. 1998; Goldsmith 1960)

$$F = k\delta + c\dot{\delta} \tag{2}$$

where k = impact stiffness; δ = indentation (or relative penetration) of colliding bodies; $\dot{\delta}$ = relative velocity; and c = impact damping, which can be obtained by

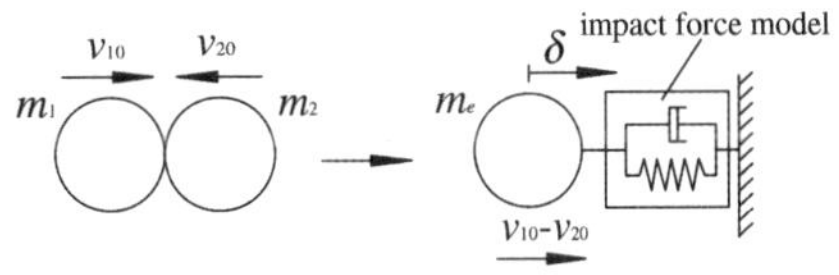

Fig. 2. Equivalent model of two colliding bodies during pounding

$$c = 2\zeta\sqrt{k\frac{m_1 m_2}{m_1 + m_2}} \quad (3)$$

where m_1 and m_2 = mass of the colliding bodies; and ζ = pounding damping, which is related to the coefficient of restitution e by the following formula (Anagnostopoulos 2004):

$$\zeta = \frac{-\ln e}{\sqrt{\pi^2 + (\ln e)^2}} \quad (4)$$

Modified Linear Viscoelastic Model

The linear viscoelastic model may produce a negative force at the end of impact, which is not consistent with experimental observations (Jankowski 2005; Goldsmith 1960). Therefore, a modified linear viscoelastic (MLV) model (Mahmoud and Jankowski 2011), which divides the pounding into two individual periods (approaching period and restitution period) and assumes that energy loss only happens in approaching period, is given by

$$F = \begin{cases} k\delta + c\dot{\delta} & \dot{\delta} \geq 0 \\ k\delta & \dot{\delta} < 0 \end{cases} \quad (5)$$

where c has the same form as in Eq. (3), and the pounding damping ζ can be calculated in a more precise relationship (Mahmoud and Jankowski 2011) between indentation δ and relative velocity $\dot{\delta}$ by

$$\zeta = \frac{1 - e^2}{e[e(\pi - 2) + 2]} \quad (6)$$

Nonlinear Viscoelastic Model

Given its simplicity, the modified linear viscoelastic model is easy to apply in numerical simulations. However, at the beginning of impact ($\delta = 0$), a sudden rise of impact force may occur as a result of the nonzero relative velocity $\dot{\delta}$. In order to overcome disadvantages of the modified linear viscoelastic model, a nonlinear viscoelastic (NV) model was introduced by Jankowski (2005, 2006), and it is expressed

$$F = \begin{cases} k\delta^{3/2} + c\dot{\delta} & \dot{\delta} \geq 0 \\ k\delta^{3/2} & \dot{\delta} < 0 \end{cases} \quad (7)$$

where c has the form

$$c = 2\zeta\sqrt{k\sqrt{\delta}\frac{m_1 m_2}{m_1 + m_2}} \quad (8)$$

where the pounding damping ζ can be calculated by

$$\zeta = \frac{9\sqrt{5}}{2}\frac{1 - e^2}{e[e(9\pi - 16) + 16]} \quad (9)$$

Hunt-Crossley Model

Given the high nonlinearity of impact problems, a fixed form of the impact force is insufficient to solve various kinds of pounding problems. Therefore, a more flexible impact force model with a variable nonlinear coefficient, called the Hunt-Crossley (HC) model, has been proposed to handle the impact process (Hunt and Crossley 1975; Quinn 2005). The energy loss of the Hunt-Crossley model accounts for the damping term with a succinct form of indentation and relative velocity. The general form of the Hunt-Crossley model is

$$F = k\delta^n + \zeta\delta^n\dot{\delta} \quad (10)$$

Different expressions of ζ have been derived by assuming different approximated relationships between the indentation and relative velocity. When the nonlinear coefficient $n = 3/2$, Lankarani and Nikravesh (1990) and Ye et al. (2009) proposed different solutions of ζ and they are, respectively, as follows:

$$\zeta = \frac{3}{4}\frac{k(1 - e^2)}{\dot{\delta}_0} \quad (11)$$

$$\zeta = \frac{8}{5}\frac{k(1 - e)}{e\dot{\delta}_0} \quad (12)$$

where $\dot{\delta}_0$ = relative velocity at the beginning of the impact.

However, Eqs. (11) and (12) have a poor performance for a small value of coefficient of restitution e. Assisted by deriving the precise relationship between indentation and relative velocity, Khatiwada et al. (2014) found an implicit expression of the pounding damping ζ with high accuracy, which is given by

$$(1 + e) = \frac{k}{\zeta\dot{\delta}_0}\ln\left|\frac{\frac{k}{\zeta\dot{\delta}_0} + 1}{\frac{k}{\zeta\dot{\delta}_0} - e}\right| \quad (13)$$

Existing viscoelastic impact force models mentioned earlier describe the impact at the beginning of the impact period, when the indentation δ is zero, and at the end of the impact period, when the indentation δ returns to zero again. However, the indentation δ right before the separation of the colliding bodies will not be zero in poundings between viscoelastic materials and steel because of the elastic aftereffect phenomenon. Therefore, the impact force model should be revised to consider the elastic aftereffect phenomenon.

Proposed Impact Force Model Considering Remaining Surface Deformation

Proposed Impact Force Model

Lankarani and Nikravesh (1994) proposed an impact force model to consider the plastic deformation for high-speed impacts. In their model, energy dissipation during the impact totally depends on the plastic deformation, and only an elastic force exists in the approaching period. Its expression is

$$F = \begin{cases} k\delta^n & \dot{\delta} \geq 0 \\ f_{\max}\left(\frac{\delta - \delta_p}{\delta_{\max} - \delta_p}\right)^n & \dot{\delta} < 0 \end{cases} \quad (14)$$

where $\delta_{\max}$ = maximum indentation during pounding; and δ_p = assumed plastic deformation.

The aforementioned model is shown to work well for poundings between hard–hard materials with low hysteretic damping. However, there are some disadvantages for the preceding force model to simulate the pounding between viscoelastic materials and steel. First of all, the damping effect of viscoelastic materials should exist in the approaching period, which means that the assumption of only an elastic force in the approaching period is unreasonable. Furthermore, for poundings between viscoelastic materials and steel happening with low relative velocity, the plastic deformation may not occur on the surfaces of the colliding bodies. In order to overcome the drawbacks in the aforementioned model and provide a reasonable explanation for the experimental

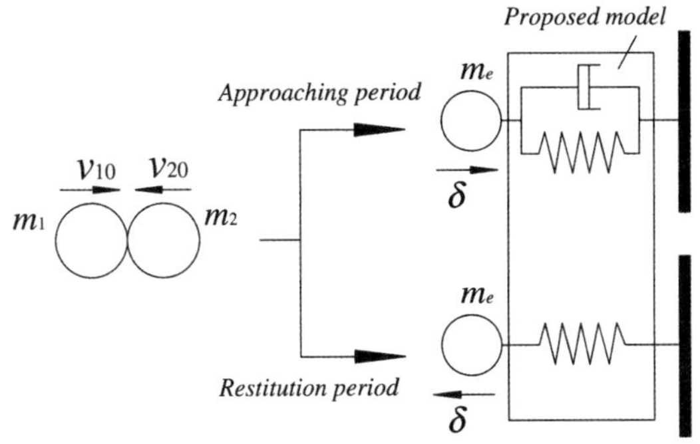

Fig. 3. Equivalent model of the proposed model for pounding between two bodies

observation of the elastic aftereffect phenomenon, a new impact force model is proposed by combining Eqs. (10) and (14).

Pounding experiments have proven that the major energy loss during the pounding took place in the approaching period (Goldsmith 1960). Therefore, the energy loss in the restitution period is ignored in the proposed model, as shown in Fig. 3. The remaining surface deformation δ_e occurring on the surface of the viscoelastic materials layer at the end of poundings is introduced in the proposed model to take the elastic aftereffect phenomenon into consideration. The impact force model proposed for low-speed pounding between viscoelastic materials and steel can be expressed

$$F=\begin{cases} k\delta^n+\zeta\delta^n\dot{\delta} & \delta_{\max}\geq\delta>0,\dot{\delta}\geq 0 \ \text{(approaching period)} \\ f_e\left(\dfrac{\delta-\delta_e}{\delta_{\max}-\delta_e}\right)^n & \delta_{\max}>\delta\geq\delta_e,\dot{\delta}<0 \ \text{(restitution period)} \\ 0 & \delta_e>\delta>0,\dot{\delta}<0 \end{cases} \tag{15}$$

where δ = indentation between the colliding bodies; $\dot{\delta}$ = relative velocity of colliding bodies; k = pounding stiffness; ζ = pounding damping ratio; $\delta_{\max}$ = maximum indentation in the process of pounding; and f_e = maximum elastic force during the pounding, which can be obtained by substituting $\delta=\delta_{\max}$ and $\dot{\delta}=0$ into the approaching period of Eq. (15). The expression of f_e is given by

$$f_e=k\delta_{\max}^n \tag{16}$$

To describe the relaxation behavior of the viscoelastic materials during poundings, the remaining surface deformation ratio e_1 can be defined by the following equation:

$$e_1=\frac{\delta_e}{\delta_{\max}} \tag{17}$$

The proposed model is used to simulate low-speed pounding involving viscoelastic materials. Consequently, the plastic deformation will be very limited during low-speed poundings, and the proposed model will not consider plastic deformation. Although Eq. (15) has a similar form to Eq. (14), the remaining surface deformation δ_e resulting from the elastic aftereffect phenomenon as well as the maximum indentation $\delta_{\max}$ in the proposed model are both elastic deformations.

Derivation of Pounding Damping ζ

The energy loss ΔE during the pounding, based on classical impact theory (Goldsmith 1960), is obtained from

$$\Delta E=\frac{1}{2}\frac{m_1m_2}{m_1+m_2}(1-e^2)\dot{\delta}_0^2 \tag{18}$$

where m_1 and m_2 = mass of two colliding bodies; e = coefficient of restitution and is determined by experiment; and $\dot{\delta}_0$ = relative velocity at the beginning of pounding.

The energy dissipation in the proposed pounding force model consists of two parts, namely the energy dissipated because of the damping force ΔE_1 and the energy dissipated because of the remaining surface deformation ΔE_2, and it is

$$\Delta E=\Delta E_1+\Delta E_2 \tag{19}$$

The energy dissipated because of the damping force ΔE_1 and the energy dissipated because of the remaining surface deformation ΔE_2 are provided by Eqs. (20) and (21), respectively

$$\Delta E_1=\int_{\delta=0,\dot{\delta}=\dot{\delta}_0}^{\delta=\delta_{\max},\dot{\delta}=0}\zeta\delta^n\dot{\delta}d\delta \tag{20}$$

$$\Delta E_2=\int_{\delta=0}^{\delta=\delta_{\max}}k\delta^n d\delta-\int_{\delta=e_1\delta_{\max},}^{\delta=\delta_{\max}}f_e\left(\frac{\delta-e_1\delta_{\max}}{\delta_{\max}-e_1\delta_{\max}}\right)^n d\delta \tag{21}$$

After applying integration by parts, Eq. (20) becomes

$$\Delta E_1=\int_{\delta=0,\dot{\delta}=\dot{\delta}_0}^{\delta=\delta_{\max},\dot{\delta}=0}\zeta\delta^n\dot{\delta}d\delta=-\zeta\int_{\delta=0,\dot{\delta}=\dot{\delta}_0}^{\delta=\delta_{\max},\dot{\delta}=0}\frac{\delta^{n+1}}{n+1}d\dot{\delta} \tag{22}$$

In order to solve Eq. (22), indentation δ shall be described in terms of the relative velocity $\dot{\delta}$. Considering the equivalent model shown in Fig. 2, the equation of motion of the equivalent system in the approaching period is

$$m_e\ddot{\delta}+\zeta\delta^n\dot{\delta}+k\delta^n=0 \tag{23}$$

where $m_e=m_1m_2/(m_1+m_2)$ = equivalent mass. With the help of $\ddot{\delta}=\dot{\delta}d\dot{\delta}/d\delta$ and integration of both sides, Eq. (23) can be transformed into

$$\int-\frac{m_e\dot{\delta}}{\zeta\dot{\delta}+k}d\dot{\delta}=\int\delta^n d\delta \tag{24}$$

Therefore, by solving Eq. (24), the relationship between indentation δ and relative velocity $\dot{\delta}$ has been given by Khatiwada et al. (2014)

$$\frac{\zeta}{m_e}\frac{\delta^{n+1}}{n+1}=\frac{k}{\zeta}\ln\left|\dot{\delta}+\frac{k}{\zeta}\right|-\dot{\delta}+C \tag{25}$$

where C = constant of integration. Substituting the indentation and relative velocity at the beginning of the impact $\delta=0$ and $\dot{\delta}=\dot{\delta}_0$ into Eq. (25), C will be

$$C=\dot{\delta}_0-\frac{k}{\zeta}\ln\left|\dot{\delta}_0+\frac{k}{\zeta}\right| \tag{26}$$

Therefore, the accurate relation of δ and $\dot{\delta}$ will be expressed

$$\frac{\delta^{n+1}}{n+1}=\frac{m_e}{\zeta}\left(\frac{k}{\zeta}\ln\left|\dot{\delta}+\frac{k}{\zeta}\right|-\dot{\delta}+\dot{\delta}_0-\frac{k}{\zeta}\ln\left|\dot{\delta}_0+\frac{k}{\zeta}\right|\right) \tag{27}$$

The energy loss attributable to damping force ΔE_1 can be obtained by substituting Eq. (27) into Eq. (22) (Khatiwada et al. 2014)

$$\Delta E_1 = m_e\left(\frac{k^2}{\zeta^2}\ln\left|\frac{\dot{\delta}_0 + \frac{k}{\zeta}}{\frac{k}{\zeta}}\right| + \frac{\dot{\delta}_0^2}{2} - \frac{k}{\zeta}\dot{\delta}_0\right) \tag{28}$$

By substituting Eq. (16) into Eq. (21), the energy dissipation from the remaining surface deformation ΔE_2 should be (Lankarani and Nikravesh 1994)

$$\Delta E_2 = \frac{ke_1}{n+1}\delta_{\max}^{n+1} \tag{29}$$

By letting $\delta = \delta_{\max}$ and $\dot{\delta} = 0$ in Eq. (27), $\delta_{\max}^{n+1}$ is obtained in the form of

$$\delta_{\max}^{n+1} = \frac{m_e(n+1)}{\zeta}\left(\frac{k}{\zeta}\ln\left|\frac{\frac{k}{\zeta}}{\dot{\delta}_0 + \frac{k}{\zeta}}\right| + \dot{\delta}_0\right) \tag{30}$$

With the help of Eq. (30), Eq. (29) is rewritten

$$\Delta E_2 = m_e e_1\left(\frac{k^2}{\zeta^2}\ln\left|\frac{\frac{k}{\zeta}}{\dot{\delta}_0 + \frac{k}{\zeta}}\right| + \frac{k}{\zeta}\dot{\delta}_0\right) \tag{31}$$

Considering that these two different forms of energy loss in Eqs. (18) and (19) should be equal to each other, by substituting Eqs. (28) and (31) into it, an explicit expression of the pounding damping ζ is obtained

$$\frac{1}{2}(1-e^2)\dot{\delta}_0^2 = (1-e_1)\left(\frac{k^2}{\zeta^2}\ln\left|\frac{\dot{\delta}_0 + \frac{k}{\zeta}}{\frac{k}{\zeta}}\right| - \frac{k}{\zeta}\dot{\delta}_0\right) + \frac{\dot{\delta}_0^2}{2} \tag{32}$$

In this model, the pounding stiffness k will be determined by equating the simulated maximum impact force to the maximum impact force observed from the impact experiment. The remaining surface deformation ratio e_1 and coefficient of restitution e should be obtained by the impact experiment. The remaining surface deformation δ_e and pounding damping ζ can be acquired through Eqs. (17) and (32). The value of the maximum elastic force f_e and maximum indentation $\delta_{\max}$ should be obtained from the numerical simulations. The detailed steps to use the proposed model in an impact simulation are

1. Pounding experiments should be performed prior to the impact simulation to obtain the coefficient of restitution e and remaining surface deformation ratio e_1 from the measured impact force and displacement of the striker. The coefficient of restitution and remaining surface deformation ratio can be calculated by Eqs. (41) and (17);
2. For an assumed value of pounding stiffness k, the pounding damping ζ can be computed from Eq. (32). In the approaching period of Eq. (15), the impact force, indentation, and relative velocity can be obtained by the fourth-order Runge-Kutta method with the initial condition of the two colliding bodies. The maximum indentation $\delta_{\max}$ is collected when the relative velocity is zero at the end of the approaching period. With the remaining surface deformation ratio e_1 and maximum indentation $\delta_{\max}$ collected from the approaching period, the remaining surface deformation δ_e can be calculated by Eq. (17). All parameters in the restitution period of Eq. (15) have now been identified and the impact force during the restitution period also can be computed by the fourth-order Runge-Kutta method; and
3. Pounding stiffness k for pounding between viscoelastic materials and steel can be obtained by equating the simulated maximum impact force to the maximum impact force observed from the pounding experiment. Moreover, the nonlinear coefficient n in Eq. (15) is determined by equating the simulated maximum indentation to the maximum indentation observed from the pounding experiment.

The polyfitting method has been used to determine the pounding stiffness k and damping ζ by minimizing the error between the simulated and measured impact force (Xue et al. 2016a). In the proposed model, the pounding stiffness k is the only variable needing to be determined when the impact velocity, e, e_1, and n are already known. The pounding damping ζ will be obtained from Eq. (32) after k is determined. Therefore, the polyfitting method is overcomplicated to determine the only parameter k in this paper.

Discussion

Because of the damping of materials during pounding, the energy loss from the remaining surface deformation will not exceed the entire energy dissipation during pounding. This restriction is given by

$$\Delta E \geq \Delta E_2 \tag{33}$$

By substituting Eqs. (18) and (29), Eq. (33) can be written

$$\frac{1}{2}m_e(1-e^2)\dot{\delta}_0^2 \geq \frac{ke_1}{n+1}\delta_{\max}^{n+1} \tag{34}$$

When the critical situation $\Delta E = \Delta E_2$ is considered, the damping term in Eq. (15) will equal zero. Therefore, the relationship between kinetic energy and potential energy in the end of the approaching period is expressed

$$\frac{1}{2}m_e\dot{\delta}_0^2 = \frac{k}{n+1}\delta_{\max}^{n+1} \tag{35}$$

Therefore, the expression of $\delta_{\max}^{n+1}$ in the critical situation is updated using Eq. (35) as follows:

$$\delta_{\max}^{n+1} = \frac{n+1}{2k}m_e\dot{\delta}_0^2 \tag{36}$$

After substituting Eq. (36) into Eq. (34), the restricted relationship of e and e_1 is obtained by the following equation:

$$e \leq \sqrt{1-e_1} \tag{37}$$

From Eq. (37), it can be seen that when e_1 is a certain value in [0, 1] and the energy loss from materials damping in Eq. (15) is assumed to be zero, the coefficient of restitution e will obtain a maximum value of $e_{\max} = \sqrt{1-e_1}$. Because of material damping during the pounding, the simulated range of coefficient of restitution e by the proposed force model will be satisfied to Eq. (37).

Stability Verification of the Proposed Model

The stability of an impact force model has been previously studied (Ye et al. 2009; Flores et al. 2011). An impact force model with good stability accurately simulates the response of the colliding bodies and ensures the equal value of the preimpact coefficient of restitution e_{pre} and postimpact coefficient of restitution e_{post}. To verify the stability of the proposed model, a numerical simulation of a dropping a ball with an initial height h_0 to a fixed plate covered with viscoelastic materials, as shown in Fig. 4, was conducted. In this verification, e_{pre} is assumed to a value in the proposed model prior to the simulation, and after the simulation, e_{post} can be obtained by following equation:

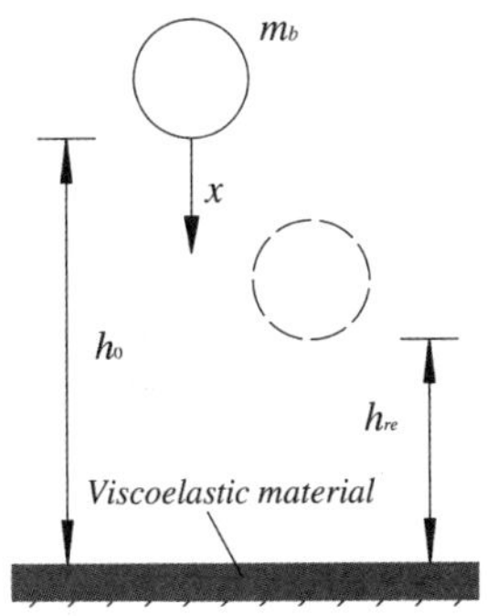

Fig. 4. Stability verification by dropping a ball to a fixed plate with VE material

$$e_{\text{post}} = \sqrt{h_{re}/h_0} \qquad (38)$$

where h_{re} = maximum rebound height after impact.

The motion equation of the impact simulation between a freely dropping ball and fixed plate with viscoelastic materials can be expressed

$$m_b \ddot{x} + F = m_b g \qquad (39)$$

where m_b = mass of the ball; $\ddot{x}$ = acceleration of the free-falling ball; g = acceleration of gravity; and F = impact force. In this simulation, $x = 0$ is defined at the initial dropping height h_0; the impact force F is 0 when $x \leq h_0$ and is defined by Eq. (15) when $x > h_0$. The expression of the indentation δ in Eq. (15) is as follows:

$$\delta = x - h_0 \qquad (40)$$

Based on a previous study on pounding involving viscoelastic materials (Zhang et al. 2013), the parameters k and n were selected to be 1.7×10^4 N/m$^{3/2}$ and 1.5, respectively. The remaining surface deformation ratios were assumed to be 0.1, 0.5, and 0.9 in different cases. The pounding damping ζ can be computed by Eq. (32) when the parameters k, n, e, e_1, and preimpact velocity $\dot{\delta}_0$ are known. The parameters used in the numerical verification are provided in Table 1. Here, e_{pre} varies from 0.1 to the maximum value of the coefficient of restitution e_{max} from Eq. (37), and the fourth-order Runge-Kutta method is used to solve Eq. (39) with a time step of 1×10^{-6} s.

The results of numerical verification with different values of the remaining surface deformation e_1 are plotted in Fig. 5. The straight line describes the relationship for $e_{\text{pre}} = e_{\text{post}}$ and can be regarded as reference. Generally, the impact force model may have a good performance and can maintain stability in impact force simulation when e_{pre} is equal to e_{post} (Flores et al. 2011). It is shown from Fig. 5 that the impact force model revised by Ye et al. (2009) is effective for high values of the coefficient of restitution e. When the coefficient of restitution is decreased to low values, however, the difference between e_{pre} and e_{post} becomes large, and this model may be totally unreliable in impact simulations. In contrast to the work carried out by Ye et al. (2009), the proposed model can keep the conservation of e_{pre} and e_{post} even at a very low value of the predefined coefficient of restitution e_{pre}. This verification not only illustrates the accuracy of the provided derivation for the proposed force model, but also shows that the proposed model can be used in the simulation of impact force with a high stability. The simulated time history of impact force in the stability verification ($e_1 = 0.9$) is also presented in Fig. 6 as a parametric study of the coefficient of restitution. It can be seen that the impact force decreases and pounding duration increases when the coefficient of restitution increases.

Table 1. Parameters Used in Stability Verification

Parameter	Value
n	1.5
k (N/m$^{3/2}$)	1.7×10^4
e_1	0.1, 0.5, and 0.9
h_0 (m)	0.5
m_b (kg)	1

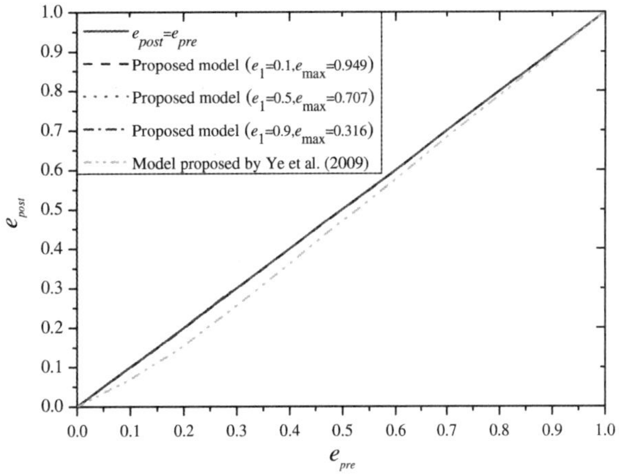

Fig. 5. Relationship between postimpact and preimpact coefficient of restitution

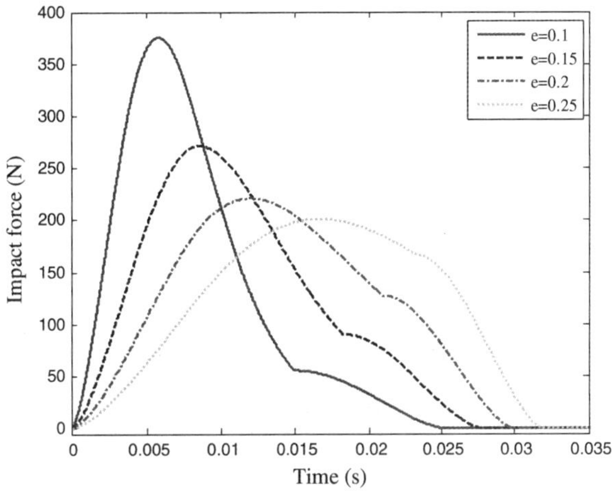

Fig. 6. Time history of impact force in stability verification when $e_1 = 0.9$

Pounding Experiment and Impact Force Prediction Using the Proposed Model

Pounding Experiment between Viscoelastic Materials and Steel

This section discusses a pounding experiment using of a steel spherical striker impacting a fixed pounding boundary covered by viscoelastic materials, as shown in Fig. 7, which was carried out to investigate the impact property and determine unknown parameters in the proposed force model. The spherical striker, shown in Fig. 8(a), was installed on a 12.5-kg mass block and connected by a rigidly articulated rod with a length $L_p = 1.0$ m. A force sensor placed behind the pounding boundary with viscoelastic materials was used to measure pounding force, and a laser displacement sensor was used to record the straight-line displacement of the striker to the viscoelastic materials layer. The sampling frequency of the force sensor and laser displacement sensor were set to 2,000 Hz. The viscoelastic materials used in the pounding experiment are called very high bond (VHB) (3M, Maplewood, Minnesota) as shown in Fig. 8(b). The thickness of the VHB layer added to the pounding boundary is 24 mm. The initially released distance between the striker and fixed boundary is 0.05 m. The pounding force and displacement of the spherical striker in the first three poundings were recorded and are plotted in Fig. 9.

From the displacement record of the first three poundings, the coefficient of restitution e for the pounding between the spherical striker and viscoelastic materials can be calculated by following equation:

$$e = \sqrt{\frac{L_p - \sqrt{L_p^2 - L_{\text{rebound}}^2}}{L_p - \sqrt{L_p^2 - L_{\text{initial}}^2}}} \tag{41}$$

where L_p = length of rigid rod; and L_{initial} and L_{rebound} = initially released length and maximum rebound length after one pounding.

The identification results of the coefficient of restitution are given in Table 2. In this experiment, even with different approaching velocities, the coefficient of restitution e remains at a relatively stable range from 0.398 to 0.412. Therefore, an average value of 0.404 was adopted as the coefficient of restitution e of the pounding between viscoelastic materials and steel in the conducted experiment.

Figs. 10–12 show the time history of the impact force and indentation during the first three poundings, with the hysteretic loop of impact force and indentation also shown. It can be seen that poundings happen when the indentation of the spherical striker δ is greater than zero, and the maximum impact force is reached before the arrival of the maximum indentation $\delta_{\max}$. In all three poundings, the remaining surface deformation δ_e resulting from the elastic aftereffect phenomenon can be observed on the surface of the VHB layer when the impact force returned to zero again, which means the spherical striker and pounding boundary covered by VHB are already separated. However, the deformed surface of the VHB layer will recover to the original surface before the next impact comes. Because the poundings happen at low speeds, the plastic deformation of the VHB layer is very limited. By using Eq. (18), the remaining surface deformation ratio was calculated,

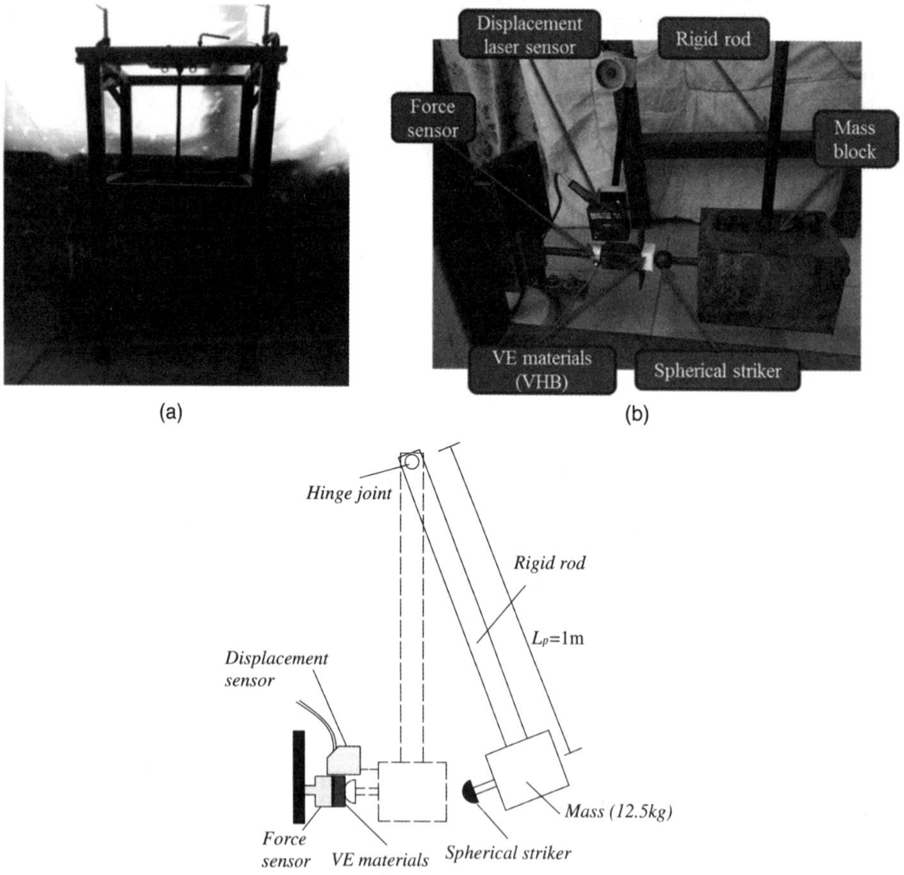

Fig. 7. Experimental setup for pounding between steel striker and VHB: (a) panoramic view; (b) close view; (c) schematic view

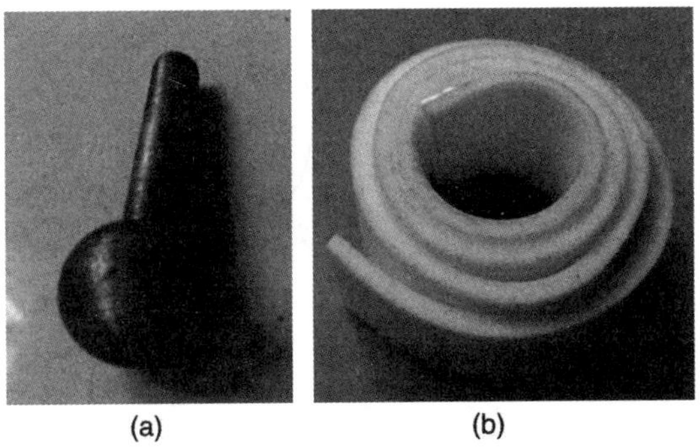

Fig. 8. Experiment device and material: (a) spherical striker; (b) viscoelastic materials (VHB)

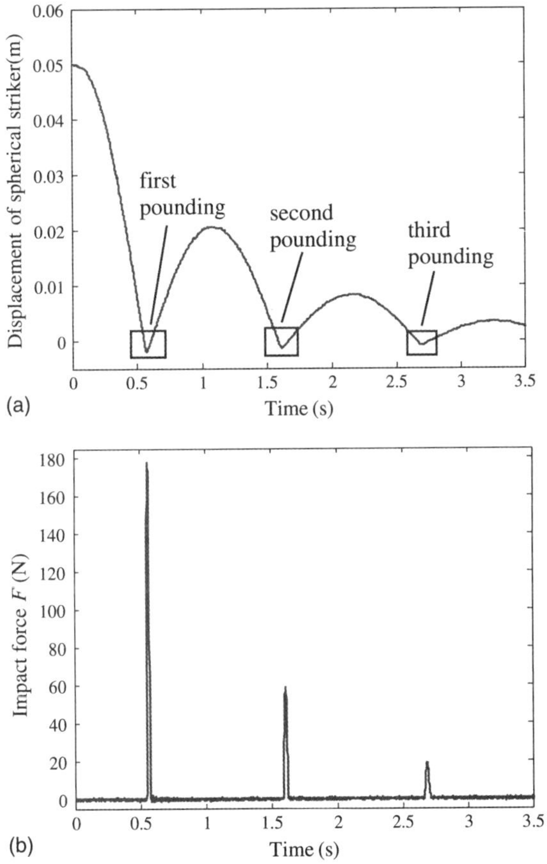

Fig. 9. Time history of displacement of the striker and impact force: (a) displacement; (b) impact force

Table 2. Coefficient of Restitution e at Different Pounding Velocities

Pounding case	L_{initial} (m)	L_{rebound} (m)	Pounding velocity (m/s)	e
First pounding	0.0500	0.0206	0.1566	0.412
Second pounding	0.0206	0.0082	0.0645	0.398
Third pounding	0.0082	0.0033	0.0257	0.402

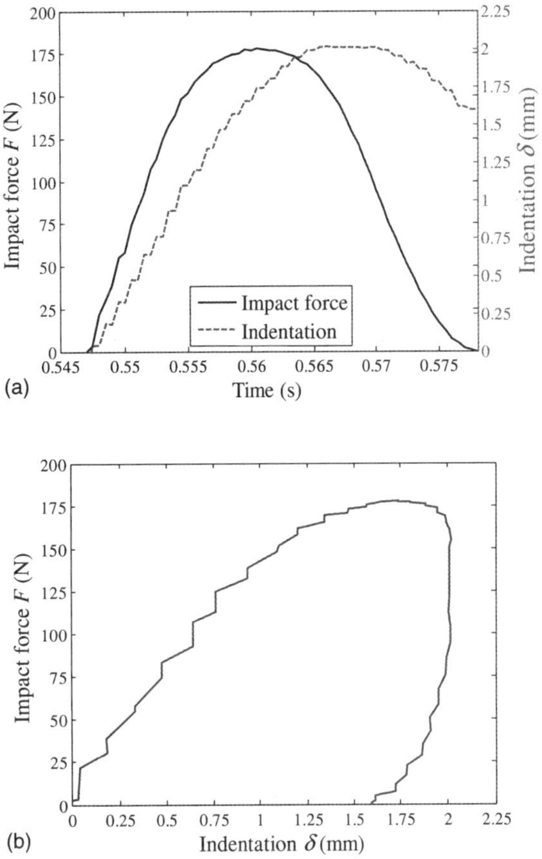

Fig. 10. Time history of the first pounding: (a) impact force and indentation; (b) hysteretic loop

and the results are given in Table 3. Even though the preimpact velocity is changed during poundings, the remaining surface deformation ratio is almost constant. Eventually, a mean value 0.785 of the remaining surface deformation ratio in the first three poundings was obtained. A previous study on the impact properties of poundings between a viscoelastic materials layer and a steel striker (Wang et al. 2017) found that the coefficient of restitution and remaining surface deformation ratio would remain at a stable value even if the poundings happen at various impact velocities.

Discussion

The critical situation $e = \sqrt{1 - e_1}$ of Eq. (37) holds when the damping term in Eq. (15) is ignored, and the proposed model formally turns into Eq. (14). In such a case, the entire energy dissipation during pounding is entirely caused by the plastic deformation, as described by Eq. (14). At first glance, if the remaining surface deformation ratio $e_1 = 0.785$ is substituted into Eq. (37), the coefficient of restitution using Eq. (14) can be calculated as $e = \sqrt{1 - e_1} = 0.464$, which is larger than the experimental result of 0.404. This implies that Eq. (14) is inadequate to describe the pounding between viscoelastic materials and steel, and materials damping always exists in impacts. The proposed model, separating energy dissipation during impact into the damping of materials and the remaining surface deformation, is more reasonable from experimental observations.

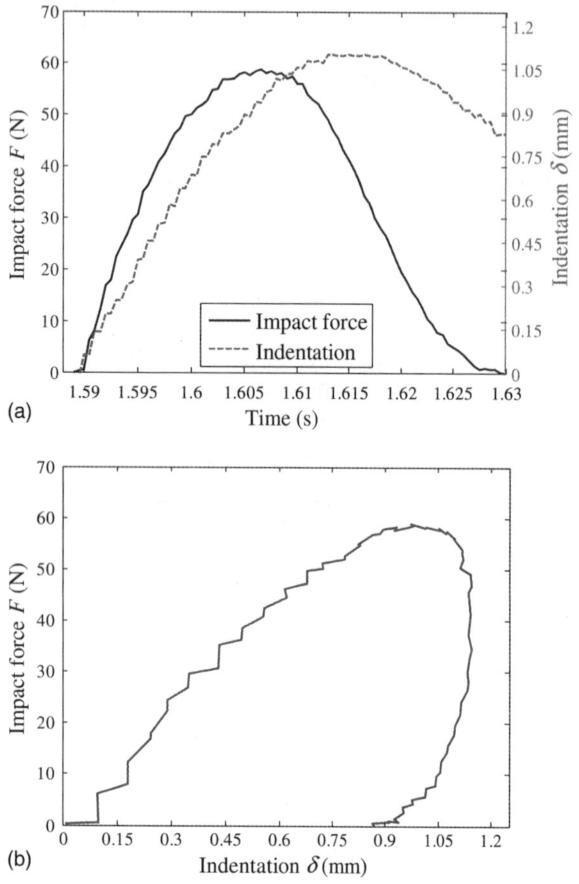

Fig. 11. Time history of the second pounding: (a) impact force and indentation; (b) hysteretic loop

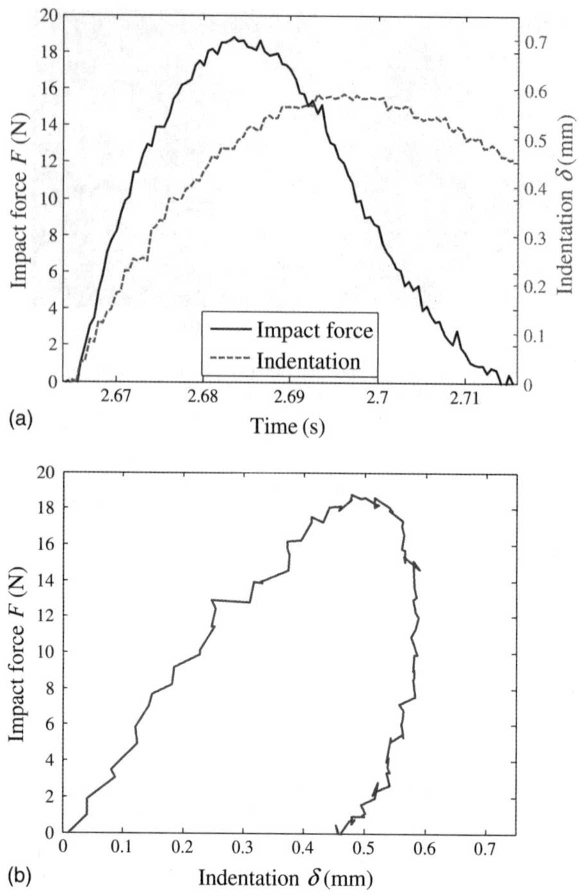

Fig. 12. Time history of the third pounding: (a) impact force and indentation; (b) hysteretic loop

Determination of Nonlinear Coefficient n in Proposed Model

With the coefficient of restitution $e = 0.404$ and remaining surface deformation ratio $e_1 = 0.785$ obtained from the pounding experiment, the proposed impact force model can be established. Fig. 13 plots the simulated impact force of the first pounding from the proposed model. In order to find the most suitable nonlinear coefficient n for the pounding between viscoelastic materials and steel, different values of n varying from 0.7 to 2.5 with a time interval of 0.2 were applied. Parameters used in impact force simulation and the simulation results are summarized in Table 4.

Table 4 indicates that both pounding stiffness and pounding damping increase with an increase of n. Moreover, from results of the impact force simulation in Table 4 demonstrate that when n increases, the pounding duration and maximum indentation of simulated impact force also grow. Specifically, the maximum indentation when $n = 1.9$ is 2.0 mm, which is extremely close to the experimental record 2.02 mm. Therefore, the best value of n is selected as 1.9 from the pounding experiment. In fact, the value of n is closely related to the geometry of the colliding bodies involved in poundings (van Mier et al. 1991). For instance, according to the Hertz contact law, n should be 1.5 for poundings between spherical and cylindrical bodies (Goldsmith 1960). Therefore, pounding experiments should be performed to determine the nonlinear coefficient n in the proposed model.

Table 3. Remaining Surface Deformation Ratio e_1 for First Three Poundings

Pounding case	δ_{max} (mm)	δ_e (mm)	e_1
First pounding	2.02	1.59	0.787
Second pounding	1.14	0.90	0.789
Third pounding	0.59	0.46	0.780

Comparative Study between the Proposed Model and Existing Viscoelastic Force Models

To illustrate the performance of the proposed model in impact simulations between a VE material and steel, a comparative study of the proposed model and existing viscoelastic impact force models was carried out. In this study, the first pounding in the pounding experiment is used as the simulation target. The pounding stiffness of each impact force model was determined by letting the maximum simulated impact force be equal to the maximum impact force 177.5 N observed from the pounding experiment. Parameters used in different impact force models are listed in Table 5. Nonlinear coefficient n in the HC model and the proposed model are 1.5 and 1.9, respectively. Fig. 14 presents the history of impact force and indentation by using each impact force model to simulate the first pounding. The normalized error ε_0 of each impact force model was calculated by Jankowski (2005)

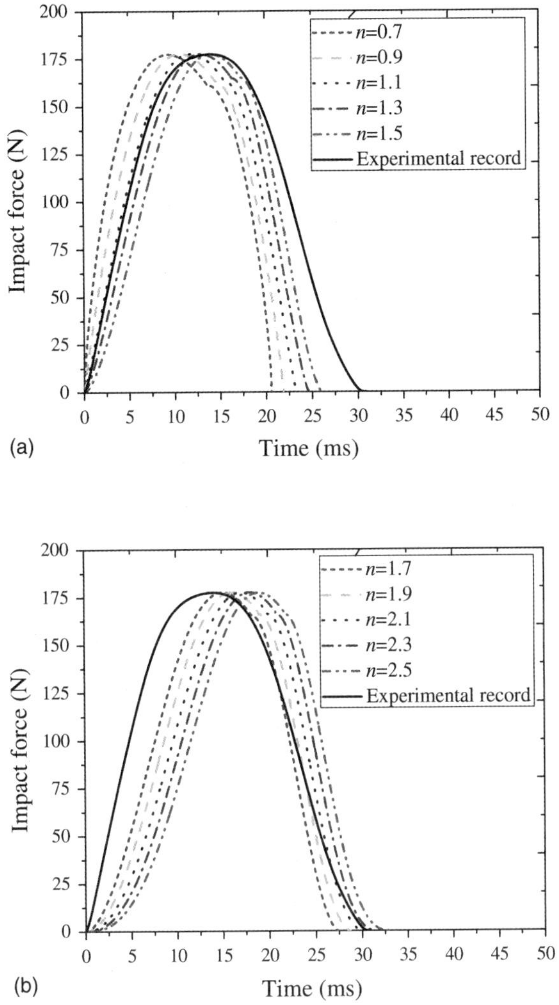

Fig. 13. Histories of simulated impact force at different values of n: (a) $n = 0.7$–1.5; (b) $n = 1.7$–2.5

$$\varepsilon_0 = \frac{\sqrt{\sum_{i=1}^{n} (F_{\exp} - F_p)^2}}{\sqrt{\sum_{i=1}^{n} (F_{\exp})^2}} \tag{42}$$

where F_p and $F_{\exp}$ = force calculated by each impact force model and force measured by an experiment, respectively. The normalized error of each impact force model is shown in Fig. 15.

Fig. 14 shows that the proposed model and HC model have a better performance than other three models in the impact force simulation. With consideration of elastic aftereffects, the indentation simulated by the proposed model concides well with the experimental results. Other viscoelastic impact force models may result in incorrect responses of the indentation. Moreover, the pounding duration from the proposed model best agrees with experimental results. From Fig. 15, the normalized error of the proposed model is 26.32%, which is the lowest value compared with those from the other models.

Even though the proposed model has a lowest error in impact simulations, it can be seen from Fig. 14 that the HC model with a nonlinear coefficient $n = 1.5$ agrees better with the experimental result at the approaching period, and the proposed model with a nonlinear coefficient $n = 1.9$ describes the impact better in the restitution period. Fig. 13 also indicates that the proposed model agrees better with the experimental record at the approaching period when the nonlinear coefficient n is set to 1.1. For the proposed model, therefore, an improvement in the accuracy of the approaching period should be made, and this will be considered in a future work.

Table 4. Parameters and Results of Impact Simulation for the First Pounding

n	Pounding stiffness k (N/m^n)	Pounding damping ζ	Pounding duration (ms)	Maximum indentation $\delta_{\max}$ (mm)
0.7	1.761×10^4	5.534×10^4	20.6	1.2
0.9	6.233×10^4	1.856×10^5	21.9	1.4
1.1	2.211×10^5	6.665×10^5	23.2	1.5
1.3	7.012×10^5	2.204×10^6	24.6	1.6
1.5	2.257×10^6	7.007×10^6	26.0	1.8
1.7	7.104×10^6	2.232×10^7	27.4	1.9
1.9	2.192×10^7	6.882×10^7	28.7	2.0
2.1	6.662×10^7	2.081×10^8	30.1	2.2
2.3	1.972×10^8	6.197×10^8	31.5	2.3
2.5	5.572×10^8	1.808×10^9	32.9	2.4

Table 5. Parameters of Each Impact Force Model

Impact force model	Pounding stiffness k	Pounding damping ζ
LV model	1.569×10^5 N/m	0.2772
MLV model	3.631×10^4 N/m	0.8416
NV model	$1.853 \times 10^6\ \text{N/m}^{1.5}$	0.9944
HC model	$1.921 \times 10^6\ \text{N/m}^{1.5}$	2.522×10^7
Proposed model	$2.192 \times 10^7\ \text{N/m}^{1.9}$	6.882×10^7

Impact Force Prediction Using the Identified Model

In order to further validate the performance of the proposed force model in impact simulation, a mathematic model of a pendulum striker impacting a pounding boundary covered by a VE material (VHB), is performed to predict the entire pounding process. The equation of motion of the pendulum striker is expressed

$$m\ddot{x} + \frac{mg}{L_p}x + F = 0 \tag{43}$$

where $\ddot{x}$, x, m, g, and L_p = horizontal acceleration, movement, mass, acceleration of gravity, and length of pendulum arm, respectively; and F = impact force, which is 0 when $x \leq 0$ and is calculated by Eq. (15) when $x > 0$ by making $x = \delta$. Both the experimental and numerical analyses start with an initial released distance of $x_0 = 0.05$ m. Parameters used in numerical prediction are taken from the first pounding and have already been provided in Tables 2, 3, and 5.

The numerical simulation of the pounding experiment was performed for the first 3.5 s, which includes three poundings. The fourth-order Runge-Kutta method was used to solve Eq. (43) with a time interval of 1×10^{-6} s. Numerical and experimental records of the displacement of the striker and impact force are shown in Fig. 16. It can be seen that the proposed force model for pounding simulation agrees well with experimental results. In contrast to the pounding force recorded during the experiments, the predicted maximum impact force, which is highly accurate, has an error that

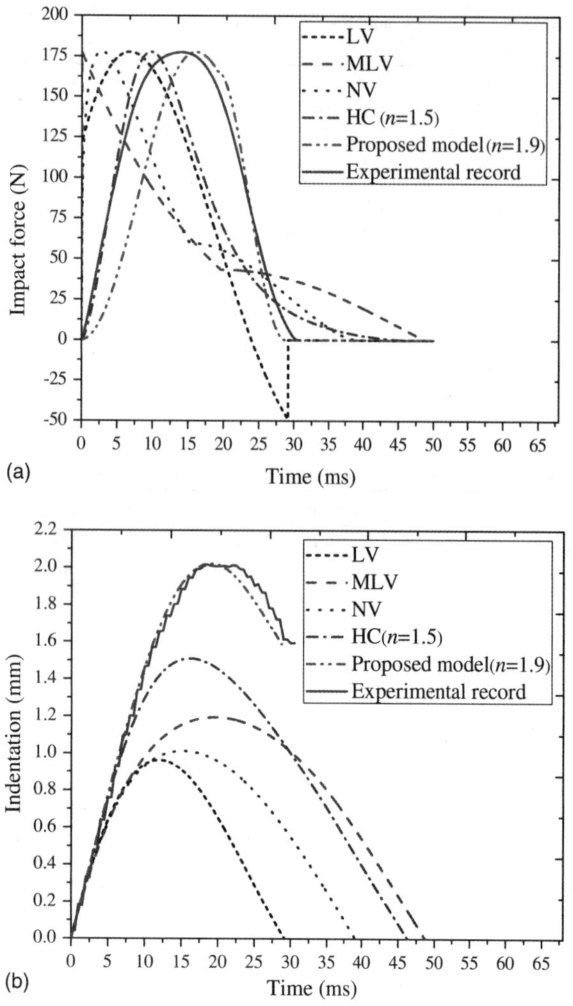

Fig. 14. Comparison of impact simulation and experiment: (a) impact force; (b) indentation

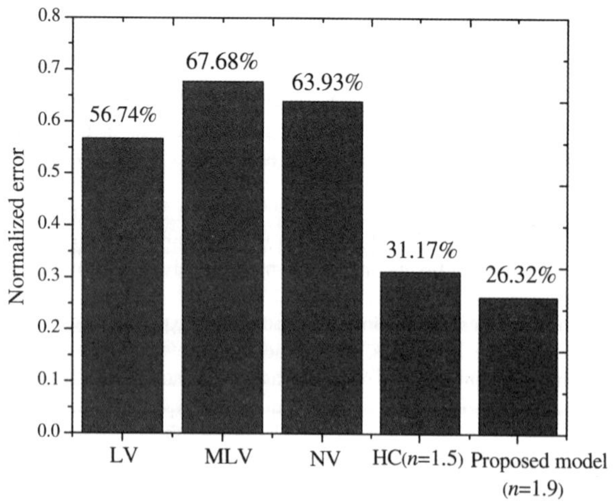

Fig. 15. Normalized error ε_0 of impact force models

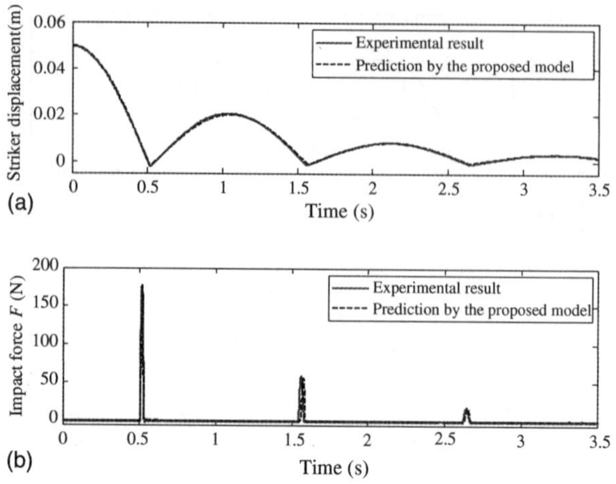

Fig. 16. Prediction of pounding experiment by the proposed impact force model ($n = 1.9$): (a) displacement of the striker; (b) impact force

is under 5%. In addition, the impact duration predicted by the proposed model is almost the same as that of the experimental results. Because the proposed model ignores the plastic deformation resulting from high-speed impacts, this model is only applicable for low impact speeds. Based on the available experimental data, the applicable range is from 0 to approximately 0.15 m/s, although the actual upper limit may be higher. Actually, the inelastic deformation on the surface of the viscoelastic materials layer is very limited; even the impact velocity is approximately 0.4 m/s (Wang et al. 2017), and poundings in impact dampers and adjacent buildings under earthquake mostly happen at a relatively low impact velocity.

Conclusions

In this paper, an advanced impact force model was developed to accurately model low-speed poundings between viscoelastic materials and steel. The relaxation behavior of the viscoelastic materials was described by the remaining surface deformation, which is produced by the elastic aftereffect phenomenon. The pounding damping in the proposed impact force model was derived theoretically. A pounding experiment with a spherical striker impacting a pounding boundary covered with VHB has been performed to study the properties of viscoelastic impact events. The unknown parameters in the proposed model were determined by the pounding experiment. The experimental results verified the numerical prediction based on the proposed impact force model. Additionally, the results of this work showed that the proposed model has good stability even for a low value of the coefficient of restitution. From the pounding experiment, the remaining surface deformation resulting from the elastic aftereffect phenomenon was observed. The proposed model, which provides a reasonable explanation for the remaining surface deformation right before the end of impact, showed better performance than other existing impact force models. By applying the proposed model, the impact force prediction for the pounding experiment demonstrated the validity of the proposed model in impact simulations.

Acknowledgments

The authors appreciate the support from the National Key Basic Research Program of China (973 Project) (Grant No. 2015CB057702) and the National Natural Foundation of China (Grant Nos. 51422806 and 51378203). In addition, the financial support sponsored by the China Scholarship Council (CSC) is acknowledged.

References

Al-Shudeifat, M. A., Wierschem, N., Quinn, D. D., Vakakis, A. F., Bergman, L. A., and Spencer, B. F., Jr. (2013). "Numerical and experimental investigation of a highly effective single-sided vibro-impact non-linear energy sink for shock mitigation." *Int. J. Non Linear Mech.*, 52, 96–109.

Anagnostopoulos, S. A. (1988). "Pounding of buildings in series during earthquakes." *Earthquake Eng. Struct. Dyn.*, 16(3), 443–456.

Anagnostopoulos, S. A. (2004). "Equivalent viscous damping for modeling inelastic impacts in earthquake pounding problems." *Earthquake Eng. Struct. Dyn.*, 33(8), 897–902.

Anagnostopoulos, S. A., and Spiliopoulos, K. V. (1992). "An investigation of earthquake induced pounding between adjacent buildings." *Earthquake Eng. Struct. Dyn.*, 21(4), 289–302.

Bayly, P. V., and Virgin, L. N. (1993). "An experimental study of an impacting pendulum." *J. Sound Vib.*, 164(2), 364–374.

Cross, W. B., and Jones, N. P. (1993). "Seismic performance of joist-pocket connections. I: Modeling." *J. Struct. Eng.*, 10.1061/(ASCE)0733-9445(1993)119:10(2986), 2986–3007.

Dimentberg, M. F., and Iourtchenko, D. V. (1999). "Towards incorporating impact losses into random vibration analyses: A model problem." *Probab. Eng. Mech.*, 14(4), 323–328.

Dimentberg, M. F., and Iourtchenko, D. V. (2004). "Random vibrations with impacts: A review." *Nonlinear Dyn.*, 36(2), 229–254.

Flores, P., Machado, M., Silva, M. T., and Martins, J. M. (2011). "On the continuous contact force models for soft materials in multibody dynamics." *Multibody Sys. Dyn.*, 25(3), 357–375.

Goldsmith, W. (1960). *Impact: The theory and physical behavior of colliding solids*, Edward Arnold, London.

Hunt, K. H., and Crossley, F. R. E. (1975). "Coefficient of restitution interpreted as damping in vibroimpact." *J. Appl. Mech.*, 42(2), 440–445.

Iourtchenko, D. V., and Song, L. L. (2006). "Numerical investigation of a response probability density function of stochastic vibroimpact systems with inelastic impacts." *Int. J. Non Linear Mech.*, 41(3), 447–455.

Jankowski, R. (2005). "Non-linear viscoelastic modelling of earthquake-induced structural pounding." *Earthquake Eng. Struct. Dyn.*, 34(6), 595–611.

Jankowski, R. (2006). "Analytical expression between the impact damping ratio and the coefficient of restitution in the non-linear viscoelastic model of structural pounding." *Earthquake Eng. Struct. Dyn.*, 35(4), 517–524.

Jankowski, R. (2010). "Experimental study on earthquake-induced pounding between structural elements made of different building materials." *Earthquake Eng. Struct. Dyn.*, 39(3), 343–354.

Jankowski, R., Wilde, K., and Fujino, Y. (1998). "Pounding of superstructure segments in isolated elevated bridge during earthquakes." *Earthquake Eng. Struct. Dyn.*, 27(5), 487–502.

Khatiwada, S., Chouw, N., and Butterworth, J. W. (2014). "A generic structural pounding model using numerically exact displacement proportional damping." *Eng. Struct.*, 62, 33–41.

Kim, S., and Shinozuka, M. (2003). "Effects of seismically induced pounding at expansion joints of concrete bridges." *J. Eng. Mech.*, 10.1061/(ASCE)0733-9399(2003)129:11(1225), 1225–1234.

Lankarani, H. M., and Nikravesh, P. E. (1990). "A contact force model with hysteresis damping for impact analysis of multibody systems." *J. Mech. Des.*, 112(3), 369–376.

Lankarani, H. M., and Nikravesh, P. E. (1994). "Continuous contact force models for impact analysis in multibody systems." *Nonlinear Dyn.*, 5(2), 193–207.

Li, H. N., Zhang, P., Song, G., Patil, D., and Mo, Y. L. (2015). "Robustness study of the pounding tuned mass damper for vibration control of subsea jumpers." *Smart Mater. Struct.*, 24(9), 095001.

Li, L., Song, G., Singla, M., and Mo, Y. L. (2013). "Vibration control of a traffic signal pole using a pounding tuned mass damper with viscoelastic materials. II: Experimental verification." *J. Vib. Control*, 21(4), 670–675.

Lu, Z., Lu, X., and Masri, S. F. (2010). "Studies of the performance of particle dampers under dynamic loads." *J. Sound Vib.*, 329(26), 5415–5433.

Lu, Z., Lu, X., and Masri, S. F. (2012). "Experimental studies of the effects of buffered particle dampers attached to a multi-degree-of-freedom system under dynamic loads." *J. Sound Vib.*, 331(9), 2007–2022.

Machado, M., Moreira, P., Flores, P., and Lankarani, H. M. (2012). "Compliant contact force models in multibody dynamics: Evolution of the Hertz contact theory." *Mech. Mach. Theory*, 53, 99–121.

Mahmoud, S., and Jankowski, R. (2011). "Modified linear viscoelastic model of earthquake-induced structural pounding." *Iran. J. Sci. Technol.*, 35(C1), 51–62.

Maison, B. F., and Kasai, K. (1990). "Analysis for a type of structural pounding." *J. Struct. Eng.*, 10.1061/(ASCE)0733-9445(1990)116:4(957), 957–977.

Malhotra, P. K. (1998). "Dynamics of seismic pounding at expansion joints of concrete bridges." *J. Eng. Mech.*, 10.1061/(ASCE)0733-9399(1998)124:7(794), 794–802.

Masri, S. F., and Caughey, T. K. (1966). "On the stability of the impact damper." *J. Appl. Mech.*, 33(3), 586–592.

Muthukumar, S., and DesRoches, R. (2006). "A Hertz contact model with non-linear damping for pounding simulation." *Earthquake Eng. Struct. Dyn.*, 35(7), 811–828.

Polycarpou, P. C., Komodromos, P., and Polycarpou, A. C. (2013). "A non-linear impact model for simulating the use of rubber shock absorbers for mitigating the effects of structural pounding during earthquakes." *Earthquake Eng. Struct. Dyn.*, 42(1), 81–100.

Quinn, D. D. (2005). "Finite duration impacts with external forces." *J. Appl. Mech.*, 72(5), 778–784.

Song, G., et al. (2016). "Vibration control of a pipeline structure using pounding tuned mass damper." *J. Eng. Mech.*, 10.1061/(ASCE)EM.1943-7889.0001078, 04016031.

Song, L. L., and Iourtchenko, D. V. (2006). "Analysis of stochastic vibroimpact systems with inelastic impacts." *Mech. Solids*, 41(2), 146–154.

van Mier, J. G. M., Pruijssers, A. F., Reinhardt, H. W., and Monnier, T. (1991). "Load-time response of colliding concrete bodies." *J. Struct. Eng.*, 10.1061/(ASCE)0733-9445(1991)117:2(354), 354–374.

Wagg, D. J. (2007). "A note on coefficient of restitution models including the effects of impact induced vibration." *J. Sound Vib.*, 300(3), 1071–1078.

Wagg, D. J., and Bishop, S. R. (2002). "Application of non-smooth modelling techniques to the dynamics of a flexible impacting beam." *J. Sound Vib.*, 256(5), 803–820.

Wang, W., Hua, X., Wang, X., Chen, Z., and Song, G. (2017). "Optimum design of a novel pounding tuned mass damper under harmonic excitation." *Smart Mater. Struct.*, 26(5), 055024.

Xu, Z., Wang, M., and Chen, T. (2005). "Particle damping for passive vibration suppression: Numerical modelling and experimental investigation." *J. Sound Vib.*, 279(3), 1097–1120.

Xue, Q., Zhang, C., He, J., Zou, G., and Zhang, J. (2016a). "An updated analytical structural pounding force model based on viscoelasticity of materials." *Shock Vib.*, 2016, 2596923.

Xue, Q., Zhang, J., He, J., and Zhang, C. (2016b). "Control performance and robustness of pounding tuned mass damper for vibration reduction in SDOF structure." *Shock Vib.*, 2016, 8021690.

Ye, K., Li, L., and Zhu, H. (2009). "A note on the Hertz contact model with nonlinear damping for pounding simulation." *Earthquake Eng. Struct. Dyn.*, 38(9), 1135–1142.

Zhang, P., Song, G., Li, H. N., and Lin, Y. (2013). "Seismic control of power transmission tower using pounding TMD." *J. Eng. Mech.*, 10.1061/(ASCE)EM.1943-7889.0000576, 1395–1406.

Received: 1 February 2018 | Revised: 7 December 2018 | Accepted: 6 June 2019
DOI: 10.1002/stc.2413

RESEARCH ARTICLE

WILEY

Optimal design of TVMD with linear and nonlinear viscous damping for SDOF systems subjected to harmonic excitation

Zhiwen Huang[1] | Xugang Hua[1] | Zhengqing Chen[1,2] | Huawei Niu[1]

[1]Key Laboratory for Wind and Bridge Engineering of Hunan Province, College of Civil Engineering, Hunan University, Changsha, Hunan, China

[2]State Key Laboratory of Advanced Design and Manufacturing for Vehicle Body, Hunan University, Changsha 410082, Hunan, China

Correspondence
Xugang Hua, Key Laboratory for Wind and Bridge Engineering of Hunan Province, College of Civil Engineering, Hunan University, Changsha 410082, Hunan, China.
Email: cexghua@hnu.edu.cn

Funding information
National Natural Science Foundation of China, Grant/Award Number: 51808210; State's Key Project of Research and Development Plan, Grant/Award Numbers: 2017YFC07036002 and 2016YFE0127900

Summary
This paper investigates the optimal design of the tuned viscous mass damper (TVMD) with linear and nonlinear viscous damping properties for a single degree-of-freedom structure subjected to harmonic excitation. The TVMD consists of a parallel-connected inerter and viscous damper, and a spring that mounts them to the structure. To examine the vibration control performance of the TVMD, both theoretical analysis and numerical optimization are conducted to obtain the optimal parameters of the TVMD with either linear or nonlinear viscous damping. The influence of the damping exponent of the nonlinear viscous damper on the optimal parameters and performance of the TVMD is analyzed. The results show that much better control effectiveness and robustness can be achieved by the TVMD when compared with tuned mass damper (TMD). Furthermore, it is also of great practical importance to show that the static deformation of the supporting spring due to the self-weight of the inerter is only a small fraction of the traditional TMD in case of vertical vibration control. Compared with the linear TVMD, the nonlinear TVMD can achieve comparable or even slightly better control performance. The optimal damping coefficient depends largely on the damping exponent, and the loading intensity fluctuation also has an influence on the performance of the nonlinear TVMD. Moreover, it is shown that that the nonlinear TVMD with damping exponent $\nu = 2$ has the best control performance on the peak displacement response of the main structure, and the TVMD with linear viscous damping optimized for the harmonic excitation can also effectively suppress the wind-induced vibration.

KEYWORDS
inerter, nonlinear damping, tuned mass dampers, tuned viscous mass dampers, vibration control

1 | INTRODUCTION

The current development of civil engineering research is aiming to achieve resilient and sustainability-informed design. The structural control technologies using active and passive dampers play a crucial role for mitigation of responses of a structure subjected to dynamic excitations. Tuned mass damper (TMD), also known as vibration absorbers, is one of the popular control devices that have been widely used all over the world for flexible and low-damped structures such as long-span bridges[1] and high-rise buildings.[2] A traditional TMD consists of a relatively small secondary mass attached to the main structure through a spring and a damper. The mechanism of a TMD is to transfer the vibration energy of the main structure to the TMD and dissipate the energy by its damper when TMD is finely tuned.

In order to obtain effective control strategy, many optimal design methods of TMD have been proposed to suppress the undesirable structural vibrations induced by winds[3,4] and earthquakes.[5-7] It is well known that the effectiveness and robustness of the TMD rely largely on its mass ratio to the host structure. However, due to the practical reasons, the TMD mass ratio is limited to several percent of the modal mass of the total structure, which attenuates the robustness of the TMD to the uncertainties in dynamic characteristics of the structure and the external excitations, making it difficult for the TMD to be used for large seismic vibration control. To improve the robustness of the traditional TMD, the concept of multiple TMDs[8] has been proposed and verified to be more robust than the TMD. However, the configuration of multiple TMDs is very complicated, and it is still constrained by the limited secondary mass. On the other hand, in the very recent years, various inerter-based vibration absorbers have been developed,[9-11] which behave like the traditional TMDs. The inerter, as first introduced by Smith in 2002,[12] is a two-terminal mechanical device; when subjected to motion, it will produce a force proportional to the relative acceleration between its two terminals. The proportionality constant is called inertance or apparent mass, which can reach several hundred times of the physical mass of the device. Therefore, the inerter-based vibration absorbers may achieve a much better effectiveness and robustness than the traditional TMD. The inerter can be physically realized by the rack-and-pinion mechanism or ball-screw mechanism.[13,14]

Currently, many inerter-based damping devices have developed by different arrangement of inerter, damper, and spring. The inerter-based vibration absorbers mainly include the tuned viscous mass damper (TVMD),[9] the tuned inerter damper (TID),[10] the tuned mass damper inerter (TMDI).[11] The TVMD consists of a viscous mass damper (an inerter and a viscous damper in parallel configuration) mounted in series with a supporting spring. It has been employed to suppress the earthquake-induced vibration for both single degree-of-freedom (SDOF) structures[9,15] and multiple degree-of-freedom structures[16,17] under different optimization criteria. The TID has a configuration similar to that of a TMD, in which the inerter is installed in series with a spring and a damper. The vibration control performances of TID for both building structures[10,18] and cables[19] have been studied. The TMDI consists of a traditional TMD and an inerter connecting the TMD mass to the supporting ground. The studies on the performances and optimal designs of the TMDI have been conducted for structural systems subjected to random ground excitations[20] and wind excitations.[21]

To the best of the authors' knowledge, nearly all of the previous studies on the inerter-based vibration absorber considered a linear viscous damper, until the very recently, Hessabi and Mercan investigated the gyro-mass damper–viscous damper–Brace (GVB) system for seismic control of an SDOF system,[22] in which a nonlinear viscous damper with a power factor of 2 was considered. The GVB system has the same mathematical model as the TVMD, and their results showed that the GVB system with nonlinear viscous damper has less requirement of the damping coefficient but can obtain nearly the same control performance as the GVB system with linear viscous damper. On the other hand, the performances of the TMD with nonlinear viscous damper attached to SDOF system or multiple degree of freedom system under white noise excitations or wind loads have been carefully studied.[23-26] The results indicate that the damping exponent of the nonlinear viscous damper has a great influence on the optimal damping coefficient.

In this study, the optimal designs of TVMD with linear and nonlinear viscous damping for SDOF structure subjected to harmonic excitation were considered. For TVMD with linear viscous damper, the optimal parameters of TVMD and the minimum dynamic amplification factor of the main structure were obtained by the fixed-point method for undamped main structure and are compared with those of the traditional TMD. The influence of the inherent damping of the main structure on the optimization was also investigated. For TVMD with nonlinear viscous damper, the equivalent linear damping ratio had been used for the linearization of the equations of motion, and an iterative method had been proposed to search for the equivalent damping ratio and the corresponding system response. Based on this, an analytical expression for the optimal damping coefficient was derived and a numerical optimization was also carried out to obtain the optimal parameters of the nonlinear TVMD.

2 | OPTIMAL DESIGN OF TVMD WITH LINEAR VISCOUS DAMPING

2.1 | The mechanism of inerter

The inerter is a key element in TVMD, and it can be realized by the ball screw mechanism as shown in Figure 1a. The inerter mainly consists of the ball screw, the ball nut, the flywheel, the external tube, and the thrust bearings. When a relative motion is given between the Node 1 and Node 2, the ball nut and the flywheel are rotated by the ball screw,

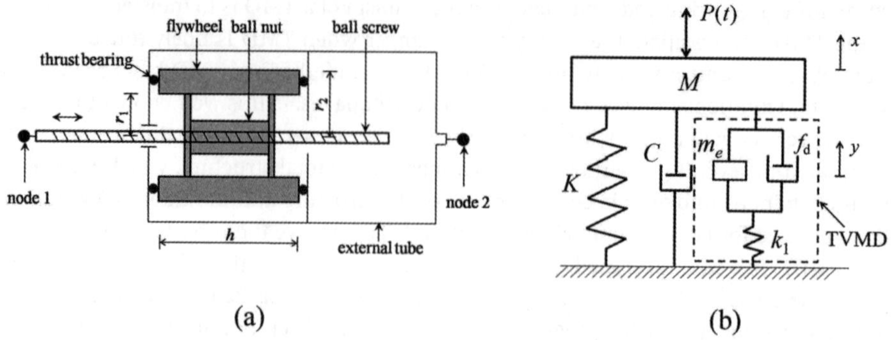

FIGURE 1 Single degree-of-freedom system with attached eddy current tuned viscous mass damper (TVMD): (a) a schematic of the inerter and (b) the structure–TVMD system under harmonic excitation

resulting in an inertial torque, which is proportional to angular acceleration of the ball nut. Through the ball screw mechanism, the inertial torque is converted to an axial inertial force applied on the ball screw and transmitted to the structure. Due to the amplifying mechanism of the ball screw,[9,27] the apparent mass of an inerter can reach hundreds or even thousands of times of the physical mass of the flywheel. Thus, when an inerter is employed in TVMD, very large mass ratio can be obtained.

2.2 | Equation of motion

Figure 1b shows a TVMD connected to an SDOF main structure. The TVMD device consists of a parallel-connected inerter and viscous damper, and a spring. When the main structure is subjected to a harmonic excitation the governing equation can be taken as

$$\begin{cases} M\ddot{x} + C\dot{x} + Kx + k_1(x-y) = P(t) \\ m_e\ddot{y} + c_1\dot{y} + k_1(y-x) = 0 \end{cases}, \tag{1}$$

where x and y denote the displacement of the main structure and deformation of the inerter or viscous damper, respectively; $q = x - y$ denotes the deformation of the supporting spring; M and m_e are the mass of the structure and the apparent mass or inertance of the inerter; respectively. K and k_1 denote the effective stiffness of the structure and the supporting spring of TVMD; the main structure is assumed to be lightly damped with a damping coefficient C; f_d denotes the damping force provided by the viscous damper in TVMD, and in this section, only linear viscous damper is considered; thus, $f_d = c_1\dot{y}$; a sinusoidal force of the form $P(t) = P_0\sin(\omega t)$ is considered.

The dynamic magnification factor (DMF), R, of the main structure can be obtained by Laplace transform of Equation (1),

$$R = \left|\frac{X}{X_0}\right| = \sqrt{\frac{A_x^2 + B_x^2}{E^2 + D^2}}, \tag{2}$$

where X and X_0 are the displacement amplitude and static deformation of the main structure. The expressions for A_x, B_x, E, and D are given below

$$\begin{aligned} A_x &= \alpha^2 - \beta^2, \quad B_x = 2\alpha\beta\zeta_1 \\ E &= (1-\beta^2)(\alpha^2-\beta^2) - 4\zeta\zeta_1\beta^2\alpha - \mu\alpha^2\beta^2 \\ D &= 2\zeta\beta(\alpha^2-\beta^2) + 2\alpha\beta\zeta_1(1-\beta^2+\mu\alpha^2) \end{aligned} \tag{3}$$

in which

$$\zeta = \frac{C}{2M\omega_n}, \quad \zeta_1 = \frac{c_1}{2m_e\omega_b}, \quad \mu = \frac{m_e}{M}, \quad \alpha = \frac{\omega_b}{\omega_n}, \quad \beta = \frac{\omega}{\omega_n}, \tag{4}$$

where $\omega_n = (K/M)^{1/2}$ and ζ are the natural frequency and damping ratio of the main structure, respectively; $\omega_b = (k_1/m_e)^{1/2}$ and ζ_1 are the natural frequency and damping ratio of the TVMD, respectively; μ and α are the mass ration and frequency ratio, respectively, between the inerter and the main structure; and β is the forced frequency ratio.

The dashpot deformation amplitude ratio R_y and the supporting spring deformation amplitude ratio R_q can also be obtained from Equation (1),

$$R_y = \left|\frac{Y}{X_0}\right| = \sqrt{\frac{\alpha^4}{E^2 + D^2}}, \tag{5}$$

$$R_q = \left|\frac{Q}{X_0}\right| = \sqrt{\frac{\beta^4 + (2\alpha\beta\zeta_1)^2}{E^2 + D^2}}, \tag{6}$$

where Y and Q are the deformation amplitudes of the viscous damper and supporting spring of the TVMD, respectively.

2.3 | Optimal design of TVMD

The optimal design parameters for a TVMD vary with the optimization criteria and the type of excitations as a TMD does.[28,29] For the optimization of TMD, many approaches have been developed in the literature. For example, Den Hartog[30] derived the optimum parameters of TMD for an undamped primary system subjected to harmonic external excitations by using the fixed-points theory. Ghosh and Basu[31] proposed alternative expressions for the optimum parameters based on the fixed-point theory for light damped structures. In this section, the optimization criterion is to minimize the maximum DMF (min max DMF) of the main structure under harmonic excitation. The procedure of fixed-point theory proposed by Liu and Liu[32] was employed to derive the optimum frequency and damping ratio of TVMD. The mass ratio is assumed as a constant parameter in the design stage in this paper, because as shown in the following section, the control efficiency of TVMD monotonically grows with the mass ratio for an SDOF system subjected to harmonic external excitation.

The derivation is long and tedious, so the detail was not given. The expression of the optimal frequency α_{opt}, the optimal damping ratio $\zeta_{1\text{opt}}$, and the corresponding maximum DMF of the main structure, R_{opt}, are given as following,

$$\alpha_{\text{opt}} = \sqrt{\frac{1}{1-\mu}}; \quad \zeta_{1\text{opt}} = \sqrt{\frac{3\mu}{4(2-\mu)}}; \quad R_{\text{opt}} = (1-\mu)\sqrt{\frac{2}{\mu}}. \tag{7}$$

Equation (7) are identical to the optimal parameters derived by Ikago et al.[9] for TMVD attached to an SDOF system subjected to harmonic ground motion, but are different from those obtained byHu et al.[33] for TVMD used as an isolator for SDOF system.

2.4 | Comparison between TVMD and traditional TMD

For comparison, the optimal parameters of TMD obtained by Den Hartog[30] are given in Equation (8),

$$\alpha_{\text{opt}}^{\text{TMD}} = \frac{1}{1+\mu_b}, \quad \xi_{1\text{opt}}^{\text{TMD}} = \sqrt{\frac{3\mu_b}{8(1+\mu_b)}}, \quad R_{\text{opt}}^{\text{TMD}} = \sqrt{1+\frac{2}{\mu_b}}, \tag{8}$$

where μ_b represents the TMD mass ratio; $\alpha_{\text{opt}}^{\text{TMD}}$ and $\zeta_{1\text{opt}}^{\text{TMD}}$ represent the optimal frequency ratio and optimal damping ratio of the TMD, respectively; and R_{opt}^{TMD} represents the corresponding maximum DMF of the main structure.

The comparison of the optimal parameters for TVMD and TMD is shown in Figure 2. It can be seen from Figure 2a,b that for the same mass ratio, the TMVD performs slightly better than the TMD. However, the mass ratio of the TVMD may be increased from 0.05 to 0.2 by employing the inerter. As a result, the DMF curve of TVMD becomes both lower and flatter, making the TVMD much more effective and robust than the TMD. From Figure 2b, it can be observed that the optimal damping ratio of TVMD is always larger than that of TMD; however, their difference is negligible if the mass ratio is less than 5%. The working strokes of the TVMD and TMD are also compared in Figure 2d, it is observed that the maximum DMF of the inerter and spring in TVMD is almost the same as that of the TMD, indicating that they have the same stroke requirement.

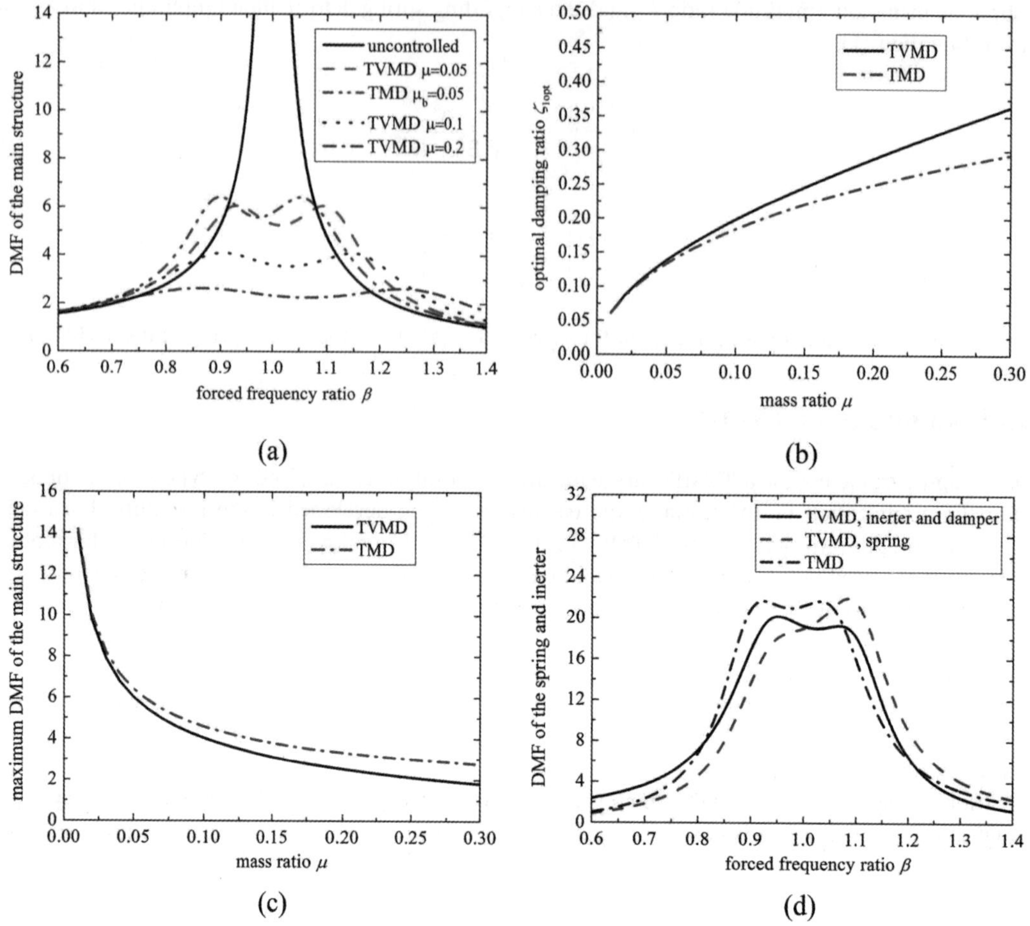

FIGURE 2 Comparison of the optimal parameters for tuned viscous mass damper (TVMD) and tuned mass damper (TMD): (a) dynamic magnification factor (DMF) curve of the main structure for different mass ratios; (b) optimal damping ratio; (c) maximum DMF of the main structure; and (d) DMF curve of the TVMD and TMD for $\mu = 0.05$

It is worth to note that in the case of vertical vibration control of low-frequency structures, the static deformation of the supporting spring of the TMD is another practical and important issue, because it usually determines the installation space requirement of the TMD.[1] For both TVMD and TMD, the vertical static deformation of the supporting spring due to self-weight of the device can be estimated by the following formula,

$$\Delta_s = \frac{mg}{k_1} = \frac{m_e g}{\lambda k_1} = \frac{g}{4\pi^2 f_b^2} \approx \frac{0.25}{\lambda f_n^2}, \tag{9}$$

where m is the secondary mass of the TMD or the total mass of the TVMD including the inerter and the viscous damper; λ is the mass amplifying coefficient for TVMD, and it equals one for TMD; and f_b and f_n are the natural frequency of the TVMD and the main structure, respectively.

It can be seen from Equation (9) that the static deformation of the supporting spring of the TVMD or TMD is determined by the natural frequency of the main structural and the mass amplifying coefficient, which can reach a value of several hundred or even several thousand for TVMD according to some recent studies.[9,12] For example, if $f_n = 0.25$ is assumed, the static deformation of the spring in TMD would be 4 m, whereas it is only 0.04 m for the supporting spring in TVMD if an inerter with $\lambda = 100$ is employed. Therefore, the TVMD provides a favorable solution to the vertical vibration suppression at very low frequencies, which was not mentioned in the previous studies.

2.5 | Influence of natural damping of main structure

The optimal formulas of the TVMD in the last section are derived for undamped main structures. In this section, the influence of the inherent structural damping on the optimal parameters of the TVMD will be discussed. In this case, the Matlab-based Nelder–Mead simplex algorithm (function FMINSEARCH) is used to search the optimal frequency ratio and damping ratio of the TVMD to minimize the DMF of the main structure.

The results of the numerical optimization are summarized in Figure 3. It can be seen that as the structural damping ratio increases, the optimal frequency ratio decreases slightly (Figure 3b), whereas the optimal damping ratio increases slightly (Figure 3c). The fixed-point method can still provide a very good approximation of the optimal parameters of the TVMD and the DMF of the main structure (Figure 3a), even the structure is lightly damped. For example, when $\mu = 0.2$, the error of the optimal frequency ratio obtained by the fixed-point method is within 1%, whereas the error of the optimal damping ratio is within 7%. Figure 3d illustrates the variation of the maximum DMF of the main structure with the structural damping for different mass ratio. It can be seen that if the mass ratio is very large, the structural damping has little influence on the control performance of the TVMD. The optimal frequency ratios of TMD are also given in Figure 3b for comparison. It can be seen that though the optimal frequencies of TVMD and TMD are distinctly different, their tendency of variation with the structural damping ratio are nearly the same.

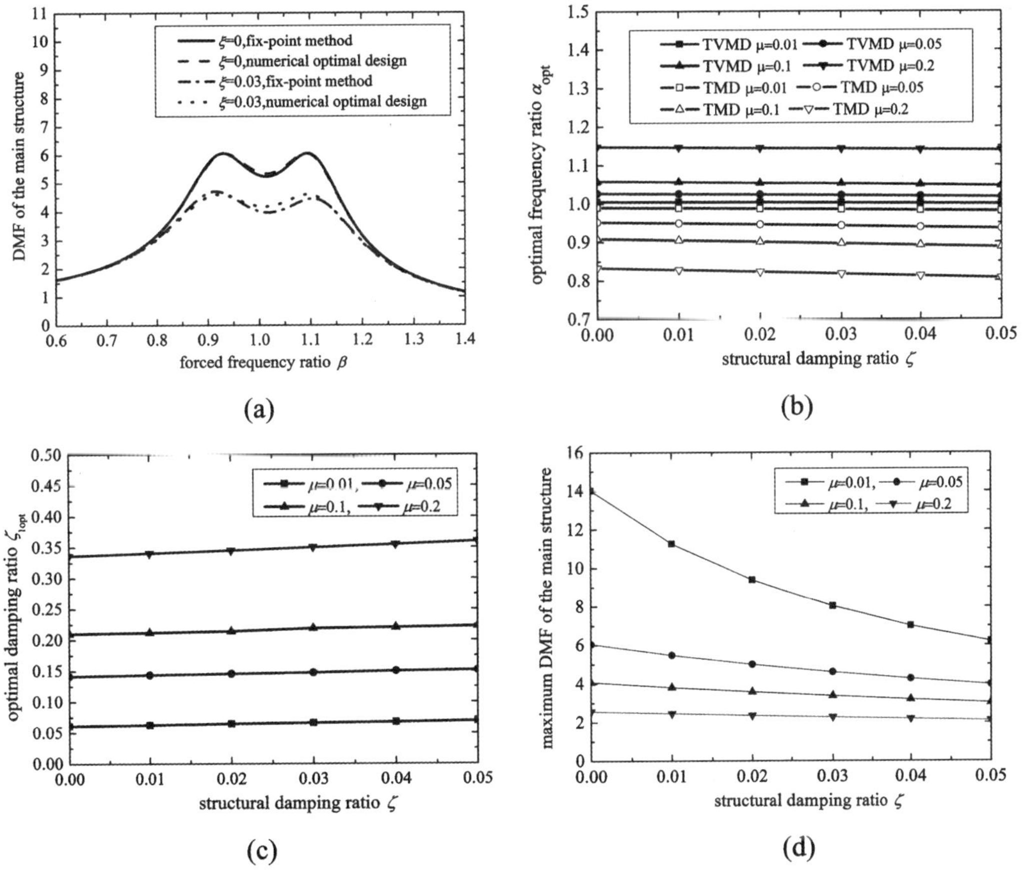

FIGURE 3 Influence of the structural damping on the optimal parameters of the tuned viscous mass damper (TVMD): (a) dynamic magnification factor curves of the main structure for various structural damping for $\mu = 0.05$; (b) optimal frequency ratio; (c) optimal damping ratio; and (d) maximum dynamic magnification factor of the main structure. TMD, tuned mass damper

3 | OPTIMAL DESIGN OF TVMD WITH NONLINEAR FLUID VISCOUS DAMPING

3.1 | Modeling of nonlinear viscous dampers

The force exerted by a nonlinear fluid viscous damper can be usually described by the following equation:

$$f_d = c_d |\dot{y}|^{\nu} \operatorname{sign}(\dot{y}), \tag{10}$$

where c_d denotes the damping coefficient of nonlinear viscous damper and ν denotes the damping exponent, which has a positive value varying from 0 and 2. Particularly, $\nu = 1$ represents a linear viscous damper, whereas $\nu = 0$ represents a friction damper.

For harmonic excitation, the nonlinear damping force can be replaced by an equivalent linear damping force with an equivalent damping coefficient c_e, whereby the energy dissipated by the nonlinear system in a steady-state vibration equals to the energy dissipated by the equivalent linear viscous system.[34] The general expression of c_e corresponding to Equation (10) is given by

$$c_e = c_d \gamma (A_d \omega)^{\nu - 1}, \tag{11}$$

$$\gamma = \frac{2\Gamma\left(\frac{1}{2}(\nu + 2)\right)}{\sqrt{\pi}\,\Gamma\left(\frac{1}{2}(\nu + 3)\right)}, \tag{12}$$

where A_d denotes the amplitude of the relative displacement of the nonlinear viscous damper. Γ denotes the function Gamma. Accordingly, the equivalent damping ratio of the TVMD is given by

$$\zeta_e = \frac{c_d \gamma (A_d \omega)^{\nu - 1}}{2 m_e \omega_1}. \tag{13}$$

In the equivalent linear structure–TVMD system, the amplitude of the relative displacement of the viscous damper is given by

$$A_d = R_y(\zeta_e) \cdot \frac{P_0}{K}, \tag{14}$$

where R_y is the DMF of the viscous damper in TVMD. It is a function of the equivalent damping ratio and has the identical form with Equation (5).

Substituting Equation (14) into Equation (13), Equation (13) can be rewritten as

$$\zeta_e = \frac{c_d \gamma (A_d \omega)^{\nu - 1}}{2 m_e \omega_1} = \frac{c_d \gamma}{2 m_e \omega_1} \left(\omega R_y \frac{P_0}{K} \right)^{\nu - 1}. \tag{15}$$

This equation suggests that the equivalent damping ratio of the TVMD depends on the amplitude and frequency of the harmonic excitation, as well as R_y, which is not known a priori. Therefore, an iterative scheme is needed to determine the equivalent damping ratio and the corresponding R_y for every prescribed excitation. Figure 4 provides a step-by-step flowchart for the mentioned iterative scheme. Moreover, optimization algorithms can be employed, together with the iterative method, to search for the optimal damping coefficient of the nonlinear viscous damping (c_d) and the optimal frequency ratio (α) of the TVMD for prescribed mass ratio μ, damping exponent ν, and the amplitude of the harmonic loading P_0. In this section, the optimization algorithm is the Matlab-based Nelder–Mead simplex algorithm (function FMINSEARCH).

On the other hand, an explicit expression for the optimal c_d in approximate sense can be obtained if we assume that the optimal frequency ratio and the equivalent damping ratio of the nonlinear TVMD are equal to those of the linear TVMD when the forced frequency is equal to the natural frequency of the main structure, that is,

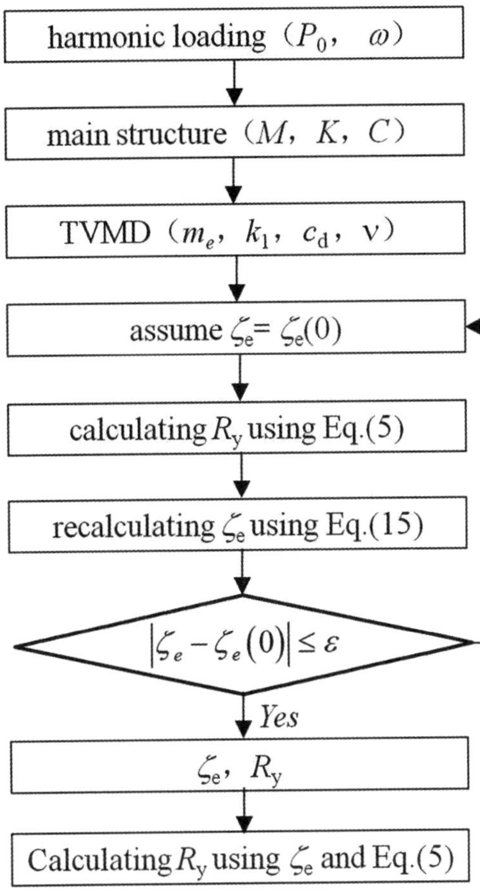

FIGURE 4 Flowchart of the iterative method for the optimization of the nonlinear tuned viscous mass damper (TMVD)

$$\left(\alpha_{\mathrm{opt}}^{\mathrm{nl}} = \alpha_{\mathrm{opt}},\quad \zeta_{\mathrm{e}} = \zeta_{1\mathrm{opt}},\quad \omega = \omega_{\mathrm{n}}\right) \Rightarrow c_{\mathrm{d}} = c_{\mathrm{dopt}}, \tag{16}$$

where $\alpha_{\mathrm{opt}}^{\mathrm{n1}}$ denotes the optimal frequency of the nonlinear TVMD and c_{dopt} denotes the optimal damping coefficient of the nonlinear viscous damper. Rearranging Equation (15), the damping coefficient is

$$c_{\mathrm{d}} = \frac{2m_e\omega_1\zeta_e}{\gamma}\left(\omega R_{\mathrm{y}}\frac{P_0}{K}\right)^{1-\nu}. \tag{17}$$

The corresponding R_{y} can be given by Equation (5)

$$R_{\mathrm{y}} = \frac{\alpha^2}{\sqrt{(4\alpha\zeta\zeta_1 + \mu\alpha^2)^2 + 4[\zeta(\alpha^2-1) + \mu\zeta_1\alpha^3]^2}}. \tag{18}$$

Then substituting Equation (16) and Equation (18) into Equation (17), c_{dopt} can be obtained,

$$c_{\mathrm{dopt}} = \frac{2M\omega_{\mathrm{n}}\mu\alpha_{\mathrm{opt}}\zeta_{1\mathrm{opt}}}{\gamma}\left(\frac{\omega_{\mathrm{n}}\alpha_{\mathrm{opt}}^2}{\sqrt{\left(4\alpha_{\mathrm{opt}}\zeta\zeta_1 + \mu\alpha_{\mathrm{opt}}^2\right)^2 + 4\left[\zeta\left(\alpha_{\mathrm{opt}}^2-1\right) + \mu\zeta_{1\mathrm{opt}}\alpha_{\mathrm{opt}}^3\right]^2}}\frac{P_0}{K}\right)^{1-\nu}. \tag{19}$$

It is worth to mention that the optimization criteria shown in Equation (16) is much weaker than that used in the case of TMD with nonlinear viscous damping for SDOF system under white noise excitation. Under harmonic loading, the equivalent damping ratio of the nonlinear viscous damper is always dependent on the forced frequency, so$\zeta_e = \zeta_{1opt}$ cannot be satisfied at every forced frequency for a prescribed damping coefficient, whereas this is not the case for white noise excitation. As a result, in order to obtain an analytical expression for c_{dopt}, only the resonant frequency $\omega = \omega_n$ is considered to satisfy $\zeta_e = \zeta_{1opt}$ for harmonic excitation.

3.2 | Optimal design of TVMD

A numerical simulation is conducted to illustrate the influence of the damping exponent of the nonlinear viscous damper on the optimal parameters of the TVMD, including the optimal frequency ratio and damping coefficient and the minimum DMF of the main structure. The mass of the main structure $M = 40$ ton, the equivalent stiffness $K = 2{,}139.6$ kN, thus the natural frequency $f_n = 1.164$ Hz, the amplitude of the harmonic load $P_0 = 6{,}000$ N and the forced frequency ranges from $0.6f_n \sim 1.4f_n$. The optimal parameters are calculated from both the optimal algorithm and the analytical expression.

In Figure 5, the maximum DMF of the main structure is shown as a function of the damping exponent for different structural damping ratio for $\mu = 0.01$ and $\mu = 0.1$. It can be seen that the analytical solution of c_{dopt} agrees well with that of the numerical optimization. The maximum DMF remains almost the same for different damping exponent when $\mu = 0.01$, whereas it grows slightly as the damping exponent increases when $\mu = 0.1$. These results are similar with those obtained by Chung et al.[3] for TMD with nonlinear viscous damping attached to wind-excited structures.

The influence of the damping exponent on the optimal parameters of the nonlinear TVMD is shown in Figure 6. It can be seen from Figure 6a that the optimal frequency ratio increases slightly with the increasing damping exponent for all the three different structural damping ratios. Though the error of the analytical expression obtained from the linear TVMD for undamped structure increases with the structural damping ratio, the biggest error is still within 2% for $\nu = 0.5$ and $\zeta = 3\%$. It can be seen from Figure 6b that the optimal damping coefficient increases with the damping exponent. Especially, as the damping exponent exceeds beyond 1.50, the optimal damping coefficient increases dramatically. When$\zeta = 1.0\%$, the optimal damping coefficient for$\nu = 2.0$ is about 7 times of that for $\nu = 1.0$ and about 32 times of that for$\nu = 0.5$. It is obvious that the structural damping has a great influence on the optimal damping coefficient for $\nu = 2.0$, which is distinctly different from the linear TVMD. Besides, the analytical expression of c_{dopt} has a good accuracy for the cases of $\nu \leq 1.0$, but its error increases significantly as the damping exponent exceeds 1.50. However, as shown in Figure 6, the minimum DMF of the main structure seems not sensitive to the damping coefficient of the nonlinear viscous damper, so the analytical expression of c_{dopt} can still be used for the cases of $\nu > 1.0$.

The DMF curves of the main structure for three different damping exponents for $\mu = 0.1$ and $\zeta = 0.01$ are obtained by the numerical optimization and shown in Figure 7a. It can be seen that they are almost identical to each other, even though $\nu = 0.5$ shows the best vibration reduction at nearly all the forced frequency ranges. Correspondingly, the

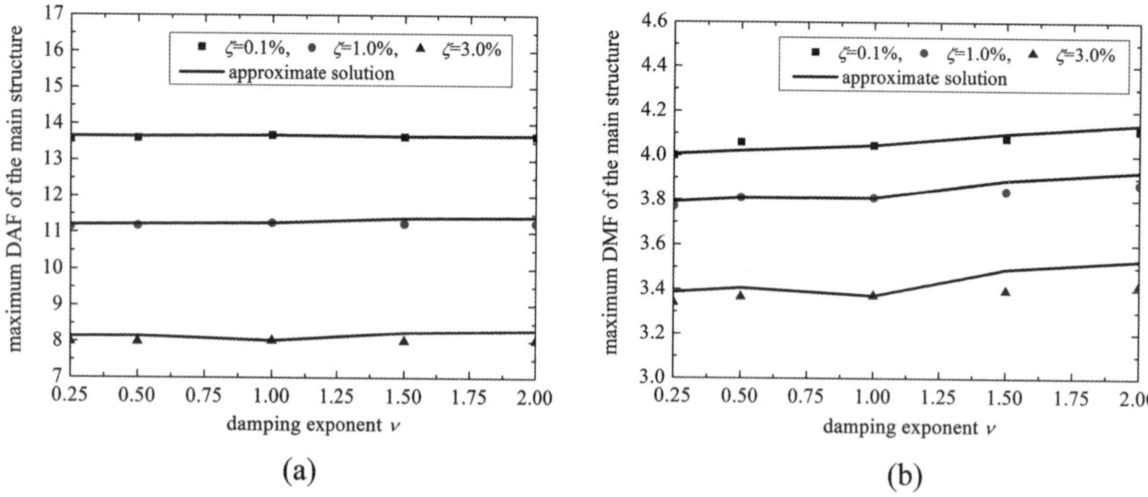

FIGURE 5 Influence of the damping exponent on the maximum dynamic magnification factor (DMF) of the main structure: (a) $\mu = 0.01$ and (b) $\mu = 0.1$

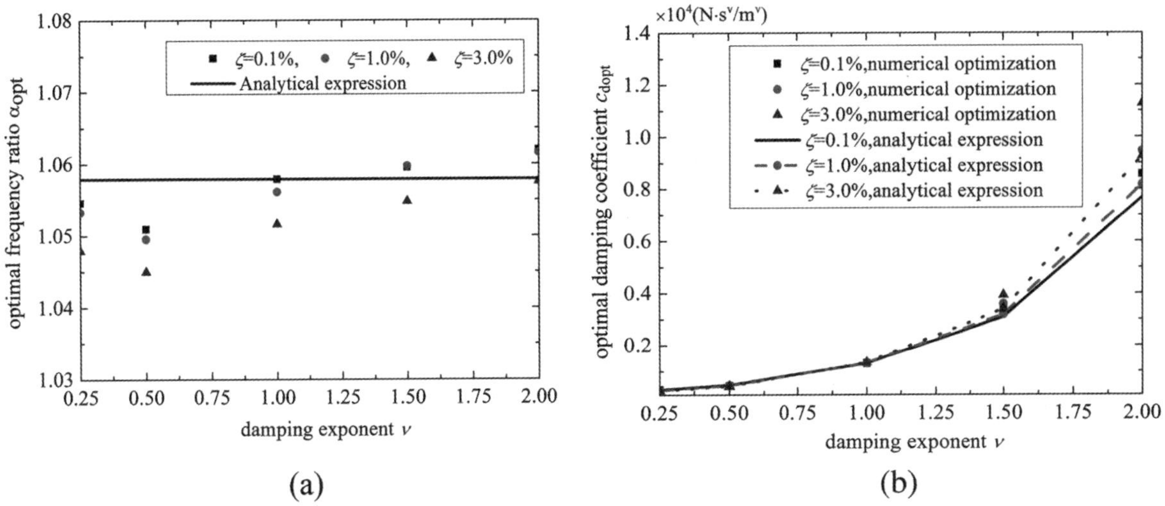

FIGURE 6 Influence of the damping exponent on the optimal parameters of the nonlinear tuned viscous mass damper for $\mu = 0.1$: (a) optimal frequency ratio and (b) optimal damping coefficient

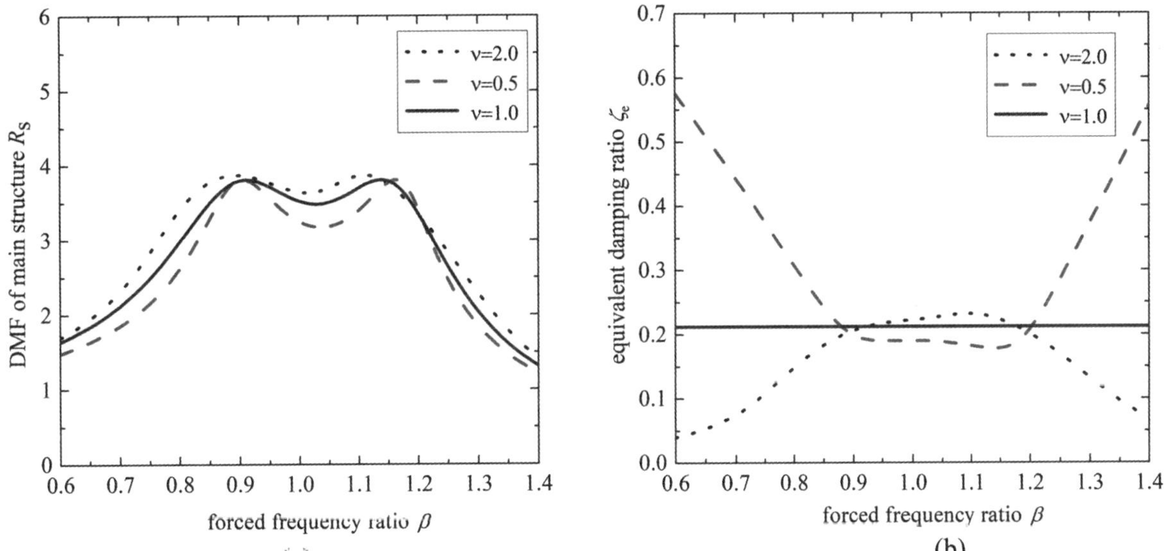

FIGURE 7 Influence of the damping exponent on the dynamic magnification factor curve of the main structure and the effective damping ratio for $\mu = 0.1$ and $\zeta = 0.01$: (a) Dynamic magnification factor curves of the main structure and (b) equivalent damping ratio of the nonlinear tuned viscous mass damper

equivalent damping ratios of the TVMD are shown in Figure 7b. It can be observed that for both $\nu = 0.5$ and $\nu = 2.0$, the equivalent damping ratios are approximately equal with the linear TVMD system in the resonant frequency range, which validates the assumptions in the derivation of the optimal damping coefficient. Outside the resonant frequency region, the nonlinear viscous damper with $\nu = 0.5$ has much larger equivalent damping ratio than the linear viscous damper, and it is contrary for that with $\nu = 2.0$.

3.3 | Sensitivity analysis

The previous section has shown that the displacement reduction of the main structure is not very sensitive to the damping coefficient of the nonlinear viscous damper employed in the TVMD, which is in agreement with the results in Chung et al.[26] for nonlinear TMD under wind excitations. In this section, the sensitivity of the maximum DMF of the main structure to the intensity of the external excitations is analyzed.

Figure 8 shows the relative increment of the maximum DMF of the main structure to its optimal value when the intensity of the harmonic excitation changes. It can be seen that for both $\nu = 0.5$ and $\nu = 2.0$, the effectiveness of the nonlinear TVMD decreases when the amplitude of the harmonic loading either increases or decreases from the design value P_0. Comparing Figure 8a with Figure 8b, it can be found that the sensitivity of the maximum DMF of the main structure to the loading intensity increases with the mass ratio and decreases with the damping exponent. For example, for $\mu = 0.1$ and $P_i/P_0 = 1.20$, the relative increment of the maximum DMF is about 6% for $\nu = 0.5$ and about 15% for $\nu = 1.0$. Therefore, the parameters design of the nonlinear TVMD requires careful evaluation of the loading intensity that a structure sustains.

3.4 | Transient response of the nonlinear TVMD system

The transient responses of the nonlinear TVMD system are further compared for three different damping exponent $\nu = 0.5$, 1, and 2 at the fixed mass ratio $\mu = 0.01$, structural damping ratio $\zeta = 0.1\%$, and harmonic loading $P_0 = 6{,}000 \cdot \sin(\omega_n t)$.

Figure 9a shows the time history of the structural displacement. The steady-state displacement responses of the main structure are 3.14, 3.33, and 3.36 cm for $\nu = 0.5$, 1.0, 2.0, respectively, whereas the peak displacement responses are 4.52, 3.80, and 3.36 cm, respectively. This indicates that the nonlinear TVMD with $\nu = 2.0$ shows the best performance on suppressing the peak response the main structure whereas the nonlinear TVMD with $\nu = 0.5$ shows the best performance on suppressing the steady-state response. To explain this result, the time histories of the damping forces of the three nonlinear viscous dampers are compared in Figure 9b. The steady-state damping forces are 610.4, 718.5, and 853.1 N for $\nu = 0.5$, 1.0, 2.0, respectively, whereas the peak damping forces are 688.1, 778.5, and 855.3 N, respectively. This indicates that in steady state, the nonlinear TVMD with $\nu = 2.0$ generates the largest damping force whereas the nonlinear TVMD with $\nu = 0.5$ generates the smallest. Moreover, it can be seen from Figure 9a that the damping force grows fastest for the case of $\nu = 2.0$ after the first 2 s, so it is able to dissipate the energy input from the external loading in a very short time and keep the main structure from a pulse-like peak response.

4 | CONTROL PERFORMANCE OF TVMD FOR WIND EXCITATION

Though the TVMD is optimized for harmonic excitations in the previous sections, it would be interesting to see how it performs under random excitations such as wind. A numerical simulation is conducted in this section to evaluate the performance of TVMD for wind excitation. Only TVMD with linear viscous damping is considered here, because the optimal parameters and performance of TVMD with nonlinear viscous damping depend on the intensity of the external loading; thus, the optimal design formulas derived for sinusoidal excitation may be not suitable for random excitations.

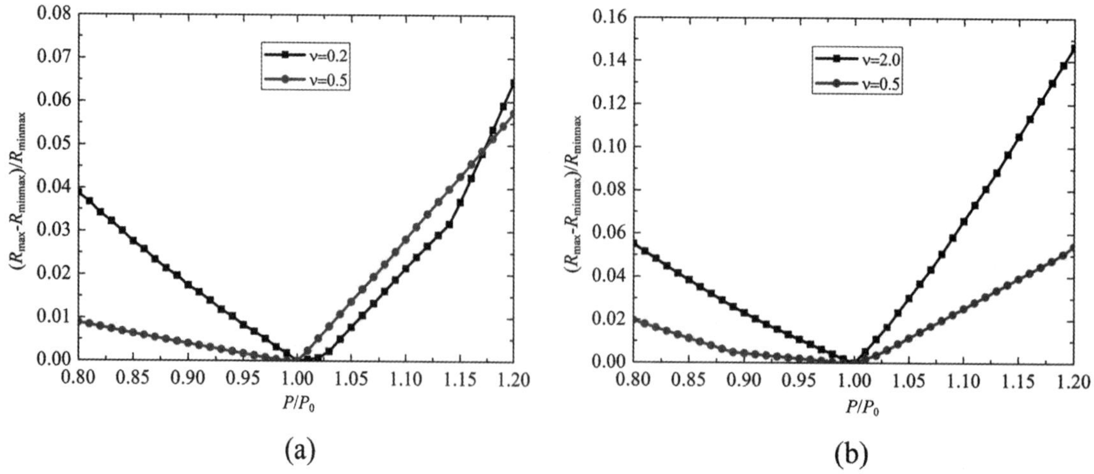

FIGURE 8 Sensitivity of the maximum dynamic magnification factor of the main structure to the intensity of the external excitation for $\zeta = 0.1\%$: (a) $\mu = 0.01$ and (b) $\mu = 0.1$

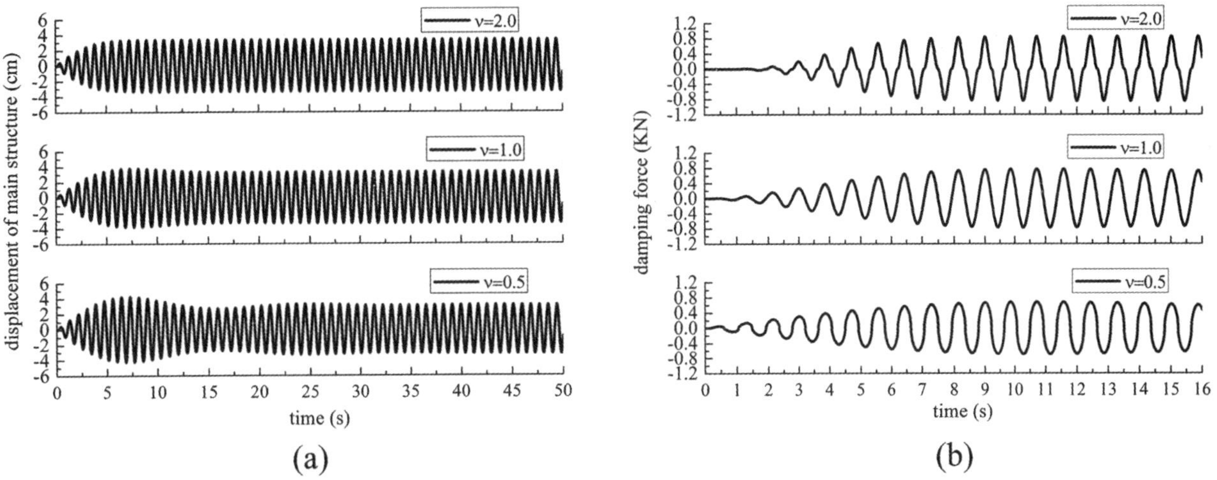

FIGURE 9 Transient response of the structure–TVMD system: (a) time history of the structural displacement and (b) time history of the damping force of the nonlinear viscous damper

The mass of the main structure is assumed to be $M = 40$ ton, whereas the natural frequency and structural damping ratio are $f_n = 0.116$ Hz and $\zeta = 1\%$, respectively. The mass ratio of TVMD is selected as $\mu = 0.01$. Substituting the mass ratio into Equation (7) yields the optimal frequency ratio α_{opt} and the optimal damping ratio ζ_{1opt}, which are 1.005 and 0.0614, respectively. Moreover, the performance of TMD is also evaluated for comparison, for which the optimum parameters are calculated from Equation (8). As shown in Figure 10a, the time-history wind force applied to the main structure is obtained based on the Kaimal wind speed power spectrum and the wave superposition method.[35] It is

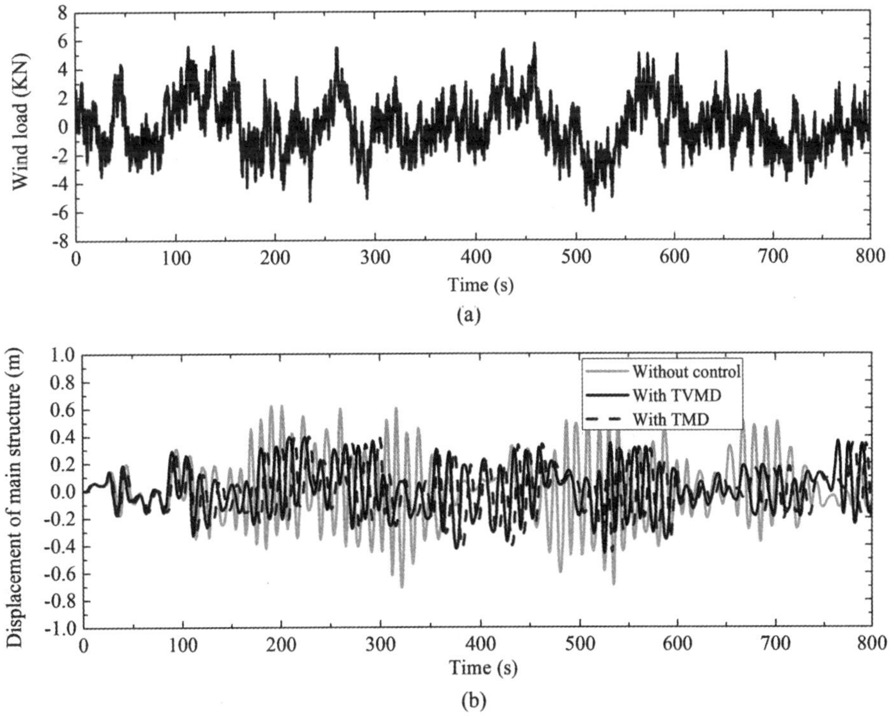

FIGURE 10 Control performance of tuned viscous mass damper (TVMD) for single degree-of-freedom structure under wind excitation: (a) time history of the wind force and (b) time history of the structural displacement with and without control. TMD, tuned mass damper

TABLE 1 Comparison of TVMD and TMD performance for wind excitation

Performance index	Without control	With TVMD	Reduction ratio	With TMD	Reduction ratio
$\lvert x\rvert_{max}$	0.710	0.458	35.5%	0.449	36.8%
$\lvert x\rvert_{RMS}$	0.219	0.161	26.5%	0.161	26.5%

Abbreviations: TMD, tuned mass damper; TVMD, tuned viscous mass damper.

obvious from this figure that the maximum instantaneous wind force has been normalized to be 6,000 N, and only the wind force contributed from the fluctuating wind gusts is considered.

Figure 10b shows the time-history displacement response of the main structure with and without control, and the results are summarized in Table.1. It is observed that the control performance of TVMD and TMD are almost the same. For TVMD, the maximum displacement of the main structure reduces by 35.5%, whereas the RMS displacement reduces by 26.5% after the installation of TVMD, which means that the TVMD optimized for harmonic excitation can also effectively suppress the structural vibration induced by wind.

5 | CONCLUSION

In this study, the optimal designs of TVMD with linear and nonlinear viscous damping for SDOF structures subjected to harmonic excitation were considered. Both theoretical derivation and numerical optimization were conducted to obtain the optimal parameters of the TVMD.

For TVMD with linear viscous damper, the optimal frequency ratio, optimal damping ratio, and the strokes of the TVMD, as well as the minimum DMF of the main structure, were obtained by the fixed-point method for undamped host structure and compared with those of the traditional TMD. The influence of the inherent damping of the main structure on the optimal parameters was also unclosed. The results show that the TVMD has a slightly better performance on the displacement reduction of the main structure, compared with the TMD having the same mass ratio. Moreover, the TVMD can take advantages of the inerter to obtain significantly large mass ratio up to 10% or even larger, so much better control effectiveness and robustness can be achieved. Another potential benefit from the employment of the inerter is that the static deformation of the supporting spring of the TVMD can be reduced to 1% or even smaller of that of the traditional TMD, when they are applied to vertical vibration control. Both the optimal frequency ratio and the optimal damping ratio of the TVMD are larger than those of the TMD, and they both increases with the increasing mass ratio. Additionally, the inherent damping of the main structure only has a little influence on optimal frequency ratio and optimal damping ratio of the TVMD. It is interesting to see from the numerical simulation that the linear TMVD designed for harmonic excitation can also effectively suppress the wind-induced vibration.

For TVMD with nonlinear viscous damper, equivalent linear damping ratio had been used for the linearization of the equations of motion, and an iterative method had been proposed to obtain the equivalent damping ratio and the corresponding system response. Then numerical optimization was carried out to obtain the optimal parameters of the nonlinear TVMD, and an analytical expression for the optimal damping coefficient was also derived based on the formulas of the optimal parameters of the linear TVMD. The results show that the damping exponent has only slightly influence on the minimum DMF and DMF curve of the main structure, as well as the optimal frequency ratio of the TVMD, indicating that the nonlinear TVMD can achieve comparable or even better control performance compared with the linear TVMD. On the other hand, the optimal damping coefficient depends largely on the damping exponent; for$\xi = 1.0\%$ and $\mu = 0.1$, the optimal damping coefficient at $\nu = 2.0$ is about 7 times the damping coefficient at $\nu = 1.0$ and about 32 times the damping coefficient at $\nu = 0.5$. The inherent damping also affects the optimal damping coefficient for nonlinear TVMD, which is very different from that of the linear TVMD. Additionally, the bias of the loading intensity from the design value will deteriorate the performance of the nonlinear TVMD. Moreover, transient response analysis was conducted by direct numerical integration, and the results show that the nonlinear TVMD with $\nu = 2.0$ are more efficiency in reducing the peak response of the main structure.

In this study, only the harmonic excitation was considered. In practical engineering, the excitation may be very different and much more complex, such as the wind excitation and base excitation due to the earthquake. This should be considered in the future study.

ACKNOWLEDGEMENTS

This study is sponsored by the National Natural Science Foundation of China with the program number 51808210 and by the State's Key Project of Research and Development Plan (2016YFE0127900 and 2017YFC07036002).

ORCID

Xugang Hua https://orcid.org/0000-0001-6150-2563

REFERENCES

1. Fujino Y, Yoshida Y. Wind-induced vibration and control of trans-Tokyo Bay crossing bridge. *J Struct Eng*. 2002;128(8):1012-1025.
2. Lu X, Zhang Q, Weng D, et al. Improving performance of a super tall building using a new eddy-current tuned mass damper. *Struct Control Health Monit*. 2016;24(3):e1882.
3. Chung LL, Wu LY, Yang CSW, Lien KH, Lin MC, Huang HH. Optimal design formulas for viscous tuned mass dampers in wind-excited structures. *Struct Control Health Monit*. 2013;20(3):320-336.
4. Morga M, Marano GC. Optimization criteria of TMD to reduce vibrations generated by the wind in a slender structure. *J Vib Control*. 2014;20(16):2404-2416.
5. Tributsch A, Adam C. Evaluation and analytical approximation of tuned mass damper performance in an earthquake environment. *Smart Struct Syst*. 2012;10(2):155-179.
6. Schmelzer B, Oberguggenberger M, Adam C. Efficiency of tuned mass dampers with uncertain parameters on the performance of structures under stochastic excitation. *Proc Inst Mech Eng Part O: J Risk Reliab*. 2010;224(4):297-308.
7. Ghosh A, Basu B. Effect of soil interaction on the performance of tuned mass dampers for seismic applications. *J Sound Vib*. 2004;3(274):1079-1090.
8. Yamaguchi H, Harnpornchai N. Fundamental characteristics of multiple tuned mass dampers for suppressing harmonically forced oscillations. *Earthq Eng Struct Dyn*. 1993;22(1):51-62.
9. Ikago K, Saito K, Inoue N. Seismic control of single-degree-of-freedom structure using tuned viscous mass damper. *Earthq Eng Struct Dyn*. 2012a;41(3):453-474.
10. Lazar IF, Neild SA, Wagg DJ. Using an inerter-based device for structural vibration suppression. *Earthq Eng Struct Dyn*. 2014;43(8):1129-1147.
11. Marian L, Giaralis A. Optimal design of a novel tuned mass-damper–inerter (TMDI) passive vibration control configuration for stochastically support-excited structural systems. *Probab Eng Eng Mech*. 2014;38:156-164.
12. Smith MC. Synthesis of mechanical networks: the inerter. *IEEE Trans Autom Control*. 2002;47(10):1648-1662.
13. Chen MZ, Papageorgiou C, Scheibe F, Wang F C, Smith MC. The missing mechanical circuit element. *IEEE Circuits Syst Mag*. 2009;9(1):10-26.
14. Chen ZQ, Huang ZW, Hua XG. Inerter damper-spring passive vibraiton control: its system reliazation with eddy-current mass dampers, In Proceedings of 14th World Conference on Seismic Isolation, Energy dissipation and active control of Structures, 2015 9–11 September 2015, San Diego, USA
15. Pan C, Zhang R, Luo H, Li C, Shen H. Demand-based optimal design of oscillator with parallel-layout viscous inerter damper. *Struct Control Health Monit*. 2018;25(1):e2051.
16. Ikago K, Sugimura Y, Saito K, Inoue N. Modal response characteristics of a multiple-degree-of-freedom structure incorporated with tuned viscous mass dampers. *J Asian Archit Build Eng*. 2012b;11(2):375-382.
17. Wen Y, Chen Z, Hua X. Design and evaluation of tuned inerter-based dampers for the seismic control of MDOF structures. *J Struct Eng*. 2016;143(4):04016207.
18. Zhang SY, Jiang JZ, Neild S. Optimal configurations for a linear vibration suppression device in a multi-storey building. *Struct Control Health Monit*. 2016;24(3):e1887.
19. Lazar IF, Neild SA, Wagg DJ. Vibration suppression of cables using tuned inerter dampers. *Eng Struct*. 2016;122:62-71.
20. Pietrosanti D, De Angelis M, Basili M. Optimal design and performance evaluation of systems with tuned mass damper inerter (TMDI). *Earthq Eng Struct Dyn*. 2017;46(8):1367-1388.
21. Giaralis A, Petrini F. Wind-induced vibration mitigation in tall buildings using the tuned mass-damper-inerter. *J Struct Eng*. 2017;143(9):04017127.
22. Mirza Hessabi R, Mercan O. Investigations of the application of gyro-mass dampers with various types of supplemental dampers for vibration control of building structures. *Eng Struct*. 2016;126:174-186.
23. Rüdinger F. Optimal vibration absorber with nonlinear viscous power law damping and white noise excitation. *J Eng Mech*. 2006;132(1):46-53.

24. Rüdinger F. Tuned mass damper with nonlinear viscous damping. *J Sound Vib.* 2007;300(3):932-948.
25. Shum KM. Tuned vibration absorbers with nonlinear viscous damping for damped structures under random load. *J Sound Vib.* 2015;346:70-80.
26. Chung L-L, Wu L-Y, Huang H-H, Chang C-H, Lien K-H. Optimal design theories of tuned mass dampers with nonlinear viscous damping. *Earthq Eng Eng Vib.* 2009;8(4):547-560.
27. Nakamura Y, Fukukita A, Tamura K, et al. Seismic response control using electromagnetic inertial mass dampers. *Earthq Eng Struct Dyn.* 2014;43(4):507-527.
28. Warburton G. Optimum absorber parameters for various combinations of response and excitation parameters. *Earthq Eng Struct Dyn.* 1982;10(3):381-401.
29. Marano GC, Greco R, Chiaia B. A comparison between different optimization criteria for tuned mass dampers design. *J Sound Vib.* 2010;329(23):4880-4890.
30. Den Hartog JP. *Mechanical Vibrations.* 4th ed. New York, NY, USA: McGraw-Hill; 1956.
31. Ghosh A, Basu B. A closed-form optimal tuning criterion for TMD in damped structures. *Struct Control Health Monit.* 2007;14(4):681-692. https://doi.org/10.1002/stc.176
32. Liu K, Liu J. The damped dynamic vibration absorbers: revisited and new result. *J Sound Vib.* 2005;284(3):1181-1189.
33. Hu Y, Chen MZ, Shu Z, Huang L. Analysis and optimisation for inerter-based isolators via fixed-point theory and algebraic solution. *J Sound Vib.* 2015;346:17-36.
34. Soong TT, sCostantinou MM. *Passive and Active Structural Vibration Control in Civil Engineering.* NewYork: Springer-Verlag; 1994.
35. Scanlan RH. *Wind Effects on Structure, Fundamentals and Alications to Design.* Third ed. New York: John Willey & Sons Inc.; 1996.

第40卷 第4期
2020年8月
地 震 工 程 与 工 程 振 动
EARTHQUAKE ENGINEERING AND ENGINEERING DYNAMICS
Vol. 40 No. 4
Aug. 2020

文章编号:1000－1301(2020)04－0118－10　　DOI:10.13197/j.eeev.2020.04.118.lianglt.015

大跨度悬索桥加劲梁纵向运动特性及其电涡流阻尼控制研究

梁龙腾[1],封周权[1,2],陈政清[1,2],牛华伟[1,2],华旭刚[1,2]

(1. 湖南大学 风工程与桥梁工程湖南省重点实验室,湖南 长沙 410082;
2. 湖南大学 汽车车身先进设计制造国家重点实验室,湖南 长沙 410082)

摘　要:大跨度悬索桥在交通荷载和环境激励下,加劲梁产生持续不断的纵向往复运动,这类振动会导致悬索桥上的连接构件如伸缩缝等出现疲劳破坏。抑制这类振动一般是在梁端安装阻尼装置,这类装置不仅能够减少运营状态下的梁端位移,还能减弱地震作用下的梁端位移响应。本文提出采用耐久性能良好的新型电涡流阻尼器(ECD)进行梁端位移控制并对其减振效果进行了评估。首先对某大跨度悬索桥的实测纵向位移数据进行时域和频域分析,通过简化建立了悬索桥纵向振动的单自由度模型(SDOF),并反演了运营状态下的等效外力时程。然后对悬索桥简化SDOF模型在运营状态等效外力时程作用下和地震激励下的响应进行了计算,并评估了ECD在运营低速状态下线性阻尼和地震高速状态下非线性阻尼的减振(震)控制效果,提出了平衡两种状态控制目标的阻尼器参数合理取值范围,为ECD在悬索桥加劲梁纵向位移控制领域的应用提供了一定的理论依据。

关键词:大跨度悬索桥,纵向振动;电涡流阻尼器;振动控制;阻尼器参数
中图分类号:TU4　　**文献标志码**:A

Longitudinal movement characteristics of long span suspension bridge girder and its control based on eddy current dampers

LIANG Longteng[1], FENG Zhouquan[1,2], CHEN Zhengqing[1,2], NIU Huawei[1,2], HUA Xugang[1,2]

(1. Key Laboratory of Wind and Bridge Engineering of Hunan Province, Hunan University, Changsha 410082, China,
2. State Key Laboratory of Advanced Design and Manufacturing for Vehicle Body, Hunan University, Changsha 410082, China)

Abstract: Under the action of traffic load and ambient excitation, the main girder of long-span suspension bridge will produce continuous longitudinal movement, which will lead to fatigue failure of connecting members on the suspension bridge, such as expansion joints. To suppress this kind of movement, damping devices are usually installed at the girder ends, which can not only reduce longitudinal displacement under the operation state, but also attenuate the longitudinal response under earthquakes. In this paper, a new type of eddy current damper (ECD) with good durability is proposed to control the girder longitudinal movement and the performance of movement reduction is evaluated. Firstly, the measured longitudinal displacement data of a long-span suspension bridge are analyzed in

收稿日期:2019－12－08;　**修订日期**:2020－02－23
基金项目:国家自然科学基金项目(51708203);湖南大学汽车车身先进设计制造国家重点实验室自主研究课题(71865006,71865004);国家重点研发计划"先进轨道交通"重点专项项目(2018YFB1201703)
Supported by: National Natural Science Foundation of China (51708203); The Independent Research Project Funds From the State Key Laboratory of Advanced Design and Manufacturing for Vehicle Body in Hunan University (71865006, 71865004); National Key R&D Program of China (2018YFB1201703)
作者简介:梁龙腾(1988－),男,博士研究生,主要从事电涡流阻尼减振方面研究. E-mail: llt_cehnu@hnu.edu.cn
通讯作者:封周权(1982－),男,助理教授,博士,主要从事风工程,结构监测与振动控制研究. E-mail: zqfeng@hnu.edu.cn

time domain and frequency domain. Secondly, a simplified equivalent single degree-of-freedom(SDOF) model is established for longitudinal movement analysis of the suspension bridge, and the time history of the equivalent external force in operation state is identified based on the SDOF model and the measured longitudinal displacement data. Thirdly, the response of the SDOF model under the action of external force in operation state is calculated and the corresponding linear damping control performance of ECD at low speed is evaluated. Fourthly, the seismic response of the SDOF model is calculated and the corresponding nonlinear damping control performance of ECD at high speed is evaluated. Lastly, a reasonable range of damper parameters for balancing control targets of two states of operation state and seismic state is proposed. It provides a theoretical basis for the application of ECD in longitudinal movement reduction of suspension bridge girders.

Key words: long span suspension bridge; longitudinal vibration; eddy current damper; vibration control; damper parameters

引言

大跨度悬索桥在环境激励下,会引起加劲梁持续不断的往复纵向运动,虽然不会引起结构毁灭性的破坏现象,但是这种频繁的纵向运动会引起加劲梁的附属构件如伸缩缝的疲劳破坏[1-2],如江阴长江大桥在通车的4年后即出现伸缩缝损坏等病害[3]。找到一种既能够有效控制纵向运动,又具有优良耐久性能抵御疲劳破坏的手段,是保证悬索桥安全平稳运行的关键。

大跨度悬索桥的日常运营状态下的纵向运动包含了受环境温度变化引起的加劲梁热胀冷缩而出现的梁端缓慢纵向位移和由车辆荷载、风荷载等环境激励引起的往复纵向位移响应。Ni 等[4]和邓扬等[5]分别对汀九大桥和润扬大桥加劲梁梁端位移与温度的关系进行了研究,其相关性近似为线性。Snæbjörnsson 等[6]基于主跨446m 的 Lysefjord 桥的实测数据分析了抖振和车辆荷载的影响,王连华等[7]基于杭瑞洞庭大桥的有限元模型分析了车辆荷载对加劲梁纵向振动的影响,并讨论了中央扣对降低加劲梁纵向振幅的效果。Larsen 等[8]通过建立有限元模型和风洞试验对由风荷载引起的纵向振动响应进行了分析。研究者们对日常运营状态下加劲梁纵向运动的各个成分分别进行了探讨,但对实测加劲梁纵向运动数据进行分解并辨识各成分构成的相关研究还相对较少。本文中引入了由 Huang 等[9]提出的经验模态分解(EMD) 方法对悬索桥的实测加劲梁纵向运动数据进行分解,将混在一起各类特性的纵向运动进行剥离,分离后的信号在频域上进行辨识。EMD 是一种自适应信号时频处理方法,它能将复杂信号分解为有限个本征模函数(Intrinsic Mode Function,简称 IMF) ,所分解出来的各 IMF 会自动按照频率从高到低进行排列,而不需要进行人为干预。这种方法是直观的、直接的、后验的和自适应的。EMD 方法在土木工程领域得到了许多应用,Yang 等[10]研究了使用 EMD 方法对建筑风致振动响应进行自振频率识别; 陈隽等[11]采用 EMD 方法分解青马大桥实测数据,获得了结构的模态响应分量。

黏滞阻尼器作为常见的耗能减振装置用于悬索桥的振动控制[12],牛建涛等[13]将黏滞阻尼器和支座结合用于降低塔梁之间的地震位移响应; 王浩等[14]探讨了控制超大跨悬索桥地震响应的各类措施,指出黏滞阻尼器的减振效果更全面; 丁幼亮等[15]应用黏滞阻尼器对嘉绍大桥的风致抖振进行振动控制,减振率达到50%左右。但黏滞阻尼器的疲劳失效问题也不容忽视,Sun 等[16]指出安装在江阴长江大桥上的黏滞阻尼器出现了油泄露并更换的情况。本文采用由陈政清等[17]开发的轴向电涡流阻尼器(ECD) 作为减振控制手段。ECD 作为一种无工作流体的全金属阻尼器,启动摩擦小,维护简单,耐久性能优越,能够有效地抵抗纵向振动导致的疲劳破坏。ECD 作为一种新型的阻尼减振(震) 装置,其力学特性不同于黏滞阻尼器,在低速状态下具有较好的线性阻尼特性,而在高速状态下则呈现出较大的非线性特征[18, 19]。对于悬索桥的纵向位移控制,ECD 不仅要满足地震作用下的减震需求,还要满足日常运营状态下的减振要求。

本文以某悬索桥未安装阻尼器情况下的实测加劲梁梁端纵向位移时程作为分析对象,进行时域和频域分析,建立等效 SDOF 模型,并反演了运营状态下的等效外力时程; 然后对悬索桥简化 SDOF 模型在运营状态等效外力时程作用下和地震激励下的响应进行了计算,并评估了 ECD 在运营低速状态下线性阻尼和地震高速状态下非线性阻尼的减振(震) 控制效果; 最后提出了平衡两种状态控制目标的阻尼器参数合理取值范围。

1　悬索桥实测纵向运动分析

以某悬索桥未安装阻尼器时加劲梁 2006 年 3 月 30 日(24 h) 实测梁端纵向运动时程曲线作为分析对象,其中不仅包含了车辆和风等激励引起的振动成分,也包含了由温度变化引起的缓慢纵向飘移[20]。因材料随着温度变化产生变形而引起的梁端纵向位移一般不作为纵向运动控制目标,所以需要将其从位移时程曲线中剥离出来。采用 EMD 方法将原信号分解为一系列的本征模函数和残差。实测数据与 EMD 分解中的残差成分进行对比如图 1(a) 所示,从图 1(a) 中可以看到,残差曲线代表了纵向位移一天内的总体趋势,而由温度引起的梁端位移变化可以视为周期是 24 h 的纵向运动,其频率为 1/86 400 Hz,残差曲线的运动特性与温度变化产生的梁端位移变化特性相符合,该趋势也与文献[20]所描述的相似。而所有本征模函数 IMF 相加的曲线是一条围绕平衡位置往复运动的曲线,即在时域中剔除了由温度变化引起的悬索桥纵向位移的其他位移成分之和的时程曲线,如图 1(b) 所示。

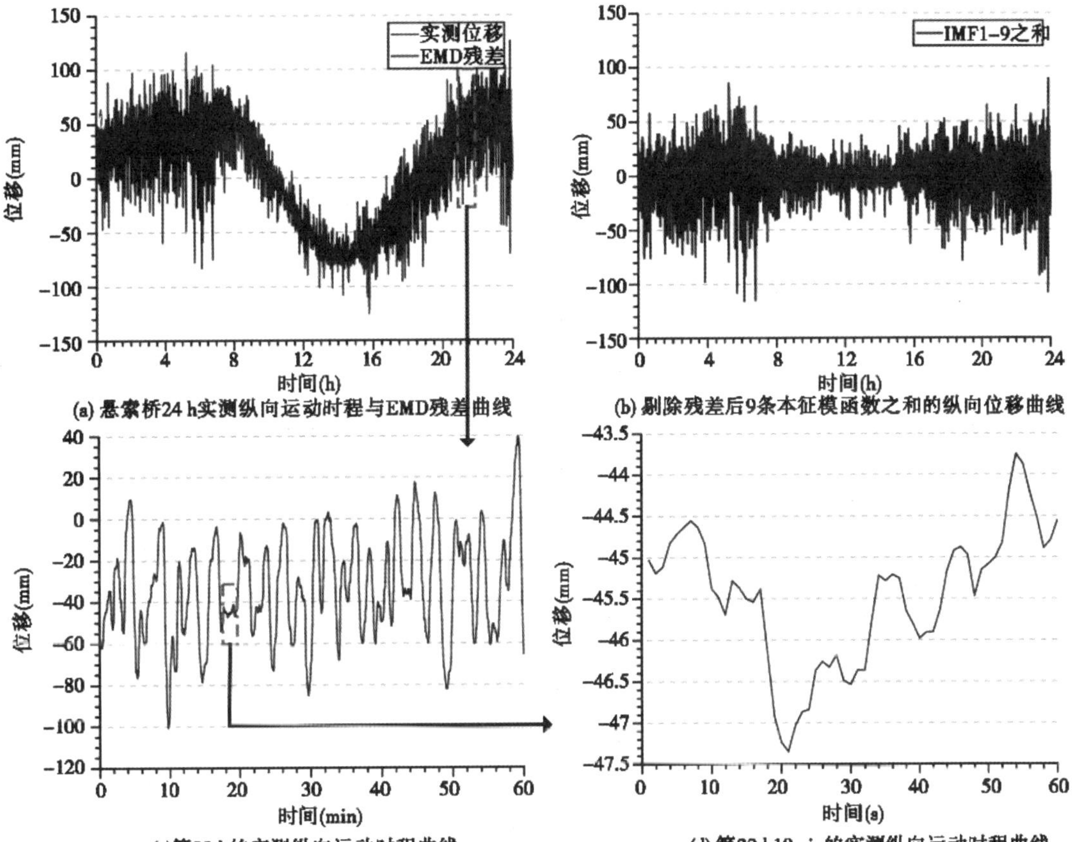

(a) 悬索桥24 h实测纵向运动时程与EMD残差曲线　(b) 剔除残差后9条本征模函数之和的纵向位移曲线

(c)第22 h的实测纵向运动时程曲线　(d) 第22 h19min的实测纵向运动时程曲线

图 1　实测悬索桥梁端纵向运动曲线

Fig. 1　Measured time histories of longitudinal movement of the suspension bridge girder ends

进一步,对实测位移时程和分解后的位移时程进行频谱分析,得到的结果如图 2 所示。剔除温度影响后的纵向位移时程曲线,即 EMD 分解出来的 IMF1-9 之和的频谱图和实测位移时程曲线图较为类似,这说明由温度变化引起的位移只会影响位移时程的总体趋势,而不会改变一天内位移时程的频谱特征。同时分别对 IMF1 与 IMF2-9 之和进行频谱分析,我们发现 IMF1 的高频成分较多,而 IMF2-9 之和基本为超低频成分。这表明,悬索桥纵向运动位移时程曲线(IMF1-9 之和) 可以大致分成两个部分: 一部分是 IMF1 对应的相对高频的纵向振动成分,其卓越频率为 0.104 Hz(周期为 9.6 s) ,与该悬索桥有限元分析的纵飘模态对应,是由风荷载和车辆动态激励引起的主要由加劲梁纵飘模态参与的共振响应,此类振动可从图 1(d) ,即 1min 的运动时程曲线上观察到; 另一个部分是由 IMF2-9 所对应的相对低频的纵向往复运动成分,其卓越频率为

0.006 2 Hz(周期为161.3 s),是由车辆荷载引起的拟静态位移,此类运动可从图1(c),即加劲梁纵向运动1 h时程曲线上观察到较为长周期的大波波形。前者是由外部激励引起的结构共振响应,可以通过增设阻尼装置加以抑制;后者是由于车辆荷载的不对称分布,引起悬索桥主缆和加劲梁的大位移变形造成的梁端拟静态位移,单靠阻尼装置较难抑制,需要辅助其他措施才能达到较好的效果。限于篇幅,本文仅讨论对加劲梁第一类位移成分(即IMF1)的振动控制,关于由车辆荷载不对称分布引起的拟静态位移控制,将在后续的研究中加以探讨。

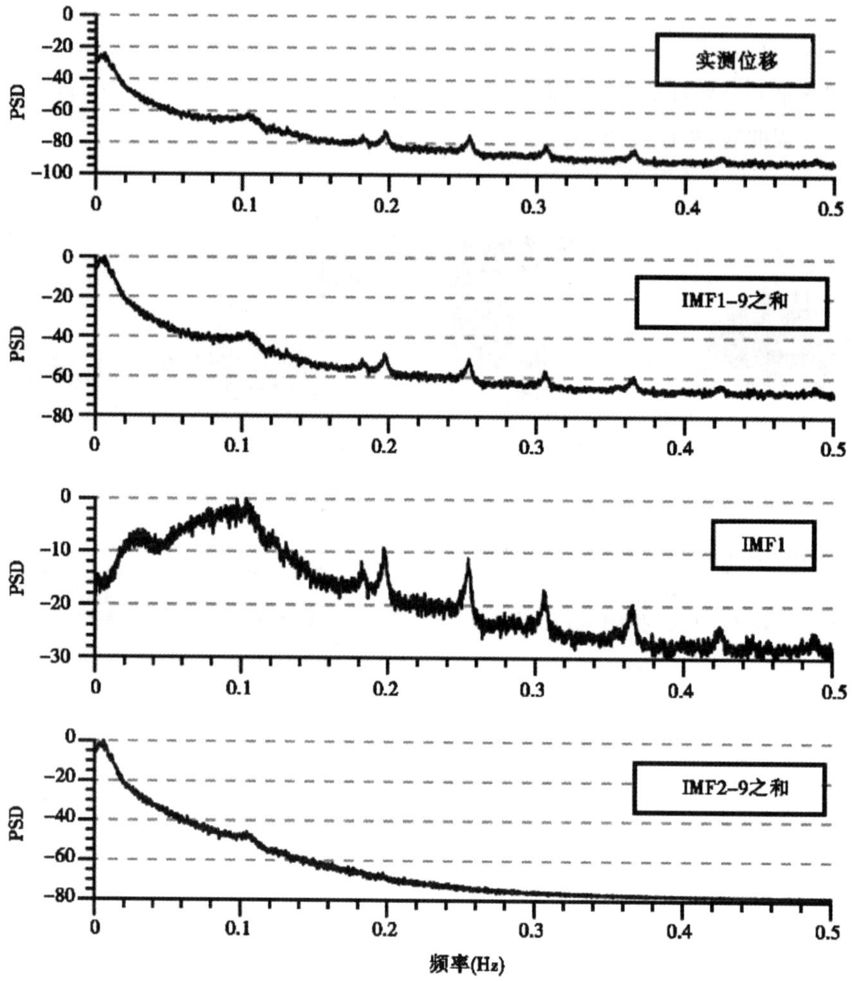

图2 位移时程曲线的频谱分析

Fig. 2 Spectral characteristics of displacement time histories

2 悬索桥SDOF简化模型

由外部动态激励引起的漂浮体系桥梁加劲梁的纵向运动可以简化为SDOF体系来进行分析。从上一节中,我们也可以看出,由外部动态激励引起的主梁纵向振动主要由悬索桥的纵飘模态参与,其他模态参与甚少,为了简化计算,将悬索桥简化为纵飘模态是合适的。SDOF体系的质量主要是来自于加劲梁,而SDOF体系的自振频率通过上节中由环境因素激起的纵向运动的卓越频率(纵飘模态)确定。而体系的刚度可以通过质量和自振频率求得。简化的模型如图3所示,其中集中质量取悬索桥加劲梁的总重m为24 930 t,体系的自振频率f取0.104 Hz。SDOF体系的圆频率和刚度分别为$\omega_n = 2\pi f = 0.65$ rad/s, $k = m\omega_n^2 =$ 10 533 kN/m。该悬索桥的加劲梁为钢箱梁,根据规范[21]取SDOF体系的结构阻尼比为0.3%,则阻尼系数为$c = 2m\omega_n\xi = 97.23$ kN·s/m。

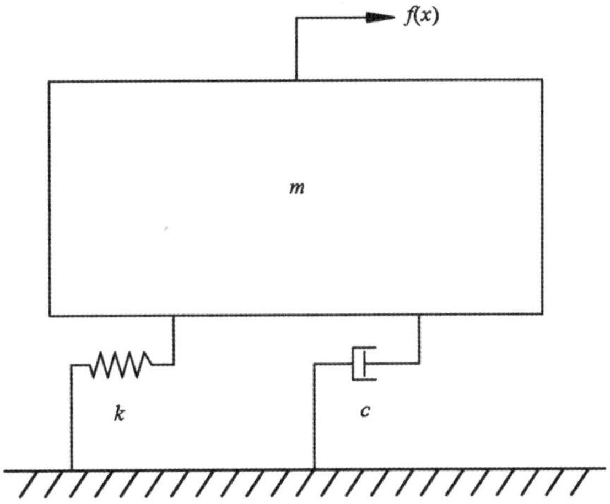

图 3　加劲梁 SDOF 体系示意图

Fig. 3　SDOF system of bridge girder

采用五点微分法[22]，对上节提取出来的由风和车辆动态激励而产生的纵向位移响应 IMF1 进行微分，可以得到相应的速度和加速度响应。结合 SDOF 体系的动力学方程式(1)进行动力反演，可以得到 SDOF 的等效外力时程曲线，如图 4 所示。

$$ma + cv + ku = f(t) \tag{1}$$

其中：m，c，k 分别是 SDOF 体系的质量、阻尼和刚度；a，v，u 分别对应 SDOF 体系的加速度，速度和位移；$f(t)$ 为外力时程曲线。

通过反演得到的等效力时程曲线，就可以进一步对建立的 SDOF 体系进行振动控制的设计和性能评估。

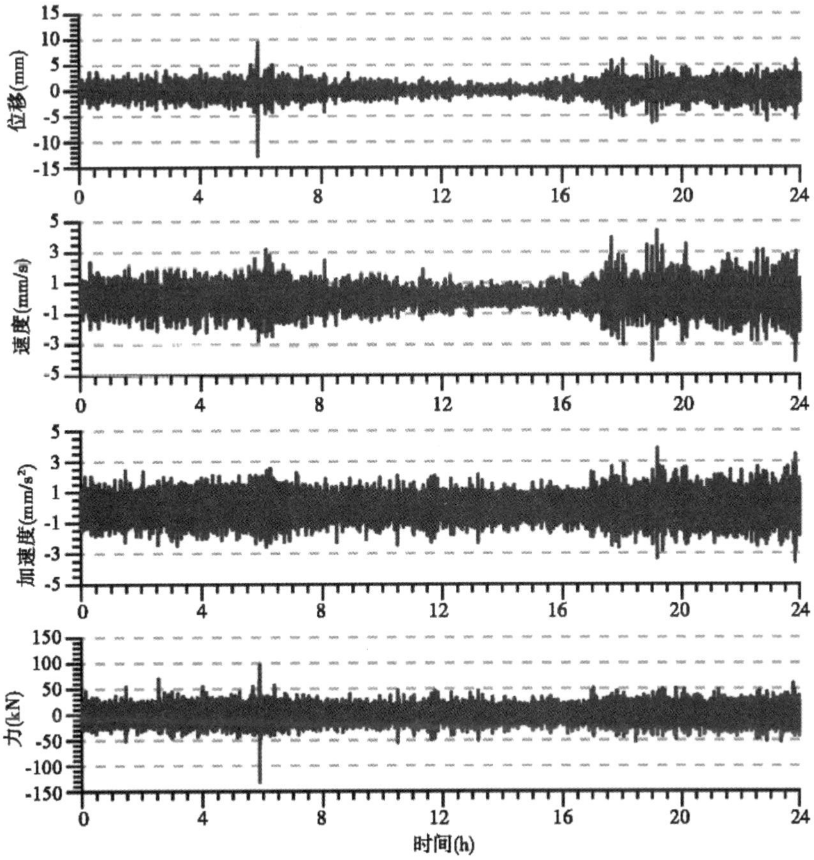

图 4　力时程曲线的反演

Fig. 4　Calculation of force time histories

3 电涡流阻尼器低速状态纵向振动控制

以悬索桥 SDOF 简化模型为基础,依据已知的质量和自振频率,将上节得到的力时程曲线作为激励输入,并以 SDOF 体系的阻尼比作为变量,定义控制率 ρ 如式(2),对位移响应的累计位移进行考察,如图 5 所示。可以发现,累计位移随着阻尼比的逐步增大而快速减小,但这个衰减的过程是先快后慢,当阻尼比 ξ 增加到 100% 后,衰减的效率变低。

$$\rho = \frac{D_{\mathrm{SUM}}(\xi)}{D_{\mathrm{SUM}}(0.3\%)} \tag{2}$$

其中: $D_{\mathrm{SUM}}(\xi)$ 是阻尼比为 ξ 时的累计位移, $D_{\mathrm{SUM}}(0.3\%)$ 是在本身结构阻尼比为 0.3% 时的累计位移。

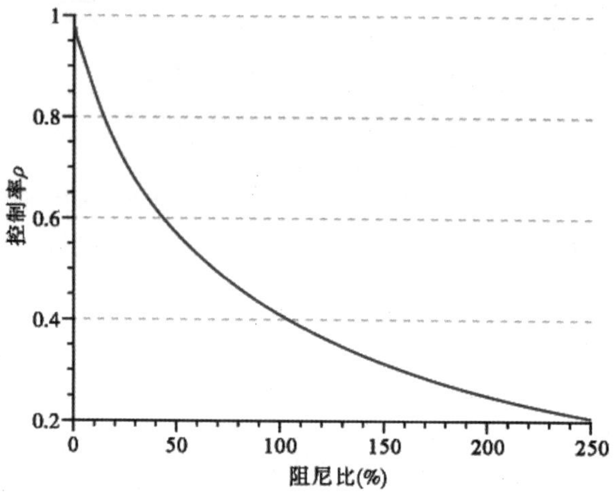

图 5 累计位移控制比

Fig. 5 Control ratio of cumulative displacement

采用耐久性能良好电涡流阻尼器进行加劲梁纵向振动控制,电涡流阻尼器的结构示意图如图 6 所示。

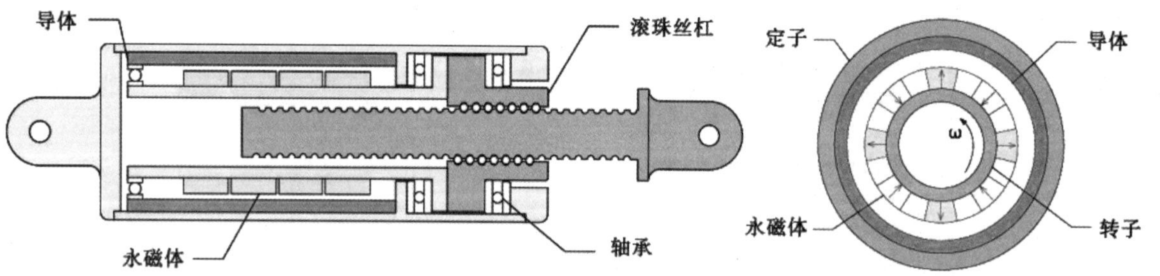

图 6 电涡流阻尼器示意图

Fig. 6 Cross sections of ECD

电涡流阻尼器采用 Wouterse 等[23] 提出的速度-阻尼力曲线模型进行分析,其本构方程和曲线分别如式(3) 和图 7 所示。

$$F_{\mathrm{ECD}} = F_{\max} \frac{2}{\dfrac{v}{v_{\mathrm{cr}}} + \dfrac{v_{\mathrm{cr}}}{v}} \tag{3}$$

其中: F_{ECD} 是电涡流阻尼器产生的阻尼力, v 对应 SDOF 体系的速度, $F_{\max}$ 是阻尼力的最大值, v_{cr} 是阻尼力最大值对应的速度,称之为临界速度。

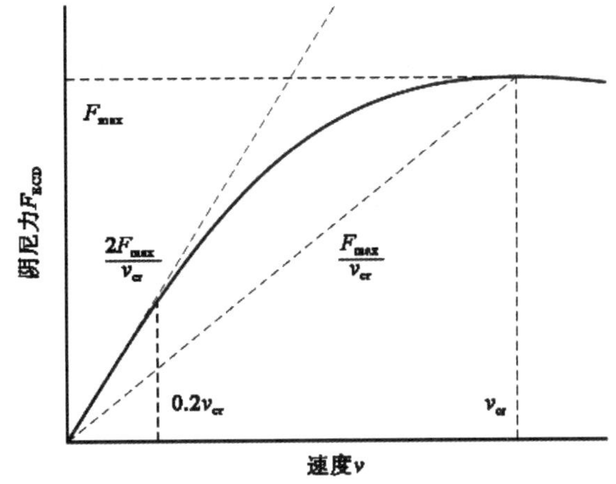

图 7　电涡流速度-阻尼力曲线

Fig. 7　Velocity-damping force curve of ECD

对 F_{ECD}求导:

$$\frac{\partial F_{ECD}}{\partial v} = 2F_{max}\frac{\frac{v_{cr}}{v^2} - \frac{1}{v_{cr}}}{\left(\frac{v}{v_{cr}} + \frac{v_{cr}}{v}\right)^2} \tag{4}$$

对阻尼力对速度的导数趋向于 0 求极限可知:

$$\lim_{n \to 0}\frac{\partial F_{ECD}}{\partial v} = \frac{2F_{max}}{v_{cr}}$$

当 $v > 0$ 时,令 $v = xv_{cr}$,其中 x 为速度与临界速度的比例系数,则式(4) 可以转化为:

$$\frac{\partial F_{ECD}}{\partial v} = \frac{2F_{max}}{v_{cr}}\frac{1 - x^2}{(1 + x^2)^2} \tag{5}$$

式(5) 中,$2F_{max}/v_{cr}$ 为常数,而 $1 - x^2/(1 + x^2)^2$ 部分当 $x \in [0,2]$ 时有如图 8(a) 所示的曲线,可知阻尼力的导数,即阻尼力-速度曲线的切线斜率随着速度的增大逐步减小,并在 $x = 1$ 时,切线斜率为零,取得阻尼力的极大值。

考察线性阻尼系数(割线斜率) c_{ECD}与速 v 的关系,

$$c_{ECD} = \frac{F_{ECD}}{v} = \frac{2F_{max}v_{cr}}{v^2 + v_{cr}^2} \tag{6}$$

令 $v = xv_{cr}$,

$$c_{ECD} = \frac{F_{ECD}}{v} = \frac{2F_{max}}{v_{cr}}\frac{1}{(x^2 + 1)} \tag{7}$$

绘制式(7) 中 $1/(x^2 + 1)$ 部分在 $x \in [0,1.5]$ 时的曲线如图 8(b) ,可知阻尼系数即阻尼力-速度曲线的割线斜率随着速度的增大逐步减小。当 $x = 0.2$ 时, $1/(x^2 + 1) \approx 0.96$,所以在 $v \in [0,0.2v_{cr}]$ 这个区间, c_{ECD} 可以近似的看作线性阻尼,其大小为 $2F_{max}/v_{cr}$ 。

所以,在低速工况下,悬索桥塔梁纵向阻尼器的纵向振动控制设计,可以用线性阻尼简化进行分析。累积位移的控制效率可以直接根据电涡流阻尼器的设计参数最大阻尼力 F_{max}和临界速度 v_{cr} 结合图 5 来进行判断。

4　电涡流阻尼器的地震纵向振动控制

采用 Runge-Kutta 法对安装了电涡流阻尼器的悬索桥 SDOF 体系进行地震分析,采用 Northridge 地震波作为输入激励,对地震波进行调幅,PGA 为 $1.5247m/s^2$,并对地震波进行频谱分析,如图 9 所示。

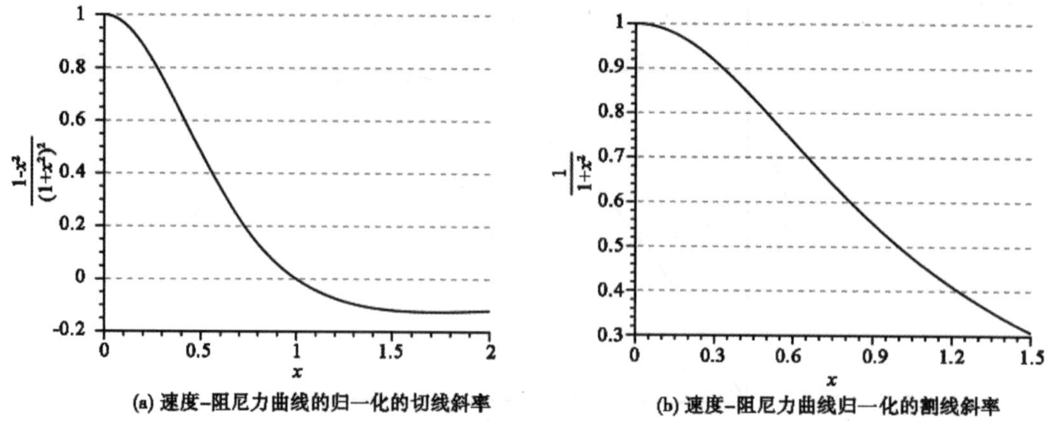

图 8　速度-阻尼力曲线的斜率

Fig. 8　Slopes of velocity-damping force curve

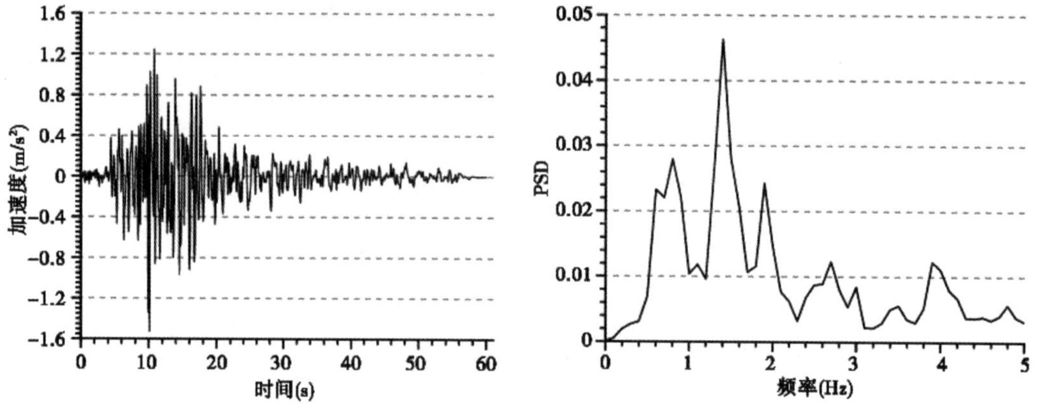

图 9　Northridge 地震波

Fig. 9　Northridge seismic wave

计算地震响应时结构本身的阻尼比按规范[21]取 2%。计算线性阻尼情况下 SDOF 体系的响应,并令电涡流阻尼器的临界速度 v_{cr} 和最大阻尼力 F_{max} 与 SDOF 体系在线性阻尼下的速度响应幅值和最大阻尼力相等,将其代入式(8),计算出对应的电涡流阻尼器下的响应,以 SDOF 体系的位移响应幅值作为评价目标。

$$ma + cv + ku + F_{max}\frac{2}{\frac{v}{v_{cr}} + \frac{v_{cr}}{v}} = -ma_g \tag{8}$$

其中: a_g 为地震加速度。

以线性阻尼比作为变量,对线性阻尼与其相应的电涡流阻尼下 SDOF 体系的位移响应幅值进行考察,如图 10 所示。可知,在相同条件下(即最大阻尼力和最大速度相等),电涡流阻尼器的位移响应幅值要小于相应的线性阻尼。用线性阻尼来预估相应电涡流阻尼器对位移响应幅值的控制水平,结果趋于保守。

5　电涡流阻尼器的合理参数取值

对于电涡流阻尼器,在最大阻尼力 F_{max} 为定值的情况下,改变临界速度 v_{cr},则其在地震下的减振性能会跟随改变,而且其低速工况下的近似线性阻尼系数 $2F_{max}/v_{cr}$ 也会随之改变。以线性阻尼比 100% 的情况为例进行讨论,计算得到线性阻尼情况下的速度幅值 $v_{max} = 0.133$ m/s,分析改变 v_{cr} 的取值范围,地震加载下的位移响应幅值和低速工况下的累计位移的变化情况,结果如图 11 所示。需要指出的是,当 v_{cr} 取值偏小时,电涡流的线性段也会相应变短,但即使取 $v_{cr} = 0.5v_{max}$ 时,根据第 3 节的结论,电涡流阻尼器在速度小于 $0.2v_{cr} = 13.3$ mm/s 的情况下,可以视为线性阻尼。从图 4 可知,在未安装阻尼器情况下,加劲梁的纵向振动速度也低于 5 mm/s,可以按线性阻尼进行计算。累积位移可以先计算线性阻尼 $2F_{max}/v_{cr}$,然后从图 5 中查到相应的数值。

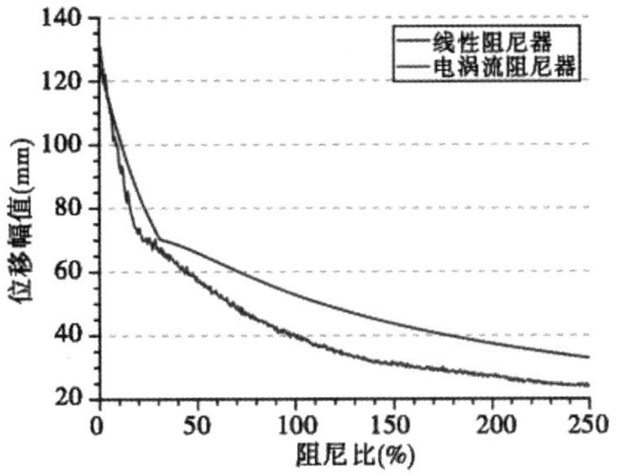

图 10　线性阻尼器与电涡流阻尼器的位移响应幅值比较

Fig. 10　Displacement response amplitude of linear damper and ECD

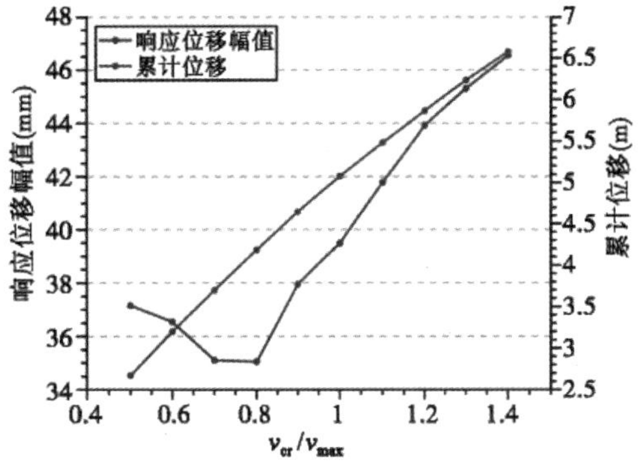

图 11　线性阻尼比为 100% 时,地震作用下位移响应幅值与运营状态累计位移随 v_{cr} 变化的曲线

Fig. 11　Seismic response amplitude and cumulative displacement vary with v_{cr} when the corresponding linear damping ration is 100%

从图 11 可以看出,随着电涡流阻尼器临界速度 v_{cr} 与最大速度 v_{max} 比值的减小,运营状态阻尼器低速运行工况下的累积位移将随之逐渐减小,而地震作用下阻尼器高速运行工况下的位移响应幅值将呈现出先减小后增大的趋势,而当 v_{cr}/v_{max} 取 0.7 ~ 0.8 时地震位移响应幅值最小。对悬索桥加劲梁纵向位移控制来说,既要考虑运营状态下的累积位移减振需求,也要考虑地震作用下的位移幅值控制,综合考虑两种控制目标,则电涡流的临界速度取为最大速度的 0.7 ~ 0.8 倍较为适宜;进一步结合运营和维护成本来考虑,累积位移过大会带来构件(如伸缩缝、支座)的疲劳破坏,可以适当降低 v_{cr}/v_{max} 以获得更好的累积位移控制效果,从图 11 可以发现 v_{cr}/v_{max} 取 0.5 时,地震响应位移幅值仅增大约 2.1mm,但是可以将日累积位移的控制率从63.6% 降低到 40.6%,这样可以进一步提高伸缩装置的疲劳寿命,从而使得运营和维护成本进一步降低。但是,过度降低 v_{cr}/v_{max} 的比值会导致生产成本的上升和安装空间的增大,这样做也是不经济的和不切实际的。

6　结论

(1) 采用 EMD 分解和频谱分析对实测悬索桥加劲梁纵向位移时程曲线进行剖析,得到了由温度变化、车辆拟静力荷载、车辆和环境动态激励等不同成因导致的悬索桥加劲梁纵向振动成分。

(2) 基于由车辆和环境动态激励引起的纵向振动建立简化等效 SDOF 模型并反演求得动态激励等效外荷载时程曲线,为悬索桥加劲梁纵向振动控制简化分析提供了一条快捷途径。

(3) 电涡流阻尼器在速度小于 $0.2\,v_{cr}$ 时,阻尼特性可以近似为线性阻尼,其值为 $2F_{max}/v_{cr}$,大大简化了低速工况下电涡流阻尼器的设计计算和性能评估。

(4) 电涡流阻尼器在地震工况下呈现出较大的非线性阻尼特性,可先通过相应线性阻尼来进行初步设计,结果趋于保守,但是计算量大为降低。

(5) 通过线性阻尼分析来确定电涡流阻尼器的参数选择时,v_{cr}/v_{max} 取 0.7 ~ 0.8 时能够平衡运营低速状态下的线性阻尼和地震高速状态下的非线性阻尼两者的减振控制目标,如考虑运营维护成本时,这个比值可进一步适当降低。

参考文献:

[1]　GUO Tong, LIU Jie, ZHANG Yufeng, et al. Displacement monitoring and analysis of expansion joints of long-span steel bridges with viscous dampers[J]. Journal of Bridge Engineering, 2015, 20(9): 04014099

[2]　丁幼亮, 周　凯, 王高新, 等. 苏通大桥斜拉桥伸缩缝位移的长期监测与分析[J]. 公路交通科技, 2014, 31(7): 60 -64,71.
DING Youliang, ZHOU Kai, WANG Gaoxin, et al. Long-term monitoring and analysis of expansion joint displacement of sutong cable-stayed

bridge[J]. Journal of Highway and Transportation Research and Development, 2014, 31(7): 60-64,71. (in Chinese)

[3] 张宇峰，陈雄飞，张立涛，等. 大跨悬索桥伸缩缝状态分析与处理措施[J]. 桥梁建设，2013，43(5)：49-54.
ZHANG Yufeng, CHEN Xiongfei, ZHANG Litao, et al. Condition analysis and handling measures for expansion joints of long span suspension bridges[J]. Bridge Construction, 2013, 43(5): 49-54. (in Chinese)

[4] NI Y. Q., HUA X. G., WONG K. Y., et al. Assessment of bridge expansion joints using long-term displacement and temperature measurement [J]. Journal of Performance of Constructed Facilities, 2007, 21(2): 143-151.

[5] 邓 扬，李爱群，丁幼亮. 大跨悬索桥梁端位移与温度的相关性研究及其应用[J]. 公路交通科技，2009，26(5)：54-58:
DENG Yang, LI Aiqun, DING Youliang. Research and appiication of correlation between beam end displacement and temperature of long-span suspension bridge [J]. Journal of Highway and Transportation Research and Development, 2009, 26(5): 54-58. (in Chinese)

[6] SNæBJöRNSSON Jónas, CHEYNET Etienne, JAKOBSEN JB. Performance evaluation of a suspension bridge excited by wind and traffic induced action[C]//Proceedings of the Proceedings of 8th European Workshop on Structural Health Monitoring (EWSHM 2016), Bilbao, Spain, F, 2016.

[7] 王连华，孙璋鸿，崔剑锋，等. 车辆激励下中央扣对悬索桥梁端位移的影响[J]. 湖南大学学报(自然科学版)，2019，46(3)：18-24.
WANG Lianghua, SUN Zhanghong, CUI Jianfeng, et al. Effects of central buckle on end displacement of suspension bridges under vehicle excitation [J]. Journal of Hunan University(Natural Sciences), 2019, 46(3): 18-24. (in Chinese)

[8] LARSEN Allan, LAROSE Guy L. Dynamic wind effects on suspension and cable-stayed bridges[J]. Journal of Sound and Vibration, 2015, 334(1): 2-28.

[9] HUANG Norden E, SHEN Zheng, LONG Steven R, et al. The empirical mode decomposition and the Hilbert spectrum for nonlinear and non-stationary time series analysis[J]. Proceedings of the Royal Society of London. Series A: Mathematical, Physical Engineering Sciences, 1998, 454(1971): 903-995.

[10] YANG Jann N, LEI Ying, LIN Silian, et al. Identification of natural frequencies and dampings of in situ tall buildings using ambient wind vibration data[J]. 2004, 130(5): 570-577.

[11] 陈 隽，徐幼麟. HHT方法在结构模态参数识别中的应用 [J]. 振动工程学报，2003，16(3)：129-134.
CHEN Juan, XU Youling. Application of HHTmethod in structural modal parameter identification[J]. Journal of Vibration Engineering, 2003, 16(3): 129-134. (in Chinese)

[12] 李延强，杜彦良. 结构控制技术在桥梁工程中的应用综述 [J]. 地震工程与工程振动，2009，29(3)：160-166.
LI Yanqiang, DU Yanliang. Application of structural control technology in bridge engineering [J]. Earthquake Engineering and Engineering Dynamics, 2009, 29(3): 160-166. (in Chinese)

[13] 牛建涛，丁 阳，石运东. 应用新型油阻尼器的斜拉桥横向减震体系[J]. 地震工程与工程振动，2019，39(2)：111-120.
NIU Jiantao, DING Yang, SHI Yundong. Transverse seismic reduction system of cable-stayed bridge introducing oil damper [J]. Earthquake Engineering and Engineering Dynamics, 2019, 39(2): 111-120. (in Chinese)

[14] 王 浩，李爱群，郭 彤. 超大跨悬索桥地震响应的综合最优控制研究[J]. 湖南大学学报(自然科学版)，2006，33(3)：6-10.
WANG Hao, LI Aiqun, GUO Tong. Compositive optimal control of the seismic response for super-long-span suspension bridges[J]. Journal of Hunan University(Natural Sciences), 2006, 33(3): 6-10. (in Chinese)

[15] 丁幼亮，耿方方，葛文浩，等. 多塔斜拉桥风致抖振响应的黏滞阻尼器控制研究[J]. 工程力学，2015，32(4)：130-137.
DING Youliang, GENG Fangfang, GE Wenhao, et al. Control of wind-induced buffeting responses of a multi-tower cable-stayed bridge using viscous dampers[J]. Engineering Mechanics, 2015, 32(4): 130-137.

[16] SUN Zhen, ZOU Zilong, ZHANG Yufeng. Utilization of structural health monitoring in long-span bridges: case studies[J]. Structural Control Health Monitoring, 2017, 24(10): e1979.

[17] 陈政清，华旭刚，张弘毅，等. 一种电涡流轴向阻尼器，CN109611486B [P/OL]. 2019-09-13.
CHEN Zhengqing, HUA Xugang, ZHANG Hongyi, et al. A type of eddy current damper, CN109611486B [P/OL]. 2019-09-13. (in Chinese)

[18] LIANG Longteng, FENG Zhouquan, CHEN Zhengqing. Seismic control of SDOF systems with nonlinear eddy current dampers[J]. Applied Sciences-Basel, 2019, 9(16): 3427.

[19] 汪志昊，田文文，王 浩，等. 非线性电涡流惯质阻尼器力学性能仿真与试验[J]. 哈尔滨工业大学学报，2019，51(6)：171-177,184.
WANG Zhihao, TIAN Wenwen, WANG Hao, et al. Numerical simulation and experimental study on mechanical properties of a nonlinear eddy-current inertial mass damper[J]. Journal of Harbin Institute of Technology, 2019, 51(6): 171-177,184. (in Chinese)

[20] MAO Jianxiao, WANG Hao, FENG Dongming, et al. Investigation of dynamic properties of long - span cable - stayed bridges based on one - year monitoring data under normal operating condition[J]. Structural Control and Health Monitoring, 2018, 25(5): e2146.

[21] JTG/T D65-05-2015 公路悬索桥设计规范[S]. 北京：人民交通出版社，2015.
JTG/T D65-05-2015 Spcifications for Design of Highway Suspension Bridge[S]. Beijing: China Communications Press, 2015. (in Chinese)

[22] 夏爱生，夏军剑，张会鹏. 五点数值微分公式及其外推算法[J]. 军事交通学院学报，2014，16(4)：93-95.
XIA Aisheng, XIA Junjian, ZHANG Huipeng, Extrapolation method of two order and five points numerical differential formulas[J]. Journal of Military Transportation University, 2014, 16(4): 93-95. (in Chinese)

[23] WOUTERSE J H. Critical torque and speed of eddy current brake with widely separated soft iron poles[J]. IEE Proceedings B (Electric Power Applications), 1991, 138(4): 153-158.

Exploration of the Nonlinear Effect of Pendulum Tuned Mass Dampers on Vibration Control

Kai Xu[1]; Xugang Hua[2]; Walter Lacarbonara[3]; Zhiwen Huang[4]; and Zhengqing Chen[5]

Abstract: Understanding the nonlinearity of pendulum tuned mass dampers (PTMDs) under large-amplitude vibration is important to enhance control performance. Here, the nonlinear effects of PTMDs on vibration control of a single-degree-of-freedom (SDOF) system under harmonic excitation are explored. The nonlinear governing equations of the linear SDOF structure coupled with the PTMD are formulated. The results obtained via the method of harmonic balance under various excitation amplitudes are compared with those obtained employing the linearized system. It is shown that the effects of PTMD nonlinearity can be ignored when the PTMD rotation angle is below about 9°. Upon increasing the excitation amplitude, remarkable differences are observed between the nonlinear and linear frequency response functions (FRFs). Large mass ratios of PTMD are found to lead to reduced PTMD motion and attenuated nonlinear effects. A parametric study is performed to find the optimal parameters of the nonlinear PTMD under various excitation amplitudes. The results indicate that the optimal tuning frequency ratio increases significantly with excitation amplitudes, whereas the optimal damping is less sensitive to excitation amplitudes. Finally, a 61-m-high steel chimney structure subjected to harmonic and random wind loadings is investigated to verify the proposed PTMD design method. **DOI: 10.1061/(ASCE)EM.1943-7889.0001961.**

Author keywords: Flexible structures; Vibration control; Nonlinear dynamics; Pendulum tuned mass damper (PTMD); Optimal design; Method of harmonic balance.

Introduction

Many excitation sources such as wind and earthquake may induce excessive dynamic responses or even severe structural damages in flexible structures. An effective means to mitigate such vibrations is addition of external damping devices, provided that the dynamic response is dominated by resonance. A tuned mass damper (TMD) is one of the simplest and most reliable vibration control devices (e.g., Fujino and Siringoringo 2013; EI-Khoury and Adeli 2013; Wen et al. 2016). TMDs usually consist of mass, spring, and damping component, where the spring attached to the primary structure is typically linear (Den Hartog 1985). When the TMD frequency is suitably tuned in resonance with the primary structure, a large part of the kinetic energy of the primary structure is transferred to the TMD and then dissipated through the TMD damping. For practical applications, Ormondroyd and Den Hartog (1928) derived a closed-form formula to optimize the TMD parameters, and the structural damping of the primary structure was neglected. Structural damping, however, plays an important role in the TMD control performance. Thus, a number of studies proposed different optimum design formulas of TMD when a damped primary structure is subjected to different types of excitations (Rathi and Chakraborty 2017; Pourzeynali et al. 2013; Leung et al. 2008; Ghosh and Basu 2007; Toshihiro and Ken 1978).

There exist various realization forms for TMDs (Elias and Matsagar 2017; Wang et al. 2018; Huang et al. 2019; Sun et al. 2019; Cao et al. 2020). Among them, pendulum tuned mass dampers (PTMDs) are preferable due to their simplicity and low cost. The PTMD consists of a pendulum rod, a tip mass, and a damper. In mechanical systems, PTMDs, also called autoparametric vibration absorbers, are often used to control the vertical vibration, where the PTMD frequency is usually tuned to about one-half the frequency of the controlled mode (Hatwal et al. 1983; Vyas and Bajaj 2001; Ertas et al. 2000). In civil engineering applications, PTMDs have been extensively used to control the horizontal vibration of primary structures, where the PTMD frequency is tuned in 1:1 resonance with the primary structure.

There are many high-rise buildings using PTMDs to control wind-induced or seismic vibrations around the world. The 102-m-tall steel antenna mast at the top of the CN Tower in Toronto, Canada, was equipped with two doughnut-shaped PTMDs to reduce the second and fourth modes of wind-induced vibration, respectively (Kwok and Samali 1995). The 305-m-tall Sydney Tower was retrofitted with a PTMD to reduce the wind-induced motion. Field measurements showed that the acceleration response of the tower was reduced by 40% to 50% (Kwok 1983; Kwok and Macdonald 1990). Other examples in high-rise buildings include those of the Crystal Tower (Kwok 1983), Chiefly Tower

[1]Ph.D. Student, Key Laboratory for Wind and Bridge Engineering of Hunan Province, College of Civil Engineering, Hunan Univ., Changsha 410082, China. Email: kaixu@hnu.edu.cn

[2]Professor, Key Laboratory for Wind and Bridge Engineering of Hunan Province, College of Civil Engineering, Hunan Univ., Changsha 410082, China (corresponding author). ORCID: https://orcid.org/0000-0001-6150-2563. Email: cexghua@hnu.edu.cn

[3]Professor, Dept. of Structural and Geotechnical Engineering, Sapienza Univ. of Rome, 00184 Rome, Italy. ORCID: https://orcid.org/0000-0002-8780-281X. Email: walter.lacarbonara@uniroma1.it

[4]Associate Professor, Key Laboratory for Wind and Bridge Engineering of Hunan Province, Hunan Univ., Changsha 410082, China. Email: zwhuang213@hnu.edu.cn

[5]Professor, Key Laboratory for Wind and Bridge Engineering of Hunan Province, College of Civil Engineering, Hunan Univ., Changsha 410082, China. Email: zqchen@hnu.edu.cn

Note. This manuscript was submitted on November 10, 2020; approved on March 22, 2021; published online on June 12, 2021. Discussion period open until November 12, 2021; separate discussions must be submitted for individual papers. This paper is part of the ***Journal of Engineering Mechanics***, © ASCE, ISSN 0733-9399.

(Nagase and Hisatoku 1992), Taipei 101 Tower (Kourakis 2007; Tuan and Shang 2014), and Shanghai Centre Tower (Lu and Chen 2011).

In addition, PTMDs can be also used to control wind-induced or seismic vibrations in other slender structures such as wind turbines (Sun and Jahangiri 2018), chimneys (Areemit and Warnitchai 2001), guyed masts (Lacarbonara and Ballerini 2009), transmission towers (Battista et al. 2003), coal-fired power plants (Shu et al. 2017, 2019), and cable-stayed bridge towers (Guo et al. 2007b, a). Moreover, new PTMD-based devices have been proposed to further increase the control performance such as pendulum-pounding TMD (Wang et al. 2019) and nontraditional PTMDs (Xiang et al. 2017).

Usually, PTMDs are treated as linear assuming small amplitude of vibrations. However, it is well known that the pendulum behaves nonlinearly at large angles (Sheheitli and Rand 2012), which may result in frequency detuning of PTMDs. Therefore, it is imperative to consider the nonlinearity of PTMDs in design and analysis of PTMDs, especially under action of strong loadings. Many works have been devoted to study nonlinear dynamics of PTMDs for a large amplitude of vibrations. Bevilacqua et al. (2000) and Battista et al. (2003) studied the nonlinear responses of a guyed tower and a transmission tower when attached with a nonlinear PTMD. Although the PTMD nonlinearity was considered and the results showed the remarkable control performance, the comparison between nonlinear and linear solutions and the parameter analysis were not discussed. Roffel et al. (2013) studied the dynamic response of a flexible structure endowed with a three-dimensional (3D) nonlinear PTMD whose motions consisted of planar and spherical motions. The results showed that the 3D PTMD best captures the negligible deviations when compared with the linearized PTMDs. The proposed approach was successfully utilized to attenuate ambient vibrations of the Apron Control Tower (Roffel and Narasimhan 2016).

Fallahpasand et al. (2015) investigated the vibration suppression of a nonlinear PTMD for a linear damped system when subjected to forced and base excitations. They pointed out that if the rotation angle of PTMD exceeds 0.2 rad, enormous errors exist for the linear solutions, resulting in control performance deterioration. However, their study was limited to three fixed excitation frequency ratios, and the maximum amplitudes of FRF were not compared. Brzeski et al. (2014) discussed the use of nonlinear PTMDs for vibration control of the Duffing oscillator.

Recently, a few studies suggested variants of the nonlinear PTMD. Sarkar and Gudmestad (2013) proposed a pendulum-type liquid column tuned mass damper. Eason et al. (2015) developed an adaptive-length pendulum tuned mass damper (APTMD), which consisted of a Duffing nonlinear TMD and a nonlinear PTMD in series configuration. Viet and Nghi (2014) studied a nonlinear single-mass two-frequency TMD, which contains one pendulum mass moving along the pendulum rod while the rod can oscillate around the fulcrum position. By taking advantage of the mass amplification effect of the inerter, Brzeski et al. (2015) investigated the dynamics of nonlinear PTMDs with an inerter attached to the pendulum.

Christie et al. (2019) utilized the novel magnetorheological materials to realize variable resonance nonlinear PTMDs. Orlando and Goncalves (2013) proposed a hybrid control system, which contains a PTMD and a nonlinear variable stiffness device based on position and velocity feedback to improve the control performance for transient responses. Ikeda (2011) investigated the nonlinear response of a single-degree-of-freedom (SDOF) system with two nonlinear PTMDs when subjected to a horizontal sinusoidal excitation. In addition, if the PTMD motion is allowed to rotate with very large rotations, it can be used as nonlinear energy sinks (NES). The details of rotary and vibroimpact NES have been described elsewhere (Saeed et al. 2019; Farid and Gendelman 2017; Sigalov et al. 2012).

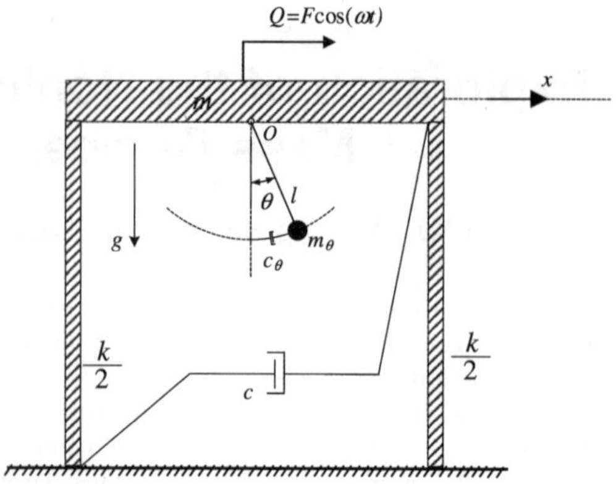

Fig. 1. Two-DOF model of a primary structure coupled with a PTMD.

Despite the aforementioned efforts, the effects of PTMD nonlinearity on vibration control performance have not been fully investigated. A detailed picture of the PTMD's nonlinear response characteristics in terms of higher-order harmonics and instantaneous frequency variation is still lacking. Moreover, a proper choice of mass ratio for mitigating the PTMD's nonlinear effects has not been reported in the literature for practical applications. This work looks more closely at the nonlinear dynamics of PTMD undergoing large-amplitude vibrations and explores the nonlinear effects of PTMD on vibration control. In addition, in order to mitigate or cancel the unfavorable effects of PTMD nonlinearity on vibration control, two approaches are developed toward proper parameter design of PTMDs coping with the nonlinearity. The outlined method allows design practitioners to directly compute the necessary parameters of a PTMD installed in a damped structure.

The paper is organized as follows. First, the nonlinear dynamics of a linear damped SDOF structure coupled with a nonlinear PTMD under harmonic excitation are studied. Then, the nonlinear frequency response functions (FRFs) are compared with those obtained from the linearized system for a series of excitation amplitudes. The time-frequency characteristics of the system responses under different excitation magnitudes are unfolded making use of the Hilbert-Huang transform. Moreover, two approaches are described to proper parameter design of PTMD in face of nonlinearity. Finally, the validity of the proposed approach is successfully demonstrated in numerical simulation of a steel chimney coupled with nonlinear PTMD under harmonic and random wind loading.

Formulation of the Theoretical Model

Consider a damped linear primary structure equipped with a PTMD under harmonic excitation, as shown in Fig. 1. The primary structure is modeled as a SDOF system with mass m, damping coefficient c, and stiffness k. The PTMD with a tip mass m_θ and damping coefficient c_θ is attached to the primary structure by a massless pendulum rod with length l. The harmonic excitation force with frequency ω and amplitude F is applied to the primary structure. The horizontal displacement of the primary structure and the rotation angle of the PTMD are denoted by x and θ, respectively. The gravity acceleration is g.

The Euler-Lagrange equations lead to the following nonlinear governing equations:

$$
\begin{aligned}
&(m+m_\theta)\ddot{x}+kx+c\dot{x}+m_\theta l(\ddot{\theta}\cos\theta-\dot{\theta}^2\sin\theta)=F\cos\omega t\\
&m_\theta l^2\ddot{\theta}+m_\theta gl\sin\theta+c_\theta l^2\dot{\theta}+m_\theta l\ddot{x}\cos\theta=0
\end{aligned}
\tag{1}
$$

For convenience, the following nondimensional parameters are introduced:

$$
\begin{aligned}
&\mu=\frac{m_\theta}{m},\quad y=\frac{x}{l},\quad \omega_0=\sqrt{\frac{k}{m}},\quad \omega_\theta=\sqrt{\frac{g}{l}},\\
&\beta=\frac{\omega_\theta}{\omega_0},\quad \zeta=\frac{c}{2m\omega_0},\quad \zeta_\theta=\frac{c_\theta}{2m_\theta\omega_\theta},\quad \tau=\omega_0 t,\\
&\alpha=\frac{\omega}{\omega_0},\quad f=\beta^2\frac{F}{mg}
\end{aligned}
\tag{2}
$$

where μ = mass ratio; y = structural displacement normalized with the pendulum length l; ω_0 and ζ = natural frequency and structural damping ratio of the primary structure; ω_θ and ζ_θ = natural frequency and damping ratio of the PTMD; β = frequency ratio of the PTMD to the primary structure; α = ratio of excitation frequency to structural natural frequency; f = dimensionless excitation amplitude; and τ = nondimensional time.

After substituting Eq. (2) into Eq. (1), the nondimensional form of Eq. (1) is obtained

$$
\begin{aligned}
&(1+\mu)\ddot{y}+y+2\zeta\dot{y}+\mu(\ddot{\theta}\cos\theta-\dot{\theta}^2\sin\theta)=f\cos\alpha\tau\\
&\ddot{\theta}+\beta^2\sin\theta+2\zeta_\theta\beta\dot{\theta}+\ddot{y}\cos\theta=0
\end{aligned}
\tag{3}
$$

where the dots represent differentiation with respect to τ.

By applying Taylor expansion and keeping the first three harmonic terms, Eq. (3) becomes

$$
\begin{aligned}
&(1+\mu)\ddot{y}+y+2\zeta\dot{y}+\mu\left[\ddot{\theta}\left(1-\frac{1}{2}\theta^2\right)-\dot{\theta}^2\left(\theta-\frac{1}{6}\theta^3\right)\right]=f\cos\alpha\tau\\
&\ddot{\theta}+\beta^2\left(\theta-\frac{1}{6}\theta^3\right)+2\zeta_\theta\beta\dot{\theta}+\ddot{y}\left(1-\frac{1}{2}\theta^2\right)=0
\end{aligned}
\tag{4}
$$

Eq. (4) can be transformed as follows:

$$
\begin{aligned}
&\ddot{y}=\frac{1}{1+\mu}\left(-y-2\zeta\dot{y}-\mu\ddot{\theta}+\frac{1}{2}\mu\ddot{\theta}\theta^2+\mu\dot{\theta}^2\theta-\frac{1}{6}\mu\dot{\theta}^2\theta^3\right)-\frac{f}{1+\mu}\cos\alpha\tau\\
&\ddot{\theta}=-\beta^2\theta-2\zeta_\theta\beta\dot{\theta}-\ddot{y}+\frac{1}{2}\theta^2\ddot{y}+\frac{1}{6}\beta^2\theta^3
\end{aligned}
\tag{5}
$$

Methodology

Method of Harmonic Balance

The general method of harmonic balance (HBM) is utilized to solve the nonlinear differential equations of Eq. (5). It is assumed that the solutions for harmonic oscillations can be expressed

$$
\begin{aligned}
&y=a(\tau)\cos\alpha\tau+b(\tau)\sin\alpha\tau=A(\tau)\cos(\alpha\tau+\varphi_1(\tau))\\
&\theta=c(\tau)\cos\alpha\tau+d(\tau)\sin\alpha\tau=B(\tau)\cos(\alpha\tau+\varphi_2(\tau))
\end{aligned}
\tag{6}
$$

where $a(\tau)$, $b(\tau)$, $c(\tau)$, and $d(\tau)$ = unknown coefficients to be determined for the corresponding harmonic terms; $A(\tau)$ and $B(\tau)$ = displacement amplitudes for the primary structure and the PTMD, respectively; and φ_1 and φ_2 = associated phase angles. The expressions of $A(\tau)$, $B(\tau)$, $\varphi_1(\tau)$, and $\varphi_2(\tau)$ are given

$$
\begin{aligned}
&A(\tau)=\sqrt{a(\tau)^2+b(\tau)^2},\quad \varphi_1(\tau)=\tan^{-1}(-b(\tau)/a(\tau))\\
&B(\tau)=\sqrt{c(\tau)^2+d(\tau)^2},\quad \varphi_2(\tau)=\tan^{-1}(-d(\tau)/c(\tau))
\end{aligned}
\tag{7}
$$

Vander Pol's method (Stoker 1950) is then used to determine the FRFs. Introducing the small parameter ε, the orders of the coefficients are assumed as follows:

$$
\begin{aligned}
&a(\tau),b(\tau),c(\tau),d(\tau)=O(1)\\
&\dot{a}(\tau),\dot{b}(\tau),\dot{c}(\tau),\dot{d}(\tau)=O(\varepsilon)\\
&\ddot{a}(\tau),\ddot{b}(\tau),\ddot{c}(\tau),\ddot{d}(\tau)=O(\varepsilon^2)\\
&\zeta,\zeta_\theta=O(\varepsilon),\quad f=O(\varepsilon)
\end{aligned}
\tag{8}
$$

After substituting Eq. (6) into Eq. (5), applying the trigonometric identities $\cos^3(\alpha\tau)\approx 3/4\cos(\alpha\tau)$ and $\sin^3(\alpha\tau)\approx 3/4\sin(\alpha\tau)$, and further taking cosine and sine terms on both sides of equations, one obtains the evolution equations within the accuracy of $O(\varepsilon)$ as follows:

$$
\begin{aligned}
&2\zeta\dot{a}+2(1+\mu)\alpha\dot{b}+\frac{1}{2}\alpha\mu\left[-cd+\frac{1}{12}(cd^3+c^3d)\right]\dot{c}+\alpha\mu\left[2-\frac{1}{4}(c^2+3d^3)+\frac{1}{24}(d^4-c^4)\right]\dot{d}\\
&\quad=\alpha^2(1+\mu)a-2\zeta\alpha b-a+\alpha^2\mu c+f+\frac{1}{8}\mu\alpha^2(-c^3-cd^2)-\frac{1}{48}\mu\alpha^2(c^5+2c^3d^2+cd^4)\\
&-2\alpha(1+\mu)\dot{a}+2\zeta\dot{b}+\alpha\mu\left[-2+\frac{1}{4}(3c^2+d^2)+\frac{1}{24}(d^4-c^4)\right]\dot{c}+\frac{1}{2}\alpha\mu\left[cd-\frac{1}{12}(cd^3+c^3d)\right]\dot{d}\\
&\quad=\alpha^2(1+\mu)b-b+2\alpha\zeta a+\alpha^2\mu d+\frac{1}{8}\mu\alpha^2(-d^3-c^2d)-\frac{1}{48}\mu\alpha^2(d^5+2c^2d^3+dc^4)\\
&\frac{1}{2}\alpha cd\dot{a}+\alpha\left[2-\frac{1}{4}(3c^2+d^2)\right]\dot{b}+2\zeta_\theta\beta\dot{c}+2\alpha\dot{d}\\
&\quad=\alpha^2(a+c)-\beta^2c-2\zeta_\theta\beta\alpha d-\frac{1}{8}\alpha^2(3ac^2+ad^2+2bcd)+\frac{1}{8}\beta^2(c^3+cd^2)\\
&\alpha\left[-2+\frac{1}{4}(c^2+3d^2)\right]\dot{a}-\frac{1}{2}\alpha cd\dot{b}-2\alpha c+2\zeta_\theta\beta\dot{d}\\
&\quad=\alpha^2(b+d)-\beta^2d+2\zeta_\theta\beta\alpha c-\frac{1}{8}\alpha^2(3bd^2+bc^2+2acd)+\frac{1}{8}\beta^2(d^3+dc^2)
\end{aligned}
\tag{9}
$$

Eq. (9) can be expressed in matrix form

$$\mathbf{H}\dot{\mathbf{X}} = \mathbf{F} \tag{10}$$

where

$$\mathbf{X} = [a,\ b,\ c,\ d]^{\mathrm{T}}, \qquad \mathbf{F} = [f_1,\ f_2,\ f_3,\ f_4]^{\mathrm{T}}$$

$$\mathbf{H} = \begin{bmatrix} 2\zeta & 2(1+\mu)\alpha & H_{13} & H_{14} \\ -2\alpha(1+\mu) & 2\zeta & H_{23} & H_{24} \\ \frac{1}{2}\alpha cd & \alpha\left[2-\frac{1}{4}(3c^2+d^2)\right] & 2\zeta_\theta\beta & 2\alpha \\ \alpha\left[-2+\frac{1}{4}(c^2+3d^2)\right] & -\frac{1}{2}\alpha cd & -2\alpha & 2\zeta_\theta\beta \end{bmatrix}$$

$$H_{13} = \frac{1}{2}\alpha\mu\left[-cd+\frac{1}{12}(cd^3+c^3d)\right], \qquad H_{14} = \alpha\mu\left[2-\frac{1}{4}(c^2+3d^2)+\frac{1}{24}(d^4-c^4)\right]$$

$$H_{23} = \alpha\mu\left[-2+\frac{1}{4}(3c^2+d^2)+\frac{1}{24}(d^4-c^4)\right], \qquad H_{24} = \frac{1}{2}\alpha\mu\left[cd-\frac{1}{12}(cd^3+c^3d)\right]$$

$$f_1 = \alpha^2(1+\mu)a - 2\zeta\alpha b - a + \alpha^2\mu c + f + \frac{1}{8}\mu\alpha^2(-c^3-cd^2) - \frac{1}{48}\mu\alpha^2(c^5+2c^3d^2+cd^4)$$

$$f_2 = \alpha^2(1+\mu)b - b + 2\alpha\zeta a + \alpha^2\mu d + \frac{1}{8}\mu\alpha^2(-d^3-c^2d) - \frac{1}{48}\mu\alpha^2(d^5+2c^2d^3+dc^4)$$

$$f_3 = \alpha^2(a+c) - \beta^2 c - 2\zeta_\theta\beta\alpha d - \frac{1}{8}\alpha^2(3ac^2+ad^2+2bcd) + \frac{1}{8}\beta^2(c^3+cd^2)$$

$$f_4 = \alpha^2(b+d) - \beta^2 d + 2\zeta_\theta\beta\alpha c - \frac{1}{8}\alpha^2(3bd^2+bc^2+2acd) + \frac{1}{8}\beta^2(d^3+dc^2) \tag{11}$$

The inverse of **H** exists on account of a nonzero determinant. Eq. (10) can be transformed into

$$\dot{\mathbf{X}} = \mathbf{H}^{-1}\mathbf{F} \tag{12}$$

where $\mathbf{H}^{-1}$ is the inverse of **H**.

Eq. (12) is a set of nonlinear ordinary differential equations. The coefficients a, b, c, and d in Eq. (12) and the displacement amplitudes A and B expressed by Eq. (7) are time-varying quantities. However, the amplitudes will ultimately approach the steady-state values. Thus, only the steady-state solution of Eq. (12) is considered, and it can be solved via the pseudo-arc-length continuation algorithm.

Linearized Two-Degrees-of-Freedom Model

The linearized two-degrees-of-freedom (2DOF) model is introduced briefly to compare with the proposed nonlinear model, whose motion equations are given as follows (Lacarbonara and Ballerini 2009):

$$\begin{aligned} &(1+\mu)\ddot{y} + y + 2\zeta\dot{y} + \mu\ddot{\theta} = f\cos\alpha\tau \\ &\ddot{\theta} + \beta^2\theta + 2\zeta_\theta\dot{\theta} + \ddot{y} = 0 \end{aligned} \tag{13}$$

The particular solutions of Eq. (13) have the following forms:

$$\begin{aligned} y &= C\cos(\alpha\tau+\phi_1) \\ \theta &= D\cos(\alpha\tau+\phi_2) \end{aligned} \tag{14}$$

where C and D = displacement amplitudes of primary structure and PTMD, respectively; and ϕ_1 and ϕ_2 = phase angles.

Substituting Eq. (14) into Eq. (13), the analytical solutions of the linearized 2DOF model can be obtained as follows:

$$C = f\sqrt{\frac{[(\alpha^2-\beta^2)^2+4\alpha^2\zeta_\theta^2]}{G}}$$

$$D = f\frac{\alpha^2}{\sqrt{G}}$$

$$\phi_1 = \tan^{-1}\left(-\frac{\alpha^6+\alpha^4[-\beta^2(2+u)-1+4\zeta_\theta^2(1+u)]+\alpha^2[\beta^4(1+u)+2\beta^2-4\zeta_\theta^2]-\beta^4}{2\alpha[\alpha^4(\zeta_\theta u+\zeta)+\alpha^2(-2\zeta\beta^2+4\zeta\zeta_\theta^2)+\zeta\beta^4]}\right)$$

$$\phi_2 = \tan^{-1}\left(\frac{-\alpha^4-\beta^2+\alpha^2[\beta^2(1+u)+1+4\zeta\zeta_\theta]}{2\alpha\zeta_\theta[\alpha^2(1+u)-1]+2\alpha\zeta(\alpha^2-\beta^2)}\right) \tag{15}$$

where

$$G = (\alpha^8 + \alpha^6[(1+u)(4\zeta_\theta^2 - 2\beta^2) + (2\zeta + 2\zeta_\theta u)^2 + 4\zeta_\theta^2 u - 2] + \alpha^4[\beta^4(1+u)^2 + 2\beta^2(2+u-4\zeta^2) + 8\zeta_\theta^2(2\zeta^2 - u - 1) + 1] + \alpha^2[2\beta^4(2\zeta^2 - u - 1) + 4\zeta_\theta^2 - 2\beta^2] + \beta^4) \quad (16)$$

Based on the expressions of C and D in Eq. (15), the displacement amplitude ratio of D/C between the PTMD and primary structure is calculated as follows:

$$\frac{D}{C} = \alpha^2\sqrt{\frac{1}{(\alpha^2-\beta^2)^2 + 4\zeta_\theta^2\alpha^2}} \quad (17)$$

From Eq. (17), the displacement amplitude ratio depends on the frequency ratio β, PTMD damping ζ_θ, and excitation frequency ratio α. It can be seen that the displacement amplitude ratio is independent of the excitation amplitude, which is different from the nonlinear case (shown subsequently).

Numerical Verification

The Runge-Kutta integral is implemented to validate the nonlinear solution and linear solution for a coupled structure-PTMD system, with its nondimensional parameters given in Table 1. Fig. 2 shows the comparison of FRF obtained with the HBM and Runge-Kutta integral. Meanwhile, the comparison of FRF between the analytical solutions of linear model obtained by Eq. (15) and the Runge-Kutta integral results is also shown in Fig. 2, where NM and LM are the abbreviations of nonlinear and linear models, respectively. It can be seen that both linear and nonlinear solutions agree well with their counterparts obtained through numerical simulation. Additionally, the obvious differences in the system responses are demonstrated between linear and nonlinear solutions. In the following section, comparison of the dynamic responses between the two models will be conducted while the PTMD is optimized by the linear model.

Table 1. Nondimensional parameters of a coupled structure-PTMD system

Parameters	Notation	Value
Mass ratio	μ	0.02
Frequency ratio of PTMD to the primary structure	β	1
Structural damping the primary structure	ζ	0.01
Damping ratio of PTMD	ζ_θ	0.02
Dimensionless amplitude of the excitation force	f	0.01
Frequency ratio of the excitation force to the primary structure	α	0.4–1.5

Comparison between Nonlinear and Linear Models

Model Parameters

PTMDs are generally optimized by the linear model assuming small vibration amplitudes and are expected to operate in linear condition. However, there is a strong nonlinearity when the PTMD experiences large oscillations. In light of this, it is necessary to make a comparison on the dynamic response between the nonlinear and linear models when the PTMD is optimized by the linear model. The optimal design formulas for TMD, fitted by Toshihiro and Ken (1978), are used to determine the parameters of a linear PTMD when attached to a damped system, expressed as follows:

$$\beta_{\text{opt-L}} = \frac{1}{1+\mu} - (0.241 + 1.74\mu - 2.6\mu^2)\zeta - (1-1.9\mu + \mu^2)\zeta^2$$

$$\zeta_{\theta,\text{opt-L}} = \sqrt{\frac{3\mu}{8(1+\mu)}} + (0.13 + 0.12\mu + 0.4\mu^2)\zeta + (0.16\text{–}2.6\mu + 5\mu^2)\zeta^2 \quad (18)$$

where $\beta_{\text{opt-L}}$ and $\zeta_{\theta,\text{opt-L}}$ = optimal frequency ratio and optimal damping ratio of a linear PTMD.

Now, the frequency ratio β and the damping ratio ζ_θ are obtained by Eq. (18), and other key parameters such as mass ratio μ, structural damping ratio ζ, and excitation amplitude ratio f are set as follows: $\mu = 0.01$–0.05, $\zeta = 0.01$–0.03, and $f = 0.001$–0.01. It is illustrated that the mass ratio is frequently less than 5% on account of installation space. The typical structural damping ratio of the design assumptions for tall buildings, especially steel structures, is lower than 3% (Rob et al. 2010; Pritchard 1984). To demonstrate the physical concept of nondimensional excitation amplitude f, its expression given by Eq. (2) can be transformed as follows:

$$f = \beta^2\frac{F}{mg} \Rightarrow F = m\frac{f}{\beta^2}g = m\kappa \quad (19)$$

where κ = equivalent acceleration exerted on the primary structure.

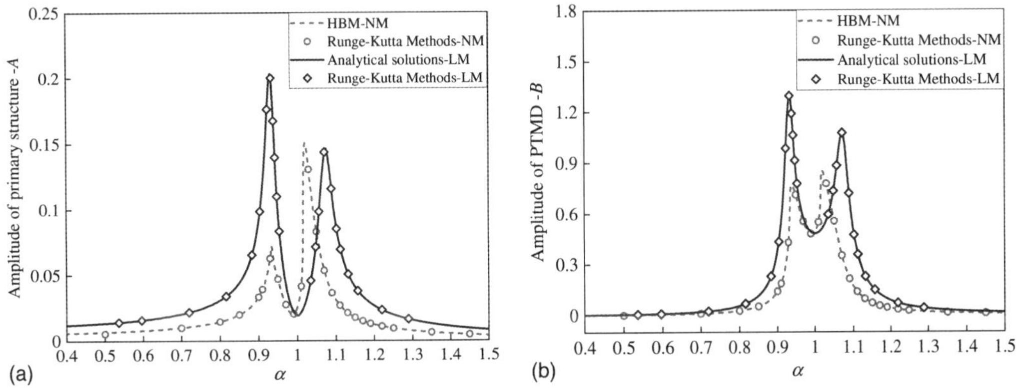

Fig. 2. Comparison of FRFs between the analytical and numerical solutions: (a) primary structure; and (b) PTMD.

Table 2. Nondimensional parameters used in analysis of FRFs for three different cases

Parameters	Notation	Case 1	Case 2	Case 3
Mass ratio	μ	0.01	0.03	0.03
Structural damping ratio	ζ	0.01	0.01	0.02
Tuned frequency ratio of PTMD [by Eq. (18)]	β	0.9874	0.9679	0.9647
Damping ratio of PTMD [by Eq. (18)]	ζ_θ	0.0623	0.1059	0.1072
Dimensionless amplitude of excitation force	f	0.002, 0.006, 0.01	0.006, 0.01	0.006, 0.01
Frequency ratio of excitation to primary structure	α	0.85–1.15	0.85–1.15	0.85–1.15

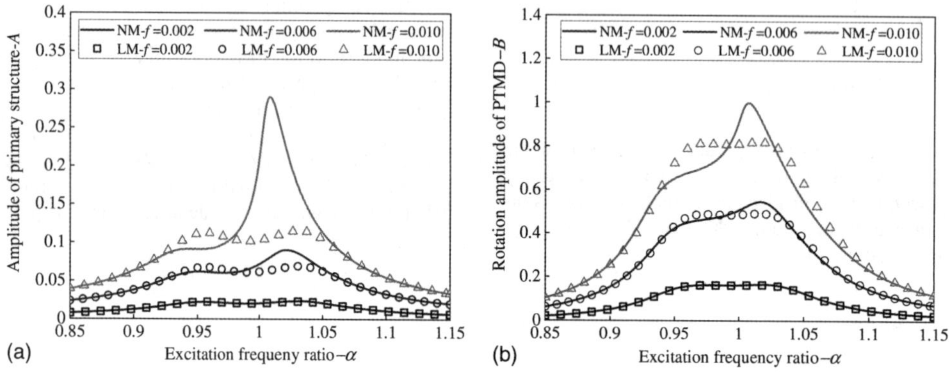

Fig. 3. Comparison of FRFs between the nonlinear and linear solutions for Case 1 under different excitation amplitudes (LM = linear model; and NM = nonlinear model): (a) primary structure; and (b) PTMD.

Because f is varied between 0.001 and 0.01, the expression of κ can be obtained as follows:

$$\kappa = \frac{f}{\beta^2} g = \frac{(0.001\text{–}0.01)}{\beta^2} g \tag{20}$$

The tuning frequency ratio of the PTMD, β, is varied from 0.9419 to 0.9874 as obtained by Eq. (18) under various mass ratios ($\mu = 0.01$–0.05) and structural damping ratios ($\zeta = 0.01$–0.03). Substituting β into Eq. (20), the range of κ is given as follows:

$$\kappa \approx (0.001\text{–}0.011)g \tag{21}$$

It is found that the range of κ is wide enough and its coefficients, that is (0.001–0.011), are very close to the values of f (0.001–0.01). The excitation amplitude f represents well the equivalent external acceleration applied to the primary structure. Moreover, the maximum of κ is not as large as the seismic loadings. The adopted range of excitation amplitude f is suitable for the simulation of relatively low-intensity excitations.

Frequency Response Functions

The excitation amplitude is varied to assess the impact of the PTMD nonlinearity on the difference in the FRFs between the nonlinear and linear models. The mass and structural damping ratios are taken to be 0.01, namely Case 1 as listed in Table 2, and the tuning frequency and damping ratios of PTMD are determined by Eq. (18).

Fig. 3(a) shows the comparison of FRFs for primary structure between the nonlinear and linear models, for three excitation amplitudes $f = 0.002, 0.006$, and 0.01. One can see that the two peaks in each FRF calculated by the linear model nearly remain equal for different excitation amplitudes, which is expected that the optimal parameters of PTMD are used. The FRF obtained by the nonlinear model is very close to that of the linear solutions at low excitation amplitude, such as $f = 0.002$. However, upon increasing the excitation amplitude, the FRF obtained from the nonlinear model shows a significant departure from the linear model. It implies that, at high forcing intensity, the PTMD nonlinearity leads to the frequency detuning of PTMD, reducing its control efficiency.

Moreover, the two peaks in FRF obtained by the nonlinear model slightly shift to the left (lower frequency side) and the maximum is located in the vicinity of 1:1 resonance with forcing frequency. The shift of FRF peaks to the lower frequency side suggests a kind of softening nonlinearity present in the PTMD. Fig. 3(b) shows the variation of FRF of PTMD rotation with excitation amplitude for both nonlinear and linear models. Similar observations can be made. However, compared with the displacement of the primary structure, the difference in PTMD rotation between the nonlinear model and linear model is less obvious.

If the PTMD operates in linear condition, the ratio of the displacement amplitude of the PTMD to that of the primary structure is constant regardless of excitation amplitude, as described in Eq. (17). Fig. 4 shows the variation of the displacement amplitude ratio under three excitation amplitudes $f = 0.002, 0.006$, and 0.01. As it can be seen, there is only one peak (in the vicinity of $\alpha = 1.0$) in each amplitude ratio curve obtained by linear model, whereas two peaks and one salient valley in the curve of the nonlinear model for $f = 0.01$.

A large mass ratio ($\mu = 0.03$) is selected in Case 2 (listed in Table 2) to assess the impact of mass ratio on nonlinearity in FRF when compared with Case 1 ($\mu = 0.01$). Fig. 5 shows the comparison of FRFs between the linear and nonlinear models for Case 2 under $f = 0.006$ and $f = 0.01$. It is found that the FRFs obtained by the nonlinear model are close to that of linear results

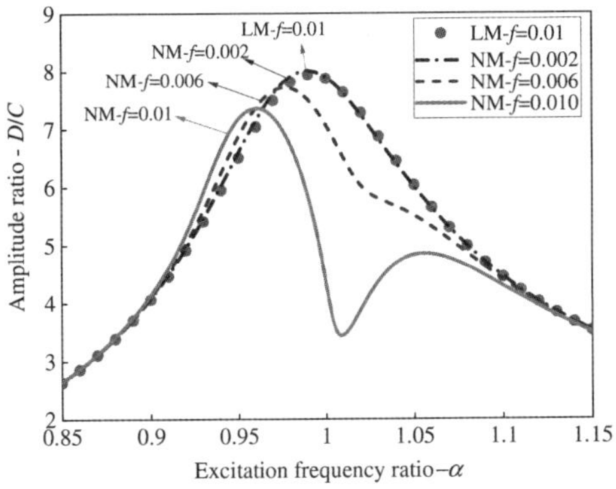

Fig. 4. Displacement amplitude ratio of PTMD to primary structure for Case 1.

even at high excitation levels (e.g., $f = 0.01$), indicating that the increase of mass ratio suppresses the nonlinear effect of PTMD. If the structural damping ratio increases to 0.02 (namely Case 3 as listed in Table 2), the difference of FRFs further reduce, as shown in Fig. 5. Increasing the structural damping also mitigates the nonlinearity effect of PTMD. In addition, the evident differences in both FRFs and the displacement amplitude ratio only exist in the relatively narrow range of excitation frequency ratio ($\alpha = 0.9$–1.1). Beyond the above range, there are no obvious differences between the nonlinear and linear models.

Time-Frequency-Domain Analysis

At a large amplitude of vibration, the frequency of the PTMD may vary with time, leading to higher-order harmonics in responses. The nonlinear solution for Case 1 is selected to analyze the frequency components of system response and the instantaneous frequency

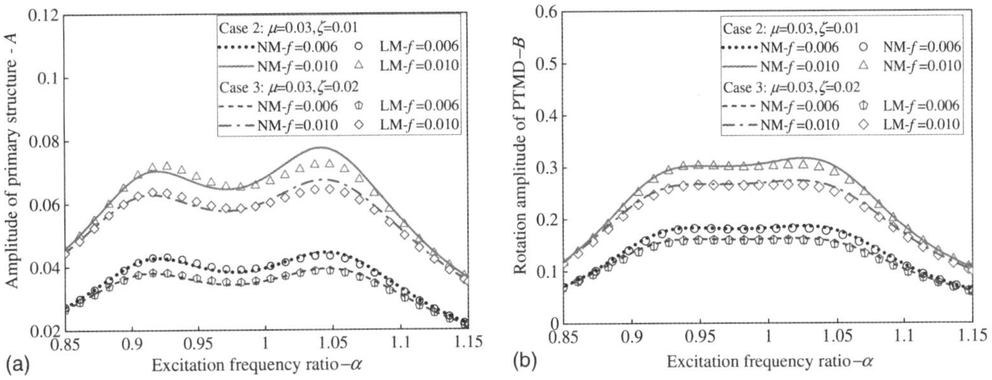

Fig. 5. Comparison of FRFs between the nonlinear and linear solutions for Cases 2 and 3 (LM = linear model; and NM = nonlinear model): (a) primary structure; and (b) PTMD.

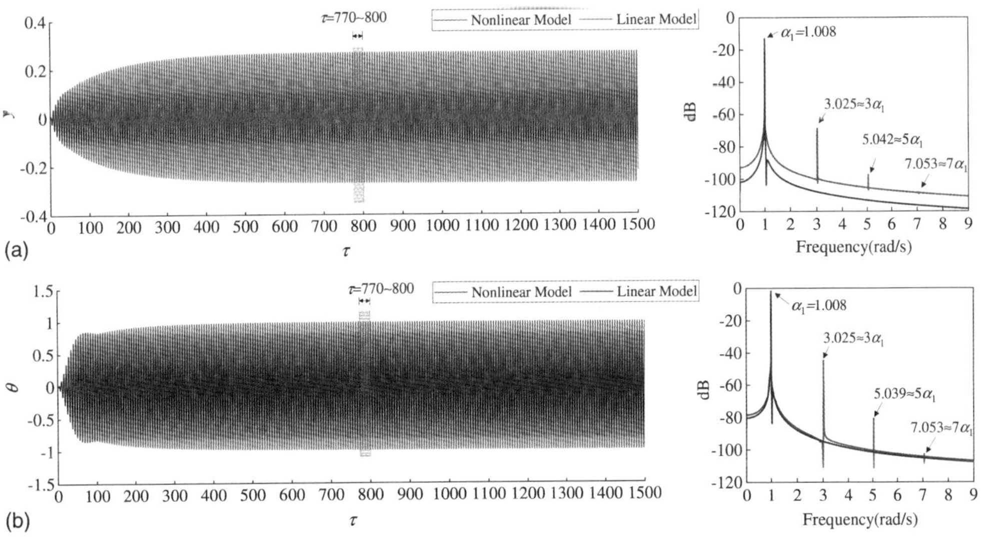

Fig. 6. Time histories of the primary structure and PTMD motions and their displacement spectrum for Case 1: (a) primary structure; and (b) PTMD.

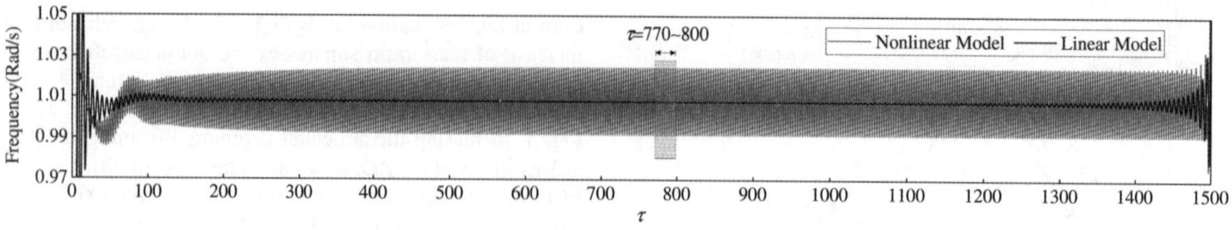

Fig. 7. Instantaneous frequency of PTMD motion for the nonlinear and linear solutions.

variation of the PTMD. Fig. 6 shows the time histories of dynamic responses and the one-sided displacement spectra for $f = 0.01$ and $\alpha_1 = 1.008$. The displacement spectra are obtained by the classical fast Fourier transform (FFT) algorithm where the sampling frequency is set as 50 rad/s, and a Hamming window is used to reduce the sidelobes of the frequency spectra. To clearly illustrate the various frequency components, the ordinate unit of decibels is utilized in the spectrum plots.

It can be seen that the displacement spectra of system displacement obtained by the linear model only shows 1:1 primary resonance with excitation frequency (that is $\alpha_1 = 1.008$). Actually, except for primary resonance, 1:3, 1:5, and 1:7 superharmonic resonances appear in the nonlinear solutions. However, due to the unit of ordinate decibels, the spectrum amplitude of the primary resonance is hundreds or thousands of times larger than that of the superharmonic resonances, indicating that the vibrational energy of the superharmonic resonances is much lower than that of the primary resonance.

Fig. 7 shows the instantaneous frequency of PTMD rotation obtained with Hilbert-Huang transform (HHT). The instantaneous frequency curve shows considerable fluctuations at the beginning and end of time due to end effects. In order to provide more details about the variation of the instantaneous frequency with displacement/velocity, the frequency variation during $\tau = 770$–800 are selected and shown in Fig. 8, along with the displacement and the corresponding velocity. One can see that the instantaneous frequency of PTMD displacement for linear solution is almost unchanged with time. Additionally, the mean value of instantaneous

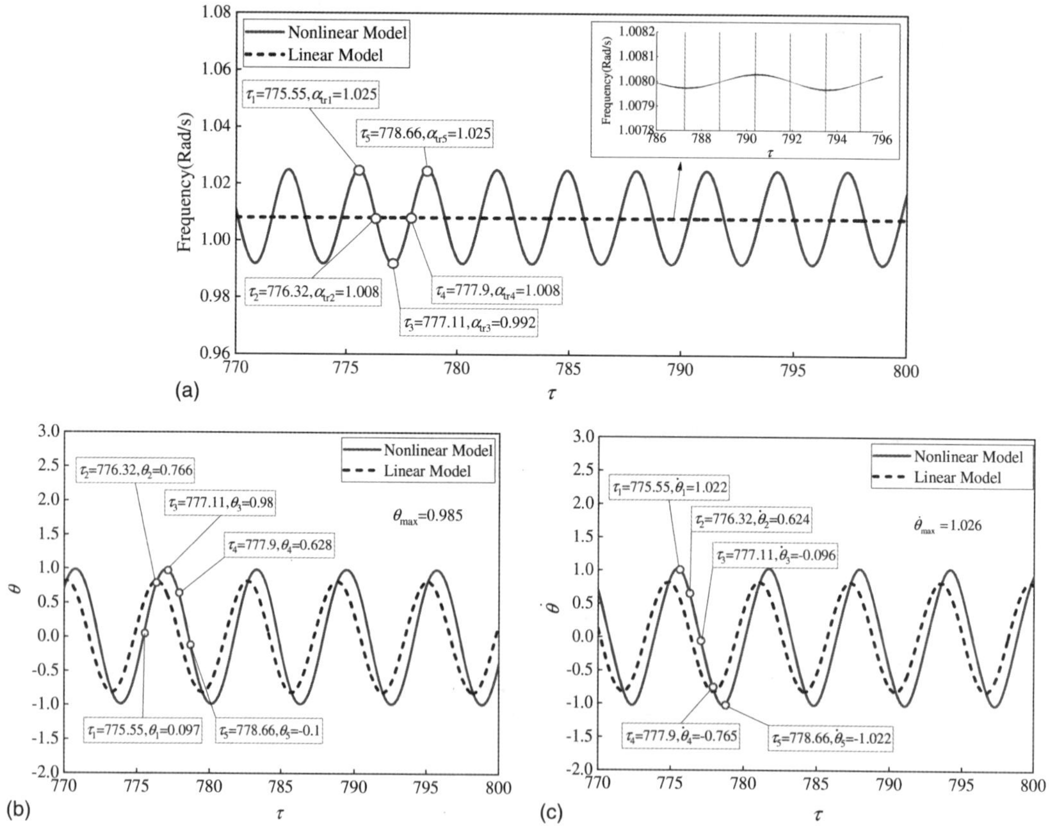

Fig. 8. Instantaneous frequency, rotation and its velocity of PTMD during $\tau = 770$–800: (a) instantaneous frequency; (b) rotation angle; and (c) rotation velocity.

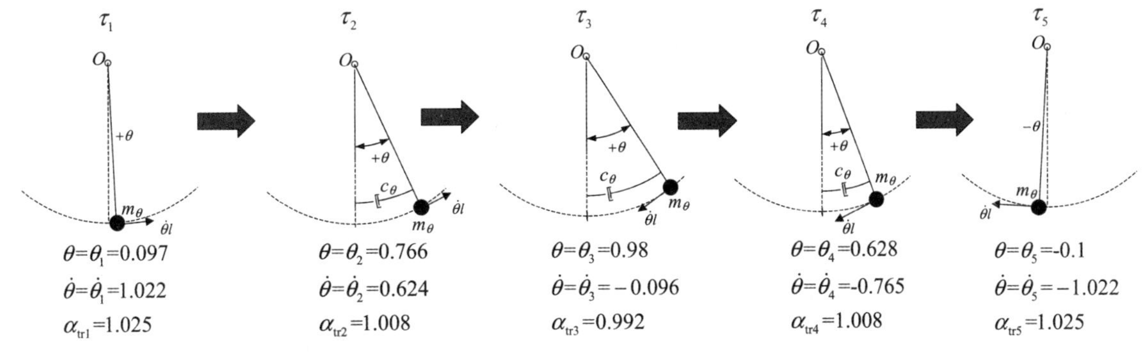

Fig. 9. Visualized PTMD motion at time instants τ_1 to τ_5.

frequency equals 1.008, namely the excitation frequency. Five time instants τ_1–τ_5 are picked out for explanation, which correspond to the maximum, minimum, or mean instantaneous frequencies. The rotation angle, velocity, and instantaneous frequency at the five instants τ_1–τ_5 are denoted by θ_1–θ_5, $\dot{\theta}_1$–$\dot{\theta}_5$, and α_{tr1}–α_{tr5}, respectively.

Fig. 9 shows the vibration trajectory of PTMD motion at the five instants. At τ_1, the PTMD passes positive displacement $\theta_1 = 0.097$, and the instantaneous frequency equals 1.025. Then, the PTMD moves to $\theta_2 = 0.766$ at τ_2 and instantaneous frequency becomes excitation frequency 1.008. At τ_3, the displacement of PTMD reaches to $\theta_3 = 0.98$ near to its maximum ($\theta_{max} = 0.985$), whereas the instantaneous frequency decreases to the minimum 0.992. At τ_5, the PTMD returns to a position close to its equilibrium ($\theta_5 = -0.1$), and the frequency becomes 1.025 again.

Fig. 10 shows the envelope curve of the instantaneous frequency of PTMD motion with rotation amplitude, obtained by increasing the excitation amplitude. The two curved lines represent the boundary of maximum and minimum steady-state instantaneous frequencies for the nonlinear model. With an increase of the rotation amplitude, the magnitude of the variation of the instantaneous frequency increases. However, the magnitude of the variation of the instantaneous frequency for the linear model (two straight lines) is not changed.

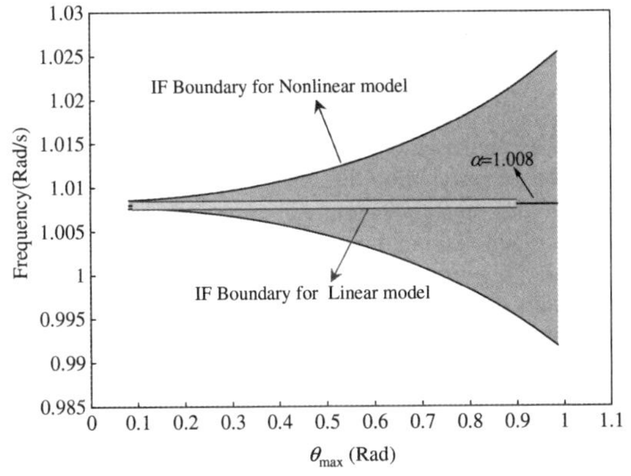

Fig. 10. Variation of instantaneous frequency with rotation amplitude of PTMD (IF = instantaneous frequency).

Comparison of Control Performance

The excitation amplitude f is varied to evaluate the influence of the PTMD nonlinearity on control performance under various mass and structural damping ratios ($\mu = 0.01$–0.05 and $\zeta = 0.01$–0.03) when the PTMD is tuned by the linear model using Eq. (18). Minimizing the maximum amplitude in the FRF of the primary structure is utilized as index of control performance. The maximum amplitudes of the primary structure and PTMD are defined as A_{max} and B_{max}, respectively. The difference in maximum amplitudes between the nonlinear and linear solutions is expressed as follows:

$$y_p = \frac{A_{max,N} - A_{max,L}}{A_{max,L}} \times 100\%$$
$$\theta_p = \frac{B_{max,N} - B_{max,L}}{B_{max,L}} \times 100\% \qquad (22)$$

where subscripts N and L = maximum amplitudes that obtained by the nonlinear and linear models, respectively.

Fig. 11 shows 3D surface plot of the maximum amplitudes of the primary structure and PTMD as functions of mass and structural damping ratios under different excitation amplitudes. As seen in Fig. 11(a), the maximum amplitudes of the primary structure increase dramatically when the excitation amplitude f increases, especially for small mass ratios or structural damping ratios. The nonlinearity of maximum amplitudes of the PTMD is less significant when the excitation amplitude is varied, as shown in Fig. 11(b). For example, for the coupled system with $\mu = 0.01$ and $\zeta = 0.01$, the maximum amplitudes of the primary structure are 0.01 and 0.29 for the excitation amplitude $f = 0.001$ and $f = 0.01$, respectively; meanwhile, those of the PTMD are 0.082 and 0.99. The maximum amplitudes of both the primary structure and PTMD increase linearly with the excitation amplitude for linear solutions, according to Eq. (15). Therefore, it can be seen that the nonlinearity in response of primary structure is more obvious than that of PTMD.

Fig. 12(a) shows the 3D surface plot of difference in the maximum amplitudes of the primary structure, y_p, as functions of mass and structural damping ratios under several excitation amplitudes. The maximum difference is less than 0.57% when the excitation amplitude equals 0.001. The amplitudes between the nonlinear and linear solutions are close to each other, indicating that the effect of PTMD nonlinearity can be ignored. When excitation amplitude increases, the difference builds up sharply up to 152% at $f = 0.01$. It means that large excitation amplitudes amplify the difference in the maximum amplitudes between the nonlinear and linear models,

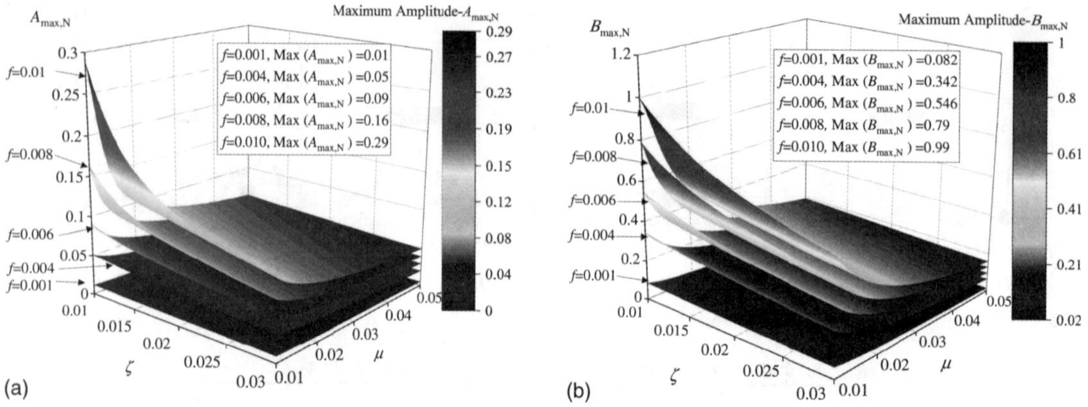

Fig. 11. Three-dimensional surface plot of maximum amplitudes as functions of mass and structural damping ratios under different excitation amplitudes: (a) primary structure; and (b) PTMD.

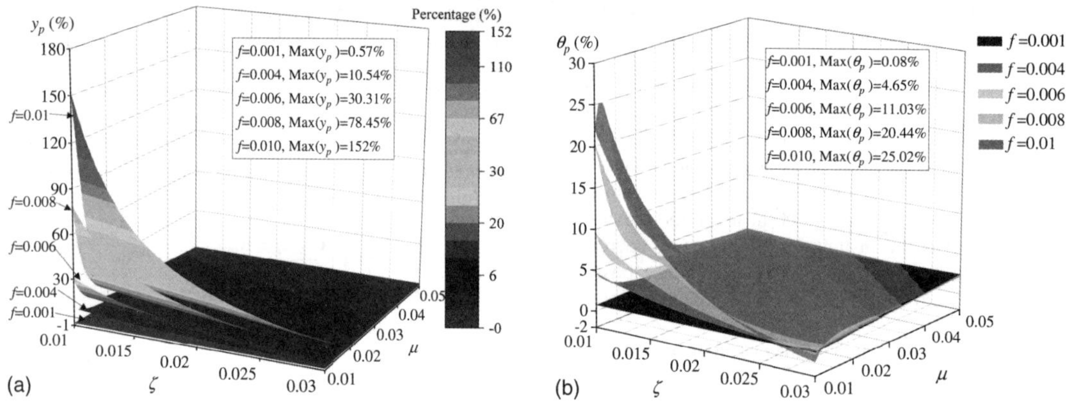

Fig. 12. Three-dimensional surface plot of difference in maximum amplitudes between nonlinear and linear solutions as functions of mass ratio and structural damping ratio under different excitation amplitudes: (a) primary structure; and (b) PTMD.

significantly attenuating the control performance of PTMD. The dynamic response of the primary structure is dramatically underestimated by the linear model. In addition, the maximum difference between the linear and nonlinear solutions occurs in the cases of small mass and structural damping ratios. Furthermore, the difference decreases with increase of mass or structural damping ratios.

Fig. 12(b) illustrates 3D surface plot of the difference in maximum amplitudes of PTMD, θ_p, as functions with mass and structural damping ratios under different excitation amplitudes. Obviously, the difference between the linear and nonlinear solutions for the PTMD is considerably smaller than the difference for the primary structure.

Fig. 13 shows the maximum amplitudes of the primary structure and the PTMD under several excitation amplitudes for $\mu = 0.01$ and $\zeta = 0.01$, along with their difference between linear and nonlinear solutions. The critical excitation amplitude and critical maximum amplitudes of the primary structure and PTMD can be obtained when the difference in maximum amplitudes for the primary structure does not exceed some specified values, such as $y_p = 2\%$, 5%, or 10%. These three values may be taken as thresholds for negligible, acceptable, and unacceptable influence of PTMD nonlinearity on control performance. For a threshold $y_p = 5\%$, the excitation amplitudes f should be smaller than 0.00286 if the nonlinear effect of the PTMD on control performance is acceptable. The corresponding maximum amplitudes of the primary structure and PTMD are 0.035 and 0.24 for $f = 0.0286$.

Likewise, if the effect of PTMD nonlinearity can be ignored ($y_p \leq 2\%$), the excitation amplitude must be less than 0.00186 or the maximum amplitude of PTMD does not exceed 0.154 rad (8.82°). Once the rotation angle of PTMD exceeds 0.334 rad (19.15°), the corresponding difference $y_p > 10\%$ may not be allowed in engineering applications. Thus, $y_p = 5\%$ is suggested because the critical maximum amplitudes of PTMD could be up to 0.24 rad (13.76°), which permits slightly larger stroke of PTMD and maintains near-optimal control performance.

Following the same procedure, Table 3 summarizes the critical excitation amplitude under various mass and structural damping ratios for $y_p = 2\%$, 5%, and 10%, and the corresponding maximum amplitudes of the primary structure and PTMD are provided as well. At the same mass ratio such as $\mu = 0.01$, the increase of structural damping improves the values of the three terms. Moreover, if the mass ratio is further increased, the critical excitation amplitude and the maximum amplitude of primary structure keep growing.

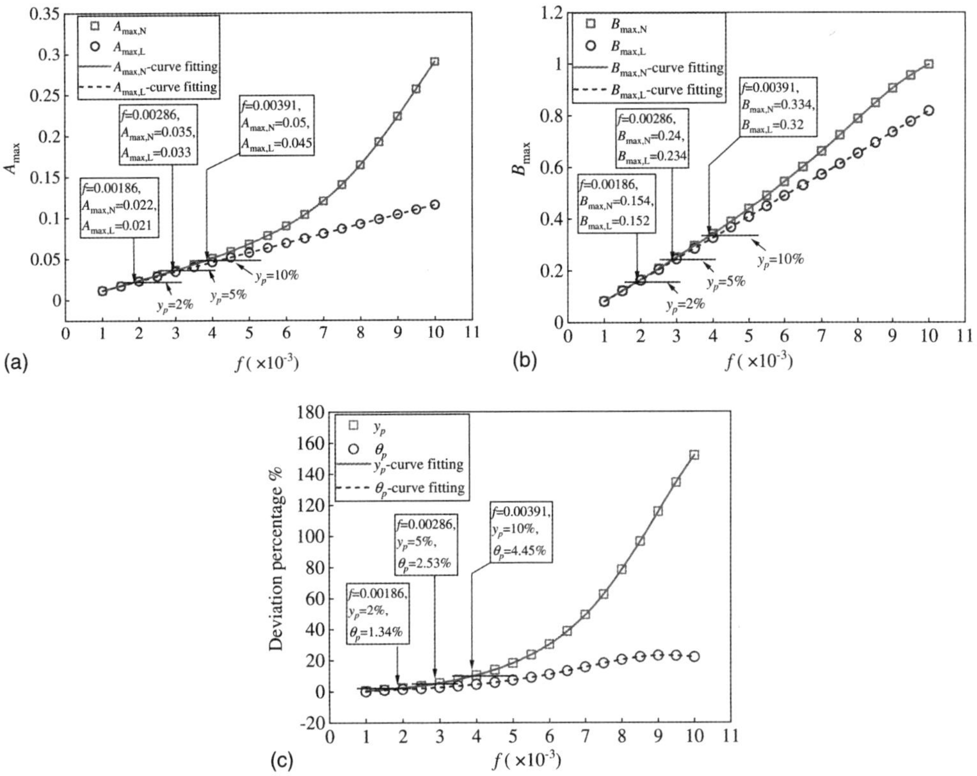

Fig. 13. Variation of linear/nonlinear maximum amplitudes and their difference for the primary structure and PTMD with excitation amplitude ($\mu = 0.01$ and $\zeta = 0.01$): (a) A_{max}; (b) B_{max}; and (c) y_p and θ_p.

However, the critical maximum amplitude of PTMD does not increase with the mass ratio if its value approaches 11.37° for $y_p = 2\%$ and 17.82° for $y_p = 5\%$.

In summary, the critical values of the three terms vary with mass and structural damping ratios. Because the effect of PTMD nonlinearity on vibration control manifests itself through frequency detuning, increasing the mass ratio leads to a control performance less sensitive to frequency detuning or PTMD nonlinearity, therefore allowing larger critical amplitudes of PTMD. Nevertheless, the critical rotation amplitude 8.82° (about 9°) is associated with the smallest $\mu = 0.01$ and $\zeta = 0.01$, which are commonly appeared in civil engineering structures. It is reasonable to take about 9° as a limitation to directly assess if PTMD nonlinearity can be ignored in practical engineering. Moreover, the obtained critical amplitudes of PTMD can be also extended to a multidegree-of-freedom (MDOF) system because its dynamic responses are mostly dominated by single resonant mode where the PTMD is tuned.

Table 3. Critical excitation amplitudes, maximum amplitudes of structure and PTMD for $y_p = 2\%$, 5%, and 10%

		Critical excitation amplitude ($\times10^{-3}$)			Critical displacement amplitude of primary structure (rad)			Critical displacement amplitude of PTMD (°)		
μ	ζ	$y_p = 2\%$	$y_p = 5\%$	$y_p = 10\%$	$y_p = 2\%$	$y_p = 5\%$	$y_p = 10\%$	$y_p = 2\%$	$y_p = 5\%$	$y_p = 10\%$
0.01	0.01	1.86	2.86	3.91	0.022	0.035	0.050	8.82	13.75	19.14
	0.02	2.59	4.05	5.57	0.026	0.041	0.059	10.04	15.76	21.92
	0.03	3.50	5.49	7.65	0.030	0.048	0.070	11.37	17.82	24.80
0.015	0.01	2.76	4.28	5.82	0.027	0.044	0.062	9.14	14.35	19.92
	0.02	3.66	5.71	7.84	0.031	0.050	0.072	10.20	16.04	22.28
	0.03	4.74	7.42	—	0.035	0.056	—	11.37	17.82	—
0.02	0.01	3.70	5.70	7.70	0.032	0.052	0.073	9.40	14.73	20.46
	0.02	4.72	7.35	—	0.036	0.058	—	10.37	16.27	—
	0.03	5.93	9.27	—	0.040	0.065	—	11.36	17.82	—
0.025	0.01	4.58	7.10	9.65	0.037	0.058	0.083	9.58	15.06	20.91
	0.02	5.76	8.96	—	0.041	0.065	—	10.49	16.44	—
	0.03	7.10	—	—	0.045	—	—	11.36	—	—

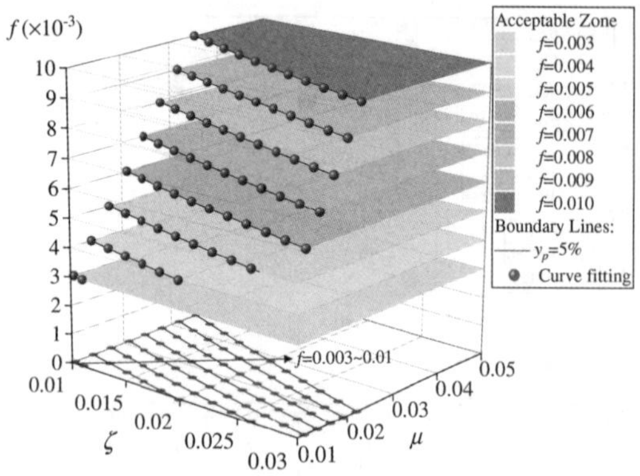

Fig. 14. Boundary lines for the difference threshold ($y_p = 5\%$) and acceptable zones ($y_p \le 5\%$) in maximum amplitude of primary structure.

PTMD Design Strategy

The preceding study indicates that the effect of PTMD nonlinearity on vibration control can be enormous. In order to attenuate the adverse effects of PTMD nonlinearity on vibration control, two approaches are proposed in this section to proper parameter design of a PTMD in face of nonlinearity. The first approach is based on increasing mass ratio, and a minimum mass ratio as function of excitation amplitude and structural damping ratio is derived. In the second approach, optimization of model parameters of the PTMD is carried out considering nonlinearity, and diagrams that describe the optimal parameters of PTMD as functions of the excitation amplitude are obtained.

Methodology of Mitigating Nonlinearity Effect of PTMD

As pointed out in the section "Comparison of Control Performance," by increasing the mass ratio, the difference in the maximum amplitude of primary structure between the nonlinear and linear models is reduced. It is possible to mitigate the nonlinear effect of PTMD by selecting an appropriate mass ratio for certain excitation amplitude. This study takes $y_p = 5\%$ as threshold for PTMD with negligible nonlinearity. Fig. 14 shows the $y_p = 5\%$ boundary lines and the acceptable zones ($y_p \le 5\%$) under various mass ratios and structural damping ratios for different excitation amplitudes. The shaded area represents that the mass and structural damping ratios are in the acceptable zone for the corresponding excitation amplitude, which guaranties $y_p \le 5\%$.

Curve-fitting is performed to evaluate the threshold $y_p = 5\%$ boundary lines as shown in Fig. 14. Additionally, the projection of boundary lines and corresponding fitted points are demonstrated in the ζ-μ plane. As can be seen, the fitted results agree well with the boundary lines through the numerical search. The expression of the fitted curves, which express the minimum mass ratio as functions of structural damping and excitation amplitude ($0.003 \le f \le 0.01$), is given by

$$\mu \ge 0.00338 + 3.8482f - 0.336\zeta + 10.15f^2 + 3.097\zeta^2 - 45.615f\zeta \quad (0.003 \le f \le 0.01) \tag{23}$$

For a certain maximum excitation amplitude f_{max} and structural damping ratio ζ, the minimum mass ratio can be determined by Eq. (23). Upon the determination of mass ratio, the tuning frequency and damping ratios of PTMD will be calculated according to Eq. (18), simply following the classical linear solutions.

Optimal Parameters of PTMD Considering Nonlinearity

Because different excitation amplitudes induce different degree of nonlinearity of PTMD, it is expected that the optimal parameters of PTMD are functions of the excitation amplitude, in addition to the mass ratio and the structural damping. Hence, the optimal tuning frequency and damping ratios of PTMD, denoted as $\beta_{opt\text{-}N}$ and $\zeta_{\theta,opt\text{-}N}$, can be obtained by the established nonlinear model through numerical search.

To illustrate the procedure of finding the optimal parameters of the PTMD, the mass ratio $\mu = 0.01$ and structural damping ratio $\zeta = 0.01$ has been considered. The maximum amplitude of the primary structure with respect to various tuning frequencies and damping ratios of the PTMD is obtained for the two intensities

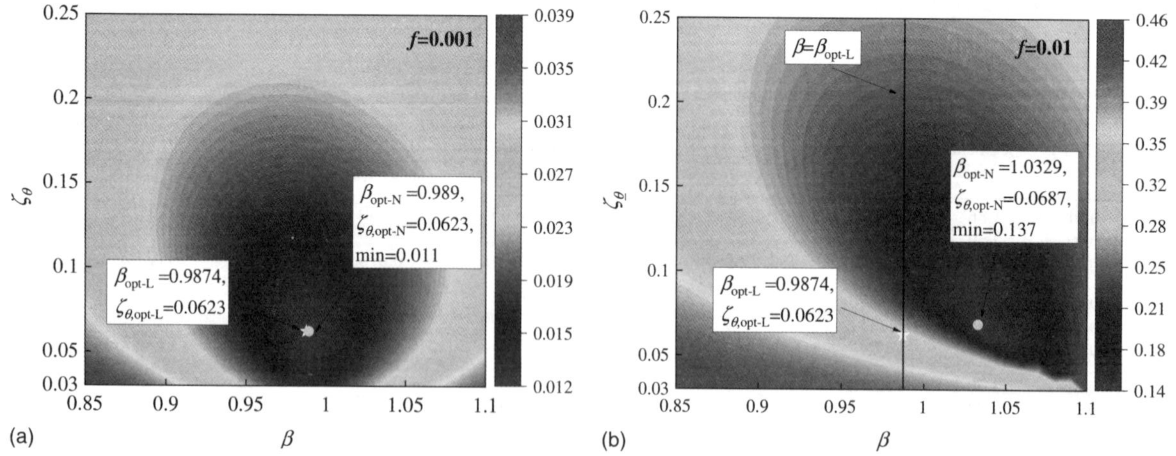

Fig. 15. Contour plot of maximum amplitude of primary structure as functions of tuning frequency and damping ratio of PTMD under $\mu = 0.01$ and $\zeta = 0.01$: (a) $f = 0.001$; and (b) $f = 0.01$.

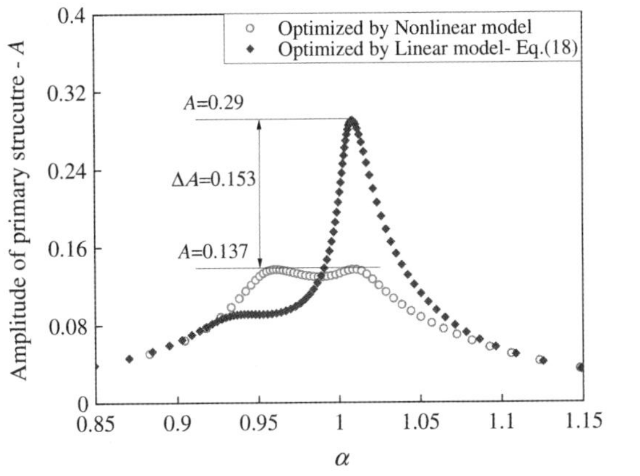

Fig. 16. Comparison of FRFs optimized by nonlinear and linear models under $\mu = 0.01$, $\zeta = 0.01$, and $f = 0.01$.

of excitations ($f = 0.001$ and $f = 0.01$), as shown in Fig. 15. The minimum of the maximum amplitude in Fig. 15 can be readily determined and the corresponding optimal parameters can be obtained. Additionally, the optimal parameters (denoted as $\beta_{\text{opt-L}}$ and $\zeta_{\theta,\text{opt-L}}$) obtained from linear model are also plotted.

It has been observed that, unlike the classical linear PTMD, the optimal parameters of nonlinear PTMD show the obvious deviation between the two excitation amplitudes. If the tuning frequency ratio is fixed at $\beta = \beta_{\text{opt-L}} = 0.9874$, the maximum amplitudes via various damping of PTMD, denoted by a vertical solid line in Fig. 15(b), are much larger than the optimal value from the nonlinear model. It indicates that by the change of damping of PTMD to achieve better performance for a large excitation amplitude is impossible while the tuning frequency ratio remains unchanged, like at a small excitation intensity ($f = 0.001$).

Fig. 16 shows the comparison of FRF curves for the primary structure, which are optimized by nonlinear and linear models, respectively. As can be seen, the two peaks are almost equal when the PTMD is retuned by the nonlinear model, which outperforms signally that tuned by the linear model.

The optimal tuning frequency ratios of PTMD under different mass and structural damping ratios are shown in Fig. 17 as function of the excitation amplitude, and the linear optimal parameters ($\beta_{\text{opt-L}}$ and $\zeta_{\theta,\text{opt-L}}$) denoted as short-dashed lines are also superimposed. It is found that the optimal tuning frequency ratio of nonlinear PTMD increases significantly with the excitation amplitude at small mass ratios. With the increase of mass and structural damping ratios, the growth of the optimal frequency ratio has a slower trend. Nevertheless, the optimal frequency ratios obtained by the nonlinear model are larger than that of the linear model.

Fig. 18 shows the variation optimal damping ratio of nonlinear PTMD with the excitation amplitude. It is observed that the optimal damping ratio is not sensitive to the excitation amplitude. Whereas the optimal parameters of the nonlinear PTMD enhance the control performance, the optimal frequency strongly depends on the excitation intensity. In practice, the excitation amplitude may vary within a range. The nonlinear optimal parameters of PTMD are difficult to implement for passive control. Therefore, the methodology of minimal PTMD nonlinearity based on proper mass ratios, proposed in previous subsection, is more preferable.

Numerical Example of a MDOF Structure Equipped with PTMD

Model Description

The primary structure is simplified as a SDOF system in the aforementioned analysis. To verify the nonlinear effects of PTMD on the control performance of MDOF structure, numerical simulation is

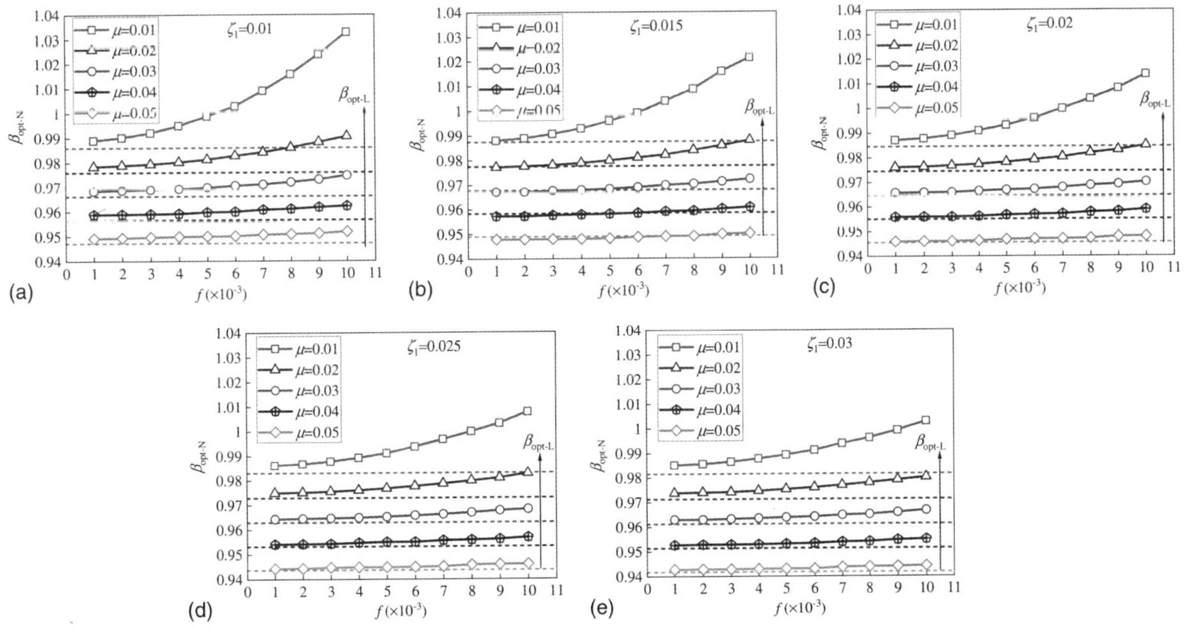

Fig. 17. Optimal frequency ratio of PTMD versus excitation amplitudes (solid lines: nonlinear model; short-dashed lines: linear model): (a) $\zeta = 0.01$; (b) $\zeta = 0.02$; (c) $\zeta = 0.03$; (d) $\zeta = 0.04$; and (e) $\zeta = 0.05$.

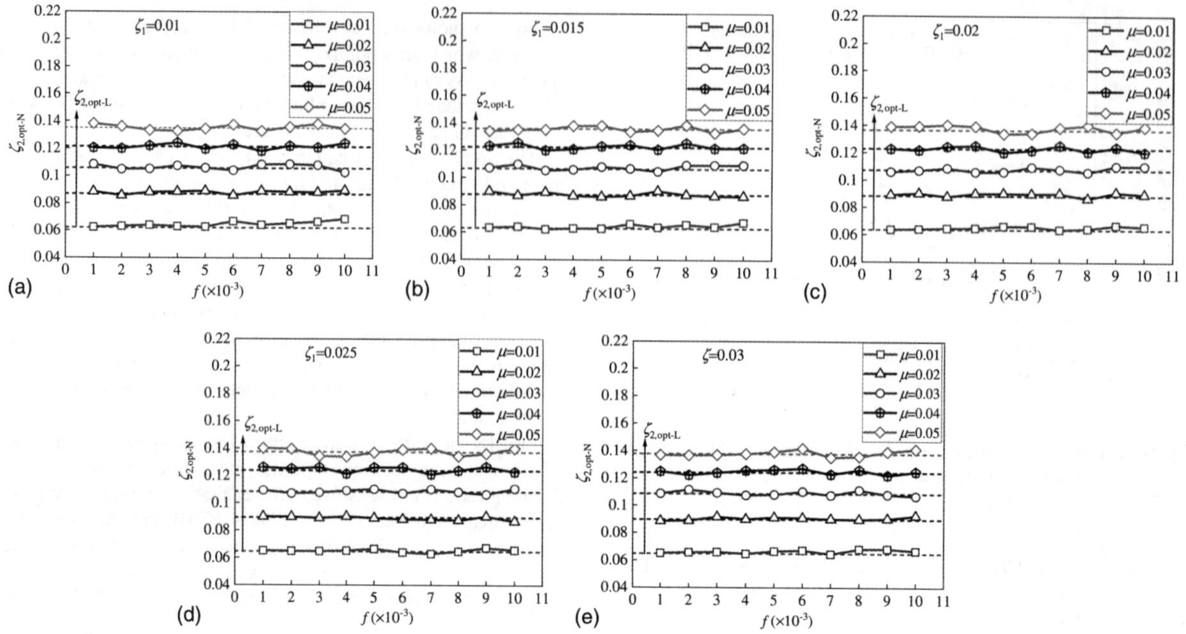

Fig. 18. Optimal damping ratio of PTMD versus excitation amplitudes (solid lines: nonlinear model; short-dashed lines: linear model): (a) $\zeta = 0.01$; (b) $\zeta = 0.02$; (c) $\zeta = 0.03$; (d) $\zeta = 0.04$; and (e) $\zeta = 0.05$.

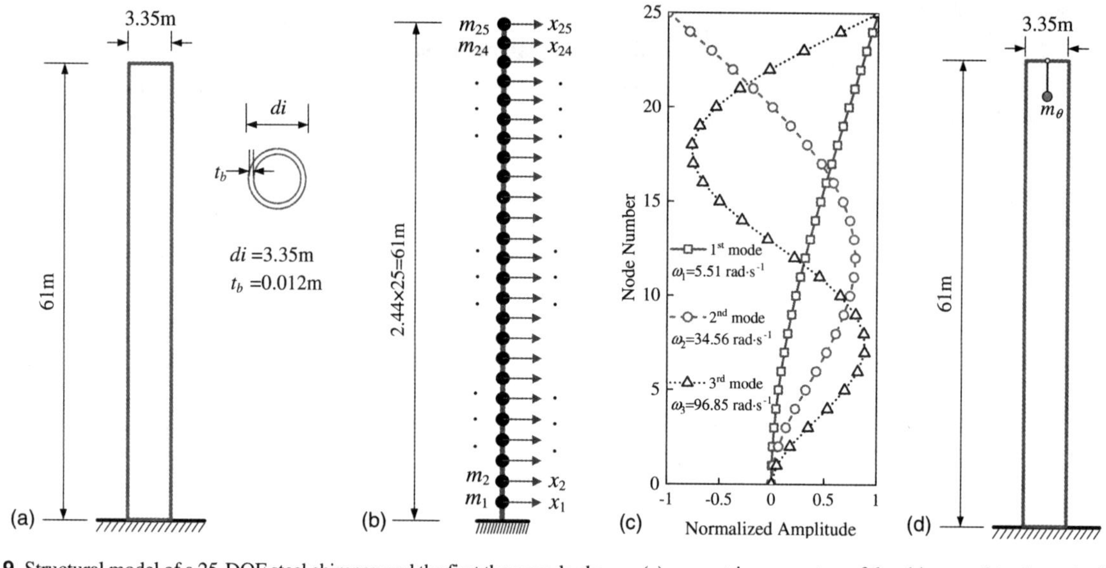

Fig. 19. Structural model of a 25-DOF steel chimney and the first three mode shapes: (a) geometric parameters of the chimney; (b) schematic diagram of the discrete model; (c) first three mode shapes without PTMD; and (d) chimney with PTMD attached.

conducted using a 25-DOF structural model, which represents a 61-m-high steel chimney as shown in Fig. 19(a). The outer diameter is 3.35 m and the shell thickness is 0.012 m.

The chimney is discretized into a 25-DOF model as shown in Fig. 19(b), and the length of each element is 2.44 m. The masses of every node, m_i, are all taken equal. The motion equations of the 25-DOF structure, subjected to excitation force $\{Q(t)\} = \{Q_1(t), Q_2(t), \ldots, Q_{25}(t)\}^{\mathrm{T}}$, is given in classical matrix form as follows:

$$\mathbf{M}\{\ddot{X}\} + \mathbf{C}\{\dot{X}\} + \mathbf{K}\{X\} = \{Q(t)\} \tag{24}$$

where $\{X\} = \{x_1, x_2, \ldots, x_{25}\}^{\mathrm{T}}$ is the vector containing the displacement of every node; **M**, **C**, and **K** = mass, damping coefficient, and stiffness matrix, respectively, which have the following form:

$$\mathbf{M} = \mathrm{diag}(m_1, m_2, \ldots, m_{25})$$

$$\mathbf{K} = \begin{bmatrix} k_{1,1} & k_{1,2} & k_{1,3} & \cdots & k_{1,25} \\ & k_{2,2} & k_{2,3} & \cdots & k_{2,25} \\ & & k_{3,3} & \cdots & k_{3,25} \\ & \text{sym} & & \ddots & \vdots \\ & & & & k_{25,25} \end{bmatrix}$$

$$\mathbf{C} = \begin{bmatrix} c_{1,1} & c_{1,2} & c_{1,3} & \cdots & c_{1,25} \\ & c_{2,2} & c_{2,3} & \cdots & c_{2,25} \\ & & c_{3,3} & \cdots & c_{3,25} \\ & \text{sym} & & \ddots & \vdots \\ & & & & c_{25,25} \end{bmatrix} \tag{25}$$

The stiffness matrix **K** is obtained by the inverse of flexibility matrix. Assume that the damping coefficient matrix **C** is proportional to the stiffness matrix and the first-order modal damping ratio is set as $\zeta_1 = 0.5\%$, which refers to the material damping of steel. Then, the first three natural frequencies of the uncontrolled chimney, ω_1–ω_3, are 5.51, 34.56, and 96.85 rad · s^{-1}, respectively, where the associated mode shapes are shown in Fig. 19(c). Additionally, the first-order modal mass is calculated as 16,117 kg.

When a PTMD with mass m_θ is attached to the top of the chimney as shown in Fig. 19(d), the equations of motion for this coupled system are given as follows:

$$\mathbf{M}^{\mathrm{N}}_{(26\times26)}\begin{Bmatrix}\{\ddot{X}\}\\ \ddot{\theta}\end{Bmatrix} + \mathbf{C}^{\mathrm{N}}_{(26\times26)}\begin{Bmatrix}\{\dot{X}\}\\ \dot{\theta}\end{Bmatrix} + \mathbf{K}^{\mathrm{N}}_{(26\times26)}\begin{Bmatrix}\{X\}\\ \theta\end{Bmatrix} + \begin{bmatrix}\mathbf{0}_{(24\times24)} & \mathbf{0}_{(24\times2)}\\ \mathbf{0}_{(2\times24)} & \mathbf{NF}_{(2\times2)}\end{bmatrix} = \begin{Bmatrix}\{Q(t)\}\\ 0\end{Bmatrix} \tag{26}$$

where $\mathbf{M}^{\mathrm{N}}_{(26\times26)}$, $\mathbf{K}^{\mathrm{N}}_{(26\times26)}$, and $\mathbf{C}^{\mathrm{N}}_{(26\times26)}$ are the mass, stiffness, and damping coefficient matrixes for the coupled system; and **NF** = nonlinear force generated by the PTMD, which are expressed

$$\mathbf{M}^{\mathrm{N}}_{(26\times26)} = \mathrm{diag}(m_1, m_2, \ldots, m_{24}, m_{25} + m_\theta, m_\theta l^2)$$

$$\mathbf{K}^{\mathrm{N}}_{(26\times26)} = \begin{bmatrix}\mathbf{K}_{(25\times25)} & \{0\}_{(25\times1)}\\ \{0\}_{(1\times25)} & 0\end{bmatrix}$$

$$\mathbf{C}^{\mathrm{N}}_{(26\times26)} = \begin{bmatrix}\mathbf{C}_{(25\times25)} & \{0\}_{(25\times1)}\\ \{0\}_{(1\times25)} & c_\theta l^2\end{bmatrix}$$

$$\mathbf{NF} = \begin{bmatrix} m_\theta l(\ddot{\theta}\cos\theta - \dot{\theta}^2\sin\theta) & 0 \\ 0 & m_\theta g l \sin\theta + m_\theta l \ddot{x}\cos\theta\end{bmatrix} \tag{27}$$

Assuming that $\sin\theta \approx \theta$ and $\cos\theta \approx 1$, and elminating the higher-order term of angular velocity ($\dot{\theta}^2$) in the nonlinear force **NF**, Eq. (26) can be linearized as follows:

$$\mathbf{M}^{\mathrm{L}}_{(26\times26)}\begin{Bmatrix}\{\ddot{X}\}\\ \ddot{\theta}\end{Bmatrix} + \mathbf{C}^{\mathrm{L}}_{(26\times26)}\begin{Bmatrix}\{\dot{X}\}\\ \dot{\theta}\end{Bmatrix} + \mathbf{K}^{\mathrm{L}}_{(25\times25)}\begin{Bmatrix}\{X\}\\ \theta\end{Bmatrix} = \begin{Bmatrix}\{Q(t)\}\\ 0\end{Bmatrix} \tag{28}$$

where

$$\mathbf{M}^{\mathrm{L}}_{(26\times26)} = \begin{bmatrix}\mathrm{diag}(m_1, \ldots, m_{24}) & \mathbf{0}_{(24\times2)} & \\ \mathbf{0}_{(2\times24)} & m_{25} + m_\theta & m_\theta l \\ & m_\theta l & m_\theta l^2\end{bmatrix}$$

$$\mathbf{K}^{\mathrm{L}}_{(25\times25)} = \begin{bmatrix}\mathbf{K}_{(25\times25)} & \{0\}_{(25\times1)}\\ \{0\}_{(1\times25)} & m_\theta g l\end{bmatrix}$$

$$\mathbf{C}^{\mathrm{L}}_{(26\times26)} = \mathbf{C}^{\mathrm{N}}_{(26\times26)} \tag{29}$$

In the following, the nonlinear effects of PTMD on the vibration control of the steel chimney under harmonic and random wind loads excitations are investigated by using the Runge-Kutta numerical method.

Response to Harmonic Excitations

The mass ratio of the PTMD to the modal mass for the fundamental mode of the 25-DOF system is set as $\mu = 0.01$, that is $m_\theta = 161.2$ kg. The tuning frequency and damping ratios are $\beta = 0.989$

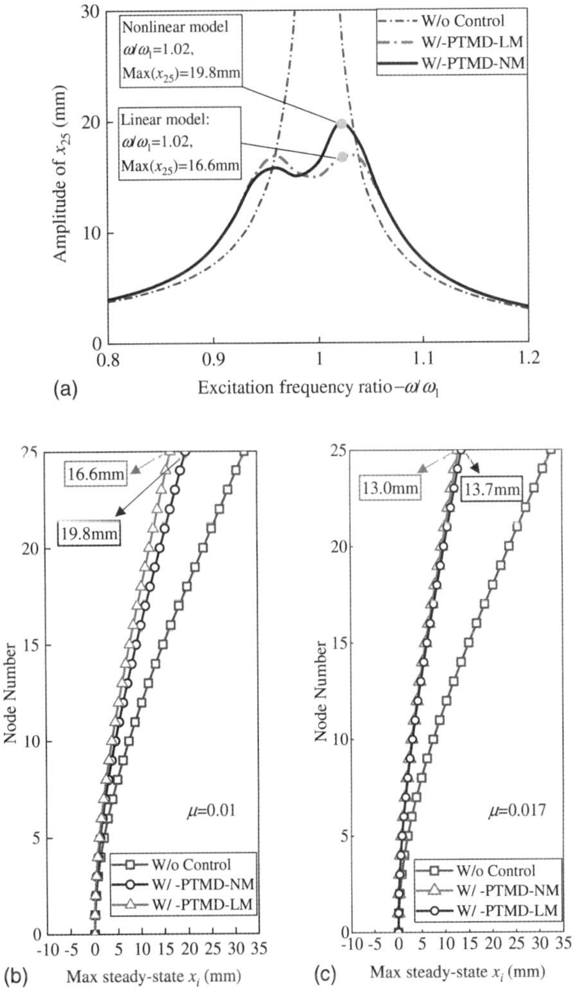

Fig. 20. FRF and maximum displacement for the chimney under harmonic excitation: (a) FRF at top node; (b) maximum displacement for $\mu = 0.01$; and (c) maximum displacement for $\mu = 0.017$.

Table 4. Comparison between nonlinear and linear solutions for the steel chimney under harmonic excitation

$\mathbf{Q}(t)$ (N)	μ	Case description	Amplitude of top node, Max x_{25} (mm)	First modal amplitude (mm)
$Q_{25}(t) = 600\cos(1.02\omega_1 t)$	0.01	Without control	32.3	220.8
		With PTMD (linear model)	16.6	133.2
		With PTMD (nonlinear model)	19.8	111.7
		Errors between the two models (%)	19.11	19.12
	0.017	With PTMD (linear model)	13.0	87.8
		With PTMD (nonlinear model)	13.7	92.0
		Errors between the two models (%)	4.80	4.81

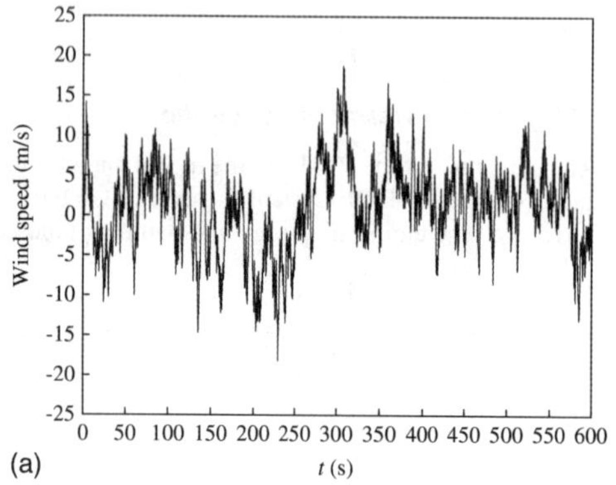

Fig. 21. Simulation results based on the Kaimal spectrum: (a) simulated wind velocity at top node; and (b) comparison of simulated and target wind spectrum.

and $\zeta_\theta = 0.0616$, respectively, according to Eq. (18). The length of the pendulum rod and damping coefficient are calculated as 0.337 m and 106.007 N · s/m (10.817 kgf · s/m), respectively.

The harmonic excitation force acting on the top node of the 25-DOF system is set as $Q_{25}(t) = 660\cos(\omega t)$ N, where ω is the excitation frequency. The excitation amplitude, $Q_{25,\max} = 660$ (N), is equivalent to $f = 0.004$ in nondimensional form. For both the linear and nonlinear models, the FRFs of the primary structure are obtained by frequency sweeping. Fig. 20(a) shows the FRFs of the top node. As can be seen, the amplitudes reach the maximum when the excitation frequency equals $1.02\omega_1$ for both nonlinear and linear models. Moreover, the FRF curves of the other nodes have the same relationship with that of the top node. Therefore, the excitation frequency is set as $1.02\omega_1$.

The steady-state displacement amplitude of every node is shown in Fig. 20(b). It can be seen that the PTMD could significantly reduce the vibration of the primary structure for the nonlinear model. However, the upper-node amplitudes obtained by the nonlinear model are much larger than that of the linear model. The error of displacement amplitude at top node is about 19.1% for the two models.

In order to design PTMD with negligible nonlinearity, a mass ratio of 0.017 is reselected according to the excitation amplitude ($f = 0.004$) and structural damping ratio ($\zeta_1 = 0.5\%$) based on Eq. (23). Fig. 20(c) shows the displacement amplitude of every node for mass ratio $\mu = 0.017$. It is evident that the difference in displacement amplitude between the linear and nonlinear solutions is very small. The PTMD properly plays a control role.

Table 4 lists the displacement amplitude of the top node and the first-order modal amplitude for uncontrolled and with PTMD

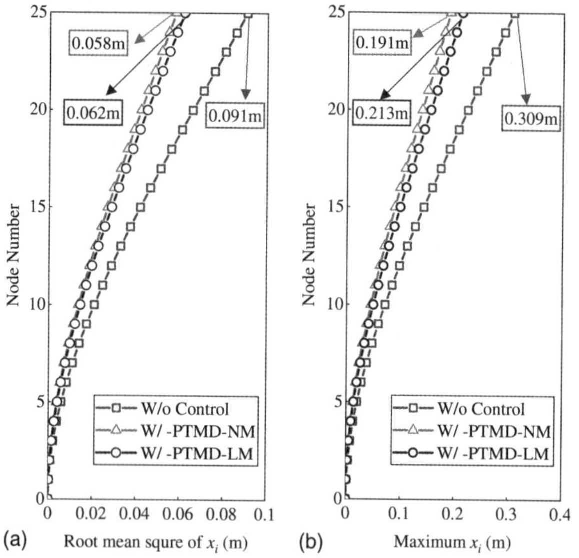

Fig. 22. RMS and maximum displacements for the chimney without and with PTMD: (a) RMS values; and (b) maximum displacement values.

controlled results. With the increased mass ratio, the errors between nonlinear and linear models are reduced significantly from 19.11% to 4.80% for the top-node amplitude and from 19.12% to 4.81% for the first-order modal amplitude. The reduced errors are very close

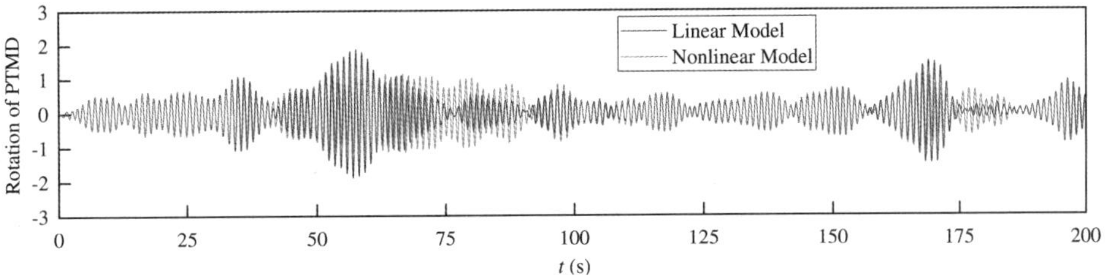

Fig. 23. Time history of the PTMD rotation angle calculated with nonlinear and linear models.

to the $y_p = 5\%$ threshold. Therefore, the feasibility of the proposed PTMD design strategy is verified.

Response to Simulated Wind Loadings

Although the nonlinearity of PTMD may attenuate the control performance when the MDOF structure subjected to harmonic excitation, it would be interesting to see how it performs under random excitation such as wind loads. The mass ratio of PTMD is selected as $\mu = 0.01$. The fluctuating wind speed is produced by the Kaimal power spectrum and the wave superposition method (Scanlan 1996). The simulated fluctuating wind on the top node is shown in Fig. 21. In Fig. 21(b), the spectrum of simulated wind speed agrees well with the Kaimal target spectrum.

Root-mean square (RMS) values of every node displacement as well as the maximum displacements are presented in Fig. 22. When PTMD is attached to the MDOF structure, the RMS difference between the nonlinear and linear models can be calculated as 6.9%, and the difference in the maximum displacement is 11.6%. The PTMD nonlinearity also leads to deterioration in control performance when the MDOF structure is subjected to fluctuating wind loads.

Additionally, Fig. 23 shows the time histories of the PTMD rotation angle for nonlinear and linear models. The amplitude of the rotation angle obtained by nonlinear model exceeds 1 rad, which is very large and may not be permitted. However, this large rotation angle is reasonable because the short length of pendulum rod. The commonly used linear model cannot predict the dynamic responses of the primary structure and PTMD accurately.

Conclusions

The nonlinear vibration problem of a linear primary structure coupled with a nonlinear PTMD subjected to harmonic excitation was addressed. The solutions were compared with the classical linear solutions under various excitation amplitudes. The nonlinearity of PTMD was shown to be negligible when the maximum PTMD rotation amplitude is below 9°. Upon increasing the excitation amplitude, remarkable differences can be observed between the nonlinear and linear FRFs. The difference in the maximum amplitudes of the primary structure reaches 152% for $f = 0.01$. For large oscillation amplitudes, the linear model cannot predict the system response accurately. Moreover, large PTMD mass ratios were found to attenuate the adverse effects of PTMD nonlinearity. The time history obtained according to the nonlinear model captures the superharmonic resonance. The instantaneous PTMD frequency actually varies with time and is dependent on the movement position and locus. Furthermore, the magnitude of instantaneous frequency variation increases with increase of the PTMD rotation amplitude.

This study also provided design formulas of the minimum mass ratio for PTMD with negligible nonlinearity effect under an acceptable tolerance 5% ($y_p < 5\%$), which allows design practitioners to directly compute the necessary parameters of a PTMD installed in a damped structure. Moreover, a parametric study was carried out to determine the optimal parameters of the PTMD considering its nonlinearity. The optimal frequency of the nonlinear PTMD was shown to increase significantly with excitation amplitudes, and the optimal damping ratio was found to be less sensitive to the excitation amplitude. The numerical simulations with a 25-DOF chimney structure coupled with the nonlinear PTMD demonstrate the importance of the PTMD nonlinearity on vibration control under both harmonic and random wind loadings. The nonlinear effects of the PTMD on vibration control can be mitigated in practice by increasing the mass ratio.

Data Availability Statement

All data, models, or codes that support the findings of this study are available from the corresponding author upon reasonable request.

Acknowledgments

The work described in this paper is supported by the National Science Fund for Distinguished Young Scholars (No. 52025082), National Natural Science Foundation of China (No. 51808210), and Hunan Provincial Innovation Foundation for Postgraduate (No. CX2018B218).

References

Areemit, N., and P. Warnitchai. 2001. "Vibration suppression of a 90-m-tall steel stack by using a high-damping tuned mass damper." In *Proc., 8th East Asia-Pacific Conf. on Structural Engineering and Construction*. Singapore: Nanyang Technological Univ.

Battista, R. C., R. S. Rodrigues, and M. S. Pfeil. 2003. "Dynamic behavior and stability of transmission line towers under wind forces." *J. Wind Eng. Ind. Aerodyn.* 91 (8): 1051–1067. https://doi.org/10.1016/S0167-6105(03)00052-7.

Bevilacqua L., R. B. Battista, and N. F. F. Ebecken. 2000. "Dynamical analysis of an offshore platform with vibration absorbers." In *Proc., IUTAM Symp. on Recent Developments in Non-linear Oscillations of Mechanical Systems*, edited by N. Van Dao and E. J. Kreuzer. Berlin: Springer.

Brzeski, P., E. Pavlovskaia, T. Kapitaniak, and P. Perlikowski. 2015. "The application of inerter in tuned mass absorber." *Int. J. Non Linear Mech.* 70 (10): 20–29. https://doi.org/10.1016/j.ijnonlinmec.2014.10.013.

Brzeski, P., P. Perlikowski, and T. Kapitaniak. 2014. "Numerical optimization of tuned mass absorbers attached to strongly nonlinear Duffing oscillator." *Commun. Nonlinear Sci. Numer. Simul.* 19 (1): 298–310. https://doi.org/10.1016/j.cnsns.2013.06.001.

Cao, L. Y., C. X. Li, and X. Chen. 2020. "Performance of multiple tuned mass dampers-inerters for structures under harmonic ground acceleration." *Smart Struct. Syst.* 26 (1): 49–61. https://doi.org/10.12989/sss.2020.26.1.049.

Christie, M. D., S. Sun, L. Deng, D. H. Ning, H. Du, S. W. Zhang, and W. H. Li. 2019. "A variable resonance magnetorheological-fluid-based pendulum tuned mass damper for seismic vibration suppression." *Mech. Syst. Sig. Process.* 116 (7): 530–544. https://doi.org/10.1016/j.ymssp.2018.07.007.

Den Hartog, J. P. 1985. *Mechanical vibrations*. New York: Dover.

Eason, R. P., C. Sun, A. J. Dick, and S. Nagarajaiah. 2015. "Steady-state response attenuation of a linear oscillator–nonlinear absorber system by using an adjustable-length pendulum in series: Numerical and experimental results." *J. Sound Vib.* 344 (1): 332–344. https://doi.org/10.1016/j.jsv.2015.01.030.

El-Khoury, O., and H. Adeli. 2013. "Recent advances on vibration control of structures under dynamic loading." *Arch. Comput. Method Eng.* 20 (4): 353–360. https://doi.org/10.1007/s11831-013-9088-2.

Elias, S., and V. Matsagar. 2017. "Research developments in vibration control of structures using passive tuned mass dampers." *Annu. Rev. Control* 44 (Jan): 129–156. https://doi.org/10.1016/j.arcontrol.2017.09.015.

Ertas, A., O. Cuvalci, and S. Ekwaro-Osire. 2000. "Performance of pendulum absorber for a non-linear system of varying orientation." *J. Sound Vib.* 229 (4): 913–933. https://doi.org/10.1006/jsvi.1999.2521.

Fallahpasand, S., M. Dardel, M. H. Pashaei, and H. R. Mohammadi Daniali. 2015. "Investigation and optimization of nonlinear pendulum vibration absorber for horizontal vibration suppression of damped system." *Struct. Design Tall Spec. Build.* 24 (14): 873–893. https://doi.org/10.1002/tal.1216.

Farid, M., and O. V. Gendelman. 2017. "Tuned pendulum as nonlinear energy sink for broad energy range." *J. Vib. Control* 23 (3): 373–388. https://doi.org/10.1177/1077546315578561.

Fujino, Y., and D. Siringoringo. 2013. "Vibration mechanisms and controls of long-span bridges: A review." *Struct. Eng. Int.* 23 (3): 248–268. https://doi.org/10.2749/101686613X13439149156886.

Ghosh, A., and B. Basu. 2007. "A closed-form optimal tuning criterion for TMD in damped structures." *Struct. Control Health Monit.* 14 (4): 681–692. https://doi.org/10.1002/stc.176.

Guo, A. X., Y. L. Xu, and H. Li. 2007a. "Dynamic performance of cable-stayed bridge tower with multi-stage pendulum mass damper under wind excitations—I: Theory." *Earthquake Eng. Eng. Vibr.* 6 (3): 295–306. https://doi.org/10.1007/s11803-007-0747-x.

Guo, A. X., Y. L. Xu, and H. Li. 2007b. "Dynamic performance of cable-stayed bridge tower with multi-stage pendulum mass damper under wind excitations—II: Experiment." *Earthquake Eng. Eng. Vibr.* 6 (4): 417–424. https://doi.org/10.1007/s11803-007-0748-9.

Hatwal, H., A. K. Mallik, and A. Ghosh. 1983. "Forced nonlinear oscillations of an autoparametric system—Part 1: Periodic responses." *J. Appl. Mech.* 50 (3): 657–662. https://doi.org/10.1115/1.3167106.

Huang, Z. W., X. G. Hua, Z. Q. Chen, and H. W. Niu. 2019. "Optimal design of TVMD with linear and nonlinear viscous damping for SDOF systems subjected to harmonic excitation." *Struct. Control Health Monit.* 26 (10): e2413. https://doi.org/10.1002/stc.2413.

Ikeda, T. 2011. "Nonlinear responses of dual-pendulum dynamic absorbers." *J. Comput. Nonlinear Dyn.* 6 (1): 11012. https://doi.org/10.1115/1.4002385.

Kourakis I. 2007. "Structural systems and tuned mass dampers of super-tall buildings: Case study of Taipei 101." Ph.D. dissertation, Dept. of Civil and Environmental Engineering, Massachusetts Institute of Technology.

Kwok, K. C. S. 1983. "Full-scale measurements of wind-induced response of Sydney Tower." *J. Wind Eng. Ind. Aerodyn.* 14 (1–3): 307–318. https://doi.org/10.1016/0167-6105(83)90033-8.

Kwok, K. C. S., and P. A. Macdonald. 1990. "Full-scale measurements of wind-induced acceleration response of Sydney Tower." *Eng. Struct.* 12 (3): 153–162. https://doi.org/10.1016/0141-0296(90)90002-A.

Kwok, K. C. S., and B. Samali. 1995. "Performance of tuned mass dampers under wind performance of tuned mass dampers under wind loads." *Eng. Struct.* 17 (9): 655–667. https://doi.org/10.1016/0141-0296(95)00035-6.

Lacarbonara, W., and S. Ballerini. 2009. "Vibration mitigation of guyed masts via tuned pendulum dampers." *Struct. Eng. Mech.* 32 (4): 517–529. https://doi.org/10.12989/sem.2009.32.4.517.

Leung, A. Y. T., H. Zhang, C. C. Cheng, and Y. Y. Lee. 2008. "Particle swarm optimization of TMD by non-stationary base excitation during earthquake." *Earthquake Eng. Struct. Dyn.* 37 (9): 1223–1246. https://doi.org/10.1002/eqe.811.

Lu, X. L., and J. R. Chen. 2011. "Parameter optimization and structural design of tuned mass damper for Shanghai Centre Tower." *Struct. Des. Tall Spec. Build.* 20 (4): 453–471. https://doi.org/10.1002/tal.649.

Nagase, T., and T. Hisatoku. 1992. "Tuned-pendulum mass damper installed in Crystal Tower." *Struct. Des. Tall Spec. Build.* 1 (1): 35–56. https://doi.org/10.1002/tal.4320010105.

Orlando, D., and P. B. Goncalves. 2013. "Hybrid nonlinear control of a tall tower with a pendulum absorber." *Struct. Eng. Mech.* 46 (2): 153–177. https://doi.org/10.12989/sem.2013.46.2.153.

Ormondroyd, J., and J. P. Den Hartog. 1928. "Theory of the dynamic vibration absorber." *J. Appl. Mech.* 50 (7): 11–22.

Pourzeynali, S., S. Salimi, and H. E. Kalesar. 2013. "Robust multi-objective optimization design of TMD control device to reduce tall building responses against earthquake excitations using genetic algorithms." *Sci. Iran.* 20 (2): 207–221. https://doi.org/10.1016/j.scient.2012.11.015.

Pritchard, B. N. 1984. "Steel chimney oscillations: A comparative study of their reported performance versus predictions using existing design techniques." *Eng. Struct.* 6 (4): 315–323. https://doi.org/10.1016/0141-0296(84)90029-4.

Rathi A. K., and A. Chakraborty. 2017. "Reliability-based performance optimization of TMD for vibration control of structures with uncertainty in parameters and excitation." *Struct. Control Health Monit.* 24 (1): e1857. https://doi.org/10.1002/stc.1857.

Rob, S., R. Merello, and M. Willford. 2010. "Intrinsic and supplementary damping in tall buildings." *Proc. Inst. Civ. Eng. Struct. Build.* 163 (2): 111–118. https://doi.org/10.1680/stbu.2010.163.2.111.

Roffel, A. J., and S. Narasimhan. 2016. "Results from a full-scale study on the condition assessment of pendulum tuned mass dampers." *J. Struct. Eng.* 142 (1): 4015096. https://doi.org/10.1061/(ASCE)ST.1943-541X.0001339.

Roffel, A. J., S. Narasimhan, and T. Haskett. 2013. "Performance of pendulum tuned mass dampers in reducing the responses of flexible structures." *J. Struct. Eng.* 139 (12): 04013019. https://doi.org/10.1061/(ASCE)ST.1943-541X.0000797.

Saeed, A. S., M. A. Al-Shudeifat, and A. F. Vakakis. 2019. "Rotary-oscillatory nonlinear energy sink of robust performance." *Int. J. Non Linear Mech.* 117 (Dec): 103249. https://doi.org/10.1016/j.ijnonlinmec.2019.103249.

Sarkar, A., and O. T. Gudmestad. 2013. "Pendulum type liquid column damper (PLCD) for controlling vibrations of a structure—Theoretical and experimental study." *Eng. Struct.* 49 (Apr): 221–233. https://doi.org/10.1016/j.engstruct.2012.10.023.

Scanlan, R. H. 1996. *Wind effects on structure, fundamentals and applications to design*. 3rd ed. New York: Wiley.

Sheheitli, H., and R. H. Rand. 2012. "Dynamics of a mass–spring–pendulum system with vastly different frequencies." *Nonlinear Dyn.* 70 (1): 25–41. https://doi.org/10.1007/s11071-012-0428-9.

Shu, Z., S. Li, X. Sun, and M. He. 2019. "Performance-based seismic design of a pendulum tuned mass damper system." *J. Earthquake Eng.* 23 (2): 334–355. https://doi.org/10.1080/13632469.2017.1323042.

Shu, Z., S. Li, J. Zhang, and M. He. 2017. "Optimum seismic design of a power plant building with pendulum tuned mass damper system by its heavy suspended buckets." *Eng. Struct.* 136 (Apr): 114–132. https://doi.org/10.1016/j.engstruct.2017.01.010.

Sigalov, G., O. V. Gendelman, M. A. AL-Shudeifat, L. I. Manevitch, A. F. Vakakis, and L. A. Bergman. 2012. "Resonance captures and targeted energy transfers in an inertially-coupled rotational nonlinear energy sink." *Nonlinear Dyn.* 69 (4): 1693–1704. https://doi.org/10.1007/s11071-012-0379-1.

Stoker, J. J. 1950. *Nonlinear vibrations*. New York: Wiley.

Sun, C., and V. Jahangiri. 2018. “Bi-directional vibration control of offshore wind turbines using a 3D pendulum tuned mass damper.” *Mech. Syst. Sig. Process.* 105 (May): 338–360. https://doi.org/10.1016/j.ymssp.2017.12.011.

Sun, H. X., L. Zuo, X. Y. Wang, J. Peng, and W. X. Wang. 2019. “Exact H2 optimal solutions to inerter-based isolation systems for building structures.” *Struct. Control Health Monit.* 26 (6): e2357. https://doi.org/10.1002/stc.2357.

Toshihiro, I. O. I., and I. Ken. 1978. “On the dynamic vibration damped absorber of the vibration system.” *Bull. JSME* 21 (151): 64–71. https://doi.org/10.1299/jsme1958.21.64.

Tuan, A. Y., and G. Q. Shang. 2014. “Vibration control in a 101-storey building using a tuned mass damper.” *J. Appl. Sci. Eng.* 17 (2): 141–156. https://doi.org/10.6180/jase.2014.17.2.05.

Viet, L. D., and N. B. Nghi. 2014. “On a nonlinear single-mass two-frequency pendulum tuned mass damper to reduce horizontal vibration.” *Eng. Struct.* 81 (Dec): 175–180. https://doi.org/10.1016/j.engstruct.2014.09.038.

Vyas, A., and A. K. Bajaj. 2001. “Dynamics of autoparametric vibration absorbers using multiple pendulums.” *J. Sound Vib.* 246 (1): 115–135. https://doi.org/10.1006/jsvi.2001.3616.

Wang, W. X., X. G. Hua, Z. Q. Chen, X. Y. Wang, and G. B. Song. 2019. “Modeling, simulation, and validation of a pendulum-pounding tuned mass damper for vibration control.” *Struct. Control Health Monit.* 26 (4): e2326. https://doi.org/10.1002/stc.2326.

Wang, W. X., X. Y. Wang, X. G. Hua, G. B. Song, and Z. Q. Chen. 2018. “Vibration control of vortex-induced vibrations of a bridge deck by a single-side pounding tuned mass damper.” *Eng. Struct.* 173 (Oct): 61–75. https://doi.org/10.1016/j.engstruct.2018.06.099.

Wen, Q., X. G. Hua, Z. Q. Chen, Y. Yang, and H. W. Niu. 2016. “Control of human-induced vibrations of a curved cable-stayed bridge: Design, implementation, and field validation.” *J. Bridge Eng.* 21 (7): 04016028. https://doi.org/10.1061/(ASCE)BE.1943-5592.0000887.

Xiang, P., A. Nishitani, and M. Wu. 2017. “Seismic vibration and damage control of high-rise structures with the implementation of a pendulum-type nontraditional tuned mass damper.” *Struct. Control Health Monit.* 24 (12): e2022. https://doi.org/10.1002/stc.2022.

International Journal of Mechanical Sciences 198 (2021) 106364

Contents lists available at ScienceDirect

International Journal of Mechanical Sciences

journal homepage: www.elsevier.com/locate/ijmecsci

Closed-form optimal calibration of a tuned liquid column damper (TLCD) for flexible structures

Bei Chen [a,b], Zili Zhang [b,*], Xugang Hua [a]

[a] Key Laboratory for Wind and Bridge Engineering, Hunan University, Changsha 410082, China
[b] Department of Engineering, Aarhus University, Aarhus 8000, Denmark

ARTICLE INFO

Keywords:
Closed-form formulas
Tuned liquid column damper
Flexible structure
Pole-placement calibration
Background flexibility

ABSTRACT

The conventional optimal calibration of a TLCD for suppressing structural vibration is based on the classic 2-degrees-of-freedom (2-DOF) model in terms of one specific structural mode (usually the resonant mode). For flexible multi-degrees-of-freedom (MDOF) structures, this implies that background flexibility contribution from other non-resonant modes is omitted, resulting in an unbalance in the frequency response of the flexible structure-TLCD coupled system. Furthermore, numerical search techniques are usually used for optimizing TLCDs, and extensive computational efforts are required. Therefore, in this paper a novel closed-form optimal calibration procedure accounting for the background flexibility contribution from the flexible structure is developed based on the pole-placement method. The background flexibility is represented by an equivalent stiffness that is derived from modal analysis. An analogue 2-DOF model accounting for background flexibility is developed, which turns out to be a generalization of the classic 2-DOF model. The root locus analysis is performed to derive the optimal frequency ratio of the TLCD by the equal modal damping ratio criterion. A straightforward approach for determining the optimal head loss coefficient is proposed based on the bifurcation point present in the root locus. It is seen from the closed-form formulas that the optimal parameters of the TLCD depend not only on the constructive parameters (such as the mass ratio and horizontal length ratio of a TLCD) but also on the structural inherent characteristics. Application to MDOF flexible structures are illustrated by a 10-story shear frame. Results demonstrates that the proposed calibration procedure leads to a balanced frequency response curve and thus improves the performance of the TLCD comparing with the calibration procedure ignoring background flexibility.

1. Introduction

Flexible structures (such as high-rise buildings, chimneys, wind turbine towers etc.) are lightly damped owing to low material damping and lack of aerodynamic damping in the cross-wind direction, making vibration inevitable [1]. Therefore, it is of crucial need to damp vibrations of these flexible structures for improving the fatigue life and the operational performance. Different types of vibration absorbers have been used for damping vibrations of flexible structures such as the tuned mass damper (TMD) [2–5], the pendulum absorber [6–9], and the tuned liquid damper (TLD) [10–13]. Among them, the tuned liquid column damper (TLCD) has several advantages over other devices including low cost, easy installation and maintenance, easy adjustment of damper geometry to the target frequency, and controllable damping property by the orifice opening. It can even serve as a water storage facility in case of fire emergency. It was shown to successfully suppress vibrations of high-rise buildings [14–16], towers [17–19], onshore or offshore wind turbines [20–23].

TLCD was first proposed by Sakai et al. [24] for suppressing horizontal vibration of building structures, and since then it has been extensively investigated [25–36]. A typical TLCD consists of a U-shaped tube with liquid inside, and an orifice in the middle of the horizontal tube acting as a damping element. The property of the TLCD is specified by a few parameters such as the total length of the liquid inside the tube, the horizontal length ratio, the head loss coefficient (mainly depends on the orifice area), the cross-sectional area of both vertical and horizontal tubes and the liquid characteristics [25]. Many investigations were performed for revealing the influences of the TLCD parameters on its vibration damping efficacy. Colwell and Basu [25] experimentally assessed the effects of three types of liquids, i.e. water, magnetorheological fluid and glycol, on the damping performance of the TLCD for a single-degree-of-freedom (SDOF) system. Gao et al. [26] investigated the influence of several TLCD parameters through a numerical procedure. Balendra et al. [27] evaluated the effectiveness of a TLCD for reducing wind-induced vibration of towers, with emphasis on the role of opening ratio of the orifice. Matteo et al. [28] showed how the TLCD

* Corresponding author.
E-mail addresses: chamber@hnu.edu.cn (B. Chen), zili_zhang@cae.au.dk (Z. Zhang), cexghua@hnu.edu.cn (X. Hua).

https://doi.org/10.1016/j.ijmecsci.2021.106364
Received 31 January 2021; Received in revised form 26 February 2021; Accepted 28 February 2021
Available online 4 March 2021

mass ratio and the structure damping ratio influence the damper performance, when the SDOF structure is subjected to earthquake excitations. Hitchcock et al. [29,30] investigated the characteristics of a TLCD having different cross-sectional areas in the horizontal and vertical tubes. Furthermore, the inherent nonlinearities of TLCD, such as the nonlinear amplitude-dependent damping mechanism [28,31] and the nonlinear air spring of a sealed TLCD [32] were also studied.

These parametric studies indicate that a proper determination of TLCD parameters is crucial to the performance of TLCD in damping structural vibrations. To facilitate calibration of the damper, optimal TLCD parameters, especially the optimal frequency ratio and the optimal head loss coefficient, were investigated by many researchers. Most of the studies were based on the classic 2-degree-of-freedom (2-DOF) structure-TLCD model, where one DOF is used for the structure representing one structural resonant mode (usually the 1st mode) and one DOF is used for the TLCD. Gao et al. [26] compared the performance of two types of TLCD (the classic U-shaped and improved V-shaped) under a harmonic excitation. Based on the classic 2-DOF model, the optimal frequency ratio and head loss coefficient were obtained by the numerical search technique, using the criterion of minimizing the maximum dynamic magnification factor of the structural displacement. Wu et al.[37] investigated the TLCD performance for suppressing a SDOF system under a white noise type of wind excitation. Due to the amplitude-related nonlinearity of the TLCD damping element, an iterative procedure together with numerical search were adopted to obtain the optimal parameters. Based on the same 2-DOF model, Debbarma et al.[38] evaluated the effect of structural parameters' uncertainties on the optimal parameters and performance of TLCD. For stiff structure subjected to earthquake excitations, Ghosh and Basu [39] proposed an alternative way of attaching the TLCD, i.e. it is connected to the 1-DOF structure model through an extra spring and a dashpot. Although the optimal parameters of the connecting spring and dashpot were analytically determined from the 'fixed-point' theory by Den Hartog [40], the optimal TLCD parameters were still obtained by numerical search. For performance enhancement of the TLCD, Al-Saif et al.[41] proposed the use of the so-called tuned liquid column ball damper (TLCBD), where a steel ball was added inside the horizontal portion of the tube. A 2-DOF model extended from the classic 2-DOF structure-TLCD model was developed without performing parametric optimization. Later, Gur et al. [42] evaluated the performance of TLCBD based on a similar 2-DOF model subjected to random earthquake excitations, and the optimal damper parameters were obtained numerically. Recently, the spring-connected TLCD model was proposed to be further enhanced by incorporating an inerter between the TLCD and the ground [43], which might be practically infeasible.

The classic 2-DOF model can be extended to accommodate the case where multiple TLCDs (MTLCDs) are used for damping one structural resonant mode (the 1st mode). The SDOF representing the structural resonant mode is unchanged as in the last paragraph. However, the frequency of each TLCD is different and distributed around the natural frequency of that mode, in order to enhance the robustness of vibration control in practical applications. By appropriately calibrating parameters of each TLCD, a lower (and wider) frequency response curve of the structure response can be obtained. Further, the MTLCDs provide more flexible installation options than a single TLCD (with the same total water mass), since each TLCD can be installed at a different location on a real structure. Gao et al. [44] assessed the damping performance of MTLCDs on a SDOF system under harmonic excitations. Optimal parameters, such as the number of TLCDs, the frequency ratio and head loss coefficient of each TLCD, were numerically optimized. Similar works were performed on the performance of MTLCDs for a SDOF system that is subjected to wind excitation [45] and earthquake excitation [46]. In the study conducted by Shum and Xu [47], the capability of MTLCDs in suppressing coupled lateral and torsional vibration of a structure were demonstrated.

All the above-mentioned studies resorted to numerical search techniques for parametric optimization of the damper, which require heavy computational efforts. The iteration needed due to the inherent nonlinear liquid damping further decreases the computational efficiency. To facilitate efficient calibration of the TLCD, Shum [48] derived explicit formulas for the optimal frequency ratio by the 'fixed-point' theory (equalized dynamic amplification at the two invariant points) [40], where the classic 2-DOF model was used as the basis. An approximate solution for the optimum head loss coefficient of TLCD was also developed by extrapolating Brock's perturbation approach [49], so that iteration was avoided. Ghosh et al. [50] presented an alternative closed-form solution for the optimal frequency ratio of the TLCD attached to a damped SDOF structure subjected to earthquake excitations. The 'fixed-point' theory was again used with good justification even though the two invariant points do not actually exist for a damped structure. Moreover, Yalla et al. [45] derived explicit formulas for the optimal frequency ratio and head loss coefficient of the TLCD attached to an undamped SDOF system subjected to white noise excitation. Unlike in [48] and [50], the derivations were based on the gradient method together with random vibration theory, with the criterion of minimizing the variance of the structural response [45]. These studies provided useful calibration formulas for a fast and simple design of the structure-mounted TLCD. However, same as the studies based on numerical optimization techniques, the classic 2-DOF model (with one DOF representing the 1st mode) was consistently employed in these works. This implies that the non-resonant modes (higher modes) of the flexibly structure were completely ignored in these studies. As a result, this might lead to non-optimal performance of the TLCD when installed to a flexible structure whose vibration involves multiple modes.

One solution is to apply semi-active TLCD [50–54] to the flexible structure, where the TLCD parameters can in principle be continuously adjusted by a proper control algorithm. In such cases, an optimal calibration of the TLCD is not necessary in advance, and the influence from all modes of the structure on the TLCD parameters is automatically accounted for during the semi-active control. For example, the head loss coefficient was continuously adjusted by an actively controlled orifice device [51]. Two different control strategies were considered, and the TLCD performance was shown to be modestly improved comparing with the passive counterpart [51]. A semi-active device similar to the one proposed by [50] was also investigated [52], where the TLCD was connected to the primary structure through an adaptive spring with adjustable stiffness. Two different control strategies (feedforward and feedback) were discussed, and the semi-active TLCD was shown to have better performance against the change of structural parameters [52]. However, the semi-active control might still be too complex and demanding for practical applications. Another alternative is to install MTLCDs into the flexible structure [45,55,56] targeting on several vibrational modes. MTLCDs were numerically investigated in suppressing the 1st and 2nd modal responses of the flexible structure [45,55], and each TLCD was independently tuned to one mode, either through numerical optimization [45] or simply setting the frequency ratio to be one [55]. Full-scale MTLCDs (two TLCDs) attached to a flexible structure were also experimentally investigated by real-time hybrid simulation (RTHS) technique [56]. One TLCD was tuned to the 1st structural mode and the other was tuned to the 2nd mode, where the classic 2-DOF model was used for parametric optimization. MTLCDs are able to suppress several modal responses of the flexible structure including the dominant resonant mode (the 1st mode) and the higher modes, but at the cost of additional amount of liquid needed. Moreover, the use of the damper targeting on higher modes is not cost effective (its vibration damping effect is not fully exploited) since these mode are not significantly excited. Therefore, a single TLCD targeting at the resonant mode (the 1st mode), with its parameters calibrated accounting for the influence from higher modes, might be preferred in many situations. Developing an efficient optimal calibration procedure for it (a single TLCD on flexible structures) becomes essential.

Therefore, the present paper focuses on the development of closed-form formulas for the optimal TLCD parameters taking into account the influence (background flexibility) from the higher modes of a flexible structure. First, an analogue 2-DOF model for the flexible structure-TLCD system is established, where the background flexibility is represented by an equivalent stiffness derived from modal analysis of the flexible structure. It is shown that this analogue 2-DOF model is an generalization of the classic 2-DOF structure-TLCD model that was extensively used in previous studies. Next, based on the characteristic equation of the analogue 2-DOF model, explicit formulas for the optimal TLCD parameters are developed by means of the pole-placement calibration approach [57]. Equal modal damping ratio (of two free vibration modes) criterion is employed for deriving the optimal frequency ratio. This is achieved by guaranteeing the two of the three roots of the characteristic equation being inverse points with respect to a reference frequency [57]. Then, the existence of a bifurcation point in the root locus is used to obtain a reference equivalent damping ratio, whereby the optimal head loss coefficient can subsequently be obtained using linearization technique for the nonlinear damping term. A calibration flow chart is presented as a ready-to-use practical tool. Finally, the proposed explicit calibration procedure is evaluated by a 10-story shear frame model. Results indicate that the proposed calibration formulas lead to a balanced frequency response curve of the flexible structure, and thus enhance the performance of the TLCD comparing with the calibration procedure ignoring background flexibility from the higher modes.

2. Mathematical modelling

When the structure is flexible, the TLCD will experience its support motion not only from the resonant mode (normally the fundamental mode), but also from non-resonant background modes (higher-frequency modes). This is represented in Fig. 1(a), where the TLCD support (attachment location) is experiencing a total displacement u_d that consists of the resonant part u_r and a displacement u_b from the non-resonant background modes. This additional displacement u_b may also be understood as being due to an additional flexibility, i.e. the background flexibility. It turns out that this is equivalent to the so-called quasi-static correction term related to truncated modal expansion in structural dynamics [58].

2.1. *Reduced-order model of the flexible structure installed with a TLCD*

As shown in Fig. 1(b), the TLCD is a U-Shaped tube with an orifice installed at the center of the horizontal tube. The horizontal length, the vertical length and the total length of the liquid inside the TLCD are denoted by B, H and L, respectively, where $L = 2H + B$. Denoting the liquid density ρ and the cross sectional area A for both the vertical and horizontal tubes (herein, only the TLCD with uniform cross section is considered), the total mass of the liquid inside the TLCD can be calculated as:

$$m_w = \rho A(2H + B) = \rho AL \tag{1}$$

The effect of an additional flexibility between the primary structure and the TLCD is shown in the modal analogue in Fig. 1(b). This effect is modelled by connecting the TLCD to the structural mass (modal mass m_r) through a spring with the stiffness k_b related to the background flexibility. The modal mass m_r and the modal stiffness k_r corresponding to the resonant mode, and the spring stiffness k_b ($1/k_b$ being the background flexibility) are obtained from the following modal analysis of the flexible structure.

The flexible structure (such as the one in Fig. 1(a)) is normally modelled by multi-degree-of-freedom (MDOF) model using for example the finite element method (FEM). The inherent structural damping is neglected ($\mathbf{C} = \mathbf{0}$) in the subsequent damper calibration procedure to facilitate deriving closed-form formulas, which is also common practice [2,57]. It is considered reasonable because structural damping is insignificant comparing with the additional damping introduced by the damping device. Moreover, it was shown that the influence of structural damping on the optimal TLCD parameters is minor [48]. Therefore, the equation of motion of the uncontrolled flexible structure can be written in a compact form as:

$$\mathbf{M}\ddot{\mathbf{u}}(t) + \mathbf{K}\mathbf{u}(t) = \mathbf{f}(t) \tag{2}$$

where $\mathbf{M}$ and $\mathbf{K}$ are the mass and stiffness matrices of the flexible structure. $\mathbf{u}(t)$ and $\mathbf{f}(t)$ represent the displacement vector and load vector, respectively. The double dots over $\mathbf{u}(t)$ denotes the second derivative with respect to time. From the generalized eigenvalue problem, the eigenvectors $\mathbf{\Phi}_1, \cdots, \mathbf{\Phi}_n$ and the corresponding eigenfrequencies $\omega_1, \cdots, \omega_n$ can be obtained. Following the standard procedure of modal analysis, Eq. (2) can be converted into n decoupled modal equation of motion:

$$\mathbf{\Phi}_i^T\mathbf{M}\mathbf{\Phi}_i\ddot{q}_i(t) + \mathbf{\Phi}_i^T\mathbf{K}\mathbf{\Phi}_i q_i(t) = \mathbf{\Phi}_i^T\mathbf{f}(t), \quad i = 1, 2 \cdots, n \tag{3}$$

where $\mathbf{\Phi}_i^T\mathbf{K}\mathbf{\Phi}_i = \omega_i^2\mathbf{\Phi}_i^T\mathbf{M}\mathbf{\Phi}_i$. $q_i(t)$ is the ith modal coordinate. The displacement vector can be expressed as the following modal expansion:

$$\mathbf{u}(t) = \sum_{i=1}^{n}\mathbf{\Phi}_i q_i(t) \tag{4}$$

In the context of vibration control using TLCD, the structural response is normally dominated by the 1st mode, i.e. the resonant mode, to which the TLCD is targeted. Quasi-static correction in truncated modal expansion [58] indicates that for the modal coordinates of the higher-frequency modes (non-resonant modes), the influence of inertial forces can be ignored, leaving only the quasi-static part. It should be noted that If a higher mode is targeted on rather than the 1st mode, the inertial forces from other non-resonant modes indeed will play a role. In the present study targeting on the 1st mode, Eq. (4) can be approximated as:

$$\begin{aligned}\mathbf{u}(t) &= \mathbf{\Phi}_1 q_1(t) + \sum_{i=2}^{n}\mathbf{\Phi}_i q_i(t)\\ &\simeq \mathbf{\Phi}_1 q_1(t) + \left(\sum_{i=2}^{n}\frac{\mathbf{\Phi}_i\mathbf{\Phi}_i^T}{\mathbf{\Phi}_i^T\mathbf{K}\mathbf{\Phi}_i}\right)\mathbf{f}(t)\\ &= \mathbf{\Phi}_1 q_1(t) + \left(\mathbf{K}^{-1} - \frac{\mathbf{\Phi}_1\mathbf{\Phi}_1^T}{\mathbf{\Phi}_1^T\mathbf{K}\mathbf{\Phi}_1}\right)\mathbf{f}(t)\end{aligned} \tag{5}$$

where the well-known expression $\mathbf{K}^{-1} = \sum_{i=1}^{n}\frac{\mathbf{\Phi}_i\mathbf{\Phi}_i^T}{\mathbf{\Phi}_i^T\mathbf{K}\mathbf{\Phi}_i}$ has been used to obtain the last line of Eq. (5).

Next, the task is to determine the displacement $u_d(t)$ at the location where the damper is attached (the damper support, as shown in Fig. 1(a) the TLCD is attached to the n_ath floor), when a force $f_d(t)$ is acting at this same location. The connectivity vector $\mathbf{w} = [0, \cdots, 1, \cdots, 0]^T$ identifies the fixture point of the damper, with a single unit entry placed at the DOF to which the damper is attached. Therefore, $u_d(t)$ and $f_d(t)$ can be related to $\mathbf{u}(t)$ and $\mathbf{f}(t)$ as:

$$u_d(t) = \mathbf{w}^T\mathbf{u}(t) \tag{6a}$$

$$\mathbf{f}(t) = \mathbf{w}f_d(t) \tag{6b}$$

Using Eq. (6) and Eq. (5), the following relationship between $u_d(t)$ and $f_d(t)$ is obtained:

$$u_d(t) = \mathbf{w}^T\mathbf{\Phi}_1 q_1(t) + \left(\mathbf{w}^T\mathbf{K}^{-1}\mathbf{w} - \frac{(\mathbf{w}^T\mathbf{\Phi}_1)^2}{\mathbf{\Phi}_1^T\mathbf{K}\mathbf{\Phi}_1}\right)f_d(t) \tag{7}$$

The loads and responses are assumed to be harmonic varying, i.e. $f_d(t) = f_{d,0}e^{i\omega t}$, $u_d(t) = u_{d,0}e^{i\omega t}$, $q_1(t) = q_{1,0}e^{i\omega t}$, where $f_{d,0}$, $u_{d,0}$ and $q_{1,0}$ are the amplitudes. By use of Eq. (3) and Eq. (7), the input-output relationship between $f_{d,0}$ and $u_{d,0}$ becomes:

$$\begin{aligned}u_{d,0} &= \left(\frac{(\mathbf{w}^T\mathbf{\Phi}_1)^2}{-\omega^2\mathbf{\Phi}_1^T\mathbf{M}\mathbf{\Phi}_1 + \mathbf{\Phi}_1^T\mathbf{K}\mathbf{\Phi}_1} + \mathbf{w}^T\mathbf{K}^{-1}\mathbf{w} - \frac{(\mathbf{w}^T\mathbf{\Phi}_1)^2}{\mathbf{\Phi}_1^T\mathbf{K}\mathbf{\Phi}_1}\right)f_{d,0}\\ &= \left(\frac{1}{k_r - \omega^2 m_r} + \mathbf{w}^T\mathbf{K}^{-1}\mathbf{w} - \frac{1}{k_r}\right)f_{d,0}\end{aligned} \tag{8}$$

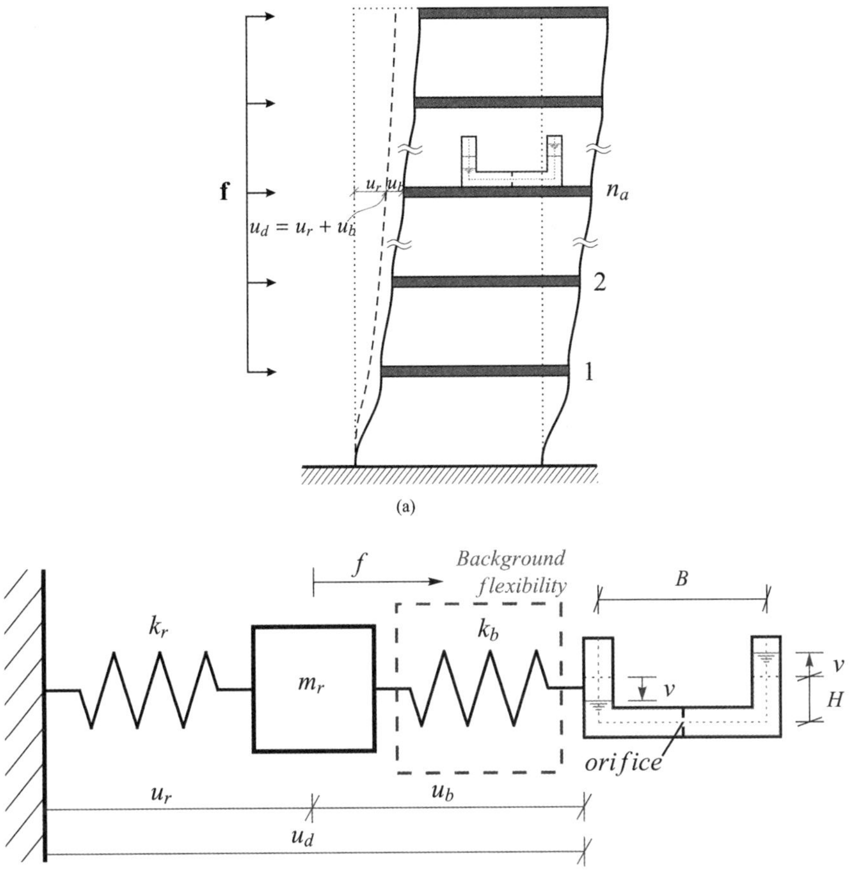

Fig. 1. Coupled TLCD-structure system under external excitation **f**. (a) The primary flexible structure instrumented with a TLCD on the n_ath floor. The total displacement of the structure u_d consists of the displacement u_r due to the resonant mode and the relative displacement u_b due to the non-resonant background modes. (b) Modal analogue (analogue 2-DOF model) of the flexible structure equipped with a TLCD under the external force f. The analogue 2-DOF model consists of the modal mass m_r and modal stiffness k_r of the resonant mode, as well as an equivalent stiffness k_b related to the non-resonant modes. B and H denote the horizontal and vertical length of liquid inside the TLCD, respectively. v is the vertical displacement of liquid in the TLCD.

where the first term inside the parentheses is related to the resonant mode (the $1st$ mode), with the corresponding modal stiffness k_r and modal mass m_r being defined as:

$$k_r = \frac{\boldsymbol{\Phi}_1^T \mathbf{K} \boldsymbol{\Phi}_1}{(\mathbf{w}^T \boldsymbol{\Phi}_1)^2}, \quad m_r = \frac{\boldsymbol{\Phi}_1^T \mathbf{M} \boldsymbol{\Phi}_1}{(\mathbf{w}^T \boldsymbol{\Phi}_1)^2} \tag{9}$$

where the normalized mode shape $\frac{\boldsymbol{\Phi}_1}{\mathbf{w}^T \boldsymbol{\Phi}_1}$ (unit modal displacement at the damper) has been used.

The remaining part inside the parentheses defines the equivalent stiffness k_b due to the background flexibility [2]:

$$\frac{1}{k_b} = \mathbf{w}^T \mathbf{K}^{-1} \mathbf{w} - \frac{1}{k_r} \tag{10}$$

2.2. Equations of motion

As shown in Fig. 1, the total displacement of the structure consist of two components: the displacement u_r due to the resonant mode and the relative displacement u_b due to the non-resonant background modes. v is the vertical displacement of liquid in the TLCD. The kinetic energy of the analogue 2-DOF model depicted in Fig. 1(b) is written as:

$$T = \frac{1}{2} m_r \dot{u}_r^2 + \frac{1}{2} \rho AB (\dot{u}_r + \dot{u}_b + \dot{v})^2 + \frac{1}{2} \rho 2AH \left((\dot{u}_r + \dot{u}_b)^2 + \dot{v}^2 \right) \tag{11}$$

The potential energy of the system is written as:

$$U = \frac{1}{2} k_r u_r^2 + \frac{1}{2} k_b u_b^2 + \rho g A v^2 \tag{12}$$

Inserting Eq. (11) and Eq. (12) into Lagrange's equation in analytical dynamics [59], one would obtain the following equations of motion (EOMs) of the system:

$$\frac{d}{dt}\left(\frac{\partial T}{\partial \dot{u}_r}\right) - \frac{\partial T}{\partial u_r} + \frac{\partial U}{\partial u_r} = f(t) \quad \Rightarrow$$
$$(m_r + m_w)\ddot{u}_r + m_w \ddot{u}_b + \rho AB \ddot{v} + k_r u_r = f(t) \tag{13a}$$

$$\frac{d}{dt}\left(\frac{\partial T}{\partial \dot{u}_b}\right) - \frac{\partial T}{\partial u_b} + \frac{\partial U}{\partial u_b} = 0 \quad \Rightarrow \\ m_w \ddot{u}_r + m_w \ddot{u}_b + \rho AB \ddot{v} + k_b u_b = 0 \tag{13b}$$

$$\frac{d}{dt}\left(\frac{\partial T}{\partial \dot{v}}\right) - \frac{\partial T}{\partial v} + \frac{\partial U}{\partial v} = f_{TLCD,nonlinear} \quad \Rightarrow \\ \rho AB \ddot{u}_r + \rho AB \ddot{u}_b + m_w \ddot{v} + 2\rho g A v + \frac{1}{2} \rho A \xi |\dot{v}| \dot{v} = 0 \tag{13c}$$

where $f(t)$ is the external force, $f_{TLCD,nonlinear} = -1/2\rho A \xi |\dot{v}| \dot{v}$ is the inherent nonlinear damping force of the TLCD; ξ is the head loss coefficient, which can be controlled by varying the orifice area [60]; g is the gravitational acceleration.

Although there are three variables in Eq. (13), only u_r and v are independent variables. Comparison of Eq. (13a) and Eq. (13b) leads to an explicit expression of u_b in terms of u_r, v and their 2nd time derivatives. Inserting this expression of u_b into any two of the three equations in Eq. (13), the following nonlinear EOMs for the analogue 2-DOF model are obtained:

$$\begin{cases} m_r \ddot{u}_r + \left(\rho AB - \frac{\rho AL^2}{B}\right)\ddot{v} - \frac{C_w L}{B} + k_r u_r - \frac{2\rho ALg}{B} v = f(t) & (14a) \\ \frac{\rho A(L^2 - B^2)}{k_b} \ddot{v} + \frac{C_w L}{k_b} \ddot{v} + B \ddot{u}_r + L\left(1 + \frac{2\rho g A}{k_b}\right)\ddot{v} + \frac{1}{2}\xi |\dot{v}| \dot{v} + 2gv = 0 & (14b) \end{cases}$$

As shown from both previous studies [25,28,48] and our preliminary time-domain simulations using Eq. (14), the nonlinearity in the damping term for the TLCD is weak and has insignificant influence on the system responses. By replacing this nonlinear damping force, $f_{TLCD,nonlinear} = -1/2\rho A\xi|\dot{v}|\dot{v}$, with the equivalent linear damping force, $f_{TLCD,linear} = -C_w\dot{v}$ (C_w being the equivalent damping coefficient of the TLCD, will be discussed in Section 3.4), the linearized EOMs becomes:

$$\begin{cases} m_r\ddot{u}_r + \left(\rho AB - \frac{\rho AL^2}{B}\right)\ddot{v} - \frac{C_wL}{B}\dot{v} + k_ru_r - \frac{2\rho ALg}{B}v = f(t) & (15a) \\ \frac{\rho A(L^2-B^2)}{k_b}\ddddot{v} + \frac{C_wL}{k_b}\dddot{v} + B\ddot{u}_r + L\left(1 + \frac{2\rho gA}{k_b}\right)\ddot{v} + \frac{C_w}{\rho A}\dot{v} + 2gv = 0 & (15b) \end{cases}$$

For conciseness of analysis and presentation, a few dimensionless parameters are introduced for Eq. (15). The linearized EOMs can thus be recast as:

$$\begin{cases} m_r\ddot{u}_r + \mu m_r\left(\gamma - \frac{1}{\gamma}\right)\ddot{v} - \frac{2\chi\mu\zeta_w\omega_rm_r}{\gamma}\dot{v} + m_r\omega_r^2u_r - \frac{\mu\chi^2m_r\omega_r^2}{\gamma}v = f(t) & (16a) \\ \frac{\mu(1-\gamma^2)}{\lambda\omega_r^2}\ddddot{v} + \frac{2\chi\mu\zeta_w}{\lambda\omega_r}\dddot{v} + \gamma\ddot{u}_r + \left(1 + \frac{\mu\chi^2}{\lambda}\right)\ddot{v} + 2\chi\zeta_w\omega_r\dot{v} + \chi^2\omega_r^2v = 0 & (16b) \end{cases}$$

in which the frequency ratio of the TLCD to the structure χ, the ratio of the modal background stiffness to the modal dynamic stiffness λ, the ratio of the horizontal length to overall length of the liquid column γ, the ratio of the liquid mass to the modal mass of the structure μ and the equivalent damping ratio of the TLCD ζ_w are defined as:

$$\chi = \frac{\omega_w}{\omega_r}, \quad \lambda = \frac{k_b}{k_r}, \quad \gamma = \frac{B}{L}, \quad \mu = \frac{m_w}{m_r}, \quad \zeta_w = \frac{C_w}{2\sqrt{k_wm_w}} \tag{17}$$

where $k_w = 2\rho Ag$ is the equivalent stiffness of the TLCD; $m_w = \rho AL$ is the total mass of the liquid, cf. Eq. (1); $\omega_r = \sqrt{k_r/m_r}$ is the circular frequency of structural resonant mode; $\omega_w = \sqrt{k_w/m_w} = \sqrt{2g/L}$ is the circular frequency of the TLCD.

Eq. (16b) turns out to be a forth-order ordinary differential equation, due to the existence of the equivalent spring k_b. As $k_b \Rightarrow \infty$ corresponding to ignoring the background flexibility, Eq. (16a) and Eq. (16b) are reduced to the following second-order differential equations:

$$(1 + \mu)m_r\ddot{u}_r + \gamma\mu m_r\ddot{v} + m_r\omega_r^2 = f(t) \tag{18a}$$

$$\gamma\mu m_r\ddot{u}_r + \mu m_r\ddot{v} + 2\chi\mu\zeta_wm_r\omega_r\dot{v} + \mu\chi^2m_r\omega_r^2v = 0 \tag{18b}$$

which are exactly the EOMs of the classic SDOF-TLCD system (classic 2-DOF system with background flexibility ignored). Therefore Eq. (16) is the generalization of Eq. (18).

Assume a harmonic external force, $f(t) = f_0e^{i\omega t} = Cm_rge^{i\omega t}$ (C is a dimensionless force intensity factor), acting on the structure with an circular frequency ω. The normalized (by the static response f_0/k_r) complex amplitudes of forced modal response, u_r, and of the liquid response, v, can be obtained from Eq. (16) as:

$$\frac{u_{r,0}}{f_0/k_r} = \frac{\frac{\mu}{\lambda}(1-\gamma^2)\varsigma^4 - \left(1 + \frac{\mu\chi^2}{\lambda}\right)\varsigma^2 + \chi^2 + 2i\zeta_w\chi\varsigma\left(1 - \frac{\mu}{\lambda}\varsigma^2\right)}{\left(\begin{array}{c}\frac{\mu(\gamma^2-1)}{\lambda}\varsigma^6 + \left((1+\frac{\mu\chi^2}{\lambda}) - \mu(\gamma^2-1) - \frac{\mu(\gamma^2-1)}{\lambda}\right)\varsigma^4 - \left(\chi^2 + \mu\chi^2 + (1+\frac{\mu\chi^2}{\lambda})\right)\varsigma^2 \\ \hline +\chi^2 + 2i\zeta_w\chi\varsigma\left(\frac{\mu}{\lambda}\varsigma^4 - (1+\mu+\frac{\mu}{\lambda})\varsigma^2 + 1\right)\end{array}\right)} \tag{19a}$$

$$\frac{v_0}{f_0/k_r} = \frac{\gamma\varsigma^2}{\left(\begin{array}{c}\frac{\mu(\gamma^2-1)}{\lambda}\varsigma^6 + \left(\left(1+\frac{\mu\chi^2}{\lambda}\right) - \mu(\gamma^2-1) - \frac{\mu(\gamma^2-1)}{\lambda}\right)\varsigma^4 - \left(\chi^2 + \mu\chi^2 + \left(1+\frac{\mu\chi^2}{\lambda}\right)\right)\varsigma^2 \\ \hline +\chi^2 + 2i\zeta_w\chi\varsigma\left(\frac{\mu}{\lambda}\varsigma^4 - \left(1+\mu+\frac{\mu}{\lambda}\right)\varsigma^2 + 1\right)\end{array}\right)} \tag{19b}$$

where $i = \sqrt{-1}$ is the imaginary unit. The non-dimensional frequency ratio, ς, employed in Eq. (19) is defined as:

$$\varsigma = \frac{\omega}{\omega_r} \tag{20}$$

For Eq. (18), the normalized complex amplitudes of u_r and v become:

$$\frac{u_{r,0}}{f_0/k_r} = \frac{(\chi^2 - \varsigma^2) + 2i\zeta_w\chi\varsigma}{(1 + \mu - \gamma^2\mu)\varsigma^4 - \left(1 + (1 + \mu)\chi^2\right)\varsigma^2 + \chi^2 + 2i\zeta_w\chi\varsigma\left(1 - (1 + \mu)\varsigma^2\right)} \tag{21a}$$

$$\frac{v_0}{f_0/k_r} = \frac{\gamma\varsigma^2}{(1 + \mu - \gamma^2\mu)\varsigma^4 - \left(1 + (1 + \mu)\chi^2\right)\varsigma^2 + \chi^2 + 2i\zeta_w\chi\varsigma\left(1 - (1 + \mu)\varsigma^2\right)} \tag{21b}$$

3. Optimal calibration of the TLCD

Fig. 2 (a) presents the dynamic amplification (the absolute value of the complex amplitude) of structural response of the analogue 2-DOF system (with background flexibility $k_b \neq \infty$ considered as shown in Fig. (1)), with the mass ratio $\mu = 0.02$, the liquid length ratio $\gamma = 0.6$ and the frequency ratio $\chi = 1$ under various values of the equivalent damping ratio ζ_w. In addition, results corresponding to the classic 2-DOF system ignoring the background flexibility ($k_b \Rightarrow \infty$) are also shown in Fig. 2(b). As seen, two fixed points (neutral frequencies, denoted by P_1 and P_2) around the resonant frequency ω_r exist in both subfigures, with their amplitudes being independent of ζ_w. The amplitude of point P_1 (0.9660, 20.805) in Fig. 2(a) is a bit lower than that in Fig. 2(b) (P_1(0.9664, 21.073)), while the amplitude of point P_2(1.0261, 13.510) in Fig. 2(a) is around 1% larger than that in Fig. 2(b) (P_2(1.0264, 13.403)). These results indicate that the fixed-points concept [40] hold for both the analogue 2-DOF systemand the classic 2-DOF system, although the specific amplitudes of P_1 and P_2 are slightly different. Therefore, the optimal frequency ratio, χ_{opt}, can in principle be derived by equalizing the dynamic amplifications at these two points [40]. For the classic 2-DOF system, the derivation of χ_{opt} is presented in Appendix C.1. However, a direct application of this procedure for the analogue 2-DOF system will lead to a ninth-degree equation in the independent-of-damping-ratio step, cf. Eq. (B.3) in Appendix B. Hence, this conventional calibration procedure is no longer applicable.

Instead, the pole-placement (root locus) calibration method [57,61] is applied herein. Such calibration procedure outlined in the following sections is concerned with the proper placement of the complex system poles obtained from the characteristic equation of the analogue 2-DOF system. A set of closed-form formulas for optimal tuning of the TLCD in flexible structures can then be derived. Since the current study focuses on deriving the closed-form formulas for optimal calibration of TLCDs accounting for the background flexibility, only the analogue 2-DOF system is presented here. Optimal calibration of the classic 2-DOF system using both the equalized dynamic amplification approach and the pole-placement approach are elaborated respectively in Appendix C.

3.1. Characteristic equation

The characteristic equation for the analogue 2-DOF system shown in Fig. 1(b) is obtained from the denominator of Eq. (19a):

$$\begin{aligned} &\varsigma^6 - \left(\frac{\lambda+\chi^2\mu+(1-\gamma^2)(1+\lambda)\mu}{\mu(1-\gamma^2)}\right)\varsigma^4 + \left(\frac{\lambda+\chi^2(\lambda+\mu+\lambda\mu)}{\mu(1-\gamma^2)}\right)\varsigma^2 - \frac{\chi^2\lambda}{\mu(1-\gamma^2)} \\ &\quad -2i\zeta_w\frac{\chi\varsigma}{1-\gamma^2}\left(\varsigma^4 - \frac{(1+\mu)\lambda+\mu}{\mu}\varsigma^2 + \frac{\lambda}{\mu}\right) = 0 \end{aligned} \tag{22}$$

As seen, for $\zeta_w = 0$ the roots (denoted by $\varsigma_{0,1}$, $\varsigma_{0,2}$ and $\varsigma_{0,3}$, as shown in Fig. 3) of the characteristic equation are determined by the cubic equation (the even-power terms and constant term in Eq. (22)). This limit is reached for vanishing damping of the TLCD. In the opposite limit $\zeta_w \Rightarrow \infty$, the roots (denoted by $\varsigma_{\infty,1}$ and $\varsigma_{\infty,2}$, see Fig. 3) are determined by the quadratic equation (the quadratic equation inside the bracket of the last term in Eq. (22)). This limit implies the orifice area inside the TLCD is totally blocked. In this case the TLCD can be treated as a concentrated mass, m_w, connect to the structural modal mass, m_r, through the stiffness k_b, whereby the two roots represent the vibration modes for the coupled 2-DOF system.

When $k_b \Rightarrow \infty$, Eq. (22) is reduced to Eq. (23), corresponding to the characteristic equation of the classic SDOF-TLCD system:

$$\varsigma^4 - \frac{1 + (1 + \mu)\chi^2}{1 + \mu - \gamma^2\mu}\varsigma^2 + \frac{\chi^2}{1 + \mu - \gamma^2\mu} - \frac{2i\zeta_w\chi}{1 + \mu - \gamma^2\mu}\varsigma\left((1 + \mu)\varsigma^2 - 1\right) = 0 \tag{23}$$

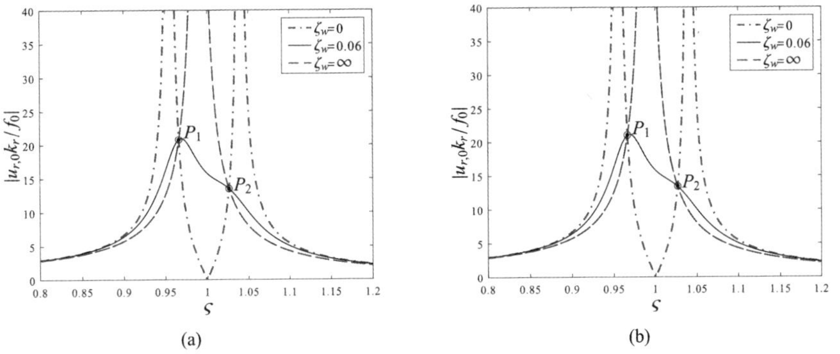

Fig. 2. Dynamic amplification curves of the 2-DOF structure-TLCD system for three different equivalent damping ratios ζ_w. $\mu = 0.02$, $\gamma = 0.6$ and $\chi = 1$. (a) The analogue 2-DOF system with background flexibility included ($\lambda = 5.58$, $k_b \neq \infty$), (b) The classic 2-DOF system with background flexibility ignored ($k_b \Rightarrow \infty$).

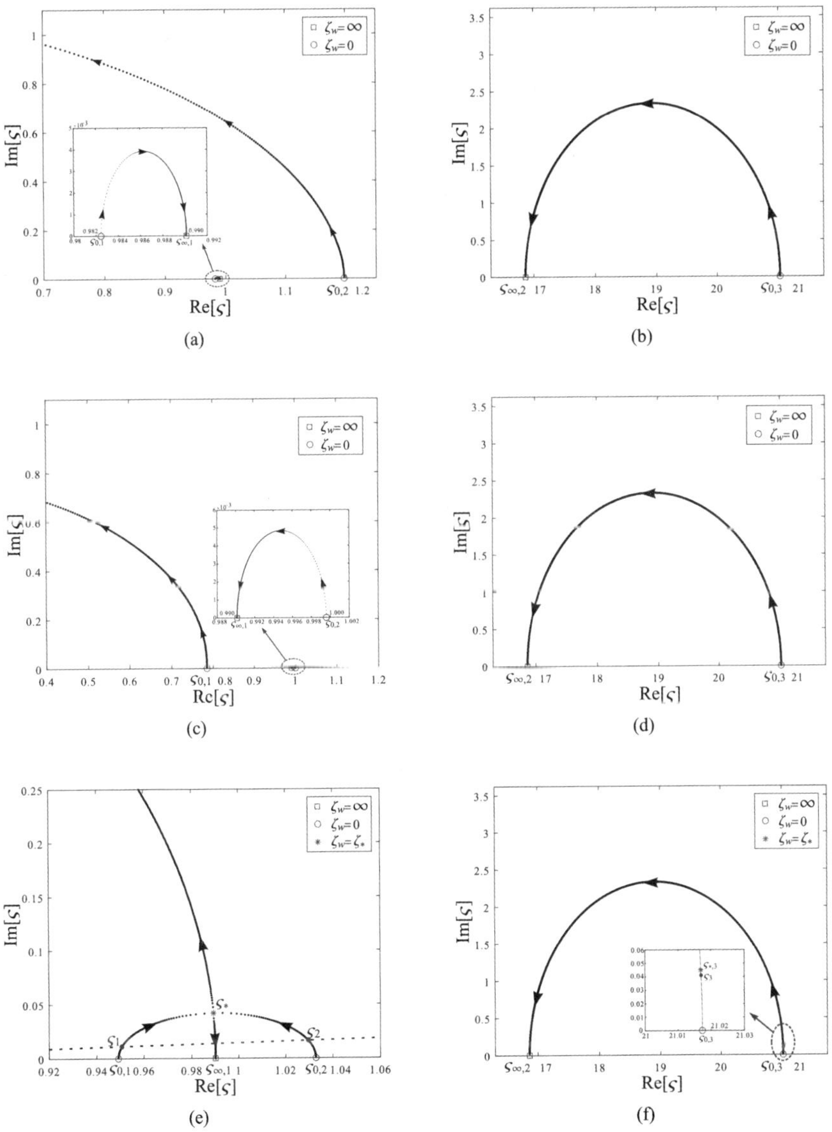

Fig. 3. Three possible types of root locus diagrams for the analogue 2-DOF system, depending on the calibration. $\mu = 0.02$, $\gamma = 0.6$ and $\lambda = 5.58$. The arrows ◄ denote the evolution directions of roots of Eq. (22) with increasing value of ζ_w. The three circles (○) on the horizontal axis denote the roots $\varsigma_{0,1}$, $\varsigma_{0,2}$ and $\varsigma_{0,3}$ of Eq. (22) obtained in the limit $\zeta_w = 0$, and the two square markers (□) on the horizontal axis denote the roots $\varsigma_{\infty,1}$ and $\varsigma_{\infty,2}$ obtained in the limit $\zeta_w \Rightarrow \infty$. (a)(c)(e) The 1st and 2nd root locus diagrams, (b)(d)(f) the 3rd root locus diagrams. (a)(b) $\chi/\chi_{opt} = 1.2$, (c)(d) $\chi/\chi_{opt} = 0.8$, (e)(f) $\chi/\chi_{opt} = 1$. The red points ● in (e)(f) denote three roots of Eq. (22) (ς_1, ς_2 and ς_3) in the first quadrant when $0 \le \zeta_w \le \zeta_*$, where ς_1 and ς_2 are inverse points with respect to the circle R. The red asterisks $*$ in (e)(f) denote two roots of Eq. (22) (ς_* and $\varsigma_{*,3}$) in the first quadrant when $\zeta_w = \zeta_*$, and ς_* is the bifurcation point.

For the limit case $\zeta_w = 0$, the roots are determined by the quadratic equation (the oven-order terms in Eq. (23)). The opposite limit $\zeta_w \Rightarrow \infty$ leads to only one root $\varsigma_\infty = 1/\sqrt{1+\mu}$, associated with an extra mass m_w added to the modal mass m_r. Further, as can be seen from Eq. (21), the classic SDOF-TLCD system is reduced to the classic SDOF-TMD system as γ trends to 1.

3.2. *Equal modal damping ratio*

Once the values of λ, χ, μ, γ and ζ_w are provided, six complex roots of the sextic characteristic equation Eq. (22) can be obtained. Let three roots be denoted by ς_1, ς_2 and ς_3, which lied in the first quadrant, and the symmetrically located roots $-\bar{\varsigma}_1$, $-\bar{\varsigma}_2$ and $-\bar{\varsigma}_3$ are located in the second quadrant, where the bar symbol (¯) denotes the complex conjugate. By varying the value of the equivalent damping ratio ζ_w from 0 to ∞, the so-called root locus diagrams for these six roots can be plotted in the complex plane. Due to symmetry characteristic of roots, only three of six roots in the first quadrant are investigated. The basic dynamic characteristics of the system can be revealed by these roots [61].

It was demonstrated in [57] that for the classic SDOF-TMD system, equalized dynamic amplification at the two invariant points leads to equal modal damping ratio of the two free vibration modes. For the present analogue 2-DOF system, the idea is that two of the three complex roots (from the sextic characteristic equation) should have equal modal damping ratio [61,62]. Similar to the pole-placement analysis of the classic SDOF-TMD system [57], the property of equal modal damping ratio for the current analogue 2-DOF system can be achieved by the two complex poles being inverse points with respect to a reference frequency [61,62].

Fig. 3 depicts three possible types of root locus diagrams for the analogue 2-DOF system, depending how the calibration is performed. The three circles (○) on the horizontal axis denote the roots $\varsigma_{0,1}$, $\varsigma_{0,2}$ and $\varsigma_{0,3}$ of Eq. (22) obtained in the limit $\zeta_w = 0$, while the two square markers (□) on the horizontal axis denote the roots $\varsigma_{\infty,1}$ and $\varsigma_{\infty,2}$ obtained in the limit $\zeta_w \Rightarrow \infty$. The specific data of Fig. 3 are obtained from the numerical example in Section 5 Fig. 3(a), 3(b) show the case of frequency ratio being slightly larger than the optimal values ($\chi/\chi_{opt} = 1.2$), where χ_{opt} denotes the optimal frequency ratio and will be derived below (as shown in Eq. (31)). In this case, one branch of the root locus forms a local closed curve starting from $\varsigma_{0,1}$ and ending at $\varsigma_{\infty,1}$ with the increasing value of the equivalent damping ratio ζ_w. Likewise, another branch (as shown in Fig. 3(b)) also forms a closed curve starting from $\varsigma_{0,3}$ and ending at $\varsigma_{\infty,2}$ with a coordinate value in the real axis far more larger than the first branch of root locus. In contrast, a non-local branch starts from $\varsigma_{0,2}$ and extends to the imaginary axis. Fig. 3(c), 3(d) present the case of frequency ratio being smaller than the optimal frequency ratio ($\chi/\chi_{opt} = 0.8$). Comparing Figs. 3(a) and 3(c) the configurations of the two branches are interchanged. On the other hand, the third branch shown in Fig. 3(d) remains the same as in Fig. 3(b).

Therefor it is anticipated that only the two local branches of the root locus will intersect at the so-called bifurcation point, ς_*, under the condition $\chi = \chi_{opt}$, as shown in Fig. 3(e). The root locus diagrams corresponding to the optimal frequency ratio condition are sketched in Fig. 3(e), 3(f). As seen, in one limit $\zeta_w = 0$ three roots ($\varsigma_{0,1}$, $\varsigma_{0,2}$ and $\varsigma_{0,3}$) appear on the positive real axis. When increasing ζ_w from 0 to a critical value ζ_*, these three roots move into the complex plane. It follows from the inverse point characteristics [61,62] that the first two roots (ς_1 and ς_2) lying on the same line containing the origin of the complex plane (as seen in Fig. 3(e)), have equal argument, corresponding to equal modal damping ratio. When $\zeta_w = \zeta_*$, the two local branches intersect at the bifurcation point ς_*. Further increasing ζ_w leads to branch off in two directions along a circle with radius of R centered at the origin point of the complex plane. As for the third root, the corresponding root locus diagram (as shown in Fig. 3(f)) seems independent of the frequency ratio, which is identical to Figs. 3(b) and 3(d). The references [61,62] present three types of possible root locus diagrams for the similar sixtic characteristic equation, cf. Fig. 2 in [61]. However, the systems investigated in [61,62] involves three DOFs, which are different with the current 2-DOF case (u_r and v). Therefore, the triple-bifurcation point (employed in references [61,62]) which involves a intersection between those three branches does not exist in the current case.

In the following derivation of the optimal frequency ratio and damping ratio of the TLCD, the root locus shown in Fig. 3(e) is aimed for, where the a bifurcation point exists and the two complex roots are inverse points (as long as $0 \le \zeta_w \le \zeta_*$).

In the limit $\zeta_w \Rightarrow \infty$, the characteristic equation Eq. (22) becomes:

$$-2i\frac{\chi\varsigma}{1-\gamma^2}\left(\varsigma^4 - \frac{(1+\mu)\lambda+\mu}{\mu}\varsigma^2 + \frac{\lambda}{\mu}\right) = 0 \tag{24}$$

from which two roots can be obtained, $\varsigma_{\infty,1}^2$ and $\varsigma_{\infty,2}^2$ ($\varsigma_{\infty,1}^2 \ll \varsigma_{\infty,2}^2$). Defining $R^2 = \varsigma_{\infty,1}^2$, the specific expression for R can be obtained from Eq. (24) as:

$$R = \sqrt{\frac{(\lambda+\mu+\lambda\mu) - \sqrt{(\lambda+\mu+\lambda\mu)^2 - 4\mu\lambda}}{2\mu}} \tag{25}$$

In the opposite limit $\zeta_w = 0$, the characteristic equation Eq. (22) becomes:

$$\varsigma^6 - \left(\frac{\lambda+\chi^2\mu+(1-\gamma^2)(1+\lambda)\mu}{\mu(1-\gamma^2)}\right)\varsigma^4 + \left(\frac{\lambda+\chi^2(\lambda+\mu+\lambda\mu)}{\mu(1-\gamma^2)}\right)\varsigma^2 - \frac{\chi^2\lambda}{\mu(1-\gamma^2)} = 0 \tag{26}$$

Based on the inverse points condition of ς_1 and ς_2 for $0 \le \zeta_w \le \zeta_*$, the following relation can thus be formulated as:

$$\varsigma_1\bar{\varsigma}_2 = \varsigma_2\bar{\varsigma}_1 = R^2 \tag{27}$$

in which, ς_1 and ς_2 are two of three roots (ς_1, ς_2 and ς_3) of Eq. (22), as seen the red dots ● in Fig. 3(e), 3(f). In the limit $\zeta_w = 0$, these three roots are denoted by $\varsigma_{0,i}$ (i.e. $\varsigma_1 = \varsigma_{0,1}$, $\varsigma_2 = \varsigma_{0,2}$ and $\varsigma_3 = \varsigma_{0,3}$ when $\zeta_w = 0$, as seen the red circles ○ in Fig. 3(e), 3(f)). A sixtic equation under the limit $\zeta_w = 0$ can thus be constructed as:

$$\begin{aligned}
&(\varsigma - \varsigma_{0,1})(\varsigma + \bar{\varsigma}_{0,1})(\varsigma - R^2/\varsigma_{0,1})(\varsigma + R^2/\bar{\varsigma}_{0,1})(\varsigma - \varsigma_{0,3})(\varsigma + \bar{\varsigma}_{0,3}) = 0 \\
&\Rightarrow \varsigma^6 - \left(\frac{R^4+\mathrm{Re}[\varsigma_{0,1}]^4+\mathrm{Re}[\varsigma_{0,1}]^2\,\mathrm{Re}[\varsigma_{0,3}]^2}{\mathrm{Re}[\varsigma_{0,1}]^2}\right)\varsigma^4 \\
&+\left(\frac{R^4(\mathrm{Re}[\varsigma_{0,1}]^2+\mathrm{Re}[\varsigma_{0,3}]^2)+\mathrm{Re}[\varsigma_{0,1}]^4\,\mathrm{Re}[\varsigma_{0,3}]^2}{\mathrm{Re}[\varsigma_{0,1}]^2}\right)\varsigma^2 - R^4\,\mathrm{Re}[\varsigma_{0,3}]^2 = 0
\end{aligned} \tag{28}$$

where Re[·] is the real part operator.

By equalizing coefficients in Eq. (26) and Eq. (28), one would obtain the following relation:

$$A_\chi\chi^4 + B_\chi\chi^2 + C_\chi = 0 \tag{29}$$

where:

$$\begin{cases}
A_\chi = \lambda(\lambda - \mu R^4) \\
B_\chi = \lambda R^4\big(\mu(1+\lambda)(\gamma^2-1) - \lambda\big) - R^8\mu(\lambda+\mu+\lambda\mu)(\gamma^2-1) \\
C_\chi = R^8\mu(1-\gamma^2)\big(R^4\mu(\gamma^2-1)+\lambda\big)
\end{cases} \tag{30}$$

The optimal frequency ratio, χ_{opt}, which enables an equal modal damping ratio of the analogue 2-DOF system can thus be obtained from Eq. (29):

$$\chi_{opt} = \sqrt{\frac{-B_\chi + \sqrt{B_\chi^2 - 4A_\chi C_\chi}}{2A_\chi}} \tag{31}$$

It is noted that, unlike the optimal frequency ratio for the classic SDOF-TMD system cf. [40,57], the optimal frequency ratio χ_{opt} for the analogue 2-DOF system is not only a function of mass ratio, μ, and liquid length ratio, γ, but also depends on the structural inherent characteristics, $\lambda = k_b/k_r$, cf.Eq. (30),(31).

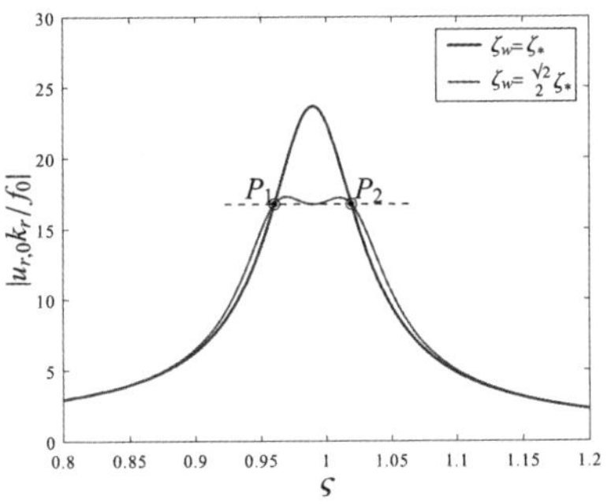

(a)

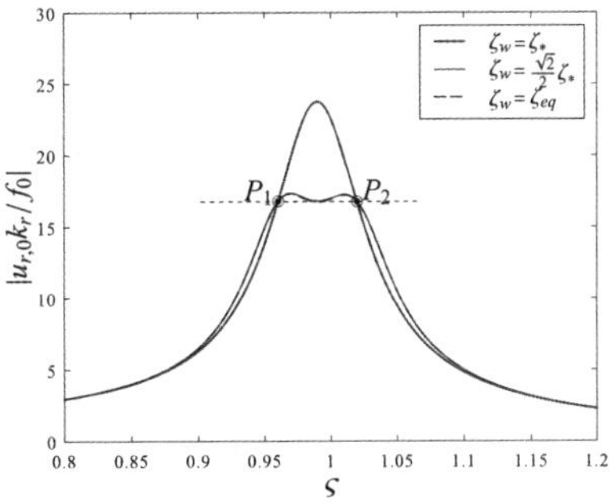

(b)

Fig. 4. Dynamic amplification curves of the 2-DOF structure-TLCD system at the optimal frequency ratio $\chi = \chi_{opt}$ for two different equivalent damping ratios ζ_w. $\mu = 0.02$ and $\gamma = 0.6$. (a) The analogue 2-DOF system with background flexibility included (i.e. $\lambda = 5.58, k_b \neq \infty$), χ_{opt} obtained from Eq. (31); (b) The classic 2-DOF system with background flexibility ignored (i.e. $k_b \Rightarrow \infty$), χ_{opt} obtained from Eq. (C.12b).

3.3. Bifurcation point

Next, the existence of the bifurcation point (see ς_* in Fig. 3(e) marked by $*$) implies that the two complex roots, ς_1 and ς_2, are equal when $\zeta_w = \zeta_*$ (i.e. $\varsigma_1 = \varsigma_2 = \varsigma_*$), whereby the sixtic polynomial equation can be formulated as:

$$(\varsigma - \varsigma_*)^2(\varsigma + \bar{\varsigma}_*)^2(\varsigma - \varsigma_{*,3})(\varsigma + \bar{\varsigma}_{*,3}) = 0 \Rightarrow$$

$$\varsigma^6 - \left(\frac{R^4 + \mathrm{Re}[\varsigma_*]^4 + \mathrm{Re}[\varsigma_*]^2\,\mathrm{Re}[\varsigma_{*,3}]^2}{\mathrm{Re}[\varsigma_*]^2}\right)\varsigma^4 + \left(\frac{R^4(\mathrm{Re}[\varsigma_*]^2 + \mathrm{Re}[\varsigma_{*,3}]^2) + \mathrm{Re}[\varsigma_*]^4\,\mathrm{Re}[\varsigma_{*,3}]^2}{\mathrm{Re}[\varsigma_*]^2}\right)\varsigma^2 - R^4\,\mathrm{Re}[\varsigma_{*,3}]^2 = 0 \tag{32}$$

where $\varsigma_{*,3}$ is the third root, ς_3, of Eq. (22) for $\zeta_w = \zeta_*$ and $\chi = \chi_{opt}$, as seen the red asterisk $*$ in Fig. 3(f).

For convenience of expression in the following analysis, the coefficients for ς in Eq. (22) are replaced by six variables, thus Eq. (22) can be manipulated into the simplified form:

$$A_6\varsigma^6 - A_4\varsigma^4 + A_2\varsigma^2 - A_0 - 2i\zeta_w\left(A_5\varsigma^5 - A_3\varsigma^3 + A_1\varsigma\right) = 0 \tag{33}$$

in which:

$$A_6 = 1,\ A_4 = \frac{\lambda+\chi^2\mu+(1-\gamma^2)(1+\lambda)\mu}{\mu(1-\gamma^2)},\ A_2 = \frac{\lambda+\chi^2(\lambda+\mu+\lambda\mu)}{\mu(1-\gamma^2)},\ A_0 = \frac{\chi^2\lambda}{\mu(1-\gamma^2)}$$
$$A_5 = \frac{\chi}{1-\gamma^2},\ A_3 = \frac{\lambda+\mu+\lambda\mu}{\mu(1-\gamma^2)},\ A_1 = \frac{\chi\lambda}{\mu(1-\gamma^2)} \tag{34}$$

By equalizing Eq. (32) with Eq. (33), the coordinates of the bifurcation point, ς_*, and the corresponding equivalent damping ratio, ζ_*, can be obtained:

$$\begin{cases} \mathrm{Re}[\varsigma_*] = \sqrt{\left|\dfrac{-5R^8+A_4R^6-A_2R^4+5A_0R^2}{-4R^6+4A_0}\right|} \\ \mathrm{Im}[\varsigma_*] = \sqrt{\left|\dfrac{-R^8+A_4R^6-A_2R^4+A_0R^2}{-4R^6+4A_0}\right|} \end{cases} \tag{35}$$

$$\zeta_* = \sqrt{\left|\frac{(R^{12} - A_2R^8 + 4A_0R^6 - 3A_4A_0R^4 + 4A_2A_0R^2 - 5A_0^2)^2}{16A_1^2R^2(-R^6 + A_0)(-R^6 + A_4R^4 - A_2R^2 + A_0)}\right|} \tag{36}$$

Fig. 4 shows the dynamic amplification of the modal response $|u_{r,0}k_r/f_0|$ of the 2-DOF structure-TLCD system at the optimal frequency ratio (with fixed $\mu = 0.02$, $\gamma = 0.6$ and $\lambda = 5.58$) for two different values of the equivalent damping ratio, ζ_w. Fig. 4(a) presents the results for the analogue 2-DOF system with background flexibility considered. The blue line corresponds to $\zeta_w = \zeta_*$, while the red line corresponds to $\zeta_w = \sqrt{2}/2\zeta_*$, which is the optimal damping ratio suggested by Krenk [57]. As seen, the amplitudes of these two lines at the invariant points (neutral frequencies), P_1 and P_2, are equal under the optimal frequency ratio, which verifies the derivation results obtained based on the pole-placement method. Further, the value suggested by Krenk, $\zeta_w = \sqrt{2}/2\zeta_*$, leads to a fairly level behavior of the curve between the neutral frequency points. On the other hand, the value ζ_* leads to a peak between points P_1 and P_2. In the following optimal procedure, this equivalent damping ratio ($\zeta_w = \sqrt{2}/2\zeta_*$) will be regarded as the optimal damping ratio. Similar conclusion can be drawn in Fig. 4(b) for the classic 2-DOF system. The optimal frequency ratio $\chi = \chi_{opt}$ employed in Fig. 4(b) is obtained based on the formulation derived in Appendix C.2. As seen, one additional black line is added in Fig. 4(b) corresponding to $\zeta_w = \zeta_{eq}$, where ζ_{eq} denotes the theoretical optimal damping ratio obtained from the conventional dynamic amplification-based calibration method, cf.Eq. (C.9) in Appendix C. The red line ($\zeta_w = \sqrt{2}/2\zeta_*$) almost merges with the black line, implying that the equivalent damping ratio value $\zeta_w = \sqrt{2}/2\zeta_*$ can be considered optimal in practical calibration of the TLCD.

3.4. Optimal head loss coefficient of the TLCD

In Eqs. (15), (16) and (18), the nonlinear damping force of the TLCD, $f_{TLCD,nonlinear} = -1/2\rho A\xi|\dot{v}|\dot{v}$, is replaced by a linearized equivalent damping force, $f_{TLCD,linear} = -C_w\dot{v}$, under a harmonic excitation. C_w is the equivalent damping coefficient, which can be represented as [26]:

$$C_w = 2\zeta_w\omega_w m_w = \frac{4}{3\pi}\rho A\xi\omega|v_0| \tag{37}$$

The corresponding equivalent damping ratio, ζ_w, can thus be obtained:

$$\zeta_w = \frac{1}{3\pi g}\xi\chi\omega_r\omega|v_0| \tag{38}$$

where $|v_0|$ is the absolute amplitude value of liquid in the vertical tube.

The optimal head loss coefficient for the TLCD, ξ_{opt}, can be obtained either by the theoretical approximations such as the perturbation-based method [48] or the numerical optimizations such as the particle-swarm optimization method [28]. In this study, a rather straightforward way to determine the head loss coefficient is adopted. As shown in Section 3.3, the optimal equivalent damping ratio is determined as $\zeta_{w,opt} = \sqrt{2}/2\zeta_*$ under $\chi = \chi_{opt}$, where ζ_* can be calculated through Eq. (36) and χ_{opt} can be obtained from Eq. (31). By inserting $\zeta_{w,opt}$ and χ_{opt} into Eq. (38), the optimal head loss coefficient is:

$$\xi_{opt} = \frac{3\pi g\zeta_{w,opt}}{\chi_{opt}\omega_r\omega|v_0|} = \frac{3\sqrt{2}\pi}{2}\frac{g\zeta_*}{\chi_{opt}\omega_r\omega|v_0|} \tag{39}$$

As seen in Eq. (39), ξ_{opt} depends on the external excitation frequency ω. In consideration of practical use, the head loss coefficient should be chosen based on the severest condition [28], i.e. $\omega = \omega_r$. Hence, the optimal head loss coefficient is given as:

$$\xi_{opt} = \frac{3\pi g\zeta_{w,opt}}{\chi_{opt}\omega_r^2|v_0|_{\omega=\omega_r}} = \frac{3\sqrt{2}\pi}{2}\frac{g\zeta_*}{\chi_{opt}\omega_r^2|v_0|_{\omega=\omega_r}} \tag{40}$$

where $|v_0|_{\omega=\omega_r}$ denotes the absolute amplitude value of liquid in the vertical tube, which can be obtained by inserting $\omega = \omega_r$, $\chi = \chi_{opt}$ and

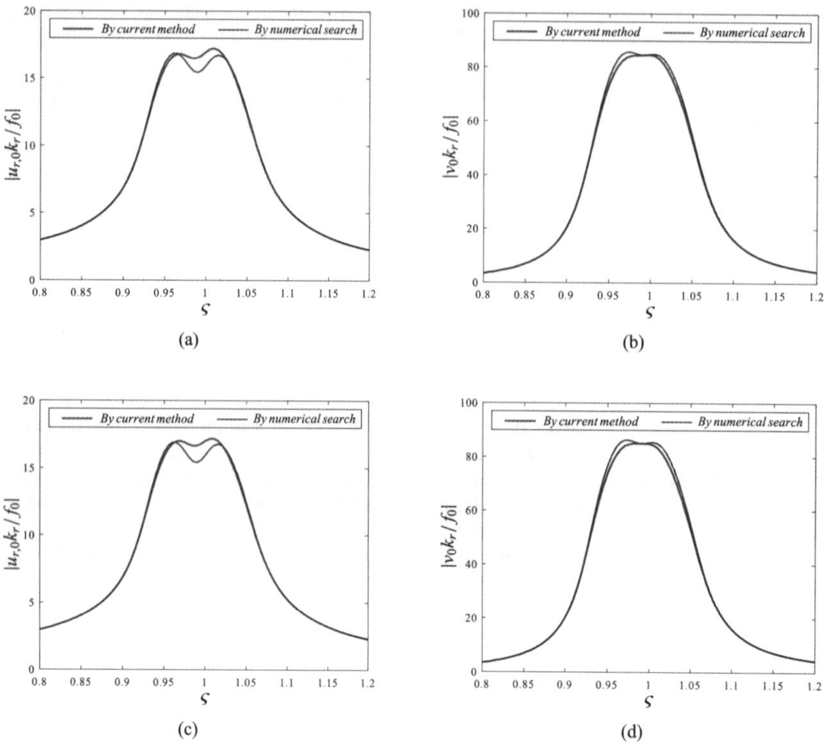

Fig. 5. Dynamic amplification curves of the 2-DOF structure-TLCD system at the optimal frequency ratio χ_{opt} and optimal head loss coefficient ξ_{opt}. $\mu = 0.02$, $\gamma = 0.6$ and $C = 0.003$. (a)(c) Structural response, (b)(d) liquid response. (a)(b) Analogue 2-DOF system with background flexibility included (i.e. $\lambda = 5.58$, $k_b \neq \infty$). The orange lines —— correspond to the results when $\chi_{opt} = 0.987$ and $\xi_{opt} = 2.24$ (obtained from Eqs. (31) and (40)); the blue lines —— correspond to the results when $\chi_{num,opt} = 0.988$ and $\xi_{num,opt} = 2.12$ (obtained by numerical search). (c)(d) Classic 2-DOF system with background flexibility ignored (i.e. $k_b \Rightarrow \infty$). The orange lines —— correspond to the results when $\chi_{opt} = 0.987$ and $\xi_{opt} = 2.23$ (obtained from Eq. (C.12b),40); the blue lines —— correspond to the results when $\chi_{num,opt} = 0.987$ and $\xi_{num,opt} = 2.06$ (obtained by numerical search).

$\zeta_w = \zeta_{w,opt} = \sqrt{2}/2\zeta_*$ into the linearized EOMs. For the flexible structure-TLCD system (as shown in Fig. 1(a)), the value of $|v_0|_{\omega=\omega_r}$ should be obtained from the EOMs of the augmented system (cf. Appendix D):

$$\hat{\mathbf{u}}_0 = \left(\hat{\mathbf{K}} + i\omega_r\hat{\mathbf{C}} - \omega_r^2\hat{\mathbf{M}}\right)^{-1}\hat{\mathbf{f}}_0 \tag{41}$$

where the augmented stiffness matrix $\hat{\mathbf{K}}$, damping matrix $\hat{\mathbf{C}}$ and mass matrix $\hat{\mathbf{M}}$ are given in Appendix D, with elements $C_w = \sqrt{2}\chi_{opt}\mu\zeta_*\omega_r m_r$ and $k_w = \mu\chi_{opt}^2 k_r$ being inserted into $\hat{\mathbf{C}}$ and $\hat{\mathbf{K}}$. The value of $|v_0|_{\omega=\omega_r}$ corresponds to the last element of $\hat{\mathbf{u}}_0$ (i.e. $|v_0|_{\omega=\omega_r} = \hat{\mathbf{u}}_{0,n+1,1}$).

Once the optimal head loss coefficient, ξ_{opt}, and the optimal frequency ratio, χ_{opt}, are determined, the dynamic amplification of structural response can be obtained through the iterative procedure in the frequency domain [48].To justify the theoretically derived ξ_{opt} and χ_{opt}, Fig. 5 compares the dynamic amplifications of the 2-DOF structure-TLCD system using ξ_{opt} and χ_{opt}, and the optimal values obtained by numerical search ($\xi_{num,opt}$, $\chi_{num,opt}$), respectively. The range of $\xi_{num,opt}$, $\chi_{num,opt}$ during the numerical search procedure are limited within 1 to 8, 0.9 to 1.0, respectively. The specific data of Fig. 5 are obtained from the numerical example in Section 5 Eq. (5). Fig. 5(a),5(b) show the comparison for the analogue 2-DOF system with background flexibility considered. The red line denotes the results corresponding to $\xi_{opt} = 2.24$ and $\chi_{opt} = 0.987$ obtained from Eq. (31),(40), while the blue line denotes the result corresponding to $\xi_{num,opt} = 2.12$ and $\chi_{num,opt} = 0.988$ obtained from the numerical search. It is seen that the current method leads to a slightly flatter behavior for the structural response curve (which is beneficial), although the numerical search results in a lower trough as expected. As for the liquid response v_0, the numerical search leads to a slightly larger amplitudes. The optimal frequency ratios from the two methods, $\chi_{num,opt} = 0.988$ and $\chi_{opt} = 0.987$, are very close to each other. As for the optimal head loss coefficient, the result obtained from current method $\xi_{opt} = 2.24$ is only around 5% higher than the numerical search result $\xi_{num,opt} = 2.12$. Hence, the current closed-form method is considered comparable with the numerical search method in terms of the accuracy. Furthermore, it is worth noticing that the computational cost for the numerical search method is much higher than the current method. For the classic 2-DOF system, similar conclusion can be drawn from Fig. 5(c),5(d).

4. Calibration procedure

For better clarity, the optimal calibration procedure for a TLCD installed on a flexible structure is summarized in Fig. 6. The equations shown in this flow chat are obtained based on the pole-placement approach, cf. Sections 3.2 and 3.4. The mass ratio, $\mu = m_w/m_r$, and the liquid length ratio, $\gamma = B/L$, are assigned by the designers according to practical constraints. The procedure should then determine the location of the TLCD with respect to the structure, i.e. to determine the location vector $\mathbf{w}$:

$$\mathbf{w} = \begin{bmatrix} 0 \\ \vdots \\ 1 \\ \vdots \\ 0 \end{bmatrix} \begin{matrix} 1 \\ \vdots \\ n_a \\ \vdots \\ n \end{matrix} \tag{42}$$

Once $\mathbf{w}$ is determined, the corresponding modal mass, m_r, modal stiffness, k_r, and background stiffness, k_b, can be calculated subsequently, based on Eq. (9),(10). The effect of the background flexibility on the TLCD calibration is accounted for via the non-dimensional stiffness ratio $\lambda = k_b/k_r$. n

The next step is to determine the optimal frequency ratio, χ_{opt}, which is related to the radius of the circular inversion R and a few constants A_χ, B_χ and C_χ. R corresponds to the smaller root of Eq. (24) when $\zeta_w \Rightarrow \infty$, and can be obtained through Eq. (25). A_χ, B_χ and C_χ are functions of R, λ, μ and γ. Hence the optimal frequency ratio, χ_{opt}, can be calculated using Eq. (31).

1.Choose the TLCD location vector: $\mathbf{w} = [0, \cdots, 1, \cdots, 0]^T$

2.Modal stiffness (both dynamic and background) and modal mass:
$$k_r = \frac{\mathbf{\Phi}_1^T \mathbf{K} \mathbf{\Phi}_1}{(\mathbf{w}^T \mathbf{\Phi}_1)^2}, \quad m_r = \frac{\mathbf{\Phi}_1^T \mathbf{M} \mathbf{\Phi}_1}{(\mathbf{w}^T \mathbf{\Phi}_1)^2}, \quad k_b = \frac{1}{\mathbf{w}^T \mathbf{K}^{-1} \mathbf{w} - 1/k_r}, \quad \lambda = k_b/k_r$$

3.Radius of the circular inversion : $R = \sqrt{\frac{(\lambda+\mu+\lambda\mu) - \sqrt{(\lambda+\mu+\lambda\mu)^2 - 4\mu\lambda}}{2\mu}}$

4.Constants:
$$A_\chi = \lambda(\lambda - \mu R^4), \quad C_\chi = R^8 \mu(1-\gamma^2)\left(R^4\mu(\gamma^2-1)+\lambda\right),$$
$$B_\chi = \lambda R^4\left(\mu(1+\lambda)(\gamma^2-1)-\lambda\right) - R^8\mu(\lambda+\mu+\lambda\mu)(\gamma^2-1)$$

5.Optimal frequency ratio: $\chi_{opt} = \sqrt{\frac{-B_\chi + \sqrt{B_\chi^2 - 4A_\chi C_\chi}}{2A_\chi}}$

6.Constants:
$$A_0 = \frac{\chi^2\lambda}{\mu(1-\gamma^2)}, \quad A_1 = \frac{\chi\lambda}{\mu(1-\gamma^2)}, \quad A_2 = \frac{\lambda+\chi^2(\lambda+\mu+\lambda\mu)}{\mu(1-\gamma^2)},$$
$$A_3 = \frac{\lambda+\mu+\lambda\mu}{\mu(1-\gamma^2)}, \quad A_4 = \frac{\lambda+\chi^2\mu+(1-\gamma^2)(1+\lambda)\mu}{\mu(1-\gamma^2)}, \quad A_5 = \frac{\chi}{1-\gamma^2}, \quad A_6 = 1$$

7.Damping ratio corresponding to the bifurcation point:
$$\zeta_* = \sqrt{\left|\frac{(-5A_0^2+4A_0R^6-3A_4A_0R^4+4A_2A_0R^2+R^{12}-A_2R^8)^2}{16A_1^2R^2(-R^6+A_0)(-R^6+A_4R^4-A_2R^2+A_0)}\right|}$$

8.Structure matrixes:
$\hat{\mathbf{M}}$, $\hat{\mathbf{K}}$ and $\hat{\mathbf{C}}$ (with element $C_w = \sqrt{2}\chi_{opt}\mu\zeta_*\omega_r m_r$ and $k_w = \mu\chi_{opt}^2 k_r$ inserted)

9.Amplitude of the augmented displacement vector:
$$\hat{\mathbf{u}}_0 = \left(\hat{\mathbf{K}} + i\omega_r\hat{\mathbf{C}} - \omega_r^2\hat{\mathbf{M}}\right)^{-1}\hat{\mathbf{f}}_0$$

10.Vertical displacement amplitude of liquid:
$|v_0|_{\omega=\omega_r} = \hat{\mathbf{u}}_{0,n+1,1}$ (the last element of $\hat{\mathbf{u}}_0$)

11.Optimal head loss coefficient:
$$\xi_{opt} = \frac{3\sqrt{2}\pi}{2}\frac{g\zeta_*}{\chi_{opt}\omega_r^2|v_0|_{\omega=\omega_r}}$$

Fig. 6. Optimal calibration procedure for a TLCD with given mass ratio μ and horizontal length ratio γ

It is shown in Eq. (40) that the optimal head loss coefficient ξ_{opt} is a function of the equivalent damping ratio ζ_w and the response amplitude of the liquid in the vertical tube $|v_0|$. The optimal damping ratio employed in Eq. (40) is taken as $\zeta_{w,opt} = \sqrt{2}/2\zeta_*$, where ζ_* can be obtained from Eq. (36), and the response amplitude of liquid $|v_0|_{\omega=\omega_r}$ corresponding to the severest working condition can be obtained from Eq. (41).

5. Numerical example

The calibration procedure for a TLCD attached to a flexible structure introduced in the previous sections is illustrated by a simple 10-story shear frame the data of which have already been used for generating Fig. 3,4,5. Fig. 1(a) presents a 10-story shear frame with floor mass m ($m_{tot} = 10 \times m$) and inter-story stiffness k. The horizontal displacements of the individual floors, $u_i(t)$, represent the 10 DOFs, and the vertical liquid displacement of the TLCD denotes the $11th$ DOF ($v(t)$). These 11 DOFs are assembled in an augmented displacement vector, $\hat{\mathbf{u}}$, as shown in Eq. (D.5).

As shown in Fig. 1(a), the TLCD is located at n_ath floor. The influence of the location of the TLCD on the optimal device parameters is investigated here, when the objective is to introduce damping into the first mode (resonant mode). The optimal parameters in terms of different values of n_a are shown in Table 1). The modal mass, m_r, corresponding to the first mode and the total mass of the TLCD, m_w, are listed in the second and third columns, which increase with the decreasing value of n_a. Column four presents the ratio of the modal background stiffness to the modal dynamic stiffness, $\lambda = k_b/k_r$. As seen, a local maximum value $\lambda = 12.0880$ is obtained near the 8th floor. It is so because the location of 8th floor is close to the nodes of 2nd and 3rd modal shapes of the structure, which means the contributions from higher modes (non-resonant modes) are small [2]. The optimal frequency ratio, χ_{opt}, the optimal equivalent damping ratio, $\zeta_{w,opt}$, and the optimal head loss coefficient, ξ_{opt}, corresponding to the calibration procedure with background flexibility included ($k_b \neq \infty$) or excluded ($k_b \Rightarrow \infty$) are present in the last six columns, respectively. According to Eq. (C.12), χ_{opt} and $\zeta_{w,opt}$ are merely dependent on μ and γ when $k_b \Rightarrow \infty$, see columns six and eight. It is seen that the values of χ_{opt} and $\zeta_{w,opt}$ are nearly increasing when the TLCD moves down, but the variations are not significant when the background flexibility is included, see columns five and seven. On the other hand, the optimal head loss coefficient, ξ_{opt}, decreases as the TLCD moves down, for both cases, see the last two columns. Furthermore, the background contribution leads to larger values of χ_{opt} and ξ_{opt}.

The effectiveness of the TLCD is evaluated by performing frequency analysis of the structure with a TLCD attained at n_ath floor. The external load is assumed to be uniformly distributed along the height of structure and has harmonic time variation with angular frequency ω, i.e. $\hat{\mathbf{f}}(t) = \hat{\mathbf{f}}_0 e^{i\omega t}$ (where $\hat{\mathbf{f}}_0 = [\mathbf{f}_0, 0]^T$, $\mathbf{f}_0 = f_0[\mathbf{1}]_{n\times 1} = Cm_r g[\mathbf{1}]_{n\times 1}$). The top floor response, u_{top}, is selected as reference and the corresponding static displacement $u_{top,static}$ ($u_{top,static} = \hat{\mathbf{u}}_{0,n,1}$, $\hat{\mathbf{u}}_0 = \hat{\mathbf{K}}^{-1}\hat{\mathbf{f}}_0$) is used for normalization. Based on the calibration results of the optimal frequency ratio, χ_{opt}, and the optimal head lose coefficient, ξ_{opt}, given in Table 1, the dynamic amplification of the Structure-TLCD coupled system can be calculated through the conventional iterative procedure. Fig. 7 presents the dynamic amplification of the structural top floor displacements and the liquid responses for a TLCD attached at the sixth, forth and second floors, respectively. The red solid line corresponds to the results accounting for the background flexibility ($k_b \neq \infty$), and the blue dashed line corresponds to the results excluding the background flexibility ($k_b \Rightarrow \infty$). It is shown that the background flexibility has modest influence on the dynamic response of the structure $u_{top,0}/u_{top,static}$ for a TLCD located at

Table 1
TLCD in the 10-story shear frame structure ($\mu = 0.02$, $\gamma = 0.6$)

n_a	$\frac{m_r}{m_{tot}}$	$\frac{m_w}{m_{tot}}$	λ	χ_{opt}		$\zeta_{w,opt}$		ξ_{opt}	
				$k_b \neq \infty$	$k_b \Rightarrow \infty$	$k_b \neq \infty$	$k_b \Rightarrow \infty$	$k_b \neq \infty$	$k_b \Rightarrow \infty$
10	0.5297	0.0106	5.5757	0.9872	0.9866	0.0596	0.0594	0.3350	0.3331
8	0.6059	0.0121	12.0880	0.9869	0.9866	0.0595	0.0594	0.3118	0.3110
6	0.8589	0.0172	6.6156	0.9871	0.9866	0.0596	0.0594	0.2624	0.2612
4	1.6544	0.0331	2.0908	0.9881	0.9866	0.0599	0.0594	0.1910	0.1882
2	6.0428	0.1209	0.5883	0.9920	0.9866	0.0610	0.0594	0.1038	0.0985

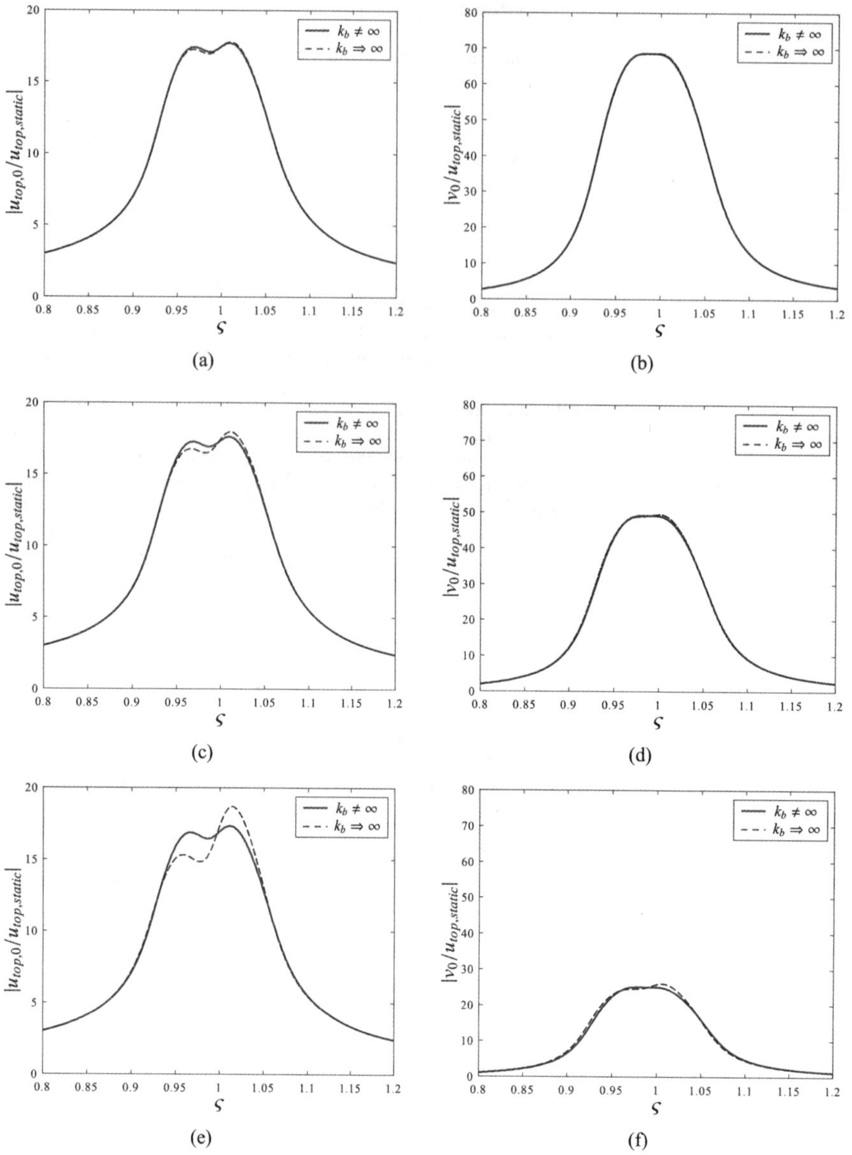

Fig. 7. Dynamic amplifications of the flexible structure-TLCD coupled system, with a TLCD located at the n_ath floor. $\mu = 0.02$, $\gamma = 0.6$ and $C = 0.003$. (a)(c)(d) Dynamic amplifications of the top floor displacement, (b)(d)(f) dynamic amplifications of the liquid response. (a)(b) $n_a = 6$, (c)(d) $n_a = 4$, (e)(f) $n_a = 2$. The orange lines (—— $k_b \neq \infty$) represent the results when the background flexibility is included; the blue lines (– – –) represent the results when the background flexibility is ignored.

the sixth and forth floors, as shown in Fig. 7(a),7(c). However, when a TLCD is located at the second floor, the dynamic amplification of the structural response becomes severely unbalanced, if neglecting the background flexibility, as shown in Eq. Fig. 7(e). Similar characteristics can be observed in the liquid response of the TLCD, v_0, as can be observed in Fig. 7(b),7(d),7(f).

Due to the implementation of a TLCD, the 1st mode (the resonant mode) of the 10-story shear frame is split into two coupled modes. Fig. 8 presents modal damping ratios corresponding to these two modes versus the external excitation frequency. The main object of the calibration procedure is to achieve equal modal damping ratio of these two modes. It can be observed in Fig. 8(a),8(c),8(e) that the object is attained when the background flexibility is included. As seen, these two curves are almost identical. On the other hand, when the background flexibility is omitted in the calibration procedure, the difference between these two modal damping ratios becomes increasingly significant as the TLCD is placed lower, as shown in Fig. 8(b),8(d),8(f). Especially for the 2*nd* floor, the maximum relative difference reaches about 25%, as shown in Fig. 8(f).

Fig. 9 shows the root locus diagram of these two coupled modes. As elaborated earlier, the equal modal damping ratio is equivalent to the existence of a local bifurcation point, which is the intersection of two branches of the root locus. Fig. 9(a),9(c),9(e) show the results when the background flexibility is considered. For $n_a = 6$, these two branches almost intersect, and the bifurcation point seems to exist (although actually not), as shown in Fig. 9(a). As n_a decreases, these two branches exhibit a very slight branch off. On the other hand, when the background flexibility is ignored in the calibration procedure, the non-intersecting property of the two branches becomes much more significant, as shown in Fig. 9(b), 9(d), Fig. 9(f). Especially for $n_a = 2$ (as seen in Fig. 9(f)), the two branches turn out to be quite far away from each other, indicating unbalanced calibration of the TLCD.

Fig. 10 shows the variation of the optimal frequency ratio χ_{opt} with respect to the mass ratio μ and the horizontal length ratio γ. The orange,

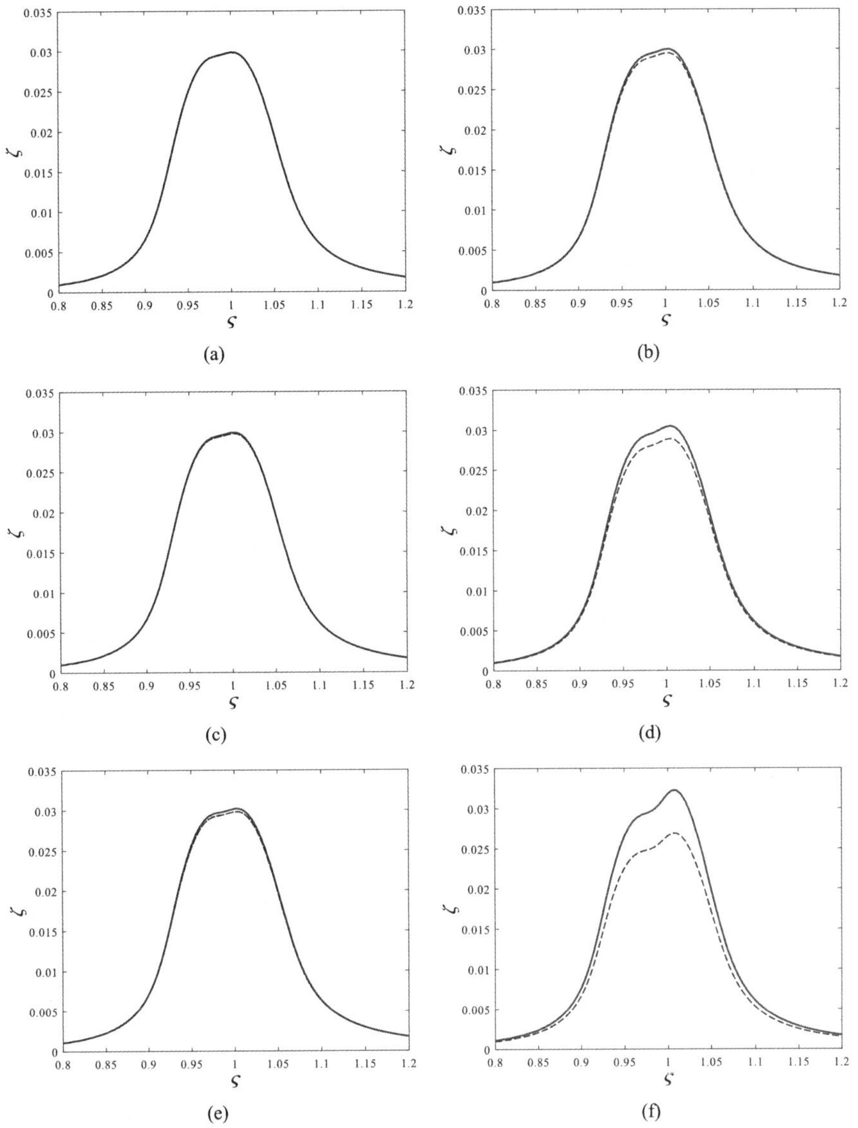

Fig. 8. Variation of modal damping ratios of the first two modes with respect to the normalized excitation frequency ς, for different values of n_a. $\mu = 0.02$, $\gamma = 0.6$ and $C = 0.003$. (a)(c)(d) Background flexibility included (i.e. $k_b \neq \infty$), (b)(d)(f) background flexibility ignored (i.e. $k_b \Rightarrow \infty$). (a)(b) $n_a = 6$, (c)(d) $n_a = 4$, (e)(f) $n_a = 2$.

purple and blue lines represent the results obtained from Eq. (31) with n_a equal to 2, 4, 6, respectively. The black line corresponds to the results obtained from Eq. (C.12b), in which χ_{opt} is independent of n_a. Similar to the results in [38,45], χ_{opt} decreases almost linearly with increasing μ for all cases, as shown in Fig. 10(a). By including the background flexibility, the value of χ_{opt} will be enlarged especially for $n_a = 2$. It is because the contribution from non-resonant modes (higher-frequency modes) make considerable contributions when the TLCD is placed lower. For the *6th* floor, the influence from the background flexibility on χ_{opt} is minor, as the black and blue lines are close to each other. Similar trend can be observed in Fig. 10(b), except for the orange line corresponding to the results of $n_a = 2$, which shows a reverse trend in comparison with others. It is obvious from the plots that γ exhibits modest influence on χ_{opt} in contrast with μ.

Fig. 11 shows the influence of the mass ratio μ, the horizontal length ratio γ and the excitation intensity factor C on the optimal head loss coefficient ξ_{opt}. The solid lines represent the results including the background flexibility ($k_b \neq \infty$), and the dashed lines represent the results ignoring the background flexibility ($k_b \Rightarrow \infty$). It can be observed from Fig. 11(a) ξ_{opt} increases as μ increases. Further, the background flexibility leads to slight enlargement of ξ_{opt}, particularly when the TLCD is located at the *2nd* floor. Similar conclusion can be drawn for γ, see Fig. 11(b). The variation of ξ_{opt} with C is depicted in Fig. 11(c). It is seen that ξ_{opt} decreases with increasing C, although the change becomes insignificant as C increases to a certain value. Similar trend has been observed for the optimal head loss coefficient versus μ and C of a TLCD given by Shum and Yalla [45,48]. It is also found that for all the relevant parameters discussed above, μ plays the most crucial role in determining ξ_{opt}.

The variations of the maximum dynamic amplifications of the structural response $|u_{top,0}/u_{top,static}|_{max}$ and the liquid response $|v_0/u_{top,static}|_{max}$ are depicted in Fig. 12. The subscript *max* denotes the peak value within the external excitation frequency interval around the resonance frequency (here ω ranges from $0.5\omega_r$ to $1.8\omega_r$). Fig. 12(a) shows the evolution of $|u_{top,0}/u_{top,static}|_{max}$ with respect to μ. It is seen that as μ increases, the maximum frequency response of the structure

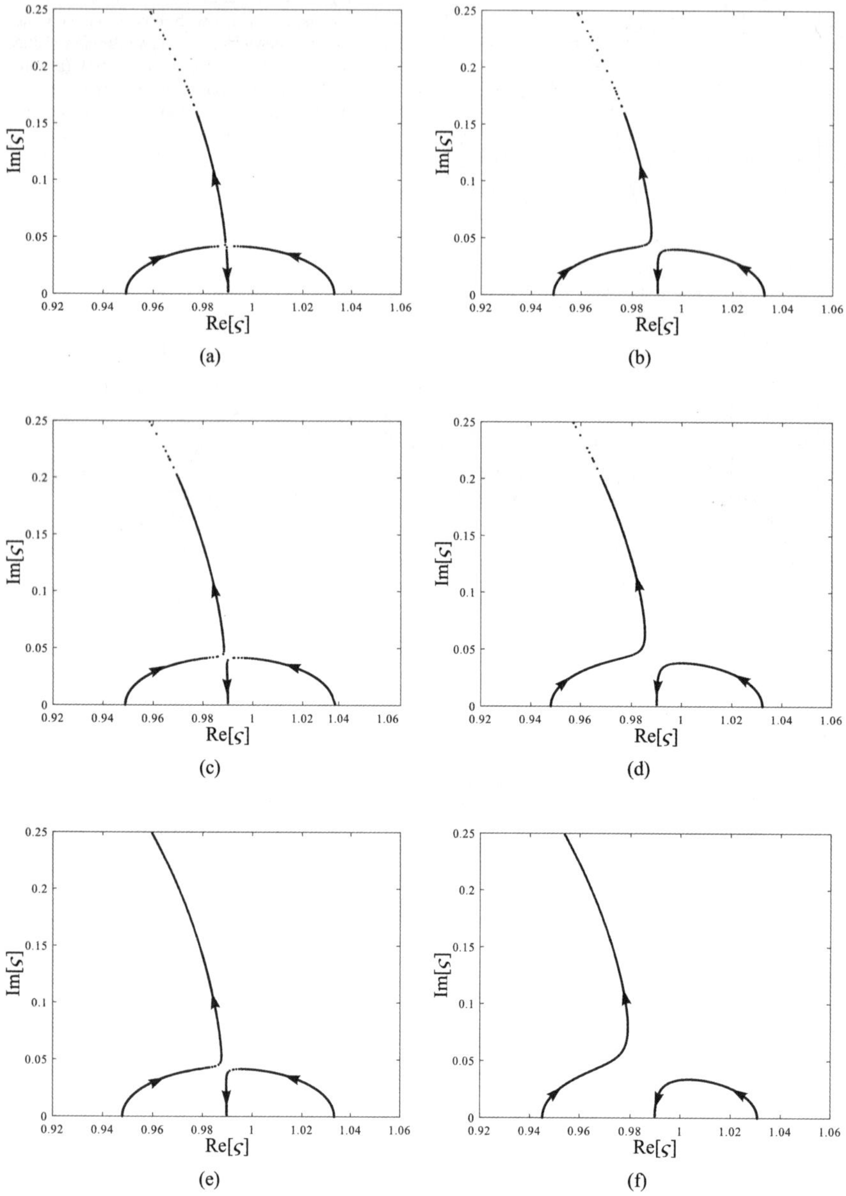

Fig. 9. Root locus diagrams corresponding to the first two modes of the flexible structure-TLCD system, for different values of n_a. $\mu = 0.02$, $\gamma = 0.6$ and $C = 0.003$. The arrows ◄ denote the evolution directions of roots corresponding to the two modes with increasing value of ζ_w. (a)(c)(d) Background flexibility included (i.e. $k_b \neq \infty$), (b)(d)(f) background flexibility ignored (i.e. $k_b \Rightarrow \infty$). (a)(b) $n_a = 6$, (c)(d) $n_a = 4$, (e)(f) $n_a = 2$.

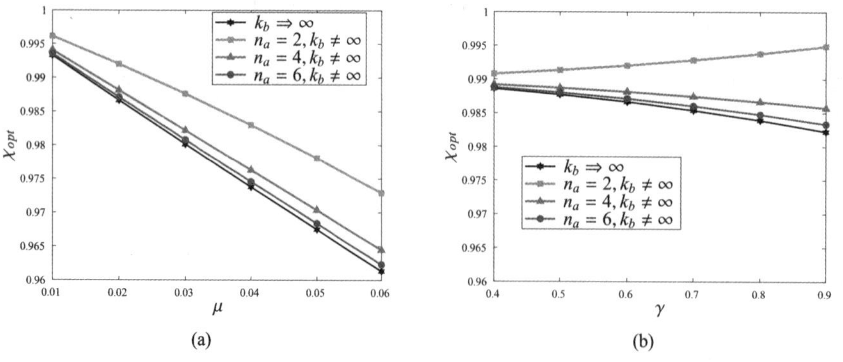

Fig. 10. Variation of the optimal frequency ratio χ_{opt} with (a) the mass ratio μ ($\gamma = 0.6$ and $C = 0.003$) and (b) the horizontal length ratio γ ($\mu = 0.02$ and $C = 0.003$), for different values of n_a. The black lines ($k_b \Rightarrow \infty$) correspond to the results when the background flexibility is ignored; the colored lines ($k_b \neq \infty$) correspond to the results when the background flexibility is included.

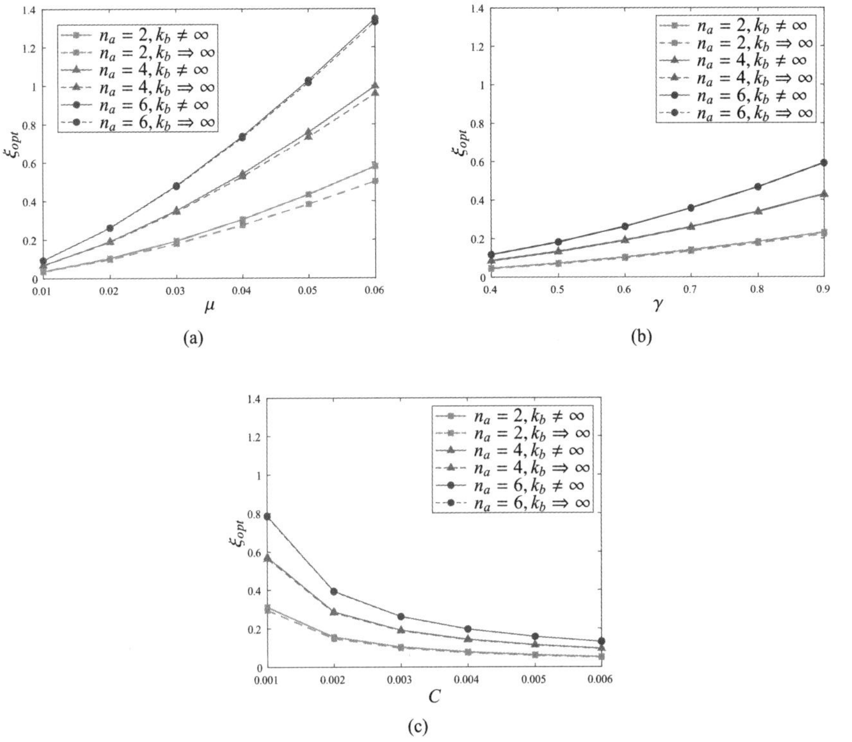

Fig. 11. Variation of the optimal head loss coefficient ξ_{opt} with (a) the mass ratio μ ($\gamma = 0.6$ and $C = 0.003$), (b) the horizontal length ratio γ ($\mu = 0.02$ and $C = 0.003$) and (c) the excitation intensity factor C ($\mu = 0.02$ and $\gamma = 0.6$) for different values of n_a. The dashed lines ($k_b \Rightarrow \infty$) correspond to the results when the background flexibility is ignored; the solid lines ($k_b \neq \infty$) correspond to the results when the background flexibility is included.

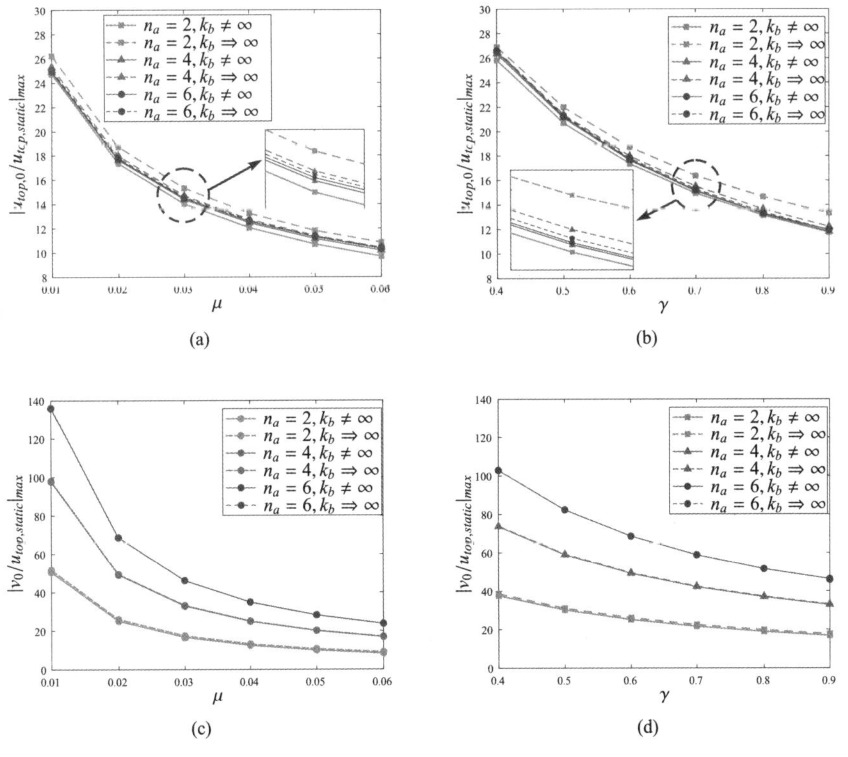

Fig. 12. Variations of the maximum dynamic amplifications for the flexible structure-TLCD system (in the external excitation frequency range from $0.5\omega_r$ to $1.8\omega_r$) with (a)(c) the mass ratio μ ($\gamma = 0.6$ and $C = 0.003$) and (b)(d) the horizontal length ratio γ ($\mu = 0.02$ and $C = 0.003$), for different values of n_a. (a)(b) Maximum dynamic amplifications of the structural top displacement, (c)(d) Maximum dynamic amplifications of the liquid response. The dashed lines ($k_b \Rightarrow \infty$) correspond to the results when the background flexibility is ignored; the solid lines ($k_b \neq \infty$) correspond to the results when the background flexibility is included.

is significantly attenuated, as well known for a vibration absorber. By accounting for the background flexibility, further reduction of the structural response can be achieved for the same n_a and μ, as revealed by the fact that the solid line is beneath the dashed line. It should also be noted that the improvement of the vibration suppression effect becomes less and less noticeable as μ increases. This is beneficial since in practice the value of μ should be limited. When the background flexibility is ignored ($k_b \Rightarrow \infty$) corresponding to the dashed lines, the best control performance is achieved when the TLCD is located at higher floor ($n_a = 6$). On the contrary, it is interesting to observe that the best control performance of the TLCD (with the same mass ratio) is achieved at $n_a = 2$ (although only slightly better than others) when the background flexibility is accounted for. This indicates the effectiveness of the proposed novel calibration procedure. As for $|u_{top,0}/u_{top,static}|_{max}$ versus γ (as shown in Fig. 12(b)), the conclusions are similar. The variations of the frequency response of liquid $|v_0/u_{top,static}|_{max}$ with μ and γ are shown in Fig. 12(c),12(d), respectively. Similarly, $|v_0/u_{top,static}|_{max}$ decreases with μ and γ. The smallest liquid response is always achieved for $n_a = 2$. It is also seen that the background flexibility has modest influence on the liquid response, and actually results in very slight reduced liquid response.

6. Concluding remarks

In this paper, a novel closed-form optimal calibration procedure for a TLCD accounting for the background flexibility contribution from the flexible structure has been proposed. The calibration is performed based on an analogue 2-DOF structure-TLCD model involving the modal mass and stiffness associated with the resonant mode, while the background flexibility from the non-resonant modes is represented in the model by an equivalent stiffness derived from modal analysis. Explicit expressions for the frequency ratio and optimal damping ratio are derived based on the pole-placement method. For obtaining the optimal frequency ratio, equal modal damping ratio criterion is employed, which also turns out to result in equal dynamic amplification at the two invariant points. It is shown from the derived formula that the optimal frequency ratio is dependent on the structure inherent characteristics (both the resonant and non-resonant modes, $\lambda = k_b/k_r$). Furthermore, a straightforward approach solution for determining the optimal head loss coefficient is proposed based on the bifurcation point in the root locus, without resorting to the conventional numerical search method that is computationally inefficient.

The developed optimal calibration procedure is evaluated by a 10-story shear frame model. It is numerically demonstrated that for all the TLCD installation locations, the optimally calibrated parameters accounting for the background flexibility lead to a fully balanced dynamic amplification curve of the structural displacement, while the curve becomes unbalanced when ignoring the background flexibility in the calibration. The effect of background contribution is more pronounced when the TLCD is located where other non-resonant modes make contribute considerably to the total displacement of structure, such as the lower floors of the shear frame. The parametric studies indicate that the background contribution leads to higher values of both the optimal frequency ratio and optimal head loss coefficient. Further, the optimal TLCD parameters varies noticeably with the changing mass ratio and horizontal length ratio. The inclusion of background contribution slightly changes the variation tendencies, especially for the optimal head loss coefficient when the TLCD is installed at lower floors. The maximal frequency responses of the structure and the liquid with optimal TLCD design also vary with the changing mass ratio and horizontal length ratio, and accounting for background flexibility leads to reduction of both responses comparing with the design ignoring background flexibility.

Declaration of Competing Interest

The authors declare that they have no known competing financial interests or personal relationships that could have appeared to influence the work reported in this paper.

CRediT authorship contribution statement

Bei Chen: Investigation, Software, Writing - original draft, Writing - review & editing. **Zili Zhang:** Conceptualization, Investigation, Software, Supervision, Writing - original draft, Writing - review & editing. **Xugang Hua:** Supervision, Writing - review & editing.

Acknowledgments

The authors gratefully acknowledge the financial support from the National Natural Science Foundation of China (No.51908209), the Hunan Provincial Natural Science Foundation of China (No.2020JJ5052), the supports of Aarhus University Research Foundation under the AUFF Assistant Professor Starting Grant (AUFF-E-2017-7-20) and the support from the Committee of Science and Technology of Shanghai, China (Grant No. 18160712800).

Appendix A. Nomenclature

$\mathbf{M}$	Mass matrix of the flexible structure
$\hat{\mathbf{M}}$	Mass matrix of the Structure-TLCD coupled system
$\mathbf{K}$	Stiffness matrix of the flexible structure
$\hat{\mathbf{K}}$	Stiffness matrix of the Structure-TLCD coupled system
$\mathbf{C}$	Damping matrix of the flexible structure
$\hat{\mathbf{C}}$	Damping matrix of the Structure-TLCD coupled system
$\mathbf{u}(t)$	Displacement vector of the flexible structure
$\hat{\mathbf{u}}(t)$	Displacement vector of the Structure-TLCD coupled system
$\mathbf{f}(t)$	Load vector of the flexible structure
$\hat{\mathbf{f}}(t)$	Load vector of the Structure-TLCD coupled system
$\mathbf{\Phi}_i$	ith eigenvector
ω_i	ith eigenfrequency
q_i	ith modal coordinate
L	Overall length of the liquid inside the TLCD
B	Horizontal length of the liquid inside the TLCD
A	Cross-sectional area of vertical and horizontal columns of the TLCD
ξ	Head loss coefficient due to the orifice
$\gamma = B/L$	Ratio of the horizontal length to overall length of the liquid column
ρ	Liquid density
m_w	Overall mass of the liquid
m_r	Modal mass of the structure
$\mu = m_w/m_r$	Ratio of the liquid mass to the modal mass of the structure
k_r	Modal dynamic stiffness of the structure
k_b	Modal background stiffness of the structure
k_w	Equivalent stiffness of the TLCD
$\lambda = k_b/k_r$	Ratio of the modal background stiffness to the modal dynamic stiffness
$\omega_r = \sqrt{k_r/m_r}$	Circular frequency of the resonant mode of structure
$\omega_w = \sqrt{k_w/m_w}$	Equivalent circular frequency of the TLCD
$\chi = \omega_w/\omega_r$	Frequency ratio of the TLCD to the structure
$\varsigma = \omega/\omega_r$	Non-dimensional frequency ratio

c_w	Equivalent damping coefficient of the TLCD
ζ_w	Equivalent damping ratio of the TLCD
$v(t)$	Displacement of the liquid in the vertical tube
v_0	Displacement amplitude of the liquid in the vertical tube
$u_r(t)$	Structure horizontal displacement
$u_{r,0}$	Structure horizontal displacement amplitude
$f(t) = f_0 e^{i\omega t}$	External harmonic force
$f_0 = Cm_r g$	Amplitudes of the external force
C	Excitation intensity factor
ω	Excitation circular frequency
g	Gravitational acceleration
$i = \sqrt{-1}$	Imaginary unit
t	Time
R	Radius of the circular inversion
$\varsigma_{0,1}, \varsigma_{0,2}, \varsigma_{0,3}$	Roots of the characteristic equation Eq. (22) for $\zeta_w = 0$
$\varsigma_{\infty,1}, \varsigma_{\infty,2}$	Roots of the characteristic equation Eq. (22) for $\zeta_w \Rightarrow \infty$
$\varsigma_1, \varsigma_2, \varsigma_3$	Roots of the characteristic equation Eq. (22)
$\varsigma_*, \varsigma_{*,3}$	Roots of the characteristic equation Eq. (22) for $\chi = \chi_{opt}$ and $\zeta_w = \zeta_*$

Appendix B. Conventional equalized dynamic amplification approach for the analogue 2-DOF system (with background flexibility included)

The normalized complex amplitude of $u_{r,0}$ (given in Eq. (19a)) can be recast in a simplified form, given as:

$$\frac{u_{r,0}}{f_0/k_r} = \frac{Y_1 + 2i\zeta_w Y_2}{Y_3 + 2i\zeta_w Y_4} \Rightarrow \left|\frac{u_{r,0}}{f_0/k_r}\right|^2 = \frac{Y_1^2 + (2\zeta_w)^2 Y_2^2}{Y_3^2 + (2\zeta_w)^2 Y_4^2} \tag{B.1}$$

By use of the amplification being independent of ζ_w at the two neutral frequencies, P_1 and P_2, as shown in Fig. 2(a), the following relations can be constructed:

$$\frac{Y_1^2}{Y_3^2} = \frac{Y_2^2}{Y_4^2} \Rightarrow \frac{Y_1}{Y_3} = \pm\frac{Y_2}{Y_4} \Rightarrow Y_1 Y_4 = \pm Y_3 Y_2 \tag{B.2}$$

The plus sign leads to the root $\varsigma = 0$, while the minus sign leads to the ninth-degree equation in P_1 and P_2:

$$\begin{aligned}&\left[-\frac{2\chi\mu^2(\gamma^2-1)}{\lambda^2}\right]\varsigma^9 + \left[-\frac{2\chi\mu(2\lambda+\mu+\lambda\mu+\chi^2\mu-\gamma^2\lambda-\gamma^2\mu-\gamma^2\lambda\mu)}{\lambda^2}\right]\varsigma^7\\&+\left[\frac{\chi(2\chi^2\lambda\mu^2+4\chi^2\lambda\mu+2\chi^2\mu^2-\gamma^2\lambda^2\mu-2\gamma^2\lambda\mu+2\lambda^2\mu+2\lambda^2+4\lambda\mu)}{\lambda^2}\right]\varsigma^5\\&+\left[-\frac{2\chi(\lambda+\chi^2\lambda+2\chi^2\mu+\chi^2\lambda\mu)}{\lambda}\right]\varsigma^3 + \left[2\chi^3\right]\varsigma = 0\end{aligned} \tag{B.3}$$

Appendix C. Optimal calibration approaches for the classic 2-DOF system (with background flexibility ignored)

C1. Conventional equalized dynamic amplification approach

The normalized amplitude of $u_{r,0}$ (given in Eq. (21a)) can be rewritten as:

$$\frac{u_{r,0}}{f_0/k_r} = \frac{Y_1 + 2i\zeta_w Y_2}{Y_3 + 2i\zeta_w Y_4} \Rightarrow \left|\frac{u_{r,0}}{f_0/k_r}\right|^2 = \frac{Y_1^2 + (2\zeta_w)^2 Y_2^2}{Y_3^2 + (2\zeta_w)^2 Y_4^2} \tag{C.1}$$

By use of the amplification being independent of ζ_w at those two neutral frequencies, P_1 and P_2, as shown in Fig. 2(b), the following relations can be obtained:

$$\frac{Y_1^2}{Y_3^2} = \frac{Y_2^2}{Y_4^2} \Rightarrow \frac{Y_1}{Y_3} = \pm\frac{Y_2}{Y_4} \Rightarrow Y_1 Y_4 = \pm Y_3 Y_2 \tag{C.2}$$

The plus sign leads to the root $\varsigma = 0$, while the minus sign leads to the quartic equation in P_1 and P_2:

$$(2 + 2\mu - \gamma^2\mu)\varsigma^4 - 2(1 + (1 + \mu)\chi)\varsigma^2 + 2\chi^2 = 0 \tag{C.3}$$

The roots of the equation above are represented by $\varsigma_{P_1}^2$ and $\varsigma_{P_2}^2$ corresponding to those two neutral frequencies, P_1 and P_2. The relation between these two roots are:

$$\varsigma_{P_1}^2 + \varsigma_{P_2}^2 = \frac{2 + 2(1 + \mu)\chi^2}{2 + 2\mu - \gamma^2\mu} \tag{C.4}$$

By inserting the roots obtained from Eq. (C.3) into Eq. (21a) and assuming $\zeta_w \Rightarrow \infty$, the normalized amplitude of the structural displacement at these two invariant points can be calculated:

$$\left|\frac{u_{r,0}k_r}{f_0}\right|_{\varsigma=\varsigma_{P_1},\varsigma_{P_2}} = \pm\frac{1}{1 - (1 + \mu)\varsigma^2} \tag{C.5}$$

Based on Eq. (C.5), the equal dynamic amplification condition at P_1 and P_2 will further leads to:

$$\varsigma_{P_1}^2 + \varsigma_{P_2}^2 = \frac{2}{1 + \mu} \tag{C.6}$$

By equalizing Eq. (C.4) with Eq. (C.6), the optimal frequency ratio can be obtained, given as:

$$\chi_{opt} = \frac{\sqrt{1 + \mu - \gamma^2\mu}}{1 + \mu} \tag{C.7}$$

When $\zeta_w \Rightarrow \infty$, the natural frequency ratio of the classic 2-DOF system is $\varsigma_\infty = 1/\sqrt{1 + \mu}$. By inserting ς_∞ and χ_{opt} into Eq. (21a), the normalized amplitude of $u_{r,0}$ at ς_∞ is:

$$\left|\frac{u_{r,0}k_r}{f_0}\right|_{\varsigma=\varsigma_\infty} = \frac{\sqrt{\gamma^4\mu^2 + (2\zeta_w)^2(1 + \mu)(1 + \mu - \gamma^2\mu)}}{\gamma^2\mu} \tag{C.8}$$

According to the optimal criterion proposed by Krenk [57], the optimal damping ratio can be obtained by equalizing Eq. (C.8) with Eq. (C.5), given as:

$$\zeta_{eq,opt} = \frac{1}{2}\frac{\gamma^2\mu(1 + \gamma\mu - \gamma^2\mu)}{(1 + \mu)(1 + \mu - \gamma^2\mu)} \tag{C.9}$$

C2. Pole-placement (root locus) calibration approach

For the classic 2-DOF system ignoring the background flexibility (i.e. $k_b \Rightarrow \infty$), the calibration procedure is similar to the classic SDOF-TMD system, which is fully elaborated in [57]. For brevity, detailed derivation is skipped in the present paper.

According to Krenk [57], by appropriate calibrating the frequency ratio χ, the two roots, ς_1 and ς_2, of Eq. (23) are inverse points with respect to a reference frequency, $\varsigma_\infty = 1/\sqrt{1 + \mu}$, obtained from the limit case $\zeta_w \Rightarrow \infty$. Hence, the following relation must be hold:

$$\varsigma_1\bar{\varsigma}_2 = \varsigma_2\bar{\varsigma}_1 = \varsigma_\infty^2 \tag{C.10}$$

For the bifurcation point ($\varsigma_1 = \varsigma_2 = \varsigma_*$), the polynomial equation can be formulated:

$$\begin{aligned}&(\varsigma - \varsigma_*)(\varsigma + \bar{\varsigma}_*)(\varsigma - \varsigma_\infty^2/\varsigma_*)(\varsigma + \varsigma_\infty^2/\bar{\varsigma}_*) = 0 \Rightarrow\\&\varsigma^4 - \left(2|\varsigma_*|^2 + 4\,\mathrm{Im}[\varsigma_*]^2\right)\varsigma^2 + |\varsigma_*|^4 - 4i\,\mathrm{Im}[\varsigma_*]\varsigma\left(\varsigma^2 + |\varsigma_*|^2\right) = 0\end{aligned} \tag{C.11}$$

where $\mathrm{Im}[\cdot]$ is the imaginary part operator. Comparing Eq. (23) and Eq. (C.11), the following results can be obtained:

$$|\varsigma_*| = \frac{1}{\sqrt{1 + \mu}} = \varsigma_\infty \tag{C.12a}$$

$$\chi_{opt} = \frac{\sqrt{1 + \mu - \gamma^2\mu}}{1 + \mu} \tag{C.12b}$$

$$\zeta_* = \sqrt{\frac{\gamma^2 \mu}{1+\mu}} \tag{C.12c}$$

where the optimal frequency ratio obtained from the pole-placement calibration method, Eq. (C.12b), is consistent with the result obtained from the conventional method, as shown in C.7, and the result from the cited reference [48].

Appendix D. Flexible structure-TLCD coupled system

The general EOM for the flexible structure-TLCD coupled system is given as:

$$\hat{\mathbf{M}}\ddot{\hat{\mathbf{u}}}(t) + \hat{\mathbf{C}}\dot{\hat{\mathbf{u}}}(t) + \hat{\mathbf{K}}\hat{\mathbf{u}}(t) = \hat{\mathbf{f}}(t) \tag{D.1}$$

where $\hat{\mathbf{M}}$, $\hat{\mathbf{C}}$ and $\hat{\mathbf{K}}$ are the augmented mass, damping and stiffness matrixes respectively with a dimension of $(n+1)\times(n+1)$; $\hat{\mathbf{u}}(t)$ and $\hat{\mathbf{f}}(t)$ are the augmented displacement vector and load vector, respectively. Those matrixes and vectors are defined as:

$$\hat{\mathbf{M}} = \begin{bmatrix} \mathbf{M}_{\mathbf{n},\mathbf{n}} & \mathbf{0}_{n,1} \\ \mathbf{0}_{1,n} & 0 \end{bmatrix} + \begin{array}{c} \begin{matrix} 1 & \cdots & n_a & \cdots & \cdots & n+1 \end{matrix} \\ \begin{bmatrix} 0 & \cdots & 0 & \cdots & \cdots & 0 \\ \vdots & \ddots & & & & \vdots \\ 0 & & m_w & \cdots & \cdots & m_w\gamma \\ \vdots & & \vdots & \ddots & & \vdots \\ \vdots & & \vdots & & \ddots & \vdots \\ 0 & \cdots & m_w\gamma & \cdots & \cdots & m_w \end{bmatrix} \end{array} \begin{matrix} 1 \\ \vdots \\ n_a \\ \vdots \\ \vdots \\ n+1 \end{matrix} \tag{D.2}$$

$$\hat{\mathbf{C}} = \begin{bmatrix} \mathbf{C}_{\mathbf{n},\mathbf{n}} & \mathbf{0}_{n,1} \\ \mathbf{0}_{1,n} & 0 \end{bmatrix} + \begin{array}{c} \begin{matrix} 1 & \cdots & & \cdots & & n+1 \end{matrix} \\ \begin{bmatrix} 0 & \cdots & 0 & \cdots & \cdots & 0 \\ \vdots & \ddots & & & & \vdots \\ 0 & & \ddots & & & 0 \\ \vdots & & & \ddots & & \vdots \\ \vdots & & & & \ddots & \vdots \\ 0 & \cdots & 0 & \cdots & \cdots & C_w \end{bmatrix} \end{array} \begin{matrix} 1 \\ \vdots \\ \\ \vdots \\ \\ n+1 \end{matrix} \tag{D.3}$$

$$\hat{\mathbf{K}} = \begin{bmatrix} \mathbf{K}_{\mathbf{n},\mathbf{n}} & \mathbf{0}_{n,1} \\ \mathbf{0}_{1,n} & 0 \end{bmatrix} + \begin{array}{c} \begin{matrix} 1 & \cdots & & \cdots & & n+1 \end{matrix} \\ \begin{bmatrix} 0 & \cdots & 0 & \cdots & \cdots & 0 \\ \vdots & \ddots & & & & \vdots \\ 0 & & \ddots & & & 0 \\ \vdots & & & \ddots & & \vdots \\ \vdots & & & & \ddots & \vdots \\ 0 & \cdots & 0 & \cdots & \cdots & k_w \end{bmatrix} \end{array} \begin{matrix} 1 \\ \vdots \\ \\ \vdots \\ \\ n+1 \end{matrix} \tag{D.4}$$

$$\hat{\mathbf{u}} = \begin{bmatrix} \mathbf{u(t)} \\ v(t) \end{bmatrix}, \quad \mathbf{u}(t) = \begin{bmatrix} u_1(t) \\ \vdots \\ u_n(t) \end{bmatrix}, \quad \hat{\mathbf{f}}(t) = \begin{bmatrix} \mathbf{f(t)} \\ 0 \end{bmatrix} \tag{D.5}$$

where $\mathbf{M}_{n,n}$ $\mathbf{C}_{n,n}$ and $\mathbf{K}_{n,n}$ are the structural mass matrix, damping matrix and stiffness matrix, respectively; $\mathbf{u}(t)$ and $\mathbf{f}(t)$ represent the displacement vector and load vector of the structure, respectively, see Eq. (2) for detailed definition. n_a is the floor location where the TLCD is attached to, see Fig. 1(a). C_w is the equivalent damping coefficient, cf.Eq. (37); $k_w = \mu\chi^2 k_r$ is the equivalent stiffness of the TLCD.

References

[1] Huertas VV, Rohal-Ilkiv B. Vibration suppression of a flexible structure. Procedia Engineering 2012;48:233–41.
[2] Krenk S, Høgsberg J. Tuned mass absorber on a flexible structure. Journal of Sound and Vibration 2014;333(6):1577–95.
[3] Krenk S, Høgsberg J. Optimal resonant control of flexible structures. Journal of Sound and Vibration 2009;323(3-5):530–54.
[4] Ghosh A, Basu B. A closed-form optimal tuning criterion for TMD in damped structures. Structural Control and Health Monitoring: The Official Journal of the International Association for Structural Control and Monitoring and of the European Association for the Control of Structures 2007;14(4):681–92.
[5] Pietrosanti D, De Angelis M, Basili M. A generalized 2-DOF model for optimal design of MDOF structures controlled by Tuned Mass Damper Inerter (TMDI). International Journal of Mechanical Sciences 2020;185:105849.
[6] Gerges RR, Vickery BJ. Optimum design of pendulum-type tuned mass dampers. The Structural Design of Tall and Special Buildings 2005;14(4):353–68.
[7] Oliveira FdS, Gomez A, Avila SM, Brito J. Design criteria for a pendulum absorber to control high building vibrations. International Journal of Innovations in Materials Science and Engineering (IMSE) 2014;1(2):82–9.
[8] Kecik K, Mitura A. Energy recovery from a pendulum tuned mass damper with two independent harvesting sources. International Journal of Mechanical Sciences 2020;174:105568.
[9] Anh N, Matsuhisa H, Viet L, Yasuda M. Vibration control of an inverted pendulum type structure by passive mass–spring-pendulum dynamic vibration absorber. Journal of Sound and Vibration 2007;307(1-2):187–201.
[10] Fujino Y, Sun L, Pacheco BM, Chaiseri P. Tuned liquid damper (TLD) for suppressing horizontal motion of structures. Journal of Engineering Mechanics 1992;118(10):2017–30.
[11] Banerji P, Murudi M, Shah AH, Popplewell N. Tuned liquid dampers for controlling earthquake response of structures. Earthquake Engineering & Structural Dynamics 2000;29(5):587–602.
[12] Zhao Z, Zhang R, Jiang Y, Pan C. A tuned liquid inerter system for vibration control. International Journal of Mechanical Sciences 2019;164:105171.
[13] Sun L, Fujino Y, Pacheco B, Chaiseri P. Modelling of tuned liquid damper (TLD). Journal of Wind Engineering and Industrial Aerodynamics 1992;43(1-3):1883–94.
[14] Kim H, Adeli H. Wind-induced motion control of 76-story benchmark building using the hybrid damper-TLCD system. Journal of Structural Engineering 2005;131(12):1794–802.
[15] Di Matteo A, Pirrotta A, Tumminelli S. Combining TMD and TLCD: analytical and experimental studies. Journal of Wind Engineering and Industrial Aerodynamics 2017;167:101–13.
[16] Diana G, Resta F, Sabato D, Tomasini G. Development of a methodology for damping of tall buildings motion using tlcd devices. Wind & structures 2013;17(6):629–46.
[17] Chen J, Liu Y, Bai X. Shaking table test and numerical analysis of offshore wind turbine tower systems controlled by TLCD. Earthquake Engineering and Engineering Vibration 2015;14(1):55–75.
[18] Sarkar S, Chakraborty A. Development of semi-active vibration control strategy for horizontal axis wind turbine tower using multiple magneto-rheological tuned liquid column dampers. Journal of Sound and Vibration 2019;457:15–36.
[19] Mensah AF, Dueñas-Osorio L. Improved reliability of wind turbine towers with tuned liquid column dampers (TLCDs). Structural Safety 2014;47:78–86.
[20] Colwell S, Basu B. Tuned liquid column dampers in offshore wind turbines for structural control. Engineering Structures 2009;31(2):358–68.
[21] Zhang Z, Basu B, Nielsen SR. Tuned liquid column dampers for mitigation of edgewise vibrations in rotating wind turbine blades. Structural Control and Health Monitoring 2015;22(3):500–17.
[22] Sarkar S, Chakraborty A. Optimal design of semiactive MR-TLCD for along-wind vibration control of horizontal axis wind turbine tower. Structural Control and Health Monitoring 2018;25(2):e2083.
[23] Zhang Z, Høeg C. Dynamics and control of spar-type floating offshore wind turbines with tuned liquid column dampers. Structural Control and Health Monitoring 2020;27(6):e2532.
[24] Sakai F. Tuned liquid column damper-new type device for suppression of building vibration. In: Proceedings of 1st International Conference on High-rise Buildings; 1989. p. 926–31.
[25] Colwell S, Basu B. Experimental and theoretical investigations of equivalent viscous damping of structures with TLCD for different fluids. Journal of Structural Engineering 2008;134(1):154–63.
[26] Gao H, Kwok K, Samali B. Optimization of tuned liquid column dampers. Engineering Structures 1997;19(6):476–86.
[27] Balendra T, Wang C, Cheong H. Effectiveness of tuned liquid column dampers for vibration control of towers. Engineering Structures 1995;17(9):668–75.
[28] Di Matteo A, Iacono FL, Navarra G, Pirrotta A. Direct evaluation of the equivalent linear damping for TLCD systems in random vibration for pre-design purposes. International Journal of Non-Linear Mechanics 2014;63:19–30.
[29] Hitchcock P, Kwok K, Watkins R, Samali B. Characteristics of liquid column vibration absorbers (LCVA)-II. Engineering Structures 1997;19(2):135–44.
[30] Hitchcock P, Kwok K, Watkins R, Samali B. Characteristics of liquid column vibration absorbers (LCVA)-I. Engineering Structures 1997;19(2):126–34.
[31] Di Matteo A, Iacono FL, Navarra G, Pirrotta A. Experimental validation of a direct pre-design formula for TLCD. Engineering Structures 2014;75:528–38.
[32] Bhattacharyya S, Ghosh AD, Basu B. Nonlinear modeling and validation of air spring effects in a sealed tuned liquid column damper for structural control. Journal of sound and vibration 2017;410:269–86.
[33] Mousavi SA, Bargi K, Zahrai SM. Optimum parameters of tuned liquid column–gas damper for mitigation of seismic-induced vibrations of offshore jacket platforms. Structural Control and Health Monitoring 2013;20(3):422–44.
[34] Shum K, Xu YL. Multiple-tuned liquid column dampers for torsional vibration control of structures: experimental investigation. Earthquake engineering & structural dynamics 2002;31(4):977–91.
[35] Di Matteo A, Lo Iacono F, Navarra G, Pirrotta A. The TLCD passive control: Numerical investigations vs. experimental results. In: ASME International Mechanical Engineering Congress and Exposition, 45202. American Society of Mechanical Engineers; 2012. p. 1283–90.
[36] Liang S, Li Q, Qu W. Control of 3-D coupled responses of wind-excited tall buildings by a spatially placed TLCD system. Wind & structures 2000;3(3):193–207.

[37] Wu J, Shih M, Lin Y, Shen Y. Design guidelines for tuned liquid column damper for structures responding to wind. Engineering Structures 2005;27(13):1893–905.
[38] Debbarma R, Chakraborty S, Ghosh SK. Optimum design of tuned liquid column dampers under stochastic earthquake load considering uncertain bounded system parameters. International Journal of Mechanical Sciences 2010;52(10):1385–93.
[39] Ghosh A, Basu B. Seismic vibration control of short period structures using the liquid column damper. Engineering Structures 2004;26(13):1905–13.
[40] Den Hartog JP. Mechanical vibrations. 4. New York: McGraw-Hill; 1956.
[41] Al-Saif K, Aldakkan K, Foda M. Modified liquid column damper for vibration control of structures. International Journal of Mechanical Sciences 2011;53(7):505–12.
[42] Gur S, Roy K, Mishra SK. Tuned liquid column ball damper for seismic vibration control. Structural Control and Health Monitoring 2015;22(11):1325–42.
[43] Wang Q, Tiwari ND, Qiao H, Wang Q. Inerter-based tuned liquid column damper for seismic vibration control of a single-degree-of-freedom structure. International Journal of Mechanical Sciences 2020:105840.
[44] Gao H, Kwok K, Samali B. Characteristics of multiple tuned liquid column dampers in suppressing structural vibration. Engineering structures 1999;21(4):316–31.
[45] Yalla SK, Kareem A. Optimum absorber parameters for tuned liquid column dampers. Journal of Structural Engineering 2000;126(8):906–15.
[46] Ahadi P, Mohebbi M, Shakeri K. Using optimal multiple tuned liquid column dampers for mitigating the seismic response of structures. International Scholarly Research Notices 2012;2012.
[47] Shum K, Xu Y. Multiple tuned liquid column dampers for reducing coupled lateral and torsional vibration of structures. Engineering Structures 2004;26(6):745–58.
[48] Shum K. Closed form optimal solution of a tuned liquid column damper for suppressing harmonic vibration of structures. Engineering Structures 2009;31(1):84–92.
[49] Brock JE. A note on the damped vibration absorber. Trans ASME, J Appl Mech 1946;13(4) A284.
[50] Ghosh A, Basu B. Alternative approach to optimal tuning parameter of liquid column damper for seismic applications. Journal of Structural Engineering 2007;133(12):1848–52.
[51] Yalla SK, Kareem A, Kantor JC. Semi-active tuned liquid column dampers for vibration control of structures. Engineering Structures 2001;23(11):1469–79.
[52] Sonmez E, Nagarajaiah S, Sun C, Basu B. A study on semi-active tuned liquid column dampers (sTLCDs) for structural response reduction under random excitations. Journal of Sound and Vibration 2016;362:1–15.
[53] Altay O, Klinkel S. A semi-active tuned liquid column damper for lateral vibration control of high-rise structures: Theory and experimental verification. Structural Control and Health Monitoring 2018;25(12):e2270.
[54] Ni Y, Ying Z, Wang J, Ko J, Spencer Jr B. Stochastic optimal control of wind-excited tall buildings using semi-active MR-TLCDs. Probabilistic Engineering Mechanics 2004;19(3):269–77.
[55] Balendra T, Wang C, Rakesh G. Vibration control of various types of buildings using TLCD. Journal of Wind Engineering and Industrial Aerodynamics 1999;83(1):197–208.
[56] Zhu F, Wang J, Jin F, Lu L. Real-time hybrid simulation of full-scale tuned liquid column dampers to control multi-order modal responses of structures. Engineering Structures 2017;138:74–90.
[57] Krenk S. Frequency analysis of the tuned mass damper. Journal of Applied Mechanics 2005;72(6):936–42.
[58] Nielsen SRK. Vibration Theory, Vol. 1A: linear vibration theory. Aalborg University Press; 1998.
[59] Pars L. A Treatise on Analytical Dynamics (Woodbridge). Ox Bow Press; 1979.
[60] Wu J. Experimental calibration and head loss prediction of tuned liquid column damper. Tamkang Journal of Science and Engineering 2005;8(4):319–25.
[61] Høgsberg J. Vibration control by piezoelectric proof-mass absorber with resistive-inductive shunt. Mechanics of Advanced Materials and Structures 2019:1–13.
[62] Zhang Z, Larsen TG. Optimal calibration of the rotational inertia double tuned mass damper (RIDTMD) for rotating wind turbine blades. Journal of Sound and Vibration 2020:115827.

Engineering Structures 318 (2024) 118674

Contents lists available at ScienceDirect

Engineering Structures

journal homepage: www.elsevier.com/locate/engstruct

Mechanical behavior and seismic control performance of a metallic torsional damper for flexible structures

Sheng Chen [a,b], Wenxi Wang [a,b,c,*], Guokun Liu [d], Chao Zhou [a,b], Bei Chen [a,b,c], Zhengqing Chen [a,b,c]

[a] State Key Laboratory of Bridge Safety and Resilience, Hunan University, Changsha 410082, China
[b] Key Laboratory for Bridge and Wind Engineering of Hunan Province, College of Civil Engineering, Hunan University, Changsha 410082, China
[c] Shock and Vibration Technology Research Center, College of Civil Engineering, Hunan University, Changsha 410082, China
[d] School of Architectural Engineering, Hunan Institute of Engineering, Xiangtan, Hunan, China 411100

ARTICLE INFO

Keywords:
Vibration control
Torsional plastic deformation
Torsional metallic dampers
Gear and rack device
Mechanical behavior
Seismic control performance

ABSTRACT

The use of metallic dampers is an effective way to reduce seismic responses and dissipate excessive vibration energy for civil structures. Conventional metallic dampers employed energy dissipation by plastic deformation under bending or shear modes are hard to design, and may cause uniform distribution of stress in the metallic components. In this paper, employing the high plastic deformation capacity of aluminum materials, a metallic torsional damper using a gear and rack device (MTDGRD) is proposed to enhance the control effectiveness for the structural response reduction under earthquakes. The static mechanical behavior and energy dissipation of aluminum metallic components under torsional deformation are experimentally studied. Moreover, the configuration and working principle of the proposed damper are introduced. The dynamic behavior of the proposed damper is experimentally studied. For precise modeling in simulations, the elastic-plastic theory of torsional deformation is used to derive the mechanical model of the proposed damper and the model accuracy is validated by experimental data. Finally, numerical simulations under different seismic ground motions controlled by a proposed MTDGRD are compared to that controlled by a conventional tuned mass damper (TMD) for a 10-floor flexible structure. The results show that aluminum 6061# of the energy dissipation capacity is higher than aluminum 1060#. The dynamical tests show that the loading frequency and operational cycles have very limited effect on mechanical behavior and energy consumption capacity of the MTDGRD. The proposed dynamic mechanical model of the MTDGRD has considerable accuracy in numerical simulations and seismic control simulations show the better performance of the proposed damper in reducing structural displacement responses than the conventional TMD with a mass ratio of 0.5 %.

1. Introduction

Flexible structures with low inherent damping are susceptible to significant vibrations under earthquakes, which can cause serious damage to the structure. Over the past few decades, there has been extensive development in structural vibration control systems aimed at mitigating vibrations resulted from by seismic events. Existing vibration control systems can be classified into passive control [1–7], semi-active control [8–14] and active control systems [15–17]. Among them, the passive control system owing to advantages of simple configuration and convenient implementation, has been widely used in practical applications. Due to their stable damping performance, superior energy dissipation capacity, low cost and easy replacement, metallic dampers, as one of the most promising passive control devices, are widely used in structural vibration control.

According to the yielding deformation mechanism, metallic dampers can be divided into axial-yielding type, flexural-yielding type, shear-yielding type and torsion-yielding type dampers. The axial-yielding type dampers realize energy dissipation through the plastic tension-compression deformation of metallic dissipation components. In practices, the most used axial-yielding type dampers are the buckling restrained brace (BRB) [18]. However, the inner core will produce multi-wave buckling when the BRB undergoes axial deformation. To reduce the occurrence of multi-wave buckling [19], inner cores with

* Correspondence to: China College of Civil Engineering, Hunan University, Changsha 410082, China.
E-mail address: wxwang@hnu.edu.cn (W. Wang).

https://doi.org/10.1016/j.engstruct.2024.118674
Received 2 February 2024; Received in revised form 11 July 2024; Accepted 23 July 2024
Available online 1 August 2024

different cross-sections have been further developed, such as rectangular cores [20], cross inner cores [21], H-shaped inner cores [22,23] and circular inner cores [24,25]. The energy consumption of the flexural-yielding type dampers is caused by the out-of-plane bending plastic deformation of the metal plate. The most common flexural-yielding type dampers are the added damping and stiffness device [26]. In order to make the bending plate buckle uniformly, the bending plate can be designed into a specific shape to improve the efficiency of energy consumption of the bending plate [27–32]. Similarly, shear-yielding type dampers consume energy through metal shear buckling. The most used shear-yielding type dampers are shear plate dampers (SPDs) [33,34]. Although SPD has strong energy dissipation characteristics, out-of-plane buckling can easily occur during the loading process. In order to overcome the buckling problem of traditional SPDs, reinforced SPDs [35,36], square tube SPDs [37,38] and corrugated SPDs [39–42] have been proposed. Meanwhile, another common shear-yield type dampers, named steel slit dampers (SSDs), have also been widely studied [43–48]. The interdependence between the strength and stiffness of SSDs can be separated due to the slit configuration, which provides enough flexibility in practical design [49].

Torsional metallic dampers dissipate energy through torsional plastic deformation. Due to the uniform torque distribution on any cross-section of a round rod in pure torsion, the torsional metallic dampers exhibit a stronger energy dissipation capacity and higher material utilization rate. Consequently, the torsional metallic dampers have increasing attention from researchers. Kelly et al. [50] first proposed the a use of torsion-yielding mechanism to dissipate energy and proposed a torsional beam damper. This kind of damper causes torsion and bending deformation of the rectangular steel plate, mainly torsion. Subsequently, Franco et al. [51] proposed a torsional steel pipe damper for seismic resistance of building structures, and conducted four full-scale device tests to study the energy dissipation capacity of the torsional steel pipe damper. Milani and Dicleli [52,53] proposed a multi-directional torsional hysteretic damper, which can convert horizontal deformations in any direction into torsional deformations at solid cylindrical steel members. Mahyari et al. [54] proposed a pure torsional buckling damper, which uses a special design to transfer inter-story shear force to the damper, causing the damping circular tube to undergo pure torsion without generating shear force and bending moment. Lie et al. [55] proposed a circular steel tube torsional yield metallic damper and conducted tests under cyclic loading to study the cyclic performance of the damper. Wang et al. [56] proposed a new type of torsional metallic damper, conducted quasi-static low-cycle reciprocating loading tests on the damper, and established a mechanical model and finite element model of the damper. The results showed that compared with the damper under combined bending-shear-torsion deformation, the damper under pure torsional deformation has higher load-bearing capacity, higher energy dissipation capacity and better ductility.

Although the torsion-yielding type dampers have a strong energy dissipation capacity, it is difficult to fully employ its mitigating potential due to the vibration amplitude in civil engineering is relatively small. In order to amplify the inter-story displacement of a structure, Chen et al. [57] use a ball screw device to propose a ball screw type displacement amplified metallic torsional damper. At the same time, some dampers using ball screws have also been developed, such as viscous dampers [58,59], electromagnetic dampers [60–62], eddy current dampers [63, 64], friction dampers [65], etc. Although the ball screw mechanism has high motion amplification efficiency and stable transmission performance. But the difficulty and cost of manufacturing the ball screw mechanism increases significantly if the maximum available force of the damper increases [66]. The gear and rack device is a mature industrial product with low cost and strong reliability. It can be used to replace ball screw mechanisms to achieve similar motion amplification. Using gear and rack devices, some novel amplifying dampers have been developed, such as gear-driven rotation amplified rubber viscoelastic dampers [67], displacement amplified fluid viscous dampers [68], gyroscopic mass dampers [69,70] and eddy current dampers [66,71].

Although many scholars have devoted themselves to the research of metallic torsional dampers, there are relatively few studies on the mechanical behavior and energy dissipation capacity of metallic torsional dampers. Moreover, failure modes of metallic torsional dampers during the torsional loading process have not yet been clarified. Additionally, the vibration amplitude of structure is relatively small in practices, making it difficult to fully employ its high plastic deformation capacity. For the purpose of amplifying the structural responses, a ball screw type metallic torsional damper was developed recently [57]. Although the control effectiveness of the damper is validated, it is difficult to apply them on a large-scale flexible structure considering the high manufacturing cost of ball screw devices. Therefore, there is an urgent need to develop a low-cost and highly reliable metallic torsional damper with displacement amplification devices. Meanwhile, very few existing studies have compared the performance of metallic dampers to other most frequently-used tuned dampers, such as the conventional TMD, which may cause difficulty in selecting damper types in the design stage of practical control systems.

This paper introduces a metallic torsional damper using a gear and rack device (MTDGRD) to improve control effectiveness in reducing structural responses during earthquakes. The gear and rack device is employed to transform the transitional displacement into angular displacement, and simultaneously amplifies the angular displacement. In this paper, the energy dissipation capacity of aluminum 6061 and aluminum 1060 are compared. The influence of the inner and outer diameter ratio β and the length and outer diameter ratio γ of the aluminum tubes on maximum energy consumption per unit mass of aluminum tubes during the torsion process is studied. Experimental investigations are conducted to analyze the dynamic behavior of the proposed damper, validating the elastic-plastic theory of torsional deformation used to simulate its performance. Finally, numerical simulations are carried out for a 10-floor flexible structure under various earthquakes. Finally, a control performance comparison is made between the proposed damper and a traditional TMD under seismic loads to solve the selection issue of damper systems in the design stage.

2. Damper configuration and working principle

As shown in the Fig. 1, the MTDGRD can be design to install between two adjacent floors. The bottom plate of the proposed damper is connected to the lower floor, and the rack is connected to the upper floor. The rack is connected to the bottom plate through sliding bearings. The function of the sliding bearing is to ensure that the rack can only move horizontally. The rack meshes with the gear, and the gear is fixed with the transmission shaft. The transmission shaft is connected through two balling bearings. The function of the balling bearings is to ensure that the gear and transmission shaft can only rotate. One end of the aluminum rod is connected to the transmission shaft, and the other end

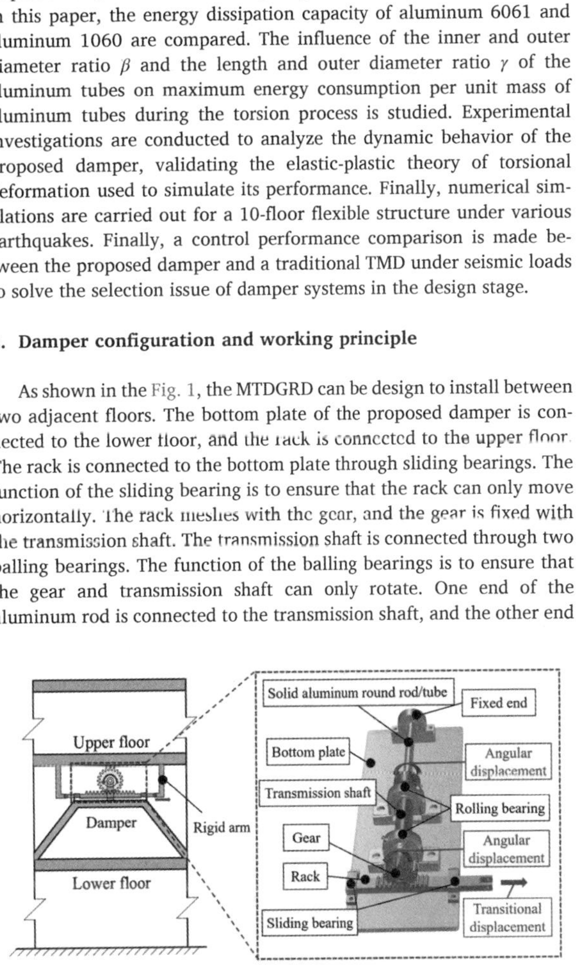

Fig. 1. Configuration of the MTDGRD.

of the aluminum rod is connected to the fixed end. The sliding bearing, ball bearing and fixed end are fixed with the bottom plate through bolts. When the structure is under dynamic load, the rack moves horizontally with the upper floor, thereby driving the rotation of the gear. The connecting shaft drives one end of the aluminum rod to rotate with the rotation of the gear, and the other end of the aluminum rod is fixed with the fixed end. The relative displacement of the adjacent floors can generate the plastic torsional deformation of the damping rod to efficiently dissipate the vibration energy. Moreover, the angular displacement amplification function can be realized by selecting the gear radius.

3. Torsional static mechanical behavior of aluminum 6061# materials

Metallic torsional dampers mainly achieve energy dissipation and seismic control through torsional plastic deformation of metal energy-dissipating materials. Aluminum as a metal energy-dissipating material has gained extensive attention in manufacturing metallic dampers due to its lower yield strain and good ductility, which can provide large energy dissipation during the yielding stage. Aluminum 6061# materials have the characteristics of high strength, light weight, and easy manufacturing. Therefore, the static behavior of aluminum 6061# materials is studied through the mechanical testing machine in this section.

3.1. Performance testing of aluminum 6061#

The experimental specimens of solid metal round rods are manufactured by aluminum 6061 materials with a diameter of 10 mm and length of 100 mm, as shown in Fig. 2. As shown in Fig. 3(a), the tension test is carried out on a electronic universal testing machine (type: WDW-100 C, manufactured by the Wuhan Gelaimo Testing Equipment Co. LTD). The test through a extensometer (type: YYU10/50-SH, manufactured by the NCS Testing Technology Co. LTD) to determine the tension displacement of the tension specimen. As shown in Fig. 3(b), a electronic torsion testing machine (type: DNJ-1000, manufactured by the Changsha Yaxing Numerical Control Technology Co. LTD) is used in the torsion test.

Axial force-displacement and torque-torsional angle curves of solid aluminum 6061# are obtained from tension and torsion experiments, as shown in Fig. 4. For solid aluminum 1060#, these curves are sourced from reference [57], as shown in Fig. 5. To compare the energy dissipation capacities of tension and torsion loading modes, the equivalent stress and strain at the specimen's outer edge is evaluated. Considering the necking phenomenon, the equivalent axial stress σ_{eq} and strain ε_{eq} of the solid round rod during the tension test can be described as [72],

$$\sigma_{eq} = \frac{4P(l_0 + \Delta l)}{\pi d^2 l_0} \tag{1}$$

$$\varepsilon_{eq} = \ln\left(1 + \frac{\Delta l}{l_0}\right) \tag{2}$$

where P is the axial tension force, l_0 is the rod's original length, Δl is the stretch length of the rod, and d is the rod's initial diameter.

The equivalent shear stress τ_{eq} and strain γ_{eq} of the outer edge on the rod during the torsion test can be expressed as [73],

$$\tau_{eq} = \frac{16T}{\pi d^3} \tag{3}$$

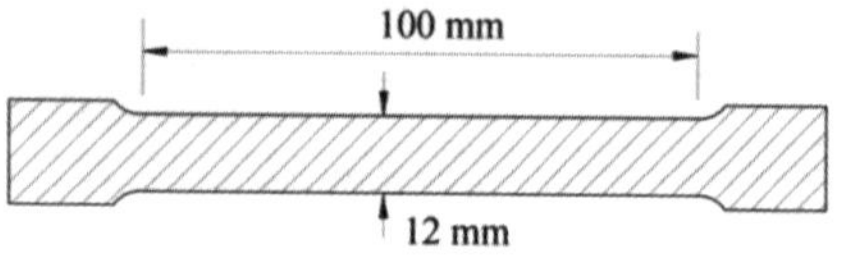

Fig. 2. Dimensions of experimental sample.

$$\gamma_{eq} = \frac{\varphi d}{2L} \tag{4}$$

where T is the loading torque, φ is the torsion angle, d is the diameter of the rod.

According to Eqs. (1), (2), (3) and (4) and the experimental data from Figs. 3 and 4, the equivalent stress-strain relations of solid aluminum 1060# and 6061# round rods can be obtained, as shown in Fig. 6. Table 1 summarize the mechanical property parameters of solid aluminum round rod. The total energy consumption in tension and torsion experiments are determined by the area enclosed by the axial force-displacement and torque-torsional angle curve with the coordinate axis from Figs. 4 and 5. It can be seen from Fig. 6 and Table 1 that the ultimate shear strain of aluminum 6061# and aluminum 1060# are much larger than the ultimate axial strain. This means the capacity of plastic deformation from torsion is higher than that from tension for aluminum 6061# and aluminum 1060# materials. The total energy consumption of aluminum 6061# during the torsion process is slightly greater than that of aluminum 1060#, which means that the energy dissipation capacity of aluminum 6061# from torsional plastic deformation is higher than that of aluminum 1060#. Therefore, aluminum 6061# is adopted in the proposed damper to manufacture a metallic dissipation component in this paper. The total energy consumption in tension experiments is larger than that in tension experiments. This is because the necking phenomenon occurred during the tension test, decreasing cross-section of the round rod at the necking point [57].

3.2. Effect of different sizes of metallic component on torsional plastic energy dissipation

The plastic energy dissipation capacity of the proposed MTDGRD can be influenced by various sizes of the metallic dissipation component. Therefore, a lot of specimens under various dimensions is used to study the static behavior of metallic component. The dimensions of the aluminum 6061# metallic component are shown in Fig. 7. D_{out} is the outer diameter, D_{in} is the inner diameter, and L is the length of the aluminum metallic component. The specific dimensions of the specimens are shown in Table 2. and the inner and outer diameter ratio $\beta = D_{in}/D_{out}$ is defined. The dimension of the aluminum 6061# metallic component is worth notable: it can be regarded as a solid rod when the inner diameter is zero and as a tube when the inner diameter is non-zero.

The torque-torsion angle relationship curves of aluminum metallic component of different inner and outer diameter ratio β are shown in Fig. 8. Table 2 summarize the energy dissipation parameters of aluminum metallic component of different β, and the maximum energy consumption per unit mass is determined by the ratio of the area of the torque-torsion angle curves and coordinate axis to mass of aluminum metallic component. The results indicate that both the maximum torque and torsional angle of the specimen decrease as β increases. Similarly, the maximum energy consumption per unit mass of the specimen also decreases with increasing β. Compared with the aluminum tubes, the maximum torque and torsion angle of the solid aluminum rod are the largest. Fig. 9 shows the typical failure patterns of four kinds of experimental specimens. Moreover, torsional buckling is observed in round tubes, and this phenomenon gradually decreases after the wall thickness increases. The buckling phenomenon that occurs during the torsion process completely disappears under the case of solid aluminum rod ($\beta = 0$). In this situation, the each cross-section of the aluminum rod has uniform deformation, and the final failure form of the aluminum rod is shearing failure. Therefore, solid aluminum rods are a better form than aluminum tube for manufacturing metallic dissipation component.

Several solid round rods are chosen to study the effects of different lengths on the plastic energy dissipation of aluminum rods. The specific dimensions of the specimens are shown in Table 3, and the length and outer diameter ratio $\gamma = L/D_{out}$ is defined.

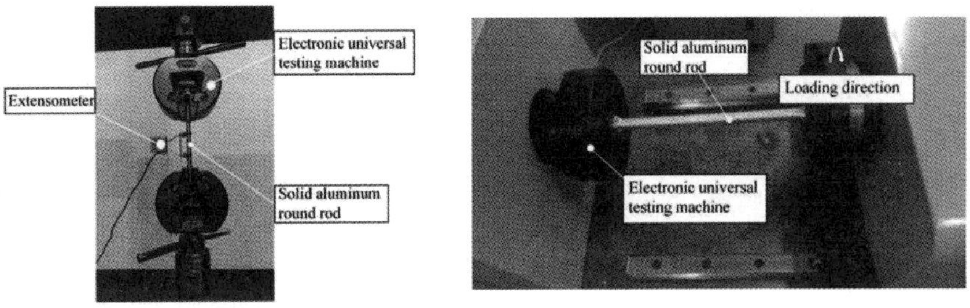

(a) Experimental setup of tension test (b) Experimental setup of torsion test

Fig. 3. Energy dissipation test of solid metal round rod.

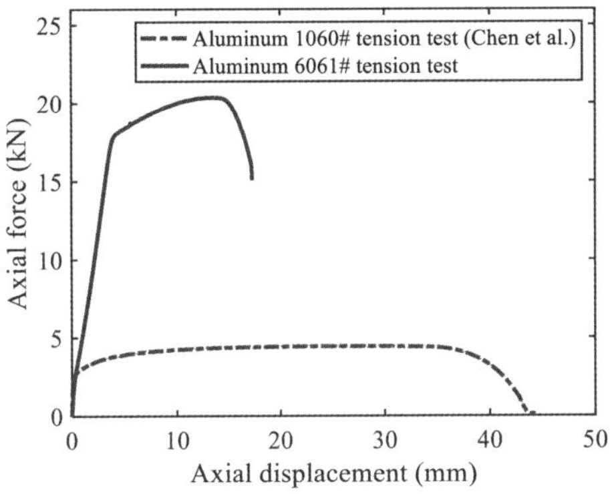

Fig. 4. Axial force and displacement curve.

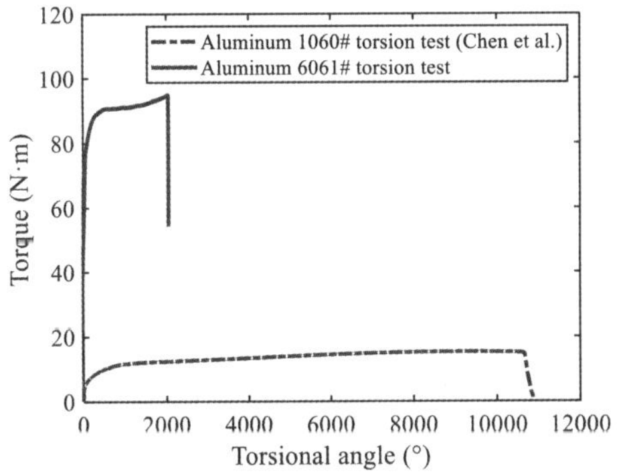

Fig. 5. Torque and torsional angle curve.

The torque-torsion angle relationship curves of aluminum rods of different lengths are shown in Fig. 10. Table 3 summarize the Energy dissipation parameters of aluminum metallic component of different γ. The results show that the maximum torque and the maximum energy consumption per unit mass of the specimen are not affected by γ. Fig. 11 provides an overview of the typical failure patterns observed in five different experimental specimens. In all cases, specimen failure occurred due to shear, without experiencing buckling. At this time, the entire cross-section of the aluminum rod consumes energy uniformly and eventually fractures in shear.

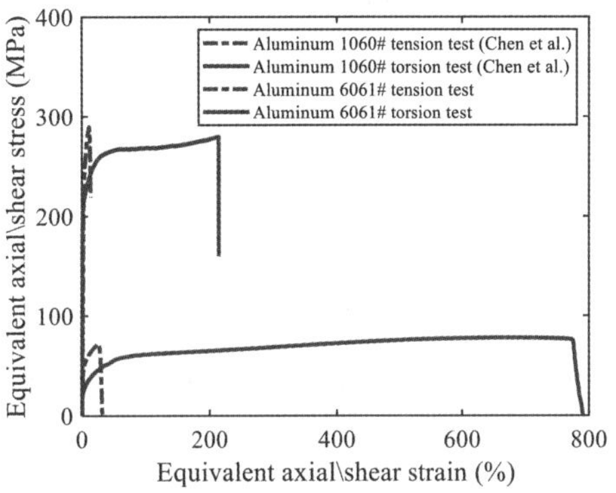

Fig. 6. Relationship of equivalent stress and strain under tension and torsion tests.

4. Configuration and dynamic mechanical behavior of the MTDGRD

4.1. Experimental setup

To evaluate its mechanical properties, the damper shown in Fig. 1 is manufactured employing the proposed configuration shown in Fig. 12. The relevant parameters of the damper are shown in Table 4. The solid round rod adopts aluminum 6061# materials, the gear and rack is manufactured by 45# steel, and the rest parts of transmission mechanism are manufactured by Q235# steel.

4.2. Hysteresis test of the MTDGRD

The MTDGRD of the hysteresis test layout is shown in Fig. 13. The test applies a displacement load to the MTDGRD through a shaking table, and the input displacement load is a sine wave. The rack of the MTDGRD is connected to the vibrating table through a connecting rod. When the vibrating table reciprocates, the dowel rod drives the rack of the MTDGRD to move horizontally, thereby driving the gear to rotate. The aluminum rod undergoes torsional plastic deformation, thereby achieving the purpose of energy consumption. Two laser displacement sensors (type: optoNCDT, manufactured by the Beijing Micro-Epsilon Measurement Co. LTD) are arranged in the test to measure the displacement at the vibration table platform and the displacement at the rack respectively. The axial force of the rack is measured by force sensors (type: GJBLS-l, manufactured by the Bengbu High Precision Sensing System Engineering Co. LTD). Finally, the data from the laser transmission sensors and the force sensor is collected through the collection system (type: DH5960, manufactured by the Donghua Testing Technology Co. LTD).

In order to study the influence of the proposed damper loading

Table 1
Mechanical property parameters of solid aluminum round rod.

Specimen	Axial stress (MPa)	Axial strain (%)	Shear stress (MPa)	Shear strain (%)	Total energy consumption in tension experiments (J)	Total energy consumption in torsion experiments (J)
Aluminum 6061#	289.2	11.46	279.3	213.9	294.9	3222.9
Aluminum 1060#	71.81	25.79	76.8	765.1	173.9	2561.0

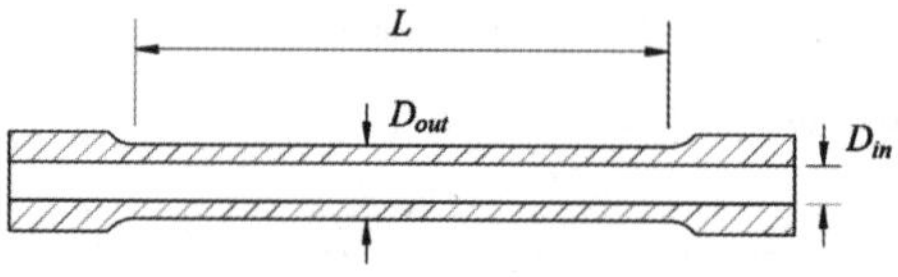

Fig. 7. Dimensions of aluminum dissipation component.

amplitude, loading frequency, aluminum rod outer diameter, aluminum rod length and hysteresis cycle number on the damping performance of the damper, a series of tests are conducted under the conditions outlined in Table 5. Due to the limitation of the shaking table device during the test, it can only produce a harmonic motion with a constant amplitude in one working condition. Therefore, the damper is subjected to constant amplitude loading in each case. Fig. 14 and Table 6 respectively show the hysteresis test curve of the damper and the energy consumption results of each cycle from working conditions T1 to T8. As shown in Fig. 14, the experimental hysteresis curve exhibits a slipping phenomenon when the damping force is zero. This slipping phenomenon is caused by the gap between the gear and rack. Additionally, the hysteresis curve is not smooth due to the squeezing and friction between the sliding bearing and the rack during the operation. These reasons cause the instability and asymmetry of the damper hysteresis curve shown in Fig. 14. As shown in Table 6, the energy dissipation of the proposed damper in the 5th cycle does not significantly differ from that in the 1st cycle under all conditions. This indicates that the proposed damper has energy dissipation stability.

Fig. 15 shows the hysteresis curves at different frequencies. It can be found in Fig. 15 that increasing the loading frequency reduces the average energy consumption of the damper by 1.02 %. This shows that the loading frequency of the damper has no effect on its mechanical properties and energy consumption characteristics. Hysteresis curve of damper after multiple cycles are shown in Fig. 16. It can be seen that the energy dissipation capacity and maximum damping force of the hysteresis curve of the 1st and 30th cycles of the damper are compared, and the results are shown in Table 7. It is found that the energy dissipation and the maximum damping force of the damper decrease to a certain extent with the increase of the action times, but the reduction rate of the maximum damping force can be controlled within 10 %. The tested specimen in T10 working condition can remain stable energy dissipation capacity after 30 cycles.

4.3. Mechanical properties of the MTDGRD

The torque balance of gear and rack in the proposed damper is shown in Fig. 17. In the figure, x is the horizontal displacement response of the rack, r_c is the radius of the gear, φ is the angle of rotation of the gear, F is the force of the rack acting on the gear, T is the torque generated by the aluminum round rod due to torsion. According to the geometric relationship, it can be obtained as,

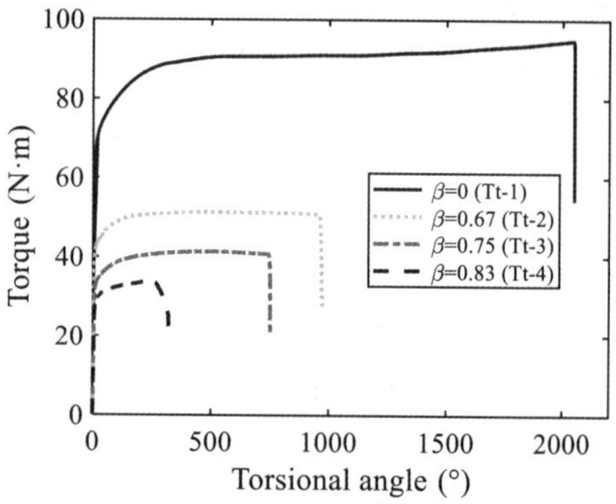

Fig. 8. Torque-torsion angle curves of aluminum tubes with different inner diameter.

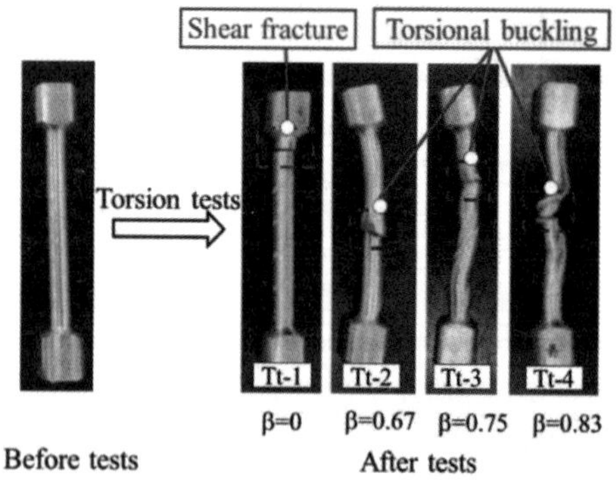

Fig. 9. Failure modes of specimens with different inner diameter.

Table 2
Energy dissipation parameters of aluminum metallic component of different β.

Number	Type	Length / L (mm)	Outer diameter / D_{out} (mm)	Inner diameter / D_i (mm)	Inner and outer diameter ratio /β	Maximum torque (N·m)	Maximum torsion angle (°)	Maximum energy consumption per unit mass (J)
Tt−1	Rod	100	12	0	0	94.86	2050	1.06×10^6
Tt−2	Tube			8	0.67	51.31	972	0.50×10^6
Tt−3	Tube			9	0.75	41.81	878	0.39×10^6
Tt−4	Tube			10	0.83	34.80	330	0.19×10^6

Table 3
Energy dissipation parameters of aluminum metallic component of different length and outer diameter ratio γ.

Specimen	Type	Length / L (mm)	Outer diameter / D_{out} (mm)	Inner diameter / D_i (mm)	Length and outer diameter ratio /γ	Maximum torque (N·m)	Maximum torsion angle (°)	Maximum energy consumption per unit mass (J)
Tl−1	Rod	20	12	0	1.7	96.36	465	6.80×10^7
Tl−2	Rod	40			3.3	96.42	927	6.86×10^7
Tl−3	Rod	80			6.7	97.15	2068	6.52×10^7
Tl−4	Rod	120			10.0	95.02	2642	6.47×10^7
Tl−5	Rod	160			13.3	97.18	4166	6.93×10^7

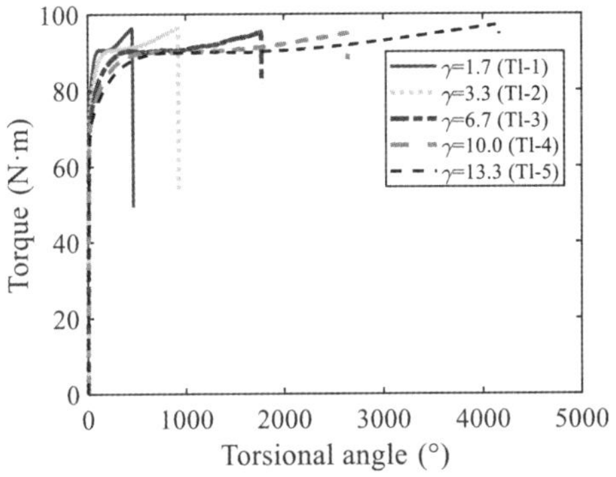

Fig. 10. Torque-torsion angle curves of aluminum tubes with different γ.

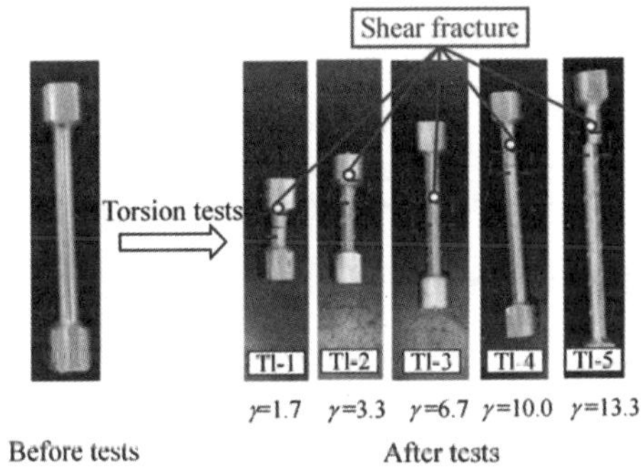

Fig. 11. Failure modes of specimens with different γ.

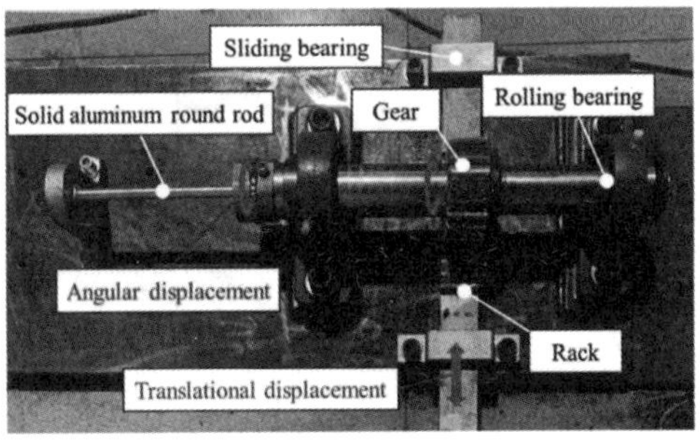

Fig. 12. The prototype damper.

Table 4
Related parameters of the MTDGRD.

Length of rack (mm)	Radius of gear (mm)	Adjustable length of aluminum rod (mm)
1500	44	1000/1600/2500

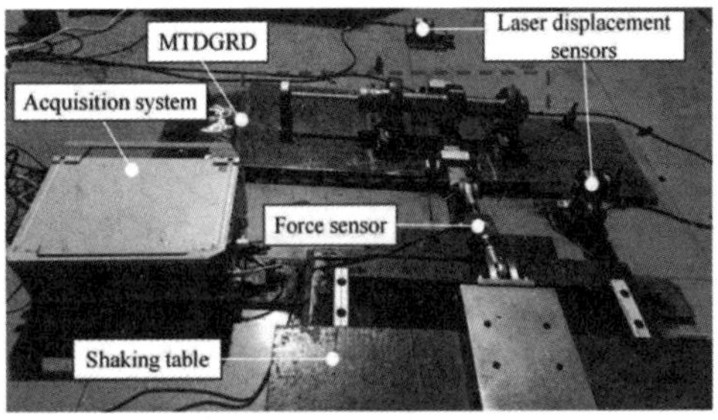

Fig. 13. Experimental setup.

Table 5
Hysteresis test conditions of the MTDGRD.

Test conditions	Length (mm)	Outer diameter (mm)	Loading amplitude (mm)	Loading frequency (Hz)	Number of cycles
T1	100	10	30	0.167	5
T2	100	10	32.5	0.167	
T3	160	10	32.5	0.167	
T4	250	10	32.5	0.167	
T5	100	10	35	0.167	
T6	160	10	35	0.167	
T7	100	8	35	0.167	
T8	100	6	35	0.167	
T9	100	10	35	0.334	
T10	100	10	35	0.167	30

$$\varphi = \frac{1}{r_c}x \tag{5}$$

$$T = Fr_c \tag{6}$$

According to the torsional deformation diagram of aluminum round rod, the shear strain γ can be expressed as,

$$\gamma = \frac{D\varphi}{2L} \tag{7}$$

Substituting Eq. (7) into Eq. (5), the relationship between the axial displacement of the damper x and the shear strain of the aluminum rod γ is as follows,

$$\gamma = \frac{Dx}{2Lr_c} \tag{8}$$

where L is the length of the aluminum round rod.

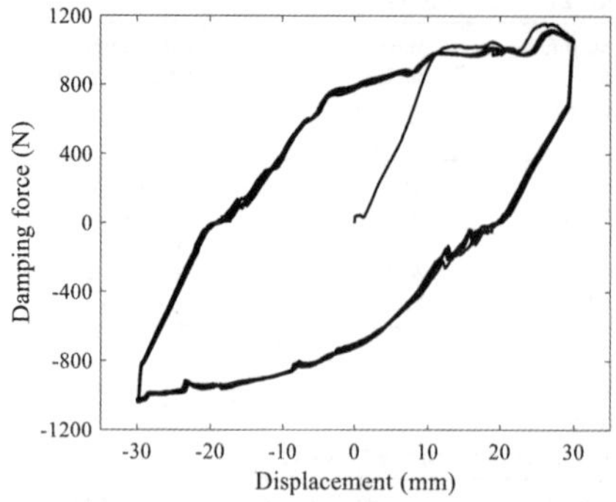

(a) T1 test condition

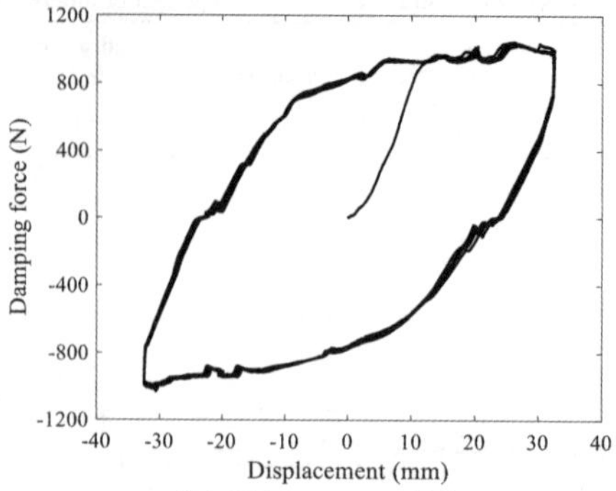

(b) T2 test condition

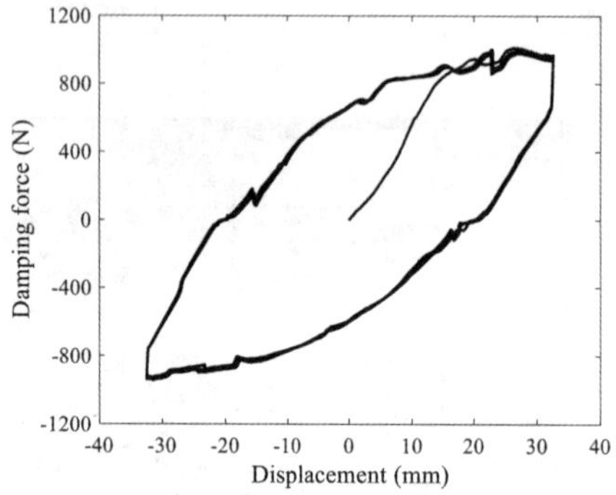

(c) T3 test condition

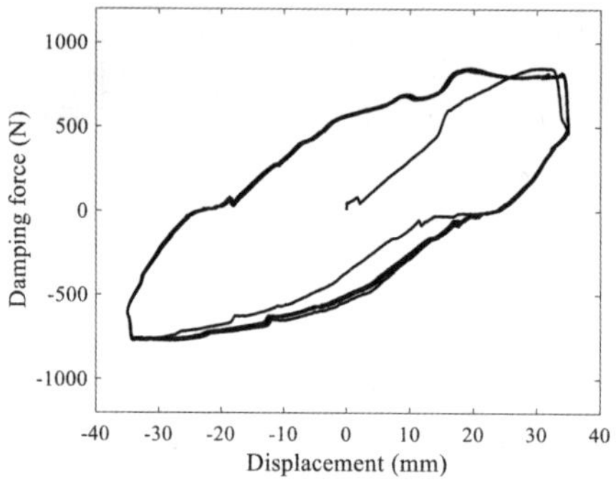

(d) T4 test condition

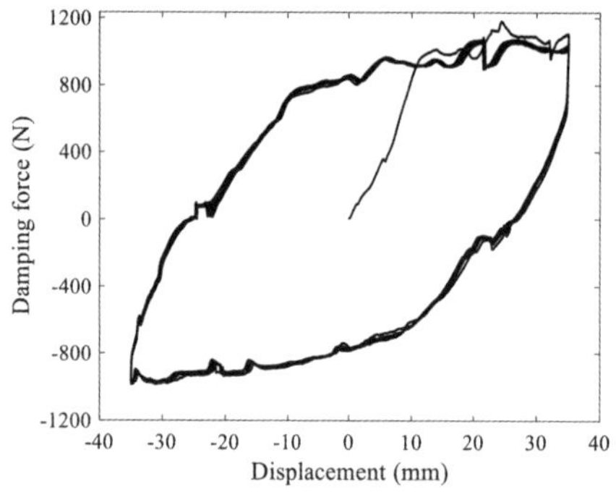

(e) T5 test condition

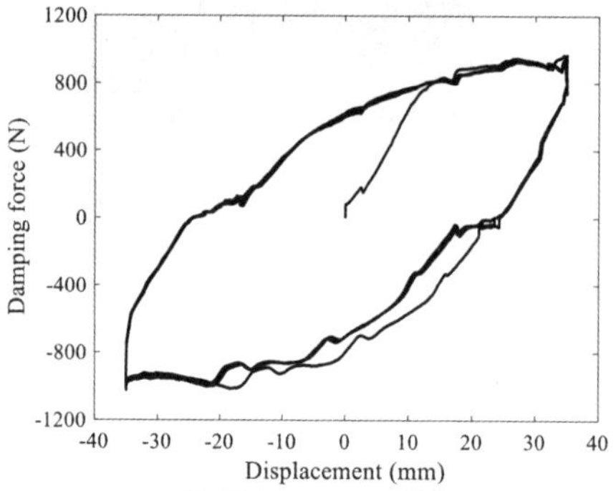

(f) T6 test condition

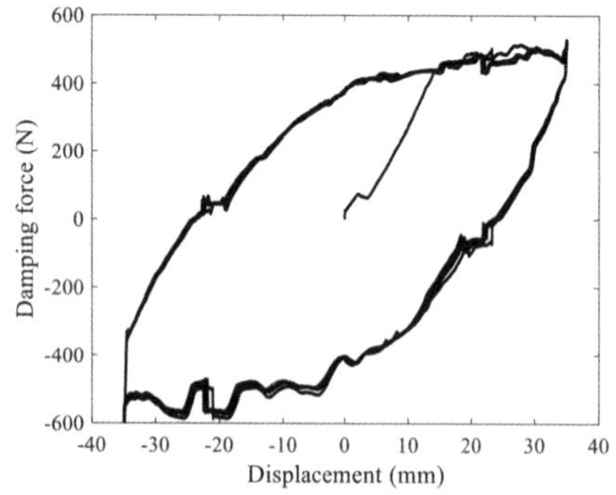

(g) T7 test condition

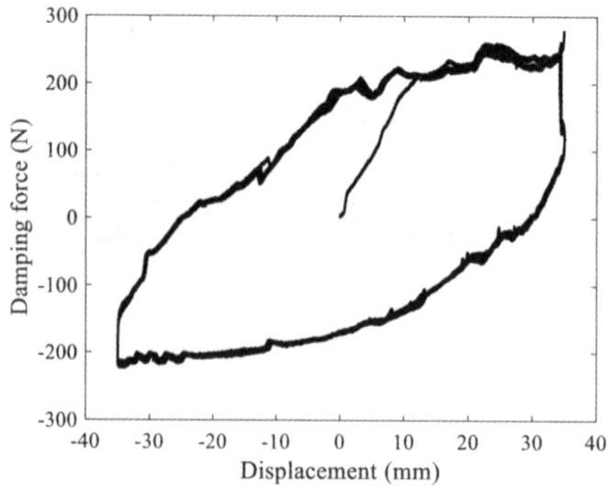

(h) T8 test condition

Fig. 14. Damping force-displacement hysteresis curves of the MTDGRD at different test conditions.

Table 6
Energy dissipation of dampers for each cycle at different test conditions.

Test conditions	Energy consumption (J)						Cycle 5/ Cycle 1 (%)
	Cycle 1	Cycle 2	Cycle 3	Cycle 4	Cycle 5	Average	
T1	59.29	62.59	62.26	61.82	61.22	61.63	103.24
T2	70.74	75.77	75.43	74.92	74.28	74.47	105.00
T3	54.40	59.53	59.23	58.89	58.51	58.52	107.55
T4	49.37	56.13	55.90	55.45	55.16	54.79	111.73
T5	83.33	87.17	86.79	86.07	85.54	85.97	102.65
T6	75.96	74.31	73.86	73.50	73.12	74.15	97.62
T7	38.06	40.47	40.01	39.86	39.49	39.65	103.76
T8	17.84	18.77	18.43	18.37	18.32	18.24	102.67
T9	77.05	82.67	81.10	79.84	78.29	80.02	101.60

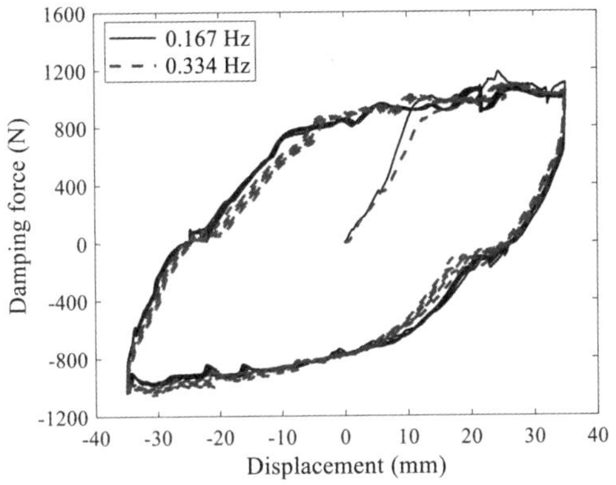

Fig. 15. Hysteresis curve of damper at different loading frequencies.

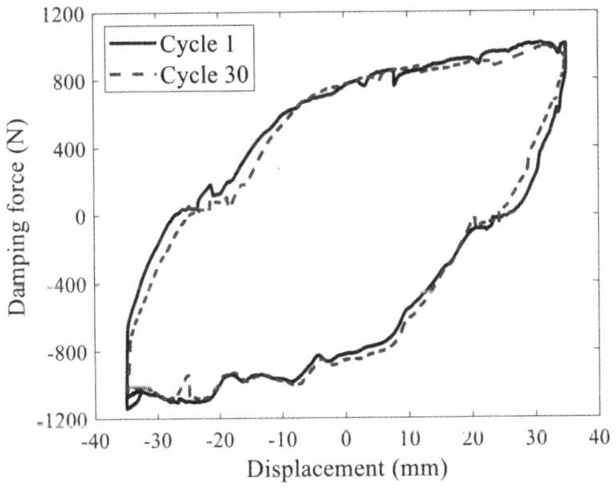

Fig. 16. Hysteresis curve of damper after multiple cycles.

Table 7
Comparison of mechanical indexes after multiple operation of dampers.

Test conditions	Evaluation Indicators	Loading Cycle		Reduction rate (%)
		Cycle 1	Cycle 30	
T10	Energy consumption (J)	83.15	78.84	5.2
	Maximum damping force (N)	1140	1020	10.5

The MTDGRD concurrently achieves the angular displacement amplification mechanism through the gear and rack and converts the transitional displacement of the rack into an angular displacement of the aluminum round rod. The angular displacement amplification effect can

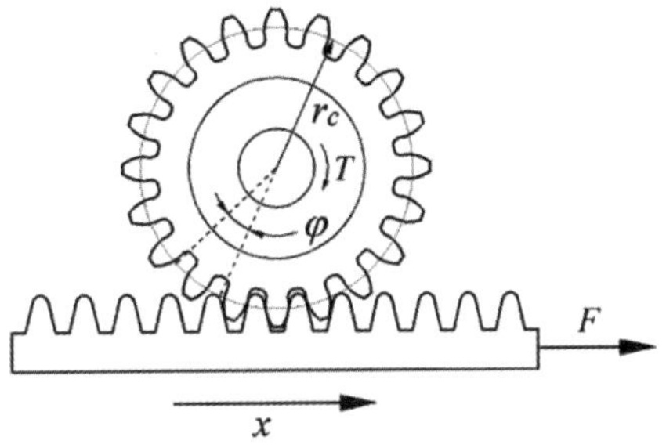

Fig. 17. Torque balance of gear and rack.

be described by a dimensionless parameter called the angular displacement amplification factor α. The displacement amplification factor can be expressed as the ratio of the axial displacement of the rack x to the rotating arc length the aluminum round rod s.

$$\alpha = \frac{s}{x} = \frac{D}{2r_c} \tag{9}$$

Based on the torsion deformation of the rod in the elastic-plastic stage, the cross-section of the rod is divided into elastic and plastic areas, as shown in Fig. 18(a), and the torque T of each cross-section can be expressed as [51,57],

$$T = \int_0^{2\pi}\int_0^{\frac{D_y}{2}} \tau r^2 drd\theta + \int_0^{2\pi}\int_{\frac{D_y}{2}}^{\frac{D}{2}} \left[\tau_y + G^*\left(\gamma - \gamma_y\right)\right] r^2 drd\theta \tag{10}$$

where τ_y is the yielding shear stress. γ_y is the yielding shear strain. D is the diameter of the aluminum round rod. D_y is the yielding diameter, as shown in Fig. 18(b), expressed as $D_y = 2\tau_y L/(G\theta)$. G is the shear elastic modulus, G^+ is the shear plastic modulus. Additionally, τ is the shear stress in the elastic region, with the following forms,

$$\tau = G\gamma \tag{11}$$

Substituting Eqs. (8) and (11) into Eq. (10) and integrating, the cross-section torque T can be obtained. Therefore, Eq. (6) can be rewritten as,

$$F = \frac{G\pi {D_y}^4}{32Lr_c^2}x + \frac{G^*\pi\left(D^4 - {D_y}^4\right)}{32Lr_c^2}x + \frac{\pi\left(\tau_y - G^*\gamma_y\right)\left(D^3 - {D_y}^3\right)}{12r_c} \tag{12}$$

When the solid cross-section aluminum round rod has a small rotation angle and each cross-section is still in an elastic torsional condition, $D_y = D$, the force of the damper can be obtained through Eq. (12) as,

$$F = \frac{G\pi D^4}{32Lr_c^2}x \tag{13}$$

When the aluminum round rod is totally plastic, which means $D_y \to 0$, the force of the damper can also be obtained through Eq. (12) as,

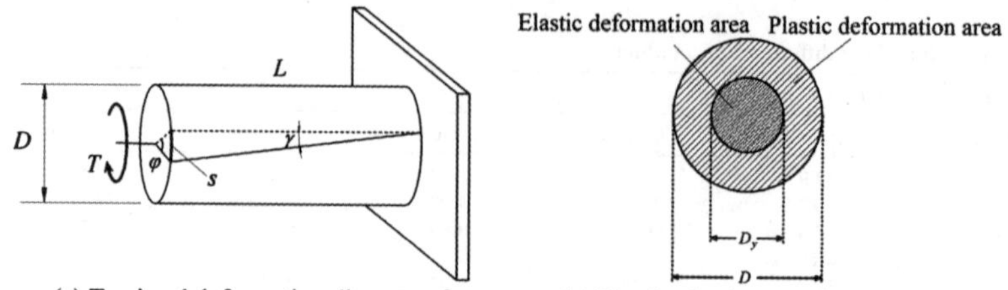

(a) Torsional deformation diagram of aluminum round rod
(b) Elastic-plastic area distribution of aluminum section

Fig. 18. Torsional deformation of aluminum round rod.

$$F=\frac{G^{*}\pi D^{4}}{32Lr_{c}^{2}}x+\frac{\pi\left(\tau_{y}-G^{*}\gamma_{y}\right)D^{3}}{12r_{c}} \tag{14}$$

According to the reference of Chen [57], the mechanical behavior of the damper is modeled using the bilinear constitutive model, which is represented by Eqs. (13) and (14), as shown in Fig. 19. The simplified constitutive relation between the damping force and the axial displacement can be expressed as,

$$F=\begin{cases}k_{p}x-b & -x_{\max}<x<-x_{ep}\\ k_{e}x & -x_{ep}<x<x_{ep}\\ k_{p}x+b & x_{ep}<x<x_{\max}\end{cases} \tag{15}$$

where k_e represents the elastic stiffness of the bilinear model. k_p represents the plastic stiffness of the bilinear model. x_{ep} represents the yielding displacement of the bilinear model. x_{max} represents the limit displacement of the bilinear model. F_{ep} represents the yielding force of the bilinear model, and these parameters are defined as follows,

$$k_{e}=\frac{G\pi D^{4}}{32Lr_{c}^{2}} \tag{16}$$

$$k_{p}=\frac{G^{*}\pi D^{4}}{32Lr_{c}^{2}} \tag{17}$$

$$x_{ep}=\frac{8Lr_{c}\left(\tau_{y}-G^{*}\gamma_{y}\right)}{3D(G-G^{*})} \tag{18}$$

$$F_{ep}=\frac{G\pi\left(\tau_{y}-G^{*}\gamma_{y}\right)D^{3}}{12r_{c}(G-G^{*})} \tag{19}$$

According to the constitutive relation of the damper in Eq. (15), a hysteretic curve model, as shown in Fig. 20, for dynamic simulations can be obtained as,

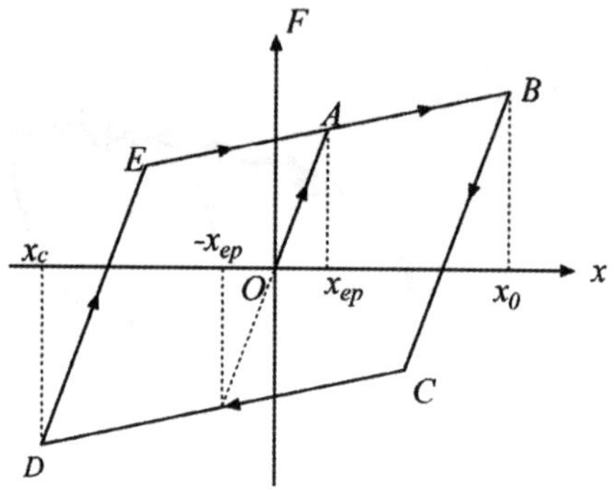

Fig. 20. Damper force hysteresis model.

$$F=\begin{cases}k_{e}x & OA\\ k_{p}x+(k_{e}-k_{p})x_{ep} & AB\\ k_{e}x-(k_{e}-k_{p})(x_{0}-x_{ep}) & BC\\ k_{p}x-(k_{e}-k_{p})x_{ep} & CD\\ k_{e}x-(k_{e}-k_{p})(x_{c}+x_{ep}) & DE\end{cases} \tag{20}$$

where x_0 and x_c are the maximum displacement amplitude in the positive and negative direction of the horizontal axial, respectively. Suppose the energy consumption of the damper is E_D, then the area of $BCDE$ in Fig. 20 is E_D,

$$E_{D}=4x_{ep}(k_{e}-k_{p})(x_{0}-x_{ep}) \tag{21}$$

Substituting Eq. (16), Eq. (17) and Eq. (18) into Eq. (21), the energy consumption of the damper E_D can be obtained. Therefore, Eq. (20) can be rewritten as,

$$E_{D}=\frac{\pi D^{3}\left(\tau_{y}-G^{*}\gamma_{y}\right)}{3r_{c}}\left[x_{0}-\frac{8Lr_{c}\left(\tau_{y}-G^{*}\gamma_{y}\right)}{3D(G-G^{*})}\right] \tag{22}$$

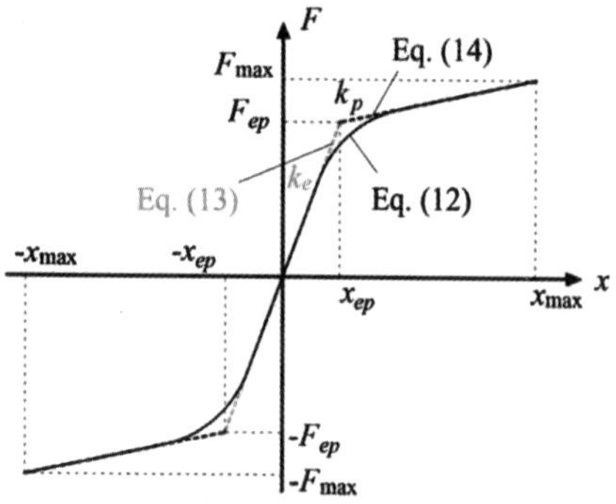

Fig. 19. Bilinear model for damper.

4.4. Theoretical hysteresis model validation

In order to verify the proposed bilinear hysteresis model of the MTDGRD, the comparison between the experimental results and the theoretical hysteresis curve results under T5 operating condition is shown in Fig. 21. The results show that the theoretical bilinear model fits well with the experimental results, and the theoretical bilinear model can be used for subsequent numerical simulations.

Fig. 22 shows the relationship between the loading amplitude and the outer diameter of the aluminum rod and the energy consumption of the proposed MTDGRD. The experimental values are basically consistent with the theoretical values from Eq. (22), which once again verifies the correctness of the elastic theory from the perspective of energy dissipation of the damper. The main reason for the lower energy consumption obtained from the experiment in Fig. 20 (a) is the presence of some

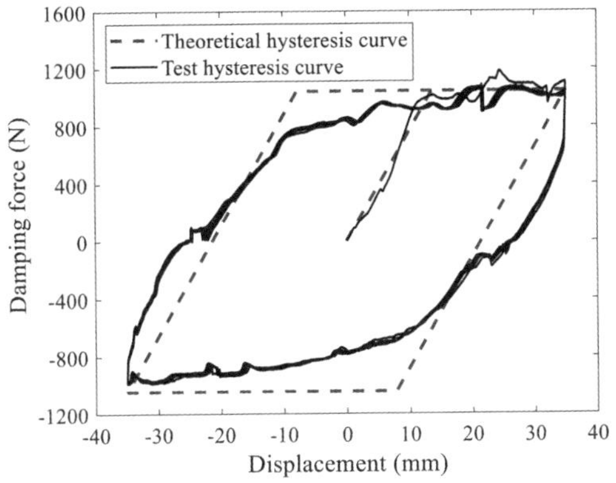

Fig. 21. Theoretical hysteresis model validation.

gaps inside the damper, which results in the actual torsion angle of the metal rod being smaller than the theoretical torsion angle.

5. Seismic performances of the MTDGRD for a super high-rise flexible structure

High-rise flexible structures are prone to large vibrations under earthquakes due to their low damping. Therefore, this section studies the seismic control performance of the proposed damper is studied in this section. To simplify the high-rise structure model, adjacent floors can be combined into a lumped mass [74,75]. Consequently, this paper employs an equivalent 10-degree-of-freedom model to simulate the high-rise structure. The mass and the inter-floor flexural stiffness of each floor of the structure is shown in Table 8. This article primarily focuses on tall flexible structures, characterized by low frequency and small damping ratios. Therefore, the Rayleigh damping matrix is used for structural damping, and the damping ratio of the first two natural modes

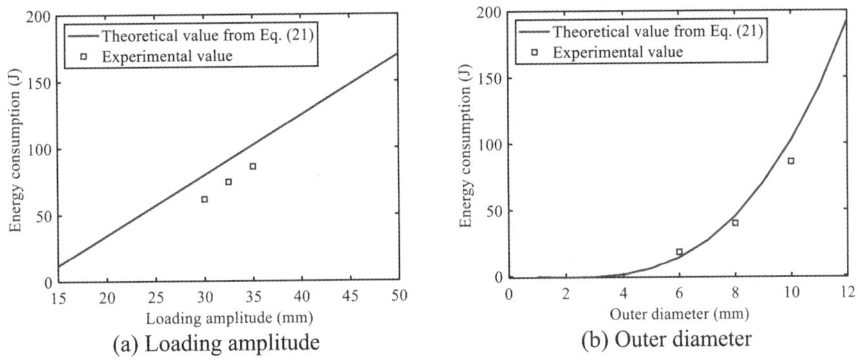

Fig. 22. Theoretical hysteresis model validation.

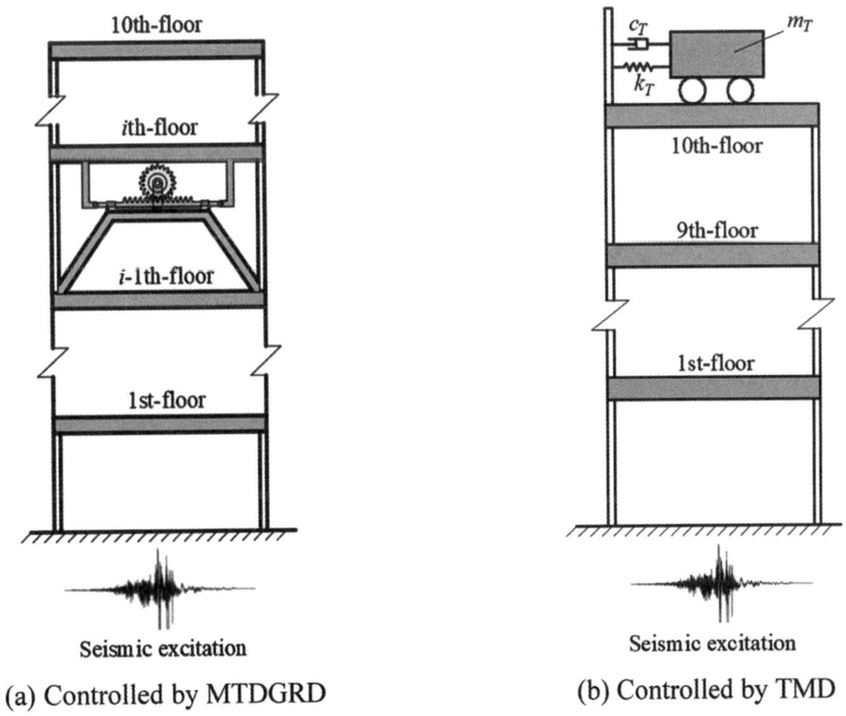

Fig. 23. High-rise flexible structure after installing dampers.

Table 8
Parameter table of 10-floor flexible structure.

Floor i	1	2	3	4	5	6	7	8	9	10
Mass m_i ($\times10^3$ kg)	3624	3600	3576	3552	3528	3504	3480	3456	3432	3408
Stiffness k_i ($\times10^6$ N/m)	64.4	64.4	54.0	54.0	51.5	45.0	38.6	32.2	25.7	19.3

is ensured to be 0.005. The eigenvalue method is used to conduct a modal analysis of the 10-floor flexible structure. After normalizing the maximum modal displacement to 1, the mode shapes of the first three modes of vibration are shown in Fig. 24. The corresponding modal frequencies, modal masses and modal damping ratios of the first three orders of vibration are summarized in Table 9.

The ground motion records of FEMA P695 far-field set [76] are used for time-history analysis to evaluate seismic performance [77,78]. Therefore, 12 seismic ground motions are selected from the FEMA P695 far-field record set to evaluate the control performance of the proposed damper. These seismic ground motions are selected from the PEER NGA database using some record selection criteria as specified in the Appendix A.7 in [76]. The peak ground acceleration (PGA) of these seismic ground motions ranges from 0.21 g to 0.82 g. The details of the 12 seismic ground motions, including the event name, station, year of occurrence, and magnitude are listed in Table 10. Response spectrum curve of the 12 seismic ground motions is also plotted in Fig. 25.

5.1. *Motion equations of a super high-rise flexible structure controlled by a MTDGRD*

As shown in the Fig. 23(a), the motion equation of the high-rise flexible structure controlled by the MTDGRD is shown below,

$$\mathbf{M}\ddot{\mathbf{X}} + \mathbf{C}\dot{\mathbf{X}} + \mathbf{K}\mathbf{X} + \mathbf{F_M} = -\mathbf{M}\mathbf{I}\ddot{x}_g \tag{23}$$

where **M**, **K** and **C** are the mass matrix, stiffness matrix, damping matrix of the system shown in Fig. 23(a). $\mathbf{F_M}$ is the damping force vector of the MTDGRD installed on the i-th story of the structure. **X** is the displacement vector of the system. $\ddot{x}_g$ is the acceleration of seismic ground motion. **I** is the unit column vector. It can be expressed as,

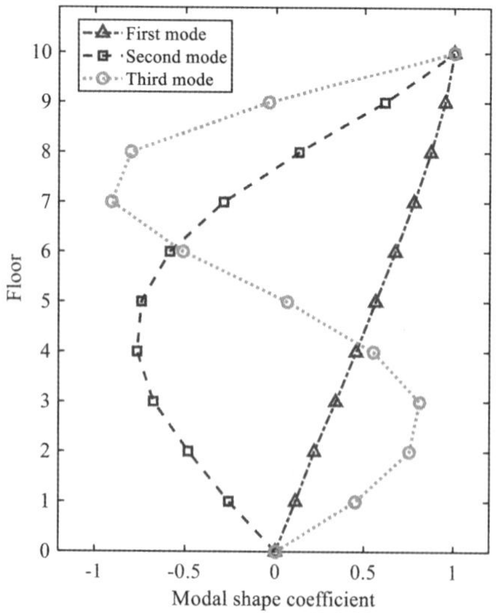

Fig. 24. Mode shape diagram of the first three modals.

Table 9
Modal parameters of the flexible structure with degrees of freedom.

Mode number i	$i=1$	$i=2$	$i=3$
Modal frequency f_{mi} (Hz)	0.29	0.75	1.22
Modal mass m_{mi} ($\times10^3$ kg)	14,739	12,317	14,769
Modal damping ratio ζ_{mi}	0.005	0.005	0.0067

Table 10
Detailed information of seismic ground motion records.

Serial number	Magnitude	Year	Event name	Station name
GM−1	7.1	1994	Duzce, Turkey	Bolu
GM−2	6.5	1979	Imperial Valley	El Centro Array #11
GM−3	7.5	1999	Kocaeli, Turkey	Duzce
GM−4	7.3	1992	Landers	Yermo Fire Station
GM−5	6.9	1989	Loma Prieta	Capitola
GM−6	7.4	1990	Manjil, Iran	Abbar
GM−7	7.0	1992	Cape Mendocino	Rio Dell Overpass
GM−8	7.6	1999	Chi-Chi, Taiwan	CHY101
GM−9	6.6	1971	San Fernando	LA-Hollywood Stor
GM−10	6.7	1994	Northridge	Beverly Hills - Mulhol
GM−11	7.1	1999	Hector Mine	Hector
GM−12	6.9	1995	Kobe, Japan	Nishi-Akashi

$$\mathbf{M} = \begin{bmatrix} m_1 & & & \\ & m_2 & & \\ & & \ddots & \\ & & & m_n \end{bmatrix}_{(10\times10)} \tag{24}$$

$$\mathbf{K} = \begin{bmatrix} k_1+k_2 & -k_2 & & \\ -k_2 & k_2 & \ddots & \\ & \ddots & \ddots & -k_n \\ & & -k_n & k_n \end{bmatrix}_{(10\times10)} \tag{25}$$

$$\mathbf{F_M} = \begin{bmatrix} 0 \\ \vdots \\ -F \\ F \\ \vdots \\ 0 \end{bmatrix} \begin{matrix} 1 \\ \vdots \\ i-1 \\ i \\ \vdots \\ 10 \end{matrix} \tag{26}$$

$$\mathbf{X} = [x_1, x_2, \ldots, x_n]^T_{(10\times1)} \tag{27}$$

where m_i, k_i, and x_i are respectively the mass, stiffness and displacement of the i-th floor. F the damping force of the MTDGRD, which can be calculated by Eq. (20).

Since the Rayleigh damping matrix is used for structural damping, the system damping matrix is expressed as follows,

$$\mathbf{C} = a\mathbf{M} + b\mathbf{K} \tag{28}$$

$$a = \frac{4\pi f_{m1} f_{m2}(\xi_{m1} f_2 - \xi_{m2} f_1)}{{f_2}^2 - {f_1}^2} \tag{29}$$

$$b = \frac{\xi_{m2} f_{m2} - \xi_{m1} f_{m1}}{\pi({f_{m2}}^2 - {f_{m1}}^2)} \tag{30}$$

To obtain the responses of the 10-floor flexible structure, the

Table 11
Parameter table of the MTDGRD.

Material	Outer diameter (m)	Length (m)	Gear radius (m)	Angular displacement amplification factor	Elastic stiffness (N/m)	Plastic stiffness (N/m)	Yielding displacement (mm)
Al 6061	0.52	0.2	0.17	1.53	9.56×10^9	7.65×10^6	2.8

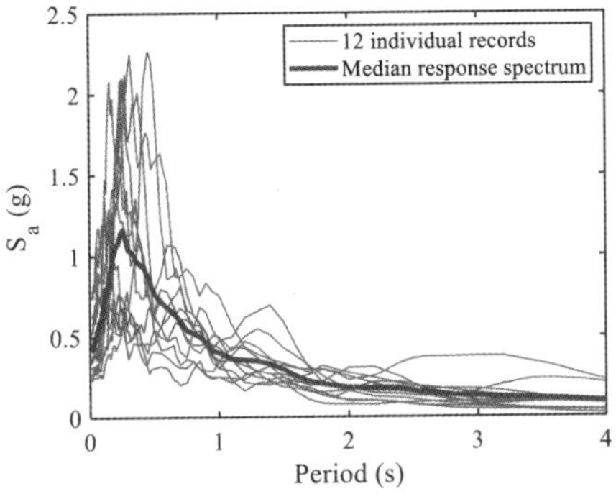

Fig. 25. Acceleration response spectra and design spectral curves.

equations of motion are solved using the Newmark-Beta method. A time step of the numerical simulations is set to 0.001 s to ensure the solution accuracy. Fig. 26 shows the design process of MTDGRD design for controlling the i-th mode of super high-rise flexible structure.

5.2. Parametric optimization

In order to evaluate the control performance of the proposed damper on the seismic response of the 10-floor flexible structure, the response reduction η is defined as follows,

$$\eta = \frac{X_0 - X}{X_0} \times 100\% \tag{31}$$

where, X_0 and X are the root mean square(RMS) of displacement of the 10-floor flexible structure without and with dampers, respectively.

In order to carry out the parametric design of the proposed damper, the seismic ground motion numbered GM-1 is selected to be input at the bottom of the 10-floor flexible structure, and the Matlab software is used to perform numerical simulations of the structural responses. Considering only one damper is installed between the inter-floor, the structure response is analyzed in time history. Fig. 27 shows root mean square (RMS) of each floor displacement. It can be seen from Fig. 27 that the maximum RMS of displacement has the best control performance than other cases when the damper is installed on the first floor. Fig. 28 shows the influence of the outer diameter of the aluminium rod on and the radius of gear of the reduction rate of the maximum RMS of

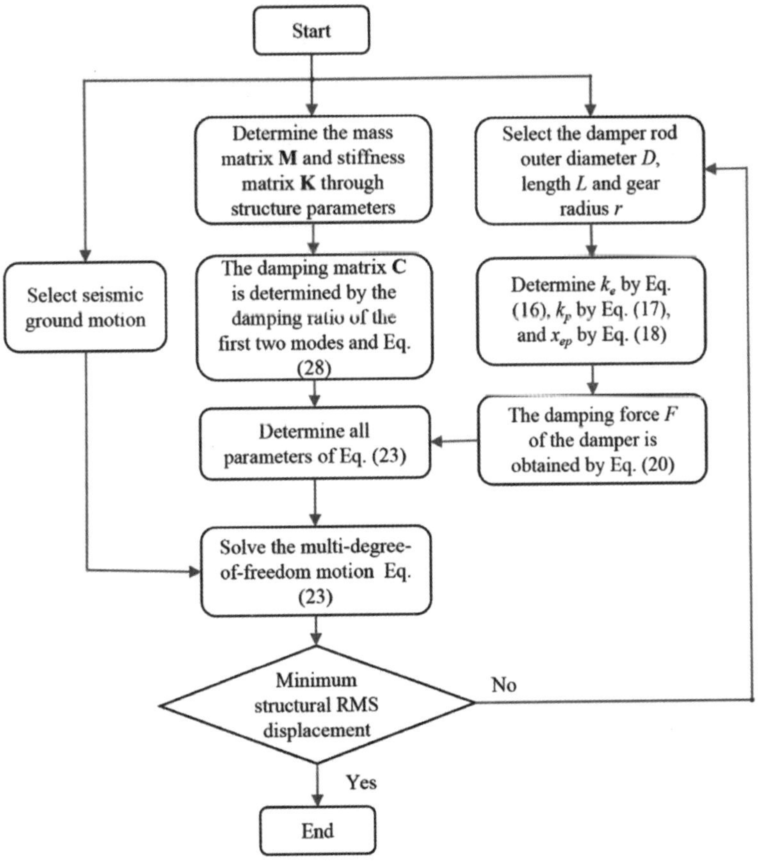

Fig. 26. The design process of MTDGRD.

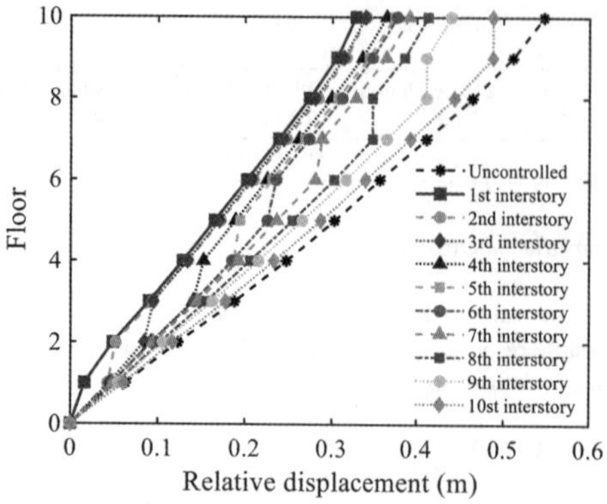

Fig. 27. Structural RMS of displacement under different positions.

displacement of the structure. Therefore, installing the damper on the first floor is the optimal installation position. As can be seen from Fig. 28, the reduction rate of the maximum RMS of displacement of the structure increases first and then decrease with the increase of the outer diameter of the round rod and the radius of gear. In this simulation, the outer diameter of the round rod is set to 0.52 m and the radius of gear is set to 0.17 m, the reduction rate of the maximum RMS of displacement of the structure is largest, and the maximum response reduction can reach about 40.13 %.

The seismic performance of the 10-floor structure is analyzed under the optimized damper parameters corresponding to the optimal installation position of the MTDGRD. Fig. 29 shows the displacement response under the seismic ground motion numbered GM-1. In contrast, the results of another damper with angular displacement amplification factor $\alpha = 1$ are also plotted in Fig. 29. It can be seen that the responses of each floor can be well controlled after installing the MTDGRD. Additionally, the seismic control performance with $\alpha = 1.53$ is superior to that with $\alpha = 1$. According to the damper displacement and its mechanical model, the hysteresis curve of the MTDGRD can be obtained in Fig. 30. It shows that the maximum damping force offered by the MTDGRD with $\alpha = 1.53$ is 42,550 kN, significantly more than the damper with $\alpha = 1$ offers. Through Eq. (5), the relationship between the rotation angle and torque of the aluminum rod under seismic ground motion numbered GM-1 can be obtained in Fig. 31. It is shown that the MTDGRD absorbs a large amount of vibration energy during the aluminum rod is in the plastic deformation stage. Moreover, the energy input by GM-1 and the energy consumed by the proposed damper are shown in Fig. 32. It can be seen from the figure that most of the energy input by GM-1 into the structure is consumed by the proposed damper. In conclusion, structural vibrations can be effectively controlled by the installation of the proposed damper, and the energy consumption of MTDGRD is enhanced after the displacement of MTDGRD is amplified.

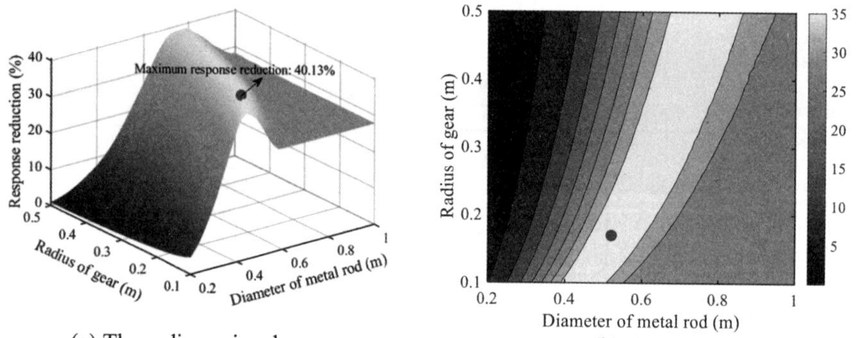

(a) Three-dimensional map (b) Top view

Fig. 28. Influence of outer diameter of round rod and radius of gear of the reduction rate of the maximum RMS of displacement of the structure.

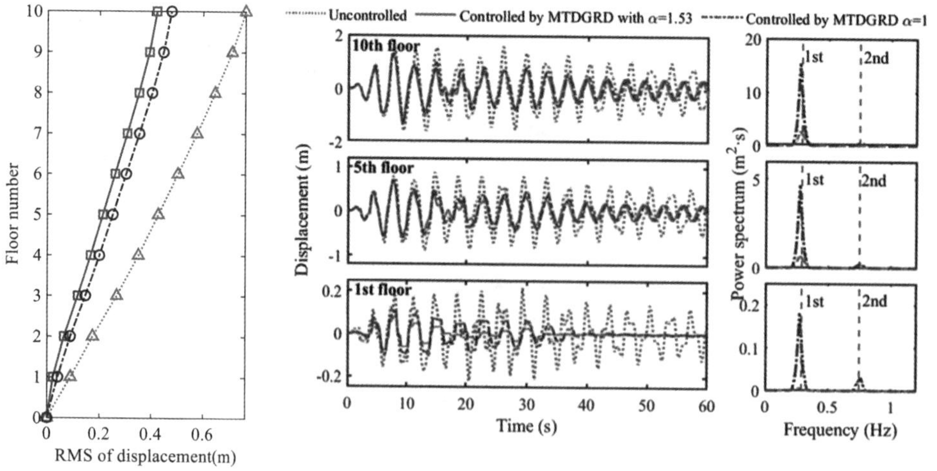

Fig. 29. Displacement response of structure under seismic ground motions numbered GM-1.

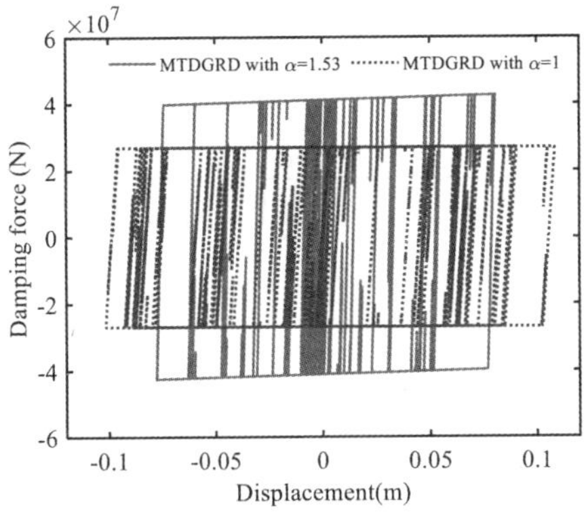

Fig. 30. The proposed damper hysteresis curve.

5.3. Comparison with conventional tuned mass damper (TMD)

Since the energy dissipation capacity of different types of metallic dampers is closely related to their size, designing another type of metallic damper for the controlled structure is hard in this paper. Therefore, it is difficult to compare the damping performance of previously developed metallic dampers with the proposed torsional damper under simulations and experiments. Meanwhile, very few existing studies have compared the performance of metallic dampers to the conventional TMD, which may cause difficulty in selecting damper types in the design stage of practical control systems. Therefore, a TMD with a mass ratio of 0.5 % under optimal parameters is selected to compare the seismic control performance with the proposed MTDGRD, as shown Fig.23(b). The TMD is installed on the top floor of the structure to control the first-order mode vibration of the structure.

The mass ratio μ, frequency ratio r and damping ratio ξ are defined as follows,

$$\mu = \frac{m_T}{m_{m1}} r = \frac{\omega_T}{\omega_{m1}} \xi = \frac{c_T}{2m_{m1}\omega_{m1}} \tag{32}$$

where

$$\omega_{m1} = 2\pi f_{m1} \omega_T = \sqrt{\frac{k_T}{m_T}} \tag{33}$$

where m_T is the mass of TMD, m_{m1} is the first-order modal mass of the structure, c_T is the damping coefficient of TMD, ω_T is the circular frequency of TMD, ω_{m1} is the first-order modal circular frequency of the structure, f_{m1} is the first-order modal frequency of the structure, k_T is the stiffness of TMD. Den Hartog's fixed-point theory [79] provides a method for the optimal parameter design of TMD, such as frequency ratio and damping ratio. Warburton [80] used the fixed point method to derive the optimal parameters of TMD under the action of base acceleration. The fixed-point theory states that the DMF of the structure under different damping ratios ζ_T of TMD always passes through two fixed-points P and Q, which indicates that P and Q points are independent with ζ_T. Therefore, the value of DMFs should be equal to each other when ζ_T is set to zero and infinity. This condition can be used to determine the optimal frequency ratio, as shown in Eq. (34). Since the gradient of the DMF at points P and Q should be zero, the optimal damping ratio also can be determined, as shown in Eq. (35). The parameters of the optimal TMD and MTDGRD are shown in Table 12.

$$r_{opt} = \frac{1}{1+\mu}\left(\sqrt{\frac{2-\mu}{2}}\right) \tag{34}$$

$$\xi_{opt} = \sqrt{\frac{3\mu}{8(1+\mu)}}\left(\sqrt{\frac{2}{2-\mu}}\right) \tag{35}$$

As shown in the Fig. 23(b), the motion equation of the high-rise flexible structure controlled by the TMD is shown below,

$$\mathbf{M_T}\ddot{\mathbf{X}}_\mathbf{T} + \mathbf{C_T}\dot{\mathbf{X}}_\mathbf{T} + \mathbf{K_T}\mathbf{X_T} = -\mathbf{M_T}\mathbf{I}\ddot{x}_g \tag{36}$$

where the $\mathbf{M_T}$, $\mathbf{K_T}$, $\mathbf{C_T}$ are the mass matrix, stiffness matrix and damping matrix of the system shown in Fig. 23(b), $\mathbf{X_T}$ is the displacement vector of the system,

$$\mathbf{M_T} = \begin{bmatrix} m_1 & & & & \\ & m_2 & & & \\ & & \ddots & & \\ & & & m_{10} & \\ & & & & m_T \end{bmatrix}_{(11\times 11)} \tag{37}$$

$$\mathbf{K_T} = \begin{bmatrix} k_1+k_2 & -k_2 & & & \\ -k_2 & k_2+k_3 & \ddots & & \\ & \ddots & \ddots & -k_{10} & \\ & & -k_{10} & k_{10}+k_T & -k_T \\ & & & -k_T & k_T \end{bmatrix}_{(11\times 11)} \tag{38}$$

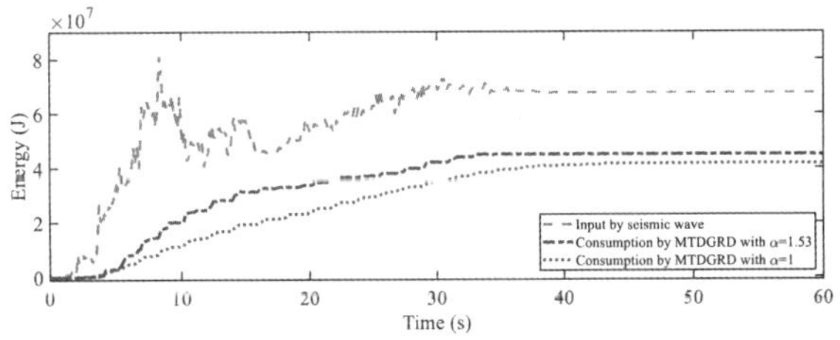

Fig. 32. Energy consumption of the MTDGRD.

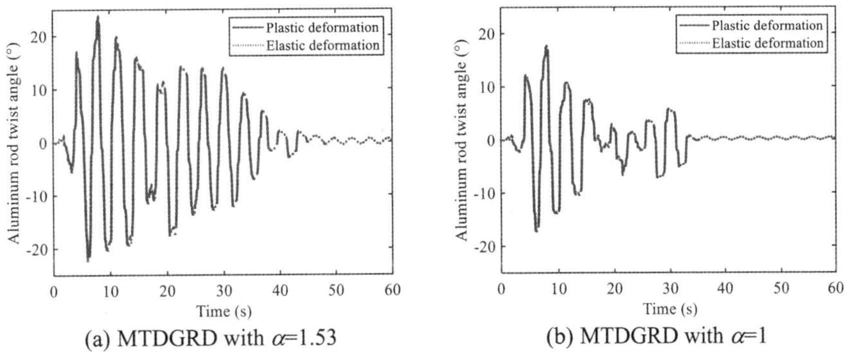

Fig. 31. Aluminum rod torsion angle time history of the MTDGRD.

Table 12
Damper parameters.

MTDGRD				TMD		
Material	Outer diameter (m)	Length (m)	Mass (kg)	Mass (kg)	Stiffness (N/m)	Damping coefficient (N·m/s)
Al6061	0.52	0.2	0.115×10^3	74×10^3	243×10^3	11.56×10^3

$$\mathbf{C}_\mathrm{T} = \begin{bmatrix} c_{11} & c_{12} & \cdots & c_{1n} & 0 \\ c_{21} & c_{22} & \cdots & c_{2n} & 0 \\ \vdots & \vdots & \ddots & \vdots & \vdots \\ c_{n1} & c_{n2} & \cdots & c_{nn}+c_T & -c_T \\ 0 & 0 & \cdots & -c_T & c_T \end{bmatrix}_{(11\times11)} \tag{39}$$

$$\mathbf{X}_\mathrm{T} = [x_1, x_2, \ldots, x_{10}, x_T]^T_{(11\times1)} \tag{40}$$

where the notation c_{ij} represents the element in the i-th row and j-th column of matrix **C**. To obtain the responses of the 10-floor flexible structure, the equations of motion are solved using the Newmark-Beta method. A time step of the numerical simulations is set to 0.001 s to ensure the solution accuracy.

Fig. 33 shows the displacement of RMS of each floor of the structure under the control of MTDGRD and TMD under the action of 12 seismic ground motions. The maximum RMS and response reduction of the structure are shown in Fig. 34 and Table 13. It can be seen from the Fig. 33 that the optimal parameters of MTDGRD based on the seismic ground motion numbered GM1 also have a better seismic performance in the case of other seismic ground motions. It can be seen from Table 13 that the average response reduction rate of MTDGRD can reach 60.79 %, while TMD can only reach 38.48 %. Compared with optimized TMD of 0.5 % mass ratio, the MTDGRD has a better seismic performance, while the mass of the aluminum rod of MTDGRD is only 0.155 % of the TMD mass. Compared with the MTDGRD, the TMD require greater mass to be installed in the structure. This requires additional reinforcement of the stiffness between structural stores, which may increase the cost of TMD.

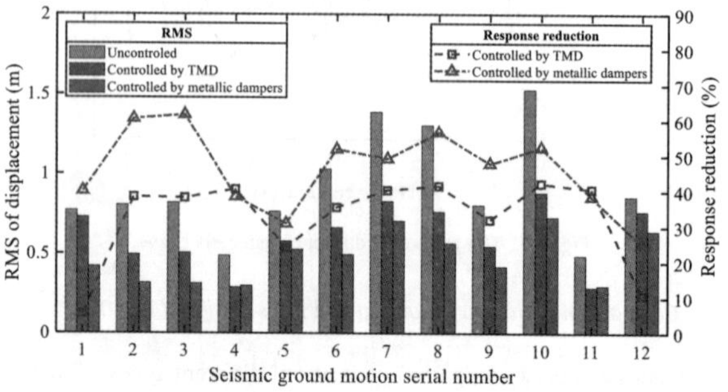

Fig. 34. RMS and response reduction under 12 seismic ground motions.

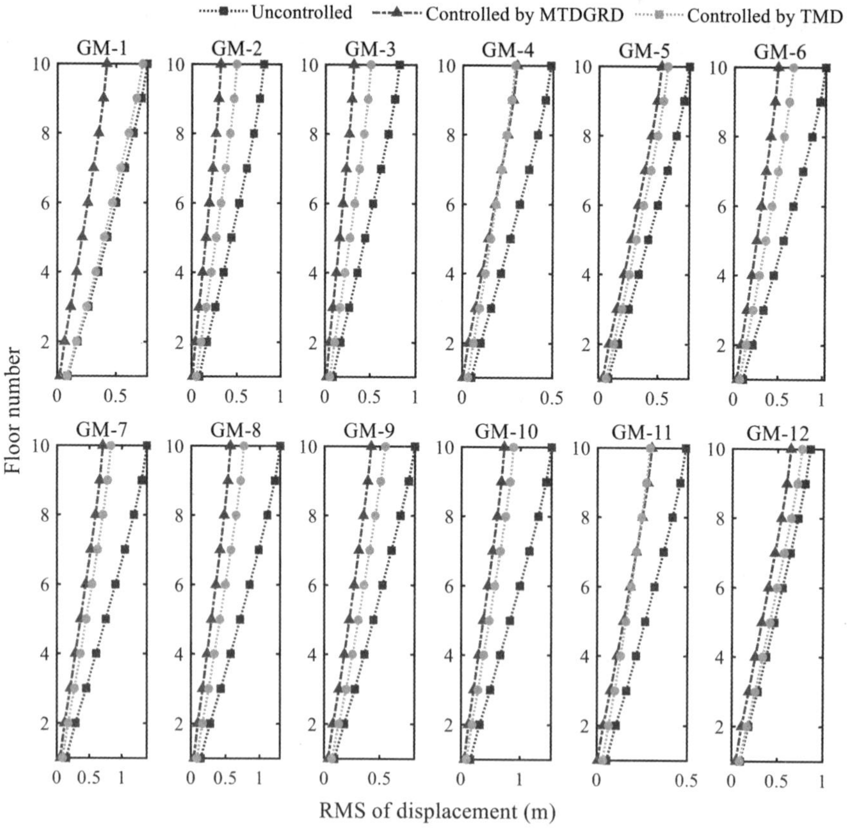

Fig. 33. RMS displacement of each floor of the flexible structure.

Table 13
RMS and response reduction under 12 seismic ground motions.

Serial number	Uncontrolled	Controlled by MTDGRD		Controlled by TMD	
	RMS (m)	RMS (m)	Response reduction (%)	RMS (m)	Response reduction (%)
GM−1	0.771	0.460	40.13	0.728	5.47
GM−2	0.805	0.316	60.79	0.495	38.48
GM−3	0.819	0.312	61.87	0.506	38.22
GM−4	0.488	0.302	38.20	0.290	40.60
GM−5	0.764	0.527	31.10	0.579	24.18
GM−6	1.030	0.496	51.82	0.664	35.52
GM−7	1.389	0.705	49.21	0.828	40.34
GM−8	1.306	0.566	56.64	0.763	41.56
GM−9	0.804	0.418	48.04	0.547	32.00
GM−10	1.529	0.730	52.23	0.883	42.25
GM−11	0.488	0.302	38.20	0.290	40.60
GM−12	0.856	0.642	24.97	0.765	10.70

Fig. 35 shows the time history and power spectrum of displacement response under the seismic ground motion numbered GM-2. From the response of the 1st and 5th floors, it is evident that the GM-2 excites several modal vibrations. The 1st mode of vibration can be effectively controlled by both MTDGRD and TMD, but the 2nd mode of vibration is more effectively controlled by MTDGRD than by TMD. For the vibrations of other modes, TMD only has limited control effect. Additionally, MTDGRD are not sensitive to the vibration frequency and have good control effects on other vibration modes. Hysteresis curve of MTDGRD and TMD is shown in Fig. 36. It can be seen from the hysteresis curves of the MTDGRD and TMD that the TMD requires a larger stroke than MTDGRD under the GM-2. This means a larger installation space for the TMD is required.

As shown in Fig. 37, the maximum response reduction of RMS displacement increases as the mass ratio increases when the TMD is in optimal tuning. TMD can achieve the same control performance as MTDGRD when the mass ratio of TMD reaches 2.7 %. The mass of the aluminum rod at this time is 0.155 % of the TMD mass. Moreover, the mass ratio needs to reach 3.8 % to achieve the response reduction of RMS of displacement when the TMD frequency is detuned by −20 %. More mass is needed to achieve the same control performance of the MTDGRD when TMD is detuned.

6. Conclusions

In this paper, the metallic torsional damper using a gear and rack device (MTDGRD) is proposed. The static mechanical behavior and energy dissipation of aluminum metallic component under torsional deformation is experimentally studied. Secondly, the configuration and working principle of the proposed damper is introduced and its dynamic behavior of the proposed damper is experimentally and theoretically studied. Finally, numerical simulations under different seismic ground motions are compared to that controlled by a conventional tuned mass damper (TMD) for a 10-floor structure. The following conclusions can be obtained,

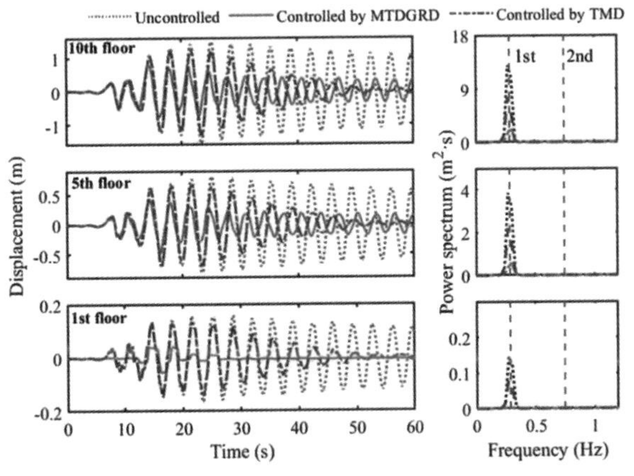

Fig. 35. Displacement response of 10-floor flexible structure under GM-2.

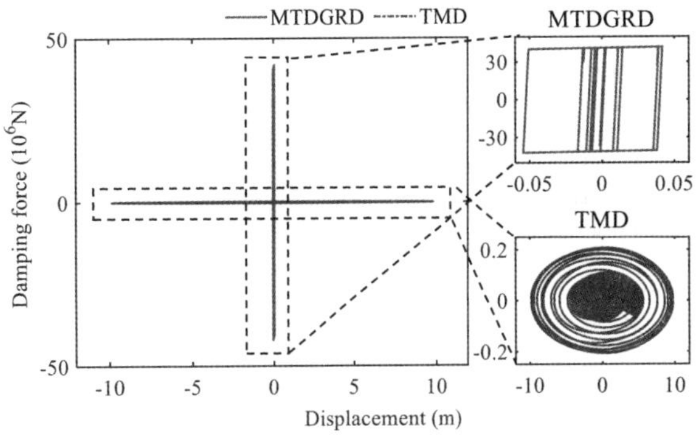

Fig. 36. Hysteresis curve of MTDGRD and TMD.

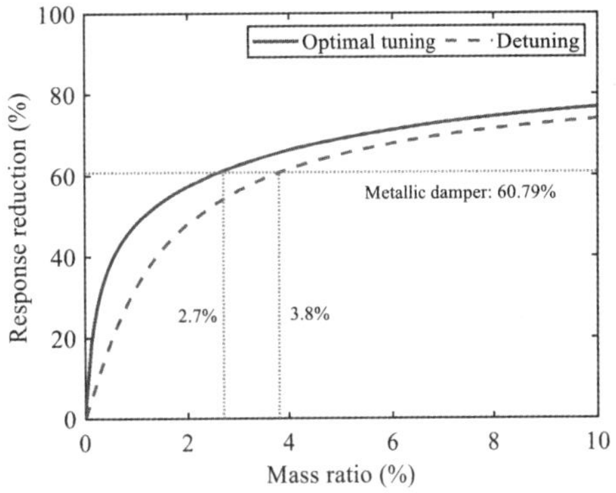

Fig. 37. The impact of TMD mass ratio on response reduction.

(1) The plastic damping capacity of aluminum 1060# and aluminum 6061# materials can be more adequately utilized under torsional plastic deformation. Aluminum 6061# of the energy dissipation capacity is higher than aluminum 1060#.
(2) The dynamical tests show that the loading frequency and operational cycles have limited effect on mechanical behavior and energy consumption capacity of the MTDGRD. The proposed dynamic mechanical model of the MTDGRD has considerable accuracy in numerical simulations.
(3) The maximum RMS of displacement of the structure has the best control performance under seismic ground motions numbered GM-1 when the MTDGRD is installed on the first floor and ground, and the maximum response reduction is 40.13 %. Additionally, the control performances of an optimized damper designed for one particular seismic ground motions also apply to other seismic ground motions with similar characteristics.
(4) The average response reduction rate of MTDGRD can reach 55.32 %, while the TMD of 0.5 % mass ratio can only reach 38.99 %. Compared with optimized TMD of 0.5 % mass ratio, the MTDGRD has a better seismic performance, while the mass of the aluminum rod of MTDGRD is only 0.115 % of the TMD mass. More mass is needed to achieve the same control performance of the MTDGRD when the TMD is detuned.

CRediT authorship contribution statement

Zhengqing Chen: Supervision, Conceptualization. **Bei Chen:** Methodology, Investigation. **Chao Zhou:** Methodology, Investigation. **Guokun Liu:** Methodology, Investigation. **Wenxi Wang:** Writing – review & editing, Supervision, Methodology, Funding acquisition, Conceptualization. **Sheng Chen:** Writing – review & editing, Writing – original draft, Methodology, Investigation.

Declaration of Competing Interest

The authors declare that they have no known competing financial interests or personal relationships that could have appeared to influence the work reported in this paper.

Data Availability

All data, models, or code that support the findings of this study are available from the corresponding author upon reasonable request.

Acknowledgments

This work was supported by the National Natural Science Foundation of China (Grant Nos. 52278304, 52378507); the Science and Technology Innovation Program of Hunan Province (Grant No. 2023RC3090); the Changsha Outstanding Innovative Youth Culturing Program (Grant No. kq2209009); the Hunan Provincial Natural Science Foundation of China (Grant No. 2023JJ30129).

References

[1] Reggio A, Angelis MD. Optimal energy-based seismic design of non-conventional Tuned Mass Damper (TMD) implemented via inter-story isolation. Earthq Eng Struct Dyn 2015;44(10):1623–42.
[2] Liu Y, Wu J, Donà M. Effectiveness of fluid-viscous dampers for improved seismic performance of inter-storey isolated buildings. Eng Struct 2018;169:276–92.
[3] Dong B, Sause R, Ricles JM. Seismic response and performance of a steel MRF building with nonlinear viscous dampers under DBE and MCE. J Struct Eng 2016; 142(6):04016023.
[4] Gong S, Zhou Y. Experimental study and numerical simulation on a new type of viscoelastic damper with strong nonlinear characteristics. Struct Control Health Monit 2017;24(4):e1897.
[5] Nasab MSE, Kim J. Fuzzy analysis of a viscoelastic damper in seismic retrofit of structures. Eng Struct 2022;250:113473.
[6] Jarrahi H, Asadi A, Khatibinia M, et al. Optimal design of rotational friction dampers for improving seismic performance of inelastic structures. J Build Eng 2020;27:100960.
[7] Javidan MM, Kim J. Seismic retrofit of soft-first-story structures using rotational friction dampers. J Struct Eng 2019;145(12):04019162.
[8] Wang L, Zhou Y, Nagarajaiah S, et al. Bi-directional semi-active tuned mass damper for torsional asymmetric structural seismic response control. Eng Struct 2023;294: 116744.
[9] Azimi M, Rasoulnia A, Lin Z, et al. Improved semi-active control algorithm for hydraulic damper-based braced buildings. Struct Control Health Monit 2017;24 (11):e1991.
[10] Oliveira F, Botto MA, Morais P, et al. Semi-active structural vibration control of base-isolated buildings using magnetorheological dampers. J Low Freq Noise, Vib Act Control 2018;37(3):565–76.
[11] YAMAZAKI I., HIRAMOTO K. Semi-active vibration control for structural systems with a variable inertia mass damper[C]//The Proceedings of the International Conference on Motion and Vibration Control 2020.15. The Japan Society of Mechanical Engineers, 2020: 10074.
[12] Wang L, Nagarajaiah S, Shi W, et al. Semi-active control of walking-induced vibrations in bridges using adaptive tuned mass damper considering human-structure-interaction. Eng Struct 2021;244:112743.
[13] Maślanka M. Optimised semi-active tuned mass damper with acceleration and relative motion feedbacks. Mech Syst Signal Process 2019;130:707–31.
[14] Shi X, Guan X, Shen W, et al. A control strategy using negative stiffness and semi-active viscous damping for fully tracking active control force for bridge cables: principles and simulations. Struct Control Health Monit 2022;29(9):e2989.
[15] Zhou K, Zhang JW, Li QS. Control performance of active tuned mass damper for mitigating wind-induced vibrations of a 600-m-tall skyscraper. J Build Eng 2022; 45:103646.
[16] Kras A, Gardonio P. Active vibration control unit with a flywheel inertial actuator. J Sound Vib 2020;464:114987.
[17] Zhang C, Wang H. Swing vibration control of suspended structures using the Active Rotary Inertia Driver system: theoretical modeling and experimental verification. Struct Control Health Monit 2020;27(6):e2543.
[18] Zhou Y, Shao H, Cao Y, et al. Application of buckling-restrained braces to earthquake-resistant design of buildings: a review. Eng Struct 2021;246:112991.
[19] Wu J, Liang RJ, Wang CL, et al. Restrained buckling behavior of core component in buckling-restrained braces. Int J Adv Steel Constr 2012;8(3):212–25.
[20] Tremblay R, Bolduc P, Neville R, et al. Seismic testing and performance of buckling-restrained bracing systems. Can J Civ Eng 2006;33(2):183–98.
[21] Murase Y, Morishita K, Inoue K, Tateyama E. Structural design method of the long brace with axial hysteresis dampers at both ends (part 1 analysis on the buckling restraint conditions). J Struct Constr Eng 2004;578. 131–8.
[22] Li W, Wu B, Ding Y, et al. Experimental performance of buckling-restrained braces with steel cores of H-section and half-wavelength evaluation of higher-order local buckling. Adv Struct Eng 2017;20(4):641–57.
[23] Hao XY, Li HN, Li G, et al. Experimental investigation of steel structure with innovative H-type steel unbuckling braces. Struct Des Tall Spec Build 2014;23(14): 1064–82.
[24] Kuwahara S, Tada M, Yoneyama T, et al. A study on stiffening capacity of double-tube members. J Struct Constr Eng 1993;445(3):151–8.
[25] Dongbin Z, Xin N, Peng P, et al. Experimental study and finite element analysis of a buckling-restrained brace consisting of three steel tubes with slotted holes in the middle tube1. J Constr Steel Res 2016;124:1–11.
[26] Whittaker AS, Bertero VV, Thompson CL, et al. Seismic testing of steel plate energy dissipation devices. Earthq Spectra 1991;7(4):563–604.
[27] Tsai KC, Chen HW, Hong CP, et al. Design of steel triangular plate energy absorbers for seismic-resistant construction. Earthq Spectra 1993;9(3):505–28.
[28] Ming-Hsiang S, Wen-pei S, Cheer-Germ GO. Investigation of newly developed added damping and stiffness device with low yield strength steel. J Zhejiang Univ-Sci A 2004;5(3):326–34.
[29] Shih MH, Sung WP. A model for hysteretic behavior of rhombic low yield strength steel added damping and stiffness. Comput Struct 2005;83(12-13):895–908.
[30] Zibasokhan H, Behnamfar F, Azhari M. Experimental study of a new pure bending yielding dissipater. Bull Earthq Eng 2019;17:4389–410.
[31] Javidana Mohammad, Mahdi, Kim Jinkoo. Steel hexagonal damper-brace system for efficient seismic protection of structures. Steel Compos Struct 45 5 2022: 683–95.
[32] Javidan Mohammad, Mahdi Seungho, Chun, Kim Jinkoo. Experimental study on steel hysteretic column dampers for seismic retrofit of structures. Steel Compos Struct 40 4 2021:495–509.
[33] Javidan MM, Kim J. A ductile steel damper-brace for low-damage framed structures. Steel Compos Struct 2022;44(3):311–23.
[34] Nakashima M, Iwai S, Iwata M, et al. Energy dissipation behaviour of shear panels made of low yield steel. Earthq Eng Struct Dyn 1994;23(12):1299–313.
[35] Abebe DY, Jeong SJ, Getahune BM, et al. Hysteretic characteristics of shear panel damper made of low yield point steel. Mater Res Innov 2015;19(S5):S5-902–5-910.
[36] Chen Z., Ge H., Usami T. Hysteretic performance of shear panel dampers. 2002, (2): 1223–1228.
[37] Chen Z, Ge H, Usami T. Hysteretic model of stiffened shear panel dampers. J Struct Eng 2006;132(3):478–83.
[38] Yao Z, Wang W, Zhu Y. Experimental evaluation and numerical simulation of low-yield-point steel shear panel dampers[J]. Eng Struct 2021;245:112860.
[39] Zhu Y, Wang W, Lu Y, et al. Finite element modeling and design recommendations for low-yield-point steel shear panel dampers. J Build Eng 2023;72:106634.
[40] Dou C, Pi YL, Gao W. Shear resistance and post-buckling behavior of corrugated panels in steel plate shear walls. Thin-Walled Struct 2018;131:816–26.
[41] Dou C, Cheng X, Zhao YY, et al. Shear resistance and design of infill panels in corrugated-plate shear walls. J Struct Eng 2021;147(11):04021179.
[42] Quan C, Wang W, Li Y. Hysteretic model and resilient application of corrugated shear panel dampers. Thin-Walled Struct 2022;178:109477.
[43] Chan RWK, Albermani F. Experimental study of steel slit damper for passive energy dissipation. Eng Struct 2008;30(4):1058–66.
[44] Ch. Lee Y, Ju J, Min S, Lho S, Kim. Non-uniform steel strip dampers subjected to cyclic loadings. Eng Struct 2015;99:192–204.
[45] Ch. Lee S, Lho D, Kim J, Oh Y, Ju. Hourglass-shaped strip damper subjected to monotonic and cyclic loadings. Eng Struct 2016;119:122–34.
[46] Moghaddam SH, Shooshtari A. Numerical and experimental investigation on seismic performance of proposed steel slit dampers. J Constr Steel Res 2023;200: 107646.
[47] Naeem A, Kim J. Seismic performance evaluation of a multi-slit damper. Eng Struct 2019;189:332–46.
[48] Naeem Asad, et al. Development and experimental verification of self-centering disc slit damper for buildings. J Constr Steel Res 2023;201:107759.
[49] Ke K, Chen Y, Zhou X, et al. Experimental and numerical study of a brace-type hybrid damper with steel slit plates enhanced by friction mechanism[J]. Thin-Walled Struct 2023;182:110249.
[50] Skinner RI, Kelly JM, Heine AJ. Hysteretic dampers for earthquake-resistant structures. Earthq Eng Struct Dyn 1974;3(3):287–96.
[51] Franco JM, Cahís X, Gracia L, et al. Experimental testing of a new anti-seismic dissipator energy device based on the plasticity of metals. Eng Struct 2010;32(9): 2672–82.
[52] Salem Milani A, Dicleli M. Systematic development of a new hysteretic damper based on torsional yielding: part I—design and development[J]. Earthq Eng Struct Dyn 2016;45(6):845–67.

[53] Salem Milani A, Dicleli M. Systematic development of a new hysteretic damper based on torsional yielding: part II—experimental phase. Earthq Eng Struct Dyn 2016;45(5):779–96.
[54] Mahyari SL, Riahi HT, Hashemi M. Investigating the analytical and experimental performance of a pure torsional yielding damper. J Constr Steel Res 2019;161: 385–99.
[55] Lie W, Wu C, Luo W, et al. Cyclic behaviour of a novel torsional steel-tube damper. J Constr Steel Res 2022;188:107010.
[56] Wang D, Zhang Y, Fan G, et al. Experimental research and theoretical analysis of mechanical behavior of torsional metallic damper. Int J Civ Eng 2023;21(5): 857–74.
[57] Chen S, Wang W, Zhou C, et al. Damping capacity and seismic performance of a torsional metallic damper using a displacement amplification mechanism. J Bridge Eng 2023;28(10):04023071.
[58] Ikago K, Saito K, Inoue N. Seismic control of single-degree-of-freedom structure using tuned viscous mass damper. Earthq Eng Struct Dyn 2012;41(3):453–74.
[59] Ma R, Bi K, Hao H. A novel rotational inertia damper for amplifying fluid resistance: Experiment and mechanical model. Mech Syst Signal Process 2021;149: 107313.
[60] Zhu H, Li Y, Shen W, et al. Mechanical and energy-harvesting model for electromagnetic inertial mass dampers. Mech Syst Signal Process 2019;120: 203–20.
[61] Wang ZH, Xu YW, Gao H, et al. Vibration control of a stay cable with a rotary electromagnetic inertial mass damper. Smart Struct Syst 2019;23(6):627–39.
[62] Luo Y, Sun H, Wang X, et al. Parametric optimization of electromagnetic tuned inerter damper for structural vibration suppression. Struct Control Health Monit 2021;28(5):e2711.
[63] Wang Z, Cheng Z, Yin G, et al. A magnetic negative stiffness eddy-current inertial mass damper for cable vibration mitigation. Mech Syst Signal Process 2023;188: 110013.
[64] Zhang HY, Chen ZQ, Hua XG, et al. Design and dynamic characterization of a large-scale eddy current damper with enhanced performance for vibration control. Mech Syst Signal Process 2020;145:106879.
[65] Zhao G, Ma Y, Yang Z, et al. Friction mechanism and experimental investigation of the response-amplified friction damper. Struct Control Health Monit 2022;29(7): e2953.
[66] Li S, Li Y, Mao W, et al. A novel 500 kN axial eddy current damper using rack and gear mechanism: Design, testing, and evaluation. Struct Control Health Monit 2022;29(10):e3022.
[67] Li HN, Huang Z, Fu X. Study on the performance of a gear-driven rotation-amplified rubber viscoelastic damper and its vibration control of a frame structure. Struct Control Health Monit 2020;27(11):e2617.
[68] Berton S, Bolander JE. Amplification system for supplemental damping devices in seismic applications. J Struct Eng 2005;131(6):979–83.
[69] Saitoh M. On the performance of gyro-mass devices for displacement mitigation in base isolation systems. Struct Control Health Monit 2012;19(2):246–59.
[70] Hessabi RM, Mercan O. Investigations of the application of gyro-mass dampers with various types of supplemental dampers for vibration control of building structures. Eng Struct 2016;126:174–86.
[71] Li Y, Li S, Wang J, et al. A new type of damper combining eddy current damping with rack and gear. J Vib Control 2021;27(9-10):1087–97.
[72] Faridmehr Iman, et al. Correlation between engineering stress-strain and true stress-strain curve. Am J Civ Eng Archit 2014;2(1):53–9.
[73] Craig Jr, Roy R., and Eric M. Taleff Mechanics of materials. John Wiley & Sons, 2020.
[74] Lu X, Li P, Guo X, et al. Vibration control using ATMD and site measurements on the Shanghai World Financial Center Tower. Struct Des Tall Spec Build 2014;23(2): 105–23.
[75] Kourakis, Ioannis. Structural systems and tuned mass dampers of super-tall buildings: case study of Taipei 101. Diss. Massachusetts Institute of Technology, 2007.
[76] Applied Technology Council. Quantification of building seismic performance factors[M]. US Department of Homeland Security, FEMA, 2009.
[77] De Domenico D, Ricciardi G, Takewaki I. Design strategies of viscous dampers for seismic protection of building structures: a review. Soil Dyn Earthq Eng 2019;118: 144–65.
[78] Anajafi H, Medina RA. Comparison of the seismic performance of a partial mass isolation technique with conventional TMD and base-isolation systems under broad-band and narrow-band excitations. Eng Struct 2018;158:110–23.
[79] Den Hartog J.P. Mechanical vibrations[M]. Courier Corporation, 1985.
[80] Warburton GB. Optimum absorber parameters for various combinations of response and excitation parameters. Earthq Eng Struct Dyn 1982;10(3):381–401.

Mechanical Systems and Signal Processing 220 (2024) 111639

Contents lists available at ScienceDirect

Mechanical Systems and Signal Processing

journal homepage: www.elsevier.com/locate/ymssp

A novel tuned hydrodynamic mass damper employing added mass and damping from interaction between heave plate and fluids

Wenxi Wang [a,b,c], Jun Zhou [a,b], Tianfu Yu [a,b], Sheng Chen [a,b,*], Yifan Luo [d], Xugang Hua [a,b,c]

[a] *State Key Laboratory of Bridge Safety and Resilience, Hunan University, Changsha 410082, China*
[b] *Key Laboratory for Bridge and Wind Engineering of Hunan Province, College of Civil Engineering, Hunan University, Changsha 410082, China*
[c] *Shock and Vibration Technology Research Center, College of Civil Engineering, Hunan University, Changsha 410082, China*
[d] *Hunan Provincial Key Laboratory of Structural Engineering for Wind Resistant and Vibration Control, Hunan University of Science and Technology, Xiangtan 411201, China*

A R T I C L E I N F O

Keywords:
Vibration control
Tuned hydrodynamic mass damper
Parameter identification
Vibration reduction experiment
Robustness analysis

A B S T R A C T

Traditional tuned mass damper (TMD) devices requires heavier masses and additional dampers to improve their vibration control performance. In this paper, a novel tuned hydrodynamic mass damper (THMD) is proposed for vibration control employing added mass and damping from the interaction between heave plate and fluids. The added mass can reduce the weight of the mass block and improve the performance of conventional TMD devices. Moreover, the added damping can replace fluid or eddy current dampers in the traditional TMD system and result in lower costs. The equations of motion of the THMD-single-degree-of-freedom (SDOF) system is established with the help of the Morison equation. The corresponding semi-analytical solutions are obtained using the harmonic balance method, and the approximate analytical solutions of optimal parameters are also derived based on the H_∞ criterion. The control performance and robustness of THMD are compared with the conventional TMD and the heave plate cases by numerical analysis. Moreover, the parametric identification of the Morison equation is performed by a forced experiment in water and the accuracy is verified through Computational Fluid Dynamics(CFD) analysis. Finally, a controlled structure and a THMD are manufactured in the laboratory where the THMD is designed by the analytically optimal solutions. The results show that the semi-analytical solutions of motion equations derived by the harmonic balance method agree well with the numerical optimization solution, and analytical solutions of optimal parameters of THMD are highly accurate. The THMD outperforms the conventional TMD and heave plate under broadband harmonic loads according to the numerical and experimental analysis. the THMD outperforms the conventional TMD and heave plate under broadband harmonic loads. When the ratio of mass μ is set to 0.045, a THMD with a added mass ratio β of 0.031 has a 32.7% reduction in peak value of DMF compared to a conventional TMD.

1. Introduction

The application of new materials and technologies lead to the increasing height and span of structures, while the stiffness of the

* Corresponding author.
E-mail address: cs2023@hnu.edu.cn (S. Chen).

https://doi.org/10.1016/j.ymssp.2024.111639
Received 7 February 2024; Received in revised form 10 June 2024; Accepted 11 June 2024
Available online 17 June 2024

structure continues to decrease. Under external loads such as winds and earthquakes, flexible structures are prone to large-amplitude vibrations [1–6], which can threaten structural safety. In recent decades, tuned mass damper (TMD), also called the dynamic vibration absorber (DVA), is introduced for vibration control of structures to safeguard human lives and property [7]. TMD consists of a damper and a spring-mass system, which has a similar natural frequency to the primary structure. Frahm [8] initially proposed undamped TMD based on the principle of anti-resonance in 1909. Subsequently, many scholars proposed some classic optimization theories. Den Hartog [9] proposed the optimization theory of undamped and damped TMD in 1928. According to fixed points theory, Den Hartog derived an analytical solution of optimization parameters for the TMD when an undamped primary structure is subjected to harmonic excitation. Based on the Den Hartog optimization criteria, Warburton [10] investigated the optimal parameters of TMD under different excitation and excitation positions. Grigoriu and Soong [11] studied various analytical solutions of TMD parameters for different minimization criterions under various white noise. De Angelis M [12] studied optimal design of structures with large mass ratio TMD.

In recent years, TMDs have been used for structural vibration control under various loads, such as wind [13–18], earthquake [19–26], human-induced excitation [27–31], etc. Kang et al. [32] conducted a study on a 39-story steel building equipped with monitoring systems to measure building and TMD vibrations during the Kompasu Typhoon in 2010. The results demonstrate that the TMD significantly reduced typhoon-induced vibrations. Morga and Marano [33] model slender structures as single-degree-of-freedom (SDOF) systems to examine the efficacy of TMD in reducing wind-induced vibrations. In that study, the authors compared two optimization criteria for TMD design and used them to design TMD parameters for vibration reduction in a lighting tower under wind load conditions. Xing et al. [34] proposed a TMD-type counterweight that utilizes the counterweight initially intended to mitigate live load-induced displacements in the central span as the mass block in the TMD. The buffeting responses of the bridge with the stationary counterweight and the TMD-type counterweight indicate that the TMD has more significant effects in reducing wind-induced vibrations. Moreover, in Domenico and Ricciardi's study [35], base isolation is combined with a basement-located TMD to improve seismic performance in a non-conventional application. The experimental and simulation results show advantages over the fixed base and conventional base isolation in terms of performance indicators such as the deformation of the isolators, the displacement demand of the structure, the base shear, the inter-story drifts, and the shear forces and bending moments on the beam-column members. Pourzeynali et al. [36] used Genetic Algorithms to optimize TMDs for earthquake vibrations, focusing on three key objectives: minimizing displacement, velocity, and acceleration of floors. The optimized TMD can reduce the building's maximum displacement by around 28 %. Wen et al. [37] investigated the effectiveness of TMD in controlling human-induced vibrations of a curved cable-stayed bridge. Field test shows that measured vertical acceleration is 40 % lower than predicted at the same pedestrian density, which indicates that the TMD increases damping ratios and prevents large-amplitude vibrations. Liu et al. [38] investigated the vibration control of a cantilevered floor using tuned mass dampers (TMDs) through dynamic testing and numerical simulations. The installation of TMDs resulted in a significant reduction in the acceleration response, achieving a maximum reduction effect of 64 %. To reduce the mass of TMDs, Marian and Giaralis [39] combined the classic TMD with an inerter, proposing the Tuned Mass Damper Inerter (TMDI). The TMDI uses the large inertial mass simulated by the inerter to enhance the performance of the classic TMD. Subsequently, mathematical models and various configurations of the TMDI have been developed [40–42].

Therefore, TMDs are widely used for structural vibration control due to their excellent performance. Viscous fluid dampers are the most commonly used damping element device in TMDs. However, viscous fluid dampers have a problem of fluid leakage [43,44], which degrades the performance of the viscous damper and affects the performance of the TMD in long-term operation. In addition, the mass of the TMD is taken to a larger value when the TMD is designed to control structural vibration under seismic loads [45–47]. Based on the TMD principle, some researchers proposed roof-garden [48,49] and swimming pools [50] as the mass of TMD to control structural vibrations, where the mass of roof-garden and swimming pool is also large. Larger mass of TMD may cause that additional reinforcement of the stiffness between structural layers is required, which increases the cost and difficulty of the construction.

To overcome the shortcomings of traditional TMDs, such as fluid leakage and the need for large mass blocks, a novel tuned hydrodynamic mass damper (THMD) is proposed for vibration control employing added mass and damper form interaction between heave plate and fluid. The added mass can reduce the weight of the mass block and improve the performance of conventional TMD.

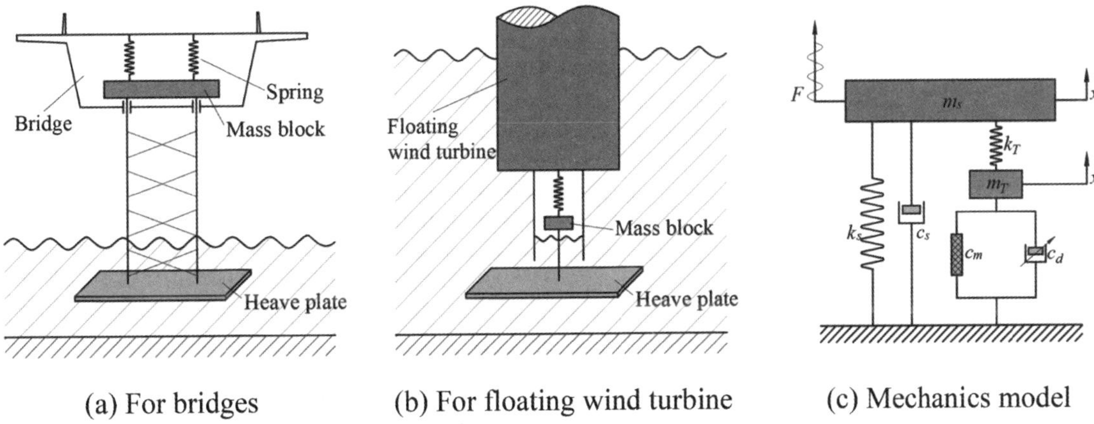

Fig. 1. Possible setup of THMD in practices and mechanics model.

Moreover, the added damping can replace traditional mechanical dampers and result in lower costs. The motion equation of the THMD-SDOF system is established with the help of the Morison equation. The harmonic balance method is used to derive the semi-analytical solutions, and the approximate analytical solutions of optimal parameters are derived based on the H_∞ criterion. The control performance and robustness of THMD are compared with conventional TMD and heave plates using numerical analysis. Moreover, the accuracy of the Morison model and parameter identification is verified through experiments. Finally, a controlled structure and a THMD are manufactured in the laboratory to verify the correctness of theoretical derivation and the control performance and robustness of THMD.

2. Parametric optimization of THMD

2.1. *Simplified model and equations of motion of THMD-SODF structure*

The THMD, which mainly consists of a mass block, a tuning spring and a heave plate, can be used to mitigate excessive vibrations for bridges and floating wind turbines, as shown in Fig. 1(a) and (b). Its simplified mechanical model is shown in Fig. 1(c). The mass block is attached to the controlled structure by a spring. The other side of the mass block is rigidly connected to the heave plate, and the heave plate is submerged in the fluid. When the controlled structure vibrates under external loads, the mass and the heave plate resonate with the controlled structure to absorb energy. At the same time, a damping force is generated by the interaction between the heave plate and the fluid to dissipate the energy.

According to the results of hydrodynamic characterization experiments on flat plates, the total force given by the fluid to the heave plate can be divided into an added damping force term and an added mass term when boundary effects are not considered. The added damping force is proportional to the square of the heave velocity, representing the viscous damping effects caused by the heave plate. On the other hand, the added mass term is proportional to the heave acceleration, representing the inertia effects due to the fluid mass surrounding the heave plate. Therefore, the forces on the heave plate during harmonic motion in the fluid can be described by the Morison equation [51,52] as shown below,

$$f_r = -\rho L^3 c_m \ddot{x}(t) - \frac{1}{2}\rho L^2 c_d \dot{x}(t)|\dot{x}(t)| \tag{1}$$

where $\ddot{x}$ and $\dot{x}$ are the acceleration and velocity of the harmonic vibration respectively. ρ is the density of the fluid, L is the characteristic length of the heave plate, and c_m and c_d are the dimensionless added mass and added damping coefficients.

Considering the controlled structures a SDOF structure in Fig. 1, and introducing the Morison equation to describe the hydrodynamic characteristics of the heave plate, the equations of motion of the THMD-SDOF system are established as,

$$\begin{cases} m_s\ddot{x}_s(t) + c_s\dot{x}_s(t) + k_s x_s(t) + k_T(x_s(t) - x_T(t)) = F(t) \\ m_T\ddot{x}_T(t) + k_T(x_T(t) - x_s(t)) + \rho L^3 c_m \ddot{x}_T(t) + \frac{1}{2}\rho L^2 c_d \dot{x}_T(t)\left|\dot{x}_T(t)\right| = 0 \end{cases} \tag{2}$$

where x_s and x_T are the displacements of the principal structure and substructure respectively, k_s and k_T are the stiffnesses of the controlled structure and the THMD, c_s is the damping coefficient of the controlled structure, $F(t)$ are the external loads, and c_m and c_d are the added mass coefficient and added damping coefficient of the THMD respectively. It is notable that the c_m and c_d are the added masse and damping generated by all components submerged in water. To convert the equations of motion to dimensionless form, the dimensionless parameters are introduced as shown in Table 1.

Assuming the external load $F(t) = F_0\cos(\omega t)$, and substituting the dimensionless parameters into Eq. (2), the dimensionless form of the equations of motion is obtained as,

Table 1
Definition of parameters.

Symbol	Definition
$\omega_s = \sqrt{k_s/m_s}$	Frequency of the primary structure
$\omega_T = \sqrt{k_T/(\rho L^3 c_m + m_T)}$	Frequency of THMD
$\mu = m_T/m_s$	Mass ratio of THMD
$\beta = \rho L^3 c_m/m_s$	Added mass ratio of THMD
$\Omega = \mu + \beta$	Total mass ratio of THMD
$\zeta_s = c_s/2m_s\omega_s$	Damping ratio of the primary structure
$\gamma_d = \rho L^2 c_d/2m_s$	Added damping ratio of the primary structure
$\nu = \omega_T/\omega_s$	Frequency ratio of THMD to the primary structure
$\lambda = \omega/\omega_s$	Frequency ratio of excitation to the primary structure
$f = F_0/k_s$	Dimensionless amplitude of excitation
$\tau = \omega t$	Dimensionless time

$$\begin{cases} \lambda^2\ddot{x}_s + 2\lambda\zeta_s\dot{x}_s + x_s + \Omega\upsilon^2(x_s - x_T) = f\cos(\tau) \\ \Omega\lambda^2\ddot{x}_T + \Omega\upsilon^2(x_T - x_s) + \gamma_d\lambda^2\dot{x}_T\left|\dot{x}_T\right| = 0 \end{cases} \tag{3}$$

2.2. Semi-analytical solution of tuning parameters

It can be seen from Eq. (3) that the damping term generated by THMD is nonlinear. To obtain an approximate solution to the nonlinear equations of motion, the equations are derived using the harmonic balance method [53–55]. For harmonic excitation, the dimensionless displacements of the main structure and substructure can be expressed as,

$$\begin{cases} x_s(\tau) = A_s\sin(\tau - \varphi_s) \\ \dot{x}_s(\tau) = A_s\cos(\tau - \varphi_s) \\ \ddot{x}_s(\tau) = -A_s\sin(\tau - \varphi_s) \end{cases} \quad \begin{cases} x_T(\tau) = A_T\sin(\tau - \varphi_T) \\ \dot{x}_T(\tau) = A_T\cos(\tau - \varphi_T) \\ \ddot{x}_T(\tau) = -A_T\sin(\tau - \varphi_T) \end{cases} \tag{4}$$

where A_s and A_T are the dimensionless steady amplitudes of the THMD displacements, and φ_s and φ_T are the phase differences between the primary structure displacements and THMD displacements with the external loads, respectively.

The added damping force is expanded in Fourier series, and the first-order term is retained, we have[56]

$$\dot{x}_T\left|\dot{x}_T\right| \cong A\cos\tau + B\sin\tau \tag{5}$$

where A and B can be calculated from the orthogonality of the trigonometric functions as shown below,

$$A = \frac{1}{\pi}\int_0^{2\pi} A_T\cos(\tau - \varphi_T)|A_T\cos(\tau - \varphi_T)|\cos\tau d\tau = \frac{8}{3\pi}A_T^2\cos\varphi_T \tag{6}$$

$$B = \frac{1}{\pi}\int_0^{2\pi} A_T\cos(\tau - \varphi_T)|A_T\cos(\tau - \varphi_T)|\sin\tau d\tau = \frac{8}{3\pi}A_T^2\sin\varphi_T$$

By substituting Eq. (6) into Eq. (5), it is obtained,

$$\dot{x}_T\left|\dot{x}_T\right| \cong \frac{8}{3\pi}A_T^2\cos(\tau - \varphi_T) \tag{7}$$

Then Eq. (5) can be derived as,

$$\dot{x}_T\left|\dot{x}_T\right| \cong \Phi A_T\dot{x}_T \tag{8}$$

where

$$\Phi = \frac{8}{3\pi} \tag{9}$$

Substituting Eqs. (4) and (8) into Eq. (3),

$$\begin{cases} [\lambda^2A_s\sin\varphi_s + 2\zeta_s\lambda A_s\cos\varphi_s - \Omega\upsilon^2(A_s\sin\varphi_s - A_T\sin\varphi_T) - A_s\sin\varphi_s]\cos(\tau) - \\ [\lambda^2A_s\cos\varphi_s - 2\zeta_s\lambda A_s\sin\varphi_s - \Omega\upsilon^2(A_s\cos\varphi_s - A_T\cos\varphi_T) - A_s\cos\varphi_s]\sin(\tau) = f\cos(\tau) \\ [\Omega\upsilon^2(A_s\sin\varphi_s - A_T\sin\varphi_T) + \Omega\lambda^2A_T\sin\varphi_T + \Phi\gamma_d\lambda^2A_T^2\cos\varphi_T]\cos(\tau) - \\ [\Omega\upsilon^2(A_s\cos\varphi_s - A_T\cos\varphi_T) + \Omega\lambda^2A_T\cos\varphi_T - \Phi\gamma_d\lambda^2A_T^2\sin\varphi_T]\sin(\tau) = 0 \end{cases} \tag{10}$$

According to the harmonic balance method, the coefficients of the same harmonic terms should be equal, the equation can be obtained as follows,

$$\begin{cases} \lambda^2A_s\sin\varphi_s + 2\zeta_s\lambda A_s\cos\varphi_s - \Omega\upsilon^2(A_s\sin\varphi_s - A_T\sin\varphi_T) - A_s\sin\varphi_s = f \\ \lambda^2A_s\cos\varphi_s - 2\zeta_s\lambda A_s\sin\varphi_s - \Omega\upsilon^2(A_s\cos\varphi_s - A_T\cos\varphi_T) - A_s\cos\varphi_s = 0 \\ \Omega\upsilon^2(A_s\sin\varphi_s - A_T\sin\varphi_T) + \Omega\lambda^2A_T\sin\varphi_T + \Phi\gamma_d\lambda^2A_T^2\cos\varphi_T = 0 \\ \Omega\upsilon^2(A_s\cos\varphi_s - A_T\cos\varphi_T) + \Omega\lambda^2A_T\cos\varphi_T - \Phi\gamma_d\lambda^2A_T^2\sin\varphi_T = 0 \end{cases} \tag{11}$$

Then Eq. (11) can be reduced to

$$\begin{cases} -\Omega\upsilon^2 A_T \sin(\varphi_s - \varphi_T) + 2\zeta_s \lambda A_s = f\cos\varphi_s \\ \Omega\upsilon^2 A_T \cos(\varphi_s - \varphi_T) + \lambda^2 A_s - A_s - \Omega\upsilon^2 A_s = f\sin\varphi_s \\ \Omega\upsilon^2 A_s \sin(\varphi_s - \varphi_T) + \Phi\gamma_d \lambda^2 A_T^2 = 0 \\ \Omega\upsilon^2 A_s \cos(\varphi_s - \varphi_T) + \Omega\lambda^2 A_T - \Omega\upsilon^2 A_T = 0 \end{cases} \tag{12}$$

Given the total mass ratio, the frequency ratio υ, the primary structure damping ratio ζ_s, the excitation frequency ratio λ, and the dimensionless excitation amplitude f, and substituting φ, the dimensionless amplitudes A_s and A_T of the primary structure and THMD, and the phases φ_s and φ_T of the primary structure and THMD can be obtained by solving Eq. (12). The dynamic amplification factor (DMF) of the primary structure can be expressed as,

$$\mathrm{DMF}_s(\lambda) = \frac{A_s(\lambda)}{f} \tag{13}$$

The excitation of different frequencies is substituted into Eq. (3) and solved using the fourth-order Runge-Kutta method. The time step is set to 0.001 to obtain the numerical solution of the DMF_s. Then the semi-analytical solution of the DMF_s is calculated using Eq. (13) derived from the harmonic balance method. The results of the comparison between the numerical solution and the semi-analytical solution are shown in Fig. 2. It can be seen from Fig. 2 that the numerical and semi-analytical solutions agree well. They are only slightly different at peaks. However, since the semi-analytical solution only requires solving the equations and does not require tedious numerical time calculations, the semi-analytical solutions can significantly reduce the computational cost.

H_∞ optimization criterion is typically used when a structure is submitted to harmonic loading, and the objective is to minimize the maximum response over the entire range of possible sinusoidal loading frequencies. H_2 optimization is typically used when the structure is subjected to random loading, and the objective is to minimize the response variance. Single-frequency dominant vibrations caused by relatively simple loads, such as vortex-induced vibrations and human-induced vibrations, are primarily considered in this paper. These vibrations can be regarded as harmonic loads acting on the structure. Therefore, H_∞ optimization criterion is chosen for this study. However, structures will also be subjected to random loads, such as buffeting from wind loads and random vibrations caused by earthquake waves and other environmental forces, where H_2 optimization would be more appropriate. Therefore, the H_2 optimization criterion for random vibrations should be studied in future.

In SDOF structural vibration control, the performance of the THMD depends on the total mass ratio Ω, structural damping ratio ζ_s, tuning frequency ratio υ and additional damping ratio γ_d. Meanwhile, the THMD performance is influenced by the external load frequency ratio f due to the existence of nonlinear terms. The maximum value of DMFs(λ) can be used to evaluate the vibration control performance of TMD. A lower maximum value of DMFs(λ) indicates superior vibration damping performance. According to the H_∞ optimization criterion, for a given Ω, ζ_s and f, there exists a set of optimal values for the parameters υ and γ_d minimizing $\mathrm{DMF}_s(\lambda)$. Therefore, in the numerical search of optimal parameters, the objective function can be expressed as,

$$\min_{r,\gamma_d}\max(\mathrm{DMF}_s(\lambda)) \tag{14}$$

The optimal parameter search function is written in MATLAB, with Ω and f as independent variables, where the range of total mass ratio Ω is set to [0.01 0.1] and the range of dimensionless excitation amplitude f is set to [0.0001 0.01]. The damping ratio ζ_s of the primary structure is set to 0, the range of tuning frequency ratio υ is set to [0.9 1.1] and the dimensionless added damping ratio γ_d is set to [0 100]. In solving the Eqs. (12), the interval of the excitation frequency ratio λ is set to 0.002. The results of the optimal parameters

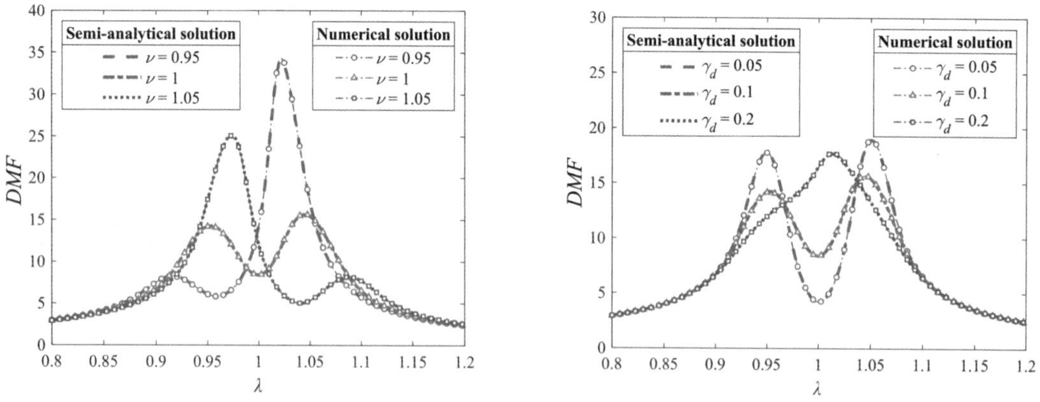

(a) Different tuning frequency ratios (γ_d=0.1) (b) Different additional damping ratios (υ=1)

Fig. 2. Comparison of DMF curves employing semi-analytical and numerical solutions.

obtained are shown in Table 2. It can be seen from Table 2 that the optimum frequency ratio υ and the added damping ratio γ_d increase with the total mass ratio Ω for all conditions. However, the dimensionless added damping ratio γ_d is not only related to the mass ratio Ω, but also to the dimensionless excitation amplitude f. The smaller the excitation amplitude, the larger the optimum added damping ratio.

The $\mathrm{DMF_s}$ curves at the optimum parameters obtained by the numerical search based on the semi-analytical solution are given in Fig. 3. It can be seen that the $\mathrm{DMF_s}$ curves have double equal peaks for different mass ratios, which indicates that the results of the numerical search are accurate. In addition, a smaller the value of DMF and better the vibration control effect can be achieved by increasing the total mass ratio when the external load amplitude ratio and the main structure damping ratio are set to a certain value.

2.3. *Approximate analytical solution of tuning parameters*

The analysis above demonstrates that the semi-analytical solution based on the harmonic balance method is accurate and can be used for parameter search to reduce the computational cost of the numerical solution. However, it is necessary to carry out the derivation of an approximate analytical solution in order to obtain a concise and explicit design formula. Assuming the existence of a linear damping factor c_T making the equations of motion equivalent to Eq. (2), the equivalent equation is given as follows,

$$\begin{cases} m_s\ddot{x}_s(t)+c_s\dot{x}_s(t)+k_sx_s(t)+k_T(x_s(t)-x_T(t))=F(t) \\ m_T\ddot{x}_T(t)+k_T(x_T(t)-x_s(t))+\rho L^3c_m\ddot{x}_T(t)+c_T\dot{x}_T(t)=0 \end{cases} \tag{15}$$

Assuming that the external load $F(t)$ is harmonic excitation, $F(t)=F_0e^{i\omega t}$ is substituted into the equations of motion. Introducing the displacement transfer functions $H_s(i\omega)$ and $H_T(i\omega)$ for the primary structure and THMD, the steady solutions for the displacement, velocity and acceleration of the primary structure and THMD are expressed as,

$$\begin{cases} x_s(t)=H_s(i\omega)F_0e^{i\omega t} \\ \dot{x}_s(t)=i\omega H_s(i\omega)F_0e^{i\omega t} \\ \ddot{x}_s(t)=-\omega^2H_s(i\omega)F_0e^{i\omega t} \end{cases} \quad \begin{cases} x_T(t)=H_T(i\omega)F_0e^{i\omega t} \\ \dot{x}_T(t)=i\omega H_T(i\omega)F_0e^{i\omega t} \\ \ddot{x}_T(t)=-\omega^2H_T(i\omega)F_0e^{i\omega t} \end{cases} \tag{16}$$

Substituting Eq. (16) into Eq. (15), and eliminating $F_0e^{i\omega t}$, the expression of $H_s(i\omega)$ is obtained as,

$$H_s(i\omega)=\frac{A_1+B_1i}{C_1+D_1i} \tag{17}$$

$$H_T(i\omega)=\frac{A_2+B_2i}{C_2+D_2i} \tag{18}$$

where

$$\begin{cases} A_1=\rho L^3c_m\omega^2-k_T+m_T\omega^2 \\ B_1=-c_T\omega \\ C_1=\rho L^3c_mk_s\omega^2-k_sk_T+\rho L^3c_mk_T\omega^2-\rho L^3c_mm_s\omega^4+k_sm_T\omega^2 \\ \qquad +k_Tm_s\omega^2+k_Tm_T\omega^2-m_sm_T\omega^4+c_sc_T\omega^2 \\ D_1=-\omega\left(c_sk_T+c_Tk_s+c_Tk_T-c_sm_T\omega^2-c_Tm_s\omega^2-\rho L^3c_mc_s\omega^2\right) \end{cases} \tag{19}$$

Table 2
Optimization parameters of THMD.

	$f=0.01$, ζ_s=0		$f=0.005$, ζ_s=0		$f=0.001$, ζ_s=0		$f=0.0005$, ζ_s=0		$f=0.0001$, ζ_s=0	
Ω	υ	γ_d	υ	γ_d	υ	γ_d	υ	γ_d	υ	γ_d
0.01	1.0051	0.0015	1.0051	0.0029	1.0051	0.0147	1.0051	0.0293	1.0051	0.1467
0.02	1.0103	0.0085	1.0103	0.0170	1.0103	0.0850	1.0102	0.1693	1.0104	0.8543
0.03	1.0158	0.0240	1.0158	0.0481	1.0159	0.2415	1.0159	0.4830	1.0159	2.4172
0.04	1.0213	0.0502	1.0217	0.1019	1.0214	0.5047	1.0214	1.0096	1.0214	5.0539
0.05	1.0270	0.0897	1.0277	0.1826	1.0275	0.9096	1.0277	1.8263	1.0275	9.0901
0.06	1.0341	0.1485	1.0336	0.2938	1.0347	1.5054	1.0336	2.9373	1.0336	14.6849
0.07	1.0404	0.2233	1.0421	0.4605	1.0404	2.2321	1.0404	4.4642	1.0404	22.3286
0.08	1.0474	0.3213	1.0474	0.6425	1.0471	3.2003	1.0474	6.4253	1.0474	32.1268
0.09	1.0546	0.4443	1.0545	0.8873	1.0545	4.4367	1.0547	8.8897	1.0545	44.3722
0.1	1.0627	0.5993	1.0624	1.1949	1.0624	5.9736	1.0627	11.9894	1.0561	54.8794

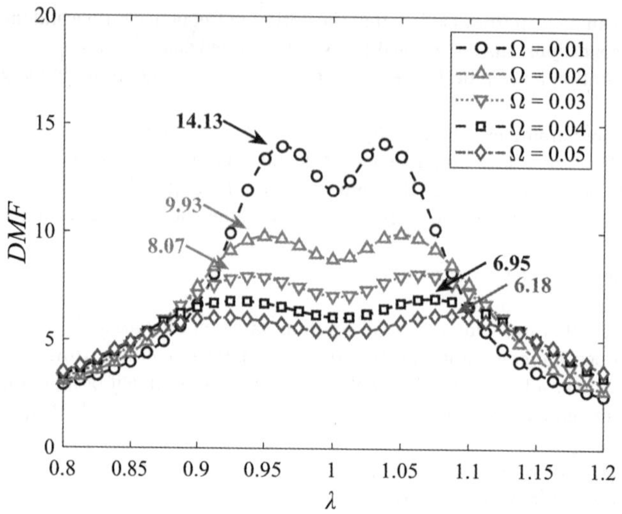

Fig. 3. Optimal DMF curves of THMD under different mass ratios ($f = 0.001$, $\zeta_s = 0$).

$$\begin{cases} A_2 = -k_T \\ B_2 = 0 \\ C_2 = \rho L^3 c_m k_s \omega^2 - k_s k_T + \rho L^3 c_m k_T \omega^2 - \rho L^3 c_m m_s \omega^4 + k_s m_T \omega^2 \\ \qquad + k_T m_s \omega^2 + k_T m_T \omega^2 - m_s m_T \omega^4 + c_s c_T \omega^2 \\ D_2 = -\omega\left(c_s k_T + c_T k_s + c_T k_T - c_s m_T \omega^2 - c_T m_s \omega^2 - \rho L^3 c_m c_s \omega^2\right) \end{cases} \tag{20}$$

Assuming the structural damping coefficient $c_s = 0$, introducing the the dimensionless parameter $\zeta_T = c_T/(2m_T\omega_T)$, and considering the dimensionless parameters in Table 1, the expression for the DMF can be obtained by multiplying k_s to Eqs. (17) and (18), as shown below,

$$\mathrm{DMF_s}(\lambda) = |H_s(i\omega)|k_s = \sqrt{\frac{A_3^2 + B_3^2\zeta_T^2}{C_3^2 + D_3^2\zeta_T^2}} \tag{21}$$

$$\mathrm{DMF_T}(\lambda) = |H_T(i\omega)|k_s = \sqrt{\frac{A_4^2 + B_4^2\zeta_T^2}{C_4^2 + D_4^2\zeta_T^2}} \tag{22}$$

where

$$\begin{cases} A_3 = -\Omega(\upsilon^2 - \lambda^2) \\ B_3 = -2\Omega\upsilon\lambda \\ C_3 = \Omega(\lambda^2 - \upsilon^2 - \lambda^4 + \upsilon^2\lambda^2 + \Omega\upsilon^2\lambda^2) \\ D_3 = -2\Omega\upsilon\lambda(\Omega\upsilon^2 - \lambda^2 + 1) \end{cases} \tag{23}$$

$$\begin{cases} A_4 = -\Omega\upsilon^2 \\ B_4 = 0 \\ C_4 = \Omega(\lambda^2 - \upsilon^2 - \lambda^4 + \upsilon^2\lambda^2 + \Omega\upsilon^2\lambda^2) \\ D_4 = -2\Omega\upsilon\lambda(\Omega\upsilon^2 - \lambda^2 + 1) \end{cases} \tag{24}$$

Den Hartog's fixed-point theory [9] provides a method for the optimal parameter design of the TMD, such as frequency ratio and damping ratio. According to the fixed-points theory, the DMF of the structure under different damping ratios ζ_T of THMD always passes through two fixed-points P and Q, which indicates that P and Q points are independent with ζ_T. Therefore, the value of $\mathrm{DMF_s}$ should be equal to each other when ζ_T is set to zero and infinity, and it leads to following relation,

$$\left|\frac{A_3}{C_3}\right| = \left|\frac{B_3}{D_3}\right| \tag{25}$$

Substituting Eq. (23) into Eq. (25). Since there is no solution when the absolute value symbol is expanded positively, the formula expanded negatively is retained as follows,

$$\frac{\lambda^2 - \upsilon^2}{\lambda^2 - \upsilon^2 - \lambda^4 + \upsilon^2\lambda^2 + \Omega\upsilon^2\lambda^2} = -\frac{1}{\Omega\upsilon^2 - \lambda^2 + 1} \tag{26}$$

Simplifying Eq. (26), it can be rewritten as an equation of λ^2,

$$2\lambda^4 - 2\Omega\upsilon^2\lambda^2 - 2\upsilon^2\lambda^2 - 2\lambda^2 + 2\upsilon^2 + \Omega\upsilon^4 = 0 \tag{27}$$

Assuming the roots of Eq. (27) are λ_P and λ_Q, the following relation can be obtained,

$$\lambda_P^2 + \lambda_Q^2 = \Omega\upsilon^2 + \upsilon^2 + 1 \tag{28}$$

Considering the optimal objective is minimizing the maximum value of the DMF_s in the frequency domain, the DMF_s value at λ_P and λ_Q should be identical. In particular, when the damping ratio of THMD ζ_T is infinite, we have,

$$\frac{1}{\Omega\upsilon^2 - \lambda_P^2 + 1} = -\frac{1}{\Omega\upsilon^2 - \lambda_Q^2 + 1} \tag{29}$$

Then Eq. (29) can be reduced to,

$$\lambda_P^2 + \lambda_Q^2 = 2\Omega\upsilon^2 + 2 \tag{30}$$

Combining Eq. (28) and Eq. (30), the optimum frequency ratio can be expressed as,

$$\upsilon_{opt} = \frac{1}{\sqrt{1-\Omega}} \tag{31}$$

The two fixed-points λ_P and λ_Q can be obtained by substituting Eq. (31) into Eq. (27), and it has following form,

$$\lambda_{P,Q}^2 = \frac{2 \pm \sqrt{2\Omega}}{2(1-\Omega)} \tag{32}$$

In this situation, the DMF_s value at λ_P and λ_Q can be found as,

$$DMF_s|_{\lambda_P,\lambda_Q} = \frac{\sqrt{2}(1-\Omega)}{\sqrt{\Omega}} \tag{33}$$

Since the response value at the two fixed-points is the maximum of DMF_s curve, the gradient of DMF_s at λ_P and λ_Q should be equal to zero, as shown below,

$$\frac{\partial}{\partial\lambda^2}|DMF_s(\lambda)|^2 = 0 \tag{34}$$

Substituting Eq. (24) into Eq. (33), the expression of the optimum damping ratio is obtained by,

$$\zeta_{P,Q}^2 = \frac{3\Omega(2 \pm \sqrt{2\Omega})}{8(2-\Omega)} \tag{35}$$

The average damping ratios corresponding to points P and Q is adopted as the optimum damping ratio, which is expressed as,

$$\zeta_{opt} = \sqrt{\frac{\zeta_P^2 + \zeta_Q^2}{2}} = \sqrt{\frac{3\Omega}{4(2-\Omega)}} \tag{36}$$

Comparing the original Eq. (2) with the linearized Eq. (15), a sufficient condition for the approximate equivalence of the two equations is

$$c_T\dot{x}_T \cong \frac{1}{2}\rho L^2 c_d \dot{x}_T\left|\dot{x}_T\right| \tag{37}$$

Using Eq. (9) obtained by the harmonic balance method and introducing $\zeta_T = c_T/(2m_T\omega_T)$ and $\gamma_d = \rho L^2 c_d/2m_s$, the relationship between non-linear and linear dimensionless damping can be obtained as shown below,

$$\gamma_d \cong \frac{2\Omega\upsilon\zeta_T}{\Phi\lambda A_T} \tag{38}$$

It can be seen from the above equation that the added damping ratio γ_d is related to the amplitude A_T of the THMD. Substituting Eqs. (31), (32) and (36) into Eq. (22), the expressions for the DMF_T corresponding to points P and Q are obtained as,

$$DMF_T|_{\lambda_P,\lambda_Q} = \frac{2}{1-\Omega}\sqrt{\frac{(1-\Omega)^4(2-\Omega)}{\Omega^2(8-\Omega \pm 3\sqrt{2\Omega})}} \tag{39}$$

Introducing the definition of the DMF in Eq. (13), the root mean square of the DMF_T at points P and Q is taken as the value of the DMF_T of the THMD, and the amplitude of the THMD is calculated as,

$$A_T = f\sqrt{\frac{DMF_T^2(\lambda_P) + DMF_T^2(\lambda_Q)}{2}} = f\sqrt{\frac{4(1-\Omega)^2(8-\Omega)}{\Omega^2(32-\Omega)}} \approx \frac{1-\Omega}{\Omega}f \tag{40}$$

Substituting Eqs. (10), (31), (36) and (40) into Eq. (38) and the υ is set to 1 since the most unfavorable condition is around the resonance frequency, we have

$$\gamma_{opt} = \frac{3\sqrt{3}\pi}{8f}\sqrt{\frac{\Omega^5}{(2-\Omega)(1-\Omega)^3}} \tag{41}$$

In summary, the Eq. (41) gives an approximate analytical solution for the optimal added damping ratio. For the optimal frequency ratio, the expression in Eq. (31) is still adopted since the damping term has a small effect on tuning frequency. The P and Q points of the DMF_S curve do not correspond to the maximum of the DMF_T curve as shown in Fig. 4, which shows that the above ζ_d is not a strict derivation, rather it is an approximate analytical solution. Such an approximate solution may cause the final DMF curve of THMD to slightly deviate from the optimum ones. Therefore, it is necessary to verify the accuracy of the approximate analytical solution.

To verify the accuracy of the approximate analytical solution, the numerical search results of the semi-analytical solution are compared with the analytical solution. The variation of the optimal tuning frequency ratio υ_{opt} and the optimal added damping ratio γ_{opt} with the total mass ratio Ω is given in Fig. 5 and compared with the results in Table 2. The comparison shows that the difference between the optimal tuning frequency ratio and the added damping ratio both increase with increasing mass ratio. The results of the semi-analytical solution and the analytical solution are complete consistency when Ω is less than 0.05, demonstrating the analytical solution is accurate. When the Ω is equal to 0.1, the difference in the υ_{opt} is only 0.9 %. The difference between numerical optimization and analytical optimum formulas for γ_{opt} is increased with the decrease of f, reaching a maximum difference of 8.2 % at $f = 0.0005$ and Ω=0.1. However, instances where Ω exceeds 0.05 are infrequent in civil engineering applications.

3. Performance analysis of the THMD

3.1. Comparison with cases of conventional TMD and heave plate

Compared to the conventional TMD, the heave plate of the THMD replaces the mechanical damper of the conventional TMD. The mechanical model of the dampers has also changed and it is necessary to compare the performance of conventional TMD and THMD. In the case of floating structures, the heave plate is a common structural heave control measure, thus it is also necessary to compare the performance of the heave plate with the proposed THMD. In this section, A comparison with the conventional TMD, the heave plate and the proposed THMD is performed to evaluate the control effectiveness of THMD. A mechanical model of a heave plate and a conventional TMD is presented in Fig. 6. In particular, the heave plate model can be simulated by setting the mass of the THMD model to zero.

THMD has nonlinear characteristics and its optimal parameters are related to the dimensionless excitation amplitude f. Therefore, the dimensionless excitation amplitude f needs to be determined in advance. Here f is set to 0.001 and the primary structural damping ratio ζ_s is taken to be 0, the mass ratio μ is set to 0.01, 0.02, 0.03 and 0.05, and the added mass ratio β is set to 0, 0.01, 0.02, 0.03 and 0.04, the optimal parameters are calculated according to the optimal Eqs. (31) and (41). For performance comparison, the mass ratio of the conventional TMD is taken to be the same as that of the THMD, and the optimal parameters are calculated according to the classical

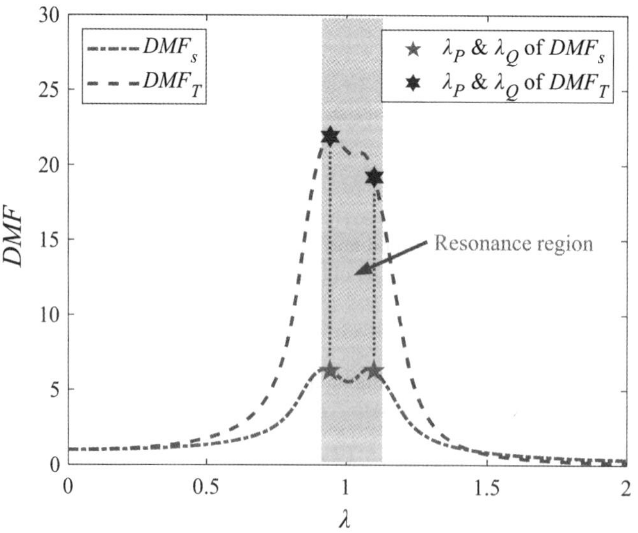

Fig. 4. P and Q points on the DMF curves.

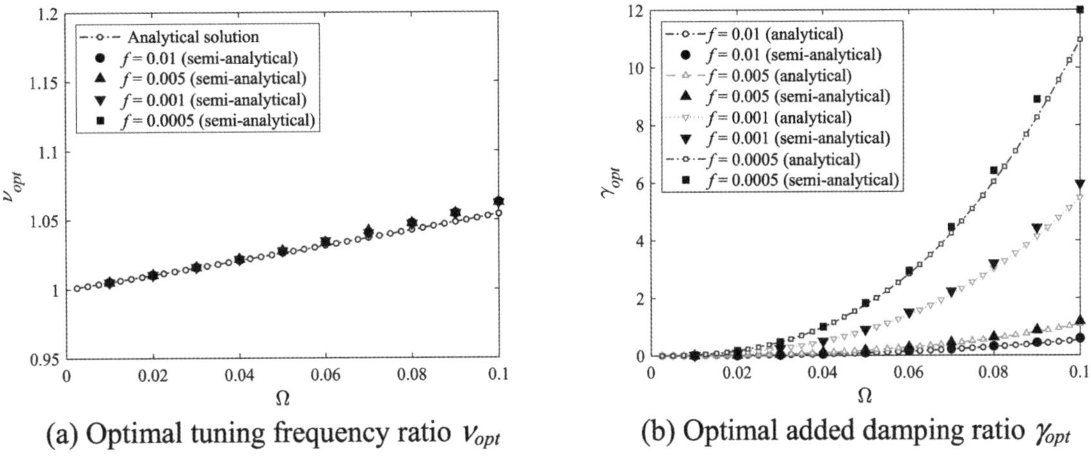

Fig. 5. The variation of the optimal parameters with the total mass ratio.

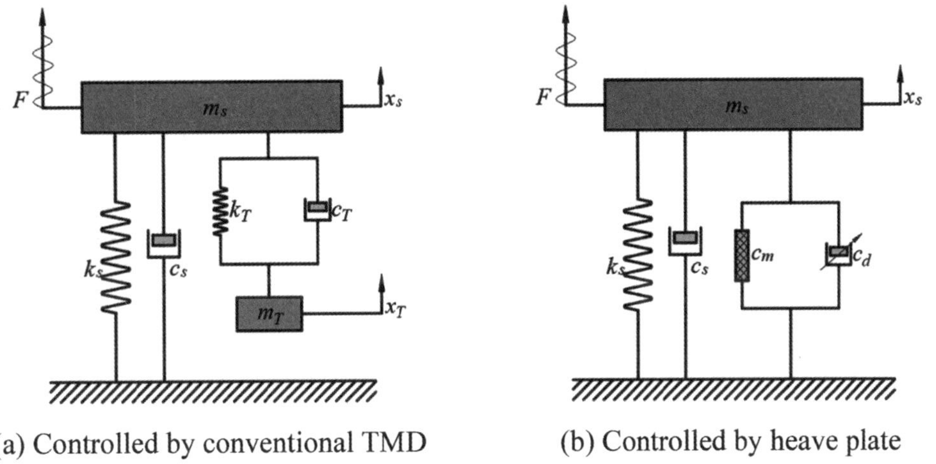

Fig. 6. Simplified mechanical model of SDOF structure controlled by heave plate and THMD.

Den Hartog formulas. The parameters for the heave plate are taken to be the same as those for the THMD. The DMF_s curves for each measure are computed according to the mechanical model in Fig. 6 using the fourth-order Runge-Kutta method with a time step of 0.001 s. The results are summarized in Fig. 7. The results in Fig. 7 show that both THMD and TMD are much more effective than the dampers at resonance. This is because the heave plate functions as a direct installation of a damper on the primary structure. It cannot achieve tuned energy absorption with the primary structure and can only dissipate vibration energy through the damping components. While the tuned damper is able to absorb much energy at resonance and dissipate it through the damping element. Therefore, the performance of THMD and TMD in the resonance region is much better than that of the heave plate.

In addition, compared with the conventional TMD, the performance of the THMD is basically the same as that of the conventional TMD when the added mass ratio β is 0. However, when the added mass ratio of the THMD increases, the performance of THMD gradually improves and eventually the performance of THMD is much better than that of the conventional TMD. Notably, both added mass and added damping are hydrodynamic properties of the heave plate. The added mass ratio and added damping ratio are not completely independent. They are related to the construction and size of the heave plate. Therefore, the hydrodynamic properties of the heave plate need to be measured and analyzed before the THMD is designed. These aspects will be discussed in Section 4.

3.2. *Robustness analysis of THMD*

In engineering, the dynamic properties of the controlled structure will change over time, and this may cause that the tuning parameters of the THMD deviate from the optimal value. Therefore, it is necessary to carry out robustness analysis to verify the performance of THMD after parameter deviations. In order to quantify the difference of performance, the maximum value of the DMF_s curve is used as a target of THMD performance, expressed as follows,

$$R_s = \max(DMF_s(\lambda)) \tag{42}$$

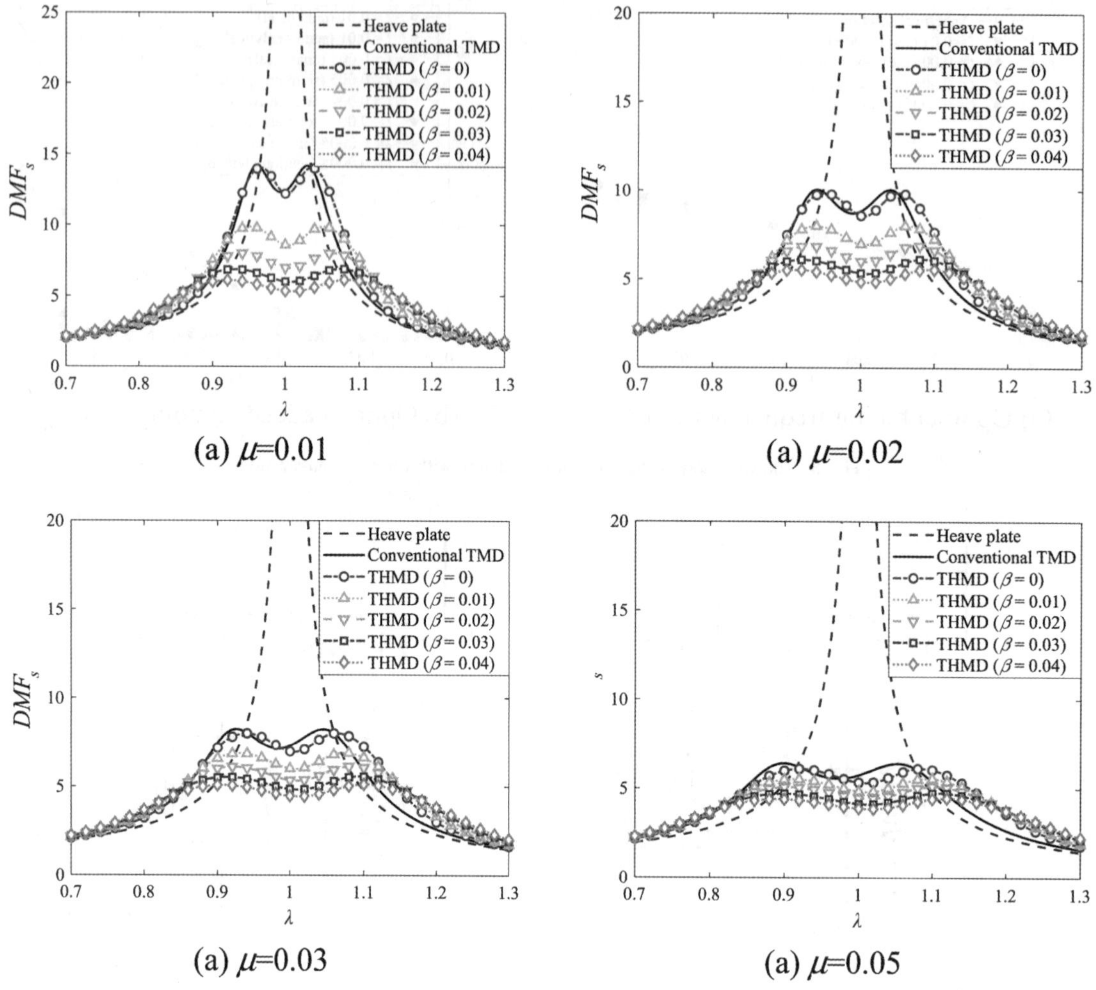

Fig. 7. *DMF* curves with different mass ratio and added mass ratio.

The performance of the THMD is influenced by the tuning frequency ratio, the added damping ratio and the dimensionless excitation amplitude. Therefore, a robustness analysis of the THMD is performed for deviations from these three parameters. The deviation ratios of these three parameters are defined as,

$$\varepsilon_\nu = \frac{\nu_e - \nu_d}{\nu_d}\ \varepsilon_\gamma = \frac{\gamma_e - \gamma_d}{\gamma_d}\varepsilon_f = \frac{f_e - f_d}{f_d} \tag{43}$$

where ε_ν, ε_γ, ε_f are the deviation ratios of tuning frequency ratio, added damping ratio and excitation amplitude, ν_e, γ_e, f_e are the tuning frequency ratio, added damping ratio and excitation amplitude after detuning, ν_d, γ_d, f_d are the design values of tuning frequency ratio, added damping ratio and excitation amplitude.

The performance of the tuned dampers is sensitive to the tuning frequency, thus the frequency deviation ratio of THMD ε_ν is set to [-10 %, 10 %]. Since the performance of the tuned dampers is not sensitive to the damping, and to better display the line details in the Fig. 8, the damping deviation ratio of THMD ε_γ is set to [-50 %, 50 %]. Similarly, the deviation ratios of excitation ratio ε_f is set to [-50 %, 50 %]. The results of the robustness analysis for a dimensionless excitation amplitude f with a design value of 0.001 and a primary structure damping ratio ζ_s of 0 are given in Fig. 8. It can be seen from Fig. 8(a) that the deviation of the tuning frequency ratio has a large impact on the performance of the THMD. The larger the deviation ratio of the tuning frequency ratio, the worse the performance. When the mass ratio Ω is set to 0.1 and the deviation rate of the frequency ratio ε_ν is set to −10 %, the performance index R_s of THMD changes from 14.04 to 61.64, indicating a 339 % increase in the maximum value of the frequency domain response. In addition, based on the results for different mass ratios μ, it can be seen that the larger the mass ratio, the better the robustness of the THMD. As can be seen in Fig. 8(b), the deviation of the added damping ratio has a limited effect on the performance of the THMD. The performance metric R_s only increases by 30 % when the mass ratio μ is set to 0.01 and the deviation ratio ε_γ is 50 %. Fig. 8(c) shows that the effect of the dimensionless excitation amplitude f is consistent with the effect of the added damping ratio.

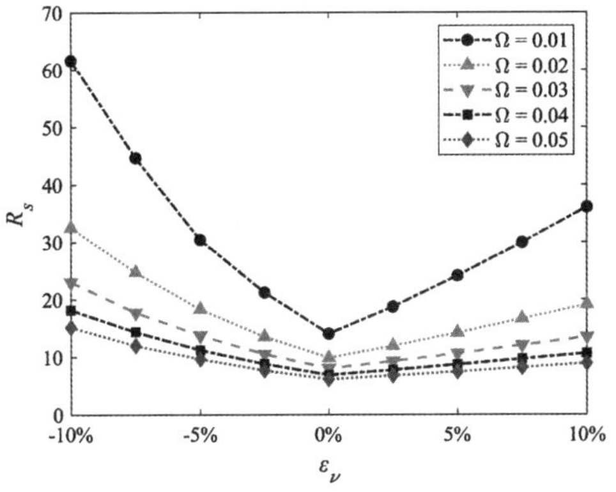

(a) Maximum value of DMFs curve under different ε_ν

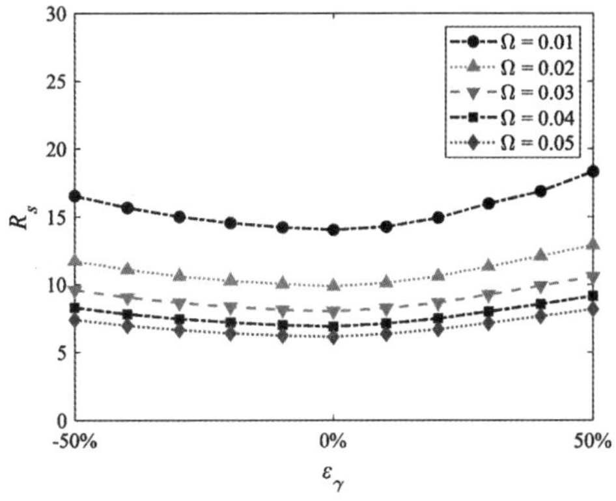

(b) Maximum value of DMFs curve under different ε_γ

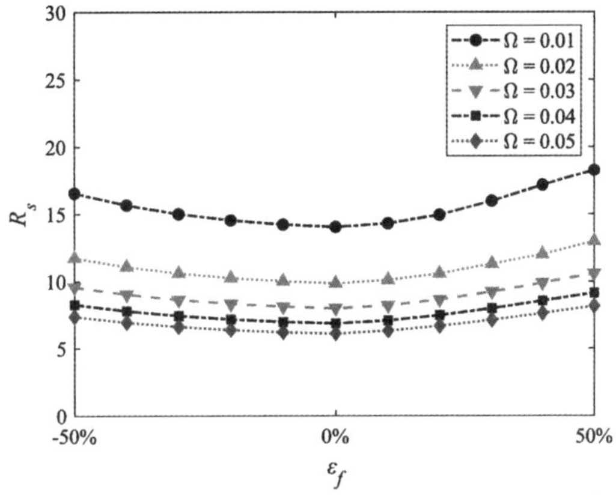

(c) Maximum value of DMFs curve under different ε_f

Fig. 8. Robustness comparison under different parameter detuning.

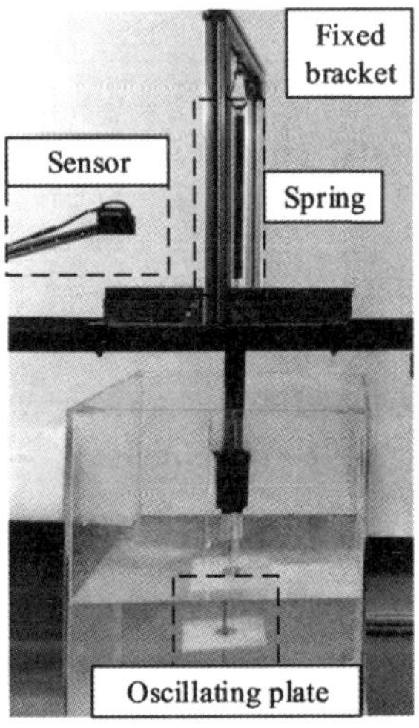

(a) Layout of the vibration system

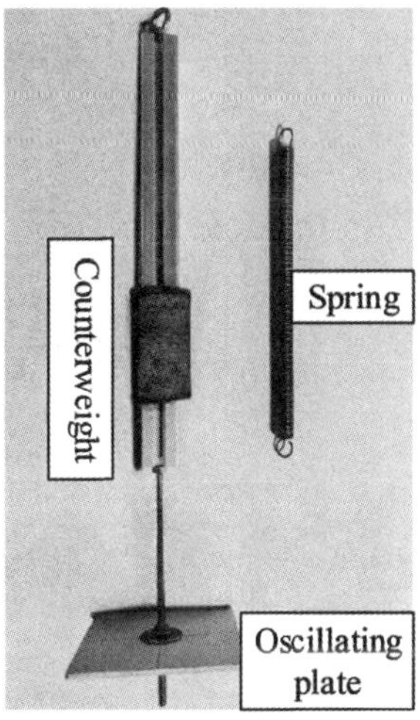

(b) Layout of the THMD

Fig. 9. Setup of free vibration experiment.

4. Experiment study of hydrodynamic characteristics

4.1. Parametric identification of Morison equation

The free vibration test is a hydrodynamic parameter identification method with easy-to-understand principles, simple processes and equipment, and is suitable for the identification of the additional mass and additional damping parameters of the oscillating plate in this article. The purpose of the test is to obtain the free vibration displacement time history of the oscillating plate during free vibration, and then identify the additional mass and additional damping parameters based on the least squares method. In order to verify the accuracy of the Morison hydrodynamic equation, a free vibration experiment study is carried out. The free vibration experiment is conducted on a SDOF system with a heave plate. The setup of the experiment and the SDOF system with the heave plate are shown in Fig. 9. It can be seen from Fig. 9 that the experimental setup consists of a fixed bracket, a water box fixed to the floor with dimensions of 50 × 50 cm, and a SDOF system with a heave plate and spring, where the spring stiffness is designed to be 200 N/m. The heave plate can be disassembled and replaced with different size plates, which are made of PVC in order to guarantee the stiffness and reduce the mass of the plate. The shape of the heave plate is chosen as a regular square with a side length of L and a thickness of 3 mm. During the test, the heave plate is submerged above the water surface for 40 cm to prevent interference from the boundary. The time history of displacements during the free vibration is collected with a laser meter and the sample frequency is set to 1000 Hz.

The hydrodynamic characteristics of the heave plate are related to the amplitude and Reynolds number, while the changes in the hydrodynamic characteristics are not significant when the amplitude and Reynolds number do not vary much. Therefore, considering the feasibility under the experimental scale, the initial amplitude of the test is taken as 10 cm. Meanwhile, the lengths L of the heave plates are taken as 15, 16, 17 and 18 cm. The free vibration displacement curves of each size of the heave plates are finally plotted in Fig. 10. As can be seen from Fig. 10, for different sizes of heave plates, their time histories of free vibration are characterized by a fast decay at the beginning and a slow decay at the end. This phenomenon is consistent with the assumption that the damping term is inversely proportional to the quadratic velocity in the numerical hydrodynamic model. In addition, it can be seen that the larger the size of the heave plate, the smaller the amplitude of the free decay curve and the lower the frequency of the free vibration. It indicates that the added mass and added damping increase as the plate size becomes larger.

The experimental free vibration equation can be expressed as,

$$m_0\ddot{x}_0(t)+\rho L^3 c_m\ddot{x}_0(t)+\frac{1}{2}\rho L^2 c_d\dot{x}_0(t)\left|\dot{x}_0(t)\right|+k_0x_0(t)=0 \tag{44}$$

where m_0, k_0 are the modal masses and spring stiffnesses of the system, ρ is the density of the liquid, L is the size of the heave plate, c_m and c_d are the added mass coefficients and added damping coefficients respectively, and $x_0(t)$ is the displacement time history of the free vibration of the heave plate.

In the free vibration experiment, m_0 is the modal mass of the vibration system, which calculated as $m_0 = 2.58 + 0.21/3 = 2.65$ kg. k_0 determined as 200 N/m, is the spring stuffiness. ρ is the density of water, which is set to 1000 kg/m^3. L is the length of heave plate, which is chosen according to the experimental conditions. Then, according to the principle of least squares, the added mass coefficient c_m and the added damping coefficient c_d are adjusted to minimize the quadratic sum of the errors between the fitted curve and the target curve. The target error function is expressed as,

$$err=\sum(x_0(t)-\dot{x}_0(t))^2 \tag{45}$$

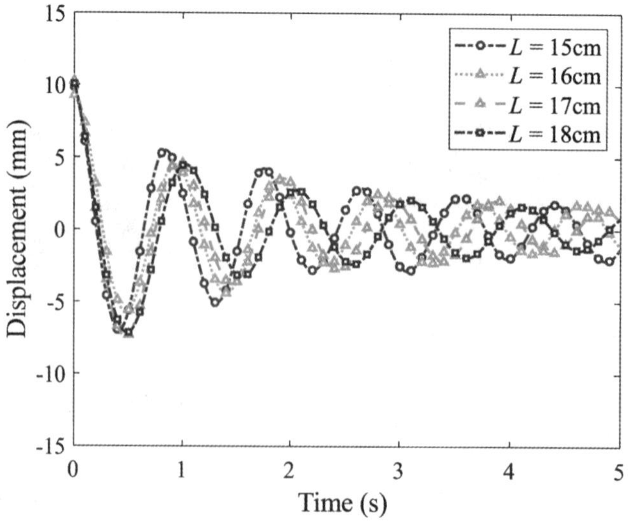

Fig. 10. Displacement time histories of different heave plates in free vibration.

where $x_0(t)$ is the displacement time history of the free vibration measured in the experiment. $\dot{x}_0(t)$ is the fitted curve obtained from the given c_m, c_d and Eq. (44).

The gradient degradation optimization algorithm is used to minimize the target error function in MATLAB. The identified added mass coefficients and added damping coefficients are obtained and the corresponding fitted curves under conditions of the $L = 16$ cm and 18 cm are plotted in Fig. 11 respectively. It can be seen that the fitted curves are in general agreement with the experimental curves, which indicates that the numerical hydrodynamic model based on the Morison equation is reliable. At the same time, it can also be proved that the hydrodynamic parameters obtained by the least squares principle are accurate.

4.2. Validation by computational fluid dynamics (CFD)

As a numerical method, CFD can ultimately reduce overall project costs by minimizing the need for extensive physical prototypes and shortening the overall design cycle. In order to prove the feasibility and accuracy of the CFD experiment, the results of the CFD experiment also need to be verified. To reduce the computational cost, a two-dimensional mesh model is used for the CFD experiment. In the CFD experiment, the heave plate undergoes forced vibration in the flow field, with a displacement amplitude of 10 cm and a frequency of 1 Hz. The size of the plate is consistent with that of the free vibration experiment. To avoid negative mesh volume, an overset mesh method is used here. The final CFD model is meshed in Fig. 12. A diagram of the mesh region is given in Fig. 12(a). The computational domain is set large enough to prevent boundary effects and is set to 2×1 m. Fig. 12(b) gives the results of the mesh, which shows that the overset mesh is divided into two component: the background mesh and the foreground mesh. The background mesh has a refinement region, the mesh size of refinement region is set to 0.002 m and the mesh size of non-refinement region is set to 0.01 m. To consider the boundary layer effect, the thickness of the first layer of boundary layer is determined as 0.00001 m, and the number of layers is set to 20. In order to verify the mesh-independence, the results of the mesh size reduction are compared with the original mesh. The computational results show that the difference is within 1 %, indicating that the mesh size no longer affects the results of the CFD method.

A condition of $L = 16$ cm is chosen for the CFD method to obtain the time history of the forces on the heave plate subjected to forced vibration in water. At the same time, the the hydrodynamic parameters identified in the free vibration experiments are substituted into the Morison model to obtain the time history of hydrodynamic force. The time history of hydrodynamic force under the experimental identified parameters compared with the results of the CFD method are plotted in Fig. 13. As can be seen from Fig. 13, the two curves are in general agreement. However, there are minor differences in the phase and peak values. The phase difference mainly depends on the added mass and the peak difference depends on the added damping. Due to the small scale in the experiments, the details of the experiments are difficult to control finely. Therefore, the CFD method will have better accuracy in the simulation for practical applications. Using the CFD method and the hydrodynamic parameter identification criterion in Eq. (45), the hydrodynamic parameters are computed for the change in side length of the heave plate from 10 cm to 18 cm. The corresponding added mass and added damping are shown in Fig. 14. It can be seen from Fig. 14 that the added mass ratio increases by 466 % as the heave plate size increases, while the added damping ratio increases by only 182 %, which indicates that the tendency for the added mass to increase is greater than the added damping. This is because the additional mass is proportional to the cube of L, and the additional damping is proportional to the square of L. According to the analytical formulas of the optimum parameters, the optimum damping ratio depends on the value of the mass ratio. The added mass and added damping provided by the heave plate do not exactly optimize the THMD performance. Therefore, the value of the mass block needs to be recalculated to ensure that the THMD is optimally tuned.

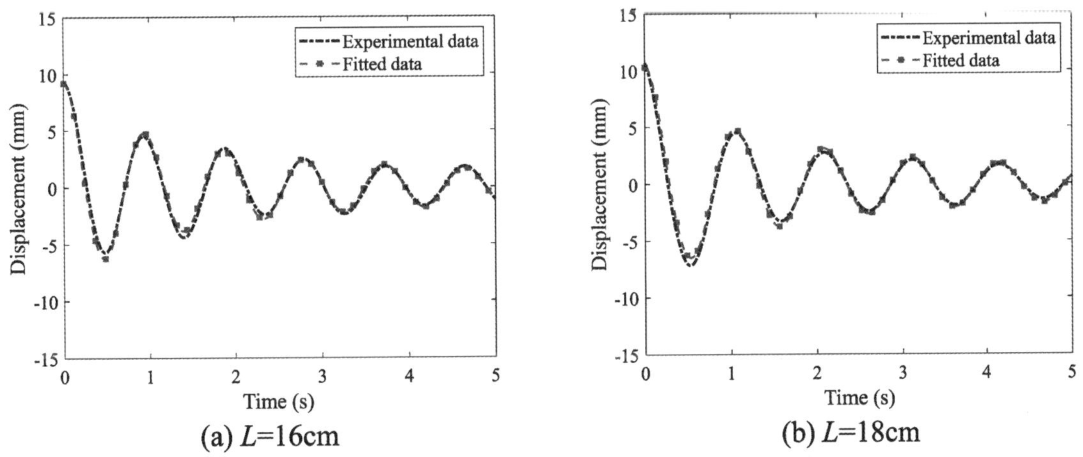

Fig. 11. Comparison of experimental and fitted curves in free vibration.

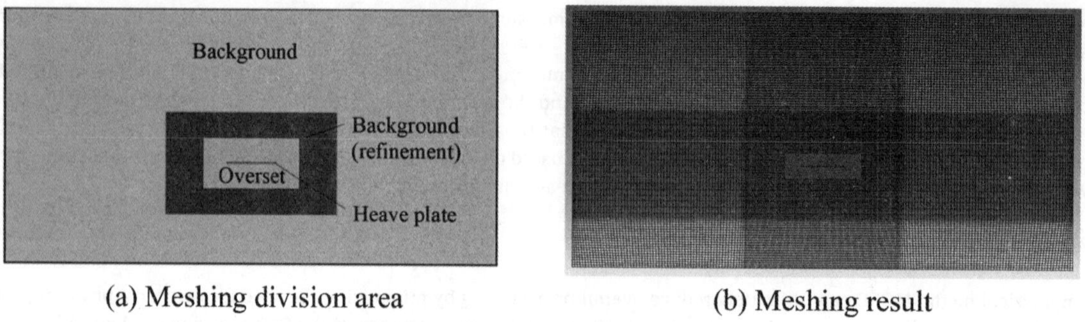

Fig. 12. CFD Model of heave plate.

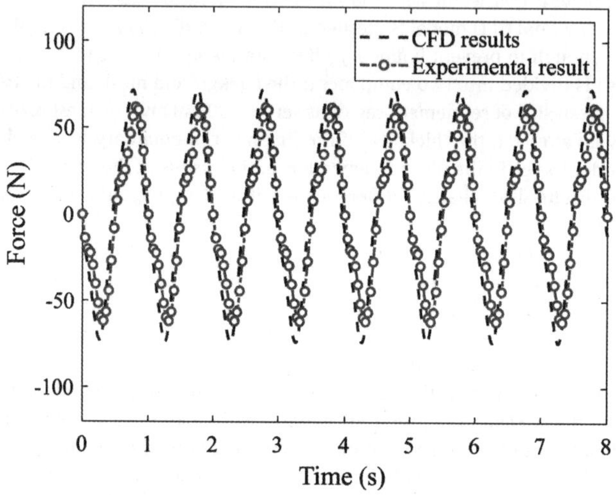

Fig. 13. Experimental and CFD simulated force curves of heave plate under forced vibration.

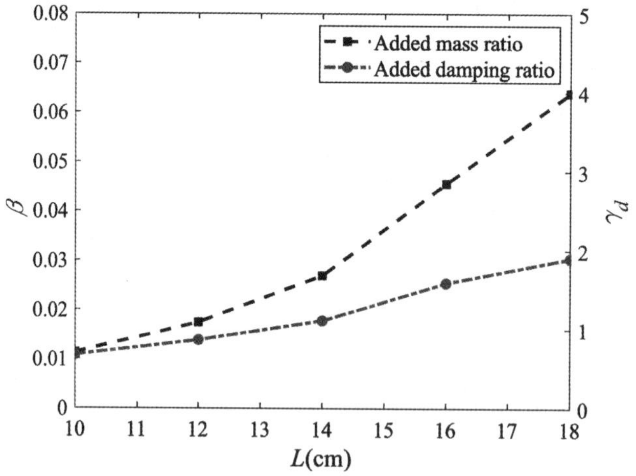

Fig. 14. Variation of added mass ratio and added damping ratio with size of heave plate.

5. Experiment study of control performance

5.1. Optimization design of the THMD in experiment

The frequency of the actual engineering structure is very low. However, considering the excessive static displacement of the low frequency structure and the limitations of the experimental site, the fundamental frequency of the primary structure is designed to be around 1 Hz in the performance experiment. The specific configuration of the controlled structure is shown in Fig. 15. As can be seen from Fig. 15(a), the controlled structure is suspended from the springs and is susceptible to vibration in the vertical direction, while the lateral direction is difficult to vibrate due to its high stiffness. The primary beam is 8 m long and weighs 43.7 kg, the counterweight on the primary beam is 50 kg and the weight of the springs is 2.6 kg. The controlled structure is modeled using the BEAM4 element in the ANSYS finite element simulation software. One end of the controlled structure is fixed, while the other end is fixed but with the longitudinal constraint released. the modal mass ms is calculated as 82.2 kg. The controlled structure is two L-shaped steel, with their cross-sectional dimensions detailed in Fig. 15(a). It can be seen from Fig. 15(b) the excitator on the controlled structure is mainly generated by a motor rotating an unbalanced mass, and this unbalanced force will produce a harmonic force in the vertical direction. The unbalanced mass weighs 0.25 kg and has a rotational radius of 0.13 m. At resonance, the amplitude of the harmonic force can be calculated as $F_0 = 1.68$ N.

The free vibration experiment is used to identify the modal parameters of the controlled structure. The controlled structure is released at a certain initial displacement, and a laser meter is used to collect the displacement time history of the free vibration, the sampling frequency is set to 1000 Hz and the experimental data is plotted in Fig. 16. The time history in the free vibration experiment is plotted in Fig. 16(a), where the peak points for each cycle are marked. The envelope curve of free decay of a SDOF system with viscous damping is also plotted. According to the least squares principle, the damping ratio ζ_s of the controlled structure is obtained as 0.22 % when the envelope is best fitted to the peak points. Fig. 16(b) gives the power spectral density curve of the free vibration of the controlled structure, which shows that the fundamental frequency of the controlled structure is 1.144 Hz. Then, the stiffness of the controlled structure $k_s = 4155.0$kN/m can be calculated.

In this experiment, the total mass ratio Ω is set to 0.045. Based on the harmonic external load amplitude F_0 and the controlled structural stiffness k_s, the dimensionless excitation amplitude f can be calculated to be 0.0004. Considering that the damping ratio of the controlled structure is negligible, the analytical Eqs. (31) and (41) of the optimal parameters are chosen for the optimal design, and the optimal frequency ratio is calculated to be 1.0233 and the optimal added damping ratio is 1.6793.

Hydrodynamic parameters can be obtained through CFD and hydrodynamic experiments. At the same time, Wendel [57] proposed an estimation formula for the additional mass coefficient as shown in Eq. (46). In summary, a comparison of the additional mass ratio calculated through CFD, experiments and Eq. (46) can be obtained in Table 3. It can be seen that the additional mass ratios obtained by the three methods are not much different, which shows that the CFD and experiments results are reliable. However, there may be some deviation in the results of the three due to the size effect and boundary effect of the water tank and plate.

$$\beta = \frac{\pi\rho L^3}{4m_s} \tag{46}$$

Based on the analysis in section 4.1, the dimensionless hydrodynamic parameters for different sizes of heave plates in the free vibration experiment are listed in Table 4. The size of the heave plate with the closest optimum value of the added damping ratio is chosen as the design size in the experiment, where $L = 16$ cm. According to the definition of dimensionless parameters, the optimal design parameters are calculated and summarized in Table 5.

The design total mass ratio of the THMD is 0.045, which means 3.7 kg. However, the actual mass block only weight 1.15 kg, which indicates that the added mass generated by the heave plate greatly reduces the physical mass of the THMD by 68.9 %. The spring stiffness of the THMD is calculated from the optimum frequency ratio to be 195.8 N/m, and a spring with $k = 200$ N/m is selected in the experiment. The heave plate is made of PVC material with a thickness of 3 mm and a side length of 16 cm.

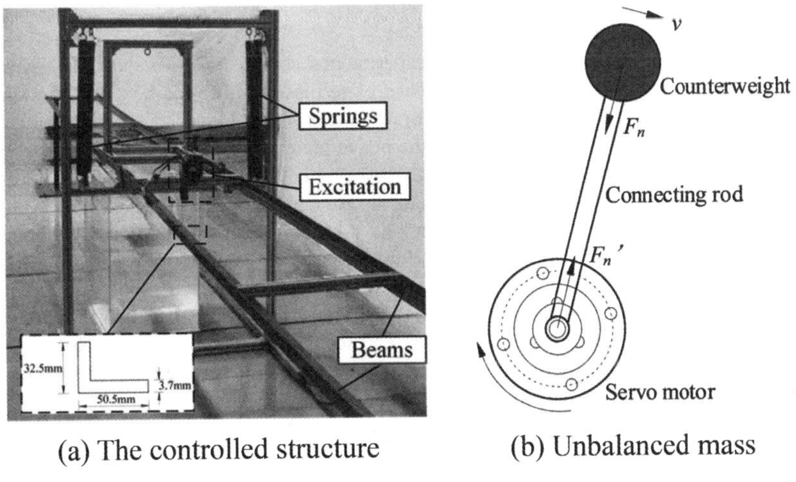

Fig. 15. Configuration of the controlled structure.

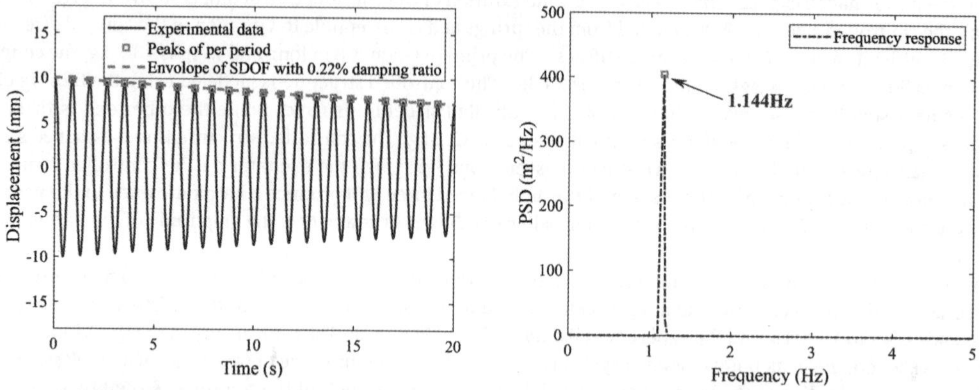

(a) Displacement time history (b) PSD of time history

Fig. 16. Results of free vibration experiment.

Table 3
Added mass ratio for different calculation method.

Heave plate length (*L*)	10 cm	12 cm	14 cm	16 cm
Added mass ratio calculated by CFD	0.0085	0.0142	0.0218	0.0310
Added mass ratio calculated by experiment	0.0112	0.0174	0.0269	0.0455
Added mass ratio calculated by Eq. (46)	0.0096	0.0165	0.0262	0.0391

Table 4
Dimensionless hydrodynamic parameters for different sizes of heave plates.

Heave plate length (*L*)	10 cm	11 cm	12 cm	13 cm	14 cm	15 cm	16 cm
Added mass ratio (β)	0.0085	0.0112	0.0142	0.0180	0.0218	0.0252	0.0310
Added damping ratio (γ_d)	0.603	0.735	0.996	1.181	1.347	1.534	1.614

Table 5
Optimal design parameters of THMD in the experiment.

Designed parameters	Value
Physical mass (m_T)	1.15 kg
Stiffness (k_T)	195.8 N/m
Heave plate length (*L*)	16 cm

5.2. *Experimental result*

Based on the design values of the optimum parameters, the experimental setup is made as shown in Fig. 17. It can be seen that a water box is arranged below the controlled structure with a heave plate placed inside the box, and the plate suspended from the controlled structure by a spring. When the controlled structure vibrates under external loads, the energy of the vibration is absorbed by the resonant auxiliary structure and dissipated in the water by the heave plate.

In a free vibration experiment, a static force is applied to the structure under control, causing a static displacement of the controlled structure and then releasing it. The displacement time histories of the controlled structure recorded by a laser meter are plotted in Fig. 18. The controlled structure controlled by the THMD is essentially stationary after about 30 s as can be seen in Fig. 18. Compared to the uncontrolled state, the control effect of the THMD on free vibrations is obvious. In addition, Fig. 18 also gives the free vibration curve obtained by fourth-order Runge-Kutta method. It can be seen that it is basically consistent with the experimental curve, indicating the accuracy of the numerical model. However, there are also minor differences between the numerical and experimental results. There are two possible reasons for this: firstly, due to the small scale of the experiment, it is difficult to accurately adjust the tuning frequency. The second is that although the numerical model has been verified for accuracy, the effects of fluid are complex so that the result inevitably differ from the actual situation.

In the forced vibration experiment, the unbalanced mass is rotated by the motor to exert a harmonic load to the controlled structure. The corresponding displacement time curves are recorded by the laser meter. Then, the speed is adjusted to obtain steady amplitudes at different excitation frequencies. To compare the control effects of THMD, experimental studies of forced vibration in the

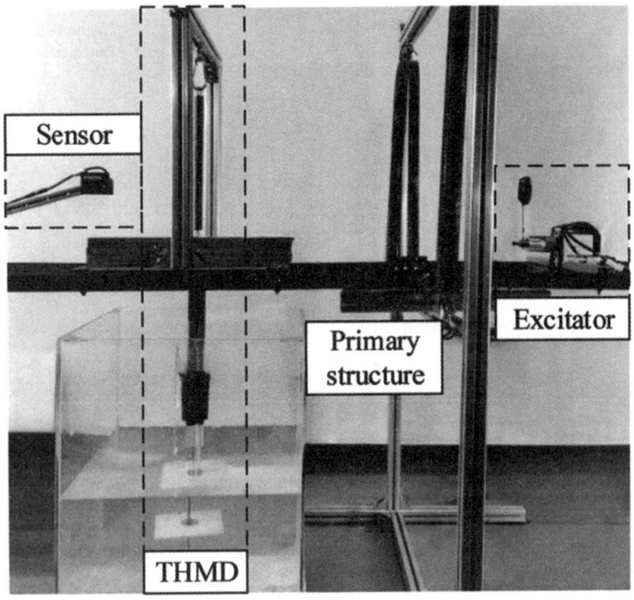

Fig. 17. Experimental layout of the structure controlled by THMD.

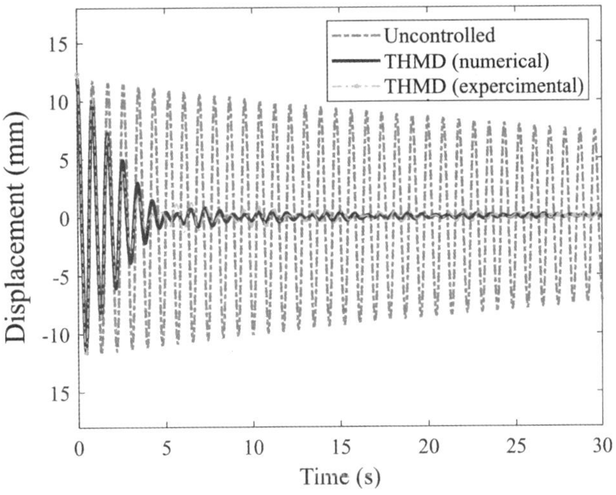

Fig. 18. Displacement time history of the primary structure in free vibration.

uncontrolled condition and in the heave plate condition are also carried out in this section. The numerical theory of conventional TMD is so mature that it is not investigated as an experimental condition. The DMF curves for the uncontrolled condition, the condition controlled by heave plate, controlled by the conventional TMD, and controlled by the THMD are plotted in Fig. 19(a), where the corresponding numerical response curves are also plotted. The THMD has a 32.7 % reduction in the peak value of DMF compared to a conventional TMD. Comparing the experimental results with the numerical results, it can be seen that the data from the experiments are generally consist with the results from the numerical simulations. However, the DMF curve of the THMD still differs from the theory due to the small scale of the experiment, which makes it difficult to tune the THMD accurately. The amplitude of the structure's response increases significantly when it resonates, which may cause structural damage. Therefore, discussing the displacement time histories at resonance is extremely important. Fig. 19(b) gives a comparison of the displacement time histories for these conditions at resonance. It can be seen that the displacement amplitude of the primary structure in the uncontrolled state is 31.8 mm, while the amplitude of the primary structure with the heave plate is 11.8 mm. The amplitude of the primary structure with the THMD is only 1.4 mm, and the corresponding damping rate is 95.6 %.

The robustness performance of the THMD is experimentally verified in experiments using heave plates of different sizes. Heave plates of different sizes are shown in Fig. 20. According to the identification results of hydrodynamic parameters in Table 2, the design parameters of THMD using heave plates of different sizes are shown in Table 6. Adopting these design parameters in the performance

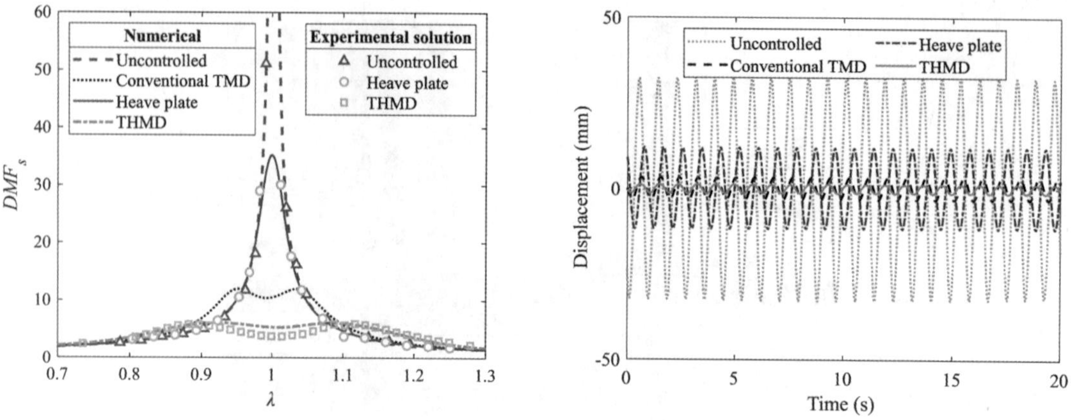

(a) DMF (b) Time history at resonance

Fig. 19. Displacement response of the primary structure in forced vibration.

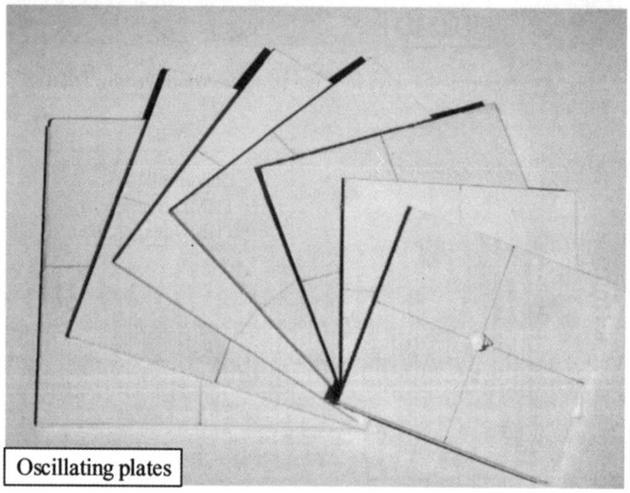

Fig. 20. Heave plates of different sizes in robustness experiment.

Table 6
Design parameters of THMD employing different heave plates.

Heave plate length L(cm)	10	11	12	13	14	15	16
Stiffness of spring k_T(N/m)	195.8	195.8	195.8	195.8	195.8	195.8	195.8
Physical mass m_T(kg)	3.00	2.78	2.53	2.22	1.91	1.63	1.15

experiments, the experimental DMF curves for different conditions are obtained as shown in Fig. 21. Fig. 21 shows that the DMF curve of the primary structure controlled by THMD have double peaks with identical height, indicating that the spring is optimal. When $L =$ 16 cm, the added damping of the heave plate is closest to the optimum value, therefore the maximum value of the DMF curve for this condition is minimum. As L becomes smaller, the added damping deviates from the optimal value and the DMF curve increases. The maximum value of the DMF curve increases to 9.7 when L decreases to 10 cm. However, this condition is still a significant advantage compared to the uncontrolled condition and the condition controlled by the heave plate.

6. Conclusion

In this paper, a novel tuned hydrodynamic mass damper employing added mass and damper form interaction between heave plate and fluid is proposed. The properties of THMD are deeply studied by numerical and experimental analysis. Therefore, the following conclusions were obtained:

(1) The motion equations of a structure controlled by a THMD are established. The semi-analytical solutions of motion equations

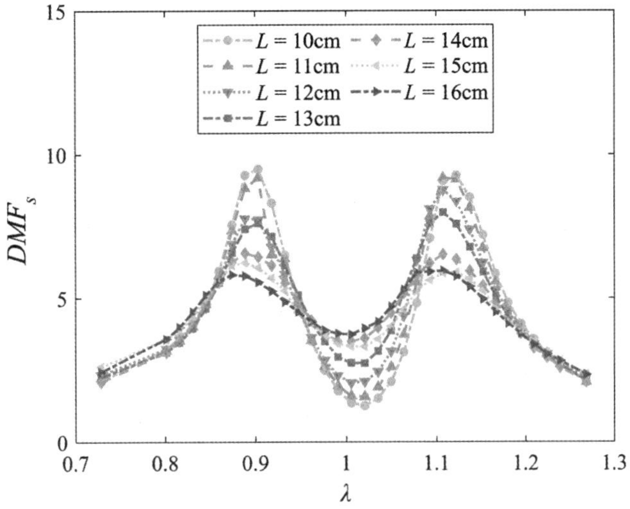

Fig. 21. DMF curves for robustness experiment.

are derived by harmonic balance method. The comparison result shows that the semi-analytical solutions agree well with the numerical solution.

(2) Base on the H_∞ optimization criterion, the approximate analytical solutions of optimal parameters of THMD are obtained by harmonic balance method. Numerical studies show that these analytical solutions are highly accurate.

(3) Using CFD simulation, the change law of the hydrodynamic properties of heave plates was investigated. The results show that the added mass of the heave plate tends to increase more than the added damping as the length increases. Therefore, the parameters of THMD needs to be recalculated to ensure that the THMD is optimally tuned.

(4) Added mass and damping of different heave plate were identified using two-dimensional CFD method and Morrison equations. From the CFD and experiment result, the accuracy of identifying hydrodynamic parameters by CFD method was verified.

(5) According to the numerical and experimental analysis, the control performance and robustness of THMD is compared to the conventional TMD and heave plate. The comparison results indicate the THMD outperforms the conventional TMD and heave plate under broadband harmonic loads. When the ratio of mass μ is set to 0.045, a THMD with a added mass ratio β of 0.031 has a 32.7 % reduction in peak value of *DMF* compared to a conventional TMD.

CRediT authorship contribution statement

Wenxi Wang: Writing – review & editing, Supervision, Methodology, Funding acquisition, Conceptualization. **Jun Zhou:** Validation, Methodology, Investigation. **Tianfu Yu:** Methodology, Investigation, Data curation. **Sheng Chen:** Writing – review & editing, Writing – original draft, Methodology, Investigation, Conceptualization. **Yifan Luo:** Methodology, Conceptualization. **Xugang Hua:** Supervision, Conceptualization.

Declaration of compcting interest

The authors declare that they have no known competing financial interests or personal relationships that could have appeared to influence the work reported in this paper.

Data availability

the explanation text will be published verbatim in your article

Acknowledgments

This work was supported by the National Natural Science Foundation of China (Grant No. 52278304); the science and technology innovation Program of Hunan Province (Grant No. 2023RC3090); the Changsha Outstanding Innovative Youth Culturing Program (Grant No. kq2209009).

References

[1] M. Jafari, F. Hou, A. Abdelkefi, Wind-induced vibration of structural cables, Nonlinear Dyn. 100 (2020) 351–421.

[2] Q. Zhu, Y.L. Xu, L.D. Zhu, et al., Vortex-induced vibration analysis of long-span bridges with twin-box decks under non-uniformly distributed turbulent winds, J. Wind Eng. Ind. Aerodyn. 172 (2018) 31–41.
[3] F. Hou, P.P. Sarkar, Aeroelastic model tests to study tall building vibration in boundary-layer and tornado winds, Eng. Struct. 207 (2020) 110259.
[4] T.J. Carey, H.B. Mason, A.R. Barbosa, et al., Multihazard earthquake and tsunami effects on soil–foundation–bridge systems, J. Bridg. Eng. 24 (4) (2019) 04019004.
[5] X. Kang, L. Jiang, Y. Bai, et al., Seismic damage evaluation of high-speed railway bridge components under different intensities of earthquake excitations, Eng. Struct. 152 (2017) 116–128.
[6] M.W. Mieler, J. Mitrani-Reiser, Review of the state of the art in assessing earthquake-induced loss of functionality in buildings, J. Struct. Eng. 144 (3) (2018) 04017218.
[7] S. Elias, V. Matsagar, Research developments in vibration control of structures using passive tuned mass dampers, Annu. Rev. Control. 44 (2017) 129–156.
[8] Frahm H. DEVICE FOR DAMPING VIBRATIONS OF BODIES.: US19090525455[P]. US989958A[2024-02-05]. DOI:US989958 A.
[9] J.P. Den Hartog, Mechanical vibrations[M], Courier Corporation, 1985.
[10] G.B. Warburton, Optimum absorber parameters for various combinations of response and excitation parameters, Earthq. Eng. Struct. Dyn. 10 (3) (1982) 381–401.
[11] T.T. Soong, M. Grigoriu, Random vibration of mechanical and structural systems, NASA STI/Recon Technical Report A 93 (1993) 14690.
[12] M. De Angelis, S. Perno, A. Reggio, Dynamic response and optimal design of structures with large mass ratio TMD, Earthq. Eng. Struct. Dyn. 41 (1) (2012) 41–60.
[13] A. Giaralis, F. Petrini, Wind-induced vibration mitigation in tall buildings using the tuned mass-damper-inerter, J. Struct. Eng. 143 (9) (2017) 04017127.
[14] S. Elias, V. Matsagar, Wind response control of tall buildings with a tuned mass damper, Journal of Building Engineering 15 (2018) 51–60.
[15] C. Sun, V. Jahangiri, Bi-directional vibration control of offshore wind turbines using a 3D pendulum tuned mass damper, Mech. Syst. Sig. Process. 105 (2018) 338–360.
[16] W. Shen, S. Zhu, Y.L. Xu, et al., Energy regenerative tuned mass dampers in high-rise buildings, Struct. Control Health Monit. 25 (2) (2018) e2072.
[17] J. Dai, Z.D. Xu, P.P. Gai, Tuned mass-damper-inerter control of wind-induced vibration of flexible structures based on inerter location, Eng. Struct. 199 (2019) 109585.
[18] Z. Sun, Z. Zou, X. Ying, et al., Tuned mass dampers for wind-induced vibration control of Chongqi Bridge, J. Bridg. Eng. 25 (1) (2020) 05019014.
[19] D. De Domenico, G. Ricciardi, An enhanced base isolation system equipped with optimal tuned mass damper inerter (TMDI), Earthq. Eng. Struct. Dyn. 47 (5) (2018) 1169–1192.
[20] A. Di Matteo, C. Masnata, A. Pirrotta, Simplified analytical solution for the optimal design of tuned mass damper inerter for base isolated structures, Mech. Syst. Sig. Process. 134 (2019) 106337.
[21] Q. Wang, N.D. Tiwari, H. Qiao, et al., Inerter-based tuned liquid column damper for seismic vibration control of a single-degree-of-freedom structure, Int. J. Mech. Sci. 184 (2020) 105840.
[22] A. Kaveh, M. Fahimi Farzam, J.H. Hojat, Statistical seismic performance assessment of tuned mass damper inerter, Struct. Control Health Monit. 27 (10) (2020) e2602.
[23] W. Wang, Z. Yang, X. Hua, et al., Evaluation of a pendulum pounding tuned mass damper for seismic control of structures, Eng. Struct. 228 (2021) 111554.
[24] L. Wang, W. Shi, Y. Zhou, Adaptive-passive tuned mass damper for structural aseismic protection including soil–structure interaction, Soil Dyn. Earthq. Eng. 158 (2022) 107298.
[25] Z. Long, W. Shen, H. Zhu, On energy dissipation or harvesting of tuned viscous mass dampers for SDOF structures under seismic excitations, Mech. Syst. Sig. Process. 189 (2023) 110087.
[26] E. Matta, Lifecycle cost optimization of tuned mass dampers for the seismic improvement of inelastic structures, Earthq. Eng. Struct. Dyn. 47 (3) (2018) 714–737.
[27] L. Wang, S. Nagarajaiah, W. Shi, et al., Semi-active control of walking-induced vibrations in bridges using adaptive tuned mass damper considering human-structure-interaction, Eng. Struct. 244 (2021) 112743.
[28] P. Wang, J. Chen, Nonlinear rotational inertial double-tuned mass damper for reducing human-induced floor vibration with frequency detuning, Mech. Syst. Sig. Process. 172 (2022) 109016.
[29] M. De Angelis, F. Petrini, D. Pietrosanti, Optimal design of the ideal grounded tuned mass damper inerter for comfort performances improvement in footbridges with practical implementation considerations, Struct. Control Health Monit. 28 (9) (2021) e2800.
[30] L. Wang, S. Nagarajaiah, Y. Zhou, et al., Experimental study on adaptive-passive tuned mass damper with variable stiffness for vertical human-induced vibration control, Eng. Struct. 280 (2023) 115714.
[31] C. Song, R. Xiao, Z. Jiang, et al., Active-learning Kriging-assisted robust design optimization of tuned mass dampers: Vibration mitigation of a steel-arch footbridge, Eng. Struct. 303 (2024) 117502.
[32] Kang N, Kim H, Choi & Seongwoo Jo S, et al. Performance evaluation of TMD under typhoon using system identification and inverse wind load estimation[J]. Computer-Aided Civil and Infrastructure Engineering, 2012, 27(6): 455-473.
[33] M. Morga, G.C. Marano, Optimization criteria of TMD to reduce vibrations generated by the wind in a slender structure, J. Vib. Control 20 (16) (2014) 2404–2416.
[34] C. Xing, H. Wang, A. Li, et al., Study on wind-induced vibration control of a long-span cable-stayed bridge using TMD-type counterweight, J. Bridg. Eng. 19 (1) (2014) 141–148.
[35] D. De Domenico, G. Ricciardi, Earthquake-resilient design of base isolated buildings with TMD at basement: Application to a case study, Soil Dyn. Earthq. Eng. 113 (2018) 503–521.
[36] S. Pourzeynali, S. Salimi, H.E. Kalesar, Robust multi-objective optimization design of TMD control device to reduce tall building responses against earthquake excitations using genetic algorithms, Sci. Iran. 20 (2) (2013) 207–221.
[37] Q. Wen, X.G. Hua, Z.Q. Chen, et al., Control of human-induced vibrations of a curved cable-stayed bridge: Design, implementation, and field validation, J. Bridg. Eng. 21 (7) (2016) 04016028.
[38] K. Liu, L. Liu, Q. Zhu, et al., Dynamic testing and numerical simulation of human-induced vibration of cantilevered floor with tuned mass dampers[C]// Structures, Elsevier 34 (2021) 1475–1488.
[39] Marian L, Giaralis A. Optimal design of inerter devices combined with TMDs for vibration control of buildings exposed to stochastic seismic excitation. Safety, Reliability, Risk and Life-Cycle Performance of Structures and Infrastructures-Proceedings of the 11th International Conference on Structural Safety and Reliability, ICOSSAR 2013. CRC Press, 2013: 1025-1032.
[40] L. Marian, A. Giaralis, Optimal design of a novel tuned mass-damper–inerter (TMDI) passive vibration control configuration for stochastically support-excited structural systems, Probab. Eng. Mech. 38 (2014) 156–164.
[41] D. Pietrosanti, M. De Angelis, M. Basili, Optimal design and performance evaluation of systems with Tuned Mass Damper Inerter (TMDI), Earthq. Eng. Struct. Dyn. 46 (8) (2017) 1367–1388.
[42] D. De Domenico, G. Ricciardi, An enhanced base isolation system equipped with optimal tuned mass damper inerter (TMDI), Earthq. Eng. Struct. Dyn. 47 (5) (2018) 1169–1192.
[43] H. Ataei, A.K. Kalbasi, A proposed structural design method considering fluid viscous damper degradations, Struct. Design Tall Spec. Build. 27 (15) (2018) e1512.
[44] J. Tu, Z. Li, J. Zhang, et al., Development, test, and mechanical model of the leak-proof magnetorheological damper, Front. Mater. 6 (2019) 118.
[45] G. Chen, J. Wu, Optimal placement of multiple tune mass dampers for seismic structures, J. Struct. Eng. 127 (9) (2001) 1054–1062.
[46] A. Giaralis, A.A. Taflanidis, Optimal tuned mass-damper-inerter (TMDI) design for seismically excited MDOF structures with model uncertainties based on reliability criteria, Struct. Control Health Monit. 25 (2) (2018) e2082.

[47] N. Hoang, Y. Fujino, P. Warnitchai, Optimal tuned mass damper for seismic applications and practical design formulas, Eng. Struct. 30 (3) (2008) 707–715.
[48] E. Matta, A. De Stefano, Seismic performance of pendulum and translational roof-garden TMDs, Mech. Syst. Sig. Process. 23 (3) (2009) 908–921.
[49] M. Rahman, A. Hasnat, Modified Roof-Top Garden as a Tuned Mass Damper for Vibration Control of Building Structure Under Earthquake Excitation. Sonargaon, University Journal 2 (2) (2018).
[50] A. Bhujel, M.R. Bhatt, P.M. Pradhan, Study of Seismic Effect on Reinforced Concrete Building Due to Swimming Pool on Roof Top[C]//International Conference on Wave Mechanics and Vibrations, Springer International Publishing, Cham, 2022, pp. 1224–1234.
[51] T. Ishihara, S. Zhang, Prediction of dynamic response of semi-submersible floating offshore wind turbine using augmented Morison's equation with frequency dependent hydrodynamic coefficients, Renew. Energy 131 (2019) 1186–1207.
[52] J.R. Morison, J.W. Johnson, S.A. Schaaf, The force exerted by surface waves on piles, J. Petrol. Tech. 2 (05) (1950) 149–154.
[53] L. Liu, J.P. Thomas, E.H. Dowell, et al., A comparison of classical and high dimensional harmonic balance approaches for a Duffing oscillator, J. Comput. Phys. 215 (1) (2006) 298–320.
[54] Z.K. Peng, Z.Q. Lang, S.A. Billings, et al., Comparisons between harmonic balance and nonlinear output frequency response function in nonlinear system analysis, J. Sound Vib. 311 (1–2) (2008) 56–73.
[55] Z.K. Peng, G. Meng, Z.Q. Lang, et al., Study of the effects of cubic nonlinear damping on vibration isolations using harmonic balance method, Int. J. Non Linear Mech. 47 (10) (2012) 1073–1080.
[56] M. Wang, Feasibility study of nonlinear tuned mass damper for machining chatter suppression, J. Sound Vib. 330 (9) (2011) 1917–1930.
[57] Wendel K. Hydrodynamic masses and hydrodynamic moments of inertia. 1956.

Engineering Structures 311 (2024) 118136

Contents lists available at ScienceDirect

Engineering Structures

journal homepage: www.elsevier.com/locate/engstruct

Active vibration control for flexible towers based on displacement observation and reduced-order controller in modal space: Theory and experiment

Xixi Wen [a,b], Hongyi Zhang [a,b,*], Zhengqing Chen [a,b], Xugang Hua [a,b], Huawei Niu [a,b], Zhaodong Xu [c]

[a] State Key Laboratory of Bridge Safety and Resilience, Hunan University, Changsha 410082, China
[b] Key Laboratory for Wind and Bridge Engineering of Hunan Province, Hunan University, Changsha 410082, China
[c] School of Civil Engineering, Southeast University, Nanjing 211189, China

ARTICLE INFO

Keywords:
Vibration control
Active control
Reduced-order controller
Wind turbine tower
Modal space
Videometrics

ABSTRACT

Active vibration control is not only suitable but also urgently needed in order to mitigate excessive vibration for dynamic systems such as wind turbine towers. However, there is a challenge of directly observing displacement and developing an efficient reduced-order control strategy. With this in view, a reduced-order controller based on displacement observation is developed for the vibration mitigation of flexible high towers, where the problems of low signal-to-noise ratio (SNR) and intricate acceleration-based control algorithms are eliminated completely. The displacement can be measured by laser or modern videometrics technology. A reduced-order model with an active mass damper (AMD) at the top of the tower is created using modal truncation approach based on modal participation factor (MPF), which successfully preserves the dominant modes of the original structure, and reduces the influence of high-frequency uncertainties. Considering structure uncertainties caused by environmental factors and aging process, an unscented Kalman filter (UKF) is introduced to identify the modal stiffness of the reduced-order model. The present study, using a 5 MW onshore wind turbine as an example, investigates active vibration control under wind and seismic loads and compares it with traditional tuned mass damper (TMD). From a dynamic perspective, the simplified model of wind turbine tower takes the form of a multi-degree-of-freedom series model. Consequently, a six-story steel frame platform has been constructed to experimentally validate the efficacy of the reduced-order controller. According to numerical simulation and laboratory testing, the control effect of the reduced-order controller with a few displacement states as feedback can be utilized as an active control strategy for high-rise flexible towers.

1. Introduction

Global wind turbine has experienced a rapid growth in recent years to meet the ever-increasing demand for carbon-free renewable energy. To better utilize high-altitude wind energy resources and expand the swept area of the blades, the wind turbine tower has consistently broken through its height, typically between 100 m and 140 m. Therefore, the wind turbine tower is a typical high-flexibility structure, with the first-order bending frequency ranging from 0.12–0.35 Hz, making them prone to cross-resonance caused by the unit. The second-order bending frequency is 0.9–3.0 Hz, making them susceptible to vortex-induced resonance [1]. Hence, flexible towers often need to stop or limit their speed, reducing power generation efficiency and impacting the fatigue life of the structure. Although wind turbines have an intricate control system that encompassing speed control and safe shutdown management, their pitch and yaw systems [2,3] alone prove inadequate for mitigating the vibration of the entire machine. Vibration-induced wind turbine tower collapse accidents caused by pitch system failure or significant vibration during yaw occasionally occur [4]. Therefore, an independent vibration control system is necessary for wind turbines [5,6].

The passive control of wind turbines has been extensively studied and applied by many scholars, primarily focusing on TMD and tuned liquid dampers (TLD). The concept of TMD was initially introduced by Den Hartog in the 1940 s and has found widespread use in high-rise

* Corresponding author at: State Key Laboratory of Bridge Safety and Resilience, Hunan University, Changsha 410082, China.
E-mail address: zhyzzb25@hnu.edu.cn (H. Zhang).

https://doi.org/10.1016/j.engstruct.2024.118136
Received 13 October 2023; Received in revised form 23 April 2024; Accepted 2 May 2024
Available online 16 May 2024

buildings, flexible bridges, and various other domains. While research in wind turbine applications started relatively late, its practical utilization gained momentum in recent years, coinciding with the widespread adoption of the flexible tower of wind turbine structures. Murtagh *et al.* [7] investigated the use of TMD for the mitigation of vibrations due to the along-wind forced vibration response of a simplified wind turbine, and theoretical calculations indicate that TMD can exhibit significant vibration reduction effects when the parameters of the wind turbine system remain constant. Lackner and Rotea [8] developed a new simulation tool based on OpenFAST to model TMD control systems in wind turbines. The simulation results indicate that TMDs can be used to improve the structural response of offshore wind turbines. Sarkar and Fitzgerald [9] introduced inertial mass into TMD system, referred to as TMDI, which improves the quality and stroke necessary for optimal TMD performance. TLD harnesses the lateral pressure generated by the oscillating liquid within the water tank to furnish damping force. Roderick [10] studied the vibration reduction efficacy of TLD in a 150 m steel-concrete composite wind turbine tower. At a mass ratio of 5%, TLD achieves a 41% reduction under small amplitudes. Nevertheless, with increasing displacement, the vibration reduction effect experiences a rapid decay. Colwell and Basu [11] found that when an offshore wind turbine is equipped with a TLCD and subjected to wind and wave forces, reductions of up to 55% in the peak response of the same system without TLCD may be achieved. However, in view of the restriction in the provision of the available "horizontal length" for a TLCD in a wind turbine, other dampers such as TMDs and TLDs may prove to be more efficient in certain cases. Shadman et al. [12] investigated response of a floating wind turbine controlled by a TLCD subjected to couple hydrodynamic and aerodynamic loads, and the results indicate that this passive type control approach can be used to improve the structural response of floating wind turbines. In essence, the optimized passive damping scheme can effectively control vibrations within a specific frequency range, yet its limitations are apparent. Firstly, passive control typically necessitates a mass ratio of 1% or even 5% to yield a discernible effect, and positioning a substantial mass block at the pinnacle of the tower is constrained not only by the limitations of tower space but also poses a rigorous challenge for both the bearing capacity and cost control; Secondly, TMD functions as a single-mode narrowband controller, inherently limiting its damping effectiveness.

To improve the effectiveness of TMD, Chang and Soong [13] introduced an AMD system based on active control theory. Three commonly used structure forms for active control systems exist. Firstly, an active control force is introduced between the TMD and the controlled structure, termed as ATMD. The second involves directly applying an active force to the controlled structure, devoid of stiffness and damping components, referred to as AMD. The third incorporates an AMD substructure, denoted as HMD, added to the TMD. Concerning active control algorithms, various methods exist, including the pole placement method [14], linear quadratic control (LQR) [15], linear quadratic Gaussian (LQG) [15], sliding mode control (SMC) [16], H∞ control [17], among others, each accompanied by relatively comprehensive theories. However, the application of active control technology in large-scale civil engineering constructions is still fraught with significant challenges. Active control systems have high environmental anti-disturbance and power supply requirements, which limit their field of applicability and make them unsuitable for controlling sudden earthquakes. Nevertheless, the power supply problem can be resolved by the wind turbine itself, creating favorable conditions for the use of active control technology. Although the high tower structure typically has the necessary power supply and equipment installation conditions for active control, two significant challenges exist. The first is an efficient control strategy for high towers, and the second is the precise identification of the displacement state.

The initial challenge lies in formulating a lower-order model that aligns more closely with the actual structure. A number of model reduction techniques have been developed and widely used. In the context of wind turbines, finite element method, modal coordinate method, and multi body dynamics method, mainly be adopted to capture the dynamics of the turbine system. Mitra [18] established a concise multi-body dynamics model of a wind turbine and presents a wavelet-based strategy for fore-aft vibration control based on the model. Sarkar [19,20] established a reduced-order model of the blade-tower system to investigate the performance of the semi-active strategies using magneto-rheological tuned liquid column dampers. Besides the multi body dynamics and modal coordinate method mentioned above, other reduction methods are also investigated by many researchers. A limited degree of freedom model is identified with synthetic data and used by Lackner [21,22] to investigate the control techniques of offshore wind turbines. For finite element model, the balanced truncation method has been demonstrated as a highly effective approach for preserving low-order dynamic properties [23–25]. Xie [26] conducted an experimental investigation of internal balance model reduction, utilizing a flexible beam as the research subject, and proposed a procedure for extracting the internal balance modal coordinates from physical sensor measurements. Chen [27] conducted a comparative analysis of various model reduction methods and concluded that the balanced truncation method is a dependable order reduction technique suitable for civil engineering applications. However, this technique involves solving two Lyapunov equations, which can be computationally demanding and lack clear physical interpretation.

Secondly, the challenge lies in accomplishing real-time identification of structural vibration states. In order to calculate the control force, active control requires accurate and real-time identification of the structural vibration state, specifically the displacement and velocity. In the past, due to the difficulty in measurement of displacement of towers, control algorithm based on acceleration observation is generally adopted, leading to a host of issues including a low SNR and intricate control algorithms. Besides, Acceleration feedback must be combined with a state observer to estimate the displacement and velocity of the structure [28], which also leads to delays. As a result, controller based on acceleration feedback significantly reduce the control accuracy and compromising the superiority of active control over passive control. Recent advancements in non-contact measuring technology, particularly videometrics technology [29,30], have made it possible to observe the real-time status of large structures, thereby enabling the direct monitoring of displacement. Videometrics technology leverages the camera as a sensor to analyze and process images, providing measurements for location, size, speed, and acceleration. Its unique feature is its ability to use sequence images as the primary object for real-time measurements of the dynamic properties of a moving target, surpassing the contact measuring technique in numerous ways. By using videometrics technology in the active control system of a high-rise structure, the structure's displacement can be directly monitored, and the displacement state can be employed as the feedback signal to calculate the control force. This approach simplifies the calculation steps, reduces processing time, and enhances accuracy, providing the necessary solution for resolving high-rise towers displacement measurement issues through the integration of modern measurement and information technologies.

In the aforementioned context, this study proposes a reduced-order displacement-based controller suitable for high flexible towers, using the active control system of a 5 MW wind turbine as an example. The measurement of displacement can be accomplished by laser or the videometrics technique which involves creating a transformation matrix to convert relative displacement of camera observations to absolute node displacement. A model reduction method called modal truncation based on MPF is introduced, which is easy to calculate and has clear physical significance. Since the high-frequency components are usually truncated, the disturbance caused by these high-frequencies becomes quite small. A reduced-order LQR controller considering time-delay is designed for the wind turbine. To assure the control performance, a UKF algorithm is employed to identify the state-space modal parameters. This reduced-order control system is exposed to turbulent wind and seismic ground motion to investigate the controller performance, respectively. A six-story steel frame model is developed to experimentally identify the controller. The results indicate that the reduced-order controller can replace the full-order controller, and that the active mass damper control system can achieve better control outcomes than traditional passive TMD control in limited space.

The following sections of the paper are organized as follows: Section 2 introduces the establishment of reduced-order control system including modal truncation method based on MPF and LQR with the consideration of time-delay effect, displacement-relay videometric network. Section 3 details the design of a reduced-order controller for a 5 MW wind turbine and presents simulation results. Section 4 provides experimental verification with a six-story steel frame model. Finally, the concluding remarks are presented in the last section.

2. Design of reduced-order control system

2.1. Basic theory of modal truncation method based on MPF

The dynamic equilibrium of a damped system with n degrees of freedom structure be:

$$\boldsymbol{M}\ddot{\boldsymbol{x}}+\boldsymbol{C}\dot{\boldsymbol{x}}+\boldsymbol{K}\boldsymbol{x}=\boldsymbol{f} \tag{1}$$

Modal transformation of the above system will be:

$$\boldsymbol{x}=\boldsymbol{\Phi}\boldsymbol{y} \tag{2}$$

where $\boldsymbol{x}$ is the n-dimensional displacement vector, $\boldsymbol{y}$ is the n-dimensional generalized modal coordinate vector and $\boldsymbol{\Phi}$ is the $n\times n$ regularized mode-shape matrix.

Assuming the damping matrix is uncoupled, one can get:

$$\boldsymbol{\Phi M\Phi}=\boldsymbol{I},\boldsymbol{\Phi K\Phi}=\boldsymbol{\Lambda},\boldsymbol{\Phi C\Phi}=\boldsymbol{\Gamma} \tag{3}$$

where I is the n-order identity matrix, $\Lambda=\mathrm{diag}\{\omega_1^2,\omega_2^2,\cdots\omega_n^2\}$ is the spectral matrix, and $\Gamma=\mathrm{diag}\{2\omega_1\xi_1,2\omega_2\xi_2,\cdots2\omega_n\xi_n\}$ is uncoupled modal damping matrix. It generally is more convenient and physically reasonable to use the damping ratio for each mode because the modal damping ratios can be determined experimentally or estimated with adequate precision in many cases.

Assuming that a total of m modes of the structure has been taken to represent the dynamic response of the structure with sufficient accuracy, then Eq. (2) is written in block matrix form:

$$\boldsymbol{x}=\begin{bmatrix}\boldsymbol{x}_o\\ \boldsymbol{x}_u\end{bmatrix}=\begin{bmatrix}\boldsymbol{\Phi}_{ro} & \boldsymbol{\Phi}_{uo}\\ \boldsymbol{\Phi}_{ru} & \boldsymbol{\Phi}_{uu}\end{bmatrix}\begin{bmatrix}\boldsymbol{y}_r\\ \boldsymbol{y}_u\end{bmatrix} \tag{4}$$

where $\boldsymbol{y}_r$ is the reduced m-dimensional modal coordinate vector , $\boldsymbol{y}_u$ is coordinate vector of the unselected modes, $\boldsymbol{x}_o$ is the observed q-dimensional displacement vector and $\boldsymbol{x}_u$ is the unobserved one. Neglecting $\boldsymbol{y}_u$ and $\boldsymbol{x}_u$, Eq. (4) is reduced to:

$$\boldsymbol{x}_o=\boldsymbol{\Phi}_{ro}\boldsymbol{y}_r \tag{5}$$

When the dimensions of $\boldsymbol{x}_o$ and $\boldsymbol{y}_r$ are equal, $\boldsymbol{y}_r$ can be derived by $\boldsymbol{y}_r=\boldsymbol{\Phi}_{ro}^{-1}\boldsymbol{x}_o$ More generally, when $q\geq m$, a pseudo inverse operation of $\boldsymbol{\Phi}_{ro}$ could be used:

$$\boldsymbol{y}_r=\boldsymbol{\Phi}_{ro}^{+}\boldsymbol{x}_o \tag{6}$$

Considering viscous damping and external load, the system is written in the form of state equation:

$$\dot{\mathbb{x}}=\boldsymbol{A}\mathbb{x}+\boldsymbol{B}\boldsymbol{u}+\boldsymbol{E}\boldsymbol{f} \tag{7}$$

where $\mathbb{x}$ is the $2n$-dimensional state vector, and $\mathbb{x}=\begin{bmatrix}\boldsymbol{x} & \dot{\boldsymbol{x}}\end{bmatrix}^{\mathrm{T}}$; $\boldsymbol{u}$ is the k-dimensional control force vector, and $\boldsymbol{u}=-\boldsymbol{G}\mathbb{x}$, $\boldsymbol{G}$ is $k\times 2n$ matrix, $\boldsymbol{f}$ is the l-dimensional external excitation load vector, $\boldsymbol{A}=\begin{bmatrix}\boldsymbol{0} & \boldsymbol{I}\\ -\boldsymbol{M}^{-1}\boldsymbol{K} & -\boldsymbol{M}^{-1}\boldsymbol{C}\end{bmatrix}$ is a $2n\times 2n$ state matrix, $\boldsymbol{B}$ is a $2n\times k$ position matrix of control force, $\boldsymbol{E}$ is a $2n\times l$ position matrix of load.

Substitute Eq. (2) into Eq. (7), and left-multiply both sides by $\boldsymbol{\Phi}^{\mathrm{T}}\boldsymbol{M}$:

$$\begin{bmatrix}\boldsymbol{\Phi}^{\mathrm{T}}\boldsymbol{M\Phi} & 0\\ 0 & \boldsymbol{\Phi}^{\mathrm{T}}\boldsymbol{M\Phi}\end{bmatrix}\dot{\mathbb{y}}=\begin{bmatrix}\boldsymbol{\Phi}^{\mathrm{T}}\boldsymbol{M\Phi} & 0\\ 0 & \boldsymbol{\Phi}^{\mathrm{T}}\boldsymbol{M\Phi}\end{bmatrix}\boldsymbol{A}\begin{bmatrix}\boldsymbol{\Phi} & 0\\ 0 & \boldsymbol{\Phi}\end{bmatrix}\mathbb{y}+\begin{bmatrix}\boldsymbol{\Phi}^{\mathrm{T}}\boldsymbol{M} & 0\\ 0 & \boldsymbol{\Phi}^{\mathrm{T}}\boldsymbol{M}\end{bmatrix}\boldsymbol{Bu}+\begin{bmatrix}\boldsymbol{\Phi}^{\mathrm{T}}\boldsymbol{M} & 0\\ 0 & \boldsymbol{\Phi}^{\mathrm{T}}\boldsymbol{M}\end{bmatrix}\boldsymbol{Ef} \tag{8}$$

where $\mathbb{y}=\begin{bmatrix}\boldsymbol{y} & \dot{\boldsymbol{y}}\end{bmatrix}^{\mathrm{T}}$.Considering Eq. (3), Eq. (8) is simplified as:

$$\dot{\mathbb{y}}=\widetilde{\boldsymbol{A}}\mathbb{y}+\widetilde{\boldsymbol{B}}\boldsymbol{u}+\widetilde{\boldsymbol{E}}\boldsymbol{f} \tag{9}$$

where $\widetilde{\boldsymbol{A}}=\begin{bmatrix}0 & \boldsymbol{I}\\ -\boldsymbol{\Lambda} & -\boldsymbol{\Gamma}\end{bmatrix}$, $\widetilde{\boldsymbol{B}}=\boldsymbol{\Phi}^{-1}\boldsymbol{B}$ and $\widetilde{\boldsymbol{E}}=\boldsymbol{\Phi}^{-1}\boldsymbol{E}$.

Rearrange $\mathbb{y}$ from $[y_r\, y_u\, \dot{y}_r\, \dot{y}_u]^T$ to $[y_r\, \dot{y}_r\, y_u\, \dot{y}_u]^T$ and neglecting the contribution of $[y_u\, \dot{y}_u]^T$, the reduced-order state equation of the modal space composed of m modes is written as:

$$\dot{\mathbb{y}}_r=\widetilde{\boldsymbol{A}}_r\mathbb{y}_r+\widetilde{\boldsymbol{B}}_r\boldsymbol{u}_r+\widetilde{\boldsymbol{E}}_r\boldsymbol{f} \tag{10}$$

where $\widetilde{\boldsymbol{A}}_r=\boldsymbol{T}_r\widetilde{\boldsymbol{A}}\boldsymbol{T}_r^{\mathrm{T}}$, $\widetilde{\boldsymbol{B}}_r=\boldsymbol{T}_r\widetilde{\boldsymbol{B}}$, $\widetilde{\boldsymbol{E}}_r=\boldsymbol{T}_r\widetilde{\boldsymbol{E}}$, and the transformation matrix $\boldsymbol{T}_r$ is expressed as:

$$\boldsymbol{T}_r=[\boldsymbol{I}_{2m}\quad 0]_{(2m\times 2n)}\cdot\begin{bmatrix}\boldsymbol{I}_m & \cdots & \cdots & 0\\ \vdots & 0 & \boldsymbol{I}_m & \vdots\\ \vdots & \boldsymbol{I}_{n-m} & 0 & \vdots\\ 0 & \cdots & \cdots & \boldsymbol{I}_{n-m}\end{bmatrix}_{(2n\times 2n)}$$

Alternatively, the selection of m reserved modes could be determined by the MPF of each mode:

$$\mu_i=\frac{(\boldsymbol{\phi}_i^{\mathrm{T}}\boldsymbol{M}\boldsymbol{e}_r)^2}{\boldsymbol{\phi}_i^{\mathrm{T}}\boldsymbol{M}\boldsymbol{\phi}_i} \tag{11}$$

where $\boldsymbol{e}_r$ is a unit vector of the actuator's main control direction, $\boldsymbol{\phi}_i$ is the i-th mode shape. The modal participating coefficient is a significant indicator in structure's seismic analysis using modal superposition method. A mode with larger participating coefficient usually contributes more to structural vibrations. Besides, the closer the sum of chosen MPF is to 1, the better the reduction model can represent the accurate model.

2.2. Basic principle of LQR algorithm

The vibration control of civil engineering structures can always be regarded as the behavior in infinite domain, so the quadratic optimal control in infinite domain time interval [0, ∞] is the theoretical basis of active structural vibration control. For linear steady-state system Eq. (7), semi-positive definite symmetric weighting matrices $\boldsymbol{Q}$ and $\boldsymbol{R}$ are selected to construct the quadratic performance control index:

$$J=\frac{1}{2}\int_{t_0}^{\infty}\left[\mathbb{x}^{\mathrm{T}}(t)\boldsymbol{Q}\mathbb{x}(t)+\boldsymbol{u}^{\mathrm{T}}(t)\boldsymbol{R}\boldsymbol{u}(t)\right]\mathrm{d}t \tag{12}$$

Then the optimal control vector $\boldsymbol{u}(t)$ satisfying the minimization of J exists and is uniquely determined by the following equation:

$$\boldsymbol{u}(t)=-\boldsymbol{G}\mathbb{x}(t)=-\boldsymbol{R}^{-1}\boldsymbol{B}^{\mathrm{T}}\boldsymbol{P}\mathbb{x}(t) \tag{13}$$

For time-delay system, the control force can be tuned by pole placement method [31],

$$\boldsymbol{u}(t)=-\boldsymbol{R}^{-1}\boldsymbol{B}^{\mathrm{T}}\boldsymbol{P}e^{\boldsymbol{A}_c\tau}\mathbb{x}(t) \tag{14}$$

where τ is time delay of control system, and $\boldsymbol{A}_c=\boldsymbol{A}-\boldsymbol{R}^{-1}\boldsymbol{B}^{\mathrm{T}}\boldsymbol{P}$.

Matrix $\boldsymbol{P}$ is the solution of the following Riccati algebraic equation:

$$\boldsymbol{PA}+\boldsymbol{A}^{\mathrm{T}}\boldsymbol{P}-\boldsymbol{PBR}^{-1}\boldsymbol{B}^{\mathrm{T}}\boldsymbol{P}+\boldsymbol{Q}=0 \tag{15}$$

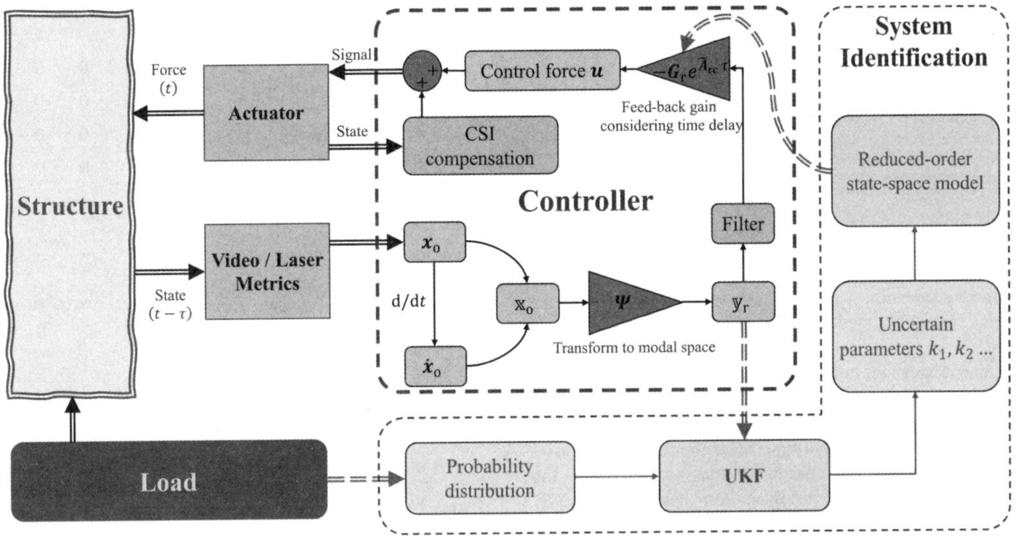

Fig. 1. The flowchart of the proposed control scheme.

As well-known, implementing the LQR control algorithm requires relatively accurate model representation to achieve effective control performance. Besides, structural uncertainties inevitably exist due to many factors, such as aerodynamic stiffness and damping, wind turbine operation conditions, structural aging process, and so forth. Therefore, the identification of structural parameters with acceptable accuracy is highly desirable. UKF is a Bayesian filter using the unscented transformation to capture the propagation of the statistical properties of state estimates through nonlinear functions. It can be used for the joint estimation of structural states and parameters.

A p-dimensional vector $\boldsymbol{\eta}$ is composed of the unknown structural parameters, saying $\boldsymbol{\eta} = [k_1, k_2, \ldots k_p]^{\mathrm{T}}$ in which k_i (i = 1,2,..., p) denotes the unknown lower order modal stiffness. Then, the implementation steps of UKF are given as follows:

(1). Combine the state vector and uncertainty vector to form a new augmented state vector:

$$\mathbb{z} = [\mathbb{y}_{\mathrm{r}} \quad \boldsymbol{\eta}]^{\mathrm{T}} \tag{16}$$

(2). Express the state transition equation of UKF in the following form:

$$\dot{\mathbb{z}} = \begin{bmatrix} \widetilde{\boldsymbol{A}}_{\mathrm{r}}(\boldsymbol{\eta}) & \mathbf{0}_{2m\times p} \\ \mathbf{0}_{p\times 2m} & \mathbf{0}_{p\times p} \end{bmatrix} \mathbb{z} + \begin{bmatrix} \widetilde{\boldsymbol{B}}_{\mathrm{r}}(\boldsymbol{\eta}) \\ \mathbf{0} \end{bmatrix} \boldsymbol{u}_{\mathrm{r}} + \begin{bmatrix} \widetilde{\boldsymbol{E}}_{\mathrm{r}}(\boldsymbol{\eta}) \\ \mathbf{0} \end{bmatrix} \boldsymbol{f} \tag{17}$$

It should be noted that the external excitation load is regarded as noise in the UKF algorithm, and only the distribution statistics such as mean value and covariance matrix are in need.

(3). After identifying the uncertainties, the state matrix of reduced-order model $\widetilde{\boldsymbol{A}}_{\mathrm{r}}(\boldsymbol{\eta})$ is updated in real time based on the identified parameters. $\widetilde{\boldsymbol{B}}_{\mathrm{r}}(\boldsymbol{\eta})$ and $\widetilde{\boldsymbol{E}}_{\mathrm{r}}(\boldsymbol{\eta})$ will also be updated if the modal shapes change. A comprehensive overview of the UKF algorithm can be found in references[32,33].

2.3. *Design of reduced-order controller based on LQR algorithm*

The accuracy of the reduced-order model is a primary requirement for the order reduction process. It is crucial that the characteristics of the reduced-order model are similar to those of the actual controlled object. However, meeting the accuracy requirements alone is not enough for order reduction in active control. In some cases, the performance of the controller designed using a slightly less accurate reduced-order model may be better than the one designed with a highly accurate reduced-order model. Therefore, it is necessary to consider the controller design along with order reduction.

Based on the reduced-order model shown in Eq. (10), the reduced-order controller designed according to Eq. (14) is:

$$\boldsymbol{u}_{\mathrm{r}} = -\boldsymbol{G}_{\mathrm{r}} e^{\tilde{\boldsymbol{A}}_{\mathrm{rc}}\tau} \mathbb{y}_{\mathrm{r}} \tag{18}$$

where $\boldsymbol{G}_{\mathrm{r}} = \widetilde{\boldsymbol{R}}^{-1} \widetilde{\boldsymbol{B}}^{\mathrm{T}} \widetilde{\boldsymbol{P}}$, $\widetilde{\boldsymbol{A}}_{\mathrm{rc}} = \widetilde{\boldsymbol{A}}_{\mathrm{r}} - \boldsymbol{R}^{-1} \widetilde{\boldsymbol{B}}_{\mathrm{r}}^{\mathrm{T}} \boldsymbol{P}$, $\boldsymbol{u}_r$ is the optimal control force.

By substituting Eqs. (6) to (18), the relationship between optimal control force and observed state vector is:

$$\boldsymbol{u}_{\mathrm{r}} = -\boldsymbol{G}_{\mathrm{r}} e^{\tilde{\boldsymbol{A}}_{\mathrm{rc}}\tau} \boldsymbol{\Psi} \mathbb{x}_{\mathrm{o}} \tag{19}$$

where

$$\boldsymbol{\Psi} = \begin{bmatrix} \boldsymbol{\Phi}_{\mathrm{ro}}^{+} & \mathbf{0} \\ \mathbf{0} & \boldsymbol{\Phi}_{\mathrm{ro}}^{+} \end{bmatrix}.$$

Fig. 1 illustrates the calculation diagram of the entire reduced-order control system. To obtain a more precise low-order state-space model of the controlled system, the UKF is employed to identify the system's modal parameters. This UKF identification system requires some measured states and the probability distribution of excitation load, and it will update the reduced-order state-space model only if the structural parameter changes. Subsequently, a controller is developed based on the identified state-space parameters. In control process, the displacement of the structure is monitored using video or laser metric technique, denoted as $\boldsymbol{x}_{\mathrm{o}}$, and the velocity, $\dot{\boldsymbol{x}}_{\mathrm{o}}$, is obtained by differentiating $\boldsymbol{x}_{\mathrm{o}}$. The combined values of $\boldsymbol{x}_{\mathrm{o}}$ and $\dot{\boldsymbol{x}}_{\mathrm{o}}$ constitute the actual feedback state, denoted as $\mathbb{x}_{\mathrm{o}}$. Subsequently, the actual states are transformed into modal displacement and modal velocity, $\mathbb{y}_{\mathrm{r}}$, by multiplying the matrix $\boldsymbol{\Psi}$. To effectively filter out high-frequency noise and interference signals from the observed data, a lowpass filter is integrated into the controller. The active control force can be determined by multiplying filtered state $\mathbb{y}_{\mathrm{r}}$ with the gain matrix $-\boldsymbol{G}_{\mathrm{r}} e^{\tilde{\boldsymbol{A}}_{\mathrm{rc}}\tau}$. If the system's time delay is negligible, the time delay τ can be set to 0, and the gain will degrade to be $-\boldsymbol{G}_{\mathrm{r}}$. Finally, considering the control-structure interaction (CSI), a composite control scheme [34,35] is introduced to reduce the CSI effect. The ultimate command signal consists of the control force and the CSI compensation force, and is then transmitted to the actuator and looped back to the structure.

Table 1
The first 4 modal parameters of different models.

Mode	Mode shape	Ref/ Hz	Simplified model/Hz	Different/ %
1	1st Tower Fore-Aft	0.3240	0.3230	0.31
2	1st Tower Side-to-Side	0.3120	0.3210	2.88
3	2nd Tower Fore-Aft	2.9003	2.8975	0.09
4	2nd Tower Side-to-Side	2.9361	2.7585	6.05

2.4. Displacement measurement network of turbine towers by displacement-relay videometric method

The displacement-to-height ratio at the top of a highly flexible tower is almost twice that of a conventional rigid tower, indicating that displacement monitoring is now a crucial indicator of structural vibration monitoring and evaluation. Real-time measurement of structural response is necessary for active control systems to determine the appropriate feedback control force using modern control theory. By utilizing displacement monitoring for active control, computational accuracy can be enhanced while calculation time can be decreased. For large and complex structures, non-contact displacement measurement technology is necessary. Various non-contact displacement measurement technologies such as GPS, total station, laser, and videometric methods have been widely used in engineering. The displacement-relay videometric method, as proposed by Yu [30,36], has numerous advantages such as multi-point measurement, real-time measurement, high degree of automation and visualization, and is suitable for measuring displacement from inside the tower. In addition, it has high anti-interference capabilities during disasters such as typhoons, and can be utilized to monitor tower displacement and deformation in real-time for an extended period.

The measurement codes for the mark points and the double-head camera are, from bottom to top, M_1, S_1, $C_1 \ldots M_i$, S_i, $C_i \ldots M_q$, S_q, C_q. The computation formula for the horizontal displacement of the two upper and lower marks captured by the double-head camera, regardless of camera rotation, is as follows:

$$\begin{cases} h_{\mathrm{M}_{i-1}}^{C_i} = k_{\mathrm{M}_{i-1}}^{C_i}\left(\Delta y_{\mathrm{M}_{i-1}} - \Delta y^{C_i} - d_{\mathrm{M}_{i-1}}^{C_i} \bullet \sin\theta^{C_i}\right) \\ h_{\mathrm{S}_{i-1}}^{C_i} = k_{\mathrm{S}_{i-1}}^{C_i}\left(\Delta y_{\mathrm{S}_{i-1}} - \Delta y^{C_i} - d_{\mathrm{S}_{i-1}}^{C_i} \bullet \sin\theta^{C_i}\right) \\ h_{\mathrm{M}_i}^{C_i} = k_{\mathrm{M}_i}^{C_i}\left(\Delta y_{\mathrm{M}_i} - \Delta y^{C_i} - d_{\mathrm{M}_i}^{C_i} \bullet \sin\theta^{C_i}\right) \\ h_{\mathrm{S}_i}^{C_i} = k_{\mathrm{M}_i}^{C_i}\left(\Delta y_{\mathrm{S}_i} - \Delta y^{C_i} - d_{\mathrm{S}_i}^{C_i} \bullet \sin\theta^{C_i}\right) \end{cases} \tag{20}$$

where superscript C_i represents the double-headed cameras, subscripts M_i, S_i represents the marking points, k is corresponding magnification, h is the displacement of the cooperation sign in the camera image, Δy is the displacement of corresponding marker, d is distance between the camera and corresponding marker, θ is the variation of the pitch angle of the camera, the values of d and k can be obtained through calibration.

Eq. (20) contains 4 n equations and 4 n + 2 unknowns. The equations can be solved if there are more than or equal to two known quantities. Assuming that Δy_{M_0} and Δy_{S_0} are fixed points, Eq. (20) can be expressed as follows:

$$\left[\Delta y^{C_1}\sin\theta^{C_1}\Delta y_{M_1}\Delta y_{s_1}\cdots\Delta y^{C_q}\sin\theta^{C_q}\Delta y_{M_q}\Delta y_{s_q}\right] = Nh \tag{21}$$

where

$$N = \begin{bmatrix} -k_{M_0}^{C_1} & -k_{M_0}^{C_1} & d_{M_0}^{C_1} & 0 & 0 & \cdots & 0 & 0 & 0 & 0 \\ -k_{S_0}^{C_1} & -k_{S_0}^{C_1} & d_{S_0}^{C_1} & 0 & 0 & \cdots & 0 & 0 & 0 & 0 \\ -k_{M_1}^{C_1} & k_{M_1}^{C_1} & d_{M_1}^{C_1} & k_{M_1}^{C_1} & 0 & \cdots & 0 & 0 & 0 & 0 \\ -k_{S_1}^{C_1} & k_{S_1}^{C_1} & d_{S_1}^{C_1} & 0 & k_{S_1}^{C_1} & \cdots & 0 & 0 & 0 & 0 \\ \vdots & \vdots & \vdots & \vdots & \vdots & \ddots & \vdots & \vdots & \vdots & \vdots \\ 0 & 0 & 0 & 0 & \cdots & -k_{M_{q-1}}^{C_q} & k_{M_{q-1}}^{C_q} & d_{M_{q-1}}^{C_q} & 0 & 0 \\ 0 & 0 & 0 & 0 & \cdots & -k_{S_{q-1}}^{C_q} & k_{S_{q-1}}^{C_q} & d_{S_{q-1}}^{C_q} & 0 & 0 \\ 0 & 0 & 0 & 0 & \cdots & -k_{M_q}^{C_q} & k_{M_q}^{C_q} & d_{M_q}^{C_q} & k_{M_q}^{C_q} & 0 \\ 0 & 0 & 0 & 0 & \cdots & -k_{S_q}^{C_q} & k_{S_q}^{C_q} & d_{S_q}^{C_q} & 0 & k_{S_q}^{C_q} \end{bmatrix}^{-1}$$

$$\boldsymbol{h} = \left[h_{\mathrm{M}_0}^{C_1} \quad h_{\mathrm{S}_0}^{C_1} \quad h_{\mathrm{M}_1}^{C_1} \quad h_{\mathrm{S}_1}^{C_1} \quad \cdots \quad h_{\mathrm{M}_{q-1}}^{C_q} \quad h_{\mathrm{S}_{q-1}}^{C_q} \quad h_{\mathrm{M}_q}^{C_q} \quad h_{\mathrm{S}_q}^{C_q}\right]^{\mathrm{T}}$$

The displacement of the mark points can be written as:

$$\begin{bmatrix} \Delta y_{\mathrm{M}_1} \\ \Delta y_{\mathrm{S}_1} \\ \vdots \\ \Delta y_{\mathrm{M}_q} \\ \Delta y_{\mathrm{S}_q} \end{bmatrix} = \begin{bmatrix} 0 & 0 & 1 & 0 & \cdots & 0 & 0 & 0 & 0 \\ 0 & 0 & 0 & 1 & \cdots & 0 & 0 & 0 & 0 \\ \vdots & \vdots & \vdots & \vdots & \ddots & \vdots & \vdots & \vdots & \vdots \\ 0 & 0 & 0 & 0 & \cdots & 0 & 0 & 1 & 0 \\ 0 & 0 & 0 & 0 & \cdots & 0 & 0 & 0 & 1 \end{bmatrix} Nh = VNh \tag{22}$$

where

$V \in \mathbb{R}^{2q\times 4q}, N \in \mathbb{R}^{4q\times 4q}, h \in \mathbb{R}^{4q\times 1}$.

Eq. (22) shows that at least $p/2$ double-head cameras are necessary when a number of the p observation points need to be placed. Assuming $q = p/2$, the observation state vector $\mathbb{x}_o$ could be derived from video metric by:

$$\mathbb{X} = \begin{bmatrix} x_o \\ \dot{x}_o \end{bmatrix} = \begin{bmatrix} VN & 0 \\ 0 & VN \end{bmatrix} \begin{bmatrix} h \\ \dot{h} \end{bmatrix} \tag{23}$$

Eq. (23) is the output equation in active control.

3. Case study

To enhance the effectiveness of the active vibration control strategy for the structure, it is crucial to simplify the structural model for the design of the controller. In this section, we employ the modal truncation technique based on MPF to reduce the number of DOFs of the 5 MW wind turbine model by transforming it from physical space to modal space, and ultimately establish a reduced-order controller.

3.1. Design of reduced-order control system for wind turbine

3.1.1. Establishment of the simplified model

The 5-MW baseline wind turbine, which is a horizontal-axis, three-bladed, upwind, variable speed, pitch-controlled turbine with a 126 m rotor diameter and a 90 m hub height, proposed by National Renewable Energy Laboratory (NREL) is taken as an example [37]. To simplify the computation, the blade, hub, and engine room are treated as a mass point [38,39], and a three-dimensional simplified model of the wind turbine is created. Table 1 shows the comparison between the first two frequencies of the tower in the Fore-Aft and Side-to-Side direction calculated in this study and that given by Jonkman [37]. Fig. 2 describes the modal shape of the simplified and OpenFAST models. Note that the simplified model considers the rotational degrees of freedom, while the mode function of OpenFAST only provides the translational degrees of freedom. The results in Table 1 and Fig. 2 indicate that the simplified model established in this paper can effectively represent the dynamic characteristics of a 5 MW onshore wind turbine.

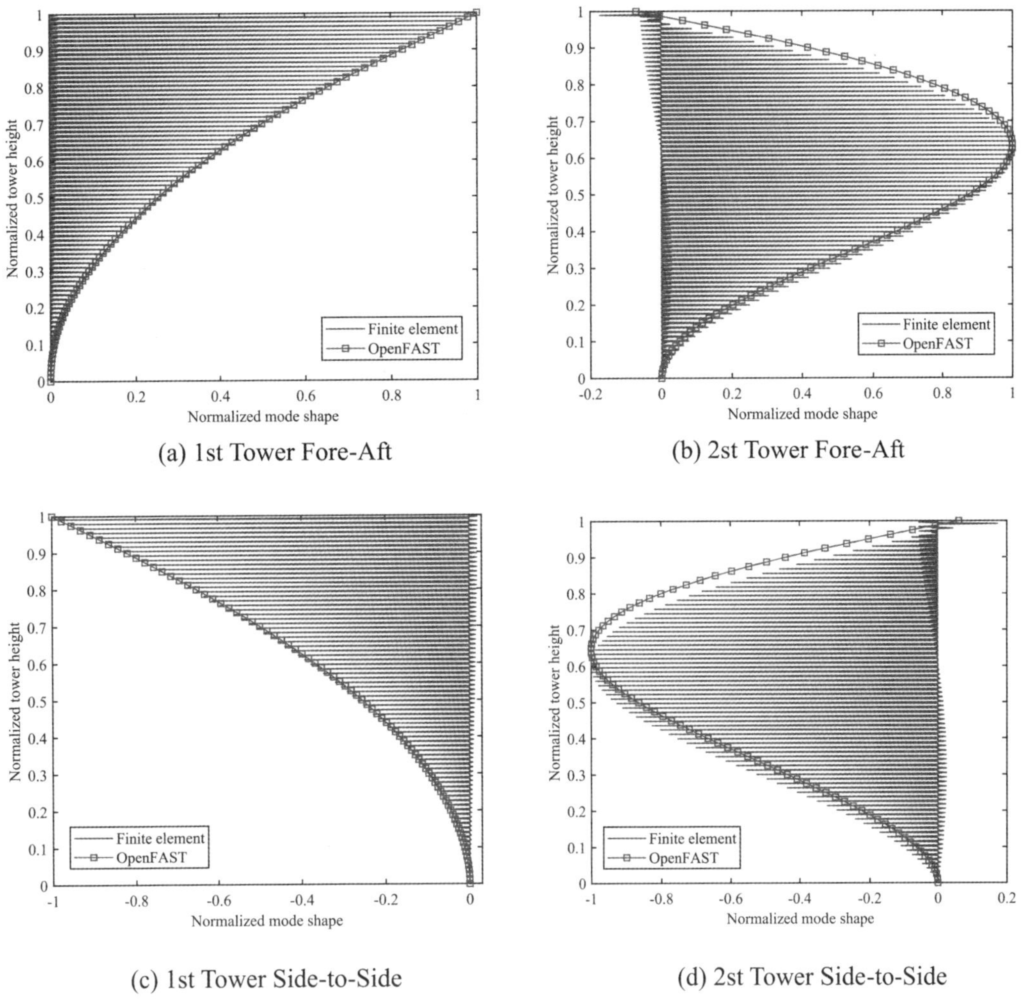

Fig. 2. Modal shape of simplified model and OpenFAST.

3.1.2. Model reduction

In practical engineering, even after simplifying the structure, there are still numerous degrees of freedom. If the simplified model is used directly to design the controller, it would result in excessive computation time for the controller. This delay may increase the required control force, decrease the control effect, and cause the structural response to diverge, resulting in the control force increasing by a single amplitude. Thus, reducing the model dimension is critical to simplify the controller design, enhance control efficiency, and prevent delays that can affect the control force.

To implement an active control system for a structure, designing a state space controller is crucial. Even without considering the damping device, the order of the tall tower in this paper's state space model is 326, making it necessary to further reduce the dimension of the simplified model for the tower structure. Typically, only a few vibration modes significantly affect the dynamic response of civil engineering structures. Therefore, this section proposes a strategy for dimension reduction in the modal space.

Fig. 3 depicts the similarity between the transfer functions of the top-level displacement and acceleration and the original system (OS) for different modal numbers. As shown in the figure, as the number of retained modes (m) increases, the curves become more closely aligned, indicating that a higher number of retained modes result in the preservation of more dynamic properties and a closer approximation of the transfer function to the original system. However, if the dimension is excessively large, the purpose of dimension reduction will not be achieved, and the controller's calculation process will still be excessively slow. Therefore, it is crucial to plan and maintain the dimension at an appropriate level to meet the required accuracy standards.

Fig. 4 illustrates a comparison of poles between the original, modal space, and reduced-order models obtained using the modal truncation method. The results indicate that the poles of the system after the modal transformation remain unchanged. The poles of the reduced-order model are also consistent with those of the original system.

3.1.3. Design of reduced-order controller based on LQR algorithm and AMD system

An AMD control system with multi-point feedback and single-point control for wind turbine tower is designed in this section. We only verify the effectiveness of the reduced-order controller. Therefore, simply one direction is chosen for control in this research for the convenience of calculation. Fig. 5 depicts a layout design for measuring tower displacement by internal videometric method. Dual-head cameras (C) are employed to monitor the displacement of marker points (M, S). These marker points are strategically positioned at 24%, 35%, 56%,62%,81%,88% and 100% of the tower height, as well as on AMD, in

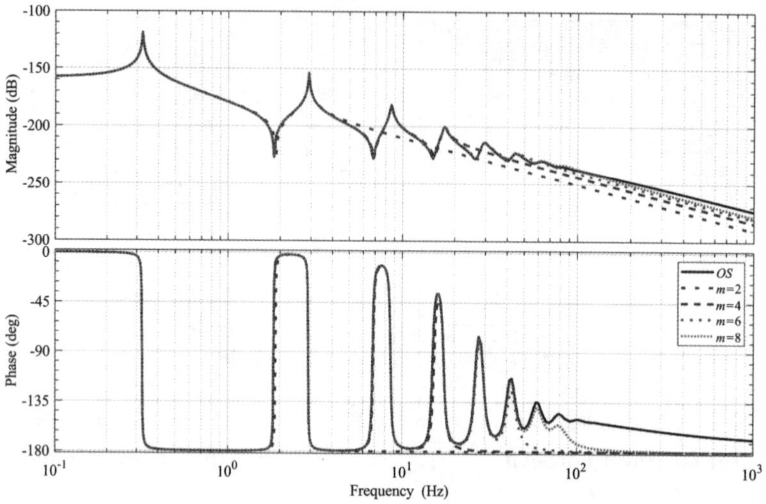

(a)Curve of displacement transfer function

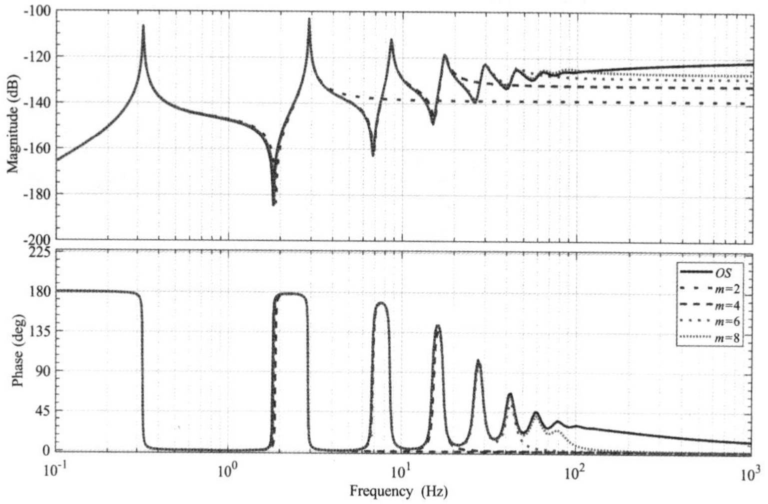

(b)Curve of acceleration transfer function

Fig. 3. Comparison of transfer functions of displacement and acceleration at nacelle of original and reduced-order model.

the subsequent simulation. Fig. 6 shows the entire AMD system, which is located in the nacelle above the top tower, acting on the Fore-Aft direction of the structure. The AMD system consists of mass blocks, actuators, guide rails, limiters, and other necessary devices.

3.2. *Control effect simulation of the 5 MW wind turbine*

This section provides a brief overview of the generation of turbulent wind field and aerodynamic load based on blade element momentum (BEM) theory. The response of the simplified wind turbine model excited by wind and seismic load under three conditions of no control, with TMD control and AMD control with a reduced-order controller, is calculated, respectively. The mass of TMD/AMD system is taken as 1% (3.99 tons) of the first modal mass. The stiffness and damping ratio of the TMD system are taken 0.94 and 0.02, respectively. The weight matrices of the reduced-order AMD control system are defined by Eq. (24). The parameters α and β in the simulation are set to 10,000 and 5.7e-6, respectively. In addition, the stiffness and mass of the AMD term in the $\boldsymbol{Q}$ matrix are multiplied by 2.2e7 and 0, respectively.

$$\boldsymbol{Q} = \alpha \begin{bmatrix} \boldsymbol{\Phi}_r^{\mathrm{T}} \boldsymbol{K} \boldsymbol{\Phi}_r & \boldsymbol{0} \\ \boldsymbol{0} & \boldsymbol{\Phi}_r^{\mathrm{T}} \boldsymbol{M} \boldsymbol{\Phi}_r \end{bmatrix}, \boldsymbol{R} = \beta \tag{24}$$

3.2.1. *Wind field and aerodynamic loads*

Wind velocity acting on the wind turbine can be divided into mean velocity $\overline{u}$ and turbulent velocity u. The mean wind component at tower height z is given by[40]:

$$\overline{u}_z = u_0 \left(\frac{z}{H_{\mathrm{hub}}} \right)^{\alpha} \tag{25}$$

where H_{hub} is the height of hub, u_0 is mean velocity at H_{hub}, which is taken as rated speed (11.4 m/s), and α is the wind shear exponent.

The turbulent velocity component is described by the IEC Kaimal

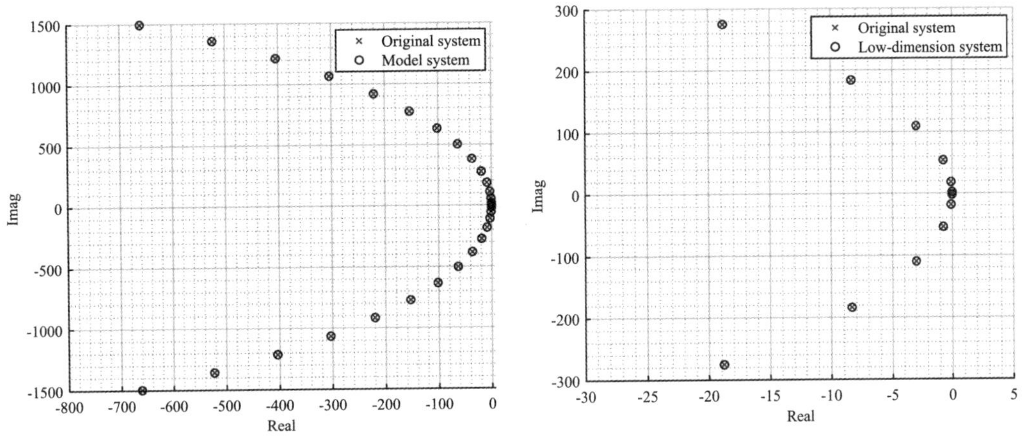

(a) Pole comparison between simplified and model system

(b)Comparison of the 8 pairs of poles between simplified and reduced-order system

Fig. 4. Comparison of poles.

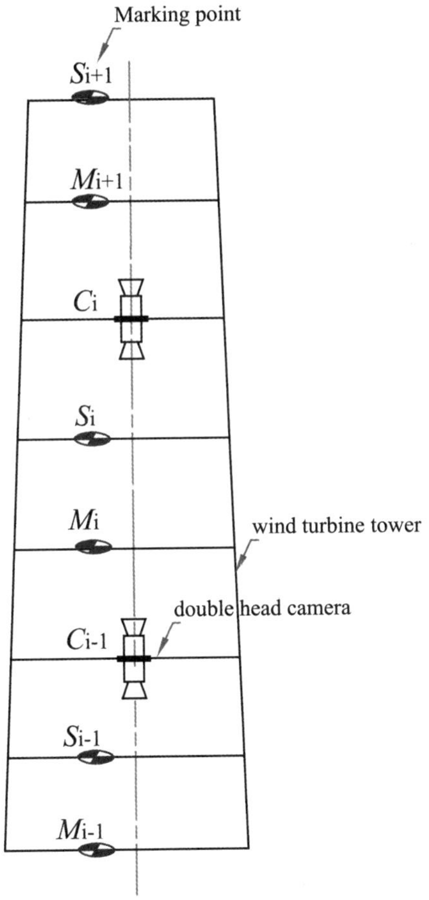

Fig. 5. Schematic diagram of measuring tower displacement by internal videometric method.

spectral model, The IEC Kaimal model is defined in IEC 61400–1[41] and assumes neutral atmospheric stability. The spectra for the wind component in Fore-Aft direction, is given by:

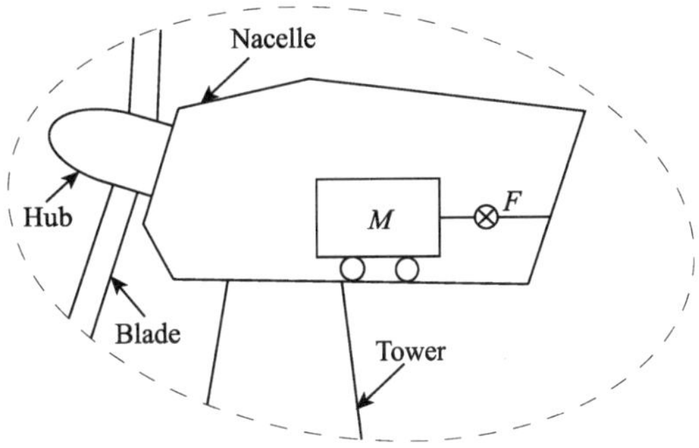

Fig. 6. Layout diagram of AMD.

$$S_u(f) = \frac{\frac{4\sigma_u^2 L_u}{\bar{u}_{hub}}}{\left(1+\frac{6fL_u}{\bar{u}_{hub}}\right)^{\frac{5}{3}}} \tag{26}$$

where f is cyclic frequency and L_u is an integral scale parameter. The IEC 61400–1 standard defines the integral scale parameter to be $L_u = 8.10\Lambda_u$,where the turbulence scale parameter $\Lambda_u = 0.7\min(30\text{m}, H_{hub})$.

In general, spatial coherence between points i and j is defined as:

$$\text{Coh}_{i,j}(f) = \frac{|S_{i,j}(f)|}{\sqrt{S_{ii}(f)S_{jj}(f)}} \tag{27}$$

where S_{ii} is the power spectral density as defined in the "Spectral Models" section, and S_{ij} is the cross-spectral density.

The IEC coherence function is given by:

$$\text{Coh}_{i,j} = \exp\left(-a\sqrt{\left(\frac{fr}{\bar{u}_{hub}}\right)^2 + \left(\frac{0.12r}{L_c}\right)^2}\right) \tag{28}$$

where r is the distance between points i and j, L_c is a coherence scale parameter, and $L_c = 2.45\min(30\text{m}, H_{hub})$ a is coherence decrement and $a = 8.8$.

Turbsim[42] is used to generate wind field based on Eqs. (25)-(28), and wind velocity at hub is shown in Fig. 7.

Combined with the shape coefficient μ_{st}, the wind load on nodes at

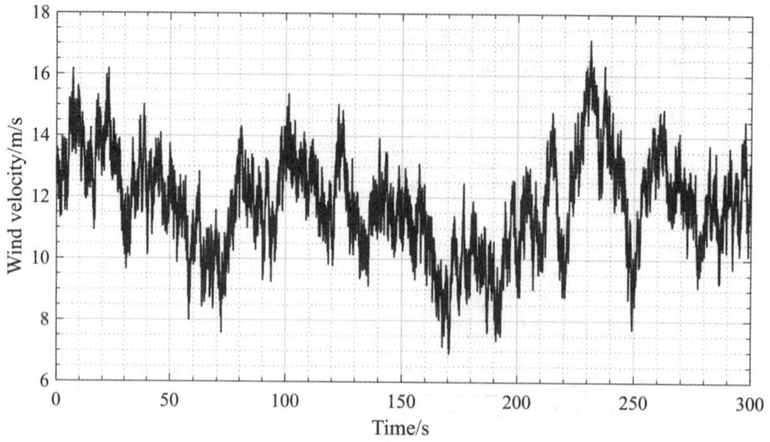

Fig. 7. Time history curve of wind velocity at hub.

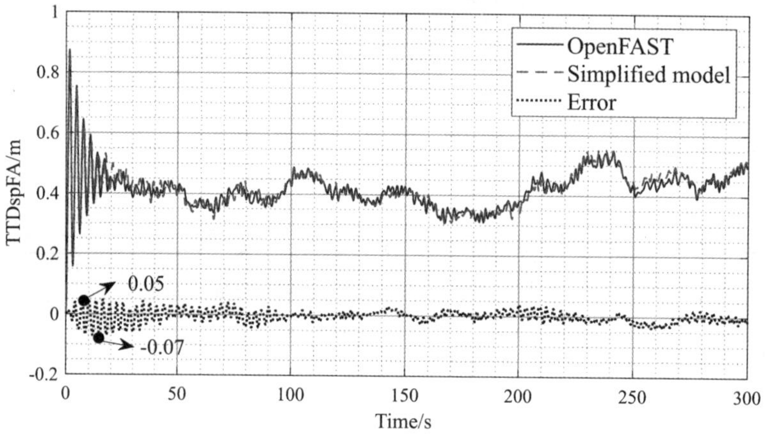

Fig. 8. Displacement of nacelle of OpenFAST and simplified model.

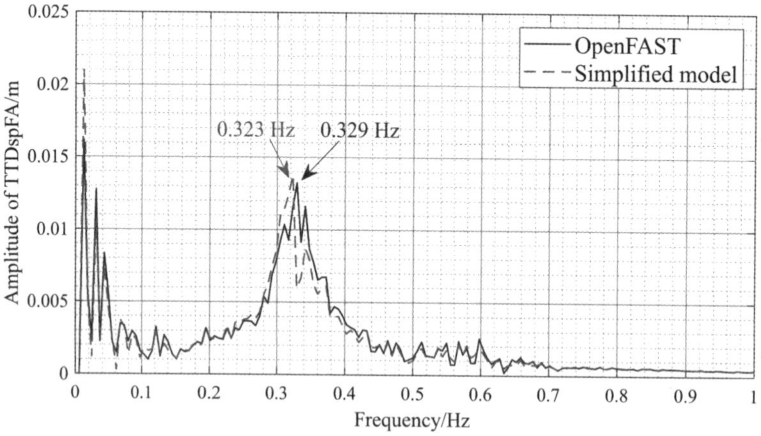

Fig. 9. Amplitude spectrum of OpenFAST and simplified model.

different heights is obtained as follows:

$$F_i(t) = 0.5\rho\mu_{st}u_i^2A_i + \rho\mu_{st}\bar{u}u_iA_i i = 1,2,\cdots \tag{29}$$

where μ_{st} is the shape coefficient of tower, in the paper, $\mu_{st} = 0.6$, ρ is the density of air, A_i is the windward area.

Under the action of wind load, the tower of onshore wind turbine is mainly affected by aerodynamic thrust. The aerodynamic thrust can be given based on momentum theory:

Table 2
Dynamic characteristics of controlled wind turbine.

Mode	TMD controlled		AMD controlled	
	Frequency/ Hz	Additional damping ratio	Frequency/ Hz	Additional damping ratio
1	0.306	0.037	0.318	0.029
	0.332	0.027		
2	2.898	0	2.863	0.105
3	8.560	0	8.530	0.131
4	17.440	0	12.861	0.905
5	29.271	0	17.489	0.076
6	43.898	0	29.328	0.027
7	61.098	0	43.931	0.001
8	80.589	0	61.107	-0.010

$$dT = 4a(1-a)\rho V_0^2 \pi r dr \tag{30}$$

where ρ is air density, a is axial induction factors, V_0 is free stream wind speed, r is span location on the rotor.

However, unless some flow state is assumed, there are not enough equations to solve for the differential rotor thrust dT. This state of flow is governed by the characteristics of the rotor blades, such as airfoil shape and twist distribution. The thrust exerted on a wind turbine that uses blade geometry properties to determine by the flow stream is given by:

$$dT = 0.5B_n \rho V_{rel}^2 (C_L \cos\varphi + C_D \sin\varphi) c dr \tag{31}$$

where B_n is number of blades, c is chord length, C_L and C_D are lift and drag coefficients, respectively, V_{rel} is the vector sum of the axial inflow velocity at the rotor plane, and $V_{rel} = \sqrt{[V_0(1-a)]^2 + [\omega r(1+a')]^2}$, where a′ is angular induction factor, ω is angular velocity of blades. φ is angle between rotation plane and relative velocity and is defined by the following equation:

$$\tan\varphi = \frac{V_0(1-a)}{\omega r(1+a')} \tag{32}$$

Iteratively solving a and a′ based on the fact that Eqs. (30) and (31) must be equivalent, the aerodynamic thrust can be obtained. More information about BEM Theory can be found in reference[43].

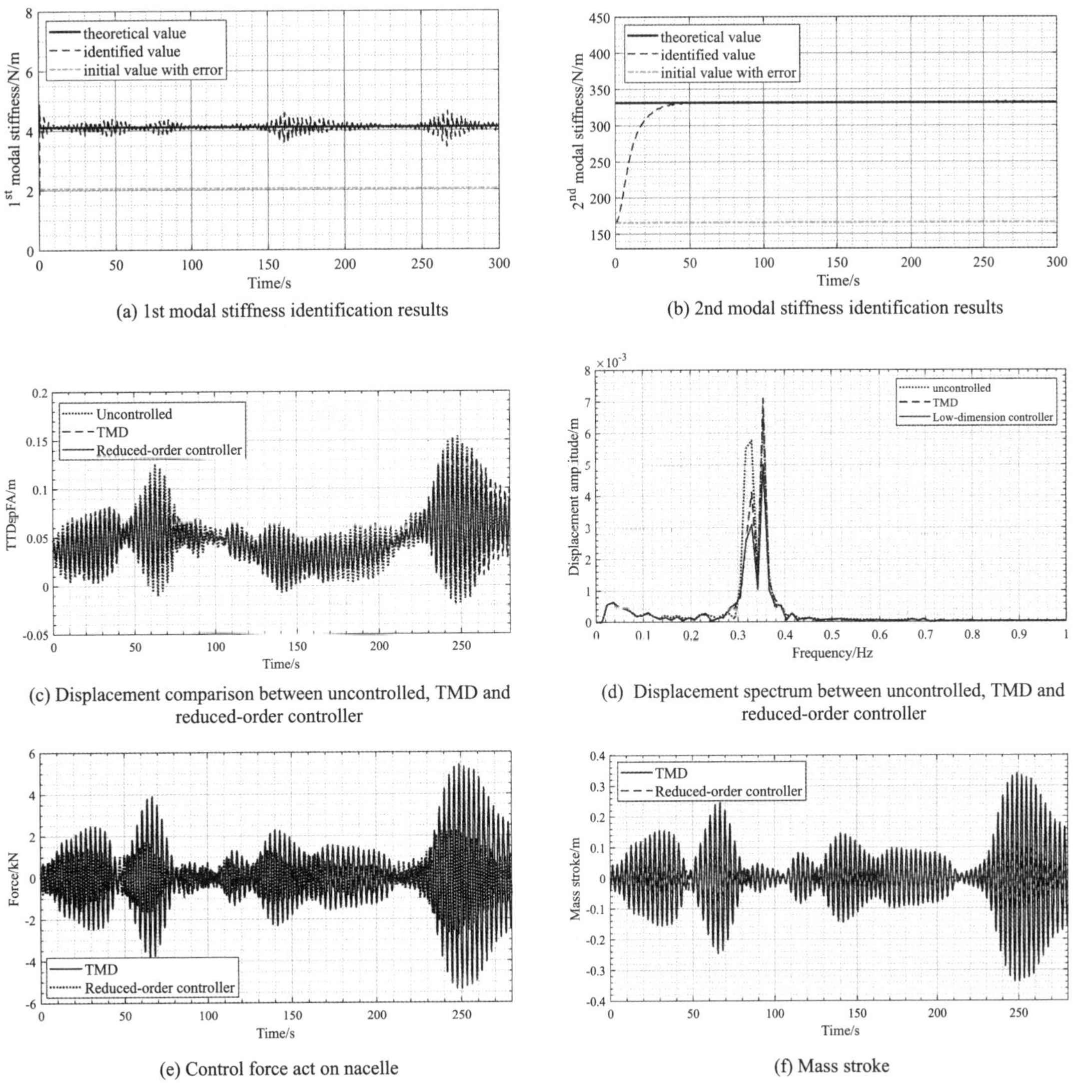

(a) 1st modal stiffness identification results

(b) 2nd modal stiffness identification results

(c) Displacement comparison between uncontrolled, TMD and reduced-order controller

(d) Displacement spectrum between uncontrolled, TMD and reduced-order controller

(e) Control force act on nacelle

(f) Mass stroke

Fig. 10. Response of tower and control parameters at case 1.

Table 3
Response of nacelle and TMD/AMD control parameters.

	Rotation speed (rpm)	Wind speed (m/s)	Index	Uncontrolled		TMD controlled		AMD controlled	
				Max	Rms	Max	Rms	Max	Rms
Case 1	6.9	3	TTDspFA (m)	0.154	0.054	0.143	0.053	0.124	0.051
			TTAccFA (m/s^2)	0.420	0.110	0.370	0.108	0.270	0.077
			Mass Stroke (m)	/	/	0.677	0.101	0.434	0.038
			Control force (kN)	/	/	5.41	1.61	2.78	0.73
Case 2	12.1	11.4	TTDspFA (m)	0.910	0.447	0.905	0.447	0.874	0.447
			TTAccFA (m/s^2)	0.350	0.090	0.370	0.093	0.370	0.089
			Mass Stroke (m)	/	/	0.632	0.06	0.258	0.15
			Control force (kN)	/	/	5.11	0.95	8.34	2.06
Case 3	12.1	25	TTDspFA (m)	0.964	0.752	0.892	0.751	0.899	0.751
			TTAccFA (m/s^2)	0.910	0.220	0.660	0.160	0.710	0.158
			Mass Stroke (m)	/	/	1.626	0.178	0.845	0.253
			Control force (kN)	/	/	13.36	2.81	16.36	3.49

Table 4
Seismic wave information.

Number	Year	Earthquake event	Station name	Magnitude	Mechanism
1	1979	Imperial Valley-06	Calipatria Fire Station	6.53	strike slip
2	1987	Superstition Hills-02	El Centro Imp. Co. Cent	6.54	strike slip
3	1992	Landers	Desert Hot Springs	7.28	strike slip
4	1966	Parkfield	Cholame - Shandon Array #12	6.19	strike slip
5	1968	Borrego Mtn	El Centro Array #9	6.63	strike slip
6	1971	San Fernando	Whittier Narrows Dam	6.61	Reverse
7	1980	Victoria, Mexico	Chihuahua	6.33	strike slip
8	1980	Irpinia, Italy-01	Bagnoli Irpinio	6.9	Normal
9	1978	Tabas	Tabas	7.35	Reverse
10	1935	Helena, Montana-01	Carroll College	6.0	strike slip
11	1992	Cape Mendocino	Petrolia	7.01	Reverse
12	1999	Kocaeli, Turkey	Izmit	7.51	strike slip
13	1940	Imperial Valley-02	El Centro Array #9	6.95	strike slip
14	1974	Hollister-03	Hollister City Hall	5.14	strike slip
15	1983	Coalinga-01	Parkfield - Vineyard Cany 6 W	6.36	Reverse
16	1986	Chalfant Valley-02	Mammoth Lakes Sheriff Subst.	6.19	strike slip
17	1986	Kalamata, Greece-02	Kalamata (bsmt) (2nd trigger)	5.4	Normal
18	1987	Whittier Narrows-01	Newhall - W Pico Canyon Rd.	5.99	Reverse Oblique
19	1989	Loma Prieta	Gilroy Array #2	6.93	Reverse Oblique
20	1994	Northridge-01	LA - Pico & Sentous	6.69	Reverse

3.2.2. *Results and analysis*

One presents response data obtained through the OpenFAST program, which further validates the accuracy of the simplified model. Fig. 8 depicts the displacement time history of the nacelle in the Fore-Aft direction under rated speed (12.1 rpm) and rated wind speed (11.4 m/s) conditions, for both the simplified model and OpenFAST. The results show that response of the simplified model is quite close to that given by OpenFAST program. Fig. 9 shows the displacement spectrum comparison after Fourier transform, revealing that the displacement peaks of the two models are located at the frequency of the tower's first bending mode.

In the design of the reduced-order controller, a total of eight modes were considered with a sum of mass participation factors reaching 90.63%, ensuring that the wind turbine's response under load is accurately reflected. The displacement monitoring points are set at the location of maximum displacement for each mode. Following the design method of controller introduced in Section 2, a structure-AMD system can be established. By calculating the poles of closed-loop system, frequency and damping ratio of each mode can be derived. The additional damping ratio of wind turbine controlled by TMD and AMD are listed in Table 2. Apparently, TMD is only valid for the 1st mode. In contrast, AMD has clear advantages since it provides damping ratio not only for the 1st mode but also for other modes.

In the analysis, three distinct normal operation cases were selected, namely, the cut-in wind speed (3 m/s) and cut-in rotational speed (6.9 rpm), rated wind speed (11.4 m/s) and rated rotational speed (12.1 rpm), and cut-out wind speed (25 m/s) and rated rotational speed (12.1 rpm). Eight modal stiffness parameters were assumed inaccurate for identification, with an initial error of 50% for each modal stiffness. The initial value variance and process variance were set to 1, with a noise variance of 1e-9. Fig. 10 (a) and (b) depict the theoretical and identified values of the first two modal stiffnesses of the tower. It is clear from the figures that the first modal stiffness quickly converges to the true value, while the second modal stiffness stabilizes around the true value within approximately 40 s, demonstrating the accuracy and effectiveness of the UKF. Table 3 presents the response of the nacelle as well as the TMD/AMD control parameters with no control, TMD control and AMD control. It can be seen from Table 3 that both TMD and AMD have a certain vibration control effect on the tower under case 2 and case 3, but the control effect is limited due to the large damping of the structure at rated rotation speed. The inclusion of aerodynamic damping in the calculation results in faster attenuation of displacement at the nacelle. Nevertheless, under these two operating conditions, the stroke controlled by AMD is half as large as that controlled by TMD, providing a space-saving advantage in installation. The displacement response under the case of uncontrolled, TMD control and AMD control with reduced-order controller at case 3 are illustrated graphically in Fig. 10 (c) and Fig. 10 (d). Combining Fig. 10 and Table 3 reveals that under case 1, the tower exhibits the most effective vibration reduction. When TMD is applied, the maximum displacement of the nacelle decreases from 154 mm to 143 mm, resulting in a vibration reduction rate of 7.1%. In contrast, when AMD is applied, the maximum displacement is reduced to 124 mm, achieving a significantly higher vibration reduction rate of 19.4%. The vibration reduction rate of AMD is notably superior, being 12.3% higher than that of TMD. Similarly, the maximum

Table 5
Displacement response of nacelle and TMD/AMD control parameters under the action of 20 seismic waves.

Seismic number	Uncontrolled		TMD controlled						AMD controlled					
	TTDspFA/m		TTDspFA/m		Mass Stroke/m		Control force/kN		TTDspFA/m		Mass Stroke/m		Control force/kN	
	Max	Rms	Max	Rms	Max	Rms	Max	Rms	Max	Rms	Max	Rms	Max	Rms
1	0.528	0.147	0.447	0.107	1.809	0.513	29.4	8.4	0.388	0.088	0.770	0.178	349.6	23.9
2	0.223	0.067	0.210	0.062	0.795	0.278	12.9	4.5	0.173	0.053	0.479	0.106	242.2	21.5
3	0.231	0.071	0.204	0.052	0.700	0.225	11.4	3.7	0.164	0.041	0.447	0.101	380.7	43.3
4	0.350	0.131	0.331	0.103	1.566	0.483	25.5	7.9	0.296	0.085	0.618	0.179	383.6	31.7
5	0.487	0.149	0.484	0.118	1.776	0.539	28.9	8.8	0.458	0.103	1.105	0.221	176.8	16.7
6	0.278	0.087	0.275	0.077	1.294	0.393	21.1	6.4	0.251	0.071	0.610	0.137	322.5	29.7
7	**0.392**	**0.134**	**0.298**	**0.090**	**1.297**	**0.422**	**21.1**	**6.9**	**0.266**	**0.079**	**0.723**	**0.164**	**451.8**	**24.3**
8	0.295	0.090	0.295	0.059	0.908	0.277	14.8	4.5	0.285	0.052	0.738	0.121	377.8	33.8
9	**0.355**	**0.118**	**0.322**	**0.073**	**1.325**	**0.385**	**21.6**	**6.3**	**0.284**	**0.064**	**0.725**	**0.135**	**310.3**	**30.3**
10	0.044	0.011	0.043	0.008	0.151	0.039	2.5	0.6	0.039	0.007	0.195	0.018	500.1	19.5
11	0.212	0.049	0.200	0.035	0.693	0.173	11.3	2.8	0.177	0.030	0.399	0.069	289.0	23.5
12	0.443	0.087	0.433	0.078	1.685	0.399	27.5	6.5	0.400	0.071	0.806	0.143	307.7	34.0
13	0.370	0.138	0.306	0.083	1.361	0.431	22.2	7.0	0.256	0.074	0.768	0.160	353.0	34.0
14	0.033	0.006	0.033	0.005	0.076	0.018	1.3	0.3	0.037	0.005	0.384	0.029	451.8	30.0
15	0.519	0.177	0.425	0.117	1.866	0.567	30.4	9.2	0.344	0.097	0.841	0.214	311.0	40.7
16	0.161	0.037	0.144	0.031	0.613	0.167	10.0	2.7	0.125	0.028	0.371	0.068	283.7	40.3
17	0.052	0.005	0.052	0.005	0.086	0.019	1.4	0.3	0.053	0.004	0.413	0.022	338.3	18.2
18	0.043	0.010	0.043	0.007	0.099	0.029	1.6	0.5	0.040	0.007	0.346	0.035	302.7	33.2
19	0.149	0.048	0.129	0.031	0.513	0.156	8.4	2.5	0.115	0.026	0.256	0.055	334.9	25.8
20	0.291	0.070	0.304	0.063	1.147	0.280	18.7	4.6	0.270	0.056	0.584	0.121	432.0	35.5

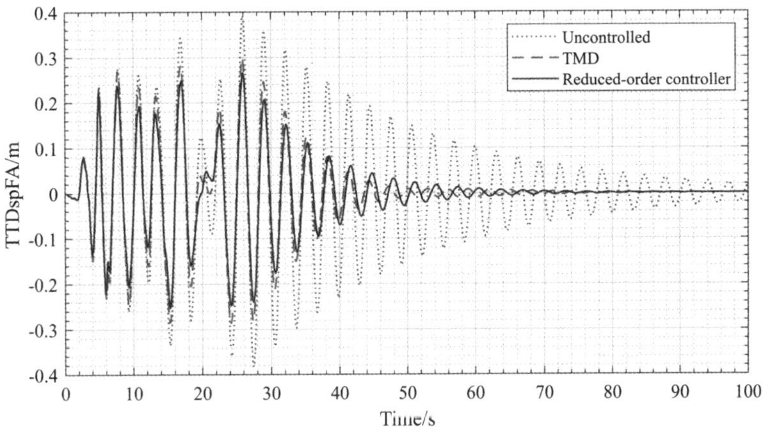

Fig. 11. Displacement comparison between uncontrolled, TMD and reduced-order controller under the action of the 7th seismic wave.

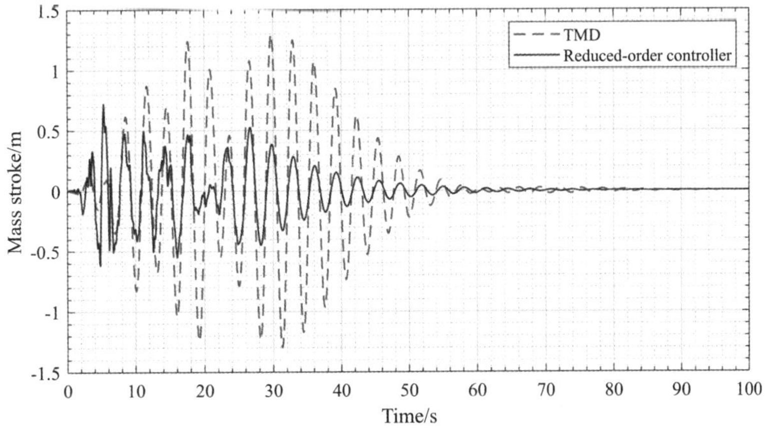

Fig. 12. Mass stroke under the action of the 7th seismic wave.

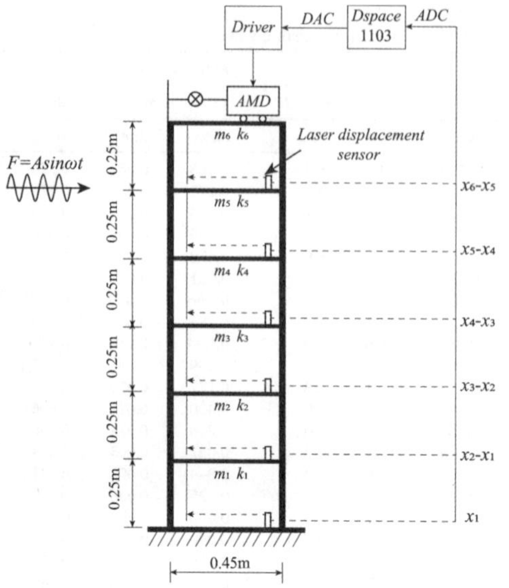

Basic parameters of frame

layer	Mass (kg)	Stiffness (kg/mm)
1	44.75	65.42
2	44.75	51.11
3	44.75	51.11
4	44.75	51.11
5	43.23	51.11
6	44.89	51.11

Fig. 13. Schematic diagram of frame test model.

(a) Side view of model (b) Front view of model

Fig. 14. Experimental set up for the frame model.

Table 6
Basic characteristics of structural model.

Order	Frequency (Hz)	Period (s)	participation mass (kg)	mass participation factor
1	1.33	0.75	226.59	84.8%
2	3.87	0.26	25.12	9.4%
3	6.18	0.16	8.95	3.35%
4	8.11	0.12	4.18	1.57%
5	9.53	0.10	1.82	0.68%
6	10.34	0.10	0.46	0.17%

acceleration damping rate of the nacelle under the action of TMD is 12%, while under AMD control, it reaches 35.7%, marking a substantial 23.7% increase compared to TMD. Fig. 10(e) depicts the control force act on the nacelle under two control modes. The maximum control force of wind-induced vibration shall not exceed 6 kN. In practical engineering, from an economic point of view, the quality of AMD can be further reduced to meet the demand for vibration reduction. It can be seen from the Fig. 10(f) that the maximum stroke of the mass of AMD is 0.434 m, which is less than 0.677 m of TMD, indicating that AMD can achieve a better control effect than TMD in a limited space.

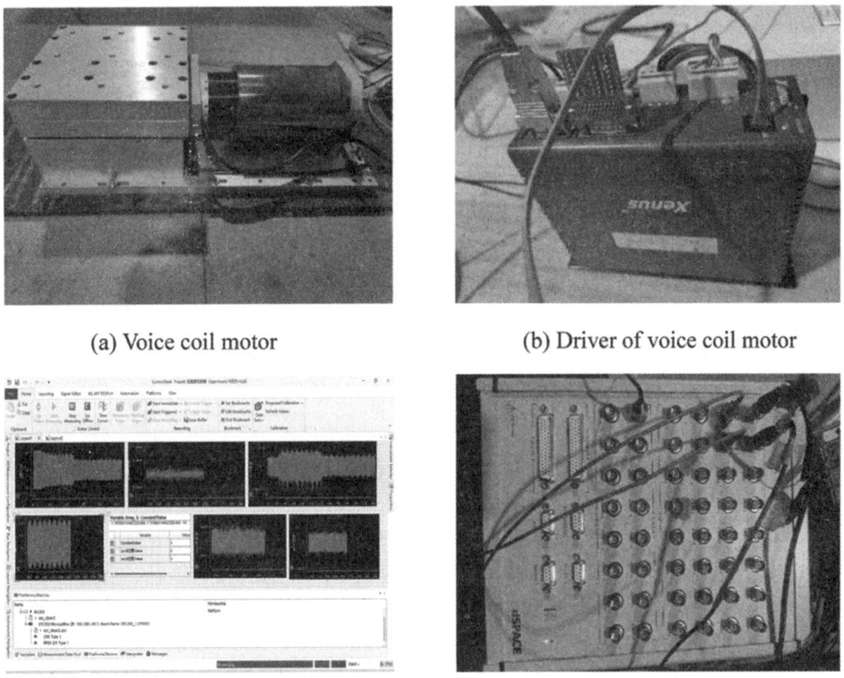

(a) Voice coil motor (b) Driver of voice coil motor

(c) Software of MicroLabBox (d) Hardware of MicroLabBox

Fig. 15. Control system.

Table 7
Basic parameters of voice coil motor.

Index	Parameter
Mover mass	6.2 kg
Stator Mass	22.8 kg
Travel range	0.1-45 mm
Peak thrust/Sustained thrust	895/265 N
Force constant	36.5 N/A

Fig. 16. Excitation system.

3.2.3. Seismic simulation

To further verify the effectiveness of the reduced-order controller, the structural responses under various seismic actions are calculated. A total of 20 seismic waves are randomly selected from the Pacific earthquake engineering research center of the University of Berkeley for simulation calculation. The basic information is displayed in Table 4. Adjust the proportion of seismic wave acceleration so that the peak value is 0.3 g. The displacement response of the nacelle and TMD/AMD control parameters under these seismic actions are presented in Table 5. Specifically, Figs. 11 and 12 illustrate the results obtained under the 7th seismic wave, which highlights the superior control effect of AMD on the maximum displacement of the nacelle.

Table 5 reveals that, under most seismic conditions, the AMD control system can achieve superior control effectiveness with a smaller mass stroke. However, only under the influence of the 14th, 17th, and 18th seismic waves, the maximum stroke of AMD is slightly greater than that of TMD. This further illustrates that the reduced-order controller, designed using the LQR algorithm, is stable and can serve as an effective substitute for the full-order controller as a control strategy for AMD wind turbine systems. Among the seismic waves considered, the 7th wave shows the best control performance of AMD on the maximum displacement of nacelle, with a vibration reduction rate of 32.1%. Moreover, the AMD shows the most outstanding control effectiveness on the root mean square value of nacelle under the 9th seismic wave, achieving a reduction rate of 45.6%.

4. Experimental verification of active control

In order to verify the effectiveness of the displacement observer and reduced-order controller in modal space, a model validation experiment is performed using a six-story steel frame model. The structural model dimensions and parameters of the theoretical design are shown in Fig. 13. The actuator used is a voice coil motor that converts electric energy into linear motion using the ampere force concept. The excitation is provided by an electric motor with a reducer, and an eccentric mass is attached to the output shaft of the motor to generate sinusoidal excitation force to the structure. During the test, the laser displacement meter is utilized in place of the camera to collect the displacement signal of each layer of the structure in real time. The physical model and experimental setting are depicted in Fig. 14.

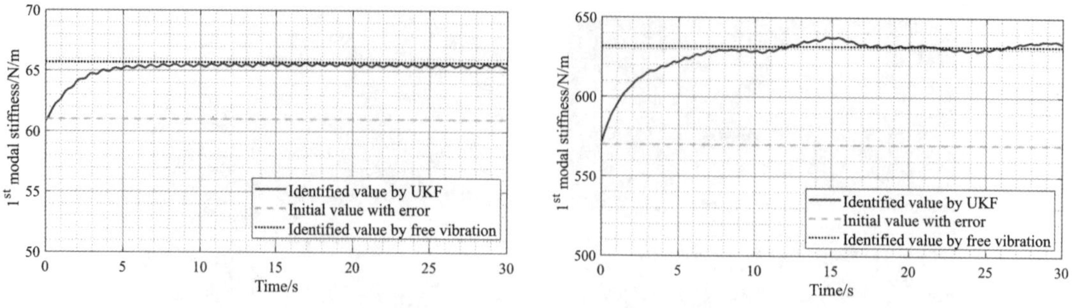

(a) (b)

Fig. 17. Identification result of 1st and 2nd modal stiffness.

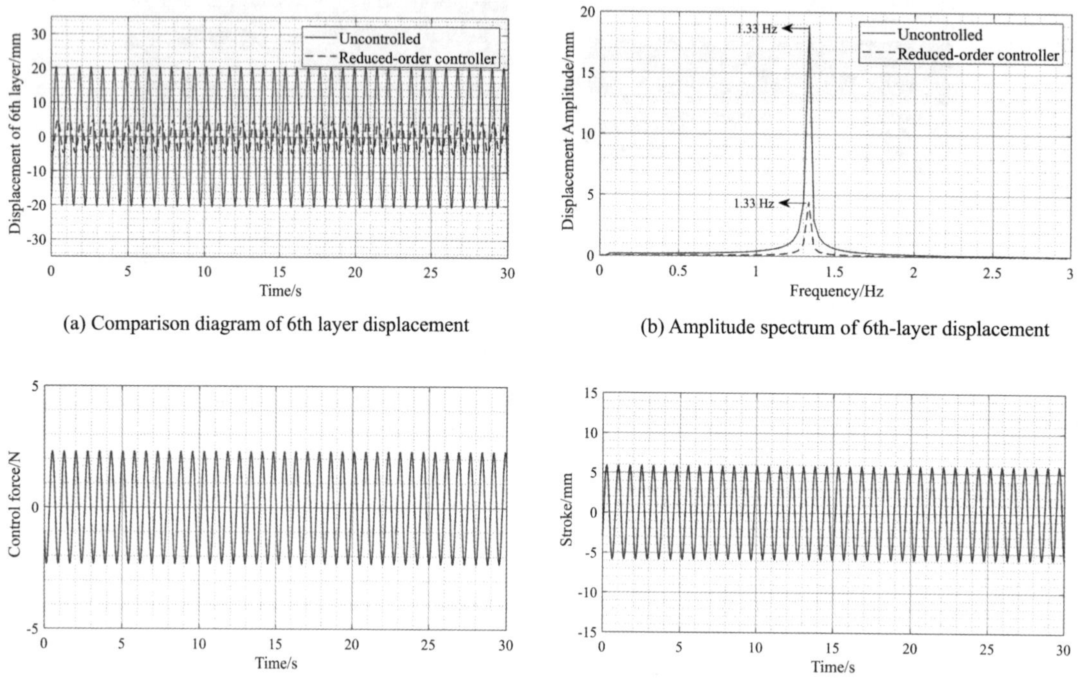

(a) Comparison diagram of 6th layer displacement

(b) Amplitude spectrum of 6th-layer displacement

(c) Control force of AMD system

(d) Stroke of AMD system

Fig. 18. Structure response and control parameters in theory (single frequency excitation).

4.1. Basic parameters of structure

The mass matrix and stiffness of the controlled structure can be estimated based on the geometric dimensions, density, and elastic modulus of the structure, which are illustrated in Fig. 14. Damping ratio of the first two modes of the structure, as determined by the free decay approach, is 0.0025 and 0.0028. Table 6 displays the essential characteristics of the structure. As can be observed, the participation mass coefficient of the first mode of the model is the largest, which means that during the design phase of reduced-order controller, retaining the first-order mode can effectively satisfy the criteria.

4.2. Control system and excitation system

Fig. 15 depicts the composition of the control system, which includes the voice coil motor, driver, and MicroLabBox (DS 1202). Time delay of the controller is measured to be about 20 ms, and the parameter τ of Eq. (10) is set to be 20 ms, although the system stays stable when set to 0 ms. The experiment employed a voice coil motor rather than a linear motor as the actuator due to the lower CSI effect associated with voice coil motors. The voice coil motor, TMEC0895–045-000, is manufactured by Kunshan Tongmao Motor Company in China, and its basic parameters are presented in Table 7. The driver employed in the system is the Xenus series produced by Copley Controls in the United States. Additionally, an incremental encoder with the model number of EXA, manufactured by Fagor in Spain, is installed on the motor, with a resolution of 1 μm. Transfer function from input voltage signal to output force of the voice motor is tested and it shows very good consistency at medium and high frequencies. There is a certain phase difference and amplitude loss at very low frequencies, but this CSI effect can be reduced through a composite control scheme[34,35]. As shown in Fig. 16, the excitation system comprises a small motor, with a speed regulating range of 0–150 rpm, an output frequency of 0–2.5hz, a reduction ratio of 1/10 and a power of 25 W. The displacement was measured utilizing a laser displacement meter (IL-300) from Keyence, boasting a linearity of 0.75 mm. Acceleration measurements were conducted using Donghua's

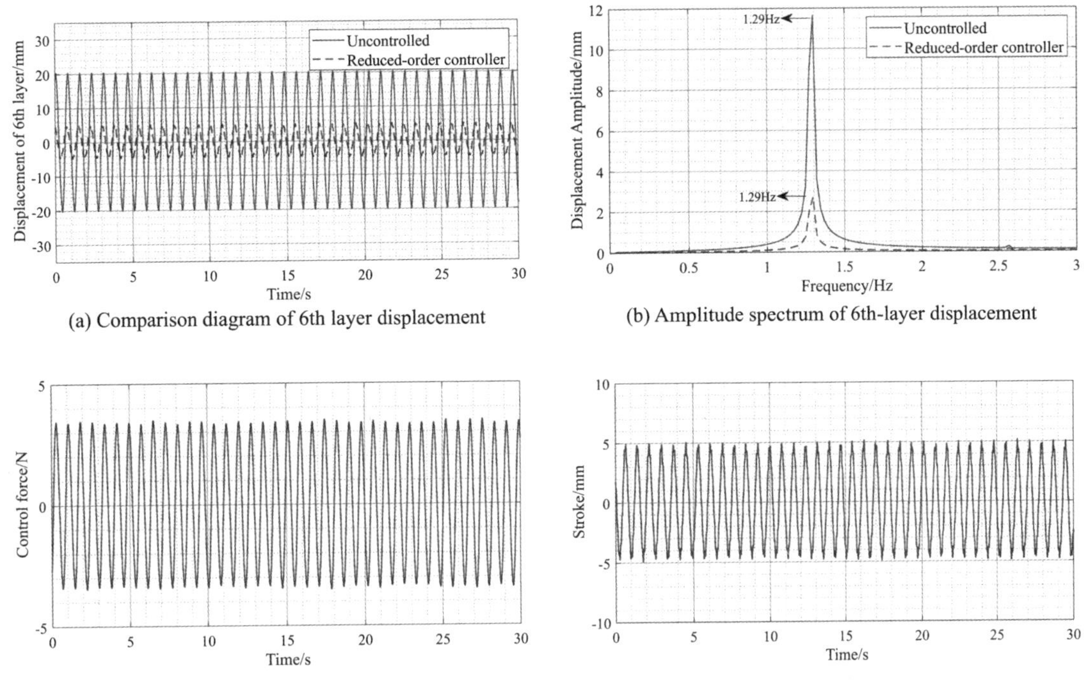

(a) Comparison diagram of 6th layer displacement

(b) Amplitude spectrum of 6th-layer displacement

(c) Control force of AMD system

(d) Stroke of AMD system

Fig. 19. Structure response and control parameters in test (single frequency excitation).

Table 8
Displacement response amplitude and control parameters of the structure (single frequency excitation).

Index		Displacement of 6th layer (mm)		Acceleration of 6th layer (m/s^2)		Control force (N)		Stroke (mm)	
		Theory	Experiment	Theory	Experiment	Theory	Experiment	Theory	Experiment
Max	Uncontrolled	20.3	20.4	1.41	1.9	\	\	\	\
	Reduced-order controller	4.73	5.29	0.31	0.49	2.3	3.2	5.6	5.1
	Vibration reduction rate (%)	77	74	78	74	\	\	\	\
Rms	Uncontrolled	14.2	13.9	0.98	1.3		\	\	
	Reduced-order controller	3.34	3.22	0.22	0.30	1.64	2.4	4.12	3.22
	Vibration reduction rate (%)	76	77	78	77	\	\	\	\

piezoresistive accelerometer (1B105).

4.3. *Result of experiment*

The maximum current of voice coil motor is set to 1 A and the frequency of the excitation motor is set to 1.34 Hz. Amplitude of the first-order harmonic excitation force is 3.2 N, and that of the second-order is 7.1 N. Feedback signals are gathered by collecting the displacement of the 6th and 2nd layers, while simultaneously monitoring the acceleration signal of the 6th layer. In the experiment, taking $\alpha = 10, \beta = 0.03$. Two modal stiffness parameters of the structure were chosen for UKF identification. The initial value of the 1st modal stiffness was set to 60 N/m with a process variance of 30 N/m, while the 2nd modal stiffness was set to 550 N/m with a process variance of 400 N/m. The measurement noise variance was set to 1e-9. The experimental data from the steady-state process of the 30 s is selected as a measure of the control effect of AMD.

As illustrated in Fig. 17, the identified first-order modal stiffness eventually stabilized at 65.08 N/m, and the second-order modal stiffness fluctuated around 613.5 N/m, which is consistent with the results measured by the free vibration method. Comparing with the initial design model, where the first-order and second-order modal stiffnesses were 69.83 N/m and 591 N/m respectively, there appears to be a 3% error between the first two frequencies of the design model and the physical model. In the controller, a reduced-order state space model is established using the identified modal parameters.

Figs. 18 and 19 visually depict the theoretical and experimental structural responses along with control parameters, respectively. Table 8 details the theoretical and experimental responses along with control parameters under both uncontrolled and reduced-order control conditions. Based on Figs. 18, 19, and Table 8, it is evident that there is minimal disparity in the response between the simulation and testing of the structure under uncontrolled condition. The maximum displacement values for the sixth layer are 20.3 mm and 20.4 mm, with mean square values of 14.2 mm and 13.9 mm, respectively. The maximum acceleration values for the sixth layer are 1.41 m/s^2 and 1.9 m/s^2, with mean square values of 0.98 m/s^2 and 1.3 m/s^2, respectively. The spectral analysis reveals identical fundamental frequencies of the structure, indicating a fundamental consistency between the theoretical model and the physical model. Under the action of a reduced-order controller, the theoretical control effect surpasses the experimental control effect. There is a difference between theoretical and practical control force, measuring 2.3 N and 3.2 N, respectively. However, the theoretical AMD mass stroke is basically consistent with the stroke observed in the experiment. This discrepancy arises from the theoretical neglect of factors such as friction between the AMD mass and the guide rail,

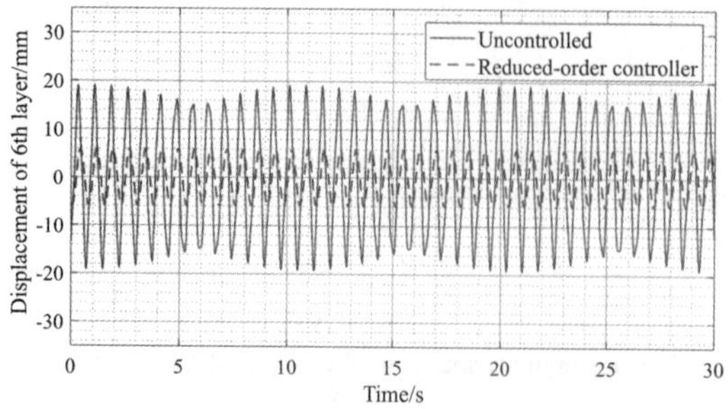

(a) Comparison diagram of 6th layer displacement

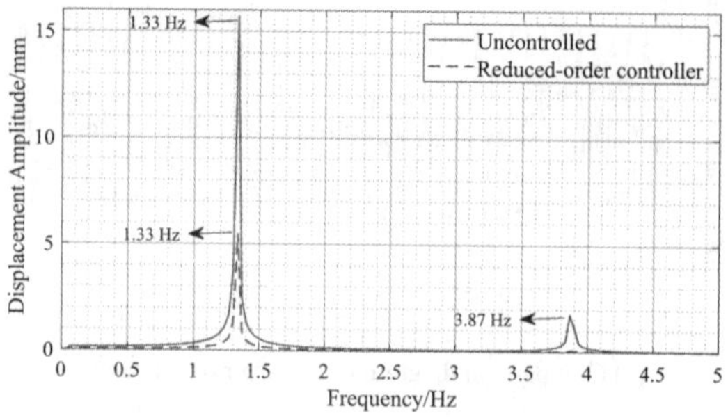

(b) Amplitude spectrum of 6th-layer displacement

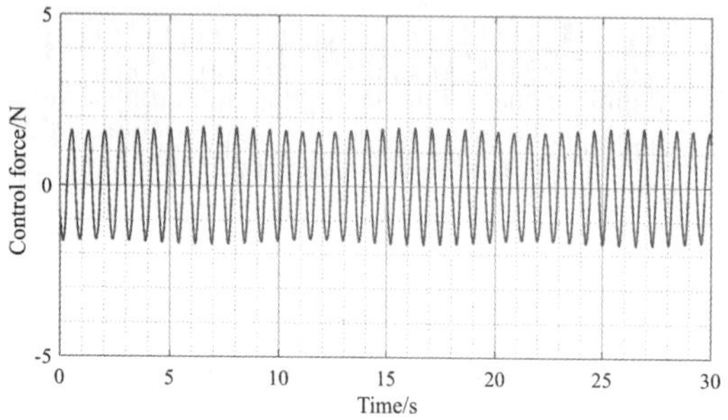

(c) Control force of AMD system

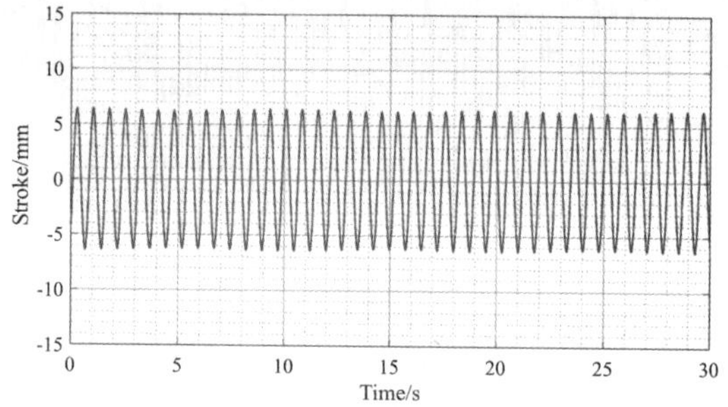

(d) Stroke of AMD system

Fig. 20. Structure response and control parameters in theory (dual frequency excitation).

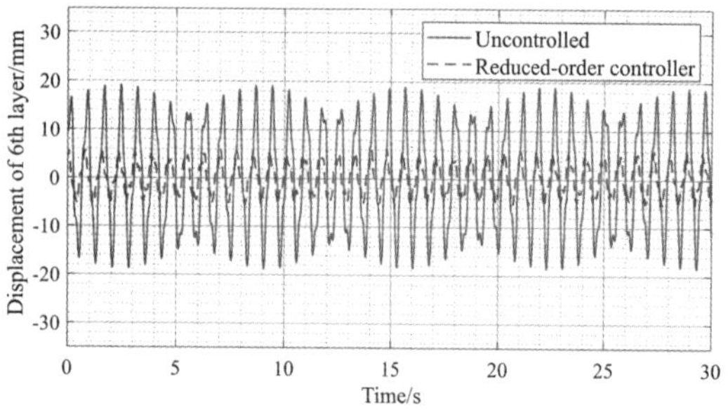

(a) Comparison diagram of 6th layer displacement

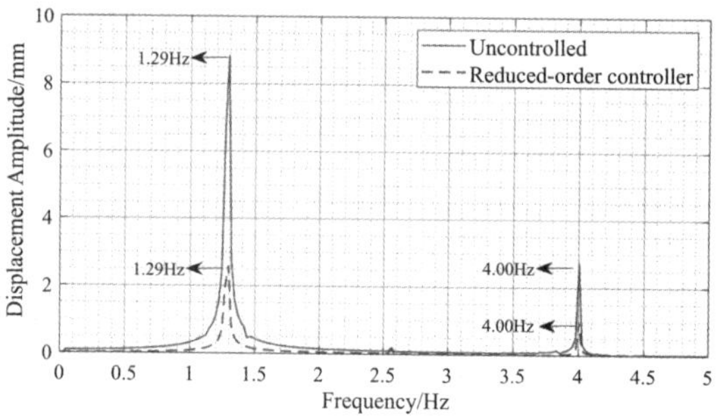

(b) Amplitude spectrum of 6th-layer displacement

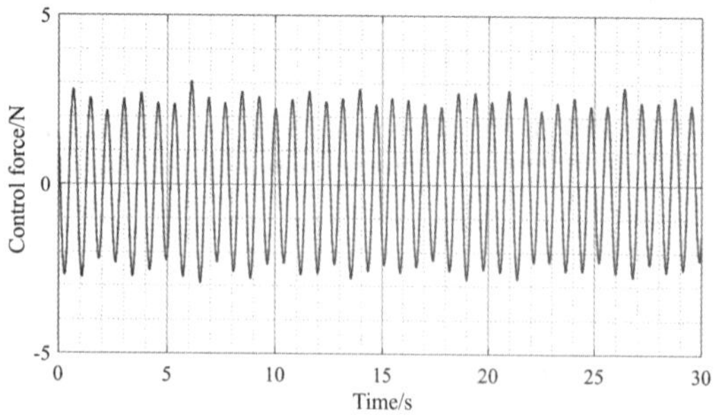

(c) Control force of AMD system

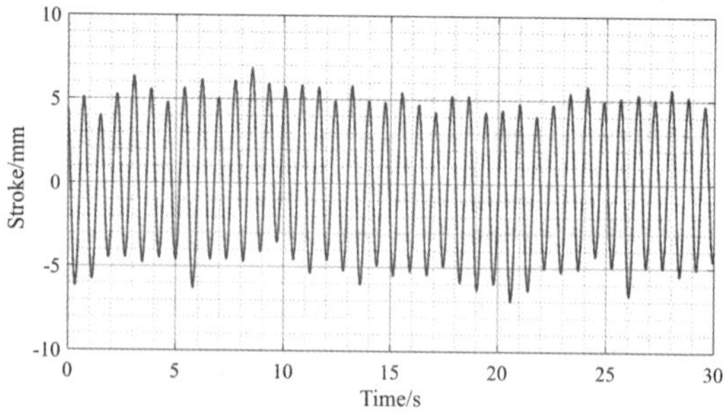

(d) Stroke of AMD system

Fig. 21. Structure response and control parameters in test (dual frequency excitation).

Table 9
Displacement response amplitude and control parameters of the structure (dual frequency excitation).

Index		Displacement of 6th layer (mm)		Acceleration of 6th layer (m/s^2)		Control force (N)		Stroke (mm)	
		Theory	Experiment	Theory	Experiment	Theory	Experiment	Theory	Experiment
Max	Uncontrolled	19.1	19.1	2.5	4.01	\		\	\
	Reduced-order controller	5.9	6.1	0.5	1.18	1.8	3.04	6.5	7
	Vibration reduction rate (%)	70	68	80	71	\	\	\	\
Rms	Uncontrolled	11.9	10.96	1.9	1.99	\	\	\	
	Reduced-order controller	4.1	3.22	0.3	0.57	1.16	1.79	4.46	3.75
	Vibration reduction rate (%)	66	71	84	71	\		\	\

environmental interference signals, and magnetic attraction between the stator and rotor of the voice coil motor. Consequently, achieving an identical control effect as the theory becomes challenging even with fundamentally consistent control forces. Additionally, due to friction and the limited range of the voice coil motor, the AMD stroke during the experiment was relatively small.

In order to further verify the multimodal vibration reduction effect of the reduced-order controller, one also conducted experiments on the structure excited by sinusoidal excitation force of different frequencies simultaneously.

Figs. 20 and 21, along with Table 9, offer a visual and tabular comparison of the theoretical and experimental responses, as well as the corresponding control parameters of the structure under various operating conditions. With the control of a reduced-order controller, the theoretical maximum displacement and mean square vibration reduction rates for the 6th layer are70% and 66%, respectively. In the experiment, the corresponding vibration reduction rates are 68% and 71%. For the 6th layer, the theoretical maximum and mean square values of acceleration exhibit damping rates of 80% and 84%, respectively, while the corresponding rates in the experiment are 71% and 71%. The vibration reduction rates in the experiment decrease by 2% to 13% when compared to the theoretical values. Spectral curves in Figs. 20 and 21 unmistakably demonstrate that the reduced-order controller exerts a pronounced vibration reduction effect on the first two modes of the structure. Under multi-frequency excitation conditions, a slight deviation is observed where the actual control force surpasses the theoretical control force, while the mass block stroke remains essentially constant. Due to the effects of friction, parameter uncertainty, etc, the control effect is slightly worse than the theory. The multifrequency excitation test demonstrates that the reduced-order controller is a broadband controller, effectively reducing vibrations in the first two modes of the structure.

5. Conclusion

In this study, a displacement feedback controller, characterized by its high SNR and a straightforward, rapid control algorithm, is implemented for active control of practical engineering applications of high towers. Theoretical, numerical, and experimental studies of active control of a 5 MW wind turbine have been carried out, yielding the following conclusions:

(1) By utilizing the modal truncation method based on MPF, the reduced-order system can effectively preserve the primary modal information of the structure while simultaneously reinstating the global dynamic characteristics of the structure to the greatest extent possible. The greater the dimension reserved in the reduced-order system, the more consistent the displacement and acceleration transfer functions are with those of the original structure. For the wind turbine studied in this paper, the sum of mass participation factor of the 8 selected modes reaches 90.63%, indicating that the dynamic characteristics of this reduction model are capable of accurately describing the real model.

(2) The reduced-order controller designed for a 5 MW wind turbine achieves good control effects. Under wind loads, the reduction in the maximum displacement of the nacelle in the fore-aft direction, as well as the stroke of the mass achieved by the AMD system, surpasses that of TMD. Under seismic loads, the peak and root mean square values of displacement are both reduced significantly, by 32.10% and 45.55%, respectively, through the utilization of the AMD system.

(3) The efficacy of the reduced-order controller is validated via a six-story steel frame test. Comparisons have been conducted between theoretical and experimental responses for the frame structure under both uncontrolled and controlled conditions. The UKF algorithm is employed to identify the structural parameters, and the identification accuracy and stability have been verified. Results reveal that the actual reduction effect is 2% ~ 13% worse than the theoretical one, which is acceptable due to the friction, structure uncertainty and other influential factors. The multifrequency excitation experiment clearly showcases the substantial vibration reduction effect of the proposed active control system.

CRediT authorship contribution statement

Zhaodong Xu: Writing – review & editing. **Hongyi Zhang:** Writing – review & editing, Software, Methodology. **Zhengqing Chen:** Writing – review & editing, Resources, Funding acquisition, Conceptualization. **Xugang Hua:** Writing – review & editing. **Huawei Niu:** Writing – review & editing. **Xixi Wen:** Writing – original draft, Software, Investigation, Formal analysis.

Declaration of Competing Interest

The authors declare that they have no known competing financial interests or personal relationships that could have appeared to influence the work reported in this paper.

Data Availability

Data will be made available on request.

Acknowledgments

The authors gratefully acknowledge the financial support received from Young Scientists Fund of the National Natural Science Foundation of China (52208478) and the Joint Fund Project of National Natural Science Foundation of China (U2141242).

References

[1] Chizfahm A, Yazdi EA, Eghtesad M. Dynamic modeling of vortex induced vibration wind turbines. Renew Energy 2018;121:632–43. https://doi.org/10.1016/j.renene.2018.01.038.

[2] Pao LY, Johnson KE. Control of wind turbines. IEEE Control Syst Mag 2011;31:44–62.

[3] Active Structural Control of Offshore Wind Turbines | Aerospace Sciences Meetings n.d. ⟨https://arc.aiaa.org/doi/10.2514/6.2010-1000⟩ (accessed May 2, 2023).

[4] Chou J-S, Ou Y-C, Lin K-Y. Collapse mechanism and risk management of wind turbine tower in strong wind. J Wind Eng Ind Aerodyn 2019;193. https://doi.org/10.1016/j.jweia.2019.103962.
[5] Ding H, Altay O, Wang J-T. Lateral vibration control of monopile supported offshore wind turbines with toroidal tuned liquid column dampers. Eng Struct 2023;286. https://doi.org/10.1016/j.engstruct.2023.116107.
[6] Xie F, Aly A-M. Structural control and vibration issues in wind turbines: a review. Eng Struct 2020;210. https://doi.org/10.1016/j.engstruct.2019.110087.
[7] Murtagh PJ, Ghosh A, Basu B, Broderick BM. Passive control of wind turbine vibrations including blade/tower interaction and rotationally sampled turbulence. Wind Energy 2008;11:305–17. https://doi.org/10.1002/we.249.
[8] Lackner MA, Rotea MA. Passive structural control of offshore wind turbines. Wind Energy 2011;14:373–88. https://doi.org/10.1002/we.426.
[9] Sarkar S, Fitzgerald B. Vibration control of spar-type floating offshore wind turbine towers using a tuned mass-damper-inerter. Struct Control Health Monit 2020;27. https://doi.org/10.1002/stc.2471.
[10] Roderick C. Vibration Reduction of Offshore Wind Turbines Using Tuned Liquid Column Dampers n.d.
[11] Colwell S, Basu B. Tuned liquid column dampers in offshore wind turbines for structural control. Eng Struct 2009;31:358–68. https://doi.org/10.1016/j.engstruct.2008.09.001.
[12] Shadman M, Akbarpour A. Utilizing TLCD (Tuned Liquid Column Damper) in Floating Wind Turbines. Vol. 7 Ocean Space Util. Ocean Renewable Energy, Rio de Janeiro. Brazil: American Society of Mechanical Engineers,; 2012. p. 241–7. https://doi.org/10.1115/OMAE2012-83330.
[13] Chang JCH, Soong TT. Structural Control Using Active Tuned Mass Dampers. J Eng Mech Div 1980;106:1091–8. https://doi.org/10.1061/JMCEA3.0002652.
[14] Soong T.T. Active structural control in civil engineering n.d.
[15] Preumont A. Vibration Control of Active Structures, vol. 246. Cham: Springer International Publishing,; 2018. https://doi.org/10.1007/978-3-319-72296-2.
[16] Rodriguez J, Collet M, Chesné S. Active Vibration Control on a Smart Composite Structure Using Modal-Shaped Sliding Mode Control. J Vib Acoust 2022;144:021013. https://doi.org/10.1115/1.4053358.
[17] Yang F, Gani M, Henrion D. Fixed-Order Robust H_{infty} Controller Design With Regional Pole Assignment. IEEE Trans Autom Control 2007;52:1959–63. ⟨https://doi.org/10.1109/TAC.2007.906242⟩.
[18] Mitra A, Giridharan Y, Chakraborty A. Wavelet linear quadratic regulator-based gain scheduling for optimal fore-aft vibration control of horizontal axis wind turbine tower using active tuned mass damper. Struct Control Health Monit 2022;29:e3055. https://doi.org/10.1002/stc.3055.
[19] Sarkar S, Chakraborty A. Development of semi-active vibration control strategy for horizontal axis wind turbine tower using multiple magneto-rheological tuned liquid column dampers. J Sound Vib 2019;457:15–36. https://doi.org/10.1016/j.jsv.2019.05.052.
[20] Sarkar S, Chakraborty A. Optimal design of semiactive MR-TLCD for along-wind vibration control of horizontal axis wind turbine tower. Struct Control Health Monit 2018;25:e2083. https://doi.org/10.1002/stc.2083.
[21] Lackner M, Rotea M, Saheba R. Active Structural Control of Offshore Wind Turbines. 48th AIAA Aerosp. Sci. Meet. New Horiz. Forum Aerosp. Expo., Orlando, Florida. American Institute of Aeronautics and Astronautics,; 2010. https://doi.org/10.2514/6.2010-1000.
[22] Lackner MA, Rotea MA. Structural control of floating wind turbines. Mechatronics 2011;21:704–19. https://doi.org/10.1016/j.mechatronics.2010.11.007.
[23] Fernando KV, Nicholson H. Minimality of SISO linear systems. Proc IEEE 1982;70:1241–2.
[24] Glover K. All optimal Hankel-norm approximations of linear multivariable systems and their L, ∞ -error bounds n.d.
[25] Jonckheere EA, Silverman LM. A new set of invariants for linear systems–Application to reduced order compensator design. IEEE Trans Autom Control 2003;28:953–64.
[26] Xie Y, Zhao T, Cai G. Model reduction and active control of flexible beam using internal balance technique. Appl Math Mech 2011;32:1009–18. https://doi.org/10.1007/s10483-011-1476-x.
[27] Chen C-J, Teng J, Li Z-H, Wu Q-G, Lin B-C. Model reduction techniques for high-rise buildings and its reduced-order controller with an improved BT method. Struct Eng Mech 2021;78:305–17. ⟨https://doi.org/10.12989/sem.2021.78.3.305⟩.
[28] Chen C-J, Li Z-H, Teng J, Hu W-H, Wang Y. An observer-based controller with a lmi-based filter against wind-induced motion for high-rise buildings. Shock Vib 2017;2017:1–18. https://doi.org/10.1155/2017/1427270.
[29] Shang Y, Yu Q, Yang Z, Xu Z, Zhang X. Displacement and deformation measurement for large structures by camera network. Opt Lasers Eng 2014;54:247–54. https://doi.org/10.1016/j.optlaseng.2013.07.009.
[30] Yu Q., Shang Y., Guan B., Lei Z., Liu X. Camera series and parallel networks for deformation measurements of large scale structures 2015.
[31] Srivastava S, Misra A, Thakur SK, Pandit VS. An optimal PID controller via LQR for standard second order plus time delay systems. ISA Trans 2016;60:244–53. https://doi.org/10.1016/j.isatra.2015.11.020.
[32] Miah MS, Chatzi EN, Weber F. Semi-active control for vibration mitigation of structural systems incorporating uncertainties. Smart Mater Struct 2015;24:055016. https://doi.org/10.1088/0964-1726/24/5/055016.
[33] Wu Y, Chen X, Wang Y. Identification of linear and nonlinear flutter derivatives of bridge decks by unscented Kalman filter approach from free vibration or stochastic buffeting response. J Wind Eng Ind Aerodyn 2021;214:104650. https://doi.org/10.1016/j.jweia.2021.104650.
[34] Yao H, Tan P, Zhou H, Zhou F. A composite control scheme for electromagnetic active mass damper against control-structure-interaction effects. Eng Struct 2023;281:115638. https://doi.org/10.1016/j.engstruct.2023.115638.
[35] Yao H, Tan P, Zhou H, Zhou F. Real-time hybrid testing for active mass driver system based on an improved control strategy against control-structure-interaction effects. J Build Eng 2023;63:105477. https://doi.org/10.1016/j.jobe.2022.105477.
[36] Yu Q, Jiang G, Shang Y, Xu Y, Liu X, Fu S, et al. A displacement-relay videometric method for surface subsidence surveillance in unstable areas. Sci China Technol Sci 2015;58:1105–11. https://doi.org/10.1007/s11431-015-5811-6.
[37] Jonkman J., Butterfield S., Musial W., Scott G. Definition of a 5-MW Reference Wind Turbine for Offshore System Development. National Renewable Energy Lab. (NREL), Golden, CO (United States); 2009. ⟨https://doi.org/10.2172/947422⟩.
[38] Bazeos N, Hatzigeorgiou GD, Hondros ID, Karamaneas H, Karabalis DL, Beskos DE. Static, seismic and stability analyses of a prototype wind turbine steel tower. Eng Struct 2002;24:1015–25. https://doi.org/10.1016/S0141-0296(02)00021-4.
[39] Adhikari S, Bhattacharya S. Dynamic Analysis of Wind Turbine Towers on Flexible Foundations. Shock Vib 2012;19:37–56. https://doi.org/10.1155/2012/408493.
[40] Liu X, Xu J, He G, Chen C. Lateral vibration mitigation of monopile offshore wind turbines with a spring pendulum pounding tuned mass damper. Ocean Eng 2022;266. https://doi.org/10.1016/j.oceaneng.2022.112954.
[41] IEC 61400–1 - Wind energy generation systems – Part 1: Design requirements | GlobalSpec n.d. ⟨https://standards.globalspec.com/std/13202600/IEC%2061400–1⟩ (accessed May 2, 2023).
[42] Jonkman BJ. Turbsim User's Guide: Version 1.50. Golden, CO (United States): National Renewable Energy Lab. (NREL); 2009. https://doi.org/10.2172/965520.
[43] Modeling of the UAE wind turbine for refinement of FAST_AD n.d. ⟨https://www.researchgate.net/publication/242718581_Modeling_of_the_UAE_wind_turbine_for_refinement_of_FAST_AD⟩ (accessed May 2, 2023).

人致振动与控制

第26卷增刊II　Vol.26 Sup.II　工　程　力　学
2009年 12月　Dec. 2009　ENGINEERING MECHANICS

文章编号：1000-4750(2009)Sup.II-0148-12

人行桥的人致振动理论与动力设计

*陈政清，刘光栋

(湖南大学风工程研究中心，湖南，长沙 410082)

摘　要：随着人行桥跨度的不断增大、新型轻质高强材料的日益运用以及对桥梁美学和城市景观的追求，人行桥的基频不断降低，其振动及带来的行走舒适性问题也相继日益突出。自英国千禧桥关闭事件以来，国内外学者对人行桥的人致振动尤其是横向振动进行了大量的研究。该文介绍了人行桥的人致振动理论及动力设计方法，重点阐述了人行桥大幅横向振动分析的主要理论：直接共振理论、人桥相互作用理论及参数共振理论。最后以国内某曲线人行斜拉桥为例，展示了人行桥的动力设计全过程。

关键词：人行桥；步行荷载；人致振动；动力设计；减振设计

中图分类号：U441^{+}.3; U448.11　**文献标识码**：A

PEDESTRIAN-INDUCED VIBRATION THEORY AND DYNAMIC DESIGN OF FOOTBRIDGES

*CHEN Zheng-qing, LIU Guang-dong

(Wind Engineering Research Center, Hunan University, Changsha, Hunan 410082, China)

Abstract: With the progress of ever-increasing span of footbridges and the application of modern light and high strength materials, as well as the pursuit of bridge aesthetic in urban areas, the fundamental frequency of modern flexible footbridges decreases significantly. Since the event of London Millennium Bridge which was closed due to excessive pedestrian-induced lateral vibration, considerable efforts have been devoted to pedestrian-induced vibration analysis, particularly the lateral vibration, and its mitigation for modern footbridges. In the present study, the fundamentals of pedestrian-induced vibration theory and dynamic design of footbridges are described. The emphasis is placed on three existing methods for analyzing the large-amplitude lateral vibration, namely the direct resonance method, the pedestrian-bridge interaction method and the parametric resonance method. Finally a curved cable-stayed footbridge is used as an example to demonstrate the dynamic design of footbridges.

Key words: footbridges; pedestrian-induced forces; pedestrian-induced vibration; dynamic design; vibration control

1　伦敦千禧桥关闭事件

1.1　千禧桥的结构特点

2000年，英国为庆祝进入21世纪的一项重要桥梁工程——伦敦千禧桥开通仅3天就不得不关闭。这一事件受到了全世界异乎寻常的关注。千禧桥(The Millennium Bridge)是英国伦敦泰晤士河上新建的一座人行桥。如图1所示，千禧桥是一座浅悬索桥，主缆垂度只有2.3m，垂跨比约为1/63，比通常的悬索桥垂跨比1/8－1/12小很多。两组4根直径120mm的主缆连接两岸桥台和河中两桥墩，形成三跨连续结构：北边跨81m，中跨144m和南边跨108m。千禧桥没有吊杆，竖向高度和横向宽度都不等的U型横梁直接固定在主缆上，使每边四

收稿日期：2009-05-09
作者简介：*陈政清(1947－)，男，湖南湘潭市人，教授，博士，湖南大学风工程试验研究中心主任，主要从事桥梁与结构抗风研究(E-mail: zqchen@hnu.cn)；
刘光栋(1935－)，男，山东日照人，教授，博导，主要从事力学与桥梁工程等研究.

根钢缆处于同一水平面上。同时，主缆水平中心距也在 10m－13.4m 之间变化。横梁每隔 8m 一道，纵梁为两根圆钢管，直接与横梁焊接。桥面板宽 4m，长 16m，是蜂窝状轻型铝板。桥面板之间为铰接滑移联接方式，直接铺设在纵横梁之上，不参与全桥结构受力。

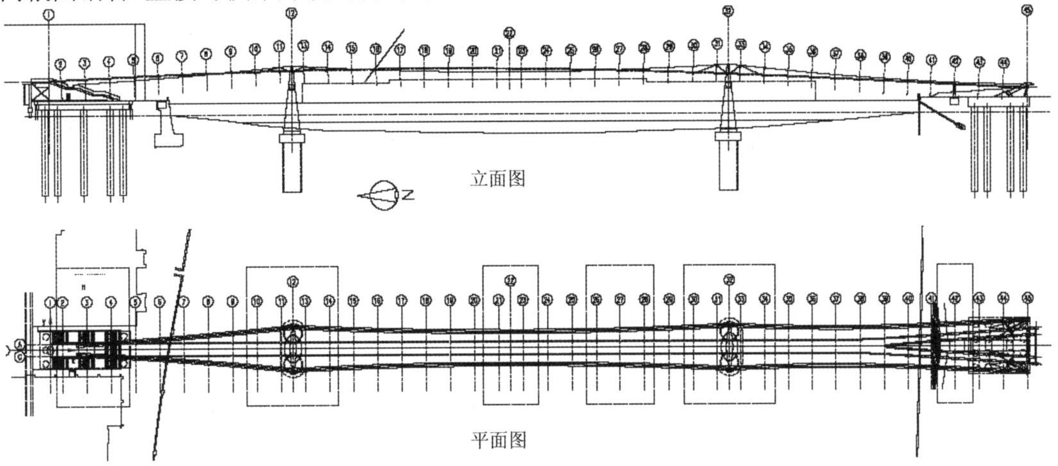

图 1　伦敦千禧桥桥型布置图

Fig.1　Layout of London Millennium Bridge

1.2　千禧桥的大幅横向振动

该桥是一座结构新颖的扁平轻巧的悬索桥，于 2000 年 6 月 10 日开通。不幸的是，桥梁开放行人当日即发生过量的横向摇摆，在减少上桥人数仍发生过量横向摇摆的情况下，6 月 12 日临时关闭该桥梁。据估计开放当天有 80000 人－100000 人通过了该桥。录像分析表明，最高峰时约有 2000 人同时在桥上，人群密度达到 1.3 人/m^2－1.5 人/m^2。其中大幅侧向振动主要发生在南跨和中跨。南跨一阶侧向频率大约为 0.8Hz，中跨的一阶侧向频率低于 0.5Hz，二阶频率低于 1.0Hz。北跨振动要小一些，其侧向振动频率略大于 1.0Hz。大幅度的振动并不是持续发生的：当大量行人在桥梁上行走时，振动就增大，而行人减少或停止行走时，振动就减小。根据日测估计南跨和中跨的最大侧向加速度为 0.200g－0.250g。在这样的加速度水平下，大量的行人行走困难，不得不停下来依靠栏杆保持身体平衡。没有观察到过大的竖向振动。

事后的实验室研究和现场研究证明，由于千禧桥的竖向及横向频率正好在步行力的频率范围之内，当行人很多时，就不可避免地会产生人桥共振效应，行人会自动调整步频和相位，以适应桥梁的振动和改善自身的行走舒适性。不幸的是，这一调整过程正好进一步加剧了桥梁的横向振动，直到行人因极度不舒适而停止前进或桥上人数减少。显然，限制过桥人数是违背建桥初衷的。因此减振措施要达到的指标被设定为 2 人/m^2 的高密度条件下桥梁振动在行人可接受的范围之内。

1.3　千禧桥的减振措施

改善桥梁特性有两种普遍的方法：第一种就是增大桥梁的刚度，使固有频率偏离行人步行力的基频范围；第二种就是提高桥梁的阻尼来减少共振响应。

通过提高刚度改善振动特性方案的最大困难在于增大中跨频率，中跨的侧向基频只有 0.49Hz。要使频率增加 3 倍，即使能做到增加刚度时不增加结构质量，理论上侧向刚度要增加 9 倍。事实上，任何涉及刚度增加的方法都会增加质量，因此刚度要求增加 10 倍以上。然而，主跨的所有刚度几乎都由主缆的几何刚度提供的。即使将桥面与两根圆钢管的纵梁连接在一起提供刚度，中跨的一阶侧向频率也仅仅只能增加几个百分点。因此如果没有相当多的附属设施，桥的刚度不可能增加到 10 倍，其中包括一些成本很高且会严重影响桥梁美观的一些设施。即使中跨的一阶侧弯频率达到 1.3Hz 以上，某些侧弯-扭转耦合模态的频率还是小于 1Hz，而这只能通过提高结构的竖向或扭转刚度，且增加的倍数与侧向刚度相同。需要增加附属设施来提高竖向或扭转刚度，而它们又会影响现有桥梁的承载力。

因此，提高桥梁刚度几乎是不可行的，减振途径只能是提高结构阻尼。要求由原结构 0.4%的阻尼比提高到 20%，以提供足够的阻尼来确保人群密度为 2 人/m^2 时结构的动力稳定性。侧向振动模态的阻尼主要由粘滞阻尼器提供，总共有 7 种不同类型的 37 个阻尼器。中跨还另外增加 4 对侧向 TMD 阻尼器来提供侧向阻尼，每个阻尼器重 2.5t。竖向阻尼主要由竖向 TMD 阻尼器提供。在桥梁的三跨上总共安装了 26 对竖向 TMD 阻尼器。这些阻尼器的质量为 1t－3t，由压缩弹簧支撑。

改造安装工程由 Cleveland Bridge UK 承包。粘滞阻尼器由美国 Taylor Devices 公司生产，TMD 阻尼器由德国 Gerb Schwingungsisolierungen 公司生产。减振工程于 2001 年 5 月初动工，2001 年底完工。2002 年元月 30 号组织了两千人过桥的现场试验，平均行人密度达到 1.5 人/m^2，是 2000 年 6 月开通当日的最高行人密度。现场试验测试结果表明，动力响应较以前下降了 40 倍，桥梁没有共振现象产生，行人也没有任何不适应的报告。2002 年 2 月 22 号，在经过 8 个多月的关闭之后，千禧桥正式向行人开放，并获得媒体一致好评。

1.4 千禧桥的教训

行人激发的桥梁过度振动现象其实很常见。1957 年 10 月 15 日，我国第一座跨长江的大桥——武汉长江大桥建成通车。五万群众“跟在汽车后面前进，他们兴奋地举起鲜花和头巾，如同碧绿般的潮水一起一伏地从武昌流向汉阳”(引自 1957 年 10 月 16 日《长江日报》)，结果激发了严重的横向摇摆振动，一度造成人群恐慌。1973 年西班牙博斯普鲁斯大桥开放当天，估计有 60000 人－100000 人涌至主跨长 1074m 的悬索桥，同时还有 2000 人/min 的人流不断从桥两端涌入。当桥开始摇摆时，桥两端的人群就被禁止进入。《工程新闻记录》记载“根据设计，除车辆外该桥可承载 30000 人，相当于 1人/m^2”。据估计那天高峰时段，桥面被人群挤满了，桥面板承载多达 5 人/m^2。新西兰奥克兰港口公路桥也发生了类似的情况。这是一座主跨 190m 的 4 车道公路桥。1975 年的一天临时封闭了一个方向的两个车道供举行示威游行的队伍通过，示威者有 2000 人到 4000 人，过桥时产生了明显的横向振动。这一事件并有电视录像记录可考。

这些事件表明：

1) 虽然生物力学和桥梁专家在人致桥梁振动的机理研究上已进行了很多研究，但这些成果还远不能满足设计要求。特别是千禧桥事件以前，横向振动研究很少。千禧桥事件激发了世界范围内对人行桥振动问题的重视，成为了当前研究的一个热点。2002 年、2005 年和 2008 年已连续召开了 3 次人行桥振动的国际会议。

2) 人桥振动的敏感频率范围是横向 1.2Hz 以下，竖向 1.5Hz－5.0Hz。大跨度的人行桥，无论取何种结构形式，其结构基频均难以避开步行力的频率范围，而且提高频率往往是很不经济的。因此对这类人行桥，减振设计应当成为结构设计的一部分。

2 桥梁人致振动的基本原理

2.1 行人是激振源

人类行进过程，两下肢交替运动，带动整个身体前进。正常行走时，脚跟先着地，然后脚尖离地迈进，每步都有一个短暂的双脚同时与地面接触的过程。奔跑的特点是始终脚尖先落地，且没有双脚同时与地接触的过程。无论是步行还是跑步，每迈一步行人的重心都会高低起伏一次。重心变化产生的加速度，使每一步的竖向荷载都具有双峰的性质。步行者没有腾空过程时程曲线是连续的，跑步的时程曲线则是一个个间断的半波。除竖向力外，行人的双腿交替运动，也导致人的重心左右摆动，因此也有横向力。由于行人重心每两步左右摆动一次，因此横向步行力的频率正好是竖向步行力的一半。为克服地面摩擦力前进，行人荷载也有纵向力分量。图 2 是典型的正常行走时的步行力时程曲线。

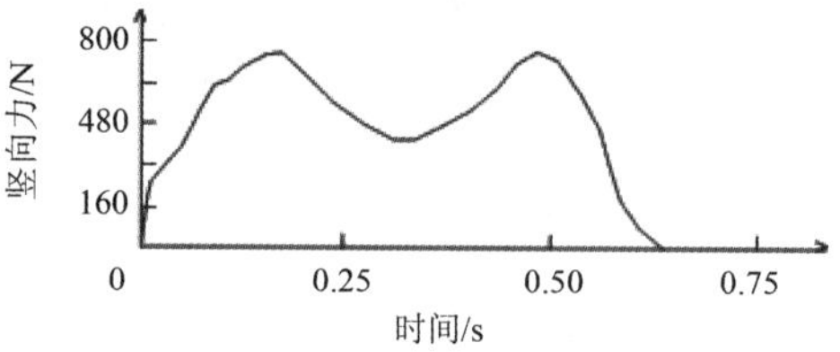

(a) 竖向力

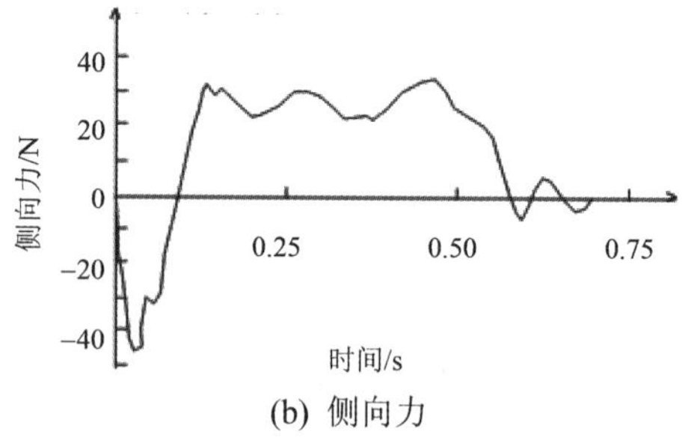

(b) 侧向力

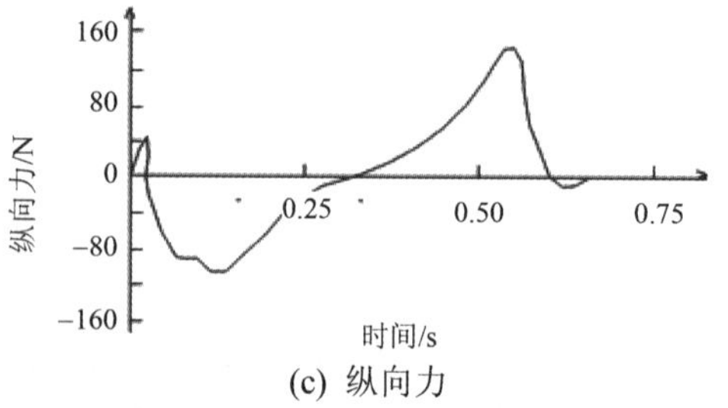

(c) 纵向力

图 2 正常行走时单步力三方向分量

Fig.2 Three components of normal working forces

与主要由路面不平顺产生的车辆动荷载不同，行人荷载具有如下特征：

1) 明显的周期性。严格来讲，即使是同一个人，每迈出一步都会和其它步有细微的区别，但总的说来，在一定近似意义上行走方式决定了人行荷载有明显的周期性。成年人正常行走时的步频(即每秒钟迈步数)在 1.6Hz－2.4Hz 之间。竖向力的双峰特征，使竖向力的前三阶谐波成分都可能激发桥梁振动。横向力则一般可只考虑一阶谐波作用。因此一般认为基频要高于 5Hz 的桥梁才不大可能出现人桥振动问题。

2) 窄带随机过程。人类在体重和行走速度方面的差别远不如车辆那么大，因此行人荷载的步频、步长均是在很窄的一个范围内随机分布，并较好地符合正态分布特征。但是，在同一桥上行走的人的步行荷载的初相位，可以认为是在 0π—2π 之间均匀随机分布。人行荷载的窄带性意味着桥上行人多到一定数目后，就会自然产生一部分频率相位非常接近同步的人，桥梁越长越宽，这一现象就会越显著，如果这一部分人的频率正好与结构的固有频率接近，就会产生明显的桥梁振动。这好比人的生日，一个小区小村同一天生日的人不多，但一个城市一个国家同一天生日的人就会很多。这也是大跨的人行桥振动特点。

3) 人桥相互作用。与车辆不同，人类是感觉很灵敏的生物，可以感知 $0.01m/s^2$(0.001g)的加速度。如果桥梁发生振动，行人会自动调整步频和相位，以适应桥梁的振动和改善自身的行走舒适性，振动学上称为“锁定”(lock in)现象。不幸的是，这一调整过程正好进一步加剧了桥梁的横向振动，直到行人因极度不舒适而停止前进或桥上人数减少，振动才会消失。

2.2 行人又是感受体

人类是对振动十分敏感的生物，不过幸运的是现有研究表明，人在行进中对振动的耐受力要比坐在建筑物内高一些。设计一座行人完全感觉不到振动的桥肯定是非常不经济的，也是没有必要的。因此合理的途径是将振动控制在容许范围内。但是，人的感觉是一个非常离散难于定量的问题，不仅不同的人在相同振动环境下的反应不一样，而且同一个人在不同时候对同样振动的反应也不一样。由于这种不同人之间的差异以及单人对振动反应的内在变异性，有关振动舒适性的问题均是通过调查大量的受试者在振动环境下的感受而定性地来划分舒适度标准。研究表明，在振动的位移、速度和加速度这 3 个要素中，影响人的生理和心理感受的主要因素是加速度。

2.3 人行桥人致振动的研究内容

人行桥人致振动问题的研究内容包括如下几个方面：

1) 单人步行力荷载的测量和数学模型的建立。由于人种的不同，即使是采用同一模式，参数也有很大差别。现有研究数据主要是依据西方人种的测量结果，亟待开展中国人步行荷载的研究。例如，笔者组织了大学生步行的频率和步长的调查研究。依据 12000 多个样本统计得出的步频平均值为 1.825Hz，步长平均值为 0.715m，两者均低于西方人种的统计结果。

2) 人群荷载模式的建立。这一部分工作难度大，目前国际上的研究资料也很少。

3) 人群荷载激励下桥梁结构的振动响应评价。特别是人行桥的横向振动还存在一种振动突然增大的“锁定”(lock in)现象，这是人行桥动力设计的一个主要环节。目前已发展了一些确定性方法和随机性方法，但都伴随着多种假定，离评估实际响应还有一定距离。

4) 人行桥动力设计的规范或指南。英国 BS5400 最早在规范中纳入人行桥设计的标准。限于当时的认识，只对竖向振动作了规定。目前国外已在总结伦敦千禧桥教训的基础上提出了竖横向振动的动力设计指南，如德国 2008 年发表的 EN03 人行桥设计指南。

3 人行桥横向振动理论

虽然人行桥也存在竖向振动问题，但已有的大跨度人行桥过度振动事件几乎无一例外都是横向振动。至伦敦千禧桥关闭事件，人们才深刻认识到人行桥在行人激励下，可能发生横向动力失稳，并将这种失稳称之为“锁定”现象。因此，自千禧桥事件后，很多学者致力于横向振动理论研究。

目前，关于人行桥发生行人激励的过量横向振动问题的解释，或者说横向动力失稳的锁定现象的机理，主要有三种理论：直接共振理论，认为行人激励频率正好与桥梁某一阶模态的频率相同而引起共振；动力相互作用理论，它基于桥梁运动与桥上行人的运动之间的某种适当的相互作用模型；内共振理论，认为是结构的非线性导致结构的不同模态之间进入内共振状态。

3.1 人行桥直接共振理论

从现有文献看，日本T桥是世界上第一个完整地报道、研究人行桥横向振动并采取了减振措施的工程实例。这一研究由日本学者 Fujino 主持[1]。根据结构动力学，人行桥的振动方程为：

$$m_s(x)\cdot\ddot{u}(x,t)+c_s(x)\cdot\dot{u}(x,t)+k_s(x)\cdot u(x,t)=q_p(x,t) \tag{1}$$

其中：$m_s(x)$、$c_s(x)$、$k_s(x)$分别为桥梁的质量分布函数、阻尼分布函数和刚度分布函数；$\ddot{u}(x,t)=\partial^2u(x,t)/\partial^2 t$、$\dot{u}(x,t)=\partial u(x,t)/\partial t$ 为加速度和位移；$q_p(x,t)$为沿桥长分布的动荷载。

令$u(x,t)=\sum_{i=1}^{N}\varphi_i(x)v_i(t)$，应用模态迭加法，由振型的正交性得到解耦后的各阶模态振动方程：

$$\ddot{v}_i+2\zeta_i\omega_i\dot{v}_i+\omega_i^2v_i=\frac{1}{M_{si}}P_i(t) \tag{2}$$

式中：v_i是第i阶模态的广义位移；M_{si}是桥梁的第i阶广义质量；ζ_i是第i阶模态阻尼比；$P_i(t)$是行人横向动荷载的第i阶模态力：

$$P_i(t)=\int_0^L q_p(x,t)\varphi_i(x)\mathrm{d}x \tag{3}$$

只考虑一阶谐波的单人横向动荷载可以表示为：

$$P_{ps}=\alpha m_p g\cdot\cos(2\pi f_p t+\varphi) \tag{4}$$

式中：α是动荷载因子；m_p是行人质量；g是重力加速度；f_p是行人横向动荷载的频率；φ是行人的初相位。

日本T桥振动的录相表明，振动时有一部分人处于完全同步状态，即频率、相位都相同，设桥长为L，均布有总数为N的行人，处于同步的行人比例为λ，那么振动时同步行人的动荷载密度为：

$$q_p=\lambda\cdot\frac{N_p}{L}\alpha m_p g\cdot\cos(2\pi f_p t) \tag{5}$$

式中略去了相位角φ，是因为λN_p个人的相位角也都相同。将式(3)和式(5)代入式(2)，可得：

$$P_i=\lambda\cdot\alpha\cdot N_p m_p\cdot g\cdot\cos(2\pi f_p t)\cdot\int_0^L\varphi(x)\mathrm{d}x \tag{6}$$

如果λN_p个行人荷载的圆频率$2\pi f_p$正好等于桥梁第i阶振型的频率ω；那么直接共振条件满足，容易得到共振时桥梁横向动挠度的表达式为：

$$u(x,t)=\frac{1}{M_{si}}\cdot\lambda\cdot\alpha\cdot N_p m_p g\cdot\frac{1}{2\zeta_i}\cdot\varphi(x)\cdot\int_0^L\varphi(x)\mathrm{d}x\cdot\cos\left(\omega_i t-\frac{\pi}{2}\right) \tag{7}$$

Fujino 用上式计算出T桥一阶横向振动最大振幅并与实测值相比较。计算时各参数的取值如表1所示，其中λ=0.2 来源于实桥观测资料，相当于有20%的桥上行人处于同步状态。

表1 T桥1阶横向共振计算参数

Table 1 Parameters for analyzing the lateral vibration of T bridge

参数	取值	参数	取值
同步因子λ	0.2	模态圆频率ω_1	0.91×2π
动载因子α	0.05	模态阻尼比ζ_1	0.01
行人平均质量m_p	70kg	模态振型$\varphi_1(x)$	$\sin(\pi x/L)$
桥上总人数N_p	2000	模态等效质量* M_{si}	800×10^3kg

*：等效质量计入了拉索振动影响，数值与$\varphi(x)$的最大值等于1相对应。

由于一阶振型为半个正弦波，所以：

$$\int_0^L\varphi_1(x)\mathrm{d}x=\int_0^L\sin\frac{\pi x}{L}\mathrm{d}x=\frac{1}{2}$$

且最大动挠度发生在跨中$L/2$处。按表1给出的参数，计算的横向最大振幅为1.5cm，与实测最大振幅1cm接近。

如果i阶振型是多个半波，例如$\varphi_i(x)=\sin\frac{2\pi x}{L}$，那么式(5)中因子$\int_0^L\varphi_i(x)\mathrm{d}x=0$显然与实际情况不符。为了克服这一缺点，可以将式(5)中因子改写为$\int_0^L|\varphi_i(x)|\mathrm{d}x$。其力学意义是处于负半波的行人动荷载的相位与正半波的相反，或者说处于不同半波上的行人动荷载都是指向荷载增大的方向。

当人行桥发生横向动力失稳时，桥上同步行人数可能因桥型而异，亦即同步因子λ对于不同的桥可以有不同的值。日本T桥分析取λ=0.2，是依据该桥现场实测数据进行分析后的结果。在设计阶段，对于一座尚未建造的桥梁，λ值取多少才能合理预测振动的幅值，是直接共振法尚未解决的问题。

由于行人横向动荷载一阶分量的频率范围为0.8Hz－1.2Hz之间，因此直接共振法只能解释在这同一频率范围内的桥梁共振现象。英国伦敦千禧桥主跨横向一阶频率为0.5Hz，并且在行人作用下产生了大幅一阶振动，就很难用直接振动法解释。

3.2 人桥相互作用理论

3.2.1 经验参数模型

1) Dallard 模型。

Dallard 等人从伦敦千禧桥现场实验数据中经

反分析得到的人桥相互作用力模型，由这一模型导出了按模态横向动力稳定判据导出的临界人行数公式。Dallard 方法的要点归纳于表 2 中[2]。

表 2　Dallard 的实验参数模型

Table 2　Experimentally-obtained empirical model parameter for Dallad formulae

单人激振力	$P_i = k\cdot\dot{v}\cdot\varphi_i$	k=300N·s/m 是源于千禧桥的实验参数；$\dot{v}$ 为模态速度。φ_i 为第 i 人所处位置的振型函数值，N 为总的行人数，M^*模态质量，ζ为模态阻尼比。
模态等效激振力	$F_p = k\dot{v}\cdot\sum_{i=1}^{N}\varphi_i^2$	
动力稳定条件	结构阻尼力=激振力	
临界人行数公式	$N<\dfrac{8\pi fM^*\cdot\zeta}{K}$	

Dallard 模型是一个动力失稳判据模型，由此判据可计算临界行人数，但不能计算振动幅值。模型的关键参数 k 是否也适合其它人行桥，是一个尚待证明的问题，但是 Dallard 模型计算方便，因此近来的法、德、日等国的人行桥设计都采用了这一公式。

2) Nakamura 模型。

Dallard 公式中经验参数 k=300N·s/m 是一个定值，这意味着激振力随着模态速度线性增长，那么人行桥的振幅也就会越来越大，这与已知的观测事实不符。实际上，一旦人行桥发生摇摆。就会有部分行人停止或放慢前进步伐，相当于激振力减少，阻尼增加，使桥的振动幅度能稳定在一定范围内。为了将这一人桥相互作用因素考虑进去，日本学者 Nakamura 提出了他的模态共振激振力模型[3]。

$$M_B\ddot{v}+C_B\dot{v}+K_Bv=F_p(t) \tag{8}$$

$$F_p=K_1K_2H(\dot{v})G(f)\cdot M_p\cdot g \tag{9}$$

$$H(\dot{v})=\frac{\dot{v}}{k_3+|\dot{v}|} \tag{10}$$

$$G(f)=1.0 \tag{11}$$

式中：v 是横向振动模态坐标；K_1 即动荷载因子 α；K_2 是处于同步状态的桥上行人的百分比(即为式(3)中的λ)，依据 Fujino 在日本 T 桥实测结果，K_2=0.2。与 Dallard 模型相比，Nakamura 模型的最大不同点是用一个分式函数 $H(\dot{v})$ 代替了 Dallard 模型中的线性因子 $\dot{v}$。

基于式(6)－式(9)，Nakamura 采用龙格-库塔法模拟研究了人行桥横向振动随各种参数变化的情况，计算以日本 T 桥的一阶横向振动为背景，主要参数取值见表 3。

Nakamura 模拟了 T 桥在行人密度为 0.8 人/m² 时的振动响应时程。在前 20s 振幅增加缓慢。20s－80s 振幅增加较快，以后逐渐进入稳定振动状态，振幅约为 10mm，速度约为 50mm/s，模拟的时程与 T 桥实测时程趋势基本一致，说明激振力参数取值较为合理。

表 3　日本 T 桥一阶横向振动模态参数和激振力参数

Table 3　Modal and exciting force parameter for the first lateral mode of T bridge

模态参数		激振力参数
结构模态质量 M_B	2.37×10³kg	K_1=0.04
结构阻尼比ζ_B	0.008	K_2=0.2
一阶模态频率 f_B	0.93Hz	K_3=0.01m/s
主跨长	174m	行人密度 0.2 人－1.5 人/m²
宽	5.25m	行人质量 70kg

3) 孙利民-袁旭斌模型。

同济大学孙利民教授和他的学生袁旭斌博士也提出了一种基于实验的人桥相互作用模型[4]。孙-袁模型的特点是不仅考虑了同步人群的自激作用，也考虑了未同步人群的强迫振动作用。另外，模型还考虑了同步人群作用力与结构振动的相位差，他们的实验在同济大学进行，设计了一套专门的实验装置，用来确定单人激振力模型。

在单自由度振动台上设置电动走步机，依据行人试验结果，孙-袁提出了同步概率、动荷载因子与振幅的关系分别为：

$$P_s(A)=\frac{A}{A+C_1}\mathrm{e}^{(-C_2(f_p-1)^2)} \tag{12}$$

$$\alpha_1(A)=1.1811A+0.0517 \tag{13}$$

式(12)由两部分的乘积组成，分式部分和指数部分。分式部分反映了同步概率与振幅的关系，其中 C_1 由试验值拟合得到，C_1=0.022。指数部分表示结构振动频率偏离行人横向摆动的中心频率而产生的同步概率折减系数。行人横向摆动的中心频率假定为 1Hz，系数 C_2=20。

孙-袁模型导出了一个单自由度变阻尼的非线性强迫振动方程，其共振解为：

$$A_0=\frac{1}{2\zeta_T}\cdot\frac{r_{\varphi r}\sqrt{(1-P_s(A_0))N}\cdot\alpha_1(A_0)m_pg}{M_B\cdot\omega^2} \tag{14}$$

振动稳定的临界状态由ζ_e=0 确定，由此式可以计算临界行人数。由于是非线性方程，需要采用迭代法求解。

袁旭斌取实验确定的参数由式(12)和式(13)计算日本 T 桥和英国伦敦桥的响应，结果都与实测不符。经分析原因后改取 C_1=0.06(试验拟合值为 0.022)，获得了较为满意的结果，由此可见孙-袁模型本质上也是一种经验参数模型，并且由单自由度振动台试验得到的参数不一定与实桥一致。

3.2.2 简化的人桥耦合振动分析法

行人本身也是一个振动系统，有自身的刚度、阻尼和质量，行人横向振动，至少也是一个单自由度振动系统。行人振动与桥梁振动由行人与桥梁之间的作用力和反作用力耦合。按照这一观点，人桥耦合振动模型的示意图应如图3所示，相应的耦合振动方程组为：

$$m_{pi}\ddot{u}_{pi}(x,t)+c_{pi}\dot{u}_{pi}(x,t)+k_{pi}u_{pi}(x,t)=-P_i(x,t),\quad i=1,2,\cdots,N \tag{15a}$$

$$m_s\ddot{u}_s(x,t)+C_s\dot{u}_s(x,t)+k_su_s(x,t)=\sum_{i=1}^{N}P_i(x,t),\quad i=1,2,\cdots,N \tag{15b}$$

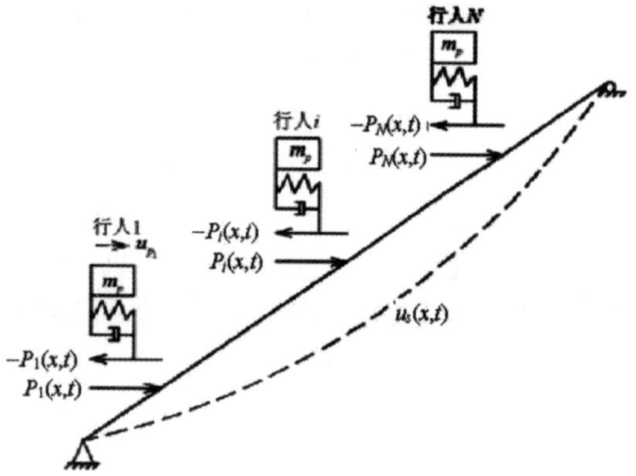

图3 人行桥耦合振动模型示意图

Fig.3 Coupled pedestrian and bridge model

此方程组由 N 个行人振动方程式(15a)和一个桥梁振动方程式(15b)组成。由于每个行人都在沿桥长方向移动且作用力 $P(x,t)$受多因素影响，要直接求解这一方程组是极其困难的。应当指出，此方程组尚未考虑桥上行人数 N 也是随时间改变的这一因素。

英国学者 Roberts T M 提出了一种简化分析方法[5]，现分步介绍如下：

1) 假定桥上维持 N 个行人不变，且沿桥长均匀分布。于是可以将 N 个行人作用力之和化为沿桥长分布的连续作用力 $P(x,t)$，相应的，N 个行人的运动简化为在均布质量密度 $N\times m_p/L$ 下的一个连续运动 $v_p(x,t)$，并且忽略人体刚度与阻尼的作用。

2) 设$\varphi_i(x)$，i=1, 2,⋯,∞，是桥梁的 i 阶模态振型函数，并设桥上有 $N_p(N_p<N)$个人进入同步状态，同步频率为ω_i，Roberts 进一步假定在这 N_p 个同步人群作用下，连续化的作用力 $P(x,t)$和行人位移 $u(x,t)$也可用桥梁振动模态加以分解，即：

$$P(x,t)=\sum_{i=1}^{\infty}\varphi_i(x)P_i\sin(\omega t),\ u_p(x,t)=\sum_{i=1}^{\infty}\varphi_i(x)v_{pi}(t) \tag{16}$$

3) 由振型的正交性可以得到分离后的桥梁第 j 阶模态广义位移运动方程：

$$M_j\cdot\ddot{v}_{sj}(t)+C_j\cdot\dot{v}_{sj}(t)+\omega_j^2M_j\cdot v_{sj}(t)=P_i\cdot\sin\omega t \tag{17}$$

由单自由度系统在简谐荷载下的受迫振动理论，可知式(17)的解为：

$$v_{sj}(t)=\frac{P_jD_j}{M_j\omega_j^2}\cdot\sin(\omega t-\psi_j)=v_{sj}\cdot\sin(\omega t-\psi_j) \tag{18}$$

式中：$D_j=\left\{\dfrac{1}{(1-r^2)^2+(2\zeta_ir)^2}\right\}^{0.5}$ 为放大系数，$\psi_j=\tan^{-1}\left(\dfrac{2\zeta_ir}{1-r^2}\right)$ 为相位差，$r=\dfrac{\omega}{\omega_i}$ 为频率比，ζ_j 为模态阻尼比。

4) 只考虑 N 中一部分同步人群 $N_p(N_p<N)$的作用，得到 j 阶模态下的行人模态位移为：

$$v_{pj}(t)=\frac{L}{N_pM_{pj}}\cdot\frac{P_j}{\omega^2}\sin\omega t=v_{pj}\cdot\sin\omega t \tag{19}$$

式中：

$$M_{pj}=\int_0^L m_p\cdot\varphi_j^2(x)\mathrm{d}x \tag{20}$$

5) Roberts 提出，当行人振幅 v_p 大于人行桥振幅 v_s 时，人行桥的 j 阶模态振动是稳定的，反之则是不稳定的，因此人行桥 j 桥模态振动稳定的临界判据为 $v_p=v_s$，于是得到确定同步人行数 N_p 的方程：

$$\frac{N_p}{L}=\frac{M_j}{M_{pj}}\cdot\frac{1}{r^2D_j} \tag{21}$$

6) 当行人只分布在人行桥的部分长度$\alpha L(\alpha<1)$时，可以证明式(21)应当修正为：

$$\frac{N_p}{\alpha L}=\frac{(1+\alpha^2)}{2\alpha}\frac{M_j}{M_{pj}}\cdot\frac{1}{r^2D_j} \tag{22}$$

7) 行人步行横向力 $P(x,t)$的频率在 0.7Hz－1.2Hz 之间，如果 j 阶横向振动模态频率ω_j 也在这一范围内，那么 $r=\omega/\omega_j$ 接近于 1，而 D_j 在 $r\to 1$ 时取极大值，从而由式(20)知 N_p 取最小值。

由于 D_j 值在 r=1 附近变化非常大，Roberts 认为假定行人频率与桥梁振动频率完全相符即 r=1 不太合乎实际，他建议取 r 在 0.8－1.2 的范围内计算 r^2D_j 的平均值。当ζ=0.0075 时，r^2D_j 在此范围内的平均值为 9.98，用这个值按式(22)计算伦敦千禧桥北边跨横向振动的临界人行数 N_p 与该桥现场实测值 165 人较为接近。

8) 当人行桥未进入横向动力失稳状态时，Roberts 也提出了一个行人横向荷载模型，可以计算 N 个行人产生的桥梁振动。沿桥均布的作用力 $P(x,t)$ 的幅值为：

$$P(x,t)=2\times\frac{\sqrt{N}}{L}\times\frac{m_p m_{up}\omega^2}{\sqrt{2}}\sin\omega t=\frac{\sqrt{2N}}{L}m_p m_{up}\omega^2\sin\omega t \tag{23}$$

式(23)是 N 个行人沿桥均布，横向振幅 u_p 的频率ω及质量 m_p 均相同，行人相位差在$(0-2\pi)$之间随机均匀条件下得到的均布横向作用力表达式。将它代替式(17)的右边项，即可用时程分析法或模态迭加法求得桥梁的稳态振动响应 $u_s(x,t)$。Roberts 没有给出具体算例结果。

Roberts 法的最大优点是分别建立人桥两个振动体系的方程，通过作用力与反作用力的耦合联立求解，但是他的方法有很多不够合理的地方。

1) 式(16)的假定与实际不符，行人作用力和横向位移不满足桥梁边界条件，不能用像桥梁位移那样用满足边界条件的函数列$\varphi_i(x)$来分解，例如在$\varphi_i(x)=0$ 的点，$P(x,t)$和 $u_p(x,t)$按此假定会等于零，而实际上即使行人走在固定地面上，$P(x,t)$和 $u_p(x,t)$也不为零。

2) 横向动力失稳的临界人数估计式(21)是基于行人模态位移幅值和桥梁模态位移幅值相等的假定，这一假定本身具有直觉的成分，缺乏严密的证明。

3) 在预测北边跨的临界人行数时取 r 在一定范围的平均值，虽然较为合理，但具体取值范围如何确定是个问题。算例取 0.8－1.2 之间的平均值，似乎覆盖了行人步频的全部范围，未免太宽了，离共振概念相关太远。若取 $0.9<r<1.1$，则 r^2D_j 的平均值为 16.1，那么预测的 N_p 值就是 88－142，低于试验值 165 人。

4) 对于伦敦桥其它模态，不论其频率为多少，Roberts 都按 $0.8<r<1.2$ 的范围的 r^2D_j 平均值 9.98 计算相应模态的临界行人数，因此导致不合理的结果：中跨一阶模态频率 0.48Hz，二阶模态频率 0.95Hz，临界行人数都相同。事实上，只有模态频率在 1Hz 附近时，取 $0.8<r<1.2$ 才合乎实际情况，因为它相当于假设行人横向频率在 0.8Hz－1.2Hz 之间。中跨一阶模态频率只有 0.48Hz，如果仍假定 r 在 0.8Hz－1.2Hz 之间，相当于假定行人横向频率在 0.384Hz－0.576Hz 之间，这样低的行人频率几乎是不可能的。

5) Roberts 法如上的一些缺点正好表明人行桥横向振动问题还有很大的研究空间。为便于阅读，本文在表达方式和数学符号上较 Roberts 原文有所改动。

3.2.3 基于运动叠加假定的分析法

在分析他人实验资料的基础上，Newland[6]假定行人运动由两部分组成：一个就是在静止的台板上行走的固有运动；另一个是由桥面运动引起的，与桥面振幅成比例，但是滞后于桥面振动。当时滞处于最差情况时，行人就产生负阻尼导致动力失稳现象。根据这一假定，稳定准则可以通过一个无量纲的参数得以表达，这个参数包含了桥梁阻尼和行人与桥梁的质量比。这也类似于用于描述风致桥梁振动的 Scruton 数。从形式上看，Newland 方法对竖向与侧向振动都可以进行分析。

行人荷载由两部分位移各自对应的加速度产生的惯性力构成：第一部分是在静止台板上行走时身体质心的自然位移；第二部分是台板运动引起的附加位移。令 $x(t)$为静止台板上的行人质心的运动，$y(t)$为台板本身的运动。当有很多行人布满桥上任何一处，并假定他们一起行走，那么可转为模态分析。假定人行激励荷载取决于人和桥的模态加速度以及行人的模态质量。包括了两个经验系数α和β，α是指行人质心运动与桥面运动的比值，β是指人行同步运动的相关系数。于是人行荷载模拟如下：

$$f(t)=m\beta\ddot{x}(t)+m\alpha\beta\ddot{y}(t-\Delta) \tag{24}$$

这里假定行人的行为在小阻尼理论下没有改变桥梁的模态特性。行人只是桥梁模态的一个荷载函数，可得振动方程：

$$\ddot{y}(t)+2\varsigma\omega_n\dot{y}+\omega_n^2 y(t)+\frac{m}{M}\alpha\beta\ddot{y}(t-\Delta)=-\frac{m}{M}\beta\ddot{x}(t) \tag{25}$$

如果行人在固定台面上的固有运动 $x(t)$已知，那么就可以求解桥梁的模态位移 $y(t)$。当然，只考虑桥梁的小振幅运动以便行人不会因为桥梁的振动而停止。滞后时间Δ考虑了身体相对脚步的延时相对于竖向振动，侧向振动时人体中心运动与桥面运动间更容易产生滞后时间Δ，这是因为侧向振动时身体容易有微笑的不同相的侧向摆动从而形成相位差。为了接下来分析的方便，质量比参数 m_r 和频率比参数定义为：

$$m_r = \frac{\alpha\beta m}{M}, \quad \bar{\omega} = \frac{\omega}{\omega_n} \tag{26}$$

这里的 m/M 是指行人与桥梁模态质量比，$\bar{\omega}$ 指激励频率与桥梁固有频率比值。由式(25)的齐次方程有非零解条件并假定 m_r 为常数，可得稳定临界频率 ω_c 应满足下式：

$$\frac{\omega_c}{\omega_n} = \frac{1}{1-m_r^2} \tag{27}$$

这时的临界阻尼比 ς_c 是：

$$\varsigma_c^2 = \frac{1}{2}(1-\sqrt{1-m_r^2}) \tag{28}$$

当质量比 m_r 很小时($m_r \ll 1$)，式(28)的右边通过多项式展开且忽略高阶部分得：

$$2\varsigma_c \approx m_r \tag{29}$$

类似的，式(27)相应的临界频率 ω_c 可转换为：

$$\frac{\omega_c}{\omega_n} \approx 1 + \frac{m_r^2}{4} \tag{30}$$

人行荷载与风荷载激励具有类似的特性。旋涡的脱落对结构激励振动通过无量纲 Scruton 数计算。McRobie 和 Morgenthal 建议将同样的方法用于人行激励振动，并将竖向与侧向振动分开以便区分桥面竖向与侧向振动时行人不同的反应。行人 Scruton 数的定义形式有一定的可选性，Newland 定义行人 Scruton 数 S_{cp} 为：

$$S_{cp} = \frac{2\varsigma M}{m} \tag{31}$$

式中：ς 为模态阻尼比；M 为模态质量；对于实桥面，就是每延米的质量；m 为行人模态质量，对于人群均匀分布的桥面，为每延米的行人质量。

对于这个定义，为了使得阻尼比超过式(29)中的最小临界阻尼比，就需要：

$$S_{cp} > \alpha\beta \tag{32}$$

跟前面定义一样，α 指行人体心运动与桥面运动的比值，β是指人行同步运动的相关系数。如果取α=2/3, β=0.4 作为稳定所需的最小 Scruton 数，那么 Newland 判据要比 Dallard 判据宽松得多。因此欧洲近年的人行桥设计指南都采用偏保守的 Dallard 判据。

3.3 Piccardo 的参数共振分析法

在伦敦千禧桥开放当日的过度横向振动中，中跨振动主要成分之一为 0.48Hz 的一阶模态振动。这一步频与行人横向摆动频率范围 0.7Hz－1.2Hz 相处甚远。上述三种理论都难以对这一现象作出合理解释。Piccardo[7]为此提出了一种参数共振分析方法。

这一方法假定行人的横向摆动为简谐运动 $\cos\Omega t$，Ω为行人横向摆动圆频率($\Omega=2\pi f_s$)。因为已有实验表明桥梁振动明显地增加了行人对桥梁的横向作用力，因此动荷载系数α假定为桥梁横向振动 $q(x, t)$的线性函数$\alpha=\alpha_0+\alpha_1 q(x,t)$。于是，作用在桥梁上的每延米行人步行力为 $f_p(x, t)$：

$$f_p(x,t) = \lambda[\alpha_0 + \alpha_1 q(x,t)]g m_p(x)\cos\Omega t \tag{33}$$

式中：λ为同步比例；$m_p(x)$是每延米行人质量，一般可假定行人沿桥长均匀分布；g 为重力加速度。采用这一荷载表达式，可得到关于 j 阶模态坐标 p_j 的振动方程：

$$\ddot{p}_j(t) + 2\omega_j\zeta_j\dot{p}_j(t) + \left[\omega_j^2 - g\lambda\alpha_1\frac{M_{p_j}}{M_{s_j}}\cos\Omega t\right]p_j(t) = \left[g\lambda x_0\int_0^L m_p(x)\varphi_j(x)\,\mathrm{d}x\right]\cos\Omega t \tag{34}$$

式中：M_{p_j}、M_{s_j} 分别是 j 阶模态下的行人模态质量及桥梁模态质量。为进行参数振动分析，引入如下无量纲位移 y 和无量纲时间 $\tilde{t}$ 参数：

$$y = \frac{p_j}{L}, \quad \tilde{t} = \frac{1}{2}\Omega t \tag{35}$$

利用式(35)的定义，可以将式(34)转化为无量纲位移 y 关于无量纲时间 $\tilde{t}$ 的方程：

$$\ddot{y} + 2\eta\dot{y} + [\delta - 2\varepsilon\cos 2t]y = \kappa\cos 2\tilde{t} \tag{36}$$

式中：

$$\eta = \frac{2\omega_j}{\Omega}\zeta_j, \quad \delta = \frac{4\omega_j^2}{\Omega^2}, \quad \varepsilon = \frac{2g\alpha_1}{\Omega^2}\frac{M_{p_j}}{M_{s_j}},$$

$$\kappa = \frac{4g\lambda x_0}{\Omega^2 M_{s_j} L}\int_0^L m_p(x)\varphi_j(x)\,\mathrm{d}x \text{。} \tag{37}$$

因为行人模态质量 M_{p_j} 总是远远小于桥梁模态质量，参数 ε 可认为是一很小的量；同理因为阻尼 ζ_j 是一个很小的量，而频率比 $2\omega_j/\Omega$ 一般在 1－2 附近变化，因此 η 也可以认为是与 ε 同阶的小量，记为：

$$\eta = \varepsilon\tilde{\eta} \tag{38}$$

将式(38)代入式(37)后可以得到：

$$\ddot{y} + 2\varepsilon\tilde{\eta}\dot{y} + [\delta - 2\varepsilon\cos 2\tilde{t}]y = \kappa\cos 2\tilde{t} \tag{39}$$

式(39)是经典的 Mathieu 型参数振动方程。依据参数的不同，式(39)的解可能是稳定的或者不稳定的。参数空间($\delta, \varepsilon, \eta$)可以用临界曲线(transition curves)区分为稳定区与非稳定区，而位于临界线上的参数可使式(39)有周期解，周期为 π 或 2π。这一临界曲线可以使用摄动法获得解析解。图4绘出了一阶近似下临界曲线随阻尼比变化趋势：在δ=1 附近，阻尼比影响很大，大的阻尼比导致大的$\varepsilon_{\lim}$值，但是偏离δ=1 后，阻尼影响迅速减小。

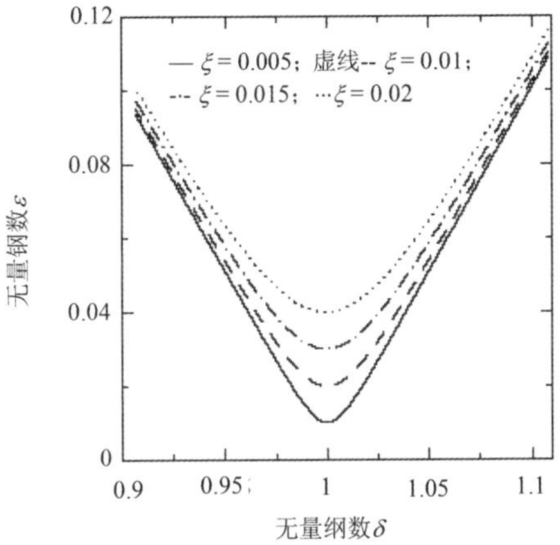

图 4　不同阻尼比下的 $\varepsilon_{\lim}$ 的一阶近似曲线

Fig.4　First-order approximation of $\varepsilon_{\lim}$ for varying damping ratio

由ε与δ的定义式(38)可以看出，去掉常数项，ε是同步人群模态质量($\lambda\alpha$, M_{pj})与桥梁模态质量之比，而δ是桥梁模态频率ω_j的 2 倍与行人步频 Ω 的比。由于伦敦千禧桥的一阶横向频率为 0.48Hz，而行人横向摆动频率接近 1Hz，所以参数共振条件得以满足，该桥出现了 0.5Hz 的大幅横向振动。用上述理论，Piccardo 对伦敦千禧桥过度振动现象作了分析。计算时所取的参数为桥质量 m_s=2000kg/m，行人平均质量 m_{ps}=70kg，α_1=2/m，λ=0.3(即有 30%的行人同步)，桥梁一阶横向振动模态频率ω_j=2πn_1 (n_1=0.48Hz)，阻尼比分别取 0.005、0.007 和 0.01 三种情况。显然，最不利情况为 n=0.96 从而δ=1 的情况，对应不同阻尼的临界行人数分别为 127、178 和 254。其中阻尼比为 0.007 为实桥实测阻尼值，而对应的临界人数 178 也正好与 Dallard 在伦敦千禧桥上组织的试验吻合。

3.4　桥上人群移动的可压缩流体模型

意大利的 Venuti 等人[8]在前人研究的基础上，将可压缩流体模型应用到人群与结构相互作用问题，提出了人群-结构相互作用的一种数学模型，包括人群动力学模型的模拟。这个模型是为了再现复杂多物理的非线性动力系统，一方面给出沿主梁人群流动的一个详细描述，另一方面，获得对工程师和设计者有用的综合结果。该模型的主要特点是把一个耦合系统通过数学和数值分解为两个物理子系统并考虑了两个子系统之间的两种相互作用方式。在水动力学建模框架中，基于质量守恒方程的一阶模型被应用到宏观描述人群动力学模型。结构系统通过一个广义单自由度模型来模拟。

4　某人行曲线斜拉桥的动力设计与振动控制

以国内在建的某人行曲线斜拉桥为例，本节阐述了人行桥在步行荷载作用下的动力设计方法。该桥为全长 400m 的三跨双层斜拉桥，上层为 S 形人行桥，下层为双向六车道车行桥，跨径布置为 100m+200m+100m。倒 Y 型桥塔塔高 98m，主体结构高 84.5m，顶部装饰高 13.5m；车行桥主梁采用大悬臂预应力混凝土箱梁，现场悬浇施工。人行桥主梁采用钢箱梁，采用分段吊装施工，梁宽 6m，高 2m。斜拉索为单索面，全桥共有 68 根拉索(其中车行桥 52 根，人行桥 16 根)。设计荷载等级为公路Ⅰ级，图 5 给出了人行桥和车行桥的效果图。

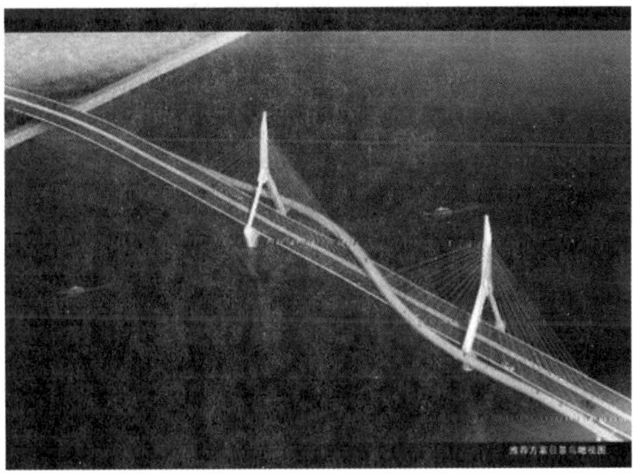

图 5　人、车行桥整体效果图

Fig.5　Artist impression of pedestrian and traffic road bridge

4.1　动力设计基本过程

由于人的荷载具有显著的周期特性和卓越频率，人行桥振动本质上是桥梁在行人简谐步行荷载作用下的多阶动力谐振响应问题。人行桥动力设计的基本过程如下[9]：

1) 第 1 步：确定敏感频率范围。结合最新德国和法国规范的敏感频率范围评价准则。依据这一准则，竖向振动及纵向振动的敏感频率范围为 1.60Hz－2.40Hz，横向振动的敏感频率范围为 0.50Hz－1.20Hz。当人行桥的自振频率不在上述敏感频率范围时，认为人行桥的人致振动问题自然满足要求。否则，需要检算人行桥在人群步行荷载作用下的加速度响应是否满足行走舒适性条件。

2) 第2步：人致动力响应分析。考虑到很多人行桥的竖向振动频率位于振动敏感区间，通过频率调整来满足人行桥的振动舒适性要求很难做到。因此第2步是通过限制动力响应值来满足行走舒适性要求。通常采用单自由度(模态)模型或者有限元三维模型，以单人和人群共振荷载情况下结构所产生的最大加速度响应来评估其振动舒适性。若最大加速度响应小于行人能够容忍的最大加速度指标，则认为人行桥的人致振动问题满足要求。否则，应该对结构进行修改或者安装阻尼减振设备。本文采用的舒适性指标见表4。

表4　人行桥的加速度舒适性指标

Table 4　Acceleration comfort class for footbridges

舒适性类别	舒适度	竖向加速度限值/(m/s^2)	侧向加速度限值/(m/s^2)
CL1	最好	<0.50	<0.10
CL2	中等	0.50－1.00	0.10－0.30
CL3	差	1.00－2.50	0.30－0.80
CL4	不可接受	>2.50	>0.80

3) 第3步：侧向同步动力失稳分析。当横向的振动加速度大于 $0.10m/s^2－0.15m/s^2$ 时，更多的人可能自然地调整步伐与桥梁振动频率一致，而进一步加剧侧向振动的程度，这一现象称为侧向同步动力失稳。所以，对于频率小于1.2Hz的侧向振动振型，还需要确定发生侧向同步动力失稳的临界行人数是否大于设计行人数，必要时增加减振措施。

4) 第4步：修改结构或安装阻尼减振设备。通过修改结构或安装阻尼减振装置，直到结构振动频率或最大加速度响应满足行走舒适性要求。

4.2　原设计方案的人致振动分析

在该桥的原设计方案中，人行桥与车行桥两桥共用桥塔，其它部位都无连接。

采用ANSYS 11.0建立全桥有限元模型，并进行了模态分析。主要结果如下：

1) 人行桥横向振动的一阶频率为 0.4588Hz，振型特点是主跨正对称侧弯。二阶频率为0.7314Hz，振型特点为主跨反对称侧弯同对称竖弯。且位于横向敏感频率范围，需要计算横向共振下的最大加速度响应及引起横向动力失稳的临界行人数。

2) 人行桥竖向振动的一阶频率为 0.4684Hz，振型特点是主跨对称竖弯。二阶频率为0.7117Hz，振型特点为主跨反对称竖弯。竖向振动频率密集。第14阶、15阶、17阶、18阶、19阶和20阶模态为人行桥的竖向振动模态，频率分别为1.620Hz、1.678Hz、1.832Hz、1.960Hz、2.392Hz和2.410Hz，正好在人行步频范围内，共振可能性也是非常大的。

3) 对只有人行桥参与振动的振型，广义质量小，阻尼也低，发生共振时容易发生大幅振动。

本桥的一阶侧向频率跟伦敦千禧桥很接近，因此极容易发生大幅侧向振动。但主梁(包括桥面等)恒载达到48.4kN/m，是千禧桥的一倍以上。因此可以推测发生大幅振动的行人数比千禧桥要高。本桥的特点是曲线梁桥，竖向振型总是伴随着横向振动，有时耦合非常严重。因此抑制横向振动和竖向振动同等重要，而且需要抑制振动的频率范围很宽。

假定结构的模态阻尼比为0.5%。以自由行走时的行人等效动荷载，计算了各模态分别共振时的最大横向和竖向加速度，得到最大竖向和最大侧向加速度分别为 $2.776m/s^2$ 和 $1.463m/s^2$，远远超过附录给出的最佳舒适加速度。对于第2阶、第5阶振型，还分别按照Dallard及Newland方法计算了当前阻尼比条件下的横向稳定临界人数，两套计算方法得到的临界人数存在很大差别。Newland方法的计算结果与系数α、系数β有关；按照Newland的建议，α=2/3, β=0.4对应临界人数的最大值，α=1, β=1对应临界人数的最小值。相对于Newland下限线(α=2/3, β=0.4)，则Dallard方法总体偏于保守。本文按照Dallard公式取值，得到第2阶及第5阶模态的临界人数分别为220人和1100人，如要在设计人群数3600时(动力设计人群密度1.5人/m^2)不发生横向动力失稳，模态阻尼比需要分别提高16.3倍和3.5倍。

所以，同时从行走舒适度和横向动力失稳考虑，第2阶、5阶、14阶、15阶、17阶、18阶、19阶、20阶模态的阻尼比分别要达到8.5%、2.0%、6.0%、7.5%、1.0%、3.0%、2.0%和3.0%以上。为达到上述阻尼要求，分别针对各敏感模态设计了TMD减振装置[10]，所有TMD装置中质量块重量共为104.1t。虽然附加给结构的重量相对较小，但减振系统的总费用较大。万一TMD失效，很少的行人即可能产生较大的振动。

4.3 设置连杆的人行桥的动力设计

针对本桥人行桥缠绕车行桥的特点，研究了在人-车行桥间设置连杆来提高整体刚度。通过比较分析，最后确定在中跨设置一对竖杆、一对斜杆、在两边跨设置一对水平杆的方案。连杆为直径 200mm 壁厚 20mm 的圆钢管。

对设置连杆的人行桥进行了动力特性分析。设置连杆后，第 4 阶模态频率(0.9860Hz)位于横向敏感频率范围，第 11 阶、13 阶和 14 阶模态频率分别为 1.647Hz、1.885Hz 和 2.054Hz，位于竖向敏感频率范围。还需要对设置连杆的人行桥进行人致振动分析及减振设计。

同样假定各模态阻尼比为 0.5%，对各模态共振时计算了横向和竖向加速度响应，并对第 4 阶模态验算了横向稳定临界人数。根据分析，同时满足横向动力稳定及行走最好舒适度，第 4 阶、11 阶、13阶和 14 阶模态的阻尼比分别要达到 4.93%、4.0%、1.0%和 2.5%以上。

根据设置连杆后人行桥的人致振动分析结果，针对第 4 阶、11 阶、13 阶和 14 阶模态设计了质量调谐阻尼器的减振方案。同样采用调谐质量阻尼器减振方案，结果见表 5。所有 TMD 装置中质量块重共为 42.49t，较原设计方案的减振系统的附加质量减少了约 60%。减振系统和连杆系统的总成本可较原方案降低一半。

表 5　控制各模态的 TMD 装置参数及设置位置

Table 5　Model parameters and installation location for all tuned mass dampers

模态号	位置*	TMD 运动方向	质量/t	刚度/(kN/m)	阻尼系数/(kN·s/m)
4	主跨跨中	水平	10.96	400.44	12.67
11	160m 处	竖向	14.16	1444.51	27.36
13	主跨跨中	竖向	13.31	1830.13	19.02
14	主跨跨中	竖向	4.06	649.38	8.80

*：以左桥塔、右桥塔的坐标分别为-100m 和 100m 为参考。

5　结论

与欧洲与日本等发达国家相比，我国的大跨人行桥建设刚刚起步。当前我国正处于城市化建设高潮之中，人行桥建设日益增多，在建人行桥最大跨度已达到 200m。而国内现有人行桥的规范要求竖向频率大于 3.0Hz，对轻柔的大跨人行桥很难做到；对更为严重的横向振动稳定问题没有涉及。国内现有规范已难以满足目前大跨人行桥建设的需要。本文介绍了人行桥的人致振动理论及动力设计方法，重点阐述了大幅横向振动分析的直接共振理论、人桥相互作用理论及内共振理论。以国内一曲线斜拉桥为例，阐述了人行桥动力设计的全过程。分析表明，不同方法得到的横向动力稳定临界人数还存在较大差别，离评估实际响应还有一定距离。

参考文献：

[1] Fujino Y, Pacheco M, Nakamura S. Synchronization of human walking observed during lateral vibration of a congested pedestrian bridge [J]. Earthquake Engineering and Structural Dynamics, 1993, 22: 741－758.

[2] Dallard P, Fitzpatrick A J, Flint A, Be Bourva S, Low A, Ridsdill R M, Willford M. The London Millennium footbridge [J]. The Structural Engineer, 2001, 79(22): 17－33.

[3] Nakamaura K. Lateral vibration of footbridge by synchronised walking [J]. Journal of Constructional Steel Research, 2006, 62: 1148－1160.

[4] 袁旭斌, 孙利民. 人行桥人致振动特性研究[D]. 上海: 同济大学, 2006.
Yuan Xubing, Sun Limin. Pedestrian-induced vibration characteristics of footbridges [D]. Shanghai: Tongji University, 2006. (in Chinese)

[5] Roberts T M. Lateral pedestrian excitation of footbridge [J]. Journal of Bridge Engineering, ASCE, 2005, 9(2): 107－112.

[6] Newland D E. Pedestrian excitation of bridges [J]. Mechanical Engineering Science, 2004, 218: 477－492.

[7] Piccardo G, Tubino F. Parametric resonance of flexible footbridge under crowd-induced lateral vibration [J]. Journal of Sound and Vibration, 2008, 311: 353－371.

[8] Venuti F F, Bruno L, Bellomo N. Crowd dynamic on a moving platform [J]. Mathematical and Computer Modelling, 2008, 45: 252－269.

[9] Hivoss. Guide line for design for footbridges [R]. Research Report, RFS2-CT-2007-0033, 2007: 1－33.

[10] Soong T T, Gargush G F. Passive energy dissipation systems in structural engineering [M]. John Wiley. New York, 1996.

注：该文在第 18 届结构工程学术会议(2009 广州)应邀作特邀报告

Control of Human-Induced Vibrations of a Curved Cable-Stayed Bridge: Design, Implementation, and Field Validation

Q. Wen[1]; X. G. Hua[2]; Z. Q. Chen, M.ASCE[3]; Y. Yang[4]; and H. W. Niu[5]

Abstract: This article describes the research work relating to the assessment and control of human-induced vibration of an unusual curved cable-stayed bridge with separate road and pedestrian decks. Dynamic simulations of human-induced vibration were performed with a mode-by-mode approach, and the results showed that a total of eight lateral and vertical modes of the bridge may suffer from excessive vibrations at the design crowd density. A hybrid control scheme was developed for control of human-induced vibration, which consists of steel braces fitted between the road bridge deck and the pedestrian deck to improve structural stiffness and eddy current tuned mass dampers to enhance damping ratios of lively modes. The modal properties of the bridge with steel braces were experimentally obtained by dynamic tests and used for parameter tuning of mass dampers. The performance of the hybrid control strategy was evaluated by group tests of up to 400 pedestrians. It is shown that the critical number of pedestrians triggering synchronous lateral excitation is generally in good agreement with the empirical model developed from measurements on the London Millennium Bridge, but the measured vertical acceleration amplitude is only about 40% of the prediction under the same pedestrian density. The damping ratios for both the vertical and lateral modes increase appreciably after installation of tuned mass dampers and no evidence of large-amplitude vibrations has been observed. **DOI: 10.1061/(ASCE)BE.1943-5592.0000887.**

Author keywords: Footbridge; Vibration; Modal identification; Damping; Vibration control; Tuned mass dampers.

Introduction

The motion of walking pedestrians exerts dynamic forces on footbridges that have vertical, lateral, and longitudinal components. The dominant forcing frequencies are 2 and 1 Hz in vertical and lateral directions, respectively, although they vary slightly depending on the individuals and on the purpose for walking. If a footbridge has a natural frequency close to the forcing frequencies, resonance can potentially be induced by pedestrians. Because of the developments in building materials, construction techniques, and design methods accompanied by aesthetic requirements for greater slenderness, the modern footbridges tend to have longer spans, are lighter, and have less damping. Accordingly, they are more susceptible to vibrations under dynamic loading caused by walking pedestrians (Bachmann 1992; Zivanovic et al. 2005; Chen and Hua 2009).

The pedestrian-induced vibrations, particularly in the lateral direction, have received wide attention in the last few decades from the earliest cases in the 1990s up to the well-known case of the London Millennium Bridge (Fujino et al. 1993; Dallard et al. 2001a, b). These vibrations often cause discomfort for the users and the temporary closure of the footbridges. Therefore, they have attracted extensive research emphasizing modeling and measurements of pedestrian-induced forces and analysis of bridge-pedestrian interactions (Zivanovic et al. 2005; Ricciardelli and Pizzimenti 2007; Racic et al. 2009; Ingolfsson et al. 2012; Racic et al. 2013).

Supplementary damping through the incorporation of energy dissipation devices into footbridges has been a popular solution to the undesirable resonant vibrations. Two typical energy dissipation devices are fluid viscous dampers and tuned mass dampers (TMDs) (Soong and Dargush 1997). The first type of damper relies on the relative movements between the two damper ends and is therefore connected between two locations of a structure. The TMDs are inertial devices attached to a single location of a structure. Fujino et al. (1993) reported the application of tuned liquid dampers for control of a lateral human-induced vibration of a cable-stayed T-bridge. The damping system for the London Millennium Bridge consists of fluid viscous dampers and TMDs for both lateral and vertical vibrations (Dallard et al. 2001a). After that, the TMDs have been widely used for the control of human-induced vibration of long-span footbridges (Occhiuzzi et al. 2008; Casado et al. 2011; Bruno et al. 2012). Caetano et al. (2010a, b) have discussed a number of practical issues relating to the implementation of TMDs for vibration control of the Pedro Ines Footbridge.

[1]Ph.D. Candidate, Key Laboratory for Wind and Bridge Engineering of Hunan Province, College of Civil Engineering, Hunan Univ., Hunan 410082, China.

[2]Professor, Key Laboratory for Wind and Bridge Engineering of Hunan Province, College of Civil Engineering, Hunan Univ., Hunan 410082, China (corresponding author). E-mail: cexghua@hotmail.com; cexghua@hnu.edu.cn

[3]Professor, Key Laboratory for Wind and Bridge Engineering of Hunan Province, College of Civil Engineering, Hunan Univ., Hunan 410082, China. E-mail: zqchen@hnu.edu.cn

[4]Professor of Engineering and Vice Dean, Guangzhou Municipal Engineering Design & Research Institute, Guangzhou 510000, China. E-mail: mail_yyy@163.com

[5]Senior Engineer, Key Laboratory for Wind and Bridge Engineering of Hunan Province, College of Civil Engineering, Hunan Univ., Hunan 410082, China. E-mail: niuhw_hd@126.com

Note. This manuscript was submitted on May 19, 2015; approved on November 4, 2015; published online on February 3, 2016. Discussion period open until July 3, 2016; separate discussions must be submitted for individual papers. This paper is part of the ***Journal of Bridge Engineering***, © ASCE, ISSN 1084-0702.

The construction of a curved cable-stayed bridge with a central span of 200 m in Mianyang City, China, has motivated the dynamic assessment of pedestrian-induced vibrations and the development of a vibration control strategy. The distinctive feature of the bridge is its separate deck systems—the straight lower deck arranged for the vehicles and the curved upper deck for pedestrians. This special configuration allows for the development of a hybrid control system for mitigating the human-induced vibrations, which consists of steel braces fitted between the road bridge deck and the footbridge deck for improving structural stiffness and TMDs for enhancing the damping ratio of lively modes. This article describes the numerical and experimental studies for the assessment and control of human-induced vibration of the footbridge using eddy current TMDs. The internal friction on the performance of eddy current TMDs was also discussed in some detail.

Description of the Mianyang First Bridge

The Mianyang First Bridge is located across the Fujiang in the Chengnan New District of Mianyang, China. It is a cable-stayed bridge with a central span of 200 m and two side spans of 100 m each. To meet the demands of high pedestrian density at this leisure area, two separate deck systems are considered for vehicles and pedestrians in the bridge design. Therefore, the bridge may be regarded as two cable-stayed bridges sharing the same pylons. The road bridge deck for vehicles has a concrete box girder 28 m wide and 3 m high, and the footbridge deck for pedestrians has a steel box girder 6 m wide and 2 m high. The footbridge deck is located above the road bridge deck in elevation, and it roughly follows a shallow letter S in the plan, such that it passes through the road bridge from one side to the opposite side in the central span. The bridge decks resting on the cross beams of pylons are supported by their individual cable systems. The inverted-Y pylons are made of steel at the anchorage zone and of concrete at the remaining parts. The longitudinal movement between the girder and tower is allowed to accommodate temperature variation, traffic, and wind loading effects, whereas the other two translational movements are restrained. Fig. 1 shows the general view of the Mianyang first cable-stayed bridge. Fig. 2 shows the elevation and plan view of the bridge.

Dynamic Assessment

Because of its large span, human-induced vibrations have been taken into consideration very early in the design process. A finite-element model is established for analyzing the modal properties and human-induced vibrations. In this model, the bridge girders and pylons are represented by the three-dimensional Timoshenko beam elements, and stay cables are modeled with the two-node cable elements. The mass of handrails and pavement on the decks are considered in finite-element modeling, but their stiffness is neglected. The soil-structure interaction of pylons is not included, because the pylons are built on the massive reinforced slabs. The material properties and cross-sectional parameters of these elements are taken from design drawings.

Modal analysis is performed to identify the lively modes easily excited by pedestrians for the lateral and vertical directions. The unity-scaled mode shape is used for calculating the modal mass (effective vibrational mass). It is clear that the modes with small modal mass are most likely to be excited by walking pedestrians. Fig. 3 presents the most relevant modes within the vertical and lateral forcing frequency ranges, in which mode shapes are dominated by the vibrations in the footbridge deck. No torsional vibration is found in the forcing frequency ranges due to the large torsional stiffness of the closed steel box girder. It is seen that the S-curved shape of the footbridge deck leads to the presence of a number of coupled vibration modes in lateral and vertical directions. This implies that the vertical pedestrian loading can also result in the lateral vibrations for coupled modes and vice visa. The natural frequency for the first lateral mode is 0.599 Hz, which falls into the sensitive frequency range of 0.5–1.2 Hz for synchronous lateral excitation, and the corresponding modal mass is 496 t.

Fig. 1. General view of Mianyang First Bridge (images by X. G. Hua)

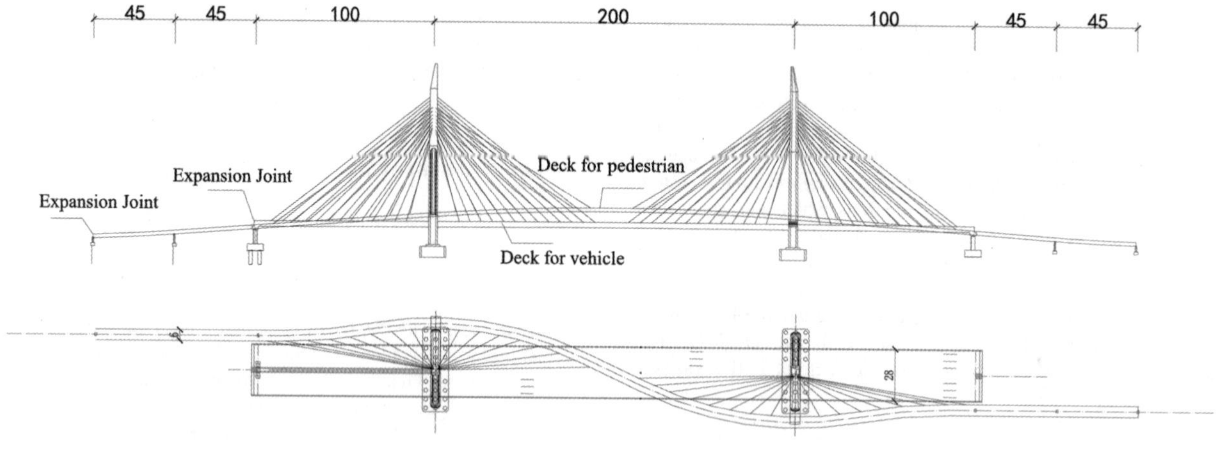

Fig. 2. Elevation and plan view of the bridge (unit: m)

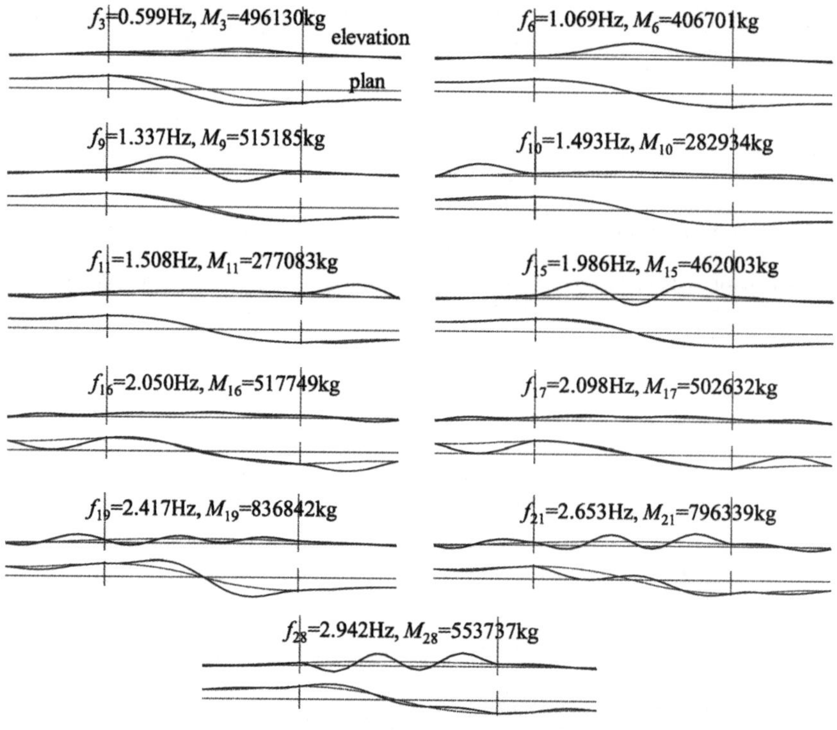

Fig. 3. Mode shapes of footbridge deck

As a comparison, the natural frequency and modal mass for the first lateral mode of the London Millennium Bridge are 0.5 Hz and 130 t (Dallard et al. 2001b), and those for the Pedro Ines Footbridge with a central span of 110 m are 0.86 Hz and 202 t (Caetano et al. 2010b). The vibrational mass per unit length in the central span is about 2.48 t/m, which is about 35% larger than that of the Pedro Ines Footbridge. Additionally, the width of the footbridge deck is larger than that of the Pedro Ines Footbridge; therefore, the footbridge is likely to be excited by the synchronous lateral excitation as observed in the London Millennium Bridge and the Pedro Ines Footbridge.

A simplified dynamic simulation with a mode-by-mode approach is performed to evaluate the maximum vertical and lateral acceleration responses. Because a university and many residential districts are located around the footbridge, an exceptional pedestrian density of 1.0 person per square meter is considered. The total length and width of the footbridge deck is 400 and 6 m, thus the number of pedestrians, n, at the design density is 2,400. This number of pedestrians with random steps is used to define an equivalent number of synchronous pedestrians with the same step frequency, which is $n' = 91$ ($=1.85 \times n^{1/2}$) based on recommendations in the design guideline for footbridges (HiVoSS 2008). A damping ratio

Table 1. Human-Induced Vertical and Lateral Acceleration Amplitude for Each Mode

Mode number	Natural frequency (Hz)	Modal mass (tons)	Vertical acceleration (m/s^2)	Lateral acceleration (m/s^2)
3	0.599	496	0.047	0.184
6	1.069	407	1.363	0.051
9	1.337	515	1.365	0.343
10	1.493	283	1.641	0.383
11	1.508	277	1.647	0.346
15	1.986	462	1.566	0.218
16	2.050	518	0.096	0.379
17	2.098	503	0.116	0.436

of 0.5% is adopted for every mode. By using the modal analysis of a single degree of freedom (SDOF), the generalized force and acceleration amplitudes for a given mode of vibration and uniform pedestrian density can be obtained as

$$q_i = \frac{P_i^*}{M_i} \times \frac{1}{2\zeta_i} \tag{1}$$

$$P_i^* = \frac{n'F}{L}\int_0^L \varphi_i(x)dx \tag{2}$$

where P_i^* = generalized force for the equivalent synchronous pedestrians; M_i and ζ_i = modal mass and damping ratio for the ith mode, respectively; and F = vertical or lateral forcing component for a single pedestrian, which is 35 N in the lateral direction for the first mode and 280 N in the vertical direction for the remaining modes (HiVoSS 2008).

Table 1 gives the maximum vertical and lateral resonance acceleration amplitudes for each mode with a frequency below 2.4 Hz. The thresholds of vertical and lateral accelerations for user comfort are taken as 0.5 and 0.1 m/s^2, respectively (HiVoSS 2008). It is seen that the vertical or lateral acceleration amplitudes for all of the eight modes exceeds the previous levels under the design pedestrian density. For Coupled Modes 9–11 and 15, vertical and lateral vibrations present simultaneously when any of these modes is excited by vertical pedestrian loading, as shown in Table 1. For Coupled Modes 16 and 17, it is observed that the pedestrian loading in vertical directions also results in large-amplitude lateral vibrations. A preliminary examination shows that a least damping ratio of 2.0–3.0% is required for these modes.

It should be noted that the synchronous lateral excitation for the lateral modes, also known as the pedestrian lateral lock-in, should be avoided as well. The synchronous lateral excitations are involved in the synchronization of footstep frequency of pedestrians with the vibration frequency of the bridge deck. It is characterized by a sudden buildup of responses, and the amplitudes achieved are greater than that calculated from the resonance model given by Eqs. (1) and (2). This phenomenon is not fully understood, and has attracted a great deal of research work with an emphasis on the discussion of excitation mechanisms and the development of appropriate mathematical models (Newland 2004; Blekherman 2005; Piccardo and Tubino 2008; Macdonald 2009). Based on the empirical model derived from the measurements on the London Millennium Bridge, the number of pedestrians, N_L, triggering pedestrian lateral lock-in can be estimated by (Dallard et al. 2001a)

$$N_L = \frac{4\pi\zeta Mf}{k}\frac{L}{\int_0^L \phi^2 dx} \tag{3}$$

where M, ζ, and f = modal mass, damping ratio, and natural frequency for the first lateral mode, respectively; k = empirical constant of 300 N/(m · s^{-1}) per person based on measurement data on the London Millennium Bridge; and the integral in the denominator is determined from the unity-scaled mode shape for the first lateral mode. With the empirical model, the critical pedestrian number for triggering pedestrian lateral lock-in is 120 for the main span, or equivalently 240 for the whole span when the damping ratio for the first lateral mode is taken as 0.5%. That corresponds to a pedestrian density of 0.1 person per square meter, and a minimum modal damping ratio of 5% is expected to avoid lateral lock-in at the pedestrian density of 1.0 person per square meter.

Design of Vibration Control System

A traditional approach to footbridge vibration control is the use of TMDs. There are a total of eight modes to be controlled, and each mode will require an independent TMD. In taking the mass ratio as 2% for vertical modes and 3% for lateral modes, the total mass for all the mass dampers would amount to 100 t. This would be prohibitive due to high cost and required maintenance of the TMD system.

A hybrid control scheme of the stiffness upgrade and damping enhancement is adopted for the present footbridge. Three pairs of steel braces are used to connect the road bridge deck and footbridge deck at the particular locations in the central span and both side spans, which improve the vertical and lateral stiffness of the bridge and reduce the number of lively modes. Then, the eddy current TMDs will be used to control the lively modes. Fig. 4 illustrates the layout of the steel brace system. A pair of vertical braces and a pair of inclined braces are installed at the central span, and a pair of horizontal braces is installed at the side spans. Fig. 5 shows the prototype of the vertical and inclined steel braces, which is located outside the concrete crash barrier of the road bridge deck. Each brace is comprised of a steel pipe with a diameter of 426 mm and a thickness of 8 mm and a high damping rubber pad 80 mm thick at both ends. The rubber pad is used in place of the traditional fixed steel joints, such that bending moments in the three directions are released, and it is also helpful to isolate the vibrations of the road bridge deck induced by traffic. The steel brace is prestressed using the high-strength strands at both ends to guarantee the rubber pads are in compression. The brace is modeled as a series system of steel pipes and two 80-mm-thick rubber pads to account for the limited axial stiffness of rubber pads. The equivalent axial stiffness is calculated as 16.6 MN/m, which is only a fraction of the stiffness of steel pipe.

The three pairs of steel braces are represented by a truss element with the equivalent axial stiffness and are incorporated in the finite-element model. The modal frequencies and modal masses of the bridge with steel braces are listed in Table 2. The stiffness of the footbridge is increased, and only five modes fall into the forcing frequency ranges. Furthermore, the critical pedestrian number for lateral lock-in will also increase with modal frequency for the lateral mode. Because the steel braces act like elastic supports for the footbridge deck, the mode shapes are similar to those of the bridge without steel braces.

As the minimum damping ratios for avoiding lateral lock-in for the lateral mode are higher than those for vertical modes, a mass ratio of 3.0% is used for the lateral mode and a mass ratio of 2% is used for the vertical modes. After determining the mass ratios, the optimum parameters for each TMD can be determined from the

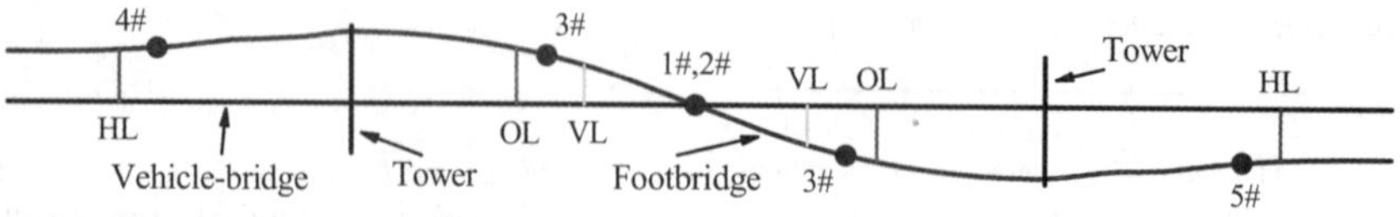

Fig. 4. Layout of steel braces; HL, horizontal; OL, oblique; VL vertical braces

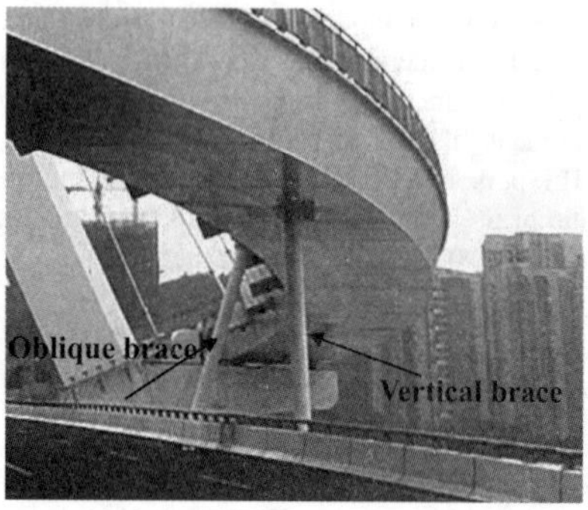

Fig. 5. Prototype of vertical and oblique braces at the central span (image by X. G. Hua)

classical theory (Soong and Dargush 1997), that is, $f_{TMD}/f_s = (1+\mu)^{-1}$ and $\zeta_{TMD} = [3\mu/8(1+\mu)]^{-1/2}$. The results are provided in Table 3.

Table 2. Main Modal Properties and Human-Induced Vibration of the Bridge with Steel Braces

Mode number	Natural frequency (Hz)	Modal mass (t)	Vertical acceleration (m/s^2)	Lateral acceleration (m/s^2)
4	0.735	470	0.01	0.190
8	1.435	242	1.692	0.032
9	1.502	289	1.741	0.271
10	1.516	283	1.752	0.245
14	1.974	603	1.289	0.130

Table 3. Optimum Parameters of TMD for Each Mode at Design Stage

Mode number	Weight sliding mass (t)	Frequency (Hz)	Damping ratio (%)	Direction	Locations
4	14.0	0.714	10.5	Horizontal	1/2 main span
8	5.0	1.406	8.6	Vertical	1/2 main span
9	6.0	1.472	8.6	Vertical	1/2 side span
10	6.0	1.486	8.6	Vertical	1/2 side span
14	12.0	1.974	8.6	Vertical	1/4 main span

Experimental Assessment of Dynamic Behavior

Performing actual tests on full-scale structures is the most reliable way to determine the modal properties of major importance, such as natural frequencies, mode shapes, and damping values. A series of dynamic tests are performed to identify modal properties of the bridge for fine-tuning of tuned mass dampers and to evaluate the lateral and vertical vibration responses to different pedestrian streams. These tests include the ambient vibration test, free vibration test, forced vibration test (FVT), and human-induced vibration test. Ambient vibration tests relying on natural wind and vehicle excitation provide a simple and most economical way for dynamic tests, because the only equipment required during the test is for data acquisition. The modal frequencies and mode shapes can be extracted with good accuracy from ambient vibration measurements; however, the damping estimates are highly scattered. In contrast to ambient vibration tests, free vibration tests can be performed by suddenly releasing the bridge from rests, and they usually produce improved estimation of damping values. Forced vibration tests are accurate but costly, because they require very heavy and expensive equipment, such as servohydraulic shakers or eccentric mass vibrators, for driving the large-scale and massive structures into steady-state resonance. Because measurements of both input and output are made for FVTs, *frequency response functions* (FRFs) can be derived and complete modal characteristics, including modal frequencies, mode shapes, damping ratios, and modal mass, will be obtained. The determination of modal mass is of particular importance for TMD design (Pavic et al. 2002).

The field dynamic tests were performed before and after installation of tuned mass dampers. The first period of dynamic testing was conducted from March 7 to 11 in 2013 shortly after installation of three pairs of steel braces between the road bridge deck and the footbridge deck; however, there are dozens of tons of temporary construction materials on the footbridge deck, which had some influence on the natural frequencies of the bridge. The dynamic tests conducted during this period were mainly ambient vibration and human-induced vibrations, and the derived results were used for the detailed parameter design of the TMDs as well as for verifying the effectiveness of the steel braces. After the final installation of TMDs, the second period of dynamic tests was performed from September 24 to 30 in 2013. Apart from performing ambient vibration and human-induced vibration tests, FVTs by means of a horizontal shaker were also used for the lateral mode. During the tests, about five to six low-frequency accelerometers (model DH610, 0.3 $V/m \cdot s^{-2}$) and a DH5937 data acquisition system (DongHua Testing Technology, Co. Ltd., Suzhou, Jiangsu, China) were used. The vertical responses were measured at 1/2 and 1/4 of the central span and 1/2 of the side spans, whereas lateral responses were measured only at the middle of central span. The sampling frequency was 100 Hz. The wind velocity during the tests was about 3–5 m/s.

Modal Parameter Identification

Many techniques are available for extracting modal properties from small-amplitude ambient vibration responses of a structure in frequency or time domains. In this study, the methods used for ambient vibration tests are the peak-picking method (PP) (Bendat and Piersol 1993), enhanced frequency domain decomposition (EFDD) (Brincker et al. 2001), and stochastic subspace identification method (SSI) (van Overschee and de Moor 1996). Fig. 6 shows an example of auto power spectral densities (PSDs) of selected hourly acceleration measurements. It is observed that the vertical acceleration

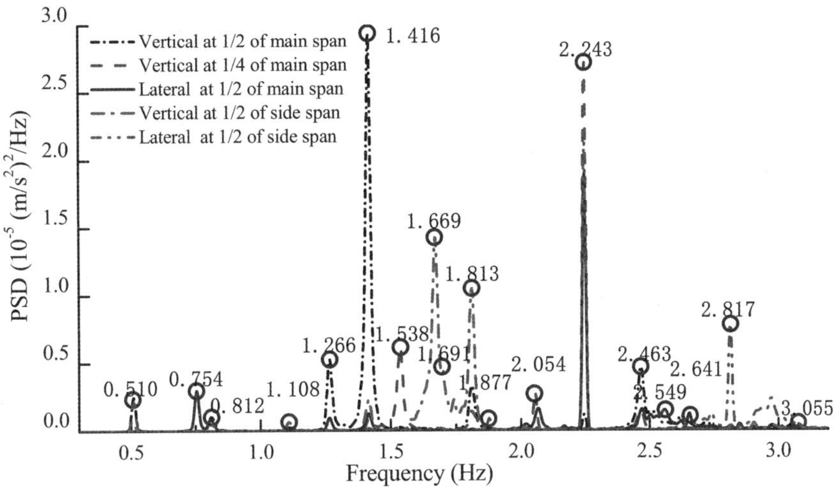

Fig. 6. Auto power spectral densities of vertical and lateral accelerations at the footbridge deck

Table 4. Summary of Identified Modal Frequencies and Damping Ratio during the First Period Tests

	Finite-element prediction (with steel braces)			PP	EFDD		SSI		
Mode	f(Hz)	Modal mass (t)	Mode shape	f(Hz)	f(Hz)	ζ (%)	f(Hz)	ζ (%)	MAC
2	0.472	8379	V-S	0.510	0.509	0.92	0.510	0.52	0.98
3	0.714	16125	V-A	0.812	0.806	0.68	0.814	0.75	0.90
4	0.734	470	L-S	0.754	0.754	0.74	0.754	0.60	0.99
5	1.033	9300	V-S	1.266	1.267	0.71	1.267	0.50	—
6	1.136	6318	L-S	1.108	1.108	0.60	1.109	0.44	—
7	1.222	11453	V-A	1.538	1.540	0.48	1.536	0.77	0.85
8	1.435	242	V-S	1.416	1.418	0.38	1.414	0.37	0.99
9	1.502	289	V-S (side)	1.669	1.669	\	1.664	1.10	0.94
10	1.516	283	V-S (side)	1.691	1.718	\	1.724	2.55	0.87
11	1.589	13291	V-S	1.813	1.810	0.57	1.810	0.41	0.96
12	1.811	5858	V-A	1.877	1.878	0.18	1.862	1.57	0.97
13	1.842	7203	V-S	2.054	2.053	0.13	2.051	0.75	0.86
14	1.975	603	V-A	2.243	2.243	0.14	2.240	0.12	0.91

Note: A = asymmetric; L = lateral; MAC = modal assurance criterion; S = symmetric; V = vertical.

and lateral acceleration have the same frequencies in the peaks of auto power spectra, which suggests the intramodal interaction between lateral and vertical directions as also found in finite-element modal analysis. Additionally, unlike the traditional footbridges in which the lower order modes usually have higher peaks in the auto power spectra, the present case exhibits the presence of higher peaks for several higher order modes with small modal mass. Compared with the analytical predictions, Table 4 summarizes the modal frequencies and damping ratios identified with the previous three methods, and the value of modal assurance criterion (MAC) between analytical and experimental mode shapes is also provided. It is shown that three methods give the consistent results of the identified modal frequencies; however, the damping ratio estimation is considerably scattered, especially for some higher order modes. Most experimentally obtained modal frequencies are larger than the analytical values. In ambient vibration tests, the small-amplitude vibration is such that the longitudinal dampers installed between girder and tower cross beams behave like a fixed support. The stiffness of the composite concrete-steel section for the towers may be less accurately reproduced in the finite-element model. These may explain the observed discrepancy to some extent. For the footbridge vibration modes without participation from the vehicle bridge, the modal frequencies are reliably identified because of their large signal-to-noise level.

Table 5. Selected Experimental Modal Frequencies and Damping during the Second Period Tests

	PP	EFDD	SSI		FVT	
Mode	f(Hz)	f(Hz)	f(Hz)	ζ (%)	f(Hz)	ζ (%)
4	0.722	0.722	0.715	0.76	0.723	0.63
8	1.372	1.371	1.362	1.81	—	—
9	1.584	1.582	1.586	1.00	1.567	1.10
10	1.640	1.640	1.637	0.62	1.618	0.30
14	2.243	2.243	2.240	0.12	—	—

Prior to the second dynamics tests, TMDs have been installed inside the box girder of the footbridge. To obtain the refined estimation of natural frequencies, the motion of the mass block is restricted such that the effect of the TMDs on natural frequencies is merely as extra weight. The previous three techniques are again applied for extracting modal properties of the footbridge from ambient vibration data, and the results are given in Table 5. As mentioned earlier, FVTs by means of a self-made horizontal shaker, as shown in Fig. 7, are performed to extract modal properties for the first lateral mode (Wen et al. 2016). The moving mass of the shaker including the mass on the platform and the mass of the platform is about 310 kg, and the stroke of mass is ±5.2 cm. The frequency range of excitation used is 0.68–0.77 Hz with a step of 0.01 Hz; thus the maximum horizontal inertia force can be precisely calculated for each test frequency. Both the excitation and response measurements are made at the middle of the central span. The steady-state lateral acceleration responses at variable excitation frequencies are measured and processed to establish the frequency response functions. Fig. 8 plots the variation of lateral acceleration amplitudes at midspan with test frequency. The least-squares curve fitting of the analytical expression of the frequency response function to the measurement data gives the estimation of the modal mass of 519 t and the damping ratio of 0.63%. As a comparison, the modal mass predicted from the finite-element model of the bridge with the extra weight of the TMD is 513 t. The close agreement between analysis and experiments suggests that the modal mass from finite-element analysis is reliable. For vertical modes, the modal frequencies and damping ratios listed in Table 5 are obtained by the measured free-decay responses after excitation from a jump of 100 persons. The modal properties obtained at the second dynamic tests permit the fine-tuning of mass dampers for the five modes.

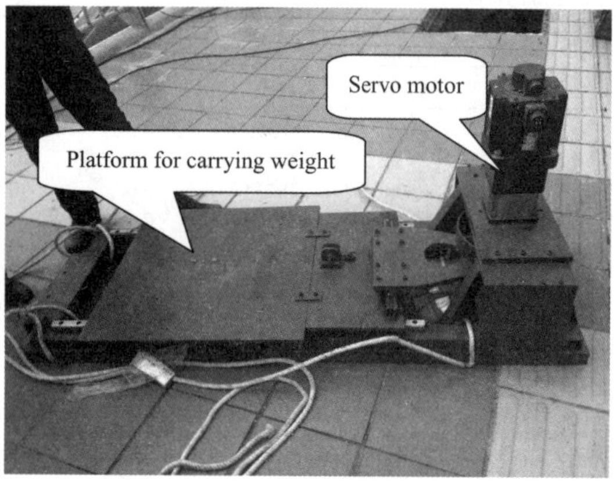

Fig. 7. Computer-controlled horizontal shaker for exciting the lateral mode (image by X. G. Hua)

Assessment of Human-Induced Vibrations

During the first period dynamic tests, the actual human-induced responses of the footbridge in lateral and vertical directions were assessed by means of pedestrian group tests of up to 400 persons. The most critical human-induced vibration is the synchronous lateral excitation, which could potentially develop for any lateral modes with a frequency less than 1.2 Hz. Therefore, the first test of human-induced vibration was performed to check the triggering number of pedestrians for the lateral lock-in. Pedestrians circulated anticlockwise around the central span with an increasing and controlled number up to a maximum of 200 persons. Fig. 9 shows the stream of the pedestrians on the footbridge. The lateral and vertical acceleration responses at the midspan were recorded, as illustrated in Fig. 10. The maximum lateral response was about 0.03 m/s^2 for 100 persons freely walking in the central span. However, it suddenly increased to 0.2 m/s^2 when the number of pedestrians is 150, suggesting the development of the synchronization lateral excitation. The maximum acceleration reached approximately 0.4 m/s^2 when there were 200 walking pedestrians. The test was forced to stop because the pedestrian paces stopped or were disordered by panic from some pedestrians. Based on the modal mass (470 t) and damping ratio (0.60%) for the first lateral mode prior to installation of the TMDs, the critical number of pedestrians based on Eq. (3) is about 170 persons. The triggering number obtained by the tests is generally in good agreement with the empirical model derived from the measurements from the London Millennium Bridge. The small deviation may be caused by the error in damping ratio. As for the vertical response, analysis indicates that the measured vertical response is only 40% of the predicted one under the same pedestrian number.

Fig. 11 shows the PSDs of the vertical and lateral accelerations induced by groups of 100, 150, and 200 persons. Before the lateral lock-in, the PSD of the vertical acceleration spreads over a wide

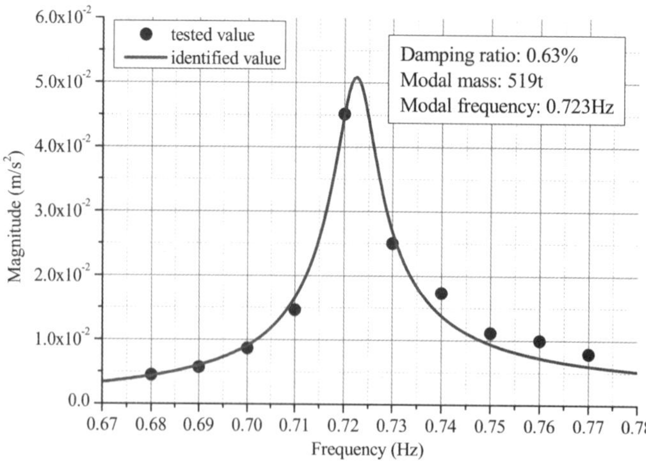

Fig. 8. Variation of acceleration amplitude with test frequency for the FVT

Fig. 9. Test of the footbridge with pedestrian streams (images by X. G. Hua)

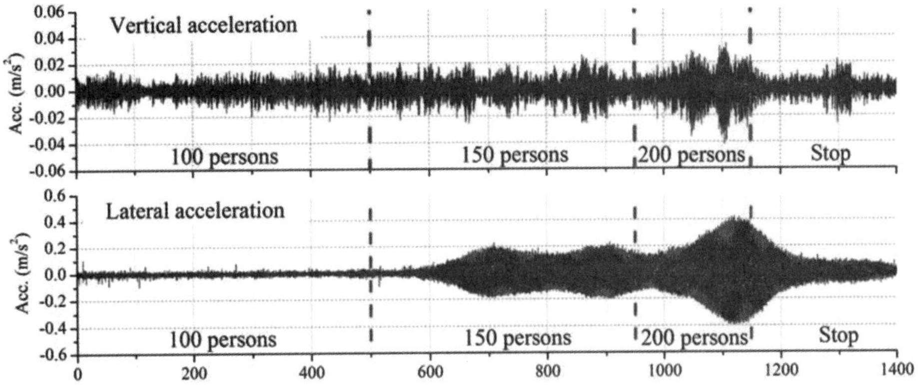

Fig. 10. Vertical and lateral acceleration responses of the midspan to pedestrian streams

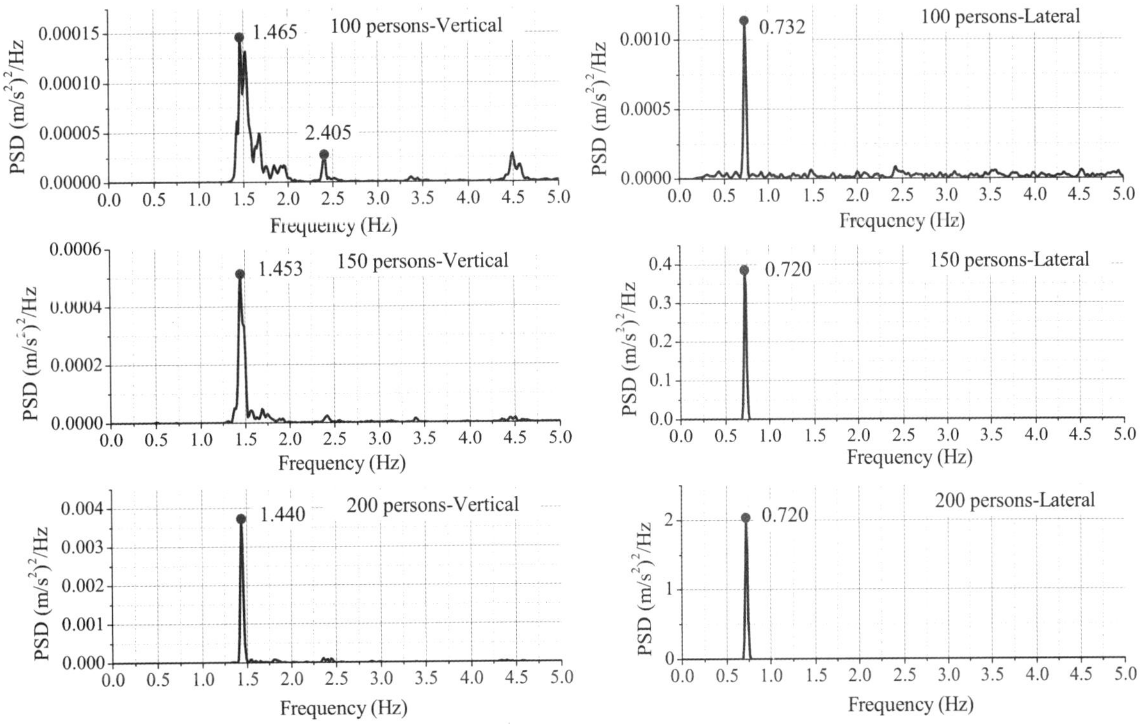

Fig. 11. Power spectral densities of vertical and lateral responses at the midspan

frequency range from 1.40 to 2.5 Hz, which is the nature of a typical random vibration. During the lateral lock-in with 200 pedestrians, only a clear peak occurs at the frequency 1.44 Hz for vertical vibrations and at the frequency 0.72 Hz for lateral vibrations, revealing that pedestrian paces are highly synchronized and their frequencies are twice the lateral resonance frequency.

Design of Eddy Current TMDs

The TMDs are mainly composed of a mass element, a stiffness element, and a damping element. The required damping in TMDs is usually provided by fluid viscous dampers, which have several disadvantages, such as oil leakage and difficulty in adjusting the damping ratio for an operating TMD. The activation of fluid viscous dampers may sometimes be problematic in small-amplitude vibrations. Alternatively, eddy current damping does not require any contact or viscous fluids to generate damping and is therefore a potential solution. Other advantages of eddy current damping are the ideal viscous forces linearly proportional to relative velocity, two-dimensional damping rather than the unidirectional damping, and accurate determination of the damping coefficient by an analytical approach. The mechanism of eddy current damping is illustrated in Fig. 12. Eddy currents develop when a conductor plate moves through a magnetic field generated by external magnets. These currents create electromagnets with a magnetic field opposite to the original external magnetic field, therefore, causing a repulsive force (damping force) proportional to the relative velocity between the magnetic field and conductor plate. The damping coefficients can be easily adjusted by simply changing the air gap between the magnets and conductive plate. As the commercially available permanent magnets are used to generate the magnetic fields, no power is needed to supply the dampers.

The free parameter central to TMD design is the mass ratio. Once the mass ratio is determined, the optimal frequency and damping ratio can be readily determined. The mechanical parameters of stiffness coefficients k_{TMD} and damping coefficients c_{TMD} will also be calculated. For the present bridge, a relatively large mass ratio, namely 3% for lateral mode and 2% for vertical modes, is used to assure the control performance of TMDs in the face of frequency mistuning in TMDs, modal identification error, and the unpredicted effect of pedestrian weight, among others.

As discussed earlier, the modal properties obtained at the second tests provide the baseline for fine-tuning of mass dampers for the five modes. However, they correspond to the bridge with the extra weight due to the locked TMDs. Another important consideration is that the best performance of TMDs should be achieved when the footbridge is loaded with a high pedestrian density rather than the empty bridge. After taking into account the above effects, the final modal properties of the footbridge used for tuning of mass dampers are summarized in Table 6. The optimal TMD parameters are also shown in Table 6. For practical reasons, each TMD is split into several TMD units, which are placed closest to the antinode for the controlled mode. For example, there are 16 TMD units for lateral TMDs and the effective mass for each TMD unit is 875 kg, totally amounting up to 14 t as required. Fig. 13 shows the vertical and the coupled horizontal and vertical TMD units. For the vertical TMDs, the inherent damping is very low, and the required damping is mainly supplied by eddy current damping; for the lateral TMDs, the required damping is contributed by both the friction-induced damping and eddy current damping, as will be discussed later.

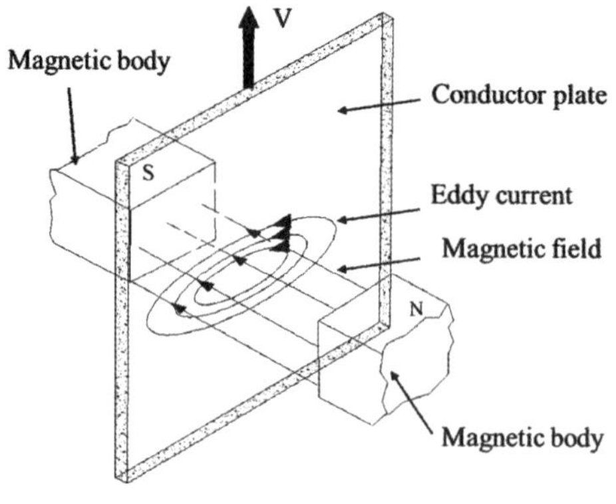

Fig. 12. Mechanism of the eddy current damping forces

Field Validation of Control Performance

Activation of the Vertical and Lateral TMDs

The TMDs will be activated when the acceleration of the TMD's mass exceeds a threshold value to overcome the inevitable friction force between the guide systems. To guarantee the footbridge meets the requirement of user comfort, the activation acceleration must be smaller than the comfort limits. For example, TMDs must take effect when bridge deck acceleration is above 0.1 m/s^2 for lateral direction, which is a threshold for triggering the synchronous lateral excitation.

Fig. 14 shows the typical acceleration response of the bridge and TMD units. For the vertical TMDs, because the mass is supported by the springs, the friction force between the guide systems is very small. Therefore, the activation acceleration is very small, which is about 0.003 m/s^2, as shown in Fig. 14(a). For the lateral TMDs, the mass is supported by two rods and it slides along the two rods. To minimize the friction force, the two rods must be installed with accurate leveling and the coefficient of static friction must be as small as possible. If the rods were installed with accurate leveling, the activation acceleration equals the coefficient of static friction multiplied by gravitational acceleration. The theoretical coefficient of friction for the rods is about 0.002–0.003. By the dynamic test, the intermittent activation of three lateral TMD units was observed, as shown in Fig. 14(b). It can be seen that the two TMD units were activated when the acceleration of the footbridge deck is larger than about 0.06 m/s^2; therefore, the measured total coefficient of friction is about 0.006. The activation acceleration for L2#TMD is 0.12 m/s^2, which may be caused by the poor leveling of the rods during the assembly of TMDs.

Table 6. Modal Parameters for the TMD

	Modal parameter of the bridge		Basic parameters of TMDs			
Mode	f_s (Hz)	M (t)	μ	f_{TMD} (Hz)	m_{TMD} (t)	ζ_{TMD} (%)
4	0.74	470	0.03	0.718	14.0	10.6
8	1.41	242	0.02	1.382	5.0	8.6
9	1.64	289	0.05	1.561	13.2	13.6
10	1.70	283	0.05	1.619	13.2	13.6
14	2.05	603	0.02	2.010	14.7	8.6

Fig. 13. Vertical and coupled lateral and vertical tuned mass dampers (images by X. G. Hua)

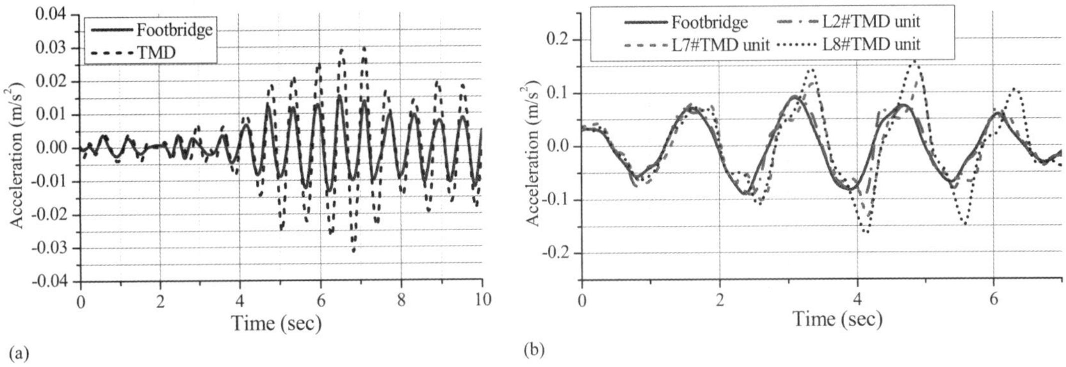

Fig. 14. Activation acceleration for vertical and lateral TMD units: (a) activation acceleration of vertical TMD; (b) activation acceleration of lateral TMD

Fig. 15. Test of the footbridge with pedestrian stream at the central span (images by X. G. Hua)

Although the static coefficient of friction for stationary bodies is larger than the kinetic one, it is used to roughly estimate the equivalent damping ratio caused by friction, as follows

$$\zeta_f = \frac{2\kappa g}{\pi \omega_{\mathrm{TMD}}^2 u_{\mathrm{TMD}}} \tag{4}$$

where κ = measured coefficient of friction; g = gravitational acceleration; and the denominator represents the relative acceleration of TMD multiplied by a constant. The equivalent damping ratio is found to be inversely linear proportional to the relative acceleration of TMD. Taking the measured coefficient of friction as 0.006, the equivalent damping ratio caused by fiction is about 9.5% for the 0.4-m/s^2 acceleration of the TMD. It reaches to 19.0% for the 0.2-m/s^{2}acceleration of the TMD, which deviates significantly from the optimal damping. In this regard, a large acceleration response of the TMD is desirable to minimize the effect of friction on an equivalent damping ratio, which may be achieved by a small mass ratio used for the TMD design.

Tests of Pedestrian Streams

Validation tests of control performance were performed after tuning TMD frequencies to their target values. The frequency was adjusted by adding or removing the mass as required, and the damping was left unchanged, which is less critical for TMD performance. After that, two dynamics tests were performed for the validation of the control performance of lateral TMDs. The first was the human-induced vibrations for pedestrians circulating around the central span with a maximum of 400 persons, as shown in Fig. 15. Fig. 16 shows the recorded vertical and lateral acceleration at the middle of the central span, and Fig. 17 shows the corresponding PSDs. From Fig. 16 the lateral acceleration response increases suddenly to 0.1 m/s^2, and then a beating phenomenon occurs with the bridge deck

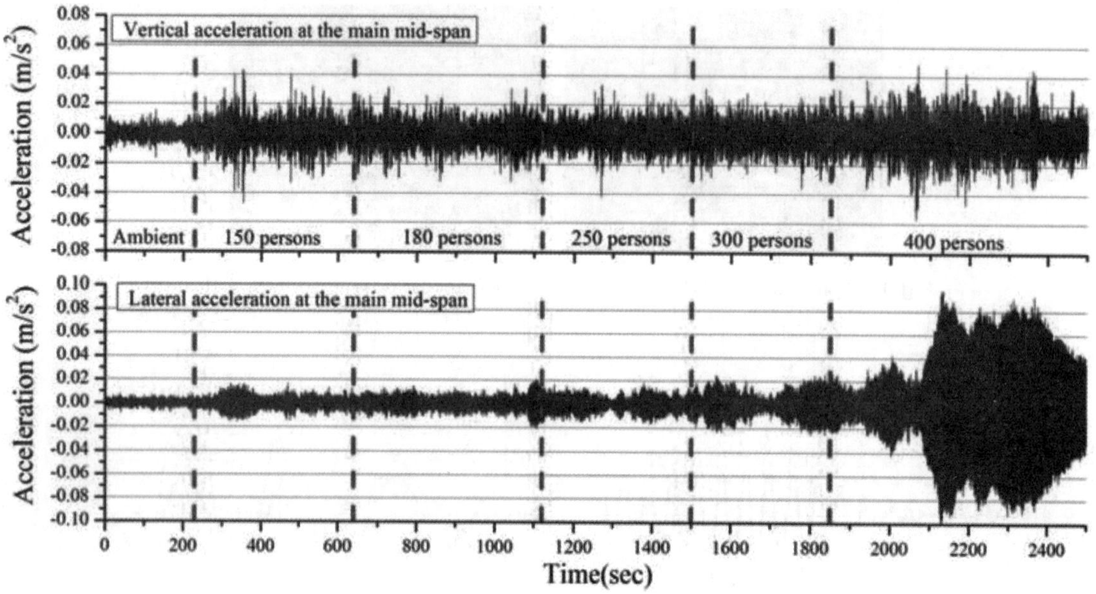

Fig. 16. Vertical and lateral accelerations at the middle of the central span with pedestrian stream

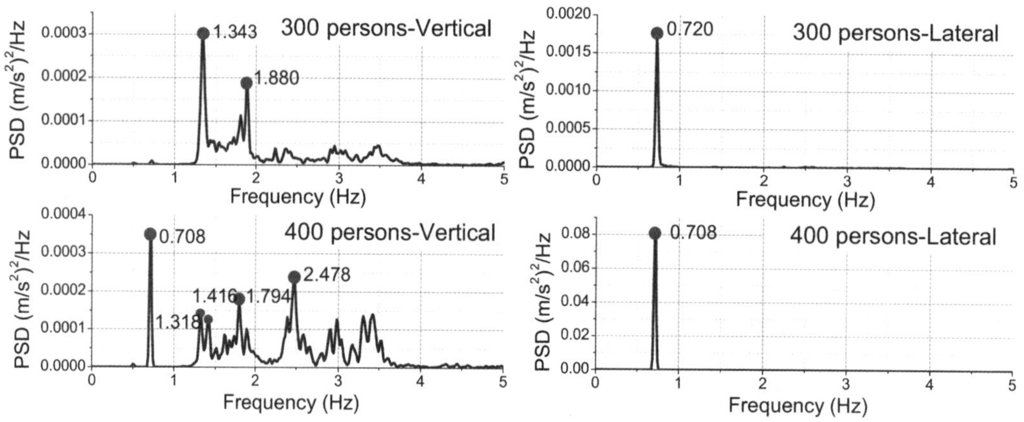

Fig. 17. Power spectral densities of the midspan vertical and lateral responses

acceleration in the 0.06–0.08-m/s^2 range when the number of pedestrians at the central span remained at 400. The PSD of vertical acceleration with 400 persons shown in Fig. 17 is different from the one with 200 persons during the lock-in. The several frequency peaks for the PSD of vertical acceleration in this test indicate that the step frequency is random and the lock-in phenomenon does not occur.

Actually, an ideal way to confirm the validity is to compare the damping ratios for the first lateral mode before and after installation of TMDs. However, the horizontal shaker previously used cannot induce sufficiently large bridge responses to activate the lateral TMDs, therefore, this method is not further pursued due to time limitations. Another alternative approach is used. First, all of the lateral TMDs are locked, and 250 pedestrians walked around the main span to trigger the lock-in. Then, all the lateral TMDs were unlocked. Fig. 18 shows the measured lateral and vertical accelerations at the main midspan. After activation of the lateral TMD, the acceleration decreases dramatically to about 0.08 m/s^2. Then the synchronous walk steps were disorganized, and the lateral lock-in phenomenon disappeared.

Fig. 19 shows the recorded lateral acceleration responses for the selected TMD units and the footbridge during the previous test. It is shown TMD Units 7 and 8 behave almost identically, whereas TMD Unit 2 does not operate properly. Because of errors in manufacture, installation, and frequency tuning, not all the TMD units will have the same characteristics, which may attenuate the TMD performance. Under the crowd excitation, the responses of TMD units are not sinusoidal, as shown in Fig. 19, which indicates the nonlinear properties of the TMD. The nonlinear properties of the TMD mainly refer to the dependence of the damping ratio on vibration amplitude, which is caused by dry friction. Because of the differences between units and the nonlinear properties of the TMD, the efficiency of the lateral TMD is reduced.

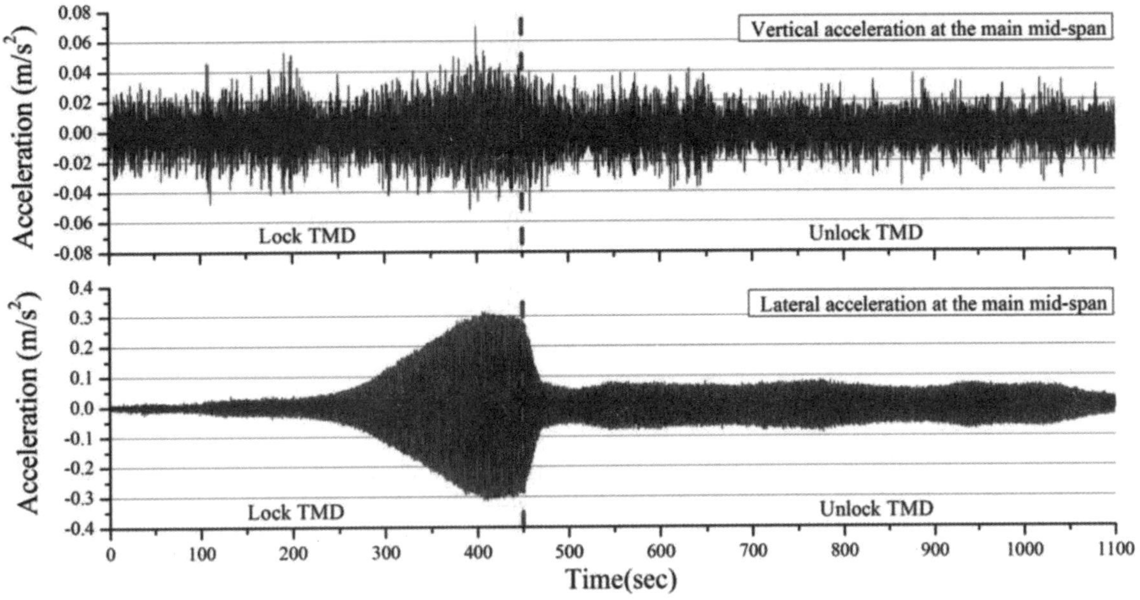

Fig. 18. Vertical and lateral acceleration at midspan before and after activation of lateral TMDs

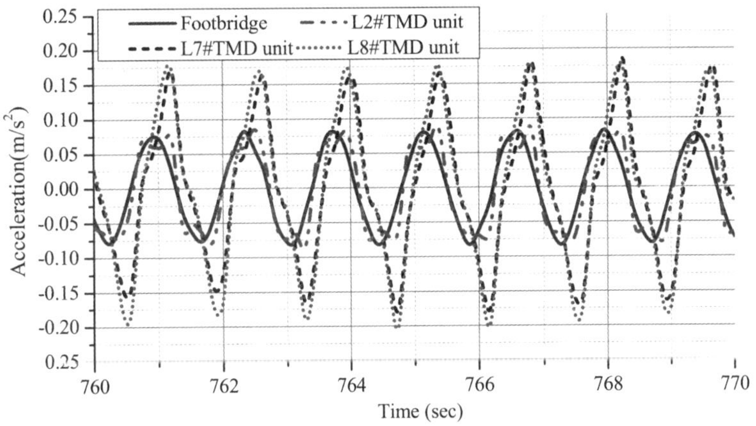

Fig. 19. Recorded TMD and footbridge accelerations induced by 250 pedestrians

Fig. 20 shows the free-decay vertical responses of both the structure and TMD mass at the middle of the side span subject to the jump of 100 persons. It is seen that the structural response lag is about 90° behind that of TMDs when the vertical acceleration exceeds 0.04 m/s^2. Below this threshold, the phase difference between the footbridge and the TMD unit deviates significantly from 90°, implying that the TMD is less effective for small-amplitude vibration. The equivalent damping ratio for the vertical mode of the side span is estimated as 5.5%, which is significantly larger than that of the uncontrolled mode.

Conclusions

The assessment and control of human-induced vibrations of an unusual curved cable-stayed bridge with separate deck systems for vehicles and pedestrians have been presented. A hybrid control scheme is pursued, which combines the steel braces to improve structural stiffness and the TMDs for enhancing the damping ratio of critical modes. Both ambient and FVTs are used to determine the modal frequencies and damping ratios of the bridge incorporated with steel braces. Numerical results showed that the use of steel braces is effective to reduce the number of lively modes. Pedestrian group tests indicated that the number of pedestrians triggering the synchronous lateral excitation generally agrees well with the empirical model derived from the measurement data on the London Millennium Bridge, whereas the measured vertical-induced responses are only about 40% of the predictions under the same pedestrian density. The damping ratios for both the vertical and lateral modes increase appreciably after installation of TMDs, and no evidence of large-amplitude vibrations has been observed. It is worth mentioning that the

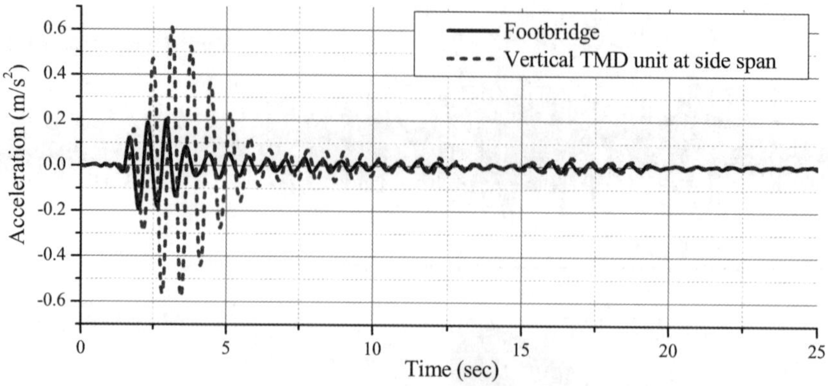

Fig. 20. Free-decay vertical acceleration at the middle of the side span subject to the jump of 100 persons

inherent damping caused by friction exceeds the required optimal damping at small relative movements of the TMD for the lateral TMDs, and in this regard a small mass ratio is desirable for relatively large movement of the TMD mass.

Acknowledgments

The authors would like to acknowledge support from the National Science Foundation of China (51278189, 51422806, and 91215302). The collaboration of the Bureau of Municipal and Rural Construction of Mianyang, in particular Dr. Xiao-Xian Kang, is of special mention. Our colleagues and Ph.D. students, Mr. J. H. Wang, Dr. Z. H. Wang, Mr. W. X. Wang, and Mr. T. F. Hu, provided great assistance during the field tests. The participation of more than 400 undergraduate students from the Southwest University of Science and Technology is also acknowledged.

References

Bachmann, H. (1992). "Case studies of structures with human-induced vibrations." *J. Struct. Eng.*, 10.1061/(ASCE)0733-9445(1992)118:3(631), 631–647.

Blekherman, A. N. (2005). "Swaying of pedestrian bridges." *J. Bridge Eng.*, 10.1061/(ASCE)1084-0702(2005)10:2(142), 142–150.

Bruno, L., Venuti, F., and Nascè V. (2012). "Pedestrian-induced torsional vibrations of suspended footbridges: proposal and evaluation of vibration countermeasures." *Eng. Struct.*, 36(1), 228–238.

Bendat, J. S., and Piersol, A. G. (1993). *Engineering applications of correlation and spectral analysis*, 2nd Ed, John Wiley & Sons, New York.

Brincker, R., Ventura, C. E., and Andersen, P. (2001). "Damping estimation by frequency domain decomposition." Proc., IMAC 19: Conf. on Structural Dynamics, Kissimmee, FL, Society for Experimental Mechanics, Bethel, CT, 698–703.

Caetano, E., Cunha, A., Magalhaes, F., and Moutinho, C. (2010a). "Studies for controlling human-induced vibration of the Pedro e Ines footbridge. Part 1: Assessment of dynamic behavior." *Eng. Struct.*, 32(4), 1069–1081.

Caetano, E., Cunha, A., Magalhaes, F., and Moutinho, C. (2010b). "Studies for controlling human-induced vibration of the Pedro e Ines footbridge. Part 2: Implementation of tuned mass dampers." *Eng. Struct.*, 32(4), 1082–1091.

Casado, C. M., Diaz, I. M., de Sebastián, J., Poncela, A. V., and Lorenzana, A. (2011). "Implementation of passive and active vibration control on an in-service footbridge." *Struct. Control Health Monit.*, 20(1), 70–87.

Chen, Z. Q., and Hua, X. G. (2009). *Dynamic design and control for footbridge vibrations*, China Communications Press, Beijing.

Dallard, P., et al. (2001a). "The London Millennium footbridge." *Struct. Eng.*, 79(22),17–33.

Dallard P., et al. (2001b). "London Millennium Bridge: Pedestrian-induced lateral vibration." *J Bridge Eng.*, 10.1061/(ASCE)1084-0702(2001)6:6(412), 412–417.

HiVoSS (Human-Induced Vibration of Steel Structures). (2008). "Design of footbridges: guideline, research report, RFS2-CT-2007-00033." ⟨http://www.stb.rwth-aachen.de/projekte/2007/HIVOSS/download.php⟩ (Aug. 26, 2008).

Fujino, Y., Pacheco, B. M., Nakamura, S. I., and Warnitchai, P. (1993). "Synchronization of human walking observed during lateral vibration of a congested pedestrian bridge." *Earthquake Eng. Struct. Dyn.*, 22(9), 741–758.

Ingolfsson, E. T., Georgakis, C. T., and Jonsson, J. (2012). "Pedestrian-induced lateral vibrations of footbridges: A literature review." *Eng. Struct.*, 45(1), 21–52.

Macdonald, J. H. G. (2009). "Lateral excitation of bridges by balancing pedestrians." *Proc. R. Soc. A*, 465(2104), 1055–1073.

Newland, D. E. (2004). "Pedestrian excitation of bridges." *J. Mech. Eng. Sci.*, 218(2), 477–492.

Occhiuzzi, A., Spizzuoco, M., and Ricciardelli, F. (2008). "Loading models and response control of footbridges excited by running pedestrians." *Struct. Control Health Monit*, 15(3), 349–368.

Pavic, A., Armitage, T., Reynolds, P., and Wright J. (2002). "Methodology for modal testing of the Millennium Bridge, London." *Proc. Inst. Civ. Eng. Struct. Build.*, 152(2), 111–121.

Piccardo, G., and Tubino, F. (2008). "Parametric resonance of flexible footbridges under crowd-induced lateral excitation." *J. Sound Vibr.*, 311(1–2), 353–371.

Racic, V., Pavic, A., and Brownjohn, J. M. W. (2009). "Experimental identification and analytical modelling of human walking forces: A literature review." *J. Sound Vibr.*, 326(1–2), 1–49.

Racic, V., Pavic, A., and Brownjohn, J. M. W. (2013). "Modern facilities for experimental measurement of dynamic loads induced by humans: A literature review." *Shock Vibr.*, 20(1), 53–67.

Ricciardelli, F., and Pizzimenti, A. D. (2007). "Lateral walking-induced forces on footbridge." *J. Bridge Eng.*, 10.1061/(ASCE)1084-0702(2007)12:6(677), 77–88.

Soong, T. T., and Dargush, G. F. (1997). *Passive energy dissipation systems in structural engineering*, John Wiley & Sons, New York.

van Overschee, P., and de Moor, B. L. (1996). *Subspace identification for linear systems: Theory, implementation and applications*, Kluwer, Dordrecht, the Netherlands.

Wen, Q., Hua, X. G., Chen, Z. Q., Wang, W. X., and Niu, H. W. (2016). "Modal parameter identification of a footbridge by forced vibration experiments." *Adv. Struct. Eng.*, in press.

Zivanovic, S., Pavic, A., and Reynolds, P. (2005). "Vibration serviceability of footbridges under human-induced excitation: a literature review." *J. Sound Vibr.*, 279(1–2), 1–74.

第30卷 第2期 中 国 公 路 学 报 Vol.30 No.2
2017年2月 China J. Highw. Transp. Feb. 2017

文章编号:1001-7372(2017)02-0098-09

基于稳态简谐激励的人行桥模态参数识别

温 青,华旭刚,陈政清,王文熙,牛华伟
(湖南大学 风工程与桥梁工程湖南省重点实验室,湖南 长沙 410082)

摘要:为了探究稳态激振现场试验方法在人行桥模态参数识别中的可行性,通过数值仿真分析和现场试验,研究了基于稳态激振的人行桥模态参数识别。该方法首先通过单点简谐激励获得结构稳态响应的频响函数(FRF),然后对实测频响函数按单模态或多模态叠加形式进行最小二乘曲线拟合得到结构的固有频率、阻尼比、模态振型和模态质量。以一频率密集分布的多点弹性支撑梁为数值算例,分析了激振器激振质量、频率间隔、激振次数、多模态参与和噪声对识别结果的影响。采用该方法识别了一座双层分离桥面曲线人行桥第1阶侧弯模态的频率、模态质量及阻尼比,为该桥的人致振动分析及减振设计提供了有效手段。结果表明:激振质量会改变结构固有模态参数,因此在满足激振条件下激振质量应尽可能小;选择合适的激振带宽和较小的激振频率间隔能获得更可靠的结果;人行桥现场稳态激振试验识别的频率和阻尼比与自由振动识别结果基本相符,识别的模态质量与有限元计算结果相符,说明该方法可行。

关键词:桥梁工程;人行桥;模态参数识别;强迫振动试验;频响函数;模态质量

中图分类号:U441.3 **文献标志码**:A

Modal Parameters Identification of Footbridges Based on Steady-state Sinusoidal Excitations

WEN Qing, HUA Xu-gang, CHEN Zheng-qing, WANG Wen-xi, NIU Hua-wei
(Key Laboratory for Wind and Bridge Engineering of Hunan Province, Hunan University, Changsha 410082, Hunan, China)

Abstract: To investigate the feasibility of field steady-state forced vibration test on modal parameters identification of footbridges, modal parameters identification of footbridges based on steady-state forced vibration was investigated by numerical simulation analysis and field tests. Firstly, frequency response functions (FRFs) of the steady-state structural responses based on sinusoidal excitation were obtained. Then, the measured FRFs were employed to estimate inherent frequency, damping ratio, modal shape, modal mass by least squares curve fitting methods in form of either a single-mode FRFs or a multi-modes FRFs. Numerical simulations of a elastically multi-supported beam with closely spaced frequency were carried out to analyze the effects of additional mass of exciter, frequency step, number of excitation, multi-mode vibration and noise on identification results. The natural frequency, modal mass and damping ratio for the first lateral mode of a double-level cable-stayed bridge with separate deck systems for pedestrians

收稿日期:2016-01-27
基金项目:国家自然科学基金项目(51278189);国家优秀青年基金项目(51422806);
国家自然科学基金重大研究计划项目(91215302)
作者简介:温 青(1985-),男,湖南桃江人,工学博士研究生,E-mail:wenqing_hn@126.com。
通讯作者:华旭刚(1978-),男,浙江义乌人,教授,博士研究生导师,工学博士,E-mail:cexghua@hotmail.com。

and vehicles were identified, which provided effective means for the analysis of pedestrian-induced vibration and the design of vibration reduction. The results show that inherent modal parameters of the structure can be changed by the exciter mass, so the exciter mass should be as small as possible under the conditions of sinusoidal forced vibrations. Choosing suitable exciting bandwidth and small vibration frequency step is important to obtain modal parameters accurately. The frequency and damping ratio of footbridge estimated by field steady-state forced vibration tests are in accordance with the results estimated by free vibration tests, and the identified modal mass is identical to the result by finite element model (FEM), which demonstrates that this method is reliable.

Key words: bridge engineering; footbridge; modal parameter identification; forced vibration test; FRF; modal mass

0 引 言

随着建筑材料、建造技术和设计方法的不断进步,以及人们对桥梁美学和城市景观的追求,许多新建人行桥造形奇特、外形美观。这些桥梁结构轻柔、基频低、阻尼小,但由此带来的人致振动和行走舒适性等问题日益突出[1],自英国伦敦千禧桥事件后,该问题受到了全世界的密切关注。为了准确预测、评估和控制人行桥的人致振动,提高行走舒适性,保证人行桥运营安全,需要准确获得人行桥的模态特征,如固有频率、阻尼比、模态质量和模态振型等。

结构动力测试是获得固有频率、振型及阻尼比等结构模态参数最为可靠的手段。常用动力测试方法有环境振动测试、自由振动测试和强迫振动测试。环境振动测试是一种即经济又方便的动力测试方法,测试结构在风荷载、车辆荷载和人群荷载以及其他荷载作用下结构的随机响应。自由振动测试是通过外荷载使结构产生较大的初始位移、速度或加速度,测试结构的自由衰减响应。这 2 种方法不需要专门的激振设备,已经在人行桥模态参数识别中得到了广泛应用[2-7]。然而,这 2 种方法都无法识别结构的模态质量,而模态质量是人行桥减振设计尤其是质量型减振设计如被动调谐质量阻尼器(TMD)所必需的一个参数。Porras 等[8]通过附加一个弹簧质量振子评估了人行桥的模态质量,该方法虽然也是采用自由振动测试,但需要专门的弹簧质量振子装置,该装置可通过无阻尼元件的 TMD 实现。强迫振动测试需要专门的激振设备,通过同步测定不同激振频率下的荷载输入和结构响应得到频响函数,从而可靠地识别出重要模态的特征参数,尤其是阻尼比和模态质量。人行桥模态质量小、阻尼比小,在较小的激振力作用下就能产生较大的响应,因此,强迫振动测试在人行桥上完全可行。Dallard 等[9]、Živanovic 等[10]和 Caetano 等[11]分别采用该方法分析了不同人行桥的模态特征。Brownjohn 等[12]使用人群激励的方法评估了人行桥的模态质量。

稳态激振振动测试技术通过测定激振力和结构响应来获得频响函数,尽管其理论已非常成熟,但在土木领域的研究与应用较少。大跨人行桥频率密集,激振器质量、激振频率间隔、激振次数、多模态参与和噪声等因素都会影响其识别精度。此外,作用在结构上的激振器会产生激振力,也会产生附加质量。因受激振器制造和设计制约,激振频率间隔和激振力大小都有限。鉴于此,本文分析了基于稳态谐振激励的人行桥模态参数识别的方法。首先,阐述结构模态参数,尤其是模态质量的定义,建立惯性式激振器单点稳态谐振激励的频响函数表达式;然后,提出模态参数识别方法,通过单点稳态谐振激励,获得结构的频响函数,采用最小二乘法对实测频响特征按单模态或多模态叠加的频响函数形式进行拟合,评估结构模态参数;采用仿真分析的方法,以 1 个多点弹性支撑梁为例,分析激振器激振质量、激振频率间隔、激振次数、多模态参与和噪声对识别结果的影响,为准确识别模态参数提供可靠依据;应用该方法识别 1 座大跨度曲线斜拉人行桥的第 1 阶侧弯模态参数,识别结果准确、可靠,可为准确预测和有效控制该桥的人致振动提供依据。

1 基本理论

1.1 模态参数

n 个自由度振动系统的运动方程可以表示为[13]

$$\boldsymbol{M}\ddot{\boldsymbol{x}}(t)+\boldsymbol{C}\dot{\boldsymbol{x}}(t)+\boldsymbol{K}\boldsymbol{x}(t)=\boldsymbol{F}(t) \tag{1}$$

式中:$\boldsymbol{M}$,$\boldsymbol{C}$ 和 $\boldsymbol{K}$ 分别为质量、阻尼和刚度矩阵;$\boldsymbol{F}(t)$ 为与时间 t 相关的激振力向量;$\boldsymbol{x}(t)$ 为与时间 t 相关

的位移向量。

假设阻尼为比例阻尼形式，则式(1)可以解耦成单自由度的运动方程，第 $r(r=1,2,\cdots,n)$ 阶模态的运动方程可以表示为[13]

$$\ddot{q}_r(t)+2\xi_r\omega_r\dot{q}_r(t)+\omega_r^2q_r(t)=p_r(t)/m_r \quad (2)$$

式中：$q_r(t)$为广义模态坐标；模态质量 $m_r=\boldsymbol{\varphi}_r^{\mathrm{T}}\boldsymbol{M}\boldsymbol{\varphi}_r$，广义模态力 $p_r(t)=\boldsymbol{\varphi}_r^{\mathrm{T}}\boldsymbol{F}(t)$，模态阻尼比 $\xi_r=\boldsymbol{\varphi}_r^{\mathrm{T}}\boldsymbol{C}\boldsymbol{\varphi}_r/(2\omega_r m_r)$，$\omega_r$ 和 $\boldsymbol{\varphi}_r$ 分别为第 r 阶模态圆频率和模态振型，通过求解公式(3)的特征值和特征向量获得，即

$$(\boldsymbol{K}-\omega^2\boldsymbol{M})\boldsymbol{x}(t)=0 \quad (3)$$

由于模态振型 $\boldsymbol{\varphi}_r$ 为任意尺度的量纲一的特征向量，不同的模态振型尺度，模态质量 m_r 和广义模态力 p_r 不同，因此，通常需要对振型进行规格化，常用的振型规格化方法有 2 种[13]：按质量归一化的振型规格化，即 $\hat{\boldsymbol{\varphi}}_r^{\mathrm{T}}\boldsymbol{M}\hat{\boldsymbol{\varphi}}_r=1$，$\hat{\boldsymbol{\varphi}}_r$ 为质量归一化模态振型，则结构的模态质量为 $m_r=1$；特定自由度幅值取 1 的振型规格化，即以某个自由度的幅值取 1，并以这个指定的值为基准确定其他自由度幅值，如指定最大自由度幅值为 1，则幅值归一化模态振型 $\boldsymbol{\varphi}_r^*=(\varphi_{1r},\varphi_{2r},\cdots,1,\cdots,\varphi_{nr})^{\mathrm{T}}$，该模态的模态质量为 $m_r=\boldsymbol{\varphi}_r^{*\mathrm{T}}\boldsymbol{M}\boldsymbol{\varphi}_r^*$。

本文采用指定最大自由度幅值为 1 的方法对振型进行规格化。

1.2 单点激励频响函数

式(1)所示的线性黏滞阻尼振动系统的频域内稳态响应 $\boldsymbol{X}(\omega)$可以表示为

$$\boldsymbol{X}(\omega)=\boldsymbol{H}(\omega)\boldsymbol{F}(\omega) \quad (4)$$

$$\boldsymbol{H}(\omega)=(\boldsymbol{K}+\mathrm{i}\omega\boldsymbol{C}-\omega^2\boldsymbol{M})^{-1} \quad (5)$$

式中：$\boldsymbol{H}(\omega)$为频响函数；ω 为激振频率；$\boldsymbol{F}(\omega)$为频域内激振力；i 为虚数单位。

假设简谐激励 $F_k(\omega)$作用于第 $k(k=1,2,\cdots,n)$个自由度，第 $j(j=1,2,\cdots,n)$个自由度的简谐位移响应为 $X_j(\omega)$，则导纳频响幅值函数 $H_{jk}(\omega)$[14]为

$$H_{jk}(\omega)=\frac{X_j(\omega)}{F_k(\omega)}=\left|\sum_{r=1}^{n}\frac{1}{m_r}\frac{\varphi_{jr}\varphi_{kr}}{\omega_r^2-\omega^2+2\mathrm{i}\xi_r\omega_r\omega}\right|=\left|\sum_{r=1}^{n}\frac{A_{jkr}}{\omega_r^2-\omega^2+2\mathrm{i}\xi_r\omega_r\omega}\right| \quad (6)$$

式中：φ_{jr} 和 φ_{kr} 分别为第 r 阶模态振型 $\boldsymbol{\varphi}_r$ 的第 j 和第 k 个自由度的幅值；模态常数 $A_{jkr}=\varphi_{jr}\varphi_{kr}/m_r$，当 $j=k$ 时，$A_{kkr}=\varphi_{kr}^2/m_r$。

导纳频响幅值函数是结构 n' 个模态响应的叠加。当结构频率分布不密集，阻尼较小时，在频率为 ω 的简谐激励作用下，导纳频响幅值函数可以近似表示为与激振频率 ω 接近的少数几个模态的叠加，远离 ω 的模态振幅很小，可以忽略。

1.3 单点激励模态试验

在单点激励模态试验中，常采用侧向[11]和竖向[10]惯性式激振器，如图 1 所示，通过伺服电机驱动质量块运动引起的惯性力来产生简谐激振力。假设激振器有效运动总质量 m_0，振幅为 A_0，激振频率为 ω，则激振力 $F(t)$为

$$F(t)=m_0A_0\omega^2\sin(\omega t) \quad (7)$$

(a) 侧向

(b) 竖向

图 1 惯性式激振器

Fig. 1 Intertial Exciter

激振力作用于第 k 个自由度位置，则激振力在频域内可以表示为

$$F_k(\omega)=m_0A_0\omega^2 \quad (8)$$

实测结构的响应通常为加速度响应，则第 j 个自由度的位移响应 $X_j(\omega)$为

$$X_j(\omega)=\ddot{X}_j(\omega)/\omega^2 \quad (9)$$

式中：$\ddot{X}_j(\omega)$为加速度响应。

通过模态试验获得的频响函数可以表示为

$$H_{jk}(\omega)=\frac{X_j(\omega)}{F_k(\omega)}=\frac{\ddot{X}_j(\omega)/\omega^2}{m_0A_0\omega^2} \quad (10)$$

由式(6)、(10)可知，第 k 个自由度激振，测试第 j 个自由度加速度的位移频响幅值函数应满足

$$H_{jk}(\omega)=\frac{\ddot{X}_j(\omega)/\omega^2}{m_0A_0\omega^2}=\left|\sum_{r=1}^{n}\frac{A_{jkr}}{\omega_r^2-\omega^2+2\mathrm{i}\xi_r\omega_r\omega}\right| \tag{11}$$

当激振器固定于结构上时,在激振器作用下,结构将发生振动,激振器质量块的加速度是相对于结构的,假设激振力作用于第 r 阶模态最大模态振型坐标位置,则第 r 阶模态的运动方程可以表示为

$$(m_r+m_0)\ddot{q}_r(t)+c_r\dot{q}_r(t)+k_rq_r=m_0A_0\omega^2\sin(\omega t) \tag{12}$$

式中:广义模态阻尼 $c_r=\boldsymbol{\varphi}_r^{\mathrm{T}}\boldsymbol{C}\boldsymbol{\varphi}_r$;广义模态刚度 $k_r=\boldsymbol{\varphi}_r^{\mathrm{T}}\boldsymbol{K}\boldsymbol{\varphi}_r$。

激振器会产生一个附加质量,进而影响模态参数的识别。

1.4 模态参数识别

为了获得结构的模态参数,在结构上作用简谐激励,激振频率为 ω,实测结构加速度响应,由此可按式(10)求得结构的实测频响特征,然后采用最小二乘法对实测频响特征按式(6)所示的函数式进行拟合,获得各未知参数的最小二乘解。每阶模态有3组未知数,固有频率 ω_r、模态阻尼比 ξ_r 和模态常数 A_{jkr}。

在实际模态试验中,激振频率带宽是有限的,$\omega\in[\omega_l,\omega_u]$,$\omega_l$ 和 ω_u 分别为激振频率的上限和下限。在此激振频率带宽内,可激起结构 n^* 次共振响应,说明结构有 n^* 阶模态频率位于激振频率内,为了减少拟合时未知参数,可以将式(11)近似成 n^* 个模态叠加的形式,即

$$H_{jk}(\omega)=\frac{\ddot{X}_j(\omega)/\omega^2}{m_0A_0\omega^2}=\left|\sum_{r=1}^{n^*}\frac{A_{jkr}}{\omega_r^2-\omega^2+2\mathrm{i}\xi_r\omega_r\omega}\right| \tag{13}$$

由模态常数 $A_{jkr}=\varphi_{jr}\varphi_{kr}/m_r$,按 $m_r=\varphi_{jr}\varphi_{kr}/A_{jkr}$ 评估模态质量,φ_{jr} 和 φ_{kr} 可根据理论模态分析结果确定,也可通过实测确定。

为了测定 φ_{jr} 和 φ_{kr},在结构上布置 N 个加速度测点,对应自由度分别为 $p_i(i=1,2,\cdots,N)$。布置测点时,测点应尽可能包括这 n^* 个模态振型幅值最大的自由度,并且必须包括激振点的第 k 个自由度,即 p_k 测点。按式(13)识别出第 r 阶模态的 N 个模态常数 $A_{p_ip_kr}$,然后按式(14)计算各测点模态振型位移 φ_{pir},获得模态振型 $\boldsymbol{\varphi}_r$,对 $\boldsymbol{\varphi}_r$ 按最大自由度幅值为1规格化,得最大幅值归一化模态振型 $\boldsymbol{\varphi}_r^*$ 第 k 个自由度的位移为 $\varphi_{p_kr}^*$,最后按式(15)评估模态质量,即

$$\varphi_{pir}=A_{p_ip_kr}/A_{p_kp_kr}\quad i=1,2,\cdots,k,\cdots,N;k\in i \tag{14}$$

$$m_r=\varphi_{p_kr}^*/A_{p_kp_kr} \tag{15}$$

上文给出了基于正弦稳态激振强迫振动测试技术的模态参数识别方法。

2 仿真分析

以一个多点弹性支撑简支梁为例,分析惯性式激振器激振质量、激振频率间隔、激振次数、多模态参与和噪声对识别结果的影响,以期为准确识别模态参数提供可靠依据。

2.1 有限元理论分析

图2为多点弹性支撑简支梁模型,该梁跨径为10 m,跨中均布19个等刚度弹簧支撑。主梁为矩形断面钢梁,梁宽0.2 m,高0.004 m,弹性模量 $E=2.06\times10^{11}$ Pa,密度 $\rho=7\ 850\ \mathrm{kg\cdot m^{-3}}$,弹簧刚度 $K=100\ \mathrm{N\cdot m^{-1}}$。采用ANSYS建立了该结构的有限元模型,通过模态分析获得了结构的模态参数,如表1所示。该结构前3阶频率接近、能有效模拟大跨人行桥频率密集的特性。在频域分析中,阻尼采用模态阻尼比形式,假定各阶模态的阻尼比均为0.5%。在研究噪声影响的时域动力响应分析中,阻尼采用Rayleigh阻尼形式,设定前2阶模态阻尼比为0.50%,反推Rayleigh阻尼常数。在仿真分析中,激振力 $F(t)=F_0\sin(\omega t)$,F_0 为激振力幅值,为常数,其值不随激振频率的改变而改变,$F_0=1$ N。

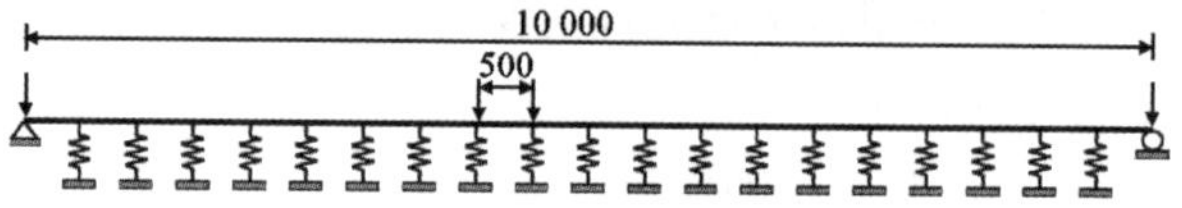

图2 多点弹性支撑梁模型(单位:mm)

Fig. 2 Elastically Multi-supported Beam (Unit: mm)

表1 有限元理论分析模态参数

Tab. 1 Modal Parameters of FEM Analysis

模态阶次	频率/Hz	阻尼比/%	模态质量/kg	振型
1	0.903	0.50	31.415	对称竖弯
2	0.971	0.50	31.446	反对称竖弯
3	1.226	0.50	31.498	对称竖弯
4	1.733	0.50	31.516	反对称竖弯

2.2 激振器质量的影响

固定于结构上的激振器以频率 ω 激振时,激振器对结构的作用可以分为2个部分:附加质量和简谐激振力。附加质量将改变结构的模态特征,从而影响评估结果。

假设激振器质量 m_0 与第1阶模态质量 m_1 之比 $\mu=m_0/m_1=0.01$,激振力作用点为1/2跨,激振频率 $f\in[0.88\ \mathrm{Hz},0.92\ \mathrm{Hz}]$,频率间隔 $\Delta f=0.004$ Hz。通过ANSYS的谐波分析获得频响函数曲线,如图3所示,将频响函数近似为单模态参与,拟合了

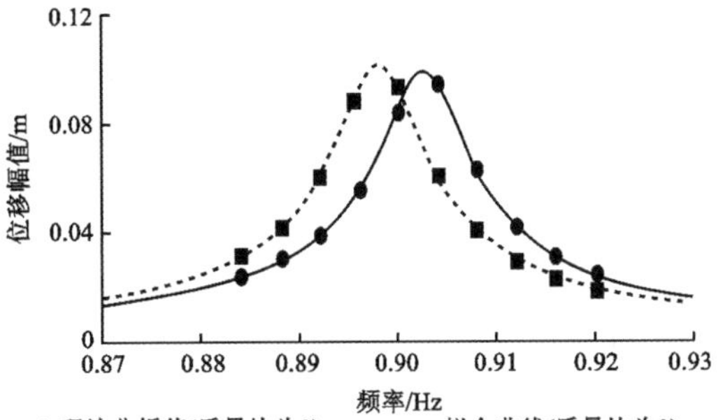

图 3 不同质量比下频响函数曲线

Fig. 3 Curves of FRFs Under Different Mass Ratios

模态参数，分析了不同质量比对模态参数识别结果的影响，结果如表 2 所示。当质量比为 0 时，识别的频率、阻尼比与理论计算结果相同，模态质量的误差仅为 0.06%，识别的模态参数精度较高。随着质量比的增大，识别的频率和模态质量均逐渐减小，与原模型的理论分析值的误差越来越大，因此，在满足激振力要求的条件下，选择较小的质量比能更准确地获得模态参数。如文献[11]中，$m_0=1\ 200$ kg，而 $m_r=165\ 880$ kg，质量比为 0.007，忽略了 m_0 对模态参数识别的影响。

表 2 不同激振器质量比识别的模态参数

Tab. 2 Identified Modal Parameters Under Mass Ratios of Different Exciters

工况	质量比	频率/Hz	阻尼比/%	模态质量/kg
1	0.000	0.903	0.50	31.435
2	0.005	0.900	0.50	31.199
3	0.010	0.898	0.50	31.109
4	0.020	0.894	0.50	30.362

2.3 激振频率间隔与激振次数

由于激振器设计和制造等因素，激振频率的最小间隔是有限的，如文献[11]中，频率间隔为 0.01 Hz。激振器频率间隔会影响模态参数的识别，较小的频率间隔能更好地表征结构频响特征。当频率间隔确定时，激振次数将决定激振频率带宽，激振次数多时，激振带宽将增大，当采用单模态参数识别时，适宜的频率间隔和激振次数带宽可以获得更准确的识别结果，见表 3。在合理的激振频率带宽内，阻尼比和模态质量的识别对频率间隔和激振次数较为敏感，较小的频率间隔和较多的激振次数能获得更准确的模态参数。

2.4 多模态参数识别

由式(6)可知，频响函数 $H_{jk}(\omega)$ 是多个模态的叠加，当模态频率分布稀疏时，采用单模态的频响函数就可以获得高精度的模态参数，如跨中加载的工况，由于第 2 阶模态振型在跨中位置的位移为 0，第 2 阶模态不参与振动，用单模态频响函数识别结果的精度较高，如表 2 所示；当模态频率分布密集时，需要采用多模态叠加的频响函数，才能准确识别各模态参数。

表 3 不同激振频率间隔与激振次数的模态参数

Tab. 3 Modal Parameters for Different Excitation Frequency Steps and Number of Excitation

工况	激振参数			识别模态参数		
	激振次数	频率间隔/Hz	频率带宽/Hz	频率/Hz	阻尼比/%	模态质量/kg
5	11	0.004	0.88～0.92	0.903	0.50	31.435
6	5	0.005	0.89～0.91	0.903	0.51	31.135
7	9	0.005	0.88～0.92	0.903	0.50	31.309
8	17	0.005	0.86～0.94	0.903	0.50	31.414
9	5	0.010	0.88～0.92	0.903	0.53	30.868
10	11	0.010	0.86～0.94	0.903	0.53	30.992
11	17	0.010	0.82～0.98	0.903	0.52	31.159

假设不考虑激振器的质量，在结构上布置 7 个测点，激振力作用点为 1/4 跨，即 k 为 p_2，激振频率 $f\in[0.88\ \text{Hz},1.30\ \text{Hz}]$，频率间隔为 0.005 Hz，通过 ANSYS 的谐波分析，获得 $H_{kk}(\omega)$ 即 $H_{p_2p_2}(\omega)$ 曲线，如图 4 所示。在激振频率带宽内，结构出现 3 次共振，分别为第 1,2 和 3 阶模态。采用 3 模态叠加的频响函数识别了模态参数，如表 4 所示，模态振型如图 5 所示。识别的模态参数与理论计算的模态参数误差很小，模态频率、阻尼比和模态质量的最大相对误差分别为 0.2%、2%和 1.4%。

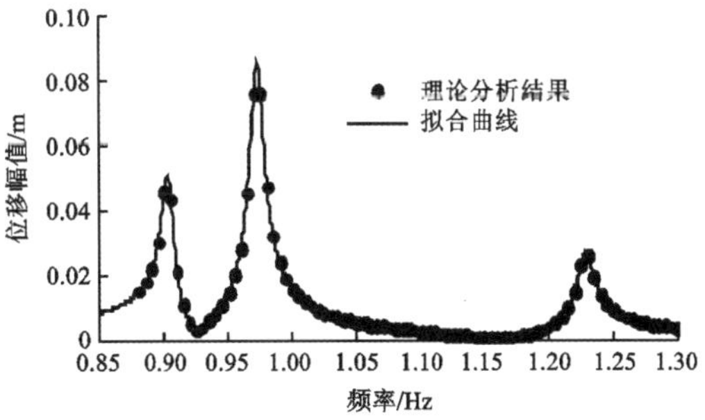

图 4 多模态频响函数曲线 $H_{kk}(\omega)$

Fig. 4 Curve of Multi-mode FRFs $H_{kk}(\omega)$

2.5 噪声的影响

在实际工程中，噪声干扰不可避免，表 5 模拟了噪声对识别结果的影响。此外，不可控的环境激励引起的结构响应实际上也可以处理为一种噪声。因此，噪声分为两部分模拟：零均值高斯白噪声激励，本文仅考虑了在 1/4,1/2 和 3/4 跨节点上作用白噪

表 4　多模态识别的模态参数

Tab. 4　Multi-mode FRFs Parameters Identification

工况	模态阶次	振型 φ_{kr}	模态常数 A_{kkr}	频率/Hz	阻尼比/%	模态质量/kg
12	1	0.699 9	0.015 8	0.903	0.49	30.974
13	2	0.995 7	0.031 7	0.972	0.50	31.232
14	3	0.709 9	−0.016 1	1.228	0.51	31.394

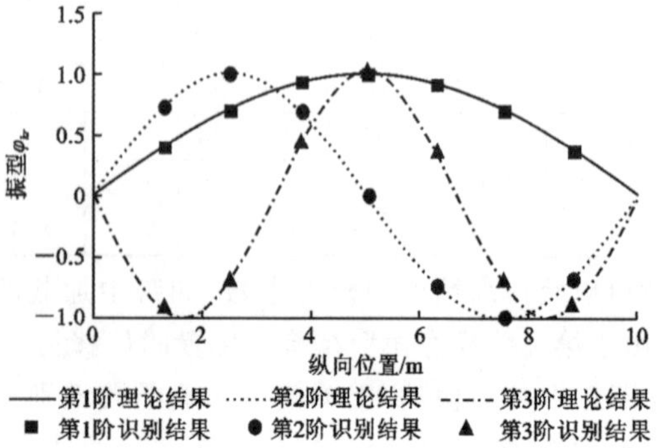

图 5　理论分析模态振型和识别振型

Fig. 5　Identified and Analytical Mode Shapes

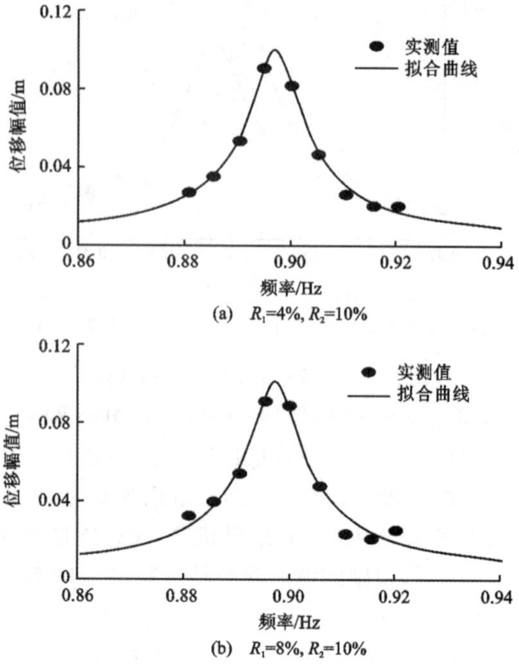

图 6　不同噪声比下频响函数曲线

Fig. 6　Curves of FRFs with Noise to Signal Ratios

声激励；测量噪声用零均值高斯白噪声模拟，叠加在结构的响应上。噪声水平由噪信比 R 确定，$R=\sigma_y/\sigma_x$。在研究激励噪声的影响时，σ_y 为激励噪声引起的响应均方根，σ_x 为最大稳态响应的均方根值，其噪声水平用 R_1 表示；在研究测量噪声的影响时，σ_y 为位移测量噪声均方根，σ_x 同样为最大稳态响应的均方根值，其噪声水平用 R_2 表示。采用 ANSYS 瞬态分析计算了简谐激励和噪声激励下结构前 200 s 的位移响应。测量噪声直接叠加在位移响应上。180 s 后，简谐激励响应基本稳定，以 180～200 s 的响应为基础计算结构的振幅。

不同噪信比对识别结果的影响如表 5 所示，部分噪信比下实测与拟合频响函数如图 6 所示。可知：随着噪声的增大，识别精度略有降低；在较大的噪声影响下，该方法仍具有很高的识别精度。

表 5　具有噪声的模态参数识别结果

Tab. 5　Modal Parameters Identification Results with Noise

模态参数	理论值	R_1=4%		R_1=8%	
		R_2=5%	R_2=10%	R_2=5%	R_2=10%
m_1/kg	31.415	31.470	31.576	30.450	30.330
f_1/Hz	0.903	0.897	0.897	0.897	0.897
ξ_1/%	0.50	0.50	0.50	0.51	0.51

3　工程应用

绵阳一号桥主桥是一座三跨双层分离桥面斜拉桥，跨径布置为 100 m＋200 m＋100 m。下层桥面为混凝土箱梁车行桥，梁高 3 m，宽 28 m；上层为钢箱梁曲线人行桥，梁高 2 m，宽 6 m，在平面上采用 S 形曲线设计，从车行桥的一侧跨越到另一侧。桥塔采用倒 Y 形桥塔，车行桥主梁通过横梁与桥塔相连，人行桥主梁通过桥塔两侧的牛腿与桥塔相连。图 7 给出了绵阳桥实桥情况。

图 7　绵阳一号桥实景

Fig. 7　General View of Mianyang First Bridge

由于大桥位于城市中心，因此动力设计时采用 1 人・m^{-2} 的高人群密度。理论与试验研究表明[15]，此人行桥极易发生人致振动。为了改善人行桥的动力特性，提出了采用连杆提高整体刚度和调谐质量阻尼器提高阻尼的联合减振方案。在主跨设置了 1 对竖杆和 1 对斜杆，在两边跨跨中分别设置了 1 对水平杆，连接人行桥主梁与车行桥主梁，有效提高了刚度，减少了易振模态数量，降低了纯调谐质量阻尼器减振方案中的 TMD 重量[15]。现场人群试验表明，横向失稳是该人行桥人致振动最显著的特征，在不到150人时就激发横向锁定现象，发生大幅横向振动，横向加速度达到0.2 $m \cdot s^{-2}$。为了提高横向失稳人数，在人行桥主跨跨中安装了1套水平TMD，用于控制1阶侧弯模态。对于竖向振动问题，共设计了4套竖向TMD分别用于控制4阶竖弯模态的振动。

模态质量是TMD优化设计的重要参数，模态质量的偏差会降低TMD的效率。因此，为了准确获得模态质量及结构阻尼比，应用伺服电机驱动的惯性质量激振器进行了简谐稳态强迫振动试验，评估1阶侧弯模态的频率、模态质量和阻尼比。激振器如图8所示。在水平活动台上增加质量块后，通过伺服电机驱动水平活动台做简谐水平运动，实现水平激励。经室内试验标定，该激振器最大水平运动质量为500 kg，最大振幅为0.055 m，最小频率间隔为0.01 Hz。

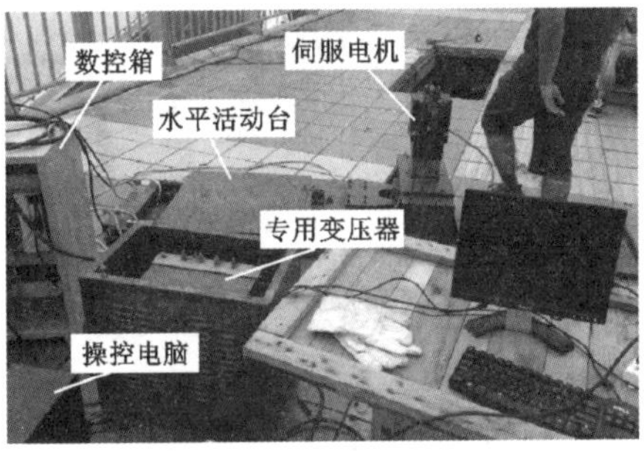

图8　水平激振器

Fig. 8　Horizontal Exciter

通过有限元分析可知，设置连杆后该桥第1阶横向振动频率为0.735 Hz，振型的最大位置在主跨跨中，如图9所示。按振型最大值为1得到的模态质量为513 t，为便于比较，上述模态质量中含有已安装在桥上的TMD重量。由于1阶对称横弯的振型最大值在跨中，因此将激振器置于主跨跨中。在强迫振动试验中，TMD全部锁定，激振器运动总质量 $m_0=310$ kg，振幅为 $A_0=0.052$ m，激振频率间隔为0.01 Hz，激振频率 $f=[0.68\ \text{Hz}, 0.77\ \text{Hz}]$。测试不同激振频率下人行桥主跨跨中加速度响应，如图10所示。从图10可以看出，当激振频率远离共振频率时，加速度响应很小，环境振动响应较为显著；当激振频率接近结构共振频率时，结构响应急速增大。

采用单模态频响函数识别了结构的模态参数，识别结果如表6和图11所示，1阶侧弯模态的频率为0.723 Hz，该频率对应的模态质量为519 t，阻尼比为0.63%，其值比设计规范中给定的0.50%略

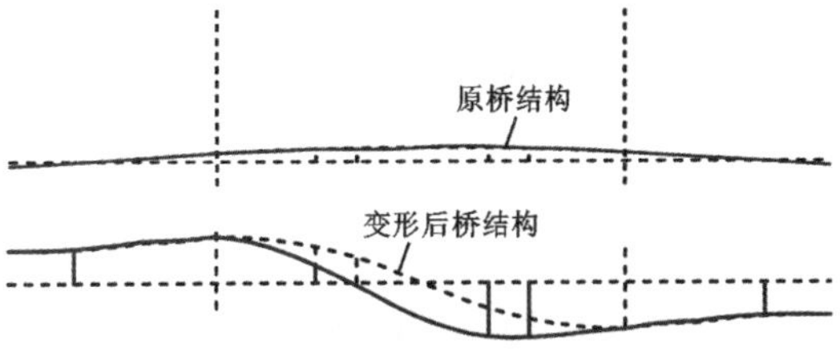

图9　第1阶侧弯模态振型

Fig. 9　Mode Shape for First Lateral Mode

大。为进一步验证识别结果，表6中也给出了由TMD激振后自由衰减响应识别的频率和阻尼比，实测自由衰减加速度响应如图12所示，但自由振动测试无法获得模态质量。在稳态激励振动测试和自由振动测试中，结构的振幅不同，频率和阻尼比具有与振幅相关的非线性特征，因此，2种方法识别的频率和阻尼比略有差异，理论分析与稳态简谐振动激励获得的模态质量相符。

4　结　语

(1)即使在较大环境噪声及测量噪声影响下，基于频响函数的模态识别方法也能够获得准确的模态参数识别结果。

(2)稳态激振器的激振质量会改变结构固有模态参数，对参数识别产生不利影响，因此，在满足激振条件下，应尽可能采用较小的激振质量，建议其不超过模态质量的1%。

(3)应用本文方法识别了绵阳一号桥侧弯模态的频率、阻尼比和模态质量，结果表明在环境振动影响下，基于正弦稳态强迫振动法识别的结果仍有很高精度，识别的模态质量与有限元分析值相符，可为质量型阻尼器的设计及精确调频提供依据。

(4)结构的模态特征具有与振幅相关的非线性特征，而本文方法未考虑结构模态特征的非线性特征，如何采用本文方法识别结构振幅相关非线性模态特征有待于进一步研究。

参考文献：

References:

[1] INGÓLFSSON E T，GEORGAKIS C T，JÖNSSON J. Pedestrian-induced Lateral Vibrations of Footbridges: A Literature Review[J]. Engineering Structures，2012 (45):21-52.

[2] BAYRAKTAR A，ALTUNISIK A，SEVIM B，et al. Modal Testing，Finite-element Model Updating，and Dynamic Analysis of an Arch Type Steel Footbridge [J]. Journal of Performance of Constructed Facilities.

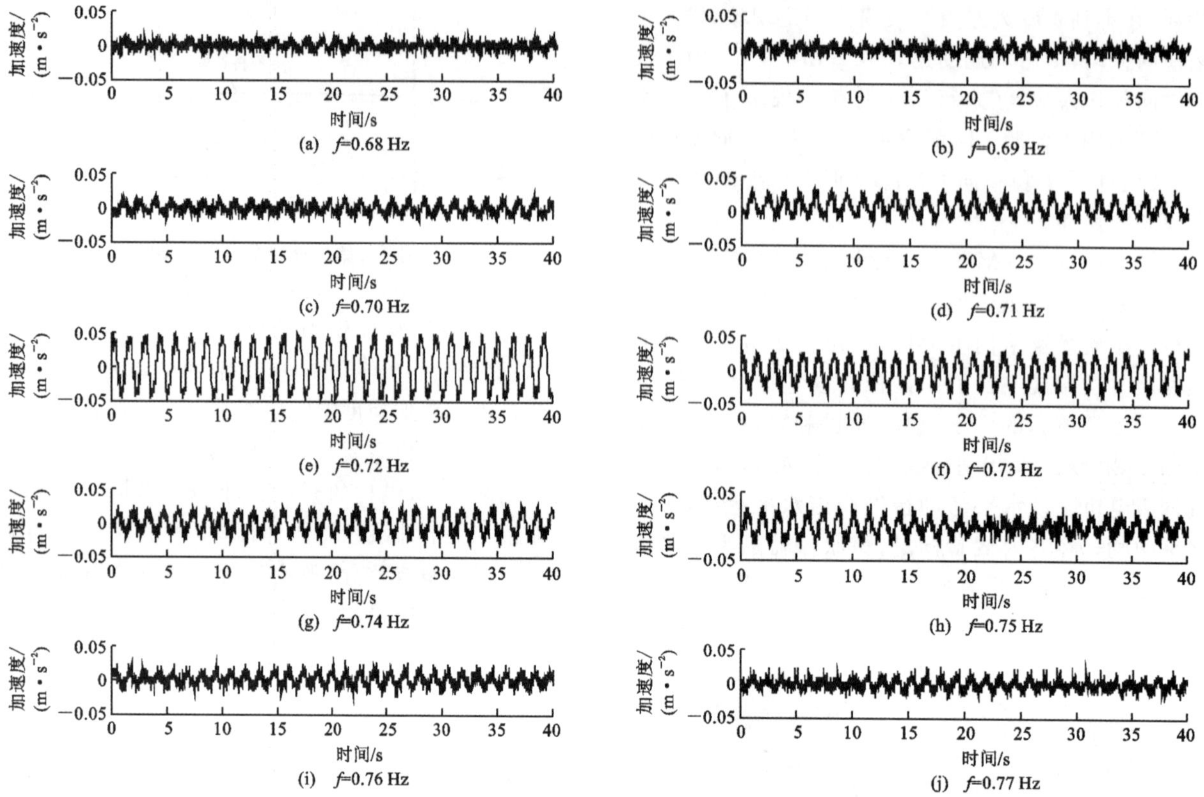

图 10 不同频率激振下主跨跨中侧向加速度响应

Fig. 10 Lateral Acceleration Responses of Main Span Under Different Excitation Frequencies

表 6 不同方法的模态参数

Tab. 6 Modal Parameters Estimated by Different Methods

分析方法	频率/Hz	阻尼比/%	模态质量/t
有限元分析	0.735		513
稳态激振法	0.723	0.63	519
自由衰减法	0.715	0.72	

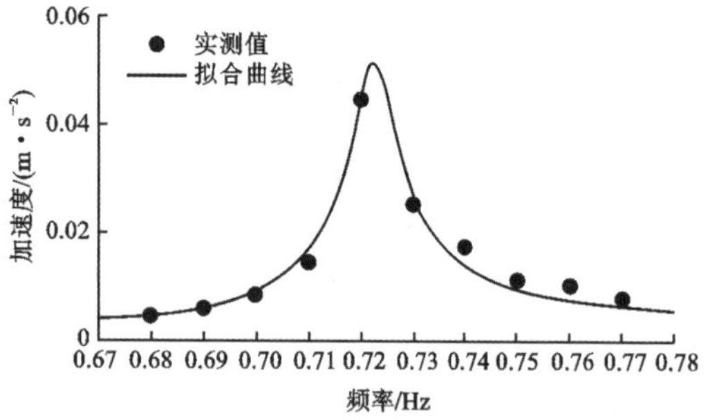

图 11 实测加速度及最小二乘拟合值比较

Fig. 11 Acceleration Amplitudes Contrast of Measured and Its Least-squares Curve Fitted Values

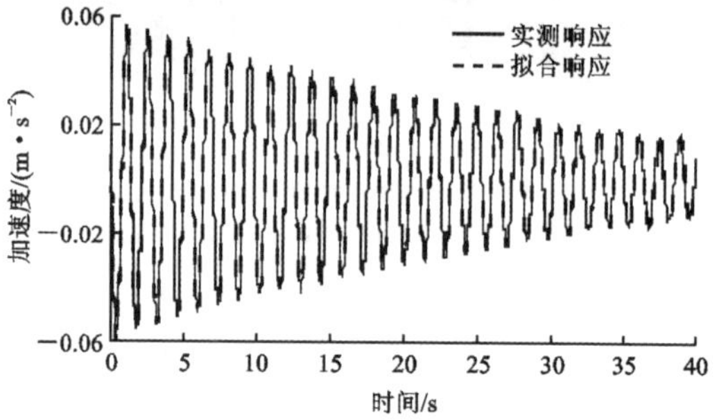

图 12 第 1 阶横向振动自由衰减加速度响应与拟合值

Fig. 12 Free Decay Acceleration Responses and Fitted Values of First Lateral Vibration

2009,23(2):81-89.

[3] 王洪涛,施卫星,韩建平,等.钢连桥人致振动及 TMD 减振效应实测与分析[J].振动、测试与诊断,2016,36(3):505-511,604-605.

WANG Hong-tao, SHI Wei-xing, HAN Jian-ping, et al. Analysis and In-situ Test of Human-induced Vibration for the Steel Footbridge with and Without TMD Devices[J]. Journal of Vibration, Measurement & Diagnosis, 2016, 36(3):505-511, 604-605.

[4] NI Y, ZHANG F, LAM H. Series of Full-scale Field Vibration Tests and Bayesian Modal Identification of a Pedestrian Bridge[J]. Journal of Bridge Engineering, 2016, 21(8):1-18.

[5] TUBINO F, CARASSALE L, PICCARDO G. Human-induced Vibrations on Two Lively Footbridges in Milan[J]. Journal of Bridge Engineering, 2016, 21(8):1-10.

[6] CAETANO E, CUNHA Á, MOUTINHO C, et al.

Studies for Controlling Human-induced Vibration of the Pedro e Inês Footbridge, Portugal. Part 1: Assessment of Dynamic Behavior[J]. Engineering Structures, 2010, 32(4): 1069-1081.

[7] 张冠华,葛耀君. 砼连续箱梁人行天桥振动特性测试与分析[J]. 振动与冲击,2009,28(2):102-106,204.
ZHANG Guan-hua, GE Yao-jun. Vibration Characteristics Test and Analysis on Concrete Box Girder Continuous Pedestrian Bridge[J], Journal of Vibration and Shock, 2009, 28(2): 102-106, 204.

[8] PORRAS J A, DE SEBASTIAN J, CASADO C M, et al. Modal Mass Estimation from Output-only Data Using Oscillator Assembly[J]. Mechanical Systems and Signal Processing, 2012, 26: 15-23.

[9] DALLARD P, FITZPATRICK A, FLINT A, et al. The London Millennium Footbridge[J]. The Structural Engineer, 2001, 79(22): 17-33.

[10] ŽIVANOVIC S, PAVIC A, REYNOLDS P. Modal Testing and FE Model Tuning of a Lively Footbridge Structure[J]. Engineering Structures, 2006, 28(6): 857-868.

[11] CAETANO E, CUNHA Á, MOUTINHO C, et al. Studies for Controlling Human-induced Vibration of the Pedro e Inês Footbridge, Portugal. Part 2: Implementation of Tuned Mass Dampers[J]. Engineering Structures, 2010, 32(4): 1082-1091.

[12] BROWNJOHN J M W, PAVIC A. Experimental Methods for Estimating Modal Mass in Footbridges Using Human-induced Dynamic Excitation[J]. Engineering Structures, 2007, 29(11): 2833-2843.

[13] CLOUGH R W, PENZIEN J. Dynamics of Structures[M]. 3rd ed. Berkeley: Computers & Structures, Inc., 2003.

[14] EWINS D J. Modal Testing: Theory and Practice[M]. London: Research Studies Press, 1985.

[15] 陈政清,刘光栋. 人行桥的人致振动理论与动力设计[J]. 工程力学,2009,26(增2):148-159.
CHEN Zheng-qing, LIU Guang-dong. Pedestrian-induced Vibration Theory and Dynamics Design of Footbridges[J]. Engineering Mechanics, 2009, 26(S2): 148-159.

健康监测与其他主题

International Journal of Structural Stability and Dynamics
Vol. 13, No. 5 (2013) 1250083 (21 pages)

DOI: 10.1142/S0219455412500836

PERFORMANCE VARIATIONS OF A CABLE-STAYED BRIDGE WITH DAMAGED CABLES

OU YANG

College of Civil Engineering, Hunan University
Changsha, China
ouyang.hnu@gmail.com

HUI LI

School of Civil Engineering
Harbin Institute of Technology, Harbin, China

JINPING OU

School of Civil Engineering
Dalian University of Technology, Dalian, China

Q. S. LI

Department of Civil and Architectural Engineering
City University of Hong Kong, Hong Kong, China

Received 16 July 2012
Accepted 18 September 2012
Published 2 April 2013

A cable-stayed bridge model was made based on the design drawings of Shandong Binzhou Yellow River Highway Bridge (BZB) and detailed model tests were conducted. In the tests, the cable damage effects on the loading behavior and dynamic performance of the bridge model were studied by considering different cable damages' locations, different numbers of damaged cables, and different damaged levels of cables etc. The experimental results revealed that the cable damages may significantly affect the internal forces of other cables and the stress distributions of the girders; however, they have less influence on the natural frequencies of the bridge model. In parallel with the model tests, a finite element model of the bridge structure has been established. Numerical analysis for the bridge model with cable damages was carried out to validate the experimental results and explain the phenomena observed in the model tests.

Keywords: Cable-stayed bridge; cable damage; model test; finite element method.

1. Introduction

Stay-cables are critical components of cable-stayed bridges, which are attached to pylons to support girders. Extensive studies have been conducted on the factors which may cause damages on cables and the means to protect cables.[1,2] For example, Sih *et al.*[3] developed a model to determine the fatigue crack growth in pre-tensioned bridge cables made of stranded steel wires. Périer *et al.*[4] investigated the cable fretting-fatigue behavior in a solution of sodium chloride. Li *et al.*[5] investigated the mechanical properties and residual fatigue life of the corroded cables of a cable-stayed bridge. Hamilton III *et al.*[6,7] analyzed the corrosion protection ability of the traditional stay-cable corrosion protection system of cement grout and high-density polyethylene sheathing. However, few studies were concentrated on the performance variations of bridges with occurrence of cable damages. Obviously, this is a crucial issue for the safe maintenance of cable-stayed bridges, which needs to be investigated in detail. To this end, a laboratorial cable-stayed bridge model was constructed in this study to investigate the loading behavior and dynamic performance of the bridge model with the presence of cable damages. Three categories of cable damages were considered in this paper: (1) one cable lost 30% of stress and fractured; (2) two cables in a same girder section lost 30% of stress and fractured; (3) two adjacent cables lost 30% of stress and fractured. Each category contained four cases with the occurrence of cable damages at different locations. A finite element model (FEM) was established and numerical analysis for the bridge model with cable damages was carried out to verify the experimental results, while explaining the phenomena observed in the model tests.

2. Introduction of the Bridge Model

A laboratorial cable-stayed bridge model was designed and manufactured based on the design drawings of Binzhou Yellow River Highway Bridge (BZB), located in the Shandong Province of China. The bridge girder consists of four spans of 84 m (side span), 300 m (main span), 300 m (main span), and 84 m (side span) in length, which has a total length of 768 m. The bridge deck is 32.8 m wide and 2.8 m deep with two triangle-box-girders on each side. The bridge has three pylons to anchor the stay-cables. The middle pylon extends 123.25 m above the base, and two side pylons extend 75.78 m above the base. All the pylons consist of two hollow columns with no connection above the deck. There are 200 cables with two parallel rows, in a modified fan layout. Each side of the middle pylon is anchored with 26 cables in one row, and each side of the side pylon is anchored with 12 cables, as shown in Fig. 1.

The laboratorial model was designed based on the similarity of girder stiffness with dimension scale of 1/40. Aluminium alloy with Young's modulus of 53.5 GP and density of 2700 kg/m^3 was utilized to manufacture the bridge model pylons and deck. The steel wires with Young's modulus of 200 GP and density of 7500 kg/m^3 were used to represent the bridge cables. The primary similarity relations used in the model design are listed in Table 1.

Because of the limitation of laboratorial space, only two main spans were considered in the present tests. The bridge model is with 15 m long in total. The deck of the bridge model has a width of 0.82 m. The main pylon and two side pylons extend

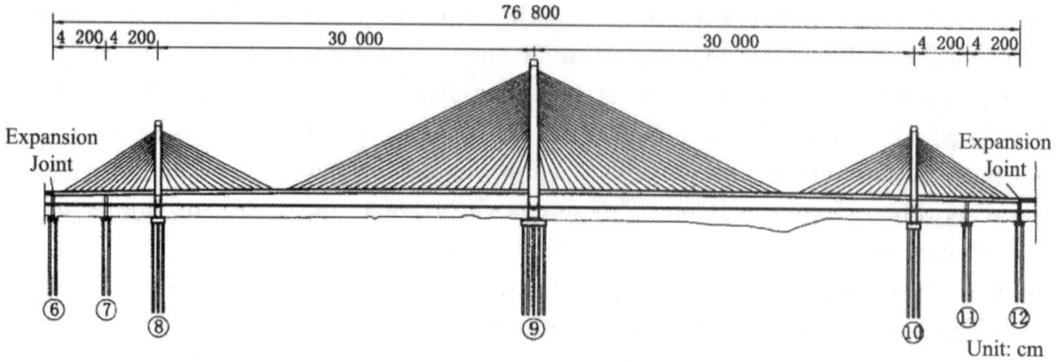

Fig. 1. Layout of BZB.

Table 1. Similarity relations.

Parameters	Dimension	Similarity Relations	Ratio of Similitude
Strain [ε]	—	$S_\varepsilon = 1$	1.000
Young's modulus [E]	$ML^{-1}T^{-2}$	S_E	1.643
Density [ρ]	ML^{-3}	S_ρ	9.456
Length [l]	L	S_l	0.025
Displacement [δ]	L	$S_\delta = S_l$	0.025
Area [A]	L^2	$S_A = S_l^2$	6.250×10^{-4}
Load [F]	MLT^{-2}	$S_F = S_E S_l^2$	1.027×10^{-3}
Frequency [ω]	T^{-1}	$S_\omega = \sqrt{S_E/S_\rho}/S_l$	16.673

3.1 m and 1.9 m above the base, respectively. Since the bridge model is very large, it is difficult to make it by casting integrally. The bridge model consists of 21 components, including five pair of symmetric deck components, five main tower components, and three components for each side tower. These components were assembled using connectors. The assembled bridge model is shown in Fig. 2.

For convenience, the number of stay-cables was reduced by keeping the vertical stiffness of the cables as constant.[8] The cables are numbered from 1 to 26 with the parallel row cables near the door labeled with U and those near the window with D,

Fig. 2. Photo of the bridge model.

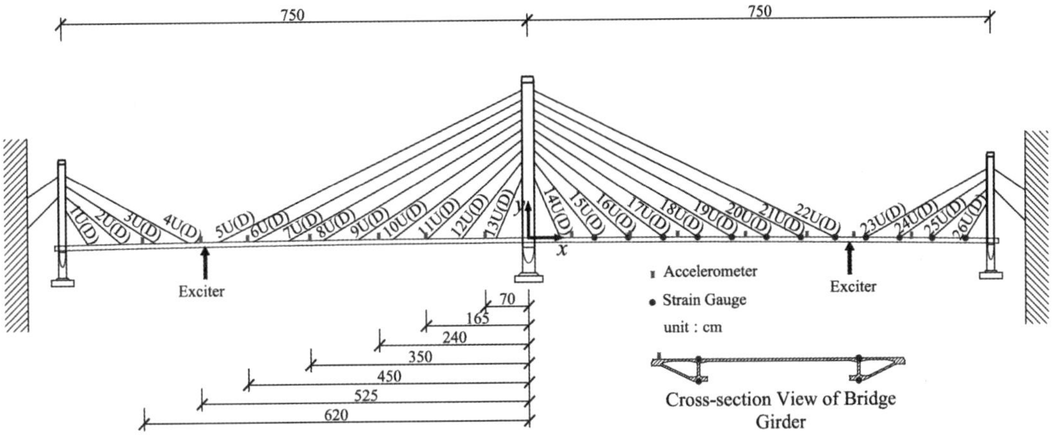

Fig. 3. Serial number of stayed-cables and measurement point layout.

as shown in Fig. 3. Every numbered cable was attached a bared Fiber Bragg Grating (FBG) sensor to record the variation of the cable force during the tests. The anchoring points at the bridge deck of cables No. 15 to 26 (including both parallel row cables) were stuck with strain gauges at both the top and bottom of the deck to monitor the variation of the girder strain when cables were damaged, as marked with blue dots in Fig. 3. A total of 14 accelerometers were mounted symmetrically along the main span, as marked with red rectangle in Fig. 3.

3. Finite Element Model

To validate the experimental results and provide more information, a FEM of the bridge model has been established using the ANSYS package. The beam-shell type element has been adopted for the bridge deck considering its characteristics, while the pylons of the bridge model are modeled using beam elements, and the bridge deck is divided to two longitudinal girders modeled by the beam elements linked by deck

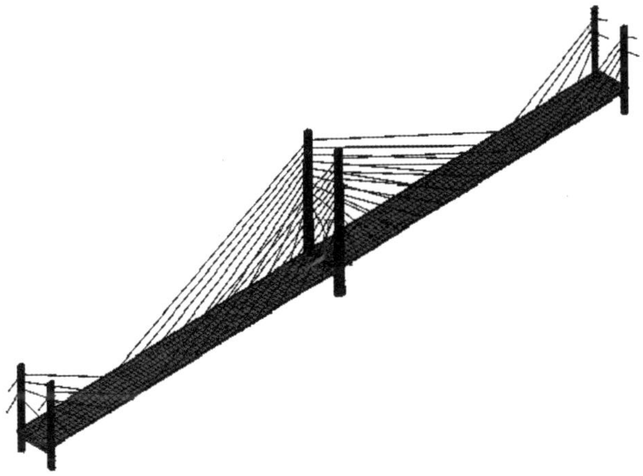

Fig. 4. 3D view of the FEM.

and transverse ribs modeled by shell elements and beam elements, respectively. The established numerical model can be used to simulate the bending stiffness of the bridge deck precisely with consideration of the warping deformation, which makes the numerical results close to the test results. The 3D view of the FEM model of the bridge is shown in Fig. 4.

4. Initial State of the Bridge Model

Before the tests were conducted, the initial state of the bridge model was recorded, including the initial cable forces and natural frequencies of the model. The initial cable forces were obtained from the optimization analysis by the finite element method, based on the equilibrium configuration defined as the cable forces that made the bending moments of the bridge girder and pylons be the minimum.[9] The influence matrix methodology[10] was adopted to adjust the cable forces of the bridge model. The cable forces after the adjustment are shown in Fig. 5.

The identified initial natural frequencies of the bridge model are listed in Table 2, where the symbols f_i and ζ_i represent the natural frequency and damping ratio of the ith mode, respectively, and f_a denotes the natural frequency calculated by the FEM.

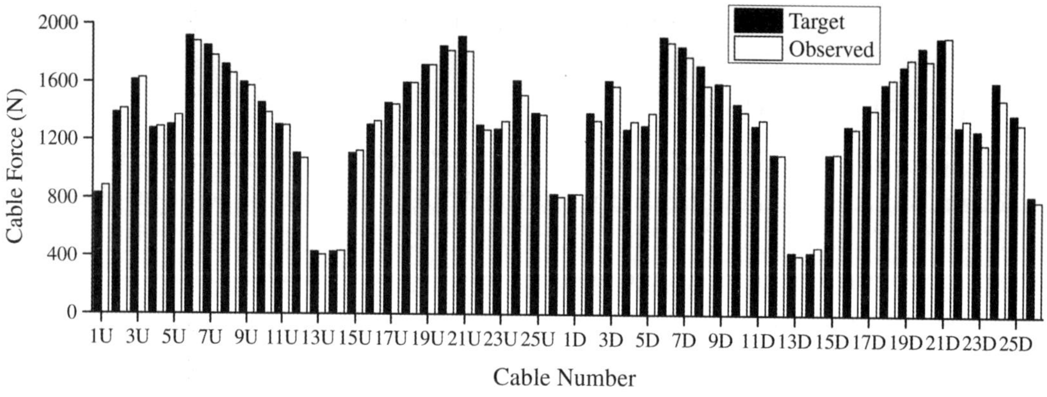

Fig. 5. Initial cable forces after adjustment.

Table 2. Modal parameter identification results of the bridge model.

Modes	f_i (Hz)	ζ_i (%)	f_a (Hz)	Error (%)
1st antisymmetric bending shape	4.014	2.71	4.254	5.97
1st symmetric bending shape	8.839	2.43	9.126	3.25
2nd antisymmetric bending shape	10.822	1.36	11.163	3.15
2nd symmetric bending shape	11.857	0.77	12.088	1.95
3rd antisymmetric bending shape	15.322	0.21	14.872	−2.93
3rd symmetric bending shape	16.344	0.81	15.992	−2.15
4th antisymmetric bending shape	21.926	0.90	22.378	2.06
4th symmetric bending shape	22.962	0.47	23.078	0.51

An approach that combines the natural excitation technique (NExT)[11,12] and the eigensystem realization algorithm (ERA)[13,14] was adopted to identify the natural frequencies of the bridge model. As shown in the table, the identified natural frequencies and calculated ones are in good agreement, illustrating that the FEM can provide accurate predictions of the dynamic characteristics of the bridge model.

5. Case Studies

In this study, the performance variation of the bridge model was investigated systemically with 12 cases of cable damages, as listed in Table 3. Each case included two different damaged levels (30% and 100%). These damage cases can be classified into three categories: (1) one cable was damaged (Case 1 to Case 4); (2) two cables anchored in the same girder section were damaged (Case 5 to Case 8); (3) two adjacent cables were damaged (Case 9 to Case 12).

Since the bridge is symmetric in geometry, only the right side cables will be considered herein. The principle for selecting the cable damages is based on the cable length and distribution. Three cases for the main pylon cables including the longest, medium and short cables, and one case for the side tower cables have been chosen in this study.

6. Results and Discussions

The performance tests of the bridge model with cable damages, including the variations of cable forces, girder strain, and natural frequencies, were conducted. To verify the test results, the numerical results by the FEM are also provided for each case. During the computational process, the reduction of cable forces was modeled by the approach of adjusting thermal stress, and the fracture of cables was simulated by the program function of "element kill" (EKILL function in ANSYS).

Table 3. Cases of cable damages.

Cases	Description
Case 1	Cable No. 24U with reduction of 30% cable force and totally fractured.
Case 2	Cable No. 22U (which is the longest cable of the bridge model) with reduction of 30% cable force and totally fractured.
Case 3	Cable No. 19U with reduction of 30% cable force and totally fractured.
Case 4	Cable No. 15U with reduction of 30% cable force and totally fractured.
Case 5	Both cables No. 24U and 24D with reduction of 30% cable forces and totally fractured.
Case 6	Both cables No. 22U and 22D with reduction of 30% cable forces and totally fractured.
Case 7	Both cables No. 19U and 19D with reduction of 30% cable forces and totally fractured.
Case 8	Both cables No. 15U and 15D with reduction of 30% cable forces and totally fractured.
Case 9	Both cables No. 24U and 25U with reduction of 30% cable forces and totally fractured.
Case 10	Both cables No. 21U and 22U with reduction of 30% cable forces and totally fractured.
Case 11	Both cables No. 18U and 19U with reduction of 30% cable forces and totally fractured.
Case 12	Both cables No. 14U and 15U with reduction of 30% cable forces and totally fractured.

For convenience, the differential value between the top and bottom strain of one girder section is defined as the bending strain. The coordinate system for the girder is defined as shown in Fig. 3, where the origin of the coordinate is set at the intersection point of the main tower and girder, with the right direction defined as positive.

The frequency change ratio (f_v) and damping ratio variance (ζ_v) are defined in Eq. (1) and Eq. (2), respectively.

$$f_v = \frac{f_c - f_o}{f_o} \times 100\%, \tag{1}$$

$$\zeta_v = \zeta_c - \zeta_o, \tag{2}$$

where f_c and f_o represent the natural frequency of the bridge model for the current damaged case and the initial state of the bridge model, respectively. ζ_c and ζ_o denote the damping ratio of the bridge model for the current damage case and the initial state of the bridge model, respectively.

6.1. *One cable damaged cases*

The variations of cable forces for the cases with one cable damaged (Cases 1−4, including cable force with reduction of 30% and cable fractured) are shown in Figs. 6− 9. The variations of the bending strain of girder section for these Cases (Cases 1−4, including cable force with reduction of 30% and cable fractured) are shown in Figs. 10−13. The dynamic performance variations for these cases (Cases 1−4, including cable force with reduction of 30% and cable fractured) are shown in Tables 4−7. These results show the same tendency of variations of the cable force and girder section bending strain for the cases of the cable force with a reduction of 30% and one cable fractured. Hence, only the fractured cable test results will be presented and discussed below.

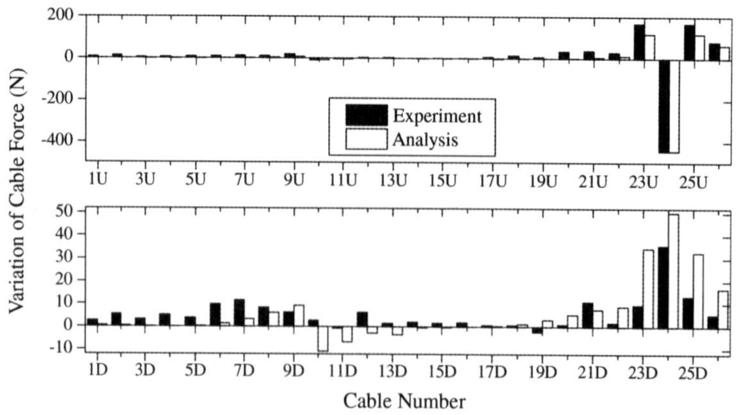

(a) Cable force reduced by 30%

Fig. 6. Variation of entire bridge cable force caused by damage of cable 24U.

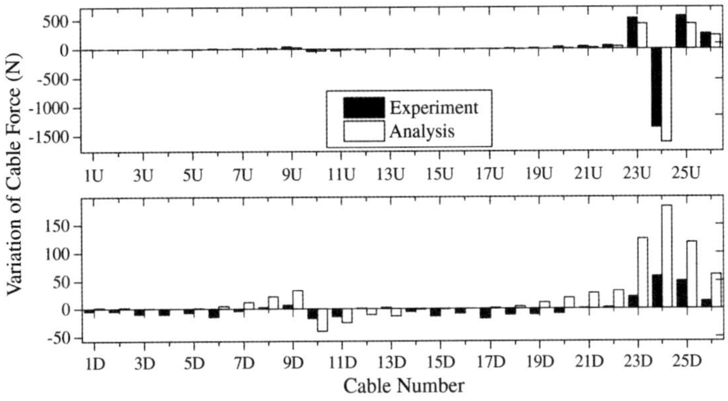

(b) Cable fractured

Fig. 6. (*Continued*)

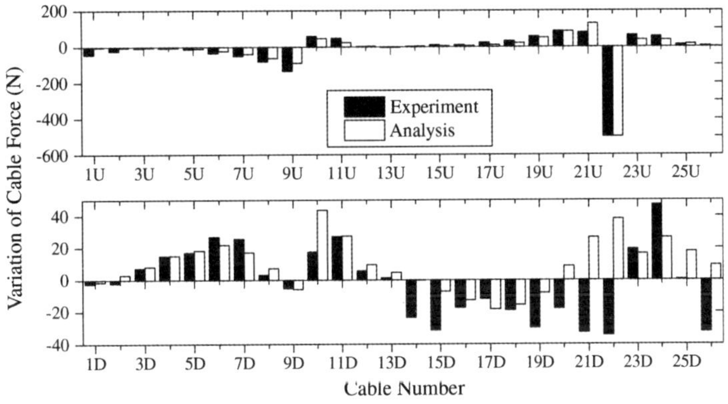

(a) Cable force reduced by 30%

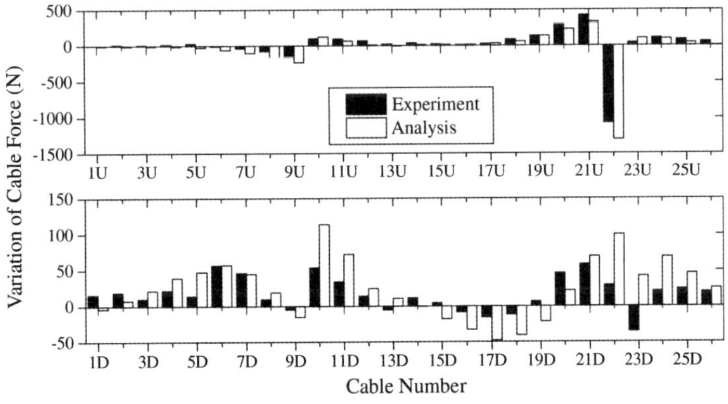

(b) Cable fractured

Fig. 7. Variation of entire bridge cable force caused by damage of cable 22U.

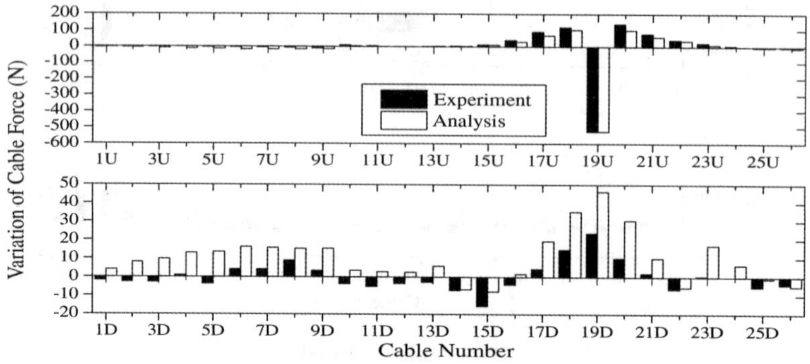

(a) Cable force reduced by 30%

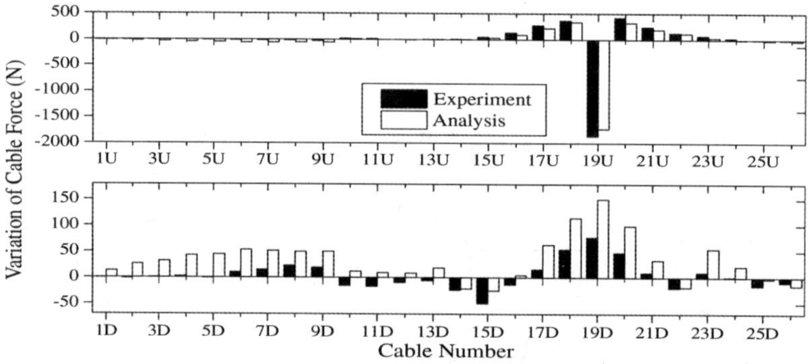

(b) Cable fractured

Fig. 8. Variation of entire bridge cable force caused by damage of cable 19U.

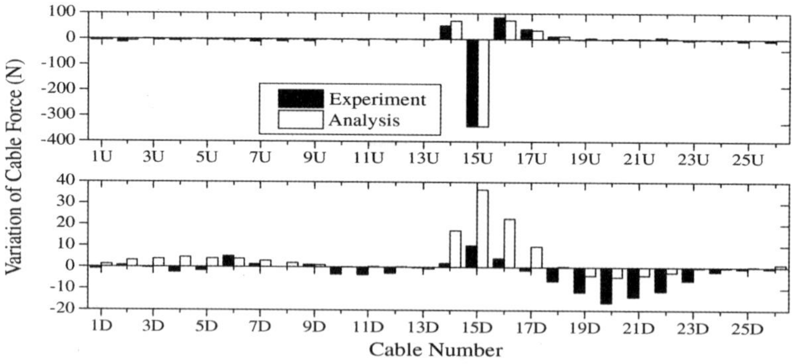

(a) Cable force reduced by 30%

Fig. 9. Variation of entire bridge cable force caused by damage of cable 15U.

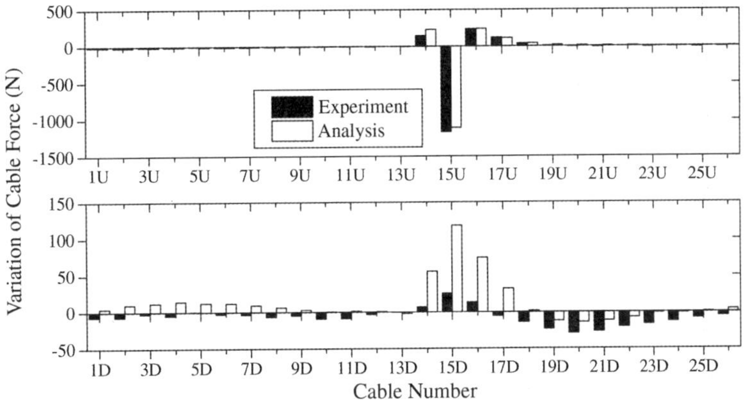

(b) Cable fractured

Fig. 9. (*Continued*)

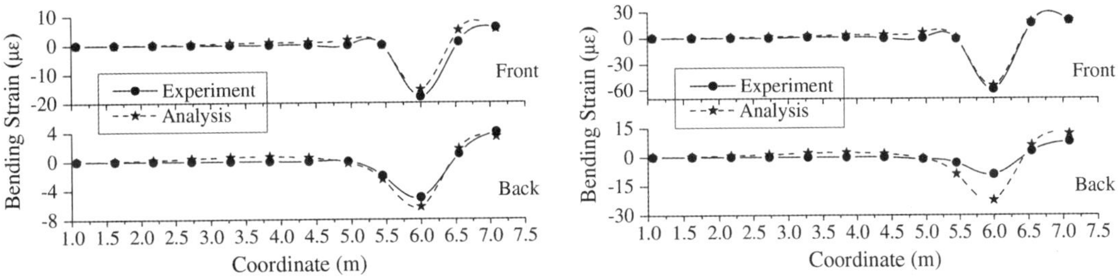

(a) Cable force reduced by 30%

(b) Cable fractured

Fig. 10. Variation of girder section bending strain caused by damage of cable 24U.

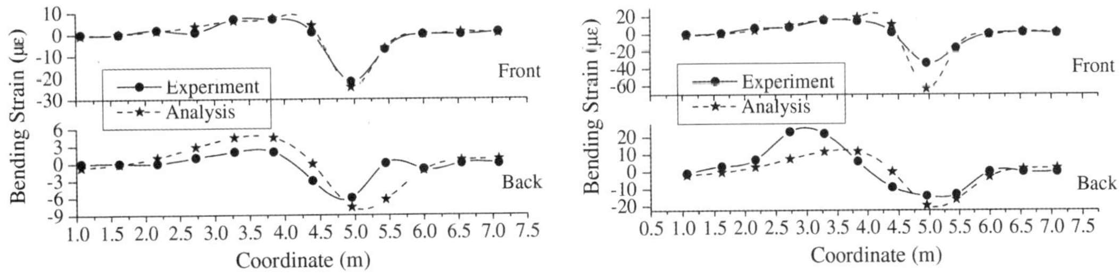

(a) Cable force reduced by 30%

(b) Cable fractured

Fig. 11. Variation of girder section bending strain caused by damage of cable 22U.

The FEM predictions and the test results are found to be in reasonably good agreement, while for the parts without cable damage, the test results are generally lower than the predictions. This is because the two triangle box girders were modeled using the beam elements, which caused the transverse span between the two box

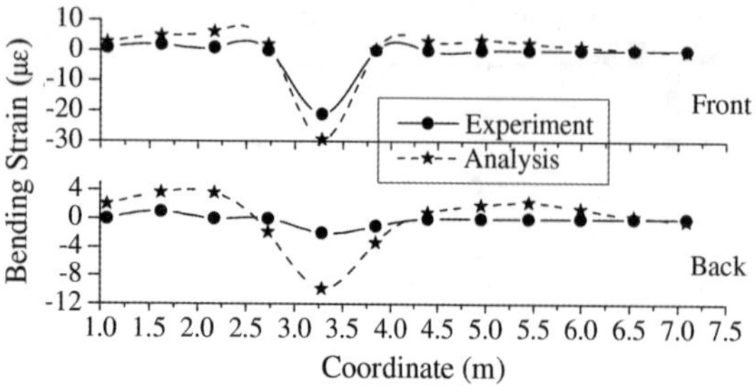

(a) Cable force reduced by 30%

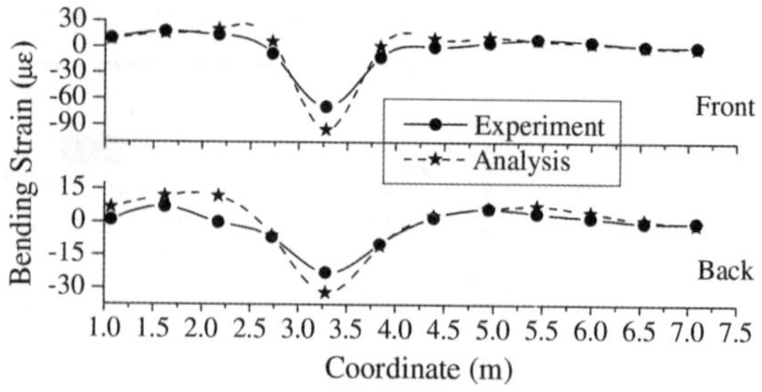

(b) Cable fractured

Fig. 12. Variation of girder section bending strain caused by damage of cable 19U.

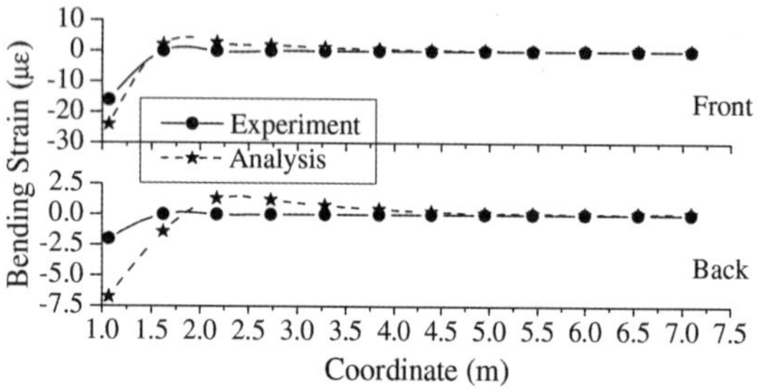

(a) Cable force reduced by 30%

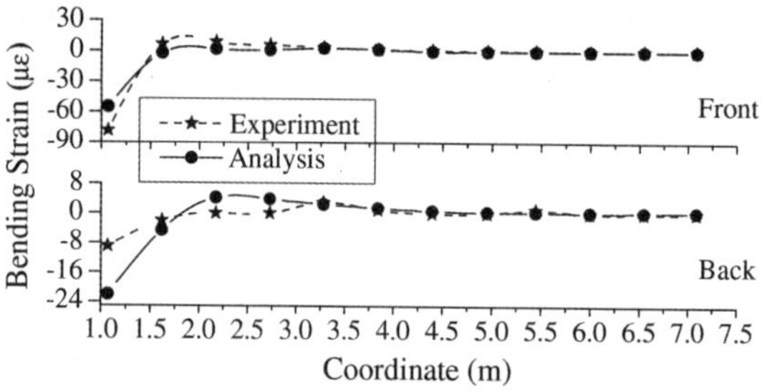

(b) Cable fractured

Fig. 13. Variation of girder section bending strain caused by damage of cable 15U.

Table 4. Variation of the dynamic performance caused by damage of cable 24U.

Description	Cable force reduced by 30%				Cable fractured			
	f (Hz)	ζ (%)	f_v (%)	ζ_v (%)	f (Hz)	ζ (%)	f_v (%)	ζ_v (%)
1st antisymmetric bending shape	4.02	1.55	0.43	−0.35	3.99	2.07	−0.24	0.16
1st symmetric bending shape	8.85	1.59	0.56	0.41	8.72	2.01	−0.84	0.83
2nd antisymmetric bending shape	10.80	1.37	0.19	0.85	10.75	0.78	−0.31	0.26
2nd symmetric bending shape	11.84	0.95	−0.02	0.46	11.72	0.91	−0.99	0.42
3rd antisymmetric bending shape	15.32	0.12	−0.02	−0.69	15.31	0.26	−0.05	−0.55
3rd symmetric bending shape	16.50	0.17	0.46	−0.22	16.41	0.26	−0.09	−0.13
4th antisymmetric bending shape	21.98	1.06	0.16	0.70	21.80	1.00	−0.68	0.64
4th symmetric bending shape	22.88	0.57	−0.07	0.04	22.92	0.45	0.11	−0.08

Table 5. Variation of the dynamic performance caused by damage of cable 22U.

Description	Cable force reduced by 30%				Cable fractured			
	f (Hz)	ζ (%)	f_v (%)	ζ_v (%)	f (Hz)	ζ (%)	f_v (%)	ζ_v (%)
1st antisymmetric bending shape	4.00	1.91	−0.39	−0.80	3.91	3.48	−2.57	0.77
1st symmetric bending shape	8.80	1.18	−0.46	−1.25	8.59	1.84	−2.87	−0.58
2nd antisymmetric bending shape	10.78	0.52	−0.40	−0.84	10.92	3.22	0.95	1.86
2nd symmetric bending shape	11.84	0.49	−0.15	−0.28	11.78	0.44	−0.65	−0.33
3rd antisymmetric bending shape	15.32	0.81	0.00	0.60	15.39	0.10	0.46	−0.11
3rd symmetric bending shape	16.42	0.39	0.49	−0.42	16.82	1.27	2.91	0.47
4th antisymmetric bending shape	21.95	0.36	0.11	−0.54	21.87	1.23	−0.27	0.34
4th symmetric bending shape	22.90	0.53	−0.27	0.06	23.00	1.02	0.17	0.56

Table 6. Variation of the dynamic performance caused by damage of cable 19#.

Description	Cable force reduced by 30%				Cable fractured			
	f (Hz)	ζ (%)	f_v (%)	ζ_v (%)	f (Hz)	ζ (%)	f_v (%)	ζ_v (%)
1st antisymmetric bending shape	3.99	2.68	−0.66	−0.03	4.02	3.46	0.03	0.75
1st symmetric bending shape	8.80	1.31	−0.46	−1.12	8.66	1.60	−1.97	−0.83
2nd antisymmetric bending shape	10.83	0.63	0.08	−0.73	10.87	0.71	0.47	−0.65
2nd symmetric bending shape	11.87	1.24	0.14	0.47	11.73	0.95	−1.10	0.18
3rd antisymmetric bending shape	15.22	0.43	0.64	0.22	15.18	0.28	−0.92	0.07
3rd symmetric bending shape	16.41	0.34	0.39	−0.47	16.42	0.13	0.50	−0.68
4th antisymmetric bending shape	21.89	1.30	−0.17	0.41	21.73	1.05	−0.91	0.15
4th symmetric bending shape	22.95	0.66	−0.06	0.20	22.88	0.40	−0.34	−0.06

girders in the finite element analysis to be larger than that of the real structure. However, the variation trends of the computational results and the test data are similar.

For the cables with tower anchoring points located near the damaged cable, their forces changed more significantly than those with their girder anchoring points

Table 7. Variation of the dynamic performance caused by damage of cable 15U.

Description	Cable force reduced by 30%				Cable fractured			
	f (Hz)	ζ (%)	f_v (%)	ζ_v (%)	f (Hz)	ζ (%)	f_v (%)	ζ_v (%)
1st antisymmetric bending shape	4.01	1.92	−0.01	−0.78	4.04	2.13	0.56	−0.58
1st symmetric bending shape	8.76	1.24	−0.87	−1.19	8.87	1.91	0.40	−0.52
2nd antisymmetric bending shape	10.83	0.98	0.05	−0.38	10.80	0.45	−0.22	−0.91
2nd symmetric bending shape	11.87	0.89	0.14	0.12	11.78	0.64	−0.63	−0.13
3rd antisymmetric bending shape	15.24	0.12	−0.54	−0.09	15.28	0.62	−0.26	0.41
3rd symmetric bending shape	16.39	0.66	0.26	−0.15	16.23	0.94	−0.70	0.13
4th antisymmetric bending shape	21.92	0.83	−0.01	−0.06	21.64	0.78	−1.31	−0.12
4th symmetric bending shape	22.89	0.40	−0.32	−0.06	22.83	0.42	−0.57	−0.05

located near the damaged cable. For example, when cable 22U was damaged (which is anchored at the main pylon), the force variation of cable 21U (which is also anchored at the main pylon) is obviously larger than that of cable 23U, whose girder anchoring point is near the damaged cable but anchored at the side pylon. The same phenomenon was observed when the damaged cable was anchored at the side pylon. The possible reason is because the stiffness of the tower is larger than that of the girder and the interval of the cable anchoring points in the pylon is much smaller than that in the girder.

The bending strain in the girder section around the damaged cables decreased significantly. Instead, the bending strain in other parts of the girder showed a little increase. The variation of the bending strain at the girder side where a damaged cable was located was larger than those of the other sides, indicating that the girder side was subjected to the bending-torsional coupling deformation.

In the four cases discussed above, several natural frequencies of the bridge model increased when the cable force was reduced by 30%. Especially when the force of cable 22U was reduced to 30%, most natural frequencies increased. The probable reason is that when one cable force was reduced, those of the other cables were increased, thereby enhancing the overall stiffness. Another reason is that the bending-torsional coupling deformation of the bridge deck enhanced its bending stiffness. The model's natural frequencies tended to decrease when one cable was fractured, but the amplitudes of such variations were less than 5%. The damping ratio of the bridge model varied randomly when a cable was damaged, because the errors of the identification method were relatively large, which covered up the variation of the damping ratio caused by the cable damage.

6.2. *Two damaged cables anchored in a same girder section*

As indicated previously, different damaged levels of a cable tend to share the same variation trend. Hence, the cable fractural test results were mainly presented and discussed in the following sections.

The variations of cable forces for the cases when two damaged cables are anchored at the same girder section (Case 5–Case 8) are shown in Fig. 14. As shown in the figure, the experimental data matched the computational results well, implying that the FEM can simulate the bending stiffness of the bridge girder very well. Moreover, the influence of the cable damages on the adjacent cables anchored at the same pylon is larger than those anchored at different pylons, which validated the observation made for the cases with one cable damaged.

The variations of the bending strain of the girder for Case 5–Case 8 are shown in Fig. 15. The results show that the bending strain varied almost identically on both sides of the girder, indicating that only bending deformation occurred in these cases.

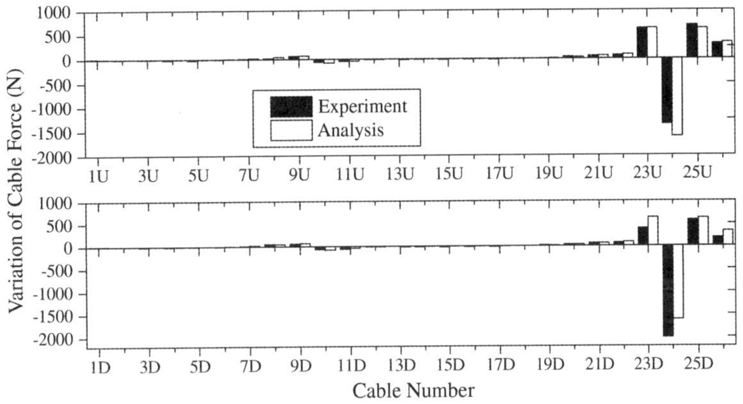

(a) Fracture of cables 24U and 24D

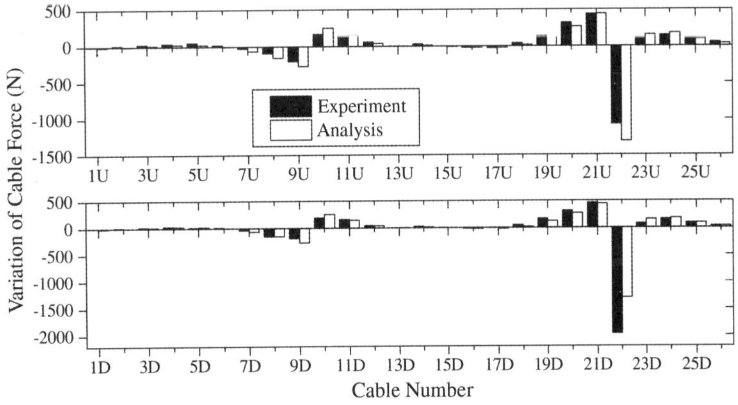

(b) Fracture of cables 22U and 22D

Fig. 14. Variation of entire bridge cable force caused by fracture of two cables on a same section.

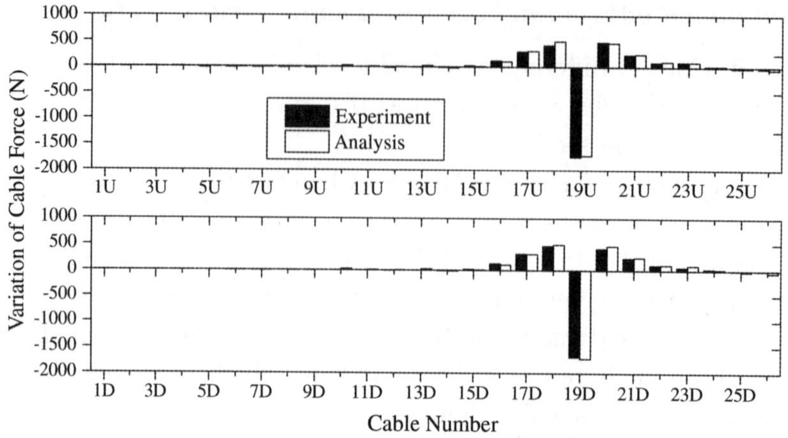

(c) Fracture of cables 19U and 19D

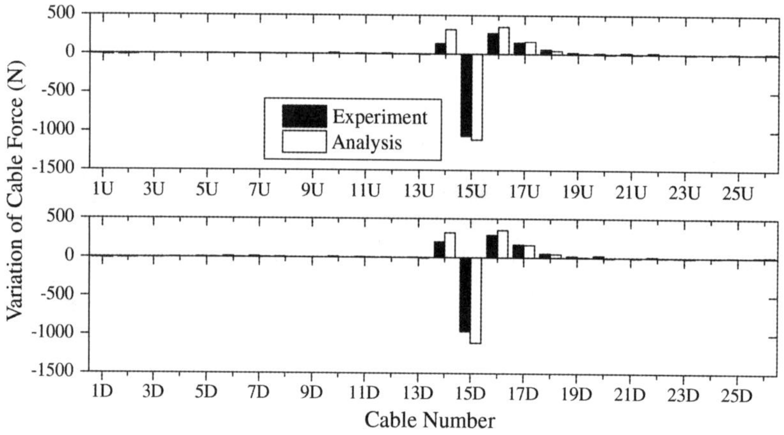

(d) Fracture of cables 15U and 15D

Fig. 14. (*Continued*)

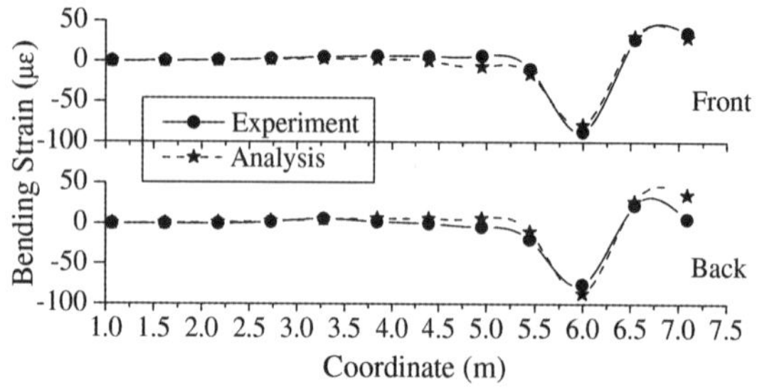

(a) Fracture of cables 24U and 24D

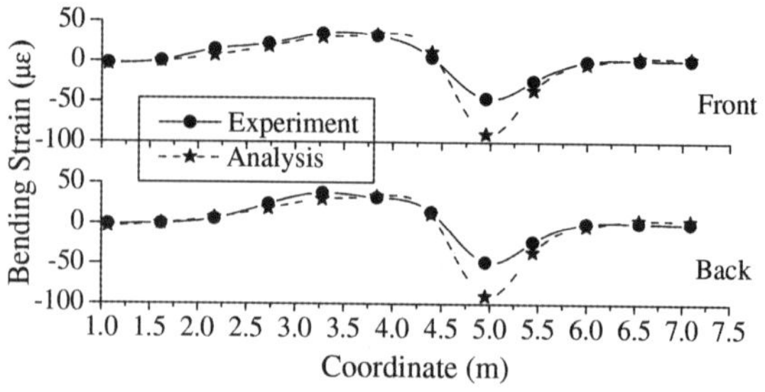

(b) Fracture of cables 22U and 22D

Fig. 15. Variation of girder section bending strain caused by fracture of two cables on a same section.

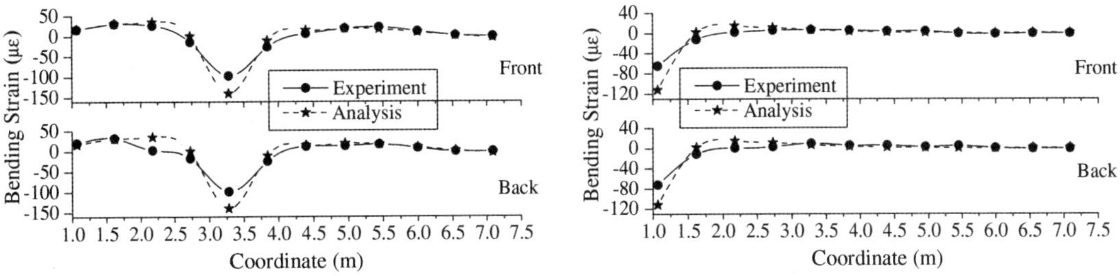

(c) Fracture of cables 19U and 19D

(d) Fracture of cables 15U and 15D

Fig. 15. (*Continued*)

Table 8. Variation of the dynamic performance caused by damage of two cables on a same section.

Description	Case 5		Case 6		Case 7		Case 8	
	f_v (%)	ζ_v (%)	f_v (%)	ζ_v (%)	f_v (%)	ζ_v (%)	f_v (%)	ζ_v (%)
1st antisymmetric bending shape	−1.98	−1.02	−3.47	−0.03	0.83	0.82	0.61	−0.28
1st symmetric bending shape	−2.24	−0.57	−3.35	−0.78	3.24	0.21	−0.16	−0.62
2nd antisymmetric bending shape	−2.11	−0.33	−0.21	−0.36	−0.56	−0.34	−0.81	−0.40
2nd symmetric bending shape	−2.07	−0.09	−1.07	−0.20	−0.12	0.12	−0.66	−0.05
3rd antisymmetric bending shape	0.17	0.81	−1.32	0.46	−0.37	0.55	−2.31	1.85
3rd symmetric bending shape	−0.18	−0.40	0.36	−0.68	0.35	−0.56	−3.59	−0.78
4th antisymmetric bending shape	0.08	0.52	−0.03	0.75	−0.82	0.00	−0.92	0.15
4th symmetric bending shape	−0.23	0.06	−0.75	0.53	−0.41	−0.18	−0.68	0.35

The variation of the model's dynamic performance for Case 5–Case 8 is shown in Table 8. The test results show that the variations of natural frequencies in these cases are larger than those of the cases with one cable damaged, but the variances are less than 5%.

6.3. *Two adjacent cables damaged cases*

The variations of cable forces when two adjacent cables were damaged (Case 9–Case 12) are shown in Fig. 16. From the results, the same conclusions as those of the previously studied cases can be obtained.

The variations of the girder bending strain for Case 9–Case 12 are shown in Fig. 17. Obviously, the variation trends of the bending strain for these cases are similar to those of the one cable damaged cases.

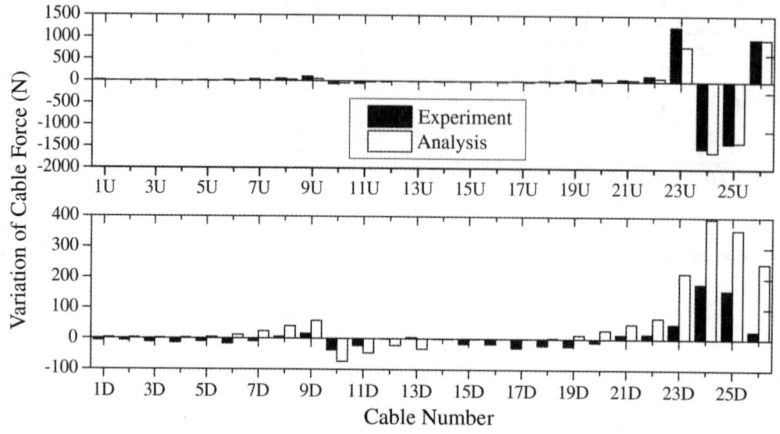

(a) Fracture of cables 24U and 25U

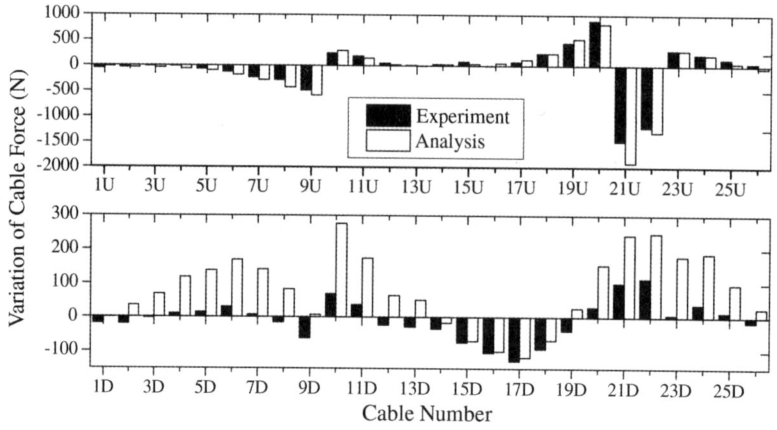

(b) Fracture of cables 21U and 22U

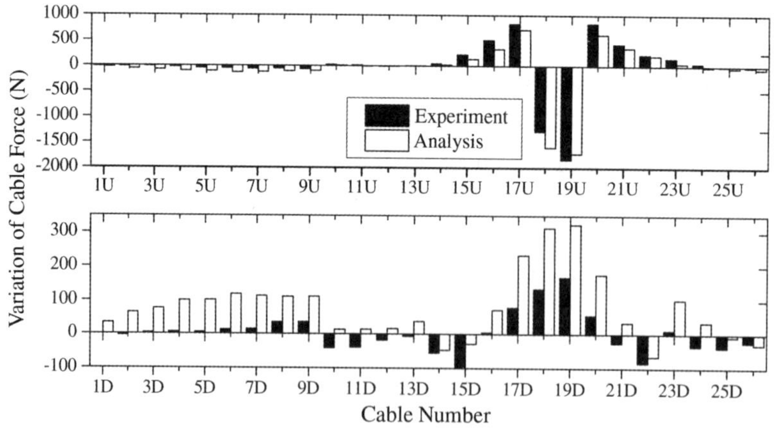

(c) Fracture of cables 18U and 19U

Fig. 16. Variation of entire bridge cable force caused by fracture of two adjacent cables.

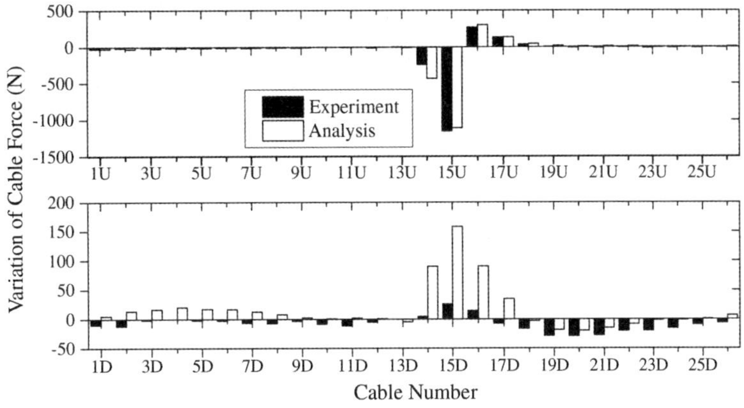

(d) Fracture of cables 14U and 15U

Fig. 16. (*Continued*)

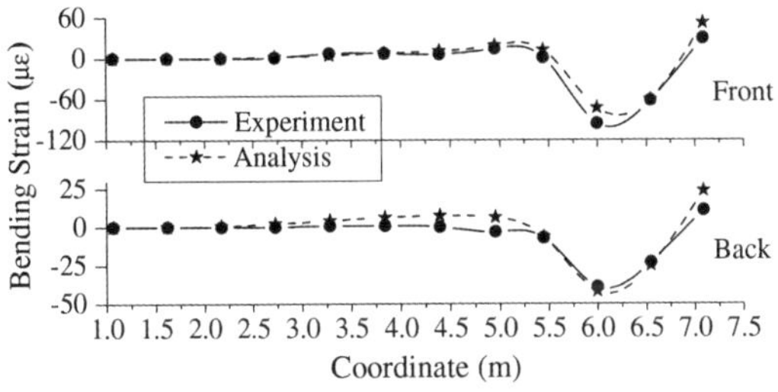

(a) Fracture of cables 24U and 25U

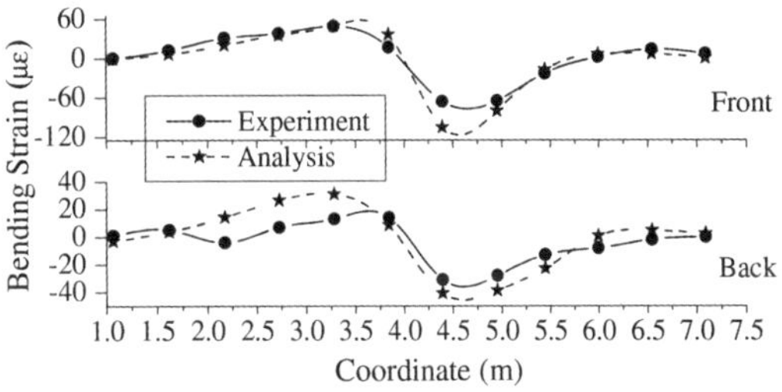

(b) Fracture of cables 21U and 22U

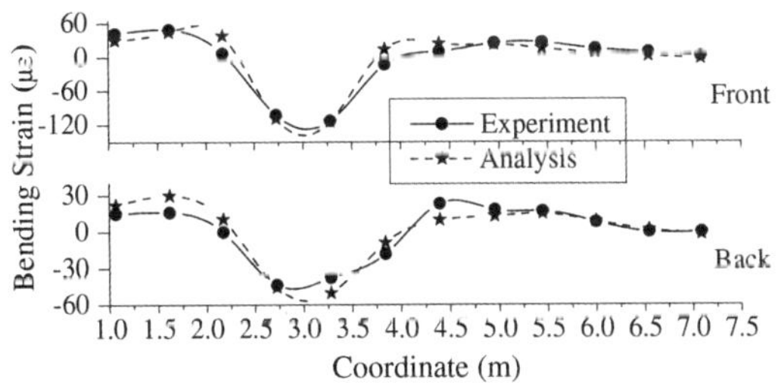

(c) Fracture of cables 18U and 19U

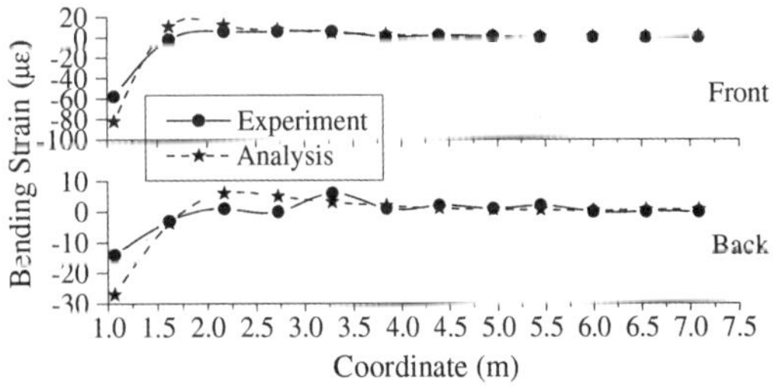

(d) Fracture of cables 14U and 15U

Fig. 17. Variation of girder section bending strain caused by fracture of two adjacent cables.

The variations of the dynamic performance for Case 9–Case 12 are shown in Table. 9. Compared with the previous test results, it is found that the influence of cable damages on the bridge model's natural frequencies is dependent on the anchoring points of the damaged cables at the bridge deck. Specifically, when a damaged cable is anchored near the peak point of the mode shape of the deck, the corresponding natural frequency varies significantly. Instead, when a damaged cable is anchored near the zero point of the mode shape of the deck, the natural frequency

Table 9. Variation of the dynamic performance caused by fracture of two adjacent cables.

Description	Case 9		Case 10		Case 11		Case 12	
	f_v (%)	ζ_v (%)	f_v (%)	ζ_v (%)	f_v (%)	ζ_v (%)	f_v (%)	ζ_v (%)
1st antisymmetric bending shape	−1.58	0.17	−2.19	−0.06	−1.23	0.39	0.01	0.15
1st symmetric bending shape	−1.66	−0.48	−5.07	−0.16	−4.34	−0.62	0.50	−1.02
2nd antisymmetric bending shape	0.63	−0.91	−2.63	−0.48	0.09	−0.27	−0.34	−0.58
2nd symmetric bending shape	−1.98	0.12	−0.82	0.00	−3.34	0.41	−0.51	−0.18
3rd antisymmetric bending shape	−2.97	1.82	0.52	0.36	−0.64	0.25	−0.08	0.36
3rd symmetric bending shape	−1.36	−0.13	1.45	0.89	0.68	−0.15	−1.94	1.84
4th antisymmetric bending shape	−0.82	0.36	−0.68	−0.08	−1.74	0.43	−1.22	0.12
4th symmetric bending shape	−0.50	0.05	−0.22	0.08	−0.48	0.00	−0.77	0.00

varies very slightly. However, the damping ratio varied randomly, since the error of the identification method for the parameter is relatively large, which covers up the variation caused by cable damages.

7. Conclusions

This paper investigated the static and dynamic performance variations of a cable-stayed bridge model caused by cable damages through the model tests and the finite element analysis. Conclusions are summarized from this combined experimental and numerical study as follows.

(1) The variations of the static performance (including the cable forces and girder section bending strain) for the cases with the cable forces reduced by 30% and cables fractured are basically identical, and the only deference is that the variation amplitude is larger when cables are fractured. This indicates that different cable damaged levels share the same static performance variation laws.
(2) The influence of a cable's damage on adjacent cables anchored at the same pylon is larger than that on cables anchored in different pylons. This is because the stiffness of the tower is larger than that of the girder, and the interval between cables, anchoring points at the tower is smaller than that at the girder.
(3) The girder section bending strain decreases significantly near the damaged cable's anchoring point on the deck. Instead, the bending strain of other parts increases, indicating that the bridge deck is subjected to downward deflection at the damaged cable's anchoring point under the dead loads.
(4) The influence of a cable's damage, even two cables, damages, on the natural frequencies of the bridge model is not significant. For most cases considered in

this paper, the frequency variation ratios are less than 5%. Moreover, the effect of cables, damages on the natural frequencies is dependent on the damaged cable's anchoring location at the bridge deck. When a damaged cable is anchored near the peak point of the mode shape of the deck, the corresponding frequency varies significantly. On the other hand, when a damaged cable is anchored near the zero point of the mode shape, the frequency varies very slightly.

(5) Only the self-weight of the bridge model has been considered in this study. The live loads, such as wind actions and vehicle loads, on the performance of the bridge model will be investigated in the future studies.

Acknowledgments

The authors gratefully acknowledge the funding supported by National Science Foundation of China (NSFC) under Grant No. 51108172 and the Chinese Post-Doctorate Science Fund 49th Batch of Surface Subsidizes.

References

1. Y. B. Yang and Jiunn-Yin Tsay, Geometric nonlinear analysis of cable structures with a two-node cable element by generalizaed displacement control method, *Int. J. Struct. Stab. Dyn.* **7**(4) (2007) 571–588.
2. N. Impollonia, G. Ricciardi and F. Staitta, Dynamics of shallow cables under tubulent wind: A nonlinear finite element approach, *Int. J. Struct. Stab. and Dyn.* **11**(4) (2011) 755–774.
3. G.C. Sih, X.S. Tang, Z.X. Li, A.Q. Li and K.K. Tang, Fatigue crack growth behavior of cables and steel wires for the cable-stayed portion of Runyang bridge: Disproportionate loosening and/or tightening of cables, *Theor. Appl. Frac. Mec.* **49**(1) (2008) 1–25.
4. V. Périer, L. Dieng, L. Gaillet, C. Tessier and S. Fouvry, Fretting-fatigue behaviour of bridge engineering cables in a solution of sodium chloride, *Wear* **267**(1–4) (2009) 308–314.
5. H. Li, C. M. Lan, Y. Ju and D. S. Li, Experimental and numerical study of the fatigue properties of corroded parallel wire cables, *J. Bridge Eng.* **17**(2) (2012) 211–220.
6. H. R. Hamilton III, J. E. Breen and K. H. Frank, Bridge stay cable corrosion protection. I: Grout injection and load testing, *J. Bridge Eng.* **3**(2) (1998) 64–71.
7. H. R. Hamilton III, J. E. Breen and K. H. Frank, Bridge stay cable corrosion protection. II: Accelerated corrosion tests, *J. Bridge Eng.* **3**(2) (1998) 72–81.
8. B. S. Smith, The single plane cable-stayed girder bridge: a method of analysis suitable for computer use, *Proc. Inst. Civ. Engrs.* **37** (1967) 183–194.
9. L. Hou, Model updating, damage diagnosis and prewarning methods based on the health monitoring of large cable stayed bridges, Ph.D thesis of Harbin Institute of Technology (2009) (in Chinese).
10. A. Cole, The influence matrix methodology: A technical report, Landcare Research Contract Report: LC0506/175 (2006).
11. G. H. James, T. G. Carne and J. P. Lauffer, The natural excitation technique for modal parameter extraction from operating wind turbines, Rep. No. SAND92-1666, UC-261, Sandia National Laboratories, Sandia, NM (1993).

12. G. H. James, T. G. Carne and R. L. Mayes, Modal parameter extraction from large operating structures using ambient excitation, in *Proc. 14th Int. Modal Analysis Conf.* (Dearborn, Michigan, 1996), pp. 260–277.
13. J. N. Juang and R. S. Pappa, An eigensystem realization algorithm for modal parameter identification and model reduction, *J. Guid. Control Dyn.* **8**(5) (1985) 620–627.
14. D. M. Siringoringo and Y. Fujino, System identification of suspension bridge from ambient vibration response, *Eng. Struct.* **30**(2) (2008) 462–477.

Materials and Structures (2016) 49:3579–3589
DOI 10.1617/s11527-015-0741-0

CrossMark

ORIGINAL ARTICLE

Influence of longitudinal bar corrosion on impact behavior of RC beams

Mingjiang Dai · Ou Yang · Yan Xiao ·
Fengchen Li

Received: 1 April 2015 / Accepted: 2 November 2015 / Published online: 19 November 2015

Abstract To investigate the influence of longitudinal bar corrosion on impact behavior of RC beams, four RC beam specimens with varied corrosion rate were tested under falling-weight, impacting the specimens at the mid-span. Four corresponding specimens were also tested under static loads to determine their static properties and behavior. Reinforcement was corroded by means of immersing the specimens in 3 % sodium chloride solution and applying a current density of 3 mA/cm^2. The test results indicate that corrosion of longitudinal bars decreases the impact force and reaction force, and increases mid-span displacement of RC beams under impact load and the beam's capacity of absorbing energy. Corrosion-induced cracks in concrete have been shown as the most relevant damages in the tested beams. All specimens, regardless of their corrosion rate, failed in shear-critical mode and presented severe diagonal shear cracks under the impact point.

Keywords Beam corrosion · Impact loads · Reaction behavior · Energy absorption

M. Dai · O. Yang (✉) · Y. Xiao
College of Civil Engineering, Hunan University, Changsha 410082, China
e-mail: ouyang.hnu@gmail.com

F. Li
School of Civil Environment Engineering, East China Institute of Technology, Fuzhou, China

1 Introduction

Reinforced concrete (RC) structures in marine environments, such as offshore installations, seawalls, and breakwaters, may suffer various types of dynamic loading in terms of impact, blasting, earthquakes, and tsunamis during their service life, all of which may reduce their serviceability. The situation is exacerbated when RC structures are subjected to chloride ion penetration, which contributes to the premature deterioration and failure of RC structures. These dynamic loadings and deterioration due to chloride ingress must be taken into account in the safety evaluation of offshore RC structures. The present study will concentrate on the impact behavior and properties of RC beams, because they are the most common type of member used in practical engineering.

Over the past few decades, a considerable number of researchers have investigated the behavior of RC beams under drop weight impact [1–6]. Some of them have proposed empirical formulas to facilitate the design of RC beams under impact loading conditions [7–12]. These studies indicate that it is possible to predict the impact-resistant capacity and the failure probability of RC beams based on static capacity and behavior. The effects of factors such as section parameters, reinforcement ratios, clear span length, impact velocity, the mass and shape of the weight, and input kinetic energy have profound influence on the impact behavior of RC beams. The ultimate load-bearing capacity, cracking stiffness, and energy absorption of RC beams were found to increase with the increment of loading rates [13–17].

Deterioration of reinforced concrete structures due to chloride ingress accompanied by reinforcement corrosion is a serious problem all over the world [18]. The corrosion of reinforcements can reduce the impact-resistant capability of the structure through the loss of cross-section, deterioration of bond strength, and formation of corrosion-induced cracks in the concrete. Although considerable researches have been conducted on the impact behavior of RC structures, little effort has been expended toward understanding the impact behavior of RC structures that suffer significant damage due to reinforcement corrosion. The objective of this study was to research the effect of different degrees of reinforcement corrosion on the loss of impact-resistance. In this study, four RC beam specimens with different corrosion rates were tested under falling-weight conditions. Histories of impact force, reaction force, and mid-span displacement were obtained from the impact test. Hysteretic loops of impact force versus displacement and reaction force versus displacement, reaction behavior, and energy absorption were used to analyze the impact behavior of the RC beams.

2 Experimental overview

2.1 Beam design

A total of 8 identical RC beams have been designed to carry out the falling-weight impact tests (Fig. 1). All the specimens showed cross sections of 120 × 200 mm with the clear cover of 25 mm and effective depth of 167 mm, and a total length of 1600 mm. The specimens were tested under simply supported conditions with a shear span of 650 mm. The shear-span ratio was 3.89.

Normal concrete was used in this study with w/c ratio 0.47. The mixing proportion ratio was 1:0.47:1.59:3.39 of cement:water:fine aggregate:-coarse aggregate. The cement used in the specimens is the ordinary Portland cement with a compressive strength of 32.5 MPa (by cylinder specimen test). The maximum size of coarse aggregate was 16 mm. The concrete specimens were cured in room temperature for 28 days. The 28 days average strength of the concrete is 19.34 MPa, which was obtained from the test results of three 150 × 150 × 150 mm cubic specimens.

Hot rolled steel bars HRB335 were used as longitudinal bars and hanger bars to acquire flexural strength and resist the inertia effects on the impact response, respectively. The main reinforcing bar ratios were 2.01 %. In order to eliminate the effect of stirrup corrosion on the properties of RC beams, only two stirrups were used to fasten steel frame in both ends of the RC beams. The dimensions and material properties of the steel bars are listed in Table 1. The yield strength, ultimate strength, and elasticity modulus of the steel bars were obtained by using standard tensile tests. The elongation was the tensile stretch of the bar over a length five times its diameter near the rupture area.

The primary variable was the magnitude of corrosion. Four specimens with different corrosion rates were tested under impact load conditions. These specimens are denoted "BI" followed by the specimen type number. Static tests with four additional specimens corresponding to the impact specimens were

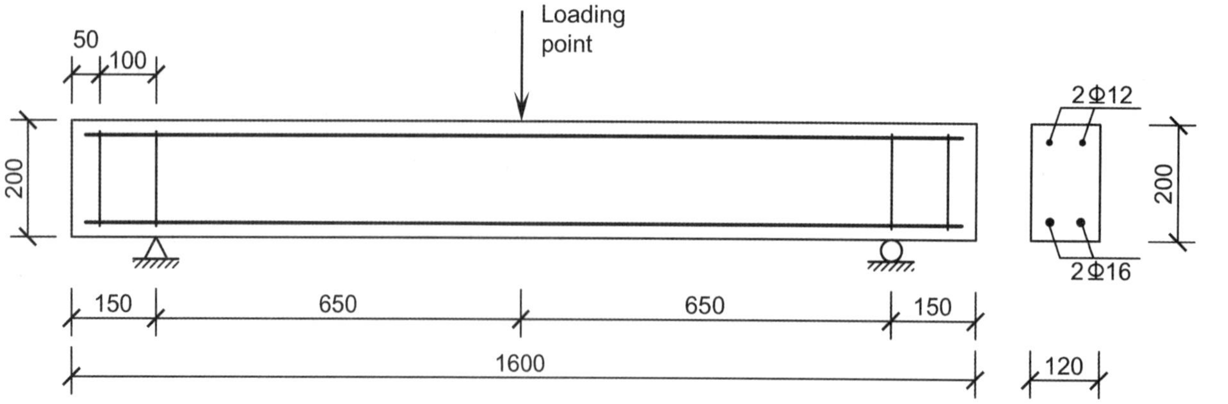

Fig. 1 Details of the beam specimen (units: mm)

Table 1 Characteristics of the reinforcing bars

Rebar grade	Diameter (mm)	Yield strength (N/mm^2)	Ultimate strength (N/mm^2)	Elasticity modulus (N/mm^2)	Elongation (%)
HRB335	12	447.20	562.61	1.8061×10^5	28.85
HRB335	16	461.40	579.22	1.8422×10^5	32.00

conducted. These specimens are denoted with "BS" followed by the specimen type number.

2.2 Corrosion experiment and evaluation index

The electrochemical corrosion setup is shown in Fig. 2. The specimens were immersed in 3 % sodium chloride (NaCl) solution. The steel bars and the copper plate were connected to the anode and cathode of a constant current generator, respectively. A current density of 3 mA/cm^2 was imposed through the longitudinal bars. The duration of the corrosion test, determined using Faraday's Law, was controlled to determine the corrosion levels of the longitudinal bars.

After the loading tests, all specimens were demolished, and the steel bars were removed from the concrete to investigate the corrosion level. The steel bars were cleaned and measured using the method described in ASTM G1-03 [19]. The relative average mass loss C (hereinafter "corrosion rate") was used to describe the corrosion level, which can be described as follows:

$$C = \frac{\Delta m}{m} \times 100\,\% \tag{1}$$

Here Δm is the average mass lost by the bars to corrosion, and m is the original mass of the bars.

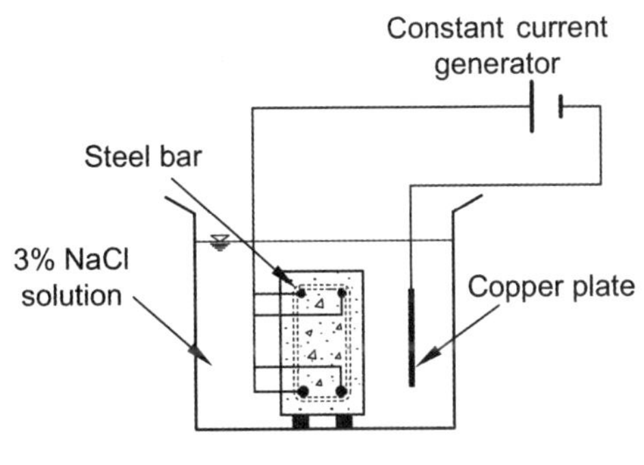

Fig. 2 Electrochemical corrosion setup

2.3 Falling-weight impact test procedure

The experimental program, carried out at Hunan University Structural Laboratories, consisted of two phases: falling-weight impact tests and static tests. Impact tests were conducted by dropping a 198 kg steel weight, along the guide rails from 400 mm height onto the mid-span of the specimen using the impact test apparatus, as shown in Fig. 3. The drop-weight, manufactured with a flat and circular bottom of radius of $r = 96$ mm was equipped with a load cell for measuring the impact load. The specimens, simply supported at the bottom with heavy steel pedestals, were clamped at each end using cross beams to prevent them from lifting off. The reaction forces were measured using two load cells equipped on the supports. A displacement meter was attached at the bottom surface of the specimens to measure mid-span displacement. A laser velocimeter was used to measure the impact velocity. The behavior of the specimens under the impact loading was captured using a high-speed camera with a performance of 1000 frames/s.

In this study, the impact velocity v of the drop-weight was 2.8 m/s and the input kinetic energy $E_k = \frac{1}{2}mv^2$ (m: mass of steel weight) was 776.2 J. The histories of the impact force P, reaction force R, and displacement δ were measured. The data collected were captured at a 1 MHz sampling rate using National Instruments (NI) digital data acquisition system. After testing, the crack patterns of the beams were sketched.

After the completion of the falling-weight impact tests, four corresponding specimens were tested to determine their static properties and behavior. One displacement meter was used to measure mid-span displacement. Another two were equipped on the top surface of the specimens at the supports to measure vertical displacements. The static tests were conducted by monotonically increasing static loads at the mid-

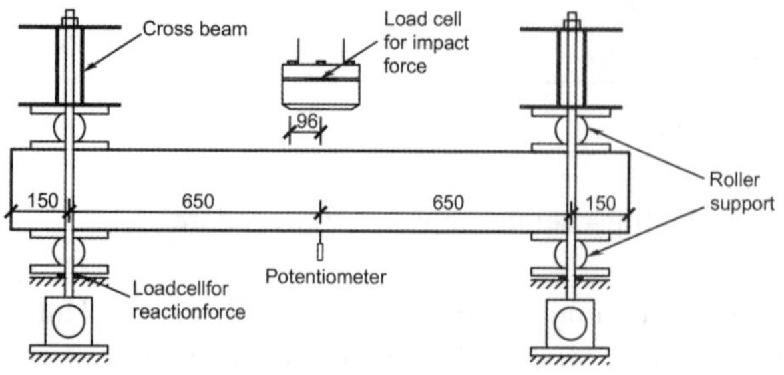

(a) Schematic drawing

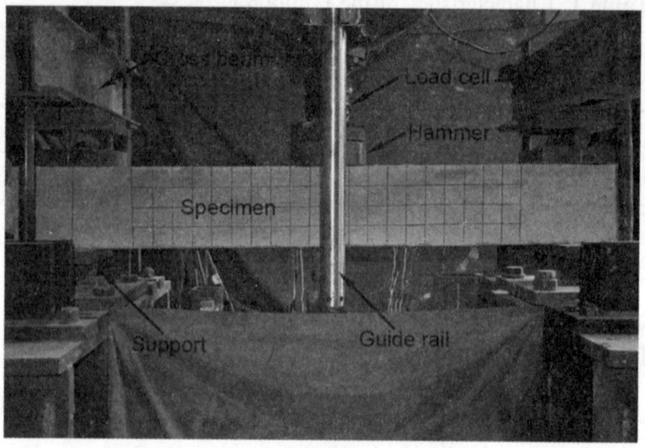

(b) Photo

Fig. 3 Experimental setup

span under support conditions identical to those used with the impact test program.

3 Experimental results and discussion

3.1 Results of the corrosion experiment

The corrosion rates of specimen BS1, BS2, BS3, and BS4 were 0, 6.7, 9.6, and 17.8 %, respectively. The corrosion rates of specimen BI1, BI2, BI3, and BI4 were 0, 7.3, 10.0, and 18.1 %, respectively. The corrosion product (rust) resides at the interface between reinforcement and concrete, involving a substantial increase in volume; that causes expansive stresses, leading to cracking of concrete cover. In this study, corrosion-induced cracks along the longitudinal bars and hanger bars were observed on the side and bottom surface in all corroded RC beams. There is no visible cracking on the top surface of the beams. The possible reasons are: (1) the lower diameter of the hanger rebars produced less corrosion product; (2) the resulting expansive stresses were mainly released by cracking on the side surface; (3) the specimens were immersed in NaCl solution and the corrosion products spread to the solution from side cracks. When the corrosion rate C was less than 7.3 %, maximum corrosion-induced cracks 1 mm appeared. The maximum width of corrosion-induced cracks increased as the corrosion rate increased until it reached 2 mm in BS4 and BI4. Observations of the side and bottom surface of the specimens showed that corrosion products were fluid and penetrated into concrete through cracks. Severe localized corrosion in the form of pitting or grooves type was observed on the steel surface for corrosion rate above 7.3 %. Such localized attacks became more severe and covered a wider area with the increase of corrosion rate and some location of the bars even had 50 % area loss in BS4 and BI4. It is also found that bars corroded more severely in the side that near the cathode. However, for the following two reasons the effects these factors were ignored. First, the corrosion pits are stochastic and hard to quantize. Second, the purpose of this study is to observe the variation trend of impact capacity with corrosion rate and a large corrosion rate interval (about 5 %) has been chosen, the corrosion pits and non-uniform corrosion style may not affect the conclusion.

3.2 Static test results

In the static test study, the external load was applied in the form of displacement increments at mid-span and the mid-span static load–displacement ($P_s - \delta$) responses observed are presented in Fig. 4. All specimens failed in a brittle shear-critical mode with the formation of a principal diagonal shear crack as depicted in Fig. 5. The static loading capacities of specimens BS1, BS2, BS3, and BS4 were 52.65, 45.98, 43.61, and 37.36 kN, respectively. The static load increased almost linearly up to the maximum value along with the increase in mid-span displacement. Both the stiffness and shear capacity of the beam decreased as the corrosion rate increased. The more heavily corroded beams showed more mid-span displacement than the lower corroded beam at the same load level due to degrading stiffness of the beams. Cover cracking and bond deterioration are believed to cause stiffness decay. The shear capacity degradation of corroded beams was primarily affected by reduction of the reinforcement in the cross area and concrete delamination attributable to cover cracking.

3.3 Crack patterns of the specimens after impact test

Figure 6 shows the crack patterns of the specimens after impact test. The horizontal cracks along the longitudinal bars and hanger bars were corrosion-induced cracks and the rest of the cracks are produced by impact load. All specimens exhibited brittle shear-critical behavior and developed severe diagonal cracks, forming shear-plugs, which are independent of corrosion rate. Diagonal shear cracks originated from the mid-height of the beam and propagated towards loading and support points for all beams. The high shear span ratio and absence of shear-resistant rebar were the reason for the shear failure. In BI2, BI3, and BI4, serious spalling occurred between the bottom surface of the beam surface and diagonal crack, which started from the supports and propagated along an angle of about 45 degree upward until to the top (Fig. 6b–d). In BI3, one shear crack started from the support and became horizontal close to the top before reaching the impact zone. No vertical cracks were induced in this impact test because of the hanger bars' resistance of the inertia effects. All these cracks were similar in nature to the typical shear cracks under static loads, and they became wider with the increase of the corrosion rate. Development and propagation of cracks around different points in reaction force-versus-time curves were portrayed in Fig. 7 for BI2.

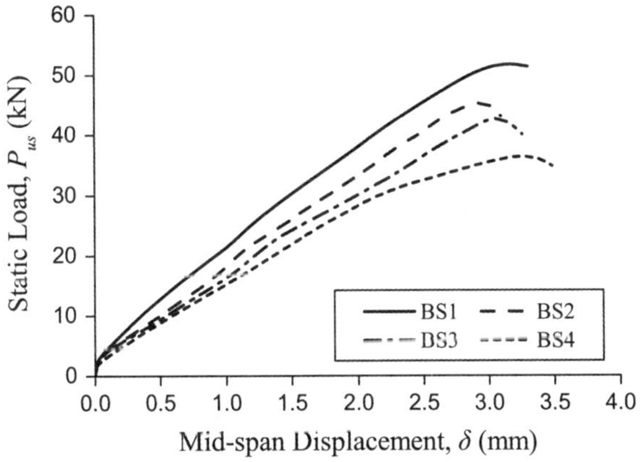

Fig. 4 Static load–displacement curve

3.4 Time histories of impact force, reaction force, and mid-span displacement

Figure 8 shows the histories of the impact force P, reaction force R, and mid-span displacement δ for the specimens in impact tests. The maximum value of each response wave is listed in Table 2. Here, the origin of the time axis was taken as the time when the weight impacts the beam. The reaction force R was calculated by adding up the values obtained from both support points.

As shown in Fig. 8a, according to the histories of the impact force, the beams behave similarly. The history exhibited two half-sine waves: (1) A triangular wave with high amplitude and short duration at the beginning of impact; and (2) a main wave with a longer duration. The maximum impact force is decreased with the increase of corrosion rate. However, the reaction force was excited with only one half-

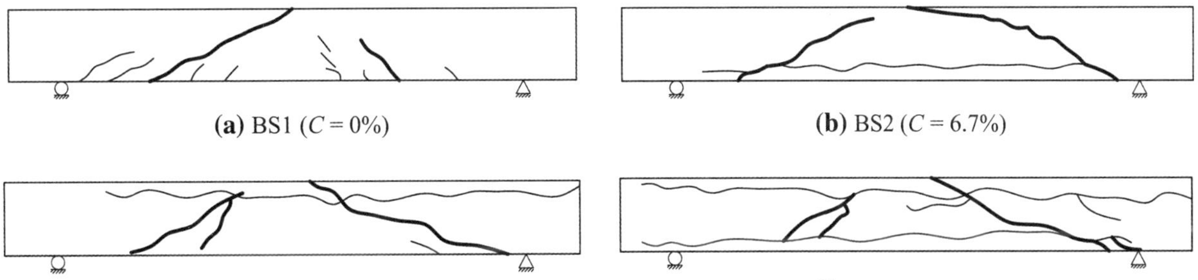

(a) BS1 ($C = 0\%$)

(b) BS2 ($C = 6.7\%$)

(c) BS3 ($C = 9.6\%$)

(d) BS4 ($C = 17.8\%$)

Fig. 5 Crack profiles for the static specimens

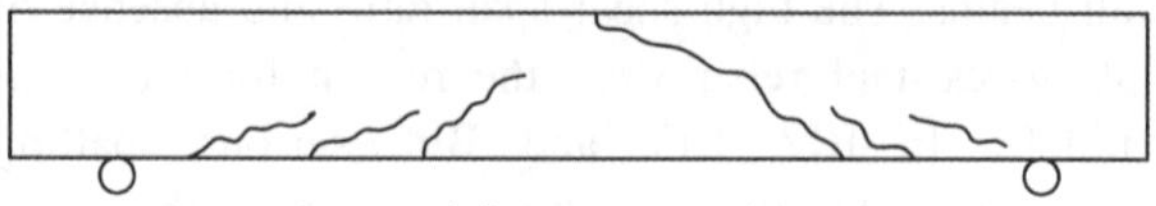

(a) BI1 ($C = 0\%$)

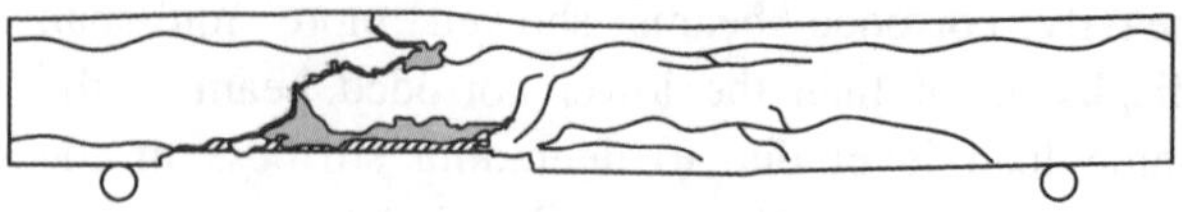

(b) BI2 ($C = 7.3\%$)

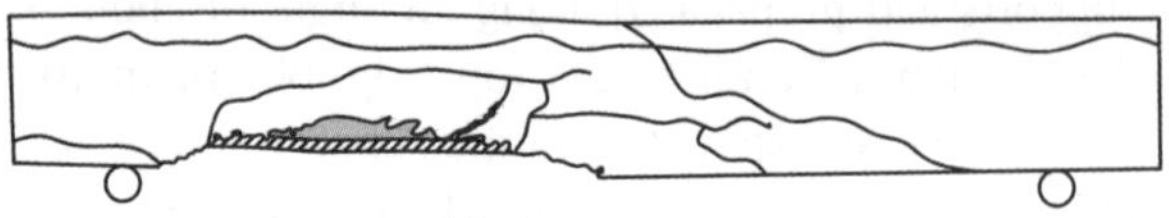

(c) BI3 ($C = 10.0\%$)

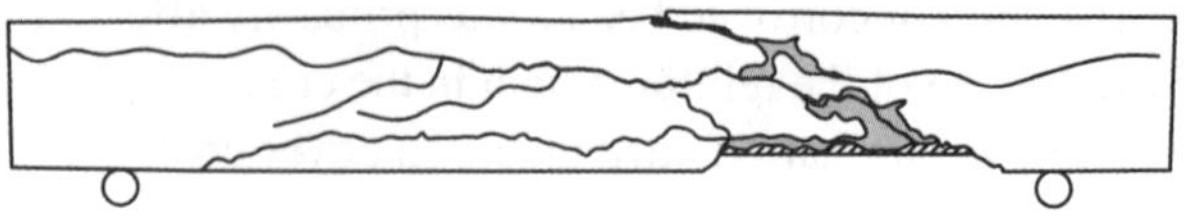

(d) BI4 ($C = 18.1\%$)

Fig. 6 Crack profiles for the impact specimens

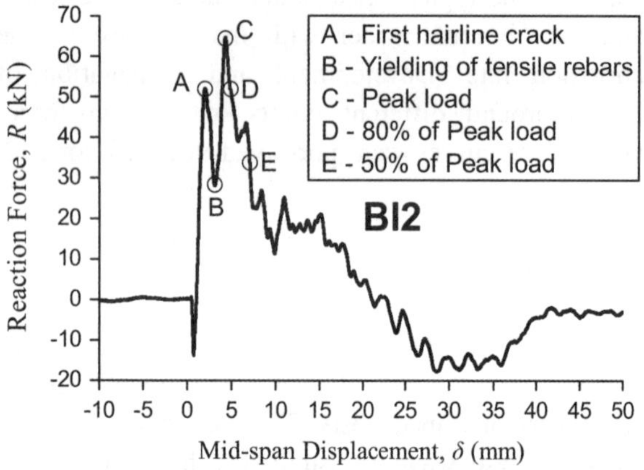

(a) Reaction force history curve of BI2

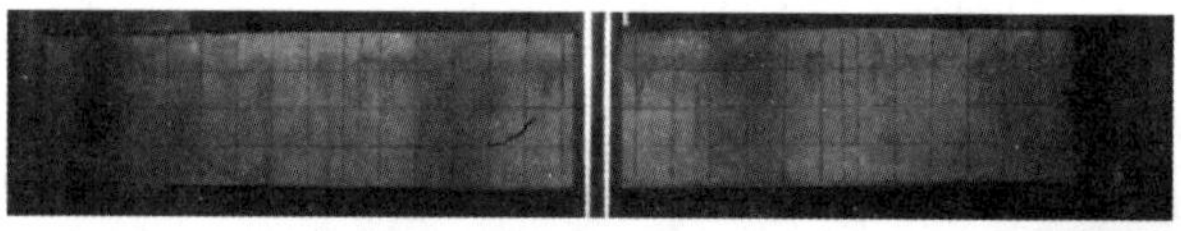

(b) First hairline crack

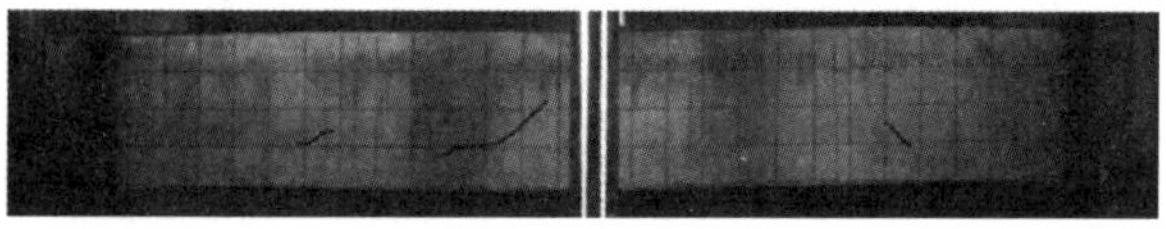

(c) Yielding of tensile rebars

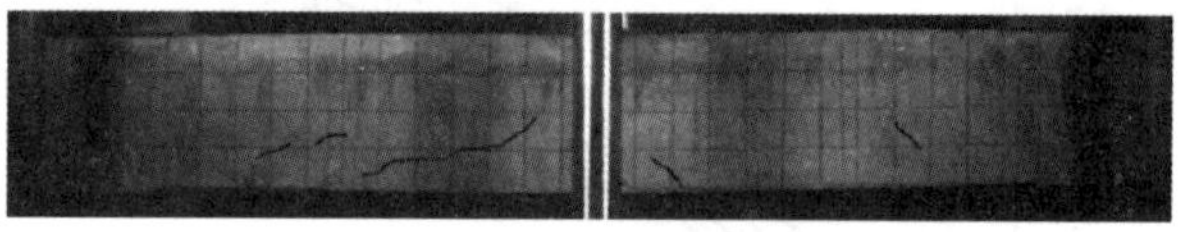

(d) Peak load

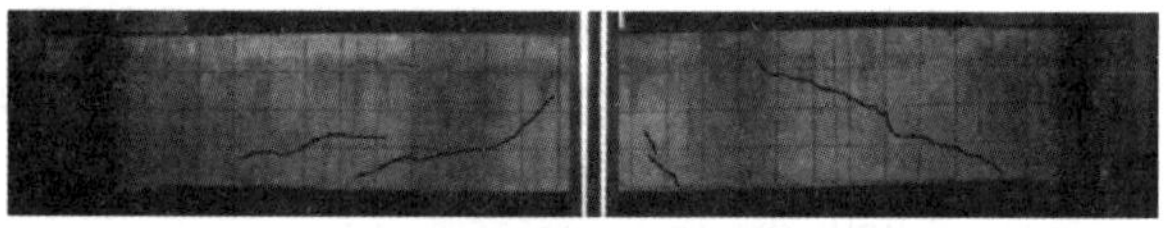

(e) 80% of peak load

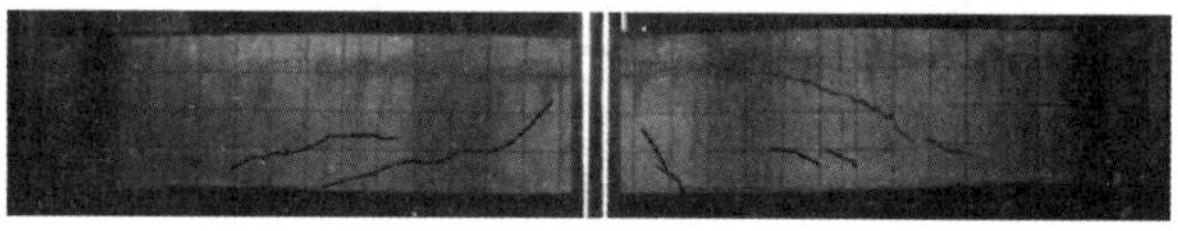

(f) 50% of peak load

Fig. 7 Development and propagation of cracks in BI2 under impact test conditions

sine wave as shown in Fig. 8b. The maximum reaction force R_{max} tended to decrease as the corrosion rate increased (Table 2). There was a negative reaction force at the beginning of impact because the beam ends had a tendency to lift off due to the rebound after the first impact, and the amplifier units of the load cells were initialized after the beam ends were tightened together with the load cells by means of the cross beams. Focusing on the mid-span displacement wave in Fig. 8c, the displacement is excited with a half-sine at first and then restrained for BI2, BI3, and BI4. This means that the beams have suffered large plastic deformation. In the case of BI1, the response appeared to drift and vibrate after the main wave for the rebound of the applied impact load. For BI1, BI2, and BI4, the maximum displacement and residual displacement showed a tendency to increase alongside increases in the corrosion rate. However, this value of BI3 decreased because the beam was less damaged than BI2 and BI4 (Fig. 6).

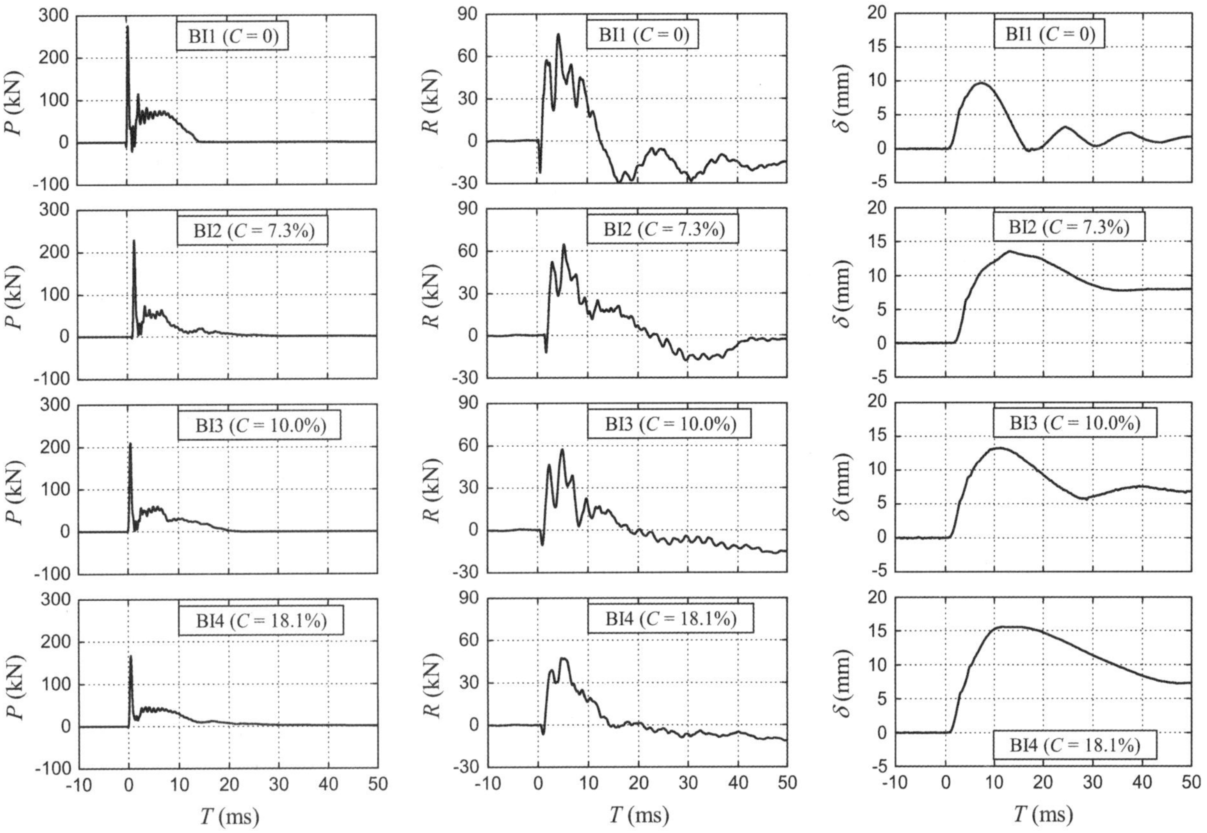

(a) Impact force **(b)** Reaction force **(c)** Mid-span displacement

Fig. 8 Histories of impact and reaction forces and displacement for the impact tests

Table 2 Maximum dynamic response values, residual displacement, and absorbed energy estimated using $P-\delta$ loop and/or $R-\delta$ loop

Specimen	C (%)	P_{max} (kN)	R_{max} (kN)	δ_{max} (mm)	δ_{rs} (mm)	E_P (J)	E_R (J)	E_R/E_k
BI1	0.0	290.8	76.1	9.67	1.72	354.4	339.6	0.438
BI2	7.3	235.6	65.1	13.60	7.98	559.9	490.8	0.632
BI3	10.0	216.7	58.1	13.32	6.89	476.7	369.0	0.475
BI4	18.1	168.6	48.7	17.20	8.53	540.4	472.1	0.608

C corrosion rate, P_{max} maximum impact force, R_{max} maximum reaction force, δ_{max} maximum displacement, δ_{rs} residual displacement, E_P absorbed energy from $P-\delta$ loop, E_R absorbed energy from $R-\delta$ loop, E_R/E_k absorbed input energy ratio

3.5 Hysteretic loops of impact force versus displacement and reaction force versus displacement

Hysteretic loops of impact force versus displacement ($P-\delta$) and reaction force versus displacement ($R-\delta$) for impact beams are presented in Fig. 9. Here, the plots of static load against displacement curves ($P_s-\delta$) are also shown for comparison. In addition, the value of the absorbed energy evaluated by integrating the loop-area for each specimen is shown in Table 2. These figures show that the values of the absorbed energy estimated from $R-\delta$ loops were within the range of 339.6–490.8 J. As shown in Fig. 9, impact force increased very rapidly at very beginning of impact and then suddenly decreased to almost zero after peaking, irrespective of the corrosion rate. After that, the impact force increased again to the second peak and then decreased to zero. Reaction forces increased with increasing displacement and then decrease until the displacement reached approximately 6.5 mm. This piece of the curve exhibited a triangular shape. A secondary triangular loop appeared after the first triangular loop formed because the longitudinal bars began to resist the shear force. In BI2, BI3, and BI4, when the mid-span displacement reached maximum value, impact force and reaction force dropped almost to zero. The primary stiffness estimated using the reaction force decreased as the corrosion rate increased and decreased faster than the primary stiffness as estimated using the static load. In this way, even though the primary stiffness estimated using the reaction force is much higher than the primary stiffness estimated using the static load in BI1, the two values are similar to each other in BI4.

The impact load cannot be considered a true bending load of the beam because of the inertial reaction of the beam. A part of the impact load was used to accelerate the beam. Load carrying capacity of RC beams under impact loading can be determined using instrumented anvil supports [16]. In the following investigations, the maximum reaction force was used to estimate the ultimate strength of RC beams under impact loading conditions.

3.6 Relationship between maximum reaction force, maximum impact force, and corrosion rate

The maximum impact force (P_{max}) and maximum reaction force (R_{max}) decreased monotonically as the corrosion rate increased (Figs. 10, 11). Corrosion-induced cracks formed along steel rebar in concrete, which are the most relevant damages in the tested beams, are the main cause of impact force and reaction force decay. The values of ratio R_{max}/P_{max} are distributed in the region from 0.26 to 0.29, with a mean value of 0.27 approximately. The values of dynamic response ratio R_{max}/P_{us} for each specimen are plotted in Fig. 12. Here, P_{us} is the ultimate load of

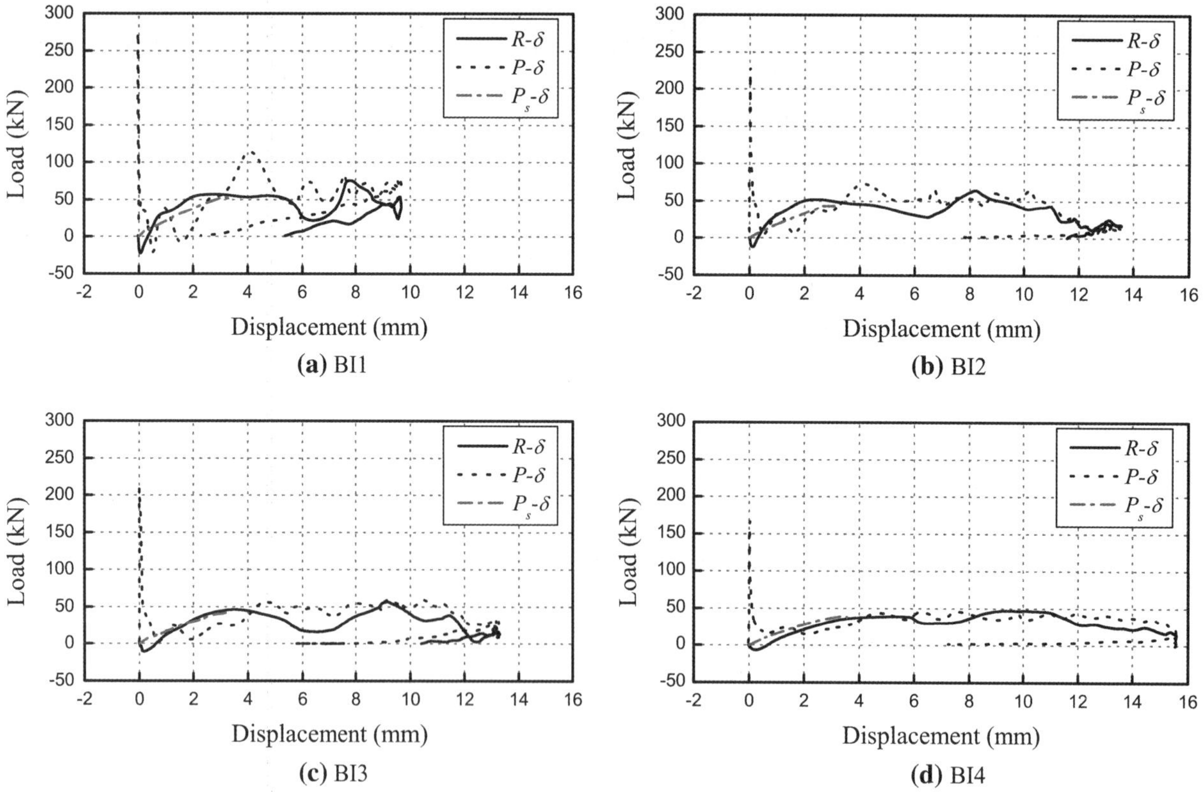

Fig. 9 Hysteretic loops of impact force versus displacement $P-\delta$ and reaction force versus displacement $R-\delta$.

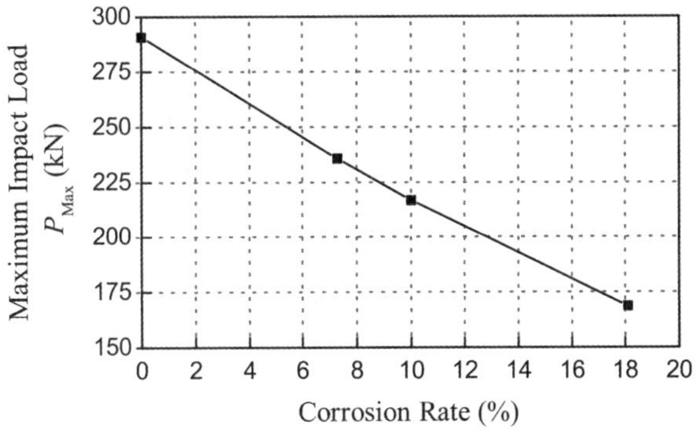

Fig. 10 Relationship between maximum impact force and corrosion rate

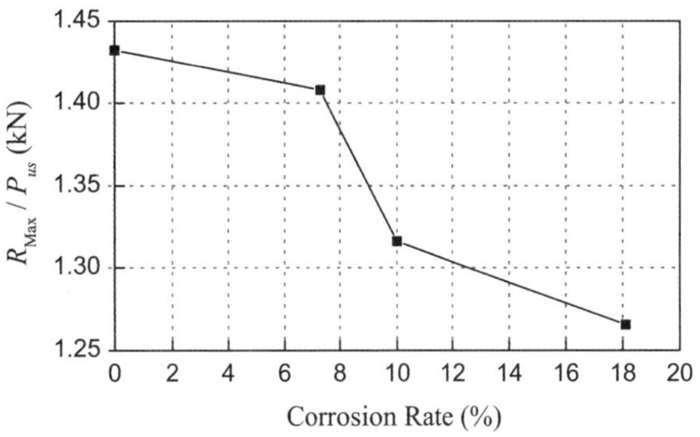

Fig. 12 Relationship between R_{max}/P_s and corrosion rate

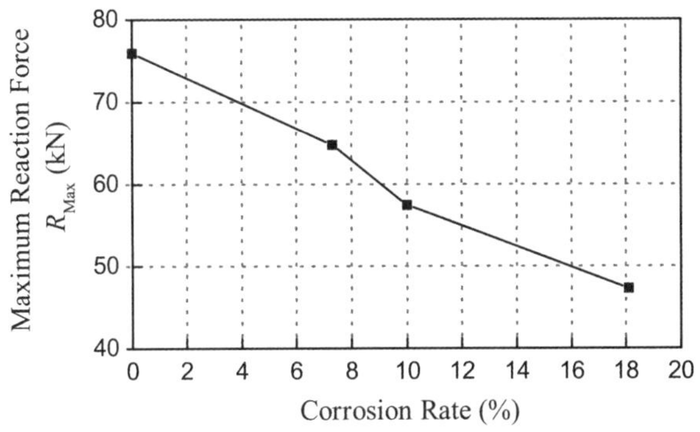

Fig. 11 Relationship between maximum reaction force and corrosion rate

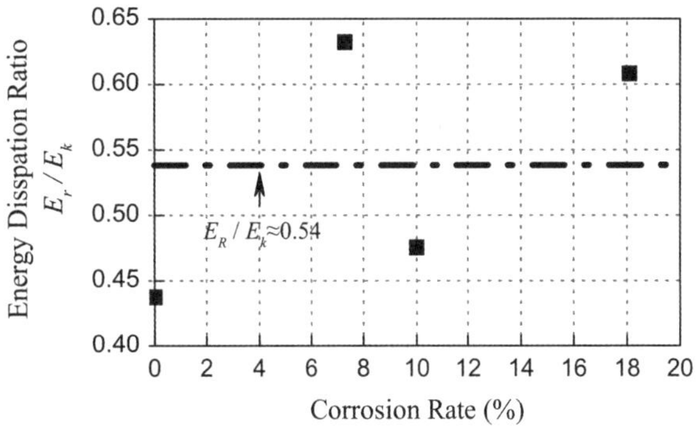

Fig. 13 Relationship between absorbed input energy ratio E_R/E_k and corrosion rate

static test. As shown, the values of R_{max}/P_{us} for all beams considered here were within the range of 1.25–1.45. This indicated that the maximum reaction force for all RC beams exceeded 1.25 times static shear capacity. This suggests that the impact shear capacity of RC beams can be estimated by 1.25 times the value of static shear capacity to ensure a margin of safety when the corrosion rate is less than 20 %.

3.7 Ratio of absorbed energy to input kinetic energy

The ratio of absorbed energy to input kinetic energy E_R/E_k for each RC beam is shown in Fig. 13 and Table 2, where E_R and E_k were absorbed energy estimated using a looped area of reaction force–displacement curve (Fig. 9). The figure shows that these values are distributed in the region from 0.438 through 0.632, with a mean value of 0.54 approximately. The ratio of absorbed energy to input kinetic energy did not decrease as the corrosion rate increased, which means that corrosion rate probably did not have much relationship with the energy dissipation capacities of RC beams. The ratio of absorbed energy to input kinetic energy in BI2 and BI4 was greater than the ratio of absorbed energy to input kinetic energy in BI1and BI3, while the degree of damage in BI2 and BI4 is more serious than BI1 and BI3 (as shown in Fig. 6). The results indicated that the energy dissipation capacities of RC beams are closely related to the extent of damage and shape.

4 Discussion

The objective of this work was to determine the effect of reinforcement corrosion on the various aspects of impact behavior of RC beams with emphasis on crack distribution and propagation, failure mechanisms, load-carrying capacity, and energy-absorbing capacity.

Comparing the crack profiles between the static specimens and impact specimens in Figs. 5 and 6, results indicate that impact load caused extensive damage such as more and wider cracks and spalling of concrete cover in corroded specimens. This was attributed to the initial corrosion damage. The application of an external impact load onto the mid-span of the specimen caused a very complex triaxial stress field to develop within the concrete medium. This was further aggravated by the propagation of cracks and the ensuing internal stress redistributions. The impact load imposed on the specimen caused stress waves, which propagated through the concrete medium away from the impact zone, as reflected in Fig. 7. The complexity of this stress field was further accentuated by the corrosion-induced cracks, which obstructed the propagation of stress waves and caused deflection. The formation of such waves affected the triaxial stress field and so caused high concentrations of stress in localized regions of the specimen, leading to local tensile failure, cracking, and even spalling.

The data presented in Fig. 8 show the time histories of impact and reaction forces and displacement for the impact specimens subjected to various corrosion levels. The results reveal that corrosion has a significant influence on the overall impact behavior of RC beams, such as lower impact force and reaction force (Figs. 10, 11) and more severe deformation. This is attributed to the degradation of stiffness and loss of strength, mainly coming from corrosion-induced cracks and delamination of concrete. The loss of cross section of reinforcement and loss of bond between concrete and reinforcement also contributed to the degradation of impact behavior. Even when there was excessive corrosion with considerable cracking of concrete, load-carrying capacity was not completely destroyed. Measurable impact force and reaction force were present even when the steel mass loss approached 20 %. In other words, even after excessive corrosion, RC beams were able to sustain considerable impact loads.

Corrosion reduces the reinforcement steel area and results in lower target stiffness and softer impact behavior. When the same energy was imparted to the specimens at various corrosion levels in impact test, the corroded specimens experienced extensive damage while the control specimen BI1, which was not corroded, suffered limited damage and no yielding of the rebars was obtained. However, the absorbing energy estimated using looped area of reaction force–displacement curve did not exhibit much relationship to the corrosion rate and may be more closely related to the degree of damage. The absorbing energy value of corroded specimens estimated using looped area of reaction force–displacement and impact force–displacement curve (as shown in Table 2) is higher than the value of the control beam BI1, which indicate that corrosion will increase the energy dissipation capacities. This is probably because a considerable part of energy was imparted through the damage of the specimen, i.e. yielding of the rebars and cracking and delamination of the concrete. Thus, corrosion-induced cracks and delamination of concrete are beneficial to the energy dissipation.

It should be noted that the corrosion rate applied in the impact test were all over 7 %, and the properties and behavior of RC structures sometimes exhibit an abnormal phenomenon at lower corrosion levels, such as the increase of the steel/concrete bond strength and load-carrying ability. The findings summarized in this project require further experimental study involving lower corrosion level to verify.

5 Conclusions

This study focused on the behavior of RC beams with varied corrosion rate under falling-weight impact load. The corroded specimens were produced in an accelerated corrosion experiment. Several conclusions were drawn based on these test results:

(1) Due to the low shear-span ratio, 3.89, and lack of shear reinforcement, RC beams failed in shear-critical mode with the formation of severe shear cracks in both static tests and impact tests, irrespective of corrosion rate.

(2) Maximum impact force and maximum reaction force decreased as the corrosion rate increased.

(3) Bending load capacity of corroded beams without shear rebar can be estimated as 1.3 times the static shear capacity with a safe margin.

(4) Corrosion increases the energy dissipation capacities of RC beams subjected to impact load. The energy dissipation capacities of RC beams are closely related to the extent of damage and shape.

Acknowledgments The research described in this paper was financially supported by National Nature Science Foundation of China project (Grant No. 51578229).

References

1. Bhatti AQ, Kishi N, Mikami H, Ando T (2009) Elasto-plastic impact response analysis of shear-failure-type RC beams with shear rebars. Mater Des 30(3):502–510
2. Fujikake K, Li B, Soeun S (2009) Impact response of reinforced concrete beam and its analytical evaluation. J Struct Eng 135(8):938–950
3. Kishi N, Ikeda K, Mikami H, Yamaguchi E (2001) Dynamic behavior of RC beams under steel weight impact loading—effects of nose-shape of steel weight. In: CONSEC'01: third international conference on concrete under severe conditions, Department of Civil Engineering, University of British Columbia, Vancouver, Canada, p 8
4. Mylrea TD (1940) Effect of impact on reinforced concrete beams. ACI J Proc 36(6):581–594
5. Saatci S, Vecchio FJ (2009) Effects of shear mechanisms on impact behavior of reinforced concrete beams. ACI Struct J 106(1):78–86
6. Koji S, Ito C, Onuma H (1994) Numerical studies of impact on reinforced concrete beam of hard missile. Nucl Eng Des 150(2–3):483–489
7. Adhikary SD, Li B, Fujikake K (2012) Dynamic behavior of reinforced concrete beams under varying rates of concentrated loading. Int J Impact Eng 47:24–38
8. Adhikary SD, Li B, Fujikake K (2013) Strength and behavior in shear of reinforced concrete deep beams under dynamic loading conditions. Nucl Eng Des 259:14–28
9. Bentur A, Mindess S, Banthia N (1986) The behaviour of concrete under impact loading: experimental procedures and method of analysis. Mater Struct 19(5):371–378
10. Kishi N, Mikami H, Matsuoka KG, Ando T (2002) Impact behavior of shear-failure-type RC beams without shear rebar. Int J Impact Eng 27(9):955–968
11. Kishi N, Mikami H (2012) Empirical formulas for designing reinforced concrete beams under impact loading. ACI Struct J 109(4):509
12. Tachibana S, Masuya H, Nakamura S (2010) Performance based design of reinforced concrete beams under impact. Nat Hazards Earth Syst Sci 10(6):1069–1078
13. Adhikary SD, Li B, Fujikake K (2014) Effects of high loading rate on reinforced concrete beams. Struct J 111(1–6):651–660
14. Banthia NP, Mindess S, Bentur A (1987) Impact behaviour of concrete beams. Mater Struct 20(4):293–302
15. Fujikake K (2014) Impact performance of ultra-high performance fiber reinforced concrete beam and its analytical evaluation. Int J Prot Struct 5(2):167–186
16. Soleimani SM, Banthia N, Mindess S (2007) Behavior of RC beams under impact loading: some new findings. Paper presented at the proceedings of the sixth international conference on fracture mechanics of concrete and concrete structures, Catania, Italy
17. Stochino F, Carta G (2014) SDOF models for reinforced concrete beams under impulsive loads accounting for strain rate effects. Nucl Eng Des 276:74–86
18. Maage M, Helland S, Poulsen E, Vennesland O, Carl JE (1996) Service life prediction of existing concrete structures exposed to marine environment. Mater J 93(6):602–608
19. ASTM G1-03 (2011) Standard practice for preparing, cleaning, and evaluating corrosion test specimens. International ASTM, West Conshohocken

Bond Performance between Slightly Corroded Steel Bar and Concrete after Exposure to High Temperature

Ou Yang[1]; Bai Zhang[2]; Guirong Yan, M.ASCE[3]; and Jun Chen[4]

Abstract: Bond-slip characteristics of slightly corroded (less than 2% mass loss) reinforcing steel bars embedded in concrete were investigated with pull-out tests conducted after exposure to high temperature and subsequent cooling. Twenty-five scenarios consisting of five corrosion levels (0.0%, 0.5%, 1.0%, 1.5%, and 2.0% corrosion loss) and five exposed temperatures (20°C, 200°C, 400°C, 600°C, and 800°C) were tested. The results indicate that the slight corrosion of the reinforcement not only affects the bond performance of concrete and steel bars but also influences their reaction to elevated temperature before the exposure temperature exceeds 400°C. The effect of slight corrosion on bond-mechanics behavior was eliminated by the dramatic decrease in concrete strength and significantly different thermal expansion between concrete and steel as the exposure temperature reached 600°C. **DOI: 10.1061/(ASCE)ST.1943-541X.0002217.**

Author keywords: Fire behavior; Pull-out test; Aging structure; Empirical model.

Introduction

Concrete structures are among the most popular building types because of their relatively lower cost, excellent bearing capacity, and adequate durability. The most common application of concrete is concrete composited with steel reinforcement because of the similar thermal expansion of steel and concrete at normal temperatures and the excellent tensile strength of steel, which can reinforce the poor tensile performance of concrete. However, a severe threat for reinforced concrete structures is the corrosion of the steel reinforcement; this is one of the major durability problems for reinforced concrete structures. Generally, concrete can provide effective protection of the reinforcing steel against corrosion. However, various situations may destroy the protection, such as carbonation of concrete reaching the reinforcement surface or chloride ions dissolving into the porous solution of concrete. It is worth noting that the corrosion process will persist after it begins, since acid ions, such as chloride and carbonate ions, are not completely consumed during the electrochemical reaction (Ahmad 2003). Reinforcement corrosion accelerates the aging of infrastructure, reduces its remaining service life, and requires expenditures for maintenance, repair, or replacement (Abaneh and Sheban 2011). Common sense dictates that the deterioration of structural performance caused by steel reinforcement corrosion is mostly due to following two reasons (Coccia et al. 2016). The first reason is section decrease of the reinforcing bars, particularly pitting corrosion, which results in decreased tensile strength of the bars; the second is longitudinal cracking of concrete because of the expansive pressure caused by corrosion products, which leads to the debonding of rebar and concrete. The latter is more dangerous, because bond failure is usually brittle in nature.

The degradation of structural performance caused by reinforcement corrosion can significantly affect building safety under extreme loads, such as earthquakes (Martinelli and Erduran 2013) and fires. In fact, the behavior of concrete and steel, and even the bond performance of these two materials under elevated temperatures, has been widely investigated. In general, the most severe damage a concrete structure suffers during a fire includes strength reduction (Fletcher et al. 2007) and cracking (even spalling) of the concrete (Lo Monte and Gambarova 2015), strength decline of the reinforcement (Kumar et al. 2013), and the degradation of the reinforcement bond strength (Ergün et al. 2016). Steel reinforcement is a homogenous material, which means its behavior in fire is relatively simple and stable (Gardner et al. 2016). Concrete properties usually initially vary with increasing temperature, since the physical and chemical changes that occur in concrete start from relatively lower temperatures (Ma et al. 2015). The damage to the reinforcement and bond interface may be more hazardous, since structural failure often occurs when the effective strength of the reinforcement is lost through heating (Kodur and Agrawal 2017). In fact, the study of the bond performance of concrete and steel bar under/after elevated temperature began in the 1960s. Since that time, many researchers have focused their efforts on this issue. Diederichs and Schneider (1981) summarized previous efforts and experimentally studied the bond performance between normal concrete and both plain round steel bar and deformed steel bar, separately, at elevated temperature. Bingöl and Gül (2009) studied the effects of cooling regimes on bond strength between concrete and steel bars after exposure to elevated temperatures. Ergün et al. (2016) investigated the effects of the material properties, such as concrete strength, rebar diameters, and rebar grades, on bond strength between concrete and steel bar after exposure to high temperature. However, previous studies of elevated temperature effect on RC structures have only concerned intact structures, and little

[1]Assistant Professor, College of Civil Engineering, Hunan Univ., Changsha 410082, P.R. China (corresponding author). ORCID: https://orcid.org/0000-0002-7267-1968. Email: ouyanghnu@hnu.edu.cn

[2]Graduate Student, College of Civil Engineering and Mechanics, Xiangtan Univ., Xiangtan 41105, P.R. China. Email: 1033751154@qq.com

[3]Assistant Professor, Dept. of Civil, Architectural, and Environmental Engineering, Missouri Univ. of Science and Technology, Rolla, MO 65409. Email: yang@mst.edu

[4]Associate Professor, College of Civil Engineering and Mechanics, Xiangtan Univ., Xiangtan 41105, P.R. China. Email: chenjun0325@126.com

Note. This manuscript was submitted on December 15, 2017; approved on June 5, 2018; published online on September 5, 2018. Discussion period open until February 5, 2019; separate discussions must be submitted for individual papers. This paper is part of the ***Journal of Structural Engineering***, © ASCE, ISSN 0733-9445.

information has been provided on the behavior of fire on an aging structure.

This research program aims to investigate the bond performance of slightly corroded steel reinforcements after elevated temperatures using pull-out tests. Specifically, a total of 75 nominally identical specimens were fabricated for 25 test scenarios comprising five alternate exposure temperatures (20°C, 200°C, 400°C, 600°C, and 800°C) and five different corrosion levels (0%, 0.5%, 1.0%, 1.5%, and 2.0% of rebar mass loss). Three specimens were tested in each of the aforementioned scenarios.

Research Significance

To understand the bond properties of corroded RC in fire is significant because corrosion is usually exhibited in aging structures, which are more vulnerable to fire because the power system and facilities are in a state of deterioration. The corrosion of reinforcement actually induces new material to form in concrete–steel composites. The physical and chemical properties of steel corrosion products vary with elevated temperature and result in the different bond performance variation laws of concrete and steel bar after exposure to high temperature. The coupled effects of steel corrosion and elevated temperature on the bond-mechanics behavior of concrete and steel bar are more evident for slightly corroded reinforcements, since the concrete, the reinforcement, and their bond interface generally remain in their original state. As a primary investigation of the coupled effect of steel corrosion and elevated temperature on the bond performance of concrete and steel bar, in this paper we present basic test results, such as the bond-slip curves and bond strength and some analytical results, such as initial bond stiffness and pull-out energy. In addition, a modified empirical model based on the fib Model Code for Concrete Structures 2010 (FIB 2013) is also proposed. These efforts may contribute to the fire-resistance evaluation of aging structures.

Experimental Procedure

Specimen Dimensions

A 550-mm-long, 20-mm-diameter deformed steel bar was embedded along the axis of a concrete cube with dimensions of 200 × 200 × 200 mm^3, as shown in Fig. 1. The embedment length between reinforcement and concrete was only 40 mm, which was two times the rebar diameter (d_s). The free 100-mm prelength and 60-mm end-length of the steel–concrete overlap were separated by PVC sleeves with a thickness of 2 mm and an outside diameter of 28 mm. To avoid errors induced during the specimen-fabrication process, the embedment length was measured manually for each specimen by comparing the relative position of the steel bar and the PVC sleeves before and after the casting procedure. The interspace between sleeve and bar was filled with polyurethane foam to prevent concrete from flowing into it during the casting procedure. In order to obtain more realistic conditions and control the failure pattern, three square stirrups with side lengths of 100 mm and diameters of 6 mm were laid out along the embedded range at intervals of 50 mm.

Specimen Materials

A type of normal-strength concrete with a compressive strength of 50 MPa was designed to cast the specimen. Ordinary portland cement with a compressive strength of 42.5 MPa and tap water were used for the cement paste mixture. The coarse aggregate was silica-based gravel with sizes ranging from 5 to 15 mm, and the fine aggregate was common silica-based river sand from Changsha City, in middle-southern China. The mix proportions of cement/water/fine aggregate/coarse aggregate were 1:0.4:0.92:1.95.

The specification for the steel bar used in this experimental study followed China National Standard No. GB 1499.2-2013 (General Administration of Quality Supervision, Inspection and Quarantine of P.R.C. and Standardization Administration of P.R.C. 2008). The reinforcing steel grade was HRB500, which was defined as hot rolled ribbed bars with a yield strength of over 500 MPa. The stirrup steel was HPB335, which was defined as hot rolled plain bars with a yield strength of over 300 MPa.

Artificial Corrosion

The specimens were corroded using an electrochemically accelerated corrosion technique. Direct electric current was impressed on the main steel bar embedded in the concrete using an integrated system that incorporated a rectifier with a built in ammeter to monitor the current and a potentiometer to control the current intensity. Two measures were adopted to ensure that the corrosion only took place in the bond range of the steel bar. First, the steel bar, apart from the bond range, was epoxy-coated to prevent the corrosion of

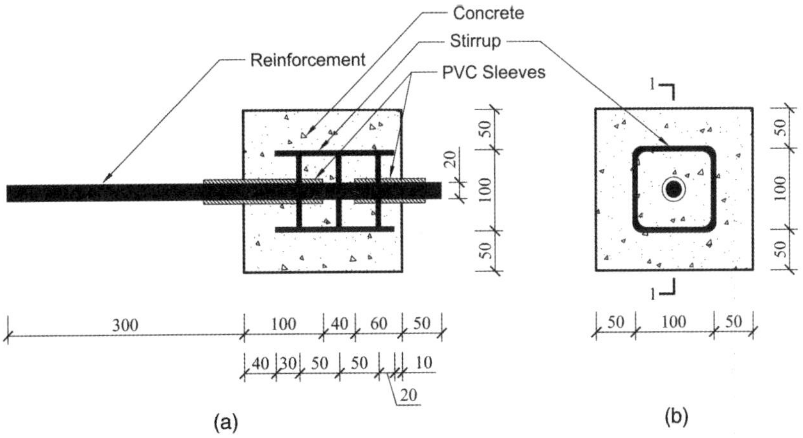

Fig. 1. Specimen dimensions (units are in millimeters): (a) 1-1 section; and (b) front view.

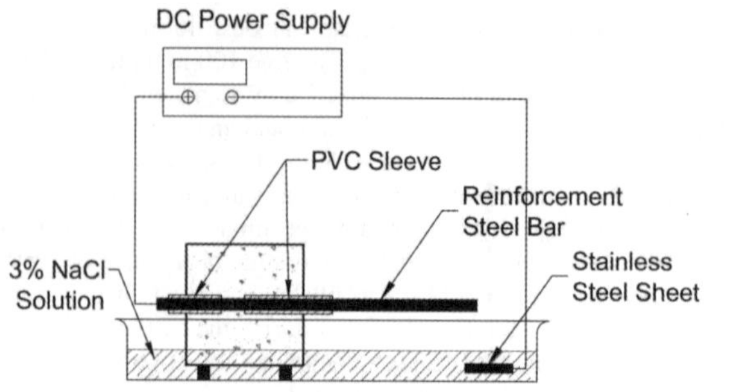

Fig. 2. Accelerated corrosion setup.

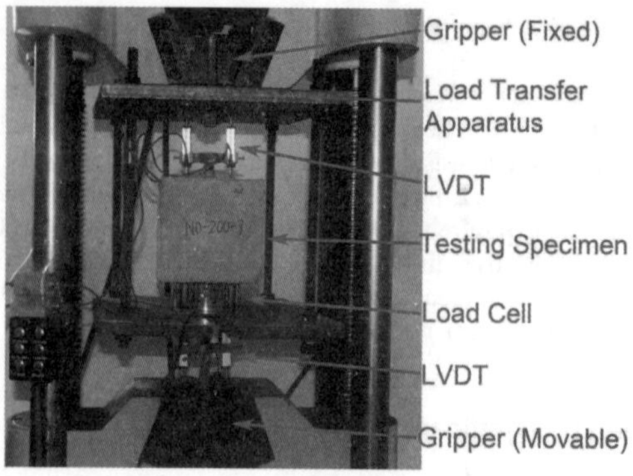

Fig. 3. Experimental setup.

the nonembedded range. Second, the portion of the specimen beneath the steel bar was immersed in a 3% solution of NaCl, as shown in Fig. 2. Because of the permeation of the solution in the concrete, only the embedded range of the steel bar was in contact with the NaCl. An average current density of 100 $\mu A/cm^2$, which corresponds to the highest corrosion rates in existing reinforced concrete structures (Caré and Raharinaivo 2007), was imposed on the surface of the corroded range of the steel bar. The duration of accelerated corrosion, determined using Faraday's law (Zhou et al. 2015), was used to control the corrosion level from 0.5% to 2.0%.

Heating Procedure

Five temperature series—20°C, 200°C, 400°C, 600°C, and 800°C—were studied in the experiment, and each series consisted of 15 specimens. The specimens were heated in an electric furnace at a rate of 10°C/min after drying in ambient conditions in the laboratory for 2 months. The temperature in the furnace was kept constant for 3 h after the specimens were heated to the predetermined temperatures (200°C, 400°C, 600°C, and 800°C). The furnace rating was 30 kW/380 V/6 phase, with a maximum operating temperature of 1,250°C, temperature control precision of ±2°C, and inner dimensions of $500 \times 500 \times 810$ mm^3 (Huo et al. 2013). After the specimens were exposed to high temperatures and cooled to room temperature (approximately 12 h), the pull-out tests were conducted.

Pull-Out Test

The pull-out tests were conducted in an electrohydraulic universal testing machine with a maximum output force of 1,000 kN. The pull-out load was first applied using load control with a loading speed of 10 kN/min, and the load was then switched to displacement control with a loading speed of 0.5 mm/min. The specimen was placed in a load-transfer apparatus, as shown in Fig. 3. The load-transfer apparatus consisted of two 20-mm-thick steel plates, with four 20-mm-diameter adjustable screws through each corner of both plates to keep the two plates at a proper position by adjusting the nuts, and a 150-mm-long, 32-mm-diameter screw fixed on the center of the top plate. A gap extended to the center of the bottom plate to facilitate easy placement of the test specimen in the load-transfer apparatus. The fixed gripper of the testing machine clamped the screw on the top plate, and the movable gripper clamped the longer cantilever steep bar to apply pull force on the specimen. Four linear variable differential transformers with a maximum stroke of 15 mm and 0.01% linear unequal observations were used to record the relative slip between the steel bar and the concrete. Two LVDTs were attached on the free side of the steel bar symmetrically with the spring trigger on the surface of the concrete cube, and two LVDTs were attached on the load side of the steel bar symmetrically with the spring trigger on the surface of the lower plate of the load-transfer apparatus. A load cell with a capacity of 30 t and a comprehensive error of 0.05% was placed between the lower side of the specimens and the steel plate of the load-transfer apparatus. An NI 9,215 data acquisition card (DAQ card) (National Instruments, Austin, Texas) was used to record the data from the LVDTs and the load cell at a sampling rate of 32 Hz.

Table 1. Mechanical properties of concrete and steel bars after exposure to different temperatures

HT (°C)	f_c (MPa)	S_{dc}	f_y (MPa)	S_{dy}	f_u (MPa)	S_{du}
20	47.60	4.33	529.07	3.49	651.42	3.12
200	41.48	1.87	520.49	1.87	646.83	1.07
400	30.16	4.11	516.94	4.23	644.04	3.93
600	17.18	1.92	512.93	2.61	640.79	3.22
800	6.97	0.82	377.74	3.02	531.48	4.88

Note: The symbol HT = exposure temperature of the specimen; f_c = compressive strength of the concrete at corresponding temperatures; f_y and f_u = yield strength and ultimate strength of the steel bar at corresponding temperatures, respectively; and S_{dc}, S_{dy}, and S_{du} = standard deviations of the compressive strength of concrete test results and of the yield strength and ultimate strength of steel bar test results, respectively.

Experimental Results and Discussion

Material Properties

Twenty-five $150 \times 150 \times 150$-mm^3 concrete cubes were cast with the specimens in the same condition for compressive-strength testing of the concrete after each elevated temperature cycle. The average compressive strength of the concrete f_c and the yield strength f_y and ultimate strength f_u of the steel bars after exposure to different temperatures are listed in Table 1. The symbol HT refers to the exposed temperature of the specimen. The strength values (f_c, f_y, and f_u) are the means of five specimen test results, and the corresponding standard deviations (S_{dc}, S_{dy}, and S_{du}) are also listed in Table 1. As shown in the table, an accelerated degradation trend appeared for the compressive strength of the concrete as the heating temperature increased, but the decrease in the strength of the steel bars, including yield strength and ultimate strength, was slight before the heating temperature exceeded 600°C. Once the heating temperature reached 800°C, the strengths of the steel bar, including yield strength and ultimate strength, decreased dramatically. These results are similar to those reported in the literature (Arel and Yazici 2014; Biolzi et al. 2008; Ergün et al. 2016; Fletcher et al. 2007; Kumar et al. 2013), in which further discussion is presented.

Corrosion-Loss Measurement and Error Analysis

Corrosion loss was defined by the percentage of steel bar mass loss. However, since the corrosion length was only 40 mm, which was less than 1/10 of the total length, using the mass loss of the entire rebar would have been inaccurate. Instead, the linear density of the bond range, which was defined as weight per unit length, was used to measure the mass loss induced by corrosion. The mass loss of the steel bars was measured as follows:

1. the weight and length of the entire steel bar before embedding into concrete was measured and the linear density ρ_0 was calculated;
2. the concrete of the specimen after pull-out testing was crushed to access the corroded steel bar;
3. the bond range of the corroded steel bar was cut off;
4. the segment of steel bar was cleaned with 5% HCl solution to remove the oxides;
5. the weight and length of the segment of steel bar was measured and the linear density ρ_c was calculated; and
6. the corrosion loss was calculated by the following equation:

$$\mathrm{CL}_a = \frac{\rho_0 - \rho_c}{\rho_0} \times 100\% \quad (1)$$

To investigate the error induced by cutting off the steel bar, an error test was conducted using the following procedure:

1. five uncorroded tested steel bars (approximately 500 mm long) were chosen randomly and their weights and lengths were measured;
2. five 40-mm-long segments on the steel bar were cut off at random positions (these segments were the same length as the bond length), and the weights and lengths of each segment were measured; and
3. the linear densities of the chosen steel bar and corresponding segments were calculated to compare the variations of their linear densities.

The test results are presented in Table 2. The maximum measurement errors of mass loss $\Delta_{\max}$ were calculated as follows:

$$\Delta_{\max} = \frac{\rho_{i,\max} - \rho_{i,\min}}{\rho_0} \times 100\% \quad (2)$$

where $\rho_{i,\max}$ and $\rho_{i,\min}$ = maximum and minimum linear density measurements of the steel segments, respectively.

As shown in the table, the measurement error of the rebar mass loss was relatively large compared to the corrosion-induced mass loss, which should be considered in measuring the corroded steel bars. The error was mostly derived from the following three aspects. First, inappropriate operation during the weighing process of the steel bar and segments could have induced error. Second, the variance of the rib along the steel bar and casting faults could also have caused error. Finally, since the steel bars were cut manually and it was difficult to ensure completely flat sections, the resulting irregular breaks could cause error.

The corrosion losses of the steel bars are listed in Table 3. As shown in the table, the average measured corrosion loss of the steel bars matches the expected values quite well. However, the corrosion loss of each specimen for the same expected corrosion level varied significantly. One of the probable reasons for this was that the insulating efficiency of the epoxy coating was not as good as was expected, which meant that the corrosion area was larger than expected and resulted in current density of the bond region under 100 $\mu\mathrm{A/cm}^2$. Measurement error may be another reason for the variation in corrosion loss. As shown in Table 2, the maximum measurement error for the linear density of a cut segment reached 0.5%, a large value compared to the corrosion-induced mass loss.

It should be pointed out that the measurements were obtained from specimens that only experienced ambient temperature because the variation of the weight of the steel bars after a high-temperature cycle would exceed the loss caused by corrosion.

Table 2. Testing errors of steel bar mass loss

Specimen ID	ρ_0 (g/mm)	ρ_i (g/mm) S1	S2	S3	S4	S5	$\Delta_{\max}$ (%)
CR01	2.343	2.335	2.337	2.339	2.341	2.341	0.254
CR02	2.375	2.373	2.373	2.376	2.377	2.381	0.361
CR03	2.344	2.340	2.340	2.342	2.342	2.343	0.120
CR04	2.343	2.339	2.340	2.341	2.346	2.351	0.506
CR05	2.367	2.351	2.358	2.359	2.362	2.364	0.516

Note: The variables ρ_0 and ρ_i = linear density of the entire steel bar and the segment of the steel bar, respectively; and $\Delta_{\max}$ = maximum measurements error of mass loss.

Table 3. Mass-loss percentage of corroded steel bars

Specimen series	CL_e (%)	CL_a (%) S1	S2	S3	S4	S5	Average
C05	0.5	0.33	0.47	0.68	0.55	0.72	0.55
C10	1.0	1.13	0.82	1.29	1.14	0.93	1.06
C15	1.5	1.24	1.48	1.31	1.76	1.48	1.45
C20	2.0	2.46	1.53	1.99	1.59	2.22	1.96

Note: The variables CL_e and CL_a = expected and actual mass-loss percentage of the steel bar, respectively.

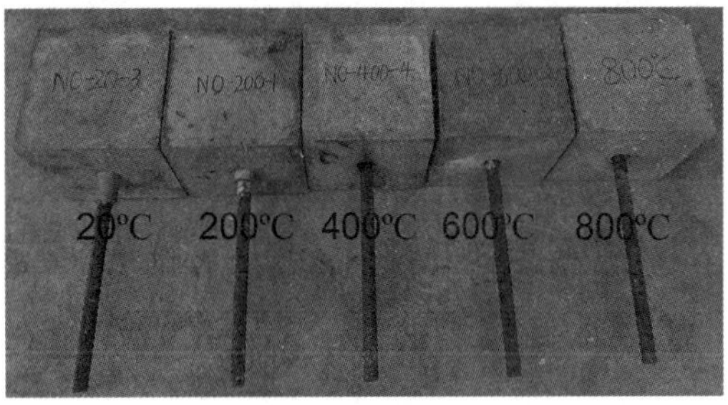

Fig. 4. Concrete after exposure to elevated temperatures.

Experimental Observations

Rust spots on the concrete surface and corrosion pits on rebar ribs appeared as the corrosion level reached 1.0% mass loss. However, cracks were not observed in all corroded specimens.

Both the concrete and steel bar of the specimens presented different visual characteristics with varying elevated temperatures, as shown in Figs. 4 and 5. The concrete generally presented a stone gray color at room temperature. As the temperature reached 200°C,

Fig. 5. Rebar after exposure to elevated temperatures.

Fig. 6. Penetrating crack on concrete cover.

the color turned muddy. Part of the concrete surface appeared dark reddish-brown and hairline cracking appeared when the heating temperature reached 400°C. Dark red color, more cracks, and partial edge and corner spalling were observed when the heating temperature was increased to 600°C. The concrete turned to gray-white and a chapped surface with obvious spalling at the edges and corners was observed when the heating temperature reached 800°C. The steel bars became more russet-colored at elevated temperature levels as they were oxidized.

Because of the presence of the restraint from the stirrup and a thick concrete cover, all of the specimens failed due to the steel bar being pulled out without the concrete cube splitting. Only a few specimens that were heated to 800°C exhibited penetrating cracks in the concrete during the pull-out test, as shown in Fig. 6.

Bond Stress and Relative Slip Relationship

The bond stress τ_b of the steel bars and concrete that were measured in this study was the average stress of the bond range, which can be expressed as

$$\tau_b = \frac{F}{\pi d L} \tag{3}$$

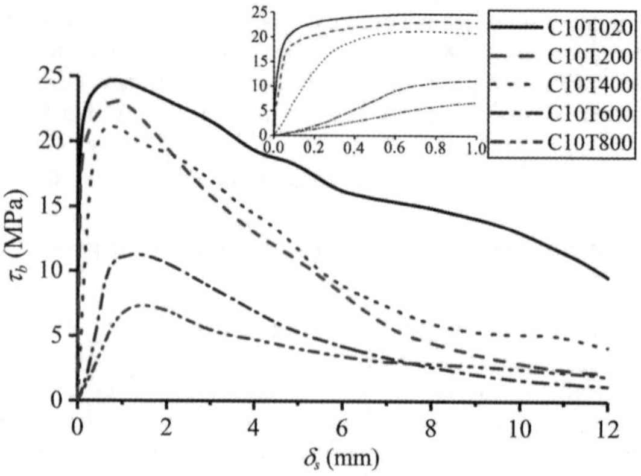

Fig. 7. Relationship of τ-s with various elevated temperature cycles.

where F = pull-out force applied on the steel bar; d = diameter of the steel bar; and L = bond length between the steel bar and the concrete.

The relative slip between rebar and concrete was measured from both sides of the rebar with two LVDTs, as shown in Fig. 3. However, the values of the relative slip adopted in this paper are the average measurement of the two LVDTs at the free end of the rebar, which is close to the fixed gripper in Fig. 3. It should be pointed out that larger rebar slip was observed at the load end for the corresponding pull force in most of the tests. The deflection of the bottom plate, deformation of the concrete, and uneven concrete surface at the load end may have resulted in the difference between the two measurements. However, since the bond length was relatively short and the pull-out force was also small (the maximum value was approximately 80 kN), the elongation of the rebar can be ignored.

Due to space limitations, only several representative bond-stress and relative slip (τ-s) curves are plotted. The five τ-s curves in Fig. 7 are for the specimens with 1.0% corrosion loss after different thermal cycles, which represent the variation law of the same corrosion level after exposure to different temperatures.

The ascent stage of the τ-s curve commonly consists of two phases for normal temperature tests (Wu 2014). The first phase is dominated by chemical adhesion and friction between the steel bar and the concrete. The second phase is dominated by the mechanical interaction between the steel bar rib and the concrete. The same phenomenon was observed for the test scenario in which the heating temperature reached 200°C. However, the chemical-adhesion-dominant phase completely disappeared after the exposure temperature of the specimens exceeded 400°C. This phenomenon could be explained by the decomposition of calcium hydroxide above 350°C, which causes damage to the chemical-adhesion interface between the steel bar and concrete (Ma et al. 2015). It should be pointed out that some specimens in the test scenario for a corrosion loss of 2.0% remained in the first phase on the τ-s curves at 400°C due to the notably increased friction caused by the expansion effect of corrosion products, as shown in Fig. 8.

The five curves in Fig. 8 are for the pull-out process of the specimens with different corrosion levels after exposure to 400°C, which represents the τ-s curve characteristics of various corrosion-level exposures to the same temperature.

As shown in the figure, the ascent stage becomes sharp with increasing corrosion loss because of the increasing friction resulting from the expansion effect of corrosion products.

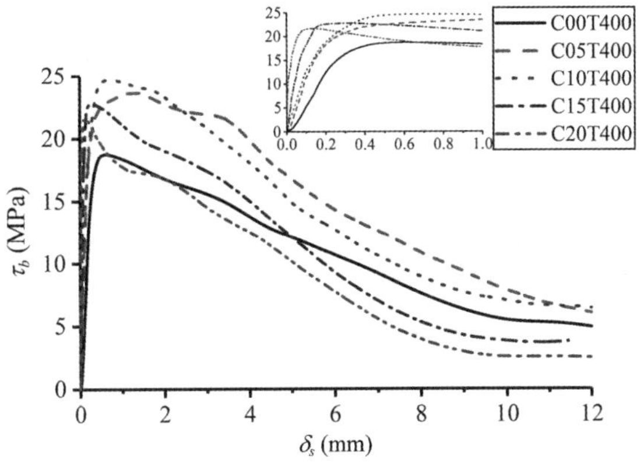

Fig. 8. Relationship of τ-s with various corrosion levels.

This phenomenon appears in the test scenario with a heating temperature under 400°C. For the test scenario in which the heating temperature exceeded 600°C, the corrosion-expansion effect was eliminated because of the significant difference in thermal expansion between steel bar and concrete (the thermal expansion of steel bar is greater than that of concrete). Thus, the τ-s curves for these test scenarios were similar and showed no observable variation law.

In summary, the τ-s curves for all corrosion levels at the same temperature are similar. Elevated exposure temperatures can destroy the chemical adhesion if the temperature exceeds 400°C, which leads to the disappearance of the chemical-adhesion-dominant phase. Slight corrosion resulted in a sharper ascent stage, but the effect only existed when the exposure temperature was lower than 400°C.

Bond Strength

Bond strength $\tau_{b,\max}$ was defined as the maximum value of the bond stress. Bond strength and the corresponding relative slip are listed in Table 4, and the average bond strength for each test scenario is shown in Fig. 9. Bond strength generally decreased with elevated temperature for all corrosion levels, except for a corrosion loss of 0.55%, whose bond strength increased (approximately 13%) as the exposure temperature increased from ambient temperature to 200°C. This abnormal phenomenon of these two corrosion levels not only appeared in bond strength, but also showed in the slope of the initial bond-slip curve and pull-out energy at the peak of the bond-slip curve, as described in the sections "Slope of Initial Bond-Slip Curve" and "Pull-Out Energy," respectively.

A possible reason for this phenomenon is that the evaporation effect leads to the corrosion product permeating into the pores of the concrete, which results in an increase in local concrete strength near the steel bar. However, with exposure temperature increasing to 400°C, the decomposition of calcium hydroxide and a significant reduction in concrete strength would eliminate this effect.

The reason for the decrease in bond strength in the scenario with a higher corrosion level when the exposure temperature was increased to 200°C can be explained as follows. With increasing corrosion loss, a large amount of iron hydroxide decomposed into iron oxides after the exposure temperature exceeded 150°C, resulting in a reduction in the volume of corrosion products (Cornell and Schwertmann 2003), and the extra hoop stress caused by corrosion

Table 4. Summary of test results

			$\tau_{b,\max}$ (MPa)				$\delta_{s,\max}$ (mm)			
Test scenario	HT (°C)	CL (%)	sp1	sp2	sp3	Average	sp1	sp2	sp3	Average
C00T020	20	0.0	29.19	27.77	21.11	26.02	1.34	0.66	0.56	0.85
C05T020		0.5	26.72	26.59	25.13	26.15	1.64	0.35	0.24	0.74
C10T020		1.0	25.82	25.20	24.68	25.23	0.22	0.33	0.83	0.46
C15T020		1.5	34.06	32.40	32.11	32.86	0.82	1.10	0.34	0.75
C20T020		2.0	35.88	34.57	34.40	34.95	0.41	0.63	0.17	0.40
C00T200	200	0.0	23.64	20.72	20.57	21.64	0.80	1.05	1.21	1.02
C05T200		0.5	32.82	28.00	27.82	29.55	1.21	0.85	0.73	0.93
C10T200		1.0	26.11	25.07	23.13	24.77	0.81	0.69	0.83	0.78
C15T200		1.5	25.13	24.37	24.34	24.61	0.48	0.58	0.50	0.52
C20T200		2.0	31.91	28.06	27.63	29.20	0.26	0.36	0.18	0.27
C00T400	400	0.0	19.68	19.36	18.74	19.26	0.72	1.15	0.61	0.83
C05T400		0.5	25.84	25.72	23.66	25.08	1.28	0.74	0.70	0.91
C10T400		1.0	24.69	21.17	21.01	22.29	0.72	0.51	0.70	0.64
C15T400		1.5	26.15	23.76	22.78	24.23	0.44	0.43	0.29	0.39
C20T400		2.0	24.39	22.14	21.67	22.73	0.14	0.40	0.61	0.38
C00T600	600	0.0	11.20	10.68	10.66	10.85	1.46	1.09	0.59	1.05
C05T600		0.5	11.14	9.85	9.42	10.14	0.92	1.15	1.20	1.09
C10T600		1.0	11.28	11.03	10.83	11.05	1.30	0.90	0.85	1.02
C15T600		1.5	13.08	10.54	10.13	11.25	0.42	0.36	0.93	0.57
C20T600		2.0	12.33	11.69	10.93	11.65	0.40	0.21	0.38	0.33
C00T800	800	0.0	8.67	7.46	6.88	7.67	1.46	1.43	1.37	1.42
C05T800		0.5	8.94	8.58	8.32	8.61	1.19	0.48	0.66	0.78
C10T800		1.0	8.70	7.51	7.34	7.85	1.79	1.49	1.17	1.48
C15T800		1.5	8.87	7.93	7.15	7.98	1.19	1.53	0.67	1.13
C20T800		2.0	8.23	7.27	7.11	7.54	1.03	0.90	0.64	0.86

Note: The variable CL = corrosion level of the specimens; sp1, sp2, and sp3 = three different specimens of each test scenario; $\tau_{b,\max}$ = bond strength of the specimen; and $\delta_{s,\max}$ = relative slip on the free end of the rebar corresponding to the bond strength.

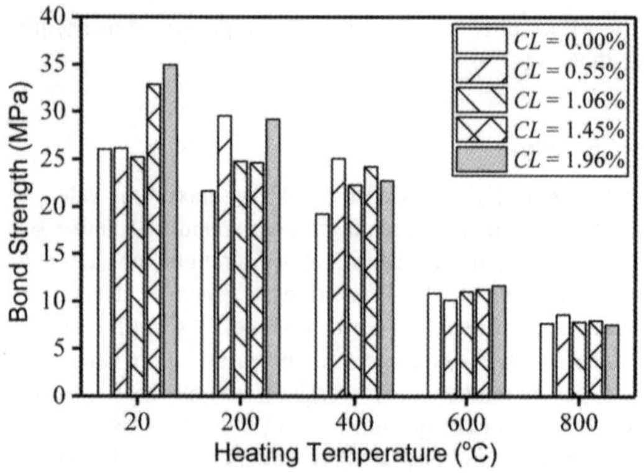

Fig. 9. Variation of average bond strength with elevated temperature.

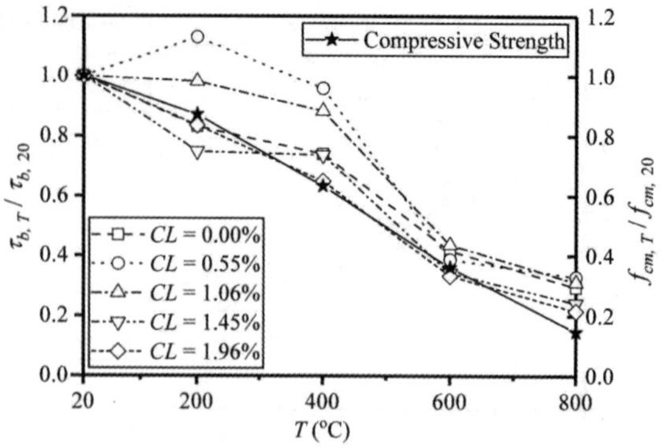

Fig. 10. Variation of normalized bond strength and compressive strength with elevated temperature.

was reduced. Thus, the bond strength of these specimens also decreased.

The bond strengths of the test scenarios with corroded specimens were generally higher than those with uncorroded specimens, but they varied randomly with increasing corrosion level at the same exposure temperature before the exposure temperature exceeded 400°C, as shown in Fig. 9. The dispersion of bond strength with varying corrosion level may be attributable to the nonuniformly distributed corrosion on the steel bar, which resulted in rust deposited locally and caused the expansion effect to exceed its corrosion level.

However, as the heating temperature exceeded 600°C, bond strength decreased dramatically and became the same for all corrosion levels. The most important reason for this occurrence was that the dramatically decreased concrete strength caused by the decomposition of the silica-based aggregate (at approximately 575°C) dominated the bond strength between the concrete and the steel bar. Other phenomena, such as elevated-temperature-induced concrete cracking that released the built-in pressure caused by corrosion expansion and the different thermal coefficient of expansion separating steel bar from concrete during the thermal process, could eliminate the extra stress caused by corrosion expansion.

The relative slip corresponding to the bond strength, $\delta_{s,\max}$, is dispersive due to the anisotropy of concrete. However, the entire development trend of $\delta_{s,\max}$ increases with increasing exposure temperature and decreases with increasing corrosion level.

To highlight the variation law of bond strength with elevated temperatures for each corrosion level, the normalized bond strength versus heating temperature are plotted in Fig. 10. The bond strength in Fig. 10 is normalized in accordance with the varying temperature series, which means that the data in each curve are the ratios of the bond strength at ambient temperature, as expressed in

$$\bar{\tau}_b(\mathrm{HT},\mathrm{CL}_i)=\frac{\tau_b(\mathrm{HT},\mathrm{CL}_i)}{\tau_b(20,\mathrm{CL}_i)} \tag{4}$$

where $\bar{\tau}_b(\mathrm{HT},\mathrm{CL}_i)$ = normalized bond strengths of each corrosion level at different heating temperatures ($\mathrm{CL}_i = 0.00\%$, 0.55%, 1.06%, 1.45%, and 1.96%); $\tau_b(\mathrm{HT},\mathrm{CL}_i)$ = corresponding bond strength; and $\tau_b(20,\mathrm{CL}_i)$ = bond strength at ambient temperature of the corresponding corrosion level.

As a reference, the normalized compressive strength, which is defined as the ratio of the compressive strength at ambient temperature, is also plotted in Fig. 10. As can be seen in the figure, the variation trend of normalized bond strength with elevated temperature for uncorroded specimens matched the normalized compressive strength well.

However, the normalized bond strength varied differently with elevated temperatures for the corroded specimens. The decrease of bond strength was relatively flat for lower corrosion levels before the heating temperature exceeded 400°C, and the bond strength of the specimens with 0.5% corrosion loss even increased by approximately 13% at 200°C. These results indicate that rebar corrosion not only affects the bond strength of concrete–rebar but also influences the reaction of the bond performance at elevated temperature of the bond performance.

Slope of Initial Bond-Slip Curve

Fig. 11 shows the slopes of the initial bond-slip curve for each tested scenario. The secant slopes of two critical points on the bond-slip curve were calculated, which were the secant slopes at relative slips of 0.01 mm [shown in Fig. 11(a)] and 0.10 mm [shown in Fig. 11(b)]. The first point is usually regarded as the initial slip of the pull-out process, and the second point is regarded as the initial crushing of concrete. For convenience, the secant slopes at relative slips of 0.01 and 0.10 mm are called the stiffness at 0.01 mm and at 0.10 mm, respectively.

The stiffnesses of the two critical points exhibited a similar trend as the exposure temperature increased. The stiffnesses at both crucial points were significantly influenced by corrosion level at exposure temperatures lower than 400°C. The stiffness of the corroded specimen test scenario was notably larger than that of the uncorroded test scenario, and the degree of enlargement generally increased with increasing corrosion level. The increment of stiffness at 0.01 mm can be explained by the expansion effect of steel corrosion products. However, the increment of stiffness at 0.10 mm should be due to the enhancement of local concrete strength as pointed out in the section "Bond Strength," since the stiffness at 0.10 mm is related to concrete crushing. The other evidence to support this point is that the stiffness values became the same as the exposure temperature increased to 400°C. This was because the permeation of the corrosion products due to the evaporation effect became homogeneous.

For the same reason as for the bond-strength variation, the stiffness-increase phenomenon disappeared as the exposure temperature reached 600°C. The stiffnesses recorded in the corroded specimen test scenarios were even lower than those recorded in the uncorroded test scenarios because of the lubrication of corrosion products at an exposure temperature of 600°C.

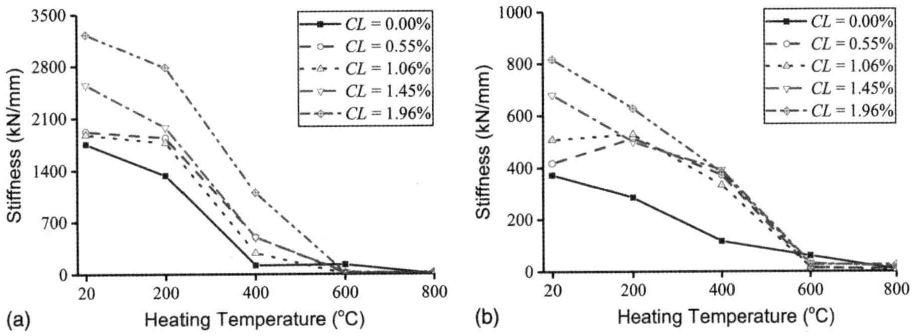

Fig. 11. Variation of bond stiffness with elevated temperature at (a) 0.01; and (b) 0.10 mm.

Pull-Out Energy

Fig. 12 shows the energy dissipation of the pull-out force, which was derived by calculating the area under the bond-force-slip curves up to its peak value. For convenience, this value is called pull-out energy.

It is clear that the pull-out energy decreased with increasing exposure temperature for all corrosion levels. Pull-out energy generally decreased with increasing corrosion level at the same exposure temperature. Compared with the opposite variation trend for bond strength and stiffness [shown in Figs. 10 and 11(b)], it can be concluded that reinforcement corrosion makes the concrete near the bond interface hard but brittle.

As with bond strength and stiffness, the trend of the enhancement of the pull-out energy was also observed in the test scenarios with corrosion losses of 0.55% and 1.06% as the exposure temperature was increased to 200°C.

Empirical Modeling

Since the bond-stress-slip curves were similar for all test scenarios, the same model can be applied to define the numerical relationship of bond stress and relative slip for all test scenarios, but using different parameters. An empirical bond-stress-slip model is proposed in this study based on the recommended expression by the fib Model Code for Concrete Structures 2010 (FIB 2013), as defined in

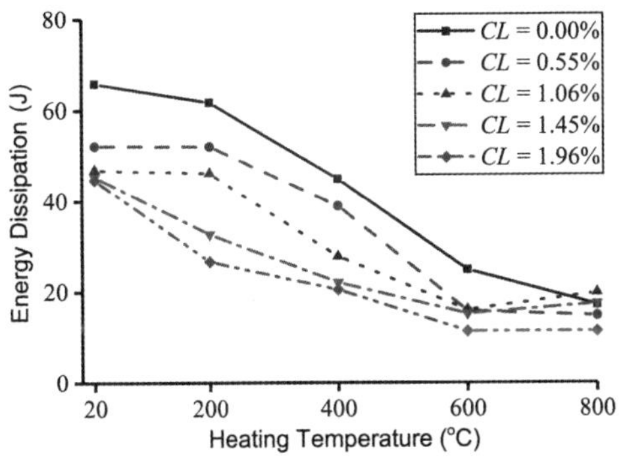

Fig. 12. Variation of pull-out energy with elevated temperature.

$$\begin{cases} \tau_b = \tau_{b,\max}(\delta/\delta_1)^{\alpha} & \text{for } 0 \leq \delta \leq \delta_1 \\ \tau_b = \tau_{b,\max} & \text{for } \delta_1 \leq \delta \leq \delta_2 \\ \tau_b = \tau_{b,\max} - (\tau_{b,\max} - \tau_{bf})(\delta - \delta_2)/(\delta_3 - \delta_2) & \text{for } \delta_2 \leq \delta \leq \delta_3 \\ \tau_b = \tau_{bf} & \text{for } \delta_3 < \delta \end{cases} \quad (5)$$

According to the corresponding relative slip of the bond strength listed in Table 4, it is reasonable to define the relative slip control points δ_1, δ_2, and δ_3 as 1 mm, 2 mm, and the clear distance between the ribs on the steel bar, respectively, following the fib code recommendations. The bond strength $\tau_{b,\max}$ and residual strength τ_{bf} are calculated using

$$\tau_{b,\max} = \beta \cdot \sqrt{f_{cu}(T)} \quad (6)$$

$$\tau_{bf} = 0.3 \cdot \tau_{b,\max} \quad (7)$$

respectively, where $f_{cu}(T)$ represents the mean cubic compressive strength at the corresponding heating temperature and can be estimated by

$$f_{cu}(T) = f_{cu}(20) - 0.05 \cdot T \quad (8)$$

In this case, the value of β can be taken as 3.5 for exposure temperatures from 20°C to 400°C and as 2.5 for those from 600°C to 800°C.

The exponent value α is calculated for each test scenario by means of nonlinear regression analysis of the average bond-stress-slip curve, as listed in Table 5. According to the curve-fitting results, a simple expression for determining the α value is proposed as follows:

$$\begin{cases} \alpha(\mathrm{CL}) = \alpha(0) - 0.10 \cdot \mathrm{CL} & \text{for } T < 600, \\ \alpha(\mathrm{CL}) = 1 & \text{for } T \geq 600 \end{cases} \quad (9)$$

Conclusions

The combined effect of slight corrosion of the reinforcement and elevated temperature on bond performance in concrete reinforced with steel bar was investigated using pull-out tests. The following conclusions can be drawn based on the experimental observations.

The bond-slip behaviors of concrete and steel bar were similar for all corrosion levels at the same exposure temperature. However,

Table 5. Regression analysis results for empirical model

Test scenario	α	R^2
C00T020	0.35	0.91
C05T020	0.28	0.95
C10T020	0.20	0.87
C15T020	0.19	0.87
C20T020	0.18	0.86
C00T200	0.30	0.95
C05T200	0.27	0.91
C10T200	0.21	0.93
C15T200	0.20	0.87
C20T200	0.21	0.90
C00T400	0.54	0.93
C05T400	0.45	0.92
C10T400	0.41	0.93
C15T400	0.32	0.88
C20T400	0.32	0.98
C00T600	0.85	0.97
C05T600	0.93	1.00
C10T600	0.88	0.96
C15T600	0.87	0.96
C20T600	0.93	0.99
C00T800	1.02	0.98
C05T800	0.96	0.99
C10T800	0.82	0.98
C15T800	0.79	0.99
C20T800	0.82	0.97

Note: The variable α = exponent of the empirical model in Eq. (5); and R^2 = adjusted R-square.

different corrosion losses exhibited diverse bond-mechanics behavior at elevated temperatures. Specifically, corrosion losses of 0.55% and 1.06% exhibited trends of increased bond strength, stiffness, and pull-out energy as the exposure temperature increased from ambient temperature to 200°C. These phenomena indicate that slight corrosion leads to local enhancement of concrete strength due to corrosion products permeating into the concrete pores, and this effect can be amplified by elevated temperature with relatively low corrosion loss.

Slight corrosion of the reinforcement caused an increase in bond strength and initial stiffness, but reduced pull-out energy when the exposure temperature was lower than 400°C. This result indicates that although the strength of concrete near the bond interface was enhanced by the permeation of the corrosion products, the deformability of the concrete in this region decreased.

The influence of slight corrosion on bond-mechanics behavior was eliminated by the dramatic decrease in concrete compressive strength because of the decomposition of aggregates and the significantly uneven thermal expansion between concrete and steel bar as the exposure temperature reached 600°C.

Based on the test data and nonlinear regression analysis, an empirical model for considering the combined effect of slight corrosion and elevated temperature was proposed by modifying the recommended bond-slip model according to the fib Model Code for Concrete Structures 2010 (FIB 2013).

A primary investigation of the combined effect of reinforcement corrosion and elevated temperature on bond-mechanics behavior was reported, and the test results could contribute to the assessment of the structural fire resistance of aging RC structures. More experimental results are needed to verify the findings presented in this paper. In addition, the influence of other factors, such as higher corrosion level, various steel and concrete types, and different cooling processes, on the bond performance of aging RC structures subjected to elevated temperatures needs to be considered in future studies.

Acknowledgments

The research described in this paper was financially supported by the National Natural Science Foundation of China (Grant No. 51578229).

References

Abaneh, A., and M. Sheban. 2011. "Impact of mechanical loading on the corrosion of steel reinforcement in concrete structures." *Mater. Struct.* 44 (6): 1123–1137. https://doi.org/10.1617/s11527-010-9688-3.

Ahmad, S. 2003. "Reinforcement corrosion in concrete structures, its monitoring and service life prediction—A review." *Cem. Concr. Compos.* 25 (4–5): 459–471. https://doi.org/10.1016/S0958-9465(02)00086-0.

Arel, H. S., and S. Yazici. 2014. "Effect of different parameters on concrete-bar bond under high temperature." *ACI Mater. J.* 111 (6): 633–639. https://doi.org/10.14359/51686992.

Bingöl, A. F., and R. Gül. 2009. "Residual bond strength between steel bars and concrete after elevated temperatures." *Fire Saf. J.* 44 (6): 854–859. https://doi.org/10.1016/j.firesaf.2009.04.001.

Biolzi, L., S. Cattaneo, and G. Rosati. 2008. "Evaluating residual properties of thermally damaged concrete." *Cem. Concr. Compos.* 30 (10): 907–916. https://doi.org/10.1016/j.cemconcomp.2008.09.005.

Caré, S., and A. Raharinaivo. 2007. "Influence of impressed current on the initiation of damage in reinforced mortar due to corrosion of embedded steel." *Cem. Concr. Res.* 37 (12): 1598–1612. https://doi.org/10.1016/j.cemconres.2007.08.022.

Coccia, S., S. Imperatore, and Z. Rinaldi. 2016. "Influence of corrosion on the bond strength of steel rebars in concrete." *Mater. Struct.* 49 (1–2): 537–551. https://doi.org/10.1617/s11527-014-0518-x.

Cornell, R. M., and U. Schwertmann. 2003. *The iron oxides: Structure, properties, reactions, occurrences and uses*. Weinheim, Germany: Wiley-VCH.

Diederichs, U., and U. Schneider. 1981. "Bond strength at high temperatures." *Mag. Concr. Res.* 33 (115): 75–84. https://doi.org/10.1680/macr.1981.33.115.75.

Ergün, A., G. Kürklü, and M. S. Başpınar. 2016. "The effects of material properties on bond strength between reinforcing bar and concrete exposed to high temperature." *Constr. Build. Mater.* 112: 691–698. https://doi.org/10.1016/j.conbuildmat.2016.02.213.

FIB (International Federation for Structural Concrete). 2013. *Fib model code for concrete structures 2010*. Berlin: Ernst & Sohn.

Fletcher, I. A., S. Welch, J. L. Torero, R. O. Carvel, and A. Usmani. 2007. "Behaviour of concrete structures in fire." *Therm. Sci.* 11 (2): 37–52. https://doi.org/10.2298/TSCI0702037F.

Gardner, L., Y. Bu, P. Francis, N. R. Baddoo, K. A. Cashell, and F. McCann. 2016. "Elevated temperature material properties of stainless steel reinforcing bar." *Constr. Build. Mater.* 114: 977–997. https://doi.org/10.1016/j.conbuildmat.2016.04.009.

General Administration of Quality Supervision, Inspection and Quarantine of P.R.C., and Standardization Administration of P.R.C. 2008. *Steel for the reinforcement of concrete—Part 2: Hot rolled ribbed bars (ISO 6935-2:1991, Steel for the reinforcement of concrete—Part 2: Ribbed bars, NEQ)*. GB 1499.2-2007. Beijing: Standards Press of China.

Huo, J. S., Y. M. He, L. P. Xiao, and B. S. Chen. 2013. "Experimental study on dynamic behaviours of concrete after exposure to high temperatures up to 700°C." *Mater. Struct.* 46 (1): 255–265. https://doi.org/10.1617/s11527-012-9899-x.

Kodur, V. K. R., and A. Agrawal. 2017. "Effect of temperature induced bond degradation on fire response of reinforced concrete beams." *Eng. Struct.* 142: 98–109. https://doi.org/10.1016/j.engstruct.2017.03.022.

Kumar, V., U. K. Sharma, B. Singh, and P. Bhargava. 2013. "Effect of temperature on mechanical properties of pre-damaged steel reinforcing bars." *Constr. Build. Mater.* 46: 19–27. https://doi.org/10.1016/j.conbuildmat.2013.03.051.

Lo Monte, F., and P. G. Gambarova. 2015. "Corner spalling and tension stiffening in heat-damaged R/C members: A preliminary investigation." *Mater. Struct.* 48 (11): 3657–3673. https://doi.org/10.1617/s11527-014-0429-x.

Ma, Q., R. Guo, Z. Zhao, Z. Lin, and K. He. 2015. "Mechanical properties of concrete at high temperature—A review." Supplement, *Constr. Build. Mater.* 93: 371–383. https://doi.org/10.1016/j.conbuildmat.2015.05.131.

Martinelli, E., and E. Erduran. 2013. "Seismic capacity design of RC frames and environment-induced degradation of materials: Any concern?" *Eng. Struct.* 52: 466–477. https://doi.org/10.1016/j.engstruct.2013.02.029.

Wu, C. 2014. *A unified bond theory, probabilistic meso-scale modeling, and experimental validation of deformed steel rebar in normal strength concrete*. Rolla, MO: Missouri Univ. of Science and Technology.

Zhou, H. J., J. L. Lu, X. Xv, B. Q. Dong, and F. Xing. 2015. "Effects of stirrup corrosion on bond-slip performance of reinforcing steel in concrete: An experimental study." *Constr. Build. Mater.* 93: 257–266. https://doi.org/10.1016/j.conbuildmat.2015.05.122.

Measurement 180 (2021) 109537

Contents lists available at ScienceDirect

Measurement

journal homepage: www.elsevier.com/locate/measurement

Substructural identification with weighted global iteration considering unknown interfacial forces and external excitation

Jia He *, Mengchen Qi, Xugang Hua, Zhengqing Chen, Ou Yang, Zhang Cao

Key Laboratory of Building Safety and Energy Efficiency of the Ministry of Education, Key Laboratory of Wind and Bridge Engineering of Hunan Province, Hunan University, Changsha, China

ARTICLE INFO

Keywords:
Substructural identification
Extended Kalman filter
Unknown interfacial forces
Unknown excitation
Weighted global iteration

ABSTRACT

Many of available substructural identification (SSI) methods are performed with the interface forces and/or the external excitation being available. However, the prior knowledge of these two forces is not always possible. Therefore, an extended Kalman filter-based substructural identification with unknown interface forces (EKF-SSI-UIF) approach is proposed in this paper for simultaneously identifying the parameters and interface forces of the concerned substructure as well as the unknown excitation applied to it. A projection matrix is introduced, and then a modified observation equation is obtained. The recursive solution for the extended substructural states is analytically derived. Data fusion technique is employed to avoid the so-called low-frequency drift in the estimated quantities. Moreover, to assure the convergence and stability of the identified results, a weighted global iteration procedure is introduced. The effectiveness of the proposed approach is numerically demonstrated via several linear substructures and a substructure with local nonlinearity.

1. Introduction

The identification of structural parameters for the purpose of structural safety assessment and remaining service life estimation becomes more and more important, particularly with relation to the increasing number of damaged or deteriorated structures caused by natural aging process, harsh environmental conditions, natural or man-made hazards, and inadequate maintenance. In this regard, various vibration-based structural identification methods have been extensively studied and developed in the frequency domain, time domain or time–frequency domain. Frequency domain methods often deal with natural frequencies, mode shapes, frequency response functions, or other transfer functions with the help of Fourier transform or Laplace transform. Time domain methods, on the other hand, directly deal with time series of the signals based on differential equations of motion or state space equation. Time-frequency methods comprise those techniques that conduct signal processing in both time and frequency domains simultaneously, using various time–frequency representations. To date, a number of literature reviews with different emphases can be found for the brief introduction of these techniques [1–4].

Since structural identification is inherently an ill-conditioned inverse problem, the numerical difficulty to assure convergence and computational efficiency increases dramatically when too many unknowns are involved. Therefore, for those large-scale structures, accurately identifying the entire structure is often time-consuming and difficult. However, if a large-size structure can be divided into several substructures whose degrees-of-freedom (DOFs) are much less than those of the whole structure, the identification procedure may be performed more efficiently and reliably. Moreover, in many practical situations, only some critical portions rather than the whole structure should be firstly taken into consideration for damage detection. From these points of view, it would be beneficial to develop substructural identification (SSI) methods to assure computational efficiency and to assess local conditions accurately. Research on SSI methods has received considerable attention in the past decades and much process has been made in this field. For example, based on genetic algorithm (GA), Koh et al. [5] proposed a progressive SSI approach for the estimation of substructural parameters in sequence. Tee et al. [6] proposed a SSI approach for the identification of large systems in a divide-and-conquer manner. By using the measured acceleration responses to determine the interface force vector, Trinh and Koh [7] presented an improved SSI strategy on the basis of multi-feature GA. Based on the virtual distortion method, Hou et al. [8,9] proposed a substructure isolation approach for local structural health monitoring. By confining each substructure to one

* Corresponding author.
E-mail address: jiahe@hnu.edu.cn (J. He).

https://doi.org/10.1016/j.measurement.2021.109537
Received 13 January 2021; Received in revised form 22 April 2021; Accepted 1 May 2021
Available online 8 May 2021
0263-2241/

DOF, Xing and Mita [10] presented an SSI approach for local damage detection of shear-type structure. Wang et al. [11] proposed a GA-based SSI approach for identifying the unknown parameters of jack-up platform in time domain and frequency domain. By analyzing the eigenparameters and measured substructural flexibility, Weng et al. [12,13] proposed a finite element model updating method for local damage detection. Based on wavelet-domain response reconstruction, Li and Hao [14] proposed a SSI method for the detection of local damages in the concerned substructures. More recently, by introducing a self-balanced virtual control system, a novel method was proposed by Zhang et al. [15] to improve the accuracy of SSI results. With the aid of auto-regressive moving average with exogenous inputs model, Mei et al. [16,17] proposed SSI-based methods to locate and quantify the structural damages. By using block pulse generalized operational matrix and GA, Ghaffarzadeh et al. [18] proposed a SSI approach for identifying the physical and modal parameters of multistory shear building. An improved objective function was presented by Hou et al. [19] for damage identification by using substructural virtual distortion method. Based on unscented Kalman filter, Su et al. [20] proposed a SSI method for model updating and parameters identification.

Although these SSI methods mentioned above are capable of accurately and efficiently identifying the substructural parameters, most of them are performed under the condition that the interfacial force are assumed to be known and treated as input (excitation) to the concerned substructure. However, it may be difficult or impossible to obtain such boundary information. Therefore, some researchers devoted to developing SSI methods without the prior knowledge of interfacial information. For example, on the basis of receptance theory and GA, Koh and Shankar [21] proposed a SSI method by eliminating interface forces through different sets of response measurements of the concerned substructure under identical excitation. Based on the prior knowledge of finite element model and an iterative model updating method, Law and Yong [22] proposed two substructural approaches for the identification of local damages and interface forces in state space. By sequentially using EKF and LSE in the concerned substructures, Lei et al. [23] proposed an approach for the identification of tall shear building under unknown seismic excitation. By combining EKF with a weighted adaptive iteration algorithm, Xu and He [24] proposed a SSI method for parameter and dynamic load identification with limited measurements. With the aid of response sensitivity matrix, Liu et al. [25] extended the explicit Newmark-β method for substructural interface force identification. By using unscented Kalman filter, Al-Hussein and Haldar [26] proposed a SSI approach for the health assessment of complex structures under unknown inputs with different nonlinear scenarios. Li and Hao [27] proposed a frequency-time domain method for the identification of substructural interface force. Based on Bayesian inference, Yuen and Huang [28] proposed a real-time SSI method for the estimation of boundary force under non-stationary responses. The aforementioned approaches are all capable of accurately identifying substructural parameters with unknown interfacial information. However, many of them are only suitable for linear structure [21–25,27,28]. Therefore, much more effort is required for the development of SSI approach with the consideration of structural nonlinearity.

The Kalman filter (KF) technique provides a recursive way for the optimal estimation of structural states, and has been actively investigated and developed by many researchers for state-input-system identification (e.g. [29–34]). Recently, by using projection matrix to derive a modified observation equation, the authors proposed a KF-based approach for multi-scale response reconstruction [35] and an EKF-based approach for parameters and loads identification [36]. As well-known, the EKF algorithm treats the unknown parameters as a part of structural states. Therefore, for large-scale structures with many parameters to be identified, the dimension of the extended state vector in EKF approaches will significantly increase resulting in computational difficulty on the calculation of derivative quantities and Jacobians as well as poor convergence. The same challenges will be encountered when the aforementioned approaches proposed by the authors are directly employed for the identification of large-size structures. Thus, to circumvent the aforementioned limitations, based on the substructural idea, an EKF-based substructural identification with unknown interface forces (EKF-SSI-UIF) approach is proposed in this study for the simultaneous identification of the parameters and interface forces of the concerned substructure as well as the unknown excitations applied to it. A projection matrix is employed, and a modified observation equation, where the unknown boundary condition of the concerned substructure and external excitations are not explicitly presented, is obtained. Then, the recursive solution for the extended state vector is analytically derived. On the basis of the estimated structural states and the measured responses, the unknown inputs and boundary forces of the concerned substructure are identified by means of LSE. Data fusion technique is introduced to avoid the so-called low-frequency drift in the estimated quantities. Moreover, to assure the convergence and stability of the identified results, a weighted global iteration procedure is employed. The aforementioned recursive estimation of the unknown quantities during the entire time history can be viewed as a local calculation process. When this local process is completed, an objective function defined as the differences between the observations and their estimated ones is evaluated. A weighted matrix is employed to scale up the covariance matrix before the subsequent global iteration. The global iteration process would continue until the minimum value of the objective function is obtained. The effectiveness of the proposed approach is numerically validated via a 20-story linear structure, a 3-span planar truss structure, and a 20-story shear building equipped with Dahl nonlinear model. Results show that the proposed EKF-SSI-UIF approach is capable of simultaneously identifying the parameters and interface forces of the concerned linear or nonlinear substructures as well as the unknown excitations applied to them.

2. Formulas of the proposed EKF-SSI-UIF approach

The equation of motion of an n-DOF structure subject to unknown excitations can be written as:

$$\mathbf{M}\ddot{\mathbf{x}}(t)+\mathbf{F}[\mathbf{x}(t),\dot{\mathbf{x}}(t),\theta]=\varphi^{u}\mathbf{f}^{u}(t) \tag{1}$$

where $\mathbf{M}$ is the mass matrix and assumed to be available herein; $\mathbf{F}[\mathbf{x}(t),\dot{\mathbf{x}}(t),\theta]$is the linear or nonlinear restoring force vector; $\mathbf{f}^u(t)$ is the unknown external excitation with its influence matrix defined as $\boldsymbol{\varphi}^u$; $\boldsymbol{\theta}$ is the structural parameter vector to be identified; $\mathbf{x}(t)$, $\dot{\mathbf{x}}(t)$and $\ddot{\mathbf{x}}(t)$are displacement, velocity, and acceleration responses, respectively.

As mentioned before, the whole structure can be divided into several substructures, and one substructure which may be vulnerable to damage or may be more important is assumed to be selected as the target structure for assessment. Based on substructural idea, the interface force can be viewed as a set of external excitation acting on the selected substructure [37], and thus the equation of motion of the concerned substructure can be given as,

$$\mathbf{M}_s\ddot{\mathbf{x}}_s(t)+\mathbf{F}_s\left[\mathbf{x}_s(t),\dot{\mathbf{x}}_s(t),\theta_s\right]=\varphi_s^{u}\mathbf{f}_s^{u}(t)+\varphi^{B}\mathbf{f}^{B}(t) \tag{2}$$

where the subscript 's' denotes the DOFs of the selected substructure; $\mathbf{f}^B(t)$ and $\boldsymbol{\varphi}^B$ are the interface force and its influence matrix, respectively. Since the interface force is assumed to be unmeasurable in this study, the two terms in the right-hand-side of the equation are both unknown quantities. Then, Eq. (2) can be rewritten as,

$$\mathbf{M}_s\ddot{\mathbf{x}}_s(t)+\mathbf{F}_s\left[\mathbf{x}_s(t),\dot{\mathbf{x}}_s(t),\theta_s\right]=\varphi'_s\mathbf{f}'_s(t) \tag{3}$$

where $\varphi'_s=\left[\varphi_s^u,\ \varphi^B\right]$; $\mathbf{f}'_s(t)=\left[\mathbf{f}_s^u(t),\ \mathbf{f}^B(t)\right]^{\mathrm{T}}$. Here, $\mathbf{f}'_s(t)$ can be called as an integrated input and composed of two parts, i.e. the external excitation applied to target substructure and its interfacial force.

In this study, the structural parameters to be identified are time invariant, i.e. $\dot{\theta} = \mathbf{0}$. Let $\mathbf{Z}(t) = \left[\mathbf{x}_s(t)^{\mathrm{T}}, \dot{\mathbf{x}}_s(t)^{\mathrm{T}}, \theta_s^{\mathrm{T}}\right]^{\mathrm{T}}$, the state equation of the target substructure can be then expressed as,

$$\dot{\mathbf{Z}}(t) = \begin{Bmatrix} \dot{\mathbf{x}}_s(t) \\ \mathbf{M}_s^{-1}\left[-\mathbf{F}_s\left[\mathbf{x}_s(t), \dot{\mathbf{x}}_s(t), \theta_s\right] + \varphi_s' \mathbf{f}_s'(t)\right] \\ \mathbf{0} \end{Bmatrix} = \mathbf{g}_s(\mathbf{Z}(t), \mathbf{f}'_s(t)) + \mathbf{w}(t) \tag{4}$$

where $\mathbf{w}(t)$ is process noise with zero mean and a covariance matrix $\mathbf{Q}(t)$.

Only using sparse noisy acceleration measurements may lead to the so-called spurious low-frequency drifts which may be caused by acceleration's insensitivity to any quasi-static component in the inputs [38,39]. Therefore, a data fusion of partially measured displacement and acceleration responses are introduced in the observation equation to solve the aforementioned problem. The discretized observation equation can be then expressed as,

$$\bar{\mathbf{y}}_k = \begin{bmatrix} \mathbf{y}_k^a \\ \mathbf{y}_k^d \end{bmatrix} = \begin{bmatrix} \mathbf{L}_a \ddot{\mathbf{x}}_{s,k} + \mathbf{v}_k^a \\ \mathbf{L}_d \mathbf{x}_{s,k} + \mathbf{v}_k^d \end{bmatrix} = \begin{Bmatrix} \mathbf{L}_a\left[-\mathbf{M}_s^{-1}\mathbf{F}_s\left(\mathbf{x}_{s,k}, \dot{\mathbf{x}}_{s,k}, \theta_s\right) + \mathbf{M}_s^{-1}\varphi_s' \mathbf{f}'_{s,k}\right] + \mathbf{v}_k^a \\ \mathbf{L}_d \mathbf{x}_{s,k} + \mathbf{v}_k^d \end{Bmatrix} \tag{5}$$

where $\bar{y}_k$ is the multi-type observations including acceleration measurements $\mathbf{y}_k^a$ and displacement measurements $\mathbf{y}_k^d$ at the time instant $t_k = k \cdot \Delta t$ with Δt being sampling interval; $\mathbf{L}_a$ and $\mathbf{L}_d$ are the matrices associated with the location of accelerometers and displacement transducers, respectively; $\mathbf{v}_k^a$ and $\mathbf{v}_k^d$ are measurement noise with zero mean and covariance matrices of $\mathbf{R}_k^a$ and $\mathbf{R}_k^d$, respectively. Given the following definition,

$$\mathbf{h}_a(\mathbf{Z}_k) = -\mathbf{L}_a \mathbf{M}_s^{-1} \mathbf{F}_s\left(\mathbf{x}_{s,k}, \dot{\mathbf{x}}_{s,k}, \theta_s\right); \ \mathbf{D} = -\mathbf{L}_a \mathbf{M}_s^{-1} \varphi'_s \tag{6}$$

The first part of Eq. (5) can be rearranged as

$$\mathbf{D}\mathbf{f}'_{s,k} = \mathbf{h}_a(\mathbf{Z}_k) - \mathbf{y}_k^a + \mathbf{v}_k^a \tag{7}$$

To ensure that matrix $\mathbf{D}$ is of full column rank, measurements should be available at the DOFs where the unknown integrated inputs are applied. Then, by using LSE, the closest solution of $\mathbf{f}'_{s,k}$ can be shown as,

$$\tilde{\mathbf{f}}'_{s,k} = (\mathbf{D}^{\mathrm{T}}\mathbf{D})^{-1}\mathbf{D}^{\mathrm{T}}\left[\mathbf{h}_a(\mathbf{Z}_k) - \mathbf{y}_k^a + \mathbf{v}_k^a\right] \tag{8}$$

Compared with Eq. (7), the error of the aforementioned solution can be calculated as,

$$e = \mathbf{D}\mathbf{f}'_{s,k} - \mathbf{D}\tilde{\mathbf{f}}'_{s,k} = \left(\mathbf{I} - \mathbf{D}(\mathbf{D}^{\mathrm{T}}\mathbf{D})^{-1}\mathbf{D}^{\mathrm{T}}\right)\left[\mathbf{h}_a(\mathbf{Z}_k) - \mathbf{y}_k^a + \mathbf{v}_k^a\right] \tag{9}$$

where $\mathbf{D}(\mathbf{D}^{\mathrm{T}}\mathbf{D})^{-1}\mathbf{D}^{\mathrm{T}}$ is known as projection matrix. Then, the following equation can be obtained because the limit of the error shown in Eq. (9) should tend to be zero.

$$\Phi \mathbf{y}_k^a = \Phi \mathbf{h}_a(\mathbf{Z}_k) + \Phi \mathbf{v}_k^a \tag{10}$$

where $\Phi = \left(\mathbf{I} - \mathbf{D}(\mathbf{D}^{\mathrm{T}}\mathbf{D})^{-1}\mathbf{D}^{\mathrm{T}}\right)$. With the combination of Eq. (5) and Eq. (10), a modified observation equation can be obtained as follows,

$$\bar{\Phi}\bar{\mathbf{y}}_k = \bar{\Phi}\bar{\mathbf{h}}(\mathbf{Z}_k) + \bar{\Phi}\bar{\mathbf{v}}_k \tag{11}$$

where

$$\bar{\Phi} = \begin{bmatrix} \Phi & \mathbf{0} \\ \mathbf{0} & \mathbf{I} \end{bmatrix}; \ \bar{\mathbf{h}}\left(\mathbf{Z}_k\right) = \begin{bmatrix} \mathbf{h}_a(\mathbf{Z}_k) \\ \mathbf{L}_d \mathbf{x}_{s,k} \end{bmatrix}; \ \bar{\mathbf{v}}_k = \begin{bmatrix} \mathbf{v}_k^a \\ \mathbf{v}_k^d \end{bmatrix} \tag{12}$$

As compared with Eq. (5), it can be seen that the unknown integrated input is not explicitly presented in Eq. (11). Therefore, the principle of EKF can be employed for the state estimation.

The priori state estimate $\widehat{\mathbf{Z}}_{k+1|k}$ can be calculated as,

$$\widehat{\mathbf{Z}}_{k+1|k} = \widehat{\mathbf{Z}}_{k|k} + \int_{k\Delta t}^{(k+1)\Delta t} \mathbf{g}_s\left(\widehat{\mathbf{Z}}_{k|k}, \widehat{\mathbf{f}}'_{s,k}, t_k\right) dt \tag{13}$$

where $\widehat{\mathbf{Z}}_{k|k}$ and $\widehat{\mathbf{f}}'_{s,k}$ are the estimates of $\mathbf{Z}_k$ and $\mathbf{f}'_{s,k}$ at the k-th time step, respectively. With proper transformation, the priori estimate error covariance matrix $\mathbf{P}_{k+1|k}$ can be then determined as,

$$\mathbf{P}_{k+1|k} = \mathbf{A}_1 \mathbf{P}_{k|k} \mathbf{A}_1^{\mathrm{T}} + \mathbf{A}_2 \mathbf{R}_k^a \mathbf{A}_2^{\mathrm{T}} + \Delta t^2 \mathbf{Q}_k \tag{14}$$

where

$$\mathbf{A}_1 = \mathbf{I} + \Delta t \left.\frac{\partial \mathbf{g}_s(\mathbf{Z}_k, \mathbf{f}'_{s,k}, t_k)}{\partial \mathbf{Z}_k}\right|_{\substack{\mathbf{Z}_k = \widehat{\mathbf{Z}}_{k|k} \\ \mathbf{f}'_{s,k} = \widehat{\mathbf{f}}'_{s,k}}} + \Delta t \begin{bmatrix} \mathbf{0} \\ -(\mathbf{L}_a^T \mathbf{L}_a)^{-1} \mathbf{L}_a^T \mathbf{H}_{k|k}^a \\ \mathbf{0} \end{bmatrix} \tag{15}$$

$$\mathbf{A}_2 = \Delta t \begin{bmatrix} \mathbf{0} \\ -(\mathbf{L}_a^{\mathrm{T}} \mathbf{L}_a)^{-1} \mathbf{L}_a^{\mathrm{T}} \\ \mathbf{0} \end{bmatrix} \tag{16}$$

$$\mathbf{H}_{k|k}^a = \left.\frac{\partial \mathbf{h}_a(\mathbf{Z}_k)}{\partial \mathbf{Z}_k}\right|_{\mathbf{Z}_k = \widehat{\mathbf{Z}}_{k|k}} \tag{17}$$

The posteriori state estimate $\widehat{\mathbf{Z}}_{k+1|k+1}$ can be determined on the basis of the priori state estimate $\widehat{\mathbf{Z}}_{k+1|k}$ shown in Eq. (13) and the modified observation equation shown in Eq. (11),

$$\widehat{\mathbf{Z}}_{k+1|k+1} = \widehat{\mathbf{Z}}_{k+1|k} + \mathbf{G}_{k+1}\left[\bar{\Phi}\bar{\mathbf{y}}_{k+1} - \bar{\Phi}\bar{\mathbf{h}}\left(\widehat{\mathbf{Z}}_{k+1|k}\right)\right] \tag{18}$$

where $\mathbf{G}_{k+1}$ is the gain matrix at the $(k+1)$-th time step and can be calculated as,

$$\mathbf{G}_{k+1} = \mathbf{P}_{k+1|k} \bar{\mathbf{H}}_{k+1|k}^{\mathrm{T}} \bar{\Phi}^{\mathrm{T}} \left[\bar{\Phi}\left(\bar{\mathbf{H}}_{k+1|k} \mathbf{P}_{k+1|k} \bar{\mathbf{H}}_{k+1|k}^{\mathrm{T}} + \bar{\mathbf{R}}_{k+1}\right)\bar{\Phi}^{\mathrm{T}}\right]^{-1} \tag{19}$$

where

$$\bar{\mathbf{R}}_{k+1} = \mathrm{E}(\bar{\mathbf{v}}_{k+1}\bar{\mathbf{v}}_{k+1}^{\mathrm{T}}); \ \bar{\mathbf{H}}_{k+1|k} = \left.\frac{\partial \bar{\mathbf{h}}(\mathbf{Z}_{k+1})}{\partial \mathbf{Z}_{k+1}}\right|_{\mathbf{Z}_{k+1} = \widehat{\mathbf{Z}}_{k+1|k}} \tag{20}$$

The posteriori estimate error covariance matrix $\mathbf{P}_{k+1|k+1}$ can be determined as,

$$\mathbf{P}_{k+1|k+1} = \left(\mathbf{I} - \mathbf{G}_{k+1}\bar{\Phi}\bar{\mathbf{H}}_{k+1|k}\right)\mathbf{P}_{k+1|k}\left(\mathbf{I} - \mathbf{G}_{k+1}\bar{\Phi}\bar{\mathbf{H}}_{k+1|k}\right)^{\mathrm{T}} + \mathbf{G}_{k+1}\bar{\Phi}\bar{\mathbf{R}}_{k+1}(\mathbf{G}_{k+1}\bar{\Phi})^{\mathrm{T}} \tag{21}$$

Some details for the determination of state estimates (i.e. $\widehat{\mathbf{Z}}_{k+1|k}$ and $\widehat{\mathbf{Z}}_{k+1|k+1}$), their error covariance matrices (i.e. $\mathbf{P}_{k+1|k}$ and $\mathbf{P}_{k+1|k+1}$), and the gain matrix $\mathbf{G}_{k+1}$ can be found in the reference [36]. Since the posteriori state estimate $\widehat{\mathbf{Z}}_{k+1|k+1}$ is obtained as shown in Eq. (18), the unknown integrated inputs at the $(k+1)$-th time step can be identified according to Eq. (8),

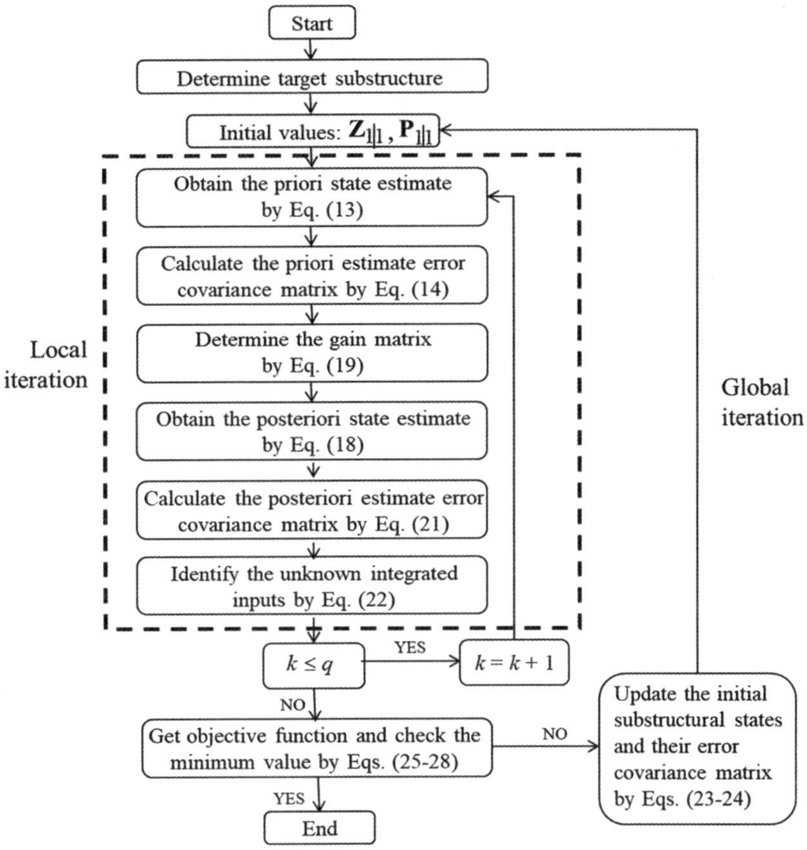

Fig. 1. Flowchart of the proposed EKF-SSI-UIF approach.

$$\widehat{\mathbf{f}}'_{s,k+1} = (\mathbf{D}^{\mathrm{T}}\mathbf{D})^{-1}\mathbf{D}^{\mathrm{T}}\left[\mathbf{h}_a\left(\widehat{\mathbf{Z}}_{k+1|k+1}\right) - \mathbf{y}^a_{k+1}\right] \tag{22}$$

The aforementioned recursive identification procedure, including the priori estimates (i.e. Eq. (13) and Eq. (14)), posteriori estimates (i.e. Eq. (18) and Eq. (21)), and load identification (i.e. Eq. (22)), are carried out for all sampling points. This process can be viewed as a local iteration. Since limited observations and arbitrarily assigned initial values are used in the local iteration, it may be doubtful or difficult to tell whether or not the estimated quantities are stable and convergent to real ones. Therefore, in order to assure the stability and convergence of the identified results, a weighted global iteration procedure is employed in this study. It was found that the error covariance matrix of the structural parameters reduced significantly during the recursive estimation, and proper amplification of this matrix at the beginning of each global iteration would be beneficial for the estimation procedure [26]. Here, a global iteration procedure as shown below is introduced and the weighted factor λ is employed to scale up the error covariance matrix.

$$\widehat{\mathbf{Z}}^{(2)}_{1|1} = \begin{bmatrix} \widehat{\mathbf{x}}^{(2)}_{s,1|1} \\ \widehat{\dot{\mathbf{x}}}^{(2)}_{s,1|1} \\ \widehat{\theta}^{(2)}_{s,1|1} \end{bmatrix} = \begin{bmatrix} \widehat{\mathbf{x}}^{(1)}_{s,1|1} \\ \widehat{\dot{\mathbf{x}}}^{(1)}_{s,1|1} \\ \widehat{\theta}^{(1)}_{s,q|q} \end{bmatrix} \tag{23}$$

$$\mathbf{P}^{(2)}_{1|1} = \begin{bmatrix} \mathbf{P}^{(1)}_{\mathbf{x},1|1} & \mathbf{0} \\ \mathbf{0} & \lambda\mathbf{P}^{(1)}_{\theta,q|q} \end{bmatrix} \tag{24}$$

where the superscripts '(1)' and '(2)' denote the previous and current global iteration, respectively; the subscript '1|1' denotes the corresponding quantities at the first time step; the subscript 'q' denotes the number of sample points.

The global iteration process mentioned above would continue until the minimum value of the following objective function is obtained [40]. Notably, since the modified observation equation is used for the identification, the differences between the observations and their estimates should be calculated on the basis of Eq. (11) in this study.

$$\delta_{jk} = \overline{\Phi}_j\overline{\mathbf{y}}_{j,k} - \overline{\Phi}_j\overline{\mathbf{h}}_j(\widehat{\mathbf{Z}}_k) \tag{25}$$

$$\gamma_j = \sum_{k=1}^{q}\delta^2_{jk} \Big/ \sum_{k=1}^{q} y^2_{j,k} \tag{26}$$

$$\overline{\gamma} = \frac{1}{B}\sum_{j=1}^{B}\gamma_j \tag{27}$$

$$\overline{\varepsilon} = \left[\sum_{j=1}^{B}\left(\gamma_j - \overline{\gamma}\right)^2\right]^{1/2} \tag{28}$$

where B is the size of the measurement vector; the subscript 'j' denotes the j-th observation; $\overline{\Phi}_j$,$\overline{\mathbf{y}}_{j,k}$, and$\overline{\mathbf{h}}_j(\widehat{\mathbf{Z}}_k)$ are the corresponding elements in $\overline{\Phi}$,$\overline{\mathbf{y}}_k$, and$\overline{\mathbf{h}}(\widehat{\mathbf{Z}}_k)$, respectively. For ease of understanding, the flowchart of the proposed EKF-SSI-UIF approach is shown in Fig. 1.

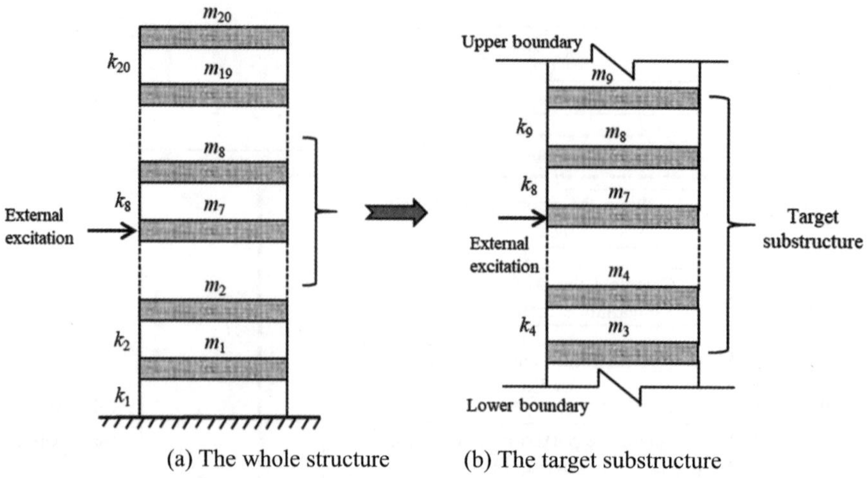

Fig. 2. The 20-story shear-type linear structure.

Table 1
The identified substructural parameters in the 1st example.

Parameters	Actual (N/m)	N.G.I. = 1		N.G.I. = 4	
		Identified (N/m)	Error (%)	Identified (N/m)	Error (%)
k_4	7.5×10^6	7.463×10^6	−0.50	7.462×10^6	−0.50
k_5	7.5×10^6	7.482×10^6	−0.25	7.472×10^6	−0.36
k_6	7.5×10^6	7.481×10^6	−0.26	7.482×10^6	−0.24
k_7	7.5×10^6	7.505×10^6	0.07	7.504×10^6	0.06
k_8	7.5×10^6	7.472×10^6	−0.37	7.468×10^6	−0.43
k_9	7.5×10^6	7.492×10^6	−0.09	7.493×10^6	−0.09
α	0.3336	0.3379	1.29	0.3377	1.22
β	2.0×10^{-3}	1.99×10^{-3}	−0.5	1.99×10^{-3}	−0.5

Note: N.G.I. denotes number of global iterations.

3. Numerical examples

To validate the effectiveness of the proposed approach, several numerical examples, including a 20-story linear structure, 3-span planar truss structure, and a 20-story building structure with a nonlinear Dahl model, are considered. The observations used for the identification are composed of partial acceleration and displacement responses. The measurement noise with 5% noise-to-signal ratio in root mean square (RMS) is considered for all of the cases. The quantities to be identified include the parameters of target substructure, the external excitation applied to it, and the corresponding interface force. Moreover, the coefficients of nonlinear model are also identified if nonlinear scenario is considered.

3.1. A 20-story shear-type linear structure

As shown in Fig. 2, a 20-story shear type linear structure is used as the first example. The structural parameters are mass of 800 kg and stiffness of 7.5×10^6N/m for each floor. The Rayleigh damping model is employed for the construction of damping matrix, and the proportional coefficients for mass matrix and stiffness matrix are set as $\alpha = 0.3336$ and $\beta = 2 \times 10^{-3}$, respectively. An external excitation of wide-banded white noise is applied to the 7th floor of the building structure, and the structural responses are calculated by state space method with the time interval of 0.001 s.

As shown in Fig. 2(b), the structure from the 3rd floor to 9th floor is considered as the target substructure. The acceleration responses of the 3rd, 4th, 5th, 6th, 7th and 9th floor and the displacement response of the 7th floor are assumed to be measured for the identification of parameters of the target substructure, the excitation applied to it, and the corresponding interface force. The initial guesses of the substructural parameters are given as 50% of the actual ones. The effect of different weights ranging from 1 to 100 was discussed by Hayashi and Saito [40] in details. It was found a weight of 100 led to accurate results with the best convergence performance. Similar description can be also found in other reference [34]. Therefore, the weighted factor used to scale up the error covariance matrix at the beginning of each global iteration is set to be 100.

By using the proposed EKF-SSI-UIF approach, the substructural parameters are identified and shown in Table 1. Obviously, the identified parameters are close to the actual ones even the number of global iteration (N.G.I.) is equal to 1. It seems that the implementation of the subsequent global iteration is unnecessary. However, focusing on the force identification, unsatisfied results are found in the first global iteration. Fig. 3 gives the identification results of the unknown excitation as well as the lower and upper boundary forces. For ease of comparison, only the time segment from 4 s to 6 s is plotted. It can be seen from Fig. 3 that significant differences between the identified forces and the actual ones are found in the first global iteration whereas much better results can be observed in the last global iteration. A normalized root mean square error (NRMSE) defined by the authors [35] is used to quantitively evaluate the difference between the actual forces and estimated ones.

$$\mathrm{NRMSE} = \frac{1}{A_m}\sqrt{\frac{1}{n_p}\sum_{i=1}^{n_p}(\chi_i^{est} - \chi_i^{act})^2} \tag{29}$$

where A_m is the force amplitude; n_p is the number of the sampling points; $\chi_i^{est}$and $\chi_i^{act}$are the estimated and actual force at the i-th sampling point, respectively. The NRMSE results of the upper boundary force, lower boundary force, and external excitation are 0.35%, 1.07% and 1.59%, receptively. It means that the identified forces are quite close to the actual ones. Moreover, to demonstrate the performance of the presented data-fusion technique, the identification procedure is conducted by solely using the aforementioned acceleration measurements. In this case, significant drifts can be found in the identified excitation, boundary forces and displacement responses as shown in Fig. 4. When the displacement of the 3th floor as mentioned before is measured and used for the identification, a good agreement can be found between the identified results and the actual ones. This means that the data fusion of displacement responses in observation equation is beneficial for circumventing the so-called low-frequency drifts.

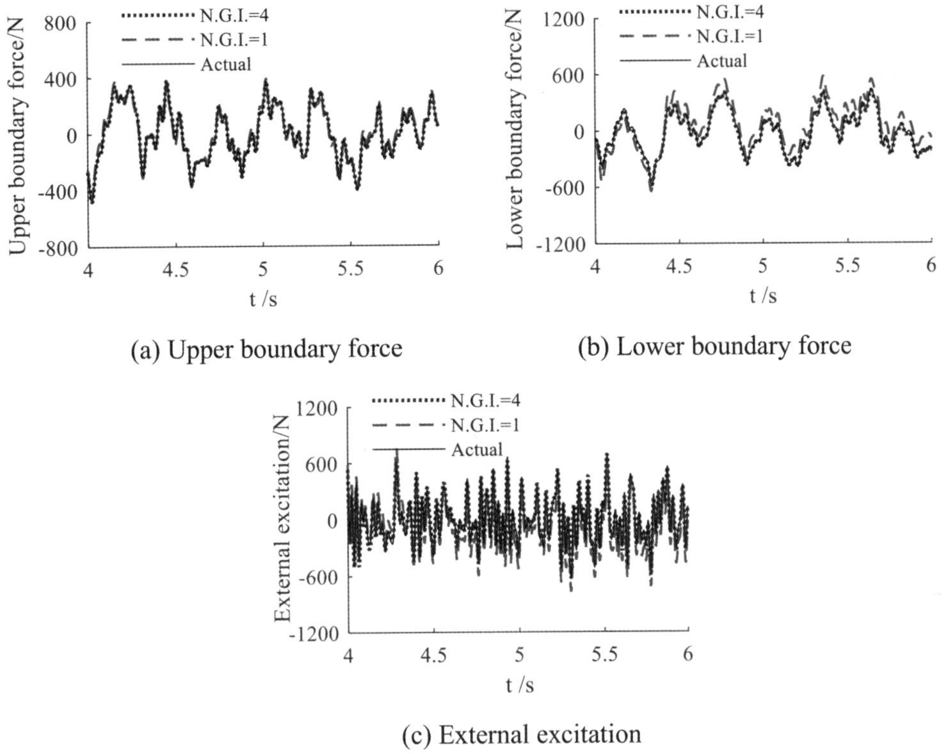

Fig. 3. The comparison of force identification with different N.G.I. in the 1st example.

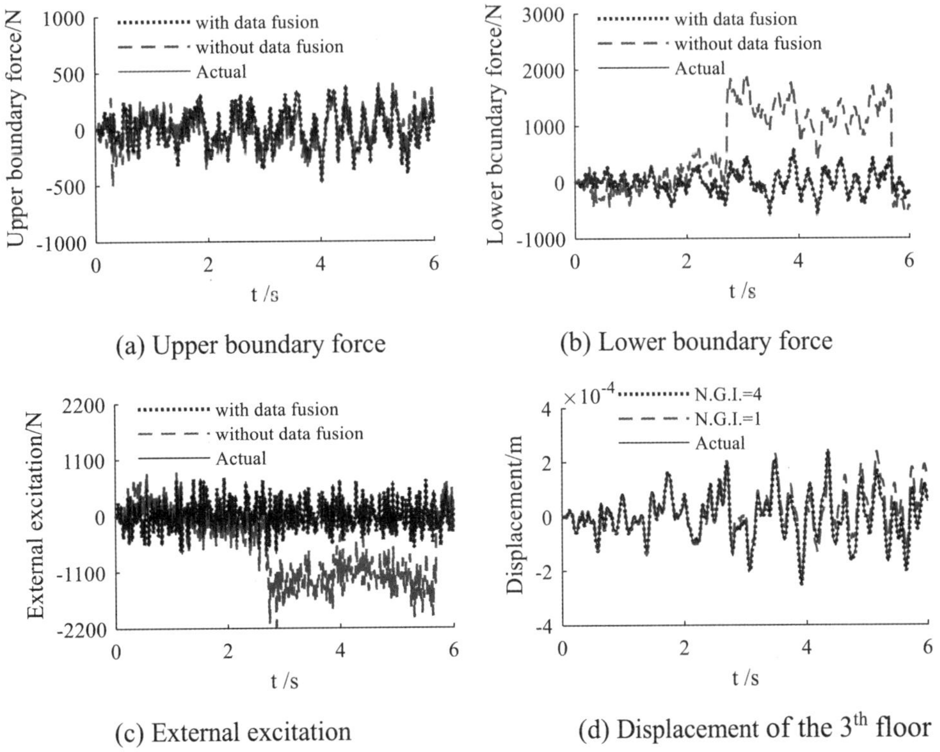

Fig. 4. The identification results with/without data fusion in the 1st example.

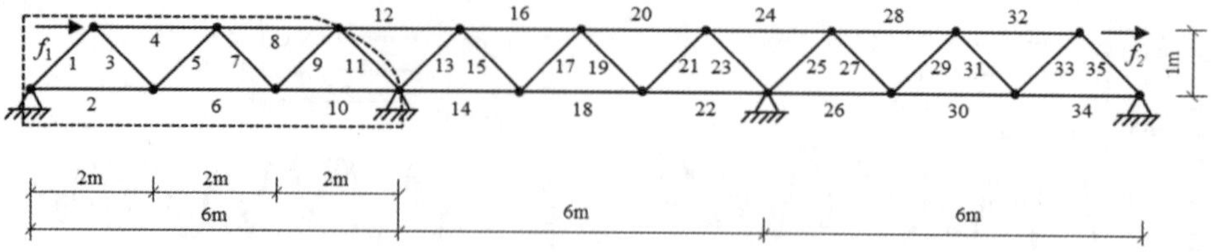

(a) The whole structure

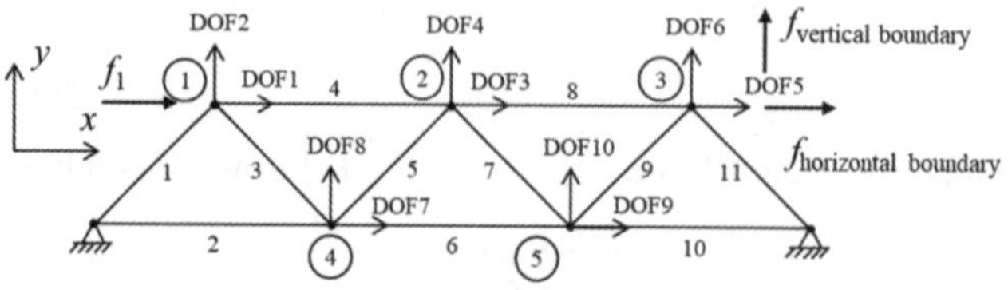

(b) The target substructure

Fig. 5. The 3-span planar truss model.

Table 2
The identified substructural parameters in the 2nd example.

Parameters	Actual (N/m)	N.G.I. = 1		N.G.I. = 3	
		Identified (N/m)	Error (%)	Identified (N/m)	Error (%)
k_1	282,843	282,019	−0.29	283,232	0.14
k_2	180,000	175,989	−2.23	1,790,245	−0.54
k_3	282,843	283,065	0.08	283,406	0.20
k_4	200,000	199,732	−0.13	200,370	0.19
k_5	282,843	277,745	−1.80	279,948	−1.02
k_6	200,000	198,439	−0.78	200,116	0.06
k_7	282,843	281,702	−0.40	282,055	−0.28
k_8	200,000	195,556	−2.22	198,203	−0.90
k_9	282,843	242,100	−14.40	281,686	−0.41
k_{10}	200,000	200,649	0.32	204,094	2.05
k_{11}	282,843	279,726	−1.10	281,001	−0.65
α	0.5449	0.4176	−23.36	0.5392	−1.03
β	0.0017	0.0017	–	0.0017	–

3.2. A 3-span planar truss structure

As shown in Fig. 5, a 3-span planar truss structure with four hinged supports is taken as another linear model to illustrate the performance of the proposed approach. The whole truss structure is assumed to be composed of 35 bars with the cross-sectional area of 2×10^{-4} m^2, Young's module of 2×10^8 Pa, and mass density of 7850 kg/m^3. The length of the inclined bar is $\sqrt{2}$m whereas the horizontal one is 2 m. The consistent mass distribution is employed for the establishment of mass matrix. The Rayleigh damping assumption is employed, and the two coefficients α and β are assumed to be 0.5449 and 1.7×10^{-3}, respectively. In order to simulate structural damage, the cross section area of the 2nd element is reduced to 90% of the intact one resulting in 10% stiffness degradation.

The external excitations are assumed to be applied at two nodes in the horizontal direction as shown in Fig. 5(a), and the corresponding structural responses are computed by state space method with the time interval of 0.001 s. The small-size truss structure as shown in Fig. 5(b) is selected as the target substructure. It should be noted that, since node 3 is considered as its boundary, the interfacial forces exist in the vertical and horizontal direction. The acceleration responses in the vertical direction of node 1, 3, 4, 5 and the horizontal direction of node 1, 2, 3, 4, 5 are assumed to be available. Moreover, to avoid the low-frequency drifts, the displacement responses in the horizontal direction of node 3 are also employed for the identification. The initial values of the substructural parameters are set to be 50% of the actual ones, and the weighted factor in the global iteration is 100.

The identified substructural parameters are listed in Table 2. The maximum error is only 2.05% which means the identification results have a good agreement with the actual ones. The external excitation applied to the concerned substructure and its boundary forces are also identified by the proposed approach as shown in Fig. 6. Significant difference between the identified boundary force and the actual one in the horizontal direction can be found when N.G.I equals to 1. However, much better results can be observed when several global iterations are conducted, i.e. N.G.I = 3. The NRMSE results of the vertical boundary force, horizontal boundary force, and external excitation are 4.51%, 2.41% and 0.83%, receptively. These small values means that the identified forces have a good agreement with the actual ones. Similarly, the comparison of the identification results with and without data fusion of displacement responses is given as shown in Fig. 7. It is clear that a significant drift exists when only the noisy acceleration measurements are used for the identification.

3.3. A 20-story building structure with a nonlinear hysteretic model

The aforementioned numerical examples are both in the linear domain. To further investigate the performance of the proposed approach, a 20-story building structure with a nonlinear hysterretic model on the 6th floor as shown in Fig. 8 is considered herein. The structural parameters are mass of 80 kg and stiffness of 5.5×10^5N/m for each story. The two coefficients of Rayleigh damping model are assumed to be $\alpha = 0.2857$ and $\beta = 2.4 \times 10^{-3}$, respectively. The external excitation is applied to the 8th floor of the building structure, and the structural responses are calculated by Runge-Kutta method with the time interval of 0.001 s.

The Dahl model located on the 6th story is employed to simulate local nonlinearity. The nonlinear restoring force provided by this model can be expressed as,

$$F_{Dahl} = k_D \Delta x + c_D \Delta \dot{x} + f_D z_D + f_0 \tag{30}$$

where k_D, c_D, f_D, and f_0 are the coefficients used for the description of the characteristics of Dahl model; Δx and $\Delta \dot{x}$are the inter-story displacement and velocity, respectively; z_D is a dimensionless coefficient and can be determined as,

$$\dot{z}_D = \sigma_D \cdot \Delta \dot{x}[1 - z_D \cdot \text{sign}(\Delta \dot{x})] \tag{31}$$

where sign(·) denotes signum function and σ_D is the coefficient used to

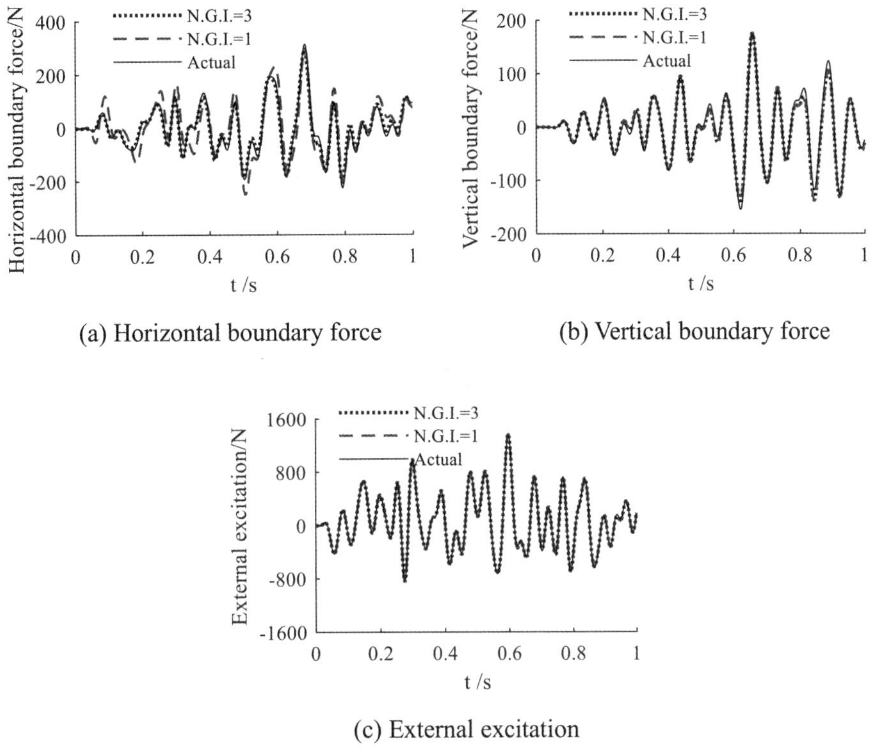

Fig. 6. The comparison of force identification with different N.G.I. in the 2nd example.

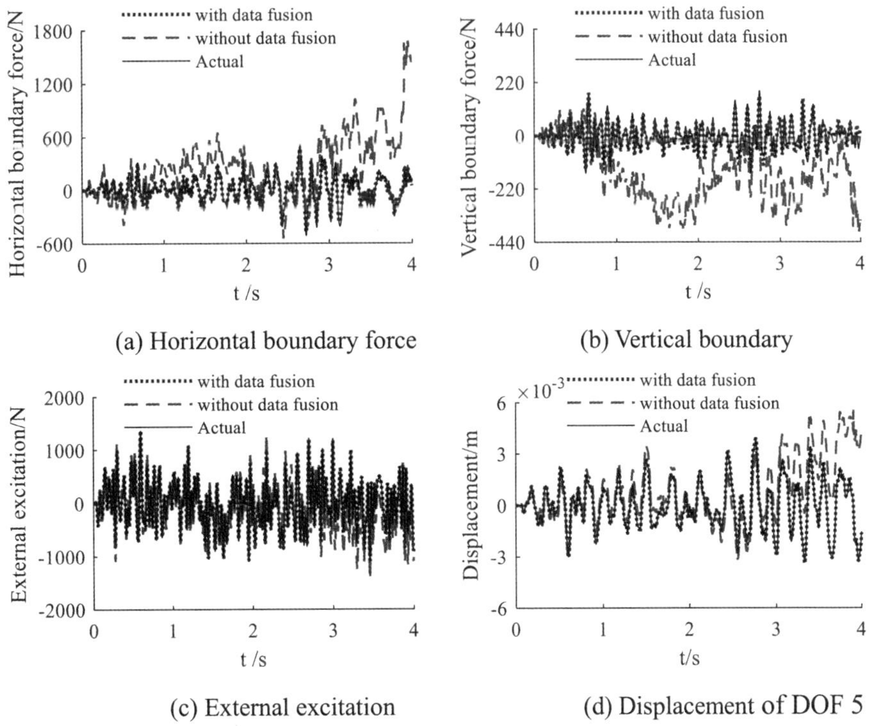

Fig. 7. The identification results with/without data fusion in the 2nd example.

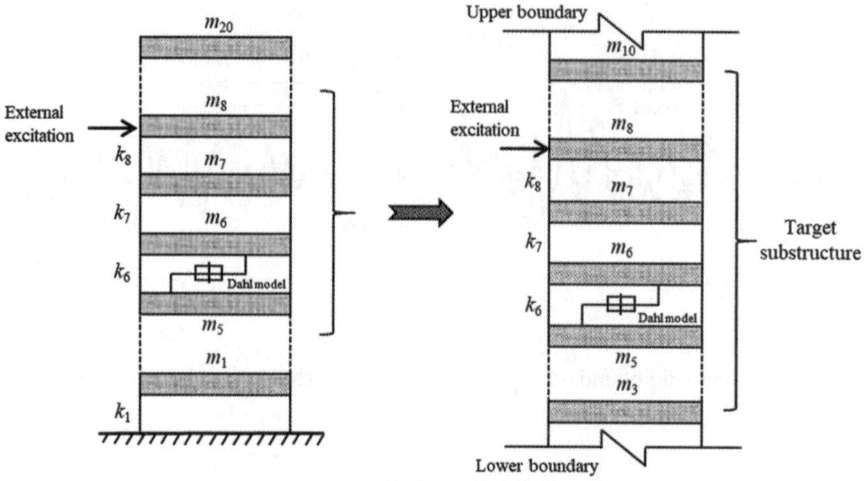

(a) The whole structure (b) The target substructure

Fig. 8. The 20-story building structure with Dahl model.

Table 3
The identified parameters in the 3rd example.

Parameters	Actual (N/m)	N.G.I. = 1		N.G.I. = 3	
		Identified (N/m)	Error (%)	Identified (N/m)	Error (%)
k_4	5.5×10^5	5.458×10^5	−0.76	5.493×10^5	−0.12
k_5	5.5×10^5	5.531×10^5	0.57	5.513×10^5	0.24
k_6	5.5×10^5	5.535×10^5	0.64	5.517×10^5	0.31
k_7	5.5×10^5	5.529×10^5	0.54	5.551×10^5	0.94
k_8	5.5×10^5	5.493×10^5	−0.13	5.529×10^5	0.53
k_9	5.5×10^5	5.511×10^5	0.20	5.540×10^5	0.73
k_{10}	5.5×10^5	5.495×10^5	−0.10	5.503×10^5	0.06
k_D	30	31	3.27	31	2.59
c_D	160	145	−9.42	158	−1.41
f_D	280	280	−0.09	2888	2.88
σ_D	1500	1431	−4.63	14,558	−2.97
α	0.2857	0.3694	29.29	0.2853	−0.12
β	2.4×10^{-3}	2.39×10^{-3}	0.42	2.39×10^{-3}	0.42

control the shape of hysteresis loop. In this example, the coefficients of Dahl model are selected as $k_D = 30$ N/m, $c_D = 160$ N·s/m, $f_D = 280$ N, $f_0 = 0$ N, and $\sigma_D = 1500$ s/m.

As shown in Fig. 8(b), the structure from the 3rd floor to 10th floor is considered as the target substructure. It can be seen that local nonlinearity exists in this substructure. The acceleration responses of the 3rd, 4th, 5th, 6th, 7th, 8th and 10th floor and the displacement response of the 8th floor are assumed to be measured for SSI. The initial guesses of the substructural parameters are given as 50% of the actual ones. The weighted factor is set to be 100.

The quantities to be identified in this example include the parameters of the nonlinear model and target substructure, the unknown external excitation, as well as the upper and lower boundary force. The identified parameters are shown in Table 3. It is obvious that the identified values are close to the real ones. Fig. 9 gives the comparison of force identification results with different global iterations. It can be found that the results in the last global iteration are much better that those in the first iteration. The NRMSE results of the upper and lower boundary forces, as well as external force are 4.39%, 1.82% and 4.78%, receptively. Although it seems larger than the linear case, the identified results are still acceptable. Fig. 10 shows the performance of the proposed approach with and without data fusion of displacement responses. Moreover, based on these identified quantities, the nonlinear restoring force provided by the Dahl model can be also estimated as shown in Fig. 11. Satisfactory results can be found for the estimation of local nonlinearity.

4. Conclusions

In this paper, a time domain EKF-SSI-UIF approach is proposed for the simultaneous identification of the parameters and interface forces of the concerned substructure as well as its unknown excitation. A modified observation equation is obtained on the basis of projection matrix, and the recursive estimation of the extended substructural states is analytically derived. With the usage of the estimated structural states in the current time step, the unknown integrated inputs including the external excitation and interfacial force are identified by LSE. In order to assure the convergence and stability of the identified results, a weighted global iteration procedure is employed. An objective function is then defined and a weighted matrix is introduced. An attractive feature of the proposed approach is the boundary forces and external excitation can be accurately identified at the same time. Moreover, different from many other SSI methods, the proposed approach can be still performed even structural nonlinearity is involved. The reliability of the proposed approach is numerically demonstrated via two linear models and a tall building structure with local nonlinearity. Results show that a good agreement between the identified quantities and actual ones can be found when several global iterations are conducted. The low-frequency drift of the estimated displacement and force can be also effectively avoided by using the data fusion technique presented in the proposed approach.

It should be noted that the parameters to be identified in this study is time-invariant. Thus, the proposed approach is not applicable if the unknown parameters are time-varying. Moreover, although the interfacial forces are identified, the interface measurements are required. Therefore, further research will be conducted to circumvent these two limitations.

CRediT authorship contribution statement

Jia He: Conceptualization, Methodology, Writing - original draft. **Mengchen Qi:** Software, Validation. **Xugang Hua:** Writing - review & editing. **Zhengqing Chen:** Supervision. **Ou Yang:** Funding acquisition. **Zhang Cao:** Investigation.

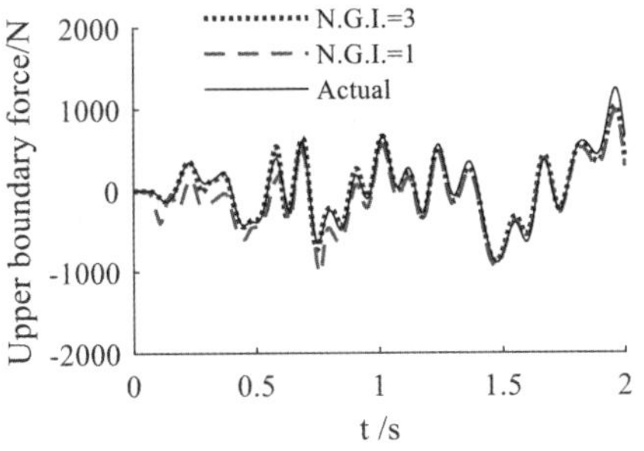

(a) Upper boundary force

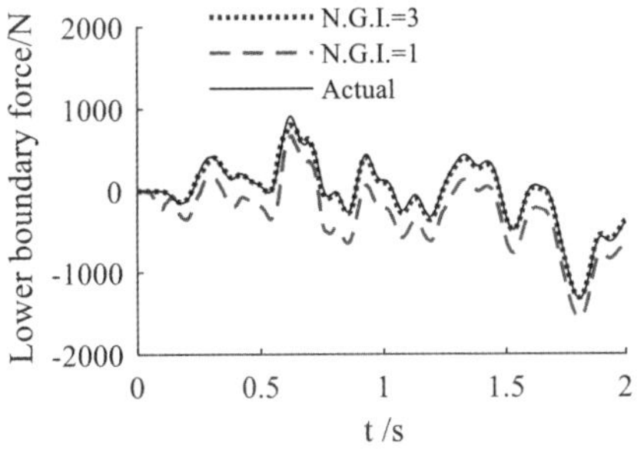

(b) Lower boundary force

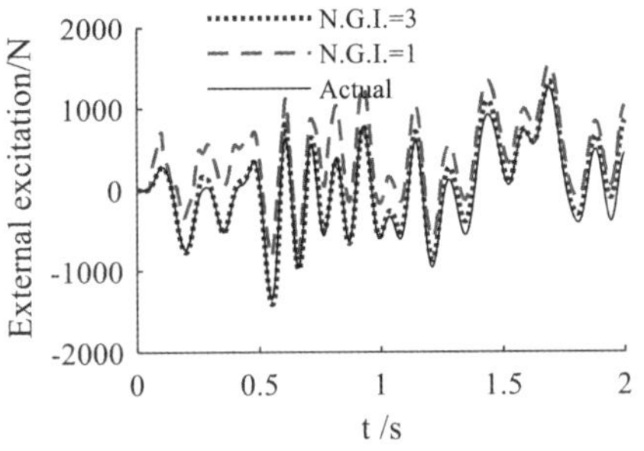

(c) External excitation

Fig. 9. The comparison of force identification with different N.G.I. in the 3rd example.

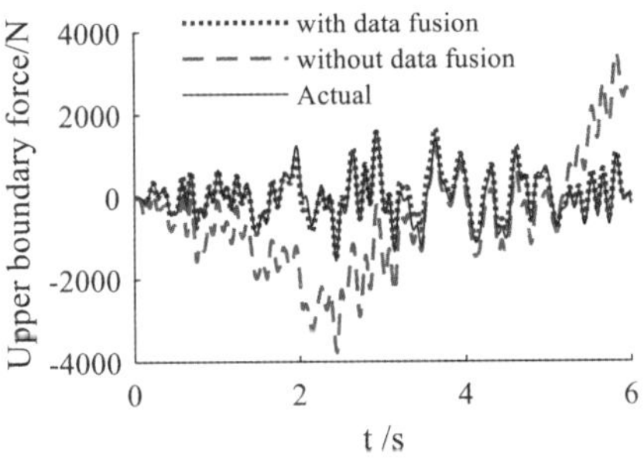

(a) Upper boundary force

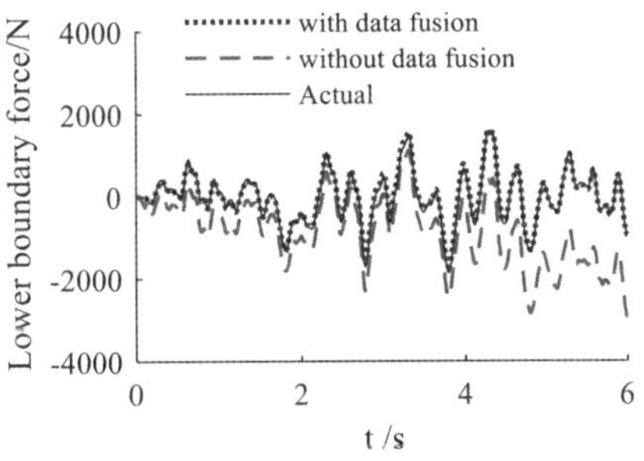

(b) Lower boundary force

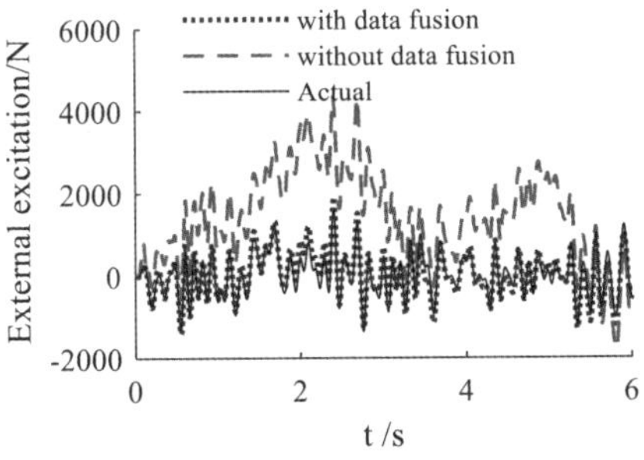

(c) External excitation

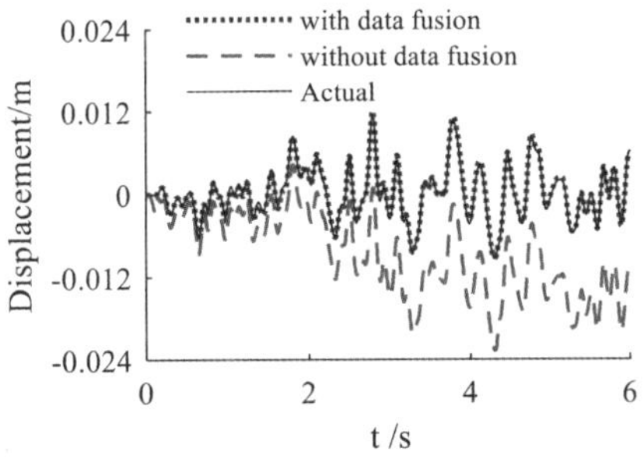

(d) Displacement of the 3th floor

Fig. 10. The identification results with/without data fusion in the 3rd example.

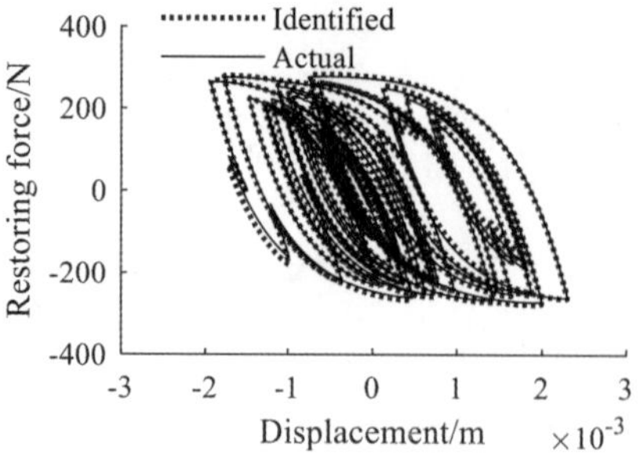

(a) Hysteresis curve

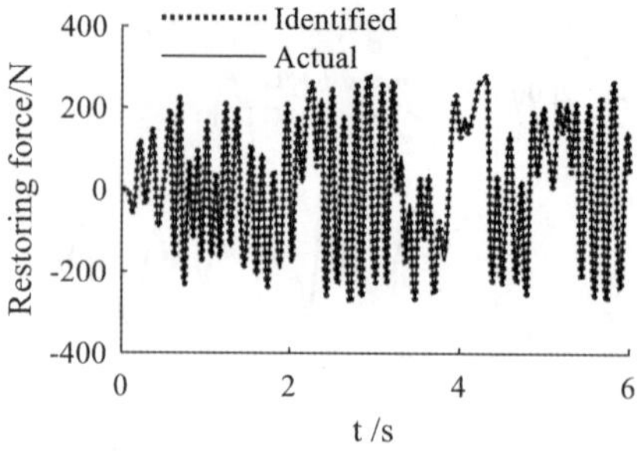

(b) Time-history curve

Fig. 11. The identified nonlinear restoring force.

Declaration of Competing Interest

The authors declare that they have no known competing financial interests or personal relationships that could have appeared to influence the work reported in this paper.

Acknowledgements

The authors gratefully acknowledge the financial support from Key R&D plan of the Ministry of Science and Technology through grant No. 2019YFC1511101. The support from Innovation Platform Open Fund project of Hunan Province (No. 19K018) is also greatly appreciated.

References

[1] T.H. Yi, H.B. Huang, H.N. Li, Development of sensor validation methodologies for structural health monitoring: A comprehensive review, Measurement 19 (2017) 200–214.
[2] Y.L. Xu, J. He, Smart Civil Structures, CRC Press, Boca Raton, 2017.
[3] C. Zhang, J. Li, K. Bi, Preface: Recent advances on structural control, health monitoring and applications in bridge engineering, Int. J. Struct. Stab. Dyn. 18 (8) (2018) 1802001.
[4] S. Weng, H.P. Zhu, Y. Xia, J. Li, W. Tian, A review on dynamic substructuring methods for model updating and damage detection of large-scale structures, Adv. Struct. Eng. 23 (3) (2020) 584–600.
[5] C.G. Koh, B. Hong, C.Y. Liaw, Substructural and progressive structural identification methods, Eng. Struct. 25 (2003) 1551–1563.
[6] K.F. Tee, C.G. Koh, S.T. Quek, Numerical and experimental studies of a substructural identification strategy, Struct. Health Monit. 8 (5) (2009) 397–410.
[7] T.N. Trinh, C.G. Koh, An improved substructural identification strategy for large structural systems, Struct. Control Health Monit. 19 (8) (2012) 686–700.
[8] J. Hou, Ł. Jankowski, J.P. Ou, A substructure isolation method for local structural health monitoring, Struct. Control Health Monit. 18 (6) (2011) 601–618.
[9] J. Hou, Ł. Jankowski, J.P. Ou, Experimental study of the substructure isolation method for local health monitoring, Struct. Control Health Monit. 19 (4) (2012) 491–510.
[10] Z. Xing, A. Mita, A substructure approach to local damage detection of shear structure, Struct. Control Health Monit. 19 (2) (2012) 309–318.
[11] X.M. Wang, C.G. Koh, J. Zhang, Substructural identification of jack-up platform in time and frequency domains, Appl. Ocean Res. 44 (2014) 53–62.
[12] S. Weng, H.P. Zhu, Y. Xia, L. Mao, Damage detection using the eigenparameter decomposition of substructural flexibility matrix, Mech. Syst. Signal Proc. 34 (1–2) (2013) 19–38.
[13] S. Weng, H.P. Zhu, P.H. Li, Y. Xia, Y. Ye, Construction of orthogonal projector for the damage identification by measured substructural flexibility, Measurement 88 (2016) 441–455.
[14] J. Li, H. Hao, Substructure damage identification based on wavelet-domain response reconstruction, Struct. Health Monit. 13 (4) (2014) 389–405.
[15] D.Y. Zhang, Y. Yang, T.Q. Wang, H. Li, Improving substructure identification accuracy of shear structures using virtual control system, Smart Mater. Struct. 27 (2) (2018), 025013.
[16] L. Mei, A. Mita, J. Zhou, An improved substructural damage detection approach of shear structure based on ARMAX model residual, Struct. Control Health Monit. 23 (2) (2016) 218–236.
[17] L. Mei, H. Li, Y. Zhou, W. Wang, F. Xing, Substructural damage detection in shear structures via ARMAX model and optimal subpattern assignment distance, Eng. Struct. 191 (2019) 625–639.
[18] H. Ghaffarzadeh, T.Y. Yang, Y.H. Ajorloo, Structural identification based on substructural technique and using generalized BPFs and GA, Struct. Eng. Mech. 67 (4) (2018) 359–368.
[19] J. Hou, S. Wang, Q. Zhang, Ł. Jankowski, An improved objective function for modal-based damage identification using substructural virtual distortion method, Appl. Sci. 9 (2019) 971, https://doi.org/10.3390/app9050971.
[20] T. Su, Z. Tang, L. Peng, Y. Bai, X. Jin, J. Kong, Model updating for real time dynamic substructures based on UKF algorithm, Earthq. Eng. Eng. Vibr. 19 (2020) 413–421.
[21] C.G. Koh, K. Shankar, Substructural identification method without interface measurement, J. Eng. Mech. 129 (7) (2003) 769–776.
[22] S.S. Law, D. Yong, Substructure methods for structural condition assessment, J. Sound Vibr. 330 (2011) 3606–3619.
[23] Y. Lei, M.Y. He, C. Liu, S.Z. Lin, Identification of tall shear buildings under unknown seismic excitation with limited output measurements, Adv. Struct. Eng. 16 (11) (2013) 1839–1849.
[24] B. Xu, J. He, Substructural parameters and dynamic loading identification with limited observations, Smart Struct. Syst. 15 (1) (2015) 169–189.
[25] K. Liu, S.S. Law, X.Q. Zhu, Substructural Condition Assessment Based on Force Identification and Interface Force Sensitivity, Int. J. Struct. Stab. Dyn. 15 (2) (2015) 1450046.
[26] A. Al-Hussein, A. Haldar, Unscented Kalman filter with unknown input and weighted global iteration for health assessment of large structural systems, Struct. Control Health Monit. 23 (1) (2016) 156–175.
[27] J. Li, H. Hao, Substructural interface force identification with limited vibration measurements, J. Civ. Struct. Health Monit. 6 (2016) 395–410.
[28] K.V. Yuen, K. Huang, Real-time substructural identification by boundary force modeling, Struct. Control Health Monit. 25 (5) (2018), e2151.
[29] H. Su, X. Yang, L. Liu, Y. Lei, Identifying nonlinear characteristics of model free MR dampers in structures with partial response data, Measurement 130 (2018) 362–371.
[30] Y. Lei, H. Qiu Hao, F. Zhang, Identification of structural element mass and stiffness changes using partial acceleration responses of chain like systems under ambient excitations, J. Sound Vibr. 488 (2020), 115678.
[31] Y. Lei, D. Xia, K. Erazo, S. Nagarajaiah, A novel unscented Kalman filter for recursive state input system identification of nonlinear systems, Mech. Syst. Signal Proc. 127 (2019) 120–135.
[32] Y. Huang, J.Q. Yu, J.L. Beck, H.P. Zhu, H. Li, Novel sparseness-inducing dual Kalman filter and its application to tracking time-varying spatially-sparse structural stiffness changes and inputs, Comput. Meth. Appl. Mech. Eng. 372 (2020), 113411.
[33] Y. Li, Y.Z. Luo, H.P. Wan, C.B. Yun, Y.B. Shen, Identification of earthquake ground motion based on limited acceleration measurements of structure using Kalman filtering technique, Struct. Control Health Monit. 27 (1) (2020), e2464.
[34] B. Xu, J. Li, S.J. Dyke, B.C. Deng, J. He, Nonparametric identification for hysteretic behavior modeled with a power series polynomial using EKF-WGI approach under limited acceleration and unknown mass, Int. J. Non Linear Mech. 119 (2020), 103324.
[35] J. He, X. Zhang, B. Xu, KF-based multi-scale response reconstruction under unknown inputs with data fusion of multi-type observations, J. Aerosp. Eng. 32 (4) (2019) 04019038.
[36] J. He, X. Zhang, B. Xu, Identification of structural parameters and unknown inputs based on revised observation equation: approach and validation, Int. J. Struct. Stab. Dyn. 19 (2) (2019) 1950156.
[37] C.G. Koh, L.M. See, T. Balendra, Estimation of structural parameters in time domain: a substructural approach, Earthq. Eng. Struct. Dyn. 20 (8) (1991) 787–801.
[38] Z. Zheng, H. Qiu, Z. Wang, S. Luo, Y. Lei, Data fusion based multi-rate Kalman filtering with unknown input for on-line estimation of dynamic displacements, Measurement 131 (2019) 211–218.
[39] L. Liu, Y. Su, J. Zhu, Y. Lei, Data fusion based EKF-UI for real-time simultaneous identification of structural systems and unknown external inputs, Measurement 88 (2016) 456–467.
[40] M. Hoshiya, E. Saito, Structural identification by extended Kalman filter, J. Eng. Mech. 110 (12) (1984) 1757–1771.

Hindawi
Structural Control and Health Monitoring
Volume 2023, Article ID 8379183, 18 pages
https://doi.org/10.1155/2023/8379183

WILEY | Hindawi

Research Article

Simultaneous Identification of Time-Varying Parameters and External Loads Based on Extended Kalman Filter: Approach and Validation

Xiaoxiong Zhang, Jia He, Xugang Hua, Zhengqing Chen, and Zhouquan Feng

Key Laboratory of Building Safety and Energy Efficiency of the Ministry of Education, Key Laboratory of Wind and Bridge Engineering of Hunan Province, College of Civil Engineering, Hunan University, Changsha, China

Correspondence should be addressed to Jia He; jiahe@hnu.edu.cn

Received 3 October 2022; Revised 2 December 2022; Accepted 20 January 2023; Published 21 February 2023

Academic Editor: E. Taciroglu

Online identification of time-variant parameters without knowledge of external loads is an important but challenging task for structural health monitoring and vibration control. In this study, a two-stage approach, named extended Kalman filter with forgetting factor matrix under unknown inputs (EKF-FFM-UI), is proposed for simultaneously identifying the time-variant parameters and external loads. In stage 1, an extended Kalman filter under unknown inputs (EKF-UI) approach previously proposed by the authors is employed for estimating the structural states and unknown loads. This EKF-UI approach is solely suitable for time-invariant system identification. Therefore, the aim of stage 2 is to improve this approach for the purpose of possessing tracking capability. In this stage, the acceleration responses are first reconstructed by using the differential equation of motion and employed for improving the accuracy of estimated structural states. A forgetting factor matrix is introduced into the priori estimation error covariance matrix to track time-varying parameters. The square errors between the measurements and the corresponding estimates are defined as an index and used for detecting the damage time instant. Then, a covariance resetting technique is employed to assure that such changes in structural parameters can be efficiently captured. A shear-type building structure without/with magneto-rheological (MR) dampers and a fixed beam structure are used as numerical examples for validating the effectiveness of the proposed approach. Experimental tests on a six-story building model are also conducted. Results show the time-varying parameters and unknown inputs can be simultaneously identified with acceptable accuracy.

1. Introduction

The occurrence and growth of structural damages (e.g., stiffness degradation) caused by severe loads or environmental effects are inevitable during the service life of infrastructure. There is no doubt that such phenomena are detrimental to structural safety and durability. Identification of these time-varying parameters by using real-time data is an important but still challenging task for structural health monitoring and vibration control. Much effort has been made in this field, and many literature reviews on parameter identification with different emphases can be found [1–3].

In general, real-time identification methods are performed in a recursive manner and enable stepwise parameter identification to detect system changes online. These methods include but do not limit to wavelet transform [4–6], recursive least-squares [7, 8], Hilbert transform [9, 10], deep learning [11, 12], and Bayesian filter [13, 14]. This paper is not intended to provide a literature review on real-time parameter identification methodologies. Instead, the following introduction will focus on the online estimation of structural systems based on the Kalman filter (KF) technique.

As known, the KF is a powerful tool for recursively estimating structural states, such as displacement and velocity. However, the classic KF is only suitable for linear systems whose parameters are assumed to be known. On this account, with the aid of a linearization procedure and Jacobian matrix, the extended Kalman filter (EKF) is developed for parameter identification by adding the unknown parameters into structural states. Considering the linearized approximation may not be accurate enough for highly nonlinear systems. The unscented Kalman filter (UKF) is further developed with the aid of unscented transformation. Both EKF and UKF can be used to identify time-invariant parameters of linear and nonlinear systems.

As mentioned before, due to various man-made or natural factors, the physical parameters of structures often display time-dependent characteristics. To cover the case of time-varying parameter identification, the EKF or UKF-based methods should be improved to possess adaptive tracking capabilities. In this regard, some attempts have been made in the past several years. For example, the tracking ability of adaptive fading KF was investigated by Loh et al. [15, 16]. By using an adaptive factor matrix, Yang et al. [17] proposed an EKF-based time-varying parameter identification approach and compared it with other tracking techniques using experimental data [18]. The derivation of this approach was complicated, and then Lei et al. [19] proposed a three-stage EKF approach to improve the computational efficiency. Based on the appropriate adjustment of the state covariance matrix, Bisht and Singh [20] proposed an adaptive UKF for tracking sudden stiffness changes. By using an innovation error and the corresponding trigger parameter, Schleiter and Altay [21] proposed a UKF-based approach for the identification of abrupt stiffness changes in controlled structures. With the combination of KF and wavelet decomposition, Chen et al. [22] proposed an approach for identifying time-variant parameters. Based on the concept of pre-estimation of system parameters, Ciolek et al. [23] proposed a decoupled KF approach for the estimation of time-varying systems. To enhance the convergence, Yu et al. [24] proposed a modified EKF for identifying time-variant Hammerstein–Wiener nonlinear systems. More recently, the KF-based online identification of cable force [25], precast segmental columns [26], or sensor fault [27] was also conducted.

Although the methods mentioned above can deal with time-varying systems, they are performed on the premise that the external loads are known for the identification. However, in many practical situations, it may be difficult or impossible to obtain the measured information of external excitation. Thus, based on the framework of KF, some effort has also been made for developing the so-called joint input-state estimation techniques. For example, Eftekhar Azam et al. [28] proposed a dual KF approach for estimating the unknown forces and structural states. This method was then validated and compared with other filters via a four-story frame structure [29]. Liu et al. [30] proposed an improved KF approach under unknown inputs by fusing acceleration and displacement together. The corresponding modal KF approach was then presented by Lei et al. [31]. Based on a modified observation equation, He et al. [32] proposed a KF-based approach for estimating nonlinear systems with unknown inputs. This approach was then extended for multiscale response reconstruction with the aid of data fusion techniques [33]. In general, the structural states to be estimated in the KF algorithm are composed of displacement and velocity. By adding the unknown inputs into the state vector, Lourens et al. [34] developed an augmented Kalman filter (AKF) for joint input-state estimation. This idea is attractive because the unknown inputs can be recursively estimated in conjunction with the states. To date, AKF is still widely investigated by many researchers. For example, under the framework of the AKF, Cumbo et al. [35] proposed an approach for optimal sensor placement. Using modal expansion, Eftekhar Azam et al. [36] proposed an AKF-based approach in the modal domain for input-state estimation of steel railway bridges. Based on sparse constraint theory, Wei et al. [37] proposed an AKF approach for improved force identification.

The KF-based force identification methods introduced before are developed based on prior knowledge of structural parameters. The simultaneous identification of structural parameters and external forces will be more complicated, especially in the case of parameters being time-variant. To the authors' knowledge, the idea of EKF under unknown inputs (EKF-UI) for adaptively capturing time-variant parameters was first presented by Yang et al. [38]. Since then, some efforts have been made. For example, with the combination of wavelet multiresolution analysis and nonlinear least-squares optimization, Lei and Yang [39] proposed a KF-based approach for identifying time-variant parameters under unknown inputs. By employing wavelet multiresolution analysis, Yang et al. [40] proposed a fading-factor UKF method for the identification of time-variant nonlinear systems under unknown inputs. Lei et al. [41] proposed a generalized EKF approach for identifying and controlling time-variant systems under unknown seismic inputs. Based on the idea of substructure and fading factor, Yang et al. [42] proposed an EKF-UI approach for time-varying parameter identification in large-scale structures. With the usage of an online tracking matrix, Zhang et al. [43] proposed an EKF-UI approach for real-time structural identification. By modelling the unknown inputs as modulated colored noise and using Bayesian methodology to recursively update the noise covariance matrices, Huang et al. [44] proposed an EKF-based real-time input-state-parameter estimation approach.

Online identification without prior knowledge of inputs is quite a challenging task. Under this circumstance, the KF-based approaches mentioned above are basically conducted with the combination of other techniques, such as wavelet transformation, least-squares estimation, or nonlinear optimization. However, these procedures may be time-consuming, resulting in the difficulty of achieving real-time identification. Therefore, by solely using the EKF principle, a two-stage approach, named extended Kalman filter with forgetting factor matrix under unknown inputs (EKF-FFM-UI), is proposed in this study for simultaneously identifying time-variant parameters and external loads. In stage 1, an EKF-UI approach previously proposed by the authors [45] is employed for estimating the structural states and unknown loads. This EKF-UI approach is solely suitable for time-invariant system identification. Therefore, the aim of stage 2 is to improve this approach for the purpose of possessing tracking capability. In this stage, the acceleration responses are first reconstructed based on the differential equation of motion and used for updating the estimated structural states. A forgetting factor matrix is introduced into the priori estimation error covariance matrix to track time-varying parameters. The square errors between the measured acceleration responses and those estimated ones are defined as an index and used for detecting the damage time instant. The abrupt increase in square errors can be viewed as an indication that variations in structural parameters are occurring. Then, a covariance resetting technique is employed at this time to assure that such changes in structural parameters can be efficiently captured.

This paper is organized as follows: First, the formulas of the proposed two-stage EKF-FFM-UI approach are introduced in Section 2. Then, a shear-type building structure without/with MR damper and a fixed beam structure are employed as numerical examples in Section 3 for validating the effectiveness of the proposed approach. In Section 4, dynamic tests on a six-story shear-type building model are conducted and used for demonstrating the effectiveness of the proposed approach for a real structure. In Section 5, some conclusions are presented.

2. The Formulas of the Proposed Two-Stage EKF-FFM-UI Approach

2.1. Stage 1: The Previously Proposed EKF-UI Approach. In this stage, the EKF-UI approach previously developed by the authors [45] is employed for estimating the structural parameters and unknown excitation. This approach is briefly introduced herein.

The equation of motion of an n degrees-of-freedom (DOFs) system can be expressed as

$$\mathbf{M}\ddot{\mathbf{x}}(t) + \mathbf{F}[\dot{\mathbf{x}}(t), \mathbf{x}(t), \boldsymbol{\theta}(t)] = \boldsymbol{\varphi}^u \mathbf{f}^u(t), \tag{1}$$

where $\ddot{\mathbf{x}}(t), \dot{\mathbf{x}}(t)$, and $\mathbf{x}(t)$ denote structural acceleration, velocity, and displacement, respectively, $\mathbf{M}$ is the mass matrix and assumed to be known, $\boldsymbol{\theta}(t)$ is the m-dimensional parameter vector to be identified, $\mathbf{F}[\dot{\mathbf{x}}(t), \mathbf{x}(t), \theta(t)]$ represents the linear/nonlinear restoring force, and $\mathbf{f}^u(t)$ and φ^u are the unknown input and the corresponding influence matrix, respectively.

An extended state vector is defined as

$$\mathbf{Z}(t) = \left[\mathbf{x}(t)^{\mathrm{T}}, \dot{\mathbf{x}}(t)^{\mathrm{T}}, \boldsymbol{\theta}(t)^{\mathrm{T}}\right]^{\mathrm{T}}. \tag{2}$$

Then, one can obtain

$$\begin{aligned}\frac{d\mathbf{Z}(t)}{dt} &= \begin{bmatrix} \dot{\mathbf{x}}(t) \\ \mathbf{M}^{-1}\left(-\mathbf{F}[\dot{\mathbf{x}}(t), \mathbf{x}(t), \boldsymbol{\theta}] + \boldsymbol{\varphi}^u \mathbf{f}^u(t)\right) \\ \dot{\boldsymbol{\theta}}(t) \end{bmatrix} \\ &= \mathbf{u}(\mathbf{Z}(t), \mathbf{f}^u(t), t) + \mathbf{w}(t),\end{aligned} \tag{3}$$

where $\mathbf{w}(t)$ is the process noise vector with zero mean and a covariance matrix $\mathbf{Q}(t)$. If the time-independent structural parameters are considered, $\dot{\theta}(t)$ in equation (3) will be zero.

Here, limited acceleration and displacement responses are measured and fused into the observation equation. The observation equation at the k-th time instant can be discretized as

$$\bar{\mathbf{y}}_k = \begin{bmatrix} \mathbf{y}_k^a \\ \mathbf{y}_k^s \end{bmatrix} = \left[\boldsymbol{\varphi}_a \ddot{\mathbf{x}}_k + \mathbf{v}_k^a \; \boldsymbol{\varphi}_s \mathbf{x}_k + \mathbf{v}_k^s\right] = \begin{bmatrix} -\boldsymbol{\varphi}_a \mathbf{M}^{-1}\left(\mathbf{F}[\dot{\mathbf{x}}_k, \mathbf{x}_k, \boldsymbol{\theta}_k] \quad \boldsymbol{\varphi}_f^u \mathbf{f}_k^u\right) + \mathbf{v}_k^a \\ \boldsymbol{\varphi}_s \mathbf{x}_k + \mathbf{v}_k^s \end{bmatrix}, \tag{4}$$

where $\mathbf{y}_k^a$ and $\mathbf{y}_k^s$ are acceleration and displacement measurements, respectively, $\boldsymbol{\varphi}_a$ and $\boldsymbol{\varphi}_s$ are associated with the placement of accelerometers and displacement transducers, respectively, and $\mathbf{v}_k^a$ and $\mathbf{v}_k^s$ are the corresponding measurement noise vectors with zero mean and covariance matrices $\mathbf{R}_k^a$ and $\mathbf{R}_k^s$, respectively.

A modified version of the observation equation can be derived with the aid of projection matrix [45],

$$\overline{\boldsymbol{\Phi}}\bar{\mathbf{y}}_k = \overline{\boldsymbol{\Phi}}\bar{\mathbf{h}}(\mathbf{Z}_k) + \overline{\boldsymbol{\Phi}}\bar{\mathbf{v}}_k, \tag{5}$$

where

$$\overline{\Phi} = \begin{bmatrix} \Phi & \mathbf{0} \\ \mathbf{0} & \mathbf{I} \end{bmatrix}; \Phi = \mathbf{I}_n - \mathbf{D}(\mathbf{D}^{\mathrm{T}}\mathbf{D})^{-1}\mathbf{D}^{\mathrm{T}}; \mathbf{D} = -\varphi_a \mathbf{M}^{-1}\varphi^u;$$

$$\bar{\mathbf{v}}_k = \begin{bmatrix} \mathbf{v}_k^a \\ \mathbf{v}_k^s \end{bmatrix};$$

$$\bar{\mathbf{h}}(\mathbf{Z}_k) = \begin{bmatrix} \mathbf{h}(\mathbf{Z}_k) \\ \boldsymbol{\varphi}_s \mathbf{x}_k \end{bmatrix}; \mathbf{h}(\mathbf{Z}_k) = -\boldsymbol{\varphi}_a \mathbf{M}^{-1}\mathbf{F}[\dot{\mathbf{x}}_k, \mathbf{x}_k, \boldsymbol{\theta}_k]. \tag{6}$$

Based on the aforementioned state equation and modified observation equation, the EKF-UI approach can be performed as follows, including priori estimation, posteriori estimation, and load identification. More details can be found in Ref. [45].

Priori estimation:

$$\widehat{\mathbf{Z}}_{k+1|k} = \widehat{\mathbf{Z}}_{k|k} + \int_{k\Delta t}^{(k+1)\Delta t} \mathbf{u}\left(\widehat{\mathbf{Z}}_{k|k}, \widehat{\mathbf{f}}_k^u, k\Delta t\right)\mathrm{d}t, \tag{7}$$

$$\mathbf{P}_{k+1|k} = \mathbf{A}_1 \mathbf{P}_{k|k} \mathbf{A}_1^{\mathrm{T}} + \mathbf{A}_2 \mathbf{R}_k^a \mathbf{A}_2^{\mathrm{T}} + \Delta t^2 \mathbf{Q}_k, \tag{8}$$

where Δt is the time interval; $\widehat{\mathbf{Z}}_{k|k}$ and $\widehat{\mathbf{f}}_k^u$ are the estimates of $\mathbf{Z}_k$ and $\mathbf{f}_k^u$, respectively, and $\widehat{\mathbf{Z}}_{k+1|k}$ and $\mathbf{P}_{k+1}|k$ are the priori state estimate and the corresponding error covariance matrix, respectively;

$$\mathbf{A}_1 = \left(\mathbf{I} + \Delta t\mathbf{U}_{k|k} - \Delta t\begin{bmatrix} \mathbf{0} \\ (\mathbf{I}-\mathbf{\Phi})\mathbf{H}_{k|k} \\ \mathbf{0}\end{bmatrix}\right); \mathbf{A}_2 = \begin{bmatrix} \mathbf{0} \\ -\Delta t(\mathbf{I}-\mathbf{\Phi}) \\ \mathbf{0}\end{bmatrix}; \mathbf{H}_{k+1|k} = \frac{\partial \mathbf{h}(\mathbf{Z}_{k+1})}{\partial \mathbf{Z}_{k+1}}\Big|_{\mathbf{Z}_{k+1}=\widehat{\mathbf{Z}}_{k+1|k}}; \tag{9}$$

$$\mathbf{U}_{k|k} = \frac{\partial \mathbf{u}(\mathbf{Z}_k, \mathbf{f}_k^u, t_k)}{\partial \mathbf{Z}_k}\Big|_{\mathbf{Z}_k=\widehat{\mathbf{Z}}_{k|k}, \mathbf{f}_k^u=\widehat{\mathbf{f}}_k^u}.$$

Posteriori estimation:

$$\widehat{\mathbf{Z}}_{k+1|k+1} = \widehat{\mathbf{Z}}_{k+1|k} + \mathbf{G}_{k+1}\left[\overline{\mathbf{\Phi}}\overline{\mathbf{y}}_{k+1} - \overline{\mathbf{\Phi}}\overline{\mathbf{h}}\left(\widehat{\mathbf{Z}}_{k+1|k}\right)\right], \tag{10}$$

$$\mathbf{P}_{k+1|k+1} = \left(\mathbf{I} - \mathbf{G}_{k+1}\overline{\mathbf{\Phi}}\overline{\mathbf{H}}_{k+1|k}\right)\mathbf{P}_{k+1|k}\left(\mathbf{I} - \mathbf{G}_{k+1}\overline{\mathbf{\Phi}}\overline{\mathbf{H}}_{k+1|k}\right)^{\mathrm{T}} + \mathbf{G}_{k+1}\overline{\mathbf{\Phi}}\overline{\mathbf{R}}_{k+1}\overline{\mathbf{\Phi}}^{\mathrm{T}}\mathbf{G}_{k+1}^{\mathrm{T}}, \tag{11}$$

$$\mathbf{G}_{k+1} = \mathbf{P}_{k+1|k}\overline{\mathbf{H}}_{k+1|k}^{\mathrm{T}}\overline{\mathbf{\Phi}}^{\mathrm{T}}\left[\overline{\mathbf{\Phi}}\left(\overline{\mathbf{H}}_{k+1|k}\mathbf{P}_{k+1|k}\overline{\mathbf{H}}_{k+1|k}^{\mathrm{T}} + \overline{\mathbf{R}}_{k+1}\right)\overline{\mathbf{\Phi}}^{\mathrm{T}}\right]^{-1}, \tag{12}$$

where $\widehat{\mathbf{Z}}_{k+1|k+1}$ and $\mathbf{P}_{k+1}|k_{+1}$ are the posteriori state estimate and the corresponding error covariance matrix, respectively, $\mathbf{G}_{k+1}$ is the gain matrix, $\overline{\mathbf{R}}_{k+1}$ is the covariance matrix of measurements noise vector $\overline{\mathbf{v}}_{k+1}$, and $\overline{\mathbf{H}}_{k+1|k} = \partial\overline{\mathbf{h}}(\mathbf{Z}_{k+1})/\partial\mathbf{Z}_{k+1}|_{\mathbf{Z}_{k+1}=\widehat{\mathbf{Z}}_{k+1|k}}$.

Load identification:

$$\widehat{\mathbf{f}}_{k+1}^u = \left(\mathbf{D}^{\mathrm{T}}\mathbf{D}\right)^{-1}\mathbf{D}^{\mathrm{T}}\left[\mathbf{h}\left(\widehat{\mathbf{Z}}_{k+1|k+1}\right) - \mathbf{y}_{k+1}^a\right]. \tag{13}$$

2.2. Stage 2: Tracking Capability Improvement. As mentioned before, the EKF-UI approach introduced in the previous subsection is only suitable for time-invariant system and unable to identify time-variant parameters. In order to capture time-varying parameters, the following procedures are developed for the improvement of tracking capability.

Based on the identified results shown in equations (10) and (13), the structural acceleration can be calculated according to equation (1).

$$\widehat{\ddot{\mathbf{x}}}_{k+1} = -\mathbf{M}^{-1}\left(\mathbf{F}\left[\widehat{\dot{\mathbf{x}}}_{k+1}, \widehat{\mathbf{x}}_{k+1}, \widehat{\mathbf{\theta}}_{k+1}\right] - \mathbf{\varphi}^u\widehat{\mathbf{f}}_{k+1}^u\right), \tag{14}$$

where $\widehat{\dot{\mathbf{x}}}_{k+1}$ and $\widehat{\mathbf{x}}_{k+1}$ are the estimates of velocity and displacement obtained from equation (10), respectively. This calculated acceleration shown in equation (14) can be divided into two parts. The first part named as $\widehat{\ddot{\mathbf{x}}}_{k+1}^a$ is associated with the DOFs where the accelerometers are installed, whereas the other part named as $\widehat{\ddot{\mathbf{x}}}_{k+1}^u$ corresponds to the unmeasured DOFs.

By assuming $\widehat{\ddot{\mathbf{x}}}_{k+1}^u$ as "fictitious measurements," the observation equation shown in equation (4) can be updated as

$$\widetilde{\mathbf{y}}_{k+1} = \begin{bmatrix}\widetilde{\mathbf{y}}_{k+1}^a \\ \mathbf{y}_{k+1}^s\end{bmatrix} + \begin{bmatrix}\widetilde{\mathbf{v}}_{k+1}^a \\ \mathbf{v}_{k+1}^s\end{bmatrix}, \tag{15}$$

where $\widetilde{\mathbf{y}}_{k+1}^a = \begin{bmatrix}\widehat{\ddot{\mathbf{x}}}_{k+1}^u \\ \mathbf{y}_{k+1}^a\end{bmatrix}$ and $\widetilde{\mathbf{v}}_{k+1}^a$ is the noise vector of $\widetilde{\mathbf{y}}_{k+1}^a$ with zero mean and covariance matrices $\widetilde{\mathbf{R}}_{k+1}^a$.

Similar to equation (5), a modified observation equation can be found as

$$\widetilde{\mathbf{\Phi}}\widetilde{\mathbf{y}}_{k+1} = \widetilde{\mathbf{\Phi}}\widetilde{\mathbf{h}}(\mathbf{Z}_{k+1}) + \widetilde{\mathbf{\Phi}}\widetilde{\mathbf{v}}_{k+1}, \tag{16}$$

where

$$\widetilde{\Phi} = \begin{bmatrix}\Phi' & \mathbf{0} \\ \mathbf{0} & \mathbf{I}\end{bmatrix}; \Phi' = \mathbf{I}_n - \widetilde{\mathbf{D}}(\widetilde{\mathbf{D}}^{\mathrm{T}}\widetilde{\mathbf{D}})^{-1}\widetilde{\mathbf{D}}^{\mathrm{T}}; \widetilde{\mathbf{D}} = -\mathbf{M}^{-1}\varphi^u;$$

$$\widetilde{\mathbf{h}}(\mathbf{Z}_{k+1}) = \begin{bmatrix}\mathbf{h}'(\mathbf{Z}_{k+1}) \\ \varphi_s\mathbf{x}_{k+1}\end{bmatrix};$$

$$\mathbf{h}'(\mathbf{Z}_{k+1}) = -\mathbf{M}^{-1}\mathbf{F}[\dot{\mathbf{x}}_{k+1}, \mathbf{x}_{k+1}, \mathbf{\theta}_{k+1}]\widetilde{\mathbf{v}}_{k+1} = \begin{bmatrix}\widetilde{\mathbf{v}}_{k+1}^a \\ \mathbf{v}_{k+1}^s\end{bmatrix}. \tag{17}$$

As can be seen from equations (15) and (16), all the acceleration responses including the measured and reconstructed ones are considered in the observation equation. The usage of these reconstructed responses is helpful to assure convergence speed and stability.

For the purpose of tracking time-variant parameters, a forgetting factor matrix $\mathbf{\Lambda}_{k+1}$ is introduced into the priori estimation error covariance matrix. The aim of using the forgetting factor is to place more importance on recent errors than older ones. As stated by Yang et al. [17], if the forgetting factor is set to be large, it has less tracking capability but is robust to noise. On the contrary, if a small value is adopted, it has better tracking capability, but is vulnerable to noise pollution. $\mathbf{\Lambda}_{k+1}$ is a $(2n+m)\times(2n+m)$ diagonal matrix. The first $2n$ diagonal elements associated with displacement and velocity are assumed to be one. The last m diagonal elements corresponding to the unknown parameters are set to be close to one.

Then, equation (8) can be updated as

$$\widetilde{\mathbf{P}}_{k+1|k} = \mathbf{\Lambda}_{k+1}\left[\widetilde{\mathbf{A}}_1\widetilde{\mathbf{P}}_{k|k}\widetilde{\mathbf{A}}_1^{\mathrm{T}}\right]\mathbf{\Lambda}_{k+1}^{\mathrm{T}} + \widetilde{\mathbf{A}}_2\widetilde{\mathbf{R}}_k^a\widetilde{\mathbf{A}}_2^{\mathrm{T}} + \Delta t^2\mathbf{Q}_k, \quad (18)$$

where $\widetilde{\mathbf{P}}_{k+1|k}$ is the updated priori estimation error covariance matrix.

$$\widetilde{\mathbf{A}}_1 = \left(\mathbf{I} + \Delta t\mathbf{U}_{k|k} - \Delta t\begin{bmatrix} \mathbf{0} \\ \left(\mathbf{I} - \mathbf{\Phi}'\right)\mathbf{H}'_{k|k} \\ \mathbf{0} \end{bmatrix}\right); \widetilde{\mathbf{A}}_2 = \begin{bmatrix} \mathbf{0} \\ -\Delta t\left(\mathbf{I} - \mathbf{\Phi}'\right) \\ \mathbf{0} \end{bmatrix}; \mathbf{H}'_{k+1|k} = \left.\frac{\partial \mathbf{h}'(\mathbf{Z}_{k+1})}{\partial \mathbf{Z}_{k+1}}\right|_{\mathbf{Z}_{k+1}=\widetilde{\mathbf{Z}}_{k+1|k}}. \quad (19)$$

According to equation (8), the updated prior state estimate $\widetilde{\mathbf{Z}}_{k+1|k}$ can be obtained as

$$\widetilde{\mathbf{Z}}_{k+1|k} = \widetilde{\mathbf{Z}}_{k|k} + \int_{k\Delta t}^{(k+1)\Delta t} \mathbf{u}\left(\widetilde{\mathbf{Z}}_{k|k}, \widetilde{\mathbf{f}}_k^u, k\Delta t\right)\mathrm{d}t, \quad (20)$$

where $\widetilde{\mathbf{f}}_k^u$ means the updated value of the identified load at the k-th time step.

Based on equation (16), the updated posteriori state estimate $\widetilde{\mathbf{Z}}_{k+1|k+1}$ can be found as

$$\widetilde{\mathbf{Z}}_{k+1|k+1} = \widetilde{\mathbf{Z}}_{k+1|k} + \widetilde{\mathbf{G}}_{k+1}\left[\widetilde{\mathbf{\Phi}}\widetilde{\mathbf{y}}_{k+1} - \widetilde{\mathbf{\Phi}}\widetilde{\mathbf{h}}\left(\widetilde{\mathbf{Z}}_{k+1|k}\right)\right], \quad (21)$$

where $\widetilde{\mathbf{G}}_{k+1}$ is the updated gain matrix.

Similarly, the updated posteriori estimation error covariance $\widetilde{\mathbf{P}}_{k+1|k+1}$ can be obtained basing on equation (11),

$$\widetilde{\mathbf{P}}_{k+1|k+1} = \left(\mathbf{I} - \widetilde{\mathbf{G}}_{k+1}\widetilde{\mathbf{\Phi}}\widetilde{\mathbf{H}}_{k+1|k}\right)\mathbf{\Lambda}_{k+1}\widetilde{\mathbf{P}}_{k+1|k}\mathbf{\Lambda}_{k+1}^{\mathrm{T}}\left(\mathbf{I} - \widetilde{\mathbf{G}}_{k+1}\widetilde{\mathbf{\Phi}}\widetilde{\mathbf{H}}_{k+1|k}\right)^{\mathrm{T}} + \widetilde{\mathbf{G}}_{k+1}\widetilde{\mathbf{\Phi}}\widetilde{\mathbf{R}}_{k+1}\widetilde{\mathbf{\Phi}}^{\mathrm{T}}\widetilde{\mathbf{G}}_{k+1}^{\mathrm{T}}, \quad (22)$$

where $\widetilde{\mathbf{R}}_{k+1}$ is the covariance matrix of noise vector $\widetilde{\mathbf{v}}_{k+1}$; $\widetilde{\mathbf{H}}_{k+1|k} = \partial\widetilde{\mathbf{h}}(\mathbf{Z}_{k+1})/\partial\mathbf{Z}_{k+1}|_{\mathbf{Z}_{k+1}=\widetilde{\mathbf{Z}}_{k+1|k}}$.

The updated gain matrix can be found as

$$\widetilde{\mathbf{G}}_{k+1} = \mathbf{\Lambda}_{k+1}\widetilde{\mathbf{P}}_{k+1|k}\mathbf{\Lambda}_{k+1}^{\mathrm{T}}\widetilde{\mathbf{H}}_{k+1|k}^{\mathrm{T}}\widetilde{\mathbf{\Phi}}^{\mathrm{T}}\left[\widetilde{\mathbf{\Phi}}\left(\widetilde{\mathbf{H}}_{k+1|k}\mathbf{\Lambda}_{k+1}\widetilde{\mathbf{P}}_{k+1|k}\mathbf{\Lambda}_{k+1}^{\mathrm{T}}\widetilde{\mathbf{H}}_{k+1|k}^{\mathrm{T}} + \widetilde{\mathbf{R}}_{k+1}\right)\widetilde{\mathbf{\Phi}}^{\mathrm{T}}\right]^{-1}. \quad (23)$$

Finally, the identified force can be updated as

$$\widetilde{\mathbf{f}}_{k+1}^u = \left(\widetilde{\mathbf{D}}^{\mathrm{T}}\widetilde{\mathbf{D}}\right)^{-1}\widetilde{\mathbf{D}}^{\mathrm{T}}\left[\mathbf{h}'\left(\widetilde{\mathbf{Z}}_{k+1|k+1}\right) - \widetilde{\mathbf{y}}_{k+1}^a\right]. \quad (24)$$

More details on the derivation of equations (16)–(24) can be found in the Appendix. Although the procedures introduced above make the EKF-UI approach possess tracking capability, the limitation of using a constant forgetting factor matrix may be especially evident for identifying abrupt changes in structural parameters. Therefore, in order to ensure the variations of structural parameters that can be efficiently captured, a covariance resetting technique is further employed.

As stated by other researchers [19, 46, 47], abrupt changes in parameters introduce immediate discontinuities in structural acceleration responses. To clearly display these discontinuities in acceleration signals, the square errors between the measured accelerations and those reconstructed ones are calculated and considered an index.

$$\Delta_{i,k+1} = \left(y_{i,k+1}^a - \widehat{\ddot{x}}_{i,k+1}^a\right)^2, \quad (25)$$

where $\Delta_{i,k+1}$ is the square error of the acceleration at the i-th DOF; $y_{i,k+1}^a$ is the measured acceleration at the i-th DOF; $\widehat{\ddot{x}}_{i,k+1}^a$ is the reconstructed acceleration as mentioned in equation (14).

A jump increase of $\Delta_{i,k+1}$ indicates the onset of parameter variations such as stiffness degradation. As this damage time instant is detected, a covariance resetting technique is used immediately. The posteriori estimate error covariance matrix is reset as follows:

$$\widetilde{\mathbf{P}} = \mathrm{diag}\left(\left[\mathbf{0}_{1\times 2n}, \mathbf{P}_{1|1}(2n+1, 2n+1), \ldots, \mathbf{P}_{1|1}(2n+m, 2n+m)\right]\right). \quad (26)$$

where $\overline{\mathbf{P}}$ is called as reset error covariance matrix; and $\mathbf{P}_{1|1}$ is the initial value of error covariance matrix which is given at the beginning. It can be seen that $\overline{\mathbf{P}}$ is a $(2n+m)\times(2n+m)$ diagonal matrix. The first $2n$ diagonal elements are set to be zero to reduce the oscillation of the estimated responses, and the last m diagonal elements are equal to the corresponding initial value of $\mathbf{P}_{1|1}$. As the usage of $\overline{\mathbf{P}}$ at the damage time instant, the error covariance of structural parameters is reset to a larger value which means more powerful estimation is conducted at this time. This makes the time-variant parameters can be captured more efficiently.

The flowchart of the proposed approach is plotted in Figure 1 for ease of understanding.

3. Numerical Investigation

To demonstrate the effectiveness of the proposed approach, a six-story shear-type building structure without/with an MR damper and a fixed beam structure are considered as numerical examples in this section. For practical considerations, the measurements are simulated by the corresponding theoretical ones superimposed with noise of 5% noise-to-signal ratio in root mean square.

3.1. A Six-Story Building Structure. A six-story shear-type building structure is considered as the first numerical example. The mass and stiffness of each story are 100 kg and 120 kN/m, respectively. The Rayleigh damping model, i.e. $\mathbf{C}=\alpha\cdot\mathbf{M}+\beta\cdot\mathbf{K}$, is used. The proportional coefficients are $\alpha=0.3739$ and $\beta=1.8227\times10^{-3}$. To mimic structural damage, the stiffness parameters of the 1st story is reduced abruptly from 120 kN/m to 72 kN/m at $t=4$s. The concentrated random excitation is assumed to be applied on the top floor, and thus the force influence matrix can be expressed as $\varphi^u=(0\ 0\ 0\ 0\ 0\ 1)^{\mathrm{T}}$. By using the state space method, the structural responses can be calculated. The time interval in this example is 0.001 s. Here, the acceleration responses of the 1st, 3rd, 5th, and 6th floors as well as the displacement response of the 6th floor are assumed to be measured and used for the identification.

The quantities to be identified include six stiffness parameters, two damping coefficients, and the external excitation. The structural parameters are initially assumed to be 50% of the actual ones. The posteriori estimate error covariance matrix is set to be $\mathbf{P}_{1|1}=\mathrm{di\,ag}([\mathbf{I}_{1\times12},\mathbf{10}^{14}_{1\times6},\mathbf{10}^{10}_{1\times2}])$ at the beginning. The first 12 and the last 8 diagonal elements of forgetting factor matrix (FFM) are set to be 1 and 1.001, respectively. The reset error covariance matrix used for the enhancement of tracking capability is set as $\overline{\mathbf{P}}=\mathrm{diag}([\mathbf{0}_{1\times12},\mathbf{10}^{14}_{1\times6},\mathbf{10}^{10}_{1\times2}])$. The proper assignment of process and measurement noise covariance matrices (i.e., $\mathbf{Q}$ and $\mathbf{R}$ in this study) plays an important role in ensuring the reliability and stability of recursive KF-based methods. As stated by Eftekhar Azam et al. [28], the $\mathbf{Q}$ matrix reflects the confidence put in the utilized structural model. A large value is usually selected if the model uncertainty is large. The $\mathbf{R}$ matrix reveals the confidence put in the observations. The larger value means the more measurement noise is considered. Research on proper adjustment of $\mathbf{Q}$ and $\mathbf{R}$ matrices has been conducted by some scholars using different techniques, such as Bayesian probabilistic algorithm [48, 49] and mathematical regularization techniques [28, 50]. However, the optimal determination of $\mathbf{Q}$ and $\mathbf{R}$ matrices is not the main purpose of this study. The constant values of these two matrices in this paper are selected based on other available literature [38, 51, 52]. Here, the $\mathbf{Q}$ and $\mathbf{R}$ matrices are set to be $10^{-4}\times\mathbf{I}$ and $\mathbf{I}$, respectively, where $\mathbf{I}$ denotes identity matrix.

The structural parameters can be identified by using the proposed approach as given in Table 1. Clearly, the identified results are quite close to the actual ones. To investigate the tracking capability, the square errors defined in equation (21) are calculated and shown in Figure 2. It is obvious that a significant increase of Δ_1 can be observed at $t=4$ s. This means the structural damage is happening at this time. Only the time histories of the identified stiffness of the 1st and 6th stories are plotted in Figure 3 due to the limitation of the paper length. Similar results can be found for the remaining stories. Moreover, for ease of understanding, two additional cases are considered in this example for demonstrating the necessity of using FFM and $\overline{\mathbf{P}}$ matrix. In case 1, only FFM is used for the online identification. In this case, the error covariance matrix will not be reset by equation (22) even though the damage instant is detected. Instead, only $\overline{\mathbf{P}}$ matrix rather than FFM is used in case 2. It is clear from Figure 3(a) that neither of these two cases can track the abrupt change of k_1. The proposed approach, however, is capable of accurately and stably capturing this sudden decrease.

Besides identifying time-variant parameters, the unknown excitation can be identified as well. As shown in Figure 4(a), the time history of the identified load is plotted by a solid line, whereas the actual one is given by a dashed line. The time segment from 4 s to 6 s is further given in Figure 4(b) for clarity of comparison. Obviously, the identified input is in good agreement with the actual one.

3.2. Six-Story Building Structure with MR Damper. The aim of this section is to investigate the performance of the proposed approach for the identification of nonlinear structure. The six-story building structure introduced in Section 3.1 is employed herein. To mimic structural nonlinearity, an MR damper described by the Bingham model is assumed to be installed on the 1st story. The restoring force provided by the MR damper is expressed as follows:

$$F_{\mathrm{MR}}=f_c\mathrm{sgn}(\Delta\dot{x}_i)+c_b\Delta\dot{x}_i+f_b, \tag{27}$$

where $\Delta\dot{x}_i$ is the interstory velocity and defined as $\Delta\dot{x}_i=\dot{x}_i-\dot{x}_{i-1}$ and f_c, c_b, and f_b are the model coefficients and set to be $f_c=1$ N, $c_b=1000$ N·s/m, and $f_b=0$.

In this section, multiple damages are considered. The stiffness parameters of the 1st and 2nd stories are reduced by 50% at $t=5$ s. A random force is applied to the top floor, and the corresponding structural responses are computed by the Runge–Kutta method. The acceleration responses of the 1st, 3rd, 5th, and 6th floors and displacement response of the 6th

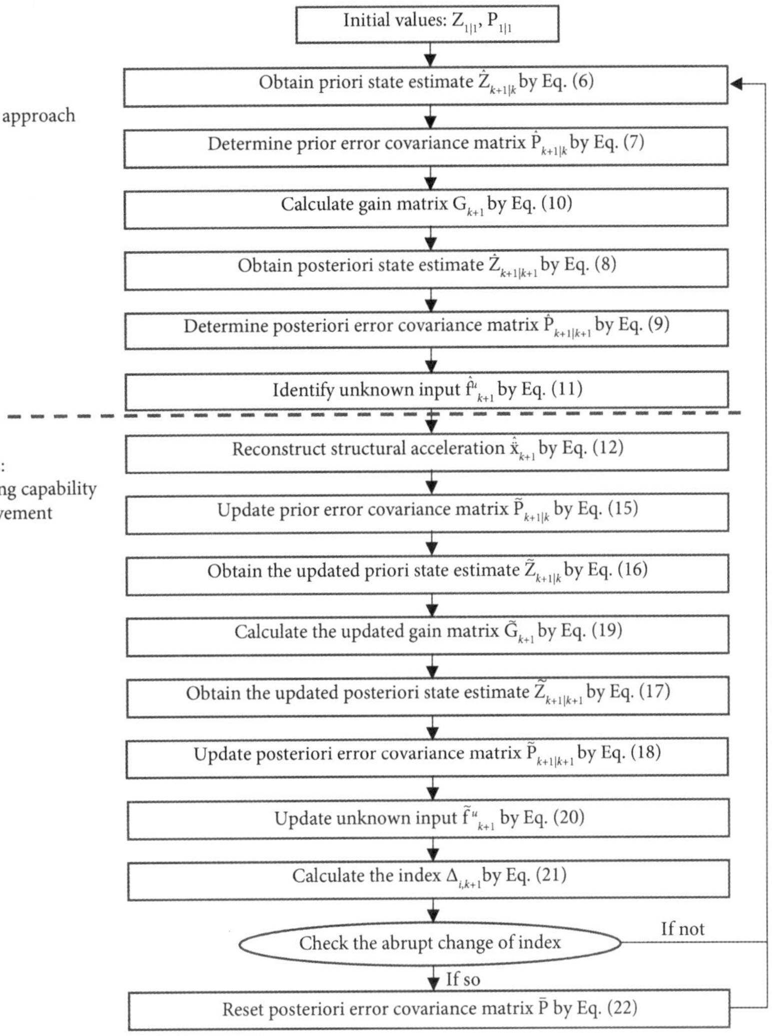

FIGURE 1: Flowchart of the proposed two-stage EKF-FFM-UI approach.

TABLE 1: The identified results of the building structure.

Parameters	Actual	Identified	Error (%)
k_1 (kN/m)	120⟶72	119.84⟶71.99	−0.13⟶−0.01
k_2 (kN/m)	120	120.44	0.36
k_3 (kN/m)	120	120.13	0.11
k_4 (kN/m)	120	120.07	0.06
k_5 (kN/m)	120	119.56	−0.37
k_6 (kN/m)	120	120.29	0.24
α	0.3739	0.3618	−3.24
$\beta \times 10^{-3}$	1.8227	1.8619	2.15

floor are employed for identification. The sampling frequency is 1000 Hz. The parameters are initially set to be 50% of the actual ones. The first 12 and the last 10 diagonal elements of the FFM are set to be 1 and 1.002, respectively. The reset error covariance matrix is $\overline{\mathbf{P}} = \text{diag}([\mathbf{0}_{1\times12}, \mathbf{10}^{12}_{1\times6}, \mathbf{10}^{2}_{1\times2}, 10^4, 10^9])$.

The parameters can be identified by using the proposed approach as listed in Table 2. Satisfactory identification results can be observed even though structural nonlinearities are involved. The square errors used for detecting damage instant are also calculated as plotted in Figure 5. Due to paper length limitations, only the square errors of the 1st and 3rd stories are given as examples. Apparently, the damage instant is at $t = 5$s. The time histories of the identified time-varying stiffness are shown in Figure 6. Similarly, two additional cases, i.e., FFM only and $\overline{\mathbf{P}}$ matrix only, are considered for comparison. It can be found that only using FFM or $\overline{\mathbf{P}}$ matrix is unable to efficiently track time-varying parameters. The proposed approach with the combination of FFM and $\overline{\mathbf{P}}$ matrix is capable of capturing time-varying parameters with good accuracy. Also, the external excitation can be simultaneously identified as shown in Figure 7. It is obvious that the identified force is close to the actual one.

3.3. A Fixed Beam Structure. In this section, a fixed beam composed of eight elements is further considered, as shown in Figure 8. The physical parameters of each element are given as follows: the area of the cross-section

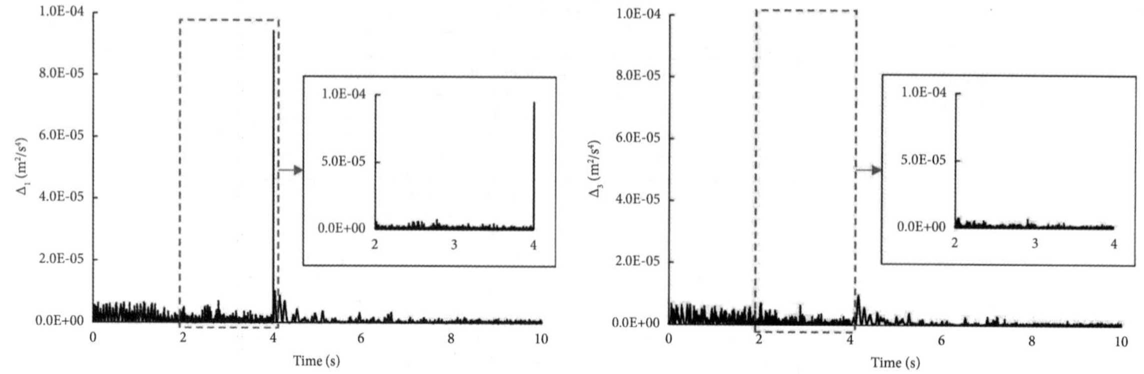

(a) (b)

FIGURE 2: Time histories of square errors in the 1st example. (a) Time history of square error of Δ_1. (b) Time history of square error of Δ_3.

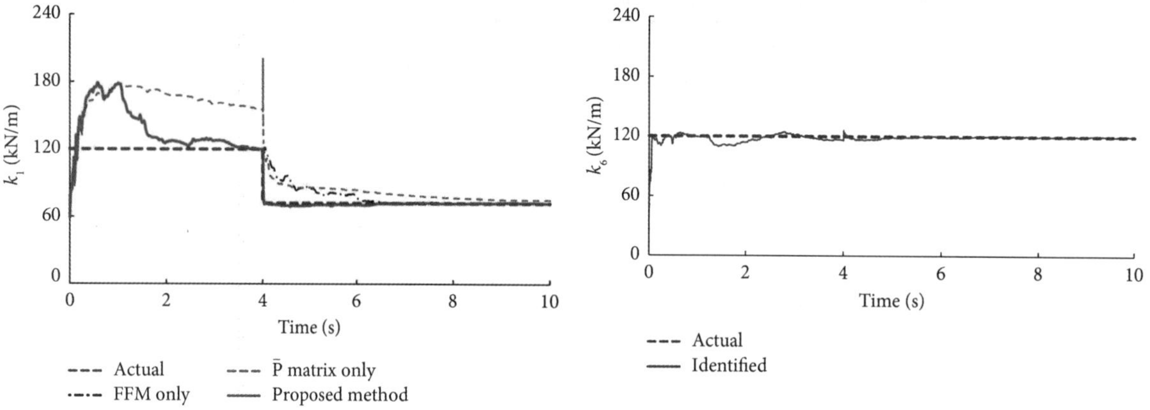

(a) (b)

FIGURE 3: The identified stiffness of building structure. (a) Stiffness of the 1st story. (b) Stiffness of the 6th story.

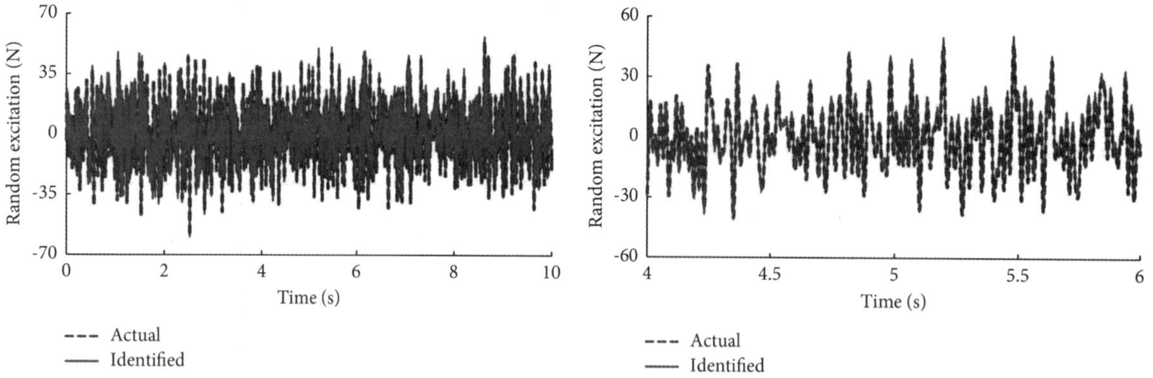

(a) (b)

FIGURE 4: The identified excitation applied to building structure. (a) The entire time history. (b) The segment from 4 s to 6 s.

$A = 2.25 \times 10^{-2}\,\text{m}^2$, the material density $\rho = 7850\,\text{kg/m}^3$, modulus of elasticity $E = 2 \times 10^8$ Pa, and the length $l = 1.5$ m. The Rayleigh damping model is adopted, and proportional coefficients are $\alpha = 0.0473$ and $\beta = 1.4866 \times 10^{-2}$. The damage scenario considered herein is an abrupt stiffness degradation of three elements. We assume that the stiffness parameters of the 1st, 2nd, and 8th elements are decreased sharply by 50%, 50%, and 40%, respectively, at the time instant $t = 5$ s.

TABLE 2: The identified results of the building structure with MR damper.

Parameter	Actual	Identified	Error (%)
k_1 (kN/m)	120⟶60	120.13⟶60.74	0.11⟶1.23
k_2 (kN/m)	120⟶60	120.73⟶60.50	0.61⟶0.83
k_3 (kN/m)	120	119.68	−0.27
k_4 (kN/m)	120	120.33	0.27
k_5 (kN/m)	120	119.72	−0.61
k_6 (kN/m)	120	119.41	−0.49
f_c (N)	1	0.9678	−3.22
c_b (N·s/m)	1000	979.24	−2.08
α	0.3739	0.3574	−4.43
$\beta\times10^{-3}$	1.8227	1.8541	1.72

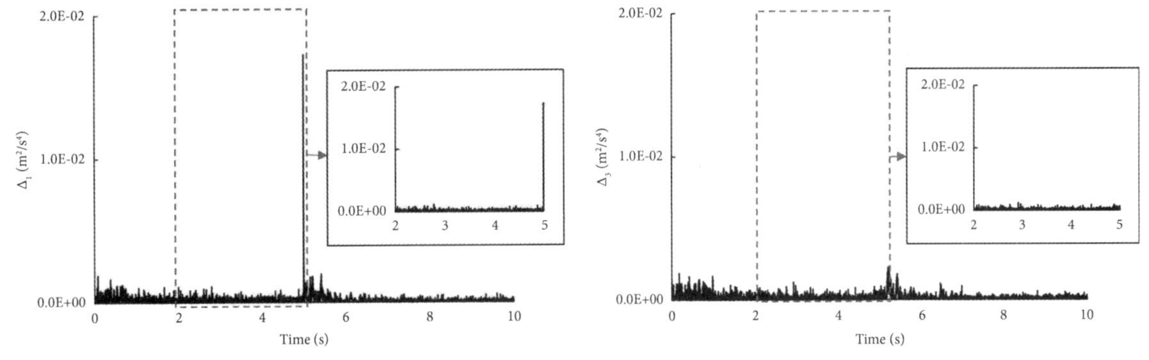

FIGURE 5: Time histories of square errors in the 2nd example. (a) Time history of square error of Δ_1. (b) Time history of square error of Δ_3.

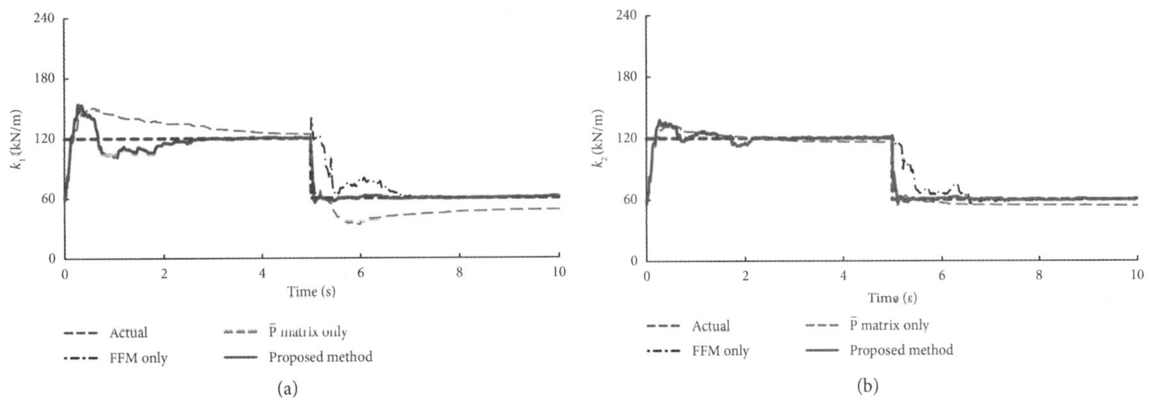

FIGURE 6: The identified stiffness of building structure with MR damper. (a) Stiffness of the 1st story. (b) Stiffness of the 2nd story.

As shown in Figure 8, there are nine nodes, and each node has two DOFs except node 1 and node 9. A random excitation is vertically applied at node 4, and the structural responses are computed numerically. Since the rotational responses are quite difficult to be measured, only the acceleration of nodes 2~8 and the displacement of node 3 in the vertical direction are used for identification. The parameters are initially set to be 50% of the actual ones. The first 28 and the last 10 diagonal elements of FFM are 1 and 1.001, respectively. The $\overline{\mathbf{P}}$ matrix is set as $\mathrm{diag}([\mathbf{0}_{1\times28}, \mathbf{10}^{14}_{1\times8}, \mathbf{10}^{6}_{1\times2}])$.

The identified results are given in Table 3. The square errors between the measurements and the corresponding estimated ones are plotted in Figure 9. Similarly, only Δ_1 and Δ_3 are given in Figure 9 as examples. The time histories of some identified stiffness are plotted in Figure 10. Two additional cases are discussed. It can be seen from these results that the proposed approach is able to satisfactorily identify the time-varying parameters of beam-type structures. Moreover, the unknown force applied to the beam can be simultaneously identified. It can be found from Figure 11 the identified force has a good agreement with the actual one.

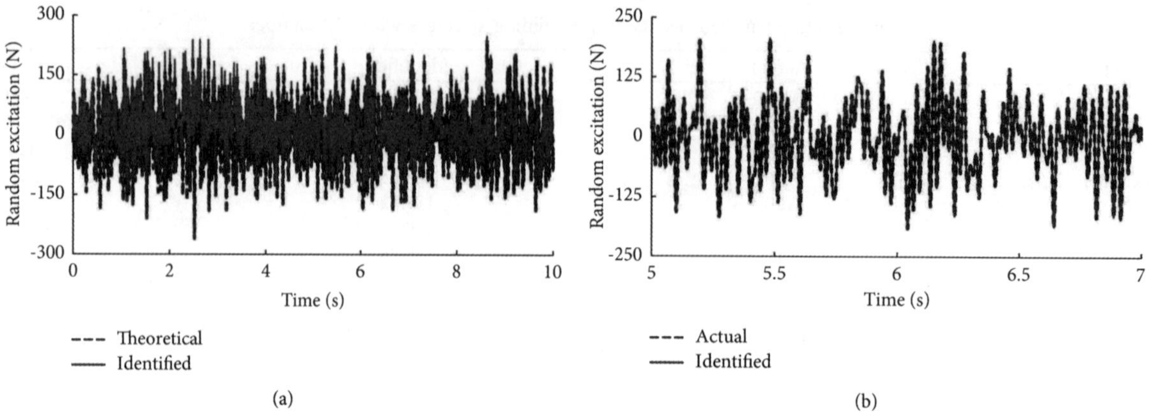

FIGURE 7: The identified excitation applied to nonlinear structure. (a) The entire time history. (b) The segment from 5 s to 7 s.

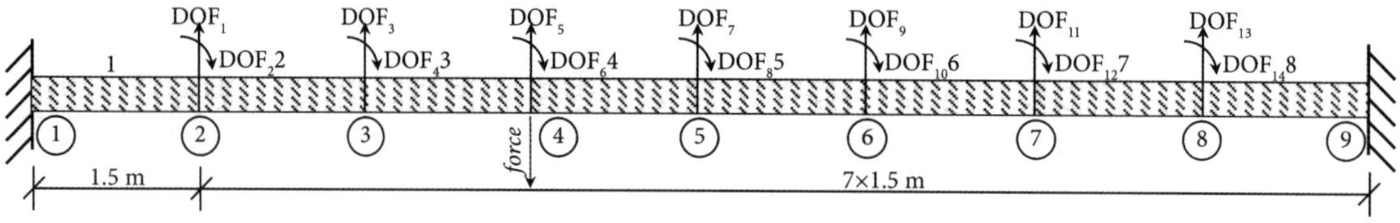

FIGURE 8: The fixed beam structure.

TABLE 3: The identified results of the beam structure.

Parameter	Actual	Identified	Error (%)
k_1 (N/m)	5625⟶2812.5	5347.0⟶2827.7	−4.94⟶2.14
k_2 (N/m)	5625⟶2812.5	5564.0⟶2810.5	−1.08⟶−0.07
k_3 (N/m)	5625	5630.9	0.10
k_4 (N/m)	5625	5583.7	−0.73
k_5 (N/m)	5625	5761.6	2.43
k_6 (N/m)	5625	5618.6	−0.11
k_7 (N/m)	5625	5680.4	0.98
k_8 (N/m)	5625⟶3375	5769.2⟶3314.7	2.56⟶−1.79
α	0.0473	0.0458	−3.16
$\beta \times 10^{-2}$	1.4866	1.4873	0.04

4. Experimental Validation

In order to validate the effectiveness of the proposed approach for real structures, a small-scale, six-story shear building model, as shown in Figure 12, is designed and built in the lab. The structure consists of aluminum columns and plates assembled firmly by bolted joints. The size of each plate is 400 mm × 400 mm, and the thickness is 10 mm. The cross-section of each column is 20 mm × 4 mm. The height of each floor is 210 mm except for the top floor. To conveniently install a lock-up system to mimic an abrupt change in structural stiffness, the height of the top floor is designed at 410 mm. An additional iron column of 400 mm × 50 mm × 1 mm fixed on the top floor plate and a powerful electromagnet device fixed on the 5th floor plate are used to construct the lock-up system. When direct current (DC) is injected into the electromagnet device by a DC power supply, a large magnetic attraction force is generated. In this case, the iron column is locked by the electromagnet device, as shown in Figure 12, providing additional stiffness to the structural system. If the DC becomes zero, the iron column is immediately released by the electromagnet device, indicating a sudden decrease in stiffness on the 6th story. In other words, structural damage is simulated by such a lock-up system in this experiment. The "lock" state indicates additional stiffness is provided, whereas the "unlock" state indicates additional stiffness is zero.

The six-story shear building model in this section can be viewed as a 6-DOFs lumped mass structural system. The weight of plates, columns, and bolts is measured and equally distributed in each story. Notably, the length of the columns of the 6th story is longer than other stories due to a lock-up system being installed on this story. Since the heavy electromagnet device is fixed to the 5th story plate, as shown in Figure 12, the concentrated mass of the 5th DOF is relatively larger than other DOFs. The stiffness of each story is determined by static tests, as schematically shown in Figure 13. Based on the weight of the calibrated mass block and the deformation of the building model recorded by displacement transducers, the structural stiffness can be determined. The mass distribution and stiffness of each story are summarized in Table 4.

The Rayleigh damping assumption is used for the consideration of damping effect. Moreover, in order to effectively identify structural frequencies, a sine sweeping-frequency excitation ranging from 0 Hz to 50 Hz is applied to the 5th floor of the model. Based on the frequency response functions (FRFs), the natural frequencies of the model are identified. The natural frequencies are 3.24 Hz, 6.95 Hz, 12.74 Hz, 20.36 Hz, 27.44 Hz, and 31.71 Hz in the lock state, and 3.14 Hz, 6.07 Hz, 12.67 Hz, 20.32 Hz, 27.43 Hz, and

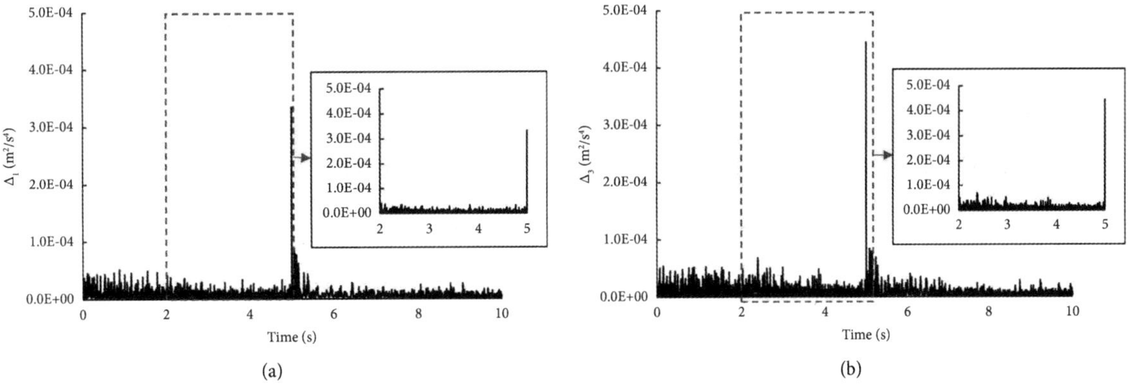

FIGURE 9: Time histories of square errors in the 3rd example. (a) Time history of square error of Δ_1. (b) Time history of square error of Δ_3.

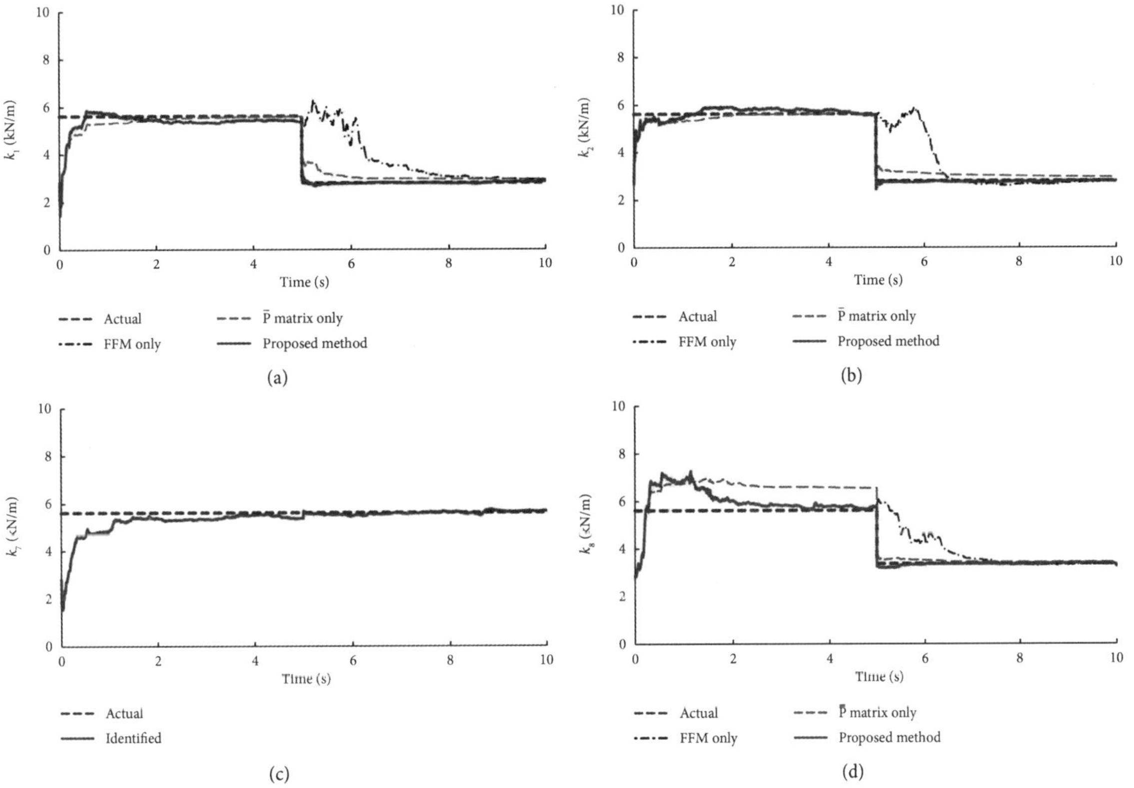

FIGURE 10: The identified stiffness of the beam structure. (a) Stiffness of the 1st element. (b) Stiffness of the 2nd element. (c) Stiffness of the 7th element. (d) Stiffness of the 8th element.

31.67 Hz in the unlock state. For clarity, only the FRF of the 1st floor is shown in Figure 14.

Dynamic tests are conducted on this six-story building model to validate the effectiveness of the proposed approach. A random force measured by a force transducer is applied to the 5th floor by an exciter. The displacement response of the 5th floor and the acceleration responses of the 1st, 3rd, 5th, and 6th floors are measured at a sampling frequency of 1000 Hz. The unknown quantities include six stiffness coefficients, two damping coefficients and the random force. The structural parameters are initially assumed to be 50% of the actual ones. The first 12 and the last 8 diagonal elements of FFM are set to be 1 and 1.001, respectively. The $\overline{\mathbf{P}}$ matrix is defined as $\mathrm{diag}([\mathbf{0}_{1\times12}, \mathbf{10}^{10}_{1\times6}, \mathbf{10}^{1}_{1\times2}])$ herein.

The structural parameters can be identified by the proposed approach, as shown in Table 5. The maximum error is 4.72%, which means the proposed approach can be employed for identifying time-varying parameters of real structure with acceptable accuracy. The square errors associated with the acceleration measurements are displayed

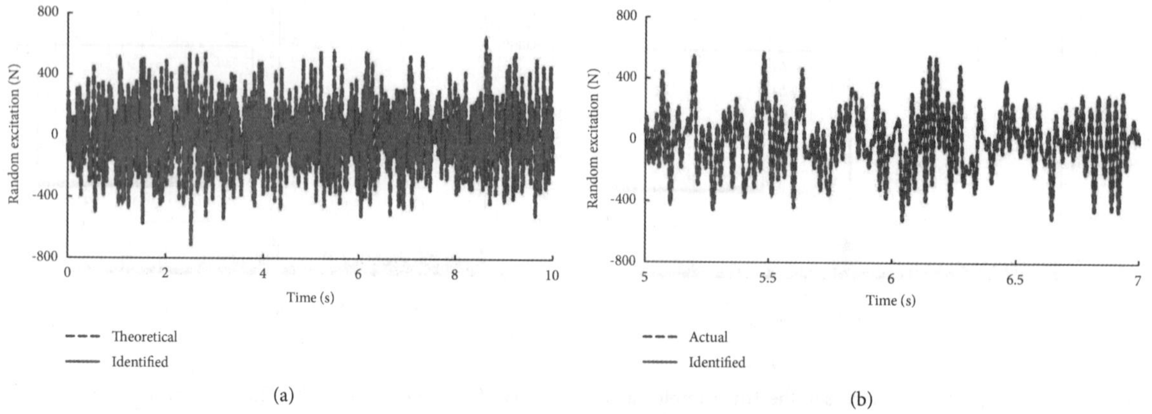

FIGURE 11: The identified excitation applied to the fixed beam. (a) The entire time history. (b) The segment from 5 s to 7 s.

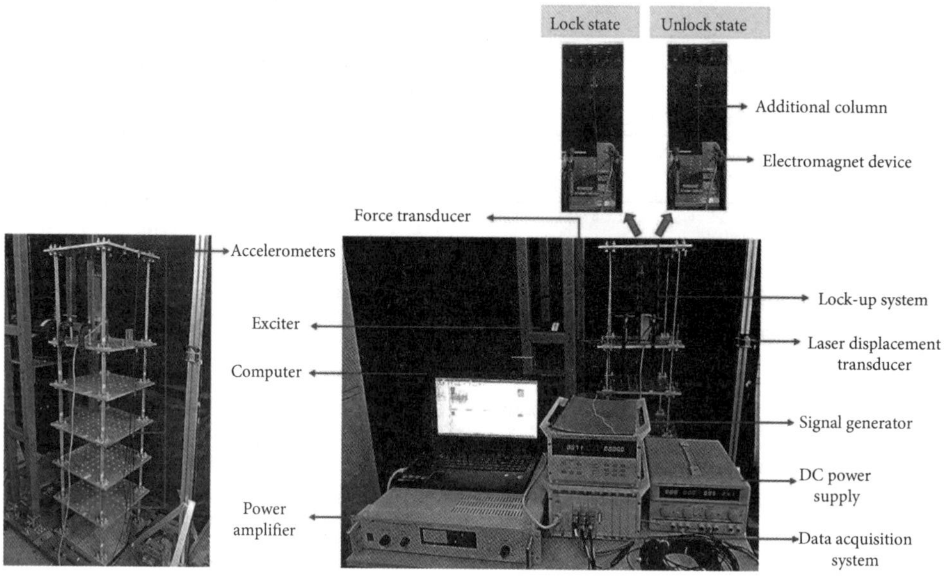

FIGURE 12: The six-story building model in the lab.

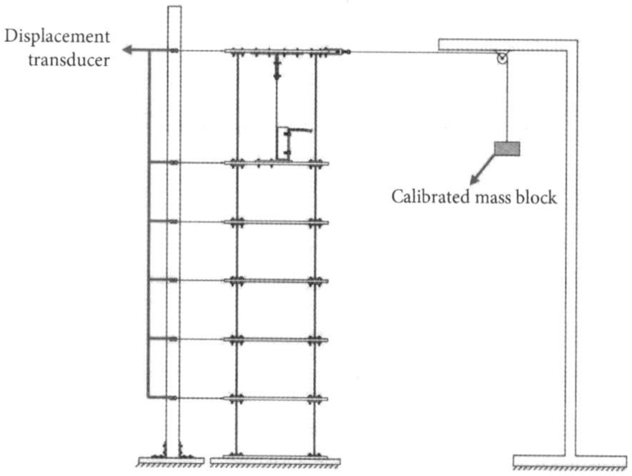

FIGURE 13: Schematic diagram of static test.

TABLE 4: The basic information of the building model.

Floor number	Mass distribution (kg)	Stiffness (kN/m)	
		Lock state	Unlock state
The 1st story	5.35	54.89	54.89
The 2nd story	5.34	56.44	56.44
The 3rd story	5.36	53.97	53.97
The 4th story	5.33	60.68	60.68
The 5th story	11.39	56.60	56.60
The 6th story	6.47	8.85	6.15

in Figure 15. The jump increases can be observed at $t = 8.13$ s indicating the significant changes of structural parameters at this time. Figure 16 gives the time histories of the identified stiffness parameters. Only the stiffness parameters of 1st and 6th story are plotted due to the limitation of paper length. It

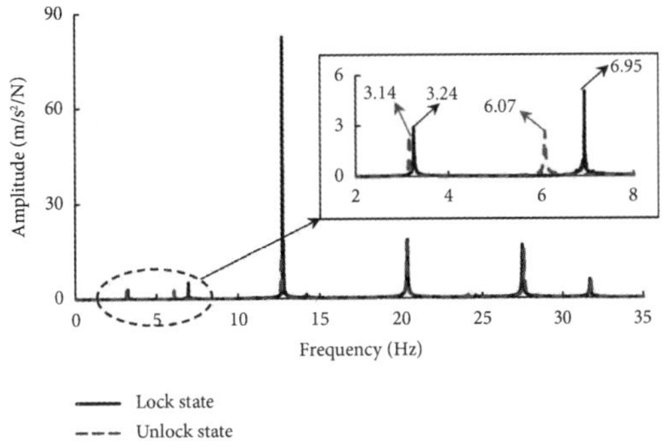

FIGURE 14: FRF of the 1st floor.

TABLE 5: The identified stiffness of the real building model.

Parameters	Actual	Identified	Error (%)
k_1 (kN/m)	54.89	54.37	−0.95
k_2 (kN/m)	56.44	57.92	2.61
k_3 (kN/m)	53.97	54.64	1.25
k_4 (kN/m)	60.68	58.78	−3.14
k_5 (kN/m)	56.60	58.16	2.75
k_6 (kN/m)	8.85⟶6.15	8.51⟶5.86	−3.84⟶−4.72

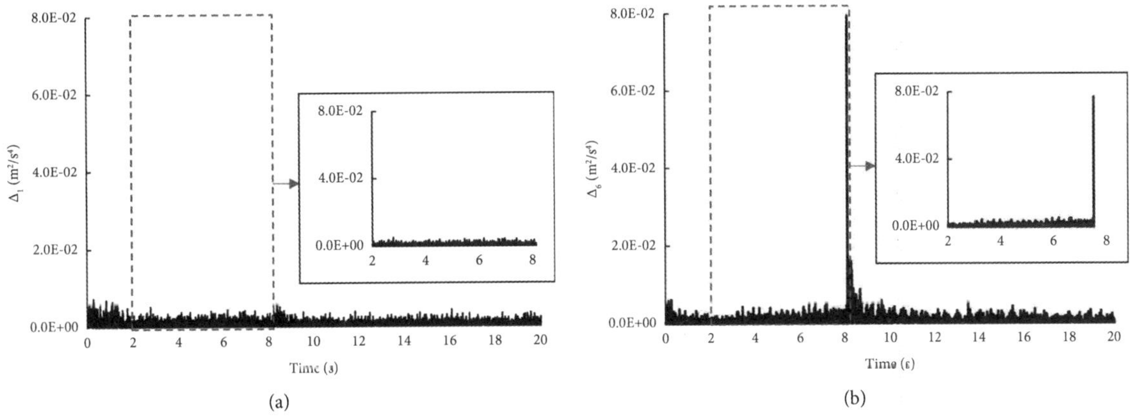

FIGURE 15: Time histories of squares errors in the experiment. (a) Time history of square error of Δ_1. (b) Time history of square error of Δ_6.

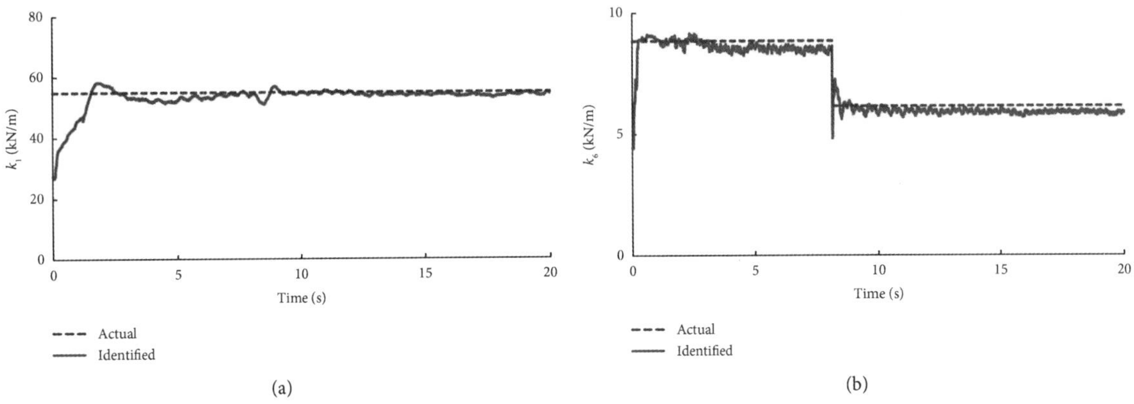

FIGURE 16: The identified stiffness of the building model in the lab. (a) Stiffness of the 1st story. (b) Stiffness of the 6th story.

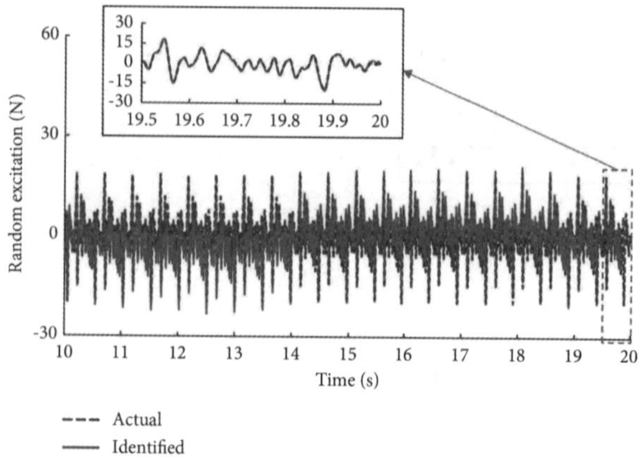

FIGURE 17: The identified excitation applied to the building model in the lab.

can be found that the proposed approach is capable of efficiently capturing abrupt changes in structural parameters. The external excitation is identified and compared with the measured one, as plotted in Figure 17. Obviously, the identified force is close to the measured one.

5. Conclusion

Online identification of time-varying parameters of linear and nonlinear structures under unknown loads is an important but challenging task. In this paper, a two-stage EKF-FFM-UI approach is proposed for simultaneously identifying time-variant parameters and external loads. In stage 1, an EKF-UI approach previously proposed by the authors is briefly introduced. At this stage, the parameters are assumed to be time-invariant. The aim of stage 2 is to improve this EKF-UI approach for the purpose of possessing tracking capability. First, based on the reconstructed acceleration, an updated observation equation is derived. Then, a forgetting factor matrix is introduced, and the formulas for the estimation of priori states, posteriori states, and unknown inputs are updated accordingly. Finally, in order to efficiently capture abrupt changes in structural parameters, the square errors associated with acceleration measurements are defined. The jump increases of these square errors indicate the occurrence of parameter variations. A covariance resetting technique is immediately used at this time instant to enhance the tracking capability. One of the merits of the proposed approach is that it is performed solely based on the EKF principle without any optimization procedures. Therefore, it is more straightforward and computationally efficient for tracking abrupt damages in real time. The effectiveness of the proposed approach is first validated via linear and nonlinear numerical examples. Experimental tests are also conducted to investigate the performance of the proposed approach for real structures. Results show that abrupt structural damages can be effectively captured and accurately identified. The unknown excitations can also be estimated at the same time. The proposed approach provides a promising way for real-time parameter identification and structural evaluation.

Appendix

For ease of understanding, the details regarding the derivation of equations (16)–(24) are given as follows:

Based on equation (1), the first part of equation (15) can be rearranged as

$$\widetilde{\mathbf{D}}\mathbf{f}^u_{k+1} = -\widetilde{\mathbf{y}}^a_{k+1} + \mathbf{h}'(\mathbf{Z}_{k+1}) + \widetilde{\mathbf{v}}^a_{k+1}, \qquad (A.1)$$

where $\widetilde{\mathbf{D}} = -\mathbf{M}^{-1}\varphi^u$; $\mathbf{h}'(\mathbf{Z}_{k+1}) = -\mathbf{M}^{-1}\mathbf{F}[\dot{\mathbf{x}}_{k+1}, \mathbf{x}_{k+1}, \theta_{k+1}]$ and $\widetilde{\mathbf{v}}^a_{k+1}$ are the noise vector of $\widetilde{\mathbf{y}}^a_{k+1}$ with zero mean and covariance matrices $\widetilde{\mathbf{R}}^a_{k+1}$.

Based on least square estimation, the closest solution of unknown inputs can be calculated as

$$\mathbf{f}^u_{k+1,LSE} = \left(\widetilde{\mathbf{D}}^{\mathrm{T}}\widetilde{\mathbf{D}}\right)^{-1}\widetilde{\mathbf{D}}^{\mathrm{T}}\left[-\widetilde{\mathbf{y}}^a_{k+1} + \mathbf{h}'(\mathbf{Z}_{k+1}) + \widetilde{\mathbf{v}}^a_{k+1}\right]. \qquad (A.2)$$

The error of the solution from equation (A.2) is given by

$$\begin{aligned}\mathbf{err} &= \widetilde{\mathbf{D}}\mathbf{f}^u_{k+1} - \widetilde{\mathbf{D}}\mathbf{f}^u_{k+1,LSE}.\\ &= \left(\mathbf{I}_n - \widetilde{\mathbf{D}}\left(\widetilde{\mathbf{D}}^{\mathrm{T}}\widetilde{\mathbf{D}}\right)^{-1}\widetilde{\mathbf{D}}^{\mathrm{T}}\right)\left[-\widetilde{\mathbf{y}}^a_{k+1} + \mathbf{h}'(\mathbf{Z}_{k+1}) + \widetilde{\mathbf{v}}^a_{k+1}\right],\end{aligned} \qquad (A.3)$$

where $\mathbf{I}$ is an identity matrix; and $\widetilde{\mathbf{D}}(\widetilde{\mathbf{D}}^{\mathrm{T}}\widetilde{\mathbf{D}})^{-1}\widetilde{\mathbf{D}}^{\mathrm{T}}$ is known as projection matrix. As a limit, the error in equation (A.3) tends to be zero, leading to

$$\mathbf{\Phi}'\widetilde{\mathbf{y}}^a_{k+1} = \mathbf{\Phi}'\mathbf{h}'(\mathbf{Z}_{k+1}) + \mathbf{\Phi}'\widetilde{\mathbf{v}}^a_{k+1}, \qquad (A.4)$$

where $\Phi' = \mathbf{I}_n - \widetilde{\mathbf{D}}(\widetilde{\mathbf{D}}^{\mathrm{T}}\widetilde{\mathbf{D}})^{-1}\widetilde{\mathbf{D}}^{\mathrm{T}}$. Then, based on equations (A.4) and (15), the modified version of the observation equation can be obtained,

$$\widetilde{\mathbf{\Phi}}\widetilde{\mathbf{y}}_{k+1} = \widetilde{\mathbf{\Phi}}\widetilde{\mathbf{h}}(\mathbf{Z}_{k+1}) + \widetilde{\mathbf{\Phi}}\widetilde{\mathbf{v}}_{k+1}, \qquad (A.5)$$

where $\widetilde{\Phi} = \begin{bmatrix}\Phi' & \mathbf{0}\\ \mathbf{0} & \mathbf{I}\end{bmatrix}$; $\widetilde{\mathbf{h}}(\mathbf{Z}_{k+1}) = \begin{bmatrix}\mathbf{h}'(\mathbf{Z}_{k+1})\\ \varphi_s\mathbf{x}_{k+1}\end{bmatrix}$; $\widetilde{\mathbf{v}}_{k+1} = \begin{bmatrix}\widetilde{\mathbf{v}}^a_{k+1}\\ \mathbf{v}^s_{k+1}\end{bmatrix}$.

Let $\widetilde{\mathbf{Z}}_{k+1|k}$ be the updated priori estimate of state $\mathbf{Z}_{k+1|k}$. Linearizing $\mathbf{h}'(\mathbf{Z}_{k+1})$ with respect to $\widetilde{\mathbf{Z}}_{k+1|k}$, one can obtain

$$\mathbf{h}'(\mathbf{Z}_{k+1}) = \mathbf{h}'\left(\widetilde{\mathbf{Z}}_{k+1|k}\right) + \mathbf{H}'_{k+1|k}\left(\mathbf{Z}_{k+1} - \widetilde{\mathbf{Z}}_{k+1|k}\right). \qquad (A.6)$$

Here, $\mathbf{H}'_{k+1|k} = \partial\mathbf{h}'(\mathbf{Z}_{k+1})/\partial\mathbf{Z}_{k+1}|_{\mathbf{Z}_{k+1}=\widetilde{\mathbf{Z}}_{k+1|k}}$.

According to the classic EKF principle, the updated priori estimation state $\widetilde{\mathbf{Z}}_{k+1|k}$ can be calculated as

$$\widetilde{\mathbf{Z}}_{k+1|k} = \widetilde{\mathbf{Z}}_{k|k} + \int_{k\Delta t}^{(k+1)\Delta t}\mathbf{u}\left(\widetilde{\mathbf{Z}}_{k|k}, \widetilde{\mathbf{f}}^u_k, k\Delta t\right)dt, \qquad (A.7)$$

where $\widetilde{\mathbf{f}}^u_k$ means the updated value of the identified load at the k-th time step.

Let $\widetilde{\mathbf{Z}}_{k|k}$ and $\widetilde{\mathbf{f}}^u_k$ be the estimates of $\mathbf{Z}_k$ and $\mathbf{f}^u_k$, respectively. Equation (3) can be linearized with respect to $\widetilde{\mathbf{Z}}_{k|k}$ and $\widetilde{\mathbf{f}}^u_k$ as follows:

$$\begin{aligned}\mathbf{u}(\mathbf{Z}_k, \mathbf{f}^u_k, t_k) &= \mathbf{u}\left(\widetilde{\mathbf{Z}}_{k|k}, \widetilde{\mathbf{f}}^u_k, t_k\right) + \mathbf{U}_{k|k}\left(\mathbf{Z}_k - \widetilde{\mathbf{Z}}_{k|k}\right)\\ &\quad + \mathbf{B}_{k|k}\left(\mathbf{f}^u_k - \widetilde{\mathbf{f}}^u_k\right) + \mathbf{w}(t),\end{aligned} \qquad (A.8)$$

where $\mathbf{U}_{k|k} = \partial \mathbf{u}(\mathbf{Z}_k, \mathbf{f}_k^u, t_k)/\partial \mathbf{Z}_k|_{\mathbf{Z}_k=\tilde{\mathbf{Z}}_{k|k}, \mathbf{f}_k^u=\tilde{\mathbf{f}}_k^u}$; $\mathbf{B}_{k|k} = \partial \mathbf{u}(\mathbf{Z}_k, \mathbf{f}_k^u, t_k)/\partial \mathbf{f}_k^u|_{\mathbf{Z}_k=\tilde{\mathbf{Z}}_{k|k}, \mathbf{f}_k^u=\tilde{\mathbf{f}}_k^u}$.

Substituting $\mathbf{Z}_{k+1} - \mathbf{Z}_k/\Delta t = d\mathbf{Z}_k/dt = \mathbf{u}(\mathbf{Z}_k, \mathbf{f}_k^u, t_k)$ into equation (A.8), one obtains

$$\mathbf{Z}_{k+1} = \mathbf{Z}_k + \Delta t\left[\mathbf{u}\left(\tilde{\mathbf{Z}}_{k|k}, \tilde{\mathbf{f}}_k^{\mathrm{u}}, t_k\right) + \mathbf{U}_{k|k}\left(\mathbf{Z}_k - \tilde{\mathbf{Z}}_{k|k}\right) + \mathbf{B}_{k|k}\left(\mathbf{f}_k^{\mathrm{u}} - \tilde{\mathbf{f}}_k^{\mathrm{u}}\right) + \mathbf{w}_k\right]. \quad (A.9)$$

The priori estimation error $\tilde{\varepsilon}_{k+1|k}$ of the unknown state vector $\mathbf{Z}_{k+1}$ is

$$\tilde{\boldsymbol{\varepsilon}}_{k+1|k} = \mathbf{Z}_{k+1} - \tilde{\mathbf{Z}}_{k+1|k} = \left(\mathbf{I} + \Delta t\mathbf{U}_{k|k}\right)\left(\mathbf{Z}_k - \tilde{\mathbf{Z}}_{k|k}\right) + \Delta t\mathbf{B}_{k|k}\left(\mathbf{f}_k^u - \tilde{\mathbf{f}}_k^u\right) + \Delta t\mathbf{w}_k. \quad (A.10)$$

The second term on the right-hand side of equation (A.10) can be simplified on the basis of equations (A.2) and (A.6) as

$$\begin{aligned}\Delta t\mathbf{B}_{k|k}\left(\mathbf{f}_k^u - \tilde{\mathbf{f}}_k^u\right) &= \Delta t\begin{bmatrix}\mathbf{0}\\ \mathbf{M}^{-1}\boldsymbol{\varphi}^u\left(\mathbf{f}_k^u - \tilde{\mathbf{f}}_k^u\right)\\ \mathbf{0}\end{bmatrix},\\ &= \Delta t\begin{bmatrix}\mathbf{0}\\ -\tilde{\mathbf{D}}\left(\tilde{\mathbf{D}}^{\mathrm{T}}\tilde{\mathbf{D}}\right)^{-1}\tilde{\mathbf{D}}^{\mathrm{T}}\left[\mathbf{h}'(\mathbf{Z}_k) - \mathbf{h}'\left(\tilde{\mathbf{Z}}_{k|k}\right) + \tilde{\mathbf{v}}_k^a\right]\\ \mathbf{0}\end{bmatrix},\\ &= -\Delta t\begin{bmatrix}\mathbf{0}\\ \left(\mathbf{I}_n - \boldsymbol{\Phi}'\right)\left[\mathbf{H}'_{k|k}\left(\mathbf{Z}_k - \tilde{\mathbf{Z}}_{k|k}\right) - \tilde{\mathbf{v}}_k^a\right]\\ \mathbf{0}\end{bmatrix}.\end{aligned} \quad (A.11)$$

Substituting equation (A.11) into equation (A.10) leads to

$$\tilde{\boldsymbol{\varepsilon}}_{k+1|k} = \mathbf{Z}_{k+1} - \tilde{\mathbf{Z}}_{k+1|k} = \tilde{\mathbf{A}}_1\tilde{\boldsymbol{\varepsilon}}_{k|k} + \tilde{\mathbf{A}}_2\tilde{\mathbf{v}}_k^a + \Delta t\mathbf{w}_k, \quad (A.12)$$

where $\tilde{\mathbf{A}}_1 = \left(\mathbf{I} + \Delta t\mathbf{U}'_{k|k} - \Delta t\begin{bmatrix}\mathbf{0}\\ (\mathbf{I} - \Phi')\mathbf{H}'_{k|k}\\ \mathbf{0}\end{bmatrix}\right)$; $\tilde{\mathbf{A}}_2 = \begin{bmatrix}\mathbf{0}\\ -\Delta t(\mathbf{I} - \Phi')\\ \mathbf{0}\end{bmatrix}$.

Then, the priori estimation error covariance matrix $\tilde{\mathbf{P}}_{k+1|k}$ for identifying time-invariant parameters can be calculated as

$$\tilde{\mathbf{P}}_{k+1|k} = E\left(\tilde{\boldsymbol{\varepsilon}}_{k+1|k}\tilde{\boldsymbol{\varepsilon}}_{k+1|k}^{\mathrm{T}}\right) = \tilde{\mathbf{A}}_1\tilde{\mathbf{P}}_{k|k}\tilde{\mathbf{A}}_1^{\mathrm{T}} + \tilde{\mathbf{A}}_2\tilde{\mathbf{R}}_k^a\tilde{\mathbf{A}}_2^{\mathrm{T}} + \Delta t^2\mathbf{Q}_k. \quad (A.13)$$

In order to track time-varying parameters, a forgetting factor matrix $\boldsymbol{\Lambda}_{k+1}$ is introduced into the priori estimation error covariance matrix. Therefore, the priori estimation error covariance matrix in equation (A.13) can be updated in the following way [27]:

$$\tilde{\mathbf{P}}_{k+1|k} = \boldsymbol{\Lambda}_{k+1}\left[\tilde{\mathbf{A}}_1\tilde{\mathbf{P}}_{k|k}\tilde{\mathbf{A}}_1^{\mathrm{T}}\right]\boldsymbol{\Lambda}_{k+1}^{\mathrm{T}} + \tilde{\mathbf{A}}_2\tilde{\mathbf{R}}_k^a\tilde{\mathbf{A}}_2^{\mathrm{T}} + \Delta t^2\mathbf{Q}_k. \quad (A.14)$$

Based on equation (A.5), the recursive solution for the updated posteriori estimation state can be given as

$$\tilde{\mathbf{Z}}_{k+1|k+1} = \tilde{\mathbf{Z}}_{k+1|k} + \tilde{\mathbf{G}}_{k+1}\left[\tilde{\boldsymbol{\Phi}}\tilde{\mathbf{y}}_{k+1} - \tilde{\boldsymbol{\Phi}}\tilde{\mathbf{h}}\left(\tilde{\mathbf{Z}}_{k+1|k}\right)\right], \quad (A.15)$$

where $\tilde{\mathbf{G}}_{k+1}$ denotes the EKF gain matrix at the $(k+1)$-th time step.

Let $\varepsilon_{k+1|k} = \Lambda_{k+1}(\mathbf{Z}_{k+1} - \tilde{\mathbf{Z}}_{k+1|k})$, the updated posteriori estimation error $\tilde{\varepsilon}_{k+1|k+1}$ of the unknown state vector $\mathbf{Z}_{k+1}$ at time $t = (k+1)\Delta t$ can then be calculated as

$$\begin{aligned}\tilde{\boldsymbol{\varepsilon}}_{k+1|k+1} &= \mathbf{Z}_{k+1} - \tilde{\mathbf{Z}}_{k+1|k+1}\\ &= \mathbf{Z}_{k+1} - \tilde{\mathbf{Z}}_{k+1|k} - \tilde{\mathbf{G}}_{k+1}\left[\tilde{\boldsymbol{\Phi}}\tilde{\mathbf{y}}_{k+1} - \tilde{\boldsymbol{\Phi}}\tilde{\mathbf{h}}\left(\tilde{\mathbf{Z}}_{k+1|k}\right)\right]\\ &= \left(\mathbf{I} - \tilde{\mathbf{G}}_{k+1}\tilde{\boldsymbol{\Phi}}\tilde{\mathbf{H}}_{k+1|k}\right)\left(\mathbf{Z}_{k+1} - \tilde{\mathbf{Z}}_{k+1|k}\right) - \tilde{\mathbf{G}}_{k+1}\tilde{\boldsymbol{\Phi}}\tilde{\mathbf{v}}_{k+1}\\ &= \left(\mathbf{I} - \tilde{\mathbf{G}}_{k+1}\tilde{\boldsymbol{\Phi}}\tilde{\mathbf{H}}_{k+1|k}\right)\boldsymbol{\Lambda}_{k+1}\tilde{\boldsymbol{\varepsilon}}_{k+1|k} - \tilde{\mathbf{G}}_{k+1}\tilde{\boldsymbol{\Phi}}\tilde{\mathbf{v}}_{k+1}.\end{aligned} \quad (A.16)$$

Similarly, the updated posteriori estimation error covariance can then be found as

$$\begin{aligned}\tilde{\mathbf{P}}_{k+1|k+1} &= E\left(\tilde{\boldsymbol{\varepsilon}}_{k+1|k+1}\tilde{\boldsymbol{\varepsilon}}_{k+1|k+1}^{\mathrm{T}}\right)\\ &= \left(\mathbf{I} - \tilde{\mathbf{G}}_{k+1}\tilde{\boldsymbol{\Phi}}\tilde{\mathbf{H}}_{k+1|k}\right)\boldsymbol{\Lambda}_{k+1}\tilde{\mathbf{P}}_{k+1|k}\boldsymbol{\Lambda}_{k+1}^{\mathrm{T}}\left(\mathbf{I} - \tilde{\mathbf{G}}_{k+1}\tilde{\boldsymbol{\Phi}}\tilde{\mathbf{H}}_{k+1|k}\right)^{\mathrm{T}} + \tilde{\mathbf{G}}_{k+1}\tilde{\boldsymbol{\Phi}}\tilde{\mathbf{R}}_{k+1}\tilde{\boldsymbol{\Phi}}^{\mathrm{T}}\tilde{\mathbf{G}}_{k+1}^{\mathrm{T}},\end{aligned} \quad (A.17)$$

where $\widetilde{\mathbf{R}}_{k+1}$ is the covariance matrix of noise vector $\widetilde{\mathbf{v}}_{k+1}$; $\widetilde{\mathbf{H}}_{k+1|k} = \partial\mathbf{h}(\mathbf{Z}_{k+1})/\partial\mathbf{Z}_{k+1}|_{\mathbf{Z}_{k+1}=\widetilde{\mathbf{Z}}_{k+1|k}}$.

To obtain the optimal value of the gain matrix $\widetilde{\mathbf{G}}_{k+1}$ that can minimize the updated estimation error covariance $\widetilde{\mathbf{P}}_{k+1|k+1}$ at time $t=(k+1)\Delta t$, the differentiation of $\widetilde{\mathbf{P}}_{k+1|k+1}$ with respect to $\widetilde{\mathbf{G}}_{k+1}$ produces

$$\begin{aligned}\frac{\partial\widetilde{\mathbf{P}}_{k+1|k+1}}{\partial\widetilde{\mathbf{G}}_{k+1}} &= 2\left(\mathbf{I}-\widetilde{\mathbf{G}}_{k+1}\widetilde{\mathbf{\Phi}}\widetilde{\mathbf{H}}_{k+1|k}\right)\mathbf{\Lambda}_{k+1}\widetilde{\mathbf{P}}_{k+1|k}\mathbf{\Lambda}_{k+1}^{\mathrm{T}}\left(-\widetilde{\mathbf{\Phi}}\widetilde{\mathbf{H}}_{k+1|k}\right)^{\mathrm{T}}+2\widetilde{\mathbf{G}}_{k+1}\widetilde{\mathbf{\Phi}}\widetilde{\mathbf{R}}_{k+1}\widetilde{\mathbf{\Phi}}^{\mathrm{T}} \\ &= 2\widetilde{\mathbf{G}}_{k+1}\widetilde{\mathbf{\Phi}}\left(\widetilde{\mathbf{H}}_{k+1|k}\mathbf{\Lambda}_{k+1}\widetilde{\mathbf{P}}_{k+1|k}\mathbf{\Lambda}_{k+1}^{\mathrm{T}}\widetilde{\mathbf{H}}_{k+1|k}^{\mathrm{T}}+\widetilde{\mathbf{R}}_{k+1}\right)\widetilde{\mathbf{\Phi}}^{\mathrm{T}}-2\mathbf{\Lambda}_{k+1}\widetilde{\mathbf{P}}_{k+1|k}\mathbf{\Lambda}_{k+1}^{\mathrm{T}}\widetilde{\mathbf{H}}_{k+1|k}^{\mathrm{T}}\widetilde{\mathbf{\Phi}}^{\mathrm{T}}.\end{aligned} \tag{A.18}$$

By setting the value of the partial derivative of equation (A.18) to zero, one can obtain

$$\widetilde{\mathbf{G}}_{k+1}=\mathbf{\Lambda}_{k+1}\widetilde{\mathbf{P}}_{k+1|k}\mathbf{\Lambda}_{k+1}^{\mathrm{T}}\widetilde{\mathbf{H}}_{k+1|k}^{\mathrm{T}}\widetilde{\mathbf{\Phi}}^{\mathrm{T}}\left[\widetilde{\mathbf{\Phi}}\left(\widetilde{\mathbf{H}}_{k+1|k}\mathbf{\Lambda}_{k+1}\widetilde{\mathbf{P}}_{k+1|k}\mathbf{\Lambda}_{k+1}^{\mathrm{T}}\widetilde{\mathbf{H}}_{k+1|k}^{\mathrm{T}}+\widetilde{\mathbf{R}}_{k+1}\right)\widetilde{\mathbf{\Phi}}^{\mathrm{T}}\right]^{-1}. \tag{A.19}$$

Since the structural states are updated as shown in equation (A.15), the unknown inputs can be also updated accordingly as

$$\widetilde{\mathbf{f}}_{k+1}^{u}=\left(\widetilde{\mathbf{D}}^{\mathrm{T}}\widetilde{\mathbf{D}}\right)^{-1}\widetilde{\mathbf{D}}^{\mathrm{T}}\left[\mathbf{h}'\left(\widetilde{\mathbf{Z}}_{k+1|k+1}\right)-\widetilde{\mathbf{y}}_{k+1}^{a}\right]. \tag{A.20}$$

Data Availability

The experimental data used to support the findings of this study were supplied by Jia He under license and so cannot be made freely available. Requests for access to these data should be made to Jia He via jiahe@hnu.edu.cn.

Conflicts of Interest

The authors declare that they have no conflicts of interest.

Acknowledgments

The authors gratefully acknowledge the financial support from the National Key Research and Development Program of China through Grant no. 2022YFC3005305, the National Natural Science Foundation of China (No. 52278305), and the Natural Science Foundation of Hunan Province (No. 2021JJ30110).

References

[1] Y. L. Xu and J. He, *Smart Civil Structures*, CRC Press, Boca Raton, FL,USA, 2017.

[2] R. Hou and Y. Xia, "Review on the new development of vibration-based damage identification for civil engineering structures: 2010–2019," *Journal of Sound and Vibration*, vol. 491, Article ID 115741, 2021.

[3] R. Liu, E. Dobriban, Z. Hou, and K. Qian, "Dynamic load identification for mechanical systems: a review," *Archives of Computational Methods in Engineering*, vol. 29, no. 2, pp. 831–863, 2022.

[4] W. C. Su, C. Y. Liu, and C. S. Huang, "Identification of instantaneous modal parameter of time-varying systems via a wavelet-based approach and its application," *Computer-Aided Civil and Infrastructure Engineering*, vol. 29, no. 4, pp. 279–298, 2014.

[5] J. L. Liu, J. Y. Zheng, Y. Q. Lin, R. H. Qiu, and Y. Luo, "Instantaneous frequency identification of time-varying structures using variational mode decomposition and synchrosqueezing wavelet transform," *Journal of Vibration and Shock*, vol. 37, no. 20, pp. 24–31, 2018.

[6] C. Wang, D. M. Ai, and W. X. Ren, "A wavelet transform and substructure algorithm for tracking the abrupt stiffness degradation of shear structureffness degradation of shear structure," *Advances in Structural Engineering*, vol. 22, no. 5, pp. 1136–1148, 2019.

[7] Z. S. Ma, L. Liu, S. D. Zhou et al., "Parametric output-only identification of time-varying structures using a kernel recursive extended least squares TARMA approach," *Mechanical Systems and Signal Processing*, vol. 98, pp. 684–701, 2018.

[8] P. Mirhoseini and M. Tabatabaei, "Bi-Loop matrix forgetting factor-based coupled recursive least squares algorithm for identification of multivariable plants," *J Control Autom Electr Syst*, vol. 29, no. 2, pp. 136–152, 2018.

[9] M. Feldman, "Hilbert transform methods for nonparametric identification of nonlinear time varying vibration systems," *Mechanical Systems and Signal Processing*, vol. 47, no. 1-2, pp. 66–77, 2014.

[10] P. H. Ni, J. Li, H. Hao et al., "Time-varying system identification using variational mode decomposition," *Structural Control and Health Monitoring*, vol. 25, no. 6, Article ID e2175, 2018.

[11] Z. J. Yue and L. Liu, "A single-mode recursive validation method for modal identification of linear time-varying structures based on prior knowledge," *Structural Control and Health Monitoring*, vol. 28, no. 12, Article ID e2845, 2021.

[12] M. Khodayar and J. H. Wang, "Probabilistic time-varying parameter identification for load modeling: a deep generative

approach," *IEEE Transactions on Industrial Informatics*, vol. 17, no. 3, pp. 1625–1636, 2021.

[13] K. V. Yuen, S. C. Kuok, and L. Dong, "Self-calibrating Bayesian real-time system identification," *Computer-Aided Civil and Infrastructure Engineering*, vol. 34, no. 9, pp. 806–821, 2019.

[14] X. Y. Yu, X. H. Li, and Y. T. Bai, "Evaluating maximum inter-story drift ratios of building structures using time-varying models and Bayesian filters," *Soil Dynamics and Earthquake Engineering*, vol. 162, Article ID 107496, 2022.

[15] C. H. Loh, C. Y. Lin, and C. C. Huang, "Time domain identification of frames under earthquake loadings," *Journal of Engineering Mechanics*, vol. 126, no. 7, pp. 693–703, 2000.

[16] C. H. Loh, J. H. Weng, Y. C. Liu, P. Y. Lin, and S. K. Huang, "Structural damage diagnosis based on on-line recursive stochastic subspace identification," *Smart Materials and Structures*, vol. 20, no. 5, Article ID 055004, 2011.

[17] J. N. Yang, S. Lin, H. W. Huang, and L. Zhou, "An adaptive extended Kalman filter for structural damage identification," *Structural Control and Health Monitoring*, vol. 13, no. 4, pp. 849–867, 2006.

[18] H. W. Huang, J. N. Yang, and L. Zhou, "Comparison of various structural damage tracking techniques based on experimental data," *Smart Structures and Systems*, vol. 6, no. 9, pp. 1057–1077, 2010.

[19] Y. Lei, H. Zhou, and Z. L. Lai, "A computationally efficient algorithm for real-time tracking the abrupt stiffness degradations of structural elements," *Computer-Aided Civil and Infrastructure Engineering*, vol. 31, no. 6, pp. 465–480, 2016.

[20] S. S. Bisht and M. P. Singh, "An adaptive unscented Kalman filter for tracking sudden stiffness changes," *Mechanical Systems and Signal Processing*, vol. 49, no. 1-2, pp. 181–195, 2014.

[21] S. Schleiter and O. Altay, "Identification of abrupt stiffness changes of structures with tuned mass dampers under sudden events," *Structural Control and Health Monitoring*, vol. 27, no. 6, Article ID e2530, 2020.

[22] S. Y. Chen, J. B. Lu, and Y. Lei, "Identification of time-varying systems with partial acceleration measurements by synthesis of wavelet decomposition and Kalman filter," *Advances in Mechanical Engineering*, vol. 12, no. 6, Article ID 168781402093046, 11 pages, 2020.

[23] M. Ciolek, M. Niedzwiecki, and A. X. Gancza, "Decoupled kalman filter based identification of time-varying FIR systems," *IEEE Access*, vol. 9, pp. 74622–74631, 2021.

[24] F. Yu, Z. Z. Mao, and D. K. He, "Identification of time-varying hammerstein-wiener systems," *IEEE Access*, vol. 8, pp. 136906–136916, 2020.

[25] N. Yang, J. Li, M. Q. Xu, and S. Q. Wang, "Real-time identification of time-varying cable force using an improved adaptive extended kalman filter," *Sensors*, vol. 22, no. 11, p. 4212, 2022.

[26] Y. Xin, J. Li, H. Hao, N. Yang, and C. Li, "Time-varying system identification of precast segmental columns subjected to seismic excitations," *Journal of Bridge Engineering*, vol. 27, no. 4, Article ID 04022013, 2022.

[27] K. Huang, K. V. Yuen, L. Wang, T. Jiang, and L. Dai, "Sensor fault detection, localization, and reconstruction for online structural identification," *Structural Control and Health Monitoring*, vol. 29, no. 4, p. 2925, 2022.

[28] S. Eftekhar Azam, E. Chatzi, and C. Papadimitriou, "A dual Kalman filter approach for state estimation via output-only acceleration measurements," *Mechanical Systems and Signal Processing*, vol. 60-61, pp. 866–886, 2015.

[29] S. Eftekhar Azam, E. Chatzi, C. Papadimitriou, and A. Smyth, "Experimental validation of the Kalman-type filters for online and real-time state and input estimation," *Journal of Vibration and Control*, vol. 23, no. 15, pp. 2494–2519, 2015.

[30] L. Liu, J. Zhu, Y. Su, and Y. Lei, "Improved Kalman filter with unknown inputs based on data fusion of partial acceleration and displacement measurements," *Smart Structures and Systems*, vol. 17, no. 6, pp. 903–915, 2016.

[31] Y. Lei, S. Luo, and Y. Su, "Data fusion based improved Kalman filter with unknown inputs and without collocated acceleration measurements," *Smart Structures and Systems*, vol. 18, no. 3, pp. 375–387, 2016.

[32] J. He, X. X. Zhang, and N. X. Dai, "An improved Kalman filter for joint estimation of structural states and unknown loadings," *Smart Structures and Systems*, vol. 24, no. 2, pp. 209–221, 2019.

[33] J. He, X. X. Zhang, and B. Xu, "KF-based multi-scale response reconstruction under unknown inputs with data fusion of multi-type observations," *Journal of Aerospace Engineering*, vol. 32, no. 4, Article ID 04019038, 2019.

[34] E. Lourens, E. Reynders, G. De Roeck, G. Degrande, and G. Lombaert, "An augmented Kalman filter for force identification in structural dynamics," *Mechanical Systems and Signal Processing*, vol. 27, pp. 446–460, 2012.

[35] R. Cumbo, L. Mazzanti, T. Tamarozzi, P. Jiranek, W. Desmet, and F. Naets, "Advanced optimal sensor placement for Kalman-based multiple-input estimation," *Mechanical Systems and Signal Processing*, vol. 160, no. 3, Article ID 107830, 2021.

[36] S. Eftekhar Azam, M. Masanes Didyk, D. Linzell, and A. Rageh, "Experimental validation and numerical investigation of virtual strain sensing methods for steel railway bridges," *Journal of Sound and Vibration*, vol. 537, Article ID 117207, 2022.

[37] D. Wei, D. Li, and J. Huang, "Improved force identification with augmented Kalman filter based on the sparse constraint," *Mechanical Systems and Signal Processing*, vol. 167, Article ID 108561, 2022.

[38] J. N. Yang, S. Pan, and H. Huang, "An adaptive extended Kalman filter for structural damage identifications II: unknown inputs," *Structural Control and Health Monitoring*, vol. 14, no. 3, pp. 497–521, 2007.

[39] Y. Lei and N. Yang, "Simultaneous identification of structural time-varying physical parameters and unknown excitations using partial measurements," *Engineering Structures*, vol. 214, Article ID 110672, 2020.

[40] N. Yang, Y. Lei, J. Li, and H. Hao, "Identification of time-varying nonlinear structural physical parameters by integrated WMA and UKF/UKF-UI," *Nonlinear Dynamics*, vol. 106, no. 1, pp. 681–706, 2021.

[41] Y. Lei, J. Lu, and J. Huang, "Synthesize identification and control for smart structures with time-varying parameters under unknown earthquake excitation," *Structural Control and Health Monitoring*, vol. 27, no. 4, Article ID e2512, 2020.

[42] N. Yang, Y. Lei, J. Li, H. Hao, and J. S. Huang, "A substructural and wavelet multiresolution approach for identifying time-varying physical parameters by partial measurements," *Journal of Sound and Vibration*, vol. 523, Article ID 116737, 2022.

[43] X. X. Zhang, J. He, X. G. Hua, Z. Q. Chen, and O. Yang, "Online identification of time-variant structural parameters under unknown inputs basing on extended Kalman filter," *Nonlinear Dynamics*, vol. 109, no. 2, pp. 963–974, 2022.

[44] K. Huang, K. V. Yuen, and L. Wang, "Real-time simultaneous input-state-parameter estimation with modulated colored noise excitation," *Mechanical Systems and Signal Processing*, vol. 165, Article ID 108378, 2022.
[45] J. He, X. X. Zhang, and B. Xu, "Identification of structural parameters and unknown inputs based on revised observation equation: approach and validation," *International Journal of Structural Stability and Dynamics*, vol. 19, no. 12, Article ID 1950156, 2019.
[46] B. Chen and Y. L. Xu, "A new damage index for detecting sudden change of structural stiffness," *Structural Engineering & Mechanics*, vol. 26, no. 3, pp. 315–341, 2007.
[47] S. S. Bisht and M. P. Singh, "Detecting sudden changes in stiffness using high-pass filters," *Structural Control and Health Monitoring*, vol. 19, no. 3, pp. 319–331, 2012.
[48] K. V. Yuen, P. F. Liang, and S. C. Kuok, "Online estimation of noise parameters for kalman filter," *Structural Engineering & Mechanics*, vol. 47, no. 3, pp. 361–381, 2013.
[49] K. V. Yuen and S. C. Kuok, "Online updating and uncertainty quantification using nonstationary output-only measurement," *Mechanical Systems and Signal Processing*, vol. 66-67, pp. 62–77, 2016.
[50] V. K. Dertimanis, E. N. Chatzi, S. Eftekhar Azam, and C. Papadimitriou, "Input-state-parameter estimation of structural systems from limited output information," *Mechanical Systems and Signal Processing*, vol. 126, pp. 711–746, 2019.
[51] H. Huang, J. N. Yang, and L. Zhou, "Adaptive quadratic sum-squares error with unknown inputs for damage identification of structures," *Structural Control and Health Monitoring*, vol. 17, no. 4, pp. 404–426, 2009.
[52] Y. Zhao, B. Xu, B. C. Deng, S. J. Dyke, J. He, and H. B. Ge, "Various damper forces and dynamic excitation nonparametric identification with a double Chebyshev polynomial using limited fused measurements," *Measurement*, vol. 193, Article ID 110940, 2022.

Smart Structures and Systems, *Vol. 31, No. 2 (2023) 131-140*
https://doi.org/10.12989/sss.2023.31.2.131

An improved extended Kalman filter for parameters and loads identification without collocated measurements

Jia He*, Mengchen Qi, Zhuohui Tong, Xugang Hua and Zhengqing Chen

Key Laboratory of Building Safety and Energy Efficiency of the Ministry of Education,
Key Laboratory of Wind and Bridge Engineering of Hunan Province, Hunan University, Changsha, China

(Received November 25, 2021, Revised August 26, 2022, Accepted October 15, 2022)

Abstract. As well-known, the extended Kalman filter (EKF) is a powerful tool for parameter identification with limited measurements. However, traditional EKF is not applicable when the external excitation is unknown. By using least-squares estimation (LSE) for force identification, an EKF with unknown input (EKF-UI) approach was recently proposed by the authors. In this approach, to ensure the influence matrix be of full column rank, the sensors have to be deployed at all the degrees-of-freedom (DOFs) corresponding to the unknown excitation, saying collocated measurements are required. However, it is not easy to guarantee that the sensors can be installed at all these locations. To circumvent this limitation, based on the idea of first-order-holder discretization (FOHD), an improved EKF with unknown input (IEKF-UI) approach is proposed in this study for the simultaneous identification of structural parameters and unknown excitation. By using projection matrix, an improved observation equation is obtained. Few displacement measurements are fused into the observation equation to avoid the so-called low-frequency drift. To avoid the ill-conditioning problem for force identification without collocated measurements, the idea of FOHD is employed. The recursive solution of the structural states and unknown loads is then analytically derived. The effectiveness of the proposed approach is validated via several numerical examples. Results show that the proposed approach is capable of satisfactorily identifying the parameters of linear and nonlinear structures and the unknown excitation applied to them.

Keywords: data fusion; first-order-holder discretization; improved EKF; parameters identification; unknown input

1. Introduction

Parameter identification plays an important role in the field of structural health monitoring (SHM). The identified parameters can be used for damage detection, condition assessment and remaining service-life prediction. To date, there are a number of methods performed in frequency domain, time domain or time frequency domain for parameter identification (Xu and He 2017, Das and Saha 2018, Hannan *et al.* 2018). The Kalman filter (KF) provides an effective way for the estimation of structural states (Xu *et al.* 2016, Lai *et al.* 2016, Zhang *et al.* 2021, Saleem and Jo 2021). However, the structural parameters in KF algorithms should be assumed to be known. The extended Kalman filter (EKF) methods treat the parameters to be identified as a part of structural states and directly identify them in a recursive manner. The structural damage including its existence, location and severity can be then estimated on the basis of these identified quantities.

The research on the development of EKF technique for the purpose of SHM has been conducted for many years. In the classic EKF methods, the external excitation should be assumed to be known. The earlier contribution to the identification of structural parameters using EKF can be traced back to the work by Yun and Shinozuka (1980). A weighted global iteration EKF (WGI-EKF) procedure was then presented by Hoshiya and Saito (1984) to ensure the convergence and stability of the EKF algorithm. In the last century, computational efficiency may be the greatest obstacle for the application of EKF technique due to the complexity of calculating Jacobian matrix. In the past two decades, as the rapid advances of computer technologies, EKF algorithm has been actively investigated and widely used for parameter identification. For example, an improved EKF approach was proposed by Lei *et al.* (2014) for impact-induced damage detection. Mu and Yuen (2015) proposed an outlier-resistant EKF approach for the online identification of structural parameters. With the combination of EKF and UKF, Liu *et al.* (2016) proposed a two-stage approach to identify the nonlinear structural parameters. A discontinuous EKF method was presented by Chatzis *et al.* (2017) for state estimation and parameters identification. Based on EKF and $l1$-norm regularization, Zhang *et al.* (2017) proposed an algorithm for detecting local damage of complex structures. More recently, Lei *et al.* (2018) proposed an EKF-based approach for damage detection under ambient excitation. Ginsberg *et al.* (2018) proposed a sparsity-constrained EKF approach for damage localization and identification. An adaptive EKF with two computational modes was proposed by Yang *et al.* (2020) for system identification under seismic inputs. Based on WGI-EKF procedure, Xu *et al.* (2020) proposed a

*Corresponding author, Associate Professor,
E-mail: jiahe@hnu.edu.cn

nonparametric approach for the identification of nonlinear hysteretic behavior. Li and Wang (2020) proposed a constraint EKF approach for parameter identification of a differentiable Bouc-Wen model. Based on unscented KF, Naranjo-Perez *et al.* (2020) proposed a multi-objective harmony search algorithm for parameter identification of the dynamic Winkler soil-structure interaction model. Xiao *et al.* (2021) proposed an adaptive EKF approach for identifying the frequencies and track irregularities of high-speed railway bridges.

In the EKF-based approaches mentioned above, the external loads are required for the estimation. Without the prior knowledge of input information, these methods are unable to track the evolution of the estimates. However, some loads may not be measured in practice due to inaccessibility of sensor locations or other practical limitations. Therefore, much effort has been made on the development of EKF under unknown input (EKF-UI) methods. For example, Yang *et al.* (2007) proposed an adaptive EKF-UI approach for the identification of time-varying parameters. By using least-squares estimation (LSE) and EKF in sequence, Lei *et al.* (2012) proposed an EKF-UI approach for damage detection. Based on WGI-EKF and an adaptive iteration procedure, Xu and He (2015) proposed an approach for the identification of substructural parameters and unknown loads. Pan *et al.* (2016) proposed a general EKF-UI approach to reduce the restrictions of the sensor location. Liu *et al.* (2018) proposed a modal EKF-UI approach for identifying large-scale structures in real time. With the combination of augmented Kalman filter and genetic algorithm, Saleem and Jo (2019, 2021) proposed an approach for impact force identification and sensor placement. Based on LSE and EKF, He *et al.* (2020) proposed a two-stage approach for parameters identification with unknown inputs. Lei *et al.* (2021) proposed an EKF-UI approach for the identification of nonlinear restoring force in a model-free manner.

The authors have also been conducted the research on EKF-UI methods in the recent year. By using projection matrix to derive a modified observation equation, an EKF-UI approach was proposed for the simultaneous identification of structural parameters and unknown loadings (He *et al.* 2019a). It was then extended for substructural identification (He *et al.* 2021). This approach is novel and not presented in the previous literature. However, in order to ensure the influence matrix be of full column rank and to obtain the modified observation equation, sensors should be placed at all of the DOFs where the unknown inputs are applied, i.e., collocated acceleration measurements at the locations of unknown inputs are required. This limitation is also mentioned in some other literatures (Lei *et al.* 2012, Xu and He 2015, Pan *et al.* 2016, Liu *et al.* 2018). As known, it is not always possible to deploy the sensors at all these desired locations. Therefore, to circumvent this limitation, based on the idea of first-order-holder discretization (FOHD), an improved EKF with unknown input (IEKF-UI) approach is proposed in this study. Data fusion technique is employed to avoid the so-called low-frequency drift, and an improved version of observation equation is obtained. The recursive solution of the proposed approach is analytically derived. Several numerical examples including linear and nonlinear structures are used for the validation of its effectiveness.

2. Formulas of the proposed IEKF-UI approach

In general, the equation of motion of an n-DOFs structure with unknown loads can be written as

$$\boldsymbol{M}\ddot{\boldsymbol{x}}(t) + \boldsymbol{F}[\boldsymbol{x}(t), \dot{\boldsymbol{x}}(t), \boldsymbol{\theta}] = \boldsymbol{\varphi}\boldsymbol{f}(t) + \boldsymbol{\varphi}_u\boldsymbol{f}_u(t) \tag{1}$$

in which $\mathbf{M}$ is mass matrix; $\mathbf{x}(t)$, $\dot{\boldsymbol{x}}(t)$ and $\ddot{\boldsymbol{x}}(t)$ are time series of displacement, velocity and acceleration, respectively; $\boldsymbol{F}[\boldsymbol{x}(t), \dot{\boldsymbol{x}}(t), \boldsymbol{\theta}]$ is restoring force vector; $\boldsymbol{\theta}$ represents the unknown parameters; $\mathbf{f}(t)$ and $\mathbf{f}_u(t)$ are the known and unknown inputs, respectively; $\boldsymbol{\varphi}$ and $\boldsymbol{\varphi}_u$ are the influence matrices associated with the known and unknown inputs, respectively.

In this study, the parameters to be identified are assumed to be time-invariant. Thus, with the definition of the extended state vector $\boldsymbol{Z}(t) = [\boldsymbol{x}(t)^T,\ \dot{\boldsymbol{x}}(t)^T,\ \boldsymbol{\theta}^T]^T$, the following state equation can be obtained

$$\begin{aligned}\dot{\boldsymbol{Z}}(t) &= \begin{bmatrix} \dot{\boldsymbol{x}}(t) \\ \boldsymbol{M}^{-1}[-\boldsymbol{F}[\boldsymbol{x}(t), \dot{\boldsymbol{x}}(t), \boldsymbol{\theta}] + \boldsymbol{\varphi}\boldsymbol{f}(t) + \boldsymbol{\varphi}_u\boldsymbol{f}_u(t)] \\ \mathbf{0} \end{bmatrix}^T \\ &= \boldsymbol{g}(\boldsymbol{Z}(t), \boldsymbol{f}(t), \boldsymbol{f}_u(t), t) + \boldsymbol{w}(t)\end{aligned} \tag{2}$$

where $\mathbf{w}(t)$ is the process noise vector with zero mean and a covariance matrix $\mathbf{Q}(t)$.

The state equation mentioned above can be linearized as

$$\begin{aligned}&\boldsymbol{g}(\boldsymbol{Z}_k, \boldsymbol{f}_k, \boldsymbol{f}_{u,k}, k\Delta t) \\ &\approx \boldsymbol{g}(\hat{\boldsymbol{Z}}_{k|k}, \boldsymbol{f}_k, \hat{\boldsymbol{f}}_{u,k}, k\Delta t) + \boldsymbol{U}_{k|k}(\boldsymbol{Z}_k - \hat{\boldsymbol{Z}}_{k|k}) \\ &\quad + \boldsymbol{W}_{k|k}(\boldsymbol{f}_{u,k} - \hat{\boldsymbol{f}}_{u,k})\end{aligned} \tag{3}$$

where $\hat{\boldsymbol{Z}}_{k|k}$ and $\hat{\boldsymbol{f}}_{u,k}$ are the estimates of $\mathbf{Z}_k$ and $\mathbf{f}_{u,k}$ at the k-th time step, respectively; Δt is the time interval; $\mathbf{U}_{k|k}$ and $\mathbf{W}_{k|k}$ can be calculated as follows

$$\begin{aligned}\boldsymbol{U}_{k|k} &= \frac{\partial \boldsymbol{g}(\boldsymbol{Z}_k, \boldsymbol{f}_k, \boldsymbol{f}_{u,k}, k\Delta t)}{\partial \boldsymbol{Z}_k}\bigg|_{\substack{\boldsymbol{Z}_k = \hat{\boldsymbol{Z}}_{k|k} \\ \boldsymbol{f}_{u,k} = \hat{\boldsymbol{f}}_{u,k}}}, \\ \boldsymbol{W}_{k|k} &= \frac{\partial \boldsymbol{g}(\boldsymbol{Z}_k, \boldsymbol{f}_k, \boldsymbol{f}_{u,k}, k\Delta t)}{\partial \boldsymbol{f}_{u,k}}\bigg|_{\substack{\boldsymbol{Z}_k = \hat{\boldsymbol{Z}}_{k|k} \\ \boldsymbol{f}_{u,k} = \hat{\boldsymbol{f}}_{u,k}}} = \begin{bmatrix} \mathbf{0} \\ \boldsymbol{M}^{-1}\boldsymbol{\varphi}_u \\ \mathbf{0} \end{bmatrix}\end{aligned} \tag{4}$$

Here, partial displacement and acceleration responses are fused together and used for the identification. The discretized observation equation can be then expressed as

$$\begin{aligned}\bar{\boldsymbol{y}}_k &= \begin{bmatrix} \boldsymbol{y}_{a,k} \\ \boldsymbol{y}_{s,k} \end{bmatrix} = \begin{bmatrix} \boldsymbol{L}_a\ddot{\boldsymbol{x}}_k + \boldsymbol{v}_{a,k} \\ \boldsymbol{L}_s\boldsymbol{x}_k + \boldsymbol{v}_{s,k} \end{bmatrix} \\ &= \begin{bmatrix} \boldsymbol{L}_a(\boldsymbol{h}(\boldsymbol{Z}_k) + \boldsymbol{M}^{-1}\boldsymbol{\varphi}_u\boldsymbol{f}_{u,k}) + \boldsymbol{v}_{a,k} \\ \boldsymbol{L}_s\boldsymbol{x}_k + \boldsymbol{v}_{s,k} \end{bmatrix}\end{aligned} \tag{5}$$

where $\boldsymbol{h}(\boldsymbol{Z}_k) = \boldsymbol{M}^{-1}[-\boldsymbol{F}[\boldsymbol{x}_k, \dot{\boldsymbol{x}}_k, \boldsymbol{\theta}] + \boldsymbol{\varphi}\boldsymbol{f}_k]$; $\mathbf{L}_a$ and $\mathbf{L}_s$ are the matrices associated with the locations of accelerometers and displacement transducers, respectively; $\mathbf{y}_{a,k}$ and $\mathbf{y}_{s,k}$ are the acceleration and displacement measurements at the k-th time step, respectively; $\mathbf{v}_{a,k}$ and $\mathbf{v}_{s,k}$ are measurement noise assumed to be Gaussian white-noise vector with zero mean and covariance matrices $\mathbf{R}_{a,k}$ and $\mathbf{R}_{s,k}$, respectively.

Focusing on the first part of the observation equation where the unknown inputs are explicitly included, the following equation regarding acceleration measurements can be obtained

$$\boldsymbol{y}_{a,k} = \boldsymbol{L}_a\left(\boldsymbol{h}(\boldsymbol{Z}_k) + \boldsymbol{M}^{-1}\boldsymbol{\varphi}_u\boldsymbol{f}_{u,k}\right) + \boldsymbol{v}_{a,k} \quad (6)$$

In this study, to investigate a general case, the unknown inputs $\mathbf{f}_{u,k}$ are considered to be composed of two parts, namely $\boldsymbol{f}^*_{u,k}$ and $\boldsymbol{f}'_{u,k}$. The former represents the unknown input vector associated with the collocated acceleration responses being measured, whereas the latter represents the unknown input vector without the collocated acceleration measurements. Then, Eq. (6) can be expressed as

$$\boldsymbol{y}_{a,k} = \boldsymbol{L}_a\boldsymbol{h}(\boldsymbol{Z}_k) - [\boldsymbol{D}^* \quad \boldsymbol{D}']\begin{bmatrix}\boldsymbol{f}^*_{u,k}\\ \boldsymbol{f}'_{u,k}\end{bmatrix} + \boldsymbol{v}_{a,k} \quad (7)$$

where $\boldsymbol{D}^* = -\boldsymbol{L}_a\boldsymbol{M}^{-1}\boldsymbol{\varphi}^*_u$; $\boldsymbol{D}' = -\boldsymbol{L}_a\boldsymbol{M}^{-1}\boldsymbol{\varphi}'_u$; $\boldsymbol{\varphi}^*_u$ and $\boldsymbol{\varphi}'_u$ are the influence matrices associated with unknown input vector $\boldsymbol{f}^*_{u,k}$ and $\boldsymbol{f}'_{u,k}$, respectively. Since the matrix $\mathbf{L}_a$ is not related to the influence matrix $\boldsymbol{\varphi}'_u$, it can be found that $\boldsymbol{D}'$ matrix is always equal to zero. Then, Eq. (7) can be rearranged as follows

$$\boldsymbol{y}_{a,k} = \boldsymbol{L}_a\boldsymbol{h}(\boldsymbol{Z}_k) - \boldsymbol{D}^*\boldsymbol{f}^*_{u,k} + \boldsymbol{v}_{a,k} \quad (8)$$

Obviously, only the unknown input vector $\boldsymbol{f}^*_{u,k}$ is explicitly presented in the observation equation. For ease of expression, $\boldsymbol{f}^*_{u,k}$ is named as explicit unknown input at the k-th time step whereas $\boldsymbol{f}'_{u,k}$ is called as implicit unknown input hereafter.

Notably, the matrix $\mathbf{D}^*$ is always of full column rank. Moreover, the measurement noise is assumed to be much smaller than the measured signal. Thus, the explicit unknown input $\boldsymbol{f}^*_{u,k}$ can be estimated by LSE as follows

$$\hat{\boldsymbol{f}}^*_{u,k} = \left(\boldsymbol{D}^{*T}\boldsymbol{D}^*\right)^{-1}\boldsymbol{D}^{*T}\left[\boldsymbol{L}_a\boldsymbol{h}(\boldsymbol{Z}_k) - \boldsymbol{y}_{a,k}\right] \quad (9)$$

Then, according to the similar procedure conducted in the previous literature (He *et al.* 2019), Eq. (8) can be re-written as

$$\boldsymbol{\Phi}\boldsymbol{y}_{a,k} = \boldsymbol{\Phi}\boldsymbol{L}_a\boldsymbol{h}(\boldsymbol{Z}_k) + \boldsymbol{\Phi}\boldsymbol{v}_{a,k} \quad (10)$$

where $\boldsymbol{\Phi} = \boldsymbol{I} - \boldsymbol{D}^*\left(\boldsymbol{D}^{*T}\boldsymbol{D}^*\right)^{-1}\boldsymbol{D}^{*T}$. Here, $\boldsymbol{D}^*\left(\boldsymbol{D}^{*T}\boldsymbol{D}^*\right)^{-1}\boldsymbol{D}^{*T}$ is known as projection matrix.

On the basis of Eq. (5) and Eq. (10), an improved observation equation can be derived as

$$\bar{\boldsymbol{\Phi}}\bar{\boldsymbol{y}}_k = \bar{\boldsymbol{\Phi}}\bar{\boldsymbol{h}}(\boldsymbol{Z}_k) + \bar{\boldsymbol{\Phi}}\bar{\boldsymbol{v}}_k \quad (11)$$

where $\bar{\boldsymbol{\Phi}} = \begin{bmatrix}\boldsymbol{\Phi} & \boldsymbol{0}\\ \boldsymbol{0} & \boldsymbol{I}\end{bmatrix}$; $\bar{\boldsymbol{h}}(\boldsymbol{Z}_k) = \begin{bmatrix}\boldsymbol{L}_a\boldsymbol{h}(\boldsymbol{Z}_k)\\ \boldsymbol{L}_s\boldsymbol{x}_k\end{bmatrix}$; $\bar{\boldsymbol{v}}_k = \begin{bmatrix}\boldsymbol{v}_{a,k}\\ \boldsymbol{v}_{s,k}\end{bmatrix}$.

According to the principle of EKF, the priori state estimate $\hat{\boldsymbol{Z}}_{k+1|k}$ can be obtained as

$$\hat{\boldsymbol{Z}}_{k+1|k} = \hat{\boldsymbol{Z}}_{k|k} + \int_{k\Delta t}^{(k+1)\Delta t} \boldsymbol{g}(\hat{\boldsymbol{Z}}_{k|k}, \boldsymbol{f}_k, \hat{\boldsymbol{f}}_{u,k}, k\Delta t)dt \quad (12)$$

The priori estimate error covariance matrix $\mathbf{P}_{k+1|k}$ can be calculated as

$$\boldsymbol{P}_{k+1|k} = \boldsymbol{A}_1\boldsymbol{P}_{k|k}\boldsymbol{A}_1 + \boldsymbol{A}_2\boldsymbol{R}_{a,k}\boldsymbol{A}_2 + \Delta t^2\boldsymbol{Q}_k \quad (13)$$

where $\boldsymbol{A}_1 = \boldsymbol{I} + \Delta t\left(\boldsymbol{U}_{k|k} + \begin{bmatrix}0\\ -\boldsymbol{H}_{k|k}\\ 0\end{bmatrix}\right)$;

$\boldsymbol{A}_2 = \Delta t\begin{bmatrix}0\\ -(\boldsymbol{L}_a^T\boldsymbol{L}_a)^{-1}\boldsymbol{L}_a^T\\ 0\end{bmatrix}$; $\boldsymbol{H}_{k|k} = \frac{\partial \boldsymbol{h}(\boldsymbol{Z}_k)}{\partial \boldsymbol{Z}_k}\big|\boldsymbol{Z}_k = \hat{\boldsymbol{Z}}_{k|k}$.

Then, the posteriori state estimate $\hat{\boldsymbol{Z}}_{k+1|k+1}$ is determined as

$$\hat{\boldsymbol{Z}}_{k+1|k+1} = \hat{\boldsymbol{Z}}_{k+1|k} + \boldsymbol{G}_{k+1}\left[\bar{\boldsymbol{\Phi}}\bar{\boldsymbol{y}}_{k+1} - \bar{\boldsymbol{\Phi}}\bar{\boldsymbol{h}}(\hat{\boldsymbol{Z}}_{k+1|k})\right] \quad (14)$$

where $\mathbf{G}_{k+1}$ represents gain matrix and can be computed as

$$\boldsymbol{G}_{k+1} = \boldsymbol{P}_{k+1|k}\bar{\boldsymbol{H}}^T_{k+1}\bar{\boldsymbol{\Phi}}^T\left[\bar{\boldsymbol{\Phi}}\left(\bar{\boldsymbol{H}}_{k+1|k}\boldsymbol{P}_{k+1|k}\bar{\boldsymbol{H}}^T_{k+1} + \bar{\boldsymbol{R}}_{k+1}\right)\bar{\boldsymbol{\Phi}}^T\right]^{-1} \quad (15)$$

in which $\bar{\boldsymbol{R}}_k = E(\bar{\boldsymbol{v}}_k\bar{\boldsymbol{v}}_k^T)$; $\bar{\boldsymbol{H}}_{k+1|k} = \frac{\partial \bar{\boldsymbol{h}}(\boldsymbol{Z}_{k+1})}{\partial \boldsymbol{Z}_{k+1}}\big|\boldsymbol{Z}_{k+1}\hat{\boldsymbol{Z}}_{k+1\,|k}$

The posteriori estimate error covariance matrix $\mathbf{P}_{k+1|k+1}$ is calculated as

$$\boldsymbol{P}_{k+1|k+1} = \left(\boldsymbol{I} - \boldsymbol{G}_{k+1}\bar{\boldsymbol{\Phi}}\bar{\boldsymbol{H}}_{k+1|k}\right)\boldsymbol{P}_{k+1|k}\left(\boldsymbol{I} - \boldsymbol{G}_{k+1}\bar{\boldsymbol{\Phi}}\bar{\boldsymbol{H}}_{k+1|k}\right)^T + \boldsymbol{G}_{k+1}\bar{\boldsymbol{\Phi}}\bar{\boldsymbol{R}}_{k+1}(\boldsymbol{G}_{k+1}\bar{\boldsymbol{\Phi}})^T \quad (16)$$

The recursive solution of the state estimation seems to be given by Eqs. (12)-(16). However, it should be noted that the unknown input in Eq. (6) is not completely determined. Only part of this input, saying the explicit unknown input $\hat{\boldsymbol{f}}^*_{u,k+1}$, can be obtained according to Eq. (9) and shown as follows.

$$\hat{\boldsymbol{f}}^*_{u,k+1} = \left(\boldsymbol{D}^{*T}\boldsymbol{D}^*\right)^{-1}\boldsymbol{D}^{*T}\left[\boldsymbol{L}_a\boldsymbol{h}(\hat{\boldsymbol{Z}}_{k+1|k+1}) - \boldsymbol{y}_{a,k+1}\right] \quad (17)$$

The remaining part, i.e., the implicit unknown input $\boldsymbol{f}'_{u,k}$ is still unknown, which means the procedure described by Eqs. (12)-(16) can't be directly implemented. In order to identify this implicit unknown input, the idea of first-order-holder discretization (FOHD) (Ding *et al.* 2013, Lei *et al.* 2020) is employed.

In the conventional EKF-based approach, the zero-order-holder discretization (ZOHD) method is used to discretize the continuous function. The unknown force in a sampling period is assumed to be constant by using ZOHD method, and thus Eq. (12) can be expressed as

$$\hat{\boldsymbol{Z}}_{k+1|k} = \hat{\boldsymbol{Z}}_{k|k} + \int_{k\Delta t}^{(k+1)\Delta t} \boldsymbol{g}(\hat{\boldsymbol{Z}}_{k|k}, \boldsymbol{f}_k, k\Delta t)dt + \begin{bmatrix}\boldsymbol{0}\\ \boldsymbol{M}^{-1}\boldsymbol{\varphi}'_u\Delta t\\ \boldsymbol{0}\end{bmatrix}\hat{\boldsymbol{f}}'_{u,k} + \begin{bmatrix}\boldsymbol{0}\\ \boldsymbol{M}^{-1}\boldsymbol{\varphi}^*_u\Delta t\\ \boldsymbol{0}\end{bmatrix}\hat{\boldsymbol{f}}^*_{u,k} \quad (18)$$

Different from ZOHD method, FOHD method treats the unknown input as a linear variable in a sampling period. For ease of distinction, the priori state estimate is called as $\bar{\boldsymbol{Z}}_{k+1|k}$ in this case, and Eq. (12) can be arranged as

$$\begin{aligned}\bar{\boldsymbol{Z}}_{k+1|k} = \hat{\boldsymbol{Z}}_{k|k} &+ \int_{k\Delta t}^{(k+1)\Delta t} \boldsymbol{g}(\hat{\boldsymbol{Z}}_{k|k}, \boldsymbol{f}_k, k\Delta t)dt \\ &+\frac{1}{2}\Delta t\left(\begin{bmatrix}\boldsymbol{0}\\ \boldsymbol{M}^{-1}\boldsymbol{\varphi}_u'\\ \boldsymbol{0}\end{bmatrix}(\hat{\boldsymbol{f}}'_{u,k}+\hat{\boldsymbol{f}}'_{u,k+1}) \right.\\ &\left. + \begin{bmatrix}\boldsymbol{0}\\ \boldsymbol{M}^{-1}\boldsymbol{\varphi}_u^*\\ \boldsymbol{0}\end{bmatrix}(\hat{\boldsymbol{f}}^*_{u,k}+\hat{\boldsymbol{f}}^*_{u,k+1})\right)\end{aligned} \tag{19}$$

By using Eqs. (18)-(19), the following relationship exists

$$\begin{aligned}&\bar{\boldsymbol{Z}}_{k+1|k} - \hat{\boldsymbol{Z}}_{k+1|k} \\ &= \frac{1}{2}\Delta t\left[\boldsymbol{B}'(\hat{\boldsymbol{f}}'_{u,k+1} - \hat{\boldsymbol{f}}'_{u,k}) + \boldsymbol{B}^*(\hat{\boldsymbol{f}}^*_{u,k+1} - \hat{\boldsymbol{f}}^*_{u,k})\right]\end{aligned} \tag{20}$$

where $\boldsymbol{B}' = \begin{bmatrix}\boldsymbol{0}\\ \boldsymbol{M}^{-1}\boldsymbol{\varphi}_u'\\ \boldsymbol{0}\end{bmatrix}; \boldsymbol{B}^* = \begin{bmatrix}\boldsymbol{0}\\ \boldsymbol{M}^{-1}\boldsymbol{\varphi}_u^*\\ \boldsymbol{0}\end{bmatrix}$.

Based on the Taylor series, Eq. (8) can be linearized as follows

$$\begin{aligned}&\boldsymbol{y}_{a,k+1} - \boldsymbol{L}_a\boldsymbol{h}(\boldsymbol{Z}_{k+1}) + \boldsymbol{D}^*\boldsymbol{f}^*_{u,k+1} = \boldsymbol{y}_{a,k+1} \\ &-\left[\boldsymbol{L}_a\boldsymbol{h}(\hat{\boldsymbol{Z}}_{k+1|k}) + \boldsymbol{L}_a\boldsymbol{H}_{k+1|k}(\bar{\boldsymbol{Z}}_{k+1|k} - \hat{\boldsymbol{Z}}_{k+1|k})\right] \\ &+\boldsymbol{D}^*\boldsymbol{f}^*_{u,k+1} \approx 0\end{aligned} \tag{21}$$

Substituting Eq. (20) into Eq. (21) leads to

$$\begin{aligned}&\boldsymbol{y}_{a,k+1} - \boldsymbol{L}_a\boldsymbol{h}(\hat{\boldsymbol{Z}}_{k+1|k}) \\ &-\frac{1}{2}\boldsymbol{L}_a\boldsymbol{H}_{k+1|k}\Delta t\left[\boldsymbol{B}'(\hat{\boldsymbol{f}}'_{u,k+1} - \hat{\boldsymbol{f}}'_{u,k}) + \boldsymbol{B}^*(\hat{\boldsymbol{f}}^*_{u,k+1} - \hat{\boldsymbol{f}}^*_{u,k})\right] \\ &+\boldsymbol{D}^*\hat{\boldsymbol{f}}^*_{u,k+1} = 0\end{aligned} \tag{22}$$

Then, with proper transformation, the implicit unknown input can be obtained as

$$\begin{aligned}\hat{\boldsymbol{f}}'_{u,k+1} = (\boldsymbol{S}_{k+1}^T\boldsymbol{S}_{k+1})^{-1}\boldsymbol{S}_{k+1}^T&\left[\boldsymbol{y}_{a,k+1} - \boldsymbol{L}_a\boldsymbol{h}(\hat{\boldsymbol{Z}}_{k+1|k})\right. \\ &-\frac{1}{2}\boldsymbol{L}_a\boldsymbol{H}_{k+1|k}\Delta t(\boldsymbol{B}^*(\hat{\boldsymbol{f}}^*_{u,k+1} - \hat{\boldsymbol{f}}^*_{u,k}) \\ &\left.-\boldsymbol{B}'\hat{\boldsymbol{f}}'_{u,k}) + \boldsymbol{D}^*\hat{\boldsymbol{f}}^*_{u,k+1}\right]\end{aligned} \tag{23}$$

where $\boldsymbol{S}_{k+1} = \frac{1}{2}\boldsymbol{L}_a\boldsymbol{H}_{k+1|k}\boldsymbol{B}'\Delta t$. It can be found that the matrix $\mathbf{S}_{k+1}$ is always of full column rank.

In summary, the equations for the proposed IEKF-UI approach fall into three groups: time update equations, measurement update equations, and force identification equations. The time update equations including Eqs. (12)-(13) are responsible for the determination of a priori estimate using the information of previous time step. The measurement update equations including Eqs. (14)-(16) are responsible for incorporating a new measurement into the priori estimate to obtain an improved posteriori estimate. These two update equations are similar to those in the conventional EKF algorithm. Moreover, the unknown inputs can be identified by Eq. (17) and Eq. (23) at the same time. It can be seen that the collocated acceleration measurements are not necessarily required for the identification indicating the restriction on the sensor locations is removed. For ease of understanding, the flowchart of the proposed IEKF-UI approach is plotted in Fig. 1.

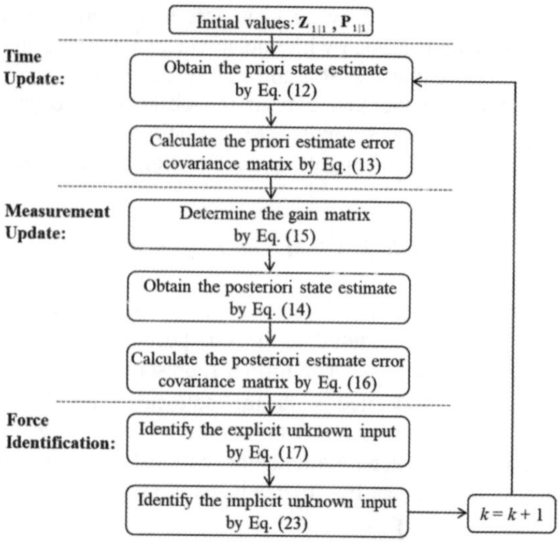

Fig. 1 The flowchart of the proposed IEKF-UI approach

3. Numerical validation

In order to validate the effectiveness of the proposed approach, several numerical examples including a shear building structure, a planar truss structure and a building structure with a nonlinear hysteretic model are considered in this section.

3.1 Eight-story shear building structure

A numerical study of an 8-story shear building structure subject to two unknown inputs is first conducted herein. The parameters of mass and stiffness are given as m_i = 180 kg and k_i = 550 kN/m (i = 1, ..., 8). The Rayleigh damping model (i.e., $\mathbf{C} = \alpha \cdot \mathbf{M} + \beta \cdot \mathbf{K}$) is employed, and the proportional coefficients of α and β are assumed to be 0.457 and 1.5×10^{-3}, respectively. The first two damping ratios are 3%, and the first three natural frequencies are 1.62 Hz, 4.82 Hz and 7.84 Hz, respectively. The random external forces are applied on 2nd floor and 8th floor of the building. The corresponding structural responses are obtained by state-space method with the time interval of 0.001 s. After the calculation of structural responses, these two external forces are considered as the unknown inputs to be identified. The system error covariance matrix $\mathbf{Q}$ in this study is set to be $10^{-6} \cdot \mathbf{I}$.

The acceleration responses of the 1st, 2nd, 3rd, 6th and 7th are assumed to be known for the identification. Since the acceleration response of the 2nd floor are available, the

Table 1 The identified structural parameters of shear building

Parameters (kN/m)	Actual	Identified	Error (%)	Parameters (kN/m)	Actual	Identified	Error (%)
k_1	550	546.05	-0.72	k_6	550	554.64	0.84
k_2	550	551.71	0.31	k_7	550	544.62	-0.98
k_3	550	544.71	-0.96	k_8	550	555.56	1.01
k_4	550	546.62	-0.61	α	0.4577	0.4650	1.59
k_5	550	543.44	-1.19	β	0.0015	0.0012	-

(a) The 1st floor (b) The 6th floor

Fig. 2 The identified stiffness of shear building structure

(a) On the 2nd floor (b) On the 8th floor

Fig. 3 The comparison of the identified excitation applied to shear structure

external force applied to this floor can be viewed as the explicit unknown input. Accordingly, the external force on the 8th floor can be treated as the implicit unknown input. Moreover, the displacement responses of the 1st and 6th floor are fused in the observation equation. The noise effect is considered by adding white noise with 5% in root mean square (RMS) to these measurements.

The extended state vector is defined as $\boldsymbol{Z} = [x_1,\dots, x_8, \dot{x}_1,\dots, \dot{x}_8, k_1,\dots, k_8, \alpha, \beta]^T$, where x_i, $\dot{x}_i$ and k_i (i = 1, …, 8) represent the displacement, velocity and stiffness of the i-th floor, respectively. The initial values of structural parameters are assumed to be 50% of the actual ones. By using the proposed approach, the structural parameters can be identified and shown in Table 1. It can be seen that the identified values are close to the actual ones. Notably, since the value of β is several orders of magnitude smaller than other variables, it is not easy to obtain the accurate identification result of such a small coefficient. Sever strategies may be helpful for improving the accuracy of the identified result, such as reducing the noise level, using more sensors, and changing the locations of sensors. Figs. 2(a)-(b) gives the convergence details of the identified parameters. It is obvious that the identified parameters are able to stably converge to the actual ones after few seconds. Only the stiffness coefficients of the 1st and 6th floor are plotted in Figs. 2(a)-(b) as examples. Similar results can be found for the remaining parameters.

In addition to the identification of structural parameters, the unknown inputs can be also estimated by the proposed approach. The comparison of the identified external forces with their real ones is plotted in Figs. 3(a)-(b). Obviously, the identified results have a good agreement with the actual ones. Moreover, a normalized root-mean-square-error (NRMSE) (He *et al.* 2019b) is used as a measure of deviation between the actual inputs and the identified ones.

$$NRMSE = \frac{1}{A}\sqrt{\frac{1}{np}\sum_{k=1}^{np}\left(\hat{f}_{u,k} - f_{u,k}\right)^2} \tag{24}$$

where A is the amplitude of force; np is the number of sampling points, $\hat{f}_{u,k}$ and $f_{u,k}$ denote the identified input and actual input at the k-th time step, respectively. In this

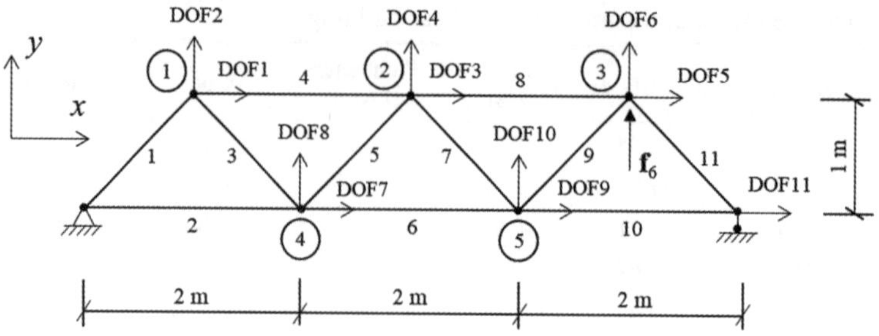

Fig. 4 The planar truss structure

example, the values of NRMSE of the unknown inputs applied on the 2nd and 8th floor are 1.02% and 1.21%, receptively, indicating that the identified forces are close to the actual ones.

3.2 Planar truss structure

A planar truss structure as shown in Fig. 4 is considered. In this example, the case of only implicit unknown input being involved is discussed. The truss structure consists of 11 members with the same cross-sectional area of 2×10^{-3} m^2. The Young's module is selected as 2×10^8 Pa, and mass density is 7850 kg/m^3. The consistent mass distribution is employed for the construction of the mass matrix. The Rayleigh damping assumption is used, and the proportional coefficients are assumed as $\alpha = 0.949$ and $\beta = 8 \times 10^{-4}$. The first two damping ratios are 3%, and the first three natural frequencies are 3.71 Hz, 7.83 Hz, 11.63 Hz, respectively.

A random force is applied at the DOF6 as shown in Fig. 4, and the corresponding structural responses are calculated

Table 2 The identified structural parameters of truss structure

Parameters (kN/m)	Actual	Identified	Error (%)	Parameters (kN/m)	Actual	Identified	Error (%)
k_1	282.84	283.12	0.10	k_8	200.00	199.28	-0.36
k_2	200.00	198.27	-0.86	k_9	282.84	282.74	-0.04
k_3	282.84	282.42	-0.15	k_{10}	200.00	199.80	-0.10
k_4	200.00	200.28	0.14	k_{11}	282.84	282.64	-0.07
k_5	282.84	282.596	-0.09	α	0.949	0.9224	-2.80
k_6	200.00	201.45	0.72	β	0.0008	0.0008	-
k_7	282.84	283.08	0.08				

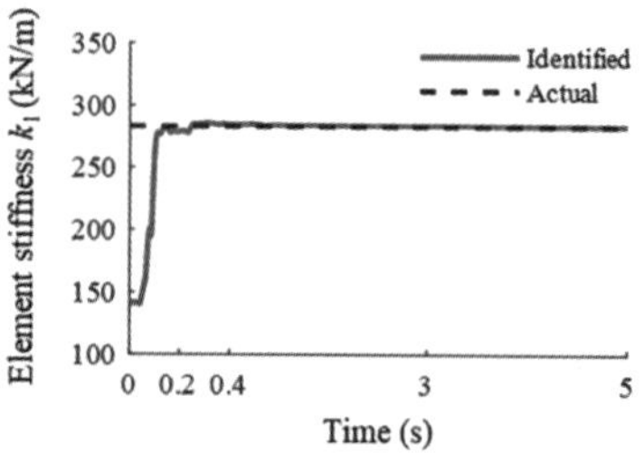

(a) k_1

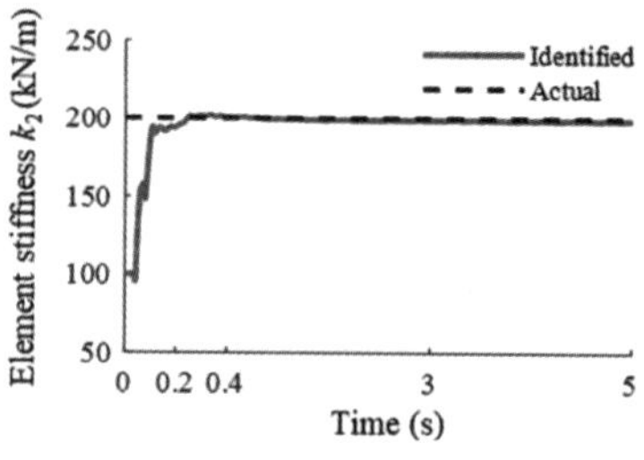

(b) k_2

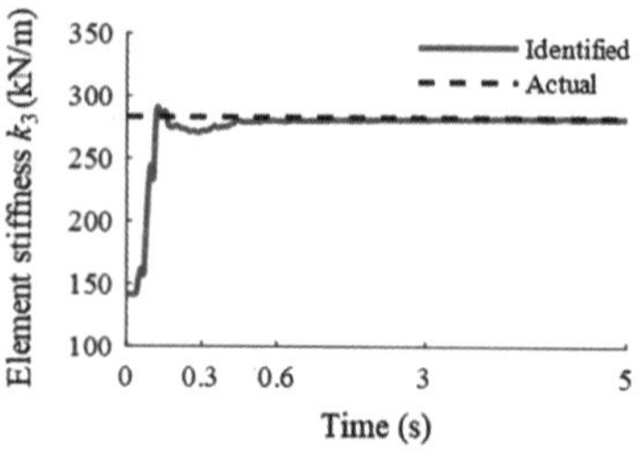

(c) k_3

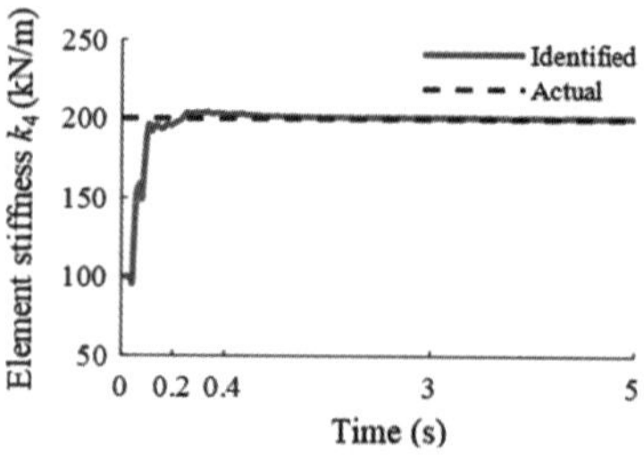

(d) k_4

Fig. 5 The identified stiffness of truss structure

by state-space method with the time interval of 0.001 s. The acceleration responses of the 1st, 2nd, 3rd, 4th, 8th, 9th, 10th and 11th DOF and the displacement responses of the 5th and 7th DOF are used for the identification. The noise level is also assumed to be 5%. The unknown parameters include the stiffness of the 11 elements and two Rayleigh damping coefficients. By using the proposed approach, the structural parameters are identified and listed in Table 2. A good agreement between the identified values and the actual ones can be observed. Fig. 5 gives the convergence procedure of some identified parameters. Due to space limitation, only the identified stiffness of the 1st element to 4th element is plotted as examples. It can be seen that the identified values can promptly and stably converge to the actual ones. Similar results can be observed for the remaining parameters.

The unknown external excitation can be also identified by the proposed approach. Notably, since the acceleration at the 6th DOF is not measured, the external force at node 3 (i.e., $\mathbf{f}_6$ as shown in Fig. 4) can be viewed as an implicit unknown input. It can be also found that there is no explicit unknown input in this example. Thus, the implicit unknown input can be determined by using Eq. (23) with the consideration of $\hat{\boldsymbol{f}}^*_{u,k+1}$ and $\hat{\boldsymbol{f}}^*_{u,k}$ being zero. The identified input is shown in Figs. 6(a)-(b) as solid line, whereas the real one is plotted as dashed line. The time segment from 4 s to 5 s is also given in Fig. 6 for clarity of comparison. The value of NRMSE defined in Eq. (23) is 0.83% in this example. It can be seen the identified input is quite close to the real one.

3.3 Shear building with a nonlinear hysteretic model

In this section, the identification of structural parameters in the nonlinear system is discussed. The same shear building introduced in section 3.1 is employed herein. The structural parameters are set as m_i = 180 kg, k_i = 550 kN/m (i = 1, …, 8), α = 0.457 and $\beta = 1.5 \times 10^{-3}$. A hysteretic Dahl model is assumed to be installed on the 5th floor to mimic local nonlinearity. The nonlinear restoring force provided by Dahl model in this case can be expressed as

$$\boldsymbol{F}_{Dahl} = k_D \Delta \boldsymbol{x}_5 + c_D \Delta \dot{\boldsymbol{x}}_5 + f_D \boldsymbol{z}_D + f_0 \tag{25}$$

where k_D and c_D are stiffness and damping coefficients of the Dahl model, respectively; f_D and f_0 are force coefficients; $\Delta\mathbf{x}_5$ and $\Delta\dot{\boldsymbol{x}}_5$ are the relative displacement and relative velocity of the 5th floor, respectively; $\mathbf{z}_D$ is a dimensionless quantity used to describe Coulomb friction as

$$\dot{\boldsymbol{z}}_D = \sigma_D \cdot \Delta \dot{\boldsymbol{x}}_5 \cdot [1 - \boldsymbol{z}_D \cdot \mathrm{sign}(\Delta \dot{\boldsymbol{x}}_5)] \tag{26}$$

where σ_D is the parameter used to control the shape of hysteresis loop; sign(·) denotes signum function. In this study, the parameters of Dahl model are set as k_D = 50 N/m, c_D = 400 N·s/m, f_D = 300 N, f_0 = 0N, and σ_D = 1000 s/m.

An external excitation is applied on the 3rd floor of the building, and the corresponding nonlinear responses are calculated by Runge-Kutta method with the time interval of 0.001 s. The acceleration responses of the 1st, 2nd, 4th, 5th, 6th and 8th floor as well as displacement response of the 2nd are used for the identification. All the measurements are assumed to be contaminated by 5% noise. The extended

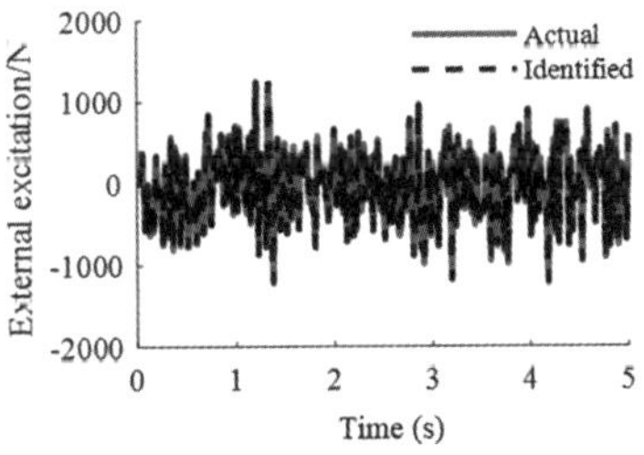

(a) The entire time series

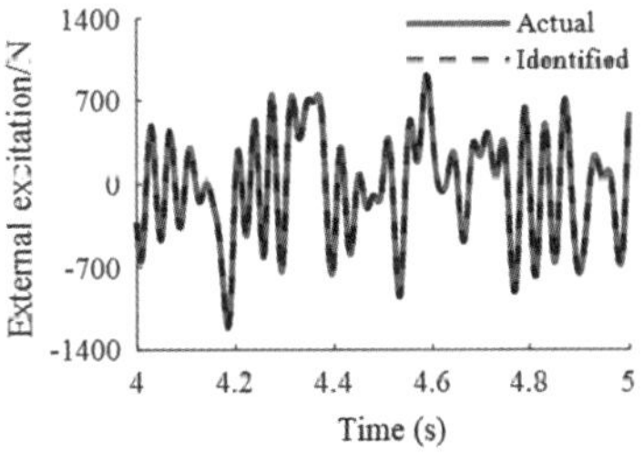

(b) Time segment from 4 s to 5 s

Fig. 6 The comparison of the identified force applied to truss structure

Table 3 The identified parameters of nonlinear system

Parameters (kN/m)	Actual	Identified	Error (%)	Parameters (kN/m)	Actual	Identified	Error (%)
k_1 (kN/m)	550	535.79	-2.58	k_8 (kN/m)	550	534.42	-2.83
k_2 (kN/m)	550	552.82	0.51	k_D (N/m)	50	49.54	-0.91
k_3 (kN/m)	550	541.83	-1.48	c_D (N·s/m)	400	396.47	-0.88
k_4 (kN/m)	550	544.50	-1.00	f_D (N)	300	294.26	-1.91
k_5 (kN/m)	550	531.00	-3.45	σ_D (s/m)	1000	989.80	-1.02
k_6 (kN/m)	550	541.24	-1.59	α	0.4577	0.4634	1.23
k_7 (kN/m)	550	557.24	1.31	β	0.0015	0.0015	-

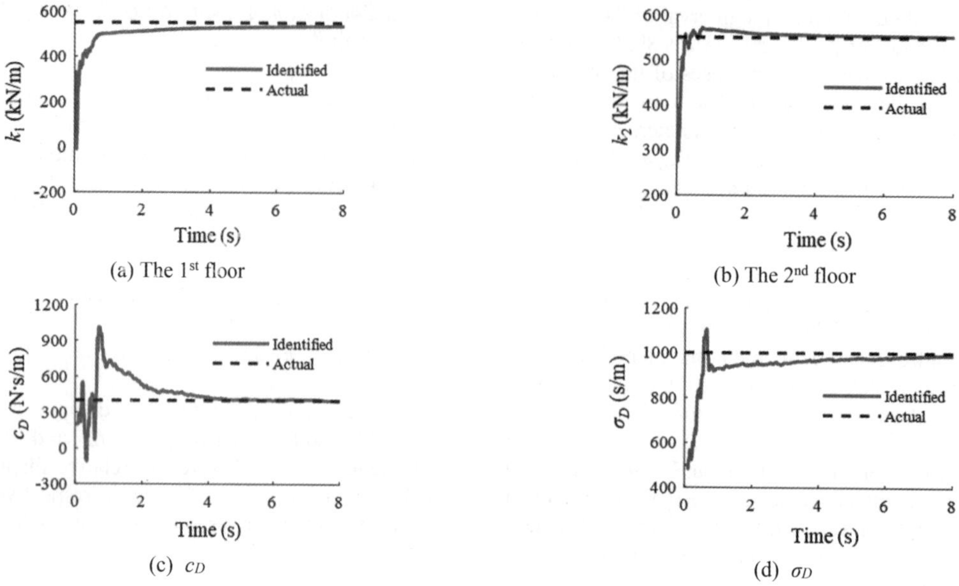

(a) The 1st floor

(b) The 2nd floor

(c) c_D

(d) σ_D

Fig. 7 The identified parameters of nonlinear system

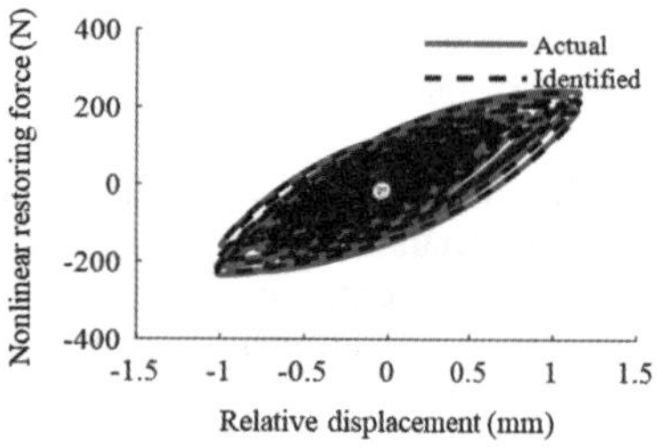

Fig. 8 The nonlinear restoring force provided by Dahl model

state vector is defined a $\boldsymbol{Z}=[x_1,\ldots,x_8,\ \dot{x}_1,\ldots,\dot{x}_8,\ k_1,\ldots,k_8,\ k_D,\ c_D,\ f_D,\ \sigma_D,\ \alpha,\ \beta]^T$. The initial values of the unknown parameters are set to be 50% of the corresponding actual ones.

Based on the aforementioned information, the unknown parameters can be identified by the proposed approach as shown in Table 3. It can be seen that as compared with the linear case, the identification errors in the nonlinear system are relatively large. Even so, the results are still acceptable with the maximum error only being 3.45%. Fig. 7 gives the time histories of some parameters to show their convergence processes. It is clear that the identified parameters can converge to their actual ones after few seconds. Similar results can be found for the remaining parameters. Based on the identified parameters of Dahl model and the estimated structural states, the nonlinear restoring force provided by Dahl model can be also obtained. The hysteretic performance of Dahl model is shown in Fig. 8 for comparison. Obviously, the estimated nonlinear restoring force matches with the theoretical one very well.

The unknown excitation can be estimated by the proposed approach as well. Notably, the force applied to the 3rd flood should be treated as implicit unknown input due to the acceleration of this floor being unmeasured. Also, there is no explicit unknown input involved in this example. The comparison of the identified excitation with the real one is depicted in Fig. 9. The time segment from 7 s to 8 s is also given for clarity of comparison. The value of NRMSE in this case is 2.24%. Although the NRMSE value is relatively larger than that of linear case, it can be concluded that the proposed approach is still capable of identifying unknown loads even structural nonlinearity involved.

(a) The entire time series

(b) Time segment from 7 s to 8 s

Fig. 9 The comparison of the identified force in nonlinear system

4. Conclusions

According to whether the external forces are presented in observation equation or not, these unknown inputs can be divided into two parts in this study, namely explicit unknown inputs and implicit unknown inputs. Many available EKF-UI methods are basically performed on the premise that collocated acceleration measurements at the locations of unknown inputs are measured. Thus, these methods are not applicable if the implicit unknown inputs are involved. In this paper, an improved extended Kalman filter with unknown input (IEKF-UI) is proposed for identifying the structural parameters and unknown inputs at the same time. An improved observation equation is obtained with the aid of projection matrix and data fusion technique. By using FOHD method for discretizing continuous function and Taylor series for linearization procedure, the implicit inputs can be directly identified without the requirement of collocated acceleration response measurements. As compared with many existing EKF-UI methods, a merit of the proposed approach is that the restriction on the sensor locations is removed. Numerical examples show that the proposed approach is capable of simultaneously identifying the parameters of linear/nonlinear structures as well as the unknown inputs applied to them. It should be noted that in order to obtain the improved observation equation, the mass matrix should be known in the proposed approach. Further research can be conducted for the parameter and force identification without the prior knowledge of mass distribution. Moreover, dynamic tests can be carried out to verify the effectiveness of the approach for real structural model.

Acknowledgments

The financial support from the National Key Research and Development Program of China (grant number 2019YFC1511101) is greatly appreciated. The support from National Natural Science Foundation of China (No. 52278305), Natural Science Foundation of Hunan Province (No. 2021JJ30110) and Innovation Platform Open Fund project of Hunan Province (No. 19K018) is also greatly appreciated.

References

Chatzis, M.N., Chatzi, E.N. and Triantafyllou, S.P. (2017), "A discontinuous extended Kalman filter for non-smooth dynamic problems", *Mech. Syst. Signal Process*, **92**, 13-29. https://doi.org/10.1016/j.ymssp.2017.01.021

Das, S. and Saha, P. (2018), "Structural health monitoring techniques implemented on IASC–ASCE benchmark problem: a review", *J. Civil Struct. Health Monit.*, **8**(4), 689-718. https://doi.org/10.1007/s13349-018-0292-5

Ding, Y., Law, S.S., Wu, B., Xu, G.S., Lin, Q., Jiang, H.B. and Miao, Q.S. (2013), "Average acceleration discrete algorithm for force identification in state space", *Eng. Struct.*, **56**, 1880-1892. https://doi.org/10.1016/j.engstruct.2013.08.004

Ginsberg, D., Fritzen, C.P. and Loffeld, O. (2018), "Sparsity-constrained extended Kalman filter concept for damage localization and identification in mechanical structures", *Smart Struct. Syst., Int. J.*, **21**(6), 741-749. https://doi.org/10.12989/sss.2018.21.6.741

Hannan, M.A., Hassan, K. and Jern, K.P. (2018), "Review on sensors and systems in structural health monitoring: current issues and challenges", *Smart Struct. Syst., Int. J.*, **22**(5), 509-525. https://doi.org/10.12989/sss.2018.22.5.509

He, J., Zhang, X.X and Xu, B. (2019a), "Identification of structural parameters and unknown inputs based on revised observation equation: approach and validation", *Int. J. Struct. Stab. Dyn.*, **19**(12), 1950156. https://doi.org/10.1142/S0219455419501566

He, J., Zhang, X.X. and Dai, N.X. (2019b), "An improved Kalman filter for joint estimation of structural states and unknown loadings", *Smart Struct. Syst., Int. J.*, **24**(2), 209-221. https://doi.org/10.12989/sss.2019.24.2.209

He, J., Zhang, X.X., Feng, Z.Q., Chen, Z.Q. and Cao, Z. (2020), "A two-stage Kalman filter for the identification of structural parameters with unknown loads", *Smart Struct. Syst., Int. J.*, **26**(6), 693-701. https://doi.org/10.12989/sss.2020.26.6.693

He, J., Qi, M.C, Hua, X.G, Chen, Z.Q, Yang, O. and Cao, Z. (2021), "Substructural identification with weighted global iteration considering unknown interfacial forces and external excitation", *Measurement*, **180**, 109537. https://doi.org/10.1016/j.measurement.2021.109537

Hoshiya, M. and Saito, E. (1984), "Structural identification by extended Kalman filter", *J. Eng. Mech.*, **110**(12), 1757-1770. https://doi.org/10.1061/(ASCE)0733-9399(1984)110:12(1757)

Lai, Z., Lei, Y., Zhu, S.Y., Xu, Y.L., Zhang, X.H. and Krishnaswamy, S. (2016), "Moving-window extended Kalman filter for structural damage detection with unknown process and measurement noises", *Measurement*, **88**, 428-440. https://doi.org/10.1016/j.measurement.2016.04.016

Lei, Y., Jiang, Y. and Xu, Z. (2012), "Structural damage detection with limited input and output measurement signals", *Mech. Syst. Signal Proc.*, **28**, 229-243. https://doi.org/10.1016/j.ymssp.2011.07.026

Lei, Y., Lai, Z., Zhu, S. and Zhang, X.H. (2014), "Experimental study on impact-induced damage detection using an improved extended Kalman filter", *Int. J. Struct. Stab. Dyn.*, **14**(5), 1440007. https://doi.org/10.1142/S0219455414400070

Lei, Y., Xia, D.D., Chen, F. and Deng, Y.M. (2018), "Synthesis of cross-correlation functions of partial responses and the extended Kalman filter approach for structural damage detection under ambient excitations", *Int. J. Struct. Stab. Dyn.*, **18**(8), 1840003. https://doi.org/10.1142/S0219455418400035

Lei, Y., Lu, J., Huang, J. and Chen, S. (2020), "A general synthesis of identification and vibration control of building structures under unknown excitations", *Mech. Syst. Signal Proc.*, **143**, 106803. https://doi.org/10.1016/j.ymssp.2020.106803

Lei, Y., Yang, X., Huang, J., Zhang, F. and Liu, L. (2021), "Identification of model-free hysteretic forces of magnetorheological dampers embedded in buildings under unknown excitations using incomplete structural responses", *Struct. Control Health Monit.*, **28**(5), e2715. https://doi.org/10.1002/stc.2715

Li, D. and Wang, Y. (2020), "Parameter identification of a differentiable Bouc-Wen model using constrained extended Kalman filter", *Struct. Health Monit.*, **20**(1), 360-378. https://doi.org/10.1177/1475921720929434

Liu, L., Lei, Y. and He, M. (2016), "A two-stage parametric identification of strong nonlinear structural systems with incomplete response measurements", *Int. J. Struct. Stab. Dyn.*, **16**(4), 1640022. https://doi.org/10.1142/S0219455416400228

Liu, L., Hua, W. and Lei, Y. (2018), "Real-time simultaneous identification of structural systems and unknown inputs without collocated acceleration measurements based on MEKF-UI",

Measurement, **122**, 545-553. https://doi.org/10.1016/j.measurement.2017.07.001

Mu, H.Q. and Yuen, K.V. (2015), "Novel outlier-resistant extended Kalman filter for robust online structural identification", *J. Eng. Mech.*, **141**(1), 04014100. https://doi.org/10.1061/(ASCE)EM.1943-7889.0000810

Naranjo-Perez, J., Jimenez-Alonso, J.F. and Saez, A. (2020), "Parameter identification of the dynamic Winkler soil-structure interaction model using a hybrid unscented Kalman filter-multi-objective harmony search algorithm", *Adv. Struct. Eng.*, **23**(12), 2653-2668. https://doi.org/10.1177/1369433220919074

Pan, S., Xiao, D., Xing, S., Law, S.S., Du, P. and Li, Y. (2016), "A general extended Kalman filter for simultaneous estimation of system and unknown inputs", *Eng. Struct.*, **109**, 85-98. https://doi.org/10.1016/j.engstruct.2015.11.014

Saleem, M.M. and Jo, H. (2019), "Impact force localization for civil infrastructure using augmented Kalman Filter optimization", *Smart Struct. Syst., Int. J.*, **23**(2), 123-139. https://doi.org/10.12989/sss.2019.23.2.123

Saleem, M.M. and Jo, H. (2021), "Multi-objective sensor placement optimization for structural response estimation under spatially varying dynamic loading of bridges", *Adv. Struct. Eng.*, **24**(10), 2255-2266. https://doi.org/10.1177/1369433221993574

Xiao, X., Xu, X. and Shen, W. (2021), "Simultaneous identification of the frequencies and track irregularities of high-speed railway bridges from vehicle vibration data", *Mech. Syst. Signal Proc.*, **152**, 107412. https://doi.org/10.1016/j.ymssp.2020.107412

Xu, B. and He, J. (2015), "Substructural parameters and dynamic loading identification with limited observations", *Smart Struct. Syst., Int. J.*, **15**(1), 169-189. https://doi.org/10.12989/sss.2015.15.1.169

Xu, Y.L. and He, J. (2017), *Smart Civil Structures*, CRC Press, Boca Raton, FL, USA.

Xu, Y.L., Zhang, X.H., Zhu, S. and Zhan, S. (2016), "Multi-type sensor placement and response reconstruction for structural health monitoring of long-span suspension bridges", *Sci. Bull.*, **61**(4), 313-329. https://doi.org/10.1007/s11434-016-1000-7

Xu, B., Li, J., Dyke, S.J., Deng, B.C. and He, J. (2020), "Nonparametric identification for hysteretic behavior modeled with a power series polynomial using EKF-WGI approach under limited acceleration and unknown mass", *Int. J. Non-Linear Mech.*, **119**, 103324. https://doi.org/10.1016/j.ijnonlinmec.2019.103324

Yang, J.N., Pan, S. and Huang, H. (2007), "An adaptive extended Kalman filter for structural damage identifications II: Unknown inputs", *Struct. Control Health Monit.*, **14**(3), 497-521. https://doi.org/10.1002/stc.171

Yang, Y., Nagayama, T. and Xue, K. (2020), "Structure system estimation under seismic excitation with an adaptive extended Kalman filter", *J. Sound Vib.*, **489**, 115690. https://doi.org/10.1016/j.jsv.2020.115690

Yun, C.B. and Shinozuka, M. (1980), "Identification of nonlinear structural dynamic systems", *J. Struct. Mech.*, **8**(2), 187-203. https://doi.org/10.1080/03601218008907359

Zhang, C., Huang, J.Z., Song, G.Q. and Chen, L. (2017), "Structural damage identification by extended Kalman filter with l1-norm regularization scheme", *Struct. Control Health Monit.*, **24**(11), 1-17. https://doi.org/10.1002/stc.1999

Zhang, X.H., Zhu, Z., Yuan, G.K. and Zhu, S.Y. (2021), "Adaptive mode selection integrating Kalman filter for dynamic response reconstruction", *J. Sound Vib.*, **515**, 116497. https://doi.org/10.1016/j.jsv.2021.116497

Mechanical Systems and Signal Processing 211 (2024) 111218

Contents lists available at ScienceDirect

Mechanical Systems and Signal Processing

journal homepage: www.elsevier.com/locate/ymssp

Identification of time-varying stiffness with unknown mass distribution based on extended Kalman filter

Xiaoxiong Zhang [a,b], Jia He [a,b,*], Xugang Hua [a,b], Zhengqing Chen [a,b]

[a] Key Laboratory of Wind and Bridge Engineering of Hunan Province, College of Civil Engineering, Hunan University, Changsha, China
[b] Key Laboratory of Building Safety and Energy Efficiency of the Ministry of Education, China

ARTICLE INFO

Communicated by Lukasz Jankowski

Keywords:
Unknown mass distribution
Time-varying parameters identification
Adaptive noise covariance matrix
Covariance resetting technique
Extended Kalman filter

ABSTRACT

Although the extended Kalman filter (EKF) provides a promising way for structural state estimation, it cannot effectively track time-varying parameters online. Besides, the structural mass distribution is usually assumed to be known in advance for many EKF-based methods, limiting their applications. In this paper, by using limited observations, an adaptive EKF with unknown mass coefficients (AEKF-UM) approach is proposed for the identification of time-variant parameters and mass distribution at the same time. A real-time updating procedure is presented for improving the process and measurement noise covariance matrices at each time step to assure the stability and accuracy of convergence results. Based on the dramatic increase of measurement noise covariance, an index is defined for determining the damage instant. A covariance resetting technique is then used to enhance the tracking capability for the purpose of effectively capturing the time-varying parameters. The unknown mass coefficients can be estimated at the same time by adding them into the extended state vector. To validate the effectiveness of the proposed approach, two numerical cases are considered, i.e. (i) the Phase I ASCE structural health monitoring benchmark building structure, and (ii) a four-story nonlinear structure equipped with a magneto-rheological (MR) damper. Experimental tests on a four-story building model subject to base excitation are also conducted to investigate the performance of the proposed approach. Results show that the proposed approach is capable of satisfactorily tracking abrupt changes of stiffness parameters with unknown mass distributions.

1. Introduction

Structural health monitoring (SHM) systems are able to provide significant information of the concerned structure during its service life [1,2]. Due to the non-destructive nature of vibration-based methodologies, assessing structural states via vibration information has significantly advanced during the past few decades [3–5]. As the saying goes, change is the only constant. Structural parameters, such as stiffness, may degrade due to the deterioration of materials or severe catastrophes. Therefore, it is highly desirable to estimate structural states through sensor networks in real-time. To date, a number of methods have been developed for tracking time-varying characteristics, including time-domain methods [6–10], frequency-domain methods [11–15], and time–frequency domain methods [16,17]. This paper is not intended to provide a comprehensive review on parameter identification methods. Instead, the introduction

* Corresponding author at: Key Laboratory of Wind and Bridge Engineering of Hunan Province, College of Civil Engineering, Hunan University, Changsha, China.
E-mail address: jiahe@hnu.edu.cn (J. He).

https://doi.org/10.1016/j.ymssp.2024.111218
Received 1 August 2023; Received in revised form 14 November 2023; Accepted 2 February 2024
Available online 7 February 2024

in this section will focus on the estimation of time-variant systems basing on Kalman filter (KF) algorithm.

As one of the most famous KF-based methods, the extended Kalman filter (EKF) provides a promising way for recursively estimating structural states. During the recursive procedure, the error covariance matrix and EKF gain matrix often decrease and become less sensitive to current data because of data saturation. This means that the classic EKF method is not suitable for time-variant system estimation. Therefore, much effort has been made to enable EKF-based methods to possess the capability of tracking time-varying parameters. A commonly used technique is the application of fading-factor related to the time-varying parameters. However, the selection of the most optimal fading-factor is often difficult. Some fading-factors were set by experience with the consideration of tracking capability and noise sensitivity [18–20]. In addition, some attempts were made to find the optimal fading-factor by updating the adaptive factor matrix [21–23], but it's time-consuming for searching the optimal solution at each time step. To improve the computational efficiency, Lei et al. [24] further developed a three-stage strategy and just solved the optimization problem at the damage instant. Besides the usage of fading-factor, some researchers also developed EKF-based approach for identifying time-variant parameters with the aid of other techniques, such as Bayesian probabilistic algorithm [25–27], wavelet multiresolution [28–31] and so forth.

Although the aforementioned approaches can deal with time-varying systems, they are carried out under the assumption that the mass coefficients are known. However, in practical situations, it may be difficult to accurately obtain the mass distribution, and the inaccuracy of the mass estimation can lead to errors in the identification results. Therefore, much effort has also been made for structural identification without the prior knowledge of mass distribution [32–35]. Focusing on the EKF-based methods, the unknown mass coefficients can be added into the extended state vector and estimated along with other parameters. Since the mass coefficients are included in the extended state vector, the number of unknowns and the extent of nonlinearity significantly increase in the EKF method. Thus, some supplementary conditions or techniques are required to ensure the accuracy and convergency of the identified results. For example, by using cross-correlation functions, Lei et al. [36] proposed an EKF-based approach for dealing with the stiffness-mass coupled coefficients. By using energy equilibrium equations to establish a relationship between structural energy and unknown parameters, Huang et al. [37] proposed a KF-based approach for simultaneously identifying the coefficients of mass, stiffness and damping. With the combination of EKF and the weighted global iteration, Xu et al. [38,39] proposed an approach for the identification of nonlinear restoring forces and unknown mass.

The studies mentioned above performed well in dealing with either time-varying stiffness or mass identification, but few of them resolved both difficulties simultaneously. To circumvent this limitation, in this paper, an adaptive EKF with unknown mass coefficients (AEKF-UM) approach is proposed for identifying time-varying stiffness and unknown mass coefficients at the same time. A real-time updating procedure is presented for improving the process and measurement noise covariance matrices at each time step to assure the stability and accuracy of convergence results. As the structural parameters are suddenly changed, a jump increase of the measurement noise covariance matrix can be detected and an index is defined to capture these changes. Then, a covariance resetting technique is employed at this damage instant for the purpose of enhancing the tracking capability. The effectiveness of the proposed approach is validated numerically and experimentally. This paper is organized as follows: First, the formulas of the proposed approach is introduced in Section 2. Then, the numerical and experimental examples are shown in Section 3 and Section 4, respectively, to demonstrate the effectiveness of the proposed approach. Finally, the main conclusions are given is Section 5.

2. Formulas of the proposed AEKF-UM approach

The motion of equation of an n degrees-of-freedom (DOFs) structure can be written as

$$\boldsymbol{M}\ddot{\boldsymbol{x}}(t)+\overline{\boldsymbol{F}}[\dot{\boldsymbol{x}}(t),\boldsymbol{x}(t),\boldsymbol{\theta}(t)]=\boldsymbol{F}(t) \tag{1}$$

where $\ddot{\boldsymbol{x}}(t)$ $\dot{\boldsymbol{x}}(t)$ and $\boldsymbol{x}$(t) are the vectors of acceleration, velocity and displacement, respectively; $\mathbf{M}$ is the mass matrix and assumed to be unknown in this study; $\overline{\boldsymbol{F}}[\dot{\boldsymbol{x}}(t),\boldsymbol{x}(t),\boldsymbol{\theta}]$ represents the linear or nonlinear restoring force; $\mathbf{F}(t)$ is the external excitation; $\boldsymbol{\theta}(t)$ is an m-dimensional vector of parameters to be identified, including the coefficients of mass, damping, stiffness and/or those in the nonlinear model.

The extended state vector is defined as

$$\boldsymbol{Z}(t)=\left[\boldsymbol{x}(t)^{\mathrm{T}},\dot{\boldsymbol{x}}(t)^{\mathrm{T}},\boldsymbol{\theta}(t)^{\mathrm{T}}\right]^{\mathrm{T}} \tag{2}$$

Then, the state equation can be given as,

$$\frac{d\boldsymbol{Z}(t)}{dt}=\begin{bmatrix}\dot{\boldsymbol{x}}(t)\\ \boldsymbol{M}^{-1}(-\overline{\boldsymbol{F}}[\dot{\boldsymbol{x}}(t),\boldsymbol{x}(t),\boldsymbol{\theta}(t)]+\boldsymbol{F}(t))\\ \dot{\boldsymbol{\theta}}(t)\end{bmatrix}=\boldsymbol{g}(\boldsymbol{Z}(t),\boldsymbol{F}(t),t) \tag{3}$$

Let $\widehat{\boldsymbol{Z}}_{k|k}$ be the estimation of $\boldsymbol{Z}_k$. Based on Tayler expansion equation, the nonlinear state equation in Eq. (3) can be linearized with respect to $\widehat{\boldsymbol{Z}}_{k|k}$ at the k-th time step as follow,

$$\boldsymbol{g}(\boldsymbol{Z}_k,\boldsymbol{F}_k,t_k)\approx\boldsymbol{g}\left(\widehat{\boldsymbol{Z}}_{k|k},\boldsymbol{F}_k,t_k\right)+\boldsymbol{U}_{k|k}\left(\boldsymbol{Z}_k-\widehat{\boldsymbol{Z}}_{k|k}\right) \tag{4}$$

where $\boldsymbol{U}_{k|k}=\left.\frac{\partial g(\boldsymbol{Z}_k,\boldsymbol{F}_k,t_k)}{\partial \boldsymbol{Z}_k}\right|_{\boldsymbol{Z}_k=\hat{\boldsymbol{Z}}_{k|k}}$.

The left side of Eq. (3) can be also expressed as $(Z_{k+1} - Z_k)/\Delta t$, where Δt is the time interval. Based on Eqs. (3–4), the following equation can be obtained,

$$\begin{aligned} \boldsymbol{Z}_{k+1} &= \boldsymbol{Z}_k + \boldsymbol{g}(\widehat{\boldsymbol{Z}}_{k|k}, \boldsymbol{F}_k, t_k)\Delta t + \boldsymbol{U}_{k|k}(\boldsymbol{Z}_k - \widehat{\boldsymbol{Z}}_{k|k})\Delta t + \boldsymbol{w}_k \\ &= \Psi(\boldsymbol{Z}_k, \widehat{\boldsymbol{Z}}_{k|k}, \boldsymbol{F}_k, t_k) + \boldsymbol{w}_k \end{aligned} \tag{5}$$

where $\mathbf{w}(t)$ is a process noise vector with a zero mean and a covariance matrix $\mathbf{Q}(t)$.

In this study, only several acceleration responses are used. Thus, the discretized observation equation at time t_k can be described as,

$$\boldsymbol{y}_k = \boldsymbol{L}\boldsymbol{M}^{-1}(\boldsymbol{F}_k - \overline{\boldsymbol{F}}[\dot{\boldsymbol{x}}_k, \boldsymbol{x}_k, \boldsymbol{\theta}_k]) + \boldsymbol{v}_k = \boldsymbol{L}\boldsymbol{h}(\boldsymbol{Z}_k, \boldsymbol{F}_k, t_{k+1}) + \boldsymbol{v}_k \tag{6}$$

in which $\mathbf{L}$ is the influence matrix determined by the locations of accelerometers; $\boldsymbol{h}(\boldsymbol{Z}_k, \boldsymbol{F}_k, t_{k+1}) = \boldsymbol{M}^{-1}(\boldsymbol{F}_k - \overline{\boldsymbol{F}}[\dot{\boldsymbol{x}}_k, \boldsymbol{x}_k, \boldsymbol{\theta}_k])$; $\mathbf{v}_k$ is the measurement noise vector with a zero mean and a covariance matrix $\mathbf{R}(t)$.

The classic EKF consists of two parts: prediction and correction. The prediction part contains the calculation of the priori estimation state $\widehat{\boldsymbol{Z}}_{k+1|k}$ and priori estimation error covariance $\boldsymbol{P}_{k+1|k}$, respectively. The correction part contains the posteriori estimation of error covariance $\boldsymbol{P}_{k+1|k+1}$ and state $\widehat{\boldsymbol{Z}}_{k+1|k+1}$. Moreover, the Kalman gain matrix $\mathbf{G}_{k+1}$, which is significantly important in EKF, is also calculated in the correction part. The equations for these two parts are shown as follows,

Prediction part:

$$\widehat{\boldsymbol{Z}}_{k+1|k} = \widehat{\boldsymbol{Z}}_{k|k} + \int_{k\Delta t}^{(k+1)\Delta t} \boldsymbol{g}(\widehat{\boldsymbol{Z}}_{k|k}, \boldsymbol{F}_k, t_k)\, \mathrm{d}t \tag{7}$$

$$\boldsymbol{P}_{k+1|k} = E\left[(\boldsymbol{Z}_{k+1} - \widehat{\boldsymbol{Z}}_{k+1|k})(\boldsymbol{Z}_{k+1} - \widehat{\boldsymbol{Z}}_{k+1|k})^T\right] = \boldsymbol{A}\boldsymbol{P}_{k|k}\boldsymbol{A}^T + \boldsymbol{Q}_k \tag{8}$$

in which $\boldsymbol{A} = \boldsymbol{I} + \Delta t\boldsymbol{U}_{k|k}$.

Correction part:

$$\boldsymbol{G}_{k+1} = \boldsymbol{P}_{k+1|k}\boldsymbol{H}_{k+1|k}^T\boldsymbol{L}^T\left(\boldsymbol{L}\boldsymbol{H}_{k+1|k}\boldsymbol{P}_{k+1|k}\boldsymbol{H}_{k+1|k}^T\boldsymbol{L}^T + \boldsymbol{R}_{k+1}\right)^{-1} \tag{9}$$

$$\widehat{\boldsymbol{Z}}_{k+1|k+1} = \widehat{\boldsymbol{Z}}_{k+1|k} + \boldsymbol{G}_{k+1}\left[\boldsymbol{y}_{k+1} - \boldsymbol{L}\boldsymbol{h}(\widehat{\boldsymbol{Z}}_{k+1|k}, \boldsymbol{F}_{k+1})\right] \tag{10}$$

$$\begin{aligned} \boldsymbol{P}_{k+1|k+1} &= E((\boldsymbol{Z}_{k+1} - \widehat{\boldsymbol{Z}}_{k+1|k+1})(\boldsymbol{Z}_{k+1} - \widehat{\boldsymbol{Z}}_{k+1|k+1})^T) \\ &= (\boldsymbol{I} - \boldsymbol{G}_{k+1}\boldsymbol{L}\boldsymbol{H}_{k+1|k})\boldsymbol{P}_{k+1|k}(\boldsymbol{I} - \boldsymbol{G}_{k+1}\boldsymbol{L}\boldsymbol{H}_{k+1|k})^T + \boldsymbol{G}_{k+1}\boldsymbol{R}_{k+1}\boldsymbol{G}_{k+1}^T \end{aligned} \tag{11}$$

in which $\boldsymbol{H}_{k+1|k} = \left.\frac{\partial \boldsymbol{h}(Z_{k+1}, F_{k+1}, t_{k+1})}{\partial Z_{k+1}}\right|_{Z_{k+1} = \hat{Z}_{k+1|k}}$.

Eqs. (7–11) are the primary process of the classical EKF. Usually, the process and measurement noise covariance matrices, saying matrices $\mathbf{Q}$ and $\mathbf{R}$, are assumed to be constant and determined by experience. However, improper choice of these two matrices may lead to unreliable or even divergent estimation of structural states. Moreover, using constant noise covariance is not adequate in achieving the best filtering performance for time-varying parameters identification. Thus, based on the idea of innovation and residual [40], a real-time updating procedure is employed in this study for improving the noise covariance matrices $\mathbf{Q}$ and $\mathbf{R}$ at each time step.

The innovation of EKF is defined as the difference between the measurement and its priori estimation. At the time instant t_k, the innovation can be written as

$$\boldsymbol{\delta}_{k+1} = \boldsymbol{y}_{k+1} - \boldsymbol{L}\boldsymbol{h}(\widehat{\boldsymbol{Z}}_{k+1|k}, \boldsymbol{F}_{k+1}) \tag{12}$$

Based on Eqs. (5) and (10), the process noise matrix at each time step can be calculated using following equation,

$$\begin{aligned} \widehat{\boldsymbol{w}}_k &= \widehat{\boldsymbol{Z}}_{k+1|k+1} - \Psi(\widehat{\boldsymbol{Z}}_{k|k}, \boldsymbol{F}_k, t_k) \\ &= \widehat{\boldsymbol{Z}}_{k+1|k+1} - \widehat{\boldsymbol{Z}}_{k+1|k} \\ &= \boldsymbol{G}_{k+1}\left[\boldsymbol{y}_{k+1} - \boldsymbol{L}\boldsymbol{h}(\widehat{\boldsymbol{Z}}_{k+1|k}, \boldsymbol{F}_{k+1})\right] \\ &= \boldsymbol{G}_{k+1}\boldsymbol{\delta}_{k+1} \end{aligned} \tag{13}$$

Thus,

$$\widehat{\boldsymbol{Q}}_k = E(\widehat{\boldsymbol{w}}_k\widehat{\boldsymbol{w}}_k^T) = \boldsymbol{G}_{k+1}(\boldsymbol{\delta}_{k+1}\boldsymbol{\delta}_{k+1}^T)\boldsymbol{G}_{k+1}^T \tag{14}$$

Different from using the moving estimation window technique [41], a weighting factor μ $(0 < \mu \le 1)$ is introduced to estimate the process noise matrix. The updated $\mathbf{Q}$ can be written as follows,

$$\boldsymbol{Q}_{k+1} = \mu\boldsymbol{Q}_k + (1 - \mu)diag(\boldsymbol{G}_{k+1}\boldsymbol{\delta}_{k+1}\boldsymbol{\delta}_{k+1}^T\boldsymbol{G}_{k+1}^T) \tag{15}$$

where $\mathbf{Q}_{k+1}$ and $\mathbf{Q}_k$ denote the process noise matrix at the current time step and the previous time step, respectively. It can be found that, the smaller value of μ is used, the more weights are put on the current estimates. When μ is equal to 1, the **Q** matrix will not update any more. Given that process noise can be assumed to be uncorrelated, the $\mathbf{Q}_{k+1}$ matrix has a diagonal form.

The residual of EKF is the difference between the measurement and its posteriori estimation as shown below,

$$\boldsymbol{\varepsilon}_{k+1} = \boldsymbol{y}_{k+1} - \boldsymbol{L}\boldsymbol{h}(\widehat{\boldsymbol{Z}}_{k+1|k+1}, \boldsymbol{F}_{k+1}) \tag{16}$$

According to Mohamed's advice, the residual sequence is more stable than the innovation sequence to calculate measurement noise covariance [41]. It is computed as,

$$\widehat{\boldsymbol{R}}_k = \boldsymbol{\varepsilon}_{k+1}\boldsymbol{\varepsilon}_{k+1}^T + \boldsymbol{L}\boldsymbol{H}_{k+1|k}\boldsymbol{P}_{k+1|k}\boldsymbol{H}_{k+1|k}^T\boldsymbol{L}^T \tag{17}$$

Similarly, the **R** matrix can be estimated as,

$$\boldsymbol{R}_{k+1} = \mu\boldsymbol{R}_k + (1-\mu)diag\left(\boldsymbol{\varepsilon}_{k+1}\boldsymbol{\varepsilon}_{k+1}^T + \boldsymbol{L}\boldsymbol{H}_{k+1|k}\boldsymbol{P}_{k+1|k}\boldsymbol{H}_{k+1|k}^T\boldsymbol{L}^T\right) \tag{18}$$

where $\mathbf{R}_{k+1}$ and $\mathbf{R}_k$ denote the measurement noise matrix at the current time step and the previous time step, respectively. Since the measurement noise is often assumed uncorrelated, the **R** matrix is in a diagonal form as well.

By using the updating procedure described above, the adaptive EKF in this study is more sensitive to structural parameter changes as compared with conventional EKF. As known, the error covariance matrix **P** will converge to some stable value and become less sensitive to the abrupt changes of structural parameters, if the recursive estimation is conducted sufficiently. To further enhance the tracking capability, a covariance resetting technique for the adjustment of matrix **P** is employed herein. Abrupt changes of parameters will result in discontinuities in measurements. A jump increase can be observed accordingly in the diagonal element of matrix **R**. For ease of capturing these abrupt changes, an index is defined by the following equation,

$$r_N = \frac{\left(\sum_{k=N_0}^{N}\text{diag}(\boldsymbol{R}_k)\right)\Big/(N-N_0)}{\left(\sum_{k=N_0}^{N-1}\text{diag}(\boldsymbol{R}_k)\right)\Big/(N-1-N_0)} \qquad (N_0+1 \leqslant N \leqslant N_s) \tag{19}$$

where diag($\mathbf{R}_k$) denotes the diagonal elements of **R** matrix at the k-th time step; N is the number of time steps used for calculating the index r_N; N_0 denotes the starting point used for the calculation of r_N; N_s is the total number of time steps. Since the initial values of structural parameters are often assumed to be quite different from the actual ones, the first few seconds are not used for calculating the index r_N. In this study, the index is computed after two seconds. The sampling frequency is 1000 Hz, and thus N_0 is set to be 2000. If there is no damage, the value of r_N should be close to 1. When structural damage is happening, an increase of r_N can be observed. If the value of r_N is larger than the predefined threshold δ, the covariance resetting technique is triggered. In this study, δ is set to be 1.1.

Once the damage instant is detected, the posteriori error covariance matrix **P** is reset as Eq. (20) to enhance the tracking capability for capturing the abrupt changes of parameters.

$$\overline{\boldsymbol{P}} = diag\left(\left[\boldsymbol{0}_{1\times 2n}, \boldsymbol{P}_{1|1}(2n+1, 2n+1), \ldots, \boldsymbol{P}_{1|1}(2n+m, 2n+m)\right]\right) \tag{20}$$

in which $\overline{\boldsymbol{P}}$ is called as reset error covariance matrix; $\mathbf{P}_{1|1}$ is the initial value of error covariance matrix which is given at the beginning. It can be seen that $\overline{\boldsymbol{P}}$ is a $(2n+m)\times(2n+m)$ diagonal matrix. The first $2n$ diagonal elements are set to be zero to reduce the oscillation of the estimated responses, and the last m diagonal elements are equal to the corresponding initial value of $\mathbf{P}_{1|1}$. At the damage instant, the error covariance matrix is reset to a large value leading to a relatively large Kalman gain matrix. This means that more weights are put on the current data, and thus the suddenly changed parameters can be captured more efficiently.

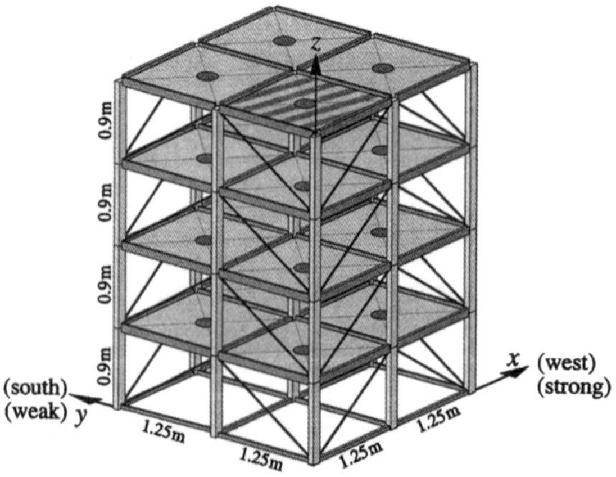

Fig. 1. The ASCE Phase I benchmark building structure [42].

3. Numerical investigation

Two numerical examples, including a linear benchmark building and a nonlinear structure, are used to demonstrate the effectiveness of the proposed approach. The external inputs and limited acceleration responses are measured for the identification. To mimic the effect of noise in real circumstances, all the measurements are polluted by a 5 % noise-to-signal ratio in the root mean square.

3.1. The ASCE Phase I benchmark building structure

The ASCE Phase I benchmark building structure as shown in Fig. 1 is a four-story steel frame. It is considered as the first example in this study. A random excitation is applied on the top floor in its weak direction. The structure can then be simplified as a four-story shear-type structure because of its symmetry [21,22]. The coefficients of stiffness and viscous damping of each story are 6.79×10^7N/m and 6.79×10^3N·s/m, respectively. The mass coefficients from the 1st floor to the 4th floor are 3.45×10^3 kg, 2.65×10^3 kg, 2.65×10^3 kg and 1.81×10^3 kg, respectively.

The parameters to be identified include four stiffness coefficients, four damping coefficients and four mass coefficients. The initial values of structural parameters are set to 50 % of the actual ones. Since the stiffness of each floor is 6.79×10^7N/m, the calculation of its covariance can reach the order of 10^{15}. Similarly, the orders of covariance for damping and mass are 10^6 and 10^6, respectively. Thus, to ensure computational efficiency and convergence rate, the posteriori error covariance matrix is initially assumed to be $\boldsymbol{P}_{1|1} = diag\left(\left[\boldsymbol{I}_{1\times8}, \mathbf{10}^{18}_{1\times4}, \mathbf{10}^{6}_{1\times4}, \mathbf{10}^{10}_{1\times4}\right]\right)$. The noise matrices are initially chosen to be $\mathbf{Q}_1 = 10^{-4} \times \mathbf{I}$ and $\mathbf{R}_1 = \mathbf{I}$, respectively, where **I** denotes identity matrix. The weighting factor μ used for updating the matrices **Q** and **R** is assumed to be 0.3.

Two different damage scenarios are discussed. Case 1: the stiffness coefficient of 1st floor reduces from 6.79×10^7N/m to 1.97×10^7N/m at the time of 10 s. Case 2: the stiffness coefficients of the 1st and the 3rd floor reduce from 6.79×10^7N/m to 1.97×10^7N/m at the time of 10 s. Single damage pattern is considered in Case 1, whereas multiple damage pattern is considered in Case 2. The acceleration responses of the 1st, 3rd and 4th are measured for the identification with the sampling frequency of 1000 Hz.

3.1.1. Case 1: Single damage pattern

As mentioned before, only the stiffness coefficient k_1 is assumed to be suddenly changed at $t = 10$ s in this case. When the damage occurs, the discontinuities will be introduced in the acceleration measurements. The index r_N defined in Eq. (19) is calculated and shown in Fig. 2. It is clear that a significant jump increase of r_N can be observed at $t = 10$ s. This indicates that structural damage is taking place at this moment. The error covariance matrix can then be reset as $\overline{\boldsymbol{P}} = diag\left(\left[\mathbf{0}_{1\times8}, \mathbf{10}^{18}_{1\times4}, \mathbf{10}^{6}_{1\times4}, \mathbf{10}^{10}_{1\times4}\right]\right)$. The identified results are listed in Table 1. It is obvious that the identified parameters are close to the actual ones.

The time histories of the identified mass coefficients are plotted in Fig. 3. The results without using weighting factor μ, saying without **Q****R** update, is also shown in Fig. 3 to illustrate the superiority of the proposed approach. Due to the limitation of paper length, only the mass coefficients of the 3rd and 4th floor are given in Fig. 3 as examples. Obviously, without updating the matrices **Q** and **R**, the convergence rates of mass coefficients are very slow and the identified results are not acceptable.

Fig. 4 gives the identified stiffness coefficients of the 1st and 3rd floor as examples. For comparison, the results of only using real-time updating procedure (i.e. **Q****R** update only) and only using covariance resetting technique (i.e. ***P*** matrix only) are plotted in Fig. 4 as well. It can be found that updating the matrices **Q** and **R** in real-time makes the convergence rate very fast at the first few seconds. However, without using covariance resetting technique, the abrupt changes of parameters cannot be efficiently captured. The proposed approach with the combination of these two techniques is capable of precisely and rapidly capturing the sudden change of k_1.

3.1.2. Case 2: Multiple damage pattern

In Case 2, the stiffness coefficients of the 1st and the 3rd floor are abruptly changed at time $t = 10$ s. Similar to Case 1, the procedures conducted before are used for the identification of structural parameters in this case. The index r_N is calculated and shown in

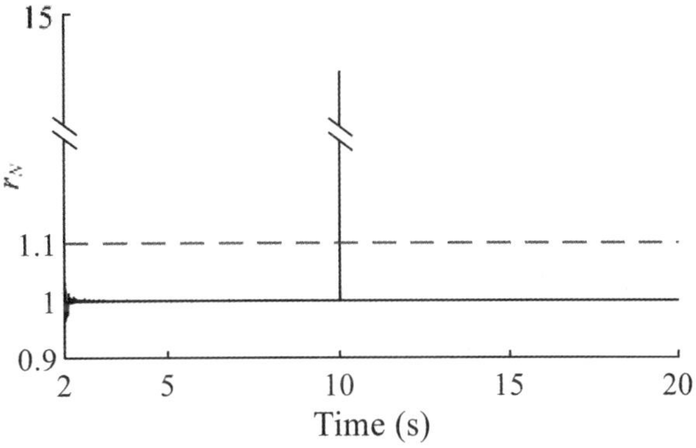

Fig. 2. The calculated index for Benchmark building structure (Case 1).

Table 1
The identified results of the benchmark building structure (Case 1).

Parameters	Actual	Identified	Error (%)
m_1 ($\times10^3$kg)	3.45	3.49	1.14
m_2 ($\times10^3$kg)	2.65	2.68	0.89
m_3 ($\times10^3$kg)	2.65	2.68	1.03
m_4 ($\times10^3$kg)	1.81	1.83	0.91
k_1 ($\times10^7$N/m)	6.79 → 1.97	6.84 → 1.99	0.74 → 0.94
k_2 ($\times10^7$N/m)	6.79	6.84	0.78
k_3 ($\times10^7$N/m)	6.79	6.85	0.93
k_4 ($\times10^7$N/m)	6.79	6.84	0.71
c_1 ($\times10^3$N·s/m)	6.79	6.80	0.16
c_2 ($\times10^3$N·s/m)	6.79	6.72	−1.09
c_3 ($\times10^3$N·s/m)	6.79	7.06	3.96
c_4 ($\times10^3$N·s/m)	6.79	6.57	−3.22

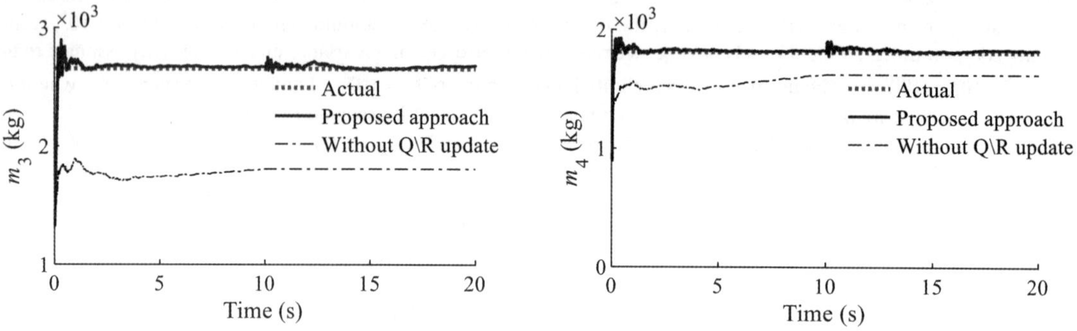

Fig. 3. The identified mass coefficients of benchmark building structure (Case 1).

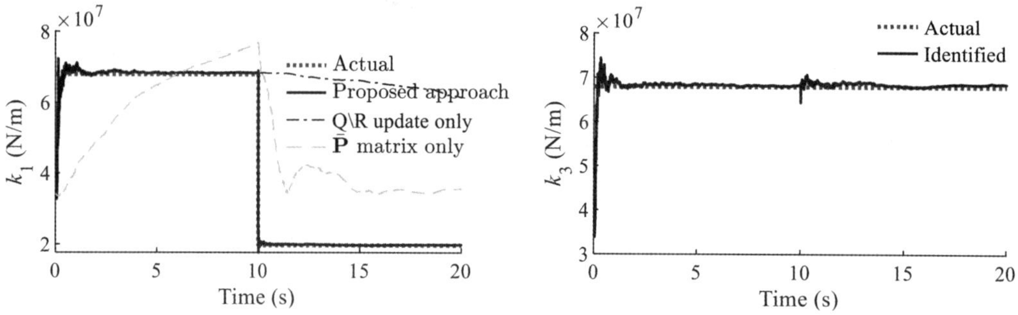

Fig. 4. The identified stiffness of benchmark building structure (Case 1).

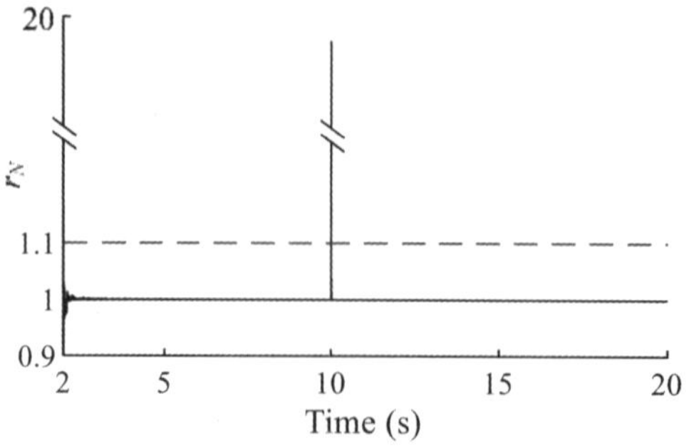

Fig. 5. The calculated index for Benchmark building structure (Case 2).

Fig. 5. A jump increase can be found at $t = 10$ s, which means structural damage is happening. The covariance resetting technique is then triggered. By using the proposed approach, satisfactory identification results can be obtained as shown in Table 2. The identified coefficients of mass and stiffness are plotted in Fig. 6 and Fig. 7, respectively. It can be seen that excellent tracking capability is exhibited by the proposed approach.

3.2. Four-story structure equipped with MR damper

To demonstrate the effectiveness of the proposed approach for the identification of structural nonlinearity, a four-story building structure equipped with MR damper as shown in Fig. 8 is considered in this section. The mass and stiffness parameters of each floor are 10 kg and 1.2×10^5N/m, respectively. The stiffness of the 2nd floor is assumed to abruptly decrease from 1.2×10^5N/m to 0.72×10^5N/m at time $t = 10$ s. The Rayleigh damping assumption is adopted. The proportional coefficients for mass and stiffness are set as $\alpha = 0.5648$ and $\beta = 1.3551 \times 10^{-4}$, respectively. The Dahl model is used to mimic to characteristic of MR damper,

$$F^{Dahl} = k_d \Delta x_i + c_d \Delta \dot{x}_i + f_d z + f_0 \tag{21}$$

in which k_d, c_d, f_d and f_0 are model coefficients and set as $k_d = 1$ N/m, $c_d = 50$ N·s/m, $f_d = 100$ N and $f_0 = 0$ N;Δx_i and $\Delta \dot{x}_i$ are relative displacement and velocity between the adjacent floors, respectively. z is a dimensionless coefficient as

$$\dot{z} = \sigma \Delta \dot{x}_i \left[1 - z \cdot sgn\left(\Delta \dot{x}_i \right) \right] \tag{22}$$

where σ is the shape coefficient of the hysteresis loop and is set to 1500 s/m; sgn(·) is signum function.

A concentrated excitation is applied on the 4th floor, and the corresponding responses are calculated by Runge-Kunta method with the time-interval of 0.001 s. The acceleration measurements of the 1st, 3rd and 4th are used for identification. The parameters to be identified include the coefficients of structural stiffness, damping, mass and Dahl model. The initial values are chosen to be 50 % of the actual ones. The process and measurement noise covariance matrices are initially set to be $10^{-4} \times \mathbf{I}$ and $\mathbf{I}$, respectively. The weighting factor μ is set as 0.3 for updating these two noise covariance matrices.

A sudden change of the index r_N can be found at $t = 10$ s as shown in Fig. 9. Then, the error covariance matrix is reset as $\overline{\boldsymbol{P}} = diag\left(\left[\mathbf{0}_{1\times9}, \mathbf{10}^{10}_{1\times4}, \mathbf{10}^{0}_{1\times2}, \mathbf{10}^{4}_{1\times4}, \mathbf{10}^{4}_{1\times4}\right]\right)$ to enhance the tracking capability. The identification results are listed in Table 3. Obviously, the identified parameters have a good agreement with the actual ones. The time histories of the mass and stiffness coefficients are shown in Fig. 10 and Fig. 11, respectively. It can be found that the abrupt changes of parameters can be efficiently captured by the proposed approach.

4. Experimental validation

To further demonstrate the effectiveness of the proposed approach, a four-story shear building model, as shown in Fig. 12, is considered in this section. The structural components, including the plates and columns, are made of aluminum alloy and firmly connected by bolts. The sizes of each plate and column are 400 mm × 400 mm × 10 mm and 20 mm × 4 mm, respectively. The height of each floor is 210 mm except for the 4th floor. A lock-up device is designed to mimic rapid changes of stiffness and installed on the 4th floor. Thus, the height of 4th floor is 410 mm. The lock-up device consists of a flexible iron column fixed to the top plate and an electromagnet fixed to the third-floor plate. When direct current (DC) is provided and a strong magnetic attraction force is generated, the electromagnet and the iron column will be attached tightly. In this locked state, the stiffness of 4th floor becomes large. As the current is not provided, the electromagnet and the iron column will be separated leading to a sudden decrease of stiffness. It is called as the unlocked state. By measuring the weight of each component, the mass distribution of the building model can be obtained as shown in Table 4. Static tests are conducted to assess the stiffness of each floor. As schematically shown in Fig. 13, the structural stiffness can be calculated using the weight of the calibrated mass block and the deformation of the building model measured by displacement

Table 2
The identified results of the benchmark building structure (Case 2).

Parameters	Actual	Identified	Error (%)
m_1 ($\times10^3$kg)	3.45	3.56	3.14
m_2 ($\times10^3$kg)	2.65	2.58	−2.73
m_3 ($\times10^3$kg)	2.65	2.68	0.87
m_4 ($\times10^3$kg)	1.81	1.81	0.26
k_1 ($\times10^7$N/m)	6.79 → 1.97	6.82 → 1.98	0.43 → 0.42
k_2 ($\times10^7$N/m)	6.79	6.73	−0.93
k_3 ($\times10^7$N/m)	6.79 → 1.97	6.84 → 1.99	0.69 → 1.14
k_4 ($\times10^7$N/m)	6.79	6.81	0.30
c_1 ($\times10^3$N·s/m)	6.79	6.75	−0.58
c_2 ($\times10^3$N·s/m)	6.79	7.11	4.75
c_3 ($\times10^3$N·s/m)	6.79	6.67	−1.78
c_4 ($\times10^3$N·s/m)	6.79	6.78	−0.12

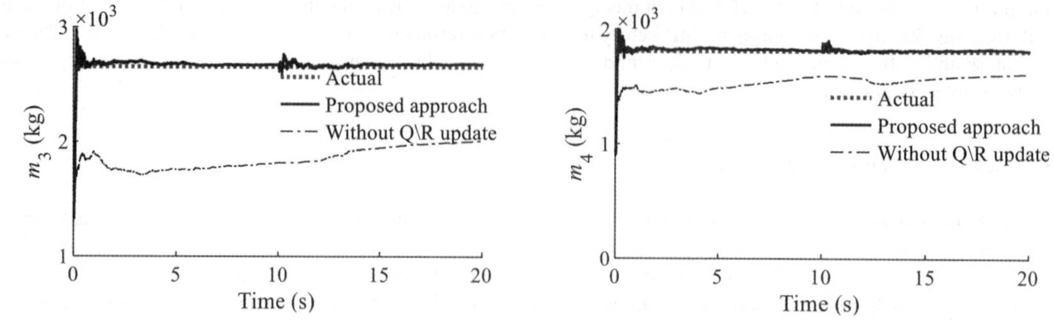

Fig. 6. The identified mass coefficients of benchmark building structure (Case 2).

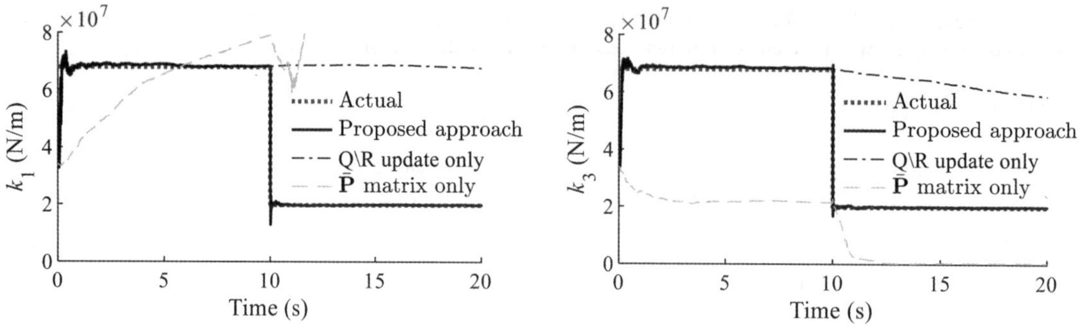

Fig. 7. The identified stiffness of benchmark building structure (Case 2).

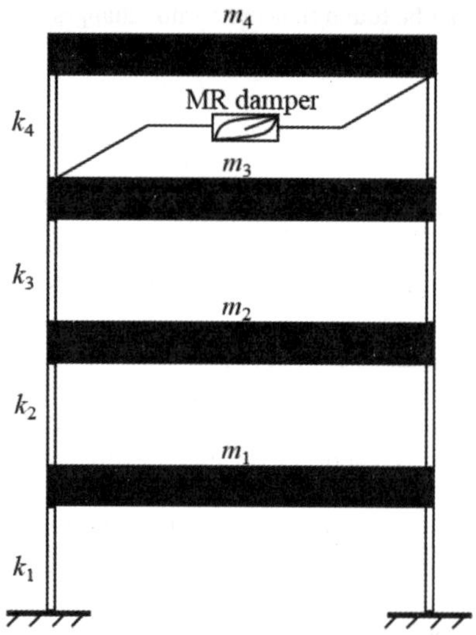

Fig. 8. Four-story building structure equipped with MR damper.

transducers. The structural stiffness in the locked and unlocked state are given in Table 4 as well. The Rayleigh damping assumption is employed for the consideration of damping effect. To obtain the natural frequencies of the model, a sine sweeping-frequency excitation with a frequency range of 0 Hz to 40 Hz is applied to the model. Based on the frequency response functions, the natural frequencies are determined as 4.10 Hz, 8.36 Hz, 18.74 Hz and 29.23 Hz in the locked state, and 3.75 Hz, 7.51 Hz, 18.71 Hz and 29.22 Hz in the unlocked state.

A random force is produced by the exciter and applied on the base floor, as shown in Fig. 12. The acceleration responses of the 1st, 2nd and 4th floors are measured with a sampling frequency of 1000 Hz. The parameters to be identified include four stiffness, two damping and four mass coefficients. The initial values are assumed to be half of the actual ones. The posteriori error covariance is initially set as $\boldsymbol{P}_{1|1} = diag([\boldsymbol{I}_{1\times8}, \mathbf{10}^7_{1\times4}, \mathbf{10}^0_{1\times2}, \mathbf{10}^4_{1\times4}])$. The matrices of $\mathbf{Q}_1$ and $\mathbf{R}_1$ are set as $10^{-4} \times \mathbf{I}$ and $\mathbf{I}$, respectively. Similar to

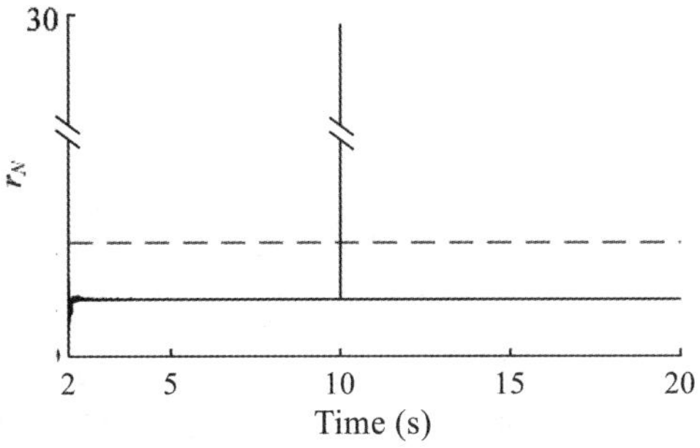

Fig. 9. The calculated index for nonlinear model.

Table 3
The identified results of the nonlinear building.

Parameters	Actual	Identified	Error (%)
m_1 (kg)	10	10.35	3.46
m_2 (kg)	10	10.26	2.59
m_3 (kg)	10	10.26	2.60
m_4 (kg)	10	10.28	2.84
k_1 ($\times 10^5$N/m)	1.20	1.23	2.64
k_2 ($\times 10^5$N/m)	1.20 → 0.72	1.23 → 0.74	2.50 → 2.78
k_3 ($\times 10^5$N/m)	1.20	1.24	3.09
k_4 ($\times 10^5$N/m)	1.20	1.18	−1.69
α	0.5648	0.5564	−1.47
$\beta \times 10^{-4}$	1.3551	1.3824	2.01
k_d(N/m)	1	1.04	4.01
c_d(N·s /m)	50	48.97	−2.06
f_d(N)	100	104.86	4.86
σ(s /m)	1500	1459.87	−2.68

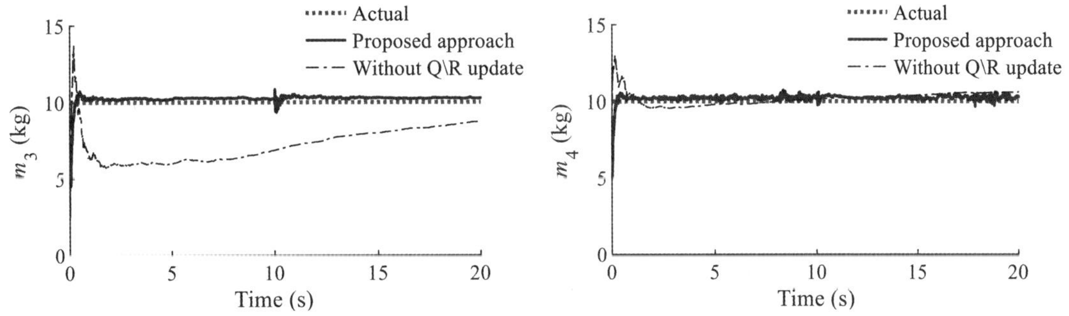

Fig. 10. The identified mass coefficients of nonlinear structure.

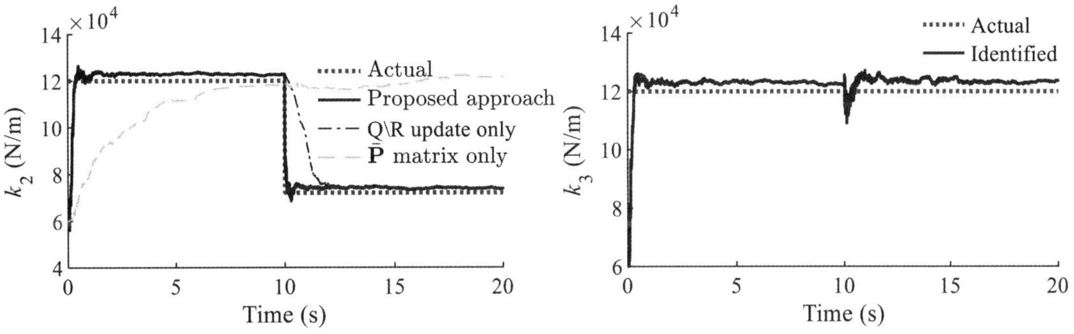

Fig. 11. The identified stiffness of nonlinear building.

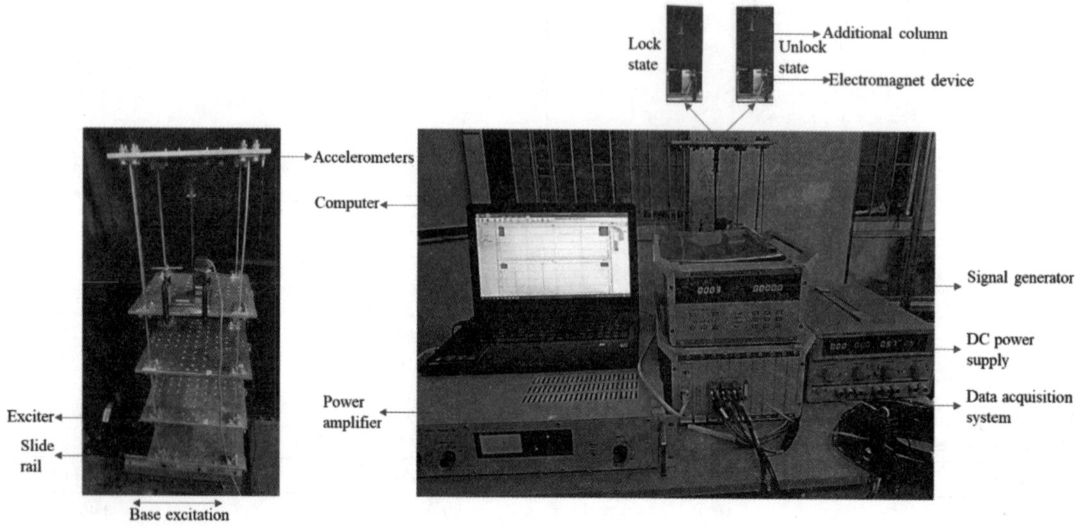

Fig. 12. The four-story experimental building model.

Table 4
The mass and stiffness of the building model.

Floor number	Mass distribution (kg)	Stiffness ($\times 10^4$N/m)	
		Locked state	Unlocked state
The 1st story	5.35	5.36	5.36
The 2nd story	5.36	5.51	5.51
The 3rd story	8.76	5.41	5.41
The 4th story	6.22	0.89	0.62

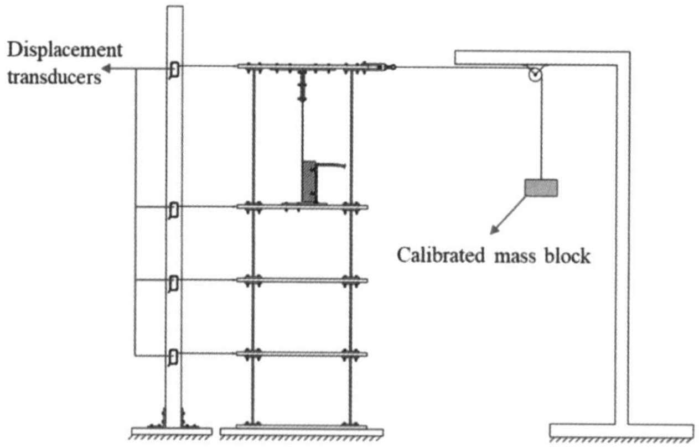

Fig. 13. Schematic diagram of static test.

numerical examples, the weighting factor is set to be 0.3.

By using the proposed approach, a sudden increase of the index r_N at $t = 41.91$ s can be observed, as shown in Fig. 14. This indicates that the variation of structural parameters is happening at this moment. Thus, the error covariance matrix is immediately reset as $\overline{\boldsymbol{P}} = diag\left(\left[\boldsymbol{0}_{1\times8}, \boldsymbol{10}^7_{1\times4}, \boldsymbol{10}^0_{1\times2}, \boldsymbol{10}^4_{1\times4}\right]\right)$ for the purpose of effectively capturing these abrupt changes. The identified parameters are listed in Table 5. The maximum errors of mass and stiffness are 3.70 % and 4.61 %, respectively, which means that the proposed approach is capable of identifying time-variant parameters of a real structure with acceptable accuracy. The time histories of the identified stiffness of 1st floor and 4th floor are further shown in Fig. 15 as examples. The comparison of the proposed approach with Yang's method [21] is also given in Fig. 15 for k_4. It can be found that the abrupt changes of stiffness can be effectively captured by both of them.

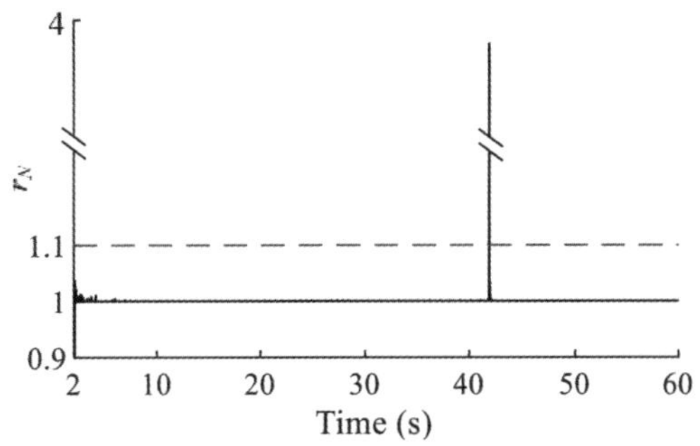

Fig. 14. The calculated index for experimental model.

Table 5
The identified results of the experimental building model.

Parameters	Actual	Identified	Error (%)
m_1 (kg)	5.35	5.21	2.31
m_2 (kg)	5.36	5.56	3.70
m_3 (kg)	8.76	8.86	2.45
m_4 (kg)	6.22	6.19	−0.41
k_1 ($\times 10^4$N/m)	5.36	5.27	−1.57
k_2 ($\times 10^4$N/m)	5.51	5.77	4.61
k_3 ($\times 10^4$N/m)	5.41	5.21	−3.64
k_4 ($\times 10^4$N/m)	0.89 → 0.62	0.87 → 0.59	−2.26 → -3.58

5. Conclusion

The research on online identification of time-varying stiffness without prior knowledge of mass distribution is relatively limited. In this paper, by using limited observations, an adaptive EKF with unknown mass coefficients (AEKF-UM) approach is proposed for the identification of time-variant parameters and mass distribution at the same time. The process and measurement noise covariance matrices, saying matrices **Q** and **R**, are updated at each time step. It can be found that a faster convergence speed is obtained by using the updated matrices **Q** and **R**. Based on the dramatic increase of measurement noise covariance, an index is defined for detecting damage instant. A covariance resetting technique is then employed to further enhance the tracking capability. Since none of optimization functions or iteration procedures are introduced in the proposed approach, it would be time-saving and more potential for real-time identification. The effectiveness of the proposed approach is first demonstrated via two numerical examples, including the Phase I ASCE SHM benchmark building and a four-story nonlinear structure equipped with MR damper. Moreover, experimental tests on a four-story building model are also conducted in the lab for the validation of the proposed approach. Results show that the abrupt changes of stiffness can be effectively captured by the proposed approach. It should be noted that the external loads are assumed to be known in this study. In practical situations, the external loads may be difficult to be measured. Thus, some attempts will be made in the near future to extend this approach for identifying time-vary parameters under unknown external inputs.

CRediT authorship contribution statement

Xiaoxiong Zhang: Methodology, Investigation, Validation, Writing – original draft. **Jia He:** Conceptualization, Methodology, Funding acquisition, Writing – review & editing. **Xugang Hua:** Writing – review & editing. **Zhengqing Chen:** Supervision.

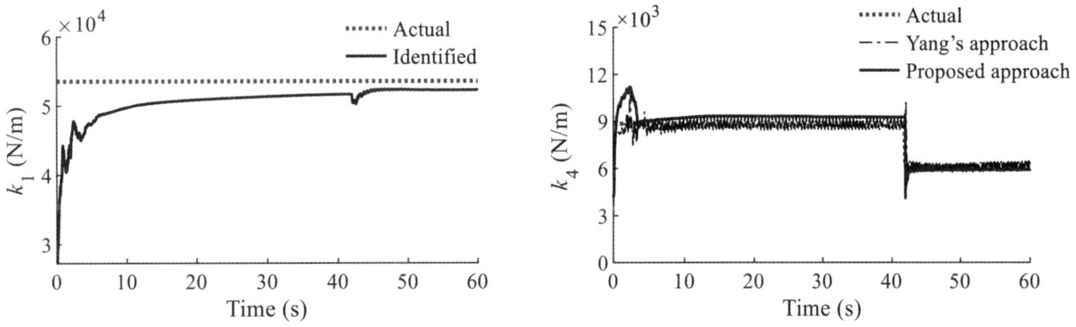

Fig. 15. The identified stiffness of experimental model.

Declaration of competing interest

The authors declare that they have no known competing financial interests or personal relationships that could have appeared to influence the work reported in this paper.

Data availability

Data will be made available on request.

Acknowledgements

The authors gratefully acknowledge the financial support from the National Key Research and Development Program of China through Grant no. 2022YFC3005305, the National Natural Science Foundation of China (No. 52278305), the Natural Science Foundation of Hunan Province (No. 2021JJ30110), and Key Laboratory for Intelligent Infrastructure and Monitoring of Fujian Province (No. IIM-01-01).

References

[1] Y.L. Xu, J. He, Smart Civil Structures, first ed., CRC Press, Boca Raton, 2017.
[2] A. Kamariotis, E. Chatzi, D. Straub, A framework for quantifying the value of vibration-based structural health monitoring, Mech. Syst. Signal Proc. 184 (2023) 109708.
[3] R. Hou, Y. Xia, Review on the new development of vibration-based damage identification for civil engineering structures: 2010–2019, J. Sound Vibr. 491 (2021) 115741.
[4] W.J. Yan, D. Chronopoulos, K.V. Yuen, Y.C. Zhu, Structural anomaly detection based on probabilistic distance measures of transmissibility function and statistical threshold selection scheme, Mech. Syst. Signal Proc. 162 (2022) 108009.
[5] K. Eltouny, M. Gomaa, X. Liang, Unsupervised learning methods for data-driven vibration-based structural health monitoring: a review, Sensors 23 (6) (2023) 3290.
[6] C. Wang, D. Ai, W.X. Ren, A wavelet transform and substructure algorithm for tracking the abrupt stiffness degradation of shear structure, Adv. Struct. Eng. 22 (5) (2018) 1136–1148.
[7] K.V. Yuen, S.C. Kuok, L. Dong, Self-calibrating Bayesian real-time system identification, Comput.-Aided Civil Infrastruct. Eng. 34 (9) (2019) 806–821.
[8] S.K. Huang, F.C. Chi, Y.T. Weng, Practical implementation of recursive subspace identification on seismically excited structures with fixed window, Appl. Sci. 12 (21) (2022) 10841.
[9] X.X. Zhang, J. He, X.G. Hua, Z.Q. Chen, O. Yang, Online identification of time-variant structural parameters under unknown inputs basing on extended Kalman filter, Nonlinear Dyn. 109 (2) (2022) 963–974.
[10] X.X. Zhang, J. He, X.G. Hua, Z.Q. Chen, Z.Q. Feng, Simultaneous identification of time-varying parameters and external loads based on extended kalman filter: approach and validation, Struct. Control. Health Monit. 2023 (2023) 1–18.
[11] C. Wang, J. Zhang, H.P. Zhu, A combined method for time-varying parameter identification based on variational mode decomposition and generalized morse wavelet, Int. J. Struct. Stab. Dyn. 20 (07) (2020) 2050077.
[12] P. Singh, M. Keyvanlou, A. Sadhu, An improved time-varying empirical mode decomposition for structural condition assessment using limited sensors, Eng. Struct. 232 (2021) 111882.
[13] W.H. Chen, W. Hseuh, J. Loh Kenneth, C.H. Loh, Damage evaluation of seismic response of structure through time-frequency analysis technique, Struct. Monitor. Maintenance Int. J. 9 (2) (2022) 107–127.
[14] R.M. Delgadillo, J.R. Casas, Bridge damage detection via improved completed ensemble empirical mode decomposition with adaptive noise and machine learning algorithms, Struct. Control. Health Monit. 29 (8) (2022) e2966.
[15] H. Qu, A. Chang, T. Li, Z. Guan, Synchro-squeezed adaptive wavelet transform-based optimized multiple analytical mode decomposition: parameter identification of cable-stayed bridge under earthquake input, Build.-Basel 12 (8) (2022) 1285.
[16] X.J. Yao, T.H. Yi, C.X. Qu, Autoregressive spectrum-guided variational mode decomposition for time-varying modal identification under nonstationary conditions, Eng. Struct. 251 (2022) 113543.
[17] C. Tan, H. Zhao, E.J. Obrien, N. Uddin, C.W. Kim, Exploring time-varying characteristics in drive-by bridge frequency extraction with the second-order synchrosqueezing transform, J. Bridge Eng. 28 (4) (2023) 04023010.
[18] T. Sato, K. Takei, Development of a Kalman filter with fading memory, In: Proceedings of Structural Safety and Reliability, ICOSSA, Balkema: Rotterdam, (1998) 387-394.
[19] J.W. Lin, R. Betti, A.W. Smyth, R.W. Longman, On-line identification of non-linear hysteretic structural systems using a variable trace approach, Earthq. Eng. Struct. Dyn. 30 (9) (2001) 1279–1303.
[20] H.Q. Mu, S.C. Kuok, K. Yuen, Stable robust extended Kalman filter, J. Aerosp. Eng. (2016). Article B4016010.
[21] J.N. Yang, S. Lin, H. Huang, L. Zhou, An adaptive extended Kalman filter for structural damage identification, Struct. Control. Health Monit. 13 (4) (2006) 849–867.
[22] J.N. Yang, S. Pan, H. Huang, An adaptive extended Kalman filter for structural damage identifications II: unknown inputs, Struct. Control. Health Monit. 14 (3) (2007) 497–521.
[23] Y.H. Park, H.B. Lee, G.W. Kim, Crack monitoring in rotating shaft using rotational speed sensor-based torsional stiffness estimation with adaptive extended kalman filters, Sensors (basel) 23 (5) (2023) 2437.
[24] Y. Lei, H. Zhou, Z.L. Lai, A computationally efficient algorithm for real-time tracking the abrupt stiffness degradations of structural elements, Comput.-Aided Civil Infrastruct. Eng. 31 (6) (2016) 465–480.
[25] K.V. Yuen, S.C. Kuok, Online updating and uncertainty quantification using nonstationary output-only measurement, Mech. Syst. Signal Proc. 66–67 (2016) 62–77.
[26] K.V. Yuen, S.C. Kuok, L. Dong, Self-calibrating Bayesian real-time system identification, Comput. Aided Civ. Infrastruct. Eng. 34 (2019) 806–821.
[27] M.E. Hassanabadi, Z. Liu, S. Eftekhar Azam, D. Dias-da-Costa, A linear Bayesian filter for input and state estimation of structural systems, Comput. Aided Civ. Infrastruct. Eng. 38 (13) (2023) 1749–1766.
[28] Y. Lei, J. Lu, J. Huang, Synthesize identification and control for smart structures with time-varying parameters under unknown earthquake excitation, Struct. Control. Health Monit. 27 (4) (2020).
[29] Y. Lei, J. Lu, J. Huang, Integration of identification and vibration control of time-varying structures subject to unknown seismic ground excitation, J. Vib. Control 26 (15–16) (2020) 1330–1344.
[30] N. Yang, Y. Lei, J. Li, H. Hao, J.S. Huang, A substructural and wavelet multiresolution approach for identifying time-varying physical parameters by partial measurements, J. Sound Vibr. 523 (2022) 116737.

[31] N. Yang, Y. Lei, J. Li, H. Hao, Identification of gradually varying physical parameters based on discrete cosine transform using partial measurements, Struct. Control. Health Monit. 29 (12) (2022) e3111.
[32] D. Zhang, H. Li, Loop substructure identification for shear structures of unknown structural mass using synthesized references, Smart Mater. Struct. 26 (8) (2017) 085046.
[33] H. Li, J. Jiang, M.S. Mohamed, Online dynamic load identification based on extended kalman filter for structures with varying parameters, Symmetry 13 (8) (2021) 1372.
[34] H. Qiu, J.S. Huang, Z.P. Zheng, Simultaneous Identification of structures and unknown seismic excitations for chain-like systems with unknown mass using partial absolute responses, Struct. Eng. Mech. 79 (6) (2021) 699–709.
[35] K.M. Jensen, I.F. Santos, L.K.H. Clemmensen, S. Theodorsen, H.J.P. Corstens, Mass estimation of ground vehicles based on longitudinal dynamics using loosely coupled integrated navigation system and CAN-bus data with model parameter estimation, Mech. Syst. Signal Proc. 171 (2022) 108925.
[36] Y. Lei, H. Qiu, F.B. Zhang, Identification of structural element mass and stiffness changes using partial acceleration responses of chain-like systems under ambient excitations, J. Sound Vibr. 488 (2020) 115678.
[37] X.H. Huang, S. Dyke, Z. Sun, Z.D. Xu, Simultaneous identification of stiffness, mass, and damping using an on-line model updating approach, Struct. Control. Health Monit. 24 (4) (2017) e1892.
[38] B. Xu, B.C. Deng, J. Li, J. He, Structural nonlinearity and mass identification with a nonparametric model using limited acceleration measurements, Adv. Struct. Eng. 22 (4) (2018) 1018–1031.
[39] B. Xu, J. Li, S.J. Dyke, B.C. Deng, J. He, Nonparametric identification for hysteretic behavior modeled with a power series polynomial using EKF-WGI approach under limited acceleration and unknown mass, Int. J. Non-Linear Mech. 119 (2020) 103324.
[40] S. Akhlaghi, N. Zhou, Z. Huang, Adaptive adjustment of noise covariance in Kalman filter for dynamic state estimation, in: 2017 IEEE Power & Energy Society General Meeting, IEEE, (2017), pp. 1-5.
[41] A.H. Mohamed, K.P. Schwarz, Adaptive Kalman Filtering for INS/GPS, J. Geodesy 9 (1) (1999) 193–203.
[42] E.A. Johnson, H.F. Lam, L.S. Katafygiotis, J.L. Beck, Phase I IASC-ASCE structural health monitoring benchmark problem using simulated data, J. Eng. Mech. 130 (1) (2004) 3–15.

Wiley
Structural Control and Health Monitoring
Volume 2024, Article ID 5137641, 20 pages
https://doi.org/10.1155/2024/5137641

Research Article

Bayesian Spectral Decomposition for Efficient Modal Identification Using Ambient Vibration

Zhouquan Feng,[1,2] **Jiren Zhang**,[1,2] **Lambros Katafygiotis**,[3] **Xugang Hua**,[1,2] **and Zhengqing Chen**[1,2]

[1]*Hunan Provincial Key Laboratory of Wind and Bridge Engineering, College of Civil Engineering, Hunan University, Changsha 410082, China*
[2]*State Key Laboratory of Bridge Safety and Resilience, Hunan University, Changsha 410082, Hunan, China*
[3]*Department of Civil and Environmental Engineering, The Hong Kong University of Science and Technology, Clear Water Bay, Kowloon, Hong Kong, China*

Correspondence should be addressed to Zhouquan Feng; zqfeng@hnu.edu.cn

Received 28 February 2024; Revised 4 April 2024; Accepted 24 May 2024

Academic Editor: Hoon Sohn

Modal parameter identification via ambient vibration is popular but faces challenges from uncertainties due to unknown inputs and low signal-to-noise ratio. Bayesian methods are gaining increasing attention for operational modal identification due to their ability to quantify uncertainties. However, improvements in computational efficiency are needed, particularly when addressing numerous modes and degrees of freedom. To address this challenge, this study proposes an innovative approach, termed the "Bayesian spectral decomposition" method (BSD), employing the decompose-and-conquer strategy. This novel method, operating within the frequency domain, identifies each mode individually by exploiting their inherent separated modal characteristics. For each mode, the response spectrum matrix undergoes an eigenvalue decomposition, yielding crucial eigenvalues (incorporating frequency and damping information) and eigenvectors (containing mode shape information). Subsequently, statistical properties of the eigenvalues and eigenvectors are utilized to establish likelihood functions for Bayesian parameter identification. By combining prior information, the posterior probability distribution functions of modal parameters are derived. The optimal solution is then obtained by resolving the maximum posterior probability distribution function problem. To further quantify the uncertainty of modal parameters, Gaussian distributions are employed to approximate the posterior probability distribution functions. The adoption of the decomposition approach circumvents the joint identification of all modal parameters, substantially reducing the parameter dimensions for optimization. Consequently, this strategy leads to decreased computational complexity and significantly improved computational stability. The effectiveness of the BSD is confirmed through simulated data generated from an 8-story shear building as well as measured data collected from both an experimental shear frame and the Canton Tower. The results demonstrate that the proposed method achieves high accuracy in identifying modal parameters, greatly improves computational efficiency, and effectively quantifies the uncertainties in modal parameters.

1. Introduction

The dynamic features of a structure can be roughly characterized by three key parameters: natural frequencies, damping ratios, and mode shapes [1]. Modal identification involves determining these modal parameters of an existing structure using measured dynamic testing data. The utilization of vibration data for modal identification has found extensive application over time, owing to its diverse uses in tasks such as model updating, damage identification [2], health monitoring, and vibration control [3]. Numerous modal identification algorithms have emerged over the past four decades. Specifically, significant emphasis has been placed on identifying modal parameters when solely response measurements, rather than input data, are at hand. These methodologies, commonly denoted as ambient modal identification, or operational modal identification, offer on-the-fly testing solutions for modal parameter determination without relying on external artificial stimuli. They utilize ambient excitations such as small tremors, vehicular movement, breezes, oceanic motions, and seismic activity. The application of operational modal identification has proven highly valuable in civil engineering, particularly in scenarios where employing large-scale exciters for buildings and bridges proves impractical. It offers an efficient alternative by using ambient vibrations during regular operation to identify modal parameters, eliminating the need for external exciters and preventing any potential interruptions or disturbances.

Commonly employed in practice, operational modal identification methods can be broadly categorized into two distinct groups: (a) frequency-domain approaches and (b) time-domain approaches. Among frequency-domain methods, the simplest one is the peak-picking method [4]. Subsequently, the frequency-domain decomposition (FDD) method [5] and modified frequency-domain decomposition (MFDD) [6] were developed, enabling the accurate identification of close modes even in the presence of strong signal noise. Several researchers also proposed frequency-domain least-squares methods [7–9], offering the advantage of extracting multiple modes across a broad frequency range. These methods achieve a least-square estimation of modal parameters by minimizing the measure of fit between the PSD matrix estimated from the measured output and the PSD matrix predicted by a model. To address the issue of non-Gaussian excitation in practical engineering, a new method based on eigensystem realization algorithm (ERA) and virtual frequency response function (FRF) has been proposed [10]. Prominent in time-domain techniques are the natural excitation technique (NExT) paired with ERA [11], the combination of random decrement technique (RDT) and the Ibrahim time-domain (ITD) method [12], and stochastic subspace identification (SSI) methods [13].

The aforementioned modal identification algorithms fall under the category of methods that rely solely on output information. Typically, these methods assume that the input follows the characteristics of Gaussian white noise, and the system exhibits time invariance. However, in practical engineering scenarios, these assumptions may deviate from reality. Moreover, due to small ambient vibration responses with low signal-to-noise ratio, the identified modal parameters, particularly the damping, become highly uncertain. To address this challenge, a promising approach is Bayesian system identification [14, 15]. This method involves evaluating the joint posterior probability density function (PDF) of the modal parameters based on the provided measured data and assumed models. The introduction of probabilistic techniques rooted in Bayesian principles for identifying modal parameters in the frequency domain through ambient data was pioneered by Katafygiotis and Yuen [16, 17]. While offering exceptional ability to identify the updated modal parameters through posterior PDF maximization and enabling uncertainty quantification, these methods face a critical challenge: the efficient computation of optimal values and uncertainty quantification. Current strategies typically involve computationally intensive nonlinear, multidimensional numerical optimization problems for optimal value determination, coupled with finite difference methods for covariance matrix calculation, hindering efficiency. This process becomes particularly challenging and computationally intensive when handling a substantial quantity of identified parameters, which is widely encountered in the modal identification of civil structures.

To enhance computational efficiency in the field of Bayesian modal identification, Au developed the fast Bayesian FFT method (FBFFT), a technique designed to quickly identify separated modes. FBFFT is an accelerated version of the Bayesian FFT method originally proposed by Yuen and Katafygiotis facilitating rapid calculation for the optimal values and their covariance matrix [18]. Li and Zhu et al. introduced the maximum expectation algorithm (EM) into the FBFFT method, further improving its computational efficiency [19, 20]. In a different approach, Yan et al. proposed a two-stage fast Bayesian spectral density method. By leveraging the assumption that the trace of the power spectral density matrix approximately follows a normal distribution, this method achieves decoupling of mode shapes from other modal parameters in cases of separated and dense modes. This significantly reduces the dimensionality of the parameter identification in optimization problem [21, 22]. Additionally, Zhang et al. also proposed a Bayesian modal identification method for free vibration [23, 24]. In addition to these developments, Sedehi et al. tackled the challenge of uncertainty assessment in identified modal parameters of linear systems by proposing a hierarchical Bayesian framework informed by multiple vibration datasets [25]. These advances in Bayesian system identification offer more efficient and accurate solutions for various modal identification challenges.

In this study, we propose a novel Bayesian algorithm designed to efficiently determine optimal values and covariance matrix in the frequency domain, offering significant computational advantages. The rest of this paper is organized as follows: Section 2 provides a brief review of two modal identification methods. Section 3 describes the formulation of the Bayesian spectral decomposition method. Section 4 presents examples using both simulated and measured data to showcase the capabilities and accuracy of BSD, including an 8-story shear building (Section 4.1) and a three-story shear frame (Section 4.2) as well as the Canton Tower (Section 4.3). Section 5 is the conclusions of this paper.

2. Revisiting of BSDA and FDD

In this section, we provide a concise review of two modal parameter identification algorithms closely related to this paper.

2.1. BSDA. Consider the acceleration response recorded at N_s degrees of freedom (DOFs) of dynamic system, denoted as $y = \{y(n) \in R^{N_s \times 1}\}$, $n = 1, \ldots, N$, with N denoting the quantity of data points of every individual measured channel. In this context, the measured response is expressed as $y(n)=x(n)+e(n)$, with $x(n)$ representing the acceleration response. The prediction error, denoted as $e(n)$, accommodates the difference amid model response and observed data, considering measurement variability and modeling inaccuracies. Utilizing the recorded y, a discrete estimator of the spectral density matrix is introduced as

$$\hat{S}(f_k)=\lim_{T\longrightarrow\infty}E\{[Y^T(f_k)]^*Y^T(f_k)\},\tag{1}$$

where $E(\bullet)$ denotes expectation operator, the superscripts $(\bullet)^T$ and $(\bullet)^*$ denote transpose and conjugate transpose, respectively, and $Y(f_k)$ represents the (scaled) Fourier transform of y at frequency f_k:

$$Y(f_k)=\sqrt{\frac{\Delta t}{N}}\sum_{n=0}^{N-1}y(n)e^{-j2\pi f_k n\Delta t},\tag{2}$$

where $j^2=-1$, Δt denotes sampling interval, $f_k=k\Delta f$, $k=1,\ldots N_1+1$, with $N_1=\mathrm{int}(N/2)$, $\Delta f=1/T$, and $T=N\times\Delta t$.

In the presence of ambient excitation (assumed to be Gaussian white noise) and i.i.d. Gaussian prediction errors, the probability density function for the spectral density matrix of the measured response has been established [16, 17]. This was subsequently employed in Bayesian modal identification. Assuming a collection of i.i.d. time histories $y^{(r)}\in R^{N_s\times N}\ (r=1,\ldots,M)$, and considering that $N\longrightarrow\infty$, the average estimate of the spectral density is as follows:

$$\hat{S}(f_k)=\frac{1}{M}\sum_{r=1}^{M}\left\{\left[\left(Y^{(r)}(f_k)\right)^T\right]^*\left(Y^{(r)}(f_k)\right)^T\right\}.\tag{3}$$

The distribution conforms to a central complex Wishart distribution of dimension N_s with M degrees of freedom [14]. The PDF of $\hat{S}(f_k)$ is given by [14]:

$$p(\hat{\mathbf{S}}(f_k))=\frac{\pi^{-N_s(N_s-1)/2}M^{N_s(M-N_s)}\left|\hat{\mathbf{S}}(f_k)\right|^{M-N_s}}{\left(\prod_{p=1}^{N_s}(M-p)!\right)\left|\mathbf{S}(f_k)\right|^{M}}\exp\left(-Mtr\left[\left(\mathbf{S}(f_k)^{-1}\hat{\mathbf{S}}(f_k)\right]\right),\tag{4}$$

where $|A|$ and tr[A] represent the determinant and the trace of matrix A, respectively. It is important to observe that equation (4) is defined only when $M\geq N_s$. Additionally, it can be demonstrated that as $N\longrightarrow\infty$, $\hat{\mathbf{S}}(f_k)$ and $\hat{\mathbf{S}}(f_l)$ are independent of each other when $k\neq l$, that is,

$$p[\hat{S}(f_k),\hat{S}(f_l)]=p[\hat{S}(f_k)]p[\hat{S}(f_l)].\tag{5}$$

Under the assumption of linearly damped vibration with N_r contributing modes and considering a relatively high sampling frequency along, the $S(f_k)$ in equation (4) is as follows:

$$S(f_k)=\Phi H_k\Phi^T+S_e,\tag{6}$$

where $\Phi\in R^{N_s\times N_r}$ represents the matrix of mode shapes limited to the observed DOFs, where one column corresponds to one mode shape; S_e denotes the constant spectrum matrix associated with the prediction error; H_k denotes the spectrum matrix associated with the response, with (i,j) element defined as

$$H_k(i,j)=\frac{S_{ij}}{\left[\left(\beta_{ik}^2-1\right)+j(2\beta_{ik}\zeta_i)\right]\left[\left(\beta_{jk}^2-1\right)-j\left(2\beta_{jk}\zeta_j\right)\right]},\tag{7}$$

where $\beta_{ik}=f_i/f_k$ and $\beta_{jk}=f_j/f_k$ are frequency ratios; f_i and ζ_i are the frequency and damping ratio pertaining to the ith mode, respectively; S_{ij} represents the cross-spectral density between the ith and j^{th} modal force.

Within the framework of modal identification utilizing the PSD matrix model outlined in equations (6) and (7), the modal parameter set θ encompasses not only frequency, damping ratio, and mode shape but also the spectrum matrices for modal forces and prediction errors. According to Bayes' theorem, the posterior PDF of θ is proportional to the likelihood function,

$$p\left(\theta\middle|\hat{S}^{k_1,k_2}\right)\propto p\left(\hat{S}^{k_1,k_2}\middle|\theta\right)\simeq c_1\prod_{k=k_1}^{k_2}\frac{\left|\hat{S}(f_k)\right|^{M-N_s}}{\left|S(f_k)\right|^{M}}\exp\left(-Mtr\left[S(f_k)^{-1}\hat{S}(f_k)\right]\right),\tag{8}$$

where c_1 serves as a normalization factor, and the multiplication is conducted across the frequencies within the designated frequency band ($f_{k_1} \sim f_{k_2}$). It is convenient to write the posterior PDF in terms of the "negative log-likelihood function" (NLLF) $L(\theta)$ as

$$p\left(\theta \middle| S^{k_1,k_2}\right) \propto \exp[-L(\theta)], \tag{9}$$

where

$$L(\theta) = M \sum_{k=k_1}^{k_2} \left[\ln\left|S(f_k)\right| + tr\left[S(f_k)^{-1}\widehat{S}(f_k)\right]\right] + c_2. \tag{10}$$

Achieving the most probable parameters $\widehat{\theta}$ involves minimizing $L(\theta)$. Additionally, when dealing with a substantial volume of data, the posterior PDF for the parameters θ can be effectively approximated as a Gaussian distribution with mean $\widehat{\theta}$ and covariance matrix $H^{-1}(\widehat{\theta})$, where $H(\widehat{\theta})$ denotes the Hessian of $L(\theta)$ calculated at $\theta = \widehat{\theta}$. Consider the approximation for $L(\theta)$

$$L(\theta) \approx L(\widehat{\theta}) + J(\widehat{\theta})(\theta - \widehat{\theta}) + \frac{1}{2}(\theta - \widehat{\theta})^T H(\widehat{\theta})(\theta - \widehat{\theta}). \tag{11}$$

Please be aware that the $L(\widehat{\theta})$ remains unchanged, and the $J(\widehat{\theta})(\theta - \widehat{\theta})$ vanishes as $J(\widehat{\theta})$ becomes zero, owing to the minimizing property. By replacing equation (11) with equation (9), one can estimate the posterior PDF through an approximation as a Gaussian distribution

$$p\left(\theta \middle| S^{k_1,k_2}\right) \propto \exp\left[-\frac{1}{2}(\theta - \widehat{\theta})^T C(\widehat{\theta})^{-1}(\theta - \widehat{\theta})\right], \tag{12}$$

where $C(\widehat{\theta}) = H^{-1}(\widehat{\theta})$.

The challenge at hand involves efficiently determining the optimal values for $\widehat{\theta}$ and $C(\widehat{\theta})$. In the BSDA, numerical optimization is employed to solve for the optimal values, and finite difference is applied to compute the Hessian matrix. Notably, the objective function presents itself as an intricate nonlinear expression of the modal parameters, resulting in computationally intensive efforts to ascertain optimal values. Moreover, the inverse of the matrix $S(f_k)$ becomes ill-conditioned within a resonance frequency range. If numerical optimization was to be employed for obtaining all optimal values, the computational burden would increase with both the number of N_s and N_r, and there are $2N_r + N_r(N_s - 1) + N_r(N_r + 1)/2 + N_s = O(N_r^2 + N_rN_s)$ parameters in need of identification [26]. In well-separated cases, every mode can be identified independently, leading to $N_r = 1$ in such instances. Also, then, the growth of the dimension with N_s becomes a major issue. The increase in dimensionality with N_s poses a significant challenge, especially in typical civil engineering scenarios where N_s may exceed 10.

2.2. FDD. Consider a linear low-damping system with white noise input, the PSD of the response can be written [5].

$$S(f_k) \approx \Phi D_k \Phi^H, \tag{13}$$

where D_k which takes the form of a diagonal matrix is the PSD matrix of input. Also, Φ is frequency response function matrix. Conversely, the singular value decomposition of the response PSD is expressed as

$$\widehat{S}(f_k) = U_k \Lambda_k U_k{}^H, \tag{14}$$

where U_k is a unitary matrix ($U_k U_k{}^H = I$) holding the eigenvector u_i of $\widehat{S}(f_k)$ and Λ_k is a diagonal matrix that holds the associated eigenvalues. And the first eigenvector associated with the largest eigenvalue at the peak frequency can be as an estimate of the mode shapes. Furthermore, the corresponding largest eigenvalues are connected to the auto-PSD function of the corresponding SDOF system. Therefore, the primary eigenvector serves as an approximation of the mode shape, and from the segment of the SDOF auto-PSD function surrounding the peak frequency, the natural frequency and the damping ratio can be obtained. When two modes exert dominance, the primary eigenvector consistently provides a reliable approximation for the stronger mode. If the two modes lack orthogonality, the potential bias in mode shape estimation for the dominant mode tends to be minor, while the bias for the mode shape estimation of the weaker mode becomes more pronounced.

3. Formulation of Bayesian Spectral Decomposition Method

Next, we embrace a decompose-and-conquer approach for the separation of modal parameters in BSDA. In the initial decompose-and-conquer tier, each mode is identified respectively, especially applicable for well-separated modes. Progressing to the second decompose-and-conquer level, the spectral parameters (contains modal frequency and damping ratios) and spatial parameters (linked to the mode shape) are addressed individually.

3.1. Identification of Modal Frequency and Damping Ratio. For a structural system featuring distinctly separated modes, the assumption is made that within a range around the peak of the resonant frequency, the response is predominantly influenced by a singular mode. Let Ω_j be the index set of the neighborhood of the frequency of the jth mode. By examining the mathematical structure of $S(f_k)(k \in \Omega_j)$ in equation (6), it can be rewritten as

$$S(f_k) = \alpha_k \phi \phi^T + S_e, \tag{15}$$

where ϕ is the mode shape of the jth mode, hereafter the modal index j is omitted for simplification; S_e is the residual spectral density of prediction error; α_k is the spectral density of dominant modal response given by

$$\alpha_k = \frac{S_f}{\left(\beta_k^2 - 1\right)^2 + \left(2\zeta\beta_k\right)^2} = S_f A_k, \tag{16}$$

where $\beta_k = f/f_k$; S_f is the spectral density of modal force; $A_k = 1/[(\beta_k^2 - 1)^2 + (2\zeta\beta_k)^2]$ represents the dynamic amplification specific to one mode.

For the well-separated modes, near the modal frequency of a specific mode, a predominant mode emerges singularly, with the spectral intensity of the forecast error typically minor compared to that of the related mode. However, the mathematical structure of the prediction error may be complicated and unknown in practice. To make the problem to be solvable, we make two assumptions: (1) S_e is a constant Hermitian matrix, as is evident from equation (15), where S_e is a Hermitian matrix since $S(f_k)$ and $\alpha_k\phi\phi^T$ are both Hermitian; the assumption "constant" implies that S_e remains flat in the vicinity of a modal frequency; (2) one of the eigenvectors of S_e is in the modal space spanned by ϕ, and this point is relatively easy to approximately satisfy because S_e is typically very smaller than that of the corresponding contributed mode $\alpha_k\phi\phi^T$.

Describe an orthogonal basis $B = \{b_i \in R^{N_s}: i = 1, 2, \ldots, N_s\}$ using the subsequent approach. Set $b_1 = \phi$ and let $\{b_i: i = 2, \ldots, N_s\}$ be an orthonormal basis in the orthogonal complement of the subspace spanned by ϕ. Using this basis, we express S_e via an eigendecomposition as

$$S_e = a_1\phi\phi^T + \sum_{i=2}^{N_s} a_i b_i b_i^T. \tag{17}$$

The eigenvalues of S_e are $a_i\,(i = 1, 2, \ldots, N_s)$, with the corresponding eigenvector $\mathbf{b}_i\,(i = 1, 2, \ldots, N_s)$. Then, equation (15) can be rewritten as

$$S(f_k) = (\alpha_k + a_1)\phi\phi^T + \sum_{i=2}^{N_s} a_i b_i b_i^T. \tag{18}$$

The eigenvalues of $S(f_k)$ are $\alpha_k + a_1, a_2, \ldots, a_{N_s}$ with the corresponding eigenvector $b_1 = \phi, b_2, \ldots, b_{N_s}$. Then, the $\alpha_k + a_1$ can be written as follows:

$$\lambda_k = \alpha_k + a_1 = \frac{S_f}{\left(\beta_k^2 - 1\right)^2 + \left(2\zeta\beta_k\right)^2} + a_1. \tag{19}$$

Subsequently, our aim is to determine the probability distributions associated with the eigenvalues of the $\widehat{S}(f_k)$ which follows a central complex Wishart distribution $W(\Sigma, M)$ of dimension N_s with the real covariance matrix Σ and M degree of freedom. Let $\{\lambda_i, i = 1, 2, \ldots, p\}$ and $\{l_i, i = 1, 2, \ldots, p\}$ represent the eigenvalues of Σ and W, respectively. Based on findings from the literature [27] (the theorem is outlined in Appendix), l_i asymptotically conforms to a normal distribution as M becomes sufficiently large. We can infer the probability distributions associated with the eigenvalues of the estimated spectrum matrix $\widehat{S}(f_k)$. Let l_k^i denote the ith eigenvalue of $\widehat{S}(f_k)$ (let the first one be the largest one), so l_k^i is normally distributed as

$$E\left(l_k^1\right) = \lambda_k, \quad Var\left(l_k^1\right) = \frac{\lambda_k^2}{M}, \tag{20a}$$

$$E\left(l_k^i\right) = a_i, \quad Var\left(l_k^i\right) = \frac{a_i^2}{M}, \quad (i = 2, \ldots, N_s), \tag{20b}$$

where M is the number of data segments when calculating the PSD estimate $\widehat{S}(f_k)$ in equation (3).

We denote the spectral parameters $\theta = \{f, \zeta, S_f, a_i, i = 1, \ldots, N_s\}$ and divide them into two groups: $\theta_1 = \{f, \zeta, S_f, a_1\}$ and $\theta_2 = \{a_2, i = 2, \ldots, N_s\}$. Using a Bayesian statistic framework [14], the posterior PDF of θ_1, given the data of the largest eigenvalues, can be written as follows:

$$p\left(\theta_1 \middle| l_{k_1\sim k_2}^1\right) \propto p\left(l_{k_1\sim k_2}^1 \middle| \theta_1\right) \simeq \prod_{k=k_1}^{k_2} \frac{\sqrt{M}}{\sqrt{2\pi}\lambda_k} \exp\left(-\frac{M\left(l_k - \lambda_k\right)^2}{2\lambda_k^2}\right). \tag{21}$$

For the other parameters $\theta_2 = \{a_i, i = 2, \ldots, N_s\}$, the posterior PDF is as follows:

$$p\left(a_i \middle| l_{k_1\sim k_2}^i\right) \propto p\left(l_{k_1\sim k_2}^i \middle| a_i\right) \simeq \prod_{k=k_1}^{k_2} \frac{\sqrt{M}}{2\sqrt{\pi}a_i} \exp\left(-\frac{M\left(l_k - a_i\right)^2}{2a_i^2}\right), \quad (i = 2, \ldots, N_s). \tag{22}$$

Integrating a substantial amount of eigenvalue information across the resonance frequency range generally enhances the overall identifiability of the problem. The optimal values $\widehat{\theta}_1$ and $\widehat{\theta}_2$ are obtained by maximizing the corresponding probability density functions as delineated in equations (21) and (22). This process is equivalent to minimizing the negative log-likelihood function, as presented in the following equation:

$$L(\theta_1) = -\ln\left[p\left(l^1_{k_1\sim k_2}|\theta_1\right)\right] = \sum_{k=k_1}^{k_2}\left[\ln\frac{\sqrt{2\pi}\lambda_k}{\sqrt{M}} + \frac{M(l_k-\lambda_k)^2}{2\lambda_k^2}\right]. \quad (23)$$

Numerical optimization algorithms or analytical derivations can be employed to solve the minimization problem. Also, the covariance matrix is equivalent to the inverse of the Hessian matrix at their most probable values, allowing for the utilization of either finite difference or analytical formulations to calculate the Hessian matrix.

3.2. Identification of Mode Shape. As indicated in equation (18), it is evident that the eigenvector ϕ associated with λ_k can be used as the optimal values of the mode shape. Due to $\widehat{S}(f_k)$ conforms to a Wishart distribution $W(\Sigma, M)$, if λ_i and e_i are eigenvalues and corresponding eigenvectors of $\Sigma, i = 1, 2, \ldots, p$. As per Ref. [27], the eigenvector u_i of $\widehat{S}(f_k)$ follows an asymptotically normal distribution, and the covariance matrix is:

$$\mathrm{cov}(u_i, u_i) = \sum_{\substack{j=1,\\ j\neq i}}^{p} \frac{\lambda_i\lambda_j}{(\lambda_i-\lambda_j)^2 M} e_j e_j^T. \quad (24)$$

So, the $\mathrm{cov}(\phi, \phi)$ can be written as follows:

$$\mathrm{cov}(\phi,\phi) = \sum_{j=2}^{N_s}\frac{(\alpha_k + a_1)a_j}{(\alpha_k + a_1 - a_j)^2 M} b_j b_j^T. \quad (25)$$

From equation (25), $\mathrm{cov}(\phi, \phi)$ is a noninvertible matrix with a rank $N_s - 1$, and the multivariate normal distribution will become degenerate. Hence, the pseudodeterminant is determined by multiplying the nonzero eigenvalues, whereas the pseudoinverse is obtained by summing the inverses of every nonzero eigenvalues multiplied by the outer product of the associated eigenvectors.

3.3. Main Procedures of BSD. The outlined four steps for the proposed method can be succinctly encapsulated as shown in Figure 1.

(i) Utilize equation (3) to compute $\widehat{S}(f_k)$

(ii) Conduct eigendecomposition on $\widehat{S}(f_k)$ in proximity to the mode targeted for identification, determining the largest eigenvalue l_k, and its associated eigenvector u_k

(iii) Determine the MPVs of spectral parameters by minimizing equation (23), and calculate their covariance matrix

(iv) Determine the MPVs of mode shape, and compute their covariance matrix using equation (25)

4. Verification and Applications

4.1. Numerical Simulation

4.1.1. Eight-Story Shear Building. An 8-story shear building was used for simulation analysis, as shown in Figure 2. The model assumes a uniform distribution of floor mass $m = 1\times10^5$ kg and interstory stiffness $k = 2.5\times10^8$ N/m, incorporating Rayleigh's damping with $\zeta_1 = \zeta_2 = 1\%$ [28]. Every floor encounters a separate stationary Gaussian white noise stimulation. Accelerations at each level are recorded at a sampling frequency of 100 Hz. Each dataset spans a time duration of 1000s across eight channels, and a total of 20 datasets (M = 20) are collected. In the computation of $\widehat{S}(f_k)$, each segment encompasses 10^5 data points, and the number of segments for spectrum averaging totals 20. The measurement noise at distinct sensors is presumed to be i.i.d. Gaussian white noise with an RMS of 1 μg.

In line with the presumption outlined in Section 3.1, the eigenvalues and eigenvectors of this estimation asymptotically conform to Gaussian distributions. In order to confirm the precision of the asymptotically Gaussian distribution for the eigenvalues, the empirical CDFs are computed by generating 500 instances, with every instance comprising M = 20 groups of measurements. Figure 3 gives the comparisons at f = 1.4685 Hz, 1.4800 Hz, 4.3555 Hz, 4.5000 Hz, 7.0941 Hz, 7.2000 Hz, 9.5912 Hz, 9.7000 Hz, 11.7617 Hz, and 11.9000 Hz with M = 20. Observably, the Gaussian distribution offers a satisfactory approximation for the largest eigenvalues when M = 20.

4.1.2. Modal Frequencies and Damping Ratios. Figure 4 illustrates the spectrum of the largest eigenvalues, revealing the prominence of the 1st to 5th modes. The BSD is employed to identify these modes. The identified outcomes, encompassing optimal values $\widehat{\theta}$, standard deviations (SD), coefficient of variance (CV), "normalized distance" δ, and the percentage of error are summarized in Table 1. The parameter δ signifies the relative deviation of the optimal values from the actual values, compared to their SD. Notably, the most probable values closely align with their respective exact values. As commonly observed, the CVs of the modal frequencies are considerably smaller than those of the damping ratios.

The oval-shaped outlines of the updated PDF for f_1 and ζ_1, which envelop the area between probabilities 0.5 and 0.9, are depicted in Figure 5. This illustration highlights that the real parameters are reasonably distant from the identified optimal parameters. Figure 5 illustrates the PDFs

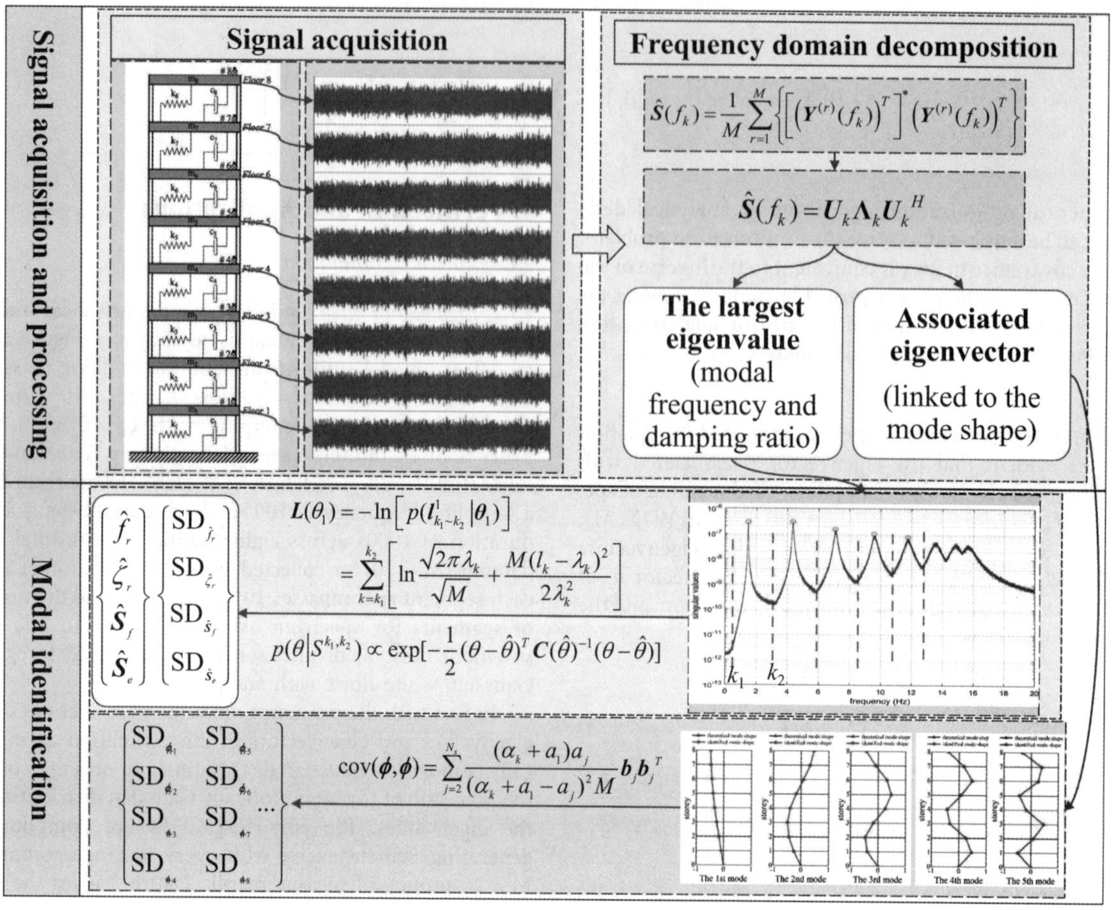

FIGURE 1: The flowchart of BSD.

conditioned on the f_1 and ζ_1, while the remaining parameters are set at their optimal values. These conditional PDFs, derived using both the BSD and the Gaussian approximation, are graphed accordingly. The results in Figure 6 suggest that the Gaussian approximation adequately represents the updated PDF.

4.1.3. Mode Shapes. The outcomes of mode shape identification, along with their respective standard deviations (SD), are presented in Table 2, while the corresponding mode shapes are depicted in Figure 7. Evaluation of the modal assurance criteria (MAC) values, depicted in Figure 8, reveals a close alignment between the identified results and theoretical values.

4.1.4. Discussion on Accuracy and Efficiency. In this subsection, the performance of the BSD method will be further compared with BSDA and automated FDD [29]. Here, we still take the eight-story shear building as the numerical example, and the results of the frequencies, and damping ratios identified by each method are listed in Table 3. It is not difficult to find that the results of the frequencies identified by the three methods are very close to the actual values. Furthermore, Bayesian methods such as BSDA and BSD can also quantify the uncertainties of the identification results, thereby aiding in the evaluation of the confidence level associated with the identified outcomes. The identified values of the modal frequencies closely approximate the actual values, with very small CVs. Concerning damping ratios, deviations are observed in the results identified by three algorithms, while the identification results of BSD are more accurate, with its CVs being lower than those of BSDA. Due to space constraints, the identification results of the mode shape are not provided. Nevertheless, the accuracy of mode shape identification for BSD and FDD is roughly equivalent and slightly better than that of BSDA.

However, in addition to identification accuracy, modal identification in practical applications also demands high computational efficiency to meet real-time requirements. Therefore, the following discussion involves analyzing and comparing the computational efficiency of the three methods. The three programs were run in MATLAB (2022b) on the same computer equipped with a 3.2 GHz Intel Core i7 processor and 64 GB of RAM. Only the time taken for the programs to identify modal parameters (including uncertainty quantification) was considered, excluding the time for generating response data. The

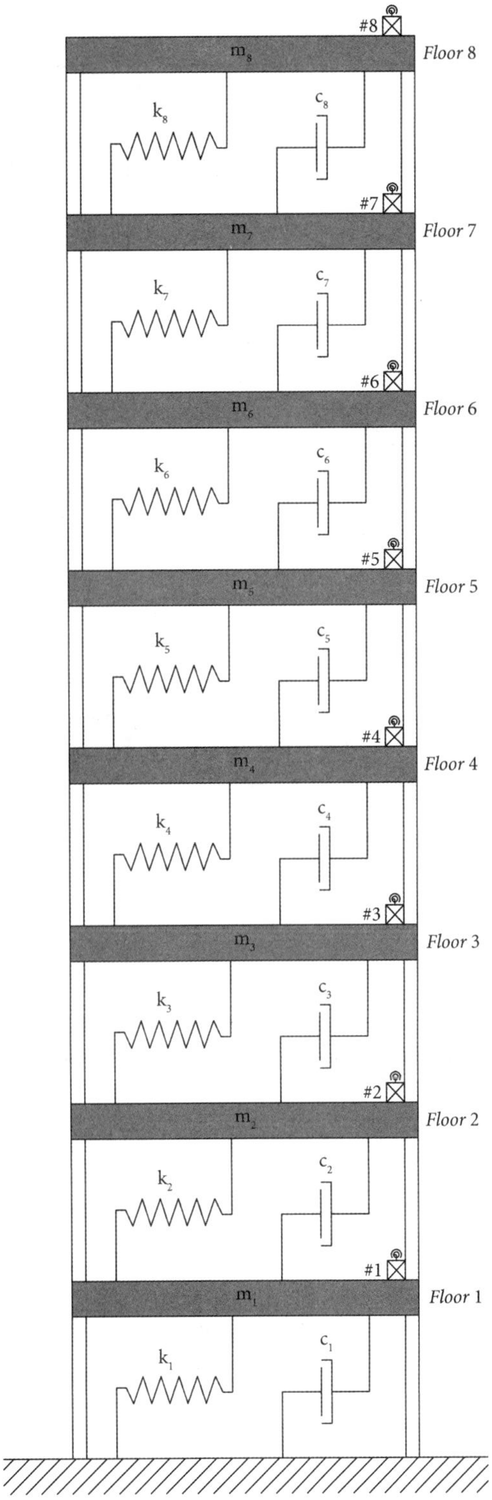

FIGURE 2: Eight-story shear building.

computational times for BSD, BSDA, and FDD were 5.31 seconds, 15.34 seconds, and 18.31 seconds, respectively. This difference can be explained by analyzing the computational processes of the algorithms. BSDA involves optimizing 51 variables (including the frequencies, damping ratios, and mode shapes with eight measured channels of the first five modes) in each iteration, whereas, although automated FDD can quickly obtain frequencies

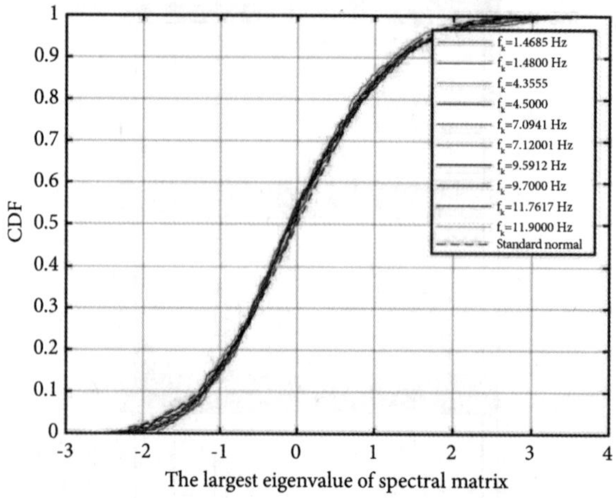

FIGURE 3: CDFs calculated from the samples of the largest eigenvalues.

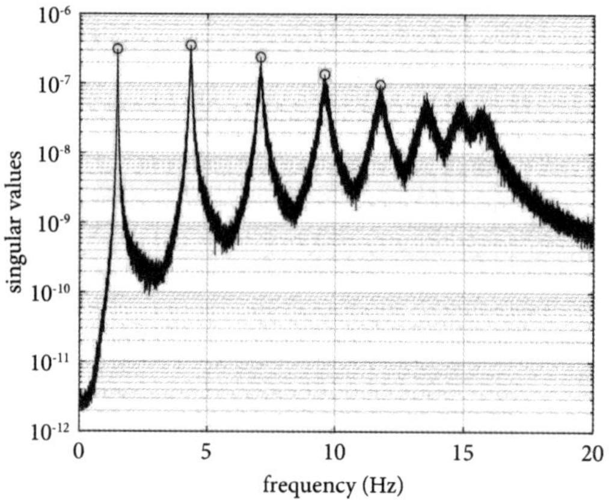

FIGURE 4: The spectrum of largest eigenvalues.

TABLE 1: The identified results of f and ζ.

Parameter	Band	Actual θ	Optimal $\widehat{\theta}$	SD (σ)	CV	$\delta = \left\|\theta - \widehat{\theta}\right\|/\sigma$	Error $= \left\|\theta - \widehat{\theta}\right\|/\theta$ (%)
f_1	[1.447, 1.81]	1.4685	1.4689	0.0007	0.0005	0.6153	0.0285
ζ_1		0.0100	0.0100	0.0002	0.0247	0.0003	0.0007
f_2	[4.325, 4.393]	4.3555	4.3556	0.0017	0.0004	0.0634	0.0025
ζ_2		0.0100	0.0102	0.0002	0.0149	1.4295	2.1813
f_3	[7.061, 7.149]	7.0941	7.0952	0.0046	0.0006	0.2360	0.0152
ζ_3		0.0137	0.0135	0.0002	0.0125	1.1586	1.4274
f_4	[9.580, 9.730]	9.5912	9.5853	0.0073	0.0008	0.8092	0.0615
ζ_4		0.0176	0.0172	0.0003	0.0175	1.2193	2.0894
f_5	[11.631, 11.795]	11.7617	11.7669	0.0111	0.0009	0.4717	0.0443
ζ_5		0.0211	0.0204	0.0003	0.0134	2.5688	3.3268

and mode shapes through singular value decomposition, it still requires additional computations using NExT and Hilbert transform to calculate damping ratios, which evidently prolongs the computational time.

4.2. Shear Frame Vibration Test. The proposed method was experimentally illustrated using a three-story shear frame, as originally made by Chang and Poon [30], as depicted in Figure 9. The frequencies and damping ratios were obtained

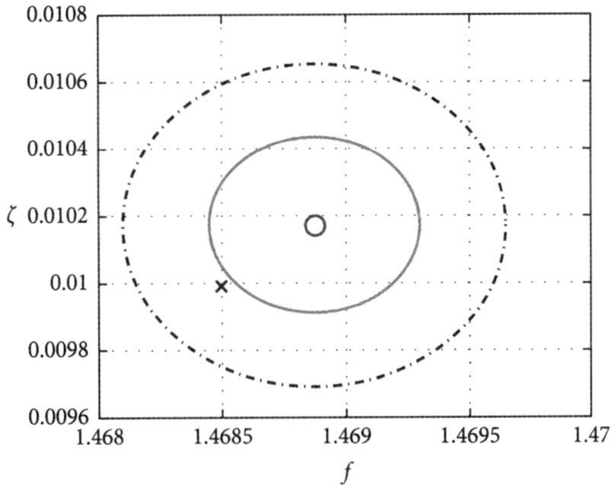

FIGURE 5: Contours of the marginal PDF of f_1 and ζ_1.

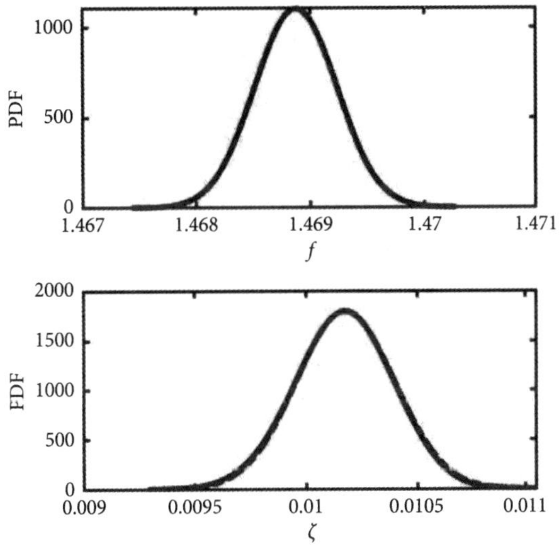

FIGURE 6: Conditional PDFs of f_1 and ζ_1.

TABLE 2: Mode shapes as identified.

Mode		DOF							
		#1	#2	#3	#4	#5	#6	#7	#8
1st	Id	−0.0900	−0.1756	−0.2551	−0.3262	−0.3864	−0.4346	−0.4667	−0.4833
	SD	0.0004	0.0004	0.0004	0.0003	0.0003	0.0003	0.0004	0.0003
2nd	Id	−0.2553	−0.4341	−0.4840	−0.3889	−0.1771	0.0862	0.3246	0.4655
	SD	0.0009	0.0010	0.0012	0.0013	0.0016	0.0017	0.0017	0.0015
3rd	Id	−0.3840	−0.4655	−0.1753	0.2534	0.4858	0.3305	−0.0882	−0.4335
	SD	0.0038	0.0040	0.0051	0.0042	0.0030	0.0045	0.0034	0.0045
4th	Id	0.4759	0.2474	−0.3208	−0.4418	0.0821	0.4772	0.1776	−0.3849
	SD	0.0096	0.0079	0.0063	0.0061	0.0074	0.0076	0.0071	0.0092
5th	Id	−0.4844	0.0846	0.4728	−0.1705	−0.4282	0.2474	0.3876	−0.3329
	SD	0.0118	0.0106	0.0091	0.0153	0.0110	0.0123	0.0117	0.0109

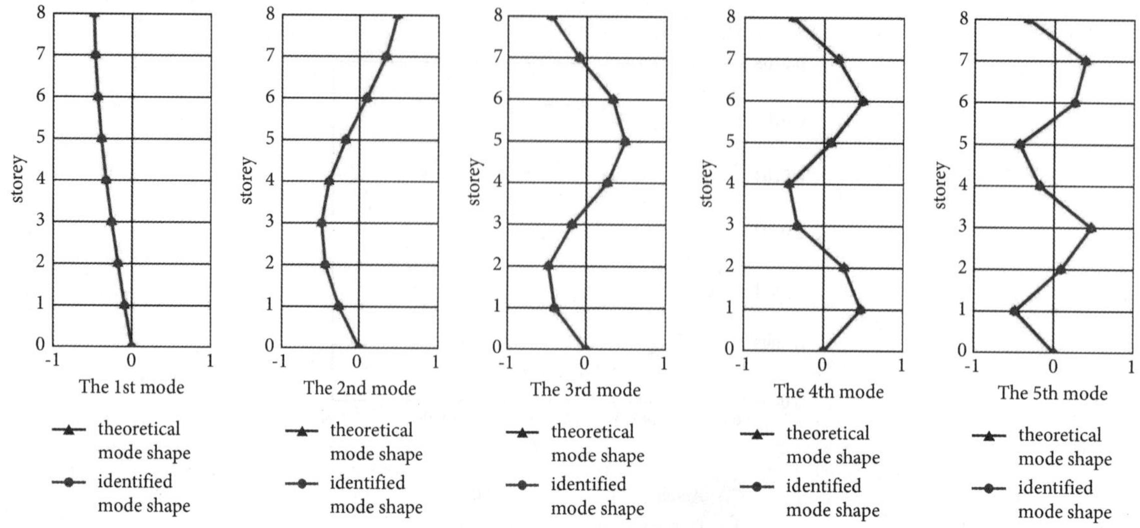

Figure 7: Theoretical and identified mode shapes.

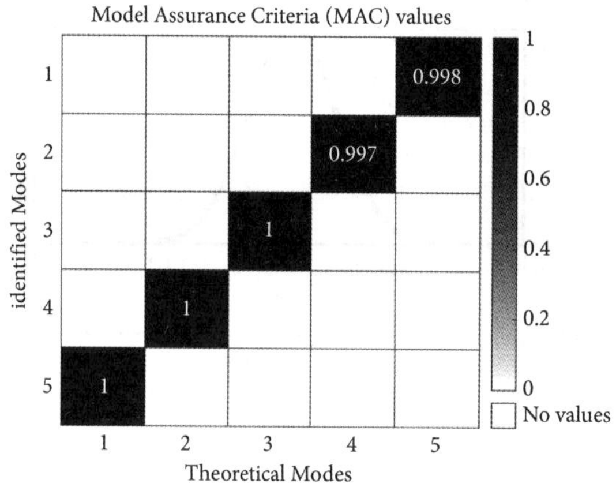

Figure 8: The modal assurance criteria (MAC) values.

Table 3: Comparison between identified results by applying BSD, BSDA, and FDD.

Mode	Parameter	Actual	BSD		BSDA		FDD
			Optimal	CV	Optimal	CV	
1	f_1	1.4685	1.4689	0.0005	1.4770	0.0011	1.4647
	ζ_1	0.0100	0.0100	0.0247	0.0101	0.1064	0.0087
2	f_2	4.3555	4.3556	0.0004	4.3608	0.0007	4.3515
	ζ_2	0.0100	0.0102	0.0149	0.0109	0.0640	0.0095
3	f_3	7.0941	7.0952	0.0006	7.0937	0.0006	7.0734
	ζ_3	0.0137	0.0135	0.0125	0.0147	0.0506	0.0132
4	f_4	9.5912	9.5853	0.0008	9.5838	0.0007	9.6611
	ζ_4	0.0176	0.0172	0.0175	0.0181	0.0489	0.0158
5	f_5	11.7617	11.7669	0.0009	11.7416	0.0008	11.8154
	ζ_5	0.0211	0.0204	0.0134	0.0223	0.0612	0.0217

by Chang and Poon [30] and as shown in Table 4. The structural design was fastened onto a vibration table and subjected to ground vibrations. Wireless sensors operating at a sampling frequency of 100 Hz were utilized to record acceleration data from each of the three levels and the ground level. Twelve sets of data were collected, each comprising 60000 points (10 minutes) per set. In Figure 10, the distribution of the primary eigenvalue's spectrum is

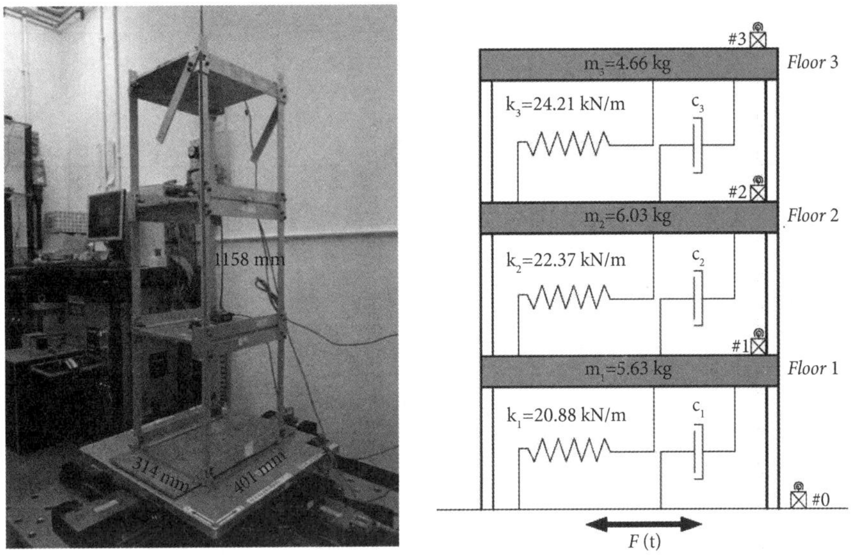

FIGURE 9: The tested shear frame.

TABLE 4: Modal frequencies and damping ratios as identified in an experimental example.

Mode	Band	Parameter	Nominal	Identified	CV
1	[4.13, 4.35]	f_1	4.5450	4.2366	0.0007
		ζ_1	0.0239	0.0077	0.0678
2	[12.69, 12.89]	f_2	13.0227	12.7754	0.0005
		ζ_2	0.0087	0.0063	0.6476
3	[18.55, 18.77]	f_3	18.2101	18.6495	0.0004
		ζ_3	0.0065	0.0039	0.0912

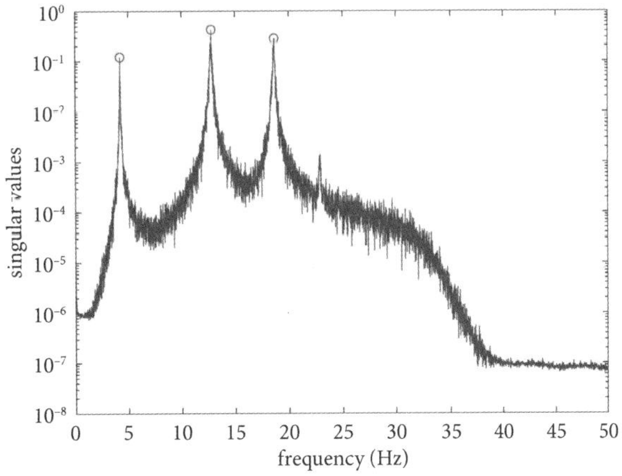

FIGURE 10: The spectrum of the largest eigenvalues from tested shear frame model.

illustrated for a single collection of recorded acceleration data, with 6 segments used for spectrum averaging and 10000 data points in each segment. Notably, all three modes were successfully excited.

Table 4 presents the outcomes of the identified modal frequencies and damping ratios for a single dataset. Examination of the results reveals a favorable correspondence between the nominal and identified values for frequency and damping ratios. As expected, CVs for damping ratios are greater than that for modal frequencies. It is important to emphasize that this experimental model exhibits some variations compared to Chang and Poon's model. Specific components underwent disassembly, and there were alterations made to the bottom fixing device upon reassembly. As

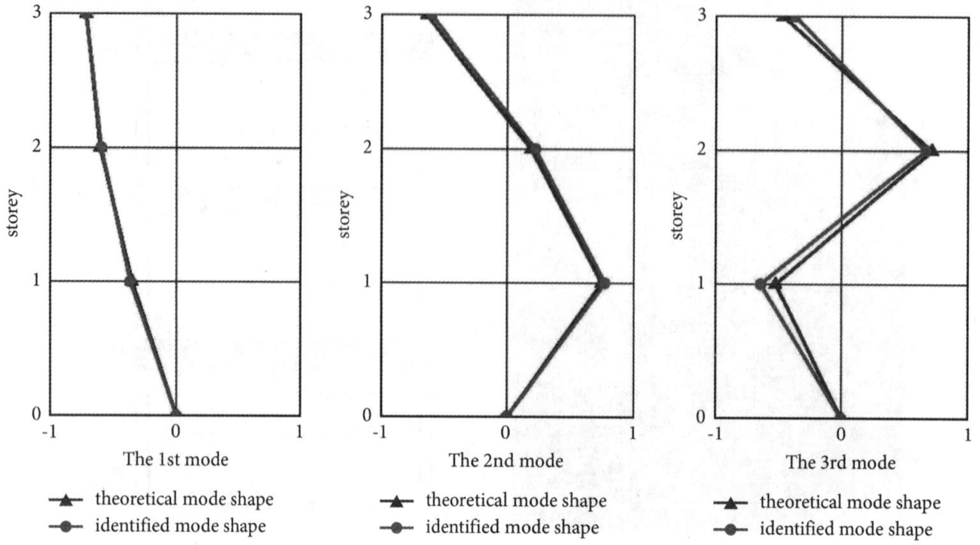

FIGURE 11: Theoretical and identified mode shapes of experimental model.

TABLE 5: Mode shapes as identified in an experimental example.

Mode	DOF	Nominal	Identified	SD	MAC
1	1	−0.3495	−0.3638	0.0018	0.9997
	2	−0.6039	−0.5935	0.0004	
	3	−0.7164	−0.7179	0.0009	
2	1	0.7450	0.7693	0.0044	0.9996
	2	0.1850	0.2187	0.0020	
	3	−0.6409	−0.6003	0.0050	
3	1	−0.5230	−0.6421	0.0124	0.9757
	2	0.7120	0.6673	0.0087	
	3	−0.4685	−0.3774	0.0053	

TABLE 6: Average figures of modal frequencies and damping ratios.

Mode	Parameter	Nominal	Identified	SD	CV
1	f_1	4.5450	4.2366	0.0026	0.0006
	ζ_1	0.0239	0.0061	0.0005	0.0749
2	f_2	13.0227	12.7834	0.0060	0.0005
	ζ_2	0.0087	0.0035	0.0011	0.2785
3	f_3	18.2101	18.6624	0.0101	0.0005
	ζ_3	0.0065	0.0053	0.0029	0.4691

TABLE 7: Average figures of mode shapes.

Mode	DOF	Nominal	Identified	SD	MAC
1	1	−0.3495	−0.3659	0.0010	0.9995
	2	−0.6039	−0.5916	0.0004	
	3	−0.7164	−0.7184	0.0005	
2	1	0.7450	0.7661	0.0033	0.9963
	2	0.1850	0.2203	0.0015	
	3	−0.6409	0.6033	0.0036	
3	1	−0.5230	−0.6513	0.0065	0.9646
	2	0.7120	0.6550	0.0049	
	3	−0.4685	−0.3735	0.0029	

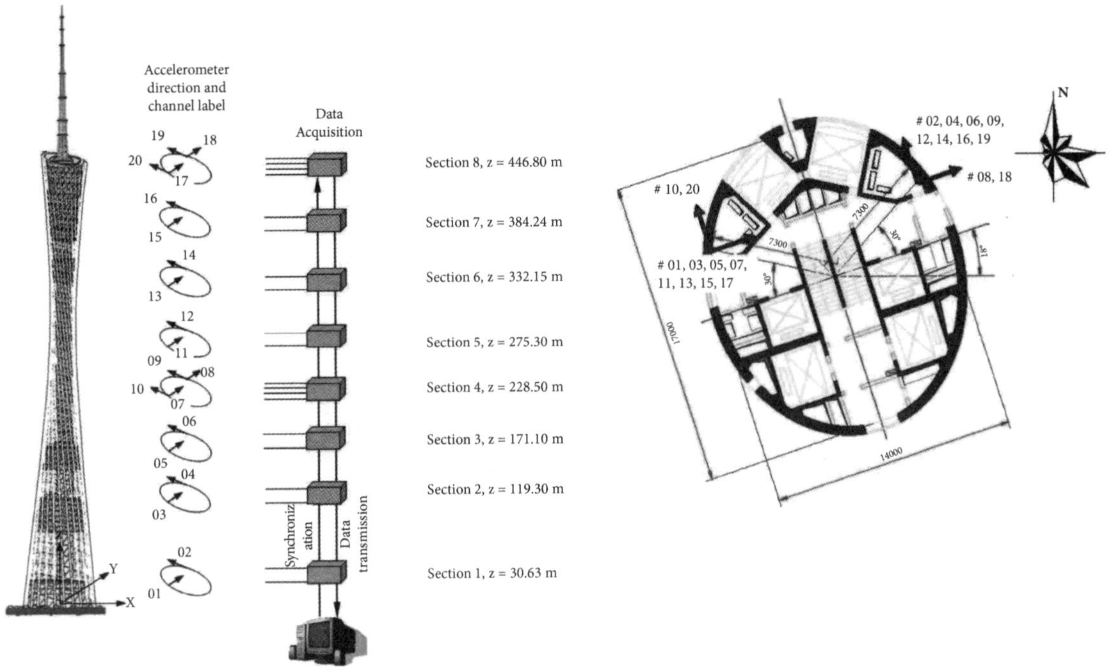

FIGURE 12: Deployment of accelerometers and data acquisition system on Canton Tower.

TABLE 8: Identified modal frequencies and damping ratios of the first ten modes.

Mode	Modal frequencies (Hz)	CV	Damping ratios (%)	CV
1	0.0937	0.0020	0.95	0.1959
2	0.1382	0.0013	0.53	0.1835
3	0.3662	0.0006	0.34	0.2126
4	0.4239	0.0005	0.26	0.1791
5	0.4752	0.0005	0.20	0.1639
6	0.5056	0.0003	0.10	0.3964
7	0.5223	0.0003	0.13	0.3023
8	0.7952	0.0006	0.37	0.0996
9	0.9668	0.0002	0.12	0.2209
10	1.1503	0.0002	0.11	0.1527

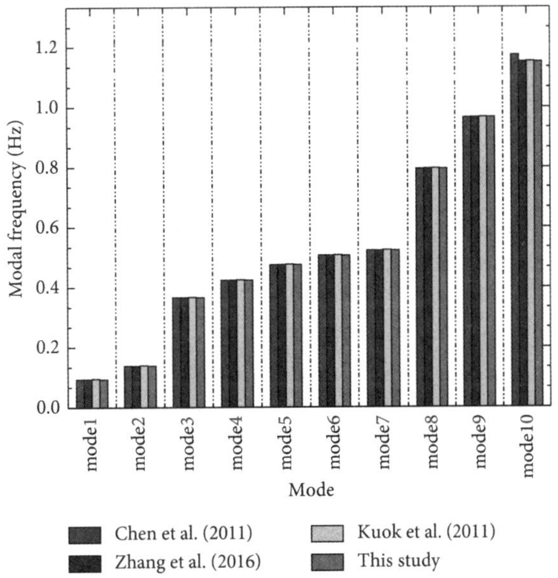

FIGURE 13: Comparison of the identified modal frequencies.

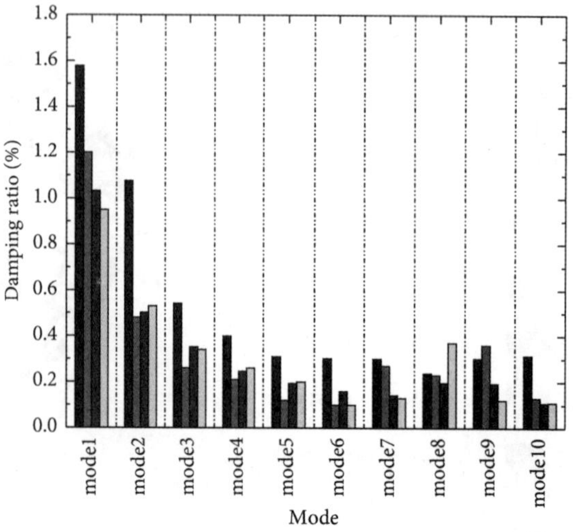

FIGURE 14: Comparison of the identified damping ratios.

TABLE 9: Comparison of the damping ratios and their respective CVs using two methods.

Mode	This study		Kuok et al. [31]	
	Damping ratios (%)	CV	Damping ratios (%)	CV
1	0.95	0.1959	1.03	0.2265
2	0.53	0.1835	0.50	0.2110
3	0.34	0.2126	0.35	0.2018
4	0.26	0.1791	0.25	0.2430
5	0.20	0.1639	0.19	0.2032
6	0.10	0.3964	0.16	0.3844
7	0.13	0.3023	0.14	0.3089
8	0.37	0.0996	0.20	0.1826
9	0.12	0.2209	0.19	0.1630
10	0.11	0.1527	0.11	0.2157

a result, these variations have resulted in minor discrepancies in their respective frequencies and significant differences in the damping ratio of the first mode. However, it is worth noting that the identification results presented in this paper closely align with those reported by Yan et al. [22].

Figure 11 shows that the theoretical and identified modal shape curves match well. Apart from the third mode, the other modes are basically consistent. This is also evident from the MAC values in Table 5, all of which are higher than 0.97. However, it cannot be denied, as mentioned earlier, that there is a certain deviation between the experimental model and the theoretical model. Therefore, model updating is often necessary in many cases.

The average figures of identified outcomes for 12 groups of data are presented in Table 6 and Table 7 correspondingly. Each group of data is divided into 6 segments for spectrum averaging, and each segment has 10,000 data points. We can find that the identified results of 12 sets of data are essentially the same as the identified results of one set of data.

4.3. Modal Identification of the Canton Tower. The Canton Tower (previously known as Guangzhou New Television Tower) stands as a tall edifice soaring to a height of 610 meters. There is a structural health monitoring system consisting of over 700 various types of sensors to capture the full-scale field measurement of the Canton Tower response under the ambient conditions [31]. The field monitoring data have been uploaded to the SHM benchmark website [32].

There are twenty uniaxial accelerometers in various locations of the tower to measure the structural ambient vibration response. The accelerometer placements and measuring orientations are depicted in Figure 12. 24-hour field measurements (recorded from 18: 00 pm on 19 January 2010 to 18: 00 pm on 20 January 2010) of structural acceleration time histories were provided. On the 4th and 8th floors, every segment features four uniaxial accelerometers: two for detecting horizontal vibrations along the structure's long axis and the remaining pair along the short axis.

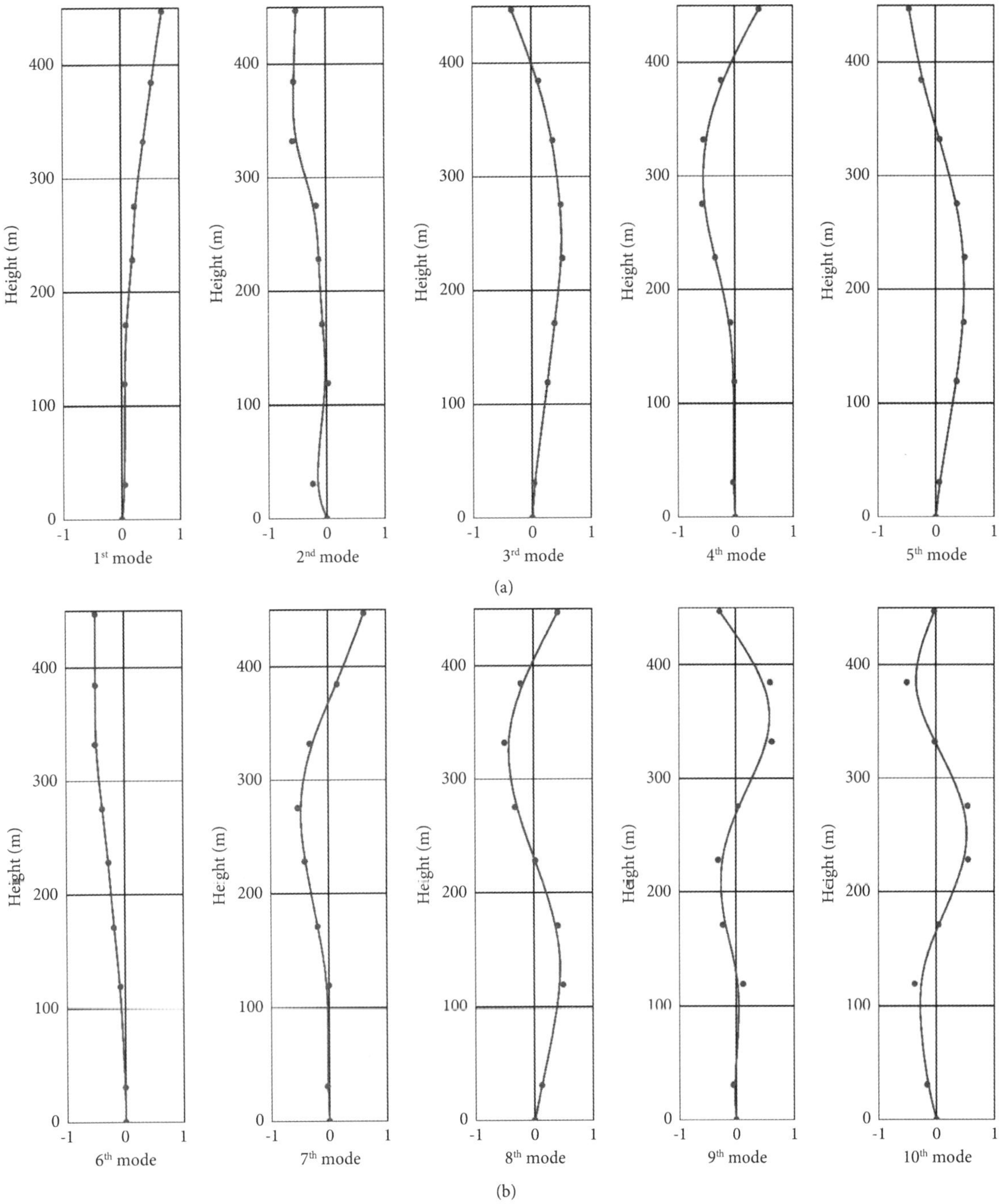

FIGURE 15: Identified mode shapes for the first 10 modes in short-axis direction. (a) 1st to 5th mode shapes. (b) 6th to 10th mode shapes.

We used the BSD to identify the modal parameters of the Canton Tower and present detailed results of modal identification using acceleration measurements (obtained from 18: 00 to 19: 00 on 19 January 2010). It is divided into ten subsets of 6 minutes each (i.e., M = 10) to calculate the averaged spectral density matrix. The identification results of the modal frequencies and damping ratios of the first ten modes are shown in Table 7. It lists the identified results and the coefficients of variation (CV). The posterior standard derivations are in the acceptable range, and the CVs of all the identified modal frequencies are less than 0.20%. However, it is not surprising to obtain large uncertainty for the damping ratios, and the CV values are in the 15%–40% range.

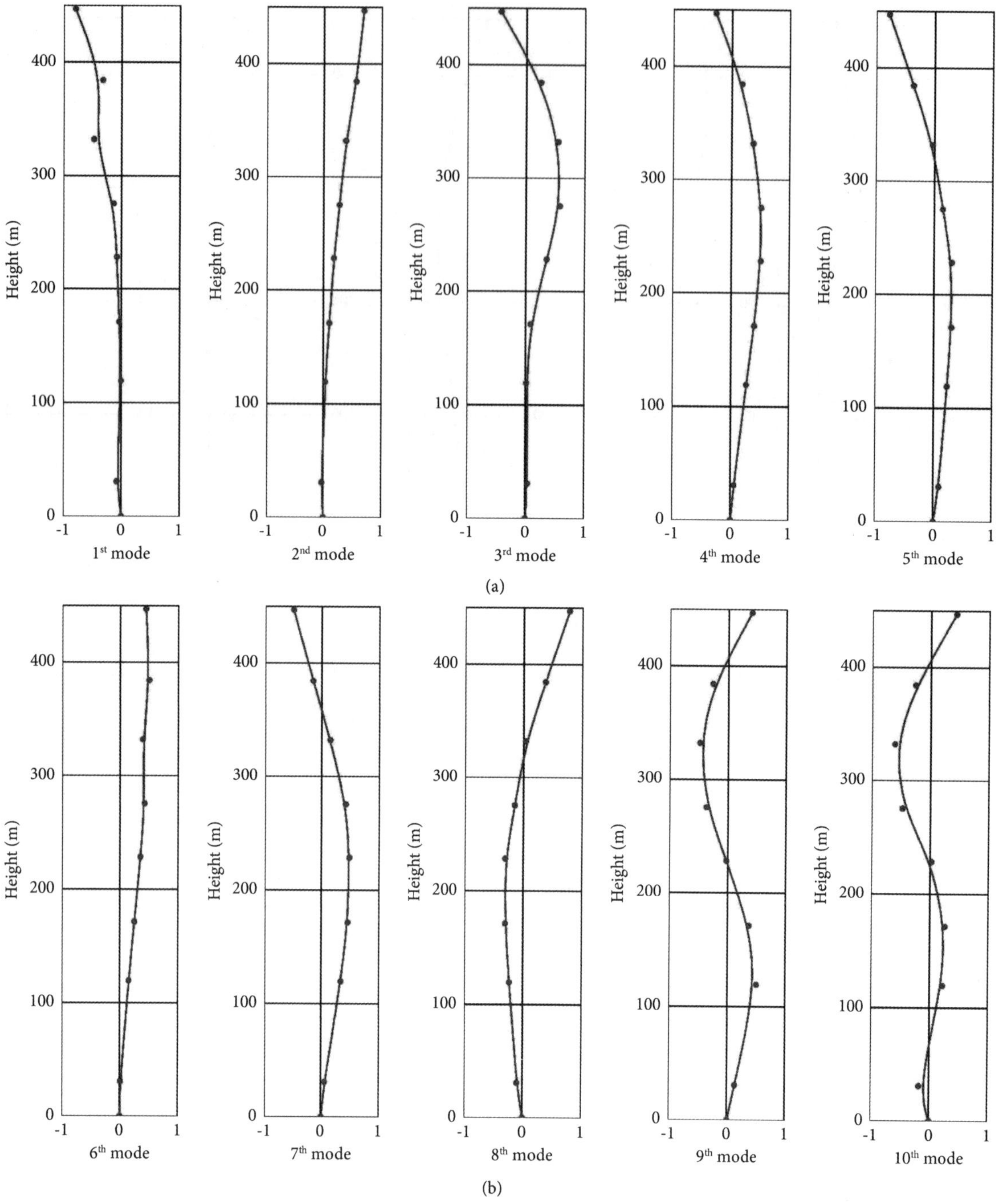

Figure 16: Identified mode shapes for the first 10 modes in long-axis direction. (a) 1st to 5th mode shapes. (b) 6th to 10th mode shapes.

Herein, the identified results in Table 8 are compared with the results of three references including Chen et al. [33], Kuok et al. [34], and Zhang et al. [35] as shown in Figure 13 and Figure 14, respectively. The identification modal frequencies of the Bayesian spectral decomposition method show good agreement with the results of the other three references. Nevertheless, the identified damping ratios of the BSD are relatively close to the results in Kuok et al. [34]. In Kuok et al. [34], the calculated CVs fall between 16% and 40%, which are also relatively close to the CVs calculated by the BSD. Table 9 presents a comparison of the damping ratios and their respective CVs.

Figures 15 and 16 show the identified mode shapes of the ten modes projected in direction along the short axis and the long axis, respectively. The results are in general agreement with the results of Kuok et al. [34].

5. Conclusions

An efficient Bayesian modal identification method based on decomposition strategy and ambient vibration data is proposed. Inspired by the BSDA and the FDD method, the BSD method capitalizes on the strengths of both while avoiding their inherent limitations. In comparison to BSDA, the BSD strategically reduces the dimensionality of identified parameters in optimization, thereby avoiding the intricacies of high-dimensional nonlinear optimization problems. Moreover, the BSD method operates within a Bayesian statistical framework, effectively addressing uncertainty—a feature absents in the original FDD method. It not only extracts optimal values for modal parameters but also captures their associated uncertainties. Importantly, estimating uncertainties does not require statistical computation from numerous datasets, and a sole dataset is adequate.

The efficacy of the BSD is validated using simulated data from an 8-story shear frame, as well as measured data from a three-story shear frame. The results from these numerical and experimental studies convincingly demonstrate that our approach excels in accurately identifying modal properties and quantifying their uncertainties, and the computation challenges of original BSDA can be well addressed. This paper also presents the results of the modal identification of the Canton Tower using acceleration measurements. The identified modal frequencies and mode shapes of the first ten modes obtained by the BSD method closely align with results from three reference sources, while there are discrepancies in the estimations of damping ratios.

Although the efficiency and accuracy of the BSD method have been well demonstrated, it should be noted that under the assumption of Gaussian white noise input, BSD is suitable for identifying modal parameters in ambient vibration with well-separated modes. Furthermore, the measurement data need to be of sufficient duration, as power spectral matrix estimation requires an adequate number of averages to reduce errors, and Fourier transform requires sufficient data length to improve frequency resolution. The influence of bandwidth and segment number on the uncertainty of identification results requires further investigation. Currently, the BSD method is only applicable to well-separated modes; for closely spaced modes, further improvements to the existing BSD methods are necessary. Considering the significant impact of operational and environmental variation on modal parameters, particularly damping ratios, developing a Bayesian method suitable for nonstationary signals or time-varying dynamic systems is valuable. Additionally, due to the characteristics of large structures having low and densely packed frequencies, enhancing BSD to make it suitable for identifying closely spaced modes is imperative. Also, proposing an indicator to evaluate the overall uncertainty of a mode shape would enhance the convenience of the BSD. In essence, the proposed BSD method stands as a robust and efficient solution for modal identification using ambient vibrations, offering both optimal estimates and a treatment of uncertainties. As uncertainty is inevitable, quantifying uncertainties (such as variance and coefficient of variation) aids in scientifically evaluating the identification results of modal parameters, which is crucial for enhancing the robustness and accuracy of structural dynamic analysis results. Additionally, uncertainty quantification helps decision-makers better understand the limitations of the model, identify key factors influencing uncertainty, and make decisions to mitigate risks or improve prediction accuracy.

Appendix

Suppose nS has the distribution $W(\Sigma, n)$, Define diagonal Λ and L and orthogonal β and B by

$$\begin{aligned} \Sigma &= \beta \Lambda \beta^T, \\ S &= BLB^T, \end{aligned} \tag{A.1}$$

$\lambda_1 > \lambda_2 > \ldots > \lambda_p$, $l_1 > l_2 > \ldots > l_p$, $\beta_{1i} \geq 0$, $b_{1i} \geq 0$, $i = 1, \ldots, p$. Define $G = \sqrt{n}(B - \beta)$ and diagonal $D = \sqrt{n}(L - \Lambda)$. Then, the limiting distribution of D and G is normal with D and G independent, and the diagonal elements of D are independent. The diagonal element d_i has the limiting distribution $N(0, 2\lambda_i^2)$. The covariance matrix of g_i in the limiting distribution of $G = (g_1, \ldots g_p)$ is

$$\operatorname{cov}(g_i) = \sum_{\substack{k=1, \\ k \neq i.}}^{p} \frac{\lambda_i \lambda_k}{(\lambda_i - \lambda_k)^2} \beta_k \beta_k^T, \tag{A.2}$$

where $\beta = (\beta_1, \ldots, \beta_p)$, the covariance matrix of g_i and g_j in the limiting distribution is

$$\operatorname{cov}(g_i, g_j) = -\frac{\lambda_i \lambda_j}{(\lambda_i - \lambda_j)^2} \beta_j \beta_i^T, i \neq j. \tag{A.3}$$

Data Availability

The data used in this study are available on request from the corresponding author.

Conflicts of Interest

The authors declare that they have no conflicts of interest with respect to the research, authorship, and/or publication of this article.

Acknowledgments

The research was funded by the Natural Science Foundation of China (52178284 and 51708203), the National Key Research and Development Program of China (2022YFC3005300), the Natural Science Foundation of Hunan Province (2021JJ30106), the Open Research Project Program of the State Key Laboratory of Internet of Things for Smart City (University of Macau) (SKL-IoTSC (UM)-2021-2023/ORP/GA09/2022), and the Graduate Research and Innovation Project of Hunan Province (CX20230443). The authors are grateful for these supports.

References

[1] F. B. Zahid, Z. C. Ong, and S. Y. Khoo, "A review of operational modal analysis techniques for in-service modal identification," *Journal of the Brazilian Society of Mechanical Sciences and Engineering*, vol. 42, no. 8, p. 398, 2020.

[2] W. Fan and P. Qiao, "Vibration-based damage identification methods: a review and comparative study," *Structural Health Monitoring*, vol. 10, no. 1, pp. 83–111, 2011.

[3] S. S. Saidin, A. Jamadin, S. Abdul Kudus, N. Mohd Amin, and M. A. Anuar, "An overview: the application of vibration-based techniques in bridge structural health monitoring," *International Journal of Concrete Structures and Materials*, vol. 16, no. 1, p. 69, 2022.

[4] A. J. Felber, *Development of a Hybrid Bridge Evaluation System*, University of British Columbia(UBC), Vancouver, Canada, 1993.

[5] R. Brincker, L. Zhang, and P. Andersen, "Modal identification of output-only systems using frequency domain decomposition," *Smart Materials and Structures*, vol. 10, no. 3, pp. 441–445, 2001.

[6] C. X. Qu, T. H. Yi, H. N. Li, and B. Chen, "Closely spaced modes identification through modified frequency domain decomposition," *Measurement*, vol. 128, pp. 388–392, 2018.

[7] P. Verboven, B. Cauberghe, and P. Guillaume, "Improved total least squares estimators for modal analysis," *Computers & Structures*, vol. 83, no. 25-26, pp. 2077–2085, 2005.

[8] B. Peeters, H. Van Der Auweraer, P. Guillaume, and J. Leuridan, "The PolyMAX frequency-domain method: a new standard for modal parameter estimation?" *Shock and Vibration*, vol. 11, no. 3-4, pp. 395–409, 2004.

[9] P. Verboven, P. Guillaume, B. Cauberghe, E. Parloo, and S. Vanlanduit, "Frequency-domain generalized total least-squares identification for modal analysis," *Journal of Sound and Vibration*, vol. 278, no. 1-2, pp. 21–38, 2004.

[10] C. Qu, T. Yi, and H. Li, "Mode identification by eigensystem realization algorithm through virtual frequency response function," *Structural Control and Health Monitoring*, vol. 26, no. 10, 2019.

[11] J. M. Caicedo, S. J. Dyke, and E. A. Johnson, "Natural excitation technique and eigensystem realization algorithm for phase I of the IASC-ASCE benchmark problem: simulated data," *Journal of Engineering Mechanics*, vol. 130, no. 1, pp. 49–60, 2004.

[12] J. C. Asmussen, R. Ibrahim, and R. Brincker, "Random decrement ldentification of structures subjected to ambient excitation," in *Proceedings of the 16th International Modal Analysis Conference, Proceedings of the 16th International Modal Analysis Conference*, Santa Barbara, CA, USA, February 1998.

[13] B. Peeters and G. De Roeck, "REFERENCE-BASED stochastic subspace identification for output-only modal analysis," *Mechanical Systems and Signal Processing*, vol. 13, no. 6, pp. 855–878, 1999.

[14] J. L. Beck and L. S. Katafygiotis, "Updating models and their uncertainties. I: bayesian statistical framework," *Journal of Engineering Mechanics*, vol. 124, no. 4, pp. 455–461, 1998.

[15] L. S. Katafygiotis and J. L. Beck, "Updating models and their uncertainties. II: model identifiability," *Journal of Engineering Mechanics*, vol. 124, no. 4, pp. 463–467, 1998.

[16] L. S. Katafygiotis and K. V. Yuen, "Bayesian spectral density approach for modal updating using ambient data," *Earthquake Engineering & Structural Dynamics*, vol. 30, no. 8, pp. 1103–1123, 2001.

[17] K. V. Yuen and L. S. Katafygiotis, "Bayesian fast fourier transform approach for modal updating using ambient data," *Advances in Structural Engineering*, vol. 6, no. 2, pp. 81–95, 2003.

[18] S. K. K. Au, "Fast bayesian FFT method for ambient modal identification with separated modes," *Journal of Engineering Mechanics*, vol. 137, no. 3, pp. 214–226, 2011.

[19] B. Li and S. K. Au, "An expectation-maximization algorithm for Bayesian operational modal analysis with multiple (possibly close) modes," *Mechanical Systems and Signal Processing*, vol. 132, pp. 490–511, 2019.

[20] Z. Zhu, S. K. Au, B. Li, and Y. L. Xie, "Bayesian operational modal analysis with multiple setups and multiple (possibly close) modes," *Mechanical Systems and Signal Processing*, vol. 150, Article ID 107261, 2021.

[21] W. J. Yan and L. S. Katafygiotis, "A two-stage fast Bayesian spectral density approach for ambient modal analysis. Part I: posterior most probable value and uncertainty," *Mechanical Systems and Signal Processing*, vol. 54-55, pp. 139–155, 2015.

[22] W. J. Yan and L. S. Katafygiotis, "A two-stage fast Bayesian spectral density approach for ambient modal analysis. Part II: mode shape assembly and case studies," *Mechanical Systems and Signal Processing*, vol. 54-55, pp. 156–171, 2015.

[23] F. L. Zhang, Y. C. Ni, S. K. Au, and H. F. Lam, "Fast Bayesian approach for modal identification using free vibration data, Part I - most probable value," *Mechanical Systems and Signal Processing*, vol. 70-71, pp. 209–220, 2016.

[24] Y. C. Ni, F. L. Zhang, H. F. Lam, and S. K. Au, "Fast Bayesian approach for modal identification using free vibration data, Part II—posterior uncertainty and application," *Mechanical Systems and Signal Processing*, vol. 70, pp. 221–244, 2016.

[25] O. Sedehi, L. S. Katafygiotis, and C. Papadimitriou, "Hierarchical Bayesian operational modal analysis: theory and computations," *Mechanical Systems and Signal Processing*, vol. 140, Article ID 106663, 2020.

[26] Z. Q. Feng, *Structural Health Monitoring Using Wireless Sensor Networks and Bayesian Probabilistic Methods*, Hong Kong University of Science and Technology, China, 2014.

[27] T. W. Anderson, *An Introduction to Multivariate Statistical Analysis*, Wiley-Interscience, Hoboken, NJ, USA, 3rd edition, 2003.

[28] Z. Q. Feng and L. S. Katafygiotis, "Bayesian spectral decomposition method for operational modal identification in wireless sensor network," in *ECCOMAS Thematic Conference-COMPDYN 2011: 3rd International Conference on Computational Methods in Structural Dynamics and Earthquake Engineering: An IACM Special Interest Conference*, Corfu, Greece, November 2011.

[29] E. Cheynet, J. B. Jakobsen, and J. Snæbjörnsson, "Damping estimation of large wind-sensitive structures," *Procedia Engineering*, vol. 199, pp. 2047–2053, 2017.

[30] C. C. Chang and C. W. Poon, "Nonlinear identification of lumped-mass buildings using empirical mode decomposition and incomplete measurement," *Journal of Engineering Mechanics*, vol. 136, no. 3, pp. 273–281, 2010.

[31] Y. Q. Ni, Y. Xia, W. Y. Liao, and J. M. Ko, "Technology innovation in developing the structural health monitoring system for Guangzhou New TV Tower," *Structural Control and Health Monitoring*, vol. 16, no. 1, pp. 73–98, 2009.

[32] Y. Q. Ni, Y. Xia, W. Lin, W. H. Chen, and J. M. Ko, "SHM benchmark for high-rise structures: a reduced-order finite element model and field measurement data," *Smart Structures and Systems*, vol. 10, no. 4_5, pp. 411–426, 2012.

[33] W. H. Chen, Z. R. Lu, W. Lin et al., "Theoretical and experimental modal analysis of the Guangzhou new TV tower," *Engineering Structures*, vol. 33, no. 12, pp. 3628–3646, 2011.
[34] S. C. Kuok and K. V. Yuen, "Structural health monitoring of Canton Tower using Bayesian framework," *Smart Structures and Systems*, vol. 10, no. 4_5, pp. 375–391, 2012.
[35] F. L. Zhang, Y. Q. Ni, Y. C. Ni, and Y. W. Wang, "Operational modal analysis of Canton Tower by a fast frequency domain Bayesian method," *Smart Structures and Systems*, vol. 17, no. 2, pp. 209–230, 2016.

Engineering Structures 315 (2024) 118477

Contents lists available at ScienceDirect

Engineering Structures

journal homepage: www.elsevier.com/locate/engstruct

Bayesian time domain approach for damping identification and uncertainty quantification in stay cables using free vibration response

Zhouquan Feng [a,b], Jiren Zhang [a,b,*], Xinyan Xuan [a,b], Yafei Wang [a,b], Xugang Hua [a,b], Zhengqing Chen [a,b], Wangji Yan [c]

[a] Hunan Provincial Key Laboratory of Wind and Bridge Engineering, College of Civil Engineering, Hunan University, Changsha 410082, China
[b] State Key Laboratory of Bridge Safety and Resilience, Hunan University, Changsha 410082, Hunan, China
[c] State Key Laboratory of Internet of Things for Smart City and Department of Civil and Environmental Engineering, University of Macau, 999078, Macao, China

ARTICLE INFO

Keywords:
Modal identification
Free vibration
Bayesian inference
Time domain
Stay Cables
Damping ratios
Uncertainty quantification

ABSTRACT

The dynamic characteristics, particularly damping, play crucial roles in the design, monitoring, and vibration control of stay cables. The free vibration test, commonly used for damping identification of stay cables, demonstrates notable technical advantages in specific mode excitation and improving the signal-to-noise ratio. Nevertheless, challenges arise from uncertainties associated with ineluctable ambient vibration influences and noise contamination in measured responses. In recent years, there has been a growing interest in Bayesian modal identification methods due to their remarkable ability to quantify uncertainties. However, further advancements are needed in the time domain, particularly when addressing the relatively short duration of free vibration response data. This is essential to minimize the impact of background ambient vibration and enhance the accuracy of modal identification. This paper proposes a novel Bayesian modal identification approach specifically crafted to address free vibration data affected by ambient vibrations, aiming to achieve high-precision modal identification and effective uncertainty quantification from free vibration response of stay cables. Additionally, a speedup algorithm is used for determining the most probable values (MPVs) to enhance computational efficiency. The efficacy of the proposed approach is validated through simulated scenarios, encompassing both a single-degree-of-freedom structure and a finite element model of a stay cable. Furthermore, the effectiveness of the proposed method is validated through its application in a laboratory test involving a scale model of the stay cable, as well as in a field test of actual stay cables of Sutong Bridge. The results demonstrate that the proposed method achieves high accuracy in identifying modal damping ratios, and effectively quantifies uncertainties in the identification results.

1. Introduction

Cable-stayed bridges play pivotal roles in modern civil infrastructure, seamlessly integrating functionality, aesthetics, and structural efficiency. The stay cables, which effectively distribute loads, are one of the essential components of cable-stayed bridges. However, long stay cables often experience significant vibrations especially in the absence of control measures. The vibration induction mechanism of stay cables is intricate, primarily attributed to their large slenderness ratio, low frequency, and minimal damping [1,2]. Damping in stay cables, a critical modal parameter, holds significant importance in addressing scientific issues related to their design, monitoring, and vibration control [3,4]. Damping ratios of stay cables can be identified from modal tests, which can be classified into three categories based on different excitation methods, including free vibration test, forced vibration test, and ambient vibration test [5]. Experimental modal analysis (EMA) and operational modal analysis (OMA) are two widely used techniques for modal identification [6].

Operational modal identification methods, commonly used, can be categorized into two groups: frequency domain approaches and time domain approaches[1,7]. Several techniques have been applied to identify modal parameters in stay cables. Among these, Stochastic Subspace Identification (SSI), a classical time-domain method in OMA, is widely used [8–10]. In these applications, Chen et al. utilized SSI to determine modal damping ratios of stay cables in four cable-stayed bridges, with damping ratios typically ranging from 0.003 % to 0.7 %

* Correspondence to: College of Civil Engineering, Hunan University, Changsha 410082, China.
E-mail address: jrzhang1994@hnu.edu.cn (J. Zhang).

https://doi.org/10.1016/j.engstruct.2024.118477
Received 4 February 2024; Received in revised form 5 June 2024; Accepted 17 June 2024
Available online 28 June 2024

[9]. High-quality vibration signals from stay cables can enhance identification accuracy. In addition to traditional modal parameter identification based on acceleration response, computer vision methods have also been utilized for displacement measurement and modal parameter identification of stay cables [11].

Although OMA has been widely implemented, certain limitations still persist. For instance, the damping ratio identified for stay cables often corresponds to low amplitude levels. This does not meet the requirements for vibration control design of ultra-long cables, which may be subject to large amplitude and high mode vibrations. Additionally, only a limited number of modes are present in the operational state for medium and short cables. Since damping represents the energy dissipation of structures, it is very convenient to obtain structural damping from the decay process of free vibration. However, compared to OMA, Experimental Modal Analysis (EMA) indeed requires more resources and time. The advantage of free vibration lies in its ability to selectively excite modes of specific orders and control their amplitudes, thereby facilitating the identification of modes that are not typically excited by ambient vibrations. Modal parameters, particularly damping, can be identified under varying amplitude conditions.

Despite the rapid decay of free vibration responses in most real applications, they usually exhibit a larger signal-to-noise ratio [12] and a higher level of response [13] compared to responses solely induced by ambient vibration. Therefore, with the ongoing advancements in excitation technology for structural dynamic testing, exciters are now capable of selectively exciting specific mode orders of stay cables, hence free vibration tests of the stay cables are becoming increasingly sophisticated and have been utilized in various studies [3,12,14–18]. Their research results indicate that free vibration response is less susceptible to noise interference, and this method can specifically excite certain modal orders, especially high-order modes that are not easily activated by ambient vibration. However, free vibration tests also face challenges in practical applications: the inevitable interference from environmental factors that cannot be eliminated. Therefore, uncertainties persist in the EMA or OMA, stemming from factors like measurement noise, modeling error, and environmental influence. When identifying damping ratios for real civil engineering structures, several factors significantly contribute to uncertainty. Key among these are the damping ratio's variation with the structural response amplitude and the impact of wind characteristics on observed damping ratios [19].

In order to deal with the uncertainty of structural dynamic system identification, Beck et al. established a basic theoretical framework for Bayesian dynamic system identification, which treats modal identification as a probabilistic inference problem where probability represents the plausibility of a statement given a model of the system and measured data [20,21]. A series of Bayesian methods focused on operational modal identification have been proposed, which can be further divided into frequency domain methods and time domain methods. Among them, frequency domain methods include the Bayesian spectral density approach [22], Bayesian Fast Fourier Transform Approach [23], Fast Bayesian FFT Method [24], Fast Bayesian spectral density approach [25, 26], Hierarchical Bayesian method [27], and Bayesian spectral decomposition method [28], among others. An et al. proposed an automated method for modal identification and uncertainty quantification of stayed cables. This method integrates the Bayesian FFT algorithm with an automatic frequency-band selection network. They validated this approach using measured data from cables on the Shanghai-Suzhou-Nantong Yangtze River Bridge [29]. There are also some Bayesian modal identification methods directly using time domain data [30,31], these studies have formed Bayesian time-domain methods. Li and Der Kiureghian presented a variational Bayes approach for modal identification using a state-space representation of a dynamical system. This approach involves neglecting the statistical correlation between latent variables and model parameters, while analytically approximating the posterior distributions [32]. Li et al. have also developed a Bayesian time domain approach for modal identification of civil structures under small or moderate seismic excitations [33]. Rather et al. presented a novel Bayesian time-domain method for identifying modal parameters of both classically and non-classically damped linear dynamic systems from ambient vibration data [34].

A predominant portion of present research on time-domain damping identification of stay cables using free vibration relies on the logarithmic decay method, with little consideration given to ambient vibration influence and evaluating the uncertainty of identification results. On the other hand, although Zhang et al. have proposed a Bayesian modal parameter identification method in frequency domain using free vibration response [35,36], potential errors caused by the Fourier transform of shortened free vibration data necessitate an alternative approach. Therefore, we propose a novel Bayesian time domain approach to achieve modal identification while simultaneously quantifying the uncertainty for free vibration data and ensuring computational efficiency through a speedup algorithm. The Bayesian time domain approach is suitable for single-channel measurement of stay cables. Simulation data, encompassing a single-degree-of-freedom structure and a finite element model of the stay cable, validates the efficacy of the Bayesian time domain method. Furthermore, its practical applicability is demonstrated through laboratory tests with a scale model and field tests on Sutong Bridge.

2. Bayesian time domain approach

2.1. Dynamics of MDOF linear system

Consider a linear dynamic system with N_s DOFs initially excited by some artificial means and then left to vibrate 'freely'. The equation of motion for dynamic system is given by:

$$\mathbf{M}\ddot{\mathbf{x}}(t) + \mathbf{C}\dot{\mathbf{x}}(t) + \mathbf{K}\mathbf{x}(t) = \mathbf{0} \tag{1}$$

where$\mathbf{M}$, $\mathbf{C}$, $\mathbf{K} \in \mathbb{R}^{N_s \times N_s}$are the mass, damping and stiffness matrix, respectively. $\mathbf{x}(t) \in \mathbb{R}^{N_s \times 1}$ is the displacement vector at time t, $\ddot{\mathbf{x}}(t)$ and $\dot{\mathbf{x}}(t)$ represent the acceleration vector and velocity vector, respectively. In the modal coordinate, Eq. (1) can be decoupled as:

$$\ddot{\mathbf{q}}_i(t) + 2\zeta_i\omega_i\dot{\mathbf{q}}_i(t) + {\omega_i}^2\mathbf{q}_i(t) = 0,\ i = 1, \ldots, N_s \tag{2}$$

where $\mathbf{q}_i$ is the modal coordinate, the ζ_i is modal damping ratio of the i-th mode, ω_i is the natural frequency of the i-th mode. The modal parameters of the stay cable are often identified from the vibration response data of a measurement channel. The measured free vibration response (displacement) at a certain location on the system of Eq. (1) can be written as [37]:

$$x(t) = \sum_{i=1}^{N_s} A_i e^{-\zeta_i\omega_i t}\cos(\omega_{di}t + \varphi_i),\ i = 1, \ldots, N_s \tag{3}$$

where $\omega_{di} = \omega_i \times \sqrt{1-{\zeta_i}^2}$, A_i and φ_i are constants determine, representing amplitude and phase respectively, noted that the information about the mode shape is already embedded in the amplitude A_i.

2.2. Damping identification and uncertainty quantification

2.2.1. Most probable value of modal parameters

Considering a stay cable excited by certain initial excitation and then vibrate freely, while also being subjected to ambient excitation during the free vibration, and the measured response (displacement) of a measurement channel at the time $t = k\Delta t$, includes the free vibration response x_{fi}, ambient vibration response x_{ai}, and prediction error ε_i, can be expressed by the following equation:

$$y_i(t) = x_{fi}(t) + x_{ai}(t) + \varepsilon_i(t),\ t = k\Delta t, k = 1, \ldots, N \tag{4}$$

where N is the number of sampling points, i is index of the excited modes in the

responses. Herein, the prediction error is modeled as zero-mean Gaussian white noise process with the mean $E[\varepsilon_i(k\Delta t)\varepsilon_i(k'\Delta t)] = \sigma_\varepsilon^2\delta_{kk'}$ and variance σ_ε^2 ,where $\delta_{tt'}$ denotes the Kronecker delta, which is given by:

$$\delta_{kk'} = \left\{ \begin{matrix} 1 & \text{if} k = k' \\ 0 & \text{if} k \neq k' \end{matrix} \right\} \tag{5}$$

Let $\boldsymbol{\theta}$ denote the set of modal parameters to be identified, which includes three groups of parameters. The first group is $\{f_i, \zeta_i : i = 1, \ldots, N_d\}$, where f_i and ζ_i denote the natural frequency and damping ratio of the i-th mode, respectively; N_d is the number of modes can be identified. The second group is $\{A_i, \varphi_i : i = 1, \ldots, N_d\}$, where A_i and φ_i are the initial amplitude and phase of the i-th modal free vibration response, respectively. The third category consists of $\{\mathbf{S}_f, \sigma_\varepsilon^2\}$, $\mathbf{S}_f$ is the Hermitian power spectral density (PSD) matrix representing modal forces in the ambient vibration response, and σ_ε^2 is the variance of prediction errors. The ambient modal force is presumed to be a consistent PSD. It is worth noting that in ambient modal identification, the input excitation is typically assumed to be white noise. Regarding the prediction error, it is reasonable to assume it as independent and identically distributed (i.i. d.) Gaussian white noise.

Let the vector $\mathbf{Y} = [y_1, y_2, \ldots, y_N]^{\mathrm{T}}$ denote a series of data points of a response measurement from time Δt to $N\Delta t$. Using Bayes' Theorem, the posterior PDF of $\boldsymbol{\theta}$ given the data is given by:

$$p(\boldsymbol{\theta}|\mathbf{Y}) \propto p(\boldsymbol{\theta})p(\mathbf{Y}|\boldsymbol{\theta}) \tag{6}$$

where$p(\boldsymbol{\theta})$ denotes the prior PDF of the parameters that reflects the plausibility of $\boldsymbol{\theta}$ and can be treated as an informative constant prior. The term $p(\boldsymbol{\theta}|\mathbf{Y})$ is directly proportional to the 'likelihood function' $p(\mathbf{Y}|\theta)$, and the term $p(\mathbf{Y}|\theta)$ reflecting the role of the measured data in shaping the posterior distribution.

For a given $\boldsymbol{\theta}$ the free vibration response x_f can be determined and hence is fixed. It then follows from Eq.(4) that the difference $y - x_f$ is referred to as the ambient vibration response with prediction error, and is then a zero-mean stationary stochastic process [35]. Therefore, for a given $\boldsymbol{\theta}$, $\mathbf{Y} - \mathbf{Y}_f$ follows Gaussian distribution and $\mathbf{Y}_f$ is the free vibration response given by Eq. (3), and covariance matrix $\boldsymbol{\Gamma}(\boldsymbol{\theta})$ determined by both the ambient response x_a and prediction error ε. The likelihood function can be given by the following equation:

$$p\left(\mathbf{Y}|\boldsymbol{\theta}\right) = (2\pi)^{-\frac{N}{2}}|\boldsymbol{\Gamma}(\boldsymbol{\theta})|^{-\frac{1}{2}}\exp\left[-\frac{1}{2}(\mathbf{Y}-\mathbf{Y}_f)^T\boldsymbol{\Gamma}(\boldsymbol{\theta})^{-1}\left(\mathbf{Y}-\mathbf{Y}_f\right)\right] \tag{7}$$

where the $\boldsymbol{\Gamma}(\boldsymbol{\theta})$ given by:

$$\Gamma^{(k,k')}(\boldsymbol{\theta}) = R[(k'-k)\Delta t|\boldsymbol{\theta}] + \sigma_\varepsilon^2\delta_{kk'} \tag{8}$$

and the autocorrelation function of the i-th mode measured data (displacement response) can be written as:

$$R_i(\tau) = \frac{\pi S_f}{2\zeta_i\omega_i^3}e^{-\zeta_i\omega_i\tau}\left[\cos\omega_{di}\tau + \frac{\zeta_i}{\sqrt{1-\zeta_i^2}}\sin\left(\omega_{di}\tau\right)\right] \tag{9}$$

2.2.2. Speedup algorithm

Although the data for free vibration may be short, the sampling frequency still affects the length of the data. Therefore, it is still necessary to shorten the runtime of the algorithm through reasonable approximation. Indeed, when dealing with a substantial volume of data samples, the direct computational cost using Eq. (7) becomes considerable. However, this increased computational effort does not necessarily lead to a commensurate improvement in identification accuracy. Recognizing this challenge, an approximate method was introduced to overcome this computational obstacle [31]. Within this theoretical framework, we have further derived new expressions specifically tailored for application in this paper.

First by using the Bayes' theorem, the likelihood function can be expanded to a product of a conditional PDF and a reduced-order PDF:

$$\begin{aligned} p(\mathbf{Y}|\boldsymbol{\theta}) &= p(y_1, y_2, \ldots, y_N|\boldsymbol{\theta}) \\ &= p(y_N|y_1, y_2, \ldots, y_{N-1}, \boldsymbol{\theta})p(y_1, y_2, \ldots, y_{N-1}|\boldsymbol{\theta}) \end{aligned} \tag{10}$$

Similarly, the process can be iterated to expand the reduced-order probability density function further. Through successive iterations, the likelihood function can be decomposed into the product of a reduced-order likelihood function and multiple conditional PDFs.

$$p\left(\mathbf{Y}|\boldsymbol{\theta}\right) = p\left(y_1, y_2, \ldots, y_{N_p}|\boldsymbol{\theta}\right)\prod_{k=N_p+1}^{N} p(y_k|y_1, y_2, \ldots, y_{k-1}, \boldsymbol{\theta}) \tag{11}$$

This expansion is exact but it does not resolve the computational difficulties encountered in the exact formulation. The computation of each of the conditional PDF $p(y_k|y_1, y_2, \ldots, y_{k-1}, \boldsymbol{\theta})$ for large k requires similar computational effort as in the computation of the likelihood function in the direct exact formulation and the right-hand side of the expansion involves many such conditional PDFs. In order to overcome this computational obstacle, the following approximation is introduced ($k > N_p$) [30]:

$$p\left(y_k|y_1, y_2, \ldots, y_{k-1}, \boldsymbol{\theta}\right) \approx p\left(y_k\middle|y_{k-N_p}, y_{k-N_p+1}, \ldots, y_{k-1}, \boldsymbol{\theta}\right) \tag{12}$$

The conditional PDFs depending on more than N_p previous data points are approximated by conditional PDFs depending on only the last N_p data points. And the likelihood function expansion in Eq. (11) can be approximated by [38]:

$$p\left(\mathbf{Y}|\boldsymbol{\theta}\right) \approx p\left(y_1, y_2, \ldots, y_{N_p}|\boldsymbol{\theta}\right)\prod_{k=N_p+1}^{N} p\left(y_k\middle|y_{k-N_p}, y_{k-N_p+1}, \ldots, y_{k-1}, \boldsymbol{\theta}\right) \tag{13}$$

Due to the vector $\mathbf{Y} - \mathbf{Y}_f$ follows Gaussian distribution with zero-mean and covariance matrix $\boldsymbol{\Gamma}_{y_{1,N_p}}(\boldsymbol{\theta})$, so the reduced-order likelihood function $p(y_1, y_2, \ldots, y_{N_p}|\theta)$ is given by:

$$\begin{aligned} &p\left(y_1, y_2, \ldots, y_{N_p}|\boldsymbol{\theta}\right) \\ &= (2\pi)^{-\frac{N_p}{2}}|\boldsymbol{\Gamma}_{y_{1,N_p}}(\boldsymbol{\theta})|^{-\frac{1}{2}}\exp\left[-\frac{1}{2}(\mathbf{Y}_{1,N_p} - \mathbf{Y}_{f,1,N_p})^T\boldsymbol{\Gamma}_{y_{1,N_p}}^{-1}(\boldsymbol{\theta})\left(\mathbf{Y}_{1,N_p} - \mathbf{Y}_{f,1,N_p}\right)\right] \end{aligned} \tag{14}$$

Then, the general expression is given for the conditional probability densities involving N_p previous points $p(y_k|y_{k-N_p}, y_{k-N_p+1}, \ldots, y_{k-1}, \boldsymbol{\theta})$ in Eq. (13). The vector $\mathbf{Y}_{k-N_p,k-1} - \mathbf{Y}_{f,k-N_p,k-1}$ has the zero mean and covariance matrix partitioned as:

$$\boldsymbol{\Gamma}_{y_{k-N_p,k-1,}} = \begin{bmatrix} \Gamma_{11} & \Gamma_{12} \\ \Gamma_{12}^T & \Gamma_{22} \end{bmatrix} \tag{15}$$

where the matrices Γ_{12} and Γ_{22} have dimensions $1 \times N_p$ and $N_p \times N_p$, respectively.

Then, the optimal estimator $\hat{y}_k$ of $y_k - y_{f,k}$ conditional on $\mathbf{Y}_{k-N_p,k-1} - \mathbf{Y}_{f,k-N_p,k-1}$ is given by:

$$\hat{y}_k \equiv E\left[y_k\middle|y_{k-N_p}, y_{k-N_p+1}, \ldots, y_{k-1}, \boldsymbol{\theta}\right] = \Gamma_{12}\Gamma_{22}^{-1}\left(\mathbf{Y}_{k-N_p,k-1} - \mathbf{Y}_{f,k-N_p,k-1}\right) \tag{16}$$

And the covariance matrix $\boldsymbol{\Gamma}_\varepsilon$ of the prediction error $\varepsilon_k = y_k - \hat{y}_k$ is given by:

$$\boldsymbol{\Gamma}_\varepsilon \equiv E\left[\varepsilon_k\varepsilon_k^T\middle|y_{k-N_p}, y_{k-N_p+1}, \ldots, y_{k-1}, \boldsymbol{\theta}\right] = \Gamma_{11} - \Gamma_{12}\Gamma_{22}^{-1}\Gamma_{12}^T \tag{17}$$

Therefore, the conditional probability density $p(y_k|y_{k-N_p}, y_{k-N_p+1}, \ldots, y_{k-1}, \boldsymbol{\theta})$ can be written as:

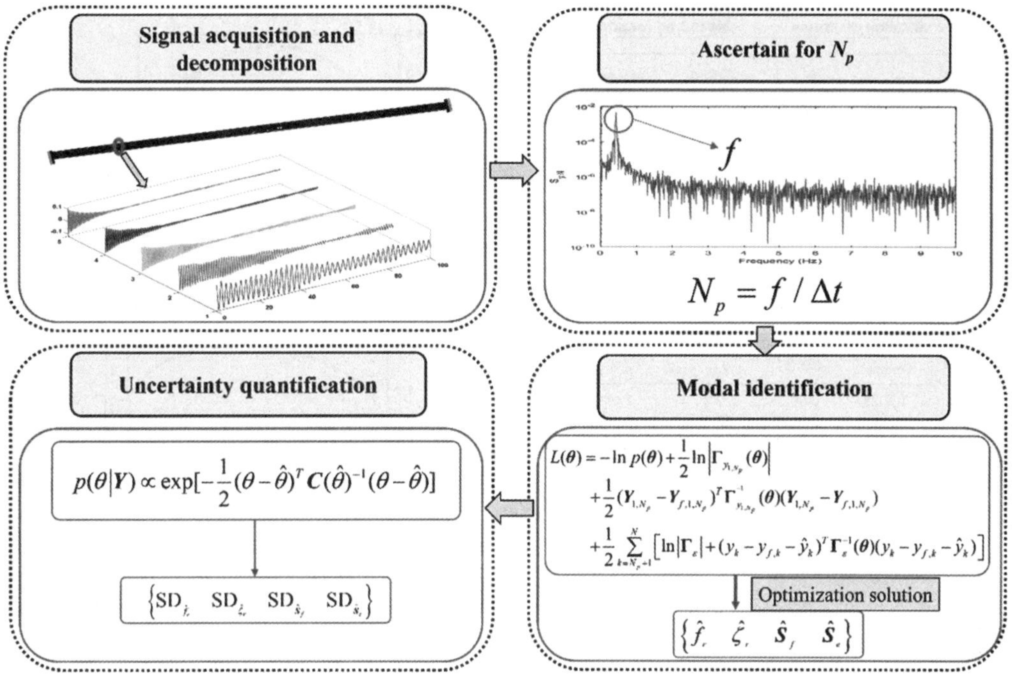

Fig. 1. The flowchart of proposed method.

$$p(y_k|y_{k-N_p}, y_{k-N_p+1}, \ldots, y_{k-1}, \boldsymbol{\theta}) = (2\pi)^{-\frac{1}{2}}|\boldsymbol{\Gamma}_\varepsilon|^{-\frac{1}{2}}\exp\left[-\frac{1}{2}\left(y_k - y_{f,k} - \widehat{y}_k\right)^T \boldsymbol{\Gamma}_\varepsilon^{-1}(\boldsymbol{\theta})\left(y_k - y_{f,k} - \widehat{y}_k\right)\right] \tag{18}$$

Then, the Eq. (13) transforms into:

$$p\left(\mathbf{Y}|\boldsymbol{\theta}\right) \approx (2\pi)^{-\frac{N_p}{2}}|\boldsymbol{\Gamma}_{y_{1,N_p}}(\boldsymbol{\theta})|^{-\frac{1}{2}}\exp\left[-\frac{1}{2}\left(\mathbf{Y}_{1,N_p} - \mathbf{Y}_{f,1,N_p}\right)^T \boldsymbol{\Gamma}_{y_{1,N_p}}^{-1}(\boldsymbol{\theta})\left(\mathbf{Y}_{1,N_p} - \mathbf{Y}_{f,1,N_p}\right)\right] \times \prod_{k=N_p+1}^{N}\left\{(2\pi)^{-\frac{1}{2}}\left|\boldsymbol{\Gamma}_\varepsilon\right|^{-\frac{1}{2}}\exp\left[-\frac{1}{2}\left(y_k - y_{f,k} - \widehat{y}_k\right)^T \boldsymbol{\Gamma}_\varepsilon^{-1}(\boldsymbol{\theta})\left(y_k - y_{f,k} - \widehat{y}_k\right)\right]\right\} \tag{19}$$

The most probable parameter vector $\widehat{\boldsymbol{\theta}}$ is obtained by minimizing the objective function which is the negative logarithm of the posterior PDF Eq. (19) (without including the terms that do not depend on $\boldsymbol{\theta}$):

$$L(\boldsymbol{\theta}) = -\ln p(\boldsymbol{\theta}) + \frac{1}{2}\ln|\boldsymbol{\Gamma}_{y_{1,N_p}}(\boldsymbol{\theta})| + \frac{1}{2}\left(\mathbf{Y}_{1,N_p} - \mathbf{Y}_{f,1,N_p}\right)^T \boldsymbol{\Gamma}_{y_{1,N_p}}^{-1}(\boldsymbol{\theta})\left(\mathbf{Y}_{1,N_p} - \mathbf{Y}_{f,1,N_p}\right) + \frac{1}{2}\sum_{k=N_p+1}^{N}\left[\ln|\boldsymbol{\Gamma}_\varepsilon| + \left(y_k - y_{f,k} - \widehat{y}_k\right)^T \boldsymbol{\Gamma}_\varepsilon^{-1}(\boldsymbol{\theta})\left(y_k - y_{f,k} - \widehat{y}_k\right)\right] \tag{20}$$

From the Eq. (19), the covariance matrix $\boldsymbol{\Gamma}_{y_{1,N_p}}^{-1}$ has a dimension of $N_p \times N_p$ and the covariance matrix $\boldsymbol{\Gamma}_\varepsilon^{-1}(\boldsymbol{\theta})$ is a number, whereas in the Eq. (7), the covariance matrix $\boldsymbol{\Gamma}(\boldsymbol{\theta})$ has a dimension of $N \times N$. The Eq. (19) clearly reduces the computational cost during the optimization process. It should be noted that since the speedup algorithm does not aim to surpass other existing methods in terms of computational efficiency, meaning it is not "faster" than other methods, but rather to overcome the low efficiency of original algorithm and achieve an "accelerated" effect, enabling the computational efficiency of the proposed method to reach a satisfactory level. It can be observed that the choice of N_p determines the trade-off between computational efficiency and accuracy. It is found that an effective value for N_p on the order of $f/\Delta t$ provides the required level of accuracy, where f is the fundamental frequency (in Hz) and Δt is the sampling time step. Although the fundamental frequency f remains an unknown parameter, a rough

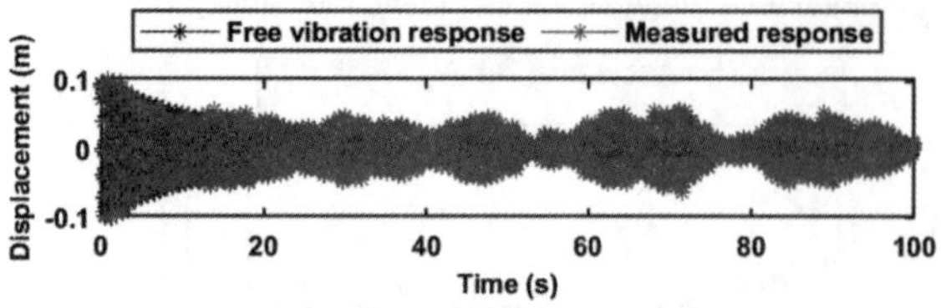

Fig. 2. Measured response and free vibration response of the SDOF dynamic model.

Table 1
Identified results of the SDOF.

Parameter	Actual θ	Optimal$\widehat{\theta}$	SD (σ)	CV ($\sigma/\widehat{\theta}$,%)	$\delta = \lvert\theta - \widehat{\theta}\rvert/\sigma$
f(Hz)	1.2995	1.3023	0.0027	0.2	1.0449
ζ(%)	0.82	0.72	0.0020	27.5	0.4931
S_f(m^{2}s^{-3})	0.0010	0.0010	0.0001	7.1	0.4527
σ_ε^2(m^2)	0.0019	0.0019	0.0000	2.0	0.4966
A(m)	0.1002	0.0971	0.0204	21.0	0.1529
φ(rad))	-0.0693	-0.0745	0.2110	283.1	0.0248

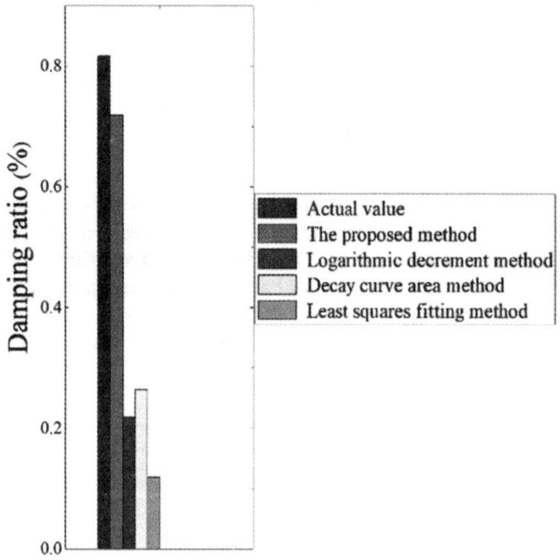

Fig. 3. Comparison of the identified damping ratios.

estimate can be derived from the first peak of the response spectrum to ascertain the value of N_p.

2.2.3. Uncertainty quantification

The approximation for $\mathbf{L}(\theta)$ can be written as [20,39]:

$$\mathbf{L}(\boldsymbol{\theta}) = \mathbf{L}(\widehat{\theta}) + \mathbf{J}(\widehat{\theta})\left(\theta - \widehat{\theta}\right) + \frac{1}{2}(\theta - \widehat{\theta})^T\mathbf{H}(\widehat{\theta})\left(\theta - \widehat{\theta}\right) \tag{21}$$

Please note that the $\mathbf{L}(\widehat{\theta})$ remains unchanged, while $\mathbf{J}(\widehat{\theta})(\theta - \widehat{\theta})$ diminishes when $\mathbf{J}(\widehat{\theta})$ approaches zero due to its minimizing property. It is found that the updated PDF of the parameter vector $\boldsymbol{\theta}$ can be well approximated by a Gaussian distribution with mean $\widehat{\boldsymbol{\theta}}$ and covariance matrix $H(\widehat{\boldsymbol{\theta}})^{-1}$.The most probable parameter vector $\widehat{\boldsymbol{\theta}}$can be obtained by minimizing $L(\widehat{\boldsymbol{\theta}})$, while the covariance matrix of the modal parameters can be obtained by taking the inverse of the Hessian matrix $H(\widehat{\boldsymbol{\theta}})$ of $L(\widehat{\boldsymbol{\theta}})$ evaluated at $\boldsymbol{\theta} = \widehat{\boldsymbol{\theta}}$. The elements of $H(\widehat{\boldsymbol{\theta}})$ can be calculated using finite difference approach.

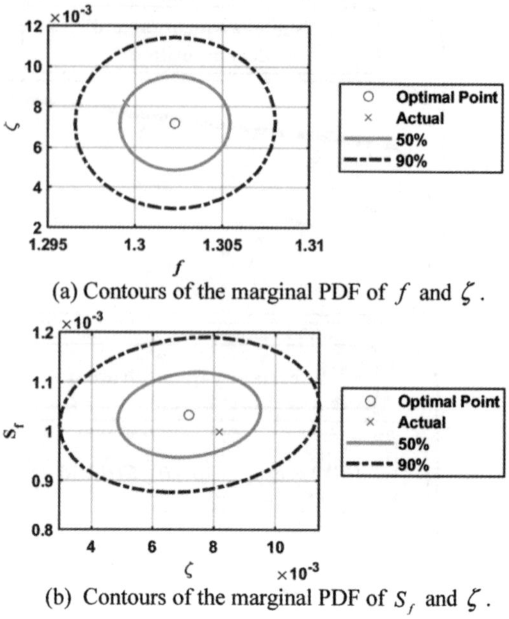

(a) Contours of the marginal PDF of f and ζ.

(b) Contours of the marginal PDF of S_f and ζ.

Fig. 4. Contours of the marginal PDF of Modal parameters.

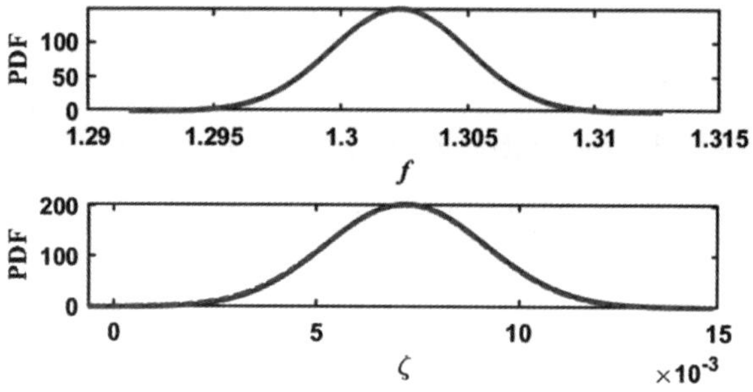

Fig. 5. Conditional PDFs of f and ζ.

Table 2
Parameters of the example cable.

Length (m)	Diameter (m)	Tension (N)	Angle (o)	Mass per length (kg/m)	Elastic modulus (Mpa)
330.4	0.127	6.403×10^6	27.8	82.1	1.9×10^5

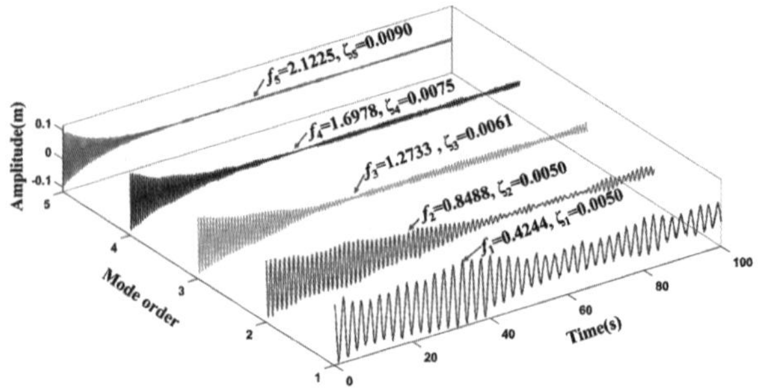

Fig. 6. Measured response of the example cable.

Table 3
Identified results of modal frequencies and damping ratios (stay cable model).

Parameter		Actual θ	Optimal$\hat{\theta}$	SD (σ)	CV ($\sigma/\hat{\theta}$,%)
Modal frequency (Hz)	f_1	0.4244	0.4234	0.0011	0.3
	f_2	0.8488	0.8511	0.0012	0.1
	f_3	1.2733	1.2720	0.0011	0.1
	f_4	1.6978	1.6976	0.0010	0.1
	f_5	2.1225	2.1224	0.0010	0.1
Damping ratio (%)	ζ_1	0.50	0.46	0.0024	51.4
	ζ_2	0.50	0.55	0.0013	24.2
	ζ_3	0.61	0.67	0.0008	12.5
	ζ_4	0.75	0.74	0.0006	7.9
	ζ_5	0.90	0.90	0.0005	5.2

Table 4
Parameters of the test cable.

Length (m)	Area (cm^2)	Tension (N)	Angle (o)	Mass per length (kg/m)	Fundamental frequency (Hz)
24.2	1.347	24.5×10^3	14.2	11.01	1.36

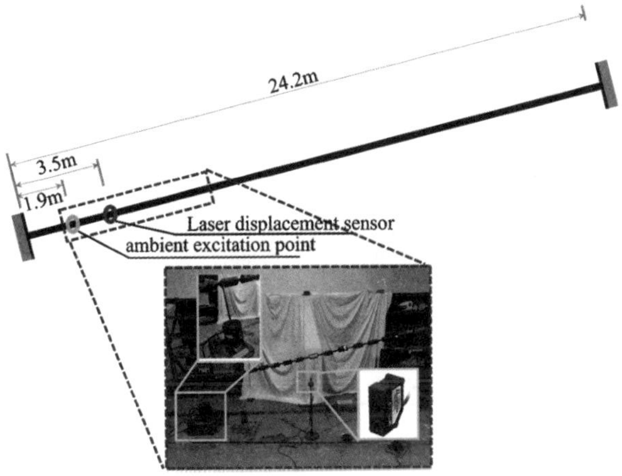

Fig. 7. Test setup in the lab on a reduced-length cable.

2.2.4. Main procedures of proposed method

The four main steps of proposed Bayesian time domain method are summarized and as shown in Fig. 1.

(i) Collect the vibration response of the stay cable and use modal decomposition technique for signal decomposition.
(ii) The fundamental frequency$f$ is roughly estimated from the first peak of the response to obtain the value of N_p using $f/\Delta t$.
(iii) Identify the most probable values of natural frequency $f$and damping ratio ζ by minimizing the objective function $L(\boldsymbol{\theta})$in Eq. (19). Optimization algorithms can be employed to solve this problem. Specifically, this paper employs the 'fminsearch', 'fmincon' and genetic algorithm to achieve optimal solutions.
(iv) The corresponding covariance matrix of the posterior distribution is also calculated by taking the inverse of the $H(\hat{\boldsymbol{\theta}})$ at the most probable values.

3. Numerical validation

3.1. Single degree of freedom dynamic model

In this example, we use simulated data from a single degree of freedom dynamic model. It is assumed that the model has a mass m= 750 kg, a stiffness 50000 N/m, and a damping ratio of 0.0082. The model starts to vibrate freely with an initial displacement of 0.1 m and an initial speed of 0.05 m/s. The model is also assumed to be subjected to a stationary Gaussian white noise excitation and measurement noise. The power spectral density of the ambient excitation is 0.001 m^{2}s^{-3}, and the value of standard deviation of measurement noise corresponds to a 10 % prediction error level, i.e. the RMS of the noise is 10 % of the RMS of the noise-free response. The sampling time step is Δt = 0.05, and the total duration is T = 100 s. Fig. 2 shows the measured response and free vibration response of the SDOF.

Table 1 shows the identified results which included the estimated optimal values $\hat{\theta}$, the calculated standard deviations (SD), coefficient of

Table 5
The identified results of damping ratios.

Mode	Parameter	Identified (%)	SD	CV (%)
5	ζ_5	0.35	0.0001	5.61
8	ζ_8	0.27	0.0000	2.11
11	ζ_{11}	0.19	0.0000	1.37

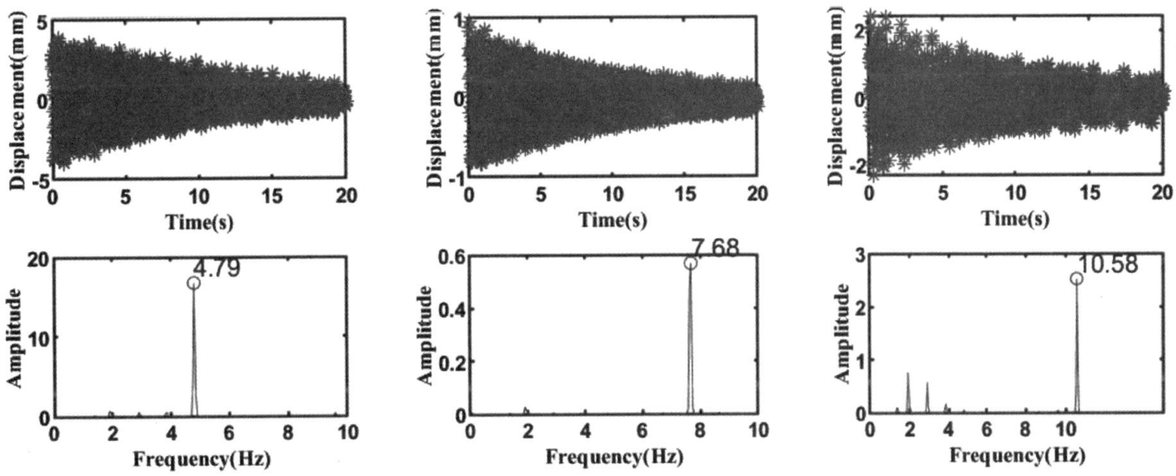

Fig. 8. Displacements recorded during free vibration of the tested cable and their power spectral density.

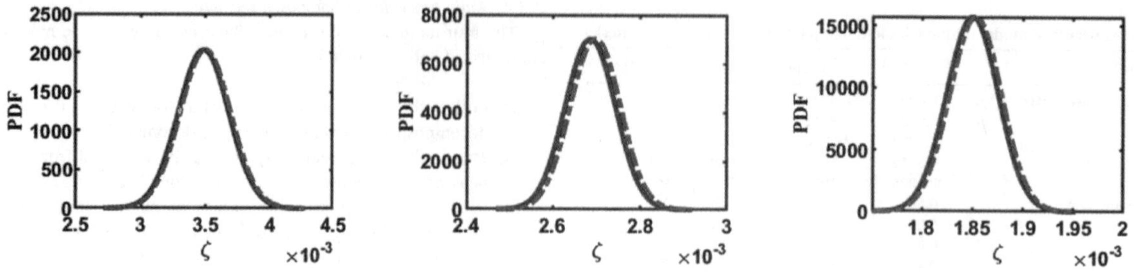

Fig. 9. Conditional PDFs of f and ζ.

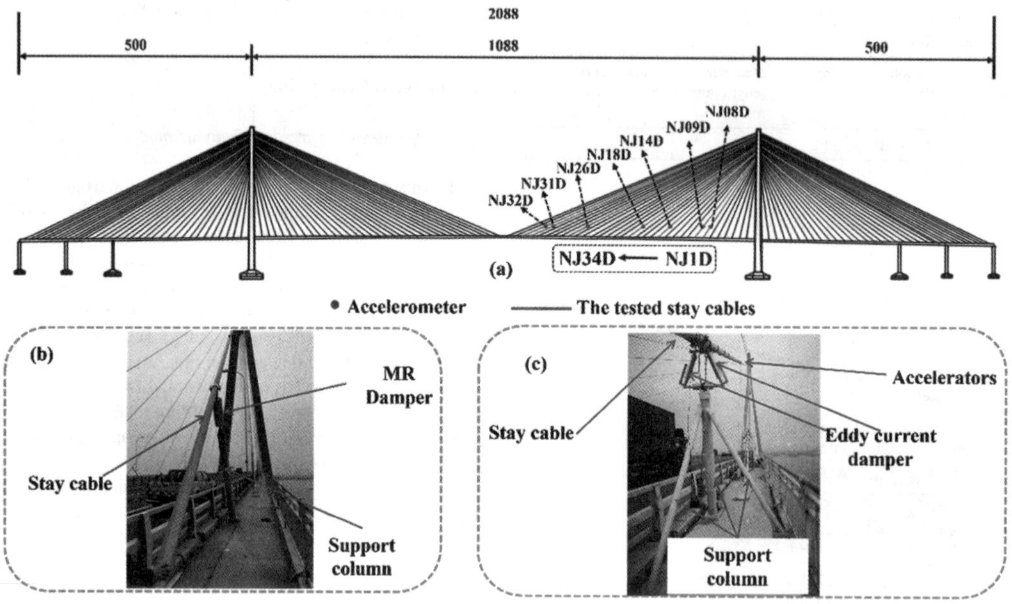

Fig. 10. Schematic diagram of the tested stay cables and the damper system of the STB. (a) The monitored stay cables and the sensors of the STB (unit: m), (b) Original stay cable damper system, and (c) The novel eddy current damper system.

Table 6
Parameters of the test stay cables.

Parameters	NJ32D	NJ31D	NJ26D	NJ18D	NJ14D	NJ9D	NJ8D
Length (m)	551.01	535.42	458.16	338.44	282.02	218.33	207.08
Diameter (mm)	152	152	142	130	127	116	116
Unit weight (Kg/m)	91.89	91.93	77.68	64.86	60.77	49.23	49.14
Tension force (ton)	579.46	564.01	501.44	408.29	382.23	317.86	300.69
Fundamental frequency (Hz) (Designed)	0.230	0.231	0.270	0.370	0.440	0.580	0.590
Fundamental frequency (Hz) (Measured)	0.238	0.231	0.270	0.375	0.475	0.588	0.610

Table 7
The identified results of modal frequencies and damping ratios.

Mode	Parameter	Identified	SD	CV (%)
5	f_5 (Hz)	1.1959	0.0000	0.0
	ζ_5(%)	1.22	0.0027	22.2
9	f_9(Hz)	2.2045	0.0049	0.2
	ζ_9(%)	1.22	0.0010	7.9
15	f_{15}(Hz)	3.6957	0.0000	0.0
	ζ_{15}(%)	0.59	0.0017	29.3
21	f_{21}(Hz)	4.8939	0.0000	0.0
	ζ_{21}(%)	0.43	0.0019	43.1

variance (CV), the value of a 'normalized distance' δ. The second column in Table 1 corresponds to the actual values used for generation of the simulated measurement data; the third and fourth columns correspond to the identified optimal parameters and the corresponding SDs, respectively; the fifth column lists the CVs for each parameter; the sixth column lists the values of δ for each parameter. The most probable values are quite close to their corresponding actual values. The SDs of the most probable values are all lower than 0.1, and the CVs are all less than 1. Consistent with common observations, the CVs of the natural frequencies are also much smaller than those of the damping ratios. and the values of δ are also within a reasonable range.

The following discussion involves analyzing the computational efficiency of speedup algorithm. The program of Bayesian time domain

Table 8
The identified results of damping ratios.

Stay cable number	Mode	Parameter	Identified (%)	SD	CV (%)
NJ08D	4	ζ_4	0.75	0.0016	21.5
	11	ζ_{11}	0.55	0.0038	67.9
NJ09D	4	ζ_4	1.15	0.0018	15.7
	12	ζ_{12}	0.67	0.0021	31.7
NJ14D	5	ζ_5	1.42	0.0016	11.0
NJ18D	6	ζ_6	1.02	0.0020	19.5
	19	ζ_{19}	0.85	0.0036	42.3
NJ26D	8	ζ_8	1.13	0.0017	15.4
	26	ζ_{26}	0.40	0.0027	67.9
NJ31D	5	ζ_5	1.03	0.0049	47.9
	15	ζ_{15}	0.67	0.0014	20.3
	21	ζ_{21}	0.45	0.0021	47.2

approach was run in MATLAB (2022b) on the computer equipped with a 3.2 GHz Intel Core i7 processor and 64 GB of RAM. Use the original algorithm (Eq. (7)) and the speedup algorithm (Eq. (19)) respectively to obtain the optimal values. Only the time taken for the programs to identify modal parameters (including uncertainty quantification) was considered, excluding the time for generating response data. The computational time for speedup algorithm is 6.60 s, but the original algorithm fails to work properly due to the excessive computational load. This illustrates the necessity of speedup algorithm, especially when the measured data is large.

We also calculated the damping ratios using three methods such as the logarithmic decrement method, the decay curve area method by Huang et al.[40] and least squares fitting method. Descriptions of the three methods are included in Appendix A, and the calculated damping ratios are shown in Fig. 3. Due to the influence of ambient excitation, the results of damping radios using the logarithmic decrement method, the decay curve area method and least squares fitting method significantly deviate from the actual values, Additionally, both the decay curve area method and the decay curve area method necessitate manual selection of calculation parameters, leading to increased uncertainty in the results.

The elliptical contours of the marginal updated PDF of f and ζ, which enclose the region between probability 0.5 and 0.9, are shown in Fig. 4 (a). Similarly, the elliptical contours of the marginal updated PDF of S_f and ζ in Fig. 4(b). The two figures show that the actual parameters are at a reasonable distance from the identified optimal parameters. The estimates of f and ζcan be considered independent, given that the ellipses exhibit horizontal major axes in Fig. 4(a). Nevertheless, the estimates of S_f and ζ are strongly correlated, so the elliptical contours have inclined major axes in Fig. 4(b).

Fig. 5 illustrates the conditional probability density functions (PDFs) of the natural frequency f and damping ratio ζ when the remaining parameters are held constant at their optimal values. The conditional PDFs, derived through both the Bayesian time domain method and Gaussian approximation, are depicted separately. The findings from Fig. 5 suggest that the Gaussian approximation adequately captures the accuracy needed to represent the updated PDF.

3.2. *Finite element model of stay cable*

In this example, we use simulated data from a finite element model of stay cable. The stay cable is the A20 cable of the Second Nanjing Yangtze River Bridge[41]. We have established a finite element model of the stay cable using the ANSYS software. The natural frequencies of the stay cable are obtained through modal analysis, and the damping for the cable is modeled using the Rayleigh damping, with a damping ratio set at 0.5 % for the first two modes. The design parameters for the stay cable are provided in Table 2 [41].

Next, we separately generate the simulated data for each mode. We initiate free vibrations of the stay cable with an initial displacement of 0.1 m and initial velocity of 0 m/s using the "original excitation". Simultaneously, we apply ambient excitation with a power spectral density of 0.0001 m^2s^{-3}, and the standard deviation of the measured noise corresponds to a 10 % level of prediction error. The sampling time step is $\Delta t = 0.05$, and the total duration is $T = 100$ s. Fig. 6 respectively depict the measured response of the example cable used for modal parameter identification.

The first five mode identified results of modal parameters are shown in Table 3. The second column in Table 3 corresponds to the actual values used for generation of the simulated measurement data; the third and fourth columns correspond to the identified optimal parameters and the corresponding SDs, respectively; the fifth column lists the CVs for each parameter. The first group of rows in the Table 3 corresponds to modal frequencies, followed by the modal damping ratios. The obtained results reveal satisfactory consistency between the identified values and the nominal values of the frequencies and damping ratios. It is

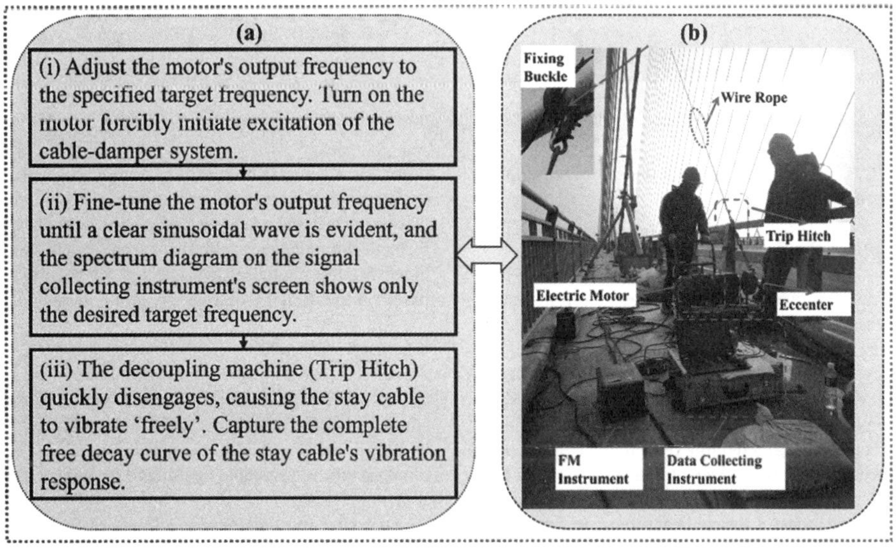

Fig. 11. Photos of the field free vibration tests, (a) free vibration test procedures (b) arrangement of the experimental apparatus.

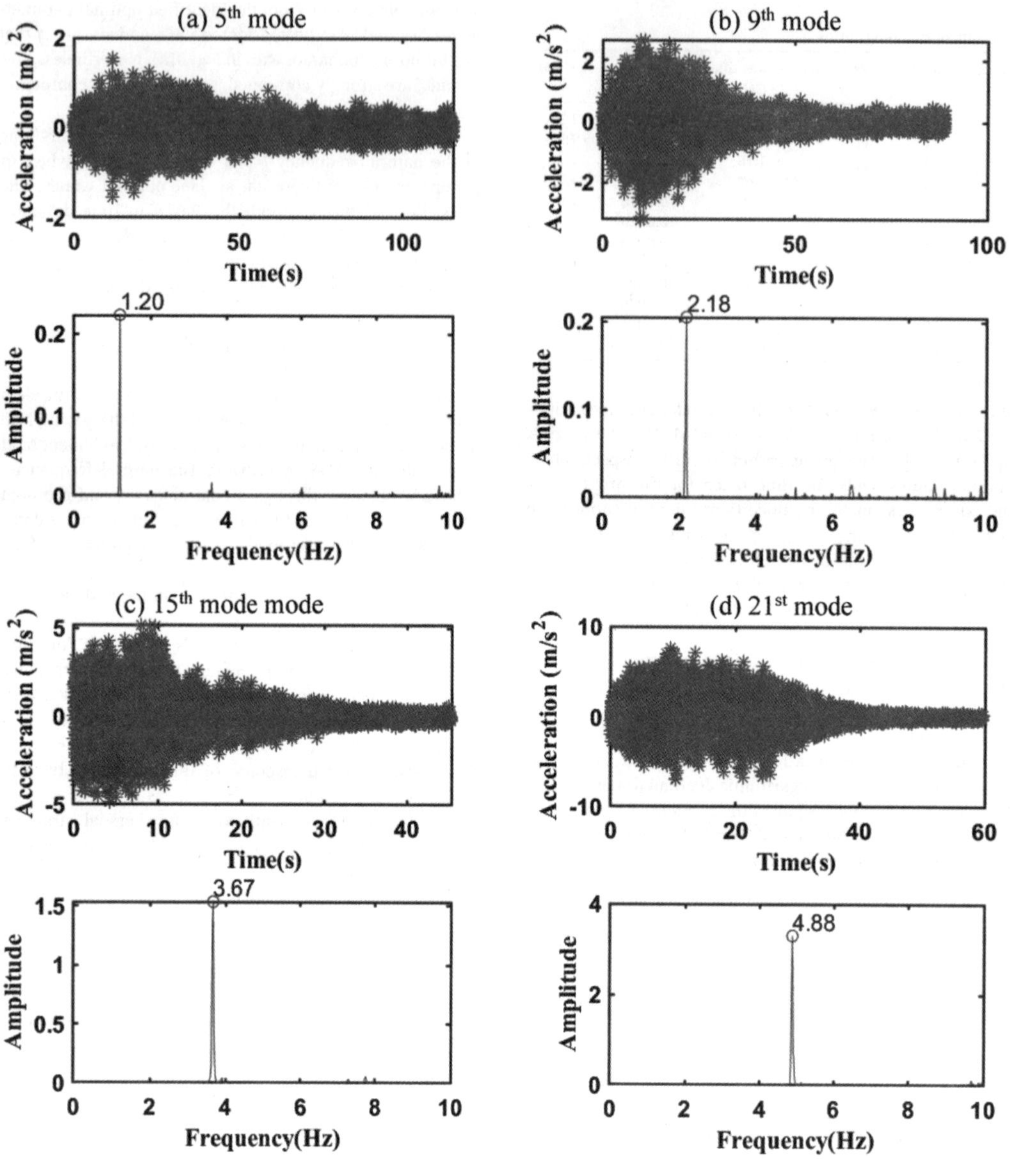

Fig. 12. Acceleration responses recorded of the tested cable and their power spectral density.

noteworthy that the CVs for damping ratios is understandably larger than that for modal frequencies.

4. Experimental applications

Unlike numerical analysis, when conducting free vibration tests of stay cables, it is necessary to obtain the targeted modal frequency in advance. Since the frequencies of stay cables are multiples of the fundamental frequency, once the fundamental frequency is received, it is easy to estimate all the natural frequencies of the stay cable. Moreover, theoretical formulas or OMA can conveniently obtain the fundamental frequency. Of course, the targeted stay cable frequencies also can be obtained in advance through finite element analysis, which is typically feasible. Additionally, to maximize the signal-to-noise ratio, the exciter only excites a single mode of the cable at a time, which implies that the number of tests will increase as the number of modes to be tested increases. Therefore, it is not necessary to calculate the damping ratios for all modal orders; they should be determined based on the design requirements of the stay cables. The following section includes a laboratory test and a field test, where the target frequencies are obtained through finite element analysis or from existing literature. It should be noted that the modal orders selected for the field test are determined based on the design requirements of the dampers.

4.1. Laboratory test

An experiment in the laboratory on a reduced-length cable was conducted. This test cable is constructed with bundled steel strands and uniformly fitted with 300 mass blocks at 16 cm intervals [42]. The parameters of the test cable are shown in Table 4 [43]. The test setup is shown in Fig. 7.

An exciter was employed to stimulate the stay cable during the tests, with an excitation frequency range spanning from 0.1 Hz to 999 Hz, and a minimum frequency interval of 0.1 Hz [42]. The exciter was placed at

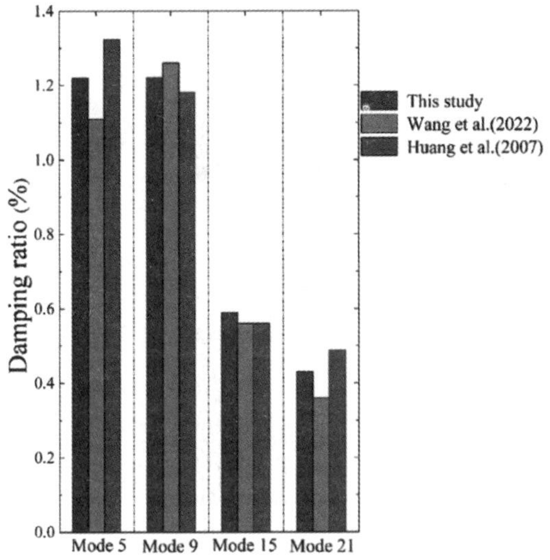

Fig. 13. Comparison of the identified damping ratios.

a distance of 1.9 m from the end of the test cable to achieve initiate excitation with a disengaging device. This disengaging device consists of a double-layered steel sleeve, springs, and bolts. The sleeve features multiple evenly spaced holes. The lower end of the disengaging device can be connected to the exciter, with bolts inserted into the holes to ensure contact between the upper end of the disengaging device and the cable, thus transmitting the excitation force. Pulling out the bolts allows for separation between the upper end of the device and the cable, at which point the cable undergoes free vibration. Adjust the exciter's output frequency to the specified target frequency. The disengaging device rapidly disengages, inducing free vibrations in the stay cable.

Moreover, the laser displacement sensor was arranged at a distance of 3.5 m from the end of the test cable to collect the displacement response. The 5th,8th and 11th modes were excited separately. The displacement response in the 5th,8th and 11th modes of the test cable were measured with a sampling frequency of 100 Hz. Three displacement responses with 2000 points (20 s) and their power spectral density were obtained and shown in Fig. 8.

Due to the obtained free vibration response signal contains vibrations from multiple modes, the Analytical Mode Decomposition (AMD) technique [44,45] is employed to decompose the vibration signal. The decomposed signals corresponding to the 5th, 8th and 11th modes are extracted separately. Subsequently, the proposed Bayesian time-domain approach is utilized for modal parameters identification. The identified results of damping ratios are shown in Table 5. It lists the identified results and their SDs and CVs. It is not surprising to obtain large uncertainty for the damping ratios and the CV values are in the 1 %– 3 % range.

Fig. 9 illustrates the conditional probability density functions (PDFs) of damping ratio ζ when the remaining parameters are held constant at their optimal values. The results in Fig. 9 indicate that the Gaussian approximation effectively achieves the required accuracy for representing the updated Probability Density Function (PDF).

4.2. Field test

The Sutong Yangtze River Highway Bridge (STB), completed in 2008, is a cable-stayed bridge featuring a main span of 1088 m and a streamlined steel closed-box girder as shown in Fig. 10 (a). The bridge has a total of $4 \times 34 \times 2 = 272$ stay cables, with the longest stay cable on the bridge measures 567.77 m. In order to mitigate wind-induced vibrations in the stay cables, a combination of dimpled surfaces along with viscous dampers and magnetorheological dampers was implemented for stay cables numbered 10 to 28 and 29 to 34 as shown in Fig. 10 (b) [3]. While in the period from Jan.1, 2019, to Nov.31,2020, seven stay cables was controlled by the novel eddy current damper system (ECDS) developed by reference [3] as shown in Fig. 10 (c). Moreover, the dampers are connected to the stay cables and the main deck via a single inclined column positioned 3.5 m above the main deck.

Seven stay cables are selected as the test stay cables, namely NJ08D, NJ09D, NJ14D, NJ18D, NJ26D, NJ31D, and NJ32D. Their parameters are shown in Table 6. During the test, the original damper system shown in Fig. 10(b) has been completely dismantled, and the test stay cables were installed with the ECDSs as shown in Fig. 10(c).

Field free vibration tests were performed to assess the actual modal damping ratios of seven stay cables following the completion of the damper system replacement project. The arrangement of the experimental apparatus and sensors is displayed in Fig. 11. Here, a wire rope connects the motor to the stay cable, transmitting the force generated by the motor. Additionally, a decoupling machine (Trip Hitch) on the wire rope is employed to disengage the wire rope from the stay cable, so that the cable can freely vibrate at a specified frequency. The two-directional accelerometers are installed on the stay cables, approximately 9 m above the bridge deck, to measure both in-plane and out-of-plane responses.

4.2.1. NJ32D cable

In this section, we first conducted free vibration tests on NJ32D cable, testing its 5th, 9th, 15th, and 21st modes. The sampling frequency of acceleration response was 200 Hz. The acceleration responses of the tested cable and their power spectral density are shown in Fig. 12. Then we performed conventional preprocessing on the signal and the Analytical Modal Decomposition (AMD) technique to improve the signal-to-noise ratio. The decomposed signals are used for damping ratios identification.

The identification results of the modal frequencies and damping ratios are shown in Table 7. It lists the identified results and their SDs and CVs. The posterior standard derivations are in acceptable range. On the other hand, the uncertainty for the damping ratios and the CV values are in the 20 %– 40 % range. The uncertainty for identification results is influenced by various factors, such as noise level, different methods of uncertainty quantification, signal length, closely-spaced or well-separated modes, among others. For cable structures, aerodynamic damping, and amplitude dependency are also significant influencing factors. Unfortunately, the damping ratios identified by the proposed method still exhibit high uncertainty. Through analysis of the examples provided in the paper, particularly numerical examples, it is preliminarily inferred that one of the reasons for this phenomenon is the significant proportion of ambient vibration data and high noise level in the free vibration data.

We also calculated the damping ratio of NJ32D cable using method proposed by Huang et al. [40], and the results obtained by the two methods and the identified damping ratios using logarithmic decay method in reference Wang et al.(2022) [3] as shown in Fig. 13. It can be seen that the damping ratios identified by the three methods are fairly close.

4.2.2. Other test cables

We also conducted free vibration tests on the remaining six cables, namely NJ08D, NJ09D, NJ14D, NJ18D, NJ26D, NJ31D, and used BTDA to identify their damping ratios for selected modal orders. The acceleration responses of the tested cable and their power spectral density are shown in Fig. 14.

Concurrently, we depict the damping ratios for the six cables in Fig. 15. In Fig. 15, the abscissa represents the modal order. Observably, the damping ratio ranges from 0.4 % to 1.42 %, with particularly high values at mode orders 4–10, followed by a decreasing trend.

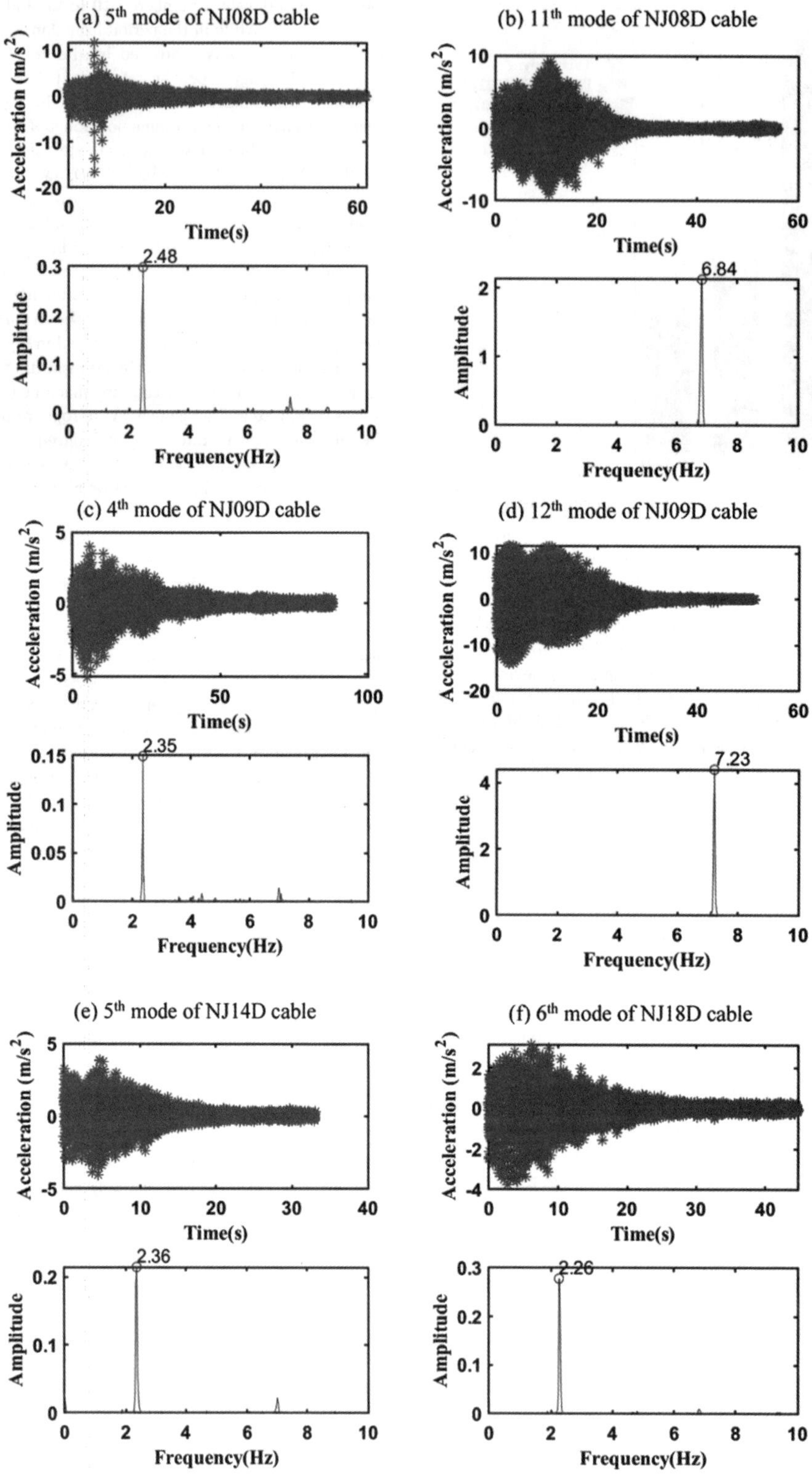

Fig. 14. Acceleration responses recorded of the other six tested cables and their power spectral density.

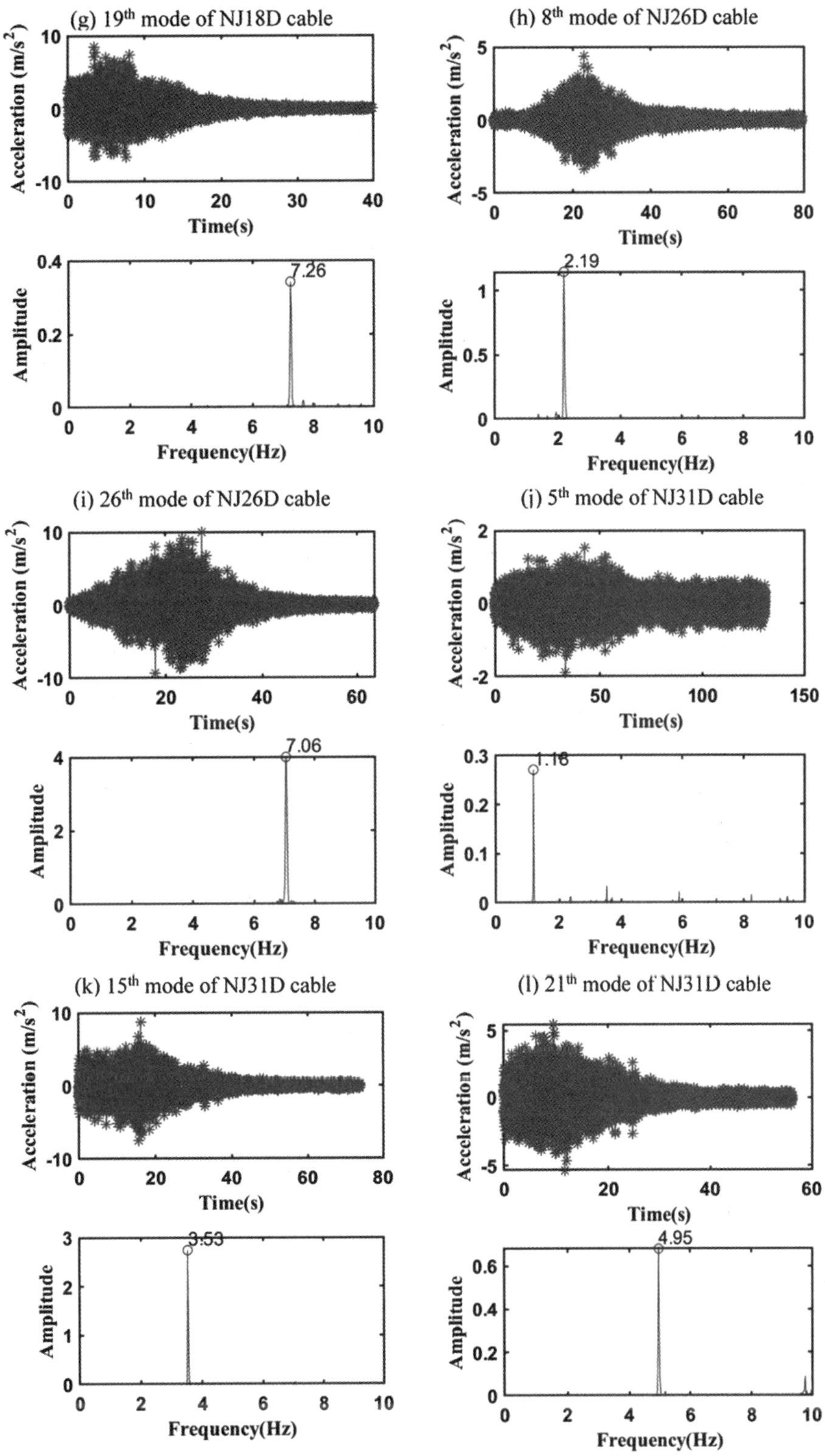

Fig. 14. (*continued*).

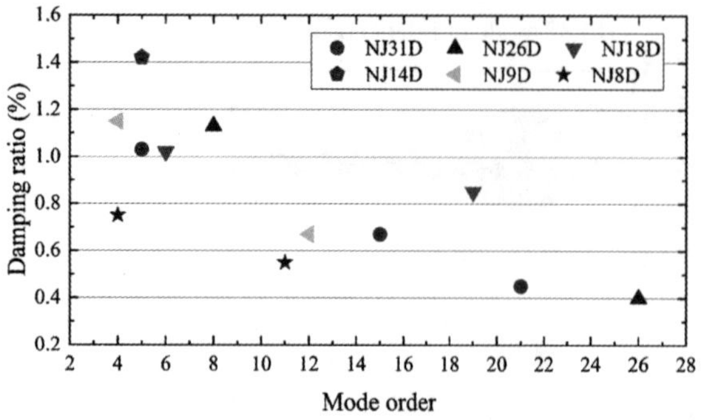

Fig. 15. The identification results of the different order damping ratios of the six cables.

5. Conclusions

A Bayesian time domain approach (BTDA) for identifying damping ratios of stay cables using free vibration data is presented. This method comprehensively accounts for the influence of potential ambient vibration on the free vibration response in the theoretical model. Moreover, time domain data is used directly to avoid the low frequency resolution caused by insufficient data length in frequency domain analysis. At the same time, it ensures computational efficiency through the speedup algorithm. By using Bayesian inference, not only the optimal values of the damping ratios are obtained, but a calculated covariance matrix also quantifies the associated uncertainties. Through numerical simulation of a SDOF model and stay cable finite element model, we verified the efficiency and accuracy of the BTDA in modal parameter identification. Laboratory test of on a reduced-length cable further confirmed its feasibility and showcased its effective handling of free vibration response data. This paper also presents the results for the damping ratios of seven stay cables of the STB by free vibration tests. The results from these numerical and experimental studies convincingly demonstrate that our approach excels in accurately identifying modal properties and quantifying their uncertainties.

Although the efficiency and accuracy of the proposed method have been well demonstrated, it should be noted that there are still many challenges that need to be addressed. Firstly, although the adopted scheme can excite specific mode orders of the stay cables, in practice, the free vibration signal does not always contain only one mode response. This necessitates the use of mode decomposition to resolve the issue. The precision of the mode decomposition still affects the identification results, making it essential to employ higher precision mode decomposition methods. Secondly, the proposed method is only applicable to structures with well-separated modes; for closely-spaced modes, further improvements are necessary. Due to the low fundamental frequency of the ultra-long cables, they can be regarded as closely-spaced modes structures. Developing time-domain identification methods suitable for closely-spaced modes can overcome the errors caused by mode decomposition. Finally, the damping ratios identified by the proposed method still exhibit high uncertainty. It is preliminarily inferred that one of the reasons for this phenomenon is the significant proportion of ambient vibration data and high noise level in the free vibration data. However, this inference has not yet been rigorously proven through theoretical derivation and extensive numerical examples. The influence of ambient vibration on the uncertainty of identification results requires further investigation.

CRediT authorship contribution statement

Zhouquan Feng: Conceptualization, Investigation, Software, Methodology, Data curation, Formal analysis, Writing–original draft, Writing–review & editing, Funding acquisition. **Jiren Zhang:** Investigation, Software, Methodology, Data curation, Formal analysis, Visualization, Writing–original draft. **Xinyan Xuan:** Methodology, Data curation, Formal analysis. **Yafei Wang:** Methodology, Data curation, Formal analysis. **Xugang Hua:** Writing –review & editing, Funding acquisition. **Zhengqing Chen:** Writing –review & editing, Supervision. **Wangji Yan:** Writing –review & editing, Funding acquisition.

Declaration of Competing Interest

The authors declare that they have no known competing financial interests or personal relationships that could have appeared to influence the work reported in this paper.

Data availability

Data will be made available on request.

Acknowledgments

The research is funded by the Natural Science Foundation of China (52178284, 51708203), the National Key Research and Development Program of China (2022YFC3005300), the Open Research Project Program of the State Key Laboratory of Internet of Things for Smart City (University of Macau) (SKL-IoTSC (UM)– 2021–2023/ORP/GA09/2022), and the Graduate Research and Innovation Project of Hunan Province (CX20230443). The authors are grateful to these supports.

Appendix A

Consider a damped SDOF model, the free vibration response of this system is

$$x(t) = Ae^{-\zeta\omega t}\cos(\omega_d t + \varphi) \tag{A.1}$$

Where ω is the natural frequency and $\omega_d = \omega \times \sqrt{1-\zeta^2}$, ζ is the modal damping ratio, A and φ respesents amplitude and phase respectively. There are three methods for calculating the damping ratio.

Method1. The logarithmic decrement method.

The logarithmic-decrement ratio can be written as:

$$\delta = \ln\frac{A_1}{A_{n+1}} = \frac{2n\pi\zeta}{\sqrt{1-\zeta^2}} \tag{A.2}$$

Where n is the number of selected periods, then ζ can be calculated by the follow-ing:

$$\zeta = \frac{\delta/2\pi}{\sqrt{1+(\delta/2\pi)^2}} \tag{A.3}$$

Method2. The decay curve area method.

Based on the logarithmic decrement method, Huang et al. introduced a novel decay curve area method to determine the damping ratio. The damping ratio is calculated using the area under the decay curve of the vibrational response, as illustrated in Fig.A.1 and Eq. (A.4). An advantage of this method is its capability to mitigate the influence of noise.

$$\zeta = \frac{1}{\sqrt{1+(\pi/E)^2}} \tag{A.4}$$

Where $E = \ln \frac{\sum S_{2n-1}}{\sum S_{2n}}$ i.e., it involves taking the natural logarithm (base e) of the ratio between the area summation of subscripts with odd values and the area summation of subscripts with even values.

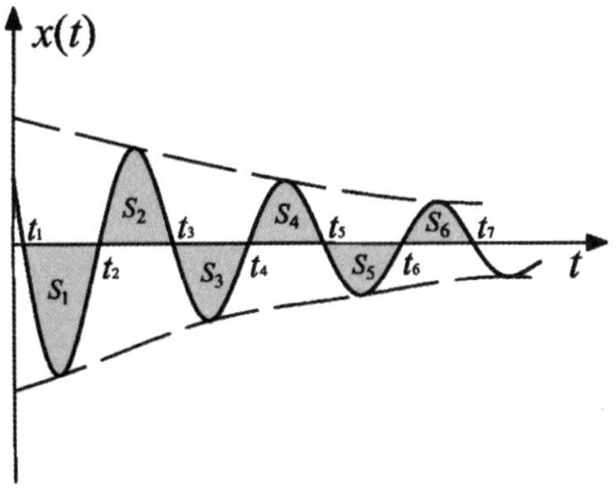

Fig. A.1. Free vibration response of the SDOF dynamic model.

Method3. The least squares fitting method.

The four main steps of least squares fitting method are summarized as follows:

(i) Define model function of four parameters such as frequencyω_d, dampingζ, amplitude $A$and phase φusing Eq. (A.1).
(ii) Select the initial value of the undetermined four parameters.
(iii) Use the measured data to fit the four parameters through least squares fitting.
(iv) Reconstruct the response data based on the identified parameters, check the correlation between the reconstructed data and the measured data.

References

[1] Kim S, Kim S, Kim H-K. High-mode vortex-induced vibration of stay cables: monitoring, cause investigation, and mitigation. J Sound Vib 2022;524:116758. https://doi.org/10.1016/j.jsv.2022.116758.
[2] FENG Z, ZHANG J, WANG Y, LIU Z, Hua X, Chen Z-Q. Bayesian approach study for identifying damping of stay cables using ambient vibration measurements. China J Highw Transp 2023;36:114–24. https://doi.org/10.19721/j.cnki.1001-7372.2023.07.009 (in Chinese).
[3] Wang Y, Chen Z, Yang C, Liu Z, He J, Feng Z. A novel eddy current damper system for multi-mode high-order vibration control of ultra-long stay cables. Eng Struct 2022;262:114319. https://doi.org/10.1016/j.engstruct.2022.114319.
[4] Chen Z, Hua X, FENG Z, CUI B, ZHANG J. Damping characteristies and identification methods for long span bridges: a review. China J Highw Transp 2023;36:1–30. https://doi.org/10.19721/j.cnki.1001-7372.2023.07.001 (in Chinese).
[5] Saidin SS, Jamadin A, Abdul Kudus S, Mohd Amin N, Anuar MA. An overview: the application of vibration-based techniques in bridge structural health monitoring. Int J Concr Struct Mater 2022;16:69. https://doi.org/10.1186/s40069-022-00557-1.
[6] Zahid FB, Ong ZC, Khoo SY. A review of operational modal analysis techniques for in-service modal identification. J Braz Soc Mech Sci Eng 2020;42:398. https://doi.org/10.1007/s40430-020-02470-8.
[7] Lardies J, Minh-Ngi T. Modal parameter identification of stay cables from output-only measurements. Mech Syst Signal Process 2011;25:133–50. https://doi.org/10.1016/j.ymssp.2010.05.020.
[8] Wu W-H, Wang S-W, Chen C-C, Lai G. Application of stochastic subspace identification for stay cables with an alternative stabilization diagram and hierarchical sifting process: SSI for Stay Cables with an Alternative Stabilization Diagram. Struct Control Health Monit 2016;23:1194–213. https://doi.org/10.1002/stc.1836.
[9] Chen C-C, Wu W-H, Yu S-T, Lai G. Investigation of modal damping ratios for stay cables based on stochastic subspace identification with ambient vibration measurements. Adv Struct Eng 2019;22:3444–60. https://doi.org/10.1177/1369433219855900.
[10] Yang X, Lei Y, Liu L, Zhu H, Shen W. Two-step online identification of in-service cable-inertial mass damper systems under nonstationary wind excitations. Mech Syst Signal Process 2023;187:109940. https://doi.org/10.1016/j.ymssp.2022.109940.
[11] Wangchuk S, Siringoringo DM, Fujino Y. Modal analysis and tension estimation of stay cables using noncontact vision-based motion magnification method. Struct Control Health Monit 2022;29. https://doi.org/10.1002/stc.2957.
[12] Mazzeo M, De Domenico D, Quaranta G, Santoro R. Automatic modal identification of bridges based on free vibration response and variational mode decomposition technique. Eng Struct 2023;280:115665. https://doi.org/10.1016/j.engstruct.2023.115665.
[13] Lorenzoni F, De Conto N, Da Porto F, Modena C. Ambient and free-vibration tests to improve the quantification and estimation of modal parameters in existing bridges. J Civ Struct Health Monit 2019;9:617–37. https://doi.org/10.1007/s13349-019-00357-4.
[14] Ko J.M., Zheng G., Chen Z.Q., Ni Y.-Q. Field vibration tests of bridge stay cables incorporated with magnetorheological (MR) dampers. In: Liu S-C, Pines DJ, editors., San Diego, CA: 2002, p. 30–40. https://doi.org/10.1117/12.472567.
[15] LI X., SUN L., CHENG W., XIA Y. Damping analysis on steel strand cables of a cable-stayed bridge based on field tests. Proc. 2014 World Congr. Adv. Civ. Environ. Mater. Res., Techno-Press; 2014.

[16] Wu W-H, Chen C-C, Liau J-A. A multiple random decrement method for modal parameter identification of stay cables based on ambient vibration signals. Adv Struct Eng 2012;15:969–82. https://doi.org/10.1260/1369-4332.15.6.969.
[17] Huang Hongwei, Liu Jiangyun, Sun Limin. Full-scale experimental verification on the vibration control of stay cable using optimally tuned MR damper. Smart Struct Syst 2015;16:1003–21. https://doi.org/10.12989/SSS.2015.16.6.1003.
[18] Li H, Liu M, Li J, Guan X, Ou J. Vibration control of stay cables of the shandong binzhou yellow river highway bridge using magnetorheological fluid dampers. J Bridge Eng 2007;12:401–9. https://doi.org/10.1061/(ASCE)1084-0702(2007)12:4(401).
[19] Magalhães F, Cunha Á, Caetano E, Brincker R. Damping estimation using free decays and ambient vibration tests. Mech Syst Signal Process 2010;24:1274–90. https://doi.org/10.1016/j.ymssp.2009.02.011.
[20] Beck JL, Katafygiotis LS. Updating models and their uncertainties. I: Bayesian statistical framework. J Eng Mech 1998;124:455–61. https://doi.org/10.1061/(ASCE)0733-9399(1998)124:4(455).
[21] Katafygiotis LS, Beck JL. Updating models and their uncertainties. II: Model identifiability. J Eng Mech 1998;124:463–7. https://doi.org/10.1061/(ASCE)0733-9399(1998)124:4(463).
[22] Katafygiotis LS, Yuen K-V. Bayesian spectral density approach for modal updating using ambient data. Earthq Eng Struct Dyn 2001;30:1103–23. https://doi.org/10.1002/eqe.53.
[23] Yuen KV, Katafygiotis LS. Bayesian fast fourier transform approach for modal updating using ambient data. Adv Struct Eng 2003;6:81–95. https://doi.org/10.1260/136943303769013183.
[24] Au SK. Fast Bayesian FFT method for ambient modal identification with separated modes. J Eng Mech 2011;137:214–26. https://doi.org/10.1061/(ASCE)EM.1943-7889.0000213.
[25] Yan WJ, Katafygiotis LS. A two-stage fast Bayesian spectral density approach for ambient modal analysis. Part I: posterior most probable value and uncertainty. 139–55 Mech Syst Signal Process 2015;54–55. https://doi.org/10.1016/j.ymssp.2014.07.027.
[26] Yan WJ, Katafygiotis LS. A two-stage fast Bayesian spectral density approach for ambient modal analysis. Part II: mode shape assembly and case studies. 156–71 Mech Syst Signal Process 2015;54–55. https://doi.org/10.1016/j.ymssp.2014.08.016.
[27] Sedehi O, Katafygiotis LS, Papadimitriou C. Hierarchical Bayesian operational modal analysis: theory and computations. Mech Syst Signal Process 2020;140:106663. https://doi.org/10.1016/j.ymssp.2020.106663.
[28] Feng Z, Zhang J, Katafygiotis LS, Hua X, Chen Z-Q. Bayesian spectral decomposition for efficient modal identification using ambient vibration. Struct Control Health Monit 2024;2024:513764. https://doi.org/10.1155/2024/5137641.
[29] An Y, Xue Z, Liao Z, Li B. Automated identification of modal parameters for cables. China Civ Eng J 2023;56:82–91. https://doi.org/10.15951/j.tmgcxb.22020140 (in Chinese).
[30] Yuen K-V, Katafygiotis LS. Bayesian time–domain approach for modal updating using ambient data. Probabilistic Eng Mech 2001;16:219–31. https://doi.org/10.1016/S0266-8920(01)00004-2.
[31] Yuen K-V, Beck JL, Katafygiotis LS. Probabilistic approach for modal identification using non-stationary noisy response measurements only. Earthq Eng Struct Dyn 2002;31:1007–23. https://doi.org/10.1002/eqe.135.
[32] Li B, Der Kiureghian A. Operational modal identification using variational Bayes. Mech Syst Signal Process 2017;88:377–98. https://doi.org/10.1016/j.ymssp.2016.11.007 (in Chinese).
[33] Li B, Der Kiureghian A, Au S. A Gibbs sampling algorithm for structural modal identification under seismic excitation. Earthq Eng Struct Dyn 2018;47:2735–55. https://doi.org/10.1002/eqe.3094.
[34] Rather S, Bansal S. Bayesian modal identification of non-classically damped systems using time-domain data. Mech Syst Signal Process 2023;197:110349. https://doi.org/10.1016/j.ymssp.2023.110349.
[35] Zhang FL, Ni YC, Au SK, Lam HF. Fast Bayesian approach for modal identification using free vibration data, Part I – most probable value. 209–20 Mech Syst Signal Process 2016;70–71. https://doi.org/10.1016/j.ymssp.2015.05.031.
[36] Ni YC, Zhang FL, Lam HF, Au SK. Fast Bayesian approach for modal identification using free vibration data, Part II—posterior uncertainty and application. 221–44 Mech Syst Signal Process 2016;70–71. https://doi.org/10.1016/j.ymssp.2015.06.009.
[37] Yu X, Dan D, Ge L. Time-domain distributed modal parameter identification based on mode decomposition of single-channel vibration response. Eng Struct 2023;289:116323. https://doi.org/10.1016/j.engstruct.2023.116323.
[38] Yuen K-V. Bayesian methods for structural dynamics and civil engineering. Singapore: Wiley; 2010.
[39] Au S-K, Xie Y-L. Calculation of Hessian under constraints with applications to Bayesian system identification. Comput Methods Appl Mech Eng 2017;323:373–88. https://doi.org/10.1016/j.cma.2017.05.021.
[40] Huang F-L, Wang X-M, Chen Z-Q, He X-H, Ni Y-Q. A new approach to identification of structural damping ratios. J Sound Vib 2007;303:144–53. https://doi.org/10.1016/j.jsv.2006.12.026.
[41] Li Guoqiang, Gu Ming, Sun Limin. Theory of Cable Vibration, Dynamic Detection and Vibration Control. Beijing: Science Press; 2014.
[42] Wang Y, Li S, Qie K, Zhang H, Chen Z. Experimental investigations on effectiveness of viscous and viscous inertial dampers on reducing the multimodal vibrations of stay cables. J Vib Control 2023;0:1–20. https://doi.org/10.1177/10775463231198641.
[43] Lai K, Fan W, Chen Z, Yang C, Liu Z, Li S. Performance of wire rope damper in vibration reduction of stay cable. Eng Struct 2023;278:115527. https://doi.org/10.1016/j.engstruct.2022.115527.
[44] Chen G, Wang Z. A signal decomposition theorem with Hilbert transform and its application to narrowband time series with closely spaced frequency components. Mech Syst Signal Process 2012;28:258–79. https://doi.org/10.1016/j.ymssp.2011.02.002.
[45] Wang Z-C, Ge B, Ren W-X, Hou J, Li D, He W-Y, et al. Discrete analytical mode decomposition with automatic bisecting frequency selection for structural dynamic response analysis and modal identification. J Sound Vib 2020;484:115520. https://doi.org/10.1016/j.jsv.2020.115520.

Engineering Structures 302 (2024) 117324

Contents lists available at ScienceDirect

Engineering Structures

journal homepage: www.elsevier.com/locate/engstruct

A machine learning based method for predicting the shear strength of Fiber-Reinforced Concrete joints in precast segmental bridges

Peng Ge [a,b], Ou Yang [a,b,*], Jia He [a,b], Kailun Zhang [a,b], Suiwen Wu [a,b]

[a] *State Key Laboratory of Bridge Safety and Resilience, Hunan University, Changsha 410082, PR China*
[b] *Laboratory for Wind and Bridge Engineering of Hunan Province, College of Civil Engineering, Hunan University, Changsha 410082, RP China*

ARTICLE INFO

Keywords:
Precast concrete segment girders
Fiber high-strength concrete
Shear behavior
Machine learning
Explainable Ensemble model
SHapley Additive exPlanations

ABSTRACT

The utilization of Fiber-reinforced concrete (FRC) in the construction of Precast segmental concrete bridges (PSCBs) is progressively increasing. The design of joints among precast members can considerably impact the overall integrity and safety of the structure. Consequently, accurately estimating the shear capacities of FRC joints has become critical for evaluating the safety of PSCBs structures. However, the effectiveness and efficiency of formulas or models intended for prediction of the shear capacities of joints with normal concrete are unknown in predicting the shear capacities of FRC joints. In this study, the machine learning (ML) algorithms are employed to predict the ultimate load of FRC joints, utilizing a database of 192 push-off test results. Three individual models (Lasso, SVM, and KNN) and three ensemble models (RF, GBDT, and XGBoost) were developed for estimating the ultimate load. The GridSearchCV technique was utilized for parameter optimization based on the training set, in conjunction with a five-fold cross-validation method. A comparative study revealed that the XGBoost ensemble model surpassed other models, with a predictive performance ranking as follows: XGBoost > GBDT > RF > KNN > SVM > Lasso. The best-performing XGBoost model achieved performance metrics of $R^2 = 0.964$, MAE = 41.318 kN, RMSE = 56.760 kN, and MAPE = 14.228%. The superior reliability of the XGBoost ensemble model was further validated through comparisons with current design codes and existing models. As ML-based models are typically black-box models, the SHapley Additive exPlanations (SHAP) method was employed to explicitly link the predicted output to the inputs. Based on SHAP results, confining stress (*CS*), depth of shear key (*D*), compressive strength of concrete (f_{cu}), total height of joints (*H*), and height of flat part (H_{sm}) recognized as the crucial parameters affecting ultimate load.

1. Introduction

Precast segmental concrete bridges (PSCBs) have become increasingly popular in bridge construction in recent decades due to their safe and economical design, rapid assembly, practical and versatile construction, and minimal impact on the environment [1]. The superstructure of precast concrete segmental bridges comprises short segments connected by prestress tendons. During design and construction, special attention must be paid to the joints between segments [2,3]. These joints represent discontinuities and significantly impact the overall behavior of the bridge by transmitting compressive and shear stresses [4,5]. Shear keys, essential components of these joints, play a pivotal role in determining the force performance and load-carrying capacity of the bridge. Therefore, investigating the mechanical properties and failure modes of shear keys under different working conditions is imperative [5,6]. The precast segmental concrete bridges (PSCBs) in Fig. 1.

In conventional precast segmental bridges, there are three types of joints in the shear key: epoxy joints, dry joints, and wet joints [7]. Extensive research has been conducted since the 1950 s on the mechanical behavior of conventional precast segmental bridges with these joints, using theoretical analysis, numerical simulations, and experimental research[2,8–10]. Previous research has shown that normal concrete joint shear keys often exhibit low tensile strength, susceptibility to brittle detachment, and poor ductility of the shear key teeth. To overcome these issues in normal concrete joint shear keys, some scholars have begun to add fibers to concrete to alter the stress performance of the joint. Experimental studies have been conducted to investigate the shear strength and deformation of Fiber-Reinforced Concrete (FRC) joints [2,4,11–13]. These studies have demonstrated that FRC joints

* Corresponding author at: State Key Laboratory of Bridge Safety and Resilience, Hunan University, Changsha 410082, PR China
E-mail address: ouyanghnu@hnu.edu.cn (O. Yang).

https://doi.org/10.1016/j.engstruct.2023.117324
Received 11 August 2023; Received in revised form 7 December 2023; Accepted 9 December 2023
Available online 27 December 2023

exhibit higher shear strength compared to normal concrete (NC) joints. Existing research essentially believes that the addition of discrete high-strength fibers provides a bridging effect to prevent crack expansion and improve post-cracking shear ductility while helping to transmit load across the cracked surface. Specifically, short fibers can efficiently control the development of microcracks, while longer fibers can increase the ultimate strength by preventing the propagation of macrocracks [14–16].

Although a great amount of research on the shear performance of joints has been carried out, the majority of studies have focused on joints made of conventional concrete, and many useful conclusions have been obtained. However, due to the high compressive strength and the existence of fibers in Fiber Reinforced Concrete (FRC), there are no clear existing design norms that apply to the design of FRC joints in the world. Existing design norms, such as AASHTO [17], adopt the compressive stress normal to the connection interface and the compression strength of grout as the main parameters to predict the shear strength of connections [13]. However, these equations were proposed based on shear key connections with ordinary strength grout, which might not be applicable to calculate the shear capacity of FRC joints [2,4,18,19]. Jiang et al. [4] conducted experiments on dry joints using steel fiber-reinforced concrete to study their shear behavior. They found that the AASHTO formula underestimated the shear strength of single-keyed joints with SFRC. Similarly, Chen et al. [20] found through both experimental and FEM analysis that there is an uneven distribution of shear stresses in multiple shear keys of Ultra-High Performance Concrete (UHPC) joints. This renders existing code equations like AASHTO ineffective for predicting the shear capacity of multi-keyed joints. Given the shortcomings of current code equations, some improved prediction methods have been proposed [18,19,21,22]. While these proposed models may yield satisfactory results when matched with researchers' experimental data, they may not provide accurate predictions for experiments conducted by other researchers. Therefore, establishing a suitable model for assessing the ultimate load of shear key specimens is necessary.

Developing a unified model that considers multiple influencing parameters simultaneously is challenging due to limitations in experimental conditions and sample size. To overcome these challenges and enhance the accuracy of ultimate load prediction, the application of artificial intelligence-based machine learning (ML) has recently garnered attention in various fields, including civil engineering [23–27]. The strength of ML lies in its capacity to circumvent errors stemming from assumptions in the mechanical model and to consider multiple influencing factors, thereby improving accuracy. ML has been successfully utilized in civil engineering to predict various properties of reinforced concrete (RC) structures, such as the prediction of FRP-concrete interfacial bond strength [24], estimation of punching shear strength in flat slabs [26], and prediction of steel section-concrete bond strength [27].

However, the black-box nature of ML models makes it difficult to understand the decision-making process due to algorithmic complexity and multi-dimensional databases [26–29]. Achieving greater interpretability of ML model results is essential for engineers to understand why specific predictions or decisions were made. One method employed to enhance interpretability is the SHAP (SHapley Additive exPlanations) method [30], which assesses the significance of features and their impact on positive and negative outcomes by computing the corresponding Shapley value for each feature. By combining SHAP model, a more comprehensive understanding of predicted ultimate load can be achieved, improving interpretability.

Given the aforementioned background, this study seeks to tackle the existing issue of the limited applicability of the ultimate load prediction models for Fiber-Reinforced Concrete joints'. To achieve this, a large database consisting of test results from 192 push-off specimens of concrete joints was established. Six robust data-driven models were employed, encompassing three individual models (Lasso, SVM, and KNN) and three ensemble models (RF, GBDT, and XGBoost). The GridSearchCV technique was utilized for parameter optimization based on the training set, along with a five-fold cross-validation method. The performance of these models was compared and evaluated based on factors such as learning capability, predictive accuracy, and residuals. Additionally, the performance of machine learning models was juxtaposed with design code provisions and various empirical fitting formulas to ascertain their efficiency and predictive capability. Finally, the SHapley Additive exPlanations (SHAP) approach was employed to elucidate the influence of various parameters on the ultimate load of shear keys. The machine learning workflow utilized for predicting the ultimate load is illustrated in Fig. 2.

2. Description of the experimental data set

2.1. Data collection

Since the 1950 s, comprehensive research has been undertaken on the mechanical behavior of precast segmental bridges with these joints, employing theoretical analysis, numerical simulations, and experimental studies [2,8–10,31–33]. Various methods have been proposed to test the shear performance of joints, with the direct shear test being a popular test method. The material properties, geometric parameters, and ultimate load are typically reported for each test. Based on practical tests and theoretical analyses, the most critical factors that influence the shear behavior of joints are the concrete strength, joint types (i.e., wet, dry, and epoxy), parameters of shear keys (e.g., shape, number, size), interface treatment, the addition of fiber, and confining stress. Machine

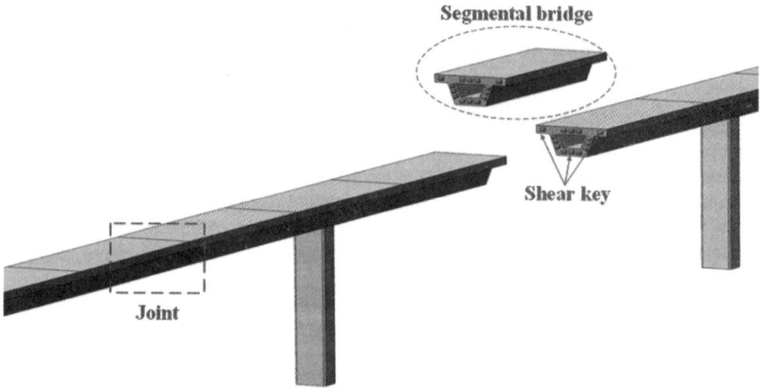

Fig. 1. Schematic of joints in precast segmental concrete bridges.

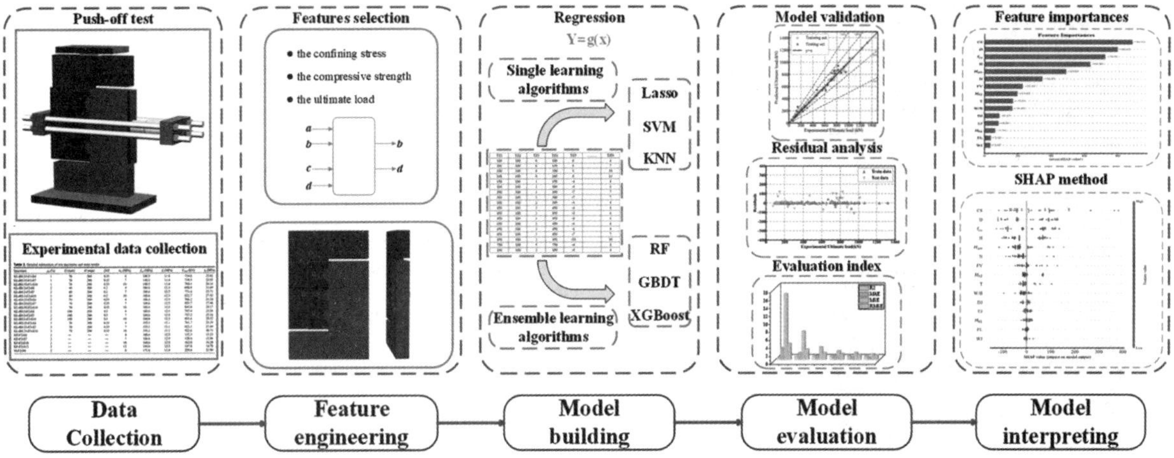

Fig. 2. Machine learning flow chart for the prediction of the ultimate load.

learning modeling offers the potential to comprehensively consider the impact of these factors.

In order to develop a data-driven model and take into account all potential influencing factors, this study collected 192 experimental data from available literature, selecting direct shear test experiments for ultimate load assessment. Detailed input and output parameters will be introduced in Section 2.2. Fig. 3 presents a schematic representation of the direct shear test.

2.2. Feature engineering

2.2.1. Selection of numerical features

12 numerical features are considered in this study, including the water-binder ratio(W/B), the compressive strength of concrete (f_{cu}, MPa), the total height of joints (H, mm); the thickness of specimen(T, mm), the number of shear keys(N), the height of flat part (H_{sm}, mm), the top width of single-keyed (H_{k1}, mm), the base width of single-keyed (H_{k2}, mm), the depth of shear key(D, mm), the confining stress (CS, MPa), the length of fiber(FL, mm), the fiber volume ratio(FV, %). The ultimate load(V_u, kN) is served as the output parameter. {{{Fig. 4}}}.

Furthermore, the compressive strength of concrete samples was primarily measured using cube specimens with dimensions of 100 mm × 100 mm× 100 mm. However, a subset of the samples in the database were obtained from specimens with varying forms and shapes, resulting in differences in the measured compressive strengths. To maintain consistency across the dataset, the compressive strength values for these samples were adjusted to correspond with equivalent values for 100 mm × 100 mm× 100 mm cube specimens, as per the relationship outlined in Eq. (1) [34,35].

$$f_{cu,100} = \frac{f_{cu,150}}{0.95} \tag{1}$$

2.2.2. Selection of categorical features

The joint type is a significant categorical feature in the database under investigation, including wet joint (WJ), dry joint (DJ), and epoxy joint (EJ), which have a considerable impact on the ultimate load and must not be disregarded [32,36–38]. However, most machine learning algorithms are unsuitable for processing categorical data and require numerical inputs and outputs. Therefore, it is essential to encode all variables in the dataset into numerical values while preserving the dataset's information as accurately as possible [39,40]. Label encoding

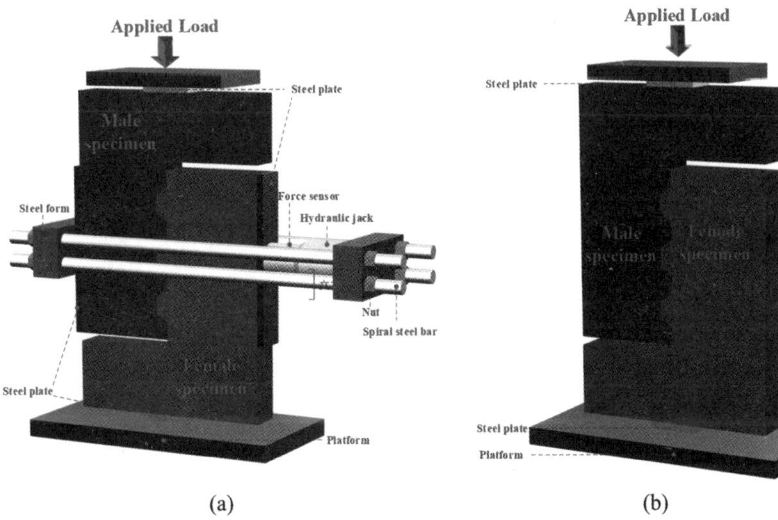

Fig. 3. Schematic drawing of the push-off test: (a) with confining stress; and (b) without confining stress.

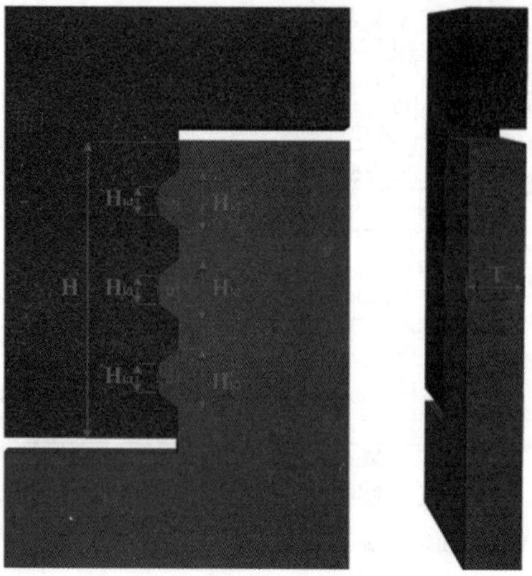

Fig. 4. Details of shear keys.

is a conventional method for converting categorical data into numerical data by assigning values to category values based on their positions in the set, where a set with N categories is numbered using consecutive integers in the interval [0, N-1][41–43]. For example, wet joint can be represented by 0, dry joint by 1 and epoxy joint by 2. However, this approach might not be appropriate for some Python-based machine learning packages like Scikit-Learn, which presume that numerical features represent algebraic quantities [41–43]. This can result in unrealistic hierarchies, such as wet joint < dry joint < epoxy joint or be mistaken for epoxy joint - dry joint = wet joint. To address this issue, one-hot encoding is a commonly employed solution that generates a new dummy feature for each category value, with each column representing a value of the nominal data. In this study, one-hot encoding was utilized to transform the joint type into three new features: wet joint, dry joint, and epoxy joint. A binary value was used to identify the type of joint by us, for example, dry joint specimens can be identified as: wet joint = 0; dry joints = 1; epoxy joints = 0. Since the majority of the values in the resulting vector are zero, one-hot encoding can be represented as a sparse matrix, thereby avoiding excessive storage requirements associated with dense matrices.

2.3. *Database description and analysis*

The experimental data encompasses a broad range of design parameters: 24 wet joint (*WJ*) records; 88 dry joint (*DJ*) records; 53 epoxy joint (*EJ*) records; and 27 monolithic specimen (MS) records. Table 1 presents a summary of the data source. Fig. 5 displays histograms representing the data distribution of each parameter used, where frequency indicates the number of times that each specific value (or range) of the variables appeared in the dataset.

As shown in Table 1 and Fig. 5, the data distribution exhibits considerable variation, indicating a highly non-linear problem. The dataset exhibits significant variations in multiple parameters, such as W/B ranging from 0.16 to 0.37, f_{cu} varying between 49.2 to 193.1 MPa, H measuring between 100 to 750 mm, T ranging from 100 to 150 mm, N varying between 0 to 5, H_{sm} ranging between 0 to 300 mm, H_{k1} varying between 0 to 200 mm, H_{k2} ranging between 0 to 300 mm, D measuring between 0 to 100 mm, CS ranging from 0 to 24 MPa, FL measuring between 0 to 40 mm, and FV varying between 0 to 3%. Due to the wide range of geometric and material properties in the specimens, V_u also varies widely from 74.1 to 1321 kN.

3. Machine learning algorithms

This study assessed the performance of six representative ML algorithms - Lasso, SVM, KNN, RF, GBDT, and XGBoost - in predicting the ultimate load. Table 2 provides a summary of the machine learning techniques employed in this manuscript.

3.1. *Model training and testing*

In machine learning applications, it is a common practice for researchers to partition the dataset into two subsets: a training set and a testing set. The training set is utilized for model training, while the testing set serves to evaluate model performance by comparing predicted and actual outcomes. This study partitioned the dataset into a training set (153 samples) and a testing set (39 samples) at an 8:2 ratio.

Table 1
Basic information of experimental data of 192 columns (numerical features).

Ref	W/B	f_{cu} /MPa	H /mm	T /mm	N	H_{sm} /mm	H_{k1} /mm	H_{k2} /mm	D /mm	CS /MPa	FL /mm	FV /%	V_u /kN
Pan et al.[44]	0.16	135.5 ~187.8	150 ~750	100	1 ~5	50 ~250	40 ~60	100	40 ~60	4 ~13	13	1 ~3	285.0 ~1321.0
Jiang et al.[1]	0.22	124.8 ~128.3	200 ~500	100	0 ~3	0 ~200	0 ~50	0 ~100	0 ~35	1 ~2	13 ~30	2	226.2 ~750.8
Liu et al.[18]	0.16 ~0.33	49.2 ~124.3	300 ~500	150	0 ~3	200 ~300	0 ~200	0 ~300	0 ~50	1.2 ~24	0 ~26.5	0 ~2	169.0 ~1229.6
Yuan et al.[11]	0.18	77.4 ~105.4	200 ~400	100	0 ~2	0 ~150	0 ~200	0 ~300	0 ~100	1 ~2	0 ~30	0 ~3	98.68 ~383.16
Pan et al.[10]	0.16	140.3 ~193.1	100 ~300	100	0 ~1	0 ~100	0 ~160	0 ~200	0 ~100	0 ~13	13	1 ~3	128.6 ~922.6
Pan et al.[37]	0.16	135.5 ~187.8	100 ~750	100	0 ~5	50 ~250	0 ~100	0 ~200	0 ~100	4 ~13	13	1 ~3	128.6 ~1321
Feng et al.[31]	0.18	112.8 ~178.5	200	100	0	0 ~200	0	0	0	0	0 ~18	0 ~3	74.1 ~494.5
Feng et al.[45]	0.18	152.6	300	150	0 ~1	0 ~300	0 ~155	0 ~100	0 ~60	0 ~8	16	2.5	382.8 ~1065.6
Jiang et al.[46]	0.37	65.1 ~71.3	200	100	0 ~1	0 ~100	0 ~50	0 ~100	0 ~35	0.5 ~2	0 ~40	0 ~1	97.4 ~259.5
Yuan et al.[47]	0.18	77.4 ~105.4	200	100	1	100	50	100	50	1	0 ~30	0 ~3	98.68 ~234.7

Note: W/B: the water-binder ratio; f_{cu}: the compressive strength of concrete; H: the total height of joints; T: the thickness of specimen; N: the number of shear keys; H_{sm}: the height of flat part; H_{k1}: the top width of single-keyed; H_{k2}: the base width of single-keyed; D: the depth of shear key; CS: the confining stress; FL: the length of fiber; FV: the fiber volume ratio; Vu: The ultimate load.

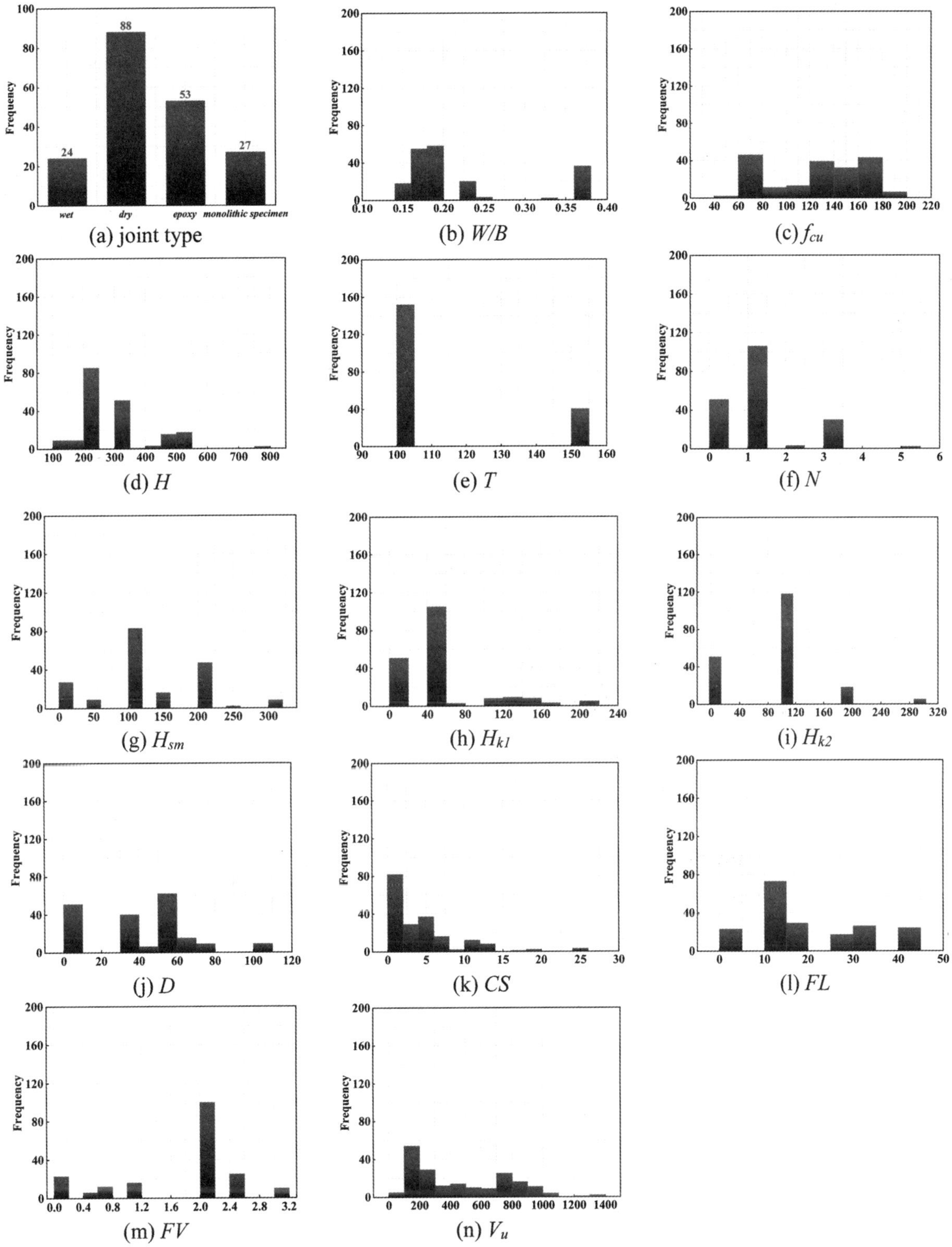

Fig. 5. Distribution histogram of characteristic parameters.

Table 2
The machine learning algorithm features.

No	Model	applications	Number of learners
1	Lasso	regression and classification algorithms	individual learning algorithms
2	SVM	regression and classification algorithms	individual learning algorithms
3	KNN	regression and classification algorithms	individual learning algorithms
4	RF	regression and classification algorithms	ensemble learning algorithms
5	GBDT	regression and classification algorithms	ensemble learning algorithms
6	XGBoost	regression and classification algorithms	ensemble learning algorithms

Subsequently, the study standardized the dataset using Eq. (2) to achieve a normal distribution with a mean of 0 and a standard deviation of 1. This standardization facilitated the parameter learning process by ensuring the same convergence effect across different dimensions. To optimize the hyperparameters for each model, this study performed grid search (function "GridSearchCV") with 5-fold cross-validations [48] on the training set. This approach enhanced the model robustness and accuracy for predicting new and unseen data. Finally, the study evaluated the trained model on the testing set. Fig. 6 depicts the flowchart of the proposed framework. The formula for Z-Score standardization is as follows.

$$X_{Z-Score} = \frac{X-\mu}{\sigma} \tag{2}$$

where $X_{Z-Score}$ denotes the eigenvalues in the original dataset; while μ and σ represent the mean and variance of each eigenvalue in the original dataset, respectively.

3.2. Model performance criteria

This study employed multiple statistical metrics to quantitatively evaluate the accuracy of the machine learning-based models. These metrics include the coefficient of determination (R^2), mean absolute error (MAE), root mean square error (RMSE), and mean absolute percent error (MAPE) [49]. The formulas for these statistical estimators are as follows:

$$R^2 = 1 - \frac{\sum_{i=1}^{n}(Y_i - \widehat{Y}_i)}{\sum_{i=1}^{n}(Y_i - \overline{Y}_i)} \tag{3}$$

$$MAE = \frac{1}{n}\sum_{i=1}^{n}|Y_i - \widehat{Y}_i| \tag{4}$$

$$RMSE = \sqrt{\frac{1}{n}\sum_{i=1}^{n}(Y_i - \widehat{Y}_i)^2} \tag{5}$$

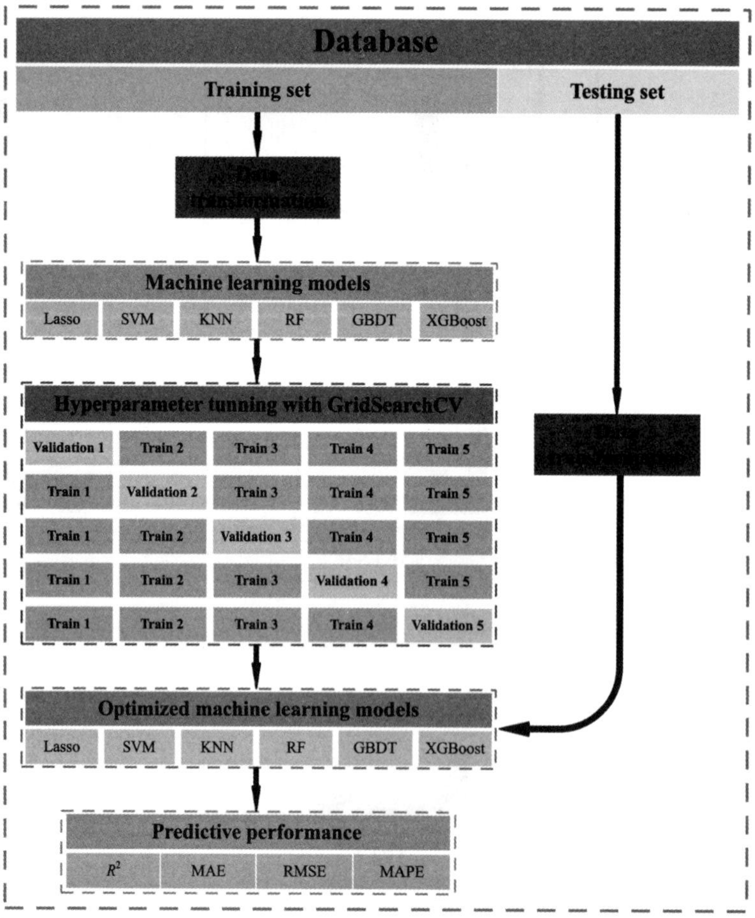

Fig. 6. Methodology flowchart of this study.

$$MAPE = \frac{1}{n}\sum_{i=1}^{n}\left|\frac{Y_i - \widehat{Y}_i}{Y_i}\right| \times 100\% \tag{6}$$

where n is the number of the sample; Y_i is the experimental value; $\widehat{Y}_i$is the predicted value; and $\overline{Y}_i$ the mean of all sample experimental values.

4. Results and discussions

This study applied various machine-learning techniques to predict the ultimate load of shear key specimens and evaluated the accuracy of these predictions by comparing them with actual experimental data. The results of this study demonstrate the effectiveness of the selected machine-learning algorithms in accurately predicting the ultimate load of shear key specimens.

4.1. Model validation

The present study contrasts the predictive performance of multiple machine learning (ML) models in estimating the ultimate load of shear key specimens. The predicted values generated by each model are compared against their corresponding experimental values, and the

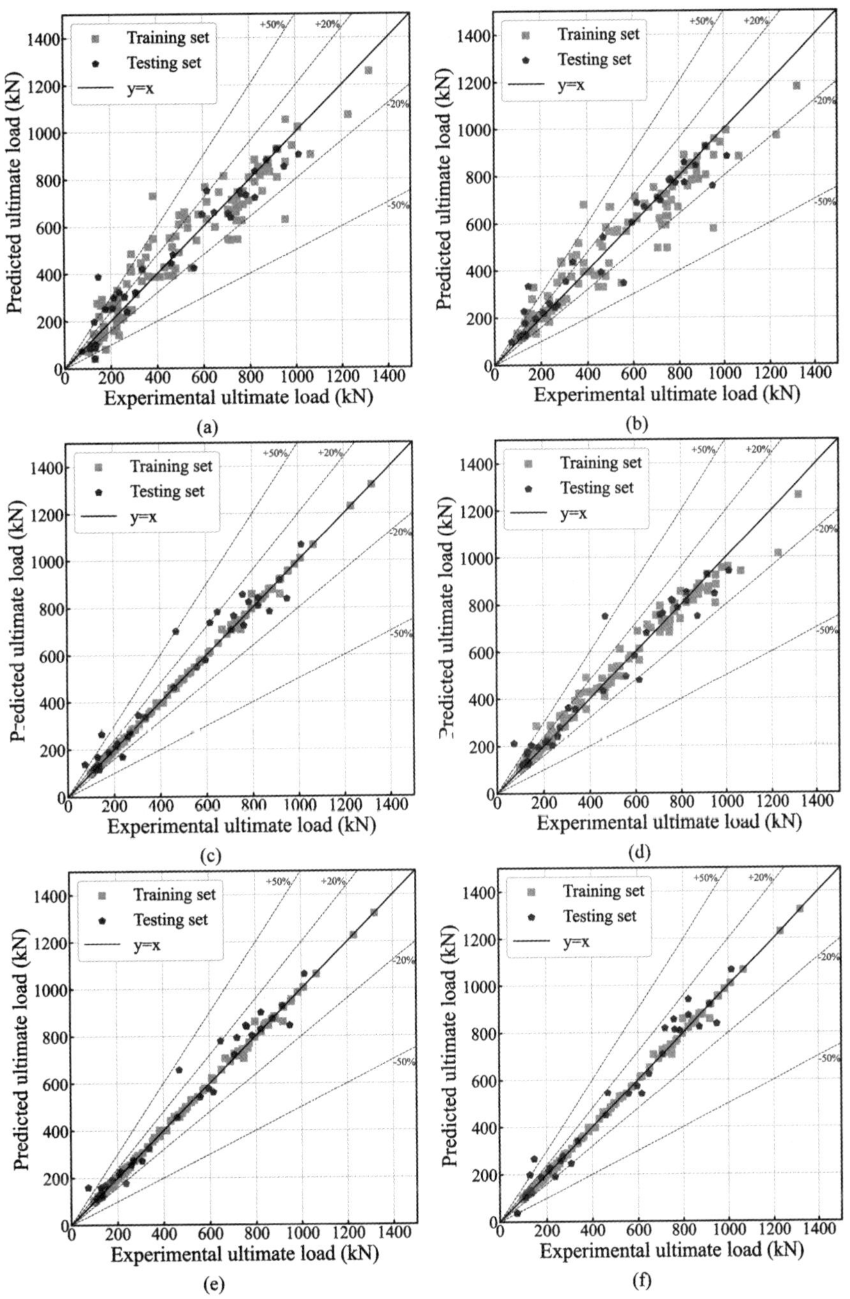

Fig. 7. Predicted the ultimate load from ML models versus experimental results: (a) Lasso; (b) SVM; (c) KNN; (d) RF; (e) GBDT; (f) XGBoost.

obtained results are presented in Fig. 7. In this figure, the line y = x represents the ideal scenario where the actual and predicted values coincide. This figure provides a comprehensive comparison of the predictive accuracy of each model on both the training and testing sets.

The scatter plots indicate that the Lasso model has the poorest forecasting results among the six models, followed by the SVM model. However, the ensemble models RF, GBDT, and XGBoost demonstrate favorable and robust predictive ability, with their training and testing predictions mostly falling within the error bounds of ± 20%. In summary, the results suggest that the evaluated ML models show potential for accurately predicting outcomes in the studied domain.

4.2. Comparison of machine learning-based prediction models

Table 3 enumerates the performance metrics of six distinct machine learning methods. The R^2 of the output results of all ML models is above 0.9, indicating that the ML models are expected to accurately predict the results of the research field. Specifically, the XGBoost algorithm offers the best predictive performance among the evaluated ML techniques, with R^2, MAE, RMSE, and MAPE values of 0.964%, 41.318%, 56.760%, and 14.228%, respectively. Conversely, the Lasso regressor exhibits the poorest performance, with R^2, MAE, RMSE, and MAPE values of 0.927%, 58.672%, 81.061%, and 25.128%, respectively. These findings substantiate the exceptional predictive performance of the XGBoost algorithm in forecasting the ultimate load of shear key specimens, while underscoring the limitations of the Lasso regressor in this context. As a result, subsequent parts of the study exclusively consider the XGBoost algorithm.

4.3. Residual analysis

The predictive capabilities of machine learning models were assessed using residual plots, which depict the discrepancy between predicted and actual values as a vertical distance from the zero line, referred to as the residual (Y_{pred}-Y_{exper}). This method simplifies the evaluation of the closeness between predicted and actual values, as residual plots closely aligned with the zero line suggest accurate predictions. Fig. 8 provides a graphical representation of the residual range and distribution of six models.

The analysis reveals that the single models (Lasso, SVM, and KNN) exhibit widely dispersed residual values, indicating poor prediction performance. Among them, the Lasso and SVM models exhibit the most extensive range of residual values, spanning from − 400 to 400 kN, while the KNN model presents a slightly narrower range, from − 150 to 300 kN. However, the KNN model's prediction performance on the test set falls short compared to its performance on the training set. The ensemble models (RF, GBDT, and XGBoost), on the other hand, demonstrate a smaller range of residual values than the single models, suggesting superior prediction performance. Specifically, The GBDT and RF models have dispersed residual values ranging from − 300 to 300 kN, In contrast, the XGBoost ensemble model exhibits a smaller range than the RF and GBDT models, with residuals scattered within a narrower interval of − 150 to 150 kN. This further suggests that the XGBoost ensemble model performed best.

Table 3
Statistical metrics of different ML models on the test set.

No	Model	R^2	MAE (kN)	RMSE (kN)	MAPE (%)
1	Lasso	0.927	58.672	81.061	25.128
2	SVM	0.935	50.157	76.507	19.210
3	KNN	0.944	46.617	70.659	15.063
4	RF	0.946	46.364	69.571	16.872
5	GBDT	0.961	39.281	59.192	11.686
6	XGBoost	0.964	41.318	56.760	14.228

4.4. Evaluation of the existing models in literature for the shear strength of joints

In this section, the effectiveness and efficiency of the existing models in literature for predicting the shear strength of joints are evaluated. For this purpose, comparison is made of the ultimate loads of the joints between the prediction results from the existing models (mainly from design specifications and empirical formulae proposed by other researchers) and the experimental results for the dataset considered in this study. Meanwhile, the results from the XGBoost are also included to demonstrate its accuracy.

4.4.1. Existing models in literature

Several typical formulae are currently in common use, including those proposed by Rombach [33], AASHTO [17], Turmo [50], and Liu [18]. These formulae are shown in Eq. (7)–(10), respectively.

In Rombach et al. [33]:

$$V_u = \frac{1}{\gamma_F}\left[0.65\sigma_n(A_{sm}+A_k)+0.14f_c^{'}A_k\right] \tag{7}$$

where V_u= calculated value of shear strength (in N); γ_F is the safety factor, 1.0 in this study; σ_n=confining stress(in MPa); $f_c^{'}$=compressive strength of concrete(in MPa); A_k=areas of the root of shear keys on the failure plane(in mm^2); A_{sm}=area of flat surfaces on the failure plane(in mm^2).

In AASHTO. [17]:

$$V_u = A_k\sqrt{6.792\times10^{-3}f_c^{'}}(2.466\sigma_n+12)+0.6A_{sm}\sigma_n \tag{8}$$

where all symbols have the same meaning as those in Eq. (7).

In Turmo et al. [50]:

$$V_u = \frac{1}{\gamma_c}\left[A_k\frac{\ln\left(1+f_c^{'}/10\right)}{100}(49\sigma_n+233)+0.6A_{sm}\sigma_n\right] \tag{9}$$

where γ_c is the safety coefficient for material properties (γ_c = 1.5 for temporary loads and γ_c =1.3 for accidental loads). γ_c = 1.0 was taken in this study to assess the accuracy of the formula; other symbols have the same meaning as those in Eq. (7).

In Liu et al.[18]:

$$V_u = A_k\sqrt{f_c^{'}}\left[0.9+\left(0.13-2.7\times10^{-4}f_c^{'}\right)\sigma_n\right]+\mu A_{sm}\sigma_n \tag{10}$$

where all symbols have the same meaning as those in Eq. (7).

4.4.2. Comparison with existing empirical models

Fig. 9 shows the comparison of the ultimate loads of joints between the above four models, the XGBoost model and the experimental results for the dataset used in this study. Table 4 shows the computed R^2, MAE, RMSE, and MAPE for the four models.

Several observations can be made from the comparison plots in Fig. 9. First, the predicted results from the four equations are almost all below the experimental results, indicating that the existing models considered in this study can conservatively estimate the shear strength of joints with FRC. Second, the level of conservatism is relatively high for the models by Turmo and Liu, as more than 50% of the predicted data points are significantly below the experimental results (i.e., less than 50% of the experimental results), indicating a high safety factor of more than 2 in these two models. However, the models by AASHTO and Rombach show a smaller conservative estimation, as most of the predicted data points fall within 50% to 100%, indicating a better estimation of shear strength of joints with FRC but less safety factor. Third, the scattering of all four models is large, with the computed RMSE being 201.145 kN, 165.366 kN, 313.456 kN, and 267.485 kN (see Table 4) for the models of AASHTO, Rombach, Turmo, and Liu, respectively, which

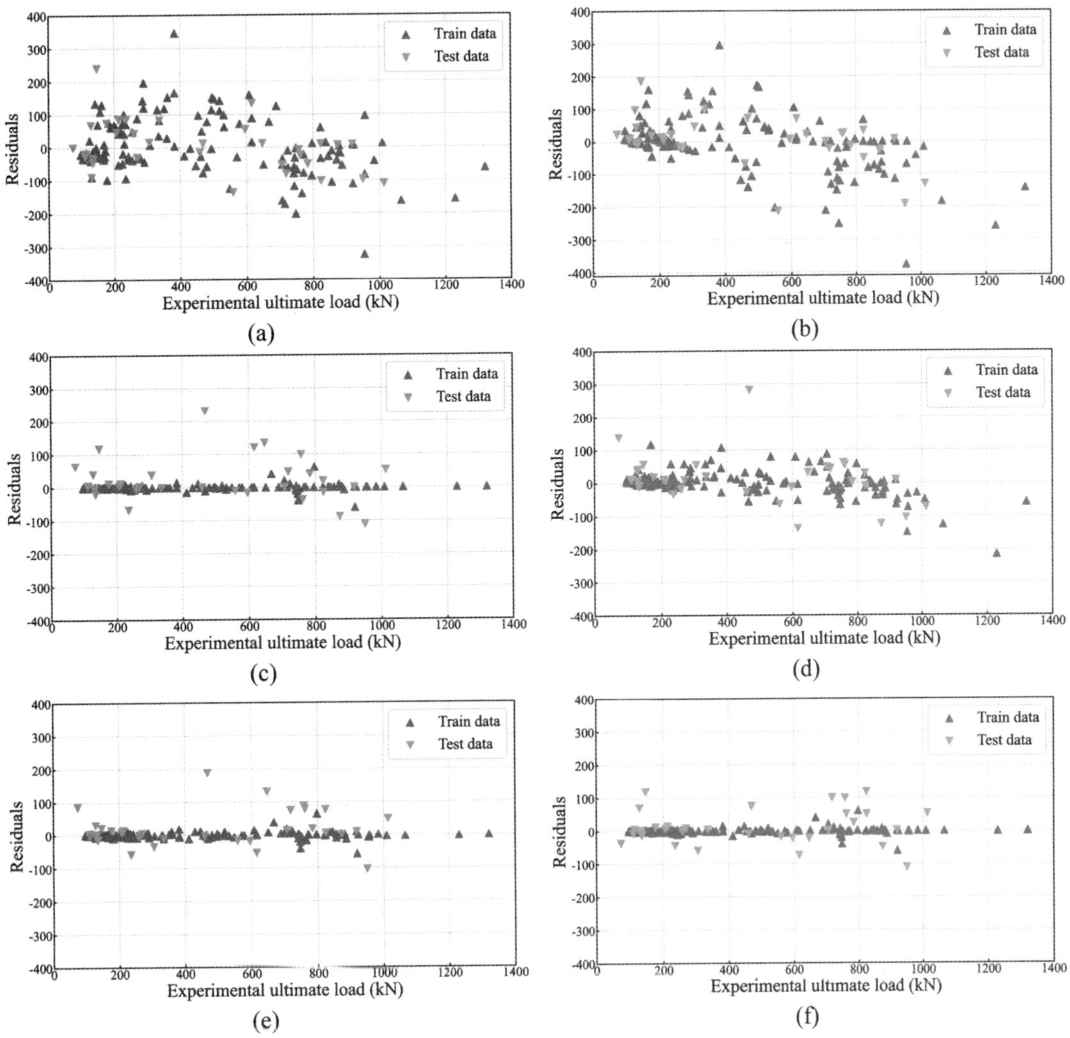

Fig. 8. Residuals of the predicted the ultimate load on training and testing data: (a) Lasso; (b) SVM; (c) KNN; (d) RF; (e) GBDT; (f) XGBoost.

are all significantly larger than that of the ML model of XGBoost (i.e., 56.760 kN). This indicates that these models might not be efficient in capturing the scattering of the shear strength of joints with FRC. Fourth, among the design specifications and empirical formulas, the models by Rombach et al. and AASHTO perform better, with the R^2 of the two empirical models being 0.714 and 0.578, respectively, which are closer to 1 and larger than those of the other two models.

Above all, the existing models in literature considered in this study can be somewhat effective but relatively conservative in estimating the shear strength of joints with FRC. Meanwhile, the scatter of these models is significantly large, indicating that these models might not be efficient in capturing the scattering of the shear strength of joints with FRC. It is noted that the empirical models offer a simplified approach to estimate the shear capacity of joints with FRC, but their scattering and conservatism may be limited by the available test results and parameters. Therefore, it is recommended to consider additional factors, such as the number of shear keys, dimensions, and fiber content, to improve the accuracy and reliability of these design specifications and empirical formulas in future applications. In addition, the ML model XGBoost can yield accurate prediction and significantly less scattering for the shear strength of joints with FRC as it considers a greater number of input variables that could affect the ultimate load of joints and was trained using part of the experimental data. However, for the design purpose, safety factors should be applied to the XGBoost model, which are the focus of the further study.

5. Model interpretations

In practical machine learning applications, understanding the rationale behind a model's prediction of the ultimate load of shear key specimens is as crucial as its predictive accuracy, given the data-driven models' black-box nature. To address this challenge, Lundberg and Lee [30] introduced the SHapley Additive exPlanations (SHAP) method, a model-agnostic game-theoretic methodology.

In this study, the best-performing XGBoost model was interpreted using the SHAP method in this study. The SHAP method quantifies the contribution of each feature to an individual prediction and provides a quantitative comprehension of the significance of each feature in predicting the specific ultimate load.

$$f(x) = \phi_0 + \sum_{i=1}^{M} \phi_i x_i \tag{11}$$

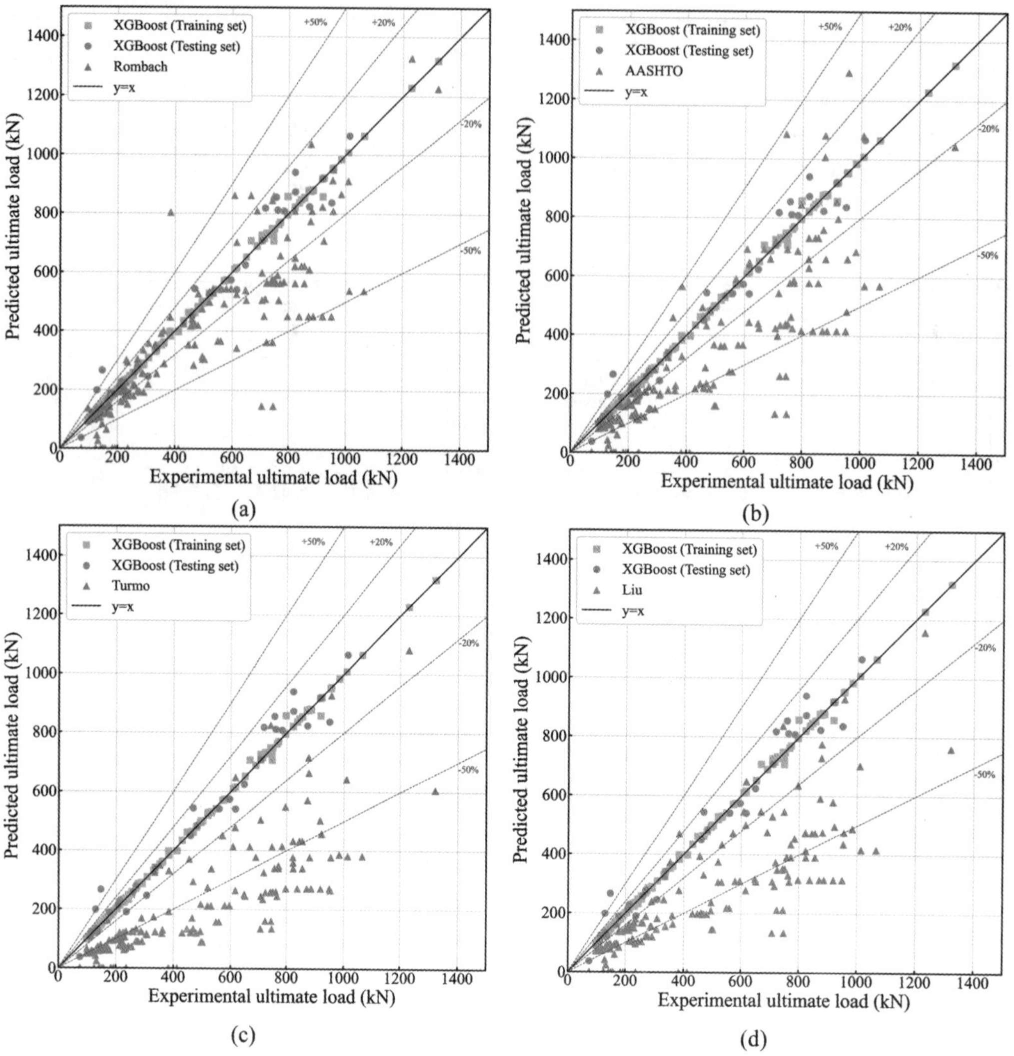

Fig. 9. Comparison between the predicted and experimental results: (a) Rombach; (b) AASHTO; (c) Turmo; (d) Liu.

Table 4
Computed statistical metrics from the empirical models.

No	Model	R^2	MAE (kN)	RMSE (kN)	MAPE (%)
1	AASHTO	0.578	149.588	201.145	34.890
2	Rombach	0.714	112.615	165.366	26.247
3	Turmo	-0.024	252.296	313.456	57.045
4	Liu	0.253	208.116	267.485	45.740

$$\phi_i = \sum_{S\subseteq N\backslash\{i\}} \frac{|S|!(M-|S|-1)!}{M!}[f(SU\{i\}) - f(S)] \tag{12}$$

where *f(x)* represents the prediction of ML model; *M* is the number of input features; $\phi_0$represents the constant base value; i.e. the mean value of all outputs; ϕ_i is the feature attribution value to i-th feature; *S* denotes a subset of the features; *N* is the set of all input features; and denotes the feature of interest.

The SHapley Additive exPlanations (SHAP) method can determine the positive or negative contribution of each input variable to a prediction. For a detailed explanation of the SHAP method, interested readers are referred to relevant references[26,28,29,51].

Fig. 10 presents the global importance factors of input variables for the ultimate load of shear key specimens. The global importance of each feature is taken to be the mean absolute Shapley values for that feature over all the given samples. Higher importance factors indicate a more significant impact of the corresponding variable on the ultimate load, facilitating a comprehensive evaluation of each variable's influence on the ultimate load. For convenience, the mean (|SHAP value|) of single feature was represented by $E(i)_{mean}$ and the sum of mean (|SHAP value|) for all features was represented by A_{mean} for convenience in this study. The relationship between E_{mean} and A_{mean} is shown in the following formula:

$$A_{mean} = \sum_{i=1}^{n} E(i)_{mean} \tag{13}$$

The results indicate that the confining stress (*CS*), depth of shear key (*D*), compressive strength of concrete (f_{cu}), total height of joints (*H*), and height of flat part (H_{sm}) are the most crucial parameters affecting the ultimate load, with mean absolute Shapley values of 88.92, 80.05, 72.51, 63.49, and 49.05, respectively. The sum of their mean absolute

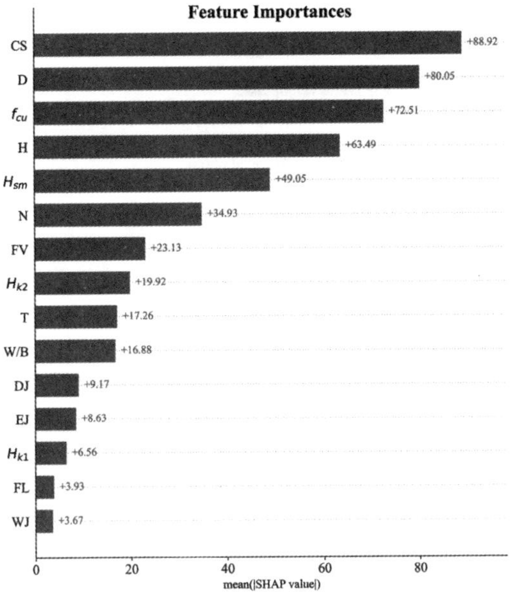

Fig. 10. Global importance factors based on SHAP.

Shapley values is 354.02, accounting for approximately 71.07% of the A_{mean}. These are followed closely by the number of shear keys (*N*), fiber volume ratio (*FV*), base width of single-keyed (H_{k2}), thickness of specimen *(T)*, and water-binder ratio (*W/B*), with mean Shapley values of 34.93, 23.13, 19.92, 17.26, and 16.88, respectively. The sum of their mean absolute Shapley values is 112.12, accounting for approximately 22.51% of the A_{mean}. The remaining features, including dry joint (*DJ*), epoxy joint (*EJ*), top width of single-keyed (H_{k1}), length of fiber (*FL*), and wet joint (*WJ*), had mean absolute Shapley values of 9.17, 8.63, 6.56, 3.93, and 3.67, respectively. The sum of their average absolute Shapley values is 31.96, accounting for approximately 6.42% of the A_{mean}.

However, it is not clear from Fig. 10 whether each parameter has a positive or negative impact on the prediction. This can be investigated using SHAP summary plots, as shown in Fig. 11. These plots further

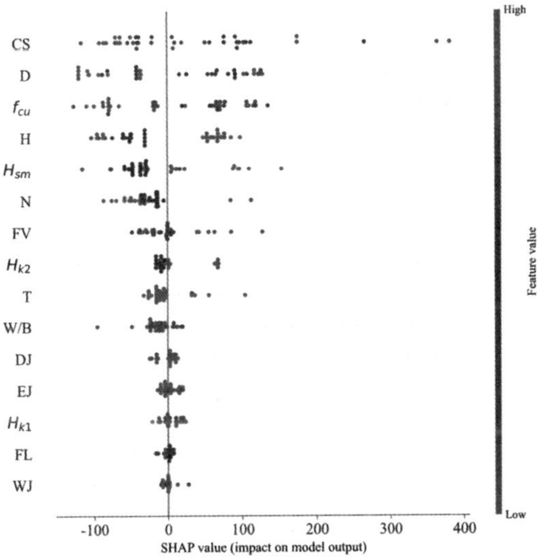

Fig. 11. SHAP summary plot.

clarify the effect of each parameter on the ultimate load and its positive or negative influence on the prediction. Each dot in Fig. 11 represents a feature's SHAP value for a sample in the test set. The sample point's color transitions from blue to red to denote changes in the corresponding parameter value from small to large. The y-axis displays input variables in descending order of importance, while the x-axis represents the actual distribution of Shapley values for each influencing parameter.

The findings depicted in Fig. 10 and Fig. 11 are consistent with current knowledge. Prior research involving shear tests on a series of shear keys has shown that factors such as confining stress (*CS*), depth of shear key (*D*), compressive strength of concrete (f_{cu}), number of shear keys (N), and fiber volume ratio (*FV*) significantly impact the ultimate load[11,12,19,38]. Further analysis in Fig. 11 reveals that increases in the most critical variable, confining stress (*CS*), result in an elevated ultimate load. This observation corroborates previous research conclusions[1], suggesting that confining stress changes the keyed joint from uniaxial to biaxial compression and effectively limits the transverse displacement of the keyed joints, thereby significantly improving the shear capacity of keyed-joints.

Additional influential parameters, such as depth of shear key (*D*), compressive strength of concrete (f_{cu}), total height of joint (*H*), number of shear keys (*N*), and fiber volume ratio (*FV*), also exhibit a positive correlation with the ultimate load. Similar results have been reported in other studies[1,11,12]. For example, composite joints with steel fibers exhibited higher ultimate loads than those without steel fibers, and the ultimate load increased as the strength of the concrete improved. Moreover, the ultimate load increased with fiber volume ratios due to the enhanced tensile performance of FRC and the bridging effect provided by steel fibers that prevent crack expansion and improve post-cracking shear ductility. Furthermore, the ultimate load increased with the number of keys, as the effective area of the interface related to shear keys increased, effectively reducing the sliding of keys along the joint surface.

Besides offering a comprehensive analysis, the SHAP method can also quantitatively determine the significance of specific features of a given sample to the output results. A typical prediction plot obtained with the XGBoost model for a specimen (Specimen KL-R0.35-F1-S10 from Pan et al. [10]) is shown in Fig. 12.

As mentioned earlier, the prediction was broken down into contributions from individual components. In Fig. 12, the base value corresponds to the average ultimate load in the database, which is 462.42 kN. The features marked in red, such as depth of shear key, confining stress, compressive strength of concrete, base width of single-keyed joint, and total height of joints, positively influenced the ultimate load, pushing it above the base value. Conversely, features marked in blue like fiber volume ratio and height of flat part negatively influenced the ultimate load, pushing it below the base value. Ultimately, when all feature actions are combined, they result in a predicted ultimate load of 809.04 kN. For this specific prediction, the output results of Specimen KL-R0.35-F1-S10 were primarily influenced by factors such as the depth of shear key, confining stress, and compressive strength of concrete.

These findings demonstrate the explanatory power of the ensemble XGBoost model and its ability to deepen our understanding of the ultimate load, especially regarding the significance of specific features of a particular sample.

6. Conclusions

This study compiles experimental results from various prior studies and constructs interpretable machine learning models to estimate the ultimate load. A total of 192 experimental results were utilized, which were randomly partitioned into a training set and a test set. Six regression methods were explored, comprising three individual models (Lasso, SVM, and KNN) and three ensemble models (RF, GBDT, and XGBoost). The GridSearchCV technique based on the training set was utilized for parameter optimization along with a five-fold cross-validation method.

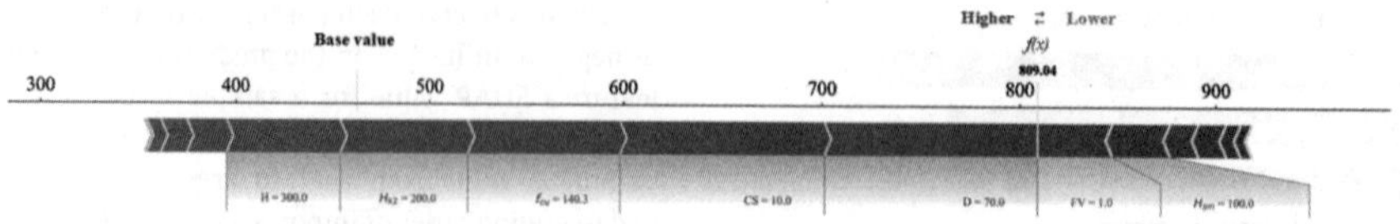

Fig. 12. Explanation of the ultimate load prediction for Specimen KL-R0.35-F1-S10 by Pan et al. [10].

The performance of these models was compared and analyzed based on factors like learning ability, predictive accuracy, and residuals. Furthermore, the XGBoost model was additionally compared with existing design code provisions and empirical models to ascertain their efficiency and predictive capability. The SHapley Additive exPlanations (SHAP) approach was employed for visual interpretation of ultimate load prediction. Based on the findings of this study, the following conclusions can be drawn.

(1) The ensemble model XGBoost outperforms other machine learning (ML) models in bond strength estimation in terms of learning ability, predictive performance, and residuals with acceptable computational time. The predictive performance of the considered models can be ranked as XGBoost > GBDT > RF > KNN > SVM > Lasso. The performance metrics R^2, MAE, RMSE, and MAPE of 0.964, 41.318 kN, 56.760 kN, and 14.228% are achieved by the best XGBoost model.

(2) Current design codes and empirical equations proposed by previous researchers were assessed using a database for predicting the ultimate load. However, it was discovered that these existing codes and equations are somewhat effective but relatively conservative in predicting the shear capacity of joints of FRC. Among the design specifications and empirical formulas, the models proposed by Rombach et al. and AASHTO performed better. However, the R^2 of these two empirical models was only 0.714 and 0.578 respectively, indicating a larger scatter. By contrast, they are far lower than that of the R^2 of XGBoost model is 0.964. Similarly, the MAE, RMSE, and MAPE values of these two empirical models were more than double those of the XGBoost model.

(3) This study further investigates the use of the SHapley Additive exPlanations (SHAP) approach to rank input variables and elucidate contributing factors to bond strength estimation. The results suggest that CS, D, f_{cu}, H, and H_{sm} are the most crucial parameters affecting ultimate load, closely followed by N, FV, H_{k2}, T, and W/B.

(4) This study also examines the changes in model output results caused by variations in factors such as CS, D, f_{cu}, N, and FV. It was discovered that increasing these factors all resulted in an improvement in the ultimate load, which is consistent with the previous research findings.

CRediT authorship contribution statement

Peng Ge: Writing - original draft, Methodology, Data curation. **Ou Yang**: Conceptualization, Writing - review & editing, Supervision, Funding acquisition. **Jia He**: Supervision, Writing - review & editing. **Kailun Zhang**: Writing - review & editing. **Suiwen Wu**: Supervision, Writing - review & editing.

Declaration of Competing Interest

The authors declare that they have no known competing financial interests or personal relationships that could have appeared to influence the work reported in this paper.

Data availability

Data will be made available on request.

Acknowledgement

This work is supported by the National Key Research and Development Program of China (2019YFC1511101).

References

[1] Jiang H, Huang C, Feng J, Gao X, Wang T, Zhong Z. Direct shear behavior of castellated dry RPC joints in precast concrete segmental bridges. Structures 2021: 4579–95.
[2] Hu Y, Qiu J, Li Z, Yao Y, Liu J, Wang J. Shear strength prediction method of the UHPC keyed dry joint considering the bridging effect of steel fibers. Eng Struct 2022;255:113937.
[3] Zhan Y, Li Z, Chen Z, Shao J, Yue F, Liu F, et al. Experimental and numerical investigations on shear performance of key tooth joints of precast concrete segmental bridge under repeated loading. Constr Build Mater 2022;351:128794.
[4] Jiang H, Wei R, John Ma Z, Li Y, Jing Y. Shear strength of steel fiber-reinforced concrete dry joints in precast segmental bridges. J Bridge Eng 2016;21:04016085.
[5] Ahmed GH, Aziz OQ. Shear behavior of dry and epoxied joints in precast concrete segmental box girder bridges under direct shear loading. Eng Struct 2019;182: 89–100.
[6] Wu J, Liu D, Chen X, Zhang J. Experimental study on shear performance of bond-tooth dry joints in prestressed assembled concrete beams. J Build Eng 2023;68: 106189.
[7] Saibabu S, Srinivas V, Sasmal S, Lakshmanan N, Iyer NR. Performance evaluation of dry and epoxy jointed segmental prestressed box girders under monotonic and cyclic loading. Constr Build Mater 2013;38:931–40.
[8] Zhou X, Mickleborough N, Li Z. Shear strength of joints in precast concrete segmental bridges. Acids Struct J 2005;102:3.
[9] Song S. Experimental study and theoretical analysis in bending and joint shear of high-speed railway precast segmental concrete box bridges. Southeast University, Dhaka, Bangladesh. 2015.
[10] Pan R, He W, Cheng L, Li C. Direct shear strength of UHPC large-keyed epoxy joint: theoretical model and experimental verification. J Bridge Eng 2022;27:04022083.
[11] Yuan A, Zhao X, Lu R. Experimental investigation on shear performance of fiber-reinforced high-strength concrete dry joints. Eng Struct 2020;223:111159.
[12] Fu J, Yuan A, Wan S. Behavior of fiber reinforced key joints in precast concrete segmental bridge: experimental and numerical analysis. J Bridge Eng 2021;26: 04021053.
[13] Hu M, Han Q, Wu S, Du X. Shear capacity of precast concrete shear keys with ultrahigh-performance concrete for connections. J Bridge Eng 2021;26:04021036.
[14] Camille C, Hewage DK, Mirza O, Mashiri F, Kirkland B, Clarke T. Performance behaviour of macro-synthetic fibre reinforced concrete subjected to static and dynamic loadings for sleeper applications. Constr Build Mater 2021;270:121469.
[15] Larsen IL, Thorstensen RT. The influence of steel fibres on compressive and tensile strength of ultra high performance concrete: a review. Constr Build Mater 2020; 256:119459.
[16] Yehia S, Douba A, Abdullahi O, Farrag S. Mechanical and durability evaluation of fiber-reinforced self-compacting concrete. Constr Build Mater 2016;121:120–33.
[17] AASHTO, Guide Specifications for Design and Construction of Segmental Concrete Bridges (Revision 2;). US-AASHTO; 2003.
[18] Liu T, Wang Z, Guo J, Wang J. Shear strength of dry joints in precast UHPC segmental bridges: experimental and theoretical research. J Bridge Eng 2019;24: 04018100.
[19] Jiang H, Chen L, Ma ZJ, Feng W. Shear behavior of dry joints with castellated keys in precast concrete segmental bridges. J Bridge Eng 2015;20:04014062.
[20] Chen L, Yan J, Xiang N, Zhong J. Shear performance of ultra-high performance concrete multi-keyed epoxy joints in precast segmental bridges. Struct: Elsevier 2022:1696–708.
[21] Gopal BA, Hejazi F, Hafezolghorani M, Voo YL. Shear strength of dry and epoxy joints for ultra-high-performance fiber-reinforced concrete. Acids Struct J 2020; 117.
[22] Yuan A, Yang C, Wang J, Chen L, Lu R. Shear behavior of epoxy resin joints in precast concrete segmental bridges. J Bridge Eng 2019;24:04019009.
[23] Li Z, Qi J, Hu Y, Wang J. Estimation of bond strength between UHPC and reinforcing bars using machine learning approaches. Eng Struct 2022;262:114311.
[24] Zhang F, Wang C, Liu J, Zou X, Sneed LH, Bao Y, et al. Prediction of FRP-concrete interfacial bond strength based on machine learning. Eng Struct 2023;274:115156.

[25] Fu B, Chen S-Z, Liu X-R, Feng D-C. A probabilistic bond strength model for corroded reinforced concrete based on weighted averaging of non-fine-tuned machine learning models. Constr Build Mater 2022;318:125767.

[26] Mangalathu S, Shin H, Choi E, Jeon J-S. Explainable machine learning models for punching shear strength estimation of flat slabs without transverse reinforcement. J Build Eng 2021;39:102300.

[27] Wang X, Chen A, Liu Y. Explainable ensemble learning model for predicting steel section-concrete bond strength. Constr Build Mater 2022;356:129239.

[28] Cavaleri L, Barkhordari MS, Repapis CC, Armaghani DJ, Ulrikh DV, Asteris PG. Convolution-based ensemble learning algorithms to estimate the bond strength of the corroded reinforced concrete. Constr Build Mater 2022;359:129504.

[29] Mei Y, Sun Y, Li F, Xu X, Zhang A, Shen J. Probabilistic prediction model of steel to concrete bond failure under high temperature by machine learning. Eng Fail Anal 2022;142:106786.

[30] Lundberg SM, Lee S-I. A unified approach to interpreting model predictions. Adv Neural Inf Process Syst 2017:30.

[31] Feng Z, Li CX, Pan RS, Zhou JL, Ke L, Ke HJ. Experimental investigation and calculation method of UHPC direct shear capacity. China J Highw Transp 2021;34 (8):78–90.

[32] Buyukozturk O, Bakhoum MM, Michael Beattie S. Shear behavior of joints in precast concrete segmental bridges. J Struct Eng 1990;116:3380–401.

[33] Rombach G. Precast segmental box girder bridges with external prestressing-design and construction. INSA Rennes 2002:1–15.

[34] Jie S, Wei L, Caijun S. Scale effect of cubic compressive strength of ultra-high performance concrete. J Chin Ceram Soc 2021;49:305–11.

[35] Abellan-Garcia J. Study of nonlinear relationships between dosage mixture design and the compressive strength of UHPC. Case Stud Constr Mater 2022;17:e01228.

[36] Turmo J, Ramos G, Aparicio ÁC. Shear strength of joints in precast concrete segmental bridges. Acids Struct J 2005;102:901.

[37] Pan R, Cheng L, He W, Zhou X, Shen X. Direct shear performance of UHPC Multi-Keyed epoxy joint. Struct: Elsevier 2022:1898–909.

[38] Kim Y-J, Chin W-J, Jeon S-J. Interface shear strength at joints of ultra-high performance concrete structures. Int J Concr Struct Mater 2018;12:1–14.

[39] Singh VK, Maurya NS, Mani A, Yadav RS. Machine learning method using position-specific mutation based classification outperforms one hot coding for disease severity prediction in haemophilia 'A. Genomics 2020;112:5122–8.

[40] Hancock JT, Khoshgoftaar TM. Survey on categorical data for neural networks. J Big Data 2020;7:1–41.

[41] VanderPlas J. Python data science handbook: essential tools for working with data. O'Reilly Media, Inc; 2016.

[42] Müller AC, Guido S. Introduction to machine learning with Python: a guide for data scientists. O'Reilly Media, Inc; 2016.

[43] Raschka S. Python machine learning. Packt publishing ltd,; 2015.

[44] Pan RS, Cheng LX, He WW, Li CX. Experimental research on direct shear performance of UHPC keyed epoxy joints. China Civ Eng J 2022;55(11):38–49. 117 (in Chinese).

[45] Feng Z, Li CX, Zhou JL, Dong CW, Ke HJ, Yoo D-Y. Direct shear test on UHPC key-wet-joints and the unified calculation formula of direct shear capacity of UHPC wet-joints. China Civ Eng J 2022;55(6):79–91.

[46] Jiang HB, Wang TL, Xiao J, Zhang DM, Cai SJ. Test on shear behavior of dry joints in precast steel fiber reinforced concrete segmental bridges. China J Highw Transp 2018;31(12):37–49.

[47] Yuan AM, Zhao XR, Lu RW. Experimental investigation on shear performance of fiber-reinforced high-strength concrete dry joints. China J Highw Transp 2020;33 (8):159 (in Chinese).

[48] Kang M-C, Yoo D-Y, Gupta R. Machine learning-based prediction for compressive and flexural strengths of steel fiber-reinforced concrete. Constr Build Mater 2021; 266:121117.

[49] Kim S, Kim H. A new metric of absolute percentage error for intermittent demand forecasts. Int J Forecast 2016;32:669–79.

[50] Turmo J, Ramos G, Aparicio J. Shear strength of match cast dry joints of precast concrete segmental bridges: proposal for Eurocode 2. Mater De Constr 2006;56:45-52.

[51] Mangalathu S, Hwang S-H, Jeon J-S. Failure mode and effects analysis of RC members based on machine-learning-based SHapley Additive exPlanations (SHAP) approach. Eng Struct 2020;219:110927.

Engineering Structures 320 (2024) 118871

Contents lists available at ScienceDirect

Engineering Structures

journal homepage: www.elsevier.com/locate/engstruct

Analytical solutions for static longitudinal displacements of suspension bridges under a moving vertical concentrated load

Haokun Jing [a,b], Zhouquan Feng [a,b,*], Zhi Chen [a,b], Guoping Huang [a,c], Xugang Hua [a,b], Zhengqing Chen [a,b], Tianbao Wan [d]

[a] Hunan Provincial Key Laboratory of Wind and Bridge Engineering, College of Civil Engineering, Hunan University, Changsha 410082, China
[b] State Key Laboratory of Bridge Safety and Resilience, Hunan University, Changsha 410082, China
[c] College of Civil Engineering, Hunan City University, Yiyang 413000, China
[d] China Railway Major Bridge Reconnaissance & Design Institute Co., Ltd., Wuhan 430056, China

ARTICLE INFO

Keywords:
Suspension bridge
Longitudinal displacement
Vertical deformation
Vertical concentrated load
Analytical solution

ABSTRACT

The longitudinal displacement at the girder ends of long-span suspension bridges is a crucial issue that affects structural safety, durability, as well as traffic safety and ride comfort. The primary cause of these reciprocating displacements is the action of moving live loads, which induces quasi-static displacements in the longitudinal direction. This paper theoretically investigates the static longitudinal displacements of suspension bridges under a moving vertical concentrated load and elucidates the underlying deformation mechanisms. Considering geometrical nonlinearity, analytical equations are established for the longitudinal deformations of the main cable and stiffening girder under a vertical concentrated load. By analyzing the geometrical deformation conditions of the suspenders, the relationship between the longitudinal displacements of the stiffening girder and the main cable is derived. Due to the coupling between longitudinal and vertical displacements, the vertical displacement of the stiffening girder must be solved to obtain its longitudinal displacement. Based on deflection theory and considering the bending stiffness of the stiffening girder, this paper derives the vertical deformations of the stiffening girder for single-span suspension bridges and those with short continuous side spans under a vertical concentrated load, which is then used to calculate the longitudinal displacement. Finite element models are used to verify the accuracy of the analytical solutions. Differences between the proposed solutions and previous solutions neglecting the bending stiffness of the stiffening girder are compared and analyzed. The effects of neglecting the longitudinal displacement at the bridge tower top and the presence of short continuous side spans on the longitudinal displacement of stiffening girder are also investigated. The results demonstrate the high accuracy of the proposed analytical solution for the longitudinal displacement at the girder ends of suspension bridges. Importantly, the bending stiffness of the stiffening girder cannot be neglected in calculating the longitudinal displacement. The longitudinal displacement at the girder ends is primarily caused by the longitudinal deformation of the main cable, while the contribution from the flexural deformation of the girder is negligible. Furthermore, the longitudinal displacement at the bridge tower top has a negligible effect on the longitudinal displacement of the girder, and the presence of short continuous side spans is beneficial for reducing the longitudinal displacement of the stiffening girder. The proposed analytical solution method provides a rapid approach for calculating the static longitudinal displacement at the girder ends of suspension bridges under passing live loads and reveals the deformation mechanism. This is beneficial for preliminary design and structural optimization, as it avoids the complex finite element modeling and solution process.

1. Introduction

As an effective way to cross expansive rivers, straits and valleys, suspension bridges have gradually been widely adopted in recent years, with spans reaching the impressive two-kilometer range. Notable examples include the 1915 Çanakkale Bridge (with a main span of 2023 m,

* Corresponding author at: Hunan Provincial Key Laboratory of Wind and Bridge Engineering, College of Civil Engineering, Hunan University, Changsha 410082, China.
E-mail address: zqfeng@hnu.edu.cn (Z. Feng).

https://doi.org/10.1016/j.engstruct.2024.118871
Received 14 April 2024; Received in revised form 23 August 2024; Accepted 24 August 2024
Available online 3 September 2024

in operation) in Turkey, the Shiziyang Bridge (with a main span of 2180 m, under construction) and Zhangjinggao Bridge (with a main span of 2300 m, under construction) in China. The suspension bridges exhibit a relatively low level of stiffness and damping, rendering them susceptible to deformation and vibration when subjected to external forces, such as temperature variations, wind, earthquakes, and vehicle loads. In recent years, particular attention has been directed towards the longitudinal movements at the girder ends of long-span suspension bridges, given their potential to cause structural damage such as the oil leakage of fluid viscous dampers [1], deterioration of expansion joints at the girder end [2], and even fatigue damage of short suspenders [3,4].

Annual Cumulative Displacement (ACD) is emerging as a key index for evaluating the effect of such longitudinal movements. Numerous studies have highlighted that the ACD of many suspension bridges exceeds the alarming threshold of 10,000 m by analyzing the monitoring data of bridges. Notable examples include the Jiangyin Yangtze River Bridge (main span of 1385 m, ACD of 25200 m) [5], Runyang Yangtze River Bridge (main span of 1490 m, ACD of 29160 m) [6], Aizhai Bridge (main span of 1176 m, ACD of 54200 m) [7,8]. In addition, scholars have carried out simulation studies aimed at building temperature-displacement models grounded in the monitoring data [9, 10], and exploring the bridge's deformation characteristics in response to various factors, such as wind and vehicle loads [11,12] as well as earthquakes [13].

Some strategies have been explored to address the challenges posed by longitudinal deformations in suspension bridges. Structural interventions, including the addition of central buckles [14,15], negative cables [16] and other supporting structures, have demonstrated effectiveness in mitigating longitudinal displacements at the girder end and reducing the relative longitudinal displacement between the main cable and the girder to improve the service lives of short suspenders. Equally, the implementation of damping measures by setting such as magneto-rheological dampers [17], fluid viscous dampers [18] or friction dampers [19,20] between the girder and the bridge tower have been proved as another effective approaches.

Recently, the realm of railway suspension bridges has also witnessed significant developments, with boasting main spans at the 1000-meter level either completed or in progress. Examples include the Wufengshan Yangtze River Bridge (with a main span of 1092 m) and the Xihoumen Railway Bridge (with a main span of 1488 m) in China. Several new railway suspension bridges are under construction in the mountainous regions of western China. These bridges incorporate short extended continuous girders for side spans from the long main span, which results in enhanced stiffness and reduced deformation and vibration when subjected to train loads. Suspension bridges with short continuous side spans differ from the traditional suspension bridges with a single span and two hinges, offering greater overall stiffness than the latter. The side spans of these bridges are not equipped with suspenders and auxiliary piers. Suspension bridges often utilize a semi-floating system, and the suspension bridge studied in this paper is no exception. This semi-floating system features vertical supports at the towers and girder ends, and employs expansion joints and longitudinal dampers to accommodate slight longitudinal displacement of the stiffening girder. However, this paper ignores the effects of longitudinal damping and friction, focusing solely on the vertical support aspects of the semi-floating system. Previous studies on the longitudinal motion of suspension bridges have mainly focused on highway suspension bridges, with limited research on railway suspension bridges. Unlike road bridges that support random traffic loads, railway bridges usually bear determined train loads, which can usually be simplified into a series of multiple vertical concentrated loads for simulation.

Regardless of whether it is a railway suspension bridge or a road suspension bridge, the severity and main cause of girder end displacements are significantly different from other bridge types (such as cable-stayed bridges, arch bridges, and beam bridges). For other bridge types, the longitudinal deformations under moving live loads are relatively small, mainly caused by vehicle braking and traction forces. However, the longitudinal displacements at the girder ends of suspension bridges are quite different. Due to the large deformation and geometric nonlinearity of the main cable in suspension bridges, the moving vertical live loads cause asymmetric deformation of the main cable. The asymmetric vertical deformation of the main cable is coupled with longitudinal deformation, which, driven by the suspenders, induces a longitudinal drift of the stiffening girder. This deformation characteristics and mechanism exhibiting a significant quasi-static nature have been verified through monitoring data [7,8] and numerical simulations [18], theoretical investigations on this subject have received relatively limited attention. Therefore, studying the longitudinal deformations of suspension bridges under moving vertical loads from a theoretical perspective holds substantial importance. Since previous studies paid little attention to longitudinal motion, existing theoretical frameworks for suspension bridges have predominantly focused on finding the shape of the suspension bridge [21–24] and analyzing the vertical deformation of the girder under live loads [25–27]. There are few studies on the longitudinal deformation of stiffening girder of suspension bridges under live loads. Huang *et al.* [28] conducted a theoretical study on the longitudinal deformation of suspension bridges under vertical live loads. However, their study has certain limitations: 1) It focused solely on single-span suspension bridges with double-hinged simply supported stiffening girders, employing an analogy method for analysis, which is not applicable to suspension bridges with short continuous side spans. 2) The study did not account for the effect of girder bending stiffness, which is considered non-negligible for railway suspension bridges [29].

Therefore, this paper proposed a theoretical method for analyzing the longitudinal displacement of the girder ends in suspension bridges under a passing vertical concentrated load. The remaining sections of this paper are organized as follows: Section 2 presents the detailed derivation process for the longitudinal and vertical deformations of suspension bridges; Section 3 summarizes the specific application procedure of the proposed method; Section 4 verifies the accuracy of the proposed method through a numerical example and provides relevant discussions and analyses; the final section draws conclusions and offers future perspectives.

2. Longitudinal and vertical deformations of suspension bridges

2.1. *Longitudinal deformations of the main cable*

Suspension bridges distinguish themselves from other bridge types. When subjected to moving vertical live loads, the suspended main cable exhibits significant vertical and longitudinal deformation, with both aspects being mutually coupled and featuring noticeable geometric nonlinear effects. The main cable is the main bearing component of the suspension bridge. This part has been confirmed in literature [28] and is used to analyze the longitudinal motion of single-span suspension bridges. For the completeness of the analysis, it is necessary to review this part.

When the live load acts on the suspension bridge, the deformation of the microelement of the main cable under the action of dead and live loads is shown in Fig. 1.

where ds, dy and dx respectively represent the geometric length and projection lengths of the initial state of the main cable microelement, and Δds, Δdy, Δdx respectively represent the deformation of the microelement under the live load. η represents the vertical displacement of the main cable under load, and $d\eta$ represents the vertical additional displacement.

When the cable transitions from the initial geometry to the final geometry, the following relations are valid for this microelement, as shown in the Eq. (1) [30].

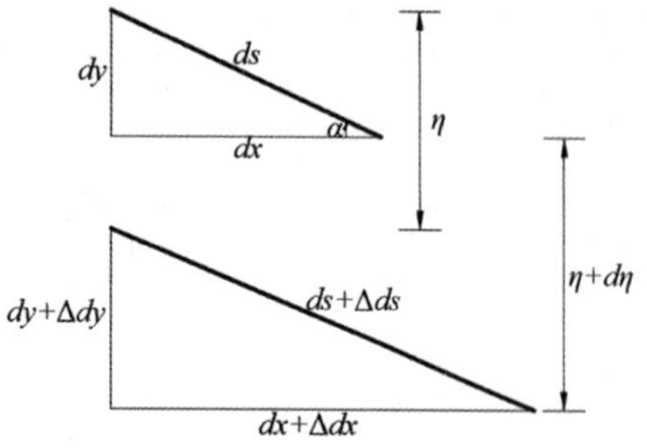

Fig. 1. The deformation of the microelement of the main cable.

$$\begin{cases}(dx)^2+(dy)^2=(ds)^2\\(dx+\Delta dx)^2+(dy+\Delta dy)^2=(ds+\Delta ds)^2\\\dfrac{ds}{dx}=\dfrac{1}{\cos\alpha}\\\Delta dy=d\eta\end{cases} \tag{1}$$

If the main cable microelement is subject to tension T and the horizontal component of tension is H_q, the following relationship can be obtained according to the basic relationship of mechanics.

$$\Delta ds=\frac{H_q ds}{EA\cos\alpha} \tag{2}$$

According to Eqs. (1) and (2), the longitudinal deformations of the main cable microelement Δdx can be obtained as follows,

$$\Delta dx=\frac{H_q dx}{EA\cos^3\alpha}-\frac{dy}{dx}d\eta \tag{3}$$

The longitudinal displacement of the main cable $u(x)$ can be obtained by integrating the longitudinal deformations of the microelement over the length of the main cable, as presented in the Eq. (4).

$$u(x)=\int_0^x\frac{H_q dx}{EA\cos^3\alpha}-\int_0^x\frac{dy}{dx}d\eta \tag{4}$$

If the longitudinal displacement of the bridge tower is not considered, the relationship $u(x=0)=0$ and $u(x=L_0)=0$ should be valid for the middle span main cable.

It can be seen from the Eq. (4) that one part of the longitudinal deformations of the main cable is the elastic deformation caused by the horizontal tension of the main cable, and the other is caused by the vertical deformation of the main cable. Generally, the elastic deformation term of the main cable is ignored, and longitudinal deformations of the main cable could be written as [28],

$$u(x)=-\int_0^x\frac{dy}{dx}d\eta \tag{5}$$

When the middle span of the suspension bridge is L_0 and the sag of the main cable is f, the shape of the main cable in the initial equilibrium state can be approximated as a parabola. Its shape can be expressed as,

$$y(x)=-\frac{4f}{L_0^2}x\left(x-L_0\right) \tag{6}$$

By combining the Eq. (5) and the Eq. (6), the final expression of the longitudinal deformations of the main cable can be obtained. As can be seen from the Eq. (7), the longitudinal displacement of the main cable includes the vertical displacement term. Therefore, before solving the longitudinal displacement $u(x)$, the vertical displacement $\eta(x)$ needs to be solved first.

$$u(x)=-\frac{4f}{L_0{}^2}\left[\eta(x)\left(L_0-2x\right)+\int_0^x 2\eta(x)dx\right] \tag{7}$$

When the suspension bridge has short continous side span without suspenders, the side span main cable can only balance the horizontal tension of the tower top under live load, and the middle span main cable is the main load-bearing component. Therefore, only the longitudinal deformations of the middle span main cable are considered in this study.

2.2. Longitudinal deformations of the girder

Under the action of live load, the longitudinal deformations of the middle span main cable cause the suspenders to incline, and the longitudinal deformations of the main cable drives the deformations of the girder, which is called the Part I. At the same time, the flexural deformation of the girder also produces the longitudinal displacement, which is called the Part II. The longitudinal displacement of the main cable and girder of the suspension bridge near the i - th suspender is shown in Fig. 2.

where, d_i and d_i' reprensent the length of the i-th suspender before and after deformation, respectively. The u_i and u_{b1i} mean the longitudinal displacement of the main cable and the girder at the position of the i-th suspender, respectively. N_i is the axial force of each suspender. N_{ix} is the force of the longitudinal force at the bottom of each suspender, and its direction is tangent to the axis of the deformed girder. According to Fig. 2, the following equilibrium conditions can be obtained ,

$$\frac{N_i}{N_{ix}}=\frac{d_i'}{u_{b1i}-u_i} \tag{8}$$

For the earth-anchored suspension bridge, the longitudinal force of the stiffening girder after construction is equal to zero, which means the resultant force of the longitudinal force at the bottom of each suspender is zero, then the longitudinal displacement (Part I) of the girder could be gained by the Eq. (9),

$$u_{b1}=\sum_{i=1}^{n}\frac{u_i}{d_i}\Big/\sum_{i=1}^{n}\frac{1}{d_i} \tag{9}$$

According to the deflection theory, the elongation of the suspender under live load is neglected, which reveals that the vertical deformation of the middle span main cable is equal to that of the girder at the same longitudinal location. Compared to the single span suspension bridge, the longitudinal displacement caused by the angle of the stiffening girder should be taken into account according to the situation, as the side span is not connected by the suspension cable. The longitudinal displacement induced by the flexural deformation of the girder (Part II) can be obtained by the Eq. (10),

$$\begin{cases}u_{b2}=\dfrac{d\eta_c}{dx}H_b\ or\ \dfrac{d\eta_b}{dx}H_b,\ \text{for middle span}\\u_{b2}=\dfrac{d\eta_b}{dx}H_b,\ \text{for extended span}\end{cases} \tag{10}$$

where H represents the height from the neutral axis of the stiffening girder to the top or bottom, subscript b represents the girder, and subscript c represents the main cable. Finally, the longitudinal

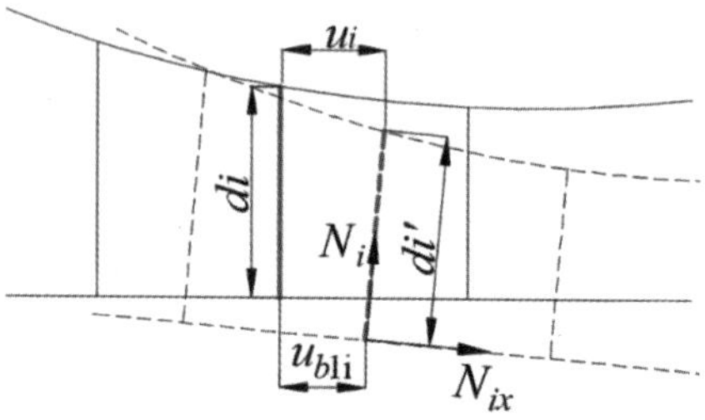

Fig. 2. The deformations of the microelements of the main cable, suspender and girder.

displacement of a certain point on a suspension bridge with short continuous side spans is equal to the sum of u_{b1} and u_{b2}.

2.3. *Vertical deformations of suspension bridges with short continuous side spans*

As can be seen in the Eqs. (7–10), the vertical deformation $\eta(x)$ of suspension bridge should be calculated before we gain the longitudinal deformations of the girder. Huang *et al.* [28] made an analogy between the cable geometry y and the bending moment diagram of the simply supported girder M. Although this method is convenient, it ignored the effects of girder bending stiffness, and can not well consider when the live load acting on the extended side span of the suspension bridge. Therefore, for suspension bridges with short continuous side spans, the vertical deformation of girder under live load should be obtained. Fig. 3 shows the force diagram of a suspension bridge with short continuous side spans in non-live load case.

where, x-y represents the Cartesian coordinate system, H_q represents the horizontal force of the main cable under dead load, w_c represents the weight of the main cable, w_b represents the weight of the girder, and $h_1(x)$ represents the force of the suspenders. In this state, we believe that the main cable bears the full load of the cable and the girder. According to the vertical force balance relationship of the main cable microelement, the following relationship can be obtained,

$$H_q\frac{d^2y(x)}{dx^2} = -\left(w_b + w_c\right) \tag{11}$$

By substituting the Eq. (6) into the Eq. (11), we can get the expression for H_q, as can be seen in the Eq. (12),

$$H_q = \frac{(w_b + w_c)L_0^2}{8f} \tag{12}$$

(1) Live load is applied to the side span

When live load P acts on the suspension bridge with short continuous side spans, at this time, the main cable and the middle span girder have the same vertical deformation η, $h_2(x)$ represents the force of the suspenders, and the horizontal tension of the main cable becomes $H_p + H_q$.

Due to the force balance of the main cable and the Eq. (11), ignoring the minimal term $H_p\frac{d^2\eta(x)}{dx^2}$, the Eq. (13) and the Eq. (14) can be established,

$$w_c + h_2(x) = -\left(H_p + H_q\right)\left[\frac{d^2y}{dx^2} + \frac{d^2\eta(x)}{dx^2}\right] \tag{13}$$

$$h_2(x) = w_b - H_p\frac{d^2y}{dx^2} - H_q\frac{d^2\eta(x)}{dx^2} \tag{14}$$

According to the condition that the horizontal projection length of the main cable remains unchanged after live load action, the Eq. (15) can be obtained, in which $L_c = \int_0^{L_0}\left(1 + \frac{dy}{dx}\right)^{\frac{3}{2}}$ [30], and after substituting it into the Eq. (14), the expression of force of suspenders under live load action can be obtained,

$$H_p = -\frac{E_cA_c}{L_c}\int_0^{L_0}\frac{d^2y}{dx^2}\eta(x)dx \tag{15}$$

$$h_2(x) = w_b - H_q\frac{d^2\eta(x)}{dx^2} - \frac{E_cA_c}{L_c}\left(\frac{d^2y}{dx^2}\right)^2\int_0^{L_0}\eta(x)dx \tag{16}$$

When live load P acts on the extended span, the loads and deformation of the main cable and stiffening girder are shown in Fig. 4.

By analyzing the micorelement of the girder, the Eq. (17) can be obtained. In where, $(EI)_b$ is the bending stiffness of the girder, x_c represents the location of the live load P.

$$\begin{cases} (EI)_b\dfrac{d\eta_1{}^4(x)}{dx} = 0, & -L_1 \le x \le x_c \\ (EI)_b\dfrac{d\eta_2{}^4(x)}{dx} = 0, & -x_c \le x \le 0 \\ (EI)_b\dfrac{d\eta_3{}^4(x)}{dx} = w_b - h_2(x), & 0 \le x \le L_0 \\ (EI)_b\dfrac{d\eta_4{}^4(x)}{dx} = 0, & L_0 \le x \le L_0 + L_2 \end{cases} \tag{17}$$

By solving the Eq. (17), we can get the vertical deformation expression of the stiffening girder at different positions, where C_1-C_{16} are unknown coefficients that needs to be determined by boundary conditions.

$$\begin{cases} \eta_1(x) = C_1x^3 + C_2x^2 + C_3x + C_4, & -L_1 \le x \le x_c \\ \eta_2(x) = C_5x^3 + C_6x^2 + C_7x + C_8, & -x_c \le x \le 0 \\ \eta_3(x) = -x^2\dfrac{3B_lb}{bL_0{}^3 - 3} + \dfrac{C_9\cosh ax + C_{10}\sinh ax}{a^2} + C_{11}x + C_{12}, & 0 \le x \le L_0 \\ \eta_4(x) = C_{13}x^3 + C_{14}x^2 + C_{15}x + C_{16}, & L_0 \le x \le L_0 + L_2 \end{cases} \tag{18}$$

where，some parameters are listed as the Eqs. (19) - (21),

$$B_l = \frac{1}{a^3}(C_9\sinh aL_0 + C_{10}\cosh aL_0 - C_{10}) + \frac{1}{2}C_{11}L_0{}^2 + C_{12}L_0 \tag{19}$$

$$b = \frac{E_cA_c}{2H_qL_c}\left(\frac{d^2y}{dx^2}\right)^2 \tag{20}$$

$$a = \sqrt{\frac{H_q}{E_bI_b}} \tag{21}$$

According to the condition that the vertical deformation of the support position is zero and the continuous displacement of the live load position, the Eq. (18) should satisfy the following expression as follow,

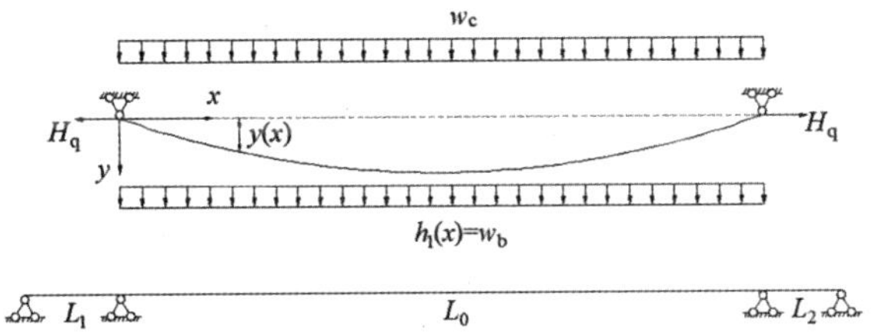

Fig. 3. Force diagram in non-live load case.

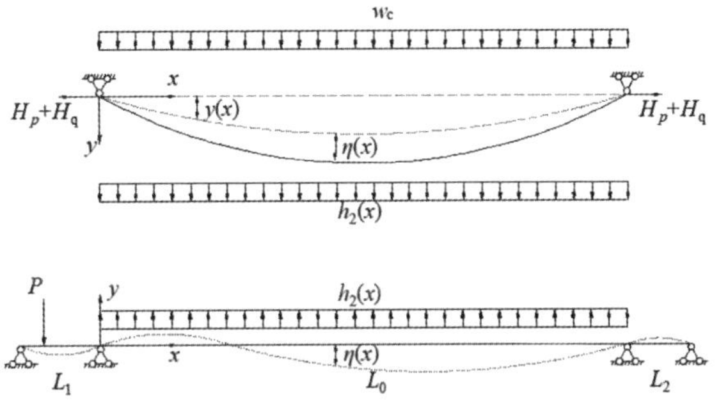

Fig. 4. Force and deformation diagram in live load case (P act on extended span).

$$\eta_1(-L_1) = 0 \tag{22}$$

$$\eta_1(x_c) - \eta_2(x_c) = 0 \tag{23}$$

$$\eta_2(0) = 0 \tag{24}$$

$$\eta_3(0) = 0 \tag{25}$$

$$\eta_3(L_0) = 0 \tag{26}$$

$$\eta_4(L_0) = 0 \tag{27}$$

$$\eta_4(L_0 + L_2) = 0 \tag{28}$$

Moreover, according to the conditions of continuous angle of support position and live load position, the Eq. (18) should meet the following expression,

$$\eta'_1(x_c) - \eta'_2(x_c) = 0 \tag{29}$$

$$\eta'_2(0) - \eta'_3(0) = 0 \tag{30}$$

$$\eta'_3(L_0) - \eta'_4(L_0) = 0 \tag{31}$$

Then, according to the conditions that the bending moment at girder end is zero, the bending moment of other support positions are continuous, and the bending moment and shear force at the live load position are continuous, the Eq. (18) should also meet the following equations,

$$\eta''_1(-L_1) = 0 \tag{32}$$

$$\eta''_1(x_c) - \eta''_2(x_c) = 0 \tag{33}$$

$$\eta''_2(0) - \eta''_3(0) = 0 \tag{34}$$

$$\eta''_3(L_0) - \eta''_4(L_0) = 0 \tag{35}$$

$$\eta''_4(L_0 + L_2) = 0 \tag{36}$$

$$\eta'''_1(x_c) - \eta'''_2(x_c) = P/(EI)_b \tag{37}$$

As can be seen from the Eq. (18), the values of coefficients C_1-C_{16} can be determined according to the Eq. (22) to the Eq. (37). Once the undetermined coefficients are determined, the vertical deformation of the girder at any position under live load P in extended span can be solved.

Similarly, when live load P is applied to the right extended span, the equation of the girder, as well as the expression for the deflection, can be seen in the Eq. (38) to the Eq. (39). The expressions of the boundary conditions are similar to the Eq. (22) to the Eq. (37).

$$\begin{cases} (EI)_b \dfrac{d\eta_1{}^4(x)}{dx} = 0, & -L_1 \le x \le 0 \\ (EI)_b \dfrac{d\eta_2{}^4(x)}{dx} = w_b - h_2(x), & 0 \le x \le L_0 \\ (EI)_b \dfrac{d\eta_3{}^4(x)}{dx} = 0, & L_0 \le x \le x_c \\ (EI)_b \dfrac{d\eta_4{}^4(x)}{dx} = 0, & x_c \le x \le L_0 + L_2 \end{cases} \tag{38}$$

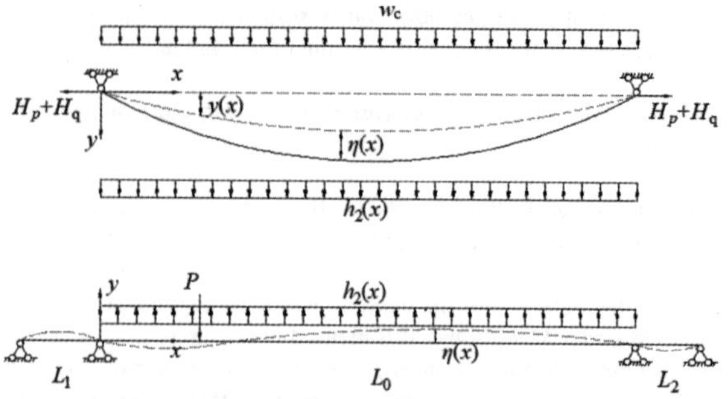

Fig. 5. Force and deformation diagram in live load case (P act on middle span).

$$\begin{cases} \eta_1(x) = C_1x^3 + C_2x^2 + C_3x + C_4, & -L_1 \le x \le 0 \\ \eta_2(x) = -x^2 \dfrac{3B_rb}{bL_0{}^3 - 3} + \dfrac{C_5 \cosh ax + C_6 \sinh ax}{a^2} + C_7x + C_8, & 0 \le x \le L_0 \\ \eta_3(x) = C_9x^3 + C_{10}x^2 + C_{11}x + C_{12}, & L_0 \le x \le x_c \\ \eta_4(x) = C_{13}x^3 + C_{14}x^2 + C_{15}x + C_{16}, & x_c \le x \le L_0 + L_2 \end{cases} \tag{39}$$

where ,

$$B_r = \frac{1}{a^3}(C_5 \sinh aL_0 + C_6 \cosh aL_0 - C_6) + \frac{1}{2}C_7L_0{}^2 + C_8L_0 \tag{40}$$

(2) Live load is applied to the main span

A slightly more complicated case, when live load P acts on the middle span, the loads and deformations of the main cable and stiffening girder are shown in Fig. 5.

By analyzing the micorelement of the girder, the Eq. (41) can be obtained.

$$\begin{cases} (EI)_b \dfrac{d\eta_1{}^4(x)}{dx} = 0, & -L_1 \le x \le 0 \\ (EI)_b \dfrac{d\eta_2{}^4(x)}{dx} = w_b - h_2(x), & 0 \le x \le x_c \\ (EI)_b \dfrac{d\eta_3{}^4(x)}{dx} = w_b - h_2(x), & x_c \le x \le L_0 \\ (EI)_b \dfrac{d\eta_4{}^4(x)}{dx} = 0, & L_0 \le x \le L_0 + L_2 \end{cases} \tag{41}$$

By solving the Eq. (41), we can get the vertical deformation expression of the stiffening girder at different positions, where C_1-C_{16} are unknown coefficients that needs to be determined by boundary conditions,

$$\begin{cases} \eta_1(x) = C_1x^3 + C_2x^2 + C_3x + C_4, & -L_1 \le x \le 0 \\ \eta_2(x) = -x^2 \dfrac{3B_mb}{bL_0{}^3 - 3} + \dfrac{C_5 \cosh ax + C_6 \sinh ax}{a^2} + C_7x + C_8, & 0 \le x \le x_c \\ \eta_3(x) = -x^2 \dfrac{3B_mb}{bL_0{}^3 - 3} + \dfrac{C_9 \cosh ax + C_{10} \sinh ax}{a^2} + C_{11}x + C_{12}, & x_c \le x \le L_0 \\ \eta_4(x) = C_{13}x^3 + C_{14}x^2 + C_{15}x + C_{16}, & L_0 \le x \le L_0 + L_2 \end{cases} \tag{42}$$

where,

$$\begin{aligned} B_m = {} & \frac{1}{a^3}[(C_5 - C_9)\sinh ax_c + (C_6 - C_{10})\cosh ax_c + C_9 \sinh aL_0 + C_{10} \cosh aL_0 - C_6] \\ & + \frac{1}{2}C_7x_c{}^2 + C_8x + \frac{1}{2}C_{11}(L_0{}^2 - x_c{}^2) + C_{12}(L_0 - x_c) \end{aligned} \tag{43}$$

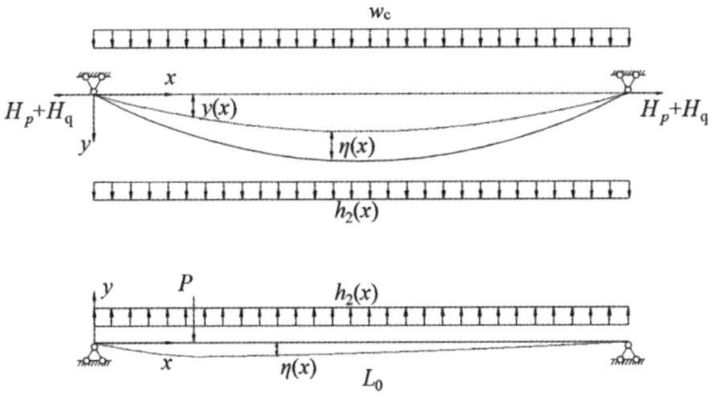

Fig. 6. Force and deformation diagram in live load case (P act on middle span).

2.4. Vertical deformations of suspension bridges with single-span stiffening girder

For the single-span suspension bridge commonly used in practice, it can be regarded as a special case ($L_1=L_2=0$), as shown in Fig. 6.

In this case, the girder equation can be obtained by simplifying the Eq. (41) ,

$$\begin{cases}(EI)_b\dfrac{d\eta_1^{\ 4}(x)}{dx}=w_b-h_2(x), & 0\le x\le x_c\\[2mm] (EI)_b\dfrac{d\eta_2^{\ 4}(x)}{dx}=w_b-h_2(x), & x_c\le x\le L_0\end{cases}\tag{44}$$

By solving the Eq. (44), we can get the vertical deformation expression of the stiffening girder at different positions, where C_1-C_8 are unknown coefficients that needs to be determined by boundary conditions,

$$\begin{cases}\eta_1(x)=-x^2\dfrac{3Bb}{bL_0^{\ 3}-3}+\dfrac{C_1\cosh ax+C_2\sinh ax}{a^2}+C_3x+C_4, & 0\le x\le x_c\\[2mm] \eta_2(x)=-x^2\dfrac{3Bb}{bL_0^{\ 3}-3}+\dfrac{C_5\cosh ax+C_6\sinh ax}{a^2}+C_7x+C_8, & x_c\le x\le L_0\end{cases}\tag{45}$$

where ,

$$B=\frac{1}{a^3}[(C_1-C_5)\sinh ax_c+(C_2-C_6)\cosh ax_c+C_5\sinh aL_0+C_6\cosh aL_0-C_2]+\frac{1}{2}C_3x_c^{\ 2}+C_4x+\frac{1}{2}C_7\left(L_0^{\ 2}-x_c^{\ 2}\right)+C_8\left(L_0-x_c\right)\tag{46}$$

3. Application procedure summary of the proposed method

The analytical method proposed for solving the longitudinal displacement of the stiffening girder needs to solve several equations simultaneously, and the result is difficult to be expressed by a single formula. In order to make the solution process more comprehensible, the solution process is shown in Fig. 7, which can be easily implemented in MATLAB.

In order to analyse a suspension bridge, it is essential to obtain the physical parameters of the stiffening girder (w_b, L_0, L_1, L_2, E_b, I_b) and the main cable (w_c, f, E_c, A_c), as well as to determine the position x_c of the moving load. Subsequently, the intermediate parameters (y, H_q, L_c, a, b) can be calculated, after which the vertical deformation of the suspension bridge can be calculated using the Eqs. (18), (39), (42) and (45). Finally, the longitudinal static deformation of the stiffening girder can be obtained. In the case of a moving train, it is necessary to ascertain a series of multiple axle weights and their positions, which can then be used to describe the vertical deformation of the girder (similar to the Eqs. (18), (39), (42) and (45)), and subsequently, the longitudinal deformation can be calculated.

Furthermore, the longitudinal motion of the stiffening girder of

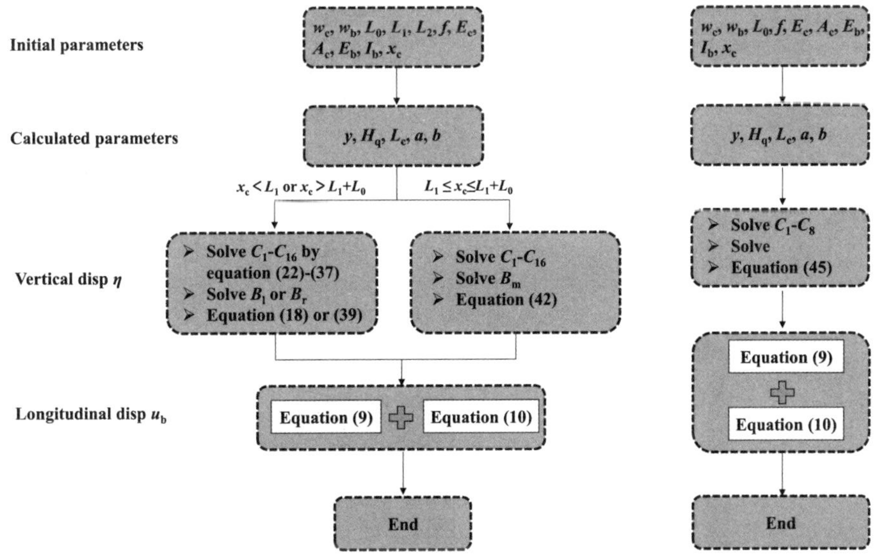

(a) with short continuous side spans (b) with single-span stiffening girder

Fig. 7. Calculation procedure for girder longitudinal displacement of suspension bridges.

suspension bridge is predominantly attributable to the ultra-low frequency quasi-static motion induced by vehicle. The static solutions obtained in this paper can be multiplied by a magnification factor of 1.1–1.2 to get the results of quasi-static motion.

4. Numerical example

4.1. Bridge model

In this section, a suspension bridge with short continuous side spans under construction in China is taken as an example to verify the accuracy of the theoretical method and conduct subsequent parameter analysis, as shown in Fig. 8. The bridge is a railway-only suspension bridge with a main span of 1060 m with side spans without suspenders and auxiliary piers. The girders have a height of 12 m, a width of 30 m and a distance of 10 m in the longitudinal direction for each segment. The towers are 262.8 m and 145.8 m high and are made of reinforced concrete. The suspension bridge's main cable has a verticality of 118 m and consist of 217 strands, each comprising 91 high-strength galvanised steel wires with a diameter of 5.68 mm. Additionally, each suspender is made up of 163 high tensile galvanised steel wires with a diameter of 5 mm. The bridge towers and girder end abutments have been fitted with both vertical and lateral bearings.

The girder and the bridge tower are simulated through the use of Beam188 element, which possess six degrees of freedom at each node. The main cables and suspenders are modelled by using Link180 element, which are set to be tension only. The Shell181 element is employed for the simulation of the bridge deck. The bearings are modelled using the CP command. The finite element model comprises a total of 18,059 elements [31]. The initial strains, material properties and real constants (i.e. cross-sectional area and moment of inertia) are calculated with great precision and assigned to the corresponding elements. In particular, the initial strain of the main cable is set to 0.0032. The initial strain of each suspender is different, with the initial strain of the suspender at the center of the span being equal to approximately 0.0023. The initial strain of the suspender close to the bridge tower was approximately 0.0025.

To validate the proposed theoretical approach, it is necessary to simplify the refined finite element model and compute the equivalent mechanical parameters of the suspension bridge, thereby establishing a simplified finite element model. Firstly, the masses and mass moments of inertia of the bridge towers, stiffening girders, and bridge deck are calculated. Then, the vertical, lateral, and torsional stiffness of each component are computed to determine the structure's stiffness. For instance, a certain length of stiffening girder is taken, with concentrated forces applied at one end in different directions while the other end is fixed. The corresponding stiffness is obtained by calculating the displacement at the loaded end, as illustrated in Fig. 9. The mechanical parameters of the bridge are summarized in Table 1. The initial strains of the main cable and suspenders are the same as in the finite element model in Fig. 8.

In the simplified model, the girder and bridge tower are simulated by Beam4 element. The main cable and suspenders are modeled by Link10 tension-only element. The simplified finite element model comprises a total of 1552 elements. In the context of finite element analysis for suspension bridges, the effect of stress stiffening is accounted for by incorporating an additional stiffness matrix, known as the geometric stiffness matrix. A preliminary nonlinear static analysis is conducted to compute this geometric stiffness matrix under dead loads. Subsequently, the dynamic response to the moving load is analyzed using transient analysis in ANSYS [32]. The two models (the simplified model and refined model) have been verified to have very close longitudinal

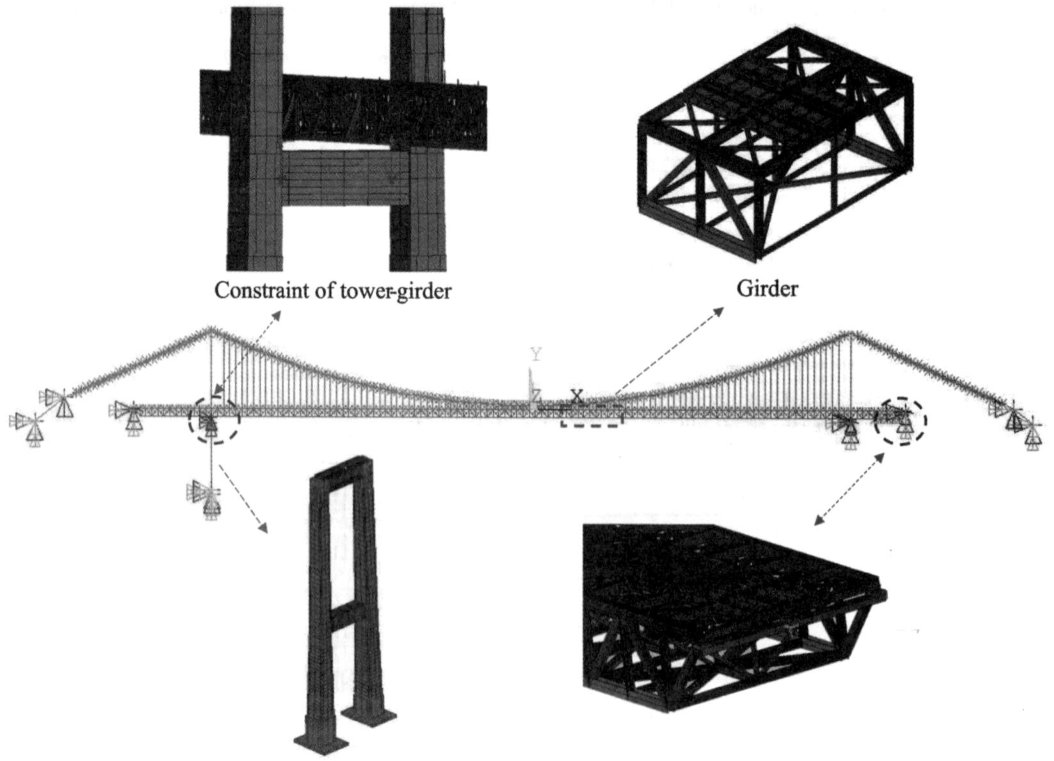

Fig. 8. The FEM of a suspension bridge with short continuous side spans.

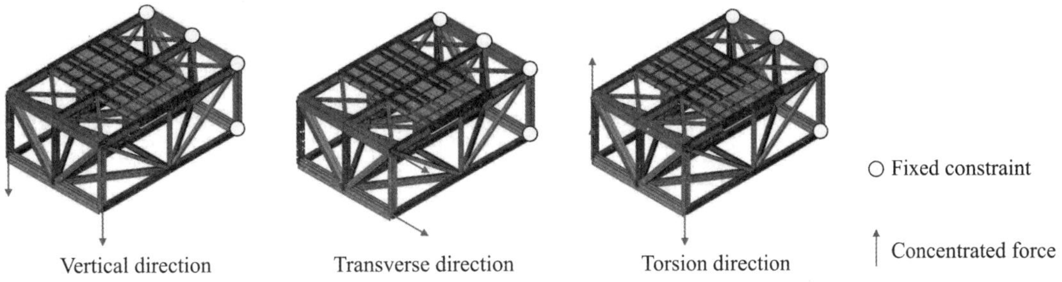

Fig. 9. The FEM of stiffening girder.

Table1
Parameters for calculation.

Parameter	Value
E_b	2.06×10^{11} N/m^2
I_b	71.8 m^4
E_c	2.06×10^{11} N/m^2
A_c	0.5 m^2
w_b	4.8×10^5 N/m
f	118 m
H_b	6 m
P	1×10^6 N

responses under moving train loads [31].

For the sake of discussion, models with constrained longitudinal displacement at the bridge tower top, and without side spans are also established. The finite element model of the bridge was conducted in ANSYS, as shown in the Fig. 10. The ratio of the length of the three spans is 0.123:1:0.085, and the side spans are extended without suspenders. Some of the parameters used in the calculations are shown in Table 1.

4.2. Comparison and discussion

Fig. 11 shows the vertical displacement results of the stiffening girder when the concentrated load P is applied to different longitudinal positions of the bridge, including the comparison of FEM results and theoretical results. In this figure, the blue dashed line distinguishes the extended spans from the main span. As can be seen from Fig. 11(a) and Fig. 11(e), when the live load P acts on the extended spans, the vertical displacement values of the whole girder are low, and the maximum value appears in the middle span. As showed in the Fig. 11(b) to the Fig. 11(d), when the load P is applied to the middle span, the maximum vertical deformation of the girder appears at the load position, and the maximum vertical deformation is 0.058 m (when $x_c = L_0/2$).

Furthermore, as showed in the Fig. 11, the results of FEM are basically consistent with those of theoretical method, which proves the accuracy of the theoretical method proposed in this paper. Based on the deflection theory, the vertical displacement of the main cable is the same as the vertical displacement of the girder, so the results of the middle span in Fig. 11 are equal to the vertical deformation of the main cable under live load.

Fig. 12 shows the longitudinal deformations results of the girder end during the whole process of live load P crossing the bridge. The distance between each time step of the live load is 5 m, and the horizontal coordinate in the Fig. 12 represents the time step.

Based on the analysis in the Section 3, it can be seen that the longitudinal deformations of the stiffening girder of the suspension bridge with Short continuous side spans is composed of the whole deformations caused by the main cable (Part I) and the local deformations caused by the flexural deformation of the girder (Part II). It can be seen from Fig. 12(a) that the longitudinal displacement (Part I) calculated by the Eq. (9) is similar to the sine curve, with a peak value of 0.009 m (when $x_c = L_0/4$).

However, the longitudinal displacement (Part II) calculated by Eq. (10) does not take the form of sine curve, and its peak value is 3.8×10^{-4} m (when $x_c = L_1/2$). For longitudinal displacement of the girder end induced by vertical live load, Part I is the primary consideration, while Part II can be ignored.

Fig 13 shows the results of the longitudinal displacement of the girder end under live load, including the comparison of FEM and theoretical solutions. It is evident that the FEM results align closely with the theoretical approach, with minimal variation observed in the total girder displacement (sum of Part I and Part II) compared to Part I alone and the finite element outcomes.

4.3. Effects of longitudinal displacement of bridge tower top

Generally, the structural bending stiffness of the suspension bridge tower is very large, so the displacement of the bridge tower is not considered in the calculation and analysis of the suspension bridge. In order to find out the influence of the bridge tower displacement on the longitudinal displacement of the girder end, the longitudinal displacement of the tower top is constrained in the finite element model.

Fig. 14 shows the longitudinal displacement results of the girder end with and without constraining the displacement of the tower top. It can be seen that constraining the tower top or not has negligible influence on the longitudinal displacement of the stiffening girder end. Therefore, the effect of the bridge tower can be ignored. The assumption of fixed tower top displacement during the derivation process is verified.

4.4. Effects of short continuous side spans

For traditional suspension bridges, there is usually only a single span with two-hinged simply supported girder. The differences between the deformation of suspension bridges with short continuous side spans and without side spans under live load are discussed.

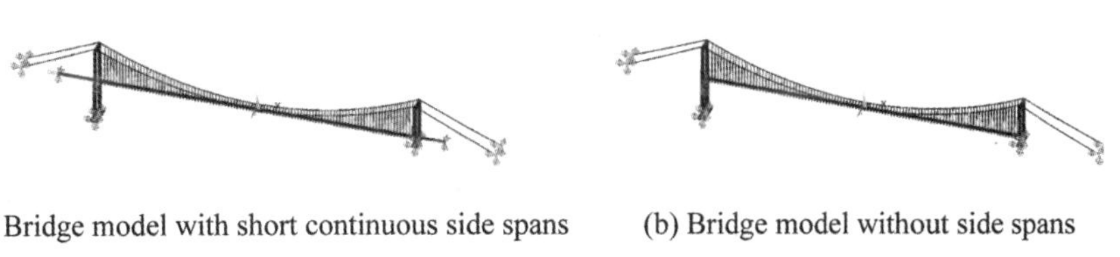

Fig. 10. The FEM of the suspension bridge.

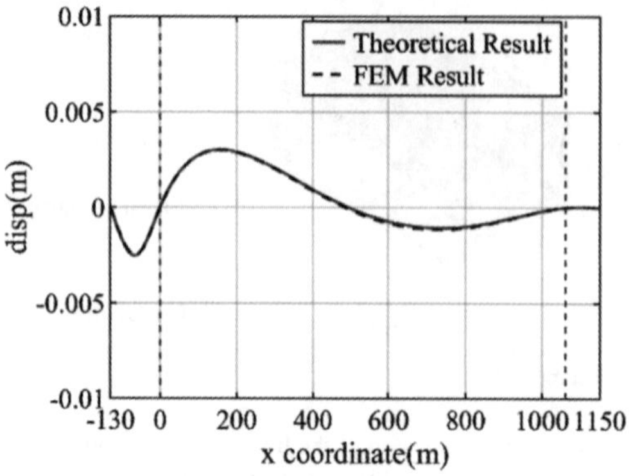

(a) $x_c=-L_1/2$

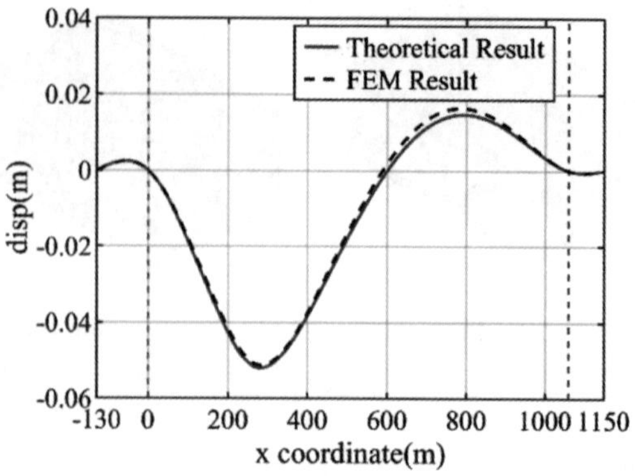

(b) $x_c=L_0/4$

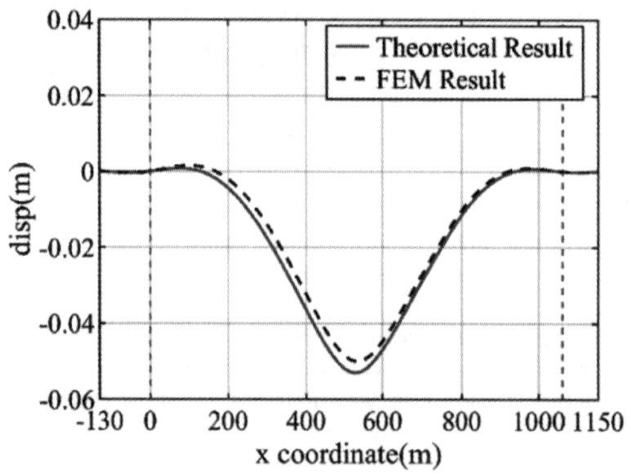

(c) $x_c=L_0/2$

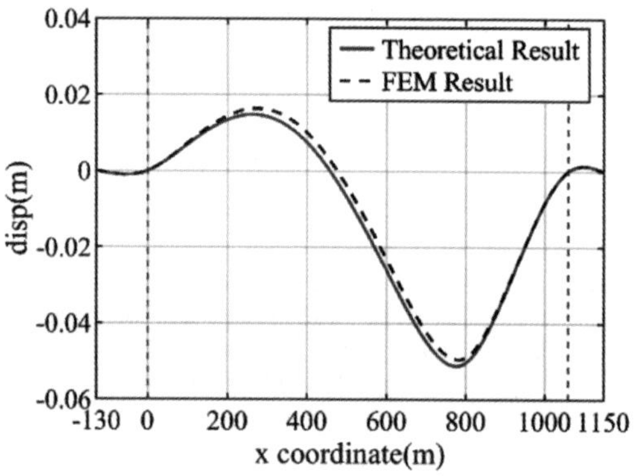

(d) $x_c=3\ L_0/4$

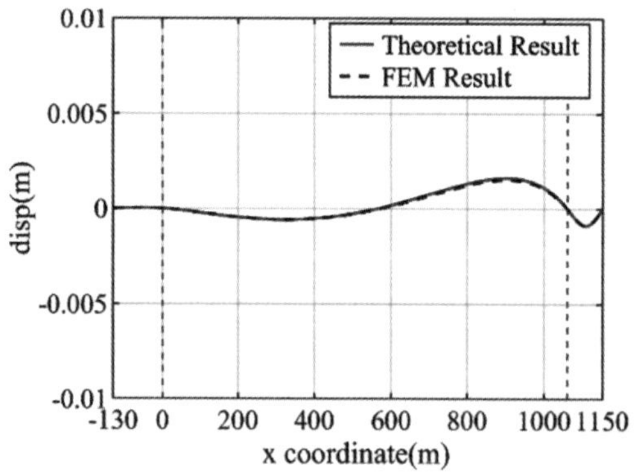

(e) $x_c=L_0+L_2/2$

Fig. 11. Comparison of vertical deformation of the girder.

Fig. 15 shows the longitudinal displacement results of the girder end before and after considering the short continuous side spans. For the case where continuous short side spans are considered, the load moves only within the main span, so both cases have the same time steps. It can be seen that the longitudinal displacement of the stiffening girder end is reduced under the effect of short continuous side spans (0.011 m to 0.007 m). This means that the short continuous side spans can significantly reduce the vertical and longitudinal displacement caused by moving loads, and proves the superiority of this new type of suspension bridge used for bridges requiring higher safety levels.

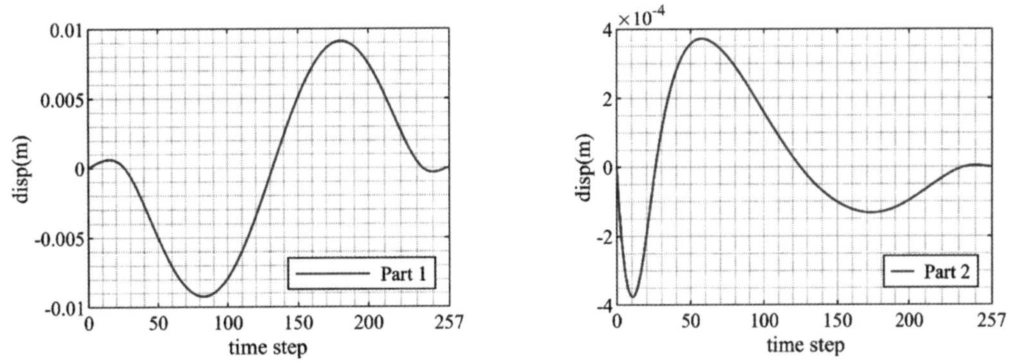

(a) The whole deformations (Part I) (b) The local deformations (Part II)

Fig. 12. The longitudinal displacements of the stiffening girder end.

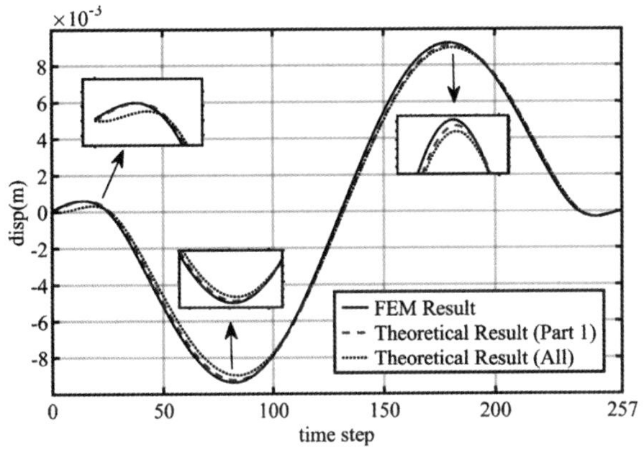

Fig. 13. Comparison of longitudinal displacements of the girder end.

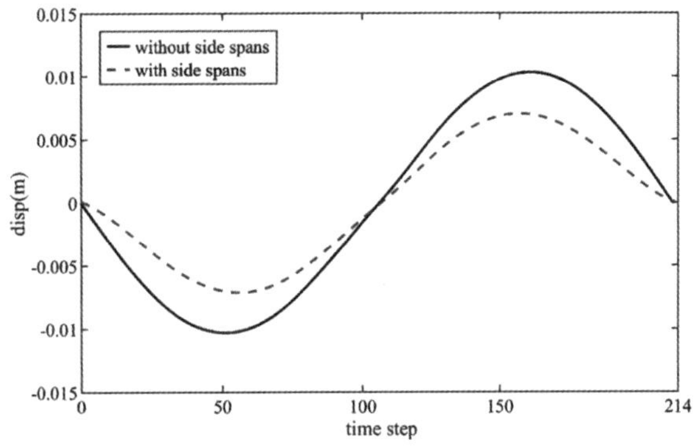

Fig. 15. Comparison of the longitudinal displacement of the girder end (Considering the effects of short continuous side spans).

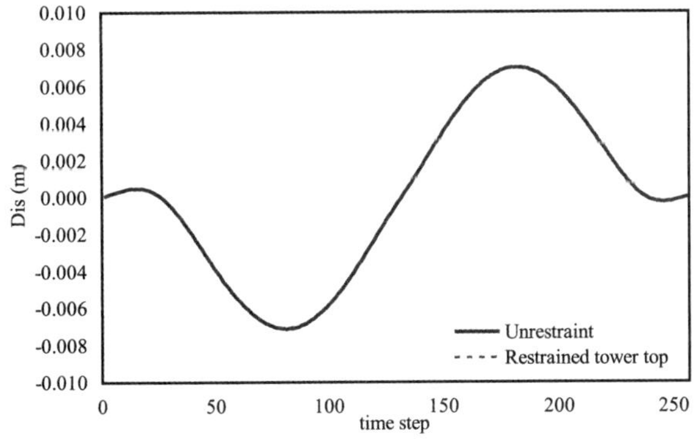

Fig. 14. Comparison of the longitudinal deformations of the girder end (Considering the effects of tower).

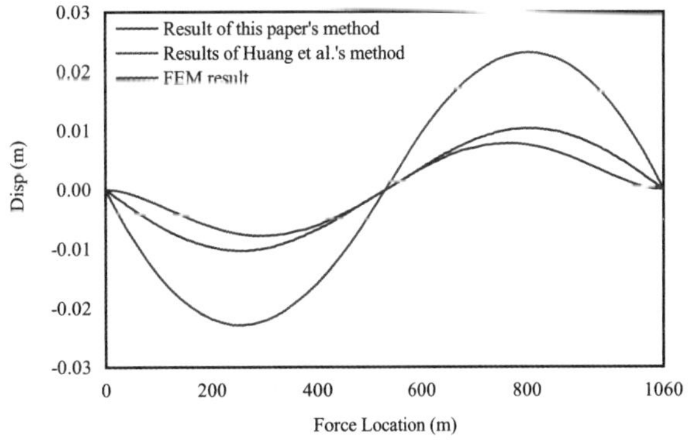

Fig. 16. Comparison of longitudinal deformation of the girder end of single-span suspension bridge.

4.5. Effects of girder bending stiffness

The bending stiffness of the stiffening girder is highly critical to the deformation of the structure under live loads for a railway suspension bridge. In this study, the bending stiffness of the girder is varied and the longitudinal displacement of the stiffening girder is compared.

First of all, for single-span suspension bridges, the method proposed by Huang *et al.* [28] and the method proposed in this paper are used for comparative analysis, and the results are shown in Fig. 16. It is evident that Huang *et al.* did not consider the bending stiffness of the stiffening girder in their analysis and adopted the main cable analogy method, so the peak longitudinal displacement is 0.023 m, which is significantly higher than the method proposed in this paper (0.007 m). Therefore, the method proposed in this paper is more accurate than the method proposed by Huang et al. in analysing the longitudinal movement of single-span suspension bridges.

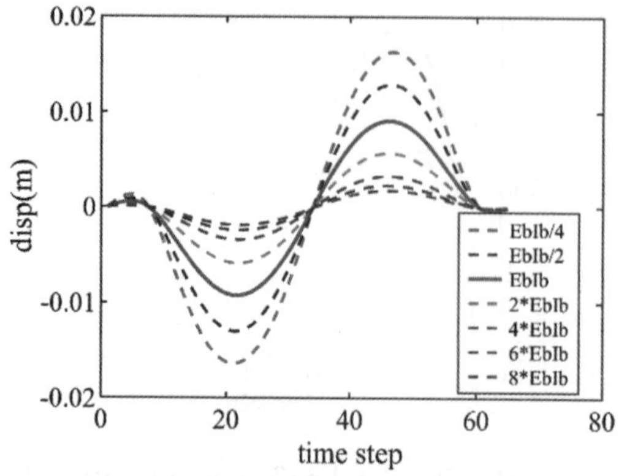

(a) The whole deformations (Part I)

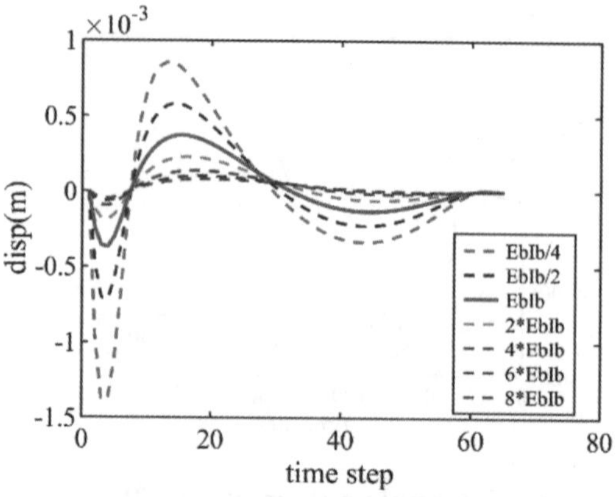

(b) The local deformations (Part II)

Fig. 17. Comparison of longitudinal deformation of the girder end with different girder bending stiffness.

Furthermore, for a suspension bridge with short continuous side spans, the girder bending stiffness increases from $(E_bI_b)/4$ to $8(E_bI_b)$, as illustrated in Fig. 17. The results indicate that as the bending stiffness increases, the longitudinal deformation, including both Part I and Part II, decreases to a certain extent. When the girder bending stiffness is $(E_bI_b)/4$ and $(E_bI_b)/2$, the peak longitudinal deformation increases by 79.12 % (0.016 m) and 41.48 % (0.013 m), respectively, compared to the original bending stiffness of E_bI_b. Conversely, when the girder bending stiffness is $2(E_bI_b)$, $4(E_bI_b)$, $6(E_bI_b)$, or $8(E_bI_b)$, the peak longitudinal deformation is reduced by 37.02 % (0.006 m), 63.95 % (0.003 m), 74.82 % (0.0023 m), and 80.68 % (0.0017 m), respectively, compared to the original bending stiffness of E_bI_b. This demonstrates that when the bending stiffness is increased to 4 times or more, the change in longitudinal deformation becomes very limited. Nevertheless, the magnitude of the local deformation Part II remains smaller than that of the overall deformation (Part I).

5. Conclusion

The longitudinal displacements at the girder ends of suspension bridges induced by moving live loads can lead to a deterioration in the performance of various structural components and auxiliary devices, impacting both traffic operations and structural safety. To address the quasi-static longitudinal motion of the girder, this paper systematically investigates analytical solutions for the static longitudinal deformation of suspension bridges under a vertical concentrated load and verifies them through finite element analyses, revealing the following key findings:

Firstly, the analytical expressions of longitudinal deformations of the stiffening girder with short continuous side spans and with single span are derived, including the deformations caused by the main cable (Part I) and the deformations caused by the flexural deformation of stiffening girder (Part II). The results show that the longitudinal displacement of the girder is coupled with the vertical displacement. The longitudinal deformations of the stiffening girder are mainly caused by the Part I, and the deformations of the Part II can be ignored.

In addition, taking a suspension bridge with short continuous side spans of 1060 m main span under construction in China as an example, a FEM was established to analyze the longitudinal deformations of the girder ends. The results showed that the vertical and longitudinal deformations of the stiffening girder obtained by the theoretical solution are consistent with the FE results.

Furthermore, using the finite element method, we analyzed the effects the bridge tower, the short continuous side spans and the bending stiffness of the girder on the longitudinal deformation of the stiffening girder. Results show that the increasing of girder bending stiffness can reduce the peak value of longitudinal deformation, the bridge tower has negligible impact, while short continuous side span significantly diminishes longitudinal displacement caused by moving live loads.

In essence, this study provides analytical solutions of static longitudinal deformations of suspension bridges, facilitating preliminary design and rapid calculation analysis. It offers an alternative to the complex finite element modeling process and significantly enhances our understanding of the longitudinal deformations at the girder ends of long-span suspension bridges. The analytical solutions presented in this paper provides a theoretical groundwork for the rapid calculation of the longitudinal dynamic response of girder ends in suspension bridges under actual train crossings.

CRediT authorship contribution statement

Haokun Jing: Writing – original draft, Visualization, Methodology, Investigation, Formal analysis, Data curation. **Zhouquan Feng:** Writing – review & editing, Supervision, Project administration, Methodology, Investigation, Funding acquisition, Conceptualization. **Zhi Chen:** Validation, Investigation, Data curation. **Guoping Huang:** Validation, Investigation, Data curation. **Xugang Hua:** Supervision, Investigation. **Zhengqing Chen:** Writing – review & editing, Supervision. **Tianbao Wan:** Supervision, Project administration.

Declaration of Competing Interest

The authors declare that they have no known competing financial interests or personal relationships that could have appeared to influence the work reported in this paper.

Data availability

Some or all data, models, or code that support the findings of this study are available from the corresponding author upon reasonable request.

Acknowledgements

This work was supported by National Natural Science Foundation of China (Grant Nos. 52178284, 51708203), Natural Science Foundation of Hunan Province (Grant No. 2021JJ30106), Natural Science Foundation of Chongqing (Grant No. CSTB2022NSCQ-MSX1465), and the Open

Research Project Programme of the State Key Laboratory of Internet of Things for Smart City (University of Macau) (Ref. No. SKL-IoTSC(UM)-2021-2023/ORP/GA09/2022).

References

[1] Hu S, Yang M, Meng D, Hu R. Damping performance of the degraded fluid viscous damper due to oil leakage. Structures 2023;2023:1609–19. https://doi.org/10.1016/j.istruc.2023.01.070.

[2] Guo T, Huang L, Liu J, Zou Y. Damage mechanism of control springs in modular expansion joints of long-span bridges. J Bridge Eng 2018;23:04018038. https://doi.org/10.1061/(ASCE)BE.1943-5592.0001255.

[3] Liu X, Han W, Yuan Y, Chen X, Xie Q. Corrosion fatigue assessment and reliability analysis of short suspender of suspension bridge depending on refined traffic and wind load condition. Eng Struct 2021;234:111950. https://doi.org/10.1016/j.engstruct.2021.111950.

[4] Liu Z, Guo T, Huang L, Pan Z. Fatigue life evaluation on short suspenders of long-span suspension bridge with central clamps. J Bridge Eng 2017;22:04017074. https://doi.org/10.1061/(ASCE)BE.1943-5592.0001097.

[5] Sun Z, Siringoringo DM, Chen S, Lu J. Cumulative displacement-based detection of damper malfunction in bridges using data-driven isolation forest algorithm. Eng Fail Anal 2023;143:106849. https://doi.org/10.1016/j.engfailanal.2022.106849.

[6] Guo T, Liu J, Huang L. Investigation and control of excessive cumulative girder movements of long-span steel suspension bridges. Eng Struct 2016;125:217–26. https://doi.org/10.1016/j.engstruct.2016.07.003.

[7] Hu J, Wang L, Song X, Sun Z, Cui J, Huang G. Field monitoring and response characteristics of longitudinal movements of expansion joints in long-span suspension bridges. Measurement 2020;162:107933. https://doi.org/10.1016/j.measurement.2020.107933.

[8] Huang G, Hu J, Liu H, Sun X. Girder longitudinal movement and its factors of suspension bridge under vehicle load. Adv Civ Eng 2021;2021:1–14. https://doi.org/10.1155/2021/1443996.

[9] Ni YQ, Hua XG, Wong KY, Ko JM. Assessment of bridge expansion joints using long-term displacement and temperature measurement. J Perform Constr Facil 2007;21:143–51. https://doi.org/10.1061/(ASCE)0887-3828(2007)21:2(143).

[10] Wang Z, Lu X, Zhang W, Fragkoulis VC, Beer M, Zhang Y. Deep learning-based reconstruction of missing long-term girder-end displacement data for suspension bridge health monitoring. Comput Struct 2023;284:107070. https://doi.org/10.1016/j.compstruc.2023.107070.

[11] Li G, Han W, Chen X, Guo T, Xie Q, Yuan Y. Wear evaluation on slide bearings in expansion joints based on cumulative displacement for long-span suspension bridge under monitored traffic flow. J Perform Constr Facil 2020;34:04019106. https://doi.org/10.1061/(ASCE)CF.1943-5509.0001388.

[12] Zhao Y, Huang P, Long G, Yuan Y, Sun Y. Influence of fluid viscous damper on the dynamic response of suspension bridge under random traffic load. Adv Civ Eng 2020;2020:1–19. https://doi.org/10.1155/2020/1857378.

[13] Hu S, Hu R, Yang M, Wang T. Seismic behavior of the combined viscous-steel damping system for a long-span suspension bridge considering wave-passage effect. J Bridge Eng 2023;28:04023034. https://doi.org/10.1061/JBENF2.BEENG-5718.

[14] Wang H, Shen R, Wen W, Dai Q, Bai L. Investigation of longitudinal movements of girder ends considering various longitudinal non-linear resistances in railway suspension bridges. J Constr Steel Res 2024;214:108446. https://doi.org/10.1016/j.jcsr.2023.108446.

[15] Gao W, Li G, Su Q, Han W. Impact of rigid central clamps on longitudinal deformation of long-span suspension bridges under vehicle excitations. Struct Infrastruct Eng 2022;18:760–74. https://doi.org/10.1080/15732479.2021.1875486.

[16] Wang H, Shen R, Bai L, Wang L, Chen W. Study on the movements and bending stresses of hangers and control measures in self-anchored rail suspension bridges. Eng Struct 2023;275:115304. https://doi.org/10.1016/j.engstruct.2022.115304.

[17] Yang M-G, Cai C. Longitudinal vibration control for a suspension bridge subjected to vehicle braking forces and earthquake excitations based on magnetorheological dampers. J Vib Control 2016;22:3659–78. https://doi.org/10.1177/1077546314564781.

[18] Wu C, Jing H, Feng Z, Song J, Wan T, Chen Z. Control of longitudinal movement response of suspension bridges induced by passing trains using low-exponent fluid viscous dampers. Structures 2024;62:106330. https://doi.org/10.1016/j.istruc.2024.106330.

[19] Xu Y, Cui B, Chen Z, Zhang T, Liu J, Li X, et al. Longitudinal displacement control of a long-span suspension bridge under random traffic loads using a polyimide friction damper. Structures 2023;55:1861–73. https://doi.org/10.1016/j.istruc.2023.05.148.

[20] Liang L, Feng Z, Zhang H, Chen Z, Liang L. An analytical study on using maxwell-coulomb friction damper to mitigate quasi-static response of SDOF systems. KSCE J Civ Eng 2023;27:4762–9. https://doi.org/10.1007/s12205-023-1365-5.

[21] Jung M-R, Min D-J, Kim M-Y. Nonlinear analysis methods based on the unstrained element length for determining initial shaping of suspension bridges under dead loads. Comput Struct 2013;128:272–85. https://doi.org/10.1016/j.compstruc.2013.06.014.

[22] Kim M-Y, Jung M-R, Attard MM. Unstrained length-based methods determining an optimized initial shape of 3-dimensional self-anchored suspension bridges. Comput Struct 2019;217:18–35. https://doi.org/10.1016/j.compstruc.2019.03.008.

[23] Zhang W, Zou H, Chang J, Liu T. Estimating design positions of suspension bridge tower saddles in the completed bridge state: an analytical approach. Eng Struct 2024;300:117178. https://doi.org/10.1016/j.engstruct.2023.117178.

[24] Tian G, Zhang W. A semi-analytical form-finding method of the 3D curved cable considering its flexural and torsional stiffnesses in suspension bridges. Appl Math Model 2023;124:806–39. https://doi.org/10.1016/j.apm.2023.08.027.

[25] Zhang W, Shen X, Chang J. Main cable shape-finding and live load response of the suspension bridge with central buckles: an analytical algorithm. J BRIDGE Eng 2023;28:04023084. https://doi.org/10.1061/JBENF2.BEENG-6350.

[26] Zhang W, Chen Y. An analytical method for assessing the live load deformation of a suspension bridge with an improved central buckle. Structures 2023;47:1341–51. https://doi.org/10.1016/j.istruc.2022.11.137.

[27] Zhang W, Lu X, Chang J, Tian G. An analytical algorithm for estimating the deck's maximum deflection and deck-end rotation angle of a suspension bridge under live load. J Bridge Eng 2022;27:04022055. https://doi.org/10.1061/(ASCE)BE.1943-5592.0001892.

[28] Huang G, Hu J, Hua X, Feng Z, Chen Z, Wan T. Analytic solution to longitudinal deformation of suspension bridges under live loads. J Bridge Eng 2023;28:04022147. https://doi.org/10.1061/JBENF2.BEENG-5835.

[29] Zhang W, Chang J, Shen X, Lu X, Liu T. Rapid assessment of suspension bridge deformation under concentrated live load considering main beam stiffness: an analytical method-Web of Science Core Collection. Struct Eng Mech 2023;88:53–65.

[30] Zhang W, Wang Z, Zhang H, Lu X, Liu Z. Analytical study on free vertical and torsional vibrations of two- and three-pylon suspension bridges via d. Struct Eng Mech 2020;76:293–310. https://doi.org/10.12989/SEM.2020.76.3.293.

[31] Jing H, Feng Z, Chen Z, Hua X, Chen Z, Wan T. Enhancing longitudinal performance in railway suspension bridges through an integrated cable restrainer system. Structures 2024;65:106769. https://doi.org/10.1016/j.istruc.2024.106769.

[32] ANSYS 19.0 - User's manual. ANSYS Inc.; 2019.

致读者

为便于专业交流，编者整理了湖南大学风工程试验研究中心成立 20 年以来公开发表的所有论文，并制作了相应的二维码，供大家扫码查阅。